Algebra for College Students

Margaret L. Lial
American River College

John Hornsby
University of New Orleans

Terry McGinnis

Vice President, Courseware Portfolio Management: Chris Hoag
Director, Courseware Portfolio Management: Michael Hirsch
Courseware Portfolio Manager: Karen Montgomery
Courseware Portfolio Assistant: Kayla Shearns
Managing Producer: Scott Disanno
Content Producer: Lauren Morse
Producers: Stacey Miller and Noelle Saligumba
Managing Producer: Vicki Dreyfus
Associate Content Producer, TestGen: Rajinder Singh
Content Managers, MathXL: Eric Gregg and Dominick Franck
Manager, Courseware QA: Mary Durnwald
Senior Product Marketing Manager: Alicia Frankel
Product Marketing Assistant: Brooke Imbornone
Senior Author Support/Technology Specialist: Joe Vetere
Full Service Vendor, Cover Design, Composition: Pearson CSC
Full Service Project Management: Pearson CSC (Carol Merrigan)
Cover Image: Borchee/E+/Getty Images

Library of Congress Cataloging-in-Publication Data

Names: Lial, Margaret L., author. | Hornsby, John, 1949- author. | McGinnis, Terry, author.
Title: Algebra for college students.
Description: 9th edition / Margaret L. Lial (American River College), John Hornsby (University of New Orleans), Terry McGinnis. | Boston : Pearson, [2020] | Includes index.
Identifiers: LCCN 2019000106 | ISBN 9780135160664 (student edition) | ISBN 0135160669 (student edition)
Subjects: LCSH: Algebra--Textbooks.
Classification: LCC QA154.3 .L53 2020 | DDC 512.9--dc23
LC record available at https://lccn.loc.gov/2019000106

8 2021

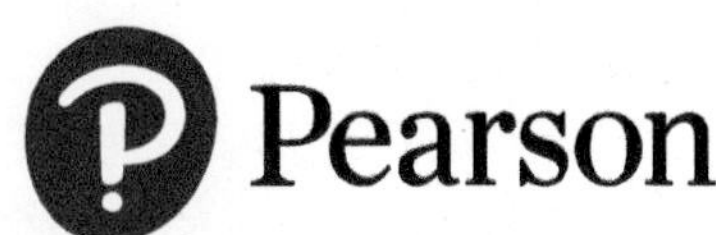

ISBN 13: 978-0-13-516066-4
ISBN 10: 0-13-516066-9

CONTENTS

3 Systems of Linear Equations 229

4 Exponents, Polynomials, and Polynomial Functions 279

5 Factoring 341

6 Rational Expressions and Functions 385

7 Roots, Radicals, and Root Functions 455

8 Quadratic Equations and Inequalities 531

9 Additional Graphs of Functions and Relations 587

10 Inverse, Exponential, and Logarithmic Functions 649

11 Polynomial and Rational Functions 715

12 Conic Sections and Nonlinear Systems 773

13 Further Topics in Algebra 819

WELCOME TO THE 9TH EDITION

The first edition of Marge Lial's *Algebra for College Students* was published in 1988, and now we are pleased to present the 9th edition—with the same successful, well-rounded framework that was established over 30 years ago and updated to meet the needs of today's students and professors. The names Lial and Miller, two faculty members from American River College in Sacramento, California, have become synonymous with excellence in Developmental Mathematics, Precalculus, Finite Mathematics, and Applications-Based Calculus.

With Chuck Miller's passing, Marge Lial was joined by a team of carefully selected coauthors who partnered with her. John Hornsby (University of New Orleans) joined Marge in this capacity in 1992, and in 1999, Terry McGinnis became part of this developmental author team. Since Marge's passing in 2012, John and Terry have dedicated themselves to carrying on the Lial/Miller legacy.

In the preface to the first edition of *Intermediate Algebra,* Marge Lial wrote

> *". . . the strongest theme . . . is a combination of readability and suitability for the book's intended audience: students who are not completely self-confident in mathematics as they come to the course, but who must be self-confident and proficient . . . by the end of the course."*

Today's Lial author team upholds these same standards. With the publication of the 9th edition of *Algebra for College Students,* we proudly present a complete course program for students who need developmental algebra. Revisions to the core text, working in concert with such innovations in the MyLab Math course as Skill Builder and Learning Catalytics, combine to provide superior learning opportunities appropriate for all types of courses (traditional, hybrid, online).

We hope you enjoy using it as much as we have enjoyed writing it. We welcome any feedback that you have as you review and use this text.

WHAT'S NEW IN THIS EDITION?

We are pleased to offer the following new features and resources in the text and MyLab.

IMPROVED STUDY SKILLS These special activities are now grouped together at the front of the text, prior to Chapter R. **Study Skills Reminders** that refer students to specific Study Skills are found liberally throughout the text. Many Study Skills now include a ***Now Try This*** section to help students implement the specific skill.

REVISED EXPOSITION With each edition of the text, we continue to polish and improve discussions and presentations of topics to increase readability and student understanding. This edition is no exception.

NEW FIGURES AND DIAGRAMS For visual learners, we have included more than 50 new mathematical figures, graphs, and diagrams, including several new "hand drawn" style graphs. These are meant to suggest what a student who is graphing with paper and pencil should obtain. We use this style when introducing a particular type of graph for the first time.

ENHANCED USE OF PEDAGOGICAL COLOR We have thoroughly reviewed the use of pedagogical color in discussions and examples and have increased its use whenever doing so would enhance concept development, emphasize important steps, or highlight key procedures.

INCREASED Concept Check AND WHAT WENT WRONG? EXERCISES The number of **Concept Check** exercises, which facilitate students' mathematical thinking and conceptual understanding, and which begin each exercise set, has been increased. We have also more than doubled the number of **WHAT WENT WRONG?** exercises that highlight common student errors.

INCREASED RELATING CONCEPTS EXERCISES We have doubled the number of these flexible groups of exercises, which are located at the end of many exercise sets. These sets of problems were specifically written to help students tie concepts together, compare and contrast ideas, identify and describe patterns, and extend concepts to new situations. They may be used by individual students or by pairs or small groups working collaboratively. All answers to these exercises appear in the student answer section.

ENHANCED MYLAB MATH RESOURCES MyLab exercise coverage in the revision has been expanded, and video coverage has also been expanded and updated to a modern format for today's students. **WHAT WENT WRONG?** problems and all **RELATING CONCEPTS** exercise sets (both even- and odd-numbered problems) are now assignable in MyLab Math.

SKILL BUILDER These exercises offer just-in-time additional adaptive practice in MyLab Math. The adaptive engine tracks student performance and delivers, to each individual, questions that adapt to his or her level of understanding. This new feature enables instructors to assign fewer questions for homework, allowing students to complete as many or as few questions as they need.

LEARNING CATALYTICS This new student response tool uses students' own devices to engage them in the learning process. Problems that draw on prerequisite skills are included at the beginning of each section to gauge student readiness for the section. Accessible through MyLab Math and customizable to instructors' specific needs, these problems can be used to generate class discussion, promote peer-to-peer learning, and provide real-time feedback to instructors. More information can be found via the Learning Catalytics link in MyLab Math. Specific exercises notated in the text can be found by searching LialACS# where the # is the chapter number.

CONTENT CHANGES

Specific content changes include the following:

- **Exercise sets** have been scrutinized and updated with a renewed focus on conceptual understanding and skill development. Even and odd pairing of the exercises, an important feature of the text, has been carefully reviewed.
- **Real world data** in all examples and exercises and their accompanying graphs has been updated.
- **An increased emphasis on fractions, decimals, and percents** appears throughout the text. **Chapter R** begins with a new section that thoroughly reviews these topics. And we have included an **all-new set of Cumulative Review Exercises,** many of which focus on fractions, decimals, and percents, at the end of Chapter 1. Sets of Cumulative Review Exercises in subsequent chapters now begin with new exercises that review skills related to these topics.
- **Solution sets of linear inequalities in Sections 1.5–1.7** are now graphed first before writing them using interval notation.
- **Scientific notation is covered in a separate section in Chapter 4.**
- **Presentations of the following topics have been enhanced and expanded,** often including new examples and exercises:

  Evaluating exponential expressions (Section R.4)

  Geometric interpretation of slope as rise/run (Section 2.2)

  Identifying functions and domains from equations (Section 2.6)

  Solving systems of linear equations in three variables (Section 3.2)

  Determining strategies for factoring polynomials (Section 5.4)

  Solving quadratic equations with double solutions (Section 5.5)

  Solving rational equations (Section 6.4)

  Concepts and relationships among real numbers, nonreal complex numbers, and imaginary numbers; simplifying powers of i (Section 7.7)

  Solving quadratic equations using the quadratic formula (Section 8.2)

  Testing for symmetry with respect to an axis or the origin (Section 9.4)

  Solving exponential and logarithmic equations (Sections 10.2, 10.3)

  Graphing polynomial and rational functions (Sections 11.3, 11.4)

  Graphing systems of linear inequalities (Section 12.4)

LIAL DEVELOPMENTAL HALLMARK FEATURES

We have enhanced the following popular features, each of which is designed to increase ease of use by students and/or instructors.

- ***Emphasis on Problem-Solving*** We introduce our six-step problem-solving method in Chapter 2 and integrate it throughout the text. The six steps, *Read, Assign a Variable, Write an Equation, Solve, State the Answer,* and *Check,* are emphasized in boldface type and repeated in examples and exercises to reinforce the problem-solving process for students. We also provide students with **PROBLEM-SOLVING HINT** boxes that feature helpful problem-solving tips and strategies.
- ***Helpful Learning Objectives*** We begin each section with clearly stated, numbered objectives, and the included material is directly keyed to these objectives so that students and instructors know exactly what is covered in each section.
- ***Cautions and Notes*** One of the most popular features of previous editions is our inclusion of information marked **! CAUTION** and **NOTE** to warn students about common errors and to emphasize important ideas throughout the exposition. The updated text design makes them easy to spot.
- ***Comprehensive Examples*** The new edition features a multitude of step-by-step, worked-out examples that include pedagogical color, helpful side comments, and special pointers. We give special attention to checking example solutions—more checks, designated using a special **CHECK** tag and ✓, are included than in past editions.

- ***More Pointers*** There are more pointers in examples and discussions throughout this edition of the text. They provide students with important on-the-spot reminders, as well as warnings about common pitfalls.
- ***Numerous Now Try Problems*** These margin exercises, with answers immediately available at the bottom of the page, have been carefully written to correspond to every example in the text. This key feature allows students to immediately practice the material in preparation for the exercise sets.
- ***Updated Figures, Photos, and Hand-Drawn Graphs*** Today's students are more visually oriented than ever. As a result, we provide detailed mathematical figures, diagrams, tables, and graphs, including a "hand-drawn" style of graphs, whenever possible. We have incorporated depictions of well-known mathematicians, as well as appealing photos to accompany applications in examples and exercises.
- ***Relevant Real-Life Applications*** We include many new or updated applications from fields such as business, pop culture, sports, technology, and the health sciences that show the relevance of algebra to daily life.
- ***Extensive and Varied Exercise Sets*** The text contains a wealth of exercises to provide students with opportunities to practice, apply, connect, review, and extend the skills they are learning. Numerous illustrations, tables, graphs, and photos help students visualize the problems they are solving. Problem types include skill building and writing exercises, as well as applications, matching, true/false, multiple-choice, and fill-in-the-blank problems. Special types of exercises include **Concept Check**, **WHAT WENT WRONG?**, **Extending Skills**, and **RELATING CONCEPTS**.
- ***Special Summary Exercises*** We include a set of these popular in-chapter exercises in every chapter. They provide students with the all-important ***mixed review problems*** they need to master topics and often include summaries of solution methods and/or additional examples.
- ***Extensive Review Opportunities*** We conclude each chapter with the following review components:

  A **Chapter Summary** that features a helpful list of **Key Terms** organized by section, **New Symbols,** a **Test Your Word Power** vocabulary quiz (with answers immediately following), and a **Quick Review** of each section's main concepts, complete with additional examples.

  A comprehensive set of **Chapter Review Exercises,** keyed to individual sections for easy student reference.

  A set of **Mixed Review Exercises** that helps students further synthesize concepts and skills.

  A **Chapter Test** that students can take under test conditions to see how well they have mastered the chapter material.

  A set of **Cumulative Review Exercises** for ongoing review that covers material going back to Chapter R.
- ***Comprehensive Glossary*** The online Glossary includes key terms and definitions (with section references) from throughout the text.

ACKNOWLEDGMENTS

The comments, criticisms, and suggestions of users, nonusers, instructors, and students have positively shaped this text over the years, and we are most grateful for the many responses we have received. The feedback gathered for this edition was particularly helpful.

We especially wish to thank the following individuals who provided invaluable suggestions.

Barbara Aaker, *Community College of Denver*
Kim Bennekin, *Georgia Perimeter College*
Dixie Blackinton, *Weber State University*
Eun Cha, *College of Southern Nevada, Charleston*
Callie Daniels, *St. Charles Community College*
Cheryl Davids, *Central Carolina Technical College*
Robert Diaz, *Fullerton College*
Chris Diorietes, *Fayetteville Technical Community College*
Sylvia Dreyfus, *Meridian Community College*
Sabine Eggleston, *Edison State College*
LaTonya Ellis, *Bishop State Community College*
Beverly Hall, *Fayetteville Technical Community College*
Loretta Hart, *NHTI, Concord's Community College*
Sandee House, *Georgia Perimeter College*
Joe Howe, *St. Charles Community College*
Lynette King, *Gadsden State Community College*
Linda Kodama, *Windward Community College*
Carlea McAvoy, *South Puget Sound Community College*
James Metz, *Kapi'olani Community College*
Jean Millen, *Georgia Perimeter College*
Molly Misko, *Gadsden State Community College*
Charles Patterson, *Louisiana Tech*
Jane Roads, *Moberly Area Community College*
Melanie Smith, *Bishop State Community College*
Erik Stubsten, *Chattanooga State Technical Community College*
Tong Wagner, *Greenville Technical College*
Rick Woodmansee, *Sacramento City College*
Sessia Wyche, *University of Texas at Brownsville*

Over the years, we have come to rely on an extensive team of experienced professionals. Our sincere thanks go to these dedicated individuals at Pearson who worked long and hard to make this revision a success.

We would like to thank Michael Hirsch, Matthew Summers, Karen Montgomery, Alicia Frankel, Lauren Morse, Vicki Dreyfus, Stacey Miller, Noelle Saligumba, Eric Gregg, and all of the Pearson math team for helping with the revision of the text.

We are especially pleased to welcome Callie Daniels, who has taught from our texts for many years, to our team. Her assistance has been invaluable. She thoroughly reviewed all chapters and helped extensively with manuscript preparation.

We are grateful to Carol Merrigan for her excellent production work. We appreciate her positive attitude, responsiveness, and expert skills. We would also like to thank Pearson CSC for their production work; Emily Keaton for her detailed help in updating real data applications; Connie Day for supplying her copyediting expertise; Pearson CSC for their photo research; and Lucie Haskins for producing another accurate, useful index. Paul Lorczak and Hal Whipple did a thorough, timely job accuracy-checking the page proofs and answers, and Sarah Sponholz checked the index.

We particularly thank the many students and instructors who have used this text over the years. You are the reason we do what we do. It is our hope that we have positively impacted your mathematics journey. We would welcome any comments or suggestions you might have via email to math@pearson.com.

John Hornsby
Terry McGinnis

DEDICATION

To Wayne and Sandra

E.J.H.

To Andrew and Tyler

Mom

Pearson MyLab

Resources for Success

Get the Most Out of MyLab Math for *Algebra for College Students,* Ninth Edition by Lial, Hornsby, McGinnis

The Lial team has helped thousands of students learn algebra with an approachable, teacherly writing style and balance of skill and concept development. With this revision, the series retains the hallmarks that have helped students succeed in math, and includes new and updated digital tools in the MyLab Math course.

Take advantage of the following resources to get the most out of your MyLab Math course.

Get Students Prepared with Integrated Review

Every student enters class with different levels of preparedness and prerequisite knowledge. To ensure students are caught up on prior skills, every Lial MyLab course now includes Integrated Review.

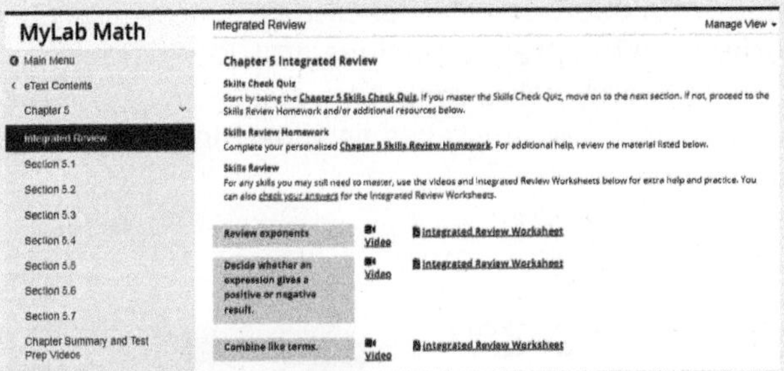

New! Integrated Review provides embedded and personalized review of prerequisite topics within relevant chapters. Students can check their prerequisite skills, and receive personalized practice on the topics they need to focus on, with study aids like worksheets and videos also available to help.

Integrated Review assignments are premade and available to assign in the Assignment Manager.

Personalize Learning

New! Skill Builder exercises offer just-in-time additional adaptive practice. The adaptive engine tracks student performance and delivers questions to each individual that adapt to his or her level of understanding. This new feature allows instructors to assign fewer questions for homework, allowing students to complete as many or as few questions as they need.

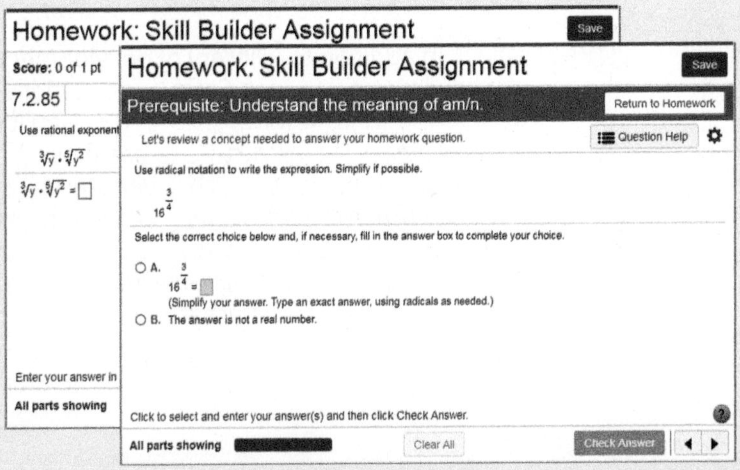

pearson.com/mylab/math

Resources for Success

Pearson MyLab

Support Students Whenever, Wherever

Updated! The **complete video program** for the Lial series includes:

- Full Section Lecture Videos
- Solution clips for select exercises
- Chapter Test Prep videos
- Short Quick Review videos that recap each section

Full Section Lecture Videos are also available as shorter, objective-level videos. No matter your students' needs—if they missed class, need help solving a problem, or want a short summary of a section's concepts—they can get support whenever they need it, wherever they need it. Much of the video series has been updated in a modern presentation format.

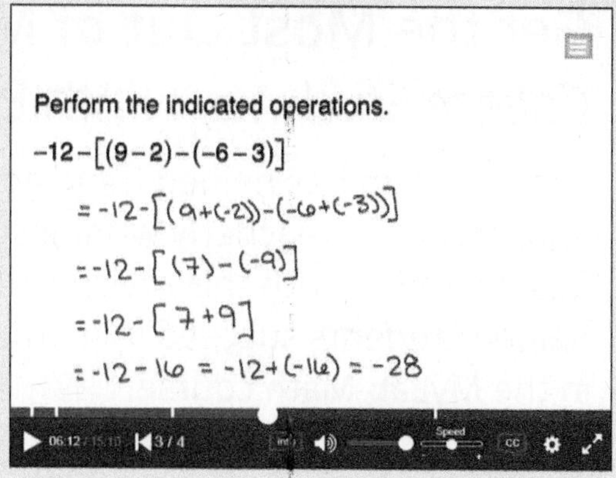

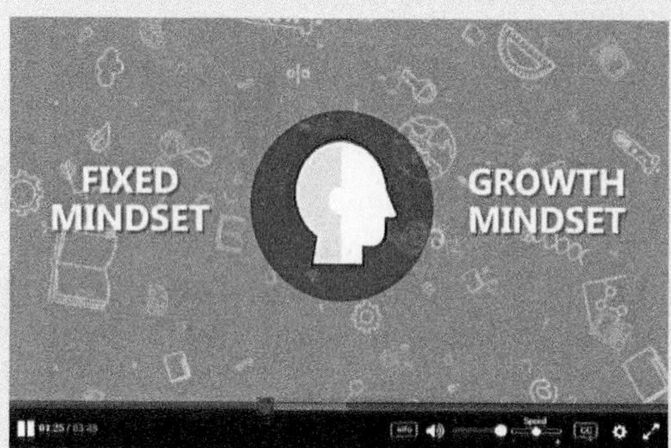

Foster a Growth Mindset

New! A **Mindset module** is available in the course, with mindset-focused videos and exercises that encourage students to maintain a positive attitude about learning, value their own ability to grow, and view mistakes as a learning opportunity.

Get Students Engaged

New! Learning Catalytics Learning Catalytics is an interactive student response tool that uses students' smartphones, tablets, or laptops to engage them in more sophisticated tasks and thinking.

In addition to a library of developmental math questions, Learning Catalytics questions created specifically for this text are pre-built to make it easy for instructors to begin using this tool! These questions, which cover prerequisite skills before each section, are noted in the margin of the Annotated Instructor's Edition, and can be found in Learning Catalytics by searching for "LialACS#", where # is the chapter number.

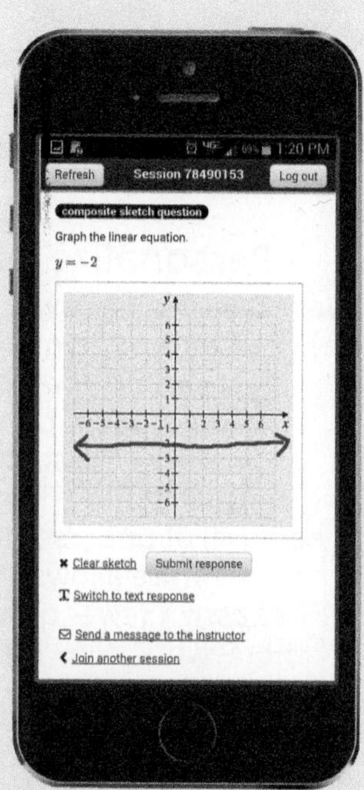

pearson.com/mylab/math

Resources for Success

Pearson MyLab

Instructor Resources

Annotated Instructor's Edition

Contains all the content found in the student edition, plus answers to even and odd exercises on the same text page, and Teaching Tips and Classroom Examples throughout the text placed at key points.

The resources below are available through Pearson's Instructor Resource Center, or from MyLab Math.

Instructor's Resource Manual with Tests

Includes mini-lectures for each text section, several forms of tests per chapter—two diagnostic pretests, four free-response and two multiple-choice test forms per chapter, and two final exams.

Instructor's Solutions Manual

Contains detailed, worked-out solutions to all exercises in the text.

TestGen®

Enables instructors to build, edit, print, and administer tests using a computerized bank of questions developed to cover all the objectives of the text. TestGen is algorithmically based, allowing instructors to create multiple but equivalent versions of the same question or test with the click of a button. Instructors can also modify test bank questions or add new questions.

PowerPoint Lecture Slides

Available for download only, these slides present key concepts and definitions from the text. Accessible versions of the PowerPoint slides are also available for students who are vision-impaired.

Student Resources

Guided Notebook

This Guided Notebook helps students keep their work organized as they work through their course. The notebook includes:

- Guided Examples that are worked out for students, plus corresponding Now Try This exercises for each text objective.
- Extra practice exercises for every section of the text, with ample space for students to show their work.
- Learning objectives and key vocabulary terms for every text section, along with vocabulary practice problems.

Student Solutions Manual

Provides completely worked-out solutions to the odd-numbered section exercises and to all exercises in the Now Trys, Relating Concepts, Chapter Reviews, Mixed Reviews, Chapter Tests, and Cumulative Reviews. Available at no additional charge in the MyLab Math course.

pearson.com/mylab/math

Using Your Math Text

Your text is a valuable resource. You will learn more if you make full use of the features it offers.

Now **TRY THIS**

General Features of This Text

Locate each feature, and complete any blanks.

- **Table of Contents** This is located at the front of the text.

  Find it and mark the chapters and sections you will cover, as noted on your course syllabus.

- **Answer Section** This is located at the back of the text.

  Tab this section so you can easily refer to it when doing homework or reviewing for tests.

- **List of Formulas** This helpful list of geometric formulas, along with review information on triangles and angles, is found at the back of the text.

  The formula for the volume of a cube is _______.

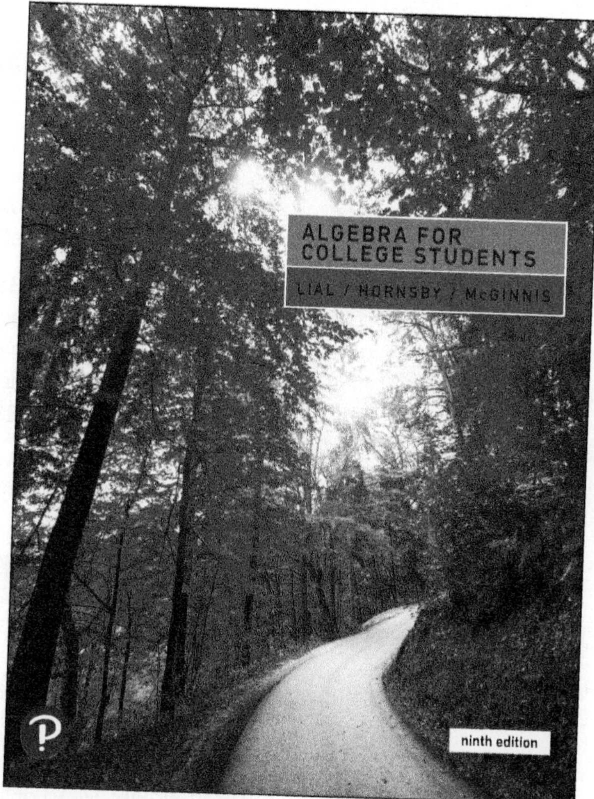

Specific Features of This Text

Look through Chapter 1 and give the number of a page that includes an example of each of the following specific features.

- **Objectives** The objectives are listed at the beginning of each section and again within the section as the corresponding material is presented. Once you finish a section, ask yourself if you have accomplished them. ***See page*** _______.
- **Vocabulary List** Important vocabulary is listed at the beginning of each section. You should be able to define these terms when you finish a section. ***See page*** _______.
- **Now Try Exercises** These margin exercises allow you to immediately practice the material covered in the examples and prepare you for the exercises. Check your results using the answers at the bottom of the page. ***See page*** _______.
- **Pointers** These small, shaded balloons provide on-the-spot warnings and reminders, point out key steps, and give other helpful tips. ***See page*** _______.
- **Cautions** These provide warnings about common errors that students often make or trouble spots to avoid. ***See page*** _______.
- **Notes** These provide additional explanations or emphasize other important ideas. ***See page*** _______.
- **Problem-Solving Hints** These boxes give helpful tips or strategies to use when you work applications. ***See page*** _______.

STUDY SKILL 2

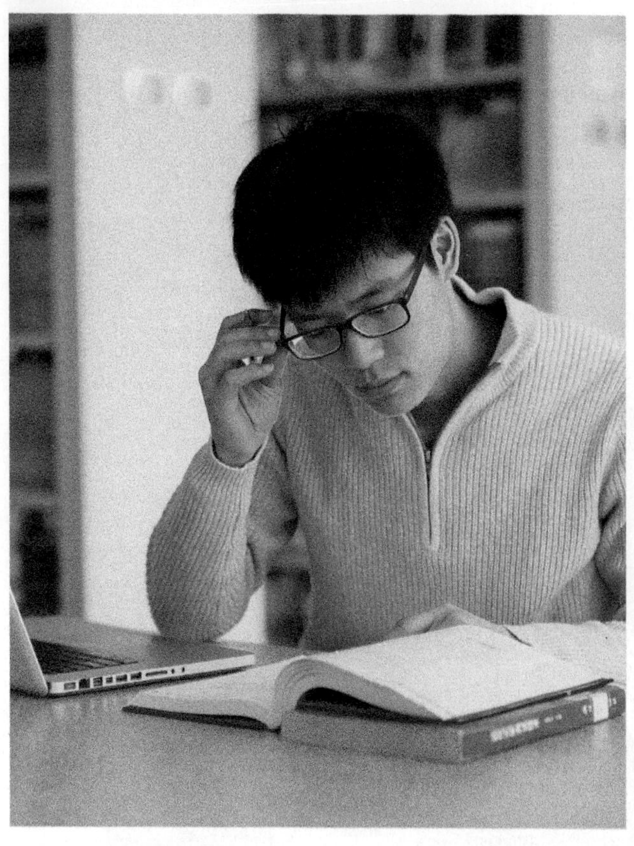

Reading Your Math Text

Take time to read each section and its examples before doing your homework. You will learn more and be better prepared to work the exercises your instructor assigns.

Approaches to Reading Your Math Text

Student A learns best by listening to her teacher explain things. She "gets it" when she sees the instructor work problems. She previews the section before the lecture, so she knows generally what to expect. **Student A carefully reads the section in her text *AFTER* she hears the classroom lecture on the topic.**

Student B learns best by reading on his own. He reads the section and works through the examples before coming to class. That way, he knows what the teacher is going to talk about and what questions he wants to ask. **Student B carefully reads the section in his text *BEFORE* he hears the classroom lecture on the topic.**

Which of these reading approaches works best for you—that of Student A or Student B?

Tips for Reading Your Math Text

- **Turn off your cell phone and the TV.** You will be able to concentrate more fully on what you are reading.
- **Survey the material.** Glance over the assigned material to get an idea of the "big picture." Look at the list of objectives to see what you will be learning.
- **Read slowly.** Read only one section—or even part of a section—at a sitting, with paper and pencil in hand.
- **Pay special attention to important information given in colored boxes or set in boldface type.** Highlight any additional information you find helpful.
- **Study the examples carefully.** Pay particular attention to the blue side comments and any pointer balloons.
- **Do the Now Try exercises in the margin on separate paper as you go.** These problems mirror the examples and prepare you for the exercise set. Check your answers with those given at the bottom of the page.
- **Make study cards as you read.** Make cards for new vocabulary, rules, procedures, formulas, and sample problems.
- **Mark anything you don't understand. *ASK QUESTIONS*** in class—everyone will benefit. Follow up with your instructor, as needed.

Now TRY THIS

Think through and answer each question.

1. Which two or three reading tips given above will you try this week?
2. Did the tips you selected improve your ability to read and understand the material? Explain.

Taking Lecture Notes

Come to class prepared.

- Bring paper, pencils, notebook, text, completed homework, and any other materials you need.
- Arrive 10–15 minutes early if possible. Use the time before class to review your notes or study cards from the last class period.
- Select a seat carefully so that you can hear and see what is going on.

Study the set of sample math notes given at the right.

- **Use a new page** for each day's lecture.
- **Include the date and title** of the day's lecture topic.
- **Skip lines and write neatly** to make reading easier.
- **Include cautions and warnings** to emphasize common errors to avoid.
- **Mark important concepts with stars, underlining, etc.**
- **Use two columns,** which allows an example and its explanation to be close together.
- **Use brackets and arrows** to clearly show steps, related material, etc.
- **Highlight any material and/or information that your instructor emphasizes.** Instructors often give "clues" about material that will definitely be on an exam.

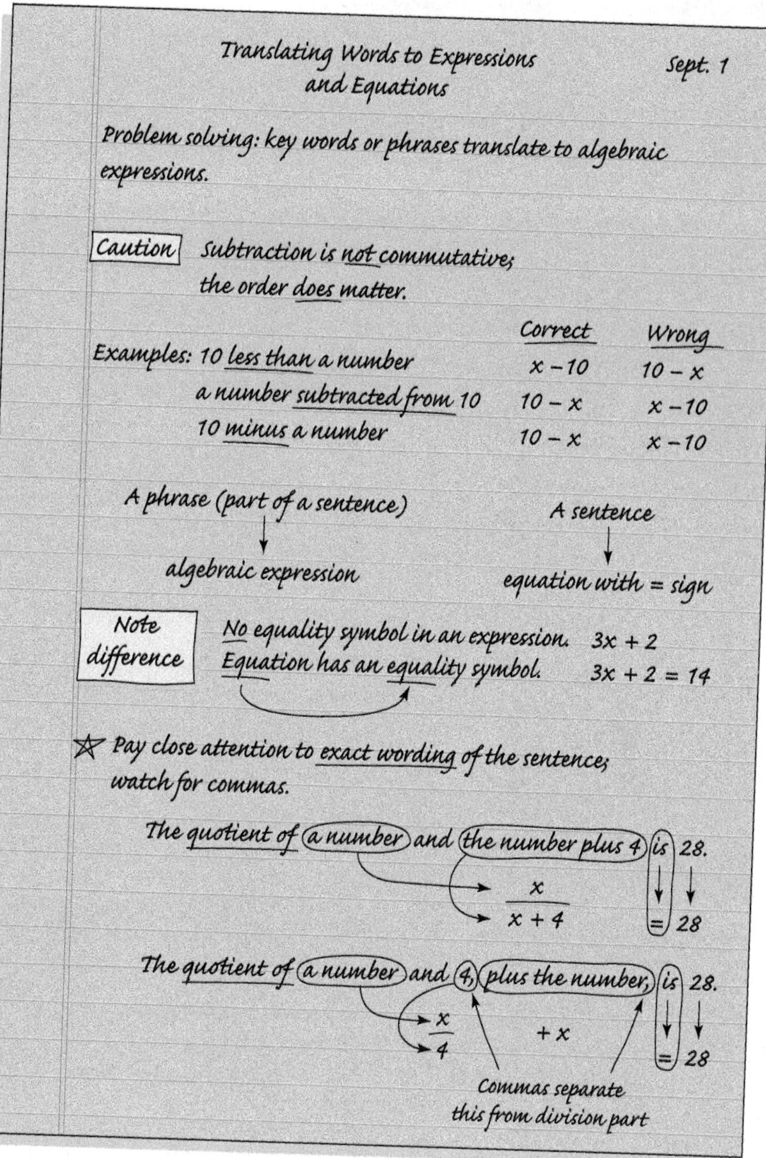

Consider using a three-ring binder to organize your notes, class handouts, and completed homework.

Now TRY THIS

With a study partner or in a small group, compare lecture notes. Then answer each question.

1. What are you doing to show main points in your notes (such as boxing, using stars, etc.)?
2. In what ways do you set off explanations from worked problems and subpoints (such as indenting, using arrows, circling, etc.)?
3. What new ideas did you learn by examining your classmates' notes?
4. What new techniques will you try when taking notes in future lectures?

STUDY SKILL 4

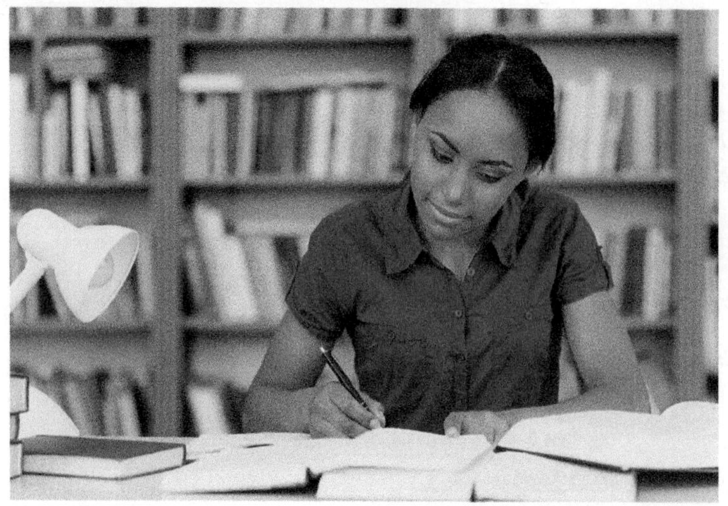

Completing Your Homework

You are ready to do your homework ***AFTER*** you have read the corresponding text section and worked through the examples and Now Try exercises.

Homework Tips

- **Keep distractions and potential interruptions to a minimum.** Turn off your cell phone and the TV. Find a quiet, comfortable place to work, away from a lot of other people, so you can concentrate on what you are doing.
- **Review your class notes.** Pay particular attention to anything your instructor emphasized during the lecture on this material.
- **Survey the exercise set.** Glance over the problems that your instructor has assigned to get a general idea of the types of exercises you will be working. Skim directions, and note any references to section examples.
- **Work problems neatly.** NEVER do your math homework in pen. Use pencil and write legibly, so others can read your work. Skip lines between steps. Clearly separate problems from each other.
- **Show all your work.** It is tempting to take shortcuts. Include ALL steps.
- **Check your work frequently to make sure you are on the right track.** It is hard to unlearn a mistake. For all odd-numbered problems, answers are given in the back of the text.
- **If you have trouble with a problem, refer to the corresponding worked example in the section.** The exercise directions will often reference specific examples to review. Pay attention to every line of the worked example to see how to get from step to step.
- **If you have trouble with an even-numbered problem, work the corresponding odd-numbered problem.** Check your answer in the back of the text, and apply the same steps to work the even-numbered problem.
- **If you have genuinely tried to work a problem but have not been able to complete it in a *reasonable* amount of time, it's ok to STOP.** Mark these problems. Ask for help at your school's tutor center or from fellow classmates, study partners, or your instructor.
- **Do some homework problems every day.** This is a good habit, even if your math class does not meet each day.

Now TRY THIS

Think through and answer each question.

1. What is your instructor's policy regarding homework?
2. Think about your current approach to doing homework. Be honest in your assessment.
   (a) What are you doing that is working well?
   (b) What improvements could you make?
3. Which one or two homework tips will you try this week?
4. In the event that you need help with homework, what resources are available? When does your instructor hold office hours?

Using Study Cards

You may have used "flash cards" in other classes. In math, "study cards" can help you remember terms and definitions, procedures, and concepts. Use study cards to

- Help you understand and learn the material;
- Quickly review when you have a few minutes;
- Review before a quiz or test.

One of the advantages of study cards is that you learn the material while you are making them.

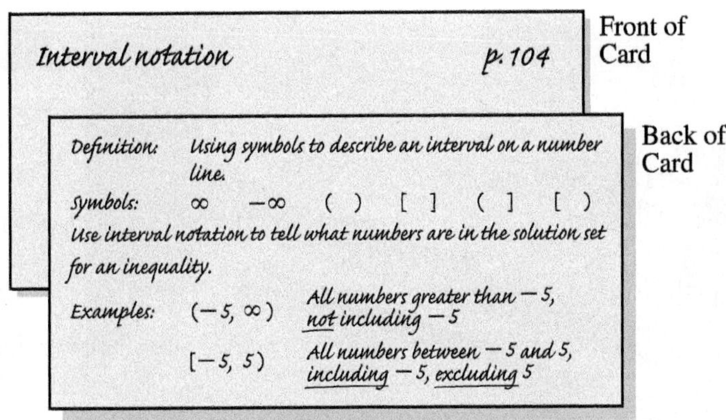

Interval notation
p. 104
Front of Card
Definition: Using symbols to describe an interval on a number line.
Symbols: ∞ −∞ ( ) [ ] ( ] [ )
Use interval notation to tell what numbers are in the solution set for an inequality.
Examples: (−5, ∞) All numbers greater than −5, not including −5
[−5, 5) All numbers between −5 and 5, including −5, excluding 5
Back of Card

Vocabulary Cards

Put the word and a page reference on the front of the card. On the back, write the definition, an example, any related words, and a sample problem (if appropriate).

Solving a Linear Inequality *p. 108*

1. *Simplify each side separately. (Clear parentheses, fractions, and decimals and combine like terms.)*
2. *Isolate variable terms on one side. (Add or subtract the same number from each side.)*
3. *Isolate the variable. (Divide each side by the same number; if dividing by a negative number, reverse direction of inequality.)*

Front of Card

Solve $-3(x+4)+2 \geq 7-x$ *and graph the solution set.*

$-3(x+4)+2 \geq 7-x$	*Clear parentheses.*
$-3x-12+2 \geq 7-x$	*Combine like terms.*
$-3x-10 \geq 7-x$	*Both sides are simplified.*
$-3x-10+x \geq 7-x+x$	*Add x to each side.*
$-2x-10 \geq 7$	*Variable term still not isolated.*
$-2x-10+10 \geq 7+10$	*Add 10 to each side.*
$\frac{-2x}{-2} \leq \frac{17}{-2}$	*Divide each side by −2; dividing by negative, reverse direction of inequality symbol.*
$x \leq -\frac{17}{2}$	

$-\frac{17}{2} = -8\frac{1}{2}$

−9 −3 0

Back of Card

Procedure ("Steps") Cards

Write the name of the procedure on the front of the card. Then write each step in words. On the back of the card, put an example showing each step.

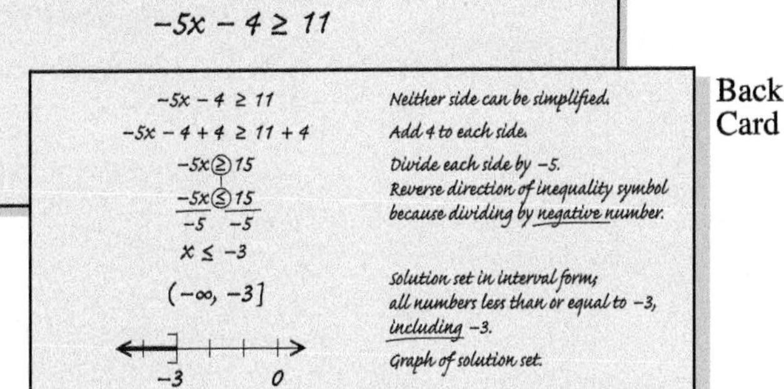

Solve this inequality. Give the solution set in both interval and graph forms.
−5x − 4 ≥ 11
Front of Card
−5x − 4 ≥ 11 Neither side can be simplified.
−5x − 4 + 4 ≥ 11 + 4 Add 4 to each side.
−5x ≥ 15 Divide each side by −5.
−5x/−5 ≤ 15/−5 Reverse direction of inequality symbol because dividing by negative number.
x ≤ −3
(−∞, −3] Solution set in interval form; all numbers less than or equal to −3, including −3.
Graph of solution set.
−3 0
Back of Card

Practice Problem Cards

Write a problem with direction words (like *solve, simplify*) on the front of the card, and work the problem on the back. Make one for each type of problem you learn.

Now TRY THIS

Make a vocabulary card, a procedure card, and a practice problem card for material that you are learning or reviewing.

STUDY SKILL 6

Managing Your Time

Many college students juggle a busy schedule and multiple responsibilities, including school, work, and family demands.

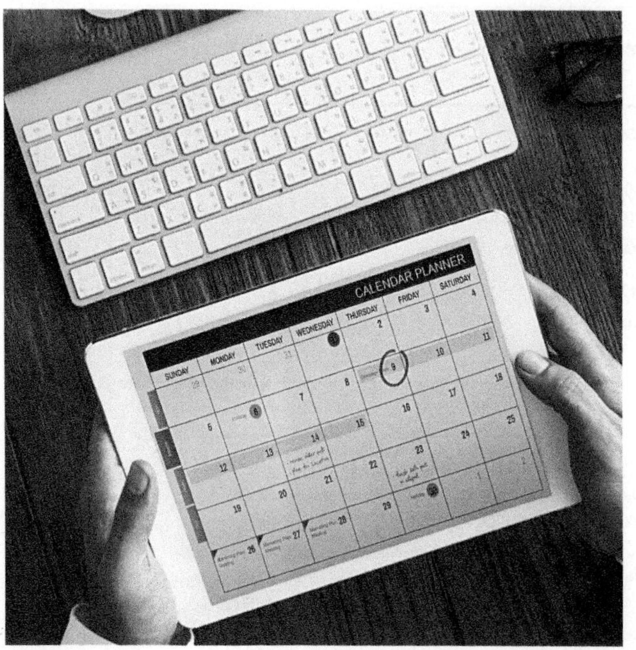

Time Management Tips

- **Read the syllabus for each class.** Understand class policies, such as attendance, late homework, and make-up tests. Find out how you are graded.
- **Make a semester or quarter calendar.** Put test dates and major due dates for *all* your classes on the *same* calendar. Try using a different color for each class.
- **Make a weekly schedule.** After you fill in your classes and other regular responsibilities, block off some study periods. Aim for 2 hours of study for each 1 hour in class.
- **Choose a regular study time and place** (such as the campus library). Routine helps.
- **Keep distractions to a minimum.** Get the most out of the time you have set aside for studying by limiting interruptions. Turn off your cell phone. Take a break from social media. Avoid studying in front of the TV.
- **Make "to-do" lists.** Number tasks in order of importance. To see your progress, cross off tasks as you complete them.
- **Break big assignments into smaller chunks.** Don't wait until the last minute to begin big assignments or to study for tests. Make deadlines for each smaller chunk so that you stay on schedule.
- **Take breaks when studying.** Do not try to study for hours at a time. Take a 10-minute break each hour or so.
- **Ask for help when you need it.** Talk with your instructor during office hours. Make use of the learning/tutoring center, counseling office, or other resources available at your school.

Now TRY THIS

Work through the following, answering any questions.

1. Evaluate when and where you are currently studying. Are these places quiet and comfortable? Are you studying when you are most alert?
2. Which of the above tips will you try this week to improve your time management?
3. Create a weekly calendar that includes your class times, study times, and other family and/or work obligations.
4. Once the week is over, evaluate how these tips worked. Did you use your calendar and stick to it? What will you do differently next week?
5. Ask classmates, friends, and/or family members for tips on how they manage their time. Try any that you think might work for you.

Reviewing a Chapter

Your text provides extensive material to help you prepare for quizzes or tests in this course. Refer to the **Chapter 1 Summary** as you read through the following techniques.

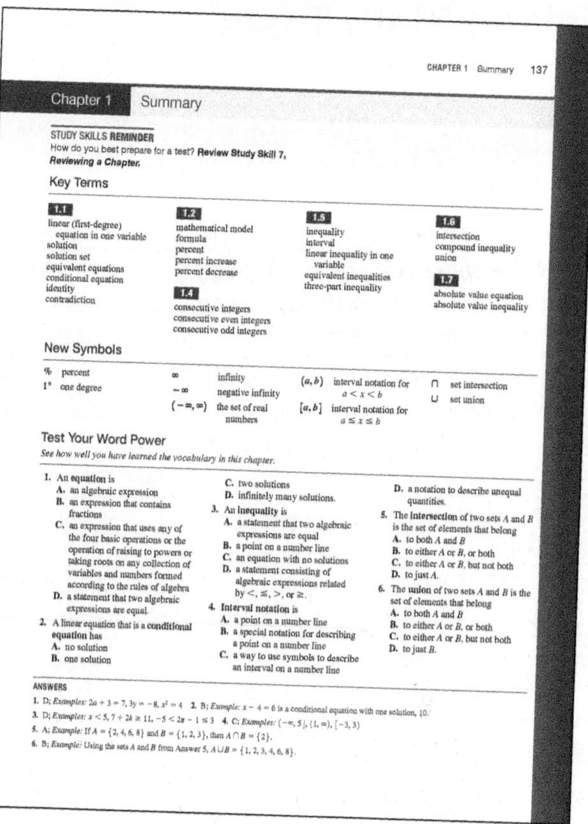

Chapter 1 Summary

STUDY SKILLS **REMINDER**
How do you best prepare for a test? **Review Study Skill 7, *Reviewing a Chapter.***

Key Terms

1.1
linear (first-degree) equation in one variable
solution
solution set
equivalent equations
conditional equation
identity
contradiction

1.2
mathematical model
formula
percent
percent increase
percent decrease

1.4
consecutive integers
consecutive even integers
consecutive odd integers

1.5
inequality
interval
linear inequality in one variable
equivalent inequalities
three-part inequality

1.6
intersection
compound inequality
union

1.7
absolute value equation
absolute value inequality

New Symbols

% percent
$1°$ one degree
∞ infinity
$-\infty$ negative infinity
$(-\infty, \infty)$ the set of real numbers
(a, b) interval notation for $a < x < b$
$[a, b]$ interval notation for $a \leq x \leq b$
$\cap$ set intersection
$\cup$ set union

Test Your Word Power

See how well you have learned the vocabulary in this chapter.

1. An **equation** is
   A. an algebraic expression
   B. an expression that contains fractions
   C. an expression that uses any of the four basic operations or the operation of raising to powers or taking roots on any collection of variables and numbers formed according to the rules of algebra
   D. a statement that two algebraic expressions are equal.
2. A linear equation that is a **conditional equation** has
   A. no solution
   B. one solution
   C. two solutions
   D. infinitely many solutions.
3. An **inequality** is
   A. a statement that two algebraic expressions are equal
   B. a point on a number line
   C. an equation with no solutions
   D. a statement consisting of algebraic expressions related by $<, \leq, >,$ or $\geq$.
4. **Interval notation** is
   A. a point on a number line
   B. a special notation for describing a point on a number line
   C. a way to use symbols to describe an interval on a number line
   D. a notation to describe unequal quantities.
5. The **intersection** of two sets A and B is the set of elements that belong
   A. to both A and B
   B. to either A or B, or both
   C. to either A or B, but not both
   D. to just A.
6. The **union** of two sets A and B is the set of elements that belong
   A. to both A and B
   B. to either A or B, or both
   C. to either A or B, but not both
   D. to just B.

ANSWERS

1. D; *Examples:* $2a + 3 = 7$, $3y = -8$, $x^2 = 4$ 2. B; *Example:* $x - 4 = 6$ is a conditional equation with one solution, 10.
3. D; *Examples:* $x < 5$, $7 + 2k \geq 11$, $-5 < 2z - 1 \leq 3$ 4. C; *Examples:* $(-\infty, 5]$, $(1, \infty)$, $[-3, 3)$
5. A; *Example:* If $A = \{2, 4, 6, 8\}$ and $B = \{1, 2, 3\}$, then $A \cap B = \{2\}$.
6. B; *Example:* Using the sets A and B from Answer 5, $A \cup B = \{1, 2, 3, 4, 6, 8\}$.

Techniques for Reviewing a Chapter

- **Review the Key Terms and any New Symbols.** Make a study card for each. Include a definition, an example, a sketch (if appropriate), and a section or page reference.
- **Take the Test Your Word Power quiz** to check your understanding of new vocabulary. The answers immediately follow.
- **Read the Quick Review.** Pay special attention to the headings. Study the explanations and examples given for each concept. Try to think about the whole chapter.
- **Reread your lecture notes.** Focus on what your instructor has emphasized in class, and review that material in your text.
- **Look over your homework.** Pay special attention to any trouble spots.
- **Work the Review Exercises.** They are grouped by section. Answers are included at the back of the text.
  - Pay attention to direction words, such as *simplify, solve,* and *evaluate*.
  - Are your answers exact and complete? Did you include the correct labels, such as \$, cm^2, ft, etc.?
  - Make study cards for difficult problems.
- **Work the Mixed Review Exercises.** They are in random order. Check your answers in the answer section at the back of the text.
- **Take the Chapter Test under test conditions.**
  - Time yourself.
  - Use a calculator or notes only if your instructor permits them on tests.
  - Take the test in one sitting.
  - Show all your work.
  - Check your answers in the answer section. Section references are provided.

Reviewing a chapter takes time. Avoid rushing through your review in one night. Use the suggestions over a few days or evenings to better understand and remember the material.

Now TRY THIS

Follow these reviewing techniques to prepare for your next test. Then answer each question.

1. How much time did you spend reviewing for your test? Was it enough?
2. Which reviewing techniques worked best for you?
3. Are you investing enough time and effort to really *know* the material and set yourself up for success? Explain.
4. What will you do differently when reviewing for your next test?

STUDY SKILL 8

Taking Math Tests

Techniques to Improve Your Test Score	Comments
Come prepared with a pencil, eraser, paper, and calculator, if allowed.	Working in pencil lets you erase, keeping your work neat.
Scan the entire test, note the point values of different problems, and plan your time accordingly.	To do 20 problems in 50 minutes, allow $50 \div 20 = 2.5$ minutes per problem. Spend less time on easier problems.
Do a "knowledge dump" when you get the test. Write important notes, such as formulas, in a corner of the test for reference.	Writing down tips and other special information that you've learned at the beginning allows you to relax as you take the test.
Read directions carefully, and circle any significant words. When you finish a problem, reread the directions. Did you do what was asked?	Pay attention to any announcements written on the board or made by your instructor. Ask if you don't understand something.
Show all your work. Many teachers give partial credit if some steps are correct, even if the final answer is wrong. ***Write neatly.***	If your teacher can't read your writing, you won't get credit for it. If you need more space to work, ask to use extra paper.
Write down anything that might help solve a problem: a formula, a diagram, etc. If necessary, circle the problem and come back to it later. Do *not* erase anything you wrote down.	If you know even a little bit about a problem, write it down. The answer may come to you as you work on it, or you may get partial credit. Don't spend too long on any one problem.
If you can't solve a problem, make a guess. Do not change it unless you find an obvious mistake.	Have a good reason for changing an answer. Your first guess is usually your best bet.
Check that the answer to an application problem is reasonable and makes sense. Reread the problem to make sure you've answered the question.	Use common sense. Can the father really be seven years old? Would a month's rent be $32,140? Remember to label your answer if needed: $, years, inches, etc.
Check for careless errors. Rework each problem without looking at your previous work. Then compare the two answers.	Reworking a problem from the beginning forces you to rethink it. If possible, use a different method to solve the problem.

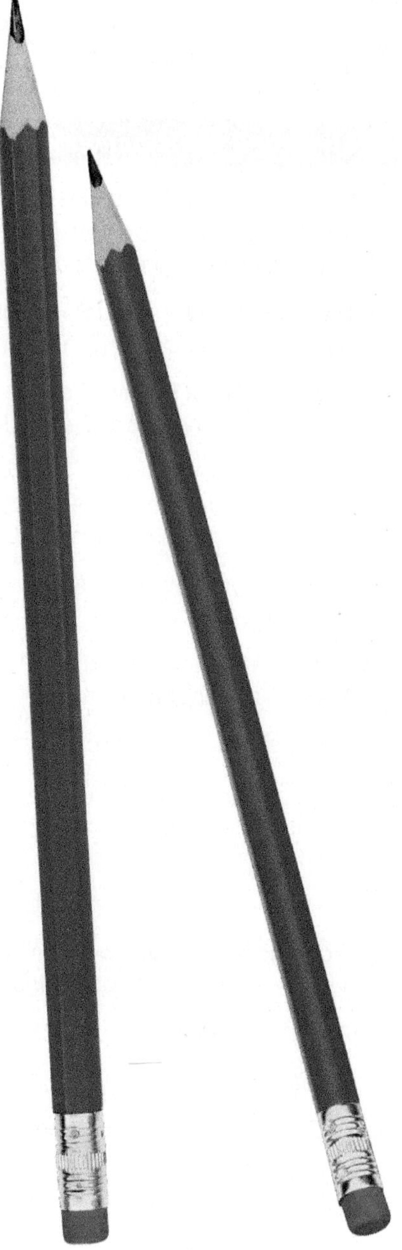

Now TRY THIS

Think through and answer each question.

1. What two or three tips will you try when you take your next math test?
2. How did the tips you selected work for you when you took your math test?
3. What will you do differently when taking your next math test?
4. Ask several classmates how they prepare for math tests. Did you learn any new preparation ideas?

Analyzing Your Test Results

An exam is a learning opportunity—learn from your mistakes. After a test is returned, do the following:

- **Note what you got wrong and why you had points deducted.**
- **Figure out how to solve the problems you missed.** Check your text or notes, or ask your instructor. Rework the problems correctly.
- **Keep all quizzes and tests that are returned to you.** Use them to study for future tests and the final exam.

Typical Reasons for Errors on Math Tests

1. You read the directions wrong.
2. You read the question wrong or skipped over something.
3. You made a computation error.
4. You made a careless error. (For example, you incorrectly copied a correct answer onto a separate answer sheet.)
5. Your answer was not complete.
6. You labeled your answer wrong. (For example, you labeled an answer "ft" instead of "ft^2.")
7. You didn't show your work.

These are test-taking errors. They are easy to correct if you read carefully, show all your work, proofread, and double-check units and labels.

8. You didn't understand a concept.
9. You were unable to set up the problem (in an application).
10. You were unable to apply a procedure.

These are test preparation errors. Be sure to practice all the kinds of problems that you will see on tests.

Now TRY THIS

Work through the following, answering any questions.

1. Use the sample charts at the right to track your test-taking progress. Refer to the tests you have taken so far in your course. For each test, check the appropriate box in the charts to indicate that you made an error in a particular category.
2. What test-taking errors did you make? Do you notice any patterns?
3. What test preparation errors did you make? Do you notice any patterns?
4. What will you do to avoid these kinds of errors on your next test?

▼ Test-Taking Errors

Test	*Read directions wrong*	*Read question wrong*	*Made computation error*	*Made careless error*	*Answer not complete*	*Answer labeled wrong*	*Didn't show work*
1							
2							
3							

▼ Test Preparation Errors

Test	*Didn't understand concept*	*Didn't set up problem correctly*	*Couldn't apply a procedure*
1			
2			
3			

STUDY SKILL 10

Preparing for Your Math Final Exam

Your math final exam is likely to be a comprehensive exam, which means it will cover material from the entire term. **One way to prepare for it now is by working a set of Cumulative Review Exercises** each time your class finishes a chapter. This continual review will help you remember concepts and procedures as you progress through the course.

Final Exam Preparation Suggestions

1. **Figure out the grade you need to earn on the final exam to get the course grade you want.** Check your course syllabus for grading policies, or ask your instructor if you are not sure.
2. **Create a final exam week plan.** Set priorities that allow you to spend extra time studying. This may mean making adjustments, in advance, in your work schedule or enlisting extra help with family responsibilities.
3. **Use the following suggestions to guide your studying.**
   - **Begin reviewing several days before the final exam.** DON'T wait until the last minute.
   - **Know exactly which chapters and sections will be covered on the exam.**
   - **Divide up the chapters.** Decide how much you will review each day.
   - **Keep returned quizzes and tests. Use them to review.**
   - **Practice all types of problems. Use the Cumulative Review Exercises** at the end of each chapter in your text beginning in Chapter 1. All answers, with section references, are given in the answer section at the back of the text.
   - **Review or rewrite your notes** to create summaries of important information.
   - **Make study cards for all types of problems.** Carry the cards with you, and review them whenever you have a few minutes.
   - **Take plenty of short breaks as you study to reduce physical and mental stress.** Exercising, listening to music, and enjoying a favorite activity are effective stress busters.

Finally, *DON'T* stay up all night the night before an exam—*get a good night's sleep.*

Now TRY THIS

Think through and answer each question.

1. How many points do you need to earn on your math final exam to get the grade you want in your course?
2. What adjustments to your usual routine or schedule do you need to make for final exam week? List two or three.
3. Which of the suggestions for studying will you use as you prepare for your math final exam? List two or three.
4. Analyze your final exam results. How will you prepare differently next time?

REVIEW OF THE REAL NUMBER SYSTEM

R.1 Fractions, Decimals, and Percents

OBJECTIVES

1. Write fractions in lowest terms.
2. Convert between improper fractions and mixed numbers.
3. Perform operations with fractions.
4. Write decimals as fractions.
5. Perform operations with decimals.
6. Write fractions as decimals.
7. Write percents as decimals and decimals as percents.
8. Write percents as fractions and fractions as percents.

Recall that **fractions** are a way to represent parts of a whole. See FIGURE 1. In a fraction, the **numerator** gives the number of parts being represented. The **denominator** gives the total number of equal parts in the whole. The fraction bar represents division $\left(\frac{a}{b} = a \div b\right)$.

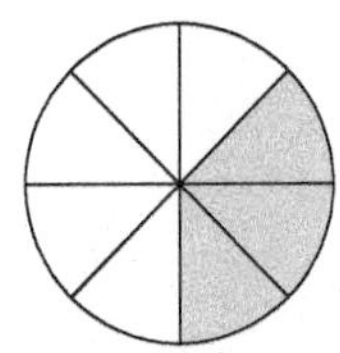

The shaded region represents $\frac{3}{8}$ of the circle.

FIGURE 1

Fraction bar → $\frac{3}{8}$ ← Numerator (3), ← Denominator (8)

A fraction is classified as either a **proper fraction** or an **improper fraction.**

Proper fractions	$\frac{1}{5}, \frac{2}{7}, \frac{9}{10}, \frac{23}{25}$	Numerator **is less than** denominator. Value is less than 1.
Improper fractions	$\frac{3}{2}, \frac{5}{5}, \frac{11}{7}, \frac{28}{4}$	Numerator **is greater than or equal to** denominator. Value is greater than or equal to 1.

OBJECTIVE 1 Write fractions in lowest terms.

A fraction is in **lowest terms** when the numerator and denominator have no factors in common (other than 1).

Writing a Fraction in Lowest Terms

Step 1 Write the numerator and denominator in factored form.

Step 2 Replace each pair of factors common to the numerator and denominator with 1.

Step 3 Multiply the remaining factors in the numerator and in the denominator.

(This procedure is sometimes called **"simplifying the fraction."**)

VOCABULARY

- ☐ fractions
- ☐ numerator
- ☐ denominator
- ☐ proper fraction
- ☐ improper fraction
- ☐ lowest terms
- ☐ mixed number
- ☐ reciprocals
- ☐ decimal
- ☐ terminating decimal
- ☐ repeating decimal
- ☐ percent

EXAMPLE 1 Writing Fractions in Lowest Terms

Write each fraction in lowest terms.

(a) $\frac{10}{15} = \frac{2 \cdot 5}{3 \cdot 5} = \frac{2}{3} \cdot \frac{5}{5} = \frac{2}{3} \cdot 1 = \frac{2}{3}$ 5 is the greatest common factor of 10 and 15.

(b) $\frac{15}{45} = \frac{15}{3 \cdot 15} = \frac{1}{3 \cdot 1} = \frac{1}{3}$ Remember to write 1 in the numerator.

(c) $\frac{150}{200} = \frac{3 \cdot 50}{4 \cdot 50} = \frac{3}{4} \cdot 1 = \frac{3}{4}$ 50 is the greatest common factor of 150 and 200.

Another strategy is to choose *any* common factor and work in stages.

$$\frac{150}{200} = \frac{15 \cdot 10}{20 \cdot 10} = \frac{3 \cdot 5 \cdot 10}{4 \cdot 5 \cdot 10} = \frac{3}{4} \cdot 1 \cdot 1 = \frac{3}{4}$$ The same answer results.

NOW TRY 

NOW TRY EXERCISE 1

Write each fraction in lowest terms.

(a) $\frac{30}{42}$ **(b)** $\frac{10}{70}$ **(c)** $\frac{72}{120}$

OBJECTIVE 2 Convert between improper fractions and mixed numbers.

A **mixed number** is a single number that represents the sum of a natural (counting) number and a proper fraction.

$$\text{Mixed number} \rightarrow 2\frac{3}{4} = 2 + \frac{3}{4}$$

NOW TRY EXERCISE 2

Write $\frac{92}{5}$ as a mixed number.

EXAMPLE 2 Converting an Improper Fraction to a Mixed Number

Write $\frac{59}{8}$ as a mixed number.

Because the fraction bar represents division $\left(\frac{a}{b} = a \div b, \text{ or } b\overline{)a}\right)$, divide the numerator of the improper fraction by the denominator.

$$\begin{array}{r} 7 \\ 8\overline{)59} \\ \underline{56} \\ 3 \end{array}$$

7 ← Quotient
Denominator of fraction → 8)59 ← Numerator of fraction
3 ← Remainder

$$\frac{59}{8} = 7\frac{3}{8}$$

NOW TRY 

NOW TRY EXERCISE 3

Write $11\frac{2}{3}$ as an improper fraction.

EXAMPLE 3 Converting a Mixed Number to an Improper Fraction

Write $6\frac{4}{7}$ as an improper fraction.

Multiply the denominator of the fraction by the natural number, and then add the numerator to obtain the numerator of the improper fraction.

$$7 \cdot 6 = 42 \quad \text{and} \quad 42 + 4 = 46$$

The denominator of the improper fraction is the same as the denominator in the mixed number, which is 7 here.

$$6\frac{4}{7} = \frac{7 \cdot 6 + 4}{7} = \frac{46}{7}$$

NOW TRY ANSWERS

1. (a) $\frac{5}{7}$ (b) $\frac{1}{7}$ (c) $\frac{3}{5}$
2. $18\frac{2}{5}$
3. $\frac{35}{3}$

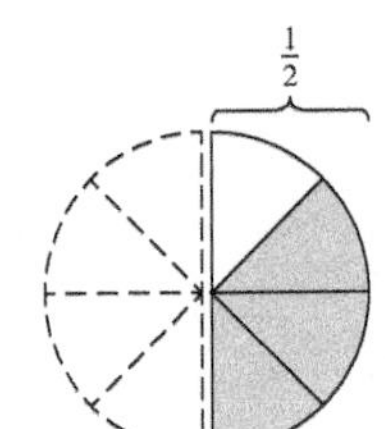

$\frac{3}{4}$ of $\frac{1}{2}$ is equivalent to $\frac{3}{4} \cdot \frac{1}{2}$, which equals $\frac{3}{8}$ of the circle.

FIGURE 2

OBJECTIVE 3 Perform operations with fractions.

FIGURE 2 illustrates multiplying fractions.

Multiplying Fractions

If $\frac{a}{b}$ and $\frac{c}{d}$ are fractions, then $\quad \frac{a}{b} \cdot \frac{c}{d} = \frac{a \cdot c}{b \cdot d}$.

That is, to multiply two fractions, multiply their numerators and then multiply their denominators.

NOW TRY EXERCISE 4

Multiply. Write the answer in lowest terms.

$$\frac{4}{7} \cdot \frac{5}{8}$$

EXAMPLE 4 Multiplying Fractions

Multiply. Write the answer in lowest terms.

$$\frac{3}{8} \cdot \frac{4}{9}$$

$= \frac{3 \cdot 4}{8 \cdot 9}$ Multiply numerators. Multiply denominators.

$= \frac{12}{72}$ Multiply.

$= \frac{1 \cdot 12}{6 \cdot 12}$ The greatest common factor of 12 and 72 is 12.

Make sure the product is in lowest terms. $= \frac{1}{6}$ $\quad \frac{1 \cdot 12}{6 \cdot 12} = \frac{1}{6} \cdot 1 = \frac{1}{6}$

NOW TRY

Two numbers are **reciprocals** of each other if their product is 1. For example, $\frac{3}{4} \cdot \frac{4}{3} = \frac{12}{12}$, or 1. Division is the inverse or opposite of multiplication, and as a result, we use reciprocals to divide fractions. **FIGURE 3** illustrates dividing fractions.

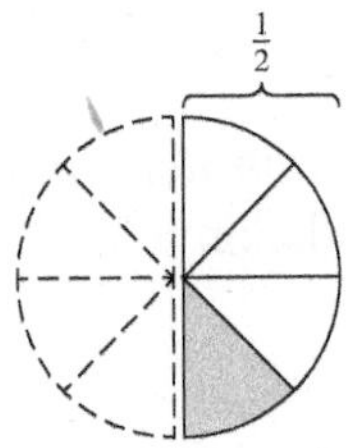

$\frac{1}{2} \div 4$ is equivalent to $\frac{1}{2} \cdot \frac{1}{4}$, which equals $\frac{1}{8}$ of the circle.

FIGURE 3

Dividing Fractions

If $\frac{a}{b}$ and $\frac{c}{d}$ are fractions, then $\quad \frac{a}{b} \div \frac{c}{d} = \frac{a}{b} \cdot \frac{d}{c}$.

That is, to divide by a fraction, multiply by its reciprocal.

EXAMPLE 5 Dividing Fractions

Divide. Write answers in lowest terms as needed.

(a) $\frac{3}{4} \div \frac{8}{5}$

$= \frac{3}{4} \cdot \frac{5}{8}$ Multiply by $\frac{5}{8}$, the reciprocal of $\frac{8}{5}$.

$= \frac{3 \cdot 5}{4 \cdot 8}$ Multiply numerators. Multiply denominators.

$= \frac{15}{32}$ Make sure the answer is in lowest terms.

(b) $\frac{5}{8} \div 10$ Think of 10 as $\frac{10}{1}$ here.

$= \frac{5}{8} \cdot \frac{1}{10}$ Multiply by $\frac{1}{10}$, the reciprocal of 10.

$= \frac{5 \cdot 1}{8 \cdot 2 \cdot 5}$ Multiply and factor.

$= \frac{1}{16}$ Remember to write 1 in the numerator.

NOW TRY ANSWER

4. $\frac{5}{14}$

NOW TRY EXERCISE 5

Divide. Write answers in lowest terms as needed.

(a) $\frac{2}{7} \div \frac{8}{9}$ **(b)** $3\frac{3}{4} \div 4\frac{2}{7}$

(c) $1\frac{2}{3} \div 4\frac{1}{2}$

$= \frac{5}{3} \div \frac{9}{2}$ Write each mixed number as an improper fraction.

$= \frac{5}{3} \cdot \frac{2}{9}$ Multiply by $\frac{2}{9}$, the reciprocal of $\frac{9}{2}$.

$= \frac{10}{27}$ Multiply. The quotient is in lowest terms.

NOW TRY  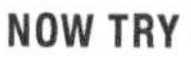

FIGURES 4 and **5** illustrate adding and subtracting fractions.

Adding Fractions

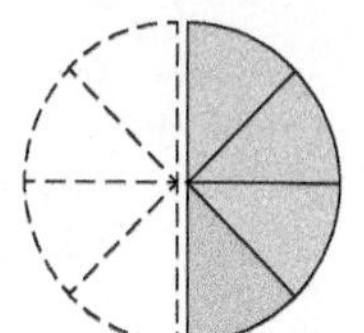

$\frac{3}{8} + \frac{1}{8} = \frac{4}{8} = \frac{1}{2}$

FIGURE 4

Subtracting Fractions

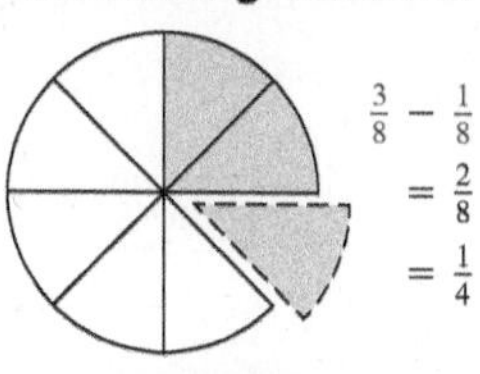

$\frac{3}{8} - \frac{1}{8} = \frac{2}{8} = \frac{1}{4}$

FIGURE 5

Adding and Subtracting Fractions

If $\frac{a}{b}$ and $\frac{c}{b}$ are fractions (where $b \neq 0$), then add or subtract as follows.

$$\frac{a}{b} + \frac{c}{b} = \frac{a + c}{b}$$

$$\frac{a}{b} - \frac{c}{b} = \frac{a - c}{b}$$

That is, to add or subtract two fractions having the same denominator, add or subtract the numerators and keep the same denominator.

If the denominators are different, first find the least common denominator (LCD). Write each fraction as an equivalent fraction with this denominator. Then add or subtract as above.

EXAMPLE 6 Adding and Subtracting Fractions

Add or subtract as indicated. Write answers in lowest terms as needed.

(a) $\frac{2}{10} + \frac{3}{10}$

$= \frac{2 + 3}{10}$ Add numerators. Keep the same denominator.

$= \frac{5}{10}$

$= \frac{1}{2}$ Write in lowest terms.

(b) $\frac{4}{15} + \frac{5}{9}$

$= \frac{4}{15} \cdot \frac{3}{3} + \frac{5}{9} \cdot \frac{5}{5}$ $15 = 3 \cdot 5$ and $45 = 3 \cdot 3 \cdot 5$, so the LCD is $3 \cdot 3 \cdot 5 = 45$.

$= \frac{12}{45} + \frac{25}{45}$ Write equivalent fractions with the common denominator.

$= \frac{37}{45}$ Add numerators. Keep the same denominator.

NOW TRY ANSWERS

5. **(a)** $\frac{9}{28}$ **(b)** $\frac{7}{8}$

NOW TRY EXERCISE 6

Add or subtract as indicated. Write answers in lowest terms as needed.

(a) $\frac{1}{8}+\frac{3}{8}$ **(b)** $\frac{5}{12}+\frac{3}{8}$

(c) $\frac{5}{11}-\frac{2}{9}$ **(d)** $4\frac{1}{3}-2\frac{5}{6}$

(c) $\frac{15}{16}-\frac{4}{9}$

$=\frac{15}{16}\cdot\frac{9}{9}-\frac{4}{9}\cdot\frac{16}{16}$ Because 16 and 9 have no common factors except 1, the LCD is $16 \cdot 9 = 144$.

$=\frac{135}{144}-\frac{64}{144}$ Write equivalent fractions with the common denominator.

$=\frac{71}{144}$ Subtract numerators. Keep the common denominator.

(d) $4\frac{1}{2}-1\frac{3}{4}$

Method 1 $4\frac{1}{2}-1\frac{3}{4}$

$=\frac{9}{2}-\frac{7}{4}$ Write each mixed number as an improper fraction.

$=\frac{18}{4}-\frac{7}{4}$ Find a common denominator. The LCD is 4.

Think: $\frac{9}{2}\cdot\frac{2}{2}=\frac{18}{4}$

$=\frac{11}{4}$, or $2\frac{3}{4}$ Subtract. Write as a mixed number.

Method 2

$$4\frac{1}{2}=4\frac{2}{4}=3\frac{6}{4}$$

The LCD is 4. $4\frac{2}{4}=3+1+\frac{2}{4}=3+\frac{4}{4}+\frac{2}{4}=3\frac{6}{4}$

$$-1\frac{3}{4}=1\frac{3}{4}=1\frac{3}{4}$$

$$2\frac{3}{4}, \text{ or } \frac{11}{4}$$

The same answer results.

NOW TRY 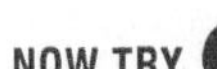

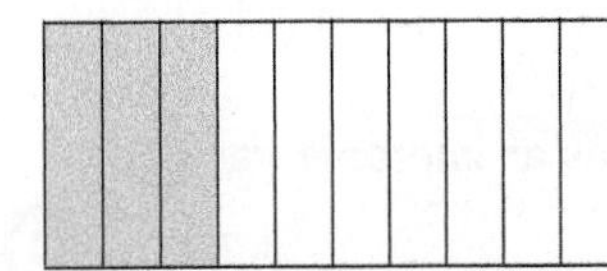

3 parts of the whole 10 are shaded. As a fraction, $\frac{3}{10}$ of the figure is shaded. As a decimal, 0.3 is shaded. Both of these numbers are read *"three-tenths."*

FIGURE 6

Fractions are one way to represent parts of a whole. Another way is with a decimal fraction or **decimal,** a number written with a decimal point.

9.25, 14.001, 0.3 Decimal numbers

See **FIGURE 6**. Each digit in a decimal number has a place value, as shown below.

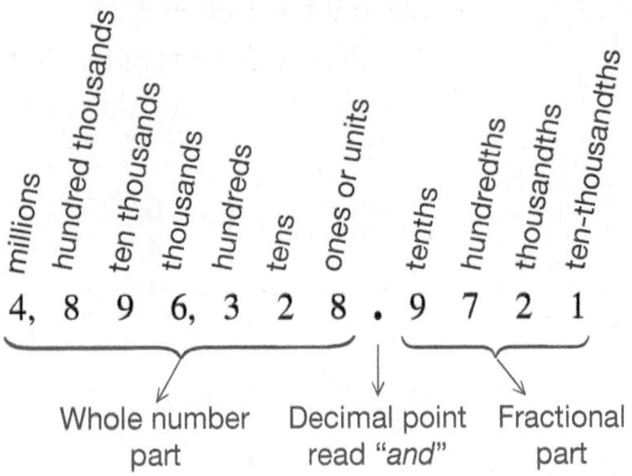

Each successive place value is ten times greater than the place value to its right and one-tenth as great as the place value to its left.

NOW TRY ANSWERS

6. (a) $\frac{1}{2}$ (b) $\frac{19}{24}$ (c) $\frac{23}{99}$
(d) $\frac{3}{2}$, or $1\frac{1}{2}$

OBJECTIVE 4 Write decimals as fractions.

Place value is used to write a decimal number as a fraction.

Converting a Decimal to a Fraction

Read the decimal using the correct place value. Write it in fractional form just as it is read.

- The numerator will be the digits to the right of the decimal point.
- The denominator will be a power of 10—that is, 10 for tenths, 100 for hundredths, and so on.

NOW TRY EXERCISE 7

Write each decimal as a fraction. (Do not write in lowest terms.)

(a) 0.8 **(b)** 0.431 **(c)** 2.58

EXAMPLE 7 Writing Decimals as Fractions

Write each decimal as a fraction. (Do not write in lowest terms.)

(a) 0.95 We read 0.95 as "*ninety-five hundredths.*"

$$0.95 = \frac{95}{100} \leftarrow \text{For hundredths}$$

(b) 0.056 We read 0.056 as "*fifty-six thousandths.*"

$$0.056 = \frac{56}{1000} \leftarrow \text{For thousandths}$$

Do not confuse **0.056** with **0.56**, read "fifty-six *hundredths*," which is the fraction $\frac{56}{100}$.

(c) 4.2095 We read this decimal number, which is greater than 1, as "*Four **and** two thousand ninety-five ten-thousandths.*"

$$4.2095 = 4\frac{2095}{10,000} \quad \text{Write the decimal number as a mixed number.}$$

$$= \frac{42,095}{10,000} \quad \text{Write the mixed number as an improper fraction.}$$

Think: $10,000 \cdot 4 + 2095$

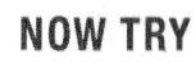

OBJECTIVE 5 Perform operations with decimals.

NOW TRY EXERCISE 8

Add or subtract as indicated.

(a) $68.9 + 42.72 + 8.973$

(b) $351.8 - 2.706$

EXAMPLE 8 Adding and Subtracting Decimals

Add or subtract as indicated.

(a) $6.92 + 14.8 + 3.217$

Place the digits of the decimal numbers in columns by place value. Attach zeros as placeholders so that there are the same number of places to the right of each decimal point.

$$\begin{array}{r} 6.92 \\ 14.8 \\ +\ 3.217 \\ \hline \end{array} \quad \text{becomes} \quad \begin{array}{r} 6.920 \\ 14.800 \\ +\ 3.217 \\ \hline 24.937 \end{array} \quad \text{Attach 0s.}$$

Be sure to line up decimal points.

6.92 is equivalent to 6.920. 14.8 is equivalent to 14.800.

(b) $47.6 - 32.509$

$$\begin{array}{r} 47.6 \\ -\ 32.509 \\ \hline \end{array} \quad \text{becomes} \quad \begin{array}{r} 47.600 \\ -\ 32.509 \\ \hline 15.091 \end{array}$$

Write the decimal numbers in columns, attaching 0s to 47.6.

NOW TRY ANSWERS

7. **(a)** $\frac{8}{10}$ **(b)** $\frac{431}{1000}$ **(c)** $\frac{258}{100}$

8. **(a)** 120.593 **(b)** 349.094

NOW TRY EXERCISE 9

Multiply or divide as indicated.

(a) 9.32×1.4

(b) 0.6×0.004

(c) $7.334 \div 1.3$ (Round the answer to two decimal places.)

CLASSROOM EXAMPLE 9

Multiply or divide as indicated.

(a) 30.2×0.052

(b) 0.06×0.12

(c) $5.476 \div 0.37$

(d) $3.76 \div 3.1$ (Round the answer to one decimal place.)

Answers: (a) 1.5704 (b) 0.0072 (c) 14.8 (d) 1.2

EXAMPLE 9 Multiplying and Dividing Decimals

Multiply or divide as indicated.

(a) 29.3×4.52

Multiply normally. Place the decimal point in the answer as shown.

$$\begin{array}{r} 29.3 \\ \times\ \ 4.52 \\ \hline 586 \\ 1465\ \ \\ 1172\ \ \ \ \\ \hline 132.436 \end{array}$$

29.3 — 1 decimal place

4.52 — 2 decimal places

1 + 2 = 3

132.436 — 3 decimal places

(b) 0.05×0.3

Here $5 \times 3 = 15$. Be careful placing the decimal point.

2 decimal places — 1 decimal place

$0.05 \times 0.3 = 0.015$

Do *not* write 0.150.

2 + 1 = 3 decimal places
Attach 0 as a placeholder in the tenths place.

(c) $8.949 \div 1.25$ (Round the answer to two decimal places.)

$1.25\overline{)8.949}$ Move each decimal point two places to the right.

$$\begin{array}{r} 7.159 \\ 125\overline{)894.900} \\ 875 \\ \hline 199 \\ 125 \\ \hline 740 \\ 625 \\ \hline 1150 \\ 1125 \\ \hline 25 \end{array}$$

Move the decimal point straight up, and divide as with whole numbers. Attach 0s as placeholders.

We carried out the division to three decimal places so that we could round to two decimal places, obtaining the answer 7.16.

NOW TRY

NOTE To round 7.159 in **Example 9(c)** to two decimal places (that is, to the nearest hundredth), we look at the digit to the *right* of the hundredths place.

- **If this digit is 5 or greater, we round up.**
- **If this digit is less than 5, we drop the digit(s) beyond the desired place.**

Hundredths place

7.159 — 9, the digit to the right of the hundredths place, is 5 or greater.

$\approx$ 7.16 — Round 5 up to 6. $\approx$ means *is approximately equal to.*

Multiplying and Dividing by Powers of 10 (Shortcuts)

- To ***multiply*** by a power of 10, ***move the decimal point to the right*** as many places as the number of zeros.
- To ***divide*** by a power of 10, ***move the decimal point to the left*** as many places as the number of zeros.

In both cases, insert 0s as placeholders if necessary.

NOW TRY ANSWERS

9. (a) 13.048 (b) 0.0024 (c) 5.64

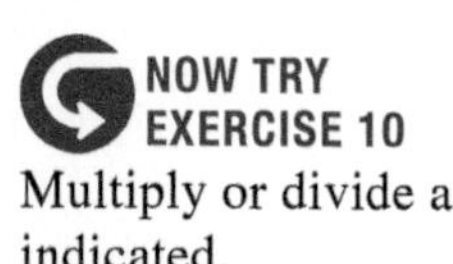

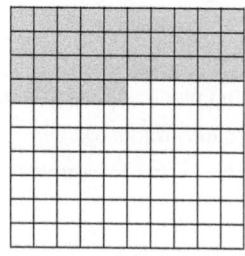

NOW TRY EXERCISE 10

Multiply or divide as indicated.

(a) 294.72×10

(b) $4.793 \div 100$

EXAMPLE 10 Multiplying and Dividing by Powers of 10

Multiply or divide as indicated.

(a) 48.731×10

$= 48.731$
$= 487.31$ Move the decimal point one place to the *right*.

(b) 48.731×1000

$= 48.731$
$= 48{,}731$ Move the decimal point three places to the *right*.

(c) $48.731 \div 10$

$= 48.731$
$= 4.8731$ Move the decimal point one place to the *left*.

(d) $48.731 \div 1000$

$= 048.731$
$= 0.048731$ Move the decimal point three places to the *left*.

NOW TRY

OBJECTIVE 6 Write fractions as decimals.

Converting a Fraction to a Decimal

Because a fraction bar indicates division, write a fraction as a decimal by dividing the numerator by the denominator.

NOW TRY EXERCISE 11

Write each fraction as a decimal. For repeating decimals, write the answer by first using bar notation and then rounding to the nearest thousandth.

(a) $\frac{17}{20}$ **(b)** $\frac{2}{9}$

EXAMPLE 11 Writing Fractions as Decimals

Write each fraction as a decimal.

(a) $\frac{19}{8}$ **(b)** $\frac{2}{3}$

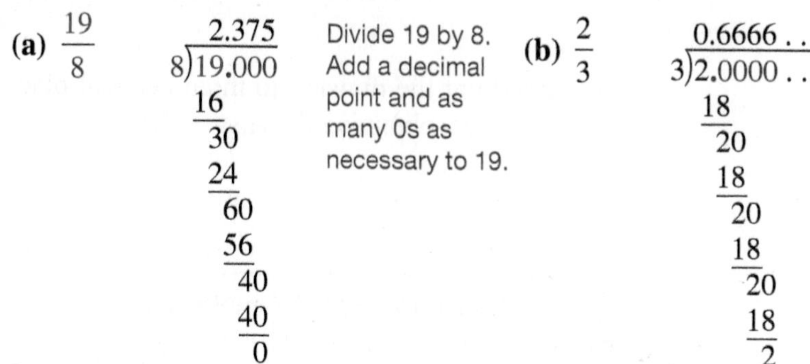

2.375
8)19.000
16
30
24
60
56
40
40
0
Divide 19 by 8. Add a decimal point and as many 0s as necessary to 19.
0.6666 . . .
3)2.0000 . . .
18
20
18
20
18
20
18
2

$\frac{19}{8} = 2.375$ ← Terminating decimal

$\frac{2}{3} = 0.6666\ldots$ ← Repeating decimal

$= 0.\overline{6}$ A bar is written over the repeating digit(s).

≈ 0.667 Nearest thousandth

NOW TRY 

OBJECTIVE 7 Write percents as decimals and decimals as percents.

The word **percent** means ***"per 100."*** Percent is written with the symbol %. ***"One percent" means "one per one hundred," or "one one-hundredth."*** See FIGURE 7.

35 of the 100 squares are shaded. That is, $\frac{35}{100}$, or 35%, of the figure is shaded.

FIGURE 7

Percent, Fraction, and Decimal Equivalents

$$1\% = \frac{1}{100} = 0.01, \quad 10\% = \frac{10}{100} = 0.10, \quad 100\% = \frac{100}{100} = 1$$

NOW TRY ANSWERS

10. **(a)** 2947.2 **(b)** 0.04793

11. **(a)** 0.85 **(b)** $0.\overline{2}$; 0.222

For example, 73% means "73 *per one hundred*."

$$73\% = \frac{73}{100} = 0.73$$

Essentially, we are dropping the % symbol from 73% and dividing 73 by 100. Doing this moves the decimal point, which is understood to be after the 3, two places to the *left*.

Writing 0.73 as a percent is the opposite process.

$$0.73 = 0.73 \cdot 100\% = 73\% \qquad 100\% = 1$$

Moving the decimal point two places to the *right* and attaching a % symbol give the same result.

Converting Percents and Decimals (Shortcuts)

- To convert a percent to a decimal, move the decimal point two places to the *left* and drop the % symbol.
- To convert a decimal to a percent, move the decimal point two places to the *right* and attach a % symbol.

Drop %. *Divide* by 100.
(Move decimal point two places *left*.)

Decimal **0.85** ← → **Percent** **85%**

Attach %. *Multiply* by 100.
(Move decimal point two places *right*.)

NOW TRY EXERCISE 12
Convert each percent to a decimal and each decimal to a percent.
(a) 52% **(b)** 2%
(c) 0.45 **(d)** 3.5

EXAMPLE 12 Converting Percents and Decimals by Moving the Decimal Point

Convert each percent to a decimal and each decimal to a percent.

(a) 45% = 0.45 **(b)** 250% = 2.50, or 2.5 **(c)** 9% = 09% = 0.09

(d) 0.57 = 57% **(e)** 1.5 = 1.50 = 150% **(f)** 0.007 = 0.7%

NOW TRY

OBJECTIVE 8 Write percents as fractions and fractions as percents.

EXAMPLE 13 Writing Percents as Fractions

Write each percent as a fraction. Give answers in lowest terms.

(a) 8%

We use the fact that *percent* means "*per one hundred*," and convert as follows.

$$8\% = \frac{8}{100} = \frac{2 \cdot 4}{25 \cdot 4} = \frac{2}{25}$$

As with converting percents to decimals, drop the % symbol and divide by 100.

Write in lowest terms.

(b) $175\% = \frac{175}{100} = \frac{7 \cdot 25}{4 \cdot 25} = \frac{7}{4}$, or $1\frac{3}{4}$ ← A number greater than 1 is more than 100%.

Write in lowest terms.

NOW TRY ANSWERS
12. **(a)** 0.52 **(b)** 0.02
(c) 45% **(d)** 350%

NOW TRY EXERCISE 13

Write each percent as a fraction. Give answers in lowest terms.

(a) 20% **(b)** 160%

(c) 13.5%

$= \frac{13.5}{100}$ Drop the % symbol. Divide by 100.

$= \frac{13.5}{100} \cdot \frac{10}{10}$ Multiply by 1 in the form $\frac{10}{10}$ to eliminate the decimal in the numerator.

$= \frac{135}{1000}$ Multiply.

$= \frac{27}{200}$ Write in lowest terms. $\frac{135}{1000} = \frac{27 \cdot 5}{200 \cdot 5} = \frac{27}{200}$

NOW TRY

We know that 100% of something is the whole thing. One way to convert a fraction to a percent is to multiply by 100%, which is equivalent to 1. This involves the same steps that are used for converting a decimal to a percent—***attach a % symbol and multiply by 100.***

NOW TRY EXERCISE 14

Write each fraction as a percent.

(a) $\frac{6}{25}$ **(b)** $\frac{7}{9}$

EXAMPLE 14 Writing Fractions as Percents

Write each fraction as a percent.

(a) $\frac{2}{5}$

$= \frac{2}{5} \cdot 100\%$ Multiply by 1 in the form 100%.

$= \frac{2}{5} \cdot \frac{100}{1}\%$

$= \frac{2 \cdot 5 \cdot 20}{5 \cdot 1}\%$ Multiply and factor.

$= \frac{2 \cdot 20}{1}\%$ Divide out the common factor.

$= 40\%$ Simplify.

(b) $\frac{1}{6}$

$= \frac{1}{6} \cdot 100\%$

$= \frac{1}{6} \cdot \frac{100}{1}\%$

$= \frac{1 \cdot 2 \cdot 50}{2 \cdot 3 \cdot 1}\%$

$= \frac{50}{3}\%$

$= 16\frac{2}{3}\%$, or $16.\overline{6}\%$

NOW TRY

NOW TRY ANSWERS

13. **(a)** $\frac{1}{5}$ **(b)** $\frac{8}{5}$, or $1\frac{3}{5}$

14. **(a)** 24% **(b)** $77.\overline{7}\%$

R.1 Exercises

MyLab Math

Video solutions for select problems available in MyLab Math

Concept Check *Decide whether each statement is* true *or* false. *If it is false, explain why.*

1. In the fraction $\frac{5}{8}$, 5 is the numerator and 8 is the denominator.

2. The mixed number equivalent of the improper fraction $\frac{31}{5}$ is $6\frac{1}{5}$.

3. The fraction $\frac{7}{7}$ is proper.

4. The reciprocal of $\frac{6}{2}$ is $\frac{3}{1}$.

Concept Check *Choose the letter of the correct response.*

5. Which choice shows the correct way to write $\frac{16}{24}$ in lowest terms?

A. $\frac{16}{24} = \frac{8+8}{8+16} = \frac{8}{16} = \frac{1}{2}$

B. $\frac{16}{24} = \frac{4 \cdot 4}{4 \cdot 6} = \frac{4}{6}$

C. $\frac{16}{24} = \frac{8 \cdot 2}{8 \cdot 3} = \frac{2}{3}$

6. Which fraction is *not* equal to $\frac{5}{9}$?

A. $\frac{15}{27}$ B. $\frac{30}{54}$ C. $\frac{40}{74}$ D. $\frac{55}{99}$

Write each fraction in lowest terms. ***See Example 1.***

7. $\frac{8}{16}$ 8. $\frac{4}{12}$ 9. $\frac{15}{18}$ 10. $\frac{16}{20}$ 11. $\frac{90}{150}$

12. $\frac{100}{140}$ 13. $\frac{18}{90}$ 14. $\frac{16}{64}$ 15. $\frac{144}{120}$ 16. $\frac{132}{77}$

Write each improper fraction as a mixed number. ***See Example 2.***

17. $\frac{12}{7}$ 18. $\frac{16}{9}$ 19. $\frac{77}{12}$ 20. $\frac{67}{13}$

Write each mixed number as an improper fraction. ***See Example 3.***

21. $2\frac{3}{5}$ 22. $5\frac{6}{7}$ 23. $12\frac{2}{3}$ 24. $10\frac{1}{5}$

Multiply or divide as indicated. Write answers in lowest terms as needed. ***See Examples 4 and 5.***

25. $\frac{4}{5} \cdot \frac{6}{7}$ 26. $\frac{5}{9} \cdot \frac{2}{7}$ 27. $\frac{2}{15} \cdot \frac{3}{8}$ 28. $\frac{3}{20} \cdot \frac{5}{21}$

29. $\frac{1}{10} \cdot \frac{12}{5}$ 30. $\frac{1}{8} \cdot \frac{10}{7}$ 31. $\frac{15}{4} \cdot \frac{8}{25}$ 32. $\frac{21}{8} \cdot \frac{4}{7}$

33. $21 \cdot \frac{3}{7}$ 34. $36 \cdot \frac{4}{9}$ 35. $3\frac{1}{4} \cdot 1\frac{2}{3}$ 36. $2\frac{2}{3} \cdot 1\frac{3}{5}$

37. $\frac{7}{9} \div \frac{3}{2}$ 38. $\frac{6}{11} \div \frac{5}{4}$ 39. $\frac{5}{4} \div \frac{3}{8}$ 40. $\frac{7}{5} \div \frac{3}{10}$

41. $\frac{32}{5} \div \frac{8}{15}$ 42. $\frac{24}{7} \div \frac{6}{21}$ 43. $\frac{3}{4} \div 12$ 44. $\frac{2}{5} \div 30$

45. $6 \div \frac{3}{5}$ 46. $8 \div \frac{4}{9}$ 47. $6\frac{3}{4} \div \frac{3}{8}$ 48. $5\frac{3}{5} \div \frac{7}{10}$

49. $2\frac{1}{2} \div 1\frac{5}{7}$ 50. $2\frac{2}{9} \div 1\frac{2}{5}$ 51. $2\frac{5}{8} \div 1\frac{15}{32}$ 52. $2\frac{3}{10} \div 1\frac{4}{5}$

Add or subtract as indicated. Write answers in lowest terms as needed. ***See Example 6.***

53. $\frac{7}{15} + \frac{4}{15}$ 54. $\frac{2}{9} + \frac{5}{9}$ 55. $\frac{7}{12} + \frac{1}{12}$ 56. $\frac{3}{16} + \frac{5}{16}$

57. $\frac{5}{9} + \frac{1}{3}$ 58. $\frac{4}{15} + \frac{1}{5}$ 59. $\frac{3}{8} + \frac{5}{6}$ 60. $\frac{5}{6} + \frac{2}{9}$

61. $\frac{5}{9} + \frac{3}{16}$ 62. $\frac{3}{4} + \frac{6}{25}$ 63. $3\frac{1}{8} + 2\frac{1}{4}$ 64. $4\frac{2}{3} + 2\frac{1}{6}$

65. $\frac{7}{9} - \frac{2}{9}$ 66. $\frac{8}{11} - \frac{3}{11}$ 67. $\frac{13}{15} - \frac{3}{15}$ 68. $\frac{11}{12} - \frac{3}{12}$

69. $\frac{7}{12} - \frac{1}{3}$ 70. $\frac{5}{6} - \frac{1}{2}$ 71. $\frac{7}{12} - \frac{1}{9}$ 72. $\frac{11}{16} - \frac{1}{12}$

73. $4\frac{3}{4} - 1\frac{2}{5}$ 74. $3\frac{4}{5} - 1\frac{4}{9}$ 75. $8\frac{2}{9} - 4\frac{2}{3}$ 76. $7\frac{5}{12} - 4\frac{5}{6}$

Work each problem involving fractions.

77. For each description, write a fraction in lowest terms that represents the region described.

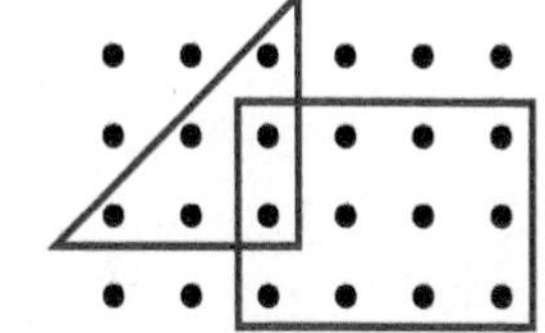

(a) The dots in the rectangle as a part of the dots in the entire figure

(b) The dots in the triangle as a part of the dots in the entire figure

(c) The dots in the overlapping region of the triangle and the rectangle as a part of the dots in the triangle alone

(d) The dots in the overlapping region of the triangle and the rectangle as a part of the dots in the rectangle alone

78. At the conclusion of the Pearson softball league season, batting statistics for five players were as shown in the table.

Player	*At-Bats*	*Hits*	*Home Runs*
Maureen	36	12	3
Christine	40	9	2
Chase	11	5	1
Joe	16	8	0
Greg	20	10	2

Use this information to answer each question. Estimate as necessary.

(a) Which player got a hit in exactly $\frac{1}{3}$ of his or her at-bats?

(b) Which player got a home run in just less than $\frac{1}{10}$ of his or her at-bats?

(c) Which player got a hit in just less than $\frac{1}{4}$ of his or her at-bats?

(d) Which two players got hits in exactly the same fractional part of their at-bats? What was that fractional part, expressed in lowest terms?

Concept Check *Provide the correct response.*

79. In the decimal number 367.9412, name the digit that has each place value.

(a) tens **(b)** tenths **(c)** thousandths **(d)** ones or units **(e)** hundredths

80. Write a decimal number that has 5 in the thousands place, 0 in the tenths place, and 4 in the ten-thousandths place.

81. For the decimal number 46.249, round to the place value indicated.

(a) hundredths **(b)** tenths **(c)** ones or units **(d)** tens

82. Round each decimal to the nearest thousandth.

(a) $0.\overline{8}$ **(b)** $0.\overline{4}$ **(c)** 0.9762 **(d)** 0.8645

Write each decimal as a fraction. (Do not write in lowest terms.) ***See Example 7.***

83. 0.4 **84.** 0.6 **85.** 0.64 **86.** 0.82 **87.** 0.138

88. 0.104 **89.** 0.043 **90.** 0.087 **91.** 3.805 **92.** 5.166

Add or subtract as indicated. ***See Example 8.***

93. $25.32 + 109.2 + 8.574$ **94.** $90.527 + 32.43 + 589.8$ **95.** $28.73 - 3.12$

96. $46.88 - 13.45$ **97.** $43.5 - 28.17$ **98.** $345.1 - 56.31$

99. 3.87 + 15 + 2.9 **100.** 8.2 + 1.09 + 12 **101.** 32.56 + 47.356 + 1.8

102. 75.2 + 123.96 + 3.897 **103.** 18 − 2.789 **104.** 29 − 8.582

Multiply or divide as indicated. ***See Example 9.***

105. 12.8 × 9.1 **106.** 34.04 × 0.56 **107.** 22.41 × 33 **108.** 55.76 × 72

109. 0.2 × 0.03 **110.** 0.07 × 0.004 **111.** 78.65 ÷ 11 **112.** 73.36 ÷ 14

113. 32.48 ÷ 11.6 **114.** 85.26 ÷ 17.4 **115.** 19.967 ÷ 9.74 **116.** 44.4788 ÷ 5.27

Multiply or divide as indicated. ***See Example 10.***

117. 123.26 × 10 **118.** 785.91 × 10 **119.** 57.116 × 100 **120.** 82.053 × 100

121. 0.094 × 1000 **122.** 0.025 × 1000 **123.** 1.62 ÷ 10 **124.** 8.04 ÷ 10

125. 124.03 ÷ 100 **126.** 490.35 ÷ 100 **127.** 23.29 ÷ 1000 **128.** 59.8 ÷ 1000

Concept Check *Complete the table of fraction, decimal, and percent equivalents.*

	Fraction in Lowest Terms (or Whole Number)	Decimal	Percent
129.	$\frac{1}{100}$	0.01	
130.	$\frac{1}{50}$		2%
131.		0.05	5%
132.	$\frac{1}{10}$		
133.	$\frac{1}{8}$	0.125	
134.			20%
135.	$\frac{1}{4}$		
136.	$\frac{1}{3}$		
137.			50%
138.	$\frac{2}{3}$		$66\frac{2}{3}$%, or $66.\overline{6}$%
139.		0.75	
140.	1	1.0	

Write each fraction as a decimal. For repeating decimals, write the answer by first using bar notation and then rounding to the nearest thousandth. ***See Example 11.***

141. $\frac{21}{5}$ **142.** $\frac{9}{5}$ **143.** $\frac{9}{4}$ **144.** $\frac{15}{4}$ **145.** $\frac{3}{8}$

146. $\frac{7}{8}$ **147.** $\frac{5}{9}$ **148.** $\frac{8}{9}$ **149.** $\frac{1}{6}$ **150.** $\frac{5}{6}$

Write each percent as a decimal. ***See Examples 12(a)–12(c).***

151. 54% **152.** 39% **153.** 7% **154.** 4% **155.** 117%

156. 189% **157.** 2.4% **158.** 3.1% **159.** $6\frac{1}{4}$% **160.** $5\frac{1}{2}$%

Write each decimal as a percent. ***See Examples 12(d)–12(f).***

161. 0.79 **162.** 0.83 **163.** 0.02 **164.** 0.08 **165.** 0.004

166. 0.005 **167.** 1.28 **168.** 2.35 **169.** 0.4 **170.** 0.6

Write each percent as a fraction. Give answers in lowest terms. **See Example 13.**

171. 51% **172.** 47% **173.** 15% **174.** 35% **175.** 2%

176. 8% **177.** 140% **178.** 180% **179.** 7.5% **180.** 2.5%

Write each fraction as a percent. **See Example 14.**

181. $\frac{4}{5}$ **182.** $\frac{3}{25}$ **183.** $\frac{7}{50}$ **184.** $\frac{9}{20}$ **185.** $\frac{2}{11}$

186. $\frac{4}{9}$ **187.** $\frac{9}{4}$ **188.** $\frac{8}{5}$ **189.** $\frac{13}{6}$ **190.** $\frac{31}{9}$

R.2 Basic Concepts from Algebra

OBJECTIVES

1. Write sets using set notation.
2. Use number lines.
3. Classify numbers.
4. Find additive inverses.
5. Use absolute value.
6. Use inequality symbols.

VOCABULARY

- ☐ set
- ☐ elements (members)
- ☐ finite set
- ☐ natural (counting) numbers
- ☐ infinite set
- ☐ whole numbers
- ☐ empty (null) set
- ☐ variable
- ☐ number line
- ☐ integers
- ☐ coordinate
- ☐ graph
- ☐ rational numbers
- ☐ irrational numbers
- ☐ real numbers
- ☐ additive inverse (opposite, negative)
- ☐ signed numbers
- ☐ absolute value
- ☐ equation
- ☐ inequality

OBJECTIVE 1 Write sets using set notation.

A **set** is a collection of objects called the **elements,** or **members**, of the set. In algebra, the elements of a set are usually numbers. Set braces, $\{\ \}$, are used to enclose the elements. For example,

$$2 \text{ is an element of the set } \{1, 2, 3\}.$$

Because we can count the number of elements in the set $\{1, 2, 3\}$ and the counting comes to an end, it is a **finite set.**

In algebra, we refer to certain sets of numbers by name. The set

$$N = \{1, 2, 3, 4, 5, 6, \ldots\} \quad \text{Natural (counting) numbers}$$

is the **natural numbers,** or the **counting numbers.** The three dots (*ellipsis* points) show that the list continues in the same pattern indefinitely. We cannot list all of the elements of the set of natural numbers, so it is an **infinite set.**

Including 0 with the set of natural numbers gives the set of **whole numbers.**

$$W = \{0, 1, 2, 3, 4, 5, 6, \ldots\} \quad \text{Whole numbers}$$

The set containing no elements is the **empty set,** or **null set,** usually written $\varnothing$. For example, the set of whole numbers less than 0 is $\varnothing$.

! CAUTION Do not write $\{\varnothing\}$ for the empty set. $\{\varnothing\}$ is a set with one element: $\varnothing$. Use the notation $\varnothing$ for the empty set.

To indicate that 2 is an element of the set $\{1, 2, 3\}$, we use the symbol $\in$, which is read "*is an element of.*"

$$2 \in \{1, 2, 3\}$$

The number 2 is also an element of the set of natural numbers N.

$$2 \in N$$

To show that 0 is *not* an element of set N, we draw a slash through the symbol $\in$.

$$0 \notin N$$

Two sets are equal if they contain exactly the same elements. For example,

$$\{1, 2\} = \{2, 1\}. \quad \text{Order does not matter.}$$

To indicate that two sets are not equal, we use the symbol $\neq$, which is read "*is not equal to*." For example,

$$\{1, 2\} \neq \{0, 1, 2\}$$

because one set contains the element 0 and the other does not.

A **variable** is a symbol, usually a letter, used to represent a number or to define a set of numbers. For example,

$$\{x \mid x \text{ is a natural number between 3 and 15}\}$$

(read "*the set of all elements x such that x is a natural number between* 3 *and* 15") defines the following set.

$$\{4, 5, 6, 7, \ldots, 14\}$$

The notation $\{x \mid x \text{ is a natural number between 3 and 15}\}$ is an example of **set-builder notation.**

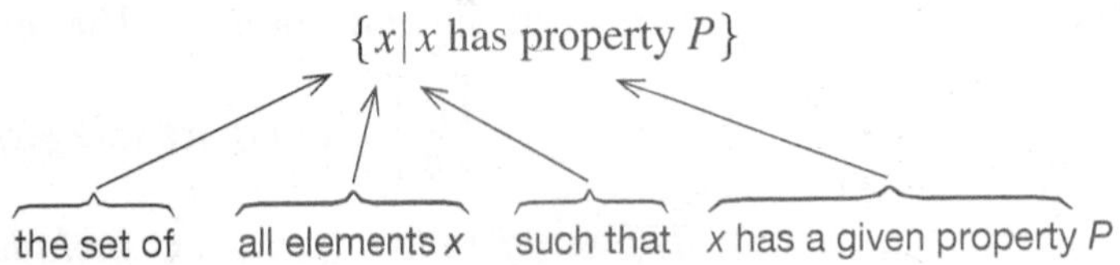

NOW TRY EXERCISE 1
List the elements in the set.

$\{p \mid p \text{ is a natural number less than } 6\}$

EXAMPLE 1 Listing the Elements in Sets

List the elements in each set.

(a) $\{x \mid x \text{ is a natural number less than } 4\}$

The natural numbers less than 4 are 1, 2, and 3. This set is $\{1, 2, 3\}$.

(b) $\{x \mid x \text{ is one of the first five even natural numbers}\}$ is $\{2, 4, 6, 8, 10\}$.

(c) $\{x \mid x \text{ is a natural number greater than or equal to } 7\}$

The set of natural numbers greater than or equal to 7 is an infinite set, which is written with ellipsis points as

$$\{7, 8, 9, 10, \ldots\}.$$

NOW TRY 

NOW TRY EXERCISE 2
Use set-builder notation to describe the set.

$\{9, 10, 11, 12\}$

EXAMPLE 2 Using Set-Builder Notation to Describe Sets

Use set-builder notation to describe each set.

(a) $\{1, 3, 5, 7, 9\}$

There are often several ways to describe a set in set-builder notation. One way to describe the given set is

$$\{x \mid x \text{ is one of the first five odd natural numbers}\}.$$

(b) $\{5, 10, 15, \ldots\}$

This set can be described as $\{x \mid x \text{ is a positive multiple of } 5\}$. NOW TRY

NOW TRY ANSWERS
1. $\{1, 2, 3, 4, 5\}$
2. One answer is $\{x \mid x \text{ is a natural number between 8 and 13}\}$.

OBJECTIVE 2 Use number lines.

A good way to picture a set of numbers is to use a **number line.** See FIGURE 8.

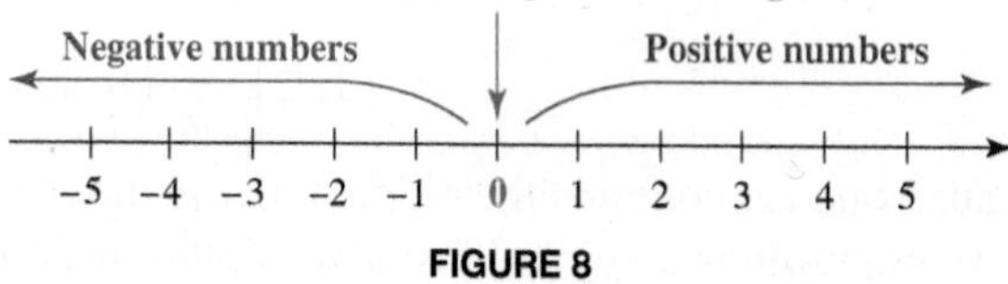

FIGURE 8

To draw a number line, choose any point on the line and label it 0. Then choose any point to the right of 0 and label it 1. Use the distance between 0 and 1 as the scale to locate, and then label, other points.

The set of numbers identified on the number line in FIGURE 8, including positive and negative numbers and 0, is part of the set of **integers.**

$$I = \{\ldots, -3, -2, -1, 0, 1, 2, 3, \ldots\} \quad \text{Integers}$$

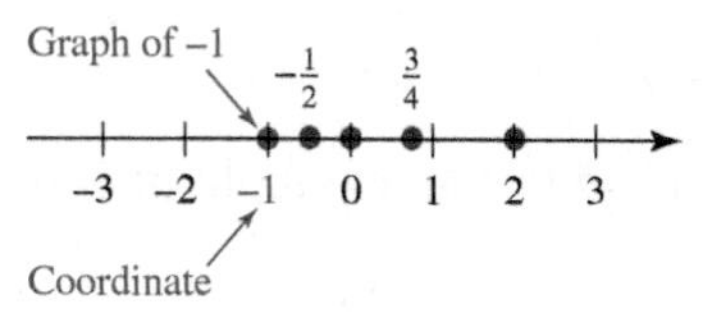

FIGURE 9

Each number on a number line is the **coordinate** of the point that it labels, while the point is the **graph** of the number. FIGURE 9 shows a number line with several points graphed on it.

The fractions $-\frac{1}{2}$ and $\frac{3}{4}$, graphed on the number line in FIGURE 9, are examples of *rational numbers*. A **rational number** can be expressed as the quotient of two integers, with denominator not 0. The set of all rational numbers is written as follows.

$$\left\{\frac{p}{q} \,\middle|\, p \text{ and } q \text{ are integers, where } q \neq 0\right\} \quad \text{Rational numbers}$$

The set of rational numbers includes the natural numbers, whole numbers, and integers because these numbers can be written as fractions.

$$14 = \frac{14}{1}, \quad -3 = \frac{-3}{1}, \quad \text{and} \quad 0 = \frac{0}{1} \quad \text{Rational numbers}$$

A rational number written as a fraction, such as $\frac{1}{8}$ or $\frac{2}{3}$, can also be expressed as a decimal by dividing the numerator by the denominator.

$8\overline{)1.000}$ = 0.125 ← Terminating decimal (rational number) 8 20 16 40 40 0 ← Remainder is 0. $\frac{1}{8} = 0.125$	$3\overline{)2.000\ldots}$ = 0.666 . . . ← Repeating decimal (rational number) 18 20 18 20 18 2 ← Remainder is never 0. $\frac{2}{3} = 0.\overline{6}$ ← A bar is written over the repeating digit(s).

Thus terminating decimals, such as $0.125 = \frac{1}{8}$, $0.8 = \frac{4}{5}$, and $2.75 = \frac{11}{4}$, and decimals that have a repeating block of digits, such as $0.\overline{6} = \frac{2}{3}$ and $0.\overline{27} = \frac{3}{11}$, are rational numbers.

Decimal numbers that neither terminate nor repeat, which include many square roots, are **irrational numbers.**

$$\sqrt{2} = 1.414213562\ldots \quad \text{and} \quad -\sqrt{7} = -2.6457513\ldots \quad \text{Irrational numbers}$$

NOTE Some square roots, such as $\sqrt{16} = 4$ and $\sqrt{\frac{9}{25}} = \frac{3}{5}$, are rational.

A decimal number such as 0.010010001 . . . has a pattern, but it is irrational because there is no fixed block of digits that repeat. Another irrational number is π. See **FIGURE 10.**

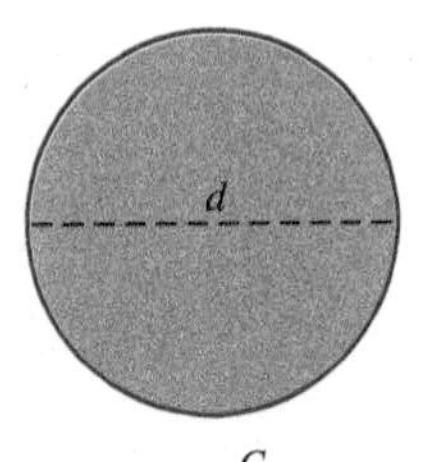

FIGURE 10

Some rational and irrational numbers are graphed on the number line in **FIGURE 11.** The rational numbers together with the irrational numbers make up the set of **real numbers.**

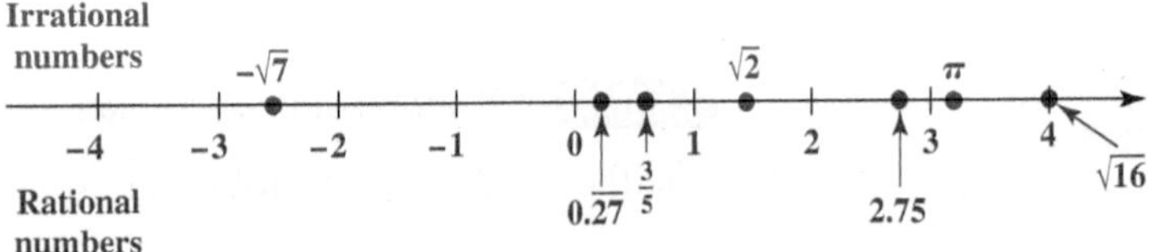

FIGURE 11

Every point on a number line corresponds to a real number, and every real number corresponds to a point on the number line.

OBJECTIVE 3 Classify numbers.

Sets of Numbers

Natural numbers, or counting numbers	$\{1, 2, 3, 4, 5, 6, \ldots\}$
Whole numbers	$\{0, 1, 2, 3, 4, 5, 6, \ldots\}$
Integers	$\{\ldots, -3, -2, -1, 0, 1, 2, 3, \ldots\}$
Rational numbers	$\left\{\frac{p}{q} \mid p \text{ and } q \text{ are integers, where } q \neq 0\right\}$ *Examples:* $\frac{4}{1}$, 1.3, $-\frac{9}{2}$ or $-4\frac{1}{2}$, $\frac{16}{8}$ or 2, $\sqrt{9}$ or 3, $0.\overline{6}$
Irrational numbers	$\{x \mid x$ **is a real number that cannot be represented by a terminating or repeating decimal**$\}$ *Examples:* $\sqrt{3}$, $-\sqrt{2}$, π, 0.010010001 . . .
Real numbers	$\{x \mid x$ **is a rational or an irrational number**$\}$*

*An example of a number that is not real is $\sqrt{-1}$. This number is part of the *complex number system* and is discussed later in the text.

FIGURE 12 shows the set of real numbers. ***Every real number is either rational or irrational.*** Notice that the integers are elements of the set of rational numbers and that the whole numbers and natural numbers are elements of the set of integers.

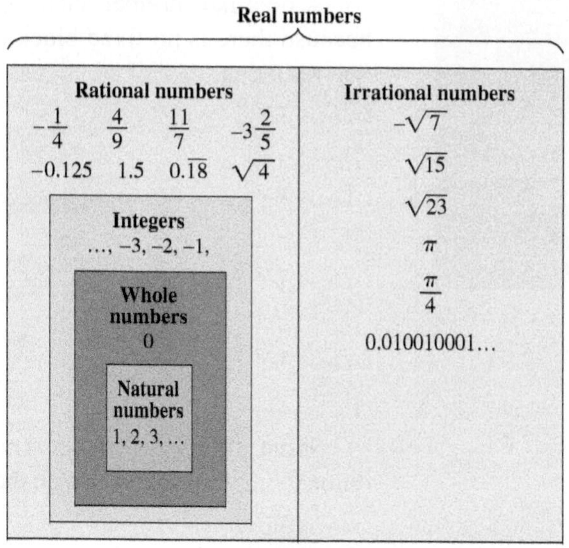

FIGURE 12

NOW TRY EXERCISE 3

List the numbers in the following set that are elements of each set.

$$\left\{-2.4, -\sqrt{1}, -\frac{1}{2}, 0, 0.\overline{3}, \sqrt{5}, \pi, 5\right\}$$

(a) Whole numbers

(b) Rational numbers

EXAMPLE 3 Identifying Examples of Number Sets

List the numbers in the following set that are elements of each set.

$$\left\{-8, -\sqrt{2}, -\frac{9}{64}, 0, 0.5, \frac{1}{3}, 1.\overline{12}, \sqrt{3}, 2, \pi\right\}$$

(a) Integers

-8, 0, and 2

(b) Rational numbers

$-8, -\frac{9}{64}, 0, 0.5, \frac{1}{3}, 1.\overline{12}$, and 2

(c) Irrational numbers

$-\sqrt{2}$, $\sqrt{3}$, and π

(d) Real numbers

All are real numbers.

NOW TRY 

NOW TRY EXERCISE 4

Determine whether each statement is *true* or *false*. If it is false, tell why.

(a) All integers are irrational numbers.

(b) Every whole number is an integer.

EXAMPLE 4 Determining Relationships between Sets of Numbers

Determine whether each statement is *true* or *false*. If it is false, tell why.

(a) All irrational numbers are real numbers.

This statement is true. As shown in FIGURE 12, the set of real numbers includes all irrational numbers.

(b) Every rational number is an integer.

This statement is false. Although some rational numbers are integers, other rational numbers, such as $\frac{2}{3}$ and $-\frac{1}{4}$, are not.

NOW TRY

NOW TRY ANSWERS

3. (a) 0, 5
(b) $-2.4, -\sqrt{1}, -\frac{1}{2}, 0, 0.\overline{3}, 5$
$(-\sqrt{1} = -1)$

4. (a) false; All integers are rational numbers.
(b) true

OBJECTIVE 4 Find additive inverses.

In FIGURE 13, for each positive number, there is a negative number on the opposite side of 0 that lies the same distance from 0. These pairs of numbers are *additive inverses, opposites,* or *negatives* of each other. For example, 3 and -3 are additive inverses.

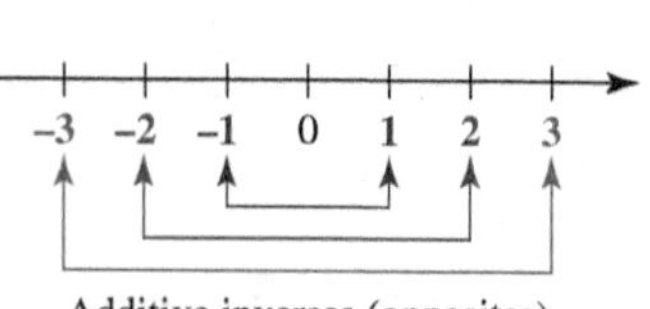

FIGURE 13

▼ **Additive Inverses of Signed Numbers**

Number	Additive Inverse
6	−6
−4	4
$\frac{2}{3}$	$-\frac{2}{3}$
−8.7	8.7
0	0

The number 0 is its own additive inverse.

Additive Inverse

For any real number a, the number $-a$ is the **additive inverse** of a.

We change the sign of a number to find its additive inverse. As we shall see later, the sum of a number and its additive inverse is always 0.

Uses of the − Symbol

The − symbol can be used to indicate any of the following.

1. A negative number, as in −9, read "*negative* 9"
2. The additive inverse of a number, as in "−4 *is the additive inverse of* 4"
3. Subtraction, as in 12 − 3, read "12 *minus* 3"

In the expression $-(-5)$, the − symbol is being used in two ways. The first − symbol indicates the additive inverse (or opposite) of −5, and the second indicates a negative number, −5. Because the additive inverse of −5 is 5,

$$-(-5) = 5.$$

−(−a)

For any real number a, $\quad \mathbf{-(-a) = a.}$

Numbers written with positive or negative signs, such as +4, +8, −9, and −5, are **signed numbers.** A positive number can be called a signed number even though the positive sign is usually left off.

OBJECTIVE 5 Use absolute value.

Geometrically, the **absolute value** of a number a, written $|a|$, is the distance on the number line from 0 to a. For example, the absolute value of 5 is the same as the absolute value of −5 because each number lies five units from 0. See **FIGURE 14.**

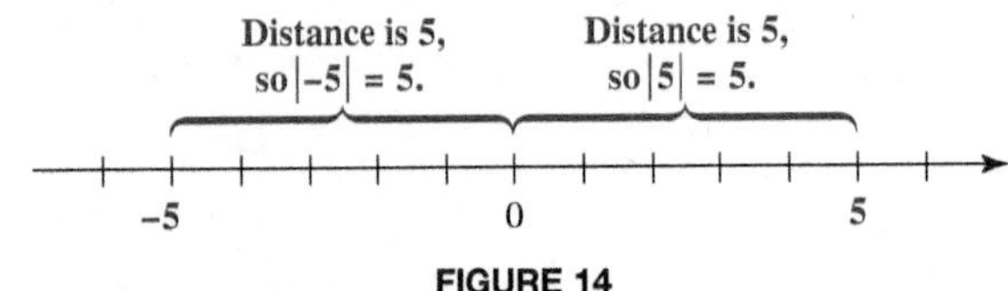

FIGURE 14

CAUTION ***Because absolute value represents undirected distance, the absolute value of a number is always positive or 0.***

The formal definition of absolute value follows.

Absolute Value

For any real number a, $\quad |a| = \begin{cases} a & \textbf{if } a \textbf{ is positive or 0} \\ -a & \textbf{if } a \textbf{ is negative.} \end{cases}$

Consider the second part of the definition of absolute value, $|a| = -a$ if a is negative. If a is a *negative* number, then $-a$, the additive inverse or opposite of a, is a positive number. Thus $|a|$ is positive. For example, let $a = -3$.

$$|a| = |-3| = -(-3) = 3 \qquad |a| = -a \text{ if } a \text{ is negative.}$$

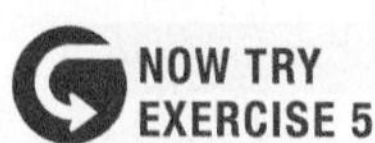
NOW TRY EXERCISE 5

Find each absolute value and simplify if needed.

(a) $|-7|$ **(b)** $-|7|$

(c) $-|-7|$ **(d)** $|4| - |-4|$

(e) $|4 - 4|$

EXAMPLE 5 Finding Absolute Value

Find each absolute value and simplify if needed.

(a) $|13| = 13$

(b) $|0| = 0$

(c) $|-2| = -(-2) = 2$

(d) $|-0.75| = 0.75$

(e) $-|8| = -(8) = -8$ Evaluate the absolute value. Then find the additive inverse.

(f) $-|-8| = -(8) = -8$ Work as in part (e); $|-8| = 8$

(g) $|5| + |-2| = 5 + 2 = 7$ Evaluate each absolute value, and then add.

(h) $-|5 - 2| = -|3| = -3$ Subtract inside the absolute value bars first.

NOW TRY EXERCISE 6

Refer to the table in **Example 6.** Of home health aides, parking enforcement workers, and locomotive firers, which occupation is expected to see the least change (without regard to sign)?

EXAMPLE 6 Comparing Rates of Change in Industries

The table shows the projected total rates of change in employment (in percent) in some of the fastest-growing and in some of the most rapidly-declining occupations from 2016 through 2026.

Occupation (2016–2026)	*Total Rate of Change (in percent)*
Wind turbine service technicians	96.1
Home health aides	46.7
Physicians assistants	37.4
Watch repairers	−28.7
Parking enforcement workers	−35.3
Locomotive firers	−78.6

Data from Bureau of Labor Statistics.

What occupation shown in the table is expected to see the greatest change? The least change?

We want the greatest change, without regard to whether the change is an increase or a decrease. Look for the number in the table with the greatest absolute value. That number is for wind turbine service technicians.

$$|96.1| = 96.1$$

Similarly, the least change is for watch repairers.

$$|-28.7| = 28.7$$

NOW TRY ANSWERS

5. **(a)** 7 **(b)** −7 **(c)** −7 **(d)** 0 **(e)** 0

6. parking enforcement workers

OBJECTIVE 6 Use inequality symbols.

The statement

$$4 + 2 = 6$$

is an **equation**—a statement that two quantities are equal. The statement

$$4 \neq 6 \quad \text{(read "4 \textit{is not equal to} 6")}$$

is an **inequality**—a statement that two quantities are *not* equal. When two numbers are not equal, one must be less than the other. When reading from left to right, the symbol $<$ means "*is less than.*"

$$8 < 9, \quad -6 < 15, \quad -6 < -1, \quad \text{and} \quad 0 < \frac{4}{3} \qquad \text{All are true.}$$

Reading from left to right, the symbol $>$ means "*is greater than.*"

$$12 > 5, \quad 9 > -2, \quad -4 > -6, \quad \text{and} \quad \frac{6}{5} > 0 \qquad \text{All are true.}$$

In each case, the symbol "points" toward the lesser number.

The number line in **FIGURE 15** shows the graphs of the numbers 4 and 9. On the graph, 4 is to the *left* of 9, so $4 < 9$. ***The lesser of two numbers is always to the left of the other on a number line.***

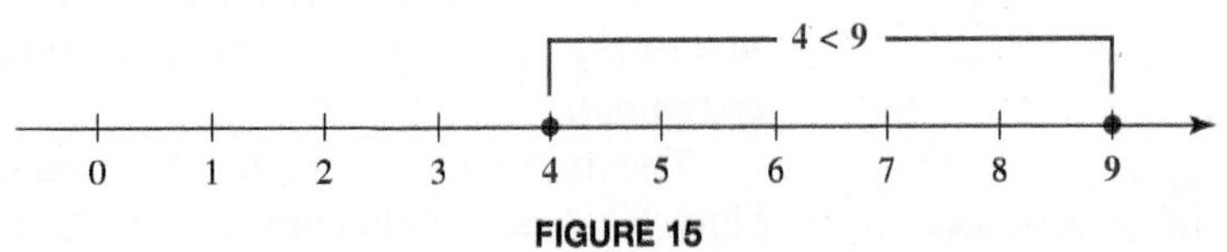

FIGURE 15

Inequalities on a Number Line

On a number line, the following hold true.

$\boldsymbol{a < b}$ if a is to the left of b. $\quad$ $\boldsymbol{a > b}$ if a is to the right of b.

NOW TRY EXERCISE 7

Use a number line to determine whether each statement is *true* or *false*.

(a) $-5 > -1$

(b) $-7 < -6$

EXAMPLE 7 Determining Order on a Number Line

Use a number line to compare -6 and 1 and to compare -5 and -2.

As shown on the number line in **FIGURE 16**, -6 is located to the left of 1. For this reason, $-6 < 1$. Also, $1 > -6$. From **FIGURE 16**, $-5 < -2$, or $-2 > -5$. In each case, the symbol points to the lesser number.

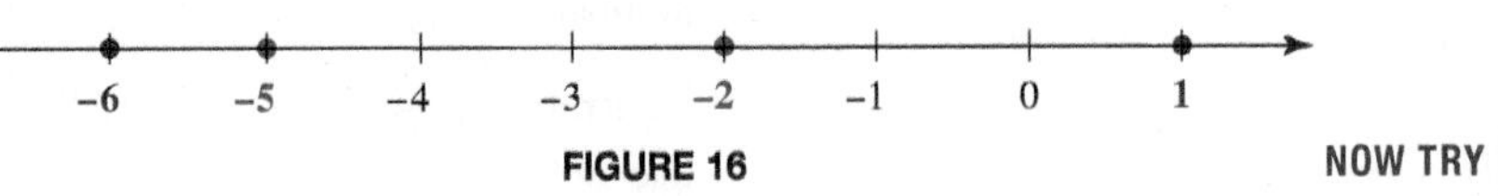

FIGURE 16

NOW TRY

The following table summarizes this discussion.

▼ **Results about Positive and Negative Numbers**

Words	*Symbols*
Every negative number is less than 0.	If a is negative, then $a < 0$.
Every positive number is greater than 0.	If a is positive, then $a > 0$.
0 is neither positive nor negative.	

NOW TRY ANSWERS

7. (a) false **(b)** true

In addition to the symbols $\neq$, $<$, and $>$, the symbols $\leq$ and $\geq$ are often used.

▼ Inequality Symbols

Symbol	*Meaning*	*Example*
$\neq$	is not equal to	$3 \neq 7$
$<$	is less than	$-4 < -1$
$>$	is greater than	$3 > -2$
$\leq$	is less than or equal to	$6 \leq 6$
$\geq$	is greater than or equal to	$-8 \geq -10$

NOW TRY EXERCISE 8

Determine whether each statement is *true* or *false*.

(a) $-5 \leq -6$

(b) $-2 \geq -10$

(c) $0.5 \leq 0.5$

(d) $10 \cdot 6 > 8(7)$

EXAMPLE 8 Using Inequality Symbols

The table shows several uses of inequalities and indicates why each is true.

Inequality	*Why It Is True*
$6 \leq 8$	$6 < 8$
$-2 \leq -2$	$-2 = -2$
$-9 \geq -12$	$-9 > -12$
$-3 \geq -3$	$-3 = -3$
$6 \cdot 4 \leq 5(5)$	$24 < 25$

Notice the reason why $-2 \leq -2$ is true. ***With the symbol $\leq$, if either the $<$ part or the $=$ part is true, then the inequality is true. This is also the case with the $\geq$ symbol.***

The dot in $6 \cdot 4$ indicates the product 6×4, or 24, and 5(5) means 5×5, or 25. Thus, $6 \cdot 4 \leq 5(5)$ becomes $24 \leq 25$, which is true.

NOW TRY

NOW TRY ANSWERS

8. **(a)** false **(b)** true
(c) true **(d)** $60 > 56$; true

R.2 Exercises

MyLab Math

Video solutions for select problems available in MyLab Math

1. Concept Check A student claimed that $\{x \mid x$ is a natural number greater than $3\}$ and $\{y \mid y$ is a natural number greater than $3\}$ actually name the same set, even though different variables are used. Was this student correct?

2. Concept Check Give a real number that satisfies each condition.

(a) An integer between 6.75 and 7.75

(b) A rational number between $\frac{1}{4}$ and $\frac{3}{4}$

(c) A whole number that is not a natural number

(d) An integer that is not a whole number

(e) An irrational number between $\sqrt{4}$ and $\sqrt{9}$

(f) An irrational number that is negative

List the elements in each set. ***See Example 1.***

3. $\{x \mid x$ is a natural number less than $6\}$

4. $\{m \mid m$ is a natural number less than $9\}$

5. $\{z \mid z$ is an integer greater than $4\}$

6. $\{y \mid y$ is an integer greater than $8\}$

7. $\{z \mid z$ is an integer less than or equal to $4\}$

8. $\{p \mid p$ is an integer less than $3\}$

9. $\{a \mid a$ is an even integer greater than $8\}$

10. $\{k \mid k$ is an odd integer less than $1\}$

11. $\{p \mid p$ is a number whose absolute value is $4\}$

12. $\{w \mid w$ is a number whose absolute value is $7\}$

13. $\{x \mid x \text{ is an irrational number that is also rational}\}$

14. $\{r \mid r \text{ is a number that is both positive and negative}\}$

Use set-builder notation to describe each set. **See Example 2.** *(More than one description is possible.)*

15. $\{2, 4, 6, 8\}$

16. $\{11, 12, 13, 14\}$

17. $\{4, 8, 12, 16, \ldots\}$

18. $\{\ldots, -6, -3, 0, 3, 6, \ldots\}$

Graph the elements of each set on a number line. **See Objective 2.**

19. $\{-4, -2, 0, 3, 5\}$

20. $\{-3, -1, 0, 4, 6\}$

21. $\left\{-\frac{6}{5}, -\frac{1}{4}, 0, \frac{5}{6}, \frac{13}{4}, 5.2, \frac{11}{2}\right\}$

22. $\left\{-\frac{2}{3}, 0, \frac{4}{5}, \frac{12}{5}, \frac{9}{2}, 4.8\right\}$

*Which elements of each set are **(a)** natural numbers, **(b)** whole numbers, **(c)** integers, **(d)** rational numbers, **(e)** irrational numbers, **(f)** real numbers?* **See Example 3.**

23. $\left\{-9, -\sqrt{6}, -0.7, 0, \frac{6}{7}, \sqrt{7}, 4.\overline{6}, 8, \frac{21}{2}, 13, \frac{75}{5}\right\}$

24. $\left\{-8, -\sqrt{5}, -0.6, 0, \frac{3}{4}, \sqrt{3}, \pi, 5, \frac{13}{2}, 17, \frac{40}{2}\right\}$

Determine whether each statement is true *or* false. *If it is false, tell why.* **See Example 4.**

25. Every integer is a whole number.

26. Every natural number is an integer.

27. Every irrational number is an integer.

28. Every integer is a rational number.

29. Every natural number is a whole number.

30. Some rational numbers are irrational.

31. Some rational numbers are whole numbers.

32. Some real numbers are integers.

33. The absolute value of any number is the same as the absolute value of its additive inverse.

34. The absolute value of any nonzero number is positive.

35. Concept Check Match each expression in parts (a)–(d) with its value in choices A–D. Choices may be used once, more than once, or not at all.

I		II	
(a) $-(-4)$	**(b)** $\lvert -4 \rvert$	**A.** 4	**B.** -4
(c) $-\lvert -4 \rvert$	**(d)** $-\lvert -(-4) \rvert$	**C.** Both A and B	**D.** Neither A nor B

36. Concept Check For what value(s) of x is $|x| = 4$ true?

*Give **(a)** the additive inverse and **(b)** the absolute value of each number.* **See Objective 4 and Example 5.**

37. 6 **38.** 9 **39.** -12 **40.** -14 **41.** $\frac{6}{5}$ **42.** 0.16

Find each absolute value and simplify if needed. **See Example 5.**

43. $|-8|$ **44.** $|-19|$ **45.** $\left|\frac{3}{2}\right|$ **46.** $\left|\frac{3}{4}\right|$

47. $-|5|$ **48.** $-|12|$ **49.** $-|-2|$ **50.** $-|-6|$

51. $-|4.5|$ **52.** $-|12.4|$ **53.** $|-2| + |3|$ **54.** $|-16| + |14|$

55. $-|10-9|$ 56. $-|12-6|$ 57. $|-9|-|-3|$

58. $|-10|-|-7|$ 59. $|-1|+|-2|-|-3|$ 60. $|-7|+|-3|-|-10|$

Solve each problem. ***See Example 6.***

61. The table shows the percent change in population from 2010 through 2016 for selected metropolitan areas.

Metropolitan Area	*Percent Change*
Detroit	−5.8
Houston	9.7
Las Vegas	8.3
New Orleans	13.9
Phoenix	11.6
Toledo	−3.0

Data from U.S. Census Bureau.

(a) Which metropolitan area had the greatest change in population? What was this change? Was it an increase or a decrease?

(b) Which metropolitan area had the least change in population? What was this change? Was it an increase or a decrease?

62. The table gives the net trade balance, in millions of U.S. dollars, for selected U.S. trade partners for October 2017.

Country	*Trade Balance (in millions of dollars)*
India	−2419
China	−35,230
Netherlands	2148
France	−1993
New Zealand	230

Data from U.S. Census Bureau.

A negative balance means that imports to the United States exceeded exports from the United States. A positive balance means that exports exceeded imports.

(a) Which country had the greatest discrepancy between exports and imports? Explain.

(b) Which country had the least discrepancy between exports and imports? Explain.

Sea level refers to the surface of the ocean. The depth of a body of water can be expressed as a negative number, representing average depth in feet below sea level. The altitude of a mountain can be expressed as a positive number, indicating its height in feet above sea level. The table gives selected depths and altitudes.

Body of Water	*Average Depth in Feet (as a negative number)*	*Mountain*	*Altitude in Feet (as a positive number)*
Pacific Ocean	−14,040	Denali	20,310
South China Sea	−4802	Point Success	14,164
Gulf of California	−2375	Matlalcueyetl	14,636
Caribbean Sea	−8448	Rainier	14,410
Indian Ocean	−12,800	Steele	16,624

Data from *The World Almanac and Book of Facts.*

63. List the bodies of water in order, starting with the deepest and ending with the shallowest.

64. List the mountains in order, starting with the shortest and ending with the tallest.

65. *True* or *false:* The absolute value of the depth of the Pacific Ocean is greater than the absolute value of the depth of the Indian Ocean.

66. *True* or *false:* The absolute value of the depth of the Gulf of California is greater than the absolute value of the depth of the Caribbean Sea.

Use a number line to determine whether each statement is true *or* false. ***See Example 7.***

67. $-6 < -1$ **68.** $-4 < -2$ **69.** $-4 > -3$ **70.** $-3 > -1$

71. $3 > -2$ **72.** $6 > -3$ **73.** $-3 > -3$ **74.** $-5 < -5$

Rewrite each statement with $>$ *so that it uses* $<$ *instead. Rewrite each statement with* $<$ *so that it uses* $>$. ***See Example 7.***

75. $6 > 2$ **76.** $5 > 1$ **77.** $-9 < 4$

78. $-6 < 1$ **79.** $-5 > -10$ **80.** $-7 > -12$

Use an inequality symbol to write each statement. ***See Example 8.***

81. 7 is greater than -1.

82. -4 is less than 10.

83. 5 is greater than or equal to 5.

84. -6 is less than or equal to -6.

85. $13 - 3$ is less than or equal to 10.

86. $5 + 14$ is greater than or equal to 19.

87. $5 + 0$ is not equal to 0.

88. $6 + 7$ is not equal to -13.

Simplify each inequality if needed. Then determine whether the statement is true *or* false. ***See Example 8.***

89. $0 \leq -5$ **90.** $-11 \geq 0$ **91.** $7 \leq 7$

92. $10 \geq 10$ **93.** $-6 < 7 + 3$ **94.** $-7 < 4 + 1$

95. $2 \cdot 5 \geq 4 + 6$ **96.** $8 + 7 \leq 3 \cdot 5$ **97.** $-|-3| \geq -3$

98. $-|-4| \leq -4$ **99.** $-8 > -|-6|$ **100.** $-10 > -|-4|$

The graph shows egg production in millions of eggs in selected states for 2015 and 2016. Use this graph to work each problem.

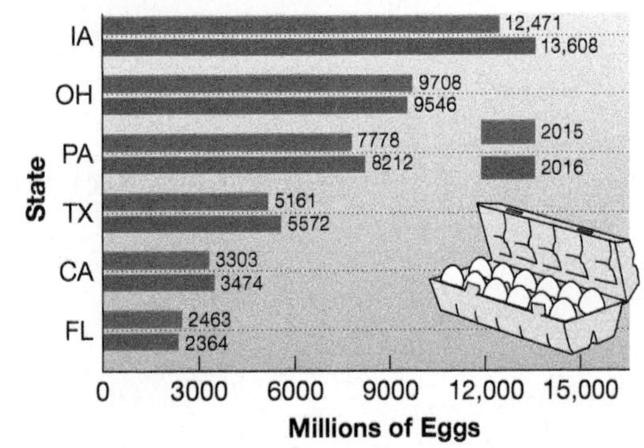

Data from U.S. Department of Agriculture.

101. In 2016, which states had production greater than 6000 million eggs?

102. In which states was 2016 egg production less than 2015 egg production?

103. If x represents 2016 egg production for Texas (TX) and y represents 2016 egg production for Ohio (OH), which is true, $x < y$ or $x > y$?

104. If x represents 2015 egg production for Iowa (IA) and y represents 2015 egg production for Pennsylvania (PA), write two inequalities, one with $>$ and one with $<$, that compare the numbers of eggs produced in these two states.

R.3 Operations on Real Numbers

OBJECTIVES

1 Add real numbers.
2 Subtract real numbers.
3 Find the distance between two points on a number line.
4 Multiply real numbers.
5 Find reciprocals and divide real numbers.

VOCABULARY

- ☐ sum
- ☐ difference
- ☐ product
- ☐ quotient
- ☐ reciprocal (multiplicative inverse)
- ☐ dividend
- ☐ divisor

NOW TRY EXERCISE 1

Find each sum.

(a) $-4 + (-9)$

(b) $-7.25 + (-3.57)$

(c) $-\frac{2}{5} + \left(-\frac{3}{10}\right)$

NOW TRY ANSWERS

1. (a) -13 **(b)** -10.82 **(c)** $-\frac{7}{10}$

OBJECTIVE 1 Add real numbers.

Recall that the answer to an addition problem is a **sum.**

Adding Real Numbers

Same sign To add two numbers with the *same* sign, add their absolute values. The sum has the same sign as the given numbers.

Examples: $2 + 7 = 9, \quad -2 + (-7) = -9$

Different signs To add two numbers with *different* signs, find the absolute values of the numbers, and subtract the lesser absolute value from the greater. The sum has the same sign as the number with the greater absolute value.

Examples: $-8 + 3 = -5, \quad 15 + (-9) = 6$

EXAMPLE 1 Adding Two Negative Real Numbers

Find each sum.

(a) $-12 + (-8)$

First find the absolute values: $|-12| = 12$ and $|-8| = 8$.

Because -12 and -8 have the *same* sign, add their absolute values.

Both numbers are negative, so the sum will be negative.

$$-12 + (-8)$$
$$= -(12 + 8) \quad \text{Add the absolute values.}$$
$$= -(20)$$
$$= -20$$

(b) $-6 + (-3)$

$$= -(|-6| + |-3|)$$
$$= -(6 + 3)$$
$$= -9$$

(c) $-1.2 + (-0.4)$

$$= -(1.2 + 0.4)$$
$$= -1.6$$

(d) $-\frac{5}{6} + \left(-\frac{1}{3}\right)$

$$= -\left(\frac{5}{6} + \frac{1}{3}\right)$$ Add the absolute values. Both numbers are negative, so the sum will be negative.

$$= -\left(\frac{5}{6} + \frac{2}{6}\right)$$ The least common denominator is 6. $\frac{1}{3} \cdot \frac{2}{2} = \frac{2}{6}$

$$= -\frac{7}{6}$$ Add numerators. Keep the same denominator.

NOW TRY

Find each sum.

(a) $-15 + 7$

(b) $-5 + 12$

(c) $4.6 + (-2.8)$

(d) $-\frac{5}{9} + \frac{2}{7}$

EXAMPLE 2 Adding Real Numbers with Different Signs

Find each sum.

(a) $-17 + 11$

First find the absolute values: $|-17| = 17$ and $|11| = 11$.

Because -17 and 11 have *different* signs, subtract their absolute values.

$$17 - 11 = 6$$

The number -17 has a greater absolute value than 11, so the answer is negative.

$$-17 + 11 = -6$$ The sum is negative because $|-17| > |11|$.

(b) $4 + (-1)$

Subtract the absolute values, 4 and 1. Because 4 has the greater absolute value, the sum must be positive.

$$4 + (-1) = 4 - 1 = 3$$ The sum is positive because $|4| > |-1|$.

(c) $-9 + 17$

$= 17 - 9$

$= 8$

(d) $-2.3 + 5.6$

$= 5.6 - 2.3$

$= 3.3$

(e) $-16 + 12$

The absolute values are 16 and 12. Subtract the absolute values.

$$-16 + 12 = -(16 - 12) = -4$$ The sum is negative because $|-16| > |12|$.

(f) $-\frac{4}{5} + \frac{2}{3}$

$= -\frac{12}{15} + \frac{10}{15}$ The least common denominator is 15. $-\frac{4}{5} \cdot \frac{3}{3} = -\frac{12}{15}$; $\frac{2}{3} \cdot \frac{5}{5} = \frac{10}{15}$

$= -\left(\frac{12}{15} - \frac{10}{15}\right)$ Subtract the absolute values. $-\frac{12}{15}$ has the greater absolute value, so the answer will be negative.

$= -\frac{2}{15}$ Subtract numerators. Keep the same denominator.

NOW TRY

OBJECTIVE 2 Subtract real numbers.

Recall that the answer to a subtraction problem is a **difference.** Compare the following two statements.

$$6 - 4 = 2$$

$$6 + (-4) = 2$$

Thus, $6 - 4 = 6 + (-4)$.

To subtract 4 from 6, we add the additive inverse of 4 to 6. This example suggests the following definition of subtraction of real numbers.

NOW TRY ANSWERS

2. (a) -8 (b) 7 (c) 1.8 (d) $-\frac{17}{63}$

Subtraction

For all real numbers a and b, the following holds.

$$a - b = a + (-b)$$

To subtract b from a, add the additive inverse (or opposite) of b to a.

Examples: $5 - 12 = 5 + (-12) = -7$, $6 - (-3) = 6 + 3 = 9$

NOW TRY EXERCISE 3

Find each difference.

(a) $9 - 15$

(b) $-4 - 11$

(c) $-5.67 - (-2.34)$

(d) $\frac{4}{9} - \frac{3}{5}$

EXAMPLE 3 Subtracting Real Numbers

Find each difference.

Change to addition.
The additive inverse of 8 is −8.

(a) $6 - 8 = 6 + (-8) = -2$

Change to addition.
The additive inverse of 4 is −4.

(b) $-12 - 4 = -12 + (-4) = -16$

(c) $-10 - (-7)$

$= -10 + 7$ The additive inverse of −7 is 7.

$= -3$

(d) $-2.4 - (-8.1)$

$= -2.4 + 8.1$

$= 5.7$

(e) $\frac{5}{6} - \left(-\frac{3}{8}\right)$

$= \frac{5}{6} + \frac{3}{8}$ To subtract $a - b$, add the additive inverse (opposite) of b to a.

$= \frac{20}{24} + \frac{9}{24}$ The least common denominator is 24. $\frac{5}{6} \cdot \frac{4}{4} = \frac{20}{24}$; $\frac{3}{8} \cdot \frac{3}{3} = \frac{9}{24}$

$= \frac{29}{24}$ Add numerators. Keep the same denominator.

NOW TRY 

When working a problem that involves both addition and subtraction, add and subtract in order from left to right. Work inside brackets or parentheses first.

EXAMPLE 4 Adding and Subtracting Real Numbers

Perform the indicated operations.

(a) $15 - (-3) - 5 - 12$

$= (15 + 3) - 5 - 12$ Work in order from left to right.

$= 18 - 5 - 12$ Add inside the parentheses.

$= 13 - 12$ Subtract from left to right.

$= 1$ Subtract.

NOW TRY ANSWERS

3. (a) -6 **(b)** -15 **(c)** -3.33 **(d)** $-\frac{7}{45}$

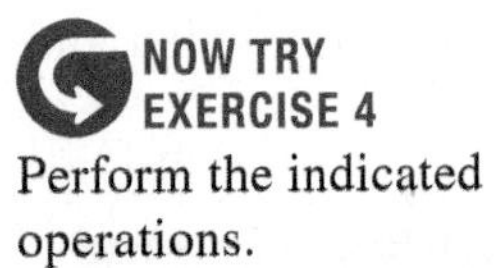

NOW TRY EXERCISE 4
Perform the indicated operations.

$$-4 - (-2 - 7) - 12$$

(b) $-10 - [-8 - (-4)] + 6$

$= -10 - [-8 + 4] + 6$ Work inside the brackets.

$= -10 - [-4] + 6$ Add.

$= -10 + 4 + 6$ Add the additive inverse.

$= -6 + 6$ Add from left to right.

$= 0$ Add. NOW TRY

OBJECTIVE 3 **Find the distance between two points on a number line.**

The number line in FIGURE 17 shows several points.

−8 −7 −6 −5 −4 −3 −2 −1 0 1 2 3 4 5 6 7 8 9 10

FIGURE 17

To find the distance between the points 4 and 7, we subtract $7 - 4 = 3$. Because distance is positive (or 0), we must be careful to subtract in such a way that the answer is positive (or 0). To avoid this problem altogether, we can find the absolute value of the difference. Then the distance between 4 and 7 is found as follows.

$|7 - 4| = |3| = 3$ or $|4 - 7| = |-3| = 3$ Distance between 4 and 7

Distance

The **distance** between two points on a number line is the absolute value of the difference of their coordinates.

NOW TRY EXERCISE 5
Find the distance between the points −7 and 10.

EXAMPLE 5 **Finding Distance between Points on a Number Line**

Find the distance between each pair of points. Refer to FIGURE 17 as needed.

(a) 8 and −4

Find the absolute value of the difference of the numbers, taken in either order.

$$|8 - (-4)| = 12, \quad \text{or} \quad |-4 - 8| = 12$$

(b) −4 and −6

$$|-4 - (-6)| = 2, \quad \text{or} \quad |-6 - (-4)| = 2$$

NOW TRY 

OBJECTIVE 4 **Multiply real numbers.**

Recall that the answer to a multiplication problem is a **product.**

Multiplying Real Numbers

Same sign The product of two numbers with the *same* sign is positive.

Examples: $4(8) = 32, \quad -4(-8) = 32$

Different signs The product of two numbers with *different* signs is negative.

Examples: $-8(9) = -72, \quad 6(-7) = -42$

NOW TRY ANSWERS
4. −7
5. 17

The **multiplication property of 0** states that the product of any real number and 0 is 0.

Multiplication Property of 0

For any real number a, the following hold.

$$\boldsymbol{a \cdot 0 = 0} \quad \text{and} \quad \boldsymbol{0 \cdot a = 0}$$

Examples: $12 \cdot 0 = 0, \quad 0 \cdot 12 = 0, \quad -12 \cdot 0 = 0, \quad 0 \cdot (-12) = 0$

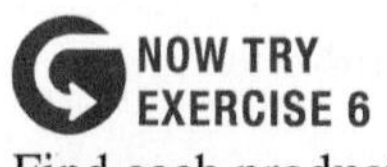
NOW TRY EXERCISE 6

Find each product.

(a) $-3(-10)$

(b) $0.7(-1.2)$

(c) $-\frac{8}{11}(33)$

(d) $14(0)$

EXAMPLE 6 Multiplying Real Numbers

Find each product.

(a) $-3(-9) = 27$ The numbers have the same sign, so the product is positive.

(b) $-0.5(-0.4) = 0.2$

(c) $-\frac{3}{4}\left(-\frac{5}{6}\right)$

$= \frac{15}{24}$ Multiply numerators. Multiply denominators.

$= \frac{5 \cdot 3}{8 \cdot 3}$ Factor to write in lowest terms.

$= \frac{5}{8}$ Divide out the common factor, 3.

(d) $-6(0) = 0$ Multiplication property of 0

(e) $6(-9) = -54$ The numbers have different signs, so the product is negative.

(f) $-0.05(0.3) = -0.015$

(g) $-\frac{3}{4}\left(\frac{2}{9}\right) = -\frac{1}{6}$

(h) $\frac{2}{3}(-6) = -4$ ($-6 = -\frac{6}{1}$)

NOW TRY

OBJECTIVE 5 Find reciprocals and divide real numbers.

The definition of division uses the concept of a **multiplicative inverse,** or *reciprocal.* Recall that two numbers are *reciprocals* if they have a product of 1.

Reciprocal

The **reciprocal** of a nonzero number a is $\frac{1}{a}$.

▼ **Reciprocals**

Number	*Reciprocal*	
$-\frac{2}{5}$	$-\frac{5}{2}$	$-\frac{2}{5}\left(-\frac{5}{2}\right) = 1$
-6, or $-\frac{6}{1}$	$-\frac{1}{6}$	$-6\left(-\frac{1}{6}\right) = 1$
$\frac{7}{11}$	$\frac{11}{7}$	$\frac{7}{11}\left(\frac{11}{7}\right) = 1$
0.05	20	$0.05(20) = 1$
0	None	

Reciprocals have a product of 1. There is no reciprocal for 0 because there is no number that can be multiplied by 0 to give a product of 1.

NOW TRY ANSWERS

6. **(a)** 30 **(b)** -0.84 **(c)** -24 **(d)** 0

CAUTION Keep the following in mind.

- ***A number and its additive inverse have* opposite *signs, such as 3 and −3.***
- ***A number and its reciprocal always have the* same *sign, such as 3 and $\frac{1}{3}$.***

The result of dividing one number by another is a **quotient.** For example, we can write the quotient of 45 and 3 as $\frac{45}{3}$, which equals 15. We obtain the same answer if we multiply 45 and $\frac{1}{3}$.

$$45 \div 3 = \frac{45}{3} = 15 \quad \text{and} \quad 45 \cdot \frac{1}{3} = 15$$

This suggests the following definition of division of real numbers.

Division

For all real numbers a and b (where $b \neq 0$), the following holds true.

$$\boldsymbol{a \div b = \frac{a}{b} = a \cdot \frac{1}{b}}$$

To divide a by b, multiply a (the **dividend**) by the reciprocal of b (the **divisor**).

There is no reciprocal for the number 0, so ***division by 0 is undefined.*** For example, $\frac{15}{0}$ is undefined and $-\frac{1}{0}$ is undefined.

CAUTION Although division by 0 is undefined, dividing 0 by a nonzero number gives the quotient 0.

Example: $\frac{6}{0}$ is undefined, but $\frac{0}{6} = 0$ (because $0 \cdot 6 = 0$).

Be careful when 0 is involved in a division problem.

Because division is defined as multiplication by the reciprocal, the rules for signs of quotients are the same as those for signs of products.

Dividing Real Numbers

Same sign The quotient of two nonzero real numbers with the *same* sign is positive.

Examples: $\frac{24}{6} = 4$, $\frac{-24}{-6} = 4$

Different signs The quotient of two nonzero real numbers with *different* signs is negative.

Examples: $\frac{-36}{3} = -12$, $\frac{36}{-3} = -12$

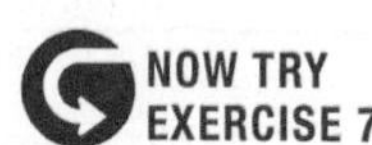

NOW TRY EXERCISE 7

Find each quotient.

(a) $\frac{-10}{-5}$ **(b)** $\frac{-1.5}{0.3}$

(c) $\frac{0}{-4}$ **(d)** $\frac{-\frac{10}{3}}{\frac{3}{8}}$

(e) $-\frac{6}{5} \div \left(-\frac{3}{7}\right)$

EXAMPLE 7 Dividing Real Numbers

Find each quotient.

(a) $\frac{-12}{4} = -3$ The numbers have opposite signs, so the quotient is negative.

(b) $\frac{6}{-3} = -2$

(c) $\frac{-3.6}{-0.4} = 9$ The numbers have the same sign, so the quotient is positive.

(d) $\frac{-9}{0}$ is undefined.

(e) $\frac{0}{-12} = 0$ This is true because $0(-12) = 0$.

(f) $\frac{-\frac{2}{3}}{-\frac{5}{9}}$ This is a *complex fraction*. A complex fraction has a fraction in the numerator, the denominator, or both.

$= -\frac{2}{3}\left(-\frac{9}{5}\right)$ $\frac{a}{b} = a \div b$; Multiply by $-\frac{9}{5}$, the reciprocal of the divisor $-\frac{5}{9}$.

$= \frac{18}{15}$ Multiply numerators. Multiply denominators.

$= \frac{6}{5}$ Write in lowest terms; $\frac{18}{15} = \frac{2 \cdot 3 \cdot 3}{3 \cdot 5} = \frac{6}{5}$

(g) $-\frac{9}{14} \div \frac{3}{7}$

$= -\frac{9}{14} \cdot \frac{7}{3}$ Multiply by $\frac{7}{3}$, the reciprocal of the divisor $\frac{3}{7}$.

$= -\frac{63}{42}$ Multiply numerators. Multiply denominators.

$= -\frac{3}{2}$ Write in lowest terms; $-\frac{63}{42} = -\frac{3 \cdot 3 \cdot 7}{2 \cdot 3 \cdot 7} = -\frac{3}{2}$

NOW TRY

Every fraction has three signs: the sign of the numerator, the sign of the denominator, and the sign of the fraction itself.

Equivalent Forms of a Fraction

The fractions $\frac{-x}{y}$, $\frac{x}{-y}$, and $-\frac{x}{y}$ are equivalent (where $y \neq 0$).

The fractions $\frac{x}{y}$ and $\frac{-x}{-y}$ are equivalent (where $y \neq 0$).

Examples: $\frac{-4}{7} = \frac{4}{-7} = -\frac{4}{7}$ and $\frac{4}{7} = \frac{-4}{-7}$

NOW TRY ANSWERS

7. (a) 2 (b) −5 (c) 0 (d) $-\frac{80}{9}$ (e) $\frac{14}{5}$

R.3 Exercises

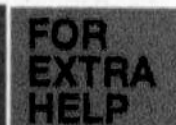

MyLab Math

Video solutions for select problems available in MyLab Math

Concept Check *Complete each statement and give an example.*

1. The sum of two positive numbers is a ______________ number.

2. The sum of two negative numbers is a ______________ number.

3. The sum of a positive number and a negative number is negative if the negative number has the ____________ absolute value.

4. The sum of a positive number and a negative number is positive if the positive number has the ____________ absolute value.

5. The difference of two positive numbers is negative if ____________.

6. The difference of two negative numbers is negative if ____________.

7. The product of two numbers with the same sign is ____________.

8. The product of two numbers with different signs is ____________.

9. The quotient formed by any nonzero number divided by 0 is ____________, and the quotient formed by 0 divided by any nonzero number is ____________.

10. The sum of a positive number and a negative number is 0 if the numbers are ____________.

Find each sum or difference. ***See Examples 1–3.***

11. $-6 + (-13)$
12. $-8 + (-16)$
13. $-15 + 6$
14. $-17 + 9$
15. $13 + (-4)$
16. $19 + (-13)$
17. $-17 + 22$
18. $-12 + 16$
19. $-\frac{7}{3} + \frac{3}{4}$
20. $-\frac{5}{6} + \frac{4}{9}$
21. $-2.8 + 4.5$
22. $-3.8 + 6.2$
23. $4 - 9$
24. $3 - 7$
25. $-6 - 5$
26. $-8 - 17$
27. $8 - (-13)$
28. $12 - (-22)$
29. $-16 - (-3)$
30. $-21 - (-6)$
31. $-12.31 - (-2.13)$
32. $-15.88 - (-9.42)$
33. $\frac{9}{10} - \left(-\frac{4}{3}\right)$
34. $\frac{3}{14} - \left(-\frac{3}{4}\right)$
35. $|-8 - 6|$
36. $|-7 - 15|$
37. $-|-4 + 9|$
38. $-|-5 + 6|$
39. $-2 - |-4|$
40. $16 - |-13|$

Perform the indicated operations. ***See Example 4.***

41. $-7 + 5 - 9$
42. $-12 + 14 - 18$
43. $6 - (-2) - 8$
44. $7 - (-4) - 11$
45. $8 - (-12) - 2 - 6$
46. $3 - (-14) - 6 - 4$
47. $-9 - 4 - (-3) + 6$
48. $-10 - 6 - (-12) + 9$
49. $-0.38 + 4 - 0.62$
50. $-2.95 + 8 - 0.05$
51. $\left(-\frac{5}{4} - \frac{2}{3}\right) + \frac{1}{6}$
52. $\left(-\frac{5}{9} - \frac{1}{6}\right) + \frac{1}{2}$
53. $-\frac{3}{4} - \left(\frac{1}{2} - \frac{3}{8}\right)$
54. $-\frac{1}{2} - \left(\frac{5}{6} - \frac{5}{12}\right)$
55. $-4 - [-3 - (-6)] + 9$
56. $-10 - [-2 - (-5)] + 16$
57. $|-11| - |-5| - |7| + |-2|$
58. $|-6| + |-3| - |4| - |-8|$

The number line has several points labeled. Find the distance between each pair of points. ***See Example 5.***

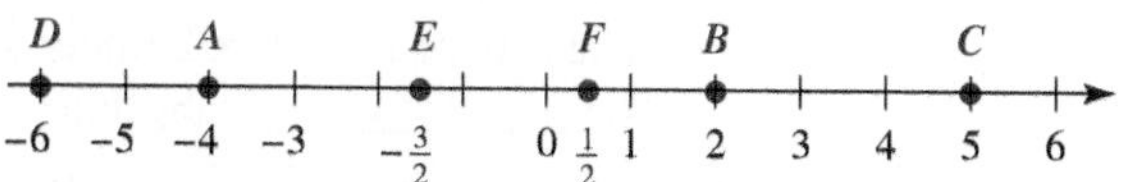

59. A and B **60.** A and C **61.** D and F **62.** E and C

63. Concept Check Consider the statement "*Two negatives give a positive.*" When is this true? When is this false? Give a more precise statement that conveys this message.

64. Concept Check Why must the reciprocal of a nonzero number have the same sign as the number?

Find each product. ***See Example 6.***

65. $-8(-5)$ **66.** $-20(-4)$ **67.** $5(-7)$ **68.** $6(-9)$

69. $4(0)$ **70.** $0(-8)$ **71.** $\frac{1}{2}(0)$ **72.** $0(-4.5)$

73. $\frac{3}{4}(-16)$ **74.** $\frac{3}{5}(-35)$ **75.** $-10\left(-\frac{1}{5}\right)$ **76.** $-18\left(-\frac{1}{2}\right)$

77. $-\frac{5}{2}\left(-\frac{12}{25}\right)$ **78.** $-\frac{9}{7}\left(-\frac{21}{36}\right)$ **79.** $-\frac{3}{8}\left(-\frac{24}{9}\right)$ **80.** $-\frac{2}{11}\left(-\frac{22}{4}\right)$

81. $-0.8(-0.5)$ **82.** $-0.5(-0.6)$ **83.** $-0.06(0.4)$ **84.** $-0.08(0.7)$

Find each quotient where possible. ***See Example 7.***

85. $\frac{-14}{2}$ **86.** $\frac{-39}{13}$ **87.** $\frac{-24}{-4}$ **88.** $\frac{-45}{-9}$ **89.** $\frac{100}{-25}$

90. $\frac{150}{-30}$ **91.** $\frac{0}{-8}$ **92.** $\frac{0}{-14}$ **93.** $\frac{5}{0}$ **94.** $\frac{13}{0}$

95. $-\frac{10}{17} \div \left(-\frac{12}{5}\right)$ **96.** $-\frac{22}{23} \div \left(-\frac{33}{5}\right)$ **97.** $\frac{\frac{12}{13}}{-\frac{4}{3}}$ **98.** $\frac{\frac{7}{6}}{-\frac{2}{3}}$

99. $\frac{-7.2}{0.8}$ **100.** $\frac{-4.5}{0.9}$ **101.** $\frac{-1.28}{-0.4}$ **102.** $\frac{-1.82}{-0.2}$

The following problems provide more practice on operations with fractions and decimals. Perform the indicated operations.

103. $\frac{1}{6} - \left(-\frac{7}{9}\right)$ **104.** $\frac{7}{10} - \left(-\frac{1}{6}\right)$ **105.** $-\frac{1}{9} + \frac{7}{12}$

106. $-\frac{1}{12} + \frac{13}{16}$ **107.** $-\frac{3}{8} - \frac{5}{12}$ **108.** $-\frac{11}{15} - \frac{4}{9}$

109. $-\frac{7}{30} + \frac{2}{45} - \frac{3}{10}$ **110.** $-\frac{8}{15} + \frac{7}{6} - \frac{3}{20}$ **111.** $\frac{8}{25}\left(-\frac{5}{12}\right)$

112. $\frac{9}{20}\left(-\frac{7}{15}\right)$ **113.** $\frac{5}{6}\left(-\frac{9}{10}\right)\left(-\frac{4}{5}\right)$ **114.** $\frac{4}{3}\left(-\frac{9}{20}\right)\left(-\frac{5}{12}\right)$

115. $\frac{8}{3} \div \left(-\frac{14}{15}\right)$ **116.** $\frac{12}{5} \div \left(-\frac{18}{25}\right)$ **117.** $\frac{\frac{2}{3}}{-2}$

118. $\dfrac{\frac{3}{4}}{-6}$

119. $\dfrac{-\frac{8}{9}}{\frac{2}{3}}$

120. $\dfrac{-\frac{15}{16}}{\frac{3}{8}}$

121. $-8.6 - 23.751$

122. $-37.8 - 13.582$

123. $-2.5(0.8)(1.5)$

124. $-1.6(0.5)(2.5)$

125. $-24.84 \div 6$

126. $-32.84 \div 8$

127. $-2496 \div (-0.52)$

128. $-1875 \div (-0.25)$

129. $\dfrac{-100}{-0.01}$

130. $\dfrac{-60}{-0.06}$

131. $-14.2 + 9.81$

132. $-89.41 + 21.325$

Solve each problem.

133. On June 27, 1915, a temperature of 100°F was recorded in Fort Yukon, Alaska. On January 23, 1971, Prospect Creek Camp, Alaska, recorded a temperature of −80°F. Express the difference between these two temperatures as a positive number. (Data from *The World Almanac and Book of Facts.*)

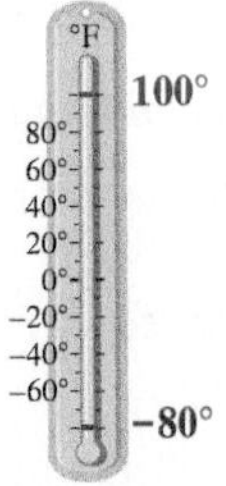

134. On August 10, 1936, a temperature of 120°F was recorded in Ozark, Arkansas. On February 13, 1905, Gravette, Arkansas, recorded a temperature of −29°F. Express the difference between these two temperatures as a positive number. (Data from *The World Almanac and Book of Facts.*)

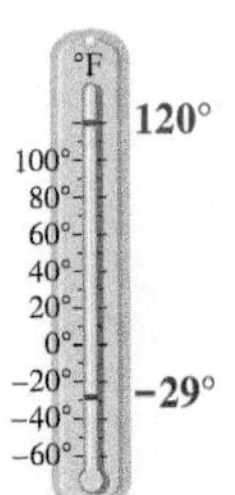

135. Andrew has \$48.35 in his checking account. He uses his debit card to make purchases of \$35.99 and \$20.00, which overdraws his account. His bank charges his account an overdraft fee of \$28.50. He then deposits his paycheck for \$66.27 from his part-time job at Arby's. What is the balance in his account?

136. Kayla has \$37.60 in her checking account. She uses her debit card to make purchases of \$25.99 and \$19.34, which overdraws her account. Her bank charges her account an overdraft fee of \$25.00. She then deposits her paycheck for \$58.66 from her part-time job at Subway. What is the balance in her account?

137. Ahmad owes \$382.45 on his Visa account. He returns two items costing \$25.10 and \$34.50 for credit. Then he makes purchases of \$45.00 and \$98.17.

(a) How much is his payment if he wants to pay off the balance on the account?

(b) Instead of paying off the balance, he makes a payment of \$300 and then incurs a finance charge of \$24.66. What is the balance on his account?

138. Charlene owes \$237.59 on her MasterCard account. She returns one item costing \$47.25 for credit and then makes two purchases of \$12.39 and \$20.00.

(a) How much is her payment if she wants to pay off the balance on the account?

(b) Instead of paying off the balance, she makes a payment of \$75.00 and incurs a finance charge of \$32.06. What is the balance on her account?

139. The graph shows annual returns in percent for shares of a charter fund.

(a) Find the sum of the percents for the years shown in the graph.

(b) Find the difference between the returns in 2017 and 2016.

(c) Find the difference between the returns in 2018 and 2017.

140. The graph shows profits and losses in thousands of dollars for a company.

(a) What was the total profit or loss for the years 2015 through 2018?

(b) Find the difference between the profit or loss in 2018 and that in 2017.

(c) Find the difference between the profit or loss in 2016 and that in 2015.

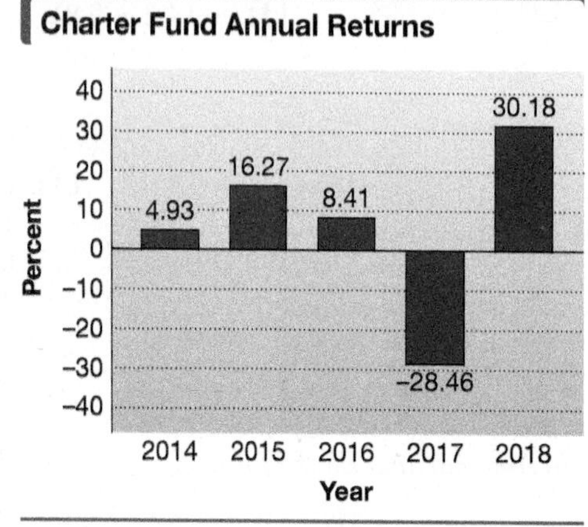

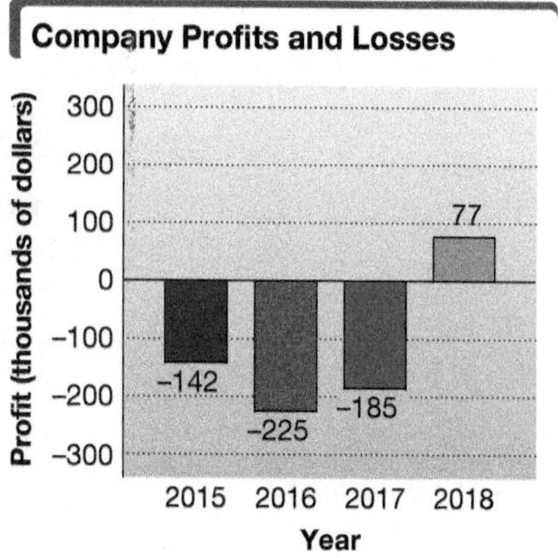

The table shows receipts (income) and outlays (spending) in billions of dollars for the U.S. government in selected years.

Fiscal Year	*Receipts*	*Outlays*
2001	1991	1863
2006	2407	2655
2011	2303	3603
2016	3268	3853

Data from Office of Management and Budget.

141. Find the difference between U.S. government receipts and outlays for each year shown in the table.

142. During which years did the budget show a surplus? A deficit? Explain the answers.

R.4 Exponents, Roots, and Order of Operations

OBJECTIVES

1 Use exponents.
2 Find square roots.
3 Use the rules for order of operations.
4 Evaluate algebraic expressions for given values of variables.

Two (or more) numbers whose product is a third number are **factors** of that third number. For example, 2 and 6 are factors of 12 because $2 \cdot 6 = 12$.

OBJECTIVE 1 Use exponents.

In algebra, we use *exponents* as a way of writing products of repeated factors. For example, we write the product $2 \cdot 2 \cdot 2 \cdot 2 \cdot 2$ as follows.

$$\underbrace{2 \cdot 2 \cdot 2 \cdot 2 \cdot 2}_{\text{5 factors of 2}} = 2^5$$

The number 5 shows that 2 is used as a factor 5 times. The number 5 is the *exponent*, and 2 is the *base*.

$$2^5 \leftarrow \text{Exponent}$$
$$\uparrow \text{Base}$$

Read 2^5 as "2 *to the fifth power*," or "2 *to the fifth*." Multiplying five 2s gives 32.

$$2^5 \quad \text{means} \quad 2 \cdot 2 \cdot 2 \cdot 2 \cdot 2, \quad \text{which equals } 32.$$

VOCABULARY
- ☐ factors
- ☐ exponent (power)
- ☐ base
- ☐ exponential expression
- ☐ square root
- ☐ positive (principal) square root
- ☐ negative square root
- ☐ constant
- ☐ algebraic expression

Exponential Expression

If a is a real number and n is a natural number, then

$$a^n = \underbrace{a \cdot a \cdot a \cdot \ldots \cdot a}_{n \text{ factors of } a},$$

where n is the **exponent,** a is the **base,** and a^n is an **exponential expression.** Exponents are also called **powers.**

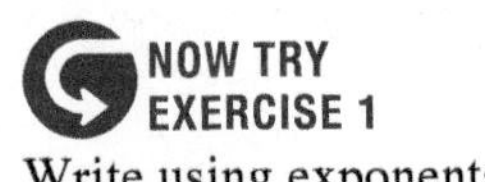
NOW TRY EXERCISE 1

Write using exponents.

(a) $(-3)(-3)(-3)$

(b) $(0.5)(0.5)$

(c) $t \cdot t \cdot t \cdot t \cdot t$

EXAMPLE 1 Using Exponential Notation

Write using exponents.

(a) $\underbrace{4 \cdot 4 \cdot 4}_{3 \text{ factors of } 4} = 4^3$

Read 4^3 as "4 ***cubed.***"

(b) $\frac{3}{5} \cdot \frac{3}{5} = \left(\frac{3}{5}\right)^2$ 2 factors of $\frac{3}{5}$

Read $\left(\frac{3}{5}\right)^2$ as "$\frac{3}{5}$ ***squared.***"

(c) $(-6)(-6)(-6)(-6) = (-6)^4$ 4 factors of -6

Read $(-6)^4$ as "-6 *to the fourth power*," or "-6 *to the fourth*."

(d) $(0.3)(0.3)(0.3)(0.3)(0.3) = (0.3)^5$ **(e)** $x \cdot x \cdot x \cdot x \cdot x \cdot x = x^6$ NOW TRY 

NOTE In **Example 1,** we used the terms *squared* and *cubed* to refer to powers of 2 and 3, respectively. The term *squared* comes from the figure of a square, which has the same measure for both length and width. See **FIGURE 18(a).** Similarly, the term *cubed* comes from the figure of a cube, where the length, width, and height have the same measure. See **FIGURE 18(b).**

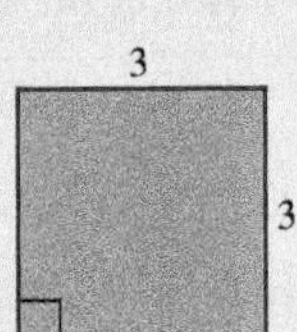

(a) $3 \cdot 3$ means 3 squared, or 3^2.

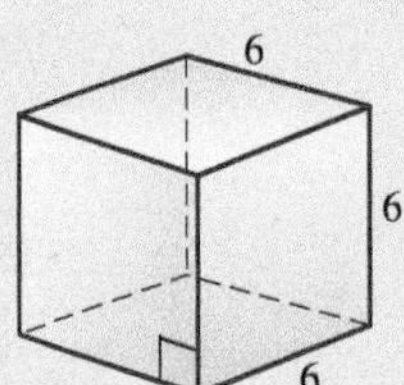

(b) $6 \cdot 6 \cdot 6$ means 6 cubed, or 6^3.

FIGURE 18

EXAMPLE 2 Evaluating Exponential Expressions

Evaluate.

(a) 5^2 means $5 \cdot 5$, which equals 25. 5 is used as a factor 2 times.

5^2 means $5 \cdot 5$, ***not*** $5 \cdot 2$.

NOW TRY ANSWERS

1. (a) $(-3)^3$ **(b)** $(0.5)^2$ **(c)** t^5

NOW TRY EXERCISE 2

Evaluate.

(a) 5^3 **(b)** $\left(\frac{2}{5}\right)^4$

(c) $(-3)^2$ **(d)** $(-3)^3$

(b) $\left(\frac{2}{3}\right)^3$ means $\frac{2}{3} \cdot \frac{2}{3} \cdot \frac{2}{3}$, which equals $\frac{8}{27}$. $\frac{2}{3}$ is used as a factor 3 times.

(c) $(0.2)^3$ means $(0.2)(0.2)(0.2)$, which equals 0.008.

(d) $(-2)^4$ means $(-2)(-2)(-2)(-2)$, which equals 16.

(e) $(-2)^5$ means $(-2)(-2)(-2)(-2)(-2)$, which equals -32. NOW TRY

Examples 2(d) and (e) suggest the following generalizations.

Sign of an Exponential Expression

The product of an *even* number of negative factors is positive.

Example: $(-2)(-2)(-2)(-2) = 16$ 4 factors of -2

The product of an *odd* number of negative factors is negative.

Example: $(-2)(-2)(-2)(-2)(-2) = -32$ 5 factors of -2

NOW TRY EXERCISE 3

Evaluate.

(a) 7^2 **(b)** $(-7)^2$ **(c)** -7^2

EXAMPLE 3 Evaluating Exponential Expressions

Evaluate.

(a) 2^6 means $2 \cdot 2 \cdot 2 \cdot 2 \cdot 2 \cdot 2$, which equals 64. The base is 2.

(b) $(-2)^6$ means $(-2)(-2)(-2)(-2)(-2)(-2)$, which equals 64. The base is -2.

(c) -2^6

There are no parentheses. The exponent 6 applies *only* to the number 2, not to -2.

-2^6 means $-(2 \cdot 2 \cdot 2 \cdot 2 \cdot 2 \cdot 2)$, which equals -64. The base is 2.

NOW TRY 

CAUTION It is important to distinguish between $-a^n$ and $(-a)^n$.

$$-a^n \text{ means } -1\underbrace{(a \cdot a \cdot a \cdot \ldots \cdot a)}_{n \text{ factors of } a} \quad \text{The base is } a.$$

$$(-a)^n \text{ means } \underbrace{(-a)(-a) \cdot \ldots \cdot (-a)}_{n \text{ factors of } -a} \quad \text{The base is } -a.$$

Be careful when evaluating an exponential expression with a negative sign.

OBJECTIVE 2 Find square roots.

As we saw in **Example 2(a),** 5 squared (or 5^2) equals 25. The opposite (inverse) of squaring a number is taking its **square root.** For example, a square root of 25 is 5. Another square root of 25 is -5 because $(-5)^2 = 25$. Thus, 25 has two square roots, 5 and -5.

We write the **positive** or **principal square root** of a number using a **radical symbol** $\sqrt{}$**.** The positive or principal square root of 25 is written

$$\sqrt{25} = 5.$$

The **negative square root** of 25 is written

$$-\sqrt{25} = -5.$$

NOW TRY ANSWERS

2. **(a)** 125 **(b)** $\frac{16}{625}$ **(c)** 9 **(d)** -27

3. **(a)** 49 **(b)** 49 **(c)** -49

Because the square of any nonzero real number is positive, the square root of a negative number, such as $\sqrt{-25}$, is not a real number.

NOW TRY EXERCISE 4

Find each square root that is a real number.

(a) $\sqrt{121}$ (b) $\sqrt{\frac{100}{9}}$

(c) $-\sqrt{121}$ (d) $\sqrt{-121}$

EXAMPLE 4 Finding Square Roots

Find each square root that is a real number.

(a) $\sqrt{36} = 6$ because $6^2 = 36$.

(b) $\sqrt{0} = 0$ because $0^2 = 0$.

(c) $\sqrt{\frac{9}{16}} = \frac{3}{4}$ because $\left(\frac{3}{4}\right)^2 = \frac{9}{16}$.

(d) $\sqrt{0.16} = 0.4$ because $(0.4)^2 = 0.16$.

(e) $\sqrt{100} = 10$ because $10^2 = 100$.

(f) $-\sqrt{100} = -10$ because the negative sign is outside the radical symbol.

(g) $\sqrt{-100}$ is not a real number because the negative sign is inside the radical symbol. No *real number* squared equals -100.

Notice that part (e) is the positive or principal square root of 100, part (f) is the negative square root of 100, and part (g) is the square root of -100, which is not a real number.

NOW TRY

OBJECTIVE 3 Use the rules for order of operations.

To simplify the following expression, what should we do first—add 5 and 2, *or* multiply 2 and 3?

$$5 + 2 \cdot 3$$

When an expression involves more than one operation symbol, we use the following rules for **order of operations.**

Order of Operations

If grouping symbols are present, work within them, innermost first (and above and below fraction bars separately), in the following order.

Step 1 Apply all **exponents.**

Step 2 Do any **multiplications** and **divisions** in order from left to right.

Step 3 Do any **additions** and **subtractions** in order from left to right.

If no grouping symbols are present, start with Step 1.

NOW TRY EXERCISE 5

Simplify.

$$15 - 3 \cdot 4 + 2$$

EXAMPLE 5 Using the Rules for Order of Operations

Simplify.

(a) $5 + 2 \cdot 3$

$= 5 + 6$ Multiply.

$= 11$ Add.

(b) $24 \div 3 \cdot 2 + 6$ Multiplications and divisions are done in order from left to right. So we divide first here.

$= 8 \cdot 2 + 6$

$= 16 + 6$ Multiply.

$= 22$ Add.

NOW TRY 

NOW TRY ANSWERS

4. (a) 11 (b) $\frac{10}{3}$ (c) -11 (d) not a real number

5. 5

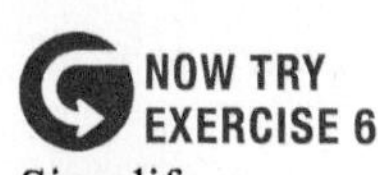

NOW TRY EXERCISE 6

Simplify.

(a) $-5^2 + 10 \div 5 - |3 - 7|$

(b) $6 + \frac{2}{3}(-9) - \frac{5}{8} \cdot 16$

EXAMPLE 6 Using the Rules for Order of Operations

Simplify.

(a)

$10 \div 5 + 2|3 - 4|$ Work inside the absolute value bars first.

$= 10 \div 5 + 2|-1|$ Subtract inside the absolute value bars.

$= 10 \div 5 + 2 \cdot 1$ Take the absolute value.

$= 2 + 2$ Divide first, and then multiply.

$= 4$ Add.

(b)

$4 - 3^2 + 7 - (2 + 8)$ Work inside the parentheses first.

$= 4 - 3^2 + 7 - 10$ Add inside the parentheses.

$= 4 - 9 + 7 - 10$ Evaluate the power.

The base is 3. 3^2 means $3 \cdot 3$, *not* $3 \cdot 2$.

$= -5 + 7 - 10$

$= 2 - 10$

$= -8$

Additions and subtractions are done in order from left to right, so subtract $4 - 9$ first here, then add $-5 + 7$.

(c)

$\frac{1}{2} \cdot 4 + (6 \div 3 - 7)$ Work inside the parentheses first.

$= \frac{1}{2} \cdot 4 + (2 - 7)$ Divide inside the parentheses.

$= \frac{1}{2} \cdot 4 + (-5)$ Subtract inside the parentheses.

$= 2 + (-5)$ Multiply.

$= -3$ Add.

NOW TRY

NOW TRY EXERCISE 7

Simplify.

$$\frac{\sqrt{36} - 4 \cdot 3^2}{-2^2 - 8 \cdot 3 + 28}$$

EXAMPLE 7 Using the Rules for Order of Operations

Simplify.

$$\frac{5 - 2^4}{6 \cdot \sqrt{9} - 9 \cdot 2}$$ Work separately above and below the fraction bar.

$$= \frac{5 - 16}{6 \cdot 3 - 9 \cdot 2}$$ Evaluate the power and the root.

$$= \frac{5 - 16}{18 - 18}$$ Multiply.

$$= \frac{-11}{0}$$ Subtract.

Because division by 0 is undefined, the given expression is undefined. NOW TRY

NOW TRY ANSWERS

6. (a) -27 (b) -10

7. undefined

OBJECTIVE 4 **Evaluate algebraic expressions for given values of variables.**

A **constant** is a fixed, unchanging number.

$$1, \quad 6, \quad -10, \quad \frac{2}{5}, \quad -3.75 \qquad \text{Constants}$$

A collection of constants, variables, operation symbols, and/or grouping symbols is an **algebraic expression.**

$$6ab, \quad 5m - 9n, \quad -2(x^2 + 4y) \qquad \text{Algebraic expressions}$$

Algebraic expressions have different numerical values for different values of the variables. We evaluate such expressions by *substituting* given values for the variables.

NOW TRY EXERCISE 8

Evaluate each expression for $x = -4$, $y = 7$, and $z = 36$.

(a) $3y - 2x$ **(b)** $\dfrac{x^2 - \sqrt{z}}{-3xy}$

EXAMPLE 8 **Evaluating Algebraic Expressions**

Evaluate each expression for $m = -4$, $n = 5$, $p = -6$, and $q = 25$.

(a) $5m - 9n$

$= 5(-4) - 9(5)$ Substitute $m = -4$ and $n = 5$.

Use parentheses around substituted values to avoid errors.

$= -20 - 45$ Multiply.

$= -65$ Subtract.

(b) $\dfrac{m + 2n}{4p}$ Work separately above and below the fraction bar.

$= \dfrac{-4 + 2(5)}{4(-6)}$ Substitute $m = -4$, $n = 5$, and $p = -6$.

$= \dfrac{-4 + 10}{-24}$ Multiply in the numerator. Multiply in the denominator.

$= \dfrac{6}{-24}$ Add in the numerator.

$= -\dfrac{1}{4}$ Write in lowest terms, and rewrite using $\frac{a}{-b} = -\frac{a}{b}$.

(c) $-3m^3 - n^2(\sqrt{q})$

$= -3(-4)^3 - (5)^2(\sqrt{25})$ Substitute $m = -4$, $n = 5$, and $q = 25$.

$= -3(-64) - 25(5)$ Evaluate the powers and the root.

$= 192 - 125$ Multiply.

$= 67$ Subtract.

NOW TRY

NOW TRY ANSWERS

8. (a) 29 **(b)** $\frac{5}{42}$

R.4 Exercises

MyLab Math

Video solutions for select problems available in MyLab Math

Concept Check *Determine whether each statement is* true *or* false. *If it is false, correct the statement so that it is true.*

1. $-7^6 = (-7)^6$

2. $-5^7 = (-5)^7$

3. $\sqrt{25}$ is a positive number.

4. $3 + 5 \cdot 8 = 3 + (5 \cdot 8)$

5. $(-6)^7$ is a negative number.

6. $(-6)^8$ is a positive number.

7. The product of 10 positive factors and 10 negative factors is positive.

8. The product of 5 positive factors and 5 negative factors is positive.

9. In the exponential expression -2^5, the base is -2.

10. $\sqrt{a}$ is positive for all positive numbers a.

Write each expression using exponents. ***See Example 1.***

11. $10 \cdot 10 \cdot 10 \cdot 10$

12. $8 \cdot 8 \cdot 8$

13. $\frac{3}{4} \cdot \frac{3}{4} \cdot \frac{3}{4} \cdot \frac{3}{4} \cdot \frac{3}{4}$

14. $\frac{1}{2} \cdot \frac{1}{2}$

15. $(-9)(-9)(-9)$

16. $(-4)(-4)(-4)(-4)$

17. $(0.8)(0.8)$

18. $(0.1)(0.1)(0.1)(0.1)(0.1)(0.1)$

19. $z \cdot z \cdot z \cdot z \cdot z \cdot z \cdot z$

20. $a \cdot a \cdot a \cdot a \cdot a$

Concept Check *Evaluate each exponential expression.*

21. (a) 8^2 **(b)** -8^2 **(c)** $(-8)^2$ **(d)** $-(-8)^2$

22. (a) 4^3 **(b)** -4^3 **(c)** $(-4)^3$ **(d)** $-(-4)^3$

Evaluate each expression. ***See Examples 2 and 3.***

23. 4^2 **24.** 6^2 **25.** $(0.3)^3$ **26.** $(0.1)^3$

27. $\left(\frac{1}{5}\right)^3$ **28.** $\left(\frac{1}{6}\right)^4$ **29.** $\left(\frac{4}{5}\right)^4$ **30.** $\left(\frac{7}{10}\right)^3$

31. $(-5)^3$ **32.** $(-3)^5$ **33.** $(-2)^8$ **34.** $(-3)^6$

35. -3^6 **36.** -4^6 **37.** -8^4 **38.** -10^3

39. Concept Check Determine whether each expression is *positive* or *negative* when evaluated. Do not actually evaluate.

(a) -7^2 **(b)** $(-7)^2$

(c) -7^3 **(d)** $(-7)^3$

(e) -7^4 **(f)** $(-7)^4$

40. Concept Check Match each square root in Column I with the appropriate value or description in Column II.

I	II
(a) $\sqrt{144}$	**A.** -12
(b) $\sqrt{-144}$	**B.** 12
(c) $-\sqrt{144}$	**C.** Not a real number

Find each square root. If it is not a real number, say so. ***See Example 4.***

41. $\sqrt{81}$ **42.** $\sqrt{64}$ **43.** $\sqrt{169}$ **44.** $\sqrt{225}$

45. $-\sqrt{400}$ **46.** $-\sqrt{900}$ **47.** $\sqrt{\frac{100}{121}}$ **48.** $\sqrt{\frac{225}{169}}$

49. $-\sqrt{0.49}$ **50.** $-\sqrt{0.64}$ **51.** $\sqrt{-36}$ **52.** $\sqrt{-121}$

53. Concept Check If a is a positive number, is $-\sqrt{-a}$ *positive, negative,* or *not a real number?*

54. Concept Check If a is a positive number, is $-\sqrt{a}$ *positive, negative,* or *not a real number?*

55. Concept Check Frank's grandson was asked to evaluate the following expression.

$$9 + 15 \div 3$$

Frank gave the answer as 8, but his grandson gave the answer as 14. The grandson explained that the answer is 14 because of the "Order of Process rule," which says that when evaluating expressions, we proceed from right to left rather than left to right. (*Note:* This is a true story.)

(a) Whose answer was correct for this expression, Frank's or his grandson's?

(b) Was the *reasoning* for the correct answer valid? Explain.

56. Concept Check Problems like this one occasionally appear on social media. Simplify this expression.

$$7 + 7 \div 7 + 7 \times 7 - 7$$

Simplify each expression. ***See Examples 5–7.***

57. $12 + 3 \cdot 4$

58. $15 + 5 \cdot 2$

59. $6 \cdot 3 - 12 \div 4$

60. $9 \cdot 4 - 8 \div 2$

61. $10 + 30 \div 2 \cdot 3$

62. $12 + 24 \div 3 \cdot 2$

63. $-3(5)^2 - (-2)(-8)$

64. $-9(2)^2 - (-3)(-2)$

65. $5 - 7 \cdot 3 - (-2)^3$

66. $-4 - 3 \cdot 5 - (-3)^3$

67. $-7(\sqrt{36}) - (-2)(-3)$

68. $-8(\sqrt{64}) - (-3)(-7)$

69. $6|4 - 5| - 24 \div 3$

70. $-4|2 - 4| + 8 \cdot 2$

71. $|-6 - 5|(-8) - 3^2$

72. $|-2 - 3|(-9) - 4^2$

73. $18 - 4^2 + 5 - (3 - 7)$

74. $10 - 2^2 + 9 - (1 - 8)$

75. $6 + \frac{2}{3}(-9) - \frac{5}{8} \cdot 16$

76. $7 - \frac{3}{4}(-8) + 12 \cdot \frac{5}{6}$

77. $-14\left(-\frac{2}{7}\right) \div (2 \cdot 6 - 10)$

78. $-12\left(-\frac{3}{4}\right) - (6 \cdot 5 \div 3)$

79. $\dfrac{(-5 + \sqrt{4}) - 2^2}{-5 - 2}$

80. $\dfrac{(-9 + \sqrt{16}) - 3^2}{-6 - 1}$

81. $\dfrac{2(-5) + (-3)(-2)}{-8 + 3^2 - 1}$

82. $\dfrac{3(-4) + (-5)(-8)}{2^3 - 2 - 6}$

Evaluate each expression for $a = -3$, $b = 64$, *and* $c = 6$. ***See Example 8.***

83. $3a + \sqrt{b}$

84. $-2a - \sqrt{b}$

85. $\sqrt{b} + c - a$

86. $\sqrt{b} - c + a$

87. $4a^3 + 2c$

88. $-3a^4 - 3c$

89. $2(a - c)^2 - ac$

90. $-4ac + (c - a)^2$

91. $\dfrac{\sqrt{b} - 4a}{c^2}$

92. $\dfrac{a^3 + 2c}{-\sqrt{b}}$

93. $\dfrac{2c + a^3}{4b + 6a}$

94. $\dfrac{3c + a^2}{2b - 6c}$

Evaluate each expression for $w = 4$, $x = -\frac{3}{4}$, $y = \frac{1}{2}$, *and* $z = 1.25$. ***See Example 8.***

95. $wy - 8x$

96. $wz - 12y$

97. $xy + y^4$

98. $xy - x^2$

99. $-w + 2x + 3y + z$

100. $w - 6x + 5y - 3z$

Residents of Linn County, Iowa, in the Cedar Rapids Community School District can use the expression

$$(v \times 0.5485 - 4850) \div 1000 \times 31.44$$

to determine their property taxes, where v is assessed home value. (Source: The Gazette.)

Use this expression to calculate, to the nearest dollar, the amount of property taxes that the owner of a home with each of the following values would pay. Follow the rules for order of operations.

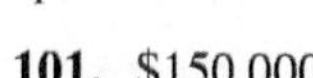

101. \$150,000 **102.** \$200,000

The Blood Alcohol Concentration (BAC) of a person who has been drinking is given by the following expression.

$$\text{number of oz} \times \%\ \text{alcohol} \times 0.075 \div \text{body weight in lb} - \text{hr of drinking} \times 0.015$$

(*Source:* Lawlor, J., *Auto Math Handbook: Calculations, Formulas, Equations and Theory for Automotive Enthusiasts,* Penguin, © 1991.)

103. Suppose a policeman stops a 190-lb man who, in 2 hr, has ingested four 12-oz beers (48 oz), each having a 3.2% alcohol content.

(a) Substitute the values into the formula, and write the expression for the man's BAC.

(b) Calculate the man's BAC to the nearest thousandth. Follow the rules for order of operations.

104. Find the BAC to the nearest thousandth for a 135-lb woman who, in 3 hr, has drunk three 12-oz beers (36 oz), each having a 4.0% alcohol content.

105. Refer to **Exercises 103 and 104.**

(a) Calculate the BACs if each person weighed 25 lb more, and the rest of the variables stayed the same. How does increased weight affect a person's BAC?

(b) Predict how decreased weight will affect the BAC of each person. Calculate the BACs if each person weighed 25 lb less and the rest of the variables stayed the same.

106. Calculate the BACs in **Exercises 103 and 104** if the time for each person decreases by 1 hr and the other variable values stay the same. How does decreased time affect a person's BAC?

Solve each problem.

107. The amount, in billions of dollars, that Americans spent on their pets from 2003 to 2016 can be approximated by substituting a given year for x in the following expression.

$$2.493x - 4962$$

(Data from American Pet Products Manufacturers Association.) Find the amount spent in each year. Round answers to the nearest tenth.

(a) 2003 **(b)** 2010 **(c)** 2016

(d) How did the amount Americans spent on their pets change from 2003 to 2016?

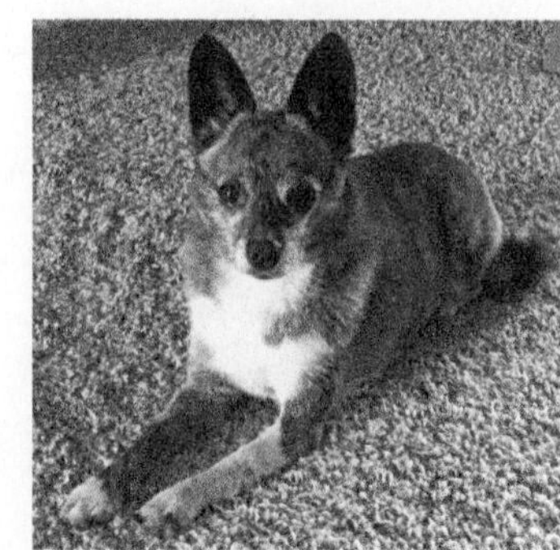

Boots

108. The average price in dollars of a movie ticket in the United States from 2000 to 2016 can be approximated using the expression

$$0.2050x - 404.6,$$

where x represents the year. (Data from National Association of Theater Owners.)

(a) Use the expression to complete the table.

(b) How much did the average price of a movie ticket in the United States increase from 2000 to 2016?

Year	Average Price (in dollars)
2000	
2005	6.43
2010	
2016	

R.5 Properties of Real Numbers

OBJECTIVES

1 Use the distributive property.
2 Use the identity properties.
3 Use the inverse properties.
4 Use the commutative and associative properties.

The basic properties of real numbers studied in this section, along with the multiplication property of zero presented earlier, reflect results that occur consistently in work with numbers. They have been generalized to apply to expressions with variables as well.

OBJECTIVE 1 Use the distributive property.

Notice the following.

$$2(3 + 5) = 2 \cdot 8 = 16$$

and $$2 \cdot 3 + 2 \cdot 5 = 6 + 10 = 16,$$

so $$2(3 + 5) = 2 \cdot 3 + 2 \cdot 5.$$

This idea is illustrated by the divided rectangle in **FIGURE 19.**

Similarly, observe these facts.

$$-4[5 + (-3)] = -4(2) = -8$$

and $$-4(5) + (-4)(-3) = -20 + 12 = -8,$$

so $$-4[5 + (-3)] = -4(5) + (-4)(-3).$$

These two examples are generalized to the set of *all* real numbers as the **distributive property of multiplication with respect to addition,** or simply the **distributive property.**

VOCABULARY

- ☐ identity element for addition (additive identity)
- ☐ identity element for multiplication (multiplicative identity)
- ☐ term
- ☐ coefficient (numerical coefficient)
- ☐ like terms
- ☐ unlike terms

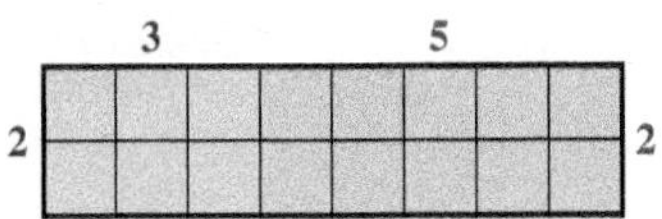

Area of left part is $2 \cdot 3 = 6$.
Area of right part is $2 \cdot 5 = 10$.
Area of total rectangle is $2(3 + 5) = 16$.

FIGURE 19

Distributive Property

For any real numbers a, b, and c, the following hold true.

$$a(b + c) = ab + ac \quad \text{and} \quad (b + c)a = ba + ca$$

Examples: $12(4 + 2) = 12 \cdot 4 + 12 \cdot 2$

$(4 + 2)12 = 4 \cdot 12 + 2 \cdot 12$

The distributive property can also be applied "in reverse."

$$ab + ac = a(b + c) \quad \text{and} \quad ba + ca = (b + c)a$$

Examples: $6 \cdot 8 + 6 \cdot 9 = 6(8 + 9)$, $8 \cdot 6 + 9 \cdot 6 = (8 + 9)6$

This property can be extended to more than two numbers as well.

$$a(b + c + d) = ab + ac + ad$$

The distributive property provides a way to rewrite a product $a(b + c)$ as a sum $ab + ac$, or a sum as a product.

NOTE When we use the distributive property to rewrite $a(b + c)$ as $ab + ac$, we refer to the process as "removing" or "clearing" parentheses.

NOW TRY EXERCISE 1

Use the distributive property to rewrite each expression.

(a) $-2(3x - y)$

(b) $4k - 12k$

(c) $3a - 5b$

EXAMPLE 1 Using the Distributive Property

Use the distributive property to rewrite each expression.

(a) $3(x + y)$

$= 3x + 3y$ Use the first form of the property to rewrite the given product as a sum.

(b) $-2(5 + k)$

$= -2(5) + (-2)(k)$ Distributive property

$= -10 - 2k$ Multiply.

(c) $4x + 8x$

$= (4 + 8)x$ Use the distributive property in reverse to rewrite the given sum as a product.

$= 12x$ Add inside the parentheses.

(d) $3r - 7r$

$= 3r + (-7r)$ Definition of subtraction

$= [3 + (-7)]r$ Distributive property in reverse

$= -4r$ Add.

(e) $5p + 7q$ This expression *cannot* be rewritten as $12pq$.

Because there is no common number or variable here, we cannot use the distributive property to rewrite the expression.

(f) $6(x + 2y - 3z)$

$= 6x + 6(2y) + 6(-3z)$ Distributive property

$= 6x + 12y - 18z$ Multiply.

NOW TRY

The distributive property can also be used for subtraction as in **Example 1(d).**

$$a(b - c) = ab - ac \qquad \textit{Example:} \quad 6(x - y) = 6x - 6y$$

OBJECTIVE 2 Use the identity properties.

The number 0 is the only number that can be added to any number to get that number, leaving the identity of the number unchanged. Thus, 0 is the **identity element for addition,** or the **additive identity.**

In a similar way, multiplying any number by 1 leaves the identity of the number unchanged. Thus, 1 is the **identity element for multiplication,** or the **multiplicative identity.** The **identity properties** summarize this discussion.

NOW TRY ANSWERS

1. **(a)** $-6x + 2y$ **(b)** $-8k$
(c) cannot be rewritten

Identity Properties

For any real number a, the following hold true.

$$a + 0 = a \quad \text{and} \quad 0 + a = a$$

Examples: $9 + 0 = 9, \quad 0 + 9 = 9$

$$a \cdot 1 = a \quad \text{and} \quad 1 \cdot a = a$$

Examples: $9 \cdot 1 = 9, \quad 1 \cdot 9 = 9$

The identity properties leave the identity of a real number unchanged. Think of a child wearing a costume on Halloween. The child's appearance is changed, but his or her identity is unchanged.

NOW TRY EXERCISE 2

Simplify each expression.

(a) $7x + x$ **(b)** $-(5p - 3q)$

EXAMPLE 2 Using the Identity Property $1 \cdot a = a$

Simplify each expression.

(a) $12m + m$

$= 12m + 1m$ Identity property

$= (12 + 1)m$ Distributive property

$= 13m$ Add inside the parentheses.

(b) $y + y$

$= 1y + 1y$

$= (1 + 1)y$

$= 2y$

(c) $-(m - 5n)$

$= -1(m - 5n)$ Identity property

$= -1(m) + (-1)(-5n)$ Distributive property

$= -m + 5n$ Multiply.

Multiply *each* term by -1. Be careful with signs.

NOW TRY

OBJECTIVE 3 Use the inverse properties.

The *additive inverse* (or *opposite*) of a number a is $-a$. Additive inverses have a sum of 0 (the additive identity).

$$5 \text{ and } -5, \quad -\frac{1}{2} \text{ and } \frac{1}{2}, \quad -34 \text{ and } 34$$

Additive inverses (sum of 0)

The *multiplicative inverse* (or *reciprocal*) of a number a is $\frac{1}{a}$ (where $a \neq 0$). Multiplicative inverses have a product of 1.

$$5 \text{ and } \frac{1}{5}, \quad -\frac{1}{2} \text{ and } -2, \quad \frac{3}{4} \text{ and } \frac{4}{3}$$

Multiplicative inverses (product of 1)

This discussion leads to the **inverse properties.**

Inverse Properties

For any real number a, the following hold true.

$$a + (-a) = 0 \quad \text{and} \quad -a + a = 0$$

Examples: $7 + (-7) = 0, \quad -7 + 7 = 0$

$$a \cdot \frac{1}{a} = 1 \quad \text{and} \quad \frac{1}{a} \cdot a = 1 \quad (\text{where } a \neq 0)$$

Examples: $7 \cdot \frac{1}{7} = 1, \quad \frac{1}{7} \cdot 7 = 1$

NOW TRY ANSWERS

2. **(a)** $8x$ **(b)** $-5p + 3q$

▼ Terms and Their Coefficients

Term	*Numerical Coefficient*
$-7y$	-7
$34r^3$	34
$-26x^5yz^4$	-26
$-k = -1k$	-1
$r = 1r$	1
$\frac{3x}{8} = \frac{3}{8}x$	$\frac{3}{8}$
$\frac{x}{3} = \frac{1x}{3} = \frac{1}{3}x$	$\frac{1}{3}$

The inverse properties "undo" addition or multiplication. Putting on your shoes when you get up in the morning and then taking them off before you go to bed at night are inverse operations that undo each other.

Expressions such as $12m$ and $5n$ from **Example 2** are examples of *terms*. A **term** is a number or the product of a number and one or more variables raised to powers. The numerical factor in a term is the **numerical coefficient,** or just the **coefficient.**

Terms with exactly the same variables raised to exactly the same powers are **like terms.** Otherwise, they are **unlike terms.**

$5p$ and $-21p$ $\quad$ $-6x^2$ and $9x^2$ $\quad$ Like terms

$3m$ and $16x$ $\quad$ $7y^3$ and $-3y^2$ $\quad$ Unlike terms

(Different variables) $\quad$ (Different exponents on the same variable)

OBJECTIVE 4 Use the commutative and associative properties.

Simplifying expressions as in **Examples 2(a) and (b)** is called **combining like terms.** ***Only like terms may be combined.*** To combine like terms in an expression such as

$$-2m + 5m + 3 - 6m + 8,$$

we need two more properties. From arithmetic, we know that the following are true.

$$3 + 9 = 12 \quad \text{and} \quad 9 + 3 = 12$$

$$3 \cdot 9 = 27 \quad \text{and} \quad 9 \cdot 3 = 27$$

The order of the numbers being added or multiplied does not matter. The same answers result. The following computations are also true.

$$(5 + 7) + 2 = 12 + 2 = 14$$

$$5 + (7 + 2) = 5 + 9 = 14$$

$$(5 \cdot 7) \cdot 2 = 35 \cdot 2 = 70$$

$$5 \cdot (7 \cdot 2) = 5 \cdot 14 = 70$$

The grouping of the numbers being added or multiplied does not matter. The same answers result.

These arithmetic examples can be extended to algebra.

Commutative and Associative Properties

For any real numbers a, b, and c, the following hold true.

$$\left.\begin{aligned} a + b &= b + a \\ ab &= ba \end{aligned}\right\} \text{ Commutative properties}$$

(The *order* of the two terms or factors changes.)

Examples: $9 + (-3) = -3 + 9$, $9(-3) = (-3)9$

$$\left.\begin{aligned} a + (b + c) &= (a + b) + c \\ a(bc) &= (ab)c \end{aligned}\right\} \text{ Associative properties}$$

(The *grouping* among the three terms or factors changes, but the order stays the same.)

Examples: $7 + (8 + 9) = (7 + 8) + 9$, $7 \cdot (8 \cdot 9) = (7 \cdot 8) \cdot 9$

The commutative properties are used to change the order of the terms or factors in an expression. Think of *commuting* from home to work and then from work to home. The associative properties are used to regroup the terms or factors of an expression. Think of *associating* the grouped terms or factors.

NOW TRY EXERCISE 3

Simplify.

$-7x + 10 - 3x - 4 + x$

EXAMPLE 3 Using the Commutative and Associative Properties

Simplify.

$$-2m + 5m + 3 - 6m + 8$$

$= (-2m + 5m) + 3 - 6m + 8$ Associative property

$= (-2 + 5)m + 3 - 6m + 8$ Distributive property

$= 3m + 3 - 6m + 8$ Add inside parentheses.

The next step would be to add $3m$ and 3, but they are unlike terms. To combine $3m$ and $-6m$, we use the associative and commutative properties, inserting parentheses and brackets according to the rules for order of operations.

$= [3m + (3 - 6m)] + 8$ Associative property

$= [3m + (-6m + 3)] + 8$ Commutative property

$= [(3m + [-6m]) + 3] + 8$ Associative property

$= (-3m + 3) + 8$ Combine like terms.

$= -3m + (3 + 8)$ Associative property

$= -3m + 11$ Add.

In practice, many of these steps are not written down, but it is important to realize that the commutative and associative properties are used whenever the terms in an expression are rearranged and regrouped to combine like terms. NOW TRY

EXAMPLE 4 Using the Properties of Real Numbers

Simplify each expression.

(a) $5y - 8y - 6y + 11y$

$= (5 - 8 - 6 + 11)y$ Distributive property

$= 2y$ Combine like terms.

(b) $3x + 4 - 5(x + 1) - 8$ Be careful with signs.

$= 3x + 4 - 5x - 5 - 8$ Distributive property

$= 3x - 5x + 4 - 5 - 8$ Commutative property

$= -2x - 9$ Combine like terms.

(c) $8 - (3m + 2)$

$= 8 - 1(3m + 2)$ Identity property

$= 8 - 3m - 2$ Distributive property

$= 6 - 3m$ Combine like terms.

NOW TRY ANSWER

3. $-9x + 6$

NOW TRY EXERCISE 4

Simplify each expression.

(a) $-3(t-4)-t+15$

(b) $7x-(4x-2)$

(c) $5x(6y)$

(d) $3(5x-7)-8(x+4)$

(d) $3x(5)(y)$

$= [3x(5)]y$ Order of operations

$= [3(x \cdot 5)]y$ Associative property

$= [3(5x)]y$ Commutative property

$= [(3 \cdot 5)x]y$ Associative property

$= (15x)y$ Multiply.

$= 15(xy)$ Associative property

$= 15xy$

As previously mentioned, many of these steps are not usually written out.

(e) $4(3x-5)-2(4x+7)$

$= 12x - 20 - 8x - 14$ Distributive property

$= 12x - 8x - 20 - 14$ Commutative property

$= 4x - 34$ Combine like terms.

Like terms may be combined by adding or subtracting the coefficients of the terms and keeping the same variable factors.

NOW TRY

CAUTION Be careful. The distributive property does not apply in **Example 4(d)** because there is no addition or subtraction involved.

$$(3x)(5)(y) \neq (3x)(5) \cdot (3x)(y)$$

NOW TRY ANSWERS

4. (a) $-4t+27$ (b) $3x+2$ (c) $30xy$ (d) $7x-53$

R.5 Exercises

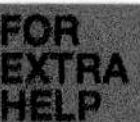

MyLab Math

Video solutions for select problems available in MyLab Math

Concept Check *Choose the correct response.*

1. The identity element for addition is

A. $-a$ **B.** 0 **C.** 1 **D.** $\frac{1}{a}$.

2. The identity element for multiplication is

A. $-a$ **B.** 0 **C.** 1 **D.** $\frac{1}{a}$.

3. The additive inverse of a is

A. $-a$ **B.** 0 **C.** 1 **D.** $\frac{1}{a}$.

4. The multiplicative inverse of a, where $a \neq 0$, is

A. $-a$ **B.** 0 **C.** 1 **D.** $\frac{1}{a}$.

Concept Check *Complete each statement.*

5. The distributive property provides a way to rewrite a product such as $a(b+c)$ as the sum _______.

6. The commutative property is used to change the _______ of two terms or factors.

7. The associative property is used to change the _______ of three terms or factors.

8. Like terms are terms with the _______ variables raised to the _______ powers.

9. When simplifying an expression, only _______ terms can be combined.

10. The numerical coefficient in the term $-7yz^2$ is _______.

Simplify each expression. **See Examples 1 and 2.**

11. $2(m + p)$
12. $3(a + b)$
13. $-12(x - y)$
14. $-10(p - q)$
15. $5k + 3k$
16. $6a + 5a$
17. $7r - 9r$
18. $4n - 6n$
19. $-8z + 4w$
20. $-12k + 3r$
21. $a + 7a$
22. $s + 9s$
23. $x + x$
24. $a + a$
25. $-(2d - f)$
26. $-(3m - n)$
27. $-(-x - y)$
28. $-(-3x - 4y)$
29. $2(x - 3y + 2z)$
30. $8(3x + y - 5z)$

Simplify each expression. **See Examples 1–4.**

31. $-12y + 4y + 3y + 2y$
32. $-5r - 9r + 8r - 5r$
33. $-6p + 5 - 4p + 6 + 11p$
34. $-8x - 12 + 3x - 5x + 9$
35. $3(k + 2) - 5k + 6 + 3$
36. $5(r - 3) + 6r - 2r + 4$
37. $10 - (4y + 8)$
38. $6 - (9y + 5)$
39. $10x(3)(y)$
40. $8x(6)(y)$
41. $-\frac{2}{3}(12w)(7z)$
42. $-\frac{5}{6}(18w)(5z)$
43. $3(m - 4) - 2(m + 1)$
44. $6(a - 5) - 4(a + 6)$
45. $0.25(8 + 4p) - 0.5(6 + 2p)$
46. $0.4(10 - 5x) - 0.8(5 + 10x)$
47. $-(2p + 5) + 3(2p + 4) - 2p$
48. $-(7m - 12) + 2(4m + 7) - 6m$
49. $2 + 3(2z - 5) - 3(4z + 6) - 8$
50. $-4 + 4(4k - 3) - 6(2k + 8) + 7$

Complete each statement so that the indicated property is illustrated. Simplify each answer if possible. **See Examples 1–4.**

51. $5x + 8x =$ ________ (distributive property)
52. $9y - 6y =$ ________ (distributive property)
53. $5(9r) =$ ________ (associative property)
54. $-4 + (12 + 8) =$ ________ (associative property)
55. $5x + 9y =$ ________ (commutative property)
56. $-5 \cdot 7 =$ ________ (commutative property)
57. $1 \cdot 7 =$ ________ (identity property)
58. $-12x + 0 =$ ________ (identity property)
59. $-\frac{1}{4}ty + \frac{1}{4}ty =$ ________ (inverse property)
60. $-\frac{9}{8}\left(-\frac{8}{9}\right) =$ ________ (inverse property)
61. $8(-4 + x) =$ ________ (distributive property)
62. $3(x - y + z) =$ ________ (distributive property)
63. $0(0.875x + 9y) =$ ________ (multiplication property of 0)
64. $0(0.35t + 12u) =$ ________ (multiplication property of 0)

65. **Concept Check** Give an "everyday" example of a commutative operation.

66. **Concept Check** Give an "everyday" example of inverse operations.

The distributive property can be used to mentally perform calculations.

$38 \cdot 17 + 38 \cdot 3$

$= 38(17 + 3)$ Distributive property

$= 38(20)$ Add inside the parentheses.

$= 760$ Multiply.

Use the distributive property to calculate each value mentally.

67. $96 \cdot 19 + 4 \cdot 19$

68. $27 \cdot 60 + 27 \cdot 40$

69. $58 \cdot \frac{3}{2} - 8 \cdot \frac{3}{2}$

70. $\frac{8}{5}(17) + \frac{8}{5}(13)$

71. $8.75(15) - 8.75(5)$

72. $4.31(69) + 4.31(31)$

RELATING CONCEPTS For Individual or Group Work (Exercises 73–78)

When simplifying an expression, we usually do some steps mentally. ***Work Exercises 73–78 in order,*** *providing the property that justifies each statement in the given simplification. (These steps could be done in other orders.)*

$3x + 4 + 2x + 7$

73. $= (3x + 4) + (2x + 7)$ ____________________

74. $= 3x + (4 + 2x) + 7$ ____________________

75. $= 3x + (2x + 4) + 7$ ____________________

76. $= (3x + 2x) + (4 + 7)$ ____________________

77. $= (3 + 2)x + (4 + 7)$ ____________________

78. $= 5x + 11$ ____________________

Chapter R Summary

Key Terms

R.1

fractions
numerator
denominator
proper fraction
improper fraction
lowest terms
mixed number
reciprocals
decimal
terminating decimal
repeating decimal
percent

R.2

set
elements (members)
finite set
natural (counting) numbers
infinite set
whole numbers
empty (null) set
variable
number line
integers
coordinate
graph
rational numbers
irrational numbers
real numbers
additive inverse (opposite, negative)
signed numbers
absolute value
equation
inequality

R.3

sum
difference
product
quotient
reciprocal (multiplicative inverse)
dividend
divisor

R.4

factors
exponent (power)
base
exponential expression
square root
positive (principal) square root
negative square root
constant
algebraic expression

R.5

identity element for addition (additive identity)
identity element for multiplication (multiplicative identity)
term
coefficient (numerical coefficient)
like terms
unlike terms

New Symbols

$0.\overline{6}$	bar notation that signifies repeating digit(s)	$\notin$	is not an element of	$>$	is greater than
%	percent	$\neq$	is not equal to	$\geq$	is greater than or equal to
$\{a, b\}$	set containing the elements a and b	$\{x \mid x$ **has property** $P\}$	set-builder notation	a^m	m factors of a
$\varnothing$	empty set	$\lvert x \rvert$	absolute value of x	$\sqrt{}$	radical symbol
$\in$	is an element of (a set)	$<$	is less than	$\sqrt{a}$	positive (principal) square root of a
		$\leq$	is less than or equal to		

Test Your Word Power

See how well you have learned the vocabulary in this chapter.

1. The **denominator** of a fraction
- **A.** is the number above the fraction bar
- **B.** gives the total number of equal parts in the whole
- **C.** gives the number of shaded parts in the whole
- **D.** is the smaller number in the fraction.

2. A **proper fraction** is a fraction that has
- **A.** numerator greater than denominator
- **B.** numerator equal to denominator
- **C.** numerator less than denominator
- **D.** denominator less than numerator.

3. The **empty set** is a set
- **A.** with 0 as its only element
- **B.** with an infinite number of elements
- **C.** with no elements
- **D.** of ideas.

4. A **variable** is
- **A.** a symbol used to represent an unknown number
- **B.** a value that makes an equation true
- **C.** a solution of an equation
- **D.** the answer in a division problem.

5. An **integer** is
- **A.** a positive or negative number
- **B.** a natural number, its opposite, or zero
- **C.** any number that can be graphed
- **D.** the quotient of two numbers.

6. The **absolute value** of a number is
- **A.** the graph of the number
- **B.** the reciprocal of the number
- **C.** the opposite of the number
- **D.** the distance between 0 and the number on a number line.

7. The **reciprocal** of a nonzero number a is

A. a **B.** $\frac{1}{a}$ **C.** $-a$ **D.** 1.

8. A **factor** is
- **A.** the answer in an addition problem
- **B.** the answer in a multiplication problem
- **C.** one of two or more numbers that are added to get another number
- **D.** any number that divides evenly into a given number.

9. An **exponent** is
- **A.** a symbol that tells how many numbers are being multiplied
- **B.** a number raised to a power
- **C.** a number that tells how many times a factor is repeated
- **D.** a number that is multiplied.

10. An **exponential expression** is
- **A.** a number that is a repeated factor in a product
- **B.** a number or a variable written with an exponent
- **C.** a number that tells how many times a factor is repeated in a product
- **D.** an expression that involves addition.

11. A **term** is
- **A.** a numerical factor
- **B.** a number or a product of a number and one or more variables raised to powers
- **C.** one of several variables with the same exponents
- **D.** a sum of numbers and variables raised to powers.

12. A **numerical coefficient** is
- **A.** the numerical factor in a term
- **B.** the number of terms in an expression
- **C.** a variable raised to a power
- **D.** the variable factor in a term.

ANSWERS

1. B; *Example:* In the fraction $\frac{3}{4}$, the denominator is 4. **2.** C; *Examples:* $\frac{1}{2}, \frac{2}{7}, \frac{5}{12}$ **3.** C; *Example:* The set of whole numbers less than 0 is the empty set, written $\varnothing$. **4.** A; *Examples:* x, y, z **5.** B; *Examples:* $-9, 0, 6$ **6.** D; *Examples:* $|2| = 2$ and $|-2| = 2$ **7.** B; *Examples:* 3 is the reciprocal of $\frac{1}{3}$; $-\frac{5}{2}$ is the reciprocal of $-\frac{2}{5}$. **8.** D; *Example:* 2 and 5 are factors of 10 because both divide evenly (without remainder) into 10. **9.** C; *Example:* In 2^3, the number 3 is the exponent (or power), so 2 is a factor three times, and $2^3 = 2 \cdot 2 \cdot 2 = 8$. **10.** B; *Examples:* 3^4 and x^{10} **11.** B; *Examples:* $6, \frac{x}{2}, -4ab^2$ **12.** A; *Examples:* The term $8z$ has numerical coefficient 8, and the term $-10x^3y$ has numerical coefficient -10.

Chapter R Test

Step-by-step test solutions are found on the Chapter Test Prep Videos available in MyLab Math.

View the complete solutions to all Chapter Test exercises in MyLab Math.

Perform the indicated operations.

1. $\frac{3}{4} + \frac{1}{6}$ **2.** $\frac{3}{7} \div \frac{9}{14}$ **3.** $13.25 - 6.417$ **4.** 0.7×0.04

Complete the table of fraction, decimal, and percent equivalents.

	Fraction in Lowest Terms	Decimal	Percent
5.			4%
6.	$\frac{5}{6}$		
7.		1.5	

8. Graph $\left\{-3, 0.75, \frac{5}{3}, 5, 6.3\right\}$ on a number line.

Let $A = \left\{-\sqrt{6}, -1, -0.5, 0, 3, \sqrt{25}, 7.5, \frac{24}{2}, \sqrt{-4}\right\}$. *Simplify the elements of A as necessary, and then list those elements of A that belong to the specified set.*

9. Whole numbers **10.** Integers **11.** Rational numbers **12.** Real numbers

Perform the indicated operations.

13. $-6 + 14 + (-11) - (-3)$ **14.** $-\frac{5}{7} - \left(-\frac{10}{9} + \frac{2}{3}\right)$

15. $10 - 4 \cdot 3 + 6(-4)$ **16.** $7 - 4^2 + 2(6) + (-4)^2$

17. $\frac{-2[3 - (-1 - 2) + 2]}{\sqrt{9}(-3) - (-2)}$ **18.** $\frac{8 \cdot 4 - 3^2 \cdot 5 - 2(-1)}{-3 \cdot 2^3 + 24}$

Find each square root. If it is not a real number, say so.

19. $\sqrt{196}$ **20.** $-\sqrt{225}$ **21.** $\sqrt{-16}$

22. Evaluate $\frac{8k + 2m^2}{r - 2}$ for $k = -3$, $m = -3$, and $r = 25$.

Simplify each expression.

23. $-3(2k - 4) + 4(3k - 5) - 2 + 4k$ **24.** $(3r + 8) - (-4r + 6)$

25. Match each statement in Column I with the appropriate property in Column II. Answers may be used more than once.

I	II
(a) $6 + (-6) = 0$	**A.** Distributive property
(b) $-2 + (3 + 6) = (-2 + 3) + 6$	**B.** Inverse property
(c) $5x + 15x = (5 + 15)x$	**C.** Identity property
(d) $13 \cdot 0 = 0$	**D.** Associative property
(e) $-9 + 0 = -9$	**E.** Commutative property
(f) $4 \cdot 1 = 4$	**F.** Multiplication property of 0
(g) $(a + b) + c = (b + a) + c$	

STUDY SKILLS **REMINDER**
You will increase your chance of success in this course if you fully utilize your text. **Review Study Skill 1, *Using Your Math Text.***

1

LINEAR EQUATIONS, INEQUALITIES, AND APPLICATIONS

The concept of balance can be applied to solving *linear equations in one variable,* a topic of this chapter.

1.1 Linear Equations in One Variable

OBJECTIVES

1. Distinguish between expressions and equations.
2. Identify linear equations.
3. Solve linear equations using the addition and multiplication properties of equality.
4. Solve linear equations using the distributive property.
5. Solve linear equations with fractions or decimals.
6. Identify conditional equations, contradictions, and identities.

VOCABULARY

- ☐ linear (first-degree) equation in one variable
- ☐ solution
- ☐ solution set
- ☐ equivalent equations
- ☐ conditional equation
- ☐ identity
- ☐ contradiction

NOW TRY EXERCISE 1

Determine whether each of the following is an *expression* or an *equation*.

(a) $2x + 17 - 3x$

(b) $2x + 17 = 3x$

NOW TRY ANSWERS

1. **(a)** expression **(b)** equation

OBJECTIVE 1 Distinguish between expressions and equations.

In our earlier work, we reviewed *algebraic expressions.*

$$8x + 9, \quad y - 4, \quad \frac{x^3y^8}{z} \qquad \text{Algebraic expressions}$$

Equations and inequalities compare algebraic expressions, just as a balance scale compares the weights of two quantities. Recall that an ***equation*** is a statement that two algebraic expressions are equal. ***An equation always contains an equality symbol, while an expression does not.***

EXAMPLE 1 Distinguishing between Expressions and Equations

Determine whether each of the following is an *expression* or an *equation*.

(a) $3x - 7 = 2$ **(b)** $3x - 7$

In part (a) we have an equation because there is an equality symbol. In part (b), there is no equality symbol, so it is an expression.

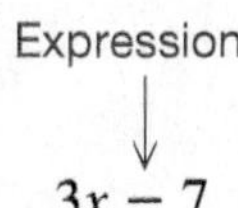

Equation: Left side $\underbrace{3x - 7}$ = Right side $\underbrace{2}$ — An equation can be *solved.*

Expression: $3x - 7$ — An expression **cannot** be solved. It can often be *evaluated* or *simplified.*

NOW TRY 

OBJECTIVE 2 Identify linear equations.

A *linear equation in one variable* involves only real numbers and one variable raised to the first power.

Linear Equation in One Variable

A **linear equation in one variable** (here x) is an equation that can be written in the form

$$ax + b = 0,$$

where a and b are real numbers and $a \neq 0$.

Examples: $x + 1 = -2, \quad x - 3 = 5, \quad 2x + 5 = 10$ Linear equations in one variable

A linear equation is a **first-degree equation** because the greatest power on the variable is 1. Some equations that are not linear equations in one variable follow.

$$4x - 5y = 6 \quad \text{(There is more than one variable.)}$$
$$x^2 + 4 = 0 \quad \text{(The exponent on the variable is not 1.)}$$
Not linear equations in one variable

If the variable in an equation can be replaced by a real number that makes the statement true, then that number is a **solution** of the equation. For example, 8 is a solution of the equation $x - 3 = 5$ because replacing x with 8 gives a true statement.

An equation is *solved* by finding its **solution set,** the set of all solutions. The solution set of the equation $x - 3 = 5$ is $\{8\}$.

Equivalent equations are related equations that have the same solution set. To solve an equation, we usually start with the given equation and replace it with a series of simpler equivalent equations. The following are all equivalent because each has solution set $\{3\}$.

$$5x + 2 = 17, \quad 5x = 15, \quad x = 3 \qquad \text{Equivalent equations}$$

OBJECTIVE 3 **Solve linear equations using the addition and multiplication properties of equality.**

We use two important properties of equality to produce equivalent equations.

Addition and Multiplication Properties of Equality

Addition Property of Equality

If a, b, and c represent real numbers, then the equations

$$a = b \quad \text{and} \quad a + c = b + c \quad \text{are equivalent.}$$

That is, the same number may be added to each side of an equation without changing the solution set.

Multiplication Property of Equality

If a, b, and c represent real numbers and $c \neq 0$, then the equations

$$a = b \quad \text{and} \quad ac = bc \quad \text{are equivalent.}$$

That is, each side of an equation may be multiplied by the same nonzero number without changing the solution set.

Because subtraction and division are defined in terms of addition and multiplication, respectively, the preceding properties can be extended.

The same number may be subtracted from each side of an equation, and each side of an equation may be divided by the same nonzero number, without changing the solution set.

EXAMPLE 2 **Solving a Linear Equation**

Solve $4x - 2x - 5 = 4 + 6x + 3$.

The goal is to isolate x on one side of the equation.

$$4x - 2x - 5 = 4 + 6x + 3$$

$$2x - 5 = 7 + 6x \qquad \text{Combine like terms.}$$

$$2x - 5 - 6x = 7 + 6x - 6x \qquad \text{Subtract } 6x \text{ from each side.}$$

$$-4x - 5 = 7 \qquad \text{Combine like terms.}$$

$$-4x - 5 + 5 = 7 + 5 \qquad \text{Add 5 to each side.}$$

$$-4x = 12 \qquad \text{Combine like terms.}$$

$$\frac{-4x}{-4} = \frac{12}{-4} \qquad \text{Divide each side by } -4.$$

$$x = -3$$

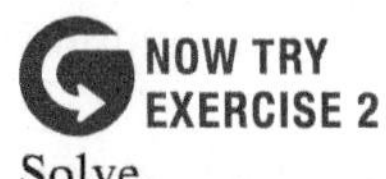
NOW TRY EXERCISE 2

Solve.

$5x + 11 = 2x - 13 - 3x$

CHECK Substitute -3 for x in the original equation.

$$4x - 2x - 5 = 4 + 6x + 3 \quad \text{Original equation}$$

$$4(-3) - 2(-3) - 5 \stackrel{?}{=} 4 + 6(-3) + 3 \quad \text{Let } x = -3.$$

$$-12 + 6 - 5 \stackrel{?}{=} 4 - 18 + 3 \quad \text{Multiply.}$$

$$-11 = -11 \checkmark \quad \text{True}$$

Use parentheses around substituted values to avoid errors.

This is *not* the solution.

The true statement indicates that $\{-3\}$ is the solution set. NOW TRY 

CAUTION In **Example 2,** the equality symbols are aligned in a column. ***Use only one equality symbol in a horizontal line of work when solving an equation.***

Solving a Linear Equation in One Variable

Step 1 **Simplify each side separately.** Use the distributive property as needed.

- Clear any parentheses.
- Clear any fractions or decimals.
- Combine like terms.

Step 2 **Isolate the variable terms on one side.** Use the addition property of equality so that all terms with variables are on one side of the equation and all constants (numbers) are on the other side.

Step 3 **Isolate the variable.** Use the multiplication property of equality to obtain an equation that has just the variable with coefficient 1 on one side.

Step 4 **Check.** Substitute the value found into the *original* equation. If a true statement results, write the solution set. If not, rework the problem.

OBJECTIVE 4 Solve linear equations using the distributive property.

EXAMPLE 3 Solving a Linear Equation

Solve $2(x - 5) + 3x = x + 6$.

Step 1 Clear parentheses using the distributive property. Then combine like terms.

Be sure to distribute over *all* terms within the parentheses.

$$2(x - 5) + 3x = x + 6$$

$$2x + 2(-5) + 3x = x + 6 \quad \text{Distributive property}$$

$$2x - 10 + 3x = x + 6 \quad \text{Multiply.}$$

$$5x - 10 = x + 6 \quad \text{Combine like terms.}$$

Step 2 Isolate the variable terms on one side and constants on the other.

$$5x - 10 - x = x + 6 - x \quad \text{Subtract } x.$$

$$4x - 10 = 6 \quad \text{Combine like terms.}$$

$$4x - 10 + 10 = 6 + 10 \quad \text{Add 10.}$$

$$4x = 16 \quad \text{Combine like terms.}$$

NOW TRY ANSWER

2. $\{-4\}$

NOW TRY EXERCISE 3

Solve.

$$5(x-4)-12=3-2x$$

Step 3 Use the multiplication property of equality to isolate x on the left.

$$\frac{4x}{4}=\frac{16}{4} \quad \text{Divide by 4.}$$

$$x=4$$

Step 4 Check by substituting 4 for x in the original equation.

CHECK

Always check your work.

$$2(x-5)+3x=x+6 \quad \text{Original equation}$$

$$2(4-5)+3(4)\stackrel{?}{=}4+6 \quad \text{Let } x=4.$$

$$2(-1)+12\stackrel{?}{=}10 \quad \text{Simplify.}$$

$$10=10 \checkmark \quad \text{True}$$

A true statement results, so the solution set is $\{4\}$. NOW TRY 

NOW TRY EXERCISE 4

Solve.

$$2-3(2+6x)=4(x+1)+36$$

EXAMPLE 4 Solving a Linear Equation

Solve $4(3x-2)=38-2(2x-1)$.

$$4(3x-2)=38-2(2x-1)$$

Step 1 $4(3x)+4(-2)=38-2(2x)-2(-1)$ Clear parentheses using the distributive property.

Be careful with signs when distributing.

$$12x-8=38-4x+2 \quad \text{Multiply.}$$

$$12x-8=40-4x \quad \text{Combine like terms.}$$

Step 2 $12x-8+4x=40-4x+4x$ Add $4x$.

$$16x-8=40 \quad \text{Combine like terms.}$$

$$16x-8+8=40+8 \quad \text{Add 8.}$$

$$16x=48 \quad \text{Combine like terms.}$$

Step 3 $\frac{16x}{16}=\frac{48}{16}$ Divide by 16.

$$x=3$$

Step 4 **CHECK** $4(3x-2)=38-2(2x-1)$ Original equation

$$4[3(3)-2]\stackrel{?}{=}38-2[2(3)-1] \quad \text{Let } x=3.$$

$$4[7]\stackrel{?}{=}38-2[5] \quad \text{Work inside the brackets.}$$

$$28=28 \checkmark \quad \text{True}$$

A true statement results, so the solution set is $\{3\}$.  

OBJECTIVE 5 Solve linear equations with fractions or decimals.

When fractions appear as coefficients in an equation, we multiply each side by the least common denominator (LCD) of all the fractions. This is an application of the multiplication property of equality.

NOW TRY ANSWERS

3. $\{5\}$

4. $\{-2\}$

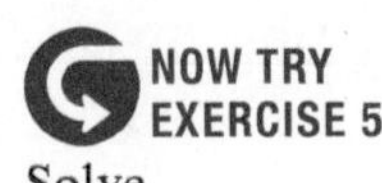
NOW TRY EXERCISE 5

Solve.

$$\frac{x-4}{4}+\frac{2x+4}{8}=5$$

EXAMPLE 5 Solving a Linear Equation with Fractions

Solve $\frac{x+7}{6}+\frac{2x-8}{2}=-4$.

$$\frac{x+7}{6}+\frac{2x-8}{2}=-4$$

Step 1 $$6\left(\frac{x+7}{6}+\frac{2x-8}{2}\right)=6(-4)$$ Eliminate the fractions. Multiply each side by the LCD, 6.

$$6\left(\frac{x+7}{6}\right)+6\left(\frac{2x-8}{2}\right)=6(-4)$$ Distributive property

$$(x+7)+3(2x-8)=-24$$ Multiply.

$$x+7+3(2x)+3(-8)=-24$$ Distributive property

$$x+7+6x-24=-24$$ Multiply.

$$7x-17=-24$$ Combine like terms.

Step 2 $$7x-17+17=-24+17$$ Add 17.

$$7x=-7$$ Combine like terms.

Step 3 $$\frac{7x}{7}=\frac{-7}{7}$$ Divide by 7.

$$x=-1$$

Step 4 **CHECK** $$\frac{x+7}{6}+\frac{2x-8}{2}=-4$$ Original equation

$$\frac{-1+7}{6}+\frac{2(-1)-8}{2}\stackrel{?}{=}-4$$ Let $x=-1$.

$$1-5\stackrel{?}{=}-4$$ Simplify each fraction.

$$-4=-4 \checkmark$$ True

A true statement results, so the solution set is $\{-1\}$. NOW TRY 

When an equation involves decimal coefficients, we can clear the decimals by multiplying by a power of 10, such as $10^1=10$, $10^2=100$, and so on, to obtain an equivalent equation with integer coefficients.

EXAMPLE 6 Solving a Linear Equation with Decimals

Solve $0.06x+0.09(15-x)=0.07(15)$.

$$0.06x+0.09(15-x)=0.07(15)$$ Clear parentheses first.

Step 1 $$0.06x+0.09(15)+0.09(-x)=0.07(15)$$ Distributive property

$$0.06x+1.35-0.09x=1.05$$ Multiply.

Now clear the decimals by multiplying by a power of 10. Because each decimal number is given in hundredths, multiply each term of the equation by 100. A number can be multiplied by 100 (that is, by 10^2) by moving the decimal point two places to the right.

NOW TRY ANSWER

5. $\{11\}$

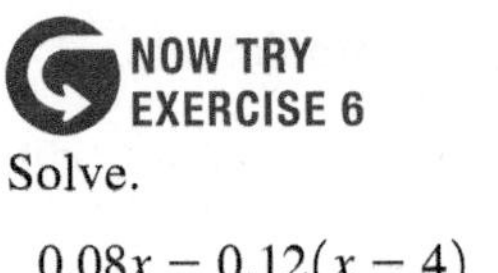

Solve.

$$0.08x - 0.12(x - 4) = 0.03(x - 5)$$

$$0.06x + 1.35 - 0.09x = 1.05 \quad \text{Multiply each term by 100.}$$

Move decimal points 2 places to the right.

$$6x + 135 - 9x = 105$$ This is an equivalent equation without decimals.

$$-3x + 135 = 105 \quad \text{Combine like terms.}$$

Step 2 $$-3x + 135 - 135 = 105 - 135 \quad \text{Subtract 135.}$$

$$-3x = -30 \quad \text{Combine like terms.}$$

Step 3 $$\frac{-3x}{-3} = \frac{-30}{-3} \quad \text{Divide by } -3.$$

$$x = 10$$

Step 4 **CHECK** $$0.06x + 0.09(15 - x) = 0.07(15) \quad \text{Original equation}$$

$$0.06(10) + 0.09(15 - 10) \stackrel{?}{=} 0.07(15) \quad \text{Let } x = 10.$$

$$0.6 + 0.09(5) \stackrel{?}{=} 1.05 \quad \text{Multiply and subtract.}$$

$$0.6 + 0.45 \stackrel{?}{=} 1.05 \quad \text{Multiply.}$$

$$1.05 = 1.05 \checkmark \quad \text{True}$$

The solution set is $\{10\}$. NOW TRY

NOTE Some students prefer to solve an equation with decimal coefficients without clearing the decimals.

$$0.06x + 0.09(15 - x) = 0.07(15) \quad \text{Equation from Example 6}$$

$$0.06x + 0.09(15) + 0.09(-x) = 0.07(15) \quad \text{Distributive property}$$

$$0.06x + 1.35 - 0.09x = 1.05 \quad \text{Multiply.}$$

Be careful with decimal points.

$$-0.03x + 1.35 = 1.05 \quad \text{Combine like terms.}$$

$$-0.03x + 1.35 - 1.35 = 1.05 - 1.35 \quad \text{Subtract 1.35.}$$

$$-0.03x = -0.3 \quad \text{Combine like terms.}$$

$$\frac{-0.03x}{-0.03} = \frac{-0.3}{-0.03} \quad \text{Divide by } -0.03.$$

$$x = 10$$ The same solution results.

OBJECTIVE 6 Identify conditional equations, contradictions, and identities.

Some equations have one solution, some have no solution, and others have an infinite number of solutions. The table gives the names of these types of equations.

▼ **Solution Sets of Equations**

Type of Linear Equation	*Number of Solutions*	*Indication When Solving*
Conditional	One solution; solution set: {a number}	Final line is $x = a$ number. (See **Examples 2–6, 7(a).**)
Identity	Infinitely many solutions; solution set: {all real numbers}	Final line is true, such as $0 = 0$. (See **Example 7(b).**)
Contradiction	No solution; solution set: $\varnothing$	Final line is false, such as $-15 = -20$. (See **Example 7(c).**)

NOW TRY ANSWER

6. $\{9\}$

Solve each equation. Determine whether it is a *conditional equation,* an *identity,* or a *contradiction.*

(a) $9x - 3(x + 4) = 6(x - 2)$

(b) $-3(2x - 1) - 2x = 3 + x$

(c) $10x - 21 = 2(x - 5) + 8x$

EXAMPLE 7 Recognizing Conditional Equations, Identities, and Contradictions

Solve each equation. Determine whether it is a *conditional equation,* an *identity,* or a *contradiction.*

(a)

$$5(2x + 6) - 2 = 7(x + 4)$$

$$10x + 30 - 2 = 7x + 28 \quad \text{Distributive property}$$

$$10x + 28 = 7x + 28 \quad \text{Combine like terms.}$$

$$10x + 28 - 7x - 28 = 7x + 28 - 7x - 28 \quad \text{Subtract } 7x. \text{ Subtract } 28.$$

$$3x = 0 \quad \text{Combine like terms.}$$

$$\frac{3x}{3} = \frac{0}{3} \quad \text{Divide by 3.}$$

$$x = 0 \quad \tfrac{0}{3} = 0$$

The last line has a variable. The number following "=" is a solution.

CHECK

$$5(2x + 6) - 2 = 7(x + 4) \quad \text{Original equation}$$

$$5[2(0) + 6] - 2 \stackrel{?}{=} 7(0 + 4) \quad \text{Let } x = 0.$$

$$5(6) - 2 \stackrel{?}{=} 7(4) \quad \text{Multiply and then add.}$$

$$28 = 28 \ \checkmark \quad \text{True}$$

The value 0 checks, so the solution set is $\{0\}$. Because the solution set has only one element, $5(2x + 6) - 2 = 7(x + 4)$ is a conditional equation.

(b)

$$5x - 15 = 5(x - 3)$$

$$5x - 15 = 5x - 15 \quad \text{Distributive property}$$

$$5x - 15 - 5x + 15 = 5x - 15 - 5x + 15 \quad \text{Subtract } 5x. \text{ Add } 15.$$

$$0 = 0 \quad \text{True}$$

The variable has "disappeared."

The *true* statement $0 = 0$ indicates that the solution set is {all real numbers}. The equation $5x - 15 = 5(x - 3)$ is an identity. Notice that the first step yielded

$$5x - 15 = 5x - 15, \quad \text{which is true for } all \text{ values of } x.$$

We could have identified the equation as an identity at that point.

(c)

$$5x - 15 = 5(x - 4)$$

$$5x - 15 = 5x - 20 \quad \text{Distributive property}$$

$$5x - 15 - 5x = 5x - 20 - 5x \quad \text{Subtract } 5x.$$

$$-15 = -20 \quad \text{False}$$

The variable has "disappeared."

Because the result, $-15 = -20$, is *false,* the equation has no solution. The solution set is the empty set $\varnothing$, so the equation $5x - 15 = 5(x - 4)$ is a contradiction.

NOW TRY

CAUTION A common error in solving an equation like that in **Example 7(a)** is to think that the equation has no solution and write the solution set as $\varnothing$. This equation has one solution, the number 0, so it is a conditional equation with solution set $\{0\}$.

NOW TRY ANSWERS

7. **(a)** {all real numbers}; identity
(b) $\{0\}$; conditional equation
(c) $\varnothing$; contradiction

1.1 Exercises

FOR EXTRA HELP  MyLab Math

Video solutions for select problems available in MyLab Math

STUDY SKILLS REMINDER
Be sure to read and work through the section material before working the exercises. **Review Study Skill 2, *Reading Your Math Text.***

Concept Check *Complete each statement. The following key terms may be used once, more than once, or not at all.*

linear equation	solution	algebraic expression	contradiction	all real numbers
solution set	identity	conditional equation	first-degree equation	empty set $\varnothing$

1. A collection of numbers, variables, operation symbols, and grouping symbols, such as $2(8x - 15)$, is a(n) ________. While an equation (*does/does not*) include an equality symbol, there (*is/is not*) an equality symbol in an algebraic expression.

2. A(n) ________ in one variable (here x) can be written in the form $ax + b\ (=/>/<)\ 0$, with $a \neq 0$. Another name for a linear equation is a(n) ________, because the greatest power on the variable is (*one/two/three*).

3. If we let $x = 2$ in the linear equation $2x + 5 = 9$, a (*true/false*) statement results. The number 2 is a(n) ________ of the equation, and $\{2\}$ is the ________.

4. A linear equation with one solution in its ________, such as $2x + 5 = 9$, is a(n) ________.

5. A linear equation with an infinite number of solutions is a(n) ________. Its solution set is {________}.

6. A linear equation with no solution is a(n) ________. Its solution set is the ________.

7. **Concept Check** Which equations are linear equations in x?

   A. $3x + x - 2 = 0$ **B.** $12 = x^2$ **C.** $9x - 4 = 9$ **D.** $3x + 2y = 6$

8. **Concept Check** Which of the equations in **Exercise 7** are nonlinear equations in x? Explain why.

Determine whether each of the following is an expression *or an* equation. ***See Example 1.***

9. $-3x + 2 - 4 = x$
10. $-3x + 2 - 4 - x = 4$
11. $4(x + 3) - 2(x + 1) - 10$
12. $4(x + 3) - 2(x + 1) + 10$
13. $-10x + 12 - 4x = -3$
14. $-10x + 12 - 4x + 3 = 0$

15. **Concept Check** This incorrect solution contains a common error.

$$8x - 2(2x - 3) = 3x + 7$$
$$8x - 4x - 6 = 3x + 7 \quad \text{Distributive property}$$
$$4x - 6 = 3x + 7 \quad \text{Combine like terms}$$
$$x = 13 \quad \text{Subtract } 3x\text{. Add 6.}$$

WHAT WENT WRONG? Give the correct solution.

16. **Concept Check** This incorrect solution contains a common error.

$$12 - 2(3x + 1) = 11$$
$$10(3x + 1) = 11 \quad \text{Subtract } 12 - 2.$$
$$30x + 10 = 11 \quad \text{Distributive property}$$
$$30x = 1 \quad \text{Subtract 10.}$$
$$x = \frac{1}{30} \quad \text{Divide by 30.}$$

WHAT WENT WRONG? Give the correct solution.

17. Concept Check Suppose we solve a linear equation and obtain, as our final result, an equation in Column I. Match each result with the solution set in Column II for the original equation.

I	II
(a) $7 = 7$	**A.** $\{0\}$
(b) $x = 0$	**B.** $\{\text{all real numbers}\}$
(c) $7 = 0$	**C.** $\varnothing$

18. Concept Check Which one of the following linear equations does *not* have solution set $\{\text{all real numbers}\}$?

A. $4x = 5x - x$ **B.** $3(x + 4) = 3x + 12$

C. $4x = 3x$ **D.** $\frac{3}{4}x = 0.75x$

E. $4(x - 2) = 2(2x - 4)$ **F.** $2x + 18x = 20x$

Solve each equation, and check the solution. If applicable, tell whether the equation is an identity *or a* contradiction. ***See Examples 2–4 and 7.***

19. $7x + 8 = 1$ **20.** $5x - 4 = 21$

21. $5x + 2 = 3x - 6$ **22.** $9x + 1 = 7x - 9$

23. $7x - 5x + 15 = x + 8$ **24.** $2x + 4 - x = 4x - 5$

25. $12w + 15w - 9 + 5 = -3w + 5 - 9$ **26.** $-4x + 5x - 8 + 4 = 6x - 4$

27. $3(2t - 4) = 20 - 2t$ **28.** $2(3 - 2x) = x - 4$

29. $-5(x + 1) + 3x + 2 = 6x + 4$ **30.** $5(x + 3) + 4x - 5 = 4 - 2x$

31. $-2x + 5x - 9 = 3(x - 4) - 5$ **32.** $-6x + 2x - 11 = -2(2x - 3) + 4$

33. $-2(x + 3) = -6(x + 7)$ **34.** $-4(x - 9) = -8(x + 3)$

35. $3(2x + 1) - 2(x - 2) = 5$ **36.** $4(x - 2) + 2(x + 3) = 6$

37. $2x + 3(x - 4) = 2(x - 3)$ **38.** $6x - 3(5x + 2) = 4(1 - x)$

39. $6x - 4(3 - 2x) = 5(x - 4) - 10$ **40.** $-2x - 3(4 - 2x) = 2(x - 3) + 2$

41. $-2(x + 3) - x - 4 = -3(x + 4) + 2$ **42.** $4(2x + 7) = 2x + 25 + 3(2x + 1)$

43. $2[x - (2x + 4) + 3] = 2(x + 1)$ **44.** $4[2x - (3 - x) + 5] = -(2 + 7x)$

45. $-[2x - (5x + 2)] = 2 + (2x + 7)$ **46.** $-[6x - (4x + 8)] = 9 + (6x + 3)$

47. $-3x + 6 - 5(x - 1) = -5x - (2x - 4) + 5$

48. $4(x + 2) - 8x - 5 = -3x + 9 - 2(x + 6)$

49. $7[2 - (3 + 4x)] - 2x = -9 + 2(1 - 15x)$

50. $4[6 - (1 + 2x)] + 10x = 2(10 - 3x) + 8x$

Concept Check *Answer each question.*

51. To solve the following linear equation, we multiply each side by the least common denominator of all the fractions in the equation. What is this LCD?

$$\frac{3}{4}x - \frac{1}{3}x = \frac{5}{6}x - 5$$

52. Suppose that in solving the following equation, we multiply each side by 12, rather than the *least* common denominator, 6. Would we obtain the correct solution? Explain.

$$\frac{1}{3}x + \frac{1}{2}x = \frac{1}{6}x$$

53. To solve a linear equation with decimals, we usually begin by multiplying by a power of 10 so that all coefficients are integers. What is the least power of 10 that will accomplish this goal in each equation?

(a) $0.05x + 0.12(x + 5000) = 940$ **(b)** $0.006(x + 2) = 0.007x + 0.009$

54. The expression $0.06(10 - x)(100)$ is equivalent to which one of the following?

A. $0.06 - 0.06x$ **B.** $60 - 6x$ **C.** $6 - 6x$ **D.** $6 - 0.06x$

Solve each equation, and check the solution. ***See Examples 5 and 6.***

55. $-\frac{5}{9}x = 2$

56. $\frac{3}{11}x = -5$

57. $\frac{6}{5}x = -1$

58. $-\frac{7}{8}x = 6$

59. $\frac{x}{2} + \frac{x}{3} = 5$

60. $\frac{x}{5} - \frac{x}{4} = 1$

61. $\frac{3x}{4} + \frac{5x}{2} = 13$

62. $\frac{8x}{3} - \frac{x}{2} = -13$

63. $\frac{x - 10}{5} + \frac{2}{5} = -\frac{x}{3}$

64. $\frac{5 - x}{6} + \frac{5}{6} = \frac{x}{54}$

65. $\frac{3x - 1}{4} + \frac{x + 3}{6} = 3$

66. $\frac{3x + 2}{7} - \frac{x + 4}{5} = 2$

67. $\frac{4x + 1}{3} = \frac{x + 5}{6} + \frac{x - 3}{6}$

68. $\frac{2x + 5}{5} = \frac{3x + 1}{2} + \frac{-x + 8}{16}$

69. $0.04x + 0.06 + 0.03x = 0.03x + 1.46$

70. $0.05x + 0.08 + 0.06x = 0.07x + 0.68$

71. $0.006x - 0.02x + 0.03 = 0.008x + 0.25$

72. $0.05x - 0.1x + 0.6 = 0.04x + 2.22$

73. $0.05x + 0.12(x + 5000) = 940$

74. $0.09x + 0.13(x + 300) = 61$

75. $0.02(50) + 0.08x = 0.04(50 + x)$

76. $0.04(90) + 0.12x = 0.06(460 + x)$

77. $0.05x + 0.10(200 - x) = 0.45x$

78. $0.08x + 0.12(260 - x) = 0.48x$

79. $0.006(x + 2) = 0.007x + 0.009$

80. $0.006(50 - x) = 0.272 - 0.004x$

1.2 Formulas and Percent

OBJECTIVES

1 Solve a formula for a specified variable.

2 Solve applied problems using formulas.

3 Solve percent problems.

4 Solve problems involving percent increase or decrease.

5 Solve problems from the health care industry.

A **mathematical model** is an equation or inequality that describes a real situation. Models for many applied problems, called *formulas,* already exist. A **formula** is an equation in which variables are used to describe a relationship. For example, the formula for finding the area $\mathcal{A}$* of a triangle is

$$\mathcal{A} = \frac{1}{2}bh.$$

Here, b is the length of the base and h is the height. See FIGURE 1. A list of formulas used in algebra is given at the back of this text.

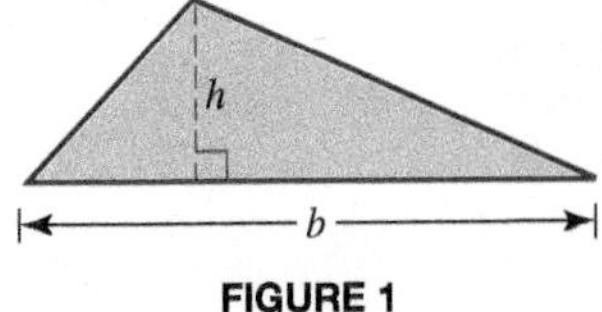

FIGURE 1

*In this text, we use $\mathcal{A}$ to denote area.

OBJECTIVE 1 Solve a formula for a specified variable.

The formula $I = prt$ says that interest on a loan or investment equals principal (amount borrowed or invested) times rate (percent) times time at interest (in years). To determine how long it will take for an investment at a stated interest rate to earn a predetermined amount of interest, it would help to first solve the formula for t. This process is called **solving for a specified variable** or **solving a literal equation.**

When solving for a specified variable, the key is to treat that variable as if it were the only one. Treat all other variables like constants (numbers).

NOW TRY EXERCISE 1
Solve the formula $I = prt$ for p.

VOCABULARY
- ☐ mathematical model
- ☐ formula
- ☐ percent
- ☐ percent increase
- ☐ percent decrease

EXAMPLE 1 Solving for a Specified Variable

Solve the formula $I = prt$ for t.

We solve this formula for t by treating I, p, and r as constants (having fixed values) and treating t as the only variable.

$$prt = I$$ Our goal is to isolate t.

$$(pr)t = I$$ Associative property

$$\frac{(pr)t}{pr} = \frac{I}{pr}$$ Divide by pr.

$$t = \frac{I}{pr}$$ The result is a formula for time t, in years.

NOW TRY

STUDY SKILLS REMINDER
Are you getting the most out of your class time? **Review Study Skill 3, *Taking Lecture Notes.***

Solving for a Specified Variable

Step 1 **Clear any parentheses and fractions.** If the equation contains parentheses, clear them using the distributive property. Clear any fractions by multiplying both sides by the LCD.

Step 2 **Isolate all terms with the specified variable.** Transform so that all terms containing the specified variable are on one side of the equation and all terms without that variable are on the other side.

Step 3 **Isolate the specified variable.** Divide each side by the factor that is the coefficient of the specified variable.

EXAMPLE 2 Solving for a Specified Variable

Solve the formula $P = 2L + 2W$ for W.

This formula gives the relationship between perimeter of a rectangle, P, length of the rectangle, L, and width of the rectangle, W. See **FIGURE 2**.

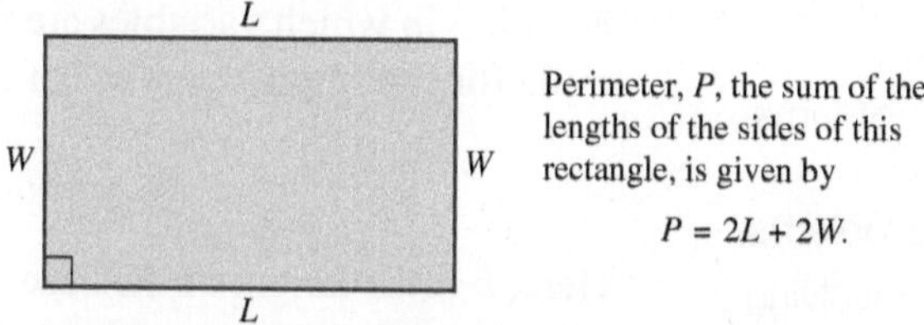

L
W
W
L
Perimeter, P, the sum of the lengths of the sides of this rectangle, is given by
P = 2L + 2W.

FIGURE 2

$$P = 2L + 2W$$ Our goal is to isolate W.

NOW TRY ANSWER
1. $p = \frac{I}{rt}$

NOW TRY EXERCISE 2
Solve the formula for b.
$$P = a + 2b + c$$

Step 1 is not needed here because there are no parentheses or fractions to clear in this formula.

$$P = 2L + 2W$$

***Step 2** $\quad P - 2L = 2L + 2W - 2L$ Subtract $2L$.

$P - 2L = 2W$ Combine like terms.

Step 3 $\quad \dfrac{P - 2L}{2} = \dfrac{2W}{2}$ Divide by 2.

$\dfrac{P - 2L}{2} = W$ $\quad \frac{2W}{2} = \frac{2}{2} \cdot W = 1 \cdot W = W$

Although W is isolated, we simplify the right side further.

$W = \dfrac{P - 2L}{2}$ Interchange sides.

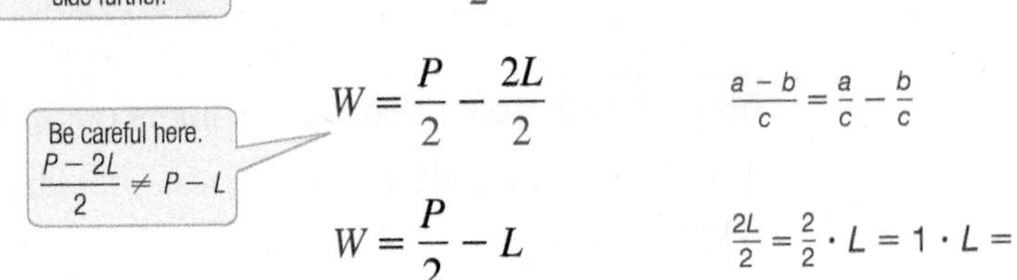

$W = \dfrac{P}{2} - \dfrac{2L}{2}$ $\quad \frac{a - b}{c} = \frac{a}{c} - \frac{b}{c}$

$W = \dfrac{P}{2} - L$ $\quad \frac{2L}{2} = \frac{2}{2} \cdot L = 1 \cdot L = L$

NOW TRY

CAUTION In Step 3 of **Example 2,** we cannot simplify the fraction $\frac{P - 2L}{2}$ by dividing 2 into the term $2L$. The fraction bar serves as a grouping symbol. Thus, the subtraction in the numerator must be done before the division.

NOW TRY EXERCISE 3
Solve $P = 2(L + W)$ for L.

EXAMPLE 3 Solving a Formula Involving Parentheses

Solve the formula $s = \frac{1}{2}(a + b + c)$ for a.

$$s = \frac{1}{2}(a + b + c)$$

Step 1 $\quad s = \dfrac{1}{2}a + \dfrac{1}{2}b + \dfrac{1}{2}c$ Distributive property

$2s = 2\left(\dfrac{1}{2}a + \dfrac{1}{2}b + \dfrac{1}{2}c\right)$ Multiply by 2 to clear the fractions.

$2s = a + b + c$ Distributive property

Step 2 $\quad 2s - b - c = a + b + c - b - c$ Subtract b and c.

Step 3 $\quad a = 2s - b - c$ Combine like terms. Interchange sides.

NOW TRY

In **Examples 1–3,** we solved formulas for specified variables. In **Example 4,** we solve an equation with two variables for one of these variables. This process will be useful later when we work with *linear equations in two variables.*

NOW TRY ANSWERS

2. $b = \frac{P - a - c}{2}$

3. $L = \frac{P - 2W}{2}$, or $L = \frac{P}{2} - W$

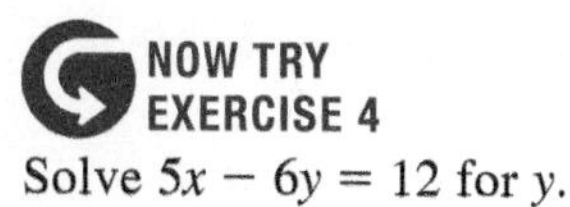

Solve $5x - 6y = 12$ for y.

EXAMPLE 4 Solving an Equation for One of the Variables

Solve $3x - 4y = 12$ for y.

Our goal is to isolate y.

$$3x - 4y = 12$$

$$3x - 4y - 3x = 12 - 3x \quad \text{Subtract } 3x.$$

$$-4y = -3x + 12 \quad \text{Combine like terms; commutative property}$$

$$\frac{-4y}{-4} = \frac{-3x + 12}{-4} \quad \text{Divide by } -4.$$

$$y = \frac{-3x}{-4} + \frac{12}{-4} \quad \frac{a+b}{c} = \frac{a}{c} + \frac{b}{c}$$

$\frac{-3x}{-4} = \frac{-3}{-4} \cdot \frac{x}{1} = \frac{3}{4}x$

$$y = \frac{3}{4}x - 3 \quad \text{Simplify the expression on the right.}$$

NOW TRY

OBJECTIVE 2 Solve applied problems using formulas.

The distance formula $d = rt$ relates d, the distance traveled, r, the rate or speed, and t, the travel time.

NOW TRY EXERCISE 5

It takes $\frac{1}{2}$ hr for Dorothy to drive 21 mi to work each day. What is her average rate?

EXAMPLE 5 Finding Average Rate

Phyllis found that on average it took her $\frac{3}{4}$ hr each day to drive a distance of 15 mi to work. What was her average rate (or speed)?

$$d = rt \quad \text{Solve the distance formula for } r.$$

$$\frac{d}{t} = \frac{rt}{t} \quad \text{Divide by } t.$$

$$\frac{d}{t} = r, \quad \text{or} \quad r = \frac{d}{t}$$

We find the desired rate by substituting the given values of d and t into this formula.

$$r = \frac{15}{\frac{3}{4}} \quad \text{Let } d = 15,\ t = \tfrac{3}{4}.$$

$$r = 15 \cdot \frac{4}{3} \quad \text{To divide by } \tfrac{3}{4}, \text{ multiply by its reciprocal, } \tfrac{4}{3}.$$

$$r = 20$$

Her average rate was 20 mph. (That is, at times she may have traveled a little faster or slower than 20 mph, but overall her rate was 20 mph.)

NOW TRY

OBJECTIVE 3 Solve percent problems.

A percent is a ratio where the second number is always 100. The percent symbol is %.

50% represents the ratio of 50 to 100—that is, $\frac{50}{100}$, or, as a decimal, 0.50.

The word **percent** means ***"per 100."*** One percent means "*one per* 100."

$$\mathbf{1\% = \frac{1}{100} = 0.01, \quad 10\% = \frac{10}{100} = 0.10, \quad 100\% = \frac{100}{100} = 1}$$

NOW TRY ANSWERS

4. $y = \frac{5}{6}x - 2$

5. 42 mph

Solving a Percent Problem

Let a represent a partial amount of b, the whole amount (or base). Then the following equation can be used to solve a percent problem.

$$\frac{\textbf{partial amount } a}{\textbf{whole amount } b} = \textbf{decimal value (which is converted to a percent)}$$

For example, if a class consists of 50 students and 32 are males, then the percent of males in the class is found as follows.

$$\frac{\text{partial amount } a}{\text{whole amount } b} = \frac{32}{50} \quad \text{Let } a = 32,\ b = 50.$$

$$= \frac{64}{100} \quad \frac{32}{50} \cdot \frac{2}{2} = \frac{64}{100}$$

$$= 0.64 \quad \text{Write as a decimal.}$$

$$= 64\% \quad \text{Convert to a percent.}$$

NOTE When interpreting the "partial amount" in the formula for solving a percent problem, understand that it may represent a quantity that is actually *larger* than the whole amount. In these cases, the percent will be greater than 100%. For example,

25 is 250% of 10.

Here 25 is the partial amount and 10 is the whole amount.

NOW TRY EXERCISE 6

Solve each problem.

(a) A 5-L mixture of water and antifreeze contains 2 L of antifreeze. What is the percent of antifreeze in the mixture?

(b) If a savings account earns 2.5% interest on a balance of \$7500 for one year, how much interest is earned?

EXAMPLE 6 Solving Percent Problems

(a) A 50-L mixture of acid and water contains 10 L of acid. What is the percent of acid in the mixture?

The given amount of the mixture is 50 L, and the part that is acid is 10 L. Let x represent the percent of acid in the mixture.

$$\begin{matrix}\text{partial amount} \rightarrow \\ \text{whole amount} \rightarrow\end{matrix} \frac{10}{50} = x$$

$$x = 0.20 \quad \text{Divide to write as a decimal. Interchange sides.}$$

$$x = 20\% \quad 0.20 = \frac{20}{100} = 20\%$$

The mixture is 20% acid.

(b) If \$4780 earns 2% interest in one year, how much interest is earned?

Let x represent the amount of interest earned (that is, the part of the whole amount invested). Because $2\% = 0.02$, the equation is written as follows.

$$\frac{x}{4780} = 0.02 \quad \frac{\text{partial amount } a}{\text{whole amount } b} = \text{decimal value}$$

$$4780 \cdot \frac{x}{4780} = 4780(0.02) \quad \text{Multiply by 4780.}$$

$$x = 95.6$$

The interest earned is \$95.60. NOW TRY

NOW TRY ANSWERS

6. **(a)** 40% **(b)** \$187.50

NOW TRY EXERCISE 7

Refer to **FIGURE 3.** How much was spent on vet care? Round the answer to the nearest tenth of a billion dollars.

EXAMPLE 7 Interpreting Percents from a Graph

In 2017, Americans spent about \$69.5 billion on their pets. Use the graph in **FIGURE 3** to determine how much of this amount was spent on pet food.

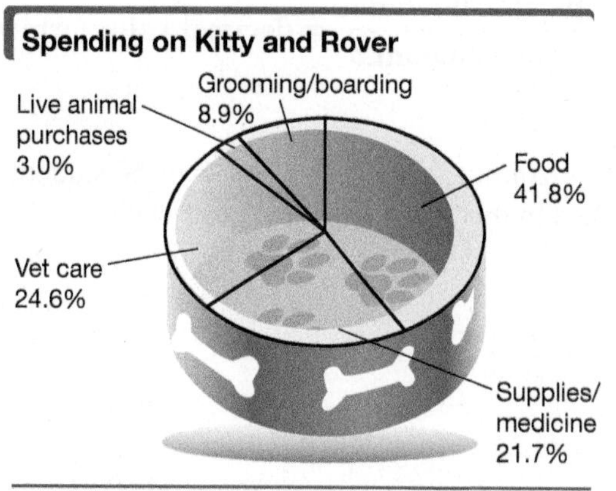

Data from American Pet Products Manufacturers Association Inc.

Bert

FIGURE 3

The graph shows that 41.8% was spent on food. Let $x =$ this amount in billions of dollars.

$$\begin{aligned} \text{partial amount} \rightarrow \quad & \frac{x}{69.5} = 0.418 && 41.8\% = 0.418 \\ \text{whole amount} \rightarrow \quad & \\ 69.5 \cdot & \frac{x}{69.5} = 69.5(0.418) && \text{Multiply by 69.5.} \\ & x \approx 29.1 && \text{Nearest tenth} \end{aligned}$$

Therefore, about \$29.1 billion was spent on pet food. NOW TRY 

OBJECTIVE 4 Solve problems involving percent increase or decrease.

Percent is often used to express a change in some quantity. Buying an item that has been marked up and getting a raise at a job are applications of **percent increase.** Buying an item on sale and finding population decline are applications of **percent decrease.**

To solve problems of this type, we use the following equation.

$$\textbf{percent change (as a decimal)} = \frac{\textbf{amount of change}}{\textbf{original amount}}$$

Subtract to find this.

EXAMPLE 8 Solving Problems about Percent Increase or Decrease

Use the percent change equation in each application.

(a) An electronics store marked up a laptop computer from their cost of \$1200 to a selling price of \$1464. What was the percent markup?

"Markup" is a name for an increase.

Let $x =$ the percent increase (as a decimal).

NOW TRY ANSWER

7. \$17.1 billion

NOW TRY EXERCISE 8

Use the percent change equation in each application.

(a) Jane bought a jacket on sale for \$56. The regular price of the jacket was \$80. What was the percent markdown?

(b) When it was time for Horatio to renew the lease on his apartment, the landlord raised his monthly rent from \$650 to \$689. What was the percent increase?

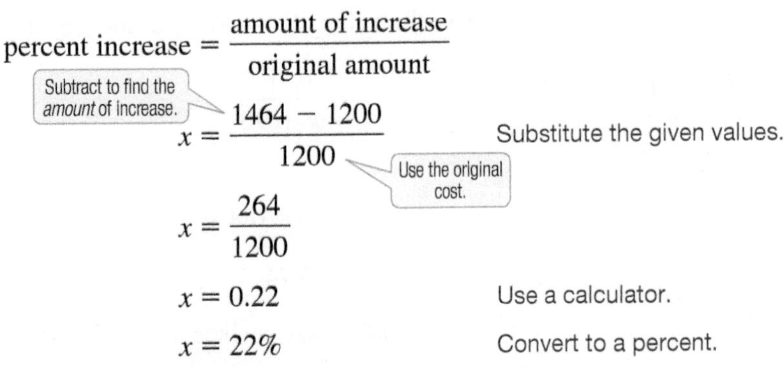
Subtract to find the *amount* of increase.
Use the original cost.

$$\text{percent increase} = \frac{\text{amount of increase}}{\text{original amount}}$$

$$x = \frac{1464 - 1200}{1200} \quad \text{Substitute the given values.}$$

$$x = \frac{264}{1200}$$

$$x = 0.22 \quad \text{Use a calculator.}$$

$$x = 22\% \quad \text{Convert to a percent.}$$

The computer was marked up 22%.

(b) The enrollment at a community college declined from 12,750 during one school year to 11,350 the following year. Find the percent decrease.

Let $x =$ the percent decrease (as a decimal).

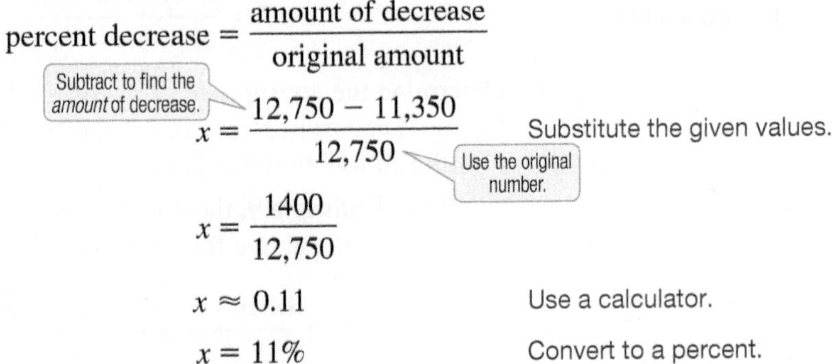
Subtract to find the *amount* of decrease.
Use the original number.

$$\text{percent decrease} = \frac{\text{amount of decrease}}{\text{original amount}}$$

$$x = \frac{12{,}750 - 11{,}350}{12{,}750} \quad \text{Substitute the given values.}$$

$$x = \frac{1400}{12{,}750}$$

$$x \approx 0.11 \quad \text{Use a calculator.}$$

$$x = 11\% \quad \text{Convert to a percent.}$$

The college enrollment decreased by about 11%. **NOW TRY** 

CAUTION When calculating a percent increase or decrease, be sure to use the original number (*before* the increase or decrease) as the denominator of the fraction, ***not*** the final number (*after* the increase or decrease).

$$\text{percent change} = \frac{\text{amount of change}}{\text{original amount}}$$

OBJECTIVE 5 Solve problems from the health care industry.

EXAMPLE 9 Determining a Child's Body Surface Area

If a child weighs k kilograms, then the child's body surface area S in square meters (m^2) is determined by the formula

$$S = \frac{4k + 7}{k + 90}.$$

What is the body surface area of a child who weighs 40 lb? (Data from Hegstad, L., and W. Hayek, *Essential Drug Dosage Calculations,* Fourth Edition, Prentice Hall.)

To convert 40 lb to kilograms, we use the identity property of multiplication and the fact that 1 lb ≈ 0.4536 kg.

(This fraction is approximately equal to 1.)

$$40 \text{ lb} \cdot \frac{0.4536 \text{ kg}}{1 \text{ lb}} \approx 18.144 \text{ kg}$$

NOW TRY ANSWERS

8. **(a)** 30% **(b)** 6%

NOW TRY EXERCISE 9

Use the formula in **Example 9** to find the body surface area, to the nearest hundredth, of a child who weighs 32 lb.

Use the value 18.144 for k in the given formula.

$$S = \frac{4k + 7}{k + 90}$$

$$S = \frac{4(18.144) + 7}{18.144 + 90} \quad \text{Let } k = 18.144.$$

$$S \approx 0.74 \quad \text{Use a calculator. Round to the nearest hundredth.}$$

The child's body surface area is 0.74 m^2.

NOW TRY EXERCISE 10

Refer to **Now Try Exercise 9.** Determine the appropriate dose, to the nearest unit, for this child if the usual adult dose is 100 mg.

EXAMPLE 10 Determining a Child's Dose of a Medication

If D represents the usual adult dose of a medication in milligrams, the corresponding child's dose C in milligrams is calculated using the formula

$$C = \frac{\text{body surface area (in square meters)}}{1.7} \times D.$$

Determine the appropriate dose for a child weighing 40 lb if the usual adult dose is 50 mg. (Data from Hegstad, L., and W. Hayek, *Essential Drug Dosage Calculations*, Fourth Edition, Prentice Hall.)

From **Example 9,** the body surface area of a child weighing 40 lb is 0.74 m^2. Use this value in the above formula to find the child's dose.

$$C = \frac{\text{body surface area (in square meters)}}{1.7} \times D$$

$$C = \frac{0.74}{1.7} \times 50 \quad \text{Body surface area} = 0.74,\ D = 50$$

$$C \approx 22 \quad \text{Use a calculator. Round to the nearest unit.}$$

The child's dose is 22 mg.

NOW TRY 

NOW TRY ANSWERS

9. 0.62 m^2
10. 36 mg

1.2 Exercises

FOR EXTRA HELP  MyLab Math

Video solutions for select problems available in MyLab Math

STUDY SKILLS REMINDER

How are you doing on your homework? **Review Study Skill 4, *Completing Your Homework.***

Concept Check *Fill in each blank with the correct response.*

1. A(n) ________ is an equation in which variables are used to describe a relationship.
2. To solve a formula for a specified variable, treat that ________ as if it were the only one and treat all other variables like ________ (numbers).

Concept Check *Work each problem to review converting between decimals and percents.*

3. Write each decimal as a percent.

   (a) 0.35 **(b)** 0.18 **(c)** 0.02 **(d)** 0.075 **(e)** 1.5

4. Write each percent as a decimal.

   (a) 60% **(b)** 37% **(c)** 8% **(d)** 3.5% **(e)** 210%

Solve each formula for the specified variable. **See Examples 1–3.**

5. $I = prt$ for r (simple interest)
6. $d = rt$ for t (distance)

7. $\mathscr{A} = LW$ (area of a rectangle)

(a) for W (b) for L

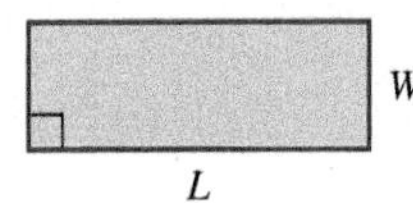

8. $\mathscr{A} = bh$ (area of a parallelogram)

(a) for b (b) for h

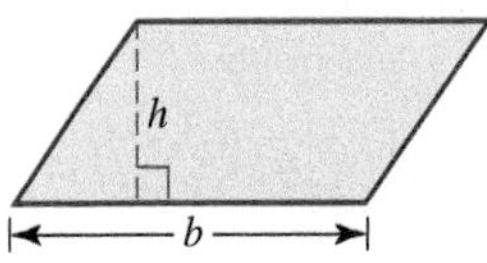

9. $P = 2L + 2W$ for L
(perimeter of a rectangle)

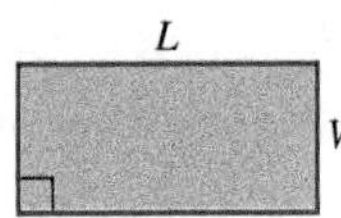

10. $P = a + b + c$ (perimeter of a triangle)

(a) for b (b) for c

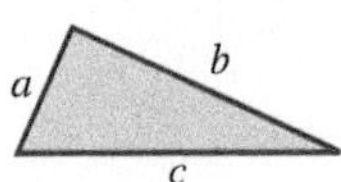

11. $V = LWH$
(volume of a rectangular solid)

(a) for W (b) for H

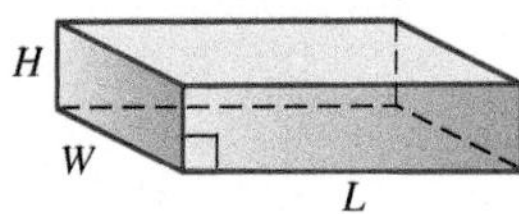

12. $\mathscr{A} = \frac{1}{2}bh$ (area of a triangle)

(a) for h (b) for b

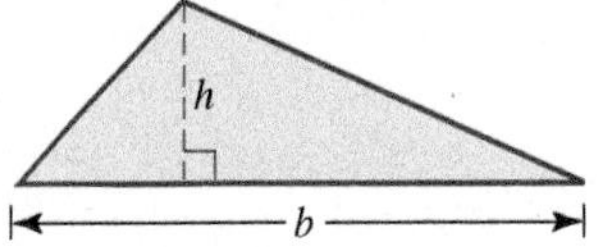

13. $C = 2\pi r$ for r
(circumference of a circle)

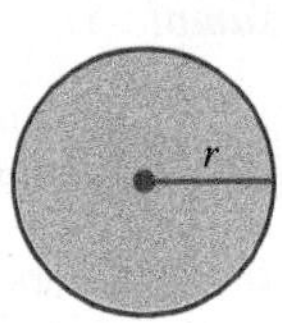

14. $V = \pi r^2 h$ for h
(volume of a right circular cylinder)

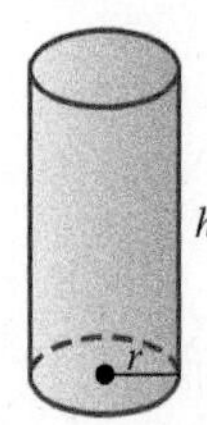

15. $\mathscr{A} = \frac{1}{2}h(b + B)$ (area of a trapezoid)

(a) for b (b) for B

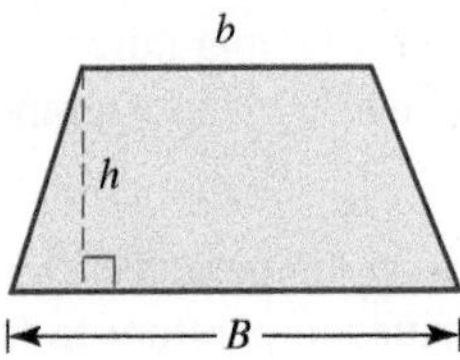

16. $V = \frac{1}{3}\pi r^2 h$ for h (volume of a cone)

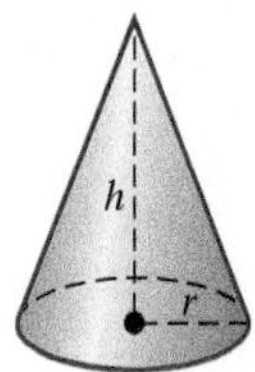

17. $F = \frac{9}{5}C + 32$ for C

(Celsius to Fahrenheit)

18. $C = \frac{5}{9}(F - 32)$ for F

(Fahrenheit to Celsius)

19. $ax + b = 0$
(linear equation in x)

(a) for x (b) for a

20. $y = mx + b$
(slope-intercept form of a linear equation)

(a) for x (b) for m

21. $A = P(1 + rt)$ for t
(future value for simple interest)

22. $M = C(1 + r)$ for r
(markup)

23. Concept Check When a formula is solved for a particular variable, several different equivalent forms may be possible. If we solve $\mathcal{A} = \frac{1}{2}bh$ for h, one correct answer is

$$h = \frac{2\mathcal{A}}{b}.$$

Which one of the following is *not* equivalent to this?

A. $h = 2\left(\frac{\mathcal{A}}{b}\right)$ **B.** $h = 2\mathcal{A}\left(\frac{1}{b}\right)$ **C.** $h = \frac{\mathcal{A}}{\frac{1}{2}b}$ **D.** $h = \frac{\frac{1}{2}\mathcal{A}}{b}$

24. Concept Check The formula $F = \frac{9}{5}C + 32$ can be solved for C to obtain $C = \frac{5}{9}(F - 32)$. Which one of the following is *not* equivalent to this?

A. $C = \frac{5}{9}F - \frac{160}{9}$ **B.** $C = \frac{5F}{9} - \frac{160}{9}$ **C.** $C = \frac{5F - 160}{9}$ **D.** $C = \frac{5}{9}F - 32$

Solve each equation for y. ***See Example 4.***

25. $4x + y = 1$ **26.** $3x + y = 9$ **27.** $x - 2y = -6$

28. $x - 5y = -20$ **29.** $4x + 9y = 11$ **30.** $7x + 8y = 11$

31. $-3x + 2y = 5$ **32.** $-5x + 3y = 12$ **33.** $6x - 5y = 7$

34. $8x - 3y = 4$ **35.** $\frac{1}{2}x - \frac{1}{3}y = 1$ **36.** $\frac{2}{3}x - \frac{2}{5}y = 2$

Solve each problem. ***See Example 5.***

37. Kurt Busch won the Daytona 500 (mile) race with a rate of 143.187 mph in 2017. Find his time to the nearest thousandth. (Data from *The World Almanac and Book of Facts.*)

38. In 2007, rain shortened the Indianapolis 500 race to 415 mi. It was won by Dario Franchitti, who averaged 151.774 mph. What was his time to the nearest thousandth? (Data from *The World Almanac and Book of Facts.*)

39. Nora traveled from Kansas City to Louisville, a distance of 520 mi, in 10 hr. Find her rate in miles per hour.

40. The distance from Melbourne to London is 10,500 mi. If a jet averages 500 mph between the two cities, what is its travel time in hours?

41. As of 2017, the highest temperature ever recorded in the state of Washington was 42°C. Find the corresponding Fahrenheit temperature to the nearest degree. (Data from National Climatic Data Center.)

42. As of 2017, the lowest temperature ever recorded in South Dakota was −41°F. Find the corresponding Celsius temperature to the nearest degree. (Data from National Climatic Data Center.)

43. The base of the Great Pyramid of Cheops is a square whose perimeter is 920 m. What is the length of each side of this square? (Data from *Atlas of Ancient Archaeology.*)

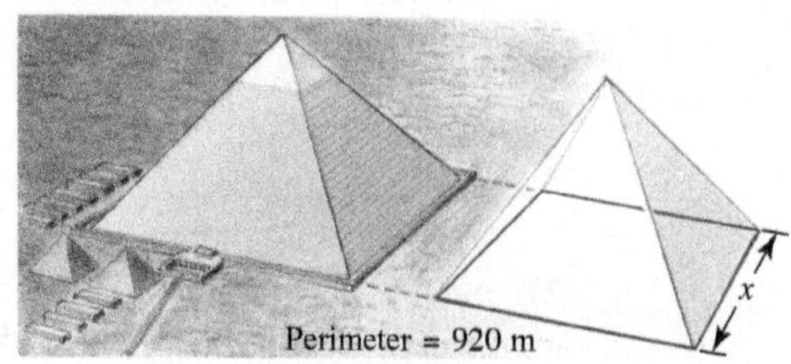

44. Marina City in Chicago is a complex of two residential towers that resemble corncobs. Each tower has a concrete cylindrical core with a 35-ft diameter and is 588 ft tall. Find the volume of the core of one of the towers to the nearest whole number. (*Hint:* Use the π key on your calculator.) (Data from www.architechgallery.com; www.aviewoncities.com)

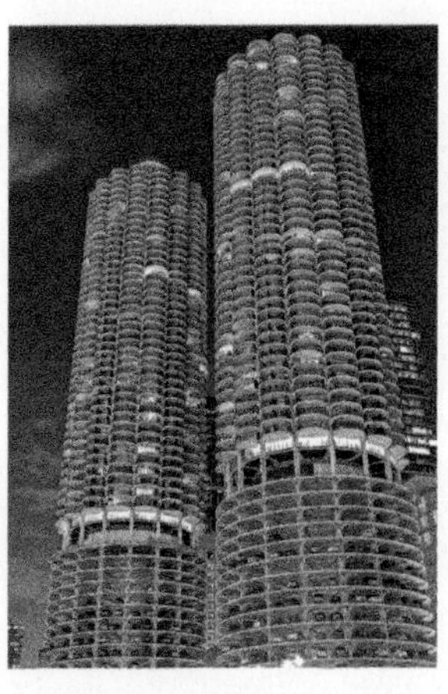

45. The circumference of a circle is 480π in.

(a) What is its radius?

(b) What is its diameter?

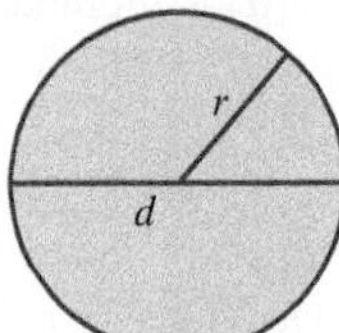

46. The radius of a circle is 2.5 in.

(a) What is its diameter?

(b) What is its circumference?

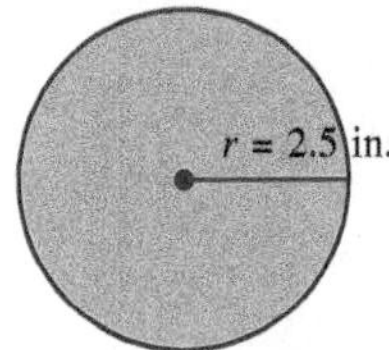

47. A sheet of standard-size copy paper measures 8.5 in. by 11 in. If a ream (500 sheets) of this paper has a volume of 187 in.^{3}, how thick is the ream?

48. Copy paper also comes in legal size, which has the same width but is longer than standard size. If a ream of legal-size paper has the same thickness as standard-size paper and a volume of 238 in.^{3}, what is the length of a sheet of legal paper?

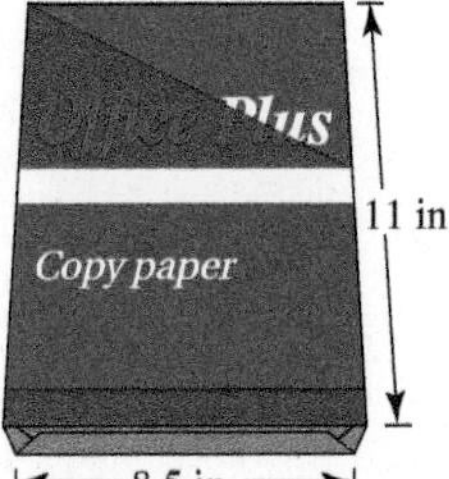

Solve each problem. ***See Example 6.***

49. 12 is what percent of 48?

50. 15 is what percent of 150?

51. A jar of 20 coins contains 4 pennies. What percent of the coins are pennies?

52. A closet containing 50 shoes has 12 that are black. What percent of the shoes are black?

53. In a group of 12 children, 3 are girls. What percent are *not* girls?

54. In a flock of 30 geese, 15 are male. What percent are *not* male?

55. A mixture of alcohol and water contains a total of 36 oz of liquid. There are 9 oz of pure alcohol in the mixture. What percent of the mixture is water? What percent is alcohol?

56. A mixture of acid and water is 35% acid. If the mixture contains a total of 40 L, how many liters of pure acid are in the mixture? How many liters of pure water are in the mixture?

57. A real-estate agent earned \$6300 commission on a property sale of \$210,000. What is her rate of commission?

58. A certificate of deposit for 1 yr pays \$25.50 simple interest on a principal of \$3400. What is the interest rate being paid on this deposit?

In baseball, winning percentage (Pct.) is commonly expressed as a decimal rounded to the nearest thousandth. To find the winning percentage of a team, divide the number of wins (W) by the total number of games played (W + L).

59. The final 2017 standings of the American League West division in Major League Baseball are shown in the table. Find the winning percentage of each team.

(a) L.A. Angels **(b)** Seattle

(c) Texas **(d)** Oakland

	W	L	Pct.
Houston	101	61	.623
L.A. Angels	80	82	
Seattle	78	84	
Texas	78	84	
Oakland	75	87	

Data from *The World Almanac and Book of Facts.*

60. The final 2017 standings of the National League Central division in Major League Baseball are shown in the table. Find the winning percentage of each team.

(a) Milwaukee **(b)** St. Louis

(c) Pittsburgh **(d)** Cincinnati

	W	L	Pct.
Chicago Cubs	92	70	.568
Milwaukee	86	76	
St. Louis	83	79	
Pittsburgh	75	87	
Cincinnati	68	94	

Data from *The World Almanac and Book of Facts.*

In June 2017, 69.5 *million U.S. households owned at least one Internet-enabled device (such as a smart TV, a video game console, or a multimedia device) that could stream video to a TV set. (Data from The Nielsen Company.) Use this information to work each problem. Round answers to the nearest percent or to the nearest tenth of a million, as needed.* ***See Example 6.***

61. About 13.4 million U.S. households owned only a smart TV in 2017. What percent of those owning at least one Internet-enabled device was this?

62. About 14.5 million U.S. households owned only a video game console in 2017. What percent of those owning at least one Internet-enabled device was this?

63. Of the households owning at least one such device in 2017, 9.4% owned all three types. How many households owned all three?

64. Of the households owning at least one Internet-enabled device in 2017, 15.8% owned both a video game console and a smart TV. How many households owned both of these?

An average middle-income family will spend \$241,080 *to raise a child from birth through age 17. The graph shows the percents spent for various categories. Use the graph to answer each question.* ***See Example 7.***

65. To the nearest dollar, how much will be spent to provide housing for the child?

66. To the nearest dollar, how much will be spent for child care and education?

67. How much will be spent on clothing?

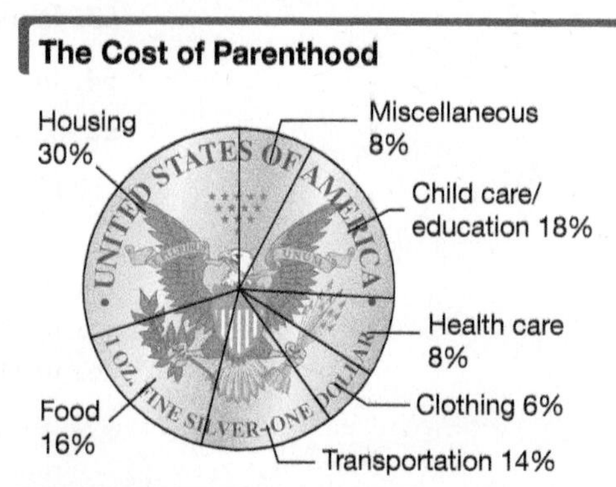

Data from U.S. Department of Agriculture.

68. About \$38,500 will be spent for food. To the nearest percent, what percent of the cost of raising a child from birth through age 17 is this? Does the answer agree with the graph?

Solve each problem. ***See Example 8.***

69. After 1 yr on the job, Grady got a raise from \$10.50 per hour to \$11.76 per hour. What was the percent increase in his hourly wage?

70. Clayton bought a ticket to a rock concert at a discount. The regular price of the ticket was \$70.00, but he paid only \$56.00. What was the percent discount?

71. Between 2000 and 2016, the estimated population of metro New Orleans, Louisiana, declined from 1,337,726 to 1,268,883. What was the percent decrease to the nearest tenth? (Data from U.S. Census.)

72. Between 2000 and 2016, the estimated population of metro Chicago, Illinois, grew from 9,098,316 to 9,512,999. What was the percent increase to the nearest tenth? (Data from U.S. Census.)

73. The movie *Transformers: The Last Knight* in DVD was on sale for \$16.17. The list price (full price) of this disc was \$17.99. To the nearest tenth, what was the percent discount? (Data from www.amazon.com)

74. The *Jurassic Park Collection* in Blu-Ray was on sale for \$29.96. The list price (full price) of this three-disc set was \$44.98. To the nearest tenth, what was the percent discount? (Data from www.amazon.com)

Apply the formulas for determining a child's body surface area and a child's dose of a medication. ***See Examples 9 and 10.***

$$S = \frac{4k + 7}{k + 90}$$

Child's body surface area S, in square meters; k = weight in kilograms

$$C = \frac{\text{body surface area (in square meters)}}{1.7} \times D$$

Child's dose of medication C, in milligrams; D = usual adult dose, in milligrams

75. Find the body surface area, to the nearest hundredth, of a child with each weight.

(a) 20 kg **(b)** 26 kg

76. Find the body surface area, to the nearest hundredth, of a child with each weight.

(a) 30 lb **(b)** 26 lb

77. If the usual adult dose of a medication is 250 mg, what is the child's dose, to the nearest unit, for a child with each weight? (*Hint:* Use the results of **Exercise 75.**)

(a) 20 kg **(b)** 26 kg

78. If the usual adult dose of a medication is 500 mg, what is the child's dose, to the nearest unit, for a child with each weight? (*Hint:* Use the results of **Exercise 76.**)

(a) 30 lb **(b)** 26 lb

RELATING CONCEPTS For Individual or Group Work **(Exercises 79–85)**

Consider the following equations.

First Equation	***Second Equation***
$\frac{7x+8}{3}=12$	$\frac{ax+k}{c}=t \quad (c \neq 0)$

Solving the second equation for x requires the same logic as solving the first equation for x. ***Work Exercises 79–85 in order,*** *to see this "parallel logic."*

79. **(a)** Clear the first equation of fractions by multiplying each side by 3.

(b) Clear the second equation of fractions by multiplying each side by c.

80. **(a)** Transform so that the term involving x is on the left side of the first equation and the constants are on the right by subtracting 8 from each side. (Do not simplify yet.)

(b) Transform so that the term involving x is on the left side of the second equation by subtracting k from each side. (Do not simplify yet.)

81. **(a)** Combine like terms in the first equation.

(b) Combine like terms in the second equation.

82. **(a)** Divide each side of the first equation by the coefficient of x. Give the solution set.

(b) Divide each side of the second equation by the coefficient of x.

83. Look at the answer for the second equation in **Exercise 82(b).** What restriction must be placed on the variables? Why is this necessary?

Now apply the concepts developed in ***Exercises 79–83*** *to work the following problems.*

84. Solve the equation. Give the solution set.

$$28=\frac{7}{2}(a+13)$$

85. Refer to the steps used to solve the equation in **Exercise 84,** and solve the following formula (from the mathematical topic of series) for a.

$$S=\frac{n}{2}(a+\ell)$$

What restriction must be placed on the variables?

1.3 Applications of Linear Equations

OBJECTIVES

1. Translate from words to mathematical expressions.
2. Write equations from given information.
3. Distinguish between simplifying expressions and solving equations.

OBJECTIVE 1 Translate from words to mathematical expressions.

Producing a mathematical model of a real situation often involves translating verbal statements into mathematical statements.

PROBLEM-SOLVING HINT Usually there are key words and phrases in a verbal problem that translate into mathematical expressions involving addition, subtraction, multiplication, and division. Translations of some commonly used expressions follow in the table on the next page.

OBJECTIVES (continued)

4 Use the six steps in solving an applied problem.

5 Solve percent problems.

6 Solve investment problems.

7 Solve mixture problems.

Use this table for reference when solving applications in this chapter.

▼ **Translating from Words to Mathematical Expressions**

Verbal Expression	*Mathematical Expression (where x and y are numbers)*
Addition	
The **sum** of a number and 7	$x + 7$
6 **more than** a number	$x + 6$
3 **plus** a number	$3 + x$
24 **added to** a number	$x + 24$
A number **increased by** 5	$x + 5$
The **sum** of two numbers	$x + y$
Subtraction	
2 **less than** a number	$x - 2$
2 **less** a number	$2 - x$
12 **minus** a number	$12 - x$
A number **decreased by** 12	$x - 12$
A number **subtracted from** 10	$10 - x$
10 **subtracted from** a number	$x - 10$
The **difference of** two numbers	$x - y$
Multiplication	
16 **times** a number	$16x$
A number **multiplied by** 6	$6x$
$\frac{2}{3}$ **of** a number (used with fractions and percent)	$\frac{2}{3}x$
$\frac{3}{4}$ **as much as** a number	$\frac{3}{4}x$
Twice (2 times) a number	$2x$
Triple (3 times) a number	$3x$
The **product** of two numbers	xy
Division	
The **quotient** of 8 and a number	$\frac{8}{x}$ $(x \neq 0)$
A number **divided by** 13	$\frac{x}{13}$
The **ratio** of two numbers x and y	$\frac{x}{y}$ $(y \neq 0)$
The **quotient** of two numbers x and y	$\frac{x}{y}$ $(y \neq 0)$

CAUTION ***Because subtraction and division are not commutative operations, it is important to correctly translate expressions involving them.***

Examples: "2 less than a number" is translated as $x - 2$, **not** $2 - x$.

"A number subtracted from 10" is translated as $10 - x$, **not** $x - 10$.

For division, the number *by which* we are dividing is the denominator, and the number *into which* we are dividing is the numerator.

Examples: "A number divided by 13" and "13 divided into x" both translate as $\frac{x}{13}$.

"The quotient of x and y" translates as $\frac{x}{y}$.

OBJECTIVE 2 Write equations from given information.

Any words that indicate equality or "sameness," such as *is*, translate as =.

NOW TRY EXERCISE 1

Translate each verbal sentence into an equation, using x as the variable.

(a) The quotient of a number and 10 is twice the number.

(b) The product of a number and 5, decreased by 7, is 0.

EXAMPLE 1 Translating Words into Equations

Translate each verbal sentence into an equation, using x as the variable.

Verbal Sentence	*Equation*
Twice a number, decreased by 3, is 42.	$2x - 3 = 42$
The product of a number and 12, decreased by 7, is 105.	$12x - 7 = 105$
The quotient of a number and the number plus 4 is 28.	$\frac{x}{x+4} = 28$
The quotient of a number and 4, plus the number, is 10.	$\frac{x}{4} + x = 10$

NOW TRY 

OBJECTIVE 3 Distinguish between simplifying expressions and solving equations.

Recall the difference between an *expression* and an *equation*.

- An expression translates as a phrase.
- An equation includes the equality symbol (=), with expressions on both sides, and translates as a sentence.

NOW TRY EXERCISE 2

Determine whether each is an *expression* or an *equation*. Simplify any expressions, and solve any equations.

(a) $3(x - 5) + 2x - 1$

(b) $3(x - 5) + 2x = 1$

EXAMPLE 2 Simplifying Expressions vs. Solving Equations

Determine whether each is an *expression* or an *equation*. Simplify any expressions, and solve any equations.

(a) $2(3 + x) - 4x + 7$

Clear parentheses and combine like terms to **simplify**.

$2(3 + x) - 4x + 7$ — There is no equality symbol. This is an **expression**.

$= 6 + 2x - 4x + 7$ — Distributive property

$= -2x + 13$ — Simplified expression

(b) $2(3 + x) - 4x + 7 = -1$

Find the value of x to **solve**.

$2(3 + x) - 4x + 7 = -1$ — There is an equality symbol. This is an **equation**.

$6 + 2x - 4x + 7 = -1$ — Distributive property

$-2x + 13 = -1$ — Combine like terms.

$-2x = -14$ — Subtract 13.

$x = 7$ — Divide by −2.

The solution set is $\{7\}$.

NOW TRY

NOW TRY ANSWERS

1. (a) $\frac{x}{10} = 2x$ (b) $5x - 7 = 0$
2. (a) expression; $5x - 16$
   (b) equation; $\left\{\frac{16}{5}\right\}$

OBJECTIVE 4 Use the six steps in solving an applied problem.

The following six steps are helpful when solving applied problems.

Solving an Applied Problem

Step 1 **Read** the problem carefully. *What information is given? What is to be found?*

Step 2 **Assign a variable** to represent the unknown value. Make a sketch, diagram, or table, as needed. If necessary, express any other unknown values in terms of the variable.

Step 3 **Write an equation** using the variable expression(s).

Step 4 **Solve** the equation.

Step 5 **State the answer.** Label it appropriately. *Does it seem reasonable?*

Step 6 **Check** the answer in the words of the *original* problem.

NOW TRY EXERCISE 3

The length of a rectangle is 2 ft more than twice the width. The perimeter is 34 ft. Find the length and the width of the rectangle.

EXAMPLE 3 Solving a Perimeter Problem

The length of a rectangle is 1 cm more than twice the width. The perimeter of the rectangle is 110 cm. Find the length and the width of the rectangle.

Step 1 **Read** the problem. *What must be found?* We must find the length and width of a rectangle. *What is given?* The length is 1 cm more than twice the width and the perimeter is 110 cm.

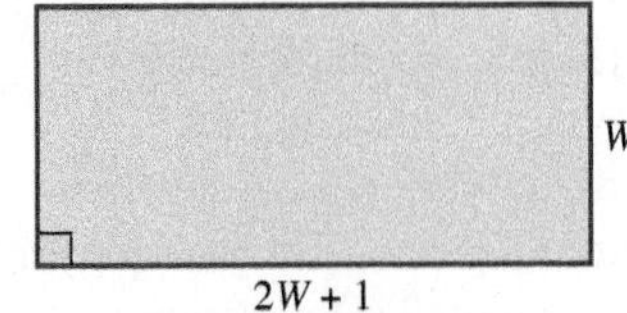

FIGURE 4

Step 2 **Assign a variable.** Make a sketch, as shown in **FIGURE 4**.

Let $W =$ the width.

Then $2W + 1 =$ the length.

Step 3 **Write an equation.** Use the formula for the perimeter of a rectangle.

$$P = 2L + 2W \quad \text{Perimeter of a rectangle}$$

$$110 = 2(2W + 1) + 2W \quad P = 110 \text{ and } L = 2W + 1$$

Step 4 **Solve.**

$$110 = 4W + 2 + 2W \quad \text{Distributive property}$$

$$110 = 6W + 2 \quad \text{Combine like terms.}$$

$$110 - 2 = 6W + 2 - 2 \quad \text{Subtract 2.}$$

$$108 = 6W \quad \text{Combine like terms.}$$

$$\frac{108}{6} = \frac{6W}{6} \quad \text{Divide by 6.}$$

$$18 = W$$

We also need to find the length.

Step 5 **State the answer.** The width of the rectangle is 18 cm and the length is

$$2(18) + 1 = 37 \text{ cm.}$$

Step 6 **Check.** The length, 37 cm, is 1 cm more than 2(18) cm (twice the width). The perimeter is $2(37) + 2(18) = 110$ cm, as required. NOW TRY

NOW TRY ANSWER

3. width: 5 ft; length: 12 ft

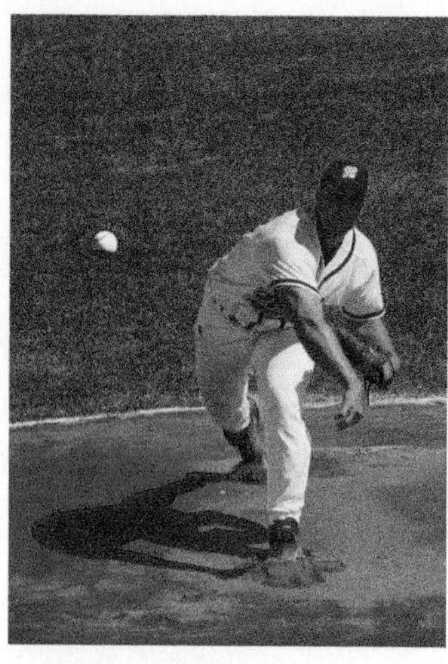

NOW TRY EXERCISE 4

During the 2017 regular NFL season, Russel Wilson of the Seattle Seahawks threw 11 more touchdown passes than Drew Brees of the New Orleans Saints. Together, these two quarterbacks completed a total of 57 touchdown passes. How many passes did each player complete? (Data from www.nfl.com)

EXAMPLE 4 Finding Unknown Numerical Quantities

During the 2017 regular Major League Baseball season, Chris Sale of the Boston Red Sox and Max Scherzer of the Washington Nationals were the top major league pitchers for strikeouts. The two pitchers had a total of 576 strikeouts. Sale had 40 more strikeouts than Scherzer. How many strikeouts did each pitcher have? (Data from www.mlb.com)

Step 1 **Read** the problem. We are asked to find the number of strikeouts each pitcher had.

Step 2 **Assign a variable** to represent the number of strikeouts for one of the men.

Let $s =$ the number of strikeouts for Max Scherzer.

We must also find the number of strikeouts for Chris Sale. Because he had 40 more strikeouts than Scherzer,

$s + 40 =$ the number of strikeouts for Sale.

Step 3 **Write an equation.** The sum of the numbers of strikeouts is 576.

Scherzer's strikeouts + Sale's strikeouts = total strikeouts.

$$s \quad + \quad (s + 40) \quad = \quad 576$$

Step 4 **Solve.**

$$s + (s + 40) = 576$$

$$2s + 40 = 576 \qquad \text{Combine like terms.}$$

$$2s + 40 - 40 = 576 - 40 \qquad \text{Subtract 40.}$$

$$2s = 536 \qquad \text{Combine like terms.}$$

$$\frac{2s}{2} = \frac{536}{2} \qquad \text{Divide by 2.}$$

$$s = 268$$

Don't stop here.

Step 5 **State the answer.** The variable s represents the number of strikeouts for Scherzer, so he had 268. Then Sale had $s + 40$, which is

$$268 + 40 = 308 \text{ strikeouts.}$$

Step 6 **Check.** 308 is 40 more than 268, and $268 + 308 = 576$. The conditions of the problem are satisfied, and our answer checks.

NOW TRY

CAUTION ***Be sure to answer all the questions asked in the problem.*** In **Example 4,** we were asked for the number of strikeouts for *each* player, so there was extra work in Step 5 in order to find Sale's number.

PROBLEM-SOLVING HINT In **Example 4,** we chose to let the variable represent the number of strikeouts for Scherzer. Students often ask, *"Can I let the variable represent the* **other** *unknown?"* The answer is yes. The equation will be different, but in the end the answers will be the same. Let s represent the number of strikeouts for Sale, and confirm this.

$s + (s - 40) = 576$ Alternative equation for **Example 4**

NOW TRY ANSWER

4. Wilson: 34; Brees: 23

OBJECTIVE 5 Solve percent problems.

Recall that percent means "*per* 100."

$$5\% \text{ means } \frac{5}{100} \text{ or } 5 \cdot \frac{1}{100} \text{ or } 5(0.01), \text{ all of which equal } 0.05.$$

$$14\% \text{ means } \frac{14}{100} \text{ or } 14 \cdot \frac{1}{100} \text{ or } 14(0.01), \text{ all of which equal } 0.14.$$

NOW TRY EXERCISE 5

In the fall of 2018, there were 96 Introductory Statistics students at a certain community college, an increase of 700% over the number of Introductory Statistics students in the fall of 2010. How many Introductory Statistics students were there in the fall of 2010?

EXAMPLE 5 Solving a Percent Problem

Total annual health expenditures in the United States were about \$3200 billion (or \$3.2 trillion) in 2015. This is an increase of 345% over the total for 1990. What were the approximate total health expenditures in billions of dollars in the United States in 1990? (Data from U.S. Centers for Medicare & Medicaid Services.)

Step 1 **Read** the problem. We are given that the total health expenditures increased by 345% from 1990 to 2015, and \$3200 billion was spent in 2015. We must find the expenditures in 1990.

Step 2 **Assign a variable.** Let x = the total health expenditures for 1990.

$$345\% = 345(0.01) = 3.45,$$

so $3.45x$ represents the increase in expenditures from 1990 to 2015.

Step 3 **Write an equation** from the given information.

The expenditures in 1990 + the increase = 3200.

$$x + 3.45x = 3200$$

Note the x in $3.45x$.

Step 4 **Solve** the equation. (We do so without clearing the decimal.)

$$1x + 3.45x = 3200 \quad \text{Identity property}$$
$$4.45x = 3200 \quad \text{Combine like terms.}$$
$$x \approx 719 \quad \text{Divide by 4.45.}$$

Step 5 **State the answer.** Total health expenditures in the United States for 1990 were about \$719 billion.

Step 6 **Check** that the increase, which is \$3200 − \$719 = \$2481 billion, is about 345% of \$719 billion.

$$3.45 \cdot 719 \approx 2481, \quad \text{as required.}$$

NOW TRY 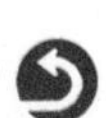

CAUTION Avoid two common errors that occur when solving problems like the one in **Example 5.**

1. Do not try to find 345% of 3200 and subtract that amount from 3200. The 345% should be applied to *the amount in 1990, not the amount in 2015.*
2. In Step 3, do not write the equation as

$$x + 3.45 = 3200. \quad \text{Incorrect}$$

The percent must be multiplied by some number. In this case, the number is the amount spent in 1990, that is, $3.45x$.

NOW TRY ANSWER

5. 12

OBJECTIVE 6 Solve investment problems.

The investment problems in this chapter deal with *simple interest*. In most real-world applications, *compound interest* (covered in a later chapter) is used.

NOW TRY EXERCISE 6

Sharon received a $20,000 inheritance from her grandfather. She invested some of the money in an account earning 3% annual interest and the remaining amount in an account earning 2.5% annual interest. If the total annual interest earned is $575, how much is invested at each rate?

EXAMPLE 6 Solving an Investment Problem

Thomas invested $40,000. He put part of the money in an account paying 2% interest and the remainder into stocks paying 3% interest. The total annual income from these investments was $1020. How much did he invest at each rate?

Step 1 **Read** the problem again. We must find the two amounts.

Step 2 **Assign a variable.**

Let $x =$ the amount invested at 2%;

$40{,}000 - x =$ the amount invested at 3%.

The formula for interest is $I = prt$. Here the time t is 1 yr. Use a table to organize the given information.

Principal (in dollars)	Rate (as a decimal)	Interest (in dollars)
x	0.02	$0.02x$
$40{,}000 - x$	0.03	$0.03(40{,}000 - x)$
40,000	XXXXX	1020 ← Total

Multiply principal, rate, and time (here, 1 yr) to find interest.

Step 3 **Write an equation.** The last column of the table gives the equation.

Interest at 2% + interest at 3% = total interest.

$$0.02x + 0.03(40{,}000 - x) = 1020$$

Step 4 **Solve** the equation. Clear the parentheses and then the decimals.

$$0.02x + 1200 - 0.03x = 1020 \quad \text{Distributive property}$$

$$100(0.02x + 1200 - 0.03x) = 100(1020) \quad \text{Multiply by 100.}$$

Move decimal points 2 places to the right.

$$2x + 120{,}000 - 3x = 102{,}000 \quad \text{Distributive property; Multiply.}$$

$$-x + 120{,}000 = 102{,}000 \quad \text{Combine like terms.}$$

$$-x = -18{,}000 \quad \text{Subtract 120,000.}$$

$$x = 18{,}000 \quad \text{Multiply by } -1.$$

Step 5 **State the answer.** Thomas invested $18,000 of the money at 2%. At 3%, he invested

$$\$40{,}000 - \$18{,}000 = \$22{,}000.$$

Step 6 **Check.** Find the annual interest at each rate. The sum of these two amounts should total $1020.

$$0.02(\$18{,}000) = \$360 \quad \text{and} \quad 0.03(\$22{,}000) = \$660$$

$$\$360 + \$660 = \$1020, \quad \text{as required.}$$

NOW TRY

NOW TRY ANSWER

6. $15,000 at 3%; $5000 at 2.5%

OBJECTIVE 7 Solve mixture problems.

Mixture problems involving rates of concentration can be solved with linear equations.

NOW TRY EXERCISE 7

How many liters of a 20% acid solution must be mixed with 5 L of a 30% acid solution to make a 24% acid solution?

EXAMPLE 7 Solving a Mixture Problem

A chemist must mix 8 L of a 40% acid solution with some 70% solution to make a 50% solution. How much of the 70% solution should be used?

Step 1 **Read** the problem. The problem asks for the amount of 70% solution to be used.

Step 2 **Assign a variable.**

Let $x =$ the number of liters of 70% solution.

The information in the problem is illustrated in **FIGURE 5**. We use it to complete a table.

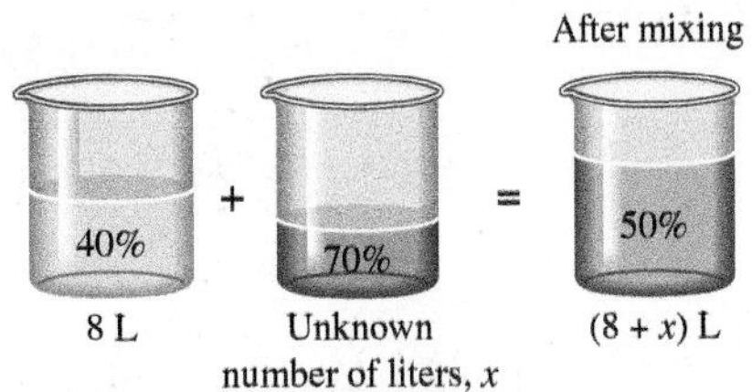

FIGURE 5

Liters of Solution	Percent (as a decimal)	Liters of Pure Acid
8	0.40	$0.40(8) = 3.2$
x	0.70	$0.70x$
$8 + x$	0.50	$0.50(8 + x)$

Sum must equal

The values in the last column were found by multiplying the strengths by the numbers of liters.

Step 3 **Write an equation** using the values in the last column of the table.

Liters of pure acid in 40% solution + liters of pure acid in 70% solution = liters of pure acid in 50% solution.

$$3.2 + 0.70x = 0.50(8 + x)$$

Step 4 **Solve.**

$$3.2 + 0.70x = 4 + 0.50x \quad \text{Distributive property}$$

Move decimal points 1 place to the right.

$$32 + 7x = 40 + 5x \quad \text{Multiply by 10 to clear the decimals.}$$

$$2x = 8 \quad \text{Subtract 32 and } 5x.$$

$$x = 4 \quad \text{Divide by 2.}$$

Step 5 **State the answer.** The chemist should use 4 L of the 70% solution.

Step 6 **Check.** 8 L of 40% solution plus 4 L of 70% solution is

$$8(0.40) + 4(0.70) = 6 \text{ L of acid.}$$

Similarly, 8 + 4 or 12 L of 50% solution has

$$12(0.50) = 6 \text{ L of acid.}$$

The total amount of pure acid is 6 L both before and after mixing, so the answer checks.

NOW TRY

NOW TRY ANSWER

7. $7\frac{1}{2}$ L

PROBLEM-SOLVING HINT When pure water is added to a solution, water is 0% of the chemical (acid, alcohol, etc.). Similarly, pure chemical is 100% chemical.

NOW TRY EXERCISE 8

How much pure antifreeze must be mixed with 3 gal of 30% antifreeze solution to obtain 40% antifreeze solution?

EXAMPLE 8 Solving a Mixture Problem When One Ingredient Is Pure

The octane rating of gasoline is a measure of its antiknock qualities. For a standard fuel, the octane rating is the percent of isooctane. How many liters of pure isooctane should be mixed with 200 L of 94% isooctane, referred to as 94 octane, to obtain a mixture that is 98% isooctane?

Step 1 **Read** the problem. We must find the amount of pure isooctane.

Step 2 **Assign a variable.** Let $x =$ the number of liters of pure (100%) isooctane. Complete a table. Recall that $100\% = 100(0.01) = 1$.

Liters of Solution	Percent (as a decimal)	Liters of Pure Isooctane
x	1	x
200	0.94	$0.94(200)$
$x + 200$	0.98	$0.98(x + 200)$

Step 3 **Write an equation** using the values in the last column of the table.

$$x + 0.94(200) = 0.98(x + 200)$$ Refer to the table.

Step 4 **Solve.**

$$x + 188 = 0.98x + 196$$ Multiply; distributive property

$$100(x + 188) = 100(0.98x + 196)$$ Multiply by 100 to clear the decimal.

$$100x + 18{,}800 = 98x + 19{,}600$$ Distributive property

$$2x = 800$$ Subtract $98x$ and 18,800.

$$x = 400$$ Divide by 2.

Step 5 **State the answer.** 400 L of isooctane is needed.

Step 6 **Check** by substituting 400 for x in the equation from Step 3.

$$400 + 0.94(200) \stackrel{?}{=} 0.98(400 + 200)$$ Let $x = 400$.

$$400 + 188 \stackrel{?}{=} 0.98(600)$$ Multiply. Add.

$$588 = 588 \checkmark$$ True

A true statement results, so the answer checks. **NOW TRY** 

NOW TRY ANSWER

8. $\frac{1}{2}$ gal

1.3 Exercises

 **MyLab Math**

Video solutions for select problems available in MyLab Math

Concept Check *In each of the following, translate part (a) as an expression and translate part (b) as an equation or inequality. Use x to represent the number.*

1. (a) 15 more than a number

(b) 15 is more than a number.

2. (a) 5 greater than a number

(b) 5 is greater than a number.

3. (a) 8 less than a number

(b) 8 is less than a number.

4. (a) 6 less than a number

(b) 6 is less than a number.

STUDY SKILLS REMINDER
You will increase your chance of success in this course if you fully utilize your text. **Review Study Skill 1, *Using Your Math Text.***

5. **Concept Check** A student translated the phrase "the difference of a number x and 7" incorrectly as $7 - x$. **WHAT WENT WRONG?** Give the correct mathematical expression.

6. **Concept Check** A student translated the phrase "the quotient of a number n and 12" incorrectly as $\frac{12}{n}$. **WHAT WENT WRONG?** Give the correct mathematical expression.

Translate each verbal phrase into a mathematical expression using x as the variable. ***See Objective 1.***

7. Twice a number, decreased by 13
8. Triple a number, decreased by 14
9. 12 increased by four times a number
10. 15 more than one-half of a number
11. The product of 8 and 16 less than a number
12. The product of 8 more than a number and 5 less than the number
13. The quotient of three times a number and 10
14. The quotient of 9 and five times a nonzero number

Translate each verbal sentence into an equation, using x as the variable. Then solve the equation. ***See Example 1.***

15. The sum of a number and 6 is -31. Find the number.
16. The sum of a number and -4 is 18. Find the number.
17. If the product of a number and -4 is subtracted from the number, the result is 9 more than the number. Find the number.
18. If the quotient of a number and 6 is added to twice the number, the result is 8 less than the number. Find the number.
19. When $\frac{2}{3}$ of a number is subtracted from 12, the result is 10. Find the number.
20. When 75% of a number is added to 6, the result is 3 more than the number. Find the number.

Determine whether each is an expression *or an* equation. *Simplify any expressions, and solve any equations.* ***See Example 2.***

21. $5(x + 3) - 8(2x - 6)$
22. $-7(x + 4) + 13(x - 6)$
23. $5(x + 3) - 8(2x - 6) = 12$
24. $-7(x + 4) + 13(x - 6) = 18$
25. $\frac{1}{2}x - \frac{1}{6}x + \frac{3}{2} - 8$
26. $\frac{1}{3}x + \frac{1}{5}x - \frac{1}{2} + 7$

Complete the six problem-solving steps to solve each problem.

27. In 2015, the corporations securing the most U.S. patents were IBM and Samsung. Together, the two corporations secured a total of 12,368 patents, with Samsung receiving 2250 fewer patents than IBM. How many patents did each corporation secure? (Data from U.S. Patent and Trademark Office.)

Step 1 **Read** the problem carefully. We are asked to find ________.

Step 2 **Assign a variable.** Let $x =$ the number of patents that IBM secured. Then $x - 2250 =$ the number of ________.

Step 3 **Write an equation.** ____ + ____ = 12,368

Step 4 **Solve** the equation. $x =$ ____

Step 5 **State the answer.** IBM secured ____ patents. Samsung secured ____ patents.

Step 6 **Check.** The number of Samsung patents was ____ fewer than the number of ________. The total number of patents was 7309 + ____ = ____.

28. In 2015, 16.2 million more U.S. residents traveled to Mexico than to Canada. There was a total of 41.2 million U.S. residents traveling to these two countries. How many traveled to each country? (Data from U.S. Department of Commerce.)

Step 1 **Read** the problem carefully. We are asked to find ______.

Step 2 **Assign a variable.** Let x = the number of travelers to Mexico (in millions). Then $x - 16.2$ = the number of ______.

Step 3 **Write an equation.** ____ + ____ = 41.2

Step 4 **Solve** the equation. x = ____

Step 5 **State the answer.** There were ____ travelers to Mexico and ____ travelers to Canada.

Step 6 **Check.** The number of ____ was ____ more than the number of travelers to ____. The total number of travelers was 28.7 + ____ = ____.

Solve each problem. ***See Examples 3 and 4.***

29. The John Hancock Center in Chicago has a rectangular base. The length of the base measures 65 ft less than twice the width. The perimeter of the base is 860 ft. What are the dimensions of the base?

30. The John Hancock Center tapers as it rises. The top floor is rectangular and has perimeter 520 ft. The width of the top floor measures 20 ft more than one-half its length. What are the dimensions of the top floor?

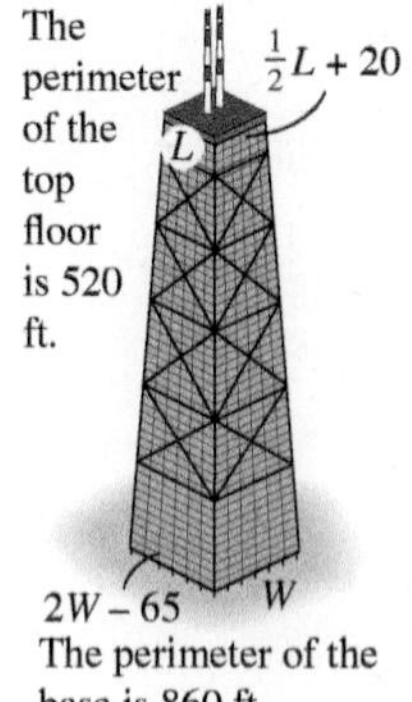

31. Grant Wood painted his most famous work, *American Gothic,* in 1930 on composition board with perimeter 108.44 in. If the rectangular painting is 5.54 in. taller than it is wide, find the dimensions of the painting. (Data from *The Gazette*.)

32. The perimeter of a certain rectangle is 16 times the width. The length is 12 cm more than the width. Find the length and width of the rectangle.

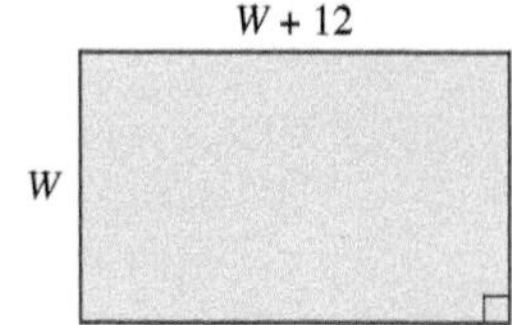

33. The Bermuda Triangle supposedly causes trouble for aircraft pilots. It has a perimeter of 3075 mi. The shortest side measures 75 mi less than the middle side, and the longest side measures 375 mi more than the middle side. Find the lengths of the three sides.

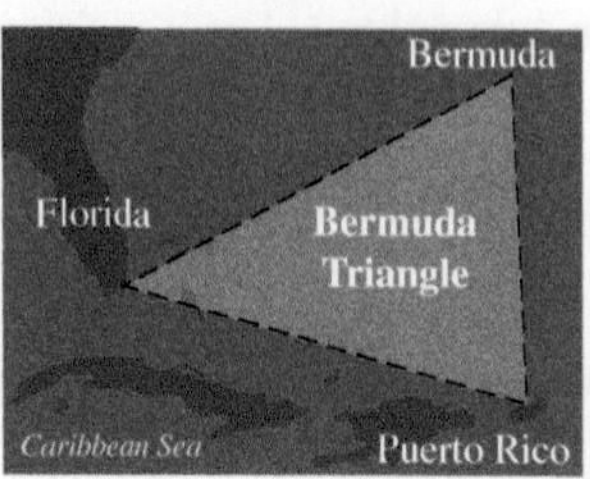

34. The Vietnam Veterans Memorial in Washington, DC, is in the shape of two sides of an isosceles triangle. If the two walls of equal length were joined by a straight line of 438 ft, the perimeter of the resulting triangle would be 931.5 ft. Find the lengths of the two walls. (Data from pamphlet obtained at Vietnam Veterans Memorial.)

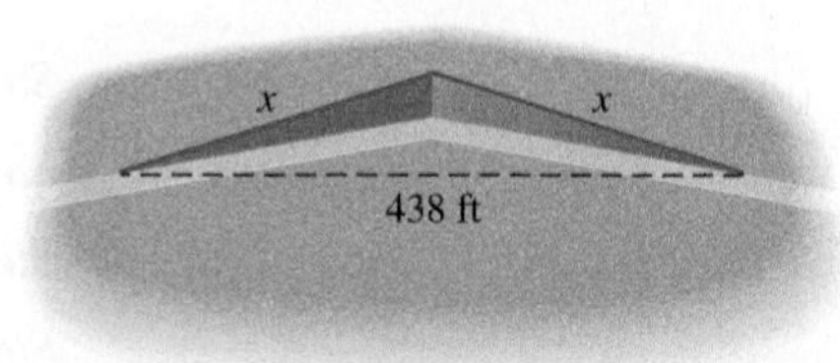

35. The two companies with top revenues in the Fortune 500 list for 2016 were Walmart and Berkshire Hathaway. Their revenues together totaled \$710.5 billion. Berkshire Hathaway revenues were \$263.3 billion less than those of Walmart. What were the revenues of each corporation? (Data from *Fortune* magazine.)

36. On a particular day in 2017, two of the longest-running Broadway shows were *The Phantom of the Opera* and *Chicago*. Together, there were 20,721 performances of these two shows during their Broadway runs. There were 3675 fewer performances of *Chicago* than of *The Phantom of the Opera*. How many performances were there of each show? (Data from The Broadway League.)

37. Galileo Galilei conducted experiments involving Italy's famous Leaning Tower of Pisa to investigate the relationship between an object's speed of fall and its weight. The Leaning Tower is 804 ft shorter than the Eiffel Tower in Paris, France. The two towers have a total height of 1164 ft. How tall is each tower? (Data from *The World Almanac and Book of Facts*.)

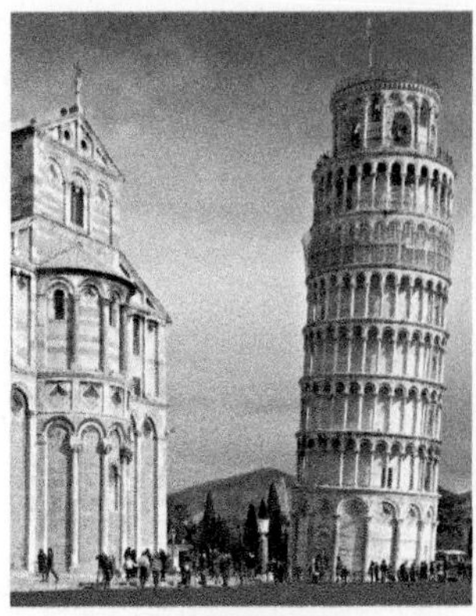

38. In 2017, the Los Angeles Dodgers and the New York Yankees had the highest payrolls in Major League Baseball. The Dodgers' payroll was \$40.6 million more than the Yankees' payroll, and the two payrolls totaled \$443.6 million. What was the payroll for each team? (Data from www.stevetheump.com)

39. In 2017, two popular brands on Instagram were National Geographic and Nike. National Geographic had 8.0 million more followers than Nike. Together, the two brands had 158.8 million followers. How many followers did each brand have? (Data from Iconosquare.)

40. Ted Williams and Rogers Hornsby were two great hitters in Major League Baseball. Together, they had 5584 hits in their careers. Hornsby had 276 more hits than Williams. How many hits did each have? (Data from Neft, D. S., and R. M. Cohen, *The Sports Encyclopedia: Baseball,* St. Martins Griffin, New York.)

Rogers Hornsby

Solve each percent problem. ***See Example 5.***

41. In 2015, the number of graduating seniors taking the ACT exam was 1,924,436. In 2011, a total of 1,617,173 graduating seniors took the exam. By what percent did the number increase over this period of time, to the nearest tenth of a percent? (Data from ACT.)

42. The U.S. population reached 321 million in 2015 and is expected to reach 347 million by the year 2025. What percent increase is expected over this time period, to the nearest tenth of a percent? (Data from www.census.gov)

43. In 2013, mobile messaging service WhatsApp had 300 million monthly active users worldwide. By 2017, this number had increased 333.3%. To the nearest million, what was the number of WhatsApp monthly active users worldwide in 2017? (Data from WhatsApp.)

44. On Major League Baseball Opening Day 2016, the payroll of the Cleveland Indians was \$87.2 million. In 2017, it had increased 44.3%. To the nearest tenth of a million, what was the opening-day payroll of the Indians for 2017? (Data from The Associated Press.)

45. In 2017, the average cost of a traditional Thanksgiving dinner for 10, featuring turkey, stuffing, cranberries, pumpkin pie, and trimmings, was \$49.12, a decrease of 1.48% from the cost in 2016. What was the average cost, to the nearest cent, of a traditional Thanksgiving dinner in 2016? (Data from American Farm Bureau.)

46. In 2016, the average cost of a Thanksgiving turkey was \$22.74. This price had decreased by 1.58% in 2017. What was the average cost of a Thanksgiving turkey in 2017? (Data from American Farm Bureau.)

47. At the end of a day, Lawrence found that the total cash register receipts at the motel where he works were \$2725. This included the 9% sales tax charged. Find the amount of tax.

48. David sold his house for \$159,000. He got this amount knowing that he would have to pay a 6% commission to his agent. What amount did he have after the agent was paid?

49. **Concept Check** Why is it impossible to mix candy worth \$4 per lb and candy worth \$5 per lb to obtain a final mixture worth \$6 per lb?

50. **Concept Check** Write an equation based on the following problem, solve the equation, and explain why the problem has no solution.

    How much 30% *acid should be mixed with* 15 L *of* 50% *acid to obtain a mixture that is* 60% *acid?*

Solve each investment problem. ***See Example 6.***

51. Mario earned \$12,000 last year giving tennis lessons. He invested part of the money at 3% simple interest and the rest at 4%. In one year, he earned a total of \$440 in interest. How much did he invest at each rate?

Principal (in dollars)	Rate (as a decimal)	Interest (in dollars)
x	0.03	
	0.04	
	XXXX	

52. Sheryl won \$60,000 on a slot machine in Las Vegas. She invested part of the money at 2% simple interest and the rest at 3%. In one year, she earned a total of \$1600 in interest. How much was invested at each rate?

Principal (in dollars)	Rate (as a decimal)	Interest (in dollars)
x	0.02	
	XXXX	

53. Michelle invested some money at 4.5% simple interest and \$1000 less than twice this amount at 3%. Her total annual income from the interest was \$1020. How much was invested at each rate?

54. Piotr invested some money at 3.5% simple interest, and \$5000 more than three times this amount at 4%. He earned \$1440 in annual interest. How much did he invest at each rate?

55. Dan has invested \$12,000 in bonds paying 6%. How much additional money should he invest in a certificate of deposit paying 3% simple interest so that the total return on the two investments will be 4%?

56. Mona received a year-end bonus of \$17,000 from her company and invested the money in an account paying 6.5%. How much additional money should she deposit in an account paying 5% so that the return on the two investments will be 6%?

Solve each problem. ***See Examples 7 and 8.***

57. How many liters of a 10% acid solution must be mixed with 10 L of a 4% solution to obtain a 6% solution?

Liters of Solution	Percent (as a decimal)	Liters of Pure Acid
10	0.04	
x	0.10	
	0.06	

58. How many liters of a 14% alcohol solution must be mixed with 20 L of a 50% solution to obtain a 30% solution?

Liters of Solution	Percent (as a decimal)	Liters of Pure Alcohol
x	0.14	
	0.50	

59. In a chemistry class, 12 L of a 12% alcohol solution must be mixed with a 20% solution to obtain a 14% solution. How many liters of the 20% solution are needed?

60. How many liters of a 10% alcohol solution must be mixed with 40 L of a 50% solution to obtain a 40% solution?

61. How much pure dye must be added to 4 gal of a 25% dye solution to increase the solution to 40%? (*Hint:* Pure dye is 100% dye.)

62. How much water must be added to 6 gal of a 4% insecticide solution to reduce the concentration to 3%? (*Hint*: Water is 0% insecticide.)

63. Randall wants to mix 50 lb of nuts worth \$2 per lb with some nuts worth \$6 per lb to make a mixture worth \$5 per lb. How many pounds of \$6 nuts must he use?

Pounds of Nuts	Cost per Pound (in dollars)	Total Cost (in dollars)

64. Lee Ann wants to mix tea worth 2¢ per oz with 100 oz of tea worth 5¢ per oz to make a mixture worth 3¢ per oz. How much 2¢ tea should be used?

Ounces of Tea	Cost per Ounce (in dollars)	Total Cost (in dollars)

RELATING CONCEPTS For Individual or Group Work (Exercises 65–68)

Consider each problem.

Problem A

Jack has \$800 invested in two accounts. One pays 3% interest per year and the other pays 6% interest per year. The amount of yearly interest is the same as he would get if the entire \$800 was invested at 5.25%. How much does he have invested at each rate?

Problem B

Jill has 800 L of acid solution. She obtained it by mixing some 3% acid with some 6% acid. Her final mixture of 800 L is 5.25% acid. How much of each of the 3% and 6% solutions did she use to obtain her final mixture?

In Problem A, let x represent the amount invested at 3% interest, and in Problem B, let y represent the amount of 3% acid used. ***Work Exercises 65–68 in order.***

65. **(a)** Write an expression in x that represents the amount of money Jack invested at 6% in Problem A.

(b) Write an expression in y that represents the amount of 6% acid solution Jill used in Problem B.

66. **(a)** Write expressions that represent the amount of interest Jack earns per year at 3% and at 6%.

(b) Write expressions that represent the amount of pure acid in Jill's 3% and 6% acid solutions.

67. **(a)** The sum of the two expressions in **Exercise 66(a)** must equal the total amount of interest earned in one year. Write an equation representing this fact.

(b) The sum of the two expressions in **Exercise 66(b)** must equal the amount of pure acid in the final mixture. Write an equation representing this fact.

68. **(a)** Solve Problem A.

(b) Solve Problem B.

(c) Explain the similarities between the processes used in solving Problems A and B.

1.4 Further Applications of Linear Equations

OBJECTIVES

1 Solve problems about different denominations of money.
2 Solve problems about uniform motion.
3 Solve problems about angles.
4 Solve problems about consecutive integers.

VOCABULARY

☐ consecutive integers
☐ consecutive even integers
☐ consecutive odd integers

OBJECTIVE 1 Solve problems about different denominations of money.

PROBLEM-SOLVING HINT In problems involving money, use the following basic fact.

number of monetary units of the same kind × denomination = total monetary value

Examples: 30 dimes have a monetary value of 30($0.10) = $3.00.
15 five-dollar bills have a value of 15($5) = $75.

EXAMPLE 1 Solving a Money Denomination Problem

For a bill totaling $5.65, a cashier received 25 coins consisting of nickels and quarters. How many of each denomination of coin did the cashier receive?

Step 1 **Read** the problem. We must find the number of nickels and the number of quarters the cashier received.

Step 2 **Assign a variable.**

Let x = the number of nickels.

Then $25 - x$ = the number of quarters.

	Conect Number of Coins	Denomination (in dollars)	Value (in dollars)
Nickels	x	0.05	$0.05x$
Quarters	$25 - x$	0.25	$0.25(25 - x)$
			5.65 ← Total

Organize the information in a table.

Step 3 **Write an equation.** Use the values from the last column of the table.

$$0.05x + 0.25(25 - x) = 5.65$$

Steven has a collection of 52 coins worth \$3.70. His collection contains only dimes and nickels. How many of each type of coin does he have?

Step 4 **Solve.** Clear the parentheses and then the decimals.

$0.05x + 0.25(25 - x) = 5.65$	Equation from Step 3
$0.05x + 6.25 - 0.25x = 5.65$	Distributive property
$100(0.05x + 6.25 - 0.25x) = 100(5.65)$	Multiply by 100.
$5x + 625 - 25x = 565$	Distributive property; Multiply.
$-20x = -60$	Subtract 625. Combine like terms.
$x = 3$	Divide by −20.

Move decimal points 2 places to the right.

Step 5 **State the answer.** There are 3 nickels and $25 - 3 = 22$ quarters.

Step 6 **Check.** The cashier has $3 + 22 = 25$ coins. The value is

$\$0.05(3) + \$0.25(22) = \$5.65,$ as required. NOW TRY

CAUTION ***Be sure that your answer is reasonable*** when working problems like **Example 1.** Because we are dealing with a number of coins, the correct answer can be neither negative nor a fraction.

OBJECTIVE 2 Solve problems about uniform motion.

PROBLEM-SOLVING HINT Uniform motion problems use the distance formula ***d = rt.*** ***When rate (or speed) is given in miles per hour, time must be given in hours. Draw a sketch*** to illustrate what is happening. ***Make a table*** to summarize given information.

EXAMPLE 2 Solving a Motion Problem (Opposite Directions)

Two cars leave the same place at the same time, one going east and the other west. The eastbound car averages 40 mph, while the westbound car averages 50 mph. In how many hours will they be 300 mi apart?

Step 1 **Read** the problem. We are looking for the time it takes for the two cars to be 300 mi apart.

Step 2 **Assign a variable.** A sketch shows what is happening in the problem. The cars are going in *opposite* directions. See **FIGURE 6.**

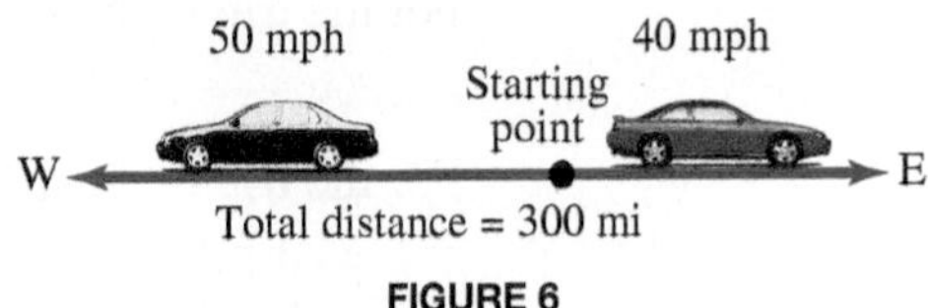

FIGURE 6

Let x = the time traveled by each car.

	Rate	Time	Distance
Eastbound Car	40	x	$40x$
Westbound Car	50	x	$50x$
			300

Fill in each distance by multiplying rate by time, using the formula $d = rt$, or $rt = d$. The sum of the two distances is 300.

NOW TRY ANSWER

1. 22 dimes; 30 nickels

NOW TRY EXERCISE 2

Two trains leave a city traveling in opposite directions. One travels at a rate of 80 km per hr and the other at a rate of 75 km per hr. How long will it take before they are 387.5 km apart?

Step 3 **Write an equation.** $40x + 50x = 300$

Step 4 **Solve.**

$90x = 300$ Combine like terms.

$x = \frac{300}{90}$ Divide by 90.

$x = \frac{10}{3}$ Lowest terms

Step 5 **State the answer.** The cars travel $\frac{10}{3} = 3\frac{1}{3}$ hr, or 3 hr, 20 min.

Step 6 **Check.** The eastbound car traveled $40\left(\frac{10}{3}\right) = \frac{400}{3}$ mi. The westbound car traveled $50\left(\frac{10}{3}\right) = \frac{500}{3}$ mi, for a total distance of $\frac{400}{3} + \frac{500}{3} = \frac{900}{3}$, or 300 mi, as required.

NOW TRY

CAUTION It is a common error to write 300 as the distance traveled by each car in **Example 2.** The ***total*** distance traveled is 300 mi.

As in **Example 2,** in general, the equation for a problem involving motion in *opposite* directions is of the follow ing form.

partial distance + partial distance = total distance

EXAMPLE 3 Solving a Motion Problem (Same Direction)

Geoff can bike to work in $\frac{3}{4}$ hr. When he takes the bus, the trip takes $\frac{1}{4}$ hr. If the bus travels 20 mph faster than Geoff rides his bike, how far is it to his workplace?

Step 1 **Read** the problem. We must find the distance between Geoff's home and his workplace.

Step 2 **Assign a variable.** Make a sketch to show what is happening. See **FIGURE 7.**

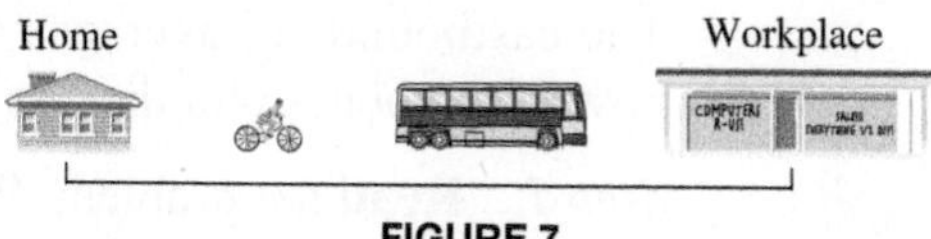

FIGURE 7

The problem asks for a *distance,* but it is easier here to let x be Geoff's *rate* when he rides his bike to work. Then the rate of the bus is $(x + 20)$ mph.

For the trip by bike, $d = rt = x \cdot \frac{3}{4} = \frac{3}{4}x.$

For the trip by bus, $d = rt = (x + 20) \cdot \frac{1}{4} = \frac{1}{4}(x + 20).$

	Rate	Time	Distance
Bike	x	$\frac{3}{4}$	$\frac{3}{4}x$
Bus	$x + 20$	$\frac{1}{4}$	$\frac{1}{4}(x + 20)$

Same distance

Step 3 **Write an equation.**

$$\frac{3}{4}x = \frac{1}{4}(x + 20)$$ The distance is the same in each case.

NOW TRY ANSWER

2. $2\frac{1}{2}$ hr

NOW TRY EXERCISE 3

Michael can drive to work in $\frac{1}{2}$ hr. When he rides his bicycle, it takes $1\frac{1}{2}$ hours. If his average rate while driving to work is 30 mph faster than his rate while bicycling to work, determine the distance that he lives from work.

Step 4 **Solve.**

$\frac{3}{4}x = \frac{1}{4}(x + 20)$	Equation from Step 3
$\frac{3}{4}x = \frac{1}{4}x + 5$	Distributive property
$4\left(\frac{3}{4}x\right) = 4\left(\frac{1}{4}x + 5\right)$	Multiply by 4 to clear the fractions.
$3x = x + 20$	Multiply; $4\left(\frac{1}{4}\right) = 1$ and $1x = x$
$2x = 20$	Subtract x.
$x = 10$	Divide by 2.

Step 5 **State the answer.** The required distance is

$$d = \frac{3}{4}x = \frac{3}{4}(10) = \frac{30}{4} = 7.5 \text{ mi.}$$

Step 6 **Check** by finding the distance using

$$d = \frac{1}{4}(x + 20) = \frac{1}{4}(10 + 20) = \frac{30}{4} = 7.5 \text{ mi.}$$

The same distance results.

NOW TRY

As in **Example 3,** the equation for a problem involving motion in the *same* direction is often of the following form.

one distance = other distance

> **PROBLEM-SOLVING HINT** In **Example 3,** it was easier to let the variable represent a quantity other than the one that we were asked to find. It takes practice to learn when this approach works best.

OBJECTIVE 3 Solve problems about angles.

An important result of Euclidean geometry (the geometry of the Greek mathematician Euclid) is as follows.

The sum of the angle measures of any triangle is 180°.

EXAMPLE 4 Finding Angle Measures

Find the value of x, and determine the measure of each angle in **FIGURE 8.**

Step 1 **Read** the problem. We are asked to find the measure of each angle.

Step 2 **Assign a variable.**

Let x = the measure of one angle.

Step 3 **Write an equation.** The sum of the three measures shown in the figure must be 180°.

$$x + (x + 20) + (210 - 3x) = 180 \quad \text{Measures are in degrees.}$$

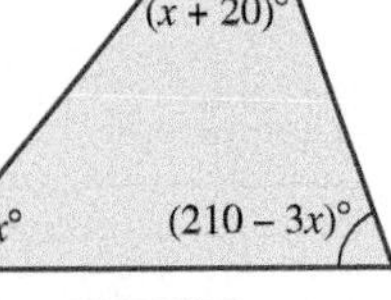

FIGURE 8

Step 4 **Solve.**

$-x + 230 = 180$	Combine like terms.
$-x = -50$	Subtract 230.
$x = 50$	Multiply by -1.

NOW TRY ANSWER
3. 22.5 mi

NOW TRY EXERCISE 4

Find the value of x, and determine the measure of each angle.

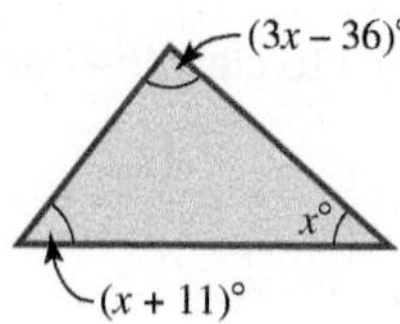

Step 5 **State the answer.** One angle measures 50°. The other two angles measure

$$x + 20 = 50 + 20 = 70^\circ$$

and

$$210 - 3x = 210 - 3(50) = 60^\circ.$$

Step 6 **Check.** Because $50^\circ + 70^\circ + 60^\circ = 180^\circ$ is true, the answer is correct.

OBJECTIVE 4 Solve problems about consecutive integers.

Recall that the set of integers is $\{\ldots, -3, -2, -1, 0, 1, 2, 3, \ldots\}$. **Consecutive integers** are integers that are next to each other on a number line. Two consecutive integers differ by 1. Some examples of consecutive integers are 4, 5 and −6, −5.

Consecutive even integers always differ by 2. Examples are 4, 6 and −4, −2. Similarly, **consecutive odd integers** also differ by 2, such as 1, 3 and −5, −3.

PROBLEM-SOLVING HINT If x = the lesser (least) integer in a consecutive integer problem, then the following apply.

- For two consecutive integers, use $x, \ x + 1.$
- For three consecutive integers, use $x, \ x + 1, \ x + 2.$
- For two consecutive even or odd integers, use $x, \ x + 2.$
- For three consecutive even or odd integers, use $x, \ x + 2, \ x + 4.$

NOW TRY EXERCISE 5

Find three consecutive integers such that the sum of the first and second is 43 less than three times the third.

EXAMPLE 5 Solving a Consecutive Integer Problem

Find three consecutive integers such that the sum of the first and third, increased by 3, is 50 more than the second.

Step 1 **Read** the problem. We are told to find three consecutive integers satisfying the given conditions.

Step 2 **Assign a variable.**

Let x = the least of the three consecutive integers.

Then $x + 1$ = the second (middle) integer,

and $x + 2$ = the greatest of the three consecutive integers.

Step 3 **Write an equation.**

Sum of the first and third,	increased by 3,	is	50 more than the second.
↓	↓	↓	↓
$x + (x + 2)$	$+ 3$	$=$	$(x + 1) + 50$

Step 4 **Solve.**

$$2x + 5 = x + 51 \quad \text{Combine like terms.}$$

$$x = 46 \quad \text{Subtract } x \text{ and 5.}$$

Step 5 **State the answer.** The solution is 46, so the first integer is $x = 46$, the second is $46 + 1 = 47$, and the third is $46 + 2 = 48$. The three integers are 46, 47, and 48.

Step 6 **Check.** The sum of the first and third is $46 + 48 = 94$. If this is increased by 3, the result is $94 + 3 = 97$, which is indeed 50 more than the second (47). The answer is correct.

NOW TRY

NOW TRY ANSWERS

4. 41°, 52°, 87°

5. 38, 39, 40

1.4 Exercises

FOR EXTRA HELP  MyLab Math

Video solutions for select problems available in MyLab Math

Concept Check *Solve each problem.*

1. What amount of money is found in a coin hoard containing 12 dimes and 18 quarters?
2. The distance between Cape Town, South Africa, and Miami is 7700 mi. If a jet averages 550 mph between the two cities, what is its travel time in hours?
3. Tri traveled from Chicago to Des Moines, a distance of 300 mi, in 10 hr. What was his rate in miles per hour?
4. A square has perimeter 160 in. What would be the perimeter of an equilateral triangle whose sides each measure the same length as the side of the square?
5. An equilateral triangle has perimeter 27 in. What would be the area of a square whose sides each measure the same length as the side of the equilateral triangle?
6. A circle has area 25π ft^2. What would be the perimeter of a square whose sides each measure the same length as the radius of the circle?
7. **Concept Check** A student was asked to find the measures of the angles of a triangle with degree measures that are consecutive integers. Letting x represent the measure of the smallest angle, the student wrote the following equation.

$$x + (x + 1) + (x + 2) = 90 \quad \text{Incorrect}$$

WHAT WENT WRONG? Write the correct equation and solve the problem.

8. **Concept Check** A student was asked to find the number of nickels and dimes that have a value of \$1.45. The total number of coins is 21. Letting x represent the number of nickels, the student wrote the following equation.

$$0.05x + 0.10(x - 21) = 1.45 \quad \text{Incorrect}$$

WHAT WENT WRONG? Write the correct equation and solve the problem.

Solve each problem. ***See Example 1.***

9. Otis has a box of coins that he uses when he plays poker with his friends. The box contains 44 coins, consisting of pennies, dimes, and quarters. The number of pennies is equal to the number of dimes. The total value is \$4.37. How many of each denomination of coin does he have in the box?

Number of Coins	Denomination (in dollars)	Value (in dollars)	
x	0.01	$0.01x$	
x			
	0.25		
XXXX	XXXXXX	4.37	← Total

10. Nana found some coins under her sofa pillows. There were equal numbers of nickels and quarters and twice as many half-dollars as quarters. If she found \$2.60 in all, how many of each denomination of coin did she find?

Number of Coins	Denomination (in dollars)	Value (in dollars)	
x	0.05	$0.05x$	
x			
$2x$	0.50		
XXXX	XXXXXX	2.60	← Total

11. In Canada, \$1 coins are called "loonies" because they have a picture of a loon on the reverse, and \$2 coins are called "toonies." When Marissa returned home to San Francisco from a trip to Vancouver, she found that she had acquired 37 of these coins, with a total value of 51 Canadian dollars. How many coins of each denomination did she have?

12. Ahmad works at an ice cream shop. At the end of his shift, he counted the bills in his cash drawer and found 119 bills with a total value of \$347. If all of the bills are \$5 bills and \$1 bills, how many of each denomination were in his cash drawer?

13. Hussein collects U.S. gold coins. He has a collection of 41 coins. Some are \$10 coins, and the rest are \$20 coins. If the face value of the coins is \$540, how many of each denomination does he have?

14. In the 19th century, the United States minted two-cent and three-cent pieces. Frances has three times as many three-cent pieces as two-cent pieces, and the face value of these coins is \$2.42. How many of each denomination does she have?

15. In 2018, general admission to the Art Institute of Chicago cost \$25 for adults and \$19 for children and seniors. If \$32,972 was collected from the sale of 1460 general admission tickets, how many adult tickets were sold? (Data from www.artic.edu)

16. For a high school production of *Hello, Dolly!*, student tickets cost \$5 each and nonstudent tickets cost \$8 each. If 480 tickets were sold and a total of \$2895 was collected, how many tickets of each type were sold?

For each event, find the rate on the basis of the information provided. Use a calculator and round answers to the nearest hundredth. All events were at the 2016 Summer Olympics in Rio de Janeiro, Brazil. (Data from The World Almanac and Book of Facts.)

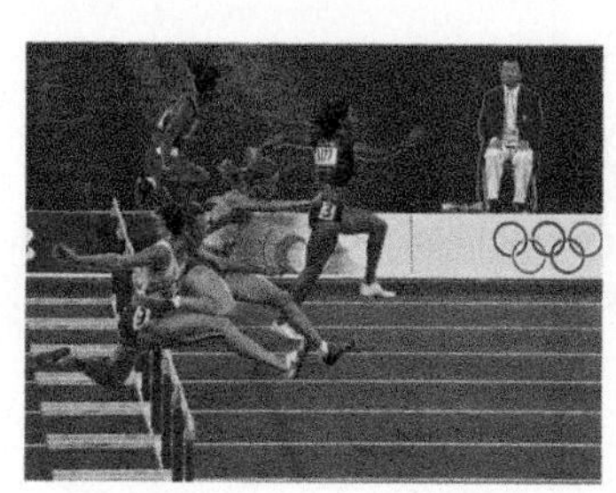

	Event	Winner	Time
17.	100-m hurdles, women	Brianna Rollins, United States	12.48 sec
18.	400-m hurdles, women	Dalilah Muhammad, United States	53.13 sec
19.	200-m run, men	Usain Bolt, Jamaica	19.78 sec
20.	400-m run, men	Wayde van Niekerk, South Africa	43.03 sec

Solve each problem. ***See Examples 2 and 3.***

21. Two steamers leave a port on a river at the same time, traveling in opposite directions. Each is traveling 22 mph. How long will it take for them to be 110 mi apart?

	Rate	Time	Distance
First Steamer		t	
Second Steamer	22		
			110

22. A train leaves Kansas City, Kansas, and travels north at 85 km per hr. Another train leaves at the same time and travels south at 95 km per hr. How long will it take before they are 315 km apart?

	Rate	Time	Distance
First Train	85	t	
Second Train			
			315

23. Mulder and Scully are driving to Georgia to investigate "Big Blue," a giant reptile reported in one of the local lakes. Mulder leaves the office at 8:30 A.M. averaging 65 mph. Scully leaves at 9:00 A.M., following the same path and averaging 68 mph. At what time will Scully catch up with Mulder?

	Rate	Time	Distance
Mulder			
Scully			

24. Lois and Clark are covering separate stories and have to travel in opposite directions. Lois leaves the *Daily Planet* building at 8:00 A.M. and travels at 35 mph. Clark leaves at 8:15 A.M. and travels at 40 mph. At what time will they be 140 mi apart?

	Rate	Time	Distance
Lois			
Clark			

25. It took Charmaine 3.6 hr to drive to her mother's house on Saturday morning for a weekend visit. On her return trip on Sunday night, traffic was heavier, so the trip took her 4 hr. Her average rate on Sunday was 5 mph slower than on Saturday. What was her average rate on Sunday?

	Rate	Time	Distance
Saturday			
Sunday			

26. Sharon commutes to her office by train. When she walks to the train station, it takes her 40 min. When she rides her bike, it takes her 12 min. Her average walking rate is 7 mph less than her average biking rate. Find the distance from her house to the train station.

	Rate	Time	Distance
Walking			
Biking			

27. Johnny leaves Memphis to visit his cousin Anne, who lives in the town of Hornsby, Tennessee, 80 mi away. He travels at an average rate of 50 mph. One-half hour later, Anne leaves to visit Johnny, traveling at an average rate of 60 mph. How long after Anne leaves will it be before they meet?

28. On an automobile trip, Laura maintained a steady rate for the first two hours. Rush-hour traffic slowed her rate by 25 mph for the last part of the trip. The entire trip, a distance of 125 mi, took $2\frac{1}{2}$ hr. What was her rate during the first part of the trip?

Find the measure of each angle in the triangles shown. ***See Example 4.***

29.

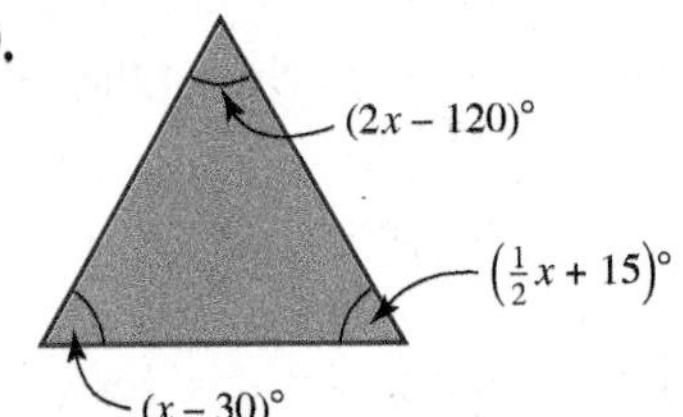

30.

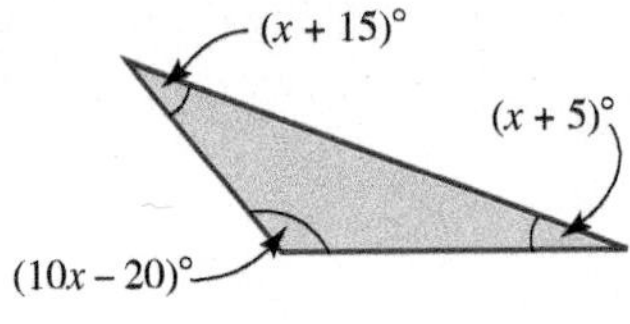

31.

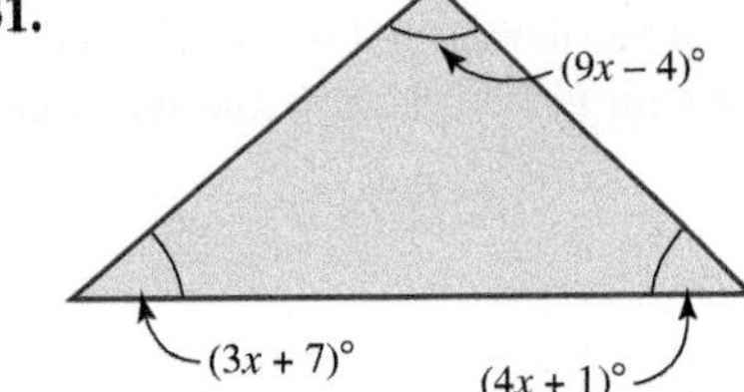

32.

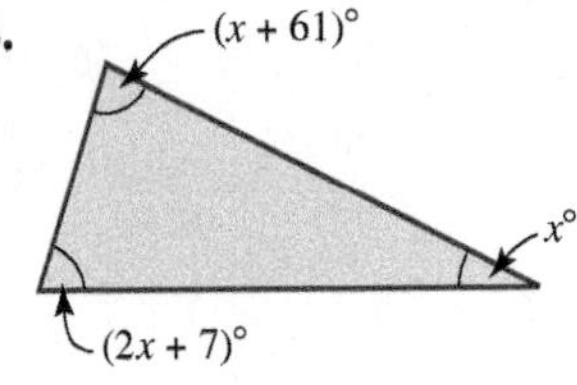

33.

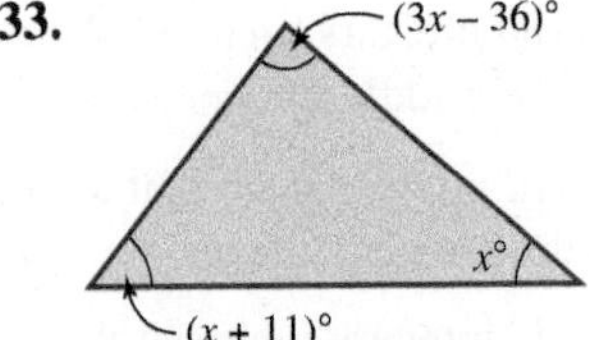

34.

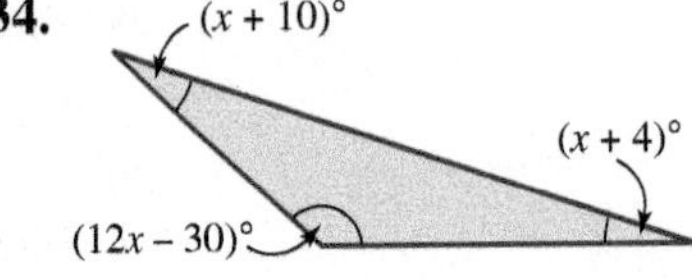

In the following problems, the angles marked with variable expressions are **vertical angles.** *It is shown in geometry that vertical angles have equal measures. Find the measure of each angle.*

35.

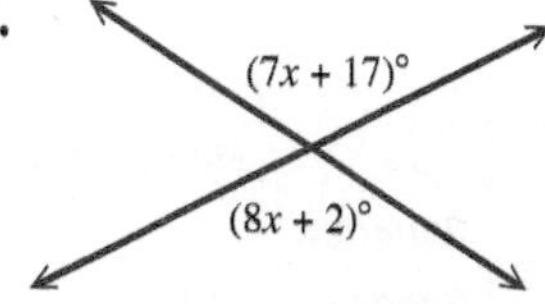

36.

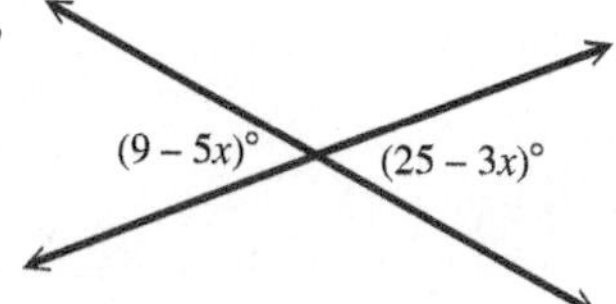

Two angles whose sum is 90° are **complementary angles.** *Find the measures of the complementary angles shown in each figure.*

37.

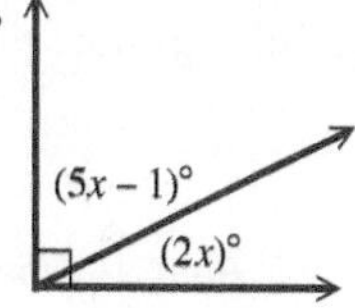

38.

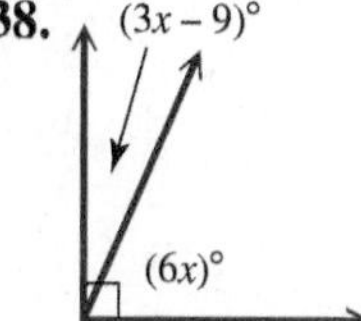

Two angles whose sum is 180° are **supplementary angles.** *Find the measures of the supplementary angles shown in each figure.*

39.

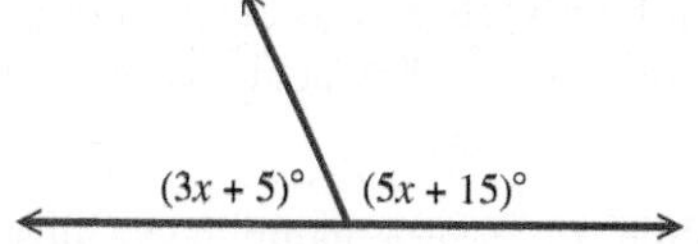

40.

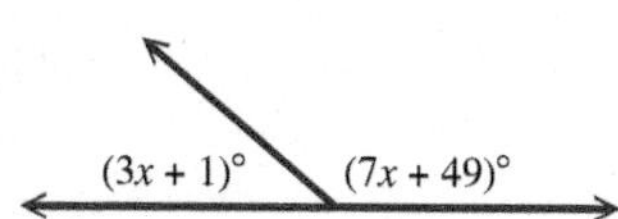

Solve each problem involving consecutive integers. ***See Example 5.***

41. Find three consecutive integers such that the sum of the first and twice the second is 17 more than twice the third.

42. Find three consecutive integers such that the sum of the first and twice the third is 39 more than twice the second.

43. Find four consecutive integers such that the sum of the first three is 54 more than the fourth.

44. Find four consecutive integers such that the sum of the last three is 86 more than the first.

45. If I add my current age to the age I will be next year on this date, the sum is 103 yr. How old will I be 10 yr from today?

46. If I add my current age to the age I will be next year on this date, the sum is 129 yr. How old will I be 5 yr from today?

October

Sun	Mon	Tue	Wed	Thu	Fri	Sat
	1	2	3	4	5	6
7	8	9	10	11	12	13
14	15	16	17	18	19	20
21	22	23	24	25	26	27
28	29	30	31			

47. Find three consecutive *even* integers such that the sum of the least integer and the middle integer is 26 more than the greatest integer.

48. Find three consecutive *even* integers such that the sum of the least integer and the greatest integer is 12 more than the middle integer.

49. Find three consecutive *odd* integers such that the sum of the least integer and the middle integer is 19 more than the greatest integer.

50. Find three consecutive *odd* integers such that the sum of the least integer and the greatest integer is 13 more than the middle integer.

RELATING CONCEPTS For Individual or Group Work **(Exercises 51–54)**

Consider the following two figures. ***Work Exercises 51–54 in order.***

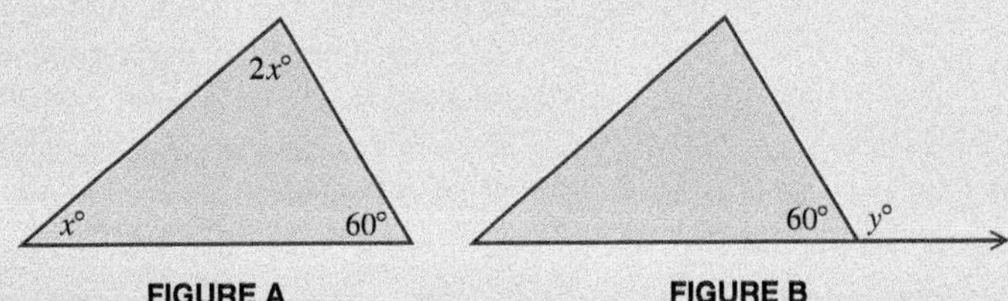

FIGURE A FIGURE B

51. Solve for the measures of the unknown angles in FIGURE A. (*Hint:* What is the sum of the angle measures of any triangle?)

52. Solve for the measure of the unknown angle marked $y°$ in FIGURE B. (*Hint:* The angles are supplementary angles.)

53. Add the measures of the two angles found in **Exercise 51.** How does the sum compare to the measure of the angle found in **Exercise 52?**

54. Based on the answers to **Exercises 51–53,** make a conjecture (an educated guess) about the relationship among the angles marked ①, ②, and ③ in the figure below.

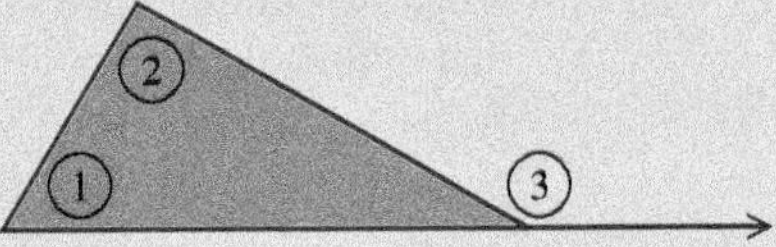

SUMMARY EXERCISES Applying Problem-Solving Techniques

Solve each problem.

1. The length of a rectangle is 3 in. more than its width. If the length were decreased by 2 in. and the width were increased by 1 in., the perimeter of the resulting rectangle would be 24 in. Find the dimensions of the original rectangle.

$x + 3$

x

2. A farmer wishes to enclose a rectangular region with 210 m of fencing in such a way that the length is twice the width and the region is divided into two equal parts, as shown in the figure. What length and width should be used?

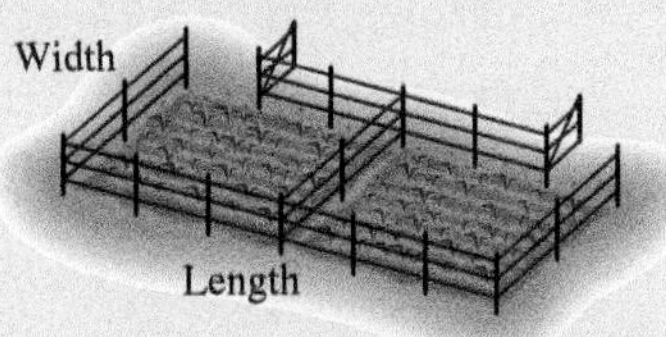

3. After a discount of 28.57%, the sale price for a Samsung Galaxy tablet computer was \$199.99. What was the regular price of the tablet computer to the nearest cent? (Data from www.amazon.com)

4. An electronics store offered a new HD television for \$255, which was the sale price after the regular price was discounted 40%. What was the regular price?

5. An amount of money is invested at 4% annual simple interest, and twice that amount is invested at 5%. The total annual interest is \$112. How much is invested at each rate?

6. An amount of money is invested at 3% annual simple interest, and \$2000 more than that amount is invested at 4%. The total annual interest is \$920. How much is invested at each rate?

7. Russell Westbrook of the Oklahoma City Thunder and James Harden of the Houston Rockets were the leading scorers in the NBA for the 2016–2017 regular season. Together they scored 4914 points, with Harden scoring 202 fewer points than Westbrook. How many points did each player score? (Data from stats.nba.com)

8. In 2016, the number of films released in U.S. and Canadian movie theaters was 161 more than the number of films released in 2009. A total of 1275 films were released in both years. How many films were released each year? (Data from comScore—Box Office Essentials.)

9. Atlanta and Cincinnati are 440 mi apart. John leaves Cincinnati, driving toward Atlanta at an average rate of 60 mph. Pat leaves Atlanta at the same time, driving toward Cincinnati in her antique auto, averaging 28 mph. How long will it take them to meet?

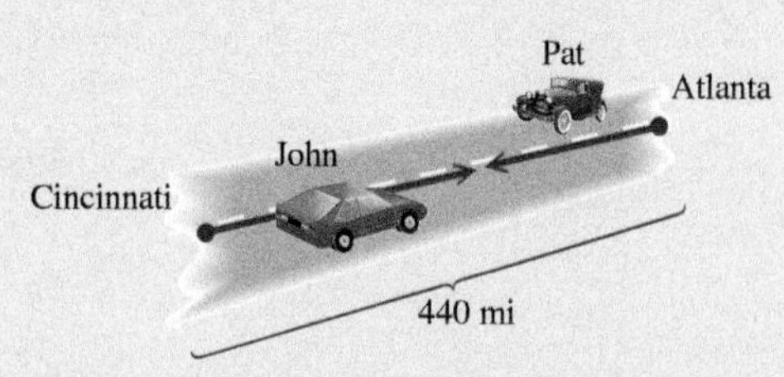

10. Joshua has a sheet of tin 12 cm by 16 cm. He plans to make a box by cutting equal squares out of each of the four corners and folding up the remaining edges. How large a square should he cut so that the finished box will have a length that is 5 cm less than twice the width?

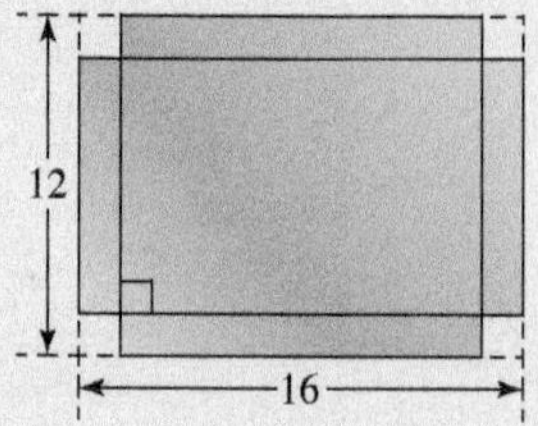

11. A pharmacist has 20 L of a 10% drug solution. How many liters of 5% solution must be added to obtain a mixture that is 8%?

12. A certain metal is 20% tin. How many kilograms of this metal must be mixed with 80 kg of a metal that is 70% tin to obtain a metal that is 50% tin?

Kilograms of Metal	*Percent (as a decimal)*	*Kilograms of Tin*
	0.20	
	0.50	

13. A cashier has a total of 126 bills in fives and tens. The total value of the money is $840. How many of each denomination of bill does he have?

14. The top-grossing domestic movie in 2016 was *Finding Dory*. On opening weekend, one theater showing this movie took in $20,520 by selling a total of 2460 tickets, some at $9 and the rest at $7. How many tickets were sold at each price? (Data from comScore, Inc.)

15. Find the measure of each angle.

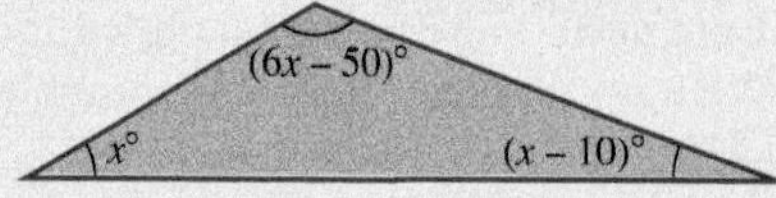

16. Find the measure of each marked angle.

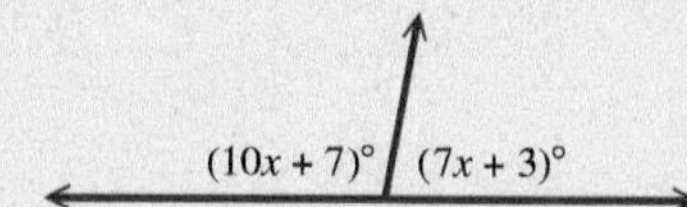

17. The sum of the least and greatest of three consecutive integers is 32 more than the middle integer. What are the three integers?

18. If the lesser of two consecutive odd integers is doubled, the result is 7 more than the greater of the two integers. Find the two integers.

19. The perimeter of a triangle is 34 in. The middle side is twice as long as the shortest side. The longest side is 2 in. less than three times the shortest side. Find the lengths of the three sides.

20. The perimeter of a rectangle is 43 in. more than the length. The width is 10 in. Find the length of the rectangle.

1.5 Linear Inequalities in One Variable

OBJECTIVES

1 Graph intervals on a number line.
2 Solve linear inequalities using the addition property.
3 Solve linear inequalities using the multiplication property.
4 Solve linear inequalities with three parts.
5 Solve applied problems using linear inequalities.

An **inequality** consists of algebraic expressions related by one of the following symbols.

$<$	"is less than"	$\leq$	"is less than or equal to"
$>$	"is greater than"	$\geq$	"is greater than or equal to"

These symbols are read as shown when the inequality is read from left to right.

We ***solve an inequality*** by finding all real number solutions for it. For example, the solution set of $x \leq 2$ includes *all* real numbers that are less than or equal to 2, not just the integers less than or equal to 2.

OBJECTIVE 1 Graph intervals on a number line.

We graph all the real numbers satisfying $\boldsymbol{x \leq 2}$ by placing a square bracket at 2 and drawing an arrow to the left to represent the fact that all numbers less than 2 are also part of the graph. See **FIGURE 9**.

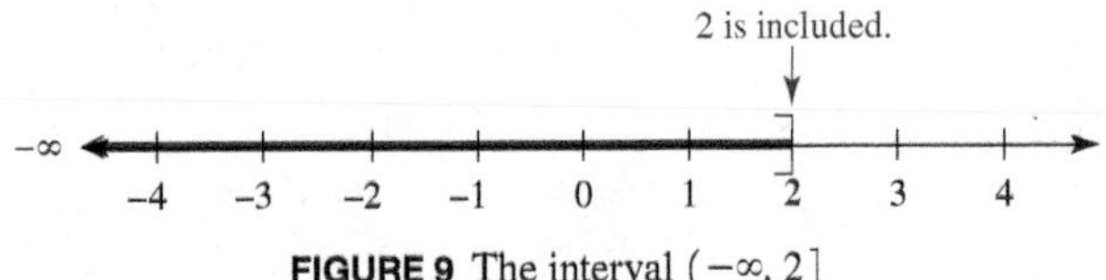

FIGURE 9 The interval $(-\infty, 2]$

VOCABULARY

☐ inequality
☐ interval
☐ linear inequality in one variable
☐ equivalent inequalities
☐ three-part inequality

The set of numbers less than or equal to 2 is an example of an **interval** on a number line. To write intervals, we use **interval notation,** which includes the **infinity symbols,** $\boldsymbol{\infty}$ or $\boldsymbol{-\infty}$. We write the interval of *all* numbers less than or equal to 2 as $\boldsymbol{(-\infty, 2]}$. The negative infinity symbol $-\infty$ does not indicate a number, but shows that the interval includes all real numbers less than 2. Both on the number line and in interval notation, the square bracket indicates that 2 is included in the solution set.

Remember the following important concepts regarding interval notation.

- A parenthesis indicates that an endpoint is *not* included.
- A square bracket indicates that an endpoint is included.
- A parenthesis is always used next to an infinity symbol, $-\infty$ or ∞.
- The set of real numbers is written in interval notation as $\boldsymbol{(-\infty, \infty)}$.

EXAMPLE 1 Using Interval Notation

Graph each inequality, and write it using interval notation.

(a) $x > -5$

This statement says that x can represent any number greater than -5, but x cannot equal -5. We graph it by placing a parenthesis at -5 and drawing an arrow to the right, as in **FIGURE 10**. The parenthesis at -5 shows that -5 is *not* part of the graph. This interval is written $(-5, \infty)$.

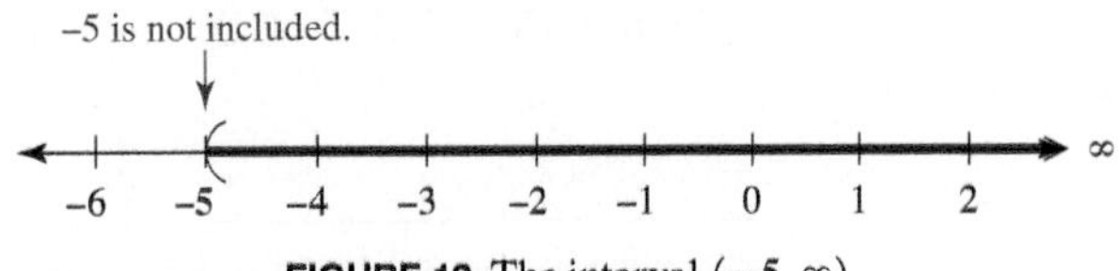

FIGURE 10 The interval $(-5, \infty)$

Graph each inequality, and write it using interval notation.

(a) $x < -1$

(b) $-4 \le x < 2$

(b) $-1 \le x < 3$

This statement is read "-1 *is less than or equal to x* ***and*** *x is less than* 3." We want the set of numbers that are *between* -1 and 3, with -1 included and 3 excluded. We use a bracket at -1 because it is part of the graph and a parenthesis at 3 because it is not part of the graph. See **FIGURE 11.** In interval notation, we write $[-1, 3)$.

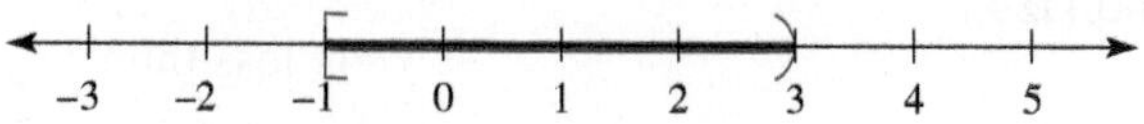

FIGURE 11 The interval $[-1, 3)$

NOW TRY

▼ Summary of Types of Intervals

Type of Interval	Set-Builder Notation	Graph	Interval Notation
Open interval	$\{x \mid a < x < b\}$	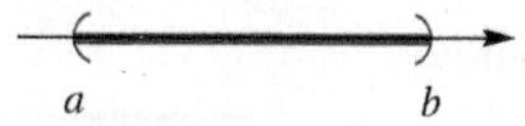 a b	(a, b)
Closed interval	$\{x \mid a \le x \le b\}$	 a b	$[a, b]$
Half-open (or half-closed) interval	$\{x \mid a \le x < b\}$	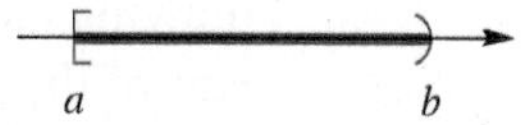 a b	$[a, b)$
	$\{x \mid a < x \le b\}$	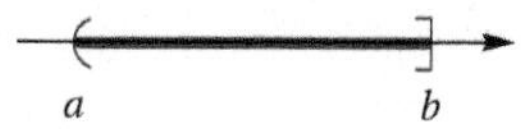 a b	$(a, b]$
Disjoint interval*	$\{x \mid x < a \text{ or } x > b\}$	 a b	$(-\infty, a) \cup (b, \infty)$
	$\{x \mid x > a\}$	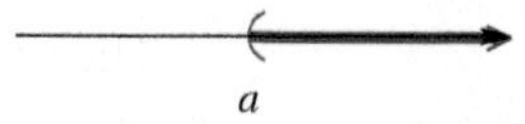 a	(a, ∞)
	$\{x \mid x \ge a\}$	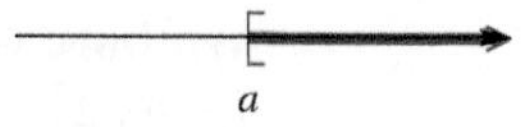 a	$[a, \infty)$
Infinite interval	$\{x \mid x < a\}$	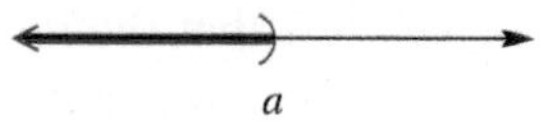 a	$(-\infty, a)$
	$\{x \mid x \le a\}$	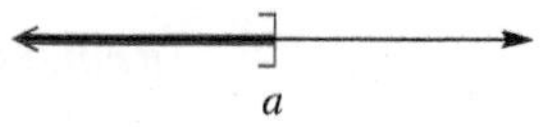 a	$(-\infty, a]$
	$\{x \mid x \text{ is a real number}\}$	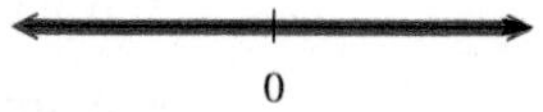 0	$(-\infty, \infty)$

CAUTION A parenthesis is *always* used next to an infinity symbol, $-\infty$ or ∞.

Solving inequalities is similar to solving equations.

Linear Inequality in One Variable

A **linear inequality in one variable** (here x) is an inequality that can be written in the form

$$ax + b < 0, \quad ax + b \le 0, \quad ax + b > 0, \quad \text{or} \quad ax + b \ge 0,$$

where a and b are real numbers and $a \neq 0$.

Examples: $x + 5 < 0$, $x - 3 \ge 5$, $2x + 5 \le 10$, $x > -1$ — Linear inequalities in one variable

NOW TRY ANSWERS

1. (a) 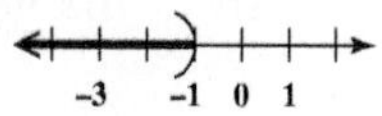 $(-\infty, -1)$

(b) 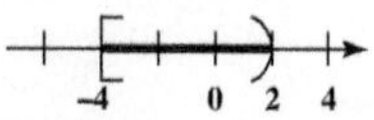 $[-4, 2)$

*We will work with disjoint intervals later when we study *set operations* and *compound inequalities.*

OBJECTIVE 2 Solve linear inequalities using the addition property.

We solve an inequality by finding all numbers that make the inequality true. Usually, an inequality has an infinite number of solutions. These solutions, like solutions of equations, are found by producing a series of simpler equivalent inequalities. **Equivalent inequalities** are inequalities with the same solution set.

We use two important properties to produce equivalent inequalities.

Addition Property of Inequality

If a, b, and c represent real numbers, then the inequalities

$$a < b \quad \text{and} \quad a + c < b + c \quad \text{are equivalent.*}$$

That is, the same number may be added to each side of an inequality without changing the solution set.

*This also applies to $a \leq b$, $a > b$, and $a \geq b$.

NOW TRY EXERCISE 2

Solve $x - 10 > -7$, and graph the solution set.

EXAMPLE 2 Using the Addition Property of Inequality

Solve $x - 7 < -12$, and graph the solution set.

$$x - 7 < -12$$

$$x - 7 + 7 < -12 + 7 \qquad \text{Add 7.}$$

$$x < -5 \qquad \text{Combine like terms.}$$

CHECK Substitute -5 for x in the *equation* $x - 7 = -12$.

$$x - 7 = -12$$

$$-5 - 7 \stackrel{?}{=} -12 \qquad \text{Let } x = -5.$$

$$-12 = -12 \quad ✓ \qquad \text{True}$$

The result, a true statement, shows that -5 is the boundary point. Now test a number on each side of -5 to verify that numbers *less than* -5 make the inequality true. We choose -6 and -4.

$$x - 7 < -12$$

$-6 - 7 \stackrel{?}{<} -12$ Let $x = -6$.	$-4 - 7 \stackrel{?}{<} -12$ Let $x = -4$.
$-13 < -12$ True ✓	$-11 < -12$ False
-6 *is* in the solution set.	-4 is *not* in the solution set.

The check confirms that the solution set is the graph shown in **FIGURE 12.**

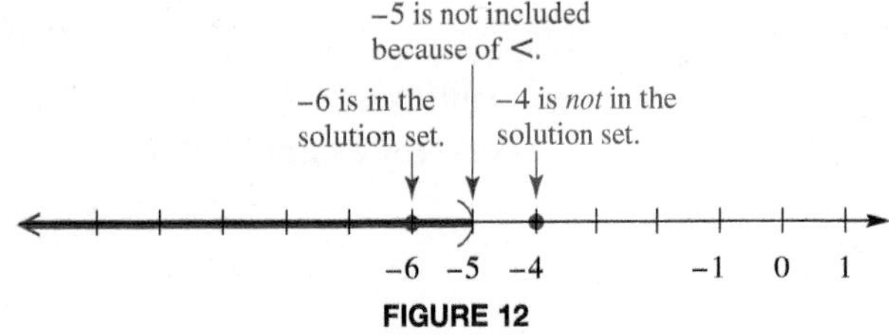

FIGURE 12

Using interval notation, the solution set is the infinite interval $(-\infty, -5)$.

NOW TRY 

As with equations, the addition property can also be used to *subtract* the same number from each side of an inequality.

NOW TRY ANSWER

2. 0 1 2 3 4 5

$(3, \infty)$

NOW TRY
EXERCISE 3

Solve $4x + 1 \geq 5x$, and graph the solution set.

EXAMPLE 3 Using the Addition Property of Inequality

Solve $14 + 2x \leq 3x$, and graph the solution set.

$$14 + 2x \leq 3x$$

$$14 + 2x - 2x \leq 3x - 2x \quad \text{Subtract } 2x.$$

$$14 \leq x \quad \text{Combine like terms.}$$

Be careful.

$$x \geq 14 \quad \text{Rewrite.}$$

The inequality $14 \leq x$ (that is, 14 is less than or equal to x) can also be written $x \geq 14$ (that is, x is greater than or equal to 14). ***Notice that in each case the inequality symbol points to the lesser number,* 14.**

CHECK

$$14 + 2x = 3x$$

$$14 + 2(14) \stackrel{?}{=} 3(14) \quad \text{Let } x = 14.$$

$$42 = 42 \ ✓ \quad \text{True}$$

So 14 satisfies the equality part of $\leq$. Choose 10 and 16 as test values.

$$14 + 2x < 3x$$

$14 + 2(10) \stackrel{?}{<} 3(10)$ Let $x = 10$.	$14 + 2(16) \stackrel{?}{<} 3(16)$ Let $x = 16$.
$34 < 30$ False	$46 < 48$ ✓ True
10 is not in the solution set.	16 is in the solution set.

The check confirms that the solution set is the graph shown in **FIGURE 13.**

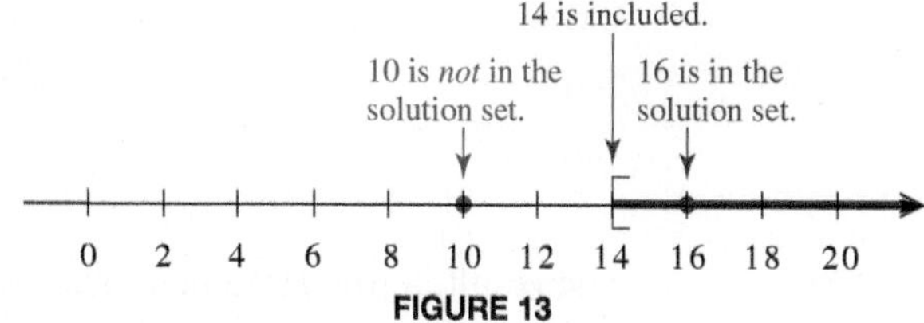
14 is included.
10 is *not* in the solution set.
16 is in the solution set.
0 2 4 6 8 10 12 14 16 18 20

FIGURE 13

Using interval notation, the solution set is the infinite interval $[14, \infty)$. **NOW TRY** 

OBJECTIVE 3 Solve linear inequalities using the multiplication property.

Consider the following true statement.

$$-2 < 5$$

Multiply each side by some positive number—for example, 8.

$$-2(8) < 5(8) \quad \text{Multiply by 8.}$$

$$-16 < 40 \quad \text{True}$$

The result is true. Start again with $-2 < 5$, and multiply each side by some negative number—for example, -8.

$$-2(-8) < 5(-8) \quad \text{Multiply by } -8.$$

$$16 < -40 \quad \text{False}$$

The result is false. To make it true, we must change the direction of the inequality symbol.

$$16 > -40 \quad \text{True}$$

As these examples suggest, multiplying each side of an inequality by a *negative* number requires reversing the direction of the inequality symbol.

NOW TRY ANSWER

3. (number line: −2 −1 0 1 2, shaded to the left through 1)

$(-\infty, 1]$

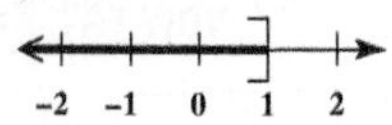

Multiplication Property of Inequality

Let a, b, and c represent real numbers, where $c \neq 0$.

(a) If c is ***positive,*** then the inequalities

$$a < b \quad \text{and} \quad ac < bc \quad \text{are equivalent.*}$$

(b) If c is ***negative,*** then the inequalities

$$a < b \quad \text{and} \quad ac > bc \quad \text{are equivalent.*}$$

That is, each side of an inequality may be multiplied (or divided) by the same *positive* number without changing the direction of the inequality symbol. ***If the multiplier is negative, we must reverse the direction of the inequality symbol.***

*This also applies to $a \leq b$, $a > b$, and $a \geq b$.

Because division is defined in terms of multiplication, the multiplication property also applies when dividing each side of an inequality by the same number.

NOW TRY EXERCISE 4

Solve each inequality, and graph the solution set.

(a) $8x \geq -40$

(b) $-20x > -60$

EXAMPLE 4 Using the Multiplication Property of Inequality

Solve each inequality, and graph the solution set.

(a) $5x \leq -30$

Divide each side by 5. ***Because $5 > 0$, do not reverse the direction of the inequality symbol.***

$$5x \leq -30$$

$$\frac{5x}{5} \leq \frac{-30}{5} \quad \text{Divide by 5.}$$

$$x \leq -6$$

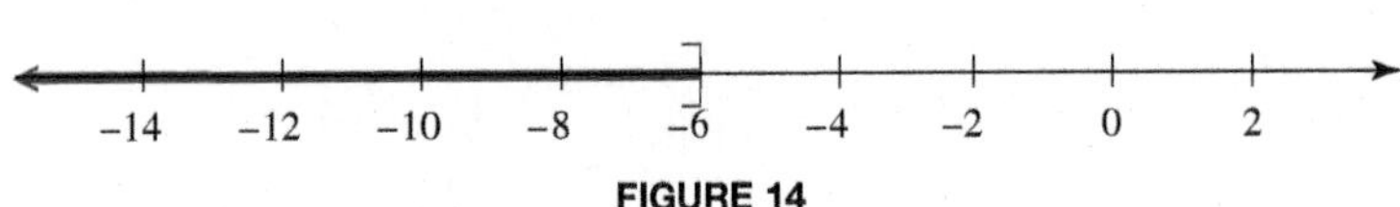

FIGURE 14

The solution set is graphed in **FIGURE 14** and written $(-\infty, -6]$.

(b) $-4x \leq 32$

Divide each side by -4. ***Because $-4 < 0$, reverse the direction of the inequality symbol.***

$$-4x \leq 32$$

$$\frac{-4x}{-4} \geq \frac{32}{-4} \quad \text{Divide by } -4. \text{ Reverse the direction of the symbol.}$$

Reverse the inequality symbol when dividing by a *negative* number.

$$x \geq -8$$

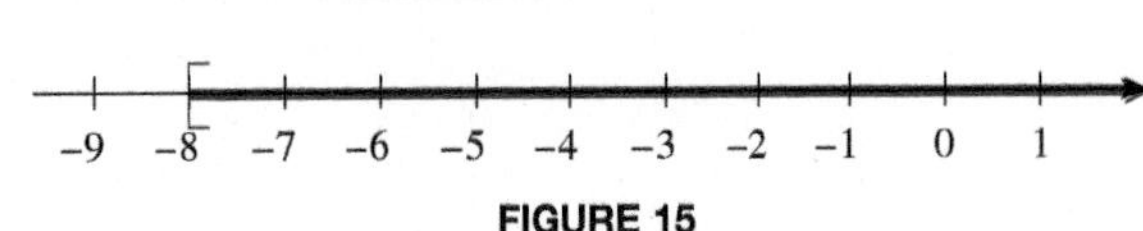

FIGURE 15

FIGURE 15 shows the graph of the solution set, $[-8, \infty)$.

NOW TRY

NOW TRY ANSWERS

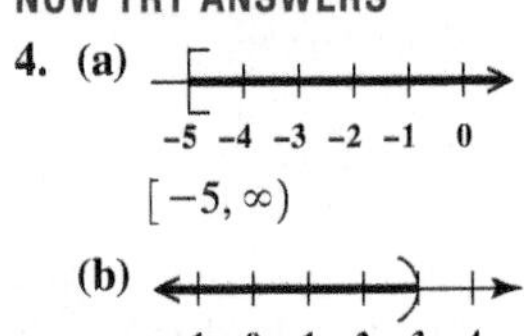

4. (a) $[-5, \infty)$

(b) $(-\infty, 3)$

To solve a linear inequality in one variable, use the following steps.

Solving a Linear Inequality in One Variable

Step 1 **Simplify each side separately.** Use the distributive property as needed.

- Clear any parentheses.
- Clear any fractions or decimals.
- Combine like terms.

Step 2 **Isolate the variable terms on one side.** Use the addition property of inequality so that all terms with variables are on one side of the inequality and all constants (numbers) are on the other side.

Step 3 **Isolate the variable.** Use the multiplication property of inequality to obtain an inequality in one of the following forms, where k is a constant (number).

$$\text{variable} < k, \quad \text{variable} \leq k, \quad \text{variable} > k, \quad \text{or} \quad \text{variable} \geq k$$

Remember: ***Reverse the direction of the inequality symbol only when multiplying or dividing each side of an inequality by a negative number.***

NOW TRY EXERCISE 5

Solve and graph the solution set.

$5 - 2(x - 4) \leq 11 - 4x$

EXAMPLE 5 **Solving a Linear Inequality**

Solve $-3(x + 4) + 2 \geq 7 - x$, and graph the solution set.

Step 1 $-3(x + 4) + 2 \geq 7 - x$

$-3x - 3(4) + 2 \geq 7 - x$ Distributive property

$-3x - 12 + 2 \geq 7 - x$ Multiply.

$-3x - 10 \geq 7 - x$ (*) Combine like terms.

Step 2 $-3x - 10 + x \geq 7 - x + x$ Add x.

$-2x - 10 \geq 7$ Combine like terms.

$-2x - 10 + 10 \geq 7 + 10$ Add 10.

$-2x \geq 17$ Combine like terms.

Step 3 $\frac{-2x}{-2} \leq \frac{17}{-2}$ Divide by -2. Change $\geq$ to $\leq$.

Be sure to reverse the direction of the inequality symbol.

$x \leq -\frac{17}{2}$

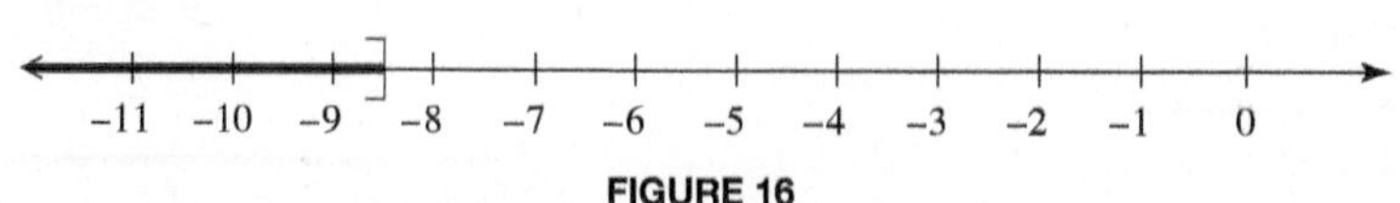

FIGURE 16

FIGURE 16 shows the graph of the solution set, $\left(-\infty, -\frac{17}{2}\right]$. NOW TRY

NOW TRY ANSWER

5. -3 -2 -1 0 1

$(-\infty, -1]$

NOTE In Step 2 of **Example 5,** if we add $3x$ (instead of x) to both sides of the inequality, we obtain the following sequence of equivalent inequalities.

$-3x - 10 \geq 7 - x$	**See (*) of Example 5.**
$-3x - 10 + 3x \geq 7 - x + 3x$	Add $3x$.
$-10 \geq 2x + 7$	Combine like terms.
$-10 - 7 \geq 2x + 7 - 7$	Subtract 7.
$-17 \geq 2x$	Combine like terms.
$-\frac{17}{2} \geq x$	Divide by 2.
$x \leq -\frac{17}{2}$	Rewrite. The same solution results.

The symbol points to x in each case.

NOW TRY EXERCISE 6
Solve and graph the solution set.

$$\frac{3}{4}(x-2)+\frac{1}{2}>\frac{1}{5}(x-8)$$

EXAMPLE 6 Solving a Linear Inequality with Fractions

Solve $-\frac{2}{3}(x-3)-\frac{1}{2}<\frac{1}{2}(5-x)$, and graph the solution set.

$$-\frac{2}{3}(x-3)-\frac{1}{2}<\frac{1}{2}(5-x)$$

Step 1 $\quad -\frac{2}{3}x+2-\frac{1}{2}<\frac{5}{2}-\frac{1}{2}x$ Clear parentheses.

$$6\left(-\frac{2}{3}x+2-\frac{1}{2}\right)<6\left(\frac{5}{2}-\frac{1}{2}x\right)$$ To clear the fractions multiply by 6, the LCD.

$$6\left(-\frac{2}{3}x\right)+6(2)+6\left(-\frac{1}{2}\right)<6\left(\frac{5}{2}\right)+6\left(-\frac{1}{2}x\right)$$ Distributive property

$$-4x+12-3<15-3x$$ Multiply.

$$-4x+9<15-3x$$ Combine like terms.

Step 2 $\quad -4x+9+3x<15-3x+3x$ Add $3x$.

$$-x+9<15$$ Combine like terms.

$$-x+9-9<15-9$$ Subtract 9.

$$-x<6$$ Combine like terms.

Step 3 $\quad -1(-x)>-1(6)$ Multiply by -1. Change $<$ to $>$.

$$x>-6$$

Reverse the inequality symbol when multiplying by a *negative* number.

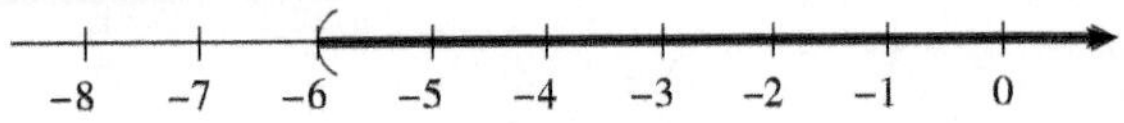

FIGURE 17

FIGURE 17 shows the graph of the solution set, $(-6, \infty)$.

NOW TRY

NOW TRY ANSWER

6. $-\frac{12}{11}$; −2 −1 0 1; $\left(-\frac{12}{11}, \infty\right)$

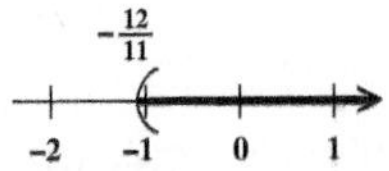

OBJECTIVE 4 Solve linear inequalities with three parts.

Some applications involve a **three-part inequality** such as

$$3 < x + 2 < 8, \quad \text{where } x + 2 \text{ is } \textit{between } 3 \text{ and } 8.$$

NOW TRY EXERCISE 7
Solve and graph the solution set.

$$-1 < x - 2 < 3$$

EXAMPLE 7 Solving a Three-Part Inequality

Solve $3 < x + 2 < 8$, and graph the solution set.

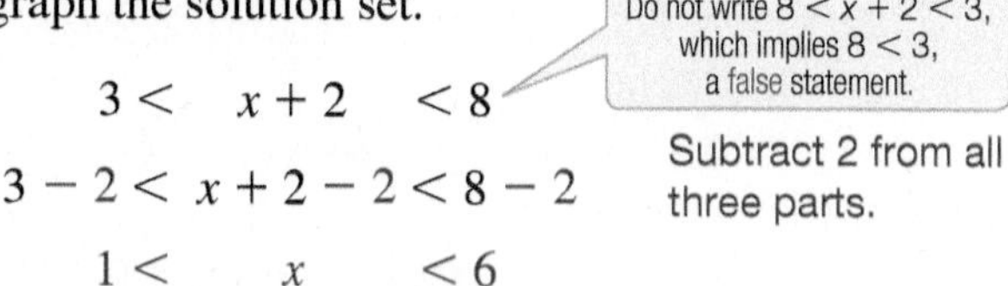

Do not write $8 < x + 2 < 3$, which implies $8 < 3$, a false statement.

$$3 < x + 2 < 8$$

$$3 - 2 < x + 2 - 2 < 8 - 2 \qquad \text{Subtract 2 from all three parts.}$$

$$1 < x < 6$$

Thus, x must be between 1 and 6 so that $x + 2$ will be between 3 and 8.

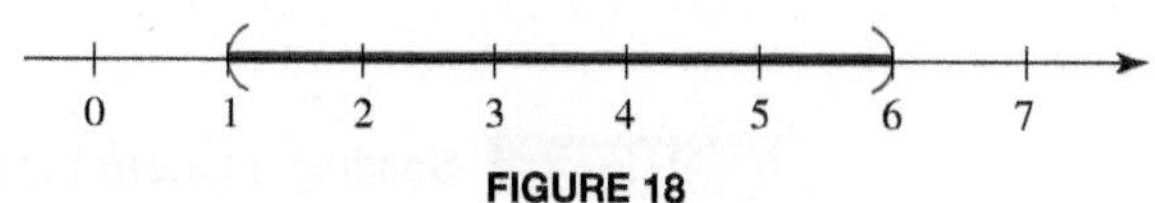

FIGURE 18

The solution set is graphed in FIGURE 18 and written as the open interval $(1, 6)$.

NOW TRY 

CAUTION *Write three-part inequalities so that the symbols point in the same direction, and both point toward the lesser number.*

NOW TRY EXERCISE 8
Solve and graph the solution set.

$$-2 < -4x - 5 \leq 7$$

EXAMPLE 8 Solving a Three-Part Inequality

Solve $-2 \leq -3x - 1 \leq 5$, and graph the solution set.

$$-2 \leq -3x - 1 \leq 5$$

$$-2 + 1 \leq -3x - 1 + 1 \leq 5 + 1 \qquad \text{Add 1 to each part.}$$

$$-1 \leq -3x \leq 6$$

$$\frac{-1}{-3} \geq \frac{-3x}{-3} \geq \frac{6}{-3} \qquad \text{Divide each part by } -3. \text{ Reverse the direction of the inequality symbols.}$$

$$\frac{1}{3} \geq x \geq -2$$

$$-2 \leq x \leq \frac{1}{3} \qquad \text{Rewrite in the order on the number line.}$$

FIGURE 19

The solution set is graphed in FIGURE 19 and written as the closed interval $\left[-2, \frac{1}{3}\right]$.

NOW TRY

NOW TRY ANSWERS

7. 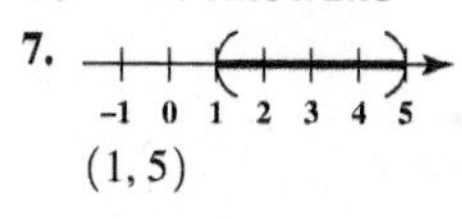

$(1, 5)$

8. 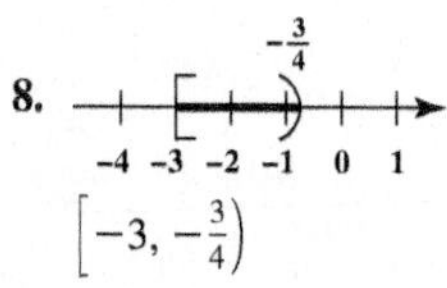

$\left[-3, -\frac{3}{4}\right)$

▼ **Solution Sets of Equations and Inequalities**

Equation or Inequality	*Typical Solution Set*	*Graph of Solution Set*
Linear equation $5x + 4 = 14$	$\{2\}$	0 2
Linear inequality $5x + 4 < 14$	$(-\infty, 2)$	0 2
$5x + 4 > 14$	$(2, \infty)$	0 2
Three-part inequality $-1 \leq 5x + 4 \leq 14$	$[-1, 2]$	−1 0 2

OBJECTIVE 5 Solve applied problems using linear inequalities.

▼ **Words and Phrases That Suggest Inequality**

Word Statement	*Interpretation*	*Example*	*Inequality*
a exceeds *b*.	$a > b$	Juan's age *j* exceeds 21 yr.	$j > 21$
a is at least *b*.	$a \geq b$	Juan is at least 21 yr old.	$j \geq 21$
a is no less than *b*.	$a \geq b$	Juan is no less than 21 yr old.	$j \geq 21$
a is at most *b*.	$a \leq b$	Mia's age *m* is at most 10 yr.	$m \leq 10$
a is no more than *b*.	$a \leq b$	Mia is no more than 10 yr old.	$m \leq 10$

In **Examples 9 and 10,** we use the six problem-solving steps from earlier, changing Step 3 from

"Write an equation" to **"Write an inequality."**

NOW TRY EXERCISE 9

A local health club charges a \$40 one-time enrollment fee, plus \$35 per month for a membership. Sara can spend no more than \$355 on this exercise expense. What is the *maximum* number of months that Sara can belong to this health club?

EXAMPLE 9 Using a Linear Inequality to Solve a Rental Problem

A rental company charges \$20 to rent a chain saw, plus \$9 per hr. Tom can spend no more than \$65 to clear some logs from his yard. What is the maximum amount of time he can use the rented saw?

Step 1 **Read** the problem again.

Step 2 **Assign a variable.** Let $x =$ the number of hours he can rent the saw.

Step 3 **Write an inequality.** He must pay \$20, plus \$9x, to rent the saw for x hours, and this amount must be *no more than* \$65.

Cost of renting	is no more than	65 dollars.
$20 + 9x$	$\leq$	65

Step 4 **Solve.**

$9x \leq 45$ Subtract 20.

$x \leq 5$ Divide by 9.

Step 5 **State the answer.** He can use the saw for a maximum of 5 hr. (Of course, he may use it for less time, as indicated by the inequality $x \leq 5$.)

Step 6 **Check.** If Tom uses the saw for 5 hr, he will spend

$$20 + 9(5) = 65 \text{ dollars, the maximum amount.}$$

NOW TRY

NOW TRY ANSWER

9. 9 months

NOW TRY EXERCISE 10

Joel has scores of 82, 97, and 93 on his first three exams. What score must he earn on the fourth exam to keep an average of at least 90?

EXAMPLE 10 Finding an Average Test Score

Martha has scores of 88, 86, and 90 on her first three algebra tests. An average score of at least 90 will earn an A in the class. What possible scores on her fourth test will earn her an A average?

Let $x =$ the score on the fourth test. Her average score must be at least 90. To find the average of four numbers, add them and then divide by 4.

Average — is at least — 90.

$$\frac{88 + 86 + 90 + x}{4} \geq 90$$

$$\frac{264 + x}{4} \geq 90 \quad \text{Add the scores.}$$

$$264 + x \geq 360 \quad \text{Multiply by 4.}$$

$$x \geq 96 \quad \text{Subtract 264.}$$

She must score 96 or more on her fourth test.

CHECK $\frac{88 + 86 + 90 + 96}{4} = \frac{360}{4} = 90$, the minimum score. ✓

A score of 96 or more will give an average of at least 90, as required. **NOW TRY**

NOW TRY ANSWER
10. at least 88

1.5 Exercises

FOR EXTRA HELP **MyLab Math**

Video solutions for select problems available in MyLab Math

STUDY SKILLS REMINDER
Study cards are a great way to learn vocabulary, procedures, and so on. **Review Study Skill 5, *Using Study Cards.***

Concept Check *Match each inequality in Column I with the correct graph or interval in Column II.*

I

1. $x \leq 3$ **2.** $x > 3$

3. $x < 3$ **4.** $x \geq 3$

5. $-3 \leq x \leq 3$

6. $-3 < x < 3$

II

A. 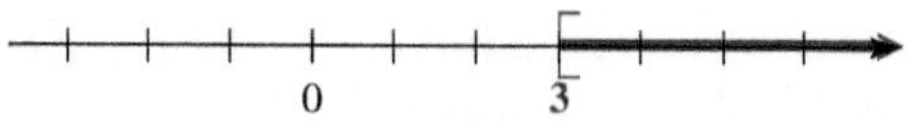

B. 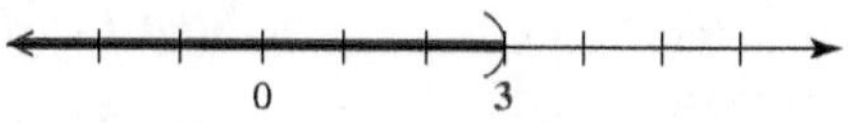

C. $(3, \infty)$ **D.** $(-\infty, 3]$

E. $(-3, 3)$ **F.** $[-3, 3]$

Concept Check *Work each problem involving inequalities.*

7. A high level of LDL cholesterol ("bad cholesterol") in the blood increases a person's risk of heart disease. The table shows how LDL levels affect risk.

If x represents the LDL cholesterol number, write a linear inequality or three-part inequality for each category. Use x as the variable.

(a) Optimal

(b) Near optimal/above optimal

(c) Borderline high

(d) High **(e)** Very high

LDL Cholesterol	*Risk Category*
Less than 100	Optimal
100–129	Near optimal/ above optimal
130–159	Borderline high
160–189	High
190 and above	Very high

Data from WebMD.

8. A high level of triglycerides in the blood also increases a person's risk of heart disease. The table shows how triglyceride levels affect risk.

If x represents the triglycerides number, write a linear inequality or three-part inequality for each category. Use x as the variable.

(a) Normal **(b)** Mildly high

(c) High **(d)** Very high

Triglycerides	*Risk Category*
Less than 100	Normal
100–199	Mildly high
200–499	High
500 or higher	Very high

Data from WebMD.

9. Concept Check A student solved the following inequality incorrectly as shown.

$$4x \geq -64$$
$$\frac{4x}{4} \leq \frac{-64}{4}$$
$$x \leq -16 \qquad \text{Solution set: } (-\infty, -16]$$

WHAT WENT WRONG? Give the correct solution set.

10. Concept Check A student solved the following inequality incorrectly as shown.

$$-2x < -18$$
$$\frac{-2x}{-2} < \frac{-18}{-2}$$
$$x < 9 \qquad \text{Solution set: } (-\infty, 9)$$

WHAT WENT WRONG? Give the correct solution set.

Solve each inequality. Graph the solution set, and write it using interval notation. ***See Examples 1–6.***

11. $x - 4 \geq 12$ **12.** $x - 3 \geq 7$ **13.** $3k + 1 > 22$

14. $5x + 6 < 76$ **15.** $4x < -16$ **16.** $2x > -10$

17. $-4x < 16$ **18.** $-5x > 25$ **19.** $-\frac{3}{4}x \geq 30$

20. $-\frac{2}{3}x \leq 12$ **21.** $-1.3x \geq -5.2$ **22.** $-2.5x \leq -1.25$

23. $5x + 2 \leq -48$ **24.** $4x + 1 \leq -31$ **25.** $\frac{3k - 1}{4} > 5$

26. $\frac{5x - 6}{8} < 8$ **27.** $\frac{2x - 5}{-4} > 5$ **28.** $\frac{3x - 2}{-5} < 6$

29. $3k + 1 < -20$ **30.** $5z + 6 > -29$ **31.** $6x - 4 \geq -2x$

32. $2x - 8 \geq -2x$ **33.** $x - 2(x - 4) \leq 3x$ **34.** $x - 3(x + 1) \leq 4x$

35. $-(4 + r) + 2 - 3r < -14$ **36.** $-(9 + x) - 5 + 4x \geq 4$

37. $-3(x - 6) > 2x - 2$ **38.** $-2(x + 4) \leq 6x + 16$

39. $\frac{2}{3}(3x - 1) \geq \frac{3}{2}(2x - 3)$ **40.** $\frac{7}{5}(10x - 1) < \frac{2}{3}(6x + 5)$

41. $-\frac{1}{4}(p + 6) + \frac{3}{2}(2p - 5) < 10$ **42.** $\frac{3}{5}(t - 2) - \frac{1}{4}(2t - 7) \leq 3$

43. $3(2x - 4) - 4x < 2x + 3$ **44.** $7(4 - x) + 5x < 2(16 - x)$

45. $8\left(\frac{1}{2}x + 3\right) < 8\left(\frac{1}{2}x - 1\right)$ **46.** $10\left(\frac{1}{5}x + 2\right) < 10\left(\frac{1}{5}x + 1\right)$

Solve each inequality. Graph the solution set, and write it using interval notation. ***See Examples 7 and 8.***

47. $-4 < x - 5 < 6$ **48.** $-1 < x + 1 < 8$ **49.** $-9 \leq x + 5 \leq 15$

50. $-4 \leq x + 3 \leq 10$ **51.** $-6 \leq 2x + 4 \leq 16$ **52.** $-15 < 3x + 6 < -12$

53. $-19 \le 3x - 5 \le 1$ **54.** $-16 < 3x + 2 < -10$ **55.** $4 \le -9x + 5 < 8$

56. $4 \le -2x + 3 < 8$ **57.** $-8 \le -4x + 2 \le 6$ **58.** $-12 \le -6x + 3 \le 15$

59. $-3 < \frac{3}{4}x < 6$ **60.** $-4 < \frac{2}{3}x < 12$ **61.** $-1 \le \frac{2x - 5}{6} \le 5$

62. $-3 \le \frac{3x + 1}{4} \le 3$ **63.** $-5 \le \frac{6 - 5x}{2} \le 0$ **64.** $-4 \le \frac{2 - 4x}{3} \le 0$

Give, in interval notation, the unknown numbers in each description.

65. A number is between 0 and 1.

66. A number is between -3 and -2.

67. Six times a number is between -12 and 12.

68. Half a number is between -3 and 2.

69. When 1 is added to twice a number, the result is greater than or equal to 7.

70. If 8 is subtracted from a number, then the result is at least 5.

71. One third of a number is added to 6, giving a result of at least 3.

72. Three times a number, minus 5, is no more than 7.

Solve each problem. ***See Examples 9 and 10.***

73. Bonnie earned scores of 90 and 82 on her first two tests in English literature. What score must she make on her third test to keep an average of 84 or greater?

74. Scott scored 92 and 96 on his first two tests in "Methods in Teaching Mathematics." What score must he make on his third test to keep an average of 90 or greater?

75. To earn a B in an algebra course requires an average of at least 80 on five tests. A student has scores of 75, 91, 82, and 74. What possible scores on the fifth test would guarantee this student a B in the class?

76. To pass Algebra II requires an average of at least 70 on four tests. A student has scores of 80, 62, and 73. What possible scores on the fourth test would guarantee this student a passing score in the class?

77. Latrice is signing up for cell phone service. She must decide between Plan A, which costs \$54.99 per month with a free phone included, and Plan B, which costs \$49.99 per month, but would require her to buy a phone for \$129. Under either plan, she does not expect to go over the included number of monthly minutes. After how many months would Plan B be a better deal?

78. Newlyweds Bryce and Lauren need to move their belongings to their new apartment. They can rent a truck from U-Haul for \$29.95 per day plus 28 cents per mile or from Budget Truck Rentals for \$34.95 per day plus 25 cents per mile. After how many miles (to the nearest mile) would the Budget rental be a better deal than the U-Haul rental?

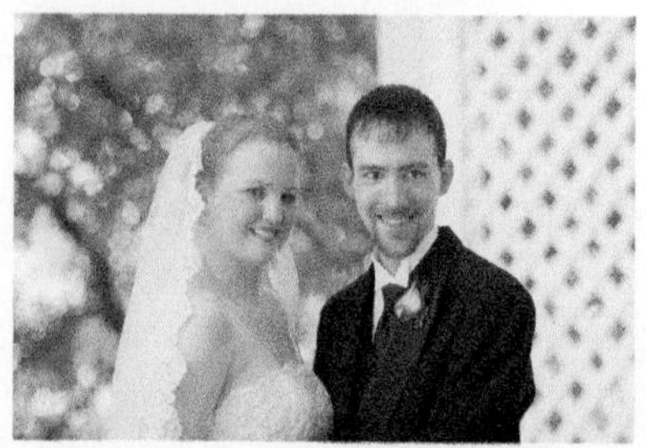

79. The average monthly precipitation for Dallas, TX, for October, November, and December is 3.47 in. If 2.88 in. falls in October and 3.13 in. falls in November, how many inches must fall in December so that the average monthly precipitation for these months exceeds 3.47 in.? (Data from www.weather.com)

80. The average monthly precipitation for Honolulu, HI, for October, November, and December is 3.11 in. If 2.98 in. falls in October and 3.05 in. falls in November, how many inches must fall in December so that the average monthly precipitation for these months exceeds 3.11 in.? (Data from www.weather.com)

81. According to some medical advisors, a body mass index (BMI) between 19 and 25 suggests a healthful weight. Use the formula

$$\text{BMI} = \frac{704 \times (\text{weight in pounds})}{(\text{height in inches})^2}$$

to find the weight range w, to the nearest pound, that gives a healthful BMI for each height.

(a) 72 in. **(b)** 63 in. **(c)** Your height in inches

82. To achieve the maximum benefit from exercising, the heart rate, in beats per minute, should be in the target heart rate (THR) zone. For a person aged A, the formula is as follows.

$$0.7(220 - A) \leq \text{THR} \leq 0.85(220 - A)$$

Find the THR to the nearest whole number for each age. (*Source:* Hockey, R. V., *Physical Fitness: The Pathway to Healthful Living,* Times Mirror/Mosby College Publishing.)

(a) 35 **(b)** 55 **(c)** Your age

A product will produce a profit only when the revenue R from selling the product exceeds the cost C of producing it. Find the least whole number of units x that must be sold for the business to show a profit for the item described.

83. Peripheral Visions, Inc., finds that the cost of producing x studio-quality DVDs is $C = 20x + 100$, while the revenue produced from them is $R = 24x$ (C and R in dollars).

84. Speedy Delivery finds that the cost of making x deliveries is $C = 3x + 2300$, while the revenue produced from them is $R = 5.50x$ (C and R in dollars).

RELATING CONCEPTS For Individual or Group Work (Exercises 85–90)

Work Exercises 85–90 in order.

85. Solve the linear equation. Graph the solution set on a number line.

$$5(x + 3) - 2(x - 4) = 2(x + 7)$$

86. Solve the linear inequality. Graph the solution set on a number line.

$$5(x + 3) - 2(x - 4) > 2(x + 7)$$

87. Solve the linear inequality. Graph the solution set on a number line.

$$5(x + 3) - 2(x - 4) < 2(x + 7)$$

88. Graph all the solution sets of the equation and inequalities in **Exercises 85–87** on the same number line. (This is the **union** of the solution sets.) What set do we obtain?

89. Complete the following:

The solution set of $-3(x + 2) = 3x + 12$ is ________.

The solution set of $-3(x + 2) < 3x + 12$ is ________.

Therefore, the solution set of $-3(x + 2) > 3x + 12$ is ________.

90. Describe the union of the three solution sets in **Exercise 89** in words, and write it in interval notation as a single interval.

1.6 Set Operations and Compound Inequalities

OBJECTIVES

1. Recognize set intersection and union.
2. Find the intersection of two sets.
3. Solve compound inequalities with the word *and*.
4. Find the union of two sets.
5. Solve compound inequalities with the word *or*.

VOCABULARY

☐ intersection
☐ compound inequality
☐ union

OBJECTIVE 1 Recognize set intersection and union.

Consider the two sets A and B defined as follows.

$$A = \{1, 2, 3\}, \qquad B = \{2, 3, 4\}$$

The set of all elements that belong to both A **and** B, called their *intersection* and symbolized $\boldsymbol{A \cap B}$, is given by

$$A \cap B = \{2, 3\}. \qquad \text{Intersection}$$

The set of all elements that belong to either A **or** B, or both, called their *union* and symbolized $\boldsymbol{A \cup B}$, is given by

$$A \cup B = \{1, 2, 3, 4\}. \qquad \text{Union}$$

OBJECTIVE 2 Find the intersection of two sets.

The intersection of two sets is defined with the word *and*.

Intersection of Sets

For any two sets A and B, the **intersection** of A and B, symbolized $A \cap B$, is defined as follows.

$$A \cap B = \{x \mid x \text{ is an element of } A \textbf{ and } x \text{ is an element of } B\}$$

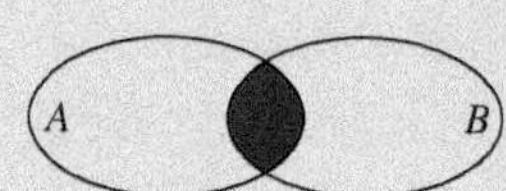

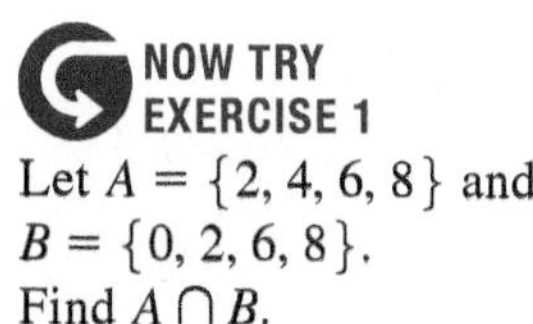
NOW TRY EXERCISE 1
Let $A = \{2, 4, 6, 8\}$ and $B = \{0, 2, 6, 8\}$. Find $A \cap B$.

EXAMPLE 1 Finding the Intersection of Two Sets

Let $A = \{1, 2, 3, 4\}$ and $B = \{2, 4, 6\}$. Find $A \cap B$.

The set $A \cap B$ contains those elements that belong to both A *and* B.

$$A \cap B = \{1, 2, 3, 4\} \cap \{2, 4, 6\}$$
$$= \{2, 4\}$$

NOW TRY 

OBJECTIVE 3 Solve compound inequalities with the word *and*.

A **compound inequality** consists of two inequalities linked by a connective word.

$$x + 1 \le 9 \quad \text{and} \quad x - 2 \ge 3$$
$$2x > 4 \quad \text{or} \quad 3x - 6 < 5 \qquad \text{Compound inequalities}$$

Solving a Compound Inequality with *and*

Step 1 Solve each inequality individually.

Step 2 The solution set of the compound inequality includes all numbers that satisfy both inequalities in Step 1—that is, the *intersection* of the solution sets.

NOW TRY ANSWER
1. $\{2, 6, 8\}$

NOW TRY EXERCISE 2

Solve the compound inequality, and graph the solution set.

$x - 2 \leq 5 \quad \text{and} \quad x + 5 \geq 9$

EXAMPLE 2 Solving a Compound Inequality with *and*

Solve the compound inequality, and graph the solution set.

$$x + 1 \leq 9 \quad \text{and} \quad x - 2 \geq 3$$

Step 1 Solve each inequality individually.

$$\begin{aligned} x + 1 &\leq 9 &\quad \text{and} \quad x - 2 &\geq 3 \\ x + 1 - 1 &\leq 9 - 1 &\quad \text{and} \quad x - 2 + 2 &\geq 3 + 2 \\ x &\leq 8 &\quad \text{and} \quad x &\geq 5 \end{aligned}$$

Step 2 The solution set will include all numbers that satisfy *both* inequalities in Step 1 at the same time. The compound inequality is true whenever $x \leq 8$ and $x \geq 5$ are both true. See the graphs in **FIGURE 20.**

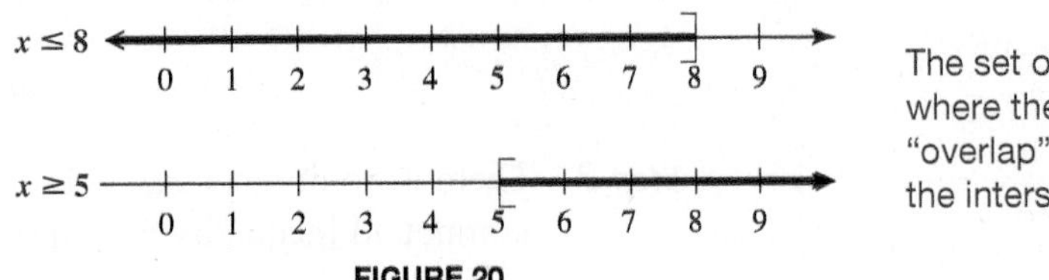

The set of points where the graphs "overlap" represents the intersection.

FIGURE 20

The intersection of the two graphs in **FIGURE 20** is the solution set. **FIGURE 21** shows the graph of the solution set, written using interval notation as the closed interval $[5, 8]$.

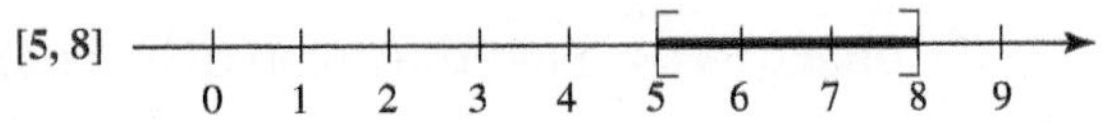

[5, 8]

FIGURE 21

NOW TRY 

NOW TRY EXERCISE 3

Solve and graph.

$-4x - 1 < 7 \quad \text{and}$

$3x + 4 \geq -5$

EXAMPLE 3 Solving a Compound Inequality with *and*

Solve the compound inequality, and graph the solution set.

$$-3x - 2 > 5 \quad \text{and} \quad 5x - 1 \leq -21$$

Step 1 Solve each inequality individually.

$$\begin{aligned} -3x - 2 &> 5 &\quad \text{and} \quad 5x - 1 &\leq -21 \\ -3x &> 7 &\quad \text{and} \quad 5x &\leq -20 \\ x &< -\frac{7}{3} &\quad \text{and} \quad x &\leq -4 \end{aligned}$$

Remember to reverse the direction of the inequality symbol.

The graphs of $x < -\frac{7}{3}$ and $x \leq -4$ are shown in **FIGURE 22.**

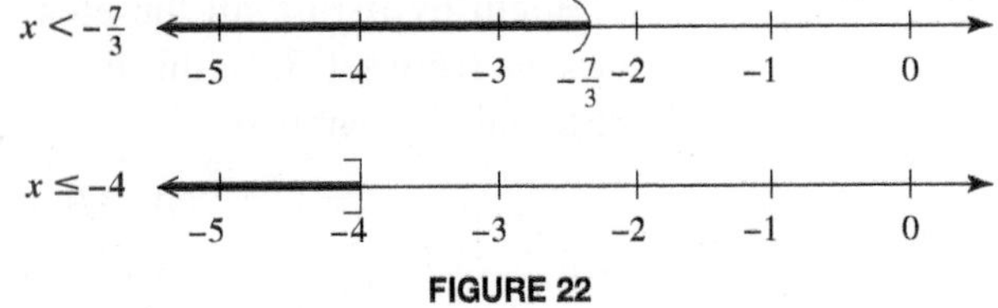

FIGURE 22

Step 2 Now find all values of x that are less than $-\frac{7}{3}$ and also less than or equal to -4. This is shown in **FIGURE 23.** The solution set is written as the infinite interval $(-\infty, -4]$.

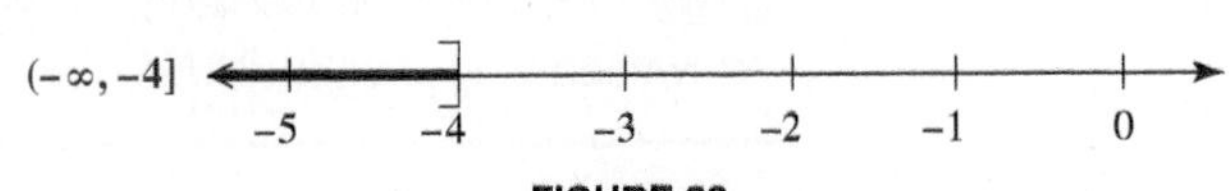

FIGURE 23

NOW TRY

NOW TRY ANSWERS

2. 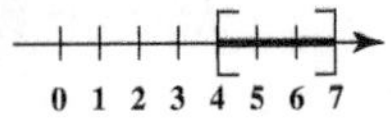 $[4, 7]$

3. 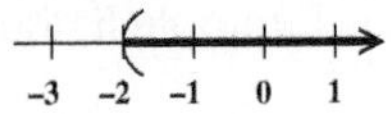 $(-2, \infty)$

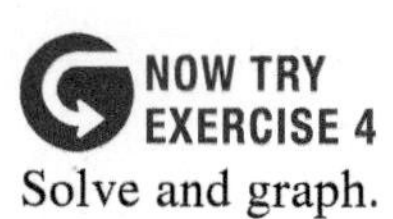
NOW TRY EXERCISE 4

Solve and graph.

$x - 7 < -12$ and $2x + 1 > 5$

EXAMPLE 4 Solving a Compound Inequality with *and*

Solve the compound inequality, and graph the solution set.

$$x + 2 < 5 \quad \text{and} \quad x - 10 > 2$$

Step 1 Solve each inequality individually.

$$x + 2 < 5 \quad \text{and} \quad x - 10 > 2$$

$$x < 3 \quad \text{and} \quad x > 12$$

The graphs of $x < 3$ and $x > 12$ are shown in **FIGURE 24.**

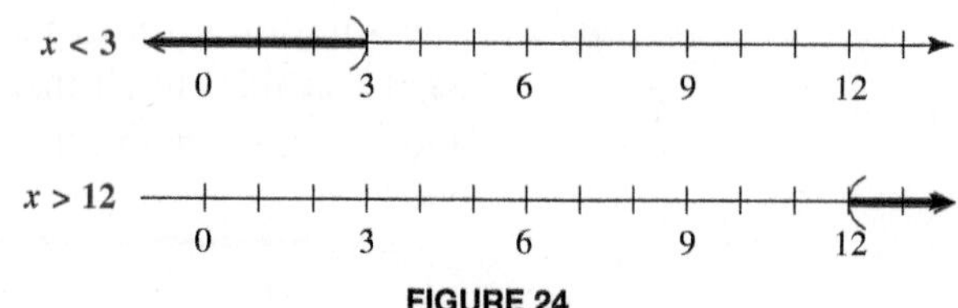

FIGURE 24

Step 2 There is no number that is both less than 3 *and* greater than 12, so the given compound inequality has no solution. See **FIGURE 25.** The solution set is $\emptyset$.

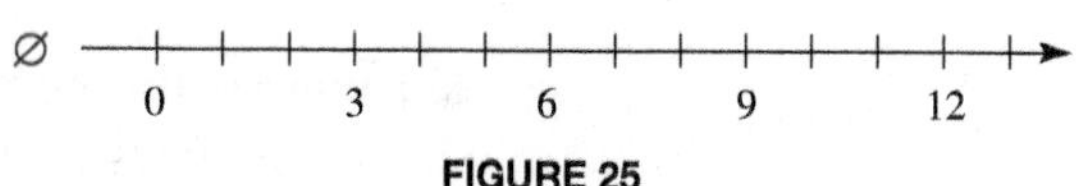

FIGURE 25

NOW TRY

OBJECTIVE 4 Find the union of two sets.

The union of two sets is defined with the word *or*.

Union of Sets

For any two sets A and B, the **union** of A and B, symbolized $A \cup B$, is defined as follows.

$$A \cup B = \{x \mid x \text{ is an element of } A \textbf{ or } x \text{ is an element of } B\}$$

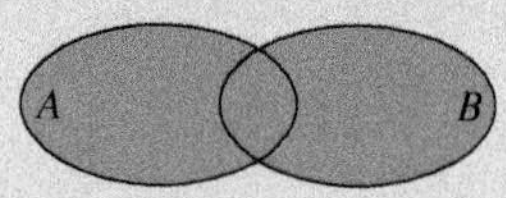

NOW TRY EXERCISE 5

Let $A = \{5, 10, 15, 20\}$ and $B = \{5, 15, 25\}$. Find $A \cup B$.

EXAMPLE 5 Finding the Union of Two Sets

Let $A = \{1, 2, 3, 4\}$ and $B = \{2, 4, 6\}$. Find $A \cup B$.

Begin by listing all the elements of set A: 1, 2, 3, 4. Then list any additional elements from set B. In this case the elements 2 and 4 are already listed, so the only additional element is 6.

$$\begin{aligned} A \cup B &= \{1, 2, 3, 4\} \cup \{2, 4, 6\} \\ &= \{1, 2, 3, 4, 6\} \end{aligned}$$

The union consists of all elements in either A *or* B (or both).

NOW TRY 

NOTE Although the elements 2 and 4 appeared in both sets A and B in **Example 5,** they are written only once in $A \cup B$.

NOW TRY ANSWERS

4. $\emptyset$

5. $\{5, 10, 15, 20, 25\}$

OBJECTIVE 5 Solve compound inequalities with the word *or*.

Solving a Compound Inequality with *or*

Step 1 Solve each inequality individually.

Step 2 The solution set of the compound inequality includes all numbers that satisfy either one or the other (or both) of the inequalities in Step 1—that is, the *union* of the solution sets.

NOW TRY EXERCISE 6

Solve and graph.

$-12x \le -24 \quad \text{or} \quad x + 9 < 8$

EXAMPLE 6 Solving a Compound Inequality with *or*

Solve the compound inequality, and graph the solution set.

$$6x - 4 < 2x \quad \text{or} \quad -3x \le -9$$

Step 1 Solve each inequality individually.

$$6x - 4 < 2x \quad \text{or} \quad -3x \le -9$$

$$4x < 4$$

$$x < 1 \quad \text{or} \quad x \ge 3$$

Remember to reverse the inequality symbol.

The graphs of these two inequalities are shown in **FIGURE 26.**

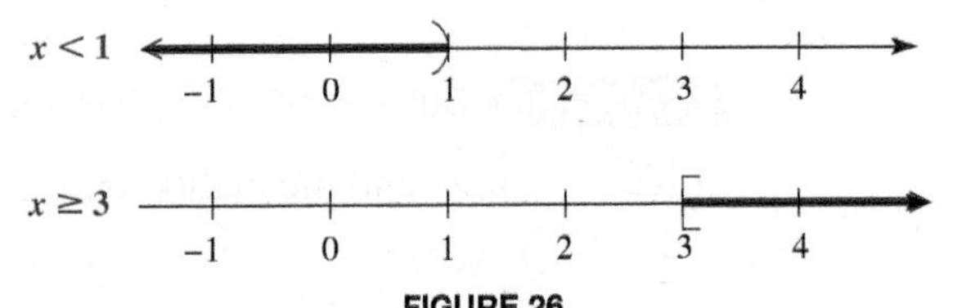

FIGURE 26

The set of points in *either* of the graphs represents the union.

Step 2 Because the inequalities are joined with *or*, find the union of the two solution sets. The union is the disjoint interval in **FIGURE 27.**

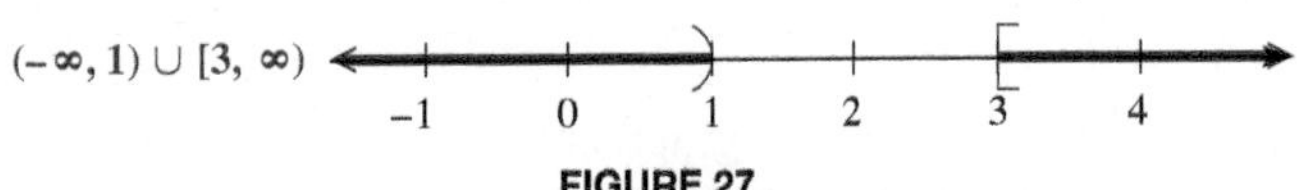

FIGURE 27

In interval notation, the solution set is $(-\infty, 1) \cup [3, \infty)$. Always pay particular attention to the end points of the solution sets and whether parentheses, brackets, or one of each should be used. **NOW TRY**

CAUTION When inequalities are used to write the solution set in **Example 6,** it *must* be written using two separate inequalities.

$$x < 1 \quad \text{or} \quad x \ge 3$$

Writing $3 \le x < 1$, which translates using *and*, would imply that

$$3 \le 1, \quad \text{which is } \textbf{\textit{FALSE.}}$$

NOW TRY ANSWER

6. -2 -1 0 1 2 3

$(-\infty, -1) \cup [2, \infty)$

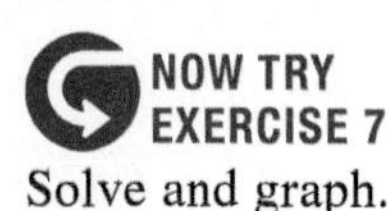

Solve and graph.

$-x + 2 < 6 \quad \text{or} \quad 6x - 8 \geq 10$

EXAMPLE 7 Solving a Compound Inequality with *or*

Solve the compound inequality, and graph the solution set.

$$-4x + 1 \geq 9 \quad \text{or} \quad 5x + 3 \leq -12$$

Step 1 Solve each inequality individually.

$$\begin{aligned} -4x + 1 &\geq 9 & \text{or} \quad 5x + 3 &\leq -12 \\ -4x &\geq 8 & \text{or} \quad 5x &\leq -15 \\ x &\leq -2 & \text{or} \quad x &\leq -3 \end{aligned}$$

The graphs of these two inequalities are shown in **FIGURE 28.**

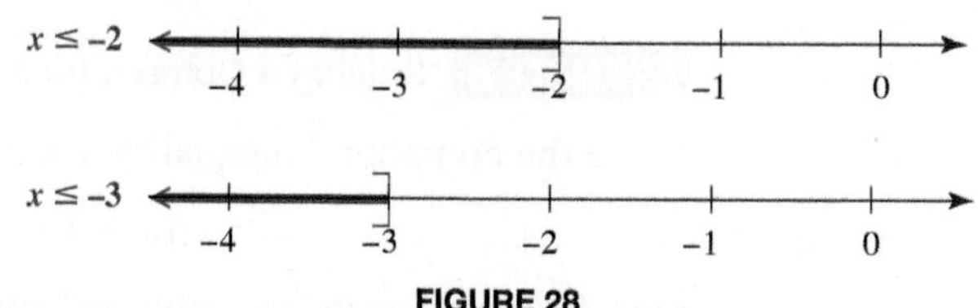

FIGURE 28

Step 2 **FIGURE 29** shows the union, the infinite interval $(-\infty, -2]$.

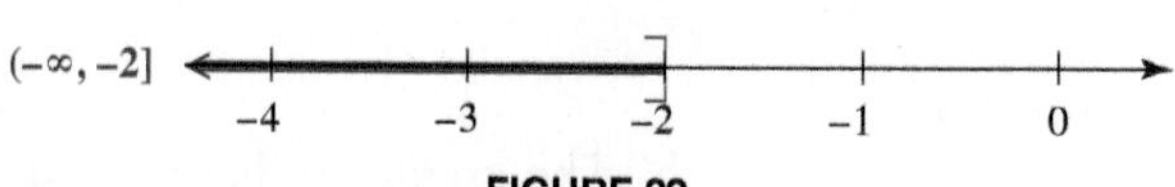

FIGURE 29

NOW TRY

NOW TRY
EXERCISE 8

Solve and graph.

$8x - 4 \geq 20 \quad \text{or}$
$-2x + 1 > -9$

EXAMPLE 8 Solving a Compound Inequality with *or*

Solve the compound inequality, and graph the solution set.

$$-2x + 5 \geq 11 \quad \text{or} \quad 4x - 7 \geq -27$$

Step 1 Solve each inequality individually.

$$\begin{aligned} -2x + 5 &\geq 11 & \text{or} \quad 4x - 7 &\geq -27 \\ -2x &\geq 6 & \text{or} \quad 4x &\geq -20 \\ x &\leq -3 & \text{or} \quad x &\geq -5 \end{aligned}$$

The graphs of these two inequalities are shown in **FIGURE 30.**

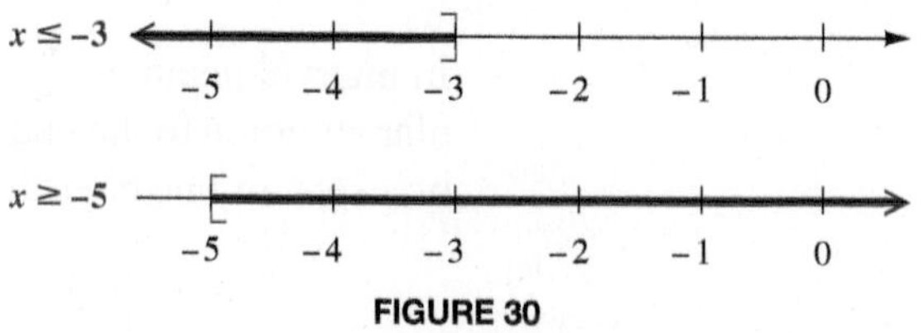

FIGURE 30

Step 2 By taking the union, we obtain every real number as a solution because every real number satisfies at least one of the two inequalities. The set of all real numbers is graphed in **FIGURE 31.** It is written in interval notation as $(-\infty, \infty)$.

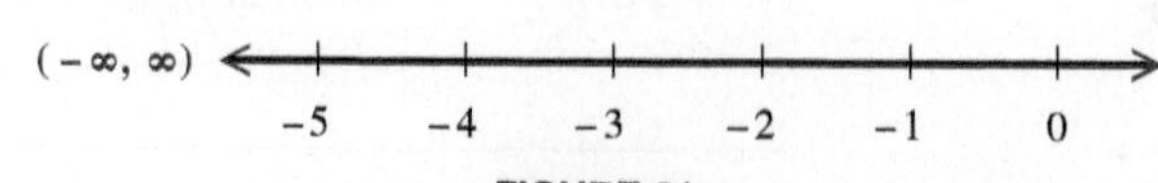

FIGURE 31

NOW TRY

NOW TRY ANSWERS

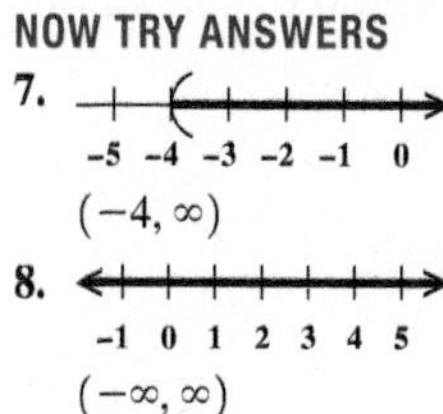

7. $(-4, \infty)$

8. $(-\infty, \infty)$

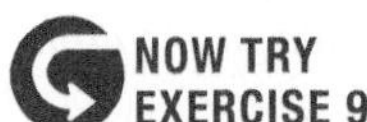

In **Example 9,** list the elements of each set.

(a) The set of countries to which exports were greater than \$100,000 million and from which imports were less than \$400,000 million

(b) The set of countries to which exports were less than \$200,000 million or from which imports were greater than \$250,000 million

EXAMPLE 9 Applying Intersection and Union

The five top U.S. trading partners for 2016 are listed in the table. Amounts are in millions of dollars.

Country	*U.S. Exports to Country*	*U.S. Imports from Country*
China	115,602	462,618
Canada	266,797	277,756
Mexico	229,702	294,056
Japan	63,236	132,046
Germany	49,363	114,099

Data from U.S. Census Bureau.

List the elements of the following sets.

(a) The set of countries to which exports were greater than \$200,000 million *and* from which imports were less than \$300,000 million

The countries that satisfy both conditions are Canada and Mexico, so the set is

$$\{\text{Canada, Mexico}\}.$$

(b) The set of countries to which exports were less than \$100,000 million *or* from which imports were greater than \$200,000 million

Here, any country that satisfies at least one of the conditions is in the set. This set includes all five countries:

$$\{\text{China, Canada, Mexico, Japan, Germany}\}.$$

NOW TRY ANSWERS

9. (a) {Canada, Mexico}
(b) {China, Canada, Mexico, Japan, Germany}

1.6 Exercises

  MyLab Math

Video solutions for select problems available in MyLab Math

STUDY SKILLS **REMINDER**

Reread your class notes before working the assigned exercises. **Review Study Skill 3, *Taking Lecture Notes.***

Concept Check *Determine whether each statement is* true *or* false. *If it is false, explain why.*

1. The union of the solution sets of $x + 1 = 6$, $x + 1 < 6$, and $x + 1 > 6$ is $(-\infty, \infty)$.
2. The intersection of the sets $\{x \mid x \geq 9\}$ and $\{x \mid x \leq 9\}$ is $\varnothing$.
3. The union of the sets $(-\infty, 7)$ and $(7, \infty)$ is $\{7\}$.
4. The intersection of the sets $(-\infty, 7]$ and $[7, \infty)$ is $\{7\}$.

Let $A = \{1, 2, 3, 4, 5, 6\}$, $B = \{1, 3, 5\}$, $C = \{1, 6\}$, *and* $D = \{4\}$. *Find each set.* ***See Examples 1 and 5.***

5. $A \cap D$ **6.** $B \cap C$ **7.** $B \cap A$ **8.** $C \cap A$ **9.** $B \cap \varnothing$

10. $A \cap \varnothing$ **11.** $A \cup B$ **12.** $B \cup D$ **13.** $B \cup C$ **14.** $C \cup D$

Concept Check *Two sets are specified by graphs. Graph the intersection of the two sets.*

15.
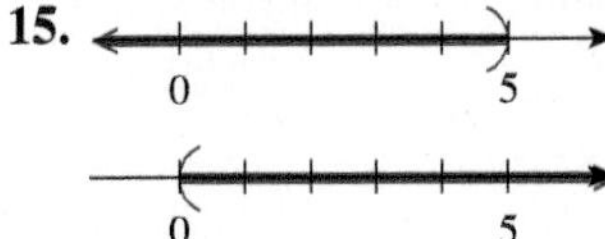

16.
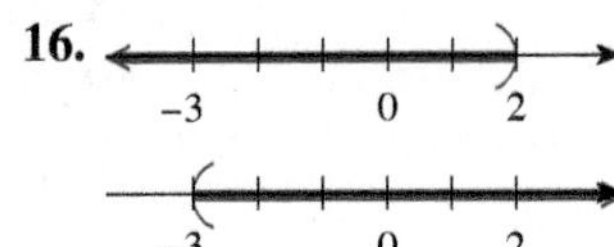

17.
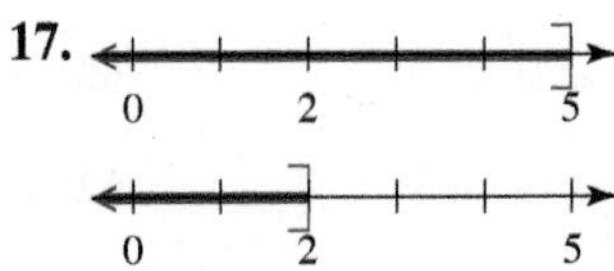

18.
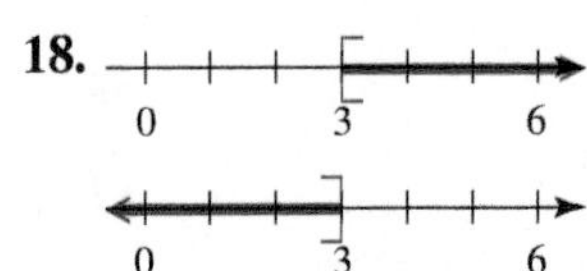

Solve each compound inequality. Graph the solution set, and write it using interval notation. ***See Examples 2–4.***

19. $x < 2$ and $x > -3$

20. $x < 5$ and $x > 0$

21. $x \leq 2$ and $x \leq 5$

22. $x \geq 3$ and $x \geq 6$

23. $x \leq 3$ and $x \geq 6$

24. $x \leq -1$ and $x \geq 3$

25. $x - 3 \leq 6$ and $x + 2 \geq 7$

26. $x + 5 \leq 11$ and $x - 3 \geq -1$

27. $-3x > 3$ and $x + 3 > 0$

28. $-3x < 3$ and $x + 2 < 6$

29. $3x - 4 \leq 8$ and $-4x + 1 \geq -15$

30. $7x + 6 \leq 48$ and $-4x + 3 \geq -21$

Concept Check *Two sets are specified by graphs. Graph the union of the two sets.*

31.

32.

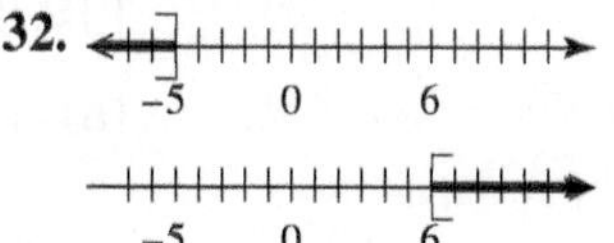

33.

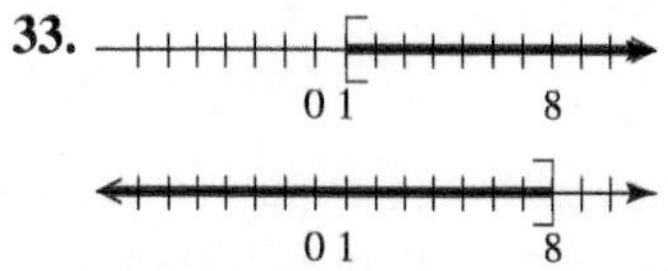

34.

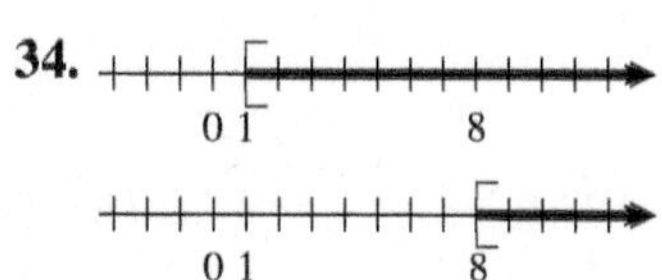

Solve each compound inequality. Graph the solution set, and write it using interval notation. ***See Examples 6–8.***

35. $x \leq 1$ or $x \leq 8$

36. $x \geq 1$ or $x \geq 8$

37. $x \geq -2$ or $x \geq 5$

38. $x \leq -2$ or $x \leq 6$

39. $x \geq -2$ or $x \leq 4$

40. $x \geq 5$ or $x \leq 7$

41. $x + 2 > 7$ or $1 - x > 6$

42. $x + 1 > 3$ or $x + 4 < 2$

43. $x + 1 > 3$ or $-4x + 1 > 5$

44. $-2x + 1 > -11$ or $x + 1 > 10$

45. $4x + 1 \geq -7$ or $-2x + 3 \geq 5$

46. $3x + 2 \leq -7$ or $-2x + 1 \leq 9$

Concept Check *Express each set in simplest interval form. (Hint: Graph each set and look for the intersection or union.)*

47. $(-\infty, -1] \cap [-4, \infty)$

48. $[-1, \infty) \cap (-\infty, 9]$

49. $(-\infty, -6] \cap [-9, \infty)$

50. $(5, 11] \cap [6, \infty)$

51. $(-\infty, 3) \cup (-\infty, -2)$

52. $[-9, 1] \cup (-\infty, -3)$

53. $[3, 6] \cup (4, 9)$

54. $[-1, 2] \cup (0, 5)$

Solve each compound inequality. Graph the solution set, and write it using interval notation. ***See Examples 2–4 and 6–8.***

55. $x < -1$ and $x > -5$

56. $x > -1$ and $x < 7$

57. $x < 4$ or $x < -2$

58. $x < 5$ or $x < -3$

59. $-3x \leq -6$ or $-3x \geq 0$

60. $-8x \leq -24$ or $-5x \geq 15$

61. $x + 1 \geq 5$ and $x - 2 \leq 10$

62. $2x - 6 \leq -18$ and $2x \geq -18$

Average expenses for full-time resident college students at 4-year institutions during the 2015–2016 academic year are shown in the table.

▼ College Expenses (in Dollars)

Type of Expense	*Public Schools (in-state)*	*Private Schools*
Tuition and fees	8778	27,951
Board rates	4561	5116
Dormitory charges	5850	6463

Data from National Center for Education Statistics.

List the elements of each set. ***See Example 9.***

63. The set of expenses that are less than \$9000 for public schools *and* are greater than \$15,000 for private schools

64. The set of expenses that are greater than \$4000 for public schools *and* are less than \$6000 for private schools

65. The set of expenses that are less than \$9000 for public schools *or* are greater than \$15,000 for private schools

66. The set of expenses that are greater than \$15,000 *or* are between \$7000 and \$8000

The figures represent the backyards of neighbors Luigi, Mario, Than, and Joe. Suppose that each resident has 150 ft *of fencing and enough sod to cover* 1400 ft^2 *of lawn.*

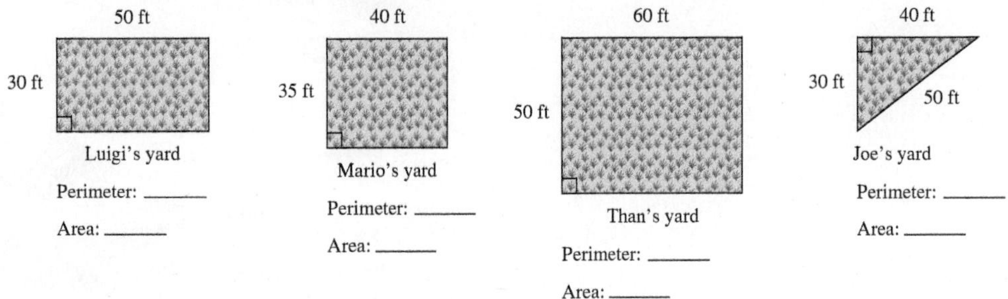

67. Determine the perimeters of the four yards.

68. Determine the areas of the four yards.

Give the name or names of the residents whose yards satisfy each description. (Hint: Use the perimeters and areas found in ***Exercises 67 and 68.****)*

69. The yard can be fenced *and* the yard can be sodded.

70. The yard can be fenced *and* the yard cannot be sodded.

71. The yard cannot be fenced *and* the yard can be sodded.

72. The yard cannot be fenced *and* the yard cannot be sodded.

73. The yard can be fenced *or* the yard can be sodded.

74. The yard cannot be fenced *or* the yard can be sodded.

RELATING CONCEPTS For Individual or Group Work (Exercises 75–78)

An intermediate algebra teacher bases final grades on points earned for activities as given in the Graded Classwork table on the left. To determine final grades, the teacher strictly adheres to the point ranges given in the Grade Distribution table on the right.

▼ **Graded Classwork**

Activity	*Points Available*
Homework and vocabulary	45
Daily activities (scaled)	55
Lab participation and completion	100
Major exams (3 at 100 points)	300
Final Exam	150
Total points	**650**

▼ **Grade Distribution**

Grade	*Points Required*
A	585–650
B	520–584
C	455–519
IP*	< 455 and active
F	< 455 and inactive

* In Progress

Use this information to ***work Exercises 75–78 in order.***

75. Suppose Lauren earns all of the homework and vocabulary points, 50 points for daily activities, and 90 points for lab participation and completion.

Let x = points to be earned on exams.

(a) Write three inequalities to find the minimum number of points she needs on exams to earn grades no lower than A, B, and C.

(b) Solve each inequality from part (a) to find the minimum number of points she needs for each grade. What "test average" (as a percent) corresponds to this number of points, given that exams account for 450 possible points? (Round up to the nearest whole number.)

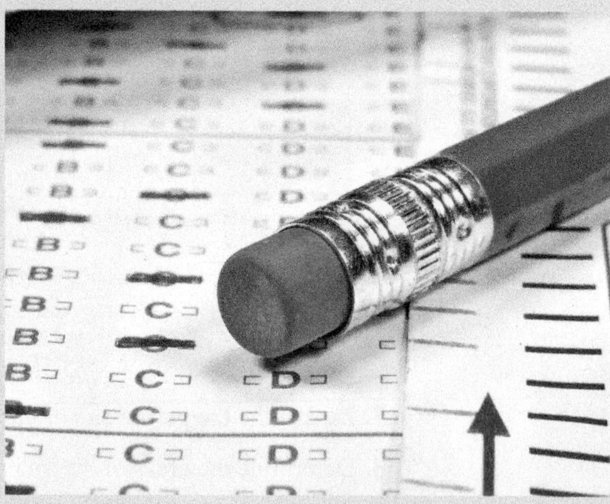

76. **See Exercise 75.** Write and solve a compound inequality to find the range of points Lauren needs in exam scores to earn a B average. What range of "test averages" (as percents) correspond to these scores, given that exams account for 450 possible points? (Round up to the nearest whole number.)

77. Suppose Mark earns only 15 points in homework and vocabulary, 40 points in daily activities, and 50 points in lab participation. Repeat **Exercise 75** using these values.

78. Repeat **Exercise 76** given that Mark wants to earn a C average. (Use his classwork points given in **Exercise 77.**)

1.7 Absolute Value Equations and Inequalities

OBJECTIVES

1. Use the distance definition of absolute value.
2. Solve equations of the form $|ax + b| = k$, for $k > 0$.
3. Solve inequalities of the form $|ax + b| < k$ and of the form $|ax + b| > k$, for $k > 0$.
4. Solve absolute value equations that involve rewriting.
5. Solve equations of the form $|ax + b| = |cx + d|$.
6. Solve special cases of absolute value equations and inequalities.
7. Solve an application involving relative error.

VOCABULARY

☐ absolute value equation
☐ absolute value inequality

Suppose the government of a country decides that it will comply with a restriction on greenhouse gas emissions *within* 3 years of 2025. This means that the *difference* between the year it will comply and 2025 is less than 3, *without regard to sign.* We state this mathematically as follows, where x represents the year in which it complies.

$$|x - 2025| < 3 \quad \text{Absolute value inequality}$$

We can intuitively reason that the year must be between 2022 and 2028, and thus values that satisfy $2022 < x < 2028$ make this inequality true.

OBJECTIVE 1 Use the distance definition of absolute value.

The **absolute value** of a number x, written $|x|$, represents the undirected distance from x to 0 on a number line. For example, the solution set of $|x| = 4$ is $\{-4, 4\}$, which means $x = -4$ or $x = 4$, as shown in **FIGURE 32**.

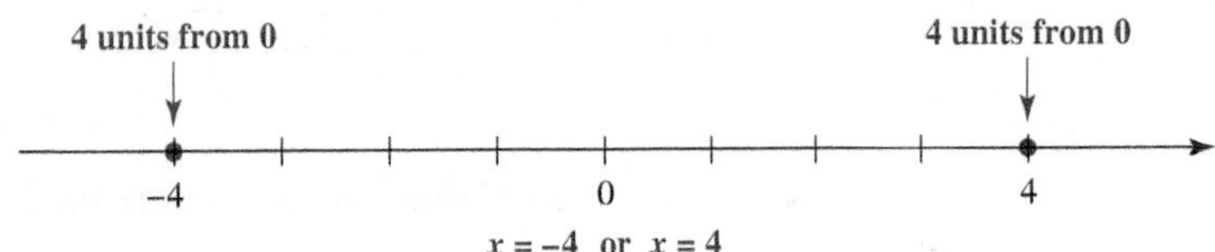

FIGURE 32

Because absolute value represents distance from 0, the solution set of $|x| > 4$ consists of all numbers that are *more* than four units from 0 on a number line. The set $(-\infty, -4) \cup (4, \infty)$ fits this description. The graph consists of disjoint intervals, which means $x < -4$ *or* $x > 4$, as shown in **FIGURE 33**.

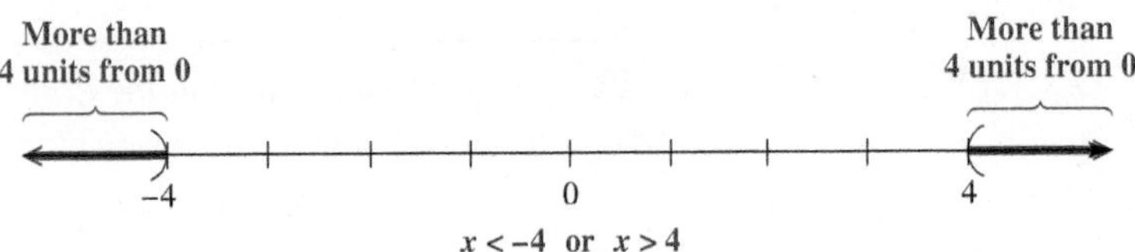

FIGURE 33

The solution set of $|x| < 4$ consists of all numbers that are *less* than 4 units from 0 on a number line. This is represented by all numbers *between* -4 and 4, which is given by the open interval $(-4, 4)$, as shown in **FIGURE 34**. Here, $-4 < x < 4$, which means $x > -4$ *and* $x < 4$.

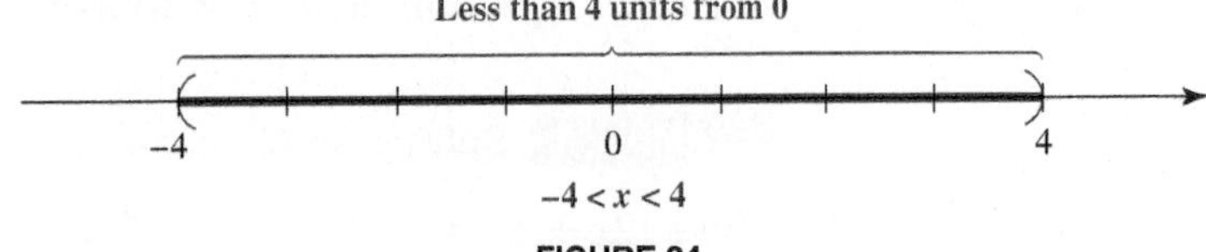

FIGURE 34

Absolute value equations and inequalities generally take the form

$$|ax + b| = k, \qquad |ax + b| > k, \qquad \text{or} \qquad |ax + b| < k,$$

where k is a positive number. From **FIGURES 32–34**, we see that

$|x| = 4$ has the same solution set as $x = -4$ or $x = 4$,

$|x| > 4$ has the same solution set as $x < -4$ or $x > 4$,

$|x| < 4$ has the same solution set as $x > -4$ and $x < 4$.

This is equivalent to $-4 < x < 4$.

Solving Absolute Value Equations and Inequalities

Let k be a positive real number, and p and q be real numbers.

Case 1 To solve $|ax + b| = k$, solve the compound equation

$$ax + b = k \quad \text{or} \quad ax + b = -k.$$

The solution set is usually of the form $\{p, q\}$, which includes two numbers.

Case 2 To solve $|ax + b| > k$,* solve the compound inequality

$$ax + b > k \quad \text{or} \quad ax + b < -k.$$

The solution set is of the form $(-\infty, p) \cup (q, \infty)$, which consists of disjoint intervals.

Case 3 To solve $|ax + b| < k$,** solve the three-part inequality

$$-k < ax + b < k.$$

The solution set is of the form (p, q), which consists of a single interval.

*This also applies to $|ax + b| \geq k$. The solution set *includes* the endpoints, using brackets rather than parentheses.

**This also applies to $|ax + b| \leq k$. The solution set *includes* the endpoints, using brackets rather than parentheses.

NOTE It is acceptable to write the compound statements in Cases 1 and 2 of the preceding box in the following equivalent forms.

$ax + b = k$ or $-(ax + b) = k$ Alternative for Case 1

$ax + b > k$ or $-(ax + b) > k$ Alternative for Case 2

OBJECTIVE 2 **Solve equations of the form $|ax + b| = k$, for $k > 0$.**

Remember that because absolute value refers to distance from the origin, an absolute value equation (with $k > 0$) will have two parts.

EXAMPLE 1 **Solving an Absolute Value Equation (Case 1)**

Solve $|2x + 1| = 7$.

For $|2x + 1|$ to equal 7, $2x + 1$ must be 7 units from 0 on a number line. This can happen only when $2x + 1 = 7$ or $2x + 1 = -7$. This is ***Case 1*** in the box above. Solve this compound equation as follows.

$$2x + 1 = 7 \quad \text{or} \quad 2x + 1 = -7$$

$$2x = 6 \quad \text{or} \quad 2x = -8 \qquad \text{Subtract 1.}$$

$$x = 3 \quad \text{or} \quad x = -4 \qquad \text{Divide by 2.}$$

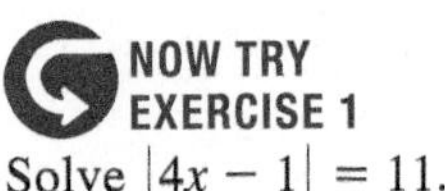

Solve $|4x - 1| = 11$.

CHECK $|2x + 1| = 7$

$|2(3) + 1| \stackrel{?}{=} 7$	Let $x = 3$.	$|2(-4) + 1| \stackrel{?}{=} 7$	Let $x = -4$.
$|6 + 1| \stackrel{?}{=} 7$		$|-8 + 1| \stackrel{?}{=} 7$	
$|7| \stackrel{?}{=} 7$		$|-7| \stackrel{?}{=} 7$	
$7 = 7$ ✓	True	$7 = 7$ ✓	True

The solution set is $\{-4, 3\}$. See **FIGURE 35**.

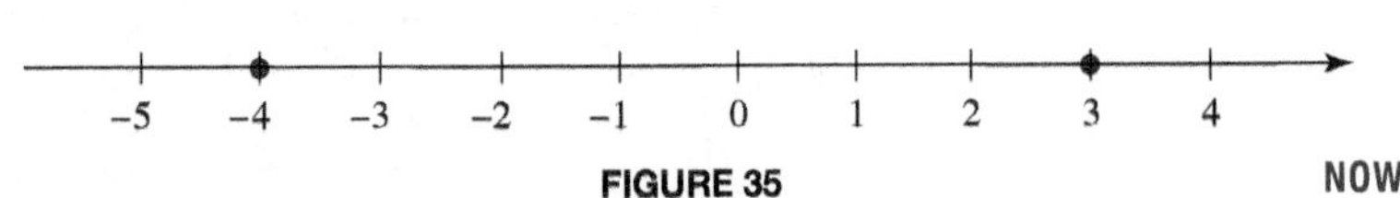

FIGURE 35

NOW TRY

OBJECTIVE 3 Solve inequalities of the form $|ax + b| < k$ and of the form $|ax + b| > k$, for $k > 0$.

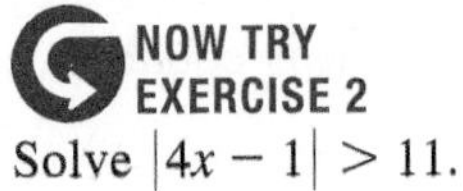

Solve $|4x - 1| > 11$.

EXAMPLE 2 Solving an Absolute Value Inequality (Case 2)

Solve $|2x + 1| > 7$.

Because $2x + 1$ must represent a number that is *more* than 7 units from 0 on either side of a number line, this absolute value inequality is rewritten as the following compound inequality. This is ***Case 2.***

$$2x + 1 > 7 \quad \text{or} \quad 2x + 1 < -7$$
$$2x > 6 \quad \text{or} \quad 2x < -8 \qquad \text{Subtract 1.}$$
$$x > 3 \quad \text{or} \quad x < -4 \qquad \text{Divide by 2.}$$

The solution set consists of the disjoint intervals shown in **FIGURE 36** and is written $(-\infty, -4) \cup (3, \infty)$.

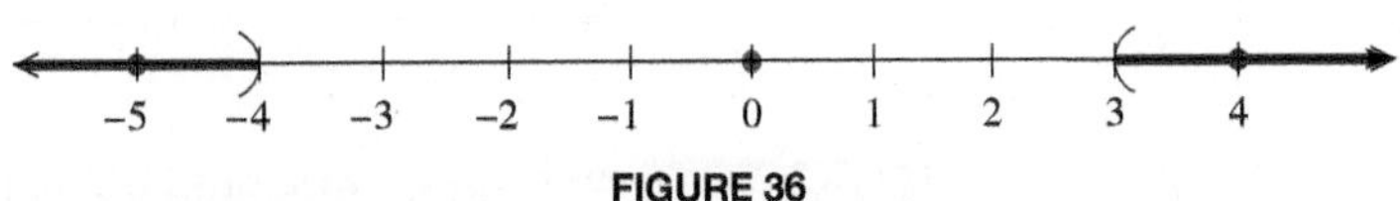

FIGURE 36

CHECK The excluded endpoints -4 and 3 are correct because from **Example 1** we know that -4 and 3 are the solutions of the related equation. Referring to **FIGURE 36**, we choose a test point in each of the three intervals $(-\infty, -4)$, $(-4, 3)$, and $(3, \infty)$.

For $(-\infty, -4)$, let $x = -5$.	For $(-4, 3)$, let $x = 0$.	For $(3, \infty)$, let $x = 4$.
$|2x + 1| > 7$	$|2x + 1| > 7$	$|2x + 1| > 7$
$|2(-5) + 1| \stackrel{?}{>} 7$	$|2(0) + 1| \stackrel{?}{>} 7$	$|2(4) + 1| \stackrel{?}{>} 7$
$|-9| \stackrel{?}{>} 7$	$|1| \stackrel{?}{>} 7$	$|9| \stackrel{?}{>} 7$
$9 > 7$ ✓ True	$1 > 7$ False	$9 > 7$ ✓ True

The check confirms that the solution set is $(-\infty, -4) \cup (3, \infty)$.

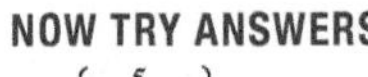

1. $\left\{-\frac{5}{2}, 3\right\}$

2. $\left(-\infty, -\frac{5}{2}\right) \cup (3, \infty)$

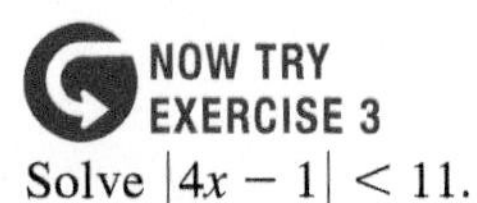
NOW TRY EXERCISE 3

Solve $|4x - 1| < 11$.

EXAMPLE 3 Solving an Absolute Value Inequality (Case 3)

Solve $|2x + 1| < 7$. Graph the solution set.

The expression $2x + 1$ must represent a number that is less than 7 units from 0 on either side of a number line. That is, $2x + 1$ must be between -7 and 7, which is written as a three-part inequality. This is ***Case 3.***

$$-7 < 2x + 1 < 7$$

$$-8 < 2x < 6 \quad \text{Subtract 1 from each part.}$$

$$-4 < x < 3 \quad \text{Divide each part by 2.}$$

The solution set consists of the open interval $(-4, 3)$ shown in **FIGURE 37**.

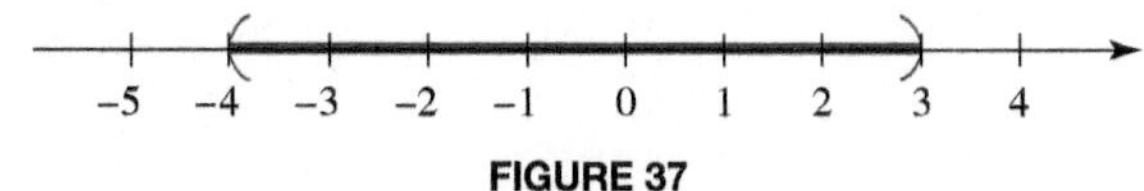

FIGURE 37

NOW TRY 

Look back at **FIGURES 35, 36, AND 37**, with the graphs of

$$|2x + 1| = 7, \quad |2x + 1| > 7, \quad \text{and} \quad |2x + 1| < 7.$$

If we find the union of the three sets, we obtain the set of all real numbers. For any value of x, $|2x + 1|$ will satisfy *one and only one* of the following: It is equal to 7, greater than 7, or less than 7.

CAUTION Remember the following important concepts.

1. The methods described apply when the constant is alone on one side of the absolute value equation or inequality and is *positive*.
2. Absolute value equations $|ax + b| = k$ and inequalities of the form $|ax + b| > k$ translate into **"or"** compound statements.
3. Absolute value inequalities of the form $|ax + b| < k$ translate into **"and"** compound statements. ***Only "and" compound statements may be written as three-part inequalities.***

NOW TRY EXERCISE 4

Solve $|7 - 4x| \geq 7$.

EXAMPLE 4 Solving an Absolute Value Inequality (Case 2, for $\geq$)

Solve $|5 - 2x| \geq 5$.

Case 2 is applied. Notice that the endpoints are included because equality is part of the symbol $\geq$.

$$5 - 2x \geq 5 \quad \text{or} \quad 5 - 2x \leq -5$$

$$-2x \geq 0 \quad \text{or} \quad -2x \leq -10 \quad \text{Subtract 5.}$$

$$x \leq 0 \quad \text{or} \quad x \geq 5 \quad \text{Divide by } -2. \text{ Reverse the direction of the inequality symbols.}$$

See the graph of the disjoint intervals in **FIGURE 38**. The solution set is written $(-\infty, 0] \cup [5, \infty)$.

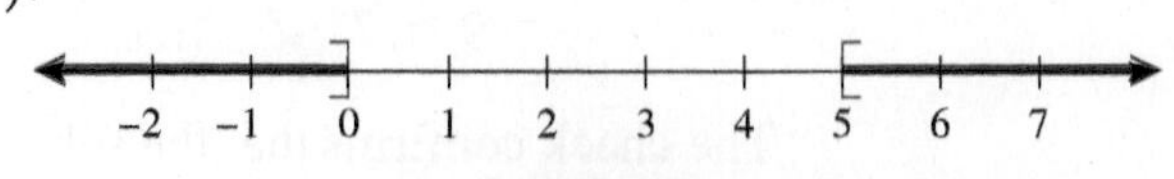

FIGURE 38

NOW TRY 

NOW TRY ANSWERS

3. $\left(-\frac{5}{2}, 3\right)$
4. $(-\infty, 0] \cup \left[\frac{7}{2}, \infty\right)$

OBJECTIVE 4 Solve absolute value equations that involve rewriting.

NOW TRY EXERCISE 5

Solve $|10x - 2| - 2 = 12$.

EXAMPLE 5 Solving an Absolute Value Equation That Requires Rewriting

Solve $|x + 3| + 5 = 12$.

Isolate the absolute value expression on one side of the equality symbol.

$$|x + 3| + 5 = 12$$

$$|x + 3| + 5 - 5 = 12 - 5 \quad \text{Subtract 5.}$$

$$|x + 3| = 7 \quad \text{Combine like terms.}$$

$$x + 3 = 7 \quad \text{or} \quad x + 3 = -7 \quad \text{Case 1}$$

$$x = 4 \quad \text{or} \quad x = -10 \quad \text{Subtract 3.}$$

CHECK $|x + 3| + 5 = 12$

$\lvert 4 + 3\rvert + 5 \stackrel{?}{=} 12$ Let $x = 4$.	$\lvert -10 + 3\rvert + 5 \stackrel{?}{=} 12$ Let $x = -10$.
$\lvert 7\rvert + 5 \stackrel{?}{=} 12$	$\lvert -7\rvert + 5 \stackrel{?}{=} 12$
$12 = 12$ ✓ True	$12 = 12$ ✓ True

The check confirms that the solution set is $\{-10, 4\}$. NOW TRY 

NOW TRY EXERCISE 6

Solve each inequality.

(a) $|x - 1| - 4 \le 2$

(b) $|x - 1| - 4 \ge 2$

EXAMPLE 6 Solving Absolute Value Inequalities That Require Rewriting

Solve each inequality.

(a) $|x + 3| + 5 \ge 12$

$$|x + 3| \ge 7 \quad \text{Case 2}$$

$$x + 3 \ge 7 \quad \text{or} \quad x + 3 \le -7$$

$$x \ge 4 \quad \text{or} \quad x \le -10$$

The solution set is $(-\infty, -10] \cup [4, \infty)$.

(b) $|x + 3| + 5 \le 12$

$$|x + 3| \le 7 \quad \text{Case 3}$$

$$-7 \le x + 3 \le 7$$

$$-10 \le x \le 4$$

The solution set is $[-10, 4]$.

NOW TRY

OBJECTIVE 5 Solve equations of the form $|ax + b| = |cx + d|$.

If two expressions have the same absolute value, they must either be equal or be negatives of each other.

Solving $|ax + b| = |cx + d|$

To solve an absolute value equation of the form

$$|ax + b| = |cx + d|,$$

solve the compound equation

$$ax + b = cx + d \quad \text{or} \quad ax + b = -(cx + d).$$

NOW TRY ANSWERS

5. $\left\{-\frac{6}{5}, \frac{8}{5}\right\}$

6. (a) $[-5, 7]$
(b) $(-\infty, -5] \cup [7, \infty)$

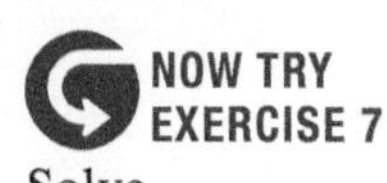
NOW TRY EXERCISE 7

Solve

$|3x - 4| = |5x + 12|.$

EXAMPLE 7 Solving an Equation Involving Two Absolute Values

Solve $|x + 6| = |2x - 3|$.

This equation is satisfied either if $x + 6$ and $2x - 3$ are equal to each other, or if $x + 6$ and $2x - 3$ are negatives of each other.

$$\begin{aligned} x + 6 &= 2x - 3 & \text{or} \quad x + 6 &= -(2x - 3) \\ x + 9 &= 2x & \text{or} \quad x + 6 &= -2x + 3 \\ 9 &= x & \text{or} \quad 3x &= -3 \\ x &= 9 & \text{or} \quad x &= -1 \end{aligned}$$

CHECK $|x + 6| = |2x - 3|$

$\lvert 9 + 6\rvert \stackrel{?}{=} \lvert 2(9) - 3\rvert$	Let $x = 9$.	$\lvert -1 + 6\rvert \stackrel{?}{=} \lvert 2(-1) - 3\rvert$	Let $x = -1$.
$\lvert 15\rvert \stackrel{?}{=} \lvert 18 - 3\rvert$		$\lvert 5\rvert \stackrel{?}{=} \lvert -2 - 3\rvert$	
$\lvert 15\rvert \stackrel{?}{=} \lvert 15\rvert$		$\lvert 5\rvert \stackrel{?}{=} \lvert -5\rvert$	
$15 = 15$ ✓	True	$5 = 5$ ✓	True

The check confirms that the solution set is $\{-1, 9\}$. NOW TRY  

OBJECTIVE 6 Solve special cases of absolute value equations and inequalities.

When an absolute value equation or inequality involves a *negative constant or 0* alone on one side, we use the following properties to solve it.

Special Properties of Absolute Value

Property 1 The absolute value of an expression can never be negative—that is, $|a| \geq 0$ for all real numbers a.

Property 2 The absolute value of an expression equals 0 only when the expression is equal to 0.

NOW TRY EXERCISE 8

Solve each equation.

(a) $|3x - 8| = -2$

(b) $|7x + 12| = 0$

EXAMPLE 8 Solving Special Cases of Absolute Value Equations

Solve each equation.

(a) $|5x - 3| = -4$

See ***Property 1*** in the preceding box. ***The absolute value of an expression can never be negative,*** so there are no solutions for this equation. The solution set is $\varnothing$.

(b) $|7x - 3| = 0$

See ***Property 2*** in the preceding box. The expression $|7x - 3|$ will equal 0 *only* if $7x - 3 = 0$.

$$7x - 3 = 0$$

$$7x = 3 \quad \text{Add 3.}$$

$$x = \frac{3}{7} \quad \text{Divide by 7.}$$

Check by substituting in the original equation.

The solution set $\left\{\frac{3}{7}\right\}$ consists of just one element. NOW TRY 

NOW TRY ANSWERS

7. $\{-8, -1\}$

8. (a) $\varnothing$ (b) $\left\{-\frac{12}{7}\right\}$
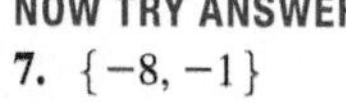

NOW TRY EXERCISE 9

Solve each inequality.

(a) $|x| > -10$

(b) $|4x + 1| + 5 < 4$

(c) $|x - 2| - 3 \leq -3$

EXAMPLE 9 Solving Special Cases of Absolute Value Inequalities

Solve each inequality.

(a) $|x| \geq -4$

The absolute value of a number is always greater than or equal to 0. (Property 1) Thus, $|x| \geq -4$ is true for *all* real numbers. The solution set is $(-\infty, \infty)$.

(b)

$$|x + 6| - 3 < -5$$

$$|x + 6| < -2 \quad \text{Add 3 to each side.}$$

There is no number whose absolute value is less than -2, so this inequality has no solution. The solution set is $\varnothing$.

(c)

$$|x - 7| + 4 \leq 4$$

$$|x - 7| \leq 0 \quad \text{Subtract 4 from each side.}$$

The value of $|x - 7|$ will never be less than 0. However, $|x - 7|$ will *equal* 0 when $x = 7$. Therefore, the solution set is $\{7\}$. NOW TRY

OBJECTIVE 7 Solve an application involving relative error.

Absolute value is used to find the **relative error,** or **tolerance,** in a measurement. If x represents the actual measurement and x_t represents the expected measurement, then taking the absolute value of the difference of x and x_t, divided by x_t, gives the relative error in x.

$$\textbf{relative error in } x = \left| \frac{x - x_t}{x_t} \right|$$

In quality control situations, the relative error must often be less than some predetermined amount.

NOW TRY EXERCISE 10

Suppose a machine filling *quart* milk cartons is set for a relative error that *is no greater than* 0.03. How many ounces may a filled carton contain?

EXAMPLE 10 Solving an Application Involving Relative Error

Suppose a machine filling *quart* milk cartons is set for a relative error that *is no greater than* 0.05. How many ounces may a filled carton contain?

Here $x_t = 32$ oz (because 1 qt $= 32$ oz) and the relative error $= 0.05$. We must find x, the actual measure of a filled carton.

$$\left| \frac{x - 32}{32} \right| \leq 0.05 \quad \text{Substitute the given values. } \textit{Is no greater than} \text{ translates as } \leq.$$

$$-0.05 \leq \frac{x - 32}{32} \leq 0.05 \quad \text{Case 3}$$

$$-1.6 \leq x - 32 \leq 1.6 \quad \text{Multiply by 32.}$$

$$30.4 \leq x \leq 33.6 \quad \text{Add 32.}$$

The filled carton may contain between 30.4 and 33.6 oz, inclusive. NOW TRY

NOW TRY ANSWERS

9. (a) $(-\infty, \infty)$ (b) $\varnothing$ (c) $\{2\}$

10. between 31.04 and 32.96 oz, inclusive

1.7 Exercises

FOR EXTRA HELP MyLab Math

Video solutions for select problems available in MyLab Math

STUDY SKILLS REMINDER
Time management can be a challenge for students. **Review Study Skill 6, *Managing Your Time.***

Concept Check *Match each absolute value equation or inequality in Column I with the graph of its solution set in Column II.*

I	II
1. $\|x\| = 5$	**A.**
$\|x\| < 5$	**B.**
$\|x\| > 5$	**C.**
$\|x\| \le 5$	**D.**
$\|x\| \ge 5$	**E.**

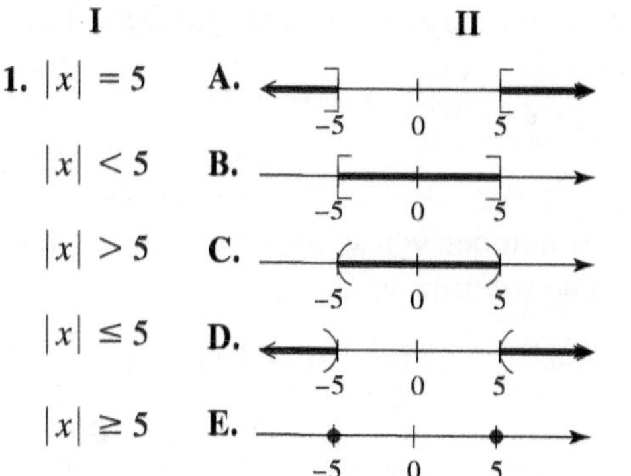

I	II
2. $\|x\| = 9$	**A.**
$\|x\| > 9$	**B.**
$\|x\| \ge 9$	**C.**
$\|x\| < 9$	**D.**
$\|x\| \le 9$	**E.**

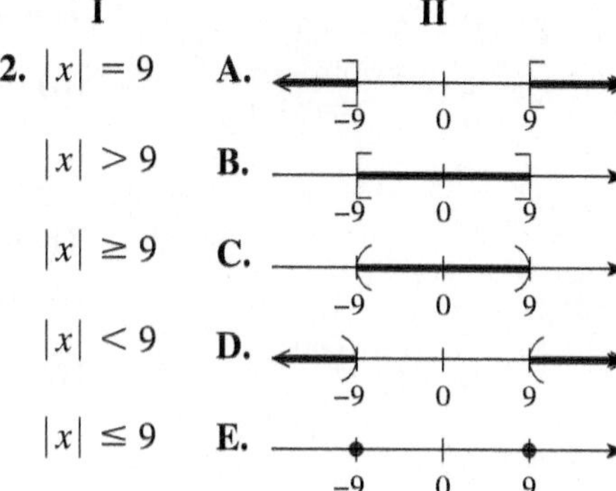

3. Concept Check How many solutions will $|ax + b| = k$ have for each situation?

(a) $k = 0$ **(b)** $k > 0$ **(c)** $k < 0$

4. Concept Check If $k < 0$, what is the solution set of each of the following?

(a) $|x - 1| < k$ **(b)** $|x - 1| > k$ **(c)** $|x - 1| = k$

Solve each equation. ***See Example 1.***

5. $|x| = 12$ **6.** $|x| = 14$ **7.** $|4x| = 20$

8. $|5x| = 30$ **9.** $|x - 3| = 9$ **10.** $|x - 5| = 13$

11. $|2x - 1| = 11$ **12.** $|2x + 3| = 19$ **13.** $|4x - 5| = 17$

14. $|5x - 1| = 21$ **15.** $|2x + 5| = 14$ **16.** $|2x - 9| = 18$

17. $|-3x + 8| = 1$ **18.** $|-6x + 5| = 4$ **19.** $\left|12 - \frac{1}{2}x\right| = 6$

20. $\left|14 - \frac{1}{3}x\right| = 8$ **21.** $|0.5x| = 6$ **22.** $|0.3x| = 9$

23. $\left|\frac{1}{2}x + 3\right| = 2$ **24.** $\left|\frac{2}{3}x - 1\right| = 5$ **25.** $\left|1 + \frac{3}{4}x\right| = 7$

26. $\left|2 - \frac{5}{2}x\right| = 14$ **27.** $|0.02x - 1| = 2.50$ **28.** $|0.04x - 3| = 5.96$

Solve each inequality. Graph the solution set, and write it using interval notation. ***See Example 2.***

29. $|x| > 3$ **30.** $|x| > 5$ **31.** $|x| \ge 4$

32. $|x| \ge 6$ **33.** $|r + 5| \ge 20$ **34.** $|x + 4| \ge 8$

35. $|5x + 2| > 10$ **36.** $|4x + 1| \ge 21$ **37.** $|3 - x| > 5$

38. $|5 - x| > 3$ **39.** $|-5x + 3| \ge 12$ **40.** $|-2x - 4| \ge 5$

41. Concept Check The graph of the solution set of $|2x + 1| = 9$ is given here.

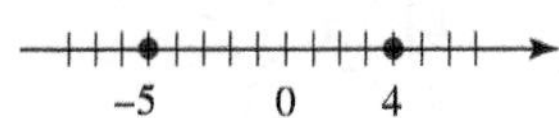

Without doing any algebraic work, graph the solution set of each inequality, referring to the graph shown.

(a) $|2x + 1| < 9$ **(b)** $|2x + 1| > 9$

42. Concept Check The graph of the solution set of $|3x - 4| < 5$ is given here.

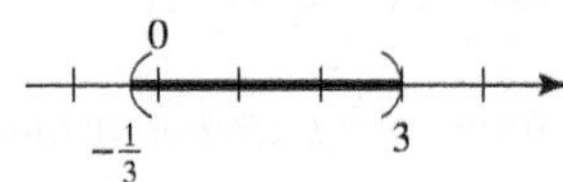

Without doing any algebraic work, graph the solution set of the following, referring to the graph shown.

(a) $|3x - 4| = 5$ **(b)** $|3x - 4| > 5$

Solve each inequality. Graph the solution set, and write it using interval notation. **See Example 3.** *(Hint: Compare the answers with those in* **Exercises 29–40.***)*

43. $|x| \le 3$ **44.** $|x| \le 5$ **45.** $|x| < 4$

46. $|x| < 6$ **47.** $|r + 5| < 20$ **48.** $|x + 4| < 8$

49. $|5x + 2| \le 10$ **50.** $|4x + 1| < 21$ **51.** $|3 - x| \le 5$

52. $|5 - x| \le 3$ **53.** $|-5x + 3| < 12$ **54.** $|-2x - 4| < 5$

Solve each equation or inequality. In Exercises 55–66, graph the solution set. **See Examples 1–4.**

55. $|-4 + x| > 9$ **56.** $|-3 + x| > 8$ **57.** $|3x + 2| < 11$

58. $|2x - 1| < 7$ **59.** $|7 + 2x| = 5$ **60.** $|9 - 3x| = 3$

61. $|3x - 1| \le 11$ **62.** $|2x - 6| \le 6$ **63.** $|-6x - 6| \le 1$

64. $|-2x - 6| \le 5$ **65.** $|-8 + x| \le 5$ **66.** $|-4 + x| \le 9$

67. $|10 - 12x| \ge 4$ **68.** $|8 - 10x| \ge 2$ **69.** $|3(x - 1)| = 8$

70. $|7(x - 2)| = 4$ **71.** $|0.1x - 1| > 3$ **72.** $|0.1x + 1| > 2$

73. $|x + 2| = 5 - 2$ **74.** $|x + 3| = 12 - 2$ **75.** $3|x - 6| = 9$

76. $5|x - 4| = 5$ **77.** $|2 - 0.2x| = 2$ **78.** $|5 - 0.5x| = 4$

Solve each equation or inequality. **See Examples 5 and 6.**

79. $|x| - 1 = 4$ **80.** $|x| + 3 = 10$

81. $|x + 4| + 1 = 2$ **82.** $|x + 5| - 2 = 12$

83. $|2x + 1| + 3 > 8$ **84.** $|6x - 1| - 2 > 6$

85. $|x + 5| - 6 \le -1$ **86.** $|x - 2| - 3 \le 4$

87. $|0.1x - 2.5| + 0.3 \ge 0.8$ **88.** $|0.5x - 3.5| + 0.2 \ge 0.6$

89. $\left|\frac{1}{2}x + \frac{1}{3}\right| + \frac{1}{4} = \frac{3}{4}$ **90.** $\left|\frac{2}{3}x + \frac{1}{6}\right| + \frac{1}{2} = \frac{5}{2}$

Solve each equation. ***See Example 7.***

91. $|3x + 1| = |2x + 4|$

92. $|7x + 12| = |x - 8|$

93. $\left|x - \frac{1}{2}\right| = \left|\frac{1}{2}x - 2\right|$

94. $\left|\frac{2}{3}x - 2\right| = \left|\frac{1}{3}x + 3\right|$

95. $|6x| = |9x + 1|$

96. $|13x| = |2x + 1|$

97. $|2x - 6| = |2x + 11|$

98. $|3x - 1| = |3x + 9|$

Solve each equation or inequality. ***See Examples 8 and 9.***

99. $|x| \geq -10$

100. $|x| \geq -15$

101. $|12t - 3| = -8$

102. $|13x + 1| = -3$

103. $|4x + 1| = 0$

104. $|6x - 2| = 0$

105. $|2x - 1| = -6$

106. $|8x + 4| = -4$

107. $|x + 5| > -9$

108. $|x + 9| > -3$

109. $|7x + 3| \leq 0$

110. $|4x - 1| \leq 0$

111. $|5x - 2| = 0$

112. $|7x + 4| = 0$

113. $|x - 2| + 3 \geq 2$

114. $|x - 4| + 5 \geq 4$

115. $|10x + 7| + 3 < 1$

116. $|4x + 1| - 2 < -5$

Determine the number of ounces a filled carton of the given size may contain for the given relative error. ***See Example 10.***

$$\left|\frac{x - x_t}{x_t}\right| = \textbf{relative error in } x$$

x represents actual measurement.
x_t represents expected measurement.

117. 64-oz carton; relative error no greater than 0.05

118. 24-oz carton; relative error no greater than 0.05

119. 32-oz carton; relative error no greater than 0.02

120. 36-oz carton; relative error no greater than 0.03

In later courses in mathematics, it is sometimes necessary to find an interval in which x must lie in order to keep y within a given difference of some number. For example, suppose

$$y = 2x + 1$$

and we want y to be within 0.01 *unit of* 4. *This criterion can be written as*

$$|y - 4| < 0.01.$$

Solving this inequality shows that x must lie in the interval (1.495, 1.505) *to satisfy the requirement.*

Find the open interval in which x must lie in order for the given condition to hold.

121. $y = 2x + 1$, and the difference of y and 1 is less than 0.1.

122. $y = 4x - 6$, and the difference of y and 2 is less than 0.02.

123. $y = 4x - 8$, and the difference of y and 3 is less than 0.001.

124. $y = 5x + 12$, and the difference of y and 4 is less than 0.0001.

Work each problem.

125. Dr. Mosely has determined that 99% of the babies he has delivered have weighed x pounds, where

$$|x - 8.3| < 1.5.$$

What range of weights corresponds to this inequality?

126. The Celsius temperatures x on Mars approximately satisfy the inequality

$$|x + 85| \le 55.$$

What range of temperatures corresponds to this inequality?

127. The recommended daily intake (RDI) of calcium for females aged 19–50 is 1000 mg. Write this statement as an absolute value inequality, with x representing the RDI, to express the RDI plus or minus 100 mg. Solve the inequality. (Data from National Academy of Sciences—Institute of Medicine.)

128. The average clotting time of blood is 7.45 sec, with a variation of plus or minus 3.6 sec. Write this statement as an absolute value inequality, with x representing the time. Solve the inequality.

RELATING CONCEPTS For Individual or Group Work (Exercises 129–132)

The 10 tallest buildings in Houston, Texas, are listed along with their heights.

Building	*Height (in feet)*
JPMorgan Chase Tower	1002
Wells Fargo Plaza	992
Williams Tower	901
Bank of America Center	780
Texaco Heritage Plaza	762
609 Main at Texas	757
Enterprise Plaza	756
Centerpoint Energy Plaza	741
1600 Smith St.	732
Fulbright Tower	725

Data from *The World Almanac and Book of Facts.*

Use this information to ***work Exercises 129–132 in order.***

129. To find the average of a group of numbers, we add the numbers and then divide by the number of numbers. Use a calculator to find the average of the heights.

130. Let k represent the average height of these buildings. If a height x satisfies the inequality

$$|x - k| < t,$$

then the height is said to be within t feet of the average. Using the result from **Exercise 129,** list the buildings that are within 50 ft of the average.

131. Repeat **Exercise 130,** but list the buildings that are within 95 ft of the average.

132. Work each of the following.

(a) Write an absolute value inequality that describes the height of a building that is *not* within 95 ft of the average.

(b) Solve the inequality from part (a).

(c) Use the result of part (b) to list the buildings that are not within 95 ft of the average.

(d) Confirm that the answer to part (c) makes sense by comparing it with the answer to **Exercise 131.**

SUMMARY EXERCISES Solving Linear and Absolute Value Equations and Inequalities

Solve each equation or inequality. Give the solution set in set notation for equations and in interval notation for inequalities.

1. $4x + 1 = 49$

2. $|x - 1| = 6$

3. $6x - 9 = 12 + 3x$

4. $3x + 7 = 9 + 8x$

5. $|x + 3| = -4$

6. $2x + 1 \leq x$

7. $8x + 2 \geq 5x$

8. $4(x - 11) + 3x = 20x - 31$

9. $2x - 1 = -7$

10. $|3x - 7| - 4 = 0$

11. $6x - 5 \leq 3x + 10$

12. $|5x - 8| + 9 \geq 7$

13. $9x - 3(x + 1) = 8x - 7$

14. $|x| \geq 8$

15. $9x - 5 \geq 9x + 3$

16. $13x - 5 > 13x - 8$

17. $|x| < 5.5$

18. $4x - 1 = 12 + x$

19. $\frac{2}{3}x + 8 = \frac{1}{4}x$

20. $-\frac{5}{8}x \geq -20$

21. $\frac{1}{4}x < -6$

22. $\frac{1}{2} \leq \frac{2}{3}x \leq \frac{5}{4}$

23. $\frac{3}{5}x - \frac{1}{10} = 2$

24. $\frac{x}{6} - \frac{3x}{5} = x - 86$

25. $x + 9 + 7x = 4(3 + 2x) - 3$

26. $6 - 3(2 - x) < 2(1 + x) + 3$

27. $-6 \leq \frac{3}{2} - x \leq 6$

28. $\frac{x}{4} - \frac{2x}{3} = -10$

29. $|5x + 1| \leq 0$

30. $5x - (3 + x) \geq 2(3x + 1)$

31. $-2 \leq 3x - 1 \leq 8$

32. $-1 \leq 6 - x \leq 5$

33. $|7x - 1| = |5x + 3|$

34. $|x + 2| = |x + 4|$

35. $|1 - 3x| \geq 4$

36. $7x - 3 + 2x = 9x - 8x$

37. $-(x + 4) + 2 = 3x + 8$

38. $|x - 1| < 7$

39. $|2x - 3| > 11$

40. $|5 - x| < 4$

41. $|x - 1| \geq -6$

42. $|2x - 5| = |x + 4|$

43. $8x - (1 - x) = 3(1 + 3x) - 4$

44. $8x - (x + 3) = -(2x + 1) - 12$

45. $|x - 5| = |x + 9|$

46. $|x + 2| < -3$

47. $2x + 1 > 5$ or $3x + 4 < 1$

48. $1 - 2x \geq 5$ and $7 + 3x \geq -2$

Chapter 1 Summary

STUDY SKILLS **REMINDER**

How do you best prepare for a test? **Review Study Skill 7, *Reviewing a Chapter.***

Key Terms

1.1

linear (first-degree) equation in one variable
solution
solution set
equivalent equations
conditional equation
identity
contradiction

1.2

mathematical model
formula
percent
percent increase
percent decrease

1.4

consecutive integers
consecutive even integers
consecutive odd integers

1.5

inequality
interval
linear inequality in one variable
equivalent inequalities
three-part inequality

1.6

intersection
compound inequality
union

1.7

absolute value equation
absolute value inequality

New Symbols

Symbol	Meaning
%	percent
1°	one degree
∞	infinity
$-\infty$	negative infinity
$(-\infty, \infty)$	the set of real numbers
(a, b)	interval notation for $a < x < b$
$[a, b]$	interval notation for $a \le x \le b$
$\cap$	set intersection
$\cup$	set union

Test Your Word Power

See how well you have learned the vocabulary in this chapter.

1. An **equation** is
   A. an algebraic expression
   B. an expression that contains fractions
   C. an expression that uses any of the four basic operations or the operation of raising to powers or taking roots on any collection of variables and numbers formed according to the rules of algebra
   D. a statement that two algebraic expressions are equal.

2. A linear equation that is a **conditional equation** has
   A. no solution
   B. one solution
   C. two solutions
   D. infinitely many solutions.

3. An **inequality** is
   A. a statement that two algebraic expressions are equal
   B. a point on a number line
   C. an equation with no solutions
   D. a statement consisting of algebraic expressions related by $<$, $\le$, $>$, or $\ge$.

4. **Interval notation** is
   A. a point on a number line
   B. a special notation for describing a point on a number line
   C. a way to use symbols to describe an interval on a number line
   D. a notation to describe unequal quantities.

5. The **intersection** of two sets A and B is the set of elements that belong
   A. to both A and B
   B. to either A or B, or both
   C. to either A or B, but not both
   D. to just A.

6. The **union** of two sets A and B is the set of elements that belong
   A. to both A and B
   B. to either A or B, or both
   C. to either A or B, but not both
   D. to just B.

ANSWERS

1. D; *Examples:* $2a + 3 = 7$, $3y = -8$, $x^2 = 4$ **2.** B; *Example:* $x - 4 = 6$ is a conditional equation with one solution, 10.

3. D; *Examples:* $x < 5$, $7 + 2k \ge 11$, $-5 < 2z - 1 \le 3$ **4.** C; *Examples:* $(-\infty, 5]$, $(1, \infty)$, $[-3, 3)$

5. A; *Example:* If $A = \{2, 4, 6, 8\}$ and $B = \{1, 2, 3\}$, then $A \cap B = \{2\}$.

6. B; *Example:* Using the sets A and B from Answer 5, $A \cup B = \{1, 2, 3, 4, 6, 8\}$.

Quick Review

CONCEPTS	EXAMPLES

1.1 Linear Equations in One Variable

Addition and Multiplication Properties of Equality

The same number may be added to (or subtracted from) each side of an equation without changing the solution set.

Similarly, each side of an equation may be multiplied (or divided) by the same nonzero number without changing the solution set.

Solving a Linear Equation in One Variable

Step 1 Simplify each side separately.

- Clear any parentheses.
- Clear any fractions or decimals.
- Combine like terms.

Step 2 Isolate the variable terms on one side.

Step 3 Isolate the variable.

Step 4 Check.

Solve. $x - 4 = 11$

$x - 4 + 4 = 11 + 4$ Add 4 to each side by the addition property of equality.

$x = 15$

Solution set: $\{15\}$

Solve. $4(8 - 3x) = 32 - 8(x + 2)$

$32 - 12x = 32 - 8x - 16$ Distributive property

$32 - 12x = 16 - 8x$ Combine like terms.

$32 - 12x + 12x = 16 - 8x + 12x$ Add $12x$.

$32 = 16 + 4x$ Combine like terms.

$32 - 16 = 16 + 4x - 16$ Subtract 16.

$16 = 4x$ Combine like terms.

$\frac{16}{4} = \frac{4x}{4}$ Divide by 4.

$4 = x$

CHECK $4(8 - 3x) = 32 - 8(x + 2)$ Original equation

$4[8 - 3(4)] \stackrel{?}{=} 32 - 8(4 + 2)$ Let $x = 4$.

$4[8 - 12] \stackrel{?}{=} 32 - 8(6)$ Multiply. Add.

$4[-4] \stackrel{?}{=} 32 - 48$ Subtract. Multiply.

$-16 = -16$ ✓ True

Solution set: $\{4\}$

1.2 Formulas and Percent

Solving a Formula for a Specified Variable (Solving a Literal Equation)

Step 1 Clear any parentheses and fractions.

Step 2 Isolate all terms with the specified variable.

Step 3 Isolate the specified variable.

Solve $\mathcal{A} = \frac{1}{2}bh$ for h.

$\mathcal{A} = \frac{1}{2}bh$

$2\mathcal{A} = 2\left(\frac{1}{2}bh\right)$ Multiply by 2.

$2\mathcal{A} = bh$ $2 \cdot \frac{1}{2} = 1$

$\frac{2\mathcal{A}}{b} = h$, or $h = \frac{2\mathcal{A}}{b}$ Divide by b.

1.3 Applications of Linear Equations

Solving an Applied Problem

Step 1 Read the problem.

Step 2 Assign a variable.

How many liters of 30% alcohol solution and 80% alcohol solution must be mixed to obtain 100 L of 50% alcohol solution?

Let x = number of liters of 30% solution needed.

Then $100 - x$ = number of liters of 80% solution needed.

Liters of Solution	Percent (as a decimal)	Liters of Pure Alcohol
x	0.30	$0.30x$
$100 - x$	0.80	$0.80(100 - x)$
100	0.50	$0.50(100)$

CONCEPTS	EXAMPLES

Step 3 Write an equation.

Step 4 Solve the equation.

Step 5 State the answer.

Step 6 Check.

From the last column of the table, write an equation.

$$0.30x + 0.80(100 - x) = 0.50(100)$$

$0.30x + 80 - 0.80x = 50$ Distributive property; Multiply.

$100(0.30x + 80 - 0.80x) = 100(50)$ Multiply by 100.

$30x + 8000 - 80x = 5000$ Distributive property; Multiply.

$-50x + 8000 = 5000$ Combine like terms.

$-50x = -3000$ Subtract 8000.

$x = 60$ Divide by -50.

The solution of the equation is 60.

60 L of 30% solution and $100 - 60 = 40$ L of 80% solution are needed.

$$0.30(60) + 0.80(100 - 60) = 50 \text{ is true.}$$

1.4 Further Applications of Linear Equations

To solve a uniform motion problem, draw a sketch and make a table. Use the distance formula.

$$d = rt, \quad \text{or} \quad rt = d$$

Two cars start from towns 400 mi apart and travel toward each other. They meet after 4 hr. Find the rate of each car if one travels 20 mph faster than the other.

Let x = rate of the slower car in miles per hour.

Then $x + 20$ = rate of the faster car.

Use the information in the problem and $d = rt$ in the form $rt = d$ to complete a table.

	Rate	Time	Distance	
Slower Car	x	4	$4x$	
Faster Car	$x + 20$	4	$4(x + 20)$	
			400	← Total

A sketch shows that the sum of the distances, $4x$ and $4(x + 20)$, must be 400.

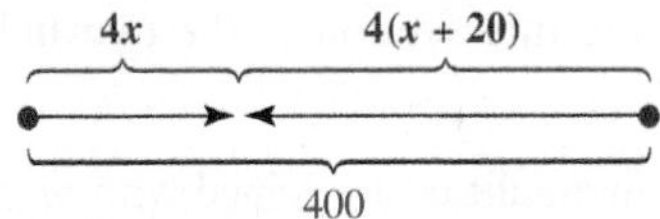

$4x + 4(x + 20) = 400$

$4x + 4x + 80 = 400$ Distributive property

$8x + 80 = 400$ Combine like terms.

$8x = 320$ Subtract 80.

$x = 40$ Divide by 8.

The slower car travels 40 mph, and the faster car travels $40 + 20 = 60$ mph.

$$4(40) + 4(40 + 20) = 400 \text{ is true.}$$

CONCEPTS	EXAMPLES

1.5 Linear Inequalities in One Variable

Solving a Linear Inequality in One Variable

Step 1 Simplify each side separately.

- Clear any parentheses.
- Clear any fractions or decimals.
- Combine like terms.

Step 2 Isolate the variable terms on one side.

Step 3 Isolate the variable (here x) to write the inequality in one of these forms.

$$x < k, \quad x \le k, \quad x > k, \quad \text{or} \quad x \ge k$$

If an inequality is multiplied or divided by a negative number, the direction of the inequality symbol must be reversed.

Solve each inequality.

$$3(x+2) - 5x \le 12$$

$3x + 6 - 5x \le 12$ Distributive property

$-2x + 6 \le 12$ Combine like terms.

$-2x + 6 - 6 \le 12 - 6$ Subtract 6.

$-2x \le 6$ Combine like terms.

$\frac{-2x}{-2} \ge \frac{6}{-2}$ Divide by −2. Change ≤ to ≥.

$$x \ge -3$$

Solution set: 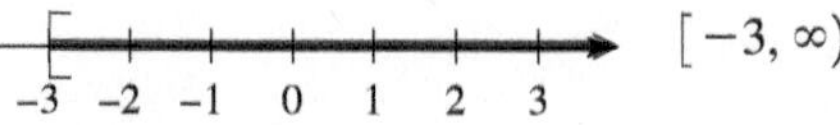 $[-3, \infty)$

To solve a three-part inequality, work with all three parts at the same time.

$$-4 < 2x + 3 \le 7$$

$-4 - 3 < 2x + 3 - 3 \le 7 - 3$ Subtract 3.

$$-7 < 2x \le 4$$

$\frac{-7}{2} < \frac{2x}{2} \le \frac{4}{2}$ Divide by 2.

$$-\frac{7}{2} < x \le 2$$

Solution set: $\left(-\frac{7}{2}, 2\right]$

1.6 Set Operations and Compound Inequalities

Solving a Compound Inequality

Step 1 Solve each inequality in the compound inequality individually.

Step 2 If the inequalities are joined with *and*, then the solution set is the intersection of the two individual solution sets.

If the inequalities are joined with *or*, then the solution set is the union of the two individual solution sets.

Solve each compound inequality.

$$x + 1 > 2 \quad \text{and} \quad 2x < 6$$

$$x > 1 \quad \text{and} \quad x < 3$$

Solution set: $(1, 3)$

$$2(x+3) - 2 \le 4 \quad \text{or} \quad -4x \le -16$$

$$2x + 6 - 2 \le 4 \quad \text{or} \quad \frac{-4x}{-4} \ge \frac{-16}{-4}$$

$$2x + 4 \le 4 \quad \text{or} \quad x \ge 4$$

$$x \le 0$$

Solution set: 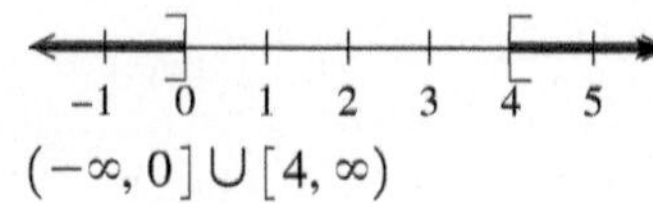

$(-\infty, 0] \cup [4, \infty)$

CONCEPTS	EXAMPLES

1.7 Absolute Value Equations and Inequalities

Solving Absolute Value Equations and Inequalities

Let k be a positive number.

Case 1 To solve $|ax + b| = k$, solve the compound equation

$$ax + b = k \quad \text{or} \quad ax + b = -k.$$

Case 2 To solve $|ax + b| > k$, solve the compound inequality

$$ax + b > k \quad \text{or} \quad ax + b < -k.$$

Case 3 To solve $|ax + b| < k$, solve the compound inequality

$$-k < ax + b < k.$$

Solve each equation or inequality.

$|x - 7| = 3$ Case 1

$x - 7 = 3$ or $x - 7 = -3$

$x = 10$ or $x = 4$ Add 7.

Solution set: (number line 0 2 4 6 8 10, points at 4 and 10) $\{4, 10\}$

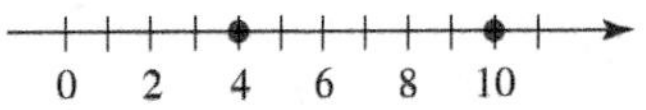

$|x - 7| > 3$ Case 2

$x - 7 > 3$ or $x - 7 < -3$

$x > 10$ or $x < 4$ Add 7.

Solution set: (number line 0 2 4 6 8 10)

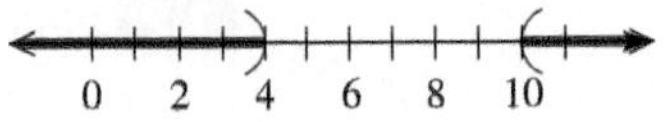

$(-\infty, 4) \cup (10, \infty)$

$|x - 7| < 3$ Case 3

$-3 < x - 7 < 3$

$4 < x < 10$ Add 7.

Solution set: (number line 0 2 4 6 8 10) $(4, 10)$

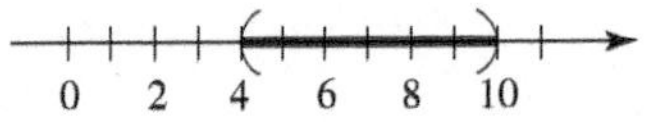

To solve an absolute value equation of the form

$$|ax + b| = |cx + d|,$$

solve the compound equation

$$ax + b = cx + d \quad \text{or} \quad ax + b = -(cx + d).$$

$|x + 2| = |2x - 6|$

$x + 2 = 2x - 6$ or $x + 2 = -(2x - 6)$

$-x = -8$ or $x + 2 = -2x + 6$

$x = 8$ or $3x = 4$

$x = \frac{4}{3}$

Solution set: $\left\{\frac{4}{3}, 8\right\}$

Special Properties of Absolute Value

Property 1 The absolute value of an expression can never be negative—that is, $|a| \geq 0$ for all real numbers a.

Property 2 The absolute value of an expression equals 0 only when the expression is equal to 0.

$|x - 2| = -4$ Solution set: $\emptyset$

$|x| \geq -1$ Solution set: $(-\infty, \infty)$

$|2x - 3| < -6$ Solution set: $\emptyset$

} Property 1

$|3x - 9| = 0$ Property 2

$3x - 9 = 0$ $|a| = 0$ implies $a = 0$.

$3x = 9$ Add 9.

$x = 3$ Divide by 3.

Solution set: $\{3\}$

Chapter 1 Review Exercises

1.1 *Solve each equation.*

1. $-(8+3x)+5=2x+6$

2. $-\frac{3}{4}x=-12$

3. $\frac{2x+1}{3}-\frac{x-1}{4}=0$

4. $5(2x-3)=6(x-1)+4x$

Solve each equation. Decide whether the equation is a conditional equation, *an* identity, *or a* contradiction.

5. $10x-3(2x-4)=4(x+3)$

6. $13x-(x+7)=12x-13$

7. $x+6(x-1)-(4-x)=-10$

8. $5+3(2x+6)=5+3(2x+6)$

1.2 *Solve for the specified variable.*

9. $V=LWH$ for L

10. $\mathcal{A}=\frac{1}{2}h(b+B)$ for b

11. $P=a+b+c+B$ for c

12. $4x+7y=9$ for y

1.2, 1.3 *Solve each problem.*

13. A rectangular solid has a volume of 180 ft^3. Its length is 6 ft and its width is 5 ft. Find its height.

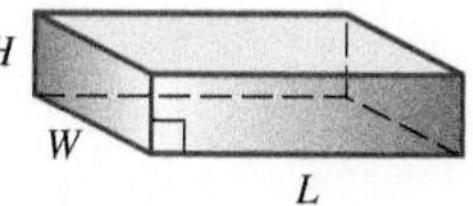

14. The number of students attending college in 2015 was 20.2 million. This number is expected to increase to 23.1 million by the year 2024. To the nearest tenth, what is the expected percent increase? (Data from National Center for Education Statistics.)

15. Find the annual simple-interest rate that Halina is earning if the principal of \$30,000 earns \$3900 interest in 4 yr.

16. If the Fahrenheit temperature is 77°, what is the corresponding Celsius temperature?

At the end of 2016, there were 395.9 *million wireless subscriptions in the United States. (Data from CTIA.) The circle graph shows market share by carrier during the fourth quarter of 2016.*

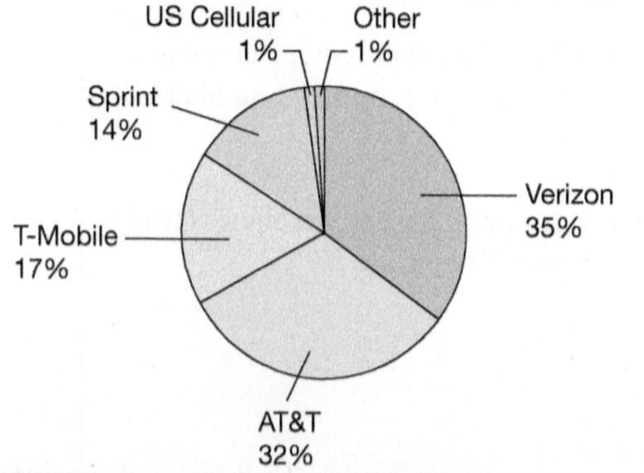

Data from Strategy Analytics; FierceWireless.

17. To the nearest tenth of a million, about how many Verizon subscriptions were there?

18. To the nearest tenth of a million, about how many T-Mobile subscriptions were there?

Translate each verbal phrase into a mathematical expression, using x as the variable.

19. One-third of a number, subtracted from 9

20. The product of 4 and a number, divided by 9 more than the number

Solve each problem.

21. The length of a rectangle is 3 m less than twice the width. The perimeter of the rectangle is 42 m. Find the length and width of the rectangle.

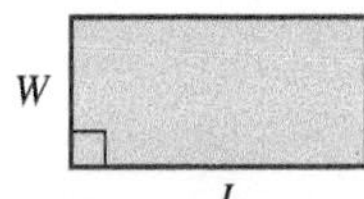

22. In a triangle with two sides of equal length, the third side measures 15 in. less than the sum of the two equal sides. The perimeter of the triangle is 53 in. Find the lengths of the three sides.

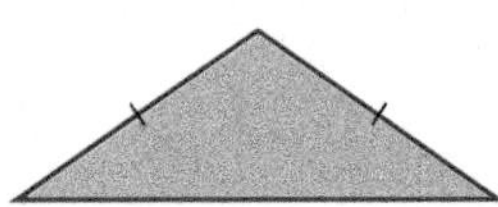

23. A candy clerk has three times as many kilograms of chocolate creams as peanut clusters. The clerk has 48 kg of the two candies altogether. How many kilograms of peanut clusters does the clerk have?

24. How many liters of a 20% solution of a chemical should be mixed with 15 L of a 50% solution to obtain a 30% mixture?

25. How much water should be added to 30 L of a 40% acid solution to reduce it to a 30% solution?

Liters of Solution	Percent (as a decimal)	Liters of Pure Acid
	0.40	
x		
	0.30	

26. Eric invested some money at 6% and $4000 less than that amount at 4%. Find the amount invested at each rate if his total annual interest income was $840.

Principal (in dollars)	Rate (as a decimal)	Interest (in dollars)
x	0.06	
	0.04	

1.4 *Solve each problem.*

27. A clerk has $3.50 in dimes and quarters in her cash drawer. The number of dimes is 1 less than twice the number of quarters. How many of each denomination are there?

28. When Jim emptied his pockets one evening, he found he had 19 nickels and dimes with a total value of $1.55. How many of each denomination did he have?

29. Which choice is the best *estimate* for the average rate for a trip of 405 mi that lasted 8.2 hr?

A. 30 mph **B.** 40 mph **C.** 50 mph **D.** 60 mph

30. A driver averaged 53 mph and took 10 hr to travel from Memphis to Chicago. What is the distance between Memphis and Chicago?

31. A small plane traveled from Warsaw to Rome, averaging 164 mph. The trip took 2 hr. What is the distance from Warsaw to Rome?

32. A passenger train and a freight train leave a town at the same time and go in opposite directions. They travel at rates of 60 mph and 75 mph, respectively. How long will it take for them to be 297 mi apart?

	Rate	Time	Distance
Passenger Train	60	x	
Freight Train	75	x	

33. Two cars leave small towns 230 km apart at the same time, traveling directly toward one another. One car travels 15 km per hr slower than the other car. They pass one another 2 hr later. What are their rates?

	Rate	Time	Distance
Faster Car	x	2	
Slower Car	$x - 15$	2	

34. An 85-mi trip to the beach took Susan 2 hr. During the second hour, a rainstorm caused her to average 7 mph less than she traveled during the first hour. Find her average rate for the first hour.

35. Find the measure of each angle in the triangle.

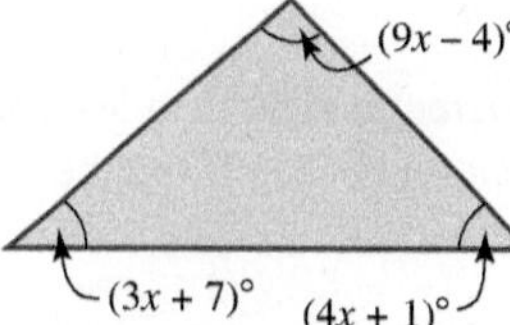

36. Find the measure of each marked angle.

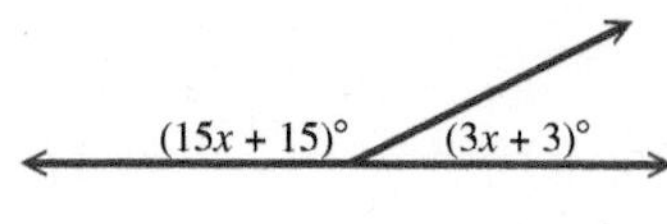

37. Find three consecutive integers such that the sum of the first and third is 49 less than three times the second.

38. Find three consecutive even integers such that the sum of the second and third is 44 more than the first.

1.5 *Solve each inequality. Give the solution set in interval form.*

39. $-\frac{2}{3}x < 6$

40. $-5x - 4 \geq 11$

41. $\frac{6x + 3}{-4} < -3$

42. $5 - (6 - 4x) \geq 2x - 7$

43. $8 \leq 3x - 1 < 14$

44. $\frac{5}{3}(x - 2) + \frac{2}{5}(x + 1) > 1$

Solve each problem.

45. Dr. Paul Donohue writes a syndicated column in which readers question him on a variety of health topics. Reader C. J. wrote, "Many people say they can weigh more because they have a large frame. How is frame size determined?" Here is Dr. Donohue's response:

"For a man, a wrist circumference between 6.75 and 7.25 in. [inclusive] indicates a medium frame. Anything above is a large frame and anything below, a small frame."

(*Source:* Dr. Donohue, © 2004. Used by permission of North America Syndicate.)

Using x to represent wrist circumference in inches, write an inequality or a three-part inequality that represents wrist circumference for a male with the following.

(a) a small frame **(b)** a medium frame **(c)** a large frame

46. The perimeter of a rectangular playground must be no greater than 120 m. One dimension of the playground must be 22 m. Find the possible lengths of the other dimension.

47. A group of college students wanted to buy tickets to attend a performance of *Motown the Musical* at the Lunt-Fontanne Theatre in New York City. They could buy student mezzanine seats for a group rate of $51 per ticket for 15 tickets or more. If they had $1600 available to spend, how many tickets could they purchase at this price? (Data from www.broadway.com)

48. To pass algebra, a student must have an average of at least 70 on five tests. On the first four tests, a student has scores of 75, 79, 64, and 71. What possible scores on the fifth test will guarantee the student a passing grade in the class?

1.6 *Let* $A = \{a, b, c, d\}$, $B = \{a, c, e, f\}$, *and* $C = \{a, e, f, g\}$. *Find each set.*

49. $A \cap B$ **50.** $A \cap C$ **51.** $B \cup C$ **52.** $A \cup C$

Solve each compound inequality. Graph the solution set, and write it using interval notation.

53. $x > 6$ and $x < 9$

54. $x + 4 > 12$ and $x - 2 < 12$

55. $x > 5$ or $x \leq -3$

56. $x \geq -2$ or $x < 2$

57. $x - 4 > 6$ and $x + 3 \leq 10$

58. $-5x + 1 \geq 11$ or $3x + 5 \geq 26$

Express each set in simplest interval form.

59. $(-3, \infty) \cap (-\infty, 4]$

60. $(-\infty, 6] \cap (-\infty, 2]$

61. $(4, \infty) \cup (9, \infty)$

62. $(1, 2) \cup (1, \infty)$

1.7 *Solve each equation or inequality.*

63. $|x| = 7$

64. $|x + 2| = 9$

65. $|3x - 7| = 8$

66. $|x - 4| = -12$

67. $|2x - 7| + 4 = 11$

68. $|4x + 2| - 7 = -3$

69. $|3x + 1| = |x + 2|$

70. $|2x - 1| = |2x + 3|$

71. $|x| < 14$

72. $|-x + 6| \leq 7$

73. $|2x + 5| \leq 1$

74. $|x + 1| \geq -3$

75. $|3 - 4x| + 7 < -4$

76. $|-8 - 3x| - 7 > -8$

Determine the number of ounces a filled carton of the given size may contain for the given relative error.

$$\left|\frac{x - x_t}{x_t}\right| = \textbf{relative error in } x$$

x represents actual measurement.
x_t represents expected measurement.

77. 48-oz carton; relative error no greater than 0.03

78. 32-oz carton; relative error no greater than 0.01

Chapter 1 Mixed Review Exercises

Solve.

1. $5 - (6 - 4x) > 2x - 5$

2. $ak + bt = 6r$ for k

3. $x + 4 < 7$ and $x + 5 \geq 3$

4. $\frac{4x + 2}{4} + \frac{3x - 1}{8} = \frac{x + 6}{16}$

5. $|3x + 6| \geq 0$

6. $-5x \geq -10$

7. $|3x + 2| + 4 = 9$

8. $0.05x + 0.03(1200 - x) = 42$

9. $|x + 3| \leq 13$

10. $\frac{3}{4}(x - 2) - \frac{1}{3}(5 - 2x) < -2$

11. $-4 < 3 - 2x < 9$

12. $-0.3x + 2.1(x - 4) \leq -6.6$

13. $|5x - 1| > 14$

14. $x \geq -2$ or $x < 4$

15. $|x - 1| = |2x + 3|$

16. $\frac{3x}{5} - \frac{x}{2} = 3$

17. $|3x - 7| = 4$

18. $5(2x - 7) = 2(5x + 3)$

19. $-5x < -30$ and $-7x > -56$

20. $-5x + 1 \geq 11$ or $3x + 5 \geq 26$

Solve each problem.

21. A newspaper recycling collection bin is in the shape of a box 1.5 ft wide and 5 ft long. If the volume of the bin is 75 ft^3, find the height.

22. The sum of the first and third of three consecutive integers is 47 more than the second integer. What are the integers?

23. To qualify for a company pension plan, an employee must average at least \$1000 per month in earnings. During the first four months of the year, an employee made \$900, \$1200, \$1040, and \$760. What possible amounts earned during the fifth month will qualify the employee for a pension plan?

24. How many liters of a 20% solution of a chemical should be mixed with 10 L of a 50% solution to obtain a 40% mixture?

Chapter 1 Test

Step-by-step test solutions are found on the Chapter Test Prep Videos available in MyLab Math.

View the complete solutions to all Chapter Test exercises in MyLab Math.

STUDY SKILLS REMINDER

Using test-taking strategies can help you improve your test scores. **Review Study Skill 8, *Taking Math Tests*.**

Solve each equation.

1. $3(2x - 2) - 4(x + 6) = 3x + 8 + x$

2. $0.08x + 0.06(x + 9) = 1.24$

3. Solve each equation. Then determine whether the equation is a *conditional equation*, an *identity*, or a *contradiction*.

(a) $3x - (2 - x) + 4x + 2 = 8x + 3$

(b) $\frac{x}{3} + 7 = \frac{5x}{6} - 2 - \frac{x}{2} + 9$

(c) $-4(2x - 6) = 5x + 24 - 7x$

(d) $\frac{x + 6}{10} + \frac{x - 4}{15} = \frac{x + 2}{6}$

4. Solve $V = \frac{1}{3}bh$ for h.

5. Solve $-3x + 2y = 6$ for y.

Solve each problem.

6. The 2017 Brickyard 400 (mile) race was won by Kasey Kahne, who averaged 114.384 mph. What was his time to the nearest thousandth of an hour? (Data from *The World Almanac and Book of Facts*.)

7. A certificate of deposit pays \$456.25 in simple interest for 1 yr on a principal of \$36,500. What is the rate of interest?

8. The final 2017 standings of the American League East division in Major League Baseball are shown in the table. Find the winning percentage of each team.

(a) NY Yankees **(b)** Tampa Bay

(c) Toronto **(d)** Baltimore

Team	W	L	Pct.
Boston	93	69	.574
NY Yankees	91	71	
Tampa Bay	80	82	
Toronto	76	86	
Baltimore	75	87	

Data from www.MLB.com

9. Tyler invested some money at 1.5% simple interest and some at 2.5% simple interest. The total amount of his investments was \$28,000, and the interest he earned during the first year was \$620. How much did he invest at each rate?

10. After working 1 yr at his new job, Kevin got a raise from \$11.25 per hour to \$12.75 per hour. What was the percent increase in his hourly wage to the nearest tenth?

11. How many liters of a 20% alcohol solution should be added to 40 L of a 50% alcohol solution to make a 30% solution?

12. Two cars leave from the same point at the same time, traveling in opposite directions. One travels 15 mph slower than the other. After 6 hr, they are 630 mi apart. Find the rate of each car.

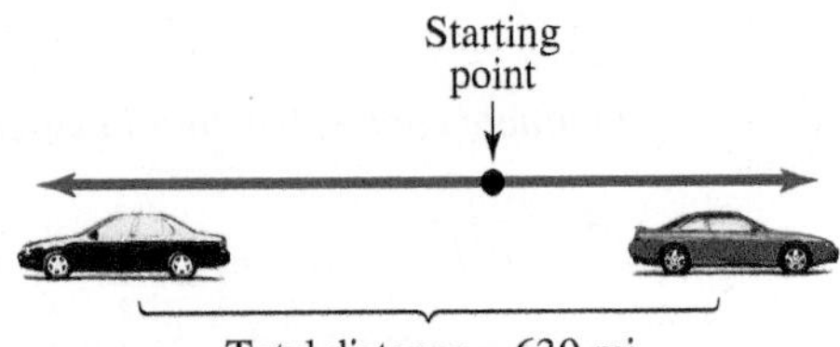

13. Find the measure of each angle in the triangle.

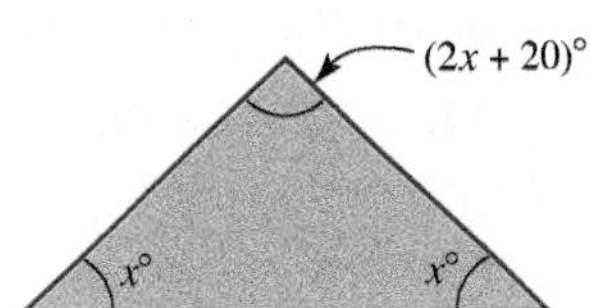

Solve each inequality. Graph the solution set, and write it using interval notation.

14. $4 - 6(x + 3) \leq -2 - 3(x + 6) + 3x$

15. $-\frac{4}{7}x > -16$

16. $-1 < 3x - 4 < 2$

17. $-6 \leq \frac{4}{3}x - 2 \leq 2$

18. Which one of the following inequalities is equivalent to $x < -3$?

A. $-3x < 9$ **B.** $-3x > -9$ **C.** $-3x > 9$ **D.** $-3x < -9$

19. Justin must have an average of at least 80 on the four tests in a course to get a B. He had scores of 83, 76, and 79 on the first three tests. What possible scores on the fourth test would guarantee him a B in the course?

20. Let $A = \{1, 2, 5, 7\}$ and $B = \{1, 5, 9, 12\}$. Find each set.

(a) $A \cap B$ **(b)** $A \cup B$

Solve each compound or absolute value equation or inequality.

21. $3x \geq 6$ and $x < 9$

22. $-4x \leq -24$ or $4x < 12$

23. $|4x + 3| \leq 7$

24. $|5 - 6x| > 12$

25. $|3x - 9| = 6$

26. $|-3x + 4| - 4 < -1$

27. $|7 - x| \leq -1$

28. $|3x - 2| + 1 = 8$

29. $|3 - 5x| = |2x + 8|$

30. If $k < 0$, what is the solution set of each of the following?

(a) $|8x - 5| < k$ **(b)** $|8x - 5| > k$ **(c)** $|8x - 5| = k$

STUDY SKILLS **REMINDER**

We can learn from the mistakes we make. **Review Study Skill 9, *Analyzing Your Test Results.***

Chapters R and 1 Cumulative Review Exercises

1. Write $\frac{7}{20}$ as **(a)** a decimal and **(b)** a percent.

2. Write 0.66 as **(a)** a percent and **(b)** a fraction in lowest terms.

Perform the indicated operations.

3. $\frac{5}{6}+\frac{1}{4}-\frac{7}{15}$

4. $0.17+3.965-12$

5. $\frac{-4(9)(-2)}{-3^2}$

6. $\sqrt{25}-5(-1)^3$

Evaluate each exponential expression.

7. $(-5)^3$

8. $\left(\frac{3}{2}\right)^4$

Evaluate each expression for $x = 2$, $y = -3$, and $z = 4$.

9. $-2y+4(x-3z)$

10. $\frac{3x^2-y^2}{4z}$

Simplify each expression.

11. $-7r+5-13r+12$

12. $-(3k+8)-2(4k-7)+3(8k+12)$

Solve each equation.

13. $\frac{2}{3}x+\frac{3}{4}x=-17$

14. $\frac{2x+3}{5}=\frac{x-4}{2}$

15. $|3m-5|=|m+2|$

16. $A=P(1+ni)$ for n

Solve each inequality. Graph the solution set, and write it using interval notation.

17. $3-2(x+7)\le -x+3$

18. $-4<5-3x\le 0$

19. $2x+1>5$ or $2-x>2$

20. $|-7k+3|\ge 4$

Solve each problem.

21. Luke invested some money at 7% interest and the same amount at 10%. His total interest for the year was \$150 less than one-tenth of the total amount he invested. How much did he invest at each rate?

22. A diet must include three foods, A, B, and C, with twice as many grams of food A as of food C, and 5 g of food B. The three foods must total at most 24 g. What is the largest amount of food C that can be used?

23. Zach has scores of 88 and 78 on his first two tests. What score must he make on his third test to earn an average of 80 or greater?

24. Telescope Peak, altitude 11,049 ft, is next to Death Valley, 282 ft below sea level. Find the difference between these altitudes. (Data from National Park Service.)

25. For a woven hanging, Janette needs three pieces of yarn, which she will cut from a 40 cm piece. The longest piece is to be 3 times as long as the middle-sized piece, and the shortest piece is to be 5 cm shorter than the middle-sized piece. What lengths should she cut?

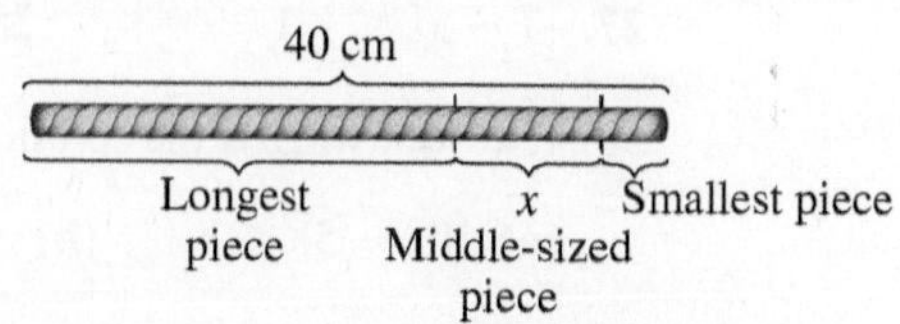

LINEAR EQUATIONS, GRAPHS, AND FUNCTIONS

The concept of steepness, or grade, is mathematically interpreted using *slope*, one of the topics of this chapter.

2.1 Linear Equations in Two Variables

OBJECTIVES

1 Interpret a line graph.
2 Plot ordered pairs.
3 Find ordered pairs that satisfy a given equation.
4 Graph lines.
5 Find x- and y-intercepts.
6 Graph equations of horizontal and vertical lines.
7 Find the midpoint of a line segment.

VOCABULARY

- ☐ ordered pair
- ☐ origin
- ☐ x-axis
- ☐ y-axis
- ☐ rectangular (Cartesian) coordinate system
- ☐ components
- ☐ plot
- ☐ coordinate
- ☐ quadrant
- ☐ table of ordered pairs (table of values)
- ☐ graph of an equation
- ☐ first-degree equation
- ☐ linear equation in two variables
- ☐ x-intercept
- ☐ y-intercept

OBJECTIVE 1 Interpret a line graph.

The line graph in **FIGURE 1** shows personal spending (in billions of dollars) on medical care in the United States from 2010 through 2016. *About how much was spent on medical care in 2015?*

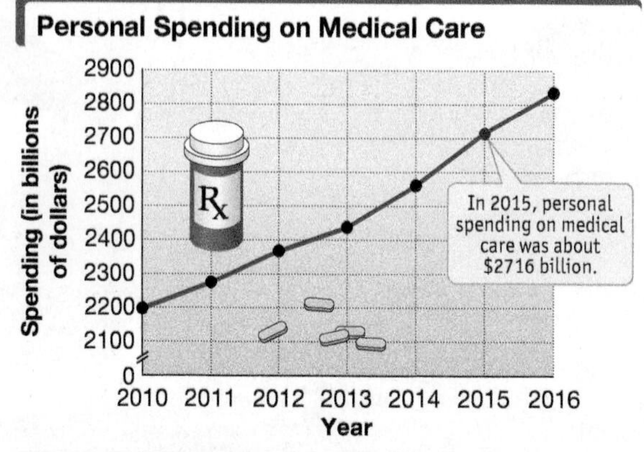

Data from Centers for Medicare & Medicaid Services.

FIGURE 1

The line graph in **FIGURE 1** presents information based on a method for locating a point in a plane developed by René Descartes, a 17th-century French mathematician. According to legend, Descartes was lying in bed ill watching a fly crawl about on the ceiling near a corner of the room. It occurred to him that the location of the fly could be described by determining its distances from the two adjacent walls, as illustrated in the figure at the right.

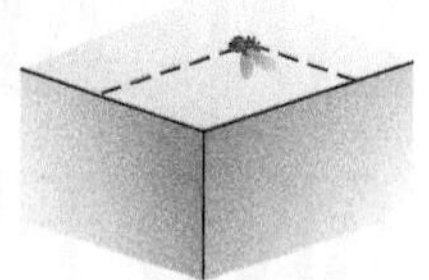

Locating a fly on a ceiling

We use this insight to plot points and graph linear equations in two variables whose graphs are straight lines.

OBJECTIVE 2 Plot ordered pairs.

Each of the pairs of numbers $(3, 2)$, $(-5, 6)$, and $(4, -1)$ is an example of an **ordered pair**—that is, a pair of numbers written within parentheses. The *order* in which the numbers are written is important. We graph an ordered pair using two perpendicular number lines that intersect at their 0 points. See **FIGURE 2**. The common 0 point is the **origin.**

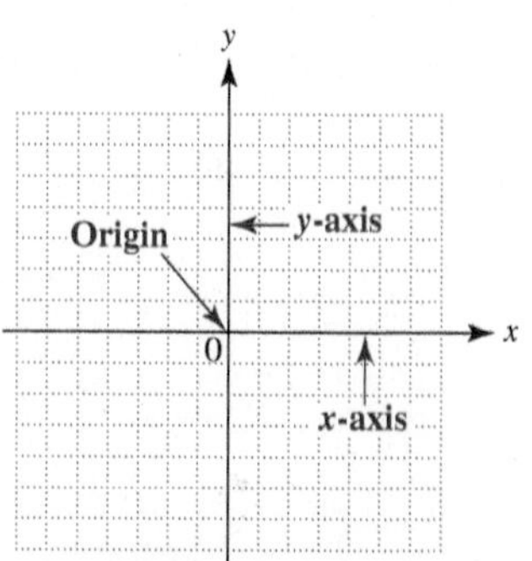

Rectangular coordinate system

FIGURE 2

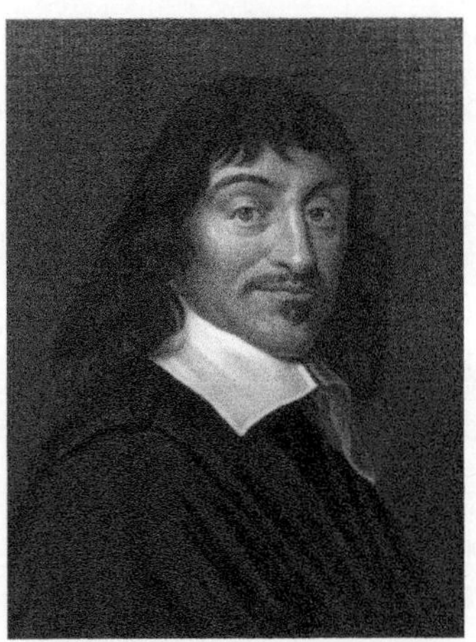
René Descartes (1596–1650)

The position of any point in this coordinate plane is determined by referring to the horizontal number line, or ***x*-axis,** and the vertical number line, or ***y*-axis.** The x-axis and the y-axis make up a **rectangular** (or **Cartesian,** for Descartes) **coordinate system.**

The numbers in an ordered pair (x, y), are its **components.** The first component indicates position relative to the x-axis, and the second component indicates position relative to the y-axis. For example, to locate, or **plot,** the point on the graph that corresponds to the ordered pair $(3, 2)$, we move three units from 0 to the right along the x-axis and then two units up parallel to the y-axis. See **FIGURE 3**. The numbers in an ordered pair are the **coordinates** of the corresponding point.

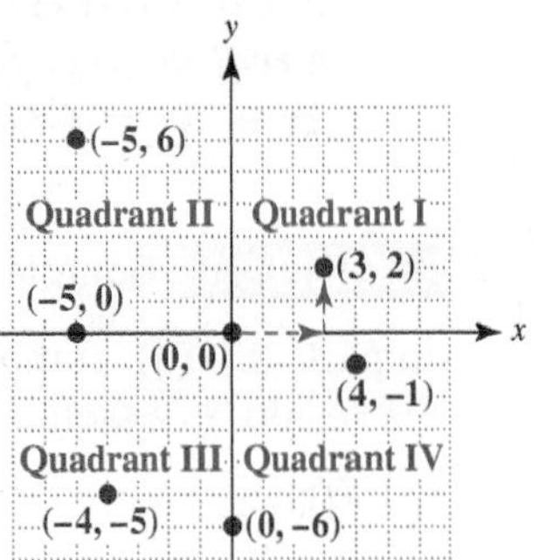

FIGURE 3

We can apply this method of locating ordered pairs to the line graph in **FIGURE 1**. We move along the horizontal axis to a year and then up parallel to the vertical axis to approximate spending for that year. Thus, we can write the ordered pair $(2015, 2716)$ to indicate that in 2015, personal spending on medical care was about \$2716 billion.

CAUTION The parentheses used to represent an ordered pair are also used to represent an open interval. The context of the discussion tells whether ordered pairs or open intervals are being represented.

The four regions of the graph shown in **FIGURE 3** are **quadrants I, II, III,** and **IV,** reading counterclockwise from the upper right quadrant. ***The points on the x-axis and y-axis do not belong to any quadrant.***

OBJECTIVE 3 Find ordered pairs that satisfy a given equation.

Each solution of an equation with two variables, such as

$$2x + 3y = 6, \quad \text{Equation with two variables } x \text{ and } y$$

will include two numbers, one for each variable. To keep track of which number goes with which variable, we write the solutions as ordered pairs. ***(If x and y are used as the variables, the x-value is given first.)***

We can show that $(6, -2)$ is a solution of $2x + 3y = 6$ by substitution.

$$2x + 3y = 6 \quad \text{Equation with two variables}$$

$$2(6) + 3(-2) \stackrel{?}{=} 6 \quad \text{Let } x = 6,\ y = -2.$$

Use parentheses to avoid errors.

$$12 - 6 \stackrel{?}{=} 6 \quad \text{Multiply.}$$

$$6 = 6 \ \checkmark \quad \text{True}$$

Because the ordered pair $(6, -2)$ makes the equation true, it is a solution.

On the other hand, (5, 1) is *not* a solution of $2x + 3y = 6$.

$$2x + 3y = 6 \quad \text{Equation with two variables}$$
$$2(5) + 3(1) \stackrel{?}{=} 6 \quad \text{Let } x = 5, y = 1.$$
$$10 + 3 \stackrel{?}{=} 6 \quad \text{Multiply.}$$
$$13 = 6 \quad \text{False}$$

To find ordered pairs that satisfy an equation, we select any number for one of the variables, substitute it into the equation for that variable, and then solve for the other variable.

Because any real number could be selected for one variable and would lead to a real number for the other variable, an equation with two variables such as $2x + 3y = 6$ has an infinite number of solutions.

NOW TRY EXERCISE 1

Complete each ordered pair for $2x - y = 4$.

(0, ___), (___, 0),

(4, ___), (___, 2)

Then write the results as a table of ordered pairs.

EXAMPLE 1 Completing Ordered Pairs and Making a Table

In parts (a)–(d), complete each ordered pair for $2x + 3y = 6$. Then, in part (e), write the results as a table of ordered pairs.

(a) (0, ___)

$$2x + 3y = 6$$
$$2(0) + 3y = 6 \quad \text{Let } x = 0.$$
$$3y = 6 \quad \text{Multiply. Add.}$$
$$y = 2 \quad \text{Divide by 3.}$$

The ordered pair is (0, 2).

(b) (___, 0)

$$2x + 3y = 6$$
$$2x + 3(0) = 6 \quad \text{Let } y = 0.$$
$$2x = 6 \quad \text{Multiply. Add.}$$
$$x = 3 \quad \text{Divide by 2.}$$

The ordered pair is (3, 0).

(c) (−3, ___)

$$2x + 3y = 6$$
$$2(-3) + 3y = 6 \quad \text{Let } x = -3.$$
$$-6 + 3y = 6 \quad \text{Multiply.}$$
$$3y = 12 \quad \text{Add 6.}$$
$$y = 4 \quad \text{Divide by 3.}$$

The ordered pair is (−3, 4).

(d) (___, −4)

$$2x + 3y = 6$$
$$2x + 3(-4) = 6 \quad \text{Let } y = -4.$$
$$2x - 12 = 6 \quad \text{Multiply.}$$
$$2x = 18 \quad \text{Add 12.}$$
$$x = 9 \quad \text{Divide by 2.}$$

The ordered pair is (9, −4).

(e) Table of ordered pairs (or table of values)

x	y	Ordered pairs
0	2	← Represents (0, 2) from part (a)
3	0	← Represents (3, 0) from part (b)
−3	4	← Represents (−3, 4) from part (c)
9	−4	← Represents (9, −4) from part (d)

NOW TRY 

NOW TRY ANSWER

1. (0, −4), (2, 0), (4, 4), (3, 2)

x	y
0	−4
2	0
4	4
3	2

OBJECTIVE 4 Graph lines.

The **graph of an equation** is the set of points corresponding to *all* ordered pairs that satisfy the equation. It gives a "picture" of the equation.

To graph $2x + 3y = 6$, we plot the ordered pairs found in **Objective 3** and **Example 1.** See **FIGURE 4(a)** on the next page. If *all* of the ordered pairs that satisfy the equation $2x + 3y = 6$ were graphed, they would form the straight line shown in **FIGURE 4(b).**

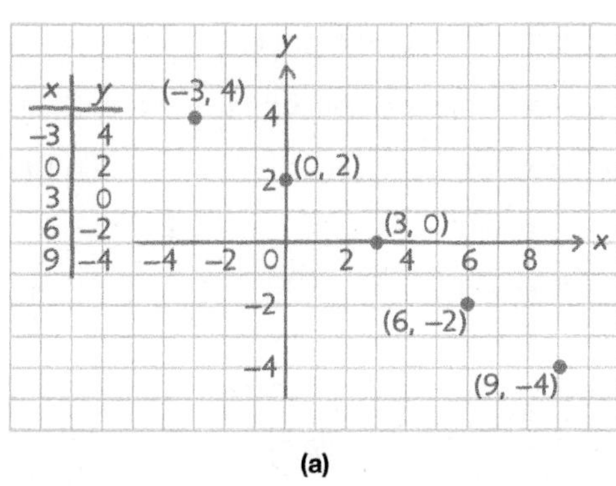

(a)

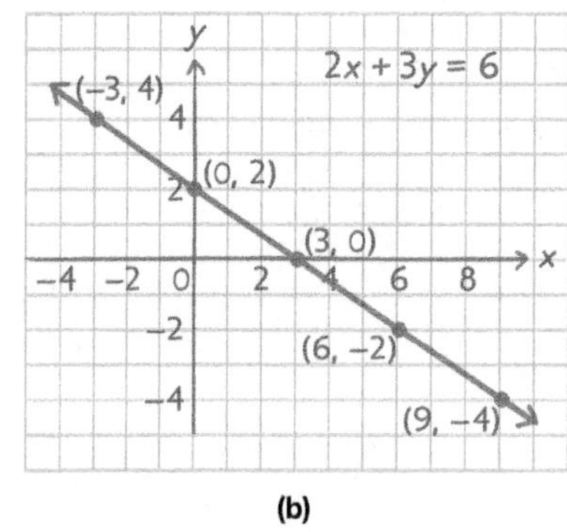

(b)

FIGURE 4

The equation $2x + 3y = 6$ is a **first-degree equation** because it has no term with a variable to a power greater than 1.

The graph of any first-degree equation in two variables is a straight line.

Because first-degree equations with two variables have straight-line graphs, they are called ***lin**ear equations in two variables.*

Linear Equation in Two Variables

A **linear equation in two variables** (here x and y) is an equation that can be written in the form

$$Ax + By = C,$$

where A, B, and C are real numbers and A and B are not both 0. This form is called **standard form.**

Examples: $3x + 4y = 9$, $x - y = 0$, $x + 2y = -8$

OBJECTIVE 5 Find x- and y-intercepts.

A straight line is determined if any two different points on the line are known. Therefore, finding two different points is sufficient to graph the line.

Two useful points for graphing are the x- and y-intercepts. The ***x*-intercept** is the point (if any) where the line intersects the x-axis. The ***y*-intercept** is the point (if any) where the line intersects the y-axis.* See FIGURE 5.

FIGURE 5

- The y-value of the point where the line intersects the x-axis is always 0.
- The x-value of the point where the line intersects the y-axis is always 0.

This suggests a method for finding the x- and y-intercepts.

Finding Intercepts

When graphing the equation of a line, find the intercepts as follows.

Let $y = 0$ to find the *x*-intercept.

Let $x = 0$ to find the *y*-intercept.

*Some texts define an intercept as a number, not a point. For example, "y-intercept $(0, 4)$" would be given as "y-intercept 4."

Find the x- and y-intercepts, and graph the equation.

$$x - 2y = 4$$

EXAMPLE 2 Finding Intercepts

Find the x- and y-intercepts of $4x - y = -3$, and graph the equation.

To find the x-intercept, let $y = 0$.

$$4x - y = -3$$

$$4x - 0 = -3 \quad \text{Let } y = 0.$$

$$4x = -3 \quad \text{Subtract.}$$

$$x = -\frac{3}{4} \quad \text{Divide by 4.}$$

The x-intercept is $\left(-\frac{3}{4}, 0\right)$.

To find the y-intercept, let $x = 0$.

$$4x - y = -3$$

$$4(0) - y = -3 \quad \text{Let } x = 0.$$

$$-y = -3 \quad \text{Multiply. Subtract.}$$

$$y = 3 \quad \text{Multiply by } -1.$$

The y-intercept is $(0, 3)$.

To guard against errors when graphing the equation, it is a good idea to find a third point. We arbitrarily choose $x = -2$, and substitute this value in the equation to find the corresponding value of y.

$$4x - y = -3$$

$$4(-2) - y = -3 \quad \text{Let } x = -2.$$

$$-8 - y = -3 \quad \text{Multiply.}$$

$$-y = 5 \quad \text{Add 8.}$$

$$y = -5 \quad \text{Multiply by } -1.$$

The ordered pair $(-2, -5)$ lies on the graph. We plot the three ordered pairs and draw a line through them. See **FIGURE 6**.

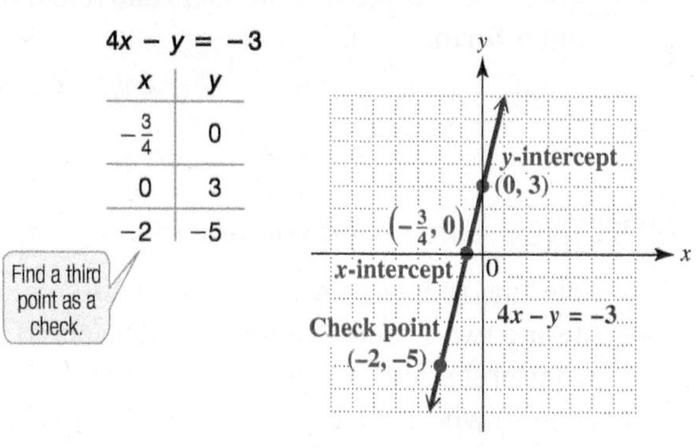

FIGURE 6

A linear equation with both x and y variables will have both x- and y-intercepts. Its graph will be a "slanted" line.

NOW TRY 

EXAMPLE 3 Graphing a Line That Passes through the Origin

Graph $x + 2y = 0$.

Find the x-intercept.

$$x + 2y = 0$$

$$x + 2(0) = 0 \quad \text{Let } y = 0.$$

$$x + 0 = 0 \quad \text{Multiply.}$$

$$x = 0 \quad x\text{-intercept is } (0, 0).$$

Find the y-intercept.

$$x + 2y = 0$$

$$0 + 2y = 0 \quad \text{Let } x = 0.$$

$$2y = 0 \quad \text{Add.}$$

$$y = 0 \quad y\text{-intercept is } (0, 0).$$

Both intercepts are the *same* point, $(0, 0)$, which means that the graph passes through the origin. To find a second point so we can graph the line, we choose any nonzero number for x or y and solve for the other variable. We arbitrarily choose $x = 4$.

NOW TRY ANSWER

2. x-intercept: $(4, 0)$; y-intercept: $(0, -2)$

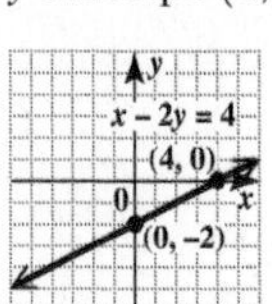

NOW TRY EXERCISE 3
Graph $2x + 3y = 0$.

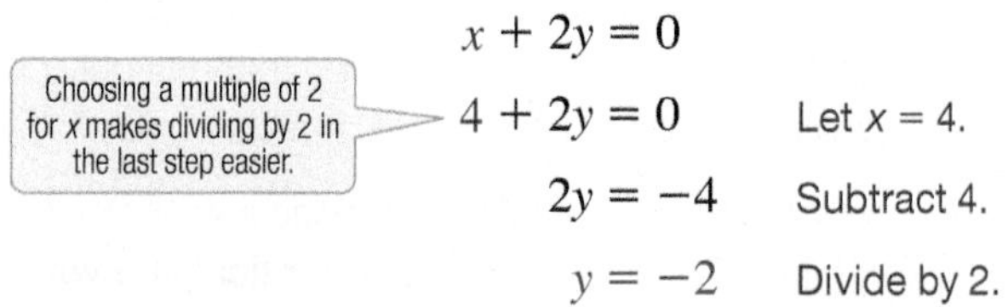

Choosing a multiple of 2 for x makes dividing by 2 in the last step easier.

$$x + 2y = 0$$
$$4 + 2y = 0 \quad \text{Let } x = 4.$$
$$2y = -4 \quad \text{Subtract 4.}$$
$$y = -2 \quad \text{Divide by 2.}$$

This gives the ordered pair $(4, -2)$. As a final check, substitute 1 for y in the equation and verify that $(-2, 1)$ also lies on the line. The graph is shown in **FIGURE 7**.

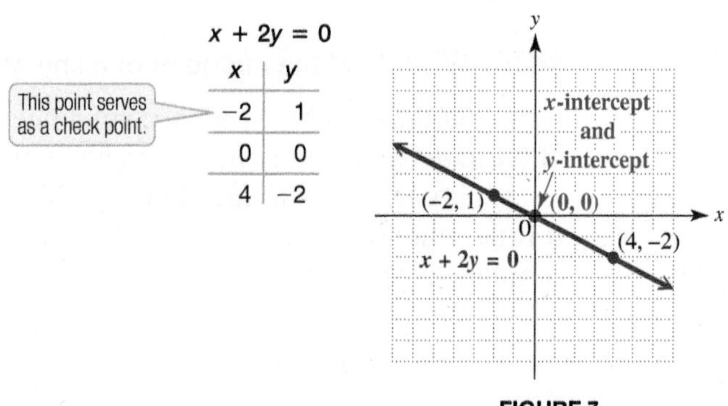

$x + 2y = 0$

x	y
−2	1
0	0
4	−2

FIGURE 7

NOW TRY

OBJECTIVE 6 Graph equations of horizontal and vertical lines.

A line parallel to the x-axis will not have an x-intercept. Similarly, a line parallel to the y-axis will not have a y-intercept. This is why we included the phrase "if any" when we introduced intercepts.

NOW TRY EXERCISE 4
Graph each equation.
(a) $y = -2$ **(b)** $x + 3 = 0$

EXAMPLE 4 Graphing Horizontal and Vertical Lines

Graph each equation.

(a) $y = 2$ (This equation can be written as $0x + y = 2$.)

Because y *always* equals 2, there is no value of x corresponding to $y = 0$, and the graph has no x-intercept. One value where $y = 2$ is on the y-axis, so the y-intercept is $(0, 2)$. Plot any two other points with y-coordinate 2, such as $(-1, 2)$ and $(3, 2)$.

The graph is shown in **FIGURE 8**. It is a horizontal line.

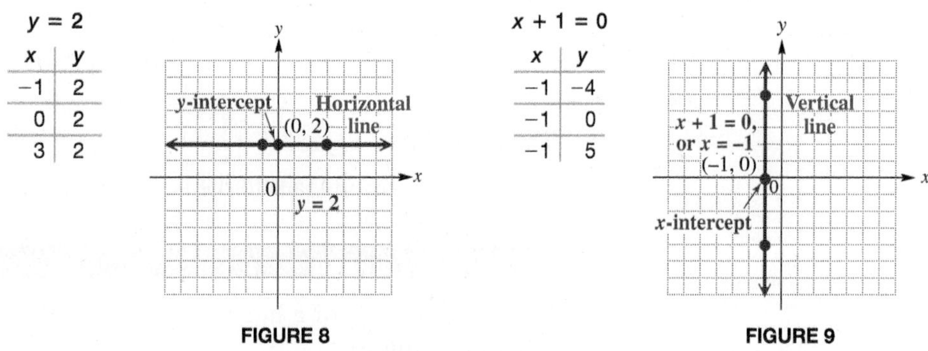

$y = 2$

x	y
−1	2
0	2
3	2

$x + 1 = 0$

x	y
−1	−4
−1	0
−1	5

FIGURE 8 **FIGURE 9**

(b) $x + 1 = 0$ (This equation can be written as $x = -1$ or $x + 0y = -1$.)

Because x *always* equals -1, there is no value of y that makes $x = 0$, and the graph has no y-intercept. One value where $x = -1$ is on the x-axis, so the x-intercept is $(-1, 0)$. Plot any two other points with x-coordinate -1, such as $(-1, -4)$ and $(-1, 5)$.

The graph is shown in **FIGURE 9**. It is a vertical line.

NOW TRY

NOW TRY ANSWERS

3.

4. (a) (b)

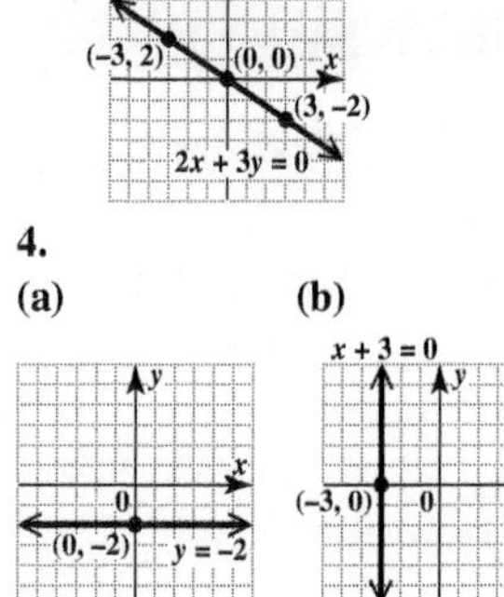

CAUTION A linear equation that has only *one* variable *x* or *y* will have a vertical or horizontal line as its graph.

1. An equation with only the variable *x* will always intersect the *x-axis* and thus will be ***vertical.*** It has the form $x = a$. **The vertical line with equation $x = 0$ is the *y*-axis.**
2. An equation with only the variable *y* will always intersect the *y-axis* and thus will be ***horizontal.*** It has the form $y = b$. **The horizontal line with equation $y = 0$ is the *x*-axis.**

OBJECTIVE 7 Find the midpoint of a line segment.

If the coordinates of the endpoints of a line segment are known, then the coordinates of the *midpoint* of the segment can be found.

FIGURE 10 shows a line segment PQ with endpoints $P(-8, 4)$ and $Q(3, -2)$. R is the point with the same x-coordinate as P and the same y-coordinate as Q. So the coordinates of R are $(-8, -2)$.

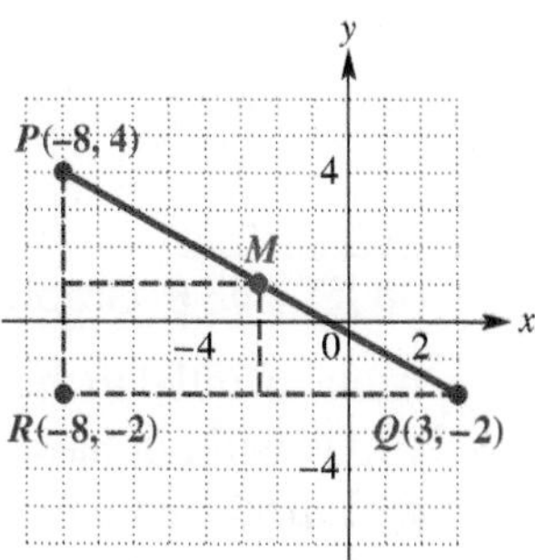

FIGURE 10

The x-coordinate of the midpoint M of PQ is the same as the x-coordinate of the midpoint of RQ. Because RQ is horizontal, the x-coordinate of its midpoint is the *average* (or *mean*) of the x-coordinates of its endpoints.

$$\frac{1}{2}(-8 + 3) = -\frac{5}{2}$$

The y-coordinate of M is the average (or *mean*) of the y-coordinates of the midpoint of PR.

$$\frac{1}{2}(4 + (-2)) = 1$$

The midpoint of PQ is $M\left(-\frac{5}{2}, 1\right)$.

This discussion leads to the *midpoint formula.*

Midpoint Formula

The midpoint M of a line segment PQ with endpoints $P(x_1, y_1)$ and $Q(x_2, y_2)$ is found as follows.

$$M = \left(\frac{x_1 + x_2}{2}, \frac{y_1 + y_2}{2}\right)$$

The two nonspecific points (x_1, y_1) and (x_2, y_2) use **subscript notation.** Read (x_1, y_1) as ***"x-sub-one, y-sub-one."***

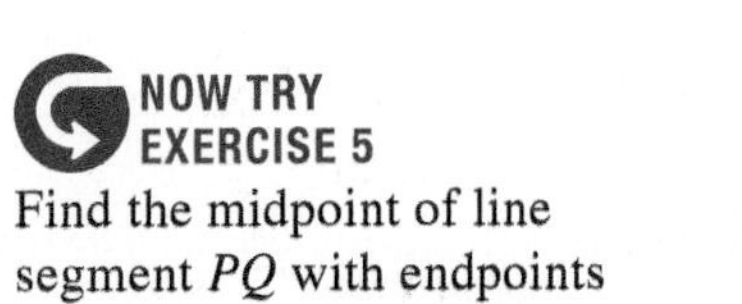

NOW TRY EXERCISE 5

Find the midpoint of line segment PQ with endpoints $P(2, -5)$ and $Q(-4, 7)$.

EXAMPLE 5 Finding the Coordinates of a Midpoint

Find the midpoint of line segment PQ with endpoints $P(4, -3)$ and $Q(6, -1)$.

$$\underset{P(4,\ -3)}{\overset{(x_1,\ y_1)}{}} \quad \text{and} \quad \underset{Q(6,\ -1)}{\overset{(x_2,\ y_2)}{}}$$ Label the points.

$$M = \left(\frac{x_1 + x_2}{2}, \frac{y_1 + y_2}{2}\right)$$ Midpoint formula

$$= \left(\frac{4+6}{2}, \frac{-3+(-1)}{2}\right)$$ Substitute.

We are finding the average of the x-coordinates and the average of the y-coordinates.

$$= \left(\frac{10}{2}, \frac{-4}{2}\right)$$ Add in the numerators.

$$= (5, -2)$$ ← Midpoint of segment PQ

NOW TRY 

NOTE When graphing with a graphing calculator, we "set up" a rectangular coordinate system. The screen in **FIGURE 11** shows the **standard viewing window.** Minimum x- and y-values are -10, and maximum x- and y-values are 10. The scale on each axis, here 1, determines the distance between the tick marks.

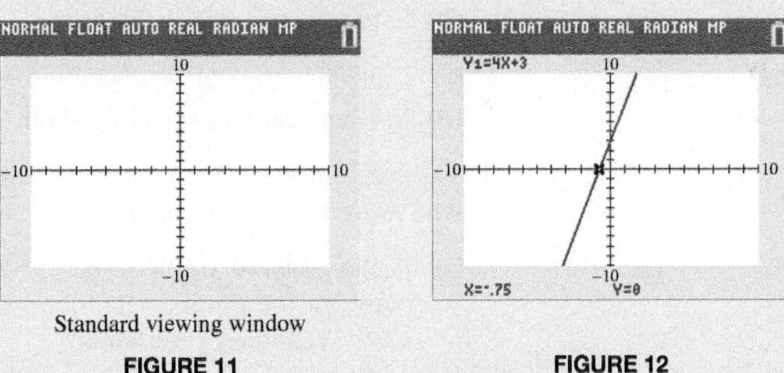

Standard viewing window

FIGURE 11

FIGURE 12

To graph an equation such as $4x - y = -3$, we use the intercepts $(-0.75, 0)$ and $(0, 3)$ to determine an appropriate window. Here, we choose the standard viewing window. We solve the equation for y to obtain $y = 4x + 3$ and enter it into the calculator. The graph in **FIGURE 12** also gives the coordinates of the x-intercept at the bottom of the screen.

NOW TRY ANSWER

5. $(-1, 1)$

2.1 Exercises

FOR EXTRA HELP  **MyLab Math**

Video solutions for select problems available in MyLab Math

STUDY SKILLS REMINDER

Are you fully utilizing the features of your text? **Review Study Skill 1, *Using Your Math Text.***

Concept Check *Fill in each blank with the correct response.*

1. The point with coordinates $(0, 0)$ is the ____ of a rectangular coordinate system.

2. For any value of x, the point $(x, 0)$ lies on the ____-axis. For any value of y, the point $(0, y)$ lies on the ____-axis.

3. To find the x-intercept of a line, we let ____ equal 0 and solve for ____. To find the y-intercept, we let ____ equal 0 and solve for ____.

4. The equation $y = 4$ has a ____ line as its graph, and $x = 4$ has a ____ line as its graph.

5. To graph a straight line, we must find a minimum of _____ points. The points (3, _____) and (_____, 4) lie on the graph of $2x - 3y = 0$.

6. The equation of the x-axis is _____. The equation of the y-axis is _____.

Concept Check *Solve each problem by locating ordered pairs on the graphs.*

7. The graph indicates personal spending in billions of dollars on medical care in the United States.

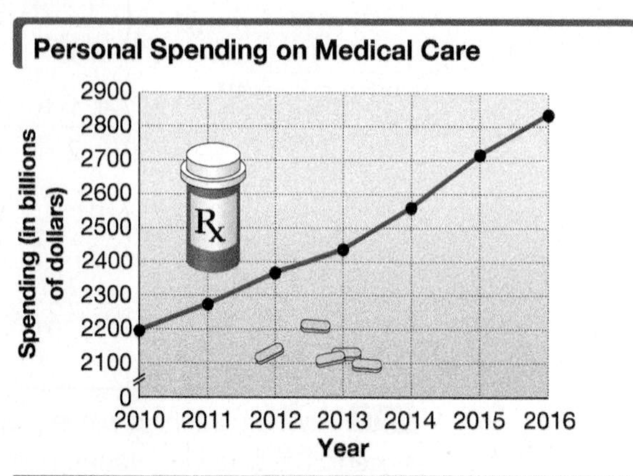

Data from Centers for Medicare & Medicaid Services.

(a) If (x, y) represents a point on the graph, what does x represent? What does y represent?

(b) Estimate spending in 2016.

(c) Write an ordered pair (x, y) that represents approximate spending in 2016.

(d) In what year was spending about $2200 billion?

8. The graph shows the percentage of Americans who moved in selected years.

(a) If the ordered pair (x, y) represents a point on the graph, what does x represent? What does y represent?

(b) Estimate the percentage of Americans who moved in 2017.

(c) Write an ordered pair (x, y) that gives the approximate percentage of Americans who moved in 2017.

(d) What does the ordered pair (1960, 20) mean in the context of this graph?

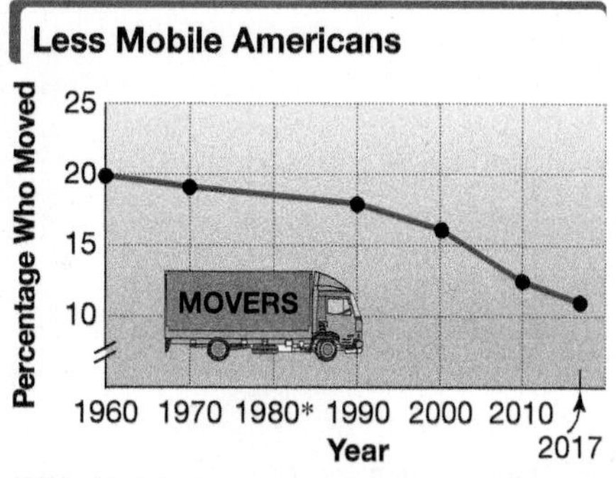

Data from U.S. Census Bureau.
* Data not collected for 1980.

Name the quadrant, if any, in which each point is located. ***See Objective 2.***

9. (a) $(1, 6)$ (b) $(-4, -2)$ (c) $(-3, 6)$ (d) $(7, -5)$ (e) $(-3, 0)$ (f) $(0, -0.5)$

10. (a) $(-2, -10)$ (b) $(4, 8)$ (c) $(-9, 12)$ (d) $(3, -9)$ (e) $(0, -8)$ (f) $(2.5, 0)$

11. **Concept Check** Use the given information to determine the quadrants in which the point (x, y) must lie. (*Hint:* Consider the signs of the coordinates in each quadrant, and the signs of their product and quotient.)

(a) $xy > 0$ (b) $xy < 0$ (c) $\frac{x}{y} < 0$ (d) $\frac{x}{y} > 0$

12. **Concept Check** What must be true about the value of at least one of the coordinates of any point that lies along an axis?

13. Concept Check A student plotted the point with coordinates $(-4, 2)$ incorrectly by moving 2 units from 0 to the right along the x-axis and then 4 units down parallel to the y-axis. **WHAT WENT WRONG?**

14. Concept Check A student incorrectly claimed that the point $(0, -4)$ lies on the x-axis because the x-coordinate is 0. **WHAT WENT WRONG?**

Plot each point in a rectangular coordinate system. ***See Objective 2.***

15. $(2, 3)$ **16.** $(-1, 2)$ **17.** $(-3, -2)$ **18.** $(1, -4)$ **19.** $(0, 5)$

20. $(-2, -4)$ **21.** $(-2, 4)$ **22.** $(3, 0)$ **23.** $(-2, 0)$ **24.** $(3, -3)$

Complete the given table for each equation and then graph the equation. ***See Example 1 and*** **FIGURE 4.**

25. $y = x - 4$

x	y
0	
1	
2	
3	
4	

26. $y = x + 3$

x	y
0	
1	
2	
3	
4	

27. $x - y = 3$

x	y
0	
	0
5	
2	

28. $x - y = 5$

x	y
0	
	0
1	
3	

29. $x + 2y = 5$

x	y
0	
	0
2	
	2

30. $x + 3y = -5$

x	y
0	
	0
1	
	−1

31. $4x - 5y = 20$

x	y
0	
	0
2	
	−3

32. $6x - 5y = 30$

x	y
0	
	0
3	
	−2

33. Concept Check Match each equation in parts (a)–(d) with its graph in choices A–D. (Coordinates of the points shown are integers.)

(a) $x + 3y = 3$ **(b)** $x - 3y = -3$ **(c)** $x - 3y = 3$ **(d)** $x + 3y = -3$

A.

y 4 0 4 x

B.

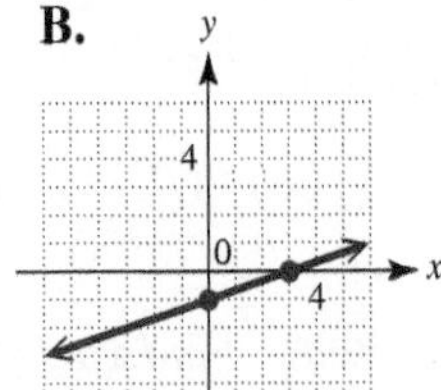

C.

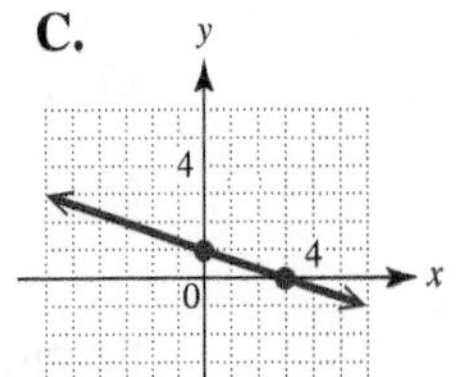

D.

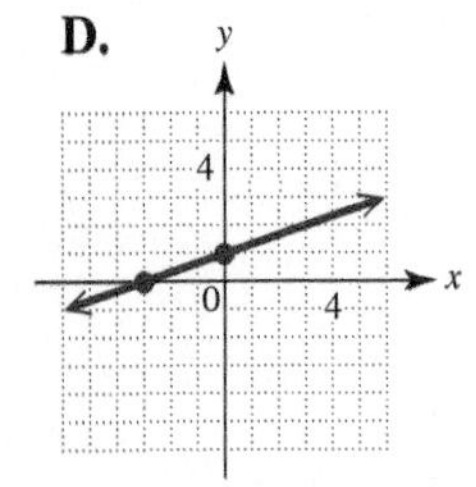

34. Concept Check Which of the following equations have a graph that is a horizontal line? A vertical line?

A. $x - 6 = 0$ **B.** $x + y = 0$ **C.** $y + 3 = 0$ **D.** $y = -10$ **E.** $x + 1 = 5$

Find the x- and y-intercepts. Then graph each equation. ***See Examples 2–4.***

35. $2x + 3y = 12$ **36.** $5x + 2y = 10$ **37.** $x - 3y = 6$

38. $x - 2y = -4$ **39.** $5x + 6y = -10$ **40.** $3x - 7y = 9$

41. $\frac{2}{3}x - 3y = 7$ **42.** $\frac{5}{7}x + \frac{6}{7}y = -2$ **43.** $x + 5y = 0$

44. $x - 3y = 0$ **45.** $2x + y = 0$ **46.** $4x - y = 0$

47. $2x = 3y$ **48.** $4y = 3x$ **49.** $-\frac{2}{3}y = x$

50. $-\frac{3}{4}y = x$ **51.** $y = 5$ **52.** $y = -3$

53. $x = 2$ **54.** $x = -3$ **55.** $x + 4 = 0$

56. $x - 4 = 0$ **57.** $y + 2 = 0$ **58.** $y - 5 = 0$

Each table of values gives several points that lie on a line.

(a) *What is the x-intercept of the line? The y-intercept?*

(b) *Which equation in choices A–D corresponds to the given table of values?*

(c) *Graph the equation.*

59.

x	y
−4	−3
−2	0
0	3
2	6

A. $3x + 2y = 6$
B. $3x - 2y = -6$
C. $3x + 2y = -6$
D. $3x - 2y = 6$

60.

x	y
−1	6
0	4
1	2
2	0

A. $2x - y = 4$
B. $2x + y = -4$
C. $2x + y = 4$
D. $2x - y = -4$

61.

x	y
−2	−1
0	−1
2	−1
4	−1

A. $y = -1$
B. $y = 1$
C. $x = 1$
D. $x = -1$

62.

x	y
6	−1
6	0
6	1
6	2

A. $x = -6$
B. $y = 0$
C. $y = 6$
D. $x = 6$

Find the midpoint of each segment with the given endpoints. ***See Example 5.***

63. $(-8, 4)$ and $(-2, -6)$ **64.** $(5, 2)$ and $(-1, 8)$

65. $(3, -6)$ and $(6, 3)$ **66.** $(-10, 4)$ and $(7, 1)$

67. $(-9, 3)$ and $(9, 8)$ **68.** $(4, -3)$ and $(-1, 3)$

69. $(2.5, 3.1)$ and $(1.7, -1.3)$ **70.** $(6.2, 5.8)$ and $(1.4, -0.6)$

Extending Skills *Find the midpoint of each segment with the given endpoints.*

71. $\left(\frac{1}{2}, \frac{1}{3}\right)$ and $\left(\frac{3}{2}, \frac{5}{3}\right)$ **72.** $\left(\frac{21}{4}, \frac{2}{5}\right)$ and $\left(\frac{7}{4}, \frac{3}{5}\right)$

73. $\left(-\frac{1}{3}, \frac{2}{7}\right)$ and $\left(-\frac{1}{2}, \frac{1}{14}\right)$ **74.** $\left(\frac{3}{5}, -\frac{1}{3}\right)$ and $\left(\frac{1}{2}, -\frac{7}{2}\right)$

Extending Skills *Segment PQ has the given coordinates for one endpoint P and for its midpoint M. Find the coordinates of the other endpoint Q. (Hint: Represent Q by (x, y) and write two equations using the midpoint formula, one involving x and the other involving y. Then solve for x and y.)*

75. $P(5, 8), M(8, 2)$ **76.** $P(7, 10), M(5, 3)$

77. $P(1.5, 1.25), M(3, 1)$ **78.** $P(2.5, 1.75), M(3, 2)$

2.2 The Slope of a Line

OBJECTIVES

1. Find the slope of a line given two points on the line.
2. Find the slope of a line given an equation of the line.
3. Graph a line given its slope and a point on the line.
4. Determine whether two lines are parallel, perpendicular, or neither using slope.
5. Solve problems involving average rate of change.

Slope (steepness) is used in many practical ways. The slope (or *grade*) of a highway is often given as a percent. For example, a 10% $\left(\text{or } \frac{10}{100} = \frac{1}{10}\right)$ slope means that the highway rises 1 unit for every 10 horizontal units. Stairs and roofs have slopes too, as shown in **FIGURE 13.**

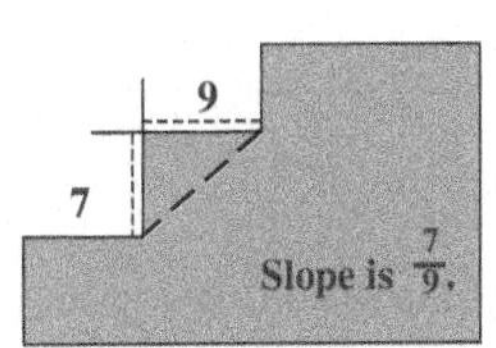

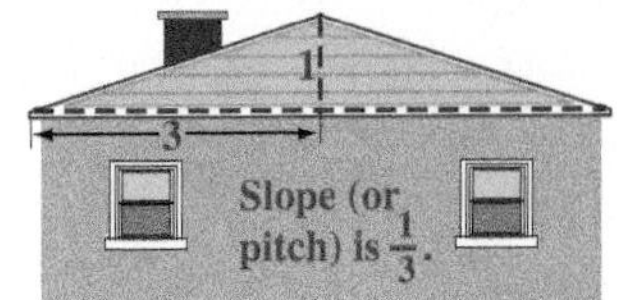

FIGURE 13

Slope is the ratio of vertical change, or **rise,** to horizontal change, or **run.** A simple way to remember this is to think, ***"Slope is rise over run."***

VOCABULARY

- ☐ rise
- ☐ run
- ☐ slope

OBJECTIVE 1 Find the slope of a line given two points on the line.

To obtain a formal definition of the slope of a line, we designate two different points on the line as (x_1, y_1) and (x_2, y_2). See **FIGURE 14.**

FIGURE 14

STUDY SKILLS REMINDER

Are you getting the most out of your class time? **Review Study Skill 3, *Taking Lecture Notes.***

As we move along the line in **FIGURE 14** from (x_1, y_1) to (x_2, y_2), the y-value changes (vertically) from y_1 to y_2, an amount equal to $y_2 - y_1$. As y changes from y_1 to y_2, the value of x changes (horizontally) from x_1 to x_2 by the amount $x_2 - x_1$.

NOTE The Greek letter **delta, Δ,** is used in mathematics to denote "change in," so Δy and Δx represent the change in y and the change in x, respectively.

The ratio of the change in y to the change in x $\left(\text{"rise over run," or } \frac{\text{rise}}{\text{run}}\right)$ is the *slope* of the line, with the letter m traditionally used for slope.

Slope Formula

The **slope m** of the line passing through the distinct points (x_1, y_1) and (x_2, y_2) is defined as follows.

$$m = \frac{\textbf{rise}}{\textbf{run}} = \frac{\textbf{change in } y}{\textbf{change in } x} = \frac{\Delta y}{\Delta x} = \frac{y_2 - y_1}{x_2 - x_1} \quad (\textbf{where } x_1 \neq x_2)$$

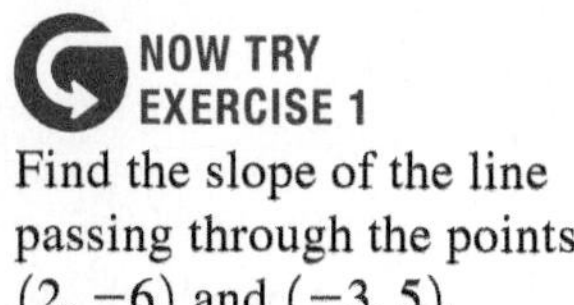

Find the slope of the line passing through the points $(2, -6)$ and $(-3, 5)$.

EXAMPLE 1 Finding the Slope of a Line

Find the slope of the line passing through the points $(2, -1)$ and $(-5, 3)$.

Label the points, and then apply the slope formula.

$$\begin{array}{ccc} (x_1, y_1) & & (x_2, y_2) \\ \downarrow \downarrow & & \downarrow \downarrow \\ (2, -1) & \text{and} & (-5, 3) \end{array}$$

$$\text{slope } m = \frac{y_2 - y_1}{x_2 - x_1} = \frac{3 - (-1)}{-5 - 2} \qquad \text{Substitute.}$$

$$= \frac{4}{-7}, \text{ or } -\frac{4}{7} \qquad \text{Subtract; } \frac{a}{-b} = -\frac{a}{b}$$

The slope is $-\frac{4}{7}$.

The same slope is obtained if we label the points in reverse order. ***It makes no difference which point is identified as (x_1, y_1) and which as (x_2, y_2).***

$$\begin{array}{ccc} (x_2, y_2) & & (x_1, y_1) \\ \downarrow \downarrow & & \downarrow \downarrow \\ (2, -1) & \text{and} & (-5, 3) \end{array}$$

$$\text{slope } m = \frac{y_2 - y_1}{x_2 - x_1} = \frac{-1 - 3}{2 - (-5)} \qquad \text{Substitute.}$$

y-values are in the numerator, *x*-values in the denominator.

$$= \frac{-4}{7}, \text{ or } -\frac{4}{7} \qquad \text{Subtract; } \frac{-a}{b} = -\frac{a}{b}$$

To confirm this slope, see **FIGURE 15**. Using the geometric interpretation of slope $\left(\text{as } \frac{\text{rise}}{\text{run}}\right)$ and beginning at the point $(2, -1)$, we move *up* 4 units and to the *left* 7 units to reach the point $(-5, 3)$.

$$\text{slope } m = \frac{\text{change in } y \text{ (rise)}}{\text{change in } x \text{ (run)}} = \frac{4}{-7}, \text{ or } -\frac{4}{7}$$

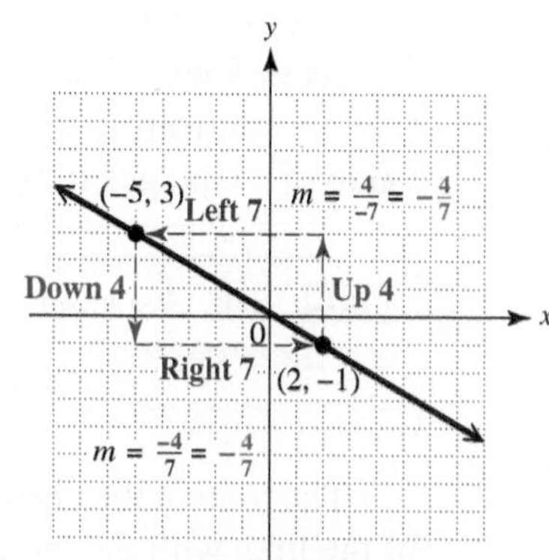

Moves *down* or to the *left* are denoted as negative numbers.

FIGURE 15

The same slope results if we begin at the point $(-5, 3)$ and move *down* 4 units and to the *right* 7 units to reach the point $(2, -1)$. See **FIGURE 15**.

$$\text{slope } m = \frac{\text{change in } y \text{ (rise)}}{\text{change in } x \text{ (run)}} = \frac{-4}{7}, \text{ or } -\frac{4}{7}$$

NOW TRY

NOW TRY ANSWER

1. $-\frac{11}{5}$

Example 1 suggests the following important ideas regarding slope.

1. The slope is the same no matter which point we consider first.
2. Using similar triangles from geometry, we can show that the slope is the same no matter which two different points on the line we choose.

CAUTION ***When calculating slope, remember that the change in y (rise) is the numerator and the change in x (run) is the denominator.***

Correct	Incorrect
$\frac{y_2 - y_1}{x_2 - x_1}$	~~$\frac{x_2 - x_1}{y_2 - y_1}$~~ or ~~$\frac{y_2 - y_1}{x_1 - x_2}$~~ or ~~$\frac{y_1 - y_2}{x_2 - x_1}$~~

Be careful to subtract the y-values and the x-values in the same order.

OBJECTIVE 2 Find the slope of a line given an equation of the line.

When an equation of a line is given, one way to find its slope is to use the definition of slope with two different points on the line.

NOW TRY EXERCISE 2
Find the slope of the line $3x - 7y = 21$.

EXAMPLE 2 Finding the Slope of a Line

Find the slope of the line $3x - y = -6$.

The intercepts can be used as the two points needed to find the slope.

Let $y = 0$ to find that the x-intercept is $(-2, 0)$.

Let $x = 0$ to find that the y-intercept is $(0, 6)$.

Use these two points in the slope formula.

$$\text{slope } m = \frac{y_2 - y_1}{x_2 - x_1} = \frac{6 - 0}{0 - (-2)} \qquad (x_1, y_1) = (-2, 0) \quad (x_2, y_2) = (0, 6)$$

$$= \frac{6}{2} \qquad \text{Subtract.}$$

$$= 3 \qquad \text{Divide.}$$

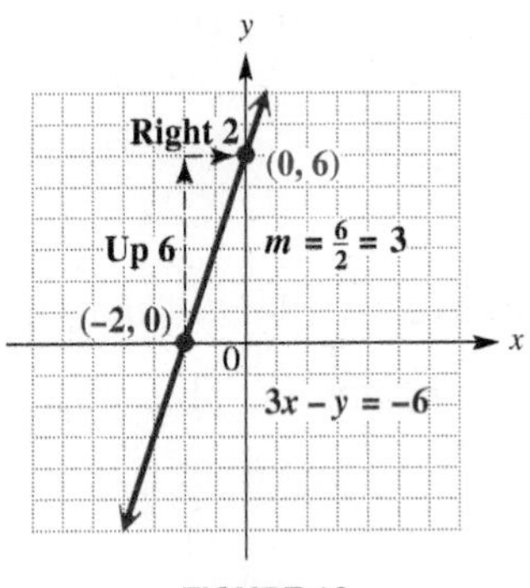

FIGURE 16

The graph of $3x - y = -6$ in **FIGURE 16** confirms that the slope of the line is 3.

NOW TRY

EXAMPLE 3 Applying the Slope Concept to Horizontal and Vertical Lines

Find the slope of each line.

(a) $y = 2$

The graph of $y = 2$ is a horizontal line. See **FIGURE 17**. To find the slope, select two points on the line, such as $(3, 2)$ and $(-1, 2)$, and apply the slope formula.

$$m = \frac{2 - 2}{-1 - 3} = \frac{0}{-4} = 0 \qquad (x_1, y_1) = (3, 2) \quad (x_2, y_2) = (-1, 2)$$

In this case, the *rise* is 0, so the slope is 0.

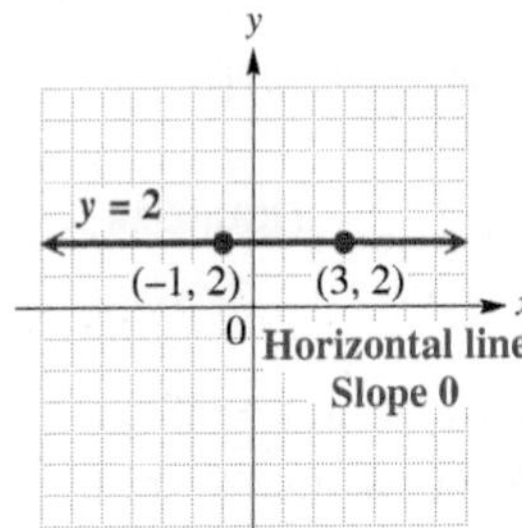

FIGURE 17

NOW TRY ANSWER
2. $\frac{3}{7}$

NOW TRY EXERCISE 3

Find the slope of each line.

(a) $x = 4$ (b) $y - 6 = 0$

(b) $x + 1 = 0$

The graph of $x + 1 = 0$, or $x = -1$, is a vertical line. See **FIGURE 18**. Two points that satisfy the equation $x = -1$ are $(-1, 5)$ and $(-1, -4)$. Use these two points and the slope formula.

$$m = \frac{-4 - 5}{-1 - (-1)} = \frac{-9}{0} \qquad (x_1, y_1) = (-1, 5) \quad (x_2, y_2) = (-1, -4)$$

Here the *run* is 0. Because division by 0 is undefined, the slope is undefined. This is why the definition of slope includes the restriction that $x_1 \neq x_2$.

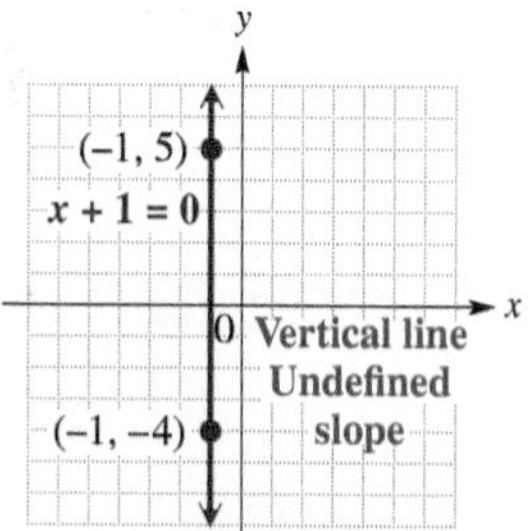

FIGURE 18

NOW TRY 

Horizontal and Vertical Lines

- An equation of the form $y = b$ always intersects the y-axis at the point $(0, b)$. A line with this equation is **horizontal** and has **slope 0.**
- An equation of the form $x = a$ always intersects the x-axis at the point $(a, 0)$. A line with this equation is **vertical** and has **undefined slope.**

The slope of a line can also be found directly from its equation. Consider the equation $3x - y = -6$ from **Example 2.** Solve this equation for y.

$$3x - y = -6 \qquad \text{Equation from Example 2}$$

$$-y = -6 - 3x \qquad \text{Subtract } 3x.$$

$$y = 3x + 6 \qquad \text{Multiply by } -1\text{; Commutative property}$$

The slope, 3, found using the slope formula in **Example 2,** is the same number as the coefficient of x in the equation $y = 3x + 6$. We will see in the next section that this is true in general, *as long as the equation is solved for y.*

NOW TRY EXERCISE 4

Find the slope of the graph of $5x - 4y = 7$.

EXAMPLE 4 Finding the Slope from an Equation

Find the slope of the graph of $3x - 5y = 8$.

Many of the ordered-pair solutions of this equation have coordinates that are fractions, including the intercepts $\left(\frac{8}{3}, 0\right)$ and $\left(0, -\frac{8}{5}\right)$. See **FIGURE 19.** This makes calculations using the slope formula more difficult.

Instead, we solve the equation for y as discussed above.

$$3x - 5y = 8 \qquad \text{Solve for } y.$$

$$-5y = 8 - 3x \qquad \text{Subtract } 3x.$$

$$\frac{-5y}{-5} = \frac{-3x + 8}{-5} \qquad \text{Commutative property; Divide each side by } -5.$$

$$y = \frac{3}{5}x - \frac{8}{5} \qquad \frac{-3x}{-5} = \frac{-3}{-5} \cdot \frac{x}{1} = \frac{3}{5}x \qquad \frac{a + b}{c} = \frac{a}{c} + \frac{b}{c}$$

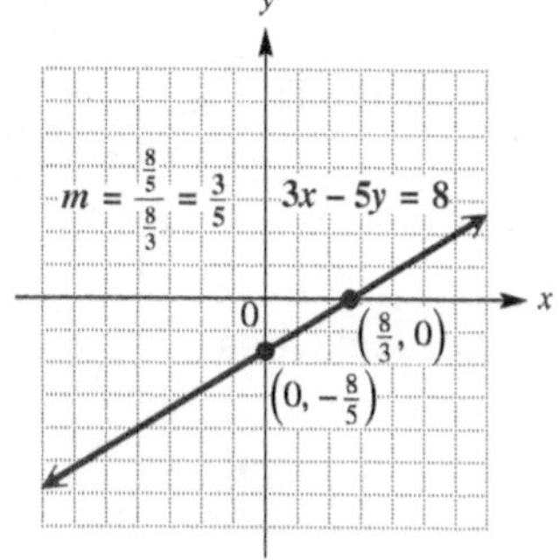

FIGURE 19

The slope, given by the coefficient of x, is $\frac{3}{5}$.

We can confirm this slope using the intercepts and the slope formula, which involves simplifying a *complex fraction*.

$$m = \frac{0 - \left(-\frac{8}{5}\right)}{\frac{8}{3} - 0} = \frac{\frac{8}{5}}{\frac{8}{3}} = \frac{8}{5} \div \frac{8}{3} = \frac{8}{5} \cdot \frac{3}{8} = \frac{3}{5} \qquad \text{The same slope results.}$$

NOW TRY

NOW TRY ANSWERS

3. (a) undefined (b) 0

4. $\frac{5}{4}$

NOTE We can solve the standard form of a linear equation $Ax + By = C$ (where $B \neq 0$) for y to show that, in general, **the slope of a line in this form is $-\frac{A}{B}$.**

$Ax + By = C$	Standard form
$By = C - Ax$	Subtract Ax.
$By = -Ax + C$	Commutative property
$y = -\frac{A}{B}x + \frac{C}{B}$	Divide each term by B.

The slope is given by the coefficient of x, $-\frac{A}{B}$. In the equation $3x - 5y = 8$ from **Example 4,** $A = 3$ and $B = -5$, so the slope is

$$-\frac{A}{B} = -\frac{3}{-5} = \frac{3}{5}. \quad \text{The same slope results.}$$

OBJECTIVE 3 Graph a line given its slope and a point on the line.

NOW TRY EXERCISE 5
Graph the line passing through $(-4, 1)$ that has slope $-\frac{2}{3}$.

EXAMPLE 5 Using the Slope and a Point to Graph Lines

Graph each line passing through the given point having the given slope.

(a) $(0, -4)$; slope $\frac{2}{3}$

Begin by plotting the point $P(0, -4)$, as shown in FIGURE 20. Then use the geometric interpretation of slope to find a second point.

$$m = \frac{\text{change in } y}{\text{change in } x} = \frac{2}{3} \begin{matrix} \leftarrow \text{rise} \\ \leftarrow \text{run} \end{matrix}$$

We move *up* 2 units from $(0, -4)$ and to the *right* 3 units to locate a second point on the graph, $R(3, -2)$. The line through $P(0, -4)$ and R is the required graph.

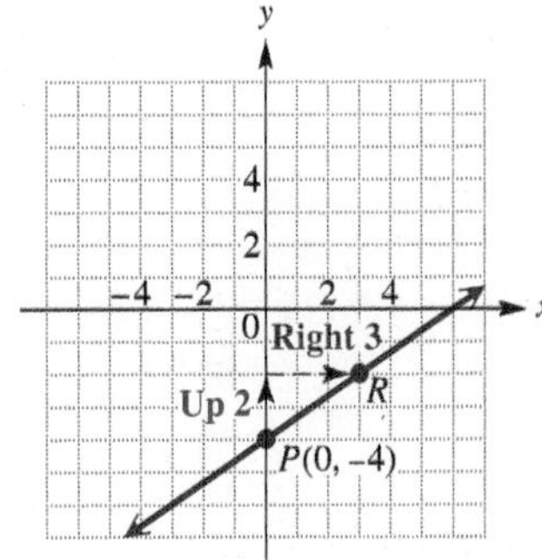

FIGURE 20

(b) $(3, 1)$; slope -4

Start by plotting the point $P(3, 1)$. See FIGURE 21. Find a second point R on the line by writing the slope -4 as $\frac{-4}{1}$ and using the geometric interpretation of slope.

$$m = \frac{\text{change in } y}{\text{change in } x} = \frac{-4}{1} \begin{matrix} \leftarrow \text{rise} \\ \leftarrow \text{run} \end{matrix}$$

We move *down* 4 units from $(3, 1)$ and to the *right* 1 unit to locate a second point, $R(4, -3)$. The line through $P(3, 1)$ and R is the required graph.

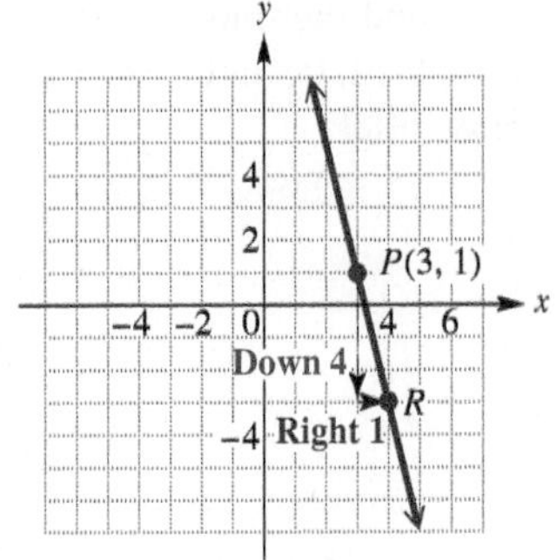

FIGURE 21

The slope -4 also could be written as

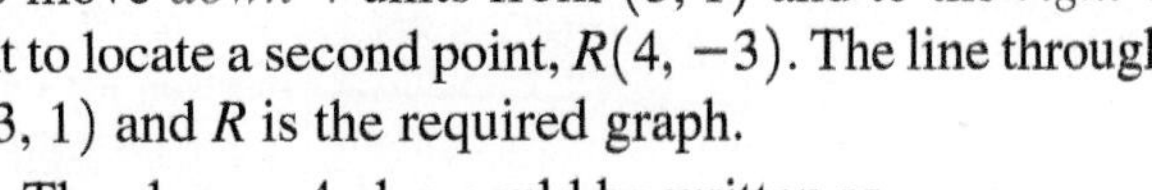

$$m = \frac{\text{change in } y}{\text{change in } x} = \frac{4}{-1}.$$

In this case, the second point R is located *up* 4 units and to the *left* 1 unit. Verify that this approach also produces the line in FIGURE 21.

NOW TRY

NOW TRY ANSWER

5.

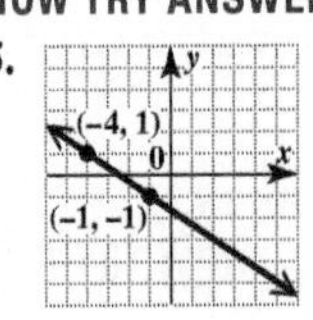

The line in **Example 5(a)** has *positive* slope $\frac{2}{3}$. The graph of the line in **FIGURE 20** *slants up (rises)* from left to right. The line in **Example 5(b)** has *negative* slope -4. As **FIGURE 21** shows, its graph *slants down (falls)* from left to right. These facts illustrate the following generalization.

Orientation of a Line in the Plane

A *positive slope* indicates that the line *slants up (rises)* from left to right.

A *negative slope* indicates that the line *slants down (falls)* from left to right.

FIGURE 22 summarizes the four cases for slopes of lines.

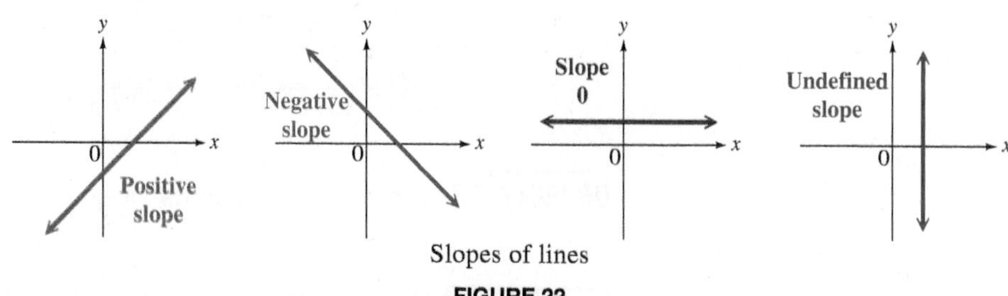

Slopes of lines

FIGURE 22

OBJECTIVE 4 Determine whether two lines are parallel, perpendicular, or neither using slope.

The slope of a line measures its steepness, and parallel lines have equal steepness.

Slopes of Parallel Lines

Two nonvertical lines with the same slope are parallel.

Two nonvertical parallel lines have the same slope.

NOW TRY EXERCISE 6

Determine whether the line passing through $(2, 5)$ and $(4, 8)$ and the line passing through $(2, 0)$ and $(-1, -2)$ are parallel.

EXAMPLE 6 Determining Whether Two Lines Are Parallel

Determine whether the lines L_1, passing through $(-2, 1)$ and $(4, 5)$, and L_2, passing through $(3, 0)$ and $(0, -2)$, are parallel.

Slope of L_1:

$$m_1 = \frac{5-1}{4-(-2)} = \frac{4}{6} = \frac{2}{3}$$

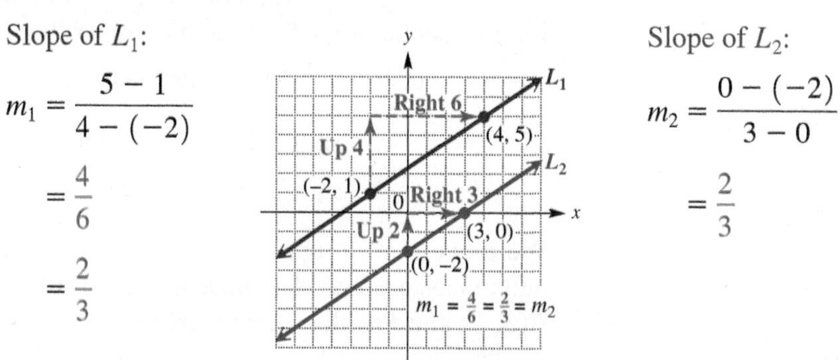

FIGURE 23

Slope of L_2:

$$m_2 = \frac{0-(-2)}{3-0} = \frac{2}{3}$$

The slopes are the *same,* both $\frac{2}{3}$, so the two lines are *parallel.* See **FIGURE 23**.

NOW TRY

NOW TRY ANSWER

6. not parallel

To see how the slopes of perpendicular lines are related, consider a nonvertical line with slope $\frac{a}{b}$. If this line is rotated 90°, the vertical change and the horizontal change are interchanged, and the slope is $-\frac{b}{a}$ because the horizontal change is now negative. See **FIGURE 24**. Thus, the slopes of perpendicular lines have product -1 and are negative reciprocals of each other.

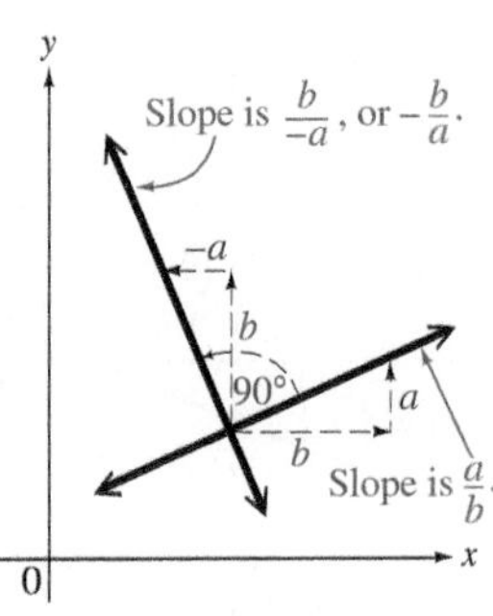

FIGURE 24

For example, if the slopes of two lines are $\frac{3}{4}$ and $-\frac{4}{3}$, then the lines are perpendicular because

$$\frac{3}{4}\left(-\frac{4}{3}\right) = -1.$$

Slopes of Perpendicular Lines

If neither is vertical, perpendicular lines have slopes that are negative reciprocals—that is, their product is -1. Also, lines with slopes that are negative reciprocals are perpendicular.

A line with slope 0 is perpendicular to a line with undefined slope.

EXAMPLE 7 Determining Whether Two Lines Are Parallel, Perpendicular, or Neither

Determine whether each pair of lines is *parallel, perpendicular,* or *neither.*

(a) $2y = 3x - 6$ and $2x + 3y = -6$

Find the slope of each line by solving each equation for y.

$$2y = 3x - 6$$

$$y = \frac{3}{2}x - 3 \quad \text{Divide by 2.}$$

↑ Slope

$$2x + 3y = -6$$

$$3y = -2x - 6 \quad \text{Subtract } 2x.$$

$$y = -\frac{2}{3}x - 2 \quad \text{Divide by 3.}$$

↑ Slope

The slopes are *negative reciprocals* because their product is $\frac{3}{2}\left(-\frac{2}{3}\right) = -1$. The two lines are *perpendicular.* See **FIGURE 25**.

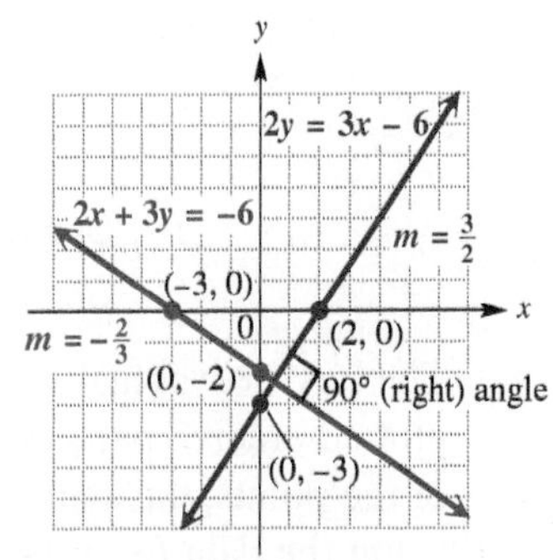

FIGURE 25

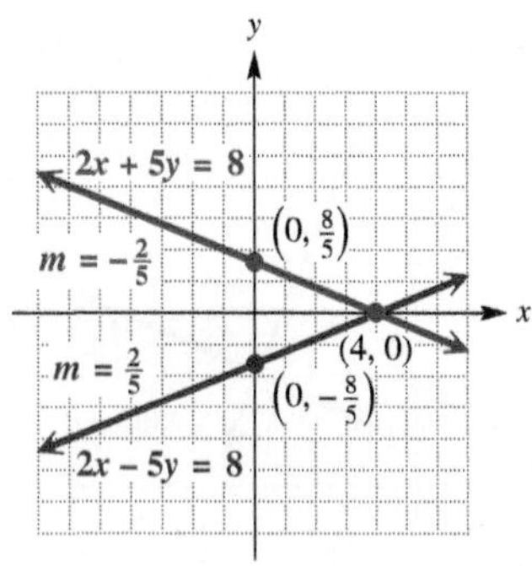

FIGURE 26

NOW TRY EXERCISE 7

Determine whether each pair of lines is *parallel, perpendicular,* or *neither.*

(a) $x + 2y = 7$ and $2x = y - 4$

(b) $2x - y = 4$ and $2x + y = 6$

(b) $2x - 5y = 8$ and $2x + 5y = 8$

$2x - 5y = 8$		$2x + 5y = 8$	
$-5y = -2x + 8$	Subtract $2x$.	$5y = -2x + 8$	Subtract $2x$.
$y = \frac{2}{5}x - \frac{8}{5}$	Divide by -5.	$y = -\frac{2}{5}x + \frac{8}{5}$	Divide by 5.
↑ Slope		↑ Slope	

The slopes, $\frac{2}{5}$ and $-\frac{2}{5}$, are not the same. They are not negative reciprocals because their product is $-\frac{4}{25}$, *not* -1. The two lines are *neither* parallel nor perpendicular. See **FIGURE 26.**

NOW TRY

NOTE In **Example 7(a),** we could have found the slope of each line using intercepts and the slope formula. The graph of $2y = 3x - 6$ has intercepts $(2, 0)$ and $(0, -3)$.

$$m = \frac{0 - (-3)}{2 - 0} = \frac{3}{2} \quad \text{The same slope found in Example 7(a) results.}$$

Find the intercepts of the graph of $2x + 3y = -6$ and use them to confirm the slope $-\frac{2}{3}$. Because the slopes are negative reciprocals, the lines are perpendicular.

To avoid fractions when using the slope formula in **Example 7(b),** we suggest identifying points with integer coordinates, such as $(4, 0)$ and $(-1, -2)$ for the equation $2x - 5y = 8$.

OBJECTIVE 5 Solve problems involving average rate of change.

The slope formula applied to any two points on a line gives the **average rate of change** in y per unit change in x.

For example, suppose the height of a boy increased from 60 to 68 in. between the ages of 12 and 16, as shown in **FIGURE 27.**

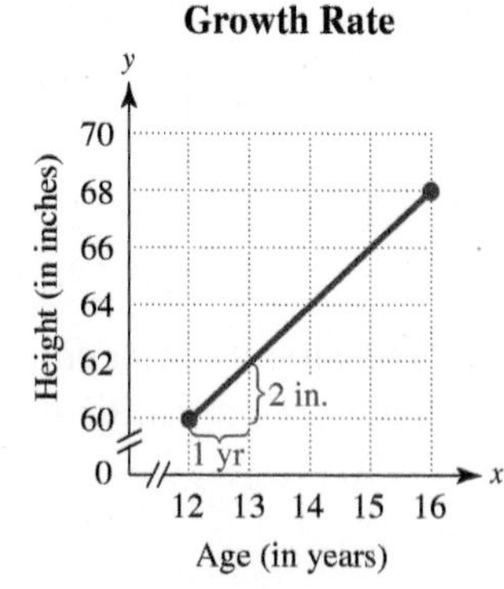

FIGURE 27

$$\frac{\text{Change in height } y \longrightarrow 68 - 60}{\text{Change in age } x \longrightarrow 16 - 12} = \frac{8}{4} = 2 \text{ in. per yr}$$

The boy may actually have grown more (or less) than 2 in. during some years. If we plotted ordered pairs (age, height) for those years and drew a line connecting any two of the points, the average rate of change would likely be slightly different. However, using the data for ages 12 and 16, the boy's *average* change in height was 2 in. per year over these years.

NOW TRY ANSWERS

7. (a) perpendicular
(b) neither

NOW TRY EXERCISE 8

There were approximately 84 million high-speed Internet subscribers in 2013. Using this number for 2013 and the number for 2017 from the graph in **FIGURE 28**, find the average rate of change per year from 2013 to 2017. How does it compare with the average rate of change found in **Example 8?**

EXAMPLE 8 Interpreting Slope as Average Rate of Change

The graph in **FIGURE 28** shows the number of high-speed Internet subscribers in the United States from 2012 to 2017. Find the average rate of change in number of subscribers per year.

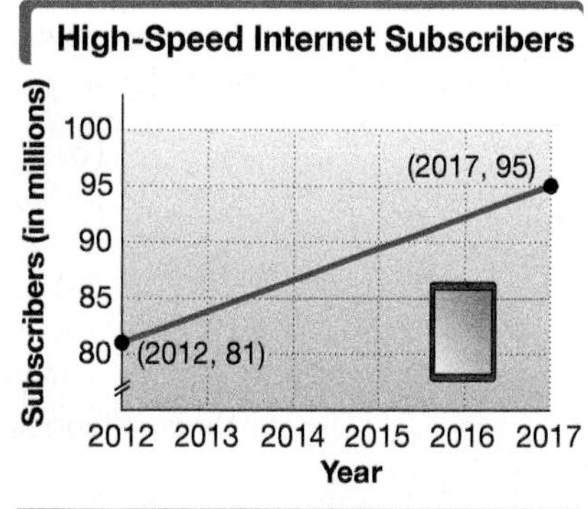

Data from Leichtman Research Group.

FIGURE 28

To find the average rate of change, we need two pairs of data. From the graph, we have the ordered pairs (2012, 81) and (2017, 95). We use the slope formula.

$$\text{average rate of change} = \frac{95 - 81}{2017 - 2012} = \frac{14}{5} = 2.8$$ A positive slope indicates an increase.

This means that the number of high-speed Internet subscribers *increased* by an average of 2.8 million subscribers per year from 2012 to 2017. NOW TRY

NOW TRY EXERCISE 9

In 2011, global sales of digital cameras totaled 33 billion euros. In 2016, sales totaled 17 billion euros. Find the average rate of change in sales of digital cameras per year. (Data from Photoindustrie–Verband.)

EXAMPLE 9 Interpreting Slope as Average Rate of Change

In 2011, there were 105 million households with cable TV in the United States. There were 96 million such households in 2017. Find the average rate of change in the number of households per year. (Data from Nielson Media Research.)

We must write two ordered pairs of the form (year, number of households) to use to find the average rate of change. From the problem, we have (2011, 105) and (2017, 96). We substitute these values in the slope formula.

$$\text{average rate of change} = \frac{96 - 105}{2017 - 2011} = \frac{-9}{6} = -1.5$$ A negative slope indicates a decrease.

The graph in **FIGURE 29** confirms that the line through the ordered pairs falls from left to right and therefore has negative slope. Thus, the number of households with cable TV *decreased* by an average of 1.5 million households per year from 2011 to 2017.

The negative sign in -1.5 denotes the *decrease*. (We say *"The number of customers decreased by 1.5 million per year."* It is ***incorrect*** to say *"The number of customers decreased by -1.5 million per year."*)

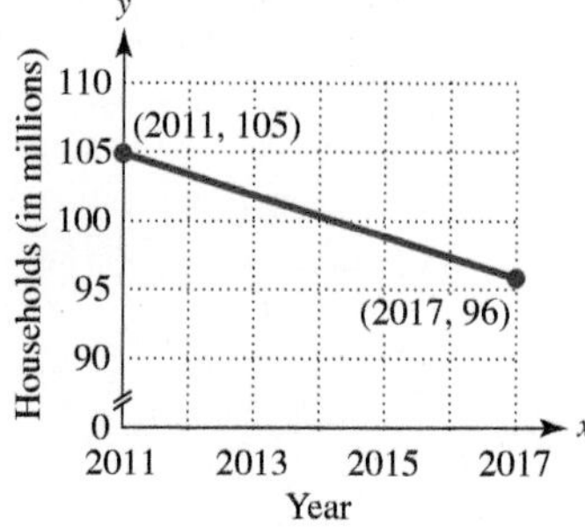

FIGURE 29

NOW TRY 

NOW TRY ANSWERS

8. 2.75 million subscribers per year; It is less than the average rate of change per year from 2012 to 2017.

9. -3.2 billion euros per year

2.2 Exercises

 **MyLab Math**

Video solutions for select problems available in MyLab Math

STUDY SKILLS REMINDER

Be sure to read and work through the section material before working the exercises. **Review Study Skill 2, *Reading Your Math Text.***

Concept Check *Answer each question.*

1. A hill rises 30 ft for every horizontal 100 ft. Which of the following express its slope (or grade)? (There are several correct choices.)

A. 0.3 **B.** $\frac{3}{10}$ **C.** $3\frac{1}{3}$

D. $\frac{30}{100}$ **E.** $\frac{10}{3}$ **F.** 30%

2. If a walkway rises 2 ft for every 24 ft on the horizontal, which of the following express its slope (or grade)? (There are several correct choices.)

A. 12% **B.** $\frac{2}{24}$ **C.** $\frac{1}{12}$

D. 12 **E.** $8.\overline{3}\%$ **F.** $\frac{24}{2}$

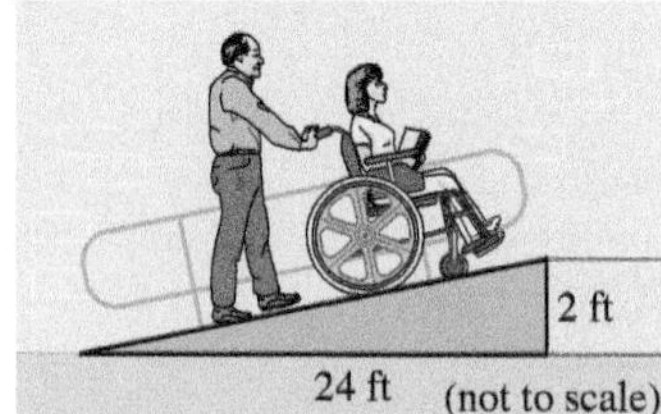

3. A ladder leaning against a wall has slope 3. How many feet in the horizontal direction correspond to a rise of 15 ft?

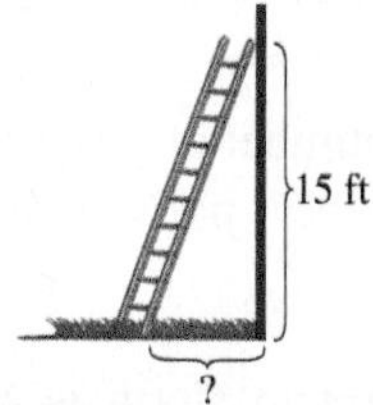

4. A hill has slope 0.05. How many feet in the vertical direction correspond to a run of 50 ft?

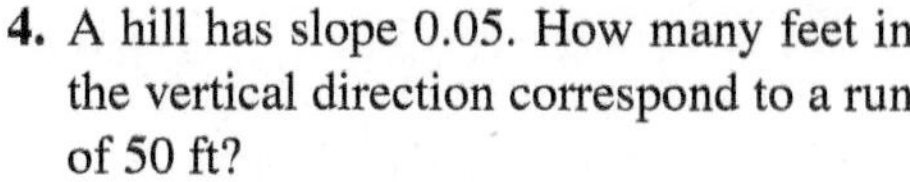

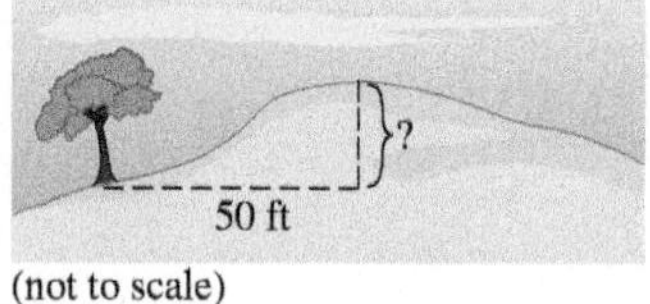

(not to scale)

5. Concept Check Match each situation in parts (a)–(d) with the most appropriate graph in choices A–D.

(a) Sales rose sharply during the first quarter, leveled off during the second quarter, and then rose slowly for the rest of the year.

(b) Sales fell sharply during the first quarter and then rose slowly during the second and third quarters before leveling off for the rest of the year.

(c) Sales rose sharply during the first quarter and then fell to the original level during the second quarter before rising steadily for the rest of the year.

(d) Sales fell during the first two quarters of the year, leveled off during the third quarter, and rose during the fourth quarter.

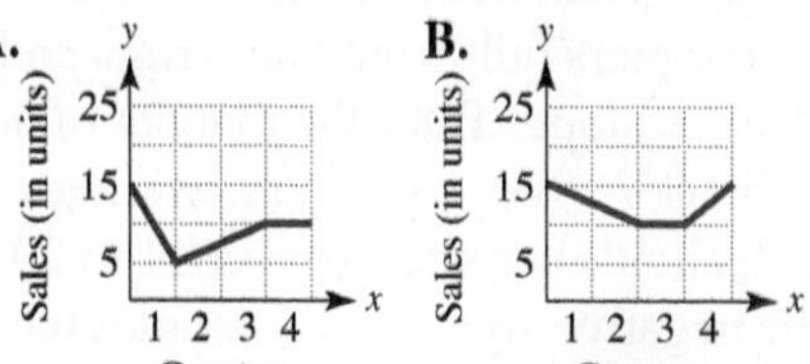

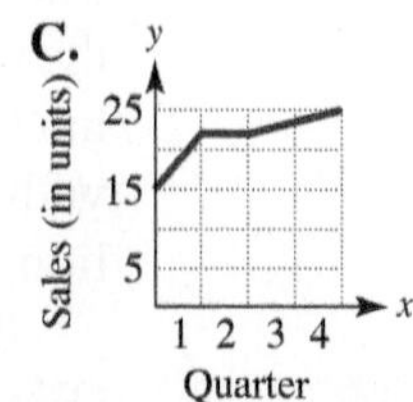

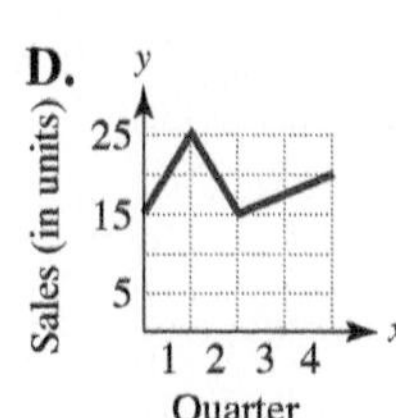

6. Concept Check Using the given axes, draw a graph that illustrates the following description.

Profits for a business were $10 million in 2013. They rose sharply from 2013 through 2015, remained constant from 2015 through 2017, and then fell slowly from 2017 through 2018.

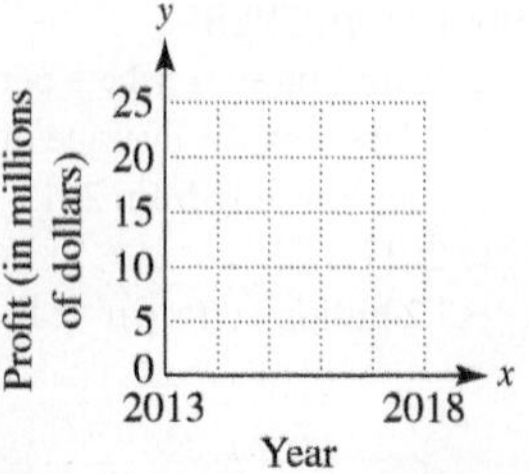

Concept Check *Determine the slope of each line segment in the given figure.*

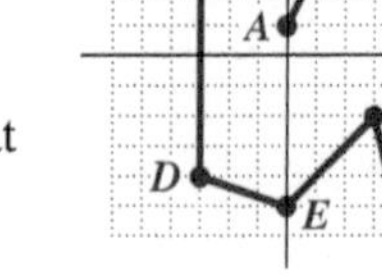

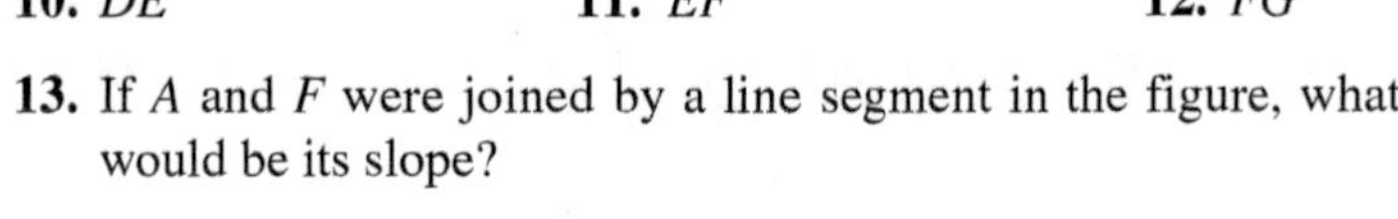

7. AB **8.** BC **9.** CD

10. DE **11.** EF **12.** FG

13. If A and F were joined by a line segment in the figure, what would be its slope?

14. If B and D were joined by a line segment in the figure, what would be its slope?

15. Concept Check Refer to the figure shown here and determine which line satisfies the given description.

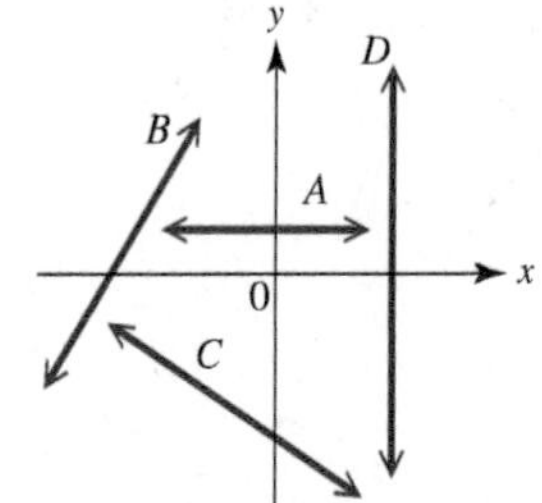

(a) The line has positive slope.

(b) The line has negative slope.

(c) The line has slope 0.

(d) The line has undefined slope.

16. Concept Check Which forms of the slope formula are correct? Explain.

A. $m = \frac{y_1 - y_2}{x_2 - x_1}$ **B.** $m = \frac{y_1 - y_2}{x_1 - x_2}$ **C.** $m = \frac{x_2 - x_1}{y_2 - y_1}$ **D.** $m = \frac{y_2 - y_1}{x_2 - x_1}$

17. Concept Check A student found the slope of the line through the points $(-4, 5)$ and $(2, 7)$ as follows.

$$m = \frac{7 - 5}{-4 - 2} = \frac{2}{-6} = -\frac{1}{3}$$

This is incorrect. **WHAT WENT WRONG?** Give the correct slope.

18. Concept Check A student found the slope of the line through the points $(-3, 4)$ and $(-5, -1)$ as follows.

$$m = \frac{-3 - (-5)}{4 - (-1)} = \frac{2}{5}$$

This is incorrect. **WHAT WENT WRONG?** Give the correct slope.

Evaluate each expression for m, applying the slope formula. ***See Example 1.***

19. $m = \frac{6 - 2}{5 - 3}$ **20.** $m = \frac{6 - 0}{0 - 3}$ **21.** $m = \frac{-5 - (-5)}{3 - 2}$

22. $m = \frac{-2 - (-2)}{4 - (-3)}$ **23.** $m = \frac{3 - 8}{-2 - (-2)}$ **24.** $m = \frac{5 - 6}{-8 - (-8)}$

In the following exercises, ***(a)*** *find the slope of the line passing through each pair of points, if possible, and* ***(b)*** *based on the slope, indicate whether the line* rises *from left to right,* falls *from left to right, is* horizontal, *or is* vertical. ***See Examples 1 and 3 and*** **FIGURE 22.**

25. $(-2, -3)$ and $(-1, 5)$ **26.** $(-4, 1)$ and $(-3, 4)$ **27.** $(-4, 1)$ and $(2, 6)$

28. $(-3, -3)$ and $(5, 6)$ **29.** $(2, 4)$ and $(-4, 4)$ **30.** $(-6, 3)$ and $(2, 3)$

31. $(-2, 2)$ and $(4, -1)$ **32.** $(-3, 1)$ and $(6, -2)$ **33.** $(5, -3)$ and $(5, 2)$

34. $(4, -1)$ and $(4, 3)$ **35.** $(1.5, 2.6)$ and $(0.5, 3.6)$ **36.** $(3.4, 4.2)$ and $(1.4, 10.2)$

Extending Skills *Find the slope of the line passing through the given pair of points.* $\left(\textit{Hint: } \dfrac{\frac{a}{b}}{\frac{c}{d}} = \dfrac{a}{b} \div \dfrac{c}{d}\right)$

37. $\left(\frac{1}{6}, \frac{1}{2}\right)$ and $\left(\frac{5}{6}, \frac{9}{2}\right)$

38. $\left(\frac{3}{4}, \frac{1}{3}\right)$ and $\left(\frac{5}{4}, \frac{10}{3}\right)$

39. $\left(-\frac{2}{9}, \frac{5}{18}\right)$ and $\left(\frac{1}{18}, -\frac{5}{9}\right)$

40. $\left(-\frac{4}{5}, \frac{9}{10}\right)$ and $\left(-\frac{3}{10}, \frac{1}{5}\right)$

Each table of values gives several points that lie on a line. Find the slope of the line.

41.

x	y
−1	8
0	6
2	2
3	0

42.

x	y
−3	6
−1	0
0	−3
2	−9

43.

x	y
−6	−4
−3	0
0	4
3	8

44.

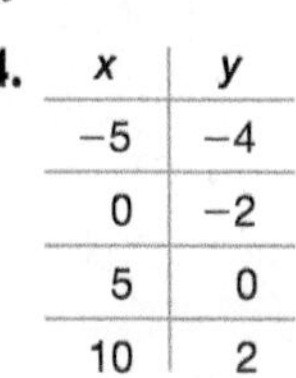

x	y
−5	−4
0	−2
5	0
10	2

Use the geometric interpretation of slope ("rise over run") *to find the slope of each line. (Coordinates of the points shown are integers.)*

45. **46.** **47.** **48.**

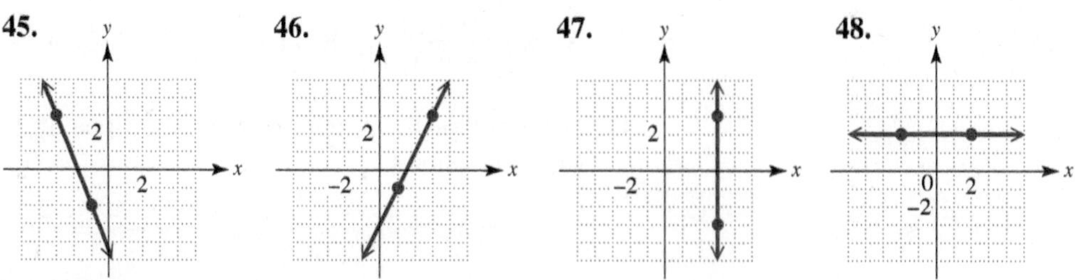

Find the slope of each line in three ways by doing the following.

(a) *Give any two points that lie on the line, and use them to determine the slope.* **See Example 2.**

(b) *Solve the equation for y, and identify the slope from the equation.* **See Example 4.**

(c) *For the form* $Ax + By = C$, *calculate* $-\frac{A}{B}$. **See the Note following Example 4.**

49. $2x - y = 8$ **50.** $3x - y = -6$ **51.** $3x + 4y = 12$

52. $6x + 5y = 30$ **53.** $x + y = -3$ **54.** $x - y = 4$

Find the slope of each line, and sketch its graph. **See Examples 1–4.**

55. $x + 2y = 4$ **56.** $x + 3y = -6$ **57.** $5x - 2y = 10$

58. $4x - y = 4$ **59.** $y = 4x$ **60.** $y = -3x$

61. $x - 3 = 0$ **62.** $x + 2 = 0$ **63.** $y = -5$

64. $y = -4$ **65.** $2y = 3$ **66.** $3x = 4$

Graph each line passing through the given point and having the given slope. **See Example 5.**

67. $(-4, 2)$; $m = \frac{1}{2}$ **68.** $(-2, -3)$; $m = \frac{5}{4}$ **69.** $(0, -2)$; $m = -\frac{2}{3}$

70. $(0, -4)$; $m = -\frac{3}{2}$ **71.** $(-1, -2)$; $m = 3$ **72.** $(-2, -4)$; $m = 4$

73. $(0, 0)$; $m = \frac{1}{5}$ **74.** $(0, 0)$; $m = \frac{5}{3}$ **75.** $(2, -5)$; $m = 0$

76. $(5, 3)$; $m = 0$ **77.** $(-3, 1)$; undefined slope **78.** $(-4, 1)$; undefined slope

79. Concept Check If a line has slope $-\frac{4}{9}$, then any line parallel to it has slope ______, and any line perpendicular to it has slope ______.

80. Concept Check If a line has slope 0.2, then any line parallel to it has slope ______, and any line perpendicular to it has slope ______.

Determine whether each pair of lines is parallel, perpendicular, *or* neither. ***See Examples 6 and 7.***

81. The line passing through $(15, 9)$ and $(12, -7)$ and the line passing through $(8, -4)$ and $(5, -20)$

82. The line passing through $(4, 6)$ and $(-8, 7)$ and the line passing through $(-5, 5)$ and $(7, 4)$

83. $x + 4y = 7$ and $4x - y = 3$

84. $2x + 5y = -7$ and $5x - 2y = 1$

85. $4x - 3y = 6$ and $3x - 4y = 2$

86. $2x + y = 6$ and $x - y = 4$

87. $x = 6$ and $6 - x = 8$

88. $3x = y$ and $2y - 6x = 5$

89. $4x + y = 0$ and $5x - 8 = 2y$

90. $2x + 5y = -8$ and $6 + 2x = 5y$

91. $4x - 3y = 8$ and $4y + 3x = 12$

92. $2x = y + 3$ and $2y + x = 3$

93. $x = 6$ and $y = 4$

94. $x + 1 = 0$ and $y - 2 = 0$

Extending Skills *Solve each problem.*

95. The upper deck at Guaranteed Rate Field in Chicago has produced, among other complaints, displeasure with its steepness. It is 160 ft from home plate to the front of the upper deck and 250 ft from home plate to the back. The top of the upper deck is 63 ft above the bottom. What is its slope?

250 ft
160 ft
63 ft

96. When designing the TD Garden in Boston, architects designed the ramps leading up to the entrances so that circus elephants would be able to walk up the ramps. The maximum grade (or slope) that an elephant will walk on is 13%. Suppose that such a ramp was constructed with a horizontal run of 150 ft. What would be the maximum vertical rise the architects could use?

Find and interpret the average rate of change illustrated in each graph. ***See Objective 5 and*** **FIGURE 27.**

97. **98.** **99.**

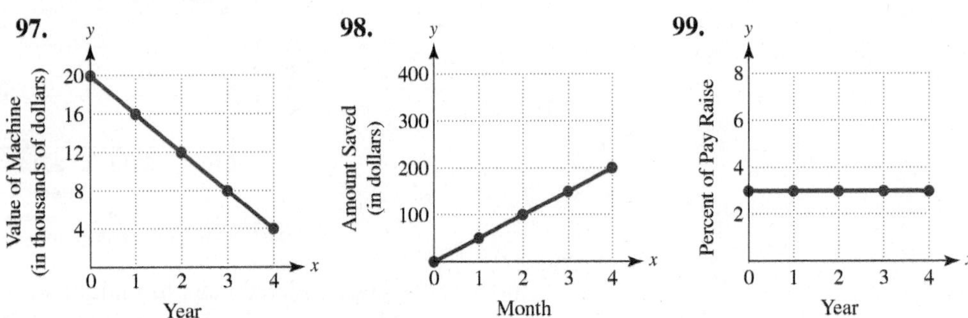

100. Concept Check If the graph of a linear equation rises from left to right, then the average rate of change is (*positive / negative*). If the graph of a linear equation falls from left to right, then the average rate of change is (*positive / negative*).

Solve each problem. ***See Examples 8 and 9.***

101. The graph shows the number of wireless subscriber connections (that is, active devices, including smartphones, feature phones, tablets, etc.) in millions in the United States for the years 2011 to 2016.

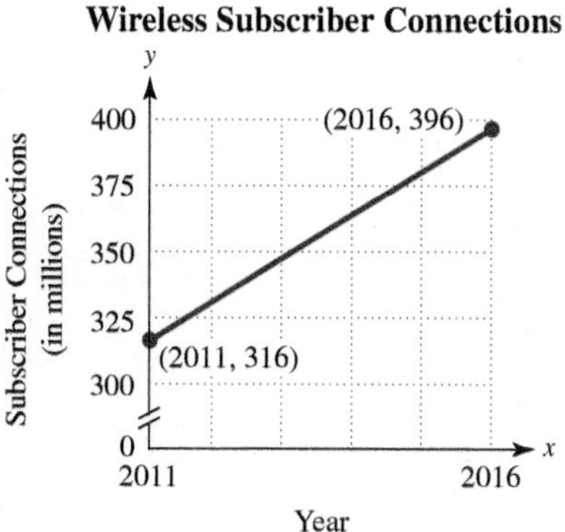

Data from CTIA.

(a) In the context of this graph, what does the ordered pair (2016, 396) mean?

(b) Use the given ordered pairs to find the slope of the line.

(c) Interpret the slope in the context of this problem.

102. The graph shows the percent of households in the United States that were wireless-only households for the years 2011 to 2016.

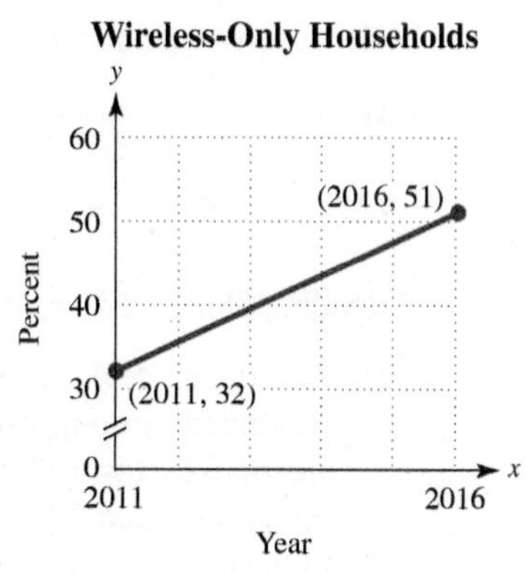

Data from CTIA.

(a) In the context of this graph, what does the ordered pair (2016, 51) mean?

(b) Use the given ordered pairs to find the slope of the line.

(c) Interpret the slope in the context of this problem.

103. The graph shows the number of drive-in movie theaters in the United States from 2010 through 2017.

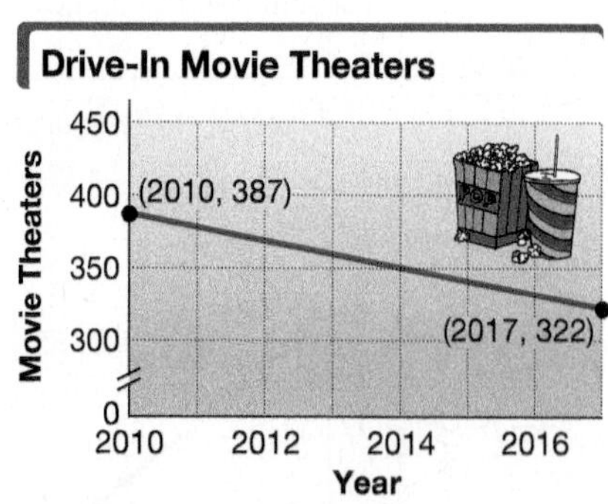

Data from www.drive-ins.com

(a) Use the given ordered pairs to find the average rate of change in the number of drive-in theaters per year during this period. Round the answer to the nearest whole number.

(b) Explain how a negative slope is interpreted in this situation.

104. The graph shows the number of U.S. travelers to Canada (in thousands) from 2000 through 2016.

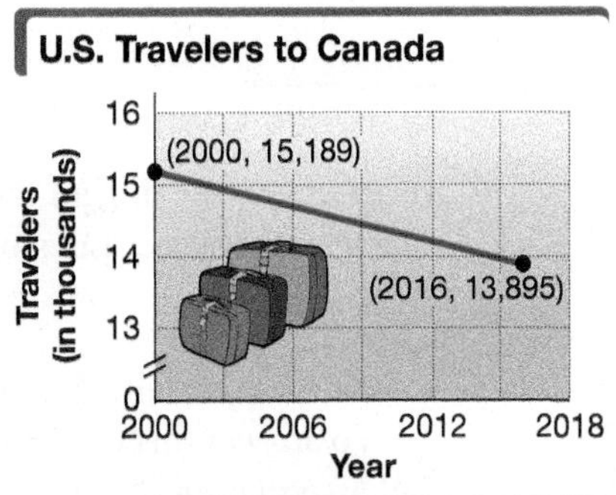

Data from U.S. Department of Commerce.

(a) Use the given ordered pairs to find the average rate of change in the number of U.S. travelers to Canada per year during this period. Round the answer to the nearest thousand.

(b) Explain how a negative slope is interpreted in this situation.

105. The average price of a gallon of unleaded gasoline in 2000 was \$1.51. In 2016, the average price was \$2.14. Find and interpret the average rate of change in the price of a gallon of gasoline per year to the nearest cent. (Data from Energy Information Administration.)

106. The average price of a movie ticket in 2004 was \$6.21. In 2016, the average price was \$8.65. Find and interpret the average rate of change in the price of a movie ticket per year to the nearest cent. (Data from Motion Picture Association of America.)

107. In 2010, the number of digital cameras shipped worldwide totaled 122 million. There were 24 million shipped in 2016. Find and interpret the average rate of change in the number of digital cameras shipped worldwide per year to the nearest million. (Data from CIPA.)

108. In 2010, worldwide shipments of desktop computers totaled 157 million units. In 2016, 103 million units were shipped. Find and interpret the average rate of change in worldwide shipments of desktop computers per year. (Data from IDC.)

Extending Skills *Use your knowledge of the slopes of parallel and perpendicular lines.*

109. Show that $(-13, -9)$, $(-11, -1)$, $(2, -2)$, and $(4, 6)$ are the vertices of a parallelogram. (*Hint:* A parallelogram is a four-sided figure with opposite sides parallel.)

110. Is the figure with vertices at $(-11, -5)$, $(-2, -19)$, $(12, -10)$, and $(3, 4)$ a parallelogram? Is it a rectangle? (*Hint:* A rectangle is a parallelogram with a right angle.)

RELATING CONCEPTS For Individual or Group Work (Exercises 111–116)

Three points that lie on the same straight line are said to be **collinear.** *Consider the points* $A(3, 1)$, $B(6, 2)$, *and* $C(9, 3)$. ***Work Exercises 111–116 in order.***

111. Find the slope of segment AB.

112. Find the slope of segment BC.

113. Find the slope of segment AC.

114. If slope of segment AB = slope of segment BC = slope of segment AC, then A, B, and C are collinear. Use the results of **Exercises 111–113** to show that this statement is satisfied.

115. Use the slope formula to determine whether the points $(1, -2)$, $(3, -1)$, and $(5, 0)$ are collinear.

116. Repeat **Exercise 115** for the points $(0, 6)$, $(4, -5)$, and $(-2, 12)$.

2.3 Writing Equations of Lines

OBJECTIVES

1 Write an equation of a line given its slope and y-intercept.
2 Graph a line using its slope and y-intercept.
3 Write an equation of a line given its slope and a point on the line.
4 Write an equation of a line given two points on the line.
5 Write equations of horizontal and vertical lines.
6 Write an equation of a line parallel or perpendicular to a given line.
7 Write an equation of a line that models real data.

OBJECTIVE 1 Write an equation of a line given its slope and y-intercept.

Recall that we can find the slope of a line from its equation by solving the equation for y. For example, the slope of the line with equation

$$y = 3x + 6$$

is 3, the coefficient of x. *What does the number 6 represent?*

To answer this question, suppose a line has slope m and y-intercept $(0, b)$. We can find an equation of this line by choosing another point (x, y) on the line, as shown in **FIGURE 30**, and applying the slope formula.

$$m = \frac{y - b}{x - 0} \quad \begin{array}{l}\leftarrow \text{Change in } y \\ \leftarrow \text{Change in } x\end{array}$$

$$m = \frac{y - b}{x} \quad \text{Subtract.}$$

$$mx = y - b \quad \text{Multiply by } x.$$

$$mx + b = y \quad \text{Add } b.$$

$$\boldsymbol{y = mx + b} \quad \text{Interchange sides.}$$

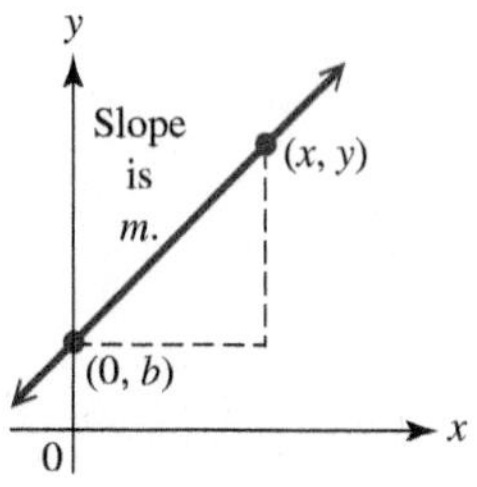

FIGURE 30

The last equation is the *slope-intercept form* of the equation of a line, because we can identify the slope m and y-intercept $(0, b)$ at a glance. In the line with equation $y = 3x + 6$, the number 6 indicates that the y-intercept is $(0, 6)$.

VOCABULARY

☐ scatter diagram

Slope-Intercept Form

The **slope-intercept form** of the equation of a line with slope m and y-intercept $(0, b)$ is

$$\boldsymbol{y = mx + b.}$$

Slope ↑ (for m); ↑ $(0, b)$ is the y-intercept.

EXAMPLE 1 Writing an Equation of a Line

Write an equation of the line with slope $-\frac{4}{5}$ and y-intercept $(0, -2)$.

Here, $m = -\frac{4}{5}$ and $b = -2$. Substitute these values into the slope-intercept form.

$$y = mx + b \quad \text{Slope-intercept form}$$

$$y = -\frac{4}{5}x + (-2) \quad \text{Let } m = -\tfrac{4}{5} \text{ and } b = -2.$$

$$y = -\frac{4}{5}x - 2 \quad \text{Definition of subtraction}$$

NOW TRY 

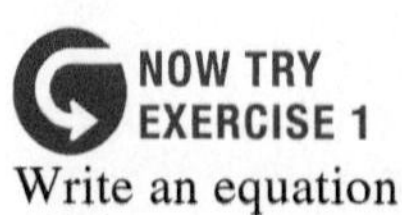
NOW TRY EXERCISE 1

Write an equation of the line with slope $\frac{2}{3}$ and y-intercept $(0, 1)$.

NOTE Every linear equation (of a nonvertical line) has a *unique* (one and only one) slope-intercept form. *Linear functions* are defined using slope-intercept form. Also, this form is used in graphing a line with a graphing calculator.

NOW TRY ANSWER

1. $y = \frac{2}{3}x + 1$

OBJECTIVE 2 Graph a line using its slope and y-intercept.

We first saw this approach in the previous section.

NOW TRY EXERCISE 2
Graph the line using its slope and y-intercept.

$$4x + 3y = 6$$

EXAMPLE 2 Graphing Lines Using Slope and y-Intercept

Graph each line using its slope and y-intercept.

(a) $y = 3x - 6$ (In slope-intercept form)

Here, $m = 3$ and $b = -6$. Plot the y-intercept $(0, -6)$. The slope 3 can be interpreted geometrically as follows.

$$m = \frac{\text{rise}}{\text{run}} = \frac{\text{change in } y}{\text{change in } x} = \frac{3}{1}$$

From $(0, -6)$, move *up* 3 units and to the *right* 1 unit, and plot a second point at $(1, -3)$. Join the two points with a straight line. See **FIGURE 31**.

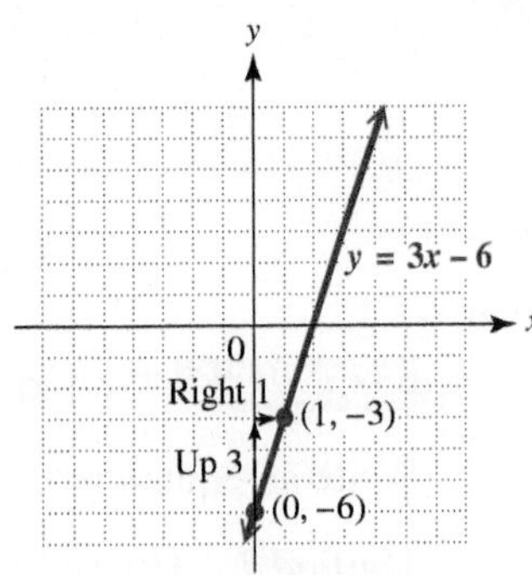

FIGURE 31

(b) $3y + 2x = 9$ (*Not* in slope-intercept form)

Write the equation in slope-intercept form by solving for y.

$$3y + 2x = 9$$

$$3y = -2x + 9 \quad \text{Subtract } 2x.$$

$$\text{Slope-intercept form} \longrightarrow y = -\frac{2}{3}x + 3 \quad \text{Divide by 3.}$$

Slope ↑ ↑ y-intercept is $(0, 3)$.

To graph this equation, plot the y-intercept $(0, 3)$. The slope can be interpreted as either $\frac{-2}{3}$ or $\frac{2}{-3}$. Using $\frac{-2}{3}$, begin at $(0, 3)$ and move *down* 2 units and to the *right* 3 units to locate the point $(3, 1)$. The line through these two points is the required graph. See **FIGURE 32**. (Verify that the point obtained using $\frac{2}{-3}$ as the slope is also on this line.)

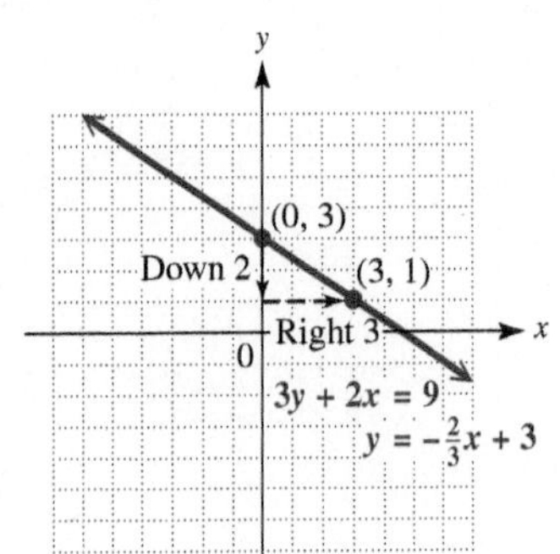

FIGURE 32

NOW TRY ANSWER

2. (graph) (0, 2), (3, −2), $4x + 3y = 6$, $y = -\frac{4}{3}x + 2$

NOW TRY 

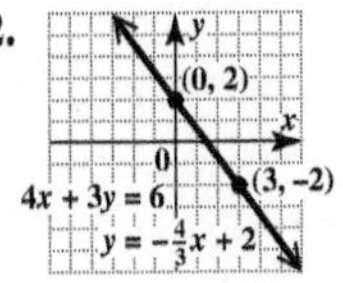

OBJECTIVE 3 Write an equation of a line given its slope and a point on the line.

Let m represent the slope of a line and (x_1, y_1) represent a given point on the line. Let (x, y) represent any other point on the line. See **FIGURE 33**.

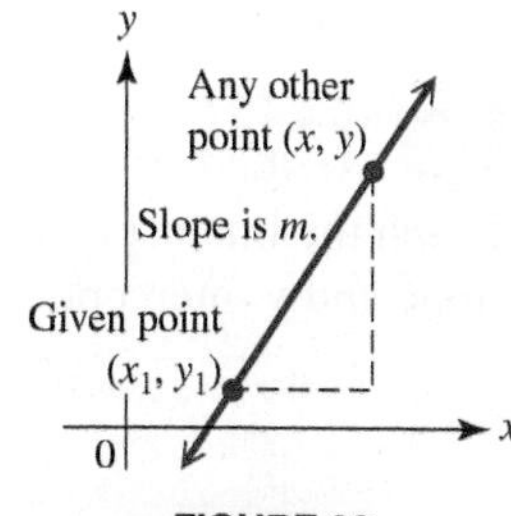

FIGURE 33

$$m = \frac{y - y_1}{x - x_1}$$ Slope formula

$$m(x - x_1) = y - y_1$$ Multiply each side by $x - x_1$.

$$\mathbf{y - y_1 = m(x - x_1)}$$ Interchange sides.

This is the *point-slope form* of the equation of a line.

Point-Slope Form

The **point-slope form** of the equation of a line with slope m passing through the point (x_1, y_1) is

Slope ↓

$$y - y_1 = m(x - x_1).$$

↑ Given point ↑

NOW TRY EXERCISE 3

Write an equation of the line with slope $-\frac{1}{5}$ passing through the point $(5, -3)$.

EXAMPLE 3 Writing an Equation of a Line Given Its Slope and a Point

Write an equation of the line with slope $\frac{1}{3}$ passing through the point $(-2, 5)$.

Method 1 **Use point-slope form.** Let $(x_1, y_1) = (-2, 5)$ and $m = \frac{1}{3}$.

$$y - y_1 = m(x - x_1)$$ Point-slope form

Leave x and y as variables.

$$y - 5 = \frac{1}{3}[x - (-2)]$$ Let $y_1 = 5$, $m = \frac{1}{3}$, and $x_1 = -2$.

$$y - 5 = \frac{1}{3}(x + 2)$$ Definition of subtraction

$$y - 5 = \frac{1}{3}x + \frac{2}{3}$$ (*) Distributive property

$$y = \frac{1}{3}x + \frac{2}{3} + \frac{15}{3}$$ Add $5 = \frac{15}{3}$.

Slope-intercept form ⟶ $$y = \frac{1}{3}x + \frac{17}{3}$$ $\frac{2}{3} + \frac{15}{3} = \frac{17}{3}$

Method 2 **Use slope-intercept form.** Let $(x, y) = (-2, 5)$ and $m = \frac{1}{3}$.

$$y = mx + b$$ Slope-intercept form

$$5 = \frac{1}{3}(-2) + b$$ Let $y = 5$, $m = \frac{1}{3}$, and $x = -2$.

Solve for b.

$$5 = -\frac{2}{3} + b$$ Multiply.

$5 = \frac{15}{3}$

$$\frac{17}{3} = b, \quad \text{or} \quad b = \frac{17}{3}$$ Add $\frac{2}{3}$; $\frac{15}{3} + \frac{2}{3} = \frac{17}{3}$

Substitute $m = \frac{1}{3}$ and $b = \frac{17}{3}$ to obtain $y = \frac{1}{3}x + \frac{17}{3}$, as above.

NOW TRY

NOW TRY ANSWER

3. $y = -\frac{1}{5}x - 2$

Previously we defined *standard form* for a linear equation.

$$Ax + By = C \quad \text{Standard form}$$

Here A, B, and C are real numbers and A and B are not both 0. ***For consistency, we give A, B, and C as integers with greatest common factor 1, and A ≥ 0.*** (If $A = 0$, then we give $B > 0$.) The equation in **Example 3** is written in standard form as follows.

$$y - 5 = \frac{1}{3}x + \frac{2}{3} \quad \text{Equation (*) from **Example 3**}$$

$$3y - 15 = x + 2 \quad \text{Multiply each term by 3.}$$

$$-x + 3y = 17 \quad \text{Subtract } x. \text{ Add 15.}$$

Standard form ⟶ $x - 3y = -17$ Multiply by -1.

NOTE "Standard form" is not standard among texts. A linear equation can be written correctly in many different ways.

Example:

$2x + 3y = 8$, (Standard form as described above)

$2x = 8 - 3y$, $3y = 8 - 2x$, $x + \frac{3}{2}y = 4$, $4x + 6y = 16$ (Equivalent forms of $2x + 3y = 8$)

OBJECTIVE 4 Write an equation of a line given two points on the line.

NOW TRY EXERCISE 4
Write an equation of the line passing through the points $(3, -4)$ and $(-2, -1)$. Give the final answer in standard form.

EXAMPLE 4 Writing an Equation of a Line Given Two Points

Write an equation of the line passing through the points $(-4, 3)$ and $(5, -7)$. Give the final answer in standard form.

First find the slope using the slope formula.

$$m = \frac{-7 - 3}{5 - (-4)} = -\frac{10}{9}$$

Use either $(-4, 3)$ or $(5, -7)$ as (x_1, y_1) in the point-slope form of the equation of a line. We choose $(-4, 3)$, so $-4 = x_1$ and $3 = y_1$.

$$y - y_1 = m(x - x_1) \quad \text{Point-slope form}$$

$$y - 3 = -\frac{10}{9}[x - (-4)] \quad \text{Let } y_1 = 3, m = -\tfrac{10}{9}, \text{ and } x_1 = -4.$$

$$y - 3 = -\frac{10}{9}(x + 4) \quad \text{Definition of subtraction}$$

$$y - 3 = -\frac{10}{9}x - \frac{40}{9} \quad \text{Distributive property}$$

$$9y - 27 = -10x - 40 \quad \text{Multiply each term by 9.}$$

Standard form ⟶ $10x + 9y = -13$ Add $10x$. Add 27.

Verify that if $(5, -7)$ were used, the same equation would result. **NOW TRY**

NOW TRY ANSWER
4. $3x + 5y = -11$

NOTE Once the slope is found in **Example 4,** the equation of the line could also be determined using slope-intercept form.

OBJECTIVE 5 Write equations of horizontal and vertical lines.

A horizontal line has slope 0. Using point-slope form, we can find the equation of a horizontal line through the point (a, b).

$$y - y_1 = m(x - x_1) \quad \text{Point-slope form}$$

$$y - b = 0(x - a) \quad y_1 = b, m = 0, x_1 = a$$

$$y - b = 0 \quad \text{Multiplication property of 0}$$

$$\text{Horizontal line} \longrightarrow \boldsymbol{y = b} \quad \text{Add } b.$$

Point-slope form does not apply to a vertical line because the slope of a vertical line is undefined. A vertical line through the point (a, b) has equation $\boldsymbol{x = a}$.

Equations of Horizontal and Vertical Lines

A **horizontal line** through the point (a, b) has equation $\boldsymbol{y = b}$.

A **vertical line** through the point (a, b) has equation $\boldsymbol{x = a}$.

NOW TRY EXERCISE 5

Write an equation of the line passing through the point $(4, -4)$ and satisfying the given condition.

(a) The line has undefined slope.

(b) The line has slope 0.

EXAMPLE 5 Writing Equations of Horizontal and Vertical Lines

Write an equation of the line passing through the point $(-3, 3)$ and satisfying the given condition.

(a) The line has slope 0.

Because the slope is 0, this is a horizontal line. A horizontal line through the point (a, b) has equation $y = b$. In $(-3, 3)$, the y-coordinate is 3, so the equation is $y = 3$.

(b) The line has undefined slope.

This is a vertical line because the slope is undefined. A vertical line through the point (a, b) has equation $x = a$. In $(-3, 3)$, the x-coordinate is -3, so the equation is $x = -3$.

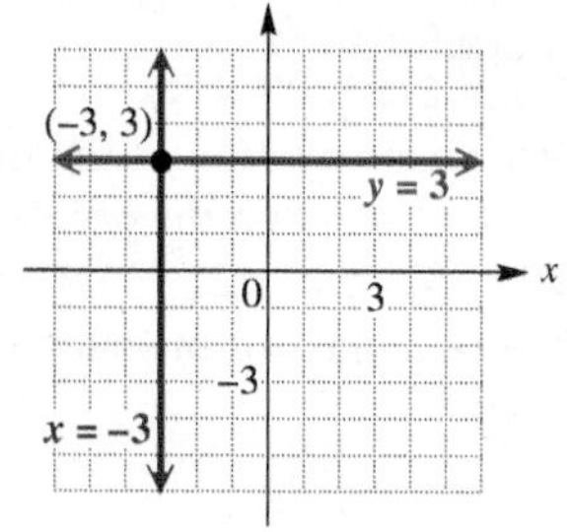

FIGURE 34

Both lines are graphed in FIGURE 34.

NOW TRY

OBJECTIVE 6 Write an equation of a line parallel or perpendicular to a given line.

EXAMPLE 6 Writing Equations of Parallel or Perpendicular Lines

Write an equation of the line passing through the point $(-3, 6)$ and satisfying the given condition. Give final answers in slope-intercept form.

(a) The line is parallel to the line $2x + 3y = 6$.

We can find the slope of the given line by solving for y.

$$2x + 3y = 6$$

$$3y = -2x + 6 \quad \text{Subtract } 2x.$$

$$y = -\frac{2}{3}x + 2 \quad \text{Divide by 3.}$$

($-\frac{2}{3}$ ← Slope)

NOW TRY ANSWERS

5. **(a)** $x = 4$ **(b)** $y = -4$

NOW TRY EXERCISE 6
Write an equation of the line passing through the point $(6, -1)$ and satisfying the given condition. Give final answers in slope-intercept form.

(a) The line is parallel to the line $3x - 5y = 7$.

(b) The line is perpendicular to the line $3x - 5y = 7$.

The slope of the line is given by the coefficient of x, so $m = -\frac{2}{3}$. See **FIGURE 35**.

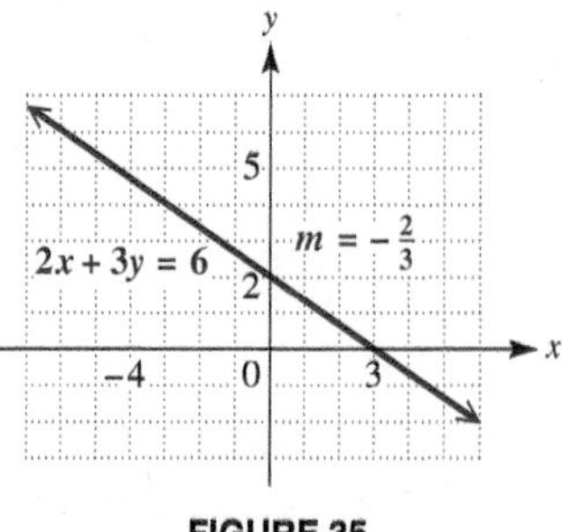

FIGURE 35

The required equation of the line through $(-3, 6)$ and *parallel* to $2x + 3y = 6$ must have the *same* slope $-\frac{2}{3}$. To find this equation, we use the point-slope form with $(x_1, y_1) = (-3, 6)$ and $m = -\frac{2}{3}$.

$$y - y_1 = m(x - x_1) \quad \text{Point-slope form}$$

$$y - 6 = -\frac{2}{3}[x - (-3)] \quad y_1 = 6, m = -\frac{2}{3}, x_1 = -3$$

$$y - 6 = -\frac{2}{3}(x + 3) \quad \text{Definition of subtraction}$$

$$y - 6 = -\frac{2}{3}x - 2 \quad \text{Distributive property}$$

$$y = -\frac{2}{3}x + 4 \quad \text{Add 6.}$$

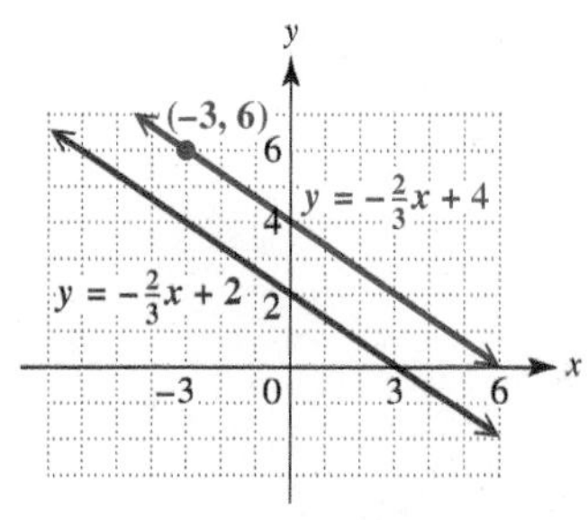

Both lines have slope $-\frac{2}{3}$. The lines are parallel.

FIGURE 36

We did not clear the fraction because we want the final equation in slope-intercept form—that is, solved for y. Both lines are shown in **FIGURE 36**.

(b) The line is perpendicular to the line $2x + 3y = 6$.

In part (a), we wrote the equation of the given line in slope-intercept form.

$$2x + 3y = 6$$

$$y = -\frac{2}{3}x + 2 \quad \text{Slope-intercept form}$$

↑ Slope

To be *perpendicular* to the line $2x + 3y = 6$, a line must have slope $\frac{3}{2}$, the *negative reciprocal* of $-\frac{2}{3}$.

We use $(-3, 6)$ and slope $\frac{3}{2}$ in the point-slope form to find the equation of the perpendicular line shown in **FIGURE 37**.

$$y - y_1 = m(x - x_1) \quad \text{Point-slope form}$$

$$y - 6 = \frac{3}{2}[x - (-3)] \quad y_1 = 6, m = \frac{3}{2}, x_1 = -3$$

$$y - 6 = \frac{3}{2}(x + 3) \quad \text{Definition of subtraction}$$

$$y - 6 = \frac{3}{2}x + \frac{9}{2} \quad \text{Distributive property}$$

$$y = \frac{3}{2}x + \frac{9}{2} + \frac{12}{2} \quad \text{Add } 6 = \frac{12}{2}.$$

$$y = \frac{3}{2}x + \frac{21}{2} \quad \frac{9}{2} + \frac{12}{2} = \frac{21}{2}$$

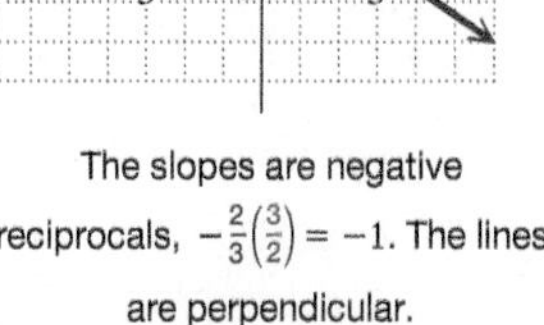

The slopes are negative reciprocals, $-\frac{2}{3}\left(\frac{3}{2}\right) = -1$. The lines are perpendicular.

FIGURE 37

NOW TRY

NOW TRY ANSWERS

6. (a) $y = \frac{3}{5}x - \frac{23}{5}$

(b) $y = -\frac{5}{3}x + 9$

▼ Summary of Forms of Linear Equations

Equation	Description	When to Use
$y = mx + b$	**Slope-Intercept Form** Slope is m. y-intercept is $(0, b)$.	The slope and y-intercept can be easily identified and used to quickly graph the equation.
$y - y_1 = m(x - x_1)$	**Point-Slope Form** Slope is m. Line passes through (x_1, y_1).	This form is ideal for finding the equation of a line if the slope and a point on the line or two points on the line are known.
$Ax + By = C$	**Standard Form** (A, B, and C integers, $A \geq 0$) Slope is $-\frac{A}{B}$ $(B \neq 0)$. x-intercept is $\left(\frac{C}{A}, 0\right)$ $(A \neq 0)$. y-intercept is $\left(0, \frac{C}{B}\right)$ $(B \neq 0)$.	The x- and y-intercepts can be found quickly and used to graph the equation. The slope must be calculated.
$y = b$	**Horizontal Line** Slope is 0. y-intercept is $(0, b)$.	If the graph intersects only the y-axis, then y is the only variable in the equation.
$x = a$	**Vertical Line** Slope is undefined. x-intercept is $(a, 0)$.	If the graph intersects only the x-axis, then x is the only variable in the equation.

OBJECTIVE 7 Write an equation of a line that models real data.

If a given set of data changes at a fairly constant rate, the data may fit a linear pattern, where the rate of change is the slope of the line.

EXAMPLE 7 Writing a Linear Equation to Describe Real Data

A local gasoline station is selling 89-octane gas for \$3.50 per gal.

(a) Write an equation that describes the cost y to buy x gallons of gas.

The total cost is determined by the number of gallons multiplied by the price per gallon (in this case, \$3.50). As the gas is pumped, two sets of numbers spin by: the number of gallons pumped and the cost of that number of gallons.

The table illustrates this situation.

Number of Gallons Pumped	Cost of This Number of Gallons
0	0(\$3.50) = \$ 0.00
1	1(\$3.50) = \$ 3.50
2	2(\$3.50) = \$ 7.00
3	3(\$3.50) = \$10.50
4	4(\$3.50) = \$14.00

If we let x denote the number of gallons pumped, then the total cost y in dollars can be found using the following linear equation.

Total cost ↘ ↙ Number of gallons

$$y = 3.50x$$

Theoretically, there are infinitely many ordered pairs (x, y) that satisfy this equation, but here we are limited to nonnegative values for x because we cannot have a negative number of gallons. In this situation, there is also a practical maximum value for x, which varies from one car to another—the size of the gas tank.

NOW TRY EXERCISE 7

A cell phone plan costs $100 for the telephone plus $85 per month for service. Write an equation that gives the cost y in dollars for x months of cell phone service using this plan.

(b) A car wash at this gas station costs an additional $3.00. Write an equation that defines the cost of gas and a car wash.

The cost will be $3.50x + 3.00$ dollars for x gallons of gas and a car wash.

$$y = 3.5x + 3 \quad \text{Final 0's need not be included.}$$

(c) Interpret the ordered pairs $(5, 20.5)$ and $(10, 38)$ in the context of the equation from part (b).

$(5, 20.5)$ indicates that 5 gal of gas and a car wash cost $20.50.

$(10, 38)$ indicates that 10 gal of gas and a car wash cost $38.00.

NOW TRY 

NOTE In **Example 7(a),** the ordered pair (0, 0) satisfied the equation $y = 3.50x$, so the linear equation has the form

$$y = mx, \quad \text{where } b = 0.$$

If a realistic situation involves an initial charge b plus a charge per unit m, as in the equation $y = 3.5x + 3$ in **Example 7(b),** then the equation has the form

$$y = mx + b, \quad \text{where } b \neq 0.$$

EXAMPLE 8 Writing an Equation of a Line That Models Data

Average annual tuition and fees for in-state students at public two-year colleges are shown in the table for selected years and graphed as ordered pairs of points in the **scatter diagram** in **FIGURE 38**, where $x = 0$ represents 2012, $x = 1$ represents 2013, and so on, and y represents the cost in dollars.

Year	*Cost (in dollars)*
2012 ($x = 0$)	3216
2013 ($x = 1$)	3264
2014 ($x = 2$)	3347
2015 ($x = 3$)	3435
2016 ($x = 4$)	3520
2017 ($x = 5$)	3570

Data from The College Board.

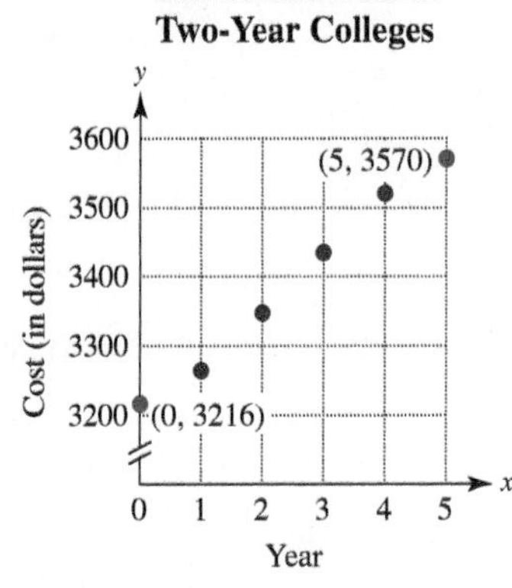

FIGURE 38

(a) Write an equation that models the data.

Because the points in **FIGURE 38** lie approximately on a straight line, we can write a linear equation that models the relationship between year x and cost y. We choose two data points, (0, 3216) and (5, 3570), to find the slope of the line.

$$m = \frac{3570 - 3216}{5 - 0} = \frac{354}{5} = 70.8$$

The slope 70.8 indicates that the cost of tuition and fees increased by about $71 per year from 2012 to 2017. We use this slope and the y-intercept (0, 3216) to write an equation of the line in slope-intercept form.

$$y = 70.8x + 3216$$

NOW TRY ANSWER

7. $y = 85x + 100$

NOW TRY EXERCISE 8

Refer to **Example 8.**

(a) Using the data values for the years 2012 and 2016, write an equation that models the data.

(b) Use the equation from part (a) to approximate the cost of tuition and fees in 2015.

(b) Use the equation from part (a) to approximate the cost of tuition and fees in 2018.

The value $x = 6$ corresponds to the year 2018.

$$y = 70.8x + 3216 \quad \text{Equation from part (a)}$$

$$y = 70.8(6) + 3216 \quad \text{Substitute 6 for } x.$$

$$y = 3640.8 \quad \text{Multiply, and then add.}$$

According to the model, average tuition and fees for in-state students at public two-year colleges in 2018 were about \$3641.

NOW TRY

NOTE Choosing different data points in **Example 8** would result in a slightly different line (particularly in regard to its slope) and, hence, a slightly different equation. However, all such equations should yield similar results. See **Now Try Exercise 8.**

EXAMPLE 9 Writing an Equation of a Line That Models Data

Hospital expenditures (in billions of dollars) in the United States are shown in the graph in **FIGURE 39.**

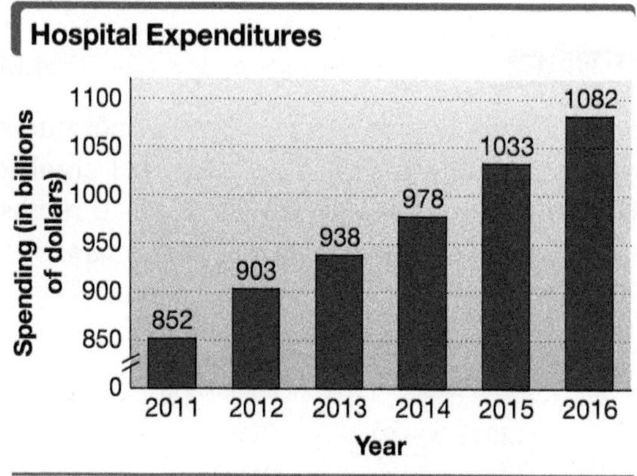

Data from Centers for Medicare & Medicaid Services.

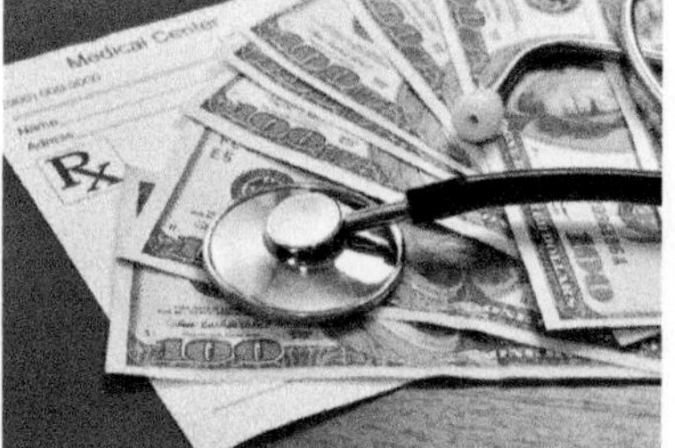

FIGURE 39

(a) Write an equation that models the data.

The data increase linearly—that is, a straight line through the tops of any two bars in the graph would be close to the top of each bar. To model the relationship between year x and hospital expenditures y, we let $x = 1$ represent 2011, $x = 2$ represent 2012, and so on. The given data for 2011 and 2016 can be written as the ordered pairs $(1, 852)$ and $(6, 1082)$.

$$m = \frac{1082 - 852}{6 - 1} = \frac{230}{5} = 46 \quad \text{Find the slope of the line through (1, 852) and (6, 1082).}$$

Thus, hospital expenditures increased by \$46 billion per year. To write an equation, we substitute this slope and the point $(1, 852)$ into the point-slope form.

$$y - y_1 = m(x - x_1) \quad \text{Point-slope form}$$

$$y - 852 = 46(x - 1) \quad (x_1, y_1) = (1, 852);\ m = 46$$

Either point can be used here. (6, 1082) provides the same answer.

$$y - 852 = 46x - 46 \quad \text{Distributive property}$$

$$y = 46x + 806 \quad \text{Add 852.}$$

NOW TRY ANSWERS

8. **(a)** $y = 76x + 3216$
(b) \$3444

NOW TRY EXERCISE 9

Refer to **Example 9.**

(a) Use the ordered pairs $(4, 978)$ and $(6, 1082)$ to write an equation that models the data.

(b) Use the equation from part (a) to estimate hospital expenditures in 2019.

(b) Use the equation from part (a) to estimate hospital expenditures in the United States in 2019. (Assume a constant rate of change.)

Because we let $x = 1$ represent 2011, $x = 9$ represents 2019.

$$y = 46x + 806 \quad \text{Equation from part (a)}$$

$$y = 46(9) + 806 \quad \text{Substitute 9 for } x.$$

$$y = 1220 \quad \text{Multiply, and then add.}$$

Hospital expenditures in the United States in 2019 were about $1220 billion.

NOW TRY ANSWERS

9. **(a)** $y = 52x + 770$
(b) $1238 billion

2.3 Exercises

FOR EXTRA HELP **MyLab Math**

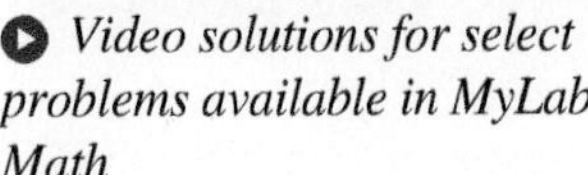

Video solutions for select problems available in MyLab Math

STUDY SKILLS REMINDER

How are you doing on your homework? **Review Study Skill 4, *Completing Your Homework.***

Concept Check *Provide the appropriate response.*

1. The following equations all represent the same line. Which one is in standard form as specified in this section?

A. $3x - 2y = 5$ **B.** $y = \frac{3}{2}x - \frac{5}{2}$ **C.** $\frac{3}{5}x - \frac{2}{5}y = 1$ **D.** $3x = 2y + 5$

2. Which equation is in point-slope form?

A. $y = 6x + 2$ **B.** $4x + y = 9$ **C.** $y - 3 = 2(x - 1)$ **D.** $2y = 3x - 7$

3. Which equation in **Exercise 2** is in slope-intercept form?

4. Write the equation $y + 2 = -3(x - 4)$ in slope-intercept form.

5. Write the equation $y + 2 = -3(x - 4)$ in standard form.

6. Write the equation $10x - 7y = 70$ in slope-intercept form.

Concept Check *Match each equation with the graph that it most closely resembles. (Hint: Determining the signs of m and b will help in each case.)*

7. $y = 2x + 3$

8. $y = -2x + 3$

9. $y = -2x - 3$

10. $y = 2x - 3$

11. $y = 2x$

12. $y = -2x$

13. $y = 3$

14. $y = -3$

A.

B.

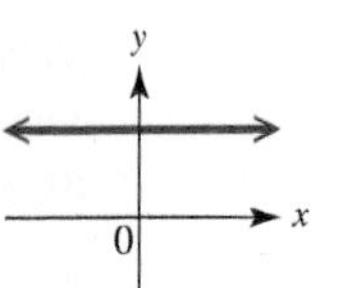

C.

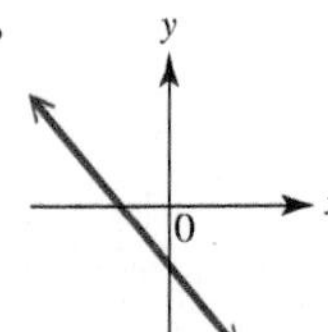

D.

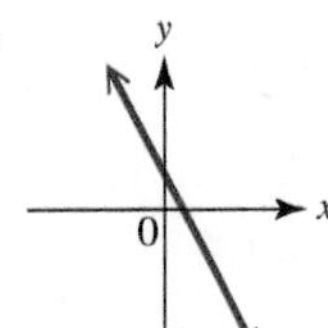

E.

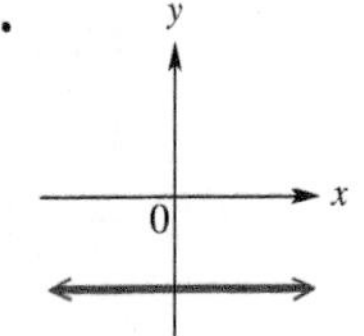

F.

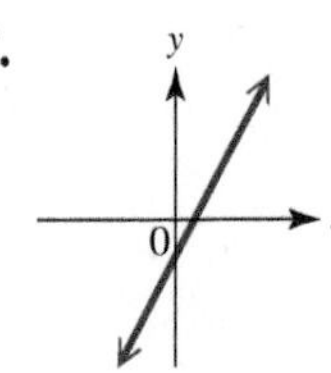

G.

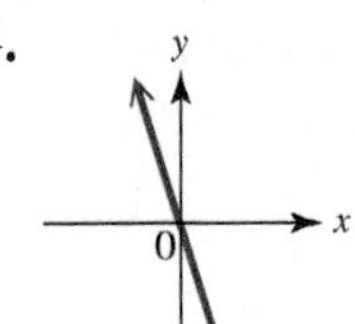

H.

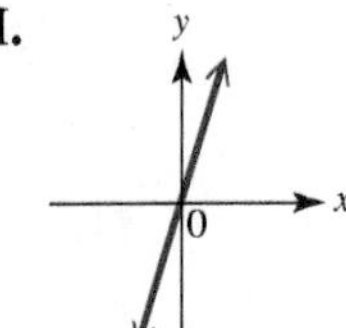

I.

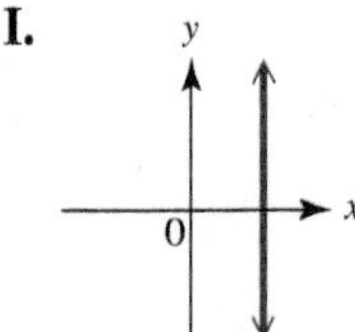

Write an equation in slope-intercept form of the line that satisfies the given conditions. ***See Example 1.***

15. $m = 5;\ b = 15$

16. $m = 2;\ b = 12$

17. $m = -\frac{2}{3};\ b = \frac{4}{5}$

18. $m = -\frac{5}{8};\ b = -\frac{1}{3}$

19. Slope 1; y-intercept $(0, -1)$

20. Slope -1; y-intercept $(0, -3)$

21. Slope $\frac{2}{5}$; y-intercept $(0, 5)$

22. Slope $-\frac{3}{4}$; y-intercept $(0, 7)$

Write an equation in slope-intercept form of the line shown in each graph. (Coordinates of the points shown are integers.)

23.

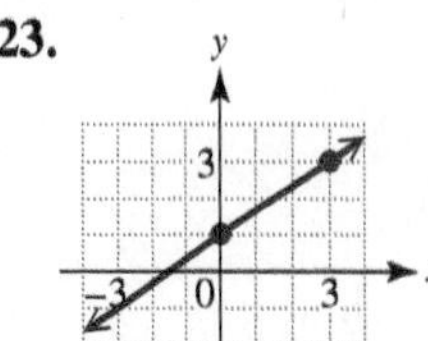

24.

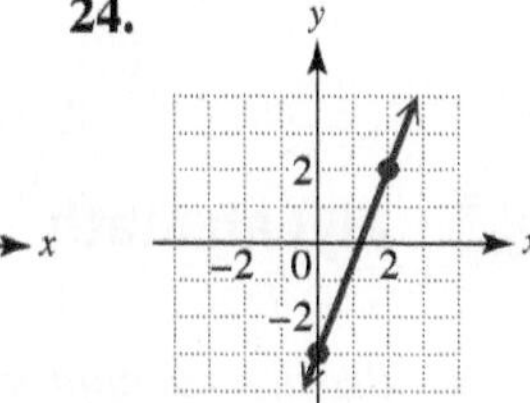

25.

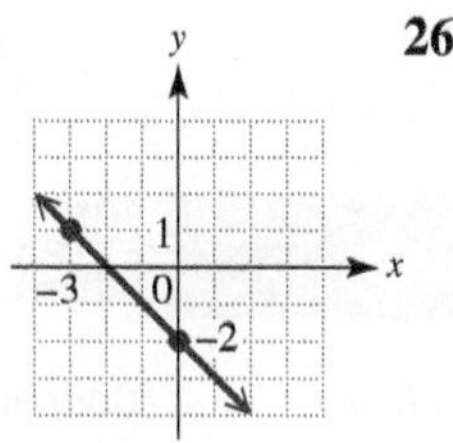

26.

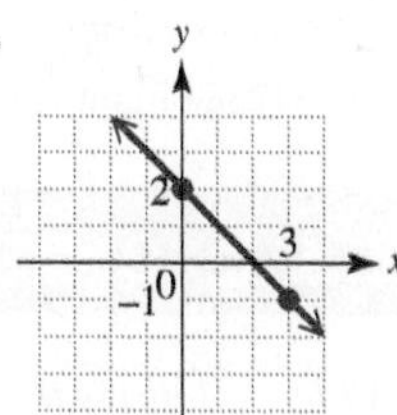

Each table of values gives several points that lie on a line. Write an equation in slope-intercept form of the line.

27.

x	y
−2	−8
0	−4
1	−2
3	2

28.

x	y
−2	−3
0	3
2	9
3	12

29.

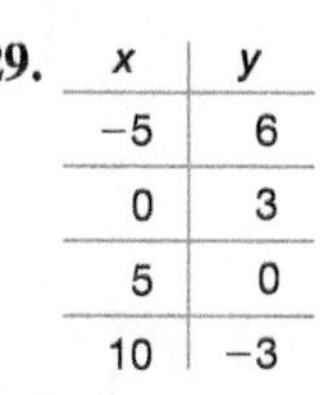

x	y
−5	6
0	3
5	0
10	−3

30.

x	y
−4	5
−2	0
0	−5
2	−10

For each equation, ***(a)*** *write it in slope-intercept form,* ***(b)*** *give the slope of the line,* ***(c)*** *give the y-intercept, and* ***(d)*** *graph the line.* ***See Example 2.***

31. $-x + y = 4$

32. $-x + y = 6$

33. $6x + 5y = 30$

34. $3x + 4y = 12$

35. $4x - 5y = 20$

36. $7x - 3y = 3$

37. $x + 2y = -4$

38. $x + 3y = -9$

Write an equation of the line passing through the given point and having the given slope. Give the equation ***(a)*** *in slope-intercept form and* ***(b)*** *in standard form.* ***See Example 3 and the discussion on standard form.***

39. $(5, 8)$; slope -2

40. $(12, 10)$; slope 1

41. $(-2, 4)$; slope $-\frac{3}{4}$

42. $(-1, 6)$; slope $-\frac{5}{6}$

43. $(-5, 4)$; slope $\frac{1}{2}$

44. $(7, -2)$; slope $\frac{1}{4}$

45. $(3, 0)$; slope 4

46. $(-2, 0)$; slope -5

47. $(2, 6.8)$; slope 1.4

48. $(6, -1.2)$; slope 0.8

Write an equation of the line passing through the given points. Give the final answer in standard form. ***See Example 4.***

49. $(3, 4)$ and $(5, 8)$

50. $(5, -2)$ and $(-3, 14)$

51. $(6, 1)$ and $(-2, 5)$

52. $(-2, 5)$ and $(-8, 1)$

53. $(2, 5)$ and $(1, 5)$

54. $(-2, 2)$ and $(4, 2)$

55. $(7, 6)$ and $(7, -8)$

56. $(13, 5)$ and $(13, -1)$

57. $\left(\frac{1}{2}, -3\right)$ and $\left(-\frac{2}{3}, -3\right)$

58. $\left(-\frac{4}{9}, -6\right)$ and $\left(\frac{12}{7}, -6\right)$

59. $\left(-\frac{2}{5}, \frac{2}{5}\right)$ and $\left(\frac{4}{3}, \frac{2}{3}\right)$

60. $\left(\frac{3}{4}, \frac{8}{3}\right)$ and $\left(\frac{2}{5}, \frac{2}{3}\right)$

Write an equation of the line passing through the given point and satisfying the given condition. ***See Example 5.***

61. $(9, 5)$; slope 0

62. $(-4, -2)$; slope 0

63. $(9, 10)$; undefined slope

64. $(-2, 8)$; undefined slope

65. $\left(-\frac{3}{4}, -\frac{3}{2}\right)$; slope 0

66. $\left(-\frac{5}{8}, -\frac{9}{2}\right)$; slope 0

67. $(-7, 8)$; horizontal

68. $(2, -7)$; horizontal

69. $(0.5, 0.2)$; vertical

70. $(0.1, 0.4)$; vertical

71. $(0, 0)$; horizontal

72. $(0, 0)$; vertical

Write an equation of the line passing through the given point and satisfying the given condition. Give the equation ***(a)*** *in slope-intercept form and* ***(b)*** *in standard form.* ***See Example 6.***

73. $(7, 2)$; parallel to $3x - y = 8$

74. $(4, 1)$; parallel to $2x + 5y = 10$

75. $(-2, -2)$; parallel to $-x + 2y = 10$

76. $(-1, 3)$; parallel to $-x + 3y = 12$

77. $(8, 5)$; perpendicular to $2x - y = 7$

78. $(2, -7)$; perpendicular to $5x + 2y = 18$

79. $(-2, 7)$; perpendicular to $x = 9$

80. $(8, 4)$; perpendicular to $x = -3$

Write an equation in the form $y = mx$ for each situation. Then give the three ordered pairs associated with the equation for x-values 0, 5, *and* 10. ***See Example 7(a).***

81. x represents the number of hours traveling at 45 mph, and y represents the distance traveled (in miles).

82. x represents the number of caps sold at \$26 each, and y represents the total cost of the caps (in dollars).

83. x represents the number of gallons of gas sold at \$3.75 per gal, and y represents the total cost of the gasoline (in dollars).

84. x represents the number of hours a bicycle is rented at \$7.50 per hour, and y represents the total charge for the rental (in dollars).

85. x represents the number of credit hours taken at a community college at \$150 per credit hour, and y represents the total tuition paid for the credit hours (in dollars).

86. x represents the number of tickets to a performance of *Hamilton* purchased at \$250 per ticket, and y represents the total paid for the tickets (in dollars).

For each situation, do the following.

(a) *Write an equation in the form $y = mx + b$.*

(b) *Find and interpret the ordered pair associated with the equation for $x = 5$.*

(c) *Answer the question posed in the problem.*

See Examples 7(b) and 7(c).

87. A ticket for Taylor Swift's Reputation Stadium Tour costs \$140. There is a ticket order delivery fee of \$18.50. Let x represent the number of tickets and y represent the cost in dollars. How much does it cost to purchase 2 tickets and have them delivered? (Data from Ticketmaster.)

88. Resident tuition at North Shore Community College is \$206 per credit hour. There is also a \$300 program fee for physical therapy. Let x represent the number of credit hours and y represent the cost in dollars. How much does it cost for a student in physical therapy to take 15 credit hours? (Data from www.northshore.edu)

89. A health club membership costs \$99, plus \$41 per month. Let x represent the number of months and y represent the cost in dollars. How much does the first year's membership cost? (Data from Midwest Athletic Club.)

90. An Executive VIP/Gold membership to a health club costs \$159, plus \$57 per month. Let x represent the number of months and y represent the cost in dollars. How much does a one-year membership cost? (Data from Midwest Athletic Club.)

91. A wireless plan includes unlimited talk, text, and data for \$90 per month. There is a \$25 activation fee. Let x represent the number of months and y represent the cost in dollars. Over two years, how much will this plan cost?

92. A wireless plan includes unlimited talk, text, and data for \$75 per month. There is a \$350 charge for a phone. Let x represent the number of months and y represent the cost in dollars. Over two years, how much will this plan cost?

93. There is a \$30 fee to rent a chain saw, plus \$6 per day. Let x represent the number of days the saw is rented and y represent the charge to the user in dollars. If the total charge is \$138, for how many days is the saw rented?

94. A rental car costs \$99, plus \$0.50 per mile. Let x represent the number of miles driven and y represent the total charge to the renter in dollars. How many miles was the car driven if the renter paid \$174.00?

Solve each problem. Give equations in slope-intercept form.
See Examples 8 and 9.

95. Total sales of e-readers in the United States (in millions of dollars) are shown in the graph, where the year 2013 corresponds to $x = 0$.

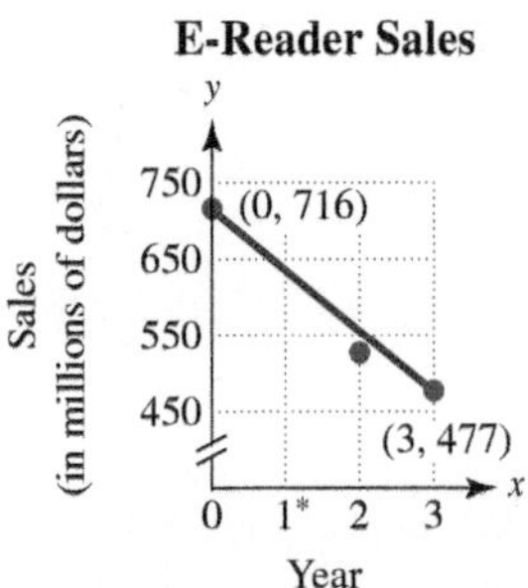

*Data for this year are unavailable.
Data from Consumer Technology Association.

(a) Use the ordered pairs from the graph to write an equation that models the data. (Round the slope to the nearest tenth.) Interpret the slope in the context of this problem.

(b) Use the equation from part (a) to approximate sales of e-readers in the United States in 2014, the year data was unavailable.

96. Total sales of smartphones in the United States (in billions of dollars) are shown in the graph, where the year 2013 corresponds to $x = 0$.

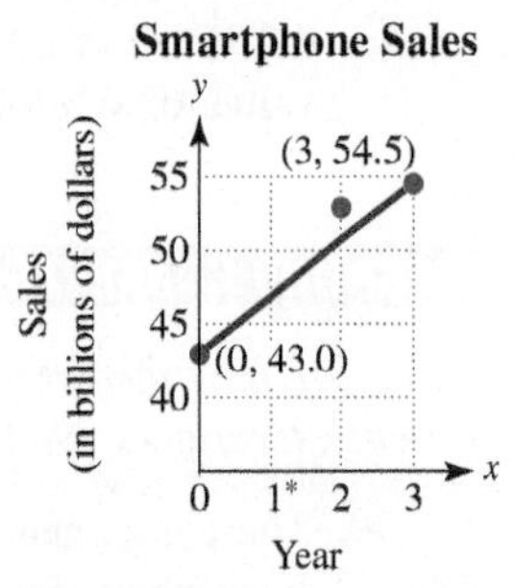

*Data for this year are unavailable.
Data from Consumer Technology Association.

(a) Use the ordered pairs from the graph to write an equation that models the data. (Round the slope to the nearest tenth.) Interpret the slope in the context of this problem.

(b) Use the equation from part (a) to approximate smartphone sales in the United States in 2014, the year data was unavailable.

97. Expenditures for home health care in the United States are shown in the graph.

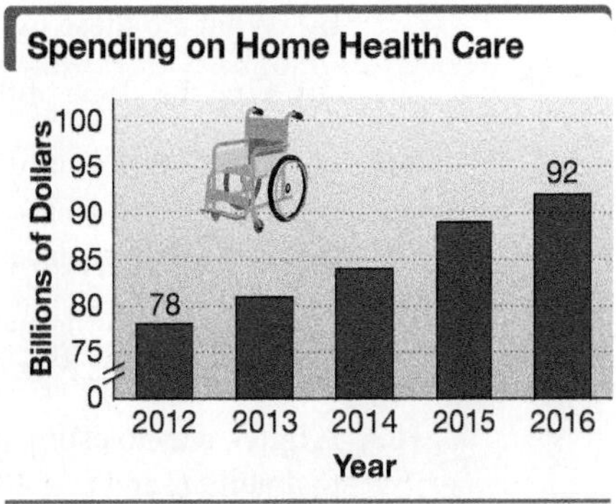

Data from Centers for Medicare & Medicaid Services.

(a) If a straight line were drawn through the tops of the bars for 2012 and 2016, would this line have positive or negative slope? Explain.

(b) Use the information given for the years 2012 and 2016, letting $x = 12$ represent 2012 and $x = 16$ represent 2016, and letting y represent spending (in billions of dollars), to write an equation that models the data.

(c) Use the equation from part (b) to approximate the amount spent on home health care in 2015. How does the result compare with the actual value, \$89 billion?

98. The number of pieces of first-class mail delivered in the United States is shown in the graph.

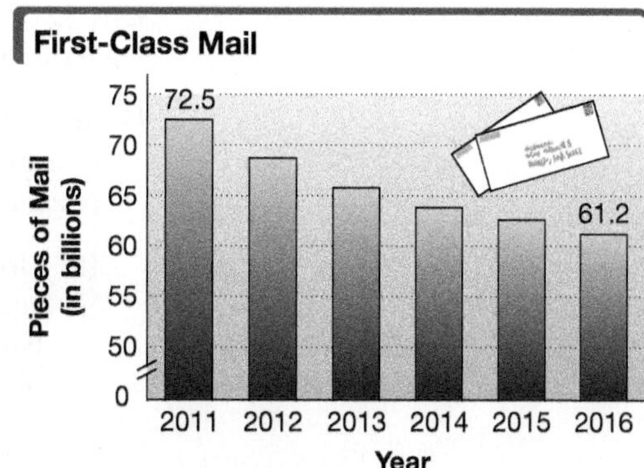

Data from U.S. Postal Service.

(a) If a straight line were drawn through the tops of the bars for 2011 and 2016, would this line have positive or negative slope? Explain.

(b) Use the information given for the years 2011 and 2016, letting $x = 11$ represent 2011 and $x = 16$ represent 2016, and letting y represent the number of pieces of mail (in billions), to write an equation that models the data.

(c) Use the equation from part (b) to approximate the number of pieces of first-class mail delivered in 2015 to the nearest tenth. How does this result compare to the actual value, 62.6 billion?

RELATING CONCEPTS For Individual or Group Work (Exercises 99–106)

To see how the formula that relates Celsius and Fahrenheit temperatures is derived, ***work Exercises 99–106 in order.***

99. There is a linear relationship between Celsius and Fahrenheit temperatures.

When $C = 0°$, $F =$ _____°.

When $C = 100°$, $F =$ _____°.

100. Think of ordered pairs of temperatures (C, F), where C and F represent corresponding Celsius and Fahrenheit temperatures. The equation that relates the two scales has a straight-line graph that contains the two points determined in **Exercise 99.**

(a) What are these two points?

(b) Find the slope of the line passing through these two points.

101. Use the slope found in **Exercise 100(b)** and one of the two points determined earlier, and write an equation that gives F in terms of C. (*Hint:* Use the point-slope form, where C replaces x and F replaces y.)

102. Use the equation found in **Exercise 101** and solve for C in terms of F. For what temperature does $F = C$? (Use the photo to confirm this temperature.)

103. A quick way to estimate Fahrenheit temperature for a given Celsius temperature is to double C and add 30. Use this method to estimate F if $C = 15$.

104. Use the equation found in **Exercise 101** to find F if $C = 15$. How does the answer compare with the answer to **Exercise 103?**

105. Use the method given in **Exercise 103** to estimate the Fahrenheit temperature given $C = 30$. Then use the equation from **Exercise 101** to find F when $C = 30$. How do the temperatures compare?

106. Explain why the method given in **Exercise 103** to estimate Fahrenheit temperature gives a good approximation of $F = \frac{9}{5}C + 32$.

SUMMARY EXERCISES Finding Slopes and Equations of Lines

Find the slope of each line passing through the given points or having the given equation.

1. $(3, -3)$ and $(8, -6)$ **2.** $(4, -5)$ and $(-1, -5)$ **3.** $y = x - 5$

4. $3x - 7y = 21$ **5.** $x - 4 = 0$ **6.** $4x + 7y = 3$

7. Concept Check Match the description in Column I with its equation in Column II.

I	II
(a) Slope -0.5, $b = -2$	**A.** $y = -\frac{1}{2}x$
(b) x-intercept $(4, 0)$, y-intercept $(0, 2)$	**B.** $y = -\frac{1}{2}x - 2$
(c) Passes through $(4, -2)$ and $(0, 0)$	**C.** $x - 2y = 2$
(d) $m = \frac{1}{2}$, passes through $(-2, -2)$	**D.** $y = 2x$
(e) $m = \frac{1}{2}$, passes through the origin	**E.** $x = 2y$
(f) Slope 2, $b = 0$	**F.** $x + 2y = 4$

8. Concept Check Which equation is written in standard form as specified in the text?

A. $y = -4x - 7$ **B.** $-3x + 4y = 12$ **C.** $x - 6y = 3$

D. $\frac{1}{2}x + y = 0$ **E.** $6x - 2y = 10$ **F.** $3y - 5x = -15$

Write the equations not written in standard form in that form.

Write an equation of the line passing through the given point(s) and satisfying the given condition. Give the equation **(a)** *in slope-intercept form and* **(b)** *in standard form.*

9. $(4, -2)$; slope -3

10. $(-3, 6)$; slope $\frac{2}{3}$

11. $(-2, 6)$ and $(4, 1)$

12. $(4, -8)$ and $(-4, 12)$

13. $(-2, 5)$ and $(1, 14)$

14. Origin; slope $-\frac{5}{2}$

15. $(5, -8)$; parallel to $y = 4$

16. $\left(\frac{3}{4}, -\frac{7}{9}\right)$; perpendicular to $x = \frac{2}{3}$

17. $(4, -2)$; perpendicular to the line passing through $(3, 7)$ and $(5, 6)$

18. $(-4, 2)$; parallel to the line passing through $(3, 9)$ and $(6, 11)$

19. $(2, -1)$; parallel to $y = \frac{1}{5}x + \frac{7}{4}$

20. $(0, -6)$; perpendicular to $y = \frac{4}{3}x + \frac{3}{8}$

2.4 Linear Inequalities in Two Variables

OBJECTIVES

1 Graph linear inequalities in two variables.
2 Graph the intersection of two linear inequalities.
3 Graph the union of two linear inequalities.

VOCABULARY

□ linear inequality in two variables
□ boundary line

OBJECTIVE 1 Graph linear inequalities in two variables.

Earlier we graphed linear inequalities in one variable on a number line. In this section, we graph linear inequalities in two variables on a rectangular coordinate system.

Linear Inequality in Two Variables

A **linear inequality in two variables** (here x and y) is an inequality that can be written in the form

$$Ax + By < C, \quad Ax + By \leq C, \quad Ax + By > C, \quad \text{or} \quad Ax + By \geq C,$$

where A, B, and C are real numbers and A and B are not both 0.

Examples:

$2x + 5y < 10,\quad x - y > 0,\quad y \geq 1$ (here $A = 0$), $\quad x \leq -6$ (here $B = 0$)

Consider the graph in **FIGURE 40**. The graph of the line $x + y = 5$ divides the points in the rectangular coordinate system into three sets of points.

1. Those points that lie ***on*** the line itself and satisfy the equation $\mathbf{x + y = 5}$, such as $(0, 5)$, $(2, 3)$, and $(5, 0)$
2. Those points that lie in the region *above* the line and satisfy the inequality $x + y > 5$, such as $(5, 3)$ and $(2, 4)$
3. Those points that lie in the region *below* the line and satisfy the inequality $x + y < 5$, such as $(0, 0)$ and $(-3, -1)$

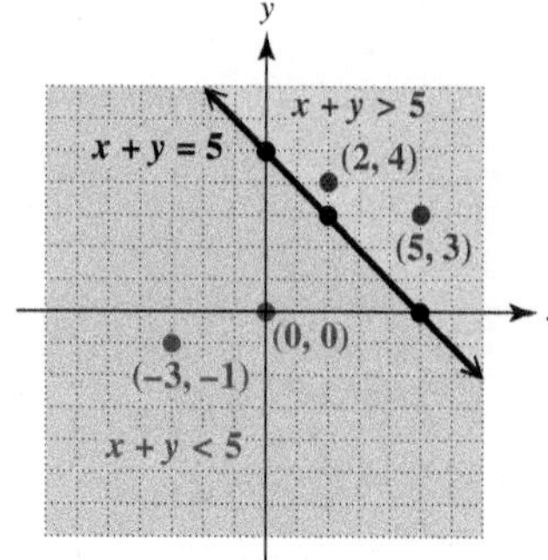

FIGURE 40

The graph of the line $x + y = 5$ is the **boundary line** for the two inequalities

$$x + y > 5 \quad \text{and} \quad x + y < 5.$$

A graph of a linear inequality in two variables is a region in the real number plane that may or may not include the boundary line.

To graph a linear inequality in two variables, follow these steps.

Graphing a Linear Inequality in Two Variables

Step 1 **Draw the graph of the straight line that is the boundary.**

- Make the line solid if the inequality involves $\leq$ or $\geq$.
- Make the line dashed if the inequality involves $<$ or $>$.

Step 2 **Choose a test point.** Choose any point not on the line, and substitute the coordinates of that point in the inequality.

Step 3 **Shade the appropriate region.** Shade the region that includes the test point if it satisfies the original inequality. Otherwise, shade the region on the other side of the boundary line.

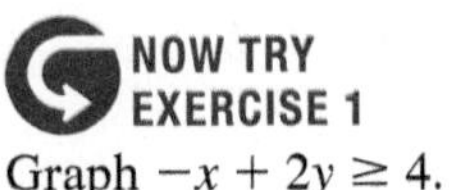

Graph $-x + 2y \geq 4$.

EXAMPLE 1 Graphing a Linear Inequality

Graph $3x + 2y \geq 6$.

Step 1 First graph the boundary line $3x + 2y = 6$, which has intercepts $(2, 0)$ and $(0, 3)$, as shown in **FIGURE 41**.

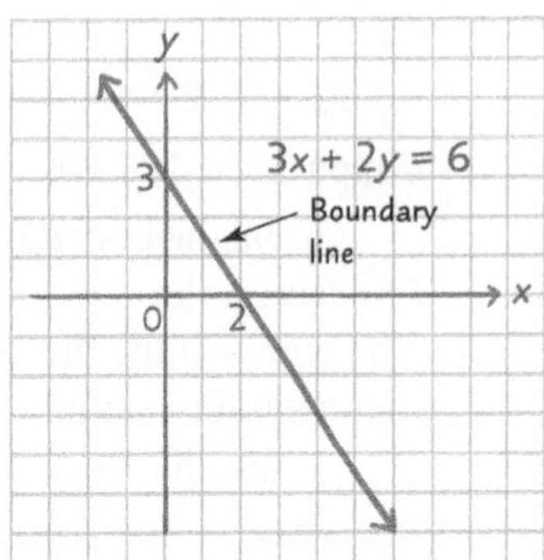

FIGURE 41

Step 2 The graph of the inequality $3x + 2y \geq 6$ includes the points of the boundary line $3x + 2y = 6$ (because the inequality symbol $\geq$ includes equality) and either the points *above* that line or the points *below* it. To decide which, select any point *not* on the boundary line to use as a test point. Substitute the values from the test point, here $(0, 0)$, for x and y in the inequality.

$$3x + 2y > 6 \quad \text{We are testing the region.}$$

$$3(0) + 2(0) \overset{?}{>} 6 \quad \text{Let } x = 0 \text{ and } y = 0.$$

(0, 0) is a convenient test point.

$$0 > 6 \quad \text{False}$$

Step 3 Because the result is false, $(0, 0)$ does *not* satisfy the inequality. The solution set includes all points in the region on the *other* side of the line. This region is shaded in **FIGURE 42**.

As a further check, select a test point in the shaded region, say $(3, 0)$.

$$3x + 2y > 6 \quad \text{Test the region.}$$

$$3(3) + 2(0) \overset{?}{>} 6 \quad \text{Let } x = 3 \text{ and } y = 0.$$

$$9 + 0 \overset{?}{>} 6 \quad \text{Multiply.}$$

$$9 > 6 \ ✓ \quad \text{True}$$

The check confirms that the correct region is graphed in **FIGURE 42**.

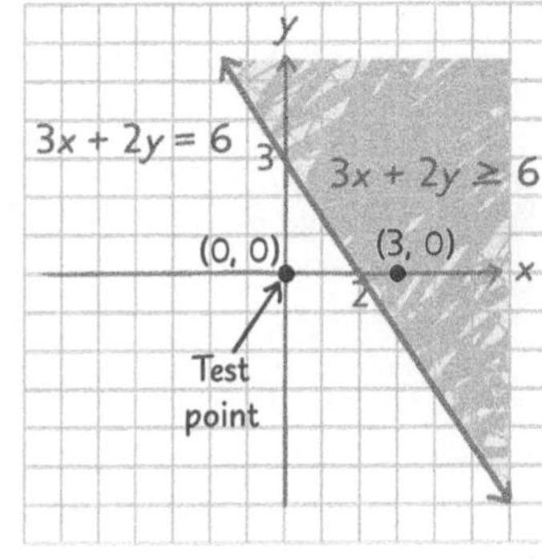

FIGURE 42

NOW TRY ANSWER

1. 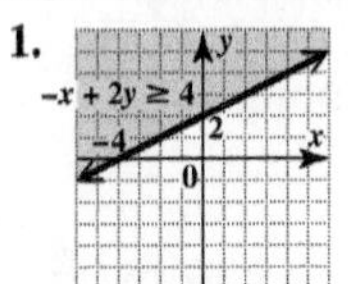

CAUTION When drawing the boundary line in Step 1, be careful to draw a solid line if the inequality includes equality ($\leq, \geq$) or a dashed line if equality is not included ($<, >$).

If an inequality is written in the form $y > mx + b$ or $y < mx + b$, then the inequality symbol indicates which region to shade.

If $y > mx + b$, then shade above the boundary line.

If $y < mx + b$, then shade below the boundary line.

This method works only if the inequality is solved for y.

NOW TRY EXERCISE 2
Graph $3x - 2y < 0$.

EXAMPLE 2 Graphing a Linear Inequality with Boundary Passing through the Origin

Graph $3x - 4y > 0$.

Graph the boundary line. The x- and y-intercepts are the same point, $(0, 0)$, so this line passes through the origin. Two other points on the line are $(4, 3)$ and $(-4, -3)$. The points of the boundary line do *not* belong to the inequality $3x - 4y > 0$ (because the inequality symbol is $>$, *not* $\geq$). For this reason, the line is dashed.

To use the method explained above, we solve the inequality for y.

$3x - 4y > 0$	Original inequality
$-4y > -3x$	Subtract $3x$.
$y < \frac{3}{4}x$	Divide by -4. Change $>$ to $<$.

Use this equivalent inequality to decide which region to shade.

Because the *is less than* symbol occurs ***when the original inequality is solved for y,*** shade the region *below* the boundary line. See **FIGURE 43**.

CHECK Choose a test point not on the line, which rules out the origin. We choose $(2, -1)$.

$3x - 4y > 0$	Original inequality
$3(2) - 4(-1) \overset{?}{>} 0$	Let $x = 2$ and $y = -1$.
$6 + 4 \overset{?}{>} 0$	Multiply.
$10 > 0$ ✓	True

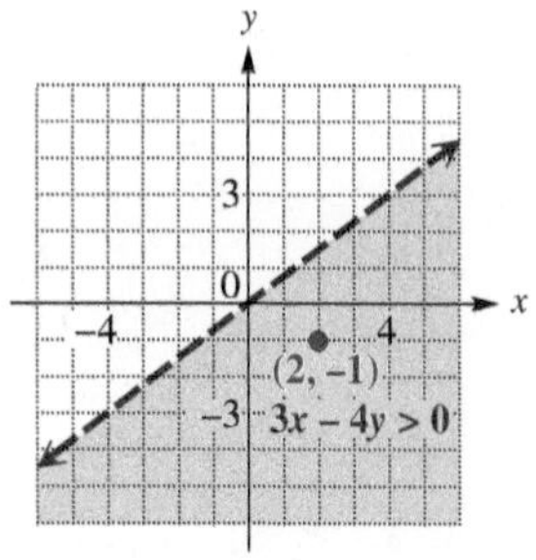

FIGURE 43

This result agrees with the decision to shade below the line. The solution set, graphed in **FIGURE 43**, includes only those points in the shaded region (and *not* those on the line).

NOW TRY EXERCISE 3
Graph $x + 2 > 0$.

EXAMPLE 3 Graphing a Linear Inequality

Graph $x - 3 < 1$.

We graph $x - 3 = 1$, which is equivalent to $x = 4$, as a dashed vertical line passing through the point $(4, 0)$. To determine which region to shade, we choose $(0, 0)$ as a test point.

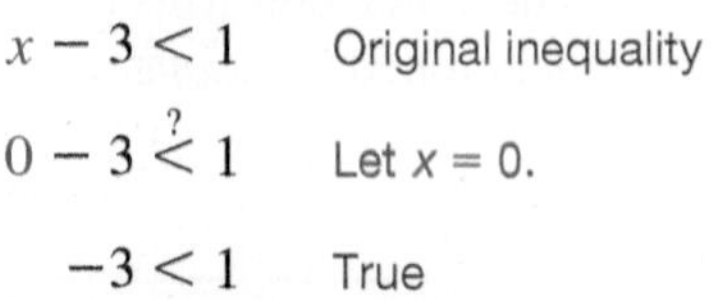

$x - 3 < 1$	Original inequality
$0 - 3 \overset{?}{<} 1$	Let $x = 0$.
$-3 < 1$	True

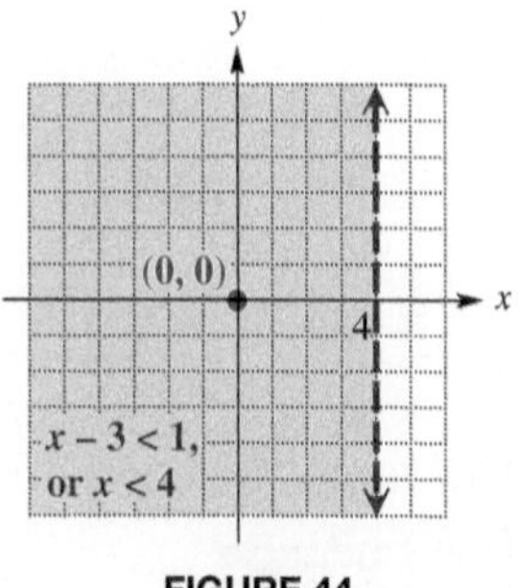

FIGURE 44

Because a true statement results, we shade the region containing $(0, 0)$. See **FIGURE 44**.

NOW TRY ANSWERS

2. $3x - 2y < 0$

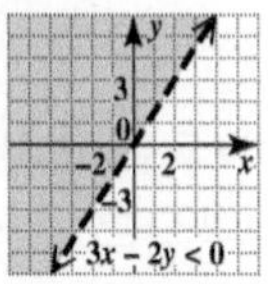

3. $x + 2 > 0$, or $x > -2$

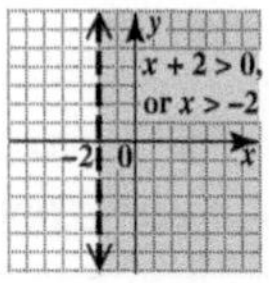

OBJECTIVE 2 Graph the intersection of two linear inequalities.

A pair of inequalities joined with the word ***and*** is interpreted as the ***intersection*** of the solution sets of the inequalities.

The graph of the intersection of two or more inequalities is the region of the plane where all points satisfy all of the inequalities at the same time.

NOW TRY EXERCISE 4
Graph $x + y < 3$ and $y \leq 2$.

EXAMPLE 4 Graphing the Intersection of Two Inequalities

Graph $2x + 4y \geq 5$ and $x \geq 1$.

To begin, we graph the solid boundary line $2x + 4y = 5$ through the intercepts $\left(\frac{5}{2}, 0\right)$ and $\left(0, \frac{5}{4}\right)$. Using $(0, 0)$ as a test point in $2x + 4y > 5$ produces a false statement, so we shade the region that does not include $(0, 0)$. See **FIGURE 45(a)**.

For the inequality $x \geq 1$, we graph the solid boundary line $x = 1$. Again, using $(0, 0)$ as a test point produces a false statement, so we shade the region to the right of the boundary line. See **FIGURE 45(b)**.

We use purple shading to identify the intersection of the graphs in **FIGURE 45(c)**.

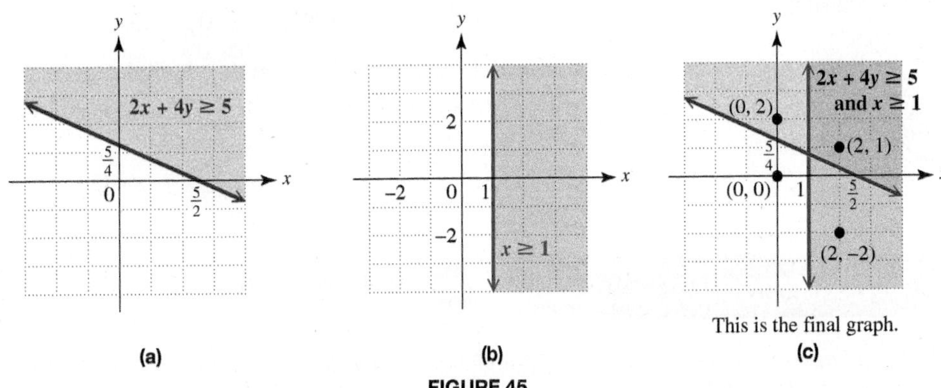

(a) (b) (c)

FIGURE 45

In practice, the graphs in **FIGURES 45(a) and (b)** are graphed on the same axes.

CHECK Using **FIGURE 45(c)**, we can choose a test point from each of the four regions formed by the intersection of the boundary lines.

$(2, 1)$,	$(0, 2)$,	$(0, 0)$,	and	$(2, -2)$	Possible test points
↑	↑	↑		↑	
Purple shaded region	Blue shaded region	Unshaded region		Red shaded region	

Verify that only ordered pairs in the purple shaded region satisfy *both* inequalities. Ordered pairs in the other regions satisfy only one of the inequalities or neither of them. ✓

NOW TRY

OBJECTIVE 3 Graph the union of two linear inequalities.

When two inequalities are joined by the word ***or***, we must find the ***union*** of the graphs of the inequalities.

The graph of the union of two inequalities includes all of the points that satisfy either inequality.

NOW TRY ANSWER

4. 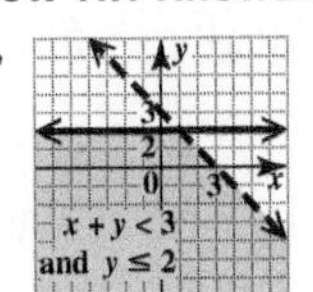

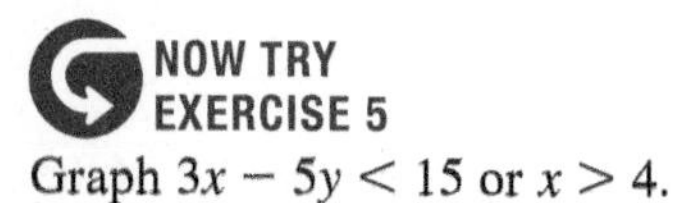

NOW TRY EXERCISE 5

Graph $3x - 5y < 15$ or $x > 4$.

EXAMPLE 5 Graphing the Union of Two Inequalities

Graph $2x + 4y \geq 5$ or $x \geq 1$.

The graphs of the two inequalities are shown in **FIGURES 45(a) and (b)** in **Example 4** on the preceding page. The graph of the union includes all points in *either* inequality, as shown in **FIGURE 46**.

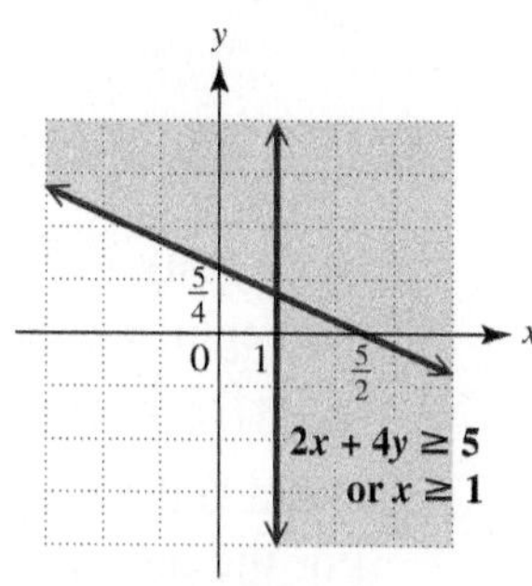

FIGURE 46

CHECK Using **FIGURE 46**, we can choose a test point from each of the four regions formed by the intersection of the boundary lines.

$$(2, 1), \quad (0, 2), \quad (0, 0), \quad \text{and} \quad (2, -2) \qquad \text{Possible test points}$$

Verify that ordered pairs from the shaded region satisfy *either* inequality—that is, they satisfy one or the other or both. (Note also that $(0, 0)$ satisfies *neither* inequality.) ✓

NOW TRY ANSWER

5.

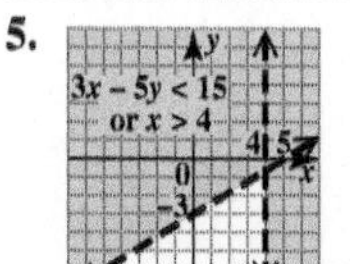

2.4 Exercises

FOR EXTRA HELP

MyLab Math

Video solutions for select problems available in MyLab Math

STUDY SKILLS **REMINDER**
Study cards are a great way to learn vocabulary, procedures, and so on. **Review Study Skill 5, *Using Study Cards.***

Concept Check *Determine whether each ordered pair is a solution of the given inequality.*

1. $x - 2y \leq 4$

(a) $(0, 0)$ **(b)** $(2, -1)$ **(c)** $(7, 1)$ **(d)** $(0, 2)$

2. $x + y > 0$

(a) $(0, 0)$ **(b)** $(-2, 1)$ **(c)** $(2, -1)$ **(d)** $(-4, 6)$

3. $x - 5 > 0$

(a) $(0, 0)$ **(b)** $(5, 0)$ **(c)** $(-1, 3)$ **(d)** $(6, 2)$

4. $y \leq 1$

(a) $(0, 0)$ **(b)** $(3, 1)$ **(c)** $(2, -1)$ **(d)** $(-3, 3)$

Concept Check *In each statement, fill in the first blank with either* solid *or* dashed. *Fill in the second blank with either* above *or* below.

5. The boundary of the graph of $y \leq -x + 2$ will be a ______ line, and the shading will be ______ the line.

6. The boundary of the graph of $y < -x + 2$ will be a ______ line, and the shading will be ______ the line.

7. The boundary of the graph of $y > -x + 2$ will be a ______ line, and the shading will be ______ the line.

8. The boundary of the graph of $y \geq -x + 2$ will be a ______ line, and the shading will be ______ the line.

Concept Check *Refer to the given graph, and complete each statement with the correct inequality symbol* $<$, $\leq$, $>$, *or* $\geq$.

9. x _____ 4

10. y _____ -3

11. y _____ $3x - 2$

12. y _____ $-x + 3$

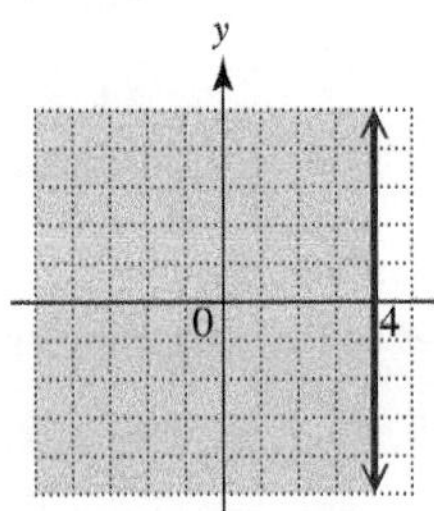

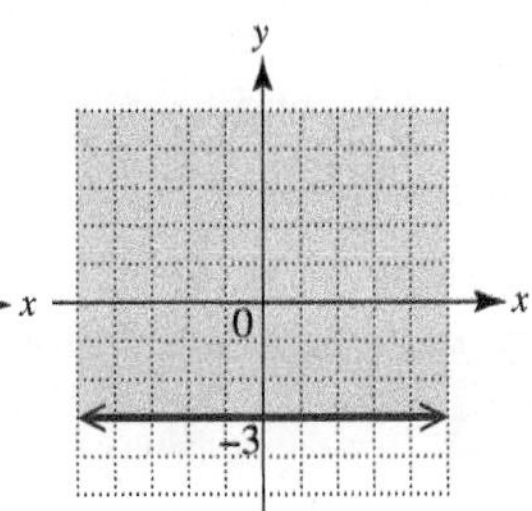

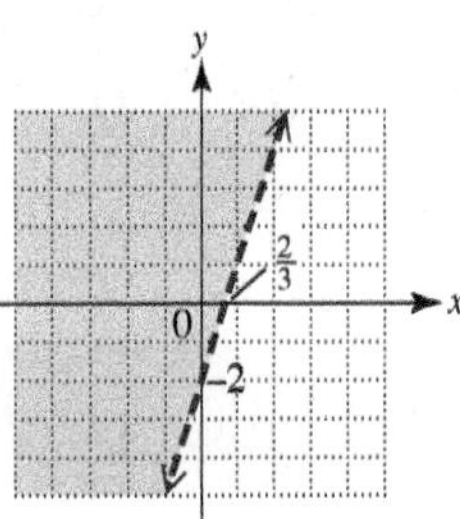

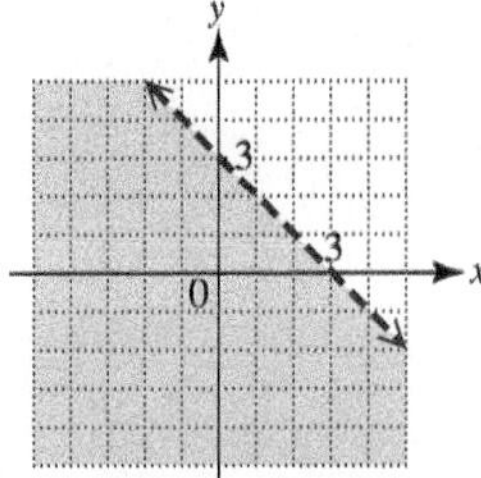

Graph each inequality. ***See Examples 1–3.***

13. $x + y \leq 2$

14. $x + y \leq -3$

15. $4x - y < 4$

16. $3x - y < 3$

17. $x + 3y \geq -2$

18. $x + 4y \geq -3$

19. $y < \frac{1}{2}x + 3$

20. $y < \frac{1}{3}x - 2$

21. $y \geq -\frac{2}{5}x + 2$

22. $y \geq -\frac{3}{2}x + 3$

23. $2x + 3y \geq 6$

24. $3x + 4y \geq 12$

25. $5x - 3y > 15$

26. $4x - 5y > 20$

27. $x + y > 0$

28. $x + 2y > 0$

29. $x - 3y \leq 0$

30. $x - 5y \leq 0$

31. $y < x$

32. $y \leq 4x$

33. $x + 3 \geq 0$

34. $x - 1 \leq 0$

35. $y + 5 < 2$

36. $y - 1 > 3$

Extending Skills *Complete each of the following to write an inequality for the graph shown.*

37. Determine the following for the boundary line.

Slope: __________

y-intercept: __________

Equation: $y =$ __________

The boundary line here is (*solid* / *dashed*), and the region (*above* / *below*) it is shaded.

The inequality symbol to indicate this is ($<$ / $\leq$ / $>$ / $\geq$).

Inequality for the graph: y __________

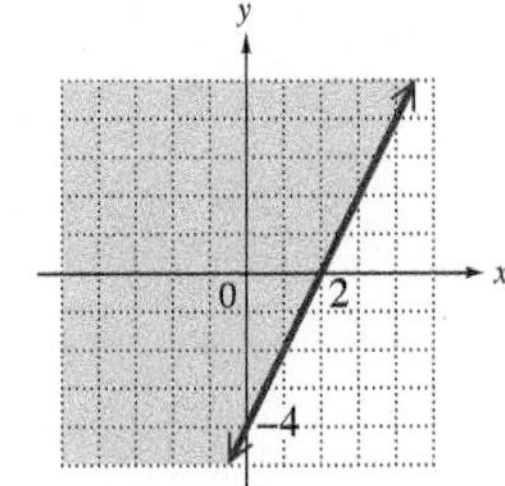

38. Determine the following for the boundary line.

Slope: __________

y-intercept: __________

Equation: $y =$ __________

The boundary line here is (*solid* / *dashed*), and the region (*above* / *below*) it is shaded.

The inequality symbol to indicate this is ($<$ / $\leq$ / $>$ / $\geq$).

Inequality for the graph: y __________

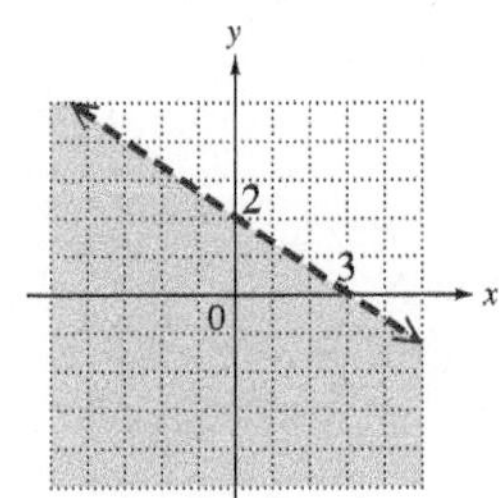

Graph the intersection of each pair of inequalities. **See Example 4.**

39. $x + y \leq 1$ and $x \geq 1$

40. $x - y \geq 2$ and $x \geq 3$

41. $2x - y \geq 2$ and $y < 4$

42. $3x - y \geq 3$ and $y < 3$

43. $x + y > -5$ and $y < -2$

44. $6x - 4y < 10$ and $y > 2$

Extending Skills *Use the definition of absolute value to write each inequality as a compound inequality, and graph its solution set in the rectangular coordinate plane.*

45. $|x| < 3$

46. $|y| < 5$

47. $|x + 1| < 2$

48. $|y - 3| < 2$

Graph the union of each pair of inequalities. **See Example 5.**

49. $x - y \geq 1$ or $y \geq 2$

50. $x + y \leq 2$ or $y \geq 3$

51. $x - 2 > y$ or $x < 1$

52. $x + 3 < y$ or $x > 3$

53. $3x + 2y < 6$ or $x - 2y > 2$

54. $x - y \geq 1$ or $x + y \leq 4$

RELATING CONCEPTS For Individual or Group Work (Exercises 55–60)

Linear programming *is a method for finding the optimal (best possible) solution that meets all the conditions for a problem such as the following.*

A factory can have no more than 200 *workers on a shift, but must have* at least 100 *and must manufacture* at least 3000 *units at minimum cost. How many workers should be on a shift in order to produce the required units at minimal cost?*

Let x represent the number of workers and y represent the number of units manufactured. ***Work Exercises 55–60 in order.***

55. Write three inequalities expressing the problem conditions.

56. Graph the inequalities from **Exercise 55** using the axes below, and shade the intersection.

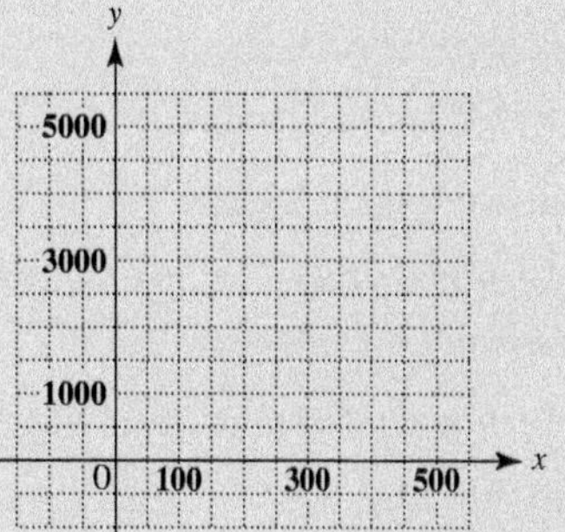

57. The cost per worker is \$50 per day and the cost to manufacture 1 unit is \$100. Write an equation in x, y, and C representing the total daily cost C.

58. Find values of x and y for several points in or on the boundary of the shaded region. Include any "corner points," where C is maximized or minimized.

59. Of the values of x and y found in **Exercise 58,** which ones give the least value when substituted in the cost equation from **Exercise 57?**

60. What does the answer in **Exercise 59** mean in terms of the given problem?

2.5 Introduction to Relations and Functions

OBJECTIVES

1 Define and identify relations and functions.
2 Find the domain and range.
3 Identify functions defined by graphs and equations.

OBJECTIVE 1 Define and identify relations and functions.

Consider the relationship illustrated in the following table between number of hours worked and paycheck amount for an hourly worker.

Number of Hours Worked	*Paycheck Amount (in dollars)*	Ordered pairs
5	50	⟶ (5, 50)
10	100	⟶ (10, 100)
20	200	⟶ (20, 200)
40	400	⟶ (40, 400)

The data from the table can be represented by a set of ordered pairs.

$$\{(5, 50), (10, 100), (20, 200), (40, 400)\}$$

Number of hours worked ↑ ↑ Paycheck amount in dollars

VOCABULARY

- ☐ relation
- ☐ function
- ☐ dependent variable
- ☐ independent variable
- ☐ domain
- ☐ range

Each first component of the ordered pairs represents a number of hours worked, and each second component represents the corresponding paycheck amount. Such a set of ordered pairs is a *relation*.

Relation

A **relation** is any set of ordered pairs.

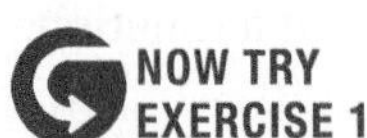
NOW TRY EXERCISE 1

Write the relation as a set of ordered pairs.

Year	*Average Gas Price per Gallon (in dollars)*
2010	2.79
2012	3.64
2014	3.37
2016	2.14

Data from Energy Information Administration.

EXAMPLE 1 Writing Ordered Pairs for a Relation

Write the relation as a set of ordered pairs.

Number of Gallons of Gas	*Cost (in dollars)*
0	0
1	3.50
2	7.00
3	10.50
4	14.00

The data in the table defines a relation between number of gallons of gas and cost and can be written as the following set of ordered pairs.

$$\{(0, 0), (1, 3.50), (2, 7.00), (3, 10.50), (4, 14.00)\}$$

Number of gallons of gas ↑ ↑ Cost in dollars

NOW TRY 

NOW TRY ANSWER

1. $\{(2010, 2.79), (2012, 3.64), (2014, 3.37), (2016, 2.14)\}$

A *function* is a special kind of relation.

Function

A **function** is a relation in which, for each distinct value of the first component of the ordered pairs, there is *exactly one* value of the second component.

NOW TRY EXERCISE 2

Determine whether each relation defines a function.

(a) $\{(1, 5), (3, 5), (5, 5)\}$

(b) $\{(-1, -3), (0, 2), (-1, 6)\}$

EXAMPLE 2 Determining Whether Relations Are Functions

Determine whether each relation defines a function.

(a) $F = \{(1, 2), (-2, 4), (3, -1)\}$

Look at the ordered pairs that define this relation. For each distinct x-value, there is *exactly one* y-value. We can show this correspondence as follows.

$$\begin{array}{ccc} \{1, & -2, & 3\} \\ \downarrow & \downarrow & \downarrow \\ \{2, & 4, & -1\} \end{array}$$

x-values of F

y-values of F

Therefore, relation F is a function.

(b) $G = \{(-2, -1), (-1, 0), (0, 1), (1, 2), (2, 2)\}$

Relation G is also a function. Although the last two ordered pairs have the same y-value (1 is paired with 2, and 2 is paired with 2), this does not violate the definition of a function.

$$\begin{array}{ccccc} \{-2, & -1, & 0, & 1, & 2\} \\ \downarrow & \downarrow & \downarrow & \downarrow\swarrow & \\ \{-1, & 0, & 1, & 2\} & \end{array}$$

x-values of G

y-values of G

The first components (x-values) are distinct, and each is paired with only one second component (y-value).

(c) $H = \{(-4, 1), (-2, 1), (-2, 0)\}$

In relation H, the last two ordered pairs have the ***same*** x-value paired with ***two different*** y-values (-2 is paired with both 1 and 0). H is a relation, but *not* a function.

$$\begin{array}{cc} \{-4, & -2\} \\ \downarrow & \swarrow\ \downarrow \\ \{1, & 0\} \end{array}$$

x-values of H

y-values of H

In a function, no two ordered pairs have the same first component and different second components.

Different y-values

Relation $H = \{(-4, 1), (-2, 1), (-2, 0)\}$ Not a function

Same x-value

NOW TRY

Relations may be defined in several different ways.

- **A relation may be defined as a set of ordered pairs.**

Relation $F = \{(1, 2), (-2, 4), (3, -1)\}$ Function **(Example 2(a))**

Relation $H = \{(-4, 1), (-2, 1), (-2, 0)\}$ Not a function **(Example 2(c))**

NOW TRY ANSWERS

2. **(a)** function

(b) not a function

- **A relation may be defined as a correspondence or *mapping*.**

See FIGURE 47. In the mapping for relation F, the arrow from 1 to 2 indicates that the ordered pair $(1, 2)$ belongs to F. Also, -2 is mapped to 4, and 3 is mapped to -1. Thus, F is a function—each first component is paired with exactly one second component.

In the mapping for relation H, which is *not* a function, the first component -2 is paired with two different second components, 1 and 0.

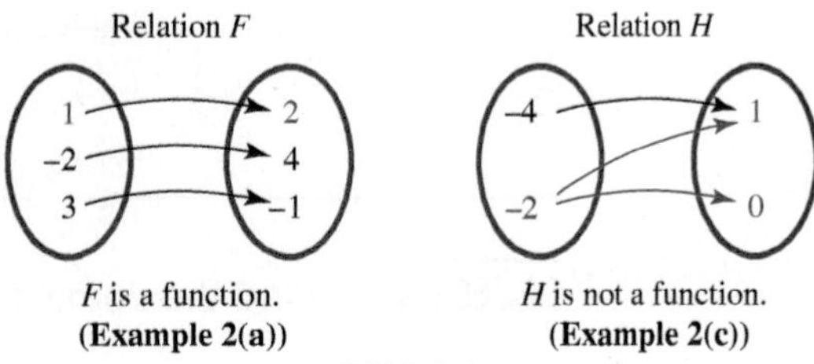

F is a function. **(Example 2(a))**

H is not a function. **(Example 2(c))**

FIGURE 47

- **A relation may be defined as a table.**
- **A relation may be defined as a graph.**

FIGURE 48 includes a table and graph for relation F, which is a function.

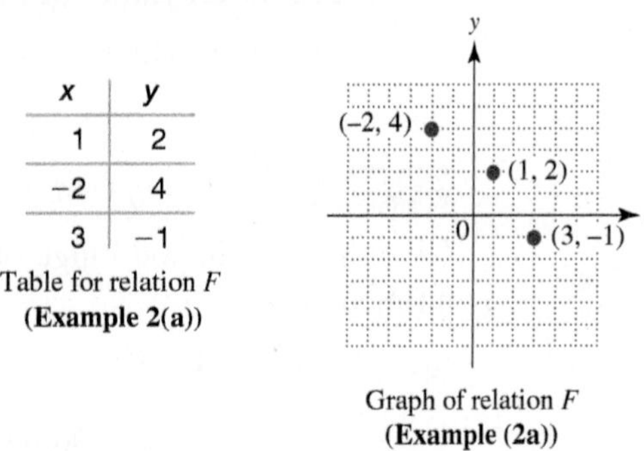

x	y
1	2
−2	4
3	−1

Table for relation F **(Example 2(a))**

Graph of relation F **(Example (2a))**

FIGURE 48

- **A relation may be defined as a rule.**

The rule may be given in words, such as "y is twice x." Usually the rule is an equation, such as

$$y = 2x.$$

The infinite number of ordered-pair solutions (x, y) can be represented by the graph in FIGURE 49.

In the equation $y = 2x$, the value of y ***depends*** on the value of x. Thus, the variable y is the **dependent variable.** The variable x is the **independent variable.**

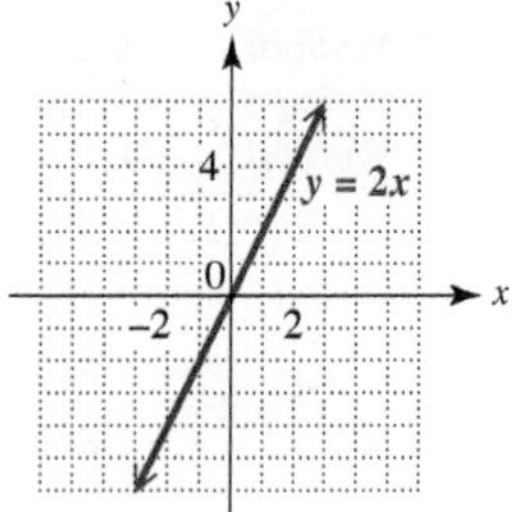

Graph of the relation $y = 2x$

FIGURE 49

Dependent variable $\rightarrow y = 2x \leftarrow$ Independent variable

An equation tells how to determine the value of the dependent variable for a specific value of the independent variable.

NOTE An equation that describes the relationship given at the beginning of this section between number of hours worked and paycheck amount is

$$y = 10x. \quad \text{10 represents the hourly rate, \$10.}$$

Paycheck amount y *depends* on number of hours worked x. Thus, *paycheck amount* is the dependent variable, and *number of hours worked* is the independent variable.

In a function, there is exactly one value of the dependent variable, the second component, for each value of the independent variable, the first component.

NOTE Another way to think of a function relationship is to visualize the independent variable as an **input** and the dependent variable as an **output.** This **input-output (function) machine** illustrates the relationship between number of hours worked and paycheck amount.

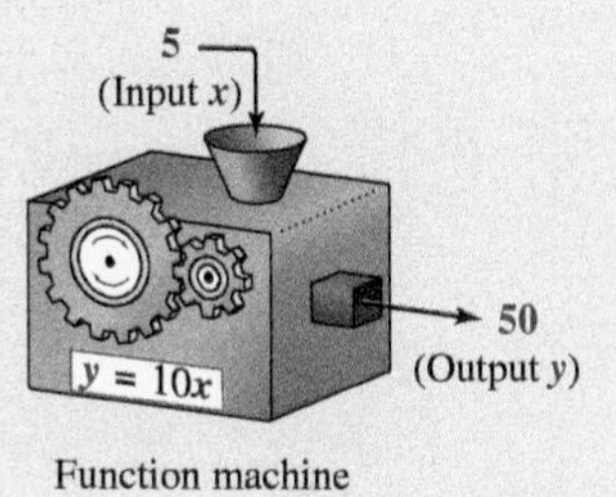

Function machine

OBJECTIVE 2 Find the domain and range.

Domain and Range

For every relation defined by a set of ordered pairs (x, y), there are two important sets of elements.

- The set of all values of the independent variable (x) is the **domain.**
- The set of all values of the dependent variable (y) is the **range.**

NOW TRY EXERCISE 3

Give the domain and range of each relation. Determine whether the relation defines a function.

(a) $\{(2, 2), (2, 5), (4, 8)\}$

(b) The table from **Objective 1**

Number of Hours Worked	Paycheck Amount (in dollars)
5	50
10	100
20	200
40	400

EXAMPLE 3 Finding Domains and Ranges of Relations

Give the domain and range of each relation. Determine whether the relation defines a function.

(a) $\{(3, -1), (4, 2), (4, 5), (6, 8)\}$ List 4 only once.

Domain: $\{3, 4, 6\}$ Set of x-values

Range: $\{-1, 2, 5, 8\}$ Set of y-values

This relation is not a function because the same x-value 4 is paired with two different y-values, 2 and 5.

(b)

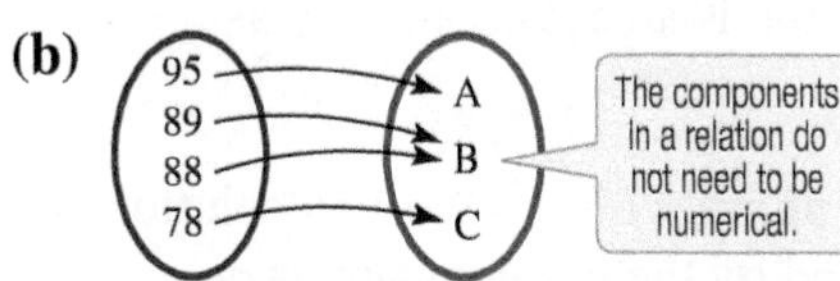

This mapping represents the following set of ordered pairs.

$\{(95, A), (89, B), (88, B), (78, C)\}$

Domain: $\{95, 89, 88, 78\}$ Set of first components

Range: $\{A, B, C\}$ Set of second components

The mapping defines a function—each domain value corresponds to exactly one range value.

(c)

x	y
-5	2
0	2
5	2

This table represents the following set of ordered pairs.

$\{(-5, 2), (0, 2), (5, 2)\}$

Domain: $\{-5, 0, 5\}$ Set of x-values

Range: $\{2\}$ Set of y-values

The table defines a function—each distinct x-value corresponds to exactly one y-value (even though it is the same y-value).

NOW TRY

NOW TRY ANSWERS

3. (a) domain: $\{2, 4\}$; range: $\{2, 5, 8\}$; not a function
(b) domain: $\{5, 10, 20, 40\}$; range: $\{50, 100, 200, 400\}$; function

A graph gives a "picture" of a relation and can be used to determine its domain and range.

> **NOTE** Pay particular attention to the use of color to interpret domain and range in **Example 4**—blue for domain and red for range.

NOW TRY EXERCISE 4
Give the domain and range of the relation.

EXAMPLE 4 Finding Domains and Ranges from Graphs

Give the domain and range of each relation.

(a)

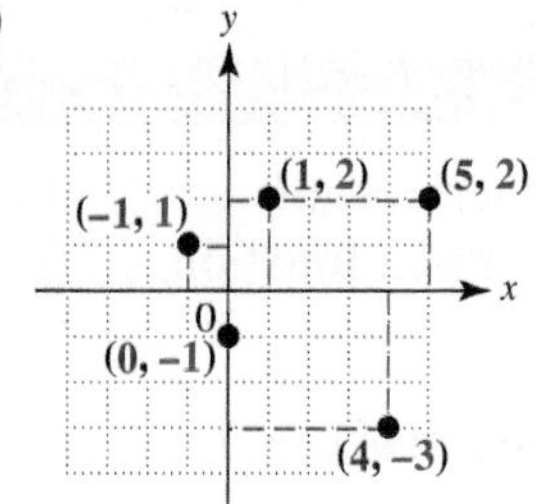

This relation includes the five ordered pairs that are graphed.

$$\{(-1, 1), (0, -1), (1, 2), (4, -3), (5, 2)\}$$

Domain: $\{-1, 0, 1, 4, 5\}$ Set of x-values

Range: $\{1, -1, 2, -3\}$ Set of y-values

List 2 only once.

(b)

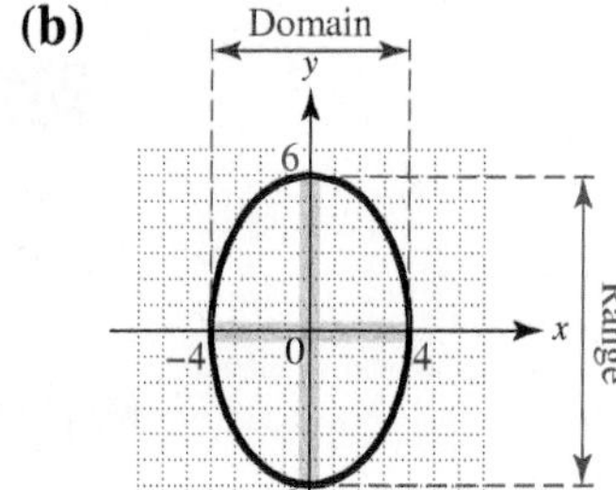

The x-values of the ordered pairs that form the graph include all numbers between -4 and 4, inclusive, as shown in blue. The y-values include all numbers between -6 and 6, inclusive, as shown in red.

Domain: $[-4, 4]$

Range: $[-6, 6]$

Use interval notation.

(c)

The arrowheads on the graphed line indicate that the line extends indefinitely left and right, as well as up and down. Therefore, both the domain (set of x-values), shown in blue, and the range (set of y-values), shown in red, include all real numbers.

Domain: $(-\infty, \infty)$ Range: $(-\infty, \infty)$

(d)

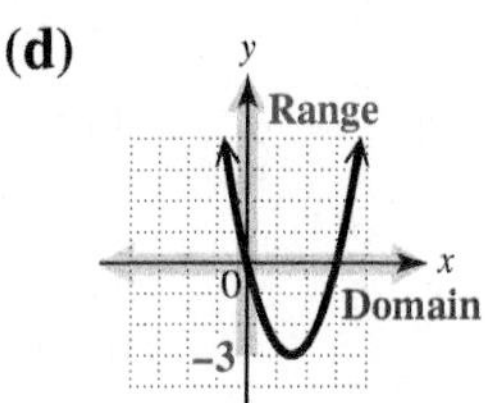

The graphed curve extends indefinitely left and right, as well as upward. The domain, shown in blue, includes all real numbers. Because there is a least y-value, -3, the range, shown in red, includes all numbers greater than or equal to -3.

Domain: $(-\infty, \infty)$ Range: $[-3, \infty)$

NOW TRY

OBJECTIVE 3 Identify functions defined by graphs and equations.

Because each value of x in a function corresponds to only one value of y, any vertical line drawn through the graph of a function must intersect the graph in at most one point. This is the *vertical line test* for a function.

FIGURE 50 on the next page illustrates this test with the graphs of two relations.

NOW TRY ANSWER

4. domain: $(-\infty, \infty)$; range: $[-2, \infty)$

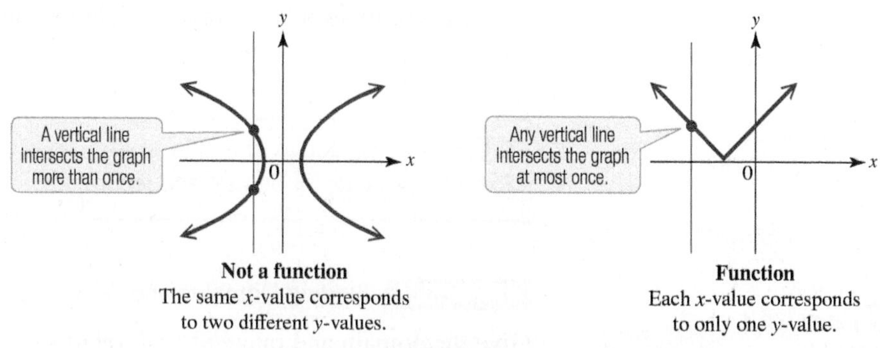

Not a function
The same x-value corresponds to two different y-values.

Function
Each x-value corresponds to only one y-value.

FIGURE 50

Vertical Line Test

If every vertical line intersects the graph of a relation in no more than one point, then the relation represents a function.

NOW TRY EXERCISE 5

Use the vertical line test to determine whether the relation is a function.

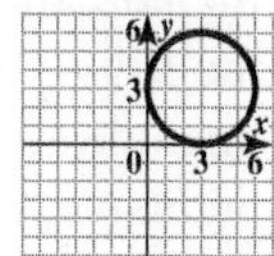

EXAMPLE 5 Using the Vertical Line Test

Use the vertical line test to determine whether each relation graphed in **Example 4** is a function. (We repeat the graphs here.)

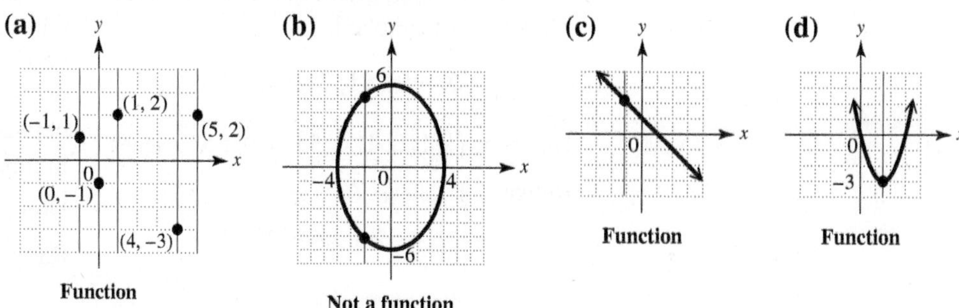

The graphs in (a), (c), and (d) satisfy the vertical line test because every vertical line intersects each graph no more than once. These graphs represent functions.

The graph in (b) fails the vertical line test because a vertical line intersects the graph more than once—that is, the same x-value corresponds to two different y-values. This is not the graph of a function.

NOW TRY

NOTE Graphs that do not represent functions are still relations. ***All equations and graphs represent relations, and all relations have a domain and range.***

If a relation is defined by an equation involving a fraction or a radical, apply these guidelines when finding its domain.

1. **Exclude from the domain any values that make the denominator of a fraction equal to 0.**

   Example: The function $y = \frac{1}{x}$ has all real numbers *except* 0 as its domain because division by 0 is undefined.

2. **Exclude from the domain any values that result in an even root of a negative number.**

   Example: The function $y = \sqrt{x}$ has all *nonnegative* real numbers as its domain because the square root of a negative number is not real.

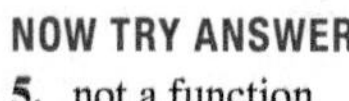

NOW TRY ANSWER

5. not a function

Agreement on Domain

Unless specified otherwise, the domain of a relation is assumed to be all real numbers that produce real numbers when substituted for the independent variable.

EXAMPLE 6 Identifying Functions and Domains from Equations

Determine whether each relation defines y as a function of x. Give the domain.

(a) $y = x + 4$

In this equation, y is found by adding 4 to x. Thus, each value of x corresponds to just one value of y, and the relation defines a function. Because x can be any real number, the domain is $(-\infty, \infty)$. The graph in **FIGURE 51** confirms this reasoning.

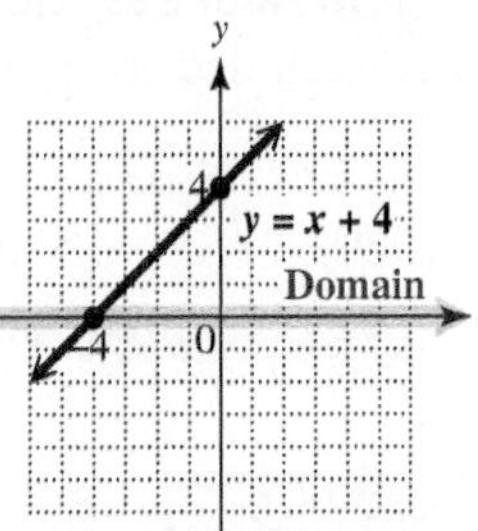

FIGURE 51

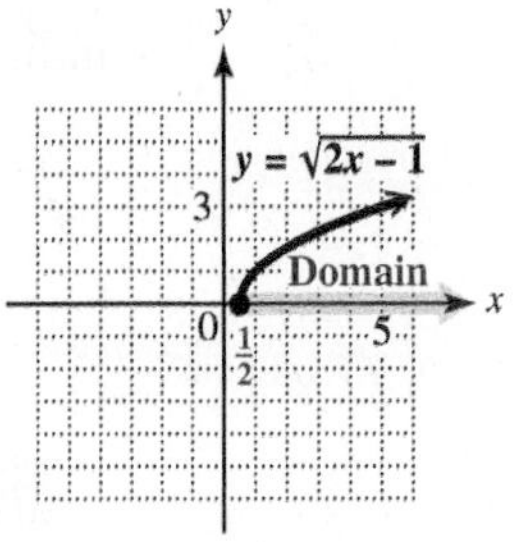

FIGURE 52

(b) $y = \sqrt{2x - 1}$

For any choice of x in the domain, there is exactly one corresponding value for y. (The radical is a single nonnegative number.) This equation defines a function. The quantity under the radical symbol cannot be negative—that is, $2x - 1$ *must be greater than or equal to* 0.

$$2x - 1 \geq 0$$

$$2x \geq 1 \quad \text{Add 1.}$$

$$x \geq \frac{1}{2} \quad \text{Divide by 2.}$$

The domain of the function is $\left[\frac{1}{2}, \infty\right)$. The graph in **FIGURE 52** confirms our work.

(c) $y^2 = x$

The ordered pairs $(4, 2)$ and $(4, -2)$ both satisfy this equation. One value of x, 4, corresponds to two values of y, 2 and -2, so this equation does not define a function. Because x is equal to the square of y, the values of x must always be nonnegative. The domain of the relation is $[0, \infty)$. See **FIGURE 53**.

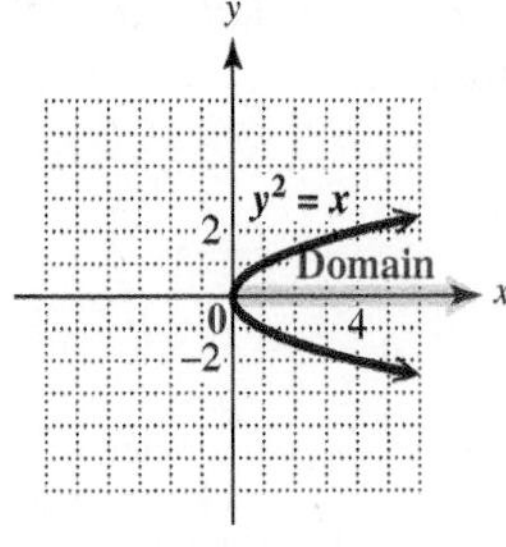

FIGURE 53

(d) $y \leq x - 1$

By definition, y is a function of x if every value of x leads to exactly one value of y. Here, a particular value of x, such as 1, corresponds to many values of y. The ordered pairs

$$(1, 0), \quad (1, -1), \quad (1, -2), \quad (1, -3), \quad \text{and so on}$$

all satisfy the inequality. This relation does not define a function. Any number can be used for x, so the domain is the set of all real numbers, $(-\infty, \infty)$. See **FIGURE 54**.

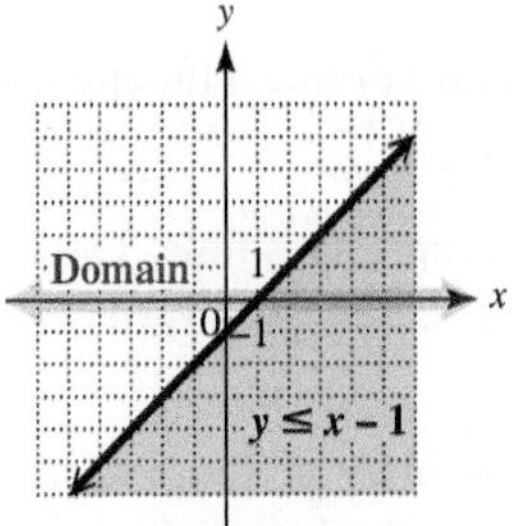

FIGURE 54

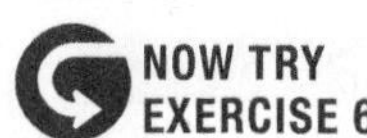

NOW TRY EXERCISE 6

Determine whether each relation defines y as a function of x. Give the domain.

(a) $y = 4x - 3$

(b) $y = \sqrt{2x - 4}$

(c) $y = \dfrac{1}{x - 2}$

(d) $y < 3x + 1$

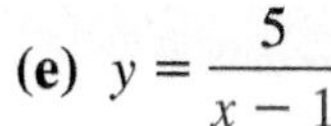

(e) $y = \dfrac{5}{x - 1}$

Given any value of x in the domain, we find y by subtracting 1 and then dividing the result into 5. This process produces exactly one value of y for each value in the domain, so the given equation defines a function.

The domain includes all real numbers except those which make the denominator 0.

$$x - 1 = 0 \quad \text{Set the denominator equal to 0.}$$

$$x = 1 \quad \text{Add 1.}$$

The domain includes all real numbers *except* 1, written $(-\infty, 1) \cup (1, \infty)$. In **FIGURE 55**, the open circle on the graph indicates that 1 is excluded from the domain.

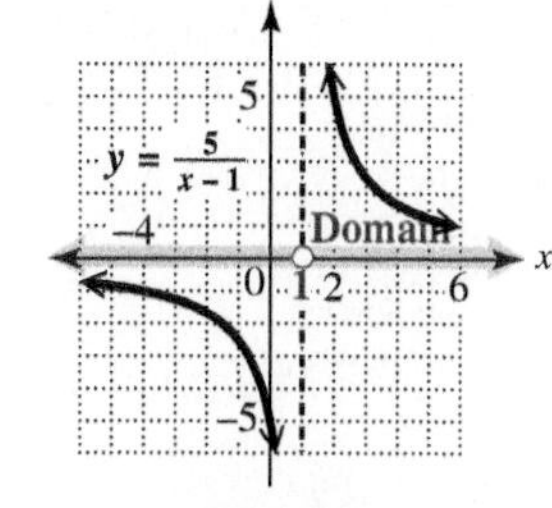

FIGURE 55

NOW TRY

In summary, we give three variations of the definition of a function.

Variations of the Definition of a Function

1. A **function** is a relation in which, for each distinct value of the first component of the ordered pairs, there is exactly one value of the second component.
2. A **function** is a set of distinct ordered pairs in which no first component is repeated.
3. A **function** is a correspondence (mapping) or an equation (rule) that assigns exactly one range value to each distinct domain value.

NOW TRY ANSWERS

6. **(a)** function; $(-\infty, \infty)$
   (b) function; $[2, \infty)$
   (c) function; $(-\infty, 2) \cup (2, \infty)$
   (d) not a function; $(-\infty, \infty)$

2.5 Exercises

FOR EXTRA HELP

MyLab Math

Video solutions for select problems available in MyLab Math

Concept Check *Complete each statement. Choices may be used more than once.*

function	independent variable	vertical line test	relation
domain	ordered pairs	dependent variable	range

1. A(n) ______ is any set of ______ $\{(x, y)\}$.
2. A(n) ______ is a relation in which, for each distinct value of the first component of the ______, there is exactly one value of the second component.
3. In a relation $\{(x, y)\}$, the ______ is the set of x-values, and the ______ is the set of y-values.
4. The relation $\{(0, -2), (2, -1), (2, -4), (5, 3)\}$ (*does / does not*) define a function. The set $\{0, 2, 5\}$ is its ______, and the set $\{-2, -1, -4, 3\}$ is its ______.
5. Consider the function $d = 50t$, where d represents distance and t represents time. The value of d depends on the value of t, so the variable t is the ______, and the variable d is the ______.
6. The ______ is used to determine whether a graph is that of a function. It says that any vertical line can intersect the graph of a(n) ______ in no more than (*zero / one / two*) point(s).

Write each relation as a set of ordered pairs. **See Example 1.**

7.

x	y
2	−2
2	0
2	1

8.

x	y
−1	−1
0	−1
1	−1

9.

Year	Average Movie Ticket Price (in dollars)
1960	0.76
1980	2.69
2000	5.39
2016	8.65

Data from Motion Picture Association of America.

10.

Year	Average ACT Composite Score
2010	21.0
2012	21.1
2014	21.0
2016	20.8

Data from ACT.

11.

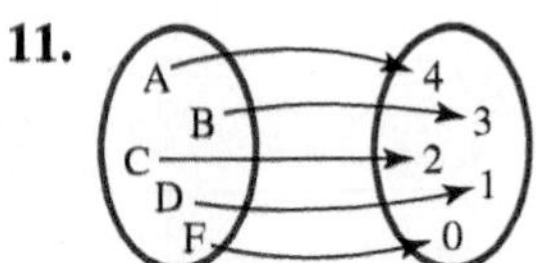

12.

A
E
I
O
U
1
5
9
15
21

Concept Check *Express each relation using a different form. (For example, if the given form is a set of ordered pairs, use a graph.) There is more than one correct way to do this.* **See Objective 1.**

13. $\{(0, 2), (2, 4), (4, 6)\}$

14. y is half of x.

15.

x	y
−1	−3
0	−1
1	1
3	3

16.

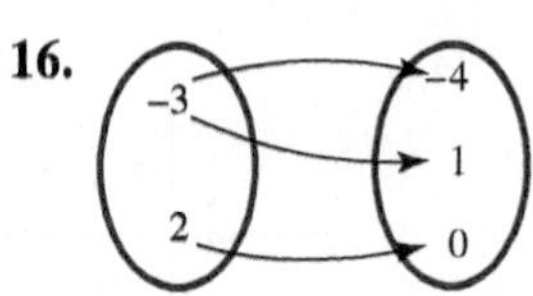

17.

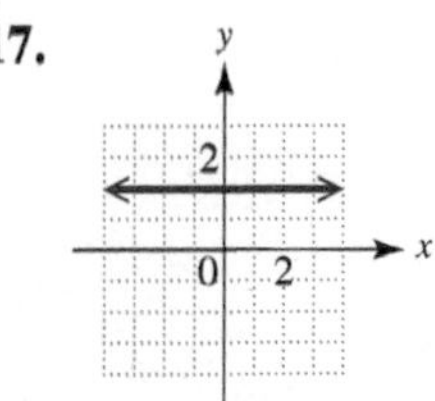

18. Concept Check Which of the relations represented in **Exercises 13–17** does *not* define a function? Explain.

Determine whether each relation defines a function, and give the domain and range. **See Examples 2–5.**

19. $\{(5, 1), (3, 2), (4, 9), (7, 6)\}$

20. $\{(8, 0), (5, 4), (9, 3), (3, 8)\}$

21. $\{(2, 4), (0, 2), (2, 5)\}$

22. $\{(9, -2), (-3, 5), (9, 2)\}$

23. $\{(-3, 1), (4, 1), (-2, 7)\}$

24. $\{(-12, 5), (-10, 3), (8, 3)\}$

25. $\{(1, 1), (1, -1), (0, 0), (2, 4), (2, -4)\}$

26. $\{(2, 5), (3, 7), (4, 9), (5, 11)\}$

27.

x	y
1	5
1	2
1	−1
1	−4

28.

x	y
−4	−4
−4	0
−4	4
−4	8

29.

x	y
4	−3
2	−3
0	−3
−2	−3

30.

x	y
−3	−6
−1	−6
1	−6
3	−6

31.
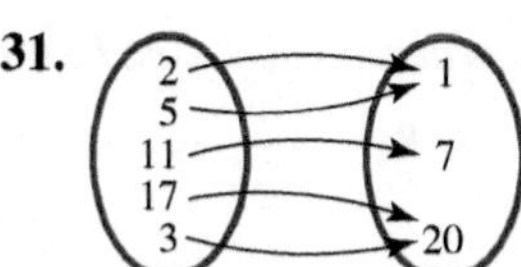

32.
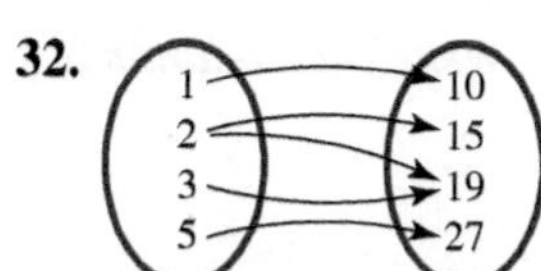

33.
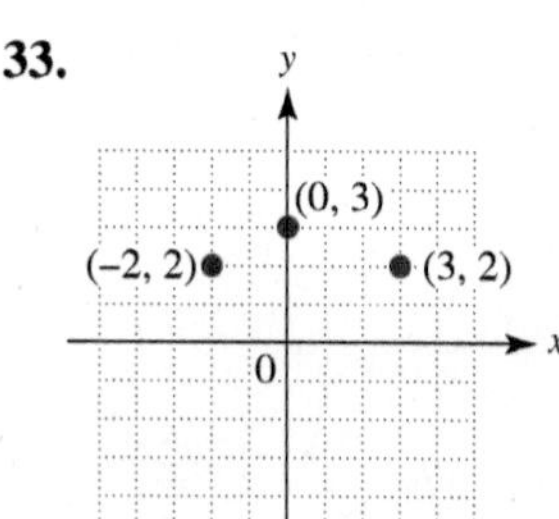

34.
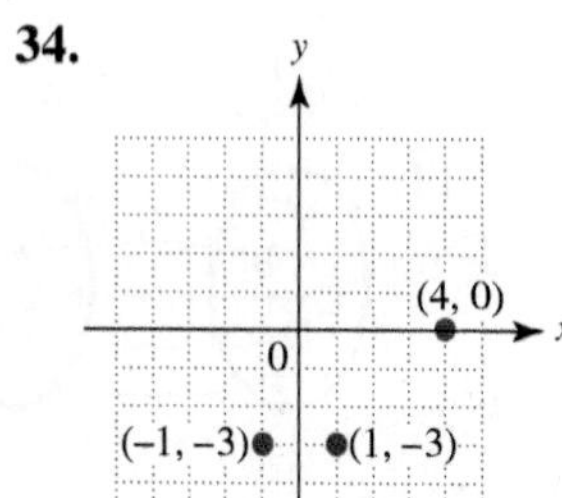

35.
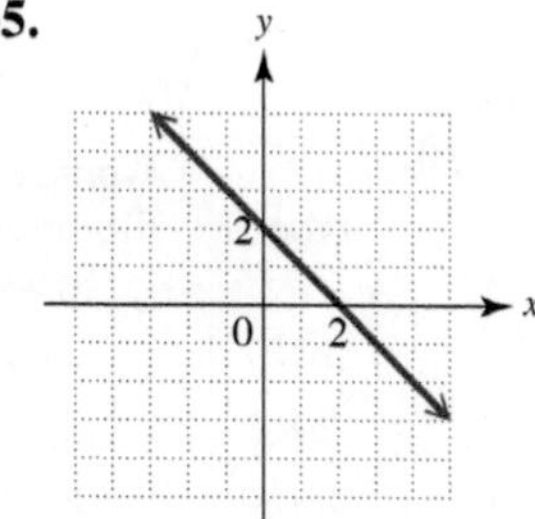

36.
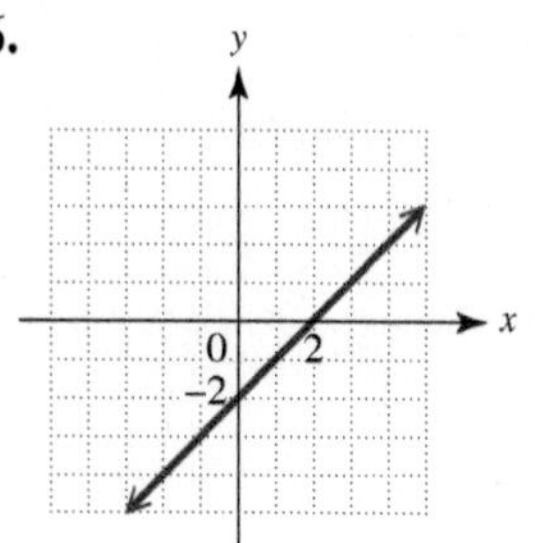

37.
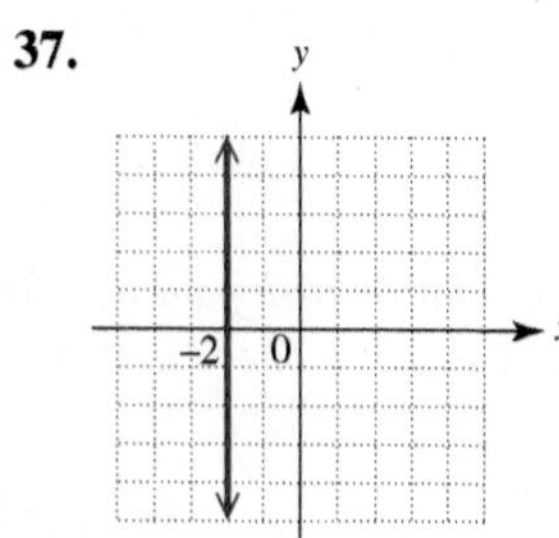

38.
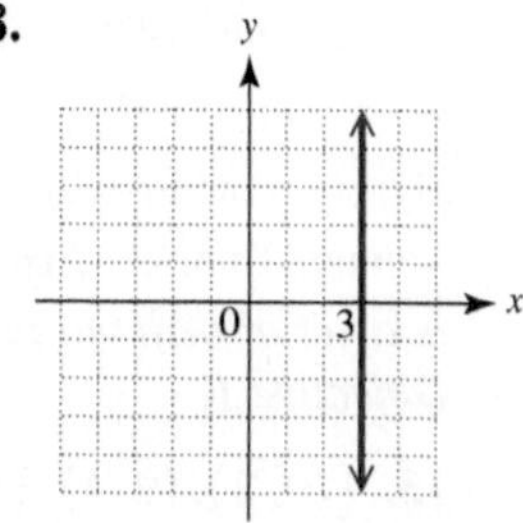

39.
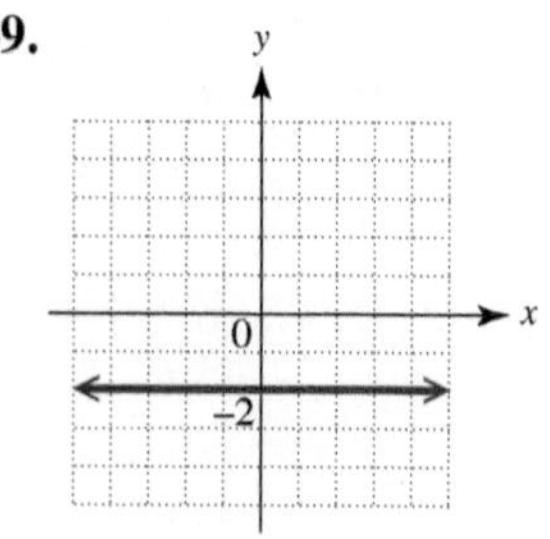

40.
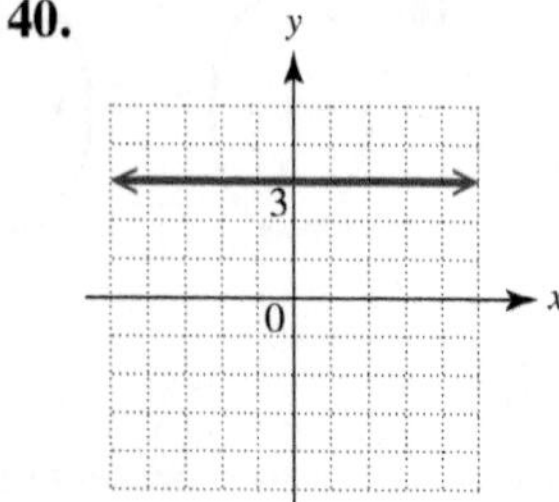

41.
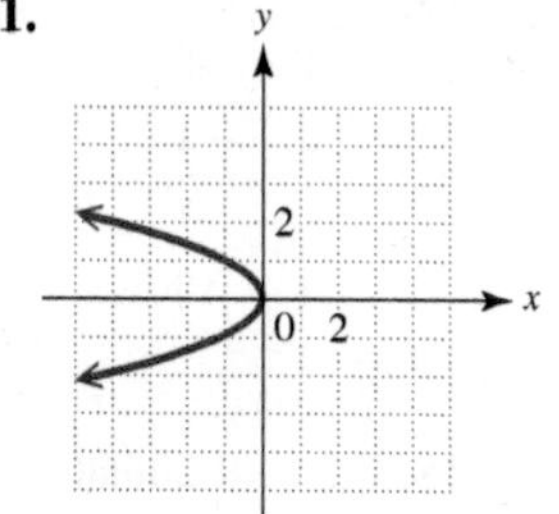

42.
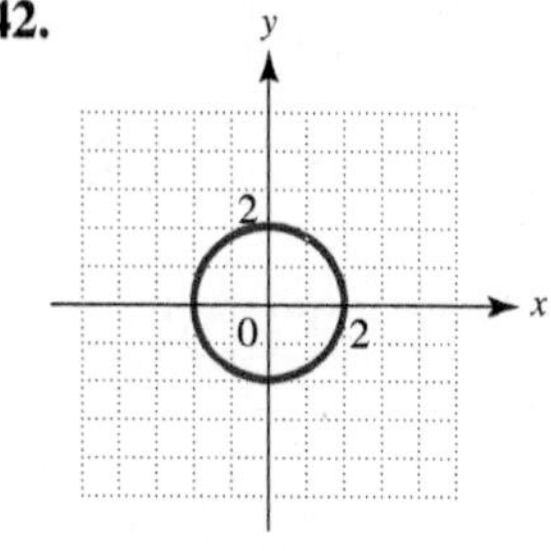

43.
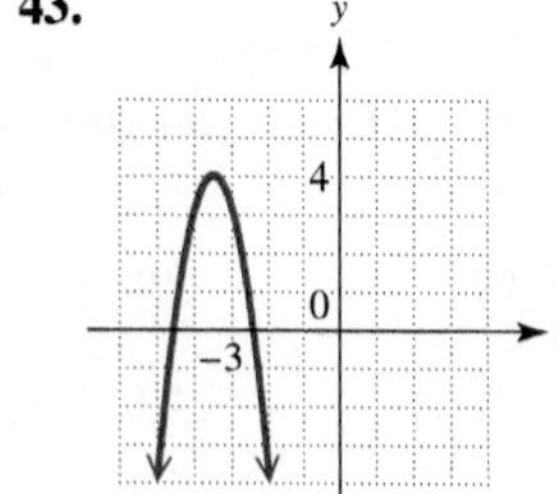

44.
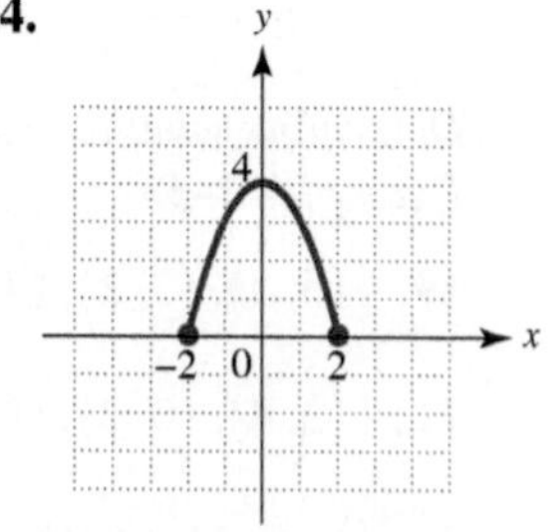

45.
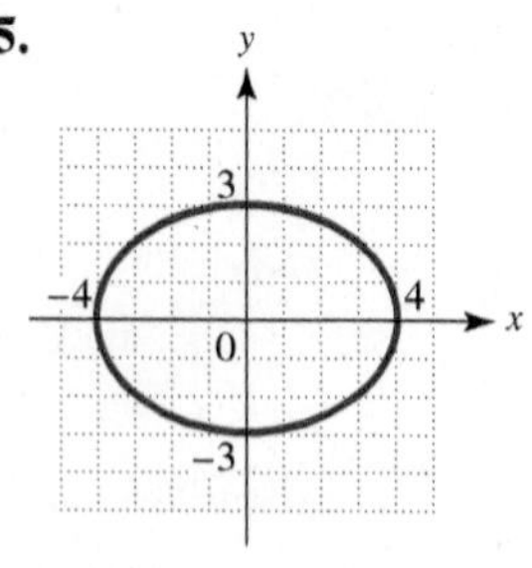

46.
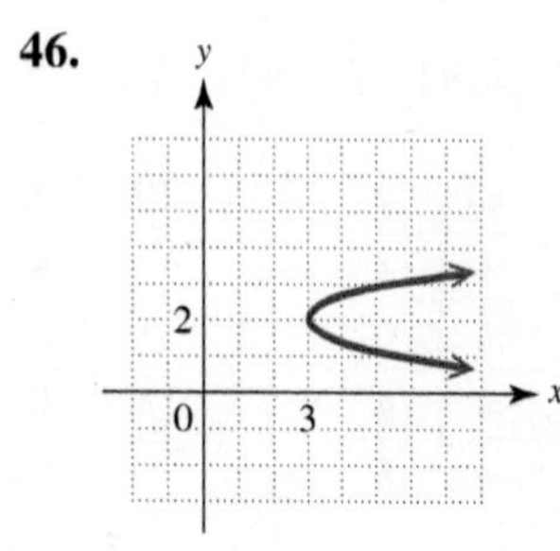

47.
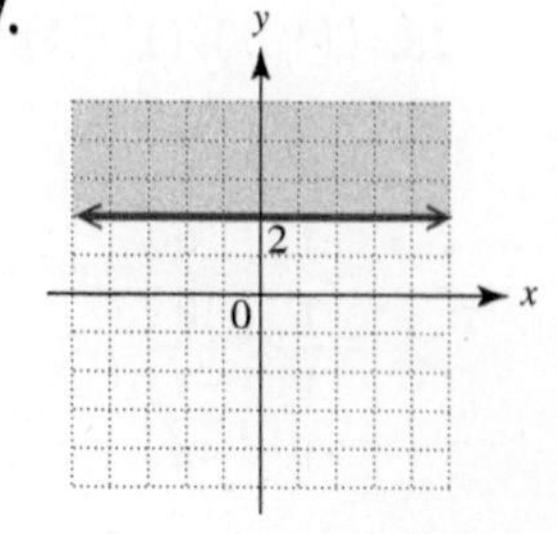

48.
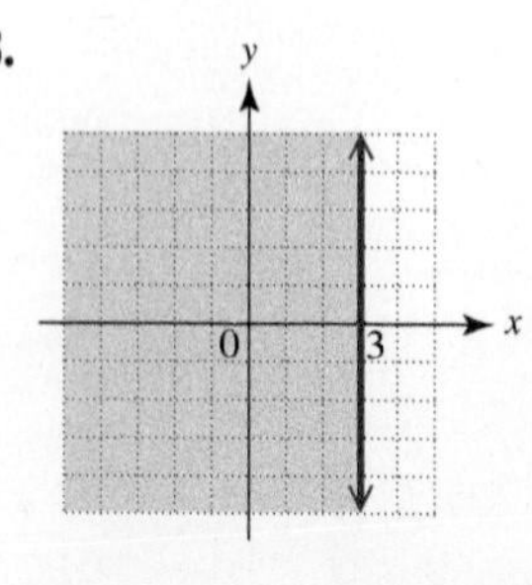

Determine whether each relation defines y as a function of x. (Solve for y first if necessary.) Give the domain. ***See Example 6.***

49. $y = -6x$

50. $y = -9x$

51. $y = 2x - 6$

52. $y = 6x + 8$

53. $y = x^2$

54. $y = x^3$

55. $x = y^6$

56. $x = y^4$

57. $x + y < 4$

58. $x - y < 3$

59. $y = \sqrt{x}$

60. $y = -\sqrt{x}$

61. $y = \sqrt{x - 3}$

62. $y = \sqrt{x - 7}$

63. $y = \sqrt{4x + 2}$

64. $y = \sqrt{2x + 9}$

65. $y = \dfrac{x + 4}{5}$

66. $y = \dfrac{x - 3}{2}$

67. $y = -\dfrac{2}{x}$

68. $y = -\dfrac{6}{x}$

69. $y = \dfrac{2}{x - 4}$

70. $y = \dfrac{7}{x - 2}$

71. $xy = 1$

72. $xy = 3$

Solve each problem.

73. The table shows the percentage of students at 4-year colleges who graduated within 5 years.

Year	*Percentage*
2013	52.8
2014	52.6
2015	52.6
2016	53.2
2017	53.7

Data from ACT.

(a) Does the table define a function?

(b) What are the domain and range?

(c) What is the range element that corresponds to 2015? The domain element that corresponds to 53.7?

(d) Call this function f. Give two ordered pairs that belong to f.

74. The table shows the percentage of persons age 12 or older who smoked cigarettes.

Year	*Percentage*
1985	38.7
2000	24.9
2005	24.9
2010	23.0
2015	19.4

Data from National Survey on Drug Use and Health.

(a) Does the table define a function?

(b) What are the domain and range?

(c) What is the range element that corresponds to 2015? The domain element that corresponds to 23.0?

(d) Call this function g. Give two ordered pairs that belong to g.

2.6 Function Notation and Linear Functions

OBJECTIVES

1 Use function notation.
2 Graph linear and constant functions.

VOCABULARY

□ linear function
□ constant function

OBJECTIVE 1 Use function notation.

When a function f is defined with a rule or an equation using x and y for the independent and dependent variables, we say, "*y is a function of x*" to emphasize that y *depends on* x. We use the notation

$$y = f(x),$$

The parentheses here do *not* indicate multiplication.

called **function notation,** to express this and read $f(x)$ as **"f of x,"** or **"f at x."** The letter f is a name for this particular function. For example, if $y = 3x - 5$, we can name this function f and write the following.

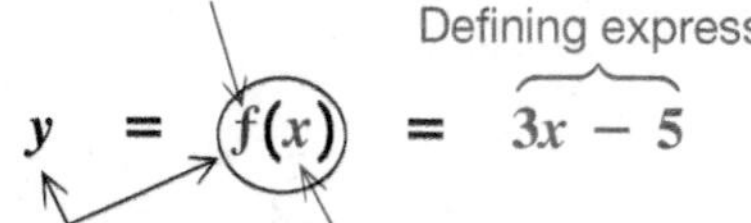

$f(x)$ is just another name for the dependent variable y.

We evaluate a function at different values of x by substituting x-values from the domain into the function.

NOW TRY EXERCISE 1

Let $f(x) = 4x + 3$. Find the value of function f for each value of x.

(a) $x = -2$ **(b)** $x = 0$

EXAMPLE 1 Evaluating a Function

Let $f(x) = 3x - 5$. Find the value of function f for each value of x.

(a) $x = 2$

$f(x) = 3x - 5$

$f(2) = 3 \cdot 2 - 5$ Replace x with 2. (Read $f(2)$ as "f of 2" or "f at 2.")

$f(2) = 6 - 5$ Multiply.

$f(2) = 1$ Subtract.

For $x = 2$, the corresponding function value (or y-value) is 1. $f(2) = 1$ symbolizes the statement

"If $x = 2$ in the function f, then $y = 1$"

and is represented by the ordered pair $(2, 1)$.

(b) $x = -1$

$f(x) = 3x - 5$

$f(-1) = 3(-1) - 5$ Replace x with -1. (Use parentheses to avoid errors.)

$f(-1) = -3 - 5$ Multiply.

$f(-1) = -8$ Subtract.

Thus, $f(-1) = -8$ and the ordered pair $(-1, -8)$ belongs to f. NOW TRY

NOW TRY ANSWERS

1. **(a)** -5 **(b)** 3

CAUTION The symbol $f(x)$ does *not* indicate "f times x," but represents the y-value associated with the indicated x-value. As shown in **Example 1(a),** $f(2)$ is the y-value that corresponds to the x-value 2 in f.

NOTE In the function $f(x) = 3x - 5$ in **Example 1,** $f(2) = 1$ and $f(-1) = -8$ correspond to the ordered pairs (2, 1) and (−1, −8). Because the domain of f is $(-\infty, \infty)$—that is, x can be any real number—this function defines an infinite set of ordered pairs whose graph is a line with slope 3 and y-intercept (0, −5). See **FIGURE 56.** This makes sense because $f(x)$ is another name for y, and

$$f(x) = 3x - 5 \quad \text{is equivalent to} \quad y = 3x - 5. \qquad f \text{ is a linear function.}$$

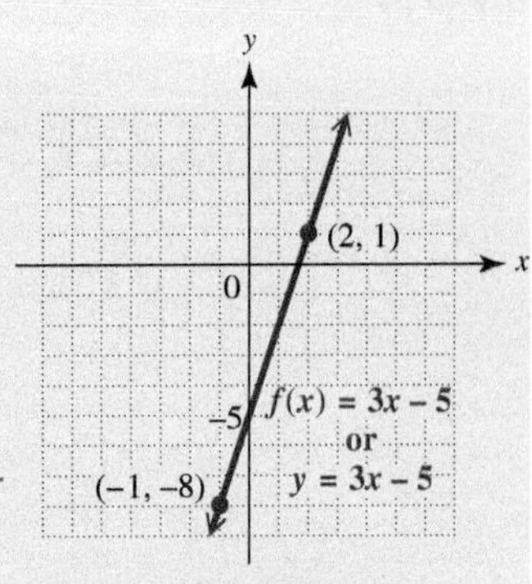

FIGURE 56

NOW TRY EXERCISE 2

Let $f(x) = 2x^2 - 4x + 1$. Find the following.

(a) $f(-2)$ **(b)** $f(a)$

EXAMPLE 2 Evaluating a Function

Let $f(x) = -x^2 + 5x - 3$. Find the following.

(a) $f(4)$

Do *not* read this as "f times 4." Read it as "f of 4," or "f at 4."

$f(x) = -x^2 + 5x - 3$ The base in $-x^2$ is x, *not* $(-x)$.

$f(4) = -4^2 + 5 \cdot 4 - 3$ Replace x with 4.

$f(4) = -16 + 20 - 3$ Apply the exponent. Multiply.

$f(4) = 1$ Add and subtract.

Thus, $f(4) = 1$, and the ordered pair (4, 1) belongs to f.

(b) $f(q)$

$f(x) = -x^2 + 5x - 3$

$f(q) = -q^2 + 5q - 3$ Replace x with q.

The replacement of one variable with another is important in later courses.

NOW TRY 

Sometimes letters other than f, such as g, h, or capital letters F, G, and H are used to name functions.

NOW TRY EXERCISE 3

Let $g(x) = 8x - 5$. Find and simplify $g(a - 2)$.

EXAMPLE 3 Evaluating a Function

Let $g(x) = 2x + 3$. Find and simplify $g(a + 1)$.

$g(x) = 2x + 3$

$g(a + 1) = 2(a + 1) + 3$ Replace x with $a + 1$.

$g(a + 1) = 2a + 2 + 3$ Distributive property

$g(a + 1) = 2a + 5$ Add.

NOW TRY ANSWERS

2. (a) 17 **(b)** $2a^2 - 4a + 1$

3. $8a - 21$

NOW TRY EXERCISE 4

For each function, find $f(-1)$.

(a) $f = \{(-5, -1), (-3, 2), (-1, 4)\}$

(b) $f(x) = x^2 - 12$

EXAMPLE 4 Evaluating Functions

For each function, find $f(3)$.

(a) $f(x) = 3x - 7$

$f(3) = 3(3) - 7$ Replace x with 3.

$f(3) = 9 - 7$ Multiply.

$f(3) = 2$ Subtract.

(b)

x	$y = f(x)$
6	−12
3	−6
0	0
−3	6

← Here, $f(3) = -6$.

(c) $f = \{(-3, 5), (0, 3), (3, 1), (6, -1)\}$

We want $f(3)$, the y-value of the ordered pair whose first component is $x = 3$. As indicated by the ordered pair $(3, 1)$, for $x = 3$, $y = 1$. Thus, $f(3) = 1$.

(d)

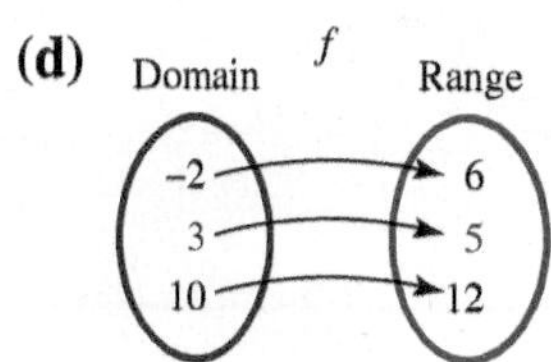

The domain element 3 is paired with 5 in the range, so

$$f(3) = 5.$$

NOW TRY 

EXAMPLE 5 Finding Function Values from a Graph

Refer to the function graphed in **FIGURE 57.**

FIGURE 57

(a) Find $f(3)$.

Locate 3 on the x-axis. See **FIGURE 58.** Moving up to the graph of f and over to the y-axis gives 4 for the corresponding y-value. Thus, $f(3) = 4$, which corresponds to the ordered pair $(3, 4)$.

(b) Find $f(0)$.

Refer to **FIGURE 58** to see that $f(0) = 1$.

NOW TRY EXERCISE 5

Refer to the function graphed in **FIGURE 57.**

(a) Find $f(-1)$.

(b) For what value of x is $f(x) = 2$?

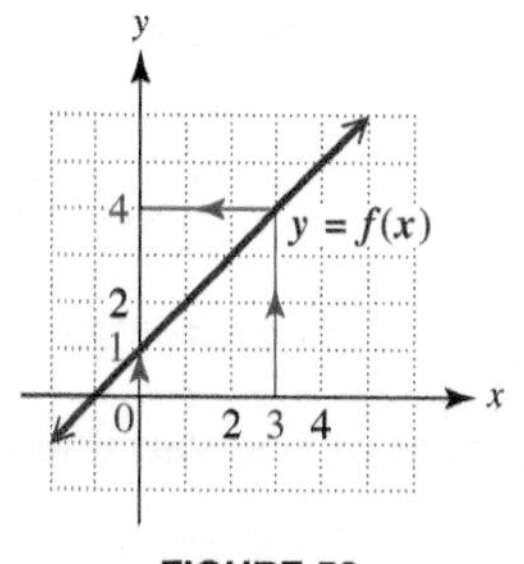

FIGURE 58

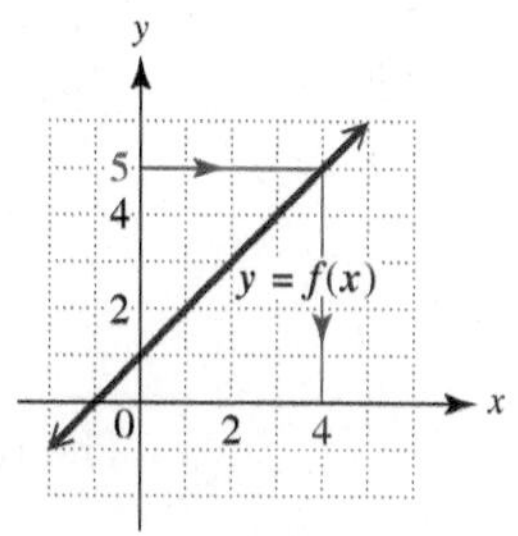

FIGURE 59

(c) For what value of x is $f(x) = 5$?

Because $f(x) = y$, we want the value of x that corresponds to $y = 5$. See **FIGURE 59,** and locate 5 on the y-axis. Moving across to the graph of f and down to the x-axis gives $x = 4$. Thus, $f(4) = 5$, which corresponds to the ordered pair $(4, 5)$. NOW TRY

If a function f is defined by an equation in x and y instead of function notation, use the following steps to find $f(x)$.

Writing an Equation Using Function Notation

Step 1 Solve the equation for y if it is not given in that form.

Step 2 Replace y with $f(x)$.

NOW TRY ANSWERS

4. (a) 4 (b) −11
5. (a) 0 (b) 1

NOW TRY EXERCISE 6

Write the equation using function notation $f(x)$. Then find $f(-3)$.

$$-4x^2 + y = 5$$

EXAMPLE 6 Writing Equations Using Function Notation

Write each equation using function notation $f(x)$. Then find $f(-2)$.

(a) $y = x^2 + 1$ ⟵ This equation is already solved for y. (Step 1)

$f(x) = x^2 + 1$ Replace y with $f(x)$. (Step 2)

To find $f(-2)$, let $x = -2$.

$$f(x) = x^2 + 1$$

$$f(-2) = (-2)^2 + 1 \quad \text{Let } x = -2.$$

$$f(-2) = 4 + 1 \quad (-2)^2 = (-2)(-2)$$

$$f(-2) = 5 \quad \text{Add.}$$

(b) $x - 4y = 5$

Step 1 $-4y = -x + 5$ Subtract x.

$$y = \frac{1}{4}x - \frac{5}{4} \quad \text{Divide by } -4.$$

Step 2 $f(x) = \frac{1}{4}x - \frac{5}{4}$ Replace y with $f(x)$.

Now find $f(-2)$.

$$f(-2) = \frac{1}{4}(-2) - \frac{5}{4} = -\frac{7}{4} \quad \text{Let } x = -2.$$

NOW TRY 

OBJECTIVE 2 Graph linear and constant functions.

Linear equations (except for vertical lines with equations of the form $x = a$) define *linear functions.*

Linear Function

A function f that can be written in the form

$$f(x) = ax + b,$$

where a and b are real numbers, is a **linear function.** The value of a is the slope m of the graph of the function. The domain of a linear function is $(-\infty, \infty)$, unless specified otherwise.

Examples:

$$f(x) = 2x + 4, \quad f(x) = -5x, \quad f(x) = -\frac{1}{2}x - \frac{5}{4}, \quad f(x) = 0.75x + 10$$

A linear function whose graph is a horizontal line has the form

$$f(x) = b \quad \text{Constant function}$$

and is a **constant function.** The range of any nonconstant linear function is $(-\infty, \infty)$, but the range of a constant function $f(x) = b$ is $\{b\}$.

NOW TRY ANSWER

6. $f(x) = 4x^2 + 5$; $f(-3) = 41$

NOW TRY EXERCISE 7

Graph each function. Give the domain and range.

(a) $f(x) = \frac{1}{3}x - 2$

(b) $f(x) = -3$

EXAMPLE 7 Graphing Linear and Constant Functions

Graph each function. Give the domain and range.

(a) $f(x) = \frac{1}{4}x - \frac{5}{4}$ (from **Example 6(b)**)

Slope ↑ ↑ y-intercept is $\left(0, -\frac{5}{4}\right)$.

To graph this function, plot the y-intercept $\left(0, -\frac{5}{4}\right)$. Use the geometric definition of slope as $\frac{\text{rise}}{\text{run}}$ to find a second point on the line. The slope is $\frac{1}{4}$, so we move up 1 unit from $\left(0, -\frac{5}{4}\right)$ and to the right 4 units to the point $\left(4, -\frac{1}{4}\right)$. Draw the straight line through these points. See **FIGURE 60.** The domain and range are both $(-\infty, \infty)$.

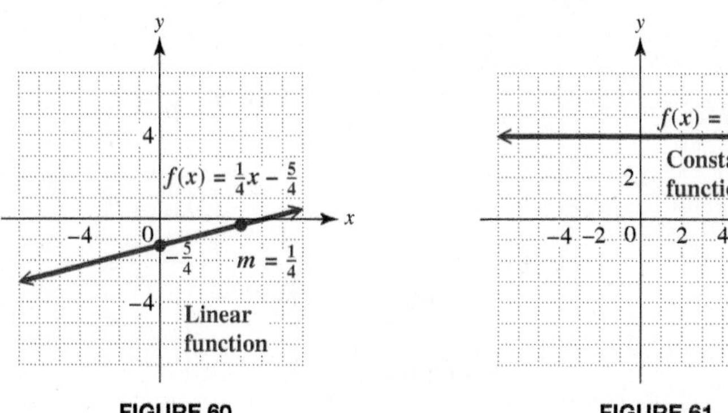

FIGURE 60 FIGURE 61

(b) $f(x) = 4$

The graph of this constant function is the horizontal line containing all points with y-coordinate 4. See **FIGURE 61.** The domain is $(-\infty, \infty)$. The value of $f(x)$—that is, y—is 4 for *every* value of x, so the range is $\{4\}$.

NOW TRY

NOW TRY ANSWERS

7. (a)

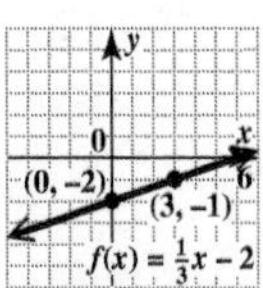

domain: $(-\infty, \infty)$; range: $(-\infty, \infty)$

(b)

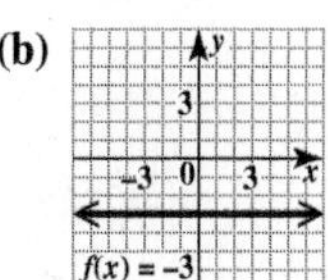

domain: $(-\infty, \infty)$; range: $\{-3\}$

2.6 Exercises

MyLab Math

Video solutions for select problems available in MyLab Math

STUDY SKILLS REMINDER

Time management can be a challenge for students. **Review Study Skill 6, *Managing Your Time.***

Concept Check *Work each problem.*

1. To emphasize that "y is a function of x" for a given function f, we use function notation and write $y =$ ______. Here, f is the name of the ______, x is a value from the ______, and $f(x)$ is the function value (or y-value) that corresponds to ______. We read $f(x)$ as "______."

2. Choose the correct response.

For a function f, the notation $f(3)$ means ______.

A. the variable f times 3, or $3f$.

B. the value of the dependent variable when the independent variable is 3.

C. the value of the independent variable when the dependent variable is 3.

D. f equals 3.

3. Fill in each blank with the correct response.

The equation $2x + y = 4$ has a straight ______ as its graph. One point that lies on the graph is (3, ____). If we solve the equation for y and use function notation, we have a(n) ______ function $f(x) =$ ______. For this function, $f(3) =$ ____, meaning that the point (____, ____) lies on the graph of the function.

4. Which of the following defines y as a linear function of x?

A. $y = \frac{1}{4}x - \frac{5}{4}$ **B.** $y = \frac{1}{x}$ **C.** $y = x^2$ **D.** $y = \sqrt{x}$

Let $f(x) = -3x + 4$ *and* $g(x) = -x^2 + 4x + 1$. *Find the following.* ***See Examples 1–3.***

5. $f(0)$ **6.** $g(0)$ **7.** $f(-3)$ **8.** $f(-5)$

9. $g(-2)$ **10.** $g(-1)$ **11.** $g(3)$ **12.** $g(10)$

13. $f(100)$ **14.** $f(-100)$ **15.** $f\left(\frac{1}{3}\right)$ **16.** $f\left(\frac{7}{3}\right)$

17. $g(0.5)$ **18.** $g(1.5)$ **19.** $f(p)$ **20.** $g(k)$

21. $f(-x)$ **22.** $g(-x)$ **23.** $f(x + 2)$ **24.** $f(x - 2)$

25. $f(2t + 1)$ **26.** $f(3t - 2)$ **27.** $g(\pi)$ **28.** $g(t)$

29. $f(x + h)$ **30.** $f(a + b)$ **31.** $g\left(\frac{p}{3}\right)$ **32.** $g\left(\frac{1}{x}\right)$

For each function, find ***(a)*** $f(2)$ *and* ***(b)*** $f(-1)$. ***See Examples 4, 5(a), and 5(b).***

33. $f = \{(-2, 2), (-1, -1), (2, -1)\}$

34. $f = \{(-1, -5), (0, 5), (2, -5)\}$

35. $f = \{(-1, 3), (4, 7), (0, 6), (2, 2)\}$

36. $f = \{(2, 5), (3, 9), (-1, 11), (5, 3)\}$

37.

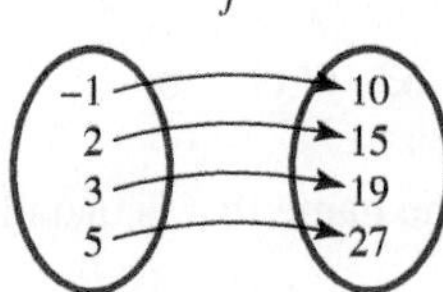

38.

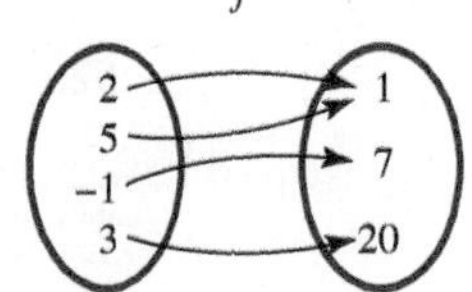

39.

x	y = f(x)
2	4
1	1
0	0
−1	1
−2	4

40.

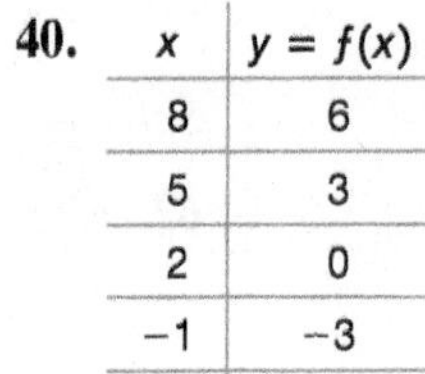

x	y = f(x)
8	6
5	3
2	0
−1	−3
−4	−6

41.

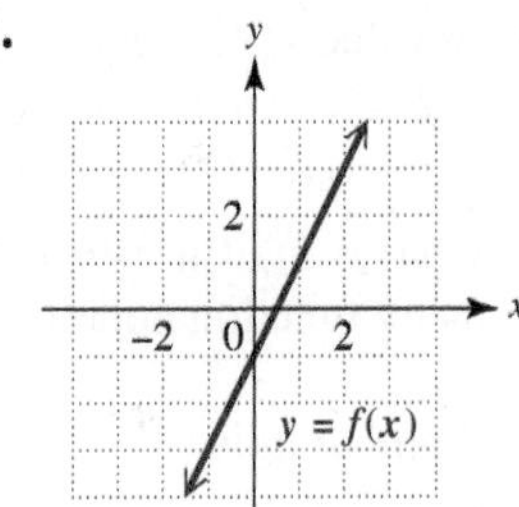

42.

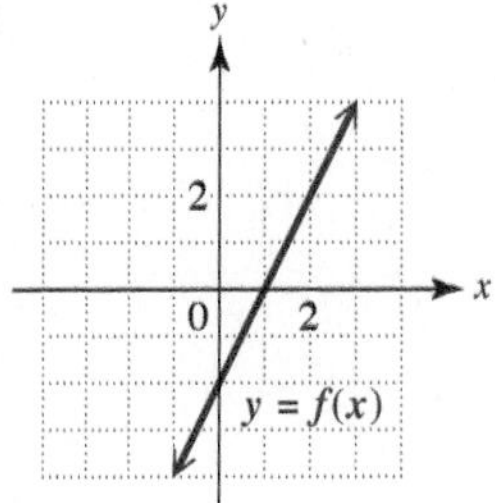

43.

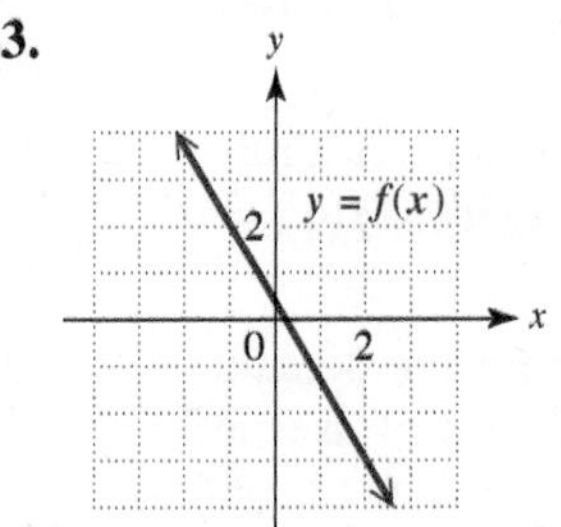

44.

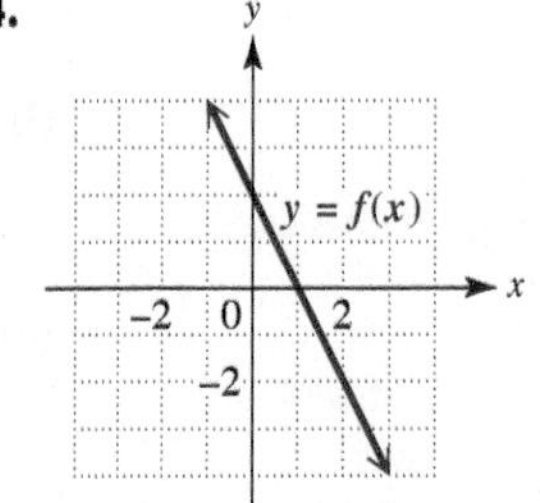

Refer to the given graph. Find the value of x for each value of $f(x)$. ***See Example 5(c).***

45. **(a)** $f(x) = 3$
(b) $f(x) = -1$
(c) $f(x) = -3$

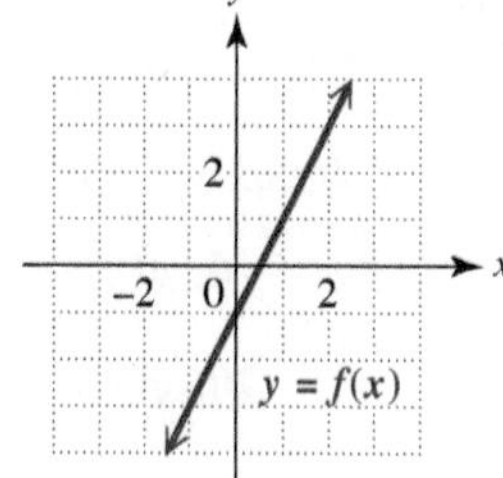

46. **(a)** $f(x) = 4$
(b) $f(x) = -2$
(c) $f(x) = 0$

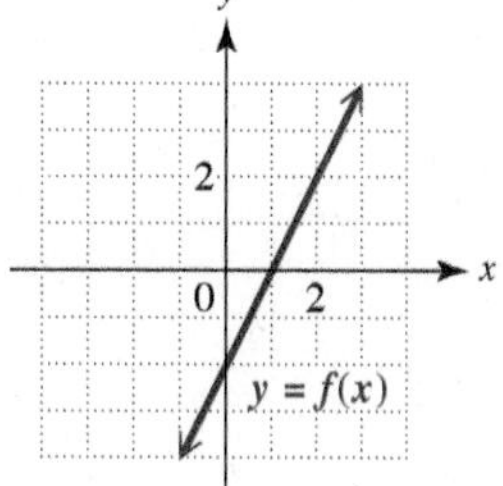

An equation that defines y as a function f of x is given. ***(a)*** *Solve for y in terms of x, and write each equation using function notation $f(x)$.* ***(b)*** *Find $f(3)$.* ***See Example 6.***

47. $x + 3y = 12$ **48.** $x - 4y = 8$ **49.** $y + 2x^2 = 3$

50. $y - 3x^2 = 2$ **51.** $4x - 3y = 8$ **52.** $-2x + 5y = 9$

Graph each linear or constant function. Give the domain and range. ***See Example 7.***

53. $f(x) = -2x + 5$ **54.** $g(x) = 4x - 1$ **55.** $h(x) = \frac{1}{2}x + 2$

56. $F(x) = -\frac{1}{4}x + 1$ **57.** $f(x) = x$ **58.** $f(x) = -x$

59. $H(x) = -3x$ **60.** $G(x) = 2x$ **61.** $g(x) = -4$

62. $f(x) = 5$ **63.** $f(x) = 0$ **64.** $f(x) = -2.5$

65. Concept Check What is the name that is usually given to the graph of $f(x) = 0$?

66. Concept Check Can the graph of a linear function have an undefined slope? Explain.

Solve each problem.

67. A package weighing x pounds costs $f(x)$ dollars to mail to a given location, where

$$f(x) = 3.75x.$$

(a) Evaluate $f(3)$.

(b) Describe what 3 and the value $f(3)$ mean in part (a), using the terms *independent variable* and *dependent variable.*

(c) How much would it cost to mail a 5-lb package? Express this situation using function notation.

68. A taxicab driver charges \$2.50 per mile.

(a) Fill in the table with the correct response for the price $f(x)$ the driver charges for a trip of x miles.

(b) The linear function that gives a rule for the amount charged is $f(x) =$ ______.

(c) Graph this function for the domain $\{0, 1, 2, 3\}$ using the set of axes at the right.

x	$f(x)$
0	
1	
2	
3	

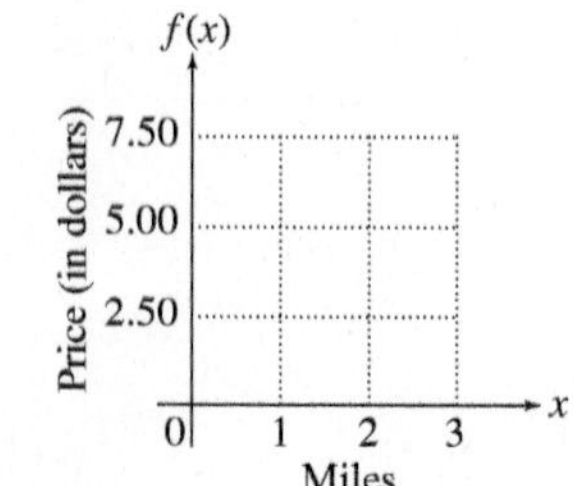

69. To print t-shirts, there is a \$100 set-up fee, plus a \$12 charge per t-shirt. Let x represent the number of t-shirts printed and $f(x)$ represent the total charge.

(a) Write a linear function that models this situation.

(b) Find $f(125)$. Interpret the answer in the context of this problem.

(c) Find the value of x if $f(x) = 1000$. Express this situation using function notation, and interpret it in the context of this problem.

70. Rental on a car is \$150, plus \$0.50 per mile. Let x represent the number of miles driven and $f(x)$ represent the total cost to rent the car.

(a) Write a linear function that models this situation.

(b) Find $f(250)$. Interpret the answer in the context of this problem.

(c) Find the value of x if $f(x) = 400$. Express this situation using function notation, and interpret it in the context of this problem.

71. The table represents a linear function.

(a) What is $f(2)$?

(b) If $f(x) = -1.3$, what is the value of x?

(c) What is the slope of the line?

(d) What is the y-intercept of the line?

(e) Using the answers from parts (c) and (d), write an equation for $f(x)$.

x	$y = f(x)$
0	3.5
1	2.3
2	1.1
3	−0.1
4	−1.3

72. The table represents a linear function.

(a) What is $f(2)$?

(b) If $f(x) = 2.1$, what is the value of x?

(c) What is the slope of the line?

(d) What is the y-intercept of the line?

(e) Using the answers from parts (c) and (d), write an equation for $f(x)$.

x	$y = f(x)$
−1	−3.9
0	−2.4
1	−0.9
2	0.6
3	2.1

73. The graph shows water in a swimming pool over time.

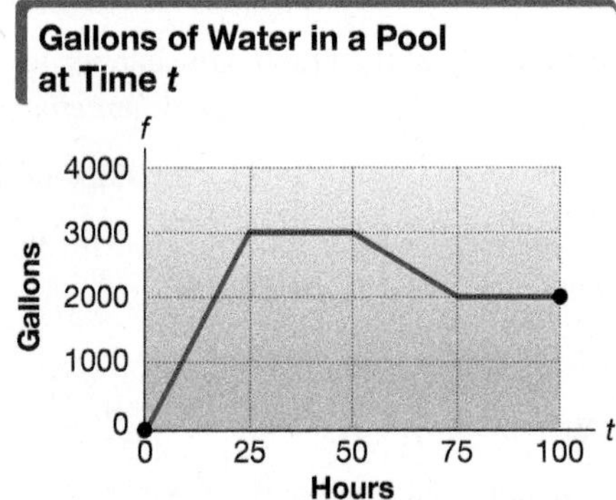

(a) What numbers are possible values of the independent variable? The dependent variable?

(b) For how long is the water level increasing? Decreasing?

(c) How many gallons of water are in the pool after 90 hr?

(d) Call this function f. What is $f(0)$? What does it mean?

(e) What is $f(25)$? What does it mean?

74. The graph shows electricity use on a summer day.

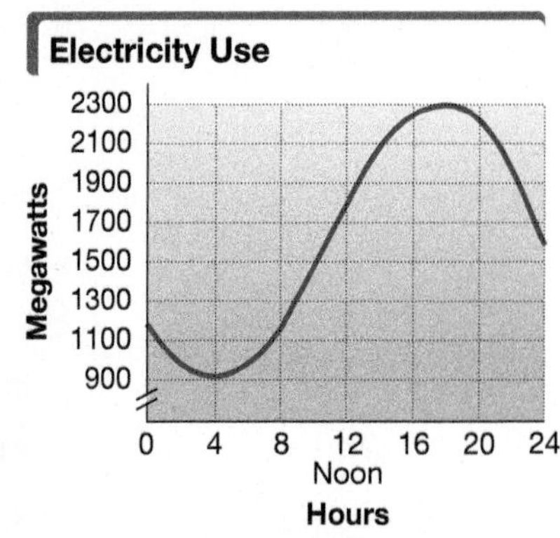

(a) Is this the graph of a function? Why or why not?

(b) What is the domain?

(c) Estimate the number of megawatts used at 8 A.M.

(d) At what time was the most electricity used? The least electricity?

(e) Call this function f. What is $f(12)$? What does it mean?

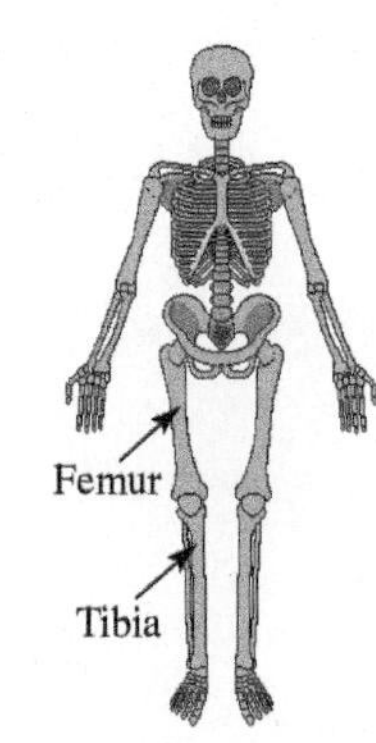

75. Forensic scientists use the lengths of certain bones to calculate the height of a person. Two such bones are the tibia (t), the bone from the ankle to the knee, and the femur (r), the bone from the knee to the hip socket. A person's height (h) in centimeters is determined from the lengths of these bones using the following functions.

For men: $h(r) = 69.09 + 2.24r$ or $h(t) = 81.69 + 2.39t$

For women: $h(r) = 61.41 + 2.32r$ or $h(t) = 72.57 + 2.53t$

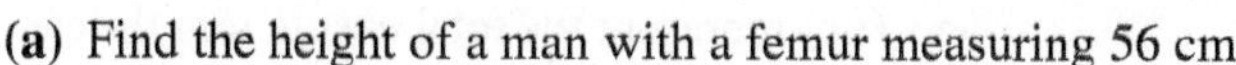

(a) Find the height of a man with a femur measuring 56 cm.

(b) Find the height of a man with a tibia measuring 40 cm.

(c) Find the height of a woman with a femur measuring 50 cm.

(d) Find the height of a woman with a tibia measuring 36 cm.

76. Based on federal regulations, a pool to house sea otters must have a volume that is "the square of the sea otter's average adult length (in meters) multiplied by 3.14 and by 0.91 meter." If x represents the sea otter's average adult length and $f(x)$ represents the volume (in cubic meters) of the corresponding pool size, this formula can be written as the function

$$f(x) = 0.91(3.14)x^2.$$

Find the volume of the pool for each adult sea otter length (in meters). Round answers to the nearest hundredth.

(a) 0.8 **(b)** 1.0 **(c)** 1.2 **(d)** 1.5

RELATING CONCEPTS For Individual or Group Work **(Exercises 77–85)**

Refer to the straight-line graph and ***work Exercises 77–85 in order.***

77. By just looking at the graph, how can we tell whether the slope is *positive, negative, 0,* or *undefined*?

78. Apply the slope formula to find the slope of the line.

79. What is the slope of any line parallel to the line shown? Perpendicular to the line shown?

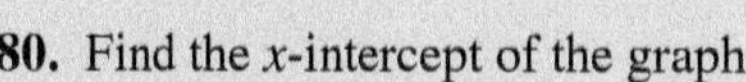

80. Find the x-intercept of the graph.

81. Find the y-intercept of the graph.

82. Use function notation to write the equation of the graphed line. Use f to designate the function.

83. Find $f(8)$.

84. If $f(x) = -8$, what is the value of x?

85. From the graph, find $f(1)$. Then confirm this value using the equation written in **Exercise 82.**

Chapter 2 Summary

STUDY SKILLS **REMINDER**

How do you best prepare for a test? **Review Study Skill 7, *Reviewing a Chapter.***

Key Terms

2.1
ordered pair
origin
x-axis
y-axis
rectangular (Cartesian) coordinate system
components
plot
coordinate
quadrant
table of ordered pairs (table of values)
graph of an equation
first-degree equation
linear equation in two variables
x-intercept
y-intercept

2.2
rise
run
slope

2.3
scatter diagram

2.4
linear inequality in two variables
boundary line

2.5
relation
function
dependent variable
independent variable
domain
range

2.6
linear function
constant function

New Symbols

(x, y) ordered pair

(x_1, y_1) subscript notation (read "*x-sub-one, y-sub-one*")

Δ Greek letter delta

m slope

$f(x)$ function notation; function of x (read "*f of x*" or "*f at x*")

Test Your Word Power

See how well you have learned the vocabulary in this chapter.

1. An **ordered pair** is a pair of numbers written
A. in numerical order within brackets
B. within parentheses or brackets
C. within parentheses in which the order of the numbers is important
D. within parentheses in which order does not matter.

2. A **linear equation in two variables** is an equation that can be written in the form
A. $Ax + By < C$ **B.** $ax + b = 0$
C. $y = x^2$ **D.** $Ax + By = C$.

3. An **intercept** is
A. the point where the x-axis and y-axis intersect
B. a pair of numbers written between parentheses in which order matters
C. one of the regions determined by a coordinate system
D. the point where a graph intersects the x-axis or the y-axis.

4. The **slope** of a line is
A. the measure of the run over the rise of the line
B. the distance between two points on the line
C. the ratio of the change in y to the change in x along the line
D. the horizontal change compared to the vertical change of two points on the line.

5. A **relation** is
A. a set of ordered pairs
B. the ratio of the change in y to the change in x along a line
C. the set of all possible values of the independent variable
D. all the second components of a set of ordered pairs.

6. A **function** is
A. the numbers in an ordered pair
B. a set of ordered pairs in which each x-value corresponds to exactly one y-value
C. a pair of numbers written within parentheses
D. the set of all ordered pairs that satisfy an equation.

7. The **domain** of a relation is
A. the set of all possible values of the dependent variable y
B. a set of ordered pairs
C. the difference of the x-values
D. the set of all possible values of the independent variable x.

8. The **range** of a relation is
A. the set of all possible values of the dependent variable y
B. a set of ordered pairs
C. the difference of the y-values
D. the set of all possible values of the independent variable x.

ANSWERS

1. C; *Examples:* $(0, 3), (3, 8), (4, 0)$ **2.** D; *Examples:* $3x + 2y = 6, x = y - 7$ **3.** D; *Example:* In the graph of the equation $2x + 3y = 6$, the x-intercept is $(3, 0)$ and the y-intercept is $(0, 2)$. **4.** C; *Example:* The line passing through the points $(3, 6)$ and $(5, 4)$ has slope $\frac{4-6}{5-3} = \frac{-2}{2} = -1$. **5.** A; *Example:* The set $\{(2, 0), (4, 3), (6, 6)\}$ defines a relation. **6.** B; *Example:* The relation given in Answer 5 is a function. **7.** D; *Example:* In the relation in Answer 5, the domain is the set of x-values, $\{2, 4, 6\}$. **8.** A; *Example:* In the relation in Answer 5, the range is the set of y-values, $\{0, 3, 6\}$.

Quick Review

CONCEPTS	EXAMPLES

2.1 Linear Equations in Two Variables

Finding Intercepts

To find the x-intercept, let $y = 0$ and solve for x.

To find the y-intercept, let $x = 0$ and solve for y.

Find the intercepts of the graph of $2x + 3y = 12$.

$2x + 3(0) = 12$ Let $y = 0$.	$2(0) + 3y = 12$ Let $x = 0$.
$2x = 12$	$3y = 12$
$x = 6$	$y = 4$
The x-intercept is $(6, 0)$.	The y-intercept is $(0, 4)$.

Midpoint Formula

The midpoint M of a line segment PQ with endpoints $P(x_1, y_1)$ and $Q(x_2, y_2)$ is found as follows.

$$M = \left(\frac{x_1 + x_2}{2}, \frac{y_1 + y_2}{2}\right)$$

Find the midpoint of the segment with endpoints $(4, -7)$ and $(-10, -13)$.

$$M = \left(\frac{4 + (-10)}{2}, \frac{-7 + (-13)}{2}\right) = (-3, -10)$$

2.2 The Slope of a Line

The slope m of the line passing through the points (x_1, y_1) and (x_2, y_2) is defined as follows.

$$\text{slope } m = \frac{\text{rise}}{\text{run}} = \frac{\text{change in } y}{\text{change in } x} = \frac{\Delta y}{\Delta x} = \frac{y_2 - y_1}{x_2 - x_1}$$

(where $x_1 \neq x_2$)

Find the slope of the graph of $2x + 3y = 12$.

Use the intercepts $(6, 0)$ and $(0, 4)$ and the slope formula.

$$m = \frac{4 - 0}{0 - 6} = \frac{4}{-6} = -\frac{2}{3} \qquad (x_1, y_1) = (6, 0) \quad (x_2, y_2) = (0, 4)$$

A horizontal line has slope 0.

A vertical line has undefined slope.

Parallel lines have the same slope.

The graph of the horizontal line $y = -5$ has slope $m = 0$.

The graph of the vertical line $x = 3$ has undefined slope.

The lines $y = 2x + 3$ and $4x - 2y = 6$ are parallel. Both have slope 2.

$y = 2x + 3$	$4x - 2y = 6$ ⟵ Solve for y.
	$-2y = -4x + 6$
	$y = 2x - 3$

Perpendicular lines, neither of which is vertical, **have slopes that are negative reciprocals** (that is, have a product of -1).

The lines $y = 3x - 1$ and $x + 3y = 4$ are perpendicular. Their slopes are negative reciprocals because $3\left(-\frac{1}{3}\right) = -1$.

$y = 3x - 1$	$x + 3y = 4$ ⟵ Solve for y.
	$3y = -x + 4$
	$y = -\frac{1}{3}x + \frac{4}{3}$

Slope gives the **average rate of change** in y per unit change in x, where the value of y depends on the value of x.

The weight of a young child increased from 30 lb to 60 lb between the ages of 3 and 8. What was the child's average change in weight per year over these years?

$$\frac{60 - 30}{8 - 3} = \frac{30}{5} = 6 \text{ lb per yr}$$

CONCEPTS	EXAMPLES
2.3 Writing Equations of Lines	
Slope-Intercept Form $y = mx + b$	$y = 2x + 3$ $m = 2$; y-intercept is $(0, 3)$.
Point-Slope Form $y - y_1 = m(x - x_1)$	$y - 3 = 4(x - 5)$ $m = 4$; The point $(5, 3)$ is on the line.
Standard Form $Ax + By = C$, where A, B, and C are real numbers and A and B are not both 0. (We give A, B, and C as integers, with greatest common factor 1, and $A \geq 0$.)	$2x - 5y = 8$ Standard form
Horizontal Line $y = b$	$y = 4$ Horizontal line
Vertical Line $x = a$	$x = -1$ Vertical line
2.4 Linear Inequalities in Two Variables	
Graphing a Linear Inequality	Graph $2x - 3y \leq 6$.
Step 1 Draw the graph of the straight line that is the boundary. • Make the line solid if the inequality involves $\leq$ or $\geq$. • Make the line dashed if the inequality involves $<$ or $>$.	Draw the graph of $2x - 3y = 6$ using the intercepts $(3, 0)$ and $(0, -2)$. Draw a solid line because of the inclusion of equality in the symbol $\leq$.
Step 2 Choose any point not on the line as a test point. Substitute the coordinates of that point in the inequality.	Choose $(0, 0)$ as a test point. $2(0) - 3(0) \overset{?}{<} 6$ $0 < 6$ True
Step 3 Shade the region that includes the test point if it satisfies the original inequality. Otherwise, shade the region on the other side of the boundary line.	Shade the region that includes (0, 0). 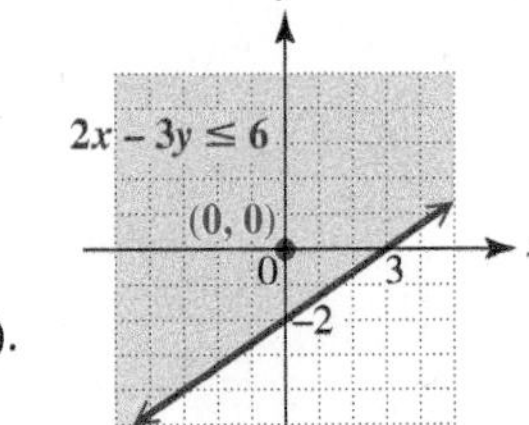
2.5 Introduction to Relations and Functions	
A **relation** is any set of ordered pairs. A **function** is a set of ordered pairs in which, for each distinct value of the first component, there is *exactly one* value of the second component.	The set of ordered pairs $\{(-1, 4), (0, 6), (1, 4)\}$ defines a function.
The set of first components (x-values) is the **domain.**	Domain: $\{-1, 0, 1\}$ Set of x-values
The set of second components (y-values) is the **range.**	Range: $\{4, 6\}$ Set of y-values The equation $y = x^2$ defines a function. Domain: $(-\infty, \infty)$ Range: $[0, \infty)$
2.6 Function Notation and Linear Functions	
To evaluate a function f, where $f(x)$ defines the range value for a given value of x in the domain, substitute the domain value wherever x appears.	Let $f(x) = x^2 - 7x + 12$. Find $f(1)$. $f(x) = x^2 - 7x + 12$ $f(1) = 1^2 - 7(1) + 12$ Let $x = 1$. $f(1) = 6$
To write an equation that defines a function f in function notation, follow these steps.	Write $2x + 3y = 12$ using function notation $f(x)$.
Step 1 Solve the equation for y if it is not given in that form.	$2x + 3y = 12$ $3y = -2x + 12$ Subtract $2x$.
Step 2 Replace y with $f(x)$.	$y = -\frac{2}{3}x + 4$ Divide by 3. $f(x) = -\frac{2}{3}x + 4$ Replace y with $f(x)$.

Chapter 2 Review Exercises

2.1 *Complete the given table of ordered pairs for each equation, and then graph the equation.*

1. $3x + 2y = 10$

x	y
0	
	0
2	
	−2

2. $x - y = 8$

x	y
2	
	−3
3	
	−2

Find the x- and y-intercepts. Then graph each equation.

3. $4x - 3y = 12$ **4.** $5x + 7y = 28$ **5.** $2x + 5y = 20$ **6.** $x - 4y = 8$

Find the midpoint of each segment with the given endpoints.

7. $(-8, -12)$ and $(8, 16)$ **8.** $(0, -5)$ and $(-9, 8)$

2.2 *Find the slope of each line. (In Exercises 17 and 18, coordinates of the points shown are integers.)*

9. Through $(-1, 2)$ and $(4, -5)$

10. Through $(0, 3)$ and $(-2, 4)$

11. Through $(-1, 5)$ and $(-1, -4)$

12. $y = 2x + 3$

13. $3x - 4y = 5$

14. $y = 4$

15. Perpendicular to $3x - y = 4$

16. Parallel to $3y = 2x + 5$

17.

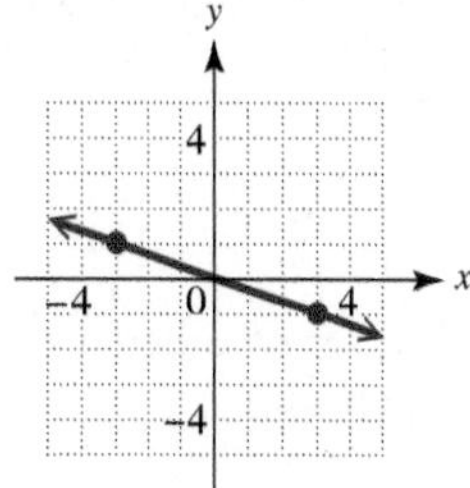

18.

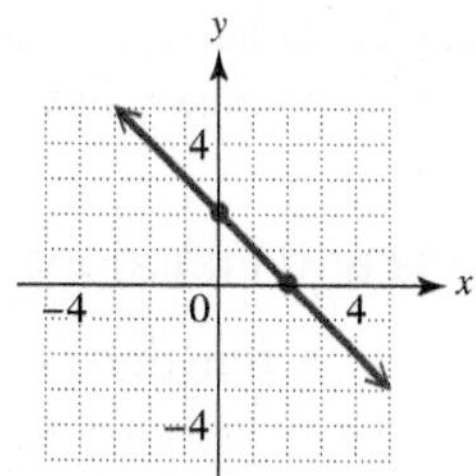

19.

x	y
−2	6
−1	3
0	0
1	−3

20. Determine whether the slope of the line is *positive, negative, 0,* or *undefined.*

(a)

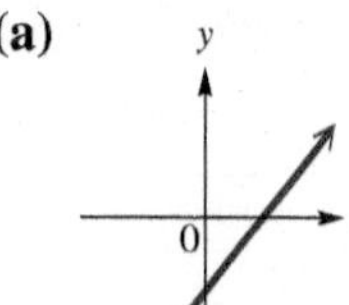

(b)

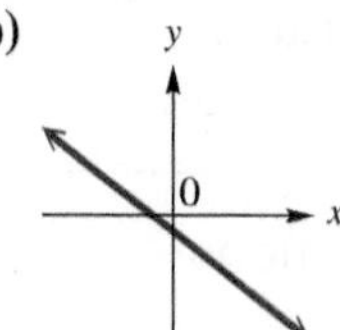

(c)

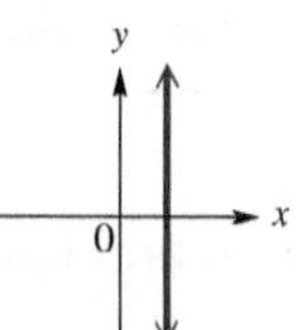

(d)

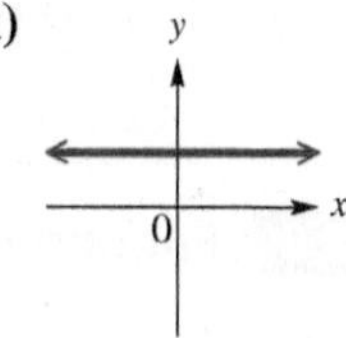

Determine whether each pair of lines is parallel, perpendicular, *or* neither.

21. $3x - y = 6$ and $x + 3y = 12$

22. $3x - y = 4$ and $6x + 12 = 2y$

Solve each problem.

23. If the pitch of a roof is $\frac{1}{4}$, how many feet in the horizontal direction correspond to a rise of 3 ft?

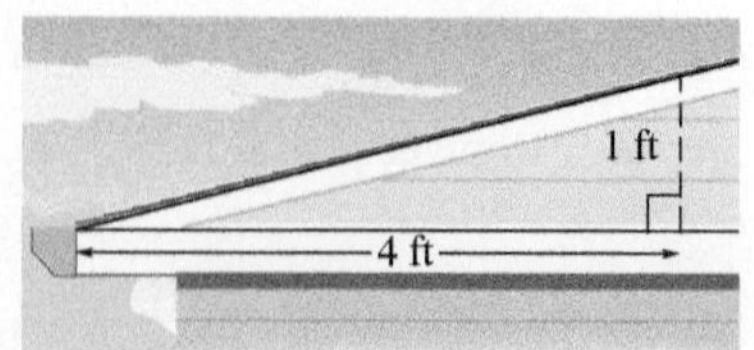

24. In 1980, the median family income in the United States was about \$21,000 per year. In 2016, it was about \$59,039 per year. Find the average rate of change in median family income per year to the nearest dollar during this period. (Data from U.S. Census Bureau.)

2.3 *Write an equation of the line passing through the given point(s) and satisfying any given condition. Give the equation **(a)** in slope-intercept form and **(b)** in standard form.*

25. $(0, -8)$; slope $\frac{3}{5}$

26. $(0, 5)$; slope $-\frac{1}{3}$

27. $(2, -5)$ and $(1, 4)$

28. $(-3, -1)$ and $(2, 6)$

29. $(6, -2)$; parallel to $4x - y = 3$

30. $(0, 1)$; perpendicular to $2x - 5y = 7$

Write an equation of the line passing through the given point and satisfying the given condition.

31. $(0, 12)$; slope 0

32. $(2, 7)$; undefined slope

33. $(0.3, 0.6)$; vertical

34. $(-1, 4)$; horizontal

Solve each problem.

35. A lease on a 2018 Chevrolet Sonic LT hatchback costs \$2829 at signing, plus \$229 per month. Let x represent the number of months since the signing and y represent the amount paid (in dollars). (Data from www.carsdirect.com)

(a) Write an equation in slope-intercept form that represents this situation.

(b) How much will be paid over the lease's full term of 39 months?

36. Average annual charges for tuition, fees, room, and board (in dollars) at private nonprofit 4-year universities in the United States are shown in the graph.

(a) Use the information given for the years 2013 and 2017, letting $x = 13$ represent 2013 and $x = 17$ represent 2017, and letting y represent cost (in dollars), to write an equation that models the data. Write the equation in slope-intercept form. Interpret the slope.

(b) Use the equation from part (a) to approximate costs at private 4-year universities in 2015.

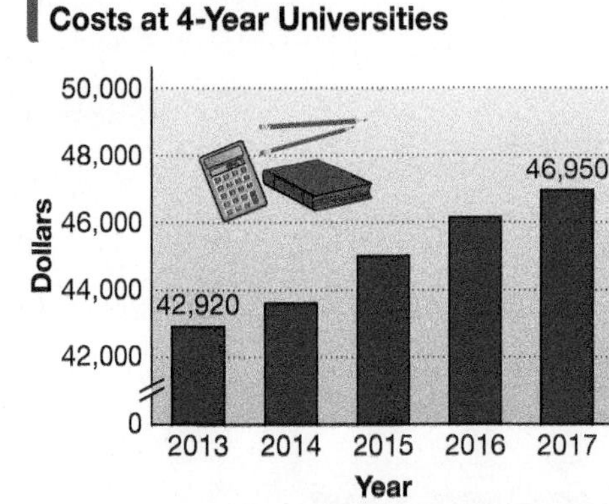

Data from The College Board.

2.4 *Graph each inequality or compound inequality.*

37. $3x - 2y \le 12$

38. $5x - y > 6$

39. $2x + y \le 1$ and $x \ge 2y$

40. $x \ge 2$ or $y \ge 2$

2.5 *Determine whether each relation defines a function, and give the domain and range.*

41. $\{(-4, 2), (-4, -2), (1, 5), (1, -5)\}$

42.

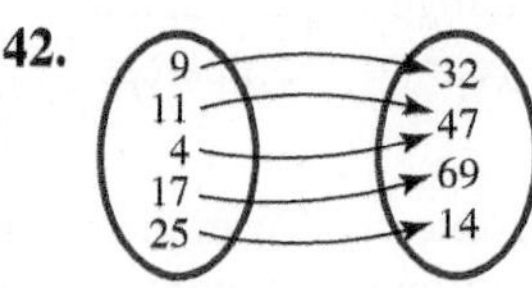

43.

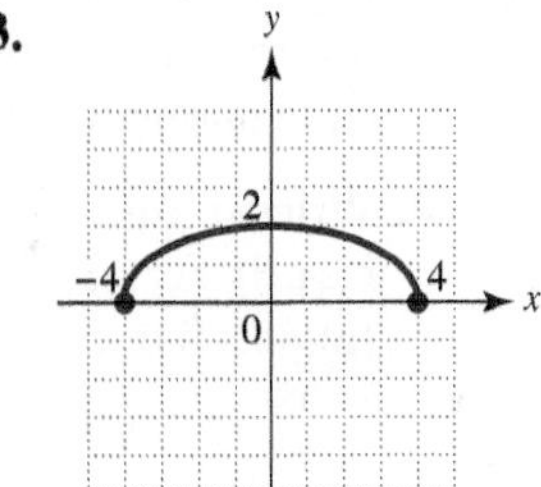

44.

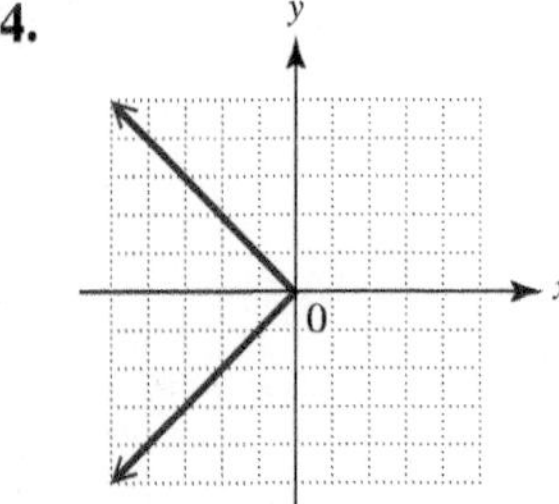

2.5, 2.6 *Determine whether each relation defines y as a function of x. Give the domain. Identify any linear functions.*

45. $y = 3x - 3$ **46.** $y < x + 2$ **47.** $y = |x|$

48. $y = \sqrt{4x + 7}$ **49.** $x = y^2$ **50.** $y = \dfrac{7}{x - 6}$

2.6 *Let* $f(x) = -2x^2 + 3x - 6$. *Find the following.*

51. $f(0)$ **52.** $f(3)$ **53.** $f\left(-\frac{1}{2}\right)$ **54.** $f(k)$

Solve each problem.

55. The equation $2x^2 - y = 0$ defines y as a function f of x. Write it using function notation $f(x)$, and find $f(3)$.

56. The linear equation $2x - 5y = 7$ defines y as a function of x. If $y = f(x)$, which of the following defines the same function?

A. $f(x) = -\frac{2}{5}x + \frac{7}{5}$ **B.** $f(x) = -\frac{2}{5}x - \frac{7}{5}$

C. $f(x) = \frac{2}{5}x - \frac{7}{5}$ **D.** $f(x) = \frac{2}{5}x + \frac{7}{5}$

57. Describe the graph of a constant function.

58. The table shows life expectancy at birth in the United States for selected years.

(a) Does the table define a function?

(b) What are the domain and range?

(c) Call this function f. Give two ordered pairs that belong to f.

(d) Find $f(1980)$. What does this mean?

(e) If $f(x) = 76.8$, what does x equal?

Year	*Life Expectancy at Birth (years)*
1960	69.7
1970	70.8
1980	73.7
1990	75.4
2000	76.8
2010	78.7
2015	78.8

Data from National Center for Health Statistics.

Chapter 2 Mixed Review Exercises

Determine whether each pair of lines is parallel, perpendicular, *or* neither.

1. $4x + 3y = 8$ and $6y = 7 - 8x$ **2.** $3x + y = 4$ and $3y = x - 6$

The graph shows per capita consumption of potatoes (in pounds) in the United States from 2008 to 2016.

3. Use the given ordered pairs to find and interpret the average rate of change in per capita potato consumption per year to the nearest tenth during this period.

4. Write an equation in slope-intercept form that models per capita consumption of potatoes y (in pounds) in year x, where $x = 0$ represents 2008.

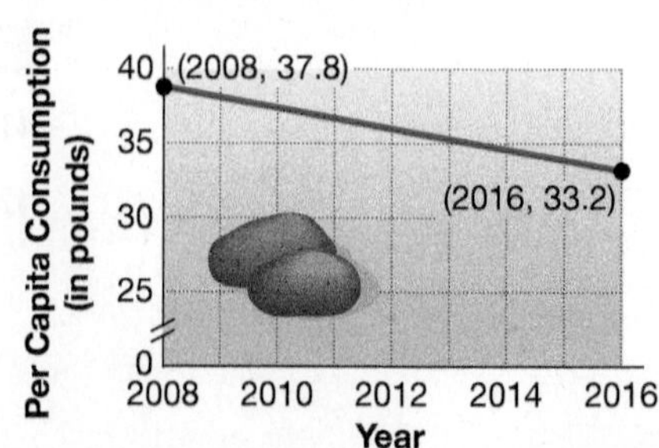

Data from U.S. Department of Agriculture.

Write an equation of a line (in the form specified, if given) passing through the given point(s) and satisfying any given conditions.

5. $(0, 0)$; $m = 3$ (slope-intercept form)

6. $(0, 3)$ and $(-2, 4)$ (standard form)

7. $(2, -3)$; perpendicular to $x = 2$

Solve each problem.

8. Which equations have a graph with just one intercept?

A. $x - 6 = 0$ **B.** $x + y = 0$ **C.** $x - y = 4$ **D.** $y = -4$

9. Which inequality has as its graph a dashed boundary line and shading below the line?

A. $y \geq 4x + 3$ **B.** $y > 4x + 3$ **C.** $y \leq 4x + 3$ **D.** $y < 4x + 3$

10. Refer to the graph of function f.

(a) Find $f(-2)$.

(b) Find $f(0)$.

(c) For what value of x is $f(x) = -3$?

(d) Give the domain and range.

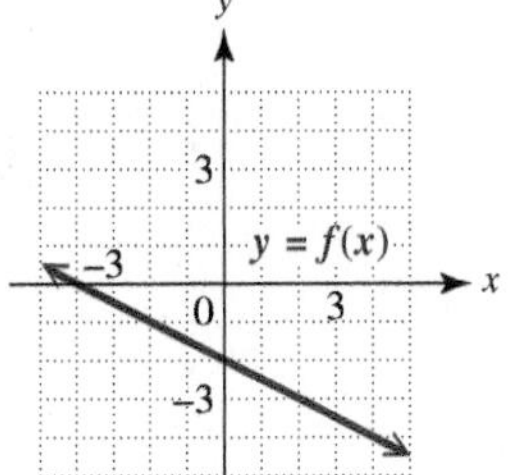

STUDY SKILLS REMINDER

Using test-taking strategies can help you improve your test scores. **Review Study Skill 8, *Taking Math Tests.***

Chapter 2 Test

Step-by-step test solutions are found on the Chapter Test Prep Videos available in MyLab Math.

View the complete solutions to all Chapter Test exercises in MyLab Math.

1. Complete the table of ordered pairs for the equation $2x - 3y = 12$.

x	y
1	
3	
	−4

Find the x- and y-intercepts. Then graph each equation.

2. $4x - 3y = -12$ **3.** $x = 2$ **4.** $y = -2x$

5. Find the slope of the line passing through the points $(6, 4)$ and $(-4, -1)$.

6. Describe how the graph of a line with undefined slope is situated in a rectangular coordinate system.

For each line, find the slope and the x- and y-intercepts.

7. $3x - 2y = 13$ **8.** $y = 5$

9. Which line has positive slope and negative y-coordinate for its y-intercept?

A.

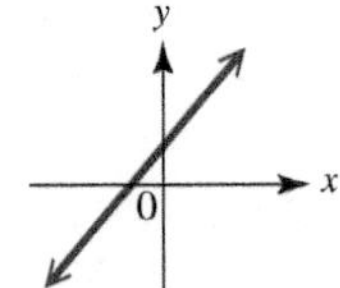

B.

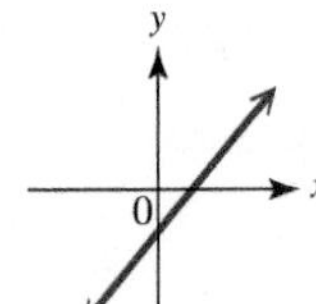

C.

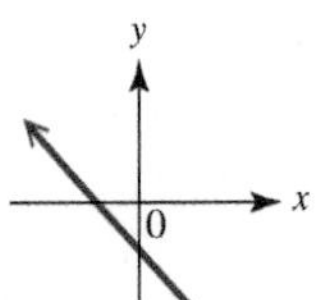

D.

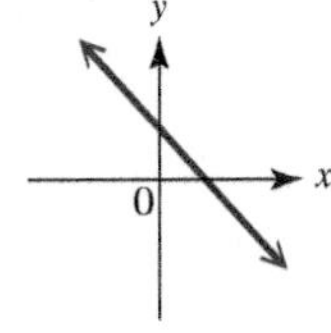

Determine whether each pair of lines is parallel, perpendicular, *or* neither.

10. $5x - y = 8$ and $5y = -x + 3$ **11.** $2y = 3x + 12$ and $3y = 2x - 5$

12. In 2000, there were 94,000 farms in Iowa. As of 2016, there were 87,000. Find and interpret the average rate of change in the number of farms per year, to the nearest whole number. (Data from U.S. Department of Agriculture.)

Write an equation of the line passing through the given point(s) and satisfying any given conditions. Give the equation **(a)** *in slope-intercept form and* **(b)** *in standard form.*

13. $(0, 3)$; slope $-\frac{2}{5}$ **14.** $(-2, 3)$ and $(6, -1)$ **15.** $(4, -1)$; $m = -5$

Write an equation of the line passing through the given point and satisfying the given condition.

16. $(-3, 14)$; horizontal **17.** $(5, -6)$; vertical **18.** $(0, 0)$; horizontal

19. Write an equation in slope-intercept form of the line passing through the point $(-7, 2)$ and satisfying the given condition.

(a) parallel to $3x + 5y = 6$ **(b)** perpendicular to $y = 2x$

20. A ticket to a rock concert costs $142.75. An advance parking pass costs $45. Let x represent the number of tickets purchased and y represent the total cost in dollars.

(a) Write an equation in slope-intercept form that represents this situation.

(b) How much does it cost for 6 tickets and a parking pass?

Graph each inequality or compound inequality.

21. $3x - 2y > 6$ **22.** $y < 2x - 1$ and $x - y < 3$

23. Which of the following is the graph of a function? Give its domain and range.

A.

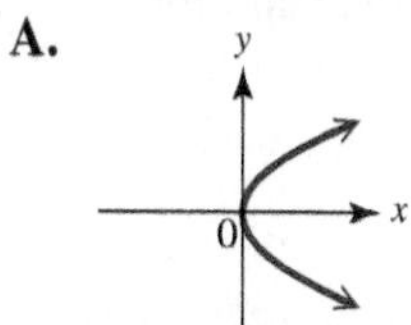

B.

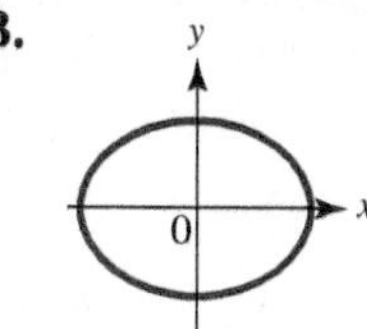

C.

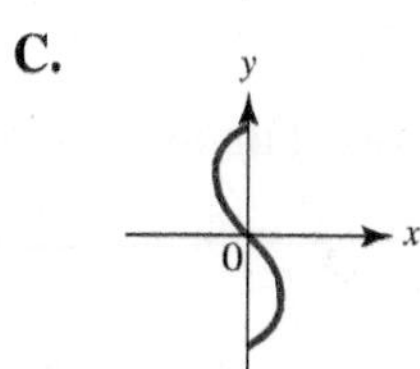

D.

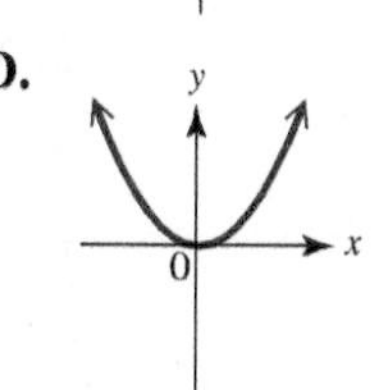

24. Which of the following does *not* define a function? Give its domain and range.

A. $\{(0, 1), (-2, 3), (4, 8)\}$

B. $y = 2x - 6$

C. $y = \sqrt{x + 2}$

D.

x	y
0	1
3	2
0	2
6	3

25. For $f(x) = -x^2 + 2x - 1$, find $f(1)$ and $f(a)$.

26. Graph the linear function $f(x) = \frac{2}{3}x - 1$. Give the domain and range.

Chapters R–2 Cumulative Review Exercises

1. Add or subtract as indicated.

(a) $\frac{1}{6} + \frac{3}{4}$ **(b)** $\frac{11}{12} - \frac{1}{4}$ **(c)** $7.5 - 2.75$

2. Multiply or divide as indicated.

(a) $6 \div \frac{3}{4}$ **(b)** 1.5×100 **(c)** $1.25 \div 10$

3. Complete the table of fraction, decimal, and percent equivalents.

	Fraction in Lowest Terms (or Whole Number)	*Decimal*	*Percent*
(a)	$\frac{1}{100}$		
(b)		0.5	
(c)			75%
(d)		1.0	

Determine whether each statement is always true, sometimes true, *or* never true. *If the statement is* sometimes true, *give examples for which it is true and for which it is false.*

4. The absolute value of a negative number equals the additive inverse of the number.

5. The sum of two negative numbers is positive.

6. The sum of a positive number and a negative number is 0.

Perform each operation.

7. $-|-2| - 4 + |-3|$ **8.** $(-0.8)^2$ **9.** -3^2 **10.** $\sqrt{-64}$

Evaluate each expression for $p = -4$, $q = \frac{1}{2}$, and $r = 16$.

11. $-3(2q - 3p)$ **12.** $\frac{\sqrt{r}}{8p + 2r}$

Simplify each expression.

13. $-(-4x + 3)$ **14.** $\sqrt{25} - 5(-1)^2$

Solve each equation.

15. $2z - 5 + 3z = 4 - (z + 2)$ **16.** $\frac{3x - 1}{5} + \frac{x + 2}{2} = -\frac{3}{10}$

Solve each problem.

17. If each side of a square were increased by 4 in., the perimeter would be 8 in. less than twice the perimeter of the original square. Find the length of a side of the original square.

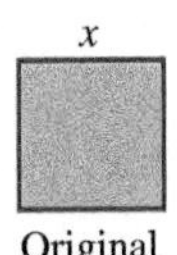

Original square

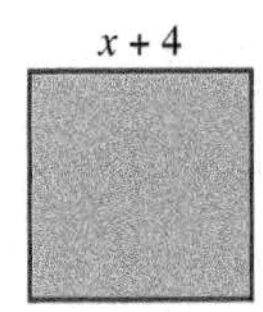

New square

18. Two planes leave the Dallas–Fort Worth airport at the same time. One travels east at 550 mph, and the other travels west at 500 mph. Assuming no wind, how long will it take for the planes to be 2100 mi apart?

West ← Airport → East

Solve.

19. $-4 < 3 - 2k < 9$

20. $-0.3x + 2.1(x - 4) \leq -6.6$

21. $\frac{1}{2}x > 3$ and $\frac{1}{3}x < \frac{8}{3}$

22. $-5x + 1 \geq 11$ or $3x + 5 > 26$

23. $|2k - 7| + 4 = 11$

24. $|3x + 6| \geq 0$

Solve each problem.

25. Find the x- and y-intercepts of the line with equation $3x + 5y = 12$, and graph the line.

26. Consider the points $A(-2, 1)$ and $B(3, -5)$.

(a) Find the slope of the line AB.

(b) Find the slope of a line perpendicular to line AB.

27. Graph the inequality $-2x + y < -6$.

*Write an equation of the line passing through the given point(s) and satisfying any given conditions. Give the equation **(a)** in slope-intercept form and **(b)** in standard form.*

28. $(0, -1)$; slope $-\frac{3}{4}$

29. $(4, -3)$ and $(1, 1)$

30. Give the domain and range of the relation. Does it define a function?

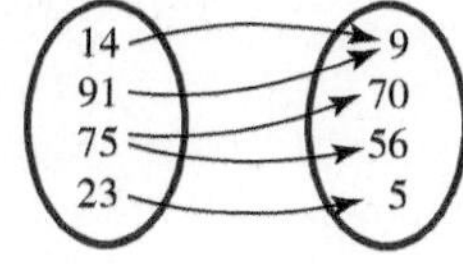

STUDY SKILLS **REMINDER**

We can learn from the mistakes we make. **Review Study Skill 9, *Analyzing Your Test Results.***

SYSTEMS OF LINEAR EQUATIONS

Just as the *intersection* of condensation trails in the sky consists of a region common to both trails, a solution of a *system of linear equations* (represented graphically by lines) is an ordered pair common to both solution sets (both lines).

3.1 Systems of Linear Equations in Two Variables

OBJECTIVES

1 Determine whether an ordered pair is a solution of a linear system.

2 Solve linear systems by graphing.

3 Solve linear systems (with two equations and two variables) by substitution.

4 Solve linear systems (with two equations and two variables) by elimination.

5 Solve special systems.

In recent years, the number of Americans who subscribe to Netflix has increased, while the number who subscribe to cable TV has decreased. These trends can be seen in the graph in **FIGURE 1**. The two straight-line graphs intersect at the time when Netflix and cable TV had the *same* number of subscribers.

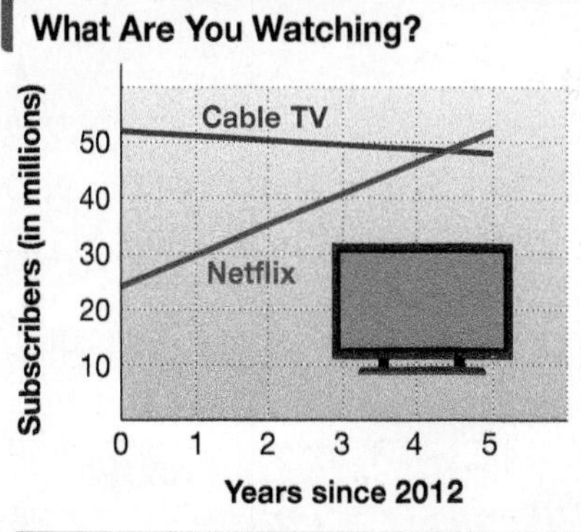

Data from Netflix, Leichtman Research Group.

FIGURE 1

$$-5.61x + y = 23.9$$
$$0.77x + y = 52.1$$
Linear system of equations

(Here, $x = 0$ represents 2012, $x = 1$ represents 2013, and so on. y represents millions of subscribers.)

As shown beside **FIGURE 1**, we can use a linear equation to model the graph of the number of Netflix subscribers (red graph) and another linear equation to model the graph of the number of cable TV subscribers (blue graph). Such a set of equations is a **system of equations**—in this case, a **system of linear equations,** or a **linear system.**

The point where the graphs in **FIGURE 1** intersect is a solution of each of the individual equations. It is also the solution of the linear system of equations.

VOCABULARY

- ☐ system of equations
- ☐ system of linear equations (linear system)
- ☐ solution set of a linear system
- ☐ consistent system
- ☐ independent equations
- ☐ inconsistent system
- ☐ dependent equations

OBJECTIVE 1 Determine whether an ordered pair is a solution of a linear system.

The **solution set of a linear system** of equations contains *all* ordered pairs that satisfy *all* the equations of the system *at the same time*.

STUDY SKILLS REMINDER

Be sure to read and work through the section material before working the exercises. **Review Study Skill 2, *Reading Your Math Text.***

EXAMPLE 1 Determining Whether an Ordered Pair Is a Solution

Determine whether the given ordered pair is a solution of the given system.

(a) $\begin{aligned} x + y &= 6 \\ 4x - y &= 14 \end{aligned}$; $(4, 2)$

Replace x with 4 and y with 2 in each equation of the system.

$x + y = 6$	$4x - y = 14$
$4 + 2 \stackrel{?}{=} 6$	$4(4) - 2 \stackrel{?}{=} 14$
$6 = 6$ ✓ True	$16 - 2 \stackrel{?}{=} 14$
	$14 = 14$ ✓ True

Because $(4, 2)$ makes *both* equations true, $(4, 2)$ is a solution of the system.

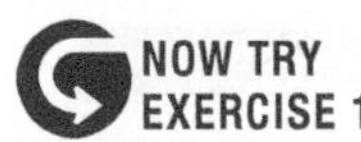
NOW TRY EXERCISE 1

Determine whether the ordered pair (2, 5) is a solution of the given system.

(a) $2x - y = -1$
$3x + 2y = 16$

(b) $-x + 2y = 8$
$3x - 2y = 0$

(b) $\begin{aligned} 3x + 2y &= 11 \\ x + 5y &= 36 \end{aligned}$; $(-1, 7)$

$3x + 2y = 11$	$x + 5y = 36$
$3(-1) + 2(7) \stackrel{?}{=} 11$	$-1 + 5(7) \stackrel{?}{=} 36$
$-3 + 14 \stackrel{?}{=} 11$	$-1 + 35 \stackrel{?}{=} 36$
$11 = 11$ ✓ True	$34 = 36$ False

The ordered pair $(-1, 7)$ is not a solution of the system—it does not make *both* equations true.

NOW TRY

OBJECTIVE 2 Solve linear systems by graphing.

One way to find the solution set of a linear system of equations is to graph each equation and find the point where the graphs intersect.

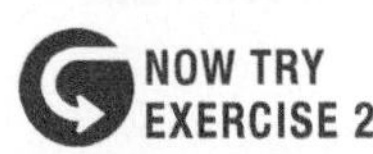
NOW TRY EXERCISE 2

Solve the system of equations by graphing.

$3x - 2y = 6$
$x - y = 1$

EXAMPLE 2 Solving a System by Graphing

Solve the system of equations by graphing.

$$x + y = 5 \quad (1)$$
$$2x - y = 4 \quad (2)$$

To graph these linear equations, we plot several points for each line.

$x + y = 5$

x	y
0	5
5	0
2	3

The intercepts are a convenient choice.

$2x - y = 4$

x	y
0	−4
2	0
4	4

Find a third ordered pair as a check.

The graph in **FIGURE 2** suggests that the point of intersection is the ordered pair (3, 2).

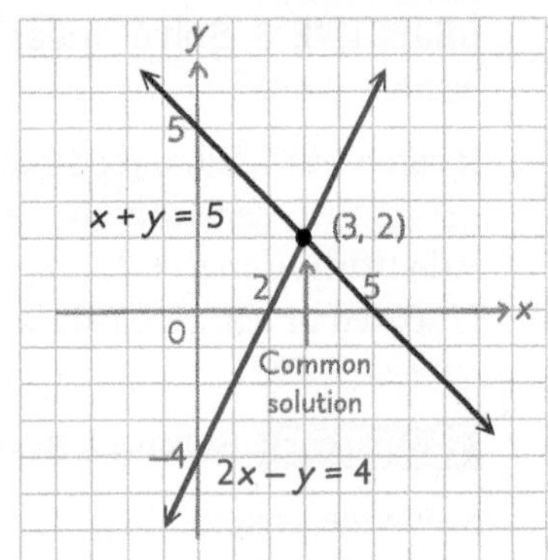

FIGURE 2

To confirm that (3, 2) is a solution of *both* equations, we check by substituting 3 for x and 2 for y in each equation.

CHECK

$x + y = 5$ (1)	$2x - y = 4$ (2)
$3 + 2 \stackrel{?}{=} 5$	$2(3) - 2 \stackrel{?}{=} 4$
$5 = 5$ ✓ True	$6 - 2 \stackrel{?}{=} 4$
	$4 = 4$ ✓ True

Because (3, 2) makes both equations true, the solution set of the system is $\{(3, 2)\}$.

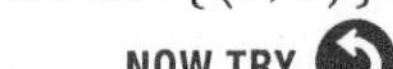
NOW TRY

NOW TRY ANSWERS

1. (a) solution (b) not a solution
2. $\{(4, 3)\}$

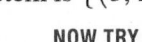

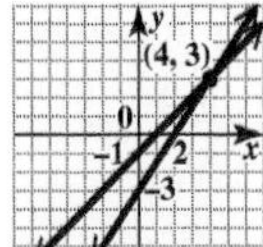

Three Cases for Solutions of Linear Systems in Two Variables

Case 1 **The two graphs intersect in a single point.**
The coordinates of this point give the only solution of the system. Because the system has a solution, it is **consistent.** The equations are *not* equivalent, so they are **independent.** See FIGURE 3(a).

Case 2 **The graphs are parallel lines.**
There is no solution common to both equations, so the solution set is $\emptyset$ and the system is **inconsistent.** Because the equations are *not* equivalent, they are **independent.** See FIGURE 3(b).

Case 3 **The graphs are the same line—that is, they coincide.**
Because any solution of one equation of the system is a solution of the other, the solution set is an infinite set of ordered pairs representing the points on the line. This type of system is **consistent** because there is a solution. The equations are equivalent, so they are **dependent.** See FIGURE 3(c).

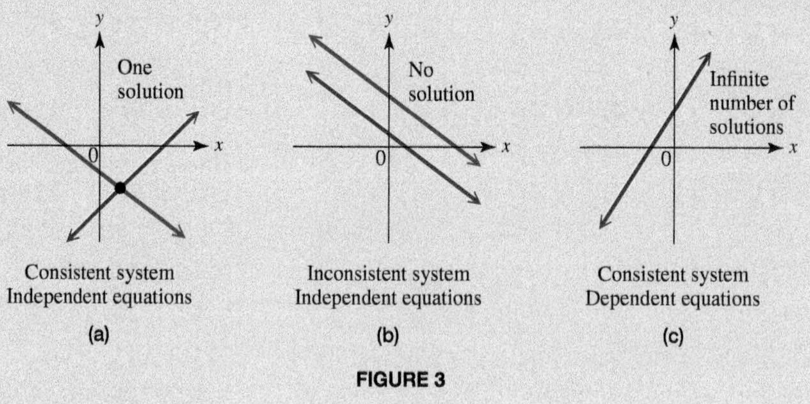

FIGURE 3

OBJECTIVE 3 **Solve linear systems (with two equations and two variables) by substitution.**

It can be difficult to read exact coordinates from a graph, especially if they are not integers, so we generally use algebraic methods to solve systems. One such method, the **substitution method,** is well suited for solving linear systems in which one equation is solved or can be easily solved for one variable in terms of the other.

EXAMPLE 3 **Solving a System by Substitution (Case 1)**

Solve the system.

$$2x - y = 6 \quad (1)$$
$$x = y + 2 \quad (2)$$

Because equation (2) is solved for x, we substitute $y + 2$ for x in equation (1).

$2x - y = 6$ (1) — Substitute in the *other* equation.

$2(y + 2) - y = 6$ Let $x = y + 2$.

$2y + 4 - y = 6$ Distributive property — Be sure to use parentheses here.

$y + 4 = 6$ Combine like terms.

$y = 2$ Subtract 4.

NOW TRY EXERCISE 3
Solve the system.

$$x = 3 + 2y$$
$$4x - 3y = 32$$

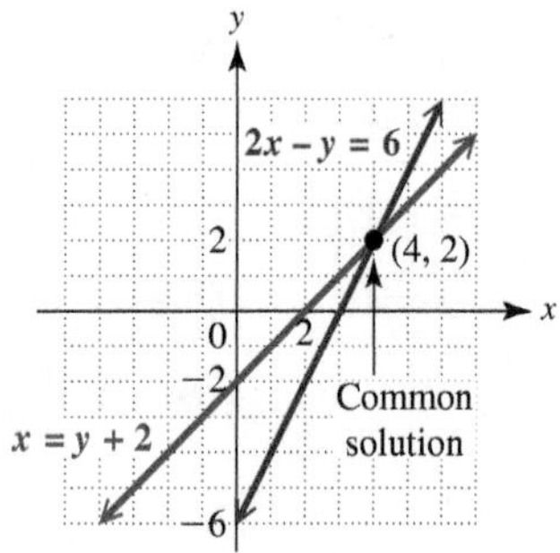

FIGURE 4

We found y. Now we solve for x by substituting 2 for y in equation (2).

$$x = y + 2 \quad (2)$$
$$x = 2 + 2 \quad \text{Let } y = 2.$$
$$x = 4 \quad \text{Add.}$$

Write the x-value first in the ordered pair.

Check the ordered pair (4, 2) in both equations of the original system.

CHECK

$2x - y = 6$ (1)	$x = y + 2$ (2)
$2(4) - 2 \stackrel{?}{=} 6$	$4 \stackrel{?}{=} 2 + 2$
$8 - 2 \stackrel{?}{=} 6$	$4 = 4$ ✓ True
$6 = 6$ ✓ True	

This check and the graph in **FIGURE 4** show that the solution set is $\{(4, 2)\}$.

NOW TRY

Solving a Linear System by Substitution

Step 1 **Solve one of the equations for either variable.** If one equation has a variable term with coefficient 1 or −1, choose it because the substitution method is usually easier.

Step 2 **Substitute** for that variable in the other equation. The result should be an equation with just one variable.

Step 3 **Solve** the equation from Step 2.

Step 4 **Find the other value.** Substitute the result from Step 3 into the equation from Step 1 and solve for the value of the other variable.

Step 5 **Check** the values in *both* of the *original* equations. Then write the solution set as a set containing an ordered pair.

EXAMPLE 4 Solving a System by Substitution (Case 1)

Solve the system.

$$3x + 2y = 13 \quad (1)$$
$$4x - y = -1 \quad (2)$$

Step 1 First solve one of the equations for either x or y. Because the coefficient of y in equation (2) is −1, it is easiest to solve for y in equation (2).

$$4x - y = -1 \quad (2)$$
$$-y = -1 - 4x \quad \text{Subtract } 4x.$$
$$y = 1 + 4x \quad \text{Multiply by } -1.$$

Step 2 Substitute $1 + 4x$ for y in equation (1).

$$3x + 2y = 13 \quad (1)$$

Substitute in the *other* equation.

$$3x + 2(1 + 4x) = 13 \quad \text{Let } y = 1 + 4x.$$

Step 3 Solve.

$$3x + 2 + 8x = 13 \quad \text{Distributive property}$$
$$11x = 11 \quad \text{Combine like terms. Subtract 2.}$$
$$x = 1 \quad \text{Divide by 11.}$$

NOW TRY ANSWER
3. $\{(11, 4)\}$

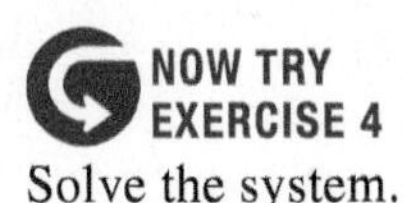

Solve the system.

$$5x + y = 7$$
$$3x - 2y = 25$$

Step 4 Now find y using the equation from Step 1.

$$y = 1 + 4x \quad \text{From Step 1}$$
$$y = 1 + 4(1) \quad \text{Let } x = 1.$$
$$y = 5 \quad \text{Multiply, and then add.}$$

Step 5 Check the ordered pair $(1, 5)$ in both equations (1) and (2).

CHECK

$3x + 2y = 13$ (1)	$4x - y = -1$ (2)
$3(1) + 2(5) \stackrel{?}{=} 13$	$4(1) - 5 \stackrel{?}{=} -1$
$3 + 10 \stackrel{?}{=} 13$	$4 - 5 \stackrel{?}{=} -1$
$13 = 13$ ✓ True	$-1 = -1$ ✓ True

The solution set is $\{(1, 5)\}$.

EXAMPLE 5 Solving a System with Fractional Coefficients (Case 1)

Solve the system.

$$\frac{2}{3}x - \frac{1}{2}y = \frac{7}{6} \quad (1)$$
$$3x - y = 6 \quad (2)$$

This system will be easier to solve if we clear the fractions in equation (1).

$$6\left(\frac{2}{3}x - \frac{1}{2}y\right) = 6\left(\frac{7}{6}\right) \quad \text{Multiply (1) by the LCD, 6.}$$

Remember to multiply *each* term by 6.

$$6 \cdot \frac{2}{3}x - 6 \cdot \frac{1}{2}y = 6 \cdot \frac{7}{6} \quad \text{Distributive property}$$
$$4x - 3y = 7 \quad (3)$$

Now the system consists of equations (2) and (3).

$$3x - y = 6 \quad (2)$$
$$4x - 3y = 7 \quad (3)$$

This equation is equivalent to equation (1).

To use the substitution method, we solve equation (2) for y.

$$3x - y = 6 \quad (2)$$
$$-y = 6 - 3x \quad \text{Subtract } 3x.$$
$$y = 3x - 6 \quad \text{Multiply by } -1. \text{ Rewrite.}$$

Substitute $3x - 6$ for y in equation (3).

$$4x - 3y = 7 \quad (3)$$
$$4x - 3(3x - 6) = 7 \quad \text{Let } y = 3x - 6.$$
$$4x - 9x + 18 = 7 \quad \text{Distributive property}$$
$$-5x + 18 = 7 \quad \text{Combine like terms.}$$

Be careful with signs.

$$-5x = -11 \quad \text{Subtract 18.}$$
$$x = \frac{11}{5} \quad \text{Divide by } -5.$$

NOW TRY ANSWER
4. $\{(3, -8)\}$

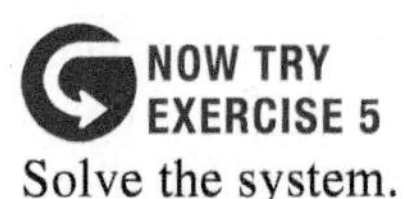

NOW TRY EXERCISE 5

Solve the system.

$$\frac{1}{10}x + \frac{3}{5}y = \frac{2}{5}$$
$$2x + y = 2$$

Now find y using $y = 3x - 6$ (equation (2) solved for y).

$$y = 3\left(\frac{11}{5}\right) - 6 \qquad \text{Let } x = \tfrac{11}{5}.$$

$$y = \frac{33}{5} - \frac{30}{5} \qquad \text{Multiply; } 6 = \tfrac{30}{5}$$

$$y = \frac{3}{5} \qquad \text{Subtract fractions.}$$

A check verifies that the solution set is $\left\{\left(\frac{11}{5}, \frac{3}{5}\right)\right\}$.

NOTE If an equation in a system contains decimal coefficients, first clear the decimals by multiplying by 10, 100, or 1000, depending on the number of decimal places. Then solve the system. For example, we multiply *each side* of the equation

$$0.5x + 0.75y = 3.25$$

by 10^2, or 100, to obtain an equivalent equation

$$50x + 75y = 325.$$

OBJECTIVE 4 Solve linear systems (with two equations and two variables) by elimination.

The **elimination method** involves combining the two equations in a system so that one variable is eliminated. This is done using the following logic.

$$\textbf{If } a = b \textbf{ and } c = d, \textbf{ then } a + c = b + d.$$

NOW TRY EXERCISE 6

Solve the system.

$$8x - 2y = 18$$
$$5x + 2y = -18$$

EXAMPLE 6 Solving a System by Elimination (Case 1)

Solve the system.

$$2x + 3y = -6 \qquad (1)$$
$$4x - 3y = 6 \qquad (2)$$

Notice that adding the equations together will eliminate the variable y.

$$2x + 3y = -6 \qquad (1)$$
$$\underline{4x - 3y = 6} \qquad (2)$$
$$6x \qquad = 0 \qquad \text{Add.}$$

Solve for x.

$$x = 0 \qquad \text{Divide by 6.}$$

To find y, substitute 0 for x in either equation (1) or equation (2).

$$2x + 3y = -6 \qquad (1)$$
$$2(0) + 3y = -6 \qquad \text{Let } x = 0.$$
$$0 + 3y = -6 \qquad \text{Multiply.}$$
$$y = -2 \qquad \text{Add, and then divide by 3.}$$

Check by substituting 0 for x and -2 for y in both equations of the original system to verify that $(0, -2)$ satisfies both equations. It does, so the solution set is $\{(0, -2)\}$.

NOW TRY 

NOW TRY ANSWERS

5. $\left\{\left(\frac{8}{11}, \frac{6}{11}\right)\right\}$
6. $\{(0, -9)\}$

From **Example 6:**

$$\begin{aligned} 2x + 3y &= -6 \quad (1) \\ 4x - 3y &= 6 \quad (2) \\ \hline 6x \quad &= 0 \end{aligned}$$

By adding the equations in **Example 6** (repeated in the margin), we eliminated the variable y because the coefficients of the y-terms were opposites. In many cases the coefficients will *not* be opposites, and we must transform one or both equations so that the coefficients of one pair of variable terms are opposites.

Solving a Linear System by Elimination

Step 1 **Write both equations in the form $Ax + By = C$.**

Step 2 **Transform the equations as needed so that the coefficients of one pair of variable terms are opposites.** Multiply one or both equations by appropriate numbers so that the sum of the coefficients of either the x- or y-terms is 0.

Step 3 **Add** the new equations to eliminate a variable. The sum should be an equation with just one variable.

Step 4 **Solve** the equation from Step 3 for the remaining variable.

Step 5 **Find the other value.** Substitute the result from Step 4 into either of the original equations and solve for the other variable.

Step 6 **Check** the values in *both* of the *original* equations. Then write the solution set as a set containing an ordered pair.

EXAMPLE 7 Solving a System by Elimination (Case 1)

Solve the system.

$$5x - 2y = 4 \quad (1)$$
$$2x + 3y = 13 \quad (2)$$

Step 1 Both equations are in $Ax + By = C$ form.

Step 2 Suppose that we wish to eliminate the variable x. One way to do this is to multiply equation (1) by 2 and equation (2) by -5.

The goal is to have *opposite* coefficients.

$$10x - 4y = 8 \quad \text{2 times each side of equation (1)}$$
$$-10x - 15y = -65 \quad \text{−5 times each side of equation (2)}$$

Step 3 Now add.

$$\begin{aligned} 10x - 4y &= 8 \\ -10x - 15y &= -65 \\ \hline -19y &= -57 \quad \text{Add.} \end{aligned}$$

Step 4 Solve for y. $\quad y = 3 \quad$ Divide by -19.

Step 5 To find x, substitute 3 for y in either equation (1) or equation (2).

$$\begin{aligned} 2x + 3y &= 13 \quad && (2) \\ 2x + 3(3) &= 13 \quad && \text{Let } y = 3. \\ 2x + 9 &= 13 \quad && \text{Multiply.} \\ 2x &= 4 \quad && \text{Subtract 9.} \\ x &= 2 \quad && \text{Divide by 2.} \end{aligned}$$

NOW TRY EXERCISE 7
Solve the system.

$$2x - 5y = 14$$
$$5x + 2y = 6$$

Step 6 To check, substitute 2 for x and 3 for y in both equations (1) and (2).

CHECK

$5x - 2y = 4$ (1)		$2x + 3y = 13$ (2)
$5(2) - 2(3) \stackrel{?}{=} 4$		$2(2) + 3(3) \stackrel{?}{=} 13$
$10 - 6 \stackrel{?}{=} 4$		$4 + 9 \stackrel{?}{=} 13$
$4 = 4$ ✓ True		$13 = 13$ ✓ True

The solution set is $\{(2, 3)\}$.

NOTE Look again at the system in **Example 7.**

$$5x - 2y = 4 \quad (1)$$
$$2x + 3y = 13 \quad (2)$$

In Step 2, we could eliminate the variable y instead of x.

$$15x - 6y = 12 \quad \text{3 times each side of equation (1)}$$
$$4x + 6y = 26 \quad \text{2 times each side of equation (2)}$$
$$19x = 38 \quad \text{Add.}$$
$$x = 2 \quad \text{Divide by 19.}$$

If we substitute 2 for x in equation (1) or (2), we find that $y = 3$. The same ordered-pair solution (2, 3) results.

OBJECTIVE 5 Solve special systems.

As we saw in **FIGURES 3(b) and (c)**, some systems of linear equations have no solution or an infinite number of solutions.

NOW TRY EXERCISE 8
Solve the system.

$$-2x + 5y = 6$$
$$6x - 15y = 4$$

EXAMPLE 8 **Solving an Inconsistent System (Case 2)**

Solve the system.

$$x + 3y = 4 \quad (1)$$
$$-2x - 6y = 3 \quad (2)$$

We multiply equation (1) by 2 and then add the result to equation (2).

$$2x + 6y = 8 \quad \text{Equation (1) multiplied by 2}$$
$$-2x - 6y = 3 \quad (2)$$
$$0 = 11 \quad \text{False (contradiction)}$$

The result of the addition step is a false statement, $0 = 11$, which indicates that the system is inconsistent. As shown in **FIGURE 5**, the graphs of the equations of the system are parallel lines.

There are no ordered pairs that satisfy both equations, so there is no solution for the system. The solution set is $\emptyset$.

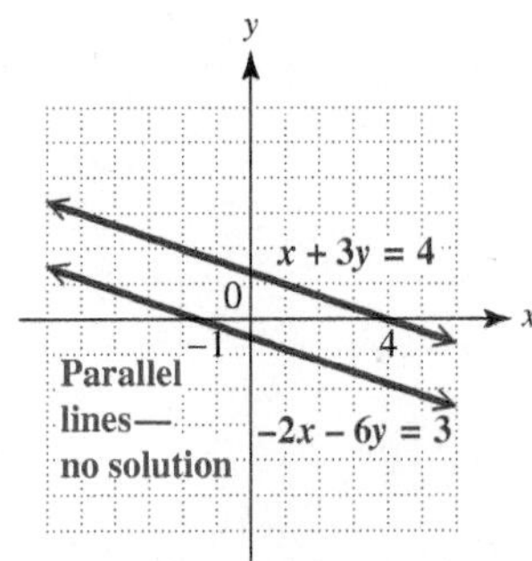

FIGURE 5

NOW TRY

NOW TRY ANSWERS
7. $\{(2, -2)\}$
8. $\emptyset$

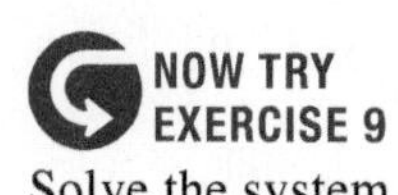
NOW TRY EXERCISE 9

Solve the system.

$$x - 3y = 7$$
$$-3x + 9y = -21$$

EXAMPLE 9 Solving a System of Dependent Equations (Case 3)

Solve the system.

$$2x - y = 3 \quad (1)$$
$$6x - 3y = 9 \quad (2)$$

We multiply equation (1) by -3 and then add the result to equation (2).

$$-6x + 3y = -9 \quad -3 \text{ times each side of equation (1)}$$
$$6x - 3y = 9 \quad (2)$$
$$0 = 0 \quad \text{True (identity)}$$

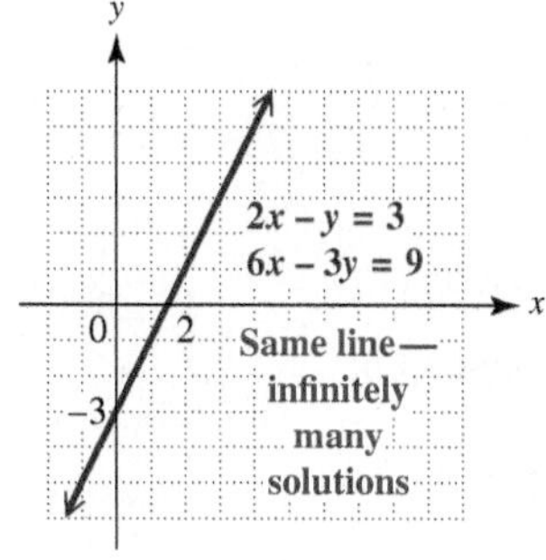

FIGURE 6

Adding these equations gives the true statement $0 = 0$. In the original system, we could obtain equation (2) from equation (1) by multiplying equation (1) by 3. Because of this, equations (1) and (2) are equivalent and have the same graph, as shown in **FIGURE 6**. The equations are dependent.

The solution set is the set of all points on the line with equation $2x - y = 3$, which is written in set-builder notation as

$$\{(x, y) \mid 2x - y = 3\}$$

and is read "*the set of all ordered pairs* (x, y), *such that* $2x - y = 3$." Either equation of the system could be used to write the solution set.

We use the equation in $Ax + By = C$ ***form with coefficients that are integers having greatest common factor 1 and positive coefficient of x.***

NOW TRY

Special Cases of Linear Systems

If both variables are eliminated when a system of linear equations is solved, then the solution sets are determined as follows.

- There is no solution if the resulting statement is *false*. **(See Example 8.)**
- There are infinitely many solutions if the resulting statement is *true*. **(See Example 9.)**

EXAMPLE 10 Using Slope-Intercept Form to Determine the Number of Solutions

Refer to **Examples 8 and 9.** Write the equations of each system in slope-intercept form, and use the results to determine how many solutions the system has.

(a)

$$x + 3y = 4 \quad (1)$$
$$-2x - 6y = 3 \quad (2)$$

Solve each equation from **Example 8** for y.

$x + 3y = 4$	(1)	$-2x - 6y = 3$	(2)
$3y = -x + 4$	Add $-x$.	$-6y = 2x + 3$	Add $2x$.
$y = -\frac{1}{3}x + \frac{4}{3}$	Divide by 3.	$y = -\frac{1}{3}x - \frac{1}{2}$	Divide by -6; $\frac{2}{-6} = -\frac{1}{3}$; $\frac{3}{-6} = -\frac{1}{2}$
↑ Slope ↑ y-intercept $\left(0, \frac{4}{3}\right)$		↑ Slope ↑ y-intercept $\left(0, -\frac{1}{2}\right)$	

The lines have the *same* slope, but *different* y-intercepts, indicating that they are parallel. Thus, the system has no solution.

NOW TRY ANSWER

9. $\{(x, y) \mid x - 3y = 7\}$

NOW TRY EXERCISE 10

Write the equations of each system in slope-intercept form, and use the results to determine how many solutions the system has.

(a) $2x - 3y = 3$
$4x - 6y = -6$

(b) $5y = -x - 4$
$-10y = 2x + 8$

(b)
$$2x - y = 3 \quad (1)$$
$$6x - 3y = 9 \quad (2)$$

Solve each equation from **Example 9** for y.

$2x - y = 3$	(1)	$6x - 3y = 9$	(2)
$-y = -2x + 3$	Add $-2x$.	$-3y = -6x + 9$	Add $-6x$.
$y = 2x - 3$	Multiply by -1.	$y = 2x - 3$	Divide by -3.
Slope, y-intercept $(0, -3)$		Slope, y-intercept $(0, -3)$	

The lines have the *same* slope and the *same* y-intercept, which means that they coincide. There are infinitely many solutions.

NOW TRY

NOTE In **Example 2**, we solved the system

$$x + y = 5 \quad (1)$$
$$2x - y = 4 \quad (2)$$

by graphing the two lines and finding their point of intersection. We can also do this with a graphing calculator, as shown in **FIGURE 7**. The two lines were graphed by solving each equation for y.

$y = 5 - x$ Equation (1) solved for y

$y = 2x - 4$ Equation (2) solved for y

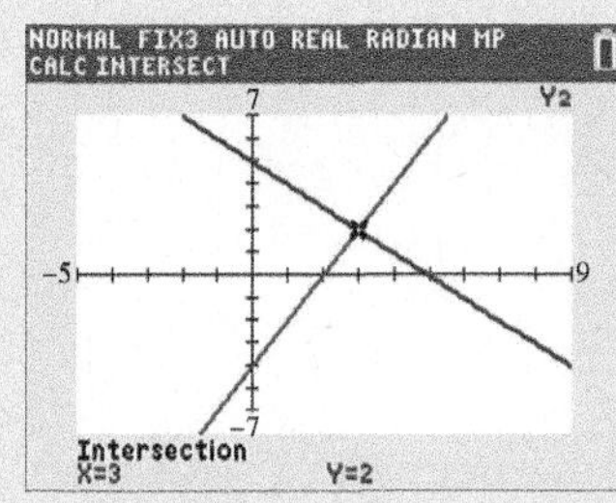

FIGURE 7

The coordinates of their point of intersection are displayed at the bottom of the screen, indicating that the solution set is $\{(3, 2)\}$. (Compare this graph with the one in **FIGURE 2**.)

NOW TRY ANSWERS

10. (a) $y = \frac{2}{3}x - 1$; $y = \frac{2}{3}x + 1$; no solution

(b) Both are $y = -\frac{1}{5}x - \frac{4}{5}$; infinitely many solutions

3.1 Exercises

FOR EXTRA HELP

MyLab Math

Video solutions for select problems available in MyLab Math

STUDY SKILLS REMINDER

Reread your class notes before working the assigned exercises. **Review Study Skill 3, *Taking Lecture Notes*.**

Concept Check *Complete each statement.*

1. If $(4, -3)$ is a solution of a linear system in two variables, then substituting ________ for x and ________ for y leads to true statements in *both* equations.

2. A solution of a system of independent linear equations in two variables is an ordered ________.

3. If solving a system leads to a false statement such as $0 = 3$, the solution set is ________.

4. If solving a system leads to a true statement such as $0 = 0$, the system has ________ equations.

5. If the two lines forming a system have the same slope and different y-intercepts, the system has (*no* / *one* / *infinitely many*) solution(s).

6. If the two lines forming a system have different slopes, the system has (*no* / *one* / *infinitely many*) solution(s).

7. **Concept Check** Which ordered pair could be a solution of the graphed system of equations? Why?

A. $(4, 4)$

B. $(-4, 4)$

C. $(-4, -4)$

D. $(4, -4)$

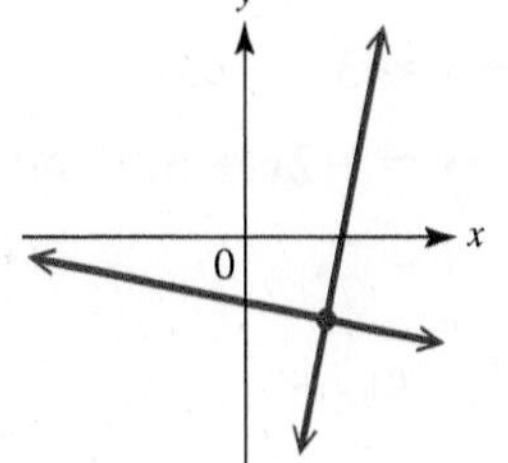

8. **Concept Check** Which ordered pair could be a solution of the graphed system of equations? Why?

A. $(4, 0)$

B. $(-4, 0)$

C. $(0, 4)$

D. $(0, -4)$

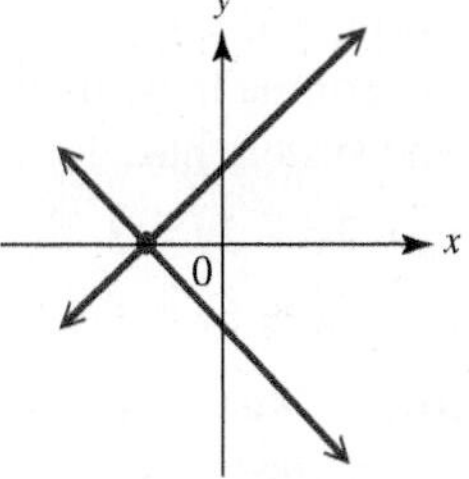

9. **Concept Check** Match each system with the correct graph.

(a) $x + y = 6$, $x - y = 0$

(b) $x + y = -6$, $x - y = 0$

(c) $x + y = 0$, $x - y = -6$

(d) $x + y = 0$, $x - y = 6$

A.

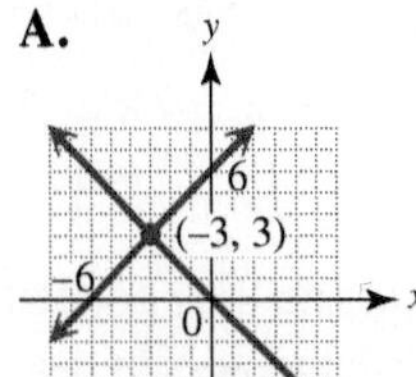

B.

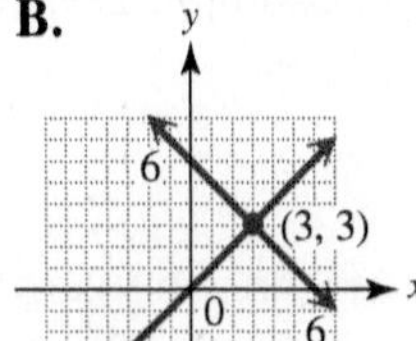

C.

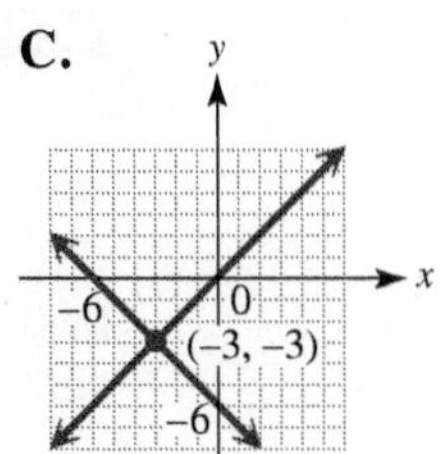

D.

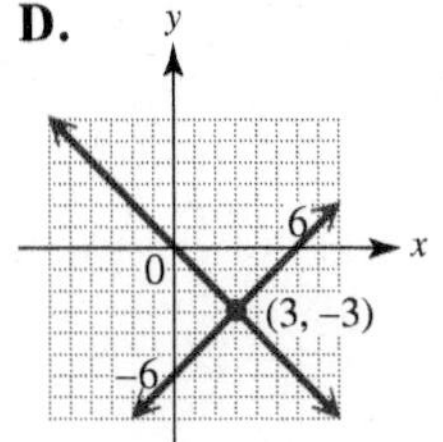

10. **Concept Check** The following systems have infinitely many solutions. Write the solution set of each system using set-builder notation and the method described in **Example 9.**

(a) $5x - 3y = 2$, $10x - 6y = 4$

(b) $4x - 2y = 8$, $2x = y + 4$

(c) $4x - 6y = 10$, $-8x + 12y = -20$

Determine whether the given ordered pair is a solution of the given system. ***See Example 1.***

11. $x - y = 17$, $x + y = -1$; $(8, -9)$

12. $x + y = 6$, $x - y = 4$; $(5, 1)$

13. $3x - 5y = -12$, $x - y = 1$; $(-1, 2)$

14. $2x - y = 8$, $3x + 2y = 20$; $(5, 2)$

Solve each system by graphing. ***See Example 2.***

15. $x + y = -5$, $-2x + y = 1$

16. $x + y = 4$, $2x - y = 2$

17. $x - 4y = -4$, $3x + y = 1$

18. $6x - y = 2$, $x - 2y = 4$

19. $2x + y = 4$, $3x - y = 6$

20. $2x + 3y = -6$, $x - 3y = -3$

Solve each system using the substitution method. If a system is inconsistent or has dependent equations, say so. ***See Examples 3–5, 8, and 9.***

21. $4x + y = 6$, $y = 2x$

22. $2x - y = 6$, $y = 5x$

23. $-x - 4y = -14$, $y = 2x - 1$

24. $-3x - 5y = -17$, $y = 4x + 8$

25. $4x - 5y = -11$, $x + 2y = 7$

26. $3x - y = 10$, $2x + 5y = 1$

27. $2x - y = 4$
$5x - 2y = 8$

28. $3x + 2y = 6$
$4x + y = 3$

29. $3x - 4y = -22$
$-3x + y = 0$

30. $-3x + y = -5$
$3x + 6y = 0$

31. $5x - 4y = 9$
$3 - 2y = -x$

32. $6x - y = -9$
$4 + 7x = -y$

33. $x = 3y + 5$
$x = \frac{3}{2}y$

34. $x = 6y - 2$
$x = \frac{3}{4}y$

35. $\frac{1}{2}x + \frac{1}{3}y = 3$
$-3x + y = 0$

36. $\frac{1}{4}x - \frac{1}{5}y = 9$
$5x - y = 0$

37. $y = 0.5x$
$1.5x - 0.5y = 5.0$

38. $y = 1.4x$
$0.5x + 1.5y = 26.0$

39. $\frac{1}{2}x - \frac{1}{4}y = -5$
$\frac{1}{8}x + \frac{1}{4}y = 0$

40. $-\frac{1}{3}x + \frac{2}{5}y = -6$
$-\frac{1}{2}x - \frac{3}{2}y = 12$

41. $y = 2x$
$4x - 2y = 0$

42. $x = 3y$
$3x - 9y = 0$

43. $x = 5y$
$5x - 25y = 5$

44. $y = -4x$
$8x + 2y = 4$

Solve each system using the elimination method. If a system is inconsistent or has dependent equations, say so. ***See Examples 6–9.***

45. $-2x + 3y = -16$
$2x - 5y = 24$

46. $6x + 5y = -7$
$-6x - 11y = 1$

47. $2x - 5y = 11$
$3x + y = 8$

48. $-2x + 3y = 1$
$-4x + y = -3$

49. $3x + 4y = -6$
$5x + 3y = 1$

50. $4x + 3y = 1$
$3x + 2y = 2$

51. $7x + 2y = 6$
$-14x - 4y = -12$

52. $x - 4y = 2$
$4x - 16y = 8$

53. $3x + 3y = 0$
$4x + 2y = 3$

54. $8x + 4y = 0$
$4x - 2y = 2$

55. $5x - 5y = 3$
$x - y = 12$

56. $2x - 3y = 7$
$-4x + 6y = 14$

57. $x + y = 0$
$2x - 2y = 0$

58. $3x + 3y = 0$
$-2x - y = 0$

59. $x - \frac{1}{2}y = 2$
$-x + \frac{2}{5}y = -\frac{8}{5}$

60. $\frac{3}{2}x + y = 3$
$\frac{2}{3}x + \frac{1}{3}y = 1$

61. $\frac{1}{2}x + \frac{1}{3}y = \frac{49}{18}$
$\frac{1}{2}x + 2y = \frac{4}{3}$

62. $\frac{1}{5}x + \frac{1}{7}y = \frac{12}{5}$
$\frac{1}{10}x + \frac{1}{3}y = \frac{5}{6}$

63. Concept Check To minimize the amount of work required, determine whether to use the substitution or the elimination method to solve each system. Explain. *Do not actually solve.*

(a) $3x + y = -7$
$x - y = -5$

(b) $6x - y = 5$
$y = 11x$

(c) $3x - 2y = 0$
$9x + 8y = 7$

64. Concept Check Explain why *System A* would be more difficult to solve by substitution than *System B*.

System A: $3x + 7y = 19$, $8x - 5y = 4$ *System B:* $x + 7y = 8$, $3x - 2y = 4$

Solve each system using the method of your choice. ***See Examples 3–9.***

65. $3x + y = -7$, $x - y = -5$

66. $6x - y = 5$, $y = 11x$

67. $3x - 2y = 0$, $9x + 8y = 7$

68. $3x - 5y = 7$, $2x + 3y = 30$

69. $2x + 3y = 10$, $-3x + y = 18$

70. $x + y = 10$, $2x - y = 5$

71. $y = -\frac{1}{5}x + 1$, $y = \frac{3}{5}x - 3$

72. $7x - 8y = 0$, $2x + 5y = 0$

73. $0.3x + 0.2y = 0.4$, $0.5x + 0.4y = 0.7$

74. $0.2x + 0.5y = 6$, $0.4x + y = 9$

75. $\frac{1}{2}x - \frac{1}{8}y = -\frac{1}{4}$, $4x - y = -2$

76. $\frac{1}{6}x + \frac{1}{3}y = 8$, $\frac{1}{4}x + \frac{1}{2}y = 12$

Write the equations of each system in slope-intercept form, and use the results to determine how many solutions the system has. Do not actually solve. ***See Example 10.***

77. $3x + 7y = 4$, $6x + 14y = 3$

78. $-x + 2y = 8$, $4x - 8y = 1$

79. $2x = -3y + 1$, $6x = -9y + 3$

80. $5x = -2y + 1$, $10x = -4y + 2$

Answer each question using the graphs provided.

81. The figure shows graphs that represent supply and demand for a certain brand of low-fat frozen yogurt at various prices per half-gallon (in dollars).

(a) At what price does supply equal demand?

(b) For how many half-gallons does supply equal demand?

(c) What are the supply and demand at a price of \$2 per half-gallon?

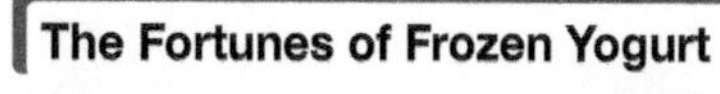

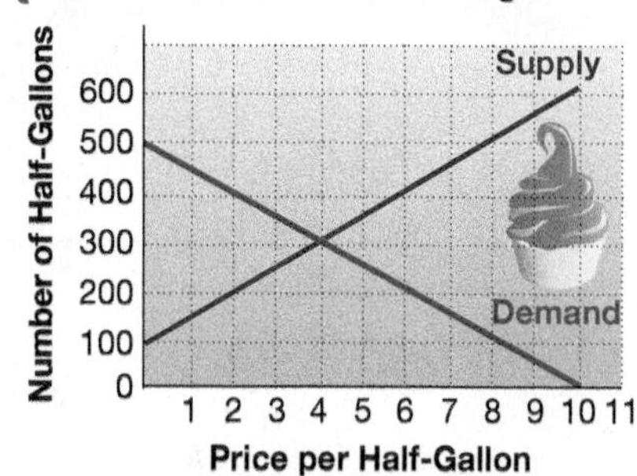

82. Sharon compared the monthly payments she would incur for two types of mortgages: fixed rate and variable rate. Her observations led to the following graphs.

(a) For which years would the monthly payment be more for the fixed-rate mortgage than for the variable-rate mortgage?

(b) In what year would the payments be the same? What would those payments be?

(c) In year 10, what are the fixed-rate and variable-rate monthly payments?

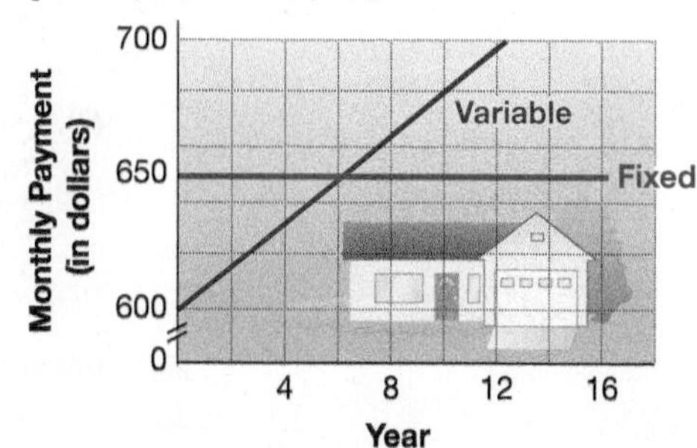

83. The graph shows college student enrollment (in thousands) for three states from 2011 through 2016.

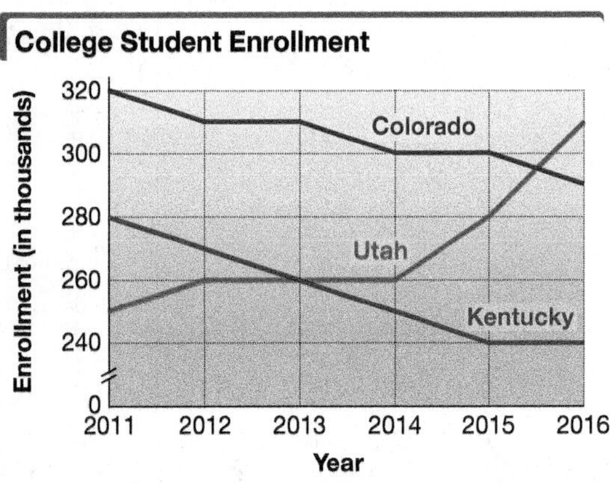

Data from National Student Clearinghouse Research Center.

(a) During what years did the number of college students enrolled in Colorado exceed that for Utah?

(b) Between what two consecutive years did Utah first have the highest enrollment?

(c) In what year were the numbers of college students enrolled in Kentucky and Utah approximately equal? Approximately how many college students were enrolled in each state during that year?

(d) Write the answer for part (c) as an ordered pair of the form (year, enrollment).

(e) Describe the trends in enrollment in Kentucky and Utah from 2012 through 2014.

84. The graph shows numbers of nuclear weapons possessed by the United States and USSR/ Russia from 1950 through 2017.

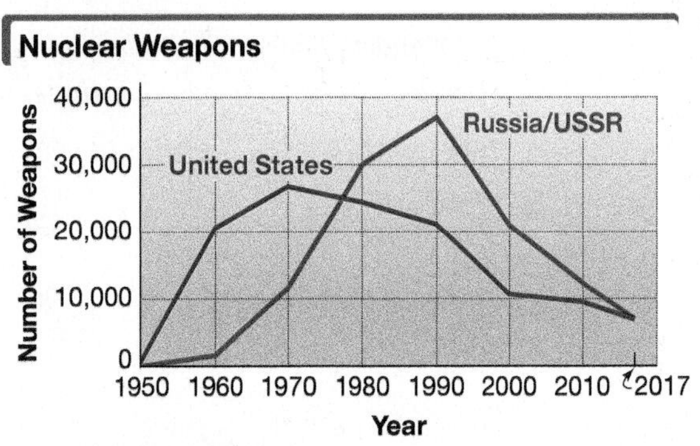

Data from *The World Almanac and Book of Facts.*

(a) In what year were the numbers of nuclear weapons approximately equal?

(b) Approximately how many nuclear weapons did each country possess in that year?

(c) Express the point of intersection of the graphs for U.S. and Russian nuclear weapons as an ordered pair of the form (year, number of nuclear weapons).

(d) Over the entire period 1950–2017, which country possessed the greatest number of nuclear weapons at any one particular time? In what year did that maximum occur, and what was the approximate number of nuclear weapons possessed?

(e) Describe the trend in number of nuclear weapons possessed by Russia from 1990 through 2017. If a straight line were used to approximate its graph, would the line have a slope that is positive, negative, or 0?

Use the graph shown in **FIGURE 1** *at the beginning of this section (and repeated here) to work each exercise.*

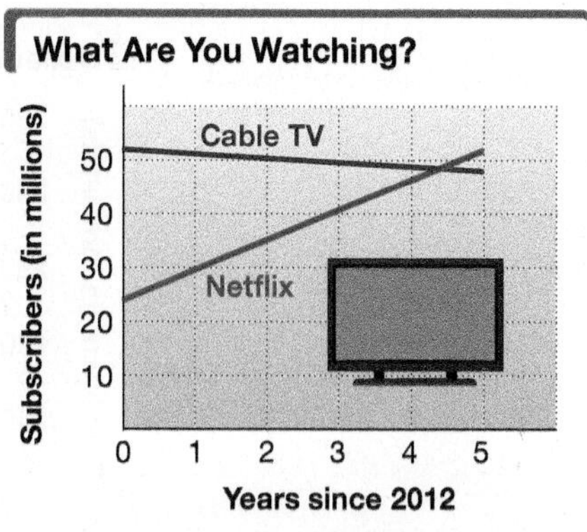

Data from Netflix, Leichtman Research Group.

85. Which type of subscription was more popular in 2017?

86. Between which two years were the numbers of Netflix and cable TV subscribers the same? About how many people subscribed to each video content provider at that point?

87. If $x = 0$ represents 2012 and $x = 5$ represents 2017, the number of subscribers y in millions can be modeled by the linear equations in the following system.

$$-5.61x + y = 23.9 \quad \text{Netflix}$$
$$0.77x + y = 52.1 \quad \text{Cable TV}$$

Solve this system. Express values as decimals rounded to the nearest tenth. Write the solution as an ordered pair of the form (year, number of subscribers).

88. Interpret the answer for **Exercise 87,** rounding down for the year. How does this compare to the estimate from **Exercise 86?**

Extending Skills *The following systems can be solved by elimination. One way to do this is to let* $p = \frac{1}{x}$ *and* $q = \frac{1}{y}$. *Substitute, solve for p and q, and then find x and y.* $\left(\right.$*For example, in Exercise 89,* $\frac{3}{x} = 3 \cdot \frac{1}{x} = 3p.$$\left.\right)$

Use this method to solve each system.

89. $\frac{3}{x} + \frac{4}{y} = \frac{5}{2}$

$\frac{5}{x} - \frac{3}{y} = \frac{7}{4}$

90. $\frac{2}{x} - \frac{5}{y} = \frac{3}{2}$

$\frac{4}{x} + \frac{1}{y} = \frac{4}{5}$

91. $\frac{2}{x} + \frac{3}{y} = \frac{11}{2}$

$-\frac{1}{x} + \frac{2}{y} = -1$

92. Extending Skills Make up a system of the following form, where a, b, c, d, e, and f are consecutive integers. Solve the system.

$$ax + by = c$$
$$dx + ey = f$$

Extending Skills *Solve by any method. Assume that a and b represent nonzero constants.*

93. $ax + by = c$

$ax - 2by = c$

94. $ax + by = 2$

$-ax + 2by = 1$

95. $2ax - y = 3$

$y = 5ax$

96. $3ax + 2y = 1$

$-ax + y = 2$

3.2 Systems of Linear Equations in Three Variables

OBJECTIVES

1 Understand the geometry of systems of three equations in three variables.
2 Solve linear systems (with three equations and three variables) by elimination.
3 Solve linear systems (with three equations and three variables) in which some of the equations have missing terms.
4 Solve special systems.

VOCABULARY

☐ ordered triple
☐ focus variable
☐ working equation

A solution of an equation in three variables, such as

$$2x + 3y - z = 4, \quad \text{Linear equation in three variables}$$

is an **ordered triple** and is written (x, y, z). For example, the ordered triple $(0, 1, -1)$ is a solution of the preceding equation, because

$$2(0) + 3(1) - (-1) = 4 \quad \text{is a true statement.}$$

Verify that another solution of this equation is $(10, -3, 7)$.

We now extend the term *linear equation* to equations of the form

$$Ax + By + Cz + \cdots + Dw = K,$$

where not all the coefficients $A, B, C, \ldots, D$ equal 0. For example,

$$2x + 3y - 5z = 7 \quad \text{and} \quad x - 2y - z + 3w = 8$$

are linear equations, the first with three variables and the second with four.

OBJECTIVE 1 Understand the geometry of systems of three equations in three variables.

Consider the solution of a system such as the following.

$$\begin{aligned} 4x + 8y + z &= 2 \\ x + 7y - 3z &= -14 \\ 2x - 3y + 2z &= 3 \end{aligned} \quad \text{System of linear equations in three variables}$$

Theoretically, a system of this type can be solved by graphing. However, the graph of a linear equation with three variables is a *plane,* not a line. Because visualizing a plane requires three-dimensional graphing, the method of graphing is not practical with these systems. However, it does illustrate the number of solutions possible for such systems, as shown in **FIGURE 8.**

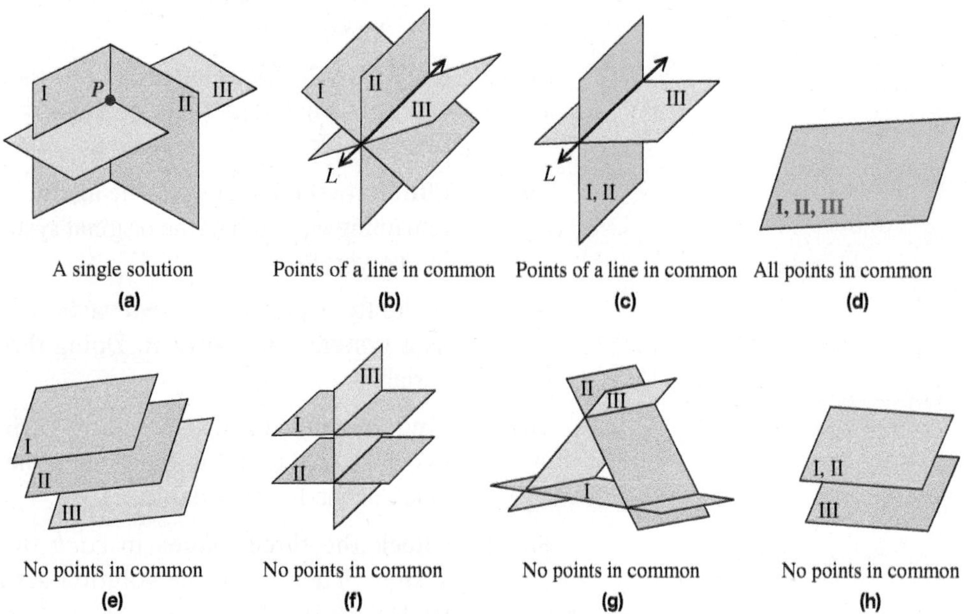

FIGURE 8

STUDY SKILLS REMINDER
Make study cards to help you learn and remember the material in this chapter. **Review Study Skill 5, *Using Study Cards*.**

FIGURE 8 illustrates the following cases.

Graphs of Linear Systems in Three Variables

Case 1 **The three planes may meet at a single, common point.** This point is the solution of the system. See FIGURE 8(a).

Case 2 **The three planes may have the points of a line in common.** The infinite set of points that satisfy the equation of the line is the solution of the system. See FIGURES 8(b) and (c).

Case 3 **The three planes may coincide.** The solution of the system is the set of all points on a plane. See FIGURE 8(d).

Case 4 **The planes may have no points common to all three.** There is no solution of the system. See FIGURES 8(e)–(h).

OBJECTIVE 2 Solve linear systems (with three equations and three variables) by elimination.

Because graphing to find the solution set of a system of three equations in three variables is impractical, these systems are solved with an extension of the elimination method.

In the steps that follow, we use the term **focus variable** to identify the first variable to be eliminated. The focus variable will always be present in the **working equation,** which will be used twice to eliminate this variable.

Solving a Linear System in Three Variables

Step 1 **Select a variable and an equation.** A good choice for the variable, which we call the *focus variable,* is one that has coefficient 1 or -1. Then select an equation, one that contains the focus variable, as the *working equation.*

Step 2 **Eliminate the focus variable.** Use the working equation and one of the other two equations of the original system. The result is an equation in two variables.

Step 3 **Eliminate the focus variable again.** Use the working equation and the remaining equation of the original system. The result is another equation in two variables.

Step 4 **Write the equations in two variables that result from Steps 2 and 3 as a system, and solve it.** Doing this gives the values of two of the variables.

Step 5 **Find the value of the remaining variable.** Substitute the values of the two variables found in Step 4 into the working equation to obtain the value of the focus variable.

Step 6 **Check** the three values in *each* of the *original* equations of the system. Then write the solution set as a set containing an ordered triple.

EXAMPLE 1 Solving a System in Three Variables

Solve the system.

$$\begin{aligned} 4x + 8y + z &= 2 && (1) \\ x + 7y - 3z &= -14 && (2) \\ 2x - 3y + 2z &= 3 && (3) \end{aligned}$$

Step 1 Because z in equation (1) has coefficient 1, we choose z as the focus variable and (1) as the working equation. (Another option would be to choose x as the focus variable—it also has coefficient 1—and use (2) as the working equation.)

Focus variable (z)

$$4x + 8y + z = 2 \qquad (1) \leftarrow \text{Working equation}$$

Step 2 Multiply working equation (1) by 3 and add the result to equation (2).

$$\begin{aligned} 12x + 24y + 3z &= 6 && \text{Multiply each side of (1) by 3.} \\ x + 7y - 3z &= -14 && (2) \\ \hline 13x + 31y \quad &= -8 && \text{Add. (4)} \end{aligned}$$

Focus variable z was eliminated.

Step 3 Multiply working equation (1) by -2 and add the result to remaining equation (3) to again eliminate focus variable z.

$$\begin{aligned} -8x - 16y - 2z &= -4 && \text{Multiply each side of (1) by } -2. \\ 2x - 3y + 2z &= 3 && (3) \\ \hline -6x - 19y \quad &= -1 && \text{Add. (5)} \end{aligned}$$

Focus variable z was eliminated.

Step 4 Write the equations in two variables that result in Steps 2 and 3 as a system.

$$\begin{aligned} 13x + 31y &= -8 && \text{(4) The result from Step 2} \\ -6x - 19y &= -1 && \text{(5) The result from Step 3} \end{aligned}$$

Make sure these equations have the same two variables.

Now solve this system. We choose to eliminate x.

$$\begin{aligned} 78x + 186y &= -48 && \text{Multiply each side of (4) by 6.} \\ -78x - 247y &= -13 && \text{Multiply each side of (5) by 13.} \\ \hline -61y &= -61 && \text{Add.} \\ y &= 1 && \text{Divide by } -61. \end{aligned}$$

Substitute 1 for y in either equation (4) or equation (5) to find x.

$$\begin{aligned} -6x - 19y &= -1 && (5) \\ -6x - 19(1) &= -1 && \text{Let } y = 1. \\ -6x - 19 &= -1 && \text{Multiply.} \\ -6x &= 18 && \text{Add 19.} \\ x &= -3 && \text{Divide by } -6. \end{aligned}$$

Step 5 Now substitute the two values we found in Step 4 in working equation (1) to find the value of the remaining variable, focus variable z.

$$\begin{aligned} 4x + 8y + z &= 2 && (1) \\ 4(-3) + 8(1) + z &= 2 && \text{Let } x = -3 \text{ and } y = 1. \\ -4 + z &= 2 && \text{Multiply, and then add.} \\ z &= 6 && \text{Add 4.} \end{aligned}$$

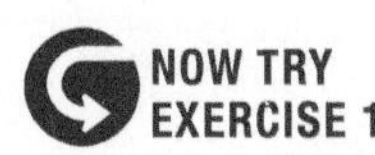

NOW TRY EXERCISE 1

Solve the system.

$$\begin{aligned} x - y + 2z &= 1 \\ 3x + 2y + 7z &= 8 \\ -3x - 4y + 9z &= -10 \end{aligned}$$

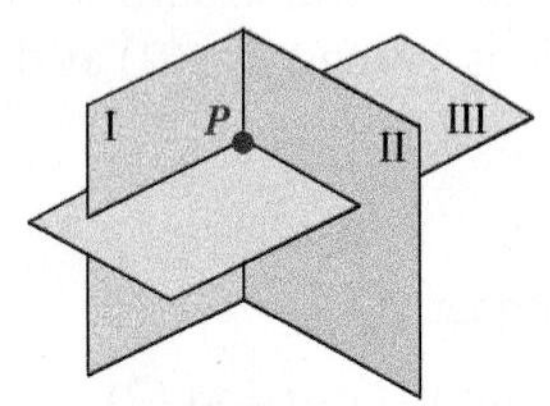

A single solution

FIGURE 8(a) (repeated)

Write the values of x, y, and z in the correct order.

Step 6 It appears that $(-3, 1, 6)$ is the only solution of the system. We must check that this ordered triple satisfies all three original equations of the system. We begin with equation (1).

CHECK

$$4x + 8y + z = 2 \quad (1)$$

$$4(-3) + 8(1) + 6 \stackrel{?}{=} 2 \quad \text{Substitute.}$$

$$-12 + 8 + 6 \stackrel{?}{=} 2 \quad \text{Multiply.}$$

$$2 = 2 \checkmark \quad \text{True}$$

Because $(-3, 1, 6)$ also satisfies equations (2) and (3), the solution set contains a single ordered triple, $\{(-3, 1, 6)\}$. This is Case 1, shown earlier in **FIGURE 8(a)**.

NOW TRY 

OBJECTIVE 3 Solve linear systems (with three equations and three variables) in which some of the equations have missing terms.

If a linear system includes an equation that is missing a term or terms, one elimination step can be omitted.

EXAMPLE 2 Solving a System of Equations with Missing Terms

Solve the system.

$$6x - 12y = -5 \quad (1) \text{ Missing } z$$

$$8y + z = 0 \quad (2) \text{ Missing } x$$

$$9x - z = 12 \quad (3) \text{ Missing } y$$

Equation (1) is missing the variable z, so one way to begin is to use z as the focus variable and eliminate z again using equations (2) and (3).

$$\begin{array}{rll} 8y + z = 0 & & (2) \\ 9x \qquad - z = 12 & & (3) \\ \hline 9x + 8y \qquad = 12 & & \text{Add. } (4) \end{array}$$

Use resulting equation (4) in x and y, together with equation (1), $6x - 12y = -5$, to eliminate x.

$$\begin{array}{rl} 18x - 36y = -15 & \text{Multiply each side of (1) by 3.} \\ -18x - 16y = -24 & \text{Multiply each side of (4) by } -2. \\ \hline -52y = -39 & \text{Add.} \end{array}$$

$$y = \frac{-39}{-52} \quad \text{Divide by } -52.$$

$$y = \frac{3}{4} \quad \text{Write in lowest terms.}$$

We can find z by substituting this value for y in equation (2).

$$8y + z = 0 \quad (2)$$

$$8\left(\frac{3}{4}\right) + z = 0 \quad \text{Let } y = \tfrac{3}{4}.$$

$$6 + z = 0 \quad \text{Multiply.}$$

$$z = -6 \quad \text{Subtract 6.}$$

NOW TRY ANSWER

1. $\{(2, 1, 0)\}$

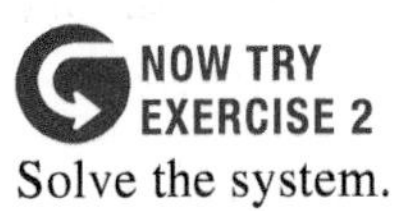

NOW TRY EXERCISE 2
Solve the system.

$$3x - z = -10$$
$$4y + 5z = 24$$
$$x - 6y = -8$$

We can find x by substituting -6 for z in equation (3).

$$9x - z = 12 \quad (3)$$
$$9x - (-6) = 12 \quad \text{Let } z = -6.$$
$$9x + 6 = 12 \quad -(-a) = a$$
$$x = \frac{6}{9} \quad \text{Subtract 6. Divide by 9.}$$
$$x = \frac{2}{3} \quad \text{Write in lowest terms.}$$

Check to verify that the solution set is $\left\{\left(\frac{2}{3}, \frac{3}{4}, -6\right)\right\}$. This is also an example of Case 1.

NOW TRY

NOTE Another way to solve the system in **Example 2** is to begin by eliminating the variable y from equations (1) and (2). The resulting equation together with equation (3) forms a system of two equations in the variables x and z. Try working **Example 2** this way to see that the same solution results. There are often multiple ways to solve a system of equations. Some ways may involve more work than others.

OBJECTIVE 4 Solve special systems.

NOW TRY EXERCISE 3
Solve the system.

$$x - 3y + 2z = 10$$
$$-2x + 6y - 4z = -20$$
$$\frac{1}{2}x - \frac{3}{2}y + z = 5$$

EXAMPLE 3 Solving a System of Dependent Equations with Three Variables

Solve the system.

$$2x - 3y + 4z = 8 \quad (1)$$
$$-x + \frac{3}{2}y - 2z = -4 \quad (2)$$
Use as the working equation, with focus variable x.
$$6x - 9y + 12z = 24 \quad (3)$$

Eliminate focus variable x using equations (1) and (2).

$$2x - 3y + 4z = 8 \quad (1)$$
$$-2x + 3y - 4z = -8 \quad \text{Multiply each side of (2) by 2.}$$
$$0 = 0 \quad \text{True (identity)}$$

Eliminating x from equations (2) and (3) gives the same result.

$$-6x + 9y - 12z = -24 \quad \text{Multiply each side of (2) by 6.}$$
$$6x - 9y + 12z = 24 \quad (3)$$
$$0 = 0 \quad \text{True (identity)}$$

All points in common
FIGURE 8(d) (repeated)

When solving a system such as this, attempting to eliminate one variable results in elimination of *all* variables. The equations are dependent—that is, they are equivalent forms of the *same* equation—and have the same graph. This is Case 3, as illustrated in **FIGURE 8(d)**. The solution set is written

$$\{(x, y, z) \mid 2x - 3y + 4z = 8\}. \quad \text{Set-builder notation}$$

Although any one of the three equations could be used to write the solution set, we use the equation with integer coefficients having greatest common factor 1 and positive coefficient of x, as in the previous section.

NOW TRY

NOW TRY ANSWERS

2. $\{(-2, 1, 4)\}$

3. $\{(x, y, z) \mid x - 3y + 2z = 10\}$

NOW TRY EXERCISE 4

Solve the system.

$$\begin{aligned} x - 5y + 2z &= 4 \\ 3x + y - z &= 6 \\ -2x + 10y - 4z &= 7 \end{aligned}$$

EXAMPLE 4 Solving an Inconsistent System with Three Variables

Solve the system.

$$2x - 4y + 6z = 5 \quad (1)$$
$$-x + 3y - 2z = -1 \quad (2)$$
$$x - 2y + 3z = 1 \quad (3)$$

Use as the working equation, with focus variable x.

Eliminate the focus variable, x, using equations (1) and (3).

$$\begin{aligned} -2x + 4y - 6z &= -2 && \text{Multiply each side of (3) by } -2. \\ 2x - 4y + 6z &= 5 && (1) \\ \hline 0 &= 3 && \text{False (contradiction)} \end{aligned}$$

The resulting false statement indicates that equations (1) and (3) have no common solution. Thus, the system is inconsistent and the solution set is $\varnothing$. The graph of this system would show two planes parallel to one another, and a third plane that intersects both, as in **FIGURE 8(f)**. This is Case 4.

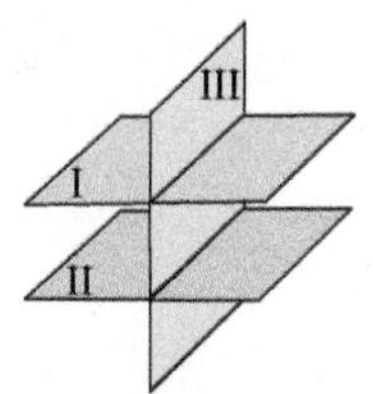

No points in common

FIGURE 8(f) (repeated)

NOW TRY 

NOTE If a false statement results when adding as in **Example 4,** it is not necessary to go any further with the solution. Because two of the three planes are parallel, it is not possible for the three planes to have any points in common.

NOW TRY EXERCISE 5

Solve the system.

$$\begin{aligned} x - 3y + 2z &= 4 \\ \frac{1}{3}x - y + \frac{2}{3}z &= 7 \\ \frac{1}{2}x - \frac{3}{2}y + z &= 2 \end{aligned}$$

EXAMPLE 5 Solving Another Special System

Solve the system.

$$2x - y + 3z = 6 \quad (1)$$
$$x - \frac{1}{2}y + \frac{3}{2}z = 3 \quad (2)$$
$$4x - 2y + 6z = 1 \quad (3)$$

Equations (1) and (2) are equivalent. If we multiply each side of equation (2) by 2, we obtain equation (1). These two equations are dependent and have the same graph.

Equations (1) and (3) are *not* equivalent, however. If we multiply each side of equation (3) by $\frac{1}{2}$, we obtain

$$2x - y + 3z = \frac{1}{2}.$$

This equation has the same coefficients as equation (1), but a different constant term. Therefore, the graphs of equations (1) and (3) have *no* points in common—that is, the planes are parallel.

Thus, this system is inconsistent and the solution set is $\varnothing$, as illustrated in **FIGURE 8(h)**. This is another example of Case 4.

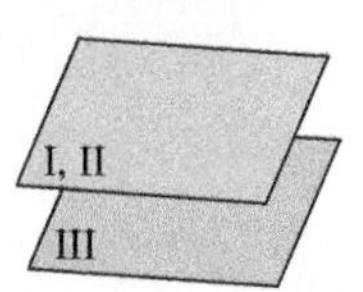

No points in common

FIGURE 8(h) (repeated)

NOW TRY 

NOW TRY ANSWERS

4. $\varnothing$

5. $\varnothing$

3.2 Exercises

MyLab Math

Video solutions for select problems available in MyLab Math

STUDY SKILLS **REMINDER**

How are you doing on your homework? **Review Study Skill 4, *Completing Your Homework.***

Concept Check *Answer each of the following.*

1. Using your immediate surroundings, give an example of three planes that satisfy the condition.

(a) They intersect in a single point.

(b) They do not intersect.

(c) They intersect in infinitely many points.

2. Suppose that a system has infinitely many ordered-triple solutions of the form (x, y, z) such that

$$x + y + 2z = 1.$$

Give three specific ordered triples that are solutions of the system.

3. Explain what the following statement means: "*The solution set of the following system is* $\{(-1, 2, 3)\}$."

$$2x + y + z = 3$$
$$3x - y + z = -2$$
$$4x - y + 2z = 0$$

4. The following two equations have a common solution of $(1, 2, 3)$.

$$x + y + z = 6$$
$$2x - y + z = 3$$

Which equation would complete a system of three linear equations in three variables having solution set $\{(1, 2, 3)\}$?

A. $3x + 2y - z = 1$ **B.** $3x + 2y - z = 4$

C. $3x + 2y - z = 5$ **D.** $3x + 2y - z = 6$

5. What constant should replace the question mark in this system so that the solution set is $\{(1, 1, 1)\}$?

$$2x - 3y + z = 0$$
$$-5x + 2y - z = -4$$
$$x + y + 2z = ?$$

6. Complete the work of **Example 1** and show that the ordered triple $(-3, 1, 6)$ is also a solution of equations (2) and (3).

$$x + 7y - 3z = -14 \quad \text{Equation (2)}$$
$$2x - 3y + 2z = 3 \quad \text{Equation (3)}$$

Solve each system. ***See Example 1.***

7. $2x - 5y + 3z = -1$
$x + 4y - 2z = 9$
$x - 2y - 4z = -5$

8. $x + 3y - 6z = 1$
$2x - y + z = 7$
$x + 2y + 2z = 14$

9. $3x + 2y + z = 8$
$2x - 3y + 2z = -16$
$x + 4y - z = 20$

10. $-3x + y - z = -10$
$-4x + 2y + 3z = -1$
$2x + 3y - 2z = -5$

11. $2x + 5y + 2z = 0$
$4x - 7y - 3z = 1$
$3x - 8y - 2z = -6$

12. $5x - 2y + 3z = -9$
$4x + 3y + 5z = 4$
$2x + 4y - 2z = 14$

13. $x + 2y + z = 4$
$2x + y - z = -1$
$x - y - z = -2$

14. $x - 2y + 5z = -7$
$-2x - 3y + 4z = -14$
$-3x + 5y - z = -7$

15. $-x + 2y + 6z = 2$
$3x + 2y + 6z = 6$
$x + 4y - 3z = 1$

16. $2x + y + 2z = 1$
$x + 2y + z = 2$
$x - y - z = 0$

17. $x + y - z = -2$
$2x - y + z = -5$
$-x + 2y - 3z = -4$

18. $x + 2y + 3z = 1$
$-x - y + 3z = 2$
$-6x + y + z = -2$

19. $\frac{1}{3}x + \frac{1}{6}y - \frac{2}{3}z = -1$
$-\frac{3}{4}x - \frac{1}{3}y - \frac{1}{4}z = 3$
$\frac{1}{2}x + \frac{3}{2}y + \frac{3}{4}z = 21$

20. $\frac{2}{3}x - \frac{1}{4}y + \frac{5}{8}z = 0$
$\frac{1}{5}x + \frac{2}{3}y - \frac{1}{4}z = -7$
$-\frac{3}{5}x + \frac{4}{3}y - \frac{7}{8}z = -5$

21. $5.5x - 2.5y + 1.6z = 11.83$
$2.2x + 5.0y - 0.1z = -5.97$
$3.3x - 7.5y + 3.2z = 21.25$

22. $6.2x - 1.4y + 2.4z = -1.80$
$3.1x + 2.8y - 0.2z = 5.68$
$9.3x - 8.4y - 4.8z = -34.20$

Solve each system. **See Example 2.**

23. $2x - 3y + 2z = -1$
$x + 2y + z = 17$
$2y - z = 7$

24. $2x - y + 3z = 6$
$x + 2y - z = 8$
$2y + z = 1$

25. $4x + 2y - 3z = 6$
$x - 4y + z = -4$
$-x + 2z = 2$

26. $2x + 3y - 4z = 4$
$x - 6y + z = -16$
$-x + 3z = 8$

27. $2x + y = 6$
$3y - 2z = -4$
$3x - 5z = -7$

28. $4x - 8y = -7$
$4y + z = 7$
$-8x + z = -4$

29. $-5x + 2y + z = 5$
$-3x - 2y - z = 3$
$-x + 6y = 1$

30. $-4x + 3y - z = 4$
$-5x - 3y + z = -4$
$-2x - 3z = 12$

31. $7x - 3z = -34$
$2y + 4z = 20$
$\frac{3}{4}x + \frac{1}{6}y = -2$

32. $5x - 2z = 8$
$4y + 3z = -9$
$\frac{1}{2}x + \frac{2}{3}y = -1$

33. $4x - z = -6$
$\frac{3}{5}y + \frac{1}{2}z = 0$
$\frac{1}{3}x + \frac{2}{3}z = -5$

34. $5x - z = 38$
$\frac{2}{3}y + \frac{1}{4}z = -17$
$\frac{1}{5}y + \frac{5}{6}z = 4$

Solve each system. If the system is inconsistent or has dependent equations, say so. **See Examples 1, 3, 4, and 5.**

35. $2x + 2y - 6z = 5$
$-3x + y - z = -2$
$-x - y + 3z = 4$

36. $-2x + 5y + z = -3$
$5x + 14y - z = -11$
$7x + 9y - 2z = -5$

37. $-5x + 5y - 20z = -40$
$x - y + 4z = 8$
$3x - 3y + 12z = 24$

38. $x + 4y - z = 3$
$-2x - 8y + 2z = -6$
$3x + 12y - 3z = 9$

39. $x + 5y - 2z = -1$
$-2x + 8y + z = -4$
$3x - y + 5z = 19$

40. $x + 3y + z = 2$
$4x + y + 2z = -4$
$5x + 2y + 3z = -2$

41. $2x + y - z = 6$
$4x + 2y - 2z = 12$
$-x - \frac{1}{2}y + \frac{1}{2}z = -3$

42. $2x - 8y + 2z = -10$
$-x + 4y - z = 5$
$\frac{1}{8}x - \frac{1}{2}y + \frac{1}{8}z = -\frac{5}{8}$

43. $x + y - 2z = 0$
$3x - y + z = 0$
$4x + 2y - z = 0$

44. $\begin{aligned} 2x + 3y - z &= 0 \\ x - 4y + 2z &= 0 \\ 3x - 5y - z &= 0 \end{aligned}$

45. $\begin{aligned} x - 2y + \frac{1}{3}z &= 4 \\ 3x - 6y + z &= 12 \\ -6x + 12y - 2z &= -3 \end{aligned}$

46. $\begin{aligned} 4x + y - 2z &= 3 \\ x + \frac{1}{4}y - \frac{1}{2}z &= \frac{3}{4} \\ 2x + \frac{1}{2}y - z &= 1 \end{aligned}$

Extending Skills *Extend the method of this section to solve each system. Express the solution in the form* (x, y, z, w).

47. $\begin{aligned} x + y + z - w &= 5 \\ 2x + y - z + w &= 3 \\ x - 2y + 3z + w &= 18 \\ -x - y + z + 2w &= 8 \end{aligned}$

48. $\begin{aligned} 3x + y - z + 2w &= 9 \\ x + y + 2z - w &= 10 \\ x - y - z + 3w &= -2 \\ -x + y - z + w &= -6 \end{aligned}$

49. $\begin{aligned} 3x + y - z + w &= -3 \\ 2x + 4y + z - w &= -7 \\ -2x + 3y - 5z + w &= 3 \\ 5x + 4y - 5z + 2w &= -7 \end{aligned}$

50. $\begin{aligned} x - 3y + 7z + w &= 11 \\ 2x + 4y + 6z - 3w &= -3 \\ 3x + 2y + z + 2w &= 19 \\ 4x + y - 3z + w &= 22 \end{aligned}$

RELATING CONCEPTS For Individual or Group Work (Exercises 51–56)

A circle *has an equation of the following form.*

$$x^2 + y^2 + ax + by + c = 0 \quad \text{Equation of a circle}$$

It is a fact from geometry that given three **non-collinear** *points—that is, points that do not all lie on the same straight line—there will be a circle that contains them. For example, the points* $(4, 2)$, $(-5, -2)$, *and* $(0, 3)$ *lie on the circle whose equation is shown in the figure.*

Work Exercises 51–56 in order, *to find an equation of the circle passing through the points*

$$(2, 1), \quad (-1, 0), \quad \text{and} \quad (3, 3).$$

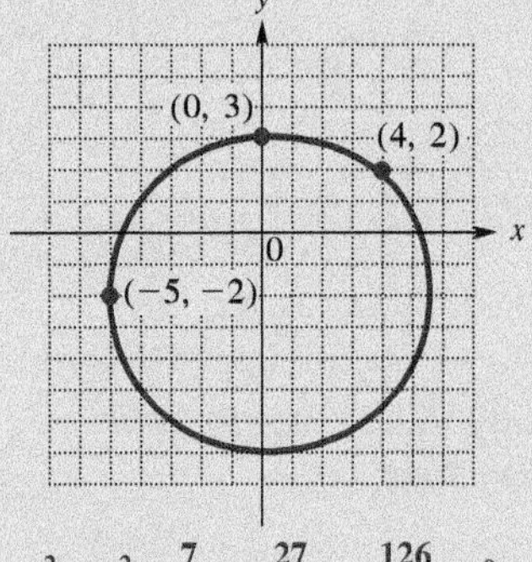

$x^2 + y^2 - \frac{7}{5}x + \frac{27}{5}y - \frac{126}{5} = 0$

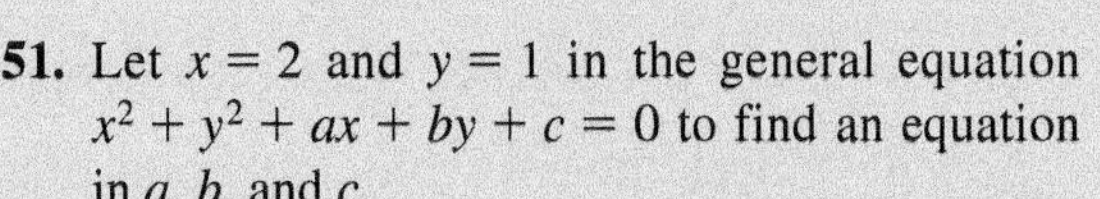

51. Let $x = 2$ and $y = 1$ in the general equation $x^2 + y^2 + ax + by + c = 0$ to find an equation in a, b, and c.

52. Let $x = -1$ and $y = 0$ to find a second equation in a, b, and c.

53. Let $x = 3$ and $y = 3$ to find a third equation in a, b, and c.

54. Form a system of three equations using the answers from **Exercises 51–53.** Solve the system to find the values of a, b, and c.

55. Use the values of a, b, and c from **Exercise 54** and the form of the equation of a circle given above to write an equation of the circle passing through the given points.

56. Explain why the relation whose graph is a circle is not a function.

3.3 Applications of Systems of Linear Equations

OBJECTIVES

1 Solve geometry problems using two variables.

2 Solve money problems using two variables.

3 Solve mixture problems using two variables.

4 Solve distance-rate-time problems using two variables.

5 Solve problems with three variables using a system of three equations.

Although some problems with two unknowns can be solved using just one variable, it is often easier to use two variables and a system of equations. The following problem, which can be solved with a system, appeared in a Hindu work that dates back to about A.D. 850. **(See Exercise 24.)**

The mixed price of 9 citrons (a lemonlike fruit) and 7 fragrant wood apples is 107; again, the mixed price of 7 citrons and 9 fragrant wood apples is 101. O you arithmetician, tell me quickly the price of a citron and the price of a wood apple here, having distinctly separated those prices well.

PROBLEM-SOLVING HINT When solving an applied problem using two variables, it is a good idea to pick letters that correspond to the descriptions of the unknown quantities. In the example above, we could choose

c to represent the number of citrons,

and w to represent the number of wood apples.

The following steps are based on the problem-solving method presented earlier in the text.

Solving an Applied Problem Using a System of Equations

Step 1 **Read** the problem carefully. *What information is given? What is to be found?*

Step 2 **Assign variables** to represent the unknown values. Write down what each variable represents. Make a sketch, diagram, or table, as needed.

Step 3 **Write a system of equations** using all the variables.

Step 4 **Solve** the system of equations.

Step 5 **State the answer**. Label it appropriately. *Does it seem reasonable?*

Step 6 **Check** the answer in the words of the *original* problem.

STUDY SKILLS REMINDER

Are you fully utilizing the features of your text? **Review Study Skill 1, *Using Your Math Text*.**

OBJECTIVE 1 Solve geometry problems using two variables.

EXAMPLE 1 Finding the Dimensions of a Soccer Field

A rectangular soccer field may have a width between 50 and 100 yd and a length between 100 and 130 yd. One particular soccer field has a perimeter of 320 yd. Its length measures 40 yd more than its width. What are the dimensions of this field? (Data from www.soccer-training-guide.com)

Step 1 **Read** the problem again. We must find the dimensions of the field.

Step 2 **Assign variables.** A sketch may be helpful. See FIGURE 9 on the next page.

Let $L =$ the length and $W =$ the width.

A rectangular parking lot has a length that is 10 ft more than twice its width. The perimeter of the parking lot is 620 ft. What are the dimensions of the parking lot?

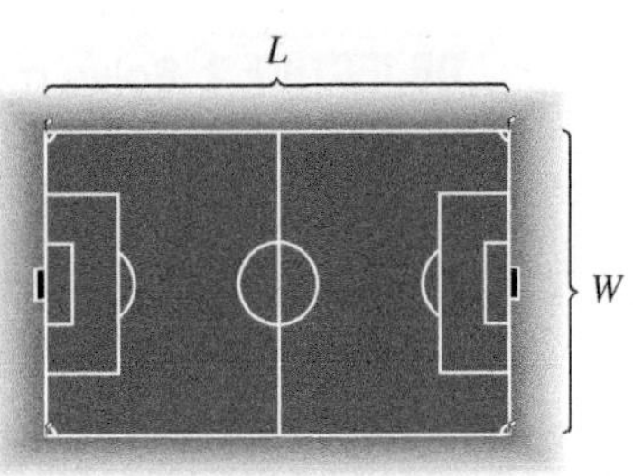

FIGURE 9

Step 3 **Write a system of equations.** Because the perimeter is 320 yd, we find one equation by using the perimeter formula.

$$2L + 2W = 320 \qquad 2L + 2W = P$$

For a second equation, use the information given about the length.

$$L = W + 40 \qquad \text{The length is 40 yd more than the width.}$$

These two equations form a system of equations.

$$2L + 2W = 320 \qquad (1)$$
$$L = W + 40 \qquad (2)$$

Step 4 **Solve** the system. Equation (2) is solved for L, so we use the substitution method, substituting $W + 40$ for L in equation (1), and solving for W.

$$2L + 2W = 320 \qquad (1)$$
$$2(W + 40) + 2W = 320 \qquad \text{Let } L = W + 40.$$
$$2W + 80 + 2W = 320 \qquad \text{Distributive property}$$
$$4W + 80 = 320 \qquad \text{Combine like terms.}$$
$$4W = 240 \qquad \text{Subtract 80.}$$
$$W = 60 \qquad \text{Divide by 4.}$$

Be sure to use parentheses around $W + 40$.

Don't stop here.

Let $W = 60$ in the equation $L = W + 40$ to find L.

$$L = 60 + 40 = 100$$

Step 5 **State the answer.** The length is 100 yd, and the width is 60 yd. Both dimensions are within the ranges given in the problem.

Step 6 **Check** using the words of the *original* problem.

$$2(100) + 2(60) = 320 \qquad \text{The perimeter is 320 yd, as required}$$
$$100 = 60 + 40 \qquad \text{Length is 40 yd more than width, as required.}$$

NOW TRY 

PROBLEM-SOLVING HINT There is often more than one way to write the equations in a system used to solve an application. In **Example 1,** we might write the second equation as

$$W = L - 40. \qquad (2)$$

In this case, we would substitute $L - 40$ for W in equation (1) to obtain

$$2L + 2(L - 40) = 320 \qquad \text{Let } W = L - 40.$$

and solve for L first (instead of W, as in Step 4 above). The *same* answer results.

NOW TRY ANSWER

1. length: 210 ft; width: 100 ft

OBJECTIVE 2 Solve money problems using two variables.

NOW TRY EXERCISE 2

Two general admission tickets and three tickets for children to a theme park cost \$239.95. One general admission ticket and four tickets for children cost \$224.95. Determine the ticket prices for general admission and for children.

EXAMPLE 2 Solving a Problem about Ticket Prices

For the 2015–2016 National Football League and National Basketball Association seasons, two football tickets and one basketball ticket purchased at their average prices cost \$241.84. One football ticket and two basketball tickets cost \$204.74. What were the average ticket prices for the two sports? (Data from Team Marketing Report.)

Step 1 **Read** the problem again. There are two unknowns.

Step 2 **Assign variables.**

Let f = the average price for a football ticket

and b = the average price for a basketball ticket.

Step 3 **Write a system of equations.** We write one equation using the fact that two football tickets and one basketball ticket cost \$241.84.

$$2f + b = 241.84$$

By similar reasoning, we can write a second equation.

$$f + 2b = 204.74$$

These two equations form a system of equations.

$$2f + b = 241.84 \quad (1)$$
$$f + 2b = 204.74 \quad (2)$$

Step 4 **Solve** the system. Either the substitution or the elimination method can be used, based on personal preference. We choose the elimination method. To eliminate f, multiply equation (2) by -2 and add.

$$2f + b = 241.84 \quad (1)$$
$$-2f - 4b = -409.48 \quad \text{Multiply each side of (2) by } -2.$$
$$-3b = -167.64 \quad \text{Add.}$$
$$b = 55.88 \quad \text{Divide by } -3.$$

To find the value of f, let $b = 55.88$ in equation (2).

$$f + 2b = 204.74 \quad (2)$$
$$f + 2(55.88) = 204.74 \quad \text{Let } b = 55.88.$$
$$f + 111.76 = 204.74 \quad \text{Multiply.}$$
$$f = 92.98 \quad \text{Subtract } 111.76.$$

Step 5 **State the answer.** The average price for one basketball ticket was \$55.88. For one football ticket, the average price was \$92.98.

Step 6 **Check** that these values satisfy the problem conditions.

$$2(\$92.98) + \$55.88 = \$241.84, \quad \text{as required.}$$
$$\$92.98 + 2(\$55.88) = \$204.74, \quad \text{as required.}$$

NOW TRY 

PROBLEM-SOLVING HINT Rework Step 4 of **Example 2** using the substitution method to confirm that the same answer results. (There are two ways to do this—either solve equation (1) for b or solve equation (2) for f.)

Solving using a different method can be a good strategy for checking your answer.

NOW TRY ANSWER

2. general admission: \$56.99; children: \$41.99

OBJECTIVE 3 Solve mixture problems using two variables.

Some mixture problems can be solved using one variable. For many mixture problems, we can use more than one variable and a system of equations.

EXAMPLE 3 Solving a Mixture Problem

How many ounces each of 5% hydrochloric acid and 20% hydrochloric acid must be combined to obtain 10 oz of solution that is 12.5% hydrochloric acid?

Step 1 **Read** the problem. Two solutions of different strengths are being mixed to obtain a specific amount of a solution with an "in-between" strength.

Step 2 **Assign variables.**

Let $x =$ the number of ounces of 5% solution

and $y =$ the number of ounces of 20% solution.

Use a table to summarize the information from the problem.

Ounces of Solution	*Percent (as a decimal)*	*Ounces of Pure Acid*
x	$5\% = 0.05$	$0.05x$
y	$20\% = 0.20$	$0.20y$
10	$12.5\% = 0.125$	$(0.125)10$

Gives equation (1) (first column); Gives equation (2) (third column)

Multiply the amount of each solution (given in the first column) by its concentration of acid (given in the second column) to find the amount of acid in that solution (given in the third column).

FIGURE 10 illustrates what is happening in the problem.

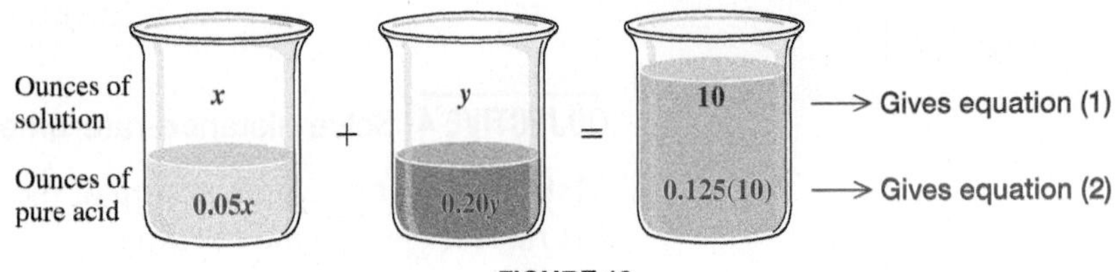

FIGURE 10

Step 3 **Write a system of equations.** When x ounces of 5% solution and y ounces of 20% solution are combined, the total number of ounces is 10.

$$x + y = 10$$

The ounces of acid in the 5% solution $(0.05x)$ added to the ounces of acid in the 20% solution $(0.20y)$ must equal the total ounces of acid in the mixture, which is $(0.125)10$, or 1.25.

$$0.05x + 0.20y = 1.25$$

***Notice that these equations can be quickly determined by reading down the table or using the labels in* FIGURE 10.**

Multiply the second equation by 100 to clear the decimals and obtain an equivalent system of equations.

$$x + y = 10 \quad (1)$$
$$5x + 20y = 125 \quad (2)$$

This is the system to solve.

NOW TRY
EXERCISE 3

How many liters each of a 15% acid solution and a 25% acid solution should be mixed to obtain 30 L of an 18% acid solution?

Step 4 **Solve** the system. Again, either the substitution or the elimination method can be used. We choose to eliminate x.

$$-5x - 5y = -50 \quad \text{Multiply each side of (1) by } -5.$$
$$5x + 20y = 125 \quad (2)$$
$$15y = 75 \quad \text{Add.}$$
$$\text{Ounces of 20\% solution} \rightarrow y = 5 \quad \text{Divide by 15.}$$

Substitute 5 for y in equation (1) to find the value of x.

$$x + y = 10 \quad (1)$$
$$x + 5 = 10 \quad \text{Let } y = 5.$$
$$\text{Ounces of 5\% solution} \rightarrow x = 5 \quad \text{Subtract 5.}$$

Step 5 **State the answer.** The desired mixture will require 5 oz of the 5% solution and 5 oz of the 20% solution.

Step 6 **Check.**

Total amount of solution: $x + y = 5 \text{ oz} + 5 \text{ oz}$
$= 10 \text{ oz}$, as required.

Total amount of acid: 5% of 5 oz + 20% of 5 oz
$= 0.05(5) + 0.20(5)$
$= 1.25 \text{ oz}$

Percent of acid in solution:

$$\frac{\text{Total acid} \rightarrow 1.25}{\text{Total solution} \rightarrow 10} = 0.125, \quad \text{or} \quad 12.5\%, \quad \text{as required.}$$

NOW TRY

OBJECTIVE 4 Solve distance-rate-time problems using two variables.

Motion problems require a form of the distance formula $d = rt$, where d is distance, r is rate (or speed), and t is time.

EXAMPLE 4 Solving a Motion Problem

A car travels 250 km in the same time that a truck travels 225 km. If the rate of the car is 8 km per hr faster than the rate of the truck, find both rates.

Step 1 **Read** the problem again. Given the distances traveled, we need to find the rate of each vehicle.

Step 2 **Assign variables.**

Let x = the rate of the car, and y = the rate of the truck.

As in **Example 3,** a table helps organize the information. Fill in the distance for each vehicle, and the variables for the unknown rates.

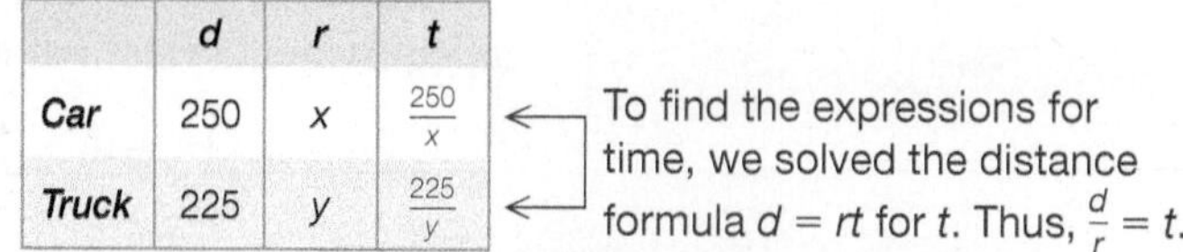

	d	r	t
Car	250	x	$\frac{250}{x}$
Truck	225	y	$\frac{225}{y}$

To find the expressions for time, we solved the distance formula $d = rt$ for t. Thus, $\frac{d}{r} = t$.

NOW TRY ANSWER

3. 25% acid solution: 9 L; 15% acid solution: 21 L

NOW TRY EXERCISE 4

On a bicycle ride, Vann can travel 50 mi in the same amount of time that Ivy can travel 40 mi. Determine each bicyclist's rate, if Vann's rate is 2 mph faster than Ivy's.

Step 3 **Write a system of equations.** The rate x of the car is 8 km per hr faster than the rate y of the truck.

Be careful writing this equation. *The car is faster.*

$$x = y + 8 \qquad (1)$$

Both vehicles travel for the *same* time, so the times must be equal.

$$\text{Time for car} \longrightarrow \frac{250}{x} = \frac{225}{y} \longleftarrow \text{Time for truck}$$

Multiply both sides by xy to obtain an equivalent equation with no variable denominators.

$$xy \cdot \frac{250}{x} = \frac{225}{y} \cdot xy \qquad \text{Multiply by the LCD, } xy.$$

$$\frac{250xy}{x} = \frac{225xy}{y} \qquad \text{Multiply; } xy = \frac{xy}{1}$$

$$250y = 225x \qquad \text{Divide out the common factors.} \quad (2)$$

We now have a system of linear equations.

$$x = y + 8 \qquad (1)$$

$$250y = 225x \qquad (2)$$

Step 4 **Solve** the system by substitution. Replace x with $y + 8$ in equation (2).

$$250y = 225x \qquad (2)$$

Be sure to use parentheses around $y + 8$.

$$250y = 225(y + 8) \qquad \text{Let } x = y + 8.$$

$$250y = 225y + 1800 \qquad \text{Distributive property}$$

$$25y = 1800 \qquad \text{Subtract } 225y.$$

$$\text{Truck's rate} \longrightarrow y = 72 \qquad \text{Divide by 25.}$$

Let $y = 72$ in the equation $x = y + 8$ to find x.

$$\text{Car's rate} \longrightarrow x = 72 + 8 = 80$$

Step 5 **State the answer.** The rate of the car is 80 km per hr, and the rate of the truck is 72 km per hr.

Step 6 **Check.**

$$\textit{Car:} \quad t = \frac{d}{r} = \frac{250}{80} = 3.125$$

$$\textit{Truck:} \quad t = \frac{d}{r} = \frac{225}{72} = 3.125$$

Times are equal, as required.

The rate of the car, 80 km per hr, is 8 km per hour greater than that of the truck, 72 km per hr, as required.

NOW TRY

PROBLEM-SOLVING HINT When one quantity in an application is compared to another (as are the rates of the car and truck in **Example 4**), be careful to translate correctly in terms of the two variables. In Step 3 of **Example 4,**

$y = x - 8$ is an alternative for equation (1).

NOW TRY ANSWER

4. Vann: 10 mph; Ivy: 8 mph

NOW TRY EXERCISE 5

In his motorboat, Ed traveled 42 mi upstream at top speed in 2.1 hr. Still at top speed, his return trip to the same spot took only 1.5 hr. Find the rate of Ed's boat in still water and the rate of the current.

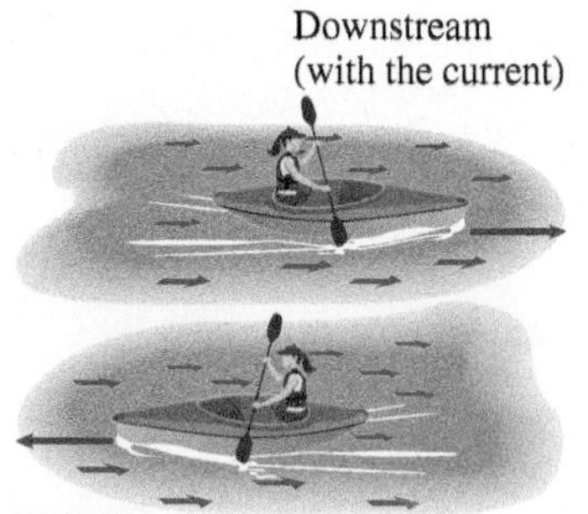

FIGURE 11

NOW TRY ANSWER

5. boat: 24 mph; current: 4 mph

EXAMPLE 5 Solving a Motion Problem

While kayaking on the Blackledge River, Rebecca traveled 9 mi upstream (against the current) in 2.25 hr. It took her only 1 hr paddling downstream (with the current) back to the spot where she started. Find Rebecca's kayaking rate in still water and the rate of the current.

Step 1 **Read** the problem. We must find two rates—Rebecca's kayaking rate in still water and the rate of the current.

Step 2 **Assign variables.**

Let x = Rebecca's kayaking rate in still water

and y = the rate of the current.

When the kayak is traveling *against* the current, the current slows it down. The rate of the kayak is the *difference* between its rate in still water and the rate of the current, which is $(x - y)$ mph.

When the kayak is traveling *with* the current, the current speeds it up. The rate of the kayak is the *sum* of its rate in still water and the rate of the current, which is $(x + y)$ mph.

Thus, $x - y$ = the rate of the kayak upstream (*against* the current),

and $x + y$ = the rate of the kayak downstream (*with* the current).

See **FIGURE 11**. Make a table. Use the formula $d = rt$, or $rt = d$.

	r	t	d
Upstream	$x - y$	2.25	$2.25(x - y)$
Downstream	$x + y$	1	$1(x + y)$

The distance is the same in each direction, 9 mi.

Step 3 **Write a system of equations.**

$$2.25(x - y) = 9 \quad \text{Upstream}$$
$$1(x + y) = 9 \quad \text{Downstream}$$

Clear parentheses in each equation, and then divide each term in the first equation by 2.25 to obtain an equivalent system.

$$x - y = 4 \quad (1)$$
$$x + y = 9 \quad (2)$$

Step 4 **Solve.**

$$2x = 13 \quad \text{Add.}$$

Rebecca's kayaking rate → $x = 6.5$ Divide by 2.

Substitute 6.5 for x in equation (2) and solve for y.

$$x + y = 9 \quad (2)$$
$$6.5 + y = 9 \quad \text{Let } x = 6.5.$$

Rate of current → $y = 2.5$ Subtract 6.5.

Step 5 **State the answer.** Rebecca's kayaking rate in still water was 6.5 mph, and the rate of the current was 2.5 mph.

Step 6 **Check.**

Distance upstream: $2.25(6.5 - 2.5) = 9$

Distance downstream: $1(6.5 + 2.5) = 9$

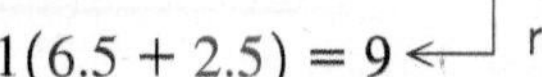

True statements result.

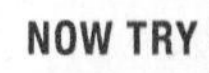

NOW TRY

OBJECTIVE 5 **Solve problems with three variables using a system of three equations.**

EXAMPLE 6 **Solving a Geometry Problem**

The sum of the measures of the angles of any triangle is 180°. The smallest angle of a certain triangle measures 36° less than the middle-sized angle. The largest angle measures 16° more than twice the smallest angle. Find the measure of each angle.

Step 1 **Read** the problem again. There are three unknowns in this problem.

Step 2 **Assign variables.** Make a sketch, as in **FIGURE 12.**

Let x = the measure of the smallest angle,

y = the measure of the middle-sized angle,

and z = the measure of the largest angle.

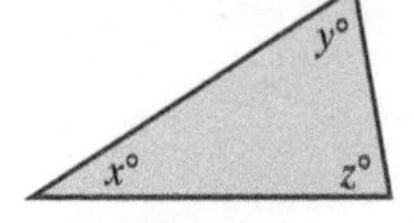

FIGURE 12

Step 3 **Write a system of three equations.** The sum of the measures of the angles of any triangle is 180°.

$$x + y + z = 180$$

We can write two more equations from the information given.

Smallest angle	measures	36° less than middle-sized angle.
↓	↓	↓
x	$=$	$y - 36$
Largest angle	measures	16° more than twice smallest angle.
↓	↓	↓
z	$=$	$16 + 2x$

These three equations form a system.

$$x + y + z = 180 \quad (1)$$
$$x = y - 36 \quad (2)$$
$$z = 16 + 2x \quad (3)$$

Step 4 **Solve** the system of equations. If we write equation (2) in terms of y, then equations (2) and (3) will give both y and z in terms of x.

$$x = y - 36 \quad (2)$$
$$y = x + 36 \quad \text{Add 36. Interchange sides.}$$

Now substitute $x + 36$ for y and $16 + 2x$ (from equation (3)) for z in equation (1).

$$x + y + z = 180 \quad (1)$$
$$x + (x + 36) + (16 + 2x) = 180 \quad \text{Let } y = x + 36 \text{ and } z = 16 + 2x.$$
$$4x + 52 = 180 \quad \text{Combine like terms.}$$
$$x = 32 \quad \text{Subtract 52. Divide by 4.}$$

Substitute 32 for x in $y = x + 36$ (equation (2) solved for y) to find y.

$$y = 32 + 36 \quad \text{Let } x = 32.$$
$$y = 68 \quad \text{Add.}$$

Substitute 32 for x in $z = 16 + 2x$ (equation (3)) to find z.

$$z = 16 + 2(32) \quad \text{Let } x = 32.$$
$$z = 80 \quad \text{Multiply, and then add.}$$

NOW TRY EXERCISE 6
The sum of the measures of the angles of any triangle is 180°. The measure of the third angle of a certain triangle is four times that of the first angle. The sum of the measures of the first and second angles is 84°. Find the measure of each angle.

Step 5 **State the answer.** The three angles measure 32°, 68°, and 80°.

Step 6 **Check.** The sum of the measures of the three angles is

$$32° + 68° + 80° = 180°, \quad \text{as required.}$$

Also, each of the following is a true statement.

$32° = 68° - 36°$ Smallest measures 36° less than middle-sized.

$80° = 16° + 2(32°)$ Largest measures 16° more than twice smallest.

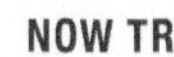

PROBLEM-SOLVING HINT In Step 4 of **Example 6,** we could also have substituted $16 + 2x$ (from equation (3)) for z in equation (1) as follows.

$x + y + z = 180$ (1)

$x + y + (16 + 2x) = 180$ Let $z = 16 + 2x$.

$3x + y = 164$ Combine like terms. Subtract 16. (4)

This gives a system of two equations in x and y.

$x - y = -36$ (2) in standard form

$3x + y = 164$ (4)

Eliminating y gives $x = 32$, which can be used to obtain $y = 68$ and $z = 80$. There is often more than one way to solve applications involving systems of three equations.

EXAMPLE 7 Solving a Problem Involving Prices

At Panera Bread, a loaf of honey wheat bread costs \$3.99, a loaf of tomato basil bread costs \$4.99, and a loaf of French bread costs \$3.19. On a recent day, three times as many loaves of honey wheat were sold as loaves of tomato basil. The number of loaves of French bread sold was 5 less than the number of loaves of honey wheat sold. Total receipts for these breads were \$90.17. How many loaves of each type of bread were sold? (Data from Panera Bread menu.)

Step 1 **Read** the problem again. There are three unknowns in this problem.

Step 2 **Assign variables** to represent the three unknowns.

Let x = the number of loaves of honey wheat bread,

y = the number of loaves of tomato basil bread,

and z = the number of loaves of French bread.

Step 3 **Write a system of three equations.** Three times as many loaves of honey wheat bread were sold as loaves of tomato basil bread.

$x = 3y$, or $x - 3y = 0$ Subtract $3y$. (1)

Also, we have the information needed for another equation.

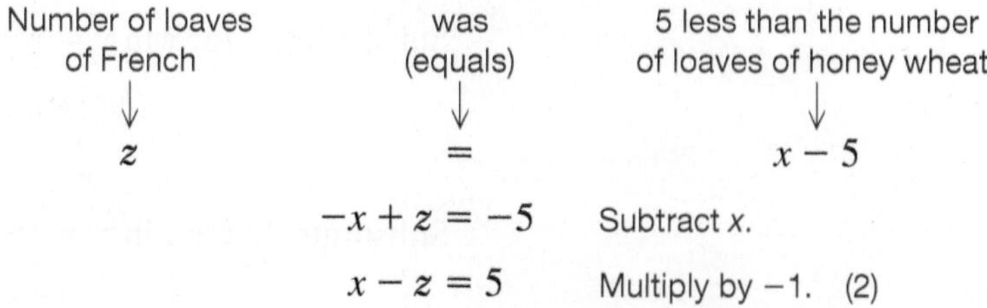

$-x + z = -5$ Subtract x.

$x - z = 5$ Multiply by −1. (2)

NOW TRY ANSWER
6. 24°, 60°, 96°

NOW TRY EXERCISE 7

At Panera Bread, a loaf of white bread costs \$3.99, a loaf of cheese bread costs \$4.79, and a loaf of sesame semolina bread costs \$8.19. On a recent day, twice as many loaves of white bread were sold as loaves of cheese bread. The number of loaves of sesame semolina sold was 3 less than the number of loaves of white sold. Total receipts for these breads were \$150.33. How many loaves of each type of bread were sold? (Data from Panera Bread menu.)

Multiplying the cost of a loaf of each kind of bread by the number of loaves of that kind sold and adding gives an equation for the total receipts.

$$3.99x + 4.99y + 3.19z = 90.17$$

$$399x + 499y + 319z = 9017 \quad \text{Multiply each term by 100 to clear decimals.} \quad (3)$$

These three equations form a system.

$$x - 3y = 0 \quad (1)$$

$$x - z = 5 \quad (2)$$

$$399x + 499y + 319z = 9017 \quad (3)$$

Step 4 **Solve** the system of equations. Equation (1) is missing the variable z, so one way to begin is to eliminate z again, using equations (2) and (3).

$$319x \qquad - 319z = 1595 \quad \text{Multiply (2) by 319.}$$

$$399x + 499y + 319z = 9017 \quad (3)$$

$$718x + 499y \qquad = 10{,}612 \quad \text{Add.} \quad (4)$$

Use resulting equation (4) in x and y, together with equation (1), $x - 3y = 0$, to eliminate x.

$$-718x + 2154y = 0 \quad \text{Multiply (1) by } -718.$$

$$718x + 499y = 10{,}612 \quad (4)$$

$$2653y = 10{,}612 \quad \text{Add.}$$

$$y = 4 \quad \text{Divide by 2653.}$$

We can find x by substituting this value for y in equation (1).

$$x - 3y = 0 \quad (1)$$

$$x - 3(4) = 0 \quad \text{Let } y = 4.$$

$$x - 12 = 0 \quad \text{Multiply.}$$

$$x = 12 \quad \text{Add 12.}$$

We can find z by substituting this value for x in equation (2).

$$x - z = 5 \quad (2)$$

$$12 - z = 5 \quad \text{Let } x = 12.$$

$$z = 7 \quad \text{Subtract 12. Multiply by } -1.$$

Step 5 **State the answer.** There were 12 loaves of honey wheat bread, 4 loaves of tomato basil bread, and 7 loaves of French bread sold.

Step 6 **Check.** Each of the following is a true statement.

$$12 = 3 \cdot 4 \quad \text{Honey wheat is three times tomato basil.}$$

$$7 = 12 - 5 \quad \text{French is 5 less than honey wheat.}$$

Multiply cost per loaf by number of loaves and add to confirm total receipts.

$$\$3.99(12) + \$4.99(4) + \$3.19(7) = \$90.17, \quad \text{as required.}$$

NOW TRY

NOW TRY ANSWER

7. white bread: 12 loaves; cheese bread: 6 loaves; sesame semolina bread: 9 loaves

3.3 Exercises

MyLab Math

Video solutions for select problems available in MyLab Math

STUDY SKILLS REMINDER

Time management can be a challenge for students. **Review Study Skill 6, *Managing Your Time*.**

Concept Check *Answer each question.*

1. If a container of liquid contains 60 oz of solution, what is the number of ounces of pure acid if the given solution contains the following acid concentrations?

   (a) 10% **(b)** 25% **(c)** 40% **(d)** 50%

2. If $5000 is invested in an account paying simple annual interest, how much interest will be earned during the first year at the following rates?

   (a) 2% **(b)** 3% **(c)** 4% **(d)** 3.5%

3. If one pound of turkey costs $1.99, how much will x pounds cost?

4. If one movie ticket costs $13.50, how much will y tickets cost?

5. If the rate of a boat in still water is 10 mph and the rate of the current of a river is x mph, what is the rate of the boat in each case?

   (a) The boat is going upstream (that is, against the current, which slows the boat down).

   (b) The boat is going downstream (that is, with the current, which speeds the boat up).

6. If the rate of a plane in still air is x mph and the rate of a steady wind is 20 mph, what is the rate of the plane in each case?

   (a) The plane is flying into the wind (that is, into a headwind, which slows the plane down).

   (b) The plane is flying with the wind (that is, with a tailwind, which speeds the plane up).

7. A bear is running at a rate of 44 ft per sec.

   (a) If the bear runs for 5 sec, what is its distance?

   (b) If the bear travels d ft, what is its time?

8. The swimming rate of a whale is 25 mph.

   (a) If the whale swims for y hours, what is its distance?

   (b) If the whale travels 10 mi, what is its time?

Solve each problem. ***See Example 1.***

9. Venus and Serena measured a tennis court and found that it was 42 ft longer than it was wide and had a perimeter of 228 ft. What were the length and the width of the tennis court?

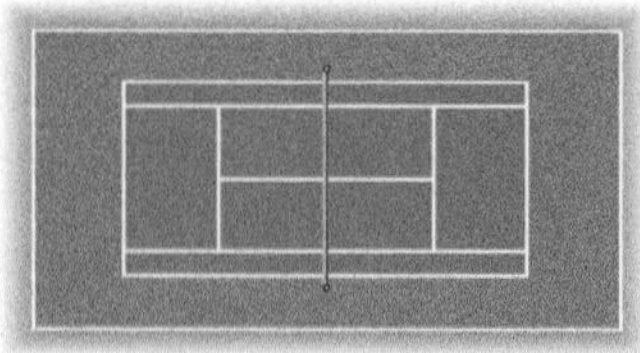

10. LeBron and Jose measured a basketball court and found that the width of the court was 44 ft less than the length. If the perimeter was 288 ft, what were the length and the width of the basketball court?

11. During the 2017 Major League Baseball season, the Cleveland Indians played 162 games. They won 42 more games than they lost. What was the team's win-loss record that year?

12. During the 2017 Major League Baseball season, the Detroit Tigers played 162 games. They lost 34 more games than they won. What was the team's win-loss record?

Team	W	L
Cleveland	___	___
Minnesota	85	77
Kansas City	80	82
Chicago White Sox	67	95
Detroit	___	___

Data from Major League Baseball.

13. In 2016, the two American telecommunication companies with the greatest revenues were AT&T and Verizon. The two companies had combined revenues of \$288.9 billion. AT&T's revenue was \$38.7 billion more than that of Verizon. What was the revenue for each company? (Data from Verizon and AT&T Annual Reports.)

14. In 2017, U.S. exports to Canada were \$39.4 billion more than exports to Mexico. Together, exports to these two countries totaled \$525.4 billion. How much were exports to each country? (Data from U.S. Census Bureau.)

Find the measures of angles x and y. Remember that **(1)** *the sum of the measures of the angles of any triangle is* 180°, **(2)** *supplementary angles have a sum of* 180°, *and* **(3)** *vertical angles have equal measures.*

15.

16.

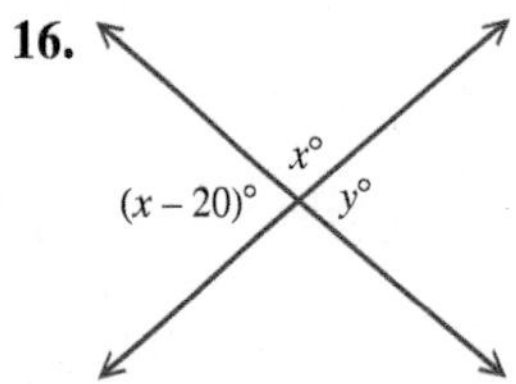

The Fan Cost Index (FCI) represents the cost of four average-price tickets to a sporting event (two adult, two child), four small soft drinks, two small beers, four hot dogs, parking for one car, two game programs, and two souvenir caps. (Data from Team Marketing Report.*)*

Use the concept of FCI to solve each problem. **See Example 2.**

17. For the 2016 Major League Baseball season, the FCI prices for the Cleveland Indians and the Boston Red Sox totaled \$540.10. The Boston FCI was \$181.22 more than that of Cleveland. What were the FCIs for these teams?

18. In 2016, the FCI prices for Major League Baseball and the National Football League totaled \$722.37. The football FCI was \$283.31 more than that of baseball. What were the FCIs for these sports?

Solve each problem. **See Example 2.**

19. Leanna is a waitress at Bonefish Grill. During one particular day she sold 15 ribeye steak dinners and 20 grilled salmon dinners, totaling \$886.50. Another day she sold 25 ribeye steak dinners and 10 grilled salmon dinners, totaling \$973.50. How much did each type of dinner cost? (Data from Bonefish Grill Menu.)

20. At a business meeting at Panera Bread, the bill for two cappuccinos and three caffe lattes was \$19.75. At another table, the bill for one cappuccino and two caffe lattes was \$11.97. How much did each type of beverage cost? (Data from Panera Bread menu.)

21. Two days at Busch Gardens (Tampa Bay) and 3 days at Universal Studios Florida (Orlando) cost \$510, while 4 days at Busch Gardens and 2 days at Universal Studios cost \$580. (Prices are based on single-day admissions.) What was the cost per day for each park? (Data from Busch Gardens and Universal Studios.)

22. On the basis of average total costs per day for business travel to New York City and Washington, DC (which include a hotel room, car rental, and three meals), 2 days in New York and 3 days in Washington cost \$2484, while 4 days in New York and 2 days in Washington cost \$3120. What was the average cost per day in each city? (Data from *Business Travel News*.)

23. Tickets to a production of *A Midsummer Night's Dream* at Broward College cost \$5 for general admission or \$4 with a student ID. If 184 people paid to see a performance and \$812 was collected, how many of each type of ticket were sold?

24. The mixed price of 9 citrons and 7 fragrant wood apples is 107; again, the mixed price of 7 citrons and 9 fragrant wood apples is 101. O you arithmetician, tell me quickly the price of a citron and the price of a wood apple here, having distinctly separated those prices well. (*Source:* Hindu work, A.D. 850.)

Solve each problem. ***See Example 3.***

25. How many gallons each of 25% alcohol and 35% alcohol should be mixed to obtain 20 gal of 32% alcohol?

Gallons of Solution	Percent (as a decimal)	Gallons of Pure Alcohol
x	25% = 0.25	
y	35% = 0.35	
20	32% =	

26. How many liters each of 15% acid and 33% acid should be mixed to obtain 120 L of 21% acid?

Liters of Solution	Percent (as a decimal)	Liters of Pure Acid
x	15% = 0.15	
y	33% =	
120	21% =	

27. A party mix is made by adding nuts that sell for \$2.50 per kg to a cereal mixture that sells for \$1 per kg. How much of each should be added to obtain 30 kg of a mix that will sell for \$1.70 per kg?

	Number of Kilograms	Price per Kilogram (in dollars)	Value (in dollars)
Nuts	x	2.50	
Cereal	y	1.00	
Mixture		1.70	

28. A fruit drink is made by mixing juices. Such a drink with 50% juice is to be mixed with a drink that is 30% juice to obtain 200 L of a drink that is 45% juice. How much of each should be used?

	Liters of Drink	Percent (as a decimal)	Liters of Pure Juice
50% Juice	x	0.50	
30% Juice	y	0.30	
Mixture		0.45	

29. A total of \$3000 is invested, part at 2% simple interest and part at 4%. If the total annual return from the two investments is \$100, how much is invested at each rate?

Principal (in dollars)	Rate (as a decimal)	Interest (in dollars)
x	0.02	$0.02x$
y	0.04	$0.04y$
3000	XXXXXX	100

30. An investor will invest a total of \$15,000 in two accounts, one paying 4% annual simple interest and the other 3%. If he wants to earn \$550 annual interest, how much should he invest at each rate?

Principal (in dollars)	Rate (as a decimal)	Interest (in dollars)
x	0.04	
y	0.03	
15,000	XXXXXX	

31. Pure acid is to be added to a 10% acid solution to obtain 54 L of a 20% acid solution. What amounts of each should be used? (*Hint:* Pure acid is 100% acid.)

32. A truck radiator holds 36 L of fluid. How much pure antifreeze must be added to a mixture that is 4% antifreeze to fill the radiator with a mixture that is 20% antifreeze?

33. How many pounds of candy that sells for \$1.75 per lb must be mixed with candy that sells for \$1.25 per lb to obtain 10 lb of a mixture that should sell for \$1.60 per lb?

34. How many pounds of candy that sells for \$2.50 per lb must be mixed with candy that sells for \$1.75 per lb to obtain 6 lb of a mixture that sells for \$2.10 per lb?

Solve each problem. ***See Examples 4 and 5.***

35. A train travels 150 km in the same time that a plane travels 400 km. If the rate of the plane is 20 km per hr less than three times the rate of the train, find both rates.

	r	t	d
Train	x		150
Plane	y		400

36. A motor scooter travels 20 mi in the same time that a bicycle travels 8 mi. If the rate of the scooter is 5 mph more than twice the rate of the bicycle, find both rates.

37. A freight train and an express train leave towns 390 km apart, traveling toward one another. The freight train travels 30 km per hr slower than the express train. They pass one another 3 hr later. What are their rates?

	r	t	d
Freight Train	x	3	
Express Train	y	3	

38. A car and a truck leave towns 230 mi apart, traveling toward each other. The car travels 15 mph faster than the truck. They pass each other 2 hr later. What are their rates?

39. In his motorboat, Bill travels upstream at top speed to his favorite fishing spot, a distance of 36 mi, in 2 hr. Returning, he finds that the trip downstream, still at top speed, takes only 1.5 hr. Find the rate of Bill's boat and the rate of the current. Let $x =$ the rate of the boat and $y =$ the rate of the current.

	r	t	d
Upstream	$x - y$	2	
Downstream	$x + y$		

40. Traveling for 3 hr into a steady headwind, a plane flies 1650 mi. The pilot determines that flying *with* the same wind for 2 hr, he could make a trip of 1300 mi. Find the rate of the plane and the wind speed.

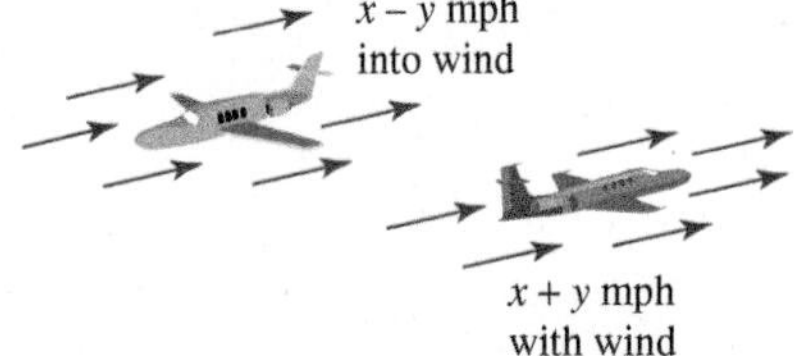

41. A plane flies 560 mi in 1.75 hr traveling with the wind. The return trip later against the same wind takes the plane 2 hr. Find the rate of the plane and the wind speed.

42. Braving blizzard conditions on the planet Hoth, Luke Skywalker sets out in his snow speeder for a rebel base 4800 mi away. He travels into a steady headwind and makes the trip in 3 hr. Returning, he finds that the trip back, now with a tailwind, takes only 2 hr. Find the rate of Luke's snow speeder and the wind speed.

Solve each problem. ***See Examples 6 and 7.***

43. In the triangle below, $z = x + 10$ and $x + y = 100$. Determine a third equation involving x, y, and z, and then find the measures of the three angles.

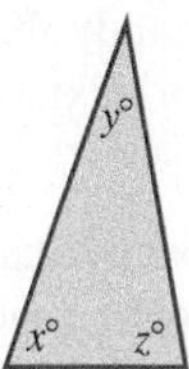

44. In the triangle below, x is 10 less than y and 20 less than z. Write a system of equations and find the measures of the three angles.

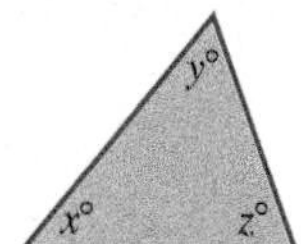

45. In a certain triangle, the measure of the second angle is 10° greater than three times the first. The third angle measure is equal to the sum of the measures of the other two. Find the measures of the three angles.

46. The measure of the largest angle of a triangle is 12° less than the sum of the measures of the other two. The smallest angle measures 58° less than the largest. Find the measures of the angles.

47. The perimeter of a triangle is 70 cm. The longest side is 4 cm less than the sum of the other two sides. Twice the shortest side is 9 cm less than the longest side. Find the length of each side of the triangle.

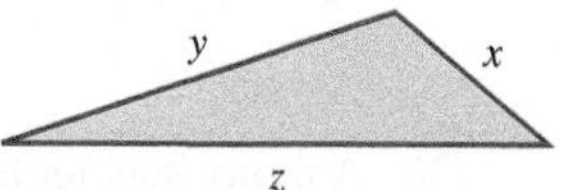

48. The perimeter of a triangle is 56 in. The longest side measures 4 in. less than the sum of the other two sides. Three times the shortest side is 4 in. more than the longest side. Find the lengths of the three sides.

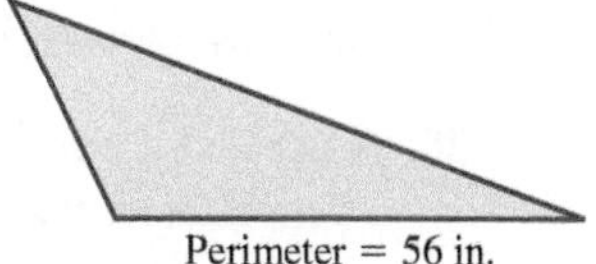

Perimeter = 56 in.

49. In the 2016 Summer Olympics, host Brazil earned 1 more gold medal than silver. The number of silver medals that Brazil earned was the same as the number of its bronze medals. Brazil earned a total of 19 medals. How many of each kind of medal did Brazil earn? (Data from *The World Almanac and Book of Facts*.)

50. In 2017, the average Facebook user in the United States spent 693 more minutes per month using Facebook than the average Instagram user spent using Instagram. The average Instagram user spent 36 fewer minutes per month using Instagram than the average Snapchat user spent using Snapchat. The total amount of time that average users of each social network spent per month using their respective networks was 1347 min. How much time did the average user spend on each social network? (Data from Verto Analytics.)

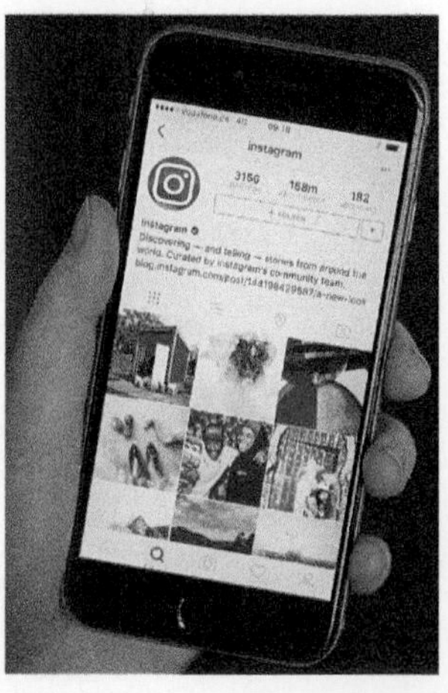

51. Tickets for a Harlem Globetrotters show cost $30 for upper level, $51 for center court, or $76 for floor seats. Nine times as many upper level tickets were sold as floor tickets, and the number of upper level tickets sold was 55 more than the sum of the number of center court tickets and floor tickets. Sales of all three kinds of tickets totaled $95,215. How many of each kind of ticket were sold? (Data from www.harlemglobetrotters.com)

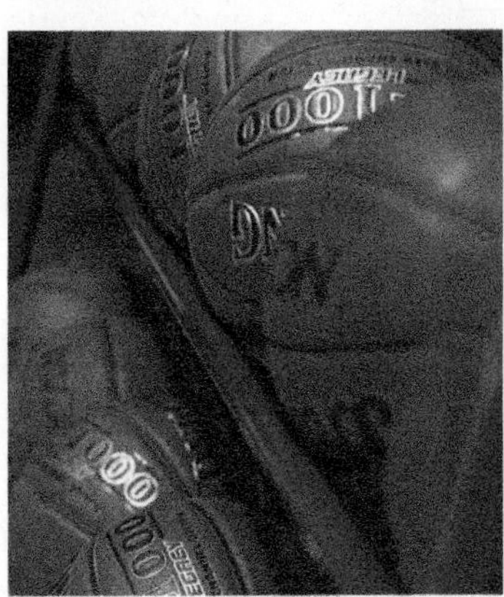

52. Three kinds of tickets are available for a rock concert: "up close," "in the middle," and "far out." "Up close" tickets cost $10 more than "in the middle" tickets. "In the middle" tickets cost $10 more than "far out" tickets. Twice the cost of an "up close" ticket is $20 more than three times the cost of a "far out" ticket. Find the price of each kind of ticket.

53. A wholesaler supplies college t-shirts to three college bookstores: A, B, and C. The wholesaler recently shipped a total of 800 t-shirts to the three bookstores. Twice as many t-shirts were shipped to bookstore B as to bookstore A, and the number shipped to bookstore C was 40 less than the sum of the numbers shipped to the other two bookstores. How many t-shirts were shipped to each bookstore?

54. An office supply store sells three models of computer desks: A, B, and C. In one month, the store sold a total of 85 computer desks. The number of model B desks was five more than the number of model C desks. The number of model A desks was four more than twice the number of model C desks. How many of each model did the store sell that month?

55. A plant food is to be made from three chemicals. The mix must include 60% of the first and second chemicals. The second and third chemicals must be in the ratio of 4 to 3 by weight. How much of each chemical is needed to make 750 kg of the plant food?

56. How many ounces of 5% hydrochloric acid, 20% hydrochloric acid, and water must be combined to obtain 10 oz of solution that is 8.5% hydrochloric acid if the amount of water used must equal the total amount of the other two solutions?

The National Hockey League uses a point system to determine team standings. A team is awarded 2 *points for a win* (W), 0 *points for a loss in regulation play* (L), *and* 1 *point for an overtime loss* (OTL). *Use this information to solve each problem.*

Team	*GP*	*W*	*L*	*OTL*	*Points*
Anaheim	82	____	____	____	105
Edmonton	82	47	26	9	103
San Jose	82	46	29	7	99
Calgary	82	45	33	4	94
Los Angeles	82	____	____	____	86

Data from www.nhl.com

57. During the 2016–2017 NHL regular season, the Anaheim Ducks played 82 games. Their wins and overtime losses resulted in a total of 105 points. They had 10 more losses in regulation play than overtimes losses. How many wins, losses, and overtime losses did they have that season?

58. During the same NHL regular season, the Los Angeles Kings also played 82 games. Their wins and overtimes losses resulted in a total of 86 points. They had 4 more total losses (in regulation play and overtime) than wins. How many wins, losses, and overtime losses did they have that season?

The following exercises are based on the "Bait Box Puzzle," featured in an "Ask Marilyn" column in Parade Magazine.

59. There are three kinds of bait boxes—those containing fish, those containing bugs, and those containing worms. Although the bait boxes look alike, each kind has a different weight. Use the information in **FIGURE A** to write a system of three equations, and solve it to determine the weight of each kind of bait box.

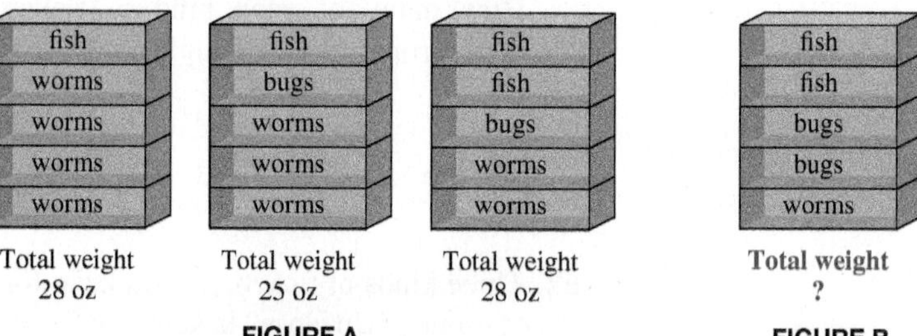

FIGURE A **FIGURE B**

60. Write an equation that describes the situation depicted in **FIGURE B**. Use the same three variables used in **Exercise 59.** Then determine the total weight of the bait boxes in **FIGURE B.**

Chapter 3 Summary

STUDY SKILLS **REMINDER**

Be prepared for your math test on this chapter. **Review Study Skills 7 and 8, *Reviewing a Chapter* and *Taking Math Tests.***

Key Terms

3.1

system of equations
system of linear equations (linear system)
solution set of a linear system
consistent system
independent equations
inconsistent system
dependent equations

3.2

ordered triple
focus variable
working equation

New Symbol

(x, y, z) ordered triple

Test Your Word Power

See how well you have learned the vocabulary in this chapter.

1. A **system of equations** consists of
A. at least two equations with different variables
B. two or more equations that have an infinite number of solutions
C. two or more equations with the same variables
D. two or more inequalities that are to be solved.

2. The **solution set of a system of equations** is
A. all ordered pairs that satisfy one equation of the system
B. all ordered pairs that satisfy all the equations of the system
C. any ordered pair that satisfies one or more equations of the system
D. the set of values that make all the equations of the system false.

3. A **consistent system** has
A. no solution
B. a solution
C. exactly two solutions
D. exactly three solutions.

4. An **inconsistent system** is a system of equations
   A. with one solution
   B. with no solution
   C. with an infinite number of solutions
   D. that have the same graph.

5. **Dependent equations**
   A. have different graphs
   B. have no solution
   C. have one solution
   D. are different forms of the same equation.

6. **Independent equations**
   A. have different graphs
   B. have no solution
   C. have graphs that are parallel lines
   D. are different forms of the same equation.

ANSWERS

1. C; *Example:* $\begin{matrix} 3x - y = 3 \\ 2x + y = 7 \end{matrix}$ 2. B; *Example:* The ordered pair $(2, 3)$ satisfies both equations of the system in Answer 1, so $\{(2, 3)\}$ is the solution set of the system. 3. B; *Example:* The system in Answer 1 is a consistent system. 4. B; *Example:* The equations of two parallel lines form an inconsistent system. Their graphs never intersect, so the system has no solution. 5. D; *Example:* The equations $4x - y = 8$ and $8x - 2y = 16$ are dependent because their graphs are the same line. 6. A; *Example:* The equations in the system in Answer 1 are not equivalent. They are independent equations with different graphs.

Quick Review

CONCEPTS	EXAMPLES
3.1 Systems of Linear Equations in Two Variables **Solving a Linear System by Substitution** ***Step 1*** Solve one of the equations for either variable. ***Step 2*** Substitute for that variable in the other equation. The result should be an equation with just one variable. ***Step 3*** Solve the equation from Step 2. ***Step 4*** Find the value of the other variable by substituting the result from Step 3 into the equation from Step 1. ***Step 5*** Check the values in *both* of the *original* equations. Then write the solution set as a set containing an ordered pair.	Solve by substitution. $4x - y = 7$ (1) $3x + 2y = 30$ (2) Solve for y in equation (1). $y = 4x - 7$ Substitute $4x - 7$ for y in equation (2), and solve for x. $3x + 2y = 30$ (2) $3x + 2(4x - 7) = 30$ Let $y = 4x - 7$. $3x + 8x - 14 = 30$ Distributive property $11x - 14 = 30$ Combine like terms. $11x = 44$ Add 14. $x = 4$ Divide by 11. Substitute 4 for x in the equation $y = 4x - 7$ to find y. $y = 4(4) - 7$ $y = 9$ Multiply. Subtract. Check to verify that $\{(4, 9)\}$ is the solution set.
Solving a Linear System by Elimination ***Step 1*** Write both equations in standard form. ***Step 2*** Transform the equations as needed so that the coefficients of one pair of variable terms are opposites. ***Step 3*** Add the new equations. The sum should be an equation with just one variable. ***Step 4*** Solve the equation from Step 3. ***Step 5*** Find the value of the other variable by substituting the result from Step 4 into either of the original equations. ***Step 6*** Check the values in *both* of the *original* equations. Then write the solution set as a set containing an ordered pair.	Solve by elimination. $5x + y = 2$ (1) $2x - 3y = 11$ (2) To eliminate y, multiply equation (1) by 3, and add the result to equation (2). $15x + 3y = 6$ 3 times equation (1) $2x - 3y = 11$ (2) $17x = 17$ Add. $x = 1$ Divide by 17.

CONCEPTS	EXAMPLES
	Let $x = 1$ in equation (1), and solve for y. $5x + y = 2$ (1) $5(1) + y = 2$ Let $x = 1$. $y = -3$ Multiply. Subtract 5. Check to verify that $\{(1, -3)\}$ is the solution set.
Solving Special Systems If the result of the addition step (Step 3) is a false statement, such as $0 = 4$, the graphs are parallel lines and *there is no solution. The solution set is* $\varnothing$. If the result is a true statement, such as $0 = 0$, the graphs are the same line, and *an infinite number of ordered pairs are solutions. The solution set is written in set-builder notation as* $\{(x, y) \mid \underline{\qquad\qquad}\}$, *where a form of the equation is written in the blank.*	Solve each system. $x - 2y = 6$ $-x + 2y = -2$ $0 = 4$ Inconsistent system Solution set: $\varnothing$ $x - 2y = 6$ $-x + 2y = -6$ $0 = 0$ Dependent equations Solution set: $\{(x, y) \mid x - 2y = 6\}$
3.2 Systems of Linear Equations in Three Variables **Solving a Linear System in Three Variables** ***Step 1*** Select a focus variable, preferably one with coefficient 1 or -1, and a working equation. ***Step 2*** Eliminate the focus variable, using the working equation and one of the equations of the system. ***Step 3*** Eliminate the focus variable again, using the working equation and the remaining equation of the system. ***Step 4*** Solve the system of two equations in two variables formed by the equations from Steps 2 and 3. ***Step 5*** Find the value of the remaining variable. ***Step 6*** Check the values in *each* of the *original* equations of the system. Then write the solution set as a set containing an ordered triple.	Solve the system. $x + 2y - z = 6$ (1) $x + y + z = 6$ (2) $2x + y - z = 7$ (3) We choose z as the focus variable and (2) as the working equation. Add equations (1) and (2). $2x + 3y = 12$ (4) Add equations (2) and (3). $3x + 2y = 13$ (5) Use equations (4) and (5) to eliminate x. $-6x - 9y = -36$ Multiply (4) by -3. $6x + 4y = 26$ Multiply (5) by 2. $-5y = -10$ Add. $y = 2$ Divide by -5. To find x, substitute 2 for y in equation (4). $2x + 3(2) = 12$ Let $y = 2$ in (4). $2x + 6 = 12$ Multiply. $2x = 6$ Subtract 6. $x = 3$ Divide by 2. Substitute 3 for x and 2 for y in working equation (2). $x + y + z = 6$ (2) $3 + 2 + z = 6$ Substitute. $z = 1$ Subtract 5. Check to verify that $\{(3, 2, 1)\}$ is the solution set.

CONCEPTS	EXAMPLES
3.3 Applications of Systems of Linear Equations Use the six-step problem-solving method. ***Step 1*** Read the problem carefully. ***Step 2*** Assign variables. ***Step 3*** Write a system of equations. ***Step 4*** Solve the system. ***Step 5*** State the answer. ***Step 6*** Check.	The perimeter of a rectangle is 18 ft. The length is 3 ft more than twice the width. What are the dimensions of the rectangle? Let x = the length and y = the width. Write a system of equations. $2x + 2y = 18$ (1) From the perimeter formula $x = 2y + 3$ (2) Length is 3 ft more than twice the width. Substitute $2y + 3$ for x in equation (1), and solve for y. $2x + 2y = 18$ (1) $2(2y + 3) + 2y = 18$ Let $x = 2y + 3$. $4y + 6 + 2y = 18$ Distributive property $6y + 6 = 18$ Combine like terms. $6y = 12$ Subtract 6. $y = 2$ Divide by 6. To find x, substitute 2 for y in equation (2). $x = 2y + 3$ (2) $x = 2(2) + 3$ Let $y = 2$. $x = 7$ Multiply, and then add. The length is 7 ft, and the width is 2 ft. The answer checks because $2(7) + 2(2) = 18$ and $2(2) + 3 = 7$, as required.

Chapter 3 Review Exercises

3.1 *Solve each system by graphing.*

1. $x + 3y = 8$
$2x - y = 2$

2. $2x + y = -8$
$x - 4y = -4$

3. $x - y = 2$
$x + 2y = -1$

4. Which ordered pair is *not* a solution of the equation $3x + 2y = 6$?

A. $(2, 0)$ **B.** $(0, 3)$ **C.** $(4, -3)$ **D.** $(3, -2)$

Solve each system using the substitution method. If a system is inconsistent or has dependent equations, say so.

5. $3x + y = -4$
$x = \frac{2}{3}y$

6. $9x - y = -4$
$y = x + 4$

7. $-5x + 2y = -2$
$x + 6y = 26$

8. $5x + y = 12$
$2x - 2y = 0$

Solve each system using the elimination method. If a system is inconsistent or has dependent equations, say so.

9. $5x - y = 14$
$4x + 2y = 0$

10. $x - 4y = -4$
$3x + y = 1$

11. $6x + 5y = 4$
$-4x + 2y = 8$

12. $\frac{1}{6}x + \frac{1}{6}y = -\frac{1}{2}$
$x - y = -9$

13. $-3x + y = 6$
$y = 6 + 3x$

14. $5x - 4y = 2$
$-10x + 8y = 7$

3.2 *Solve each system. If a system is inconsistent or has dependent equations, say so.*

15. $2x + 3y - z = -16$
$x + 2y + 2z = -3$
$-3x + y + z = -5$

16. $3x - y - z = -8$
$4x + 2y + 3z = 15$
$-6x + 2y + 2z = 10$

17. $4x - y = 2$
$3y + z = 9$
$x + 2z = 7$

18. $5x - y = 26$
$4y + 3z = -4$
$x + z = 5$

19. $3x - 4y + z = 8$
$-6x + 8y - 2z = -16$
$\frac{3}{2}x - 2y + \frac{1}{2}z = 4$

20. $2x - y + 3z = 0$
$5x + y - z = 0$
$-2x + 3y + 4z = 0$

3.3 *Solve each problem.*

21. A regulation National Hockey League ice rink has perimeter 570 ft. The length of the rink is 30 ft longer than twice the width. What are the dimensions of an NHL ice rink? (Data from www.nhl.com)

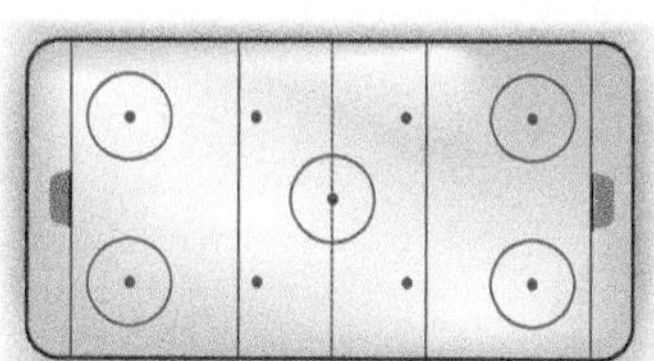

22. A plane flies 560 mi in 1.75 hr traveling with the wind. The return trip later against the same wind takes the plane 2 hr. Find the rate of the plane and the speed of the wind. Let $x =$ the rate of the plane and $y =$ the speed of the wind.

	r	*t*	*d*
With Wind	$x + y$	1.75	
Against Wind		2	

23. In the 2016 Summer Olympics in Rio de Janeiro, Japan earned 9 fewer gold medals than bronze. The number of silver medals earned was 34 less than twice the number of bronze medals. Japan earned a total of 41 medals. How many of each kind of medal did Japan earn? (Data from *The World Almanac and Book of Facts.*)

24. For Valentine's Day, Ms. Sweet will mix some \$6-per-lb nuts with some \$3-per-lb chocolate candy to obtain 100 lb of mix, which she will sell at \$3.90 per lb. How many pounds of each should she use?

	Number of Pounds	*Price per Pound (in dollars)*	*Value (in dollars)*
Nuts	x		
Chocolate	y		
Mixture	100		

25. How many liters each of 8% and 20% hydrogen peroxide should be mixed together to obtain 8 L of 12.5% solution?

26. The sum of the measures of the angles of any triangle is 180°. The largest angle measures 10° less than the sum of the other two. The measure of the middle-sized angle is the average of the other two. Find the measures of the three angles.

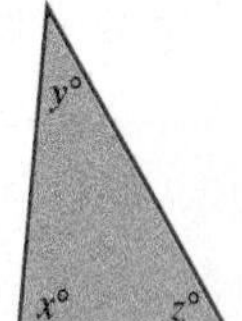

27. Noemi sells real estate. On three recent sales, she made 10% commission, 6% commission, and 5% commission. Her total commissions on these sales were \$17,000, and she sold property worth \$280,000. If the 5% sale amounted to the sum of the other two, what were the three sales prices?

28. In the great baseball year of 1961, Yankee teammates Mickey Mantle, Roger Maris, and Yogi Berra combined for 137 home runs. Mantle hit 7 fewer than Maris, and Maris hit 39 more than Berra. What were the home run totals for each player? (Data from www.mlb.com)

Chapter 3 Mixed Review Exercises

1. Suppose that two linear equations are graphed on the same set of coordinate axes. Sketch what the graph might look like if the system has the given description.
   - **(a)** The system has a single solution.
   - **(b)** The system has no solution.
   - **(c)** The system has infinitely many solutions.

2. Without doing any algebraic work, but answering on the basis of knowledge of the graphs of the two lines, explain why the following system has $\emptyset$ as its solution set.

$$y = 3x + 2$$
$$y = 3x - 4$$

Solve each system.

3. $-7x + 3y = 12$
   $5x + 2y = 8$

4. $2x - 5y = 8$
   $3x + 4y = 10$

5. $x + 2y = 48$
   $\frac{1}{4}x + \frac{1}{2}y = 12$

6. $x + 4y = 17$
   $-3x + 2y = -9$

7. $\frac{2}{3}x + \frac{1}{6}y = \frac{19}{2}$
   $\frac{1}{3}x - \frac{2}{9}y = 2$

8. $x = 7y + 10$
   $2x + 3y = 3$

9. $x + y - z = 0$
   $2y - z = 1$
   $2x + 3y - 4z = -4$

10. $x + 3y - 6z = 7$
    $2x - y + z = 1$
    $x + 2y + 2z = -1$

11. $2x + 5y - z = 12$
    $-x + y - 4z = -10$
    $-8x - 20y + 4z = 31$

Solve each problem.

12. Which system would be easier to solve using the substitution method? Why?

System A: $5x - 3y = 7$, $2x + 8y = 3$ *System B:* $7x + 2y = 4$, $y = -3x + 1$

13. To make a 10% acid solution, Jeffrey wants to mix some 5% solution with 10 L of 20% solution. How many liters of 5% solution should he use?

14. In the 2016 Summer Olympics, China, the United States, and Great Britain won a combined total of 258 medals. China won 51 fewer medals than the United States. Great Britain won 54 fewer medals than the United States. How many medals did each country win? (Data from *The World Almanac and Book of Facts.*)

Chapter 3 Test

Step-by-step test solutions are found on the Chapter Test Prep Videos available in MyLab Math.

View the complete solutions to all Chapter Test exercises in MyLab Math.

Assuming growth rates continue, the populations of Houston, Phoenix, Dallas, and Philadelphia will follow the trends indicated in the graph. Use the graph to work each problem.

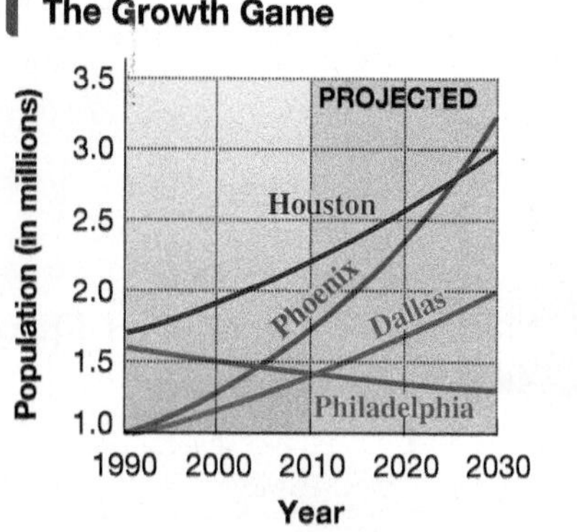

Data from U.S. Census Bureau, *Chronicle* research.

1. (a) Which of these cities will experience population growth?

(b) Which city will experience population decline?

(c) Rank the city populations from least to greatest for the year 2000.

2. (a) In which year did the population of Dallas equal that of Philadelphia? About what was this population?

(b) Write as an ordered pair (year, population in millions) the point at which Houston and Phoenix will have the same population.

3. Solve the system by graphing. $\begin{aligned} x + y &= 7 \\ x - y &= 5 \end{aligned}$

Solve each system. If a system is inconsistent or has dependent equations, say so.

4. $\begin{aligned} 2x - 3y &= 24 \\ y &= -\frac{2}{3}x \end{aligned}$

5. $\begin{aligned} 3x - y &= -8 \\ 2x + 6y &= 3 \end{aligned}$

6. $\begin{aligned} 12x - 5y &= 8 \\ 3x &= \frac{5}{4}y + 2 \end{aligned}$

7. $\begin{aligned} 3x + y &= 12 \\ 2x - y &= 3 \end{aligned}$

8. $\begin{aligned} -5x + 2y &= -4 \\ 6x + 3y &= -6 \end{aligned}$

9. $\begin{aligned} 3x + 4y &= 8 \\ 8y &= 7 - 6x \end{aligned}$

10. $\begin{aligned} 3x + 5y + 3z &= 2 \\ 6x + 5y + z &= 0 \\ 3x + 10y - 2z &= 6 \end{aligned}$

11. $\begin{aligned} 4x + y + z &= 11 \\ x - y - z &= 4 \\ y + 2z &= 0 \end{aligned}$

12. $\begin{aligned} -2x + 4y + 10z &= 3 \\ x - 2y - 5z &= 1 \\ x + 2y + z &= -1 \end{aligned}$

Solve each problem.

13. Two top-grossing super hero films, *Captain America: Civil War* and *Deadpool,* earned \$771.2 million together. If *Deadpool* grossed \$45 million less than *Captain America,* how much did each film gross? (Data from comScore, Inc.)

14. Two cars start from points 420 mi apart and travel toward each other. They meet after 3.5 hr. Find the average rate of each car if one travels 30 mph slower than the other.

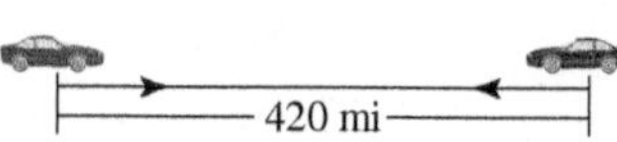

15. A chemist needs 12 L of a 40% alcohol solution. She must mix a 20% solution and a 50% solution. How many liters of each will be required to obtain what she needs?

Liters of Solution	Percent (as a decimal)	Liters of Pure Alcohol

16. A local electronics store will sell seven AC adaptors and two rechargeable flashlights for \$86, or three AC adaptors and four rechargeable flashlights for \$84. What is the price of a single AC adaptor and a single rechargeable flashlight?

17. The largest angle of a triangle measures 10° less than twice the middle-sized angle. The middle-sized angle measures 5° more than twice the smallest angle. The sum of the measures of the angles of any triangle is 180°. Find the measure of each angle.

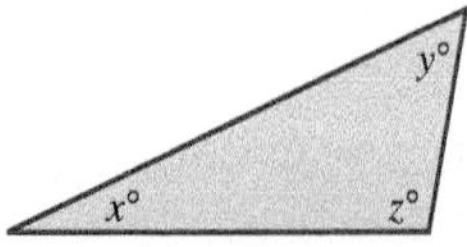

18. The owner of a tea shop wants to mix three kinds of tea to make 100 oz of a mixture that will sell for \$0.83 per oz. He uses Orange Pekoe, which sells for \$0.80 per oz, Irish Breakfast, for \$0.85 per oz, and Earl Grey, for \$0.95 per oz. If he wants to use twice as much Orange Pekoe as Irish Breakfast, how much of each kind of tea should he use?

Chapters R–3 Cumulative Review Exercises

Evaluate each expression if possible.

1. $(-3)^4$ **2.** -3^4 **3.** $-(-3)^4$

4. $\sqrt{0.49}$ **5.** $-\sqrt{0.49}$ **6.** $\sqrt{-0.49}$

Perform the indicated operations.

7. $-15.42 - (-8.25)$ **8.** $-\frac{3}{4} - \left(\frac{2}{3} + \frac{1}{8}\right)$ **9.** $\frac{7}{6} \div \left(-\frac{11}{8}\right)$

10. Simplify $\frac{(-23 + \sqrt{64})(-4^2)}{-6 - 2}$.

Evaluate for $x = -4$, $y = 3$, and $z = 6$.

11. $|2x| + 3y - z^3$ **12.** $-5(x^3 - y^3)$

Solve each equation.

13. $7(2x + 3) - 4(2x + 1) = 2(x + 1)$ **14.** $|6x - 8| = 4$

15. $ax + by = d$ for x **16.** $0.04x + 0.06(x - 1) = 1.04$

Solve each inequality.

17. $\frac{2}{3}x + \frac{5}{12}x \le 20$ **18.** $|3x + 2| \le 4$

19. $|12t + 7| \ge 0$ **20.** $2x + 3 > 5$ or $x - 1 \le 6$

Solve each problem.

21. A jar contains only pennies, nickels, and dimes. The number of dimes is one more than the number of nickels, and the number of pennies is six more than the number of nickels. How many of each denomination are in the jar, if the total value is \$4.80?

22. Two angles of a triangle have the same measure. The measure of the third angle is 4° less than twice the measure of each of the equal angles. Find the measures of the three angles.

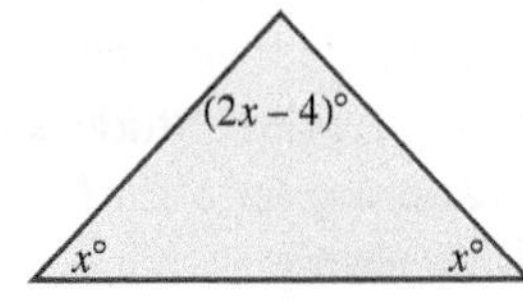

23. A survey measured public recognition of some popular contemporary advertising slogans. Complete the results shown in the table if 2500 people were surveyed.

Slogan (product or company)	*Percent Recognition (nearest tenth of a percent)*	*Actual Number Who Recognized Slogan (nearest whole number)*
Please Don't Squeeze the . . . (Charmin)	80.4%	
The Breakfast of Champions (Wheaties)	72.5%	
The King of Beers (Budweiser)		1570
Like a Good Neighbor (State Farm)		1430

Data from Department of Integrated Marketing Communications, Northwestern University.

Point A has coordinates $(-2, 6)$ *and point B has coordinates* $(4, -2)$.

24. What is the equation of the horizontal line through A?

25. What is the equation of the vertical line through B?

26. What is the slope of line AB?

27. What is the slope of a line perpendicular to line AB?

28. What is the standard form of the equation of line AB?

Solve each problem.

29. Graph the line having slope $\frac{2}{3}$ and passing through the point $(-1, -3)$.

30. Graph the inequality $-3x - 2y \leq 6$.

31. Given that $f(x) = x^2 + 3x - 6$, find **(a)** $f(-3)$ and **(b)** $f(a)$.

Solve each system.

32. $-2x + 3y = -15$
$4x - y = 15$

33. $x - 3y = 7$
$2x - 6y = 14$

34. $x + y + z = 10$
$x - y - z = 0$
$-x + y - z = -4$

35. The graph shows a company's costs to produce computer parts and the revenue from the sale of computer parts.

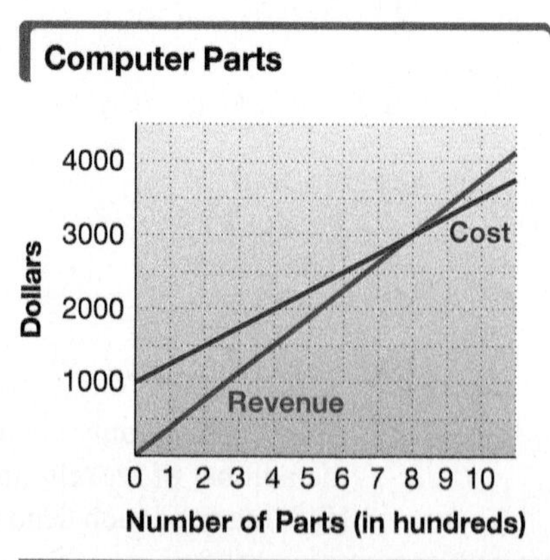

(a) At what production level does the cost equal the revenue? What is the revenue at that point?

(b) Profit is revenue less cost. Estimate the profit on the sale of 1100 parts.

STUDY SKILLS **REMINDER**
It is not too soon to begin preparing for your final exam. **Review Study Skill 10, *Preparing for Your Math Final Exam.***

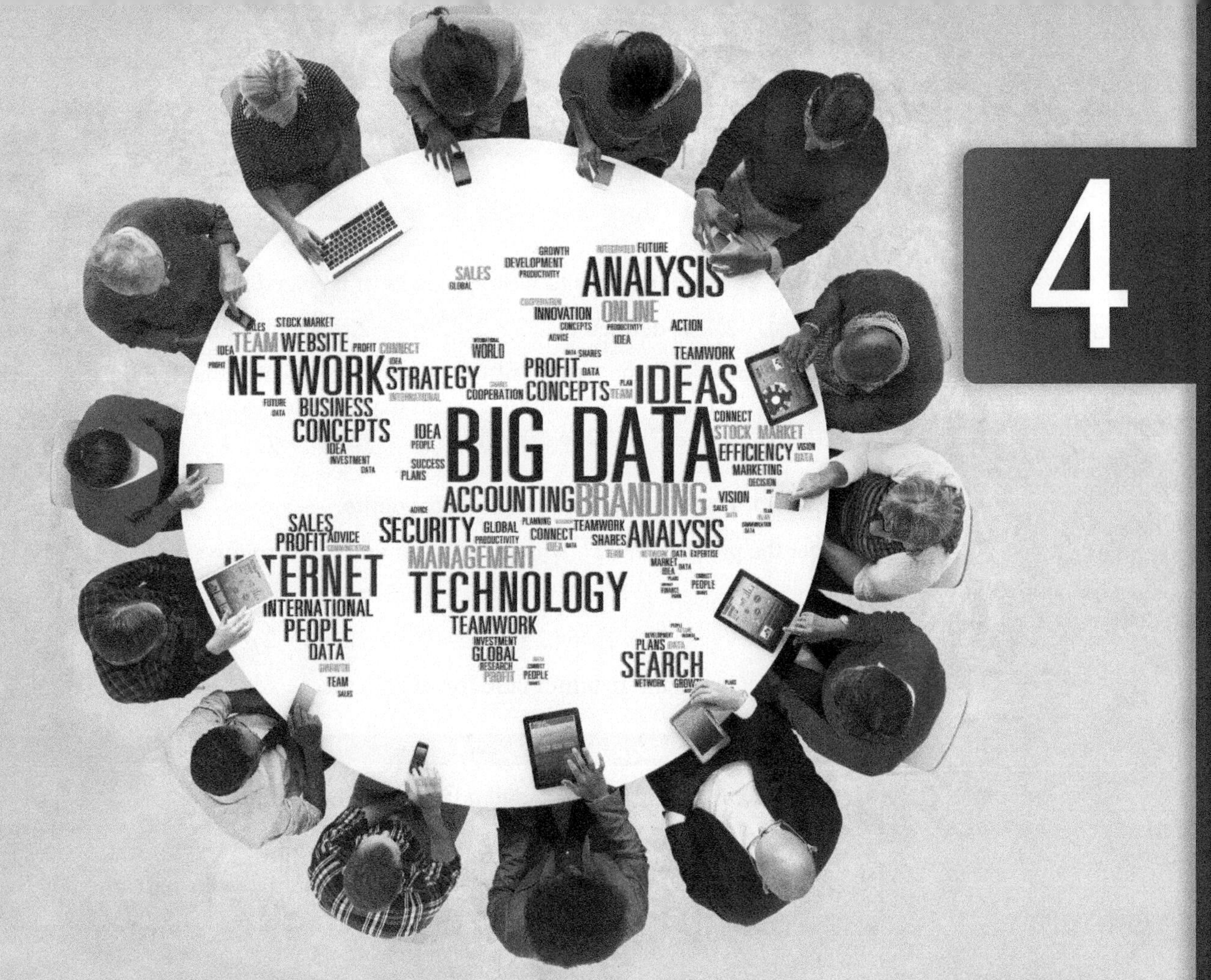

4

EXPONENTS, POLYNOMIALS, AND POLYNOMIAL FUNCTIONS

"Big Data" describes the massive amounts of information collected, analyzed, and applied worldwide today. Large numbers like Big Data numbers can be written using *scientific notation*, one of the topics of this chapter.

4.1 Integer Exponents

OBJECTIVES

1. Use the product rule for exponents.
2. Define 0 and negative exponents.
3. Use the quotient rule for exponents.
4. Use the power rules for exponents.
5. Simplify exponential expressions.

VOCABULARY

☐ exponent (power)
☐ base
☐ exponential expression

STUDY SKILLS REMINDER
Make study cards to help you learn and remember the material in this chapter. **Review Study Skill 5, *Using Study Cards.***

Recall that we use exponents to write products of repeated factors.

$$2^5 \text{ is defined as } 2 \cdot 2 \cdot 2 \cdot 2 \cdot 2, \text{ which equals } 32.$$

In 2^5, the number 5 is the **exponent** (or **power**). It indicates that the **base** 2 appears as a factor five times. The quantity 2^5 is an **exponential expression.** We read 2^5 as ***"2 to the fifth power"*** or ***"2 to the fifth."***

OBJECTIVE 1 Use the product rule for exponents.

Consider the product $2^5 \cdot 2^3$, which can be simplified as follows.

$$2^5 \cdot 2^3 = (2 \cdot 2 \cdot 2 \cdot 2 \cdot 2)(2 \cdot 2 \cdot 2) = 2^8 \quad (5 + 3 = 8)$$

This result suggests the **product rule for exponents.**

Product Rule for Exponents

If m and n are natural numbers and a is any real number, then

$$a^m \cdot a^n = a^{m+n}.$$

That is, when multiplying powers of like bases, keep the same base and add the exponents.

To see that the product rule is true, use the definition of an exponent.

$$a^m = \underbrace{a \cdot a \cdot a \cdot \ldots \cdot a}_{a \text{ appears as a factor } m \text{ times.}} \qquad a^n = \underbrace{a \cdot a \cdot a \cdot \ldots \cdot a}_{a \text{ appears as a factor } n \text{ times.}}$$

From this,

$$a^m \cdot a^n = \underbrace{a \cdot a \cdot a \cdot \ldots \cdot a}_{m \text{ factors}} \cdot \underbrace{a \cdot a \cdot a \cdot \ldots \cdot a}_{n \text{ factors}}$$

$$= \underbrace{a \cdot a \cdot a \cdot \ldots \cdot a}_{(m+n) \text{ factors}}$$

$$a^m \cdot a^n = a^{m+n}.$$

NOW TRY EXERCISE 1
Apply the product rule, if possible.

(a) $8^5 \cdot 8^4$ (b) $4 \cdot 4^2$
(c) $r^7 \cdot r \cdot r^9$ (d) $p^2 \cdot q^2$
(e) $(5x^4y^7)(-7xy^3)$

EXAMPLE 1 Using the Product Rule for Exponents

Apply the product rule for exponents, if possible.

(a) $3^4 \cdot 3^7$
$= 3^{4+7}$ (Keep the same base.)
$= 3^{11}$

(b) $5^3 \cdot 5$
$= 5^3 \cdot 5^1$
$= 5^{3+1}$
$= 5^4$

(c) $y^3 \cdot y^8 \cdot y^2$
$= y^{3+8+2}$
$= y^{13}$

(d) $x^2 \cdot y^4$ The product rule does not apply because the bases are not the same.

(e) $(5y^2)(-3y^4)$
$= 5(-3)y^2y^4$ Commutative property
$= -15y^{2+4}$ Multiply; product rule
$= -15y^6$

(f) $(7p^3q)(2p^5q^2)$
$= 7(2)p^3p^5q^1q^2$
$= 14p^{3+5}q^{1+2}$
$= 14p^8q^3$

NOW TRY 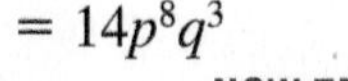

NOW TRY ANSWERS
1. (a) 8^9 (b) 4^3 (c) r^{17}
(d) The product rule does not apply.
(e) $-35x^5y^{10}$

CAUTION Do *not* multiply the bases.

$3^4 \cdot 3^7 = 3^{11}$, ***not*** 9^{11}. See Example 1(a).

Keep the same base and add the exponents.

OBJECTIVE 2 Define 0 and negative exponents.

Consider the following, where the product rule is extended to whole numbers.

$$4^2 \cdot 4^0 = 4^{2+0} = 4^2$$

For the product rule to hold, 4^0 must equal 1, so we define a^0 this way for any nonzero real number a.

Zero Exponent

If a is any nonzero real number, then

$$a^0 = 1.$$

The expression 0^0 is undefined.*

NOW TRY EXERCISE 2

Evaluate.

(a) 5^0 **(b)** -5^0 **(c)** $(-5)^0$

(d) $(-5x)^0$ $(x \neq 0)$

(e) $10^0 - 9^0$

EXAMPLE 2 Using 0 as an Exponent

Evaluate.

(a) $6^0 = 1$

(b) $(-6)^0 = 1$ Here the base is −6.

(c) $-6^0 = -(6)^0 = -1$ The base is 6, not −6.

(d) $-(-6)^0 = -1$

(e) $(8k)^0 = 1$ $(k \neq 0)$

Any nonzero quantity raised to the zero power equals 1.

(f) $5^0 + 12^0$

$= 1 + 1$

$= 2$

NOW TRY

To define a negative exponent, we extend the product rule, as follows.

$$8^2 \cdot 8^{-2} = 8^{2+(-2)} = 8^0 = 1$$

Because their product is 1, it appears that 8^{-2} is the *reciprocal* of 8^2. But we know that $\frac{1}{8^2}$ is the reciprocal of 8^2, and a number can have only one reciprocal. Therefore, **8^{-2} must equal $\frac{1}{8^2}$.**

We can generalize this result.

Negative Exponent

For any natural number n and any nonzero real number a,

$$a^{-n} = \frac{1}{a^n}.$$

With this definition, the expression a^n is meaningful for any integer exponent n and any nonzero real number a.

*In advanced courses, 0^0 is called an *indeterminate form.*

NOW TRY ANSWERS

2. (a) 1 (b) −1 (c) 1 (d) 1 (e) 0

CAUTION ***A negative exponent does not indicate that an expression represents a negative number. Negative exponents lead to reciprocals.***

$$3^{-2} = \frac{1}{3^2} = \frac{1}{9} \quad \text{Not negative} \qquad -3^{-2} = -\frac{1}{3^2} = -\frac{1}{9} \quad \text{Negative}$$

The base is 3 in both cases. 3^{-2} represents the reciprocal of 3^2.

NOW TRY EXERCISE 3

Write using only positive exponents.

(a) 9^{-4} (b) 2^{-1}

(c) $(3y)^{-6}$ $(y \neq 0)$

(d) $-4k^{-3}$ $(k \neq 0)$

EXAMPLE 3 Using Negative Exponents

Write using only positive exponents.

(a) $2^{-3} = \frac{1}{2^3}$ This is the reciprocal of 2^3.

(b) $6^{-1} = \frac{1}{6^1} = \frac{1}{6}$ This is the *reciprocal* of 6^1, or 6.

(c) $(5z)^{-3} = \frac{1}{(5z)^3}$ $(z \neq 0)$

The base is $5z$. (Note the parentheses.)

(d) $5z^{-3} = 5\left(\frac{1}{z^3}\right) = \frac{5}{z^3}$ $(z \neq 0)$

The base is z, *not* $5z$.

(e) $-m^{-2} = -\frac{1}{m^2}$ $(m \neq 0)$

(What is the base here?)

(f) $(-m)^{-2} = \frac{1}{(-m)^2}$, or $\frac{1}{m^2}$ $(m \neq 0)$

(What is the base here?)

NOW TRY EXERCISE 4

Evaluate.

(a) $4^{-1} + 6^{-1}$ (b) $\frac{1}{5^{-3}}$

(c) $\frac{10^{-2}}{2^{-5}}$

EXAMPLE 4 Using Negative Exponents

Evaluate.

(a) $3^{-1} + 4^{-1}$

$= \frac{1}{3} + \frac{1}{4}$ Definition of negative exponent

$= \frac{4}{12} + \frac{3}{12}$ $\frac{1}{3} \cdot \frac{4}{4} = \frac{4}{12}$; $\frac{1}{4} \cdot \frac{3}{3} = \frac{3}{12}$

$= \frac{7}{12}$ $\frac{a}{c} + \frac{b}{c} = \frac{a+b}{c}$

(b) $5^{-1} - 2^{-1}$

$= \frac{1}{5} - \frac{1}{2}$ Definition of negative exponent

$= \frac{2}{10} - \frac{5}{10}$ Find a common denominator.

$= -\frac{3}{10}$ $\frac{a}{c} - \frac{b}{c} = \frac{a-b}{c}$

(c) $$\frac{1}{2^{-3}} = \frac{1}{\frac{1}{2^3}} = 1 \div \frac{1}{2^3} = 1 \cdot \frac{2^3}{1} = 2^3 = 8$$

Multiply by the reciprocal of the divisor.

$\frac{1}{2^{-3}}$ represents the reciprocal of 2^{-3}. Because $2^{-3} = \frac{1}{8}$, the reciprocal is 8, which agrees with the final answer.

(d) $$\frac{2^{-3}}{3^{-2}} = \frac{\frac{1}{2^3}}{\frac{1}{3^2}} = \frac{1}{2^3} \div \frac{1}{3^2} = \frac{1}{2^3} \cdot \frac{3^2}{1} = \frac{3^2}{2^3} = \frac{9}{8}$$

CAUTION In **Example 4(a),** $3^{-1} + 4^{-1} \neq (3 + 4)^{-1}$. The expression $3^{-1} + 4^{-1}$ simplifies to $\frac{7}{12}$, while the expression $(3 + 4)^{-1}$ simplifies to 7^{-1}, which equals $\frac{1}{7}$. Similar reasoning can be applied in **Example 4(b).**

NOW TRY ANSWERS

3. **(a)** $\frac{1}{9^4}$ **(b)** $\frac{1}{2}$ **(c)** $\frac{1}{(3y)^6}$ **(d)** $-\frac{4}{k^3}$

4. **(a)** $\frac{5}{12}$ **(b)** 125 **(c)** $\frac{8}{25}$

Example 4 suggests the following generalizations.

Special Rules for Negative Exponents

If $a \neq 0$ and $b \neq 0$, then

$$\frac{1}{a^{-n}} = a^n \quad \text{and} \quad \frac{a^{-n}}{b^{-m}} = \frac{b^m}{a^n}.$$

OBJECTIVE 3 Use the quotient rule for exponents.

We simplify a quotient, such as $\frac{a^8}{a^3}$, in much the same way as a product. (In all quotients of this type, assume that the denominator is not 0.) Consider this example.

$$\frac{a^8}{a^3} = \frac{a \cdot a \cdot a \cdot a \cdot a \cdot a \cdot a \cdot a}{a \cdot a \cdot a} = a \cdot a \cdot a \cdot a \cdot a = a^5$$

Notice that $8 - 3 = 5$. In the same way, we simplify $\frac{a^3}{a^8}$.

$$\frac{a^3}{a^8} = \frac{a \cdot a \cdot a}{a \cdot a \cdot a \cdot a \cdot a \cdot a \cdot a \cdot a} = \frac{1}{a^5} = a^{-5}$$

Here, $3 - 8 = -5$. These examples suggest the **quotient rule for exponents.**

Quotient Rule for Exponents

If a is any nonzero real number and m and n are integers, then

$$\frac{a^m}{a^n} = a^{m-n}.$$

That is, when dividing powers of like bases, keep the same base and subtract the exponent of the denominator from the exponent of the numerator.

NOW TRY EXERCISE 5

Apply the quotient rule, if possible, and write each result using only positive exponents.

(a) $\frac{t^8}{t^2}$ $(t \neq 0)$ **(b)** $\frac{4^5}{4^{-2}}$

(c) $\frac{2^3}{2^7}$ **(d)** $\frac{m^4}{n^3}$ $(n \neq 0)$

EXAMPLE 5 Using the Quotient Rule for Exponents

Apply the quotient rule for exponents, if possible, and write each result using only positive exponents.

Numerator exponent — Denominator exponent

(a) $\frac{3^7}{3^2} = 3^{7-2} = 3^5$

Subtraction symbol

(b) $\frac{p^6}{p^2} = p^{6-2} = p^4 \quad (p \neq 0)$

(c) $\frac{k^7}{k^{12}} = k^{7-12} = k^{-5} = \frac{1}{k^5} \quad (k \neq 0)$

(d) $\frac{2^7}{2^{-3}} = 2^{7-(-3)} = 2^{7+3} = 2^{10}$ (Use parentheses to avoid errors.)

(e) $\frac{8^{-2}}{8^5} = 8^{-2-5} = 8^{-7} = \frac{1}{8^7}$

(f) $\frac{6}{6^{-1}} = \frac{6^1}{6^{-1}} = 6^{1-(-1)} = 6^2$

(g) $\frac{z^{-5}}{z^{-8}} = z^{-5-(-8)} = z^3 \quad (z \neq 0)$ (Be careful with signs.)

(h) $\frac{a^3}{b^4} \quad (b \neq 0)$ This expression cannot be simplified further.

The quotient rule does not apply because the bases are different.

NOW TRY

NOW TRY ANSWERS

5. **(a)** t^6 **(b)** 4^7 **(c)** $\frac{1}{2^4}$
(d) The quotient rule does not apply.

OBJECTIVE 4 Use the power rules for exponents.

We can simplify $(3^4)^2$ as follows.

$$(3^4)^2 = 3^4 \cdot 3^4 = 3^{4+4} = 3^8$$

Notice that $4 \cdot 2 = 8$. This example suggests the first **power rule for exponents.** The other two power rules can be demonstrated similarly.

Power Rules for Exponents

If a and b are real numbers and m and n are integers, then

$$\textbf{(a)}\ (a^m)^n = a^{mn}, \quad \textbf{(b)}\ (ab)^m = a^m b^m, \quad \text{and} \quad \textbf{(c)}\ \left(\frac{a}{b}\right)^m = \frac{a^m}{b^m} \quad (b \neq 0).$$

That is,

(a) To raise a power to a power, multiply exponents.

(b) To raise a product to a power, raise each factor to that power.

(c) To raise a quotient to a power, raise the numerator and the denominator to that power.

NOW TRY EXERCISE 6

Simplify using the power rules.

(a) $(x^4)^2$ **(b)** $(9x)^3$

(c) $(-2m^3)^4$

(d) $\left(\frac{3x^2}{y^3}\right)^3 \quad (y \neq 0)$

EXAMPLE 6 Using the Power Rules for Exponents

Simplify using the power rules.

(a) $(p^8)^3$

$= p^{8 \cdot 3}$ Power rule (a)

$= p^{24}$

(b) $(3y)^4$

$= 3^4y^4$ Power rule (b)

$= 81y^4$

(c) $\left(\frac{2}{3}\right)^4$

$= \frac{2^4}{3^4}$ Power rule (c)

$= \frac{16}{81}$

(d) $(6p^7)^2$

$= 6^2(p^7)^2$ Power rule (b)

$= 6^2p^{7 \cdot 2}$ Power rule (a)

$= 36p^{14}$ Square 6. Multiply exponents.

(e) $\left(\frac{-2m^5}{z}\right)^3$

$= \frac{(-2m^5)^3}{z^3}$ Power rule (c)

$= \frac{(-2)^3m^{5 \cdot 3}}{z^3}$ Power rule (b)

$= -\frac{8m^{15}}{z^3} \quad (z \neq 0)$ Simplify.

NOW TRY

The reciprocal of a^n is $\frac{1}{a^n}$, which equals $\left(\frac{1}{a}\right)^n$. Also, a^n and a^{-n} are reciprocals.

$$a^n \cdot a^{-n} = a^n \cdot \frac{1}{a^n} = 1 \qquad \text{Reciprocals have product 1.}$$

Thus, because both $\left(\frac{1}{a}\right)^n$ and a^{-n} are reciprocals of a^n, the following is true.

$$a^{-n} = \left(\frac{1}{a}\right)^n \qquad a^{-n} \text{ represents the reciprocal of } a^n.$$

NOW TRY ANSWERS

6. (a) x^8 **(b)** $729x^3$ **(c)** $16m^{12}$ **(d)** $\frac{27x^6}{y^9}$

Special Rules for Negative Exponents, Continued

If $a \neq 0$ and $b \neq 0$ and n is an integer, then

$$a^{-n} = \left(\frac{1}{a}\right)^n \quad \text{and} \quad \left(\frac{a}{b}\right)^{-n} = \left(\frac{b}{a}\right)^n.$$

That is, any nonzero number raised to the negative nth power is equal to the reciprocal of that number raised to the nth power.

NOW TRY EXERCISE 7

Write using only positive exponents and then evaluate.

(a) $\left(\frac{1}{3}\right)^{-2}$ **(b)** $\left(\frac{5}{3}\right)^{-3}$

(c) $\left(\frac{3}{2x}\right)^{-4}$ $(x \neq 0)$

EXAMPLE 7 Using Negative Exponents with Fractions

Write using only positive exponents and then evaluate.

(a) $\left(\frac{1}{2}\right)^{-4}$

$= 2^4$ $\quad \left(\frac{1}{a}\right)^n = a^{-n}$

$= 16$

Applying the negative exponent involves changing the base to its reciprocal and changing the sign of the exponent.

(b) $\left(\frac{3}{7}\right)^{-2}$

$= \left(\frac{7}{3}\right)^2$ $\quad \left(\frac{a}{b}\right)^{-n} = \left(\frac{b}{a}\right)^n$

$= \frac{7^2}{3^2}$ $\quad$ Power rule (c)

$= \frac{49}{9}$ $\quad$ Apply exponents.

(c) $\left(\frac{4x}{5}\right)^{-3}$ $(x \neq 0)$

$= \left(\frac{5}{4x}\right)^3$ $\quad \left(\frac{a}{b}\right)^{-n} = \left(\frac{b}{a}\right)^n$

$= \frac{5^3}{4^3x^3}$ $\quad$ Power rules (c) and (b)

$= \frac{125}{64x^3}$ $\quad$ Apply exponents.

NOW TRY 

Summary of Definitions and Rules for Exponents

For all integers m and n and all real numbers a and b for which the following are defined, these rules hold true.

		Examples
Product Rule	$a^m \cdot a^n = a^{m+n}$	$5^2 \cdot 5^4 = 5^{2+4} = 5^6$
Quotient Rule	$\frac{a^m}{a^n} = a^{m-n}$	$\frac{4^9}{4^2} = 4^{9-2} = 4^7$
Zero Exponent	$a^0 = 1$	$(-2)^0 = 1$
Negative Exponent	$a^{-n} = \frac{1}{a^n}$	$3^{-4} = \frac{1}{3^4}$
Power Rules	(a) $(a^m)^n = a^{mn}$	$(3^2)^5 = 3^{2 \cdot 5} = 3^{10}$
	(b) $(ab)^m = a^m b^m$	$(3x)^2 = 3^2x^2$
	(c) $\left(\frac{a}{b}\right)^m = \frac{a^m}{b^m}$	$\left(\frac{2}{3}\right)^5 = \frac{2^5}{3^5}$

Special Rules for Negative Exponents

$\frac{1}{a^{-n}} = a^n$ $\qquad \frac{1}{4^{-3}} = 4^3$ $\qquad \frac{a^{-n}}{b^{-m}} = \frac{b^m}{a^n}$ $\qquad \frac{9^{-2}}{3^{-4}} = \frac{3^4}{9^2}$

$a^{-n} = \left(\frac{1}{a}\right)^n$ $\qquad 6^{-2} = \left(\frac{1}{6}\right)^2$ $\qquad \left(\frac{a}{b}\right)^{-n} = \left(\frac{b}{a}\right)^n$ $\qquad \left(\frac{4}{7}\right)^{-3} = \left(\frac{7}{4}\right)^3$

NOW TRY ANSWERS

7. **(a)** 9 **(b)** $\frac{27}{125}$ **(c)** $\frac{16x^4}{81}$

OBJECTIVE 5 Simplify exponential expressions.

NOW TRY EXERCISE 8

Simplify. Assume that all variables represent nonzero real numbers.

(a) $x^{-8} \cdot x \cdot x^4$

(b) $(5^{-3})^2$

(c) $\dfrac{p^2q^{-4}}{p^{-2}q^{-1}}$

(d) $\left(\dfrac{2x^2}{y^2}\right)^3\left(\dfrac{-5x^{-2}}{y}\right)^{-2}$

EXAMPLE 8 Using the Definitions and Rules for Exponents

Simplify. Assume that all variables represent nonzero real numbers.

(a) $3^2 \cdot 3^{-5}$

$= 3^{2+(-5)}$ Product rule

$= 3^{-3}$ Add exponents.

$= \dfrac{1}{3^3}$ $a^{-n} = \frac{1}{a^n}$

$= \dfrac{1}{27}$ $3^3 = 27$

(b) $x^{-3}x^{-4}x^2$

$= x^{-3+(-4)+2}$ Product rule

$= x^{-5}$ Add exponents.

$= \dfrac{1}{x^5}$ $a^{-n} = \frac{1}{a^n}$

(c) $(4^{-2})^{-5}$

$= 4^{(-2)(-5)}$ Power rule (a)

$= 4^{10}$ Multiply exponents.

(d) $(x^{-4})^6$

$= x^{(-4)6}$ Power rule (a)

$= x^{-24}$ Multiply exponents.

$= \dfrac{1}{x^{24}}$ $a^{-n} = \frac{1}{a^n}$

(e) $\dfrac{x^{-4}y^2}{x^2y^{-5}}$

$= \dfrac{x^{-4}}{x^2} \cdot \dfrac{y^2}{y^{-5}}$ $\frac{ab}{cd} = \frac{a}{c} \cdot \frac{b}{d}$

$= x^{-4-2} \cdot y^{2-(-5)}$ Quotient rule

$= x^{-6}y^7$ Subtract exponents.

$= \dfrac{y^7}{x^6}$ $a^{-n} = \frac{1}{a^n}$

(f) $(2^3x^{-2})^{-2}$

$= (2^3)^{-2} \cdot (x^{-2})^{-2}$ Power rule (b)

$= 2^{-6}x^4$ Power rule (a)

$= \dfrac{x^4}{2^6}$ $a^{-n} = \frac{1}{a^n}$

$= \dfrac{x^4}{64}$ $2^6 = 64$

(g) $\left(\dfrac{3x^2}{y}\right)^2\left(\dfrac{4x^3}{y^{-2}}\right)^{-1}$

$= \dfrac{3^2(x^2)^2}{y^2} \cdot \dfrac{y^{-2}}{4x^3}$ Combination of rules

$= \dfrac{9x^4}{y^2} \cdot \dfrac{y^{-2}}{4x^3}$ $3^2 = 9$; Power rule (a)

$= \dfrac{9}{4}x^{4-3}y^{-2-2}$ Quotient rule

$= \dfrac{9}{4}x^1y^{-4}$ Subtract exponents.

$= \dfrac{9x}{4y^4}$ $a^{-n} = \frac{1}{a^n}$

(h) $\left(\dfrac{-4m^5n^4}{24mn^{-7}}\right)^{-2}$

$= \left(\dfrac{m^{5-1}n^{4-(-7)}}{-6}\right)^{-2}$ Quotient rule; Divide coefficients.

$= \left(\dfrac{m^4n^{11}}{-6}\right)^{-2}$ Subtract exponents.

$= \dfrac{(m^4)^{-2}(n^{11})^{-2}}{(-6)^{-2}}$ Power rules (b) and (c)

$= \dfrac{m^{-8}n^{-22}}{(-6)^{-2}}$ Power rule (a)

$= \dfrac{(-6)^2}{m^8n^{22}}$ $\frac{a^{-n}}{b^{-m}} = \frac{b^m}{a^n}$

The sign on −6 does *not* change in this step.

$= \dfrac{36}{m^8n^{22}}$ $(-6)^2 = 36$

NOW TRY

NOW TRY ANSWERS

8. (a) $\dfrac{1}{x^3}$ (b) $\dfrac{1}{5^6}$ (c) $\dfrac{p^4}{q^3}$ (d) $\dfrac{8x^{10}}{25y^4}$

4.1 Exercises

FOR EXTRA HELP 

Video solutions for select problems available in MyLab Math

STUDY SKILLS REMINDER

We can learn from the mistakes we make. **Review Study Skill 9, *Analyzing Your Test Results.***

Concept Check *Determine whether each statement is* true *or* false. *If false, correct the right-hand side of the statement.*

1. $(ab)^2 = ab^2$

2. $(5x)^3 = 5^3x^3$

3. $\left(\frac{4}{a}\right)^3 = \frac{4^3}{a} \quad (a \neq 0)$

4. $x^3 \cdot x^4 = x^7$

5. $xy^0 = 0 \quad (y \neq 0)$

6. $(5^2)^3 = 5^6$

7. $-(-10)^0 = 1$

8. $3^{-1} = -\frac{1}{3}$

9. Concept Check A friend incorrectly simplified

$$4^5 \cdot 4^2 \quad \text{as} \quad 16^7.$$

WHAT WENT WRONG? Give the correct answer.

10. Concept Check A student incorrectly simplified

$$\frac{6^5}{3^2} \quad \text{as} \quad 2^3.$$

WHAT WENT WRONG? Give the correct answer.

Apply the product rule for exponents, if possible. ***See Example 1.***

11. $13^4 \cdot 13^8$

12. $11^6 \cdot 11^4$

13. $8^9 \cdot 8$

14. $12 \cdot 12^6$

15. $x^3 \cdot x^5 \cdot x^9$

16. $y^4 \cdot y^5 \cdot y^6$

17. $r^2 \cdot s^4$

18. $p^3 \cdot q^2$

19. $(-3w^5)(9w^3)$

20. $(-5x^2)(3x^4)$

21. $(2x^2y^5)(9xy^3)$

22. $(8s^4t)(3s^3t^5)$

Match the expression in Column I with its equivalent expression in Column II. Choices may be used once, more than once, or not at all. ***See Example 2.***

23.

I	II
(a) 9^0	A. 0
(b) -9^0	B. 1
(c) $(-9)^0$	C. -1
(d) $-(-9)^0$	D. 9
	E. -9

24.

I	II
(a) $2x^0$	A. 0
(b) $-2x^0$	B. 1
(c) $(2x)^0$	C. -1
(d) $(-2x)^0$	D. 2
(*Note:* $x \neq 0$)	E. -2

Evaluate. Assume that all variables represent nonzero real numbers. ***See Example 2.***

25. 15^0

26. 19^0

27. -8^0

28. -10^0

29. $(-25)^0$

30. $(-30)^0$

31. $-(-1)^0$

32. $-(-2)^0$

33. x^0

34. y^0

35. $3^0 + (-3)^0$

36. $5^0 + (-5)^0$

37. $-3^0 + 3^0$

38. $-5^0 + 5^0$

39. $-4^0 - m^0$

40. $-8^0 - k^0$

Match the expression in Column I with its equivalent expression in Column II. Choices may be used once, more than once, or not at all. ***See Example 3.***

41.

I	II
(a) 5^{-2}	A. 25
(b) -5^{-2}	B. $\frac{1}{25}$
(c) $(-5)^{-2}$	C. -25
(d) $-(-5)^{-2}$	D. $-\frac{1}{25}$

42.

I	II
(a) 4^{-3}	A. 64
(b) -4^{-3}	B. -64
(c) $(-4)^{-3}$	C. $\frac{1}{64}$
(d) $-(-4)^{-3}$	D. $-\frac{1}{64}$

Write each expression using only positive exponents. Assume that all variables represent nonzero real numbers. **See Example 3.**

43. 5^{-4} **44.** 7^{-2} **45.** 3^{-5} **46.** 2^{-3}

47. 9^{-1} **48.** 4^{-1} **49.** $(4x)^{-2}$ **50.** $(5t)^{-3}$

51. $4x^{-2}$ **52.** $5t^{-3}$ **53.** $-a^{-3}$ **54.** $-b^{-4}$

55. $(-a)^{-4}$ **56.** $(-b)^{-6}$ **57.** $(-3x)^{-3}$ **58.** $(-2x)^{-5}$

Evaluate each expression. **See Example 4.**

59. $5^{-1}+6^{-1}$ **60.** $2^{-1}+8^{-1}$ **61.** $4^{-1}+5^{-1}$ **62.** $6^{-1}+7^{-1}$

63. $8^{-1}-3^{-1}$ **64.** $6^{-1}-4^{-1}$ **65.** $\frac{1}{4^{-2}}$ **66.** $\frac{1}{3^{-3}}$

67. $\frac{2^{-2}}{3^{-3}}$ **68.** $\frac{3^{-3}}{2^{-2}}$ **69.** $\frac{5^{-2}}{2^{-4}}$ **70.** $\frac{3^{-3}}{7^{-2}}$

Apply the quotient rule for exponents, if possible, and write each result using only positive exponents. Assume that all variables represent nonzero real numbers. **See Example 5.**

71. $\frac{4^8}{4^6}$ **72.** $\frac{5^9}{5^7}$ **73.** $\frac{x^{12}}{x^8}$ **74.** $\frac{y^{14}}{y^{10}}$

75. $\frac{r^7}{r^{10}}$ **76.** $\frac{y^8}{y^{12}}$ **77.** $\frac{6^4}{6^{-2}}$ **78.** $\frac{7^5}{7^{-3}}$

79. $\frac{6^{-3}}{6^7}$ **80.** $\frac{5^{-4}}{5^2}$ **81.** $\frac{7}{7^{-1}}$ **82.** $\frac{8}{8^{-1}}$

83. $\frac{r^{-3}}{r^{-6}}$ **84.** $\frac{s^{-4}}{s^{-8}}$ **85.** $\frac{x^3}{y^2}$ **86.** $\frac{y^5}{t^3}$

Simplify using the power rules. Assume that all variables represent nonzero real numbers. **See Example 6.**

87. $(x^3)^6$ **88.** $(y^5)^4$ **89.** $\left(\frac{3}{5}\right)^3$ **90.** $\left(\frac{4}{3}\right)^2$

91. $(4t)^3$ **92.** $(5t)^4$ **93.** $(6x^2)^3$ **94.** $(2x^5)^5$

95. $\left(\frac{-4m^2}{t}\right)^3$ **96.** $\left(\frac{-5n^4}{r^2}\right)^3$ **97.** $\left(\frac{-s^3}{t^5}\right)^4$ **98.** $\left(\frac{-2a^4}{b^5}\right)^6$

Match the expression in Column I with its equivalent expression in Column II. Choices may be used once, more than once, or not at all. **See Example 7.**

	I	II
99.	**(a)** $\left(\frac{1}{3}\right)^{-1}$	**A.** $\frac{1}{3}$
	(b) $\left(-\frac{1}{3}\right)^{-1}$	**B.** 3
	(c) $-\left(\frac{1}{3}\right)^{-1}$	**C.** $-\frac{1}{3}$
	(d) $-\left(-\frac{1}{3}\right)^{-1}$	**D.** -3

	I	II
100.	**(a)** $\left(\frac{2}{5}\right)^{-2}$	**A.** $\frac{25}{4}$
	(b) $\left(-\frac{2}{5}\right)^{-2}$	**B.** $-\frac{25}{4}$
	(c) $-\left(\frac{2}{5}\right)^{-2}$	**C.** $\frac{4}{25}$
	(d) $-\left(-\frac{2}{5}\right)^{-2}$	**D.** $-\frac{4}{25}$

Write using only positive exponents and then evaluate. Assume that all variables represent nonzero real numbers. ***See Example 7.***

101. $\left(\frac{1}{4}\right)^{-3}$ **102.** $\left(\frac{1}{5}\right)^{-2}$ **103.** $\left(\frac{2}{3}\right)^{-3}$ **104.** $\left(\frac{3}{2}\right)^{-3}$

105. $\left(\frac{4}{5}\right)^{-2}$ **106.** $\left(\frac{5}{4}\right)^{-2}$ **107.** $\left(\frac{2t}{3}\right)^{-4}$ **108.** $\left(\frac{3z}{4}\right)^{-3}$

109. $\left(\frac{1}{4x}\right)^{-2}$ **110.** $\left(\frac{1}{5x}\right)^{-3}$ **111.** $\left(\frac{x}{2}\right)^{-5}$ **112.** $\left(\frac{t}{4}\right)^{-4}$

Simplify each expression. Assume that all variables represent nonzero real numbers. ***See Examples 1–8.***

113. $3^5 \cdot 3^{-6}$ **114.** $4^4 \cdot 4^{-6}$ **115.** $a^{-3}a^2a^{-4}$

116. $k^{-5}k^{-3}k^4$ **117.** $(5^{-3})^{-2}$ **118.** $(4^{-4})^{-2}$

119. $(x^{-4})^3$ **120.** $(x^{-3})^6$ **121.** $(k^2)^{-3}k^4$

122. $(x^3)^{-4}x^5$ **123.** $-4r^{-2}(r^4)^2$ **124.** $-2m^{-1}(m^3)^2$

125. $(5a^{-1})^4(a^2)^{-3}$ **126.** $(3p^{-4})^2(p^3)^{-1}$ **127.** $(z^{-4}x^3)^{-1}$

128. $(y^{-2}z^4)^{-3}$ **129.** $\frac{m^2n^{-8}}{m^{-6}n^3}$ **130.** $\frac{a^{-9}b^6}{a^3b^{-4}}$

131. $\frac{(p^{-2})^3}{5p^4}$ **132.** $\frac{(m^4)^{-1}}{9m^3}$ **133.** $\frac{(3pq)q^2}{6p^2q^4}$

134. $\frac{(-8xy)y^3}{4x^5y^4}$ **135.** $\frac{4a^5(a^{-1})^3}{(a^{-2})^{-2}}$ **136.** $\frac{12k^{-2}(k^{-3})^{-4}}{6k^5}$

137. $\frac{(-y^{-4})^2}{6(y^{-5})^{-1}}$ **138.** $\frac{2(-m^{-1})^{-4}}{9(m^{-3})^2}$ **139.** $\frac{(2k)^2m^{-5}}{(km)^{-3}}$

140. $\frac{(3rs)^{-2}}{3^2r^2s^{-4}}$ **141.** $\left(\frac{2p}{q^2}\right)^3\left(\frac{3p^4}{q^{-4}}\right)^{-1}$ **142.** $\left(\frac{5z^3}{2a^2}\right)^{-3}\left(\frac{8a^{-1}}{15z^{-2}}\right)^{-3}$

143. $\frac{2^2y^4(y^{-3})^{-1}}{2^5y^{-2}}$ **144.** $\frac{3^{-1}m^4(m^2)^{-1}}{3^2m^{-2}}$ **145.** $\left(\frac{5m^4n^{-3}}{m^{-5}n^2}\right)^{-2}$

146. $\left(\frac{8x^{-6}y^3}{x^4y^{-4}}\right)^{-2}$ **147.** $\left(\frac{-3x^4y^6}{15x^{-6}y^7}\right)^{-3}$ **148.** $\left(\frac{-4a^3b^2}{12a^5b^{-4}}\right)^{-3}$

Extending Skills *Simplify each expression. Assume that all variables represent nonzero real numbers.*

149. $\frac{(2k)^2k^3}{k^{-1}k^{-5}}(5k^{-2})^{-3}$ **150.** $\frac{(3r^2)^2r^{-5}}{r^{-2}r^3}(2r^{-6})^2$

151. $\frac{(2m^2p^3)^2(4m^2p)^{-2}}{(-3mp^4)^{-1}(2m^3p^4)^3}$ **152.** $\frac{(-5y^3z^4)^2(2yz^5)^{-2}}{10(y^4z)^3(3y^3z^2)^{-1}}$

4.2 Scientific Notation

OBJECTIVES

1 Write numbers in scientific notation.
2 Convert numbers in scientific notation to standard notation.
3 Use scientific notation in calculations.

OBJECTIVE 1 Write numbers in scientific notation.

The number of one-celled organisms that will sustain a whale for a few hours is 400,000,000,000,000, and the shortest wavelength of visible light is approximately 0.0000004 m. It is often simpler to write such numbers in *scientific notation.*

Scientific Notation

A number is written in **scientific notation** when it is expressed in the form

$$a \times 10^n, \quad \text{where } 1 \le |a| < 10 \text{ and } n \text{ is an integer.}$$

$1 \le a < 10$

It is customary to use × rather than • for multiplication.

Examples:

$400{,}000{,}000{,}000{,}000 = 4 \times 10^{14}$

$8500 = 8.5 \times 10^3$

$0.039 = 3.9 \times 10^{-2}$

$0.0000004 = 4 \times 10^{-7}$

Scientific notation

CAUTION Although a number such as 8500 can be written using exponents in multiple ways, these forms are *not* considered scientific notation.

0.85×10^4 or 85×10^2, *Not* in scientific notation

$a < 1$ $\quad$ $a > 10$

In scientific notation, a number is written with the decimal point after the first nonzero digit and then multiplied by a power of 10.

Converting a Positive Number to Scientific Notation

Step 1 **Position the decimal point.** Place a caret ^ to the right of the first nonzero digit, where the decimal point will be placed.

Step 2 **Determine the numeral for the exponent.** Count the number of digits from the decimal point to the caret. This number gives the absolute value of the exponent on 10.

Step 3 **Determine the sign for the exponent.** Decide whether multiplying by 10^n should make the result of Step 1 greater or less.

- The exponent should be positive to make the result greater.
- The exponent should be negative to make the result less.

It is helpful to remember the following.

$$\textbf{For } n \ge 1, \quad 10^{-n} < 1 \quad \textbf{and} \quad 10^n \ge 10.$$

To convert a negative number to scientific notation, temporarily ignore the negative sign and go through the steps in the box above. Then attach a negative sign to the result.

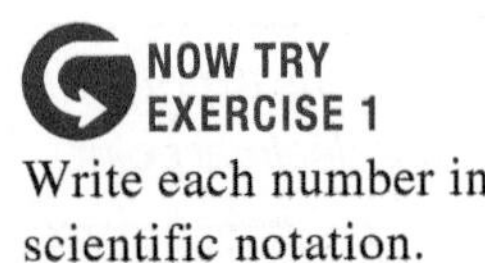

NOW TRY EXERCISE 1

Write each number in scientific notation.

(a) 7,560,000,000

(b) −0.000000245

EXAMPLE 1 Writing Numbers in Scientific Notation

Write each number in scientific notation.

(a) 820,000

Step 1 Place a caret to the right of the 8 (the first nonzero digit) to mark the new location of the decimal point.

$$8_{\wedge}20{,}000$$

Step 2 Count from the decimal point, which is understood to be after the last 0, to the caret.

8.20,000. ← Decimal point

Count 5 places.

Step 3 Because 8.2 is to be made greater, the exponent on 10 is positive.

$$820{,}000 = 8.2 \times 10^5$$

(b) 0.0000072

$0.000007_{\wedge}2$ Place a caret to the right of the first nonzero digit.

0.000007.2 Count from left to right.

6 places

Because 7.2 is to be made less, the exponent on 10 is negative.

$$0.0000072 = 7.2 \times 10^{-6}$$

(c) $-0.0004_{\wedge}62 = -4.62 \times 10^{-4}$ Remember the negative sign.

Count 4 places.

OBJECTIVE 2 Convert numbers in scientific notation to standard notation.

Converting a Positive Number from Scientific Notation

Multiplying a positive number by a positive power of 10 makes the number greater, so move the decimal point to the right if n is positive in 10^n.

Multiplying a positive number by a negative power of 10 makes the number less, so move the decimal point to the left if n is negative in 10^n.

If n is 0, leave the decimal point where it is in 10^n.

NOW TRY EXERCISE 2

Write each number in standard notation.

(a) 4.45×10^{10}

(b) -5.9×10^{-5}

EXAMPLE 2 Converting from Scientific Notation to Standard Notation

Write each number in standard notation.

(a) 6.93×10^7 6.9300000 Attach 0's as necessary.

7 places

We moved the decimal point 7 places to the right. (We had to attach five 0's.)

$$6.93 \times 10^7 = 69{,}300{,}000$$ Insert commas.

↑ Standard notation

(b) 4.7×10^{-3} .004.7 Move the decimal point 3 places to the left. Attach 0's as necessary.

3 places

$$4.7 \times 10^{-3} = 0.0047$$ Attach a leading zero.

(c) $-1.083 \times 10^0 = -1.083 \times 1 = -1.083$

NOW TRY ANSWERS

1. (a) 7.56×10^9
   (b) -2.45×10^{-7}
2. (a) 44,500,000,000
   (b) −0.000059

NOTE *When converting from scientific notation to standard notation, use the exponent to determine the number of places and the direction in which to move the decimal point.*

OBJECTIVE 3 Use scientific notation in calculations.

NOW TRY EXERCISE 3
Evaluate.

$$\frac{0.00063 \times 400{,}000}{1400 \times 0.000003}$$

EXAMPLE 3 Using Scientific Notation in Computation

Evaluate.

$$\frac{1{,}800{,}000 \times 0.0015}{0.00003 \times 45{,}000}$$

$= \dfrac{1.8 \times 10^6 \times 1.5 \times 10^{-3}}{3 \times 10^{-5} \times 4.5 \times 10^4}$ Express all numbers in scientific notation.

$= \dfrac{1.8 \times 1.5 \times 10^6 \times 10^{-3}}{3 \times 4.5 \times 10^{-5} \times 10^4}$ Commutative property

$= \dfrac{1.8 \times 1.5 \times 10^3}{3 \times 4.5 \times 10^{-1}}$ Product rule

$= \dfrac{1.8 \times 1.5}{3 \times 4.5} \times 10^4$ Quotient rule

$= 0.2 \times 10^4$ Simplify.

Don't stop here.

$= (2 \times 10^{-1}) \times 10^4$ Write 0.2 in scientific notation.

$= 2 \times 10^3$ Scientific notation

$= 2000$ Standard notation

NOW TRY

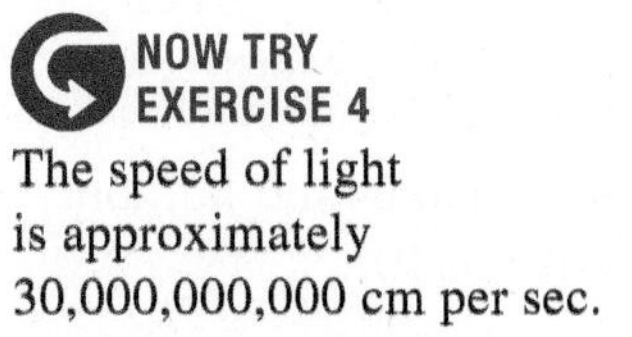

NOW TRY EXERCISE 4
The speed of light is approximately 30,000,000,000 cm per sec. How long will it take light to travel 1.2×10^{15} cm?

EXAMPLE 4 Using Scientific Notation to Solve Problems

In 1990, national health care expenditures in the United States were \$724 billion. By 2016, health care expenditures were 4.5 times this amount. (Data from Centers for Medicare & Medicaid Services.)

(a) Write the amount of health care expenditures in 1990 using scientific notation.

724 billion

$= 724 \times 10^9$ 1 billion = 1,000,000,000 = 10^9

$= (7.24 \times 10^2) \times 10^9$ Write 724 in scientific notation.

$= 7.24 \times 10^{11}$ Product rule; $10^2 \times 10^9 = 10^{2+9}$

In 1990, health care expenditures were \$$7.24 \times 10^{11}$.

(b) What were health care expenditures in 2016?

$(7.24 \times 10^{11}) \times 4.5$ Multiply the result in part (a) by 4.5.

$= (4.5 \times 7.24) \times 10^{11}$ Commutative and associative properties

$= 32.58 \times 10^{11}$ Multiply.

$= (3.258 \times 10^1) \times 10^{11}$ Write 32.58 in scientific notation.

$= 3.258 \times 10^{12}$ Product rule; $10^1 \times 10^{11} = 10^{12}$

Expenditures in 2016 were about \$3,258,000,000,000 (over \$3 trillion).

NOW TRY

NOW TRY ANSWERS

3. 6×10^4, or 60,000

4. 4.0×10^4 sec, or 40,000 sec

4.2 Exercises

FOR EXTRA HELP  MyLab Math

Video solutions for select problems available in MyLab Math

STUDY SKILLS REMINDER

Reread your class notes before working the assigned exercises. **Review Study Skill 3, *Taking Lecture Notes.***

1. **Concept Check** In scientific notation, a number is written with a decimal point (*before / after*) the first nonzero digit and multiplied by a ________ of 10. A number written in scientific notation is expressed in the form ________ × ________, where $1 \leq |a| < 10$ and n is an integer.

2. **Concept Check** Match each number written in scientific notation in (a)–(d) in Column I with the equivalent number written in standard notation in A–F in Column II. Not all choices will be used.

I		II		
(a) 1.5×10^4	**(b)** 1.5×10^5	**A.** 150,000	**B.** 1500	**C.** 0.00015
(c) 1.5×10^{-4}	**(d)** 1.5×10^{-5}	**D.** 15,000	**E.** 0.0015	**F.** 0.000015

3. **Concept Check** When asked to write 9,275,000 in scientific notation, a student incorrectly wrote

$$92.75 \times 10^5.$$

WHAT WENT WRONG? What is the correct answer?

4. **Concept Check** A student incorrectly claimed that the following number is written in scientific notation.

$$0.2 \times 10^{-2}$$

WHAT WENT WRONG? Write this number in scientific notation.

Concept Check *The term "Big Data," originally coined by NASA researchers, includes all data streaming from sources such as cell phones, tablets, satellites, Amazon, Google, Facebook, and Twitter. Write each "Big Data" number in scientific notation. (Data from* Thrivent Magazine.*)*

5. 1,000,000 bytes (a *megabyte*)
6. 1,000,000,000 bytes (a *gigabyte*)
7. 1,000,000,000,000 bytes (a *terabyte*)
8. 1,000,000,000,000,000,000 bytes (an *exabyte*)

Write each number in scientific notation. ***See Example 1.***

9. 530
10. 1600
11. 0.830
12. 0.0072
13. 0.00000692
14. 0.875
15. −38,500
16. −976,000,000

Write the boldfaced numbers in each problem in scientific notation. ***See Example 1.***

17. The annual U.S. budget first passed **\$1,000,000,000** in 1917. In 1987, it exceeded **\$1,000,000,000,000** for the first time. The budget request for fiscal-year 2018 was **\$4,094,000,000,000.** If stacked in dollar bills, this amount would stretch **277,843** mi. (Data from Office of Management and Budget.)

18. The largest of the **50** United States is Alaska, with land area of about **365,482,000** acres, while the smallest is Rhode Island, with land area of about **677,000** acres. The total land area of the United States is about **2,271,343,000** acres. (Data from General Services Administration.)

Write each number in standard notation. ***See Example 2.***

19. 7.2×10^4
20. 8.91×10^2
21. 2.54×10^{-3}
22. 5.42×10^{-4}
23. -6×10^4
24. -9×10^3
25. 1.2×10^{-5}
26. 2.7×10^{-6}
27. -2.89×10^{-2}
28. -7.68×10^{-4}
29. 8.761×10^0
30. 5.92×10^0

Evaluate. Express answers in standard notation. ***See Example 3.***

31. $\frac{12 \times 10^4}{2 \times 10^6}$

32. $\frac{16 \times 10^5}{4 \times 10^8}$

33. $\frac{3 \times 10^{-2}}{12 \times 10^3}$

34. $\frac{5 \times 10^{-3}}{25 \times 10^2}$

35. $\frac{18 \times 10^{-3} \times 24 \times 10^4}{8 \times 10^5 \times 2 \times 10^{-9}}$

36. $\frac{15 \times 10^{-4} \times 12 \times 10^5}{4 \times 10^3 \times 5 \times 10^{-8}}$

37. $\frac{0.05 \times 1600}{0.0004}$

38. $\frac{0.003 \times 40{,}000}{0.00012}$

39. $\frac{15{,}000 \times 0.25}{0.0075 \times 50{,}000}$

40. $\frac{400{,}000 \times 0.05}{0.008 \times 25{,}000}$

41. $\frac{20{,}000 \times 0.018}{300 \times 0.0004}$

42. $\frac{840{,}000 \times 0.03}{0.00021 \times 600}$

Solve each problem. ***See Example 4.***

43. In 2016, the population of the United States was approximately 323.1 million. (Data from U.S. Census Bureau.)

(a) Write the population using scientific notation.

(b) Write \$1 trillion—that is, \$1,000,000,000,000—using scientific notation.

(c) Using the answers from parts (a) and (b), calculate how much each person would have had to contribute in order to make someone a trillionaire. Write this amount in standard notation to the nearest dollar.

44. In 2016, the national debt of the U.S. government was about \$19.32 trillion. The U.S. population at that time was approximately 323.1 million. (Data from U.S. Census Bureau, U.S. Department of the Treasury.)

(a) Write the population using scientific notation.

(b) Write the amount of the national debt using scientific notation.

(c) Using the answers for parts (a) and (b), calculate how much debt this is per American. Write this amount in standard notation to the nearest dollar.

45. In a lottery, a player must choose five numbers from 1 through 50 and one number from 1 through 40. There are about 8.5×10^7 different ways to do this. If 2000 people purchase tickets, at \$1.00 each, for all these numbers, how much will each person pay?

LOTTERY

Choose 5	Choose 1
1 2 3 4 5 6 7 8 9 10	1 2 3 4 5 6 7 8 9 10
11 12 13 14 15 16 17 18 19 20	11 12 13 14 15 16 17 18 19 20
21 22 23 24 25 26 27 28 29 30	21 22 23 24 25 26 27 28 29 30
31 32 33 34 35 36 37 38 39 40	31 32 33 34 35 36 37 38 39 40
41 42 43 44 45 46 47 48 49 50	

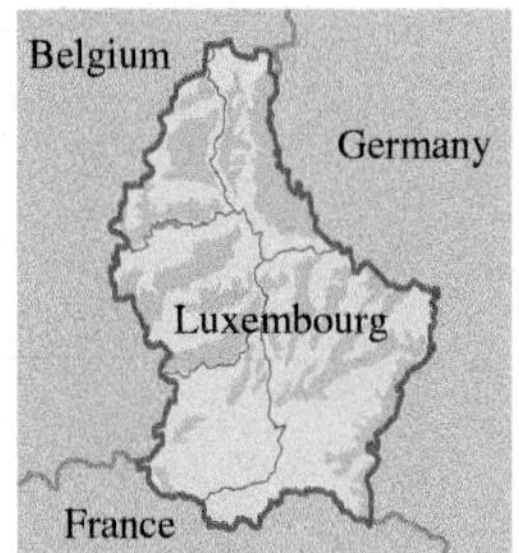

46. In 2017, the population of India was about 1.28×10^9, which was 40,000 times the population of Monaco. What was the population of Monaco? (Data from *The World Almanac and Book of Facts.*)

47. In 2017, the population of Luxembourg was approximately 5.94×10^5. The population density was 595 people per square mile. (Data from *The World Almanac and Book of Facts.*)

(a) Write the population density in scientific notation.

(b) To the nearest square mile, what is the area of Luxembourg?

48. In 2017, the population of Costa Rica was approximately 4.93×10^6. The population density was 96.6 people per square kilometer. (Data from *The World Almanac and Book of Facts*.)

(a) Write the population density in scientific notation.

(b) To the nearest square kilometer, what is the area of Costa Rica?

49. The speed of light is approximately 3×10^{10} cm per sec. How long will it take light to travel 9×10^{12} cm?

50. The average distance from Earth to the sun is 9.3×10^7 mi. How long would it take a rocket, traveling at 2.9×10^3 mph, to reach the sun? Round to the nearest thousand hours.

51. A *light-year* is the distance that light travels in one year. Find the number of miles in a light-year if light travels 1.86×10^5 mi per sec.

52. Use the information given in the previous two exercises to find the number of minutes necessary for light from the sun to reach Earth.

Calculators can express numbers in scientific notation. The displays often use notation such as

2.5E5 *to represent* 2.5×10^5.

Similarly, 2.5E⁻3 *represents* 2.5×10^{-3}. *They can also perform operations with numbers entered in scientific notation.*

Predict the result the calculator will give for each screen. (Use the customary scientific notation to write the answers.)

53.
```
NORMAL FLOAT AUTO REAL RADIAN MP
(1.5E12)*(5E-3)
```

54.
```
NORMAL FLOAT AUTO REAL RADIAN MP
(3.2E-5)*(3E12)
```

55.
```
NORMAL FLOAT AUTO REAL RADIAN MP
(8.4E14)/(2.1E-3)
```

56.
```
NORMAL FLOAT AUTO REAL RADIAN MP
(2.5E10)/(2E-3)
```

RELATING CONCEPTS For Individual or Group Work (Exercises 57–60)

The ***intensity*** *of an earthquake is measured relative to the intensity of a standard* ***zero-level*** *earthquake of intensity I_0. The relationship is equivalent to $I = I_0 \times 10^R$, where R is the* ***Richter scale*** *measure. For example, if an earthquake has magnitude 5.0 on the Richter scale, then its intensity is calculated as*

$$I = I_0 \times 10^{5.0} = I_0 \times 100{,}000,$$

which is 100,000 times as intense as a zero-level earthquake.

To compare an earthquake that measures 8.1 on the Richter scale to one that measures 5.2, find the ratio of the intensities.

$$\frac{\text{intensity } 8.1}{\text{intensity } 5.2} = \frac{I_0 \times 10^{8.1}}{I_0 \times 10^{5.2}} = \frac{10^{8.1}}{10^{5.2}} = 10^{8.1-5.2} = 10^{2.9} \approx 794 \quad \text{Use a calculator.}$$

Therefore, an earthquake that measures 8.1 on the Richter scale is almost 800 times as intense as one that measures 5.2.

Use the information in the table to **work Exercises 57–60 in order.**

Earthquake	Richter Scale Measurement
1960 Chile: Valdivia–Puerto Montt area	9.5
2002 Alaska: Slana, Mentasta Lake, Fairbanks	7.9
2005 Montana: Dillon, Silver Star, Twin Bridges	5.6
2010 Haiti: Port-au-Prince	7.0
2018 Mexico: Santiago Jamiltepec	5.9

Data from www.earthquake.usgs.gov

57. Compare the intensity of the 2002 Alaska earthquake to that of the 2018 Mexico earthquake.

58. Compare the intensity of the 1960 Chile earthquake to that of the 2010 Haiti earthquake.

59. Compare the intensity of the 2010 Haiti earthquake to that of the 2005 Montana earthquake.

60. Suppose an earthquake measures a value of x on the Richter scale. How does the intensity of a second earthquake compare if its Richter scale measure is $x + 3.0$? If its Richter scale measure is $x - 1.0$?

4.3 Adding and Subtracting Polynomials

OBJECTIVES

1. Define and classify polynomials.
2. Add and subtract polynomials.

VOCABULARY

- ☐ term
- ☐ algebraic expression
- ☐ polynomial
- ☐ numerical coefficient (coefficient)
- ☐ degree of a term
- ☐ polynomial in x
- ☐ descending powers
- ☐ leading term
- ☐ leading coefficient
- ☐ trinomial
- ☐ binomial
- ☐ monomial
- ☐ degree of a polynomial
- ☐ like terms
- ☐ negative of a polynomial

OBJECTIVE 1 Define and classify polynomials.

Recall that a **term** is a number (constant), a variable, or the product of a number and one or more variables raised to powers.

Examples: $4x,\quad \frac{1}{2}m^5 \left(\text{or } \frac{m^5}{2}\right),\quad -7z^9,\quad 6x^2z,\quad \frac{5}{3x^2},\quad 3\sqrt{x},\quad 9$ Terms

A term or a sum of two or more terms is an **algebraic expression.** The simplest kind of algebraic expression is a *polynomial.*

Polynomial

A **polynomial** is a term or a finite sum of terms of the form ax^n, where a is a real number, x is a variable, and the exponent n is a whole number.

Examples: $12x^9,\quad 3t - 5,\quad 4m^3 - 5m^2 + 8$ Polynomials in x, t, and m

Even though an expression such as $3t - 5$ involves subtraction, it can be written as a sum of terms, here $3t + (-5)$.

Some algebraic expressions are *not* polynomials.

Examples: $x^{-1} + 3x^{-2},\quad \sqrt{9 - x},\quad \frac{1}{x}$ Not polynomials

The first has negative integer exponents, the second involves a variable under a radical, and the third has a variable in the denominator.

For each term ax^n of a polynomial, the factor a is the **numerical coefficient** of x^n, or just the **coefficient** of x^n, and the exponent n is the **degree of the term.**

▼ Terms and Their Coefficients and Degrees

Term ax^n	*Numerical Coefficient*	*Degree*
$12x^9$	12	9
$3x$, or $3x^1$	3	1
-6, or $-6x^0$	-6	0
$-x^4$, or $-1x^4$	-1	4
$\frac{x^2}{3} = \frac{1x^2}{3} = \frac{1}{3}x^2$	$\frac{1}{3}$	2

Any nonzero constant has degree 0.

NOTE The number 0 has no degree because 0 times a variable to any power is 0.

A polynomial containing only the variable x is a **polynomial in x.** A polynomial in one variable is written in **descending powers** of the variable if the exponents on the variable in the terms decrease from left to right.

$$x^5 - 6x^2 + 12x - 5$$

Descending powers of x

Think of $12x$ as $12x^1$ and -5 as $-5x^0$.

When a polynomial is written in descending powers of the variable, the greatest-degree term is written first and is the **leading term** of the polynomial. Its coefficient is the **leading coefficient.**

NOW TRY EXERCISE 1
Write the polynomial in descending powers of the variable. Then give the leading term and the leading coefficient.

$-2x^3 - 2x^5 + 4x^2 + 7 - x$

EXAMPLE 1 Writing Polynomials in Descending Powers

Write each polynomial in descending powers of the variable. Then give the leading term and the leading coefficient.

(a) $y - 6y^3 + 8y^5 - 9y^4 + 12$ is written as $8y^5 - 9y^4 - 6y^3 + y + 12$.

(b) $-2 + m + 6m^2 - m^3$ is written as $-m^3 + 6m^2 + m - 2$.

Each leading term is shown in color. In part (a), the leading coefficient is 8, and in part (b) it is -1 (because $-m^3 = -1m^3$). NOW TRY

Some polynomials with a specific number of terms are given special names.

- A polynomial with exactly three terms is a **trinomial.**
- A two-term polynomial is a **binomial.**
- A single-term polynomial is a **monomial.**

Although many polynomials contain only one variable, polynomials may have more than one variable. The degree of a term with more than one variable is the sum of the exponents on the variables. The **degree of a polynomial** is the greatest degree of all of its terms.

▼ Polynomials and Their Degrees

Type of Polynomial	*Example*	*Degree*
Monomial	7	0 $(7 = 7x^0)$
	$5x^3y^7$	10 $(3 + 7 = 10)$
Binomial	$6 + 2x^3$	3
	$11y + 8$	1 $(y = y^1)$
Trinomial	$t^2 + 11t + 4$	2
	$-3 + 2k^5 + 9z^4$	5
	$x^3y^9 + 12xy^4 + 7xy$	12 (The terms have degrees 12, 5, and 2, and 12 is the greatest.)

NOW TRY ANSWER

1. $-2x^5 - 2x^3 + 4x^2 - x + 7$; $-2x^5$; -2

NOTE If a polynomial in a single variable is written in descending powers of that variable, the degree of the polynomial will be the degree of the leading term.

NOW TRY EXERCISE 2

Identify each polynomial as a *monomial, binomial, trinomial,* or *none of these.* Give the degree.

(a) $-4x^3 + 10x^5 + 7$

(b) $-6m^2n^5$

EXAMPLE 2 Classifying Polynomials

Identify each polynomial as a *monomial, binomial, trinomial,* or *none of these.* Give the degree.

(a) $-x^2 + 5x + 1$ This is a trinomial of degree 2.

(b) $\frac{3}{4}xy^4 \left(\text{or } \frac{3}{4}x^1y^4\right)$ This is a monomial of degree 5 (because $1 + 4 = 5$).

(c) $7m^9 + 18m^{14}$ This is a binomial of degree 14.

(d) $p^4 - p^2 - 6p - 5$ Polynomials of four terms or more do not have special names, so *none of these* is the answer that applies here. This polynomial has degree 4. NOW TRY

OBJECTIVE 2 Add and subtract polynomials.

We use the distributive property to simplify polynomials by combining like terms.

$$x^3 + 4x^2 + 5x^2 - 1$$
$$= x^3 + (4 + 5)x^2 - 1 \quad \text{Distributive property}$$
$$= x^3 + 9x^2 - 1 \quad \text{Add.}$$

The terms in a polynomial such as $4x + 5x^2$ cannot be combined. ***Only terms containing exactly the same variables raised to the same powers may be combined.*** Recall that such terms are **like terms.**

CAUTION *Only like terms can be combined.*

NOW TRY EXERCISE 3

Combine like terms.

(a) $2x^2 - 8x^2 + x^2$

(b) $3p^3 - 2q + p^3 - 5q$

(c) $-x^2t + 4x^2t + 3xt^2 - 7xt^2$

EXAMPLE 3 Combining Like Terms

Combine like terms.

(a) $-5y^3 + 8y^3 - y^3$

$$= (-5 + 8 - 1)y^3 \quad -y^3 = -1y^3\text{; Distributive property}$$
$$= 2y^3 \quad \text{Add and subtract.}$$

(b) $6x + 5y - 9x + 2y$

$$= 6x - 9x + 5y + 2y \quad \text{Commutative property}$$
$$= -3x + 7y \quad \text{Combine like terms.}$$

Because $-3x$ and $7y$ are unlike terms, no further simplification is possible.

(c) $5x^2y - 6xy^2 + 9x^2y + 13xy^2$

$$= 5x^2y + 9x^2y - 6xy^2 + 13xy^2 \quad \text{Commutative property}$$
$$= 14x^2y + 7xy^2 \quad \text{Combine like terms.}$$

NOW TRY 

NOW TRY ANSWERS

2. (a) trinomial; 5
   (b) monomial; 7
3. (a) $-5x^2$ (b) $4p^3 - 7q$
   (c) $3x^2t - 4xt^2$

Adding Polynomials

To find the sum of two polynomials, combine like terms.

NOW TRY EXERCISE 4

Add.

$(7x^2 - 9x + 4) + (x^3 - 3x^2 - 5)$

EXAMPLE 4 Adding Polynomials

Add $(3a^5 - 9a^3 + 4a^2) + (-8a^5 + 8a^3 + 2)$.

$$(3a^5 - 9a^3 + 4a^2) + (-8a^5 + 8a^3 + 2)$$

$$= 3a^5 - 8a^5 - 9a^3 + 8a^3 + 4a^2 + 2 \quad \text{Commutative and associative properties}$$

$$= -5a^5 - a^3 + 4a^2 + 2 \quad \text{Combine like terms.}$$

Alternatively, we can add these two polynomials vertically.

$$\begin{array}{r} 3a^5 - 9a^3 + 4a^2 + 0 \\ -8a^5 + 8a^3 + 0a^2 + 2 \\ \hline -5a^5 - a^3 + 4a^2 + 2 \end{array}$$

Place like terms in columns. Using placeholders may help.

Add in columns.

Add only the coefficients.

NOW TRY

NOTE The vertical addition in **Example 4** could be written as follows.

$$\begin{array}{r} 3a^5 - 9a^3 + 4a^2 + 0 \\ + \underline{(-8a^5 + 8a^3 + 0a^2 + 2)} \\ -5a^5 - a^3 + 4a^2 + 2 \end{array}$$

Recall that subtraction of real numbers is defined as follows.

$$a - b = a + (-b)$$

That is, we add the first number and the negative (or opposite) of the second. We define the **negative of a polynomial** as that polynomial with the sign of every coefficient changed.

Subtracting Polynomials

To find the difference of two polynomials, add the first polynomial (minuend) and the negative (or opposite) of the *second* polynomial (subtrahend).

EXAMPLE 5 Subtracting Polynomials

Subtract $(-6m^2 - 8m + 5) - (-5m^2 + 7m - 8)$.

Change every sign in the second polynomial (subtrahend) and add.

$$(-6m^2 - 8m + 5) - (-5m^2 + 7m - 8)$$

$$= -6m^2 - 8m + 5 + 5m^2 - 7m + 8 \quad \text{Definition of subtraction}$$

$$= -6m^2 + 5m^2 - 8m - 7m + 5 + 8 \quad \text{Rearrange terms.}$$

$$= -m^2 - 15m + 13 \quad \text{Combine like terms.}$$

CHECK $(-m^2 - 15m + 13) + (-5m^2 + 7m - 8) = -6m^2 - 8m + 5$ ✓

↑ Difference ↑ Subtrahend ↑ Minuend

NOW TRY ANSWER

4. $x^3 + 4x^2 - 9x - 1$

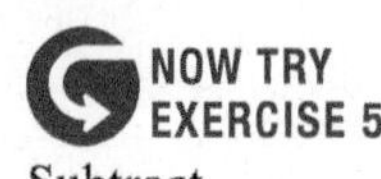
NOW TRY EXERCISE 5

Subtract.

$(2y^2 - 7y - 4) - (8y^2 - 2y + 10)$

Alternatively, we can subtract these two polynomials vertically.

$$\begin{array}{r} -6m^2 - 8m + 5 \\ \underline{-5m^2 + 7m - 8} \end{array}$$

Write the subtrahend below the minuend, lining up like terms in columns.

Change all the signs in the subtrahend and add.

$$\begin{array}{r} -6m^2 - 8m + 5 \\ \underline{+5m^2 - 7m + 8} \\ -m^2 - 15m + 13 \end{array}$$

Change all signs.

Add in columns.

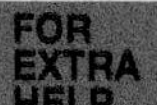

The difference is the same.

NOW TRY 

NOTE The vertical subtraction in **Example 5** could be written as follows.

$$\begin{array}{r} -6m^2 - 8m + 5 \\ \underline{-\ (-5m^2 + 7m - 8)} \end{array} \quad \text{Change each sign and add.} \quad \begin{array}{r} -6m^2 - 8m + 5 \\ \underline{+\ (+5m^2 - 7m + 8)} \\ -m^2 - 15m + 13 \end{array}$$

NOW TRY ANSWER

5. $-6y^2 - 5y - 14$

4.3 Exercises

FOR EXTRA HELP

MyLab Math

Video solutions for select problems available in MyLab Math

STUDY SKILLS REMINDER

Time management can be a challenge for students. **Review Study Skill 6, *Managing Your Time.***

Concept Check *Work each problem involving the vocabulary of polynomials.*

1. Match each description in Column I with the correct polynomial in Column II. Choices in Column II may be used once, more than once, or not at all.

I	II
(a) Monomial of degree 2	**A.** $x^5 + 5x^4 - 5x$
(b) Trinomial of degree 5	**B.** 5
(c) Polynomial with leading coefficient 1	**C.** $5 + x - x^2$
(d) Binomial in descending powers	**D.** $3ab$
(e) Term with degree 0	**E.** $-m + 5$

2. Which polynomial is a trinomial in descending powers that has degree 6?

A. $5x^6 - 4x^5 + 12$ **B.** $6x^5 - x^6 + 4$

C. $2x + 4x^2 - x^6$ **D.** $4x^6 - 6x^4 + 9x^2 - 8$

3. Give an example of a polynomial of four terms in the variable x that is of degree 5, is written in descending powers, and lacks a fourth-degree term.

4. Give an example of a trinomial in the variable y that is of degree 3, has leading coefficient -1, and lacks a term with degree 0.

5. Concept Check A student incorrectly gave 2 as the leading coefficient of the following polynomial.

$$2x^2 - 5x^3 - 8x^4$$

WHAT WENT WRONG? Give the correct leading coefficient.

6. Concept Check A student incorrectly gave 6 as the degree of the following polynomial.

$$2xy^6 + 3xy^5 + 4xy^4$$

WHAT WENT WRONG? Give the degree of this polynomial.

Give the numerical coefficient and the degree of each term. ***See Objective 1.***

7. $7z$ **8.** $3r$ **9.** $-15p^2$ **10.** $-27k^3$ **11.** x^4 **12.** y^6

13. $\frac{t}{6}$ **14.** $\frac{m}{4}$ **15.** 8 **16.** 2 **17.** $-x^3$ **18.** $-y^9$

Write each polynomial in descending powers of the variable. Then give the leading term and the leading coefficient. ***See Example 1.***

19. $2x^3 + x - 3x^2 + 4$

20. $q^2 + 3q^4 - 2q + 1$

21. $4p^3 - 8p^5 + p^7$

22. $3y^2 + y^4 - 2y^3$

23. $10 - m^3 - 3m^4$

24. $4 - x - 8x^2$

Identify each polynomial as a monomial, binomial, trinomial, *or* none of these. *Give the degree.* ***See Example 2.***

25. 25

26. 15

27. $7m - 22$

28. $6x + 15$

29. $-7y^6 + 11y^8$

30. $12k^2 - 9k^5$

31. $-mn^5$

32. $-a^3b$

33. $-5m^3 + 6m - 9m^2$

34. $4z^2 - 11z + 2$

35. $-6p^4q - 3p^3q^2 + 2pq^3 - q^4$

36. $8s^3t - 4s^2t^2 + 2st^3 + 9$

Combine like terms. ***See Example 3.***

37. $5z^4 + 3z^4$

38. $8r^5 - 2r^5$

39. $-m^3 + 2m^3 + 6m^3$

40. $3p^4 + 5p^4 - 2p^4$

41. $x + x + x + x + x$

42. $z - z - z + z$

43. $m^4 - 3m^2 + m$

44. $5a^5 + 2a^4 - 9a^3$

45. $5t + 4s - 6t + 9s$

46. $8p - 9q - 3p + q$

47. $2k + 3k^2 + 5k^2 - 7$

48. $4x^2 + 2x - 6x^2 - 6$

49. $n^4 - 2n^3 + n^2 - 3n^4 + n^3$

50. $2q^3 + 3q^2 - 4q - q^3 + 5q^2$

51. $3ab^2 + 7a^2b - 5ab^2 + 13a^2b$

52. $6m^2n - 8mn^2 + 3mn^2 - 7m^2n$

53. Concept Check A student added two polynomials vertically as follows.

$$\begin{array}{r} 4x^2 - 7x + 4 \\ \underline{2x^2 + 3x - 5} \\ 6x^4 - 4x^2 - 1 \end{array} \quad \text{Incorrect}$$

WHAT WENT WRONG? Give the correct sum.

54. Concept Check A student subtracted $(15x^2 + 12) - (7x^2 - 3)$ vertically as follows.

$$\begin{array}{r} 15x^2 + 12 \\ \underline{7x^2 - 3} \\ 8x^2 + 9 \end{array} \quad \text{Incorrect}$$

WHAT WENT WRONG? Give the correct difference.

Add or subtract as indicated. ***See Examples 4 and 5.***

55. $(5x^2 + 7x - 4) + (3x^2 - 6x + 2)$

56. $(4k^3 + k^2 + k) + (2k^3 - 4k^2 - 3k)$

57. $(6t^2 - 4t^4 - t) + (3t^4 - 4t^2 + 5)$

58. $(3p^2 + 2p - 5) + (7p^2 - 4p^3 + 3p)$

59. $(y^3 + 3y + 2) + (4y^3 - 3y^2 + 2y - 1)$

60. $(2x^5 - 2x^4 + x^3 - 1) + (x^4 - 3x^3 + 2)$

61. $(3r + 8) - (2r - 5)$

62. $(2d + 7) - (3d - 1)$

63. $(2a^2 + 3a - 1) - (4a^2 + 5a + 6)$

64. $(q^4 - 2q^2 + 10) - (3q^4 + 5q^2 - 5)$

65. $(z^5 + 3z^2 + 2z) - (4z^5 + 2z^2 - 5z)$

66. $(5t^3 - 3t^2 + 2t) - (4t^3 + 2t^2 + 3t)$

67. Add.
$$\begin{array}{r} 21p - 8 \\ \underline{-9p + 4} \end{array}$$

68. Add.
$$\begin{array}{r} 15m - 9 \\ \underline{4m + 12} \end{array}$$

69. Add.
$$\begin{array}{r} -12p^2 + 4p - 1 \\ \underline{3p^2 + 7p - 8} \end{array}$$

70. Add.
$$\begin{array}{r} -6y^3 + 8y + 5 \\ \underline{9y^3 + 4y - 6} \end{array}$$

71. Subtract.
$$\begin{array}{r} 12a + 15 \\ \underline{7a - 3} \end{array}$$

72. Subtract.
$$\begin{array}{r} -3b + 6 \\ \underline{2b - 8} \end{array}$$

73. Subtract.
$$\begin{array}{r} 6m^2 - 11m + 5 \\ \underline{-8m^2 + 2m - 1} \end{array}$$

74. Subtract.
$$\begin{array}{r} -4z^2 + 2z - 1 \\ \underline{3z^2 - 5z + 2} \end{array}$$

75. Add.

$$\begin{array}{r} 12z^2 - 11z + 8 \\ 5z^2 + 16z - 2 \\ \underline{-4z^2 + 5z - 9} \end{array}$$

76. Add.

$$\begin{array}{r} -6m^3 + 2m^2 + 5m \\ 8m^3 + 4m^2 - 6m \\ \underline{-3m^3 + 2m^2 - 7m} \end{array}$$

77. Add.

$$\begin{array}{rrrr} 6y^3 & -\ 9y^2 & & +\ 8 \\ \underline{4y^3} & \underline{+\ 2y^2} & \underline{+\ 5y} & \underline{} \end{array}$$

78. Add.

$$\begin{array}{rrrr} -7r^8 & +\ 2r^6 & -\ r^5 & \\ \underline{} & \underline{3r^6} & \underline{} & \underline{+\ 5} \end{array}$$

79. Subtract.

$$\begin{array}{rrrr} -5a^4 & & +\ 8a^2 & -\ 9 \\ \underline{} & \underline{6a^3} & \underline{-\ a^2} & \underline{+\ 2} \end{array}$$

80. Subtract.

$$\begin{array}{rrrr} & -\ 2m^3 & +\ 8m^2 & \\ \underline{m^4} & \underline{-\ m^3} & \underline{} & \underline{+\ 2m} \end{array}$$

Extending Skills *Perform the indicated operations.*

81. Subtract $4y^2 - 2y + 3$ from $7y^2 - 6y + 5$.

82. Subtract $-(-4x + 2z^2 + 3m)$ from $[(2z^2 - 3x + m) + (z^2 - 2m)]$.

83. $(-4m^2 + 3n^2 - 5n) - [(3m^2 - 5n^2 + 2n) + (-3m^2) + 4n^2]$

84. $[-(4m^2 - 8m + 4m^3) - (3m^2 + 2m + 5m^3)] + m^2$

85. $[-(y^4 - y^2 + 1) - (y^4 + 2y^2 + 1)] + (3y^4 - 3y^2 - 2)$

86. $[2p - (3p - 6)] - [(5p - (8 - 9p)) + 4p]$

87. $-[3z^2 + 5z - (2z^2 - 6z)] + [(8z^2 - [5z - z^2]) + 2z^2]$

88. $5k - (5k - [2k - (4k - 8k)]) + 11k - (9k - 12k)$

Find the perimeter of each figure. Express it as a polynomial in descending powers of the variable x.

89.

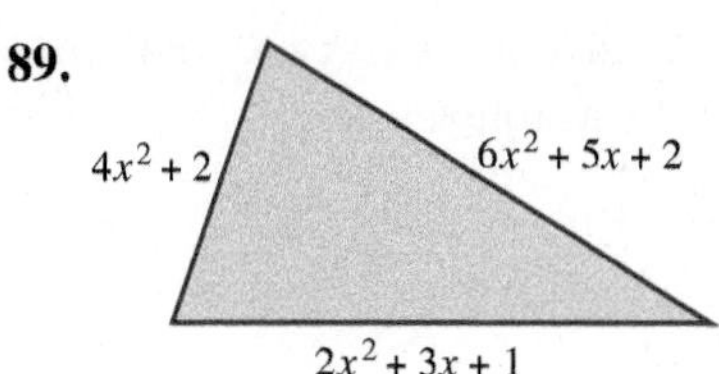

90.

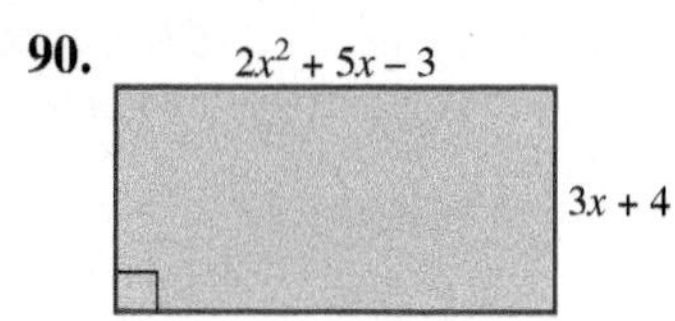

4.4 Polynomial Functions, Graphs, and Composition

OBJECTIVES

1. Recognize and evaluate polynomial functions.
2. Use a polynomial function to model data.
3. Add and subtract polynomial functions.
4. Graph basic polynomial functions.
5. Find the composition of functions.

OBJECTIVE 1 **Recognize and evaluate polynomial functions.**

We have studied linear (first-degree polynomial) functions $f(x) = ax + b$. Now we consider more general polynomial functions.

Polynomial Function

A **polynomial function of degree *n*** is defined by

$$f(x) = a_n x^n + a_{n-1}x^{n-1} + \cdots + a_1 x + a_0,$$

for real numbers $a_n, a_{n-1}, \ldots, a_1$, and a_0, where $a_n \neq 0$ and n is a whole number.

Examples: $f(x) = 2$, $f(x) = -6x + 1$, $f(x) = 5x^2 - x + 4$,
$f(x) = x^3 + 2x^2 - 5x - 3$

Another way of describing a polynomial function is to say that it is a function defined by a polynomial in one variable, consisting of one or more terms. It is usually written in descending powers of the variable, and its degree is the degree of the polynomial that defines it.

VOCABULARY

- ☐ polynomial function
- ☐ identity function
- ☐ squaring function
- ☐ cubing function
- ☐ composite function (composition of functions)

NOW TRY EXERCISE 1

Let $f(x) = x^3 - 2x^2 + 7$. Find $f(-3)$.

We can evaluate a polynomial function $f(x)$ at different values of the variable x.

EXAMPLE 1 Evaluating Polynomial Functions

Let $f(x) = 4x^3 - x^2 + 5$. Find each value.

(a) $f(3)$

Read this as "f of 3," not "f times 3."

$f(x) = 4x^3 - x^2 + 5$ Given function

$f(3) = 4(3)^3 - 3^2 + 5$ Substitute 3 for x.

$f(3) = 4(27) - 9 + 5$ Apply the exponents.

$f(3) = 108 - 9 + 5$ Multiply.

$f(3) = 104$ Subtract, and then add.

Thus, $f(3) = 104$ and the ordered pair $(3, 104)$ belongs to f.

(b) $f(-4)$

$f(x) = 4x^3 - x^2 + 5$ Use parentheses.

$f(-4) = 4(-4)^3 - (-4)^2 + 5$ Let $x = -4$.

$f(-4) = 4(-64) - 16 + 5$ Be careful with signs.

$f(-4) = -256 - 16 + 5$ Multiply.

$f(-4) = -267$ Subtract, and then add.

So, $f(-4) = -267$. The ordered pair $(-4, -267)$ belongs to f. NOW TRY 

The capital letter P is sometimes used for polynomial functions. The function

$$P(x) = 4x^3 - x^2 + 5$$

yields the same ordered pairs as the function f in **Example 1.**

OBJECTIVE 2 Use a polynomial function to model data.

NOW TRY EXERCISE 2

Use the function in **Example 2** to approximate the number of public school students in the United States in 2025 (to two decimal places).

EXAMPLE 2 Using a Polynomial Model to Approximate Data

The number of public school students (grades pre-K–12) in the United States from 1990 projected through 2025 can be modeled by the polynomial function

$$P(x) = -0.0002221x^2 + 0.1392x + 46.80,$$

where $x = 0$ corresponds to the year 1990, $x = 1$ corresponds to 1991, and so on, and $P(x)$ is in millions. Use this function to approximate the number of public school students in 2020. (Data from Department of Education.)

Here $x = 30$ corresponds to 2020, so we find $P(30)$.

$P(x) = -0.0002221x^2 + 0.1392x + 46.80$ Given function

$P(30) = -0.0002221(30)^2 + 0.1392(30) + 46.80$ Let $x = 30$.

$P(30) \approx 50.78$ Evaluate.

The model suggests that there will be approximately 50.78 million public school students in 2020.

NOW TRY

NOW TRY ANSWERS

1. -38
2. 51.40 million students

OBJECTIVE 3 Add and subtract polynomial functions.

The graph in **FIGURE 1** shows dollars (in billions) spent for general science and for space/other technologies over a 20-year period.

$G(x)$ represents dollars spent for general science.

$S(x)$ represents dollars spent for space/other technologies.

$T(x)$ represents total expenditures for these two categories.

Science and Space Spending

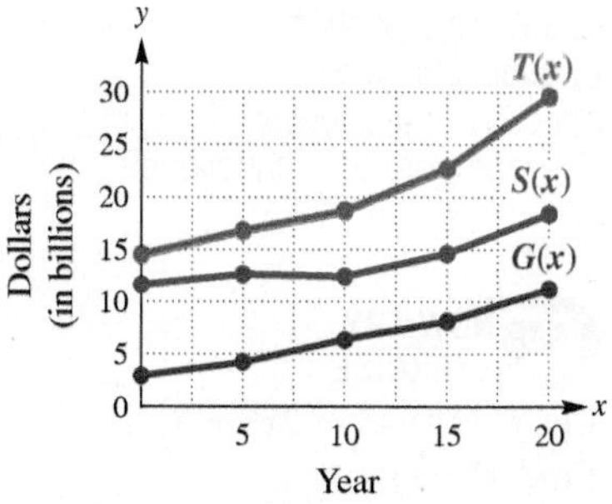

Data from U.S. Office of Management and Budget.

FIGURE 1

The total expenditures function can be found by *adding* the spending functions for the two individual categories.

$$T(x) = G(x) + S(x)$$

As another example, businesses use the equation "profit equals revenue minus cost," which can be written using function notation.

$$\underset{\text{Profit function}}{P(x)} = \underset{\text{Revenue function}}{R(x)} - \underset{\text{Cost function}}{C(x)}$$

x is the number of items produced and sold.

The profit function is found by *subtracting* the cost function from the revenue function.

We define the following **operations on functions.**

Adding and Subtracting Functions

If $f(x)$ and $g(x)$ define functions, then

$$(f + g)(x) = f(x) + g(x) \quad \text{Sum function}$$

and

$$(f - g)(x) = f(x) - g(x). \quad \text{Difference function}$$

In each case, the domain of the new function is the intersection of the domains of $f(x)$ and $g(x)$.

NOW TRY EXERCISE 3

For $f(x) = x^3 - 3x^2 + 4$ and $g(x) = -2x^3 + x^2 - 12$, find each of the following.

(a) $(f + g)(x)$

(b) $(f - g)(x)$

EXAMPLE 3 Adding and Subtracting Functions

Find each of the following for the polynomial functions f and g as defined.

$$f(x) = x^2 - 3x + 7 \quad \text{and} \quad g(x) = -3x^2 - 7x + 7$$

(a) $(f + g)(x)$ — This notation does *not* indicate the distributive property.

$= f(x) + g(x)$ — Sum function

$= (x^2 - 3x + 7) + (-3x^2 - 7x + 7)$ — Substitute.

$= -2x^2 - 10x + 14$ — Add the polynomials.

(b) $(f - g)(x)$

$= f(x) - g(x)$ — Difference function

$= (x^2 - 3x + 7) - (-3x^2 - 7x + 7)$ — Substitute.

$= x^2 - 3x + 7 + 3x^2 + 7x - 7$ — Change subtraction to addition.

$= 4x^2 + 4x$ — Add the polynomials. **NOW TRY**

NOW TRY ANSWERS

3. (a) $-x^3 - 2x^2 - 8$

(b) $3x^3 - 4x^2 + 16$

NOW TRY EXERCISE 4

For $f(x) = x^2 - 4$ and $g(x) = -6x^2$, find each of the following.

(a) $(f + g)(x)$

(b) $(f - g)(-4)$

EXAMPLE 4 Adding and Subtracting Functions

Find each of the following for the polynomial functions f and g as defined.

$$f(x) = 10x^2 - 2x \quad \text{and} \quad g(x) = 2x$$

(a) $(f + g)(2)$

$= f(2) + g(2)$ — Sum function

$= \underbrace{[10(2)^2 - 2(2)]}_{f(x) = 10x^2 - 2x} + \underbrace{2(2)}_{g(x) = 2x}$ — Substitute. (This is a key step.)

$= [40 - 4] + 4$ — Order of operations

$= 40$ — Subtract, and then add.

Alternative method: $(f + g)(x)$ — Find $(f + g)(x)$.

$= f(x) + g(x)$ — Sum function

$= (10x^2 - 2x) + 2x$ — Substitute.

$= 10x^2$ — Combine like terms.

$(f + g)(2)$ — Now find $(f + g)(2)$.

$= 10(2)^2$ — $(f + g)(x) = 10x^2$; Substitute.

$= 40$ — Apply the exponent. Multiply. (The result is the same.)

(b) $(f - g)(x)$ and $(f - g)(1)$

$(f - g)(x)$

$= f(x) - g(x)$ — Difference function

$= (10x^2 - 2x) - 2x$ — Substitute.

$= 10x^2 - 4x$ — Combine like terms.

$(f - g)(1)$ — Now find $(f - g)(1)$.

$= 10(1)^2 - 4(1)$ — $(f - g)(x) = 10x^2 - 4x$; Substitute.

$= 6$ — Perform the operations. **NOW TRY** 

(Confirm that $f(1) - g(1)$ gives the same result.)

OBJECTIVE 4 Graph basic polynomial functions.

Recall that each input (or x-value) of a function results in one output (or y-value). The set of input values (for x) defines the domain of the function, and the set of output values (for y) defines the range.

The simplest polynomial function is the **identity function $f(x) = x$,** graphed in **FIGURE 2.** This function pairs each real number with itself.

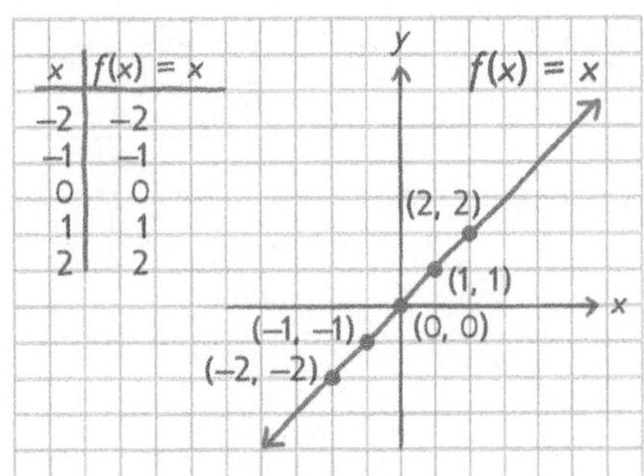

FIGURE 2

Identity function

$f(x) = x$

Domain: $(-\infty, \infty)$

Range: $(-\infty, \infty)$

NOW TRY ANSWERS

4. (a) $-5x^2 - 4$
(b) 108

NOTE The identity function $f(x) = x$ shown in **FIGURE 2** is a *linear function* of the form $f(x) = ax + b$, where the slope a is 1 and the y-value of the y-intercept b is 0.

Another polynomial function, the **squaring function** $f(x) = x^2$, is graphed in **FIGURE 3**. For this function, every real number is paired with its square. The graph of the squaring function is a *parabola.*

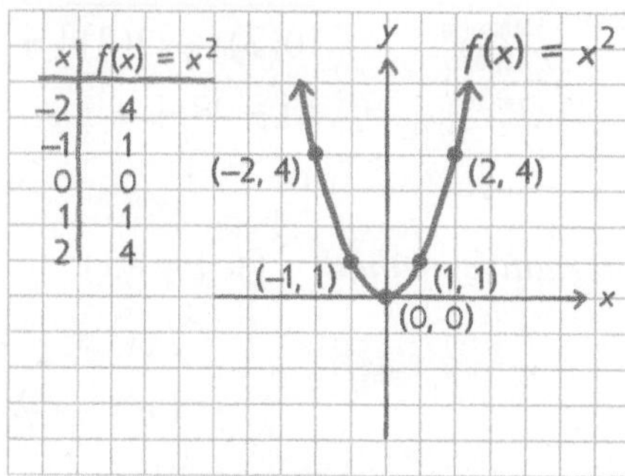

Squaring function

$f(x) = x^2$

Domain: $(-\infty, \infty)$

Range: $[0, \infty)$

FIGURE 3

The **cubing function** $f(x) = x^3$ is graphed in **FIGURE 4**. This function pairs every real number with its cube.

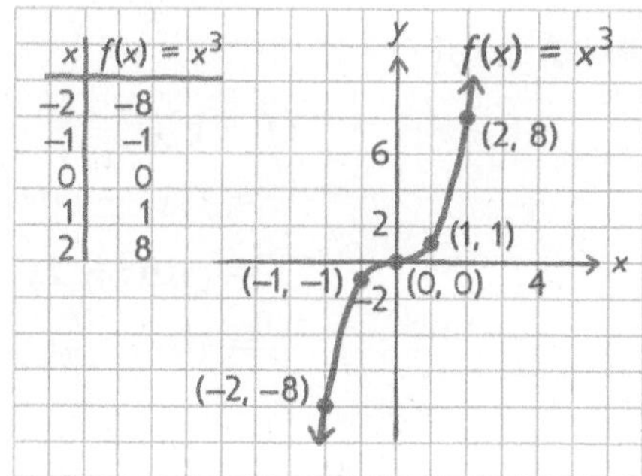

Cubing function

$f(x) = x^3$

Domain: $(-\infty, \infty)$

Range: $(-\infty, \infty)$

FIGURE 4

EXAMPLE 5 Graphing Variations of Polynomial Functions

Graph each function by creating a table of ordered pairs. Give the domain and range of each function by observing its graph.

(a) $f(x) = 2x$

To find each range value, multiply the domain value by 2. Plot the points and join them with a straight line. See **FIGURE 5**. Both the domain and the range are $(-\infty, \infty)$.

x	$f(x) = 2x$
−2	−4
−1	−2
0	0
1	2
2	4

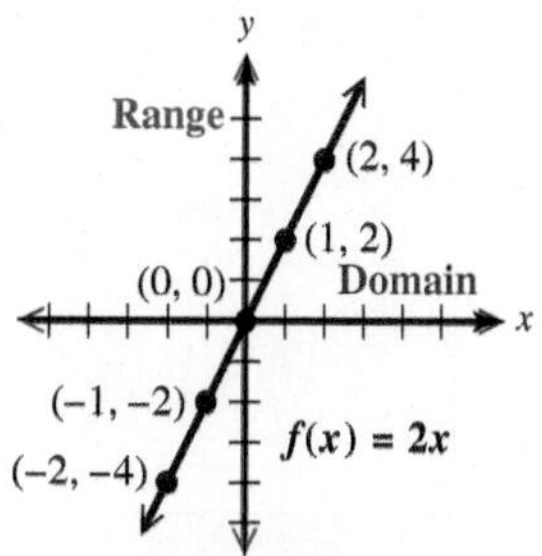

FIGURE 5

NOW TRY EXERCISE 5

Graph $f(x) = x^2 - 4$. Give the domain and range.

(b) $f(x) = -x^2$

For each input x, square it and then take its opposite. Plotting and joining the points gives a parabola that opens down. It is a *reflection* of the graph of the squaring function across the x-axis. See the table and **FIGURE 6**. The domain is $(-\infty, \infty)$ and the range is $(-\infty, 0]$.

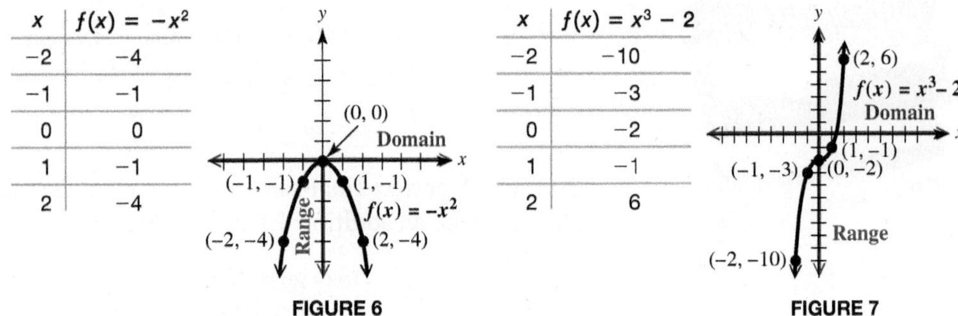

x	$f(x) = -x^2$
−2	−4
−1	−1
0	0
1	−1
2	−4

FIGURE 6

x	$f(x) = x^3 - 2$
−2	−10
−1	−3
0	−2
1	−1
2	6

FIGURE 7

(c) $f(x) = x^3 - 2$

For this function, cube the input and then subtract 2 from the result. The graph is that of the cubing function *shifted* down 2 units. See the table and **FIGURE 7**. The domain and range are both $(-\infty, \infty)$.

NOW TRY

OBJECTIVE 5 Find the composition of functions.

The diagram in **FIGURE 8** shows a function g that assigns, to each element x of set X, some element y of set Y. Suppose that a function f takes each element of set Y and assigns a value z of set Z. Then f and g together assign an element x in X to an element z in Z.

The result of this process is a new function h that takes an element x in X and assigns it an element z in Z.

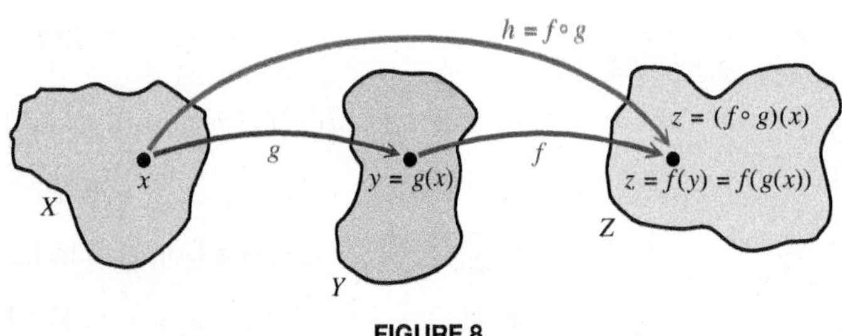

FIGURE 8

Function h is the *composition* of functions f and g, written $\boldsymbol{f \circ g}$.

Composition of Functions

The **composite function,** or **composition,** of functions f and g is defined by

$$(f \circ g)(x) = f(g(x)),$$

for all x in the domain of g such that $g(x)$ is in the domain of f.

Read $f \circ g$ as "*f of g*" (or "*f compose g*").

NOW TRY ANSWER

5.

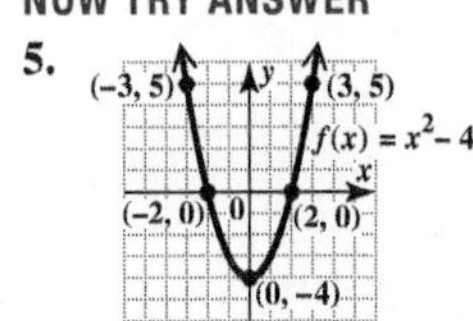

domain: $(-\infty, \infty)$;
range: $[-4, \infty)$

As a real-life example of how composite functions occur, consider the following retail situation.

A \$40 pair of blue jeans is on sale for 25% off. If we purchase the jeans before noon, the retailer offers an additional 10% off. What is the final sale price of the blue jeans?

We might be tempted to say that the jeans are 35% off and calculate as follows.

$$
\begin{aligned}
&\$40 - 0.35(\$40) \quad && \text{Original price} - \text{Discount} = \text{Sale price} \\
&= \$40 - \$14 \\
&= \$26 && \leftarrow \text{This is not correct.}
\end{aligned}
$$

To find the correct final sale price, we must first find the price after taking 25% off, and then take an additional 10% off *that* price.

Take 25% off the original price.	Take an additional 10% off.
$\$40 - 0.25(\$40)$	$\rightarrow \$30 - 0.10(\$30)$
$= \$40 - \$10$	$= \$30 - \$3$
$= \$30$	$= \$27$

This is the idea behind composition of functions.

As another real-life example of composition, suppose an oil well off the coast is leaking, with the leak spreading oil in a circular layer over the surface. See **FIGURE 9**.

At any time t, in minutes, after the beginning of the leak, the radius of the circular oil slick is given by $r(t) = 5t$ feet. Because $\mathcal{A}(r) = \pi r^2$ gives the area of a circle of radius r, the area can be expressed as a function of time.

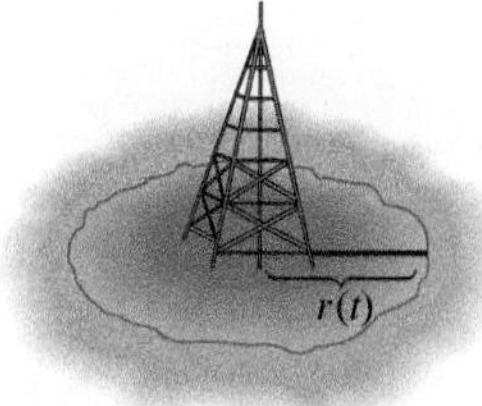

FIGURE 9

$$\mathcal{A}(r) = \pi r^2 \quad \text{Area of a circle}$$

$$\mathcal{A}(r(t)) = \pi(5t)^2 \quad \text{Substitute } 5t \text{ for } r.$$

$$\mathcal{A}(r(t)) = 25\pi t^2 \quad (5t)^2 = 5^2t^2 = 25t^2; \text{ Commutative property}$$

The function $\mathcal{A}(r(t))$ is a composite function of the functions $\mathcal{A}$ and r.

NOW TRY EXERCISE 6

Let $f(x) = 3x + 7$ and $g(x) = x - 2$. Find $(f \circ g)(7)$.

EXAMPLE 6 Finding a Composite Function

Let $f(x) = x^2$ and $g(x) = x + 3$. Find $(f \circ g)(4)$.

$(f \circ g)(4)$ — Evaluate the "inside" function value first.

$$
\begin{aligned}
&= f(g(4)) \quad && \text{Definition of composition} \\
&= f(4 + 3) && \text{Use the rule for } g(x);\ g(4) = 4 + 3 \\
&= f(7) && \text{Add.} \\
&= 7^2 && \text{Use the rule for } f(x);\ f(7) = 7^2 \\
&= 49 && \text{Square 7.}
\end{aligned}
$$

Now evaluate the "outside" function.

In this composition, g is the innermost "operation" and acts on x (here 4) first. Then the output value of g (here 7) becomes the input (domain) value of f. NOW TRY

NOW TRY ANSWER

6. 22

If we interchange the order of functions f and g, the composition $g \circ f$, read "g *of* f" (or "g *compose* f"), is defined by

$$(g \circ f)(x) = g(f(x)),$$ for all x in the domain of f such that $f(x)$ is in the domain of g.

Let $f(x) = 3x + 7$ and $g(x) = x - 2$ as in **Now Try Exercise 6.** Find $(g \circ f)(7)$.

EXAMPLE 7 Finding a Composite Function

Let $f(x) = x^2$ and $g(x) = x + 3$ as in **Example 6.** Find $(g \circ f)(4)$.

$(g \circ f)(4)$ — Evaluate the "inside" function value first.

$= g(f(4))$ — Definition of composition

$= g(4^2)$ — Use the rule for $f(x)$; $f(4) = 4^2$

$= g(16)$ — Square 4.

$= 16 + 3$ — Use the rule for $g(x)$; $g(16) = 16 + 3$ (Now evaluate the "outside" function.)

$= 19$ — Add.

In this composition, f is the innermost "operation" and acts on x (again 4) first. Then the output value of f (here 16) becomes the input (domain) value of g. NOW TRY

We see in **Examples 6 and 7** that $(f \circ g)(4) \neq (g \circ f)(4)$ because $49 \neq 19$. In general,

$$(f \circ g)(x) \neq (g \circ f)(x).$$

NOW TRY EXERCISE 8

Let $f(x) = x - 5$ and $g(x) = -x^2 + 2$. Find each of the following.

(a) $(g \circ f)(-1)$

(b) $(f \circ g)(x)$

EXAMPLE 8 Finding Composite Functions

Let $f(x) = 4x - 1$ and $g(x) = x^2 + 5$. Find each of the following.

(a) $(f \circ g)(2)$

$= f(g(2))$ — Definition of composition

$= f(2^2 + 5)$ — $g(x) = x^2 + 5$

$= f(9)$ — Work inside the parentheses.

$= 4(9) - 1$ — $f(x) = 4x - 1$

$= 35$ — Multiply, and then subtract.

(b) $(f \circ g)(x)$

$= f(g(x))$ — Use $g(x)$ as the input for function f.

$= 4(g(x)) - 1$ — Use the rule for $f(x)$; $f(x) = 4x - 1$

$= 4(x^2 + 5) - 1$ — $g(x) = x^2 + 5$

$= 4x^2 + 20 - 1$ — Distributive property

$= 4x^2 + 19$ — Combine like terms.

(c) Find $(f \circ g)(2)$ again, this time using the rule obtained in part (b).

$(f \circ g)(x) = 4x^2 + 19$ — From part (b)

$(f \circ g)(2) = 4(2)^2 + 19$ — Let $x = 2$.

$= 4(4) + 19$ — Square 2.

$= 16 + 19$ — Multiply.

Same result as in part (a) → $= 35$ — Add. NOW TRY

NOW TRY ANSWERS

7. 26

8. **(a)** -34 **(b)** $-x^2 - 3$

4.4 Exercises

MyLab Math

Video solutions for select problems available in MyLab Math

STUDY SKILLS REMINDER
How are you doing on your homework? **Review Study Skill 4, *Completing Your Homework*.**

Concept Check *Work each problem involving polynomial functions and function notation.*

1. A polynomial function is a function defined by a ______ in (*one / two / three*) variable(s), consisting of one or more (*factors / terms*) and usually written in descending ______ of the variable.

2. Which of the following are *not* polynomial functions?

A. $P(x) = x^{-2} - 2x$ **B.** $f(x) = \frac{1}{2}x^2 + x - 1$

C. $g(x) = -4x + 1.5$ **D.** $p(x) = x^3 - x^2 - \frac{5}{x}$

3. For a function f, the notation $f(5)$ means ______.

A. the variable f times 5, or $5f$ **B.** f equals 5

C. the range value when the domain value is 5 **D.** the domain value when the range value is 5

4. If $f(x) = 2x$, then $f(5) =$ ______. Here $f(5) = 10$ is an abbreviation for the statement "If $x =$ ______ in the function f, then $y =$ ______." The ordered pair ______ belongs to f.

5. For $f(x) = x$, find $f(0)$, $f(1)$, and $f(2)$. Use the results to write three ordered pairs that belong to f.

6. For $f(x) = x^2$, find $f(-2)$, $f(0)$, and $f(2)$. Use the results to write three ordered pairs that belong to f.

7. Concept Check For the function

$$f(x) = -x^2 + 4,$$

a student claimed that $f(-2) = 8$. This is incorrect. **WHAT WENT WRONG?** Find the correct value of $f(-2)$.

8. Concept Check For the function

$$f(x) = x^2 - 1,$$

a student claimed that $f(-1) = -2$. This is incorrect. **WHAT WENT WRONG?** Find the correct value of $f(-1)$.

For each polynomial function, find **(a)** $f(-1)$, **(b)** $f(2)$, *and* **(c)** $f(0)$. **See Example 1.**

9. $f(x) = 6x - 4$ **10.** $f(x) = -2x + 5$ **11.** $f(x) = x^2 - 7x$

12. $f(x) = x^2 + 5x$ **13.** $f(x) = x^2 - 3x + 4$ **14.** $f(x) = x^2 - 5x - 4$

15. $f(x) = 2x^2 - 4x + 1$ **16.** $f(x) = 3x^2 + x - 5$ **17.** $f(x) = 5x^4 - 3x^2 + 6$

18. $f(x) = 4x^4 + 2x^2 - 1$ **19.** $f(x) = -x^2 + 2x^3 - 8$ **20.** $f(x) = -x^2 - x^3 + 11$

Solve each problem. **See Example 2.**

21. Imports of Fair Trade Certified™ coffee into the United States during the years 2007 through 2016 can be modeled by the polynomial function

$$P(x) = -2.169x^2 + 60.28x - 256.6,$$

where $x = 7$ represents 2007, $x = 8$ represents 2008, and so on, and $P(x)$ is in millions of pounds. Use this function to approximate the amount of Fair Trade coffee imported into the United States (to the nearest hundredth) in each given year. (Data from Fair Trade USA.)

(a) 2007 **(b)** 2014 **(c)** 2016

22. The percent of the U.S. population that was foreign-born during the years 1930 through 2016 can be modeled by the polynomial function

$$P(x) = 0.0039x^2 - 0.3001x + 11.33,$$

where $x = 0$ represents 1930, $x = 10$ represents 1940, and so on, and $P(x)$ is percent. Use this function to approximate the percent of the U.S. population that was foreign-born (to the nearest hundredth as needed) in each given year. (Data from U.S. Census Bureau.)

(a) 1930 **(b)** 1970 **(c)** 2016

23. The amount spent by international visitors to the United States during the years 1990 through 2016 can be modeled by the polynomial function

$$P(x) = 0.01287x^3 - 0.3514x^2 + 4.979x + 45.24,$$

where $x = 0$ represents 1990, $x = 1$ represents 1991, and so on, and $P(x)$ is in billions of dollars. Use this function to approximate the amount spent by international visitors to the United States (to the nearest tenth) in each given year. (Data from U.S. Travel Association.)

(a) 1990 **(b)** 2005 **(c)** 2016

24. World tourism receipts during the years 2000 through 2016 can be modeled by the polynomial function

$$P(x) = -0.2869x^3 + 6.139x^2 + 22.84x + 465.3,$$

where $x = 0$ represents 2000, $x = 1$ represents 2001, and so on, and $P(x)$ is in billions of dollars. Use this function to approximate world tourism receipts (to the nearest tenth as needed) in each given year. (Data from World Tourism Association.)

(a) 2000 **(b)** 2008 **(c)** 2016

For each pair of functions, find **(a)** $(f + g)(x)$ *and* **(b)** $(f - g)(x)$. ***See Example 3.***

25. $f(x) = 5x - 10, \quad g(x) = 3x + 7$

26. $f(x) = -4x + 1, \quad g(x) = 6x + 2$

27. $f(x) = 4x^2 + 8x - 3, \quad g(x) = -5x^2 + 4x - 9$

28. $f(x) = 3x^2 - 9x + 10, \quad g(x) = -4x^2 + 2x + 12$

Concept Check *Find two polynomial functions defined by* $f(x)$ *and* $g(x)$ *such that each statement is true.*

29. $(f + g)(x) = 3x^3 - x + 3$

30. $(f - g)(x) = -x^2 + x - 5$

Let $f(x) = x^2 - 9$, $g(x) = 2x$, *and* $h(x) = x - 3$. *Find each of the following.* ***See Example 4.***

31. $(f + g)(x)$

32. $(f - g)(x)$

33. $(f + g)(3)$

34. $(f - g)(-3)$

35. $(f - h)(x)$

36. $(f + h)(x)$

37. $(f - h)(-3)$

38. $(f + h)(-2)$

39. $(g + h)(-10)$

40. $(g - h)(10)$

41. $(g - h)(-3)$

42. $(g + h)(1)$

43. $(g + h)\left(\frac{1}{4}\right)$

44. $(g + h)\left(\frac{1}{3}\right)$

45. $(g + h)\left(-\frac{1}{2}\right)$

46. $(g + h)\left(-\frac{1}{4}\right)$

Solve each problem. ***See Objective 3.***

47. The cost in dollars to produce x t-shirts is $C(x) = 2.5x + 50$. The revenue in dollars from sales of x t-shirts is $R(x) = 10.99x$.

(a) Write and simplify a function P that gives profit in terms of x.

(b) Find the profit if 100 t-shirts are produced and sold.

48. The cost in dollars to produce x youth baseball caps is $C(x) = 4.3x + 75$. The revenue in dollars from sales of x caps is $R(x) = 25x$.

(a) Write and simplify a function P that gives profit in terms of x.

(b) Find the profit if 50 caps are produced and sold.

Graph each polynomial function. Give the domain and range. ***See Example 5.***

49. $f(x) = 3x$ **50.** $f(x) = -4x$ **51.** $f(x) = -2x + 1$

52. $f(x) = 3x + 2$ **53.** $f(x) = -3x^2$ **54.** $f(x) = \frac{1}{2}x^2$

55. $f(x) = x^2 - 2$ **56.** $f(x) = -x^2 + 2$ **57.** $f(x) = x^3 + 1$

58. $f(x) = -x^3 + 2$ **59.** $f(x) = -2x^3 - 1$ **60.** $f(x) = \frac{1}{2}x^3 + 3$

Concept Check *Let $f(x) = x^2$ and $g(x) = 2x - 1$. Match each expression in Column I with the description of how to evaluate it in Column II.*

I	II
61. $(f \circ g)(5)$	**A.** Square 5. Take the result and square it.
62. $(g \circ f)(5)$	**B.** Double 5 and subtract 1. Take the result and square it.
63. $(f \circ f)(5)$	**C.** Double 5 and subtract 1. Take the result, double it, and subtract 1.
64. $(g \circ g)(5)$	**D.** Square 5. Take the result, double it, and subtract 1.

65. Concept Check For $f(x) = x - 8$ and $g(x) = 2$, a student found $(f \circ g)(x)$ incorrectly as follows.

$$(f \circ g)(x) = 2(x - 8)$$
$$= 2x - 16$$

WHAT WENT WRONG? Give the correct composition.

66. Concept Check For $f(x) = 2x + 3$ and $g(x) = x + 5$, a student found $(f \circ g)(x)$ correctly as

$$(f \circ g)(x) = 2x + 13.$$

When asked to find $(g \circ f)(x)$, he gave the same result. **WHAT WENT WRONG?** Give the correct composition.

Let $f(x) = x^2 + 4$, $g(x) = 2x + 3$, and $h(x) = x - 5$. Find each of the following. ***See Examples 6–8.***

67. $(h \circ g)(4)$ **68.** $(f \circ g)(4)$ **69.** $(g \circ f)(6)$ **70.** $(h \circ f)(6)$

71. $(f \circ h)(-2)$ **72.** $(h \circ g)(-2)$ **73.** $(f \circ g)(0)$ **74.** $(f \circ h)(0)$

75. $(g \circ f)(x)$ **76.** $(g \circ h)(x)$ **77.** $(h \circ g)(x)$ **78.** $(h \circ f)(x)$

79. $(f \circ h)\left(\frac{1}{2}\right)$ **80.** $(h \circ f)\left(\frac{1}{2}\right)$ **81.** $(f \circ g)\left(-\frac{1}{2}\right)$ **82.** $(g \circ f)\left(-\frac{1}{2}\right)$

Extending Skills *The tables give some selected ordered pairs for functions f and g.*

x	3	4	6	8
$f(x)$	1	3	9	2

x	2	7	1	9
$g(x)$	3	6	9	12

Tables like these can be used to evaluate composite functions. For example, to evaluate $(g \circ f)(6)$, use the first table to find $f(6) = 9$. Then use the second table to find

$$(g \circ f)(6) = g(f(6)) = g(9) = 12.$$

Find each of the following.

83. $(f \circ g)(2)$ **84.** $(f \circ g)(7)$ **85.** $(g \circ f)(3)$

86. $(g \circ f)(8)$ **87.** $(f \circ f)(4)$ **88.** $(g \circ g)(1)$

Solve each problem. ***See Objective 5.***

89. The function $f(x) = 12x$ computes the number of inches in x feet, and the function $g(x) = 5280x$ computes the number of feet in x miles. Find and simplify $(f \circ g)(x)$. What does it compute?

90. The function $f(x) = 60x$ computes the number of minutes in x hours, and the function $g(x) = 24x$ computes the number of hours in x days. Find and simplify $(f \circ g)(x)$. What does it compute?

91. The perimeter x of a square with sides of length s is given by the formula $x = 4s$.

(a) Solve for s in terms of x.

(b) If y represents the area of this square, write y as a function of the perimeter x.

(c) Use the composite function of part (b) to find the area of a square with perimeter 6.

92. The perimeter x of an equilateral triangle with sides of length s is given by the formula $x = 3s$.

(a) Solve for s in terms of x.

(b) The area y of an equilateral triangle with sides of length s is given by the formula $y = \frac{s^2\sqrt{3}}{4}$. Write y as a function of the perimeter x.

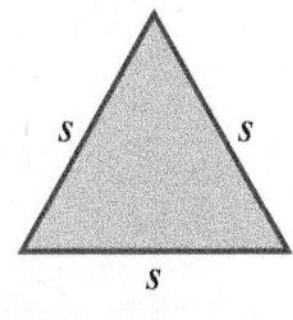

(c) Use the composite function of part (b) to find the area of an equilateral triangle with perimeter 12.

93. In a sale room at a clothing store, every item is on sale for half its original price, plus $1.

(a) Write a function g that finds half of x.

(b) Write a function f that adds 1 to x.

(c) Write and simplify the function $(f \circ g)(x)$.

(d) Use the function from part (c) to find the sale price of a shirt that has an original price of $60.

94. A pair of shoes is marked 50% off. A customer has a coupon for an additional $10 off.

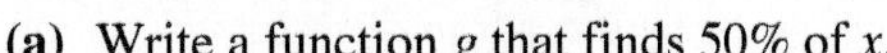

(a) Write a function g that finds 50% of x.

(b) Write a function f that subtracts 10 from x.

(c) Write and simplify the function $(f \circ g)(x)$.

(d) Use the function from part (c) to find the sale price of a pair of shoes that has an original price of $100.

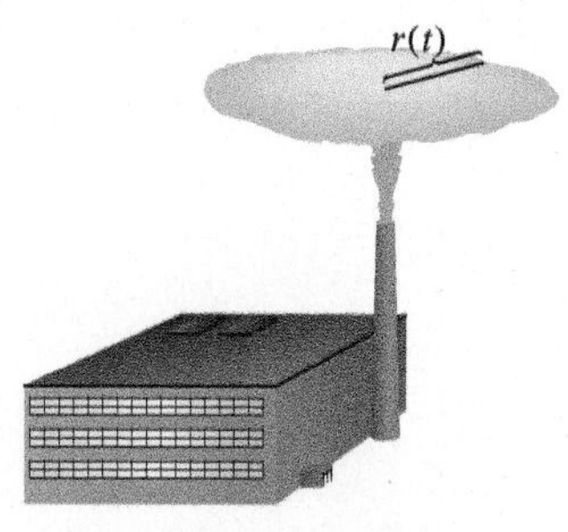

95. When a thermal inversion layer is over a city (as happens often in Los Angeles), pollutants cannot rise vertically, but are trapped below the layer and must disperse horizontally.

Assume that a factory smokestack begins emitting a pollutant at 8 A.M. and that the pollutant disperses horizontally over a circular area. Suppose that t represents the time, in hours, since the factory began emitting pollutants ($t = 0$ represents 8 A.M.), and assume that the radius of the circle of pollution is $r(t) = 2t$ miles. Let $\mathscr{A}(r) = \pi r^2$ represent the area of a circle of radius r. Find and interpret $(\mathscr{A} \circ r)(t)$.

96. An oil well is leaking, with the leak spreading oil over the surface as a circle. At any time t, in minutes, after the beginning of the leak, the radius of the circular oil slick on the surface is $r(t) = 4t$ feet. Let $\mathscr{A}(r) = \pi r^2$ represent the area of a circle of radius r. Find and interpret $(\mathscr{A} \circ r)(t)$.

RELATING CONCEPTS For Individual or Group Work (Exercises 97–100)

The polynomial function

$$f(x) = \frac{1}{24}x^4 - \frac{1}{4}x^3 + \frac{23}{24}x^2 - \frac{3}{4}x + 1$$

will give the maximum number of interior regions formed in a circle if x points on the circumference are joined by all possible chords. For $x = 1, 2, 3, 4,$ *and* 5, *see* **FIGURES A–E.**

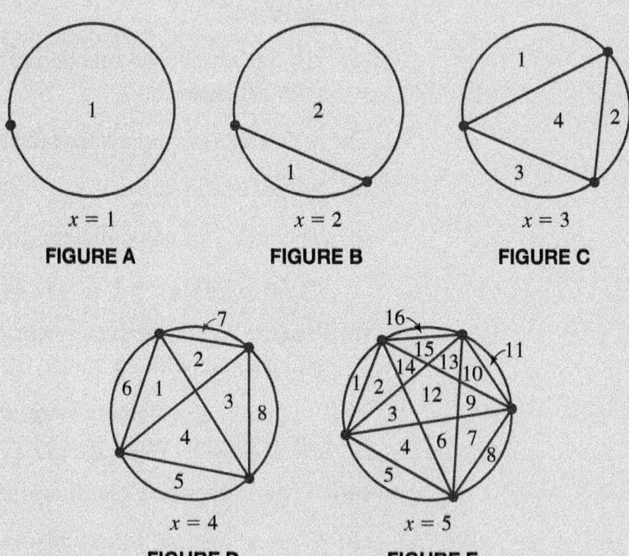

For example, in **FIGURE A** *we have 1 point, and because no chords can be drawn, we have only 1 interior region. In* **FIGURE B** *there are 2 points, and 2 interior regions are formed.*

Apply this information and ***work Exercises 97–100 in order.***

97. Use the given polynomial function to verify that $f(1) = 1$ and $f(2) = 2$.

98. Based on the appropriate figure alone, what is a logical prediction for the value of $f(3)$? $f(4)$? $f(5)$?

99. Verify the answers in **Exercise 98** by evaluating $f(3)$, $f(4)$, and $f(5)$.

100. Observe a pattern in the results of **Exercises 97–99.** Use the pattern to make a logical prediction for the value of $f(6)$. Does the prediction equal $f(6)$?

4.5 Multiplying Polynomials

OBJECTIVES

1. Multiply terms.
2. Multiply any two polynomials.
3. Multiply binomials.
4. Find the product of a sum and difference of two terms.
5. Find the square of a binomial.
6. Multiply polynomial functions.

OBJECTIVE 1 Multiply terms.

EXAMPLE 1 Multiplying Monomials

Find each product.

(a) $3x^4(5x^3)$

$= 3(5)x^4 \cdot x^3$ Commutative and associative properties

$= 15x^{4+3}$ Multiply; product rule for exponents

$= 15x^7$ Add the exponents.

(b) $-4a^3(3a^5)$

$= -4(3)a^3 \cdot a^5$

$= -12a^8$

(c) $2m^2z^4(8m^3z^2)$

$= 2(8)m^2 \cdot m^3 \cdot z^4 \cdot z^2$

$= 16m^5z^6$

NOW TRY 

NOW TRY EXERCISE 1

Find the product.

$-3s^2t(15s^3t^4)$

OBJECTIVE 2 Multiply any two polynomials.

EXAMPLE 2 Multiplying Polynomials

Find each product.

(a) $-2(8x^3 - 9x^2)$

$= -2(8x^3) - 2(-9x^2)$ Distributive property (Be careful with signs.)

$= -16x^3 + 18x^2$ Multiply.

(b) $5x^2(-4x^2 + 3x - 2)$

$= 5x^2(-4x^2) + 5x^2(3x) + 5x^2(-2)$ Distributive property

$= -20x^4 + 15x^3 - 10x^2$ Multiply.

(c) $(5x + 4)(2x^2 + x)$ Treat $2x^2 + x$ as a single expression. Distributive property; Multiply each term of $5x + 4$ by $2x^2 + x$.

$= 5x(2x^2 + x) + 4(2x^2 + x)$

$= 10x^3 + 5x^2 + 8x^2 + 4x$ Distributive property

$= 10x^3 + 13x^2 + 4x$ Combine like terms.

(d) $2x^2(x + 1)(x - 3)$

$= 2x^2[x(x - 3) + 1(x - 3)]$ Distributive property

$= 2x^2[x^2 - 3x + x - 3]$ Distributive property

$= 2x^2(x^2 - 2x - 3)$ Combine like terms.

$= 2x^4 - 4x^3 - 6x^2$ Distributive property

NOW TRY

NOW TRY EXERCISE 2

Find each product.

(a) $3k^3(-2k^5 + 3k^2 - 4)$

(b) $5x(2x - 1)(x + 4)$

NOW TRY ANSWERS

1. $-45s^5t^5$
2. (a) $-6k^8 + 9k^5 - 12k^3$
   (b) $10x^3 + 35x^2 - 20x$

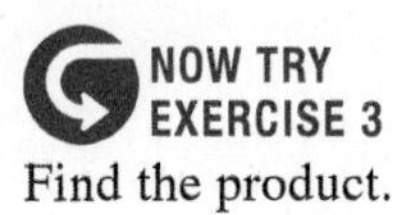

Find the product.

$$\begin{array}{r} 3t^2 - 5t + 4 \\ t - 3 \\ \hline \end{array}$$

EXAMPLE 3 Multiplying Polynomials Vertically

Find each product.

(a) $(5a - 2b)(3a + b)$

$$\begin{array}{rrl} & 5a - 2b & \\ & 3a + b & \text{Write the factors vertically.} \\ \hline & 5ab - 2b^2 & \leftarrow \text{Multiply } b(5a - 2b). \\ 15a^2 - 6ab & & \leftarrow \text{Multiply } 3a(5a - 2b). \\ \hline 15a^2 - ab - 2b^2 & & \text{Combine like terms.} \end{array}$$

(b) $(3m^3 - 2m^2 + 4)(3m - 5)$

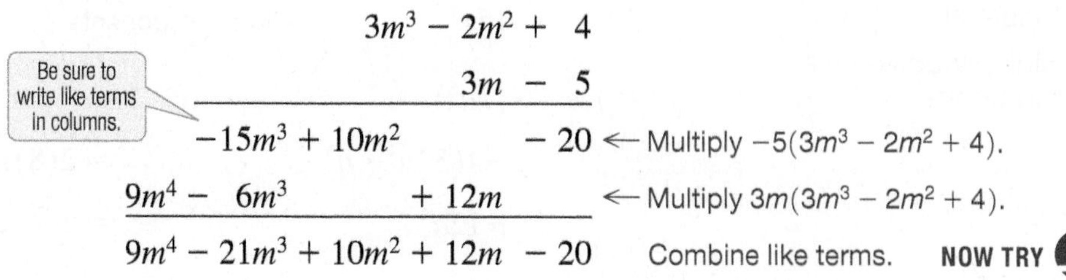

Be sure to write like terms in columns.

$$\begin{array}{rl} 3m^3 - 2m^2 + 4 & \\ 3m - 5 & \\ \hline -15m^3 + 10m^2 \quad - 20 & \leftarrow \text{Multiply } -5(3m^3 - 2m^2 + 4). \\ 9m^4 - 6m^3 \quad + 12m \quad & \leftarrow \text{Multiply } 3m(3m^3 - 2m^2 + 4). \\ \hline 9m^4 - 21m^3 + 10m^2 + 12m - 20 & \text{Combine like terms.} \end{array}$$

NOW TRY

NOTE We can use a rectangle to model polynomial multiplication.

$(5a - 2b)(3a + b)$ **See Example 3(a).**

First, we label a rectangle with each term, as shown below on the left. Then we write the product of each pair of monomials in the appropriate box, as shown on the right.

	$3a$	b
$5a$		
$-2b$		

	$3a$	b
$5a$	$15a^2$	$5ab$
$-2b$	$-6ab$	$-2b^2$

$(5a - 2b)(3a + b)$

$= 15a^2 + 5ab - 6ab - 2b^2$ Add the four monomial products.

$= 15a^2 - ab - 2b^2$ The result is the same as in **Example 3(a).**

OBJECTIVE 3 Multiply binomials.

There is a shortcut method for finding the product of two binomials. In **Example 2,** we found such products using the distributive property.

$(3x - 4)(2x + 3)$

$= 3x(2x + 3) - 4(2x + 3)$ Distributive property

$= 3x(2x) + 3x(3) - 4(2x) - 4(3)$ Distributive property again

$= 6x^2 + 9x - 8x - 12$ Multiply.

Before combining like terms to find the simplest form of the answer, we check the origin of each of the four terms in the sum

$$6x^2 + 9x - 8x - 12.$$

NOW TRY ANSWER

3. $3t^3 - 14t^2 + 19t - 12$

From the discussion on the preceding page, multiplying

$(3x - 4)(2x + 3)$

using the distributive property gave

$6x^2 + 9x - 8x - 12.$

The term $6x^2$ is the product of the two *first* terms of the binomials.

$(3x - 4)(2x + 3)$ $\quad 3x(2x) = 6x^2$ $\quad$ First terms

To obtain $9x$, the *outer* terms are multiplied.

$(3x - 4)(2x + 3)$ $\quad 3x(3) = 9x$ $\quad$ Outer terms

The term $-8x$ comes from the *inner* terms.

$(3x - 4)(2x + 3)$ $\quad -4(2x) = -8x$ $\quad$ Inner terms

Finally, -12 comes from the *last* terms.

$(3x - 4)(2x + 3)$ $\quad -4(3) = -12$ $\quad$ Last terms

The product is found by adding these four results.

$$
\begin{aligned}
&(3x - 4)(2x + 3) \\
&= 6x^2 + 9x - 8x - 12 \quad \text{FOIL method} \\
&= 6x^2 + x - 12 \quad \text{Combine like terms.}
\end{aligned}
$$

To keep track of the order of multiplying terms, we use the initials FOIL, where the letters refer to the positions of the terms (First, Outer, Inner, Last). The above steps can be done as follows.

Compact form: $(3x - 4)(2x + 3)$ — First, Last, Inner, Outer $\qquad (3x - 4)(2x + 3)$ — $6x^2$, -12, $-8x$, $9x$; x Add.

Try to do as many of these steps as possible mentally.

CAUTION ***The FOIL method applies only to multiplying two binomials.***

EXAMPLE 4 Using the FOIL Method

Use the FOIL method to find each product.

(a) $(4m - 5)(3m + 1)$

First terms $\quad (4m - 5)(3m + 1) \quad 4m(3m) = 12m^2$

Outer terms $\quad (4m - 5)(3m + 1) \quad 4m(1) = 4m$

Inner terms $\quad (4m - 5)(3m + 1) \quad -5(3m) = -15m$

Last terms $\quad (4m - 5)(3m + 1) \quad -5(1) = -5$

$$
\begin{aligned}
&(4m - 5)(3m + 1) \\
&= 12m^2 + 4m - 15m - 5 \quad \text{FOIL method} \\
&= 12m^2 - 11m - 5 \quad \text{Combine like terms.}
\end{aligned}
$$

NOW TRY EXERCISE 4

Use the FOIL method to find each product.

(a) $(2x + 1)(7x + 2)$

(b) $(3p - k)(5p + 4k)$

Compact form: $(4m - 5)(3m + 1)$ — products $12m^2$, -5, $-15m$, $4m$; $-15m + 4m = -11m$ Add.

Add these results to obtain

$$12m^2 - 11m - 5.$$

(b) $(5x - 4)(2x - 5)$

First, Outer, Inner, Last

$= 5x(2x) + 5x(-5) - 4(2x) - 4(-5)$ FOIL method

$= 10x^2 - 25x - 8x + 20$ Multiply.

$= 10x^2 - 33x + 20$ Combine like terms.

(c) $(6a + 5b)(3a - 4b)$

First, Outer, Inner, Last

$= 18a^2 - 24ab + 15ab - 20b^2$ FOIL method

$= 18a^2 - 9ab - 20b^2$ Combine like terms.

(d) $(2k + 3z)(5k + 3z)$

$= 10k^2 + 6kz + 15kz + 9z^2$ FOIL method

$= 10k^2 + 21kz + 9z^2$ Combine like terms.

NOW TRY

OBJECTIVE 4 Find the product of a sum and difference of two terms.

The product of a sum and difference of the same two terms occurs frequently.

$(x + y)(x - y)$

$= x^2 - xy + xy - y^2$ FOIL method

$= x^2 - y^2$ Combine like terms.

Product of a Sum and Difference of Two Terms

The **product of the sum and difference of the two terms x and y** is the difference of the squares of the terms.

$$(x + y)(x - y) = x^2 - y^2$$

NOW TRY EXERCISE 5

Find each product.

(a) $(x + 8)(x - 8)$

(b) $(3x - 7y)(3x + 7y)$

(c) $5k(2k - 3)(2k + 3)$

EXAMPLE 5 Multiplying a Sum and Difference of Two Terms

Find each product.

(a) $(p + 7)(p - 7)$

$= p^2 - 7^2$

$= p^2 - 49$

(b) $(2r + 5)(2r - 5)$

$= (2r)^2 - 5^2$ Be careful. $(ab)^2 = a^2b^2$

$= 4r^2 - 25$ $(2r)^2 = 2^2r^2$

(c) $(6m + 5n)(6m - 5n)$

$= (6m)^2 - (5n)^2$ $(6m)^2 = 6^2m^2$; $(5n)^2 = 5^2n^2$

$= 36m^2 - 25n^2$

(d) $2x^3(x + 3)(x - 3)$

$= 2x^3(x^2 - 9)$

$= 2x^5 - 18x^3$

NOW TRY

NOW TRY ANSWERS

4. (a) $14x^2 + 11x + 2$
(b) $15p^2 + 7kp - 4k^2$

5. (a) $x^2 - 64$
(b) $9x^2 - 49y^2$
(c) $20k^3 - 45k$

OBJECTIVE 5 Find the square of a binomial.

To find the square of a binomial sum $x + y$—that is, $(x + y)^2$—multiply $x + y$ by itself.

$$(x + y)^2$$

$$= (x + y)(x + y) \qquad a^2 = a \cdot a$$

$$= x^2 + xy + xy + y^2 \qquad \text{FOIL method}$$

$$= x^2 + 2xy + y^2 \qquad \text{Combine like terms.}$$

A similar result is true for the square of a binomial difference.

Square of a Binomial

The **square of a binomial** is the sum of the square of the first term, twice the product of the two terms, and the square of the last term.

$$(x + y)^2 = x^2 + 2xy + y^2$$

$$(x - y)^2 = x^2 - 2xy + y^2$$

NOW TRY EXERCISE 6

Find each product.

(a) $(y - 10)^2$

(b) $(4x + 5y)^2$

EXAMPLE 6 Squaring Binomials

Find each product.

(a) $(m + 7)^2$

$$= m^2 + 2 \cdot m \cdot 7 + 7^2 \qquad (x + y)^2 = x^2 + 2xy + y^2$$

$$= m^2 + 14m + 49 \qquad \text{Multiply. Apply the exponent.}$$

(b) $(p - 4)^2$

$$= p^2 - 2 \cdot p \cdot 4 + 4^2 \qquad (x - y)^2 = x^2 - 2xy + y^2$$

$$= p^2 - 8p + 16 \qquad \text{Multiply. Apply the exponent.}$$

(c) $(2p + 3v)^2$

$$= (2p)^2 + 2(2p)(3v) + (3v)^2$$

$$= 4p^2 + 12pv + 9v^2$$

(d) $(3r - 5s)^2$

$$= (3r)^2 - 2(3r)(5s) + (5s)^2$$

$$= 9r^2 - 30rs + 25s^2$$

NOW TRY

CAUTION As the products in the formula for the square of a binomial show,

$$(x + y)^2 \neq x^2 + y^2.$$

In general for $n \neq 1$,

$$(x + y)^n \neq x^n + y^n.$$

EXAMPLE 7 Multiplying More Complicated Binomials

Find each product.

(a) $[(3p - 2) + 5q][(3p - 2) - 5q]$

$$= (3p - 2)^2 - (5q)^2 \qquad \text{Product of a sum and difference of terms}$$

$$= 9p^2 - 12p + 4 - 25q^2 \qquad \text{Square both quantities.}$$

(b) $[(2z + r) + 1]^2$

$$= (2z + r)^2 + 2(2z + r)(1) + 1^2 \qquad \text{Square of a binomial}$$

$$= 4z^2 + 4zr + r^2 + 4z + 2r + 1 \qquad \text{Square again; distributive property}$$

NOW TRY ANSWERS

6. **(a)** $y^2 - 20y + 100$

(b) $16x^2 + 40xy + 25y^2$

NOW TRY EXERCISE 7
Find each product.
(a) $[(4x - y) + 2][(4x - y) - 2]$
(b) $(y - 3)^4$

(c) $(x + y)^3$

This does *not* equal $x^3 + y^3$.

$= (x + y)(x + y)^2$ — $a^3 = a \cdot a^2$

$= (x + y)(x^2 + 2xy + y^2)$ — Square $x + y$.

$= x(x^2 + 2xy + y^2) + y(x^2 + 2xy + y^2)$ — Distributive property

$= x^3 + 2x^2y + xy^2 + x^2y + 2xy^2 + y^3$ — Distributive property

$= x^3 + 3x^2y + 3xy^2 + y^3$ — Combine like terms.

(d) $(2a + b)^4$

$= (2a + b)^2(2a + b)^2$ — $a^4 = a^2 \cdot a^2$

$= (4a^2 + 4ab + b^2)(4a^2 + 4ab + b^2)$ — Square $2a + b$ twice.

$= 4a^2(4a^2 + 4ab + b^2) + 4ab(4a^2 + 4ab + b^2) + b^2(4a^2 + 4ab + b^2)$ — Distributive property

$= 16a^4 + 16a^3b + 4a^2b^2 + 16a^3b + 16a^2b^2 + 4ab^3 + 4a^2b^2 + 4ab^3 + b^4$ — Distributive property again

$= 16a^4 + 32a^3b + 24a^2b^2 + 8ab^3 + b^4$ — Combine like terms.

NOW TRY

OBJECTIVE 6 Multiply polynomial functions.

Previously, we added and subtracted functions. They can also be multiplied.

Multiplying Functions

If $f(x)$ and $g(x)$ define functions, then

$$(fg)(x) = f(x) \cdot g(x). \quad \text{Product function}$$

The domain of the product function is the intersection of the domains of $f(x)$ and $g(x)$.

NOW TRY EXERCISE 8
For $f(x) = 3x^2 - 1$
and $g(x) = 8x + 7$,
find $(fg)(x)$ and $(fg)(-2)$.

EXAMPLE 8 Multiplying Polynomial Functions

For $f(x) = 3x + 4$ and $g(x) = 2x^2 + x$, find $(fg)(x)$ and $(fg)(-1)$.

$(fg)(x)$

This notation indicates function multiplication.

$= f(x) \cdot g(x)$ — Use the definition.

$= (3x + 4)(2x^2 + x)$ — Substitute.

$= 6x^3 + 3x^2 + 8x^2 + 4x$ — FOIL method

$= 6x^3 + 11x^2 + 4x$ — Combine like terms.

$(fg)(-1)$ — $(fg)(x) = 6x^3 + 11x^2 + 4x$

$= 6(-1)^3 + 11(-1)^2 + 4(-1)$ — Let $x = -1$ in $(fg)(x)$.

Be careful with signs.

$= -6 + 11 - 4$ — Apply the exponents. Multiply.

$= 1$ — Add and subtract.

An alternative method for finding $(fg)(-1)$ is to find $f(-1)$ and $g(-1)$ and then multiply the results. Verify that $f(-1) \cdot g(-1) = 1$. This follows from the definition.

NOW TRY

NOW TRY ANSWERS
7. (a) $16x^2 - 8xy + y^2 - 4$
(b) $y^4 - 12y^3 + 54y^2 - 108y + 81$
8. $24x^3 + 21x^2 - 8x - 7$; -99

CAUTION Write the product $f(x) \cdot g(x)$ as $(fg)(x)$, ***not*** $f(g(x))$, which indicates the composition of functions f and g.

4.5 Exercises

FOR EXTRA HELP MyLab Math

Video solutions for select problems available in MyLab Math

Concept Check *Match each product in Column I with the correct polynomial in Column II.*

I	II
1. $(2x-5)(3x+4)$	**A.** $6x^2+23x+20$
2. $(2x+5)(3x+4)$	**B.** $6x^2+7x-20$
3. $(2x-5)(3x-4)$	**C.** $6x^2-7x-20$
4. $(2x+5)(3x-4)$	**D.** $6x^2-23x+20$

Find each product. ***See Examples 1–3.***

5. $-8m^3(3m^2)$

6. $-4p^2(5p^4)$

7. $14x^2y^3(-2x^5y)$

8. $5m^3n^4(-4m^2n^5)$

9. $3x(-2x+5)$

10. $5y(-6y+1)$

11. $-q^3(2+3q)$

12. $-3a^4(4+a)$

13. $6k^2(3k^2+2k+1)$

14. $5r^3(2r^2+3r+4)$

15. $(5x+1)(3x^2+x)$

16. $(9a+2)(4a^2+3a)$

17. $(2y+3)(3y-4)$

18. $(2m+6)(5m-3)$

19. $m(m+5)(m-8)$

20. $p(p+4)(p-6)$

21. $4z(2z+1)(3z-4)$

22. $2y(2y+1)(8y-3)$

23. $4x^3(x-3)(x+2)$

24. $2y^5(y-8)(y+2)$

25. $(2t+3)(3t^2-4t-1)$

26. $(4z+2)(z^2-3z-5)$

27. $\begin{array}{r} 5m-3n \\ \underline{5m+3n} \end{array}$

28. $\begin{array}{r} 2k+6q \\ \underline{2k-6q} \end{array}$

29. $\begin{array}{r} -b^2+3b+3 \\ \underline{2b+4} \end{array}$

30. $\begin{array}{r} -r^2-4r+8 \\ \underline{3r-2} \end{array}$

31. $\begin{array}{r} 2p^2+3p+6 \\ \underline{3p^2-4p-1} \end{array}$

32. $\begin{array}{r} 5y^2-2y+4 \\ \underline{2y^2+\ y+3} \end{array}$

33. $\begin{array}{r} 2z^3-5z^2+8z-1 \\ \underline{4z+3} \end{array}$

34. $\begin{array}{r} 3z^4-2z^3+\ z-5 \\ \underline{2z-5} \end{array}$

Use the FOIL method to find each product. ***See Example 4.***

35. $(m+5)(m-8)$

36. $(p+4)(p-6)$

37. $(4k+3)(3k-2)$

38. $(5w+2)(2w-5)$

39. $(3x-2)(5x-1)$

40. $(4x-1)(6x-7)$

41. $(5x+2y)(4x+3y)$

42. $(3x+4y)(2x+3y)$

43. $(z-w)(3z+4w)$

44. $(s-t)(2s+5t)$

45. $(6c-d)(2c+3d)$

46. $(2m-n)(3m+5n)$

47. Concept Check A student multiplied incorrectly as follows.

$$(x+4)^2=x^2+16$$

WHAT WENT WRONG? Give the correct product.

48. Concept Check A student multiplied incorrectly as follows.

$$(x-9)(x+9)=x^2+81$$

WHAT WENT WRONG? Give the correct product.

Find each product. **See Example 5.**

49. $(x+9)(x-9)$ **50.** $(z+6)(z-6)$ **51.** $(2p-3)(2p+3)$

52. $(3x-8)(3x+8)$ **53.** $(5m-1)(5m+1)$ **54.** $(6y-3)(6y+3)$

55. $(3a+2c)(3a-2c)$ **56.** $(5r+4s)(5r-4s)$ **57.** $(4m+7n^2)(4m-7n^2)$

58. $(2k^2+6h)(2k^2-6h)$ **59.** $3y(5y^3+2)(5y^3-2)$ **60.** $4x(3x^3+4)(3x^3-4)$

Find each product. **See Example 6.**

61. $(y-5)^2$ **62.** $(a-3)^2$ **63.** $(x+1)^2$ **64.** $(t+2)^2$

65. $(2p+7)^2$ **66.** $(3z+8)^2$ **67.** $(4n-3m)^2$ **68.** $(5r-7s)^2$

Extending Skills *The factors in the following exercises involve fractions or decimals. Apply the methods of this section, and find each product.*

69. $\left(k-\frac{5}{7}p\right)^2$ **70.** $\left(q-\frac{3}{4}r\right)^2$ **71.** $(0.2x-1.4y)^2$

72. $(0.3x-1.6y)^2$ **73.** $\left(4x-\frac{2}{3}\right)\left(4x+\frac{2}{3}\right)$ **74.** $\left(3t-\frac{5}{4}\right)\left(3t+\frac{5}{4}\right)$

75. $(0.2x+1.3)(0.5x-0.1)$ **76.** $(0.1y+2.1)(0.5y-0.4)$

77. $\left(3w+\frac{1}{4}z\right)(w-2z)$ **78.** $\left(5r+\frac{2}{3}y\right)(r-5y)$

Find each product. **See Example 7.**

79. $[(5x+1)+6y]^2$ **80.** $[(3m-2)+p]^2$

81. $[(2a+b)-3]^2$ **82.** $[(4k+h)-4]^2$

83. $[(2a+b)-3][(2a+b)+3]$ **84.** $[(m+p)-5][(m+p)+5]$

85. $[(2h-k)+j][(2h-k)-j]$ **86.** $[(3m-y)+z][(3m-y)-z]$

87. $(y+2)^3$ **88.** $(z-3)^3$ **89.** $(5r-s)^3$

90. $(x+3y)^3$ **91.** $(q-2)^4$ **92.** $(r+3)^4$

Extending Skills *Find each product.*

93. $(2a+b)(3a^2+2ab+b^2)$ **94.** $(m-5p)(m^2-2mp+3p^2)$

95. $(4z-x)(z^3-4z^2x+2zx^2-x^3)$ **96.** $(3r+2s)(r^3+2r^2s-rs^2+2s^3)$

97. $(m^2-2mp+p^2)(m^2+2mp-p^2)$ **98.** $(3+x+y)(-3+x-y)$

99. $ab(a+b)(a+2b)(a-3b)$ **100.** $mp(m-p)(m-2p)(2m+p)$

Find the area of each figure. Express it as a polynomial in descending powers of the variable x. Refer to the formulas at the back of this text if necessary.

101.

$3x-2y$

$3x+2y$

102.

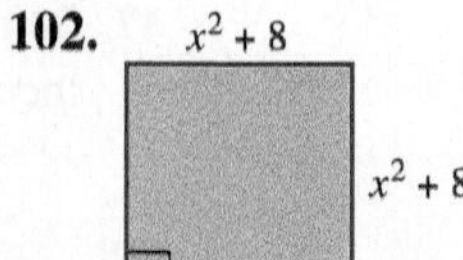

103.

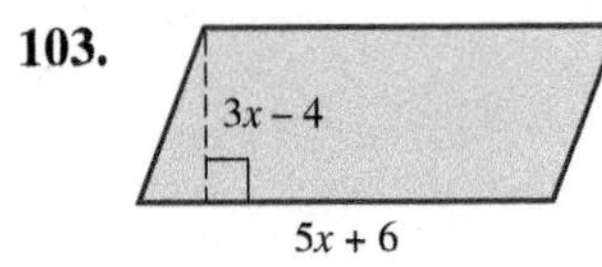

104.

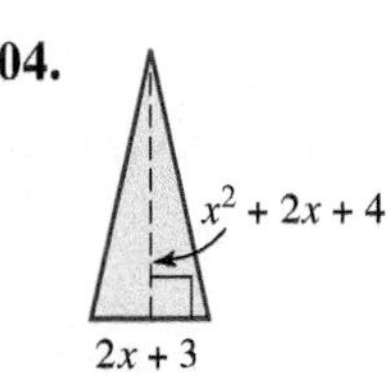

For each pair of functions, find $(fg)(x)$. ***See Example 8.***

105. $f(x) = 2x, \quad g(x) = 5x - 1$

106. $f(x) = 3x, \quad g(x) = 6x - 8$

107. $f(x) = x + 1, \quad g(x) = 2x - 3$

108. $f(x) = x - 7, \quad g(x) = 4x + 5$

109. $f(x) = 2x - 3, \quad g(x) = 4x^2 + 6x + 9$

110. $f(x) = 3x + 4, \quad g(x) = 9x^2 - 12x + 16$

Let $f(x) = x^2 - 9$, $g(x) = 2x$, *and* $h(x) = x - 3$. *Find each of the following.* ***See Example 8.***

111. $(fg)(x)$

112. $(fh)(x)$

113. $(fg)(2)$

114. $(fh)(1)$

115. $(fh)(-1)$

116. $(gh)(-2)$

117. $(gh)(-3)$

118. $(fg)(-2)$

119. $(fg)(0)$

120. $(fh)(0)$

121. $(fg)\left(-\frac{1}{2}\right)$

122. $(fg)\left(-\frac{1}{3}\right)$

RELATING CONCEPTS For Individual or Group Work (Exercises 123–130)

Consider the figure. ***Work Exercises 123–130 in order.***

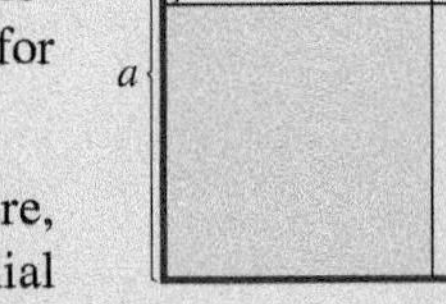

123. What is the length of each side of the blue square in terms of a and b?

124. What is the formula for the area of a square? Use the formula to write an expression, in the form of a product, for the area of the blue square.

125. Each green rectangle has an area of ______. Therefore, the total area in green is represented by the polynomial ______.

126. The yellow square has an area of ______.

127. The area of the entire colored region is represented by ______, because each side of the entire colored region has length ______.

128. The area of the blue square is equal to the area of the entire colored region, minus the total area of the green squares, minus the area of the yellow square. Write this as a simplified polynomial in a and b.

129. **(a)** What must be true about the expressions for the area of the blue square found in **Exercises 124 and 128?**

(b) Write an equation based on the answer in part (a).

130. Draw a figure and give a similar proof for $(a + b)^2 = a^2 + 2ab + b^2$.

4.6 Dividing Polynomials

OBJECTIVES

1. Divide a polynomial by a monomial.
2. Divide a polynomial by a polynomial of two or more terms.
3. Divide polynomial functions.

OBJECTIVE 1 **Divide a polynomial by a monomial.**

Recall that a monomial is a single term, such as 3, $5m^2$, or x^2y^2.

Dividing a Polynomial by a Monomial

To divide a polynomial by a monomial, divide each term of the polynomial by the monomial.

$$\frac{a+b}{c} = \frac{a}{c} + \frac{b}{c} \quad \textbf{(where } c \neq 0\textbf{)}$$

Then write each quotient in lowest terms.

EXAMPLE 1 **Dividing Polynomials by Monomials**

Divide.

(a) $\dfrac{15x^2 - 12x + 3}{3}$ — Do *not* divide out the 3's.

$= \dfrac{15x^2}{3} - \dfrac{12x}{3} + \dfrac{3}{3}$ — Divide *each* term by 3.

$= 5x^2 - 4x + 1$ — Write in lowest terms.

CHECK $3(5x^2 - 4x + 1) = 15x^2 - 12x + 3$ ✓

(Divisor: 3; Quotient: $5x^2 - 4x + 1$; Original polynomial (Dividend): $15x^2 - 12x + 3$)

(b) $\dfrac{5m^3 - 9m^2 + 10m}{5m^2}$

$= \dfrac{5m^3}{5m^2} - \dfrac{9m^2}{5m^2} + \dfrac{10m}{5m^2}$ — Divide each term by $5m^2$.

Think: $\frac{10m}{5m^2} = \frac{10}{5}m^{1-2} = 2m^{-1} = \frac{2}{m}$

$= m - \dfrac{9}{5} + \dfrac{2}{m}$ — Simplify each term. Use the quotient rule for exponents.

CHECK $5m^2\left(m - \dfrac{9}{5} + \dfrac{2}{m}\right) = 5m^3 - 9m^2 + 10m$ ✓ Divisor × Quotient = Original polynomial

The result $m - \frac{9}{5} + \frac{2}{m}$ is not a polynomial because the last term has a variable in its denominator. The quotient of two polynomials need not be a polynomial.

(c) $\dfrac{8xy^2 - 9x^2y + 6x^2y^2}{x^2y^2}$

$= \dfrac{8xy^2}{x^2y^2} - \dfrac{9x^2y}{x^2y^2} + \dfrac{6x^2y^2}{x^2y^2}$ — Divide each term by x^2y^2.

$= \dfrac{8}{x} - \dfrac{9}{y} + 6$ — $\frac{a^m}{a^n} = a^{m-n}$

NOW TRY

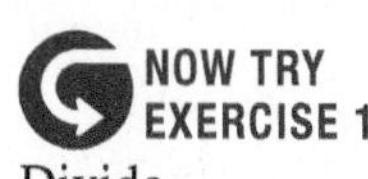

NOW TRY EXERCISE 1

Divide.

(a) $\dfrac{16x^3 - 8x^2 - 4}{4}$

(b) $\dfrac{81y^4 - 54y^3 + 18y}{9y^3}$

NOW TRY ANSWERS

1. **(a)** $4x^3 - 2x^2 - 1$
   (b) $9y - 6 + \dfrac{2}{y^2}$

OBJECTIVE 2 **Divide a polynomial by a polynomial of two or more terms.**

This process is similar to that for dividing whole numbers.

$$\begin{array}{r} 23 \\ 25\overline{)581} \\ -50 \\ \hline 81 \\ -75 \\ \hline 6 \end{array} \qquad 581 \div 25 = 23\frac{6}{25}$$

CHECK $25 \times 23 + 6 = 581$ ✓

(Divisor, Quotient, Remainder, Dividend labels: 25 is the Divisor, 23 the Quotient, 6 the Remainder, 581 the Dividend.)

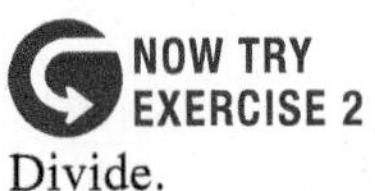

NOW TRY EXERCISE 2

Divide.

$$\frac{3x^2 - 4x - 15}{x - 3}$$

EXAMPLE 2 **Dividing a Polynomial by a Polynomial**

Divide $\dfrac{2x^2 + x - 10}{x - 2}$.

Make sure each polynomial is written in descending powers of the variables.

$$x - 2\overline{)2x^2 + x - 10}$$ ← Write as if dividing whole numbers.

Divide the first term of the dividend $2x^2 + x - 10$ by the first term of the divisor $x - 2$. Here $\frac{2x^2}{x} = 2x$.

$$\begin{array}{r} 2x \qquad\quad \\ x - 2\overline{)2x^2 + x - 10} \end{array}$$ ← Result of $\frac{2x^2}{x}$

Multiply $x - 2$ and $2x$, and write the result below $2x^2 + x - 10$.

$$\begin{array}{r} 2x \qquad\quad \\ x - 2\overline{)2x^2 + x - 10} \\ 2x^2 - 4x \qquad \end{array}$$ ← $2x(x - 2) = 2x^2 - 4x$

Now subtract by mentally changing the signs on $2x^2 - 4x$ and *adding*.

$$\begin{array}{r} 2x \qquad\quad \\ x - 2\overline{)2x^2 + x - 10} \\ \underline{2x^2 - 4x} \qquad \\ 5x \qquad \end{array}$$

Subtract. $2x^2 + x$ $-$ $(2x^2 - 4x)$

← Subtract. The difference is $5x$.

Bring down -10 and continue by dividing $5x$ by x.

$$\begin{array}{r} 2x + 5 \qquad \\ x - 2\overline{)2x^2 + x - 10} \\ \underline{2x^2 - 4x} \qquad \\ 5x - 10 \\ \underline{5x - 10} \\ 0 \end{array}$$

← $\frac{5x}{x} = 5$

← Bring down -10.

← $5(x - 2) = 5x - 10$

← Subtract. The difference is 0.

Think: $5x - 10$ $-$ $(5x - 10)$

CHECK Multiply $x - 2$ (the divisor) and $2x + 5$ (the quotient).

$$\begin{aligned} &(x - 2)(2x + 5) \\ &= 2x^2 + 5x - 4x - 10 && \text{FOIL method} \\ &= 2x^2 + x - 10 && \text{Combine like terms.} \end{aligned}$$

The result is $2x^2 + x - 10$ (the dividend). ✓

NOW TRY

NOW TRY ANSWER

2. $3x + 5$

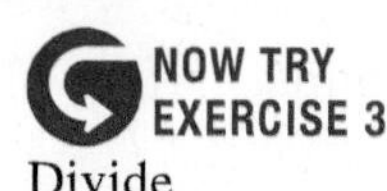

NOW TRY EXERCISE 3

Divide

$2x^3 - 12x - 10$ by $x - 4$.

EXAMPLE 3 Dividing a Polynomial with a Missing Term

Divide $3x^3 - 2x + 5$ by $x - 3$.

Add a term with 0 coefficient as a placeholder for the missing x^2-term.

$$x - 3\overline{)\,3x^3 + 0x^2 - 2x + 5}$$

← Missing term ($0x^2$)

← Descending powers of the variable

Start with $\frac{3x^3}{x} = 3x^2$.

$$\begin{array}{r} 3x^2 \phantom{{}+0x^2-2x+5} \\ x - 3\overline{)\,3x^3 + 0x^2 - 2x + 5} \\ \underline{3x^3 - 9x^2} \phantom{{}-2x+5} \end{array}$$

← $\frac{3x^3}{x} = 3x^2$

← $3x^2(x - 3)$

Subtract by mentally changing the signs on $3x^3 - 9x^2$ and adding.

$$\begin{array}{r} 3x^2 \phantom{{}+0x^2-2x+5} \\ x - 3\overline{)\,3x^3 + 0x^2 - 2x + 5} \\ \underline{3x^3 - 9x^2} \phantom{{}-2x+5} \\ 9x^2 \phantom{{}-2x+5} \end{array}$$

Subtract. $3x^3 + 0x^2$ $-\underline{(3x^3 - 9x^2)}$

← Subtract.

Bring down the next term.

$$\begin{array}{r} 3x^2 \phantom{{}+0x^2-2x+5} \\ x - 3\overline{)\,3x^3 + 0x^2 - 2x + 5} \\ \underline{3x^3 - 9x^2} \phantom{{}-2x+5} \\ 9x^2 - 2x \phantom{{}+5} \end{array}$$

← Bring down $-2x$.

In the next step, $\frac{9x^2}{x} = 9x$.

$$\begin{array}{r} 3x^2 + 9x \phantom{{}+5} \\ x - 3\overline{)\,3x^3 + 0x^2 - 2x + 5} \\ \underline{3x^3 - 9x^2} \phantom{{}-2x+5} \\ 9x^2 - 2x \phantom{{}+5} \\ \underline{9x^2 - 27x} \phantom{{}+5} \\ 25x + 5 \end{array}$$

← $\frac{9x^2}{x} = 9x$

Think: $9x^2 - 2x$ $-\underline{(9x^2 - 27x)}$

← $9x(x - 3)$

← Subtract. Bring down 5.

Finally, $\frac{25x}{x} = 25$.

$$\begin{array}{r} 3x^2 + 9x + 25 \\ x - 3\overline{)\,3x^3 + 0x^2 - 2x + 5} \\ \underline{3x^3 - 9x^2} \phantom{{}-2x+5} \\ 9x^2 - 2x \phantom{{}+5} \\ \underline{9x^2 - 27x} \phantom{{}+5} \\ 25x + 5 \\ \underline{25x - 75} \\ 80 \end{array}$$

← $\frac{25x}{x} = 25$

Think: $25x + 5$ $-\underline{(25x - 75)}$

← $25(x - 3)$

← Remainder

Write the remainder, 80, as the numerator of a fraction, $\frac{80}{x - 3}$.

$$3x^2 + 9x + 25 + \frac{80}{x - 3}$$

Be sure to add $\frac{\text{remainder}}{\text{divisor}}$. Don't forget the + sign.

CHECK Multiply $x - 3$ (the divisor) and $3x^2 + 9x + 25$ (the quotient), and then add 80 (the remainder). The result is $3x^3 - 2x + 5$ (the dividend). ✓

NOW TRY

NOW TRY ANSWER

3. $2x^2 + 8x + 20 + \frac{70}{x - 4}$

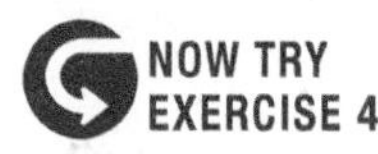

NOW TRY EXERCISE 4

Divide

$$2x^4 + 8x^3 + 2x^2 - 5x - 3$$

by $2x^2 - 2$.

EXAMPLE 4 Dividing by a Polynomial with a Missing Term

Divide $6r^4 + 9r^3 + 2r^2 - 8r + 7$ by $3r^2 - 2$.

$$\begin{array}{r@{}l}
 & 2r^2 + 3r + 2 \\
3r^2 + 0r - 2 \overline{)} & 6r^4 + 9r^3 + 2r^2 - 8r + 7 \\
 & \underline{6r^4 + 0r^3 - 4r^2} \\
 & \quad 9r^3 + 6r^2 - 8r \\
 & \quad \underline{9r^3 + 0r^2 - 6r} \\
 & \qquad 6r^2 - 2r + 7 \\
 & \qquad \underline{6r^2 + 0r - 4} \\
 & \qquad\quad -2r + 11 \leftarrow \text{Remainder}
\end{array}$$

Missing term (the $0r$ in the divisor)

Stop when the degree of the remainder is less than the degree of the divisor.

The degree of the remainder, $-2r + 11$, is less than the degree of the divisor, $3r^2 - 2$, so the division process is finished. The result is written as follows.

$$2r^2 + 3r + 2 + \frac{-2r + 11}{3r^2 - 2} \qquad \text{Quotient} + \frac{\text{remainder}}{\text{divisor}}$$

CHECK $(3r^2 - 2)(2r^2 + 3r + 2) + (-2r + 11)$

Divisor × Quotient + Remainder

$$= 6r^4 + 9r^3 + 6r^2 - 4r^2 - 6r - 4 - 2r + 11$$

$$= 6r^4 + 9r^3 + 2r^2 - 8r + 7 \checkmark \qquad \text{Original polynomial (Dividend)}$$

NOW TRY 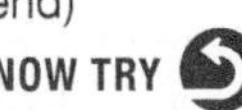

CAUTION When dividing a polynomial by a polynomial of two or more terms:

1. Be sure the terms in both polynomials are in descending powers.
2. Write any missing terms with 0 placeholders.

NOW TRY EXERCISE 5

Divide

$$6m^3 - 8m^2 - 5m - 6$$

by $3m - 6$.

EXAMPLE 5 Finding a Quotient with a Fractional Coefficient

Divide $2p^3 + 5p^2 + p - 2$ by $2p + 2$.

$\frac{3p^2}{2p} = \frac{3}{2}p$

$$\begin{array}{r@{}l}
 & p^2 + \frac{3}{2}p - 1 \\
2p + 2 \overline{)} & 2p^3 + 5p^2 + p - 2 \\
 & \underline{2p^3 + 2p^2} \\
 & \quad 3p^2 + p \\
 & \quad \underline{3p^2 + 3p} \\
 & \qquad -2p - 2 \\
 & \qquad \underline{-2p - 2} \\
 & \qquad\qquad 0
\end{array}$$

The remainder is 0, so the quotient is $p^2 + \frac{3}{2}p - 1$.

NOW TRY

NOW TRY ANSWERS

4. $x^2 + 4x + 2 + \frac{3x + 1}{2x^2 - 2}$
5. $2m^2 + \frac{4}{3}m + 1$

OBJECTIVE 3 **Divide polynomial functions.**

Dividing Functions

If $f(x)$ and $g(x)$ define functions, then

$$\left(\frac{f}{g}\right)(x) = \frac{f(x)}{g(x)}. \quad \text{Quotient function}$$

The domain of the quotient function is the intersection of the domains of $f(x)$ and $g(x)$, excluding any values of x for which $g(x) = 0$.

NOW TRY EXERCISE 6

For $f(x) = 8x^2 + 2x - 3$ and $g(x) = 2x - 1$, find $\left(\frac{f}{g}\right)(x)$ and $\left(\frac{f}{g}\right)(8)$.

EXAMPLE 6 **Dividing Polynomial Functions**

For $f(x) = 2x^2 + x - 10$ and $g(x) = x - 2$, find $\left(\frac{f}{g}\right)(x)$ and $\left(\frac{f}{g}\right)(-3)$.

$$\left(\frac{f}{g}\right)(x) = \frac{f(x)}{g(x)} = \frac{2x^2 + x - 10}{x - 2}$$

This quotient, found in **Example 2,** is $2x + 5$. Thus,

$$\left(\frac{f}{g}\right)(x) = 2x + 5, \quad x \neq 2. \quad \leftarrow \text{2 is not in the domain. It causes denominator } g(x) = x - 2 \text{ to equal 0.}$$

$$\left(\frac{f}{g}\right)(-3) = 2(-3) + 5 = -1 \quad \text{Let } x = -3 \text{ in } \left(\frac{f}{g}\right)(x) = 2x + 5.$$

Alternative method: The same result is found by evaluating $\frac{f(-3)}{g(-3)}$.

$$\left(\frac{f}{g}\right)(-3)$$

$$= \frac{f(-3)}{g(-3)}$$

$$= \frac{2(-3)^2 + (-3) - 10}{-3 - 2} \quad \begin{matrix} \leftarrow f(x) = 2x^2 + x - 10 \\ \leftarrow g(x) = x - 2 \end{matrix}$$

$$= \frac{5}{-5}$$

$$= -1 \quad \text{The result is the same.}$$

NOW TRY 

NOW TRY ANSWER

6. $4x + 3$, $x \neq \frac{1}{2}$; 35

4.6 Exercises

 MyLab Math

Video solutions for select problems available in MyLab Math

Concept Check *Complete each statement.*

1. We find the quotient of two monomials using the ______ rule for ______.
2. When dividing polynomials that are not monomials, first write them in ______ powers.
3. If a polynomial in a division problem has a missing term, insert a term with coefficient equal to ______ as a placeholder.

4. To check a division problem, multiply the _____ by the quotient. Then add the _____.

Divide. ***See Example 1.***

5. $\dfrac{15x^3 - 10x^2 + 5}{5}$

6. $\dfrac{27m^4 - 18m^3 - 9}{9}$

7. $\dfrac{9y^2 + 12y - 15}{3y}$

8. $\dfrac{80r^2 - 40r + 10}{10r}$

9. $\dfrac{15m^3 + 25m^2 + 30m}{5m^3}$

10. $\dfrac{64x^3 - 72x^2 + 12x}{8x^3}$

11. $\dfrac{4m^2n^2 - 21mn^3 + 18mn^2}{14m^2n^3}$

12. $\dfrac{24h^2k + 56hk^2 - 28hk}{16h^2k^2}$

13. $\dfrac{8wxy^2 + 3wx^2y + 12w^2xy}{4wx^2y}$

14. $\dfrac{12ab^2c + 10a^2bc + 18abc^2}{6a^2bc}$

Complete the division. ***See Example 2.***

15. $3r - 1\overline{)3r^3 - 22r^2 + 25r - 6}$ with quotient r^2; $\underline{3r^3 - r^2}$; $-21r^2$

16. $2b - 5\overline{)6b^3 - 7b^2 - 4b - 40}$ with quotient $3b^2$; $\underline{6b^3 - 15b^2}$; $8b^2$

Divide. ***See Examples 2–5.***

17. $\dfrac{y^2 + y - 20}{y + 5}$

18. $\dfrac{y^2 + 3y - 18}{y + 6}$

19. $\dfrac{q^2 + 4q - 32}{q - 4}$

20. $\dfrac{q^2 + 2q - 35}{q - 5}$

21. $\dfrac{3t^2 + 17t + 10}{3t + 2}$

22. $\dfrac{2k^2 - 3k - 20}{2k + 5}$

23. $\dfrac{p^2 + 2p + 20}{p + 6}$

24. $\dfrac{x^2 + 11x + 16}{x + 8}$

25. $\dfrac{3m^3 + 5m^2 - 5m + 1}{3m - 1}$

26. $\dfrac{8z^3 - 6z^2 - 5z + 3}{4z + 3}$

27. $\dfrac{4x^3 + 9x^2 - 10x - 6}{4x + 1}$

28. $\dfrac{10z^3 - 26z^2 + 17z - 13}{5z - 3}$

29. $\dfrac{6x^3 - 19x^2 + 14x - 15}{3x^2 - 2x + 4}$

30. $\dfrac{8m^3 - 18m^2 + 37m - 13}{2m^2 - 3m + 6}$

31. $\dfrac{m^3 - 2m^2 - 9}{m - 3}$

32. $\dfrac{p^3 + 3p^2 - 4}{p + 2}$

33. $(x^3 + 2x - 3) \div (x - 1)$

34. $(x^3 + 5x^2 - 18) \div (x + 3)$

35. $(2x^3 - 11x^2 + 25) \div (x - 5)$

36. $(2x^3 + 3x^2 - 5) \div (x - 1)$

37. $(3x^3 - x + 4) \div (x - 2)$

38. $(3k^3 + 9k - 14) \div (k - 2)$

39. $\dfrac{4k^4 + 6k^3 + 3k - 1}{2k^2 + 1}$

40. $\dfrac{9k^4 + 12k^3 - 4k - 1}{3k^2 - 1}$

41. $(2z^3 - 5z^2 + 6z - 15) \div (2z - 5)$

42. $(3p^3 + p^2 + 18p + 6) \div (3p + 1)$

43. $\dfrac{6y^4 + 4y^3 + 4y - 6}{3y^2 + 2y - 3}$

44. $\dfrac{8t^4 + 6t^3 + 12t - 32}{4t^2 + 3t - 8}$

45. $\dfrac{p^3 - 1}{p - 1}$

46. $\dfrac{8a^3 + 1}{2a + 1}$

47. $(x^4 - 4x^3 + 5x^2 - 3x + 2) \div (x^2 + 3)$

48. $(3t^4 + 5t^3 - 8t^2 - 13t + 2) \div (t^2 - 5)$

49. $(2p^3 + 7p^2 + 9p + 3) \div (2p + 2)$

50. $(3x^3 + 4x^2 + 7x + 4) \div (3x + 3)$

51. $(3a^2 - 11a + 17) \div (2a + 6)$

52. $(5t^2 + 19t + 7) \div (4t + 12)$

Extending Skills *Divide.*

53. $\left(2x^2 - \dfrac{7}{3}x - 1\right) \div (3x + 1)$

54. $\left(m^2 + \dfrac{7}{2}m + 3\right) \div (2m + 3)$

55. $\left(3a^2 - \dfrac{23}{4}a - 5\right) \div (4a + 3)$

56. $\left(3q^2 + \dfrac{19}{5}q - 3\right) \div (5q - 2)$

Solve each problem.

57. Suppose that the volume of a box is

$$(2p^3 + 15p^2 + 28p) \text{ cubic feet.}$$

The height is p feet and the length is $(p + 4)$ feet. Find an expression in p that represents the width.

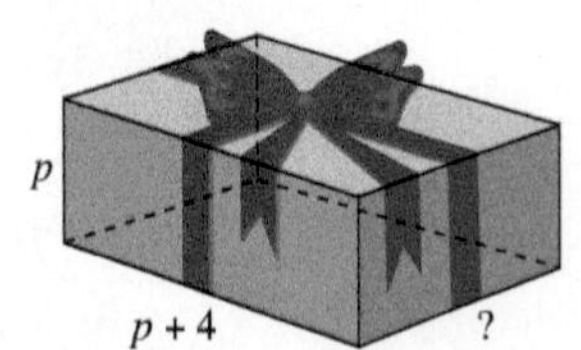

58. Suppose that a minivan travels

$$(2m^3 + 15m^2 + 35m + 36) \text{ miles}$$

in $(2m + 9)$ hours. Find an expression in m that represents the rate of the van in miles per hour (mph).

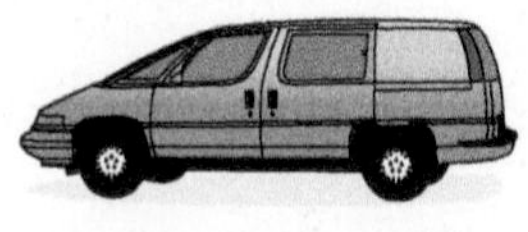

59. **Concept Check** Let $P(x) = 4x^3 - 8x^2 + 13x - 2$ and $D(x) = 2x - 1$. Use division to find polynomials $Q(x)$ and $R(x)$ such that

$$P(x) = Q(x) \cdot D(x) + R(x).$$

60. For $P(x) = x^3 - 4x^2 + 3x - 5$, find $P(-1)$. Then divide $P(x)$ by $D(x) = x + 1$. Compare the remainder with $P(-1)$. What do these results suggest?

For each pair of functions, find $\left(\frac{f}{g}\right)(x)$ *and give any x-values that are not in the domain of the quotient function.* ***See Example 6.***

61. $f(x) = 10x^2 - 2x, \quad g(x) = 2x$

62. $f(x) = 18x^2 - 24x, \quad g(x) = 3x$

63. $f(x) = 2x^2 - x - 3, \quad g(x) = x + 1$

64. $f(x) = 4x^2 - 23x - 35, \quad g(x) = x - 7$

65. $f(x) = 8x^3 - 27, \quad g(x) = 2x - 3$

66. $f(x) = 27x^3 + 64, \quad g(x) = 3x + 4$

Let $f(x) = x^2 - 9$, $g(x) = 2x$, *and* $h(x) = x - 3$. *Find each of the following.* ***See Example 6.***

67. $\left(\dfrac{f}{g}\right)(x)$

68. $\left(\dfrac{f}{h}\right)(x)$

69. $\left(\dfrac{f}{g}\right)(2)$

70. $\left(\dfrac{f}{h}\right)(1)$

71. $\left(\dfrac{h}{g}\right)(x)$

72. $\left(\dfrac{g}{h}\right)(x)$

73. $\left(\dfrac{h}{g}\right)(3)$

74. $\left(\dfrac{g}{h}\right)(-1)$

75. $\left(\dfrac{f}{g}\right)\left(\dfrac{1}{2}\right)$

76. $\left(\dfrac{f}{g}\right)\left(\dfrac{3}{2}\right)$

77. $\left(\dfrac{h}{g}\right)\left(-\dfrac{1}{2}\right)$

78. $\left(\dfrac{h}{g}\right)\left(-\dfrac{3}{2}\right)$

Chapter 4 Summary

STUDY SKILLS **REMINDER**

Be prepared for your math test over this chapter. **Review Study Skill 7, *Reviewing a Chapter,* and Study Skill 8, *Taking Math Tests.***

Key Terms

4.1

exponent (power)
base
exponential expression
polynomial
numerical coefficient (coefficient)
degree of a term
polynomial in x
descending powers
leading term
leading coefficient
trinomial
binomial
monomial
degree of a polynomial
like terms
negative of a polynomial

4.3

term
algebraic expression

4.4

polynomial function
identity function
squaring function
cubing function
composite function (composition of functions)

New Symbols

$(f \circ g)(x) = f(g(x))$ composite function

Test Your Word Power

See how well you have learned the vocabulary in this chapter.

1. A **polynomial** is an algebraic expression made up of
   A. a term or a finite product of terms with positive coefficients and exponents
   B. the sum of two or more terms with whole number coefficients and exponents
   C. the product of two or more terms
   D. a term or a finite sum of terms with real number coefficients and whole number exponents.

2. A **monomial** is a polynomial with
   A. only one term
   B. exactly two terms
   C. exactly three terms
   D. more than three terms.

3. A **binomial** is a polynomial with
   A. only one term
   B. exactly two terms
   C. exactly three terms
   D. more than three terms.

4. A **trinomial** is a polynomial with
   A. only one term
   B. exactly two terms
   C. exactly three terms
   D. more than three terms.

5. The **FOIL** method is used to
   A. add two binomials
   B. add two trinomials
   C. multiply two binomials
   D. multiply two trinomials.

ANSWERS

1. D; *Example:* $5x^3 + 2x^2 - 7$ **2.** A; *Examples:* $-4, 2x^3, 15a^2b$ **3.** B; *Example:* $3t^3 + 5t$ **4.** C; *Example:* $2a^2 - 3ab + b^2$
5. C; *Example:* $(m + 4)(m - 3) = \underset{\text{F}}{m(m)} \underset{\text{O}}{- 3m} \underset{\text{I}}{+ 4m} \underset{\text{L}}{+ 4(-3)} = m^2 + m - 12$

Quick Review

CONCEPTS	EXAMPLES
4.1 Integer Exponents	
Definitions and Rules for Exponents For all integers m and n and all real numbers a and b for which the following are defined, these rules hold true.	Apply the rules for exponents.
Product Rule $a^m \cdot a^n = a^{m+n}$	$3^4 \cdot 3^2 = 3^6$
Quotient Rule $\frac{a^m}{a^n} = a^{m-n}$	$\frac{2^5}{2^3} = 2^2$
Zero Exponent $a^0 = 1$	$27^0 = 1, \quad (-5)^0 = 1$
Negative Exponent $a^{-n} = \frac{1}{a^n}$	$5^{-2} = \frac{1}{5^2}$
Power Rules (a) $(a^m)^n = a^{mn}$ (b) $(ab)^m = a^m b^m$	$(6^3)^4 = 6^{12}, \quad (5p)^4 = 5^4p^4$
(c) $\left(\frac{a}{b}\right)^n = \frac{a^n}{b^n}$	$\left(\frac{2}{3}\right)^5 = \frac{2^5}{3^5}$
Special Rules for Negative Exponents $\frac{1}{a^{-n}} = a^n \qquad \frac{a^{-n}}{b^{-m}} = \frac{b^m}{a^n}$	$\frac{1}{x^{-3}} = x^3 \qquad \frac{r^{-3}}{t^{-6}} = \frac{t^6}{r^3}$
$a^{-n} = \left(\frac{1}{a}\right)^n \qquad \left(\frac{a}{b}\right)^{-n} = \left(\frac{b}{a}\right)^n$	$4^{-3} = \left(\frac{1}{4}\right)^3 \qquad \left(\frac{4}{7}\right)^{-2} = \left(\frac{7}{4}\right)^2$
4.2 Scientific Notation	
Scientific Notation A number is written in scientific notation when it is expressed in the form $a \times 10^n$, where $1 \le \lvert a \rvert < 10$ and n is an integer.	Write 23,500,000,000 in scientific notation. $23{,}500{,}000{,}000 = 2.35 \times 10^{10}$ Write 4.3×10^{-6} in standard notation. $4.3 \times 10^{-6} = 0.0000043$
4.3 Adding and Subtracting Polynomials Add or subtract polynomials by combining like terms.	Add. $\begin{array}{r} 13x^2 - 3x + 1 \\ -4x^2 + 5x - 7 \\ \hline 9x^2 + 2x - 6 \end{array}$ Subtract. $(5x^4 + 3x^2) - (7x^4 + x^2 - x)$ $= 5x^4 + 3x^2 - 7x^4 - x^2 + x$ $= -2x^4 + 2x^2 + x$
4.4 Polynomial Functions, Graphs, and Composition	
Adding and Subtracting Functions If $f(x)$ and $g(x)$ define functions, then $(f+g)(x) = f(x) + g(x)$ and $(f-g)(x) = f(x) - g(x)$.	Let $f(x) = x^2$ and $g(x) = 2x + 1$. Find $(f+g)(x)$ and $(f-g)(x)$. $(f+g)(x)$ $= f(x) + g(x)$ $= x^2 + 2x + 1$; $(f-g)(x)$ $= f(x) - g(x)$ $= x^2 - (2x + 1)$ $= x^2 - 2x - 1$

CONCEPTS	EXAMPLES

Graphs of Basic Polynomial Functions

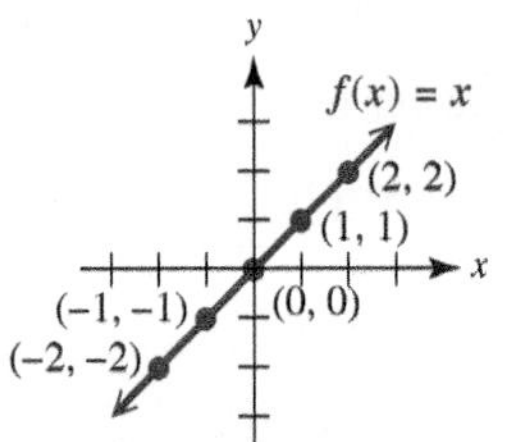

Identity function
$f(x) = x$
Domain: $(-\infty, \infty)$
Range: $(-\infty, \infty)$

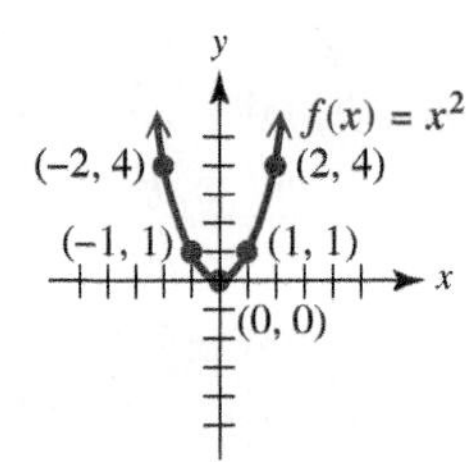

Squaring function
$f(x) = x^2$
Domain: $(-\infty, \infty)$
Range: $[0, \infty)$

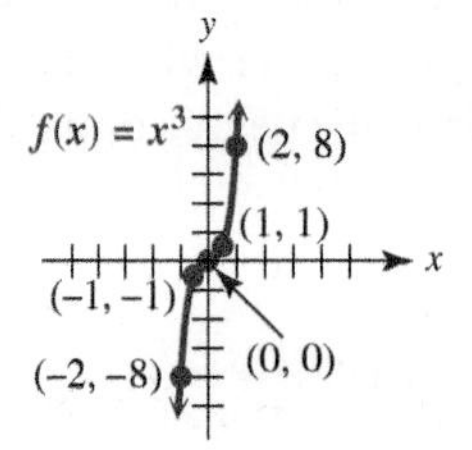

Cubing function
$f(x) = x^3$
Domain: $(-\infty, \infty)$
Range: $(-\infty, \infty)$

Composition of f and g

$$(f \circ g)(x) = f(g(x))$$
$$(g \circ f)(x) = g(f(x))$$

Let $f(x) = x^2$ and $g(x) = 2x + 1$. Find $(f \circ g)(x)$ and $(g \circ f)(x)$.

$$(f \circ g)(x) = f(g(x)) = f(2x + 1) = (2x + 1)^2$$

$$(g \circ f)(x) = g(f(x)) = g(x^2) = 2x^2 + 1$$

4.5 Multiplying Polynomials

To multiply two polynomials, multiply each term of one by each term of the other.

Multiply.

$$(x^3 + 3x)(4x^2 - 5x + 2)$$
$$= x^3(4x^2 - 5x + 2) + 3x(4x^2 - 5x + 2)$$
$$= 4x^5 - 5x^4 + 2x^3 + 12x^3 - 15x^2 + 6x$$
$$= 4x^5 - 5x^4 + 14x^3 - 15x^2 + 6x$$

To multiply two binomials, use the **FOIL method.** Multiply the **First** terms, the **Outer** terms, the **Inner** terms, and the **Last** terms. Then add these products.

$(2x + 3)(x - 7)$

$= 2x(x) + 2x(-7) + 3x + 3(-7)$ FOIL method

$= 2x^2 - 14x + 3x - 21$ Multiply.

$= 2x^2 - 11x - 21$ Combine like terms.

Special Products

$$(x + y)(x - y) = x^2 - y^2$$
$$(x + y)^2 = x^2 + 2xy + y^2$$
$$(x - y)^2 = x^2 - 2xy + y^2$$

$$(3m + 8)(3m - 8) = 9m^2 - 64$$

$$(5a + 3b)^2 = 25a^2 + 30ab + 9b^2$$

$$(2k - 1)^2 = 4k^2 - 4k + 1$$

Multiplying Functions
If $f(x)$ and $g(x)$ define functions, then

$$(fg)(x) = f(x) \cdot g(x).$$

Let $f(x) = x^2$ and $g(x) = 2x + 1$. Find $(fg)(x)$.

$$(fg)(x) = f(x) \cdot g(x) = x^2(2x + 1) = 2x^3 + x^2$$

4.6 Dividing Polynomials

Dividing by a Monomial
To divide a polynomial by a monomial, divide each term in the polynomial by the monomial.

$$\frac{a + b}{c} = \frac{a}{c} + \frac{b}{c} \quad \text{(where } c \neq 0\text{)}$$

Then write each quotient in lowest terms.

Divide. $\dfrac{2x^3 - 4x^2 + 6x - 8}{2x}$

$$= \frac{2x^3}{2x} - \frac{4x^2}{2x} + \frac{6x}{2x} - \frac{8}{2x}$$

$$= x^2 - 2x + 3 - \frac{4}{x}$$

CONCEPTS	EXAMPLES
Dividing by a Polynomial Use the "long division" process. The process ends when the remainder is 0 or when the degree of the remainder is less than the degree of the divisor.	Divide $m^3 - m^2 + 2m + 5$ by $m + 1$. $$\begin{array}{r} m^2 - 2m + 4 \\ m+1\overline{)\, m^3 - m^2 + 2m + 5} \\ \underline{m^3 + m^2 } \\ -2m^2 + 2m \\ \underline{-2m^2 - 2m } \\ 4m + 5 \\ \underline{4m + 4} \\ 1 \end{array}$$ ← Remainder The answer is $m^2 - 2m + 4 + \frac{1}{m+1}$.
Dividing Functions If $f(x)$ and $g(x)$ define functions, then $\left(\frac{f}{g}\right)(x) = \frac{f(x)}{g(x)}$, **(where $g(x) \neq 0$).**	Let $f(x) = x^2$ and $g(x) = 2x + 1$. Find $\left(\frac{f}{g}\right)(x)$. $\left(\frac{f}{g}\right)(x)$ $= \frac{f(x)}{g(x)}$ $= \frac{x^2}{2x+1}, \quad x \neq -\frac{1}{2}$

Chapter 4 Review Exercises

4.1 *Simplify each expression. Assume that all variables represent nonzero real numbers.*

1. 4^3

2. $\left(\frac{1}{3}\right)^4$

3. $(-5)^3$

4. $\frac{2}{(-3)^{-2}}$

5. $\left(\frac{2}{3}\right)^{-4}$

6. $\left(\frac{5}{4}\right)^{-2}$

7. $5^{-1} - 6^{-1}$

8. $2^{-1} + 4^{-1}$

9. $-3^0 + 3^0$

10. $(3^{-4})^2$

11. $(x^{-4})^{-2}$

12. $(xy^{-3})^{-2}$

13. $(z^{-3})^3z^{-6}$

14. $(5m^{-3})^2(m^4)^{-3}$

15. $\frac{12k^{-3}(k^{-2})^{-5}}{3k^{-4}}$

16. $\left(\frac{5z^{-3}}{z^{-1}}\right)\left(\frac{5}{z^2}\right)$

17. $\left(\frac{6m^{-4}}{m^{-9}}\right)^{-1}\left(\frac{m^{-2}}{16}\right)$

18. $\left(\frac{3r^5}{5r^{-3}}\right)^{-2}\left(\frac{9r^{-1}}{2r^{-5}}\right)^3$

19. $(-3x^4y^3)(4x^{-2}y^5)$

20. $\frac{6m^{-4}n^3}{-3mn^2}$

21. $\frac{(5p^{-2}q)(4p^5q^{-3})}{2p^{-5}q^5}$

22. Explain the difference between the expressions $(-6)^0$ and -6^0.

4.2 *Write each number in scientific notation.*

23. 13,450

24. 0.0000000765

25. -0.138

26. The total population of the United States was recently estimated at **321,400,000.** Of this amount, about **72,000** Americans were centenarians, that is, age **100** or older. Write the three boldfaced numbers using scientific notation. (Data from U.S. Census Bureau, Pew Research Center.)

Write each number in standard notation.

27. 1.21×10^6

28. -2.67×10^8

29. 5.8×10^{-3}

Evaluate. Express answers in both scientific notation and standard notation.

30. $\dfrac{16 \times 10^4}{8 \times 10^8}$

31. $\dfrac{6 \times 10^{-2}}{4 \times 10^{-5}}$

32. $\dfrac{0.0000000164}{0.0004}$

33. $\dfrac{0.0009 \times 12{,}000{,}000}{400{,}000}$

34. The planet Mercury has an average distance from the sun of 3.6×10^7 mi, and the average distance from Venus to the sun is 6.7×10^7 mi.

(a) How long would it take a spacecraft traveling at 1.55×10^3 mph to travel from Venus to Mercury? Give the answer in hours, in standard notation.

(b) Use the answer from part (a) to find the number of days it would take the spacecraft to travel from Venus to Mercury, to the nearest whole number.

4.3 *Give the numerical coefficient and the degree of each term.*

35. $14p^5$

36. $-z^2$

37. $\dfrac{x}{10}$

38. $504p^3r^5$

*For each polynomial, **(a)** write in descending powers, **(b)** identify as a* monomial, binomial, trinomial, *or* none of these, *and **(c)** give the degree.*

39. $9k + 11k^3 - 3k^2$

40. $14m^6 + 9m^7$

41. $-5y^4 + 3y^3 + 7y^2 - 2y$

42. $-7q^5r^3$

Add or subtract as indicated.

43. Add.

$$\begin{array}{r} 3x^2 - 5x + 6 \\ \underline{-4x^2 + 2x - 5} \end{array}$$

44. Subtract.

$$\begin{array}{rr} -5y^3 & + 8y - 3 \\ \hline & 4y^2 + 2y + 9 \end{array}$$

45. $(4a^3 - 9a + 15) - (-2a^3 + 4a^2 + 7a)$

46. $(3y^2 + 2y - 1) + (5y^2 - 11y + 6)$

47. Find the perimeter of the triangle.

48. Give an example of a polynomial in the variable x that is of degree 5, is written in descending powers, and lacks a third-degree term.

4.4 *Work each problem.*

49. For the polynomial function $f(x) = -2x^2 + 5x + 7$, find each value.

(a) $f(-2)$ **(b)** $f(3)$ **(c)** $f(0)$

50. The number of twin births in the United States during the years 2005 through 2014 can be modeled by the polynomial function

$$P(x) = 140.33x^3 - 4034.71x^2 + 36{,}125x + 35{,}677,$$

where $x = 5$ represents 2005, $x = 6$ represents 2006, and so on. Use this function to approximate the number of twin births (to the nearest whole number as needed) in each given year. (Data from National Center for Health Statistics.)

(a) 2005 **(b)** 2010 **(c)** 2014

51. Let $f(x) = 2x + 3$ and $g(x) = 5x^2 - 3x + 2$. Find each of the following.

(a) $(f + g)(x)$ **(b)** $(f - g)(x)$ **(c)** $(f + g)(-1)$ **(d)** $(f - g)(-1)$

Graph each polynomial function. Give the domain and range.

52. $f(x) = -2x + 5$ **53.** $f(x) = x^2 - 6$ **54.** $f(x) = -x^3 + 1$

55. Let $f(x) = 3x^2 + 2x - 1$ and $g(x) = 5x + 7$. Find each of the following.

(a) $(g \circ f)(3)$ **(b)** $(f \circ g)(3)$ **(c)** $(f \circ g)(-2)$

(d) $(g \circ f)(-2)$ **(e)** $(f \circ g)(x)$ **(f)** $(g \circ f)(x)$

4.5 *Find each product.*

56. $-6k(2k^2 + 7)$ **57.** $2x(x + 4)(x + 7)$ **58.** $(3m - 2)(5m + 1)$

59. $(3w - 2t)(2w - 3t)$ **60.** $(2p^2 + 6p)(5p^2 - 4)$ **61.** $(3q^2 + 2q - 4)(q - 5)$

62. $(6r^2 - 1)(6r^2 + 1)$ **63.** $(4m + 3)^2$ **64.** $(3t - 2s)^2$

4.6 *Divide.*

65. $\dfrac{4y^3 - 12y^2 + 5y}{4y}$ **66.** $\dfrac{x^3 - 9x^2 + 26x - 30}{x - 5}$

67. $\dfrac{2p^3 + 9p^2 + 27}{2p - 3}$ **68.** $\dfrac{5p^4 + 15p^3 - 33p^2 - 9p + 18}{5p^2 - 3}$

4.5, 4.6 *Let $f(x) = 12x^2 - 3x$ and $g(x) = 3x$. Find each of the following.*

69. **(a)** $(fg)(x)$ **(b)** $(fg)(-1)$ **70.** **(a)** $\left(\dfrac{f}{g}\right)(x)$ **(b)** $\left(\dfrac{f}{g}\right)(2)$

Chapter 4 Mixed Review Exercises

Solve each problem.

1. Match each expression (a)–(j) in Column I with its equivalent expression A–J in Column II. Choices may be used once, more than once, or not at all.

I		II	
(a) 4^{-2}	**(b)** -4^2	**A.** $\frac{1}{16}$	**B.** 0
(c) 4^0	**(d)** $(-4)^0$	**C.** 1	**D.** $-\frac{1}{16}$
(e) $(-4)^{-2}$	**(f)** -4^0	**E.** -1	**F.** $\frac{5}{16}$
(g) $-4^0 + 4^0$	**(h)** $-4^0 - 4^0$	**G.** -16	**H.** -2
(i) $4^{-2} + 4^{-1}$	**(j)** 4^2	**I.** 16	**J.** none of these

2. In 2017, the estimated population of New Zealand was 4.510×10^6. The population density was 44.16 people per square mile. What is the area of New Zealand to the nearest square mile? (Data from *The World Almanac and Book of Facts.*)

Perform the indicated operations and then simplify. Assume that all variables represent nonzero real numbers.

3. $\dfrac{6^{-1}y^3(y^2)^{-2}}{6y^{-4}(y^{-1})}$

4. 5^{-3}

5. $-(-3)^2$

6. $\dfrac{(-z^{-2})^3}{5(z^{-3})^{-1}}$

7. $7p^5(3p^4 + p^3 + 2p^2)$

8. $(2x - 9)^2$

9. $(y^6)^{-5}(2y^{-3})^{-4}$

10. $\left(\dfrac{3w^{-2}z^4}{-6wz^{-5}}\right)^{-2}$

11. $(4x + 1)(2x - 3)$

12. $\dfrac{8x^2 - 23x + 2}{x - 3}$

13. $[(3m - 5n) + p][(3m - 5n) - p]$

14. $\dfrac{20y^3x^3 + 15y^4x + 25yx^4}{10yx^2}$

15. $(2k - 1) - (3k^2 - 2k + 6)$

Chapter 4 Test

Step-by-step test solutions are found on the Chapter Test Prep Videos available in MyLab Math.

View the complete solutions to all Chapter Test exercises in MyLab Math.

1. Match each expression (a)–(j) in Column I with its equivalent expression A–J in Column II. Choices may be used once, more than once, or not at all.

I		II	
(a) 7^{-2}	**(b)** 7^0	**A.** 1	**B.** $\frac{1}{9}$
(c) -7^0	**(d)** $(-7)^0$	**C.** $\frac{1}{49}$	**D.** -1
(e) -7^2	**(f)** $7^{-1} + 2^{-1}$	**E.** -49	**F.** $\frac{9}{14}$
(g) $(7 + 2)^{-1}$	**(h)** $\frac{7^{-1}}{2^{-1}}$	**G.** $\frac{2}{7}$	**H.** 0
(i) 7^2	**(j)** $(-7)^{-2}$	**I.** 49	**J.** none of these

Simplify each expression. Assume that all variables represent nonzero real numbers.

2. $(3x^{-2}y^3)^{-2}(4x^3y^{-4})$

3. $\dfrac{36r^{-4}(r^2)^{-3}}{6r^4}$

4. $\left(\dfrac{4p^2}{q^4}\right)^3\left(\dfrac{6p^8}{q^{-8}}\right)^{-2}$

5. $(-2x^4y^{-3})^0(-4x^{-3}y^{-8})^2$

6. Write 9.1×10^{-7} in standard notation.

7. Evaluate. Express the answer in both scientific notation and standard notation.

$$\frac{2{,}500{,}000 \times 0.00003}{0.05 \times 5{,}000{,}000}$$

8. Let $f(x) = -2x^2 + 5x - 6$ and $g(x) = 7x - 3$. Find each of the following.

(a) $f(4)$ **(b)** $(f + g)(x)$ **(c)** $(f - g)(x)$ **(d)** $(f - g)(-2)$

Graph each polynomial function. Give the domain and range.

9. $f(x) = -2x^2 + 3$

10. $f(x) = -x^3 + 3$

11. Let $f(x) = 3x + 5$ and $g(x) = x^2 + 2$. Find each of the following.

(a) $(f \circ g)(-2)$ **(b)** $(f \circ g)(x)$ **(c)** $(g \circ f)(x)$

Perform the indicated operations.

12. $(5x - 3)(2x + 1)$

13. $(2m - 5)(3m^2 + 4m - 5)$

14. $(6x + y)(6x - y)$

15. $(3k + q)^2$

16. $[2y + (3z - x)][2y - (3z - x)]$

17. $\dfrac{16p^3 - 32p^2 + 24p}{4p^2}$

18. $(x^3 + 3x^2 - 4) \div (x - 1)$

19. $(4x^3 - 3x^2 + 2x - 5) - (3x^3 + 11x + 8) + (x^2 - x)$

20. Let $f(x) = x^2 + 3x + 2$ and $g(x) = x + 1$. Find each of the following.

(a) $(fg)(x)$ **(b)** $(fg)(-2)$ **(c)** $\left(\dfrac{f}{g}\right)(x)$ **(d)** $\left(\dfrac{f}{g}\right)(-2)$

Chapters R–4 Cumulative Review Exercises

1. Match each number in Column I with the equivalent number(s) in Column II. Choices may be used more than once.

I		II		
(a) $\dfrac{1}{100}$	**(b)** $\dfrac{1}{10}$	**A.** 0.5	**B.** 0.25	**C.** 0.1
(c) 1	**(d)** $\dfrac{1}{2}$	**D.** 0.01	**E.** 125%	**F.** 50%
(e) $\dfrac{1}{4}$	**(f)** $\dfrac{5}{4}$	**G.** 100%	**H.** 10%	**I.** 1%

2. Match each number in Column I with the choice or choices of sets of numbers in Column II to which the number belongs.

I		II	
(a) 34	**(b)** 0	**A.** Natural numbers	**B.** Whole numbers
(c) 2.16	**(d)** $-\sqrt{36}$	**C.** Integers	**D.** Rational numbers
(e) $\sqrt{13}$	**(f)** $-\frac{4}{5}$	**E.** Irrational numbers	**F.** Real numbers

Evaluate.

3. $9 \cdot 4 - 16 \div 4$

4. $\left(\frac{1}{3}\right)^2 - \left(\frac{1}{2}\right)^3$

5. $-|8 - 13| - |-4| + |-9|$

6. Simplify $0.5(8x - 10) - 0.4(5 + 10x)$.

Solve.

7. $-5(8 - 2z) + 4(7 - z) = 7(8 + z) - 3$

8. $3(x + 2) - 5(x + 2) = -2x - 4$

9. $A = p + prt$ for t

10. $2(m + 5) - 3m + 1 > 5$

11. $|3x - 1| = 2$

12. $|3z + 1| \geq 7$

13. A survey polled Internet users age 12 and older about the Internet activities they engaged in daily. Complete the results shown in the table if 5000 such users were surveyed.

Daily Internet Activity	*Percent*	*Actual Number*
Browse the web	70%	
Download/listen to music	39%	
Download/watch videos		1800
Play games		1550

Data from the 2017 Center for the Digital Future report.

14. Find the measure of each angle of the triangle.

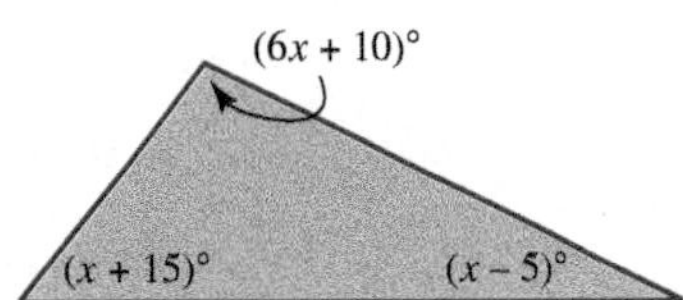

Find the slope of each line described.

15. Through $(-4, 5)$ and $(2, -3)$

16. Through $(4, 5)$; horizontal

Write an equation of each line that satisfies the given conditions. Give the equation **(a)** *in slope-intercept form and* **(b)** *in standard form.*

17. Through $(4, -1)$; $m = -4$

18. Through $(0, 0)$ and $(1, 4)$

Graph each equation or inequality.

19. $-3x + 4y = 12$

20. $y \leq 2x - 6$

21. $3x + 2y < 0$

22. Per capita consumption of whole milk in the United States (in gallons) is shown in the graph, where $x = 0$ represents 1970.

(a) Use the given ordered pairs to find the average rate of change in per capita consumption of whole milk (in gallons) per year during this period. Interpret the answer.

(b) Use the answer from part (a) to write an equation of the line in slope-intercept form that models per capita consumption of whole milk y (in gallons).

(c) Use the equation from part (b) to approximate per capita consumption of whole milk in 2000.

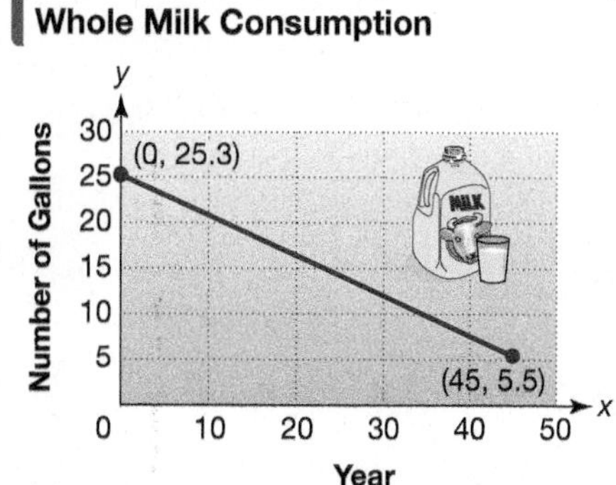

Data from U.S. Department of Agriculture.

23. Give the domain and range of the relation

$$\{(-4, -2), (-1, 0), (2, 0), (5, 2)\}.$$

Does this relation define a function?

24. If $g(x) = -x^2 - 2x + 6$, find $g(3)$.

Solve each system.

25. $3x - 4y = 1$
$2x + 3y = 12$

26. $3x - 2y = 4$
$-6x + 4y = 7$

27. $x + 3y - 6z = 7$
$2x - y + z = 1$
$x + 2y + 2z = -1$

Solve each problem.

28. The Star-Spangled Banner that flew over Fort McHenry during the War of 1812 had a perimeter of 144 ft. Its length measured 12 ft more than its width. Find the dimensions of this flag, which is displayed in the Smithsonian Institution's Museum of American History in Washington, DC. (Data from National Park Service brochure.)

29. Agbe needs 9 L of a 20% solution of alcohol. Agbe has a 15% solution on hand, as well as a 30% solution. How many liters of the 15% solution and how many liters of the 30% solution should Agbe mix to obtain the 20% solution needed?

30. Simplify $\dfrac{x^{-6}y^3 z^{-1}}{x^7y^{-4}z}$ $(x, y, z \neq 0)$. Write the answer with only positive exponents.

Perform the indicated operations.

31. $(3x^2 - 8x + 1) - (x^2 - 3x - 9)$

32. $(x + 2y)(x^2 - 2xy + 4y^2)$

33. $(3x + 2y)(5x - y)$

34. $\dfrac{16x^3y^5 - 8x^2y^2 + 4}{4x^2y}$

35. $\dfrac{m^3 - 3m^2 + 5m - 3}{m - 1}$

STUDY SKILLS **REMINDER**

It is not too soon to begin preparing for your final exam. **Review Study Skill 10, *Preparing for Your Math Final Exam.***

5

FACTORING

Determining the dimensions for a garden of fixed area is accomplished by solving a *quadratic equation,* one of the topics covered in this chapter on *factoring*.

5.1 Greatest Common Factors and Factoring by Grouping

OBJECTIVES

1 Factor out the greatest common factor.

2 Factor by grouping.

Writing a polynomial as the product of two or more simpler polynomials is called **factoring** the polynomial. For example, consider the following.

$$3x(5x - 2) = 15x^2 - 6x \quad \text{Multiplying}$$

$$15x^2 - 6x = 3x(5x - 2) \quad \text{Factoring}$$

Notice that both multiplying and factoring use the distributive property, but in opposite directions. ***Factoring "undoes," or reverses, multiplying.***

VOCABULARY

□ factoring

□ greatest common factor (GCF)

OBJECTIVE 1 Factor out the greatest common factor.

The first step in factoring a polynomial is to find the *greatest common factor* for the terms of the polynomial. The product of the greatest common numerical factor and each variable factor of least degree common to every term in a polynomial is the **greatest common factor (GCF)** of the terms of the polynomial.

For example, the greatest common factor for $8x + 12$ is 4 because 4 is the greatest factor that *divides into* both $8x$ and 12.

$$8x + 12$$

$$= 4(2x) + 4(3) \quad \text{Factor 4 from each term.}$$

$$= 4(2x + 3) \quad \text{Distributive property, } ab + ac = a(b + c)$$

CHECK Multiply $4(2x + 3)$ to obtain $8x + 12$. ✓ Original polynomial

Using the distributive property in this way is called **factoring out the greatest common factor.**

NOW TRY EXERCISE 1

Factor out the greatest common factor.

(a) $54m - 45$

(b) $2k - 7$

(c) $18 + 54r$

EXAMPLE 1 Factoring Out the Greatest Common Factor

Factor out the greatest common factor.

(a) $9z - 18$

$$= 9 \cdot z - 9 \cdot 2 \quad \text{GCF = 9; Factor 9 from each term.}$$

$$= 9(z - 2) \quad \text{Distributive property}$$

CHECK Multiply $9(z - 2)$ to obtain $9z - 18$. ✓ Original polynomial

(b) $56m + 35p$

$$= 7(8m + 5p) \quad \text{GCF = 7}$$

(c) $2y + 5$

There is no common factor other than 1.

(d) $12 + 24z$

$$= 12 \cdot 1 + 12 \cdot 2z \quad \text{Identity property; 12 is the GCF.}$$

Remember to write the 1.

$$= 12(1 + 2z) \quad \text{Distributive property}$$

CHECK $12(1 + 2z)$

$$= 12(1) + 12(2z) \quad \text{Distributive property}$$

$$= 12 + 24z \; ✓ \quad \text{Original polynomial}$$

NOW TRY

NOW TRY ANSWERS

1. **(a)** $9(6m - 5)$
(b) There is no common factor other than 1.
(c) $18(1 + 3r)$

NOW TRY EXERCISE 2

Factor out the greatest common factor.

(a) $16y^4 + 8y^3$

(b) $14p^2 - 9p^3 + 6p^4$

(c) $9x^3y^2 - 3x^2y^2 + 6x^2y^3$

EXAMPLE 2 Factoring Out the Greatest Common Factor

Factor out the greatest common factor.

(a) $9x^2 + 12x^3$

The numerical part of the GCF is 3, the greatest integer that divides into both 9 and 12. The least exponent that appears on x is 2. The GCF is $3x^2$.

$$9x^2 + 12x^3$$
$$= 3x^2(3) + 3x^2(4x) \quad \text{GCF} = 3x^2$$
$$= 3x^2(3 + 4x) \quad \text{Distributive property}$$

(b) $32p^4 - 24p^3 + 40p^5$

$$= 8p^3(4p) + 8p^3(-3) + 8p^3(5p^2) \quad \text{GCF} = 8p^3$$
$$= 8p^3(4p - 3 + 5p^2) \quad \text{Distributive property}$$

(c) $3k^4 - 15k^7 + 24k^9$

Remember the 1.

$$= 3k^4(1 - 5k^3 + 8k^5) \quad \text{GCF} = 3k^4$$

(d) $24m^3n^2 - 18m^2n + 6m^4n^3$

$$= 6m^2n(4mn) + 6m^2n(-3) + 6m^2n(m^2n^2) \quad \text{GCF} = 6m^2n$$
$$= 6m^2n(4mn - 3 + m^2n^2) \quad \text{Distributive property}$$

(e) $25x^2y^3 + 30y^5 - 15x^4y^7$

$$= 5y^3(5x^2 + 6y^2 - 3x^4y^4)$$

In each case, check the factored form by multiplying. NOW TRY

NOW TRY EXERCISE 3

Factor out the greatest common factor.

(a) $(3x - 2)(x + 1) + (3x - 2)(2x - 5)$

(b) $z(a - b)^3 - 2w(a - b)^2$

EXAMPLE 3 Factoring Out a Binomial Factor

Factor out the greatest common factor.

(a) $(x - 5)(x + 6) + (x - 5)(2x + 5)$ The greatest common factor is $x - 5$.

$$= (x - 5)[(x + 6) + (2x + 5)] \quad \text{Factor out } x - 5.$$
$$= (x - 5)(3x + 11) \quad \text{Combine like terms inside the brackets.}$$

(b) $z^2(m + n) + x^2(m + n)$

$$= (m + n)(z^2 + x^2) \quad \text{Factor out the common factor } (m + n).$$

(c) $p(r + 2s)^2 - q(r + 2s)^3$ $\quad (r + 2s)^3 = (r + 2s)^2(r + 2s)$

$$= (r + 2s)^2[p - q(r + 2s)] \quad \text{Factor out the common factor } (r + 2s)^2.$$
$$= (r + 2s)^2(p - qr - 2qs)$$

Be careful with signs.

(d) $(p - 5)(p + 2) - (p - 5)(3p + 4)$

$$= (p - 5)[(p + 2) - (3p + 4)] \quad \text{Factor out the common factor } p - 5.$$

Don't stop here.

$$= (p - 5)[p + 2 - 3p - 4] \quad \text{Distributive property}$$
$$= (p - 5)[-2p - 2] \quad \text{Combine like terms.}$$
$$= (p - 5)[-2(p + 1)] \quad \text{Look for a common factor.}$$
$$= -2(p - 5)(p + 1) \quad \text{Commutative property}$$

NOW TRY 

NOW TRY ANSWERS

2. **(a)** $8y^3(2y + 1)$
   (b) $p^2(14 - 9p + 6p^2)$
   (c) $3x^2y^2(3x - 1 + 2y)$
3. **(a)** $(3x - 2)(3x - 4)$
   (b) $(a - b)^2(za - zb - 2w)$

NOW TRY EXERCISE 4

Factor $-4y^5 - 3y^3 + 8y$ in two ways.

EXAMPLE 4 Factoring Out a Negative Common Factor

Factor $-a^3 + 3a^2 - 5a$ in two ways.

First, a could be used as the common factor.

$$
\begin{aligned}
&-a^3 + 3a^2 - 5a \\
&= a(-a^2) + a(3a) + a(-5) && \text{Factor out } a. \\
&= a(-a^2 + 3a - 5) && \text{Distributive property}
\end{aligned}
$$

CHECK Multiply $a(-a^2 + 3a - 5)$ to obtain $-a^3 + 3a^2 - 5a$. ✓

Because of the leading negative sign, $-a$ could also be used as the common factor.

$$
\begin{aligned}
&-a^3 + 3a^2 - 5a \\
&= -a(a^2) + (-a)(-3a) + (-a)(5) && \text{Factor out } -a. \\
&= -a(a^2 - 3a + 5) && \text{Distributive property}
\end{aligned}
$$

CHECK Multiply $-a(a^2 - 3a + 5)$ to obtain $-a^3 + 3a^2 - 5a$. ✓

Sometimes there may be a reason to prefer one of these forms, but either is correct.

NOW TRY

NOTE In cases where the leading coefficient is negative (as in **Example 4**), the answer section will usually give the factored form that has the common factor with a negative coefficient.

OBJECTIVE 2 Factor by grouping.

Sometimes the *individual terms* of a polynomial have greatest common factor of 1, but it still may be possible to factor the polynomial using a process called **factoring by grouping.**

We usually factor by grouping when a polynomial has more than three terms.

NOW TRY EXERCISE 5

Factor.

$3m - 3n + xm - xn$

EXAMPLE 5 Factoring by Grouping

Factor $ax - ay + bx - by$.

Group the terms in pairs so that each pair has a common factor.

$$ax - ay + bx - by$$

Terms with common factor a — Terms with common factor b

$$
\begin{aligned}
&= (ax - ay) + (bx - by) \\
&= a(x - y) + b(x - y) && \text{Factor each group.} \\
&= (x - y)(a + b) && \text{The common factor is } x - y.
\end{aligned}
$$

By the commutative property, $(a + b)(x - y)$ is also correct.

CHECK $(x - y)(a + b)$

$$
\begin{aligned}
&= xa + xb - ya - yb && \text{Multiply using the FOIL method.} \\
&= ax + bx - ay - by && \text{Commutative property} \\
&= ax - ay + bx - by \ ✓ && \text{Original polynomial}
\end{aligned}
$$

NOW TRY

NOW TRY ANSWERS

4. $y(-4y^4 - 3y^2 + 8)$; $-y(4y^4 + 3y^2 - 8)$

5. $(m - n)(3 + x)$

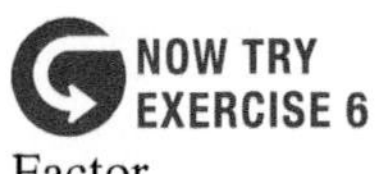

NOW TRY EXERCISE 6

Factor.

$ab - 7a - 5b + 35$

EXAMPLE 6 Factoring by Grouping

Factor $3x - 3y - ax + ay$.

$$3x - 3y - ax + ay \quad \text{Pay close attention here.}$$
$$= (3x - 3y) + (-ax + ay) \quad (*) \text{ Group the terms.}$$
$$= 3(x - y) + a(-x + y) \quad \text{Factor out 3, and factor out } a.$$

The factors $(x - y)$ and $(-x + y)$ are opposites. If we factor out $-a$ instead of a in the second group, we obtain the common binomial factor $(x - y)$. So we start over.

$$(3x - 3y) + (-ax + ay) \quad \text{Equation } (*) \text{ from above}$$
$$= 3(x - y) - a(x - y) \quad \text{Be careful with signs.}$$
$$= (x - y)(3 - a) \quad \text{Factor out } x - y.$$

CHECK $(x - y)(3 - a)$

$$= 3x - ax - 3y + ay \quad \text{Multiply using the FOIL method.}$$
$$= 3x - 3y - ax + ay \checkmark \quad \text{Original polynomial}$$

NOW TRY

NOTE In **Example 6,** a different grouping would lead to the factored form $(a - 3)(y - x)$. Verify by multiplying that this is also correct.

Factoring by Grouping

Step 1 **Group terms.** Collect the terms into groups so that each group has a common factor.

Step 2 **Factor within the groups.** Factor out the common factor in each group.

Step 3 **Factor the entire polynomial.** If each group now has a common factor, factor it out. If not, try a different grouping.

Always check the factored form by multiplying.

NOW TRY EXERCISE 7

Factor.

$3ax - 6xy - a + 2y$

EXAMPLE 7 Factoring by Grouping

Factor $6ax + 12bx + a + 2b$.

$$6ax + 12bx + a + 2b$$
$$= (6ax + 12bx) + (a + 2b) \quad \text{Group the terms.}$$

Now factor $6x$ from the first group, and use the identity property of multiplication to introduce the factor 1 in the second group.

Remember to write the 1.

$$= 6x(a + 2b) + 1(a + 2b) \quad \text{Factor each group.}$$
$$= (a + 2b)(6x + 1) \quad \text{Factor out } a + 2b.$$

CHECK $(a + 2b)(6x + 1)$

$$= 6ax + a + 12bx + 2b \quad \text{Multiply using the FOIL method.}$$
$$= 6ax + 12bx + a + 2b \checkmark \quad \text{Original polynomial}$$

NOW TRY

NOW TRY ANSWERS

6. $(b - 7)(a - 5)$
7. $(a - 2y)(3x - 1)$

NOW TRY EXERCISE 8
Factor.
$$2kp^2 + 6 - 3p^2 - 4k$$

EXAMPLE 8 **Rearranging Terms before Factoring by Grouping**

Factor $p^2q^2 - 10 - 2q^2 + 5p^2$.

Neither the first two terms nor the last two terms have a common factor except 1. We can rearrange and group the terms as follows.

$p^2q^2 - 10 - 2q^2 + 5p^2$

$= (p^2q^2 - 2q^2) + (5p^2 - 10)$ Rearrange and group the terms.

$= q^2(p^2 - 2) + 5(p^2 - 2)$ Factor out the common factors.

Don't stop here.

$= (p^2 - 2)(q^2 + 5)$ Factor out $p^2 - 2$. Use parentheses.

CHECK $(p^2 - 2)(q^2 + 5)$

$= p^2q^2 + 5p^2 - 2q^2 - 10$ Multiply using the FOIL method.

$= p^2q^2 - 10 - 2q^2 + 5p^2$ ✓ Original polynomial **NOW TRY**

CAUTION In **Example 8,** do not stop at the step

$$q^2(p^2 - 2) + 5(p^2 - 2).$$

This expression is *not in factored form* because it is a *sum* of two terms, $q^2(p^2 - 2)$ and $5(p^2 - 2)$, not a *product*.

NOW TRY EXERCISE 9
Factor.
$$12wy - 24xy + 4wz - 8xz$$

EXAMPLE 9 **Factoring Out a GCF before Factoring by Grouping**

Factor $10ax - 5ay + 10bx - 5by$.

Always start by factoring out any greatest common factor from the terms.

$10ax - 5ay + 10bx - 5by$

$= 5(2ax - ay + 2bx - by)$ Factor out the GCF, 5.

$= 5[(2ax - ay) + (2bx - by)]$ Group the terms inside the brackets.

$= 5[a(2x - y) + b(2x - y)]$ Factor out the common factors.

$= 5[(2x - y)(a + b)]$ Factor out $2x - y$.

$= 5(2x - y)(a + b)$ Write without the brackets.

CHECK $5(2x - y)(a + b)$

$= 5(2ax + 2bx - ay - by)$ Use the FOIL method.

$= 10ax + 10bx - 5ay - 5by$ Distributive property

$= 10ax - 5ay + 10bx - 5by$ ✓ Original polynomial **NOW TRY**

NOTE Verify that the result in **Example 9** can also be obtained by grouping the terms of the polynomial

$$2ax - ay + 2bx - by \quad \text{as} \quad (2ax + 2bx) + (-ay - by).$$

NOW TRY ANSWERS

8. $(p^2 - 2)(2k - 3)$

9. $4(w - 2x)(3y + z)$

5.1 Exercises

  MyLab Math

Video solutions for select problems available in MyLab Math

Concept Check *Choose the letter of the correct response.*

1. Which choice is an example of a polynomial in factored form?

A. $3x^2y^3 + 6x^2(2x + y)$ **B.** $5(x + y)^2 - 10(x + y)^3$

C. $(-2 + 3x)(5y^2 + 4y + 3)$ **D.** $(3x + 4)(5x - y) - (3x + 4)(2x - 1)$

2. Which choice is the correct completely factored form of $8 + 24r$?

A. $8(3r)$ **B.** $-8(-3r)$ **C.** $8(1 + 3r)$ **D.** $4(2 + 6r)$

3. Concept Check A student factored the following polynomial as shown.

$$\begin{aligned} &2tw^2 - 16t + 5w^2 - 40 \\ &= (2tw^2 - 16t) + (5w^2 - 40) \\ &= 2t(w^2 - 8) + 5(w^2 - 8) \end{aligned}$$

The teacher did not give the student full credit. **WHAT WENT WRONG?** Give the correct answer.

4. Concept Check A student factored the following polynomial as shown.

$$\begin{aligned} &4x^2y^5 - 8xy^3 \\ &= 2xy^3(2xy^2 - 4) \end{aligned}$$

The teacher did not give the student full credit. **WHAT WENT WRONG?** Give the correct answer.

Concept Check *Find the greatest common factor for each list of terms.*

5. $9m^3,\ 3m^2,\ 15m$

6. $4a^2,\ 6a,\ 2a^3$

7. $16xy^3,\ 24x^2y^2,\ 8x^2y$

8. $10m^2n^2,\ 25mn^3,\ 50m^2n$

9. $6m(r + t)^2,\ 3p(r + t)^4$

10. $7z^2(m + n)^4,\ 9z^3(m + n)^5$

Factor out the greatest common factor. ***See Examples 1–4.***

11. $12m - 60$

12. $15r - 45$

13. $8s + 16t$

14. $35p + 70q$

15. $4 + 20z$

16. $9 + 27x$

17. $8y - 15$

18. $7x - 40$

19. $8k^3 + 24k$

20. $9z^4 + 81z$

21. $-4p^3q^4 - 2p^2q^5$

22. $-3z^5w^2 - 18z^3w^4$

23. $7x^3 + 35x^4 - 14x^5$

24. $6k^3 - 36k^4 - 48k^5$

25. $10t^5 - 2t^3 - 4t^4$

26. $6p^3 - 3p^2 - 9p^4$

27. $15a^2c^3 - 25ac^2 + 5ac$

28. $15y^3z^3 - 27y^2z^4 + 3yz^3$

29. $16z^2n^6 + 64zn^7 - 32z^3n^3$

30. $5r^3s^5 + 10r^2s^2 - 15r^4s^2$

31. $14a^3b^2 + 7a^2b - 21a^5b^3 + 42ab^4$

32. $12km^3 - 24k^3m^2 + 36k^2m^4 - 60k^4m^3$

33. $(m - 4)(m + 2) + (m - 4)(m + 3)$

34. $(z - 5)(z + 7) + (z - 5)(z + 10)$

35. $(2z - 1)(z + 6) - (2z - 1)(z - 5)$

36. $(3x + 2)(x - 4) - (3x + 2)(x + 8)$

37. $5(2 - x)^2 - 2(2 - x)^3$

38. $2(5 - x)^3 - 3(5 - x)^2$

39. $4(3 - x)^2 - (3 - x)^3 + 3(3 - x)$

40. $2(t - s) + 4(t - s)^2 - (t - s)^3$

41. $15(2z + 1)^3 + 10(2z + 1)^2 - 25(2z + 1)$

42. $6(a + 2b)^2 - 4(a + 2b)^3 + 12(a + 2b)^4$

43. $5(m + p)^3 - 10(m + p)^2 - 15(m + p)^4$

44. $-9a^2(p + q) - 3a^3(p + q)^2 + 6a(p + q)^3$

Factor each polynomial in two ways. First use a common factor with a positive coefficient, and then use a common factor with a negative coefficient. ***See Example 4.***

45. $-r^3 + 3r^2 + 5r$

46. $-t^4 + 8t^3 - 12t$

47. $-12s^5 + 48s^4$

48. $-16y^4 + 64y^3$

49. $-2x^5 + 6x^3 + 4x^2$

50. $-5a^3 + 10a^4 - 15a^5$

Factor by grouping. ***See Examples 5–9.***

51. $mx + qx + my + qy$

52. $2k + 2h + jk + jh$

53. $10m + 2n + 5mk + nk$

54. $3ma + 3mb + 2ab + 2b^2$

55. $4 - 2q - 6p + 3pq$

56. $20 + 5m + 12n + 3mn$

57. $p^2 - 4zq + pq - 4pz$

58. $r^2 - 9tw + 3rw - 3rt$

59. $2xy + 3y + 2x + 3$

60. $7ab + 35bc + a + 5c$

61. $m^3 + 4m^2 - 6m - 24$

62. $2a^3 + a^2 - 14a - 7$

63. $-3a^3 - 3ab^2 + 2a^2b + 2b^3$

64. $-16m^3 + 4m^2p^2 - 4mp + p^3$

65. $4 + xy - 2y - 2x$

66. $10ab - 21 - 6b + 35a$

67. $8 + 9y^4 - 6y^3 - 12y$

68. $x^3y^2 - 3 - 3y^2 + x^3$

69. $1 - a + ab - b$

70. $2ab^2 - 8b^2 + a - 4$

71. $2mx - 6qx + 2my - 6qy$

72. $12 - 6q - 18p + 9pq$

73. $4a^3 - 4a^2b^2 + 8ab - 8b^3$

74. $5x^3 + 15x^2y^2 - 5xy - 15y^3$

75. $2x^3y^2 + x^2y^2 - 14xy^2 - 7y^2$

76. $3m^2n^3 + 15m^2n - 2m^2n^2 - 10m^2$

Extending Skills *Factor out the variable that is raised to the lesser exponent. (For example, in* ***Exercise 77****, factor out* m^{-5}*.)*

77. $3m^{-5} + m^{-3}$

78. $k^{-2} + 2k^{-4}$

79. $3p^{-3} + 2p^{-2}$

80. $-5q^{-3} + 8q^{-2}$

5.2 Factoring Trinomials

OBJECTIVES

1. Factor trinomials when the coefficient of the second-degree term is 1.
2. Factor trinomials by grouping when the coefficient of the second-degree term is not 1.
3. Factor trinomials using the FOIL method when the coefficient of the second-degree term is not 1.
4. Factor using substitution.

VOCABULARY

□ prime polynomial

OBJECTIVE 1 Factor trinomials when the coefficient of the second-degree term is 1.

We begin by finding the product of $x + 3$ and $x - 5$.

$$(x + 3)(x - 5)$$

$$= x^2 - 5x + 3x - 15 \quad \text{FOIL method}$$

$$= x^2 - 2x - 15 \quad \text{Combine like terms.}$$

By this result, the factored form of $x^2 - 2x - 15$ is $(x + 3)(x - 5)$.

$$\text{Factored form} \rightarrow (x + 3)(x - 5) \underset{\text{Factoring}}{\overset{\text{Multiplication}}{=}} x^2 - 2x - 15 \leftarrow \text{Product}$$

Because multiplying and factoring are operations that "undo" each other, factoring trinomials involves using the FOIL method in reverse. As shown here, the x^2-term comes from multiplying x and x, and -15 comes from multiplying 3 and -5.

Product of x and x is x^2.

$$(x+3)(x-5) = x^2 - 2x - 15$$

Product of 3 and −5 is −15.

We find the term $-2x$ in $x^2 - 2x - 15$ by multiplying the outer terms, multiplying the inner terms, and adding the results.

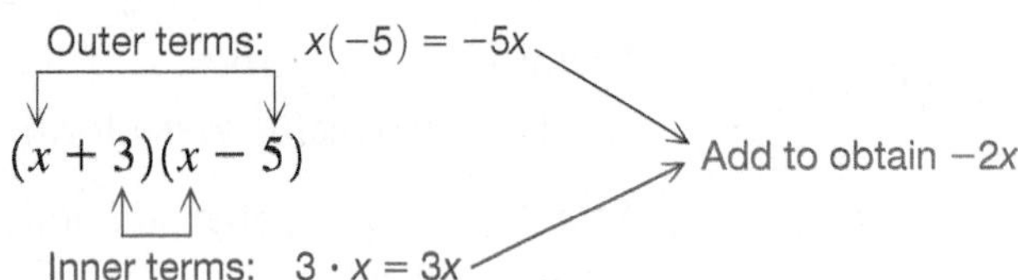

STUDY SKILLS **REMINDER**
Are you getting the most out of your class time?
Review Study Skill 3, *Taking Lecture Notes.*

Factoring $x^2 + bx + c$ (Coefficient of Second-Degree Term = 1)

Step 1 **Find pairs of integers whose product is c.** Write all pairs whose product is c, the third term of the trinomial.

Step 2 **Find the sum of each pair.** Choose the pair whose sum is b, the coefficient of the middle term.

Step 3 **Factor the trinomial as the product of two binomials.** Use the pair of integers identified in Step 2.

If there is no pair of integers in Step 2 whose sum is b, then the polynomial cannot be factored. A polynomial that cannot be factored with integer coefficients is a **prime polynomial.**

Examples: $x^2 + x + 2$, $x^2 - x - 1$, $2x^2 + x + 7$ Prime polynomials

NOW TRY EXERCISE 1
Factor each trinomial.
(a) $t^2 - t - 30$
(b) $w^2 + 12w + 32$

EXAMPLE 1 Factoring Trinomials in $x^2 + bx + c$ Form

Factor each trinomial.

(a) $x^2 + 2x - 35$

Step 1 Find pairs of integers whose product is −35.	*Step 2* Write sums of those pairs of integers.
$35(-1)$	$35 + (-1) = 34$
$-35(1)$	$-35 + 1 = -34$
$7(-5)$	$7 + (-5) = 2 \leftarrow$ Coefficient of the middle term
$-7(5)$	$-7 + 5 = -2$

Step 3 The integers 7 and −5 have the necessary product and sum.

$x^2 + 2x - 35$ factors as $(x+7)(x-5)$. Multiply to check.

(b) $r^2 + 8r + 12$

Look for two integers with a product of 12 and a sum of 8. Of all pairs having a product of 12, only the pair 6 and 2 has a sum of 8.

$r^2 + 8r + 12$ factors as $(r+6)(r+2)$. Multiply to check.

By the commutative property, it is equally correct to write $(r+2)(r+6)$.

NOW TRY

NOW TRY ANSWERS
1. **(a)** $(t-6)(t+5)$
(b) $(w+4)(w+8)$

NOW TRY EXERCISE 2

Factor $m^2 + 12m - 11$.

EXAMPLE 2 Recognizing a Prime Polynomial

Factor $m^2 + 6m + 7$.

Look for two integers whose product is 7 and whose sum is 6. Only 7 and 1 and -7 and -1 give a product of 7. Neither pair has a sum of 6, so $m^2 + 6m + 7$ cannot be factored with integer coefficients and is prime. NOW TRY

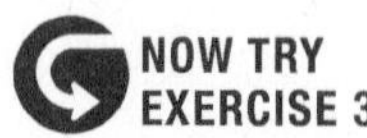
NOW TRY EXERCISE 3

Factor $a^2 + ab - 20b^2$.

EXAMPLE 3 Factoring a Multivariable Trinomial

Factor $x^2 + 6ax - 16a^2$.

This trinomial is in the form $x^2 + bx + c$, where $b = 6a$ and $c = -16a^2$.

Step 1 Find pairs of expressions whose product is $-16a^2$.	***Step 2*** Write sums of those pairs of expressions.
$16a(-a)$	$16a + (-a) = 15a$
$-16a(a)$	$-16a + a = -15a$
$8a(-2a)$	$8a + (-2a) = 6a$ ← Coefficient of the middle term
$-8a(2a)$	$-8a + 2a = -6a$
$-4a(4a)$	$-4a + 4a = 0$

Step 3 The expressions $8a$ and $-2a$ have the necessary product and sum.

$$x^2 + 6ax - 16a^2 \quad \text{factors as} \quad (x + 8a)(x - 2a).$$

CHECK $(x + 8a)(x - 2a)$

$= x^2 - 2ax + 8ax - 16a^2$ FOIL method

$= x^2 + 6ax - 16a^2$ ✓ Original polynomial NOW TRY

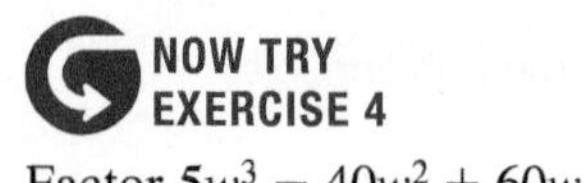
NOW TRY EXERCISE 4

Factor $5w^3 - 40w^2 + 60w$.

EXAMPLE 4 Factoring a Trinomial (Terms Have a Common Factor)

Factor $16y^3 - 32y^2 - 48y$.

$16y^3 - 32y^2 - 48y$

$= 16y(y^2 - 2y - 3)$ Factor out the GCF, $16y$.

To factor $y^2 - 2y - 3$, look for two integers whose product is -3 and whose sum is -2. The necessary integers are -3 and 1.

Remember to include the GCF, $16y$. $= 16y(y - 3)(y + 1)$ Factor the trinomial. NOW TRY 

CAUTION ***When factoring, always look for a common factor first. Remember to write the common factor as part of the answer.***

OBJECTIVE 2 Factor trinomials by grouping when the coefficient of the second-degree term is not 1.

We can use a generalization of the method shown in **Objective 1** to factor a trinomial of the form $ax^2 + bx + c$, where $a \neq 1$. To factor $3x^2 + 7x + 2$, for example, we first identify the values a, b, and c.

$$\begin{matrix} ax^2 + bx + c \\ \swarrow \quad \downarrow \quad \downarrow \\ 3x^2 + 7x + 2, \quad \text{so} \quad a = 3, \quad b = 7, \quad c = 2 \end{matrix}$$

NOW TRY ANSWERS

2. prime

3. $(a + 5b)(a - 4b)$

4. $5w(w - 6)(w - 2)$

The product ac is $3 \cdot 2 = 6$, so we must find integers having a product of 6 and a sum of 7 (because the middle term has coefficient 7). The necessary integers are 1 and 6.

$$3x^2 + 7x + 2$$

$$= 3x^2 + \underbrace{x + 6x}_{x + 6x = 7x} + 2 \quad \text{Write } 7x \text{ as } 1x + 6x, \text{ or } x + 6x.$$

$$= (3x^2 + x) + (6x + 2) \quad \text{Group the terms.}$$

$$= x(3x + 1) + 2(3x + 1) \quad \text{Factor by grouping.}$$

Check by multiplying.

$$= (3x + 1)(x + 2) \quad \text{Factor out the common factor.}$$

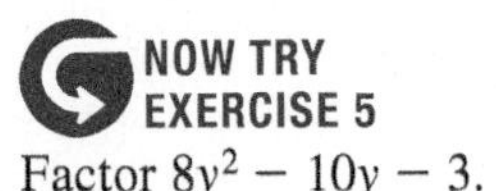
NOW TRY EXERCISE 5
Factor $8y^2 - 10y - 3$.

EXAMPLE 5 Factoring a Trinomial in $ax^2 + bx + c$ Form

Factor $12r^2 - 5r - 2$.

Here $a = 12$, $b = -5$, and $c = -2$, so the product ac is -24. The two integers whose product is -24 and whose sum b is -5 are 3 and -8.

$$12r^2 - 5r - 2$$

$$= 12r^2 + 3r - 8r - 2 \quad \text{Write } -5r \text{ as } 3r - 8r.$$

$$= (12r^2 + 3r) + (-8r - 2) \quad \text{Group the terms.}$$

$$= 3r(4r + 1) - 2(4r + 1) \quad \text{Factor by grouping.}$$

Check by multiplying.

$$= (4r + 1)(3r - 2) \quad \text{Factor out the common factor.}$$

NOW TRY

OBJECTIVE 3 Factor trinomials using the FOIL method when the coefficient of the second-degree term is not 1.

This method involves trying repeated combinations and using the FOIL method.

EXAMPLE 6 Factoring Trinomials in $ax^2 + bx + c$ Form

Factor each trinomial.

(a) $3x^2 + 7x + 2$

The goal is to find the correct numbers to put in the blanks.

$$(\underline{\quad\quad}x + \underline{\quad\quad})(\underline{\quad\quad}x + \underline{\quad\quad})$$

Addition signs are used because all the signs in the polynomial indicate addition. The first two expressions have a product of $3x^2$, so they must be $3x$ and x.

$$(3x + \underline{\quad\quad})(x + \underline{\quad\quad})$$

The product of the two last terms must be 2, so the numbers must be 2 and 1. There is a choice. The 2 could be placed with the $3x$ or with the x. Only one of these choices will give the correct middle term, $7x$.

$(3x + 2)(x + 1)$	$(3x + 1)(x + 2)$
Outer: $3x$; Inner: $2x$	Outer: $6x$; Inner: x
$3x + 2x = 5x$	$6x + x = 7x$
Wrong middle term	Correct middle term

Use the FOIL method to check each factored form.

Therefore, $3x^2 + 7x + 2$ factors as $(3x + 1)(x + 2)$.

NOW TRY ANSWER
5. $(2y - 3)(4y + 1)$

NOW TRY EXERCISE 6

Factor $10r^2 + 19r + 6$.

(b) $12r^2 - 5r - 2$

To reduce the number of trials, we note that the terms of the trinomial have greatest common factor 1. This means that neither of its factors can have a common factor except 1. We try 4 and 3 for the two first terms.

$$(4r \underline{\quad\quad})(3r \underline{\quad\quad})$$

The factors of -2 are -2 and 1 or -1 and 2. We try both possibilities.

$(4r - 2)(3r + 1)$	$(4r - 1)(3r + 2)$ (outer: $8r$; inner: $-3r$)
Wrong: The terms of $4r - 2$ have a common factor of 2. This cannot be correct because 2 is not a factor of $12r^2 - 5r - 2$.	$8r - 3r = 5r$ Wrong middle term

The middle term on the right is $5r$, instead of the $-5r$ that is needed. We obtain $-5r$ by interchanging the signs of the second terms in the factors.

$$(4r + 1)(3r - 2)$$

(outer: $-8r$; inner: $3r$)

$-8r + 3r = -5r$

Correct middle term

Thus, $12r^2 - 5r - 2$ factors as $(4r + 1)(3r - 2)$. (Compare to **Example 5.**)

CHECK $(4r + 1)(3r - 2)$

$= 12r^2 - 8r + 3r - 2$ FOIL method

$= 12r^2 - 5r - 2$ ✓ Original polynomial

NOW TRY 

NOTE As shown in **Example 6(b),** if the terms of a polynomial have greatest common factor 1, then none of the terms of its factors can have a common factor (except 1). Remembering this will eliminate some potential factors.

Factoring $ax^2 + bx + c$ (GCF of $a, b, c = 1$)

Step 1 **Find pairs of integers whose product is *a*.** Write all pairs of integer factors of a, the coefficient of the second-degree term.

Step 2 **Find pairs of integers whose product is *c*.** Write all pairs of integer factors of c, the last term.

Step 3 **Choose inner and outer terms.** Use the FOIL method, and try various combinations of the factors from Steps 1 and 2 until the necessary middle term is obtained.

If no such combinations exist, the trinomial is prime.

NOW TRY ANSWER

6. $(2r + 3)(5r + 2)$

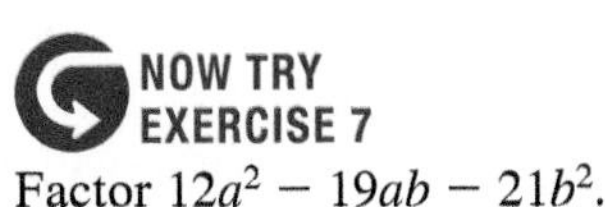

Factor $12a^2 - 19ab - 21b^2$.

EXAMPLE 7 Factoring a Multivariable Trinomial

Factor $18m^2 - 19mx - 12x^2$.

Follow the steps in the preceding box to factor the trinomial. There are many possible factors of both 18 and -12. Try 6 and 3 for 18 and -3 and 4 for -12.

$(6m - 3x)(3m + 4x)$	$(6m + 4x)(3m - 3x)$
Wrong: common factor	Wrong: common factors

Because 6 and 3 do not work as factors of 18, try 9 and 2 instead, with 3 and -4 as factors of -12.

$(9m + 3x)(2m - 4x)$	$27mx$ $(9m - 4x)(2m + 3x)$ $-8mx$
Wrong: common factors	$27mx + (-8mx) = 19mx$ Wrong middle term

The result on the right differs from the correct middle term only in sign, so interchange the signs of the second terms in the factors.

$$18m^2 - 19mx - 12x^2 \quad \text{factors as} \quad (9m + 4x)(2m - 3x).$$

NOW TRY

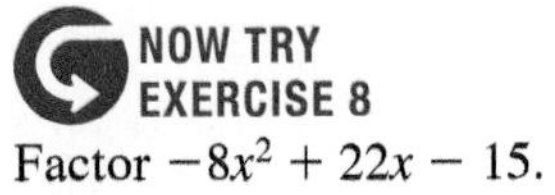

Factor $-8x^2 + 22x - 15$.

EXAMPLE 8 Factoring a Trinomial in $ax^2 + bx + c$ Form ($a < 0$)

Factor $-3x^2 + 16x + 12$.

While we could factor directly, it is helpful to first factor out -1 so that the coefficient of the x^2-term is positive.

$$-3x^2 + 16x + 12$$

$$= -1(3x^2 - 16x - 12) \quad \text{Factor out } -1.$$

$$= -1(3x + 2)(x - 6) \quad \text{Factor the trinomial.}$$

$$= -(3x + 2)(x - 6) \quad -1a = -a$$

NOW TRY

NOTE The factored form in **Example 8** can be written in other ways, such as

$$(-3x - 2)(x - 6) \quad \text{and} \quad (3x + 2)(-x + 6).$$

Verify that these both give the original trinomial when multiplied.

NOW TRY EXERCISE 9

Factor $12y^3 + 33y^2 - 9y$.

EXAMPLE 9 Factoring a Trinomial (Terms Have a Common Factor)

Factor $16y^3 + 24y^2 - 16y$.

$$16y^3 + 24y^2 - 16y$$

$$= 8y(2y^2 + 3y - 2) \quad \text{GCF} = 8y$$

$$= 8y(2y - 1)(y + 2) \quad \text{Factor the trinomial.}$$

Remember the common factor.

NOW TRY

NOW TRY ANSWERS

7. $(4a + 3b)(3a - 7b)$
8. $-(4x - 5)(2x - 3)$
9. $3y(4y - 1)(y + 3)$

OBJECTIVE 4 Factor using substitution.

NOW TRY EXERCISE 10
Factor.

$3(a+2)^2 - 11(a+2) - 4$

EXAMPLE 10 Factoring a Polynomial Using Substitution

Factor $2(x+3)^2 + 5(x+3) - 12$.

The binomial $x + 3$ appears to powers 2 and 1, so we let a substitution variable represent $x + 3$. We may choose any letter we wish except x. We choose t.

$$2(x+3)^2 + 5(x+3) - 12$$

$$= 2t^2 + 5t - 12 \qquad \text{Let } t = x + 3.$$

Don't stop here.

$$= (2t-3)(t+4) \qquad \text{Factor.}$$

Be sure to make the final substitution.

$$= [2(x+3) - 3][(x+3) + 4] \qquad \text{Replace } t \text{ with } x + 3.$$

$$= (2x + 6 - 3)(x + 7) \qquad \text{Simplify.}$$

$$= (2x+3)(x+7) \qquad \text{Combine like terms.}$$

NOW TRY

NOW TRY EXERCISE 11
Factor $6x^4 + 11x^2 + 3$.

EXAMPLE 11 Factoring a Trinomial in $ax^4 + bx^2 + c$ Form

Factor $6y^4 + 7y^2 - 20$.

The variable y appears to powers in which the greater exponent is twice the lesser exponent. We can let a substitution variable represent the variable to the lesser power.

$$6y^4 + 7y^2 - 20$$

$$= 6(y^2)^2 + 7y^2 - 20 \qquad y^4 = (y^2)^2$$

$$= 6t^2 + 7t - 20 \qquad \text{Let } t = y^2.$$

Don't stop here. Replace t with y^2.

$$= (3t-4)(2t+5) \qquad \text{Factor.}$$

$$= (3y^2-4)(2y^2+5) \qquad \text{Replace } t \text{ with } y^2.$$

NOW TRY

NOTE Some students prefer to factor polynomials like the one in **Example 11** directly.

$$6y^4 + 7y^2 - 20$$

Try 3 and 2 as factors of 6 and −4 and 5 as factors of −20.

$$15y^2$$

$$= (3y^2 - 4)(2y^2 + 5)$$

$$-8y^2$$

$3y^2 \cdot 2y^2 = 6y^4$

$-4(5) = 20$

$15y^2 - 8y^2 = 7y^2$

Correct middle term

NOW TRY ANSWERS

10. $(3a+7)(a-2)$

11. $(3x^2+1)(2x^2+3)$

5.2 Exercises

MyLab Math

Video solutions for select problems available in MyLab Math

Concept Check *Choose the letter of the correct response.*

1. Which is *not* a valid way of starting the process of factoring the trinomial $12x^2 + 29x + 10$?

A. $(12x \quad\;)(x \quad\;)$ **B.** $(4x \quad\;)(3x \quad\;)$

C. $(6x \quad\;)(2x \quad\;)$ **D.** $(8x \quad\;)(4x \quad\;)$

2. Which is the completely factored form of the trinomial $2x^6 - 5x^5 - 3x^4$?

A. $x^4(2x + 1)(x - 3)$ B. $x^4(2x - 1)(x + 3)$
C. $(2x^5 + x^4)(x - 3)$ D. $x^3(2x^2 + x)(x - 3)$

3. Which is the completely factored form of the trinomial $4x^2 - 4x - 24$?

A. $4(x - 2)(x + 3)$ B. $4(x + 2)(x + 3)$
C. $4(x + 2)(x - 3)$ D. $4(x - 2)(x - 3)$

4. Which is *not* a factored form of the trinomial $-x^2 + 16x - 60$?

A. $(x - 10)(-x + 6)$ B. $(-x - 10)(x + 6)$
C. $(-x + 10)(x - 6)$ D. $-1(x - 10)(x - 6)$

5. **Concept Check** When a student was given the polynomial $4x^2 + 2x - 20$ to factor completely on a test, the student lost some credit when her answer was

$$(4x + 10)(x - 2).$$

WHAT WENT WRONG? Give the correct answer.

6. **Concept Check** When asked to factor $x^2y^2 - 6x^2 + 5y^2 - 30$, a student gave the following incorrect answer.

$$x^2(y^2 - 6) + 5(y^2 - 6)$$

WHAT WENT WRONG? What is the correct answer?

Complete each factoring. ***See Examples 1 and 3–8.***

7. $x^2 + 8x + 15 = (x + 5)(\_\_\_\_)$
8. $y^2 + 11y + 18 = (y + 2)(\_\_\_\_)$
9. $m^2 - 10m + 21 = (m - 3)(\_\_\_\_)$
10. $n^2 - 14n + 48 = (n - 6)(\_\_\_\_)$
11. $r^2 - r - 20 = (r + 4)(\_\_\_\_)$
12. $x^2 + 4x - 32 = (x - 4)(\_\_\_\_)$
13. $x^2 + ax - 6a^2 = (x + 3a)(\_\_\_\_)$
14. $m^2 - 3mn - 10n^2 = (m + 2n)(\_\_\_\_)$
15. $4x^2 - 4x - 3 = (2x + 1)(\_\_\_\_)$
16. $6z^2 - 11z + 4 = (3z - 4)(\_\_\_\_)$
17. $12u^2 + 10uv + 2v^2 = 2(3u + v)(\_\_\_\_)$
18. $16p^2 - 4pq - 2q^2 = 2(2p - q)(\_\_\_\_)$

Factor each trinomial. ***See Examples 1–9.***

19. $y^2 + 7y - 30$
20. $z^2 + 2z - 24$
21. $p^2 + 15p + 56$
22. $k^2 - 11k + 30$
23. $m^2 - 11m + 60$
24. $p^2 - 12p - 27$
25. $a^2 - 2ab - 35b^2$
26. $z^2 + 8zw + 15w^2$
27. $a^2 - 9ab + 18b^2$
28. $k^2 - 11hk + 28h^2$
29. $x^2y^2 + 12xy + 18$
30. $p^2q^2 - 5pq - 18$
31. $-6m^2 - 13m + 15$
32. $-15y^2 + 17y + 18$
33. $10x^2 + 3x - 18$
34. $8k^2 + 34k + 35$
35. $20k^2 + 47k + 24$
36. $27z^2 + 42z - 5$
37. $15a^2 - 22ab + 8b^2$
38. $14c^2 - 17cd - 6d^2$
39. $36m^2 - 60m + 25$
40. $25r^2 - 90r + 81$
41. $40x^2 + xy + 6y^2$
42. $15p^2 + 24pq + 8q^2$
43. $6x^2z^2 + 5xz - 4$
44. $8m^2n^2 - 10mn + 3$
45. $24x^2 + 42x + 15$
46. $36x^2 + 18x - 4$
47. $-15a^2 - 70a + 120$
48. $-12a^2 - 10a + 42$
49. $-11x^3 + 110x^2 - 264x$
50. $-9k^3 - 36k^2 + 189k$
51. $2x^3y^3 - 48x^2y^4 + 288xy^5$
52. $6m^3n^2 - 60m^2n^3 + 150mn^4$
53. $6a^3 + 12a^2 - 90a$
54. $3m^4 + 6m^3 - 72m^2$
55. $13y^3 + 39y^2 - 52y$
56. $4p^3 + 24p^2 - 64p$

57. $12p^3 - 12p^2 + 3p$

58. $45t^3 + 60t^2 + 20t$

Factor each trinomial. ***See Example 10.***

59. $12p^6 - 32p^3r + 5r^2$

60. $2y^6 + 7xy^3 + 6x^2$

61. $10(k+1)^2 - 7(k+1) + 1$

62. $4(m-5)^2 - 4(m-5) - 15$

63. $3(m+p)^2 - 7(m+p) - 20$

64. $4(x-y)^2 - 23(x-y) - 6$

Factor each trinomial. ***See Example 11.***

65. $p^4 - 10p^2 + 16$

66. $k^4 + 10k^2 + 9$

67. $2x^4 - 9x^2 - 18$

68. $6z^4 + z^2 - 1$

69. $16x^4 + 16x^2 + 3$

70. $9r^4 + 9r^2 + 2$

Extending Skills *Factor each trinomial. (Hint: Factor out the GCF first.)*

71. $z^2(z-x) - zx(x-z) - 2x^2(z-x)$

72. $r^2(r-s) - 5rs(s-r) - 6s^2(r-s)$

73. $a^2(a+b)^2 - ab(a+b)^2 - 6b^2(a+b)^2$

74. $m^2(m-p)^2 - mp(m-p)^2 - 12p^2(m-p)^2$

75. $p^2(p+q) + 4pq(p+q) + 3q^2(p+q)$

76. $2k^2(5-y) - 7k(5-y) + 5(5-y)$

RELATING CONCEPTS For Individual or Group Work (Exercises 77–82)

The trinomial $6x^2 + 7x - 20$ *may be factored using substitution.* ***To see this, work Exercises 77–82 in order.***

77. Multiply each term of $6x^2 + 7x - 20$ by 6 (the leading coefficient). Then compensate by dividing the result by 6 (that is, write it as a fraction with denominator 6).

78. The first term in the numerator of the answer from **Exercise 77** is now a perfect square. Let $t = 6x$ (the square root of $36x^2$), and rewrite the expression.

79. Factor the numerator of the expression from **Exercise 78** in terms of the variable t.

80. Replace t with $6x$ in the answer from **Exercise 79,** and simplify to obtain the correct factorization of the trinomial.

81. Factor $3x^2 - 4x - 32$ using the method just described. (Multiply each term and then divide the result by the leading coefficient 3.)

82. Factor $12x^2 + 35x + 8$ using the method just described. (Multiply and divide by 12.)

5.3 Special Factoring

OBJECTIVES

1. Factor a difference of squares.
2. Factor a perfect square trinomial.
3. Factor a difference of cubes.
4. Factor a sum of cubes.

OBJECTIVE 1 Factor a difference of squares.

The special products introduced earlier are used in reverse when factoring. Recall that the product of the sum and difference of two terms leads to a **difference of squares.**

Factoring a Difference of Squares

$$x^2 - y^2 = (x+y)(x-y)$$

VOCABULARY

☐ difference of squares
☐ perfect square trinomial
☐ difference of cubes
☐ sum of cubes

NOW TRY EXERCISE 1

Factor each polynomial.

(a) $4m^2 - 25n^2$

(b) $9x^2 - 729$

(c) $(a + b)^2 - 25$

(d) $v^4 - 1$

EXAMPLE 1 Factoring Differences of Squares

Factor each polynomial.

(a) $t^2 - 36$

$= t^2 - 6^2$ $\quad$ $36 = 6^2$

$= (t + 6)(t - 6)$ $\quad$ Factor the difference of squares.

(b) $4a^2 - 64$ $\quad$ Always check for a common factor first.

$= 4(a^2 - 16)$ $\quad$ Factor out the common factor, 4.

$= 4(a + 4)(a - 4)$ $\quad$ Factor the difference of squares.

(c) $16m^2 - 49p^2 = (4m)^2 - (7p)^2 = (4m + 7p)(4m - 7p)$ $\quad$ ($x^2 - y^2 = (x + y)(x - y)$)

(d) $81k^2 - (a + 2)^2 = (9k)^2 - (a + 2)^2 = (9k + a + 2)(9k - (a + 2))$ $\quad$ ($x^2 - y^2 = (x + y)(x - y)$)

$= (9k + a + 2)(9k - a - 2)$

We could have used the method of substitution here.

(e) $x^4 - 81$ $\quad$ Don't stop here.

$= (x^2 + 9)(x^2 - 9)$ $\quad$ Factor the difference of squares.

$= (x^2 + 9)(x + 3)(x - 3)$ $\quad$ Factor the difference of squares again. $\quad$ **NOW TRY**

This sum of squares cannot be factored.

Sum of Squares

If x and y have greatest common factor 1, then the sum of squares $x^2 + y^2$ cannot be factored using real numbers and is prime.

Examples: $x^2 + 9$, $t^2 + 36$, $4m^2 + 49$ $\quad$ Prime polynomials

A sum of squares can be factored *only* if the terms have a common factor.

Example: $4x^2 + 36 = 4(x^2 + 9)$ $\quad$ Factor out the GCF, 4.

OBJECTIVE 2 Factor a perfect square trinomial.

Two other special products lead to the following rules for factoring.

Factoring a Perfect Square Trinomial

$$x^2 + 2xy + y^2 = (x + y)^2$$
$$x^2 - 2xy + y^2 = (x - y)^2$$

Because the trinomial $x^2 + 2xy + y^2$ is the square of $x + y$, it is a **perfect square trinomial.** Both the first and the last terms of the trinomial must be perfect squares.

NOW TRY ANSWERS

1. **(a)** $(2m + 5n)(2m - 5n)$
(b) $9(x + 9)(x - 9)$
(c) $(a + b + 5)(a + b - 5)$
(d) $(v^2 + 1)(v + 1)(v - 1)$

In the factored form $(x + y)^2$, twice the product of the first and the last terms must give the middle term of the trinomial.

$4m^2 + 20m + 25$	$p^2 - 8p + 64$
$4m^2 = (2m)^2$, $25 = 5^2$, and $2(2m)(5) = 20m$. **Perfect square trinomial**	p^2 and $64 = 8^2$ are perfect squares, but the middle term would have to be $16p$ or $-16p$. **Not a perfect square trinomial**

NOW TRY EXERCISE 2

Factor each polynomial.

(a) $a^2 + 12a + 36$

(b) $16x^2 - 56xy + 49y^2$

(c) $y^2 - 16y + 64 - z^2$

EXAMPLE 2 Factoring Perfect Square Trinomials

Factor each polynomial.

(a) $144p^2 - 120p + 25$

Here, $144p^2 = (12p)^2$ and $25 = 5^2$. The sign on the middle term is $-$, so if $144p^2 - 120p + 25$ is a perfect square trinomial, the factored form will have to be

$$(12p - 5)^2.$$

Determine twice the product of the two terms to see whether this is correct.

$$2(12p)(-5) = -120p \leftarrow \text{Desired middle term}$$

This is the middle term of the given trinomial.

$$144p^2 - 120p + 25 \quad \text{factors as} \quad (12p - 5)^2.$$

(b) $4m^2 + 20mn + 49n^2$

If this is a perfect square trinomial, it will equal $(2m + 7n)^2$. If multiplied out, this squared binomial has a middle term of

$$2(2m)(7n) = 28mn, \quad \textit{which does not equal} \quad 20mn.$$

This trinomial cannot be factored by any other method either. It is prime.

(c) $(r + 5)^2 + 6(r + 5) + 9$ $\qquad 9 = 3^2$

$= [(r + 5) + 3]^2$ $\qquad 2(r + 5)(3) = 6(r + 5)$ is the middle term.

$= (r + 8)^2$ $\qquad$ Add.

(d) $m^2 - 8m + 16 - p^2$ $\qquad$ The first three terms here form a perfect square trinomial.

$= (m^2 - 8m + 16) - p^2$ $\qquad$ Group the terms of the perfect square trinomial.

$= (m - 4)^2 - p^2$ $\qquad$ Factor the perfect square trinomial.

$= (m - 4 + p)(m - 4 - p)$ $\qquad$ Factor the difference of squares. **NOW TRY**

NOTE Perfect square trinomials can be factored by the general methods for other trinomials. The patterns given here provide "shortcuts."

Example: $\qquad 4m^2 + \mathbf{20m} + 25$

$= (2m + 5)(2m + 5)$ $\qquad$ (outer product $10m$, inner product $10m$) $\qquad 10m + 10m = \mathbf{20m}$

$= (2m + 5)^2$ $\qquad a \cdot a = a^2$

NOW TRY ANSWERS

2. **(a)** $(a + 6)^2$
(b) $(4x - 7y)^2$
(c) $(y - 8 + z)(y - 8 - z)$

OBJECTIVE 3 Factor a difference of cubes.

A **difference of cubes,** $x^3 - y^3$, can be factored as follows.

Factoring a Difference of Cubes

$$x^3 - y^3 = (x - y)(x^2 + xy + y^2)$$

NOW TRY EXERCISE 3
Factor each polynomial.
(a) $t^3 - 1$
(b) $125a^3 - 8b^3$

EXAMPLE 3 Factoring Differences of Cubes

Factor each polynomial.

$$x^3 - y^3 = (x - y)\,(x^2 + x \cdot y + y^2)$$

(a) $m^3 - 8 = m^3 - 2^3 = (m - 2)(m^2 + m \cdot 2 + 2^2)$

$$= (m - 2)(m^2 + 2m + 4)$$

CHECK $(m - 2)(m^2 + 2m + 4)$

$= m^3 + 2m^2 + 4m - 2m^2 - 4m - 8$ Distributive property

$= m^3 - 8$ ✓ Combine like terms.

(b) $27x^3 - 8y^3$

$= (3x)^3 - (2y)^3$ Difference of cubes

$= (3x - 2y)[(3x)^2 + (3x)(2y) + (2y)^2]$ Factor.

$= (3x - 2y)(9x^2 + 6xy + 4y^2)$

$(3x)^2 = 3^2x^2$, *not* $3x^2$.
$(2y)^2 = 2^2y^2$, *not* $2y^2$.

(c) $1000k^3 - 27n^3$

$= (10k)^3 - (3n)^3$ Difference of cubes

$= (10k - 3n)[(10k)^2 + (10k)(3n) + (3n)^2]$ Factor.

$= (10k - 3n)(100k^2 + 30kn + 9n^2)$ Multiply.

NOW TRY

OBJECTIVE 4 Factor a sum of cubes.

Although the binomial $x^2 + y^2$ (a sum of *squares*) cannot be factored with real numbers, a **sum of cubes,** such as $x^3 + y^3$, can be factored as follows.

Factoring a Sum of Cubes

$$x^3 + y^3 = (x + y)(x^2 - xy + y^2)$$

Compare the rules for factoring a *difference* of cubes and a *sum* of cubes.

$$x^3 - y^3 = (x - y)(x^2 + xy + y^2)$$

Same sign (between $x^3 - y^3$ and $x - y$); Opposite sign (+ xy); Always positive (x^2 and y^2 terms)

$$x^3 + y^3 = (x + y)(x^2 - xy + y^2)$$

Same sign (between $x^3 + y^3$ and $x + y$); Opposite sign ($- xy$); Always positive (x^2 and y^2 terms)

NOW TRY ANSWERS
3. **(a)** $(t - 1)(t^2 + t + 1)$
(b) $(5a - 2b)(25a^2 + 10ab + 4b^2)$

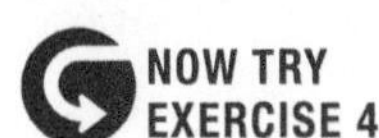

Factor each polynomial.

(a) $1000 + z^3$

(b) $81a^6 + 3b^3$

(c) $(x - 3)^3 + y^3$

EXAMPLE 4 Factoring Sums of Cubes

Factor each polynomial.

(a) $r^3 + 27$

$= r^3 + 3^3$ Sum of cubes

$= (r + 3)(r^2 - 3r + 3^2)$ Factor.

$= (r + 3)(r^2 - 3r + 9)$ This trinomial cannot be factored further.

(b) $27z^3 + 125$

$= (3z)^3 + 5^3$ Sum of cubes

$= (3z + 5)[(3z)^2 - (3z)(5) + 5^2]$ Factor.

$= (3z + 5)(9z^2 - 15z + 25)$ Multiply; $(3z)^2 = 3^2z^2$

(c) $125t^3 + 216s^6$

$= (5t)^3 + (6s^2)^3$ Sum of cubes

$= (5t + 6s^2)[(5t)^2 - (5t)(6s^2) + (6s^2)^2]$ Factor.

$= (5t + 6s^2)(25t^2 - 30ts^2 + 36s^4)$ Multiply.

(d) $3x^2 + 192$

$= 3(x^3 + 64)$ Factor out the common factor.

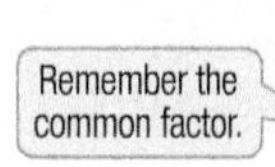

$= 3(x^3 + 4^3)$ Write as a sum of cubes.

$= 3(x + 4)(x^2 - 4x + 16)$ Factor.

(e) $(x + 2)^3 + t^3$

$= [(x + 2) + t][(x + 2)^2 - (x + 2)t + t^2]$ Sum of cubes

$= (x + 2 + t)(x^2 + 4x + 4 - xt - 2t + t^2)$ Multiply.

CAUTION A common error when factoring $x^3 + y^3$ or $x^3 - y^3$ is to think that the xy-term has a coefficient of 2. Because there is no coefficient of 2, expressions of the form $x^2 + xy + y^2$ and $x^2 - xy + y^2$ usually cannot be factored further.

STUDY SKILLS **REMINDER**
Make study cards to help you learn and remember the material in this chapter. **Review Study Skill 5, *Using Study Cards.***

Summary of Special Types of Factoring

Difference of Squares	$x^2 - y^2 = (x + y)(x - y)$
Perfect Square Trinomial	$x^2 + 2xy + y^2 = (x + y)^2$
	$x^2 - 2xy + y^2 = (x - y)^2$
Difference of Cubes	$x^3 - y^3 = (x - y)(x^2 + xy + y^2)$
Sum of Cubes	$x^3 + y^3 = (x + y)(x^2 - xy + y^2)$

NOW TRY ANSWERS

4. (a) $(10 + z) \cdot (100 - 10z + z^2)$

(b) $3(3a^2 + b) \cdot (9a^4 - 3a^2b + b^2)$

(c) $(x - 3 + y) \cdot (x^2 - 6x + 9 - xy + 3y + y^2)$

5.3 Exercises

FOR EXTRA HELP — MyLab Math

Video solutions for select problems available in MyLab Math

Concept Check *Choose the letter(s) of the correct response.*

1. Which of the following binomials are differences of squares?

A. $64 - k^2$ **B.** $2x^2 - 25$ **C.** $k^2 + 9$ **D.** $4z^4 - 49$

2. Which of the following binomials are sums or differences of cubes?

A. $64 + r^3$ **B.** $125 - p^6$ **C.** $9x^3 + 125$ **D.** $(x + y)^3 - 1$

3. Which of the following trinomials are perfect squares?

A. $x^2 - 8x - 16$ **B.** $4m^2 + 20m + 25$

C. $9z^4 + 30z^2 + 25$ **D.** $25p^2 - 45p + 81$

4. Which sum of squares can be factored? Factor it.

A. $x^2 + 100$ **B.** $9m^2 + 25$ **C.** $t^2 + 16$ **D.** $4x^2 + 64$

5. Concept Check A student incorrectly factored

$$x^2 + 4 \text{ as } (x + 2)^2.$$

WHAT WENT WRONG? Explain this error.

6. Concept Check A student incorrectly factored

$$x^3 + 8 \text{ as } (x + 2)^3.$$

WHAT WENT WRONG? Explain this error.

Factor each polynomial. **See Examples 1–4.**

7. $p^2 - 16$ **8.** $k^2 - 9$ **9.** $25x^2 - 4$

10. $36m^2 - 25$ **11.** $18a^2 - 98b^2$ **12.** $32c^2 - 98d^2$

13. $64m^4 - 4y^4$ **14.** $243x^4 - 3t^4$ **15.** $(y + z)^2 - 81$

16. $(h + k)^2 - 9$ **17.** $16 - (x + 3y)^2$ **18.** $64 - (r + 2t)^2$

19. $p^4 - 256$ **20.** $a^4 - 625$ **21.** $k^2 - 6k + 9$

22. $x^2 + 10x + 25$ **23.** $4z^2 + 4zw + w^2$ **24.** $9y^2 + 6yz + z^2$

25. $16m^2 - 8m + 1 - n^2$ **26.** $25c^2 - 20c + 4 - d^2$ **27.** $4r^2 - 12r + 9 - s^2$

28. $9a^2 - 24a + 16 - b^2$ **29.** $x^2 - y^2 + 2y - 1$ **30.** $-k^2 - h^2 + 2kh + 4$

31. $98m^2 + 84mn + 18n^2$ **32.** $80z^2 - 40zw + 5w^2$

33. $(p + q)^2 + 2(p + q) + 1$ **34.** $(x + y)^2 + 6(x + y) + 9$

35. $(a - b)^2 + 8(a - b) + 16$ **36.** $(m - n)^2 + 4(m - n) + 4$

37. $x^3 - 27$ **38.** $y^3 - 64$ **39.** $216 - t^3$

40. $512 - m^3$ **41.** $x^3 + 64$ **42.** $r^3 + 343$

43. $1000 + y^3$ **44.** $729 + x^3$ **45.** $8x^3 + 1$

46. $27y^3 + 1$ **47.** $125x^3 - 216$ **48.** $8w^3 - 125$

49. $x^3 - 8y^3$ **50.** $z^3 - 125p^3$ **51.** $64g^3 - 27h^3$

52. $27a^3 - 8b^3$ **53.** $343p^3 + 125q^3$ **54.** $512t^3 + 27s^3$

55. $24n^3 + 81p^3$ **56.** $250x^3 + 16y^3$ **57.** $(y + z)^3 + 64$

58. $(p - q)^3 + 125$ **59.** $m^6 - 125$ **60.** $x^6 - 216$

61. $27 - 1000x^9$ **62.** $64 - 729p^9$ **63.** $125y^6 + z^3$

64. $64y^9 + z^6$ **65.** $125k - 64k^4$ **66.** $216k^2 + 125k^5$

Extending Skills *In some cases, the method of factoring by grouping can be combined with the methods of special factoring discussed in this section. Consider this example.*

$8x^3 + 4x^2 + 27y^3 - 9y^2$

$= (8x^3 + 27y^3) + (4x^2 - 9y^2)$ Associative and commutative properties

$= (2x + 3y)(4x^2 - 6xy + 9y^2) + (2x + 3y)(2x - 3y)$ Factor within groups.

$= (2x + 3y)[(4x^2 - 6xy + 9y^2) + (2x - 3y)]$ Factor out the greatest common factor, $2x + 3y$.

$= (2x + 3y)(4x^2 - 6xy + 9y^2 + 2x - 3y)$ Combine like terms.

In problems such as this, how we choose to group in the first step is essential to factoring correctly. If we reach a "dead end," then we should group differently and try again.

Use the method just described to factor each polynomial.

67. $125p^3 + 25p^2 + 8q^3 - 4q^2$ **68.** $27x^3 + 9x^2 + y^3 - y^2$

69. $27a^3 + 15a - 64b^3 - 20b$ **70.** $1000k^3 + 20k - m^3 - 2m$

71. $8t^4 - 24t^3 + t - 3$ **72.** $y^4 + y^3 + y + 1$

73. $64m^2 - 512m^3 - 81n^2 + 729n^3$ **74.** $10x^2 + 5x^3 - 10y^2 + 5y^3$

RELATING CONCEPTS For Individual or Group Work (Exercises 75–80)

The binomial $x^6 - y^6$ may be considered as either a difference of squares or a difference of cubes. ***Work Exercises 75–80 in order.***

75. Factor $x^6 - y^6$ by first factoring as a difference of squares. Then factor further by considering one of the factors as a sum of cubes and the other factor as a difference of cubes.

76. Based on the answer in **Exercise 75,** fill in the blank with the correct factors so that $x^6 - y^6$ is factored completely.

$$x^6 - y^6 = (x - y)(x + y)\_\_\_\_\_\_\_\_\_\_$$

77. Factor $x^6 - y^6$ by first factoring as a difference of cubes. Then factor further by considering one of the factors as a difference of squares.

78. Based on the answer in **Exercise 77,** fill in the blank with the correct factor so that $x^6 - y^6$ is factored.

$$x^6 - y^6 = (x - y)(x + y)\_\_\_\_\_\_\_\_\_\_$$

79. Notice that the factor written in the blank in **Exercise 78** is a fourth-degree polynomial, while the two factors written in the blank in **Exercise 76** are both second-degree polynomials. What must be true about the product of the two factors written in the blank in **Exercise 76?** Verify this.

80. If we have a choice of factoring as a difference of squares or a difference of cubes, how should we start to more easily obtain the completely factored form of the polynomial? Base the answer on the results in **Exercises 75–79.**

5.4 A General Approach to Factoring

OBJECTIVE

1 Factor any polynomial.

OBJECTIVE 1 Factor any polynomial.

A polynomial is *completely factored* when it is written as a product of prime polynomials with integer coefficients.

Factoring a Polynomial

Step 1 **Factor out any common factor.**

Step 2 **How many terms are in the polynomial?**

If the polynomial is a binomial, check to see whether it is a difference of squares, a difference of cubes, or a sum of cubes.

If the polynomial is a trinomial, check to see whether it is a perfect square trinomial. If it is not, use one of the following methods.

- To factor $x^2 + bx + c$, find two integers whose product is c and whose sum is b, the coefficient of the middle term.
- To factor $ax^2 + bx + c$, find two integers having product ac and sum b. Use these integers to rewrite the middle term, and factor by grouping.

  Alternatively, use the FOIL method and try various combinations of the factors until the correct middle term is found.

If the polynomial has more than three terms, try to factor it by grouping.

Step 3 **If any of the factors can be factored further, do so.**

Step 4 **Check the factored form by multiplying.**

NOW TRY EXERCISE 1

Factor each polynomial.

(a) $13x + 39$

(b) $21x^3y^2 - 27x^2y^4$

(c) $8y(m - n) - 5(m - n)$

(d) $(y - 1)(2y + 1) - (y - 1)(y + 4)$

EXAMPLE 1 Factoring Out a Common Factor

Factor each polynomial.

(a) $9p + 45$

$= 9(p + 5)$ GCF = 9

(b) $8m^2p^2 + 4mp$

$= 4mp(2mp + 1)$ GCF = $4mp$

(c) $5x(a + b) - y(a + b)$

$= (a + b)(5x - y)$ Factor out $(a + b)$.

(d) $(x - 4)(x + 2) + (x - 4)(2x - 1)$

$= (x - 4)[(x + 2) + (2x - 1)]$ Factor out $x - 4$.

$= (x - 4)(3x + 1)$ Combine like terms inside the brackets.

NOW TRY

Factoring a Binomial

For a **binomial** (two terms), check for the following patterns.

Difference of squares $x^2 - y^2 = (x + y)(x - y)$

Difference of cubes $x^3 - y^3 = (x - y)(x^2 + xy + y^2)$

Sum of cubes $x^3 + y^3 = (x + y)(x^2 - xy + y^2)$

NOW TRY ANSWERS

1. **(a)** $13(x + 3)$
   (b) $3x^2y^2(7x - 9y^2)$
   (c) $(m - n)(8y - 5)$
   (d) $(y - 1)(y - 3)$

NOW TRY EXERCISE 2

Factor each binomial if possible.

(a) $4a^2 - 49b^2$

(b) $9x^2 + 100$

(c) $27v^3 - 1000$

EXAMPLE 2 Factoring Binomials

Factor each binomial if possible.

(a) $64m^2 - 9n^2$

$= (8m)^2 - (3n)^2$ Difference of squares

$= (8m + 3n)(8m - 3n)$ $x^2 - y^2 = (x + y)(x - y)$

(b) $8p^3 - 27$

$= (2p)^3 - 3^3$ Difference of cubes

$= (2p - 3)[(2p)^2 + (2p)(3) + 3^2]$ $x^3 - y^3 = (x - y)(x^2 + xy + y^2)$

$= (2p - 3)(4p^2 + 6p + 9)$ $(2p)^2 = 2^2p^2 = 4p^2$

(c) $1000m^3 + 1$

$= (10m)^3 + 1^3$ Sum of cubes

$= (10m + 1)[(10m)^2 - (10m)(1) + 1^2]$ $x^3 + y^3 = (x + y)(x^2 - xy + y^2)$

$= (10m + 1)(100m^2 - 10m + 1)$ $(10m)^2 = 10^2m^2 = 100m^2$

(d) $25m^2 + 121$

This *sum* of squares is prime. There is no common factor (except 1).

NOTE Although the binomial $25m^2 + 625$ is a sum of squares, it can be factored because the greatest common factor of the terms is *not* 1.

$25m^2 + 625 = 25(m^2 + 25)$ Factor out the common factor 25.

This sum of squares cannot be factored further.

Factoring a Trinomial

For a **trinomial** (three terms), decide whether it is a perfect square trinomial of either of these forms.

$$\mathbf{x^2 + 2xy + y^2 = (x + y)^2} \quad \text{or} \quad \mathbf{x^2 - 2xy + y^2 = (x - y)^2}$$

If not, use the methods shown in **Examples 3(c)–(f).**

EXAMPLE 3 Factoring Trinomials

Factor each trinomial.

(a) $p^2 + 10p + 25$ Perfect square trinomial

$= (p + 5)^2$

(b) $49z^2 - 42z + 9$ Perfect square trinomial

$= (7z - 3)^2$

(c) $y^2 - 5y - 6$

$= (y - 6)(y + 1)$ The numbers -6 and 1 have a product of -6 and a sum of -5.

NOW TRY ANSWERS

2. (a) $(2a + 7b)(2a - 7b)$
(b) prime
(c) $(3v - 10)(9v^2 + 30v + 100)$

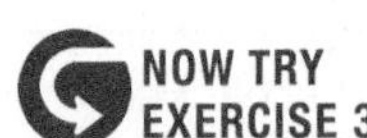
NOW TRY EXERCISE 3

Factor each trinomial.

(a) $25x^2 - 90x + 81$

(b) $7x^2 - 7xy - 84y^2$

(c) $12m^2 + 5m - 28$

(d) $12x^2 + 28xy - 5y^2$

(d) $2k^2 - k - 6 = (2k + 3)(k - 2)$

$-4k$ (outer), $3k$ (inner): $-4k + 3k = -k$

Use the FOIL method and try various combinations of the factors until the correct middle term is found.

(e) $10x^2 + xy - 3y^2$

$= 10x^2 - 5xy + 6xy - 3y^2$ ($-5xy + 6xy = xy$) The integers -5 and 6 have a product of $10(-3) = -30$ and a sum of 1, the coefficient of the middle term.

$= (10x^2 - 5xy) + (6xy - 3y^2)$ Group the terms.

$= 5x(2x - y) + 3y(2x - y)$ Factor each group.

$= (2x - y)(5x + 3y)$ Factor out the common factor.

(f) $28z^2 + 6z - 10$

$= 2(14z^2 + 3z - 5)$ Factor out the common factor.

$= 2(7z + 5)(2z - 1)$ Factor the trinomial.

Remember the common factor.

NOW TRY

If a polynomial has more than three terms, consider factoring by grouping.

NOW TRY EXERCISE 4

Factor each polynomial.

(a) $5a^3 + 5a^2b - ab^2 - b^3$

(b) $9u^2 - 48u + 64 - v^2$

(c) $x^3 - 9y^2 - 27y^3 + x^2$

EXAMPLE 4 Factoring Polynomials with More Than Three Terms

Factor each polynomial.

(a) $xy^2 - y^3 + x^3 - x^2y$

$= (xy^2 - y^3) + (x^3 - x^2y)$ Group the terms.

$= y^2(x - y) + x^2(x - y)$ Factor each group.

$= (x - y)(y^2 + x^2)$ $x - y$ is a common factor.

(b) $20k^3 + 4k^2 - 45k - 9$

$= (20k^3 + 4k^2) + (-45k - 9)$ Group the terms. Be careful with signs.

$= 4k^2(5k + 1) - 9(5k + 1)$ Factor each group.

$= (5k + 1)(4k^2 - 9)$ $5k + 1$ is a common factor.

$= (5k + 1)(2k + 3)(2k - 3)$ Factor the difference of squares.

(c) $4a^2 + 4a + 1 - b^2$

$= (4a^2 + 4a + 1) - b^2$ The first three terms form a perfect square trinomial. $4a^2 = (2a)^2$, $1 = 1^2$, and $2(2a)(1) = 4a$, as required.

$= (2a + 1)^2 - b^2$ Factor the perfect square trinomial.

$= (2a + 1 + b)(2a + 1 - b)$ Factor the difference of squares.

(d) $8m^3 + 4m^2 - n^3 - n^2$

$= (8m^3 - n^3) + (4m^2 - n^2)$ Rearrange and group the terms. (Difference of cubes; Difference of squares)

$= (2m - n)(4m^2 + 2mn + n^2) + (2m - n)(2m + n)$

Factor the difference of cubes and the difference of squares.

$= (2m - n)(4m^2 + 2mn + n^2 + 2m + n)$ Factor out the common factor $2m - n$.

NOW TRY 

NOW TRY ANSWERS

3. (a) $(5x - 9)^2$
(b) $7(x + 3y)(x - 4y)$
(c) $(4m + 7)(3m - 4)$
(d) $(6x - y)(2x + 5y)$

4. (a) $(a + b)(5a^2 - b^2)$
(b) $(3u - 8 + v)(3u - 8 - v)$
(c) $(x - 3y)(x^2 + 3xy + 9y^2 + x + 3y)$

 Observe the final two lines in **Example 4(d).** Students often ask, *"What happened to the other* $(2m - n)$*?"* Think of the variable a in the expression $ax + ay$, which is factored as $a(x + y)$. The binomial $(2m - n)$ is handled just like a in this simpler example.

5.4 Exercises

FOR EXTRA HELP MyLab Math

Video solutions for select problems available in MyLab Math

STUDY SKILLS REMINDER
How are you doing on your homework? **Review Study Skill 4, *Completing Your Homework.***

Concept Check *Match each polynomial in Column I with the method or methods for factoring it in Column II. The choices in Column II may be used once, more than once, or not at all.*

	I	II
1.	**(a)** $49x^2 - 81y^2$	**A.** Factor out the GCF.
	(b) $125z^6 + 1$	**B.** Factor a difference of squares.
	(c) $88r^2 - 55s^2$	**C.** Factor a difference of cubes.
	(d) $64a^3 - 8b^9$	**D.** Factor a sum of cubes.
	(e) $50x^2 - 128y^4$	**E.** The polynomial is prime.

	I	II
2.	**(a)** $ab - 5a + 3b - 15$	**A.** Factor out the GCF.
	(b) $z^2 - 3z + 6$	**B.** Factor a perfect square trinomial.
	(c) $25x^2 + 100$	**C.** Factor by grouping.
	(d) $r^2 - 24r + 144$	**D.** Factor into two distinct binomials.
	(e) $2y^2 + 36y + 162$	**E.** The polynomial is prime.

The following exercises are of mixed variety. Factor each polynomial. **See Examples 1–4.**

3. $3p^4 - 3p^3 - 90p^2$

4. $k^4 - 16$

5. $3a^2pq + 3abpq - 90b^2pq$

6. $49z^2 - 16$

7. $225p^2 + 256$

8. $18m^3n + 3m^2n^2 - 6mn^3$

9. $6b^2 - 17b - 3$

10. $k^2 - 6k - 16$

11. $x^3 - 1000$

12. $6t^2 + 19tu - 77u^2$

13. $4(p + 2) + m(p + 2)$

14. $40p - 32r$

15. $9m^2 - 45m + 18m^3$

16. $4k^2 + 28kr + 49r^2$

17. $54m^3 - 2000$

18. $mn - 2n + 5m - 10$

19. $9m^2 - 30mn + 25n^2$

20. $2a^2 - 7a - 4$

21. $kq - 9q + kr - 9r$

22. $56k^3 - 875$

23. $16z^3x^2 - 32z^2x$

24. $9r^2 + 100$

25. $(r + 4)(3r - 1) - (r + 4)(r + 2)$

26. $9 - a^2 + 2ab - b^2$

27. $625 - x^4$

28. $2m^2 - mn - 15n^2$

29. $p^3 + 1$

30. $(t + 8)(4t + 1) - (t + 8)(3t - 11)$

31. $64m^2 - 625$

32. $14z^2 - 3zk - 2k^2$

33. $12z^3 - 6z^2 + 18z$

34. $225k^2 - 36r^2$

35. $256b^2 - 400c^2$

36. $z^2 - zp - 20p^2$

37. $512 + 1000z^3$

38. $64m^2 - 25n^2$

39. $10r^2 + 23rs - 5s^2$

40. $12k^2 - 17kq - 5q^2$

41. $24p^3q + 52p^2q^2 + 20pq^3$

42. $32x^2 + 16x^3 - 24x^5$

43. $48k^4 - 243$

44. $14x^2 - 25xq - 25q^2$

45. $m^3 + m^2 - n^3 - n^2$

46. $64x^3 + y^3 - 16x^2 + y^2$

47. $x^2 - 4m^2 - 4mn - n^2$

48. $4r^2 - s^2 - 2st - t^2$

49. $18p^5 - 24p^3 + 12p^6$

50. $k^2 - 6k + 16$

51. $2x^2 - 2x - 40$

52. $27x^3 - 3y^3$

53. $(2m + n)^2 - (2m - n)^2$

54. $(3k + 5)^2 - 4(3k + 5) + 4$

55. $50p^2 - 162$

56. $y^2 + 3y - 10$

57. $12m^2rx + 4mnrx + 40n^2rx$

58. $18p^2 + 53pr - 35r^2$

59. $21a^2 - 5ab - 4b^2$

60. $x^2 - 2xy + y^2 - 4$

61. $x^2 - y^2 - 4$

62. $(5r + 2s)^2 - 6(5r + 2s) + 9$

63. $(p + 8q)^2 - 10(p + 8q) + 25$

64. $z^4 - 9z^2 + 20$

65. $21m^4 - 32m^2 - 5$

66. $(x - y)^3 - (27 - y)^3$

67. $(r + 2t)^3 + (r - 3t)^3$

68. $16x^3 + 32x^2 - 9x - 18$

69. $x^5 + 3x^4 - x - 3$

70. $1 - x^{16}$

71. $m^2 - 4m + 4 - n^2 + 6n - 9$

72. $x^2 + 4 + x^2y + 4y$

5.5 Solving Quadratic Equations Using the Zero-Factor Property

OBJECTIVES

1 Use the zero-factor property.

2 Solve applied problems that require the zero-factor property.

3 Solve a formula for a specified variable, where factoring is necessary.

We have solved linear, or first-degree, equations. Solving higher-degree polynomial equations requires other methods, one of which involves factoring.

OBJECTIVE 1 Use the zero-factor property.

The **zero-factor property** is a special property of the number 0.

Zero-Factor Property

If two numbers have a product of 0, then at least one of the numbers must be 0.

If $ab = 0$, then either $a = 0$ or $b = 0$ (or both).

VOCABULARY

☐ quadratic equation
☐ double solution

To prove the zero-factor property, we first assume that $a \neq 0$. (If $a = 0$, then the property is proved already.) If $a \neq 0$, then $\frac{1}{a}$ exists.

$$ab = 0$$

$$\frac{1}{a} \cdot ab = \frac{1}{a} \cdot 0 \quad \text{Multiply each side by } \tfrac{1}{a}.$$

$$b = 0 \quad \text{Multiply.}$$

Thus, if $a \neq 0$, then $b = 0$, and the property is proved.

CAUTION If $ab = 0$, then $a = 0$ or $b = 0$. However, if $ab = 6$, for example, it is not necessarily true that $a = 6$ or $b = 6$. In fact, it is very likely that *neither* $a = 6$ *nor* $b = 6$. ***The zero-factor property works only for a product equal to 0.***

NOW TRY EXERCISE 1

Solve.

$(x + 5)(4x - 7) = 0$

EXAMPLE 1 Using the Zero-Factor Property to Solve an Equation

Solve $(x + 6)(2x - 3) = 0$.

The product of $x + 6$ and $2x - 3$ is 0. The zero-factor property applies.

$$x + 6 = 0 \quad \text{or} \quad 2x - 3 = 0 \quad \text{Zero-factor property}$$

$$x = -6 \quad \text{or} \quad 2x = 3 \quad \text{Solve each of these equations.}$$

$$x = \frac{3}{2}$$

CHECK

$$(x + 6)(2x - 3) = 0$$

$$(-6 + 6)[2(-6) - 3] \stackrel{?}{=} 0 \quad \text{Let } x = -6.$$

$$0(-15) \stackrel{?}{=} 0$$

$$0 = 0 \ ✓ \ \text{True}$$

$$(x + 6)(2x - 3) = 0$$

$$\left(\frac{3}{2} + 6\right)\left(2 \cdot \frac{3}{2} - 3\right) \stackrel{?}{=} 0 \quad \text{Let } x = \tfrac{3}{2}.$$

$$\frac{15}{2}(0) \stackrel{?}{=} 0$$

$$0 = 0 \ ✓ \ \text{True}$$

Both values check, so the solution set is $\left\{-6, \frac{3}{2}\right\}$. NOW TRY

Because the product $(x + 6)(2x - 3)$ equals $2x^2 + 9x - 18$, the equation of **Example 1** has a term with a squared variable and is an example of a *quadratic equation. **A quadratic equation has degree 2.***

Quadratic Equation

A **quadratic equation** (in x here) is an equation that can be written in the form

$$ax^2 + bx + c = 0,$$

where a, b, and c are real numbers, and $a \neq 0$. The given form is called **standard form.**

Examples: $3x^2 - 7x + 2 = 0$, $x^2 - 4 = 0$, $t^2 + 2t = 0$ Quadratic equations

NOW TRY ANSWER

1. $\left\{-5, \frac{7}{4}\right\}$

Solving a Quadratic Equation Using the Zero-Factor Property

Step 1 **Write the equation in standard form**—that is, with all terms on one side of the equality symbol in descending powers of the variable and with 0 on the other side.

Step 2 **Factor** completely.

Step 3 **Apply the zero-factor property.** Set each factor with a variable equal to 0.

Step 4 **Solve** the resulting equations.

Step 5 **Check** each value in the *original* equation. Write the solution set.

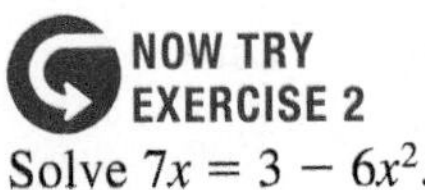

NOW TRY EXERCISE 2

Solve $7x = 3 - 6x^2$.

EXAMPLE 2 Solving a Quadratic Equation

Solve $2x^2 + 3x = 2$.

$$2x^2 + 3x = 2$$

Step 1 $2x^2 + 3x - 2 = 0$ Standard form

Step 2 $(2x - 1)(x + 2) = 0$ Factor.

Step 3 $2x - 1 = 0 \quad \text{or} \quad x + 2 = 0$ Zero-factor property

Step 4 $2x = 1 \quad \text{or} \quad x = -2$ Solve each equation.

$$x = \frac{1}{2}$$

Step 5 *Check* each value in the original equation.

CHECK $2x^2 + 3x = 2$

$$2\left(\frac{1}{2}\right)^2 + 3\left(\frac{1}{2}\right) \stackrel{?}{=} 2 \qquad \text{Let } x = \tfrac{1}{2}.$$

$$2\left(\frac{1}{4}\right) + \frac{3}{2} \stackrel{?}{=} 2$$

$$\frac{1}{2} + \frac{3}{2} \stackrel{?}{=} 2$$

$$2 = 2 \ \checkmark \ \text{True}$$

$$2x^2 + 3x = 2$$

$$2(-2)^2 + 3(-2) \stackrel{?}{=} 2 \qquad \text{Let } x = -2.$$

$$2(4) - 6 \stackrel{?}{=} 2$$

$$8 - 6 \stackrel{?}{=} 2$$

$$2 = 2 \ \checkmark \ \text{True}$$

We give solutions from least to greatest.

Because both values check, the solution set is $\left\{-2, \frac{1}{2}\right\}$. **NOW TRY**

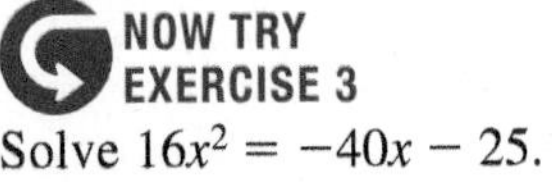

NOW TRY EXERCISE 3

Solve $16x^2 = -40x - 25$.

EXAMPLE 3 Solving a Quadratic Equation (Double Solution)

Solve $4x^2 = 4x - 1$.

$$4x^2 = 4x - 1$$

$4x^2 - 4x + 1 = 0$ Standard form

$(2x - 1)^2 = 0$ Factor.

We could factor as $(2x - 1)(2x - 1)$. The same solution results.

$2x - 1 = 0$ Zero-factor property

$x = \frac{1}{2}$ Add 1. Divide by 2.

There is only one *distinct* solution, called a **double solution,** because the trinomial $4x^2 - 4x + 1$ is a perfect square. The solution set is $\left\{\frac{1}{2}\right\}$. **NOW TRY**

NOW TRY ANSWERS

2. $\left\{-\frac{3}{2}, \frac{1}{3}\right\}$

3. $\left\{-\frac{5}{4}\right\}$

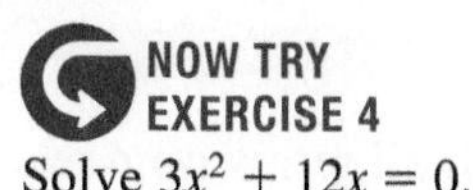
Solve $3x^2 + 12x = 0$.

EXAMPLE 4 Solving a Quadratic Equation (Missing Constant Term)

Solve $4x^2 - 20x = 0$.

This quadratic equation has a missing constant term. Comparing it with the standard form $ax^2 + bx + c = 0$ shows that $c = 0$. The zero-factor property can still be used.

$$4x^2 - 20x = 0$$

$$4x(x - 5) = 0 \quad \text{Factor.}$$

Set each *variable* factor equal to 0.

$$4x = 0 \quad \text{or} \quad x - 5 = 0 \quad \text{Zero-factor property}$$

$$x = 0 \quad \text{or} \quad x = 5 \quad \text{Solve each equation.}$$

CHECK

$$4x^2 - 20x = 0$$
$$4(0)^2 - 20(0) \stackrel{?}{=} 0 \quad \text{Let } x = 0.$$
$$0 - 0 = 0 \checkmark \quad \text{True}$$

$$4x^2 - 20x = 0$$
$$4(5)^2 - 20(5) \stackrel{?}{=} 0 \quad \text{Let } x = 5.$$
$$100 - 100 = 0 \checkmark \quad \text{True}$$

The solution set is $\{0, 5\}$. Remember to include the solution 0.

NOW TRY

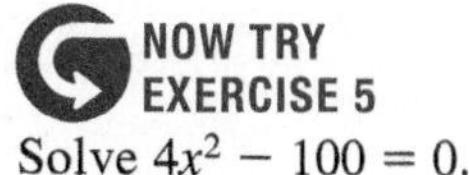
Solve $4x^2 - 100 = 0$.

EXAMPLE 5 Solving a Quadratic Equation (Missing Linear Term)

Solve $3x^2 - 108 = 0$.

$$3x^2 - 108 = 0$$

$$3(x^2 - 36) = 0 \quad \text{Factor out 3.}$$

The factor 3 does *not* lead to a solution.

$$3(x + 6)(x - 6) = 0 \quad \text{Factor } x^2 - 36.$$

$$x + 6 = 0 \quad \text{or} \quad x - 6 = 0 \quad \text{Zero-factor property}$$

$$x = -6 \quad \text{or} \quad x = 6 \quad \text{Solve each equation.}$$

Check that the solution set is $\{-6, 6\}$.

NOW TRY

CAUTION The factor $4x$ in **Example 4** is a ***variable factor*** and leads to the solution 0. The factor 3 in **Example 5** is ***not a variable factor,*** so it does *not* lead to a solution of the equation.

EXAMPLE 6 Solving an Equation That Requires Rewriting

Solve $(2x + 1)(x + 1) = 2(1 - x) + 6$.

$$(2x + 1)(x + 1) = 2(1 - x) + 6$$

$$2x^2 + 3x + 1 = 2 - 2x + 6 \quad \text{Multiply on each side.}$$

$$2x^2 + 3x + 1 = 8 - 2x \quad \text{Add on the right.}$$

Write in standard form.

$$2x^2 + 5x - 7 = 0 \quad \text{Add } 2x. \text{ Subtract 8.}$$

$$(2x + 7)(x - 1) = 0 \quad \text{Factor.}$$

$$2x + 7 = 0 \quad \text{or} \quad x - 1 = 0 \quad \text{Zero-factor property}$$

$$x = -\frac{7}{2} \quad \text{or} \quad x = 1 \quad \text{Solve each equation.}$$

NOW TRY ANSWERS

4. $\{0, -4\}$
5. $\{-5, 5\}$

NOW TRY EXERCISE 6

Solve.

$$(x+3)(2x-1) = 4(x+4)-4$$

CHECK $\quad (2x+1)(x+1) = 2(1-x)+6$

$$\left[2\left(-\frac{7}{2}\right)+1\right]\left(-\frac{7}{2}+1\right) \stackrel{?}{=} 2\left[1-\left(-\frac{7}{2}\right)\right]+6 \quad \text{Let } x = -\tfrac{7}{2}.$$

$$(-7+1)\left(-\frac{5}{2}\right) \stackrel{?}{=} 2\left(\frac{9}{2}\right)+6 \quad \text{Simplify; } 1 = \tfrac{2}{2}.$$

$$(-6)\left(-\frac{5}{2}\right) \stackrel{?}{=} 9+6$$

$$15 = 15 \ \checkmark \quad \text{True}$$

Check that 1 is also a solution. The solution set is $\left\{-\frac{7}{2}, 1\right\}$.

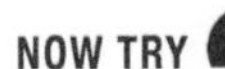

The zero-factor property can be extended to solve certain polynomial equations of degree 3 or greater, as shown in the next example.

NOW TRY EXERCISE 7

Solve $12x = 2x^3 + 5x^2$.

EXAMPLE 7 **Solving an Equation of Degree 3**

Solve $-x^3 + x^2 = -6x$.

$$-x^3 + x^2 = -6x$$

$$-x^3 + x^2 + 6x = 0 \quad \text{Add } 6x \text{ to each side.}$$

$$x^3 - x^2 - 6x = 0 \quad \text{Multiply each side by } -1.$$

$$x(x^2 - x - 6) = 0 \quad \text{Factor out } x.$$

$$x(x-3)(x+2) = 0 \quad \text{Factor the trinomial.}$$

Remember to set x equal to 0.

$$x = 0 \quad \text{or} \quad x-3=0 \quad \text{or} \quad x+2=0 \quad \text{Extend the zero-factor property to the three \textit{variable} factors.}$$

$$x = 3 \quad \text{or} \quad x = -2$$

Check that the solution set is $\{-2, 0, 3\}$. **NOW TRY**

OBJECTIVE 2 **Solve applied problems that require the zero-factor property.**

An application may lead to a quadratic equation.

EXAMPLE 8 **Using a Quadratic Equation in an Application**

A piece of sheet metal is in the shape of a parallelogram. The longer sides of the parallelogram are each 8 m longer than the distance between them. The area of the piece is 48 m^2. Find the length of the longer sides and the distance between them.

Step 1 **Read** the problem again. There will be two answers.

Step 2 **Assign a variable.**

Let $\quad x =$ the distance between the two longer sides

and $\quad x + 8 =$ the length of each of the longer sides. (See **FIGURE 1**.)

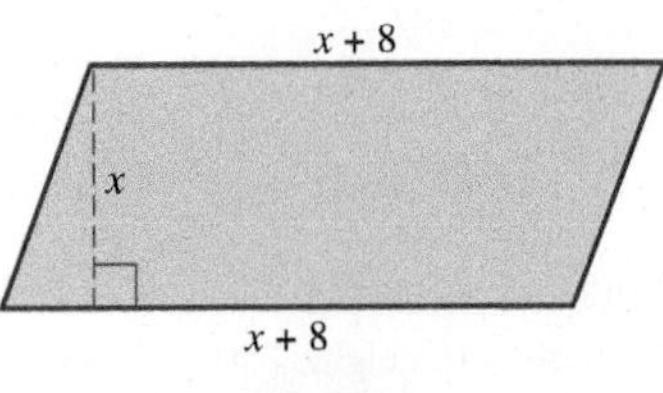

FIGURE 1

NOW TRY ANSWERS

6. $\left\{-3, \frac{5}{2}\right\}$

7. $\left\{-4, 0, \frac{3}{2}\right\}$

NOW TRY EXERCISE 8
The height of a triangle is 1 ft less than twice the length of the base. The area is 14 ft^2. What are the measures of the base and the height?

Step 3 **Write an equation.** The area of a parallelogram is given by $\mathcal{A} = bh$, where b is the length of the longer side and h is the distance between the longer sides. Here, $b = x + 8$ and $h = x$.

$$\mathcal{A} = bh \quad \text{Formula for area of a parallelogram}$$

$$48 = (x + 8)x \quad \text{Let } \mathcal{A} = 48,\ b = x + 8,\ h = x.$$

Step 4 **Solve.**

$$48 = x^2 + 8x \quad \text{Distributive property}$$

$$x^2 + 8x - 48 = 0 \quad \text{Standard form}$$

$$(x + 12)(x - 4) = 0 \quad \text{Factor.}$$

$$x + 12 = 0 \quad \text{or} \quad x - 4 = 0 \quad \text{Zero-factor property}$$

$$x = -12 \quad \text{or} \quad x = 4 \quad \text{Solve each equation.}$$

Step 5 **State the answer.** ***The distance between the longer sides cannot be negative, so reject*** -12 ***as an answer.*** The only possible answer is 4, so the required distance is 4 m. The length of the longer sides is $4 + 8 = 12$ m.

Step 6 **Check.** The length of the longer sides is 8 m more than the distance between them, and the area is

$$4 \cdot 12 = 48 \text{ m}^2, \quad \text{as required.}$$

NOW TRY 

CAUTION When an application leads to a quadratic equation, a solution of the equation may not satisfy the physical requirements of the problem, as in **Example 8.** Reject such solutions as answers.

A function defined by a quadratic polynomial is a ***quadratic function.*** The next example uses such a function.

NOW TRY EXERCISE 9
Refer to **Example 9.** After how many seconds will the rocket be 192 ft above the ground?

EXAMPLE 9 Using a Quadratic Function in an Application

If a small rocket is launched vertically upward from ground level with an initial velocity of 128 ft per sec, then its height in feet after t seconds (if air resistance is neglected) can be modeled by the function

$$h(t) = -16t^2 + 128t.$$

After how many seconds will the rocket be 220 ft above the ground?

We let $h(t) = 220$ and solve for t.

$$220 = -16t^2 + 128t \quad \text{Let } h(t) = 220.$$

$$16t^2 - 128t + 220 = 0 \quad \text{Standard form}$$

$$4t^2 - 32t + 55 = 0 \quad \text{Divide by 4.}$$

$$(2t - 5)(2t - 11) = 0 \quad \text{Factor.}$$

$$2t - 5 = 0 \quad \text{or} \quad 2t - 11 = 0 \quad \text{Zero-factor property}$$

$$t = 2.5 \quad \text{or} \quad t = 5.5 \quad \text{Solve each equation.}$$

The rocket will reach a height of 220 ft twice: on its way up at 2.5 sec and again on its way down at 5.5 sec.

NOW TRY

NOW TRY ANSWERS
8. base: 4 ft; height: 7 ft
9. 2 sec and 6 sec

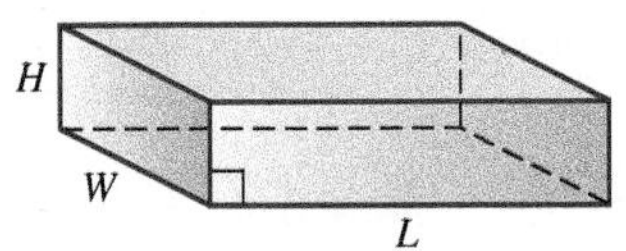

Rectangular solid
$S = 2HW + 2LW + 2LH$

FIGURE 2

OBJECTIVE 3 Solve a formula for a specified variable, where factoring is necessary.

A rectangular solid has the shape of a box, but is solid. See **FIGURE 2**. The surface area of any solid three-dimensional figure is the total area of its surface. For a rectangular solid, the surface area S is

$$S = 2HW + 2LW + 2LH.$$ H, W, and L represent height, width, and length.

NOW TRY EXERCISE 10
Solve the formula for H.
$$S = 2HW + 2LW + 2LH$$

EXAMPLE 10 Solving for a Specified Variable

Solve the formula $S = 2HW + 2LW + 2LH$ for L.

To solve for L, treat L as the only variable and treat H and W as constants.

$$S = 2HW + 2LW + 2LH$$ We must isolate the L-terms.

$$S - 2HW = 2LW + 2LH$$ Subtract $2HW$.

This is a key step. $$S - 2HW = L(2W + 2H)$$ Factor out L.

$$\frac{S - 2HW}{2W + 2H} = L, \quad \text{or} \quad L = \frac{S - 2HW}{2W + 2H}$$ Divide by $2W + 2H$. NOW TRY 

CAUTION *In* **Example 10,** ***we must write the expression so that the specified variable is a factor. Then we can divide by its coefficient in the final step.***

NOTE Earlier we saw that the graph of $f(x) = x^2$ is a parabola. In general, the graph of

$$f(x) = ax^2 + bx + c \quad (\text{where } a \neq 0)$$

is a parabola, and the x-intercepts of its graph give the real number solutions of the equation $ax^2 + bx + c = 0$. A graphing calculator can locate these x-intercepts (called **zeros** of the function).

Consider $f(x) = 2x^2 + 3x - 2$. Notice that this quadratic expression was found on the left side of the equation in Step 1 in **Example 2** earlier in this section, where the equation was written in standard form. The x-intercepts (zeros) given with the graphs in **FIGURE 3** are the same as the solutions found in **Example 2.**

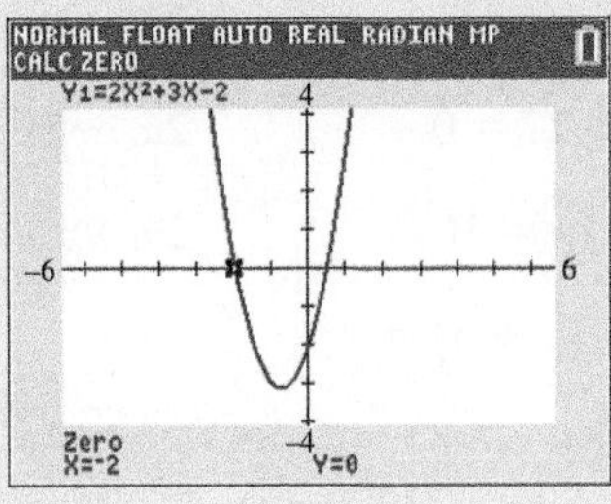

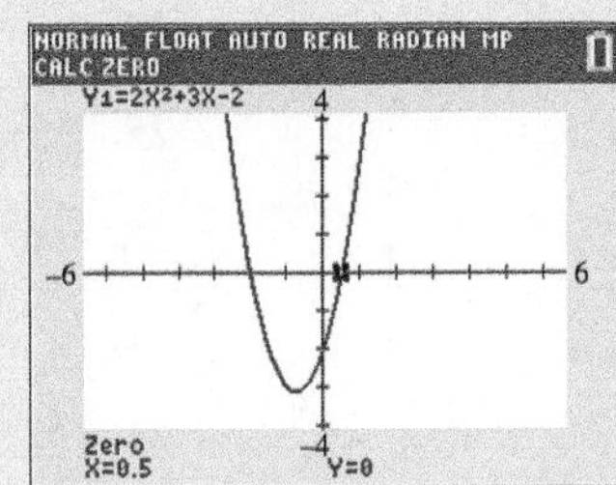

FIGURE 3

NOW TRY ANSWER
10. $H = \frac{S - 2LW}{2W + 2L}$

5.5 Exercises

FOR EXTRA HELP

 MyLab Math

Video solutions for select problems available in MyLab Math

Concept Check *Choose the letter(s) of the correct response.*

1. Which of the following are quadratic equations?

A. $2x + 3 = 6$ **B.** $x^2 - 3x + 2 = 0$ **C.** $5t^2 + 3t = 2$

D. $2x^3 + x^2 - 6x = 0$ **E.** $x^2 + 16 = 0$ **F.** $3x^2 - x = 0$

STUDY SKILLS **REMINDER**

Time management can be a challenge for students. **Review Study Skill 6, *Managing Your Time.***

2. Which of the following can be solved using the zero-factor property?

A. $25x^2 + 20x + 4 = 0$ **B.** $3x^2 = 15x$

C. $2x^2 - 128 = 0$ **D.** $x(4x - 15) - 30 = 0$

3. Concept Check A student solved the following equation incorrectly as shown.

$$(x-3)(x+2) = 6$$
$$x - 3 = 6 \quad \text{or} \quad x + 2 = 6$$
$$x = 9 \quad \text{or} \quad x = 4$$

The student gave the solution set as $\{4, 9\}$. **WHAT WENT WRONG?** Give the correct solution set.

4. Concept Check A student solved the following equation incorrectly as shown.

$$7(x+4)(x-3) = 0$$
$$x = 7 \quad \text{or} \quad x + 4 = 0 \quad \text{or} \quad x - 3 = 0$$
$$x = 7 \quad \text{or} \quad x = -4 \quad \text{or} \quad x = 3$$

The student gave the solution set as $\{-4, 3, 7\}$. **WHAT WENT WRONG?** Give the correct solution set.

5. Concept Check A student solved the following equation incorrectly as shown.

$$3x^2 + 12x = 0$$
$$3x(x+4) = 0 \quad \text{Factor.}$$
$$x + 4 = 0 \quad \text{Divide by } 3x.$$
$$x = -4$$

The student gave $\{-4\}$ as the solution set. **WHAT WENT WRONG?** Give the correct solution set.

6. Concept Check A student solved the following formula incorrectly for L as shown.

$$S = 2HW + 2LW + 2LH$$
$$S - 2LW - 2HW = 2LH$$
$$\frac{S - 2LW - 2HW}{2H} = L$$

WHAT WENT WRONG?

Solve each equation. ***See Examples 1–6.***

7. $(x+10)(x-5) = 0$
8. $(x+7)(x+3) = 0$
9. $(3k+8)(2k-5) = 0$
10. $(3q-4)(2q+5) = 0$
11. $x^2 - 3x - 10 = 0$
12. $x^2 + x - 12 = 0$
13. $x^2 + 9x + 18 = 0$
14. $x^2 - 18x + 80 = 0$
15. $2x^2 = 7x + 4$
16. $2x^2 = 3 - x$
17. $15x^2 - 7x = 4$
18. $12x^2 + 4x = 5$
19. $4p^2 + 16p = 0$
20. $2t^2 + 8t = 0$
21. $6x^2 - 36x = 0$
22. $3x^2 - 27x = 0$
23. $4p^2 - 16 = 0$
24. $9z^2 - 81 = 0$
25. $-3x^2 + 27 = 0$
26. $-2x^2 + 8 = 0$
27. $-x^2 = 9 - 6x$
28. $-x^2 - 8x = 16$
29. $9x^2 + 24x + 16 = 0$
30. $4x^2 - 20x + 25 = 0$
31. $(x-3)(x+5) = -7$
32. $(x+8)(x-2) = -21$
33. $(2x+1)(x-3) = 6x + 3$
34. $(3x+2)(x-3) = 7x - 1$
35. $2x^2 - 12 - 4x = x^2 - 3x$
36. $3x^2 + 9x + 30 = 2x^2 - 2x$
37. $(5x+1)(x+3) = -2(5x+1)$
38. $(3x+1)(x-3) = 2 + 3(x+5)$
39. $(x+3)(x-6) = (2x+2)(x-6)$
40. $(2x+1)(x+5) = (x+11)(x+3)$

Solve each equation. ***See Example 7.***

41. $2x^3 - 9x^2 - 5x = 0$
42. $6x^3 - 13x^2 - 5x = 0$
43. $x^3 - 2x^2 = 3x$
44. $x^3 - 6x^2 = -8x$
45. $9x^3 = 16x$
46. $25x^3 = 64x$

47. $2x^3 + 5x^2 - 2x - 5 = 0$

48. $2x^3 + x^2 - 98x - 49 = 0$

49. $x^3 - 6x^2 - 9x + 54 = 0$

50. $x^3 - 3x^2 - 4x + 12 = 0$

Extending Skills *Solve each equation. (Hint: In Exercises 51–54, use the substitution of variable method.)*

51. $2(x - 1)^2 - 7(x - 1) - 15 = 0$

52. $4(2x + 3)^2 - (2x + 3) - 3 = 0$

53. $5(3x - 1)^2 + 3 = -16(3x - 1)$

54. $2(x + 3)^2 = 5(x + 3) - 2$

55. $(2x - 3)^2 = 16x^2$

56. $9x^2 = (5x + 2)^2$

Solve each problem. ***See Examples 8 and 9.***

57. A garden has an area of 320 ft². Its length is 4 ft more than its width. What are the dimensions of the garden?

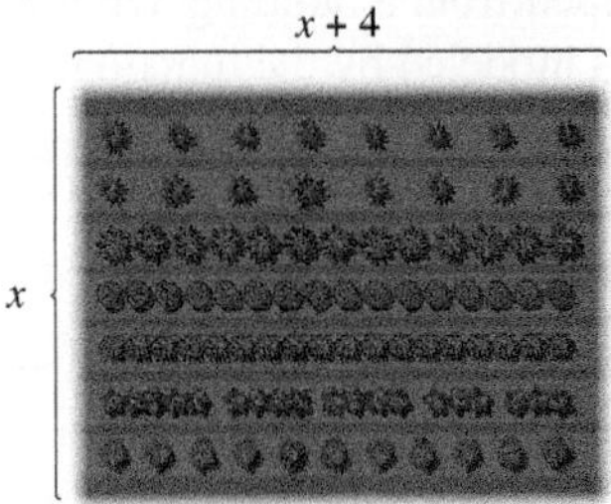

58. A square mirror has sides measuring 2 ft less than the sides of a square painting. If the difference between their areas is 32 ft², find the lengths of the sides of the mirror and the painting.

59. The base of a parallelogram is 7 ft more than the height. If the area of the parallelogram is 60 ft², what are the measures of the base and the height?

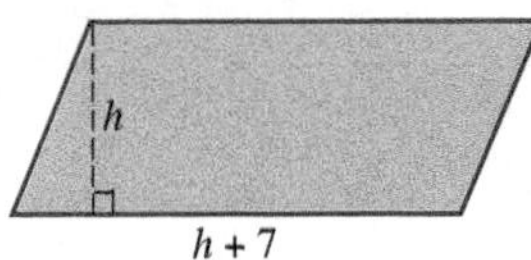

60. A sign has the shape of a triangle. The length of the base is 3 m less than the height. What are the measures of the base and the height if the area is 44 m²?

61. A farmer has 300 ft of fencing and wants to enclose a rectangular area of 5000 ft². What dimensions should she use?

62. A rectangular landfill has an area of 30,000 ft². Its length is 200 ft more than its width. What are the dimensions of the landfill?

63. Find two consecutive integers such that the sum of their squares is 61.

64. Find two consecutive integers such that their product is 72.

65. A box with no top is to be constructed from a piece of cardboard whose length measures 6 in. more than its width. The box is to be formed by cutting squares that measure 2 in. on each side from the four corners and then folding up the sides. If the volume of the box will be 110 in.³, what are the dimensions of the piece of cardboard?

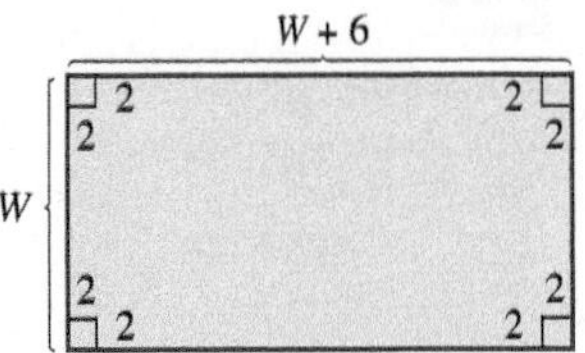

66. The surface area of the box with open top shown in the figure is 161 in.^{2}. Find the dimensions of the base. (*Hint:* The surface area of the box is modeled by the function $S(x) = x^2 + 16x$.)

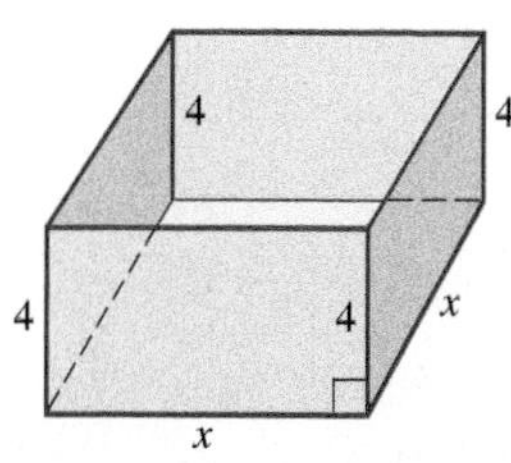

67. If an object is projected upward with an initial velocity of 64 ft per sec from a height of 80 ft, then its height in feet t seconds after it is projected is modeled by the function

$$f(t) = -16t^2 + 64t + 80.$$

How long after it is projected will it hit the ground? (*Hint:* When it hits the ground, its height is 0 ft.)

68. Refer to **Example 9.** After how many seconds will the rocket be

(a) 240 ft above the ground? **(b)** 112 ft above the ground?

69. If a rock is dropped from a building 576 ft high, then its distance in feet from the ground t seconds later is modeled by the function

$$f(t) = -16t^2 + 576.$$

How long after it is dropped will it hit the ground?

70. If a baseball is dropped from a helicopter 625 ft above the ground, then its distance in feet from the ground t seconds later is modeled by the function

$$f(t) = -16t^2 + 625.$$

How long after it is dropped will it hit the ground?

Solve each equation for the specified variable. **See Example 10.**

71. $k = dF - DF$ for F

72. $Mv = mv - Vm$ for m

73. $A = P + Prt$ for P

74. $P = kR + R - u$ for R

75. $2k + ar = r - 3y$ for r

76. $4s + 7p = tp - 7$ for p

77. $w = \dfrac{3y - x}{y}$ for y

78. $c = \dfrac{-2t + 4}{t}$ for t

Chapter 5 Summary

STUDY SKILLS REMINDER

Be prepared for your math test over this chapter. **Review Study Skill 7, *Reviewing a Chapter,* and Study Skill 8, *Taking Math Tests.***

Key Terms

5.1
factoring
greatest common factor (GCF)

5.2
prime polynomial

5.3
difference of squares
perfect square trinomial
difference of cubes
sum of cubes

5.5
quadratic equation
double solution

Test Your Word Power

See how well you have learned the vocabulary in this chapter.

1. **Factoring** is
   A. a method of multiplying polynomials
   B. the process of writing a polynomial as a product of prime factors
   C. the answer in a multiplication problem
   D. a way to add the terms of a polynomial.

2. A **difference of squares** is a binomial
   A. that can be factored as the difference of two cubes
   B. that cannot be factored
   C. that is squared
   D. that can be factored as the product of the sum and difference of two terms.

3. A **perfect square trinomial** is a trinomial
   A. that can be factored as the square of a binomial
   B. that cannot be factored
   C. that is multiplied by a binomial
   D. where all terms are perfect squares.

4. A **quadratic equation** is a polynomial equation of
   A. degree 1
   B. degree 2
   C. degree 3
   D. degree 4.

5. The **zero-factor property** is used to
   A. factor a perfect square trinomial
   B. factor by grouping
   C. solve a polynomial equation of degree 2 or more
   D. solve a linear equation.

ANSWERS

1. B; *Example:* $x^2 - 5x - 14$ factors as $(x - 7)(x + 2)$. **2.** D; *Example:* $b^2 - 49$ is the difference of the squares b^2 and 7^2. It can be factored as $(b + 7)(b - 7)$. **3.** A; *Example:* $a^2 + 2a + 1$ is a perfect square trinomial. Its factored form is $(a + 1)^2$. **4.** B; *Examples:* $x^2 - 3x + 2 = 0$, $x^2 - 9 = 0$, and $2x^2 = 6x + 8$ **5.** C; *Example:* Use the zero-factor property to write $(x + 4)(x - 2) = 0$ as $x + 4 = 0$ or $x - 2 = 0$, and then solve each linear equation to find the solution set $\{-4, 2\}$.

Quick Review

CONCEPTS	EXAMPLES

5.1 Greatest Common Factors and Factoring by Grouping

Factoring Out the Greatest Common Factor
Use the distributive property to write the given polynomial as a product of two factors, one of which is the greatest common factor of the terms of the polynomial.

Factor $4x^2y - 50xy^2$.

$4x^2y - 50xy^2$

$= 2xy(2x - 25y)$ The greatest common factor is $2xy$.

Factoring by Grouping

Step 1 Group the terms so that each group has a common factor.

Step 2 Factor out the common factor in each group.

Step 3 If each group now has a common factor, factor it out. If not, try a different grouping.

Check the factored form by multiplying.

Factor by grouping.

$5a - 5b - ax + bx$

$= (5a - 5b) + (-ax + bx)$ Group the terms.

$= 5(a - b) - x(a - b)$ Factor out 5 and $-x$.

$= (a - b)(5 - x)$ Factor out $a - b$.

5.2 Factoring Trinomials

Factoring $x^2 + bx + c$

Step 1 Find pairs of integers whose product is c.

Step 2 Choose the pair whose sum is b.

Step 3 Factor the trinomial as the product of two binomials using the pair of integers identified in Step 2.

Check the factored form by multiplying.

Factor $x^2 - 6x - 27$.

Pairs of integers with product -27	Sum of each pair
$27(-1)$	26
$-27(1)$	-26
$9(-3)$	6
$-9(3)$	-6 ⟸ Coefficient b

The integers -9 and 3 have the necessary product and sum.

$x^2 - 6x - 27$ factors as $(x - 9)(x + 3)$.

CONCEPTS	EXAMPLES
Factoring $ax^2 + bx + c$ Use one of the following methods. • Identify the values of a, b, and c. Find two integers having product ac and sum b. Use these integers to rewrite the middle term, and factor by grouping.	Factor $15x^2 + 14x - 8$. $a = 15$, $b = 14$, and $c = -8$, so $ac = 15(-8) = -120$. We must find two integers with product -120 and sum 14. $20(-6) = -120$ and $20 + (-6) = 14$ The necessary integers are 20 and -6. $15x^2 + 14x - 8$ $= 15x^2 + 20x - 6x - 8$ $20x + (-6x) = 14x$ $= (15x^2 + 20x) + (-6x - 8)$ Group the terms. $= 5x(3x + 4) - 2(3x + 4)$ Factor each group. $= (3x + 4)(5x - 2)$ Factor out $3x + 4$.
• Choose factors of the first term and factors of the last term. Write them in a pair of parentheses of this form. ( )( ) Use the FOIL method and try various combinations of the factors until the correct middle term is found. ***Check the factored form by multiplying.***	Factor $15x^2 + 14x - 8$. The positive factors of 15 are 5 and 3, and 15 and 1. The factors of -8 are -4 and 2, 4 and -2, -1 and 8, and 1 and -8. Various combinations of these factors lead to the following. $15x^2 + 14x - 8$ $= (5x - 2)(3x + 4)$ (outer: $20x$; inner: $-6x$) $20x - 6x = 14x$, the correct middle term.
5.3 Special Factoring	
Difference of Squares $x^2 - y^2 = (x + y)(x - y)$	Factor. $4m^2 - 25n^2$ $= (2m)^2 - (5n)^2$ $= (2m + 5n)(2m - 5n)$
Perfect Square Trinomials $x^2 + 2xy + y^2 = (x + y)^2$ $x^2 - 2xy + y^2 = (x - y)^2$	Factor. $9y^2 + 6y + 1$ $= (3y + 1)^2$ $16p^2 - 56p + 49$ $= (4p - 7)^2$
Difference of Cubes $x^3 - y^3 = (x - y)(x^2 + xy + y^2)$	Factor. $8 - 27a^3$ $= (2 - 3a)(4 + 6a + 9a^2)$
Sum of Cubes $x^3 + y^3 = (x + y)(x^2 - xy + y^2)$	$64z^3 + 1$ $= (4z + 1)(16z^2 - 4z + 1)$
5.4 A General Approach to Factoring See this section for guidelines and examples.	

CONCEPTS	EXAMPLES
5.5 Solving Quadratic Equations Using the Zero-Factor Property	
Zero-Factor Property If two numbers have a product of 0, then at least one of the numbers must be 0. **If $ab = 0$, then $a = 0$ or $b = 0$ (or both).**	Solve. $2x^2 + 5x = 3$ $2x^2 + 5x - 3 = 0$ Standard form $(x + 3)(2x - 1) = 0$ Factor. $x + 3 = 0$ or $2x - 1 = 0$ Zero-factor property $x = -3$ or $x = \frac{1}{2}$ Solve each equation.
Solving a Quadratic Equation Using the Zero-Factor Property *Step 1* Write the equation in standard form. *Step 2* Factor completely. *Step 3* Apply the zero-factor property. *Step 4* Solve the resulting equations. *Step 5* Check each value in the original equation. Write the solution set.	**CHECK** $2x^2 + 5x = 3$ $2(-3)^2 + 5(-3) \stackrel{?}{=} 3$ Let $x = -3$. $2(9) - 15 \stackrel{?}{=} 3$ $3 = 3$ ✓ True A check of $\frac{1}{2}$ is done similarly. Solution set: $\left\{-3, \frac{1}{2}\right\}$

Chapter 5 Review Exercises

5.1 *Factor out the greatest common factor.*

1. $12p^2 - 6p$

2. $21x^2 + 35x$

3. $12q^2b + 8qb^2 - 20q^3b^2$

4. $6r^3t - 30r^2t^2 + 18rt^3$

5. $(x + 3)(4x - 1) - (x + 3)(3x + 2)$

6. $(z + 1)(z - 4) + (z + 1)(2z + 3)$

Factor by grouping.

7. $4m + nq + mn + 4q$

8. $x^2 + 5y + 5x + xy$

9. $2m + 6 - am - 3a$

10. $x^2 + 3x - 3y - xy$

5.2 *Factor completely.*

11. $3p^2 - p - 4$

12. $6k^2 + 11k - 10$

13. $12r^2 - 5r - 3$

14. $10m^2 + 37m + 30$

15. $10k^2 - 11kh + 3h^2$

16. $9x^2 + 4xy - 2y^2$

17. $24x - 2x^2 - 2x^3$

18. $6b^3 - 9b^2 - 15b$

19. $y^4 + 2y^2 - 8$

20. $2k^4 - 5k^2 - 3$

21. $p^2(p + 2)^2 + p(p + 2)^2 - 6(p + 2)^2$

22. $3(r + 5)^2 - 11(r + 5) - 4$

23. $24x^8 - 126x^6 + 81x^4$

24. $30y^9 - 155y^6 - 55y^3$

5.3 *Factor completely.*

25. $16x^2 - 25$

26. $9t^2 - 49$

27. $36m^2 - 25n^2$

28. $x^2 + 14x + 49$

29. $9k^2 - 12k + 4$

30. $r^3 + 27$

31. $125x^3 - 1$

32. $m^6 - 1$

33. $x^8 - 1$

34. $x^2 + 6x + 9 - 25y^2$

35. $(a + b)^3 - (a - b)^3$

36. $x^5 - x^3 - 8x^2 + 8$

5.5 *Solve each equation.*

37. $x^2 - 8x + 16 = 0$

38. $(5x + 2)(x + 1) = 0$

39. $x^2 - 5x + 6 = 0$

40. $x^2 + 2x = 8$

41. $6x^2 = 5x + 50$

42. $6x^2 + 7x = 3$

43. $8x^2 + 14x + 3 = 0$

44. $-4x^2 + 36 = 0$

45. $6x^2 + 9x = 0$

46. $(2x + 1)(x - 2) = -3$

47. $(x - 1)(x + 3) = (2x - 1)(x - 1)$

48. $2x^3 - x^2 - 28x = 0$

49. $-x^3 - 3x^2 + 4x + 12 = 0$

50. $(x + 2)(5x^2 - 9x - 18) = 0$

Solve each problem.

51. A triangular wall brace has the shape of a right triangle. One of the perpendicular sides is 1 ft longer than twice the other. The area enclosed by the triangle is 10.5 ft². Find the shorter of the perpendicular sides.

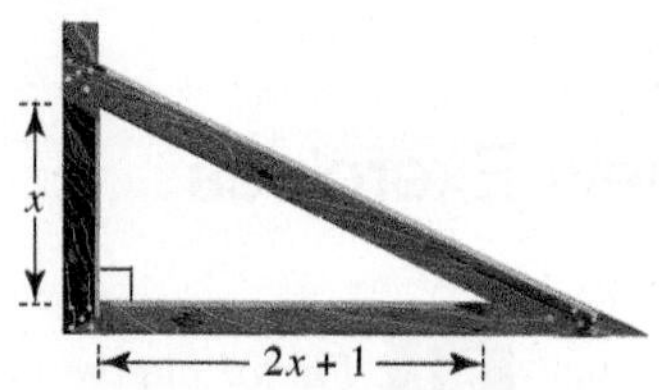

The area is 10.5 ft².

52. A rectangular parking lot has a length 20 ft more than its width. Its area is 2400 ft². What are the dimensions of the lot?

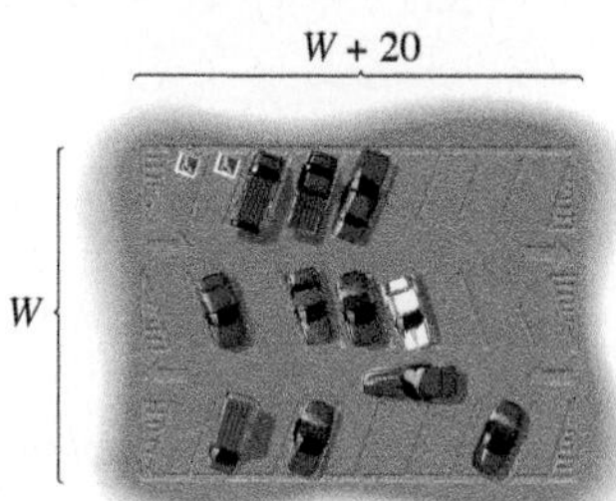

The area is 2400 ft².

A rock is projected directly upward from ground level. After t seconds, its height (if air resistance is neglected) is modeled by the function

$$f(t) = -16t^2 + 256t.$$

53. After how many seconds will the rock return to the ground?

54. After how many seconds will it be 240 ft above the ground?

55. After how many seconds does the rock reach its maximum height, 1024 ft?

56. Why does the question in **Exercise 54** have two answers?

Solve each equation for the specified variable.

57. $3s + bk = k - 2t$ for k

58. $z = \dfrac{3w + 7}{w}$ for w

Chapter 5 Mixed Review Exercises

Factor completely.

1. $x^2 + 5y + 5x + xy$
2. $-x^2 + 3x + 10$
3. $16x^2 + 144$
4. $30a + am - am^2$
5. $8 - a^3$
6. $81k^2 - 16$
7. $9x^2 + 13xy - 3y^2$
8. $15y^3 + 20y^2$
9. $25z^2 - 30zm + 9m^2$
10. Which equation is *not* in proper form for using the zero-factor property? Why?

    A. $(x + 2)(x - 6) = 0$
    B. $x(3x - 7) = 0$
    C. $3t(t + 8)(t - 9) = 0$
    D. $y(y - 3) + 6(y - 3) = 0$

Solve.

11. $5x^2 - 17x - 12 = 0$
12. $x^3 - x = 0$
13. $3m^2 - 9m = 0$
14. $2\mathcal{A} - hb = hB$ for h
15. $S = 2HW + 2LW + 2LH$ for H
16. The rectangular floor area of a typical Native American longhouse was about 2750 ft^2. The length was 85 ft greater than the width. What were the dimensions of the floor?

17. The length of a rectangular picture frame is 2 in. longer than its width. The area enclosed by the frame is 48 in.^2. What is the width?

18. An arrow is shot upward from a platform 40 ft high with an initial velocity of 224 ft per sec. Its height h in feet after t seconds is modeled by the function

$$h(t) = -16t^2 + 224t + 40.$$

After how many seconds will the arrow be 760 ft above the ground?

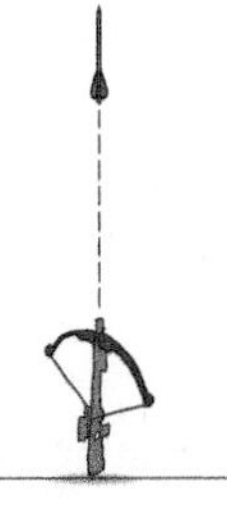

STUDY SKILLS **REMINDER**

We can learn from the mistakes we make. **Review Study Skill 9, *Analyzing Your Test Results.***

Chapter 5 Test

Step-by-step test solutions are found on the Chapter Test Prep Videos available in MyLab Math.

View the complete solutions to all Chapter Test exercises in MyLab Math.

Factor.

1. $11z^2 - 44z$

2. $10x^2y^5 - 5x^2y^3 - 25x^5y^3$

3. $3x + by + bx + 3y$

4. $-2x^2 - x + 36$

5. $6x^2 + 11x - 35$

6. $4p^2 + 3pq - q^2$

7. $16a^2 + 40ab + 25b^2$

8. $x^2 + 2x + 1 - 4z^2$

9. $a^3 + 2a^2 - ab^2 - 2b^2$

10. $9k^2 - 121j^2$

11. $y^3 - 216$

12. $6k^4 - k^2 - 35$

13. $27x^6 + 1$

14. $(t^2 + 3)^2 + 4(t^2 + 3) - 5$

15. $16x^2 + 9$

16. $-x^2 + x + 30$

17. $125 - 216y^6$

18. Which one of the following is *not* a factored form of $-x^2 - x + 12$?

A. $(3 - x)(x + 4)$ **B.** $-(x - 3)(x + 4)$

C. $(-x + 3)(x + 4)$ **D.** $(x - 3)(-x + 4)$

Solve each equation.

19. $3x^2 + 8x = -4$

20. $3x^2 - 5x = 0$

21. $5m(m - 1) = 2(1 - m)$

22. $ar + 2 = 3r - 6t$ for r

Solve each problem.

23. The area of the rectangle shown is 40 in.^{2}. Find the length and the width of the rectangle.

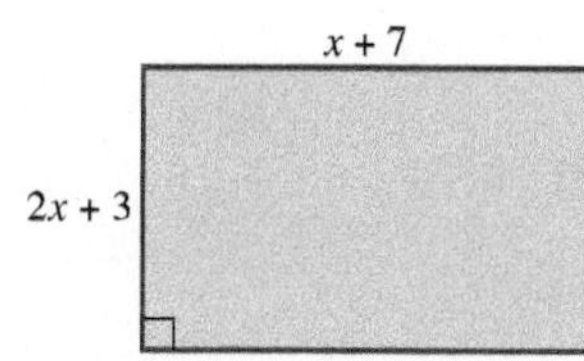

The area is 40 in.^{2}.

24. A ball is projected upward from ground level. After t seconds, its height in feet is modeled by the function

$$f(t) = -16t^2 + 96t.$$

After how many seconds will it reach a height of 128 ft?

Chapters R–5 Cumulative Review Exercises

Perform the indicated operations.

1. $\frac{5}{6} + \frac{7}{12} - \frac{1}{3}$

2. $2.5(6.5) - 0.15$

Simplify.

3. $-2(m - 3)$

4. $3x^2 - 4x + 4 + 9x - x^2$

Evaluate for $p = -4$, $q = -2$, and $r = 5$.

5. $\frac{5p + 6r^2}{p^2 + q - 1}$

6. $\frac{\sqrt{r}}{-p + 2q}$

Solve.

7. $2x - 5 + 3x = 4 - (x + 2)$

8. $\frac{3x - 1}{5} + \frac{x + 2}{2} = -\frac{3}{10}$

9. $3 - 2(x + 3) < 4x$

10. $2x + 4 < 10$ and $3x - 1 > 5$

11. $2x + 4 > 10$ or $3x - 1 < 5$

12. $|5x + 3| - 10 = 3$

13. $|x + 2| < 9$

14. $|2x - 5| \geq 9$

15. Two planes leave the Dallas–Fort Worth airport at the same time. One travels east at 550 mph, and the other travels west at 500 mph. Assuming no wind, how long will it take for the planes to be 2100 mi apart?

Plane	r	t	d
Eastbound	550	x	
Westbound	500	x	
			← Total

16. What is the slope of the line shown here?

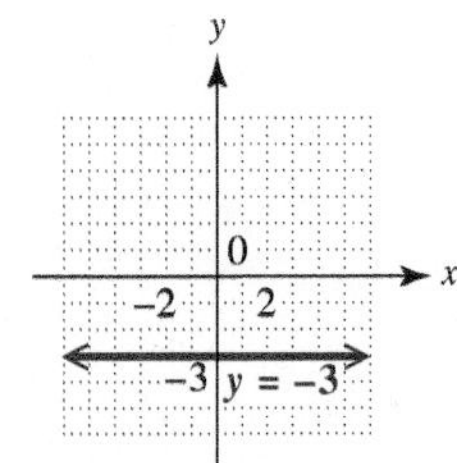

17. Find the slope of the line passing through the points $(-4, 8)$ and $(-2, 6)$.

18. Graph $4x + 2y = -8$.

Use the function $f(x) = 2x + 7$ to find each of the following.

19. $f(-4)$

20. The x-intercept of its graph

21. The y-intercept of its graph

Solve each system.

22. $3x - 2y = -7$
$2x + 3y = 17$

23. $y = 5x + 6$
$-10x = -2y - 12$

24. $2x + 3y - 6z = 5$
$8x - y + 3z = 7$
$3x + 4y - 3z = 7$

Perform the indicated operations.

25. $(7x + 3y)^2$

26. $\frac{2x^4 + x^3 + 7x^2 + 2x + 6}{x^2 + 2}$

27. $(3x^3 + 4x^2 - 7) - (2x^3 - 8x^2 + 3x)$

28. **(a)** Write 0.0004638 using scientific notation.

(b) Write 5.66×10^5 in standard notation.

Factor.

29. $16w^2 + 50wz - 21z^2$

30. $4x^2 - 4x + 1 - y^2$

31. $100x^4 - 81$

32. $8p^3 + 27$

Solve.

33. $(x - 1)(2x + 3)(x + 4) = 0$

34. $9x^2 = 6x - 1$

35. A sign is to have the shape of a triangle with a height 3 ft greater than the length of the base. How long should the base be if the area is to be 14 ft^2?

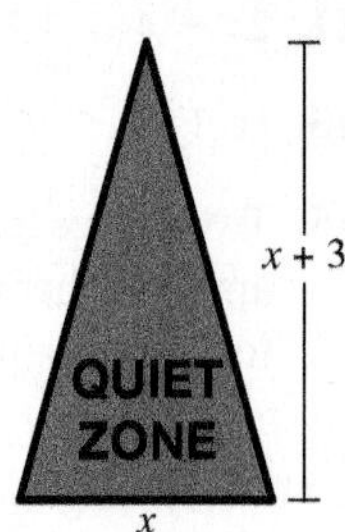

36. A game board has the shape of a rectangle. The longer sides are each 2 in. longer than the distance between them. The area of the board is 288 in.^2. Find the length of the longer sides and the distance between them.

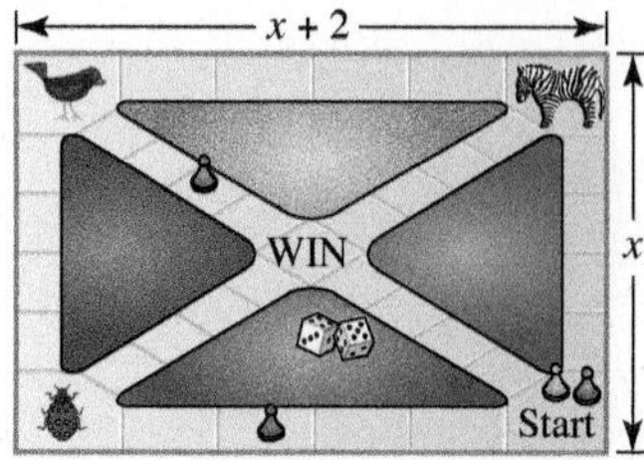

STUDY SKILLS **REMINDER**

It is not too soon to begin preparing for your final exam. **Review Study Skill 10, *Preparing for Your Math Final Exam.***

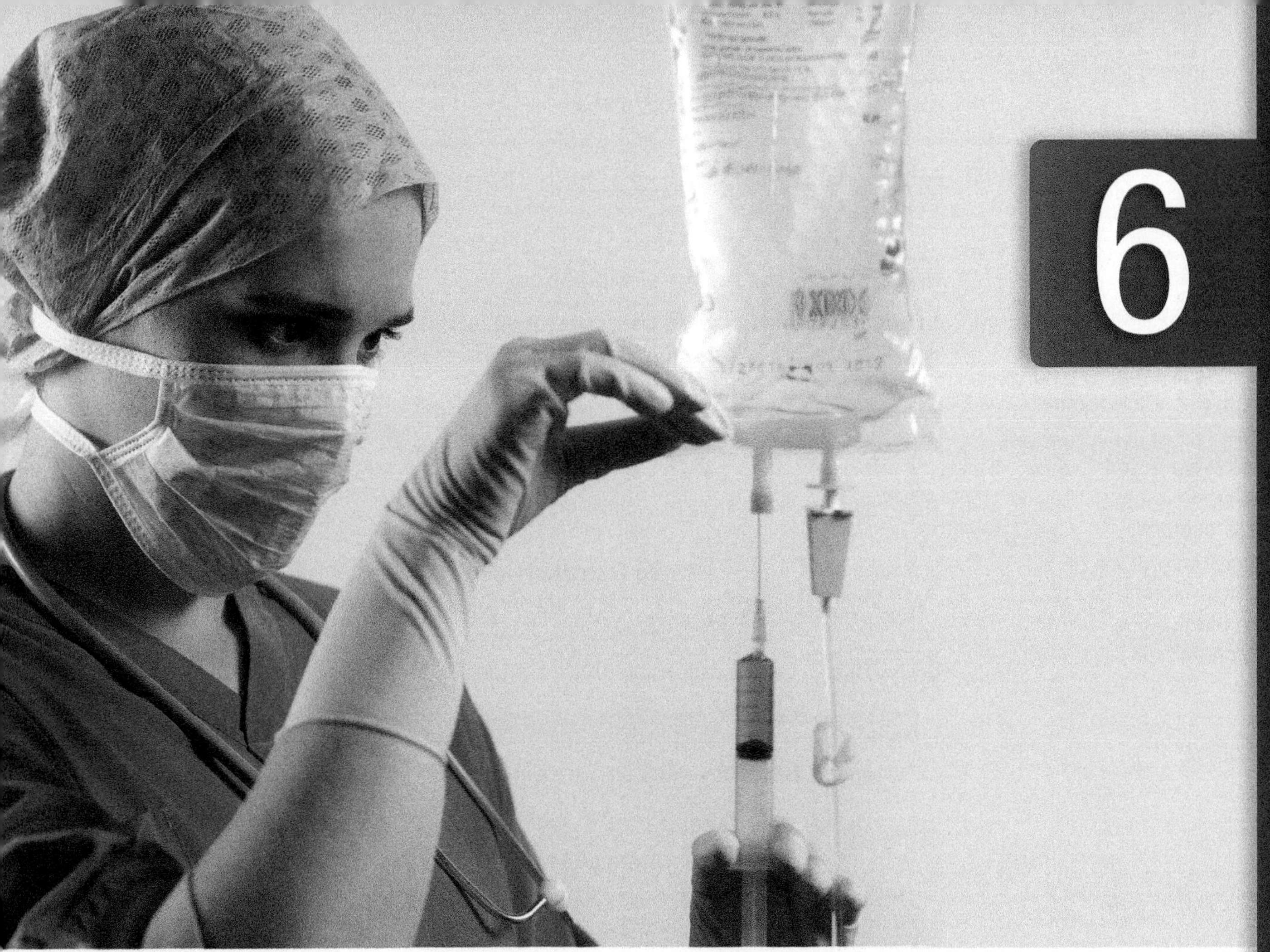

6

RATIONAL EXPRESSIONS AND FUNCTIONS

Ratios and *proportions,* topics covered in this chapter and used in health care to determine medication dosages, involve *rational expressions*.

6.1 Rational Expressions and Functions; Multiplying and Dividing

OBJECTIVES

1 Define rational expressions.
2 Define rational functions and give their domains.
3 Write rational expressions in lowest terms.
4 Multiply rational expressions.
5 Find reciprocals of rational expressions.
6 Divide rational expressions.

VOCABULARY

☐ rational expression
☐ rational function
☐ reciprocal

OBJECTIVE 1 Define rational expressions.

In arithmetic, a rational number is the quotient of two integers, with the denominator not 0. In algebra, a **rational expression,** or *algebraic fraction,* is the quotient of two polynomials, again with the denominator not 0.

$$\frac{x}{y},\quad \frac{-a}{4},\quad \frac{m+4}{m-2},\quad \frac{8x^2-2x+5}{4x^2+5x},\quad x^5 \left(\text{or } \frac{x^5}{1}\right) \qquad \text{Rational expressions}$$

Rational expressions are elements of the set

$$\left\{\frac{P}{Q} \,\middle|\, \textbf{P and Q are polynomials, where } Q \neq 0\right\}.$$

OBJECTIVE 2 Define rational functions and give their domains.

Rational Function

A **rational function** is defined by a quotient of polynomials and has the form

$$f(x) = \frac{P(x)}{Q(x)}, \quad \textbf{where } Q(x) \neq 0.$$

The domain of a rational function includes all real numbers except those that make $Q(x)$—that is, the denominator—equal to 0.

For example, the domain of the rational function

$$f(x) = \frac{2}{x-5} \leftarrow \text{Cannot equal 0}$$

includes all real numbers except 5 because 5 would make the denominator equal to 0. See **FIGURE 1**. The graph does not exist when $x = 5$. (It does not intersect the dashed vertical line with equation $x = 5$. This line is an *asymptote*.)

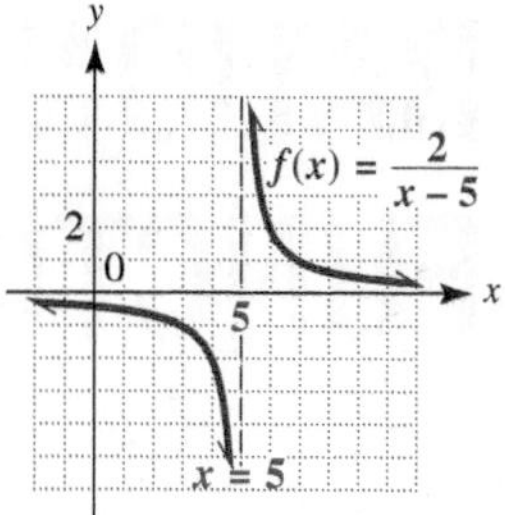

FIGURE 1

EXAMPLE 1 Finding Domains of Rational Functions

Give the domain of each rational function.

(a) $f(x) = \frac{3}{7x-14}$

Values of x that make the denominator 0 must be *excluded* from the domain. To find these values, set the denominator equal to 0 and solve.

$$7x - 14 = 0$$

$$7x = 14 \qquad \text{Add 14.}$$

$$x = 2 \qquad \text{Divide by 7.}$$

The number 2 cannot be used as a replacement for x. Therefore, the domain of f includes all real numbers *except* 2. It can be written using set-builder or interval notation.

Set-builder notation: $\{x \mid x \text{ is a real number}, x \neq 2\}$

Interval notation: $(-\infty, 2) \cup (2, \infty)$

NOW TRY EXERCISE 1

Give the domain of each rational function using set-builder notation and interval notation.

(a) $\frac{1}{x+9}$

(b) $\frac{2x-1}{x^2-4x-5}$

(c) $\frac{15}{2x^2+1}$

(b) $g(x) = \frac{3+x}{x^2-4x+3}$ Values that make the denominator 0 must be *excluded.*

$$x^2 - 4x + 3 = 0 \quad \text{Set the denominator equal to 0.}$$

$$(x-1)(x-3) = 0 \quad \text{Factor.}$$

$$x - 1 = 0 \quad \text{or} \quad x - 3 = 0 \quad \text{Zero-factor property}$$

$$x = 1 \quad \text{or} \quad x = 3 \quad \text{Solve each equation.}$$

Domain: $\{x \mid x \text{ is a real number}, x \neq 1, 3\}$ Set-builder notation

$(-\infty, 1) \cup (1, 3) \cup (3, \infty)$ Interval notation

(c) $h(x) = \frac{8x+2}{3}$ The denominator, 3, can never be 0, so the domain of h includes all real numbers.

Domain: $\{x \mid x \text{ is a real number}\}$ Set-builder notation

$(-\infty, \infty)$ Interval notation

(d) $f(x) = \frac{2}{x^2+4}$

Setting the denominator $x^2 + 4$ equal to 0 leads to $x^2 = -4$. There is no real number whose square is -4, so any real number can be used as a replacement for x.

Domain: $\{x \mid x \text{ is a real number}\}$ Set-builder notation

$(-\infty, \infty)$ Interval notation

NOW TRY 

OBJECTIVE 3 Write rational expressions in lowest terms.

In arithmetic, we use the **fundamental property of rational numbers** to write a fraction in lowest terms by dividing out any common factors in the numerator and denominator.

Fundamental Property of Rational Numbers

If $\frac{a}{b}$ is a rational number and if c is any nonzero real number, then

$$\frac{a}{b} = \frac{ac}{bc}.$$

That is, the numerator and denominator of a rational number may either be multiplied or divided by the same *nonzero number* without changing the value of the rational number.

Example: $\frac{15}{20} = \frac{3 \cdot 5}{4 \cdot 5} = \frac{3}{4} \cdot 1 = \frac{3}{4}$

Because $\frac{c}{c}$ is equivalent to 1, the fundamental property is based on the identity property of multiplication.

We write a rational expression in lowest terms similarly.

Writing a Rational Expression in Lowest Terms

Step 1 **Factor** both numerator and denominator to find their greatest common factor (GCF).

Step 2 **Apply the fundamental property.** Divide out common factors.

NOW TRY ANSWERS

1. (a) $\{x \mid x \text{ is a real number}, x \neq -9\}$; $(-\infty, -9) \cup (-9, \infty)$
   (b) $\{x \mid x \text{ is a real number}, x \neq -1, 5\}$; $(-\infty, -1) \cup (-1, 5) \cup (5, \infty)$
   (c) $\{x \mid x \text{ is a real number}\}$; $(-\infty, \infty)$

NOW TRY EXERCISE 2

Write each rational expression in lowest terms.

(a) $\dfrac{3a^2 - 7a + 2}{a^2 + 2a - 8}$

(b) $\dfrac{t^3 + 8}{t + 2}$

(c) $\dfrac{am - bm + an - bn}{am + bm + an + bn}$

EXAMPLE 2 Writing Rational Expressions in Lowest Terms

Write each rational expression in lowest terms.

(a) $\dfrac{(x+5)(x+2)}{(x+2)(x-3)}$

$\dfrac{x+2}{x+2} = 1$

$= \dfrac{(x+5)(x+2)}{(x-3)(x+2)}$ Commutative property

$= \dfrac{x+5}{x-3}$ Fundamental property

(b) $\dfrac{a^2 - a - 6}{a^2 + 5a + 6}$

$= \dfrac{(a-3)(a+2)}{(a+3)(a+2)}$ Factor the numerator. Factor the denominator.

$= \dfrac{a-3}{a+3}$ $\frac{a+2}{a+2} = 1$; Fundamental property

(c) $\dfrac{y^2 - 4}{2y + 4}$

$= \dfrac{(y+2)(y-2)}{2(y+2)}$ Factor the difference of squares in the numerator. Factor the denominator.

$= \dfrac{y-2}{2}$ $\frac{y+2}{y+2} = 1$; Fundamental property

(d) $\dfrac{x^3 - 27}{x - 3}$

$= \dfrac{(x-3)(x^2+3x+9)}{x-3}$ Factor the difference of cubes in the numerator.

$= x^2 + 3x + 9$ Fundamental property

(e) $\dfrac{pr + qr + ps + qs}{pr + qr - ps - qs}$

$= \dfrac{(pr+qr) + (ps+qs)}{(pr+qr) - (ps+qs)}$ Group the terms.

Be careful with signs.

$= \dfrac{r(p+q) + s(p+q)}{r(p+q) - s(p+q)}$ Factor within the groups.

$= \dfrac{(p+q)(r+s)}{(p+q)(r-s)}$ Factor by grouping.

$= \dfrac{r+s}{r-s}$ Fundamental property

(f) $\dfrac{8+k}{16}$

Be careful. The numerator cannot be factored.

This expression cannot be simplified further and is in lowest terms. NOW TRY

NOW TRY ANSWERS

2. (a) $\frac{3a-1}{a+4}$

(b) $t^2 - 2t + 4$

(c) $\frac{a-b}{a+b}$

CAUTION ***Only common factors may be divided out.***

Examples: $\frac{8+k}{16} \neq \frac{k}{2}$ and $\frac{8+k}{16} \neq \frac{1}{2}+k$ The 8 in $8+k$ is ***not*** a *factor* of the numerator.

The expression $\frac{8+k}{16}$ **(Example 2(f))** indicates that the *entire* numerator is being divided by 16, not just one term. It is already in lowest terms, although it could be written in the equivalent form $\frac{8}{16}+\frac{k}{16}$, or $\frac{1}{2}+\frac{k}{16}$.

Compare this to the expression $\frac{8k}{16}$, which *can* be written in lowest terms.

$$\frac{8k}{16} = \frac{8 \cdot k}{8 \cdot 2} = 1 \cdot \frac{k}{2} = \frac{k}{2}$$ ***Factor*** before writing a fraction in lowest terms.

Look again at the rational expression from **Example 2(b).**

$$\frac{a^2-a-6}{a^2+5a+6}, \quad \text{or} \quad \frac{(a-3)(a+2)}{(a+3)(a+2)}$$

In this expression, a can take any value *except* -3 or -2, because these values make the denominator 0. In the simplified expression $\frac{a-3}{a+3}$, a cannot equal -3. Thus,

$$\frac{a^2-a-6}{a^2+5a+6} = \frac{a-3}{a+3}, \quad \text{for all values of } a \text{ except } -3 \text{ or } -2.$$

From now on, such statements of equality will be made with the understanding that they apply only to those real numbers that make neither denominator equal 0. We will no longer state such restrictions.

NOW TRY EXERCISE 3
Write each rational expression in lowest terms.

(a) $\frac{a-10}{10-a}$ **(b)** $\frac{81-y^2}{y-9}$

EXAMPLE 3 Writing Rational Expressions in Lowest Terms

Write each rational expression in lowest terms.

(a) $\frac{m-3}{3-m}$ Here, the numerator and denominator are opposites.

To write in lowest terms, we write the denominator as $-1(m-3)$, as shown on the left. Alternatively, we could write the numerator as $-1(3-m)$ and obtain the same result, as shown on the right.

$$\frac{m-3}{3-m} = \frac{m-3}{-1(m-3)} = \frac{1}{-1} = -1$$

Factor out -1 in the denominator.

$$\frac{m-3}{3-m} = \frac{-1(3-m)}{3-m} = \frac{-1}{1} = -1$$

Factor out -1 in the numerator.

(b) $\frac{r^2-16}{4-r}$

$= \frac{(r+4)(r-4)}{4-r}$ Factor the difference of squares in the numerator.

$= \frac{(r+4)(r-4)}{-1(r-4)}$ Factor out -1 in the denominator to write $4-r$ as $-1(r-4)$.

Distribute to check.
$-1(r-4)$
$= -1 \cdot r - 1(-4)$
$= -r+4$, or $4-r$

$= \frac{r+4}{-1}$ $\frac{r-4}{r-4} = 1$; Fundamental property

$= -(r+4)$, or $-r-4$ Lowest terms

NOW TRY

NOW TRY ANSWERS
3. (a) -1
(b) $-(y+9)$, or $-y-9$

As shown in **Example 3,** the quotient $\frac{a}{-a}$ (where $a \neq 0$) can be simplified.

$$\frac{a}{-a} = \frac{a}{-1(a)} = \frac{1}{-1} = -1$$

The quotient $\frac{-a}{a}$ (again where $a \neq 0$) can be simplified similarly.

Quotient of Opposites

In general, if the numerator and the denominator of a rational expression are opposites, then the expression equals -1.

Examples: $\frac{q-7}{7-q} = -1$ and $\frac{-5a+2b}{5a-2b} = -1$ Numerator and denominator in each expression are opposites.

However, the following expression *cannot* be simplified further.

$\frac{r-2}{r+2}$ Numerator and denominator are *not* opposites.

OBJECTIVE 4 Multiply rational expressions.

Recall that if $\frac{a}{b}$ and $\frac{c}{d}$ are rational numbers (where $b \neq 0$ and $d \neq 0$), then

$$\frac{\mathbf{a}}{\mathbf{b}} \cdot \frac{\mathbf{c}}{\mathbf{d}} = \frac{\mathbf{ac}}{\mathbf{bd}}.$$ Multiply numerators. Multiply denominators.

Examples: $\frac{2}{3} \cdot \frac{4}{5} = \frac{2 \cdot 4}{3 \cdot 5} = \frac{8}{15}$ and $\frac{7}{10} \cdot \frac{5}{6} = \frac{7}{2 \cdot 5} \cdot \frac{5}{2 \cdot 3} = \frac{7}{12}$

Multiplying Rational Expressions

Step 1 **Factor** all numerators and denominators as completely as possible.

Step 2 **Apply the fundamental property**—that is, $\frac{ac}{bc} = \frac{a}{b}$ (where $b, c \neq 0$).

Step 3 **Multiply** remaining factors in the numerators and remaining factors in the denominators. The final expression can be left in factored form.

Step 4 **Check** to be sure the product is in lowest terms.

EXAMPLE 4 Multiplying Rational Expressions

Multiply.

(a) $\frac{5p-5}{p} \cdot \frac{3p^2}{10p-10}$

$= \frac{5(p-1)}{p} \cdot \frac{3p \cdot p}{2 \cdot 5(p-1)}$ Factor.

$= \frac{5(p-1)}{5(p-1)} \cdot \frac{p}{p} \cdot \frac{3p}{2}$ Commutative property

$= 1 \cdot 1 \cdot 1 \cdot \frac{3p}{2}$ Fundamental property

In practice, this step is usually done mentally.

$= \frac{3p}{2}$ Lowest terms

NOW TRY EXERCISE 4

Multiply.

(a) $\frac{8t^2}{t^2-4} \cdot \frac{3t+6}{9t}$

(b) $\frac{m^2+2m-15}{m^2-5m+6} \cdot \frac{m^2-4}{m^2+5m}$

(b) $\frac{k^2+2k-15}{k^2-4k+3} \cdot \frac{k^2-k}{k^2+k-20}$

$= \frac{(k+5)(k-3)}{(k-3)(k-1)} \cdot \frac{k(k-1)}{(k+5)(k-4)}$ Factor.

$= \frac{k}{k-4}$ Fundamental property; Multiply.

(c) $(p-4) \cdot \frac{3}{5p-20}$

$= \frac{p-4}{1} \cdot \frac{3}{5p-20}$ Write $p-4$ as $\frac{p-4}{1}$.

$= \frac{p-4}{1} \cdot \frac{3}{5(p-4)}$ Factor.

$= \frac{3}{5}$ Fundamental property; Multiply.

(d) $\frac{x^2+2x}{x+1} \cdot \frac{x^2-1}{x^3+x^2}$

$= \frac{x(x+2)}{x+1} \cdot \frac{(x+1)(x-1)}{x^2(x+1)}$ Factor.

$= \frac{(x+2)(x-1)}{x(x+1)}$ Fundamental property; Multiply.

The final expression can be left in factored form. It is in lowest terms.

(e) $\frac{x-6}{x^2-12x+36} \cdot \frac{x^2-3x-18}{x^2+7x+12}$

$= \frac{x-6}{(x-6)^2} \cdot \frac{(x+3)(x-6)}{(x+3)(x+4)}$ Factor.

Remember to include 1 in the numerator when all other factors are eliminated.

$= \frac{1}{x+4}$ Fundamental property; Multiply. NOW TRY

▼ Reciprocals of Rational Expressions

Rational Expression	*Reciprocal*
3, or $\frac{3}{1}$	$\frac{1}{3}$
$\frac{5}{k}$	$\frac{k}{5}$
$\frac{m^2-9m}{2}$	$\frac{2}{m^2-9m}$
$\frac{0}{4}$	undefined

Reciprocals have a product of 1. Recall that 0 has no reciprocal.

OBJECTIVE 5 Find reciprocals of rational expressions.

When dividing rational numbers, we use reciprocals. The rational numbers $\frac{a}{b}$ and $\frac{c}{d}$ are reciprocals of each other if they have a product of 1.

Example: $\frac{3}{4} \cdot \frac{4}{3} = 1$, so $\frac{3}{4}$ and $\frac{4}{3}$ are reciprocals.

The **reciprocal** of a rational expression is defined in the same way. ***Two rational expressions are reciprocals of each other if they have a product of 1.***

Finding the Reciprocal

To find the reciprocal of a nonzero rational expression, interchange the numerator and denominator of the expression. (See the table.)

NOW TRY ANSWERS

4. (a) $\frac{8t}{3(t-2)}$ (b) $\frac{m+2}{m}$

OBJECTIVE 6 Divide rational expressions.

Recall that if $\frac{a}{b}$ and $\frac{c}{d}$ are rational numbers (where $b \neq 0$, $c \neq 0$, and $d \neq 0$), then

$$\frac{a}{b} \div \frac{c}{d} = \frac{a}{b} \cdot \frac{d}{c} = \frac{ad}{bc}.$$ To divide by a fraction, multiply by its reciprocal.

Examples: $\frac{3}{4} \div \frac{2}{3} = \frac{3}{4} \cdot \frac{3}{2} = \frac{9}{8}$ and $\frac{6}{7} \div \frac{4}{3} = \frac{6}{7} \cdot \frac{3}{4} = \frac{2 \cdot 3}{7} \cdot \frac{3}{2 \cdot 2} = \frac{9}{14}$

Dividing rational expressions is similar to dividing rational numbers.

Dividing Rational Expressions

To divide two rational expressions, *multiply* the first expression (the *dividend*) by the reciprocal of the second expression (the *divisor*).

NOW TRY EXERCISE 5

Divide.

(a) $\frac{16k^2}{5} \div \frac{3k}{10}$

(b) $\frac{3k^2 + 5k - 2}{9k^2 - 1} \div \frac{4k^2 + 8k}{k^2 - 7k}$

EXAMPLE 5 Dividing Rational Expressions

Divide.

(a) $\frac{2z}{9} \div \frac{5z^2}{18}$

$= \frac{2z}{9} \cdot \frac{18}{5z^2}$ Multiply by the reciprocal of the divisor.

$= \frac{2z}{9} \cdot \frac{2 \cdot 9}{5z \cdot z}$ Factor.

$= \frac{4}{5z}$ Fundamental property; Multiply.

(b) $\frac{8k - 16}{3k} \div \frac{3k - 6}{4k^2}$

$= \frac{8k - 16}{3k} \cdot \frac{4k^2}{3k - 6}$ Multiply by the reciprocal.

$= \frac{8(k - 2)}{3k} \cdot \frac{4k \cdot k}{3(k - 2)}$ Factor.

$= \frac{32k}{9}$ Fundamental property; Multiply.

(c) $\frac{5m^2 + 17m - 12}{3m^2 + 7m - 20} \div \frac{5m^2 + 2m - 3}{15m^2 - 34m + 15}$

$= \frac{5m^2 + 17m - 12}{3m^2 + 7m - 20} \cdot \frac{15m^2 - 34m + 15}{5m^2 + 2m - 3}$ Definition of division

$= \frac{(5m - 3)(m + 4)}{(3m - 5)(m + 4)} \cdot \frac{(5m - 3)(3m - 5)}{(5m - 3)(m + 1)}$ Factor.

$= \frac{5m - 3}{m + 1}$ Fundamental property; Multiply.

NOW TRY

NOW TRY ANSWERS

5. **(a)** $\frac{32k}{3}$ **(b)** $\frac{k - 7}{4(3k + 1)}$

6.1 Exercises

FOR EXTRA HELP  MyLab Math

Video solutions for select problems available in MyLab Math

STUDY SKILLS REMINDER

We can learn from the mistakes we make. **Review Study Skill 9, *Analyzing Your Test Results.***

Concept Check *Complete each statement.*

1. A rational number such as $\frac{3}{4}$ is the quotient of two ______, with denominator not ____.

A ______ expression such as $\frac{2x^2}{2x+1}$ is the quotient of two ______, with denominator not ____.

2. A function of the form $f(x) = \frac{P(x)}{\_\_\_\_}$, where $P(x)$ and $Q(x)$ are polynomials and $Q(x) \neq$ ____, is a ______ function. The domain includes all ______ except any values of x that make the (*numerator / denominator*) equal to 0.

*Give the domain of each rational function using **(a)** set-builder notation and **(b)** interval notation.* ***See Example 1.***

3. $f(x) = \frac{x}{x-7}$

4. $f(x) = \frac{x}{x+3}$

5. $f(x) = \frac{6x-5}{7x+1}$

6. $f(x) = \frac{8x-3}{2x+7}$

7. $f(x) = \frac{12x+3}{x}$

8. $f(x) = \frac{9x+8}{x}$

9. $f(x) = \frac{3x+1}{2x^2+x-6}$

10. $f(x) = \frac{2x+4}{3x^2+11x-42}$

11. $f(x) = \frac{x+2}{14}$

12. $f(x) = \frac{x-9}{26}$

13. $f(x) = \frac{2x^2-3x+4}{3x^2+8}$

14. $f(x) = \frac{9x^2-8x+3}{4x^2+1}$

Concept Check *As review, multiply or divide the rational numbers as indicated. Write answers in lowest terms.*

15. $\frac{4}{21} \cdot \frac{7}{10}$

16. $\frac{5}{9} \cdot \frac{12}{25}$

17. $\frac{3}{8} \div \frac{5}{12}$

18. $\frac{5}{6} \div \frac{14}{15}$

19. $\frac{2}{3} \div \frac{8}{9}$

20. $\frac{3}{8} \div \frac{9}{14}$

Concept Check *Answer each question.*

21. Which rational expressions are equivalent to $-\frac{x}{y}$?

A. $\frac{-x}{-y}$ **B.** $\frac{x}{-y}$ **C.** $\frac{x}{y}$ **D.** $-\frac{x}{-y}$ **E.** $\frac{-x}{y}$ **F.** $-\frac{-x}{-y}$

22. Which rational expression can be simplified?

A. $\frac{x^2+2}{x^2}$ **B.** $\frac{x^2+2}{2}$ **C.** $\frac{x^2+y^2}{y^2}$ **D.** $\frac{x^2-5x}{x}$

23. Which rational expression is *not* equivalent to $\frac{x-3}{4-x}$?

A. $\frac{3-x}{x-4}$ **B.** $\frac{x+3}{4+x}$ **C.** $-\frac{3-x}{4-x}$ **D.** $-\frac{x-3}{x-4}$

24. Which two rational expressions are equivalent to -1?

A. $\frac{2x+3}{2x-3}$ **B.** $\frac{2x-3}{3-2x}$ **C.** $\frac{2x+3}{3+2x}$ **D.** $\frac{2x+3}{-2x-3}$

25. Concept Check Consider the expressions $\frac{x-2}{4}$ and $\frac{2x}{4}$. Only one of these expressions is not in lowest terms. Which one is it and why? Write it in lowest terms.

26. Concept Check Identify the two *terms* in the numerator and the two *terms* in the denominator of the rational expression $\frac{x^2+4x}{x+4}$, and write it in lowest terms.

27. Concept Check A student was asked to simplify the rational expression

$$\frac{2x+y}{4x+2y}$$

and wrote the incorrect answer $\frac{x+y}{4x+y}$.

WHAT WENT WRONG? What is the correct answer?

28. Concept Check A student was asked to simplify the rational expression

$$\frac{a+b}{a-b}$$

and wrote the incorrect answer $\frac{+1}{-1}$.

WHAT WENT WRONG? What is the correct answer?

Write each rational expression in lowest terms. ***See Example 2.***

29. $\frac{x^2(x+1)}{x(x+1)}$ **30.** $\frac{y^3(y-4)}{y^2(y-4)}$ **31.** $\frac{(x+4)(x-3)}{(x+5)(x+4)}$

32. $\frac{(2x+7)(x-1)}{(2x+3)(2x+7)}$ **33.** $\frac{4x(x+3)}{8x^2(x-3)}$ **34.** $\frac{5y^2(y+8)}{15y(y-8)}$

35. $\frac{3x+7}{3}$ **36.** $\frac{4x-9}{4}$ **37.** $\frac{6m+18}{7m+21}$

38. $\frac{5r-20}{3r-12}$ **39.** $\frac{3z^2+z}{18z+6}$ **40.** $\frac{2x^2-5x}{16x-40}$

41. $\frac{t^2-9}{3t+9}$ **42.** $\frac{m^2-25}{4m-20}$ **43.** $\frac{2t+6}{t^2-9}$

44. $\frac{5s-25}{s^2-25}$ **45.** $\frac{x^2+2x-15}{x^2+6x+5}$ **46.** $\frac{y^2-5y-14}{y^2+y-2}$

47. $\frac{8x^2-10x-3}{8x^2-6x-9}$ **48.** $\frac{12x^2-4x-5}{8x^2-6x-5}$ **49.** $\frac{a^3+b^3}{a+b}$

50. $\frac{r^3-s^3}{r-s}$ **51.** $\frac{2c^2+2cd-60d^2}{2c^2-12cd+10d^2}$ **52.** $\frac{3s^2-9st-54t^2}{3s^2-6st-72t^2}$

53. $\frac{ac-ad+bc-bd}{ac-ad-bc+bd}$ **54.** $\frac{2xy+2xw+y+w}{2xy+y-2xw-w}$

Write each rational expression in lowest terms. ***See Example 3.***

55. $\frac{7-b}{b-7}$ **56.** $\frac{r-13}{13-r}$ **57.** $\frac{x^2-y^2}{y-x}$

58. $\frac{m^2-n^2}{n-m}$ **59.** $\frac{x^2-4}{2-x}$ **60.** $\frac{x^2-81}{9-x}$

61. $\frac{(a-3)(x+y)}{(3-a)(x-y)}$ **62.** $\frac{(8-p)(x+2)}{(p-8)(x-2)}$ **63.** $\frac{5k-10}{20-10k}$

64. $\frac{7x-21}{63-21x}$ **65.** $\frac{a^2-b^2}{a^2+b^2}$ **66.** $\frac{p^2+q^2}{p^2-q^2}$

Multiply or divide as indicated. ***See Examples 4 and 5.***

67. $\dfrac{(x+2)(x+1)}{(x+3)(x-2)} \cdot \dfrac{(x+3)(x+4)}{(x+2)(x+1)}$

68. $\dfrac{(x+3)(x-6)}{(x-4)(x+2)} \cdot \dfrac{(x+5)(x-4)}{(x+3)(x-6)}$

69. $\dfrac{(2x+3)(x-4)}{(x+8)(x-4)} \div \dfrac{(x-4)(x+2)}{(x-4)(x+8)}$

70. $\dfrac{(6x+5)(x-3)}{(x-1)(x-3)} \div \dfrac{(2x+7)(x+9)}{(x-1)(x+9)}$

71. $\dfrac{4x}{8x+4} \cdot \dfrac{14x+7}{6}$

72. $\dfrac{12x-20}{5x} \cdot \dfrac{6}{9x-15}$

73. $\dfrac{p^2-25}{4p} \cdot \dfrac{2}{5-p}$

74. $\dfrac{a^2-1}{4a} \cdot \dfrac{2}{1-a}$

75. $(7k+7) \div \dfrac{4k+4}{5}$

76. $(8y-16) \div \dfrac{3y-6}{10}$

77. $(z^2-1) \cdot \dfrac{1}{1-z}$

78. $(y^2-4) \cdot \dfrac{8}{2-y}$

79. $\dfrac{4x-20}{5x} \div \dfrac{2x-10}{7x^3}$

80. $\dfrac{3x+12}{4x} \div \dfrac{4x+16}{6x^2}$

81. $\dfrac{m^2-49}{m+1} \div \dfrac{7-m}{m}$

82. $\dfrac{k^2-4}{3k^2} \div \dfrac{2-k}{11k}$

83. $\dfrac{12x-10y}{3x+2y} \cdot \dfrac{6x+4y}{10y-12x}$

84. $\dfrac{9s-12t}{2s+2t} \cdot \dfrac{3s+3t}{4t-3s}$

85. $\dfrac{x^2-25}{x^2+x-20} \cdot \dfrac{x^2+7x+12}{x^2-2x-15}$

86. $\dfrac{t^2-49}{t^2+4t-21} \cdot \dfrac{t^2+8t+15}{t^2-2t-35}$

87. $\dfrac{a^3-b^3}{a^2-b^2} \div \dfrac{2a-2b}{2a+2b}$

88. $\dfrac{x^3+y^3}{2x+2y} \div \dfrac{x^2-y^2}{2x-2y}$

89. $\dfrac{8x^3-27}{2x^2-18} \cdot \dfrac{2x+6}{8x^2+12x+18}$

90. $\dfrac{64x^3+1}{4x^2-100} \cdot \dfrac{4x+20}{64x^2-16x+4}$

91. $\dfrac{a^3-8b^3}{a^2-ab-6b^2} \cdot \dfrac{a^2+ab-12b^2}{a^2+2ab-8b^2}$

92. $\dfrac{p^3-27q^3}{p^2+pq-12q^2} \cdot \dfrac{p^2-2pq-24q^2}{p^2-5pq-6q^2}$

93. $\dfrac{6x^2+5x-6}{12x^2-11x+2} \div \dfrac{4x^2-12x+9}{8x^2-14x+3}$

94. $\dfrac{8a^2-6a-9}{6a^2-5a-6} \div \dfrac{4a^2+11a+6}{9a^2+12a+4}$

Extending Skills *Multiply or divide as indicated.*

95. $\dfrac{(-3mn)^2 \cdot 64(m^2n)^3}{16m^2n^4(mn^2)^3} \div \dfrac{24(m^2n^2)^4}{(3m^2n^3)^2}$

96. $\dfrac{(-4a^2b^3)^2 \cdot 9(a^2b^4)^2}{(2a^2b^3)^4 \cdot (3a^3b)^2} \div \dfrac{(ab)^4}{(a^2b^3)^2}$

97. $\dfrac{3k^2+17kp+10p^2}{6k^2+13kp-5p^2} \div \dfrac{6k^2+kp-2p^2}{6k^2-5kp+p^2}$

98. $\dfrac{16c^2+24cd+9d^2}{16c^2-16cd+3d^2} \div \dfrac{16c^2-9d^2}{16c^2-24cd+9d^2}$

99. $\left(\dfrac{6k^2-13k-5}{k^2+7k} \div \dfrac{2k-5}{k^3+6k^2-7k}\right) \cdot \dfrac{k^2-5k+6}{3k^2-8k-3}$

100. $\left(\dfrac{2x^3+3x^2-2x}{3x-15} \div \dfrac{2x^3-x^2}{x^2-3x-10}\right) \cdot \dfrac{5x^2-10x}{3x^2+12x+12}$

6.2 Adding and Subtracting Rational Expressions

OBJECTIVES

1. Add and subtract rational expressions with the same denominator.
2. Find a least common denominator.
3. Add and subtract rational expressions with different denominators.

VOCABULARY

☐ least common denominator (LCD)

OBJECTIVE 1 **Add and subtract rational expressions with the same denominator.**

Recall that if $\frac{a}{b}$ and $\frac{c}{b}$ are rational numbers (where $b \neq 0$), then

$$\frac{a}{b} + \frac{c}{b} = \frac{a+c}{b} \quad \text{and} \quad \frac{a}{b} - \frac{c}{b} = \frac{a-c}{b}.$$

Add or subtract the numerators.
Keep the same denominator.

Examples: $\frac{4}{7} + \frac{1}{7} = \frac{5}{7}$, $\frac{4}{7} - \frac{1}{7} = \frac{3}{7}$, and $\frac{3}{10} - \frac{1}{10} = \frac{2}{10} = \frac{2 \cdot 1}{2 \cdot 5} = \frac{1}{5}$

Adding and subtracting rational expressions is similar.

Adding or Subtracting Rational Expressions

Step 1 **If the denominators are the same,** add or subtract the numerators. Place the result over the common denominator.

If the denominators are different, find the least common denominator and write all rational expressions with this LCD. Add or subtract the numerators. Place the result over the common denominator.

Step 2 **Simplify.** Write all answers in lowest terms.

EXAMPLE 1 **Adding and Subtracting Rational Expressions (Same Denominators)**

Add or subtract as indicated.

(a) $\frac{3y}{5} + \frac{x}{5}$

$= \frac{3y + x}{5}$ ← Add the numerators. ← Keep the common denominator.

(b) $\frac{7}{2r^2} - \frac{11}{2r^2}$

$= \frac{7-11}{2r^2}$ Subtract the numerators. Keep the common denominator.

$= \frac{-4}{2r^2}$

$= -\frac{2}{r^2}$ $\frac{-4}{2r^2} = \frac{-2 \cdot 2}{2 \cdot r^2}$; Write in lowest terms.

(c) $\frac{m}{m^2 - p^2} + \frac{p}{m^2 - p^2}$

$= \frac{m+p}{m^2 - p^2}$ Add the numerators. Keep the common denominator.

$= \frac{m+p}{(m+p)(m-p)}$ Factor.

$= \frac{1}{m-p}$ Fundamental property

Remember to write 1 in the numerator.

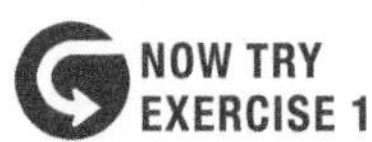

NOW TRY EXERCISE 1

Add or subtract as indicated.

(a) $\frac{5}{3x} + \frac{2}{3x}$

(b) $\frac{x^2}{x-3} - \frac{9}{x-3}$

(c) $\frac{2}{x^2+x-2} + \frac{x}{x^2+x-2}$

(d) $\frac{4}{x^2+2x-8} + \frac{x}{x^2+2x-8}$

$= \frac{4+x}{x^2+2x-8}$ Add.

$= \frac{4+x}{(x-2)(x+4)}$ Factor.

$= \frac{1}{x-2}$ Fundamental property

NOW TRY 

OBJECTIVE 2 Find a least common denominator.

We add or subtract rational numbers with different denominators by first writing them with a common denominator, usually the **least common denominator (LCD).**

Finding the Least Common Denominator

Step 1 **Factor** each denominator.

Step 2 **Find the least common denominator.** The LCD is the product of all of the different factors from each denominator, with each factor raised to the *greatest* power that occurs in any denominator.

Example: To find the LCD for $\frac{7}{15}$ and $\frac{5}{12}$, we factor each denominator.

$$15 = 3 \cdot 5 \quad \text{and} \quad 12 = 2 \cdot 2 \cdot 3 = 2^2 \cdot 3$$

$$\text{LCD} = 2^2 \cdot 3 \cdot 5 = 60$$ Find the product of the factors, each raised to its *greatest* power.

We use a similar procedure to find the LCD of two or more rational expressions.

EXAMPLE 2 Finding Least Common Denominators

Suppose that the given expressions are denominators of rational expressions. Find the LCD for each group of denominators.

(a) $5xy^2$, $2x^3y$

$$5xy^2 = 5 \cdot x \cdot y^2$$
$$2x^3y = 2 \cdot x^3 \cdot y$$
Each denominator is already factored.

$$\text{LCD} = 5 \cdot 2 \cdot x^3 \cdot y^2$$ ← Greatest exponent on y is 2.

$$= 10x^3y^2$$ Greatest exponent on x is 3.

(b) $k-3$, k The LCD must be divisible by *both* $k-3$ and k.

$$\text{LCD} = k(k-3)$$ Don't forget the factor k.

It is usually best to leave a least common denominator in factored form.

(c) y^2-2y-8, y^2+3y+2

$$y^2-2y-8 = (y-4)(y+2)$$
$$y^2+3y+2 = (y+2)(y+1)$$
Factor.

$$\text{LCD} = (y-4)(y+2)(y+1)$$

NOW TRY ANSWERS

1. (a) $\frac{7}{3x}$ (b) $x+3$ (c) $\frac{1}{x-1}$

NOW TRY EXERCISE 2

Find the LCD for each group of denominators.

(a) $15m^3n,\ 10m^2n$

(b) $t,\ t-8$

(c) $3x^2+9x-30,\ x^2-4,\ x^2+10x+25$

(d) $8z-24, \quad 5z^2-15z$

$$\left.\begin{aligned} 8z-24 &= 8(z-3) \\ 5z^2-15z &= 5z(z-3) \end{aligned}\right\} \text{ Factor.}$$

$$\begin{aligned} \text{LCD} &= 8 \cdot 5z \cdot (z-3) \\ &= 40z(z-3) \end{aligned}$$

(e) $m^2+5m+6, \quad m^2+4m+4, \quad 2m^2+4m-6$

$$\left.\begin{aligned} m^2+5m+6 &= (m+3)(m+2) \\ m^2+4m+4 &= (m+2)^2 \\ 2m^2+4m-6 &= 2(m+3)(m-1) \end{aligned}\right\} \text{ Factor. Notice that } (m+2) \text{ is squared in the second denominator.}$$

Use each factor, raised to its *greatest* power in any denominator. This means that $(m+2)^2$ must be used.

$$\text{LCD} = 2(m+3)(m+2)^2(m-1)$$

NOW TRY 

OBJECTIVE 3 **Add and subtract rational expressions with different denominators.**

Recall how we add fractions with different denominators.

Example: $\dfrac{7}{15}+\dfrac{5}{12}$ The LCD for 15 and 12 is 60. (See Objective 2.)

$= \dfrac{7 \cdot 4}{15 \cdot 4} + \dfrac{5 \cdot 5}{12 \cdot 5}$ Fundamental property

($\frac{4}{4}$ and $\frac{5}{5}$ are forms of 1, the identity element of multiplication.)

$= \dfrac{28}{60} + \dfrac{25}{60}$ Write each fraction with the common denominator.

$= \dfrac{28+25}{60}$ Add the numerators. Keep the common denominator.

$= \dfrac{53}{60}$

We add and subtract rational expressions with different denominators similarly.

EXAMPLE 3 **Adding and Subtracting Rational Expressions (Different Denominators)**

Add or subtract as indicated.

(a) $\dfrac{5}{2p}+\dfrac{3}{8p}$ The LCD for $2p$ and $8p$ is $8p$.

$= \dfrac{5 \cdot 4}{2p \cdot 4} + \dfrac{3}{8p}$ Fundamental property

$= \dfrac{20}{8p} + \dfrac{3}{8p}$ Write the first fraction with the common denominator.

$= \dfrac{23}{8p}$ Add the numerators. Keep the common denominator.

NOW TRY ANSWERS

2\. **(a)** $30m^3n$ **(b)** $t(t-8)$ **(c)** $3(x-2)(x+2)(x+5)^2$

NOW TRY EXERCISE 3

Add or subtract as indicated.

(a) $\frac{2}{3x}+\frac{7}{4x}$

(b) $\frac{3}{z}-\frac{6}{z-5}$

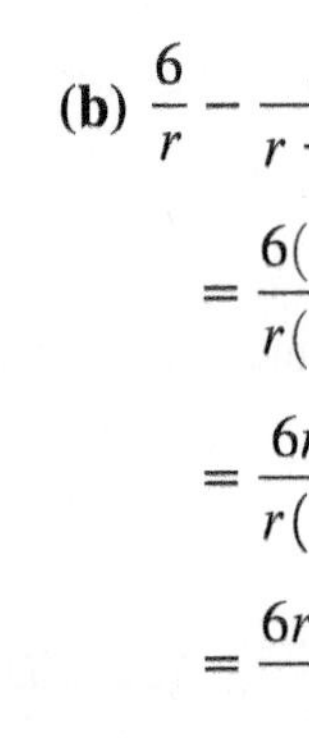

(b) $\frac{6}{r}-\frac{5}{r-3}$ The LCD is $r(r-3)$.

$$=\frac{6(r-3)}{r(r-3)}-\frac{r\cdot 5}{r(r-3)}$$ Fundamental property

$$=\frac{6r-18}{r(r-3)}-\frac{5r}{r(r-3)}$$ Distributive and commutative properties

$$=\frac{6r-18-5r}{r(r-3)}$$ Subtract the numerators. Keep the common denominator.

$$=\frac{r-18}{r(r-3)}$$ Combine like terms in the numerator. NOW TRY

CAUTION Sign errors occur easily when a rational expression with two or more terms in the numerator is being subtracted. ***The subtraction sign must be distributed to every term in the numerator of the fraction that follows it.*** **Study Example 4 carefully.**

NOW TRY EXERCISE 4

Subtract.

(a) $\frac{18x+7}{4x+5}-\frac{2x-13}{4x+5}$

(b) $\frac{5}{x-3}-\frac{5}{x+3}$

EXAMPLE 4 Subtracting Rational Expressions

Subtract.

(a) $\frac{7x}{3x+1}-\frac{x-2}{3x+1}$ The denominators are the same. ***The subtraction sign must be applied to both terms in the numerator of the second rational expression.***

$$\frac{7x}{3x+1}-\frac{x-2}{3x+1}$$ Use parentheses to avoid errors.

$$=\frac{7x-(x-2)}{3x+1}$$ Subtract the numerators. Keep the common denominator.

Be careful with signs.

$$=\frac{7x-x+2}{3x+1}$$ Distributive property

$$=\frac{6x+2}{3x+1}$$ Combine like terms in the numerator.

$$=\frac{2(3x+1)}{3x+1}$$ Factor the numerator.

$$=2$$ Fundamental property

(b) $\frac{1}{q-1}-\frac{1}{q+1}$ The LCD is $(q-1)(q+1)$.

$$=\frac{1(q+1)}{(q-1)(q+1)}-\frac{1(q-1)}{(q+1)(q-1)}$$ Fundamental property

$$=\frac{(q+1)-(q-1)}{(q-1)(q+1)}$$ Subtract the numerators.

$$=\frac{q+1-q+1}{(q-1)(q+1)}$$ Be careful with signs. Distributive property

$$=\frac{2}{(q-1)(q+1)}$$ Combine like terms in the numerator. NOW TRY

NOW TRY ANSWERS

3. (a) $\frac{29}{12x}$ **(b)** $\frac{-3z-15}{z(z-5)}$

4. (a) 4 **(b)** $\frac{30}{(x-3)(x+3)}$

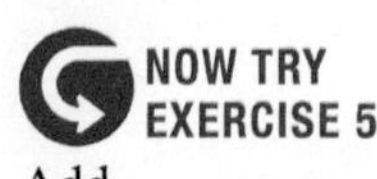

NOW TRY EXERCISE 5

Add.

$$\frac{t}{t-9}+\frac{2}{9-t}$$

EXAMPLE 5 Adding Rational Expressions (Denominators Are Opposites)

Add.

$$\frac{y}{y-2}+\frac{8}{2-y} \leftarrow \text{Denominators are opposites.}$$

$$=\frac{y}{y-2}+\frac{8(-1)}{(2-y)(-1)}$$ Multiply the second expression by 1 in the form $\frac{-1}{-1}$.

$$=\frac{y}{y-2}+\frac{-8}{y-2}$$ The LCD is $y-2$.

$$=\frac{y-8}{y-2}$$ Add the numerators.

We could use $2-y$ as the common denominator and rewrite the first expression.

$$\frac{y}{y-2}+\frac{8}{2-y}$$

$$=\frac{y(-1)}{(y-2)(-1)}+\frac{8}{2-y}$$ Multiply the first expression by 1 in the form $\frac{-1}{-1}$.

$$=\frac{-y}{2-y}+\frac{8}{2-y}$$ The LCD is $2-y$.

$$=\frac{-y+8}{2-y}$$ Add the numerators.

$$=\frac{8-y}{2-y}$$ This is an equivalent form of the answer in red above. Multiply it by $\frac{-1}{-1}$ to see this.

NOW TRY

NOW TRY EXERCISE 6

Add and subtract as indicated.

$$\frac{6}{y}-\frac{1}{y-3}+\frac{3}{y^2-3y}$$

EXAMPLE 6 Adding and Subtracting Three Rational Expressions

Add and subtract as indicated.

$$\frac{3}{x-2}+\frac{5}{x}-\frac{6}{x^2-2x}$$

$$=\frac{3}{x-2}+\frac{5}{x}-\frac{6}{x(x-2)}$$ Factor the third denominator.

$$=\frac{3x}{x(x-2)}+\frac{5(x-2)}{x(x-2)}-\frac{6}{x(x-2)}$$ The LCD is $x(x-2)$; fundamental property

$$=\frac{3x+5(x-2)-6}{x(x-2)}$$ Add and subtract the numerators.

$$=\frac{3x+5x-10-6}{x(x-2)}$$ Distributive property

$$=\frac{8x-16}{x(x-2)}$$ Combine like terms in the numerator.

$$=\frac{8(x-2)}{x(x-2)}$$ Factor the numerator.

$$=\frac{8}{x}$$ Fundamental property

NOW TRY

NOW TRY ANSWERS

5. $\frac{t-2}{t-9}$, or $\frac{2-t}{9-t}$

6. $\frac{5}{y}$

NOW TRY EXERCISE 7

Subtract.

$$\frac{t-1}{t^2-2t-8}-\frac{2t+3}{t^2+3t+2}$$

EXAMPLE 7 Subtracting Rational Expressions

Subtract.

$$\frac{m+4}{m^2-2m-3}-\frac{2m-3}{m^2-5m+6}$$

$$=\frac{m+4}{(m-3)(m+1)}-\frac{2m-3}{(m-3)(m-2)}$$ Factor each denominator.

$$=\frac{(m+4)(m-2)}{(m-3)(m+1)(m-2)}-\frac{(2m-3)(m+1)}{(m-3)(m-2)(m+1)}$$ Fundamental property

The LCD is $(m-3)(m+1)(m-2)$.

$$=\frac{(m+4)(m-2)-(2m-3)(m+1)}{(m-3)(m+1)(m-2)}$$ Subtract the numerators.

$$=\frac{m^2+2m-8-(2m^2-m-3)}{(m-3)(m+1)(m-2)}$$ Note the careful use of parentheses. Multiply in the numerator.

$$=\frac{m^2+2m-8-2m^2+m+3}{(m-3)(m+1)(m-2)}$$ Be careful with signs. Distributive property

$$=\frac{-m^2+3m-5}{(m-3)(m+1)(m-2)}$$ Combine like terms in the numerator.

If we try to factor the numerator, we find that this expression is in lowest terms.

NOW TRY 

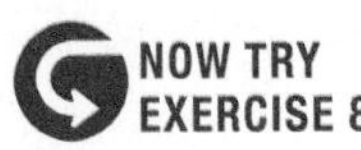

NOW TRY EXERCISE 8

Add.

$$\frac{2}{m^2-6m+9}+\frac{4}{m^2+m-12}$$

EXAMPLE 8 Adding Rational Expressions

Add.

$$\frac{5}{x^2+10x+25}+\frac{2}{x^2+7x+10}$$

$$=\frac{5}{(x+5)^2}+\frac{2}{(x+5)(x+2)}$$ Factor each denominator.

$$=\frac{5(x+2)}{(x+5)^2(x+2)}+\frac{2(x+5)}{(x+5)(x+2)(x+5)}$$ The LCD is $(x+5)^2(x+2)$; fundamental property

$$=\frac{5(x+2)+2(x+5)}{(x+5)^2(x+2)}$$ Add the numerators.

$$=\frac{5x+10+2x+10}{(x+5)^2(x+2)}$$ Distributive property

$$=\frac{7x+20}{(x+5)^2(x+2)}$$ Combine like terms in the numerator.

NOW TRY 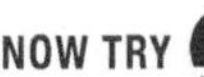

NOW TRY ANSWERS

7. $\frac{-t^2+5t+11}{(t+2)(t-4)(t+1)}$

8. $\frac{6m-4}{(m-3)^2(m+4)}$

6.2 Exercises

FOR EXTRA HELP  MyLab Math

Video solutions for select problems available in MyLab Math

STUDY SKILLS REMINDER
Make study cards to help you learn and remember the material in this chapter. **Review Study Skill 5, *Using Study Cards.***

1. **Concept Check** The least common denominator of two fractions is the (*sum/product*) of all the different (*terms/factors*) from each denominator, with each factor raised to the (*least/greatest*) power that occurs in any (*numerator/denominator*).

2. **Concept Check** Find the missing numerator so that the two fractions are equivalent.

(a) $\frac{4}{5} = \frac{?}{25}$ **(b)** $\frac{7}{x} = \frac{?}{10x}$ **(c)** $\frac{3}{2-x} = \frac{?}{x-2}$

3. **Concept Check** What is the LCD for the expressions $\frac{1}{x}$ and $-\frac{1}{9x}$?

A. 9 **B.** x **C.** $9x$ **D.** $9x^2$

4. **Concept Check** What is the LCD for the expressions $\frac{4}{a}$ and $\frac{2}{a+6}$?

A. a **B.** $a+6$ **C.** $2a+6$ **D.** $a(a+6)$

Concept Check *As review, add or subtract the rational numbers as indicated. Write answers in lowest terms.*

5. $\frac{8}{15} + \frac{4}{15}$
6. $\frac{5}{16} + \frac{9}{16}$
7. $\frac{5}{6} - \frac{8}{9}$
8. $\frac{3}{4} - \frac{5}{6}$
9. $\frac{5}{18} + \frac{7}{12}$
10. $\frac{3}{10} + \frac{7}{15}$

Add or subtract as indicated. ***See Example 1.***

11. $\frac{7}{t} + \frac{2}{t}$
12. $\frac{5}{r} + \frac{9}{r}$
13. $\frac{6x}{7} + \frac{y}{7}$
14. $\frac{12t}{5} + \frac{s}{5}$
15. $\frac{11}{5x} - \frac{1}{5x}$
16. $\frac{7}{4y} - \frac{3}{4y}$
17. $\frac{9}{4x^3} - \frac{17}{4x^3}$
18. $\frac{6}{5y^4} - \frac{21}{5y^4}$
19. $\frac{5x+4}{6x+5} + \frac{x+1}{6x+5}$
20. $\frac{6y+12}{4y+3} + \frac{2y-6}{4y+3}$
21. $\frac{x^2}{x+5} - \frac{25}{x+5}$
22. $\frac{y^2}{y+6} - \frac{36}{y+6}$
23. $\frac{4}{p^2+7p+12} + \frac{p}{p^2+7p+12}$
24. $\frac{5}{x^2+x-20} + \frac{x}{x^2+x-20}$
25. $\frac{-3x+7}{x^2+2x-8} + \frac{8x+13}{x^2+2x-8}$
26. $\frac{5x+6}{x^2+11x+30} + \frac{4-3x}{x^2+11x+30}$
27. $\frac{a^3}{a^2+ab+b^2} - \frac{b^3}{a^2+ab+b^2}$
28. $\frac{p^3}{p^2-pq+q^2} + \frac{q^3}{p^2-pq+q^2}$

Suppose that the given expressions are denominators of rational expressions. Find the least common denominator (LCD) for each group of denominators. ***See Example 2.***

29. $18x^2y^3$, $24x^4y^5$
30. $24a^3b^4$, $18a^5b^2$
31. $z-2$, z
32. $x+3$, x
33. $2y+8$, $y+4$
34. $3r-21$, $r-7$

35. $x^2 - 81,\quad x^2 + 18x + 81$

36. $y^2 - 16,\quad y^2 - 8y + 16$

37. $m + n,\quad m - n,\quad m^2 - n^2$

38. $r + s,\quad r - s,\quad r^2 - s^2$

39. $x^2 - 3x - 4,\quad x + x^2$

40. $y^2 - 8y + 12,\quad y^2 - 6y$

41. $2t^2 + 7t - 15,\quad t^2 + 3t - 10$

42. $s^2 - 3s - 4,\quad 3s^2 + s - 2$

43. $2y + 6,\quad y^2 - 9,\quad y$

44. $9x + 18,\quad x^2 - 4,\quad x$

45. $2x - 6,\quad x^2 - x - 6,\quad (x + 2)^2$

46. $3a - 3b,\quad a^2 + ab - 2b^2,\quad (a - b)^2$

47. **Concept Check** Consider the following incorrect work.

$$\frac{x}{x+2} - \frac{4x-1}{x+2}$$
$$= \frac{x - 4x - 1}{x+2}$$
$$= \frac{-3x - 1}{x+2}$$

WHAT WENT WRONG? Give the correct answer.

48. **Concept Check** On a test a student wrote the answer

$$\frac{-1}{6-x}.$$

His friend claimed that this is wrong and that the correct answer is

$$\frac{1}{x-6}.$$

The friend's assertion is incorrect. **WHAT WENT WRONG?**

Add or subtract as indicated. ***See Examples 3–8.***

49. $\frac{8}{t} + \frac{7}{3t}$

50. $\frac{5}{x} + \frac{9}{4x}$

51. $\frac{5}{12x^2y} - \frac{11}{6xy}$

52. $\frac{7}{18a^3b^2} - \frac{2}{9ab}$

53. $\frac{4}{15a^4b^5} + \frac{3}{20a^2b^6}$

54. $\frac{5}{12x^5y^2} + \frac{5}{18x^4y^5}$

55. $\frac{2r}{7p^3q^4} + \frac{3s}{14p^4q}$

56. $\frac{4t}{9a^8b^7} + \frac{5s}{27a^4b^3}$

57. $\frac{1}{a^3b^2} - \frac{2}{a^4b} + \frac{3}{a^5b^7}$

58. $\frac{5}{t^4u^7} - \frac{3}{t^5u^9} + \frac{6}{t^{10}u}$

59. $\frac{1}{x-1} - \frac{1}{x}$

60. $\frac{3}{x-3} - \frac{1}{x}$

61. $\frac{3a}{a+1} + \frac{2a}{a-3}$

62. $\frac{2x}{x+4} + \frac{3x}{x-7}$

63. $\frac{11x-13}{2x-3} - \frac{3x-1}{2x-3}$

64. $\frac{13x-5}{4x-1} - \frac{x-2}{4x-1}$

65. $\frac{17y+3}{9y+7} - \frac{-10y-18}{9y+7}$

66. $\frac{4x+1}{3x+2} - \frac{-2x-3}{3x+2}$

67. $\frac{2}{4-x} + \frac{5}{x-4}$

68. $\frac{3}{2-t} + \frac{1}{t-2}$

69. $\frac{w}{w-z} - \frac{z}{z-w}$

70. $\frac{a}{a-b} - \frac{b}{b-a}$

71. $\frac{1}{x+1} - \frac{1}{x-1}$

72. $\frac{-2}{x-1} + \frac{2}{x+1}$

73. $\frac{4x}{x-1} - \frac{2}{x+1} - \frac{4}{x^2-1}$

74. $\frac{4}{x+3} - \frac{x}{x-3} - \frac{18}{x^2-9}$

75. $\frac{15}{y^2+3y} + \frac{2}{y} + \frac{5}{y+3}$

76. $\frac{7}{t-2} - \frac{6}{t^2-2t} - \frac{3}{t}$

77. $\frac{5}{x-2} + \frac{1}{x} + \frac{2}{x^2-2x}$

78. $\frac{5x}{x-3} + \frac{2}{x} + \frac{6}{x^2-3x}$

79. $\frac{3x}{x+1} + \frac{4}{x-1} - \frac{6}{x^2-1}$

80. $\frac{5x}{x+3} + \frac{x+2}{x} - \frac{6}{x^2+3x}$

81. $\dfrac{4}{x+1}+\dfrac{1}{x^2-x+1}-\dfrac{12}{x^3+1}$

82. $\dfrac{5}{x+2}+\dfrac{2}{x^2-2x+4}-\dfrac{60}{x^3+8}$

83. $\dfrac{2x+4}{x+3}+\dfrac{3}{x}-\dfrac{6}{x^2+3x}$

84. $\dfrac{4x+1}{x+5}-\dfrac{2}{x}+\dfrac{10}{x^2+5x}$

85. $\dfrac{3}{(p-2)^2}-\dfrac{5}{p-2}+4$

86. $\dfrac{8}{(3r-1)^2}+\dfrac{2}{3r-1}-6$

87. $\dfrac{3}{x^2-5x+6}-\dfrac{2}{x^2-4x+4}$

88. $\dfrac{2}{m^2-4m+4}+\dfrac{3}{m^2+m-6}$

89. $\dfrac{3}{x^2+4x+4}+\dfrac{7}{x^2+5x+6}$

90. $\dfrac{5}{x^2+6x+9}-\dfrac{2}{x^2+4x+3}$

91. $\dfrac{5x}{x^2+xy-2y^2}-\dfrac{3x}{x^2+5xy-6y^2}$

92. $\dfrac{6x}{6x^2+5xy-4y^2}-\dfrac{2y}{9x^2-16y^2}$

93. $\dfrac{5x-y}{x^2+xy-2y^2}-\dfrac{3x+2y}{x^2+5xy-6y^2}$

94. $\dfrac{6x+5y}{6x^2+5xy-4y^2}-\dfrac{x+2y}{9x^2-16y^2}$

95. $\dfrac{r+s}{3r^2+2rs-s^2}-\dfrac{s-r}{6r^2-5rs+s^2}$

96. $\dfrac{3y}{y^2+yz-2z^2}+\dfrac{4y-1}{y^2-z^2}$

Work each problem.

97. A **Concours d'Elegance** is a competition in which a maximum of 100 points is awarded to a car on the basis of its general attractiveness. The function defined by the rational expression

$$C(x)=\frac{9010}{49(101-x)}-\frac{10}{49}$$

approximates the cost, in thousands of dollars, of restoring a car so that it will win x points.

(a) Simplify the expression for $C(x)$ by performing the indicated subtraction.

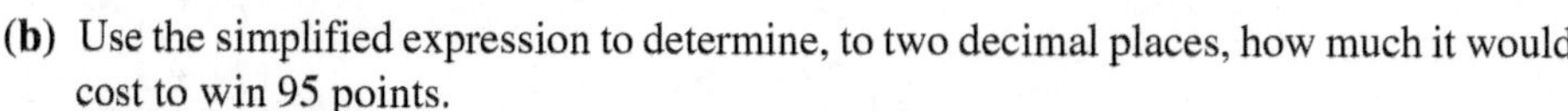

(b) Use the simplified expression to determine, to two decimal places, how much it would cost to win 95 points.

98. A **cost-benefit model** expresses cost of an undertaking in terms of benefits received. One such model gives the cost, in thousands of dollars, to remove x percent of a pollutant as

$$C(x)=\frac{6.7x}{100-x}.$$

Another model produces the relationship

$$C(x)=\frac{6.5x}{102-x}.$$

(a) What is the cost found by averaging the two models? (*Hint:* The average of two quantities is half their sum.)

(b) Using the two given models and the answer to part (a), find the cost to the nearest thousand dollars to remove 95% ($x=95$) of the pollutant.

(c) Average the two costs in part (b) from the given models, to the nearest thousand dollars. Compare this result with the cost obtained using the average of the two models.

6.3 Complex Fractions

OBJECTIVES

1 Simplify complex fractions by simplifying the numerator and denominator (Method 1).
2 Simplify complex fractions by multiplying by a common denominator (Method 2).
3 Compare the two methods of simplifying complex fractions.
4 Simplify rational expressions with negative exponents.

VOCABULARY

☐ complex fraction

A **complex fraction** is a quotient having a fraction in the numerator, denominator, or both.

$$\frac{\frac{2}{3}}{\frac{4}{9}}, \quad \frac{1+\frac{1}{x}}{2}, \quad \frac{\frac{4}{y}}{6-\frac{3}{y}}, \quad \text{and} \quad \frac{\frac{m^2-9}{m+1}}{\frac{m+3}{m^2-1}} \qquad \text{Complex fractions}$$

OBJECTIVE 1 Simplify complex fractions by simplifying the numerator and denominator (Method 1).

Simplifying a Complex Fraction (Method 1)

Step 1 Simplify the numerator and denominator separately.

Step 2 Divide by multiplying the numerator by the reciprocal of the denominator.

Step 3 Simplify the resulting fraction if possible.

In Step 2, we are treating the complex fraction as a quotient of two rational expressions and dividing. ***In order for us to perform this step, both the numerator and the denominator must be single fractions.***

$$\frac{\frac{q-5}{8}}{\frac{q+5}{3}} \begin{matrix} \leftarrow \text{Single fraction} \\ \leftarrow \text{Single fraction} \end{matrix} \qquad \frac{\frac{1}{x}+x}{\frac{x^2+1}{8}} \begin{matrix} \leftarrow \text{Not a single fraction} \\ \leftarrow \text{Single fraction} \end{matrix} \qquad \frac{6+\frac{3}{x}}{\frac{x}{4}+\frac{7}{8}} \begin{matrix} \leftarrow \text{Not a single fraction} \\ \leftarrow \text{Not a single fraction} \end{matrix}$$

EXAMPLE 1 Simplifying Complex Fractions (Method 1)

Use Method 1 to simplify each complex fraction.

(a) $\dfrac{\frac{x+1}{x}}{\frac{x-1}{2x}}$ Both the numerator and the denominator are single fractions. Each is already simplified. (Step 1)

$= \dfrac{x+1}{x} \div \dfrac{x-1}{2x}$ Write as a division problem.

$= \dfrac{x+1}{x} \cdot \dfrac{2x}{x-1}$ Multiply by the reciprocal of $\frac{x-1}{2x}$. (Step 2)

$= \dfrac{2x(x+1)}{x(x-1)}$ Multiply.

$= \dfrac{2(x+1)}{x-1}$ Simplify. (Step 3)

NOW TRY EXERCISE 1

Use Method 1 to simplify each complex fraction.

(a) $\dfrac{\frac{t+4}{3t}}{\frac{2t+1}{9t}}$ **(b)** $\dfrac{5-\frac{2}{y}}{4+\frac{1}{y}}$

(b) $\dfrac{2+\frac{1}{y}}{3-\frac{2}{y}}$ The numerator and denominator are ***not*** single fractions. Simplify them separately. (Step 1)

$= \dfrac{\frac{2y}{y}+\frac{1}{y}}{\frac{3y}{y}-\frac{2}{y}}$ Prepare to write the numerator and denominator as single fractions.

$= \dfrac{\frac{2y+1}{y}}{\frac{3y-2}{y}}$ In the numerator, $\frac{a}{c}+\frac{b}{c}=\frac{a+b}{c}$. In the denominator, $\frac{a}{c}-\frac{b}{c}=\frac{a-b}{c}$.

The numerator is a single fraction. So is the denominator.

$= \dfrac{2y+1}{y} \div \dfrac{3y-2}{y}$ Write as a division problem.

$= \dfrac{2y+1}{y} \cdot \dfrac{y}{3y-2}$ Multiply by the reciprocal of $\frac{3y-2}{y}$. (Step 2)

$= \dfrac{2y+1}{3y-2}$ Multiply and simplify. (Step 3)

NOW TRY

OBJECTIVE 2 Simplify complex fractions by multiplying by a common denominator (Method 2).

This method uses the identity property for multiplication.

Simplifying a Complex Fraction (Method 2)

Step 1 Multiply the numerator and denominator of the complex fraction by the least common denominator of the fractions in the numerator and the fractions in the denominator of the complex fraction.

Step 2 Simplify the resulting fraction if possible.

EXAMPLE 2 Simplifying Complex Fractions (Method 2)

Use Method 2 to simplify each complex fraction.

(a) $\dfrac{2+\frac{1}{y}}{3-\frac{2}{y}}$

This is the same fraction as in **Example 1(b).** Compare the solution methods.

$= \dfrac{\left(2+\frac{1}{y}\right)\cdot y}{\left(3-\frac{2}{y}\right)\cdot y}$ The LCD of all the fractions is y. Multiply the numerator and denominator by y because $\frac{y}{y}=1$. (Step 1)

$= \dfrac{2\cdot y+\frac{1}{y}\cdot y}{3\cdot y-\frac{2}{y}\cdot y}$ Distributive property (Step 2)

$= \dfrac{2y+1}{3y-2}$ Multiply. We obtain the same answer as in **Example 1(b).**

NOW TRY ANSWERS

1. (a) $\frac{3(t+4)}{2t+1}$ (b) $\frac{5y-2}{4y+1}$

NOW TRY EXERCISE 2

Use Method 2 to simplify each complex fraction.

(a) $\dfrac{5-\frac{2}{y}}{4+\frac{1}{y}}$ **(b)** $\dfrac{x+\frac{5}{x}}{4x-\frac{1}{x-2}}$

(b) $$\frac{2p+\frac{5}{p-1}}{3p-\frac{2}{p}}$$

$$=\frac{\left(2p+\frac{5}{p-1}\right)\cdot p(p-1)}{\left(3p-\frac{2}{p}\right)\cdot p(p-1)}$$ Multiply the numerator and denominator by the LCD, $p(p-1)$. (Step 1)

$$=\frac{2p[p(p-1)]+\frac{5}{p-1}\cdot p(p-1)}{3p[p(p-1)]-\frac{2}{p}\cdot p(p-1)}$$ Distributive property (Step 2)

$$=\frac{2p[p(p-1)]+5p}{3p[p(p-1)]-2(p-1)}$$ Multiply.

$$=\frac{2p[p^2-p]+5p}{3p[p^2-p]-2p+2}$$ Distributive property

This rational expression is in lowest terms.

$$=\frac{2p^3-2p^2+5p}{3p^3-3p^2-2p+2}$$ Distributive property again

NOW TRY

OBJECTIVE 3 Compare the two methods of simplifying complex fractions.

Some students prefer one method over the other, while other students feel comfortable with both methods and rely on practice with many examples to determine which method they will use in a particular problem.

EXAMPLE 3 Simplifying Complex Fractions (Both Methods)

Use both Method 1 and Method 2 to simplify each complex fraction.

Method 1

(a) $$\frac{\frac{2}{x-3}}{\frac{5}{x^2-9}}$$

$$=\frac{\frac{2}{x-3}}{\frac{5}{(x-3)(x+3)}}$$

$$=\frac{2}{x-3}\div\frac{5}{(x-3)(x+3)}$$

$$=\frac{2}{x-3}\cdot\frac{(x-3)(x+3)}{5}$$

$$=\frac{2(x+3)}{5}$$

Method 2

(a) $$\frac{\frac{2}{x-3}}{\frac{5}{x^2-9}}$$

$$=\frac{\frac{2}{x-3}}{\frac{5}{(x-3)(x+3)}}$$

$$=\frac{\frac{2}{x-3}\cdot(x-3)(x+3)}{\frac{5}{(x-3)(x+3)}\cdot(x-3)(x+3)}$$

$$=\frac{2(x+3)}{5}$$

NOW TRY ANSWERS

2. **(a)** $\dfrac{5y-2}{4y+1}$

(b) $\dfrac{x^3-2x^2+5x-10}{4x^3-8x^2-x}$

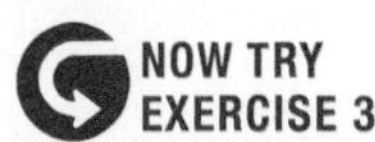

NOW TRY EXERCISE 3

Use both Method 1 and Method 2 to simplify each complex fraction.

(a) $\dfrac{\frac{1}{p-6}}{\frac{5}{p^2-36}}$ **(b)** $\dfrac{\frac{1}{m^2}-\frac{1}{n^2}}{\frac{1}{m}+\frac{1}{n}}$

Method 1

(b) $\dfrac{\frac{1}{x}+\frac{1}{y}}{\frac{1}{x^2}-\frac{1}{y^2}}$

$$= \frac{\frac{y}{xy}+\frac{x}{xy}}{\frac{y^2}{x^2y^2}-\frac{x^2}{x^2y^2}}$$

$$= \frac{\frac{y+x}{xy}}{\frac{y^2-x^2}{x^2y^2}}$$

$$= \frac{y+x}{xy} \div \frac{y^2-x^2}{x^2y^2}$$

$$= \frac{y+x}{xy} \cdot \frac{x^2y^2}{(y-x)(y+x)}$$

$$= \frac{xy}{y-x}$$

Method 2

(b) $\dfrac{\frac{1}{x}+\frac{1}{y}}{\frac{1}{x^2}-\frac{1}{y^2}}$

$$= \frac{\left(\frac{1}{x}+\frac{1}{y}\right)\cdot x^2y^2}{\left(\frac{1}{x^2}-\frac{1}{y^2}\right)\cdot x^2y^2}$$

$$= \frac{\left(\frac{1}{x}\right)x^2y^2+\left(\frac{1}{y}\right)x^2y^2}{\left(\frac{1}{x^2}\right)x^2y^2-\left(\frac{1}{y^2}\right)x^2y^2}$$

$$= \frac{xy^2+x^2y}{y^2-x^2}$$

$$= \frac{xy(y+x)}{(y+x)(y-x)}$$

$$= \frac{xy}{y-x}$$

NOW TRY

OBJECTIVE 4 Simplify rational expressions with negative exponents.

We begin by rewriting the expressions with only positive exponents.

EXAMPLE 4 Simplifying Rational Expressions with Negative Exponents

Simplify each expression, using only positive exponents in the answer.

(a) $\dfrac{m^{-1}+p^{-2}}{2m^{-2}-p^{-1}}$ $\quad a^{-n}=\frac{1}{a^n}$

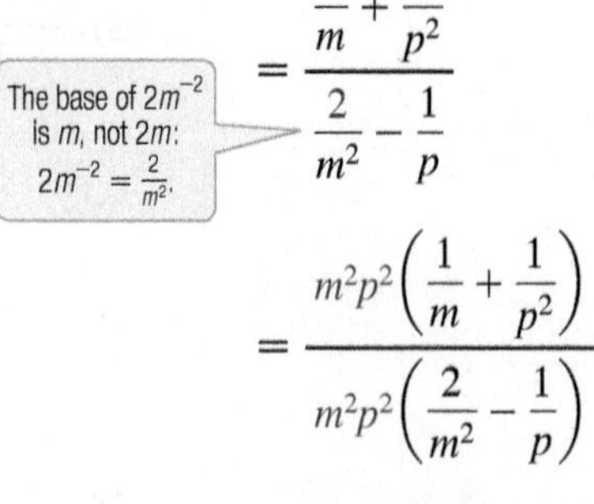

The base of $2m^{-2}$ is m, not $2m$: $2m^{-2}=\frac{2}{m^2}$.

$$= \frac{\frac{1}{m}+\frac{1}{p^2}}{\frac{2}{m^2}-\frac{1}{p}}$$

Write with positive exponents.
$2m^{-2}=2\cdot m^{-2}=\frac{2}{1}\cdot\frac{1}{m^2}=\frac{2}{m^2}$

$$= \frac{m^2p^2\left(\frac{1}{m}+\frac{1}{p^2}\right)}{m^2p^2\left(\frac{2}{m^2}-\frac{1}{p}\right)}$$

Simplify by Method 2. Multiply the numerator and denominator by the LCD, m^2p^2.

$$= \frac{m^2p^2\cdot\frac{1}{m}+m^2p^2\cdot\frac{1}{p^2}}{m^2p^2\cdot\frac{2}{m^2}-m^2p^2\cdot\frac{1}{p}}$$

Distributive property

$$= \frac{mp^2+m^2}{2p^2-m^2p}$$

Write in lowest terms.

NOW TRY ANSWERS

3. (a) $\frac{p+6}{5}$ (b) $\frac{n-m}{mn}$

NOW TRY EXERCISE 4

Simplify each expression, using only positive exponents in the answer.

(a) $\dfrac{r^{-2} - s^{-1}}{4r^{-1} + s^{-2}}$

(b) $\dfrac{2y^{-1} - 3y^{-2}}{y^{-2} + 3x^{-1}}$

(b) $\dfrac{x^{-2} - 2y^{-1}}{y - 2x^2}$

$= \dfrac{\frac{1}{x^2} - \frac{2}{y}}{y - 2x^2}$ Write with positive exponents. (The 2 does *not* go in the denominator of this fraction.)

$= \dfrac{\left(\frac{1}{x^2} - \frac{2}{y}\right)x^2y}{(y - 2x^2)x^2y}$ Use Method 2. Multiply by the LCD, x^2y.

$= \dfrac{y - 2x^2}{(y - 2x^2)x^2y}$ Use the distributive property in the numerator.

$= \dfrac{1}{x^2y}$ Write in lowest terms. (Remember to write 1 in the numerator.)

NOW TRY

NOW TRY ANSWERS

4. (a) $\dfrac{s^2 - r^2s}{4rs^2 + r^2}$ (b) $\dfrac{2xy - 3x}{x + 3y^2}$

6.3 Exercises

FOR EXTRA HELP

MyLab Math

Video solutions for select problems available in MyLab Math

1. Concept Check A(n) ________ fraction is a quotient that has a fraction in the ________, denominator, or ________.

2. Concept Check Find the slope of the line that passes through each pair of points. (*Hint:* This will involve simplifying complex fractions. Recall that slope $m = \frac{y_2 - y_1}{x_2 - x_1}$.)

(a) $\left(-\frac{5}{2}, \frac{1}{6}\right)$ and $\left(\frac{5}{3}, \frac{3}{8}\right)$ **(b)** $\left(-\frac{5}{6}, -\frac{1}{2}\right)$ and $\left(-\frac{1}{3}, -\frac{3}{2}\right)$

Concept Check *Simplify.*

3. $\dfrac{\frac{2}{3}}{4}$ **4.** $\dfrac{\frac{3}{5}}{6}$ **5.** $\dfrac{\frac{3}{4}}{\frac{5}{12}}$ **6.** $\dfrac{\frac{5}{6}}{\frac{4}{9}}$

7. $\dfrac{\frac{5}{9} - \frac{1}{3}}{\frac{2}{3} + \frac{1}{6}}$ **8.** $\dfrac{\frac{7}{8} - \frac{3}{2}}{-\frac{1}{4} - \frac{3}{8}}$ **9.** $\dfrac{2 - \frac{1}{4}}{\frac{5}{4} + 3}$ **10.** $\dfrac{\frac{4}{3} - 2}{1 - \frac{3}{8}}$

Use either method to simplify each complex fraction. ***See Examples 1–3.***

11. $\dfrac{\frac{12}{x-1}}{\frac{6}{x}}$ **12.** $\dfrac{\frac{24}{t+4}}{\frac{6}{t}}$ **13.** $\dfrac{\frac{k+1}{2k}}{\frac{3k-1}{4k}}$ **14.** $\dfrac{\frac{1-r}{4r}}{\frac{1+r}{8r}}$

15. $\dfrac{\frac{4z^2x^4}{9}}{\frac{12x^2z^5}{15}}$ **16.** $\dfrac{\frac{3y^2x^3}{8}}{\frac{9y^3x^4}{16}}$ **17.** $\dfrac{\frac{1}{x} + 1}{-\frac{1}{x} + 1}$ **18.** $\dfrac{\frac{2}{k} - 1}{\frac{2}{k} + 1}$

19. $\dfrac{6 + \frac{1}{x}}{7 - \frac{3}{x}}$ **20.** $\dfrac{4 - \frac{1}{p}}{9 + \frac{5}{p}}$ **21.** $\dfrac{\frac{3}{x} + \frac{3}{y}}{\frac{3}{x} - \frac{3}{y}}$ **22.** $\dfrac{\frac{4}{t} - \frac{4}{s}}{\frac{4}{t} + \frac{4}{s}}$

23. $\dfrac{\frac{8x-24y}{10}}{\frac{x-3y}{5x}}$ **24.** $\dfrac{\frac{20x-10y}{12y}}{\frac{2x-y}{6y^2}}$ **25.** $\dfrac{\frac{6}{y-4}}{\frac{12}{y^2-16}}$ **26.** $\dfrac{\frac{8}{t+7}}{\frac{24}{t^2-49}}$

27. $\dfrac{\frac{x^2-16y^2}{xy}}{\frac{1}{y}-\frac{4}{x}}$ **28.** $\dfrac{\frac{4t^2-9s^2}{st}}{\frac{2}{s}-\frac{3}{t}}$ **29.** $\dfrac{\frac{1}{b^2}-\frac{1}{a^2}}{\frac{1}{b}-\frac{1}{a}}$ **30.** $\dfrac{\frac{1}{x^2}-\frac{1}{y^2}}{\frac{1}{x}+\frac{1}{y}}$

31. $\dfrac{x+y}{\frac{1}{y}+\frac{1}{x}}$ **32.** $\dfrac{s-r}{\frac{1}{r}-\frac{1}{s}}$ **33.** $\dfrac{y-\frac{y-3}{3}}{\frac{4}{9}+\frac{2}{3y}}$ **34.** $\dfrac{p-\frac{p+2}{4}}{\frac{3}{4}-\frac{5}{2p}}$

35. $\dfrac{\frac{1}{x^3}-\frac{1}{y^3}}{\frac{1}{x^2}-\frac{1}{y^2}}$ **36.** $\dfrac{\frac{1}{y^3}+\frac{1}{x^3}}{\frac{1}{y^2}-\frac{1}{x^2}}$ **37.** $\dfrac{\frac{x+2}{x}+\frac{1}{x+2}}{\frac{5}{x}+\frac{x}{x+2}}$ **38.** $\dfrac{\frac{y+3}{y}-\frac{4}{y-1}}{\frac{y}{y-1}+\frac{1}{y}}$

39. Concept Check When simplifying the expression

$$\frac{1}{x^{-2}+y^{-2}},$$

a student noted the negative exponents in the denominator and wrote his answer as x^2+y^2. This is incorrect. **WHAT WENT WRONG?**

40. Concept Check A student simplifying the expression

$$\frac{a^{-2}-4b^{-2}}{3b-6a} \quad \text{wrote} \quad \frac{\frac{1}{a^2}-\frac{1}{4b^2}}{3b-6a}$$

as her first step. This is a common error. **WHAT WENT WRONG?**

Simplify each expression, using only positive exponents in the answer. ***See Example 4.***

41. $\dfrac{1}{x^{-2}+y^{-2}}$ **42.** $\dfrac{1}{p^{-2}-q^{-2}}$ **43.** $\dfrac{x^{-2}+y^{-2}}{x^{-1}+y^{-1}}$ **44.** $\dfrac{x^{-1}-y^{-1}}{x^{-2}-y^{-2}}$

45. $\dfrac{k^{-1}+p^{-2}}{k^{-1}-3p^{-2}}$ **46.** $\dfrac{x^{-2}-y^{-1}}{x^{-2}+4y^{-1}}$ **47.** $\dfrac{x^{-1}+2y^{-1}}{2y+4x}$ **48.** $\dfrac{a^{-2}-4b^{-2}}{3b-6a}$

RELATING CONCEPTS For Individual or Group Work (Exercises 49–54)

Simplifying a complex fraction by Method 1 is a good way to review the methods of adding, subtracting, multiplying, and dividing rational expressions. Method 2 gives a good review of the fundamental property of rational numbers.

Refer to the following complex fraction, and ***work Exercises 49–54 in order.***

$$\frac{\frac{4}{m}+\frac{m+2}{m-1}}{\frac{m+2}{m}-\frac{2}{m-1}}$$

49. Add the fractions in the numerator.

50. Subtract as indicated in the denominator.

51. Divide the answer from **Exercise 49** by the answer from **Exercise 50.**

52. Go back to the original complex fraction and find the LCD of all denominators.

53. Multiply the numerator and denominator of the complex fraction by the answer from **Exercise 52.**

54. The answers for **Exercises 51 and 53** should be the same. Which method do you prefer? Explain why.

6.4 Equations with Rational Expressions and Graphs

OBJECTIVES

1 Determine the domain of the variable in a rational equation.
2 Solve rational equations.
3 Recognize the graph of a rational function.

VOCABULARY

- ☐ domain of the variable in a rational equation
- ☐ proposed solution
- ☐ extraneous solution
- ☐ discontinuous
- ☐ reciprocal function
- ☐ vertical asymptote
- ☐ horizontal asymptote

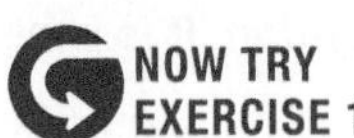

Give the domain of the variable in each equation.

(a) $\frac{1}{3x} - \frac{3}{4x} = \frac{1}{3}$

(b) $\frac{1}{x-7} + \frac{2}{x+7} = \frac{14}{x^2-49}$

OBJECTIVE 1 Determine the domain of the variable in a rational equation.

The **domain of the variable in a rational equation** is the intersection (overlap) of the domains of the rational expressions in the equation. When we solve rational equations, knowing the domain of the variable enables us to determine whether a **proposed solution**—that is, a value of the variable that *appears* to be a solution—must be discarded.

EXAMPLE 1 Determining Domains of Variables

Give the domain of the variable in each equation.

(a) $\frac{2}{x} - \frac{3}{2} = \frac{7}{2x}$

The domains of the expressions $\frac{2}{x}$, $\frac{3}{2}$, and $\frac{7}{2x}$ in the equation are, in order,

$\{x \mid x \text{ is a real number}, x \neq 0\}$,

$\{x \mid x \text{ is a real number}\}$,

and $\{x \mid x \text{ is a real number}, x \neq 0\}$.

Set-builder notation (We could use interval notation.)

The intersection of these three domains is $\{x \mid x \text{ is a real number}, x \neq 0\}$.

(b) $\frac{2}{x-3} - \frac{3}{x+3} = \frac{12}{x^2-9}$

The domains of the three expressions are, respectively,

$\{x \mid x \text{ is a real number}, x \neq 3\}$,

$\{x \mid x \text{ is a real number}, x \neq -3\}$,

and $\{x \mid x \text{ is a real number}, x \neq \pm 3\}$.

$\pm$ is read **"positive or negative,"** or **"plus or minus."**

The domain of the variable is the intersection of the three domains, all real numbers except 3 and -3.

$$\{x \mid x \text{ is a real number}, x \neq \pm 3\}$$

NOW TRY 

OBJECTIVE 2 Solve rational equations.

To solve rational equations, we usually multiply all terms by the least common denominator to clear the fractions. ***We can do this only with an equation*** because we want to solve for a variable (not with an expression, where we are carrying out operations).

Solving an Equation with Rational Expressions

Step 1 **Determine the domain of the variable.**

Step 2 **Multiply each side of the equation by the LCD** to clear the fractions.

Step 3 **Solve** the resulting equation.

Step 4 **Check** that each proposed solution is in the domain, and discard any values that are not. Check the remaining proposed solution(s) in the original equation.

NOW TRY ANSWERS

1. (a) $\{x \mid x \text{ is a real number}, x \neq 0\}$
   (b) $\{x \mid x \text{ is a real number}, x \neq \pm 7\}$

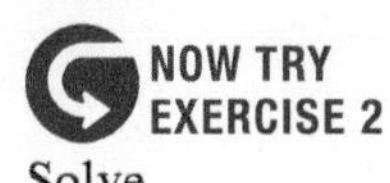
NOW TRY EXERCISE 2

Solve.

$$\frac{1}{3x}-\frac{3}{4x}=\frac{1}{3}$$

EXAMPLE 2 Solving a Rational Equation

Solve $\frac{2}{x}-\frac{3}{2}=\frac{7}{2x}$.

Step 1 The domain, which excludes 0, was found in **Example 1(a).**

Step 2 $2x\left(\frac{2}{x}-\frac{3}{2}\right)=2x\left(\frac{7}{2x}\right)$ Multiply by the LCD, $2x$.

Step 3 $2x\left(\frac{2}{x}\right)-2x\left(\frac{3}{2}\right)=2x\left(\frac{7}{2x}\right)$ Distributive property

$4-3x=7$ Multiply.

$-3x=3$ Subtract 4.

$x=-1$ Divide by -3. (This proposed solution is in the domain.)

Step 4 **CHECK** (Don't forget this step.)

$\frac{2}{x}-\frac{3}{2}=\frac{7}{2x}$ Original equation

$\frac{2}{-1}-\frac{3}{2}\stackrel{?}{=}\frac{7}{2(-1)}$ Let $x=-1$.

$-\frac{7}{2}=-\frac{7}{2}$ ✓ True

A true statement results, so the solution set is $\{-1\}$. **NOW TRY** 

CAUTION When each side of an equation is multiplied by a *variable* expression, the resulting "solutions" may not satisfy the original equation. ***We must either determine and observe the domain or check all proposed solutions in the original equation. It is wise to do both.***

EXAMPLE 3 Solving a Rational Equation with No Solution

Solve $\frac{2}{x-3}-\frac{3}{x+3}=\frac{12}{x^2-9}$.

Step 1 The domain, which excludes ± 3, was found in **Example 1(b).**

Step 2 Factor x^2-9. Then multiply each side by the LCD, $(x+3)(x-3)$.

$$(x+3)(x-3)\left(\frac{2}{x-3}-\frac{3}{x+3}\right)=(x+3)(x-3)\left[\frac{12}{(x+3)(x-3)}\right]$$

Step 3 $(x+3)(x-3)\left(\frac{2}{x-3}\right)-(x+3)(x-3)\left(\frac{3}{x+3}\right)$

$=(x+3)(x-3)\left[\frac{12}{(x+3)(x-3)}\right]$ Distributive property

$(x+3)(2)-(x-3)(3)=12$ Multiply.

$2x+6-3x+9=12$ Distributive property

$-x+15=12$ Combine like terms.

Proposed solution $\rightarrow x=3$ Subtract 15. Multiply by -1.

NOW TRY ANSWER

2. $\left\{-\frac{5}{4}\right\}$

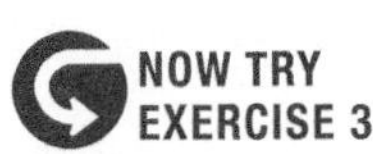
NOW TRY EXERCISE 3

Solve.

$$\frac{1}{t-7}+\frac{2}{t+7}=\frac{14}{t^2-49}$$

Step 4 The proposed solution, 3, is not in the domain. Such solutions are **extraneous** and must be rejected. Substituting 3 into the original equation shows why.

CHECK
$$\frac{2}{x-3}-\frac{3}{x+3}=\frac{12}{x^2-9} \quad \text{Original equation}$$
$$\frac{2}{3-3}-\frac{3}{3+3}\stackrel{?}{=}\frac{12}{3^2-9} \quad \text{Let } x=3.$$
$$\frac{2}{0}-\frac{3}{6}\stackrel{?}{=}\frac{12}{0} \quad \text{Division by 0 is undefined.}$$

The equation has no solution. The solution set is $\varnothing$.

NOW TRY  

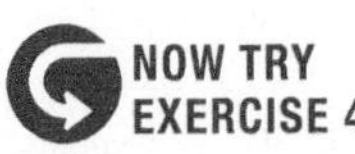
NOW TRY EXERCISE 4

Solve.

$$\frac{2}{t^2-2t-8}-\frac{4}{t^2+6t+8}=\frac{2}{t^2-16}$$

EXAMPLE 4 Solving a Rational Equation

Solve $\dfrac{3}{p^2+p-2}-\dfrac{1}{p^2-1}=\dfrac{7}{2(p^2+3p+2)}$.

Factor each denominator to find the domain and the LCD.

$$\frac{3}{(p-1)(p+2)}-\frac{1}{(p+1)(p-1)}=\frac{7}{2(p+2)(p+1)} \quad \text{Factor the denominators.}$$

The domain excludes 1, -2, and -1. Multiply each side of the equation by the LCD, $2(p-1)(p+2)(p+1)$.

$$2(p-1)(p+2)(p+1)\left[\frac{3}{(p-1)(p+2))}-\frac{1}{(p+1)(p-1)}\right]$$
$$=2(p-1)(p+2)(p+1)\left[\frac{7}{2(p+2)(p+1)}\right]$$

$$2(p+1)(3)-2(p+2)=(p-1)7 \quad \text{Distributive property}$$
$$6p+6-2p-4=7p-7 \quad 2\cdot 3=6;\ \text{Distributive property}$$
$$4p+2=7p-7 \quad \text{Combine like terms.}$$
$$9=3p \quad \text{Subtract } 4p. \text{ Add } 7.$$
$$\text{Proposed solution} \rightarrow 3=p \quad \text{Divide by 3.}$$

CHECK 3 is in the domain. Substitute it in the original equation to check.

$$\frac{3}{p^2+p-2}-\frac{1}{p^2-1}=\frac{7}{2(p^2+3p+2)} \quad \text{Original equation}$$
$$\frac{3}{3^2+3-2}-\frac{1}{3^2-1}\stackrel{?}{=}\frac{7}{2(3^2+3\cdot 3+2)} \quad \text{Let } p=3.$$
$$\frac{3}{10}-\frac{1}{8}\stackrel{?}{=}\frac{7}{40} \quad \text{Work in the denominators.}$$

$\frac{3}{10}-\frac{1}{8}=\frac{12}{40}-\frac{5}{40}$

$$\frac{7}{40}=\frac{7}{40}\ \checkmark \quad \text{True}$$

The solution set is $\{3\}$.

NOW TRY

NOW TRY ANSWERS

3. $\varnothing$

4. $\{5\}$

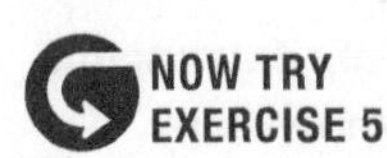

NOW TRY EXERCISE 5

Solve.

$$\frac{2x}{x-2}=\frac{-3}{x}+\frac{4}{x-2}$$

EXAMPLE 5 Solving a Rational Equation

Solve $\frac{2}{3x+1}=\frac{1}{x}-\frac{6x}{3x+1}$. (We must exclude $-\frac{1}{3}$ and 0 from the domain.)

$$x(3x+1)\left(\frac{2}{3x+1}\right)=x(3x+1)\left(\frac{1}{x}-\frac{6x}{3x+1}\right)$$ Multiply by the LCD, $x(3x+1)$.

$$x(3x+1)\left(\frac{2}{3x+1}\right)=x(3x+1)\left(\frac{1}{x}\right)-x(3x+1)\left(\frac{6x}{3x+1}\right)$$ Distributive property

(This is a quadratic equation.) $2x=3x+1-6x^2$ Multiply.

$6x^2-x-1=0$ Standard form

$(3x+1)(2x-1)=0$ Factor.

$3x+1=0$ or $2x-1=0$ Zero-factor property

Proposed solutions → $x=-\frac{1}{3}$ or $x=\frac{1}{2}$ Solve each equation.

Because $-\frac{1}{3}$ is not in the domain, it is not a solution. Check the proposed solution $\frac{1}{2}$ in the original equation. The solution set is $\left\{\frac{1}{2}\right\}$.

NOW TRY

NOW TRY EXERCISE 6

Solve.

$$\frac{4}{x}+\frac{3}{x-6}=2$$

EXAMPLE 6 Solving a Rational Equation

Solve $\frac{5}{x}+\frac{6}{x+2}=1$. (We must exclude 0 and -2 from the domain.)

$$x(x+2)\left(\frac{5}{x}+\frac{6}{x+2}\right)=x(x+2)(1)$$ Multiply by the LCD, $x(x+2)$.

$$x(x+2)\left(\frac{5}{x}\right)+x(x+2)\left(\frac{6}{x+2}\right)=x(x+2)(1)$$ Distributive property

$(x+2)(5)+6x=x^2+2x$ Multiply.

$5x+10+6x=x^2+2x$ Distributive property

$x^2-9x-10=0$ Standard form

$(x+1)(x-10)=0$ Factor.

$x+1=0$ or $x-10=0$ Zero-factor property

$x=-1$ or $x=10$ Solve each equation.

Both -1 and 10 are in the domain. Check that the solution set is $\{-1, 10\}$.

NOW TRY

OBJECTIVE 3 Recognize the graph of a rational function.

A function defined by a quotient of polynomials is a **rational function.** Because one or more values of x may be excluded from the domain of a rational function, its graph is often **discontinuous**—that is, there will be one or more breaks in the graph.

One simple rational function is the **reciprocal function** $f(x)=\frac{1}{x}$**.** The domain of this function includes all real numbers except 0. Thus, this function pairs every real number except 0 with its reciprocal.

NOW TRY ANSWERS

5. $\left\{-\frac{3}{2}\right\}$

6. $\left\{\frac{3}{2}, 8\right\}$

Some ordered pairs that belong to the function $f(x) = \frac{1}{x}$ are listed in the table.

The closer negative values of x are to 0, the less ("more negative") y is. → ← The closer positive values of x are to 0, the greater y is.

x	−3	−2	−1	−0.5	−0.25	−0.1	0.1	0.25	0.5	1	2	3
y	$-\frac{1}{3}$	$-\frac{1}{2}$	−1	−2	−4	−10	10	4	2	1	$\frac{1}{2}$	$\frac{1}{3}$

Plotting the points from the table, we obtain the graph in **FIGURE 2**.

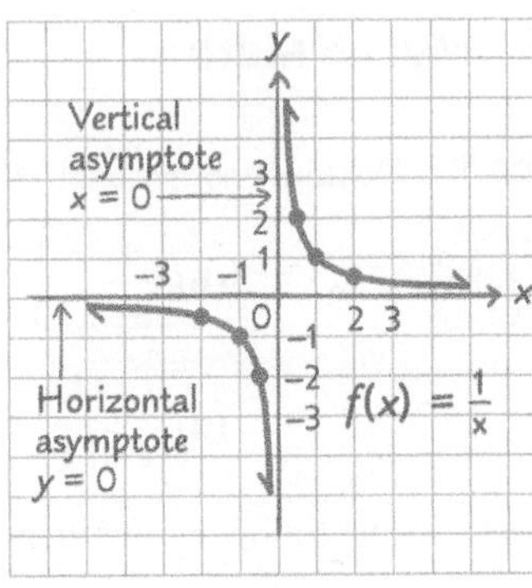

FIGURE 2

Reciprocal function

$$f(x) = \frac{1}{x}$$

Domain: $\{x \mid x \text{ is a real number}, x \neq 0\}$

Range: $\{y \mid y \text{ is a real number}, y \neq 0\}$

Because the domain of $f(x) = \frac{1}{x}$ includes all real numbers except 0, there is no point on the graph where $x = 0$. The vertical line with equation $x = 0$ is a **vertical asymptote.** The horizontal line with equation $y = 0$ is a **horizontal asymptote.** The graph approaches these asymptotes but does not cross them.

NOW TRY EXERCISE 7

Graph the rational function. Give the equations of the vertical and horizontal asymptotes.

$$f(x) = \frac{1}{x+1}$$

EXAMPLE 7 Graphing a Rational Function

Graph the rational function. Give the equations of the vertical and horizontal asymptotes.

$$g(x) = \frac{-2}{x-3}$$

Some ordered pairs that belong to the function are listed in the table.

x	−2	−1	0	1	2	2.5	2.75	3.25	3.5	4	5	6	7
y	$\frac{2}{5}$	$\frac{1}{2}$	$\frac{2}{3}$	1	2	4	8	−8	−4	−2	−1	$-\frac{2}{3}$	$-\frac{1}{2}$

There is no point on the graph, shown in **FIGURE 3**, for $x = 3$, because 3 is excluded from the domain. The dashed line

$$x = 3 \text{ represents the vertical asymptote.}$$

It is not part of the graph. The graph gets closer to the vertical asymptote as the x-values get closer to 3. Again,

$$y = 0 \text{ is a horizontal asymptote.}$$

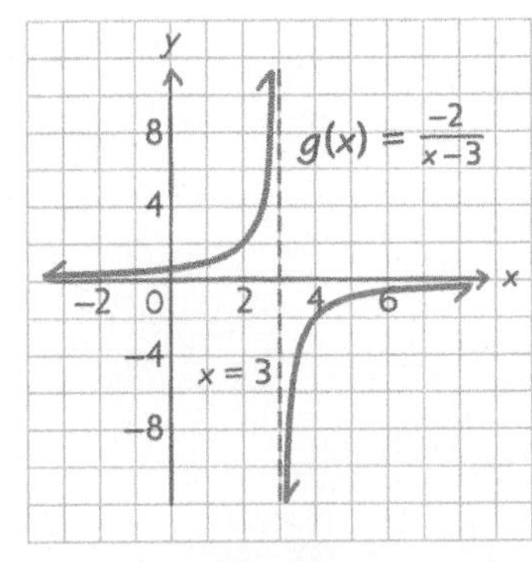

FIGURE 3

NOW TRY

NOW TRY ANSWER

7.

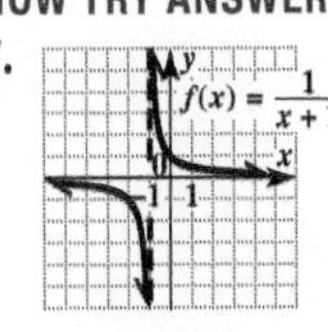

vertical asymptote: $x = -1$; horizontal asymptote: $y = 0$

NOTE In general, the following statements are true of a rational function.

1. If the y-values approach ∞ or $-\infty$ as the x-values approach a real number a, the **vertical line $x = a$ is a vertical asymptote** of the graph.
2. If the y-values approach a real number b as $|x|$ increases without bound, the **horizontal line $y = b$ is a horizontal asymptote** of the graph.

6.4 Exercises

MyLab Math

Video solutions for select problems available in MyLab Math

1. Concept Check Decide whether each of the following is an *expression* or an *equation*.

(a) $\frac{2}{x} = \frac{4}{3x} + \frac{1}{3}$ **(b)** $\frac{4}{3x} + \frac{1}{3}$

(c) $\frac{5}{x+1} - \frac{2}{x-1}$ **(d)** $\frac{5}{x+1} - \frac{2}{x-1} = \frac{4}{x^2-1}$

2. Concept Check The following instruction is incorrect.

"Solve $\frac{2x+1}{3x-4} + \frac{1}{2x+3}$."

WHAT WENT WRONG?

3. CONCEPT CHECK A student solved

$$\frac{3}{x-3} - \frac{2}{x-2} = \frac{3}{x^2-5x+6},$$

performing all algebraic steps correctly, and obtained $x = 3$ as the final step. She gave the solution set $\{3\}$, but this is incorrect. **WHAT WENT WRONG?** Give the correct solution set.

4. Concept Check Consider the following incorrect "solution."

$$x - \frac{9}{x} = -8 \quad \text{Multiply by the LCD, } x.$$
$$x - 9 = -8x$$
$$9x = 9$$
$$x = 1 \quad \text{Solution set: } \{1\}$$

WHAT WENT WRONG? Give the correct solution set.

Give the domain of the variable in each equation. ***See Example 1.***

5. $\frac{1}{3x} + \frac{1}{2x} = \frac{x}{3}$

6. $\frac{5}{6x} - \frac{8}{2x} = \frac{x}{4}$

7. $\frac{1}{x+1} - \frac{1}{x-2} = 0$

8. $\frac{3}{x+4} - \frac{2}{x-9} = 0$

9. $\frac{3x+1}{x-4} = \frac{6x+5}{2x-7}$

10. $\frac{4x-1}{2x+3} = \frac{12x-25}{6x-2}$

11. $\frac{6}{4x+7} - \frac{3}{x} = \frac{5}{6x-13}$

12. $\frac{4}{3x-5} + \frac{2}{x} = \frac{9}{4x+13}$

13. $\frac{1}{x^2-16} - \frac{2}{x-4} = \frac{1}{x+4}$

14. $\frac{2}{x^2-25} - \frac{1}{x+5} = \frac{1}{x-5}$

15. $\frac{2}{x^2-x} + \frac{1}{x+3} = \frac{4}{x-2}$

16. $\frac{3}{x^2+x} - \frac{1}{x+5} = \frac{2}{x-7}$

Solve each equation. ***See Examples 2–6.***

17. $\frac{3}{4x}=\frac{5}{2x}-\frac{7}{4}$

18. $\frac{2}{3x}=\frac{6}{5x}+\frac{8}{45}$

19. $x-\frac{24}{x}=-2$

20. $p+\frac{15}{p}=-8$

21. $\frac{x}{4}-\frac{21}{4x}=-1$

22. $\frac{x}{2}-\frac{12}{x}=1$

23. $\frac{x-4}{x+6}=\frac{2x+3}{2x-1}$

24. $\frac{5x-3}{x+2}=\frac{5x-1}{x+3}$

25. $\frac{3x+1}{x-4}=\frac{6x+5}{2x-7}$

26. $\frac{4x-1}{2x+3}=\frac{12x-25}{6x-2}$

27. $\frac{1}{y-1}+\frac{5}{12}=\frac{-2}{3y-3}$

28. $\frac{4}{m+2}-\frac{11}{9}=\frac{1}{3m+6}$

29. $\frac{-2}{3t-6}-\frac{1}{36}=\frac{-3}{4t-8}$

30. $\frac{6}{t+1}-\frac{34}{15}=\frac{-4}{5t+5}$

31. $\frac{7}{6x+3}-\frac{1}{3}=\frac{2}{2x+1}$

32. $\frac{3}{4m+2}=\frac{17}{2}-\frac{7}{2m+1}$

33. $\frac{9}{x}+\frac{4}{6x-3}=\frac{2}{6x-3}$

34. $\frac{5}{n}+\frac{4}{6-3n}=\frac{2}{6-3n}$

35. $\frac{6}{x-4}+\frac{5}{x}=\frac{-20}{x^2-4x}$

36. $\frac{7}{x-4}+\frac{3}{x}=\frac{-12}{x^2-4x}$

37. $\frac{3}{x+2}-\frac{2}{x^2-4}=\frac{1}{x-2}$

38. $\frac{3}{x-2}+\frac{21}{x^2-4}=\frac{14}{x+2}$

39. $\frac{1}{y+2}+\frac{3}{y+7}=\frac{5}{y^2+9y+14}$

40. $\frac{1}{t+3}+\frac{4}{t+5}=\frac{2}{t^2+8t+15}$

41. $\frac{6}{w+3}+\frac{-7}{w-5}=\frac{-48}{w^2-2w-15}$

42. $\frac{2}{r-5}+\frac{3}{2r+1}=\frac{22}{2r^2-9r-5}$

43. $\frac{2}{x}+\frac{4}{x+9}=1$

44. $\frac{3}{x}-\frac{9}{x-8}=-2$

45. $\frac{2}{4x+7}+\frac{x}{3}=\frac{6}{12x+21}$

46. $\frac{3}{2x+5}+\frac{x}{2}=\frac{9}{6x+15}$

47. $\frac{1}{x-2}+\frac{1}{4}=\frac{1}{4(x^2-4)}$

48. $\frac{1}{x+4}+\frac{1}{3}=\frac{-10}{3(x^2-16)}$

49. $\frac{x}{x-3}+\frac{4}{x+3}=\frac{18}{x^2-9}$

50. $\frac{2x}{x-3}+\frac{4}{x+3}=\frac{-24}{x^2-9}$

51. $\frac{1}{x+4}+\frac{x}{x-4}=\frac{-8}{x^2-16}$

52. $\frac{5}{x-4}-\frac{3}{x-1}=\frac{x^2-1}{x^2-5x+4}$

53. $\frac{2}{k^2+k-6}+\frac{1}{k^2-k-2}=\frac{4}{k^2+4k+3}$

54. $\frac{5}{p^2+3p+2}-\frac{3}{p^2-4}=\frac{1}{p^2-p-2}$

55. $\frac{5x+14}{x^2-9}=\frac{-2x^2-5x+2}{x^2-9}+\frac{2x+4}{x-3}$

56. $\frac{4x-7}{4x^2-9}=\frac{-2x^2+5x-4}{4x^2-9}+\frac{x+1}{2x+3}$

57. Concept Check Match each function in Column I with its type in Column II.

I	II
(a) $f(x) = 3x - 4$	**A.** quadratic function
(b) $f(x) = \frac{10}{x}$	**B.** cubing function
(c) $f(x) = x^2 + 2x - 8$	**C.** rational function
(d) $f(x) = x^3$	**D.** linear function

58. Concept Check Which graph has vertical and horizontal asymptotes? What are their equations?

A.

B.

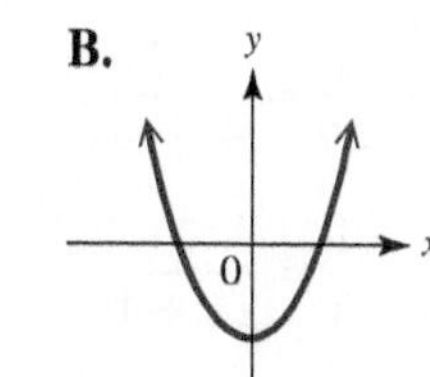

C.

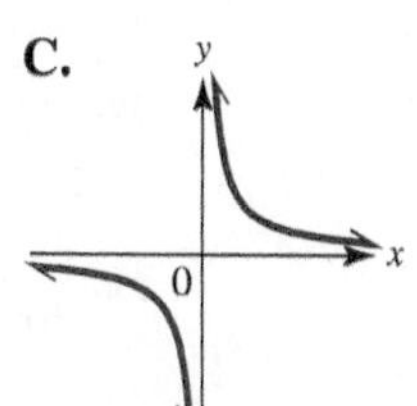

D.

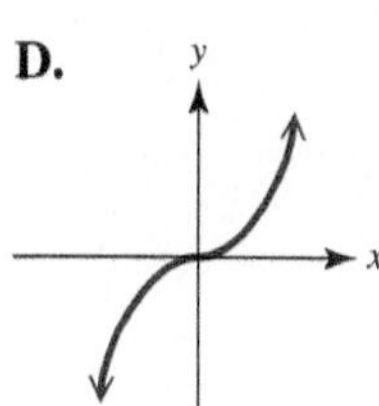

Graph each rational function. Give the equations of the vertical and horizontal asymptotes. ***See Example 7.***

59. $f(x) = \frac{2}{x}$ **60.** $f(x) = \frac{3}{x}$ **61.** $g(x) = -\frac{1}{x}$ **62.** $g(x) = -\frac{2}{x}$

63. $f(x) = \frac{1}{x - 2}$ **64.** $f(x) = \frac{1}{x + 2}$ **65.** $g(x) = \frac{-2}{x - 1}$ **66.** $g(x) = \frac{-1}{x + 3}$

Solve each problem.

67. The average number of vehicles waiting in line to enter a parking area is modeled by the function

$$W(x) = \frac{x^2}{2(1 - x)},$$

where x is a quantity between 0 and 1 known as the **traffic intensity.** (Data from Mannering, F., and W. Kilareski, *Principles of Highway Engineering and Traffic Control,* John Wiley and Sons.) For each traffic intensity, find the average number of vehicles waiting (to the nearest tenth).

(a) 0.1 **(b)** 0.8 **(c)** 0.9

(d) What happens to waiting time as traffic intensity increases?

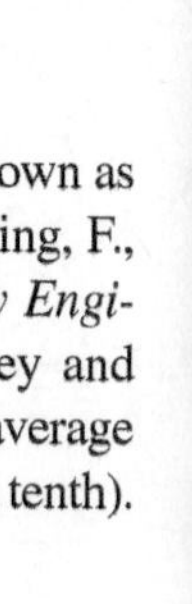

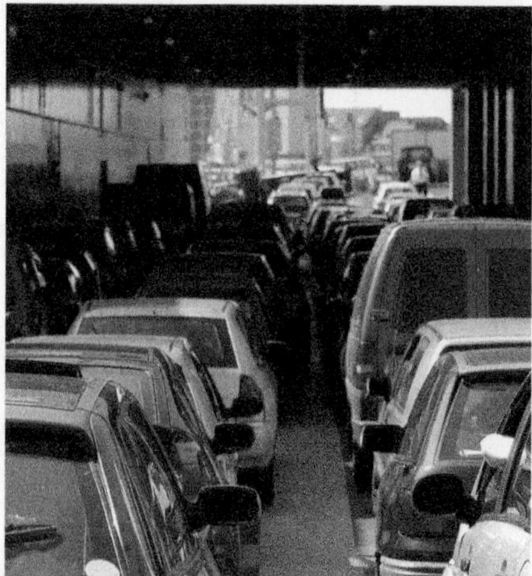

68. The percent of deaths caused by smoking is modeled by the rational function

$$P(x) = \frac{x - 1}{x},$$

where x is the number of times more likely a smoker is to die of lung cancer than a nonsmoker. This is the **incidence rate.** (Data from Walker, A., *Observation and Inference: An Introduction to the Methods of Epidemiology,* University of Michigan.) For example, $x = 10$ means that a smoker is 10 times more likely than a nonsmoker to die of lung cancer.

(a) Find $P(x)$ if $x = 10$.

(b) For what values of x is $P(x) = 80\%$? (*Hint:* Change 80% to a decimal.)

(c) Can the incidence rate equal 0? Explain.

69. The force required to keep a 2000-lb car that is going 30 mph from skidding on a curve, where r is the radius of the curve in feet, is given by

$$F(r) = \frac{225{,}000}{r}.$$

(a) What radius must a curve have if a force of 450 lb is needed to keep the car from skidding?

(b) As the radius of the curve is lengthened, how is the force affected?

70. The amount of heating oil produced (in gallons per day) by an oil refinery is modeled by the rational function

$$f(x) = \frac{125{,}000 - 25x}{125 + 2x},$$

where x is the amount of gasoline produced (in hundreds of gallons per day). Suppose the refinery must produce 300 gal of heating oil per day to meet the needs of its customers.

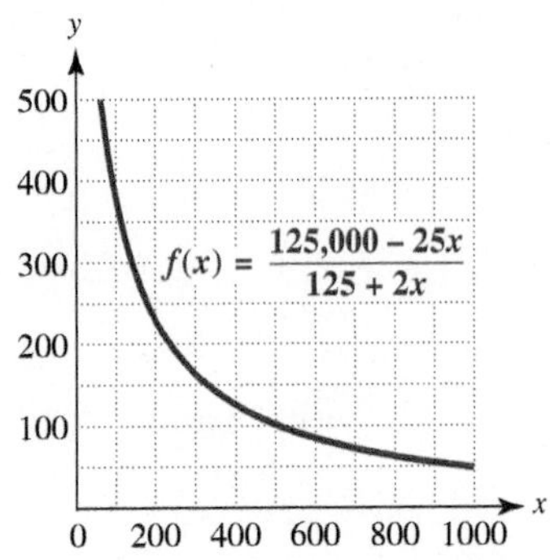

(a) How much gasoline will be produced per day?

(b) A graph of f, for $x > 0$, is shown in the figure. Use it to decide what happens to the amount of gasoline (x) produced as the amount of heating oil (y) produced increases.

SUMMARY EXERCISES Simplifying Rational Expressions vs. Solving Rational Equations

A common student error is to confuse an equation, such as $\frac{x}{2} + \frac{x}{3} = -5$, with an expression involving an operation, such as $\frac{x}{2} + \frac{x}{3}$. ***Equations are solved for a numerical answer, while problems involving operations result in simplified expressions.***

Solving an Equation	**Simplifying an Expression Involving an Operation**
Solve: $\frac{x}{2} + \frac{x}{3} = -5$ Look for the equality symbol.	***Add:*** $\frac{x}{2} + \frac{x}{3}$ There is no equality symbol.
Multiply each side by the LCD, 6.	Write both fractions with the LCD, 6.
$6\left(\frac{x}{2} + \frac{x}{3}\right) = 6(-5)$	$\frac{x}{2} + \frac{x}{3}$
$6\left(\frac{x}{2}\right) + 6\left(\frac{x}{3}\right) = 6(-5)$	$= \frac{x \cdot 3}{2 \cdot 3} + \frac{x \cdot 2}{3 \cdot 2}$
$3x + 2x = -30$	$= \frac{3x}{6} + \frac{2x}{6}$
$5x = -30$	$= \frac{3x + 2x}{6}$
$x = -6$	
Check that the solution set is $\{-6\}$.	$= \frac{5x}{6}$

Identify each exercise as an expression *or an* equation. *Then either simplify the expression by performing the indicated operation, or solve the equation, as appropriate.*

1. $\frac{x}{2} - \frac{x}{4} = 5$

2. $\frac{4x - 20}{x^2 - 25} \cdot \frac{(x + 5)^2}{10}$

3. $\frac{6}{7x} - \frac{4}{x}$

4. $\frac{\frac{1}{x} + \frac{1}{y}}{\frac{1}{x} - \frac{1}{y}}$

5. $\frac{5}{7t} = \frac{52}{7} - \frac{3}{t}$

6. $\frac{x - 5}{3} + \frac{1}{3} = \frac{x - 2}{5}$

7. $\frac{7}{6x} + \frac{5}{8x}$

8. $\frac{4}{x} - \frac{8}{x + 1} = 0$

9. $\frac{\frac{6}{x + 1} - \frac{1}{x}}{\frac{2}{x} - \frac{4}{x + 1}}$

10. $\frac{8}{r + 2} - \frac{7}{4r + 8}$

11. $\frac{x}{x + y} + \frac{2y}{x - y}$

12. $\frac{3p^2 - 6p}{p + 5} \div \frac{p^2 - 4}{8p + 40}$

13. $\frac{x - 2}{9} \cdot \frac{5}{8 - 4x}$

14. $\frac{a - 4}{3} + \frac{11}{6} = \frac{a + 1}{2}$

15. $\frac{5}{x^2 - 2x} - \frac{3}{x^2 - 4}$

16. $\frac{3}{t - 1} + \frac{1}{t} = \frac{7}{2}$

17. $\frac{\frac{5}{x} - \frac{3}{y}}{\frac{9x^2 - 25y^2}{x^2y}}$

18. $\frac{\frac{t}{4} - \frac{1}{t}}{1 + \frac{t + 4}{t}}$

19. $\frac{3r}{r - 2} = 1 + \frac{6}{r - 2}$

20. $\frac{7}{2x^2 - 8x} + \frac{3}{x^2 - 16}$

21. $\frac{b^2 + b - 6}{b^2 + 2b - 8} \cdot \frac{b^2 + 8b + 16}{3b + 12}$

22. $\frac{10z^2 - 5z}{3z^3 - 6z^2} \div \frac{2z^2 + 5z - 3}{z^2 + z - 6}$

23. $\frac{4y^2 - 13y + 3}{2y^2 - 9y + 9} \div \frac{4y^2 + 11y - 3}{6y^2 - 5y - 6}$

24. $\frac{-2}{a^2 + 2a - 3} - \frac{5}{3 - 3a} = \frac{4}{3a + 9}$

25. $\frac{-1}{3 - x} - \frac{2}{x - 3}$

26. $\frac{8}{3k + 9} - \frac{8}{15} = \frac{2}{5k + 15}$

27. $\frac{2}{y + 1} - \frac{3}{y^2 - y - 2} = \frac{3}{y - 2}$

28. $\frac{6z^2 - 5z - 6}{6z^2 + 5z - 6} \cdot \frac{12z^2 - 17z + 6}{12z^2 - z - 6}$

29. $\frac{3}{y - 3} - \frac{3}{y^2 - 5y + 6} = \frac{2}{y - 2}$

30. $\frac{2k + \frac{5}{k - 1}}{3k - \frac{2}{k}}$

6.5 Applications of Rational Expressions

OBJECTIVES

1 Find the value of an unknown variable in a formula.
2 Solve a formula for a specified variable.
3 Solve applications using proportions.
4 Solve applications about distance, rate, and time.
5 Solve applications about work rates.

VOCABULARY

☐ ratio
☐ proportion
☐ cross products of a proportion

NOW TRY EXERCISE 1

Use the formula in **Example 1** to find f if $p = 50$ cm and $q = 10$ cm.

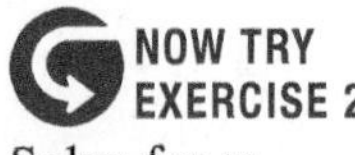

NOW TRY EXERCISE 2

Solve for m.

$$\frac{1}{m} - \frac{2}{n} = 5$$

NOW TRY ANSWERS

1. $\frac{25}{3}$ cm
2. $m = \frac{n}{5n + 2}$

OBJECTIVE 1 Find the value of an unknown variable in a formula.

Formulas may contain rational expressions, such as $t = \frac{d}{r}$ and $\frac{1}{f} = \frac{1}{p} + \frac{1}{q}$.

EXAMPLE 1 Finding the Value of a Variable in a Formula

In physics, the focal length f of a lens is given by the formula

$$\frac{1}{f} = \frac{1}{p} + \frac{1}{q}.$$

In the formula, p is the distance from the object to the lens and q is the distance from the lens to the image. See **FIGURE 4**. Find q if $p = 20$ cm and $f = 10$ cm.

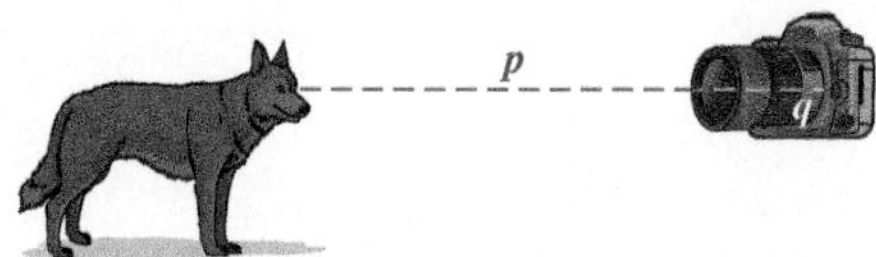

Focal Length of Camera Lens

FIGURE 4

$$\frac{1}{f} = \frac{1}{p} + \frac{1}{q}$$

$$\frac{1}{10} = \frac{1}{20} + \frac{1}{q} \qquad \text{Let } f = 10,\ p = 20.$$

(Solve this equation for q.)

$$20q \cdot \frac{1}{10} = 20q\left(\frac{1}{20} + \frac{1}{q}\right) \qquad \text{Multiply by the LCD, } 20q.$$

$$20q \cdot \frac{1}{10} = 20q\left(\frac{1}{20}\right) + 20q\left(\frac{1}{q}\right) \qquad \text{Distributive property}$$

$$2q = q + 20 \qquad \text{Multiply.}$$

$$q = 20 \qquad \text{Subtract } q.$$

The distance from the lens to the image is 20 cm. **NOW TRY** 

OBJECTIVE 2 Solve a formula for a specified variable.

We solve for a specified variable by isolating it on one side of the equality symbol.

EXAMPLE 2 Solving a Formula for a Specified Variable

Solve $\frac{1}{f} = \frac{1}{p} + \frac{1}{q}$ for p.

$$\frac{1}{f} = \frac{1}{p} + \frac{1}{q}$$

$$fpq \cdot \frac{1}{f} = fpq\left(\frac{1}{p} + \frac{1}{q}\right) \qquad \text{To clear the fractions, multiply by the LCD, } fpq.$$

$$pq = fq + fp \qquad \text{Multiply; distributive property}$$

(We want the two terms with p on the same side.)

$$pq - fp = fq \qquad \text{Subtract } fp.$$

(This is a key step.)

$$p(q - f) = fq \qquad \text{Factor out } p.$$

$$p = \frac{fq}{q - f} \qquad \text{Divide by } q - f.$$

NOW TRY

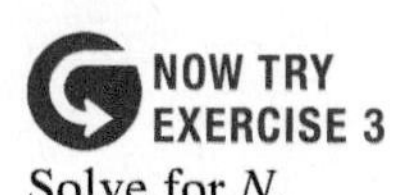
NOW TRY EXERCISE 3

Solve for N.

$$\frac{NR}{N-n} = t$$

EXAMPLE 3 Solving a Formula for a Specified Variable

Solve $I = \frac{nE}{R + nr}$ for n.

$$I = \frac{nE}{R + nr}$$

$$(R + nr)I = (R + nr)\frac{nE}{R + nr}$$ To clear the fraction, multiply by $R + nr$.

$$RI + nrI = nE$$ Distributive property on the left

Write the n-terms on the ***same*** side in preparation for factoring.

$$RI = nE - nrI$$ Subtract nrI.

$$RI = n(E - rI)$$ Factor out n.

$$\frac{RI}{E - rI} = n, \quad \text{or} \quad n = \frac{RI}{E - rI}$$ Divide by $E - rI$. Interchange sides.

NOW TRY 

CAUTION Refer to the steps in **Examples 2 and 3** that factor out the desired variable. ***This variable must be a factor on only one side of the equation.*** Then each side can be divided by the remaining factor in the last step.

OBJECTIVE 3 Solve applications using proportions.

A **ratio** is a comparison of two quantities. The ratio of a to b may be written in any of the following ways.

$$a \text{ to } b, \quad a:b, \quad \text{or} \quad \frac{a}{b}$$ Ratio of a to b

Ratios are usually written as quotients in algebra. A **proportion** is a statement that two ratios are equal.

$$\frac{a}{b} = \frac{c}{d}$$ Proportion

EXAMPLE 4 Solving a Proportion

In 2016, about 9 of every 100 Americans had no health insurance. The population at that time was about 323 million. How many million Americans had no health insurance? (Data from U.S. Census Bureau.)

Step 1 **Read** the problem.

Step 2 **Assign a variable.**

Let x = the number (in millions) who had no health insurance.

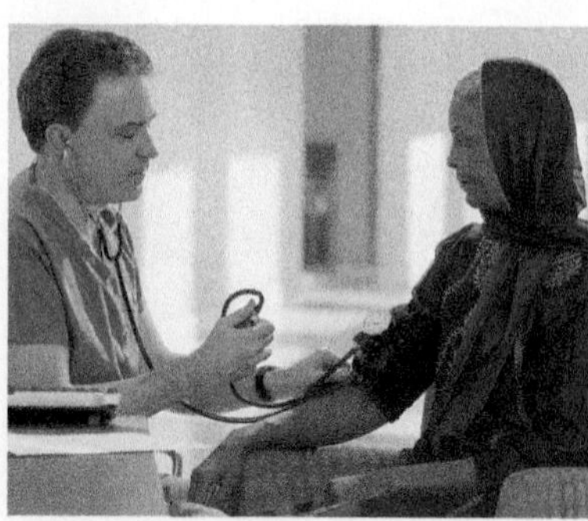

Step 3 **Write an equation.** We set up a proportion. The ratio 9 to 100 should equal the ratio x to 323.

$$\frac{9}{100} = \frac{x}{323}$$ Write a proportion.

NOW TRY ANSWER

3. $N = \frac{nt}{t - R}$

NOW TRY EXERCISE 4

In 2016, 18 of every 100 West Virginians lived in poverty. West Virginia's population was about 1,831,000 at that time. How many West Virginians lived in poverty? (Data from U.S. Census Bureau.)

Step 4 **Solve.** $32{,}300\left(\frac{9}{100}\right) = 32{,}300\left(\frac{x}{323}\right)$ Multiply by a common denominator, $100 \cdot 323 = 32{,}300$.

$2907 = 100x$ Simplify.

$29.07 = x$ Divide by 100.

Step 5 **State the answer.** 29.07 million Americans had no health insurance.

Step 6 **Check** that the ratio of 29.07 million to 323 million equals $\frac{9}{100}$.

$$\frac{29.07}{323} = \frac{9}{100}$$ Use a calculator to divide 29.07 by 323. A true statement results.

NOW TRY

NOW TRY EXERCISE 5

Clayton's SUV uses 28 gal of gasoline to drive 500 mi. He has 10 gal of gasoline in the SUV, and he still needs to drive 400 mi. If the car continues to use gasoline at the same rate, how many more gallons will he need?

EXAMPLE 5 Solving a Proportion Involving Rates

Jodie's car uses 10 gal of gasoline to travel 210 mi. She has 5 gal of gasoline in the car. She still needs to drive 640 mi. If the car continues to use gasoline at the same rate, how many more gallons will she need?

Step 1 **Read** the problem.

Step 2 **Assign a variable.**

Let x = the additional number of gallons of gas needed.

Step 3 **Write an equation.** We set up a proportion.

$$\frac{\text{gallons} \rightarrow 10}{\text{miles} \rightarrow 210} = \frac{5 + x \leftarrow \text{gallons}}{640 \leftarrow \text{miles}}$$

Step 4 **Solve.** We could multiply each side by the LCD $10 \cdot 21 \cdot 64$. Instead we use an alternative method that involves **cross products of a proportion.**

For $\frac{a}{b} = \frac{c}{d}$ to be true, the cross products ad and bc must be equal.

$a \cdot d = b \cdot c$

$\downarrow \quad \downarrow \quad \downarrow \quad \downarrow$

$10 \cdot 640 = 210(5 + x)$ If $\frac{a}{b} = \frac{c}{d}$, then $ad = bc$.

$6400 = 1050 + 210x$ Multiply; distributive property

$5350 = 210x$ Subtract 1050.

$25.5 = x$ Divide by 210. Round to the nearest tenth.

Step 5 **State the answer.** Jodie will need 25.5 more gallons of gas.

Step 6 **Check.** 25.5 more gallons plus the 5 that she has equals 30.5 gal. From Step 3, the ratio of 10 to 210 must equal the ratio of 30.5 to 640.

$$\frac{10}{210} \approx 0.048 \quad \text{and} \quad \frac{30.5}{640} \approx 0.048$$ Use a calculator.

Because the ratios are approximately equal, the answer is correct.

NOW TRY 

OBJECTIVE 4 Solve applications about distance, rate, and time.

When the distance formula $d = rt$ is solved for r or t, rational expressions result.

Rate is the ratio of distance to time, or $r = \frac{d}{t}$.

Time is the ratio of distance to rate, or $t = \frac{d}{r}$.

NOW TRY ANSWERS

4. 329,580 West Virginians
5. 12.4 more gallons

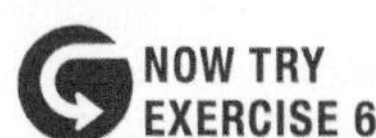
NOW TRY EXERCISE 6

A small fishing boat travels 36 mi against the current in a river in the same time that it travels 44 mi with the current. If the rate of the boat in still water is 20 mph, find the rate of the current.

EXAMPLE 6 Solving a Distance, Rate, Time Problem

A paddle wheeler travels 10 mi against the current in a river in the same time that it travels 15 mi with the current. If the rate of the current is 3 mph, find the rate of the boat in still water.

Step 1 **Read** the problem.

Step 2 **Assign a variable.** Let $x =$ the rate of the boat in still water.

Traveling ***against*** the current slows the boat down, so the rate of the boat is the ***difference*** between its rate in still water and the rate of the current—that is, $(x - 3)$ mph.

Traveling ***with*** the current speeds the boat up, so the rate of the boat is the ***sum*** of its rate in still water and the rate of the current—that is, $(x + 3)$ mph.

Thus, $x - 3 =$ the rate of the boat *against* the current,

and $x + 3 =$ the rate of the boat *with* the current.

Because the time is the same going against the current as with the current, find time in terms of distance and rate for each situation. Against the current, the distance is 10 mi and the rate is $(x - 3)$ mph.

$$t = \frac{d}{r} = \frac{10}{x - 3} \quad \text{Time } \textit{against} \text{ the current}$$

With the current, the distance is 15 mi and the rate is $(x + 3)$ mph.

$$t = \frac{d}{r} = \frac{15}{x + 3} \quad \text{Time } \textit{with} \text{ the current}$$

	Distance	Rate	Time
Against Current	10	$x - 3$	$\frac{10}{x - 3}$
With Current	15	$x + 3$	$\frac{15}{x + 3}$

Summarize the information in a table.

Times are equal.

Step 3 **Write an equation.** Use the fact that the times are equal.

Time against current equals time with current.

$$\frac{10}{x - 3} = \frac{15}{x + 3}$$

To solve, we can multiply by the LCD or use cross products.

Step 4 **Solve.**

$$(x + 3)(x - 3)\left(\frac{10}{x - 3}\right) = (x + 3)(x - 3)\left(\frac{15}{x + 3}\right) \quad \text{Multiply by the LCD, } (x + 3)(x - 3).$$

$$(x + 3)10 = (x - 3)15 \quad \text{Multiply.}$$

$$10x + 30 = 15x - 45 \quad \text{Distributive property}$$

$$75 = 5x \quad \text{Subtract } 10x. \text{ Add } 45.$$

$$15 = x \quad \text{Divide by } 5.$$

Step 5 **State the answer.** The rate of the boat in still water is 15 mph.

Step 6 **Check** the answer: $\frac{10}{15 - 3} = \frac{15}{15 + 3}$ is true.

NOW TRY

NOW TRY ANSWER

6. 2 mph

EXAMPLE 7 Solving a Distance, Rate, Time Problem

At O'Hare International Airport in Chicago, Cheryl and Bill are walking to the gate at the same rate to catch their flight to Denver. Bill wants a window seat, so he steps onto the moving sidewalk and continues to walk while Cheryl uses the stationary sidewalk. If the sidewalk moves at 1 m per sec and Bill saves 50 sec covering the 300-m distance, what is their walking rate?

Step 1 **Read** the problem. We must find their walking rate.

Step 2 **Assign a variable.**

Let $x =$ their walking rate in meters per second.

Thus, Cheryl travels at a rate of x meters per second and Bill travels at a rate of $(x + 1)$ meters per second. Express their times in terms of the known distances and the variable rates, as in **Example 6.** Cheryl travels 300 m at a rate of x meters per second.

$$t = \frac{d}{r} = \frac{300}{x} \quad \text{Cheryl's time}$$

Bill travels 300 m at a rate of $(x + 1)$ meters per second.

$$t = \frac{d}{r} = \frac{300}{x+1} \quad \text{Bill's time}$$

	Distance	*Rate*	*Time*
Cheryl	300	x	$\frac{300}{x}$
Bill	300	$x + 1$	$\frac{300}{x+1}$

← Bill's time is 50 sec less than Cheryl's time.

Step 3 **Write an equation.** Use the times from the table.

Bill's time is Cheryl's time less 50 sec

$$\frac{300}{x+1} = \frac{300}{x} - 50$$

We ***cannot*** solve using cross products because there are two terms on the right.

Step 4 **Solve.**

$$x(x+1)\left(\frac{300}{x+1}\right) = x(x+1)\left(\frac{300}{x} - 50\right) \quad \text{Multiply by the LCD, } x(x+1).$$

$$x(x+1)\left(\frac{300}{x+1}\right) = x(x+1)\left(\frac{300}{x}\right) - x(x+1)(50) \quad \text{Distributive property}$$

$$300x = 300(x+1) - 50x(x+1) \quad \text{Multiply.}$$

$$300x = 300x + 300 - 50x^2 - 50x \quad \text{Distributive property}$$

$$50x^2 + 50x - 300 = 0 \quad \text{Standard form}$$

$$x^2 + x - 6 = 0 \quad \text{Divide by 50.}$$

$$(x+3)(x-2) = 0 \quad \text{Factor.}$$

$$x + 3 = 0 \quad \text{or} \quad x - 2 = 0 \quad \text{Zero-factor property}$$

$$x = -3 \quad \text{or} \quad x = 2 \quad \text{Solve each equation.}$$

Discard the negative solution because rate (speed) cannot be negative.

NOW TRY EXERCISE 7

James and Pat are driving from Atlanta to Jacksonville, a distance of 310 mi. James, whose average rate is 5 mph faster than Pat's, will drive the first 130 mi of the trip, and then Pat will drive the rest of the way to their destination. If the total driving time is 5 hr, determine the average rate of each driver.

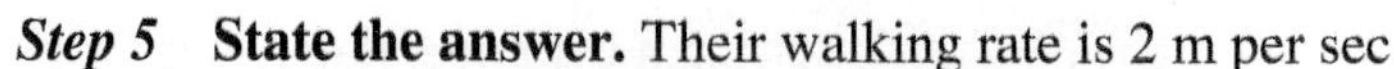

Step 5 **State the answer.** Their walking rate is 2 m per sec.

Step 6 **Check** the solution in the words of the original problem. NOW TRY 

CAUTION ***We can solve a rational equation using cross products only when there is a single rational expression on each side, such as in the equation in*** **Example 6.**

$$\frac{10}{x-3} = \frac{15}{x+3}$$ To solve, we can multiply by the LCD *or* use cross products of the proportion.

We *cannot* solve the rational equation in **Example 7** using cross products.

$$\frac{300}{x+1} = \frac{300}{x} - 50$$ There is more than one term on the right. To solve, we must multiply by the LCD.

OBJECTIVE 5 Solve applications about work rates.

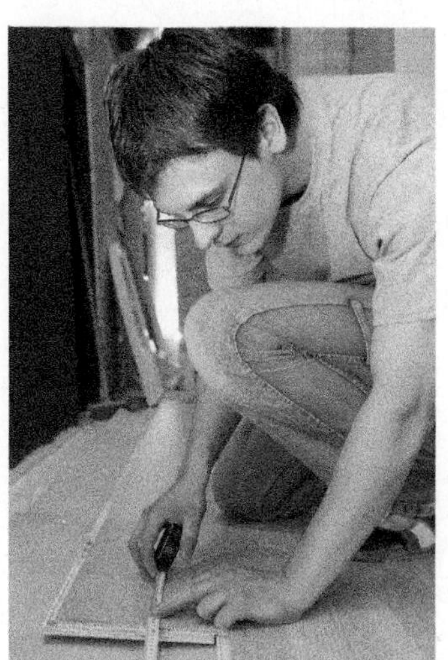

PROBLEM-SOLVING HINT If the letters r, t, and A represent the rate at which work is done, the time required, and the amount of work accomplished, respectively, then $\boldsymbol{A = rt}$. Notice the similarity to the distance formula, $d = rt$.

Amount of work can be measured in terms of jobs accomplished. Thus, if 1 job is completed, then $A = 1$, and the formula gives the rate as

$$1 = rt, \quad \text{or} \quad r = \frac{1}{t}.$$

To solve a work problem, we use the following fact to express all rates of work.

Rate of Work

If 1 complete job can be accomplished in t units of time, then the rate of work is

$$\frac{1}{t} \text{ job per unit of time.}$$

EXAMPLE 8 Solving a Work Problem

Letitia and Kareem are working on a neighborhood cleanup. Kareem can clean up all the trash in the area in 7 hr, while Letitia can do the same job in 5 hr. How long will it take them if they work together?

Steps 1 and 2 **Read** the problem, and **assign a variable.**

Let $x =$ the number of hours it will take the two people working together.

We use the formula $A = rt$. Because $A = 1$, the rate for each person will be $\frac{1}{t}$, where t is the time it takes the person to complete the job alone. Kareem can clean up all the trash in 7 hr, so his rate is $\frac{1}{7}$ of the job per hour. Letitia's rate is $\frac{1}{5}$ of the job per hour.

	Rate	*Time Working Together*	*Fractional Part of the Job Done*
Kareem	$\frac{1}{7}$	x	$\frac{1}{7}x$
Letitia	$\frac{1}{5}$	x	$\frac{1}{5}x$

NOW TRY ANSWER

7. 60 mph; 65 mph

NOW TRY EXERCISE 8

Using her zero-turn riding lawn mower, Tracy can mow her lawn in 2 hr. Using a traditional mower, Matt can mow the same lawn in 3 hr. How long will it take them to mow the lawn working together?

Step 3 **Write an equation.**

Part done by Kareem	+	part done by Letitia	is	1 whole job.	
$\frac{1}{7}x$	+	$\frac{1}{5}x$	=	1	Together they complete 1 job. The sum of the fractional parts must equal 1.

Step 4 **Solve.**

$$35\left(\frac{1}{7}x + \frac{1}{5}x\right) = 35 \cdot 1 \quad \text{Multiply by the LCD, 35.}$$

$$5x + 7x = 35 \quad (*) \quad \text{Distributive property}$$

$$12x = 35 \quad \text{Combine like terms.}$$

$$x = \frac{35}{12} \quad \text{Divide by 12.}$$

Step 5 **State the answer.** Kareem and Letitia can do the entire job working together in $\frac{35}{12}$ hr, or 2 hr, 55 min.

$\frac{35}{12} = 2\frac{11}{12} = 2\frac{11}{12} \cdot \frac{5}{5} = 2\frac{55}{60}$, or 2 hr, 55 min

Step 6 **Check** this result in the original problem.

NOW TRY 

NOTE There is another way to approach problems about rates of work. In **Example 8,** x represents the number of hours it will take the two people working together to complete the entire job. In 1 hr, $\frac{1}{x}$ of the entire job will be completed, so in 1 hr, Kareem completes $\frac{1}{7}$ of the job and Letitia completes $\frac{1}{5}$ of the job.

$$\frac{1}{7} + \frac{1}{5} = \frac{1}{x} \quad \text{The sum of their rates equals } \frac{1}{x}.$$

Multiplying each side of this equation by the LCD $35x$ gives $5x + 7x = 35$. This is equation $(*)$ in **Example 8.** The same solution results using either approach.

NOW TRY ANSWER

8. $\frac{6}{5}$ hr, or 1 hr, 12 min

6.5 Exercises

FOR EXTRA HELP

MyLab Math

 *Video solutions for select problems available in MyLab Math*

Concept Check *Each exercise presents a familiar formula. Give the letter of the choice that is an equivalent form of the formula.*

1. $p = br$ (percent)

A. $b = \frac{p}{r}$ **B.** $r = \frac{b}{p}$ **C.** $b = \frac{r}{p}$ **D.** $p = \frac{r}{b}$

2. $V = LWH$ (geometry)

A. $H = \frac{LW}{V}$ **B.** $L = \frac{V}{WH}$ **C.** $L = \frac{WH}{V}$ **D.** $W = \frac{H}{VL}$

3. $m = \frac{F}{a}$ (physics)

A. $a = mF$ **B.** $F = \frac{m}{a}$ **C.** $F = \frac{a}{m}$ **D.** $F = ma$

4. $I = \frac{E}{R}$ (electricity)

A. $R = \frac{I}{E}$ **B.** $R = IE$ **C.** $E = \frac{I}{R}$ **D.** $E = RI$

Solve each problem. ***See Example 1.***

5. In work with electric circuits, the formula

$$\frac{1}{a}=\frac{1}{b}+\frac{1}{c}$$

occurs. Find b if $a = 8$ and $c = 12$.

6. A gas law in chemistry says that

$$\frac{PV}{T}=\frac{pv}{t}.$$

Suppose that $T = 300$, $t = 350$, $V = 9$, $P = 50$, and $v = 8$. Find p.

7. A formula from anthropology says that

$$c=\frac{100b}{L}.$$

Find L if $c = 80$ and $b = 5$.

8. The gravitational force between two masses is given by

$$F=\frac{GMm}{d^2}.$$

Find M to the nearest thousandth if $F = 10$, $G = 6.67 \times 10^{-11}$, $m = 1$, and $d = 3 \times 10^{-6}$.

9. Concept Check To solve the equation

$$m=\frac{ab}{a-b} \text{ for } a,$$

a student multiplied by $a - b$ and divided by b to obtain

$$\frac{m(a-b)}{b}=a.$$

WHAT WENT WRONG? Give the correct answer.

10. Concept Check To solve the equation

$$rp - rq = p + q \text{ for } r,$$

a student began by adding rq and then dividing by p to obtain

$$r=\frac{p+q+rq}{p}.$$

WHAT WENT WRONG? Give the correct answer.

Solve each formula for the specified variable. ***See Examples 2 and 3.***

11. $F=\frac{GMm}{d^2}$ for G (physics)

12. $F=\frac{GMm}{d^2}$ for M (physics)

13. $\frac{1}{a}=\frac{1}{b}+\frac{1}{c}$ for a (electricity)

14. $\frac{1}{a}=\frac{1}{b}+\frac{1}{c}$ for b (electricity)

15. $\frac{PV}{T}=\frac{pv}{t}$ for v (chemistry)

16. $\frac{PV}{T}=\frac{pv}{t}$ for T (chemistry)

17. $I=\frac{nE}{R+nr}$ for r (engineering)

18. $a=\frac{V-v}{t}$ for V (physics)

19. $\mathcal{A}=\frac{1}{2}h(b+B)$ for b (mathematics)

20. $S=\frac{n}{2}(a+\ell)d$ for n (mathematics)

21. $\frac{E}{e}=\frac{R+r}{r}$ for r (engineering)

22. $y=\frac{x+z}{a-x}$ for x (mathematics)

23. $D=\frac{R}{1+RT}$ for R (banking)

24. $R=\frac{D}{1-DT}$ for D (banking)

Concept Check *Use proportions to solve each problem mentally.*

25. In a mathematics class, 3 of every 4 students are girls. If there are 28 students in the class, how many are girls? How many are boys?

26. In a certain southern state, sales tax on a purchase of \$1.50 is \$0.12. What is the sales tax on a purchase of \$9.00?

The water content of snow is affected by the temperature, wind speed, and other factors present when the snow is falling. The average snow-to-liquid ratio is 10 in. *of snow to* 1 in. *of liquid precipitation. This means that if* 10 in. *of snow fell and was melted, it would produce* 1 in. *of liquid precipitation in a rain gauge.* (Data from www.theweatherprediction.com)

Use a proportion to solve each problem. ***See Examples 4 and 5.***

27. A dry snow might have a snow-to-liquid ratio of 18 to 1. Using this ratio, how much liquid precipitation would be produced by 31.5 in. of snow?

28. A wet, sticky snow good for making a snow man might have a snow-to-liquid ratio of 5 to 1. How many inches of fresh snow would produce 3.25 in. of liquid precipitation using this ratio?

Solve each problem. Give answers to the nearest tenth if an approximation is needed (unless specified otherwise). ***See Examples 4 and 5.***

29. On a U.S. map, the distance between Seattle and Durango is 4.125 in. The two cities are actually 1238 mi apart. On this same map, what would be the distance between Chicago and El Paso, two cities that are actually 1606 mi apart? (Data from Universal Map Atlas.)

30. On a U.S. map, the distance between Reno and Phoenix is 2.5 in. The two cities are actually 768 mi apart. On this same map, what would be the distance between St. Louis and Jacksonville, two cities that are actually 919 mi apart? (Data from Universal Map Atlas.)

31. On a world globe, the distance between New York and Cairo, two cities that are actually 5619 mi apart, is 8.5 in. On this same globe, how far apart are Madrid and Rio de Janeiro, two cities that are actually 5045 mi apart? (Data from *The World Almanac and Book of Facts*.)

32. On a world globe, the distance between San Francisco and Melbourne, two cities that are actually 7856 mi apart, is 11.875 in. On this same globe, how far apart are Mexico City and Singapore, two cities that are actually 10,327 mi apart? (Data from *The World Almanac and Book of Facts*.)

33. In 2016, the average ratio of teachers to students in public elementary and secondary schools was approximately 1 to 16. If a public school had 896 students, how many teachers would be at the school according to this ratio? (Data from U.S. National Center for Education Statistics.)

34. At one point during the 2017–2018 NBA Season, the Indiana Pacers had won 19 of their 38 regular season games. If the team continued to win the same fraction of its games, how many games would the Pacers win for the complete 82-game season? Round the answer to the nearest whole number. (Data from www.nba.com)

35. To estimate the deer population of a forest preserve, wildlife biologists caught, tagged, and then released 42 deer. A month later, they returned and caught a sample of 75 deer and found that 15 of them were tagged. Based on this experiment, how many deer lived in the forest preserve?

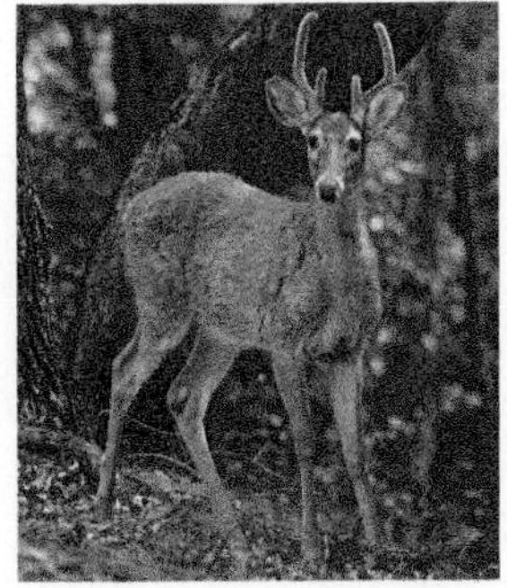

36. Suppose that in the experiment in **Exercise 35,** only 5 of the previously tagged deer were collected in the sample of 75. What would be the estimate of the deer population?

37. Biologists tagged 500 fish in a lake on January 1. On February 1, they returned and collected a random sample of 400 fish, 8 of which had been previously tagged. On the basis of this experiment, how many fish does the lake have?

38. Suppose that in the experiment of **Exercise 37,** 10 of the previously tagged fish were collected on February 1. What would be the estimate of the fish population?

39. Bruce Johnston's Shelby Cobra uses 5 gal of gasoline to drive 156 mi. He has 3 gal of gasoline in the car, and he still needs to drive 300 mi. If the car continues to use gasoline at the same rate, how many more gallons will he need?

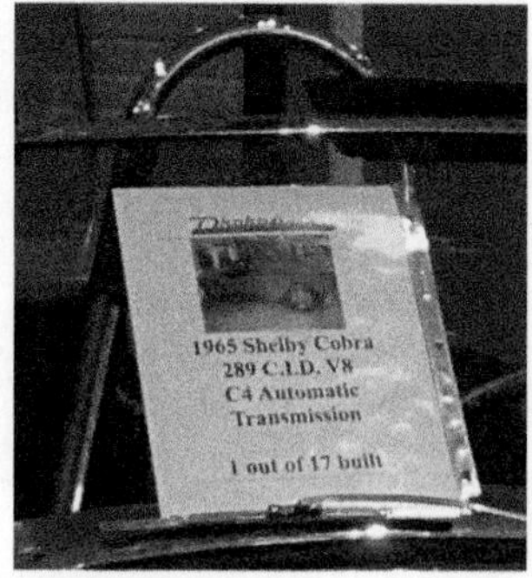

40. Mike Love's T-bird uses 6 gal of gasoline to drive 141 mi. He has 4 gal of gasoline in the car, and he still needs to drive 275 mi. If the car continues to use gasoline at the same rate, how many more gallons will he need?

In geometry, two triangles with corresponding angle measures equal, called ***similar triangles,*** *have corresponding sides proportional. For example, in the figure, angle A = angle D, angle B = angle E, and angle C = angle F, so the triangles are similar. Then the following ratios of corresponding sides are equal.*

$$\frac{4}{6} = \frac{6}{9} = \frac{2x+1}{2x+5}$$

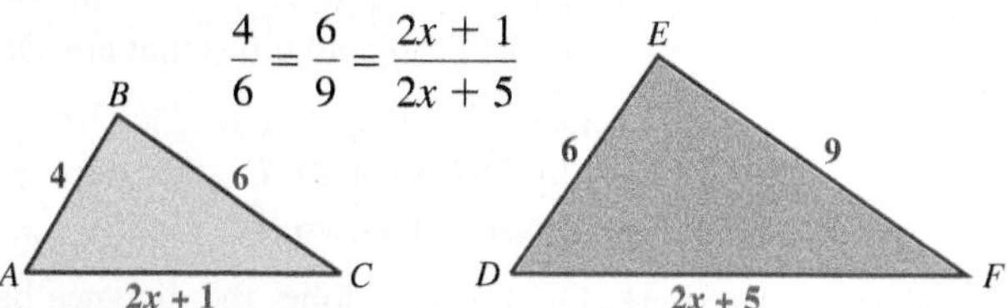

41. Solve for x, using the given proportion to find the lengths of the third sides of the triangles.

42. Suppose the following triangles are similar. Find y and the lengths of the two unknown sides of each triangle.

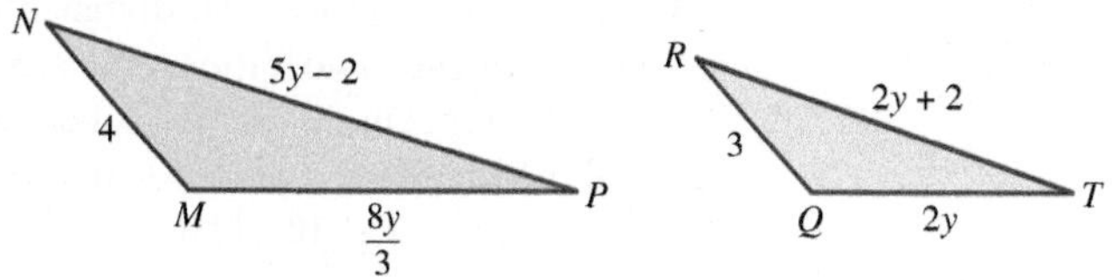

Nurses use proportions to determine the amount of a drug to administer when the dose is measured in milligrams but the drug is packaged in a diluted form in milliliters. (Data from Hoyles, C., R. Noss, and S. Pozzi, "Proportional Reasoning in Nursing Practice," *Journal for Research in Mathematics Education.)*

For example, to find the number of milliliters of fluid needed to administer 300 mg *of a drug that comes packaged as* 120 mg *in* 2 mL *of fluid, a nurse sets up the proportion*

$$\frac{120 \text{ mg}}{2 \text{ mL}} = \frac{300 \text{ mg}}{x \text{ mL}},$$

where x represents the amount to administer in milliliters. Use this method to find the correct dose for each prescription.

43. 120 mg of amikacin packaged as 100 mg in 2-mL vials

44. 1.5 mg of morphine packaged as 20-mg ampules diluted in 10 mL of fluid

Concept Check *Solve each problem mentally.*

45. If Marin can mow her yard in 3 hr, what is her rate (in terms of the proportion of the job per hour)?

46. A van traveling from Atlanta to Detroit averages 50 mph and takes 14 hr to make the trip. How far is it from Atlanta to Detroit?

47. Concept Check Greg paddles his canoe at a rate of 4 mph. If the rate of a stream is x mph, what is an expression that represents his rate in miles per hour when traveling each direction?

(a) Upstream **(b)** Downstream

48. Concept Check Which equation can be solved using cross products of the proportion? Explain. Then solve each equation.

Equation A: $\frac{12}{x} - 1 = \frac{5}{x - 1}$

Equation B: $\frac{12}{x + 3} = \frac{8}{x - 3}$

Solve each problem. ***See Examples 6 and 7.***

49. Kellen's boat travels 12 mph. Find the rate of the current of the river if she can travel 6 mi upstream in the same amount of time she can go 10 mi downstream. (Let x = the rate of the current.)

	Distance	Rate	Time
Downstream	10	$12 + x$	
Upstream	6	$12 - x$	

50. Kasey can travel 8 mi upstream in the same time it takes her to go 12 mi downstream. Her boat goes 15 mph in still water. What is the rate of the current? (Let x = the rate of the current.)

	Distance	Rate	Time
Downstream			
Upstream		$15 - x$	

51. In his boat, Sheldon can travel 30 mi downstream in the same time that it takes to travel 10 mi upstream. If the rate of the current is 5 mph, find the rate of the boat in still water.

52. In his boat, Leonard can travel 24 mi upstream in the same time that it takes to travel 36 mi downstream. If the rate of the current is 2 mph, find the rate of the boat in still water.

53. On his drive from Montpelier, Vermont, to Columbia, South Carolina, Victor averaged 51 mph. If he had been able to average 60 mph, he would have reached his destination 3 hr earlier. What is the driving distance between Montpelier and Columbia?

	Distance	Rate	Time
Actual Trip			
Alternative Trip	x	60	

54. Kelli lives in an off-campus apartment. When she rides her bike to campus, she arrives 36 min faster than when she walks. If her average walking rate is 3 mph and her average biking rate is 12 mph, how far is it from her apartment to campus?

	Distance	Rate	Time
Bike	x	12	
Walk			

55. A plane averaged 500 mph on a trip going east, but only 350 mph on the return trip. The total flying time in both directions was 8.5 hr. What was the one-way distance?

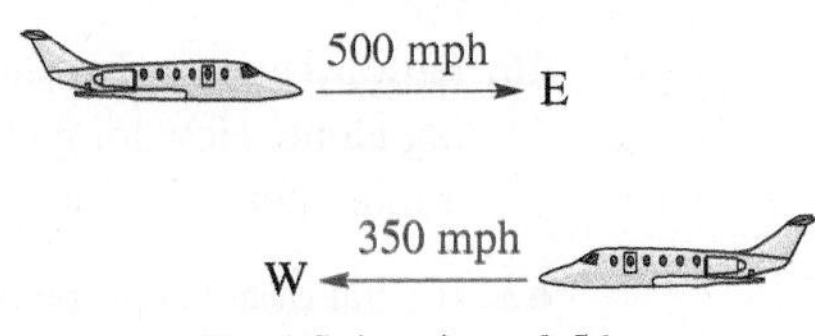

56. Mary drove from her apartment to her parents' house. She averaged 60 mph because traffic was light. On the return trip, which took 1.5 hr longer, she averaged only 45 mph on the same route. What is the distance between Mary's apartment and her parents' house?

57. On the first part of a trip to Carmel traveling on the freeway, Marge averaged 60 mph. On the rest of the trip, which was 10 mi longer than the first part, she averaged 50 mph. Find the total distance to Carmel if the second part of the trip took 30 min more than the first part.

58. During the first part of a trip on the highway, Jim and Annie averaged 60 mph. In Houston, traffic caused them to average only 30 mph. The distance they drove in Houston was 100 mi less than their distance on the highway. What was their total driving distance if they spent 50 min more on the highway than they did in Houston?

Solve each problem. ***See Example 8.***

59. A chef can finish a catering job in 6 hr. His student can do the same job in 12 hr. How long will it take them to do the job if they work together?

60. A printer can complete a large job in 8 hr using a new machine. Using an older machine requires 24 hr to do the same job. How long would it take to complete the job using both machines?

61. Butch and Peggy want to pick up the mess that their grandson, Grant, has made in his playroom. Butch could do it in 15 min working alone. Peggy, working alone, could clean it in 12 min. How long will it take them if they work together?

	Rate	*Time Working Together*	*Fractional Part of the Job Done*
Butch	$\frac{1}{15}$	x	
Peggy	$\frac{1}{12}$	x	

62. Lou can groom a customer's dogs in 8 hr, but it takes his business partner, Janet, only 5 hr to groom the same dogs. How long will it take them to groom the dogs if they work together?

	Rate	*Time Working Together*	*Fractional Part of the Job Done*
Lou	$\frac{1}{8}$	x	
Janet	$\frac{1}{5}$	x	

63. Jerry and Kuba are laying a hardwood floor. Working alone, Jerry can do the job in 20 hr. If the two of them work together, they can complete the job in 12 hr. How long would it take Kuba to lay the floor working alone? (Let $x =$ the time it would take Kuba working alone.)

	Rate	*Time Working Together*	*Fractional Part of the Job Done*
Jerry		12	
Kuba		12	

64. Mrs. Disher can grade a set of tests in 5 hr working alone. If her student teacher Mr. Howes helps her, it will take 3 hr to grade the tests. How long would it take Mr. Howes to grade the tests if he worked alone? (Let $x =$ the time it would take Mr. Howes working alone.)

	Rate	*Time Working Together*	*Fractional Part of the Job Done*
Mrs. Disher		3	
Mr. Howes		3	

65. Dixie can paint a room in 3 hr working alone. Trixie can paint the same room in 6 hr working alone. How long after Dixie starts to paint the room will it be finished if Trixie joins her 1 hr later?

66. Howard can wash a car in 30 min working alone. Raj can do the same job in 45 min working alone. How long after Howard starts to wash the car will it be finished if Raj joins him 5 min later?

67. If a vat of acid can be filled by an inlet pipe in 10 hr and emptied by an outlet pipe in 20 hr, how long will it take to fill the vat if both pipes are open?

68. A winery has a vat to hold Zinfandel. An inlet pipe can fill the vat in 9 hr, and an outlet pipe can empty it in 12 hr. How long will it take to fill the vat if both the outlet and the inlet pipes are open?

69. Suppose that Hortense and Mort can clean their entire house in 7 hr, while their toddler, Mimi, can mess it up in only 2 hr. If Hortense and Mort clean the house while Mimi is at her grandma's and then start cleaning up after Mimi the minute she gets home, how long does it take from the time Mimi gets home until the whole place is a shambles?

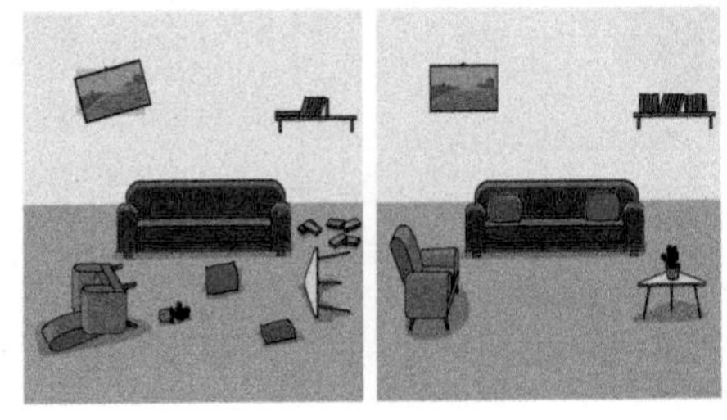

70. An inlet pipe can fill an artificial lily pond in 60 min, while an outlet pipe can empty it in 80 min. Through an error, both pipes are left open. How long will it take for the pond to fill?

6.6 Variation

OBJECTIVES

1 Write an equation expressing direct variation.
2 Find the constant of variation, and solve direct variation problems.
3 Solve inverse variation problems.
4 Solve joint variation problems.
5 Solve combined variation problems.

VOCABULARY

- ☐ direct variation
- ☐ constant of variation
- ☐ inverse variation
- ☐ joint variation
- ☐ combined variation

OBJECTIVE 1 Write an equation expressing direct variation.

The circumference of a circle is given by the formula $C = 2\pi r$, where r is the radius of the circle. See **FIGURE 5**. The circumference is always a constant multiple of the radius—that is, C is always found by multiplying r by the constant 2π.

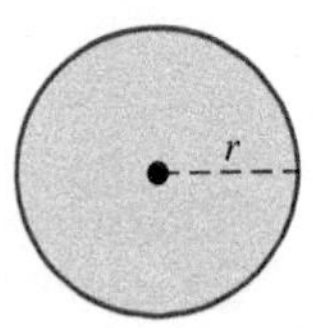

$C = 2\pi r$

FIGURE 5

As the ***radius increases,*** the ***circumference increases.***

As the ***radius decreases,*** the ***circumference decreases.***

Because of these relationships, the circumference is said to *vary directly* as the radius.

Direct Variation

y* varies directly as *x if there exists a real number k such that

$$y = kx.$$

Stated another way, ***y* is *directly proportional to x.*** The number k is the **constant of variation.**

In direct variation, for k > 0, as the value of x increases, the value of y increases. Similarly, as x decreases, y decreases.

OBJECTIVE 2 Find the constant of variation, and solve direct variation problems.

The direct variation equation y = kx defines a linear function, where the constant of variation k is the slope of the line. For example, the following equation describes the cost y to buy x gallons of gasoline.

$$y = 3.50x$$

The cost varies directly as the number of gallons purchased.

As the *number of gallons increases,* the *cost increases.*

As the *number of gallons decreases,* the *cost decreases.*

The constant of variation k is 3.50, the cost of 1 gal of gasoline.

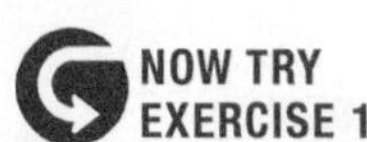

NOW TRY EXERCISE 1

One week Morgan sold 8 dozen eggs for \$20. How much does she charge for one dozen eggs?

EXAMPLE 1 Solving a Direct Variation Problem

Eva is paid an hourly wage. One week she worked 43 hr and was paid \$795.50. How much does she earn per hour?

Let h = the number of hours she works

and P = her corresponding pay.

Write a variation equation.

k represents Eva's hourly wage.	$P = kh$	P varies directly as h.
	$795.50 = k \cdot 43$	Let $P = 795.50$ and $h = 43$.
This is the constant of variation.	$k = 18.50$	Use a calculator.

Her hourly wage is \$18.50, and P and h are related by the equation

$$P = 18.50h.$$

We can use this equation to find her pay for any number of hours worked.

NOW TRY  

NOW TRY EXERCISE 2

For a constant height, the area of a parallelogram is directly proportional to its base. If the area is 20 cm^2 when the base is 4 cm, find the area when the base is 7 cm.

EXAMPLE 2 Solving a Direct Variation Problem

Hooke's law for an elastic spring states that the distance a spring stretches is directly proportional to the force applied. If a force of 150 newtons* stretches a certain spring 8 cm, how much will a force of 400 newtons stretch the spring? See **FIGURE 6**.

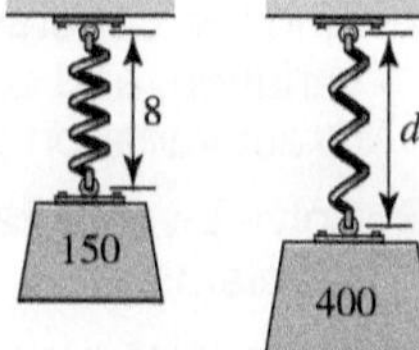

FIGURE 6

Let d = the distance the spring stretches

and f = the force applied.

Then $d = kf$ for some constant k. A force of 150 newtons stretches the spring 8 cm, so we use these values to find k.

	$d = kf$	Variation equation
Solve for *k*.	$8 = k \cdot 150$	Let $d = 8$ and $f = 150$.
	$k = \frac{8}{150}$	Solve for k.
	$k = \frac{4}{75}$	Write in lowest terms.

Now we rewrite the variation equation $d = kf$ using $\frac{4}{75}$ for k.

$$d = \frac{4}{75}f \quad \text{Let } k = \tfrac{4}{75}.$$

For a force of 400 newtons, substitute 400 for f.

$$d = \frac{4}{75}(400) = \frac{64}{3} \quad \text{Let } f = 400.$$

The spring will stretch $\frac{64}{3}$ cm, or $21\frac{1}{3}$ cm, if a force of 400 newtons is applied.

NOW TRY  

NOW TRY ANSWERS

1. \$2.50
2. 35 cm^2

*A newton is a unit of measure of force used in physics.

Solving a Variation Problem

Step 1 Write a variation equation.

Step 2 Substitute the initial values and solve for k.

Step 3 Rewrite the variation equation with the value of k from Step 2.

Step 4 Substitute the remaining values, solve for the unknown, and find the required answer.

One variable can be directly proportional to a power of another variable.

Direct Variation as a Power

y varies directly as the nth power of x if there exists a real number k such that

$$y = kx^n.$$

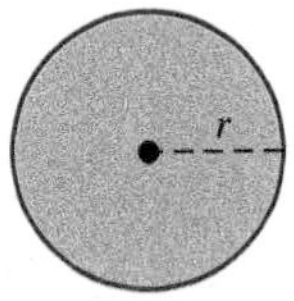

$\mathcal{A} = \pi r^2$

FIGURE 7

An example of direct variation as a power is the formula for the area of a circle, $\mathcal{A} = \pi r^2$. See FIGURE 7. Here, π is the constant of variation, and the area $\mathcal{A}$ varies directly as the *square* of the radius r.

NOW TRY EXERCISE 3

Suppose y varies directly as the square of x, and $y = 200$ when $x = 5$. Find y when $x = 7$.

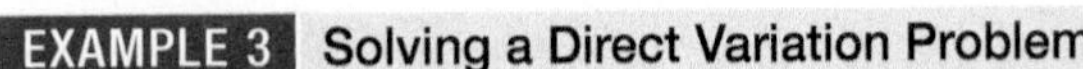

EXAMPLE 3 Solving a Direct Variation Problem

The distance a body falls from rest varies directly as the square of the time it falls (disregarding air resistance). If a skydiver falls 64 ft in 2 sec, how far will she fall in 8 sec?

Step 1 Let d = the distance the skydiver falls

and t = the time it takes to fall.

Then d is a function of t for some constant k.

$$d = kt^2 \quad d \text{ varies directly as the square of } t.$$

Step 2 To find the value of k, use the fact that the skydiver falls 64 ft in 2 sec.

$$d = kt^2 \quad \text{Variation equation}$$

$$64 = k(2)^2 \quad \text{Let } d = 64 \text{ and } t = 2.$$

$$k = 16 \quad \text{Find } k.$$

Step 3 Now we rewrite the variation equation $d = kt^2$ using 16 for k.

$$d = 16t^2 \quad \text{Let } k = 16.$$

Step 4 Let $t = 8$ to find the number of feet the skydiver will fall in 8 sec.

$$d = 16(8)^2 = 1024 \quad \text{Let } t = 8.$$

The skydiver will fall 1024 ft in 8 sec.

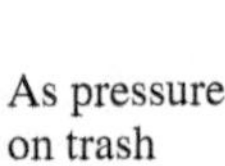
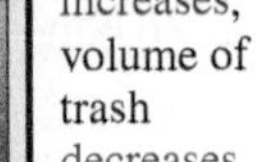

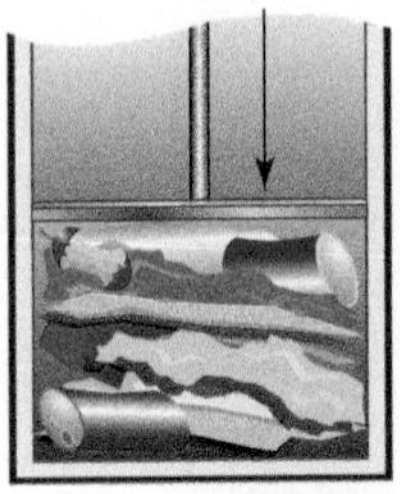

As pressure on trash increases, volume of trash decreases.

FIGURE 8

OBJECTIVE 3 Solve inverse variation problems.

Another type of variation is *inverse variation*. ***With inverse variation, for k > 0, as one variable increases, the other variable decreases.***

For example, in a closed space, volume decreases as pressure increases, which can be illustrated by a trash compactor. See FIGURE 8. As the compactor presses down, the pressure on the trash increases, and in turn, the trash occupies a smaller space.

NOW TRY ANSWER

3. 392

Inverse Variation

y* varies inversely as *x if there exists a real number k such that

$$y = \frac{k}{x}.$$

y* varies inversely as the *n*th power of *x if there exists a real number k such that

$$y = \frac{k}{x^n}.$$

The inverse variation equation $y = \frac{k}{x}$ defines a rational function.

Another example of inverse variation comes from the distance formula.

$$d = rt \quad \text{Distance formula}$$

$$t = \frac{d}{r} \quad \text{Divide each side by } r.$$

In the form $t = \frac{d}{r}$, t (time) varies inversely as r (rate or speed), with d (distance) serving as the constant of variation. For example, if the distance between Chicago and Des Moines is 300 mi, then

$$t = \frac{300}{r}.$$

The values of r and t might be any of the following.

$$\left.\begin{array}{l} r = 50, t = 6 \\ r = 60, t = 5 \\ r = 75, t = 4 \end{array}\right\} \text{As } r \text{ increases, } t \text{ decreases.} \qquad \left.\begin{array}{l} r = 30, t = 10 \\ r = 25, t = 12 \\ r = 20, t = 15 \end{array}\right\} \text{As } r \text{ decreases, } t \text{ increases.}$$

If we *increase* the rate (speed) at which we drive, time *decreases*. If we *decrease* the rate (speed) at which we drive, time *increases*.

NOW TRY EXERCISE 4

For a constant area, the height of a triangle varies inversely as the base. If the height is 7 cm when the base is 8 cm, find the height when the base is 14 cm.

EXAMPLE 4 Solving an Inverse Variation Problem

In the manufacture of a phone-charging device, the cost of producing the device varies inversely as the number produced. If 10,000 units are produced, the cost is \$2 per unit. Find the cost per unit to produce 25,000 units.

Let x = the number of units produced

and c = the cost per unit.

Here, as production increases, cost decreases, and as production decreases, cost increases. We write a variation equation using the variables c and x and the constant k.

$$c = \frac{k}{x} \quad c \text{ varies inversely as } x.$$

$$2 = \frac{k}{10{,}000} \quad \text{Let } c = 2 \text{ and } x = 10{,}000.$$

$$20{,}000 = k \quad \text{Multiply by } 10{,}000.$$

Thus, the variation equation $c = \frac{k}{x}$ becomes $c = \frac{20{,}000}{x}$.

$$c = \frac{20{,}000}{25{,}000} = 0.80 \quad \text{Let } x = 25{,}000.$$

The cost per unit to make 25,000 units is \$0.80.

NOW TRY

NOW TRY ANSWER

4. 4 cm

The weight of an object above Earth varies inversely as the square of its distance from the center of Earth. If an object weighs 150 lb on the surface of Earth, and the radius of Earth is about 3960 mi, how much does it weigh when it is 1000 mi above Earth's surface? Round to the nearest pound.

EXAMPLE 5 Solving an Inverse Variation Problem

The weight of an object above Earth varies inversely as the square of its distance from the center of Earth. A space shuttle in an elliptical orbit has a maximum distance from the center of Earth (**apogee**) of 6700 mi. Its minimum distance from the center of Earth (**perigee**) is 4090 mi. See **FIGURE 9**. If an astronaut in the shuttle weighs 57 lb at its apogee, what does the astronaut weigh at its perigee?

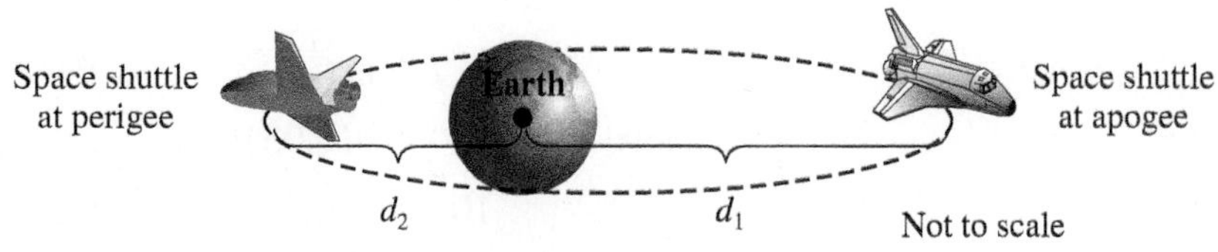

FIGURE 9

Let w = the astronaut's weight and d = the distance from the center of Earth, for some constant k. We write a variation equation using these variables.

$$w = \frac{k}{d^2} \quad w \text{ varies inversely as the square of } d.$$

At the apogee, the astronaut weighs 57 lb, and the distance from the center of Earth is 6700 mi. Use these values to find k.

$$57 = \frac{k}{(6700)^2} \quad \text{Let } w = 57 \text{ and } d = 6700.$$

$$k = 57(6700)^2 \quad \text{Solve for } k.$$

Substitute $k = 57(6700)^2$ and $d = 4090$ in the variation equation $w = \frac{k}{d^2}$ to find the astronaut's weight at the perigee.

$$w = \frac{57(6700)^2}{(4090)^2} = 153 \text{ lb} \quad \text{Use a calculator. Round to the nearest pound.}$$

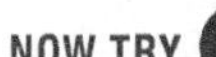

OBJECTIVE 4 Solve joint variation problems.

If one variable varies directly as the *product* of several other variables (perhaps raised to powers), the first variable is said to *vary jointly* as the others.

Joint Variation

y varies jointly as x and z if there exists a real number k such that

$$y = kxz.$$

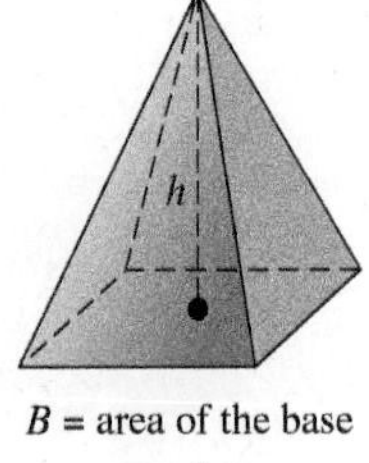

B = area of the base

$V = \frac{1}{3}Bh$

FIGURE 10

An example of joint variation is the formula for the volume of a right pyramid, $V = \frac{1}{3}Bh$. See **FIGURE 10**. Here, $\frac{1}{3}$ is the constant of variation, and the volume V varies jointly as the area of the base B and the height h.

CAUTION Note that *and* in the expression "y varies jointly as x *and* z" translates as a product in the variation equation $y = kxz$. The word *and* does **not** indicate addition here.

NOW TRY ANSWER

5. 96 lb

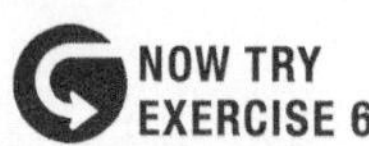

NOW TRY EXERCISE 6

The volume of a right pyramid varies jointly as the height and the area of the base. If the volume is 100 ft^3 when the area of the base is 30 ft^2 and the height is 10 ft, find the volume when the area of the base is 90 ft^2 and the height is 20 ft.

EXAMPLE 6 Solving a Joint Variation Problem

The interest on a loan or an investment is given by the formula $I = prt$. Here, for a given principal p, the interest earned I varies jointly as the interest rate r and the time t the principal is left earning interest. If an investment earns \$100 interest at 5% for 2 yr, how much interest will the same principal earn at 4.5% for 3 yr?

We use the formula $I = prt$, where p is the constant of variation because it is the same for both investments.

$$I = prt \quad \text{Here, } p \text{ is the constant of variation.}$$

Solve for p.

$$100 = p(0.05)(2) \quad \text{Let } I = 100,\ r = 5\% = 0.05, \text{ and } t = 2.$$

$$100 = 0.1p \quad \text{Multiply.}$$

$$p = 1000 \quad \text{Divide by 0.1. Rewrite.}$$

Now we find I when $p = 1000$, $r = 4.5\% = 0.045$, and $t = 3$.

$$I = 1000(0.045)(3) = 135 \quad \text{Let } p = 1000,\ r = 4.5\% = 0.045, \text{ and } t = 3.$$

The interest will be \$135.

NOW TRY

OBJECTIVE 5 Solve combined variation problems.

There are combinations of direct and inverse variation, called **combined variation.**

NOW TRY EXERCISE 7

In statistics, the sample size used to estimate a population mean varies directly as the variance and inversely as the square of the maximum error of the estimate. If the sample size is 200 when the variance is 25 m^2 and the maximum error of the estimate is 0.5 m, find the sample size when the variance is 25 m^2 and the maximum error of the estimate is 0.1 m.

EXAMPLE 7 Solving a Combined Variation Problem

Body mass index (BMI) is used to assess whether a person's weight is healthy. A BMI from 19 through 25 is considered desirable. BMI varies directly as an individual's weight in pounds and inversely as the square of his or her height in inches.

A person who weighs 118 lb and is 64 in. tall has a BMI of 20. (BMI is rounded to the nearest whole number.) Find the BMI of a man who weighs 165 lb and is 70 in. tall. (Data from *Washington Post*.)

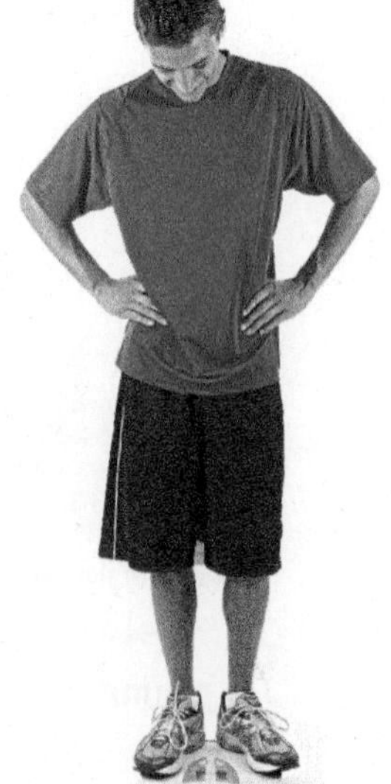

Let $B =$ BMI, $w =$ weight, and $h =$ height. Write a variation equation.

$$B = \frac{kw}{h^2}$$

kw ← BMI varies directly as the weight.

h^2 ← BMI varies inversely as the square of the height.

To find k, let $B = 20$, $w = 118$, and $h = 64$.

$$20 = \frac{k(118)}{64^2} \quad B = \frac{kw}{h^2}$$

$$k = \frac{20(64^2)}{118} \quad \text{Multiply by } 64^2. \text{ Divide by 118.}$$

$$k = 694 \quad \text{Use a calculator. Round to the nearest whole number.}$$

Now find B when $k = 694$, $w = 165$, and $h = 70$.

$$B = \frac{694(165)}{70^2} = 23 \quad \text{Round to the nearest whole number.}$$

The man's BMI is 23.

NOW TRY

NOW TRY ANSWERS

6. 600 ft^3

7. 5000

6.6 Exercises

MyLab Math

Video solutions for select problems available in MyLab Math

Concept Check *Fill in each blank with the correct response.*

1. For $k > 0$, if y varies directly as x, then when x increases, y ________, and when x decreases, y ________.

2. For $k > 0$, if y varies inversely as x, then when x increases, y ________, and when x decreases, y ________.

Concept Check *Use personal experience or intuition to determine whether the situation suggests* direct *or* inverse *variation.*

3. The number of movie tickets purchased and the total price for the tickets

4. The rate and the distance traveled by a pickup truck in 3 hr

5. The amount of pressure put on the accelerator of a car and the speed of the car

6. The percentage off an item that is on sale and the price of the item

7. Your age and the probability that you believe in the tooth fairy

8. The surface area of a balloon and its diameter

9. The demand for an item and the price of the item

10. The number of hours worked by an hourly worker and the amount of money earned

Concept Check *Determine whether each equation represents* direct, inverse, joint, *or* combined *variation.*

11. $y = \frac{3}{x}$

12. $y = \frac{8}{x}$

13. $y = 10x^2$

14. $y = 2x^3$

15. $y = 3xz^4$

16. $y = 6x^3z^2$

17. $y = \frac{4x}{wz}$

18. $y = \frac{6x}{st}$

Concept Check *Write each formula using the "language" of variation. For example, the formula for the circumference of a circle, $C = 2\pi r$, can be written as*

"The circumference of a circle varies directly as the length of its radius."

19. $P = 4s$, where P is the perimeter of a square with side of length s

20. $d = 2r$, where d is the diameter of a circle with radius r

21. $S = 4\pi r^2$, where S is the surface area of a sphere with radius r

22. $V = \frac{4}{3}\pi r^3$, where V is the volume of a sphere with radius r

23. $\mathcal{A} = \frac{1}{2}bh$, where $\mathcal{A}$ is the area of a triangle with base b and height h

24. $V = \frac{1}{3}\pi r^2 h$, where V is the volume of a cone with radius r and height h

25. Concept Check What is the constant of variation in each of the variation equations in **Exercises 19–24?**

26. Concept Check What is meant by the constant of variation in a direct variation problem? If we were to graph the linear equation $y = kx$ for some nonnegative constant k, what role would k play in the graph?

Write a variation equation for each situation. Use k as the constant of variation. ***See Examples 1–6.***

27. A varies directly as b.

28. W varies directly as f.

29. h varies inversely as t.

30. p varies inversely as s.

31. M varies directly as the square of d.

32. P varies inversely as the cube of x.

33. I varies jointly as g and h.

34. C varies jointly as a and the square of b.

Solve each problem. ***See Examples 1–7.***

35. If x varies directly as y, and $x = 9$ when $y = 3$, find x when $y = 12$.

36. If x varies directly as y, and $x = 10$ when $y = 7$, find y when $x = 50$.

37. If a varies directly as the square of b, and $a = 4$ when $b = 3$, find a when $b = 2$.

38. If h varies directly as the square of m, and $h = 15$ when $m = 5$, find h when $m = 7$.

39. If z varies inversely as w, and $z = 10$ when $w = 0.5$, find z when $w = 8$.

40. If t varies inversely as s, and $t = 3$ when $s = 5$, find s when $t = 5$.

41. If m varies inversely as the square of p, and $m = 20$ when $p = 2$, find m when $p = 5$.

42. If a varies inversely as the square of b, and $a = 48$ when $b = 4$, find a when $b = 7$.

43. p varies jointly as q and the square of r, and $p = 200$ when $q = 2$ and $r = 3$. Find p when $q = 5$ and $r = 2$.

44. f varies jointly as h and the square of g, and $f = 50$ when $h = 2$ and $g = 4$. Find f when $h = 6$ and $g = 3$.

Solve each problem. ***See Examples 1–7.***

45. Ben bought 8.5 gal of gasoline and paid $33.32. What is the price of gasoline per gallon?

46. Sara gives horseback rides at Shadow Mountain Ranch. A 2.5-hr ride costs $50.00. What is the price per hour?

47. The weight of an object on Earth is directly proportional to the weight of that same object on the moon. A 200-lb astronaut would weigh 32 lb on the moon. How much would a 50-lb dog weigh on the moon?

48. The pressure exerted by a certain liquid at a given point is directly proportional to the depth of the point beneath the surface of the liquid. The pressure at 30 m is 80 newtons. What pressure is exerted at 50 m?

49. The volume of a can of tomatoes is directly proportional to the height of the can. If the volume of the can is 300 cm^3 when its height is 10.62 cm, find the volume to the nearest whole number of a can with height 15.92 cm.

50. The force required to compress a spring is directly proportional to the change in length of the spring. If a force of 20 newtons is required to compress a certain spring 2 cm, how much force is required to compress the spring from 20 cm to 8 cm?

20 newtons
20 cm
18 cm
?
8 cm

51. For a body falling freely from rest (disregarding air resistance), the distance the body falls varies directly as the square of the time. If an object is dropped from the top of a tower 576 ft high and hits the ground in 6 sec, how far did it fall in the first 4 sec?

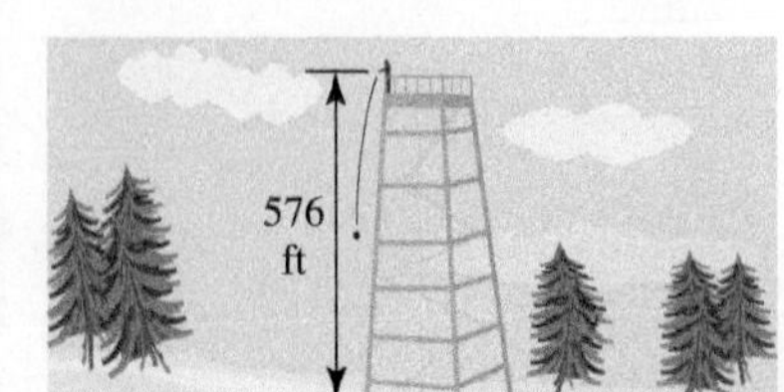

52. The amount of water emptied by a pipe varies directly as the square of the diameter of the pipe. For a certain constant water flow, a pipe emptying into a canal will allow 200 gal of water to escape in an hour. The diameter of the pipe is 6 in. How much water would a 12-in. pipe empty into the canal in an hour, assuming the same water flow?

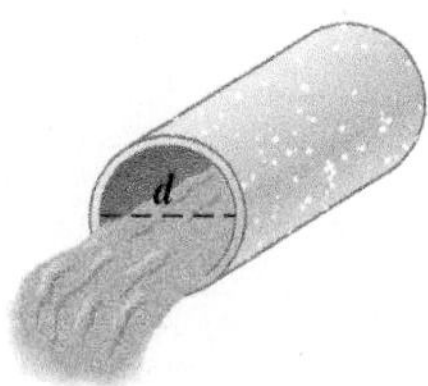

53. Over a specified distance, rate varies inversely as time. If a Dodge Viper on a test track goes a certain distance in one-half minute at 160 mph, what rate is needed to go the same distance in three-fourths minute?

54. For a constant area, the length of a rectangle varies inversely as the width. The length of a rectangle is 27 ft when the width is 10 ft. Find the width of a rectangle with the same area if the length is 18 ft.

55. The frequency of a vibrating string varies inversely as its length. That is, a longer string vibrates fewer times in a second than a shorter string. Suppose a piano string 2 ft long vibrates at 250 cycles per sec. What frequency would a string 5 ft long have?

56. The current in a simple electrical circuit varies inversely as the resistance. If the current is 20 amps when the resistance is 5 ohms, find the current when the resistance is 7.5 ohms.

57. The amount of light (measured in foot-candles) produced by a light source varies inversely as the square of the distance from the source. If the illumination produced 1 m from a light source is 768 foot-candles, find the illumination produced 6 m from the same source.

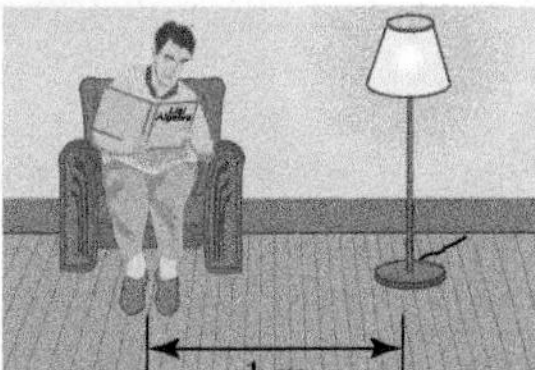

58. The force with which Earth attracts an object above Earth's surface varies inversely as the square of the distance of the object from the center of Earth. If an object 4000 mi from the center of Earth is attracted with a force of 160 lb, find the force of attraction if the object were 6000 mi from the center of Earth.

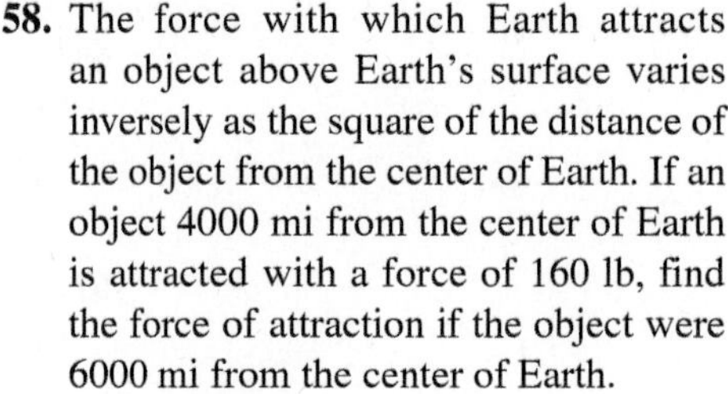

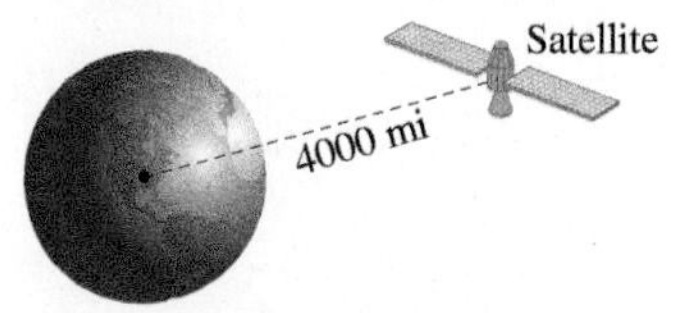

59. For a given interest rate, simple interest varies jointly as principal and time. If \$2000 left in an account for 4 yr earned interest of \$280, how much interest would be earned in 6 yr?

60. The collision impact of an automobile varies jointly as its mass and the square of its speed. Suppose a 2000-lb car traveling at 55 mph has a collision impact of 6.1. What is the collision impact (to the nearest tenth) of the same car at 65 mph?

61. The weight of a bass varies jointly as its girth and the square of its length. (**Girth** is the distance around the body of the fish.) A prize-winning bass weighed in at 22.7 lb and measured 36 in. long with a 21-in. girth. How much (to the nearest tenth of a pound) would a bass 28 in. long with an 18-in. girth weigh?

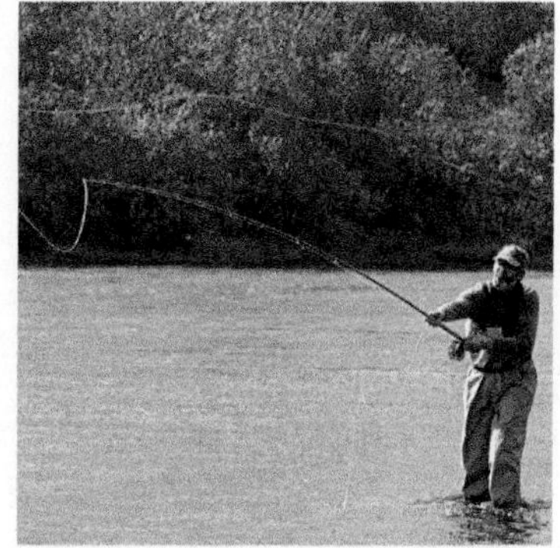

62. The weight of a trout varies jointly as its length and the square of its girth. One angler caught a trout that weighed 10.5 lb and measured 26 in. long with an 18-in. girth. Find the weight (to the nearest tenth of a pound) of a trout that is 22 in. long with a 15-in. girth.

63. The force needed to keep a car from skidding on a curve varies inversely as the radius of the curve and jointly as the weight of the car and the square of the speed. If 242 lb of force keeps a 2000-lb car from skidding on a curve of radius 500 ft at 30 mph, what force (to the nearest tenth of a pound) would keep the same car from skidding on a curve of radius 750 ft at 50 mph?

64. The maximum load that a cylindrical column with a circular cross section can hold varies directly as the fourth power of the diameter of the cross section and inversely as the square of the height. A 9-m column 1 m in diameter will support 8 metric tons. How many metric tons can be supported by a column 12 m high and $\frac{2}{3}$ m in diameter?

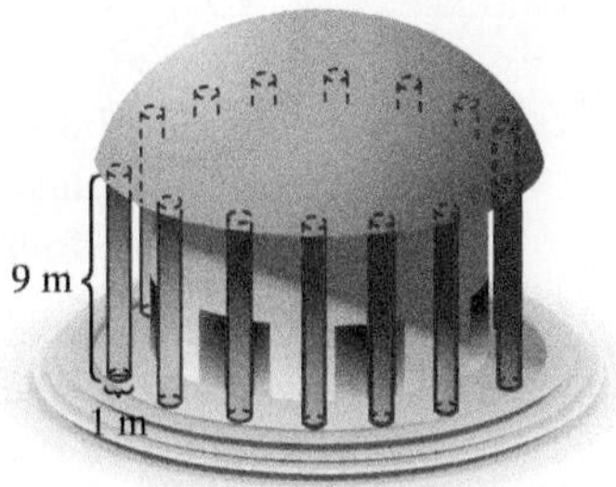

Load = 8 metric tons

65. The number of long-distance phone calls between two cities during a certain period varies jointly as the populations of the cities, p_1 and p_2, and inversely as the distance between them, in miles. If 80,000 calls are made between two cities 400 mi apart, with populations of 70,000 and 100,000, how many calls (to the nearest hundred) are made between cities with populations of 50,000 and 75,000 that are 250 mi apart?

66. The volume of gas varies inversely as the pressure and directly as the temperature. (Temperature must be measured in *kelvins* (K), a unit of measurement used in physics.) If a certain gas occupies a volume of 1.3 L at 300 K and a pressure of 18 newtons, find the volume at 340 K and a pressure of 24 newtons.

67. A body mass index from 27 through 29 carries a slight risk of weight-related health problems, while a BMI of 30 or more indicates a great increase in risk. Use your own height and weight and the information in **Example 7** to determine your BMI and whether you are at risk.

68. The maximum load of a horizontal beam that is supported at both ends varies directly as the width and the square of the height and inversely as the length between the supports. A beam 6 m long, 0.1 m wide, and 0.06 m high supports a load of 360 kg. What is the maximum load supported by a beam 16 m long, 0.2 m wide, and 0.08 m high?

RELATING CONCEPTS For Individual or Group Work **(Exercises 69–74)**

A routine activity such as pumping gasoline can be related to many of the concepts studied in this text. Suppose that premium unleaded costs \$3.75 per gal. ***Work Exercises 69–74 in order.***

69. 0 gal of gasoline cost \$0.00, while 1 gal costs \$3.75. Represent these two pieces of information as ordered pairs of the form (gallons, price).

70. Use the information from **Exercise 69** to find the slope of the line on which the two points lie.

71. Write the slope-intercept form of the equation of the line on which the two points lie.

72. Using function notation, if $f(x) = ax + b$ represents the line from **Exercise 71**, what are the values of a and b?

73. How is the value of a from **Exercise 72** related to gasoline in this situation? With relationship to the line, what do we call this number?

74. Why does the equation from **Exercise 72** satisfy the conditions for direct variation? In the context of variation, what do we call the value of a?

Chapter 6 Summary

Key Terms

6.1
rational expression
rational function
reciprocal

6.2
least common denominator (LCD)

6.3
complex fraction

6.4
domain of the variable in a rational equation
proposed solution
extraneous solution
discontinuous
reciprocal function
vertical asymptote
horizontal asymptote

6.5
ratio
proportion
cross products of a proportion

6.6
direct variation
constant of variation
inverse variation
joint variation
combined variation

Test Your Word Power

See how well you have learned the vocabulary in this chapter.

1. A **rational expression** is
   - **A.** an algebraic expression made up of a term or the sum of a finite number of terms with real coefficients and integer exponents
   - **B.** a polynomial equation of degree 2
   - **C.** a quotient with one or more fractions in the numerator, denominator, or both
   - **D.** a quotient of two polynomials with denominator not 0.

2. In a given set of fractions, the **least common denominator** is
   - **A.** the smallest denominator of all the denominators
   - **B.** the product of the factors from each denominator, with each factor raised to the greatest power that occurs
   - **C.** the largest integer that evenly divides the numerator and denominator of all the fractions
   - **D.** the sum of the factors raised to the greatest power that occurs in each denominator.

3. A **complex fraction** is
   - **A.** an algebraic expression made up of a term or the sum of a finite number of terms with real coefficients and integer exponents
   - **B.** a polynomial equation of degree 2
   - **C.** a quotient with one or more fractions in the numerator, denominator, or both
   - **D.** a quotient of two polynomials with denominator not 0.

4. A **ratio**
   - **A.** compares two quantities using a quotient
   - **B.** says that two quotients are equal
   - **C.** is a product of two quantities
   - **D.** is a difference of two quantities.

5. A **proportion**
   - **A.** compares two quantities using a quotient
   - **B.** says that two quotients are equal
   - **C.** is a product of two quantities
   - **D.** is a difference of two quantities.

ANSWERS

1. D; *Examples:* $-\frac{3}{4y^2}, \frac{5x^3}{x+2}, \frac{a+3}{a^2-4a-5}$ **2.** B; *Example:* The LCD of $\frac{1}{x}, \frac{2}{3}$, and $\frac{5}{x+1}$ is $3x(x+1)$. **3.** C; *Examples:* $\frac{\frac{2}{3}}{\frac{4}{7}}, \frac{x-\frac{1}{x}}{x+\frac{1}{y}}, \frac{\frac{2}{a+1}}{a^2-1}$

4. A; *Example:* $\frac{7 \text{ in.}}{12 \text{ in.}}$ compares two quantities. **5.** B; *Example:* The proportion $\frac{2}{3} = \frac{8}{12}$ states that the two ratios are equal.

Quick Review

CONCEPTS	EXAMPLES
6.1 Rational Expressions and Functions; Multiplying and Dividing	
Rational Function A rational function is defined by a quotient of polynomials and has the form $$f(x)=\frac{P(x)}{Q(x)},\quad \text{where } Q(x)\neq 0.$$ Its domain includes all real numbers except those that make $Q(x)$ equal to 0.	Give the domain of the rational function. $$f(x)=\frac{2x+1}{3x+6}$$ Solve $3x+6=0$ to find $x=-2$. This value must be excluded from the domain, which can be written $$\{x \mid x \text{ is a real number}, x\neq -2\} \quad \text{or} \quad (-\infty,-2)\cup(-2,\infty).$$
Fundamental Property of Rational Numbers If $\frac{a}{b}$ is a rational number and if c is any nonzero real number, then $$\frac{a}{b}=\frac{ac}{bc}.$$	Write in lowest terms. $\frac{15}{20}=\frac{3\cdot 5}{4\cdot 5}=\frac{3}{4}$ $\frac{15}{20}$ and $\frac{3}{4}$ are equivalent.
Writing a Rational Expression in Lowest Terms ***Step 1*** Factor both numerator and denominator to find their greatest common factor (GCF). ***Step 2*** Apply the fundamental property. Divide out common factors.	$\frac{2x+8}{x^2-16}$ $=\frac{2(x+4)}{(x-4)(x+4)}$ Factor. $=\frac{2}{x-4}$ Fundamental property
Multiplying Rational Expressions ***Step 1*** Factor numerators and denominators. ***Step 2*** Apply the fundamental property. ***Step 3*** Multiply remaining factors in the numerators and in the denominators. The final expression can be left in factored form. ***Step 4*** Check that the product is in lowest terms.	Multiply or divide. $\frac{x^2+2x+1}{x^2-1}\cdot\frac{5}{3x+3}$ $=\frac{(x+1)^2}{(x-1)(x+1)}\cdot\frac{5}{3(x+1)}$ Factor. $=\frac{5}{3(x-1)}$ Fundamental property; Multiply.
Dividing Rational Expressions Multiply the first rational expression (the dividend) by the reciprocal of the second expression (the divisor).	$\frac{2x+5}{x-3}\div\frac{2x^2+3x-5}{x^2-9}$ $=\frac{2x+5}{x-3}\cdot\frac{x^2-9}{2x^2+3x-5}$ Multiply by the reciprocal. $=\frac{2x+5}{x-3}\cdot\frac{(x+3)(x-3)}{(2x+5)(x-1)}$ Factor. $=\frac{x+3}{x-1}$ Fundamental property; Multiply.

CONCEPTS	EXAMPLES

6.2 Adding and Subtracting Rational Expressions

Adding or Subtracting Rational Expressions

Step 1 **If the denominators are the same,** add or subtract the numerators. Place the result over the common denominator.

If the denominators are different, write all rational expressions with the LCD. Then add or subtract the numerators, and place the result over the common denominator.

Step 2 Make sure that the answer is in lowest terms.

Subtract.

$$\frac{1}{x+6}-\frac{3}{x+2} \qquad \text{The LCD is } (x+6)(x+2).$$

$$=\frac{x+2}{(x+6)(x+2)}-\frac{3(x+6)}{(x+6)(x+2)}$$

$$=\frac{x+2-3(x+6)}{(x+6)(x+2)} \qquad \text{Subtract the numerators.}$$

$$=\frac{x+2-3x-18}{(x+6)(x+2)} \qquad \text{Distributive property}$$

$$=\frac{-2x-16}{(x+6)(x+2)} \qquad \text{Combine like terms.}$$

The numerator factors as $-2(x+8)$, so there is no common factor in the numerator and denominator. The answer is in lowest terms.

6.3 Complex Fractions

Simplifying a Complex Fraction

Method 1

Step 1 Simplify the numerator and denominator separately.

Step 2 Divide by multiplying the numerator by the reciprocal of the denominator.

Step 3 Simplify the resulting fraction if possible.

Method 2

Step 1 Multiply the numerator and denominator of the complex fraction by the least common denominator of all the fractions appearing in the complex fraction.

Step 2 Simplify the resulting fraction if possible.

Simplify using both methods.

Method 1

$$\frac{\frac{1}{x^2}-\frac{1}{y^2}}{\frac{1}{x}+\frac{1}{y}}$$

$$=\frac{\frac{y^2}{x^2y^2}-\frac{x^2}{x^2y^2}}{\frac{y}{xy}+\frac{x}{xy}}$$

$$=\frac{\frac{y^2-x^2}{x^2y^2}}{\frac{y+x}{xy}}$$

$$=\frac{y^2-x^2}{x^2y^2}\div\frac{y+x}{xy}$$

$$=\frac{(y+x)(y-x)}{x^2y^2}\cdot\frac{xy}{y+x}$$

$$=\frac{y-x}{xy}$$

Method 2

$$\frac{\frac{1}{x^2}-\frac{1}{y^2}}{\frac{1}{x}+\frac{1}{y}}$$

$$=\frac{x^2y^2\left(\frac{1}{x^2}-\frac{1}{y^2}\right)}{x^2y^2\left(\frac{1}{x}+\frac{1}{y}\right)}$$

$$=\frac{y^2-x^2}{xy^2+x^2y}$$

$$=\frac{(y-x)(y+x)}{xy(y+x)}$$

$$=\frac{y-x}{xy}$$

CONCEPTS | **EXAMPLES**

6.4 Equations with Rational Expressions and Graphs

Solving an Equation with Rational Expressions

Step 1 Determine the domain of the variable.

Step 2 Multiply each side of the equation by the LCD to clear the fractions.

Step 3 Solve the resulting equation.

Step 4 Check that each proposed solution is in the domain, and discard any values that are not. Check the remaining proposed solution(s) in the original equation.

Solve.

$$\frac{3x+2}{x-2} + \frac{2}{x(x-2)} = \frac{-1}{x}$$ Note that 0 and 2 are excluded from the domain.

$$x(3x+2) + 2 = -(x-2)$$ Multiply by the LCD, $x(x-2)$.

$$3x^2 + 2x + 2 = -x + 2$$ Distributive property

$$3x^2 + 3x = 0$$ Add x. Subtract 2.

$$3x(x+1) = 0$$ Factor.

$$3x = 0 \quad \text{or} \quad x + 1 = 0$$ Zero-factor property

$$x = 0 \quad \text{or} \quad x = -1$$ Solve each equation.

Of the two proposed solutions, 0 must be discarded because it is not in the domain. A check confirms that the solution set is $\{-1\}$.

Graphing a Rational Function

The graph of a rational function (written in lowest terms) may have one or more breaks. At such points, the graph will approach an asymptote.

Graph $f(x) = \frac{1}{x+2}$.

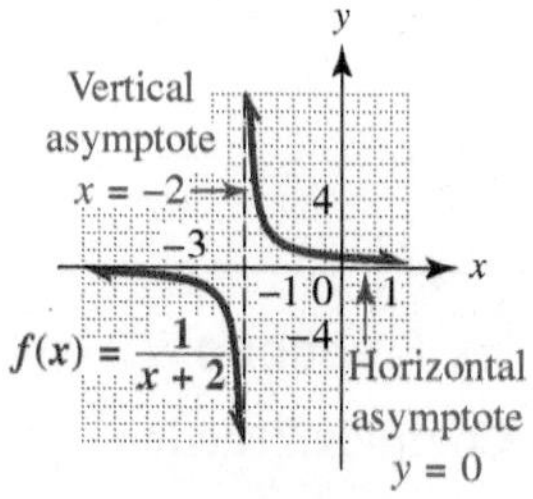

6.5 Applications of Rational Expressions

Solving a Distance, Rate, Time Problem

Use the distance formula

$$d = rt$$

or one of its equivalents,

$$t = \frac{d}{r} \quad \text{or} \quad r = \frac{d}{t}.$$

Solving a Work Problem

Use the fact that if 1 complete job is done in t units of time, then the rate of work is $\frac{1}{t}$ job per unit of time.

$$1 = rt, \quad \text{or} \quad r = \frac{1}{t}$$

Solve.

A canal has a current of 2 mph. Find the rate of Amy's boat in still water if it travels 11 mi downstream in the same time that it travels 8 mi upstream.

Let $x =$ the rate of the boat in still water.

	Distance	Rate	Time
Downstream	11	$x+2$	$\frac{11}{x+2}$
Upstream	8	$x-2$	$\frac{8}{x-2}$

The times are equal.

$$\frac{11}{x+2} = \frac{8}{x-2}$$ Write an equation. Use $t = \frac{d}{r}$.

$$11(x-2) = 8(x+2)$$ If $\frac{a}{b} = \frac{c}{d}$, then $ad = bc$.

$$11x - 22 = 8x + 16$$ Distributive property

$$3x = 38$$ Subtract $8x$. Add 22.

$$x = \frac{38}{3}, \quad \text{or} \quad 12\frac{2}{3}$$ Divide by 3.

The rate in still water is $12\frac{2}{3}$ mph.

CONCEPTS	EXAMPLES
6.6 Variation	
Let k be a real number.	
If $y = kx$, then y varies directly as x.	The diameter of a circle varies directly as the radius. $d = kr$ Here, $k = 2$.
If $y = kx^n$, then y varies directly as x^n.	The area of a circle varies directly as the square of the radius. $\mathcal{A} = kr^2$ Here, $k = \pi$.
If $y = \frac{k}{x}$, then y varies inversely as x.	Pressure varies inversely as volume. $p = \frac{k}{V}$
If $y = kxz$, then y varies jointly as x and z.	For a given principal, interest varies jointly as interest rate and time. $I = krt$ k is the given principal.

Chapter 6 Review Exercises

6.1 *Give the domain of each rational function using **(a)** set-builder notation and **(b)** interval notation.*

1. $f(x) = \frac{-7}{3x + 18}$ **2.** $f(x) = \frac{5x + 17}{x^2 - 7x + 10}$ **3.** $f(x) = \frac{9}{x^2 - 18x + 81}$

Write each rational expression in lowest terms.

4. $\frac{12x^2 + 6x}{24x + 12}$ **5.** $\frac{25m^2 - n^2}{25m^2 - 10mn + n^2}$ **6.** $\frac{r - 2}{4 - r^2}$

Multiply or divide as indicated.

7. $\frac{(2y + 3)^2}{5y} \cdot \frac{15y^3}{4y^2 - 9}$ **8.** $\frac{w^2 - 16}{w} \cdot \frac{3}{4 - w}$

9. $\frac{x^2 - 4}{x - 2} \div \frac{x + 2}{x - 4}$ **10.** $\frac{z^2 - z - 6}{z - 6} \cdot \frac{z^2 - 6z}{z^2 + 2z - 15}$

11. $\frac{8 - x}{8 + x} \cdot \frac{16 + 2x}{x^2 - 64}$ **12.** $\frac{m^3 - n^3}{m^2 - n^2} \div \frac{m^2 + mn + n^2}{m + n}$

6.2 *Suppose that the given expressions are denominators of rational expressions. Find the least common denominator (LCD) for each group of denominators.*

13. $32b^3$, $24b^5$ **14.** $9r^2$, $3r + 1$, 9

15. $6x^2 + 13x - 5$, $9x^2 + 9x - 4$ **16.** $3x - 12$, $x^2 - 2x - 8$, $x^2 - 8x + 16$

Add or subtract as indicated.

17. $\frac{8}{z} - \frac{3}{2z^2}$ **18.** $\frac{5y + 13}{y + 1} - \frac{1 - 7y}{y + 1}$

19. $\frac{6}{5a + 10} + \frac{7}{6a + 12}$ **20.** $\frac{3r}{10r^2 - 3rs - s^2} + \frac{2r}{2r^2 + rs - s^2}$

6.3 *Simplify each complex fraction.*

21. $\dfrac{\frac{3}{t}+2}{\frac{4}{t}-7}$ **22.** $\dfrac{\frac{2}{m-3n}}{\frac{1}{3n-m}}$ **23.** $\dfrac{\frac{3}{p}-\frac{2}{q}}{\frac{9q^2-4p^2}{qp}}$ **24.** $\dfrac{x^{-2}-y^{-2}}{x^{-1}-y^{-1}}$

6.4 *Solve each equation.*

25. $\dfrac{1}{t+4}+\dfrac{1}{2}=\dfrac{3}{2t+8}$

26. $\dfrac{-5m}{m+1}+\dfrac{m}{3m+3}=\dfrac{56}{6m+6}$

27. $\dfrac{4x+3}{5x+5}=\dfrac{8x+2}{10x+5}$

28. $\dfrac{2}{k-1}-\dfrac{4k+1}{k^2-1}=\dfrac{-1}{k+1}$

29. $\dfrac{r}{r-2}-\dfrac{8}{r^2-4}=\dfrac{-1}{r+2}$

30. $\dfrac{5}{x+2}+\dfrac{3}{x+3}=\dfrac{x}{x^2+5x+6}$

31. Decide whether each of the following is an *expression* or an *equation*. Simplify the one that is an expression, and solve the one that is an equation.

(a) $\dfrac{4}{x}+\dfrac{1}{2}=\dfrac{1}{3}$ **(b)** $\dfrac{4}{x}+\dfrac{1}{2}-\dfrac{1}{3}$

32. Graph the rational function $f(x)=\frac{2}{x+1}$. Give the equations of the vertical and horizontal asymptotes.

6.5 *Solve each problem.*

33. According to a law from physics,

$$\frac{1}{A}=\frac{1}{B}+\frac{1}{C}.$$

Find A if $B=30$ and $C=10$.

34. In banking and finance, the formula

$$P=\frac{M}{1+RT}$$

gives present value at simple interest. If $P=600$, $M=750$, and $R=0.125$, find T.

Solve each formula for the specified variable.

35. $F=\dfrac{GMm}{d^2}$ for m (physics)

36. $\mu=\dfrac{Mv}{M+m}$ for M (electronics)

Solve each problem.

37. An article in *Scientific American* predicted that in the year 2050, about 23,200 of the 58,000 passenger-km per day in North America will be provided by high-speed trains. If the traffic volume in a typical region of North America is 15,000, how many passenger-kilometers per day will high-speed trains provide there? (Data from Schafer, A., and D. Victor, "The Past and Future of Global Mobility," *Scientific American*.)

38. Johnny's car uses 15 gal of gasoline to travel 495 mi. He has 6 gal of gasoline in the car. He still needs to drive 600 mi. If the car continues to use gasoline at the same rate, how many more gallons will he need (to the nearest tenth)?

39. A river has a current of 4 km per hr. Find the rate of Herby's boat in still water if it travels 40 km downstream in the same time that it takes to travel 24 km upstream.

	d	r	t
Upstream	24	$x - 4$	
Downstream	40		

40. A sink can be filled by a cold-water tap in 8 min and by a hot-water tap in 12 min. How long would it take to fill the sink with both taps open?

	Rate	Time Working Together	Fractional Part of the Job Done
Cold		x	
Hot		x	

41. Jane and Jessica need to sort a pile of bottles at the recycling center. Working alone, Jane could do the entire job in 9 hr, while Jessica could do the entire job in 6 hr. How long will it take them if they work together?

6.6

42. In which one of the following does y vary inversely as x?

A. $y = 2x$ **B.** $y = \frac{x}{3}$ **C.** $y = \frac{3}{x}$ **D.** $y = x^2$

Solve each problem.

43. For a particular camera, the viewing distance varies directly as the amount of enlargement. A picture that is taken with this camera and enlarged 5 times should be viewed from a distance of 250 mm. Suppose a print 8.6 times the size of the negative is made. From what distance should it be viewed?

44. The frequency (number of vibrations per second) of a vibrating guitar string varies inversely as its length. That is, a longer string vibrates fewer times in a second than a shorter string. Suppose a guitar string 0.65 m long vibrates 4.3 times per sec. What frequency would a string 0.5 m long have?

45. The volume of a rectangular box of a given height is jointly proportional to its width and length. A box with width 2 ft and length 4 ft has volume 12 ft^3. Find the volume of a box with the same height that is 3 ft wide and 5 ft long.

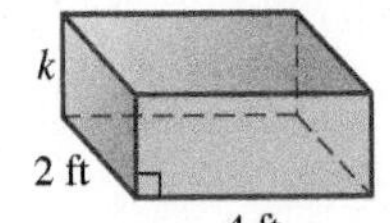

Volume: 12 ft^3

Chapter 6 Mixed Review Exercises

Write each rational expression in lowest terms.

1. $\frac{x + 2y}{x^2 - 4y^2}$

2. $\frac{x^2 + 2x - 15}{x^2 - x - 6}$

Perform the indicated operations.

3. $\frac{2}{m} + \frac{5}{3m^2}$

4. $\frac{9}{3 - x} - \frac{2}{x - 3}$

5. $\frac{\frac{-3}{x} + \frac{x}{2}}{1 + \frac{x + 1}{x}}$

6. $\frac{\frac{3}{x} - 5}{6 + \frac{1}{x}}$

7. $\frac{4y + 16}{30} \div \frac{2y + 8}{5}$

8. $\frac{t^{-2} + s^{-2}}{t^{-1} - s^{-1}}$

9. $\frac{k^2 - 6k + 9}{1 - 216k^3} \cdot \frac{6k^2 + 17k - 3}{9 - k^2}$

10. $\frac{9x^2 + 46x + 5}{3x^2 - 2x - 1} \div \frac{x^2 + 11x + 30}{x^3 + 5x^2 - 6x}$

11. $\frac{4a}{a^2 - ab - 2b^2} - \frac{6b - a}{a^2 + 4ab + 3b^2}$

12. $\frac{a}{b} + \frac{b}{c} + \frac{c}{d}$

Solve each equation.

13. $\frac{x + 3}{x^2 - 5x + 4} - \frac{1}{x} = \frac{2}{x^2 - 4x}$

14. $\frac{3x}{x - 4} + \frac{2}{x} = \frac{48}{x^2 - 4x}$

15. $1 - \frac{5}{r} = \frac{-4}{r^2}$

16. $A = \frac{Rr}{R + r}$ for r

Solve each problem.

17. The strength of a contact lens is given in units called diopters and also in millimeters of arc. As the diopters increase, the millimeters of arc decrease. The rational function

$$f(x) = \frac{337}{x}$$

relates the arc measurement $f(x)$ to the diopter measurement x. (Data from Bausch and Lomb.)

(a) To the nearest hundredth, what arc measurement will correspond to 40.5-diopter lenses?

(b) A lens with an arc measurement of 7.51 mm will provide what diopter strength, to the nearest hundredth?

18. At a certain gasoline station, 3 gal of unleaded gasoline cost $12.27. How much would 13 gal of the same gasoline cost?

19. If Dr. Dawson rides his bike to his office, he averages 12 mph. If he drives his car, he averages 36 mph. His time driving is $\frac{1}{4}$ hr less than his time riding his bike. How far is his office from home?

20. The area of a triangle varies jointly as the lengths of the base and height. A triangle with base 10 ft and height 4 ft has area 20 ft^2. Find the area of a triangle with base 3 ft and height 8 ft.

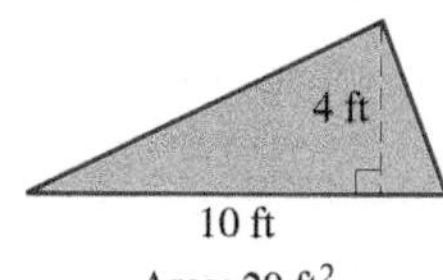

Area: 20 ft^2

Chapter 6 Test

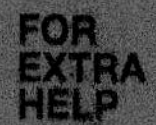

Step-by-step test solutions are found on the Chapter Test Prep Videos available in MyLab Math.

View the complete solutions to all Chapter Test exercises in MyLab Math.

1. Give the domain of the following rational function using **(a)** set-builder notation and **(b)** interval notation.

$$f(x) = \frac{x + 3}{3x^2 + 2x - 8}$$

2. Write $\frac{6x^2 - 13x - 5}{9x^3 - x}$ in lowest terms.

Multiply or divide as indicated.

3. $\frac{(x + 3)^2}{4} \cdot \frac{6}{2x + 6}$

4. $\frac{y^2 - 16}{y^2 - 25} \cdot \frac{y^2 + 2y - 15}{y^2 - 7y + 12}$

5. $\frac{3 - t}{5} \div \frac{t - 3}{10}$

6. $\frac{x^2 - 9}{x^3 + 3x^2} \div \frac{x^2 + x - 12}{x^3 + 9x^2 + 20x}$

7. Find the least common denominator for the group of denominators.

$$t^2 + t - 6, \quad t^2 + 3t, \quad t^2$$

Add or subtract as indicated.

8. $\frac{7}{6t^2} - \frac{1}{3t}$

9. $\frac{9}{x-7} + \frac{4}{x+7}$

10. $\frac{9}{x^2 - 6x + 9} + \frac{2}{x^2 - 9}$

11. $\frac{6}{x+4} + \frac{1}{x+2} - \frac{3x}{x^2 + 6x + 8}$

Simplify each expression.

12. $\frac{\frac{12}{r+4}}{\frac{11}{6r+24}}$

13. $\frac{\frac{1}{a} - \frac{1}{b}}{\frac{a}{b} - \frac{b}{a}}$

14. $\frac{2x^{-2} + y^{-2}}{x^{-1} - y^{-1}}$

15. Decide whether each of the following is an *expression* or an *equation*. Simplify the one that is an expression, and solve the one that is an equation.

(a) $\frac{2x}{3} + \frac{x}{4} - \frac{11}{2}$

(b) $\frac{2x}{3} + \frac{x}{4} = \frac{11}{2}$

Solve each equation.

16. $\frac{1}{x} - \frac{4}{3x} = \frac{1}{x-2}$

17. $\frac{y}{y+2} - \frac{1}{y-2} = \frac{8}{y^2 - 4}$

18. Solve for the variable ℓ in this formula from mathematics: $S = \frac{n}{2}(a + \ell)$.

19. Graph the rational function $f(x) = \frac{-2}{x+1}$. Give the equations of the vertical and horizontal asymptotes.

Solve each problem.

20. Chris can do a job in 9 hr, while Dana can do the same job in 5 hr. How long would it take them to do the job if they worked together?

21. The rate of the current in a stream is 3 mph. Nana's boat can travel 36 mi downstream in the same time that it takes to travel 24 mi upstream. Find the rate of her boat in still water.

22. Biologists collected a sample of 600 fish from West Okoboji Lake on May 1 and tagged each of them. When they returned on June 1, a new sample of 800 fish was collected and 10 of these had been previously tagged. Use this experiment to determine the approximate fish population of West Okoboji Lake.

23. In biology, the rational function

$$g(x) = \frac{5x}{2 + x}$$

gives the growth rate g of a population for x units of available food. (Data from Smith, J. Maynard, *Models in Ecology,* Cambridge University Press.)

(a) What amount of food (in appropriate units) would produce a growth rate of 3 units of growth per unit of food?

(b) What is the growth rate if no food is available?

24. The current in a simple electrical circuit is inversely proportional to the resistance. If the current is 80 amps when the resistance is 30 ohms, find the current when the resistance is 12 ohms.

25. The force of the wind blowing on a vertical surface varies jointly as the area of the surface and the square of the velocity. If a wind blowing at 40 mph exerts a force of 50 lb on a surface of 500 ft^2, how much force will a wind of 80 mph exert on a surface of 2 ft^2?

Chapters R–6 Cumulative Review Exercises

Perform the indicated operations.

1. $-\frac{2}{3}-\frac{1}{4}$

2. $-\frac{3}{4}\left(-\frac{5}{6}\right)$

3. $-\frac{3}{5}\div\frac{1}{2}$

4. $-8.75+3.08$

5. $5.25\div 100$

6. 43.6×100

Evaluate.

7. (a) 5^2 **(b)** -5^2 **(c)** $(-5)^2$ **(d)** $\sqrt{25}$ **(e)** $-\sqrt{25}$ **(f)** $\sqrt{-25}$

8. Evaluate $|2x|+3y-z^3$ for $x=-4$, $y=3$, and $z=6$.

Solve each equation or inequality.

9. $7(2x+3)-4(2x+1)=2(x+1)$

10. $|6x-8|-4=0$

11. $\frac{2}{3}y+\frac{5}{12}y\leq 20$

12. $|3x+2|\geq 4$

Solve each problem.

13. Abushieba invested some money at 4% interest and twice as much at 3% interest. His interest for the first year was \$400. How much did he invest at each rate?

14. A triangle has area 42 m^2. The base is 14 m long. Find the height of the triangle.

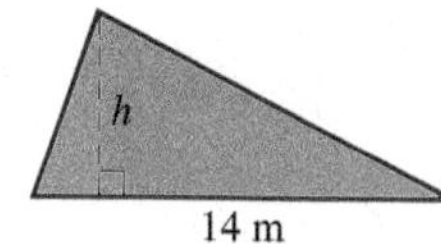

15. Graph $-4x+2y=8$ and give the intercepts.

Find the slope of each line described.

16. Passing through $(-5, 8)$ and $(-1, 2)$

17. Perpendicular to $4x-3y=12$

18. Write an equation of the line in **Exercise 16.** Give the equation in the form $y=mx+b$.

Graph each inequality or compound inequality.

19. $2x+5y>10$

20. $x-y\geq 3$ and $3x+4y\leq 12$

21. Consider the equation $5x-3y=8$.

(a) Write y as a function f of x, using function notation $f(x)$.

(b) Find $f(1)$.

22. Consider the relation $y=-\sqrt{x+2}$.

(a) Does it define a function?

(b) Give its domain and range.

Solve each system.

23. $4x - y = -7$
$5x + 2y = 1$

24. $x + y - 2z = -1$
$2x - y + z = -6$
$3x + 2y - 3z = -3$

25. $x + 2y + z = 5$
$x - y + z = 3$
$2x + 4y + 2z = 11$

26. Simplify $\left(\frac{m^{-4}n^2}{m^2n^{-3}}\right) \cdot \left(\frac{m^5n^{-1}}{m^{-2}n^5}\right)$. Write the answer with only positive exponents. Assume that all variables represent nonzero real numbers.

Perform the indicated operations.

27. $(3y^2 - 2y + 6) - (-y^2 + 5y + 12)$

28. $(4f + 3)(3f - 1)$

29. $\left(\frac{1}{4}x + 5\right)^2$

30. $(3x^3 + 13x^2 - 17x - 7) \div (3x + 1)$

31. Find each of the following for the polynomial functions

$$f(x) = x^2 + 2x - 3, \quad g(x) = 2x^3 - 3x^2 + 4x - 1, \quad \text{and} \quad h(x) = x^2.$$

(a) $(f + g)(x)$ **(b)** $(g - f)(x)$ **(c)** $(f + g)(-1)$ **(d)** $(f \circ h)(x)$

Factor each polynomial completely.

32. $2x^2 - 13x - 45$

33. $100t^4 - 25$

34. $8p^3 + 125$

Perform the indicated operations. Express answers in lowest terms.

35. $\frac{2a^2}{a + b} \cdot \frac{a - b}{4a}$

36. $\frac{x^2 - 9}{2x + 4} \div \frac{x^3 - 27}{4}$

37. $\frac{x + 4}{x - 2} + \frac{2x - 10}{x - 2}$

Solve.

38. $3x^2 + 4x = 7$

39. $\frac{-3x}{x + 1} + \frac{4x + 1}{x} = \frac{-3}{x^2 + x}$

40. Machine A can complete a certain job in 2 hr. To speed up the work, Machine B, which can complete the job alone in 3 hr, is brought in to help. How long will it take the two machines to complete the job working together?

STUDY SKILLS **REMINDER**

Have you begun to prepare for your final exam? **Review Study Skill 10, *Preparing for Your Math Final Exam.***

7

ROOTS, RADICALS, AND ROOT FUNCTIONS

The formula for calculating the distance one can see to the horizon from the top of a tall building involves a *square root radical,* one of the topics covered in this chapter.

7.1 Radical Expressions and Graphs

OBJECTIVES

1 Find roots of numbers.
2 Find principal roots.
3 Graph functions defined by radical expressions.
4 Find *n*th roots of *n*th powers.
5 Use a calculator to find roots.

VOCABULARY

- ☐ radicand
- ☐ index (order)
- ☐ radical
- ☐ principal root
- ☐ radical expression
- ☐ square root function
- ☐ cube root function

OBJECTIVE 1 Find roots of numbers.

Recall that $6^2 = 36$. We say that "6 *squared* equals 36." The opposite (or inverse) of *squaring* a number is taking its *square root.*

It is customary to write $\sqrt{\ }$ rather than $\sqrt[2]{\ }$.

$$\sqrt{36} = 6, \quad \text{because} \quad 6^2 = 36.$$

We extend this discussion to *cube roots* $\sqrt[3]{\ }$, *fourth roots* $\sqrt[4]{\ }$, and higher roots.

Meaning of $\sqrt[n]{a}$

The *n*th root of *a*, written $\sqrt[n]{a}$, is a number whose *n*th power equals *a*. That is,

$$\sqrt[n]{a} = b \quad \textbf{means} \quad b^n = a.$$

The number *a* is the **radicand,** *n* is the **index,** or **order,** and the expression $\sqrt[n]{a}$ is a **radical.**

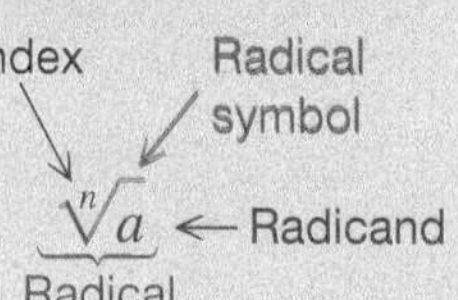

EXAMPLE 1 Simplifying Higher Roots

Find each root.

(a) $\sqrt[3]{64} = 4$, because $4^3 = 64$.

(b) $\sqrt[3]{125} = 5$, because $5^3 = 125$.

(c) $\sqrt[4]{16} = 2$, because $2^4 = 16$.

(d) $\sqrt[5]{32} = 2$, because $2^5 = 32$.

(e) $\sqrt[3]{\frac{8}{27}} = \frac{2}{3}$, because $\left(\frac{2}{3}\right)^3 = \frac{8}{27}$.

(f) $\sqrt[4]{0.0016} = 0.2$, because $(0.2)^4 = 0.0016$.

NOW TRY

NOW TRY EXERCISE 1

Find each root.

(a) $\sqrt[3]{1000}$ **(b)** $\sqrt[4]{625}$

(c) $\sqrt[4]{\frac{1}{256}}$ **(d)** $\sqrt[3]{0.027}$

OBJECTIVE 2 Find principal roots.

If *n* is even, positive numbers have two *n*th roots. For example, both 4 and -4 are square roots of 16, and 2 and -2 are fourth roots of 16. For $a > 0$, $\sqrt[n]{a}$ represents the positive root, called the **principal root,** and $-\sqrt[n]{a}$ represents the negative root. For all *n*, $\sqrt[n]{0} = 0$.

***n*th Root**

Case 1 If *n* is ***even*** and *a* is ***positive or 0,*** then

$\sqrt[n]{a}$ represents the **principal *n*th root** of *a*,

and $-\sqrt[n]{a}$ represents the **negative *n*th root** of *a*.

Case 2 If *n* is ***even*** and *a* is ***negative,*** then $\sqrt[n]{a}$ is not a real number.

Case 3 If *n* is ***odd,*** then there is exactly one real *n*th root of *a*, written $\sqrt[n]{a}$.

If *n* is even (Case 1), then the two *n*th roots of *a* are often written together as $\pm\sqrt[n]{a}$, with $\pm$ read ***"positive or negative"*** or ***"plus or minus."***

NOW TRY ANSWERS

1. (a) 10 (b) 5 (c) $\frac{1}{4}$ (d) 0.3

NOW TRY EXERCISE 2

Find each root.

(a) $-\sqrt{25}$ **(b)** $\sqrt[4]{-625}$

(c) $\sqrt[5]{-32}$ **(d)** $-\sqrt[3]{64}$

EXAMPLE 2 Finding Roots

Find each root.

(a) $\sqrt{100} = 10$ (Case 1)

Because the radicand, 100, is *positive,* there are two square roots: 10 and -10. We want the principal square root, which is 10.

(b) $-\sqrt{100} = -10$ (Case 1)

Here, we want the negative square root, -10.

(c) $\sqrt[4]{81} = 3$ Principal 4th root (Case 1)

(d) $-\sqrt[4]{81} = -3$ Negative 4th root (Case 1)

(e) $\sqrt[4]{-81}$ (Case 2)

The index is *even* and the radicand is *negative.* Any real number raised to an *even* power is *positive*—here $3^4 = 81$ and $(-3)^4 = 81$. Thus,

$\sqrt[4]{-81}$ is not a real number.

(f) $\sqrt[3]{8} = 2$, because $2^3 = 8$. (Case 3)

(g) $\sqrt[3]{-8} = -2$, because $(-2)^3 = -8$. (Case 3)

In parts (f) and (g), the index is *odd.* Each radical represents exactly one *n*th root (regardless of whether the radicand is positive, negative, or 0). NOW TRY

OBJECTIVE 3 Graph functions defined by radical expressions.

A **radical expression** is an algebraic expression that contains radicals.

$$3 - \sqrt{x}, \quad \sqrt[3]{x}, \quad \sqrt{2x - 1} \qquad \text{Radical expressions}$$

In earlier chapters, we graphed functions defined by polynomial and rational expressions. Now we graph functions defined by basic radical expressions, such as $f(x) = \sqrt{x}$ and $f(x) = \sqrt[3]{x}$.

FIGURE 1 shows the graph of the **square root function,**

$$f(x) = \sqrt{x},$$

together with a table of selected points. Only nonnegative values can be used for x, so the domain is $[0, \infty)$. Because $\sqrt{x}$ is the principal square root of x, it always has a nonnegative value, so the range is also $[0, \infty)$.

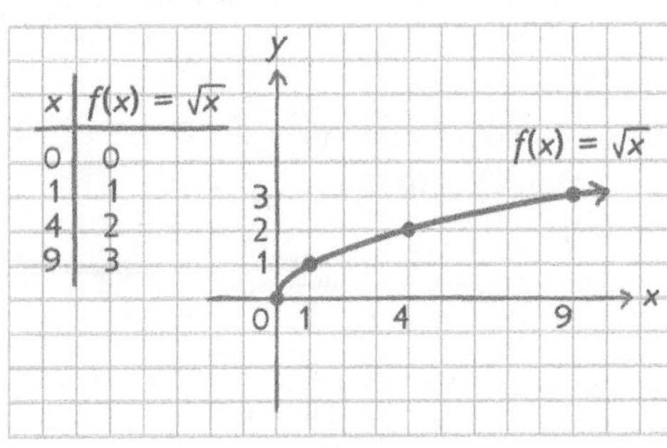

Square root function

$f(x) = \sqrt{x}$

Domain: $[0, \infty)$

Range: $[0, \infty)$

FIGURE 1

NOW TRY ANSWERS

2. **(a)** -5

(b) It is not a real number.

(c) -2 **(d)** -4

FIGURE 2 shows the graph of the **cube root function**

$$f(x) = \sqrt[3]{x}.$$

Any real number (positive, negative, or 0) can be used for x in the cube root function, so $\sqrt[3]{x}$ can be positive, negative, or 0. Thus, both the domain and the range of the cube root function are $(-\infty, \infty)$.

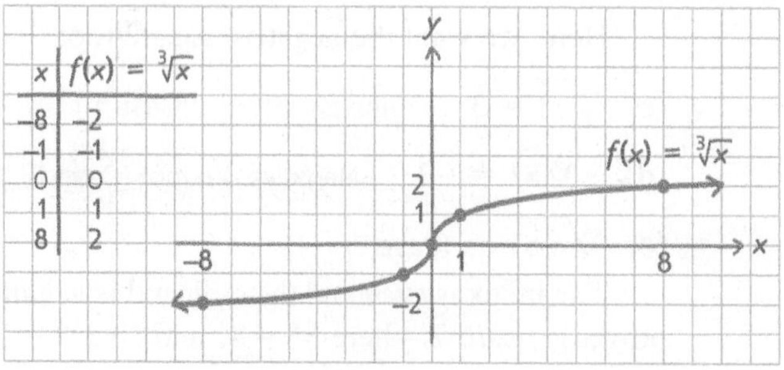

Cube root function

$f(x) = \sqrt[3]{x}$

Domain: $(-\infty, \infty)$

Range: $(-\infty, \infty)$

FIGURE 2

NOW TRY EXERCISE 3

Graph each function, and give its domain and range.

(a) $f(x) = \sqrt{x+1}$

(b) $f(x) = \sqrt[3]{x} - 1$

EXAMPLE 3 Graphing Functions Defined with Radicals

Graph each function, and give its domain and range.

(a) $f(x) = \sqrt{x-3}$

Create a table of values as given with the graph in **FIGURE 3**. The x-values were chosen in such a way that the function values are all integers. For the radicand to be nonnegative, we must have

$$x - 3 \geq 0, \quad \text{or} \quad x \geq 3.$$

Therefore, the domain of this function is $[3, \infty)$. Function values are positive or 0, so the range is $[0, \infty)$.

x	$f(x) = \sqrt{x-3}$
3	$\sqrt{3-3} = 0$
4	$\sqrt{4-3} = 1$
7	$\sqrt{7-3} = 2$

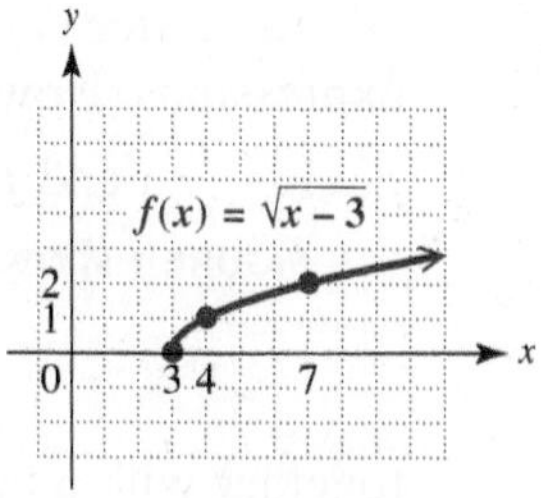

This graph is shifted to the right 3 units compared to the graph of $y = \sqrt{x}$.

FIGURE 3

(b) $f(x) = \sqrt[3]{x} + 2$

See **FIGURE 4**. Both the domain and the range are $(-\infty, \infty)$.

x	$f(x) = \sqrt[3]{x} + 2$
−8	$\sqrt[3]{-8} + 2 = 0$
−1	$\sqrt[3]{-1} + 2 = 1$
0	$\sqrt[3]{0} + 2 = 2$
1	$\sqrt[3]{1} + 2 = 3$
8	$\sqrt[3]{8} + 2 = 4$

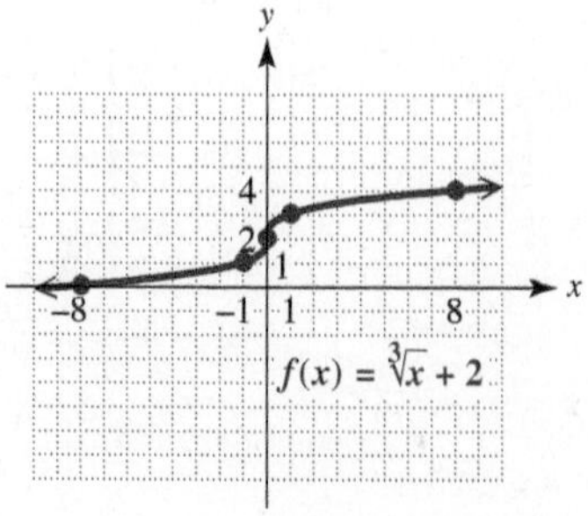

This graph is shifted up 2 units compared to the graph of $y = \sqrt[3]{x}$.

FIGURE 4

NOW TRY

NOW TRY ANSWERS

3. (a)

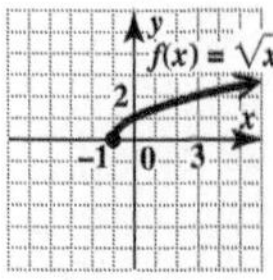

domain: $[-1, \infty)$; range: $[0, \infty)$

(b)

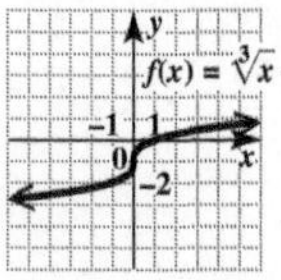

domain: $(-\infty, \infty)$; range: $(-\infty, \infty)$

OBJECTIVE 4 Find *n*th roots of *n*th powers.

Consider the expression $\sqrt{a^2}$. At first glance, we might think that it is equivalent to a. However, this is not necessarily true. For example, consider the following.

If $a = 6$, then $\sqrt{a^2} = \sqrt{6^2} = \sqrt{36} = 6$.

If $a = -6$, then $\sqrt{a^2} = \sqrt{(-6)^2} = \sqrt{36} = 6$. ← Instead of −6, we get 6, the *absolute value* of −6.

The symbol $\sqrt{a^2}$ represents the *nonnegative* square root, so we express $\sqrt{a^2}$ with absolute value bars as $|a|$ because a may be a negative number.

Meaning of $\sqrt{a^2}$

For any real number a, $\sqrt{a^2} = |a|$.

That is, the principal square root of a^2 is the absolute value of a.

NOW TRY EXERCISE 4

Find each square root.

(a) $\sqrt{11^2}$ (b) $\sqrt{(-11)^2}$

(c) $\sqrt{z^2}$ (d) $\sqrt{(-z)^2}$

EXAMPLE 4 Simplifying Square Roots Using Absolute Value

Find each square root.

(a) $\sqrt{7^2} = |7| = 7$ **(b)** $\sqrt{(-7)^2} = |-7| = 7$

(c) $\sqrt{k^2} = |k|$ **(d)** $\sqrt{(-k)^2} = |-k| = |k|$ NOW TRY

We can generalize this idea to any *n*th root.

Meaning of $\sqrt[n]{a^n}$

If ***n*** **is an *even* positive integer,** then $\sqrt[n]{a^n} = |a|$.

If ***n*** **is an *odd* positive integer,** then $\sqrt[n]{a^n} = a$.

That is, use the absolute value symbol when n is even.

NOW TRY EXERCISE 5

Simplify each root.

(a) $\sqrt[8]{(-2)^8}$ (b) $\sqrt[3]{(-9)^3}$

(c) $-\sqrt[4]{(-10)^4}$ (d) $-\sqrt{m^8}$

(e) $\sqrt[3]{x^{18}}$ (f) $\sqrt[4]{t^{20}}$

EXAMPLE 5 Simplifying Higher Roots Using Absolute Value

Simplify each root.

(a) $\sqrt[6]{(-3)^6} = |-3| = 3$ *n* is even. Use absolute value.

(b) $\sqrt[5]{(-4)^5} = -4$ *n* is odd. Absolute value is not necessary.

(c) $-\sqrt[4]{(-9)^4} = -|-9| = -9$ *n* is even. Use absolute value.

(d) $-\sqrt{m^4} = -|m^2| = -m^2$ For all m, $|m^2| = m^2$.

No absolute value bars are needed here, because m^2 is nonnegative for any real number value of m.

(e) $\sqrt[3]{a^{12}} = a^4$, because $a^{12} = (a^4)^3$.

(f) $\sqrt[4]{x^{12}} = |x^3|$

We use absolute value to guarantee that the result is not negative (because x^3 is negative when x is negative). If desired, $|x^3|$ can be written as $x^2 \cdot |x|$. NOW TRY

NOW TRY ANSWERS

4. (a) 11 (b) 11 (c) $|z|$ (d) $|z|$

5. (a) 2 (b) −9 (c) −10 (d) $-m^4$ (e) x^6 (f) $|t^5|$

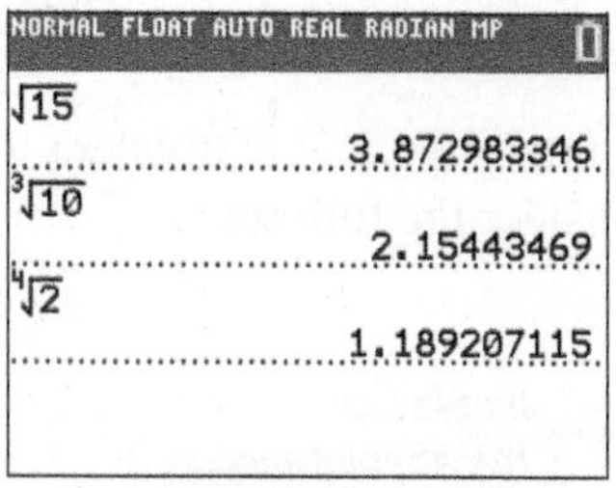

NORMAL FLOAT AUTO REAL RADIAN MP

$\sqrt{15}$ 3.872983346

$\sqrt[3]{10}$ 2.15443469

$\sqrt[4]{2}$ 1.189207115

(a)

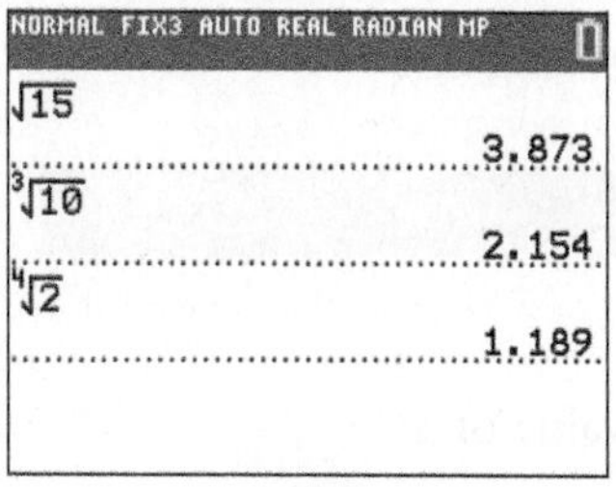

NORMAL FIX3 AUTO REAL RADIAN MP

$\sqrt{15}$ 3.873

$\sqrt[3]{10}$ 2.154

$\sqrt[4]{2}$ 1.189

(b)

FIGURE 5

OBJECTIVE 5 Use a calculator to find roots.

While numbers such as $\sqrt{9}$ and $\sqrt[3]{-8}$ are rational, radicals are often irrational numbers. To find approximations of such radicals, we usually use a calculator. For example,

$$\sqrt{15} \approx 3.872983346, \quad \sqrt[3]{10} \approx 2.15443469, \quad \text{and} \quad \sqrt[4]{2} \approx 1.189207115,$$

where the symbol $\approx$ means "is approximately equal to." In this text, we often show approximations rounded to three decimal places. Thus,

$$\sqrt{15} \approx 3.873, \quad \sqrt[3]{10} \approx 2.154, \quad \text{and} \quad \sqrt[4]{2} \approx 1.189.$$

FIGURE 5 shows how the preceding approximations are displayed on a TI-83/84 Plus C graphing calculator. In **FIGURE 5(a)**, eight or nine decimal places are shown, while in **FIGURE 5(b)**, the number of decimal places is fixed at three.

There is a simple way to check that a calculator approximation is "in the ballpark." For example, because 16 is a little larger than 15, $\sqrt{16} = 4$ should be a little larger than $\sqrt{15}$. Thus, 3.873 is reasonable as an approximation for $\sqrt{15}$.

NOTE Methods for finding approximations differ among makes and models of calculators. ***Consult your owner's manual for keystroke instructions.*** Be aware that graphing calculators often differ from scientific calculators in the order in which keystrokes are made.

NOW TRY EXERCISE 6
Use a calculator to approximate each radical to three decimal places.

(a) $\sqrt{73}$ (b) $-\sqrt{92}$

(c) $\sqrt[4]{92}$ (d) $\sqrt[5]{33}$

EXAMPLE 6 Finding Approximations for Roots

Use a calculator to approximate each radical to three decimal places.

(a) $\sqrt{39} \approx 6.245$ (b) $-\sqrt{72} \approx -8.485$

(c) $\sqrt[3]{93} \approx 4.531$ (d) $\sqrt[4]{39} \approx 2.499$

NOW TRY

NOW TRY EXERCISE 7
Use the formula in **Example 7** to approximate f to the nearest thousand if

$L = 7 \times 10^{-5}$

and $C = 3 \times 10^{-9}$.

EXAMPLE 7 Using Roots to Calculate Resonant Frequency

In electronics, the resonant frequency f of a circuit may be found using the formula

$$f = \frac{1}{2\pi\sqrt{LC}}, \quad \text{Electronics formula}$$

where f is in cycles per second, L is in henrys, and C is in farads. (Henrys and farads are units of measure in electronics.) Find the resonant frequency f if $L = 5 \times 10^{-4}$ henry and $C = 3 \times 10^{-10}$ farad. Give the answer to the nearest thousand.

Find the value of f when $L = 5 \times 10^{-4}$ and $C = 3 \times 10^{-10}$.

$$f = \frac{1}{2\pi\sqrt{LC}} \quad \text{Given formula}$$

$$f = \frac{1}{2\pi\sqrt{(5 \times 10^{-4})(3 \times 10^{-10})}} \quad \text{Substitute for } L \text{ and } C.$$

$$f \approx 411{,}000 \quad \text{Use a calculator.}$$

The resonant frequency f is approximately 411,000 cycles per sec. NOW TRY

NOW TRY ANSWERS

6. (a) 8.544 (b) −9.592 (c) 3.097 (d) 2.012

7. 347,000 cycles per sec

NOTE The expression in the second-to-last line of **Example 7** can be difficult to compute on a calculator. There are several ways to approach it, but here is one way that works nicely. Use a calculator to verify the following steps.

Step 1 Calculate the product under the radical in the denominator. The result is

$$1.5 \times 10^{-13}.$$

Step 2 Use the square root function ($\sqrt{x}$) to find the square root of this number. The result is approximately

$$3.872983346 \times 10^{-7}.$$

Step 3 Multiply by 2π, using the π function of the calculator. The result is approximately

$$2.433467206 \times 10^{-6}.$$

Step 4 Now use the reciprocal function (labeled $\frac{1}{x}$ or x^{-1}) of the calculator. The result should be 410936.296, which, rounded to the nearest thousand, is 411,000.

If the numerator had not been 1, and perhaps an expression to evaluate, one approach would be to save the result found in Step 3 in the calculator memory, evaluate the numerator, and then divide by the number saved in the memory.

7.1 Exercises

MyLab Math

Video solutions for select problems available in MyLab Math

STUDY SKILLS **REMINDER**

Make study cards to help you learn and remember the material in this chapter. ***Review Study Skill 5, Using Study Cards.***

Concept Check *Find each square root.*

1. (a) $\sqrt{25}$ **(b)** $\sqrt{81}$ **(c)** $\sqrt{144}$

2. (a) $\sqrt{0.25}$ **(b)** $\sqrt{0.81}$ **(c)** $\sqrt{1.44}$

3. Concept Check A student incorrectly claimed that $\sqrt{16} = 8$. **WHAT WENT WRONG?** Evaluate $\sqrt{16}$ correctly.

4. Concept Check When approximating $\sqrt{19}$ to three decimal places, a student incorrectly rounded as follows.

$$\sqrt{19} \approx 4.358898944$$

$$\sqrt{19} \approx 4.358$$

WHAT WENT WRONG? Give the correct approximation.

Concept Check *Match each expression in Column I with the equivalent choice in Column II. Answers may be used once, more than once, or not at all.*

I		II	
5. $-\sqrt{16}$	**6.** $\sqrt{-16}$	**A.** 3	**B.** -2
7. $\sqrt[3]{-27}$	**8.** $\sqrt[5]{-32}$	**C.** 2	**D.** -3
9. $\sqrt[4]{16}$	**10.** $-\sqrt[3]{64}$	**E.** -4	**F.** Not a real number

Concept Check *Choose the closest approximation of each square root. Do not use a calculator.*

11. $\sqrt{123}$

A. 9 **B.** 10 **C.** 11 **D.** 12

12. $\sqrt{67}$

A. 7 **B.** 8 **C.** 9 **D.** 10

Concept Check *Refer to the figure to answer each question.*

$\sqrt{98}$

$\sqrt{26}$

13. Which one of the following is the best estimate of its area?

A. 50 **B.** 100 **C.** 250 **D.** 2500

14. Which one of the following is the best estimate of its perimeter?

A. 15 **B.** 30 **C.** 100 **D.** 250

15. **Concept Check** Consider the expression $-\sqrt{-a}$. Decide whether it is *positive, negative, 0,* or *not a real number* in each case.

(a) $a > 0$ **(b)** $a < 0$ **(c)** $a = 0$

16. **Concept Check** If n is odd, under what conditions is $\sqrt[n]{a}$ the following?

(a) positive **(b)** negative **(c)** 0

Find each root. **See Examples 1 and 2.**

17. $-\sqrt{81}$ 18. $-\sqrt{121}$ 19. $\sqrt[3]{216}$ 20. $\sqrt[3]{343}$

21. $\sqrt[3]{-64}$ 22. $\sqrt[3]{-125}$ 23. $-\sqrt[3]{512}$ 24. $-\sqrt[3]{1000}$

25. $\sqrt[4]{1296}$ 26. $\sqrt[4]{625}$ 27. $-\sqrt[4]{16}$ 28. $-\sqrt[4]{256}$

29. $\sqrt[4]{-625}$ 30. $\sqrt[4]{-256}$ 31. $\sqrt[6]{64}$ 32. $\sqrt[6]{729}$

33. $\sqrt[6]{-32}$ 34. $\sqrt[8]{-1}$ 35. $\sqrt{\frac{64}{81}}$ 36. $\sqrt{\frac{100}{9}}$

37. $\sqrt[3]{\frac{64}{27}}$ 38. $\sqrt[4]{\frac{81}{16}}$ 39. $-\sqrt[6]{\frac{1}{64}}$ 40. $-\sqrt[5]{\frac{1}{32}}$

41. $-\sqrt[3]{-27}$ 42. $-\sqrt[3]{-64}$ 43. $\sqrt{0.25}$ 44. $\sqrt{0.36}$

45. $-\sqrt{0.49}$ 46. $-\sqrt{0.81}$ 47. $\sqrt[3]{0.001}$ 48. $\sqrt[3]{0.125}$

Graph each function, and give its domain and range. **See Example 3.**

49. $f(x) = \sqrt{x+3}$ 50. $f(x) = \sqrt{x-5}$ 51. $f(x) = \sqrt{x} - 2$

52. $f(x) = \sqrt{x} + 4$ 53. $f(x) = \sqrt[3]{x} - 3$ 54. $f(x) = \sqrt[3]{x} + 1$

55. $f(x) = \sqrt[3]{x-3}$ 56. $f(x) = \sqrt[3]{x+1}$

Simplify each root. **See Examples 4 and 5.**

57. $\sqrt{12^2}$ 58. $\sqrt{19^2}$ 59. $\sqrt{(-10)^2}$ 60. $\sqrt{(-13)^2}$

61. $\sqrt[6]{(-2)^6}$ 62. $\sqrt[6]{(-4)^6}$ 63. $\sqrt[5]{(-9)^5}$ 64. $\sqrt[5]{(-8)^5}$

65. $-\sqrt[6]{(-5)^6}$ 66. $-\sqrt[6]{(-7)^6}$ 67. $\sqrt{x^2}$ 68. $-\sqrt{x^2}$

69. $\sqrt{(-z)^2}$ 70. $\sqrt{(-q)^2}$ 71. $\sqrt[3]{x^3}$ 72. $-\sqrt[3]{x^3}$

73. $\sqrt[3]{x^{15}}$ 74. $\sqrt[3]{m^9}$ 75. $\sqrt[6]{x^{30}}$ 76. $\sqrt[4]{k^{20}}$

Use a calculator to approximate each radical to three decimal places. ***See Example 6.***

77. $\sqrt{9483}$ **78.** $\sqrt{6825}$ **79.** $\sqrt{284.361}$ **80.** $\sqrt{846.104}$

81. $-\sqrt{82}$ **82.** $-\sqrt{91}$ **83.** $\sqrt[3]{423}$ **84.** $\sqrt[3]{555}$

85. $\sqrt[4]{100}$ **86.** $\sqrt[4]{250}$ **87.** $\sqrt[5]{23.8}$ **88.** $\sqrt[5]{98.4}$

Solve each problem. ***See Example 7.***

89. Use the electronics formula

$$f = \frac{1}{2\pi\sqrt{LC}}$$

to calculate the resonant frequency f of a circuit, in cycles per second, to the nearest thousand for the following values of L and C.

(a) $L = 7.237 \times 10^{-5}$ and $C = 2.5 \times 10^{-10}$

(b) $L = 5.582 \times 10^{-4}$ and $C = 3.245 \times 10^{-9}$

90. The threshold weight T for a person is the weight above which the risk of death increases greatly. The threshold weight in pounds for men aged 40–49 is related to height h in inches by the formula

$$h = 12.3\sqrt[3]{T}.$$

What height corresponds to a threshold weight of 216 lb for a 43-year-old man? Round the answer to the nearest inch and then to the nearest tenth of a foot.

91. According to an article in *The World Scanner Report,* the distance D, in miles, to the horizon from an observer's point of view over water or "flat" earth is given by

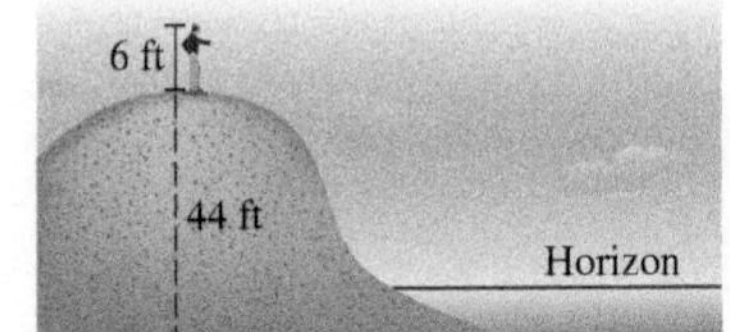

$$D = \sqrt{2H},$$

where H is the height of the point of view, in feet. If a person whose eyes are 6 ft above ground level is standing at the top of a hill 44 ft above "flat" earth, approximately how far to the horizon will she be able to see?

92. The time for one complete swing of a simple pendulum is given by

$$t = 2\pi\sqrt{\frac{L}{g}},$$

2 ft

where t is time in seconds, L is the length of the pendulum in feet, and g, the force due to gravity, is about 32 ft per sec^2. Find the time of a complete swing of a 2-ft pendulum to the nearest tenth of a second.

The coefficient of self-induction L (in henrys), the energy P stored in an electronic circuit (in joules), and the current I (in amps) are related by the formula

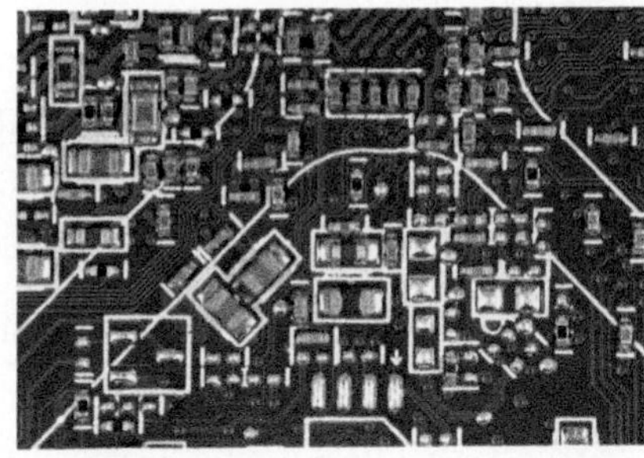

$$I = \sqrt{\frac{2P}{L}}.$$

93. Find I if $P = 120$ and $L = 80$.

94. Find I if $P = 100$ and $L = 40$.

Heron's formula *gives a method of finding the area of a triangle if the lengths of its sides are known. Suppose that a, b, and c are the lengths of the sides. Let s denote one-half of the perimeter of the triangle (called the* ***semiperimeter***)*—that is,*

$$s = \frac{1}{2}(a + b + c).$$

Then the area of the triangle is given by

$$\mathcal{A} = \sqrt{s(s - a)(s - b)(s - c)}.$$

Use Heron's formula to solve each problem.

95. Find the area of the Bermuda Triangle, to the nearest thousand square miles, if the "sides" of this triangle measure approximately 960 mi, 1030 mi, and 1030 mi.

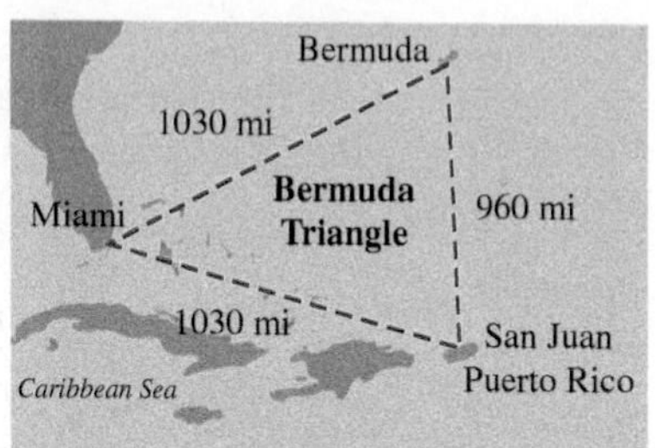

96. The Vietnam Veterans Memorial in Washington, DC, is in the shape of an unenclosed isosceles triangle with equal sides of length 246.75 ft. If the triangle were enclosed, the third side would have length 438.14 ft. Find the area of this enclosure to the nearest hundred square feet. (Data from information pamphlet obtained at the Vietnam Veterans Memorial.)

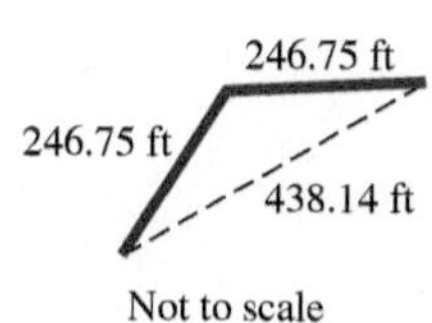

97. Find the area of a triangle with sides of lengths $a = 11$ m, $b = 60$ m, and $c = 61$ m.

98. Find the area of a triangle with sides of lengths $a = 20$ ft, $b = 34$ ft, and $c = 42$ ft.

7.2 Rational Exponents

OBJECTIVES

1 Use exponential notation for *n*th roots.

2 Define and use expressions of the form $a^{m/n}$.

3 Convert between radicals and rational exponents.

4 Use the rules for exponents with rational exponents.

OBJECTIVE 1 Use exponential notation for *n*th roots.

Consider the expression $(3^{1/2})^2$. We can simplify it as follows.

$$\begin{aligned} &(3^{1/2})^2 \\ &= 3^{1/2} \cdot 3^{1/2} && a^2 = a \cdot a \\ &= 3^{1/2+1/2} && \text{Product rule: } a^m \cdot a^n = a^{m+n} \\ &= 3^1 && \text{Add exponents.} \\ &= 3 && a^1 = a \end{aligned}$$

Also, by definition,

$$(\sqrt{3})^2 = \sqrt{3} \cdot \sqrt{3} = 3.$$

Because both $(3^{1/2})^2$ and $(\sqrt{3})^2$ are equal to 3, it seems reasonable to define

$$3^{1/2} = \sqrt{3}.$$

This suggests the following generalization.

Meaning of $a^{1/n}$

If $\sqrt[n]{a}$ is a real number, then $\quad a^{1/n} = \sqrt[n]{a}.$

Examples: $4^{1/2} = \sqrt{4}$, $8^{1/3} = \sqrt[3]{8}$, $16^{1/4} = \sqrt[4]{16}$

Notice that the denominator of the rational exponent is the index of the radical.

NOW TRY EXERCISE 1

Evaluate each exponential.

(a) $81^{1/2}$ **(b)** $125^{1/3}$

(c) $-625^{1/4}$ **(d)** $(-625)^{1/4}$

(e) $(-125)^{1/3}$ **(f)** $\left(\frac{1}{16}\right)^{1/4}$

EXAMPLE 1 **Evaluating Exponentials of the Form $a^{1/n}$**

Evaluate each exponential.

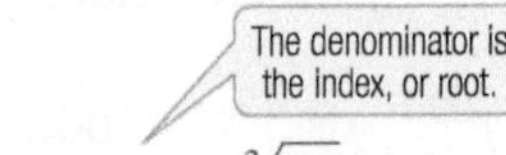
The denominator is
the index, or root.

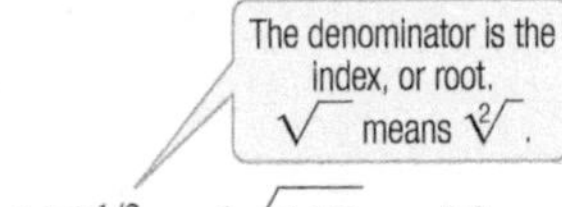
The denominator is the
index, or root.
√ means ²√ .

(a) $64^{1/3} = \sqrt[3]{64} = 4$ **(b)** $100^{1/2} = \sqrt{100} = 10$

(c) $-256^{1/4} = -\sqrt[4]{256} = -4$

(d) $(-256)^{1/4} = \sqrt[4]{-256}$ is not a real number because the radicand, -256, is negative and the index is even.

(e) $(-32)^{1/5} = \sqrt[5]{-32} = -2$ **(f)** $\left(\frac{1}{8}\right)^{1/3} = \sqrt[3]{\frac{1}{8}} = \frac{1}{2}$

NOW TRY 

CAUTION Notice the distinction between **Examples 1(c) and 1(d).** The radical in part (c) is the ***negative fourth root of a positive number,*** while the radical in part (d) is the ***principal fourth root of a negative number, which is not a real number.***

OBJECTIVE 2 Define and use expressions of the form $a^{m/n}$.

We know that $8^{1/3} = \sqrt[3]{8}$. Now we can define a number like $8^{2/3}$, where the numerator of the exponent is not 1. For past rules of exponents to be valid,

$$8^{2/3} = 8^{(1/3)2} = (8^{1/3})^2.$$

Because $8^{1/3} = \sqrt[3]{8}$,

$$8^{2/3} = \left(\sqrt[3]{8}\right)^2 = 2^2 = 4.$$

Generalizing from this example, we define $a^{m/n}$ as follows.

Meaning of $a^{m/n}$

If m and n are positive integers with m/n in lowest terms, then

$$a^{m/n} = \left(a^{1/n}\right)^m,$$

provided that $a^{1/n}$ is a real number. If $a^{1/n}$ is not a real number, then $a^{m/n}$ is not a real number.

EXAMPLE 2 **Evaluating Exponentials of the Form $a^{m/n}$**

Evaluate each exponential.

Think: $36^{1/2} = \sqrt{36} = 6$

Think: $125^{1/3} = \sqrt[3]{125} = 5$

(a) $36^{3/2} = (36^{1/2})^3 = 6^3 = 216$ **(b)** $125^{2/3} = (125^{1/3})^2 = 5^2 = 25$

NOW TRY ANSWERS

1. **(a)** 9 **(b)** 5 **(c)** -5 **(d)** It is not a real number. **(e)** -5 **(f)** $\frac{1}{2}$

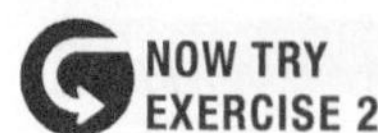

NOW TRY EXERCISE 2

Evaluate each exponential.

(a) $32^{2/5}$ **(b)** $8^{5/3}$

(c) $-100^{3/2}$ **(d)** $(-125)^{4/3}$

(e) $(-121)^{3/2}$

Be careful. The base is 4.

(c) $-4^{5/2} = -(4^{5/2}) = -(4^{1/2})^5 = -(2)^5 = -32$

Because the base is 4, the negative sign is *not* affected by the exponent.

(d) $(-27)^{2/3} = [(-27)^{1/3}]^2 = (-3)^2 = 9$

Notice that in part (c) we first evaluate the exponential and then find its negative. In part (d), the $-$ sign is part of the base, -27.

(e) $(-100)^{3/2} = [(-100)^{1/2}]^3$, which is not a real number, because

$(-100)^{1/2}$, or $\sqrt{-100}$, is not a real number. NOW TRY

Recall that for any natural number n,

$$a^{-n} = \frac{1}{a^n} \quad (\text{where } a \neq 0). \qquad \text{Definition of negative exponent}$$

When a rational exponent is negative, this earlier interpretation is applied.

Meaning of $a^{-m/n}$

If $a^{m/n}$ is a real number, then

$$a^{-m/n} = \frac{1}{a^{m/n}} \quad (\text{where } a \neq 0).$$

NOW TRY EXERCISE 3

Evaluate each exponential.

(a) $243^{-3/5}$ **(b)** $4^{-5/2}$

(c) $\left(\frac{216}{125}\right)^{-2/3}$

EXAMPLE 3 Evaluating Exponentials of the Form $a^{-m/n}$

Evaluate each exponential.

(a) $16^{-3/4} = \frac{1}{16^{3/4}} = \frac{1}{(16^{1/4})^3} = \frac{1}{(\sqrt[4]{16})^3} = \frac{1}{2^3} = \frac{1}{8}$

The denominator of 3/4 is the index and the numerator is the exponent.

(b) $25^{-3/2} = \frac{1}{25^{3/2}} = \frac{1}{(25^{1/2})^3} = \frac{1}{(\sqrt{25})^3} = \frac{1}{5^3} = \frac{1}{125}$

(c) $\left(\frac{8}{27}\right)^{-2/3} = \frac{1}{\left(\frac{8}{27}\right)^{2/3}} = \frac{1}{\left(\sqrt[3]{\frac{8}{27}}\right)^2} = \frac{1}{\left(\frac{2}{3}\right)^2} = \frac{1}{\frac{4}{9}} = \frac{9}{4}$

$\frac{1}{\frac{4}{9}} = 1 \div \frac{4}{9} = 1 \cdot \frac{9}{4}$

We can also use the rule $\left(\frac{b}{a}\right)^{-m} = \left(\frac{a}{b}\right)^m$ here, as follows.

$$\left(\frac{8}{27}\right)^{-2/3} = \left(\frac{27}{8}\right)^{2/3} = \left(\sqrt[3]{\frac{27}{8}}\right)^2 = \left(\frac{3}{2}\right)^2 = \frac{9}{4} \qquad \text{The result is the same.}$$

Take the reciprocal of only the base, ***not*** the exponent.

NOW TRY

NOW TRY ANSWERS

2. **(a)** 4 **(b)** 32 **(c)** -1000 **(d)** 625 **(e)** It is not a real number.

3. **(a)** $\frac{1}{27}$ **(b)** $\frac{1}{32}$ **(c)** $\frac{25}{36}$

CAUTION Be careful to distinguish between exponential expressions like the following.

$16^{-1/4}$, which equals $\frac{1}{2}$, $-16^{1/4}$, which equals -2, and $-16^{-1/4}$, which equals $-\frac{1}{2}$

A negative exponent does not necessarily lead to a negative result. Negative exponents lead to reciprocals, which may be positive.

We obtain an alternative definition of $a^{m/n}$ by applying the power rule for exponents differently than in the earlier definition.

Alternative Meaning of $a^{m/n}$

If all indicated roots are real numbers, then

$$a^{m/n} = (a^{1/n})^m = (a^m)^{1/n}.$$

As a result, we can evaluate an expression such as $27^{2/3}$ in two ways.

$$27^{2/3} = (27^{1/3})^2 = 3^2 = 9$$

or

$$27^{2/3} = (27^2)^{1/3} = 729^{1/3} = 9$$

The result is the same.

In most cases, it is easier to use $(a^{1/n})^m$.

Radical Form of $a^{m/n}$

If all indicated roots are real numbers, then

$$a^{m/n} = \sqrt[n]{a^m} = \left(\sqrt[n]{a}\right)^m.$$

That is, raise a to the mth power and then take the nth root, or take the nth root of a and then raise it to the mth power.

For example, $8^{2/3} = \sqrt[3]{8^2} = \sqrt[3]{64} = 4$, and $8^{2/3} = \left(\sqrt[3]{8}\right)^2 = 2^2 = 4$,

so

$$8^{2/3} = \sqrt[3]{8^2} = \left(\sqrt[3]{8}\right)^2.$$

NOW TRY EXERCISE 4

Write each exponential as a radical. Assume that all variables represent positive real numbers. Use the definition that takes the root first.

(a) $21^{1/2}$ **(b)** $17^{5/4}$

(c) $4t^{3/5} + (4t)^{2/3}$

(d) $w^{-2/5}$ **(e)** $(a^2 - b^2)^{1/4}$

NOW TRY ANSWERS

4. **(a)** $\sqrt{21}$ **(b)** $\left(\sqrt[4]{17}\right)^5$

(c) $4\left(\sqrt[5]{t}\right)^3 + \left(\sqrt[3]{4t}\right)^2$

(d) $\frac{1}{\left(\sqrt[5]{w}\right)^2}$ **(e)** $\sqrt[4]{a^2 - b^2}$

OBJECTIVE 3 Convert between radicals and rational exponents.

Using the definition of rational exponents, we can simplify many problems involving radicals by converting the radicals to numbers with rational exponents. After simplifying, we can convert the answer back to radical form if required.

EXAMPLE 4 Converting Exponentials to Radicals

Write each exponential as a radical. Assume that all variables represent positive real numbers. Use the definition that takes the root first.

(a) $13^{1/2} = \sqrt{13}$ **(b)** $6^{3/4} = \left(\sqrt[4]{6}\right)^3$ **(c)** $9m^{5/8} = 9\left(\sqrt[8]{m}\right)^5$

(d) $6x^{2/3} - (4x)^{3/5} = 6\left(\sqrt[3]{x}\right)^2 - \left(\sqrt[5]{4x}\right)^3$

(e) $r^{-2/3} = \frac{1}{r^{2/3}} = \frac{1}{\left(\sqrt[3]{r}\right)^2}$

(f) $(a^2 + b^2)^{1/2} = \sqrt{a^2 + b^2}$ $\sqrt{a^2 + b^2} \neq a + b$

NOW TRY EXERCISE 5

Write each radical as an exponential and simplify. Assume that all variables represent positive real numbers.

(a) $\sqrt[3]{15}$ **(b)** $\sqrt[4]{4^2}$

(c) $\sqrt[4]{x^4}$

EXAMPLE 5 Simplifying Radicals Using Rational Exponents

Write each radical as an exponential and simplify. Assume that all variables represent positive real numbers.

(a) $\sqrt{10} = 10^{1/2}$

(b) $\sqrt{5^4} = 5^{4/2} = 5^2 = 25$

(c) $\sqrt[4]{3^8} = 3^{8/4} = 3^2 = 9$

(d) $\sqrt[6]{z^6} = z^{6/6} = z^1 = z$, because z is positive. NOW TRY 

NOTE In **Example 5(d),** it is not necessary to use absolute value bars because the directions specifically state that the variable represents a positive real number. The absolute value of the positive real number z is z itself, so the answer is simply z.

OBJECTIVE 4 Use the rules for exponents with rational exponents.

The definition of rational exponents allows us to apply the rules for exponents.

Rules for Rational Exponents

Let r and s be rational numbers. For all real numbers a and b for which the following are defined, these rules hold true.

Rules	Examples	Rules	Examples
$a^r \cdot a^s = a^{r+s}$	$5^{2/3} \cdot 5^{5/3} = 5^{2/3+5/3} = 5^{7/3}$	$\left(\frac{a}{b}\right)^r = \frac{a^r}{b^r}$	$\left(\frac{5}{7}\right)^{2/3} = \frac{5^{2/3}}{7^{2/3}}$
$\frac{a^r}{a^s} = a^{r-s}$	$\frac{4^{5/3}}{4^{4/3}} = 4^{5/3-4/3} = 4^{1/3}$	$a^{-r} = \frac{1}{a^r}$	$4^{-3/7} = \frac{1}{4^{3/7}}$
$(a^r)^s = a^{rs}$	$(5^{1/2})^{2/5} = 5^{1/2 \cdot 2/5} = 5^{1/5}$	$\frac{a^{-r}}{b^{-s}} = \frac{b^s}{a^r}$	$\frac{5^{-1/2}}{3^{-3/4}} = \frac{3^{3/4}}{5^{1/2}}$
$(ab)^r = a^r b^r$	$(3x)^{1/2} = 3^{1/2}x^{1/2}$		

EXAMPLE 6 Applying Rules for Rational Exponents

Simplify each expression. Assume that all variables represent positive real numbers.

(a) $2^{1/2} \cdot 2^{1/4}$

$= 2^{1/2+1/4}$ Product rule

$= 2^{2/4+1/4}$ Write exponents with a common denominator.

$= 2^{3/4}$ Add exponents.

(b) $\frac{5^{2/3}}{5^{7/3}}$

$= 5^{2/3-7/3}$ Quotient rule

$= 5^{-5/3}$ Subtract exponents.

$= \frac{1}{5^{5/3}}$ $a^{-r} = \frac{1}{a^r}$

NOW TRY ANSWERS

5. **(a)** $15^{1/3}$ **(b)** 2 **(c)** x

NOW TRY EXERCISE 6

Simplify each expression. Assume that all variables represent positive real numbers.

(a) $5^{1/4} \cdot 5^{2/3}$ **(b)** $\frac{9^{3/5}}{9^{7/5}}$

(c) $\frac{(r^{2/3}t^{1/4})^8}{t}$

(d) $\left(\frac{2x^{1/2}y^{-2/3}}{x^{-3/5}y^{-1/5}}\right)^{-3}$

(e) $y^{2/3}(y^{1/3} + y^{5/3})$

(c) $\frac{(x^{1/2}y^{2/3})^4}{y}$

$= \frac{(x^{1/2})^4(y^{2/3})^4}{y}$ Power rule

$= \frac{x^2y^{8/3}}{y^1}$ Power rule; $y = y^1$

$= x^2y^{8/3-1}$ Quotient rule

$= x^2y^{5/3}$ $\frac{8}{3} - 1 = \frac{8}{3} - \frac{3}{3} = \frac{5}{3}$

(d) $\left(\frac{x^4y^{-6}}{x^{-2}y^{1/3}}\right)^{-2/3}$

$= \frac{(x^4)^{-2/3}(y^{-6})^{-2/3}}{(x^{-2})^{-2/3}(y^{1/3})^{-2/3}}$ Power rule

$= \frac{x^{-8/3}y^4}{x^{4/3}y^{-2/9}}$ Power rule

$= x^{-8/3-4/3}y^{4-(-2/9)}$ Quotient rule

$= x^{-4}y^{38/9}$ $4 - \left(-\frac{2}{9}\right) = \frac{36}{9} + \frac{2}{9} = \frac{38}{9}$

Use parentheses to avoid errors.

$= \frac{y^{38/9}}{x^4}$ Definition of negative exponent

Alternative method: We can simplify within the parentheses first, as follows.

$\left(\frac{x^4y^{-6}}{x^{-2}y^{1/3}}\right)^{-2/3}$

$= (x^{4-(-2)}y^{-6-1/3})^{-2/3}$ Quotient rule

$= (x^6y^{-19/3})^{-2/3}$ $-6 - \frac{1}{3} = -\frac{18}{3} - \frac{1}{3} = -\frac{19}{3}$

$= (x^6)^{-2/3}(y^{-19/3})^{-2/3}$ Power rule

$= x^{-4}y^{38/9}$ Power rule

$= \frac{y^{38/9}}{x^4}$ Definition of negative exponent; The result is the same.

(e) $m^{3/4}(m^{5/4} - m^{1/4})$

$= m^{3/4}(m^{5/4}) - m^{3/4}(m^{1/4})$ Distributive property

$= m^{3/4+5/4} - m^{3/4+1/4}$ Product rule

Do not make the common mistake of multiplying exponents in the first step.

$= m^{8/4} - m^{4/4}$ Add exponents.

$= m^2 - m$ Write the exponents in lowest terms.

NOW TRY 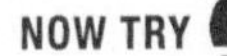

CAUTION Use the rules of exponents in problems like those in **Example 6.** Do not convert the expressions to radical form.

NOW TRY ANSWERS

6. (a) $5^{11/12}$ (b) $\frac{1}{9^{4/5}}$
(c) $r^{16/3}t$ (d) $\frac{y^{7/5}}{8x^{33/10}}$
(e) $y + y^{7/3}$

NOW TRY EXERCISE 7

Write each radical as an exponential and simplify. Leave answers in exponential form. Assume that all variables represent positive real numbers.

(a) $\sqrt[5]{y^3} \cdot \sqrt[3]{y}$ **(b)** $\dfrac{\sqrt[4]{y^3}}{\sqrt{y^5}}$

(c) $\sqrt{\sqrt[3]{y}}$

EXAMPLE 7 Applying Rules for Rational Exponents

Write each radical as an exponential and simplify. Leave answers in exponential form. Assume that all variables represent positive real numbers.

(a) $\sqrt[3]{x^2} \cdot \sqrt[4]{x}$

$= x^{2/3} \cdot x^{1/4}$ Convert to rational exponents.

$= x^{2/3+1/4}$ Product rule

$= x^{8/12+3/12}$ Write exponents with a common denominator.

$= x^{11/12}$ Add exponents.

(b) $\dfrac{\sqrt{x^3}}{\sqrt[3]{x^2}}$

$= \dfrac{x^{3/2}}{x^{2/3}}$ Convert to rational exponents.

$= x^{3/2-2/3}$ Quotient rule

$= x^{9/6-4/6}$ Write exponents with a common denominator.

$= x^{5/6}$ Subtract exponents.

(c) $\sqrt{\sqrt[4]{z}}$

$= \sqrt{z^{1/4}}$ Convert the inside radical to a rational exponent.

$= (z^{1/4})^{1/2}$ Convert the square root to a rational exponent.

$= z^{1/8}$ Power rule

NOW TRY

NOW TRY ANSWERS

7. (a) $y^{14/15}$ **(b)** $\dfrac{1}{y^{7/4}}$ **(c)** $y^{1/6}$

7.2 Exercises

FOR EXTRA HELP  **MyLab Math**

Video solutions for select problems available in MyLab Math

Concept Check *Match each expression in Column I with the equivalent choice in Column II.*

I		II	
1. $3^{1/2}$	**2.** $(-27)^{1/3}$	**A.** -4	**B.** 8
3. $-16^{1/2}$	**4.** $(-25)^{1/2}$	**C.** $\sqrt{3}$	**D.** $-\sqrt{6}$
5. $(-32)^{1/5}$	**6.** $(-32)^{2/5}$	**E.** -3	**F.** $\sqrt{6}$
7. $4^{3/2}$	**8.** $6^{2/4}$	**G.** 4	**H.** -2
9. $-6^{2/4}$	**10.** $36^{0.5}$	**I.** 6	**J.** Not a real number

11. Concept Check A student incorrectly evaluated $27^{1/3}$ as 9. **WHAT WENT WRONG?** Evaluate $27^{1/3}$ correctly.

12. Concept Check A student evaluating $16^{-3/2}$ incorrectly suggested that the result is a negative number because the exponent is negative. **WHAT WENT WRONG?** Evaluate $16^{-3/2}$ correctly.

Evaluate each exponential. **See Examples 1–3.**

13. $169^{1/2}$ **14.** $121^{1/2}$ **15.** $729^{1/3}$ **16.** $512^{1/3}$

17. $16^{1/4}$ **18.** $625^{1/4}$ **19.** $\left(\dfrac{64}{81}\right)^{1/2}$ **20.** $\left(\dfrac{8}{27}\right)^{1/3}$

21. $(-27)^{1/3}$ **22.** $(-32)^{1/5}$ **23.** $(-144)^{1/2}$ **24.** $(-36)^{1/2}$

25. $100^{3/2}$ **26.** $64^{3/2}$ **27.** $81^{3/4}$ **28.** $216^{2/3}$

29. $-16^{5/2}$ **30.** $-32^{3/5}$ **31.** $(-8)^{4/3}$ **32.** $(-243)^{2/5}$

33. $32^{-3/5}$ 34. $27^{-4/3}$ 35. $64^{-3/2}$ 36. $81^{-3/2}$

37. $\left(\frac{125}{27}\right)^{-2/3}$ 38. $\left(\frac{64}{125}\right)^{-2/3}$ 39. $\left(\frac{16}{81}\right)^{-3/4}$ 40. $\left(\frac{729}{64}\right)^{-5/6}$

Write each exponential as a radical. Assume that all variables represent positive real numbers. Use the definition that takes the root first. **See Example 4.**

41. $10^{1/2}$ 42. $3^{1/2}$ 43. $8^{3/4}$

44. $7^{2/3}$ 45. $5x^{2/3}$ 46. $8x^{3/4}$

47. $9q^{5/8} - (2x)^{2/3}$ 48. $(3p)^{3/4} + 4x^{1/3}$ 49. $x^{-3/5}$

50. $z^{-4/9}$ 51. $(2y + x)^{2/3}$ 52. $(r + 2z)^{3/2}$

Write each radical as an exponential and simplify. Assume that all variables represent positive real numbers. **See Example 5.**

53. $\sqrt{15}$ 54. $\sqrt{26}$ 55. $\sqrt{2^{12}}$ 56. $\sqrt{5^{10}}$ 57. $\sqrt[3]{4^9}$

58. $\sqrt[4]{6^8}$ 59. $\sqrt[8]{x^8}$ 60. $\sqrt{t^2}$ 61. $\sqrt{x^{20}}$ 62. $\sqrt{r^{50}}$

Simplify each expression. Assume that all variables represent positive real numbers. **See Example 6.**

63. $3^{1/2} \cdot 3^{3/2}$ 64. $6^{4/3} \cdot 6^{2/3}$ 65. $\frac{64^{5/3}}{64^{4/3}}$

66. $\frac{125^{7/3}}{125^{5/3}}$ 67. $y^{7/3} \cdot y^{-4/3}$ 68. $r^{-8/9} \cdot r^{17/9}$

69. $x^{2/3} \cdot x^{-1/4}$ 70. $x^{2/5} \cdot x^{-1/3}$ 71. $\frac{k^{1/3}}{k^{2/3} \cdot k^{-1}}$

72. $\frac{z^{3/4}}{z^{5/4} \cdot z^{-2}}$ 73. $\frac{(x^{1/4}y^{2/5})^{20}}{x^2}$ 74. $\frac{(r^{1/5}s^{2/3})^{15}}{r^2}$

75. $\frac{(x^{2/3})^2}{(x^2)^{7/3}}$ 76. $\frac{(p^3)^{1/4}}{(p^{5/4})^2}$ 77. $\frac{m^{3/4}n^{-1/4}}{(m^2n)^{1/2}}$

78. $\frac{a^{-1/2}b^{-5/4}}{(a^{-3}b^2)^{1/6}}$ 79. $\left(\frac{x^8y^{-8}}{x^{-4}y^{1/2}}\right)^{-2/3}$ 80. $\left(\frac{x^{2/3}y^{-6}}{x^{-1/12}y^{1/4}}\right)^{-4/5}$

81. $\left(\frac{b^{-3/2}}{c^{-5/3}}\right)^2(b^{-1/4}c^{-1/3})^{-1}$ 82. $\left(\frac{m^{-2/3}}{a^{-3/4}}\right)^4(m^{-3/8}a^{1/4})^{-2}$ 83. $\left(\frac{p^{-1/4}q^{-3/2}}{3^{-1}p^{-2}q^{-2/3}}\right)^{-2}$

84. $\left(\frac{2^{-2}w^{-3/4}x^{-5/8}}{w^{3/4}x^{-1/2}}\right)^{-3}$ 85. $p^{2/3}(p^{1/3} + 2p^{4/3})$ 86. $z^{5/8}(3z^{5/8} + 5z^{11/8})$

87. $k^{1/4}(k^{3/2} - k^{1/2})$ 88. $r^{3/5}(r^{1/2} + r^{3/4})$ 89. $6a^{7/4}(a^{-7/4} + 3a^{-3/4})$

90. $4m^{5/3}(m^{-2/3} - 4m^{-5/3})$ 91. $-5x^{7/6}(x^{5/6} - x^{-1/6})$ 92. $-8y^{11/7}(y^{3/7} - y^{-4/7})$

Write each radical as an exponential and simplify. Leave answers in exponential form. Assume that all variables represent positive numbers. **See Example 7.**

93. $\sqrt[5]{x^3} \cdot \sqrt[4]{x}$ 94. $\sqrt[6]{y^5} \cdot \sqrt[3]{y^2}$ 95. $\frac{\sqrt[3]{t^4}}{\sqrt[5]{t^4}}$ 96. $\frac{\sqrt[4]{w^3}}{\sqrt[6]{w}}$

97. $\frac{\sqrt{x^5}}{\sqrt{x^8}}$ 98. $\frac{\sqrt[3]{k^5}}{\sqrt[3]{k^7}}$ 99. $\sqrt{y} \cdot \sqrt[3]{yz}$ 100. $\sqrt[3]{xz} \cdot \sqrt{z}$

101. $\sqrt[4]{\sqrt[3]{m}}$ **102.** $\sqrt[3]{\sqrt{k}}$ **103.** $\sqrt{\sqrt[3]{\sqrt[4]{x}}}$ **104.** $\sqrt[3]{\sqrt[5]{\sqrt{y}}}$

Concept Check *Work each problem.*

105. Replace a with 3 and b with 4 to show that, in general,

$$\sqrt{a^2 + b^2} \neq a + b.$$

106. Suppose someone claims that $\sqrt[n]{a^n + b^n}$ must equal $a + b$, because when $a = 1$ and $b = 0$, a true statement results:

$$\sqrt[n]{a^n + b^n} = \sqrt[n]{1^n + 0^n} = \sqrt[n]{1^n} = 1 = 1 + 0 = a + b.$$

Explain why this is faulty reasoning.

Solve each problem.

107. Meteorologists can determine the duration of a storm using the function

$$T(d) = 0.07d^{3/2},$$

where d is the diameter of the storm in miles and T is the time in hours. Find the duration of a storm with a diameter of 16 mi. Round the answer to the nearest tenth of an hour.

108. The threshold weight t, in pounds, for a person is the weight above which the risk of death increases greatly. The threshold weight in pounds for men aged 40–49 is related to height H in inches by the function

$$H(t) = (1860.867t)^{1/3}.$$

What height corresponds to a threshold weight of 200 lb for a 46-yr-old man? Round the answer to the nearest inch and then to the nearest tenth of a foot.

The ***windchill factor*** *is a measure of the cooling effect that the wind has on a person's skin. It calculates the equivalent cooling temperature if there were no wind. The National Weather Service uses the formula*

$$\text{Windchill temperature} = 35.74 + 0.6215T - 35.75V^{4/25} + 0.4275TV^{4/25},$$

where T is the temperature in °F and V is the wind speed in miles per hour, to calculate windchill. The table gives the windchill factor for various wind speeds and temperatures at which frostbite is a risk, and how quickly it may occur.

Temperature (°F)

Wind speed (mph) / Calm	40	30	20	10	0	−10	−20	−30	−40
5	36	25	13	1	−11	−22	−34	−46	−57
10	34	21	9	−4	−16	−28	−41	−53	−66
15	32	19	6	−7	−19	−32	−45	−58	−71
20	30	17	4	−9	−22	−35	−48	−61	−74
25	29	16	3	−11	−24	−37	−51	−64	−78
30	28	15	1	−12	−26	−39	−53	−67	−80
35	28	14	0	−14	−27	−41	−55	−69	−82
40	27	13	−1	−15	−29	−43	−57	−71	−84

Frostbites times: 30 minutes 10 minutes 5 minutes

Data from National Oceanic and Atmospheric Administration, National Weather Service.

Use the formula to determine the windchill temperature to the nearest tenth of a degree, given the following conditions. Compare answers with the appropriate entries in the table.

109. 30°F, 15-mph wind

110. 10°F, 30-mph wind

111. 20°F, 20-mph wind

112. 40°F, 10-mph wind

7.3 Simplifying Radicals, the Distance Formula, and Circles

OBJECTIVES

1. Use the product rule for radicals.
2. Use the quotient rule for radicals.
3. Simplify radicals.
4. Simplify products and quotients of radicals with different indexes.
5. Use the Pythagorean theorem.
6. Use the distance formula.
7. Find an equation of a circle given its center and radius.

OBJECTIVE 1 Use the product rule for radicals.

Consider the expressions $\sqrt{36 \cdot 4}$ and $\sqrt{36} \cdot \sqrt{4}$.

$$\sqrt{36 \cdot 4} = \sqrt{144} = 12$$
$$\sqrt{36} \cdot \sqrt{4} = 6 \cdot 2 = 12$$

The result is the same.

This is an example of the **product rule for radicals.**

Product Rule for Radicals

If n is a natural number and $\sqrt[n]{a}$ and $\sqrt[n]{b}$ are real numbers, then the following holds true.

$$\sqrt[n]{a} \cdot \sqrt[n]{b} = \sqrt[n]{ab}$$

That is, the product of two nth roots is the nth root of the product.

We justify the product rule using the rules for rational exponents. Because $\sqrt[n]{a} = a^{1/n}$ and $\sqrt[n]{b} = b^{1/n}$,

$$\sqrt[n]{a} \cdot \sqrt[n]{b} = a^{1/n} \cdot b^{1/n} = (ab)^{1/n} = \sqrt[n]{ab}.$$

CAUTION *Use the product rule only when the radicals have the same index.*

NOW TRY EXERCISE 1

Multiply. Assume that all variables represent positive real numbers.

(a) $\sqrt{7} \cdot \sqrt{11}$

(b) $\sqrt{2mn} \cdot \sqrt{15}$

EXAMPLE 1 Using the Product Rule

Multiply. Assume that all variables represent positive real numbers.

(a) $\sqrt{5} \cdot \sqrt{7}$
$= \sqrt{5 \cdot 7}$
$= \sqrt{35}$

(b) $\sqrt{11} \cdot \sqrt{p}$
$= \sqrt{11p}$

(c) $\sqrt{7} \cdot \sqrt{11xyz}$
$= \sqrt{77xyz}$

NOW TRY

NOW TRY EXERCISE 2

Multiply. Assume that all variables represent positive real numbers.

(a) $\sqrt[3]{4} \cdot \sqrt[3]{5}$ **(b)** $\sqrt[4]{5t} \cdot \sqrt[4]{6r^3}$

(c) $\sqrt[7]{20x} \cdot \sqrt[7]{3xy^3}$

(d) $\sqrt[3]{5} \cdot \sqrt[4]{9}$

EXAMPLE 2 Using the Product Rule

Multiply. Assume that all variables represent positive real numbers.

(a) $\sqrt[3]{3} \cdot \sqrt[3]{12}$
$= \sqrt[3]{3 \cdot 12}$
$= \sqrt[3]{36}$

(b) $\sqrt[4]{8y} \cdot \sqrt[4]{3r^2}$
$= \sqrt[4]{24yr^2}$

(c) $\sqrt[6]{10m^4} \cdot \sqrt[6]{5m}$
$= \sqrt[6]{50m^5}$

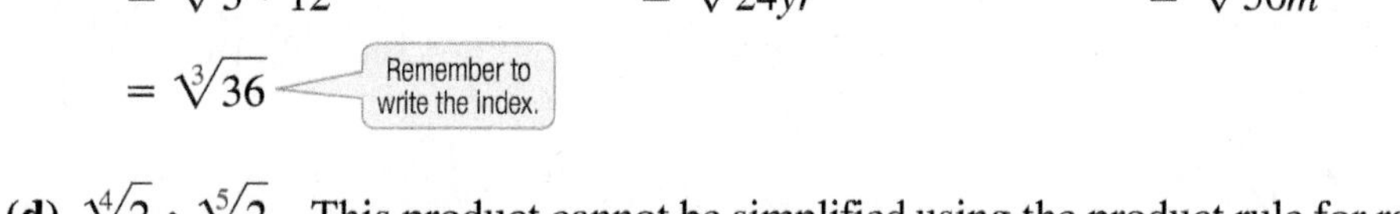
Remember to write the index.

(d) $\sqrt[4]{2} \cdot \sqrt[5]{2}$ This product cannot be simplified using the product rule for radicals because the indexes (4 and 5) are different.

NOW TRY

NOW TRY ANSWERS

1. (a) $\sqrt{77}$ (b) $\sqrt{30mn}$
2. (a) $\sqrt[3]{20}$ (b) $\sqrt[4]{30tr^3}$
   (c) $\sqrt[7]{60x^2y^3}$
   (d) This expression cannot be simplified using the product rule.

VOCABULARY

- ☐ hypotenuse
- ☐ legs (of a right triangle)
- ☐ circle
- ☐ center
- ☐ radius

OBJECTIVE 2 Use the quotient rule for radicals.

The **quotient rule for radicals** is similar to the product rule.

Quotient Rule for Radicals

If n is a natural number and $\sqrt[n]{a}$ and $\sqrt[n]{b}$ are real numbers, then the following holds true.

$$\sqrt[n]{\frac{a}{b}} = \frac{\sqrt[n]{a}}{\sqrt[n]{b}} \quad \textbf{(where } b \neq 0\textbf{)}$$

That is, the nth root of a quotient is the quotient of the nth roots.

NOW TRY EXERCISE 3

Simplify. Assume that all variables represent positive real numbers.

(a) $\sqrt{\frac{49}{36}}$ (b) $\sqrt{\frac{5}{144}}$

(c) $\sqrt[3]{-\frac{27}{1000}}$ (d) $\sqrt[4]{\frac{t}{16}}$

(e) $-\sqrt[5]{\frac{m^{15}}{243}}$

EXAMPLE 3 Using the Quotient Rule

Simplify. Assume that all variables represent positive real numbers.

(a) $\sqrt{\frac{16}{25}} = \frac{\sqrt{16}}{\sqrt{25}} = \frac{4}{5}$

(b) $\sqrt{\frac{7}{36}} = \frac{\sqrt{7}}{\sqrt{36}} = \frac{\sqrt{7}}{6}$

(c) $\sqrt[3]{-\frac{8}{125}} = \sqrt[3]{\frac{-8}{125}} = \frac{\sqrt[3]{-8}}{\sqrt[3]{125}} = \frac{-2}{5} = -\frac{2}{5}$ $\quad \frac{-a}{b} = -\frac{a}{b}$

(d) $\sqrt[3]{\frac{7}{216}} = \frac{\sqrt[3]{7}}{\sqrt[3]{216}} = \frac{\sqrt[3]{7}}{6}$

(e) $\sqrt[5]{\frac{x}{32}} = \frac{\sqrt[5]{x}}{\sqrt[5]{32}} = \frac{\sqrt[5]{x}}{2}$

(f) $-\sqrt[3]{\frac{m^6}{125}} = -\frac{\sqrt[3]{m^6}}{\sqrt[3]{125}} = -\frac{m^2}{5}$ Think: $\sqrt[3]{m^6} = m^{6/3} = m^2$

NOW TRY 

OBJECTIVE 3 Simplify radicals.

We use the product and quotient rules to simplify radicals. A radical is **simplified** if the following four conditions are met.

Conditions for a Simplified Radical

1. The radicand has no factor raised to a power greater than or equal to the index.
2. The radicand has no fractions.
3. No denominator contains a radical.
4. Exponents in the radicand and the index of the radical have greatest common factor 1.

Examples: $\sqrt{22}$, $\sqrt{15xy}$, $\sqrt[3]{18}$, $\frac{\sqrt[4]{m^3}}{m}$ These radicals are simplified.

$\sqrt{28}$, $\sqrt[3]{\frac{3}{5}}$, $\frac{7}{\sqrt{7}}$, $\sqrt[3]{r^{12}}$ These radicals are not simplified. Each violates one of the above conditions.

NOW TRY ANSWERS

3. (a) $\frac{7}{6}$ (b) $\frac{\sqrt{5}}{12}$ (c) $-\frac{3}{10}$ (d) $\frac{\sqrt[4]{t}}{2}$ (e) $-\frac{m^3}{3}$

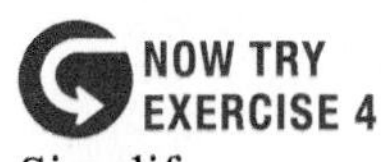

Simplify.

(a) $\sqrt{50}$ **(b)** $\sqrt{192}$

(c) $\sqrt{42}$ **(d)** $\sqrt[3]{108}$

(e) $-\sqrt[4]{80}$

EXAMPLE 4 Simplifying Roots of Numbers

Simplify.

(a) $\sqrt{24}$

Check to see whether 24 is divisible by a perfect square (the square of a natural number) such as 4, 9, 16, ... The greatest perfect square that divides into 24 is 4.

$$\sqrt{24}$$
$$= \sqrt{4 \cdot 6} \quad \text{Factor. 4 is a perfect square.}$$
$$= \sqrt{4} \cdot \sqrt{6} \quad \text{Product rule}$$
$$= 2\sqrt{6} \quad \sqrt{4} = 2$$

(b) $\sqrt{108}$

As shown on the left, the number 108 is divisible by the perfect square 36. If this perfect square is not immediately clear, try factoring 108 into its prime factors, as shown on the right.

$$\sqrt{108}$$
$$= \sqrt{36 \cdot 3} \quad \text{Factor.}$$
$$= \sqrt{36} \cdot \sqrt{3} \quad \text{Product rule}$$
$$= 6\sqrt{3} \quad \sqrt{36} = 6$$

$$\sqrt{108}$$
$$= \sqrt{2^2 \cdot 3^3} \quad \text{Factor into prime factors.}$$
$$= \sqrt{2^2 \cdot 3^2 \cdot 3} \quad a^3 = a^2 \cdot a$$
$$= \sqrt{2^2} \cdot \sqrt{3^2} \cdot \sqrt{3} \quad \text{Product rule}$$
$$= 2 \cdot 3 \cdot \sqrt{3} \quad \sqrt{2^2} = 2,\ \sqrt{3^2} = 3$$
$$= 6\sqrt{3} \quad \text{Multiply.}$$

(c) $\sqrt{10}$ No perfect square (other than 1) divides into 10, so $\sqrt{10}$ cannot be simplified further.

(d) $\sqrt[3]{16}$

$$\sqrt[3]{16} \quad \text{Look for the greatest perfect } \textit{cube} \text{ that divides into 16.}$$
$$= \sqrt[3]{8 \cdot 2} \quad \text{Factor. 8 is a perfect cube.}$$
$$= \sqrt[3]{8} \cdot \sqrt[3]{2} \quad \text{Product rule}$$
$$= 2\sqrt[3]{2} \quad \sqrt[3]{8} = 2$$

Remember to write the index.

(e)

$$-\sqrt[4]{162}$$
$$= -\sqrt[4]{81 \cdot 2} \quad \text{81 is a perfect fourth power.}$$
$$= -\sqrt[4]{81} \cdot \sqrt[4]{2} \quad \text{Product rule}$$
$$= -3\sqrt[4]{2} \quad \sqrt[4]{81} = 3$$

Remember the negative sign in each line.

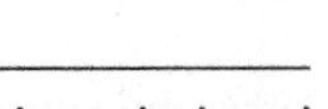

CAUTION ***Be careful with which factors belong outside the radical symbol and which belong inside.*** In **Example 4(b),** the $2 \cdot 3$ is written ***outside*** because $\sqrt{2^2} = 2$ and $\sqrt{3^2} = 3$. The remaining 3 is left inside the radical.

NOW TRY ANSWERS

4. **(a)** $5\sqrt{2}$ **(b)** $8\sqrt{3}$

(c) $\sqrt{42}$ cannot be simplified further.

(d) $3\sqrt[3]{4}$ **(e)** $-2\sqrt[4]{5}$

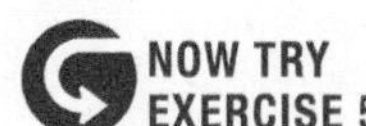

NOW TRY EXERCISE 5

Simplify. Assume that all variables represent positive real numbers.

(a) $\sqrt{36x^5}$ (b) $\sqrt{32m^5n^4}$

(c) $\sqrt[3]{-125k^3p^7}$

(d) $-\sqrt[4]{162x^7y^8}$

EXAMPLE 5 Simplifying Radicals Involving Variables

Simplify. Assume that all variables represent positive real numbers.

(a) $\sqrt{16m^3}$

$= \sqrt{16m^2 \cdot m}$ Factor.

$= \sqrt{16m^2} \cdot \sqrt{m}$ Product rule

$= 4m\sqrt{m}$ Take the square root.

Absolute value bars are not needed around the m in color because all the variables represent *positive* real numbers.

(b) $\sqrt{200k^7q^8}$

$= \sqrt{10^2 \cdot 2 \cdot (k^3)^2 \cdot k \cdot (q^4)^2}$ Factor into perfect squares.

$= 10k^3q^4\sqrt{2k}$ Take the square root.

(c) $\sqrt[3]{-8x^4y^5}$

$= \sqrt[3]{(-8x^3y^3)(xy^2)}$ Choose $-8x^3y^3$ as the perfect cube that divides into $-8x^4y^5$.

$= \sqrt[3]{-8x^3y^3} \cdot \sqrt[3]{xy^2}$ Product rule

$= -2xy\sqrt[3]{xy^2}$ Take the cube root.

(d) $-\sqrt[4]{32y^9}$

$= -\sqrt[4]{(16y^8)(2y)}$ $16y^8$ is the greatest fourth power that divides into $32y^9$.

$= -\sqrt[4]{16y^8} \cdot \sqrt[4]{2y}$ Product rule

$= -2y^2\sqrt[4]{2y}$ Take the fourth root. NOW TRY 

NOTE From **Example 5,** we see that if a variable is raised to a power with an exponent divisible by 2, it is a perfect square. If it is raised to a power with an exponent divisible by 3, it is a perfect cube. ***In general, if it is raised to a power with an exponent divisible by n, it is a perfect nth power.***

NOW TRY EXERCISE 6

Simplify. Assume that all variables represent positive real numbers.

(a) $\sqrt[6]{7^2}$ (b) $\sqrt[6]{y^4}$

(c) $\sqrt[4]{x^{10}}$

EXAMPLE 6 Simplifying Radicals Using Lesser Indexes

Simplify. Assume that all variables represent positive real numbers.

(a) $\sqrt[9]{5^6}$ Exponents in the radicand and the index must have GCF 1.

We write this radical using rational exponents and then write the exponent in lowest terms. We then express the answer as a radical.

$$\sqrt[9]{5^6} = (5^6)^{1/9} = 5^{6/9} = 5^{2/3} = \sqrt[3]{5^2} = \sqrt[3]{25}$$

(b) $\sqrt[4]{p^2} = (p^2)^{1/4} = p^{2/4} = p^{1/2} = \sqrt{p}$ (Recall the assumption that $p > 0$.)

(c) $\sqrt[4]{x^{18}} = (x^{18})^{1/4} = x^{18/4} = x^{9/2} = \sqrt{x^9} = \sqrt{x^8 \cdot x} = \sqrt{x^8} \cdot \sqrt{x} = x^4\sqrt{x}$

NOW TRY

NOW TRY ANSWERS

5. (a) $6x^2\sqrt{x}$ (b) $4m^2n^2\sqrt{2m}$ (c) $-5kp^2\sqrt[3]{p}$ (d) $-3xy^2\sqrt[4]{2x^3}$

6. (a) $\sqrt[3]{7}$ (b) $\sqrt[3]{y^2}$ (c) $x^2\sqrt{x}$

These examples suggest the following rule.

Meaning of $\sqrt[kn]{a^{km}}$

If m is an integer, n and k are natural numbers, and all indicated roots exist, then the following holds true.

$$\sqrt[kn]{a^{km}} = \sqrt[n]{a^m}$$

OBJECTIVE 4 Simplify products and quotients of radicals with different indexes.

We multiply and divide radicals with different indexes using rational exponents.

NOW TRY EXERCISE 7

Simplify $\sqrt[3]{3} \cdot \sqrt{6}$.

EXAMPLE 7 Multiplying Radicals with Different Indexes

Simplify $\sqrt{7} \cdot \sqrt[3]{2}$.

The indexes, 2 and 3, have a least common multiple of 6, so we use rational exponents to write each radical as a sixth root.

$$\sqrt{7} = 7^{1/2} = 7^{3/6} = \sqrt[6]{7^3} = \sqrt[6]{343}$$

$$\sqrt[3]{2} = 2^{1/3} = 2^{2/6} = \sqrt[6]{2^2} = \sqrt[6]{4}$$

Now we can multiply.

$$\sqrt{7} \cdot \sqrt[3]{2}$$

$$= \sqrt[6]{343} \cdot \sqrt[6]{4} \quad \text{Substitute; } \sqrt{7} = \sqrt[6]{343},\ \sqrt[3]{2} = \sqrt[6]{4}$$

$$= \sqrt[6]{1372} \quad \text{Product rule}$$

NOW TRY  

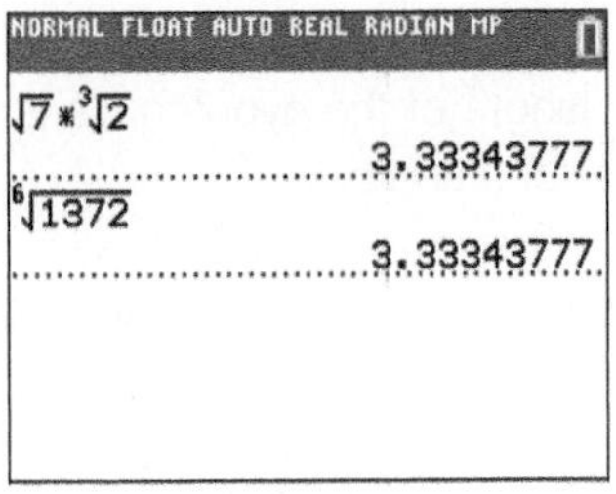

NORMAL FLOAT AUTO REAL RADIAN MP
$\sqrt{7}*\sqrt[3]{2}$
3.33343777
$\sqrt[6]{1372}$
3.33343777

FIGURE 6

Results such as the one in **Example 7** can be supported with a calculator, as shown in **FIGURE 6**. Notice that the calculator gives the same *approximation* for the initial product and the final radical that we obtained.

CAUTION The computation in **FIGURE 6** is not *proof* that the two expressions are equal. The algebra in **Example 7,** however, is valid proof of their equality.

OBJECTIVE 5 Use the Pythagorean theorem.

The **Pythagorean theorem** provides an equation that relates the lengths of the three sides of a right triangle.

Pythagorean Theorem

If a and b are the lengths of the shorter sides of a right triangle and c is the length of the longest side, then the following holds true.

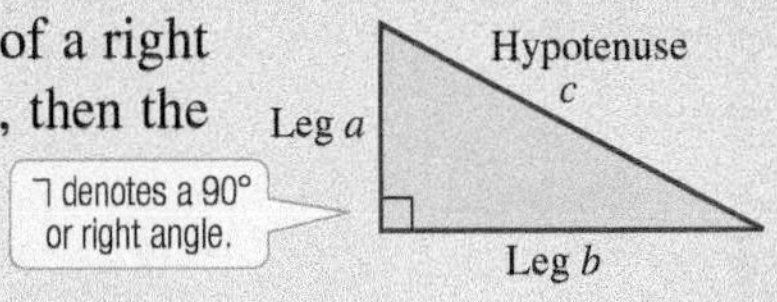

$$a^2 + b^2 = c^2$$

The two shorter sides are the **legs** of the triangle, and the longest side is the **hypotenuse.** The hypotenuse is the side opposite the right angle.

$$\textbf{leg}^2 + \textbf{leg}^2 = \textbf{hypotenuse}^2$$

NOW TRY ANSWER

7. $\sqrt[6]{1944}$

Later we will see that an equation such as $x^2 = 7$ has two solutions: $\sqrt{7}$ (the principal, or positive, square root of 7) and $-\sqrt{7}$. Similarly, $c^2 = 15$ has two solutions: $\pm\sqrt{15}$. In applications we often choose only the principal (positive) square root.

NOW TRY EXERCISE 8

Find the length of the unknown side in each right triangle.

(a)

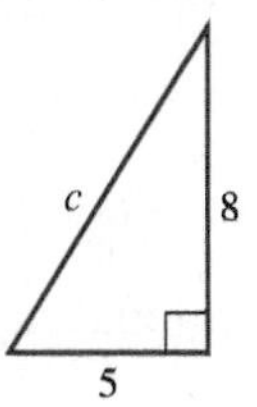

(b)

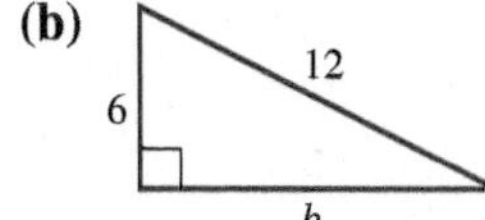

EXAMPLE 8 Using the Pythagorean Theorem

Find the length of the unknown side in the right triangle in **FIGURE 7**.

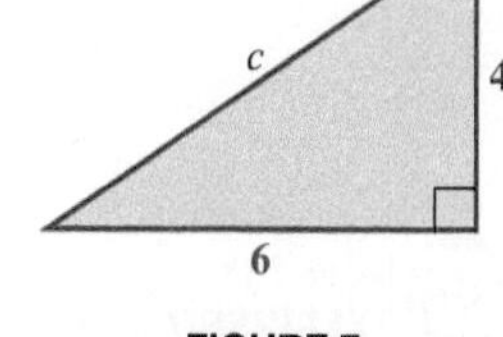

FIGURE 7

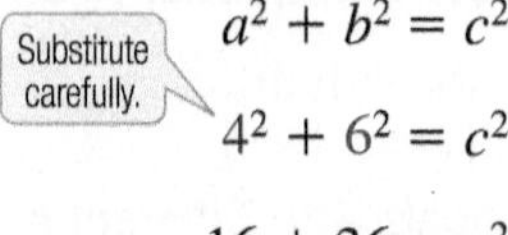

$$a^2 + b^2 = c^2 \quad \text{Pythagorean theorem}$$
$$4^2 + 6^2 = c^2 \quad \text{Let } a = 4 \text{ and } b = 6.$$
$$16 + 36 = c^2 \quad \text{Apply the exponents.}$$
$$c^2 = 52 \quad \text{Add. Interchange sides.}$$
$$c = \sqrt{52} \quad \text{Choose the principal root.}$$
$$c = \sqrt{4 \cdot 13} \quad \text{Factor.}$$
$$c = \sqrt{4} \cdot \sqrt{13} \quad \text{Product rule}$$
$$c = 2\sqrt{13} \quad \text{Simplify.}$$

The length of the hypotenuse is $2\sqrt{13}$.

NOW TRY 

CAUTION In the equation $a^2 + b^2 = c^2$, be sure that the length of the hypotenuse is substituted for c and that the lengths of the legs are substituted for a and b.

OBJECTIVE 6 Use the distance formula.

The *distance formula* enables us to find the distance between two points in the coordinate plane, or the length of the line segment joining those two points.

FIGURE 8 shows the points $(3, -4)$ and $(-5, 3)$. The vertical line through $(-5, 3)$ and the horizontal line through $(3, -4)$ intersect at the point $(-5, -4)$. Thus, the point $(-5, -4)$ becomes the vertex of the right angle in a right triangle.

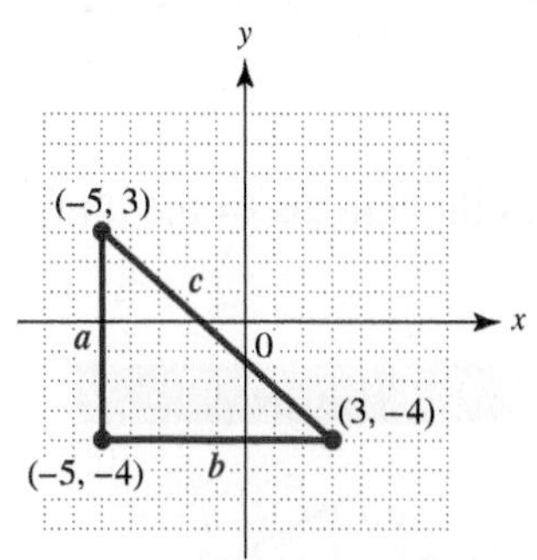

FIGURE 8

By the Pythagorean theorem, the sum of the squares of the lengths of the two legs a and b of the right triangle in **FIGURE 8** is equal to the square of the length of the hypotenuse, c.

$$a^2 + b^2 = c^2, \quad \text{or} \quad c^2 = a^2 + b^2.$$

The length a is the difference of the y-coordinates of the endpoints. The x-coordinate of both points in **FIGURE 8** is -5, so the side is vertical, and we can find a by finding the difference of the y-coordinates. We subtract -4 from 3 to obtain a positive value for a.

$$a = 3 - (-4) = 7$$

Similarly, we find b by subtracting -5 from 3.

$$b = 3 - (-5) = 8$$

NOW TRY ANSWERS

8. (a) $\sqrt{89}$ (b) $6\sqrt{3}$

Now substitute these values into the equation.

$$c^2 = a^2 + b^2 \quad \text{Pythagorean theorem}$$

$$c^2 = 7^2 + 8^2 \quad \text{Let } a = 7 \text{ and } b = 8.$$

$$c^2 = 49 + 64 \quad \text{Apply the exponents.}$$

$$c^2 = 113 \quad \text{Add.}$$

$$c = \sqrt{113} \quad \text{Choose the principal root.}$$

We choose the principal root because the distance cannot be negative. Therefore, the distance between $(-5, 3)$ and $(3, -4)$ is $\sqrt{113}$.

NOTE It is customary to leave the distance in simplified radical form. Do not use a calculator to find an approximation unless specifically directed to do so.

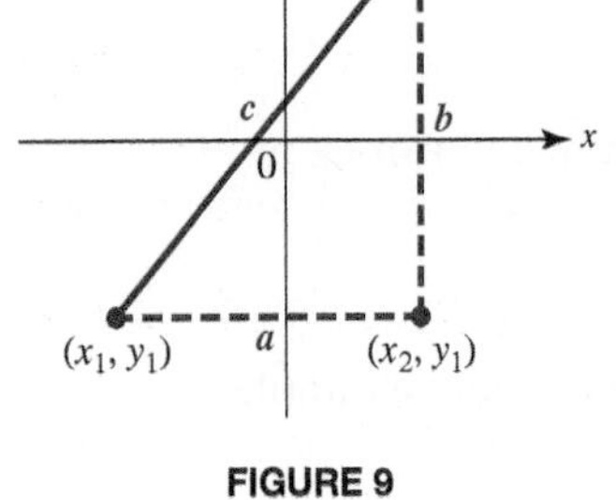

FIGURE 9

This work can be generalized. **FIGURE 9** shows the two points (x_1, y_1) and (x_2, y_2). The distance a between (x_1, y_1) and (x_2, y_1) is given by

$$a = |x_2 - x_1|,$$

and the distance b between (x_2, y_2) and (x_2, y_1) is given by

$$b = |y_2 - y_1|.$$

From the Pythagorean theorem, we obtain the following.

$$c^2 = a^2 + b^2$$

$$c^2 = (x_2 - x_1)^2 + (y_2 - y_1)^2$$

For all real numbers a, $|a|^2 = a^2$.

Choosing the principal square root gives the **distance formula.** In this formula, we use d (to denote distance) rather than c.

Distance Formula

The distance d between the points (x_1, y_1) and (x_2, y_2) is given by the following.

$$d = \sqrt{(x_2 - x_1)^2 + (y_2 - y_1)^2}$$

NOW TRY EXERCISE 9
Find the distance between the points $(-4, -3)$ and $(-8, 6)$.

EXAMPLE 9 Using the Distance Formula

Find the distance between the points $(-3, 5)$ and $(6, 4)$.

We arbitrarily choose to let $(x_1, y_1) = (-3, 5)$ and $(x_2, y_2) = (6, 4)$.

$$d = \sqrt{(x_2 - x_1)^2 + (y_2 - y_1)^2} \quad \text{Distance formula}$$

$$d = \sqrt{[6 - (-3)]^2 + (4 - 5)^2} \quad \text{Let } x_2 = 6, y_2 = 4, x_1 = -3, y_1 = 5.$$

Substitute carefully.

$$d = \sqrt{9^2 + (-1)^2}$$

$$d = \sqrt{82}$$

Leave in radical form. The distance is $\sqrt{82}$.

NOW TRY

NOW TRY ANSWER
9. $\sqrt{97}$

OBJECTIVE 7 Find an equation of a circle given its center and radius.

A **circle** is the set of all points in a plane that lie a fixed distance from a fixed point. The fixed point is the **center,** and the fixed distance is the **radius.** We use the distance formula to find an equation of a circle.

NOW TRY EXERCISE 10
Find an equation of the circle with center $(0, 0)$ and radius 6, and graph it.

EXAMPLE 10 Finding an Equation of a Circle and Graphing It

Find an equation of the circle with center $(0, 0)$ and radius 3, and graph it.

If the point (x, y) is on the circle, then the distance from (x, y) to the center $(0, 0)$ is the radius 3.

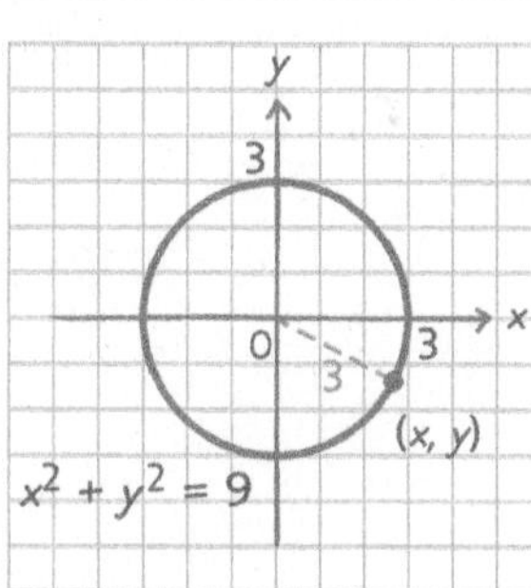

FIGURE 10

$$\sqrt{(x_2 - x_1)^2 + (y_2 - y_1)^2} = d \quad \text{Distance formula}$$

$$\sqrt{(x - 0)^2 + (y - 0)^2} = 3 \quad \text{Let } x_1 = 0, y_1 = 0, \text{ and } d = 3.$$

$$\sqrt{x^2 + y^2} = 3 \quad \text{Simplify.}$$

$$\left(\sqrt{x^2 + y^2}\right)^2 = 3^2 \quad \text{Square each side.}$$

$$x^2 + y^2 = 9 \quad \left(\sqrt{a}\right)^2 = a;\ 3^2 = 9$$

An equation of this circle is $x^2 + y^2 = 9$. The graph is shown in **FIGURE 10.**

NOW TRY 

A circle may not be centered at the origin, as seen in the next example.

NOW TRY EXERCISE 11
Find an equation of the circle with center $(-2, 2)$ and radius 3, and graph it.

EXAMPLE 11 Finding an Equation of a Circle and Graphing It

Find an equation of the circle with center $(4, -3)$ and radius 5, and graph it.

$$\sqrt{(x_2 - x_1)^2 + (y_2 - y_1)^2} = d \quad \text{Distance formula}$$

$$\sqrt{(x - 4)^2 + [y - (-3)]^2} = 5 \quad \text{Let } x_1 = 4, y_1 = -3, \text{ and } d = 5.$$

$$(x - 4)^2 + (y + 3)^2 = 25 \quad \text{Square each side.}$$

To graph the circle

$$(x - 4)^2 + (y + 3)^2 = 25,$$

plot the center $(4, -3)$, and then, because the radius is 5, move right, left, up, and down 5 units from the center, plotting the points

$$(9, -3), \quad (-1, -3), \quad (4, 2), \quad \text{and} \quad (4, -8).$$

Draw a smooth curve through these four points. When graphing by hand, it is helpful to sketch one quarter of the circle at a time. See **FIGURE 11.** (Note that the center is not part of the actual graph, but provides help in drawing a more accurate graph.)

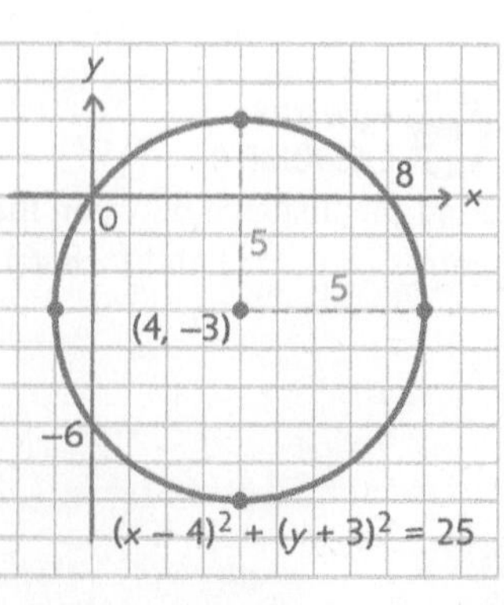

FIGURE 11

NOW TRY

NOW TRY ANSWERS

10. $x^2 + y^2 = 36$

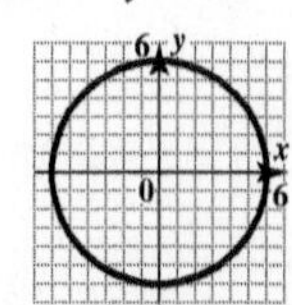

11. $(x + 2)^2 + (y - 2)^2 = 9$

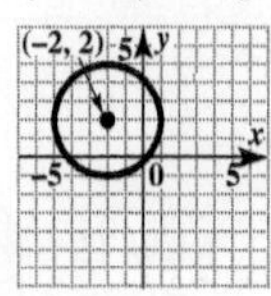

Examples 10 and 11 suggest the form of an equation of a circle with radius r and center (h, k). If (x, y) is a point on the circle, then the distance from the center (h, k) to the point (x, y) is r. See **FIGURE 12**. By the distance formula,

$$\sqrt{(x-h)^2+(y-k)^2}=r.$$

Squaring each side gives the **center-radius form** of the equation of a circle.

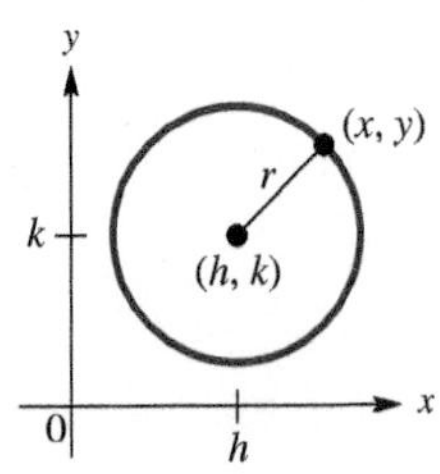

FIGURE 12

Equation of a Circle (Center-Radius Form)

A circle with center (h, k) and radius $r > 0$ has an equation of the form

$$(x-h)^2+(y-k)^2=r^2.$$

As a special case, a circle with center at the origin $(0, 0)$ and radius $r > 0$ has the following equation.

$$x^2+y^2=r^2$$

NOW TRY EXERCISE 12
Find an equation of the circle with center $(-5, 4)$ and radius $\sqrt{6}$.

NOW TRY ANSWER
12. $(x+5)^2+(y-4)^2=6$

EXAMPLE 12 **Using the Center-Radius Form of the Equation of a Circle**

Find an equation of the circle with center $(-1, 2)$ and radius $\sqrt{7}$.

$(x-h)^2+(y-k)^2=r^2$ Center-radius form

$[x-(-1)]^2+(y-2)^2=(\sqrt{7})^2$ Let $h=-1$, $k=2$, and $r=\sqrt{7}$.

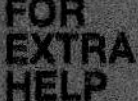

Pay attention to signs here.

$(x+1)^2+(y-2)^2=7$ Simplify; $(\sqrt{a})^2=a$ **NOW TRY**

7.3 Exercises

FOR EXTRA HELP **MyLab Math**

Video solutions for select problems available in MyLab Math

Concept Check *Choose the correct response.*

1. Which is the greatest perfect square factor of 128?
   A. 12 B. 16 C. 32 D. 64

2. Which is the greatest perfect cube factor of $81a^7$?
   A. $8a^3$ B. $27a^3$ C. $81a^6$ D. $27a^6$

3. Which radical can be simplified?
   A. $\sqrt{21}$ B. $\sqrt{48}$ C. $\sqrt[3]{12}$ D. $\sqrt[4]{10}$

4. Which radical *cannot* be simplified?
   A. $\sqrt[3]{30}$ B. $\sqrt[3]{27a^2b}$ C. $\sqrt{\frac{25}{81}}$ D. $\frac{2}{\sqrt{7}}$

5. Which one of the following is *not* equal to $\sqrt{\frac{1}{2}}$? (Do not use calculator approximations.)
   A. $\sqrt{0.5}$ B. $\sqrt{\frac{2}{4}}$ C. $\sqrt{\frac{3}{6}}$ D. $\frac{\sqrt{4}}{\sqrt{16}}$

6. Which one of the following is *not* equal to $\sqrt[3]{\frac{2}{5}}$? (Do not use calculator approximations.)

A. $\sqrt[3]{\frac{6}{15}}$ B. $\frac{\sqrt[3]{50}}{5}$ C. $\frac{\sqrt[3]{10}}{\sqrt[3]{25}}$ D. $\frac{\sqrt[3]{10}}{5}$

7. **Concept Check** A student multiplied incorrectly as follows.

$\sqrt[3]{13} \cdot \sqrt[3]{5}$

$= \sqrt{13 \cdot 5}$ Product rule

$= \sqrt{65}$ Multiply.

WHAT WENT WRONG? Give the correct product.

8. **Concept Check** A student multiplied incorrectly as follows.

$\sqrt[3]{x} \cdot \sqrt[3]{x} = x$

WHAT WENT WRONG? Give the correct product.

Multiply. Assume that all variables represent positive real numbers. **See Examples 1 and 2.**

9. $\sqrt{3} \cdot \sqrt{3}$
10. $\sqrt{5} \cdot \sqrt{5}$
11. $\sqrt{18} \cdot \sqrt{2}$
12. $\sqrt{12} \cdot \sqrt{3}$
13. $\sqrt{5} \cdot \sqrt{6}$
14. $\sqrt{10} \cdot \sqrt{3}$
15. $\sqrt{14} \cdot \sqrt{x}$
16. $\sqrt{23} \cdot \sqrt{t}$
17. $\sqrt{14} \cdot \sqrt{3pqr}$
18. $\sqrt{7} \cdot \sqrt{5xt}$
19. $\sqrt[3]{2} \cdot \sqrt[3]{5}$
20. $\sqrt[3]{3} \cdot \sqrt[3]{6}$
21. $\sqrt[3]{7x} \cdot \sqrt[3]{2y}$
22. $\sqrt[3]{9x} \cdot \sqrt[3]{4y}$
23. $\sqrt[4]{11} \cdot \sqrt[4]{3}$
24. $\sqrt[4]{6} \cdot \sqrt[4]{9}$
25. $\sqrt[4]{2x} \cdot \sqrt[4]{3y^2}$
26. $\sqrt[4]{3y^2} \cdot \sqrt[4]{6yz}$
27. $\sqrt[3]{7} \cdot \sqrt[4]{3}$
28. $\sqrt[5]{8} \cdot \sqrt[6]{12}$

Simplify. Assume that all variables represent positive real numbers. **See Example 3.**

29. $\sqrt{\frac{64}{121}}$
30. $\sqrt{\frac{16}{49}}$
31. $\sqrt{\frac{3}{25}}$
32. $\sqrt{\frac{13}{49}}$
33. $\sqrt{\frac{x}{25}}$
34. $\sqrt{\frac{k}{100}}$
35. $\sqrt{\frac{p^6}{81}}$
36. $\sqrt{\frac{w^{10}}{36}}$
37. $\sqrt[3]{-\frac{27}{64}}$
38. $\sqrt[3]{-\frac{216}{125}}$
39. $\sqrt[3]{\frac{r^2}{8}}$
40. $\sqrt[3]{\frac{t}{125}}$
41. $-\sqrt[4]{\frac{81}{x^4}}$
42. $-\sqrt[4]{\frac{625}{y^4}}$
43. $\sqrt[5]{\frac{1}{x^{15}}}$
44. $\sqrt[5]{\frac{32}{y^{20}}}$

45. **Concept Check** A student simplified $\sqrt{48}$ as follows and did not receive credit for his answer.

$\sqrt{48}$

$= \sqrt{4 \cdot 12}$

$= \sqrt{4} \cdot \sqrt{12}$

$= 2\sqrt{12}$

WHAT WENT WRONG? Give the correct simplified form.

46. **Concept Check** A student incorrectly claimed that the following radical is in simplified form.

$\sqrt[3]{k^4}$

WHAT WENT WRONG? Give the correct simplified form.

Simplify. ***See Example 4.***

47. $\sqrt{12}$
48. $\sqrt{18}$
49. $\sqrt{288}$
50. $\sqrt{72}$
51. $-\sqrt{32}$
52. $-\sqrt{48}$
53. $-\sqrt{28}$
54. $-\sqrt{24}$
55. $\sqrt{30}$
56. $\sqrt{46}$
57. $\sqrt[3]{128}$
58. $\sqrt[3]{24}$
59. $\sqrt[3]{40}$
60. $\sqrt[3]{375}$
61. $\sqrt[3]{-16}$
62. $\sqrt[3]{-250}$
63. $-\sqrt[4]{512}$
64. $-\sqrt[4]{1250}$
65. $\sqrt[5]{64}$
66. $\sqrt[5]{128}$
67. $-\sqrt[5]{486}$
68. $-\sqrt[5]{2048}$
69. $\sqrt[6]{128}$
70. $\sqrt[6]{1458}$

Simplify. Assume that all variables represent positive real numbers. ***See Example 5.***

71. $\sqrt{72k^2}$
72. $\sqrt{18m^2}$
73. $\sqrt{144x^3y^9}$
74. $\sqrt{169s^5t^{10}}$
75. $\sqrt{121x^6}$
76. $\sqrt{256z^{12}}$
77. $-\sqrt[3]{27t^{12}}$
78. $-\sqrt[3]{64y^{18}}$
79. $-\sqrt{100m^8z^4}$
80. $-\sqrt{25t^6s^{20}}$
81. $-\sqrt[3]{-125a^6b^9c^{12}}$
82. $-\sqrt[3]{-216y^{15}x^6z^3}$
83. $\sqrt[4]{\frac{1}{16}r^8t^{20}}$
84. $\sqrt[4]{\frac{81}{256}t^{12}u^8}$
85. $\sqrt{50x^3}$
86. $\sqrt{300z^3}$
87. $-\sqrt{500r^{11}}$
88. $-\sqrt{200p^{13}}$
89. $\sqrt{13x^7y^8}$
90. $\sqrt{23k^9p^{14}}$
91. $\sqrt[3]{8z^6w^9}$
92. $\sqrt[3]{64a^{15}b^{12}}$
93. $\sqrt[3]{-16z^5t^7}$
94. $\sqrt[3]{-81m^4n^{10}}$
95. $\sqrt[4]{81x^{12}y^{16}}$
96. $\sqrt[4]{81t^8u^{28}}$
97. $-\sqrt[4]{162r^{15}s^9}$
98. $-\sqrt[4]{32k^5m^{11}}$
99. $\sqrt{\frac{y^{11}}{36}}$
100. $\sqrt{\frac{v^{13}}{49}}$
101. $\sqrt[3]{\frac{x^{16}}{27}}$
102. $\sqrt[3]{\frac{y^{17}}{125}}$

Simplify. Assume that $x \geq 0$. ***See Example 6.***

103. $\sqrt[4]{48^2}$
104. $\sqrt[4]{50^2}$
105. $\sqrt[4]{25}$
106. $\sqrt[6]{8}$
107. $\sqrt[10]{x^{25}}$
108. $\sqrt[12]{x^{44}}$
109. $\sqrt[10]{x^{16}}$
110. $\sqrt[12]{x^{38}}$

Simplify. Assume that all variables represent positive real numbers. ***See Example 7.***

111. $\sqrt[3]{4} \cdot \sqrt{3}$
112. $\sqrt[3]{5} \cdot \sqrt{6}$
113. $\sqrt[4]{3} \cdot \sqrt[3]{4}$
114. $\sqrt[3]{2} \cdot \sqrt[5]{3}$
115. $\sqrt{x} \cdot \sqrt[3]{x}$
116. $\sqrt[3]{y} \cdot \sqrt[4]{y}$

Find the length of the unknown side in each right triangle. Simplify answers if possible. ***See Example 8.***

117. 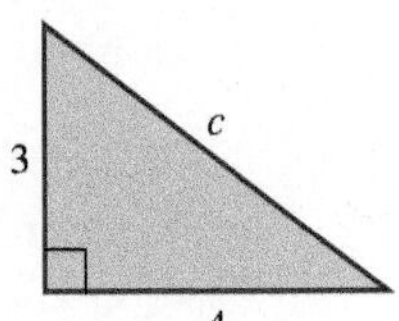

118. 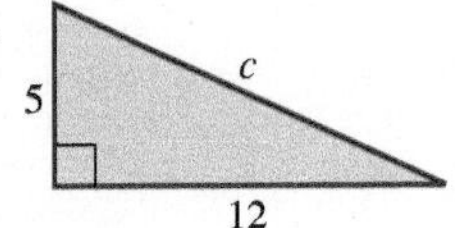

119.

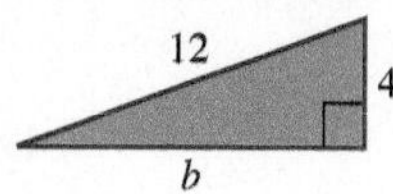

120.

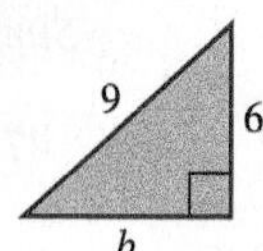

121.

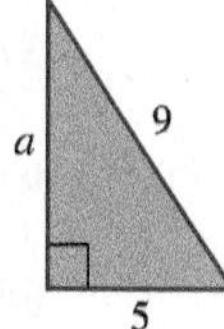

122.

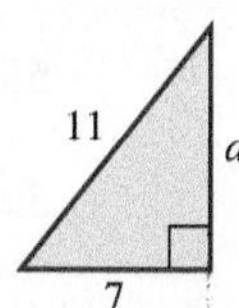

Find the distance between each pair of points. **See Example 9.**

123. $(6, 13)$ and $(1, 1)$

124. $(8, 13)$ and $(2, 5)$

125. $(-6, 5)$ and $(3, -4)$

126. $(-1, 5)$ and $(-7, 7)$

127. $(-8, 2)$ and $(-4, 1)$

128. $(-1, 2)$ and $(5, 3)$

129. $(4.7, 2.3)$ and $(1.7, -1.7)$

130. $(-2.9, 18.2)$ and $(2.1, 6.2)$

131. $(\sqrt{2}, \sqrt{6})$ and $(-2\sqrt{2}, 4\sqrt{6})$

132. $(\sqrt{7}, 9\sqrt{3})$ and $(-\sqrt{7}, 4\sqrt{3})$

133. $(x + y, y)$ and $(x - y, x)$

134. $(c, c - d)$ and $(d, c + d)$

Concept Check *Work each problem.*

135. Match each equation with the correct graph.

(a) $(x - 3)^2 + (y - 2)^2 = 25$

(b) $(x - 3)^2 + (y + 2)^2 = 25$

(c) $(x + 3)^2 + (y - 2)^2 = 25$

(d) $(x + 3)^2 + (y + 2)^2 = 25$

A. **B.**

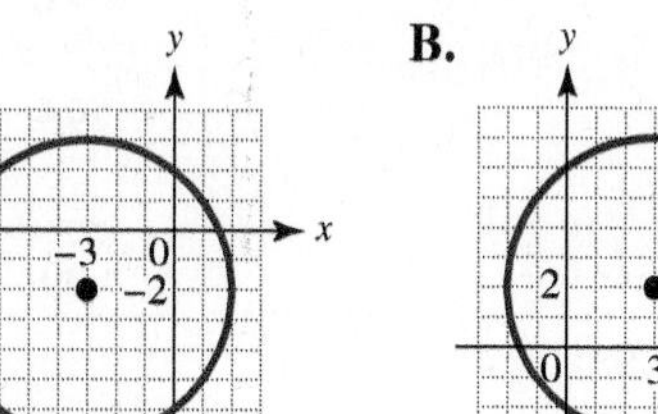

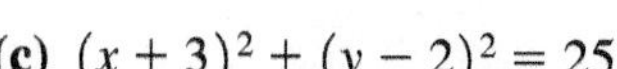

C. **D.**

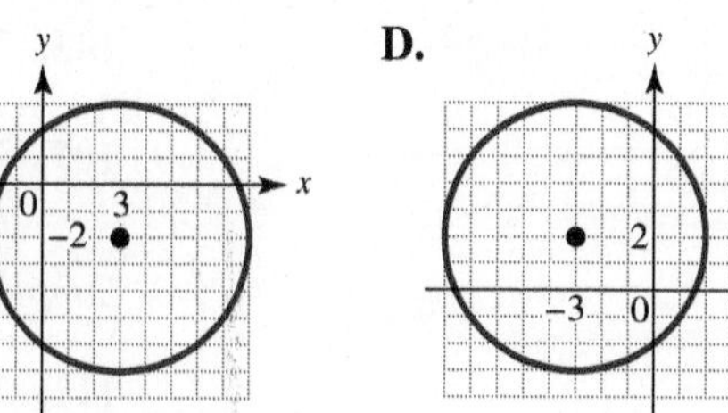

136. A circle can be drawn on a piece of posterboard by fastening one end of a string with a thumbtack, pulling the string taut with a pencil, and tracing a curve, as shown in the figure. Explain why this method works.

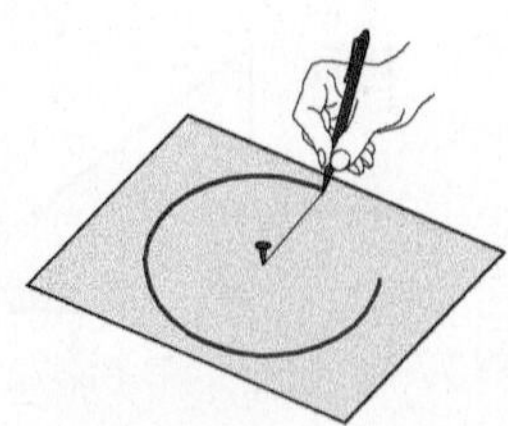

Find the equation of a circle satisfying the given conditions. **See Examples 10–12.**

137. Center: $(0, 0)$; radius: 12

138. Center: $(0, 0)$; radius: 9

139. Center: $(-4, 3)$; radius: 2

140. Center: $(5, -2)$; radius: 4

141. Center: $(-8, -5)$; radius: $\sqrt{5}$

142. Center: $(-12, 13)$; radius: $\sqrt{7}$

Graph each circle. Identify the center and the radius. **See Examples 10–12.**

143. $x^2 + y^2 = 9$

144. $x^2 + y^2 = 4$

145. $x^2 + y^2 = 16$

146. $x^2 + y^2 = 25$

147. $(x + 3)^2 + (y - 2)^2 = 9$

148. $(x - 1)^2 + (y + 3)^2 = 16$

149. $(x - 2)^2 + (y - 3)^2 = 4$

150. $(x + 4)^2 + (y + 1)^2 = 25$

Solve each problem.

151. A Panasonic Smart Viera E50 LCD HDTV has a rectangular screen with a 36.5-in. width. Its height is 20.8 in. What is the length of the diagonal of the screen to the nearest tenth of an inch? (Data from measurements of the author's television.)

152. The length of the diagonal of a box is given by

$$D = \sqrt{L^2 + W^2 + H^2},$$

where L, W, and H are, respectively, the length, width, and height of the box. Find the length of the diagonal D of a box that is 4 ft long, 2 ft wide, and 3 ft high. Give the exact value, and then round to the nearest tenth of a foot.

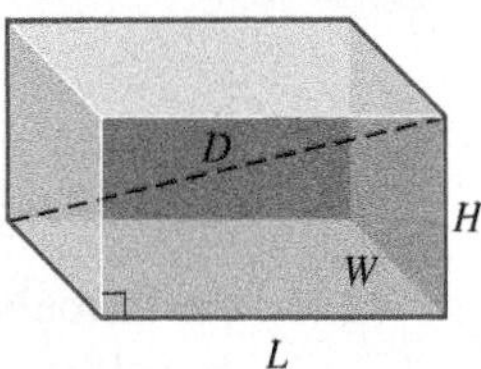

153. In the study of sound, one version of the law of tensions is

$$f_1 = f_2\sqrt{\frac{F_1}{F_2}}.$$

If $F_1 = 300$, $F_2 = 60$, and $f_2 = 260$, find f_1 to the nearest unit.

154. The illumination I, in foot-candles, produced by a light source is related to the distance d, in feet, from the light source by the equation

$$d = \sqrt{\frac{k}{I}},$$

where k is a constant. If $k = 640$, how far from the light source will the illumination be 2 foot-candles? Give the exact value, and then round to the nearest tenth of a foot.

Tom owns a condominium in a high rise building on the shore of Lake Michigan, and has a beautiful view of the lake from his window. He discovered that he can find the number of miles to the horizon by multiplying 1.224 by the square root of his eye level in feet from the ground.

155. Use Tom's discovery to do the following.

(a) Write a formula that could be used to calculate the distance d in miles to the horizon from a height h in feet from the ground.

(b) Tom lives on the 14th floor, which is 150 ft above the ground. His eyes are 6 ft above his floor. Use the formula from part (a) to calculate the distance, to the nearest tenth of a mile, that Tom can see to the horizon from his condominium window.

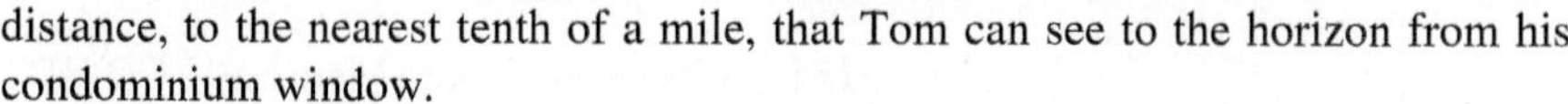

156. Tom's neighbor Sheri lives on a floor that is 100 ft above the ground. Assuming that her eyes are 5 ft above the ground, to the nearest tenth of a mile, how far can she see to the horizon?

RELATING CONCEPTS For Individual or Group Work (Exercises 157–162)

Heron's formula for finding the area $\mathcal{A}$ of a triangle with sides a, b, and c is

$$\mathcal{A} = \sqrt{s(s-a)(s-b)(s-c)},$$

where s is the semiperimeter, and $s = \frac{1}{2}(a + b + c)$.

Consider the triangle, and **work Exercises 157–162** *in order.*

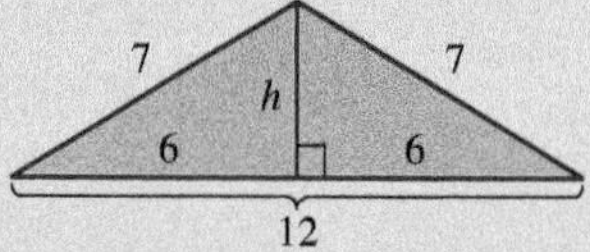

157. The lengths of the sides of the entire triangle shown are 7, 7, and 12. Find the semiperimeter s.

158. Use Heron's formula to find the area of the entire triangle. Write it as a simplified radical.

159. Find the height h of the triangle using the Pythagorean theorem.

160. Find the area of each congruent right triangle forming the entire triangle, using the common area formula, $\mathcal{A} = \frac{1}{2}bh$.

161. Double the result from **Exercise 160** to determine the area of the entire triangle.

162. How do the answers in **Exercises 158 and 161** compare?

7.4 Adding and Subtracting Radical Expressions

OBJECTIVE

1 Simplify radical expressions involving addition and subtraction.

OBJECTIVE 1 Simplify radical expressions involving addition and subtraction.

We do so using the distributive property,

$$ac + bc = (a + b)c.$$

EXAMPLE 1 Adding and Subtracting Radicals

Add or subtract to simplify each radical expression.

(a) $4\sqrt{2} + 3\sqrt{2}$

$= (4 + 3)\sqrt{2}$

$= 7\sqrt{2}$

This is similar to simplifying $4x + 3x$ as $7x$.

(b) $2\sqrt{3} - 5\sqrt{3}$

$= (2 - 5)\sqrt{3}$

$= -3\sqrt{3}$

This is similar to simplifying $2x - 5x$ as $-3x$.

(c) $3\sqrt{24} + \sqrt{54}$ — Simplify each individual radical.

$= 3\sqrt{4 \cdot 6} + \sqrt{9 \cdot 6}$ — Factor the radicands so that one factor is a perfect square.

$= 3\sqrt{4} \cdot \sqrt{6} + \sqrt{9} \cdot \sqrt{6}$ — Product rule

$= 3 \cdot 2\sqrt{6} + 3\sqrt{6}$ — Find the square roots.

$= 6\sqrt{6} + 3\sqrt{6}$ — Multiply.

$= (6 + 3)\sqrt{6}$ — Distributive property

$= 9\sqrt{6}$ — Add.

(d) $2\sqrt{20x} - \sqrt{45x}, \quad x \geq 0$

$= 2\sqrt{4} \cdot \sqrt{5x} - \sqrt{9} \cdot \sqrt{5x}$ — Product rule

$= 2 \cdot 2\sqrt{5x} - 3\sqrt{5x}$ — Find the square roots.

$= 4\sqrt{5x} - 3\sqrt{5x}$ — Multiply.

$= (4 - 3)\sqrt{5x}$ — Distributive property

$= \sqrt{5x}$ — $1\sqrt{5x} = \sqrt{5x}$ — Subtract.

(e) $2\sqrt{3} - 4\sqrt{5}$

The radicands differ and are already simplified, so this expression cannot be simplified further.

NOW TRY

NOW TRY EXERCISE 1

Add or subtract to simplify each radical expression.

(a) $3\sqrt{5} + 7\sqrt{5}$

(b) $2\sqrt{7} - 3\sqrt{7}$

(c) $\sqrt{12} + \sqrt{75}$

(d) $-\sqrt{63t} + 3\sqrt{28t}, \quad t \geq 0$

(e) $6\sqrt{7} - 2\sqrt{3}$

CAUTION ***Only radical expressions with the same index and the same radicand may be combined.***

NOW TRY ANSWERS

1. **(a)** $10\sqrt{5}$ **(b)** $-\sqrt{7}$
(c) $7\sqrt{3}$ **(d)** $3\sqrt{7t}$
(e) The expression cannot be simplified further.

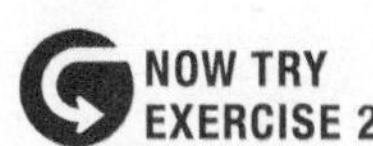

Add or subtract to simplify each radical expression. Assume that all variables represent positive real numbers.

(a) $3\sqrt[3]{2000} - 4\sqrt[3]{128}$

(b) $5\sqrt[4]{a^5b^3} + \sqrt[4]{81ab^7}$

(c) $\sqrt[3]{128t^4} - 2\sqrt{72t^3}$

EXAMPLE 2 Adding and Subtracting Radicals with Higher Indexes

Add or subtract to simplify each radical expression. Assume that all variables represent positive real numbers.

(a) $2\sqrt[3]{16} - 5\sqrt[3]{54}$ Remember to write the index with each radical.

$= 2\sqrt[3]{8 \cdot 2} - 5\sqrt[3]{27 \cdot 2}$ Factor.

$= 2\sqrt[3]{8} \cdot \sqrt[3]{2} - 5\sqrt[3]{27} \cdot \sqrt[3]{2}$ Product rule

$= 2 \cdot 2 \cdot \sqrt[3]{2} - 5 \cdot 3 \cdot \sqrt[3]{2}$ Find the cube roots.

$= 4\sqrt[3]{2} - 15\sqrt[3]{2}$ Multiply.

$= (4 - 15)\sqrt[3]{2}$ Distributive property

$= -11\sqrt[3]{2}$ Combine like terms.

In practice, the step indicating $(4 - 15)\sqrt[3]{2}$ can be done mentally, giving the final answer $-11\sqrt[3]{2}$ directly.

(b) $2\sqrt[3]{x^2y} + \sqrt[3]{8x^5y^4}$

$= 2\sqrt[3]{x^2y} + \sqrt[3]{8x^3y^3 \cdot x^2y}$ Factor.

$= 2\sqrt[3]{x^2y} + \sqrt[3]{8x^3y^3} \cdot \sqrt[3]{x^2y}$ Product rule

$= 2\sqrt[3]{x^2y} + 2xy\sqrt[3]{x^2y}$ Find the cube root.

$= (2 + 2xy)\sqrt[3]{x^2y}$ Distributive property

This result cannot be simplified further.

Although we were able to use the distributive property in the last step, 2 and $2xy$ are not like terms and cannot be combined into a single term.

(c) $5\sqrt{4x^3} + 3\sqrt[3]{64x^4}$ Be careful. The indexes are different.

$= 5\sqrt{4x^2 \cdot x} + 3\sqrt[3]{64x^3 \cdot x}$ Factor.

$= 5\sqrt{4x^2} \cdot \sqrt{x} + 3\sqrt[3]{64x^3} \cdot \sqrt[3]{x}$ Product rule

$= 5 \cdot 2x\sqrt{x} + 3 \cdot 4x\sqrt[3]{x}$ Keep track of the indexes.

$= 10x\sqrt{x} + 12x\sqrt[3]{x}$ Multiply.

The two terms in the final expression cannot be combined into a single term—the indexes are different.

NOW TRY

CAUTION *The root of a sum does not equal the sum of the roots.*

Example: $\sqrt{9 + 16} \neq \sqrt{9} + \sqrt{16}$

because $\sqrt{9 + 16} = \sqrt{25} = 5$, but $\sqrt{9} + \sqrt{16} = 3 + 4 = 7$.

NOW TRY ANSWERS

2. (a) $14\sqrt[3]{2}$

(b) $(5a + 3b)\sqrt[4]{ab^3}$

(c) $4t\sqrt[3]{2t} - 12t\sqrt{2t}$

NOW TRY EXERCISE 3

Perform the indicated operations. Assume that all variables represent positive real numbers.

(a) $5\frac{\sqrt{5}}{\sqrt{45}} - 4\sqrt{\frac{28}{9}}$

(b) $6\sqrt[3]{\frac{16}{x^{12}}} + 7\sqrt[3]{\frac{9}{x^9}}$

EXAMPLE 3 Adding and Subtracting Radicals with Fractions

Perform the indicated operations. Assume that all variables represent positive real numbers.

(a) $2\sqrt{\frac{75}{16}} + 4\frac{\sqrt{8}}{\sqrt{32}}$

$= 2\frac{\sqrt{25 \cdot 3}}{\sqrt{16}} + 4\frac{\sqrt{4 \cdot 2}}{\sqrt{16 \cdot 2}}$ Quotient rule; Factor.

$= 2\left(\frac{5\sqrt{3}}{4}\right) + 4\left(\frac{2\sqrt{2}}{4\sqrt{2}}\right)$ Product rule; Find the square roots.

$= \frac{5\sqrt{3}}{2} + 2$ Multiply; $\frac{\sqrt{2}}{\sqrt{2}} = 1$

$= \frac{5\sqrt{3}}{2} + \frac{4}{2}$ Write with a common denominator; $2 = \frac{4}{2}$

$= \frac{5\sqrt{3} + 4}{2}$ $\frac{a}{c} + \frac{b}{c} = \frac{a+b}{c}$

(b) $10\sqrt[3]{\frac{5}{x^6}} - 3\sqrt[3]{\frac{4}{x^9}}$

$= 10\frac{\sqrt[3]{5}}{\sqrt[3]{x^6}} - 3\frac{\sqrt[3]{4}}{\sqrt[3]{x^9}}$ Quotient rule

$= \frac{10\sqrt[3]{5}}{x^2} - \frac{3\sqrt[3]{4}}{x^3}$ Simplify denominators.

This equals x^3, so there is a common denominator.

$= \frac{10\sqrt[3]{5} \cdot x}{x^2 \cdot x} - \frac{3\sqrt[3]{4}}{x^3}$ Write with a common denominator.

$= \frac{10x\sqrt[3]{5} - 3\sqrt[3]{4}}{x^3}$ $\frac{a}{c} - \frac{b}{c} = \frac{a-b}{c}$

NOW TRY

NOW TRY ANSWERS

3. (a) $\frac{5 - 8\sqrt{7}}{3}$

(b) $\frac{12\sqrt[3]{2} + 7x\sqrt[3]{9}}{x^4}$

7.4 Exercises

FOR EXTRA HELP MyLab Math

Video solutions for select problems available in MyLab Math

Concept Check *Choose the correct response.*

1. Which sum can be simplified without first simplifying the individual radical expressions?

A. $\sqrt{50} + \sqrt{32}$ **B.** $3\sqrt{6} + 9\sqrt{6}$

C. $\sqrt[3]{32} - \sqrt[3]{108}$ **D.** $\sqrt[5]{6} - \sqrt[5]{192}$

2. Which difference can be simplified without first simplifying the individual radical expressions?

A. $\sqrt{81} - \sqrt{18}$ **B.** $\sqrt[3]{8} - \sqrt[3]{16}$

C. $4\sqrt[3]{7} - 9\sqrt[3]{7}$ **D.** $\sqrt{75} - \sqrt{12}$

3. Concept Check A student incorrectly simplified

$$(3 + 3xy)\sqrt[3]{xy^2} \quad \text{as} \quad 6xy\sqrt[3]{xy^2}.$$

His teacher did not give him any credit for this answer. **WHAT WENT WRONG?**

4. Concept Check A student incorrectly gave the difference

$$28 - 4\sqrt{2} \quad \text{as} \quad 24\sqrt{2}.$$

Her teacher did not give her any credit for this answer. **WHAT WENT WRONG?**

Add or subtract to simplify each radical expression. Assume that all variables represent positive real numbers. ***See Examples 1 and 2.***

5. $\sqrt{36} - \sqrt{100}$ **6.** $\sqrt{25} - \sqrt{81}$ **7.** $6\sqrt{10} + 2\sqrt{10}$

8. $5\sqrt{6} + 4\sqrt{6}$ **9.** $6\sqrt{5} - 7\sqrt{5}$ **10.** $3\sqrt{2} - 4\sqrt{2}$

11. $-2\sqrt{48} + 3\sqrt{75}$ **12.** $4\sqrt{32} - 2\sqrt{8}$ **13.** $5\sqrt{6} + 2\sqrt{10}$

14. $3\sqrt{11} - 5\sqrt{13}$ **15.** $\sqrt[3]{16} + 4\sqrt[3]{54}$ **16.** $3\sqrt[3]{24} - 2\sqrt[3]{192}$

17. $6\sqrt{18} - \sqrt{32} + 2\sqrt{50}$ **18.** $5\sqrt{8} + 3\sqrt{72} - 3\sqrt{50}$

19. $2\sqrt{5} + 3\sqrt{20} + 4\sqrt{45}$ **20.** $5\sqrt{54} - 2\sqrt{24} - 2\sqrt{96}$

21. $\sqrt{72x} - \sqrt{8x}$ **22.** $\sqrt{18k} - \sqrt{72k}$

23. $3\sqrt{72m^2} - 5\sqrt{32m^2} - 3\sqrt{18m^2}$ **24.** $9\sqrt{27p^2} - 14\sqrt{108p^2} + 2\sqrt{48p^2}$

25. $\sqrt[4]{32} + 3\sqrt[4]{2}$ **26.** $\sqrt[4]{405} - 2\sqrt[4]{5}$

27. $2\sqrt[3]{16} + \sqrt[3]{54}$ **28.** $15\sqrt[3]{81} + 4\sqrt[3]{24}$

29. $2\sqrt[3]{27x} - 2\sqrt[3]{8x}$ **30.** $6\sqrt[3]{128m} - 3\sqrt[3]{16m}$

31. $3\sqrt[3]{x^2y} - 5\sqrt[3]{8x^2y}$ **32.** $3\sqrt[3]{x^2y^2} - 2\sqrt[3]{64x^2y^2}$

33. $3x\sqrt[3]{xy^2} - 2\sqrt[3]{8x^4y^2}$ **34.** $6q^2\sqrt[3]{5q} - 2q\sqrt[3]{40q^4}$

35. $5\sqrt[4]{32} + 3\sqrt[4]{162}$ **36.** $2\sqrt[4]{512} + 4\sqrt[4]{32}$

37. $3\sqrt[4]{x^5y} - 2x\sqrt[4]{xy}$ **38.** $2\sqrt[4]{m^9p^6} - 3m^2p\sqrt[4]{mp^2}$

39. $2\sqrt[4]{32a^3} + 5\sqrt[4]{2a^3}$ **40.** $5\sqrt[4]{243x^3} + 2\sqrt[4]{3x^3}$

41. $\sqrt[3]{64xy^2} + \sqrt[3]{27x^4y^5}$ **42.** $\sqrt[4]{625s^3t} + \sqrt[4]{81s^7t^5}$

43. $\sqrt[3]{192st^4} - \sqrt{27s^3t}$ **44.** $\sqrt{125a^5b^5} + \sqrt[3]{125a^4b^4}$

45. $2\sqrt[3]{8x^4} + 3\sqrt[4]{16x^5}$ **46.** $3\sqrt[3]{64m^4} + 5\sqrt[4]{81m^5}$

Perform the indicated operations. Assume that all variables represent positive real numbers. ***See Example 3.***

47. $\sqrt{8} - \dfrac{\sqrt{64}}{\sqrt{16}}$ **48.** $\sqrt{48} - \dfrac{\sqrt{81}}{\sqrt{9}}$ **49.** $\dfrac{2\sqrt{5}}{3} + \dfrac{\sqrt{5}}{6}$

50. $\dfrac{4\sqrt{3}}{3} + \dfrac{2\sqrt{3}}{9}$ **51.** $\sqrt{\dfrac{8}{9}} + \sqrt{\dfrac{18}{36}}$ **52.** $\sqrt{\dfrac{12}{16}} + \sqrt{\dfrac{48}{64}}$

53. $\dfrac{\sqrt{32}}{3}+\dfrac{2\sqrt{2}}{3}-\dfrac{\sqrt{2}}{\sqrt{9}}$

54. $\dfrac{\sqrt{27}}{2}-\dfrac{3\sqrt{3}}{2}+\dfrac{\sqrt{3}}{\sqrt{4}}$

55. $3\sqrt{\dfrac{50}{9}}+8\dfrac{\sqrt{2}}{\sqrt{8}}$

56. $5\sqrt{\dfrac{288}{25}}+21\dfrac{\sqrt{2}}{\sqrt{18}}$

57. $\sqrt{\dfrac{25}{x^8}}+\sqrt{\dfrac{9}{x^6}}$

58. $\sqrt{\dfrac{100}{y^4}}+\sqrt{\dfrac{81}{y^{10}}}$

59. $3\sqrt{\dfrac{50}{49}}-\dfrac{\sqrt{27}}{\sqrt{12}}$

60. $9\sqrt{\dfrac{48}{25}}-2\dfrac{\sqrt{2}}{\sqrt{98}}$

61. $3\sqrt[3]{\dfrac{m^5}{27}}-2m\sqrt[3]{\dfrac{m^2}{64}}$

62. $2a\sqrt[4]{\dfrac{a}{16}}-5a\sqrt[4]{\dfrac{a}{81}}$

63. $3\sqrt[3]{\dfrac{2}{x^6}}-4\sqrt[3]{\dfrac{5}{x^9}}$

64. $-4\sqrt[3]{\dfrac{4}{t^9}}+3\sqrt[3]{\dfrac{9}{t^{12}}}$

65. Concept Check Consider the expression

$$\sqrt{63}+\sqrt{112}-\sqrt{252}.$$

(a) Simplify this expression using the methods of this section.

(b) Use a calculator to approximate the given expression.

(c) Use a calculator to approximate the simplified expression in part (a).

(d) Complete the following: Assuming the work in part (a) is correct, the approximations in parts (b) and (c) should be (*equal* / *unequal*).

66. Concept Check Let $a = 1$ and let $b = 64$.

(a) Evaluate $\sqrt{a}+\sqrt{b}$. Then find $\sqrt{a+b}$. Are they equal?

(b) Evaluate $\sqrt[3]{a}+\sqrt[3]{b}$. Then find $\sqrt[3]{a+b}$. Are they equal?

(c) Complete the following: In general,

$$\sqrt[n]{a}+\sqrt[n]{b}\neq \underline{\qquad\qquad\qquad},$$

based on the observations in parts (a) and (b) of this exercise.

Solve each problem.

67. A rectangular yard has a length of $\sqrt{192}$ m and a width of $\sqrt{48}$ m. Choose the best estimate of its dimensions. Then estimate the perimeter.

A. 14 m by 7 m **B.** 5 m by 7 m **C.** 14 m by 8 m **D.** 15 m by 8 m

68. If the sides of a triangle are $\sqrt{65}$ in., $\sqrt{35}$ in., and $\sqrt{26}$ in., which one of the following is the best estimate of its perimeter?

A. 19 in. **B.** 20 in. **C.** 24 in. **D.** 26 in.

Find the perimeter of each figure. Give answers as simplified radical expressions.

69.

$3\sqrt{20}$ in.

$2\sqrt{45}$ in.

$\sqrt{75}$ in.

70.

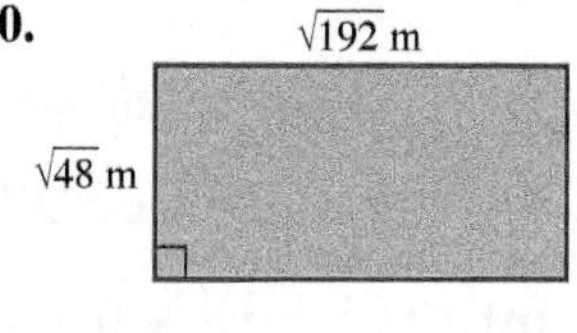

71.

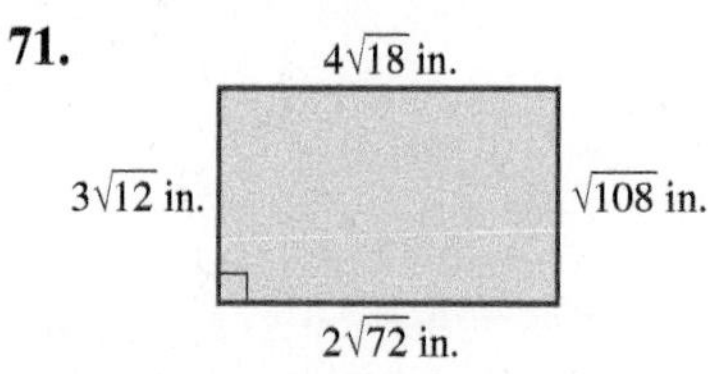

72. Find the area of the trapezoid. Give the answer as a simplified radical.

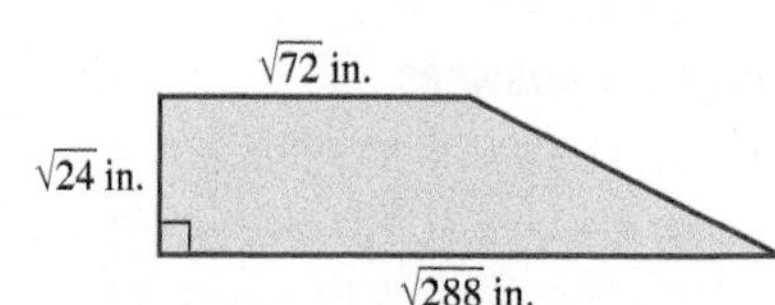

Extending Skills *Find the perimeter of each triangle.*

73.

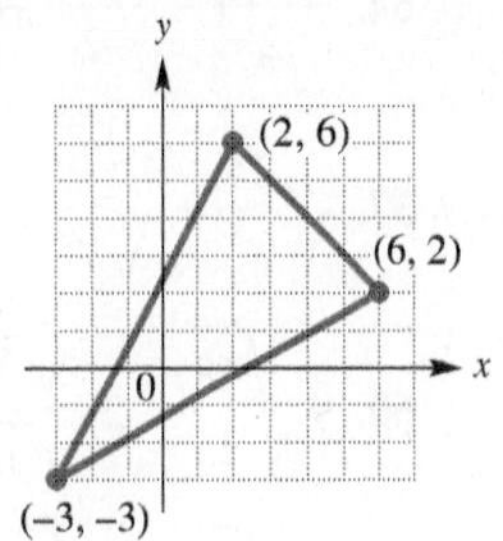

74.

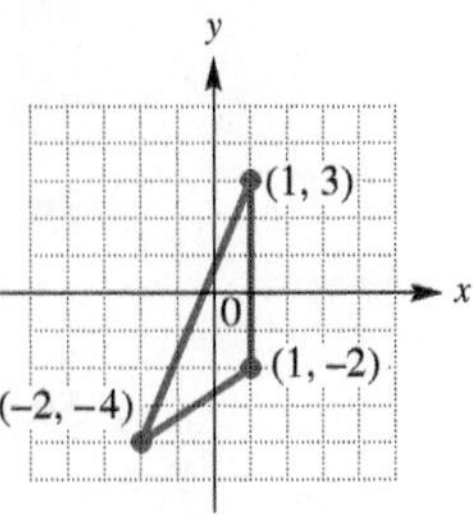

7.5 Multiplying and Dividing Radical Expressions

OBJECTIVES

1. Multiply radical expressions.
2. Rationalize denominators with one radical term.
3. Rationalize denominators with binomials involving radicals.
4. Write radical quotients in lowest terms.

VOCABULARY

☐ conjugates

OBJECTIVE 1 Multiply radical expressions.

The distributive property may be used when multiplying radical expressions.

EXAMPLE 1 Using the Distributive Property with Radicals

Multiply.

(a) $\sqrt{5}(2+\sqrt{6})$

$= \sqrt{5}\cdot 2 + \sqrt{5}\cdot\sqrt{6}$ Distributive property: $a(b+c) = ab + ac$

$= 2\sqrt{5} + \sqrt{30}$ Commutative property; product rule

(b) $4(\sqrt{12}-\sqrt{27})$

$= 4\sqrt{12} - 4\sqrt{27}$ Distributive property

$= 4\sqrt{4\cdot 3} - 4\sqrt{9\cdot 3}$ Factor the radicands so that one factor is a perfect square.

$= 4\cdot 2\sqrt{3} - 4\cdot 3\sqrt{3}$ $\sqrt{4}=2$; $\sqrt{9}=3$

$= 8\sqrt{3} - 12\sqrt{3}$ Multiply.

$= -4\sqrt{3}$ $8\sqrt{3}-12\sqrt{3} = (8-12)\sqrt{3}$ NOW TRY

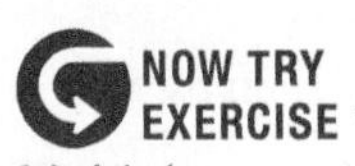

NOW TRY EXERCISE 1

Multiply.

(a) $\sqrt{10}(4+\sqrt{7})$

(b) $3(\sqrt{20}-\sqrt{45})$

We multiply binomial expressions involving radicals using the FOIL method. Recall that the acronym **FOIL** refers to the positions of the terms. We multiply the **F**irst terms, **O**uter terms, **I**nner terms, and **L**ast terms of the binomials.

EXAMPLE 2 Multiplying Binomials Involving Radical Expressions

Multiply, using the FOIL method.

(a) $(\sqrt{5}+3)(\sqrt{6}+1)$

$= \underbrace{\sqrt{5}\cdot\sqrt{6}}_{\text{First}} + \underbrace{\sqrt{5}\cdot 1}_{\text{Outer}} + \underbrace{3\cdot\sqrt{6}}_{\text{Inner}} + \underbrace{3\cdot 1}_{\text{Last}}$ FOIL method

$= \sqrt{30} + \sqrt{5} + 3\sqrt{6} + 3$ This result cannot be simplified further.

NOW TRY ANSWERS

1. (a) $4\sqrt{10}+\sqrt{70}$
   (b) $-3\sqrt{5}$

NOW TRY EXERCISE 2

Multiply, using the FOIL method.

(a) $(8-\sqrt{5})(9-\sqrt{2})$

(b) $(\sqrt{7}+\sqrt{5})(\sqrt{7}-\sqrt{5})$

(c) $(\sqrt{15}-4)^2$

(d) $(8+\sqrt[3]{5})(8-\sqrt[3]{5})$

(e) $(\sqrt{m}-\sqrt{n})(\sqrt{m}+\sqrt{n})$, $m \geq 0$ and $n \geq 0$

(b) $(7-\sqrt{3})(\sqrt{5}+\sqrt{2})$

F O I L

$= 7\sqrt{5}+7\sqrt{2}-\sqrt{3}\cdot\sqrt{5}-\sqrt{3}\cdot\sqrt{2}$ FOIL method

$= 7\sqrt{5}+7\sqrt{2}-\sqrt{15}-\sqrt{6}$ Product rule

(c) $(\sqrt{10}+\sqrt{3})(\sqrt{10}-\sqrt{3})$

$= \sqrt{10}\cdot\sqrt{10}-\sqrt{10}\cdot\sqrt{3}+\sqrt{3}\cdot\sqrt{10}-\sqrt{3}\cdot\sqrt{3}$ FOIL method

$= 10-3$ Product rule; $-\sqrt{30}+\sqrt{30}=0$

$= 7$ Subtract.

A product such as $(\sqrt{10}+\sqrt{3})(\sqrt{10}-\sqrt{3}) = (\sqrt{10})^2-(\sqrt{3})^2$ is a difference of squares.

$$(x+y)(x-y) = x^2-y^2$$ Here, $x=\sqrt{10}$ and $y=\sqrt{3}$.

(d) $(\sqrt{7}-3)^2$

$= (\sqrt{7}-3)(\sqrt{7}-3)$ $a^2 = a\cdot a$

$= \sqrt{7}\cdot\sqrt{7}-3\sqrt{7}-3\sqrt{7}+3\cdot 3$ FOIL method

$= 7-6\sqrt{7}+9$ Multiply. Combine like terms.

$= 16-6\sqrt{7}$ Add.

Be careful. These terms cannot be combined.

(e) $(5-\sqrt[3]{3})(5+\sqrt[3]{3})$

Remember to write the index 3 in *each* radical.

$= 5\cdot 5+5\sqrt[3]{3}-5\sqrt[3]{3}-\sqrt[3]{3}\cdot\sqrt[3]{3}$ FOIL method

$= 25-\sqrt[3]{3^2}$ Multiply. Combine like terms.

$= 25-\sqrt[3]{9}$ Apply the exponent.

(f) $(\sqrt{k}+\sqrt{y})(\sqrt{k}-\sqrt{y})$, $k \geq 0$ and $y \geq 0$

$= (\sqrt{k})^2-(\sqrt{y})^2$ Difference of squares

$= k-y$

NOW TRY 

NOTE In **Example 2(d),** we could have used the formula for the square of a binomial to obtain the same result.

$(\sqrt{7}-3)^2$

$= (\sqrt{7})^2-2(\sqrt{7})(3)+3^2$ $(x-y)^2 = x^2-2xy+y^2$

$= 7-6\sqrt{7}+9$ Apply the exponents. Multiply.

$= 16-6\sqrt{7}$ Add.

NOW TRY ANSWERS

2. (a) $72-8\sqrt{2}-9\sqrt{5}+\sqrt{10}$
(b) 2 (c) $31-8\sqrt{15}$
(d) $64-\sqrt[3]{25}$ (e) $m-n$

OBJECTIVE 2 **Rationalize denominators with one radical term.**

A simplified radical expression has no radical in the denominator. The origin of this agreement no doubt occurred before the days of high-speed calculation, when computation was a tedious process performed by hand.

Consider the expression $\frac{1}{\sqrt{2}}$. To find a decimal approximation by hand, it is necessary to divide 1 by a decimal approximation for $\sqrt{2}$, such as 1.414. It is much easier if the divisor is a whole number. This can be accomplished by multiplying $\frac{1}{\sqrt{2}}$ by 1 in the form $\frac{\sqrt{2}}{\sqrt{2}}$. ***Multiplying by 1 in any form does not change the value of the original expression.***

$$\frac{1}{\sqrt{2}} \cdot \frac{\sqrt{2}}{\sqrt{2}} = \frac{\sqrt{2}}{2} \qquad \text{Multiply by 1; } \frac{\sqrt{2}}{\sqrt{2}} = 1$$

Now the computation requires dividing 1.414 by 2 to obtain 0.707, which is easier.

With current technology, either form $\frac{1}{\sqrt{2}}$ or $\frac{\sqrt{2}}{2}$ can be approximated with the same number of keystrokes. See **FIGURE 13**, which shows that a calculator gives the same approximation for both forms of the expression.

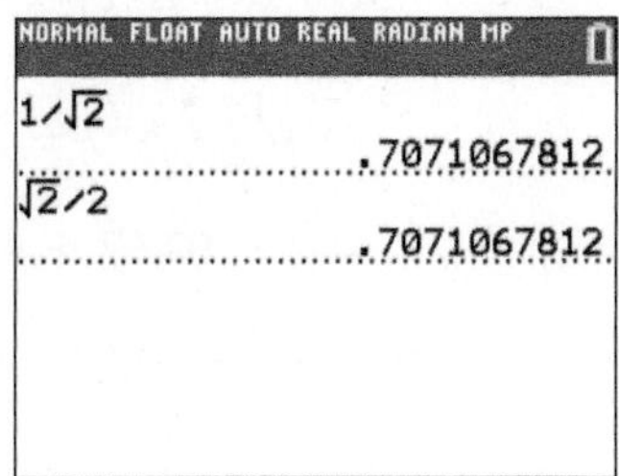

FIGURE 13

Rationalizing the Denominator

The process of removing radicals from a denominator so that the denominator contains only rational numbers is called **rationalizing the denominator.** This is done by multiplying by a form of 1.

EXAMPLE 3 **Rationalizing Denominators with Square Roots**

Rationalize each denominator.

(a) $\frac{3}{\sqrt{7}}$

Multiply by $\frac{\sqrt{7}}{\sqrt{7}}$. This is an application of the identity property of multiplication—in effect, we are multiplying by 1.

$$\frac{3}{\sqrt{7}} = \frac{3 \cdot \sqrt{7}}{\sqrt{7} \cdot \sqrt{7}} = \frac{3\sqrt{7}}{7}$$

In the denominator, $\sqrt{7} \cdot \sqrt{7} = \sqrt{7 \cdot 7} = \sqrt{49} = 7$. The final denominator is now a rational number.

(b) $\frac{5\sqrt{2}}{\sqrt{5}} = \frac{5\sqrt{2} \cdot \sqrt{5}}{\sqrt{5} \cdot \sqrt{5}} = \frac{5\sqrt{10}}{5} = \sqrt{10}$ Divide out the common factor, 5.

NOW TRY EXERCISE 3

Rationalize each denominator.

(a) $\dfrac{8}{\sqrt{13}}$ **(b)** $\dfrac{9\sqrt{7}}{\sqrt{3}}$

(c) $\dfrac{-10}{\sqrt{20}}$

(c) $\dfrac{-6}{\sqrt{12}}$

Less work is involved if the radical in the denominator is simplified first.

$$\frac{-6}{\sqrt{12}} = \frac{-6}{\sqrt{4 \cdot 3}} = \frac{-6}{2\sqrt{3}} = \frac{-3}{\sqrt{3}}$$

Now we rationalize the denominator.

$$\frac{-3}{\sqrt{3}} = \frac{-3 \cdot \sqrt{3}}{\sqrt{3} \cdot \sqrt{3}} = \frac{-3\sqrt{3}}{3} = -\sqrt{3}$$

NOW TRY 

NOW TRY EXERCISE 4

Simplify.

(a) $-\sqrt{\dfrac{27}{80}}$

(b) $\sqrt{\dfrac{48x^8}{y^3}}, \quad y > 0$

EXAMPLE 4 Rationalizing Denominators in Roots of Fractions

Simplify.

(a) $-\sqrt{\dfrac{18}{125}}$

$= -\dfrac{\sqrt{18}}{\sqrt{125}}$ Quotient rule

$= -\dfrac{\sqrt{9 \cdot 2}}{\sqrt{25 \cdot 5}}$ Factor.

$= -\dfrac{3\sqrt{2}}{5\sqrt{5}}$ Product rule

$= -\dfrac{3\sqrt{2} \cdot \sqrt{5}}{5\sqrt{5} \cdot \sqrt{5}}$ Multiply by $\dfrac{\sqrt{5}}{\sqrt{5}}$.

$= -\dfrac{3\sqrt{10}}{5 \cdot 5}$ Product rule

$= -\dfrac{3\sqrt{10}}{25}$ Multiply.

(b) $\sqrt{\dfrac{50m^4}{p^5}}, \quad p > 0$

$= \dfrac{\sqrt{50m^4}}{\sqrt{p^5}}$ Quotient rule

$= \dfrac{\sqrt{25m^4 \cdot 2}}{\sqrt{p^4 \cdot p}}$ Factor.

$= \dfrac{5m^2\sqrt{2}}{p^2\sqrt{p}}$ Product rule

$= \dfrac{5m^2\sqrt{2} \cdot \sqrt{p}}{p^2\sqrt{p} \cdot \sqrt{p}}$ Multiply by $\dfrac{\sqrt{p}}{\sqrt{p}}$.

$= \dfrac{5m^2\sqrt{2p}}{p^2 \cdot p}$ Product rule

$= \dfrac{5m^2\sqrt{2p}}{p^3}$ Multiply.

NOW TRY

EXAMPLE 5 Rationalizing Denominators with Higher Roots

Simplify.

(a) $\sqrt[3]{\dfrac{27}{16}}$

Use the quotient rule, and simplify the numerator and denominator.

$$\sqrt[3]{\frac{27}{16}} = \frac{\sqrt[3]{27}}{\sqrt[3]{16}} = \frac{3}{\sqrt[3]{8} \cdot \sqrt[3]{2}} = \frac{3}{2\sqrt[3]{2}}$$

NOW TRY ANSWERS

3. (a) $\dfrac{8\sqrt{13}}{13}$ (b) $3\sqrt{21}$
   (c) $-\sqrt{5}$

4. (a) $-\dfrac{3\sqrt{15}}{20}$ (b) $\dfrac{4x^4\sqrt{3y}}{y^2}$

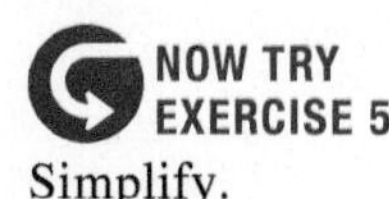

NOW TRY EXERCISE 5

Simplify.

(a) $\sqrt[3]{\dfrac{8}{81}}$

(b) $\sqrt[4]{\dfrac{7x}{y}}$, $x \geq 0, y > 0$

Because $2 \cdot 4 = 8$ is a perfect cube, multiply the numerator and denominator of the simplified expression by $\sqrt[3]{4}$.

$$\frac{3}{2\sqrt[3]{2}} \qquad \sqrt[3]{\frac{27}{16}} = \frac{3}{2\sqrt[3]{2}} \text{ from above}$$

$$= \frac{3 \cdot \sqrt[3]{4}}{2\sqrt[3]{2} \cdot \sqrt[3]{4}} \qquad \text{Multiply by } \sqrt[3]{4} \text{ in the numerator and denominator. This will give } \sqrt[3]{8} = 2 \text{ in the denominator.}$$

$$= \frac{3\sqrt[3]{4}}{2\sqrt[3]{8}} \qquad \text{Multiply.}$$

$$= \frac{3\sqrt[3]{4}}{2 \cdot 2} \qquad \sqrt[3]{8} = 2$$

$$= \frac{3\sqrt[3]{4}}{4} \qquad \text{Multiply.}$$

(b) $\sqrt[4]{\dfrac{5x}{z}}$, $x \geq 0, z > 0$

$$= \frac{\sqrt[4]{5x}}{\sqrt[4]{z}} \qquad \text{Quotient rule}$$

$$= \frac{\sqrt[4]{5x} \cdot \sqrt[4]{z^3}}{\sqrt[4]{z} \cdot \sqrt[4]{z^3}} \qquad \text{Multiply by } \sqrt[4]{z^3} \text{ in the numerator and denominator.}$$

$\sqrt[4]{z} \cdot \sqrt[4]{z^3}$ will give $\sqrt[4]{z^4}$.

$$= \frac{\sqrt[4]{5xz^3}}{\sqrt[4]{z^4}} \qquad \text{Product rule}$$

$$= \frac{\sqrt[4]{5xz^3}}{z}$$

NOW TRY 

CAUTION In **Example 5(a),** a typical error is to multiply the numerator and denominator by $\sqrt[3]{2}$, forgetting that

$$\sqrt[3]{2} \cdot \sqrt[3]{2} = \sqrt[3]{2^2}, \quad \text{which does } \textbf{\textit{not}} \text{ equal } \quad 2.$$

We need ***three*** factors of 2 to obtain 2^3 under the radical.

$$\sqrt[3]{2} \cdot \sqrt[3]{2} \cdot \sqrt[3]{2} = \sqrt[3]{2^3}, \quad \text{which does equal} \quad 2.$$

OBJECTIVE 3 Rationalize denominators with binomials involving radicals.

To rationalize a denominator that contains a binomial expression (one that contains exactly two terms) involving radicals, such as

$$\frac{3}{1 + \sqrt{2}},$$

we use the special product $(x + y)(x - y) = x^2 - y^2$ and the concept of *conjugates.* The conjugate of $1 + \sqrt{2}$ is $1 - \sqrt{2}$.

NOW TRY ANSWERS

5. (a) $\dfrac{2\sqrt[3]{9}}{9}$ (b) $\dfrac{\sqrt[4]{7xy^3}}{y}$

In general, $x + y$ and $x - y$ are **conjugates.** Specifically, if x and y represent nonnegative rational numbers, the product

$$(\sqrt{x} + \sqrt{y})(\sqrt{x} - \sqrt{y}) \quad \text{produces the rational number} \quad x - y.$$

Rationalizing a Binomial Denominator

Whenever a radical expression has a sum or difference with square root radicals in the denominator, rationalize the denominator by multiplying both the numerator and denominator by the conjugate of the denominator.

EXAMPLE 6 Rationalizing Binomial Denominators

Rationalize each denominator.

(a) $\dfrac{3}{1 + \sqrt{2}}$

Again, we are multiplying by a form of 1.

$= \dfrac{3(1 - \sqrt{2})}{(1 + \sqrt{2})(1 - \sqrt{2})}$ Multiply the numerator and denominator by $1 - \sqrt{2}$, the conjugate of the denominator.

$(1 + \sqrt{2})(1 - \sqrt{2})$
$= 1^2 - (\sqrt{2})^2$
$= 1 - 2$
$= -1$

$= \dfrac{3(1 - \sqrt{2})}{-1}$ The denominator is now a rational number.

$= \dfrac{3}{-1}(1 - \sqrt{2})$ $\quad \dfrac{a \cdot b}{c} = \dfrac{a}{c} \cdot b$

$= -3(1 - \sqrt{2})$ $\quad \dfrac{a}{-1} = -a$

$= -3 + 3\sqrt{2}$ Distributive property

Either of the forms in the last two lines can be given as the answer.

(b) $\dfrac{5}{4 - \sqrt{3}}$

$= \dfrac{5(4 + \sqrt{3})}{(4 - \sqrt{3})(4 + \sqrt{3})}$ Multiply the numerator and denominator by $4 + \sqrt{3}$.

$= \dfrac{5(4 + \sqrt{3})}{16 - 3}$ Multiply in the denominator.

$= \dfrac{5(4 + \sqrt{3})}{13}$ Subtract in the denominator.

We leave the numerator in factored form. This makes it easier to determine whether the expression is written in lowest terms.

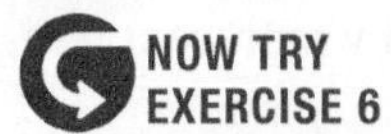

NOW TRY EXERCISE 6

Rationalize each denominator.

(a) $\frac{4}{1+\sqrt{3}}$ (b) $\frac{4}{5-\sqrt{7}}$

(c) $\frac{\sqrt{3}+\sqrt{7}}{\sqrt{5}-\sqrt{2}}$

(d) $\frac{8}{\sqrt{3x}+\sqrt{y}}$, $3x \neq y, x > 0, y > 0$

(c) $\frac{\sqrt{2}-\sqrt{3}}{\sqrt{5}+\sqrt{3}}$

$= \frac{(\sqrt{2}-\sqrt{3})(\sqrt{5}-\sqrt{3})}{(\sqrt{5}+\sqrt{3})(\sqrt{5}-\sqrt{3})}$ Multiply the numerator and denominator by $\sqrt{5}-\sqrt{3}$.

$= \frac{\sqrt{10}-\sqrt{6}-\sqrt{15}+3}{5-3}$ Multiply.

$= \frac{\sqrt{10}-\sqrt{6}-\sqrt{15}+3}{2}$ Subtract in the denominator.

(d) $\frac{3}{\sqrt{5m}-\sqrt{p}}$, $5m \neq p, m > 0, p > 0$

$= \frac{3(\sqrt{5m}+\sqrt{p})}{(\sqrt{5m}-\sqrt{p})(\sqrt{5m}+\sqrt{p})}$ Multiply the numerator and denominator by $\sqrt{5m}+\sqrt{p}$.

$= \frac{3(\sqrt{5m}+\sqrt{p})}{5m-p}$ Multiply in the denominator. NOW TRY 

OBJECTIVE 4 Write radical quotients in lowest terms.

NOW TRY EXERCISE 7

Write each quotient in lowest terms.

(a) $\frac{15-6\sqrt{2}}{18}$

(b) $\frac{15k+\sqrt{50k^2}}{20k}$, $k > 0$

EXAMPLE 7 Writing Radical Quotients in Lowest Terms

Write each quotient in lowest terms.

(a) $\frac{6+2\sqrt{5}}{4}$

$= \frac{2(3+\sqrt{5})}{2 \cdot 2}$ Factor the numerator and denominator. (This is a key step.)

$= \frac{3+\sqrt{5}}{2}$ Divide out the common factor.

Alternative method: $\frac{6+2\sqrt{5}}{4} = \frac{6}{4} + \frac{2\sqrt{5}}{4} = \frac{3}{2} + \frac{\sqrt{5}}{2} = \frac{3+\sqrt{5}}{2}$

(b) $\frac{5y-\sqrt{8y^2}}{6y}$, $y > 0$

$= \frac{5y-2y\sqrt{2}}{6y}$ $\sqrt{8y^2} = \sqrt{4y^2 \cdot 2} = 2y\sqrt{2}$

$= \frac{y(5-2\sqrt{2})}{6y}$ Factor the numerator.

$= \frac{5-2\sqrt{2}}{6}$ Divide out the common factor. NOW TRY 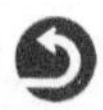

CAUTION *Be careful to factor before writing a quotient in lowest terms.*

NOW TRY ANSWERS

6. (a) $-2+2\sqrt{3}$ (b) $\frac{2(5+\sqrt{7})}{9}$ (c) $\frac{\sqrt{15}+\sqrt{6}+\sqrt{35}+\sqrt{14}}{3}$ (d) $\frac{8(\sqrt{3x}-\sqrt{y})}{3x-y}$

7. (a) $\frac{5-2\sqrt{2}}{6}$ (b) $\frac{3+\sqrt{2}}{4}$

7.5 Exercises

MyLab Math

Video solutions for select problems available in MyLab Math

Concept Check *Match each part of a rule for a special product in Column I with the part it equals in Column II. Assume that A and B represent positive real numbers.*

I	II
1. $(x+\sqrt{y})(x-\sqrt{y})$	**A.** $x-y$
2. $(\sqrt{x}+y)(\sqrt{x}-y)$	**B.** $x+2y\sqrt{x}+y^2$
3. $(\sqrt{x}+\sqrt{y})(\sqrt{x}-\sqrt{y})$	**C.** $x-y^2$
4. $(\sqrt{x}+\sqrt{y})^2$	**D.** $x-2\sqrt{xy}+y$
5. $(\sqrt{x}-\sqrt{y})^2$	**E.** x^2-y
6. $(\sqrt{x}+y)^2$	**F.** $x+2\sqrt{xy}+y$

Multiply, and then simplify each product. Assume that all variables represent positive real numbers. ***See Examples 1 and 2.***

7. $\sqrt{6}(3+\sqrt{2})$

8. $\sqrt{10}(5+\sqrt{3})$

9. $\sqrt{3}(\sqrt{12}-4)$

10. $\sqrt{5}(\sqrt{125}-6)$

11. $5(\sqrt{72}-\sqrt{8})$

12. $7(\sqrt{50}-\sqrt{18})$

13. $\sqrt{2}(\sqrt{18}-\sqrt{3})$

14. $\sqrt{5}(\sqrt{15}+\sqrt{5})$

15. $(\sqrt{2}+1)(\sqrt{3}+1)$

16. $(\sqrt{3}+3)(\sqrt{5}+2)$

17. $(\sqrt{2}-\sqrt{3})(\sqrt{2}+\sqrt{3})$

18. $(\sqrt{7}+\sqrt{14})(\sqrt{7}-\sqrt{14})$

19. $(\sqrt{8}-\sqrt{2})(\sqrt{8}+\sqrt{2})$

20. $(\sqrt{20}-\sqrt{5})(\sqrt{20}+\sqrt{5})$

21. $(\sqrt{11}-\sqrt{7})(\sqrt{2}+\sqrt{5})$

22. $(\sqrt{13}-\sqrt{7})(\sqrt{3}+\sqrt{11})$

23. $(2\sqrt{3}+\sqrt{5})(3\sqrt{3}-2\sqrt{5})$

24. $(\sqrt{7}-\sqrt{11})(2\sqrt{7}+3\sqrt{11})$

25. $(\sqrt{5}+2)^2$

26. $(\sqrt{11}-1)^2$

27. $(\sqrt{21}-\sqrt{5})^2$

28. $(\sqrt{6}-\sqrt{2})^2$

29. $(2+\sqrt[3]{6})(2-\sqrt[3]{6})$

30. $(\sqrt[3]{3}+6)(\sqrt[3]{3}-6)$

31. $(2+\sqrt[3]{2})(4-2\sqrt[3]{2}+\sqrt[3]{4})$

32. $(\sqrt[3]{3}-1)(\sqrt[3]{9}+\sqrt[3]{3}+1)$

33. $(3\sqrt{x}-\sqrt{5})(2\sqrt{x}+1)$

34. $(4\sqrt{p}+\sqrt{7})(\sqrt{p}-9)$

35. $(3\sqrt{r}-\sqrt{s})(3\sqrt{r}+\sqrt{s})$

36. $(\sqrt{k}+4\sqrt{m})(\sqrt{k}-4\sqrt{m})$

37. $(\sqrt[3]{2y}-5)(4\sqrt[3]{2y}+1)$

38. $(\sqrt[3]{9z}-2)(5\sqrt[3]{9z}+7)$

39. $(\sqrt{3x}+2)(\sqrt{3x}-2)$

40. $(\sqrt{6y}-4)(\sqrt{6y}+4)$

41. Concept Check A student incorrectly simplified the radical expression

$$6 - 4\sqrt{3} \quad \text{as} \quad 2\sqrt{3}.$$

WHAT WENT WRONG?

42. Concept Check A student rationalized the following denominator as shown.

$$\frac{5}{\sqrt[3]{2}} = \frac{5 \cdot \sqrt[3]{2}}{\sqrt[3]{2} \cdot \sqrt[3]{2}} = \frac{5\sqrt[3]{2}}{2} \quad \text{Incorrect}$$

WHAT WENT WRONG? Give the correct answer.

Rationalize each denominator. Assume that all variables represent positive real numbers. ***See Examples 3 and 4.***

43. $\frac{7}{\sqrt{7}}$ **44.** $\frac{11}{\sqrt{11}}$ **45.** $\frac{15}{\sqrt{3}}$ **46.** $\frac{12}{\sqrt{6}}$

47. $\frac{\sqrt{3}}{\sqrt{2}}$ **48.** $\frac{\sqrt{7}}{\sqrt{6}}$ **49.** $\frac{9\sqrt{3}}{\sqrt{5}}$ **50.** $\frac{3\sqrt{2}}{\sqrt{11}}$

51. $\frac{-7}{\sqrt{48}}$ **52.** $\frac{-5}{\sqrt{24}}$ **53.** $\sqrt{\frac{7}{2}}$ **54.** $\sqrt{\frac{10}{3}}$

55. $-\sqrt{\frac{7}{50}}$ **56.** $-\sqrt{\frac{13}{75}}$ **57.** $\sqrt{\frac{24}{x}}$ **58.** $\sqrt{\frac{52}{y}}$

59. $\frac{-8\sqrt{3}}{\sqrt{k}}$ **60.** $\frac{-4\sqrt{13}}{\sqrt{m}}$ **61.** $-\sqrt{\frac{150m^5}{n^3}}$ **62.** $-\sqrt{\frac{98r^3}{s^5}}$

63. $\sqrt{\frac{288x^7}{y^9}}$ **64.** $\sqrt{\frac{242t^9}{u^{11}}}$ **65.** $\frac{5\sqrt{2m}}{\sqrt{y^3}}$

66. $\frac{2\sqrt{5r}}{\sqrt{m^3}}$ **67.** $-\sqrt{\frac{48k^2}{z}}$ **68.** $-\sqrt{\frac{75m^3}{p}}$

Simplify. Assume that all variables represent positive real numbers. ***See Example 5.***

69. $\sqrt[3]{\frac{2}{3}}$ **70.** $\sqrt[3]{\frac{4}{5}}$ **71.** $\sqrt[3]{\frac{4}{9}}$ **72.** $\sqrt[3]{\frac{5}{16}}$ **73.** $\sqrt[3]{\frac{9}{32}}$

74. $\sqrt[3]{\frac{10}{9}}$ **75.** $-\sqrt[3]{\frac{2p}{r^2}}$ **76.** $-\sqrt[3]{\frac{6x}{y^2}}$ **77.** $\sqrt[3]{\frac{x^6}{y}}$ **78.** $\sqrt[3]{\frac{m^9}{q}}$

79. $\sqrt[4]{\frac{16}{x}}$ **80.** $\sqrt[4]{\frac{81}{y}}$ **81.** $\sqrt[4]{\frac{2y}{z}}$ **82.** $\sqrt[4]{\frac{7t}{s^2}}$

Rationalize each denominator. Assume that all variables represent positive real numbers and that no denominators are 0. ***See Example 6.***

83. $\frac{3}{4 + \sqrt{5}}$ **84.** $\frac{4}{5 + \sqrt{6}}$ **85.** $\frac{6}{\sqrt{5} + \sqrt{3}}$

86. $\frac{12}{\sqrt{6} + \sqrt{3}}$ **87.** $\frac{\sqrt{8}}{3 - \sqrt{2}}$ **88.** $\frac{\sqrt{27}}{3 - \sqrt{3}}$

89. $\dfrac{2}{3\sqrt{5}+2\sqrt{3}}$

90. $\dfrac{-1}{3\sqrt{2}-2\sqrt{7}}$

91. $\dfrac{\sqrt{2}-\sqrt{3}}{\sqrt{6}-\sqrt{5}}$

92. $\dfrac{\sqrt{5}+\sqrt{6}}{\sqrt{3}-\sqrt{2}}$

93. $\dfrac{m-4}{\sqrt{m}+2}$

94. $\dfrac{r-9}{\sqrt{r}-3}$

95. $\dfrac{4}{\sqrt{x}-2\sqrt{y}}$

96. $\dfrac{5}{3\sqrt{r}+\sqrt{s}}$

97. $\dfrac{\sqrt{x}-\sqrt{y}}{\sqrt{x}+\sqrt{y}}$

98. $\dfrac{\sqrt{a}+\sqrt{b}}{\sqrt{a}-\sqrt{b}}$

99. $\dfrac{5\sqrt{k}}{2\sqrt{k}+\sqrt{q}}$

100. $\dfrac{3\sqrt{x}}{\sqrt{x}-2\sqrt{y}}$

Write each quotient in lowest terms. Assume that all variables represent positive real numbers. ***See Example 7.***

101. $\dfrac{30+20\sqrt{6}}{10}$

102. $\dfrac{24+12\sqrt{5}}{12}$

103. $\dfrac{3-3\sqrt{5}}{3}$

104. $\dfrac{-5+5\sqrt{2}}{5}$

105. $\dfrac{16-4\sqrt{8}}{12}$

106. $\dfrac{12-9\sqrt{72}}{18}$

107. $\dfrac{6p+\sqrt{24p^3}}{3p}$

108. $\dfrac{11y-\sqrt{242y^5}}{22y}$

Extending Skills *Rationalize each denominator. Assume that all radicals represent real numbers and that no denominators are* 0.

109. $\dfrac{3}{\sqrt{x+y}}$

110. $\dfrac{5}{\sqrt{m-n}}$

111. $\dfrac{p}{\sqrt{p+2}}$

112. $\dfrac{q}{\sqrt{5+q}}$

Solve each problem.

113. The following expression occurs in a standard problem in trigonometry.

$$\frac{1}{\sqrt{2}}\cdot\frac{\sqrt{3}}{2}-\frac{1}{\sqrt{2}}\cdot\frac{1}{2}$$

Show that it simplifies to $\frac{\sqrt{6}-\sqrt{2}}{4}$. Then verify, using a calculator approximation.

114. The following expression occurs in a standard problem in trigonometry.

$$\frac{\sqrt{3}+1}{1-\sqrt{3}}$$

Show that it simplifies to $-2-\sqrt{3}$. Then verify, using a calculator approximation.

RELATING CONCEPTS For Individual or Group Work (Exercises 115–118)

In calculus, it is sometimes desirable to rationalize the numerator. To rationalize a numerator, we multiply the numerator and the denominator by the conjugate of the numerator. For example,

$$\frac{6-\sqrt{2}}{4}=\frac{(6-\sqrt{2})(6+\sqrt{2})}{4(6+\sqrt{2})}=\frac{36-2}{4(6+\sqrt{2})}=\frac{34}{4(6+\sqrt{2})}=\frac{17}{2(6+\sqrt{2})}.$$

Rationalize each numerator. Assume that all variables represent positive real numbers.

115. $\dfrac{6-\sqrt{3}}{8}$

116. $\dfrac{2\sqrt{5}-3}{2}$

117. $\dfrac{2\sqrt{x}-\sqrt{y}}{3x}$

118. $\dfrac{\sqrt{p}-3\sqrt{q}}{4q}$

SUMMARY EXERCISES Performing Operations with Radicals and Rational Exponents

Conditions for a Simplified Radical

1. The radicand has no factor raised to a power greater than or equal to the index.
2. The radicand has no fractions.
3. No denominator contains a radical.
4. Exponents in the radicand and the index of the radical have greatest common factor 1.

Concept Check *Give the reason why each radical is not simplified.*

1. $\sqrt{\frac{2}{5}}$
2. $\sqrt[15]{x^5}$
3. $\frac{5}{\sqrt[3]{10}}$
4. $\sqrt[3]{x^5y^6}$

Perform the indicated operations, and express each answer in simplest form. Assume that all variables represent positive real numbers.

5. $6\sqrt{10} - 12\sqrt{10}$
6. $\sqrt{7}(\sqrt{7} - \sqrt{2})$
7. $(1 - \sqrt{3})(2 + \sqrt{6})$
8. $\sqrt{50} - \sqrt{98} + \sqrt{72}$
9. $(3\sqrt{5} + 2\sqrt{7})^2$
10. $\frac{-3}{\sqrt{6}}$
11. $\frac{8}{\sqrt{7} + \sqrt{5}}$
12. $\frac{1 - \sqrt{2}}{1 + \sqrt{2}}$
13. $(\sqrt{5} + 7)(\sqrt{5} - 7)$
14. $\frac{1}{\sqrt{x} - \sqrt{5}}, \quad x \neq 5$
15. $\sqrt[3]{8a^3b^5c^9}$
16. $\frac{15}{\sqrt[3]{9}}$
17. $\frac{3}{\sqrt{5} + 2}$
18. $\sqrt{\frac{3}{5x}}$
19. $\frac{16\sqrt{3}}{5\sqrt{12}}$
20. $\frac{2\sqrt{25}}{8\sqrt{50}}$
21. $\frac{-10}{\sqrt[3]{10}}$
22. $\frac{\sqrt{6} + \sqrt{5}}{\sqrt{6} - \sqrt{5}}$
23. $\sqrt{12x} - \sqrt{75x}$
24. $(5 - 3\sqrt{3})^2$
25. $\sqrt[3]{\frac{13}{81}}$
26. $\frac{\sqrt{3} + \sqrt{7}}{\sqrt{6} - \sqrt{5}}$
27. $\frac{6}{\sqrt[4]{3}}$
28. $\sqrt[3]{\frac{x^2y}{x^{-3}y^4}}$
29. $\sqrt{12} - \sqrt{108} - \sqrt[3]{27}$
30. $\frac{4^{1/2} + 3^{1/2}}{4^{1/2} - 3^{1/2}}$
31. $\sqrt[3]{16x^2} - \sqrt[3]{54x^2} + \sqrt[3]{128x^2}$
32. $(1 - \sqrt[3]{3})(1 + \sqrt[3]{3} + \sqrt[3]{9})$

Simplify each expression. Write answers with positive exponents. Assume that all variables represent positive real numbers.

33. $3^{1/2} \cdot 3^{1/3}$
34. $\left(\frac{x^2y}{x^{-3}y^4}\right)^{1/3}$
35. $\frac{x^{-2/3}y^{4/5}}{x^{-5/3}y^{-2/5}}$
36. $\left(\frac{x^{3/4}y^{2/3}}{x^{1/3}y^{5/8}}\right)^{24}$
37. $(125x^3)^{-2/3}$
38. $(3x^{-2/3}y^{1/2})(-2x^{5/8}y^{-1/3})$

7.6 Solving Equations with Radicals

OBJECTIVES

1. Solve radical equations using the power rule.
2. Solve radical equations that require additional steps.
3. Solve radical equations with indexes greater than 2.
4. Use the power rule to solve a formula for a specified variable.

OBJECTIVE 1 Solve radical equations using the power rule.

An equation that includes one or more radical expressions with a variable in a radicand is a **radical equation.**

$$\sqrt{x-4} = 8, \quad \sqrt{5x+12} = 3\sqrt{2x-1}, \quad \sqrt[3]{6+x} = 27 \qquad \text{Radical equations}$$

Solving radical equations involves a process that we have not yet seen, and it requires careful application. Notice that the equation $x = 1$ has only one solution. Its solution set is $\{1\}$. If we square both sides of this equation, we obtain $x^2 = 1$. This new equation has *two* solutions: -1 and 1. The solution of the original equation is also a solution of the "squared" equation. However, that equation has another solution, -1, that is *not* a solution of the original equation.

When solving equations with radicals, we use this idea of raising both sides to a power, which is an application of the **power rule.**

Power Rule for Solving a Radical Equation

If both sides of an equation are raised to the same power, all solutions of the original equation are also solutions of the new equation.

VOCABULARY

- ☐ radical equation
- ☐ proposed solution
- ☐ extraneous solution

The power rule does not say that all solutions of the new equation are solutions of the original equation. They may or may not be. A value of the variable that appears to be a solution is a **proposed solution.** Such values that do not satisfy the original equation are **extraneous solutions.** They must be rejected.

CAUTION ***When the power rule is used to solve an equation, every solution of the new equation must be checked in the original equation.***

NOW TRY EXERCISE 1

Solve $\sqrt{9x+7} = 5$.

EXAMPLE 1 Using the Power Rule

Solve $\sqrt{3x+4} = 8$.

$(\sqrt{a})^2 = \sqrt{a} \cdot \sqrt{a} = a$ for $a \geq 0$.

$(\sqrt{3x+4})^2 = 8^2$	Use the power rule and square each side.
$3x + 4 = 64$	Apply the exponents.
$3x = 60$	Subtract 4.
$x = 20$	Divide by 3.

CHECK

$\sqrt{3x+4} = 8$	Original equation
$\sqrt{3 \cdot 20 + 4} \stackrel{?}{=} 8$	Let $x = 20$.
$\sqrt{64} \stackrel{?}{=} 8$	Simplify.
$8 = 8$ ✓	True

Because 20 satisfies the *original* equation, the solution set is $\{20\}$. NOW TRY 

NOW TRY ANSWER

1. $\{2\}$

Solving an Equation with Radicals

Step 1 **Isolate the radical.** Make sure that one radical term is alone on one side of the equation.

Step 2 **Apply the power rule.** Raise each side of the equation to a power that is the same as the index of the radical.

Step 3 **Solve** the resulting equation. If it still contains a radical, repeat Steps 1 and 2.

Step 4 **Check** all proposed solutions in the *original* equation. Discard any values that are not solutions of the original equation.

CAUTION Remember to check all proposed solutions (Step 4) or an incorrect solution set may result.

NOW TRY EXERCISE 2

Solve $\sqrt{3x+4}+5=0$.

EXAMPLE 2 **Using the Power Rule**

Solve $\sqrt{5x-1}+3=0$.

Step 1 $\sqrt{5x-1}=-3$ To isolate the radical on one side, subtract 3 from each side.

Step 2 $(\sqrt{5x-1})^2=(-3)^2$ Square each side.

Step 3 $5x-1=9$ Apply the exponents.

$5x=10$ Add 1.

$x=2$ Divide by 5.

Step 4 **CHECK** $\sqrt{5x-1}+3=0$ Original equation

$\sqrt{5\cdot 2-1}+3\stackrel{?}{=}0$ Let $x=2$.

Be sure to check the proposed solution.

$\sqrt{9}+3\stackrel{?}{=}0$ Evaluate the radicand.

$3+3\stackrel{?}{=}0$ Take the square root.

$6=0$ False

This false result shows that the *proposed* solution 2 is *not* a solution of the original equation. It is extraneous. The solution set is $\varnothing$. NOW TRY

NOTE We could have determined after Step 1 that the equation in **Example 2** has no solution because the expression on the left cannot equal a negative number.

$\sqrt{5x-1}=-3$ A square root radical cannot be negative.

OBJECTIVE 2 **Solve radical equations that require additional steps.**

Recall the following rule for squaring a binomial.

$$(x+y)^2=x^2+2xy+y^2$$

NOW TRY ANSWER

2. $\varnothing$

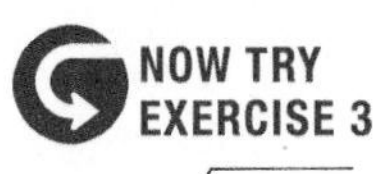

NOW TRY EXERCISE 3

Solve $\sqrt{16 - x} = x + 4$.

EXAMPLE 3 Using the Power Rule (Squaring a Binomial)

Solve $\sqrt{4 - x} = x + 2$.

Step 1 The radical is isolated on the left side of the equation.

Step 2 Square each side. On the right, $(x + 2)^2 = x^2 + 2(x)(2) + 2^2$.

$$\left(\sqrt{4 - x}\right)^2 = (x + 2)^2$$

Remember the middle term.

$$4 - x = x^2 + 4x + 4$$

Twice the product of 2 and x

Step 3 The new equation is quadratic, so write it in standard form.

$x^2 + 5x = 0$ Subtract 4, add x, and interchange sides.

$x(x + 5) = 0$ Factor.

Set *each* factor equal to 0. $x = 0$ or $x + 5 = 0$ Zero-factor property

$x = -5$ Solve.

Step 4 Check each proposed solution in the original equation.

CHECK

$\sqrt{4 - x} = x + 2$		$\sqrt{4 - x} = x + 2$	
$\sqrt{4 - 0} \stackrel{?}{=} 0 + 2$	Let $x = 0$.	$\sqrt{4 - (-5)} \stackrel{?}{=} -5 + 2$	Let $x = -5$.
$\sqrt{4} \stackrel{?}{=} 2$		$\sqrt{9} \stackrel{?}{=} -3$	
$2 = 2$ ✓	True	$3 = -3$	False

The solution set is $\{0\}$. The proposed solution -5 is extraneous.

NOW TRY 

NOW TRY EXERCISE 4

Solve

$\sqrt{x^2 - 3x + 18} = x + 3$.

EXAMPLE 4 Using the Power Rule (Squaring a Binomial)

Solve $\sqrt{x^2 - 4x + 9} = x - 1$.

Square each side. On the right, $(x - 1)^2 = x^2 - 2(x)(1) + 1^2$.

$$\left(\sqrt{x^2 - 4x + 9}\right)^2 = (x - 1)^2$$

Remember the middle term.

$$x^2 - 4x + 9 = x^2 - 2x + 1$$

Twice the product of x and -1

$-2x = -8$ Subtract x^2 and 9. Add $2x$.

$x = 4$ Divide by -2.

CHECK

$\sqrt{x^2 - 4x + 9} = x - 1$ Original equation

$\sqrt{4^2 - 4 \cdot 4 + 9} \stackrel{?}{=} 4 - 1$ Let $x = 4$.

$\sqrt{9} \stackrel{?}{=} 4 - 1$ Evaluate the radicand.

$3 = 3$ ✓ True

The solution set is $\{4\}$.

NOW TRY 

NOW TRY ANSWERS

3. $\{0\}$

4. $\{1\}$

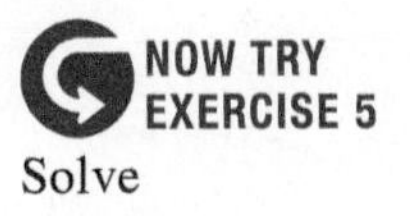

Solve

$\sqrt{3x+1} - \sqrt{x+4} = 1.$

EXAMPLE 5 Using the Power Rule (Squaring Twice)

Solve $\sqrt{5x+6} + \sqrt{3x+4} = 2$.

Isolate one radical on one side by subtracting $\sqrt{3x+4}$ from each side.

$$\sqrt{5x+6} = 2 - \sqrt{3x+4} \quad \text{Subtract } \sqrt{3x+4}.$$

$$\left(\sqrt{5x+6}\right)^2 = \left(2 - \sqrt{3x+4}\right)^2 \quad \text{Square each side.}$$

$$5x + 6 = 4 - 4\sqrt{3x+4} + (3x+4)$$

Be careful here.

Remember the middle term.

Twice the product of 2 and $-\sqrt{3x+4}$

The equation still contains a radical, so isolate the radical term on the right and square both sides again.

$$5x + 6 = 8 - 4\sqrt{3x+4} + 3x \quad \text{Combine like terms.}$$

$$2x - 2 = -4\sqrt{3x+4} \quad \text{Subtract 8 and } 3x.$$

Divide *each* term by 2.

$$x - 1 = -2\sqrt{3x+4} \quad \text{Divide each term by 2 to obtain smaller numbers.}$$

$$(x-1)^2 = \left(-2\sqrt{3x+4}\right)^2 \quad \text{Square each side again.}$$

$$x^2 - 2x + 1 = (-2)^2\left(\sqrt{3x+4}\right)^2 \quad \text{On the right, } (ab)^2 = a^2b^2.$$

$$x^2 - 2x + 1 = 4(3x+4) \quad \text{Apply the exponents.}$$

$$x^2 - 2x + 1 = 12x + 16 \quad \text{Distributive property}$$

$$x^2 - 14x - 15 = 0 \quad \text{Standard form}$$

$$(x-15)(x+1) = 0 \quad \text{Factor.}$$

$$x - 15 = 0 \quad \text{or} \quad x + 1 = 0 \quad \text{Zero-factor property}$$

$$x = 15 \quad \text{or} \quad x = -1 \quad \text{Solve each equation.}$$

CHECK First check the proposed solution 15. Then check -1.

$$\sqrt{5x+6} + \sqrt{3x+4} = 2 \quad \text{Original equation}$$

$$\sqrt{5(15)+6} + \sqrt{3(15)+4} \stackrel{?}{=} 2 \quad \text{Let } x = 15.$$

$$\sqrt{81} + \sqrt{49} \stackrel{?}{=} 2 \quad \text{Evaluate the radicands.}$$

$$9 + 7 \stackrel{?}{=} 2 \quad \text{Take square roots.}$$

$$16 = 2 \quad \text{False}$$

$$\sqrt{5x+6} + \sqrt{3x+4} = 2 \quad \text{Original equation}$$

$$\sqrt{5(-1)+6} + \sqrt{3(-1)+4} \stackrel{?}{=} 2 \quad \text{Let } x = -1.$$

$$\sqrt{1} + \sqrt{1} \stackrel{?}{=} 2 \quad \text{Evaluate the radicands.}$$

$$1 + 1 \stackrel{?}{=} 2 \quad \text{Take square roots.}$$

$$2 = 2 \ \checkmark \quad \text{True}$$

The proposed solution 15 is extraneous and must be rejected, but -1 is valid. Thus, the solution set is $\{-1\}$.

NOW TRY

NOW TRY ANSWER

5. $\{5\}$

OBJECTIVE 3 Solve radical equations with indexes greater than 2.

NOW TRY EXERCISE 6
Solve $\sqrt[3]{4x-5} = \sqrt[3]{3x+2}$.

EXAMPLE 6 Using the Power Rule for a Power Greater Than 2

Solve $\sqrt[3]{z+5} = \sqrt[3]{2z-6}$.

$$\left(\sqrt[3]{z+5}\right)^3 = \left(\sqrt[3]{2z-6}\right)^3 \quad \text{Cube each side.}$$

$$z + 5 = 2z - 6 \quad (\sqrt[3]{a})^3 = a$$

$$11 = z \quad \text{Subtract } z. \text{ Add 6.}$$

CHECK

$$\sqrt[3]{z+5} = \sqrt[3]{2z-6} \quad \text{Original equation}$$

$$\sqrt[3]{11+5} \stackrel{?}{=} \sqrt[3]{2 \cdot 11 - 6} \quad \text{Let } z = 11.$$

$$\sqrt[3]{16} = \sqrt[3]{16} \checkmark \quad \text{True}$$

The solution set is $\{11\}$.

OBJECTIVE 4 Use the power rule to solve a formula for a specified variable.

NOW TRY EXERCISE 7
Solve the formula for a.

$$x = \sqrt{\frac{y+2}{a}}$$

EXAMPLE 7 Solving a Formula from Electronics for a Variable

An important property of a radio-frequency transmission line is its **characteristic impedance,** represented by Z and measured in ohms. If L and C are the inductance and capacitance, respectively, per unit of length of the line, then these quantities are related by the following formula. Solve this formula for C.

$$Z = \sqrt{\frac{L}{C}}$$

Our goal is to isolate C on one side of the equality symbol.

$$Z^2 = \left(\sqrt{\frac{L}{C}}\right)^2 \quad \text{Square each side.}$$

$$Z^2 = \frac{L}{C} \quad (\sqrt{a})^2 = a$$

$$CZ^2 = L \quad \text{Multiply by } C.$$

$$C = \frac{L}{Z^2} \quad \text{Divide by } Z^2.$$

NOW TRY

NOW TRY ANSWERS
6. $\{7\}$
7. $a = \frac{y+2}{x^2}$

7.6 Exercises

 FOR EXTRA HELP

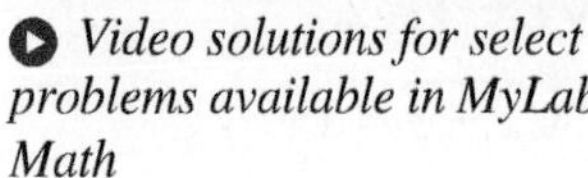 MyLab Math

Video solutions for select problems available in MyLab Math

Concept Check *Check each equation to see if the given value for x is a solution.*

1. $\sqrt{3x+18} - x = 0$
   (a) 6 **(b)** -3

2. $\sqrt{3x-3} - x + 1 = 0$
   (a) 1 **(b)** 4

3. $\sqrt{x+2} - \sqrt{9x-2} = -2\sqrt{x-1}$
   (a) 2 **(b)** 7

4. $\sqrt{8x-3} = 2x$
   (a) $\frac{3}{2}$ **(b)** $\frac{1}{2}$

5. **Concept Check** A student solved the following equation and claimed that 9 is a solution of the equation.

$$\sqrt{x} = -3$$

He received no credit for his answer. **WHAT WENT WRONG?** Give the correct solution set.

6. **Concept Check** A student solved the following equation and obtained the proposed solutions $x = -3$ and $x = 6$.

$$\sqrt{3x + 18} = x$$

She gave $\{-3, 6\}$ as the solution set. **WHAT WENT WRONG?** Give the correct solution set.

Solve each equation. ***See Examples 1–4.***

7. $\sqrt{x - 2} = 3$
8. $\sqrt{x + 1} = 7$
9. $\sqrt{6k - 1} = 1$
10. $\sqrt{7x - 3} = 6$
11. $\sqrt{4r + 3} + 1 = 0$
12. $\sqrt{5k - 3} + 2 = 0$
13. $\sqrt{3x + 1} - 4 = 0$
14. $\sqrt{5x + 1} - 11 = 0$
15. $4 - \sqrt{x - 2} = 0$
16. $9 - \sqrt{4x + 1} = 0$
17. $\sqrt{9x - 4} = \sqrt{8x + 1}$
18. $\sqrt{4x - 2} = \sqrt{3x + 5}$
19. $2\sqrt{x} = \sqrt{3x + 4}$
20. $2\sqrt{x} = \sqrt{5x - 16}$
21. $3\sqrt{x - 1} = 2\sqrt{2x + 2}$
22. $5\sqrt{4x + 1} = 3\sqrt{10x + 25}$
23. $x = \sqrt{x^2 + 4x - 20}$
24. $x = \sqrt{x^2 - 3x + 18}$
25. $x = \sqrt{x^2 + 3x + 9}$
26. $x = \sqrt{x^2 - 4x - 8}$
27. $\sqrt{9 - x} = x + 3$
28. $\sqrt{36 - x} = x + 6$
29. $\sqrt{5 - x} = x + 1$
30. $\sqrt{3 - x} = x + 3$
31. $\sqrt{6x + 7} = x + 2$
32. $\sqrt{4x + 13} = x + 4$
33. $\sqrt{k^2 + 2k + 9} = k + 3$
34. $\sqrt{x^2 - 3x + 3} = x - 1$
35. $\sqrt{z^2 + 12z - 4} + 4 - z = 0$
36. $\sqrt{x^2 - 15x + 15} + 5 - x = 0$
37. $\sqrt{r^2 + 9r + 15} - r - 4 = 0$
38. $\sqrt{m^2 + 3m + 12} - m - 2 = 0$

39. **Concept Check** When solving the equation $\sqrt{3x + 4} = 8 - x$, a student wrote the following as her first step.

$$3x + 4 = 64 + x^2$$

WHAT WENT WRONG? Solve the given equation correctly.

40. **Concept Check** When solving the equation $\sqrt{5x + 6} - \sqrt{x + 3} = 3$, a student wrote the following as his first step.

$$(5x + 6) + (x + 3) = 9$$

WHAT WENT WRONG? Solve the given equation correctly.

Solve each equation. ***See Example 5.***

41. $\sqrt{k + 2} - \sqrt{k - 3} = 1$
42. $\sqrt{r + 6} - \sqrt{r - 2} = 2$
43. $\sqrt{2r + 11} - \sqrt{5r + 1} = -1$
44. $\sqrt{3x - 2} - \sqrt{x + 3} = 1$
45. $\sqrt{3p + 4} - \sqrt{2p - 4} = 2$
46. $\sqrt{4x + 5} - \sqrt{2x + 2} = 1$
47. $\sqrt{3 - 3p} - 3 = \sqrt{3p + 2}$
48. $\sqrt{4x + 7} - 4 = \sqrt{4x - 1}$
49. $\sqrt{2\sqrt{x + 11}} = \sqrt{4x + 2}$
50. $\sqrt{1 + \sqrt{24 - 10x}} = \sqrt{3x + 5}$

Solve each equation. ***See Example 6.***

51. $\sqrt[3]{p-1}=2$

52. $\sqrt[3]{x+8}=3$

53. $\sqrt[3]{2x+5}=\sqrt[3]{6x+1}$

54. $\sqrt[3]{p+5}=\sqrt[3]{2p-4}$

55. $\sqrt[3]{2m-1}=\sqrt[3]{m+13}$

56. $\sqrt[3]{2k-11}=\sqrt[3]{5k+1}$

57. $\sqrt[3]{x^2+5x+1}=\sqrt[3]{x^2+4x}$

58. $\sqrt[3]{r^2+2r+8}=\sqrt[3]{r^2+3r+12}$

59. $\sqrt[4]{x+12}=\sqrt[4]{3x-4}$

60. $\sqrt[4]{z+11}=\sqrt[4]{2z+6}$

61. $\sqrt[3]{x-8}+2=0$

62. $\sqrt[3]{r+1}+1=0$

63. $\sqrt[4]{2k-5}+4=0$

64. $\sqrt[4]{8z-3}+2=0$

65. $\sqrt[3]{r^2+2r+8}=\sqrt[3]{r^2+3r+12}$

66. $\sqrt[3]{x^2+7x+2}=\sqrt[3]{x^2+6x+1}$

Extending Skills *For each equation, write the expressions with rational exponents as radical expressions, and then solve, using the procedures explained in this section.*

67. $(2x-9)^{1/2}=2+(x-8)^{1/2}$

68. $(3w+7)^{1/2}=1+(w+2)^{1/2}$

69. $(2w-1)^{2/3}-w^{1/3}=0$

70. $(x^2-2x)^{1/3}-x^{1/3}=0$

Solve each formula for the specified variable. ***See Example 7.*** *(Data from Cooke, N., and J. Orleans,* Mathematics Essential to Electricity and Radio, *McGraw-Hill.)*

71. $Z=\sqrt{\frac{L}{C}}$ for L

72. $r=\sqrt{\frac{\mathcal{A}}{\pi}}$ for $\mathcal{A}$

73. $V=\sqrt{\frac{2K}{m}}$ for K

74. $V=\sqrt{\frac{2K}{m}}$ for m

75. $r=\sqrt{\frac{Mm}{F}}$ for M

76. $r=\sqrt{\frac{Mm}{F}}$ for F

To find the rotational rate N of a space station, the formula

$$N=\frac{1}{2\pi}\sqrt{\frac{a}{r}}$$

can be used. Here, a is the acceleration and r represents the radius of the space station in meters. To find the value of r that will make N simulate the effect of gravity on Earth, the equation must be solved for r, using the required value of N. (Data from Kastner, B., Space Mathematics, *NASA.)*

77. Solve the equation for the indicated variable.

(a) for r **(b)** for a

78. If $a=9.8$ m per sec^2, find the value of r (to the nearest tenth) using each value of N.

(a) $N=0.063$ rotation per sec **(b)** $N=0.04$ rotation per sec

7.7 Complex Numbers

OBJECTIVES

1 Simplify numbers of the form $\sqrt{-b}$, where $b > 0$.
2 Identify subsets of the complex numbers.
3 Add and subtract complex numbers.
4 Multiply complex numbers.
5 Divide complex numbers.
6 Simplify powers of i.

Recall that the set of real numbers includes many other number sets—the rational numbers, integers, and natural numbers, for example. In this section, we introduce a new set of numbers that includes the set of real numbers, as well as numbers that are even roots of negative numbers, like $\sqrt{-2}$.

OBJECTIVE 1 Simplify numbers of the form $\sqrt{-b}$, where $b > 0$.

The equation $x^2 + 1 = 0$ has no real number solution because any solution must be a number whose square is -1. In the set of real numbers, all squares are *nonnegative* numbers because the product of two positive numbers or two negative numbers is positive and $0^2 = 0$. To provide a solution of the equation

$$x^2 + 1 = 0,$$

we introduce a new number i.

VOCABULARY

- ☐ complex number
- ☐ real part
- ☐ imaginary part
- ☐ nonreal complex number
- ☐ pure imaginary number
- ☐ complex conjugates

Imaginary Unit i

The **imaginary unit i** is defined as follows.

$$\boldsymbol{i = \sqrt{-1}}, \quad \text{and thus} \quad \boldsymbol{i^2 = -1}$$

That is, i is the principal square root of -1.

We can use this definition to define any square root of a negative real number.

Meaning of $\sqrt{-b}$

For any positive real number b, $\quad \boldsymbol{\sqrt{-b} = i\sqrt{b}}$.

NOW TRY EXERCISE 1

Write each number as a product of a real number and i.

(a) $\sqrt{-49}$ (b) $-\sqrt{-121}$

(c) $\sqrt{-3}$ (d) $\sqrt{-32}$

EXAMPLE 1 Simplifying Square Roots of Negative Numbers

Write each number as a product of a real number and i.

(a) $\sqrt{-100} = i\sqrt{100} = 10i$

(b) $-\sqrt{-36} = -i\sqrt{36} = -6i$

(c) $\sqrt{-2} = i\sqrt{2}$

(d) $\sqrt{-54} = i\sqrt{54} = i\sqrt{9 \cdot 6} = 3i\sqrt{6}$

NOW TRY

CAUTION It is easy to mistake $\sqrt{2}i$ for $\sqrt{2i}$ with the i under the radical. For this reason, we usually write $\sqrt{2}i$ as $i\sqrt{2}$, as in the definition of $\sqrt{-b}$.

When finding a product such as $\sqrt{-4} \cdot \sqrt{-9}$, we cannot use the product rule for radicals because it applies only to *nonnegative* radicands.

For this reason, we change $\sqrt{-b}$ to the form $i\sqrt{b}$ before performing any multiplications or divisions.

NOW TRY ANSWERS

1. (a) $7i$ (b) $-11i$ (c) $i\sqrt{3}$ (d) $4i\sqrt{2}$

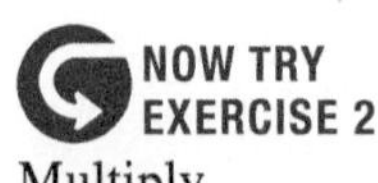
NOW TRY EXERCISE 2

Multiply.

(a) $\sqrt{-4} \cdot \sqrt{-16}$

(b) $\sqrt{-5} \cdot \sqrt{-11}$

(c) $\sqrt{-3} \cdot \sqrt{-12}$

(d) $\sqrt{13} \cdot \sqrt{-2}$

EXAMPLE 2 Multiplying Square Roots of Negative Numbers

Multiply.

(a) $\sqrt{-4} \cdot \sqrt{-9}$

$= i\sqrt{4} \cdot i\sqrt{9}$ — $\sqrt{-b} = i\sqrt{b}$ (First write all square roots in terms of i.)

$= i \cdot 2 \cdot i \cdot 3$ — Take square roots.

$= 6i^2$ — Multiply.

$= 6(-1)$ — Substitute -1 for i^2.

$= -6$

(b) $\sqrt{-3} \cdot \sqrt{-7}$

$= i\sqrt{3} \cdot i\sqrt{7}$ — $\sqrt{-b} = i\sqrt{b}$ (First write all square roots in terms of i.)

$= i^2\sqrt{3 \cdot 7}$ — Product rule

$= (-1)\sqrt{21}$ — Substitute -1 for i^2.

$= -\sqrt{21}$ — $(-1)a = -a$

(c) $\sqrt{-2} \cdot \sqrt{-8}$

$= i\sqrt{2} \cdot i\sqrt{8}$ — $\sqrt{-b} = i\sqrt{b}$

$= i^2\sqrt{2 \cdot 8}$ — Product rule

$= (-1)\sqrt{16}$ — $i^2 = -1$

$= -4$ — Take the square root.

(d) $\sqrt{-5} \cdot \sqrt{6}$

$= i\sqrt{5} \cdot \sqrt{6}$

$= i\sqrt{30}$

NOW TRY

CAUTION Using the product rule for radicals *before* using the definition of $\sqrt{-b}$ gives an *incorrect* answer. **Example 2(a)** shows that

$\sqrt{-4} \cdot \sqrt{-9} = -6,$ — Correct (**Example 2(a)**)

but $\sqrt{-4(-9)} = \sqrt{36} = 6.$ — Incorrect

Thus, $\sqrt{-4} \cdot \sqrt{-9} \neq \sqrt{-4(-9)}.$

NOW TRY EXERCISE 3

Divide.

(a) $\dfrac{\sqrt{-72}}{\sqrt{-8}}$ (b) $\dfrac{\sqrt{-48}}{\sqrt{3}}$

EXAMPLE 3 Dividing Square Roots of Negative Numbers

Divide.

(a) $\dfrac{\sqrt{-75}}{\sqrt{-3}}$

$= \dfrac{i\sqrt{75}}{i\sqrt{3}}$ (First write all square roots in terms of i.)

$= \sqrt{\dfrac{75}{3}}$ — $\frac{i}{i} = 1$; Quotient rule

$= \sqrt{25}$ — Divide.

$= 5$

(b) $\dfrac{\sqrt{-32}}{\sqrt{8}}$

$= \dfrac{i\sqrt{32}}{\sqrt{8}}$ — $\sqrt{-32} = i\sqrt{32}$

$= i\sqrt{\dfrac{32}{8}}$ — Quotient rule

$= i\sqrt{4}$ — Divide.

$= 2i$

NOW TRY

NOW TRY ANSWERS

2. (a) -8 (b) $-\sqrt{55}$ (c) -6 (d) $i\sqrt{26}$

3. (a) 3 (b) $4i$

OBJECTIVE 2 Identify subsets of the complex numbers.

A new set of numbers, the *complex numbers,* is defined as follows.

Complex Number

If a and b are real numbers, then any number of the form

$$a + bi$$

↑ Real part ↑ Imaginary part

is a **complex number.** In the complex number $a + bi$, the number a is the **real part** and b is the **imaginary part.***

The following important concepts apply to a complex number $a + bi$.

1. If $b = 0$, then $a + bi = a$, which is a real number.

   Thus, the set of real numbers is a subset of the set of complex numbers. See FIGURE 14.

2. If $b \neq 0$, then $a + bi$ is a **nonreal complex number.**

   Examples: $7 + 2i, \quad -1 - i$

3. If $a = 0$ and $b \neq 0$, then the nonreal complex number is a **pure imaginary number.**

   Examples: $3i, \quad -16i$

A complex number written in the form $a + bi$ is in **standard form.** In this section, most answers will be given in standard form, but if $a = 0$ or $b = 0$, we consider answers such as a or bi to be in standard form.

The relationships among the various sets of numbers are shown in FIGURE 14.

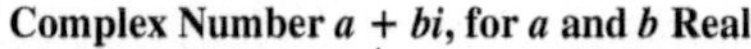

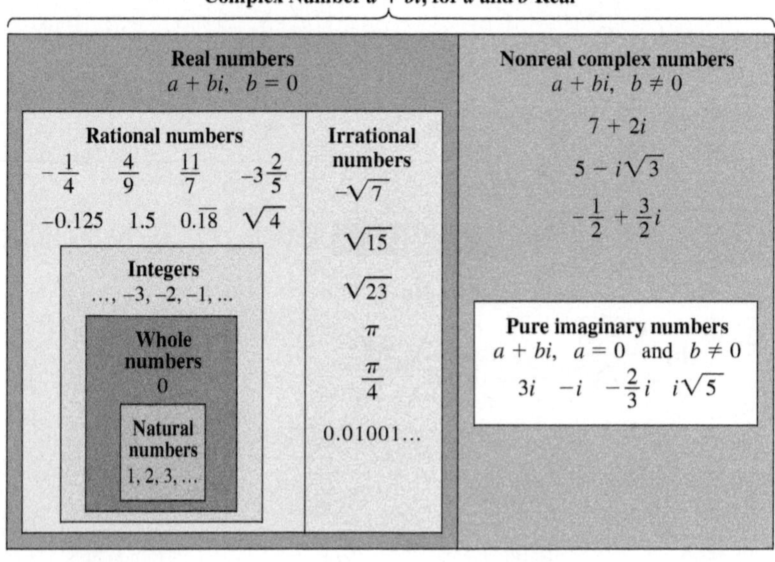

FIGURE 14

*Some texts define bi as the imaginary part of the complex number $a + bi$.

OBJECTIVE 3 Add and subtract complex numbers.

The commutative, associative, and distributive properties for real numbers are also valid for complex numbers.

To add complex numbers, we add their real parts and add their imaginary parts.

NOW TRY EXERCISE 4

Add.

(a) $(-3 + 2i) + (4 + 7i)$

(b) $(5 - i) + (-3 + 3i) + (6 - 4i)$

EXAMPLE 4 Adding Complex Numbers

Add.

(a) $(2 + 3i) + (6 + 4i)$

$= (2 + 6) + (3 + 4)i$ Commutative, associative, and distributive properties

$= 8 + 7i$ Add real parts. Add imaginary parts.

(b) $(4 + 2i) + (3 - i) + (-6 + 3i)$

$= [4 + 3 + (-6)] + [2 + (-1) + 3]i$ Commutative, associative, and distributive properties

$= 1 + 4i$ Add real parts. Add imaginary parts.

NOW TRY

To subtract complex numbers, we subtract their real parts and subtract their imaginary parts.

NOW TRY EXERCISE 5

Subtract.

(a) $(7 + 10i) - (3 + 5i)$

(b) $(5 - 2i) - (9 - 7i)$

(c) $(-1 + 12i) - (-1 - i)$

EXAMPLE 5 Subtracting Complex Numbers

Subtract.

(a) $(6 + 5i) - (3 + 2i)$

$= (6 - 3) + (5 - 2)i$ Commutative, associative, and distributive properties

$= 3 + 3i$ Subtract real parts. Subtract imaginary parts.

(b) $(7 - 3i) - (8 - 6i)$

$= (7 - 8) + [-3 - (-6)]i$

$= -1 + 3i$

(c) $(-9 + 4i) - (-9 + 8i)$

$= (-9 + 9) + (4 - 8)i$ Be careful.

$= 0 - 4i$

$= -4i$

NOW TRY

OBJECTIVE 4 Multiply complex numbers.

We multiply complex numbers in the same way that we multiply polynomials.

EXAMPLE 6 Multiplying Complex Numbers

Multiply.

(a) $4i(2 + 3i)$

$= 4i(2) + 4i(3i)$ Distributive property

$= 8i + 12i^2$ Multiply.

$= 8i + 12(-1)$ Substitute -1 for i^2.

$= -12 + 8i$ Standard form

NOW TRY ANSWERS

4. (a) $1 + 9i$ (b) $8 - 2i$
5. (a) $4 + 5i$ (b) $-4 + 5i$ (c) $13i$

NOW TRY EXERCISE 6
Multiply.
(a) $8i(3 - 5i)$
(b) $(7 - 2i)(4 + 3i)$

(b) $(3 + 5i)(4 - 2i)$

$$= \underbrace{3(4)}_{\text{First}} + \underbrace{3(-2i)}_{\text{Outer}} + \underbrace{5i(4)}_{\text{Inner}} + \underbrace{5i(-2i)}_{\text{Last}}$$ Use the FOIL method.

$= 12 - 6i + 20i - 10i^2$ Multiply.

$= 12 + 14i - 10(-1)$ Add imaginary parts; $i^2 = -1$

$= 12 + 14i + 10$ Multiply.

$= 22 + 14i$ Add real parts.

(c) $(2 + 3i)(1 - 5i)$

$= 2(1) + 2(-5i) + 3i(1) + 3i(-5i)$ FOIL method

$= 2 - 10i + 3i - 15i^2$ Multiply.

$= 2 - 7i - 15(-1)$ Add imaginary parts; $i^2 = -1$ (Use parentheses around -1 to avoid errors.)

$= 2 - 7i + 15$ Multiply.

$= 17 - 7i$ Add real parts. **NOW TRY**

The two complex numbers $a + bi$ and $a - bi$ are **complex conjugates,** or simply *conjugates,* of each other. ***The product of a complex number and its conjugate is always a real number,*** as shown here.

$$(a + bi)(a - bi)$$

$= a^2 - abi + abi - b^2i^2$ FOIL method

$= a^2 - b^2(-1)$ Combine like terms; $i^2 = -1$

$= a^2 + b^2$ (The product eliminates i.)

Example: $(3 + 7i)(3 - 7i) = 3^2 + 7^2 = 9 + 49 = 58$

OBJECTIVE 5 Divide complex numbers.

EXAMPLE 7 Dividing Complex Numbers

Divide.

(a) $\dfrac{8 + 9i}{5 + 2i}$ ($\frac{5 - 2i}{5 - 2i} = 1$)

$= \dfrac{(8 + 9i)(5 - 2i)}{(5 + 2i)(5 - 2i)}$ Multiply numerator and denominator by $5 - 2i$, the conjugate of the denominator.

$= \dfrac{40 - 16i + 45i - 18i^2}{5^2 + 2^2}$ In the denominator, $(a + bi)(a - bi) = a^2 + b^2$.

$= \dfrac{40 + 29i - 18(-1)}{25 + 4}$ In the numerator, add imaginary parts; $i^2 = -1$

$= \dfrac{58 + 29i}{29}$ Multiply. Add real parts. Add in the denominator.

$= \dfrac{29(2 + i)}{29}$ Factor the numerator. (Factor first. Then divide out the common factor.)

$= 2 + i$ Lowest terms

NOW TRY ANSWERS
6. (a) $40 + 24i$ (b) $34 + 13i$

NOW TRY EXERCISE 7

Find each quotient.

(a) $\frac{4+2i}{1+3i}$ **(b)** $\frac{5-4i}{i}$

(b) $\frac{1+i}{i}$

$$= \frac{(1+i)(-i)}{i(-i)}$$ Multiply numerator and denominator by $-i$, the conjugate of i.

$$= \frac{-i-i^2}{-i^2}$$ Use the distributive property in the numerator. Multiply in the denominator.

$$= \frac{-i-(-1)}{-(-1)}$$ Substitute -1 for i^2. Use parentheses to avoid errors.

$$= \frac{-i+1}{1}$$

$$= 1-i$$ $\frac{a}{1} = a$

NOW TRY 

OBJECTIVE 6 Simplify powers of i.

Powers of i can be simplified using the facts

$$i^2 = -1 \quad \text{and} \quad i^4 = (i^2)^2 = (-1)^2 = 1.$$

Consider the following powers of i.

$$i^1 = i \qquad i^5 = i^4 \cdot i = 1 \cdot i = i$$

$$i^2 = -1 \qquad i^6 = i^4 \cdot i^2 = 1(-1) = -1$$

$$i^3 = i^2 \cdot i = (-1) \cdot i = -i \qquad i^7 = i^4 \cdot i^3 = 1 \cdot (-i) = -i$$

$$i^4 = i^2 \cdot i^2 = (-1)(-1) = 1 \qquad i^8 = i^4 \cdot i^4 = 1 \cdot 1 = 1, \quad \text{and so on.}$$

Powers of i cycle through the same four outcomes

$$i, \quad -1, \quad -i, \quad \text{and} \quad 1$$

because i^4 has the same multiplicative property as 1. Also, any power of i with an exponent that is a multiple of 4 has value 1. As with real numbers, $i^0 = 1$.

NOW TRY EXERCISE 8

Find each power of i.

(a) i^{16} **(b)** i^{21}

(c) i^{-6} **(d)** i^{-13}

EXAMPLE 8 Simplifying Powers of i

Find each power of i.

(a) $i^{12} = (i^4)^3 = 1^3 = 1 \qquad i^4 = 1$

(b) $i^{39} = i^{36} \cdot i^3 = (i^4)^9 \cdot i^3 = 1^9 \cdot (-i) = 1 \cdot (-i) = -i$

(c) $i^{-2} = \frac{1}{i^2} = \frac{1}{-1} = -1$

(d) $i^{-15} = \frac{1}{i^{15}} = \frac{1}{(i^4)^3 \cdot i^3} = \frac{1}{1^3 \cdot (-i)} = \frac{1}{-i}$

We divide by multiplying the numerator and denominator by i, the conjugate of $-i$.

$$\frac{1}{-i} = \frac{1(i)}{-i(i)} = \frac{i}{-i^2} = \frac{i}{-(-1)} = \frac{i}{1} = i$$

NOW TRY 

NOTE In **Example 8(d)**, we could also multiply by a power of i that is a multiple of 4 (because $i^4 = 1$).

$$i^{-15} = i^{-15} \cdot i^{16} = i^1 = i \qquad i^{16} = 1$$

NOW TRY ANSWERS

7. (a) $1-i$ (b) $-4-5i$

8. (a) 1 (b) i (c) -1 (d) $-i$

7.7 Exercises

  MyLab Math

Video solutions for select problems available in MyLab Math

Concept Check *List all of the following sets to which each number belongs. A number may belong to more than one set.*

real numbers pure imaginary numbers nonreal complex numbers complex numbers

1. $3 + 5i$ **2.** $-7i$ **3.** $\sqrt{2}$

4. $\frac{13}{3}$ **5.** $\sqrt{-49}$ **6.** $-\sqrt{-8}$

Concept Check *Decide whether each expression is equal to 1, −1, i, or −i.*

7. $\sqrt{-1}$ **8.** $-\sqrt{-1}$ **9.** i^2

10. $-i^2$ **11.** $\frac{1}{i}$ **12.** $(-i)^2$

Write each number as a product of a real number and i. Simplify all radical expressions. ***See Example 1.***

13. $\sqrt{-169}$ **14.** $\sqrt{-225}$ **15.** $-\sqrt{-144}$ **16.** $-\sqrt{-196}$

17. $\sqrt{-5}$ **18.** $\sqrt{-21}$ **19.** $\sqrt{-48}$ **20.** $\sqrt{-96}$

21. Concept Check A student incorrectly multiplied as follows.

$$\sqrt{-9} \cdot \sqrt{-25} = \sqrt{225} = 15$$

WHAT WENT WRONG? Find the product correctly.

22. Concept Check A student incorrectly simplified $-\sqrt{-16}$ as follows.

$$-\sqrt{-16} = -(-4) = 4$$

WHAT WENT WRONG? Simplify the expression correctly.

Multiply or divide as indicated. ***See Examples 2 and 3.***

23. $\sqrt{-15} \cdot \sqrt{-15}$ **24.** $\sqrt{-19} \cdot \sqrt{-19}$ **25.** $\sqrt{-7} \cdot \sqrt{-15}$ **26.** $\sqrt{-3} \cdot \sqrt{-19}$

27. $\sqrt{-4} \cdot \sqrt{-25}$ **28.** $\sqrt{-9} \cdot \sqrt{-81}$ **29.** $\sqrt{-3} \cdot \sqrt{11}$ **30.** $\sqrt{-5} \cdot \sqrt{13}$

31. $\sqrt{5} \cdot \sqrt{-30}$ **32.** $\sqrt{-10} \cdot \sqrt{2}$ **33.** $\frac{\sqrt{-300}}{\sqrt{-100}}$ **34.** $\frac{\sqrt{-40}}{\sqrt{-10}}$

35. $\frac{\sqrt{-75}}{\sqrt{3}}$ **36.** $\frac{\sqrt{-160}}{\sqrt{10}}$ **37.** $\frac{-\sqrt{-64}}{\sqrt{-16}}$ **38.** $\frac{-\sqrt{-100}}{\sqrt{-25}}$

Add or subtract as indicated. Give answers in standard form. ***See Examples 4 and 5.***

39. $(3 + 2i) + (-4 + 5i)$ **40.** $(7 + 15i) + (-11 + 14i)$

41. $(5 - i) + (-5 + i)$ **42.** $(-2 + 6i) + (2 - 6i)$

43. $(4 + i) - (-3 - 2i)$ **44.** $(9 + i) - (3 + 2i)$

45. $(-3 - 4i) - (-1 - 4i)$ **46.** $(-2 - 3i) - (-5 - 3i)$

47. $(-4 + 11i) + (-2 - 4i) + (7 + 6i)$ **48.** $(-1 + i) + (2 + 5i) + (3 + 2i)$

49. $[(7 + 3i) - (4 - 2i)] + (3 + i)$ **50.** $[(7 + 2i) + (-4 - i)] - (2 + 5i)$

Concept Check *Fill in the blank with the correct response.*

51. Because $(4 + 2i) - (3 + i) = 1 + i$, using the definition of subtraction we can check this to find that

$$(1 + i) + (3 + i) = \_\_\_\_.$$

52. Because $\frac{-5}{2 - i} = -2 - i$, using the definition of division we can check this to find that

$$(-2 - i)(2 - i) = \_\_\_\_.$$

Multiply. Give answers in standard form. ***See Example 6.***

53. $(3i)(27i)$ **54.** $(5i)(125i)$ **55.** $(-8i)(-2i)$

56. $(-32i)(-2i)$ **57.** $5i(-6 + 2i)$ **58.** $3i(4 + 9i)$

59. $(4 + 3i)(1 - 2i)$ **60.** $(7 - 2i)(3 + i)$ **61.** $(4 + 5i)^2$

62. $(3 + 2i)^2$ **63.** $2i(-4 - i)^2$ **64.** $3i(-3 - i)^2$

65. $(12 + 3i)(12 - 3i)$ **66.** $(6 + 7i)(6 - 7i)$ **67.** $(4 + 9i)(4 - 9i)$

68. $(7 + 2i)(7 - 2i)$ **69.** $(1 + i)^2(1 - i)^2$ **70.** $(2 - i)^2(2 + i)^2$

Concept Check *Answer each of the following.*

71. Let a and b represent real numbers.

(a) What is the conjugate of $a + bi$?

(b) If we multiply $a + bi$ by its conjugate, we obtain ____ + ____, which is always a real number.

72. By what complex number should we multiply the numerator and denominator of $\frac{2 + i\sqrt{2}}{2 - i\sqrt{2}}$ to write the quotient in standard form?

A. $\sqrt{2}$ **B.** $i\sqrt{2}$ **C.** $2 + i\sqrt{2}$ **D.** $2 - i\sqrt{2}$

Divide. Give answers in standard form. ***See Example 7.***

73. $\frac{2}{1 - i}$ **74.** $\frac{2}{1 + i}$ **75.** $\frac{8i}{2 + 2i}$ **76.** $\frac{-8i}{1 + i}$

77. $\frac{-7 + 4i}{3 + 2i}$ **78.** $\frac{-38 - 8i}{7 + 3i}$ **79.** $\frac{2 - 3i}{2 + 3i}$ **80.** $\frac{-1 + 5i}{3 + 2i}$

81. $\frac{3 + i}{i}$ **82.** $\frac{5 - i}{i}$ **83.** $\frac{3 - i}{-i}$ **84.** $\frac{5 + i}{-i}$

Find each power of i. ***See Example 8.***

85. i^{18} **86.** i^{26} **87.** i^{89} **88.** i^{48}

89. i^{38} **90.** i^{102} **91.** i^{43} **92.** i^{83}

93. i^{-5} **94.** i^{-17} **95.** i^{-20} **96.** i^{-27}

Concept Check *Provide a short response for each problem.*

97. A student simplified i^{-18} as shown. Provide justification for this correct work.

$$i^{-18} = i^{-18} \cdot i^{20} = i^{-18+20} = i^2 = -1$$

98. Explain why the following two expressions must be equal. (Do not actually perform the computation.)

$$(46 + 25i)(3 - 6i) \quad \text{and} \quad (46 + 25i)(3 - 6i)i^{12}$$

Ohm's law *for the current I in a circuit with voltage E, resistance R, capacitive reactance X_c, and inductive reactance X_L is*

$$I = \frac{E}{R + (X_L - X_c)i}.$$

Use this law to work each exercise.

99. Find I if $E = 2 + 3i$, $R = 5$, $X_L = 4$, and $X_c = 3$.

100. Find E if $I = 1 - i$, $R = 2$, $X_L = 3$, and $X_c = 1$.

RELATING CONCEPTS For Individual or Group Work (Exercises 101–105)

Some equations have nonreal complex solutions. ***Work Exercises 101–105 in order,*** *to see how these nonreal complex solutions are related.*

101. Show that $1 + 5i$ is a solution of $x^2 - 2x + 26 = 0$.

102. Show that $1 - 5i$ is a solution of $x^2 - 2x + 26 = 0$.

103. From **Exercises 101 and 102,** the nonreal complex solutions of the equation

$$x^2 - 2x + 26 = 0$$

are $1 + 5i$ and $1 - 5i$. What do we call two complex numbers $a + bi$ and $a - bi$?

104. Show that $3 + 2i$ is a solution of the equation $x^2 - 6x + 13 = 0$.

105. Using the results of **Exercises 101–104,** make a conjecture about another nonreal complex solution of the equation $x^2 - 6x + 13 = 0$, and verify it.

Chapter 7 Summary

Key Terms

7.1

radicand
index (order)
radical
principal root
radical expression
square root function
cube root function

7.3

hypotenuse
legs (of a right triangle)
circle
center
radius

7.5

conjugates

7.6

radical equation
proposed solution
extraneous solution

7.7

complex number
real part
imaginary part
nonreal complex number
pure imaginary number
complex conjugates

New Symbols

$\sqrt{\ }$ radical symbol

$\sqrt[n]{a}$ radical; principal nth root of a

$\pm$ *"positive or negative"* or *"plus or minus"*

$\approx$ is approximately equal to

$a^{1/n}$ a to the power $\frac{1}{n}$

$a^{m/n}$ a to the power $\frac{m}{n}$

i imaginary unit

Test Your Word Power

See how well you have learned the vocabulary in this chapter.

1. A **radicand** is
   A. the index of a radical
   B. the number or expression under the radical symbol
   C. the positive root of a number
   D. the radical symbol.

2. The **Pythagorean theorem** states that, in a right triangle,
   A. the sum of the measures of the angles is 180°
   B. the sum of the lengths of the two shorter sides equals the length of the longest side
   C. the longest side is opposite the right angle
   D. the square of the length of the longest side equals the sum of the squares of the lengths of the two shorter sides.

3. A **hypotenuse** is
   A. either of the two shorter sides of a triangle
   B. the shortest side of a triangle
   C. the side opposite the right angle in a triangle
   D. the longest side in any triangle.

4. **Rationalizing the denominator** is the process of
   A. eliminating fractions from a radical expression
   B. changing the denominator of a fraction from a radical expression to a rational number
   C. clearing a radical expression of radicals
   D. multiplying radical expressions.

5. An **extraneous solution** is a value
   A. that does not satisfy the original equation
   B. that makes an equation true
   C. that makes an expression equal 0
   D. that checks in the original equation.

6. A **complex number** is
   A. a real number that includes a complex fraction
   B. a zero multiple of i
   C. a number of the form $a + bi$, where a and b are real numbers
   D. the square root of -1.

ANSWERS

1. B; *Example:* In $\sqrt{3xy}$, $3xy$ is the radicand. **2.** D; *Example:* In a right triangle where $a = 6$, $b = 8$, and $c = 10$, $6^2 + 8^2 = 10^2$.
3. C; *Example:* In a right triangle where the sides measure 9, 12, and 15 units, the hypotenuse is the side opposite the right angle, with measure 15 units.
4. B; *Example:* To rationalize the denominator of $\frac{5}{\sqrt{3}+1}$, multiply both numerator and denominator by $\sqrt{3} - 1$ to obtain $\frac{5(\sqrt{3}-1)}{2}$.
5. A; *Example:* The proposed solution 2 is extraneous in $\sqrt{5x - 1} + 3 = 0$. **6.** C; *Examples:* -5 (or $-5 + 0i$), $7i$ (or $0 + 7i$), $\sqrt{2} - 4i$

Quick Review

CONCEPTS	EXAMPLES

7.1 Radical Expressions and Graphs

CONCEPTS

$\sqrt[n]{a} = b$ **means** $b^n = a$.

$\sqrt[n]{a}$ is the **principal nth root** of a.

$\sqrt[n]{a^n} = |a|$ if n is even. $\quad \sqrt[n]{a^n} = a$ if n is odd.

EXAMPLES

Find each root.

$\sqrt{64} = 8$ Principal square root

$-\sqrt{64} = -8$ $\quad \sqrt{-64}$ is not a real number.

$\sqrt[4]{(-2)^4} = |-2| = 2$ $\quad \sqrt[3]{-27} = -3$

Functions Defined by Radical Expressions
The square root function is

$$f(x) = \sqrt{x}.$$

The cube root function is

$$f(x) = \sqrt[3]{x}.$$

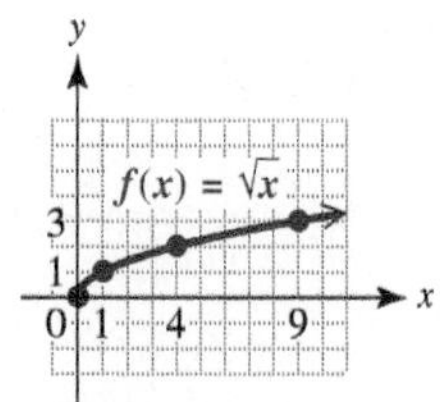

Square root function
Domain and range: $[0, \infty)$

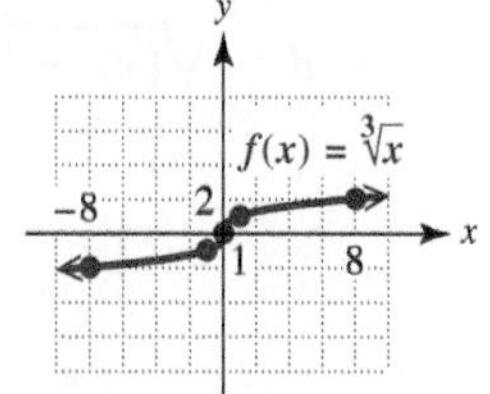

Cube root function
Domain and range: $(-\infty, \infty)$

CONCEPTS | **EXAMPLES**

7.2 Rational Exponents

$a^{1/n} = \sqrt[n]{a}$ whenever $\sqrt[n]{a}$ exists.

If m and n are positive integers with m/n in lowest terms, then

$$a^{m/n} = (a^{1/n})^m, \text{ provided that } a^{1/n} \text{ is a real number.}$$

All of the usual definitions and rules for exponents are valid for rational exponents.

Apply the rules for rational exponents.

$$81^{1/2} = \sqrt{81} = 9 \qquad -64^{1/3} = -\sqrt[3]{64} = -4$$

$$8^{5/3} = (8^{1/3})^5 = 2^5 = 32 \qquad (y^{2/5})^{10} = y^4$$

Write with positive exponents.

$$5^{-1/2} \cdot 5^{1/4} = 5^{-1/2+1/4} = 5^{-1/4} = \frac{1}{5^{1/4}}$$

$$\frac{x^{-1/3}}{x^{-1/2}} = x^{-1/3-(-1/2)} = x^{-1/3+1/2} = x^{1/6}, \quad x > 0$$

7.3 Simplifying Radicals, the Distance Formula, and Circles

Product and Quotient Rules for Radicals

If n is a natural number and $\sqrt[n]{a}$ and $\sqrt[n]{b}$ are real numbers, then the following hold true.

$$\sqrt[n]{a} \cdot \sqrt[n]{b} = \sqrt[n]{ab} \quad \text{and} \quad \sqrt[n]{\frac{a}{b}} = \frac{\sqrt[n]{a}}{\sqrt[n]{b}} \quad (\text{where } b \neq 0)$$

Simplify.

$$\sqrt{3} \cdot \sqrt{7} = \sqrt{21} \qquad \sqrt[5]{x^3y} \cdot \sqrt[5]{xy^2} = \sqrt[5]{x^4y^3}$$

$$\frac{\sqrt{x^5}}{\sqrt{x^4}} = \sqrt{\frac{x^5}{x^4}} = \sqrt{x}, \quad x > 0$$

Conditions for a Simplified Radical

1. The radicand has no factor raised to a power greater than or equal to the index.
2. The radicand has no fractions.
3. No denominator contains a radical.
4. Exponents in the radicand and the index of the radical have greatest common factor 1.

$$\sqrt{18} = \sqrt{9 \cdot 2} = \sqrt{9} \cdot \sqrt{2} = 3\sqrt{2}$$

$$\sqrt[3]{54x^5y^3} = \sqrt[3]{27x^3y^3 \cdot 2x^2} = \sqrt[3]{27x^3y^3} \cdot \sqrt[3]{2x^2} = 3xy\sqrt[3]{2x^2}$$

$$\sqrt{\frac{7}{4}} = \frac{\sqrt{7}}{\sqrt{4}} = \frac{\sqrt{7}}{2}$$

$$\sqrt[9]{x^3} = x^{3/9} = x^{1/3}, \quad \text{or} \quad \sqrt[3]{x}$$

Pythagorean Theorem

If a and b are the lengths of the shorter sides of a right triangle and c is the length of the longest side, then the following holds.

$$a^2 + b^2 = c^2$$

Find the length of the unknown side in the right triangle.

$$10^2 + b^2 = (2\sqrt{61})^2$$
$$100 + b^2 = 4(61)$$
$$100 + b^2 = 244$$
$$b^2 = 144$$
$$b = 12$$

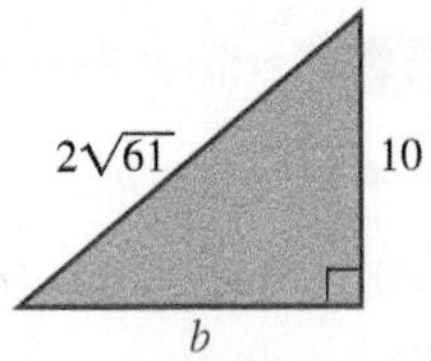

Distance Formula

The distance d between the points (x_1, y_1) and (x_2, y_2) is given by the following.

$$d = \sqrt{(x_2 - x_1)^2 + (y_2 - y_1)^2}$$

Find the distance between $(3, -2)$ and $(-1, 1)$.

$d = \sqrt{(-1-3)^2 + [1-(-2)]^2}$	Substitute.
$d = \sqrt{(-4)^2 + 3^2}$	Subtract.
$d = \sqrt{16 + 9}$	Square.
$d = \sqrt{25}$	Add.
$d = 5$	Take the square root.

CONCEPTS	EXAMPLES
Equation of a Circle A circle with center (h, k) and radius $r > 0$ has an equation of the form $$(x - h)^2 + (y - k)^2 = r^2.$$ A circle with center at the origin $(0, 0)$ and radius $r > 0$ has the following equation. $$x^2 + y^2 = r^2$$	Graph $(x + 2)^2 + (y - 3)^2 = 25$. This equation, which can be written $$[x - (-2)]^2 + (y - 3)^2 = 5^2,$$ is a circle with center $(-2, 3)$ and radius 5. 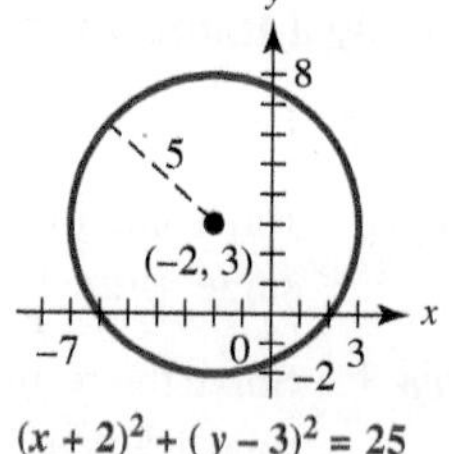 $(x + 2)^2 + (y - 3)^2 = 25$
7.4 Adding and Subtracting Radical Expressions Add or subtract radical expressions using the distributive property. $$ac + bc = (a + b)c$$ ***Only radical expressions with the same index and the same radicand may be combined.***	Simplify. $$\begin{aligned}&2\sqrt{28} - 3\sqrt{63} + 8\sqrt{112}\\&= 2\sqrt{4 \cdot 7} - 3\sqrt{9 \cdot 7} + 8\sqrt{16 \cdot 7}\\&= 2 \cdot 2\sqrt{7} - 3 \cdot 3\sqrt{7} + 8 \cdot 4\sqrt{7}\\&= 4\sqrt{7} - 9\sqrt{7} + 32\sqrt{7}\\&= (4 - 9 + 32)\sqrt{7}\\&= 27\sqrt{7}\end{aligned}$$ $\sqrt{15} + \sqrt{30}$, $\sqrt{3} + \sqrt[3]{9}$ } Cannot be simplified further
7.5 Multiplying and Dividing Radical Expressions Multiply binomial radical expressions using the FOIL method. Special product rules may apply. $$(x + y)(x - y) = x^2 - y^2$$ $$(x + y)^2 = x^2 + 2xy + y^2$$ $$(x - y)^2 = x^2 - 2xy + y^2$$	Perform the operations and simplify. $(\sqrt{2} + \sqrt{7})(\sqrt{3} - \sqrt{6})$ F O I L $= \sqrt{6} - 2\sqrt{3} + \sqrt{21} - \sqrt{42}$ $\quad \sqrt{12} = 2\sqrt{3}$ $(\sqrt{5} - \sqrt{10})(\sqrt{5} + \sqrt{10})$ $= 5 - 10$ $= -5$ $(\sqrt{3} - \sqrt{2})^2$ $= 3 - 2\sqrt{3} \cdot \sqrt{2} + 2$ $= 5 - 2\sqrt{6}$
Rationalizing a Denominator Rationalize the denominator by multiplying both the numerator and the denominator by the same expression, one that will yield a rational number in the final denominator.	$$\frac{\sqrt{7}}{\sqrt{5}} = \frac{\sqrt{7} \cdot \sqrt{5}}{\sqrt{5} \cdot \sqrt{5}} = \frac{\sqrt{35}}{5}$$ $$\frac{4}{\sqrt{5} - \sqrt{2}} = \frac{4(\sqrt{5} + \sqrt{2})}{(\sqrt{5} - \sqrt{2})(\sqrt{5} + \sqrt{2})} = \frac{4(\sqrt{5} + \sqrt{2})}{5 - 2} = \frac{4(\sqrt{5} + \sqrt{2})}{3}$$
To write a radical quotient in lowest terms, factor the numerator and denominator. Then divide out any common factor(s).	$$\frac{5 + 15\sqrt{6}}{10} = \frac{5(1 + 3\sqrt{6})}{5 \cdot 2} = \frac{1 + 3\sqrt{6}}{2}$$ Factor first, then divide.

CONCEPTS	EXAMPLES
7.6 Solving Equations with Radicals **Solving a Radical Equation** *Step 1* Isolate one radical on one side of the equation. *Step 2* Raise each side of the equation to a power that is the same as the index of the radical. *Step 3* Solve the resulting equation. If it still contains a radical, repeat Steps 1 and 2. *Step 4* Check all proposed solutions in the *original* equation. Discard any values that are not solutions of the original equation.	Solve. $\sqrt{2x+3}-x=0$ $\sqrt{2x+3}=x$ Add x. $(\sqrt{2x+3})^2=x^2$ Square each side. $2x+3=x^2$ Apply the exponents. $x^2-2x-3=0$ Standard form $(x-3)(x+1)=0$ Factor. $x-3=0$ or $x+1=0$ Zero-factor property $x=3$ or $x=-1$ Solve each equation. A check shows that 3 is a solution, but -1 is extraneous (because it leads to $2=0$, a false statement). The solution set is $\{3\}$.
7.7 Complex Numbers $i=\sqrt{-1}$, and thus $i^2=-1$. For any positive real number b, $\sqrt{-b}=i\sqrt{b}$. ***To multiply radicals with negative radicands, first change each factor to the form $i\sqrt{b}$ and then multiply. The same procedure applies to quotients.***	Simplify. $\sqrt{-25}=i\sqrt{25}=5i$ $\sqrt{-3}\cdot\sqrt{-27}$ $=i\sqrt{3}\cdot i\sqrt{27}$ $\sqrt{-b}=i\sqrt{b}$ $=i^2\sqrt{81}$ Product rule $=-1\cdot 9$ $i^2=-1$; Take the square root. $=-9$ Multiply. $\dfrac{\sqrt{-18}}{\sqrt{-2}}=\dfrac{i\sqrt{18}}{i\sqrt{2}}=\sqrt{\dfrac{18}{2}}=\sqrt{9}=3$
Adding and Subtracting Complex Numbers Add (or subtract) the real parts and add (or subtract) the imaginary parts.	Perform the operations. $(5+3i)+(8-7i)$ $=13-4i$ $(5+3i)-(8-7i)$ $=-3+10i$
Multiplying Complex Numbers Multiply complex numbers using the FOIL method.	Multiply. $(2+i)(5-3i)$ $=10-6i+5i-3i^2$ Use the FOIL method. $=10-i-3(-1)$ Add imaginary parts; $i^2=-1$ $=13-i$ Multiply. Add real parts.
Dividing Complex Numbers Divide complex numbers by multiplying the numerator and the denominator by the conjugate of the denominator.	Divide. $\dfrac{20}{3+i}$ $=\dfrac{20(3-i)}{(3+i)(3-i)}$ Multiply both numerator and denominator by the conjugate of the denominator. $=\dfrac{20(3-i)}{9-i^2}$ $(x+y)(x-y)=x^2-y^2$ $=\dfrac{20(3-i)}{10}$ $i^2=-1$ and $9-(-1)=10$. $=2(3-i)$ Divide out the common factor, 10. $=6-2i$ Distributive property

Chapter 7 Review Exercises

7.1 *Find each root.*

1. $\sqrt{1764}$ **2.** $-\sqrt{289}$ **3.** $\sqrt[3]{216}$

4. $\sqrt[3]{-125}$ **5.** $-\sqrt[3]{27}$ **6.** $\sqrt[5]{-32}$

7. $\sqrt{x^2}$ **8.** $\sqrt[3]{x^3}$ **9.** $\sqrt[4]{x^{20}}$

10. Under what conditions is $\sqrt[n]{a}$ not a real number?

Graph each function, and give its domain and range.

11. $f(x) = \sqrt{x - 1}$ **12.** $f(x) = \sqrt[3]{x} + 4$

Use a calculator to approximate each radical to three decimal places.

13. $-\sqrt{47}$ **14.** $\sqrt[3]{-129}$ **15.** $\sqrt[4]{605}$

16. $\sqrt[4]{500}$ **17.** $-\sqrt[3]{500}$ **18.** $-\sqrt{28}$

Solve each problem.

19. The time t in seconds for one complete swing of a pendulum is given by

$$t = 2\pi\sqrt{\frac{L}{g}},$$

where L is the length of the pendulum in feet, and g, the force due to gravity, is about 32 ft per sec^2. Find the time of a complete swing of a 3-ft pendulum to the nearest tenth of a second.

20. Use Heron's formula

$$\mathcal{A} = \sqrt{s(s - a)(s - b)(s - c)},$$

where $s = \frac{1}{2}(a + b + c)$, to find the area of a triangle with sides of lengths 11 in., 13 in., and 20 in.

7.2 *Evaluate each exponential.*

21. $49^{1/2}$ **22.** $-121^{1/2}$ **23.** $16^{5/4}$ **24.** $-8^{2/3}$

25. $-\left(\frac{36}{25}\right)^{3/2}$ **26.** $\left(-\frac{1}{8}\right)^{-5/3}$ **27.** $\left(\frac{81}{10{,}000}\right)^{-3/4}$ **28.** $(-16)^{3/4}$

Write each exponential as a radical. Assume that all variables represent positive real numbers. Use the definition that takes the root first.

29. $10x^{1/2}$ **30.** $(3a + b)^{-5/3}$

Write each radical as an exponential and simplify. Assume that all variables represent positive real numbers.

31. $\sqrt{7^9}$ **32.** $\sqrt{11^6}$ **33.** $\sqrt[3]{r^{12}}$ **34.** $\sqrt[4]{z^4}$

Simplify each expression. Assume that all variables represent positive real numbers.

35. $5^{1/4} \cdot 5^{7/4}$ **36.** $\dfrac{96^{2/3}}{96^{-1/3}}$ **37.** $\dfrac{(a^{1/3})^4}{a^{2/3}}$

38. $\dfrac{y^{-1/3} \cdot y^{5/6}}{y}$ **39.** $\left(\dfrac{z^{-1}x^{-3/5}}{2^{-2}z^{-1/2}x}\right)^{-1}$ **40.** $r^{-1/2}(r + r^{3/2})$

Write each radical as an exponential and simplify. Leave answers in exponential form. Assume that all variables represent positive real numbers.

41. $\sqrt[5]{y} \cdot \sqrt[3]{y}$ **42.** $\dfrac{\sqrt{x^3}}{\sqrt{x^4}}$ **43.** $\dfrac{\sqrt{p^5}}{p^2}$

44. $\sqrt[4]{k^3} \cdot \sqrt{k^3}$ **45.** $\sqrt[3]{m^5} \cdot \sqrt[3]{m^8}$ **46.** $\dfrac{\sqrt[3]{t^5}}{\sqrt[6]{t}}$

47. $\sqrt{\sqrt{\sqrt{x}}}$ **48.** $\sqrt[3]{\sqrt[5]{x}}$ **49.** $\sqrt{\sqrt[6]{\sqrt[3]{x}}}$

50. The product rule does not apply to $3^{1/4} \cdot 2^{1/5}$. Why?

7.3 *Simplify. Assume that all variables represent positive real numbers.*

51. $\sqrt{6} \cdot \sqrt{11}$ **52.** $\sqrt{5} \cdot \sqrt{r}$ **53.** $\sqrt[3]{6} \cdot \sqrt[3]{5}$ **54.** $\sqrt[4]{7} \cdot \sqrt[4]{3}$

55. $\sqrt{20}$ **56.** $\sqrt{75}$ **57.** $-\sqrt{125}$ **58.** $\sqrt[3]{-108}$

59. $\sqrt{100y^7}$ **60.** $\sqrt[3]{64p^4q^6}$ **61.** $\sqrt[3]{108a^8b^5}$ **62.** $\sqrt[3]{632r^8t^4}$

63. $\sqrt{\dfrac{y^3}{144}}$ **64.** $\sqrt[3]{\dfrac{m^{15}}{27}}$ **65.** $\sqrt[3]{\dfrac{r^2}{8}}$ **66.** $\sqrt[4]{\dfrac{a^9}{81}}$

67. $\sqrt[6]{15^3}$ **68.** $\sqrt[4]{p^6}$ **69.** $\sqrt[3]{2} \cdot \sqrt[4]{5}$ **70.** $\sqrt{x} \cdot \sqrt[5]{x}$

71. Find the length c of the hypotenuse of the right triangle shown.

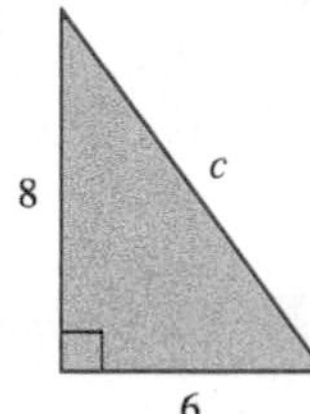

72. Find the distance between the points $(-4, 7)$ and $(10, 6)$.

Find the equation of a circle satisfying the given conditions.

73. Center $(0, 0)$, $r = 11$ **74.** Center $(-2, 4)$, $r = 3$

75. Center $(-1, -3)$, $r = 5$ **76.** Center $(4, 2)$, $r = 6$

Graph each circle. Identify the center and the radius.

77. $x^2 + y^2 = 25$ **78.** $(x + 3)^2 + (y - 3)^2 = 9$ **79.** $(x - 2)^2 + (y + 5)^2 = 9$

80. Why does the equation $x^2 + y^2 = -1$ have no points on its graph?

7.4 *Perform the indicated operations. Assume that all variables represent positive real numbers.*

81. $2\sqrt{8} - 3\sqrt{50}$

82. $8\sqrt{80} - 3\sqrt{45}$

83. $-\sqrt{27y} + 2\sqrt{75y}$

84. $2\sqrt{54m^3} + 5\sqrt{96m^3}$

85. $3\sqrt[3]{54} + 5\sqrt[3]{16}$

86. $-6\sqrt[4]{32} + \sqrt[4]{512}$

7.5 *Multiply, and then simplify.*

87. $(\sqrt{3} + 1)(\sqrt{3} - 2)$

88. $(\sqrt{7} + \sqrt{5})(\sqrt{7} - \sqrt{5})$

89. $(3\sqrt{2} + 1)(2\sqrt{2} - 3)$

90. $(\sqrt{13} - \sqrt{2})^2$

91. $(\sqrt[3]{2} + 3)(\sqrt[3]{4} - 3\sqrt[3]{2} + 9)$

92. $(\sqrt[3]{4y} - 1)(\sqrt[3]{4y} + 3)$

Rationalize each denominator. Assume that all variables represent positive real numbers.

93. $\dfrac{\sqrt{6}}{\sqrt{5}}$

94. $\dfrac{-6\sqrt{3}}{\sqrt{2}}$

95. $\dfrac{3\sqrt{7p}}{\sqrt{y}}$

96. $\sqrt{\dfrac{11}{8}}$

97. $-\sqrt[3]{\dfrac{9}{25}}$

98. $\sqrt[3]{\dfrac{108m^3}{n^5}}$

99. $\dfrac{1}{\sqrt{2} + \sqrt{7}}$

100. $\dfrac{-5}{\sqrt{6} - 3}$

Write each quotient in lowest terms.

101. $\dfrac{2 - 2\sqrt{5}}{8}$

102. $\dfrac{-18 + \sqrt{27}}{6}$

7.6 *Solve each equation.*

103. $\sqrt{8x + 9} = 5$

104. $\sqrt{2x - 3} - 3 = 0$

105. $\sqrt{7x + 1} = x + 1$

106. $3\sqrt{x} = \sqrt{10x - 9}$

107. $\sqrt{x^2 + 3x + 7} = x + 2$

108. $\sqrt{x + 2} - \sqrt{x - 3} = 1$

109. $\sqrt[3]{5x - 1} = \sqrt[3]{3x - 2}$

110. $\sqrt[3]{1 - 2x} - \sqrt[3]{-x - 13} = 0$

111. $\sqrt[4]{x - 1} + 2 = 0$

112. $\sqrt[4]{x + 7} = \sqrt[4]{2x}$

Carpenters stabilize wall frames with a diagonal brace, as shown in the figure. The length of the brace is given by

$$L = \sqrt{H^2 + W^2}.$$

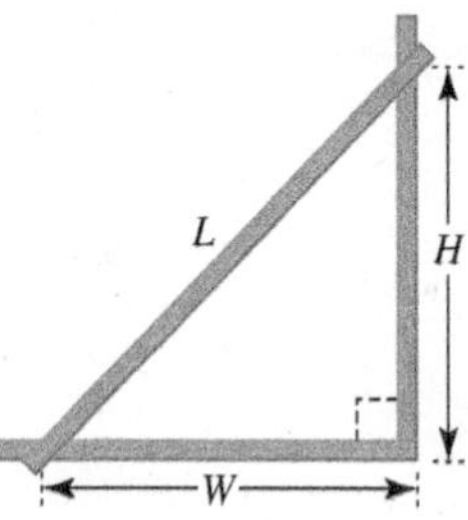

113. Solve this formula for H.

114. If the bottom of the brace is attached 9 ft from the corner and the brace is 12 ft long, how far up the corner post should it be nailed? Give the answer to the nearest tenth of a foot.

7.7 *Write each number as a product of a real number and i. Simplify.*

115. $\sqrt{-16}$

116. $\sqrt{-200}$

Perform the indicated operations. Give answers in standard form.

117. $(-2 + 5i) + (-8 - 7i)$

118. $(5 + 4i) - (-9 - 3i)$

119. $\sqrt{-5} \cdot \sqrt{-7}$

120. $\sqrt{-25} \cdot \sqrt{-81}$

121. $\dfrac{\sqrt{-72}}{\sqrt{-8}}$

122. $(2 + 3i)(1 - i)$

123. $(6 - 2i)^2$

124. $\dfrac{3 - i}{2 + i}$

Find each power of i.

125. i^{11}

126. i^{36}

127. i^{-10}

128. i^{-8}

Chapter 7 Mixed Review Exercises

Simplify. Assume that all variables represent positive real numbers.

1. $-\sqrt{169a^2b^4}$

2. $1000^{-2/3}$

3. $\dfrac{z^{-1/5} \cdot z^{3/10}}{z^{7/10}}$

4. $\sqrt[3]{54z^9t^8}$

5. $\sqrt{-49}$

6. $\dfrac{-1}{\sqrt{12}}$

7. $\sqrt[3]{\dfrac{12}{25}}$

8. i^{-1000}

9. $-5\sqrt{18} + 12\sqrt{72}$

10. $(4 - 9i) - (-1 + 2i)$

11. $\dfrac{\sqrt{50}}{\sqrt{-2}}$

12. $\dfrac{3 + \sqrt{54}}{6}$

13. $(3 + 2i)^2$

14. $8\sqrt[3]{x^3y^2} - 2x\sqrt[3]{y^2}$

15. $5i(3 - 7i)$

16. $\sqrt[3]{2} \cdot \sqrt[4]{5}$

17. $\dfrac{2\sqrt{z}}{\sqrt{z} - 2}, \quad z \neq 4$

18. Find the perimeter of a triangular electronic highway road sign having the dimensions shown in the figure.

All
Traffic
Must Exit
Iowa
Highway 64
$\sqrt{108}$ ft
$2\sqrt{27}$ ft
$\sqrt{50}$ ft

Solve each equation.

19. $\sqrt{x + 4} = x - 2$

20. $\sqrt[3]{2x - 9} = \sqrt[3]{5x + 3}$

21. $\sqrt{6 + 2x} - 1 = \sqrt{7 - 2x}$

22. $\sqrt{11 + 2x} + 1 = \sqrt{5x + 1}$

Chapter 7 Test

Step-by-step test solutions are found on the Chapter Test Prep Videos available in MyLab Math.

View the complete solutions to all Chapter Test exercises in MyLab Math.

Evaluate.

1. $-\sqrt{841}$ **2.** $\sqrt[3]{-512}$ **3.** $125^{1/3}$

Use a calculator to approximate each radical to three decimal places.

4. $\sqrt{478}$ **5.** $\sqrt[3]{-832}$

6. Graph the function $f(x) = \sqrt{x+6}$, and give its domain and range.

Simplify each expression. Leave answers in exponential form. Assume that all variables represent positive real numbers.

7. $\left(\frac{16}{25}\right)^{-3/2}$ **8.** $(-64)^{-4/3}$ **9.** $\frac{3^{2/5}x^{-1/4}y^{2/5}}{3^{-8/5}x^{7/4}y^{1/10}}$

10. $\left(\frac{x^{-4}y^{-6}}{x^{-2}y^{3}}\right)^{-2/3}$ **11.** $7^{3/4} \cdot 7^{-1/4}$ **12.** $\sqrt[3]{a^4} \cdot \sqrt[3]{a^7}$

13. Find the exact length of side b in the figure.

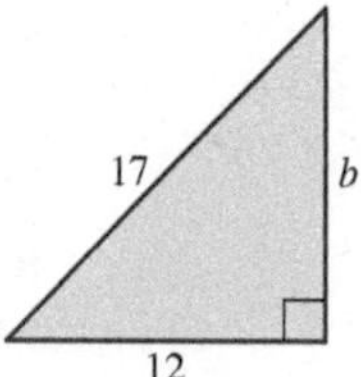

14. Find the distance between the points $(-4, 2)$ and $(2, 10)$.

15. Write an equation of a circle with center $(-4, 6)$ and radius 5.

16. Graph the circle $(x-2)^2 + (y+3)^2 = 16$. Identify the center and the radius.

Simplify. Assume that all variables represent positive real numbers.

17. $\sqrt{54x^5y^6}$ **18.** $\sqrt[4]{32a^7b^{13}}$

19. $\sqrt{2} \cdot \sqrt[3]{5}$ **20.** $3\sqrt{20} - 5\sqrt{80} + 4\sqrt{500}$

21. $\sqrt[3]{16t^3s^5} - \sqrt[3]{54t^6s^2}$ **22.** $(7\sqrt{5}+4)(2\sqrt{5}-1)$

23. $(\sqrt{3} - 2\sqrt{5})^2$ **24.** $\frac{-5}{\sqrt{40}}$ **25.** $\frac{2}{\sqrt[3]{5}}$ **26.** $\frac{-4}{\sqrt{7}+\sqrt{5}}$

27. Write $\frac{6+\sqrt{24}}{2}$ in lowest terms.

28. The following formula from physics relates the velocity V of sound to the temperature T.

$$V = \frac{V_0}{\sqrt{1-kT}}$$

(a) Approximate V to the nearest tenth if $V_0 = 50$, $k = 0.01$, and $T = 30$.

(b) Solve the formula for T.

Solve each equation.

29. $\sqrt[3]{5x} = \sqrt[3]{2x-3}$ **30.** $\sqrt{x+6} = 9 - 2x$

31. $\sqrt{x+4} - \sqrt{1-x} = -1$

Perform the indicated operations. Give the answers in standard form.

32. $(-2 + 5i) - (3 + 6i) - 7i$ **33.** $(1 + 5i)(3 + i)$ **34.** $\frac{7 + i}{1 - i}$

35. Simplify i^{37}.

36. Answer *true* or *false* to each of the following.

(a) $i^2 = -1$ **(b)** $i = \sqrt{-1}$ **(c)** $i = -1$ **(d)** $\sqrt{-3} = i\sqrt{3}$

Chapters R–7 Cumulative Review Exercises

Perform the indicated operations.

1. $-4|2 - 7| - (-3)(-9)$ **2.** $\frac{8(-4) - 3^2 \cdot 6}{-4 \cdot 2^3 + 12}$ **3.** $\left(-\frac{5}{8} + \frac{1}{6}\right) - \left(-\frac{2}{3}\right)$

Solve each equation or inequality.

4. $3(x + 2) - 4(2x + 3) = -3x + 2$

5. $\frac{1}{3}x + \frac{1}{4}(x + 8) = x + 7$

6. $0.04x + 0.06(100 - x) = 5.88$

7. $|6x + 7| = 13$

8. $-5 - 3(x - 2) < 11 - 2(x + 2)$

9. $-2 < 1 - 3x < 7$

Solve each problem.

10. A piggy bank has 100 coins, all of which are nickels and quarters. The total value of the money is \$17.80. How many of each denomination are there in the bank?

11. How many liters of pure alcohol must be mixed with 40 L of 18% alcohol to obtain a 22% alcohol solution?

12. Graph the equation $4x - 3y = 12$.

13. Find the slope of the line passing through the points $(-4, 6)$ and $(2, -3)$. Then find the equation of the line and write it in the form $y = mx + b$.

14. If $f(x) = 3x - 7$, find $f(-10)$.

Solve each system.

15. $3x - y = 23$
$2x + 3y = 8$

16. $x + y + z = 1$
$x - y - z = -3$
$x + y - z = -1$

17. In 2017, sending five 2-oz letters and three 3-oz letters by first-class mail would have cost \$6.23. Sending three 2-oz letters and five 3-oz letters would have cost \$6.65. What was the rate for one 2-oz letter and one 3-oz letter? (Data from U.S. Postal Service.)

Perform the indicated operations.

18. $(3k^3 - 5k^2 + 8k - 2) - (4k^3 + 11k + 7) + (2k^2 - 5k)$

19. $(8x - 7)(x + 3)$

20. $\dfrac{6y^4 - 3y^3 + 5y^2 + 6y - 9}{2y + 1}$

Factor each polynomial completely.

21. $2p^2 - 5pq + 3q^2$

22. $3k^4 + k^2 - 4$

23. $x^3 + 512$

24. What is the domain of $f(x) = \dfrac{4}{x^2 - 9}$? Express it using set-builder notation.

Perform each operation, and express the answer in lowest terms.

25. $\dfrac{y^2 + y - 12}{y^3 + 9y^2 + 20y} \div \dfrac{y^2 - 9}{y^3 + 3y^2}$

26. $\dfrac{1}{x + y} + \dfrac{3}{x - y}$

Simplify.

27. $\dfrac{\frac{-6}{x - 2}}{\frac{8}{3x - 6}}$

28. $\dfrac{\frac{1}{a} - \frac{1}{b}}{\frac{a}{b} - \frac{b}{a}}$

29. $\dfrac{x^{-1}}{y - x^{-1}}$

Solve.

30. $2x^2 + 11x + 15 = 0$

31. $5x(x - 1) = 2(1 - x)$

32. $\dfrac{x + 1}{x - 3} = \dfrac{4}{x - 3} + 6$

33. $\sqrt{3x - 8} = x - 2$

34. Danielle can ride her bike 4 mph faster than her husband, Richard. If Danielle can ride 48 mi in the same time that Richard can ride 24 mi, what are their rates?

Simplify. Assume that all variables represent positive real numbers.

35. $27^{-2/3}$

36. $\sqrt[3]{16x^2y} \cdot \sqrt[3]{3x^3y}$

37. $\sqrt{50} + \sqrt{8}$

38. $\dfrac{1}{\sqrt{10} - \sqrt{8}}$

39. Find the distance between the points $(-4, 4)$ and $(-2, 9)$.

40. Express $\frac{6 - 2i}{1 - i}$ in standard form.

STUDY SKILLS **REMINDER**

Have you begun to prepare for your final exam? **Review Study Skill 10, *Preparing for Your Math Final Exam.***

QUADRATIC EQUATIONS AND INEQUALITIES

The trajectory of a water spout follows a *parabolic* path, which can be described by a *quadratic* (second-degree) *equation*, the topic of this chapter.

8.1 The Square Root Property and Completing the Square

OBJECTIVES

1. Review the zero-factor property.
2. Learn the square root property.
3. Solve quadratic equations of the form $(ax + b)^2 = c$ by extending the square root property.
4. Solve quadratic equations by completing the square.
5. Solve quadratic equations with nonreal complex solutions.

Recall that a *quadratic equation* is defined as follows.

Quadratic Equation

A **quadratic equation** (in x here) is an equation that can be written in the form

$$ax^2 + bx + c = 0,$$

where a, b, and c are real numbers and $a \neq 0$. The given form is called **standard form.**

Examples: $4x^2 + 4x - 5 = 0$, $3x^2 = 4x - 8$ Quadratic equations (The first equation is in standard form.)

A quadratic equation is a **second-degree equation**—that is, an equation with a squared variable term and no terms of greater degree.

OBJECTIVE 1 Review the zero-factor property.

We have used the zero-factor property to solve quadratic equations.

Zero-Factor Property

If two numbers have a product of 0, then at least one of the numbers must be 0. **That is, if $ab = 0$, then $a = 0$ or $b = 0$.**

EXAMPLE 1 Using the Zero-Factor Property

Solve $3x^2 - 5x - 28 = 0$.

$3x^2 - 5x - 28 = 0$ This trinomial can be factored. The quadratic equation must be in standard form.

$(3x + 7)(x - 4) = 0$ Factor.

$3x + 7 = 0$ or $x - 4 = 0$ Zero-factor property

$3x = -7$ or $x = 4$ Solve each equation.

$x = -\frac{7}{3}$

To check, substitute each value in the original equation. The solution set is $\left\{-\frac{7}{3}, 4\right\}$.

NOW TRY

VOCABULARY

☐ quadratic equation
☐ second-degree equation

NOW TRY EXERCISE 1

Solve $2x^2 + 5x - 12 = 0$.

OBJECTIVE 2 Learn the square root property.

Not every quadratic equation can be solved using the zero-factor property. Other methods of solving quadratic equations are based on the following property.

Square Root Property

If x and k are complex numbers and $x^2 = k$, then

$$x = \sqrt{k} \quad \text{or} \quad x = -\sqrt{k}.$$

NOW TRY ANSWER

1. $\left\{-4, \frac{3}{2}\right\}$

These steps justify the square root property.

$$x^2 = k$$

$$x^2 - k = 0 \quad \text{Subtract } k.$$

$$(x - \sqrt{k})(x + \sqrt{k}) = 0 \quad \text{Factor.}$$

$$x - \sqrt{k} = 0 \quad \text{or} \quad x + \sqrt{k} = 0 \quad \text{Zero-factor property}$$

$$\boldsymbol{x = \sqrt{k}} \quad \text{or} \quad \boldsymbol{x = -\sqrt{k}} \quad \text{Solve each equation.}$$

CAUTION If $k \neq 0$, then using the square root property always produces *two* square roots, one positive and one negative.

NOW TRY EXERCISE 2

Solve each equation.

(a) $t^2 = 10$

(b) $2x^2 - 90 = 0$

EXAMPLE 2 Using the Square Root Property

Solve each equation.

(a) $x^2 = 5$

By the square root property, if $x^2 = 5$, then

$$x = \sqrt{5} \quad \text{or} \quad x = -\sqrt{5}.$$

Don't forget the negative solution.

These solutions are irrational numbers. The solution set is $\{\sqrt{5}, -\sqrt{5}\}$.

(b)

$$4x^2 - 48 = 0$$

$$4x^2 = 48 \quad \text{Add 48.}$$

$$x^2 = 12 \quad \text{Divide by 4.}$$

$$x = \sqrt{12} \quad \text{or} \quad x = -\sqrt{12} \quad \text{Square root property}$$

Don't stop here. Simplify the radicals.

$$x = 2\sqrt{3} \quad \text{or} \quad x = -2\sqrt{3} \quad \sqrt{12} = \sqrt{4} \cdot \sqrt{3} = 2\sqrt{3}$$

CHECK $4x^2 - 48 = 0$ Original equation

$4(2\sqrt{3})^2 - 48 \stackrel{?}{=} 0$ Let $x = 2\sqrt{3}$.	$4(-2\sqrt{3})^2 - 48 \stackrel{?}{=} 0$ Let $x = -2\sqrt{3}$.
$4(12) - 48 \stackrel{?}{=} 0$	$4(12) - 48 \stackrel{?}{=} 0$
$48 - 48 \stackrel{?}{=} 0$	$48 - 48 \stackrel{?}{=} 0$
$0 = 0$ ✓ True	$0 = 0$ ✓ True

$(2\sqrt{3})^2 = 2^2 \cdot (\sqrt{3})^2$

The solution set is $\{2\sqrt{3}, -2\sqrt{3}\}$.

(c)

$$3x^2 + 5 = 11$$

$$3x^2 = 6 \quad \text{Subtract 5.}$$

$$x^2 = 2 \quad \text{Divide by 3.}$$

$$x = \sqrt{2} \quad \text{or} \quad x = -\sqrt{2} \quad \text{Square root property}$$

The solution set is $\{\sqrt{2}, -\sqrt{2}\}$.

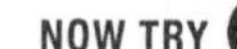

NOTE Using the symbol ± (read "*positive or negative*," or "*plus or minus*"), the solutions in **Example 2** could be written $\pm\sqrt{5}$, $\pm 2\sqrt{3}$, and $\pm\sqrt{2}$.

NOW TRY ANSWERS

2. (a) $\{\sqrt{10}, -\sqrt{10}\}$

(b) $\{3\sqrt{5}, -3\sqrt{5}\}$

NOW TRY EXERCISE 3
Tim is dropping roofing nails from the top of a roof 25 ft high into a large bucket on the ground. Use the formula in **Example 3** to determine how long it will take a nail dropped from 25 ft to hit the bottom of the bucket.

EXAMPLE 3 Using the Square Root Property in an Application

Galileo Galilei developed a formula for freely falling objects described by

$$d = 16t^2,$$

where d is the distance in feet that an object falls (disregarding air resistance) in t seconds, regardless of weight.

If the Leaning Tower of Pisa is about 180 ft tall, use Galileo's formula to determine how long it would take an object dropped from the top of the tower to fall to the ground. (Data from www.brittanica.com)

Galileo Galilei (1564–1642)

$$d = 16t^2 \quad \text{Galileo's formula}$$

$$180 = 16t^2 \quad \text{Let } d = 180.$$

$$11.25 = t^2 \quad \text{Divide by 16.}$$

$$t = \sqrt{11.25} \quad \text{or} \quad t = -\sqrt{11.25} \quad \text{Square root property}$$

Time cannot be negative, so we discard $-\sqrt{11.25}$. Using a calculator,

$$\sqrt{11.25} \approx 3.4 \quad \text{so} \quad t \approx 3.4. \quad \text{Round to the nearest tenth.}$$

The object would fall to the ground in 3.4 sec.

NOW TRY 

CAUTION When solving applications, a solution of a quadratic equation may not satisfy the physical requirements, as in **Example 3.** Reject such solutions as answers.

OBJECTIVE 3 Solve quadratic equations of the form $(ax + b)^2 = c$ by extending the square root property.

NOW TRY EXERCISE 4
Solve $(x + 3)^2 = 25$.

EXAMPLE 4 Extending the Square Root Property

Solve $(x - 5)^2 = 36$.

$$x^2 = k$$

$$(x - 5)^2 = 36 \quad \text{Substitute } (x-5)^2 \text{ for } x^2 \text{ and 36 for } k \text{ in the square root property.}$$

$$x - 5 = \sqrt{36} \quad \text{or} \quad x - 5 = -\sqrt{36} \quad \text{Square root property}$$

$$x - 5 = 6 \quad \text{or} \quad x - 5 = -6 \quad \text{Take square roots.}$$

$$x = 11 \quad \text{or} \quad x = -1 \quad \text{Add 5.}$$

CHECK $(x - 5)^2 = 36$ Original equation

Let $x = 11$.	Let $x = -1$.
$(11 - 5)^2 \stackrel{?}{=} 36$	$(-1 - 5)^2 \stackrel{?}{=} 36$
$6^2 \stackrel{?}{=} 36$	$(-6)^2 \stackrel{?}{=} 36$
$36 = 36$ ✓ True	$36 = 36$ ✓ True

Both values satisfy the original equation. The solution set is $\{-1, 11\}$.

NOW TRY

NOW TRY ANSWERS
3. 1.25 sec
4. $\{-8, 2\}$

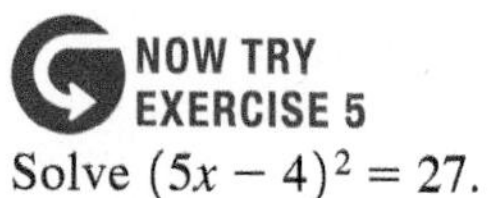

Solve $(5x-4)^2 = 27$.

EXAMPLE 5 **Extending the Square Root Property**

Solve $(2x-3)^2 = 18$.

$$2x - 3 = \sqrt{18} \quad \text{or} \quad 2x - 3 = -\sqrt{18} \qquad \text{Square root property}$$

$$2x = 3 + \sqrt{18} \quad \text{or} \quad 2x = 3 - \sqrt{18} \qquad \text{Add 3.}$$

$$x = \frac{3+\sqrt{18}}{2} \quad \text{or} \quad x = \frac{3-\sqrt{18}}{2} \qquad \text{Divide by 2.}$$

$$x = \frac{3+3\sqrt{2}}{2} \quad \text{or} \quad x = \frac{3-3\sqrt{2}}{2} \qquad \sqrt{18} = \sqrt{9 \cdot 2} = 3\sqrt{2}$$

We show the check for the first value. The check for the other value is similar.

CHECK

$$(2x-3)^2 = 18 \qquad \text{Original equation}$$

$$\left[2\left(\frac{3+3\sqrt{2}}{2}\right) - 3\right]^2 \stackrel{?}{=} 18 \qquad \text{Let } x = \frac{3+3\sqrt{2}}{2}.$$

$$\left(3 + 3\sqrt{2} - 3\right)^2 \stackrel{?}{=} 18 \qquad \text{Multiply.}$$

$$\left(3\sqrt{2}\right)^2 \stackrel{?}{=} 18 \qquad \text{Simplify.}$$

$$18 = 18 \checkmark \qquad \text{True}$$

The ± symbol denotes two solutions.

The solution set is $\left\{\frac{3+3\sqrt{2}}{2}, \frac{3-3\sqrt{2}}{2}\right\}$, abbreviated $\left\{\frac{3 \pm 3\sqrt{2}}{2}\right\}$. **NOW TRY**

OBJECTIVE 4 **Solve quadratic equations by completing the square.**

We can use the square root property to solve *any* quadratic equation by writing it in the form

$$\text{Square of a binomial} \rightarrow (x+k)^2 = n. \leftarrow \text{Constant}$$

That is, we must write the left side of the equation as a perfect square trinomial that can be factored as $(x+k)^2$, the square of a binomial, and the right side must be a constant. This process is called **completing the square.**

Recall that the perfect square trinomial

$$x^2 + 10x + 25 \quad \text{can be factored as} \quad (x+5)^2.$$

In the trinomial, the coefficient of x (the first-degree term) is 10 and the constant term is 25. If we take half of 10 and square it, we obtain the constant term, 25.

Coefficient of x ↓ Constant ↓

$$\left[\frac{1}{2}(10)\right]^2 = 5^2 = 25$$

Similarly, in $x^2 + 12x + 36$, $\left[\frac{1}{2}(12)\right]^2 = 6^2 = 36$,

and in $m^2 - 6m + 9$, $\left[\frac{1}{2}(-6)\right]^2 = (-3)^2 = 9$.

This relationship is true in general and is the idea behind completing the square.

NOW TRY ANSWER

5. $\left\{\frac{4 \pm 3\sqrt{3}}{5}\right\}$

NOW TRY EXERCISE 6
Solve $x^2 + 6x - 2 = 0$.

EXAMPLE 6 Solving a Quadratic Equation by Completing the Square ($a = 1$)

Solve $x^2 + 8x + 10 = 0$.

The trinomial on the left is nonfactorable, so this quadratic equation cannot be solved using the zero-factor property. It is not in the correct form to solve using the square root property. We can solve it by completing the square.

$$x^2 + 8x + 10 = 0 \quad \text{Original equation}$$

Only terms with variables remain of the left side.

$$x^2 + 8x = -10 \quad \text{Subtract 10.}$$

$$\underbrace{x^2 + 8x + \underline{\ ?\ }}_{\text{Needs to be a perfect square trinomial}} = -10 \quad \text{We must add a constant.}$$

Take half the coefficient of the first-degree term, $8x$, and square the result.

$$\left[\frac{1}{2}(8)\right]^2 = 4^2 = 16 \longleftarrow \text{Desired constant}$$

We add this constant, 16, to *each* side of the equation and continue solving as shown.

$$x^2 + 8x + 16 = -10 + 16 \quad \text{Add 16 to each side.}$$

This is a key step.

$$(x + 4)^2 = 6 \quad \text{Factor on the left. Add on the right.}$$

$$x + 4 = \sqrt{6} \quad \text{or} \quad x + 4 = -\sqrt{6} \quad \text{Square root property}$$

$$x = -4 + \sqrt{6} \quad \text{or} \quad x = -4 - \sqrt{6} \quad \text{Add } -4.$$

CHECK

$$x^2 + 8x + 10 = 0 \quad \text{Original equation}$$

$$\left(-4 + \sqrt{6}\right)^2 + 8\left(-4 + \sqrt{6}\right) + 10 \stackrel{?}{=} 0 \quad \text{Let } x = -4 + \sqrt{6}.$$

Remember the middle term when squaring $-4 + \sqrt{6}$.

$$16 - 8\sqrt{6} + 6 - 32 + 8\sqrt{6} + 10 \stackrel{?}{=} 0 \quad \text{Multiply.}$$

$$0 = 0 \ \checkmark \quad \text{True}$$

The check of $-4 - \sqrt{6}$ is similar. The solution set is

$$\left\{-4 + \sqrt{6}, -4 - \sqrt{6}\right\}, \quad \text{or} \quad \left\{-4 \pm \sqrt{6}\right\}.$$

NOW TRY

STUDY SKILLS REMINDER
Make study cards to help you learn and remember the material in this chapter. **Review Study Skill 5, *Using Study Cards*.**

Completing the Square to Solve $ax^2 + bx + c = 0$ (Where $a \neq 0$)

Step 1 **Be sure the second-degree term has coefficient 1.**
- If the coefficient of the second-degree term is 1, go to Step 2.
- If it is not 1, but some other nonzero number a, divide each side of the equation by a.

Step 2 **Write the equation in correct form.** Make sure that all variable terms are on one side of the equality symbol and the constant term is on the other side.

Step 3 **Complete the square.**
- Take half the coefficient of the first-degree term, and square it.
- Add the square to each side of the equation.
- Factor the variable side, which should be a perfect square trinomial, as the square of a binomial. Combine terms on the other side.

Step 4 **Solve** the equation using the square root property.

NOW TRY ANSWER
6. $\{-3 \pm \sqrt{11}\}$

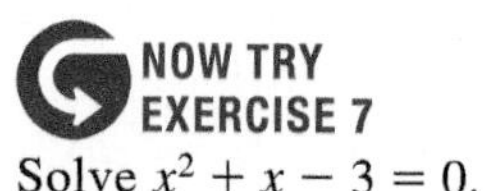

Solve $x^2 + x - 3 = 0$.

EXAMPLE 7 Solving a Quadratic Equation by Completing the Square ($a = 1$)

Solve $x^2 + 5x - 1 = 0$.

The coefficient of the second-degree term is 1, so we begin with Step 2.

Step 2 $x^2 + 5x = 1$ — Add 1 to each side.

Step 3 $\left[\frac{1}{2}(5)\right]^2 = \left(\frac{5}{2}\right)^2 = \frac{25}{4}$ — Take half the coefficient of the first-degree term and square the result.

$$x^2 + 5x + \frac{25}{4} = 1 + \frac{25}{4}$$ Add the square to each side of the equation.

$$\left(x + \frac{5}{2}\right)^2 = \frac{29}{4}$$ Factor on the left. Add on the right.

Step 4 $x + \frac{5}{2} = \sqrt{\frac{29}{4}}$ or $x + \frac{5}{2} = -\sqrt{\frac{29}{4}}$ — Square root property

$x + \frac{5}{2} = \frac{\sqrt{29}}{2}$ or $x + \frac{5}{2} = -\frac{\sqrt{29}}{2}$ — $\sqrt{\frac{a}{b}} = \frac{\sqrt{a}}{\sqrt{b}}$

$x = -\frac{5}{2} + \frac{\sqrt{29}}{2}$ or $x = -\frac{5}{2} - \frac{\sqrt{29}}{2}$ — Add $-\frac{5}{2}$.

$x = \frac{-5 + \sqrt{29}}{2}$ or $x = \frac{-5 - \sqrt{29}}{2}$ — $\frac{a}{c} \pm \frac{b}{c} = \frac{a \pm b}{c}$

The solution set is $\left\{\frac{-5 \pm \sqrt{29}}{2}\right\}$.

EXAMPLE 8 Solving a Quadratic Equation by Completing the Square ($a \neq 1$)

Solve $2x^2 - 4x - 5 = 0$.

Divide each side by 2 to obtain 1 as the coefficient of the second-degree term.

$$x^2 - 2x - \frac{5}{2} = 0$$ Divide by 2. (Step 1)

$$x^2 - 2x = \frac{5}{2}$$ Add $\frac{5}{2}$. (Step 2)

$$\left[\frac{1}{2}(-2)\right]^2 = (-1)^2 = 1$$ Complete the square. (Step 3)

$$x^2 - 2x + 1 = \frac{5}{2} + 1$$ Add 1 to each side.

$$(x - 1)^2 = \frac{7}{2}$$ Factor on the left. Add on the right.

$x - 1 = \sqrt{\frac{7}{2}}$ or $x - 1 = -\sqrt{\frac{7}{2}}$ — Square root property (Step 4)

$x = 1 + \sqrt{\frac{7}{2}}$ or $x = 1 - \sqrt{\frac{7}{2}}$ — Add 1.

$x = 1 + \frac{\sqrt{14}}{2}$ or $x = 1 - \frac{\sqrt{14}}{2}$ — $\sqrt{\frac{7}{2}} = \frac{\sqrt{7}}{\sqrt{2}} = \frac{\sqrt{7}}{\sqrt{2}} \cdot \frac{\sqrt{2}}{\sqrt{2}} = \frac{\sqrt{14}}{2}$

NOW TRY ANSWER

7. $\left\{\frac{-1 \pm \sqrt{13}}{2}\right\}$

NOW TRY EXERCISE 8
Solve $3x^2 + 12x - 5 = 0$.

Add the two terms in each solution as follows.

$$1 + \frac{\sqrt{14}}{2} = \frac{2}{2} + \frac{\sqrt{14}}{2} = \frac{2 + \sqrt{14}}{2}$$

$$1 - \frac{\sqrt{14}}{2} = \frac{2}{2} - \frac{\sqrt{14}}{2} = \frac{2 - \sqrt{14}}{2}$$

$1 = \frac{2}{2}$

The solution set is $\left\{\frac{2 \pm \sqrt{14}}{2}\right\}$. NOW TRY

OBJECTIVE 5 **Solve quadratic equations with nonreal complex solutions.**

If $k < 0$ in the equation $x^2 = k$, then there will be two nonreal complex solutions.

NOW TRY EXERCISE 9
Solve each equation.
(a) $t^2 = -24$
(b) $(x + 4)^2 = -36$
(c) $x^2 + 8x + 21 = 0$

EXAMPLE 9 **Solving Quadratic Equations (Nonreal Complex Solutions)**

Solve each equation.

(a)
$$x^2 = -15$$
$$x = \sqrt{-15} \quad \text{or} \quad x = -\sqrt{-15} \qquad \text{Square root property}$$
$$x = i\sqrt{15} \quad \text{or} \quad x = -i\sqrt{15} \qquad \sqrt{-a} = i\sqrt{a}$$

The solution set is $\{i\sqrt{15}, -i\sqrt{15}\}$, or $\{\pm i\sqrt{15}\}$.

(b)
$$(x + 2)^2 = -16$$
$$x + 2 = \sqrt{-16} \quad \text{or} \quad x + 2 = -\sqrt{-16} \qquad \text{Square root property}$$
$$x + 2 = 4i \quad \text{or} \quad x + 2 = -4i \qquad \sqrt{-16} = 4i$$
$$x = -2 + 4i \quad \text{or} \quad x = -2 - 4i \qquad \text{Add } -2.$$

The solution set is $\{-2 + 4i, -2 - 4i\}$, or $\{-2 \pm 4i\}$.

(c)
$$x^2 + 2x + 7 = 0$$
$$x^2 + 2x = -7 \qquad \text{Subtract 7.}$$
$$x^2 + 2x + 1 = -7 + 1 \qquad \left[\tfrac{1}{2}(2)\right]^2 = 1; \text{ Add 1 to each side.}$$
$$(x + 1)^2 = -6 \qquad \text{Factor on the left. Add on the right.}$$
$$x + 1 = \sqrt{-6} \quad \text{or} \quad x + 1 = -\sqrt{-6} \qquad \text{Square root property}$$
$$x + 1 = i\sqrt{6} \quad \text{or} \quad x + 1 = -i\sqrt{6} \qquad \sqrt{-6} = i\sqrt{6}$$
$$x = -1 + i\sqrt{6} \quad \text{or} \quad x = -1 - i\sqrt{6} \qquad \text{Add } -1.$$

The solution set is $\{-1 + i\sqrt{6}, -1 - i\sqrt{6}\}$, or $\{-1 \pm i\sqrt{6}\}$.  

NOW TRY ANSWERS

8. $\left\{\frac{-6 \pm \sqrt{51}}{3}\right\}$

9. **(a)** $\{\pm 2i\sqrt{6}\}$
(b) $\{-4 \pm 6i\}$
(c) $\{-4 \pm i\sqrt{5}\}$

8.1 Exercises

FOR EXTRA HELP  MyLab Math

Video solutions for select problems available in MyLab Math

Concept Check *Fill in each blank with the correct response.*

1. An equation in the form $ax^2 + bx + c = 0$, where a, b, and c are real numbers and $a \neq 0$, is a(n) ________ equation, also called a(n) ________ degree equation. The greatest degree of the variable is ________.

2. Which of the following are quadratic equations?

A. $x + y = 0$ **B.** $x^2 - 4x + 4 = 0$ **C.** $x^3 + x^2 + 8 = 0$ **D.** $2t^2 - 7t = -4$

3. Concept Check A student incorrectly solved the following equation as shown.

$$x^2 - x - 2 = 4$$

$$(x - 2)(x + 1) = 4 \quad \text{Factor.}$$

$$x - 2 = 4 \quad \text{or} \quad x + 1 = 4 \quad \text{Zero-factor property}$$

$$x = 6 \quad \text{or} \quad x = 3 \quad \text{Solve each equation.}$$

WHAT WENT WRONG? Solve correctly and give the solution set.

4. Concept Check A student was asked to solve the quadratic equation

$$x^2 = 16$$

and did not get full credit for the solution set $\{4\}$. **WHAT WENT WRONG?** Give the correct solution set.

Use the zero-factor property to solve each equation. (Hint: In Exercises 11–14, write the equation in standard form first.) ***See Example 1.***

5. $x^2 + 3x + 2 = 0$ **6.** $x^2 + 8x + 15 = 0$ **7.** $3x^2 + 8x - 3 = 0$

8. $2x^2 + x - 6 = 0$ **9.** $5x^2 - 7x - 6 = 0$ **10.** $2x^2 - 5x - 3 = 0$

11. $x^2 - 6x = -8$ **12.** $x^2 - 10x = -24$ **13.** $2x^2 = 9x - 4$

14. $5x^2 = 11x - 2$ **15.** $8x^2 + 14x + 3 = 0$ **16.** $12x^2 + 19x + 5 = 0$

Use the square root property to solve each equation. ***See Examples 2, 4, and 5.***

17. $x^2 = 81$ **18.** $x^2 = 225$ **19.** $x^2 = 17$ **20.** $x^2 = 19$

21. $x^2 = 32$ **22.** $x^2 = 54$ **23.** $x^2 - 20 = 0$ **24.** $p^2 - 50 = 0$

25. $3x^2 - 72 = 0$ **26.** $5z^2 - 200 = 0$ **27.** $2x^2 + 7 = 61$ **28.** $3x^2 + 8 = 80$

29. $3x^2 - 10 = 86$ **30.** $4x^2 - 7 = 65$ **31.** $(x + 2)^2 = 25$ **32.** $(t + 8)^2 = 9$

33. $(x - 6)^2 = 49$ **34.** $(x - 4)^2 = 64$ **35.** $(x - 4)^2 = 3$

36. $(x + 3)^2 = 11$ **37.** $(t + 5)^2 = 48$ **38.** $(m - 6)^2 = 27$

39. $(3x - 1)^2 = 7$ **40.** $(2x - 5)^2 = 10$ **41.** $(4p + 1)^2 = 24$

42. $(5t + 2)^2 = 12$ **43.** $(2 - 5t)^2 = 12$ **44.** $(1 - 4p)^2 = 24$

Use Galileo's formula to solve each problem. Round answers to the nearest tenth. **See Example 3.**

45. The sculpture of American presidents at Mount Rushmore National Memorial is 500 ft above the valley floor. How long would it take a rock dropped from the top of the sculpture to fall to the ground? (Data from www.travelsd.com)

46. The Gateway Arch in St. Louis, Missouri, is 630 ft tall. How long would it take an object dropped from the top of the arch to fall to the ground? (Data from www.gatewayarch.com)

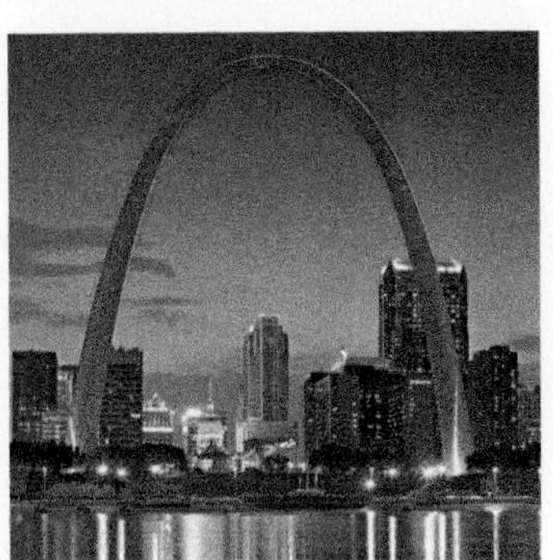

47. Concept Check Which one of the two equations

$$(2x + 1)^2 = 5 \quad \text{and} \quad x^2 + 4x = 12$$

is more suitable for solving by the square root property? By completing the square?

48. Concept Check According to the procedure described in this section, what would be the first step in solving $2x^2 + 8x = 9$ by completing the square?

Concept Check *Find the constant that must be added to make each expression a perfect square trinomial. Then factor the trinomial.*

49. $x^2 + 6x +$ ____ It factors as ________.

50. $x^2 + 14x +$ ____ It factors as ________.

51. $p^2 - 12p +$ ____ It factors as ________.

52. $x^2 - 20x +$ ____ It factors as ________.

53. $q^2 + 9q +$ ____ It factors as ________.

54. $t^2 + 3t +$ ____ It factors as ________.

Solve each equation by completing the square. **See Examples 6–8.**

55. $x^2 - 2x - 24 = 0$

56. $m^2 - 4m - 32 = 0$

57. $x^2 + 4x - 2 = 0$

58. $t^2 + 2t - 1 = 0$

59. $x^2 + 10x + 18 = 0$

60. $x^2 + 8x + 11 = 0$

61. $3w^2 - w = 24$

62. $4z^2 - z = 39$

63. $x^2 + 7x - 1 = 0$

64. $x^2 + 13x - 3 = 0$

65. $2k^2 + 5k - 2 = 0$

66. $3r^2 + 2r - 2 = 0$

67. $5x^2 - 10x + 2 = 0$

68. $2x^2 - 16x + 25 = 0$

69. $9x^2 - 24x = -13$

70. $25n^2 - 20n = 1$

71. $z^2 - \frac{4}{3}z = -\frac{1}{9}$

72. $p^2 - \frac{8}{3}p = -1$

73. $0.1x^2 - 0.2x - 0.1 = 0$ (*Hint:* First clear the decimals.)

74. $0.1p^2 - 0.4p + 0.1 = 0$ (*Hint:* First clear the decimals.)

Solve each equation. (All solutions for these equations are nonreal complex numbers.) ***See Example 9.***

75. $x^2 = -100$ **76.** $x^2 = -64$ **77.** $x^2 = -12$

78. $x^2 = -18$ **79.** $(x + 3)^2 = -4$ **80.** $(x - 5)^2 = -36$

81. $(r - 5)^2 = -3$ **82.** $(t + 6)^2 = -5$ **83.** $(6k - 1)^2 = -8$

84. $(4m - 7)^2 = -27$ **85.** $m^2 + 4m + 13 = 0$ **86.** $t^2 + 6t + 10 = 0$

87. $m^2 + 6m + 12 = 0$ **88.** $x^2 + 10x + 27 = 0$ **89.** $3r^2 + 4r + 4 = 0$

90. $4x^2 + 5x + 5 = 0$ **91.** $-k^2 - 5k - 10 = 0$ **92.** $-x^2 - 3x - 8 = 0$

Extending Skills *Solve for x. Assume that a and b represent positive real numbers.*

93. $x^2 - b = 0$ **94.** $x^2 = 4b$ **95.** $4x^2 = b^2 + 16$

96. $9x^2 - 25a = 0$ **97.** $(5x - 2b)^2 = 3a$ **98.** $x^2 - a^2 - 36 = 0$

RELATING CONCEPTS For Individual or Group Work (Exercises 99–104)

The Greeks had a method of completing the square geometrically in which they literally changed a figure into a square. For example, to complete the square for $x^2 + 6x$, we begin with a square of side x, as in the figure on the top. We add three rectangles of width 1 to the right side and the bottom to get a region with area $x^2 + 6x$. To fill in the corner (complete the square), we must add nine 1-by-1 squares as shown.

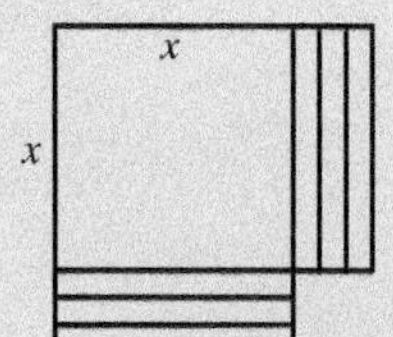

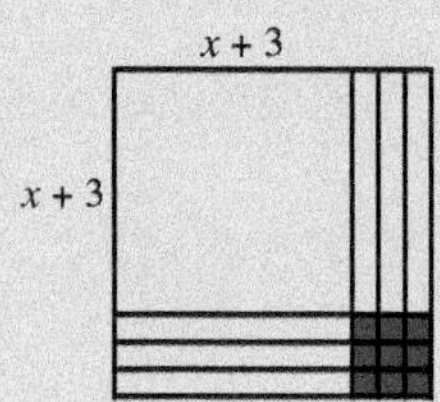

Work Exercises 99–104 in order.

99. What is the area of the original square?

100. What is the area of each strip?

101. What is the total area of the six strips?

102. What is the area of each small square in the corner of the second figure?

103. What is the total area of the small squares?

104. What is the area of the new "complete" square?

8.2 The Quadratic Formula

OBJECTIVES

1. Derive the quadratic formula.
2. Solve quadratic equations using the quadratic formula.
3. Use the discriminant to determine number and type of solutions.

In this section, we complete the square to solve the general quadratic equation

$$ax^2 + bx + c = 0,$$

where a, b, and c are complex numbers and $a \neq 0$. The solution of this general equation gives a formula for finding the solution of *any* specific quadratic equation.

OBJECTIVE 1 Derive the quadratic formula.

VOCABULARY

☐ quadratic formula
☐ discriminant

We solve $ax^2 + bx + c = 0$ by completing the square (where $a > 0$) as follows.

$$ax^2 + bx + c = 0$$

$$x^2 + \frac{b}{a}x + \frac{c}{a} = 0 \qquad \text{Divide by } a. \text{ (Step 1)}$$

$$x^2 + \frac{b}{a}x = -\frac{c}{a} \qquad \text{Subtract } \tfrac{c}{a}. \text{ (Step 2)}$$

$$\left[\frac{1}{2}\left(\frac{b}{a}\right)\right]^2 = \left(\frac{b}{2a}\right)^2 = \frac{b^2}{4a^2} \qquad \text{Complete the square. (Step 3)}$$

$$x^2 + \frac{b}{a}x + \frac{b^2}{4a^2} = -\frac{c}{a} + \frac{b^2}{4a^2} \qquad \text{Add } \tfrac{b^2}{4a^2} \text{ to each side.}$$

$$\left(x + \frac{b}{2a}\right)^2 = \frac{b^2}{4a^2} + \frac{-c}{a} \qquad \text{Factor on the left. Rearrange the terms on the right.}$$

$$\left(x + \frac{b}{2a}\right)^2 = \frac{b^2}{4a^2} + \frac{-4ac}{4a^2} \qquad \text{Write with a common denominator.}$$

$$\left(x + \frac{b}{2a}\right)^2 = \frac{b^2 - 4ac}{4a^2} \qquad \text{Add fractions.}$$

$$x + \frac{b}{2a} = \sqrt{\frac{b^2 - 4ac}{4a^2}} \quad \text{or} \quad x + \frac{b}{2a} = -\sqrt{\frac{b^2 - 4ac}{4a^2}} \qquad \text{Square root property (Step 4)}$$

We can simplify $\sqrt{\frac{b^2 - 4ac}{4a^2}}$ as $\frac{\sqrt{b^2 - 4ac}}{\sqrt{4a^2}}$, or $\frac{\sqrt{b^2 - 4ac}}{2a}$.

The right side of each equation can be expressed as follows.

$$x + \frac{b}{2a} = \frac{\sqrt{b^2 - 4ac}}{2a} \quad \text{or} \quad x + \frac{b}{2a} = \frac{-\sqrt{b^2 - 4ac}}{2a}$$

$$x = \frac{-b}{2a} + \frac{\sqrt{b^2 - 4ac}}{2a} \quad \text{or} \quad x = \frac{-b}{2a} - \frac{\sqrt{b^2 - 4ac}}{2a}$$

$$\boldsymbol{x = \frac{-b + \sqrt{b^2 - 4ac}}{2a}} \quad \text{or} \quad \boldsymbol{x = \frac{-b - \sqrt{b^2 - 4ac}}{2a}}$$

If $a < 0$, the same two solutions are obtained.

This result is the **quadratic formula,** which is abbreviated as follows.

Quadratic Formula

The solutions of the equation $ax^2 + bx + c = 0$ (where $a \neq 0$) are given by

$$\boldsymbol{x = \frac{-b \pm \sqrt{b^2 - 4ac}}{2a}}.$$

CAUTION ***In the quadratic formula, the square root is added to or subtracted from the value of $-b$ before dividing by $2a$.***

OBJECTIVE 2 Solve quadratic equations using the quadratic formula.

NOW TRY EXERCISE 1
Solve $2x^2 + 3x - 20 = 0$.

EXAMPLE 1 Using the Quadratic Formula (Two Rational Solutions)

Solve $6x^2 - 5x - 4 = 0$.

This equation is in standard form, so we identify the values of a, b, and c. Here a, the coefficient of the second-degree term, is 6, and b, the coefficient of the first-degree term, is -5. The constant c is -4. Now substitute into the quadratic formula.

$$x = \frac{-b \pm \sqrt{b^2 - 4ac}}{2a} \qquad \text{Quadratic formula}$$

$$x = \frac{-(-5) \pm \sqrt{(-5)^2 - 4(6)(-4)}}{2(6)} \qquad a = 6,\ b = -5,\ c = -4$$

Use parentheses and substitute carefully to avoid errors.

$$x = \frac{5 \pm \sqrt{25 + 96}}{12}$$

$$x = \frac{5 \pm \sqrt{121}}{12} \qquad \text{Add under the radical.}$$

$$x = \frac{5 \pm 11}{12} \qquad \text{Take the square root.}$$

There are two values represented, one from the + sign and one from the − sign.

$$x = \frac{5 + 11}{12} = \frac{16}{12} = \frac{4}{3} \quad \text{or} \quad x = \frac{5 - 11}{12} = \frac{-6}{12} = -\frac{1}{2}$$

Check each value in the original equation. The solution set is $\left\{-\frac{1}{2}, \frac{4}{3}\right\}$. NOW TRY

NOTE We could have factored the trinomial and then used the zero-factor property to solve the equation in **Example 1.**

$$6x^2 - 5x - 4 = 0 \qquad \text{See Example 1.}$$

$$(3x - 4)(2x + 1) = 0 \qquad \text{Factor.}$$

$$3x - 4 = 0 \quad \text{or} \quad 2x + 1 = 0 \qquad \text{Zero-factor property}$$

$$3x = 4 \quad \text{or} \quad 2x = -1 \qquad \text{Solve each equation.}$$

$$x = \frac{4}{3} \quad \text{or} \quad x = -\frac{1}{2} \qquad \text{Same solutions as in Example 1}$$

When solving a quadratic equation, it is a good idea to try to factor the quadratic expression first. If it can be factored, then apply the zero-factor property. If it cannot be factored or if factoring is difficult, then use the quadratic formula.

EXAMPLE 2 Using the Quadratic Formula (One Rational Solution)

Solve $9x^2 + 12x + 4 = 0$.

The trinomial on the left side of the equality symbol can be factored, so this equation could be solved using the zero-factor property. However, if we did not recognize this and solved using the quadratic formula, our work might look similar to that shown on the next page.

NOW TRY ANSWER

1. $\left\{-4, \frac{5}{2}\right\}$

NOW TRY EXERCISE 2
Solve $4x^2 - 20x + 25 = 0$.

The equation $9x^2 + 12x + 4 = 0$ is in standard form, so $a = 9$, $b = 12$, and $c = 4$.

$$x = \frac{-b \pm \sqrt{b^2 - 4ac}}{2a} \quad \text{Quadratic formula}$$

$$x = \frac{-12 \pm \sqrt{12^2 - 4(9)(4)}}{2(9)} \quad a = 9,\ b = 12,\ c = 4$$

$$x = \frac{-12 \pm \sqrt{144 - 144}}{18} \quad \text{Simplify.}$$

$$x = \frac{-12 \pm 0}{18} \quad \sqrt{0} = 0$$

$$x = -\frac{2}{3} \quad \text{Write } \frac{-12}{18} \text{ in lowest terms.}$$

In this case, $b^2 - 4ac = 0$ because $9x^2 + 12x + 4$ is a perfect square trinomial. There is one *distinct* solution, $-\frac{2}{3}$. The solution set is $\left\{-\frac{2}{3}\right\}$. **NOW TRY**

NOTE Solve the equation in **Example 2** by factoring $9x^2 + 12x + 4$ and using the zero-factor property to confirm that the same solution, $-\frac{2}{3}$, results.

NOW TRY EXERCISE 3
Solve $3x^2 + 1 = -5x$.

EXAMPLE 3 Using the Quadratic Formula (Two Irrational Solutions)

Solve $4x^2 = 8x - 1$.

Write the equation in standard form as $4x^2 - 8x + 1 = 0$. ← This is a key step.

$$x = \frac{-b \pm \sqrt{b^2 - 4ac}}{2a} \quad \text{Quadratic formula}$$

$$x = \frac{-(-8) \pm \sqrt{(-8)^2 - 4(4)(1)}}{2(4)} \quad a = 4,\ b = -8,\ c = 1$$

$$x = \frac{8 \pm \sqrt{64 - 16}}{8} \quad \text{Simplify in the numerator and denominator.}$$

$$x = \frac{8 \pm \sqrt{48}}{8} \quad \text{Subtract under the radical.}$$

$$x = \frac{8 \pm 4\sqrt{3}}{8} \quad \sqrt{48} = \sqrt{16} \cdot \sqrt{3} = 4\sqrt{3}$$

$$x = \frac{4(2 \pm \sqrt{3})}{4(2)} \quad \text{Factor.}$$

$$x = \frac{2 \pm \sqrt{3}}{2} \quad \text{Divide out the common factor 4 to write in lowest terms.}$$

The solution set is $\left\{\frac{2 \pm \sqrt{3}}{2}\right\}$. **NOW TRY**

NOW TRY ANSWERS
2. $\left\{\frac{5}{2}\right\}$
3. $\left\{\frac{-5 \pm \sqrt{13}}{6}\right\}$

CAUTION

1. ***Before solving, every quadratic equation must be expressed in standard form*** $ax^2 + bx + c = 0$, whether we use the zero-factor property or the quadratic formula.
2. ***When writing solutions in lowest terms, be sure to factor first. Then divide out the common factor.*** See the last two steps in **Example 3.**

NOW TRY EXERCISE 4
Solve $(x + 5)(x - 1) = -18$.

EXAMPLE 4 Using the Quadratic Formula (Two Nonreal Complex Solutions)

Solve $(9x + 3)(x - 1) = -8$.

This is a quadratic equation—when the first terms $9x$ and x are multiplied, we obtain a second-degree term, $9x^2$. We must write the equation in standard form.

$$(9x + 3)(x - 1) = -8$$

$$9x^2 - 6x - 3 = -8 \quad \text{Multiply.}$$

$$\text{Standard form} \rightarrow 9x^2 - 6x + 5 = 0 \quad \text{Add 8.}$$

From the equation $9x^2 - 6x + 5 = 0$, we identify $a = 9$, $b = -6$, and $c = 5$.

$$x = \frac{-b \pm \sqrt{b^2 - 4ac}}{2a} \quad \text{Quadratic formula}$$

$$x = \frac{-(-6) \pm \sqrt{(-6)^2 - 4(9)(5)}}{2(9)} \quad \text{Substitute.}$$

$$x = \frac{6 \pm \sqrt{-144}}{18} \quad \text{Simplify.}$$

$$x = \frac{6 \pm 12i}{18} \quad \sqrt{-144} = 12i$$

$$x = \frac{6(1 \pm 2i)}{6(3)} \quad \text{Factor.}$$

Factor first. Then divide out the common factor.

$$x = \frac{1 \pm 2i}{3} \quad \text{Divide out the common factor 6 to write in lowest terms.}$$

$$x = \frac{1}{3} \pm \frac{2}{3}i \quad \text{Standard form } a + bi \text{ for a complex number}$$

The solution set is $\left\{\frac{1}{3} \pm \frac{2}{3}i\right\}$.

NOW TRY 

OBJECTIVE 3 Use the discriminant to determine number and type of solutions.

The solutions of the quadratic equation $ax^2 + bx + c = 0$ are given by

$$x = \frac{-b \pm \sqrt{b^2 - 4ac}}{2a}. \leftarrow \text{Discriminant}$$

The expression under the radical symbol, $\boldsymbol{b^2 - 4ac}$, is called the **discriminant** because it distinguishes among the number of solutions—one or two—and the type of solutions—rational, irrational, or nonreal complex—of a quadratic equation.

NOW TRY ANSWER
4. $\{-2 \pm 3i\}$

Using the Discriminant

If a, b, and c are integers in a quadratic equation $ax^2 + bx + c = 0$, then the discriminant $b^2 - 4ac$ can be used to determine the number and type of solutions of the equation as follows.

Discriminant $b^2 - 4ac$	*Number and Type of Solutions*
Positive, and the square of an integer	Two rational solutions
Positive, but not the square of an integer	Two irrational solutions
Zero	One rational solution
Negative	Two nonreal complex solutions

We can also use the discriminant to help decide how to solve a quadratic equation.

If a, b, and c are integers and the discriminant is a perfect square (including 0), then the equation can be solved using the zero-factor property. Otherwise, the quadratic formula should be used.

EXAMPLE 5 Using the Discriminant

Find the discriminant. Use it to predict the number and type of solutions for each equation. Tell whether the equation can be solved using the zero-factor property, or if the quadratic formula should be used instead.

(a) $6x^2 - x - 15 = 0$

First identify the values of a, b, and c. Because $-x = -1x$, the value of b is -1. We find the discriminant by evaluating $b^2 - 4ac$.

$b^2 - 4ac$ — Use parentheses and substitute carefully.

$= (-1)^2 - 4(6)(-15)$ — $a = 6, b = -1, c = -15$ (all integers)

$= 1 + 360$ — Apply the exponent. Multiply.

$= 361$ — Add.

$= 19^2$, which is a perfect square.

The discriminant 361 is a perfect square, so referring to the table we see that there will be two rational solutions. We can solve using the zero-factor property.

(b) $3x^2 - 4x = 5$

Write in standard form as $3x^2 - 4x - 5 = 0$.

$b^2 - 4ac$ — Discriminant

$= (-4)^2 - 4(3)(-5)$ — $a = 3, b = -4, c = -5$ (all integers)

$= 16 + 60$ — Apply the exponent. Multiply.

$= 76$ — Add.

Because 76 is positive but *not* the square of an integer, the equation will have two irrational solutions. We solve using the quadratic formula.

NOW TRY EXERCISE 5

Find the discriminant. Use it to predict the number and type of solutions for each equation. Tell whether the equation can be solved using the zero-factor property, or if the quadratic formula should be used instead.

(a) $8x^2 - 6x - 5 = 0$

(b) $9x^2 = 24x - 16$

(c) $3x^2 + 2x = -1$

NOW TRY ANSWERS

5. (a) 196; two rational solutions; zero-factor property
(b) 0; one rational solution; zero-factor property
(c) -8; two nonreal complex solutions; quadratic formula

(c) $4x^2 + x + 1 = 0$

$x = 1x$, so $b = 1$.

$b^2 - 4ac$	Discriminant
$= 1^2 - 4(4)(1)$	$a = 4, b = 1, c = 1$ (all integers)
$= 1 - 16$	Apply the exponent. Multiply.
$= -15$	Subtract.

Because the discriminant is negative, there will be two nonreal complex solutions. We solve using the quadratic formula.

(d) $4x^2 + 9 = 12x$ Write in standard form as $4x^2 - 12x + 9 = 0$.

$b^2 - 4ac$	Discriminant
$= (-12)^2 - 4(4)(9)$	$a = 4, b = -12, c = 9$ (all integers)
$= 144 - 144$	Apply the exponent. Multiply.
$= 0$	Subtract.

The discriminant is 0, so there is only one rational solution. We solve using the zero-factor property.

NOW TRY

8.2 Exercises

MyLab Math

Video solutions for select problems available in MyLab Math

Concept Check *Answer each question.*

1. The documentation for an early version of Microsoft *Word* for Windows used the following for the quadratic formula. Was this correct? If not, correct it.

$$x = -b \pm \frac{\sqrt{b^2 - 4ac}}{2a} \qquad \text{Correct or incorrect?}$$

2. One patron wrote the quadratic formula, as shown here, on a wall at the Cadillac Bar in Houston, Texas. Was this correct? If not, correct it.

$$x = \frac{-b\sqrt{b^2 - 4ac}}{2a} \qquad \text{Correct or incorrect?}$$

3. A student solved $5x^2 - 5x + 1 = 0$ incorrectly as follows.

$$x = \frac{-(-5) \pm \sqrt{(-5)^2 - 4(5)(1)}}{2(5)} \qquad a = 5, b = -5, c = 1$$

$$x = \frac{5 \pm \sqrt{5}}{10} \qquad \text{Simplify.}$$

$$x = \frac{1}{2} \pm \sqrt{5} \qquad \text{Write in lowest terms.}$$

WHAT WENT WRONG? Give the correct solution set.

4. A student incorrectly claimed that the equation $2x^2 - 5 = 0$ cannot be solved using the quadratic formula because there is no first-degree x-term. **WHAT WENT WRONG?** Give the values of a, b, and c for this equation.

Use the quadratic formula to solve each equation. (All solutions for these equations are real numbers.) ***See Examples 1–3.***

5. $x^2 - 8x + 15 = 0$ **6.** $x^2 + 3x - 28 = 0$ **7.** $6x^2 + 11x - 10 = 0$

8. $8x^2 + 10x - 3 = 0$ **9.** $4x^2 + 12x + 9 = 0$ **10.** $16x^2 + 40x + 25 = 0$

11. $36x^2 - 12x + 1 = 0$ **12.** $9x^2 - 6x + 1 = 0$ **13.** $2x^2 + 4x + 1 = 0$

14. $2x^2 + 3x - 1 = 0$ **15.** $2x^2 - 2x = 1$ **16.** $9x^2 + 6x = 1$

17. $x^2 + 18 = 10x$ **18.** $x^2 - 4 = 2x$ **19.** $4x^2 + 4x - 1 = 0$

20. $4r^2 - 4r - 19 = 0$ **21.** $2 - 2x = 3x^2$ **22.** $26r - 2 = 3r^2$

23. $\frac{x^2}{4} - \frac{x}{2} = 1$ **24.** $p^2 + \frac{p}{3} = \frac{1}{6}$ **25.** $-2t(t + 2) = -3$

26. $-3x(x + 2) = -4$ **27.** $(r - 3)(r + 5) = 2$ **28.** $(x + 1)(x - 7) = 1$

29. $(x + 2)(x - 3) = 1$ **30.** $(x - 5)(x + 2) = 6$ **31.** $p = \frac{5(5 - p)}{3(p + 1)}$

32. $x = \frac{2(x + 3)}{x + 5}$ **33.** $(2x + 1)^2 = x + 4$ **34.** $(2x - 1)^2 = x + 2$

Use the quadratic formula to solve each equation. (All solutions for these equations are nonreal complex numbers.) ***See Example 4.***

35. $x^2 - 3x + 6 = 0$ **36.** $x^2 - 5x + 20 = 0$ **37.** $r^2 - 6r + 14 = 0$

38. $t^2 + 4t + 11 = 0$ **39.** $4x^2 - 4x = -7$ **40.** $9x^2 - 6x = -7$

41. $x(3x + 4) = -2$ **42.** $z(2z + 3) = -2$

43. $(2x - 1)(8x - 4) = -1$ **44.** $(x - 1)(9x - 3) = -2$

45. $(6x + 1)(x - 2) = (x + 2)(x - 5)$ **46.** $(4x + 3)(x - 1) = (x + 3)(x - 6)$

Find the discriminant. Use it to determine whether the solutions for each equation are

A. *two rational numbers* **B.** *one rational number*

C. *two irrational numbers* **D.** *two nonreal complex numbers.*

Tell whether the equation can be solved using the zero-factor property, or if the quadratic formula should be used instead. Do not actually solve. ***See Example 5.***

47. $25x^2 + 70x + 49 = 0$ **48.** $4x^2 - 28x + 49 = 0$ **49.** $x^2 + 4x + 2 = 0$

50. $9x^2 - 12x - 1 = 0$ **51.** $3x^2 = 5x + 2$ **52.** $4x^2 = 4x + 3$

53. $3m^2 - 10m + 15 = 0$ **54.** $18x^2 + 60x + 82 = 0$

55. Find the discriminant for each quadratic equation. Use it to tell whether the equation can be solved using the zero-factor property, or the quadratic formula should be used instead. Then solve each equation.

(a) $3x^2 + 13x = -12$ **(b)** $2x^2 + 19 = 14x$

56. Refer to the answers in **Exercises 47–54,** and solve the equation given in each exercise.

(a) $25x^2 + 70x + 49 = 0$ **(Exercise 47)** **(b)** $4x^2 - 28x + 49 = 0$ **(Exercise 48)**

(c) $3x^2 = 5x + 2$ **(Exercise 51)** **(d)** $4x^2 = 4x + 3$ **(Exercise 52)**

Extending Skills *Find the value of a, b, or c so that each equation will have exactly one rational solution. (Hint: The discriminant must equal 0 for an equation to have one rational solution.)*

57. $p^2 + bp + 25 = 0$

58. $r^2 - br + 49 = 0$

59. $am^2 + 8m + 1 = 0$

60. $at^2 + 24t + 16 = 0$

61. $9x^2 - 30x + c = 0$

62. $4m^2 + 12m + c = 0$

63. One solution of $4x^2 + bx - 3 = 0$ is $-\frac{5}{2}$. Find b and the other solution.

64. One solution of $3x^2 - 7x + c = 0$ is $\frac{1}{3}$. Find c and the other solution.

8.3 Equations That Lead to Quadratic Methods

OBJECTIVES

1 Solve rational equations that lead to quadratic equations.

2 Solve applied problems involving quadratic equations.

3 Solve radical equations that lead to quadratic equations.

4 Solve equations that are quadratic in form.

OBJECTIVE 1 Solve rational equations that lead to quadratic equations.

A variety of nonquadratic equations can be written in the form of a quadratic equation and solved using the methods of this chapter.

EXAMPLE 1 Solving a Rational Equation That Leads to a Quadratic Equation

Solve $\frac{1}{x} + \frac{1}{x-1} = \frac{7}{12}$.

Clear fractions by multiplying each side by the least common denominator, $12x(x-1)$. (The domain is $\{x \mid x \text{ is a real number}, x \neq 0, 1\}$.)

$$12x(x-1)\left(\frac{1}{x} + \frac{1}{x-1}\right) = 12x(x-1)\left(\frac{7}{12}\right) \quad \text{Multiply by the LCD.}$$

$$12x(x-1)\frac{1}{x} + 12x(x-1)\frac{1}{x-1} = 12x(x-1)\frac{7}{12} \quad \text{Distributive property}$$

$$12(x-1) + 12x = x(x-1) \cdot 7 \quad \text{Multiply.}$$

$$12x - 12 + 12x = 7x^2 - 7x \quad \text{Distributive property}$$

$$24x - 12 = 7x^2 - 7x \quad \text{Combine like terms.}$$

This trinomial is factorable.

$$7x^2 - 31x + 12 = 0 \quad \text{Standard form}$$

$$(7x-3)(x-4) = 0 \quad \text{Factor.}$$

$$7x - 3 = 0 \quad \text{or} \quad x - 4 = 0 \quad \text{Zero-factor property}$$

$$x = \frac{3}{7} \quad \text{or} \quad x = 4 \quad \text{Solve each equation.}$$

These values are in the domain. Check them in the original equation. The solution set is $\left\{\frac{3}{7}, 4\right\}$.

NOW TRY

VOCABULARY

☐ quadratic in form

NOW TRY EXERCISE 1

Solve $\frac{2}{x} + \frac{3}{x+2} = 1$.

OBJECTIVE 2 Solve applied problems involving quadratic equations.

Some distance-rate-time (or motion) problems lead to quadratic equations. We use a form of the distance formula $d = rt$ to solve them.

NOW TRY ANSWER

1. $\{-1, 4\}$

NOW TRY EXERCISE 2
A small fishing boat averages 18 mph in still water. It takes the boat $\frac{9}{10}$ hr to travel 8 mi upstream and return. Find the rate of the current.

EXAMPLE 2 Solving a Motion Problem

A riverboat for tourists averages 12 mph in still water. It takes the boat 1 hr, 4 min to travel 6 mi upstream and return. Find the rate of the current.

Step 1 **Read** the problem carefully.

Step 2 **Assign a variable.** Let $x =$ the rate of the current.

The current slows down the boat as it travels upstream, so the rate of the boat traveling upstream is its rate in still water *less* the rate of the current, or $(12 - x)$ mph. See **FIGURE 1**.

Similarly, the current speeds up the boat as it travels downstream, so its rate downstream is $(12 + x)$ mph. Thus,

$12 - x =$ the rate upstream in miles per hour,

and $12 + x =$ the rate downstream in miles per hour.

Riverboat traveling upstream—the current slows it down.
FIGURE 1

	d	r	t
Upstream	6	$12 - x$	$\frac{6}{12 - x}$
Downstream	6	$12 + x$	$\frac{6}{12 + x}$

Complete a table. Use the distance formula, $d = rt$, solved for time t, $t = \frac{d}{r}$, to write expressions for t.

Step 3 **Write an equation.** We use the total time, 1 hr, 4 min, written as a fraction.

$$1 + \frac{4}{60} = 1 + \frac{1}{15} = \frac{16}{15} \text{ hr} \qquad \text{Total time}$$

The time upstream plus the time downstream equals $\frac{16}{15}$ hr.

Time upstream + Time downstream = Total time

$$\frac{6}{12 - x} + \frac{6}{12 + x} = \frac{16}{15}$$

Step 4 **Solve** the equation. The LCD is $15(12 - x)(12 + x)$.

$$15(12 - x)(12 + x)\left(\frac{6}{12 - x} + \frac{6}{12 + x}\right) = 15(12 - x)(12 + x)\left(\frac{16}{15}\right)$$

Multiply by the LCD.

$$15(12 + x) \cdot 6 + 15(12 - x) \cdot 6 = (12 - x)(12 + x)16$$

Distributive property; Multiply.

$$90(12 + x) + 90(12 - x) = 16(144 - x^2) \qquad \text{Multiply.}$$

$$1080 + 90x + 1080 - 90x = 2304 - 16x^2 \qquad \text{Distributive property}$$

$$2160 = 2304 - 16x^2 \qquad \text{Combine like terms.}$$

$$16x^2 = 144 \qquad \text{Add } 16x^2\text{. Subtract 2160.}$$

$$x^2 = 9 \qquad \text{Divide by 16.}$$

$$x = 3 \quad \text{or} \quad x = -3 \qquad \text{Square root property}$$

Step 5 **State the answer.** The current rate cannot be -3, so the answer is 3 mph.

Step 6 **Check** that this value satisfies the original problem. NOW TRY 

NOW TRY ANSWER
2. 2 mph

PROBLEM-SOLVING HINT Recall from earlier work that a person's work rate is $\frac{1}{t}$ part of the job per hour, where t is the time in hours required to do the complete job. Thus, the part of the job the person will do in x hours is $\frac{1}{t}x$.

EXAMPLE 3 Solving a Work Problem

It takes two carpet layers 4 hr to carpet a room. If each worked alone, one of them could do the job in 1 hr less time than the other. How long would it take each carpet layer to complete the job alone?

Step 1 **Read** the problem again. There will be two answers.

Step 2 **Assign a variable.**

Let x = the number of hours for the slower carpet layer to complete the job.

Then $x - 1$ = the number of hours for the faster carpet layer to complete the job.

The slower worker's rate is $\frac{1}{x}$, and the faster worker's rate is $\frac{1}{x-1}$. Together they can do the job in 4 hr. Complete a table as shown.

	Rate	Time Working Together	Fractional Part of the Job Done
Slower Worker	$\frac{1}{x}$	4	$\frac{1}{x}(4)$
Faster Worker	$\frac{1}{x-1}$	4	$\frac{1}{x-1}(4)$

Sum is 1 whole job.

Step 3 **Write an equation.**

Part done by slower worker + Part done by faster worker = 1 whole job

$$\frac{4}{x} + \frac{4}{x-1} = 1$$

Step 4 **Solve** the equation. The LCD is $x(x-1)$.

$$x(x-1)\left(\frac{4}{x} + \frac{4}{x-1}\right) = x(x-1)(1) \quad \text{Multiply by the LCD.}$$

$$(x-1) \cdot 4 + 4x = x(x-1) \quad \text{Distributive property}$$

$$4x - 4 + 4x = x^2 - x \quad \text{Distributive property}$$

$$x^2 - 9x + 4 = 0 \quad \text{Standard form}$$

The trinomial on the left cannot be factored, so the equation cannot be solved using the zero-factor property. We use the quadratic formula.

$$x = \frac{-b \pm \sqrt{b^2 - 4ac}}{2a} \quad \text{Quadratic formula}$$

$$x = \frac{-(-9) \pm \sqrt{(-9)^2 - 4(1)(4)}}{2(1)} \quad a = 1, b = -9, c = 4$$

$$x = \frac{9 \pm \sqrt{65}}{2} \quad \text{Simplify.}$$

$$x = \frac{9 + \sqrt{65}}{2} = 8.5 \quad \text{or} \quad x = \frac{9 - \sqrt{65}}{2} = 0.5 \quad \text{Use a calculator. Round to the nearest tenth.}$$

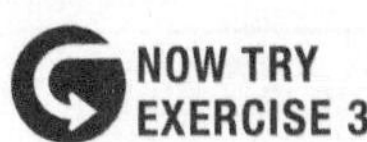
NOW TRY EXERCISE 3

Two electricians are running wire to finish a basement. One electrician could finish the job in 2 hr less time than the other. Together, they complete the job in 6 hr. How long (to the nearest tenth) would it take the slower electrician to complete the job alone?

NOW TRY EXERCISE 4

Solve each equation.

(a) $x = \sqrt{9x - 20}$

(b) $x + \sqrt{x} = 20$

Step 5 **State the answer.** Only the solution 8.5 makes sense in the original problem.

If $x = 0.5$, then $x - 1 = 0.5 - 1 = -0.5$, Time cannot be negative.

which cannot represent the time for the faster worker. The slower worker could do the job in 8.5 hr and the faster in $8.5 - 1 = 7.5$ hr.

Step 6 **Check** that these results satisfy the original problem. NOW TRY 

OBJECTIVE 3 Solve radical equations that lead to quadratic equations.

EXAMPLE 4 Solving Radical Equations That Lead to Quadratic Equations

Solve each equation.

(a) $x = \sqrt{6x - 8}$

This equation is not quadratic. However, squaring each side of the equation gives a quadratic equation that can be solved using the zero-factor property.

$$x^2 = \left(\sqrt{6x - 8}\right)^2 \quad \text{Square each side.}$$

$$x^2 = 6x - 8 \quad (\sqrt{a})^2 = a$$

This trinomial is factorable.

$$x^2 - 6x + 8 = 0 \quad \text{Standard form}$$

$$(x - 4)(x - 2) = 0 \quad \text{Factor.}$$

$$x - 4 = 0 \quad \text{or} \quad x - 2 = 0 \quad \text{Zero-factor property}$$

$$x = 4 \quad \text{or} \quad x = 2 \quad \text{Proposed solutions}$$

Squaring each side of a radical equation can introduce extraneous solutions. ***All proposed solutions must be checked in the original (not the squared) equation.***

CHECK

$x = \sqrt{6x - 8}$		$x = \sqrt{6x - 8}$	
$4 \stackrel{?}{=} \sqrt{6(4) - 8}$	Let $x = 4$.	$2 \stackrel{?}{=} \sqrt{6(2) - 8}$	Let $x = 2$.
$4 \stackrel{?}{=} \sqrt{16}$		$2 \stackrel{?}{=} \sqrt{4}$	
$4 = 4$ ✓	True	$2 = 2$ ✓	True

Both proposed solutions check, so the solution set is $\{2, 4\}$.

(b)

$$x + \sqrt{x} = 6$$

$$\sqrt{x} = 6 - x \quad \text{Isolate the radical on one side.}$$

$$\left(\sqrt{x}\right)^2 = (6 - x)^2 \quad \text{Square each side.}$$

$$x = 36 - 12x + x^2 \quad (x - y)^2 = x^2 - 2xy + y^2$$

$$x^2 - 13x + 36 = 0 \quad \text{Write in standard form.}$$

$$(x - 4)(x - 9) = 0 \quad \text{Factor.}$$

$$x - 4 = 0 \quad \text{or} \quad x - 9 = 0 \quad \text{Zero-factor property}$$

$$x = 4 \quad \text{or} \quad x = 9 \quad \text{Proposed solutions}$$

CHECK

$x + \sqrt{x} = 6$		$x + \sqrt{x} = 6$	
$4 + \sqrt{4} \stackrel{?}{=} 6$	Let $x = 4$.	$9 + \sqrt{9} \stackrel{?}{=} 6$	Let $x = 9$.
$6 = 6$ ✓	True	$12 = 6$	False

Only the proposed solution 4 checks, so the solution set is $\{4\}$. NOW TRY 

NOW TRY ANSWERS

3. 13.1 hr

4. (a) $\{4, 5\}$ **(b)** $\{16\}$

OBJECTIVE 4 Solve equations that are quadratic in form.

A nonquadratic equation that can be written in the form

$$au^2 + bu + c = 0,$$

for $a \neq 0$ and an algebraic expression u, is **quadratic in form.**

Many equations that are quadratic in form can be solved more easily by defining and substituting a "temporary" variable u for an expression involving the variable in the original equation.

NOW TRY EXERCISE 5

Define a variable u in terms of x, and write each equation in the quadratic form

$$au^2 + bu + c = 0.$$

(a) $x^4 - 10x^2 + 9 = 0$

(b) $6(x+2)^2 - 11(x+2) + 4 = 0$

EXAMPLE 5 Defining Substitution Variables

Define a variable u in terms of x, and write each equation in the quadratic form $au^2 + bu + c = 0$.

(a) $x^4 - 13x^2 + 36 = 0$

Look at the two terms involving the variable x, ignoring their coefficients. Try to find one variable expression that is the square of the other. Here $x^4 = (x^2)^2$, so we can define $u = x^2$, and rewrite the original equation as a quadratic equation in u.

$$u^2 - 13u + 36 = 0 \quad \text{Here, } u = x^2.$$

(b) $2(4x-3)^2 + 7(4x-3) + 5 = 0$

Because this equation involves both $(4x-3)^2$ and $(4x-3)$, we let $u = 4x - 3$.

$$2u^2 + 7u + 5 = 0 \quad \text{Here, } u = 4x - 3.$$

(c) $2x^{2/3} - 11x^{1/3} + 12 = 0$

We apply a power rule for exponents, $(a^m)^n = a^{mn}$. Because $(x^{1/3})^2 = x^{2/3}$, we define $u = x^{1/3}$ and write the original equation as follows.

$$2u^2 - 11u + 12 = 0 \quad \text{Here, } u = x^{1/3}.$$

NOW TRY 

EXAMPLE 6 Solving Equations That Are Quadratic in Form

Solve each equation.

(a)

		$x^4 - 13x^2 + 36 = 0$	See Example 5(a).
		$(x^2)^2 - 13x^2 + 36 = 0$	$x^4 = (x^2)^2$
		Quadratic in form → $u^2 - 13u + 36 = 0$	Let $u = x^2$.
		$(u-4)(u-9) = 0$	Factor.
$u - 4 = 0$	or	$u - 9 = 0$	Zero-factor property
Don't stop here. → $u = 4$	or	$u = 9$	Solve.
$x^2 = 4$	or	$x^2 = 9$	Substitute x^2 for u.
$x = \pm 2$	or	$x = \pm 3$	Square root property

Each value can be verified by substituting it into the original equation for x. The equation $x^4 - 13x^2 + 36 = 0$ is a fourth-degree equation and has four solutions, $-3, -2, 2, 3$.* The solution set is abbreviated $\{\pm 2, \pm 3\}$.

NOW TRY ANSWERS

5. **(a)** $u = x^2$; $u^2 - 10u + 9 = 0$
(b) $u = x + 2$; $6u^2 - 11u + 4 = 0$

*In general, an equation in which an nth-degree polynomial equals 0 has n complex solutions, although some of them may be repeated.

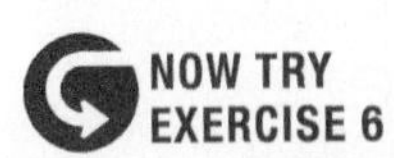

Solve each equation.

(a) $x^4 - 17x^2 + 16 = 0$

(b) $x^4 + 4 = 8x^2$

(b)

$$4x^4 + 1 = 5x^2$$

$$4x^4 - 5x^2 + 1 = 0 \quad \text{Standard form}$$

$$4(x^2)^2 - 5x^2 + 1 = 0 \quad x^4 = (x^2)^2$$

$$4u^2 - 5u + 1 = 0 \quad \text{Let } u = x^2.$$

$$(4u - 1)(u - 1) = 0 \quad \text{Factor.}$$

$$4u - 1 = 0 \quad \text{or} \quad u - 1 = 0 \quad \text{Zero-factor property}$$

$$u = \frac{1}{4} \quad \text{or} \quad u = 1 \quad \text{Solve.}$$

This is a key step.

$$x^2 = \frac{1}{4} \quad \text{or} \quad x^2 = 1 \quad \text{Substitute } x^2 \text{ for } u.$$

$$x = \pm\frac{1}{2} \quad \text{or} \quad x = \pm 1 \quad \text{Square root property}$$

Check that the solution set is $\left\{\pm\frac{1}{2}, \pm 1\right\}$.

(c)

$$x^4 = 6x^2 - 3$$

$$x^4 - 6x^2 + 3 = 0 \quad \text{Standard form}$$

$$(x^2)^2 - 6x^2 + 3 = 0 \quad x^4 = (x^2)^2$$

$$u^2 - 6u + 3 = 0 \quad \text{Let } u = x^2.$$

The trinomial on the left is nonfactorable, so we cannot solve the equation using the zero-factor property. To solve, we use the quadratic formula.

$$u = \frac{-(-6) \pm \sqrt{(-6)^2 - 4(1)(3)}}{2(1)} \quad a = 1, b = -6, c = 3$$

$$u = \frac{6 \pm \sqrt{24}}{2} \quad \text{Simplify.}$$

$$u = \frac{6 \pm 2\sqrt{6}}{2} \quad \sqrt{24} = \sqrt{4} \cdot \sqrt{6} = 2\sqrt{6}$$

$$u = \frac{2(3 \pm \sqrt{6})}{2} \quad \text{Factor.}$$

$$u = 3 \pm \sqrt{6} \quad \text{Divide out the common factor 2.}$$

$$x^2 = 3 + \sqrt{6} \quad \text{or} \quad x^2 = 3 - \sqrt{6} \quad \text{Substitute } x^2 \text{ for } u.$$

Find *both* square roots in each case.

$$x = \pm\sqrt{3 + \sqrt{6}} \quad \text{or} \quad x = \pm\sqrt{3 - \sqrt{6}} \quad \text{Square root property}$$

The solution set contains four numbers and is written as follows.

$$\left\{\pm\sqrt{3 + \sqrt{6}}, \pm\sqrt{3 - \sqrt{6}}\right\}$$

NOW TRY

NOW TRY ANSWERS

6. (a) $\{\pm 1, \pm 4\}$

(b) $\left\{\pm\sqrt{4 + 2\sqrt{3}}, \pm\sqrt{4 - 2\sqrt{3}}\right\}$

NOTE Expressions in equations like those in **Examples 6(a) and (b)** can be factored directly.

$$x^4 - 13x^2 + 36 = 0 \quad \text{Example 6(a) equation}$$

$$(x^2 - 9)(x^2 - 4) = 0 \quad \text{Factor.}$$

$$(x + 3)(x - 3)(x + 2)(x - 2) = 0 \quad \text{Factor again.}$$

Using the zero-factor property gives the same solutions that we obtained in **Example 6(a).** Equations that include nonfactorable quadratic expressions (as in **Example 6(c)**) must be solved using substitution and the quadratic formula.

Solving an Equation That Is Quadratic in Form by Substitution

Step 1 **Define a temporary variable u,** based on the relationship between the variable expressions in the given equation. Substitute u in the original equation and rewrite the equation in the form

$$au^2 + bu + c = 0.$$

Step 2 **Solve the quadratic equation obtained in Step 1** either by factoring the trinomial and applying the zero-factor property or by using the quadratic formula.

Step 3 **Replace u with the expression it defined in Step 1.**

Step 4 **Solve the resulting equations for the original variable.**

Step 5 **Check** all solutions by substituting them in the original equation. Write the solution set.

EXAMPLE 7 Solving Equations That Are Quadratic in Form

Solve each equation.

(a) $2(4x - 3)^2 + 7(4x - 3) + 5 = 0$

Step 1 Because of the repeated quantity $4x - 3$, substitute u for $4x - 3$.

$$2(4x - 3)^2 + 7(4x - 3) + 5 = 0 \quad \text{See Example 5(b).}$$

$$2u^2 + 7u + 5 = 0 \quad \text{Let } u = 4x - 3.$$

Step 2

$$(2u + 5)(u + 1) = 0 \quad \text{Factor.}$$

$$2u + 5 = 0 \quad \text{or} \quad u + 1 = 0 \quad \text{Zero-factor property}$$

Don't stop here.

$$u = -\frac{5}{2} \quad \text{or} \quad u = -1 \quad \text{Solve for } u.$$

Step 3

$$4x - 3 = -\frac{5}{2} \quad \text{or} \quad 4x - 3 = -1 \quad \text{Substitute } 4x - 3 \text{ for } u.$$

Step 4

$$4x = \frac{1}{2} \quad \text{or} \quad 4x = 2 \quad \text{Solve for } x.$$

$$x = \frac{1}{8} \quad \text{or} \quad x = \frac{1}{2}$$

Step 5 Check that the solution set of the original equation is $\left\{\frac{1}{8}, \frac{1}{2}\right\}$.

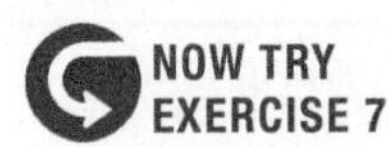

NOW TRY EXERCISE 7

Solve each equation.

(a) $6(x-4)^2 + 11(x-4) - 10 = 0$

(b) $2x^{2/3} - 7x^{1/3} + 3 = 0$

(b) $2x^{2/3} - 11x^{1/3} + 12 = 0$

Step 1 Because $x^{2/3} = (x^{1/3})^2$, we substitute u for $x^{1/3}$.

$$2x^{2/3} - 11x^{1/3} + 12 = 0 \quad \text{See Example 5(c).}$$

$$2(x^{1/3})^2 - 11x^{1/3} + 12 = 0 \quad x^{2/3} = (x^{1/3})^2$$

$$2u^2 - 11u + 12 = 0 \quad \text{Let } u = x^{1/3}.$$

Step 2

$$(2u-3)(u-4) = 0 \quad \text{Factor.}$$

$$2u - 3 = 0 \quad \text{or} \quad u - 4 = 0 \quad \text{Zero-factor property}$$

$$u = \frac{3}{2} \quad \text{or} \quad u = 4 \quad \text{Solve for } u.$$

Step 3

$$x^{1/3} = \frac{3}{2} \quad \text{or} \quad x^{1/3} = 4 \quad \text{Substitute } x^{1/3} \text{ for } u.$$

Step 4

$$(x^{1/3})^3 = \left(\frac{3}{2}\right)^3 \quad \text{or} \quad (x^{1/3})^3 = 4^3 \quad \text{Cube each side.}$$

$$x = \frac{27}{8} \quad \text{or} \quad x = 64 \quad \text{Apply the exponents.}$$

Step 5 Because the original equation involves variables with rational exponents, check that neither of these solutions is extraneous. The solution set is $\left\{\frac{27}{8}, 64\right\}$.

NOW TRY

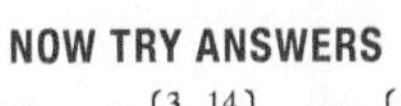

CAUTION A common error when solving problems like those in **Examples 6 and 7** is to stop too soon. ***Once we have solved for u, we must remember to substitute and solve for the values of the original variable.***

NOW TRY ANSWERS

7. (a) $\left\{\frac{3}{2}, \frac{14}{3}\right\}$ (b) $\left\{\frac{1}{8}, 27\right\}$

8.3 Exercises

FOR EXTRA HELP

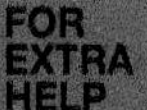

MyLab Math

Video solutions for select problems available in MyLab Math

Concept Check *Based on the discussion and examples of this section, give the first step to solve each equation. Do not actually solve.*

1. $\frac{14}{x} = x - 5$

2. $\sqrt{1+x} + x = 5$

3. $(x^2+x)^2 - 8(x^2+x) + 12 = 0$

4. $3x = \sqrt{16 - 10x}$

5. Concept Check Study this incorrect "solution."

$$x = \sqrt{3x+4} \quad \text{Square each side.}$$

$$x^2 = 3x + 4$$

$$x^2 - 3x - 4 = 0$$

$$(x-4)(x+1) = 0$$

$$x - 4 = 0 \quad \text{or} \quad x + 1 = 0$$

$$x = 4 \quad \text{or} \quad x = -1$$

Solution set: $\{4, -1\}$

WHAT WENT WRONG? Give the correct solution set.

6. Concept Check Study this incorrect "solution."

$$2(x-1)^2 - 3(x-1) + 1 = 0 \quad \text{Let}$$

$$2u^2 - 3u + 1 = 0 \quad u = x - 1.$$

$$(2u-1)(u-1) = 0$$

$$2u - 1 = 0 \quad \text{or} \quad u - 1 = 0$$

$$u = \frac{1}{2} \quad \text{or} \quad u = 1$$

Solution set: $\left\{\frac{1}{2}, 1\right\}$

WHAT WENT WRONG? Give the correct solution set.

Solve each equation. Check the solutions. ***See Example 1.***

7. $\frac{14}{x} = x - 5$

8. $\frac{-12}{x} = x + 8$

9. $1 - \frac{3}{x} - \frac{28}{x^2} = 0$

10. $4 - \frac{7}{r} - \frac{2}{r^2} = 0$

11. $3 - \frac{1}{t} = \frac{2}{t^2}$

12. $1 + \frac{2}{x} = \frac{3}{x^2}$

13. $\frac{1}{x} + \frac{2}{x+2} = \frac{17}{35}$

14. $\frac{2}{m} + \frac{3}{m+9} = \frac{11}{4}$

15. $\frac{2}{x+1} + \frac{3}{x+2} = \frac{7}{2}$

16. $\frac{4}{3-p} + \frac{2}{5-p} = \frac{26}{15}$

17. $\frac{3}{2x} - \frac{1}{2(x+2)} = 1$

18. $\frac{4}{3x} - \frac{1}{2(x+1)} = 1$

19. $3 = \frac{1}{t+2} + \frac{2}{(t+2)^2}$

20. $1 + \frac{2}{3z+2} = \frac{15}{(3z+2)^2}$

21. $\frac{6}{p} = 2 + \frac{p}{p+1}$

22. $\frac{x}{2-x} + \frac{2}{x} = 5$

23. $1 - \frac{1}{2x+1} - \frac{1}{(2x+1)^2} = 0$

24. $1 - \frac{1}{3x-2} - \frac{1}{(3x-2)^2} = 0$

Concept Check *Answer each question.*

25. A boat travels 20 mph in still water, and the rate of the current is t mph.

(a) What is the rate of the boat when it travels upstream?

(b) What is the rate of the boat when it travels downstream?

26. It takes m hours to grade a set of papers.

(a) What is the grader's rate (in job per hour)?

(b) How much of the job will the grader do in 2 hr?

Solve each problem. Round answers to the nearest tenth as needed. ***See Examples 2 and 3.***

27. In 4 hr, Kerrie can travel 15 mi upriver and come back. The rate of the current is 5 mph. Find the rate of her boat in still water.

Let $x =$ __________.

The rate traveling upriver (*against* the current) is ____ mph.

The rate traveling back downriver (*with* the current) is ____ mph.

Complete the table.

	d	r	t
Up			
Down			

Write an equation, and complete the solution.

28. Carlos can complete a certain lab test in 2 hr less time than Jaime can. If they can finish the job together in 2 hr, how long would it take each of them working alone?

Let $x =$ Jaime's time alone (in hours).

Then ____ = Carlos' time alone (in hours).

Complete the table.

	Rate	Time Working Together	Fractional Part of the Job Done
Carlos			
Jaime			

Write an equation, and complete the solution.

29. On a windy day William found that he could travel 16 mi downstream and then 4 mi back upstream at top speed in a total of 48 min. What was the top speed of William's boat if the rate of the current was 15 mph? (Let x represent the rate of the boat in still water.)

	d	r	t
Upstream	4	$x - 15$	
Downstream	16		

30. Vera flew for 6 hr at a constant rate. She traveled 810 mi with the wind, then turned around and traveled 720 mi against the wind. The wind speed was a constant 15 mph. Find the rate of the plane.

	d	r	t
With Wind	810		
Against Wind	720		

31. The distance from Jackson to Lodi is about 40 mi, as is the distance from Lodi to Manteca. Adrian drove from Jackson to Lodi, stopped in Lodi for a high-energy drink, and then drove on to Manteca at 10 mph faster. Driving time for the entire trip was 88 min. Find her rate from Jackson to Lodi. (Data from *State Farm Road Atlas.*)

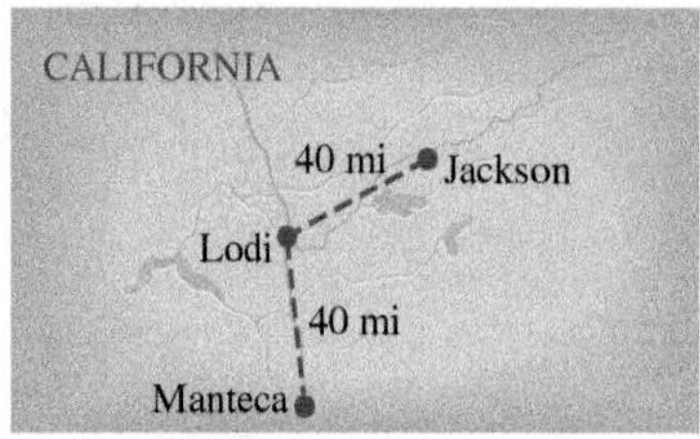

32. Medicine Hat and Cranbrook are 300 km apart. Steve rides his Harley 20 km per hr faster than Mohammad rides his Yamaha. Find Steve's average rate if he travels from Cranbrook to Medicine Hat in $1\frac{1}{4}$ hr less time than Mohammad. (Data from *State Farm Road Atlas.*)

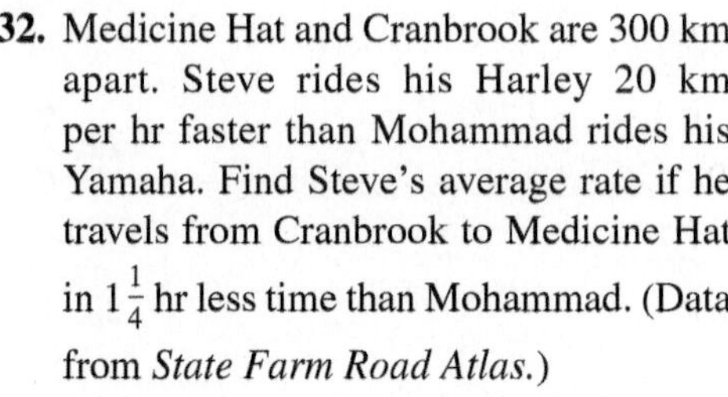

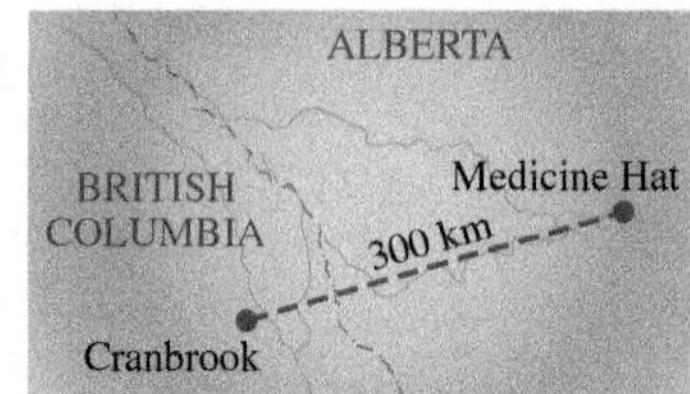

33. Working together, two people can cut a large lawn in 2 hr. One person can do the job alone in 1 hr less time than the other. How long would it take the faster worker to do the job? (Let x represent the time of the faster worker.)

	Rate	Time Working Together	Fractional Part of the Job Done
Faster Worker	$\frac{1}{x}$	2	
Slower Worker		2	

34. Working together, two people can clean an office building in 5 hr. One person takes 2 hr longer than the other to clean the building alone. How long would it take the slower worker to clean the building alone? (Let x represent the time of the slower worker.)

	Rate	Time Working Together	Fractional Part of the Job Done
Faster Worker			
Slower Worker	$\frac{1}{x}$		

35. Rusty and Nancy are planting flowers. Working alone, Rusty would take 2 hr longer than Nancy to plant the flowers. Working together, they do the job in 12 hr. How long would it have taken each person working alone?

36. Joel can work through a stack of invoices in 1 hr less time than Noel can. Working together they take $1\frac{1}{2}$ hr. How long would it take each person working alone?

37. A washing machine can be filled in 6 min if both the hot water and the cold water taps are fully opened. Filling the washer with hot water alone takes 9 min longer than filling it with cold water alone. How long does it take to fill the washer with cold water?

38. Two pipes together can fill a tank in 2 hr. One of the pipes, used alone, takes 3 hr longer than the other to fill the tank. How long would each pipe take to fill the tank alone?

Solve each equation. Check the solutions. ***See Example 4.***

39. $x = \sqrt{7x - 10}$

40. $z = \sqrt{5z - 4}$

41. $2x = \sqrt{11x + 3}$

42. $4x = \sqrt{6x + 1}$

43. $3x = \sqrt{16 - 10x}$

44. $4t = \sqrt{8t + 3}$

45. $t + \sqrt{t} = 12$

46. $p - 2\sqrt{p} = 8$

47. $x = \sqrt{\frac{6 - 13x}{5}}$

48. $r = \sqrt{\frac{20 - 19r}{6}}$

49. $-x = \sqrt{\frac{8 - 2x}{3}}$

50. $-x = \sqrt{\frac{3x + 7}{4}}$

Solve each equation. Check the solutions. ***See Examples 5–7.***

51. $x^4 - 29x^2 + 100 = 0$

52. $x^4 - 37x^2 + 36 = 0$

53. $4q^4 - 13q^2 + 9 = 0$

54. $9x^4 - 25x^2 + 16 = 0$

55. $x^4 + 48 = 16x^2$

56. $z^4 + 72 = 17z^2$

57. $(x + 3)^2 + 5(x + 3) + 6 = 0$

58. $(x - 4)^2 + (x - 4) - 20 = 0$

59. $3(m + 4)^2 - 8 = 2(m + 4)$

60. $(t + 5)^2 + 6 = 7(t + 5)$

61. $x^{2/3} + x^{1/3} - 2 = 0$

62. $x^{2/3} - 2x^{1/3} - 3 = 0$

63. $r^{2/3} + r^{1/3} - 12 = 0$

64. $3x^{2/3} - x^{1/3} - 24 = 0$

65. $4x^{4/3} - 13x^{2/3} + 9 = 0$

66. $9t^{4/3} - 25t^{2/3} + 16 = 0$

67. $2 + \frac{5}{3x - 1} = \frac{-2}{(3x - 1)^2}$

68. $3 - \frac{7}{2p + 2} = \frac{6}{(2p + 2)^2}$

69. $2 - 6(z - 1)^{-2} = (z - 1)^{-1}$

70. $3 - 2(x - 1)^{-1} = (x - 1)^{-2}$

The following exercises are not grouped by type. Solve each equation. (Exercises 83 and 84 require knowledge of complex numbers.) ***See Examples 1 and 4–7.***

71. $12x^4 - 11x^2 + 2 = 0$

72. $\left(x - \frac{1}{2}\right)^2 + 5\left(x - \frac{1}{2}\right) - 4 = 0$

73. $\sqrt{2x + 3} = 2 + \sqrt{x - 2}$

74. $\sqrt{m + 1} = -1 + \sqrt{2m}$

75. $2\left(1 + \sqrt{r}\right)^2 = 13\left(1 + \sqrt{r}\right) - 6$

76. $(x^2 + x)^2 + 12 = 8(x^2 + x)$

77. $2m^6 + 11m^3 + 5 = 0$

78. $8x^6 + 513x^3 + 64 = 0$

79. $6 = 7(2w - 3)^{-1} + 3(2w - 3)^{-2}$

80. $x^6 - 10x^3 = -9$

81. $2x^4 - 9x^2 = -2$

82. $8x^4 + 1 = 11x^2$

83. $2x^4 + x^2 - 3 = 0$

84. $4x^4 + 5x^2 + 1 = 0$

SUMMARY EXERCISES Applying Methods for Solving Quadratic Equations

We have introduced four methods for solving quadratic equations written in standard form $ax^2 + bx + c = 0$.

Method	*Advantages*	*Disadvantages*
Zero-factor property	This is usually the fastest method.	Not all polynomials are factorable. Some factorable polynomials are difficult to factor.
Square root property	This is the simplest method for solving equations of the form $(ax + b)^2 = c$.	Few equations are given in this form.
Completing the square	This method can always be used.	It requires more steps than other methods.
Quadratic formula	This method can always be used.	Sign errors may occur when evaluating $\sqrt{b^2 - 4ac}$.

Concept Check *Decide whether the* zero-factor property, *the* square root property, *or the* quadratic formula *is most appropriate for solving each quadratic equation. Do not actually solve.*

1. $(2x + 3)^2 = 4$

2. $4x^2 - 3x = 1$

3. $x^2 + 5x - 8 = 0$

4. $2x^2 + 3x = 1$

5. $3x^2 = 2 - 5x$

6. $x^2 = 5$

Solve each quadratic equation by the method of your choice.

7. $p^2 = 7$

8. $6x^2 - x - 15 = 0$

9. $n^2 + 6n + 4 = 0$

10. $(x - 3)^2 = 25$

11. $\frac{5}{x} + \frac{12}{x^2} = 2$

12. $3x^2 = 3 - 8x$

13. $2r^2 - 4r + 1 = 0$

*__14.__ $x^2 = -12$

15. $x\sqrt{2} = \sqrt{5x - 2}$

16. $x^4 - 10x^2 + 9 = 0$

17. $(2x + 3)^2 = 8$

18. $\frac{2}{x} + \frac{1}{x - 2} = \frac{5}{3}$

19. $t^4 + 14 = 9t^2$

20. $8x^2 - 4x = 2$

*__21.__ $z^2 + z + 1 = 0$

22. $5x^6 + 2x^3 - 7 = 0$

23. $4t^2 - 12t + 9 = 0$

24. $x\sqrt{3} = \sqrt{2 - x}$

25. $r^2 - 72 = 0$

26. $-3x^2 + 4x = -4$

27. $x^2 - 5x - 36 = 0$

28. $w^2 = 169$

*__29.__ $3p^2 = 6p - 4$

30. $z = \sqrt{\frac{5z + 3}{2}}$

*__31.__ $\frac{4}{r^2} + 3 = \frac{1}{r}$

32. $2(3x - 1)^2 + 5(3x - 1) = -2$

---

*This exercise requires knowledge of complex numbers.

8.4 Formulas and Further Applications

OBJECTIVES

1. Solve formulas involving squares and square roots for specified variables.
2. Solve applied problems using the Pythagorean theorem.
3. Solve applied problems using area formulas.
4. Solve applied problems using quadratic functions as models.

OBJECTIVE 1 Solve formulas involving squares and square roots for specified variables.

EXAMPLE 1 Solving for Specified Variables

Solve each formula for the specified variable. Keep $\pm$ in the answer in part (a).

(a) $w = \dfrac{kFr}{v^2}$ for v — The goal is to isolate v on one side.

$v^2 w = kFr$ — Multiply by v^2.

$v^2 = \dfrac{kFr}{w}$ — Divide by w.

$v = \pm\sqrt{\dfrac{kFr}{w}}$ — Square root property (Include both positive and negative roots.)

$v = \dfrac{\pm\sqrt{kFr}}{\sqrt{w}} \cdot \dfrac{\sqrt{w}}{\sqrt{w}}$ — Rationalize the denominator.

$v = \dfrac{\pm\sqrt{kFrw}}{w}$ — $\sqrt{a}\cdot\sqrt{b} = \sqrt{ab}$; $\sqrt{a}\cdot\sqrt{a} = a$

(b) $d = \sqrt{\dfrac{4\mathscr{A}}{\pi}}$ for $\mathscr{A}$ — The goal is to isolate $\mathscr{A}$ on one side.

$d^2 = \dfrac{4\mathscr{A}}{\pi}$ — Square each side.

$\pi d^2 = 4\mathscr{A}$ — Multiply by π.

$\dfrac{\pi d^2}{4} = \mathscr{A}$, or $\mathscr{A} = \dfrac{\pi d^2}{4}$ — Divide by 4. Interchange sides.

NOW TRY

NOW TRY EXERCISE 1

Solve each formula for the specified variable. Keep $\pm$ in the answer in part (a).

(a) $n = \dfrac{ab}{E^2}$ for E

(b) $S = \sqrt{\dfrac{pq}{n}}$ for p

EXAMPLE 2 Solving for a Specified Variable

Solve $s = 2t^2 + kt$ for t.

Because the given equation has terms with t^2 and t, write it in standard form $ax^2 + bx + c = 0$, with t as the variable instead of x.

$s = 2t^2 + kt$

$0 = 2t^2 + kt - s$ — Subtract s.

$2t^2 + kt - s = 0$ — Standard form

We can solve this equation using the quadratic formula.

NOW TRY ANSWERS

1. (a) $E = \dfrac{\pm\sqrt{abn}}{n}$ (b) $p = \dfrac{nS^2}{q}$

NOW TRY EXERCISE 2
Solve for r.

$$r^2 + 9r = -c$$

In the equation $2t^2 + kt - s = 0$, we have $a = 2$, $b = k$, and $c = -s$.

$$t = \frac{-b \pm \sqrt{b^2 - 4ac}}{2a} \quad \text{Quadratic formula}$$

$$t = \frac{-k \pm \sqrt{k^2 - 4(2)(-s)}}{2(2)} \quad \text{In } 2t^2 + kt - s = 0,\ a = 2,\ b = k, \text{ and } c = -s.$$

$$t = \frac{-k \pm \sqrt{k^2 + 8s}}{4} \quad \text{Simplify.}$$

The solutions are $t = \frac{-k + \sqrt{k^2 + 8s}}{4}$ and $t = \frac{-k - \sqrt{k^2 + 8s}}{4}$. NOW TRY

OBJECTIVE 2 Solve applied problems using the Pythagorean theorem.

The Pythagorean theorem is represented by the equation

$$a^2 + b^2 = c^2. \quad \text{See FIGURE 2.}$$

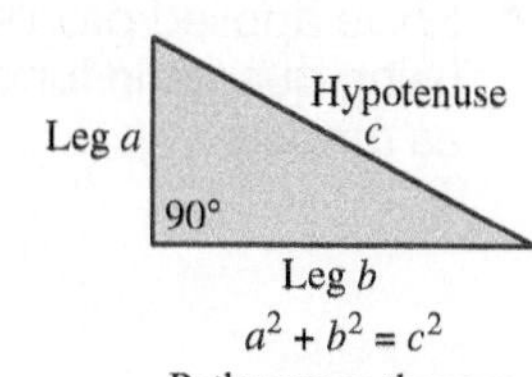

FIGURE 2

NOW TRY EXERCISE 3
Matt is building a new barn, with length 10 ft more than width. While determining the footprint of the barn, he measured the diagonal as 50 ft. What will be the dimensions of the barn?

EXAMPLE 3 Using the Pythagorean Theorem

Two cars left an intersection at the same time, one heading due north, the other due west. Some time later, they were exactly 100 mi apart. The car headed north had gone 20 mi farther than the car headed west. How far had each car traveled?

Step 1 **Read** the problem carefully.

Step 2 **Assign a variable.**

Let x = the distance traveled by the car headed west.

Then $x + 20$ = the distance traveled by the car headed north.

See **FIGURE 3**. The cars are 100 mi apart, so the hypotenuse of the right triangle equals 100.

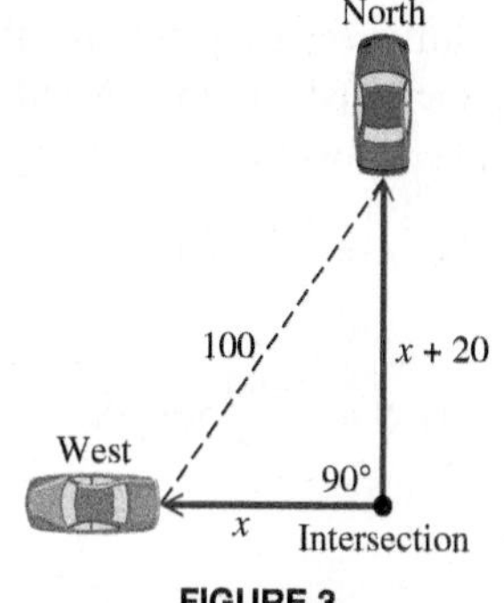

FIGURE 3

Step 3 **Write an equation.** Use the Pythagorean theorem.

$$a^2 + b^2 = c^2 \quad \text{Pythagorean theorem}$$

$$x^2 + (x + 20)^2 = 100^2 \quad \text{See FIGURE 3.}$$

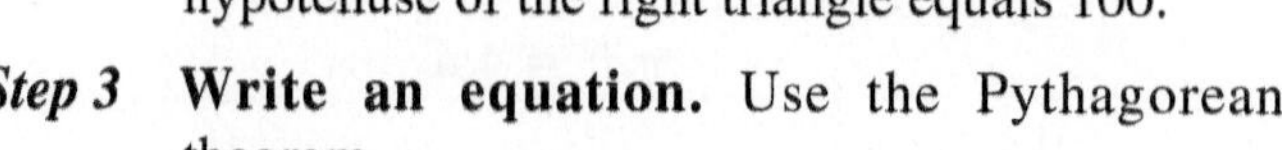

Step 4 **Solve.**

$$x^2 + x^2 + 40x + 400 = 10{,}000 \quad \text{Square the binomial.}$$

$$2x^2 + 40x - 9600 = 0 \quad \text{Standard form}$$

$$x^2 + 20x - 4800 = 0 \quad \text{Divide by 2.}$$

$$(x + 80)(x - 60) = 0 \quad \text{Factor.}$$

$$x + 80 = 0 \quad \text{or} \quad x - 60 = 0 \quad \text{Zero-factor property}$$

$$x = -80 \quad \text{or} \quad x = 60 \quad \text{Solve for } x.$$

Step 5 **State the answer.** Distance cannot be negative, so discard the negative solution. The required distances are 60 mi and 60 + 20 = 80 mi.

Step 6 **Check.** Here $60^2 + 80^2 = 100^2$, as required.

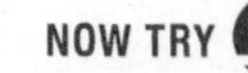

NOW TRY ANSWERS

2. $r = \frac{-9 \pm \sqrt{81 - 4c}}{2}$

3. 30 ft by 40 ft

OBJECTIVE 3 Solve applied problems using area formulas.

NOW TRY EXERCISE 4

A football practice field is 30 yd wide and 40 yd long. A strip of grass sod of uniform width is to be placed around the perimeter of the practice field. There is enough money budgeted for 296 sq yd of sod. How wide will the strip be?

EXAMPLE 4 Solving an Area Problem

A rectangular reflecting pool in a park is 20 ft wide and 30 ft long. The gardener wants to plant a strip of grass of uniform width around the edge of the pool. She has enough seed to cover 336 ft². How wide will the strip be?

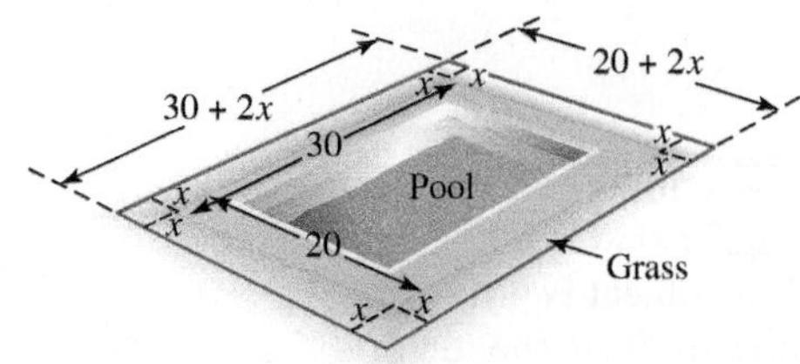

FIGURE 4

Step 1 **Read** the problem carefully.

Step 2 **Assign a variable.** The pool is shown in FIGURE 4.

Let x = the unknown width of the grass strip.

Then $20 + 2x$ = the width of the large rectangle (the width of the pool plus two grass strips),

and $30 + 2x$ = the length of the large rectangle.

Step 3 **Write an equation.** Refer to FIGURE 4.

$(30 + 2x)(20 + 2x)$ Area of large rectangle (length · width)

$30 \cdot 20$, or 600 Area of pool (in square feet)

The area of the large rectangle minus the area of the pool should equal 336 ft², the area of the grass strip.

Area of large rectangle − area of pool = area of grass.

$$(30 + 2x)(20 + 2x) - 600 = 336$$

Step 4 **Solve.**

$$600 + 60x + 40x + 4x^2 - 600 = 336 \quad \text{Multiply.}$$

$$4x^2 + 100x - 336 = 0 \quad \text{Standard form}$$

$$x^2 + 25x - 84 = 0 \quad \text{Divide by 4.}$$

$$(x + 28)(x - 3) = 0 \quad \text{Factor.}$$

$$x + 28 = 0 \quad \text{or} \quad x - 3 = 0 \quad \text{Zero-factor property}$$

$$x = -28 \quad \text{or} \quad x = 3 \quad \text{Solve for } x.$$

Step 5 **State the answer.** The width cannot be −28 ft, so the grass strip should be 3 ft wide.

Step 6 **Check.** If $x = 3$, we can find the area of the large rectangle (which includes the grass strip).

$$(30 + 2 \cdot 3)(20 + 2 \cdot 3) = 36 \cdot 26 = 936 \text{ ft}^2 \quad \text{Area of pool and strip}$$

The area of the pool is $30 \cdot 20 = 600$ ft², so the area of the grass strip is

$$936 - 600 = 336 \text{ ft}^2, \quad \text{as required.}$$

NOW TRY

NOW TRY ANSWER

4. 2 yd

OBJECTIVE 4 **Solve applied problems using quadratic functions as models.**

Some applied problems can be modeled by *quadratic functions,* which for real numbers a, b, and c, can be written in the form

$$f(x) = ax^2 + bx + c \quad (\text{where } a \neq 0).$$

NOW TRY EXERCISE 5
If an object is projected upward from the top of a 120-ft building at 60 ft per second, its position (in feet above the ground) is given by

$$s(t) = -16t^2 + 60t + 120,$$

where t is time in seconds after it was projected. When does it hit the ground (to the nearest tenth)?

EXAMPLE 5 **Solving an Applied Problem Using a Quadratic Function**

If an object is projected upward from the top of a 144-ft building at 112 ft per second, its position (in feet above the ground) is given by

$$s(t) = -16t^2 + 112t + 144,$$

where t is time in seconds after it was projected. When does it hit the ground?

When the object hits the ground, its distance above the ground is 0. We must find the value of t that makes $s(t) = 0$.

$s(t) = -16t^2 + 112t + 144$ Given model

$0 = -16t^2 + 112t + 144$ Let $s(t) = 0$.

$0 = t^2 - 7t - 9$ Divide by -16.

$t = \frac{-b \pm \sqrt{b^2 - 4ac}}{2a}$ Quadratic formula

$t = \frac{-(-7) \pm \sqrt{(-7)^2 - 4(1)(-9)}}{2(1)}$ Let $a = 1$, $b = -7$, and $c = -9$.

$t = \frac{7 \pm \sqrt{85}}{2}$ Simplify.

$t \approx \frac{7 \pm 9.2}{2}$ Approximate the square root to the nearest tenth.

Time cannot be negative, so discard -1.1.

$t \approx 8.1$ or $t \approx -1.1$ Find the two solutions.

The object will hit the ground about 8.1 sec after it is projected. NOW TRY 

EXAMPLE 6 **Using a Quadratic Function to Model the CPI**

The Consumer Price Index (CPI) is used to measure trends in prices for a "basket" of goods purchased by typical American families. This index uses a base period of 1982–1984, which means that the index number for that period is 100. The quadratic function

$$f(x) = -0.00833x^2 + 4.745x + 83.10$$

approximates the CPI for the years 1980–2016, where x is the number of years that have elapsed since 1980. (Data from Bureau of Labor Statistics.)

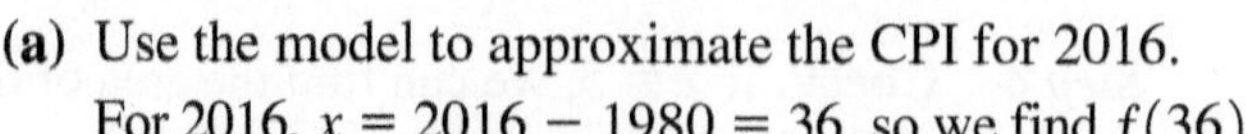

(a) Use the model to approximate the CPI for 2016.

For 2016, $x = 2016 - 1980 = 36$, so we find $f(36)$.

$f(x) = -0.00833x^2 + 4.745x + 83.10$ Given model

$f(36) = -0.00833(36)^2 + 4.745(36) + 83.10$ Let $x = 36$.

$f(36) = 243$ Nearest whole number

According to the model, the CPI for 2016 was 243.

NOW TRY ANSWER
5. 5.2 sec after it is projected

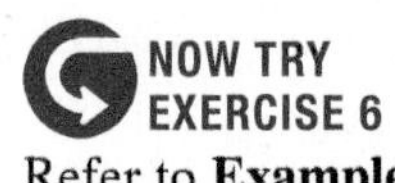

NOW TRY EXERCISE 6

Refer to **Example 6.**

(a) Use the model to approximate the CPI for 2010, to the nearest whole number.

(b) In what year did the CPI reach 175? (Round down for the year.)

NOW TRY ANSWERS

6. (a) 218 **(b)** 2000

(b) In what year did the CPI reach 200?

Find the value of x that makes $f(x) = 200$.

$$f(x) = -0.00833x^2 + 4.745x + 83.10 \quad \text{Given model}$$

$$200 = -0.00833x^2 + 4.745x + 83.10 \quad \text{Let } f(x) = 200.$$

$$0 = -0.00833x^2 + 4.745x - 116.9 \quad \text{Subtract 200.}$$

$$x = \frac{-4.745 \pm \sqrt{4.745^2 - 4(-0.00833)(-116.9)}}{2(-0.00833)} \quad \text{Use } a = -0.00833,\ b = 4.745, \text{ and } c = -116.9 \text{ in the quadratic formula.}$$

$$x = 25.8 \quad \text{or} \quad x = 543.8 \quad \text{Use a calculator. Round to the nearest tenth.}$$

Rounding the first solution 25.8 down, the CPI first reached 200 in

$$1980 + 25 = 2005.$$

(Reject the solution $x = 543.8$, as this corresponds to a totally unreasonable year.)

NOW TRY 

8.4 Exercises

FOR EXTRA HELP  MyLab Math

Video solutions for select problems available in MyLab Math

Concept Check *Answer each question.*

1. In solving a formula that has the specified variable in the denominator, what is the first step?

2. What is the first step in solving a formula like $gw^2 = 2r$ for w?

3. What is the first step in solving a formula like $gw^2 = kw + 24$ for w?

4. Why is it particularly important to check all proposed solutions to an applied problem against the information in the original problem?

For each triangle, solve for m in terms of the other variables (where $m > 0$).

5.

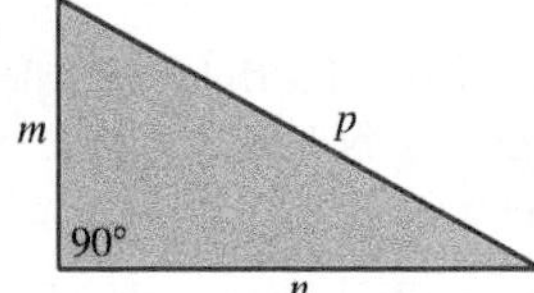

6.

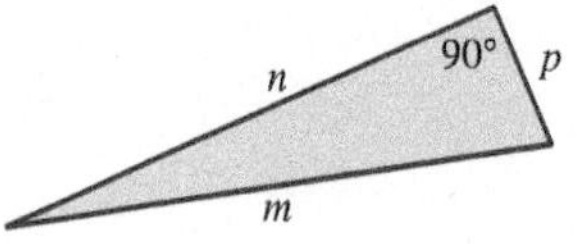

Solve each formula for the specified variable. (Leave ± in the answers as needed.) **See Examples 1 and 2.**

7. $d = kt^2$ for t **8.** $S = 6e^2$ for e **9.** $S = 4\pi r^2$ for r **10.** $s = kwd^2$ for d

11. $I = \dfrac{ks}{d^2}$ for d **12.** $R = \dfrac{k}{d^2}$ for d **13.** $F = \dfrac{kA}{v^2}$ for v

14. $L = \dfrac{kd^4}{h^2}$ for h **15.** $V = \dfrac{1}{3}\pi r^2 h$ for r **16.** $V = \pi r^2 h$ for r

17. $At^2 + Bt = -C$ for t **18.** $S = 2\pi rh + \pi r^2$ for r **19.** $D = \sqrt{kh}$ for h

20. $F = \dfrac{k}{\sqrt{d}}$ for d **21.** $p = \sqrt{\dfrac{k\ell}{g}}$ for ℓ **22.** $p = \sqrt{\dfrac{k\ell}{g}}$ for g

Extending Skills *Solve each equation for the specified variable. (Leave ± in the answers.)*

23. $p = \dfrac{E^2R}{(r + R)^2}$ for R (where $E > 0$)

24. $S(6S - t) = t^2$ for S

25. $10p^2c^2 + 7pcr = 12r^2$ for r

26. $S = vt + \dfrac{1}{2}gt^2$ for t

27. $LI^2 + RI + \dfrac{1}{c} = 0$ for I

28. $P = EI - RI^2$ for I

Solve each problem. When appropriate, round answers to the nearest tenth. ***See Example 3.***

29. Find the lengths of the sides of the triangle.

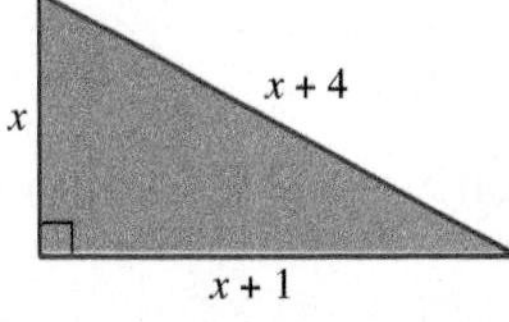

30. Find the lengths of the sides of the triangle.

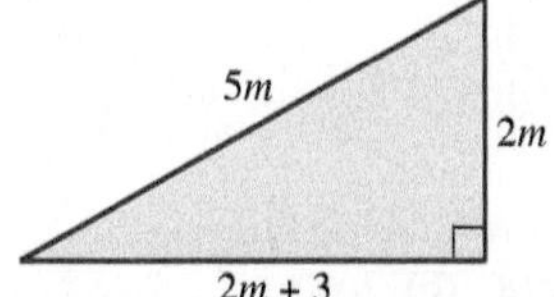

31. Two ships leave port at the same time, one heading due south and the other heading due east. Several hours later, they are 170 mi apart. If the ship traveling south traveled 70 mi farther than the other ship, how many miles did they each travel?

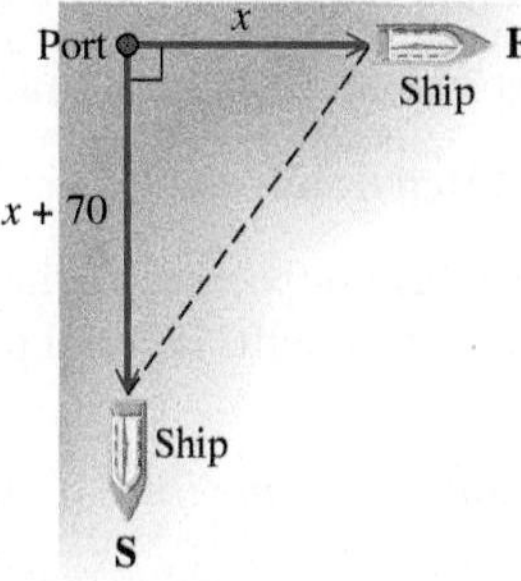

32. Deborah is flying a kite that is 30 ft farther above her hand than its horizontal distance from her. The string from her hand to the kite is 150 ft long. How high is the kite?

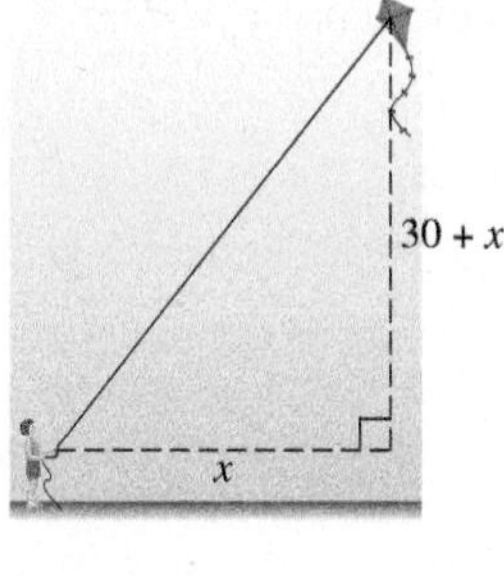

33. A game board is in the shape of a right triangle. The hypotenuse is 2 in. longer than the longer leg, and the longer leg is 1 in. less than twice as long as the shorter leg. How long is each side of the game board?

34. Manuel is planting a vegetable garden in the shape of a right triangle. The longer leg is 3 ft longer than the shorter leg, and the hypotenuse is 3 ft longer than the longer leg. Find the lengths of the three sides of the garden.

35. The diagonal of a rectangular rug measures 26 ft, and the length is 4 ft more than twice the width. Find the length and width of the rug.

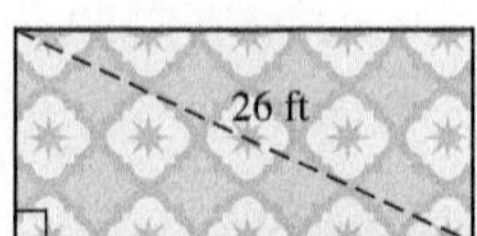

36. A 13-ft ladder is leaning against a house. The distance from the bottom of the ladder to the house is 7 ft less than the distance from the top of the ladder to the ground. How far is the bottom of the ladder from the house?

Solve each problem. ***See Example 4.***

37. A club swimming pool is 30 ft wide and 40 ft long. The club members want an exposed aggregate border in a strip of uniform width around the pool. They have enough material for 296 ft^2. How wide can the strip be?

38. Lyudmila wants to buy a rug for a room that is 20 ft long and 15 ft wide. She wants to leave an even strip of flooring uncovered around the edges of the room. How wide a strip will she have if she buys a rug with an area of 234 ft^2?

30 ft
Pool
40 ft

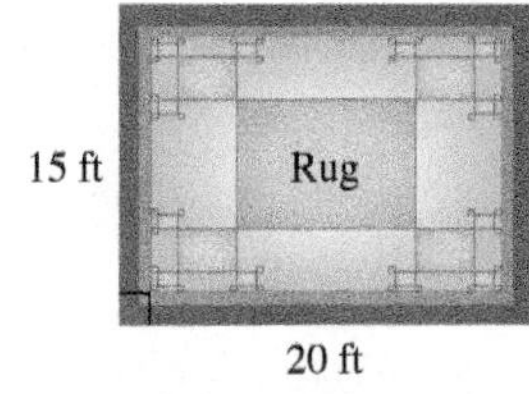

39. A rectangle has a length 2 m less than twice its width. When 5 m are added to the width, the resulting figure is a square with an area of 144 m^2. Find the dimensions of the original rectangle.

40. Mariana's backyard measures 20 m by 30 m. She wants to put a flower garden in the middle of the yard, leaving a strip of grass of uniform width around the flower garden. Mariana must have 184 m^2 of grass. Under these conditions, what will the length and width of the garden be?

41. A rectangular piece of sheet metal has a length that is 4 in. less than twice the width. A square piece 2 in. on a side is cut from each corner. The sides are then turned up to form an uncovered box of volume 256 in.^{3}. Find the length and width of the original piece of metal.

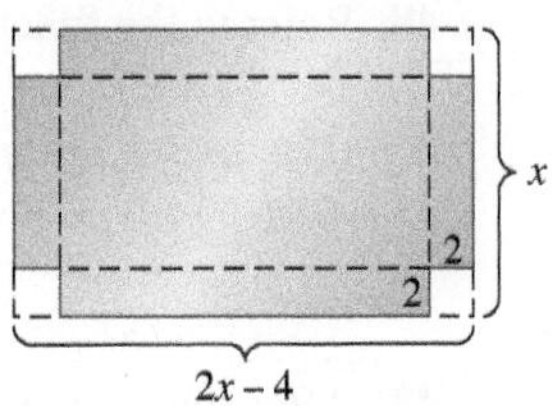

42. A rectangular piece of cardboard is 2 in. longer than it is wide. A square piece 3 in. on a side is cut from each corner. The sides are then turned up to form an uncovered box of volume 765 in.^{3}. Find the dimensions of the original piece of cardboard.

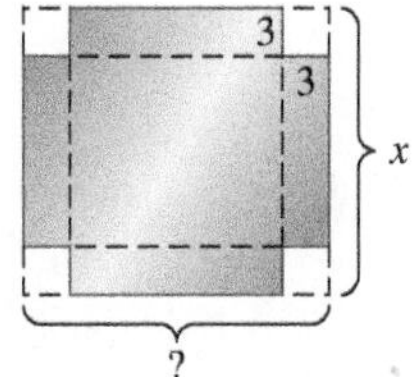

Solve each problem. When appropriate, round answers to the nearest tenth. ***See Example 5.***

43. An object is projected directly upward from the ground. After t seconds its distance in feet above the ground is

$$s(t) = 144t - 16t^2.$$

After how many seconds will the object be 128 ft above the ground? (*Hint:* Look for a common factor before solving the equation.)

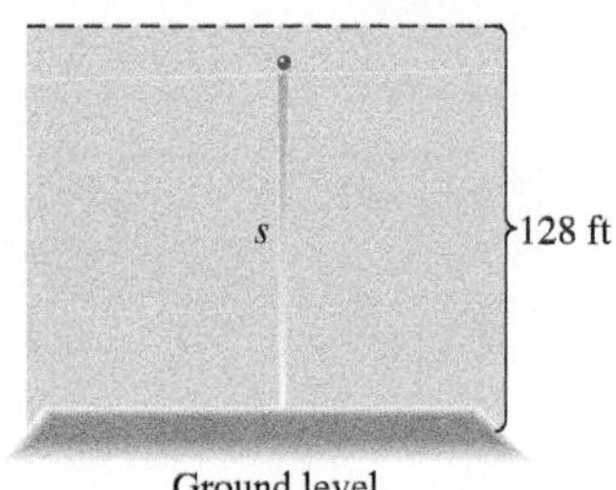

44. When does the object in **Exercise 43** strike the ground?

45. A ball is projected upward from the ground. Its distance in feet from the ground in t seconds is given by

$$s(t) = -16t^2 + 128t.$$

At what times will the ball be 213 ft from the ground?

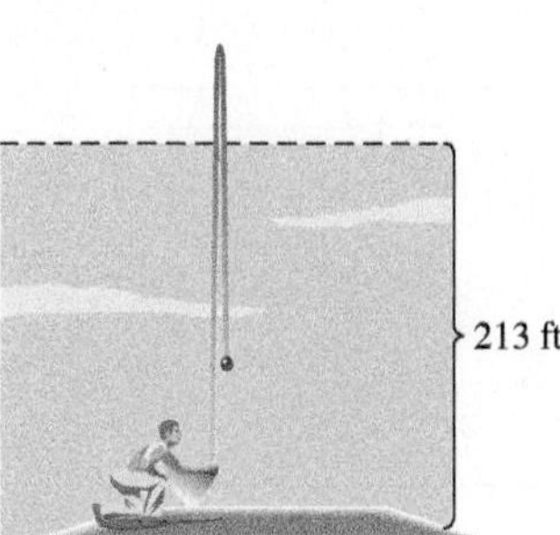

46. A toy rocket is launched from ground level. Its distance in feet from the ground in t seconds is given by

$$s(t) = -16t^2 + 208t.$$

At what times will the rocket be 550 ft from the ground?

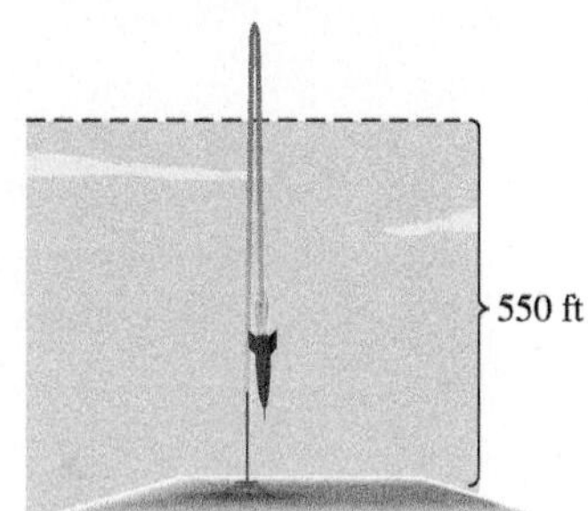

47. The following function gives the distance in feet a car going approximately 68 mph will skid in t seconds.

$$D(t) = 13t^2 - 100t$$

Find the time it would take for the car to skid 180 ft.

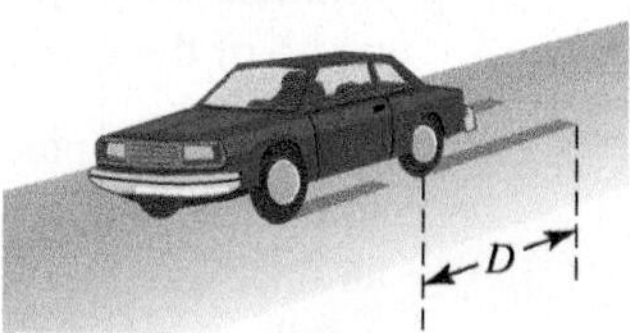

48. Refer to the function in **Exercise 47.** Find the time it would take for the car to skid 500 ft.

A ball is projected upward from ground level, and its distance in feet from the ground in t seconds is given by

$$s(t) = -16t^2 + 160t.$$

49. After how many seconds does the ball reach a height of 400 ft? Describe in words its position at this height.

50. After how many seconds does the ball reach a height of 425 ft? Interpret the mathematical result here.

Extending Skills *Solve each problem using a quadratic equation.*

51. A certain bakery has found that the daily demand for blueberry muffins is $\frac{6000}{p}$, where p is the price of a muffin in cents. The daily supply is $3p - 410$. Find the price at which supply and demand are equal.

52. In one area the demand for Blu-ray discs is $\frac{1900}{P}$ per day, where P is the price in dollars per disc. The supply is $5P - 1$ per day. At what price, to the nearest cent, does supply equal demand?

53. The formula $A = P(1 + r)^2$ gives the amount A in dollars that P dollars will grow to in 2 yr at interest rate r (where r is given as a decimal), using compound interest. What interest rate will cause \$2000 to grow to \$2142.45 in 2 yr?

54. Use the formula $A = P(1 + r)^2$ to find the interest rate r at which a principal P of \$10,000 will increase to \$10,920.25 in 2 yr.

*William Froude was a 19th century naval architect who used the following expression, known as the **Froude number,** in shipbuilding.*

$$\frac{v^2}{g\ell}$$

This expression was also used by R. McNeill Alexander in his research on dinosaurs. (Data from "How Dinosaurs Ran," Scientific American.)

Use this expression to find the value of v (in meters per second), given $g = 9.8$ m per sec^2. (Round to the nearest tenth.)

55. Rhinoceros: $\ell = 1.2$;
Froude number $= 2.57$

56. Triceratops: $\ell = 2.8$;
Froude number $= 0.16$

Recall that corresponding sides of similar triangles are proportional. Use this fact to find the lengths of the indicated sides of each pair of similar triangles. Check all possible solutions in both triangles. Sides of a triangle cannot be negative (and are not drawn to scale here).

57. Side AC

58. Side RQ

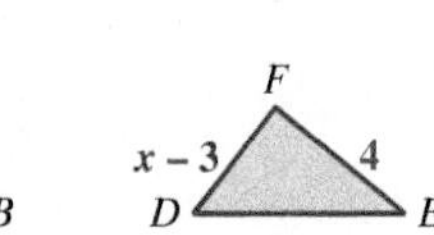

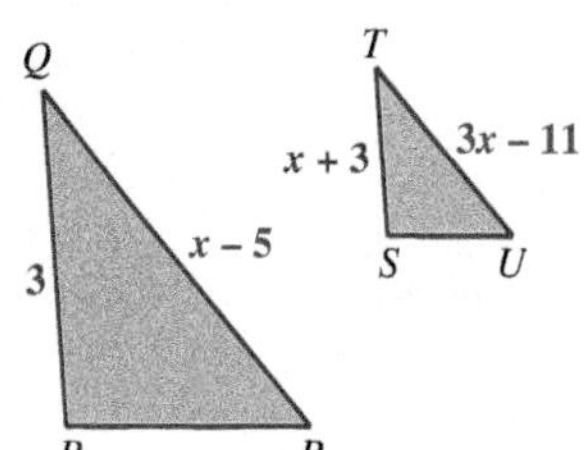

Total spending (in billions of dollars) in the United States from all sources on physician and clinical services for the years 2000–2016 are shown in the bar graph. One model for the data is the quadratic function

$$f(x) = -0.2901x^2 + 25.90x + 291.6.$$

Here, $x = 0$ *represents 2000,* $x = 2$ *represents 2002, and so on. Use the graph and the model to work each of the following exercises. **See Example 6.***

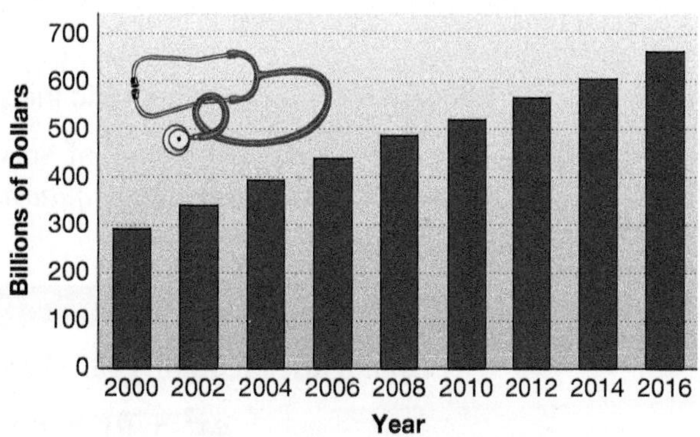

Data from Centers for Medicare and Medicaid Services.

59. Approximate spending on physician and clinical services in 2014 to the nearest $10 billion using **(a)** the graph and **(b)** the model. How do the two approximations compare?

60. Repeat **Exercise 59** for the year 2008.

61. According to the model, in what year did spending on physician and clinical services first exceed \$500 billion? (Round down for the year.)

62. Repeat **Exercise 61** for \$400 billion.

RELATING CONCEPTS For Individual or Group Work **(Exercises 63–66)**

In the 1939 classic movie The Wizard of Oz, *Ray Bolger's character, the Scarecrow, wants a brain. When the Wizard grants him his "Th.D." (Doctor of Thinkology), the Scarecrow replies with the following statement.*

Scarecrow: The sum of the square roots of any two sides of an isosceles triangle is equal to the square root of the remaining side.

His statement sounds like the formula for the Pythagorean theorem. To see why it is incorrect, ***work Exercises 63–66 in order.***

63. To what kind of triangle does the Scarecrow refer in his statement? To what kind of triangle does the Pythagorean theorem actually refer?

64. In the Scarecrow's statement, he refers to square roots. In applying the formula for the Pythagorean theorem, do we find square roots of the sides? If not, what do we find?

65. An isosceles triangle has two sides of equal length. Draw an isosceles triangle with two sides of length 9 units and remaining side of length 4 units. Show that this triangle does not satisfy the Scarecrow's statement. (This is a *counterexample* and is sufficient to show that his statement is false in general.)

66. Use wording similar to that of the Scarecrow, but state the Pythagorean theorem correctly.

8.5 Polynomial and Rational Inequalities

OBJECTIVES

1 Solve quadratic inequalities.
2 Solve polynomial inequalities of degree 3 or greater.
3 Solve rational inequalities.

OBJECTIVE 1 Solve quadratic inequalities.

We can combine methods of solving linear inequalities and methods of solving quadratic equations to solve *quadratic inequalities.*

Quadratic Inequality

A **quadratic inequality** (in x here) is an inequality that can be written in the form

$$ax^2 + bx + c < 0, \qquad ax^2 + bx + c > 0,$$
$$ax^2 + bx + c \le 0, \quad \text{or} \quad ax^2 + bx + c \ge 0,$$

where a, b, and c are real numbers and $a \neq 0$.

Examples: $x^2 - 3x - 10 < 0$, $6x^2 + x \ge 1$ Quadratic inequalities

VOCABULARY

- ☐ quadratic inequality
- ☐ rational inequality

EXAMPLE 1 Solving a Quadratic Inequality Using Test Values

Solve and graph the solution set of $x^2 - x - 12 > 0$.

Solve the quadratic equation $x^2 - x - 12 = 0$.

$$x^2 - x - 12 = 0 \quad \text{Let } f(x) = 0.$$
$$(x - 4)(x + 3) = 0 \quad \text{Factor.}$$
$$x - 4 = 0 \quad \text{or} \quad x + 3 = 0 \quad \text{Zero-factor property}$$
$$x = 4 \quad \text{or} \quad x = -3 \quad \text{Solve each equation.}$$

The numbers 4 and -3 divide a number line into Intervals A, B, and C, as shown in **FIGURE 5**. ***Be careful to put the lesser number on the left.***

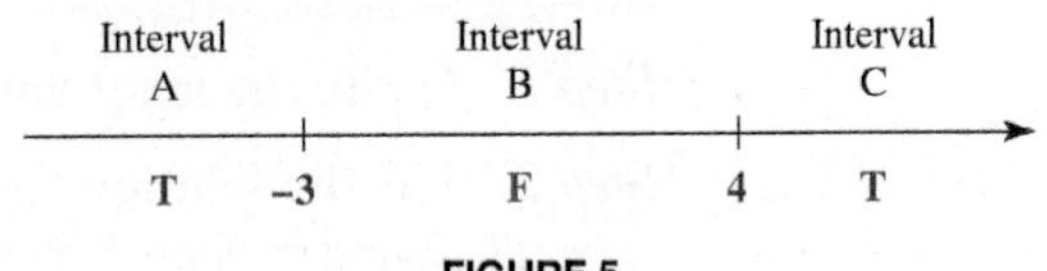

FIGURE 5

The numbers 4 and -3 are the only values that make the quadratic expression $x^2 - x - 12$ equal to 0. All other numbers make the expression either positive or negative. The sign of the expression can change from positive to negative or from negative to positive only at a number that makes it 0.

Therefore, if one number in an interval satisfies the inequality, then all numbers in that interval will satisfy the inequality.

To see if the numbers in Interval A satisfy the inequality, choose any number from Interval A in **FIGURE 5** (that is, any number less than -3). We choose -5. Substitute this test value for x in the original inequality $x^2 - x - 12 > 0$.

$$x^2 - x - 12 > 0 \quad \text{Original inequality}$$
$$(-5)^2 - (-5) - 12 \overset{?}{>} 0 \quad \text{Let } x = -5.$$
$$25 + 5 - 12 \overset{?}{>} 0 \quad \text{Simplify.}$$
$$18 > 0 \quad \text{True}$$

Use parentheses to avoid sign errors.

Because -5 satisfies the inequality, *all* numbers from Interval A are solutions, indicated by T in **FIGURE 5**.

Now try 0 from Interval B.

$$x^2 - x - 12 > 0 \quad \text{Original inequality}$$
$$0^2 - 0 - 12 \overset{?}{>} 0 \quad \text{Let } x = 0.$$
$$-12 > 0 \quad \text{False}$$

The numbers in Interval B are *not* solutions, indicated by F in **FIGURE 5**.

Now try 5 from Interval C.

$$x^2 - x - 12 > 0 \quad \text{Original inequality}$$
$$5^2 - 5 - 12 \overset{?}{>} 0 \quad \text{Let } x = 5.$$
$$8 > 0 \quad \text{True}$$

All numbers from Interval C are solutions, indicated by T in **FIGURE 5**.

NOW TRY EXERCISE 1

Solve and graph the solution set.

$$x^2 + 2x - 8 > 0$$

Based on these results (shown by the colored letters in **FIGURE 5**), the solution set includes all numbers in Intervals A and C, as shown in **FIGURE 6**.

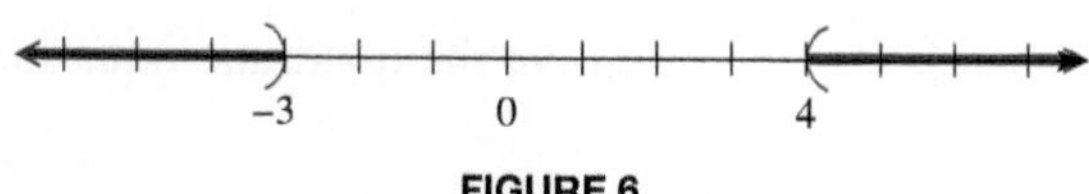

FIGURE 6

The solution set is written in interval notation as

$$(-\infty, -3) \cup (4, \infty).$$

−3 and 4 are ***not*** included because the symbol > does not include equality.

NOW TRY

Solving a Quadratic Inequality

Step 1 **Write the inequality as an equation and solve it.**

Step 2 **Use the solutions from Step 1 to determine intervals.** Graph the values found in Step 1 on a number line. These values divide the number line into intervals.

Step 3 **Find the intervals that satisfy the inequality.** Substitute a test value from each interval into the original inequality to determine the intervals that satisfy the inequality. All numbers in those intervals are in the solution set. A graph of the solution set will usually look like one of these.

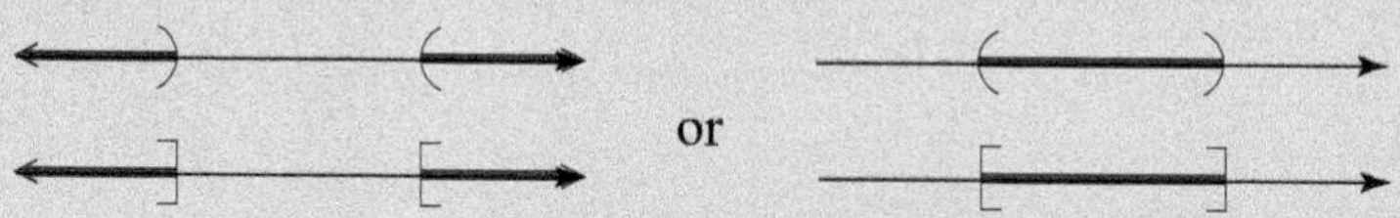

Step 4 **Consider the endpoints separately.** The values from Step 1 are included in the solution set if the inequality symbol is $\leq$ or $\geq$. They are not included if it is $<$ or $>$.

EXAMPLE 2 Solving a Quadratic Inequality

Step 1 Solve and graph the solution set of $2x^2 + 5x \leq 12$.

$2x^2 + 5x = 12$	Related quadratic equation
$2x^2 + 5x - 12 = 0$	Standard form
$(2x - 3)(x + 4) = 0$	Factor.
$2x - 3 = 0 \quad \text{or} \quad x + 4 = 0$	Zero-factor property
$x = \frac{3}{2} \quad \text{or} \quad x = -4$	Solve each equation.

Step 2 The numbers $\frac{3}{2}$ and -4 divide a number line into three intervals. See **FIGURE 7**.

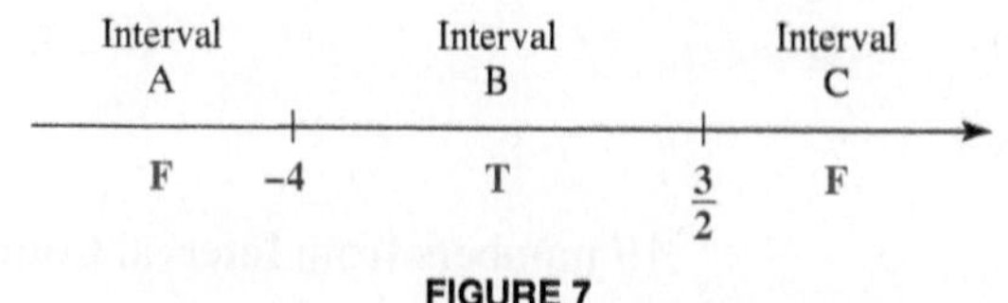

FIGURE 7

NOW TRY ANSWER

1. $(-\infty, -4) \cup (2, \infty)$

Solve and graph the solution set.

$$3x^2 - 11x \le 4$$

Steps 3 and 4 Substitute a test value from each interval in the *original* inequality $2x^2 + 5x \le 12$ to determine which intervals satisfy the inequality.

Interval	*Test Value*	*Test of Inequality*	*True or False?*
A	−5	$25 \le 12$	F
B	0	$0 \le 12$	T
C	2	$18 \le 12$	F

We use a table to organize this information. (Verify it.)

The numbers in Interval B are solutions. See **FIGURE 8**.

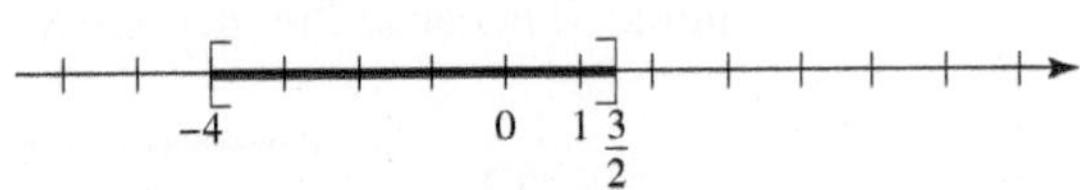

FIGURE 8

The solution set is the interval $\left[-4, \frac{3}{2}\right]$.

-4 and $\frac{3}{2}$ are included because the symbol $\le$ includes equality.

NOW TRY

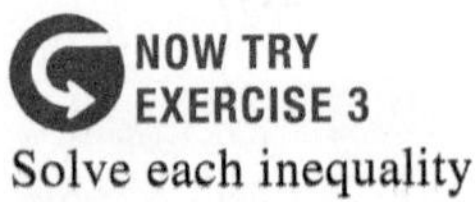

Solve each inequality.

(a) $(4x - 1)^2 > -3$

(b) $(4x - 1)^2 < -3$

EXAMPLE 3 Solving Special Cases

Solve each inequality.

(a) $(2x - 3)^2 > -1$

Because $(2x - 3)^2$ is never negative, it is always greater than -1. Thus, the solution set for $(2x - 3)^2 > -1$ is the set of all real numbers, $(-\infty, \infty)$.

(b) $(2x - 3)^2 < -1$

Using similar reasoning as in part (a), there is no solution for this inequality. The solution set is $\emptyset$.

NOW TRY

OBJECTIVE 2 Solve polynomial inequalities of degree 3 or greater.

EXAMPLE 4 Solving a Third-Degree Polynomial Inequality

Solve and graph the solution set of $(x - 1)(x + 2)(x - 4) \le 0$.

This *cubic* (third-degree) inequality can be solved by extending the zero-factor property to more than two factors. (Step 1)

$$(x - 1)(x + 2)(x - 4) = 0 \qquad \text{Set the factored polynomial } \textit{equal} \text{ to 0.}$$

$$x - 1 = 0 \quad \text{or} \quad x + 2 = 0 \quad \text{or} \quad x - 4 = 0 \qquad \text{Zero-factor property}$$

$$x = 1 \quad \text{or} \quad x = -2 \quad \text{or} \quad x = 4 \qquad \text{Solve each equation.}$$

Locate the numbers -2, 1, and 4 on a number line, as in **FIGURE 9**, to determine the Intervals A, B, C, and D. (Step 2)

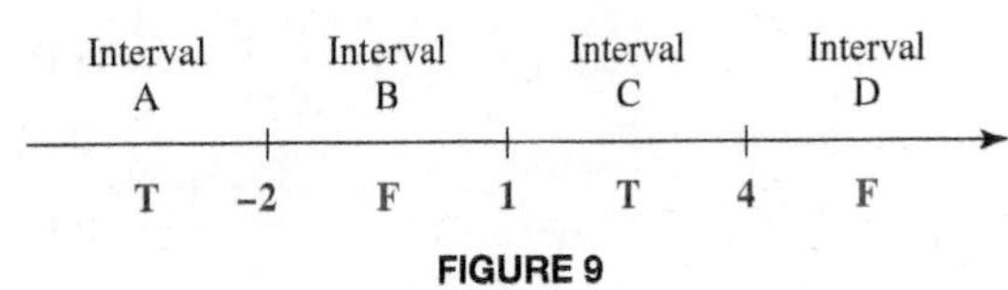

FIGURE 9

NOW TRY ANSWERS

2.

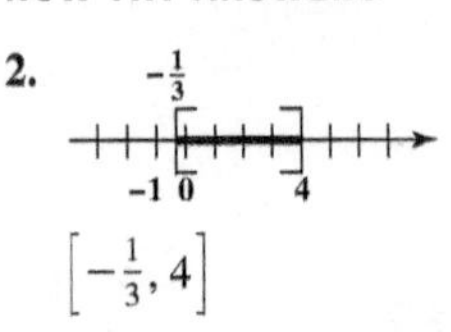

$\left[-\frac{1}{3}, 4\right]$

3. **(a)** $(-\infty, \infty)$ **(b)** $\emptyset$

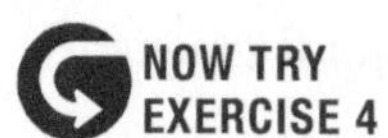

NOW TRY EXERCISE 4

Solve and graph the solution set.

$(x+4)(x-3)(2x+1) \le 0$

Substitute a test value from each interval in the *original* inequality to determine which intervals satisfy $(x-1)(x+2)(x-4) \le 0$. (Step 3)

Interval	*Test Value*	*Test of Inequality*	*True or False?*
A	−3	$-28 \le 0$	T
B	0	$8 \le 0$	F
C	2	$-8 \le 0$	T
D	5	$28 \le 0$	F

The numbers in Intervals A and C are in the solution set. The three endpoints are included because the inequality symbol $\le$ includes equality. (Step 4) See **FIGURE 10**.

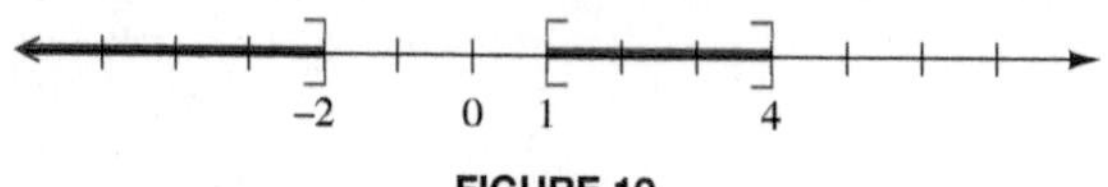

FIGURE 10

The solution set is the interval $(-\infty, -2] \cup [1, 4]$. **NOW TRY** 

OBJECTIVE 3 Solve rational inequalities.

Rational inequalities involve rational expressions and are solved similarly.

Solving a Rational Inequality

Step 1 **Write the inequality so that 0 is on one side** and there is a single fraction on the other side.

Step 2 **Determine the values that make the numerator or denominator equal to 0.**

Step 3 **Divide a number line into intervals.** Use the values from Step 2.

Step 4 **Find the intervals that satisfy the inequality.** Test a value from each interval by substituting it into the *original* inequality.

Step 5 **Consider the endpoints separately.** Exclude any values that make the denominator 0.

EXAMPLE 5 Solving a Rational Inequality

Solve and graph the solution set of $\dfrac{-1}{x-3} > 1$.

Write the inequality so that 0 is on one side. (Step 1)

$$\frac{-1}{x-3} - 1 > 0 \quad \text{Subtract 1.}$$

$$\frac{-1}{x-3} - \frac{x-3}{x-3} > 0 \quad \text{Use } x-3 \text{ as the common denominator.}$$

$$\frac{-1-(x-3)}{x-3} > 0 \quad \text{Write the left side as a single fraction.}$$

Be careful with signs.

$$\frac{-1-x+3}{x-3} > 0 \quad \text{Distributive property}$$

$$\frac{-x+2}{x-3} > 0 \quad \text{Combine like terms in the numerator.}$$

NOW TRY ANSWER

4. [number line: −4, $-\frac{1}{2}$, 0, 3]

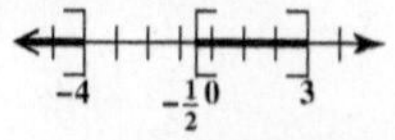

$(-\infty, -4] \cup \left[-\frac{1}{2}, 3\right]$

NOW TRY EXERCISE 5
Solve and graph the solution set.

$$\frac{3}{x+1} > 4$$

The sign of $\frac{-x+2}{x-3}$ will change from positive to negative or negative to positive only at those values that make the numerator or denominator 0. The number 2 makes the numerator 0, and 3 makes the denominator 0. (Step 2) These two numbers, 2 and 3, divide a number line into three intervals. See **FIGURE 11**. (Step 3)

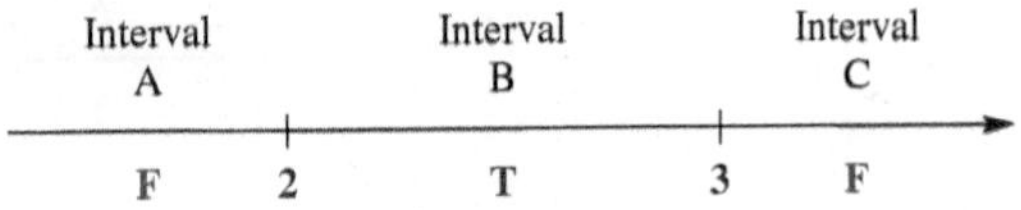

FIGURE 11

Substituting a test value from each interval in the *original* inequality, $\frac{-1}{x-3} > 1$, gives the results shown in the table. (Step 4)

Interval	*Test Value*	*Test of Inequality*	*True or False?*
A	0	$\frac{1}{3} > 1$	F
B	2.5	$2 > 1$	T
C	4	$-1 > 1$	F

The numbers in Interval B are solutions. This interval does not include 3 because it makes the denominator in the original inequality 0. The number 2 is not included either because the inequality symbol > does not include equality. (Step 5) See **FIGURE 12**.

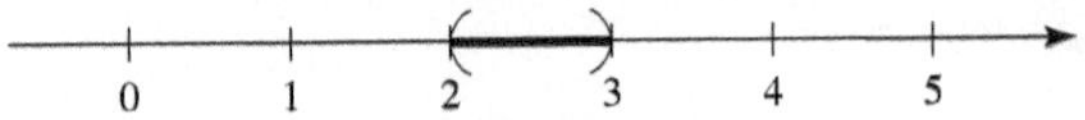

FIGURE 12

The solution set is the interval $(2, 3)$. **NOW TRY**

CAUTION ***When solving a rational inequality, any number that makes the denominator 0 must be excluded from the solution set.***

EXAMPLE 6 Solving a Rational Inequality

Solve and graph the solution set of $\frac{x-2}{x+2} \le 2$.

Write the inequality so that 0 is on one side. (Step 1)

$$\frac{x-2}{x+2} - 2 \le 0 \quad \text{Subtract 2.}$$

$$\frac{x-2}{x+2} - \frac{2(x+2)}{x+2} \le 0 \quad \text{Use } x+2 \text{ as the common denominator.}$$

$$\frac{x-2-2(x+2)}{x+2} \le 0 \quad \text{Write as a single fraction.}$$

Be careful with signs.

$$\frac{x-2-2x-4}{x+2} \le 0 \quad \text{Distributive property}$$

$$\frac{-x-6}{x+2} \le 0 \quad \text{Combine like terms in the numerator.}$$

NOW TRY ANSWER

5. 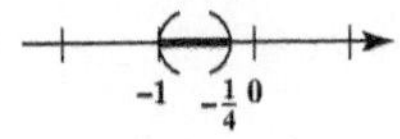

$\left(-1, -\frac{1}{4}\right)$

NOW TRY EXERCISE 6

Solve and graph the solution set.

$$\frac{x-3}{x+3} \le 2$$

In the inequality $\frac{-x-6}{x+2} \le 0$, the number -6 makes the numerator 0, and -2 makes the denominator 0. (Step 2) These two numbers determine three intervals on a number line. See **FIGURE 13.** (Step 3)

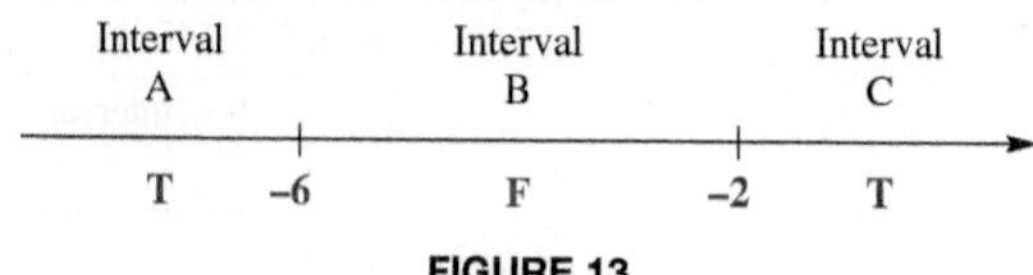

FIGURE 13

Substitute a test value from each interval in the *original* inequality $\frac{x-2}{x+2} \le 2$. (Step 4)

Interval	*Test Value*	*Test of Inequality*	*True or False?*
A	-8	$\frac{5}{3} \le 2$	T
B	-4	$3 \le 2$	F
C	0	$-1 \le 2$	T

The numbers in Intervals A and C are solutions. The number -6 satisfies the original inequality, but -2 does not because it makes the denominator 0. (Step 5) See **FIGURE 14.**

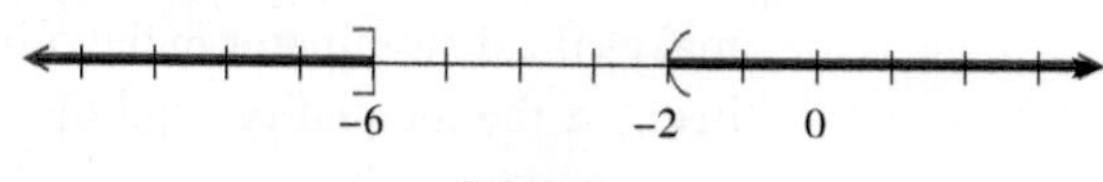

FIGURE 14

NOW TRY ANSWER

6. 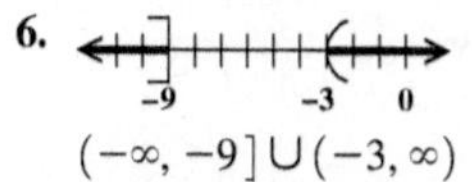

$(-\infty, -9] \cup (-3, \infty)$

The solution set is the interval $(-\infty, -6] \cup (-2, \infty)$.

NOW TRY 

8.5 Exercises

 MyLab Math

Video solutions for select problems available in MyLab Math

1. **Concept Check** Which of the following are quadratic inequalities?

   A. $2x - 6 \le 3$ **B.** $(3x + 1)^2 > -1$ **C.** $x^2 - 4x + 3 \ge 0$ **D.** $x^2 - 8x + 16 = 0$

2. Explain how to determine whether to include or to exclude endpoints when solving a quadratic or higher-degree inequality.

Concept Check *Choose a test number from the given interval and decide whether the inequality is* true *or* false.

3. $x^2 - 2x - 3 > 0$; interval: $(-1, 3)$
4. $x^2 + 2x - 3 < 0$; interval: $(-\infty, -3)$
5. $2x^2 - 7x - 15 > 0$; interval: $(5, \infty)$
6. $2x^2 + x - 10 < 0$; interval: $\left(-\frac{5}{2}, 2\right)$

Solve each inequality, and graph the solution set. **See Examples 1 and 2.** *(Hint: In Exercises 23 and 24, use the quadratic formula.)*

7. $(x + 1)(x - 5) > 0$
8. $(x + 6)(x - 2) > 0$
9. $(x + 4)(x - 6) < 0$
10. $(x + 4)(x - 8) < 0$
11. $x^2 - 4x + 3 \ge 0$
12. $x^2 - 3x - 10 \ge 0$
13. $10x^2 + 9x \ge 9$
14. $3x^2 + 10x \ge 8$
15. $4x^2 - 9 \le 0$
16. $9x^2 - 25 \le 0$
17. $6x^2 + x \ge 1$
18. $4x^2 + 7x \ge -3$
19. $z^2 - 4z \ge 0$
20. $x^2 + 2x < 0$
21. $3x^2 - 5x \le 0$
22. $2z^2 + 3z > 0$
23. $x^2 - 6x + 6 \ge 0$
24. $3x^2 - 6x + 2 \le 0$

Solve each inequality. ***See Example 3.***

25. $(4 - 3x)^2 \geq -2$

26. $(7 - 6x)^2 \geq -1$

27. $(3x + 5)^2 \leq -4$

28. $(8x + 5)^2 \leq -5$

29. $(2x + 5)^2 < 0$

30. $(3x - 7)^2 < 0$

31. $(5x - 1)^2 \geq 0$

32. $(4x + 1)^2 \geq 0$

Solve each inequality, and graph the solution set. ***See Example 4.***

33. $(x - 1)(x - 2)(x - 4) < 0$

34. $(2x + 1)(3x - 2)(4x + 7) < 0$

35. $(x - 4)(2x + 3)(3x - 1) \geq 0$

36. $(x + 2)(4x - 3)(2x + 7) \geq 0$

Solve each inequality, and graph the solution set. ***See Examples 5 and 6.***

37. $\frac{x - 1}{x - 4} > 0$

38. $\frac{x + 1}{x - 5} > 0$

39. $\frac{2x + 3}{x - 5} \leq 0$

40. $\frac{3x + 7}{x - 3} \leq 0$

41. $\frac{8}{x - 2} \geq 2$

42. $\frac{20}{x - 1} \geq 1$

43. $\frac{3}{2x - 1} < 2$

44. $\frac{6}{x - 1} < 1$

45. $\frac{x - 3}{x + 2} \geq 2$

46. $\frac{m + 4}{m + 5} \geq 2$

47. $\frac{x - 8}{x - 4} < 3$

48. $\frac{2t - 3}{t + 1} > 4$

49. $\frac{4k}{2k - 1} < k$

50. $\frac{r}{r + 2} < 2r$

51. $\frac{2x - 3}{x^2 + 1} \geq 0$

52. $\frac{9x - 8}{4x^2 + 25} < 0$

53. $\frac{(3x - 5)^2}{x + 2} > 0$

54. $\frac{(5x - 3)^2}{2x + 1} \leq 0$

RELATING CONCEPTS For Individual or Group Work (Exercises 55–58)

A model rocket is projected vertically upward from the ground. Its distance s in feet above the ground after t seconds is given by the quadratic function

$$s(t) = -16t^2 + 256t.$$

Work Exercises 55–58 in order, *to see how quadratic equations and inequalities are related.*

55. At what times will the rocket be 624 ft above the ground? (*Hint:* Let $s(t) = 624$ and solve the quadratic *equation.*)

56. At what times will the rocket be more than 624 ft above the ground? (*Hint:* Let $s(t) > 624$ and solve the quadratic *inequality.*)

57. At what times will the rocket be at ground level? (*Hint:* Let $s(t) = 0$ and solve the quadratic *equation.*)

58. At what times will the rocket be less than 624 ft above the ground? (*Hint:* Let $s(t) < 624$, solve the quadratic *inequality,* and observe the solutions in **Exercises 56 and 57** to determine the least and greatest possible values of t.)

Chapter 8 Summary

Key Terms

8.1
quadratic equation
second-degree equation

8.2
quadratic formula
discriminant

8.3
quadratic in form

8.5
quadratic inequality
rational inequality

Test Your Word Power

See how well you have learned the vocabulary in this chapter.

1. A **quadratic equation** is an equation that can be written in the form
- **A.** $y = mx + b$, for real numbers m and b
- **B.** $ax + b = 0$, for real numbers a and b $(a \neq 0)$
- **C.** $ax^2 + bx + c = 0$, for real numbers a, b, and c $(a \neq 0)$
- **D.** $Ax + By = C$, for real numbers A, B, and C (A and B not both 0).

2. The **quadratic formula** is
- **A.** a formula to find the number of solutions of a quadratic equation
- **B.** a formula to find the type of solutions of a quadratic equation
- **C.** the standard form of a quadratic equation
- **D.** a general formula for solving any quadratic equation.

3. The **discriminant** is
- **A.** the quantity under the radical in the quadratic formula
- **B.** the quantity in the denominator in the quadratic formula
- **C.** the solution set of a quadratic equation
- **D.** the result of using the quadratic formula.

4. A **quadratic inequality** is a polynomial inequality of
- **A.** degree one
- **B.** degree two
- **C.** degree three
- **D.** degree four.

5. A **rational inequality** is an inequality that
- **A.** has a second-degree term
- **B.** has three factors
- **C.** involves a fraction with a variable factor in the denominator
- **D.** involves a radical.

ANSWERS

1. C; *Examples:* $x^2 - 4x + 5 = 0$, $2x^2 = 9x - 4$, $x^2 = 25$ **2.** D; *Example:* The solutions of $ax^2 + bx + c = 0$ $(a \neq 0)$ are given by $x = \frac{-b \pm \sqrt{b^2 - 4ac}}{2a}$. **3.** A; *Example:* In the quadratic formula, the discriminant is $b^2 - 4ac$. **4.** B; *Examples:* $(x + 1)(2x - 5) > 4$, $2x^2 + 5x < 0$, $x^2 - 4x \leq 4$ **5.** C; *Examples:* $\frac{x - 1}{x + 3} < 0$, $\frac{5}{y - 2} > 3$, $\frac{x}{2x - 1} \geq 3x$

Quick Review

CONCEPTS	EXAMPLES
8.1 The Square Root Property and Completing the Square **Square Root Property** If x and k are complex numbers and $x^2 = k$, then $x = \sqrt{k}$ or $x = -\sqrt{k}$.	Solve $(x - 1)^2 = 8$. $x - 1 = \sqrt{8}$ or $x - 1 = -\sqrt{8}$ $x = 1 + 2\sqrt{2}$ or $x = 1 - 2\sqrt{2}$ Solution set: $\{1 + 2\sqrt{2}, 1 - 2\sqrt{2}\}$, or $\{1 \pm 2\sqrt{2}\}$

CONCEPTS	EXAMPLES

Completing the Square
To solve $ax^2 + bx + c = 0$ (where $a \neq 0$), follow these steps.

Step 1 If $a \neq 1$, divide each side by a.

Step 2 Write the equation with the variable terms on one side and the constant on the other.

Step 3 Complete the square.

- Take half the coefficient of x and square it.
- Add the square to each side.
- Factor the perfect square trinomial, and write it as the square of a binomial. Combine terms on the other side.

Step 4 Use the square root property to solve.

Solve $2x^2 - 4x - 18 = 0$.

$x^2 - 2x - 9 = 0$ Divide by 2.

$x^2 - 2x = 9$ Add 9.

$\left[\frac{1}{2}(-2)\right]^2 = (-1)^2 = 1$

$x^2 - 2x + 1 = 9 + 1$ Add 1.

$(x - 1)^2 = 10$ Factor. Add.

$x - 1 = \sqrt{10}$ or $x - 1 = -\sqrt{10}$ Square root property

$x = 1 + \sqrt{10}$ or $x = 1 - \sqrt{10}$

Solution set: $\{1 + \sqrt{10}, 1 - \sqrt{10}\}$, or $\{1 \pm \sqrt{10}\}$

8.2 The Quadratic Formula

Quadratic Formula
The solutions of $ax^2 + bx + c = 0$ (where $a \neq 0$) are given by

$$x = \frac{-b \pm \sqrt{b^2 - 4ac}}{2a}.$$

Solve $3x^2 + 5x + 2 = 0$.

$x = \frac{-5 \pm \sqrt{5^2 - 4(3)(2)}}{2(3)}$ $a = 3, b = 5, c = 2$

$x = \frac{-5 \pm 1}{6}$ Simplify.

$x = -\frac{2}{3}$ or $x = -1$ Two solutions, one from + and one from −

Solution set: $\left\{-1, -\frac{2}{3}\right\}$

Using the Discriminant
The discriminant $b^2 - 4ac$ of $ax^2 + bx + c = 0$ (where a, b, and c are integers) can be used to determine the number and type of solutions.

Discriminant $b^2 - 4ac$	*Number and Type of Solutions*
Positive, and the square of an integer	Two rational solutions
Positive, but not the square of an integer	Two irrational solutions
Zero	One rational solution
Negative	Two nonreal complex solutions

For $x^2 + 3x - 10 = 0$, the discriminant is as follows.

$b^2 - 4ac$

$= 3^2 - 4(1)(-10)$ $a = 1, b = 3, c = -10$

$= 49$ Simplify.

$= 7^2$ There are two rational solutions.

Because the discriminant is a perfect square, the quadratic equation can be solved using the zero-factor property.

CONCEPTS	EXAMPLES
8.3 Equations That Lead to Quadratic Methods A nonquadratic equation that can be written in the form $$au^2 + bu + c = 0,$$ for $a \neq 0$ and an algebraic expression u, is quadratic in form. **Solving an Equation Quadratic in Form by Substitution** ***Step 1*** Define a temporary variable u. ***Step 2*** Solve the quadratic equation obtained in Step 1. ***Step 3*** Replace u with the expression it defined. ***Step 4*** Solve the resulting equations for the original variable. ***Step 5*** Check all solutions in the original equation.	Solve $3(x+5)^2 + 7(x+5) + 2 = 0$. $3u^2 + 7u + 2 = 0$ Let $u = x + 5$. $(3u+1)(u+2) = 0$ Factor. $3u + 1 = 0$ or $u + 2 = 0$ Zero-factor property $u = -\frac{1}{3}$ or $u = -2$ Solve for u. $x + 5 = -\frac{1}{3}$ or $x + 5 = -2$ Replace u with $x + 5$. $x = -\frac{16}{3}$ or $x = -7$ Subtract 5. Solution set: $\left\{-7, -\frac{16}{3}\right\}$
8.4 Formulas and Further Applications **Solving a Formula for a Squared Variable** ***Case 1*** **If the variable appears only to the second power:** Isolate the second-degree variable on one side of the equation, and then use the square root property. ***Case 2*** **If the variable appears to the first and second powers:** Write the equation in standard form, and then use the quadratic formula.	Solve $A = \frac{2mp}{r^2}$ for r. ***(Case 1)*** $r^2 = \frac{2mp}{A}$ Multiply by r^2. Divide by A. $r = \pm\sqrt{\frac{2mp}{A}}$ Square root property $r = \frac{\pm\sqrt{2mpA}}{A}$ Rationalize denominator. Solve $x^2 + rx - t = 0$ for x. ***(Case 2)*** $x = \frac{-r \pm \sqrt{r^2 - 4(1)(-t)}}{2(1)}$ $a = 1$, $b = r$, $c = -t$ $x = \frac{-r \pm \sqrt{r^2 + 4t}}{2}$ Simplify.
8.5 Polynomial and Rational Inequalities **Solving a Quadratic (or Higher-Degree Polynomial) Inequality** ***Step 1*** Write the inequality as an equation and solve it. ***Step 2*** Use the values found in Step 1 to divide a number line into intervals. ***Step 3*** Substitute a test value from each interval into the *original* inequality to determine the intervals that belong to the solution set. ***Step 4*** Consider the endpoints separately.	Solve $2x^2 + 5x + 2 < 0$. $2x^2 + 5x + 2 = 0$ Related equation $(2x+1)(x+2) = 0$ Factor. $2x + 1 = 0$ or $x + 2 = 0$ Zero-factor property $x = -\frac{1}{2}$ or $x = -2$ Solve each equation. Interval A, Interval B, Interval C; number line marked -2 and $-\frac{1}{2}$; F, T, F Intervals: $(-\infty, -2)$, $\left(-2, -\frac{1}{2}\right)$, $\left(-\frac{1}{2}, \infty\right)$ Test values: -3 (Interval A), -1 (Interval B), 0 (Interval C); $x = -3$ makes the original inequality false, $x = -1$ makes it true, and $x = 0$ makes it false. Solution set: $\left(-2, -\frac{1}{2}\right)$ (Endpoints are not included because $<$ does not include equality.)

CONCEPTS	EXAMPLES
Solving a Rational Inequality	Solve $\frac{x}{x+2} \geq 4$.
Step 1 Write the inequality so that 0 is on one side and there is a single fraction on the other side.	$\frac{x}{x+2} - 4 \geq 0$ Subtract 4.
Step 2 Determine the values that make the numerator or denominator 0.	$\frac{x}{x+2} - \frac{4(x+2)}{x+2} \geq 0$ Write with a common denominator.
Step 3 Use the values from Step 2 to divide a number line into intervals.	$\frac{x - 4x - 8}{x+2} \geq 0$ Write as a single fraction; distributive property
Step 4 Substitute a test value from each interval into the *original* inequality to determine the intervals that belong to the solution set.	$\frac{-3x-8}{x+2} \geq 0$ Subtract fractions.
Step 5 Consider the endpoints separately. Exclude any values that make the denominator 0.	$-\frac{8}{3}$ makes the numerator 0, and -2 makes the denominator 0. Interval A Interval B Interval C; F $-\frac{8}{3}$ T -2 F Intervals: $\left(-\infty, -\frac{8}{3}\right)$, $\left(-\frac{8}{3}, -2\right)$, $(-2, \infty)$ Test values: -4 from Interval A makes the original inequality false, $-\frac{7}{3}$ from Interval B makes it true, and 0 from Interval C makes it false. Solution set: $\left[-\frac{8}{3}, -2\right)$ (The endpoint -2 is not included because it makes the denominator 0.)

Chapter 8 Review Exercises

8.1 *Solve each equation using the square root property or completing the square.*

1. $t^2 = 121$

2. $p^2 = 3$

3. $(2x + 5)^2 = 100$

*__4.__ $(3x - 2)^2 = -25$

5. $x^2 + 4x = 15$

6. $2x^2 - 3x = -1$

Solve each problem. Round answers to the nearest tenth of a second.

7. The height of the Eiffel Tower is 984 ft. Use Galileo's formula

$$d = 16t^2 \quad \text{(where } d \text{ is in feet)}$$

to find how long it would take a phone dropped from the top of the Eiffel Tower to reach the ground. (Data from www.wonderopolis.org)

8. The High Roller observation wheel in Las Vegas has a height of about 168 m. Use the metric version of Galileo's formula,

$$d = 4.9t^2 \quad \text{(where } d \text{ is in meters),}$$

to find how long it would take a wallet dropped from the top of the High Roller to reach the ground. (Data from www.caesars.com)

---

*This exercise requires knowledge of complex numbers.

8.2 *Solve each equation using the quadratic formula.*

9. $2x^2 + x - 21 = 0$ **10.** $x^2 + 5x = 7$ **11.** $(t + 3)(t - 4) = -2$

***12.** $2x^2 + 3x + 4 = 0$ ***13.** $3p^2 = 2(2p - 1)$ **14.** $x(2x - 7) = 3x^2 + 3$

Find the discriminant and use it to predict whether the solutions to each equation are

A. *two rational numbers* **B.** *one rational number*

C. *two irrational numbers* **D.** *two nonreal complex numbers.*

Tell whether each equation can be solved using the zero-factor property, or the quadratic formula should be used instead. Do not actually solve.

15. $x^2 + 5x + 2 = 0$ **16.** $4t^2 = 3 - 4t$

17. $4x^2 = 6x - 8$ **18.** $9z^2 + 30z + 25 = 0$

8.3 *Solve each equation. Check the solutions.*

19. $\frac{15}{x} = 2x - 1$ **20.** $\frac{1}{n} + \frac{2}{n + 1} = 2$

21. $-2r = \sqrt{\frac{48 - 20r}{2}}$ **22.** $8(3x + 5)^2 + 2(3x + 5) - 1 = 0$

23. $2x^{2/3} - x^{1/3} - 28 = 0$ **24.** $p^4 - 10p^2 + 9 = 0$

Solve each problem. Round answers to the nearest tenth, as necessary.

25. Bahaa paddled a canoe 20 mi upstream, then paddled back. If the rate of the current was 3 mph and the total trip took 7 hr, what was Bahaa's rate?

26. Carol Ann drove 8 mi to pick up a friend, and then drove 11 mi to a mall at a rate 15 mph faster. If Carol Ann's total travel time was 24 min, what was her rate on the trip to pick up her friend?

27. An old machine processes a batch of checks in 1 hr more time than a new one. How long would it take the old machine to process a batch of checks that the two machines together process in 2 hr?

28. Zoran can process a stack of invoices 1 hr faster than Claude can. Working together, they take 1.5 hr. How long would the job take each person working alone?

	Rate	Time Working Together	Fractional Part of the Job Done
Zoran			
Claude			

8.4 *Solve each formula for the specified variable. (Leave ± in the answers as needed.)*

29. $k = \frac{rF}{wv^2}$ for v **30.** $p = \sqrt{\frac{yz}{6}}$ for y **31.** $mt^2 = 3mt + 6$ for t

Solve each problem.

32. A large machine requires a part in the shape of a right triangle with a hypotenuse 9 ft less than twice the length of the longer leg. The shorter leg must be $\frac{3}{4}$ the length of the longer leg. Find the lengths of the three sides of the part.

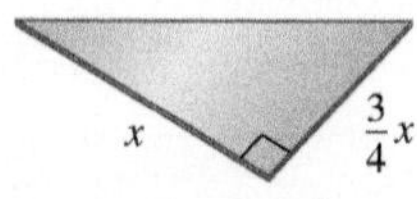

*This exercise requires knowledge of complex numbers.

33. A square has an area of 256 cm^2. If the same amount is removed from one dimension and added to the other, the resulting rectangle has an area 16 cm^2 less. Find the dimensions of the rectangle.

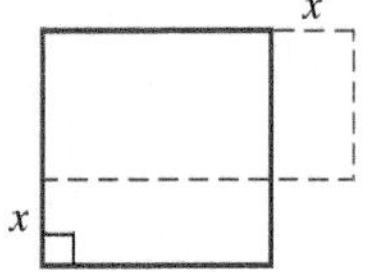

34. Allen wants to buy a mat for a photograph that measures 14 in. by 20 in. He wants to have an even border around the picture when it is mounted on the mat. If the area of the mat he chooses is 352 in.^{2}, how wide will the border be?

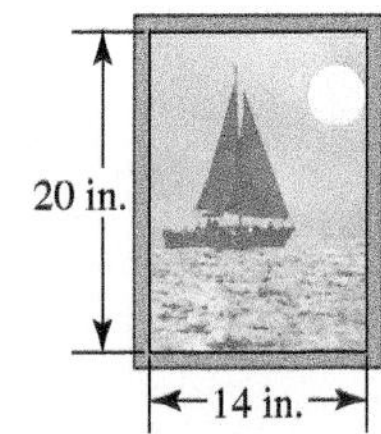

35. If a square piece of cardboard has 3-in. squares cut from its corners and then has the flaps folded up to form an open-top box, the volume of the box is given by the formula

$$V = 3(x - 6)^2,$$

where x is the length of each side of the original piece of cardboard in inches. What original length would yield a box with volume 432 in.^{3}?

36. A searchlight moves horizontally back and forth along a wall. The distance of the light from a starting point at t minutes is given by the quadratic function

$$f(t) = 100t^2 - 300t.$$

How long will it take before the light returns to the starting point?

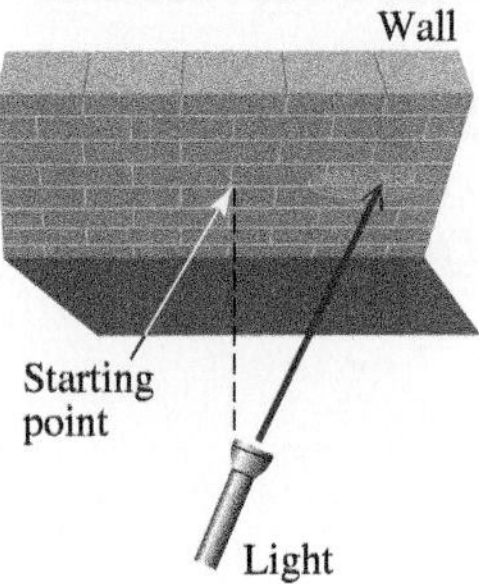

8.5 *Solve each inequality, and graph the solution set.*

37. $(x - 4)(2x + 3) > 0$

38. $x^2 + x \leq 12$

39. $(x + 2)(x - 3)(x + 5) \leq 0$

40. $(4x + 3)^2 \leq -4$

41. $\dfrac{6}{2z - 1} < 2$

42. $\dfrac{3t + 4}{t - 2} \leq 1$

Chapter 8 Mixed Review Exercises

Solve each equation or inequality.

1. $V = r^2 + R^2h$ for R

***2.** $3t^2 - 6t = -4$

3. $(3x + 11)^2 = 7$

4. $S = \dfrac{Id^2}{k}$ for d

5. $(8x - 7)^2 \geq -1$

6. $2x - \sqrt{x} = 6$

7. $x^4 - 8x^2 = -1$

8. $\dfrac{-2}{x + 5} \leq -5$

9. $6 + \dfrac{15}{s^2} = -\dfrac{19}{s}$

10. $(x^2 - 2x)^2 = 11(x^2 - 2x) - 24$

11. $(r - 1)(2r + 3)(r + 6) < 0$

*This exercise requires knowledge of complex numbers.

Solve each problem.

12. In 4 hr, Rajeed can travel 15 mi upriver and come back. The rate of the current is 5 mph. Find the rate of the boat in still water.

13. Two pieces of a large wooden puzzle fit together to form a rectangle with length 1 cm less than twice the width. The diagonal, where the two pieces meet, is 2.5 cm in length. Find the length and width of the rectangle.

14. Wachovia Center Tower in Raleigh, North Carolina, is 400 ft high. Suppose that a ball is projected upward from the top of the tower, and its position in feet above the ground is given by the quadratic function

$$f(t) = -16t^2 + 45t + 400,$$

where t is the number of seconds elapsed. How long will it take for the ball to reach a height of 200 ft above the ground? (Data from *World Almanac and Book of Facts.*)

Chapter 8 Test

Step-by-step test solutions are found on the Chapter Test Prep Videos available in MyLab Math.

View the complete solutions to all Chapter Test exercises in MyLab Math.

1. Solve $5x^2 + 13x = 6$ using the zero-factor property.

Solve each equation using the square root property or completing the square.

2. $t^2 = 54$ **3.** $(7x + 3)^2 = 25$ **4.** $x^2 + 2x = 4$

Solve each equation using the quadratic formula.

5. $2x^2 - 3x - 1 = 0$ ***6.** $3t^2 - 4t = -5$ **7.** $3x = \sqrt{\frac{9x + 2}{2}}$

***8.** If k is a negative number, then which one of the following equations will have two nonreal complex solutions?

A. $x^2 = 4k$ **B.** $x^2 = -4k$ **C.** $(x + 2)^2 = -k$ **D.** $x^2 + k = 0$

9. What is the discriminant for $2x^2 - 8x - 3 = 0$? How many and what type of solutions does this equation have? (Do not actually solve.)

Solve each equation by any method.

10. $3 - \frac{16}{x} - \frac{12}{x^2} = 0$ **11.** $4x^2 + 7x - 3 = 0$

12. $9x^4 + 4 = 37x^2$ **13.** $12 = (2n + 1)^2 + (2n + 1)$

14. Solve $S = 4\pi r^2$ for r. (Leave $\pm$ in your answer.)

Solve each problem.

15. Terry and Callie do word processing. For a certain prospectus, Callie can prepare it 2 hr faster than Terry can. If they work together, they can do the entire prospectus in 5 hr. How long will it take each of them working alone to prepare the prospectus? Round answers to the nearest tenth of an hour.

16. Qihong paddled a canoe 10 mi upstream and then paddled back to the starting point. If the rate of the current was 3 mph and the entire trip took $3\frac{1}{2}$ hr, what was Qihong's rate?

*This exercise requires knowledge of complex numbers.

17. Endre has a pool 24 ft long and 10 ft wide. He wants to construct a concrete walk around the pool. If he plans for the walk to be of uniform width and cover 152 ft^2, what will the width of the walk be?

18. At a point 30 m from the base of a tower, the distance to the top of the tower is 2 m more than twice the height of the tower. Find the height of the tower.

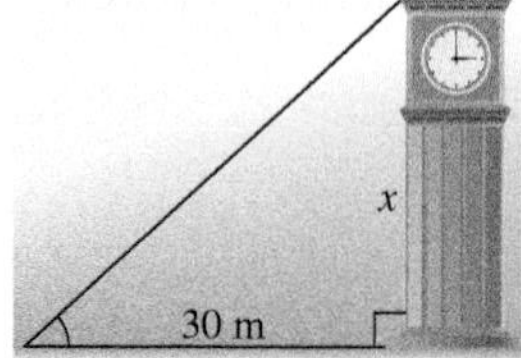

Solve each inequality, and graph the solution set.

19. $2x^2 + 7x > 15$

20. $\dfrac{5}{t-4} \le 1$

Chapters R–8 Cumulative Review Exercises

Perform the indicated operations.

1. $-8.84 - (-3.46)$

2. $\dfrac{7}{5} - \left(\dfrac{9}{10} - \dfrac{3}{2}\right)$

3. $|-12| + |13|$

4. Find 6% of 12.

5. Simplify $3 - 6(4^2 - 8)$.

6. Let $S = \left\{-\frac{7}{3}, -2, -\sqrt{3}, 0, 0.7, \sqrt{12}, \sqrt{-8}, 7, \frac{32}{3}\right\}$. List the elements of S that are elements of each set.

(a) Integers (b) Rational numbers (c) Real numbers (d) Complex numbers

Solve each equation or inequality.

7. $7 - (4 + 3t) + 2t = -6(t - 2) - 5$

8. $|6x - 9| = |-4x + 2|$

9. $2x = \sqrt{\dfrac{5x + 2}{3}}$

10. $\dfrac{3}{x-3} - \dfrac{2}{x-2} = \dfrac{3}{x^2 - 5x + 6}$

11. $(r - 5)(2r + 3) = 1$

12. $x^4 - 5x^2 + 4 = 0$

13. $-2x + 4 \le -x + 3$

14. $|3x - 7| \le 1$

15. $x^2 - 4x + 3 < 0$

16. $\dfrac{3}{p+2} > 1$

Graph each relation. Decide whether or not y can be expressed as a function f of x, and if so, give its domain and range, and write using function notation.

17. $4x - 5y = 15$

18. $4x - 5y < 15$

19. Find the slope and intercepts of the line with equation $-2x + 7y = 16$.

20. Write an equation for the specified line. Express each equation in slope-intercept form.

(a) Passing through $(2, -3)$ and parallel to the line with equation $5x + 2y = 6$

(b) Passing through $(-4, 1)$ and perpendicular to the line with equation $5x + 2y = 6$

Solve each system of equations.

21. $2x - 4y = 10$
$9x + 3y = 3$

22. $x + 2y = 5$
$4y = -2x + 8$

23. $x + y + 2z = 3$
$-x + y + z = -5$
$2x + 3y - z = -8$

24. The two top-grossing films of 2017 were *Star Wars: The Last Jedi* and *Beauty and the Beast*. The two films together grossed \$1049 million. *Beauty and the Beast* grossed \$41 million less than *Star Wars: The Last Jedi*. How much did each film gross? (Data from Box Office Mojo.)

Write with positive exponents only. Assume that variables represent positive real numbers.

25. $\left(\dfrac{x^{-3}y^2}{x^5y^{-2}}\right)^{-1}$

26. $\dfrac{(4x^{-2})^2(2y^3)}{8x^{-3}y^5}$

Perform the indicated operations.

27. $\left(\dfrac{2}{3}t + 9\right)^2$

28. Divide $4x^3 + 2x^2 - x + 26$ by $x + 2$.

Factor completely.

29. $24m^2 + 2m - 15$

30. $4t^2 - 100$

31. $8x^3 + 27y^3$

32. $9x^2 - 30xy + 25y^2$

Simplify. Express each answer in lowest terms. Assume denominators are nonzero.

33. $\dfrac{5x + 2}{-6} \div \dfrac{15x + 6}{5}$

34. $\dfrac{3}{2 - x} - \dfrac{5}{x} + \dfrac{6}{x^2 - 2x}$

35. $\dfrac{\frac{r}{s} - \frac{s}{r}}{\frac{r}{s} + 1}$

36. Two cars left an intersection at the same time, one heading due south and the other due east. Later they were exactly 95 mi apart. The car heading east had gone 38 mi less than twice as far as the car heading south. How far had each car traveled?

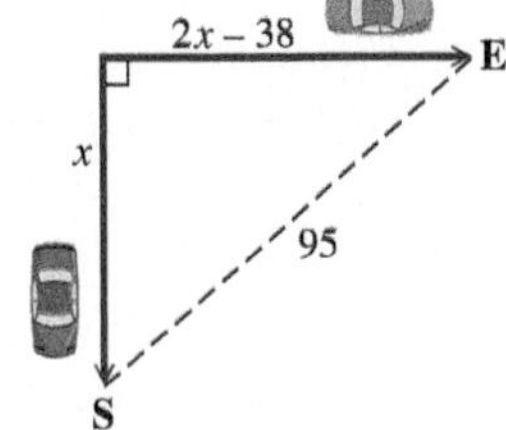

Simplify each radical expression.

37. $\sqrt[3]{\dfrac{27}{16}}$

38. $\dfrac{2}{\sqrt{7} - \sqrt{5}}$

STUDY SKILLS **REMINDER**

Have you begun to prepare for your final exam? **Review Study Skill 10, *Preparing for Your Math Final Exam*.**

9

ADDITIONAL GRAPHS OF FUNCTIONS AND RELATIONS

Quadratic functions, one of the topics of this chapter, have graphs that are *parabolas*. Cross sections of telescopes, satellite dishes, and automobile headlights form parabolas, as do the cables that support suspension bridges.

9.1 Review of Operations and Composition

OBJECTIVES

1 Review operations on functions.
2 Find a difference quotient.
3 Form composite functions and find their domains.

VOCABULARY

☐ difference quotient
☐ composite function (composition of functions)

OBJECTIVE 1 Review operations on functions.

Recall that we can add, subtract, multiply, and divide functions to create other functions. For example, in business, "profit equals revenue minus cost." Using function notation, this can be written

$P(x) = R(x) - C(x)$, where x is the number of items produced and sold.

The profit function is found by *subtracting* the cost function from the revenue function.

Operations on Functions

If f and g are functions, then for all values of x for which both $f(x)$ and $g(x)$ exist, the following operations are defined.

$$(f+g)(x) = f(x) + g(x)$$ Sum function

$$(f-g)(x) = f(x) - g(x)$$ Difference function

$$(fg)(x) = f(x) \cdot g(x)$$ Product function

$$\left(\frac{f}{g}\right)(x) = \frac{f(x)}{g(x)}, \quad g(x) \neq 0$$ Quotient function

NOTE The condition $g(x) \neq 0$ in the definition of the quotient function means that the domain of $\left(\frac{f}{g}\right)(x)$ consists of all values of x for which $f(x)$ and $g(x)$ are defined and $g(x)$ is not 0.

EXAMPLE 1 Using Operations on Functions

Let $f(x) = x^2 + 1$ and $g(x) = 3x + 5$. Find each of the following.

(a) $(f+g)(1)$

First evaluate $f(1) = 1^2 + 1 = 2$ and $g(1) = 3(1) + 5 = 8$.

$(f+g)(1)$

$= f(1) + g(1)$ Use the definition of the sum function.

$= 2 + 8$ Substitute for $f(1)$ and $g(1)$.

$= 10$ Add.

(b) $(f-g)(-3)$

$= f(-3) - g(-3)$ Use the definition of the difference function.

$= \underbrace{[(-3)^2 + 1]}_{f(x) = x^2+1} - \underbrace{[3(-3) + 5]}_{g(x) = 3x+5}$ Substitute.

$= 10 - (-4)$ Work inside the brackets. (This is a key step.)

$= 14$ Subtract.

NOW TRY EXERCISE 1

Let $f(x) = 3x - 4$

and $g(x) = 2x^2 + 5$.

Find each of the following.

(a) $(f + g)(2)$

(b) $(f - g)(-1)$

(c) $(fg)(0)$

(d) $\left(\frac{f}{g}\right)(3)$

(c) $(fg)(5)$

$= f(5) \cdot g(5)$ Use the definition of the product function.

$= [5^2 + 1] \cdot [3(5) + 5]$ Substitute; $f(x) = x^2 + 1$, $g(x) = 3x + 5$

$= 26 \cdot 20$ Work inside the brackets.

$= 520$ Multiply.

(d) $\left(\frac{f}{g}\right)(0)$

$= \frac{f(0)}{g(0)}$ Use the definition of the quotient function.

$= \frac{0^2 + 1}{3(0) + 5}$ $f(x) = x^2 + 1$, $g(x) = 3x + 5$

$= \frac{1}{5}$ Simplify. **NOW TRY**

STUDY SKILLS REMINDER

Make study cards to help you learn and remember the material in this chapter. **Review Study Skill 5, *Using Study Cards.***

Domains of $f + g$, $f - g$, fg, $\frac{f}{g}$

For functions f and g, the following hold true.

- The domains of $f + g$, $f - g$, and fg include all real numbers in the intersection of the domains of f and g.
- The domain of $\frac{f}{g}$ includes those real numbers in the intersection of the domains of f and g for which $g(x) \neq 0$.

EXAMPLE 2 Using Operations on Functions

Let $f(x) = 8x - 9$ and $g(x) = \sqrt{2x - 1}$. Find each of the following, and give their domains.

(a) $(f + g)(x)$

$= f(x) + g(x)$

$= 8x - 9 + \sqrt{2x - 1}$

(b) $(f - g)(x)$

$= f(x) - g(x)$

$= 8x - 9 - \sqrt{2x - 1}$

(c) $(fg)(x)$

$= f(x) \cdot g(x)$

$= (8x - 9)\sqrt{2x - 1}$

(d) $\left(\frac{f}{g}\right)(x)$

$= \frac{f(x)}{g(x)}$

$= \frac{8x - 9}{\sqrt{2x - 1}}$

To find the domains of the functions in parts (a)–(d), we first find the domains of f and g.

The domain of f is the set of all real numbers, $(-\infty, \infty)$.

NOW TRY ANSWERS

1. (a) 15 **(b)** -14 **(c)** -20 **(d)** $\frac{5}{23}$

NOW TRY EXERCISE 2

Let $f(x) = 4x - 2$ and $g(x) = \sqrt{3x + 12}$. Find each of the following, and give their domains.

(a) $(f + g)(x)$

(b) $(f - g)(x)$

(c) $(fg)(x)$

(d) $\left(\frac{f}{g}\right)(x)$

Because g is defined by a square root radical, the radicand must be nonnegative (that is, greater than or equal to 0).

$$g(x) = \sqrt{2x - 1} \qquad \text{Rule for } g(x)$$

$$2x - 1 \geq 0 \qquad \text{The radicand must be nonnegative.}$$

$$x \geq \frac{1}{2} \qquad \text{Add 1. Divide by 2.}$$

Thus, the domain of g is $\left[\frac{1}{2}, \infty\right)$.

The domains of $f + g$, $f - g$, and fg are the intersection of the domains of f and g.

$$(-\infty, \infty) \cap \left[\frac{1}{2}, \infty\right) = \left[\frac{1}{2}, \infty\right)$$

The intersection of two sets is the set of all elements belonging to *both* sets.

The domain of $\frac{f}{g}$ includes those real numbers in the intersection of f and g above for which $g(x) = \sqrt{2x - 1} \neq 0$—that is, the domain of $\frac{f}{g}$ is $\left(\frac{1}{2}, \infty\right)$. NOW TRY

OBJECTIVE 2 Find a difference quotient.

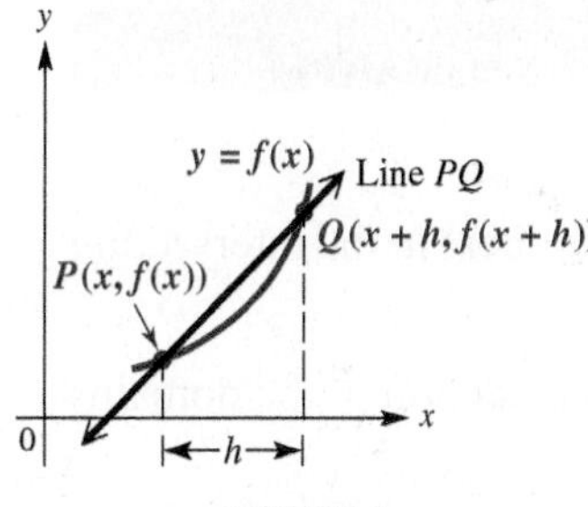

FIGURE 1

Suppose a point P lies on the graph of $y = f(x)$ as in **FIGURE 1**, and suppose h is a positive number. If we let $(x, f(x))$ denote the coordinates of P and $(x + h, f(x + h))$ denote the coordinates of Q, then the line joining P and Q has slope as follows.

$$m = \frac{f(x + h) - f(x)}{(x + h) - x} \qquad \begin{array}{l}\leftarrow \text{Change in } y \\ \leftarrow \text{Change in } x\end{array}$$

$$m = \frac{\mathbf{f(x + h) - f(x)}}{\mathbf{h}}, \quad h \neq 0 \qquad \text{Difference quotient}$$

The boldface expression is the **difference quotient.** It is important in calculus.

EXAMPLE 3 Finding a Difference Quotient

For $f(x) = x^2 - 2x$, find the following and simplify each expression.

(a) $f(x + h) - f(x)$ (This is the numerator of the difference quotient.)

First find $f(x + h)$.

$$f(x) = x^2 - 2x \qquad \text{Given function}$$

$$f(x + h) = (x + h)^2 - 2(x + h) \qquad \text{Replace } x \text{ with } x + h.$$

$$f(x + h) = x^2 + 2xh + h^2 - 2x - 2h \qquad \text{Square the binomial; distributive property}$$

Remember the middle term when squaring.

Now subtract $f(x)$ from this expression for $f(x + h)$.

$$f(x + h) - f(x)$$

$$= (x^2 + 2xh + h^2 - 2x - 2h) - (x^2 - 2x) \qquad \text{Substitute for } f(x + h) \text{ and } f(x).$$

$$= x^2 + 2xh + h^2 - 2x - 2h - x^2 + 2x \qquad \text{Distributive property}$$

$$= 2xh + h^2 - 2h \qquad \text{Combine like terms.}$$

NOW TRY ANSWERS

2. (a) $4x - 2 + \sqrt{3x + 12}$
(b) $4x - 2 - \sqrt{3x + 12}$
(c) $(4x - 2)\sqrt{3x + 12}$
(d) $\frac{4x - 2}{\sqrt{3x + 12}}$

In parts (a), (b), and (c), the domain is $[-4, \infty)$. In part (d), the domain is $(-4, \infty)$.

For $f(x) = 2x^2 + 3$, find and simplify the expression for the difference quotient.

$$\frac{f(x+h) - f(x)}{h}$$

(b) $\dfrac{f(x+h) - f(x)}{h}$ This is the difference quotient.

$= \dfrac{2xh + h^2 - 2h}{h}$ Substitute the result from part (a) for the numerator.

$= \dfrac{h(2x + h - 2)}{h}$ Factor out h in the numerator.

$= 2x + h - 2$ Divide out the common factor. NOW TRY

OBJECTIVE 3 Form composite functions and find their domains.

The diagram in **FIGURE 2** shows a function g that assigns to each x in its domain a value $g(x)$. Then another function f assigns to each $g(x)$ in its domain a value $f(g(x))$. This two-step process takes an element x and produces a corresponding element $f(g(x))$.

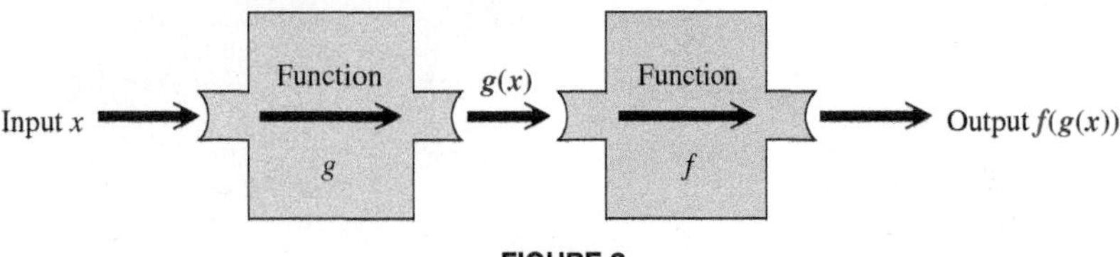

FIGURE 2

The function with y-values $f(g(x))$ is the *composition* of functions f and g, written $f \circ g$ and read **"f of g" (or "f compose g").**

Composition of Functions

If f and g are functions, then the **composite function,** or **composition,** of f and g is

$$(f \circ g)(x) = f(g(x)).$$

The domain of $f \circ g$ is the set of all numbers x in the domain of g such that $g(x)$ is in the domain of f.

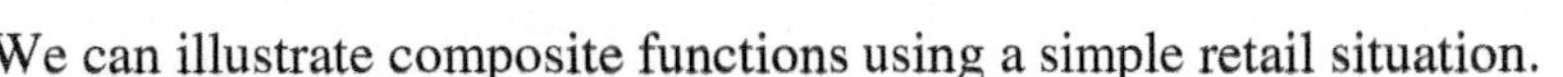

We can illustrate composite functions using a simple retail situation.

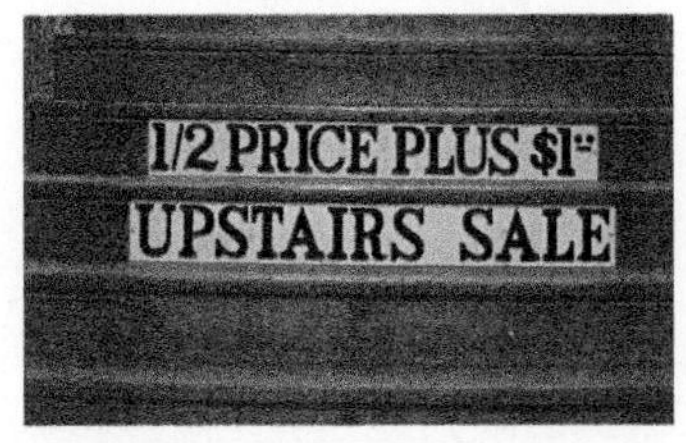

In the sale room at a clothing store, every item is half price, plus \$1. *How much would a* \$60 *shirt from the sale room cost?*

See **FIGURE 3.**

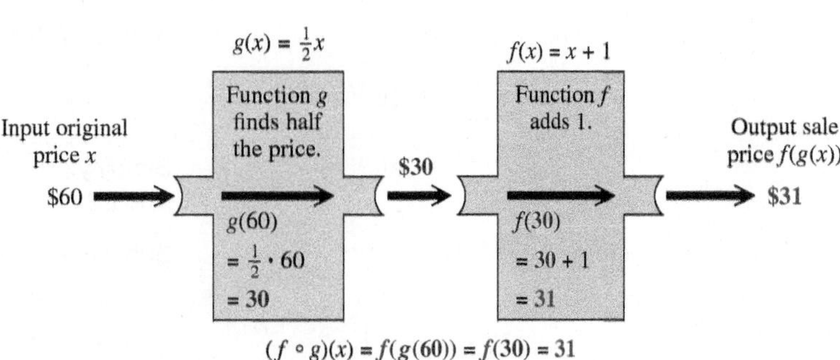

FIGURE 3

NOW TRY ANSWER

3. $4x + 2h$

As another real-life example of how composite functions occur, suppose an oil well off the coast is leaking, with the leak spreading oil in a circular layer over the surface. See FIGURE 4. At any time t, in minutes, after the beginning of the leak, the radius of the circular oil slick is given by $r(t) = 5t$ feet. Because $\mathscr{A}(r) = \pi r^2$ gives the area of a circle of radius r, the area can be expressed as a function of time.

$$\mathscr{A}(r) = \pi r^2 \qquad \text{Area of a circle}$$

$$\mathscr{A}(r(t)) = \pi(5t)^2 \qquad \text{Substitute } 5t \text{ for } r.$$

See the table. The function $\mathscr{A}(r(t))$ is a composite function of the functions $\mathscr{A}$ and r.

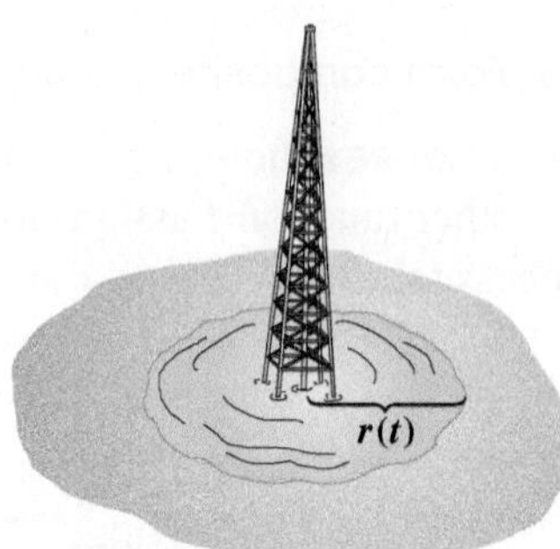

FIGURE 4

t	$r(t)$	$\mathscr{A}(r(t))$
1	5	$\pi(5)^2$, or 25π
2	10	$\pi(10)^2$, or 100π
3	15	$\pi(15)^2$, or 225π
t	$5t$	$\pi(5t)^2$, or $25\pi t^2$

NOW TRY EXERCISE 4

Let $f(x) = \frac{5}{x+2}$ and $g(x) = x - 4$.

Find each of the following.

(a) $(f \circ g)(-3)$

(b) $(g \circ f)(-3)$

EXAMPLE 4 Evaluating Composite Functions

Let $f(x) = 2x - 1$ and $g(x) = \frac{4}{x-1}$. Find each of the following.

(a) $(f \circ g)(2)$

By definition, $(f \circ g)(2) = f(g(2))$. We first find $g(2)$ using $g(x) = \frac{4}{x-1}$.

$$g(2) = \frac{4}{2-1} = \frac{4}{1} = 4$$

Now we find $(f \circ g)(2) = f(g(2)) = f(4)$ using $f(x) = 2x - 1$.

$$f(g(2))$$

$$= f(4) \qquad g(2) = 4 \text{ from above}$$

$$= 2(4) - 1 \qquad \text{Substitute 4 for } x \text{ in } f(x) = 2x - 1.$$

$$= 7 \qquad \text{Multiply. Subtract.}$$

(b) $(g \circ f)(-3)$ — Don't confuse composition with multiplication.

$$= g(f(-3)) \qquad \text{Use the definition.}$$

$$= g(2(-3) - 1) \qquad \text{Evaluate } f(-3) \text{ first. Substitute } -3 \text{ for } x \text{ in } f(x) = 2x - 1.$$

$$= g(-7) \qquad \text{Simplify.}$$

$$= \frac{4}{-7-1} \qquad \text{Substitute } -7 \text{ for } x \text{ in } g(x) = \frac{4}{x-1}.$$

$$= -\frac{1}{2} \qquad \text{Subtract in the denominator. Write in lowest terms.}$$

NOW TRY

NOW TRY ANSWERS

4. (a) -1 (b) -9

NOW TRY EXERCISE 5

Let $f(x) = 2x^2 - x$ and $g(x) = x - 4$.

Find each of the following, and give their domains.

(a) $(f \circ g)(x)$ **(b)** $(g \circ f)(x)$

EXAMPLE 5 Finding Composite Functions

Let $f(x) = 4x + 1$ and $g(x) = 2x^2 + 5x$. Find each of the following, and give their domains.

(a)

$(f \circ g)(x)$	By definition, $(f \circ g)(x) = f(g(x))$.
$= f(g(x))$	
$= f(2x^2 + 5x)$	$g(x) = 2x^2 + 5x$
$= 4(2x^2 + 5x) + 1$	$f(x) = 4x + 1$
$= 8x^2 + 20x + 1$	Distributive property

Substitute $2x^2 + 5x$ for x in the rule for f.

(b)

$(g \circ f)(x)$	By definition, $(g \circ f)(x) = g(f(x))$.
$= g(f(x))$	
$= g(4x + 1)$	$f(x) = 4x + 1$
$= 2(4x + 1)^2 + 5(4x + 1)$	$g(x) = 2x^2 + 5x$
$= 2(16x^2 + 8x + 1) + 20x + 5$	Multiply.
$= 32x^2 + 16x + 2 + 20x + 5$	Distributive property
$= 32x^2 + 36x + 7$	Combine like terms.

Substitute $4x + 1$ for x in the rule for g.

The domain of both composite functions is the set of all real numbers, $(-\infty, \infty)$.

NOW TRY

In **Example 5,** $\underbrace{8x^2 + 20x + 1}_{(f \circ g)(x) \text{ from part (a)}} \neq \underbrace{32x^2 + 36x + 7}_{(g \circ f)(x) \text{ from part (b)}}$.

In general, $$(f \circ g)(x) \neq (g \circ f)(x).$$

CAUTION ***The composite function $f \circ g$ is not the same as the product fg.***

Example: $(f \circ g)(x) = 8x^2 + 20x + 1$ — f and g defined as in **Example 5**

but $(fg)(x) = (4x + 1)(2x^2 + 5x)$

$= 8x^3 + 22x^2 + 5x$

EXAMPLE 6 Finding Composite Functions and Their Domains

Let $f(x) = \frac{1}{x}$ and $g(x) = \sqrt{3 - x}$. Find each of the following, and give their domains.

(a)

$(f \circ g)(x)$	
$= f(g(x))$	Use the definition.
$= f(\sqrt{3 - x})$	Substitute; $g(x) = \sqrt{3 - x}$
$= \dfrac{1}{\sqrt{3 - x}}$	$f(x) = \frac{1}{x}$

NOW TRY ANSWERS

5. (a) $2x^2 - 17x + 36$; $(-\infty, \infty)$

(b) $2x^2 - x - 4$; $(-\infty, \infty)$

NOW TRY EXERCISE 6

Let $f(x) = \sqrt{x+5}$ and $g(x) = 2x - 1$.
Find each of the following, and give their domains.
(a) $(f \circ g)(x)$ **(b)** $(g \circ f)(x)$

For $(f \circ g)(x) = \dfrac{1}{\sqrt{3-x}}$, the radical expression $\sqrt{3-x}$ must be a positive real number.

$$3 - x > 0$$

$$-x > -3 \quad \text{Subtract 3.}$$

$$x < 3 \quad \text{Multiply by } -1. \text{ Change } > \text{ to } <.$$

Thus, the domain of $f \circ g$ is the interval $(-\infty, 3)$.

(b) $(g \circ f)(x)$

$$= g(f(x)) \quad \text{Use the definition.}$$

$$= g\left(\frac{1}{x}\right) \quad \text{Substitute; } f(x) = \tfrac{1}{x}$$

$$= \sqrt{3 - \frac{1}{x}} \quad g(x) = \sqrt{3-x}$$

$$= \sqrt{\frac{3x-1}{x}} \quad \text{Write the radicand as a single fraction.}$$

The domain of $g \circ f$ is the set of all real numbers x such that

$$x \neq 0 \quad \left(\text{because } \tfrac{1}{x} \text{ will be undefined if } x = 0\right)$$

and

$$3 - f(x) \geq 0 \quad g(x) = \sqrt{3-x} \geq 0, \text{ because of the square root radical.}$$

$$3 - \frac{1}{x} \geq 0 \quad f(x) = \tfrac{1}{x}$$

$$\frac{3x-1}{x} \geq 0. \quad \text{Write as a single fraction.}$$

The number $\frac{1}{3}$ makes the numerator of this rational inequality 0, and 0 makes the denominator 0. Using earlier methods, we consider test values in the intervals $(-\infty, 0)$, $\left(0, \frac{1}{3}\right)$, and $\left(\frac{1}{3}, \infty\right)$. Doing so gives the results shown in the table.

Interval	*Test Value*	*Test of Inequality*	*True or False?*
$(-\infty, 0)$	-1	$4 \geq 0$	T
$\left(0, \frac{1}{3}\right)$	$\frac{1}{6}$	$-3 \geq 0$	F
$\left(\frac{1}{3}, \infty\right)$	1	$2 \geq 0$	T

The numbers in the intervals $(-\infty, 0)$ and $\left(\frac{1}{3}, \infty\right)$ are solutions. Considering the endpoints separately, the domain of $g \circ f$ is

$$(-\infty, 0) \cup \left[\frac{1}{3}, \infty\right).$$

NOW TRY

NOW TRY ANSWERS

6. (a) $\sqrt{2x+4}$; $[-2, \infty)$
(b) $2\sqrt{x+5} - 1$; $[-5, \infty)$

NOW TRY EXERCISE 7

Find functions f and g such that

$$(f \circ g)(x) = 3\sqrt{x+4} - 1.$$

NOW TRY ANSWER

7. $f(x) = 3\sqrt{x} - 1$ and $g(x) = x + 4$ is one possible answer.

EXAMPLE 7 **Finding Functions That Form a Given Composite**

Find functions f and g such that

$$(f \circ g)(x) = (x^2 - 5)^3 - 4(x^2 - 5) + 3.$$

Note the repeated quantity $x^2 - 5$. If $g(x) = x^2 - 5$ and $f(x) = x^3 - 4x + 3$, then $(f \circ g)(x)$ can be written as follows.

$(f \circ g)(x)$

$= f(g(x))$ Use the definition.

$= f(x^2 - 5)$ Replace $g(x)$ with $x^2 - 5$.

$= (x^2 - 5)^3 - 4(x^2 - 5) + 3$ In $f(x)$, replace x with $x^2 - 5$.

There are other pairs of functions f and g that also work. For instance,

$$f(x) = (x - 5)^3 - 4(x - 5) + 3 \quad \text{and} \quad g(x) = x^2.$$

NOW TRY 

9.1 Exercises

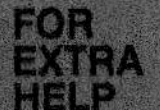
FOR EXTRA HELP

MyLab Math

Video solutions for select problems available in MyLab Math

STUDY SKILLS REMINDER

We can learn from the mistakes we make. **Review Study Skill 9, *Analyzing Your Test Results.***

Concept Check *The graph shows dollars (in billions) spent for general science and for space/other technologies over a 20-year period.*

$G(x)$ *represents dollars spent for general science.*

$S(x)$ *represents dollars spent for space/other technologies.*

$T(x)$ *represents total expenditures for these two categories.*

Science and Space Spending

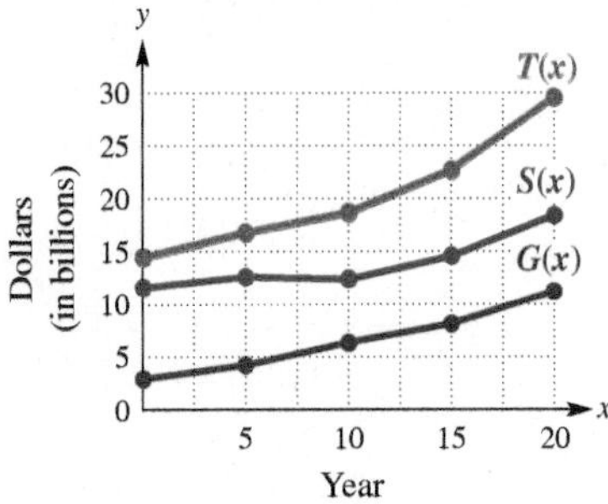

Data from U.S. Office of Management and Budget.

1. Estimate $G(15)$ and $S(15)$ and use the results to estimate $T(15)$.

2. Estimate $G(10)$ and $S(10)$ and use the results to estimate $T(10)$.

3. Estimate $(T - G)(20)$. What does this function represent?

4. Estimate $(S - G)(5)$. What does this function represent?

The figure shows the situation for a company that manufactures cellular phones. The two lines are the graphs of the linear functions for revenue and cost,

$$R(x) = 168x \quad \textit{and} \quad C(x) = 118x + 800.$$

Here, x is the number of cellular phones produced and sold, and x, $R(x)$, and $C(x)$ are in thousands. ***See Objective 1.***

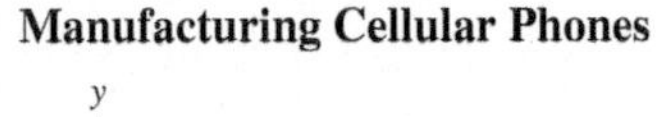
Manufacturing Cellular Phones

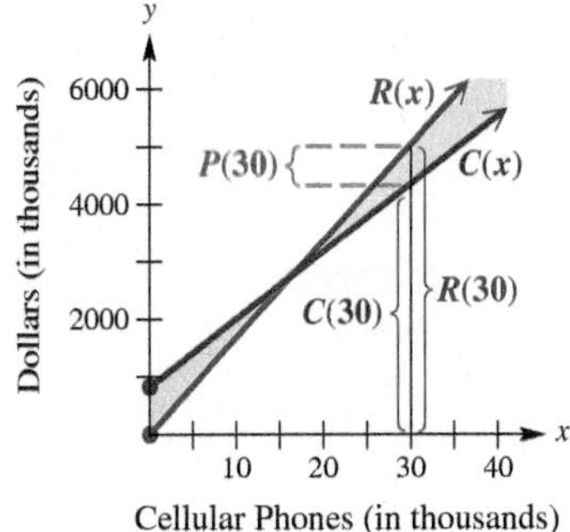

5. Write and simplify a function P that gives profit in terms of x.

6. Find the profit if 30,000 cellular phones are produced and sold. (*Hint:* What is the value of x?)

Let $f(x) = 2x^2 - 4$ and let $g(x) = 3x + 1$. Find each of the following. **See Examples 1 and 4.**

7. $(f+g)(3)$

8. $(f+g)(4)$

9. $(f-g)(1)$

10. $(f-g)(-5)$

11. $(f+g)\left(\frac{1}{2}\right)$

12. $(f+g)\left(\frac{1}{3}\right)$

13. $(g-f)\left(-\frac{1}{2}\right)$

14. $(g-f)\left(-\frac{1}{3}\right)$

15. $(fg)(4)$

16. $(fg)(2)$

17. $(gf)(-3)$

18. $(gf)(-4)$

19. $\left(\frac{g}{f}\right)(-1)$

20. $\left(\frac{g}{f}\right)(3)$

21. $\left(\frac{f}{g}\right)(4)$

22. $\left(\frac{f}{g}\right)(5)$

23. $(f \circ g)(2)$

24. $(f \circ g)(-5)$

25. $(g \circ f)(2)$

26. $(g \circ f)(-5)$

27. $(f \circ g)(-2)$

28. $(f \circ g)(-6)$

For each pair of functions f and g, find **(a)** $f+g$, **(b)** $f-g$, **(c)** fg, *and* **(d)** $\frac{f}{g}$. *Give the domain for each.* **See Example 2.**

29. $f(x) = 4x - 1, \quad g(x) = 6x + 3$

30. $f(x) = -2x + 9, \quad g(x) = -5x + 2$

31. $f(x) = 3x^2 - 2x, \quad g(x) = x^2 - 2x + 1$

32. $f(x) = 6x^2 - 11x, \quad g(x) = x^2 - 4x - 5$

33. $f(x) = 2x + 5, \quad g(x) = \sqrt{4x + 3}$

34. $f(x) = 11x - 3, \quad g(x) = \sqrt{2x - 5}$

Extending Skills *Use the table to evaluate each expression in parts (a)–(d), if possible.*

(a) $(f+g)(2)$ **(b)** $(f-g)(4)$ **(c)** $(fg)(-2)$ **(d)** $\left(\frac{f}{g}\right)(0)$

35.

x	$f(x)$	$g(x)$
−2	0	6
0	5	0
2	7	−2
4	10	5

36.

x	$f(x)$	$g(x)$
−2	−4	2
0	8	−1
2	5	4
4	0	0

For each function, find and simplify **(a)** $f(x+h) - f(x)$ *and* **(b)** $\frac{f(x+h) - f(x)}{h}$. **See Example 3.**

37. $f(x) = 6x + 2$

38. $f(x) = 4x + 11$

39. $f(x) = x^2$

40. $f(x) = -x^2$

41. $f(x) = 2x^2 - 1$

42. $f(x) = 3x^2 + 2$

43. $f(x) = x^2 + 4x$

44. $f(x) = x^2 - 5x$

Let $f(x) = 2x - 3$ and $g(x) = -x + 3$. Find each of the following. **See Example 4.**

45. $(f \circ g)(4)$

46. $(f \circ g)(2)$

47. $(f \circ g)(-2)$

48. $(g \circ f)(3)$

49. $(g \circ f)(0)$

50. $(g \circ f)(-2)$

51. $(f \circ f)(2)$

52. $(g \circ g)(-2)$

Find $(f \circ g)(x)$ and $(g \circ f)(x)$ and their domains for each pair of functions. ***See Examples 5 and 6.***

53. $f(x) = -6x + 9, \quad g(x) = 5x + 7$

54. $f(x) = 8x + 12, \quad g(x) = 3x - 1$

55. $f(x) = 5x + 3, \quad g(x) = -x^2 + 4x + 3$

56. $f(x) = 4x^2 + 2x + 8, \quad g(x) = x + 5$

57. $f(x) = \sqrt{x}, \quad g(x) = x + 3$

58. $f(x) = \sqrt{x}, \quad g(x) = x - 1$

59. $f(x) = \frac{1}{x}, \quad g(x) = x^2$

60. $f(x) = \frac{2}{x^4}, \quad g(x) = 2 - x$

61. $f(x) = \sqrt{x + 2}, \quad g(x) = 8x - 6$

62. $f(x) = 9x - 11, \quad g(x) = 2\sqrt{x + 2}$

63. $f(x) = \frac{1}{x - 5}, \quad g(x) = \frac{2}{x}$

64. $f(x) = \frac{8}{x - 6}, \quad g(x) = \frac{4}{3x}$

Extending Skills *The graphs of functions f and g are shown. Use these graphs to find each indicated value. For example, to find $(f + g)(0)$ we can determine from the graphs that $f(0) = 1$ and $g(0) = 4$, so*

$$(f + g)(0) = 1 + 4 = 5.$$

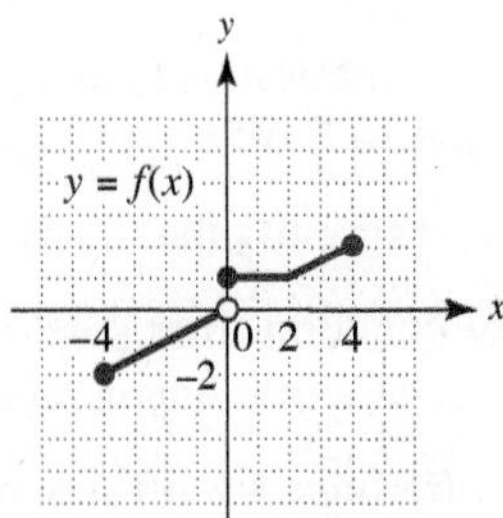

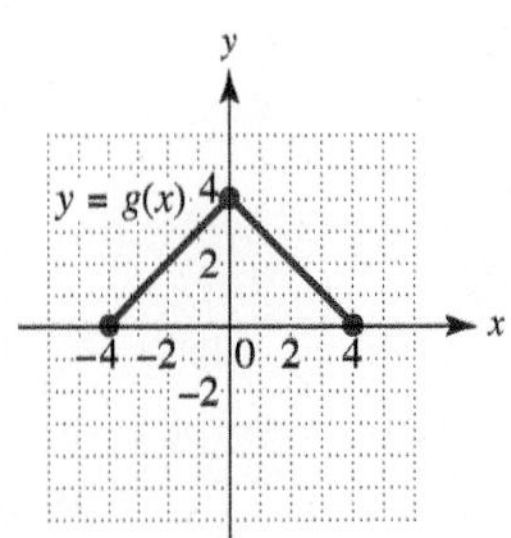

65. $(f + g)(1)$

66. $(f + g)(4)$

67. $(f - g)(0)$

68. $(f - g)(-4)$

69. $f(-2) \cdot g(4)$

70. $f(1) \cdot g(4)$

71. $(f \circ g)(2)$

72. $(f \circ g)(-2)$

73. $(g \circ f)(2)$

74. $(g \circ f)(-4)$

75. $[f(0)]^6$

76. $[g(4)]^8$

Extending Skills *The tables give some selected ordered pairs for functions f and g.*

x	f(x)
−1	1
2	−1
5	9

x	g(x)
2	−1
7	5
1	9
9	20

Find each of the following.

77. $(f \circ g)(2)$

78. $(f \circ g)(7)$

79. $(g \circ f)(-1)$

80. $(g \circ f)(5)$

81. $(f \circ f)(2)$

82. $(g \circ g)(1)$

83. Why can we not determine $(f \circ g)(1)$ given the information in the tables for **Exercises 77–82?**

84. Using the tables for **Exercises 77–82,** extend the concept of composition of functions to evaluate $(g \circ (f \circ g))(7)$.

A function h is given. Find functions f and g such that $h(x) = (f \circ g)(x)$. Many such pairs of functions exist. ***See Example 7.***

85. $h(x) = (6x - 2)^2$

86. $h(x) = (2x - 3)^3$

87. $h(x) = \sqrt{x^2 + 3}$

88. $h(x) = \sqrt{x^2 - 1}$

89. $h(x) = \dfrac{1}{x^2 + 2}$

90. $h(x) = \dfrac{1}{x^3 - 3}$

91. $h(x) = 4(2x - 3)^3 + (2x - 3) + 5$

92. $h(x) = (x + 2)^3 - 3(x + 2)^2$

Solve each problem. ***See Objective 3.***

93. The function $f(x) = 3x$ computes the number of feet in x yards, and the function $g(x) = 1760x$ computes the number of yards in x miles. What is $(f \circ g)(x)$ and what does it compute?

94. The function $f(x) = 24x$ computes the number of hours in x days, and the function $g(x) = 7x$ computes the number of days in x weeks. Find and simplify $(f \circ g)(x)$. What does it compute?

95. A store offers 50% off the regular price. This price is marked down an additional 10% during a special sale.

(a) Write a function $g(x)$ that finds 50% of x.

(b) Write a function $f(x)$ that subtracts an additional 10% off x.

(c) Write and simplify the function $(f \circ g)(x)$.

(d) Use the function from part (c) to find the sale price of a coat that is regularly \$80.

96. A store is offering 25% off a purchase. A gift card takes an additional \$20 off.

(a) Write a function $g(x)$ that finds 25% of x.

(b) Write a function $f(x)$ that adds 20 to x.

(c) Write and simplify the function $(f \circ g)(x)$.

(d) Use the function from part (c) to find the amount saved on a \$100 purchase.

97. Suppose the demand for a certain brand of vacuum cleaner is given by

$$D(p) = \frac{-p^2}{100} + 500,$$

where p is the price in dollars. If the price in terms of the cost, c, is expressed as

$$p(c) = 2c - 10,$$

find an expression for $D(c)$, the demand in terms of the cost.

98. Suppose the population P of a certain species of fish depends on the number x (in hundreds) of a smaller kind of fish that serves as its food supply, where

$$P(x) = 2x^2 + 1.$$

Suppose, also, that the number x (in hundreds) of the smaller species of fish depends on the amount a (in appropriate units) of its food supply, a kind of plankton, where

$$x = f(a) = 3a + 2.$$

Find an expression for $(P \circ f)(a)$, the relationship between the population P of the large fish and the amount a of plankton available.

RELATING CONCEPTS For Individual or Group Work (Exercises 99–106)

Work Exercises 99–106 in order, *to see how important properties of operations with real numbers are related to similar properties of composition of functions. Fill in the blanks when appropriate.**

99. Because ____ is the identity element for addition, $a +$ ____ $=$ ____ $+ a = a$ for all real numbers a.

100. Because ____ is the identity element for multiplication, $a \cdot$ ____ $=$ ____ $\cdot a = a$ for all real numbers a.

101. Consider the following function.

$$f(x) = x$$

Choose any function g, and find $(f \circ g)(x)$. Then find $(g \circ f)(x)$. How do the two results compare?

102. Based on **Exercise 101,** what might the function $f(x) = x$ be called (with respect to composition of functions)?

103. The inverse property of addition says that for every real number a, there exists a unique real number ____ such that $a +$ ____ $=$ ____ $+ a = 0$.

104. The inverse property of multiplication says that for every nonzero real number a, there exists a unique real number ____ such that $a \cdot$ ____ $=$ ____ $\cdot a = 1$.

105. Consider the following functions.

$$f(x) = x^3 + 2 \quad \text{and} \quad g(x) = \sqrt[3]{x - 2}$$

Find $(f \circ g)(x)$ and then find $(g \circ f)(x)$. How do the results compare?

106. Based on **Exercise 105,** what might the functions f and g be called with respect to each other (regarding composition of functions)?

9.2 Graphs of Quadratic Functions

OBJECTIVES

1. Graph a quadratic function.
2. Graph parabolas with horizontal and vertical shifts.
3. Use the coefficient of x^2 to predict the shape and direction in which a parabola opens.
4. Find a quadratic function to model data.

OBJECTIVE 1 Graph a quadratic function.

FIGURE 5 gives a graph of the simplest *quadratic function* $y = x^2$. This graph is a **parabola.** The point $(0, 0)$, the lowest point on the curve, is the **vertex** of the parabola. The vertical line through the vertex is the **axis of symmetry,** or simply the **axis,** of the parabola. Here its equation is $x = 0$. A parabola is **symmetric about its axis**—that is, if the graph were folded along the axis, the two portions of the curve would coincide.

$y = x^2$

x	y
−2	4
−1	1
0	0
1	1
2	4

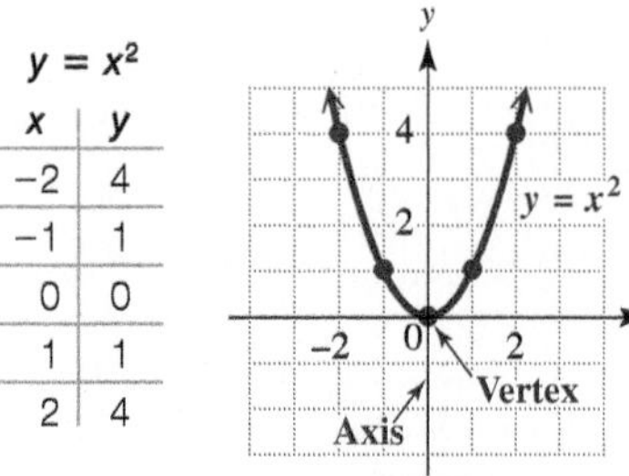

FIGURE 5

As **FIGURE 5** suggests, x can be any real number, so the domain of the function $y = x^2$, written in interval notation, is $(-\infty, \infty)$. Values of y are always nonnegative, so the range is $[0, \infty)$.

*The ideas developed in these Relating Concepts exercises are more fully developed in a later chapter.

VOCABULARY

- ☐ parabola
- ☐ vertex
- ☐ axis of symmetry (axis)
- ☐ quadratic function

Quadratic Function

A function that can be written in the form

$$f(x) = ax^2 + bx + c$$

for real numbers a, b, and c, where $a \neq 0$, is a **quadratic function.**

Examples: $f(x) = x^2$, $f(x) = x^2 - 4x + 4$, $f(x) = -\frac{1}{2}x^2 + 3$

The graph of any quadratic function is a parabola with a vertical axis.

NOTE We use the variable y and function notation $f(x)$ interchangeably. Although we use the letter f most often to name quadratic functions, other letters can be used. We use the capital letter F to distinguish between different parabolas graphed on the same coordinate axes.

Parabolas have a special reflecting property that makes them useful in the design of telescopes, radar equipment, solar furnaces, and automobile headlights. (See **FIGURE 6**.)

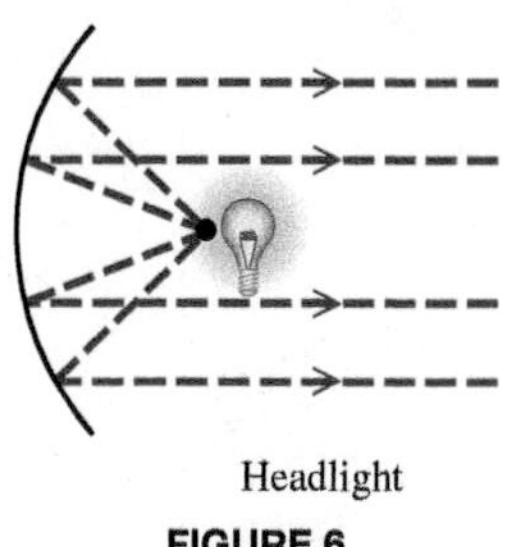

FIGURE 6

OBJECTIVE 2 Graph parabolas with horizontal and vertical shifts.

Parabolas need not have their vertices at the origin, as does the graph of $f(x) = x^2$.

NOW TRY EXERCISE 1

Graph $f(x) = x^2 - 3$. Give the vertex, axis, domain, and range.

EXAMPLE 1 Graphing a Parabola (Vertical Shift)

Graph $F(x) = x^2 - 2$.

If $x = 0$, then $F(x) = -2$, which gives the vertex $(0, -2)$. The graph of $F(x) = x^2 - 2$ has the same shape as that of $f(x) = x^2$ but is *shifted,* or *translated, down* 2 units. Every function value is 2 less than the corresponding function value of $f(x) = x^2$. Plotting points on both sides of the vertex gives the graph in **FIGURE 7**.

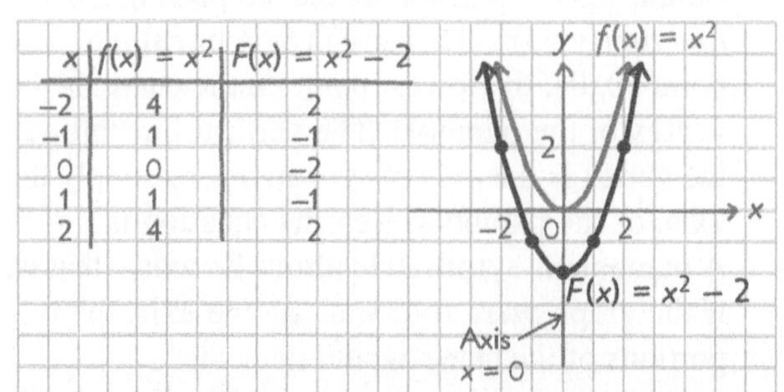

x	$f(x) = x^2$	$F(x) = x^2 - 2$
−2	4	2
−1	1	−1
0	0	−2
1	1	−1
2	4	2

$F(x) = x^2 - 2$
Vertex: $(0, -2)$
Axis of symmetry: $x = 0$
Domain: $(-\infty, \infty)$
Range: $[-2, \infty)$
The graph of $f(x) = x^2$ is shown for comparison.

FIGURE 7

This parabola is symmetric about its axis $x = 0$, so the plotted points are "mirror images" of each other. Because x can be any real number, the domain is $(-\infty, \infty)$. The value of y (or $F(x)$) is always greater than or equal to -2, so the range is $[-2, \infty)$.

NOW TRY

NOW TRY ANSWER

1\.

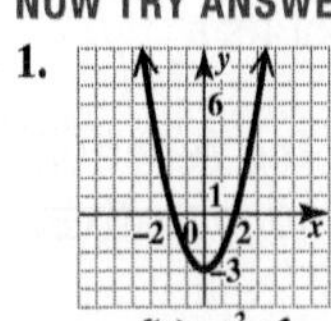

vertex: $(0, -3)$; axis: $x = 0$; domain: $(-\infty, \infty)$; range: $[-3, \infty)$

Parabola with a Vertical Shift

The graph of $F(x) = x^2 + k$ is a parabola.

- The graph has the same shape as the graph of $f(x) = x^2$.
- The parabola is shifted up k units if $k > 0$, and down $|k|$ units if $k < 0$.
- The vertex of the parabola is $(0, k)$.

NOW TRY EXERCISE 2

Graph $f(x) = (x + 1)^2$. Give the vertex, axis, domain, and range.

EXAMPLE 2 Graphing a Parabola (Horizontal Shift)

Graph $F(x) = (x - 2)^2$.

If $x = 2$, then $F(x) = 0$, giving the vertex $(2, 0)$. The graph of

$$F(x) = (x - 2)^2$$

has the same shape as that of $f(x) = x^2$ but is shifted *to the right* 2 units. We plot several points on one side of the vertex and then use symmetry about the axis $x = 2$ to find corresponding points on the other side of the vertex.

If $x = 0$, for example, then

$$F(0) = (0 - 2)^2 = 4,$$

and the point $(0, 4)$ lies on the graph. The corresponding point two units on the "other" side of the axis $x = 2$ is the point $(4, 4)$, which also lies on the graph. See **FIGURE 8**.

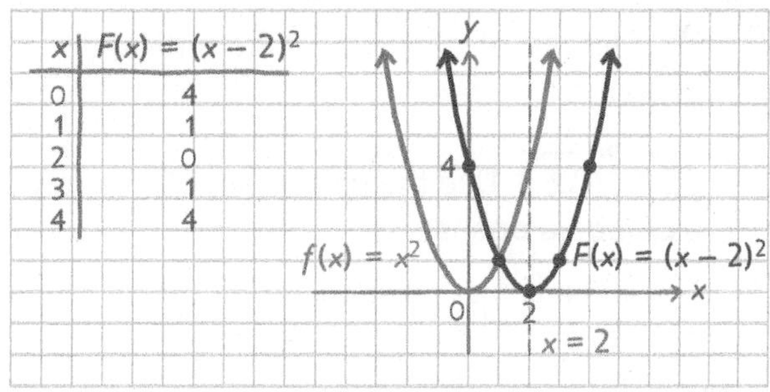

$F(x) = (x - 2)^2$
Vertex: $(2, 0)$
Axis of symmetry: $x = 2$
Domain: $(-\infty, \infty)$
Range: $[0, \infty)$

FIGURE 8

NOW TRY

Parabola with a Horizontal Shift

The graph of $F(x) = (x - h)^2$ is a parabola.

- The graph has the same shape as the graph of $f(x) = x^2$.
- The parabola is shifted to the right h units if $h > 0$, and to the left $|h|$ units if $h < 0$.
- The vertex of the parabola is $(h, 0)$.

CAUTION ***Errors frequently occur when horizontal shifts are involved.*** To determine the direction and magnitude of a horizontal shift, find the value that causes the expression $x - h$ to equal 0, as shown below.

$F(x) = (x - 5)^2$	$F(x) = (x + 5)^2$
Because **+5** causes $x - 5$ to equal 0, the graph of $F(x)$ illustrates a shift **to the right 5 units.**	Because **−5** causes $x + 5$ to equal 0, the graph of $F(x)$ illustrates a shift **to the left 5 units.**

NOW TRY ANSWER

2. 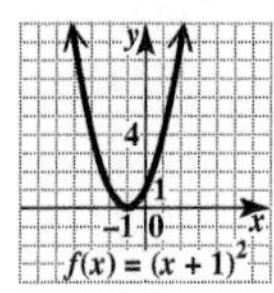

vertex: $(-1, 0)$; axis: $x = -1$; domain: $(-\infty, \infty)$; range: $[0, \infty)$

NOW TRY EXERCISE 3

Graph $f(x) = (x + 1)^2 - 2$. Give the vertex, axis, domain, and range.

EXAMPLE 3 Graphing a Parabola (Horizontal and Vertical Shifts)

Graph $F(x) = (x + 3)^2 - 2$.

This graph has the same shape as that of $f(x) = x^2$, but is shifted *to the left* 3 units (because $x + 3 = 0$ if $x = -3$) and *down* 2 units (because of the negative sign in -2). This gives the vertex $(-3, -2)$. Find and plot several ordered pairs, using symmetry as needed, to obtain the graph in **FIGURE 9**.

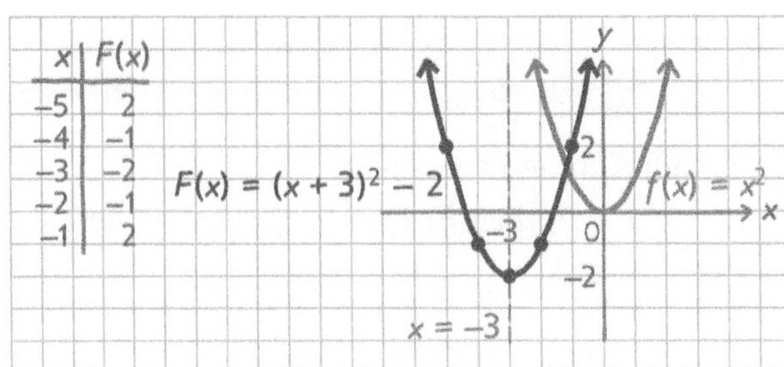

x	F(x)
−5	2
−4	−1
−3	−2
−2	−1
−1	2

$F(x) = (x + 3)^2 - 2$
Vertex: $(-3, -2)$
Axis of symmetry: $x = -3$
Domain: $(-\infty, \infty)$
Range: $[-2, \infty)$

FIGURE 9

NOW TRY 

Parabola with Horizontal and Vertical Shifts

The graph of $F(x) = (x - h)^2 + k$ is a parabola.

- The graph has the same shape as the graph of $f(x) = x^2$.
- The vertex of the parabola is (h, k).
- The axis of symmetry is the vertical line $x = h$.

OBJECTIVE 3 Use the coefficient of x^2 to predict the shape and direction in which a parabola opens.

Not all parabolas open up, and not all parabolas have the same shape as the graph of $f(x) = x^2$.

NOW TRY EXERCISE 4

Graph $f(x) = -3x^2$. Give the vertex, axis, domain, and range.

EXAMPLE 4 Graphing a Parabola That Opens Down

Graph $f(x) = -\frac{1}{2}x^2$.

This parabola is shown in **FIGURE 10**. The coefficient of x^2, $-\frac{1}{2}$, affects the shape of the graph—the $\frac{1}{2}$ makes the parabola wider $\left(\text{because the values of } \frac{1}{2}x^2 \text{ increase more slowly than those of } x^2\right)$, and the negative sign makes the parabola open down.

The graph is not shifted in any direction, so the vertex is $(0, 0)$. Unlike the parabolas graphed in **Examples 1–3,** the vertex $(0, 0)$ has the *greatest* function value of any point on the graph.

$f(x) = -\frac{1}{2}x^2$

x	$f(x)$
−2	−2
−1	$-\frac{1}{2}$
0	0
1	$-\frac{1}{2}$
2	−2

$f(x) = -\frac{1}{2}x^2$
Vertex: $(0, 0)$
Axis of symmetry: $x = 0$
Domain: $(-\infty, \infty)$
Range: $(-\infty, 0]$

FIGURE 10

NOW TRY

NOW TRY ANSWERS

3.

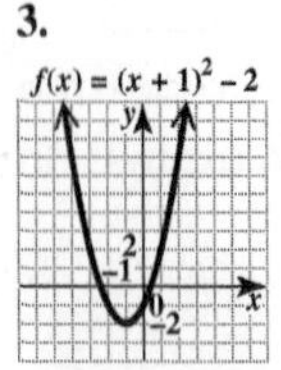

vertex: $(-1, -2)$; axis: $x = -1$; domain: $(-\infty, \infty)$; range: $[-2, \infty)$

4. $f(x) = -3x^2$

vertex: $(0, 0)$; axis: $x = 0$; domain: $(-\infty, \infty)$; range: $(-\infty, 0]$

General Characteristics of the Graph of a Vertical Parabola

The graph of the quadratic function $F(x) = a(x - h)^2 + k$ (where $a \neq 0$) is a parabola.

- The vertex of the parabola is (h, k).
- The axis of symmetry is the vertical line $x = h$.
- The graph opens up if $a > 0$ and down if $a < 0$.
- The graph is wider than that of $f(x) = x^2$ if $0 < |a| < 1$. The graph is narrower than that of $f(x) = x^2$ if $|a| > 1$.

NOW TRY EXERCISE 5

Graph $f(x) = 2(x - 1)^2 + 2$. Give the vertex, axis, domain, and range.

EXAMPLE 5 Using the General Characteristics to Graph a Parabola

Graph $F(x) = -2(x + 3)^2 + 4$.

The parabola opens down (because $a = -2$ and $-2 < 0$) and is narrower than the graph of $f(x) = x^2$ (because $|-2| = 2$ and $2 > 1$). This causes values of $F(x)$ to decrease more quickly than those of $f(x) = -x^2$.

The graph is shifted *to the left* 3 units (because $x + 3 = 0$ when $x = -3$) and *up* 4 units (because of the $+4$), which gives the vertex $(-3, 4)$. To complete the graph, we plotted the ordered pairs $(-4, 2)$ and, by symmetry, $(-2, 2)$. Symmetry can be used to find additional ordered pairs that satisfy the equation. See **FIGURE 11**.

$F(x) = -2(x + 3)^2 + 4$

x	$F(x)$
-5	-4
-4	2
-3	4
-2	2
-1	-4

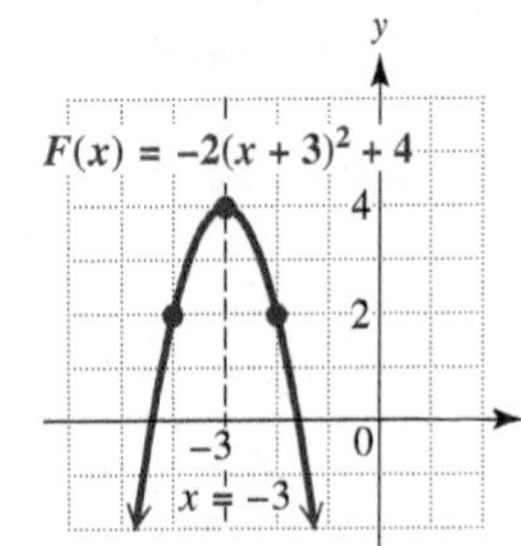

$F(x) = -2(x + 3)^2 + 4$
Vertex: $(-3, 4)$
Axis of symmetry: $x = -3$
Domain: $(-\infty, \infty)$
Range: $(-\infty, 4]$

FIGURE 11

NOW TRY

OBJECTIVE 4 Find a quadratic function to model data.

EXAMPLE 6 Modeling the Number of Multiple Births

The number of higher-order multiple births (triplets or more) in the United States is shown in the table. Here, x represents the number of years since 1996, and y represents the number of higher-order multiple births (to the nearest hundred).

Year	x	y
1996	0	5900
2000	4	7300
2002	6	7400
2004	8	7300
2006	10	6500
2008	12	6300
2010	14	5500
2012	16	4900
2014	18	4500

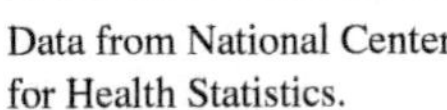

Data from National Center for Health Statistics.

NOW TRY ANSWER

5.

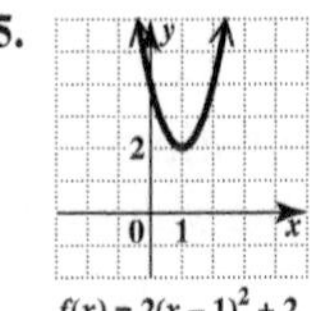

vertex: $(1, 2)$; axis: $x = 1$; domain: $(-\infty, \infty)$; range: $[2, \infty)$

NOW TRY EXERCISE 6

Using the points $(0, 5900)$, $(4, 7300)$, and $(12, 6300)$, find another quadratic model for the data on higher-order multiple births in **Example 6.** (Round values of a and b to the nearest tenth.)

Find a quadratic function that models the data in the table on the previous page.

A scatter diagram of the ordered pairs (x, y) is shown in **FIGURE 12**. The general shape suggested by the scatter diagram indicates that a parabola should approximate these points, as shown by the dashed curve in **FIGURE 13**. The equation for such a parabola would have a negative coefficient for x^2 because the graph opens down.

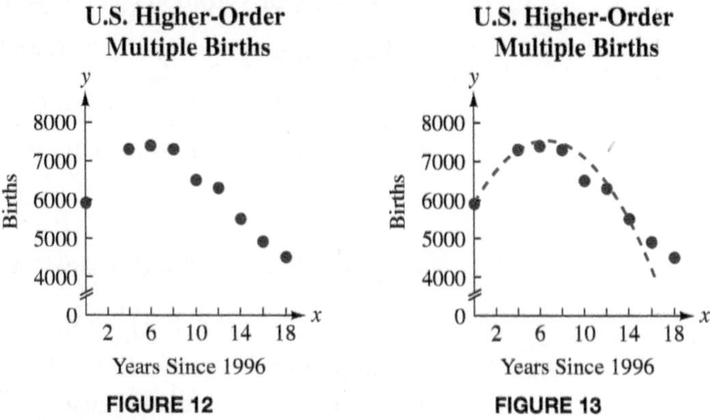

FIGURE 12 **FIGURE 13**

To find a quadratic function of the form

$$y = ax^2 + bx + c$$

that models, or *fits,* these data, we choose three representative ordered pairs from the table and use them to write a system of three equations.

$$(0, 5900), \quad (6, 7400), \quad \text{and} \quad (12, 6300) \qquad \text{Three ordered pairs } (x, y)$$

We substitute the x- and y-values from each ordered pair into the quadratic form $y = ax^2 + bx + c$ to obtain three equations.

$$a(0)^2 + b(0) + c = 5900 \xrightarrow{\text{Simplify.}} c = 5900 \quad (1)$$
$$a(6)^2 + b(6) + c = 7400 \longrightarrow 36a + 6b + c = 7400 \quad (2)$$
$$a(12)^2 + b(12) + c = 6300 \longrightarrow 144a + 12b + c = 6300 \quad (3)$$

To find the values of a, b, and c, we solve this system of three equations in three variables. From equation (1), $c = 5900$, so we substitute 5900 for c in equations (2) and (3) to obtain two equations in two variables.

$$36a + 6b + 5900 = 7400 \longrightarrow 36a + 6b = 1500 \quad (4)$$
$$144a + 12b + 5900 = 6300 \xrightarrow{\text{Subtract 5900.}} 144a + 12b = 400 \quad (5)$$

We eliminate b from this system of equations in two variables by multiplying equation (4) by -2 and adding the results to equation (5).

$$-72a - 12b = -3000 \quad \text{Multiply equation (4) by } -2.$$
$$144a + 12b = 400 \quad (5)$$
$$72a = -2600 \quad \text{Add.}$$
$$a \approx -36.1 \quad \text{Use a calculator. Round to one decimal place.}$$

We substitute -36.1 for a in equation (4) to find that $b \approx 466.6$. (Substituting in equation (5) will give $b \approx 466.5$ due to rounding procedures.) Using the values we found for a, b, and c, the model is

$$y = \underset{a}{-36.1}x^2 + \underset{b}{466.6}x + \underset{c}{5900}.$$

NOW TRY 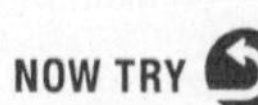

NOW TRY ANSWER

6. $y = -39.6x^2 + 508.4x + 5900$ (Answers may vary slightly due to rounding.)

NOTE If we had chosen three different ordered pairs of data in **Example 6,** a slightly different, though similar, model would have resulted. (See **Now Try Exercise 6.**)

The *quadratic regression* feature on a graphing calculator can also be used to generate the quadratic model that best fits given data. See your owner's manual for details.

9.2 Exercises

FOR EXTRA HELP  **MyLab Math**

Video solutions for select problems available in MyLab Math

Concept Check *Match each quadratic function in parts (a)–(d) with its graph from choices A–D.*

1. (a) $f(x) = (x+2)^2 - 1$

(b) $f(x) = (x+2)^2 + 1$

(c) $f(x) = (x-2)^2 - 1$

(d) $f(x) = (x-2)^2 + 1$

A.

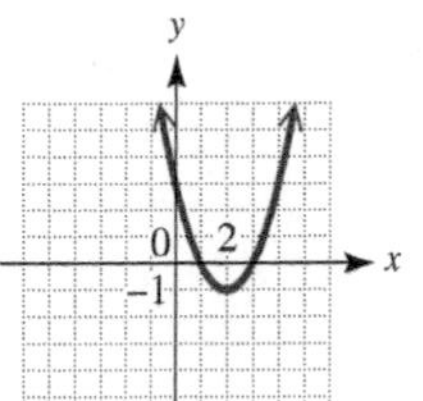

B.

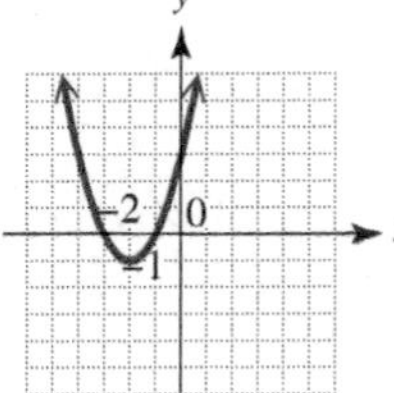

C.

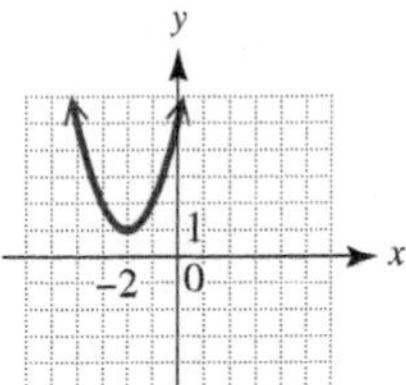

D.

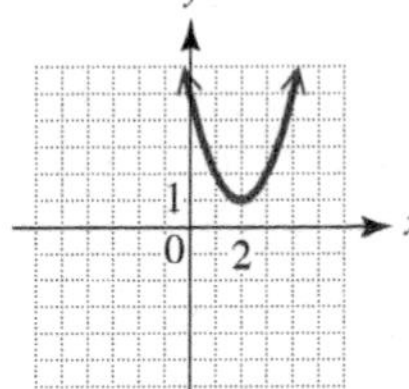

2. (a) $f(x) = -x^2 + 2$

(b) $f(x) = -x^2 - 2$

(c) $f(x) = -(x+2)^2$

(d) $f(x) = -(x-2)^2$

A.

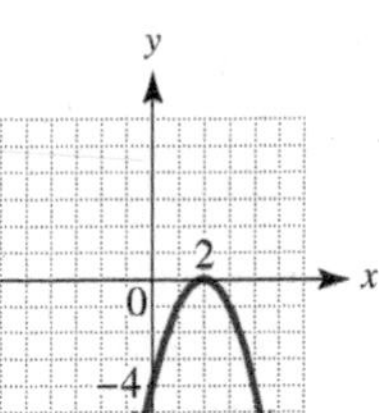

B.

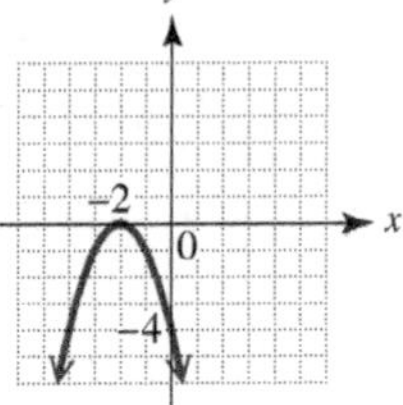

C.

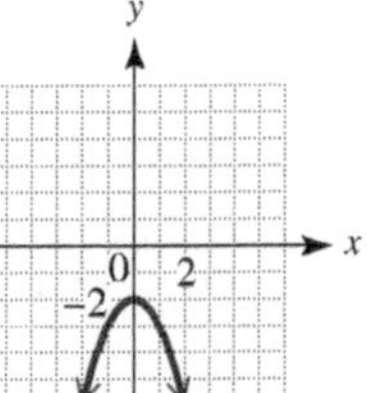

D.

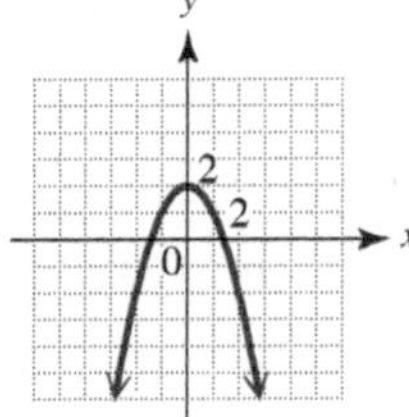

3. CONCEPT CHECK Match each quadratic function in Column I with the description of the parabola that is its graph in Column II.

I	II
(a) $f(x) = (x-4)^2 - 2$	**A.** Vertex $(2, -4)$, opens down
(b) $f(x) = (x-2)^2 - 4$	**B.** Vertex $(2, -4)$, opens up
(c) $f(x) = (x+4)^2 + 2$	**C.** Vertex $(4, -2)$, opens down
(d) $f(x) = -(x-4)^2 - 2$	**D.** Vertex $(4, -2)$, opens up
(e) $f(x) = -(x-2)^2 - 4$	**E.** Vertex $(4, 2)$, opens down
(f) $f(x) = -(x-4)^2 + 2$	**F.** Vertex $(-4, 2)$, opens up

4. Concept Check For the quadratic function $f(x) = a(x - h)^2 + k$, in what quadrant is the vertex if the values of h and k are as follows?

(a) $h > 0, k > 0$ **(b)** $h > 0, k < 0$

(c) $h < 0, k > 0$ **(d)** $h < 0, k < 0$

Consider the value of a, and make the correct choice.

(e) If $a > 0$, then the graph opens (*up / down*).

(f) If $a < 0$, then the graph opens (*up / down*).

(g) If $|a| > 1$, then the graph is (*narrower / wider*) than the graph of $f(x) = x^2$.

(h) If $0 < |a| < 1$, then the graph is (*narrower / wider*) than the graph of $f(x) = x^2$.

Identify the vertex of each parabola. **See Examples 1–4.**

5. $f(x) = -3x^2$ **6.** $f(x) = -4x^2$ **7.** $f(x) = \frac{1}{3}x^2$

8. $f(x) = \frac{1}{2}x^2$ **9.** $f(x) = x^2 + 4$ **10.** $f(x) = x^2 - 4$

11. $f(x) = (x - 1)^2$ **12.** $f(x) = (x + 3)^2$ **13.** $f(x) = (x + 3)^2 - 4$

14. $f(x) = (x + 5)^2 - 8$ **15.** $f(x) = -(x - 5)^2 + 6$ **16.** $f(x) = -(x - 2)^2 + 1$

For each quadratic function, tell whether the graph opens up *or* down *and whether the graph is* wider, narrower, *or the* same shape *as the graph of* $f(x) = x^2$. **See Examples 4 and 5.**

17. $f(x) = -\frac{2}{5}x^2$ **18.** $f(x) = -2x^2$

19. $f(x) = 3x^2 + 1$ **20.** $f(x) = \frac{2}{3}x^2 - 4$

21. $f(x) = -4(x + 2)^2 + 5$ **22.** $f(x) = -\frac{1}{3}(x + 6)^2 + 3$

Graph each parabola. Give the vertex, axis of symmetry, domain, and range. **See Examples 1–5.**

23. $f(x) = 3x^2$ **24.** $f(x) = \frac{1}{2}x^2$

25. $f(x) = -2x^2$ **26.** $f(x) = -\frac{1}{3}x^2$

27. $f(x) = x^2 - 1$ **28.** $f(x) = x^2 + 3$

29. $f(x) = -x^2 + 2$ **30.** $f(x) = -x^2 - 2$

31. $f(x) = (x - 4)^2$ **32.** $f(x) = (x + 1)^2$

33. $f(x) = (x + 2)^2 - 1$ **34.** $f(x) = (x - 1)^2 + 2$

35. $f(x) = (x - 1)^2 - 3$ **36.** $f(x) = (x + 1)^2 + 1$

37. $f(x) = 2(x - 2)^2 - 4$ **38.** $f(x) = 3(x - 2)^2 + 1$

39. $f(x) = \frac{1}{2}(x - 2)^2 - 3$ **40.** $f(x) = \frac{4}{3}(x - 3)^2 - 2$

41. $f(x) = -2(x + 3)^2 + 4$ **42.** $f(x) = -2(x - 2)^2 - 3$

43. $f(x) = -\frac{1}{2}(x + 1)^2 + 2$ **44.** $f(x) = -\frac{2}{3}(x + 2)^2 + 1$

Determine whether a linear function *or a* quadratic function *would be a more appropriate model for each set of graphed data. If linear, tell whether the slope should be* positive *or* negative. *If quadratic, tell whether the coefficient of* x^2 *should be* positive *or* negative. ***See Example 6.***

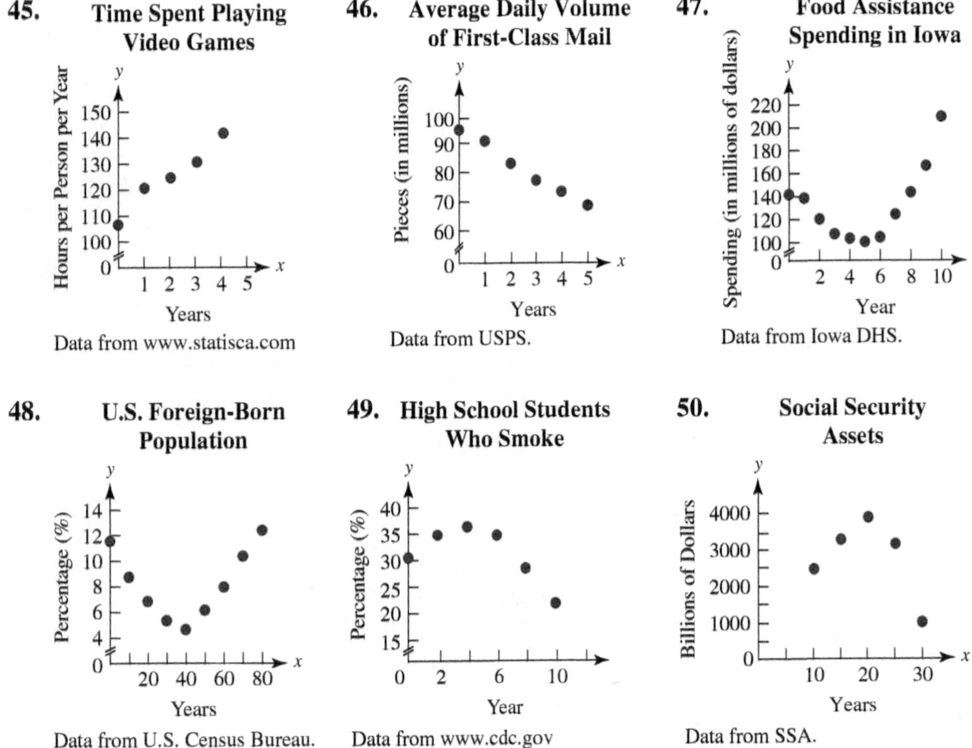

Solve each problem. ***See Example 6.***

51. The number of U.S. households (in millions) with cable television service is shown in the table, where x represents the number of years since 2007 and y represents households.

Year	x	y
2007	0	95
2009	2	103
2011	4	105
2013	6	103
2015	8	100
2017	10	96

Data from Nielsen Media Research.

(a) Use the ordered pairs (x, y) to make a scatter diagram of the data.

(b) Would a linear or a quadratic function better model the data?

(c) Should the coefficient a of x^2 in a quadratic model $y = ax^2 + bx + c$ be positive or negative?

(d) Use the ordered pairs $(0, 95)$, $(4, 105)$, and $(10, 96)$ to find a quadratic function that models the data.

(e) Use the model from part (d) to approximate the number of U.S. households (in millions) with cable television service in 2009 and 2015. How well does the model approximate the actual data from the table?

52. The number (in thousands) of new, privately owned housing units completed in the United States is shown in the table. Here x represents the number of years since 2006, and y represents total housing units completed.

Year	x	y
2006	0	1980
2008	2	1120
2010	4	650
2012	6	650
2014	8	880
2016	10	1060

Data from U.S. Census Bureau.

(a) Use the ordered pairs (x, y) to make a scatter diagram of the data.

(b) Would a linear or a quadratic function better model the data?

(c) Should the coefficient a of x^2 in a quadratic model $y = ax^2 + bx + c$ be positive or negative?

(d) Use the ordered pairs (0, 1980), (4, 650), and (8, 880) to find a quadratic function that models the data.

(e) Use the model from part (d) to approximate the number of such housing units (in thousands) completed in 2008 and 2012. How well does the model approximate the actual data from the table?

9.3 More about Parabolas and Their Applications

OBJECTIVES

1 Find the vertex of a vertical parabola.
2 Graph a quadratic function.
3 Use the discriminant to find the number of x-intercepts of a parabola with a vertical axis.
4 Use quadratic functions to solve problems involving maximum or minimum value.
5 Graph parabolas with horizontal axes.

OBJECTIVE 1 Find the vertex of a vertical parabola.

When the equation of a parabola is given in the form $f(x) = ax^2 + bx + c$, there are two ways to locate the vertex.

1. Complete the square. (See **Examples 1 and 2.**)
2. Use a formula derived by completing the square. (See **Example 3.**)

EXAMPLE 1 Completing the Square to Find the Vertex ($a = 1$)

Find the vertex of the graph of $f(x) = x^2 - 4x + 5$.

We can express $x^2 - 4x + 5$ in the form $(x - h)^2 + k$ by completing the square, which was introduced in an earlier chapter. The process is slightly different here because we want to keep $f(x)$ alone on one side of the equation. Instead of adding the appropriate number to each side, we *add and subtract* it on the right.

$f(x) = x^2 - 4x + 5$

$f(x) = (x^2 - 4x \qquad) + 5$ — Group the variable terms.

$\left[\frac{1}{2}(-4)\right]^2 = (-2)^2 = 4$ — Square half the coefficient of the first-degree term.

This is equivalent to adding 0.

$f(x) = (x^2 - 4x + 4 - 4) + 5$ — Add and subtract 4.

$f(x) = (x^2 - 4x + 4) - 4 + 5$ — Bring −4 outside the parentheses.

$f(x) = (x - 2)^2 + 1$ — Factor. Combine like terms.

The vertex of this parabola is $(2, 1)$.

NOW TRY 

NOW TRY EXERCISE 1

Find the vertex of the graph of

$f(x) = x^2 + 2x - 8.$

NOW TRY ANSWER

1. $(-1, -9)$

NOW TRY EXERCISE 2

Find the vertex of the graph of

$f(x) = -4x^2 + 16x - 10.$

EXAMPLE 2 **Completing the Square to Find the Vertex ($a \neq 1$)**

Find the vertex of the graph of $f(x) = -3x^2 + 6x - 1$.

Because the x^2-term has a coefficient other than 1, we factor that coefficient out of the first two terms before completing the square.

$$f(x) = -3x^2 + 6x - 1$$

$$f(x) = (-3x^2 + 6x) - 1 \quad \text{Group the variable terms.}$$

$$f(x) = -3(x^2 - 2x) - 1 \quad \text{Factor out } -3.$$

$$f(x) = -3(x^2 - 2x \qquad) - 1 \quad \text{Prepare to complete the square.}$$

$$\left[\tfrac{1}{2}(-2)\right]^2 = (-1)^2 = 1 \quad \text{Square half the coefficient of the first-degree term.}$$

$$f(x) = -3(x^2 - 2x + 1 - 1) - 1 \quad \text{Add and subtract 1.}$$

Now bring −1 outside the parentheses. Be sure to multiply it by −3.

This is a key step.

$$f(x) = -3(x^2 - 2x + 1) + (-3)(-1) - 1 \quad \text{Distributive property}$$

$$f(x) = -3(x^2 - 2x + 1) + 3 - 1 \quad \text{Multiply.}$$

$$f(x) = -3(x - 1)^2 + 2 \quad \text{Factor. Combine like terms.}$$

The vertex is $(1, 2)$. **NOW TRY**

We can complete the square to derive a formula for the vertex of the graph of the quadratic function $f(x) = ax^2 + bx + c$ (where $a \neq 0$).

$$f(x) = ax^2 + bx + c \quad \text{Standard form}$$

$$f(x) = (ax^2 + bx) + c \quad \text{Group the terms with } x.$$

$$f(x) = a\left(x^2 + \frac{b}{a}x \qquad \right) + c \quad \text{Factor } a \text{ from the first two terms.}$$

$$\left[\frac{1}{2}\left(\frac{b}{a}\right)\right]^2 = \left(\frac{b}{2a}\right)^2 = \frac{b^2}{4a^2} \quad \text{Square half the coefficient of the first-degree term.}$$

$$f(x) = a\left(x^2 + \frac{b}{a}x + \frac{b^2}{4a^2} - \frac{b^2}{4a^2}\right) + c \quad \text{Add and subtract } \frac{b^2}{4a^2}.$$

$$f(x) = a\left(x^2 + \frac{b}{a}x + \frac{b^2}{4a^2}\right) + a\left(-\frac{b^2}{4a^2}\right) + c \quad \text{Distributive property}$$

$$f(x) = a\left(x^2 + \frac{b}{a}x + \frac{b^2}{4a^2}\right) - \frac{b^2}{4a} + c \quad -\frac{ab^2}{4a^2} = -\frac{b^2}{4a}$$

$$f(x) = a\left(x + \frac{b}{2a}\right)^2 + \frac{4ac - b^2}{4a} \quad \text{Factor. Rewrite terms with a common denominator.}$$

$$f(x) = a\Bigg[x - \underbrace{\left(\frac{-b}{2a}\right)}_{h}\Bigg]^2 + \underbrace{\frac{4ac - b^2}{4a}}_{k} \quad f(x) = a(x-h)^2 + k; \text{ The vertex } (h, k) \text{ can be expressed in terms of } a, b, \text{ and } c.$$

The expression for k can be found by replacing x with $\frac{-b}{2a}$. Using function notation, if $y = f(x)$, then the y-value of the vertex is $f\left(\frac{-b}{2a}\right)$.

NOW TRY ANSWER

2. $(2, 6)$

Vertex Formula

The graph of the quadratic function $f(x) = ax^2 + bx + c$ (where $a \neq 0$) has vertex

$$\left(\frac{-b}{2a}, f\left(\frac{-b}{2a}\right)\right).$$

The axis of symmetry of the parabola is the line having equation

$$x = \frac{-b}{2a}.$$

NOW TRY EXERCISE 3
Use the vertex formula to find the vertex of the graph of

$$f(x) = 3x^2 - 2x + 8.$$

EXAMPLE 3 Using the Formula to Find the Vertex

Use the vertex formula to find the vertex of the graph of $f(x) = x^2 - x - 6$.

The x-coordinate of the vertex of the parabola is given by $\frac{-b}{2a}$.

$$\frac{-b}{2a} = \frac{-(-1)}{2(1)} = \frac{1}{2} \leftarrow \text{x-coordinate of vertex}$$

$a = 1$, $b = -1$, and $c = -6$

The y-coordinate of the vertex of $f(x) = x^2 - x - 6$ is $f\left(\frac{-b}{2a}\right) = f\left(\frac{1}{2}\right)$.

$$f\left(\frac{1}{2}\right) = \left(\frac{1}{2}\right)^2 - \frac{1}{2} - 6$$

$$= \frac{1}{4} - \frac{1}{2} - 6$$

$$= -\frac{25}{4} \leftarrow \text{y-coordinate of vertex}$$

$\frac{1}{4} - \frac{1}{2} - 6 = \frac{1}{4} - \frac{2}{4} - \frac{24}{4}$

The vertex is $\left(\frac{1}{2}, -\frac{25}{4}\right)$.

NOW TRY

OBJECTIVE 2 Graph a quadratic function.

Graphing a Quadratic Function $y = f(x)$

Step 1 **Determine whether the graph opens up or down.**

- If $a > 0$, then the parabola opens up.
- If $a < 0$, then it opens down.

Step 2 **Find the vertex.** Use the vertex formula or complete the square.

Step 3 **Find any intercepts.**

- To find the x-intercepts (if any), solve $f(x) = 0$.
- To find the y-intercept, evaluate $f(0)$.

Step 4 **Complete the graph.** Plot the points found so far. Find and plot additional points as needed, using symmetry about the axis.

NOW TRY ANSWER

3. $\left(\frac{1}{3}, \frac{23}{3}\right)$

Graph the quadratic function

$$f(x) = x^2 + 2x - 3.$$

Give the vertex, axis, domain, and range.

EXAMPLE 4 Graphing a Quadratic Function

Graph the quadratic function $f(x) = x^2 - x - 6$.

Step 1 From the equation, $a = 1$, so the graph of the function opens up.

Step 2 The vertex, $\left(\frac{1}{2}, -\frac{25}{4}\right)$, was found in **Example 3** using the vertex formula.

Step 3 Find any intercepts. The vertex, $\left(\frac{1}{2}, -\frac{25}{4}\right)$, is in quadrant IV and the graph opens up, so there will be two x-intercepts. Let $f(x) = 0$ and solve to find them.

$$f(x) = x^2 - x - 6$$

$$0 = x^2 - x - 6 \qquad \text{Let } f(x) = 0.$$

$$0 = (x - 3)(x + 2) \qquad \text{Factor.}$$

$$x - 3 = 0 \quad \text{or} \quad x + 2 = 0 \qquad \text{Zero-factor property}$$

$$x = 3 \quad \text{or} \quad x = -2 \qquad \text{Solve each equation.}$$

The x-intercepts are $(3, 0)$ and $(-2, 0)$.

Find the y-intercept by evaluating $f(0)$.

$$f(x) = x^2 - x - 6$$

$$f(0) = 0^2 - 0 - 6 \qquad \text{Let } x = 0.$$

$$f(0) = -6 \qquad \text{Apply the exponent. Subtract.}$$

The y-intercept is $(0, -6)$.

Step 4 Plot the points found so far and additional points as needed using symmetry about the axis, $x = \frac{1}{2}$. The graph is shown in **FIGURE 14**.

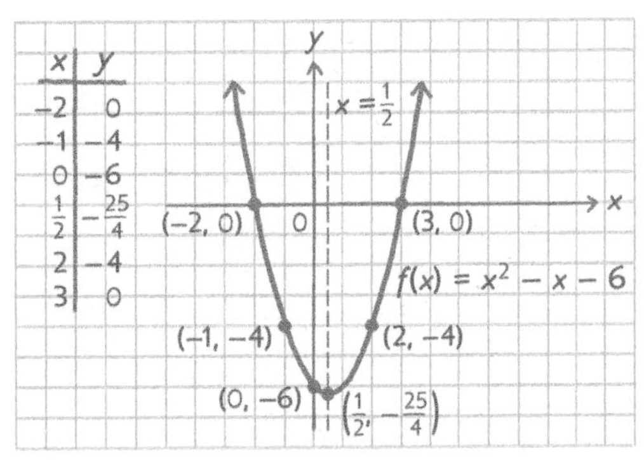

$f(x) = x^2 - x - 6$

Vertex: $\left(\frac{1}{2}, -\frac{25}{4}\right)$

Axis of symmetry: $x = \frac{1}{2}$

Domain: $(-\infty, \infty)$

Range: $\left[-\frac{25}{4}, \infty\right)$

FIGURE 14

NOW TRY 

NOW TRY ANSWER

4. 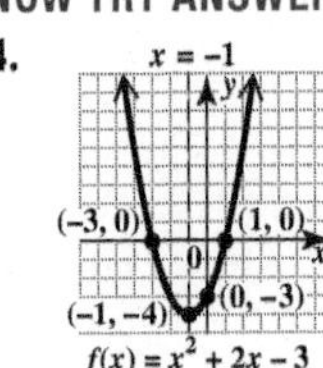

vertex: $(-1, -4)$; axis: $x = -1$; domain: $(-\infty, \infty)$; range: $[-4, \infty)$

OBJECTIVE 3 Use the discriminant to find the number of x-intercepts of a parabola with a vertical axis.

Recall that the expression under the radical in $x = \frac{-b \pm \sqrt{b^2 - 4ac}}{2a}$, that is,

$$b^2 - 4ac, \qquad \text{Discriminant}$$

is the *discriminant* of the quadratic equation $ax^2 + bx + c = 0$ and that we can use it to determine the number of real solutions of a quadratic *equation.*

In a similar way, we can use the discriminant of a quadratic *function* to determine the number of x-intercepts of its graph. The three possibilities are shown in **FIGURE 15.**

1. If the discriminant is positive, the parabola will have two x-intercepts.
2. If the discriminant is 0, there will be only one x-intercept, and it will be the vertex of the parabola.
3. If the discriminant is negative, the graph will have no x-intercepts.

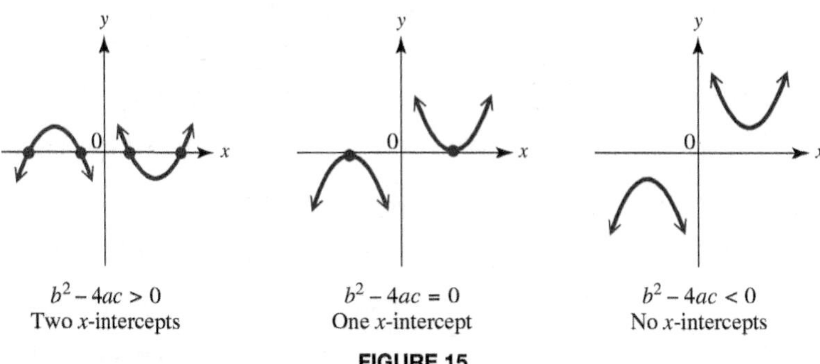

FIGURE 15

NOW TRY EXERCISE 5

Find the discriminant and use it to determine the number of x-intercepts of the graph of each quadratic function.

(a) $f(x) = -2x^2 + 3x - 2$

(b) $f(x) = 3x^2 + 2x - 1$

(c) $f(x) = 4x^2 - 12x + 9$

EXAMPLE 5 Using the Discriminant to Determine Number of x-Intercepts

Find the discriminant and use it to determine the number of x-intercepts of the graph of each quadratic function.

(a) $f(x) = 2x^2 + 3x - 5$

$b^2 - 4ac$	Discriminant
$= 3^2 - 4(2)(-5)$	$a = 2$, $b = 3$, $c = -5$
$= 9 - (-40)$	Apply the exponent. Multiply.
$= 49$	Subtract.

Because the discriminant is positive, the parabola has two x-intercepts.

(b) $f(x) = -3x^2 - 1$ (which can be written as $f(x) = -3x^2 + 0x - 1$)

$b^2 - 4ac$	
$= 0^2 - 4(-3)(-1)$	$a = -3$, $b = 0$, $c = -1$
$= 0 - 12$	Apply the exponent. Multiply.
$= -12$	Subtract.

The discriminant is negative, so the graph has no x-intercepts.

(c) $f(x) = 9x^2 + 6x + 1$

$b^2 - 4ac$	
$= 6^2 - 4(9)(1)$	$a = 9$, $b = 6$, $c = 1$
$= 36 - 36$	Apply the exponent. Multiply.
$= 0$	Subtract.

Because the discriminant is 0, the parabola has only one x-intercept (its vertex).

NOW TRY

NOW TRY ANSWERS

5. (a) -7; none **(b)** 16; two **(c)** 0; one

OBJECTIVE 4 Use quadratic functions to solve problems involving maximum or minimum value.

The vertex of the graph of a quadratic function is either the highest or the lowest point on the parabola. It provides the following information.

1. The y-value of the vertex gives the maximum or minimum value of y.
2. The x-value tells where the maximum or minimum occurs.

NOW TRY EXERCISE 6

Solve the problem in **Example 6** if the farmer has only 80 ft of fencing.

EXAMPLE 6 Finding Maximum Area

A farmer has 120 ft of fencing to enclose a rectangular area next to a building. (See **FIGURE 16**.) Find the maximum area he can enclose and the dimensions of the field when the area is maximized.

FIGURE 16

Let x = the width of the rectangle.

$$x + x + \text{length} = 120 \quad \text{Sum of the sides is 120 ft.}$$

$$2x + \text{length} = 120 \quad \text{Combine like terms.}$$

$$\text{length} = 120 - 2x \quad \text{Subtract } 2x.$$

The area $\mathcal{A}(x)$ is given by the product of the length and width.

$$\mathcal{A}(x) = (120 - 2x)x \quad \text{Area = length} \cdot \text{width}$$

$$\mathcal{A}(x) = 120x - 2x^2 \quad \text{Distributive property}$$

$$\mathcal{A}(x) = -2x^2 + 120x \quad \text{Standard form}$$

The graph of $\mathcal{A}(x) = -2x^2 + 120x$ is a parabola that opens down. To determine the maximum area, use the vertex formula to find the vertex of the parabola.

$$x = \frac{-b}{2a} = \frac{-120}{2(-2)} = \frac{-120}{-4} = 30 \quad \text{Vertex formula; } a = -2,\ b = 120,\ c = 0$$

$$\mathcal{A}(30) = -2(30)^2 + 120(30) = -2(900) + 3600 = 1800$$

The vertex is $(30, 1800)$. The maximum area will be 1800 ft^2 when x, the width, is 30 ft and the length is $120 - 2(30) = 60$ ft.

NOW TRY

EXAMPLE 7 Finding Maximum Height

If air resistance is neglected, a projectile on Earth shot straight upward with an initial velocity of 40 m per sec will be at a height s in meters given by

$$s(t) = -4.9t^2 + 40t,$$

where t is the number of seconds elapsed after projection. After how many seconds will it reach its maximum height, and what is this maximum height?

For this function, $a = -4.9$, $b = 40$, and $c = 0$. Use the vertex formula.

$$t = \frac{-b}{2a} = \frac{-40}{2(-4.9)} \approx 4.1 \quad \text{Use a calculator. Round to the nearest tenth.}$$

This indicates that the maximum height is attained at 4.1 sec.

NOW TRY ANSWER

6. The field should be 20 ft by 40 ft with maximum area 800 ft^2.

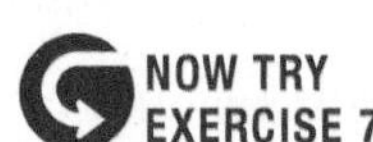

NOW TRY EXERCISE 7

A stomp rocket is launched from the ground with an initial velocity of 48 ft per sec so that its distance in feet above the ground after t seconds is

$$s(t) = -16t^2 + 48t.$$

Find the maximum height attained by the rocket and the number of seconds it takes to reach that height.

To find the maximum height at 4.1 sec, calculate $s(4.1)$.

$$s(t) = -4.9t^2 + 40t$$

$$s(4.1) = -4.9(4.1)^2 + 40(4.1) \quad \text{Let } t = 4.1.$$

$$s(4.1) \approx 81.6 \quad \text{Use a calculator. Round to the nearest tenth.}$$

The projectile will attain a maximum height of 81.6 m at 4.1 sec. **NOW TRY** 

CAUTION ***Be careful when interpreting the meanings of the coordinates of the vertex.*** The first coordinate, x, gives the value for which the *function value,* y or $f(x)$, is a maximum or a minimum.

OBJECTIVE 5 Graph parabolas with horizontal axes.

If x and y are interchanged in the equation

$$y = ax^2 + bx + c, \quad \text{the equation becomes} \quad x = ay^2 + by + c.$$

Because of the interchange of the roles of x and y, these parabolas are horizontal (with horizontal lines as axes of symmetry).

General Characteristics of the Graph of a Horizontal Parabola

The graph of an equation of the form

$$x = ay^2 + by + c \quad \text{or} \quad x = a(y - k)^2 + h$$

is a horizontal parabola.

- The vertex of the parabola is (h, k).
- The axis of symmetry is the horizontal line $y = k$.
- The graph opens to the right if $a > 0$ and to the left if $a < 0$.

NOW TRY EXERCISE 8

Graph $x = (y + 2)^2 - 1$. Give the vertex, axis, domain, and range.

EXAMPLE 8 Graphing a Horizontal Parabola ($a = 1$)

Graph $x = (y - 2)^2 - 3$. Give the vertex, axis, domain, and range.

This graph has its vertex at $(-3, 2)$ because the roles of x and y are interchanged. It opens to the right (the positive x-direction) because $a = 1$ and $1 > 0$, and has the same shape as $y = x^2$ (but situated horizontally).

To find additional points to plot, it is easiest to substitute a value for y and find the corresponding value for x. For example, let $y = 3$. Then

$$x = (3 - 2)^2 - 3 = -2, \quad \text{giving the point} \quad (-2, 3).$$

Write the x-value first.

Using symmetry, we can locate the point $(-2, 1)$. See **FIGURE 17**.

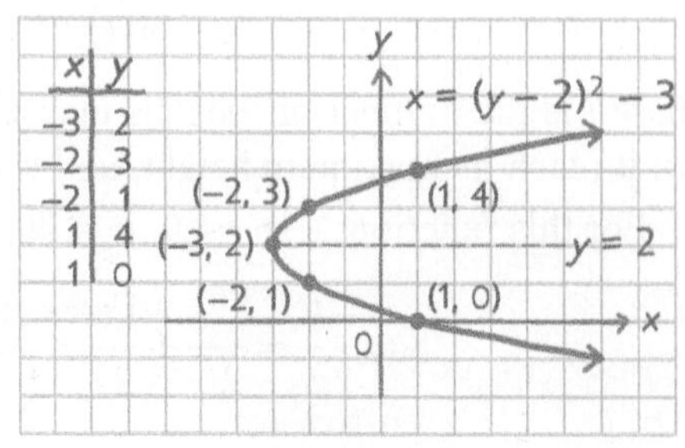

$x = (y - 2)^2 - 3$
Vertex: $(-3, 2)$
Axis of symmetry: $y = 2$
Domain: $[-3, \infty)$
Range: $(-\infty, \infty)$

FIGURE 17

NOW TRY

NOW TRY ANSWERS

7. 36 ft; 1.5 sec

8. $x = (y + 2)^2 - 1$; $(-1, -2)$

vertex: $(-1, -2)$; axis: $y = -2$; domain: $[-1, \infty)$; range: $(-\infty, \infty)$

Graph $x = -3y^2 - 6y - 5$. Give the vertex, axis, domain, and range.

EXAMPLE 9 Graphing a Horizontal Parabola ($a \neq 1$)

Graph $x = -2y^2 + 4y - 3$. Give the vertex, axis, domain, and range.

$$x = -2y^2 + 4y - 3$$

$$x = (-2y^2 + 4y) - 3 \qquad \text{Group the variable terms.}$$

$$x = -2(y^2 - 2y) - 3 \qquad \text{Factor out } -2.$$

$$x = -2(y^2 - 2y \qquad) - 3 \qquad \text{Prepare to complete the square.}$$

$$\left[\tfrac{1}{2}(-2)\right]^2 = (-1)^2 = 1 \qquad \text{Square half the coefficient of the first-degree term.}$$

$$x = -2(y^2 - 2y + 1 - 1) - 3 \qquad \text{Complete the square within the parentheses. Add and subtract 1.}$$

$$x = -2(y^2 - 2y + 1) + (-2)(-1) - 3 \qquad \text{Distributive property}$$

Be careful here.

$$x = -2(y - 1)^2 - 1 \qquad \text{Factor. Simplify.}$$

The vertex is $(-1, 1)$. Because of the negative coefficient -2 in $x = -2(y - 1)^2 - 1$, the graph opens to the left (the negative x-direction). The graph is narrower than the graph of $y = x^2$ because $|-2| = 2$, and $2 > 1$. See **FIGURE 18**.

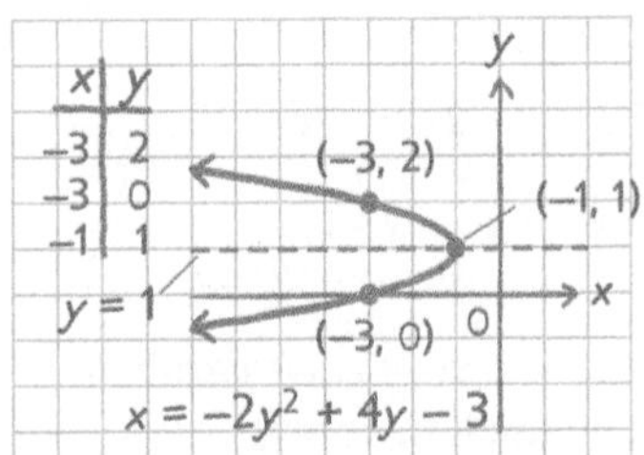

$x = -2y^2 + 4y - 3$
Vertex: $(-1, 1)$
Axis of symmetry: $y = 1$
Domain: $(-\infty, -1]$
Range: $(-\infty, \infty)$

FIGURE 18

NOW TRY 

CAUTION ***Only quadratic equations that are solved for y (whose graphs are vertical parabolas) are functions.*** The horizontal parabolas in **Examples 8 and 9** are ***not*** graphs of functions because they do not satisfy the conditions of the vertical line test.

▼ Summary of Graphs of Parabolas

Equation	Graph
$y = ax^2 + bx + c$ or $y = a(x - h)^2 + k$	(h, k), $a > 0$; (h, k), $a < 0$. These graphs represent functions.
$x = ay^2 + by + c$ or $x = a(y - k)^2 + h$	(h, k), $a > 0$; (h, k), $a < 0$. These graphs are not graphs of functions.

NOW TRY ANSWER

9. 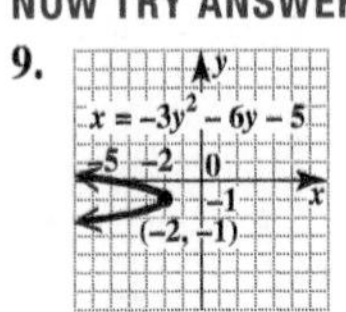

vertex: $(-2, -1)$; axis: $y = -1$; domain: $(-\infty, -2]$; range: $(-\infty, \infty)$

9.3 Exercises

FOR EXTRA HELP  MyLab Math

Video solutions for select problems available in MyLab Math

Concept Check *Answer each question.*

1. How can we determine just by looking at the equation of a parabola whether it has a vertical or a horizontal axis?

2. Why can't the graph of a quadratic function be a parabola with a horizontal axis?

3. How can we determine the number of x-intercepts of the graph of a quadratic function without graphing the function?

4. If the vertex of the graph of a quadratic function is $(1, -3)$, and the graph opens down, how many x-intercepts does the graph have?

5. Which equations have a graph that is a vertical parabola? A horizontal parabola?

A. $y = -x^2 + 20x + 80$ **B.** $x = 2y^2 + 6y + 5$

C. $x + 1 = (y + 2)^2$ **D.** $f(x) = (x - 4)^2$

6. Which of the equations in **Exercise 5** represent functions?

Find the vertex of each parabola. ***See Examples 1–3.***

7. $f(x) = x^2 + 8x + 10$

8. $f(x) = x^2 + 10x + 23$

9. $f(x) = -2x^2 + 4x - 5$

10. $f(x) = -3x^2 + 12x - 8$

11. $f(x) = x^2 + x - 7$

12. $f(x) = x^2 - x + 5$

Find the vertex of each parabola. For each equation, decide whether the graph opens up, down, to the left, *or* to the right, *and whether it is* wider, narrower, *or the* same shape *as the graph of $y = x^2$. If it is a parabola with a vertical axis of symmetry, find the discriminant and use it to determine the number of x-intercepts.* ***See Examples 1–3, 5, 8, and 9.***

13. $f(x) = 2x^2 + 4x + 5$

14. $f(x) = 3x^2 - 6x + 4$

15. $f(x) = -x^2 + 5x + 3$

16. $f(x) = -x^2 + 7x + 2$

17. $x = \frac{1}{3}y^2 + 6y + 24$

18. $x = \frac{1}{2}y^2 + 10y - 5$

Concept Check *Match each equation with its graph in choices A–F.*

19. $y = 2x^2 + 4x - 3$

20. $y = -x^2 + 3x + 5$

21. $y = -\frac{1}{2}x^2 - x + 1$

22. $x = y^2 + 6y + 3$

23. $x = -y^2 - 2y + 4$

24. $x = 3y^2 + 6y + 5$

A.

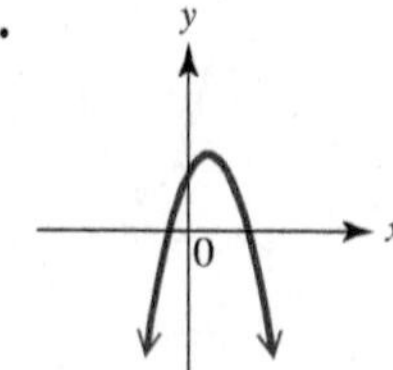

B.

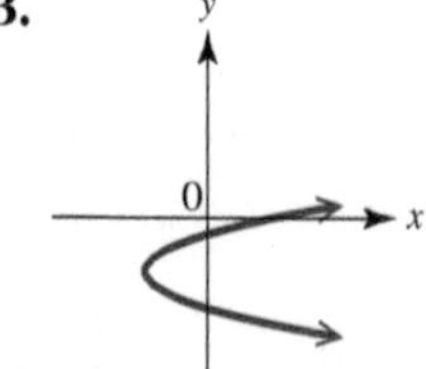

C.

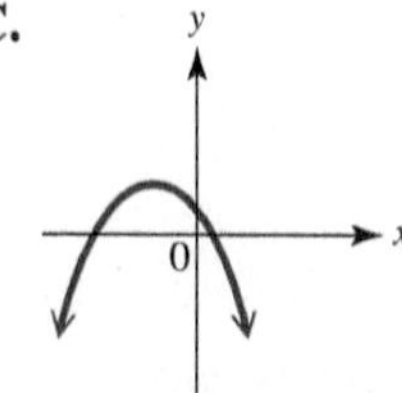

D.

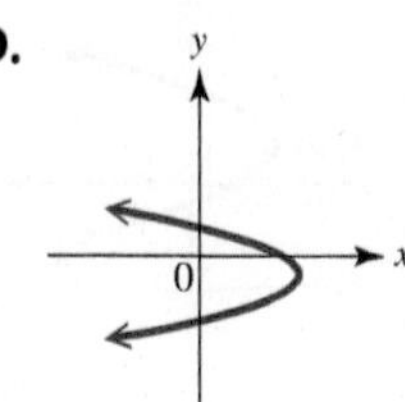

E.

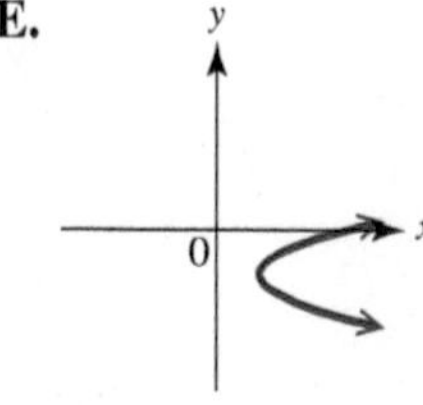

F.

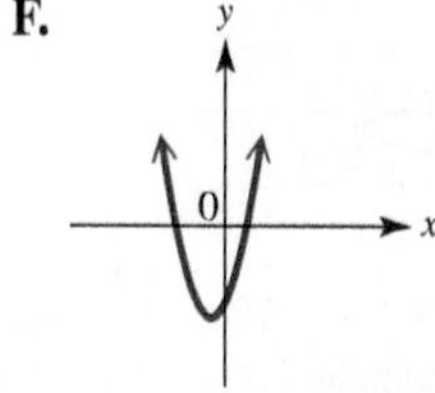

Graph each parabola. Give the vertex, axis of symmetry, domain, and range. ***See Examples 4, 8, and 9.***

25. $f(x) = x^2 + 8x + 10$

26. $f(x) = x^2 + 10x + 23$

27. $f(x) = x^2 + 4x + 3$

28. $f(x) = x^2 + 2x - 2$

29. $f(x) = -2x^2 + 4x - 5$

30. $f(x) = -3x^2 + 12x - 8$

31. $f(x) = -3x^2 - 6x + 2$

32. $f(x) = -2x^2 + 12x - 13$

33. $x = (y + 2)^2 + 1$

34. $x = (y + 3)^2 - 2$

35. $x = -(y - 3)^2 - 1$

36. $x = -(y - 2)^2 + 4$

37. $x = -\frac{1}{5}y^2 + 2y - 4$

38. $x = -\frac{1}{2}y^2 - 4y - 6$

39. $x = 3y^2 + 12y + 5$

40. $x = 4y^2 + 16y + 11$

Solve each problem. ***See Examples 6 and 7.***

41. Find the pair of numbers whose sum is 40 and whose product is a maximum. (*Hint:* Let x and $40 - x$ represent the two numbers.)

42. Find the pair of numbers whose sum is 60 and whose product is a maximum.

43. Polk Community College wants to construct a rectangular parking lot on land bordered on one side by a highway. It has 280 ft of fencing that is to be used to fence off the other three sides. What should be the dimensions of the lot if the enclosed area is to be a maximum? What is the maximum area?

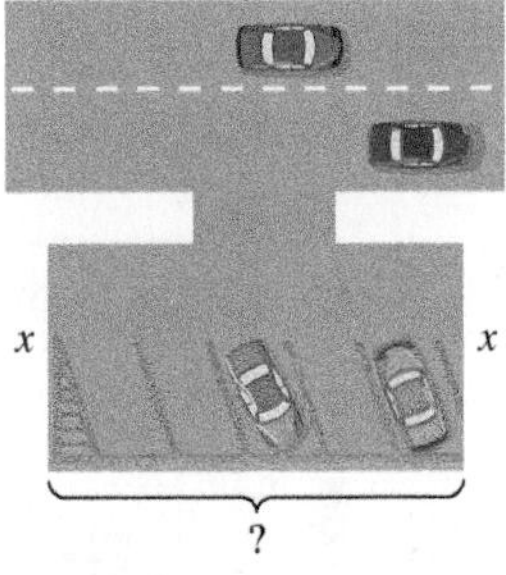

44. Bonnie has 100 ft of fencing material to enclose a rectangular exercise run for her dog. One side of the run will border her house, so she will only need to fence three sides. What dimensions will give the enclosure the maximum area? What is the maximum area?

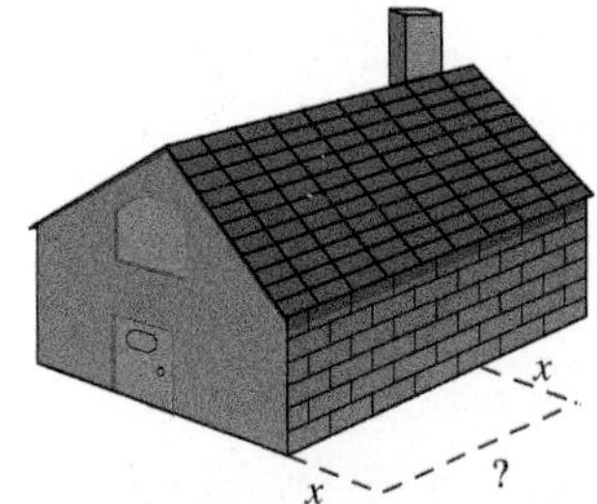

45. Two physics students from American River College find that when a bottle of California sparkling wine is shaken several times, held upright, and uncorked, its cork travels according to the function

$$s(t) = -16t^2 + 64t + 1,$$

where s is its height in feet above the ground t seconds after being released. After how many seconds will it reach its maximum height? What is the maximum height?

46. Professor Barbu has found that the number of students attending his intermediate algebra class is approximated by

$$S(x) = -x^2 + 20x + 80,$$

where x is the number of hours that the Campus Center is open daily. Find the number of hours that the center should be open so that the number of students attending class is a maximum. What is this maximum number of students?

47. Klaus has a taco stand. He has found that his daily costs are approximated by

$$C(x) = x^2 - 40x + 610,$$

where $C(x)$ is the cost, in dollars, to sell x units of tacos. Find the number of units of tacos he should sell to minimize his costs. What is the minimum cost?

48. Mohammad has a frozen yogurt cart. His daily costs are approximated by

$$C(x) = x^2 - 70x + 1500,$$

where $C(x)$ is the cost, in dollars, to sell x units of frozen yogurt. Find the number of units of frozen yogurt he must sell to minimize his costs. What is the minimum cost?

49. If an object on Earth is projected upward with an initial velocity of 32 ft per sec, then its height after t seconds is given by

$$s(t) = -16t^2 + 32t.$$

Find the maximum height attained by the object and the number of seconds it takes to hit the ground.

50. A projectile on Earth is fired straight upward so that its distance (in feet) above the ground t seconds after firing is given by

$$s(t) = -16t^2 + 400t.$$

Find the maximum height it reaches and the number of seconds it takes to reach that height.

The graph shows how Social Security trust fund assets are expected to change and suggests that a quadratic function would be a good fit to the data.

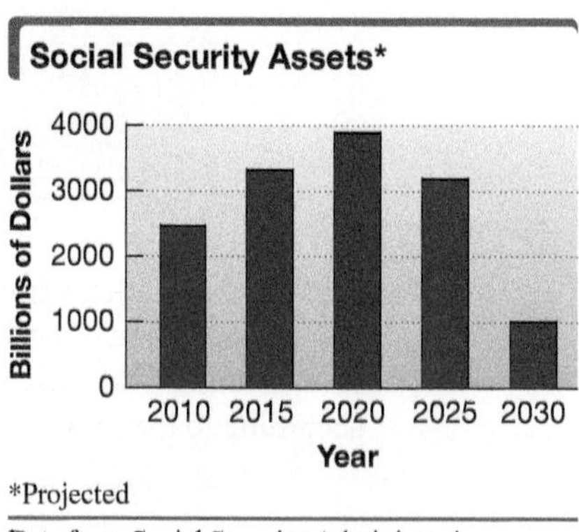

*Projected

Data from Social Security Administration.

The data are approximated by the function

$$f(x) = -20.57x^2 + 758.9x - 3140.$$

In the model, $x = 10$ *represents 2010,* $x = 15$ *represents 2015, and so on, and* $f(x)$ *is in billions of dollars.*

51. How could we have predicted that this quadratic model would have a negative coefficient for x^2, based only on the graph shown?

52. Algebraically determine the vertex of the graph. Express the x-coordinate to the nearest hundredth and the y-coordinate to the nearest whole number. Interpret the answer as it applies to this application.

Extending Skills *In each problem, find the following.*

(a) A function $R(x)$ that describes the total revenue received

(b) The graph of the function from part (a)

(c) The number of unsold seats that will produce the maximum revenue

(d) The maximum revenue

53. A charter flight charges a fare of \$200 per person, plus \$4 per person for each unsold seat on the plane. The plane holds 100 passengers. Let x represent the number of unsold seats. (*Hint:* To find $R(x)$, multiply the number of people flying, $100 - x$, by the price per ticket, $200 + 4x$.)

54. A charter bus charges a fare of \$48 per person, plus \$2 per person for each unsold seat on the bus. The bus has 42 seats. Let x represent the number of unsold seats. (*Hint:* To find $R(x)$, multiply the number riding, $42 - x$, by the price per ticket, $48 + 2x$.)

9.4 Symmetry; Increasing and Decreasing Functions

OBJECTIVES

1. Determine how multiplying a function by a real number a affects its graph.
2. Test for symmetry with respect to an axis.
3. Test for symmetry with respect to the origin.
4. Determine whether a function is increasing, decreasing, or constant on an open interval.

OBJECTIVE 1 Determine how multiplying a function by a real number *a* affects its graph.

Recall that the value of a affects the graph of $g(x) = ax^2$ in several ways.

- If $a > 0$, then the graph opens up.
- If $a < 0$, then the graph opens down.
- If $0 < |a| < 1$, then the graph is wider than the graph of $f(x) = x^2$.
- If $|a| > 1$, then the graph is narrower than the graph of $f(x) = x^2$.

FIGURE 19 illustrates these effects with the graphs of $f(x) = x^2$ and $g(x) = -\frac{1}{2}x^2$.

x	$f(x) = x^2$	$g(x) = -\frac{1}{2}x^2$
−2	4	−2
−1	1	$-\frac{1}{2}$
0	0	0
1	1	$-\frac{1}{2}$
2	4	−2

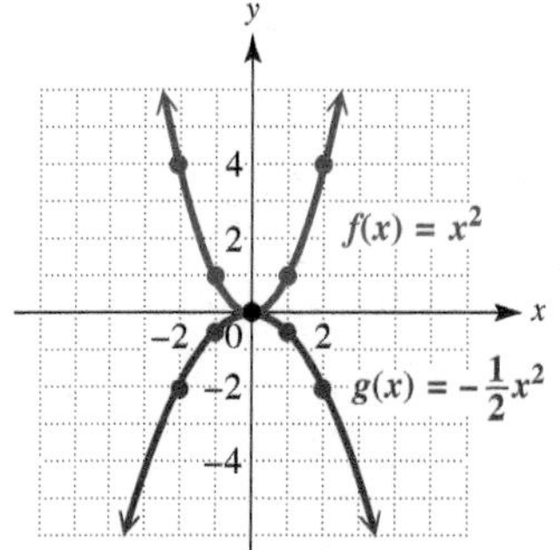

FIGURE 19

In $g(x) = -\frac{1}{2}x^2$, we have $a = -\frac{1}{2}$. Because $a < 0$, the graph of $g(x)$ opens down. And because $0 < |a| < 1$, the graph of $g(x)$ is wider than the graph of $f(x)$. As shown in the table, each y-value of $g(x)$ is $-\frac{1}{2}$ times the corresponding y-value of $f(x)$.

The same effects are true with the graphs of other functions.

VOCABULARY

- ☐ symmetric with respect to the y-axis
- ☐ symmetric with respect to the x-axis
- ☐ symmetric with respect to the origin
- ☐ increasing function
- ☐ decreasing function
- ☐ constant function

Reflection, Stretching, and Shrinking

The graph of $g(x) = a \cdot f(x)$ has the same general shape as the graph of $f(x)$.

- If $a < 0$, then the graph is **reflected** across the x-axis.
- If $|a| > 1$, then the graph is **stretched vertically** compared to the graph of $f(x)$.
- If $0 < |a| < 1$, then the graph is **shrunken vertically** compared to the graph of $f(x)$.

NOW TRY EXERCISE 1

Graph $g(x) = 3\sqrt{x}$. How does it compare with the graph of $f(x) = \sqrt{x}$?

EXAMPLE 1 Comparing the Graph of $g(x) = a \cdot f(x)$ with the Graph of $f(x)$

Graph each function. How does it compare with the graph of $f(x) = \sqrt{x}$?

(a) $g(x) = 2\sqrt{x}$

In $g(x) = 2\sqrt{x}$, we have $a = 2$. Because the coefficient 2 is positive and greater than 1, the graph of $g(x) = 2\sqrt{x}$ will have the same general shape as the graph of $f(x) = \sqrt{x}$ but is stretched vertically by a factor of 2. Thus, every y-value of $g(x)$ is two times the corresponding y-value of $f(x)$. See **FIGURE 20.**

x	$f(x) = \sqrt{x}$	$g(x) = 2\sqrt{x}$
0	0	0
1	1	2
4	2	4
9	3	6

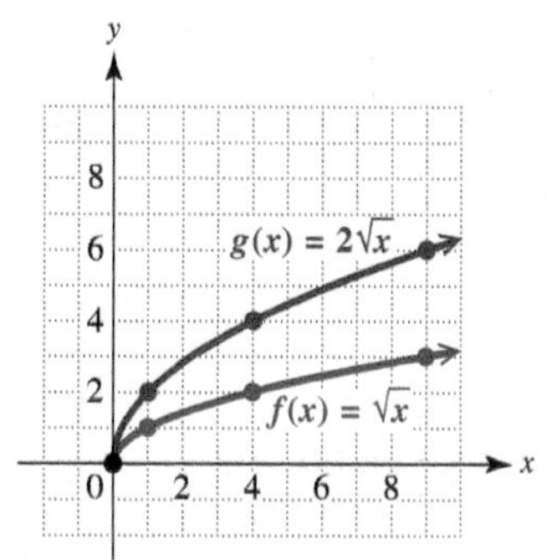

FIGURE 20

(b) $h(x) = -\sqrt{x}$

In $h(x) = -\sqrt{x}$, $a = -1$. Because $a < 0$, the graph of $h(x) = -\sqrt{x}$ will have the same general shape as the graph of $f(x) = \sqrt{x}$ but is reflected across the x-axis. The negative sign can be interpreted as multiplication by -1. Thus, every y-value of $h(x)$ is the negative of the corresponding y-value of $f(x)$. See **FIGURE 21.**

x	$f(x) = \sqrt{x}$	$h(x) = -\sqrt{x}$
0	0	0
1	1	−1
4	2	−2
9	3	−3

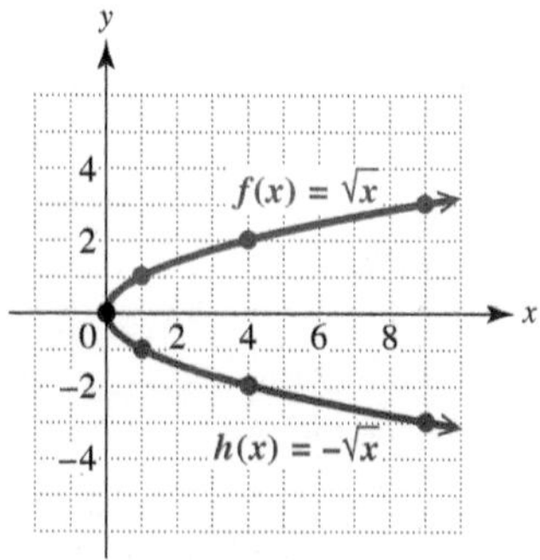

FIGURE 21

NOW TRY 

The parabolas graphed in the previous two sections were symmetric with respect to the axis of the parabola. The graphs of many other relations exhibit symmetry with respect to a line or a point.

NOW TRY ANSWER

1. The graph of $g(x) = 3\sqrt{x}$ is stretched vertically compared with the graph of $f(x) = \sqrt{x}$. Every y-value of $g(x)$ is three times the y-value of $f(x)$.

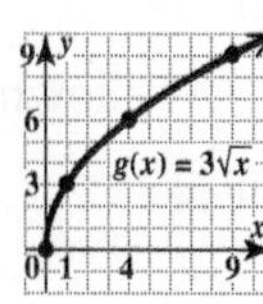

OBJECTIVE 2 Test for symmetry with respect to an axis.

The graph in **FIGURE 22(a)** is cut in half by the y-axis with each half the mirror image of the other half. Such a graph is *symmetric with respect to the y-axis.*

In general, for a graph to be symmetric with respect to the y-axis, the point $(-x, y)$ must be on the graph whenever the point (x, y) is on the graph.

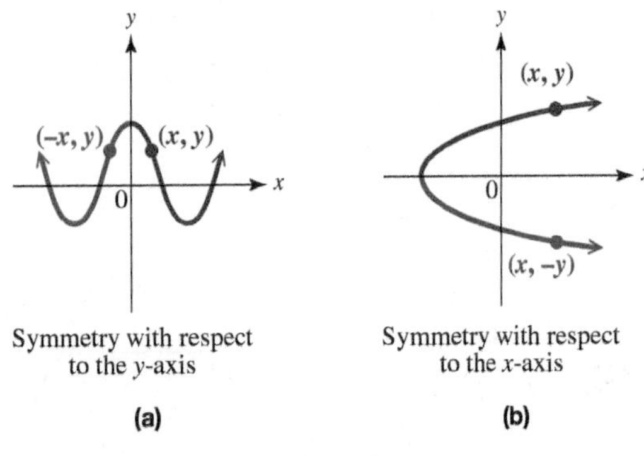

Symmetry with respect to the y-axis

(a)

Symmetry with respect to the x-axis

(b)

FIGURE 22

Similarly, if the graph in **FIGURE 22(b)** were folded in half along the x-axis, the portion from the top would exactly match the portion from the bottom. Such a graph is *symmetric with respect to the x-axis.*

In general, for a graph to be symmetric with respect to the x-axis, the point $(x, -y)$ must be on the graph whenever the point (x, y) is on the graph.

Tests for Symmetry with Respect to an Axis

The graph of a relation is **symmetric with respect to the y-axis** if the replacement of x with $-x$ results in an equivalent equation.

The graph of a relation is **symmetric with respect to the x-axis** if the replacement of y with $-y$ results in an equivalent equation.

EXAMPLE 2 Testing for Symmetry with Respect to an Axis

Test for symmetry with respect to the x-axis and the y-axis.

(a) $y = x^2 + 4$

Replace x with $-x$ to test for symmetry with respect to the y-axis, as shown on the left below. The result is equivalent to the original equation, so the graph, shown in **FIGURE 23**, is symmetric with respect to the y-axis. The y-axis cuts the graph in half—the two halves are mirror images.

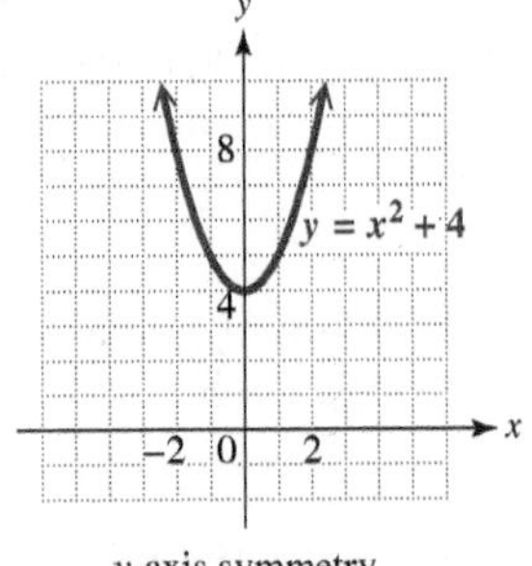

y-axis symmetry

FIGURE 23

$$y = x^2 + 4$$
$$y = (-x)^2 + 4 \quad \text{Equivalent}$$
$$y = x^2 + 4$$

Use parentheses around $-x$.

$$y = x^2 + 4$$
$$-y = x^2 + 4 \quad \text{Not equivalent}$$
$$y = -x^2 - 4$$

Now replace y with $-y$ to test for symmetry with respect to the x-axis, as shown on the right above. The result is *not* equivalent to the original equation, so the graph is *not* symmetric with respect to the x-axis. The graph in **FIGURE 23** confirms this.

NOW TRY EXERCISE 2

Test for symmetry with respect to the x-axis and the y-axis.

(a) $x = y^2 + 1$

(b) $x^2 + y^2 = 4$

(b) $2x + y = 4$

Replace x with $-x$ and then replace y with $-y$.

$2x + y = 4$ $2(-x) + y = 4$ $-2x + y = 4$ — Not equivalent	$2x + y = 4$ $2x + (-y) = 4$ $2x - y = 4$ — Not equivalent

The graph of $2x + y = 4$, which is a straight line, is not symmetric with respect to the x-axis *or* the y-axis. See **FIGURE 24**. Some straight lines exhibit symmetry. This one does not.

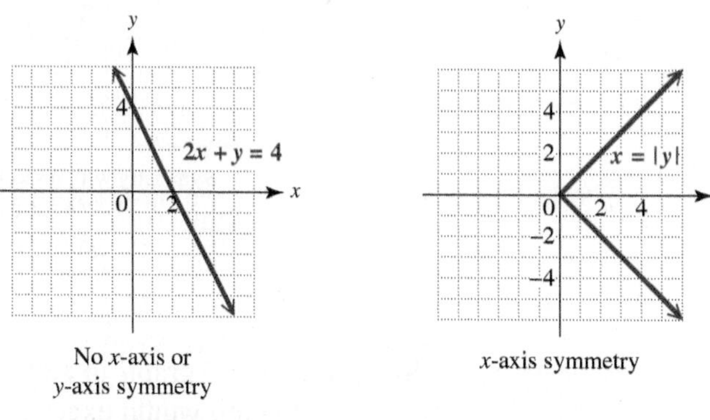

No x-axis or y-axis symmetry

FIGURE 24

x-axis symmetry

FIGURE 25

(c) $x = |y|$

Replacing x with $-x$ gives the following.

$$x = |y|$$
$$-x = |y|$$
Not equivalent

The graph is not symmetric with respect to the y-axis. Confirm this in **FIGURE 25**.

Replacing y with $-y$ gives the following.

$$x = |y|$$
$$x = |-y|$$
$$x = |y|$$
Equivalent

$|-y| = |y|$ for all y.

The graph is symmetric with respect to the x-axis. Confirm this in **FIGURE 25**.

(d) $x^2 + y^2 = 16$

Substitute $-x$ for x and then $-y$ for y in $x^2 + y^2 = 16$.

$$(-x)^2 + y^2 = 16 \quad \text{and} \quad x^2 + (-y)^2 = 16$$

Both simplify to the original equation, $x^2 + y^2 = 16$. The graph, which is a circle of radius 4 centered at the origin, is symmetric with respect to *both* axes. See **FIGURE 26**.

y, $x^2 + y^2 = 16$, 4, −4, 0, 4, x, −4

x-axis and y-axis symmetry

FIGURE 26

NOW TRY 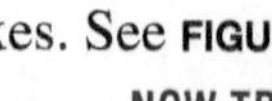

OBJECTIVE 3 Test for symmetry with respect to the origin.

Another kind of symmetry occurs when a graph can be rotated 180° about the origin and have the result coincide exactly with the original graph. Such a graph is *symmetric with respect to the origin.*

In general, for a graph to be symmetric with respect to the origin, the point $(-x, -y)$ is on the graph whenever the point (x, y) is on the graph.

FIGURE 27 on the next page shows two such graphs.

NOW TRY ANSWERS

2. (a) The graph is symmetric with respect to the x-axis but not the y-axis.
   (b) The graph is symmetric with respect to the x-axis and the y-axis.

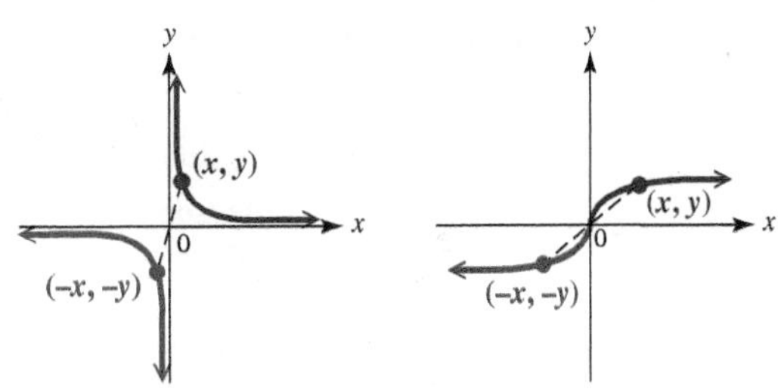

Symmetry with respect to the origin

FIGURE 27

Test for Symmetry with Respect to the Origin

The graph of a relation is **symmetric with respect to the origin** if the replacement of both x with $-x$ and y with $-y$ at the same time results in an equivalent equation.

NOW TRY EXERCISE 3

Decide whether the graph of each relation is symmetric with respect to the origin.

(a) $x = y^2 - 2$

(b) $y = 2x$

EXAMPLE 3 Testing for Symmetry with Respect to the Origin

Decide whether the graph of each relation is symmetric with respect to the origin.

(a) $3x = 5y$

Replace x with $-x$ and y with $-y$ in the equation.

$$3x = 5y$$
$$3(-x) = 5(-y)$$
$$-3x = -5y$$
$$3x = 5y$$

Use parentheses here. — Equivalent

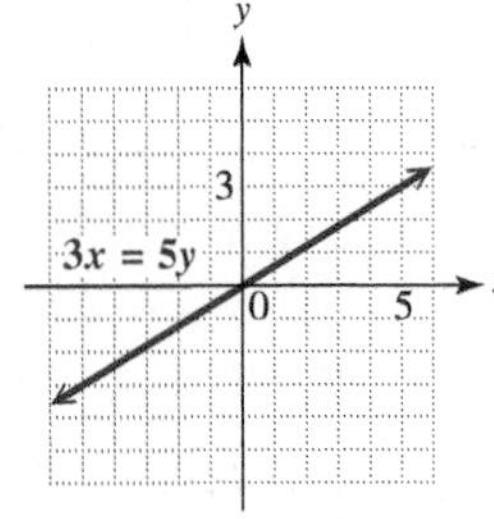

Origin symmetry

FIGURE 28

The graph is symmetric with respect to the origin. See **FIGURE 28**.

(b) $2y = 3x^2$

Substituting $-x$ for x and $-y$ for y gives the following.

$$2y = 3x^2$$
$$2(-y) = 3(-x)^2$$
$$-2y = 3x^2$$

Not equivalent

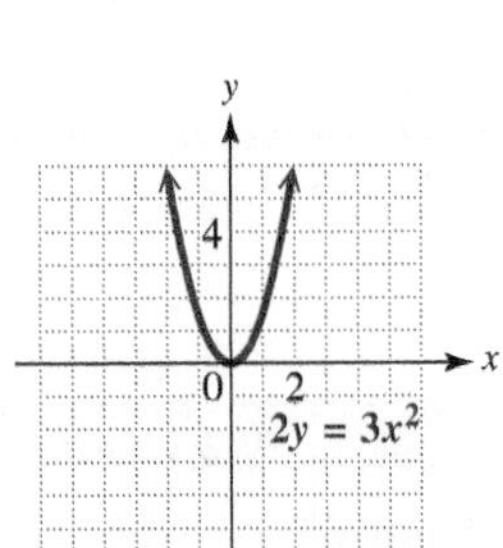

No origin symmetry

FIGURE 29

The graph, shown in **FIGURE 29**, is not symmetric with respect to the origin.

(c) $x^2 + y^2 = 16$

Replace x with $-x$ and y with $-y$.

$$x^2 + y^2 = 16$$
$$(-x)^2 + (-y)^2 = 16$$
$$x^2 + y^2 = 16$$

Equivalent

The graph, which is the circle shown in **FIGURE 26** in **Example 2(d)**, is symmetric with respect to the origin.

NOW TRY

Notice the following important concepts regarding symmetry:

- A graph symmetric with respect to both the x- and y-axes is automatically symmetric with respect to the origin. (See **FIGURE 26**.)
- A graph symmetric with respect to the origin need *not* be symmetric with respect to either axis. (See **FIGURE 28**.)
- A graph possessing any two types of symmetry must also exhibit the third type.

NOW TRY ANSWERS

3. **(a)** The graph is not symmetric with respect to the origin.
   (b) The graph is symmetric with respect to the origin.

▼ Tests for Symmetry

	Symmetric with Respect to		
	x-axis	*y-axis*	*Origin*
Test	*Replace y with $-y$.*	*Replace x with $-x$.*	*Replace x with $-x$ and replace y with $-y$.*
Example			

The symmetry exists if the replacement leads to an equation equivalent to the original one.

OBJECTIVE 4 **Determine whether a function is increasing, decreasing, or constant on an open interval.**

Informally speaking, a function *increases* on an open interval of its domain if its graph rises from left to right on the interval. It *decreases* on an open interval of its domain if its graph falls from left to right on the interval. It is *constant* on an open interval of its domain if its graph is horizontal on the interval.

For example, consider **FIGURE 30**.

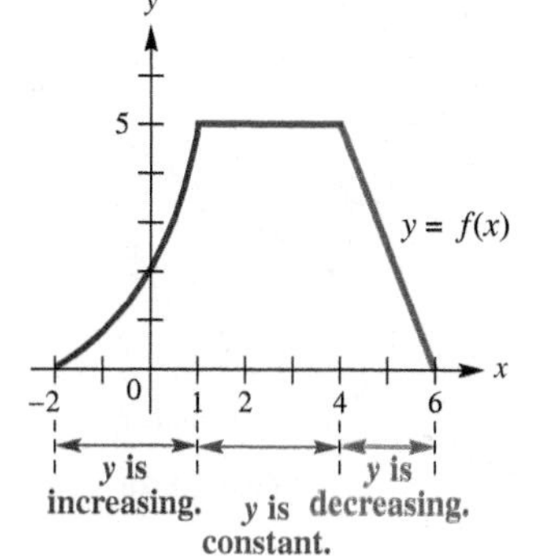

FIGURE 30

- The function increases on the open interval $(-2, 1)$ because the y-values continue to get larger for x-values in that interval.
- The function is constant on the open interval $(1, 4)$ because the y-values are always 5 for all x-values there.
- The function decreases on the open interval $(4, 6)$ because there the y-values continuously get smaller.

The intervals refer to the x-values where the y-values either increase, decrease, or are constant.

The formal definitions of these concepts follow.

Increasing, Decreasing, and Constant Functions

Suppose that a function f is defined over an *open* interval I and x_1 and x_2 are in I.

(a) f **increases** on I if, whenever $x_1 < x_2$, $f(x_1) < f(x_2)$.

(b) f **decreases** on I if, whenever $x_1 < x_2$, $f(x_1) > f(x_2)$.

(c) f is **constant** on I if, for every x_1 and x_2, $f(x_1) = f(x_2)$.

FIGURE 31 illustrates these ideas.

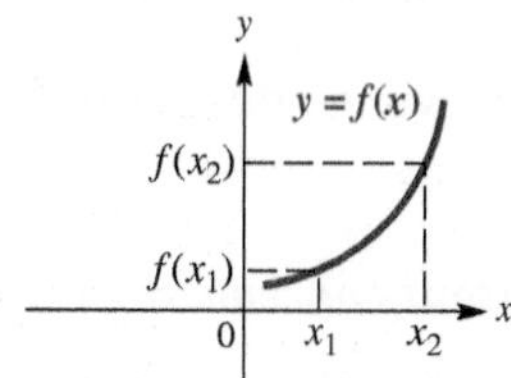

Whenever $x_1 < x_2$, and $f(x_1) < f(x_2)$, f is increasing.

(a)

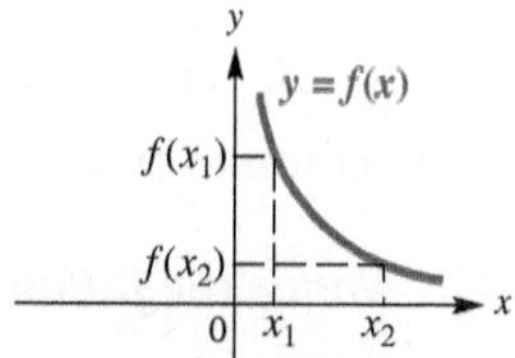

Whenever $x_1 < x_2$, and $f(x_1) > f(x_2)$, f is decreasing.

(b)

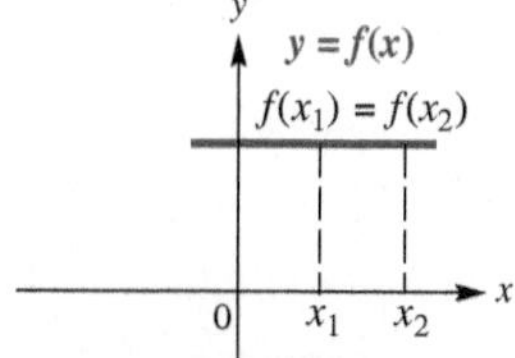

For every x_1 and x_2, if $f(x_1) = f(x_2)$, then f is constant.

(c)

FIGURE 31

NOTE To decide whether a function is increasing, decreasing, or constant over an open interval, ask, ***"What does y do as x goes from left to right?" Consider open intervals of the domain, not individual points.***

NOW TRY EXERCISE 4

Determine the open intervals over which the function is increasing, decreasing, or constant.

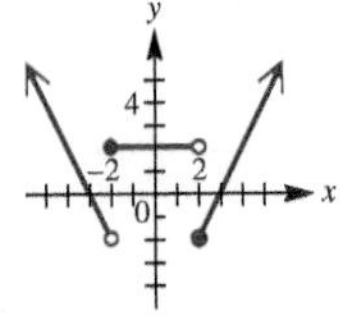

EXAMPLE 4 Determining Increasing, Decreasing, or Constant Intervals

Determine the open intervals over which the function in **FIGURE 32** is increasing, decreasing, or constant.

Ask, "*What is happening to the y-values as the x-values are getting larger?*" Moving from left to right on the graph, we see the following:

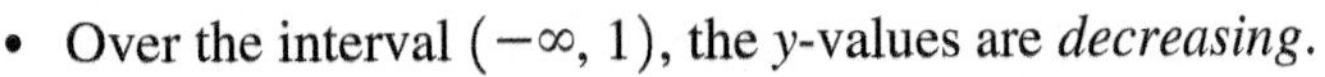

- Over the interval $(-\infty, 1)$, the y-values are *decreasing*.
- Over the interval $(1, 3)$, the y-values are *increasing*.
- Over the interval $(3, \infty)$, the y-values are *constant* (and equal to 6).

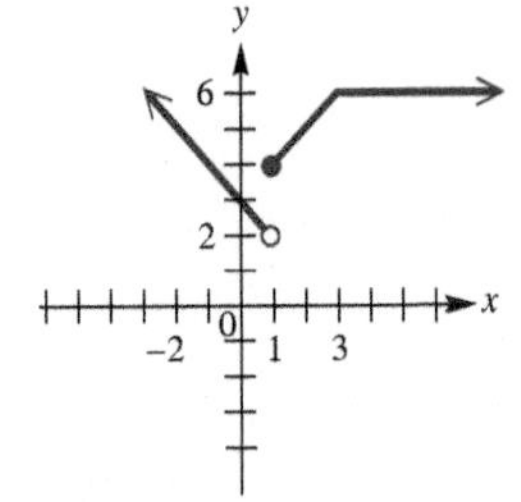

FIGURE 32

The function is decreasing on $(-\infty, 1)$, increasing on $(1, 3)$, and constant on $(3, \infty)$.

NOW TRY

CAUTION ***When identifying open intervals over which a function is increasing, decreasing, or constant, remember that we are interested in identifying domain intervals. Range values do not appear in these stated intervals.***

NOW TRY ANSWER

4. decreasing on $(-\infty, -2)$; constant on $(-2, 2)$; increasing on $(2, \infty)$

9.4 Exercises

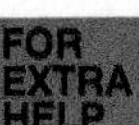

MyLab Math

Video solutions for select problems available in MyLab Math

Concept Check *Complete each statement.*

1. Compared to the graph of $f(x) = \sqrt{x}$, the graph of $g(x) = \frac{1}{2}\sqrt{x}$ is (*stretched / shrunken*) by a factor of ____. Some points on the graph of $f(x)$ are $(0, 0)$, $(4, 2)$, and $(16, 4)$. Corresponding points on the graph of $g(x)$ are $(0, 0)$, $(4, \_\_\_)$, and $(16, \_\_\_)$.

2. Compared to the graph of $f(x) = \sqrt{x}$, the graph of $g(x) = 4\sqrt{x}$ is (*stretched / shrunken*) by a factor of ____. Some points on the graph of $f(x)$ are $(0, 0)$, $(1, 1)$, and $(4, 2)$. Corresponding points on the graph of $g(x)$ are $(0, 0)$, $(1, \_\_\_)$, and $(4, \_\_\_)$.

3. Compared to the graph of $f(x) = \sqrt{x}$, the graph of $g(x) = -2\sqrt{x}$ is ____ across the (*x-axis / y-axis*) and (*stretched / shrunken*) by a factor of ____. Some points on the graph of $f(x)$ are $(0, 0)$, $(1, 1)$, and $(4, 2)$. Corresponding points on the graph of $g(x)$ are $(0, 0)$, $(1, \_\_\_)$, and $(4, \_\_\_)$.

4. Compared to the graph of $f(x) = \sqrt{x}$, the graph of $g(x) = -\frac{1}{2}\sqrt{x}$ is ____ across the (*x-axis / y-axis*) and (*stretched / shrunken*) by a factor of ____. Some points on the graph of $f(x)$ are $(0, 0)$, $(4, 2)$, and $(16, 4)$. Corresponding points on the graph of $g(x)$ are $(0, 0)$, $(4, \_\_\_)$, and $(16, \_\_\_)$.

Sketch each graph. ***See Example 1 and Exercises 1–4.***

5. $g(x) = \frac{1}{2}\sqrt{x}$ 6. $g(x) = 4\sqrt{x}$ 7. $g(x) = -2\sqrt{x}$ 8. $g(x) = -\frac{1}{2}\sqrt{x}$

Work each problem. **See Example 1.**

9. Given the graph of $y = f(x)$ in the figure, sketch the graph of each function, and describe how it is related to the graph of $y = f(x)$.

(a) $y = -f(x)$ **(b)** $y = 2f(x)$

(c) $y = \frac{1}{2}f(x)$

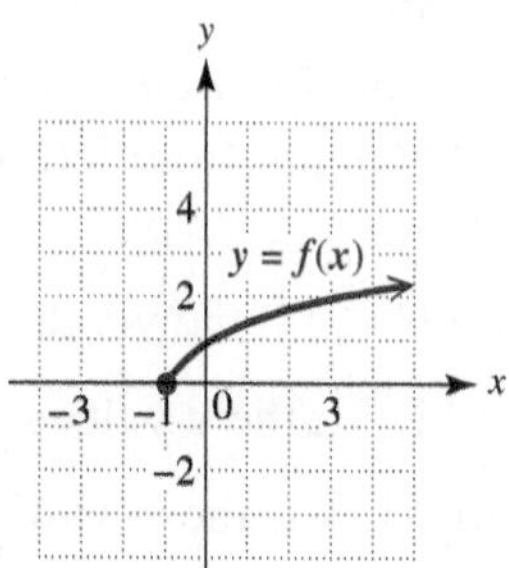

10. Given the graph of $y = g(x)$ in the figure, sketch the graph of each function, and describe how it is related to the graph of $y = g(x)$.

(a) $y = \frac{1}{2}g(x)$ **(b)** $y = -g(x)$

(c) $y = 2g(x)$

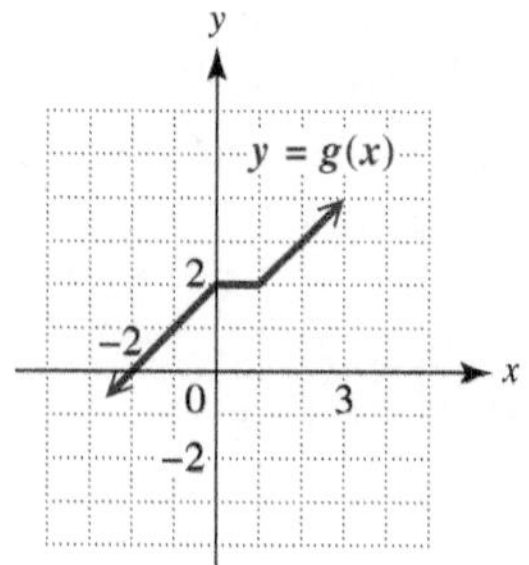

Concept Check *Decide whether each figure is symmetric with respect to* **(a)** *the given line, and* **(b)** *the given point.*

11.

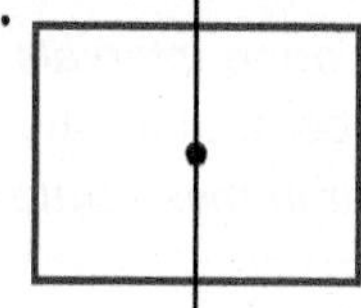

12.

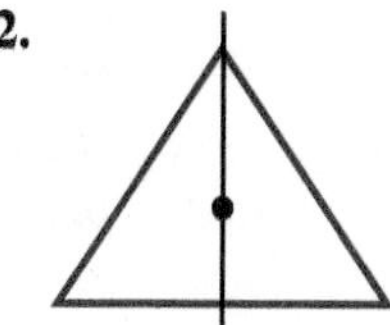

13.

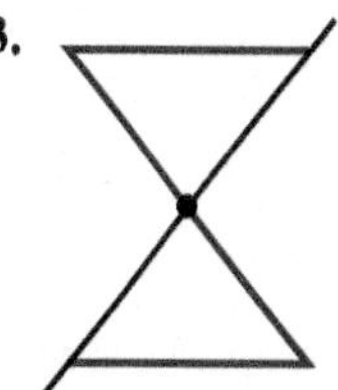

14.

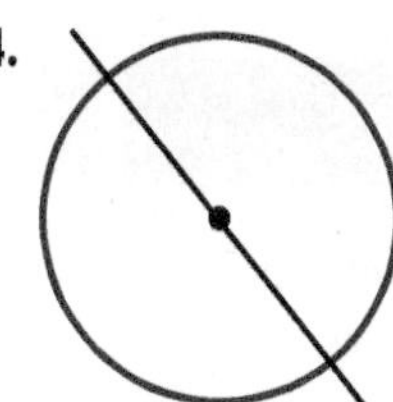

Concept Check *Plot each point, and then use the same axes to plot the points that are symmetric to the given point with respect to the following:* **(a)** *x-axis,* **(b)** *y-axis,* **(c)** *origin.*

15. $(-4, -2)$ **16.** $(-8, 3)$ **17.** $(-8, 0)$ **18.** $(0, -3)$

Use the tests for symmetry to decide whether the graph of each relation is symmetric with respect to the x-axis, the y-axis, or the origin. More than one of these symmetries, or none of them, may apply. **See Examples 2 and 3.** *(Do not graph.)*

19. $x = y^2 + 3$ **20.** $x = y^2 - 5$ **21.** $-2x = y$ **22.** $y = 5x$

23. $x^2 + y^2 = 5$ **24.** $y^2 = 1 - x^2$ **25.** $y = x^2 - 8x$ **26.** $y = 4x - x^2$

27. $y = |x|$ **28.** $y = |x| + 1$ **29.** $y = x^3$ **30.** $y = -x^3$

31. $x + y = -3$ **32.** $x - y = 6$ **33.** $x = y^4$ **34.** $x = y^2$

35. $f(x) = \frac{1}{1 + x^2}$ **36.** $f(x) = \frac{-1}{x^2 + 9}$ **37.** $xy = 2$ **38.** $xy = -6$

39. Explain why the graph of a function cannot be symmetric with respect to the x-axis.

40. A graph that is symmetric with respect to both the x-axis and the y-axis is also symmetric with respect to the origin. Explain why.

Extending Skills *Assume that for $y = f(x)$, $f(2) = 3$. For each given statement, find another value for the function.*

41. The graph of $y = f(x)$ is symmetric with respect to the origin.

42. The graph of $y = f(x)$ is symmetric with respect to the y-axis.

43. The graph of $y = f(x)$ is symmetric with respect to the line $x = 3$.

44. The graph of $y = f(x)$ is symmetric with respect to the line $x = -1$.

*Determine the open intervals over which each function is **(a)** increasing, **(b)** decreasing, and **(c)** constant.* ***See Example 4.***

45.

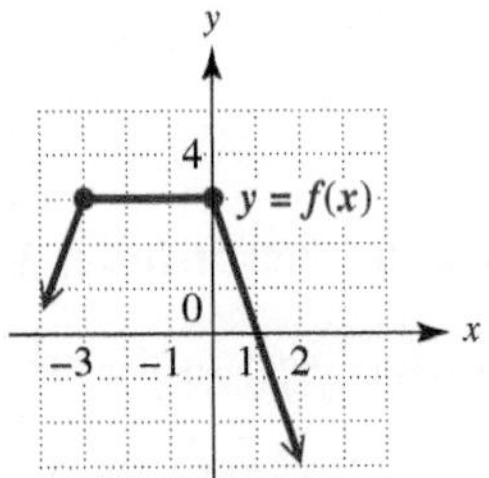

46.

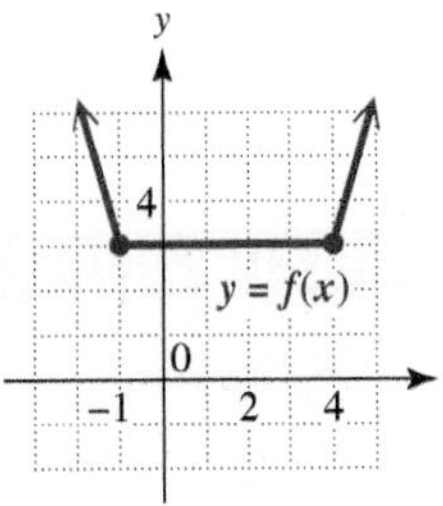

47.

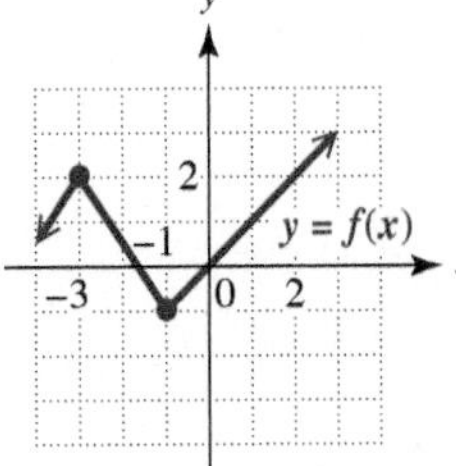

48.

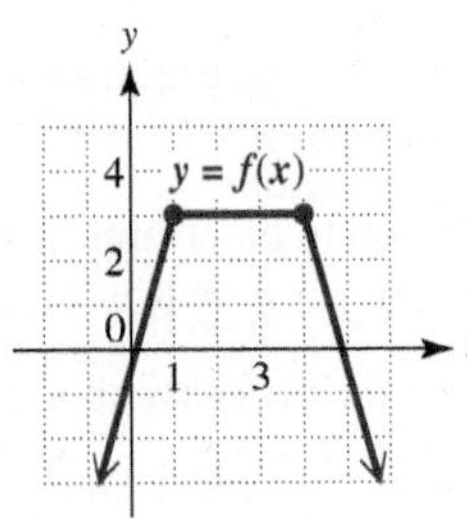

49.

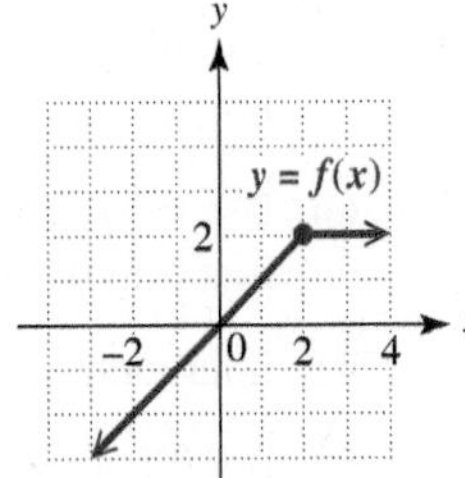

50.

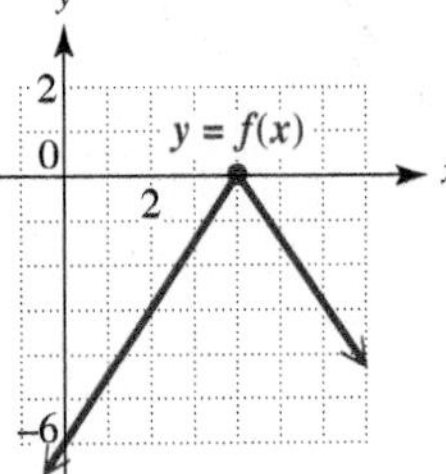

51.

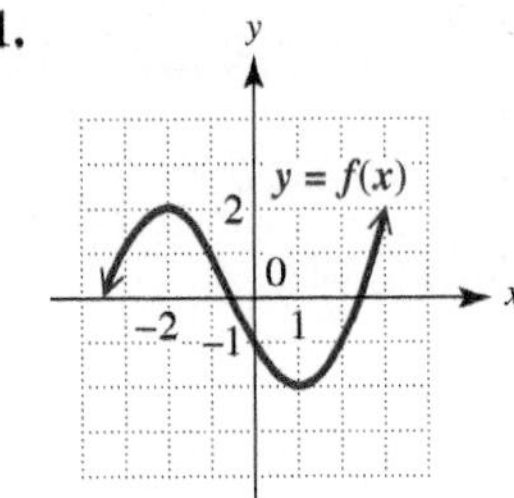

52.

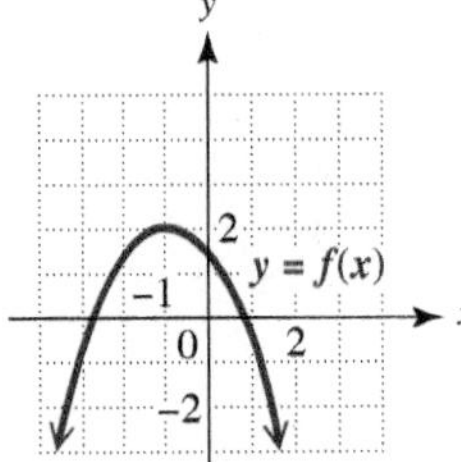

53.

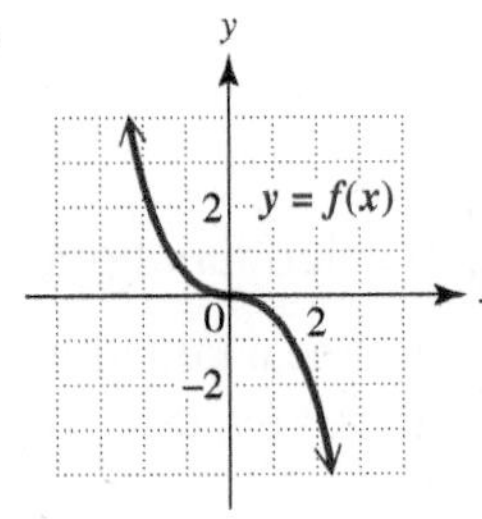

54.

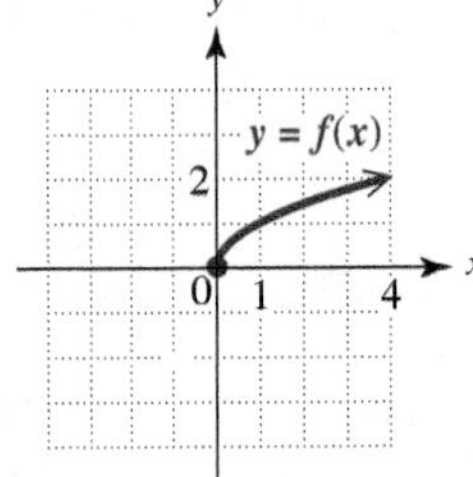

55.

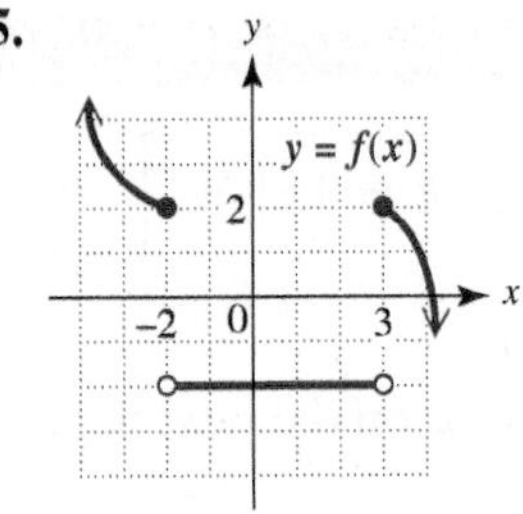

56.

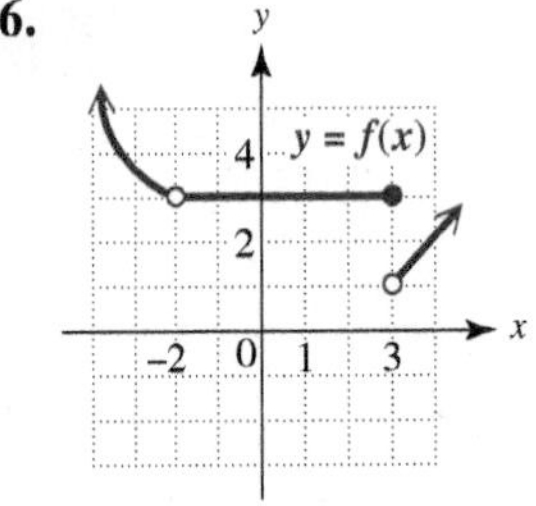

The graph shows a function illustrating average prices for a gallon of regular unleaded gasoline in the United States for the years 2008 through 2016. Refer to the graph and answer the following questions.

57. Over what interval (s), in years, is the function increasing?

58. Over what interval (s), in years, is the function decreasing?

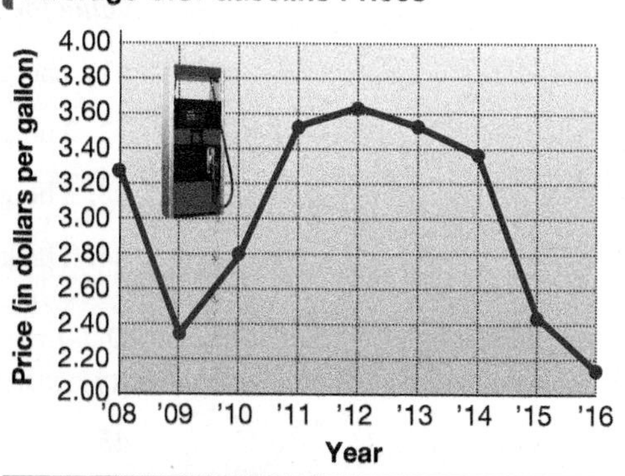

Data from U.S. Department of Energy.

RELATING CONCEPTS For Individual or Group Work (Exercises 59–64)

*A function f is an **even function** if $f(-x) = f(x)$ for all x in the domain of f.*

*A function f is an **odd function** if $f(-x) = -f(x)$ for all x in the domain of f.*

*To see how these ideas relate to symmetry, **work Exercises 59–64 in order.***

59. Use the preceding definition to determine whether the function $f(x) = x^n$ is an even function or an odd function for $n = 2$, $n = 4$, and $n = 6$.

60. Use the preceding definition to determine whether the function $f(x) = x^n$ is an even function or an odd function for $n = 1$, $n = 3$, and $n = 5$.

61. If a function is even, what do we know about its symmetry?

62. If a function is odd, what do we know about its symmetry?

*Complete the left half of the graph of $y = f(x)$ based on the given assumption and the answers to **Exercises 61 and 62.***

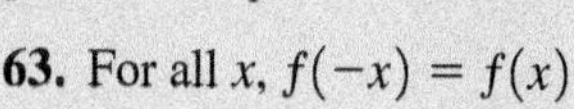

63. For all x, $f(-x) = f(x)$.

64. For all x, $f(-x) = -f(x)$.

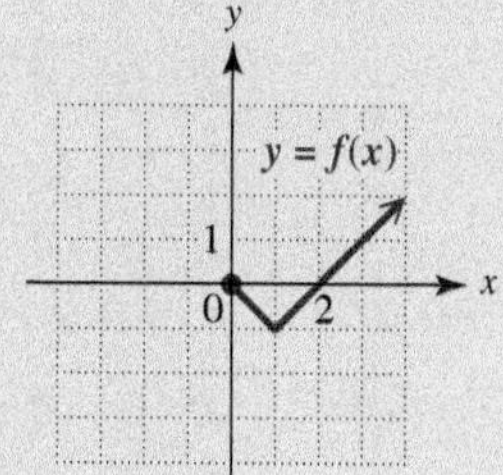

9.5 Piecewise Linear Functions

OBJECTIVES

1 Graph absolute value functions.

2 Graph other piecewise linear functions.

3 Graph step functions.

OBJECTIVE 1 Graph absolute value functions.

A function defined by different linear equations over different intervals of its domain is a **piecewise linear function.** An example of such a function is the **absolute value function,** $f(x) = |x|$.

Absolute Value Function

$$f(x) = |x| = \begin{cases} x & \text{if } x \geq 0 \\ -x & \text{if } x < 0 \end{cases}$$

Its graph is shown in **FIGURE 33** on the next page.

VOCABULARY
- ☐ piecewise linear function
- ☐ absolute value function
- ☐ greatest integer function
- ☐ step function

The absolute value function pairs each real number with its absolute value.

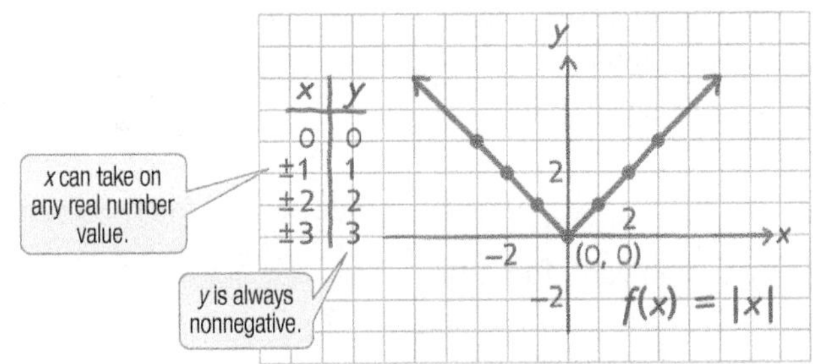

FIGURE 33

Absolute value function
$f(x) = |x|$
Axis of symmetry: $x = 0$
Domain: $(-\infty, \infty)$
Range: $[0, \infty)$

NOW TRY EXERCISE 1
Graph $f(x) = 2|x + 1| - 3$.

EXAMPLE 1 Graphing Absolute Value Functions

Graph each function.

(a) $f(x) = -|x|$

The negative sign indicates that the graph of $f(x)$ is a reflection of the graph of $y = |x|$ across the x-axis. The domain is $(-\infty, \infty)$, and the range is $(-\infty, 0]$. As shown in **FIGURE 34**, on the interval $(-\infty, 0]$ the graph is the same as the graph of $y = x$. On the interval $(0, \infty)$, it is the graph of $y = -x$.

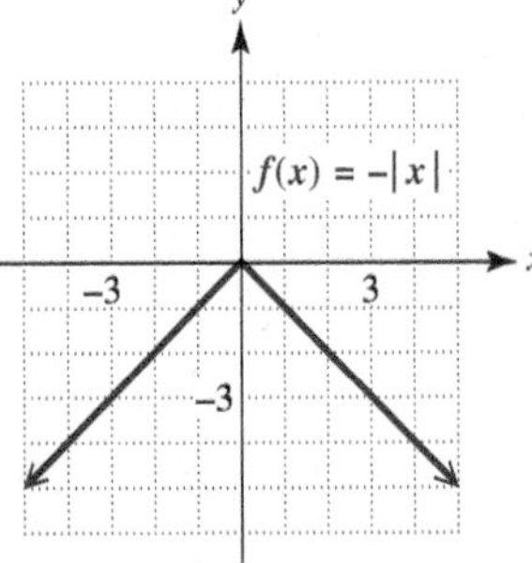

FIGURE 34

(b) $f(x) = |3x + 4| + 1$

Recall that the graph of $y = a(x - h)^2 + k$ is a parabola that, depending on the absolute value of a, is stretched or shrunken compared with the graph of $y = x^2$. It is shifted h units horizontally and k units vertically. The same idea applies here. If we write

$$y = |3x + 4| + 1 \quad \text{in the form} \quad \boldsymbol{y = a|x - h| + k},$$

its graph will compare similarly with the graph of $y = |x|$.

$$y = |3x + 4| + 1 \qquad \text{Given function}$$

$$y = \left|3\left(x + \frac{4}{3}\right)\right| + 1 \qquad \text{Factor 3 from the absolute value expression.}$$

$$y = |3|\left|x + \frac{4}{3}\right| + 1 \qquad |ab| = |a| \cdot |b|\text{; Write each factor using absolute value bars.}$$

$$y = 3\left|x - \left(-\frac{4}{3}\right)\right| + 1 \qquad |3| = 3;\ x + h = x - (-h)$$

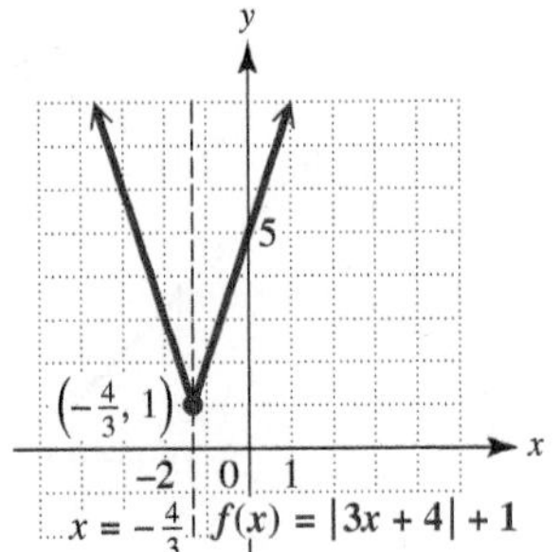

FIGURE 35

In this form, we see that the graph is narrower than the graph of $y = |x|$ (that is, stretched vertically), with the "vertex" at the point $\left(-\frac{4}{3}, 1\right)$. See **FIGURE 35**. The axis of symmetry has equation $x = -\frac{4}{3}$.

The coefficient of x, which is 3, determines the slopes of the two partial lines that form the graph. One has slope 3, and the other has slope -3. Because the absolute value of the slopes is greater than 1, the lines are steeper than the lines that form the graph of $y = |x|$.

NOW TRY

NOW TRY ANSWER

1. 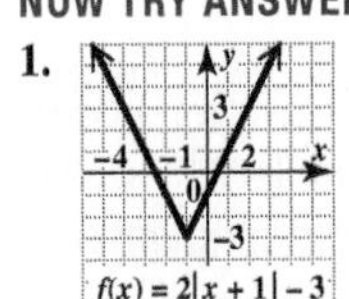

OBJECTIVE 2 Graph other piecewise linear functions.

The different parts of piecewise linear functions may have completely different equations.

NOW TRY EXERCISE 2
Graph the function.

$$f(x)=\begin{cases}\frac{1}{2}x+2 & \text{if } x \geq -2\\ -x-3 & \text{if } x<-2\end{cases}$$

EXAMPLE 2 Graphing Piecewise Linear Functions

Graph each function.

(a) $f(x)=\begin{cases}x+1 & \text{if } x \leq 2\\ -2x+7 & \text{if } x>2\end{cases}$

We graph the function over each interval of the domain separately. We begin with $f(x)=x+1$ over the interval $(-\infty, 2]$. This portion of the graph has an endpoint at $x=2$. We find the y-value by substituting 2 for x in

$$y=x+1 \quad \text{to obtain} \quad y=2+1=3. \qquad \text{Gives the endpoint } (2, 3)$$

Another point is needed to complete this portion of the graph. We choose an x-value less than 2, such as -1. Substitute -1 for x in

$$y=x+1 \quad \text{to obtain} \quad y=-1+1=0. \qquad \text{Gives } (-1, 0)$$

We draw the graph through endpoint $(2, 3)$ and the point $(-1, 0)$ as a partial line.

Now we graph the function similarly over the interval $(2, \infty)$, using $y=-2x+7$. This partial line will have an open endpoint when $x=2$ and

$$y=-2(2)+7=3. \qquad \text{Gives the open endpoint } (2, 3)$$

Choosing an x-value greater than 2, such as 4, we substitute and obtain

$$y=-2(4)+7=-1. \qquad \text{Gives } (4, -1)$$

The partial line through the open endpoint $(2, 3)$ and the point $(4, -1)$ completes the graphs. The two parts "meet" at $(2, 3)$. See **FIGURE 36.**

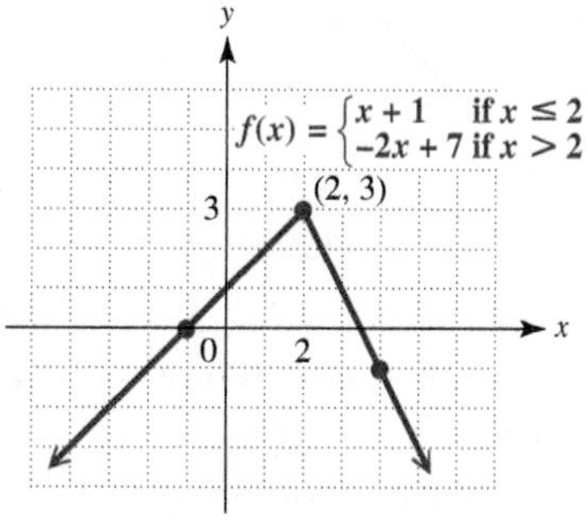

$f(x)=\begin{cases}x+1 & \text{if } x \leq 2\\ -2x+7 & \text{if } x>2\end{cases}$

FIGURE 36

(b) $f(x)=\begin{cases}2x+3 & \text{if } x \leq 1\\ -x+4 & \text{if } x>1\end{cases}$

Graph the function over each interval of the domain separately.

For $(-\infty, 1]$, the graph has an endpoint at $x=1$. Substitute 1 for x in $y=2x+3$ to obtain the ordered pair $(1, 5)$. For another point on this portion of the graph, we choose a number less than 1, such as $x=-2$. This gives the ordered pair $(-2, -1)$. Draw the partial line through these points with an endpoint at $(1, 5)$.

We graph the function over the interval $(1, \infty)$ similarly, using $y=-x+4$. This line has an open endpoint at $(1, 3)$ and passes through $(4, 0)$.

The completed graph is shown in **FIGURE 37.**

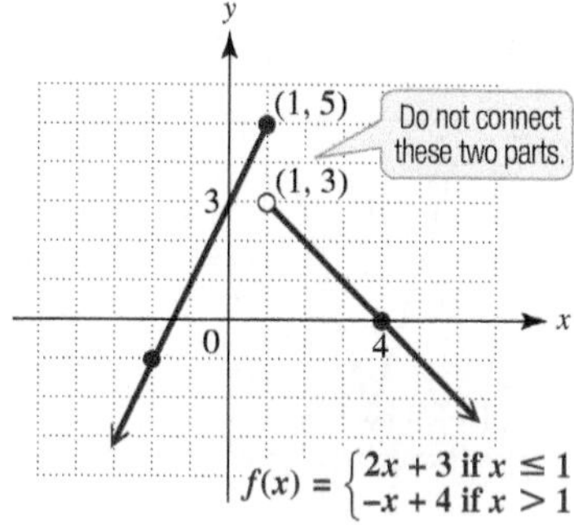

FIGURE 37

NOW TRY 

NOW TRY ANSWER

2.

$f(x)=\begin{cases}\frac{1}{2}x+2 & \text{if } x \geq -2\\ -x-3 & \text{if } x<-2\end{cases}$

CAUTION In **Example 2,** we did not graph the entire lines but only those portions with domain intervals as given. Graphs of these functions should *not* be two intersecting lines.

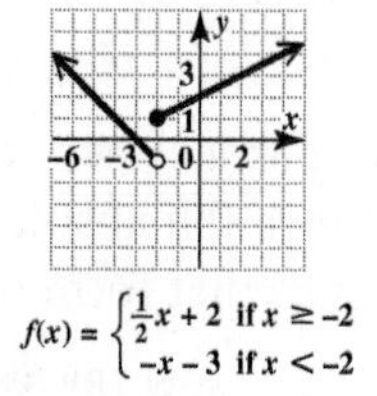

EXAMPLE 3 Applying a Piecewise Linear Function

The table and the graph in **FIGURE 38** show the number of country format commercial radio stations in the United States from 2007 through 2017.

Year	x	y
2007	0	2027
2011	4	1987
2017	10	2121

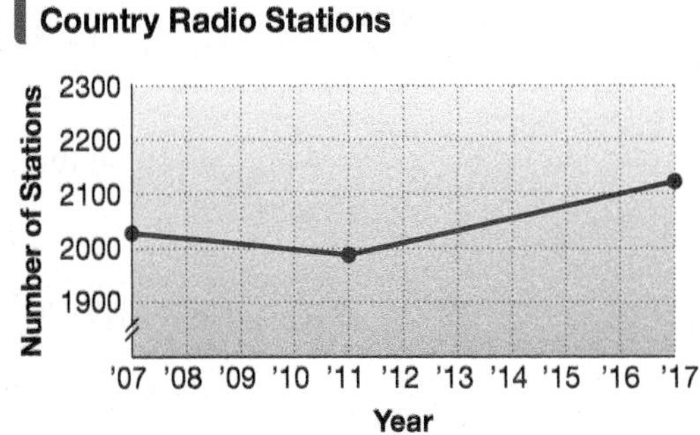

Data from www.insideradio.com

FIGURE 38

(a) Write equations for each part of the graph, and use them to find a piecewise linear function f that models the number of country radio stations over these years. Let $x = 0$ represent 2007, $x = 1$ represent 2008, and so on, and y represent the number of stations.

The data in the table can be used to write the ordered pairs $(0, 2027)$ and $(4, 1987)$. The slope of the line through these two points is

$$m = \frac{1987 - 2027}{4 - 0} = \frac{-40}{4} = -10.$$

Using the ordered pair $(0, 2027)$ and $m = -10$ in the point-slope form of the equation of a line gives an equation of the line for $0 \le x \le 4$—that is, for 2007 through 2011.

$$y - y_1 = m(x - x_1) \quad \text{Point-slope form}$$

Using the point (4, 1987) will result in the same equation.

$$y - 2027 = -10(x - 0) \quad \text{Substitute for } y_1, m, \text{ and } x_1.$$

$$y = -10x + 2027 \quad \text{Solve for } y.$$

The slope of the "other" line is found using (4, 1987) and (10, 2121).

$$m = \frac{2121 - 1987}{10 - 4} = \frac{134}{6} = 22.3 \quad \text{Round to the nearest tenth.}$$

Using the ordered pair $(10, 2121)$ and $m = 22.3$, we find an equation of the line for $4 < x \le 10$—that is, for 2011 through 2017.

$$y - y_1 = m(x - x_1) \quad \text{Point-slope form}$$

$$y - 2121 = 22.3(x - 10) \quad \text{Substitute for } y_1, m, \text{ and } x_1.$$

$$y - 2121 = 22.3x - 223 \quad \text{Distributive property}$$

$$y = 22.3x + 1898 \quad \text{Add 2121.}$$

The number of country stations can be modeled by the piecewise linear function

$$f(x) = \begin{cases} -10x + 2027 & \text{if } 0 \le x \le 4 \\ 22.3x + 1898 & \text{if } 4 < x \le 10. \end{cases}$$

NOW TRY EXERCISE 3

Use the model from **Example 3(a)** to find the number of country stations in 2010 and 2016. Round to the nearest whole number as needed.

(b) Use the model from part (a) to find the number of country stations (to the nearest whole number) in 2015.

For 2015, $x = 2015 - 2007 = 8$. We must find $f(8)$.

$$f(x) = 22.3x + 1898 \quad \text{For } 4 < x \leq 10 \text{, use the second equation.}$$

$$f(8) = 22.3(8) + 1898 \quad \text{Let } x = 8.$$

$$f(8) = 2076 \quad \text{Multiply, and then add.}$$

Based on the model, there were 2076 country stations in 2015.

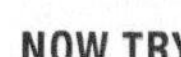

OBJECTIVE 3 Graph step functions.

The greatest integer function is defined as follows.

$f(x) = [\![x]\!]$

The **greatest integer function**

$$f(x) = [\![x]\!]$$

pairs every real number x with the greatest integer less than or equal to x.

NOW TRY EXERCISE 4

Evaluate.

(a) $[\![5]\!]$ (b) $[\![-6]\!]$

(c) $[\![3.5]\!]$ (d) $[\![-4.1]\!]$

EXAMPLE 4 Finding the Greatest Integer

Evaluate.

(a) $[\![8]\!] = 8$ **(b)** $[\![-1]\!] = -1$ **(c)** $[\![0]\!] = 0$ If x is an integer, then $[\![x]\!] = x$.

(d) $[\![7.45]\!] = 7$ The greatest integer *less than or equal to* 7.45 is 7. This is like "rounding down."

(e) $[\![-2.6]\!] = -3$ Think of a number line with -2.6 graphed on it. Because -3 is to the *left of* (and therefore *less than*) -2.6, the greatest integer less than or equal to -2.6 is -3, ***not*** -2.

-2.6
-3 -2 -1 0

NOW TRY

EXAMPLE 5 Graphing the Greatest Integer Function

Graph $f(x) = [\![x]\!]$. Give the domain and range.

$$\begin{aligned}
\text{For } [\![x]\!],\ &\text{if } -1 \leq x < 0, && \text{then } [\![x]\!] = -1;\\
&\text{if } 0 \leq x < 1, && \text{then } [\![x]\!] = 0;\\
&\text{if } 1 \leq x < 2, && \text{then } [\![x]\!] = 1;\\
&\text{if } 2 \leq x < 3, && \text{then } [\![x]\!] = 2;\\
&\text{if } 3 \leq x < 4, && \text{then } [\![x]\!] = 3, \quad \text{and so on.}
\end{aligned}$$

The graph, shown in **FIGURE 39** on the next page, consists of a series of horizontal line segments. In each segment, the left endpoint is included and the right endpoint is excluded. These segments continue infinitely following this pattern to the left and right. The appearance of the graph is the reason why this function is an example of a **step function.**

NOW TRY ANSWERS

3. 2010: 1997 country stations; 2016: 2099 country stations
4. (a) 5 (b) −6 (c) 3 (d) −5

NOW TRY EXERCISE 5

Graph $f(x) = [\![x - 1]\!]$. Give the domain and range.

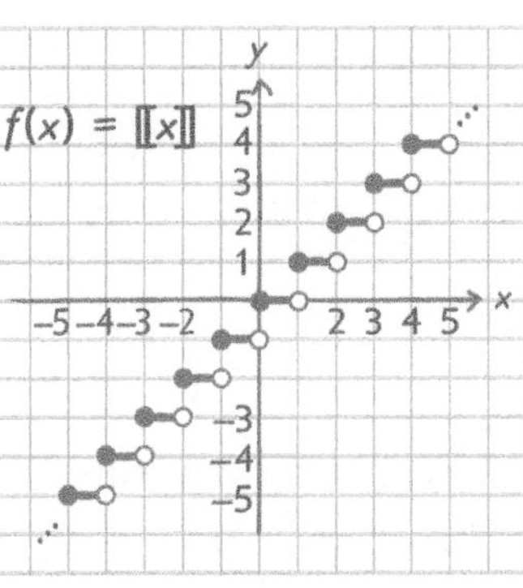

FIGURE 39

Greatest integer function

$$f(x) = [\![x]\!]$$

Domain: $(-\infty, \infty)$

Range: $\{\ldots, -3, -2, -1, 0, 1, 2, 3, \ldots\}$ (the set of integers)

The ellipsis points indicate that the graph continues indefinitely in the same pattern.

The graph of a step function also may be shifted. For example, the graph of

$h(x) = [\![x - 2]\!]$ is the graph of $f(x) = [\![x]\!]$ shifted to the right 2 units.

Similarly, the graph of

$g(x) = [\![x]\!] + 2$ is the graph of $f(x)$ shifted up 2 units. NOW TRY 

NOW TRY EXERCISE 6

Graph $f(x) = \left[\!\left[\frac{1}{3}x - 1\right]\!\right]$.

EXAMPLE 6 Graphing a Greatest Integer Function $f(x) = [\![ax + b]\!]$

Graph $f(x) = \left[\!\left[\frac{1}{2}x + 1\right]\!\right]$.

Some sample ordered pairs are given in the table.

x	0	$\frac{1}{2}$	1	$\frac{3}{2}$	2	3	4	−1	−2	−3
y	1	1	1	1	2	2	3	0	0	−1

For x in the interval $[0, 2)$, $y = 1$.

For x in the interval $[2, 4)$, $y = 2$.

For x in the interval $[-2, 0)$, $y = 0$.

For x in the interval $[-4, -2)$, $y = -1$.

The graph is shown in **FIGURE 40**. Again, the domain of the function is $(-\infty, \infty)$. The range is the set of integers, $\{\ldots, -2, -1, 0, 1, 2, \ldots\}$.

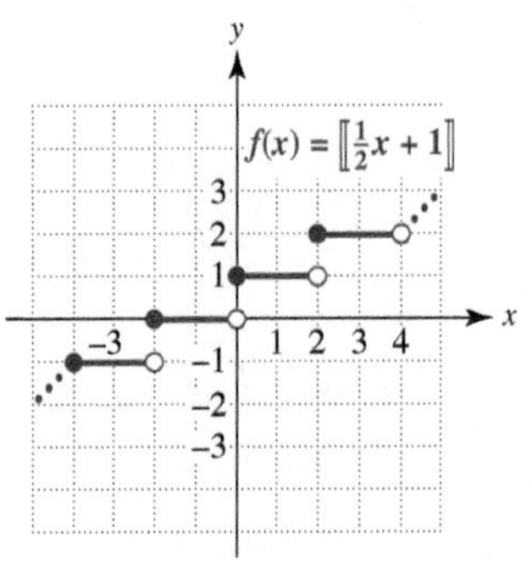

FIGURE 40

NOW TRY 

NOW TRY ANSWERS

5. 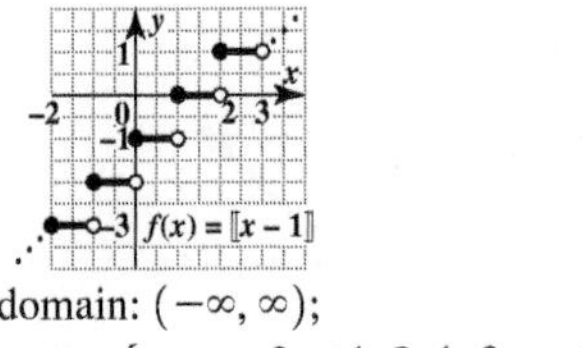

domain: $(-\infty, \infty)$;
range: $\{\ldots, -2, -1, 0, 1, 2, \ldots\}$

6. 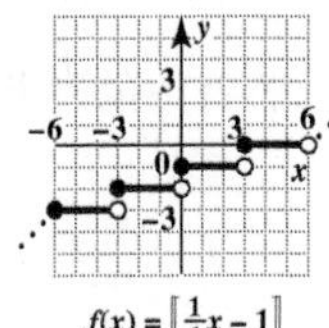

NOW TRY EXERCISE 7

The cost of parking a car at an airport hourly parking lot is \$4 for the first hour and \$2 for each additional hour or fraction of an hour. Let $y = f(x)$ represent the cost of parking a car for x hours. Graph $f(x)$ for x in the interval $(0, 5]$.

EXAMPLE 7 Applying a Greatest Integer Function

An overnight delivery service charges \$25 for a package weighing up to 2 lb. For each additional pound or fraction of a pound, there is an additional charge of \$3. Let $y = D(x)$ represent the cost to send a package weighing x pounds. Graph $D(x)$ for x in the interval $(0, 6]$.

For x in the interval $(0, 2]$, $y = 25$.

For x in the interval $(2, 3]$, $y = 25 + 3 = 28$.

For x in the interval $(3, 4]$, $y = 28 + 3 = 31$.

For x in the interval $(4, 5]$, $y = 31 + 3 = 34$.

For x in the interval $(5, 6]$, $y = 34 + 3 = 37$.

The graph, which is that of a step function, is shown in **FIGURE 41**.

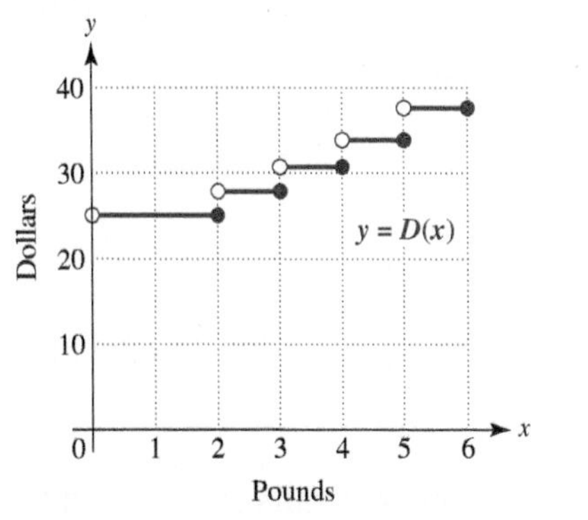

FIGURE 41

NOW TRY ANSWER

7. Dollars, Hours, $y = f(x)$

NOW TRY 

9.5 Exercises

FOR EXTRA HELP  MyLab Math

Video solutions for select problems available in MyLab Math

Concept Check *Refer to the basic graphs in choices A–F, and answer each question in Exercises 1–6.*

A.

B.

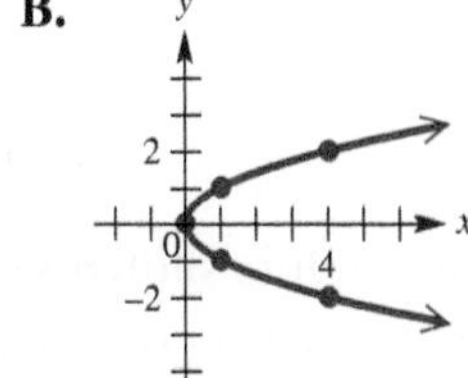

C.

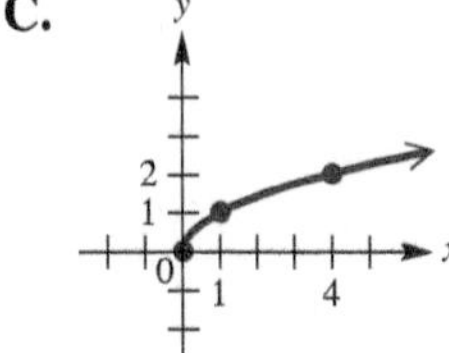

D.

E.

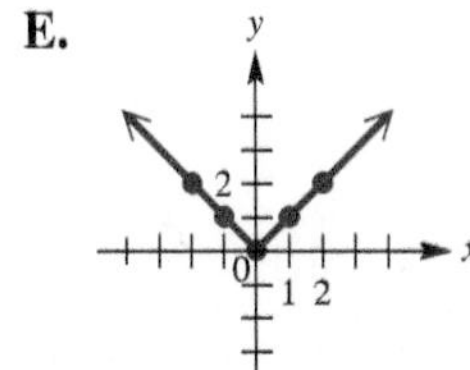

F.

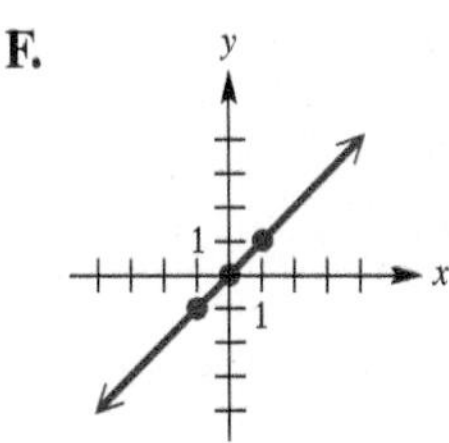

1. Which is the graph of $f(x) = |x|$? What is the largest open interval over which it is increasing?

2. Which is the graph of $f(x) = x^2$? Give the domain and range.

3. Which is the graph of $f(x) = [\![x]\!]$? What is the value of y when $x = 1.5$?

4. Which is the graph of $f(x) = \sqrt{x}$? Give the domain and range.

5. Which is not the graph of a function? Why?

6. Which graph is (or graphs are) symmetric with respect to the x-axis? The y-axis?

Concept Check *Without actually plotting points, match each function with its graph from choices A–D.*

7. $f(x) = |x - 2| + 2$

8. $f(x) = |x + 2| + 2$

9. $f(x) = |x - 2| - 2$

10. $f(x) = |x + 2| - 2$

A.

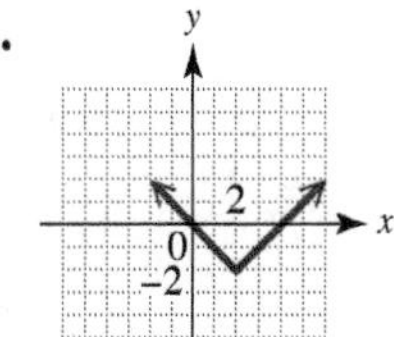

B.

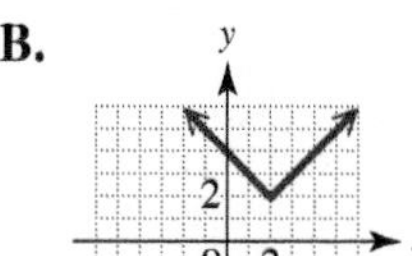

C.

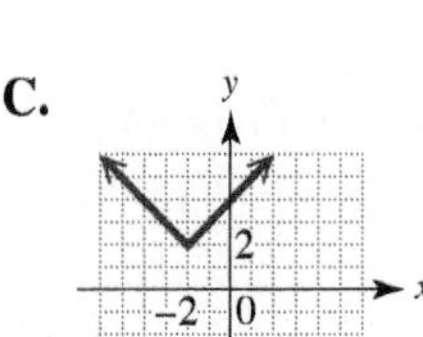

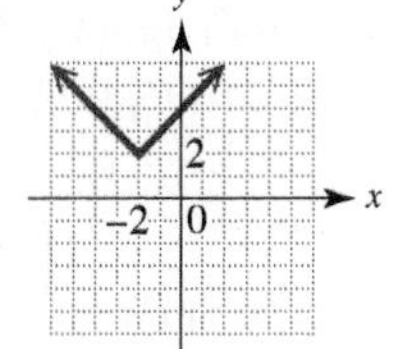

D.

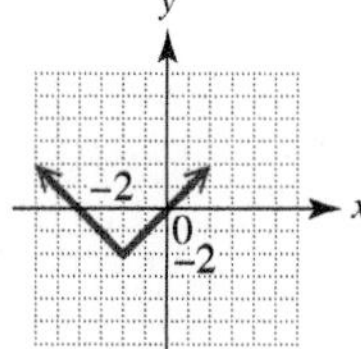

Graph each absolute value function. ***See Example 1.***

11. $f(x) = |x + 1|$

12. $f(x) = |x - 1|$

13. $f(x) = |2 - x|$

14. $f(x) = |-3 - x|$

15. $y = |x| + 4$

16. $y = |x| + 3$

17. $y = 3|x - 2| - 1$

18. $y = \frac{1}{2}|x + 3| + 1$

For each piecewise linear function, find ***(a)*** $f(-5)$, ***(b)*** $f(-1)$, ***(c)*** $f(0)$, ***(d)*** $f(3)$, *and* ***(e)*** $f(5)$. ***See Example 2.***

19. $f(x) = \begin{cases} 2x & \text{if } x \leq -1 \\ x - 1 & \text{if } x > -1 \end{cases}$

20. $f(x) = \begin{cases} 3x + 5 & \text{if } x \leq 0 \\ x & \text{if } x > 0 \end{cases}$

21. $f(x) = \begin{cases} 2 & \text{if } x \leq 0 \\ -6 & \text{if } x > 0 \end{cases}$

22. $f(x) = \begin{cases} 8 & \text{if } x < 0 \\ 10 & \text{if } x \geq 0 \end{cases}$

Graph each piecewise linear function. ***See Example 2.***

23. $f(x) = \begin{cases} 4 - x & \text{if } x < 2 \\ 1 + 2x & \text{if } x \geq 2 \end{cases}$

24. $f(x) = \begin{cases} 8 - x & \text{if } x \leq 3 \\ 3x - 6 & \text{if } x > 3 \end{cases}$

25. $f(x) = \begin{cases} x - 1 & \text{if } x \leq 3 \\ 2 & \text{if } x > 3 \end{cases}$

26. $f(x) = \begin{cases} x + 2 & \text{if } x \geq 1 \\ 3 & \text{if } x < 1 \end{cases}$

27. $f(x) = \begin{cases} 2x + 1 & \text{if } x \geq 0 \\ x & \text{if } x < 0 \end{cases}$

28. $f(x) = \begin{cases} 5x - 4 & \text{if } x \geq 1 \\ x & \text{if } x < 1 \end{cases}$

29. $f(x) = \begin{cases} -3 & \text{if } x \leq 1 \\ -1 & \text{if } x > 1 \end{cases}$

30. $f(x) = \begin{cases} -2 & \text{if } x \leq 1 \\ 2 & \text{if } x > 1 \end{cases}$

31. $f(x) = \begin{cases} 2 + x & \text{if } x < -4 \\ -x & \text{if } -4 \le x \le 5 \\ 3x & \text{if } x > 5 \end{cases}$

32. $f(x) = \begin{cases} -2x & \text{if } x < -3 \\ 3x - 1 & \text{if } -3 \le x \le 2 \\ -4x & \text{if } x > 2 \end{cases}$

Extending Skills *Graph each piecewise function. (Hint: At least one part is nonlinear.)*

33. $f(x) = \begin{cases} 2 + x & \text{if } x < -4 \\ -x^2 & \text{if } x \ge -4 \end{cases}$

34. $f(x) = \begin{cases} -2x & \text{if } x \le 2 \\ -x^2 & \text{if } x > 2 \end{cases}$

35. $f(x) = \begin{cases} |x| & \text{if } x > -2 \\ x^2 - 2 & \text{if } x \le -2 \end{cases}$

36. $f(x) = \begin{cases} |x| - 1 & \text{if } x > -1 \\ x^2 - 1 & \text{if } x \le -1 \end{cases}$

The snow depth in a particular location varies throughout the winter. In a typical winter, the snow depth in inches might be approximated by the following function.

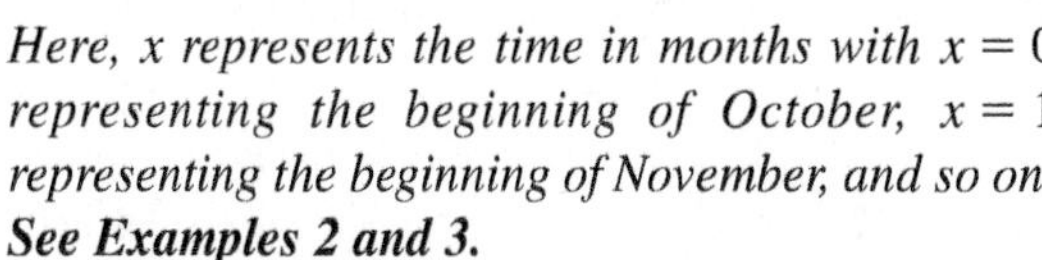

$$f(x) = \begin{cases} 6.5x & \text{if } 0 \le x \le 4 \\ -5.5x + 48 & \text{if } 4 < x \le 6 \\ -30x + 195 & \text{if } 6 < x \le 6.5 \end{cases}$$

Here, x represents the time in months with $x = 0$ representing the beginning of October, $x = 1$ representing the beginning of November, and so on. ***See Examples 2 and 3.***

37. Graph $y = f(x)$.

38. In what month is the snow deepest? What is the deepest snow depth?

39. What is the snow depth at the beginning of January? At the beginning of April?

40. In what months does the snow begin and end?

Solve each problem. ***See Example 3.***

41. The graph shows the number of news/talk radio stations in the United States over the years 2007 through 2017.

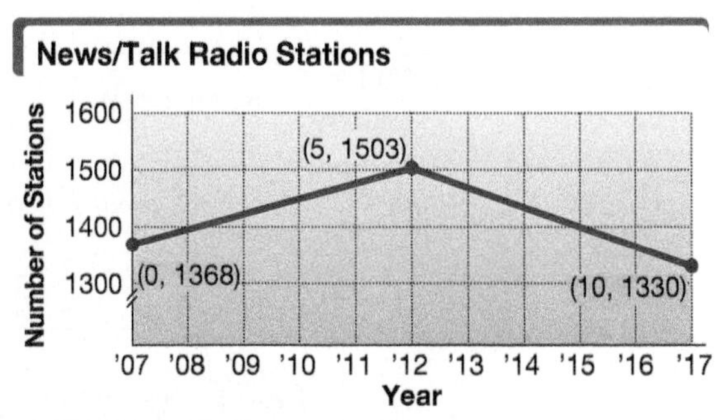

Data from www.insideradio.com

(a) Write a piecewise linear function f that models the number of news/talk stations over these years. Let $x = 0$ represent 2007, $x = 1$ represent 2008, and so on, and y represent the number of stations.

(b) Use the model from part (a) to find the number of news/talk stations (to the nearest whole number) in 2010 and 2016.

42. The graph shows the unemployment rate (percent) in the United States over the years 2007 through 2016.

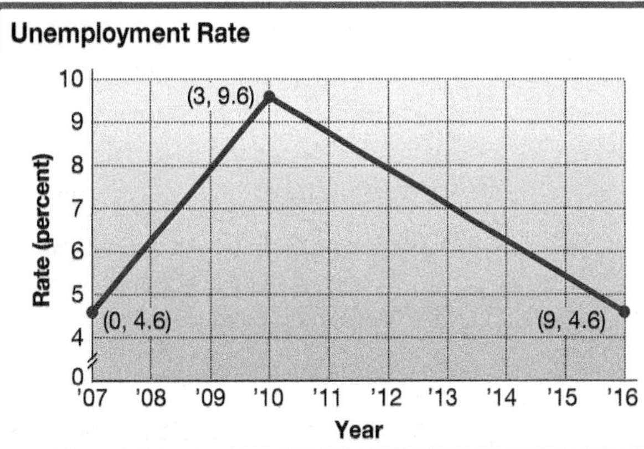

Data from Bureau of Labor Statistics.

(a) Write a piecewise linear function f that models the unemployment rate over these years. Let $x = 0$ represent 2007, $x = 1$ represent 2008, and so on, and y represent the percent of unemployment. (Round to the nearest tenth, as needed.)

(b) Use the model from part (a) to find the unemployment rate in 2009 and 2015. (Round to the nearest percent, as needed.)

43. Concept Check A student incorrectly evaluated $[\![-5.1]\!]$ as -5. **WHAT WENT WRONG?** Evaluate correctly.

44. Concept Check A student incorrectly evaluated $[\![3.75]\!]$ as 4. **WHAT WENT WRONG?** Evaluate correctly.

Evaluate each expression. ***See Example 4.***

45. $[\![3]\!]$ **46.** $[\![18]\!]$ **47.** $[\![4.5]\!]$

48. $[\![8.7]\!]$ **49.** $\left[\!\!\left[\frac{1}{2}\right]\!\!\right]$ **50.** $\left[\!\!\left[\frac{3}{4}\right]\!\!\right]$

51. $\left[\!\!\left[\frac{8}{3}\right]\!\!\right]$ **52.** $\left[\!\!\left[\frac{5}{2}\right]\!\!\right]$ **53.** $[\![-14]\!]$

54. $[\![-5]\!]$ **55.** $[\![-10.1]\!]$ **56.** $[\![-6.9]\!]$

Graph each step function. ***See Examples 5 and 6.***

57. $f(x) = [\![x]\!] - 1$ **58.** $f(x) = [\![x]\!] + 1$

59. $f(x) = [\![x - 3]\!]$ **60.** $f(x) = [\![x + 2]\!]$

61. $f(x) = [\![-x]\!]$ **62.** $f(x) = [\![2x]\!]$

63. $f(x) = [\![2x - 1]\!]$ **64.** $f(x) = [\![3x + 1]\!]$

65. $f(x) = [\![3x]\!]$ **66.** $f(x) = [\![3x]\!] + 1$

Work each problem. ***See Example 7.***

67. Suppose a chain-saw rental costs a fixed \$4 sharpening fee, plus \$7 per day or fraction of a day. Let $S(x)$ represent the cost of renting a saw for x days. Find each of the following.

(a) $S(1)$ **(b)** $S(1.25)$ **(c)** $S(3.5)$ **(d)** Graph $y = S(x)$.

68. To rent a midsized car costs \$30 per day or fraction of a day. If we pick up the car in Lansing and drop it in West Lafayette, there is a fixed \$50 dropoff charge. Let $C(x)$ represent the cost of renting the car for x days, taking it from Lansing to West Lafayette. Find each of the following.

(a) $C\left(\frac{3}{4}\right)$ **(b)** $C\left(\frac{9}{10}\right)$ **(c)** $C(1)$ **(d)** $C\left(1\frac{5}{8}\right)$

(e) $C(2.4)$ **(f)** Graph $y = C(x)$.

69. Suppose that postage rates are \$0.55 for the first ounce, plus \$0.24 for each additional ounce, and that each letter carries one \$0.55 stamp and as many \$0.24 stamps as necessary. Graph the function

$$y = p(x) = \text{the number of stamps}$$

on a letter weighing x ounces. Use the interval $(0, 5]$.

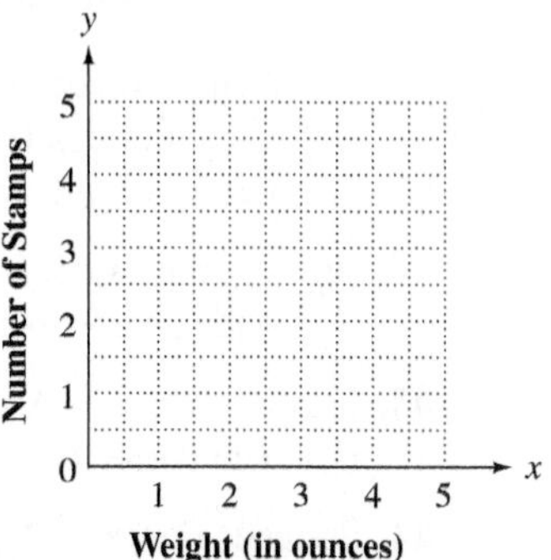

70. The cost of parking a car at an airport hourly parking lot is \$3 for the first half-hour and \$2 for each additional half-hour or fraction thereof. Graph the function

$$y = f(x) = \text{the cost of parking a car}$$

for x hours. Use the interval $(0, 2]$.

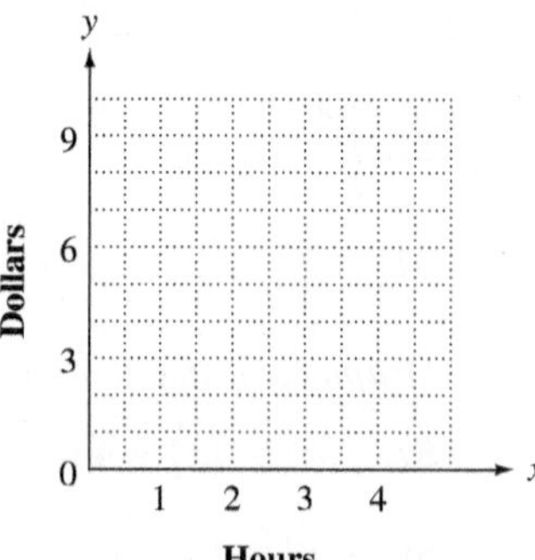

71. A certain long-distance carrier provides service between Podunk and Nowhereville. If x represents the number of minutes for a call, where $x > 0$, then the function

$$f(x) = 0.40[\![x]\!] + 0.75$$

gives the total cost of the call in dollars. Find the cost of a 5.5-minute call.

72. Total rental cost in dollars for a power washer, where x represents the number of hours with $x > 0$, can be represented by the function

$$f(x) = 12[\![x]\!] + 25.$$

Find the cost of a $7\frac{1}{2}$-hr rental.

Chapter 9 Summary

STUDY SKILLS **REMINDER**

Be prepared for your math test over this chapter. **Review Study Skills 7 and 8, *Reviewing a Chapter* and *Taking Math Tests.***

Key Terms

9.1

difference quotient
composite function (composition of functions)

9.2

parabola
vertex
axis of symmetry (axis)
quadratic function

9.4

symmetric with respect to the y-axis
symmetric with respect to the x-axis
symmetric with respect to the origin
increasing function
decreasing function
constant function

9.5

piecewise linear function
absolute value function
greatest integer function
step function

New Symbols

$(f \circ g)(x) = f(g(x))$ composite function

$[\![x]\!]$ greatest integer function

Test Your Word Power

See how well you have learned the vocabulary in this chapter.

1. A **quadratic function** is a function that can be written in the form
   A. $f(x) = mx + b$ for real numbers m and b
   B. $f(x) = \frac{P(x)}{Q(x)}$, where $Q(x) \neq 0$
   C. $f(x) = ax^2 + bx + c$ for real numbers a, b, and c, where $a \neq 0$
   D. $f(x) = \sqrt{x}$ for $x \geq 0$.

2. A **parabola** is the graph of
   A. any equation in two variables
   B. a linear equation
   C. an equation of degree 3
   D. a quadratic equation in two variables.

3. The **vertex** of a parabola is
   A. the point where the graph intersects the y-axis
   B. the point where the graph intersects the x-axis
   C. the lowest point on a parabola that opens up or the highest point on a parabola that opens down
   D. the origin.

4. The **axis** of a parabola is
   A. either the x-axis or the y-axis
   B. the vertical line (of a vertical parabola) or the horizontal line (of a horizontal parabola) through the vertex
   C. the lowest or highest point on the graph of a parabola
   D. a line through the origin.

5. A parabola is **symmetric about its axis** because
   A. its graph is near the axis
   B. its graph is identical on each side of the axis
   C. its graph looks different on each side of the axis
   D. its graph intersects the axis.

6. An **absolute value function** is a function that can be written in the form
   A. $f(x) = ax^2 + bx + c$, where $a \neq 0$
   B. $f(x) = a(x - h)^2 + k$, where $a \neq 0$
   C. $f(x) = a|x - h| + k$, where $a \neq 0$
   D. $f(x) = [\![x - h]\!] + k$.

ANSWERS

1. C; *Examples:* $f(x) = x^2 - 2$, $f(x) = (x + 4)^2 + 1$, $f(x) = x^2 - 4x + 5$ **2.** D; *Example:* The quadratic equation $y = x^2$ has a parabola as its graph. **3.** C; *Example:* The graph of $y = (x + 3)^2$ has vertex $(-3, 0)$, which is the lowest point on the graph. **4.** B; *Example:* The axis of $y = (x + 3)^2$ is the vertical line $x = -3$. **5.** B; *Example:* Because the graph of $y = (x + 3)^2$ is symmetric with respect to its axis $x = -3$, the points $(-2, 1)$ and $(-4, 1)$ lie on the graph. **6.** C; *Examples:* $f(x) = |x - 1|$, $f(x) = 2|x| + 5$, $f(x) = -|x|$

Quick Review

CONCEPTS	EXAMPLES
9.1 Review of Operations and Composition **Operations on Functions** $(f+g)(x) = f(x) + g(x)$ Sum $(f-g)(x) = f(x) - g(x)$ Difference $(fg)(x) = f(x) \cdot g(x)$ Product $\left(\frac{f}{g}\right)(x) = \frac{f(x)}{g(x)}, \; g(x) \neq 0$ Quotient	Let $f(x) = \sqrt{x}$ and $g(x) = 2x - 4$. $(f+g)(x) = \sqrt{x} + 2x - 4$ $(f-g)(x) = \sqrt{x} - 2x + 4$ $(fg)(x) = \sqrt{x}(2x-4)$ The domain is $[0, \infty)$. $\left(\frac{f}{g}\right)(x) = \frac{\sqrt{x}}{2x-4}$ The domain is $[0, 2) \cup (2, \infty)$.
Composition of Functions If f and g are functions, then the composite function of f and g is $(f \circ g)(x) = f(g(x))$. The domain of $f \circ g$ is the set of all x in the domain of g such that $g(x)$ is in the domain of f.	Using f and g as defined above, $(f \circ g)(x) = f(g(x)) = \sqrt{2x-4}$. The domain is all x such that $2x - 4 \geq 0$. Solving this inequality gives the interval $[2, \infty)$.
9.2 Graphs of Quadratic Functions The graph of the quadratic function $F(x) = a(x-h)^2 + k$ **(where** $a \neq 0$**)** is a parabola. • The vertex of the parabola is (h, k). • The axis of symmetry is the vertical line $x = h$. • The graph opens up if $a > 0$ and down if $a < 0$. • The graph is wider than the graph of $f(x) = x^2$ if $0 < \lvert a \rvert < 1$ and narrower if $\lvert a \rvert > 1$.	Graph $f(x) = -(x+3)^2 + 1$. 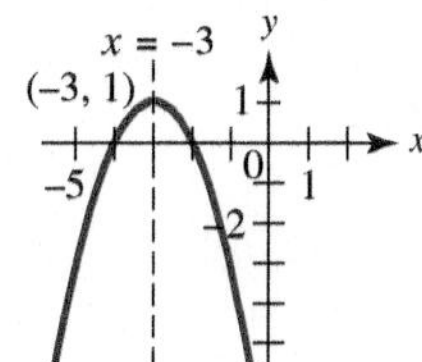 The graph opens down because $a < 0$. Vertex: $(-3, 1)$ Axis of symmetry: $x = -3$ Domain: $(-\infty, \infty)$ Range: $(-\infty, 1]$
9.3 More about Parabolas and Their Applications The vertex of the graph of $f(x) = ax^2 + bx + c$ (where $a \neq 0$) may be found by completing the square or using the vertex formula $\left(\frac{-b}{2a}, f\left(\frac{-b}{2a}\right)\right)$. **Graphing a Quadratic Function** *Step 1* Determine whether the graph opens up or down. *Step 2* Find the vertex. *Step 3* Find any intercepts. *Step 4* Find and plot additional points as needed.	Graph $f(x) = x^2 + 4x + 3$. 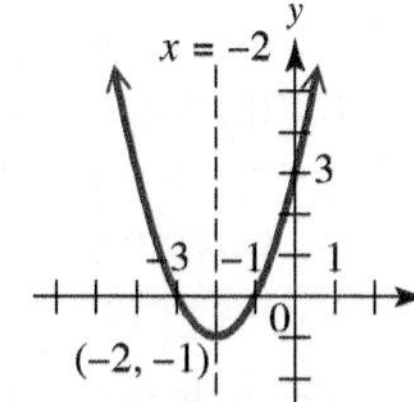 The graph opens up because $a > 0$. Vertex: $(-2, -1)$ The solutions of $x^2 + 4x + 3 = 0$ are -1 and -3, so the x-intercepts are $(-1, 0)$ and $(-3, 0)$. $f(0) = 3$, so the y-intercept is $(0, 3)$. Axis of symmetry: $x = -2$ Domain: $(-\infty, \infty)$ Range: $[-1, \infty)$
Horizontal Parabolas The graph of $x = ay^2 + by + c$ or $x = a(y-k)^2 + h$ is a horizontal parabola. • The vertex of the parabola is (h, k) • The axis of symmetry is the horizontal line $y = k$. • The graph opens to the right if $a > 0$ and to the left if $a < 0$. ***Horizontal parabolas do not represent functions.***	Graph $x = 2y^2 + 6y + 5$. 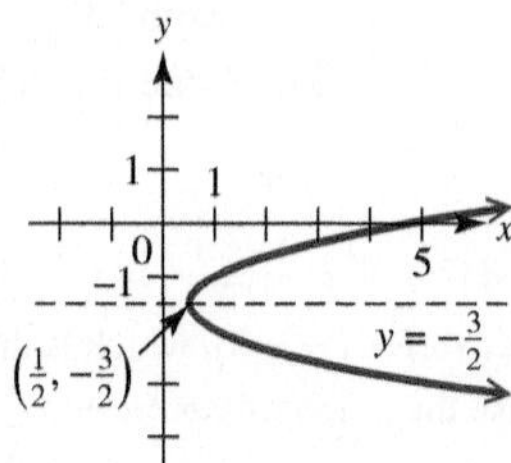 The graph opens to the right because $a > 0$. Vertex: $\left(\frac{1}{2}, -\frac{3}{2}\right)$ Axis of symmetry: $y = -\frac{3}{2}$ Domain: $\left[\frac{1}{2}, \infty\right)$ Range: $(-\infty, \infty)$

CONCEPTS	EXAMPLES

9.4 Symmetry; Increasing and Decreasing Functions

Reflections, Stretching, and Shrinking
The graph of $g(x) = a \cdot f(x)$ has the same general shape as the graph of $f(x)$.

- If $a < 0$, then the graph is reflected across the x-axis.
- If $|a| > 1$, then the graph is stretched vertically compared to the graph of $f(x)$.
- If $0 < |a| < 1$, then the graph is shrunken vertically compared to the graph of $f(x)$.

Graph $g(x) = -\frac{1}{2}\sqrt{x}$.

x	$f(x) = \sqrt{x}$	$g(x) = -\frac{1}{2}\sqrt{x}$
0	0	0
1	1	$-\frac{1}{2}$
4	2	-1
9	3	$-\frac{3}{2}$

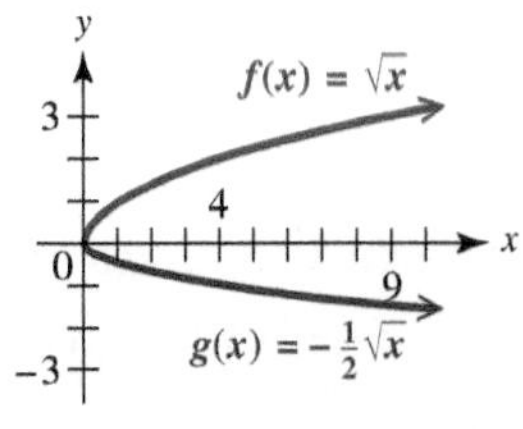

The graph of $g(x)$ has the same general shape as the graph of $f(x) = \sqrt{x}$.

The graph is reflected across the x-axis because $a = -\frac{1}{2}$ and $a < 0$.

The graph is shrunken vertically by a factor of $\frac{1}{2}$.

Symmetry
Perform the indicated test to decide whether the graph of a relation is symmetric with respect to the following.

Test each relation for symmetry.

(a) The x-axis
Replace y with $-y$.

(a)
$$x = y^2 - 5$$
$$x = (-y)^2 - 5 \quad \text{Equivalent}$$
$$x = y^2 - 5$$

The graph of $x = y^2 - 5$ is symmetric with respect to the x-axis.

(b) The y-axis
Replace x with $-x$.

(b)
$$y = -2x^2 + 1$$
$$y = -2(-x)^2 + 1 \quad \text{Equivalent}$$
$$y = -2x^2 + 1$$

The graph of $y = -2x^2 + 1$ is symmetric with respect to the y-axis.

(c) The origin
Replace x with $-x$ and y with $-y$ at the same time.

(c)
$$x^2 + y^2 = 1$$
$$(-x)^2 + (-y)^2 = 1 \quad \text{Equivalent}$$
$$x^2 + y^2 = 1$$

The symmetry exists if the replacement leads to an equation equivalent to the original one.

The graph of $x^2 + y^2 = 4$ is symmetric with respect to the origin (and to the x-axis and y-axis).

Increasing, Decreasing, and Constant Functions
Suppose that a function f is defined over an *open* interval I, and x_1 and x_2 are in I.

Determine the open intervals over which the function is increasing, decreasing, or constant.

A function f is **increasing** on I
if $f(x_1) < f(x_2)$ whenever $x_1 < x_2$.

f is increasing on $(-\infty, a)$.

A function f is **decreasing** on I
if $f(x_1) > f(x_2)$ whenever $x_1 < x_2$.

f is decreasing on (a, b).

A function f is **constant** on I
if $f(x_1) = f(x_2)$ for all x_1 and x_2 in I.

f is constant on (b, ∞).

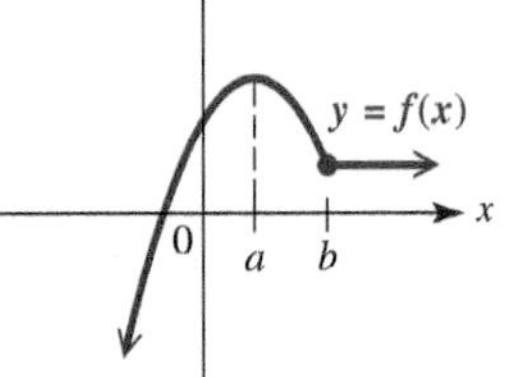

CONCEPTS	EXAMPLES

9.5 Piecewise Linear Functions

Absolute Value Function

$$f(x) = |x| = \begin{cases} x & \text{if } x \geq 0 \\ -x & \text{if } x < 0 \end{cases}$$

Compared to the graph of $f(x) = |x|$, the graph of

$$g(x) = a|x - h| + k$$

is shifted h units horizontally and k units vertically. The value of a affects whether the graph is reflected, stretched, or shrunken.

Graph $f(x) = 2|x - 1| + 3$.

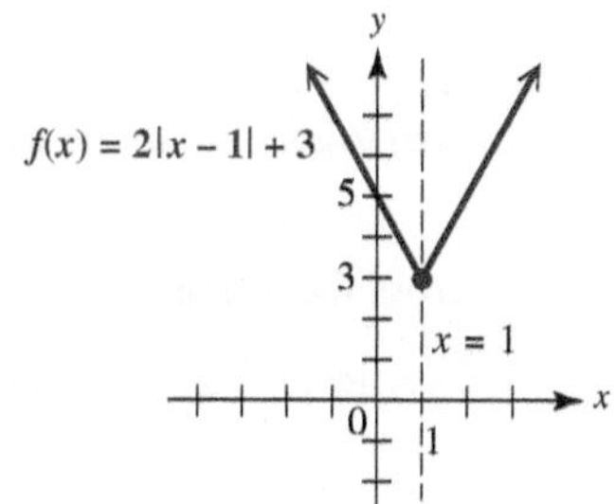

The graph opens up because $a > 0$.

The graph is stretched vertically because $a > 1$.

The graph is shifted to the right 1 unit and up 3 units.

Axis of symmetry: $x = 1$

Other Piecewise Linear Functions

Graph the function over each interval of the domain separately. Use open or closed endpoints as appropriate.

Greatest Integer Function

$$f(x) = [\![x]\!]$$

$[\![x]\!]$ is the greatest integer less than or equal to x.

Graph each function.

$$f(x) = \begin{cases} x - 2 & \text{if } x \geq 1 \\ 3x & \text{if } x < 1 \end{cases}$$

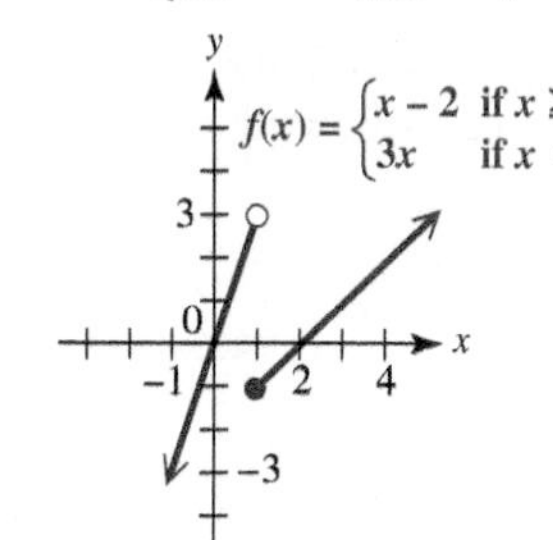

$$f(x) = [\![2x - 1]\!]$$

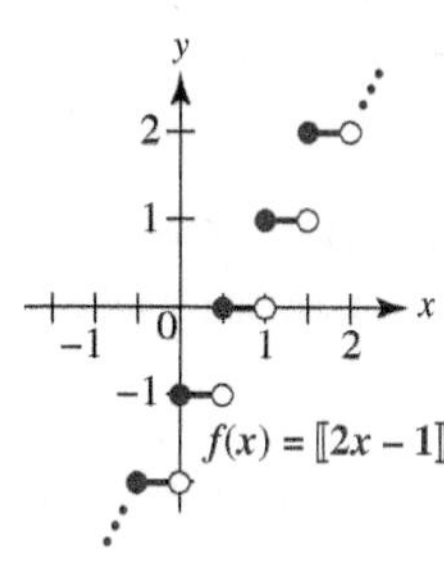

Chapter 9 Review Exercises

9.1 *Let $f(x) = x^2 - 2x$ and $g(x) = 5x + 3$. Find the following. Give the domain of each.*

1. $(f + g)(x)$ **2.** $(f - g)(x)$ **3.** $(fg)(x)$

4. $\left(\frac{f}{g}\right)(x)$ **5.** $(g \circ f)(x)$ **6.** $(f \circ g)(x)$

Let $f(x) = 2x - 3$ and $g(x) = \sqrt{x}$. Find each of the following.

7. $(f - g)(4)$ **8.** $\left(\frac{f}{g}\right)(9)$ **9.** $(fg)(5)$

10. $(f + g)(5)$ **11.** $(f \circ g)(1)$ **12.** $(g \circ f)(2)$

13. Concept Check After working **Exercise 12,** find $(f \circ g)(2)$. Are the answers equal? Is composition of functions a commutative operation?

14. Why is 5 not in the domain of $f(x) = \sqrt{9 - 2x}$?

9.2, 9.3 *Identify the vertex of each parabola.*

15. $f(x) = -(x - 1)^2$ **16.** $f(x) = (x - 3)^2 + 7$

17. $x = (y - 3)^2 - 4$ **18.** $y = -3x^2 + 4x - 2$

Graph each parabola. Give the vertex, axis of symmetry, domain, and range.

19. $y = 2(x - 2)^2 - 3$

20. $f(x) = -2x^2 + 8x - 5$

21. $x = 2(y + 3)^2 - 4$

22. $x = -\frac{1}{2}y^2 + 6y - 14$

Solve each problem.

23. The height (in feet) of a projectile t seconds after being fired from Earth into the air is given by

$$f(t) = -16t^2 + 160t.$$

Find the number of seconds required for the projectile to reach maximum height. What is the maximum height?

24. Find the length and width of a rectangle having a perimeter of 200 m if the area is to be a maximum. What is the maximum area?

9.4 *Use the tests for symmetry to decide whether the graph of each relation is symmetric with respect to the x-axis, the y-axis, or the origin. More than one of these symmetries, or none of them, may apply.*

25. $2x^2 - y^2 = 4$

26. $3x^2 + 4y^2 = 12$

27. $2x - y^2 = 8$

28. $y = 2x^2 + 3$

29. $y = 2\sqrt{x} - 4$

30. $y = \frac{1}{x^2}$

31. Suppose that a circle has its center at the origin. Is it symmetric with respect to **(a)** the x-axis, **(b)** the y-axis, **(c)** the origin?

32. Suppose that a linear function in the form $f(x) = mx + b$ has $m < 0$. Is it increasing or decreasing over all real numbers?

*Determine the open intervals over which each function is **(a)** increasing, **(b)** decreasing, and **(c)** constant.*

33.

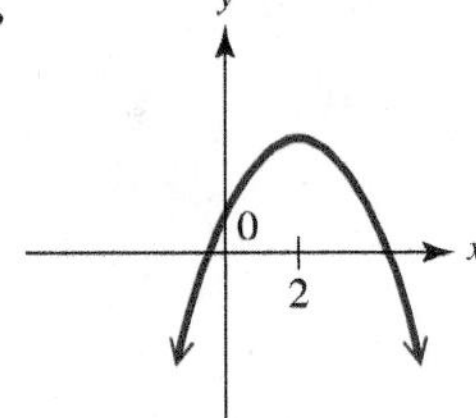

34.

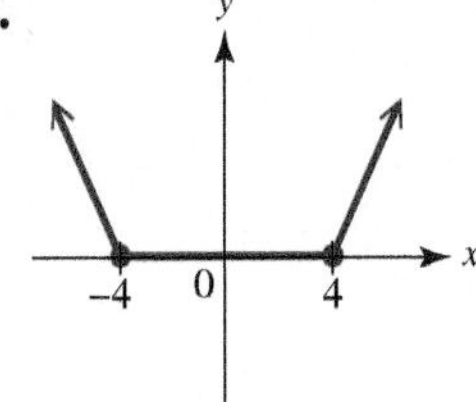

35.

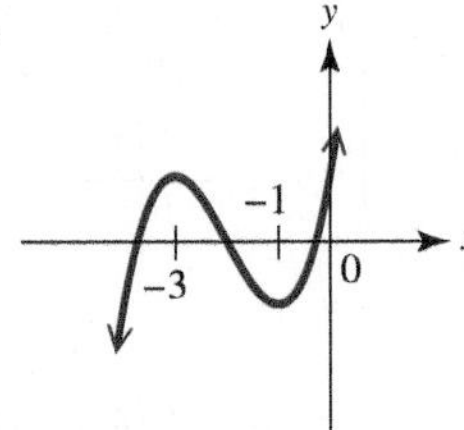

36.

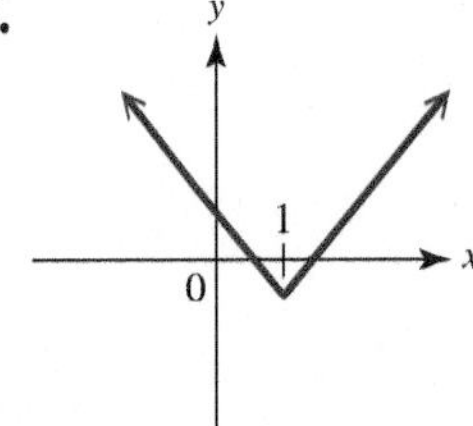

9.5 *Graph each function.*

37. $f(x) = |x - 2|$

38. $f(x) = -|x - 1|$

39. $f(x) = \begin{cases} 2x + 1 & \text{if } x \leq -1 \\ x + 3 & \text{if } x > -1 \end{cases}$

40. $f(x) = \begin{cases} 3 & \text{if } x < -2 \\ 2 - \frac{1}{2}x & \text{if } x \geq -2 \end{cases}$

41. $f(x) = -[\![x]\!]$

42. $f(x) = [\![x + 1]\!]$

Work each problem.

43. Describe how the graph of $y = 2|x + 4| - 3$ can be obtained from the graph of $y = |x|$.

44. Taxi rates in a small town are \$0.90 for the first $\frac{1}{9}$ mi and \$0.10 for each additional $\frac{1}{9}$ mi or fraction of $\frac{1}{9}$ mi. Let $y = C(x)$ represent the cost in dollars for a taxi ride of $\frac{x}{9}$ mile(s). Find each of the following.

(a) $C(1)$ **(b)** $C(2.3)$ **(c)** $C(8)$

(d) Graph $y = C(x)$. Use the interval $(0, 4]$.

Chapter 9 Mixed Review Exercises

Let $f(x) = 2x^2 - 3x + 2$ and $g(x) = -2x + 1$. Find each of the following. Simplify the expressions when possible.

1. (a) $(f - g)(x)$ **(b)** $\left(\frac{f}{g}\right)(x)$

(c) the domain of $\frac{f}{g}$ **(d)** $\frac{f(x + h) - f(x)}{h} \quad (h \neq 0)$

2. (a) $(f + g)(1)$ **(b)** $(fg)(2)$ **(c)** $(f \circ g)(0)$

Work each problem.

3. Match each equation in parts (a)–(f) with the figure that most closely resembles its graph in choices A–F.

(a) $g(x) = x^2 - 5$ **(b)** $h(x) = -x^2 + 4$ **(c)** $F(x) = (x - 1)^2$

(d) $G(x) = (x + 1)^2$ **(e)** $H(x) = (x - 1)^2 + 1$ **(f)** $K(x) = (x + 1)^2 + 1$

A.

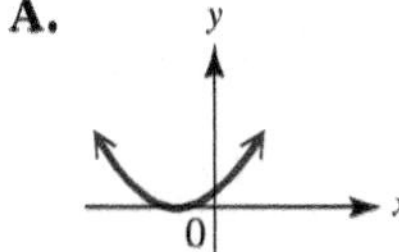

B.

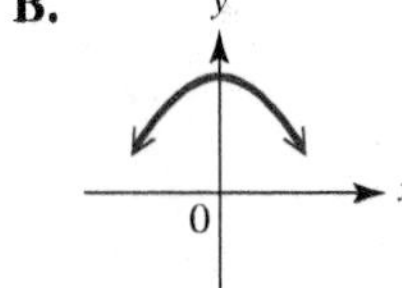

C.

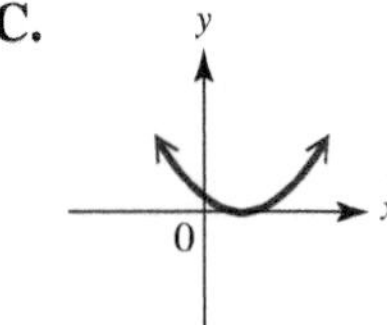

D.

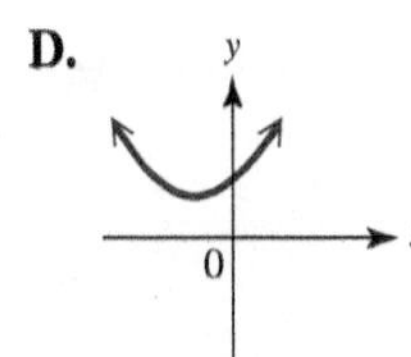

E.

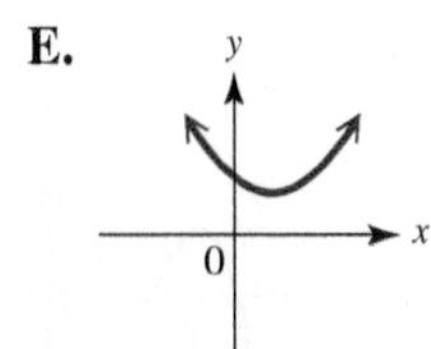

F.

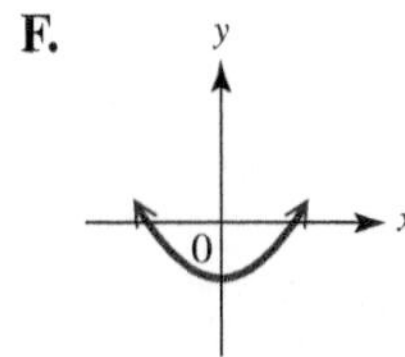

4. Graph $f(x) = 4x^2 + 4x - 2$. Give the vertex, axis of symmetry, domain, and range.

5. Determine the open interval(s) over which the function graphed here is

(a) increasing,

(b) decreasing,

(c) constant.

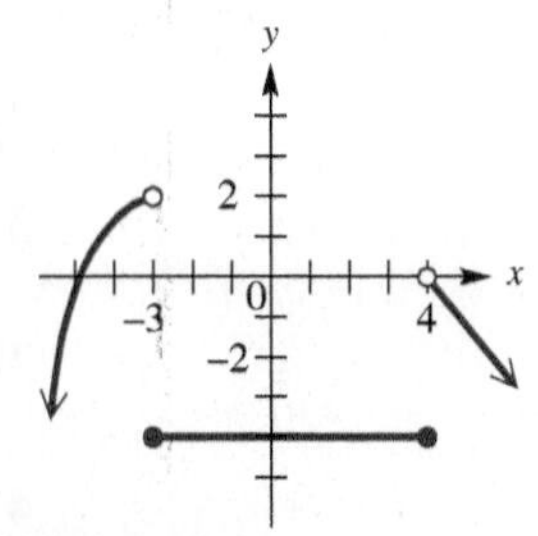

6. A car rental costs \$37 for one day, which includes 50 free miles. Each additional 25 mi or portion of 25 mi costs \$10. Let $y = C(x)$ represent the cost in dollars to drive x miles. Graph $C(x)$. Use the interval $(0, 150]$.

Graph each relation.

7. $f(x) = -3\sqrt{x}$

8. $f(x) = |2x + 1|$

9. $y^2 = x - 1$

10. $f(x) = [\![x]\!] - 2$

Chapter 9 Test

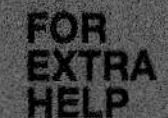

Step-by-step test solutions are found on the Chapter Test Prep Videos available in MyLab Math.

View the complete solutions to all Chapter Test exercises in MyLab Math.

Let $f(x) = 4x + 2$ and $g(x) = -x^2 + 3$. Find each of the following.

1. $g(1)$

2. $(f + g)(-2)$

3. $\left(\frac{f}{g}\right)(3)$

4. $(f \circ g)(2)$

5. $(g \circ f)(x)$

6. $(f - g)(x)$; Give its domain.

7. Which figure most closely resembles the graph of $f(x) = a(x - h)^2 + k$ if $a < 0$, $h > 0$, and $k < 0$?

A.

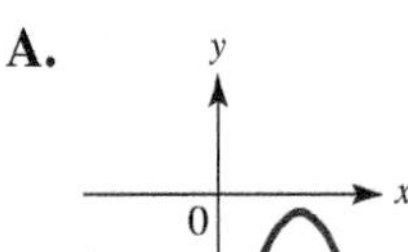

B.

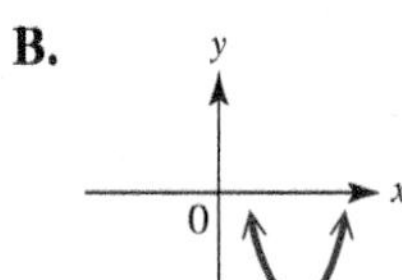

C.

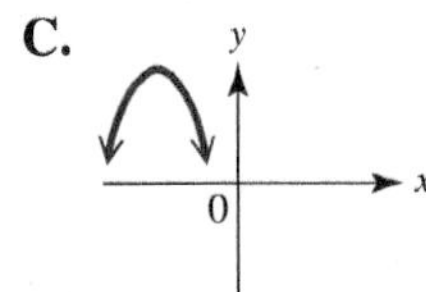

D.

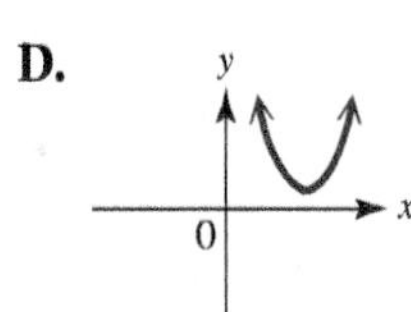

Graph each parabola. Identify the vertex, axis of symmetry, domain, and range.

8. $f(x) = \frac{1}{2}x^2 - 2$

9. $f(x) = -x^2 + 4x - 1$

10. $x = -(y - 2)^2 + 2$

11. Houston Community College is planning to construct a rectangular parking lot on land bordered on one side by a highway. The plan is to use 640 ft of fencing to fence off the other three sides. What should the dimensions of the lot be if the enclosed area is to be a maximum? What is the maximum area?

Use the tests for symmetry to decide whether the graph of each relation is symmetric with respect to the x-axis, the y-axis, or the origin. More than one of these symmetries, or none of them, may apply.

12. $f(x) = -x^2 + 1$

13. $x = y^2 + 7$

14. $x^2 + y^2 = 4$

15. Match each function in parts (a)–(d) with its graph from choices A–D.

(a) $f(x) = |x - 2|$ **(b)** $f(x) = |x + 2|$ **(c)** $f(x) = |x| + 2$ **(d)** $f(x) = |x| - 2$

A.

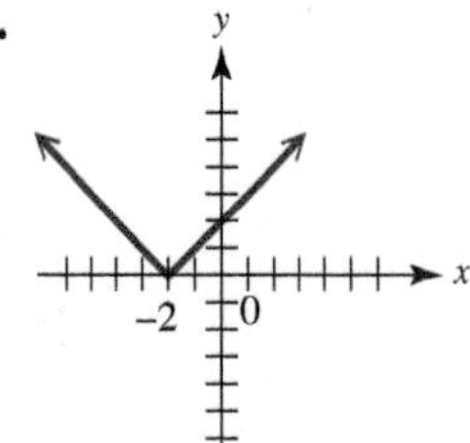

B.

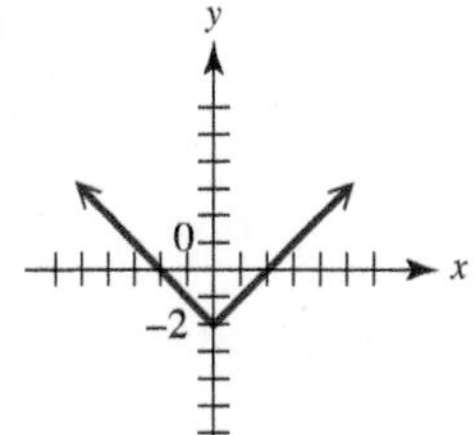

C.

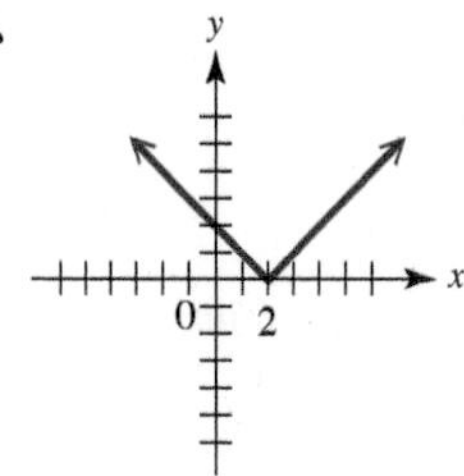

D.

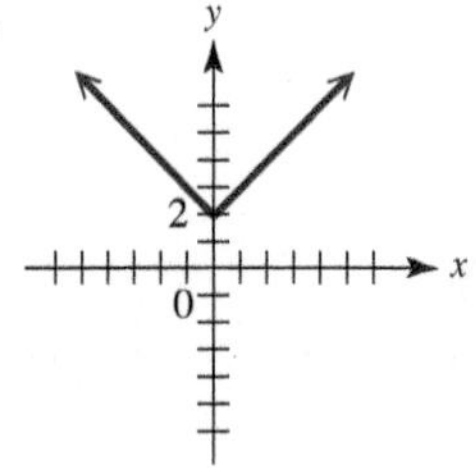

16. Determine the open intervals over which the function graphed here is

(a) increasing,

(b) decreasing,

(c) constant.

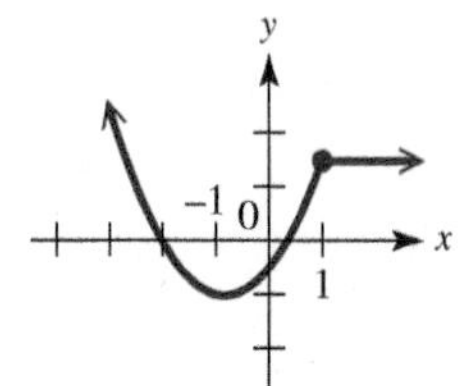

Graph each relation.

17. $f(x) = |x - 3| + 4$

18. $f(x) = [\![2x]\!]$

19. $f(x) = \begin{cases} -x & \text{if } x \le 2 \\ x - 4 & \text{if } x > 2 \end{cases}$

20. The range of $f(x) = [\![x]\!]$, the greatest integer function, is ________.

Chapters R–9 Cumulative Review Exercises

1. Write $\frac{1}{100}$ as a decimal and as a percent.

2. Simplify $-|-1| - 5 + |4 - 10|$.

3. Match each number in Column I with the set (or sets) of numbers in Column II to which the number belongs.

I		II	
(a) 1	**(b)** 0	**A.** Natural numbers	**B.** Whole numbers
(c) $-\frac{2}{3}$	**(d)** π	**C.** Integers	**D.** Rational numbers
(e) $-\sqrt{49}$	**(f)** 2.75	**E.** Irrational numbers	**F.** Real numbers

4. Evaluate each expression in the real number system.

(a) 6^2 **(b)** -6^2 **(c)** $(-6)^2$ **(d)** $\sqrt{36}$ **(e)** $-\sqrt{36}$ **(f)** $\sqrt{-36}$

Solve.

5. $2(3x - 1) = -(4 - x) - 28$

6. $3p = q(p + q)$ for p

7. $2x + 3 \leq 5 - (x - 4)$

8. $2x + 1 > 5$ or $2 - x \geq 2$

9. $|5 - 3x| = 12$

10. $|12x + 7| \geq 0$

Find the slope of each line.

11. Through $(-2, 1)$ and $(5, -4)$

12. With equation $2x - 3y = 6$

Solve.

13. Find the equation of the line passing through the point $(1, 6)$, perpendicular to the graph of $y = -\frac{1}{2}x + 3$. Give it in **(a)** slope-intercept form and **(b)** standard form.

14. What is the domain of $f(x) = \sqrt{1 - 4x}$?

15. Find $f(-3)$ if $f(x) = x^2 - 1$.

Solve each system.

16. $$\begin{aligned} x + 2y &= 10 \\ 3x - y &= 9 \end{aligned}$$

17. $$\begin{aligned} x + y + z &= 2 \\ 2x + y - z &= 5 \\ x - y + z &= -2 \end{aligned}$$

Perform the indicated operations.

18. $(4x^3 - 3x^2 + 2x - 3) \div (x - 1)$

19. $\dfrac{3p}{p + 1} + \dfrac{4}{p - 1} - \dfrac{6}{p^2 - 1}$

20. $\dfrac{6x - 10}{3x} \cdot \dfrac{6}{18x - 30}$

Factor completely.

21. $x^6 - y^6$

22. $2k^4 + k^2 - 3$

23. Write $\dfrac{2(m^{-1})^{-4}}{(3m^{-3})^2}$ with only positive exponents.

24. Write $\sqrt[3]{k^2} \cdot \sqrt{k}$ with rational exponents and simplify.

Solve.

25. $2x^3 - x^2 - 28x = 0$

26. $\dfrac{4}{x + 2} - \dfrac{11}{9} = \dfrac{5}{3x + 6}$

27. $\sqrt{x + 1} - \sqrt{x - 2} = 1$

28. $5x^2 + 3x = 3$

29. $S = 2\pi rh + \pi r^2$ for r

30. $3x^2 - 8 > 10x$

31. A toy rocket is projected directly upward. Its height in feet after t seconds is given by

$$h(t) = -16t^2 + 256t$$

(neglecting air resistance).

(a) When will it return to the ground?

(b) After how many seconds will it be 768 ft above the ground?

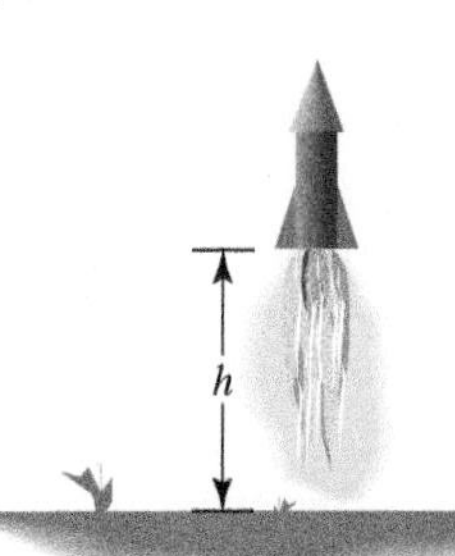

Graph.

32. $2x - 3y = 6$

33. $x + 2y \leq 4$

34. $f(x) = -2x^2 + 5x + 3$

35. $f(x) = |x + 1|$

Use the tests for symmetry to decide whether the graph of each relation exhibits any symmetries.

36. $y^2 = x + 3$

37. $x^2 - 6y^2 = 18$

38. If $f(x) = [\![x + 2]\!]$, find **(a)** $f(3)$ and **(b)** $f(-3.1)$.

STUDY SKILLS **REMINDER**

Have you begun to prepare for your final exam? **Review Study Skill 10, *Preparing for Your Math Final Exam.***

10

INVERSE, EXPONENTIAL, AND LOGARITHMIC FUNCTIONS

Compound interest earned on money, intensities of sounds, and population growth and decay are some examples of applications of *exponential* and *logarithmic functions*.

10.1 Inverse Functions

OBJECTIVES

1 Decide whether a function is one-to-one and, if it is, find its inverse.

2 Use the horizontal line test to determine whether a function is one-to-one.

3 Find the equation of the inverse of a function.

4 Graph f^{-1}, given the graph of f.

In this chapter we study two important types of functions, *exponential* and *logarithmic*. These functions are related: They are *inverses* of one another.

OBJECTIVE 1 Decide whether a function is one-to-one and, if it is, find its inverse.

Suppose we define the function

$$G = \{(-2, 2), (-1, 1), (0, 0), (1, 3), (2, 5)\}.$$

We can form another set of ordered pairs from G by interchanging the x- and y-values of each pair in G. We can call this set F, so

$$F = \{(2, -2), (1, -1), (0, 0), (3, 1), (5, 2)\}.$$

To show that these two sets are related as just described, F is called the *inverse* of G. For a function f to have an inverse, f must be a *one-to-one function.*

One-to-One Function

In a **one-to-one function,** each x-value corresponds to only one y-value, and each y-value corresponds to only one x-value.

VOCABULARY

☐ one-to-one function
☐ inverse of a function

The function in **FIGURE 1(a)** is one-to-one. The function shown in **FIGURE 1(b)** is not one-to-one because the y-value 7 corresponds to *two* x-values, 2 and 3. That is, the ordered pairs $(2, 7)$ and $(3, 7)$ both belong to the function.

STUDY SKILLS REMINDER

Make study cards to help you learn and remember the material in this chapter. **Review Study Skill 5, *Using Study Cards.***

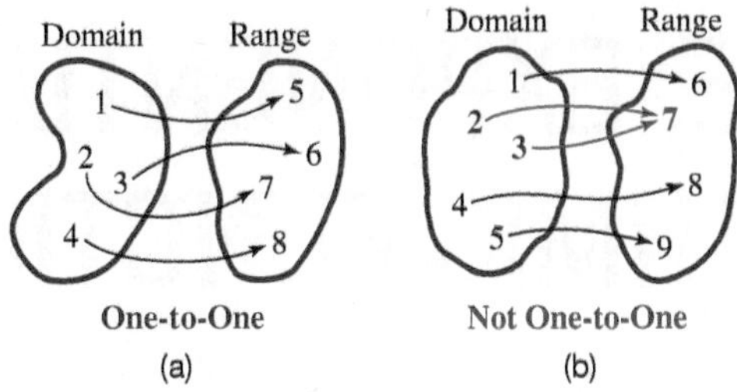

FIGURE 1

The *inverse* of any one-to-one function f is found by interchanging the components of the ordered pairs of f. The inverse of f is written $\boldsymbol{f^{-1}}$. Read f^{-1} as ***"the inverse of f"*** or ***"f-inverse."***

CAUTION The symbol $f^{-1}(x)$ does ***not*** represent $\frac{1}{f(x)}$.

Inverse of a Function

The **inverse** of a one-to-one function f, written f^{-1}, is the set of all ordered pairs of the form (y, x), where (x, y) belongs to f. ***The inverse is formed by interchanging x and y, so the domain of f becomes the range of f^{-1} and the range of f becomes the domain of f^{-1}.***

For inverses f and f^{-1}, it follows that for all x in their domains,

$$(f \circ f^{-1})(x) = x \quad \text{and} \quad (f^{-1} \circ f)(x) = x.$$

NOW TRY EXERCISE 1

Determine whether each function is one-to-one. If it is, find the inverse.

(a) $F = \{(0, 0), (1, 1), (4, 2), (9, 3)\}$

(b) $G = \{(-1, -2), (0, 0), (1, -2), (2, -8)\}$

(c) A Norwegian physiologist has developed a rule for predicting running times based on the time to run 5 km (5K). An example for one runner is shown here.

Distance	Time
1.5K	4:22
3K	9:18
5K	16:00
10K	33:40

Data from Stephen Seiler, Agder College, Kristiansand, Norway.

EXAMPLE 1 Finding Inverses of One-to-One Functions

Determine whether each function is one-to-one. If it is, find the inverse.

(a) $F = \{(3, 1), (0, 2), (2, 3), (4, 0)\}$

Every x-value in F corresponds to only one y-value, and every y-value corresponds to only one x-value, so F is a one-to-one function. The inverse function is found by interchanging the x- and y-values in each ordered pair.

$$F^{-1} = \{(1, 3), (2, 0), (3, 2), (0, 4)\}$$

The domain and range of F become the range and domain, respectively, of F^{-1}.

(b) $G = \{(-2, 1), (-1, 0), (0, 1), (1, 2), (2, 2)\}$

Each x-value in G corresponds to just one y-value. However, the y-value 1 corresponds to two x-values, -2 and 0. Also, the y-value 2 corresponds to both 1 and 2. Because some y-values correspond to more than one x-value, G is not one-to-one and does not have an inverse.

(c) The table shows the number of days in which the air in the Los Angeles–Long Beach–Anaheim metropolitan area exceeded air-quality standards in recent years.

Year	Number of Days Exceeding Standards
2010	4
2011	14
2012	37
2013	25
2014	28
2015	38
2016	25

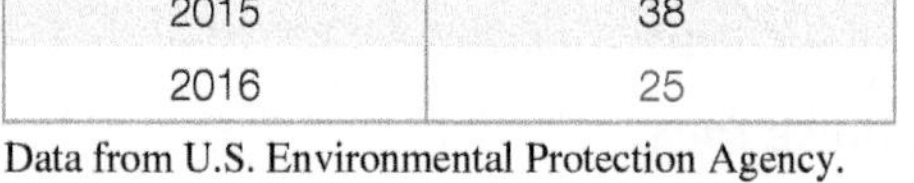

Data from U.S. Environmental Protection Agency.

Let f be the function defined in the table, with the years forming the domain and the number of days exceeding air-quality standards forming the range. Then f is not one-to-one because in two different years (2013 and 2016) the number of days exceeding air-quality standards was the same, 25.

NOW TRY

OBJECTIVE 2 Use the horizontal line test to determine whether a function is one-to-one.

By graphing a function and observing the graph, we can use the *horizontal line test* to determine whether the function is one-to-one.

Horizontal Line Test

A function is one-to-one if every horizontal line intersects the graph of the function at most once.

The horizontal line test follows from the definition of a one-to-one function. Any two points that lie on the same horizontal line have the same y-coordinate. No two ordered pairs that belong to a one-to-one function may have the same y-coordinate. Therefore, no horizontal line will intersect the graph of a one-to-one function more than once.

NOW TRY ANSWERS

1. **(a)** one-to-one; $F^{-1} = \{(0, 0), (1, 1), (2, 4), (3, 9)\}$

(b) not one-to-one

(c) one-to-one

Time	Distance
4:22	1.5K
9:18	3K
16:00	5K
33:40	10K

NOW TRY EXERCISE 2

Use the horizontal line test to determine whether each graph is the graph of a one-to-one function.

(a)

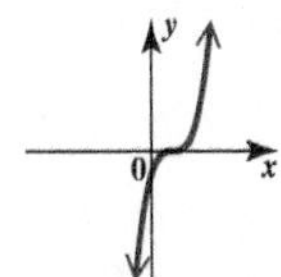

(b)

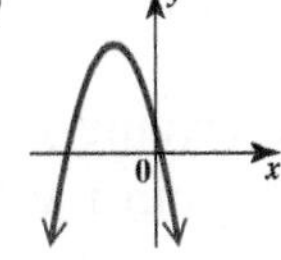

EXAMPLE 2 Using the Horizontal Line Test

Use the horizontal line test to determine whether each graph is the graph of a one-to-one function.

(a)

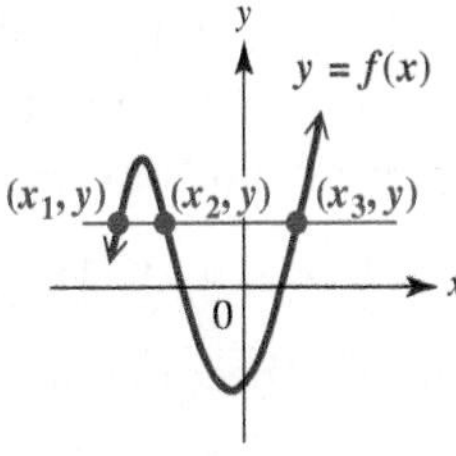

FIGURE 2

A horizontal line intersects the graph in FIGURE 2 in more than one point. This function is not one-to-one.

(b)

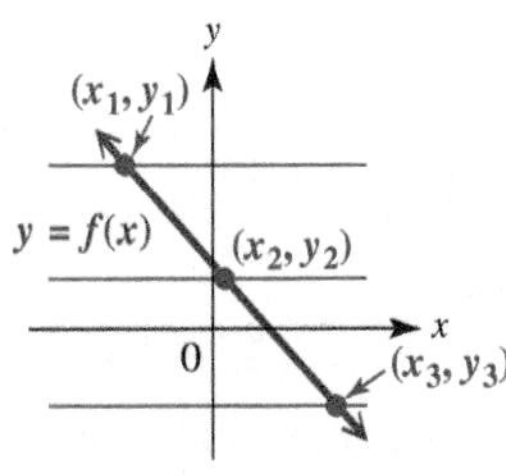

FIGURE 3

Every horizontal line will intersect the graph in FIGURE 3 in exactly one point. This function is one-to-one. NOW TRY

OBJECTIVE 3 Find the equation of the inverse of a function.

The inverse of a one-to-one function is found by interchanging the x- and y-values of each of its ordered pairs. The equation of the inverse of a function $y = f(x)$ is found in the same way.

Finding the Equation of the Inverse of $y = f(x)$

For a one-to-one function f defined by an equation $y = f(x)$, find the defining equation of the inverse function f^{-1} as follows.

Step 1 Interchange x and y.

Step 2 Solve for y.

Step 3 Replace y with $f^{-1}(x)$.

EXAMPLE 3 Finding Equations of Inverses

Determine whether each equation defines a one-to-one function. If so, find the equation that defines the inverse.

(a) $f(x) = 2x + 5$

The graph of $y = 2x + 5$ is a nonvertical line, so by the horizontal line test, f is a one-to-one function. Find the inverse as follows.

$$f(x) = 2x + 5$$

$$y = 2x + 5 \quad \text{Let } y = f(x).$$

Step 1 $$x = 2y + 5 \quad \text{Interchange } x \text{ and } y.$$

Step 2 $$2y = x - 5 \quad \text{Subtract 5. Interchange sides.}$$

$$y = \frac{x-5}{2} \quad \text{Divide by 2.}$$

(Subtract 5, interchange sides, and divide by 2: Solve for y.)

Step 3 $$f^{-1}(x) = \frac{x-5}{2} \quad \text{Replace } y \text{ with } f^{-1}(x).$$

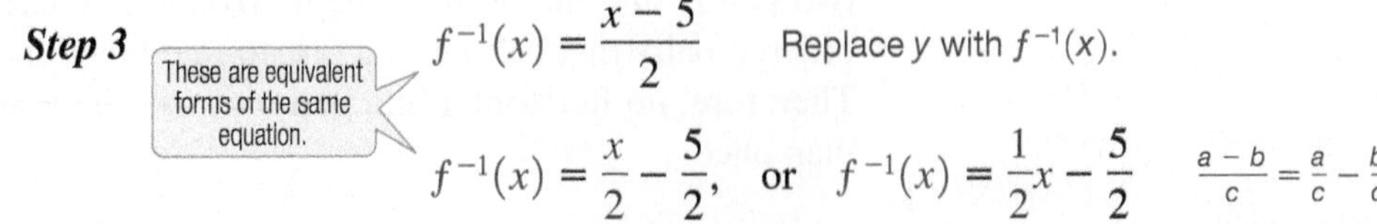

These are equivalent forms of the same equation.

$$f^{-1}(x) = \frac{x}{2} - \frac{5}{2}, \quad \text{or} \quad f^{-1}(x) = \frac{1}{2}x - \frac{5}{2} \qquad \frac{a-b}{c} = \frac{a}{c} - \frac{b}{c}$$

NOW TRY ANSWERS

2. (a) one-to-one function
(b) not a one-to-one function

Determine whether each equation defines a one-to-one function. If so, find the equation that defines the inverse.

(a) $f(x) = 5x - 7$

(b) $f(x) = (x + 1)^2$

(c) $f(x) = x^3 - 4$

Thus, f^{-1} is a linear function. In the given function

$$f(x) = 2x + 5,$$

we start with a value of x, *multiply* by 2, and *add* 5. In the equation for the inverse

$$f^{-1}(x) = \frac{x - 5}{2}, \quad \text{One form of } f^{-1}(x)$$

we *subtract* 5, and then *divide* by 2. This shows how an inverse is used to "undo" what a function does to the variable x.

(b) $y = x^2 + 2$

This equation has a vertical parabola as its graph, so some horizontal lines will intersect the graph at two points. For example, both $x = 1$ and $x = -1$ correspond to $y = 3$. See **FIGURE 4**.

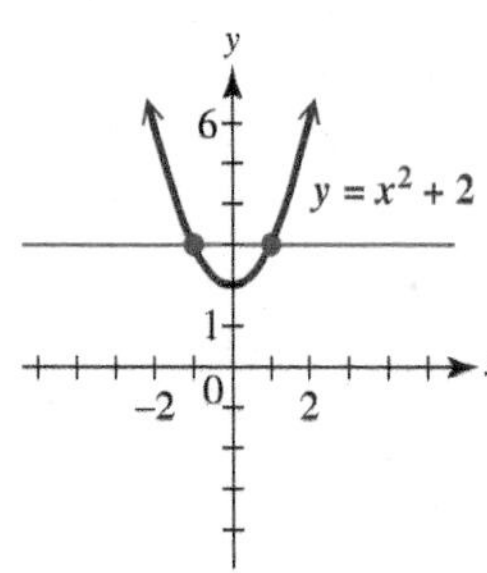

FIGURE 4

Because of the x^2-term in $y = x^2 + 2$, there are many pairs of x-values that correspond to the same y-value. The function is not one-to-one and does not have an inverse.

If we try to find the equation of an inverse, we obtain the following.

$$y = x^2 + 2$$

$$x = y^2 + 2 \quad \text{Interchange } x \text{ and } y.$$

$$y^2 = x - 2 \quad \text{Solve for } y.$$

$$y = \pm\sqrt{x - 2} \quad \text{Square root property}$$

Remember both roots.

The last step shows that there are two y-values for each choice of x in $(2, \infty)$—that is, for $x > 2$—so the function is not one-to-one. It does not have an inverse.

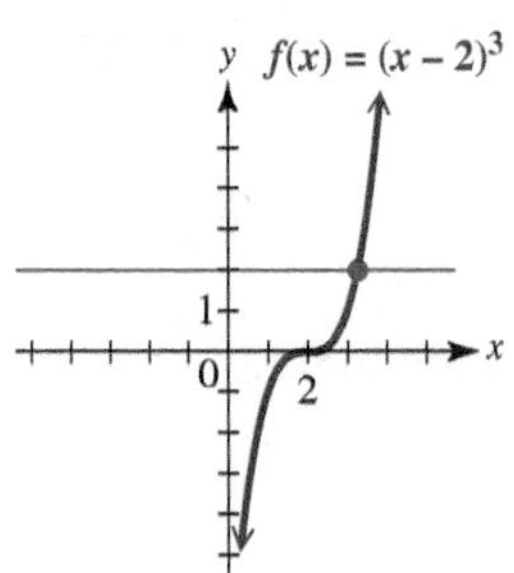

FIGURE 5

(c) $f(x) = (x - 2)^3$

A cubing function like this is one-to-one. See the graph in **FIGURE 5**.

$$f(x) = (x - 2)^3$$

$$y = (x - 2)^3 \quad \text{Replace } f(x) \text{ with } y.$$

Step 1 $\quad x = (y - 2)^3 \quad$ Interchange x and y.

Step 2 $\quad \sqrt[3]{x} = \sqrt[3]{(y - 2)^3} \quad$ Take the cube root on each side.

$\sqrt[3]{x} = y - 2 \quad \sqrt[3]{a^3} = a$

$y = \sqrt[3]{x} + 2 \quad$ Add 2. Interchange sides.

(Step 2 lines: Solve for y.)

Step 3 $\quad f^{-1}(x) = \sqrt[3]{x} + 2 \quad$ Replace y with $f^{-1}(x)$.

NOW TRY

NOW TRY ANSWERS

3. **(a)** one-to-one function; $f^{-1}(x) = \frac{x + 7}{5}$, or $f^{-1}(x) = \frac{1}{5}x + \frac{7}{5}$

(b) not a one-to-one function

(c) one-to-one function; $f^{-1}(x) = \sqrt[3]{x + 4}$

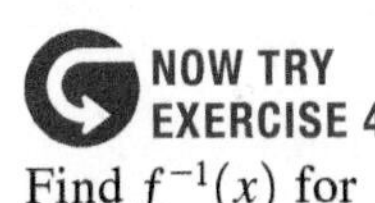
NOW TRY EXERCISE 4

Find $f^{-1}(x)$ for

$$f(x) = \frac{x+3}{x-4}, \quad x \neq 4.$$

EXAMPLE 4 Using Factoring to Find an Inverse

It is shown in standard college algebra texts that the following function is one-to-one.

$$f(x) = \frac{x+1}{x-2}, \quad x \neq 2$$

Find $f^{-1}(x)$.

$$f(x) = \frac{x+1}{x-2}, \quad x \neq 2 \qquad \text{Given equation}$$

$$y = \frac{x+1}{x-2}, \quad x \neq 2 \qquad \text{Replace } f(x) \text{ with } y.$$

Step 1 $$x = \frac{y+1}{y-2}, \quad y \neq 2 \qquad \text{Interchange } x \text{ and } y.$$

Step 2

$$x(y-2) = y+1 \qquad \text{Multiply by } y-2.$$

$$xy - 2x = y + 1 \qquad \text{Distributive property}$$

$$xy - y = 2x + 1 \qquad \text{Subtract } y. \text{ Add } 2x.$$

$$y(x-1) = 2x + 1 \qquad \text{Factor out } y.$$

$$y = \frac{2x+1}{x-1} \qquad \text{Divide by } x-1.$$

(Solve for y.)

Step 3 $$f^{-1}(x) = \frac{2x+1}{x-1}, \quad x \neq 1 \qquad \text{Replace } y \text{ with } f^{-1}(x). \text{ Note the restriction.}$$

NOW TRY 

OBJECTIVE 4 Graph f^{-1}, given the graph of f.

One way to graph the inverse of a function f whose equation is given follows.

Graphing the Inverse

Step 1 Find several ordered pairs that belong to f.

Step 2 Interchange x and y to obtain ordered pairs that belong to f^{-1}.

Step 3 Plot those points, and sketch the graph of f^{-1} through them.

We can also select points on the graph of f and *use symmetry* to find corresponding points on the graph of f^{-1}.

For example, suppose the point (a, b) shown in **FIGURE 6** belongs to a one-to-one function f. Then the point (b, a) belongs to f^{-1}. The line segment connecting the points (a, b) and (b, a) is perpendicular to, and cut in half by, the line $y = x$. The points (a, b) and (b, a) are "mirror images" of each other with respect to $y = x$.

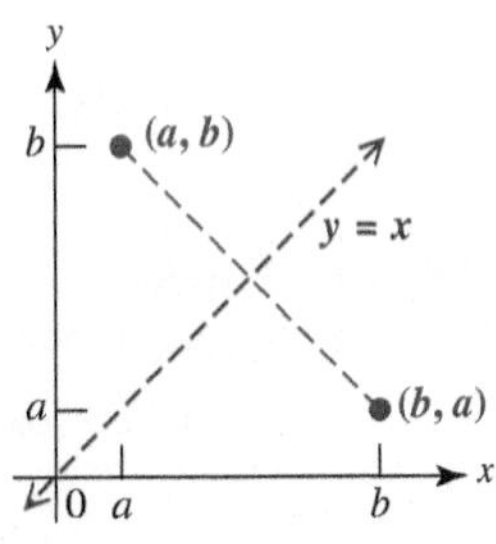

FIGURE 6

We can find the graph of f^{-1} from the graph of f by locating the mirror image of each point in f with respect to the line $y = x$.

NOW TRY ANSWER

4. $f^{-1}(x) = \frac{4x+3}{x-1}, \quad x \neq 1$

NOW TRY EXERCISE 5

Graph the inverse of the function labeled f in the figure.

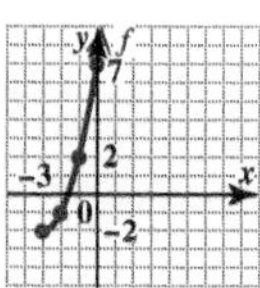

EXAMPLE 5 Graphing Inverses of Functions

Graph the inverse of each function labeled f in the figures.

(a) FIGURE 7(a) shows the graph of a one-to-one function f. The points

$$\left(-1, \frac{1}{2}\right), (0, 1), (1, 2), \text{ and } (2, 4) \qquad \text{Points on } f$$

lie on its graph. Interchange x and y to obtain ordered pairs that belong to f^{-1}.

$$\left(\frac{1}{2}, -1\right), (1, 0), (2, 1), \text{ and } (4, 2) \qquad \text{Points on } f^{-1}$$

Plot these points, and sketch the graph of f^{-1} through them. See FIGURE 7(b).

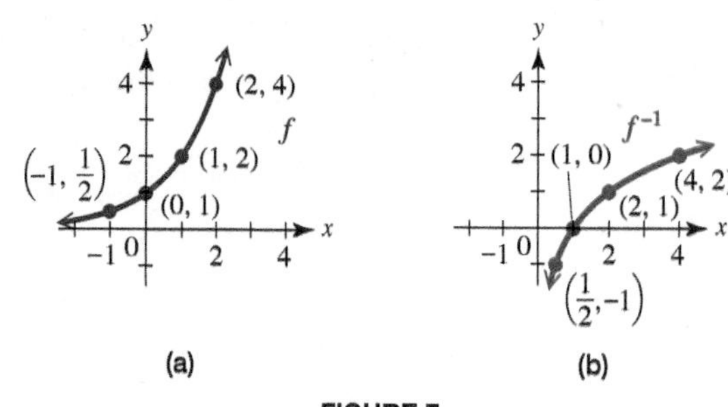

FIGURE 7

(b) Each function f (shown in blue) in FIGURES 8 and 9 is a one-to-one function.

Each inverse f^{-1} is shown in red. In both cases, the graph of f^{-1} is a reflection of the graph of f across the line $y = x$.

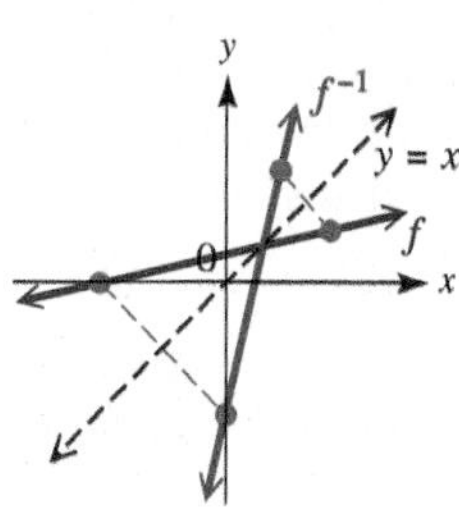

FIGURE 8

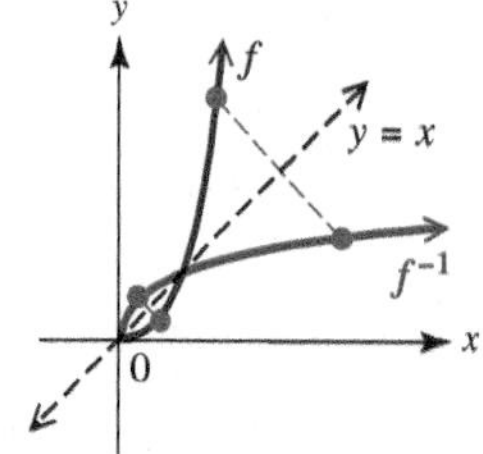

FIGURE 9

NOW TRY

NOW TRY ANSWER

5.

10.1 Exercises

MyLab Math

Video solutions for select problems available in MyLab Math

Concept Check *Choose the correct response.*

1. If a function is made up of ordered pairs in such a way that the same y-value appears in a correspondence with two different x-values, then

   A. the function is one-to-one
   B. the function is not one-to-one
   C. its graph does not pass the vertical line test
   D. it has an inverse function associated with it.

2. Which equation defines a one-to-one function? Explain why the others do not, using specific examples.

   A. $f(x) = x$
   B. $f(x) = x^2$
   C. $f(x) = |x|$
   D. $f(x) = -x^2 + 2x - 1$

3. Only one of the graphs illustrates a one-to-one function. Which one is it? **(See Example 2.)**

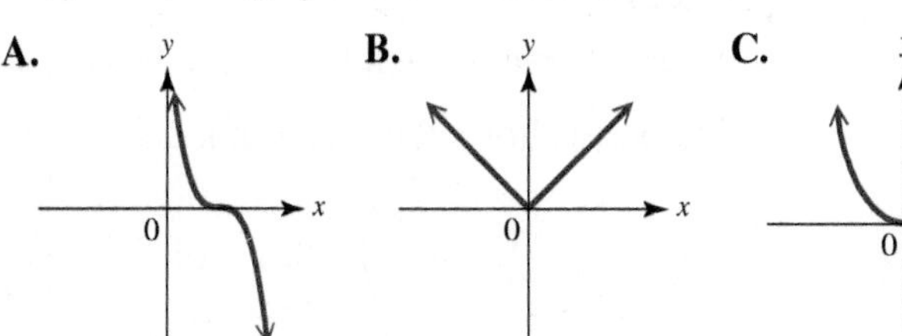

4. If a function f is one-to-one and the point (p, q) lies on the graph of f, then which point *must* lie on the graph of f^{-1}?

A. $(-p, q)$ **B.** $(-q, -p)$ **C.** $(p, -q)$ **D.** (q, p)

Answer each question.

5. The table shows fat content of various menu items at McDonald's. If the set of menu items is the domain and the set of fat contents is the range of a function, is it one-to-one? Why or why not?

Menu Item	*Fat Content (in grams)*
Chicken McNuggets (20 pieces)	60
Quarter Pounder with Cheese	43
Big Breakfast with Egg Whites	41
Bacon Clubhouse Burger	41
McFlurry with Reese's	37
Sausage Biscuit with Egg	33

Data from www.thebalancesmb.com

6. The table shows the most-visited social networking web sites, by number of visitors in millions, in June 2017. If the set of web sites is the domain and the set of numbers of visitors is the range of a function, is it one-to-one? Why or why not?

Social Networking Web Site	*Number of Visitors (in millions)*
Facebook	202
Instagram	121
Twitter	110
LinkedIn	103
Snapchat	95
Pinterest	86

Data from comScore, Inc.

7. The road mileage between Denver, Colorado, and several selected U.S. cities is shown in the table. If we consider this a function that pairs each city with a distance, is it one-to-one? How could we change the answer to this question by adding 1 mile to one of the distances shown?

City	*Distance to Denver (in miles)*
Atlanta	1398
Dallas	781
Indianapolis	1058
Kansas City, MO	600
Los Angeles	1059

8. The table lists caffeine amounts in several popular 12-oz sodas. If the set of sodas is the domain and the set of caffeine amounts is the range of a function, is it one-to-one? Why or why not?

Soda	*Caffeine (in mg)*
Mountain Dew	54
Diet Coke	46
Sunkist Orange Soda	41
Diet Pepsi-Cola	34
Coca-Cola Classic	34

Data from www.caffeineinformer.com

Determine whether each function is one-to-one. If it is, find the inverse. **See Examples 1 and 3.**

9. $\{(3, 6), (2, 10), (5, 12)\}$

10. $\left\{(-1, 3), (0, 5), (5, 0), \left(7, -\frac{1}{2}\right)\right\}$

11. $\{(-1, 3), (2, 7), (4, 3), (5, 8)\}$

12. $\{(-8, 6), (-4, 3), (0, 6), (5, 10)\}$

13. $\{(0, 4.5), (2, 8.6), (4, 12.7)\}$

14. $\{(1, 5.8), (2, 8.8), (3, 8.5)\}$

15. $f(x) = x + 3$

16. $f(x) = x + 8$

17. $f(x) = -\frac{1}{2}x - 2$

18. $f(x) = -\frac{1}{4}x - 8$

19. $f(x) = 2x + 4$

20. $f(x) = 3x + 1$

21. $g(x) = -4x + 3$

22. $g(x) = -6x - 8$

23. $f(x) = 5$

24. $f(x) = -7$

25. $f(x) = \sqrt{x - 3}, \quad x \geq 3$

26. $f(x) = \sqrt{x + 2}, \quad x \geq -2$

27. $f(x) = \sqrt{x + 6}, \quad x \geq -6$

28. $f(x) = \sqrt{x - 4}, \quad x \geq 4$

29. $f(x) = 3x^2 + 2$

30. $f(x) = 4x^2 - 1$

31. $g(x) = (x + 1)^3$

32. $g(x) = (x - 4)^3$

33. $f(x) = x^3 - 4$

34. $f(x) = x^3 + 5$

Each function is one-to-one. Find its inverse. **See Example 4.**

35. $f(x) = \frac{x + 4}{x + 2}, \quad x \neq -2$

36. $f(x) = \frac{x + 3}{x + 5}, \quad x \neq -5$

37. $f(x) = \frac{4x - 2}{x + 5}, \quad x \neq -5$

38. $f(x) = \frac{5x - 10}{x + 4}, \quad x \neq -4$

39. $f(x) = \frac{-2x + 1}{2x - 5}, \quad x \neq \frac{5}{2}$

40. $f(x) = \frac{-3x + 2}{3x - 4}, \quad x \neq \frac{4}{3}$

Concept Check *Let $f(x) = 2^x$. This function is one-to-one. Find each value.*

41. **(a)** $f(3)$ **(b)** $f^{-1}(8)$

42. **(a)** $f(4)$ **(b)** $f^{-1}(16)$

43. **(a)** $f(0)$ **(b)** $f^{-1}(1)$

44. **(a)** $f(-2)$ **(b)** $f^{-1}\left(\frac{1}{4}\right)$

Graphs of selected functions are given in the following exercises.

(a) Use the horizontal line test to determine whether each function graphed is one-to-one. **See Example 2.**

(b) If the function is one-to-one, graph its inverse. **See Example 5.**

45.

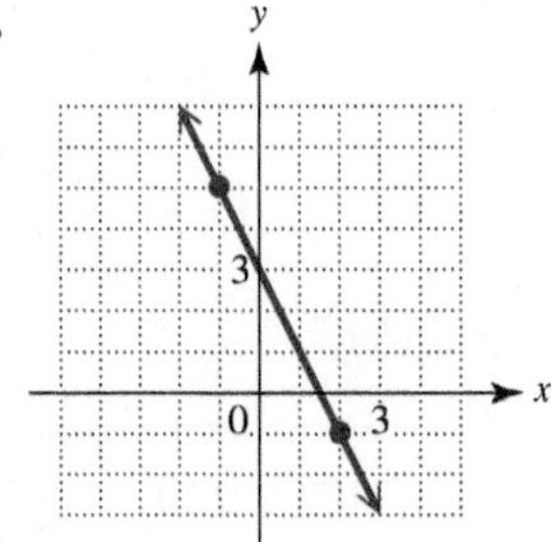

46.

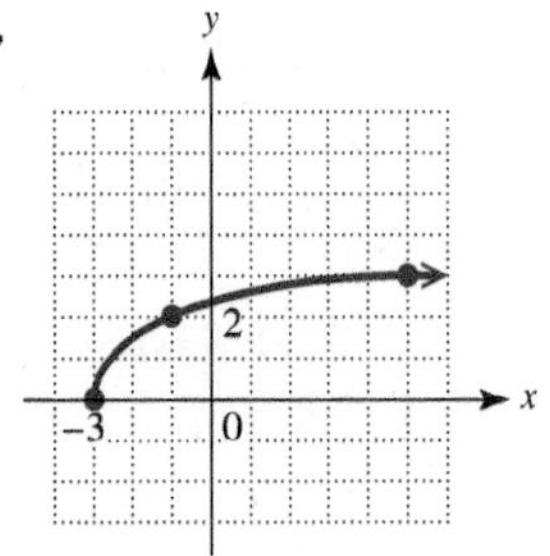

47.

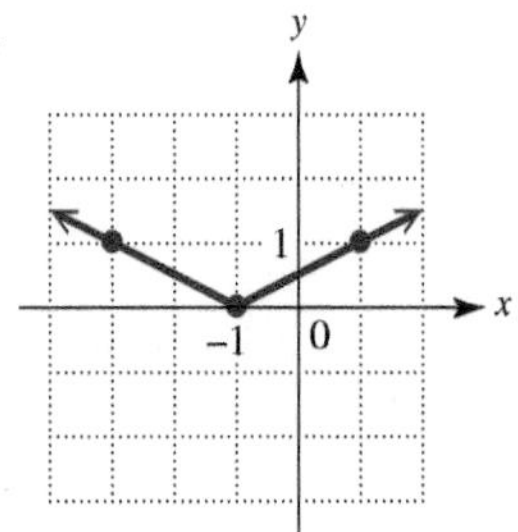

48.

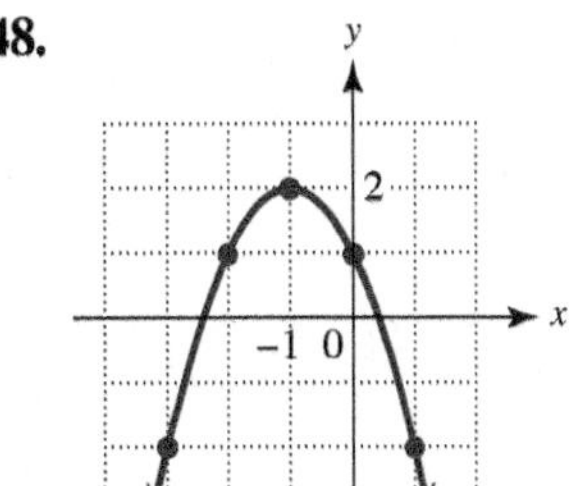

49.

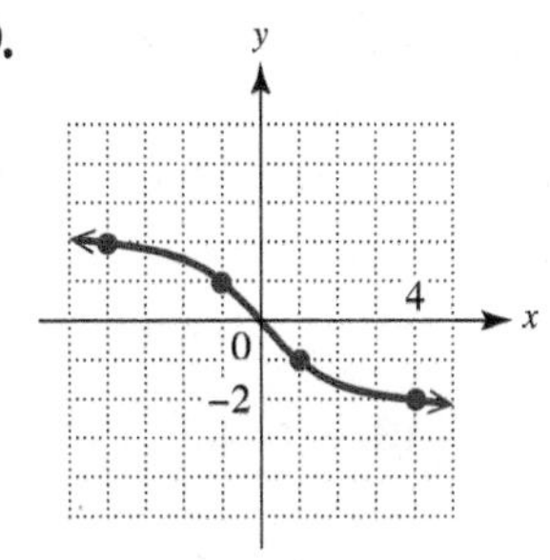

50.

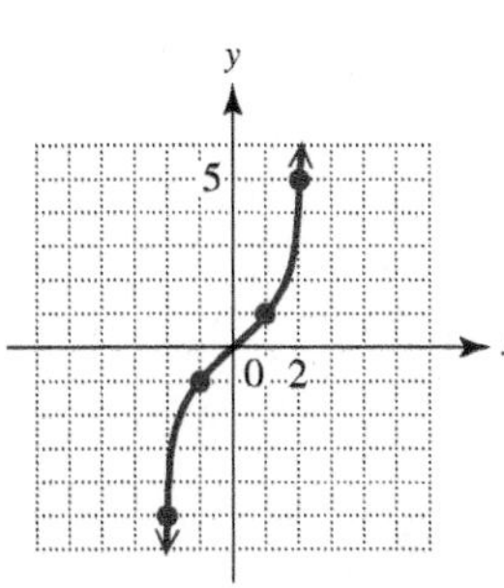

Each of the following functions is one-to-one. Graph the function as a solid line (or curve), and then graph its inverse on the same set of axes as a dashed line (or curve). Complete any tables to help graph the functions. ***See Example 5.***

51. $f(x) = 2x - 1$ **52.** $f(x) = 2x + 3$ **53.** $f(x) = -4x$ **54.** $f(x) = -2x$

55. $f(x) = \sqrt{x}$, $x \geq 0$

x	$f(x)$
0	
1	
4	

56. $f(x) = -\sqrt{x}$, $x \geq 0$

x	$f(x)$
0	
1	
4	

57. $f(x) = x^3 - 2$

x	$f(x)$
−1	
0	
1	
2	

58. $f(x) = x^3 + 3$

x	$f(x)$
−2	
−1	
0	
1	

RELATING CONCEPTS For Individual or Group Work **(Exercises 59–62)**

Inverse functions can be used to send and receive coded information. A simple example might use the function

$f(x) = 2x + 5$. *(Note that it is one-to-one.)*

Suppose that each letter of the alphabet is assigned a numerical value according to its position, as follows.

A	1	**G**	7	**L**	12	**Q**	17	**V**	22
B	2	**H**	8	**M**	13	**R**	18	**W**	23
C	3	**I**	9	**N**	14	**S**	19	**X**	24
D	4	**J**	10	**O**	15	**T**	20	**Y**	25
E	5	**K**	11	**P**	16	**U**	21	**Z**	26
F	6								

This is an Enigma machine used by the Germans in World War II to send coded messages.

Using the function, the word ALGEBRA *would be encoded as*

7 29 19 15 9 41 7,

because

$$f(\text{A}) = f(1) = 2(1) + 5 = 7, \quad f(\text{L}) = f(12) = 2(12) + 5 = 29, \quad \textit{and so on.}$$

The message would then be decoded using the inverse of f, *which is* $f^{-1}(x) = \frac{x-5}{2}$.

$$f^{-1}(7) = \frac{7-5}{2} = 1 = \text{A}, \quad f^{-1}(29) = \frac{29-5}{2} = 12 = \text{L}, \quad \textit{and so on.}$$

Work Exercises 59–62 in order.

59. Suppose that you are an agent for a detective agency. Today's encoding function is $f(x) = 4x - 5$. Find the rule for f^{-1} algebraically.

60. You receive the following coded message today. (Read across from left to right.)

47 95 7 −1 43 7 79 43 −1 75 55 67 31 71 75 27
15 23 67 15 −1 75 15 71 75 75 27 31 51 23 71
31 51 7 15 71 43 31 7 15 11 3 67 15 −1 11

Use the letter/number assignment described on the previous page to decode the message.

61. Why is a one-to-one function essential in this encoding/decoding process?

62. Use $f(x) = x^3 + 4$ to encode your name, using the above letter/number assignment.

10.2 Exponential Functions

OBJECTIVES

1 Evaluate exponential expressions using a calculator.
2 Define and graph exponential functions.
3 Solve exponential equations of the form $a^x = a^k$ for x.
4 Use exponential functions in applications involving growth or decay.

VOCABULARY

- ☐ exponential function with base *a*
- ☐ asymptote
- ☐ exponential equation

OBJECTIVE 1 Evaluate exponential expressions using a calculator.

Consider the exponential expression 2^x for rational values of x.

$$2^3 = 8, \quad 2^{-1} = \frac{1}{2}, \quad 2^{1/2} = \sqrt{2}, \quad 2^{3/4} = \sqrt[4]{2^3} = \sqrt[4]{8}$$ Examples of 2^x for rational x

In more advanced courses it is shown that 2^x exists for all real number values of x, both rational and irrational. We can use a calculator to find approximations of exponential expressions that are not easily determined.

EXAMPLE 1 Evaluating Exponential Expressions

Use a calculator to approximate each exponential expression to three decimal places.

(a) $2^{1.6}$ **(b)** $2^{-1.3}$ **(c)** $2^{1/3}$

FIGURE 10 shows how a TI-84 Plus calculator approximates these values. The display shows more decimal places than we usually need, so we round to three decimal places as directed.

$$2^{1.6} \approx 3.031, \quad 2^{-1.3} \approx 0.406, \quad 2^{1/3} \approx 1.260$$

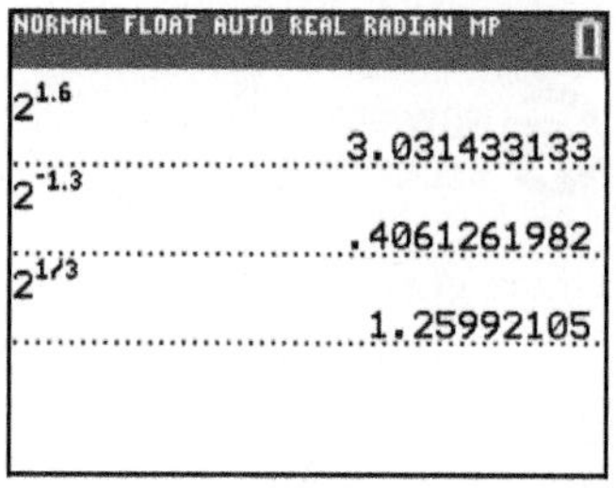

FIGURE 10

NOW TRY

NOW TRY EXERCISE 1

Use a calculator to approximate each exponential expression to three decimal places.

(a) $2^{3.1}$ **(b)** $2^{-1.7}$ **(c)** $2^{1/4}$

OBJECTIVE 2 Define and graph exponential functions.

The definition of an exponential function assumes that a^x exists for all real numbers x.

Exponential Function

For $a > 0$, $a \neq 1$, and all real numbers x,

$$f(x) = a^x$$

defines the **exponential function with base *a*.**

NOW TRY ANSWERS

1. **(a)** 8.574 **(b)** 0.308 **(c)** 1.189

When graphing an exponential function of the form $f(x) = a^x$, pay particular attention to whether $a > 1$ or $0 < a < 1$.

NOW TRY EXERCISE 2
Graph $f(x) = 4^x$.

EXAMPLE 2 Graphing an Exponential Function ($a > 1$)

Graph $f(x) = 2^x$. Then compare it to the graph of $F(x) = 5^x$.

Choose some values of x, and find the corresponding values of $f(x) = 2^x$. Plotting these points and drawing a smooth curve through them gives the darker graph shown in **FIGURE 11**. This graph is typical of the graph of an exponential function of the form $f(x) = a^x$, where $a > 1$.

The larger the value of a, the faster the graph rises.

To see this, compare the graph of $F(x) = 5^x$ with the graph of $f(x) = 2^x$ in **FIGURE 11**. When graphing such functions, be sure to plot a sufficient number of points to see how rapidly the graph rises.

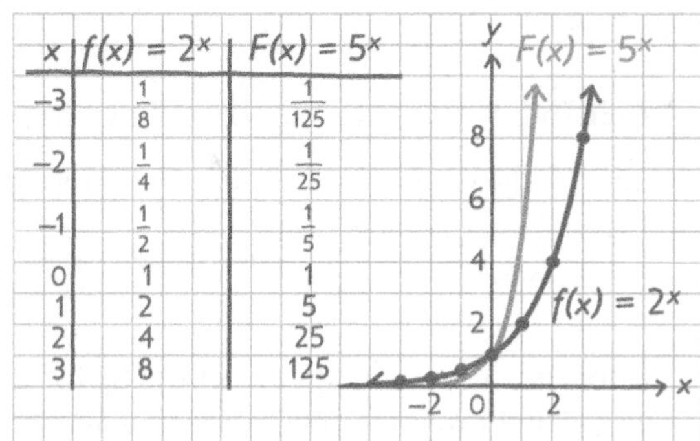

x	$f(x) = 2^x$	$F(x) = 5^x$
-3	$\frac{1}{8}$	$\frac{1}{125}$
-2	$\frac{1}{4}$	$\frac{1}{25}$
-1	$\frac{1}{2}$	$\frac{1}{5}$
0	1	1
1	2	5
2	4	25
3	8	125

Exponential function with base $a > 1$

Domain: $(-\infty, \infty)$

Range: $(0, \infty)$

y-intercept: $(0, 1)$

The function is one-to-one, and its graph rises from left to right.

FIGURE 11

The vertical line test assures us that the graphs in **FIGURE 11** represent functions. These graphs show an important characteristic of exponential functions where $a > 1$.

As x gets larger, y increases at a faster and faster rate. NOW TRY 

NOW TRY EXERCISE 3
Graph $g(x) = \left(\frac{1}{10}\right)^x$.

EXAMPLE 3 Graphing an Exponential Function ($0 < a < 1$)

Graph $g(x) = \left(\frac{1}{2}\right)^x$.

Find and plot some points on the graph. The graph in **FIGURE 12** is similar to that of $f(x) = 2^x$ (**FIGURE 11**) with the same domain and range, except that here ***as x gets larger, y decreases.*** This graph is typical of the graph of an exponential function of the form $f(x) = a^x$, where $0 < a < 1$.

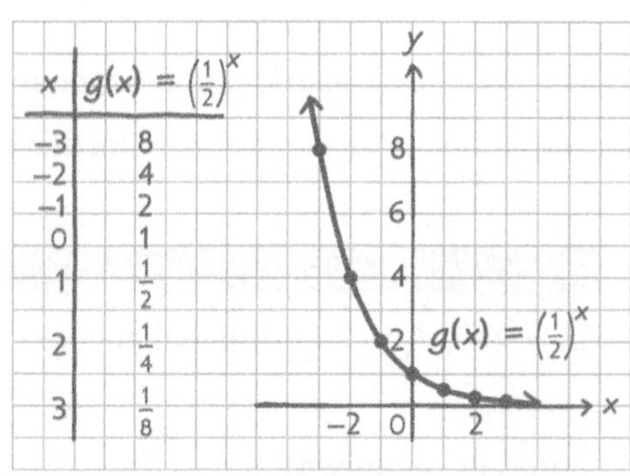

x	$g(x) = \left(\frac{1}{2}\right)^x$
-3	8
-2	4
-1	2
0	1
1	$\frac{1}{2}$
2	$\frac{1}{4}$
3	$\frac{1}{8}$

Exponential function with base $0 < a < 1$

Domain: $(-\infty, \infty)$

Range: $(0, \infty)$

y-intercept: $(0, 1)$

The function is one-to-one, and its graph falls from left to right.

FIGURE 12

NOW TRY 

NOW TRY ANSWERS

2. $f(x) = 4^x$

3. $g(x) = \left(\frac{1}{10}\right)^x$

CAUTION The graph of an exponential function of the form

$$f(x) = a^x$$

approaches the *x*-axis, but does ***not*** touch it. Recall that such a line is called an **asymptote.**

Characteristics of the Graph of $f(x) = a^x$

1. The graph contains the point $(0, 1)$, which is its y-intercept.
2. The function is one-to-one.
   - When $a > 1$, the graph *rises* from left to right. (See **FIGURE 11**.)
   - When $0 < a < 1$, the graph *falls* from left to right. (See **FIGURE 12**.)

   In both cases, the graph goes from the second quadrant to the first.
3. The graph approaches the x-axis but never touches it—that is, the x-axis is an asymptote.
4. The domain is $(-\infty, \infty)$, and the range is $(0, \infty)$.

NOW TRY EXERCISE 4
Graph $f(x) = 4^{2x-1}$.

EXAMPLE 4 **Graphing a More Complicated Exponential Function**

Graph $f(x) = 3^{2x-4}$.

Find some ordered pairs. We let $x = 0$ and $x = 2$ and find values of $f(x)$, or y.

$f(x) = 3^{2x-4}$		$f(x) = 3^{2x-4}$	
$y = 3^{2(0)-4}$	Let $x = 0$.	$y = 3^{2(2)-4}$	Let $x = 2$.
$y = 3^{-4}$		$y = 3^0$	
$y = \frac{1}{81}$	$a^{-n} = \frac{1}{a^n}$	$y = 1$	$a^0 = 1$

These ordered pairs, $\left(0, \frac{1}{81}\right)$ and $(2, 1)$, along with the other ordered pairs shown in the table, lead to the graph in **FIGURE 13**.

$f(x) = 3^{2x-4}$

x	y
0	$\frac{1}{81}$
1	$\frac{1}{9}$
2	1
3	9

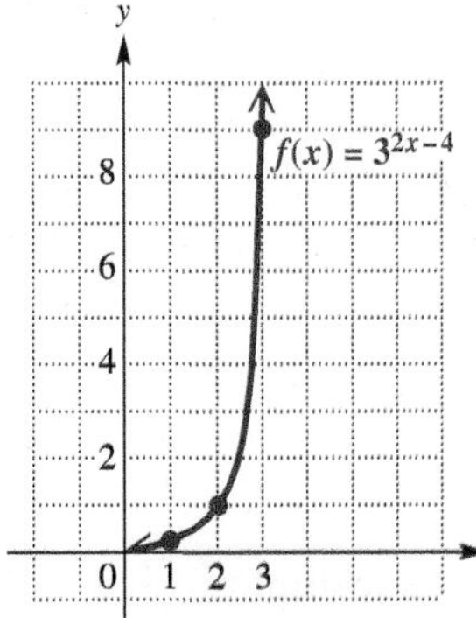

FIGURE 13

The graph of $f(x) = 3^{2x-4}$ is similar to the graph of $f(x) = 3^x$ except that it is shifted to the right and rises more rapidly.

NOW TRY

NOW TRY ANSWER

4. 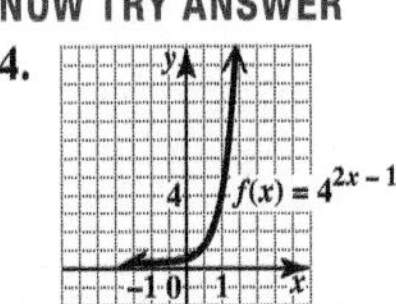

NOTE ***The two restrictions on the value of a in the definition of an exponential function $f(x) = a^x$ are important.***

1. The restriction $\mathbf{a > 0}$ is necessary so that the function can be defined for *all* real numbers x. Letting a be negative, such as $a = -2$, and letting $x = \frac{1}{2}$ gives the expression $(-2)^{1/2}$, which is not a real number.
2. The restriction $\mathbf{a \neq 1}$ is necessary because 1 raised to *any* power is equal to 1, resulting in the linear function $f(x) = 1$.

OBJECTIVE 3 Solve exponential equations of the form $a^x = a^k$ for x.

Up to this point, we have solved only equations that had the variable as a base, like $x^2 = 8$. In these equations, all exponents have been constants. An **exponential equation** is an equation that has a variable in an exponent, such as

$$9^x = 27.$$

We can use the following property to solve many exponential equations.

Property for Solving an Exponential Equation

For $a > 0$ and $a \neq 1$, if $\quad \boldsymbol{a^x = a^y} \quad$ then $\quad \boldsymbol{x = y}.$

This property would not necessarily be true if $a = 1$.

Solving an Exponential Equation

Step 1 **Both sides must have the same base.** If the two sides of the equation do not have the same base, express each as a power of the same base if possible.

Step 2 **Simplify exponents** if necessary, using the rules of exponents.

Step 3 **Set exponents equal** using the property given in this section.

Step 4 **Solve** the equation obtained in Step 3.

NOW TRY EXERCISE 5
Solve $8^x = 16$.

EXAMPLE 5 **Solving an Exponential Equation**

Solve $9^x = 27$.

	$9^x = 27$	
Step 1	$(3^2)^x = 3^3$	Write with the same base; $9 = 3^2$ and $27 = 3^3$.
Step 2	$3^{2x} = 3^3$	Power rule for exponents
Step 3	$2x = 3$	If $a^x = a^y$, then $x = y$.
Step 4	$x = \frac{3}{2}$	Solve for x.

CHECK Substitute $\frac{3}{2}$ for x: $\quad 9^x = 9^{3/2} = (9^{1/2})^3 = 3^3 = 27. \checkmark \quad$ True

The solution set is $\left\{\frac{3}{2}\right\}$.

NOW TRY

NOW TRY ANSWER
5. $\left\{\frac{4}{3}\right\}$

Solve each equation.

(a) $3^{2x-1} = 27^{x+4}$

(b) $5^x = \frac{1}{625}$

(c) $\left(\frac{2}{7}\right)^x = \frac{343}{8}$

EXAMPLE 6 **Solving Exponential Equations**

Solve each equation.

(a) $4^{3x-1} = 16^{x+2}$

$4^{3x-1} = (4^2)^{x+2}$ Write with the same base; $16 = 4^2$ (Step 1)

Be careful multiplying the exponents.

$4^{3x-1} = 4^{2x+4}$ Power rule for exponents (Step 2)

$3x - 1 = 2x + 4$ Set the exponents equal. (Step 3)

$x = 5$ Subtract $2x$. Add 1. (Step 4)

CHECK $4^{3x-1} = 16^{x+2}$

$4^{3(5)-1} \stackrel{?}{=} 16^{5+2}$ Substitute. Let $x = 5$.

$4^{14} \stackrel{?}{=} 16^7$ Perform the operations in the exponents.

$4^{14} \stackrel{?}{=} (4^2)^7$ $16 = 4^2$

$4^{14} = 4^{14}$ ✓ True

The solution set is $\{5\}$.

(b) $6^x = \frac{1}{216}$

$6^x = \frac{1}{6^3}$ $216 = 6^3$

$6^x = 6^{-3}$ Write with the same base; $\frac{1}{6^3} = 6^{-3}$.

$x = -3$ Set exponents equal.

CHECK Substitute -3 for x.

$$6^x = 6^{-3} = \frac{1}{6^3} = \frac{1}{216} \quad ✓ \text{ True}$$

The solution set is $\{-3\}$.

(c) $\left(\frac{2}{3}\right)^x = \frac{9}{4}$

$\left(\frac{2}{3}\right)^x = \left(\frac{4}{9}\right)^{-1}$ $\frac{9}{4} = \left(\frac{4}{9}\right)^{-1}$

$\left(\frac{2}{3}\right)^x = \left[\left(\frac{2}{3}\right)^2\right]^{-1}$ Write with the same base.

$\left(\frac{2}{3}\right)^x = \left(\frac{2}{3}\right)^{-2}$ Power rule for exponents

$x = -2$ Set exponents equal.

Check that the solution set is $\{-2\}$. **NOW TRY**

NOW TRY ANSWERS

6. **(a)** $\{-13\}$ **(b)** $\{-4\}$ **(c)** $\{-3\}$

OBJECTIVE 4 Use exponential functions in applications involving growth or decay.

NOW TRY EXERCISE 7

Use the function in **Example 7** to approximate average annual carbon dioxide concentration in 2010, to the nearest unit.

EXAMPLE 7 Applying an Exponential Growth Function

The graph in **FIGURE 14** shows average annual concentration of carbon dioxide (in parts per million) in the air. This concentration is increasing exponentially.

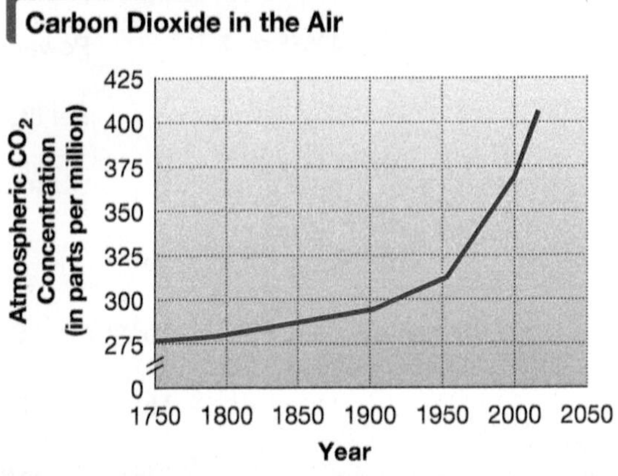

Data from National Oceanic and Atmospheric Administration.

FIGURE 14

The data in **FIGURE 14** are approximated by the exponential function

$$f(x) = 275.8 + 1.5172(1.0161)^x,$$

where x is number of years since 1750. Use this function to approximate average annual concentration of carbon dioxide in parts per million, to the nearest unit, for each year.

(a) 1900

Because x represents number of years since 1750, $x = 1900 - 1750 = 150$.

$$f(x) = 275.8 + 1.5172(1.0161)^x \quad \text{Given function}$$

$$f(150) = 275.8 + 1.5172(1.0161)^{150} \quad \text{Let } x = 150.$$

$$f(150) \approx 292 \quad \text{Evaluate with a calculator.}$$

The concentration in 1900 was approximately 292 parts per million.

(b) 2000

$$f(x) = 275.8 + 1.5172(1.0161)^x \quad \text{Given function}$$

$$f(250) = 275.8 + 1.5172(1.0161)^{250} \quad x = 2000 - 1750 = 250$$

$$f(250) \approx 358 \quad \text{Evaluate with a calculator.}$$

The concentration in 2000 was approximately 358 parts per million. NOW TRY 

EXAMPLE 8 Applying an Exponential Decay Function

Atmospheric pressure (in millibars) at a given altitude x, in meters, can be approximated by the exponential function

$$f(x) = 1038(1.000134)^{-x}, \quad \text{for values of } x \text{ between 0 and 10,000.}$$

Because the base is greater than 1 and the coefficient of x in the exponent is negative, function values decrease as x increases. This means that as altitude increases, atmospheric pressure decreases. (Data from Miller, A. and J. Thompson, *Elements of Meteorology*, Fourth Edition, Charles E. Merrill Publishing Company.)

NOW TRY ANSWER

7. 372 parts per million

NOW TRY EXERCISE 8
Use the function in **Example 8** to approximate the pressure at 6000 m, to the nearest unit.

(a) According to this function, what is the pressure at ground level?

At ground level, $x = 0$.

$f(x) = 1038(1.000134)^{-x}$	Given function
$f(0) = 1038(1.000134)^{-0}$	Let $x = 0$.
$f(0) = 1038(1)$	$a^0 = 1$
$f(0) = 1038$	Identity property

The pressure is 1038 millibars.

(b) What is the pressure at 5000 m, to the nearest unit?

$f(x) = 1038(1.000134)^{-x}$	Given function
$f(5000) = 1038(1.000134)^{-5000}$	Let $x = 5000$.
$f(5000) \approx 531$	Evaluate with a calculator.

The pressure is approximately 531 millibars.  

NOTE The function in **Example 8** is equivalent to

$$f(x) = 1038\left(\frac{1}{1.000134}\right)^x.$$

In this form, the base a satisfies the condition $0 < a < 1$.

NOW TRY ANSWER
8. 465 millibars

10.2 Exercises

  **MyLab Math**

Video solutions for select problems available in MyLab Math

Concept Check *Choose the correct response.*

1. For an exponential function $f(x) = a^x$, if $a > 1$, then the graph (*rises / falls*) from left to right.

2. For an exponential function $f(x) = a^x$, if $0 < a < 1$, then the graph (*rises / falls*) from left to right.

3. Which point lies on the graph of $f(x) = 3^x$?

A. $(1, 0)$ **B.** $(3, 1)$ **C.** $(0, 1)$ **D.** $\left(\sqrt{3}, \frac{1}{3}\right)$

4. The asymptote of the graph of $f(x) = a^x$

A. is the x-axis
B. is the y-axis
C. has equation $x = 1$
D. has equation $y = 1$.

5. Which statement is true?

A. The point $\left(\frac{1}{2}, \sqrt{5}\right)$ lies on the graph of $f(x) = 5^x$.

B. $f(x) = 5^x$ is not a one-to-one function.

C. The y-intercept of the graph of $f(x) = 5^x$ is $(0, 5)$.

D. The graph of $y = 5^x$ rises at a faster rate than the graph of $y = 10^x$.

6. Which statement is false?

A. The domain of the function $f(x) = \left(\frac{1}{4}\right)^x$ is $(-\infty, \infty)$.

B. The graph of the function $f(x) = \left(\frac{1}{4}\right)^x$ has one x-intercept.

C. The range of the function $f(x) = \left(\frac{1}{4}\right)^x$ is $(0, \infty)$.

D. The point $(-2, 16)$ lies on the graph of $f(x) = \left(\frac{1}{4}\right)^x$.

Use a calculator to approximate each exponential expression to three decimal places. ***See Example 1.***

7. $2^{1.9}$ **8.** $2^{2.7}$ **9.** $2^{-1.54}$ **10.** $2^{-1.88}$

11. $10^{0.3}$ **12.** $10^{0.5}$ **13.** $4^{1/3}$ **14.** $6^{1/5}$

15. $\left(\frac{1}{3}\right)^{1.5}$ **16.** $\left(\frac{1}{3}\right)^{2.4}$ **17.** $\left(\frac{1}{4}\right)^{-3.1}$ **18.** $\left(\frac{1}{4}\right)^{-1.4}$

Graph each exponential function. ***See Examples 2–4.***

19. $f(x) = 3^x$ **20.** $f(x) = 5^x$ **21.** $g(x) = \left(\frac{1}{3}\right)^x$ **22.** $g(x) = \left(\frac{1}{5}\right)^x$

23. $f(x) = 4^{-x}$ **24.** $f(x) = 6^{-x}$ **25.** $f(x) = 2^{2x-2}$ **26.** $f(x) = 2^{2x+1}$

27. Concept Check A student incorrectly solved the following equation as shown.

$2^x = 32$ Given equation

$\frac{2^x}{2} = \frac{32}{2}$ Divide by 2.

$x = 16$

WHAT WENT WRONG? Give the correct solution set.

28. Concept Check A student incorrectly solved the following equation as shown.

$3^x = 81$ Given equation

$3^x - 3 = 81 - 3$ Subtract 3.

$x = 78$

WHAT WENT WRONG? Give the correct solution set.

Solve each equation. ***See Examples 5 and 6.***

29. $6^x = 36$ **30.** $8^x = 64$ **31.** $100^x = 1000$ **32.** $8^x = 4$

33. $16^x = 64$ **34.** $8^x = 32$ **35.** $4^{x-5} = 64^{2x}$ **36.** $125^{3x} = 5^{2x-7}$

37. $16^{2x+1} = 64^{x+3}$ **38.** $9^{2x-8} = 27^{x-4}$ **39.** $5^x = \frac{1}{125}$

40. $3^x = \frac{1}{81}$ **41.** $9^x = \frac{1}{27}$ **42.** $8^x = \frac{1}{32}$

43. $5^x = 0.2$ **44.** $10^x = 0.1$ **45.** $\left(\frac{3}{2}\right)^x = \frac{8}{27}$

46. $\left(\frac{4}{3}\right)^x = \frac{27}{64}$ **47.** $\left(\frac{5}{4}\right)^x = \frac{16}{25}$ **48.** $\left(\frac{3}{2}\right)^x = \frac{16}{81}$

The amount of radioactive material in an ore sample is given by the exponential function

$$A(t) = 100(3.2)^{-0.5t},$$

where $A(t)$ is the amount present, in grams, of the sample t months after the initial measurement.

49. How much radioactive material was present at the initial measurement? (*Hint:* $t = 0$.)

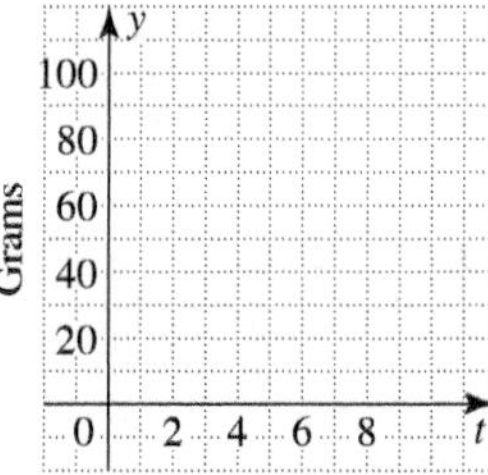

50. How much, to the nearest hundredth, was present 2 months later?

51. How much, to the nearest hundredth, was present 10 months later?

52. Graph the function on the axes as shown.

A major scientific periodical published an article in 1990 dealing with the problem of global warming. The article was accompanied by a graph that illustrated two possible scenarios.

(a) *The warming might be modeled by an exponential function of the form*

$$f(x) = (1.046 \times 10^{-38})(1.0444^x).$$

(b) *The warming might be modeled by a linear function of the form*

$$g(x) = 0.009x - 17.67.$$

In both cases, x represents the year, and the function value represents the increase in degrees Celsius due to the warming. Use these functions to approximate the increase in temperature for each year, to the nearest tenth of a degree.

53. 2000 **54.** 2010 **55.** 2020 **56.** 2040

Solve each problem. ***See Examples 7 and 8.***

57. The estimated number of monthly active Snapchat users (in millions) from 2013 to 2016 can be modeled by the exponential function

$$f(x) = 39.154(2.0585)^x,$$

where $x = 0$ represents 2013, $x = 1$ represents 2014, and so on. Use this model to approximate the number of monthly active Snapchat users in each year, to the nearest thousandth. (Data from Activate.)

(a) 2014 **(b)** 2015 **(c)** 2016

58. The number of paid music subscriptions (in millions) in the United States from 2010 to 2016 can be modeled by the exponential function

$$f(x) = 1.365(1.565)^x,$$

where $x = 0$ represents 2010, $x = 1$ represents 2011, and so on. Use this model to approximate the number of paid music subscriptions in each year, to the nearest thousandth. (Data from RIAA.)

(a) 2010 **(b)** 2013 **(c)** 2016

59. A small business estimates that the value $V(t)$ of a copy machine is decreasing according to the exponential function

$$V(t) = 5000(2)^{-0.15t},$$

where t is the number of years that have elapsed since the machine was purchased, and $V(t)$ is in dollars.

(a) What was the original value of the machine?

(b) What is the value of the machine 5 yr after purchase, to the nearest dollar?

(c) What is the value of the machine 10 yr after purchase, to the nearest dollar?

(d) Graph the function.

60. Refer to the exponential function in **Exercise 59.**

(a) When will the value of the machine be \$2500? (*Hint:* Let $V(t) = 2500$, divide both sides by 5000, and use the method of **Example 5.**)

(b) When will the value of the machine be \$1250?

10.3 Logarithmic Functions

OBJECTIVES

1 Define a logarithm.
2 Convert between exponential and logarithmic forms, and evaluate logarithms.
3 Solve logarithmic equations of the form $\log_a b = k$ for a, b, or k.
4 Use the definition of logarithm to simplify logarithmic expressions.
5 Define and graph logarithmic functions.
6 Use logarithmic functions in applications involving growth or decay.

OBJECTIVE 1 Define a logarithm.

The graph of $y = 2^x$ is the blue curve in **FIGURE 15.** Because $y = 2^x$ defines a one-to-one function, it has an inverse. Interchanging x and y gives

$$x = 2^y, \quad \text{the inverse of} \quad y = 2^x.$$ Roles of x and y are interchanged.

The graph of the inverse is found by reflecting the graph of $y = 2^x$ across the line $y = x$. The graph of the inverse $x = 2^y$ is the red curve in **FIGURE 15.**

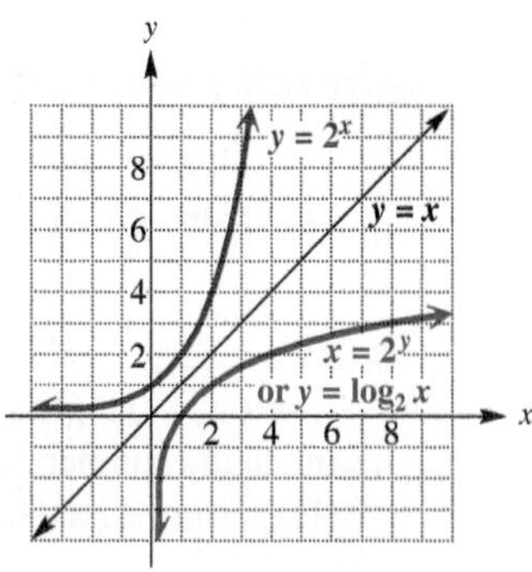

FIGURE 15

We can also write the equation of the red curve using a new notation that involves the concept of *logarithm*.

VOCABULARY

☐ logarithm
☐ logarithmic equation
☐ logarithmic function with base *a*

Logarithm

For all positive numbers a, where $a \neq 1$, and all positive real numbers x,

$$y = \log_a x \quad \text{is equivalent to} \quad x = a^y.$$

The abbreviation **log** is used for the word **logarithm.** Read $\log_a x$ as "***the logarithm of x with base a***" or "***the base a logarithm of x.***" To remember the location of the base and the exponent in each form, refer to the following diagrams.

Meaning of $\log_a x$

A logarithm is an exponent. ***The expression $\log_a x$ represents the exponent to which the base a must be raised to obtain x.***

OBJECTIVE 2 Convert between exponential and logarithmic forms, and evaluate logarithms.

We can use the definition of logarithm to carry out these conversions.

NOW TRY EXERCISE 1

(a) Write $6^3 = 216$ in logarithmic form.

(b) Write $\log_{64} 4 = \frac{1}{3}$ in exponential form.

EXAMPLE 1 Converting between Exponential and Logarithmic Forms

The table shows several pairs of equivalent forms.

Exponential Form	*Logarithmic Form*
$3^2 = 9$	$\log_3 9 = 2$
$\left(\frac{1}{5}\right)^{-2} = 25$	$\log_{1/5} 25 = -2$
$10^5 = 100{,}000$	$\log_{10} 100{,}000 = 5$
$4^{-3} = \frac{1}{64}$	$\log_4 \frac{1}{64} = -3$

$y = \log_a x$ is equivalent to $x = a^y$.

NOW TRY

NOW TRY EXERCISE 2

Use a calculator to approximate each logarithm to four decimal places.

(a) $\log_2 7$ **(b)** $\log_5 8$

(c) $\log_{1/3} 12$ **(d)** $\log_{10} 18$

EXAMPLE 2 Evaluating Logarithms

Use a calculator to approximate each logarithm to four decimal places.

(a) $\log_2 5$ **(b)** $\log_3 12$ **(c)** $\log_{1/2} 12$ **(d)** $\log_{10} 20$

FIGURE 16 shows how a TI-84 Plus calculator approximates the logarithms in parts (a)–(c).

NORMAL FIX4 AUTO REAL RADIAN MP
$\log_2(5)$
2.3219
$\log_3(12)$
2.2619
$\log_{1/2}(12)$
-3.5850

FIGURE 16

FIGURE 17

FIGURE 17 shows the approximation for the expression $\log_{10} 20$ in part (d). Notice that the second display, which indicates log 20 (with no base shown), gives the same result. We shall see in a later section that when no base is indicated, the base is understood to be 10. A base 10 logarithm is a *common logarithm*.

NOW TRY

NOW TRY ANSWERS

1. (a) $\log_6 216 = 3$
   (b) $64^{1/3} = 4$
2. (a) 2.8074 (b) 1.2920
   (c) -2.2619 (d) 1.2553

NOTE In a later section, we introduce another method of calculating logarithms like those in **Example 2(a)–(c)** using the *change-of-base rule.*

OBJECTIVE 3 Solve logarithmic equations of the form $\log_a b = k$ for a, b, or k.

A **logarithmic equation** is an equation with a logarithm in at least one term.

EXAMPLE 3 Solving Logarithmic Equations

Solve each equation.

(a)

$$\log_4 x = -2 \qquad \log_a x = y \text{ is equivalent to } x = a^y.$$

$$x = 4^{-2}$$

$$x = \frac{1}{4^2} \qquad \text{Definition of negative exponent}$$

$$x = \frac{1}{16} \qquad \text{Apply the exponent.}$$

CHECK $\log_4 \frac{1}{16} = -2$ because $4^{-2} = \frac{1}{16}$. ✓

The solution set is $\left\{\frac{1}{16}\right\}$.

(b)

$$\log_{1/2}(3x + 1) = 2$$

This is a key step.

$$3x + 1 = \left(\frac{1}{2}\right)^2 \qquad \text{Write in exponential form.}$$

$$3x + 1 = \frac{1}{4} \qquad \text{Apply the exponent.}$$

$$12x + 4 = 1 \qquad \text{Multiply each term by 4.}$$

$$12x = -3 \qquad \text{Subtract 4.}$$

$$x = -\frac{1}{4} \qquad \text{Divide by 12. Write in lowest terms.}$$

CHECK

$$\log_{1/2}(3x + 1) = 2$$

$$\log_{1/2}\left[3\left(-\frac{1}{4}\right) + 1\right] \stackrel{?}{=} 2 \qquad \text{Let } x = -\tfrac{1}{4}.$$

$$\log_{1/2} \frac{1}{4} \stackrel{?}{=} 2 \qquad \text{Simplify within parentheses.}$$

$$\left(\frac{1}{2}\right)^2 \stackrel{?}{=} \frac{1}{4} \qquad \text{Write in exponential form.}$$

$$\frac{1}{4} = \frac{1}{4} \ ✓ \qquad \text{True}$$

The solution set is $\left\{-\frac{1}{4}\right\}$.

Solve each equation.

(a) $\log_2 x = -5$

(b) $\log_{3/2} (2x - 1) = 3$

(c) $\log_x 10 = 2$

(d) $\log_{125} \sqrt[3]{5} = x$

(c)

$$\log_x 3 = 2$$

$$x^2 = 3 \quad \text{Write in exponential form.}$$

$$x = \pm\sqrt{3} \quad \text{Take square roots.}$$

Be careful here. $-\sqrt{3}$ is extraneous.

Only the *principal* square root $\sqrt{3}$ satisfies the equation because the base must be a positive number.

CHECK

$$\log_x 3 = 2$$

$$\log_{\sqrt{3}} 3 \stackrel{?}{=} 2 \quad \text{Let } x = \sqrt{3}.$$

$$\left(\sqrt{3}\right)^2 \stackrel{?}{=} 3 \quad \text{Write in exponential form.}$$

$$3 = 3 \ \checkmark \quad \text{True}$$

The solution set is $\left\{\sqrt{3}\right\}$.

(d)

$$\log_{49} \sqrt[3]{7} = x$$

$$49^x = \sqrt[3]{7} \quad \text{Write in exponential form.}$$

$$(7^2)^x = 7^{1/3} \quad \text{Write with the same base.}$$

$$7^{2x} = 7^{1/3} \quad \text{Power rule for exponents}$$

$$2x = \frac{1}{3} \quad \text{Set the exponents equal.}$$

$$x = \frac{1}{6} \quad \text{Divide by 2 (which is the same as multiplying by } \tfrac{1}{2}\text{).}$$

Check to verify that the solution set is $\left\{\frac{1}{6}\right\}$.

NOW TRY 

OBJECTIVE 4 **Use the definition of logarithm to simplify logarithmic expressions.**

The definition of logarithm enables us to state several special properties.

Special Properties of Logarithms

For any positive real number b, where $b \neq 1$, the following hold true.

$$\log_b b = 1 \qquad \log_b 1 = 0$$

$$\log_b b^r = r \ \ (r \text{ is real.}) \qquad b^{\log_b r} = r \ \ (r > 0)$$

To prove the last statement, let $x = \log_b r$.

$$x = \log_b r$$

$$b^x = r \quad \text{Write in exponential form.}$$

$$b^{\log_b r} = r \quad \text{Replace } x \text{ with } \log_b r.$$

This is the statement to be proved.

NOW TRY ANSWERS

3. (a) $\left\{\frac{1}{32}\right\}$ (b) $\left\{\frac{35}{16}\right\}$

(c) $\left\{\sqrt{10}\right\}$ (d) $\left\{\frac{1}{9}\right\}$

NOW TRY EXERCISE 4

Use the special properties to evaluate each expression.

(a) $\log_{10} 10$ **(b)** $\log_8 1$

(c) $\log_{0.1} 1$ **(d)** $\log_3 3^9$

(e) $5^{\log_5 3}$ **(f)** $\log_3 81$

EXAMPLE 4 **Using Special Properties of Logarithms**

Use the special properties to evaluate.

(a) $\log_7 7 = 1$ $\quad\log_b b = 1$

(b) $\log_{\sqrt{2}} \sqrt{2} = 1$ $\quad\log_b b = 1$

(c) $\log_9 1 = 0$ $\quad\log_b 1 = 0$

(d) $\log_{0.2} 1 = 0$ $\quad\log_b 1 = 0$

(e) $\log_2 2^6 = 6$ $\quad\log_b b^r = r$

(f) $\log_3 3^{-2.5} = -2.5$ $\quad\log_b b^r = r$

(g) $4^{\log_4 9} = 9$ $\quad b^{\log_b r} = r$

(h) $10^{\log_{10} 13} = 13$ $\quad b^{\log_b r} = r$

(i) $\log_2 32$

$= \log_2 2^5$ $\quad 32 = 2^5$

$= 5$ $\quad\log_b b^r = r$

(j) $\log_3 \frac{1}{3}$

$= \log_3 3^{-1}$ $\quad$ Definition of negative exponent

$= -1$ $\quad\log_b b^r = r$

NOW TRY

OBJECTIVE 5 Define and graph logarithmic functions.

Logarithmic Function

If a and x are positive real numbers, where $a \neq 1$, then

$$g(x) = \log_a x$$

defines the **logarithmic function with base a.**

NOW TRY EXERCISE 5

Graph $f(x) = \log_6 x$.

EXAMPLE 5 **Graphing a Logarithmic Function ($a > 1$)**

Graph $f(x) = \log_2 x$.

By writing $y = f(x) = \log_2 x$ in exponential form as $x = 2^y$, we can identify ordered pairs that satisfy the equation. It is easier to choose values for y and find the corresponding values of x. Plotting the points in the table of ordered pairs and connecting them with a smooth curve gives the graph in **FIGURE 18**. This graph is typical of logarithmic functions with base $a > 1$.

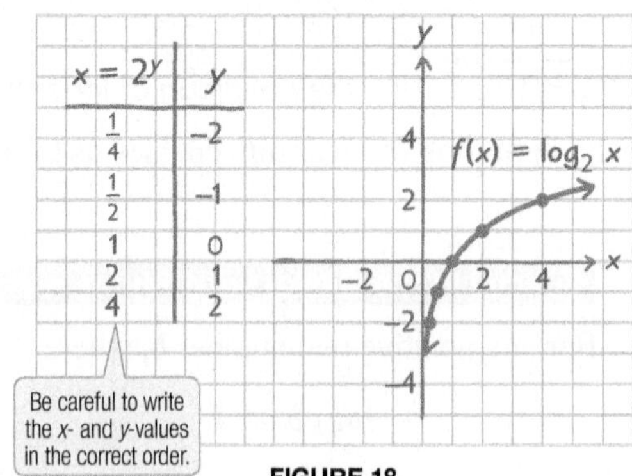

FIGURE 18

Logarithmic function with base $a > 1$

Domain: $(0, \infty)$

Range: $(-\infty, \infty)$

x-intercept: $(1, 0)$

The function is one-to-one, and its graph rises from left to right.

NOW TRY 

EXAMPLE 6 **Graphing a Logarithmic Function ($0 < a < 1$)**

Graph $g(x) = \log_{1/2} x$.

We write $y = g(x) = \log_{1/2} x$ in exponential form as

$$x = \left(\frac{1}{2}\right)^y,$$

and then choose values for y and find the corresponding values of x. Plotting these points and connecting them with a smooth curve gives the graph in **FIGURE 19**. This graph is typical of logarithmic functions with base $0 < a < 1$.

NOW TRY ANSWERS

4. (a) 1 **(b)** 0 **(c)** 0 **(d)** 9 **(e)** 3 **(f)** 4

5. $f(x) = \log_6 x$

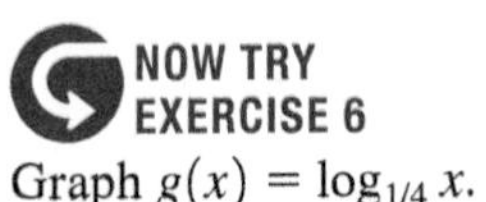
NOW TRY EXERCISE 6

Graph $g(x) = \log_{1/4} x$.

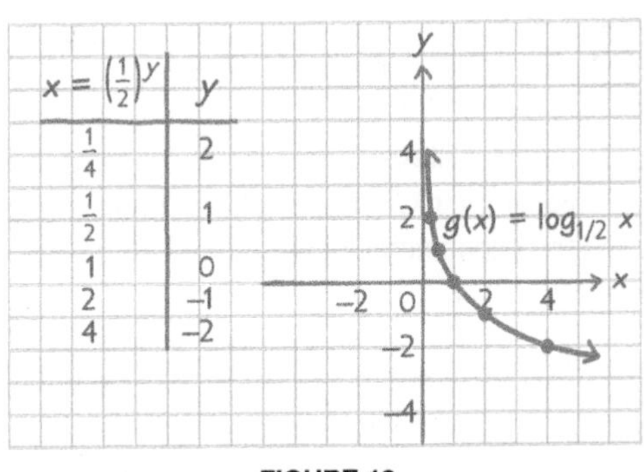

$x = \left(\frac{1}{2}\right)^y$	y
$\frac{1}{4}$	2
$\frac{1}{2}$	1
1	0
2	−1
4	−2

FIGURE 19

Logarithmic function with base $0 < a < 1$

Domain: $(0, \infty)$

Range: $(-\infty, \infty)$

x-intercept: $(1, 0)$

The function is one-to-one, and its graph falls from left to right.

NOW TRY

Characteristics of the Graph of $g(x) = \log_a x$

1. The graph contains the point $(1, 0)$, which is its x-intercept.
2. The function is one-to-one.
   - When $a > 1$, the graph *rises* from left to right, from the fourth quadrant to the first. (See **FIGURE 18**.)
   - When $0 < a < 1$, the graph *falls* from left to right, from the first quadrant to the fourth. (See **FIGURE 19**.)
3. The graph approaches the y-axis, but never touches it—that is, the y-axis is an asymptote.
4. The domain is $(0, \infty)$, and the range is $(-\infty, \infty)$.

OBJECTIVE 6 Use logarithmic functions in applications involving growth or decay.

NOW TRY EXERCISE 7

Suppose the gross national product (GNP) of a small country (in millions of dollars) is approximated by the logarithmic function

$$G(t) = 15.0 + 2.00 \log_{10} t,$$

where t is time in years since 2013. Approximate to the nearest tenth the GNP for each value of t.

(a) $t = 1$ **(b)** $t = 10$

EXAMPLE 7 Applying a Logarithmic Function

Barometric pressure in inches of mercury at a distance of x miles from the eye of a typical hurricane can be approximated by the logarithmic function

$$f(x) = 27 + 1.105 \log_{10} (x + 1).$$

(Data from Miller, A. and R. Anthes, *Meteorology*, Fifth Edition, Charles E. Merrill Publishing Company.)

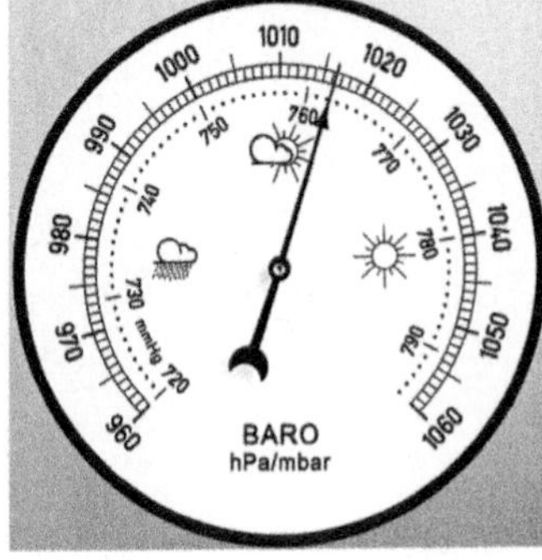

Approximate the pressure 9 mi from the eye of the hurricane.

$$f(x) = 27 + 1.105 \log_{10} (x + 1)$$

$$f(9) = 27 + 1.105 \log_{10} (9 + 1) \quad \text{Let } x = 9.$$

$$f(9) = 27 + 1.105 \log_{10} 10 \quad \text{Add inside parentheses.}$$

$$f(9) = 27 + 1.105(1) \quad \log_{10} 10 = 1$$

$$f(9) = 28.105 \quad \text{Add.}$$

The pressure 9 mi from the eye of the hurricane is 28.105 in.

NOW TRY

NOW TRY ANSWERS

6. 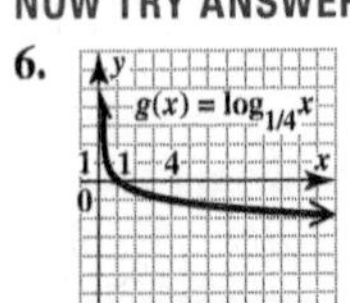

7. (a) \$15.0 million
   (b) \$17.0 million

10.3 Exercises

MyLab Math

Video solutions for select problems available in MyLab Math

1. Concept Check Match each logarithmic equation in Column I with the corresponding exponential equation in Column II.

I	II
(a) $\log_{1/3} 3 = -1$	A. $8^{1/3} = \sqrt[3]{8}$
(b) $\log_5 1 = 0$	B. $\left(\frac{1}{3}\right)^{-1} = 3$
(c) $\log_2 \sqrt{2} = \frac{1}{2}$	C. $4^1 = 4$
(d) $\log_{10} 1000 = 3$	D. $2^{1/2} = \sqrt{2}$
(e) $\log_8 \sqrt[3]{8} = \frac{1}{3}$	E. $5^0 = 1$
(f) $\log_4 4 = 1$	F. $10^3 = 1000$

2. Concept Check Match each logarithm in Column I with its corresponding value in Column II.

I	II
(a) $\log_4 16$	A. -2
(b) $\log_3 81$	B. -1
(c) $\log_3 \left(\frac{1}{3}\right)$	C. 2
(d) $\log_{10} 0.01$	D. 0
(e) $\log_5 \sqrt{5}$	E. $\frac{1}{2}$
(f) $\log_{13} 1$	F. 4

3. Concept Check The domain of $f(x) = a^x$ is $(-\infty, \infty)$, while the range is $(0, \infty)$. Therefore because $g(x) = \log_a x$ is the inverse of f, the domain of g is ________, while the range of g is ________.

4. Concept Check The graphs of both $f(x) = 3^x$ and $g(x) = \log_3 x$ rise from left to right. Which one rises at a faster rate as x gets large?

Write in logarithmic form. ***See Example 1.***

5. $4^5 = 1024$ **6.** $3^6 = 729$ **7.** $\left(\frac{1}{2}\right)^{-3} = 8$ **8.** $\left(\frac{1}{6}\right)^{-3} = 216$

9. $10^{-3} = 0.001$ **10.** $36^{1/2} = 6$ **11.** $\sqrt[4]{625} = 5$ **12.** $\sqrt[3]{343} = 7$

13. $8^{-2/3} = \frac{1}{4}$ **14.** $16^{-3/4} = \frac{1}{8}$ **15.** $5^0 = 1$ **16.** $7^0 = 1$

Write in exponential form. ***See Example 1.***

17. $\log_4 64 = 3$ **18.** $\log_2 512 = 9$ **19.** $\log_{12} 12 = 1$

20. $\log_{100} 100 = 1$ **21.** $\log_6 1 = 0$ **22.** $\log_\pi 1 = 0$

23. $\log_9 3 = \frac{1}{2}$ **24.** $\log_{64} 2 = \frac{1}{6}$ **25.** $\log_{1/4} \frac{1}{2} = \frac{1}{2}$

26. $\log_{1/8} \frac{1}{2} = \frac{1}{3}$ **27.** $\log_5 5^{-1} = -1$ **28.** $\log_{10} 10^{-2} = -2$

29. Concept Check Match each logarithm in Column I with its value in Column II.

I	II
(a) $\log_8 8$	A. -1
(b) $\log_{16} 1$	B. 0
(c) $\log_{0.3} 1$	C. 1
(d) $\log_{\sqrt{7}} \sqrt{7}$	D. 0.1

30. Concept Check When a student asked his teacher to explain how to evaluate

$$\log_9 3$$

without showing any work, his teacher told him to "*Think radically.*" Explain what the teacher meant by this hint.

Use a calculator to approximate each logarithm to four decimal places. **See Example 2.**

31. $\log_2 9$ **32.** $\log_2 15$ **33.** $\log_5 18$ **34.** $\log_5 26$

35. $\log_{1/4} 12$ **36.** $\log_{1/5} 27$ **37.** $\log_2 \frac{1}{3}$ **38.** $\log_2 \frac{1}{7}$

39. $\log_{10} 84$ **40.** $\log_{10} 126$ **41.** $\log 50$ **42.** $\log 90$

Solve each equation. **See Example 3.**

43. $x = \log_{27} 3$ **44.** $x = \log_{125} 5$ **45.** $\log_5 x = -3$

46. $\log_{10} x = -2$ **47.** $\log_x 9 = \frac{1}{2}$ **48.** $\log_x 5 = \frac{1}{2}$

49. $\log_x 125 = -3$ **50.** $\log_x 64 = -6$ **51.** $\log_{12} x = 0$

52. $\log_4 x = 0$ **53.** $\log_x x = 1$ **54.** $\log_x 1 = 0$

55. $\log_x \frac{1}{25} = -2$ **56.** $\log_x \frac{1}{10} = -1$ **57.** $\log_8 32 = x$

58. $\log_{81} 27 = x$ **59.** $\log_\pi \pi^4 = x$ **60.** $\log_{\sqrt{2}} (\sqrt{2})^9 = x$

61. $\log_6 \sqrt{216} = x$ **62.** $\log_4 \sqrt{64} = x$ **63.** $\log_4 (2x + 4) = 3$

64. $\log_3 (2x + 7) = 4$ **65.** $\log_{1/3} (x - 4) = 2$ **66.** $\log_{1/2} (2x - 1) = 3$

Use the special properties of logarithms to evaluate each expression. **See Example 4.**

67. $\log_3 3$ **68.** $\log_8 8$ **69.** $\log_5 1$ **70.** $\log_{12} 1$

71. $\log_4 4^9$ **72.** $\log_5 5^6$ **73.** $\log_2 2^{-1}$ **74.** $\log_4 4^{-6}$

75. $6^{\log_6 9}$ **76.** $12^{\log_{12} 3}$ **77.** $8^{\log_8 5}$ **78.** $5^{\log_5 11}$

79. $\log_2 64$ **80.** $\log_2 128$ **81.** $\log_3 81$ **82.** $\log_3 27$

83. $\log_4 \frac{1}{4}$ **84.** $\log_6 \frac{1}{6}$ **85.** $\log_6 \sqrt[3]{6}$ **86.** $\log_9 \sqrt[3]{9}$

Graph each logarithmic function. **See Examples 5 and 6.**

87. $g(x) = \log_3 x$ **88.** $g(x) = \log_5 x$ **89.** $f(x) = \log_4 x$ **90.** $f(x) = \log_6 x$

91. $f(x) = \log_{1/3} x$ **92.** $f(x) = \log_{1/5} x$ **93.** $g(x) = \log_{1/4} x$ **94.** $g(x) = \log_{1/6} x$

Use the graph at the right to predict the value of $f(t)$ for the given value of t.

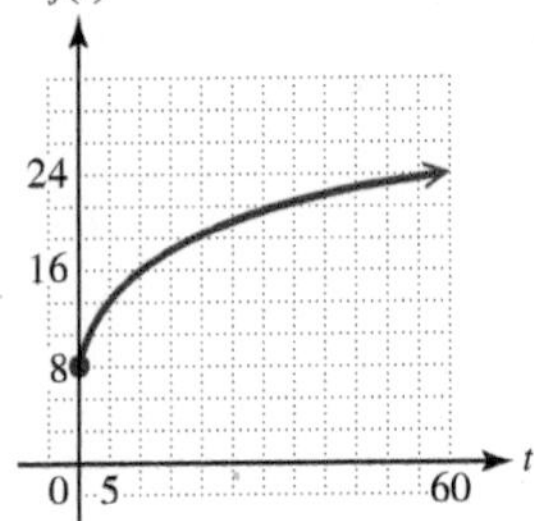

95. $t = 0$ **96.** $t = 10$ **97.** $t = 60$

98. Show that the points determined in **Exercises 95–97** lie on the graph of

$$f(t) = 8 \log_5 (2t + 5).$$

Concept Check *Answer each question.*

99. Why is 1 not allowed as a base for a logarithmic function?

100. Why is a negative number not allowed as a base for a logarithmic function?

101. Why is $\log_a 1 = 0$ true for any value of a that is allowed as the base of a logarithm? Use a rule of exponents introduced earlier in the explanation.

102. Why is $\log_a a = 1$ true for any value of a that is allowed as the base of a logarithm?

Solve each problem. ***See Example 7.***

103. Sales (in thousands of units) of a new product are approximated by the logarithmic function

$$S(t) = 100 + 30 \log_3 (2t + 1),$$

where t is the number of years after the product is introduced.

(a) What were the sales, to the nearest unit, after 1 yr?

(b) What were the sales, to the nearest unit, after 13 yr?

(c) Graph $y = S(t)$.

104. A study showed that the number of mice in an old abandoned house was approximated by the logarithmic function

$$M(t) = 6 \log_4 (2t + 4),$$

where t is measured in months and $t = 0$ corresponds to January 2018. Find the number of mice in the house for each month.

(a) January 2018 **(b)** July 2018 **(c)** July 2020 **(d)** Graph $y = M(t)$.

105. An online sales company finds that its sales (in millions of dollars) are approximated by the logarithmic function

$$S(x) = \log_2 (3x + 1),$$

where x is the number of advertisements placed on a popular website. How many advertisements must be placed to earn sales of \$4 million?

106. The population of deer (in thousands) in a certain area is approximated by the logarithmic function

$$f(x) = \log_5 (100x - 75),$$

where x is the number of years since 2017. During what year is the population expected to be 4 thousand deer?

RELATING CONCEPTS For Individual or Group Work (Exercises 107–110)

To see how exponential and logarithmic functions are related, ***work Exercises 107–110 in order.***

107. Complete the table of values, and sketch the graph of $y = 10^x$. Give the domain and range of the function.

x	y
−2	
−1	
0	
1	
2	

108. Complete the table of values, and sketch the graph of $y = \log_{10} x$. Give the domain and range of the function.

x	y
$\frac{1}{100}$	
$\frac{1}{10}$	
1	
10	
100	

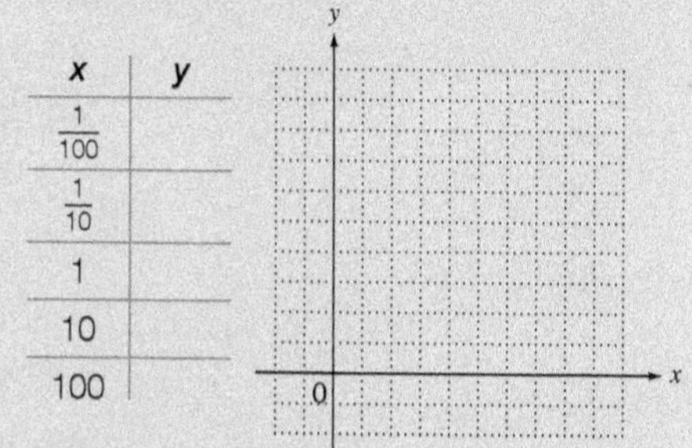

109. Describe the symmetry between the graphs in **Exercises 107 and 108.**

110. What can we conclude about the functions

$$y = f(x) = 10^x \quad \text{and} \quad y = g(x) = \log_{10} x?$$

10.4 Properties of Logarithms

OBJECTIVES

1 Use the product rule for logarithms.
2 Use the quotient rule for logarithms.
3 Use the power rule for logarithms.
4 Use properties to write alternative forms of logarithmic expressions.

Logarithms were used as an aid to numerical calculation for several hundred years. Today the widespread use of calculators has made the use of logarithms for calculation obsolete. However, logarithms are still very important in applications and in further work in mathematics.

OBJECTIVE 1 Use the product rule for logarithms.

One way in which logarithms simplify problems is by changing a problem of multiplication into one of addition. For example, we know that

$$\log_2 4 = 2, \quad \log_2 8 = 3, \quad \text{and} \quad \log_2 32 = 5.$$

Therefore, we can make the following statements.

$$\log_2 32 = \log_2 4 + \log_2 8 \qquad 5 = 2 + 3$$

$$\log_2 (4 \cdot 8) = \log_2 4 + \log_2 8 \qquad 32 = 4 \cdot 8$$

This is an example of the product rule for logarithms.

Product Rule for Logarithms

If x, y, and b are positive real numbers, where $b \neq 1$, then the following holds true.

$$\mathbf{\log_b xy = \log_b x + \log_b y}$$

That is, the logarithm of a product is the sum of the logarithms of the factors.

Examples: $\log_3 (4 \cdot 7) = \log_3 4 + \log_3 7$, $\log_{10} 8 + \log_{10} 9 = \log_{10} (8 \cdot 9)$

To prove this rule, let $m = \log_b x$ and $n = \log_b y$, and recall that

$\log_b x = m$ is equivalent to $b^m = x$ and $\log_b y = n$ is equivalent to $b^n = y$.

Now consider the product xy.

$$xy = b^m \cdot b^n \qquad \text{Substitute.}$$

$$xy = b^{m+n} \qquad \text{Product rule for exponents}$$

$$\log_b xy = m + n \qquad \text{Write in logarithmic form.}$$

$$\log_b xy = \log_b x + \log_b y \qquad \text{Substitute for } m \text{ and } n.$$

The last statement is the result we wished to prove.

NOTE The word statement of the product rule can be restated by replacing the word "logarithm" with the word "exponent." The rule then becomes the familiar rule for multiplying exponential expressions:

The *exponent* of a product is *equal* to the sum of the *exponents* of the factors.

NOW TRY EXERCISE 1

Use the product rule to rewrite each logarithm.

(a) $\log_{10}(7 \cdot 9)$

(b) $\log_5 11 + \log_5 8$

(c) $\log_5(5x), \quad x > 0$

(d) $\log_2 t^3, \quad t > 0$

EXAMPLE 1 Using the Product Rule

Use the product rule to rewrite each logarithm. Assume $x > 0$.

(a) $\log_5(6 \cdot 9)$

$= \log_5 6 + \log_5 9$ Product rule

(b) $\log_7 8 + \log_7 12$

$= \log_7(8 \cdot 12)$ Product rule

$= \log_7 96$ Multiply.

(c) $\log_3(3x)$

$= \log_3 3 + \log_3 x$ Product rule

$= 1 + \log_3 x$ $\log_b b = 1$

(d) $\log_4 x^3$

$= \log_4(x \cdot x \cdot x)$ $x^3 = x \cdot x \cdot x$

$= \log_4 x + \log_4 x + \log_4 x$ Product rule

$= 3 \log_4 x$ Combine like terms.

NOW TRY 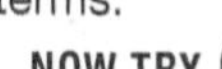

OBJECTIVE 2 Use the quotient rule for logarithms.

The rule for division is similar to the rule for multiplication.

Quotient Rule for Logarithms

If x, y, and b are positive real numbers, where $b \neq 1$, then the following holds true.

$$\log_b \frac{x}{y} = \log_b x - \log_b y$$

That is, the logarithm of a quotient is the difference of the logarithm of the numerator and the logarithm of the denominator.

Examples: $\log_5 \frac{2}{3} = \log_5 2 - \log_5 3$, $\log_7 3 - \log_7 5 = \log_7 \frac{3}{5}$

NOW TRY EXERCISE 2

Use the quotient rule to rewrite each logarithm.

(a) $\log_{10} \frac{7}{9}$

(b) $\log_4 x - \log_4 12, \quad x > 0$

(c) $\log_5 \frac{25}{27}$

(d) $\log_2 15 - \log_2 3$

The proof of this rule is similar to the proof of the product rule.

EXAMPLE 2 Using the Quotient Rule

Use the quotient rule to rewrite each logarithm. Assume $x > 0$.

(a) $\log_4 \frac{7}{9}$

$= \log_4 7 - \log_4 9$ Quotient rule

(b) $\log_5 6 - \log_5 x$

$= \log_5 \frac{6}{x}$ Quotient rule

(c) $\log_3 \frac{27}{5}$

$= \log_3 27 - \log_3 5$ Quotient rule

$= 3 - \log_3 5$ $\log_3 27 = 3$

(d) $\log_6 28 - \log_6 7$

$= \log_6 \frac{28}{7}$ Quotient rule

$= \log_6 4$ $\frac{28}{7} = 4$

NOW TRY 

NOW TRY ANSWERS

1. **(a)** $\log_{10} 7 + \log_{10} 9$
   (b) $\log_5 88$
   (c) $1 + \log_5 x$
   (d) $3 \log_2 t$
2. **(a)** $\log_{10} 7 - \log_{10} 9$
   (b) $\log_4 \frac{x}{12}$
   (c) $2 - \log_5 27$
   (d) $\log_2 5$

CAUTION *There is no property of logarithms to rewrite the logarithm of a sum.*

$$\log_b (x + y) \neq \log_b x + \log_b y$$

Also, $\log_b x \cdot \log_b y \neq \log_b xy$, and $\dfrac{\log_b x}{\log_b y} \neq \log_b \dfrac{x}{y}$.

OBJECTIVE 3 Use the power rule for logarithms.

Consider the exponential expression 2^3.

2^3 means $2 \cdot 2 \cdot 2$. The base 2 is used as a factor 3 times.

Similarly, the product rule can be extended to rewrite the logarithm of a power as the product of the exponent and the logarithm of the base.

$\log_5 2^3$	$\log_2 7^4$
$= \log_5 (2 \cdot 2 \cdot 2)$	$= \log_2 (7 \cdot 7 \cdot 7 \cdot 7)$
$= \log_5 2 + \log_5 2 + \log_5 2$	$= \log_2 7 + \log_2 7 + \log_2 7 + \log_2 7$
$= 3 \log_5 2$	$= 4 \log_2 7$

Furthermore, we saw in **Example 1(d)** that $\log_4 x^3 = 3 \log_4 x$. These examples suggest the following rule.

Power Rule for Logarithms

If x and b are positive real numbers, where $b \neq 1$, and if r is any real number, then the following holds true.

$$\mathbf{\log_b x^r = r \log_b x}$$

That is, the logarithm of a number to a power equals the exponent times the logarithm of the number.

Examples: $\log_b m^5 = 5 \log_b m$, $\log_3 5^4 = 4 \log_3 5$

To prove the power rule, let $\log_b x = m$.

$b^m = x$	Write in exponential form.
$(b^m)^r = x^r$	Raise each side to the power r.
$b^{mr} = x^r$	Power rule for exponents
$\log_b x^r = mr$	Write in logarithmic form.
$\log_b x^r = rm$	Commutative property
$\log_b x^r = r \log_b x$	$m = \log_b x$ from above

This is the statement to be proved.

As a special case of the power rule, let $r = \frac{1}{p}$, so

$$\log_b \sqrt[p]{x} = \log_b x^{1/p} = \frac{1}{p} \log_b x.$$

Examples: $\log_b \sqrt[5]{x} = \log_b x^{1/5} = \frac{1}{5} \log_b x$, $\log_b \sqrt[3]{x^4} = \log_b x^{4/3} = \frac{4}{3} \log_b x \ (x > 0)$

Another special case is

$$\log_b \frac{1}{x} = \log_b x^{-1} = -\log_b x. \qquad -a = -1 \cdot a$$

Example: $\log_9 \frac{1}{5} = \log_9 5^{-1} = -\log_9 5$

NOW TRY EXERCISE 3

Use the power rule to rewrite each logarithm. Assume $a > 0, x > 0$, and $a \neq 1$.

(a) $\log_7 5^3$ **(b)** $\log_a \sqrt{10}$

(c) $\log_3 \sqrt[4]{x^3}$ **(d)** $\log_4 \frac{1}{x^5}$

EXAMPLE 3 Using the Power Rule

Use the power rule to rewrite each logarithm. Assume $b > 0, x > 0$, and $b \neq 1$.

(a) $\log_5 4^2$

$= 2 \log_5 4$ Power rule

(b) $\log_b x^5$

$= 5 \log_b x$ Power rule

(c) $\log_b \sqrt{7}$

$= \log_b 7^{1/2}$ $\sqrt{x} = x^{1/2}$

$= \frac{1}{2} \log_b 7$ Power rule

(d) $\log_2 \sqrt[5]{x^2}$

$= \log_2 x^{2/5}$ $\sqrt[5]{x^2} = x^{2/5}$

$= \frac{2}{5} \log_2 x$ Power rule

(e) $\log_3 \frac{1}{x^4}$

$= \log_3 x^{-4}$ Definition of negative exponent

$= -4 \log_3 x$ Power rule

NOW TRY 

We summarize the properties of logarithms from the previous section and this one.

Properties of Logarithms

If x, y, and b are positive real numbers, where $b \neq 1$, and r is any real number, then the following hold true.

Special Properties $\log_b b = 1$ $\log_b 1 = 0$

$\log_b b^r = r$ $b^{\log_b r} = r \quad (r > 0)$

Product Rule $\log_b xy = \log_b x + \log_b y$

Quotient Rule $\log_b \frac{x}{y} = \log_b x - \log_b y$

Power Rule $\log_b x^r = r \log_b x$

OBJECTIVE 4 Use properties to write alternative forms of logarithmic expressions.

EXAMPLE 4 Writing Logarithms in Alternative Forms

Use the properties of logarithms to rewrite each expression if possible. Assume that all variables represent positive real numbers.

(a) $\log_4 4x^3$

$= \log_4 4 + \log_4 x^3$ Product rule

$= 1 + 3 \log_4 x$ $\log_b b = 1$; Power rule

NOW TRY ANSWERS

3. **(a)** $3 \log_7 5$ **(b)** $\frac{1}{2} \log_a 10$

(c) $\frac{3}{4} \log_3 x$ **(d)** $-5 \log_4 x$

NOW TRY EXERCISE 4

Use the properties of logarithms to rewrite each expression if possible. Assume that all variables represent positive real numbers.

(a) $\log_3 9z^4$

(b) $\log_6 \sqrt{\frac{n}{3m}}$

(c) $\log_2 x + 3\log_2 y - \log_2 z$

(d) $\log_5 (x+10) + \log_5 (x-10) - \frac{3}{5}\log_5 x, \quad x > 10$

(e) $\log_7 (49 + 2x)$

(b) $\log_7 \sqrt{\frac{m}{n}}$

$= \log_7 \left(\frac{m}{n}\right)^{1/2}$ Write the radical expression with a rational exponent.

$= \frac{1}{2}\log_7 \frac{m}{n}$ Power rule

$= \frac{1}{2}(\log_7 m - \log_7 n)$ Quotient rule

(c) $\log_5 \frac{a^2}{bc}$

$= \log_5 a^2 - \log_5 bc$ Quotient rule

$= 2\log_5 a - \log_5 bc$ Power rule

$= 2\log_5 a - (\log_5 b + \log_5 c)$ Product rule

$= 2\log_5 a - \log_5 b - \log_5 c$ Parentheses are necessary here.

(d) $4\log_b m - \log_b n, \quad b \neq 1$

$= \log_b m^4 - \log_b n$ Power rule

$= \log_b \frac{m^4}{n}$ Quotient rule

(e) $\log_b (x+1) + \log_b (2x+1) - \frac{2}{3}\log_b x, \quad b \neq 1$

$= \log_b (x+1) + \log_b (2x+1) - \log_b x^{2/3}$ Power rule

$= \log_b \frac{(x+1)(2x+1)}{x^{2/3}}$ Product and quotient rules

$= \log_b \frac{2x^2+3x+1}{x^{2/3}}$ Multiply in the numerator.

(f) $\log_8 (2p + 3r)$ cannot be rewritten using the properties of logarithms. ***There is no property of logarithms to rewrite the logarithm of a sum.*** NOW TRY

In the next example, we use numerical values for $\log_2 5$ and $\log_2 3$. While we use the equality symbol to give these values, they are actually approximations because most logarithms of this type are irrational numbers. ***We use = with the understanding that the values are correct to four decimal places.***

EXAMPLE 5 Using the Properties of Logarithms with Numerical Values

Given that $\log_2 5 = 2.3219$ and $\log_2 3 = 1.5850$, use these values and the properties of logarithms to evaluate each expression.

(a) $\log_2 15$

$= \log_2 (3 \cdot 5)$ Factor 15.

$= \log_2 3 + \log_2 5$ Product rule

$= 1.5850 + 2.3219$ Substitute the given values.

$= 3.9069$ Add.

NOW TRY ANSWERS

4. (a) $2 + 4\log_3 z$
(b) $\frac{1}{2}(\log_6 n - \log_6 3 - \log_6 m)$
(c) $\log_2 \frac{xy^3}{z}$ (d) $\log_5 \frac{x^2-100}{x^{3/5}}$
(e) cannot be rewritten

NOW TRY EXERCISE 5

Given that $\log_2 7 = 2.8074$ and $\log_2 10 = 3.3219$, use these values and the properties of logarithms to evaluate each expression.

(a) $\log_2 70$ **(b)** $\log_2 0.7$

(c) $\log_2 49$

(b) $\log_2 0.6$

$= \log_2 \frac{3}{5}$	$0.6 = \frac{6}{10} = \frac{3}{5}$
$= \log_2 3 - \log_2 5$	Quotient rule
$= 1.5850 - 2.3219$	Substitute the given values.
$= -0.7369$	Subtract.

(c) $\log_2 27$

$= \log_2 3^3$	Write 27 as a power of 3.
$= 3 \log_2 3$	Power rule
$= 3(1.5850)$	Substitute the given value.
$= 4.7550$	Multiply.

NOW TRY

NOW TRY EXERCISE 6

Determine whether each statement is *true* or *false*.

(a) $\log_2 16 + \log_2 16 = \log_2 32$

(b) $(\log_2 4)(\log_3 9) = \log_6 6^4$

EXAMPLE 6 Determining Whether Statements about Logarithms Are True

Determine whether each statement is *true* or *false*.

(a) $\log_2 8 - \log_2 4 = \log_2 4$

Evaluate each side.

$\log_2 8 - \log_2 4$	Left side	$\log_2 4$	Right side
$= \log_2 2^3 - \log_2 2^2$	Write 8 and 4 as powers of 2.	$= \log_2 2^2$	Write 4 as a power of 2.
$= 3 - 2$	$\log_b b^r = r$	$= 2$	$\log_b b^r = r$
$= 1$	Subtract.		

The statement is false because $1 \neq 2$.

(b) $\log_3 (\log_2 8) = \dfrac{\log_7 49}{\log_8 64}$

$\log_3 (\log_2 8)$	Left side	$\dfrac{\log_7 49}{\log_8 64}$	Right side
$= \log_3 (\log_2 2^3)$	Write 8 as a power of 2.	$= \dfrac{\log_7 7^2}{\log_8 8^2}$	Write 49 and 64 using exponents.
$= \log_3 3$	$\log_b b^r = r$	$= \dfrac{2}{2}$	$\log_b b^r = r$
$= 1$	$\log_b b = 1$	$= 1$	Simplify.

The statement is true because $1 = 1$.

NOW TRY 

NOW TRY ANSWERS

5. **(a)** 6.1293 **(b)** −0.5145 **(c)** 5.6148

6. **(a)** false **(b)** true

10.4 Exercises

MyLab Math

Video solutions for select problems available in MyLab Math

Concept Check *Determine whether each statement of a logarithmic property is* true *or* false*. If it is false, correct it by changing the right side of the equation.*

1. $\log_b x + \log_b y = \log_b (x + y)$

2. $\log_b \frac{x}{y} = \log_b x - \log_b y$

3. $\log_b xy = \log_b x + \log_b y$

4. $b^{\log_b r} = 1$

5. $\log_b b^r = r$

6. $\log_b x^r = \log_b rx$

7. Concept Check A student erroneously wrote $\log_a (x + y) = \log_a x + \log_a y$. When his teacher explained that this was indeed wrong, the student claimed that he had used the distributive property. **WHAT WENT WRONG?**

8. Concept Check Consider the following "proof" that $\log_2 16$ does not exist.

$$\begin{aligned} &\log_2 16 \\ &= \log_2 (-4)(-4) \\ &= \log_2 (-4) + \log_2 (-4) \end{aligned}$$

The logarithm of a negative number is not defined, so the final step cannot be evaluated. Thus $\log_2 16$ does not exist. **WHAT WENT WRONG?**

Concept Check *Use the indicated rule of logarithms to complete each equation.*

9. $\log_{10} (7 \cdot 8) =$ ________________ (product rule)

10. $\log_{10} \frac{7}{8} =$ ________________ (quotient rule)

11. $3^{\log_3 4} =$ ________________ (special property)

12. $\log_{10} 3^6 =$ ________________ (power rule)

13. $\log_3 3^9 =$ ________________ (special property)

14. $\log_3 9^2 =$ ________________ (special property)

Use the properties of logarithms to express each logarithm as a sum or difference of logarithms, or as a single logarithm if possible. Assume that all variables represent positive real numbers. ***See Examples 1–4.***

15. $\log_7 (4 \cdot 5)$

16. $\log_8 (9 \cdot 11)$

17. $\log_5 \frac{8}{3}$

18. $\log_3 \frac{7}{5}$

19. $\log_4 6^2$

20. $\log_5 7^4$

21. $\log_3 \frac{\sqrt[3]{4}}{x^2 y}$

22. $\log_7 \frac{\sqrt[3]{13}}{pq^2}$

23. $\log_3 \sqrt{\frac{xy}{5}}$

24. $\log_6 \sqrt{\frac{pq}{7}}$

25. $\log_2 \frac{\sqrt[3]{x} \cdot \sqrt[5]{y}}{r^2}$

26. $\log_4 \frac{\sqrt[4]{z} \cdot \sqrt[5]{w}}{s^2}$

Use the properties of logarithms to write each expression as a single logarithm. Assume that all variables are defined in such a way that the variable expressions are positive, and bases are positive numbers not equal to 1. ***See Examples 1–4.***

27. $\log_b x + \log_b y$

28. $\log_b w + \log_b z$

29. $\log_a m - \log_a n$

30. $\log_b x - \log_b y$

31. $(\log_a r - \log_a s) + 3 \log_a t$

32. $(\log_a p - \log_a q) + 2 \log_a r$

33. $3 \log_a 5 - 4 \log_a 3$

34. $3 \log_a 5 - \frac{1}{2} \log_a 9$

35. $\log_{10} (x + 3) + \log_{10} (x - 3)$

36. $\log_{10} (y + 4) + \log_{10} (y - 4)$

37. $\log_{10} (x + 3) + \log_{10} (x + 5)$

38. $\log_{10} (x + 4) + \log_{10} (x + 6)$

39. $3 \log_p x + \frac{1}{2} \log_p y - \frac{3}{2} \log_p z - 3 \log_p a$

40. $\frac{1}{3} \log_b x + \frac{2}{3} \log_b y - \frac{3}{4} \log_b s - \frac{2}{3} \log_b t$

To four decimal places, the values of $\log_{10} 2$ *and* $\log_{10} 9$ *are*

$$\log_{10} 2 = 0.3010 \quad \text{and} \quad \log_{10} 9 = 0.9542.$$

Use these values and the properties of logarithms to evaluate each expression. DO NOT USE A CALCULATOR. ***See Example 5.***

41. $\log_{10} 18$

42. $\log_{10} 4$

43. $\log_{10} \frac{2}{9}$

44. $\log_{10} \frac{9}{2}$

45. $\log_{10} 36$

46. $\log_{10} 162$

47. $\log_{10} \sqrt[4]{9}$

48. $\log_{10} \sqrt[5]{2}$

49. $\log_{10} 3$

50. $\log_{10} \frac{1}{9}$

51. $\log_{10} 9^5$

52. $\log_{10} 2^{19}$

Determine whether each statement is true *or* false. ***See Example 6.***

53. $\log_2 (8 + 32) = \log_2 8 + \log_2 32$

54. $\log_2 (64 - 16) = \log_2 64 - \log_2 16$

55. $\log_3 7 + \log_3 7^{-1} = 0$

56. $\log_3 49 + \log_3 49^{-1} = 0$

57. $\log_6 60 - \log_6 10 = 1$

58. $\log_3 8 + \log_3 \frac{1}{8} = 0$

59. $\frac{\log_{10} 7}{\log_{10} 14} = \frac{1}{2}$

60. $\frac{\log_{10} 10}{\log_{10} 100} = \frac{1}{10}$

10.5 Common and Natural Logarithms

OBJECTIVES

1. Evaluate common logarithms using a calculator.
2. Use common logarithms in applications.
3. Evaluate natural logarithms using a calculator.
4. Use natural logarithms in applications.
5. Use the change-of-base rule.

Logarithms are important in many applications in biology, engineering, economics, and social science. In this section we find numerical approximations for logarithms. Traditionally, base 10 logarithms were used most often because our number system is base 10. Logarithms to base 10 are **common logarithms,** and

$$\mathbf{\log_{10} x} \text{ is abbreviated as } \mathbf{\log x},$$

where the base is understood to be 10.

OBJECTIVE 1 Evaluate common logarithms using a calculator.

We evaluate common logarithms using a calculator with a (LOG) key.

EXAMPLE 1 Evaluating Common Logarithms

Evaluate each logarithm to four decimal places using a calculator as needed.

(a) $\log 327.1 \approx 2.5147$

(b) $\log 437{,}000 \approx 5.6405$

(c) $\log 0.0615 \approx -1.2111$

(d) $\log 10^{6.1988} = 6.1988$

Special property $\log_b b^r = r$

VOCABULARY

- ☐ common logarithm
- ☐ natural logarithm

NOW TRY EXERCISE 1

Evaluate each logarithm to four decimal places using a calculator as needed.

(a) log 115

(b) log 539,000

(c) log 0.023

(d) $\log 10^{12.2139}$

In part (c), $\log 0.0615 \approx -1.2111$, which is a negative result. ***The common logarithm of a number between 0 and 1 is always negative*** because the logarithm is the exponent on 10 that produces the number. In this case, we have

$$10^{-1.2111} \approx 0.0615.$$

If the exponent (the logarithm) were positive, the result would be greater than 1 because $10^0 = 1$. The graph in **FIGURE 20** illustrates these concepts.

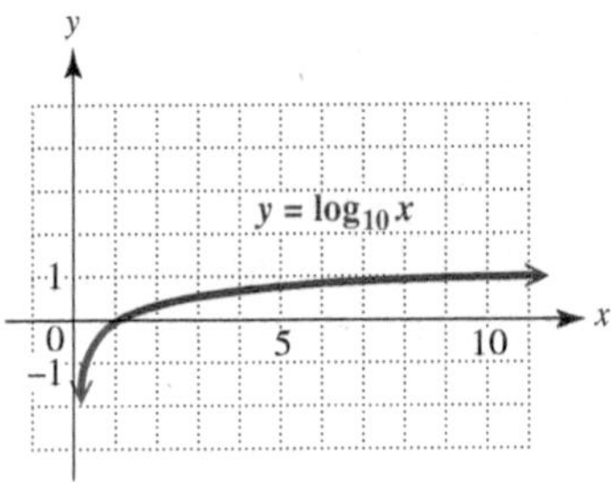

FIGURE 20

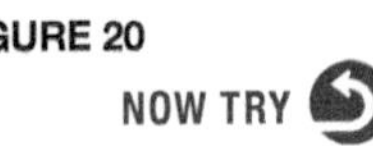

NOW TRY

OBJECTIVE 2 Use common logarithms in applications.

In chemistry, pH is a measure of the acidity or alkalinity of a solution. Water, for example, has pH 7. In general, acids have pH numbers less than 7, and alkaline solutions have pH values greater than 7. **FIGURE 21** illustrates the pH scale.

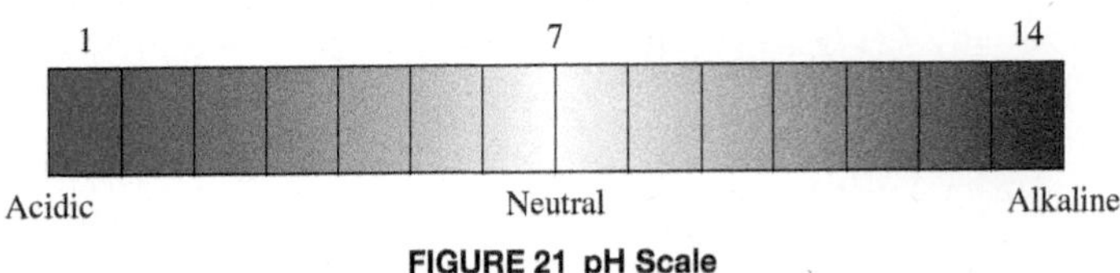

FIGURE 21 pH Scale

The **pH** of a solution is defined as

$$\mathbf{pH = -\log\,[H_3O^+]},$$

where $[H_3O^+]$ is the hydronium ion concentration in moles per liter. It is customary to round pH values to the nearest tenth.

NOW TRY EXERCISE 2

Water taken from a wetland has a hydronium ion concentration of

3.4×10^{-5} mole per liter.

Find the pH value for the water and classify the wetland as a rich fen, a poor fen, or a bog.

EXAMPLE 2 Using pH in an Application

Wetlands are classified as *bogs, fens, marshes,* and *swamps,* on the basis of pH values. A pH value between 6.0 and 7.5, such as that of Summerby Swamp in Michigan's Hiawatha National Forest, indicates that the wetland is a "rich fen." When the pH is between 3.0 and 6.0, the wetland is a "poor fen," and if the pH falls to 3.0 or less, it is a "bog." (Data from Mohlenbrock, R., "Summerby Swamp, Michigan," *Natural History.*)

Suppose that the hydronium ion concentration of a sample of water from a wetland is 6.3×10^{-3} mole per liter. How would this wetland be classified?

$\text{pH} = -\log\,[H_3O^+]$	Definition of pH
$\text{pH} = -\log\,(6.3 \times 10^{-3})$	Let $[H_3O^+] = 6.3 \times 10^{-3}$.
$\text{pH} = -(\log 6.3 + \log 10^{-3})$	Product rule
$\text{pH} \approx -[0.7993 - 3]$	Use a calculator to approximate log 6.3; $\log_b b^r = r$
$\text{pH} = -0.7993 + 3$	Distributive property
$\text{pH} \approx 2.2$	Add.

The pH is less than 3.0, so the wetland is a bog.

NOW TRY

NOW TRY ANSWERS

1. **(a)** 2.0607 **(b)** 5.7316 **(c)** −1.6383 **(d)** 12.2139

2. 4.5; poor fen

Find the hydronium ion concentration of a solution with pH 2.6.

EXAMPLE 3 Finding Hydronium Ion Concentration

Find the hydronium ion concentration of drinking water with pH 6.5.

$$\text{pH} = -\log\left[\text{H}_3\text{O}^+\right] \quad \text{Definition of pH}$$

$$6.5 = -\log\left[\text{H}_3\text{O}^+\right] \quad \text{Let pH} = 6.5.$$

$$\log\left[\text{H}_3\text{O}^+\right] = -6.5 \quad \text{Multiply by } -1\text{. Interchange sides.}$$

$$\left[\text{H}_3\text{O}^+\right] = 10^{-6.5} \quad \text{Write in exponential form, base 10.}$$

$$\left[\text{H}_3\text{O}^+\right] \approx 3.2 \times 10^{-7} \quad \text{Evaluate with a calculator.}$$

The hydronium ion concentration of drinking water is approximately 3.2×10^{-7} mole per liter.

NOW TRY

The loudness of sound is measured in a unit called a **decibel,** abbreviated **dB.** To measure with this unit, we first assign an intensity of I_0 to a very faint sound, called the **threshold sound.** If a particular sound has intensity I, then the decibel level of this louder sound is

$$\boldsymbol{D = 10 \log\left(\frac{I}{I_0}\right).}$$

The loudness of some common sounds are shown in the table. Any sound over 85 dB exceeds what hearing experts consider safe. Permanent hearing damage can be suffered at levels above 150 dB.

▼ Loudness of Common Sounds

Decibel Level	*Example*
60	Normal conversation
90	Rush hour traffic, lawn mower
100	Garbage truck, chain saw, pneumatic drill
120	Rock concert, thunderclap
140	Gunshot blast, jet engine
180	Rocket launching pad

Data from Deafness Research Foundation.

NOW TRY EXERCISE 4

Find the decibel level to the nearest whole number of the sound from a jet engine with intensity I of

$$(6.312 \times 10^{13})I_0.$$

EXAMPLE 4 Measuring the Loudness of Sound

If music delivered through Bluetooth headphones has intensity I of $(3.162 \times 10^9)I_0$, find the average decibel level.

$$D = 10 \log\left(\frac{I}{I_0}\right)$$

$$D = 10 \log\left(\frac{(3.162 \times 10^9)I_0}{I_0}\right) \quad \text{Substitute the given value for } I.$$

$$D = 10 \log\left(3.162 \times 10^9\right)$$

$$D \approx 95 \text{ dB} \quad \text{Evaluate with a calculator.}$$

NOW TRY

NOW TRY ANSWERS

3. 2.5×10^{-3}
4. 138 dB

Leonhard Euler (1707–1783)

The number e is named after Euler.

OBJECTIVE 3 Evaluate natural logarithms using a calculator.

Logarithms used in applications are often **natural logarithms,** which have as base the number e. The letter e was chosen to honor Leonhard Euler, who published extensive results on the number in 1748. It is an irrational number, so its decimal expansion never terminates and never repeats.

One way to see how e appears in an exponential situation involves calculating the values of $\left(1 + \frac{1}{x}\right)^x$ as x gets larger without bound. See the table.

x	$\left(1 + \frac{1}{x}\right)^x$
1	2
10	2.59374246
100	2.704813829
1000	2.716923932
10,000	2.718145927
100,000	2.718268237

These approximations are found using a calculator.

It appears that as x gets larger without bound, $\left(1 + \frac{1}{x}\right)^x$ approaches some number. This number is e.

Approximation for e

$$e \approx 2.718281828459$$

A calculator with an e^x key can approximate powers of e.

$$e^2 \approx 7.389056099, \quad e^3 \approx 20.08553692, \quad \text{and} \quad e^{0.6} \approx 1.8221188$$

Logarithms with base e are called natural logarithms because they occur in natural situations that involve growth or decay.

The base e logarithm of x is written **ln *x*** (read "***el en x***").

A graph of $y = \ln x$ is given in **FIGURE 22**.

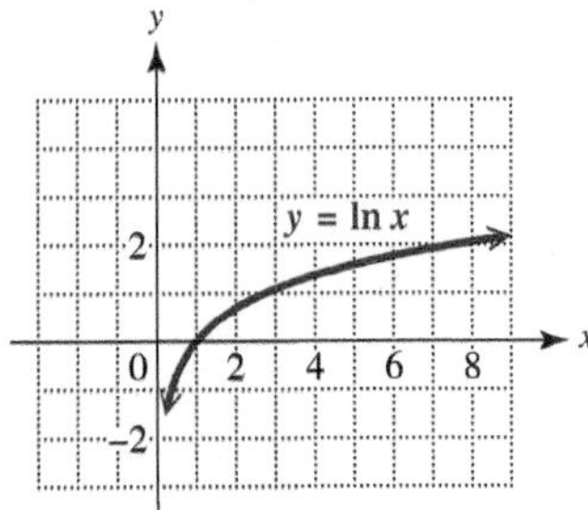

FIGURE 22

A calculator key labeled LN is used to evaluate natural logarithms.

NOW TRY EXERCISE 5

Evaluate each logarithm to four decimal places using a calculator as needed.

(a) $\ln 0.26$ **(b)** $\ln 12$

(c) $\ln 150$ **(d)** $\ln e^{5.8321}$

EXAMPLE 5 Evaluating Natural Logarithms

Evaluate each logarithm to four decimal places using a calculator as needed.

(a) $\ln 0.5841 \approx -0.5377$ **(b)** $\ln 0.9215 \approx -0.0818$

As with common logarithms, a number between 0 and 1 has a negative natural logarithm.

(c) $\ln 192.7 \approx 5.2611$ **(d)** $\ln e^{4.6832} = 4.6832$

Special property $\log_b b^r = r$

NOW TRY

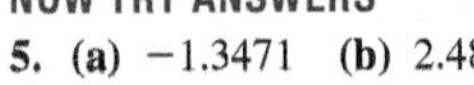

NOW TRY ANSWERS

5. (a) −1.3471 (b) 2.4849
(c) 5.0106 (d) 5.8321

OBJECTIVE 4 Use natural logarithms in applications.

NOW TRY EXERCISE 6
Use the natural logarithmic function in **Example 6** to approximate the altitude when atmospheric pressure is 600 millibars. Round to the nearest hundred.

EXAMPLE 6 Applying a Natural Logarithmic Function

Altitude in meters that corresponds to an atmospheric pressure of x millibars can be approximated by the natural logarithmic function

$$f(x) = 51{,}600 - 7457 \ln x.$$

(Data from Miller, A. and J. Thompson, *Elements of Meteorology,* Fourth Edition, Charles E. Merrill Publishing Company.)

Use this function to find the altitude when atmospheric pressure is 400 millibars. Round to the nearest hundred.

$$f(x) = 51{,}600 - 7457 \ln x$$

$$f(400) = 51{,}600 - 7457 \ln 400 \quad \text{Let } x = 400.$$

$$f(400) \approx 6900 \quad \text{Evaluate with a calculator.}$$

Atmospheric pressure is 400 millibars at 6900 m.

NOW TRY 

NOTE In **Example 6,** the final answer was obtained using a calculator *without* rounding the intermediate values. In general, it is best to wait until the final step to round the answer. Otherwise, a buildup of round-off error may cause the final answer to have an incorrect final decimal place digit or digits.

OBJECTIVE 5 Use the change-of-base rule.

The change-of-base rule enables us to convert logarithms from one base to another.

Change-of-Base Rule

If $a > 0$, $a \neq 1$, $b > 0$, $b \neq 1$, and $x > 0$, then the following holds true.

$$\log_a x = \frac{\log_b x}{\log_b a}$$

To derive the change-of-base rule, let $\log_a x = m$.

$$\log_a x = m$$

$$a^m = x \quad \text{Write in exponential form.}$$

$$\log_b a^m = \log_b x \quad \text{Take the logarithm on each side.}$$

$$m \log_b a = \log_b x \quad \text{Power rule}$$

$$(\log_a x)(\log_b a) = \log_b x \quad \text{Substitute for } m.$$

$$\log_a x = \frac{\log_b x}{\log_b a} \quad \text{Divide by } \log_b a.$$

This last statement is the change-of-base rule.

NOW TRY ANSWER
6. 3900 m

NOTE Any positive number other than 1 can be used for base b in the change-of-base rule. Usually the only practical bases are e and 10, because calculators give logarithms for these two bases.

NOW TRY EXERCISE 7
Use the change-of-base rule to approximate each logarithm to four decimal places.

(a) $\log_2 7$ **(b)** $\log_{1/3} 12$

NOW TRY ANSWERS
7. (a) 2.8074 **(b)** −2.2619

EXAMPLE 7 Using the Change-of-Base Rule

Use the change-of-base rule to approximate $\log_5 12$ to four decimal places.

$$\log_5 12 = \frac{\log 12}{\log 5} \approx 1.5440$$

$$\log_5 12 = \frac{\ln 12}{\ln 5} \approx 1.5440$$

Either common or natural logarithms can be used.

NOW TRY  

10.5 Exercises

  MyLab Math

Video solutions for select problems available in MyLab Math

Concept Check *Choose the correct response.*

1. What is the base in the expression $\log x$?

A. x **B.** 1 **C.** 10 **D.** e

2. What is the base in the expression $\ln x$?

A. x **B.** 1 **C.** 10 **D.** e

3. Given that $10^0 = 1$ and $10^1 = 10$, between what two consecutive integers is the value of $\log 6.3$?

A. −1 and 0 **B.** 0 and 1 **C.** 6 and 7 **D.** 10 and 11

4. Given that $e^1 \approx 2.718$ and $e^2 \approx 7.389$, between what two consecutive integers is the value of $\ln 6.3$?

A. 0 and 1 **B.** 1 and 2 **C.** 2 and 3 **D.** 6 and 7

Concept Check *Without using a calculator, give the value of each expression.*

5. $\log 10^{9.6421}$ **6.** $\ln e^{\sqrt{2}}$ **7.** $10^{\log \sqrt{3}}$

8. $e^{\ln 75.2}$ **9.** $\ln e^{-11.4007}$ **10.** $10 \ln e^4$

Use a calculator for most of the remaining exercises in this set.

Evaluate each logarithm to four decimal places. ***See Examples 1 and 5.***

11. $\log 43$ **12.** $\log 98$ **13.** $\log 328.4$

14. $\log 457.2$ **15.** $\log 0.0326$ **16.** $\log 0.1741$

17. $\log (4.76 \times 10^9)$ **18.** $\log (2.13 \times 10^4)$ **19.** $\ln 7.84$

20. $\ln 8.32$ **21.** $\ln 0.0556$ **22.** $\ln 0.0217$

23. $\ln 388.1$ **24.** $\ln 942.6$ **25.** $\ln (8.59 \times e^2)$

26. $\ln (7.46 \times e^3)$ **27.** $\ln 10$ **28.** $\log e$

29. Concept Check Use a calculator to find approximations of each logarithm.

(a) $\log 356.8$ **(b)** $\log 35.68$ **(c)** $\log 3.568$

(d) Observe the answers and make a conjecture concerning the decimal values of the common logarithms of numbers greater than 1 that have the same digits.

30. Concept Check Let k represent the number of letters in your last name.

(a) Use a calculator to find $\log k$.

(b) Raise 10 to the power indicated by the number found in part (a). What is the result?

(c) Explain why we obtained the answer found in part (b). Would it matter what number we used for k to observe the same result?

Suppose that water from a wetland area is sampled and found to have the given hydronium ion concentration. Is the wetland a rich fen, *a* poor fen, *or* a bog? **See *Example 2*.**

31. 3.1×10^{-5} **32.** 2.5×10^{-5} **33.** 2.5×10^{-2}

34. 3.6×10^{-2} **35.** 2.7×10^{-7} **36.** 2.5×10^{-7}

Find the pH *(to the nearest tenth) of the substance with the given hydronium ion concentration.* **See *Example 2*.**

37. Ammonia, 2.5×10^{-12}

38. Egg white, 1.6×10^{-8}

39. Sodium bicarbonate, 4.0×10^{-9}

40. Tuna, 1.3×10^{-6}

41. Grapes, 5.0×10^{-5}

42. Grapefruit, 6.3×10^{-4}

Find the hydronium ion concentration of the substance with the given pH. **See *Example 3*.**

43. Human blood plasma, 7.4

44. Milk, 6.4

45. Human gastric contents, 2.0

46. Spinach, 5.4

47. Bananas, 4.6

48. Milk of magnesia, 10.5

Solve each problem. **See *Examples 4 and 6*.**

49. Managements of sports stadiums and arenas often encourage fans to make as much noise as possible. Find the average decibel level

$$D = 10 \log \left(\frac{I}{I_0}\right)$$

for each venue with the given intensity I.

(a) NFL fans, Kansas City Chiefs at Arrowhead Stadium: $I = (1.58 \times 10^{14})I_0$

(b) NBA fans, Sacramento Kings at Sleep Train Arena: $I = (3.9 \times 10^{12})I_0$

(c) MLB fans, Baltimore Orioles at Camden Yards: $I = (1.1 \times 10^{12})I_0$

(Data from www.baltimoresportsreport.com, www.guinessworldrecords.com)

50. Find the decibel level of each sound.

(a) noisy restaurant: $I = 10^8 I_0$

(b) farm tractor: $I = (5.340 \times 10^9)I_0$

(c) snowmobile: $I = 31{,}622{,}776{,}600 I_0$

(Data from The Canadian Society of Otolaryngology.)

51. The number of years, $N(x)$, since two independently evolving languages split off from a common ancestral language is approximated by

$$N(x) = -5000 \ln x,$$

where x is the percent of words (in decimal form) from the ancestral language common to both languages now. Find the number of years (to the nearest hundred years) since the split for each percent of common words.

(a) 85% (or 0.85) **(b)** 35% (or 0.35) **(c)** 10% (or 0.10)

52. The time t in years for an amount of money invested at an interest rate r (in decimal form) to double is given by

$$t(r) = \frac{\ln 2}{\ln (1 + r)}.$$

This is the **doubling time.** Find the doubling time to the nearest tenth for an investment at each interest rate.

(a) 2% (or 0.02) **(b)** 5% (or 0.05) **(c)** 8% (or 0.08)

53. The number of monthly active Twitter users (in millions) worldwide during the third quarter of each year from 2010 to 2017 is approximated by

$$f(x) = 25.6829 + 149.1368 \ln x,$$

where $x = 1$ represents 2010, $x = 2$ represents 2011, and so on. (Data from Twitter.)

(a) What does this model give for the number of monthly active Twitter users in 2012?

(b) According to this model, when did the number of monthly active Twitter users reach 300 million? (*Hint:* Substitute for $f(x)$, and then write the equation in exponential form to solve it.)

54. In the central Sierra Nevada of California, the percent of moisture that falls as snow rather than rain is approximated by

$$f(x) = 86.3 \ln x - 680,$$

where x is the altitude in feet.

(a) What percent of the moisture at 5000 ft falls as snow?

(b) What percent at 7500 ft falls as snow?

55. The approximate tax $T(x)$, in dollars per ton, that would result in an x% (in decimal form) reduction in carbon dioxide emissions is approximated by the **cost-benefit equation**

$$T(x) = -0.642 - 189 \ln (1 - x).$$

(a) What tax will reduce emissions 25%?

(b) Explain why the equation is not valid for $x = 0$ or $x = 1$.

56. The age in years of a female blue whale of length x in feet is approximated by

$$f(x) = -2.57 \ln \left(\frac{87 - x}{63}\right).$$

(a) How old is a female blue whale that measures 80 ft?

(b) The equation that defines this function has domain $24 < x < 87$. Explain why.

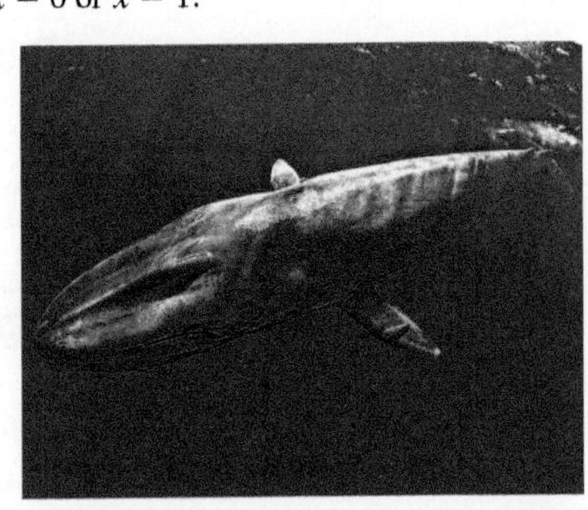

Use the change-of-base rule (with either common or natural logarithms) to approximate each logarithm to four decimal places. ***See Example 7.***

57. $\log_3 12$ **58.** $\log_4 18$ **59.** $\log_5 3$ **60.** $\log_7 4$

61. $\log_{21} 0.7496$ **62.** $\log_{19} 0.8325$ **63.** $\log_{1/2} 5$ **64.** $\log_{1/3} 7$

65. $\log_3 \sqrt{2}$ **66.** $\log_6 \sqrt[3]{5}$ **67.** $\log_\pi e$ **68.** $\log_\pi 10$

10.6 Exponential and Logarithmic Equations; Further Applications

OBJECTIVES

1 Solve equations involving variables in the exponents.
2 Solve equations involving logarithms.
3 Solve applications of compound interest.
4 Solve applications involving base *e* exponential growth and decay.

General methods for solving these equations depend on the following properties.

Properties for Solving Exponential and Logarithmic Equations

For all real numbers $b > 0$, $b \neq 1$, and any real numbers x and y, the following hold true.

1. If $x = y$, then $b^x = b^y$.

2. If $b^x = b^y$, then $x = y$.

3. If $x = y$, and $x > 0$, $y > 0$, then $\log_b x = \log_b y$.

4. If $x > 0$, $y > 0$, and $\log_b x = \log_b y$, then $x = y$.

OBJECTIVE 1 **Solve equations involving variables in the exponents.**

VOCABULARY

☐ compound interest
☐ continuous compounding

EXAMPLE 1 **Solving an Exponential Equation**

Solve $3^x = 12$. Approximate the solution to three decimal places.

$$3^x = 12$$

$$\log 3^x = \log 12 \quad \text{Property 3 (common logarithms): If } x = y\text{, then } \log_b x = \log_b y.$$

$$x \log 3 = \log 12 \quad \text{Power rule}$$

Exact solution ⟶ $x = \dfrac{\log 12}{\log 3}$ Divide by log 3.

Decimal approximation ⟶ $x \approx 2.262$ Evaluate with a calculator.

CHECK $3^x = 3^{2.262} \approx 12$ ✓ True

The solution set is $\{2.262\}$. **NOW TRY**

NOW TRY EXERCISE 1
Solve $5^x = 20$. Approximate the solution to three decimal places.

CAUTION Be careful: $\frac{\log 12}{\log 3}$ is ***not*** equal to log 4.

$$\frac{\log 12}{\log 3} \approx 2.262, \quad \text{but} \quad \log 4 \approx 0.6021.$$

NOTE In **Example 1,** we could have used Property 3 with natural logarithms.

$$\frac{\ln 12}{\ln 3} = \frac{\log 12}{\log 3} \approx 2.262 \quad \text{(Verify this.)}$$

NOW TRY ANSWER
1. $\{1.861\}$

NOW TRY EXERCISE 2
Solve $e^{0.12x} = 10$. Approximate the solution to three decimal places.

EXAMPLE 2 **Solving an Exponential Equation (Base e)**

Solve $e^{0.003x} = 40$. Approximate the solution to three decimal places.

$$e^{0.003x} = 40$$

$$\ln e^{0.003x} = \ln 40 \qquad \text{Property 3 (natural logarithms): If } x = y \text{, then } \ln x = \ln y.$$

$$0.003x \ln e = \ln 40 \qquad \text{Power rule}$$

$$0.003x = \ln 40 \qquad \ln e = \ln e^1 = 1$$

$$x = \frac{\ln 40}{0.003} \qquad \text{Divide by 0.003.}$$

$$x \approx 1229.626 \qquad \text{Evaluate with a calculator.}$$

CHECK $e^{0.003x} = e^{0.003(1229.626)} \approx 40$ ✓ True

The solution set is $\{1229.626\}$. **NOW TRY** 

NOW TRY EXERCISE 3
Solve $3^{-x+5} = 8$. Approximate the solution to three decimal places.

EXAMPLE 3 **Solving an Exponential Equation**

Solve $7^{-x+4} = 17$. Approximate the solution to three decimal places.

$$7^{-x+4} = 17$$

$$\log 7^{-x+4} = \log 17 \qquad \text{Property 3 (common logarithms): If } x = y \text{, then } \log_b x = \log_b y.$$

$$(-x + 4) \log 7 = \log 17 \qquad \text{Power rule}$$

$$-x \log 7 + 4 \log 7 = \log 17 \qquad \text{Distributive property}$$

$$-x \log 7 = \log 17 - 4 \log 7 \qquad \text{Subtract 4 log 7.}$$

$$x = \frac{\log 17 - 4 \log 7}{-\log 7} \qquad \text{Divide by } -\log 7.$$

$$x \approx 2.544 \qquad \text{Evaluate with a calculator.}$$

CHECK $7^{-x+4} = 7^{-2.544+4} \approx 17$ ✓ True

The solution set is $\{2.544\}$. **NOW TRY** 

General Method for Solving an Exponential Equation

Take logarithms having the same base on both sides and then use the power rule of logarithms or the special property $\log_b b^x = x$. **(See Examples 1–3.)**

As a special case, if both sides can be written as exponentials with the same base, do so, and set the exponents equal.

OBJECTIVE 2 Solve equations involving logarithms.

EXAMPLE 4 **Solving a Logarithmic Equation**

Solve $\log_3 (4x + 1) = 4$.

$$\log_3 (4x + 1) = 4$$

$$4x + 1 = 3^4 \qquad \text{Write in exponential form.}$$

$$4x + 1 = 81 \qquad 3^4 = 81$$

$$4x = 80 \qquad \text{Subtract 1.}$$

$$x = 20 \qquad \text{Divide by 4.}$$

NOW TRY ANSWERS
2. $\{19.188\}$
3. $\{3.107\}$

NOW TRY EXERCISE 4
Solve $\log_6 (2x + 4) = 2$.

CHECK

$$\log_3 (4x + 1) = 4$$
$$\log_3 [4(20) + 1] \stackrel{?}{=} 4 \quad \text{Let } x = 20.$$
$$\log_3 81 \stackrel{?}{=} 4 \quad \text{Work inside the brackets.}$$
$$4 = 4 \ \checkmark \quad \text{True; } \log_3 81 = 4$$

A true statement results, so the solution set is $\{20\}$. NOW TRY 

NOW TRY EXERCISE 5
Solve $\log_5 (x - 1)^3 = 2$. Give the exact solution.

EXAMPLE 5 Solving a Logarithmic Equation

Solve $\log_2 (x + 5)^3 = 4$. Give the exact solution.

$$\log_2 (x + 5)^3 = 4$$
$$(x + 5)^3 = 2^4 \quad \text{Write in exponential form.}$$
$$(x + 5)^3 = 16 \quad 2^4 = 16$$
$$x + 5 = \sqrt[3]{16} \quad \text{Take the cube root on each side.}$$
$$x = -5 + \sqrt[3]{16} \quad \text{Add } -5.$$
$$x = -5 + 2\sqrt[3]{2} \quad \sqrt[3]{16} = \sqrt[3]{8 \cdot 2} = \sqrt[3]{8} \cdot \sqrt[3]{2} = 2\sqrt[3]{2}$$

CHECK

$$\log_2 (x + 5)^3 = 4 \quad \text{Original equation}$$
$$\log_2 \left(-5 + 2\sqrt[3]{2} + 5\right)^3 \stackrel{?}{=} 4 \quad \text{Let } x = -5 + 2\sqrt[3]{2}.$$
$$\log_2 \left(2\sqrt[3]{2}\right)^3 \stackrel{?}{=} 4 \quad \text{Work inside the parentheses.}$$
$$\log_2 16 \stackrel{?}{=} 4 \quad \left(2\sqrt[3]{2}\right)^3 = 2^3\left(\sqrt[3]{2}\right)^3 = 8 \cdot 2 = 16$$
$$4 = 4 \ \checkmark \quad \text{True; } \log_2 16 = 4$$

A true statement results, so the solution set is $\left\{-5 + 2\sqrt[3]{2}\right\}$. NOW TRY 

CAUTION Recall that the domain of $f(x) = \log_b x$ is $(0, \infty)$. ***For this reason, always check that each proposed solution of an equation with logarithms yields only logarithms of positive numbers in the original equation.***

EXAMPLE 6 Solving a Logarithmic Equation

Solve $\log_2 (x + 1) - \log_2 x = \log_2 7$.

$$\log_2 (x + 1) - \log_2 x = \log_2 7$$

Transform the left side to an expression with only *one* logarithm.

$$\log_2 \frac{x + 1}{x} = \log_2 7 \quad \text{Quotient rule}$$
$$\frac{x + 1}{x} = 7 \quad \text{Property 4: If } \log_b x = \log_b y, \text{ then } x = y.$$
$$x + 1 = 7x \quad \text{Multiply by } x.$$
$$1 = 6x \quad \text{Subtract } x.$$
$$\frac{1}{6} = x \quad \text{Divide by 6.}$$

This proposed solution must be checked.

We cannot take the logarithm of a *nonpositive* number, so both $x + 1$ and x must be positive here. If $x = \frac{1}{6}$, then this condition is satisfied.

NOW TRY ANSWERS
4. $\{16\}$
5. $\{1 + \sqrt[3]{25}\}$

NOW TRY EXERCISE 6
Solve.

$\log_4 (2x + 13) - \log_4 (x + 1) = \log_4 10$

CHECK

$\log_2 (x + 1) - \log_2 x = \log_2 7$	Original equation
$\log_2 \left(\frac{1}{6} + 1\right) - \log_2 \frac{1}{6} \stackrel{?}{=} \log_2 7$	Let $x = \frac{1}{6}$.
$\log_2 \frac{7}{6} - \log_2 \frac{1}{6} \stackrel{?}{=} \log_2 7$	Add.
$\log_2 \frac{\frac{7}{6}}{\frac{1}{6}} \stackrel{?}{=} \log_2 7$	Quotient rule
$\log_2 7 = \log_2 7$ ✓	True

$\frac{\frac{7}{6}}{\frac{1}{6}} = \frac{7}{6} \div \frac{1}{6} = \frac{7}{6} \cdot \frac{6}{1} = 7$

A true statement results, so the solution set is $\left\{\frac{1}{6}\right\}$. **NOW TRY**

NOW TRY EXERCISE 7
Solve.

$\log_4 (x + 2) + \log_4 2x = 2$

EXAMPLE 7 **Solving a Logarithmic Equation**

Solve $\log x + \log (x - 21) = 2$.

$\log x + \log (x - 21) = 2$	
$\log x(x - 21) = 2$	Product rule
$x(x - 21) = 10^2$	Write in exponential form.
$x^2 - 21x = 100$	Distributive property; $10^2 = 100$
$x^2 - 21x - 100 = 0$	Standard form
$(x - 25)(x + 4) = 0$	Factor.
$x - 25 = 0$ or $x + 4 = 0$	Zero-factor property
$x = 25$ or $x = -4$	Solve each equation.

The base is 10.

The value -4 must be rejected as a solution because it leads to the logarithm of a negative number in the original equation.

$$\log (-4) + \log (-4 - 21) = 2 \qquad \text{The left side is undefined.}$$

Check that the only solution is 25, so the solution set is $\{25\}$. **NOW TRY** 

CAUTION *Do not reject a potential solution just because it is nonpositive. Reject any value that leads to the logarithm of a nonpositive number.*

Solving a Logarithmic Equation

Step 1 **Transform the equation so that a single logarithm appears on one side.** Use the product rule or quotient rule of logarithms to do this.

Step 2 **Do one of the following.**

(a) Use Property 4.

If $\log_b x = \log_b y$, then $x = y$. **(See Example 6.)**

(b) Write the equation in exponential form.

If $\log_b x = k$, then $x = b^k$. **(See Examples 4, 5, and 7.)**

NOW TRY ANSWERS

6. $\left\{\frac{3}{8}\right\}$
7. $\{2\}$

OBJECTIVE 3 Solve applications of compound interest.

We have solved simple interest problems using the formula $I = prt$. In most cases, interest paid or charged is **compound interest** (interest paid on both principal and interest). The formula for compound interest is an application of exponential functions. ***In this text, monetary amounts are given to the nearest cent.***

Compound Interest Formula (for a Finite Number of Periods)

If a principal of P dollars is deposited at an annual rate of interest r compounded (paid) n times per year, then the account will contain

$$A = P\left(1 + \frac{r}{n}\right)^{nt}$$

dollars after t years. (In this formula, r is expressed as a decimal.)

NOW TRY EXERCISE 8
How much money will there be in an account at the end of 10 yr if $10,000 is deposited at 2.5% compounded monthly?

EXAMPLE 8 Solving a Compound Interest Problem for A

How much money will there be in an account at the end of 5 yr if $1000 is deposited at 3% compounded quarterly? (Assume no withdrawals are made.)

Because interest is compounded quarterly, $n = 4$.

$$A = P\left(1 + \frac{r}{n}\right)^{nt} \qquad \text{Compound interest formula}$$

$$A = 1000\left(1 + \frac{0.03}{4}\right)^{4 \cdot 5} \qquad \text{Substitute } P = 1000,\ r = 0.03 \text{ (because } 3\% = 0.03),\ n = 4, \text{ and } t = 5.$$

$$A = 1000(1.0075)^{20} \qquad \text{Simplify.}$$

$$A = 1161.18 \qquad \text{Evaluate with a calculator.}$$

The account will contain $1161.18. NOW TRY

NOW TRY EXERCISE 9
Find the number of years, to the nearest hundredth, it will take for money deposited in an account paying 2% interest compounded quarterly to double.

EXAMPLE 9 Solving a Compound Interest Problem for t

Suppose inflation is averaging 3% per year. To the nearest hundredth of a year, how long will it take for prices to double? (This is the **doubling time** of the money.)

We want to find the number of years t for P dollars to grow to $2P$ dollars.

$$A = P\left(1 + \frac{r}{n}\right)^{nt} \qquad \text{Compound interest formula}$$

$$2P = P\left(1 + \frac{0.03}{1}\right)^{1t} \qquad \text{Let } A = 2P,\ r = 0.03, \text{ and } n = 1.$$

$$2 = (1.03)^t \qquad \text{Divide by } P. \text{ Simplify.}$$

$$\log 2 = \log (1.03)^t \qquad \text{Property 3}$$

$$\log 2 = t \log (1.03) \qquad \text{Power rule}$$

$$t = \frac{\log 2}{\log 1.03} \qquad \text{Interchange sides. Divide by log 1.03.}$$

$$t \approx 23.45 \qquad \text{Evaluate with a calculator.}$$

Prices will double in 23.45 yr. To check, verify that $1.03^{23.45} \approx 2$. NOW TRY 

NOW TRY ANSWERS
8. $12,836.92
9. 34.74 yr

Interest can be compounded over various time periods per year, including

annually, semiannually, quarterly, daily, and so on.

If the value of n increases without bound, we have an example of **continuous compounding.**

Continuous Compound Interest Formula

If a principal of P dollars is deposited at an annual rate of interest r compounded continuously for t years, the final amount A on deposit is given by

$$A = Pe^{rt}.$$

NOW TRY EXERCISE 10

Suppose that \$4000 is invested at 3% interest for 2 yr.

(a) How much will the investment grow to if it is compounded continuously?

(b) How long, to the nearest tenth of a year, would it take for the original investment to double?

EXAMPLE 10 Solving a Continuous Compound Interest Problem

In **Example 8** we found that \$1000 invested for 5 yr at 3% interest compounded quarterly would grow to \$1161.18.

(a) How much would this investment grow to if it is compounded continuously?

$A = Pe^{rt}$ Continuous compounding formula

$A = 1000e^{0.03(5)}$ Let $P = 1000$, $r = 0.03$, and $t = 5$.

$A = 1000e^{0.15}$ Multiply in the exponent.

$A = 1161.83$ Evaluate with a calculator.

The investment will grow to \$1161.83 (which is \$0.65 more than the amount in **Example 8** when interest was compounded quarterly).

(b) How long, to the nearest tenth of a year, would it take for the initial investment to triple?

We must find the value of t that will cause A to be $3(\$1000) = \$3000$.

$A = Pe^{rt}$ Continuous compounding formula

$3000 = 1000e^{0.03t}$ Let $A = 3P = 3000$, $P = 1000$, and $r = 0.03$.

$3 = e^{0.03t}$ Divide by 1000.

$\ln 3 = \ln e^{0.03t}$ Take natural logarithms.

$\ln 3 = 0.03t$ $\ln e^k = k$

$t = \frac{\ln 3}{0.03}$ Divide by 0.03. Interchange sides.

$t \approx 36.6$ Evaluate with a calculator.

It would take 36.6 yr for the original investment to triple.  

OBJECTIVE 4 Solve applications involving base e exponential growth and decay.

When situations involve growth or decay of a population, the amount or number of some quantity present at time t can be approximated by

$$f(t) = y_0 e^{kt}.$$

In this equation, y_0 is the amount or number present at time $t = 0$ and k is a constant.

NOW TRY ANSWERS

10. (a) \$4247.35
(b) 23.1 yr

NOW TRY EXERCISE 11

Radium-226 decays according to the function

$$f(t) = y_0e^{-0.00043t},$$

where t is time in years.

(a) If an initial sample contains $y_0 = 4.5$ g of radium-226, how many grams, to the nearest tenth, will be present after 150 yr?

(b) What is the half-life of radium-226? Round to the nearest year.

The continuous compounding of money is an example of exponential growth. In **Example 11,** we investigate exponential decay.

EXAMPLE 11 Applying an Exponential Decay Function

After a plant or animal dies, the amount of radioactive carbon-14 that is present disintegrates according to the natural logarithmic function

$$f(t) = y_0e^{-0.000121t},$$

where t is time in years, $f(t)$ is the amount of the sample at time t, and y_0 is the initial amount present at $t = 0$.

(a) If an initial sample contains $y_0 = 10$ g of carbon-14, how many grams, to the nearest hundredth, will be present after 3000 yr?

$f(3000) = 10e^{-0.000121(3000)} \approx 6.96$ g   Let $y_0 = 10$ and $t = 3000$ in the formula.

(b) How long would it take to the nearest year for the initial sample to decay to half of its original amount? (This is the **half-life** of the sample.)

$f(t) = y_0e^{-0.000121t}$ Exponential decay formula

$5 = 10e^{-0.000121t}$ Let $f(t) = \frac{1}{2}(10) = 5$ and $y_0 = 10$.

$\frac{1}{2} = e^{-0.000121t}$ Divide by 10.

$\ln \frac{1}{2} = -0.000121t$ Take natural logarithms; $\ln e^k = k$

$t = \dfrac{\ln \frac{1}{2}}{-0.000121}$ Divide by -0.000121. Interchange sides.

$t \approx 5728$ Evaluate with a calculator.

The half-life is 5728 yr.

NOW TRY

NOW TRY ANSWERS

11. (a) 4.2 g **(b)** 1612 yr

10.6 Exercises

FOR EXTRA HELP MyLab Math

Video solutions for select problems available in MyLab Math

Many of the problems in these exercises require a calculator.

Concept Check *Determine whether common logarithms or natural logarithms would be a better choice to use for solving each equation. Do not actually solve.*

1. $10^{0.0025x} = 75$

2. $10^{3x+1} = 13$

3. $e^{x-2} = 24$

4. $e^{-0.28x} = 30$

Solve each equation. Approximate solutions to three decimal places. **See Examples 1 and 3.**

5. $7^x = 5$

6. $4^x = 3$

7. $9^{-x+2} = 13$

8. $6^{-x+1} = 22$

9. $3^{2x} = 14$

10. $5^{3x} = 11$

11. $2^{x+3} = 5^x$

12. $6^{x+3} = 4^x$

13. $2^{x+3} = 3^{x-4}$

14. $4^{x-2} = 5^{3x+2}$

15. $4^{2x+3} = 6^{x-1}$

16. $3^{2x+1} = 5^{x-1}$

Solve each equation. Use natural logarithms. Approximate solutions to three decimal places when appropriate. ***See Example 2.***

17. $e^{0.012x} = 23$

18. $e^{0.006x} = 30$

19. $e^{-0.205x} = 9$

20. $e^{-0.103x} = 7$

21. $\ln e^{3x} = 9$

22. $\ln e^{2x} = 4$

23. $\ln e^{0.45x} = \sqrt{7}$

24. $\ln e^{0.04x} = \sqrt{3}$

25. $\ln e^{-x} = \pi$

26. $\ln e^{2x} = \pi$

27. $e^{\ln 2x} = e^{\ln (x+1)}$

28. $e^{\ln (6-x)} = e^{\ln (4+2x)}$

Solve each equation. Give exact solutions. ***See Examples 4 and 5.***

29. $\log_4 (2x + 8) = 2$

30. $\log_5 (5x + 10) = 3$

31. $\log_3 (6x + 5) = 2$

32. $\log_5 (12x - 8) = 3$

33. $\log_2 (2x - 1) = 5$

34. $\log_6 (4x + 2) = 2$

35. $\log_7 (x + 1)^3 = 2$

36. $\log_4 (x - 3)^3 = 4$

37. $\log_2 (x^2 + 7) = 4$

38. $\log_6 (x^2 + 11) = 2$

39. Concept Check Suppose that in solving a logarithmic equation having the term $\log (x - 3)$, we obtain the proposed solution 2. We know that our algebraic work is correct, so we give $\{2\}$ as the solution set. **WHAT WENT WRONG?**

40. Concept Check Suppose that in solving a logarithmic equation having the term $\log (3 - x)$, we obtain the proposed solution -4. We know that our algebraic work is correct, so we reject -4 and give $\varnothing$ as the solution set. **WHAT WENT WRONG?**

Solve each equation. Give exact solutions. ***See Examples 6 and 7.***

41. $\log (6x + 1) = \log 3$

42. $\log (7 - 2x) = \log 4$

43. $\log_5 (3t + 2) - \log_5 t = \log_5 4$

44. $\log_2 (t + 5) - \log_2 (t - 1) = \log_2 3$

45. $\log 4x - \log (x - 3) = \log 2$

46. $\log (-x) + \log 3 = \log (2x - 15)$

47. $\log_2 x + \log_2 (x - 7) = 3$

48. $\log (2x - 1) + \log 10x = \log 10$

49. $\log 5x - \log (2x - 1) = \log 4$

50. $\log_3 x + \log_3 (2x + 5) = 1$

51. $\log_2 x + \log_2 (x - 6) = 4$

52. $\log_2 x + \log_2 (x + 4) = 5$

Solve each problem. ***See Examples 8–10.***

53. Suppose that \$2000 is deposited at 4% compounded quarterly.

(a) How much money will be in the account at the end of 6 yr? (Assume no withdrawals are made.)

(b) To one decimal place, how long will it take for the account to grow to \$3000?

54. Suppose that \$3000 is deposited at 3.5% compounded quarterly.

(a) How much money will be in the account at the end of 7 yr? (Assume no withdrawals are made.)

(b) To one decimal place, how long will it take for the account to grow to \$5000?

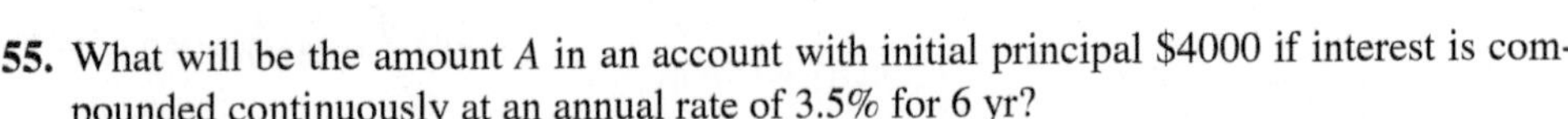

55. What will be the amount A in an account with initial principal \$4000 if interest is compounded continuously at an annual rate of 3.5% for 6 yr?

56. What will be the amount A in an account with initial principal \$10,000 if interest is compounded continuously at an annual rate of 2.5% for 5 yr?

57. How long, to the nearest hundredth of a year, would it take an initial principal P to double if it were invested at 2.5% compounded continuously?

58. How long, to the nearest hundredth of a year, would it take \$4000 to double at 3.25% compounded continuously?

59. Find the amount of money in an account after 12 yr if \$5000 is deposited at 7% annual interest compounded as follows.

(a) Annually **(b)** Semiannually **(c)** Quarterly

(d) Daily (Use $n = 365$.) **(e)** Continuously

60. Find the amount of money in an account after 8 yr if \$4500 is deposited at 6% annual interest compounded as follows.

(a) Annually **(b)** Semiannually **(c)** Quarterly

(d) Daily (Use $n = 365$.) **(e)** Continuously

61. How much money must be deposited today to amount to \$1850 in 40 yr at 6.5% compounded continuously?

62. How much money must be deposited today to amount to \$1000 in 10 yr at 5% compounded continuously?

Solve each problem. ***See Example 11.***

63. Revenues of software publishers in the United States for the years 2004–2016 can be modeled by the function

$$S(x) = 91.412e^{0.05195x},$$

where $x = 4$ represents 2004, $x = 5$ represents 2005, and so on, and $S(x)$ is in billions of dollars. Approximate, to the nearest unit, revenue for 2016. (Data from U.S. Census Bureau.)

64. Based on selected figures obtained during the years 1970–2015, the total number of bachelor's degrees earned in the United States can be modeled by the function

$$D(x) = 792{,}377e^{0.01798x},$$

where $x = 0$ corresponds to 1970, $x = 5$ corresponds to 1975, and so on. Approximate, to the nearest unit, the number of bachelor's degrees earned in 2015. (Data from U.S. National Center for Education Statistics.)

65. Suppose that the amount, in grams, of plutonium-241 present in a given sample is determined by the function

$$A(t) = 2.00e^{-0.053t},$$

where t is measured in years. Approximate the amount present, to the nearest hundredth, in the sample after the given number of years.

(a) 4 **(b)** 10 **(c)** 20 **(d)** What was the initial amount present?

66. Suppose that the amount, in grams, of radium-226 present in a given sample is determined by the function

$$A(t) = 3.25e^{-0.00043t},$$

where t is measured in years. Approximate the amount present, to the nearest hundredth, in the sample after the given number of years.

(a) 20 **(b)** 100 **(c)** 500 **(d)** What was the initial amount present?

67. A sample of 400 g of lead-210 decays to polonium-210 according to the function

$$A(t) = 400e^{-0.032t},$$

where t is time in years. Approximate answers to the nearest hundredth.

(a) How much lead will be left in the sample after 25 yr?

(b) How long will it take the initial sample to decay to half of its original amount?

68. The concentration of a drug in a person's system decreases according to the function

$$C(t) = 2e^{-0.125t},$$

where $C(t)$ is in appropriate units, and t is in hours. Approximate answers to the nearest hundredth.

(a) How much of the drug will be in the system after 1 hr?

(b) How long will it take for the concentration to be half of its original amount?

RELATING CONCEPTS For Individual or Group Work (Exercises 69–72)

Previously, we solved an equation such as $5^x = 125$ *as follows.*

$5^x = 125$	Original equation
$5^x = 5^3$	$125 = 5^3$
$x = 3$	Set exponents equal.

Solution set: $\{3\}$

The method described in this section can also be used to solve this equation. ***Work Exercises 69–72 in order,*** *to see how this is done.*

69. Take common logarithms on both sides, and write this equation.

70. Apply the power rule for logarithms on the left.

71. Write the equation so that x is alone on the left.

72. Use a calculator to find the decimal form of the solution. What is the solution set?

Chapter 10 Summary

Key Terms

10.1

one-to-one function
inverse of a function

10.2

exponential function with base a
asymptote
exponential equation

10.3

logarithm
logarithmic equation
logarithmic function with base a

10.5

common logarithm
natural logarithm

10.6

compound interest
continuous compounding

New Symbols

$f^{-1}(x)$ inverse of $f(x)$

$\log_a x$ logarithm of x with base a

$\log x$ common (base 10) logarithm of x

$\ln x$ natural (base e) logarithm of x

e a constant, approximately 2.718281828459

Test Your Word Power

See how well you have learned the vocabulary in this chapter.

1. In a **one-to-one function**
   A. each x-value corresponds to only one y-value
   B. each x-value corresponds to one or more y-values
   C. each x-value is the same as each y-value
   D. each x-value corresponds to only one y-value and each y-value corresponds to only one x-value.

2. If f is a one-to-one function, then the **inverse** of f is
   A. the set of all solutions of f
   B. the set of all ordered pairs formed by interchanging the coordinates of the ordered pairs of f
   C. the set of all ordered pairs that are the opposite (negative) of the coordinates of the ordered pairs of f
   D. an equation involving an exponential expression.

3. An **exponential function** is a function defined by an expression of the form
   A. $f(x) = ax^2 + bx + c$ for real numbers a, b, c $(a \neq 0)$
   B. $f(x) = \log_a x$ for positive numbers a and x $(a \neq 1)$
   C. $f(x) = a^x$ for all real numbers x $(a > 0, a \neq 1)$
   D. $f(x) = \sqrt{x}$ for $x \geq 0$.

4. An **asymptote** is
   A. a line that a graph intersects just once
   B. a line that the graph of a function more and more closely approaches as the x-values increase or decrease without bound
   C. the x-axis or y-axis
   D. a line about which a graph is symmetric.

5. A **logarithmic function** is a function defined by an expression of the form
   A. $f(x) = ax^2 + bx + c$ for real numbers a, b, c $(a \neq 0)$
   B. $f(x) = \log_a x$ for positive numbers a and x $(a \neq 1)$
   C. $f(x) = a^x$ for all real numbers x $(a > 0, a \neq 1)$
   D. $f(x) = \sqrt{x}$ for $x \geq 0$.

6. A **logarithm** is
   A. an exponent
   B. a base
   C. an equation
   D. a polynomial.

ANSWERS

1. D; *Example:* The function $f = \{(0, 2), (1, -1), (3, 5), (-2, 3)\}$ is one-to-one. **2.** B; *Example:* The inverse of the one-to-one function f defined in Answer 1 is $f^{-1} = \{(2, 0), (-1, 1), (5, 3), (3, -2)\}$. **3.** C; *Examples:* $f(x) = 4^x$, $g(x) = \left(\frac{1}{2}\right)^x$ **4.** B; *Example:* The graph of $f(x) = 2^x$ has the x-axis $(y = 0)$ as an asymptote. **5.** B; *Examples:* $f(x) = \log_3 x$, $g(x) = \log_{1/3} x$ **6.** A; *Example:* $\log_a x$ is the exponent to which a must be raised to obtain x; $\log_3 9 = 2$ because $3^2 = 9$.

Quick Review

CONCEPTS	EXAMPLES
10.1 Inverse Functions	
Horizontal Line Test A function is one-to-one if every horizontal line intersects the graph of the function at most once.	Find f^{-1} if $f(x) = 2x - 3$. The graph of f is a nonhorizontal (slanted) straight line, so f is one-to-one by the horizontal line test and has an inverse.
Inverse Functions For a one-to-one function f defined by an equation $y = f(x)$, find the defining equation of the inverse function f^{-1} as follows. ***Step 1*** Interchange x and y. ***Step 2*** Solve for y. ***Step 3*** Replace y with $f^{-1}(x)$.	$f(x) = 2x - 3$ $y = 2x - 3$ Let $y = f(x)$. $x = 2y - 3$ Interchange x and y. $x + 3 = 2y$ Add 3. (Solve for y.) $y = \frac{x+3}{2}$ Divide by 2. Interchange sides. (Solve for y.) $f^{-1}(x) = \frac{1}{2}x + \frac{3}{2}$ Replace y with $f^{-1}(x)$; $\frac{a+b}{c} = \frac{a}{c} + \frac{b}{c}$

CONCEPTS	EXAMPLES

In general, the graph of f^{-1} is the mirror image of the graph of f with respect to the line $y = x$. If the point (a, b) lies on the graph of f, then the point (b, a) lies on the graph of f^{-1}.

The graphs of a function f and its inverse f^{-1} are shown here.

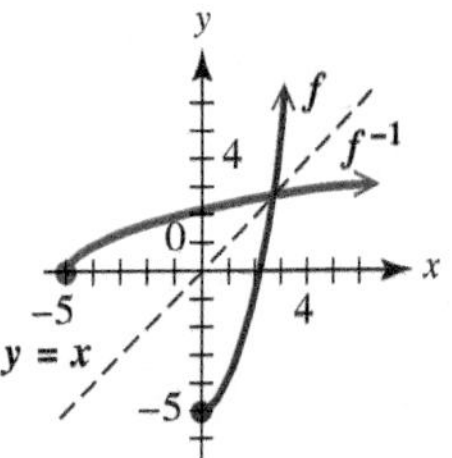

10.2 Exponential Functions

For $a > 0$, $a \neq 1$, and all real numbers x,

$$f(x) = a^x$$

defines the **exponential function with base a.**

Graph of $f(x) = a^x$

1. The graph contains the point $(0, 1)$, which is its y-intercept.
2. When $a > 1$, the graph rises from left to right.
   When $0 < a < 1$, the graph falls from left to right.
3. The x-axis is an asymptote.
4. The domain is $(-\infty, \infty)$, and the range is $(0, \infty)$.

$f(x) = 2^x$ defines the exponential function with base 2.

$F(x) = \left(\frac{1}{2}\right)^x$ defines the exponential function with base $\frac{1}{2}$.

x	$f(x) = 2^x$	$F(x) = \left(\frac{1}{2}\right)^x$
-2	$\frac{1}{4}$	4
-1	$\frac{1}{2}$	2
0	1	1
1	2	$\frac{1}{2}$
2	4	$\frac{1}{4}$

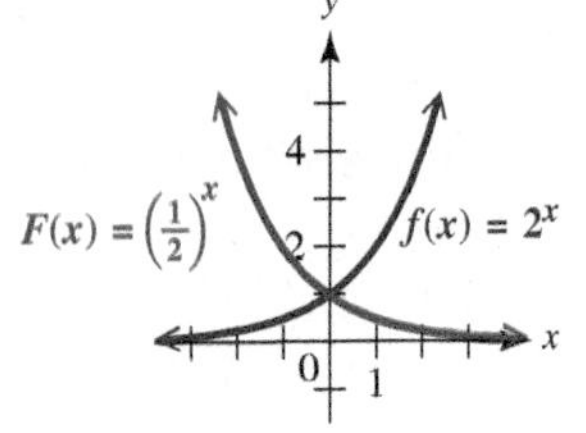

10.3 Logarithmic Functions

For all positive real numbers a, where $a \neq 1$, and all positive real numbers x,

$$y = \log_a x \quad \text{is equivalent to} \quad x = a^y.$$

$y = \log_2 x$ is equivalent to $x = 2^y$.

Special Properties of Logarithms

For $b > 0$, where $b \neq 1$, the following hold true.

$$\log_b b = 1 \qquad \log_b 1 = 0$$

$$\log_b b^r = r \quad (r \text{ is real.}) \qquad b^{\log_b r} = r \quad (r > 0)$$

Evaluate.

$$\log_3 3 = 1 \qquad \log_5 1 = 0$$

$$\log_5 5^6 = 6 \qquad 4^{\log_4 6} = 6$$

If a and x are positive real numbers, where $a \neq 1$, then

$$g(x) = \log_a x$$

defines the **logarithmic function with base a.**

Graph of $g(x) = \log_a x$

1. The graph contains the point $(1, 0)$, which is its x-intercept.
2. When $a > 1$, the graph rises from left to right.
   When $0 < a < 1$, the graph falls from left to right.
3. The y-axis is an asymptote.
4. The domain is $(0, \infty)$, and the range is $(-\infty, \infty)$.

$g(x) = \log_2 x$ defines the logarithmic function with base 2.

$G(x) = \log_{1/2} x$ defines the logarithmic function with base $\frac{1}{2}$.

x	$g(x) = \log_2 x$	$G(x) = \log_{1/2} x$
$\frac{1}{4}$	-2	2
$\frac{1}{2}$	-1	1
1	0	0
2	1	-1
4	2	-2

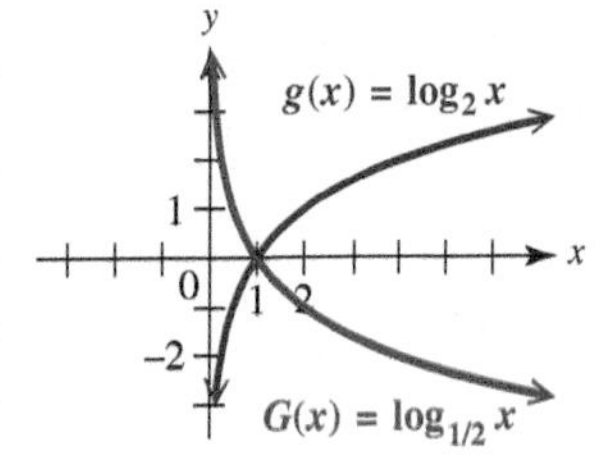

CONCEPTS	EXAMPLES
10.4 Properties of Logarithms	
If x, y, and b are positive real numbers, where $b \neq 1$, and r is any real number, then the following hold true.	Rewrite each logarithm.
Product Rule $\log_b xy = \log_b x + \log_b y$	$\log_2 3m = \log_2 3 + \log_2 m$ Product rule
Quotient Rule $\log_b \frac{x}{y} = \log_b x - \log_b y$	$\log_5 \frac{9}{4} = \log_5 9 - \log_5 4$ Quotient rule
Power Rule $\log_b x^r = r \log_b x$	$\log_{10} 2^3 = 3 \log_{10} 2$ Power rule
10.5 Common and Natural Logarithms	
Common logarithms (base 10) are used in applications such as pH and loudness of sound. Use the (LOG) key of a calculator to evaluate common logarithms.	Use the formula $\mathbf{pH = -\log [H_3O^+]}$ to find the pH of grapes with hydronium ion concentration 5.0×10^{-5}. $\text{pH} = -\log (5.0 \times 10^{-5})$ Substitute. $\text{pH} = -(\log 5.0 + \log 10^{-5})$ Property of logarithms $\text{pH} \approx 4.3$ Evaluate with a calculator.
Natural logarithms (base e) are most often used in applications of growth and decay, such as continuous compounding of money, decay of chemical compounds, and biological growth. Use the (LN) key of a calculator to evaluate natural logarithms.	Use the following formula to find doubling time, to the nearest hundredth of a year, for an amount of money invested at 4%. $t(r) = \frac{\ln 2}{\ln (1 + r)}$ Formula for doubling time (in years) $t(0.04) = \frac{\ln 2}{\ln (1 + 0.04)}$ Let $r = 0.04$. $t(0.04) \approx 17.67$ Evaluate with a calculator. The doubling time is 17.67 yr.
Change-of-Base Rule If $a > 0$, $a \neq 1$, $b > 0$, $b \neq 1$, and $x > 0$, then the following holds true. $\log_a x = \frac{\log_b x}{\log_b a}$	Use the change-of-base rule to approximate $\log_3 17$. $\log_3 17 = \frac{\ln 17}{\ln 3} = \frac{\log 17}{\log 3} \approx 2.5789$
10.6 Exponential and Logarithmic Equations; Further Applications	
To solve exponential equations, use these properties (where $b > 0$, $b \neq 1$). **1. If $b^x = b^y$, then $x = y$.** **2. If $x = y$ and $x > 0$, $y > 0$, then $\log_b x = \log_b y$.**	Solve each equation. $2^{3x} = 2^5$ $3x = 5$ Set the exponents equal. $x = \frac{5}{3}$ Divide by 3. Solution set: $\left\{\frac{5}{3}\right\}$ $5^x = 8$ $\log 5^x = \log 8$ Take common logarithms. $x \log 5 = \log 8$ Power rule $x = \frac{\log 8}{\log 5}$ Divide by log 5. $x \approx 1.292$ Evaluate with a calculator. Solution set: $\{1.292\}$

CONCEPTS	EXAMPLES
To solve logarithmic equations, use these properties (where $b > 0, b \neq 1, x > 0$, and $y > 0$). First use the product rule, quotient rule, power rule, or special properties as necessary to write the equation in the proper form. **1. If $\log_b x = \log_b y$, then $x = y$.** **2. If $\log_b x = k$, then $x = b^k$.** ***Always check that each proposed solution of a logarithmic equation yields only logarithms of positive numbers in the original equation.***	Solve each equation. $\log_3 2x = \log_3 (x+1)$ $2x = x + 1$ Property 1 $x = 1$ Subtract x. A true statement results when 1 is substituted for x. Solution set: $\{1\}$ $\log_2 (3x - 1) = 4$ $3x - 1 = 2^4$ Write in exponential form. $3x - 1 = 16$ Apply the exponent. $3x = 17$ Add 1. $x = \frac{17}{3}$ Divide by 3. A true statement results when $\frac{17}{3}$ is substituted for x. Solution set: $\left\{\frac{17}{3}\right\}$

Chapter 10 Review Exercises

10.1 *Determine whether each graph is the graph of a one-to-one function.*

1.

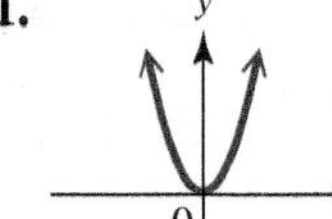

2.

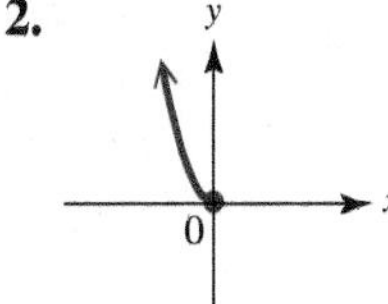

Determine whether each function is one-to-one. If it is, find its inverse.

3. $\{(-2, 4), (-1, 1), (0, 0), (1, 1), (2, 4)\}$

4. $\{(-2, -8), (-1, -1), (0, 0), (1, 1), (2, 8)\}$

5. $f(x) = -3x + 7$

6. $f(x) = \sqrt[3]{6x - 4}$

7. $f(x) = -x^2 + 3$

8. The table lists basic minimum wages in several states. If the set of states is the domain and the set of wage amounts is the range of a function, is it one-to-one? Why or why not?

State	Minimum Wage (in dollars)
Washington	11.50
Massachusetts	11.00
California	10.50
Arizona	10.50
Ohio	8.30
Iowa	7.25

Data from U.S. Department of Labor.

Each function graphed is one-to-one. Graph its inverse.

9.

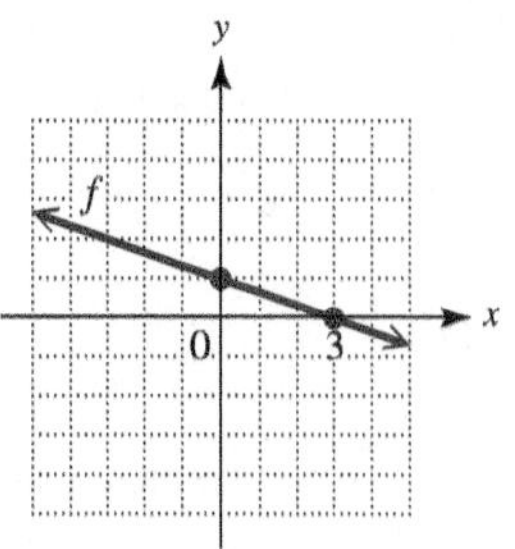

10.

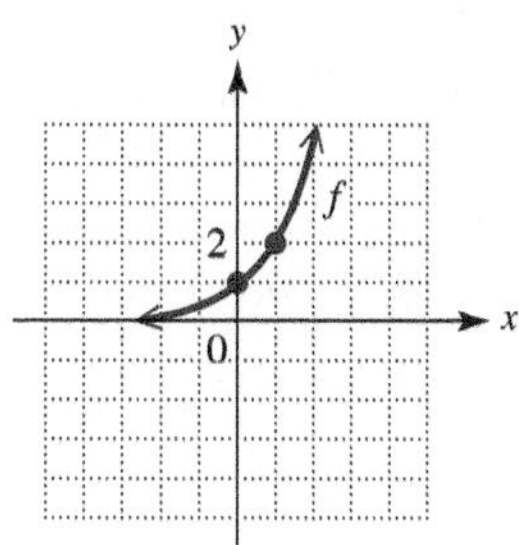

10.2 *Use a calculator to approximate each exponential expression to three decimal places.*

11. $5^{3.2}$

12. $\left(\frac{1}{5}\right)^{2.1}$

13. $8.3^{-1.2}$

Graph each exponential function.

14. $f(x) = 3^x$

15. $f(x) = \left(\frac{1}{3}\right)^x$

16. $f(x) = 2^{2x+3}$

Solve each equation.

17. $5^{2x+1} = 25$

18. $4^{3x} = 8^{x+4}$

19. $\left(\frac{1}{27}\right)^{x-1} = 9^{2x}$

20. The U.S. Hispanic population (in millions) can be approximated by

$$f(x) = 46.9 \cdot 2^{0.035811x},$$

where x represents the number of years since 2008. Use this function to approximate, to the nearest tenth, the Hispanic population in each year. (Data from U.S. Census Bureau.)

(a) 2015 **(b)** 2030

10.3 *Work each problem.*

21. Convert each equation to the indicated form.

(a) Write in exponential form: $\log_5 625 = 4$.

(b) Write in logarithmic form: $5^{-2} = 0.04$.

22. Fill in each blank with the correct response: The value of $\log_3 81$ is ____. This means that if we raise ____ to the ____ power, the result is ____.

Use a calculator to approximate each logarithm to four decimal places.

23. $\log_2 6$

24. $\log_7 3$

25. $\log_{10} 55$

Graph each logarithmic function.

26. $g(x) = \log_3 x$

27. $g(x) = \log_{1/3} x$

28. Use the special properties of logarithms to evaluate each expression.

(a) $4^{\log_4 12}$ **(b)** $\log_9 9^{13}$ **(c)** $\log_5 625$

Solve each equation.

29. $\log_8 64 = x$ **30.** $\log_2 \sqrt{8} = x$ **31.** $\log_x \left(\frac{1}{49}\right) = -2$

32. $\log_4 x = \frac{3}{2}$ **33.** $\log_k 4 = 1$ **34.** $\log_6 x = -2$

A company has found that total sales, in thousands of dollars, are given by the function

$$S(x) = 100 \log_2 (x + 2),$$

where x is the number of weeks after a major advertising campaign was introduced.

35. What were total sales 6 weeks after the campaign was introduced?

36. Graph the function.

10.4 *Use the properties of logarithms to express each logarithm as a sum or difference of logarithms. Assume that all variables represent positive real numbers.*

37. $\log_4 3x^2$ **38.** $\log_5 \frac{a^3b^2}{c^4}$

39. $\log_4 \frac{\sqrt{x} \cdot w^2}{z}$ **40.** $\log_2 \frac{p^2 r}{\sqrt{z}}$

Use the properties of logarithms to rewrite each expression as a single logarithm. Assume that all variables are defined in such a way that the variable expressions are positive, and bases are positive numbers not equal to 1.

41. $2 \log_a 7 - 4 \log_a 2$ **42.** $3 \log_a 5 + \frac{1}{3} \log_a 8$

43. $\log_b 3 + \log_b x - 2 \log_b y$ **44.** $\log_3 (x + 7) - \log_3 (4x + 6)$

10.5 *Evaluate each logarithm to four decimal places.*

45. $\log 28.9$ **46.** $\log 0.257$ **47.** $\ln 28.9$ **48.** $\ln 0.257$

Use the change-of-base rule (with either common or natural logarithms) to approximate each logarithm to four decimal places.

49. $\log_{16} 13$ **50.** $\log_4 12$ **51.** $\log_{1/4} 17$ **52.** $\log_\pi \sqrt{2}$

Find the pH *(to the nearest tenth) of the substance with the given hydronium ion concentration.*

53. Milk, 4.0×10^{-7} **54.** Crackers, 3.8×10^{-9}

Solve each problem.

55. If orange juice has pH 4.6, what is its hydronium ion concentration?

56. The magnitude of a star is given by the equation

$$M = 6 - 2.5 \log \frac{I}{I_0},$$

where I_0 is the measure of the faintest star and I is the actual intensity of the star being measured. The dimmest stars are of magnitude 6, and the brightest are of magnitude 1. Determine the ratio of intensities between stars of magnitude 1 and 3.

57. The concentration of a drug injected into the bloodstream decreases with time. The intervals of time T when the drug should be administered are given by

$$T = \frac{1}{k} \ln \frac{C_2}{C_1},$$

where k is a constant determined by the drug in use, C_2 is the concentration at which the drug is harmful, and C_1 is the concentration below which the drug is ineffective. (Data from Horelick, B. and S. Koont, "Applications of Calculus to Medicine: Prescribing Safe and Effective Dosage," *UMAP Module 202.*) Thus, if $T = 4$, the drug should be administered every 4 hr.

For a certain drug, $k = \frac{1}{3}$, $C_2 = 5$, and $C_1 = 2$. How often should the drug be administered? (*Hint:* Round down.)

10.6 *Solve each equation. Approximate solutions to three decimal places.*

58. $3^x = 9.42$ **59.** $2^{x-1} = 15$ **60.** $e^{0.06x} = 3$

Solve each equation. Give exact solutions.

61. $\log_3 (9x + 8) = 2$

62. $\log_5 (x + 6)^3 = 2$

63. $\log_3 (x + 2) - \log_3 x = \log_3 2$

64. $\log (2x + 3) = 1 + \log x$

65. $\log_4 x + \log_4 (8 - x) = 2$

66. $\log_2 x + \log_2 (x + 15) = \log_2 16$

Solve each problem.

67. Suppose \$20,000 is deposited at 4% annual interest compounded quarterly. How much will be in the account at the end of 5 yr? (Assume no withdrawals are made.)

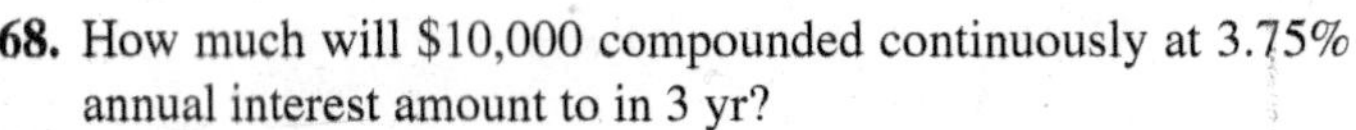

68. How much will \$10,000 compounded continuously at 3.75% annual interest amount to in 3 yr?

69. Which plan is better? How much more would it pay?

Plan A: Invest \$1000 at 4% compounded quarterly for 3 yr

Plan B: Invest \$1000 at 3.9% compounded monthly for 3 yr

70. Find the half-life of a radioactive substance that decays according to the function

$Q(t) = A_0 e^{-0.05t}$, where t is in days, to the nearest tenth.

Chapter 10 Mixed Review Exercises

Evaluate.

1. $\log_2 128$ **2.** $\log_{12} 1$ **3.** $\log_{2/3} \frac{27}{8}$ **4.** $5^{\log_5 36}$

5. $e^{\ln 4}$ **6.** $10^{\log e}$ **7.** $\log_3 3^{-5}$ **8.** $\ln e^{5.4}$

Solve.

9. $\log_3 (x + 9) = 4$

10. $\log_2 32 = x$

11. $\log_x \frac{1}{81} = 2$

12. $27^x = 81$

13. $2^{2x-3} = 8$

14. $5^{x+2} = 25^{2x+1}$

15. $\log_3 (x + 1) - \log_3 x = 2$

16. $\log (3x - 1) = \log 10$

17. $\log_4 (x + 2) - \log_4 x = 3$

18. $\ln (x^2 + 3x + 4) = \ln 2$

19. A machine purchased for business use **depreciates,** or loses value, over a period of years. The value of the machine at the end of its useful life is its **scrap value.** By one method of depreciation, the scrap value, S, is given by

$$S = C(1 - r)^n,$$

where C is original cost, n is useful life in years, and r is the constant percent of depreciation.

(a) Find the scrap value of a machine costing \$30,000, having a useful life of 12 yr and a constant annual rate of depreciation of 15%. Round to the nearest dollar.

(b) A machine has a "half-life" of 6 yr. Find the constant annual rate of depreciation, to the nearest percent.

20. One measure of the diversity of the species in an ecological community is the **index of diversity,** given by the logarithmic expression

$$-(p_1 \ln p_1 + p_2 \ln p_2 + \cdots + p_n \ln p_n),$$

where $p_1, p_2, \ldots, p_n$ are the proportions of a sample belonging to each of n species in the sample. (Data from Ludwig, J. and J. Reynolds, *Statistical Ecology: A Primer on Methods and Computing,* New York, John Wiley and Sons.)

Approximate the index of diversity to the nearest thousandth if a sample of 100 from a community produces the following numbers.

(a) 90 of one species, 10 of another

(b) 60 of one species, 40 of another

21. To solve the equation $5^x = 7$, we must find the exponent to which 5 must be raised in order to obtain 7. This is $\log_5 7$.

(a) Use the change-of-base rule and a calculator to find $\log_5 7$.

(b) Raise 5 to the number found in part (a). What is the result?

(c) Using as many decimal places as the calculator gives, write the solution set of $5^x = 7$.

22. Let m be the number of letters in your first name, and let n be the number of letters in your last name.

(a) Explain what $\log_m n$ means.

(b) Use a calculator to find $\log_m n$.

(c) Raise m to the power indicated by the number found in part (b). What is the result?

Chapter 10 Test

Step-by-step test solutions are found on the Chapter Test Prep Videos available in MyLab Math.

View the complete solutions to all Chapter Test exercises in MyLab Math.

1. Decide whether each function is one-to-one.

(a) $f(x) = x^2 + 9$

(b)

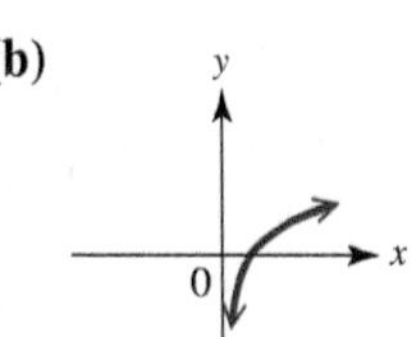

2. Find $f^{-1}(x)$ for the one-to-one function

$$f(x) = \sqrt[3]{x + 7}.$$

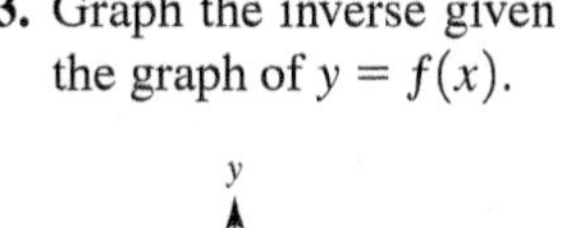

3. Graph the inverse given the graph of $y = f(x)$.

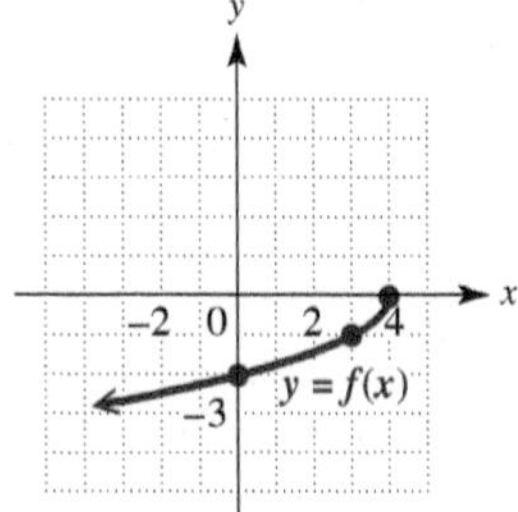

Graph each function.

4. $f(x) = 6^x$

5. $g(x) = \log_6 x$

6. Explain how the graph of the function in **Exercise 5** can be obtained from the graph of the function in **Exercise 4.**

Use the special properties of logarithms to evaluate each expression.

7. $7^{\log_7 9}$

8. $\log_3 3^6$

9. $\log_5 1$

Solve each equation. Give exact solutions.

10. $5^x = \dfrac{1}{625}$

11. $2^{3x-7} = 8^{2x+2}$

12. The atmospheric pressure (in millibars) at a given altitude x (in meters) is approximated by

$$f(x) = 1013e^{-0.0001341x}.$$

Use this function to approximate the atmospheric pressure, to the nearest unit, at each altitude.

(a) 2000 m **(b)** 10,000 m

13. Write in logarithmic form: $4^{-2} = 0.0625$.

14. Write in exponential form: $\log_7 49 = 2$.

Solve each equation.

15. $\log_{1/2} x = -5$

16. $x = \log_9 3$

17. $\log_x 16 = 4$

18. Fill in each blank with the correct response: The value of $\log_2 32$ is ______. This means that if we raise ______ to the ______ power, the result is ______.

Use the properties of logarithms to write each expression as a sum or difference of logarithms. Assume that variables represent positive real numbers.

19. $\log_3 x^2y$

20. $\log_5 \left(\dfrac{\sqrt{x}}{yz}\right)$

Use the properties of logarithms to rewrite each expression as a single logarithm. Assume that all variables represent positive real numbers, and that bases are positive numbers not equal to 1.

21. $3 \log_b s - \log_b t$

22. $\dfrac{1}{4} \log_b r + 2 \log_b s - \dfrac{2}{3} \log_b t$

Evaluate each logarithm to four decimal places.

23. $\log 23.1$

24. $\ln 0.82$

25. $\log_6 45$

Work each problem.

26. Use the change-of-base rule to express $\log_3 19$ as described.

(a) in terms of common logarithms **(b)** in terms of natural logarithms

(c) approximated to four decimal places

27. Solve $3^x = 78$, giving the solution to three decimal places.

28. Solve $\log_8 (x + 5) + \log_8 (x - 2) = 1$.

29. Suppose that \$10,000 is invested at 3.5% annual interest, compounded quarterly.

(a) How much will be in the account in 5 yr if no money is withdrawn?

(b) How long, to the nearest tenth of a year, will it take for the initial principal to double?

30. Suppose that \$15,000 is invested at 3% annual interest, compounded continuously.

(a) How much will be in the account in 5 yr if no money is withdrawn?

(b) How long, to the nearest tenth of a year, will it take for the initial principal to double?

Chapters R–10 Cumulative Review Exercises

1. Write each fraction as a decimal and a percent.

(a) $\frac{1}{20}$ **(b)** $\frac{5}{4}$

2. Perform the indicated operations. Give answers in lowest terms.

(a) $-\frac{7}{10} - \frac{1}{2}$ **(b)** $-\frac{8}{15} \div \frac{2}{3}$

3. Multiply or divide as indicated.

(a) 37.5×100 **(b)** $37.5 \div 1000$

4. Let $S = \left\{-\frac{9}{4}, -2, -\sqrt{2}, 0, 0.6, \sqrt{11}, \sqrt{-8}, 6, \frac{30}{3}\right\}$. List the elements of S that are members of each set.

(a) Integers **(b)** Rational numbers **(c)** Irrational numbers

Simplify each expression.

5. $|-8| + 6 - |-2| - (-6 + 2)$

6. $2(-5) + (-8)(4) - (-3)$

Solve each equation or inequality.

7. $7 - (3 + 4x) + 2x = -5(x - 1) - 3$

8. $2x + 2 \le 5x - 1$

9. $|2x - 5| = 9$

10. $|4x + 2| > 10$

Graph.

11. $5x + 2y = 10$

12. $-4x + y \le 5$

13. The graph projects that the number of international travelers to the United States will increase from 51.2 million in 2000 to 90.3 million in 2020.

(a) Is this the graph of a function?

(b) What is the slope of the line in the graph? Interpret the slope in the context of international travelers to the United States.

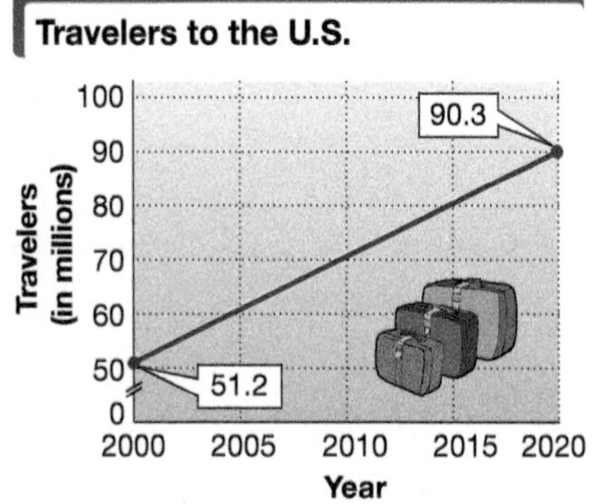

Data from U.S. Department of Commerce.

14. Find an equation of the line passing through $(5, -1)$ and parallel to the line with equation $3x - 4y = 12$. Write the equation in slope-intercept form.

Solve each system.

15. $5x - 3y = 14$
$2x + 5y = 18$

16. $3x - 2y = 3$
$x = \frac{2}{3}y + 1$

17. $x + 2y + 3z = 11$
$3x - y + z = 8$
$2x + 2y - 3z = -12$

18. Candy worth \$1.00 per lb is to be mixed with 10 lb of candy worth \$1.96 per lb to obtain a mixture that will be sold for \$1.60 per lb. How many pounds of the \$1.00 candy should be used?

Number of Pounds	Price per Pound (in dollars)	Value (in dollars)
x	1.00	$1x$
	1.60	

Perform the indicated operations.

19. $(2p + 3)(3p - 1)$

20. $(4k - 3)^2$

21. $(3m^3 + 2m^2 - 5m) - (8m^3 + 2m - 4)$

22. Divide $15x^3 - x^2 + 22x + 8$ by $3x + 1$.

Factor.

23. $8x + x^3$

24. $24y^2 - 7y - 6$

25. $5z^3 - 19z^2 - 4z$

26. $16a^2 - 25b^4$

27. $8c^3 + d^3$

28. $16r^2 + 56rq + 49q^2$

Perform the indicated operations.

29. $\frac{(5p^3)^4(-3p^7)}{2p^2(4p^4)}$

30. $\frac{x^2 - 9}{x^2 + 7x + 12} \div \frac{x - 3}{x + 5}$

31. $\frac{2}{k + 3} - \frac{5}{k - 2}$

Simplify.

32. $\sqrt{288}$

33. $2\sqrt{32} - 5\sqrt{98}$

34. $(5 + 4i)(5 - 4i)$

Solve each equation or inequality.

35. $\sqrt{2x + 1} - \sqrt{x} = 1$

36. $3x^2 - x - 1 = 0$

37. $x^2 + 2x - 8 > 0$

38. $x^4 - 5x^2 + 4 = 0$

39. $5^{x+3} = \left(\frac{1}{25}\right)^{3x+2}$

40. $\log_5 x + \log_5 (x + 4) = 1$

Graph.

41. $f(x) = \frac{1}{3}(x - 1)^2 + 2$ **42.** $f(x) = 2^x$ **43.** $f(x) = \log_3 x$

Without graphing, decide whether the graph of each relation is symmetric with respect to the x-axis, the y-axis, or the origin. More than one of these symmetries may apply.

44. $x^2 + 2y^2 = 6$ **45.** $|x| = y$

Work each problem.

46. If $f(x) = 3x$ and $g(x) = x - 1$, find each of the following. Give the domain for each.

(a) $f(x) - g(x)$ **(b)** $(fg)(x)$ **(c)** $\left(\frac{f}{g}\right)(x)$

47. Use properties of logarithms to write the following as a sum or difference of logarithms. Assume that variables represent positive real numbers.

$$\log \frac{x^3\sqrt{y}}{z}$$

48. Let the number of bacteria present in a certain culture be given by

$$B(t) = 25{,}000e^{0.2t},$$

where t is time measured in hours, and $t = 0$ corresponds to noon. Approximate, to the nearest hundred, the number of bacteria present at each time.

(a) noon **(b)** 1 P.M. **(c)** 2 P.M.

(d) When will the population double?

STUDY SKILLS **REMINDER**

Have you begun to prepare for your final exam? **Review Study Skill 10, *Preparing for Your Math Final Exam.***

11

POLYNOMIAL AND RATIONAL FUNCTIONS

Under certain conditions, if traffic intensity increases even slightly, congestion and waiting time increase dramatically. A *rational function* can model this phenomenon.

11.1 Zeros of Polynomial Functions (I)

OBJECTIVES

1. Interpret the division algorithm for two polynomials.
2. Use synthetic division to divide by a polynomial of the form $x - k$.
3. Use the remainder theorem to evaluate a polynomial.
4. Determine whether a given number is a zero of a polynomial function.

VOCABULARY

- ☐ polynomial function of degree n
- ☐ zero (of a function)
- ☐ root (solution)

STUDY SKILLS REMINDER
Make study cards to help you learn and remember the material in this chapter. **Review Study Skill 5, *Using Study Cards.***

OBJECTIVE 1 Interpret the division algorithm for two polynomials.

The following **division algorithm** will be used in this chapter.

Division Algorithm

Let $f(x)$ and $g(x)$ be polynomials with $g(x)$ of lesser degree than $f(x)$, and $g(x)$ of degree 1 or more. There exist unique polynomials $q(x)$ and $r(x)$ such that

$$f(x) = g(x) \cdot q(x) + r(x),$$

where either $r(x) = 0$ or the degree of $r(x)$ is less than the degree of $g(x)$.

In a previous chapter we learned how to perform long division of two polynomials. When

$$6x^4 + 9x^3 + 2x^2 - 8x + 7 \quad \text{is divided by} \quad 3x^2 - 2,$$

we can show that the following is true. This illustrates the division algorithm.

$$\underbrace{6x^4 + 9x^3 + 2x^2 - 8x + 7}_{\substack{f(x)\\ \text{Dividend}\\ \text{(Original polynomial)}}} = \underbrace{(3x^2 - 2)}_{\substack{g(x)\\ \text{Divisor}}} \cdot \underbrace{(2x^2 + 3x + 2)}_{\substack{q(x)\\ \text{Quotient}}} + \underbrace{(-2x + 11)}_{\substack{r(x)\\ \text{Remainder}}}$$

OBJECTIVE 2 Use synthetic division to divide by a polynomial of the form $x - k$.

When one polynomial is divided by a second and the second polynomial has the form $x - k$, where the coefficient of the x-term is 1, there is a shortcut method for performing the division. Look first below left, where we show the division of

$$3x^3 - 2x + 5 \quad \text{by} \quad x - 3.$$

Notice that we inserted 0 for the missing x^2-term.

Polynomial Division

$$\begin{array}{r}
3x^2 + 9x + 25 \\
x - 3\overline{)3x^3 + 0x^2 - 2x + 5} \\
\underline{3x^3 - 9x^2} \qquad\qquad \\
9x^2 - 2x \qquad \\
\underline{9x^2 - 27x} \qquad \\
25x + 5 \\
\underline{25x - 75} \\
80
\end{array}$$

Synthetic Division

$$\begin{array}{rrrrr}
 & & 3 & 9 & 25 \\
1 - 3\overline{)} & 3 & 0 & -2 & 5 \\
 & \underline{3} & \underline{-9} & & \\
 & & 9 & -2 & \\
 & & \underline{9} & \underline{-27} & \\
 & & & 25 & 5 \\
 & & & \underline{25} & \underline{-75} \\
 & & & & 80
\end{array}$$

On the right, exactly the same division is shown written without the variables. This is why it is *essential* to use 0 as a placeholder. All the numbers in color on the right are repetitions of the numbers directly above them, so we omit them to condense our work, as shown on the left at the top of the next page.

$$\begin{array}{rrrrr}
 & 3 & 9 & 25 & \\
\hline
1-3)\,3 & 0 & -2 & 5 & \\
-9 & & & & \\
\hline
9 & -2 & & & \\
-27 & & & & \\
\hline
25 & 5 & & & \\
-75 & & & & \\
\hline
80 & & & &
\end{array}
\qquad\Bigg|\qquad
\begin{array}{rrrr}
 & 3 & 9 & 25 \\
\hline
1-3)\,3 & 0 & -2 & 5 \\
-9 & & & \\
\hline
9 & & & \\
-27 & & & \\
\hline
25 & & & \\
-75 & & & \\
\hline
80 & & &
\end{array}$$

The numbers in color on the left are again repetitions of the numbers directly above them. They too are omitted, as shown on the right above. If we bring the 3 in the dividend down to the beginning of the bottom row, the top row can be omitted because it duplicates the bottom row.

$$\begin{array}{r|rrrr}
1-3 & 3 & 0 & -2 & 5 \\
 & & -9 & -27 & -75 \\
\hline
 & 3 & 9 & 25 & 80
\end{array}$$

We omit the 1 at the upper left—it represents $1x$, which will always be the first term in the divisor. Also, we replace subtraction in the second row by addition. To compensate for this, we change the -3 at the upper left to its additive inverse, 3.

$$\begin{array}{r|rrrrl}
\text{Additive inverse of } -3 \rightarrow 3 & 3 & 0 & -2 & 5 & \\
 & & 9 & 27 & 75 & \leftarrow \text{Signs changed} \\
\hline
 & 3 & 9 & 25 & 80 & \leftarrow \text{Remainder}
\end{array}$$

The quotient is read from the bottom row.

$$3x^2 + 9x + 25 + \frac{80}{x-3}$$

Remember to add $\frac{\text{remainder}}{\text{divisor}}$.

The quotient polynomial has degree one less than the degree of the dividend.

Synthetic Division

Synthetic division is a shortcut procedure for polynomial division that eliminates writing the variable factors. It is used only when dividing a polynomial by a binomial of the form $x - k$.

EXAMPLE 1 Using Synthetic Division

Use synthetic division to divide.

$$\frac{5x^2 + 16x + 15}{x+2}$$

Remember that a fraction bar means division.

We change $x + 2$ into the form $x - k$ by writing it as

$$x + 2 = x - (-2), \quad \text{where } k = -2.$$

Now write the coefficients of $5x^2 + 16x + 15$, placing -2 to the left.

$$\begin{array}{r|rrrl}
x+2 \text{ leads to } -2. \rightarrow -2 & 5 & 16 & 15 & \leftarrow \text{Coefficients}
\end{array}$$

$$\begin{array}{r|rrr}
-2 & 5 & 16 & 15 \\
 & \downarrow & -10 & \\
\hline
 & 5 & &
\end{array}$$

Bring down the 5, and multiply: $-2 \cdot 5 = -10$.

NOW TRY EXERCISE 1
Use synthetic division to divide.

$$\frac{3x^2 - 8x - 10}{x - 4}$$

$$\begin{array}{r|rrr} -2 & 5 & 16 & 15 \\ & & -10 & -12 \\ \hline & 5 & 6 & \end{array}$$

Add 16 and −10, obtaining 6, and multiply −2 and 6 to obtain −12.

$$\begin{array}{r|rrr} -2 & 5 & 16 & 15 \\ & & -10 & -12 \\ \hline & 5 & 6 & 3 \end{array}$$

Add 15 and −12, obtaining 3.

Read the result from the bottom row. ← Remainder

$$\frac{5x^2 + 16x + 15}{x + 2} = 5x + 6 + \frac{3}{x + 2}$$

Add $\frac{\text{remainder}}{\text{divisor}}$.

NOW TRY

By multiplying both sides by the denominator $x + 2$, we can write the result of the division in **Example 1** as

$$\underbrace{5x^2 + 16x + 15}_{f(x)} = \underbrace{(x + 2)}_{(x-k)} \cdot \underbrace{(5x + 6)}_{q(x)} + \underbrace{3}_{r}.$$

The following theorem is a special case of the division algorithm given earlier. Here $g(x)$ is the first-degree polynomial $x - k$.

Division Algorithm for Divisor $x - k$

For any polynomial $f(x)$ and any complex number k, there exist a unique polynomial $q(x)$ and number r such that the following holds.

$$f(x) = (x - k)q(x) + r$$

NOW TRY EXERCISE 2
Use synthetic division to divide.

$$\frac{2x^4 + 5x^3 + x - 10}{x + 3}$$

EXAMPLE 2 Using Synthetic Division with a Missing Term

Use synthetic division to divide.

$$\frac{-4x^5 + x^4 + 6x^3 + 2x^2 + 50}{x - 2}$$

This quotient can be written as follows.

$$(-4x^5 + x^4 + 6x^3 + 2x^2 + 50) \div (x - 2)$$

In long division form, the procedure is set up as follows.

$$x - 2\overline{)-4x^5 + x^4 + 6x^3 + 2x^2 + 50}$$

Now use synthetic division.

$$\begin{array}{r|rrrrrr} 2 & -4 & 1 & 6 & 2 & 0 & 50 \\ & & -8 & -14 & -16 & -28 & -56 \\ \hline & -4 & -7 & -8 & -14 & -28 & -6 \end{array}$$

Use the steps given above, first inserting a 0 for the missing x-term.

Read the result from the bottom row.

$$\frac{-4x^5 + x^4 + 6x^3 + 2x^2 + 50}{x - 2} = -4x^4 - 7x^3 - 8x^2 - 14x - 28 + \frac{-6}{x - 2}$$

NOW TRY

NOW TRY ANSWERS

1. $3x + 4 + \frac{6}{x - 4}$
2. $2x^3 - x^2 + 3x - 8 + \frac{14}{x + 3}$

NOW TRY EXERCISE 3

Express the polynomial function

$$f(x) = 2x^3 + 5x^2 - x + 1$$

in the form

$$f(x) = (x - k)q(x) + r$$

for $k = 3$.

EXAMPLE 3 Applying the Division Algorithm

Express the polynomial function

$$f(x) = 2x^3 - 3x^2 + x + 7$$

in the form $f(x) = (x - k)q(x) + r$ for $k = 2$.

Begin by using synthetic division to find the quotient and the remainder when $f(x)$ is divided by $x - 2$.

$$\begin{array}{r|rrrr} 2 & 2 & -3 & 1 & 7 \\ & & 4 & 2 & 6 \\ \hline & 2 & 1 & 3 & 13 \end{array}$$

13 ← Remainder; 2 1 3: Coefficients of $q(x)$

The quotient has degree one less than that of $f(x)$, namely 2. The remainder r is 13. Now use the division algorithm equation.

$$2x^3 - 3x^2 + x + 7 = (x - 2)(2x^2 + 1x + 3) + 13$$

$$f(x) = (x - 2)(2x^2 + x + 3) + 13$$

NOW TRY 

OBJECTIVE 3 Use the remainder theorem to evaluate a polynomial.

We can use synthetic division to evaluate polynomials. For example, in the synthetic division of **Example 2,** where the polynomial was divided by $x - 2$, the remainder was -6.

Replacing x in the polynomial with 2 gives the following.

$$-4x^5 + x^4 + 6x^3 + 2x^2 + 50 \quad \text{Dividend in Example 2}$$

$$= -4 \cdot 2^5 + 2^4 + 6 \cdot 2^3 + 2 \cdot 2^2 + 50 \quad \text{Let } x = 2.$$

$$= -4 \cdot 32 + 16 + 6 \cdot 8 + 2 \cdot 4 + 50 \quad \text{Evaluate the powers.}$$

$$= -128 + 16 + 48 + 8 + 50 \quad \text{Multiply.}$$

$$= -6 \quad \text{Add.}$$

This number, -6, is the same number as the remainder. Dividing by $x - 2$ produced a remainder equal to the result when x is replaced with 2.

To prove that this is true in general, consider the division algorithm,

$$f(x) = (x - k)q(x) + r.$$

This equality is true for all complex values of x, so it is true for $x = k$.

$$f(k) = (k - k)q(k) + r \quad \text{Let } x = k.$$

$$f(k) = r \quad k - k = 0$$

This proves the **remainder theorem,** which gives an alternative method of evaluating functions defined by polynomials.

Remainder Theorem

If the polynomial $f(x)$ is divided by $x - k$, then the remainder is equal to $f(k)$.

NOW TRY ANSWER

3. $f(x) = (x - 3)(2x^2 + 11x + 32) + 97$

NOW TRY EXERCISE 4

Let $f(x) = 2x^3 - x^2 - 2x + 40$. Use the remainder theorem to find $f(-3)$.

EXAMPLE 4 Using the Remainder Theorem

Let $f(x) = 2x^3 - 5x^2 - 3x + 11$. Use the remainder theorem to find $f(-2)$.

Divide $f(x)$ by $x - (-2)$ using synthetic division.

$$\begin{array}{r|rrrr} \text{Value of } k \to -2 & 2 & -5 & -3 & 11 \\ & & -4 & 18 & -30 \\ \hline & 2 & -9 & 15 & -19 \leftarrow \text{Remainder} \end{array}$$

Thus, $f(-2) = -19$.

NOW TRY  

OBJECTIVE 4 Determine whether a given number is a zero of a polynomial function.

The function $f(x) = 2x^3 - 5x^2 - 3x + 11$ in **Example 4** is an example of a *polynomial function* with real coefficients. Here, we extend the definition of a polynomial function to include complex numbers as coefficients.

Polynomial Function

A **polynomial function of degree** n (in x here) is a function of the form

$$f(x) = a_nx^n + a_{n-1}x^{n-1} + \cdots + a_1x + a_0,$$

for complex numbers $a_n, a_{n-1}, \ldots, a_1$, and a_0, where $a_n \neq 0$.

A **zero** of a polynomial function $f(x)$ is a number k such that $f(k) = 0$. ***Real number zeros are the x-values of the x-intercepts of the graph of the function.***

The remainder theorem gives a quick way to decide whether a number k is a zero of a polynomial function $f(x)$, as follows.

1. Use synthetic division to find $f(k)$.
2. If the remainder is 0, then $f(k) = 0$, and k is a zero of $f(x)$. If the remainder is not 0, then k is not a zero of $f(x)$.

A zero of $f(x)$ is a **root,** or **solution,** of the equation $f(x) = 0$.

EXAMPLE 5 Determining Whether a Number Is a Zero

Use synthetic division to determine whether the given number is a zero of the polynomial function.

(a) -5; $f(x) = 2x^4 + 12x^3 + 6x^2 - 5x + 75$

$$\begin{array}{r|rrrrr} \text{Proposed zero} \to -5 & 2 & 12 & 6 & -5 & 75 \\ & & -10 & -10 & 20 & -75 \\ \hline & 2 & 2 & -4 & 15 & 0 \leftarrow \text{Remainder} \end{array}$$

Use synthetic division and the remainder theorem.

Because the remainder is 0, $f(-5) = 0$, and -5 is a zero—that is, a solution of $f(x) = 0$—for $f(x) = 2x^4 + 12x^3 + 6x^2 - 5x + 75$.

(b) -4; $f(x) = x^4 + x^2 - 3x + 1$

$$\begin{array}{r|rrrrr} -4 & 1 & 0 & 1 & -3 & 1 \\ & & -4 & 16 & -68 & 284 \\ \hline & 1 & -4 & 17 & -71 & 285 \leftarrow \text{Remainder} \end{array}$$

Use 0 as coefficient for the missing x^3-term.

The remainder is not 0, so -4 is *not* a zero of $f(x) = x^4 + x^2 - 3x + 1$. In fact, $f(-4) = 285$.

NOW TRY ANSWER

4. -17

NOW TRY EXERCISE 5

Use synthetic division to determine whether the given number is a zero of

$f(x) = x^4 - 5x^3 + 10x^2 - 10x + 4.$

(a) 1 **(b)** -1 **(c)** $1 - i$

(c) $1 + 2i$; $f(x) = x^4 - 2x^3 + 4x^2 + 2x - 5$

$$\begin{array}{r|rrrrr} 1+2i & 1 & -2 & 4 & 2 & -5 \\ & & 1+2i & -5 & -1-2i & 5 \\ \hline & 1 & -1+2i & -1 & 1-2i & 0 \end{array}$$

Use synthetic division and operations with complex numbers; $i^2 = -1$

0 ← Remainder

$(1 + 2i)(-1 + 2i)$
$= -1 + 4i^2$
$= -5$

The remainder is 0, so $1 + 2i$ is a zero of the given polynomial function.

NOW TRY

The synthetic division in **Example 5(a)** shows that $x - (-5)$ divides the polynomial with 0 remainder. Thus $x - (-5) = x + 5$ is a *factor* of the polynomial.

$$2x^4 + 12x^3 + 6x^2 - 5x + 75 = (x + 5)(2x^3 + 2x^2 - 4x + 15)$$

The coefficients of the second factor are found in the last row of the synthetic division.

NOW TRY ANSWERS

5. (a) yes (b) no (c) yes

11.1 Exercises

MyLab Math

Video solutions for select problems available in MyLab Math

Concept Check *Choose the letter of the correct setup to perform synthetic division on the indicated quotient.*

1. $\dfrac{x^2 + 3x - 6}{x - 2}$

A. $-2\overline{)1\ \ 3\ \ -6}$ **B.** $-2\overline{)-1\ \ -3\ \ 6}$

C. $2\overline{)1\ \ 3\ \ -6}$ **D.** $2\overline{)-1\ \ -3\ \ 6}$

2. $\dfrac{x^3 - 3x^2 + 2}{x - 1}$

A. $1\overline{)1\ \ -3\ \ 2}$ **B.** $-1\overline{)1\ \ -3\ \ 2}$

C. $1\overline{)1\ \ -3\ \ 0\ \ 2}$ **D.** $1\overline{)-1\ \ 3\ \ 0\ \ -2}$

Concept Check *Fill in each blank with the appropriate response.*

3. In arithmetic, the result of the division

$$\begin{array}{r} 3 \\ 5\overline{)19} \\ \underline{15} \\ 4 \end{array}$$

can be written

$19 = 5 \cdot$ _____ + _____.

4. In algebra, the result of the division

$$\begin{array}{r} x + 3 \\ x - 1\overline{)x^2 + 2x + 3} \\ \underline{x^2 - x} \\ 3x + 3 \\ \underline{3x - 3} \\ 6 \end{array}$$

can be written $x^2 + 2x + 3 =$

$(x - 1)($_____$) +$ _____.

5. To perform the division

$$x - 3\overline{)x^3 + 6x^2 + 2x}$$

using synthetic division, we begin by writing the following.

____$\overline{)1\ \ ___\ \ 2\ \ ___}$

6. Consider the following function.

$$f(x) = 2x^4 + 6x^3 - 5x^2 + 3x + 8$$
$$f(x) = (x - 2)(2x^3 + 10x^2 + 15x + 33) + 74$$

By inspection, we can state that $f(2) =$ _____.

7. Concept Check A student attempted to divide

$$4x^3 + 2x^2 + 6 \quad \text{by} \quad x + 2$$

synthetically by setting up the division as follows.

$$-2\overline{)4 \quad 2 \quad 6}$$

This is incorrect. **WHAT WENT WRONG?** Give the correct setup and the answer.

8. Concept Check A student attempted to divide

$$4x^3 + 2x^2 + 6x \quad \text{by} \quad x + 2$$

synthetically by setting up the division as follows.

$$-2\overline{)4 \quad 2 \quad 6}$$

This is incorrect. **WHAT WENT WRONG?** Give the correct setup and the answer.

Use synthetic division to divide. ***See Examples 1 and 2.***

9. $\dfrac{x^2 - 6x + 5}{x - 1}$

10. $\dfrac{x^2 + 4x - 21}{x - 3}$

11. $\dfrac{4x^2 + 19x - 5}{x + 5}$

12. $\dfrac{3x^2 + 5x - 12}{x + 3}$

13. $\dfrac{2x^2 + 8x + 13}{x + 2}$

14. $\dfrac{4x^2 - 5x - 20}{x - 4}$

15. $\dfrac{4x^3 - 3x^2 + 2x - 3}{x - 1}$

16. $\dfrac{5x^3 - 6x^2 + 3x + 14}{x + 1}$

17. $\dfrac{x^3 + 2x^2 - 4x + 3}{x - 4}$

18. $\dfrac{x^3 - 3x^2 + 5x - 1}{x - 5}$

19. $\dfrac{2x^5 - 2x^3 + 3x^2 - 24x - 2}{x - 2}$

20. $\dfrac{3x^5 + x^4 - 84x^2 - 12x + 3}{x - 3}$

21. $\dfrac{-3x^5 - 3x^4 + 5x^3 - 6x^2 + 3}{x + 1}$

22. $\dfrac{-3x^5 + 2x^4 - 5x^3 - 6x^2 - 1}{x + 2}$

23. $\dfrac{x^5 + x^4 + x^3 + x^2 + x + 3}{x + 1}$

24. $\dfrac{x^5 - x^4 + x^3 - x^2 + x - 2}{x - 1}$

Express each polynomial function in the form $f(x) = (x - k)q(x) + r$ for the given value of k. ***See Example 3.***

25. $f(x) = 2x^3 + x^2 + x - 8; \quad k = -1$

26. $f(x) = 2x^3 + 3x^2 - 6x + 1; \quad k = -4$

27. $f(x) = -x^3 + 2x^2 + 4; \quad k = -2$

28. $f(x) = -2x^3 + 6x^2 + 5; \quad k = 2$

29. $f(x) = 4x^4 - 3x^3 - 20x^2 - x; \quad k = 3$

30. $f(x) = 2x^4 + x^3 - 15x^2 + 3x; \quad k = -3$

For each polynomial function, use the remainder theorem and synthetic division to find $f(k)$. ***See Example 4.***

31. $f(x) = x^2 - 4x + 5; \quad k = 3$

32. $f(x) = x^2 + 5x + 6; \quad k = -2$

33. $f(x) = 2x^2 - 3x - 3; \quad k = 2$

34. $f(x) = -x^3 + 8x^2 + 63; \quad k = 4$

35. $f(x) = x^3 - 4x^2 + 2x + 1; \quad k = -1$

36. $f(x) = 2x^3 - 3x^2 - 5x + 4; \quad k = 2$

37. $f(x) = 2x^5 - 10x^3 - 19x^2; \quad k = 3$

38. $f(x) = x^4 - 6x^3 + 9x^2 - 3x; \quad k = 4$

39. $f(x) = x^2 - 5x + 1; \quad k = 2 + i$

40. $f(x) = x^2 - x + 3; \quad k = 3 - 2i$

41. $f(x) = 9x^3 - 6x^2 + x; \quad k = \dfrac{1}{3}$

42. $f(x) = 6x^3 - 31x^2 - 15x; \quad k = -\dfrac{1}{2}$

Use synthetic division to determine whether the given number is a zero of the polynomial function. **See Example 5.**

43. $3;\quad f(x) = 2x^3 - 6x^2 - 9x + 27$

44. $-6;\quad f(x) = 2x^3 + 9x^2 - 16x + 12$

45. $-5;\quad f(x) = x^3 + 7x^2 + 10x$

46. $-2;\quad f(x) = x^3 - 7x^2 - 18x$

47. $\frac{2}{5};\quad f(x) = 5x^4 + 2x^3 - x + 15$

48. $\frac{1}{2};\quad f(x) = 2x^4 - 3x^2 + 4$

49. $2 - i;\quad f(x) = x^2 + 3x + 4$

50. $1 - 2i;\quad f(x) = x^2 - 3x + 5$

RELATING CONCEPTS For Individual or Group Work (Exercises 51–56)

We have seen the close connection between polynomial division and writing a quotient of polynomials in lowest terms after factoring the numerator. We can also show a connection between dividing one polynomial by another and factoring the first polynomial. ***Work Exercises 51–56 in order,*** *letting*

$$f(x) = 2x^2 + 5x - 12.$$

51. Factor $f(x)$.

52. Solve $f(x) = 0$.

53. Evaluate $f(-4)$.

54. Evaluate $f\left(\frac{3}{2}\right)$.

55. Complete the following sentence: If $f(a) = 0$, then $x -$ ______ is a factor of $f(x)$.

56. Use the conclusion reached in **Exercise 55** to determine whether $x - 3$ is a factor of $g(x) = 3x^3 - 4x^2 - 17x + 6$. If so, factor $g(x)$ completely.

11.2 Zeros of Polynomial Functions (II)

OBJECTIVES

1. Use the factor theorem.
2. Use the rational zeros theorem.
3. Find polynomial functions that satisfy given conditions.
4. Apply the conjugate zeros theorem.

VOCABULARY

- ☐ multiplicity of a zero
- ☐ complex conjugate

OBJECTIVE 1 Use the factor theorem.

By the remainder theorem, if $f(k) = 0$, then the remainder when $f(x)$ is divided by $x - k$ is 0. This means that $x - k$ is a factor of $f(x)$. Conversely, if $x - k$ is a factor of $f(x)$, then $f(k)$ must equal 0. This is summarized as the **factor theorem.**

Factor Theorem

The polynomial $x - k$ is a factor of the polynomial $f(x)$ if and only if $f(k) = 0$.

EXAMPLE 1 Deciding Whether $x - k$ Is a Factor of $f(x)$

Determine whether $x - 1$ is a factor of $f(x)$.

(a) $f(x) = 2x^4 + 3x^2 - 5x + 7$

By the factor theorem, $x - 1$ will be a factor of $f(x)$ only if $f(1) = 0$. Use synthetic division and the remainder theorem to evaluate $f(1)$.

$$\begin{array}{r|rrrrr} 1 & 2 & 0 & 3 & -5 & 7 \\ & & 2 & 2 & 5 & 0 \\ \hline & 2 & 2 & 5 & 0 & 7 \end{array}$$

Use 0 as coefficient for the missing x^3-term.

$\leftarrow$ Remainder $= f(1) \neq 0$

Because the remainder is 7 (not 0), $x - 1$ is not a factor of $f(x)$.

NOW TRY EXERCISE 1

Determine the following.

(a) Is $x + 2$ a factor of $f(x) = x^3 - x^2 + x + 14$?

(b) Is $x - 4$ a factor of $f(x) = x^3 + x^2 - x + 1$?

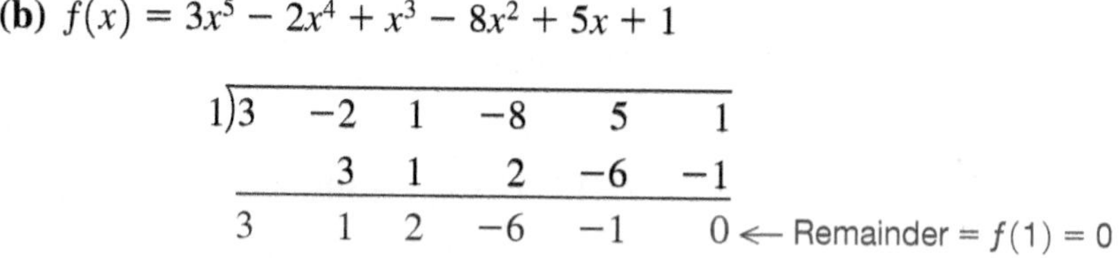

(b) $f(x) = 3x^5 - 2x^4 + x^3 - 8x^2 + 5x + 1$

$$\begin{array}{r|rrrrrr} 1 & 3 & -2 & 1 & -8 & 5 & 1 \\ & & 3 & 1 & 2 & -6 & -1 \\ \hline & 3 & 1 & 2 & -6 & -1 & 0 \end{array} \leftarrow \text{Remainder} = f(1) = 0$$

Because the remainder is 0, $x - 1$ is a factor. We can determine from the coefficients in the bottom row that the other factor is

$$3x^4 + x^3 + 2x^2 - 6x - 1. \qquad 1x^3 = x^3$$

NOW TRY

NOW TRY EXERCISE 2

Factor $f(x)$ into linear factors, given that -2 is a zero of f.

$f(x) = 6x^3 + 17x^2 + 6x - 8$

EXAMPLE 2 Factoring a Polynomial Given a Zero

Factor $f(x)$ into linear factors, given that -3 is a zero of f.

$$f(x) = 6x^3 + 19x^2 + 2x - 3$$

Because -3 is a zero of f, $x - (-3) = x + 3$ is a factor.

$$\begin{array}{r|rrrr} -3 & 6 & 19 & 2 & -3 \\ & & -18 & -3 & 3 \\ \hline & 6 & 1 & -1 & 0 \end{array} \leftarrow \text{Remainder} = f(-3) = 0$$

Use synthetic division to divide $f(x)$ by $x + 3$.

The remainder is 0, so $f(-3) = 0$, and $x + 3$ is a factor, as expected. The quotient is $6x^2 + x - 1$. It is factored using earlier methods.

$$f(x) = (x + 3)(6x^2 + x - 1)$$

$$f(x) = (x + 3)(2x + 1)(3x - 1) \qquad \text{Factor } 6x^2 + x - 1.$$

These factors are all linear.

NOW TRY

OBJECTIVE 2 Use the rational zeros theorem.

The **rational zeros theorem** gives a method to determine all possible candidates for rational zeros of a polynomial function with integer coefficients.

Rational Zeros Theorem

For $a_n \neq 0$, let

$$f(x) = a_n x^n + a_{n-1}x^{n-1} + \cdots + a_1 x + a_0$$

define a polynomial function with integer coefficients. If $\frac{p}{q}$ is a rational number written in lowest terms, and if $\frac{p}{q}$ is a zero of f, then p is a factor of the constant term a_0, and q is a factor of the leading coefficient a_n.

PROOF $f\left(\frac{p}{q}\right) = 0$ because $\frac{p}{q}$ is a zero of $f(x)$.

$$a_n\left(\frac{p}{q}\right)^n + a_{n-1}\left(\frac{p}{q}\right)^{n-1} + \cdots + a_1\left(\frac{p}{q}\right) + a_0 = 0 \qquad \text{Definition of zero of } f$$

$$a_n\left(\frac{p^n}{q^n}\right) + a_{n-1}\left(\frac{p^{n-1}}{q^{n-1}}\right) + \cdots + a_1\left(\frac{p}{q}\right) + a_0 = 0 \qquad \text{Power rule for exponents}$$

$$a_n p^n + a_{n-1}p^{n-1}q + \cdots + a_1 pq^{n-1} = -a_0 q^n \qquad \text{Multiply by } q^n. \text{ Add } -a_0q^n \text{ to each side.}$$

$$p(a_n p^{n-1} + a_{n-1}p^{n-2}q + \cdots + a_1 q^{n-1}) = -a_0 q^n \qquad \text{Factor out } p.$$

NOW TRY ANSWERS

1. (a) yes (b) no
2. $f(x) = (x + 2)(2x - 1)(3x + 4)$

Thus, $-a_0 q^n$ equals the product of the two factors,

$$p \quad \text{and} \quad (a_n p^{n-1} + \cdots + a_1 q^{n-1}).$$

For this reason, p must be a factor of $-a_0 q^n$. Because it was assumed that $\frac{p}{q}$ is written in lowest terms, p and q have no common factor other than 1. It follows that p is not a factor of q^n. Thus, p must be a factor of a_0. In a similar way, q is a factor of a_n.

NOW TRY EXERCISE 3

Consider the following polynomial function.

$$f(x) = 8x^4 + 10x^3 - 11x^2 - 10x + 3$$

(a) List all possible rational zeros.

(b) Find all rational zeros, and factor $f(x)$ into linear factors.

EXAMPLE 3 Using the Rational Zeros Theorem

Consider the following polynomial function.

$$f(x) = 6x^4 + 7x^3 - 12x^2 - 3x + 2$$

(a) List all possible rational zeros.

For a rational number $\frac{p}{q}$ to be a zero, p must be a factor of $a_0 = 2$, and q must be a factor of $a_4 = 6$. Thus, p can be ± 1 or ± 2, and q can be ± 1, ± 2, ± 3, or ± 6. The possible rational zeros $\frac{p}{q}$ are

$$\pm 1, \quad \pm 2, \quad \pm \frac{1}{2}, \quad \pm \frac{1}{3}, \quad \pm \frac{1}{6}, \quad \pm \frac{2}{3}.$$

(b) Find all rational zeros, and factor $f(x)$ into linear factors.

Use the remainder theorem to show that 1 and -2 are zeros.

Use "trial and error" to find zeros.

$$\begin{array}{r|rrrrr} 1 & 6 & 7 & -12 & -3 & 2 \\ & & 6 & 13 & 1 & -2 \\ \hline & 6 & 13 & 1 & -2 & 0 \end{array} \leftarrow f(1) = 0$$

The 0 remainder shows that 1 is a zero. The quotient is $6x^3 + 13x^2 + x - 2$.

$$f(x) = (x - 1)(6x^3 + 13x^2 + x - 2) \quad \text{Begin factoring } f(x).$$

Now, use the quotient polynomial and synthetic division to find that -2 is a zero.

$$\begin{array}{r|rrrr} -2 & 6 & 13 & 1 & -2 \\ & & -12 & -2 & 2 \\ \hline & 6 & 1 & -1 & 0 \end{array} \leftarrow f(-2) = 0$$

The new quotient polynomial is $6x^2 + x - 1$. Therefore, $f(x)$ can now be completely factored as follows.

$$f(x) = (x - 1)(x + 2)(6x^2 + x - 1)$$
$$f(x) = (x - 1)(x + 2)(3x - 1)(2x + 1)$$

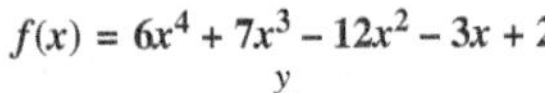
$f(x) = 6x^4 + 7x^3 - 12x^2 - 3x + 2$

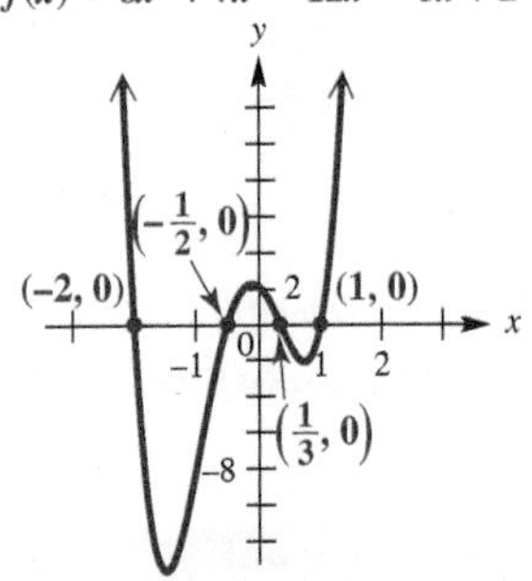

FIGURE 1

Setting $3x - 1 = 0$ and $2x + 1 = 0$ yields the zeros $\frac{1}{3}$ and $-\frac{1}{2}$. In summary, the rational zeros are $1, -2, \frac{1}{3}$, and $-\frac{1}{2}$. These zeros correspond to the x-intercepts of the graph of $f(x)$ in **FIGURE 1**. The linear factorization of $f(x)$ is as follows.

$$f(x) = 6x^4 + 7x^3 - 12x^2 - 3x + 2$$
$$f(x) = (x - 1)(x + 2)(3x - 1)(2x + 1)$$

Check by multiplying these factors.

NOW TRY

NOW TRY ANSWERS

3. (a) $\pm 1, \pm 3, \pm \frac{1}{2}, \pm \frac{1}{4}, \pm \frac{1}{8}, \pm \frac{3}{2}, \pm \frac{3}{4}, \pm \frac{3}{8}$

(b) $-\frac{3}{2}, -1, \frac{1}{4}, 1$; $f(x) = (2x + 3)(x + 1)(4x - 1)(x - 1)$

CAUTION **The rational zeros theorem gives *possible* rational zeros. It does not tell us whether these rational numbers are *actual* zeros.** Furthermore, the function must have integer coefficients.

To apply the rational zeros theorem to a polynomial with fractional coefficients, multiply by the least common denominator of all the fractions. In the example that follows, any rational zeros of $f(x)$ will also be rational zeros of $g(x)$.

$$f(x) = x^4 - \frac{1}{6}x^3 + \frac{2}{3}x^2 - \frac{1}{6}x - \frac{1}{3}$$

$$g(x) = 6x^4 - x^3 + 4x^2 - x - 2 \qquad \text{Multiply each term of } f(x) \text{ by 6.}$$

Carl Friederich Gauss (1777–1855)

The **fundamental theorem of algebra** was first proved by Carl Friederich Gauss in his doctoral thesis in 1799, when he was 22 years old.

OBJECTIVE 3 Find polynomial functions that satisfy given conditions.

The **fundamental theorem of algebra** says that every function defined by a polynomial of degree 1 or more has a zero, which means that every such polynomial can be factored.

Fundamental Theorem of Algebra

Every polynomial function of degree 1 or more has at least one complex zero.

From the fundamental theorem, if $f(x)$ is of degree 1 or greater, then there is some number k_1 such that $f(k_1) = 0$. By the factor theorem,

$$f(x) = (x - k_1) \cdot q_1(x), \quad \text{for some polynomial } q_1(x).$$

If $q_1(x)$ is of degree 1 or greater, the fundamental theorem and the factor theorem can be used to factor $q_1(x)$ in the same way. There is some number k_2 such that $q_1(k_2) = 0$, so that

$$q_1(x) = (x - k_2)q_2(x)$$

and

$$f(x) = (x - k_1)(x - k_2)q_2(x).$$

Assuming that $f(x)$ has degree n and repeating this process n times give

$$f(x) = a(x - k_1)(x - k_2) \cdots (x - k_n),$$

where a is the leading coefficient of $f(x)$. Each factor leads to a zero of $f(x)$, so $f(x)$ has the n zeros $k_1, k_2, \ldots, k_n$. This suggests the **number of zeros theorem.**

Number of Zeros Theorem

A polynomial function of degree n has at most n distinct zeros.

The theorem says that there exist *at most n* distinct zeros. For example, consider $f(x) = x^3 + 3x^2 + 3x + 1$, which can be written in factored form as $f(x) = (x + 1)^3$. Although this is a polynomial function of degree 3, it has only *one* distinct zero, -1. Actually, the zero -1 occurs three times because there are three factors of $x + 1$. The number of times a zero occurs is the **multiplicity of the zero.**

NOW TRY EXERCISE 4

Find a polynomial function $f(x)$ of degree 3 that satisfies the given conditions:

Zeros of -1, 2, and 5; $f(8) = 54$.

EXAMPLE 4 Finding a Polynomial Function That Satisfies Given Conditions (Real Zeros)

Find a polynomial function $f(x)$ of degree 3 that satisfies the given conditions.

(a) Zeros of -1, 2, and 4; $f(1) = 3$

Given these three zeros, the factors of $f(x)$ are

$$x - (-1) \text{ or } x + 1, \quad x - 2, \quad \text{and} \quad x - 4.$$

Because $f(x)$ is to be of degree 3, these are the only possible variable factors by the theorem just stated. Therefore, $f(x)$ has the form

$$f(x) = a(x + 1)(x - 2)(x - 4), \quad \text{for some real number } a.$$

To find a, use the fact that $f(1) = 3$.

$$f(1) = a(1 + 1)(1 - 2)(1 - 4) \quad \text{Let } x = 1.$$

$$3 = a(2)(-1)(-3) \quad f(1) = 3 \text{ is given.}$$

$$a = \frac{1}{2} \quad \text{Solve for } a.$$

Thus, $$f(x) = \frac{1}{2}(x + 1)(x - 2)(x - 4), \quad \text{Let } a = \tfrac{1}{2}.$$

or $$f(x) = \frac{1}{2}x^3 - \frac{5}{2}x^2 + x + 4. \quad \text{Multiply.}$$

(b) -2 is a zero of multiplicity 3; $f(-1) = 4$

The polynomial function $f(x)$ has the following form.

$$f(x) = a(x + 2)(x + 2)(x + 2)$$

$$f(x) = a(x + 2)^3$$

Because $f(-1) = 4$, we can find a.

$$f(-1) = a(-1 + 2)^3 \quad \text{Let } x = -1.$$

$$4 = a(1)^3 \quad f(-1) = 4 \text{ is given.}$$

$$a = 4 \quad \text{Solve for } a.$$

Remember $(x + 2)^3 \neq x^3 + 2^3$

Thus, $f(x) = 4(x + 2)^3$, or $f(x) = 4x^3 + 24x^2 + 48x + 32$. NOW TRY

NOTE In **Example 4(a),** we cannot clear the denominators in $f(x)$ by multiplying each side by 2 because the result would equal $2 \cdot f(x)$, not $f(x)$.

OBJECTIVE 4 Apply the conjugate zeros theorem.

If $z = a + bi$, then the **complex conjugate** of z is written $\overline{z}$, where $\overline{z} = a - bi$.

Example: If $z = -5 + 2i$, then $\overline{z} = -5 - 2i$.

The following properties of complex conjugates are needed to prove the **conjugate zeros theorem.**

NOW TRY ANSWER

4. $f(x) = \frac{1}{3}(x + 1)(x - 2)(x - 5)$, or $f(x) = \frac{1}{3}x^3 - 2x^2 + x + \frac{10}{3}$

Properties of Complex Conjugates

For any complex numbers c and d,

$$\overline{c+d} = \overline{c} + \overline{d}, \quad \overline{c \cdot d} = \overline{c} \cdot \overline{d}, \quad \text{and} \quad \overline{c^n} = (\overline{c})^n.$$

We can show that if the complex number z is a zero of $f(x)$, then the conjugate of z (that is, $\overline{z}$) is also a zero of $f(x)$.

PROOF Start with the polynomial function

$$f(x) = a_n x^n + a_{n-1}x^{n-1} + \cdots + a_1 x + a_0,$$

where all coefficients are real numbers. If $z = a + bi$ is a zero of $f(x)$, then

$$f(z) = a_n z^n + a_{n-1}z^{n-1} + \cdots + a_1 z + a_0 = 0.$$

Taking the conjugate of each side of this equation gives the following.

$$\overline{a_n z^n + a_{n-1}z^{n-1} + \cdots + a_1 z + a_0} = \overline{0}$$

$$\overline{a_n z^n} + \overline{a_{n-1}z^{n-1}} + \cdots + \overline{a_1 z} + \overline{a_0} = \overline{0}$$

$$\overline{a_n}\,\overline{z^n} + \overline{a_{n-1}}\,\overline{z^{n-1}} + \cdots + \overline{a_1}\,\overline{z} + \overline{a_0} = \overline{0}$$

Properties of complex conjugates

Now use the third property and the fact that for any real number a, $\overline{a} = a$.

$$a_n(\overline{z})^n + a_{n-1}(\overline{z})^{n-1} + \cdots + a_1(\overline{z}) + a_0 = 0$$

Thus, $\overline{z}$ is also a zero of $f(x)$, completing the proof of the **conjugate zeros theorem.**

Conjugate Zeros Theorem

If $f(x)$ is a polynomial function *having only real coefficients* and if $a + bi$ is a zero of $f(x)$, where a and b are real numbers, then $a - bi$ is also a zero of $f(x)$.

In applying this theorem, it is essential that a polynomial function have only real coefficients. For example, $f(x) = x - (1 + i)$ has $1 + i$ as a zero, but the conjugate $1 - i$ is *not* a zero.

NOW TRY EXERCISE 5

Find a polynomial function $f(x)$ of least possible degree having only real coefficients and zeros 4 and $3 + 2i$.

EXAMPLE 5 Finding a Polynomial Function That Satisfies Given Conditions (Complex Zeros)

Find a polynomial function $f(x)$ of least possible degree having only real coefficients and zeros 3 and $2 + i$.

The complex number $2 - i$ also must be a zero, so the polynomial function has at least three zeros: 3, $2 + i$, and $2 - i$. For the polynomial to be of least possible degree, these must be the only zeros. There must be three variable factors,

$$x - 3, \quad x - (2 + i), \quad \text{and} \quad x - (2 - i). \qquad \text{Factor theorem}$$

One such polynomial function of least possible degree is found as follows.

$$f(x) = (x - 3)[x - (2 + i)][x - (2 - i)]$$

$$f(x) = (x - 3)(x - 2 - i)(x - 2 + i) \qquad \text{Distribute negative signs.}$$

$$f(x) = (x - 3)(x^2 - 4x + 5) \qquad \text{Multiply and combine like terms; } i^2 = -1$$

$$f(x) = x^3 - 7x^2 + 17x - 15 \qquad \text{Multiply again.}$$

Any nonzero constant multiple of $x^3 - 7x^2 + 17x - 15$ also satisfies the given conditions on zeros. The information on zeros given in the problem is not sufficient to give a specific value for the leading coefficient.

NOW TRY

NOW TRY ANSWER

5. $f(x) = x^3 - 10x^2 + 37x - 52$ (There are others.)

The conjugate zeros theorem can help predict the number of real zeros of polynomial functions with real coefficients.

- A polynomial function with real coefficients of *odd* degree n, where $n \geq 1$, must have at least one real zero (because zeros of the form $a + bi$, where $b \neq 0$, occur in conjugate pairs).
- A polynomial function with real coefficients of *even* degree n may have no real zeros.

NOW TRY EXERCISE 6
Find all zeros of

$$f(x) = x^4 - x^3 + 6x^2 + 14x - 20,$$

given that $1 + 3i$ is a zero.

EXAMPLE 6 Finding All Zeros of a Polynomial Function Given One Zero

Find all zeros of $f(x) = x^4 - 7x^3 + 18x^2 - 22x + 12$, given that $1 - i$ is a zero.

Because the polynomial function has only real coefficients and $1 - i$ is a zero, by the conjugate zeros theorem $1 + i$ is also a zero. To find the remaining zeros, we first divide the original polynomial by $x - (1 - i)$.

$(1 - i)(-6 - i) = -7 + 5i$

$$\begin{array}{r|rrrrr} 1-i & 1 & -7 & 18 & -22 & 12 \\ & & 1-i & -7+5i & 16-6i & -12 \\ \hline & 1 & -6-i & 11+5i & -6-6i & 0 \end{array}$$

By the factor theorem, because $x = 1 - i$ is a zero of $f(x)$, $x - (1 - i)$ is a factor. Using the bottom row of the synthetic division to obtain coefficients of the other factor, we can write $f(x)$ as

$$f(x) = [x - (1 - i)][x^3 + (-6 - i)x^2 + (11 + 5i)x + (-6 - 6i)].$$

We know that $x = 1 + i$ is also a zero of $f(x)$, so

$$f(x) = [x - (1 - i)][x - (1 + i)]q(x) \quad \text{for some polynomial } q(x).$$

Thus,

$$x^3 + (-6 - i)x^2 + (11 + 5i)x + (-6 - 6i) = [x - (1 + i)]q(x).$$

We use synthetic division to find $q(x)$.

$$\begin{array}{r|rrrr} 1+i & 1 & -6-i & 11+5i & -6-6i \\ & & 1+i & -5-5i & 6+6i \\ \hline & 1 & -5 & 6 & 0 \end{array}$$

Because $q(x) = x^2 - 5x + 6$, $f(x)$ can be written as follows.

$$f(x) = [x - (1 - i)][x - (1 + i)](x^2 - 5x + 6)$$

$$f(x) = [x - (1 - i)][x - (1 + i)](x - 2)(x - 3) \quad \text{Factor.}$$

The remaining zeros are 2 and 3. Thus, the four zeros of $f(x)$ are

$$1 - i, \quad 1 + i, \quad 2, \quad \text{and} \quad 3.$$

NOW TRY 

NOW TRY ANSWER
6. $1 + 3i, 1 - 3i, 1, -2$

11.2 Exercises

MyLab Math

Video solutions for select problems available in MyLab Math

Concept Check *Determine whether each statement is* true *or* false.

1. Given that $x - 1$ is a factor of $f(x) = x^6 - x^4 + 2x^2 - 2$, we are assured that $f(1) = 0$.
2. Given that $f(1) = 0$ for $f(x) = x^6 - x^4 + 2x^2 - 2$, we are assured that $x - 1$ is a factor of $f(x)$.

3. For the function $f(x) = (x + 2)^4(x - 3)$, 2 is a zero of multiplicity 4.

4. Given that $2 + 3i$ is a zero of $f(x) = x^2 - 4x + 13$, we are assured that $-2 + 3i$ is also a zero.

Determine whether the second polynomial is a factor of the first. ***See Example 1.***

5. $4x^2 + 2x + 54$; $x - 4$

6. $5x^2 - 14x + 10$; $x + 2$

7. $x^3 + 2x^2 - 3$; $x - 1$

8. $2x^3 + x + 3$; $x + 1$

9. $2x^4 + 5x^3 - 2x^2 + 5x + 6$; $x + 3$

10. $5x^4 + 16x^3 - 15x^2 + 8x + 16$; $x + 4$

Factor $f(x)$ into linear factors, given that k is a zero of f. ***See Example 2.***

11. $f(x) = 2x^3 - 3x^2 - 17x + 30$; $k = 2$

12. $f(x) = 2x^3 - 3x^2 - 5x + 6$; $k = 1$

13. $f(x) = 6x^3 + 13x^2 - 14x + 3$; $k = -3$

14. $f(x) = 6x^3 + 17x^2 - 63x + 10$; $k = -5$

For each polynomial function, one zero is given. Find all others. ***See Examples 2 and 6.***

15. $f(x) = x^3 - x^2 - 4x - 6$; 3

16. $f(x) = x^3 + 2x^2 + 32x - 80$; 2

17. $f(x) = x^3 - 7x^2 + 17x - 15$; $2 - i$

18. $f(x) = x^3 - 10x^2 + 34x - 40$; $3 + i$

19. $f(x) = x^4 - 14x^3 + 51x^2 - 14x + 50$; i

20. $f(x) = x^4 + 10x^3 + 27x^2 + 10x + 26$; i

For each polynomial function ***(a)*** *list all possible rational zeros,* ***(b)*** *find all rational zeros, and* ***(c)*** *factor $f(x)$ into linear factors.* ***See Example 3.***

21. $f(x) = x^3 - 2x^2 - 13x - 10$

22. $f(x) = x^3 + 5x^2 + 2x - 8$

23. $f(x) = x^3 + 6x^2 - x - 30$

24. $f(x) = x^3 - x^2 - 10x - 8$

25. $f(x) = 6x^3 + 17x^2 - 31x - 12$

26. $f(x) = 15x^3 + 61x^2 + 2x - 8$

27. $f(x) = 12x^3 + 20x^2 - x - 6$

28. $f(x) = 12x^3 + 40x^2 + 41x + 12$

For each polynomial function, find all zeros and their multiplicities. ***See the discussion preceding Example 4.***

29. $f(x) = (x + 4)^2(x^2 - 7)(x + 1)^4$

30. $f(x) = (x + 1)^2(x - 1)^3(x^2 - 10)$

31. $f(x) = 3x^3(x - 2)(x + 3)(x^2 - 1)$

32. $f(x) = 5x^2(x + 6)(x - 5)(x^2 - 4)$

33. $f(x) = (9x + 7)^2(x^2 + 16)^2$

34. $f(x) = (7x - 2)^3(x^2 + 9)^2$

Find a polynomial function $f(x)$ of least possible degree with only real coefficients and having the given zeros. ***See Examples 4 and 5.***

35. $3 + i$ and $3 - i$

36. $7 - 2i$ and $7 + 2i$

37. $1 + \sqrt{2}$, $1 - \sqrt{2}$, and 3

38. $1 - \sqrt{3}$, $1 + \sqrt{3}$, and 1

39. $-2 + i$, 3, and -3

40. $3 + 2i$, -1, and 2

41. 2 and $3i$

42. -1 and $5i$

43. $1 + 2i$, 2 (multiplicity 2)

44. $2 + i$, -3 (multiplicity 2)

Find a polynomial function $f(x)$ of degree 3 with only real coefficients that satisfies the given conditions. ***See Examples 4 and 5.***

45. Zeros of -3, 1, and 4; $f(2) = 30$

46. Zeros of -2, -1, and 4; $f(2) = 48$

47. Zeros of -2, 1, and 0; $f(-1) = -1$

48. Zeros of 2, -3, and 0; $f(-1) = -3$

49. Zeros of 5, i, and $-i$; $f(-1) = 48$

50. Zeros of -2, i, and $-i$; $f(-3) = 30$

Concept Check *Work each problem.*

51. Show that -2 is a zero of multiplicity 2 of

$$f(x) = x^4 + 2x^3 - 7x^2 - 20x - 12,$$

and find all other complex zeros. Then write $f(x)$ in factored form.

52. Show that -1 is a zero of multiplicity 3 of

$$f(x) = x^5 - 4x^3 - 2x^2 + 3x + 2,$$

and find all other complex zeros. Then write $f(x)$ in factored form.

Descartes' rule of signs can help determine the number of positive and the number of negative real zeros of a polynomial function.

Descartes' Rule of Signs

Let $f(x)$ define a polynomial function with real coefficients and a nonzero constant term, with terms in descending powers of x.

(a) The number of positive real zeros of f either equals the number of variations in sign occurring in the coefficients of $f(x)$, or is less than the number of variations by a positive even integer.

(b) The number of negative real zeros of f either equals the number of variations in sign occurring in the coefficients of $f(-x)$, or is less than the number of variations by a positive even integer.

In the theorem, a *variation in sign* is a change from positive to negative or negative to positive in successive terms of the polynomial. Missing terms (those with 0 coefficients) are counted as no change in sign and can be ignored. For example, in the polynomial function $f(x) = x^4 - 6x^3 + 8x^2 + 2x - 1$, $f(x)$ has three variations in sign.

$$+x^4 - 6x^3 + 8x^2 + 2x - 1$$

1 2 3

Thus, by Descartes' rule of signs, f has either 3 or $3 - 2 = 1$ positive real zeros. Because

$$f(-x) = (-x)^4 - 6(-x)^3 + 8(-x)^2 + 2(-x) - 1$$

$$f(-x) = x^4 + 6x^3 + 8x^2 - 2x - 1$$

1

has only one variation in sign, f has only one negative real zero.

Use Descartes' rule of signs to determine the possible number of positive real zeros and the possible number of negative real zeros for each function.

53. $f(x) = 2x^3 - 4x^2 + 2x + 7$

54. $f(x) = x^3 + 2x^2 + x - 10$

55. $f(x) = 5x^4 + 3x^2 + 2x - 9$

56. $f(x) = 3x^4 + 2x^3 - 8x^2 - 10x - 1$

57. $f(x) = x^5 + 3x^4 - x^3 + 2x + 3$

58. $f(x) = 2x^5 - x^4 + x^3 - x^2 + x + 5$

Use a graphing calculator to find (or approximate) the real zeros of each function $f(x)$. Express decimal approximations to the nearest hundredth.

59. $f(x) = 0.86x^3 - 5.24x^2 + 3.55x + 7.84$

60. $f(x) = -2.47x^3 - 6.58x^2 - 3.33x + 0.14$

61. $f(x) = 2.45x^4 - 3.22x^3 + 0.47x^2 - 6.54x + 3$

62. $f(x) = 4x^4 + 8x^3 - 4x^2 + 4x + 1$

63. $f(x) = -\sqrt{7}x^3 + \sqrt{5}x + \sqrt{17}$

64. $f(x) = \sqrt{10}x^3 - \sqrt{11}x - \sqrt{8}$

11.3 Graphs and Applications of Polynomial Functions

OBJECTIVES

1. Graph functions of the form $f(x) = a(x - h)^n + k$.
2. Graph general polynomial functions.
3. Use the intermediate value and boundedness theorems.
4. Approximate real zeros of polynomial functions.
5. Use a polynomial function to model data.

OBJECTIVE 1 Graph functions of the form $f(x) = a(x - h)^n + k$.

EXAMPLE 1 Graphing Functions of the Form $f(x) = ax^n$

Graph each function.

(a) $f(x) = x^3$

Choose several values for x and find the corresponding values of $f(x)$, or y. Plot the resulting ordered pairs and connect the points with a smooth curve. The graph of $f(x) = x^3$ is the darker graph shown in **FIGURE 2**.

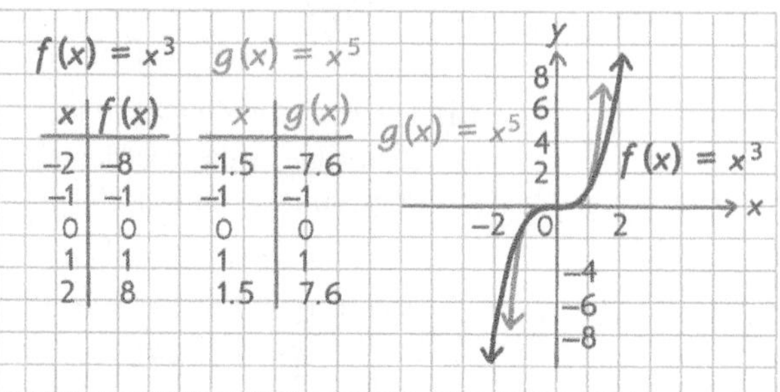

FIGURE 2

VOCABULARY

- ☐ turning points
- ☐ end behavior
- ☐ dominating term

(b) $g(x) = x^5$

Work as in part (a) to obtain the lighter graph shown in **FIGURE 2**. The graphs of $f(x) = x^3$ and $g(x) = x^5$ are both symmetric with respect to the origin.

NOW TRY EXERCISE 1

Graph $f(x) = -x^3$.

(c) $f(x) = x^4$, $g(x) = x^6$

Some typical ordered pairs for the graphs of $f(x) = x^4$ and $g(x) = x^6$ are given in the tables in **FIGURE 3**. These graphs are symmetric with respect to the y-axis.

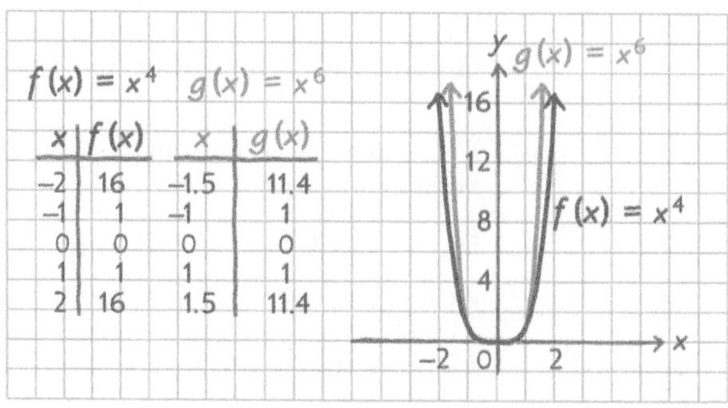

FIGURE 3

NOW TRY 

NOW TRY ANSWER

1. $f(x) = -x^3$; (−2, 8), (0, 0), (2, −8)

NOTE The ZOOM feature of a graphing calculator is useful with graphs like those in **Example 1** to compare the graphs of $y = x^3$ and $y = x^5$ and those of $y = x^4$ and $y = x^6$ for values of x in the interval $[-1.5, 1.5]$. See **FIGURE 4** on the next page.

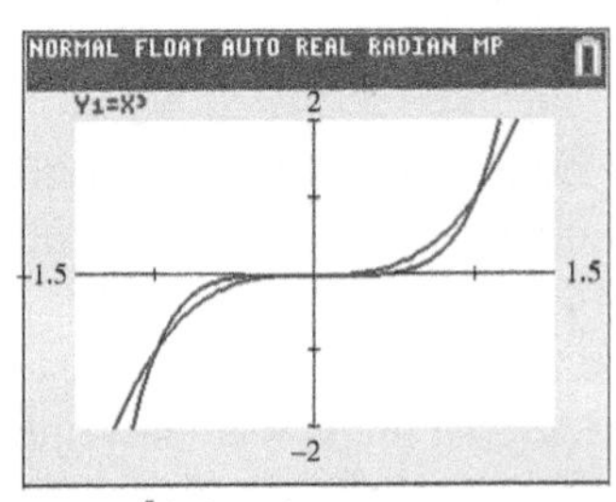

$Y_2 = X^5$ is in red.

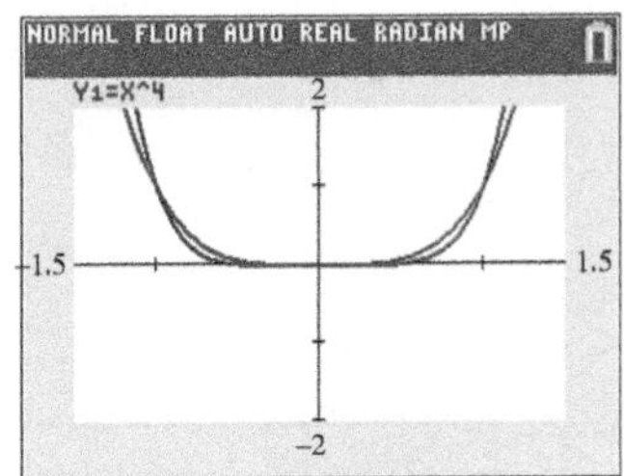

$Y_2 = X^6$ is in red.

FIGURE 4

NOW TRY EXERCISE 2

Graph $f(x) = \frac{1}{2}(x + 1)^4 - 2$.

Characteristics of the Graph of $f(x) = ax^n$

The value of a in $f(x) = ax^n$ determines the width of the graph.

- When $|a| > 1$, the graph is stretched vertically, making it narrower.
- When $0 < |a| < 1$, the graph is shrunk or compressed vertically, making it broader.

Consider the graph of $\boldsymbol{f(x) = ax^n}$. Other graphs can be obtained from it.

- The graph of $\boldsymbol{f(x) = -ax^n}$ is reflected across the x-axis.
- The graph of $\boldsymbol{f(x) = ax^n + k}$ is translated (shifted) up k units if $k > 0$ and down $|k|$ units if $k < 0$.
- The graph of $\boldsymbol{f(x) = a(x - h)^n}$ is translated to the right h units if $h > 0$ and to the left $|h|$ units if $h < 0$.
- The graph of $\boldsymbol{f(x) = a(x - h)^n + k}$ shows a combination of translations (as with quadratic functions).

EXAMPLE 2 Examining Vertical and Horizontal Translations

Graph each function.

(a) $f(x) = x^5 - 2$

The graph will be the same as that of $f(x) = x^5$, but translated down 2 units. See **FIGURE 5**.

(b) $f(x) = (x + 1)^6$

This function f has a graph the same as that of $f(x) = x^6$, but because

$$x + 1 = x - (-1),$$

it is translated to the left 1 unit, as shown in **FIGURE 6**.

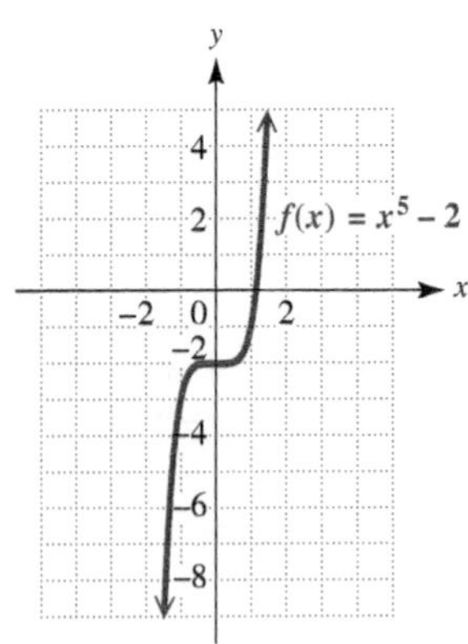

FIGURE 5

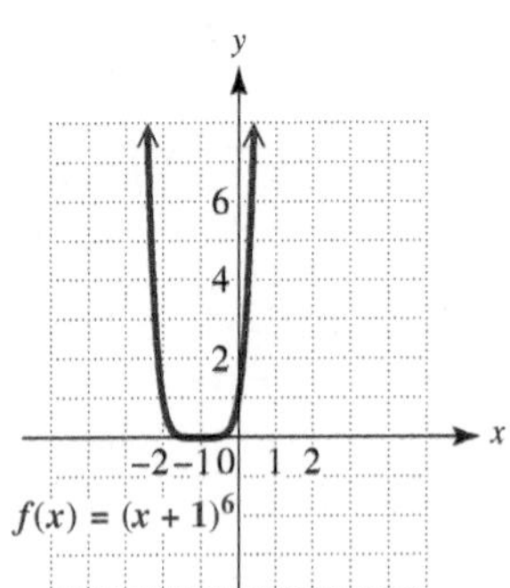

FIGURE 6

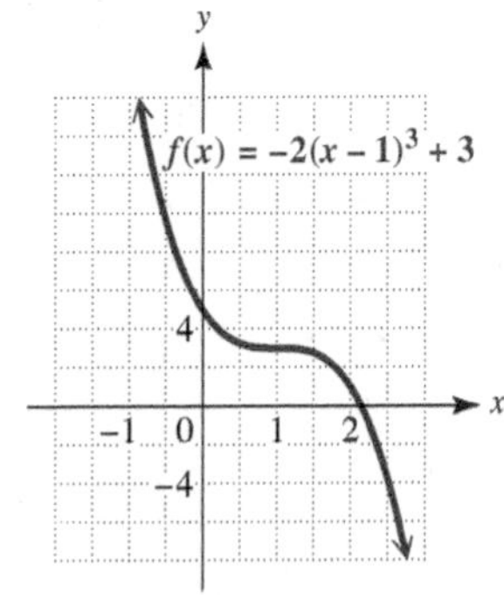

FIGURE 7

(c) $f(x) = -2(x - 1)^3 + 3$

The negative sign in -2 causes the graph to be reflected across the x-axis as compared with the graph of $f(x) = x^3$. Because $|-2| > 1$, the graph is stretched vertically as compared with the graph of $f(x) = x^3$. As shown in **FIGURE 7**, the graph is also translated to the right 1 unit and up 3 units.

NOW TRY

NOW TRY ANSWER

2.

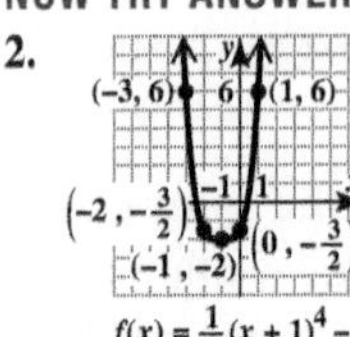

OBJECTIVE 2 Graph general polynomial functions.

The domain of every polynomial function (unless stated otherwise) is the set of all real numbers. The range of a polynomial function of odd degree is also the set of all real numbers.

FIGURE 8 shows three typical graphs of polynomial functions of odd degree. These graphs suggest that for every polynomial function f of odd degree there is at least one real value of x that satisfies $f(x) = 0$. The real zeros are located at the x-intercepts of the graph and can be determined by inspecting the factored form of each polynomial.

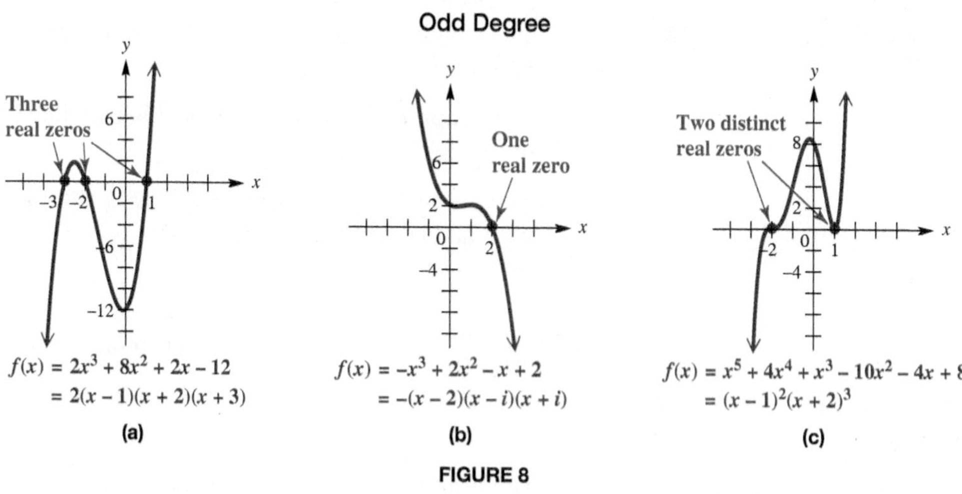

FIGURE 8

A polynomial function of even degree has a range of the form $(-\infty, k]$ or $[k, \infty)$, for some real number k. **FIGURE 9** shows two typical graphs.

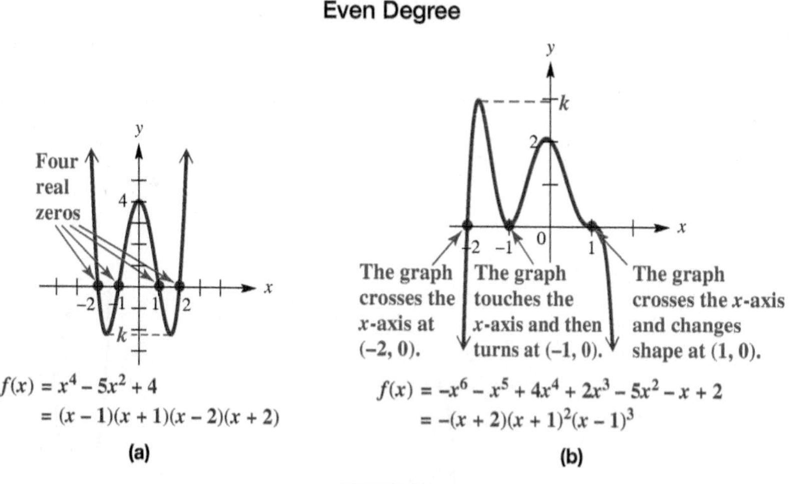

FIGURE 9

A zero k of a polynomial function has as multiplicity the exponent of the factor $x - k$. Determining the multiplicity of a zero helps sketch the graph near that zero.

- If the zero has multiplicity one, the graph crosses the x-axis at the corresponding x-intercept as seen in **FIGURE 10(a)** on the next page.
- If the zero has even multiplicity, the graph is tangent to the x-axis at the corresponding x-intercept (that is, it touches but does not cross the x-axis there). See **FIGURE 10(b)**.
- If the zero has odd multiplicity greater than one, the graph crosses the x-axis ***and*** is tangent to the x-axis at the corresponding x-intercept. This causes a change in concavity, or shape, at the x-intercept and the graph wiggles there. See **FIGURE 10(c)**.

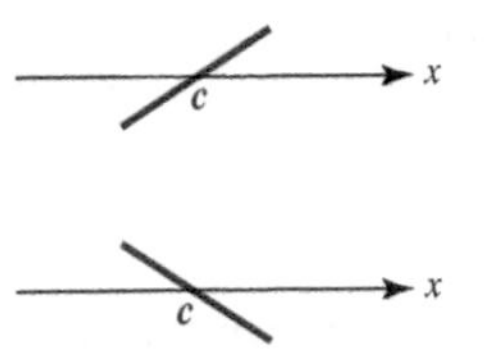

The graph crosses the x-axis at $(c, 0)$ if c is a zero of multiplicity one.

(a)

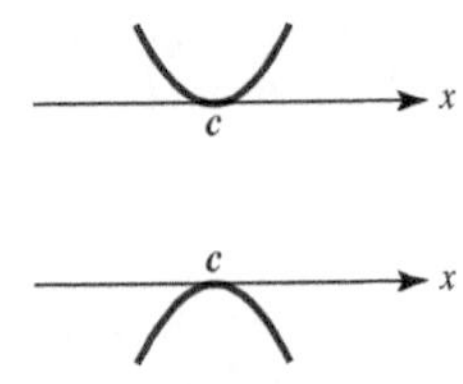

The graph is tangent to the x-axis at $(c, 0)$ if c is a zero of even multiplicity. The graph bounces, or turns, at c.

(b)

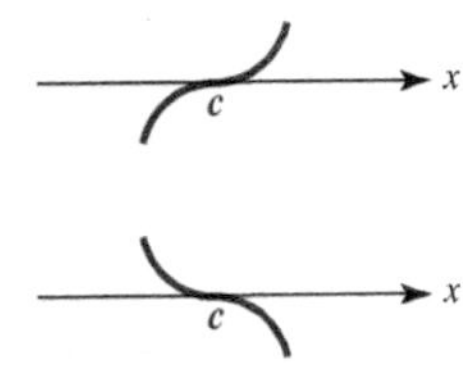

The graph crosses ***and*** is tangent to the x-axis at $(c, 0)$ if c is a zero of odd multiplicity greater than one. The graph wiggles at c.

(c)

FIGURE 10

The graphs in **FIGURES 8** and **9** on the previous page show that polynomial functions often have **turning points** where the function changes from increasing to decreasing or from decreasing to increasing.

Turning Points

A polynomial function of degree n has at most $n - 1$ turning points, with at least one turning point between each pair of successive zeros.

The **end behavior** of a polynomial graph is determined by the **dominating term**—that is, the term of greatest degree. The graph of a polynomial function of the form

$$f(x) = a_nx^n + a_{n-1}x^{n-1} + \cdots + a_0$$

has the same end behavior as that of $f(x) = a_nx^n$. For example,

$$f(x) = 2x^3 - 8x^2 + 9$$

has the same end behavior as $f(x) = 2x^3$. It is large and positive for large positive values of x and large and negative for negative values of x with large absolute value. The arrows at the ends of the graph look like those of the graph in **FIGURE 8(a)**. It rises to the right and falls to the left.

FIGURE 8(a) shows that as x increases without bound, y does also. For the same graph, as x decreases without bound, y does also.

$$\text{As} \quad x \rightarrow \infty, \quad y \rightarrow \infty \quad \text{and} \quad \text{as} \quad x \rightarrow -\infty, \quad y \rightarrow -\infty.$$

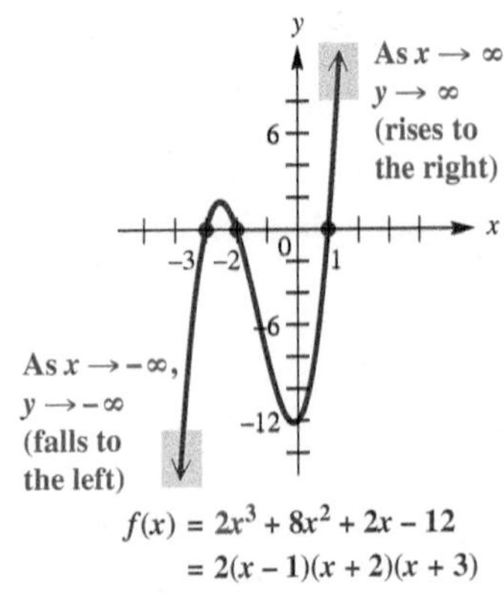

FIGURE 8(a) (repeated)

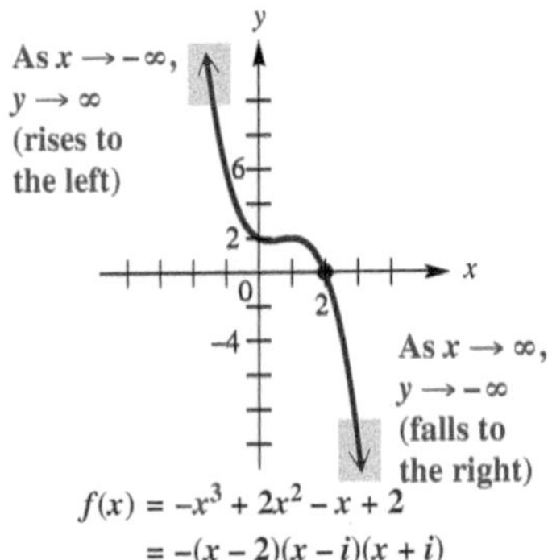

FIGURE 8(b) (repeated)

The graph in **FIGURE 8(b)** has the same end behavior as $f(x) = -x^3$.

$$\text{As} \quad x \rightarrow \infty, \quad y \rightarrow -\infty \quad \text{and} \quad \text{as} \quad x \rightarrow -\infty, \quad y \rightarrow \infty.$$

End Behavior of Graphs of Polynomial Functions

Suppose that ax^n is the dominating term of a polynomial function f of ***odd degree.***

1. If $a > 0$, then as $x \to \infty$, $f(x) \to \infty$, and as $x \to -\infty$, $f(x) \to -\infty$. Therefore, the end behavior of the graph is of the type shown in **FIGURE 11(a)**. We symbolize it as .
2. If $a < 0$, then as $x \to \infty$, $f(x) \to -\infty$, and as $x \to -\infty$, $f(x) \to \infty$. Therefore, the end behavior of the graph is of the type shown in **FIGURE 11(b)**. We symbolize it as .

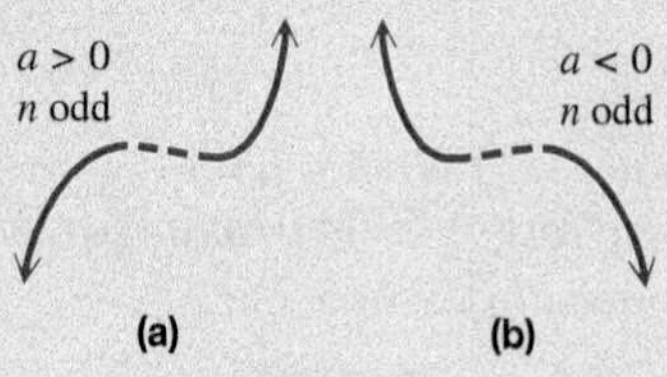

FIGURE 11

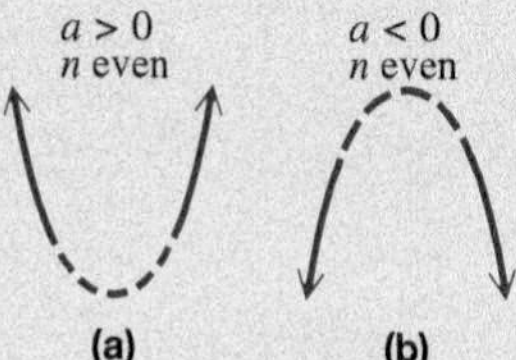

FIGURE 12

Suppose that ax^n is the dominating term of a polynomial function f of ***even degree.***

1. If $a > 0$, then as $|x| \to \infty$, $f(x) \to \infty$. Therefore, the end behavior of the graph is of the type shown in **FIGURE 12(a)**. We symbolize it as .
2. If $a < 0$, then as $|x| \to \infty$, $f(x) \to -\infty$. Therefore, the end behavior of the graph is of the type shown in **FIGURE 12(b)**. We symbolize it as .

Graphing a Polynomial Function

Let $f(x) = a_n x^n + a_{n-1}x^{n-1} + \cdots + a_1 x + a_0$, where $a_n \neq 0$, be a polynomial function of degree n. Sketch its graph as follows.

Step 1 Find the real zeros of f. Plot the corresponding x-intercepts.

Step 2 Find $f(0) = a_0$. Plot the corresponding y-intercept.

Step 3 Use end behavior, whether the graph crosses, bounces on, or wiggles through the x-axis at the x-intercepts, and selected points as necessary to complete the graph.

EXAMPLE 3 Graphing a Polynomial Function

Graph $f(x) = 2x^3 + 5x^2 - x - 6$.

Step 1 The possible rational zeros are ± 1, ± 2, ± 3, ± 6, $\pm \frac{1}{2}$, and $\pm \frac{3}{2}$. Use synthetic division to show that 1 is a zero.

$$\begin{array}{r|rrrr} 1 & 2 & 5 & -1 & -6 \\ & & 2 & 7 & 6 \\ \hline & 2 & 7 & 6 & 0 \end{array} \leftarrow f(1) = 0$$

We use the results of synthetic division to factor as follows.

$$f(x) = (x - 1)(2x^2 + 7x + 6)$$

$$f(x) = (x - 1)(2x + 3)(x + 2) \qquad \text{Factor } 2x^2 + 7x + 6.$$

Set each linear factor equal to 0, and solve for x to find zeros. The three zeros of f are 1, $-\frac{3}{2}$, and -2, so plot the corresponding x-intercepts. See **FIGURE 13** on the next page.

Graph

$f(x) = x^3 + 3x^2 - 6x - 8.$

Step 2 For $f(x) = 2x^3 + 5x^2 - x - 6$, we find $f(0) = -6$ and plot the y-intercept $(0, -6)$. See **FIGURE 13.**

Step 3 The dominating term of $f(x)$ is $2x^3$, so the graph will have end behavior similar to that of $f(x) = x^3$.

It will rise to the right and fall to the left as .

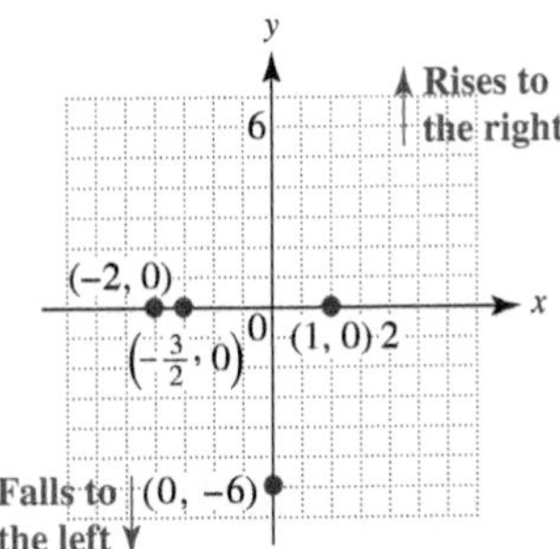

FIGURE 13

See **FIGURE 13.** Each zero of $f(x)$ occurs with multiplicity one, meaning that the graph of $f(x)$ will cross the x-axis at each of its zeros. Because the graph of a polynomial function has no breaks, gaps, or sudden jumps, we now have sufficient information to sketch the graph of $f(x)$.

Begin sketching at either end of the graph with the appropriate end behavior, and draw a smooth curve that crosses the x-axis at each zero, has a turning point between successive zeros, and passes through the y-intercept as shown in **FIGURE 14.**

Additional points may be used to verify whether the graph is above or below the x-axis between the zeros and to add detail to the sketch of the graph. The zeros divide the x-axis into four intervals:

$$(-\infty, -2), \quad \left(-2, -\frac{3}{2}\right), \quad \left(-\frac{3}{2}, 1\right), \quad \text{and} \quad (1, \infty).$$

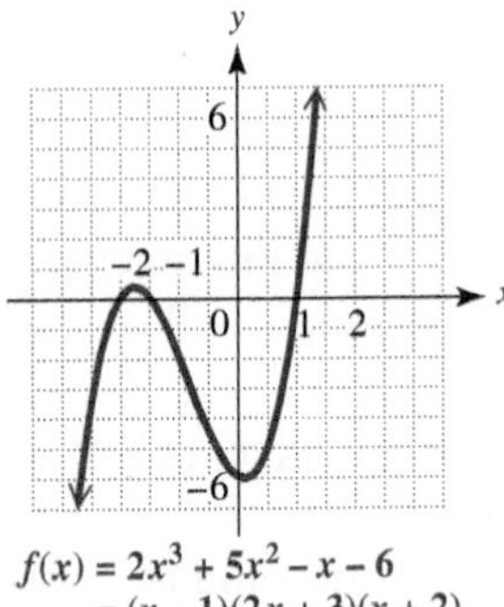

$f(x) = 2x^3 + 5x^2 - x - 6$
$= (x - 1)(2x + 3)(x + 2)$

FIGURE 14

Select an x-value as a test point in each interval, and substitute it into the equation for $f(x)$ to determine additional points on the graph. We test -3 from the interval $(-\infty, -2)$.

$$f(x) = (x - 1)(2x + 3)(x + 2) \qquad \text{Factored form of } f(x)$$

$$f(-3) = (-3 - 1)(2(-3) + 3)(-3 + 2) \qquad \text{Let } x = -3 \text{ from the interval } (-\infty, -2).$$

$$f(-3) = -4(-3)(-1) \qquad \text{Simplify in parentheses.}$$

$$f(-3) = -12 \qquad \text{Multiply.}$$

$(-3, -12)$ lies on the graph. $\leftarrow f(x)$ is negative. Graph is below x-axis.

A typical selection of test points and the results of the tests are shown in the table.

Interval	Test Point x	Value of $f(x)$	Sign of $f(x)$	Graph Above or Below x-Axis
$(-\infty, -2)$	-3	-12	Negative	Below
$\left(-2, -\frac{3}{2}\right)$	$-\frac{7}{4}$	$\frac{11}{32}$	Positive	Above
$\left(-\frac{3}{2}, 1\right)$	0	-6	Negative	Below
$(1, \infty)$	2	28	Positive	Above

The graph in **FIGURE 14** shows that this function has two turning points, the maximum number for a third-degree polynomial function. The sketch could be improved by plotting the points found in each interval in the table.

NOW TRY

NOW TRY ANSWER

3.

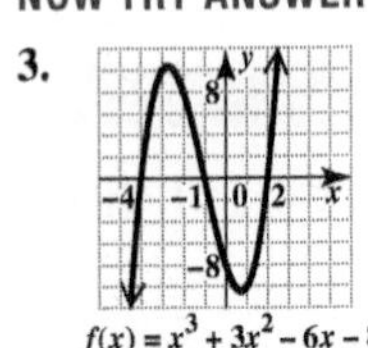

NOW TRY EXERCISE 4

Graph

$f(x) = -(x-2)(x+3)(x+1)^2$.

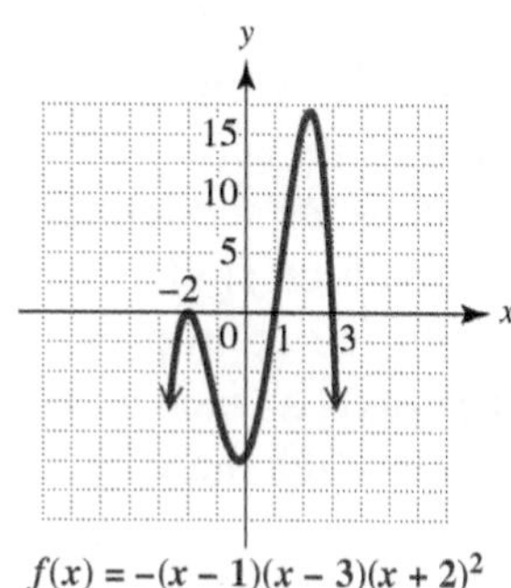

$f(x) = -(x-1)(x-3)(x+2)^2$

FIGURE 15

EXAMPLE 4 Graphing a Polynomial Function

Graph $f(x) = -(x-1)(x-3)(x+2)^2$.

Step 1 Because the polynomial is given in factored form, the zeros can be determined by inspection. They are 1, 3, and -2. Plot the corresponding x-intercepts of the graph of $f(x)$. See **FIGURE 15.**

Step 2

$$f(x) = -(x-1)(x-3)(x+2)^2$$

$$f(0) = -(0-1)(0-3)(0+2)^2 \quad \text{Find } f(0).$$

$$f(0) = -(-1)(-3)(2)^2 \quad \text{Simplify in parentheses.}$$

$$f(0) = -12 \leftarrow \text{The } y\text{-intercept is } (0, -12).$$

Plot the y-intercept $(0, -12)$. See **FIGURE 15.**

Step 3 The dominating term of $f(x)$ can be found by multiplying the factors and identifying the term of greatest degree. Here it is $-(x)(x)(x)^2 = -x^4$, indicating that the end behavior of the graph is ↙ ↘. Because 1 and 3 are zeros of multiplicity one, the graph will cross the x-axis at these zeros. The graph of $f(x)$ will touch the x-axis at -2 and then turn and change direction because it is a zero of even multiplicity.

Begin at either end of the graph with the appropriate end behavior and draw a smooth curve that crosses the x-axis at 1 and 3 and that touches the x-axis at -2, then turns and changes direction. The graph will also pass through the y-intercept $(0, -12)$. See **FIGURE 15.**

Using test points within intervals formed by the x-intercepts is a good way to add detail to the graph and verify the accuracy of the sketch. A typical selection of test points is $(-3, -24)$, $(-1, -8)$, $(2, 16)$, and $(4, -108)$.

NOW TRY 

In summary, there are important relationships among the following ideas.

- The x-intercepts of the graph of $y = f(x)$
- The zeros of the function f
- The solutions of the equation $f(x) = 0$
- The linear factors of $f(x)$

For example, the graph of the function from **Example 3,**

$$f(x) = 2x^3 + 5x^2 - x - 6$$

$$f(x) = (x-1)(2x+3)(x+2), \quad \text{Factored form}$$

has x-intercepts $(1, 0)$, $\left(-\frac{3}{2}, 0\right)$, and $(-2, 0)$, as shown in **FIGURE 14** on the previous page. Because $1, -\frac{3}{2}$, and -2 are the x-values for which the function is 0, they are the zeros of f. Futhermore, $1, -\frac{3}{2}$, and -2 are solutions of the equation

$$2x^3 + 5x^2 - x - 6 = 0.$$

This discussion is summarized as follows.

NOW TRY ANSWER

4.

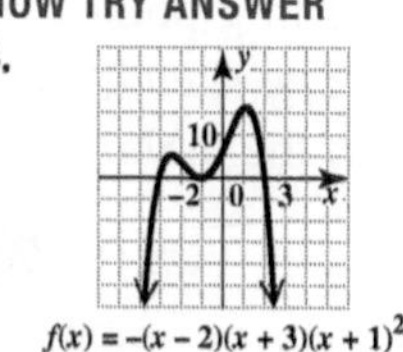

$f(x) = -(x-2)(x+3)(x+1)^2$

Relationships among x-Intercepts, Zeros, Solutions, and Factors

If f is a polynomial function and $(c, 0)$ is an x-intercept of the graph of $y = f(x)$, then

c is a zero of f, c is solution of $f(x) = 0$,

and **$x - c$ is a factor of $f(x)$.**

NOTE It is possible to reverse the process of **Example 4** and write the polynomial function from its graph if the zeros and any other point on the graph are known. Suppose that we are asked to find a polynomial function of least degree having the graph shown in **FIGURE 15.** Because the graph crosses the x-axis at 1 and 3 and bounces at -2, we know that the factored form of the function is as follows.

Multiplicity one — Multiplicity two

$$f(x) = a(x-1)^1(x-3)^1(x+2)^2$$

Now find the value of a by substituting the x- and y-values of any other point on the graph, say $(0, -12)$, into this function and solving for a.

$$f(x) = a(x-1)(x-3)(x+2)^2$$

$$-12 = a(0-1)(0-3)(0+2)^2 \quad \text{Let } x = 0 \text{ and } y = -12.$$

$$-12 = a(12) \quad \text{Simplify.}$$

$$a = -1 \quad \text{Divide by 12. Interchange sides.}$$

Verify in **Example 4** that the polynomial function is

$$f(x) = -(x-1)(x-3)(x+2)^2.$$

Exercises of this type are labeled **Extending Skills** in the exercise set for this section.

OBJECTIVE 3 Use the intermediate value and boundedness theorems.

As **Examples 3** and **4** show, one key to graphing a polynomial function is locating its zeros. In the special case where the zeros are rational numbers, the zeros can be found by the rational zeros theorem. Occasionally, irrational zeros can be found by inspection. For instance, $f(x) = x^3 - 2$ has the irrational zero $\sqrt[3]{2}$.

Two theorems presented in this section apply to the zeros of every polynomial function with real coefficients. The **intermediate value theorem** uses the fact that graphs of polynomial functions are unbroken curves, with no gaps or sudden jumps. The proof requires advanced methods, so it is not given here. **FIGURE 16** illustrates the theorem.

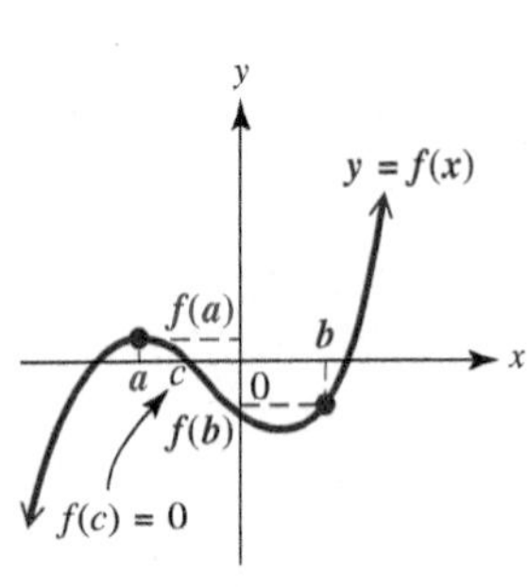

FIGURE 16

Intermediate Value Theorem (as applied to locating zeros)

If $f(x)$ is a polynomial function *with only real coefficients,* and if for real numbers a and b the values $f(a)$ and $f(b)$ are opposite in sign, then there exists at least one real zero between a and b. See **FIGURE 16.**

This theorem helps to identify intervals where zeros of polynomial functions are located. If $f(a)$ and $f(b)$ are opposite in sign, then 0 is between $f(a)$ and $f(b)$, and there must be a number c between a and b where $f(c) = 0$.

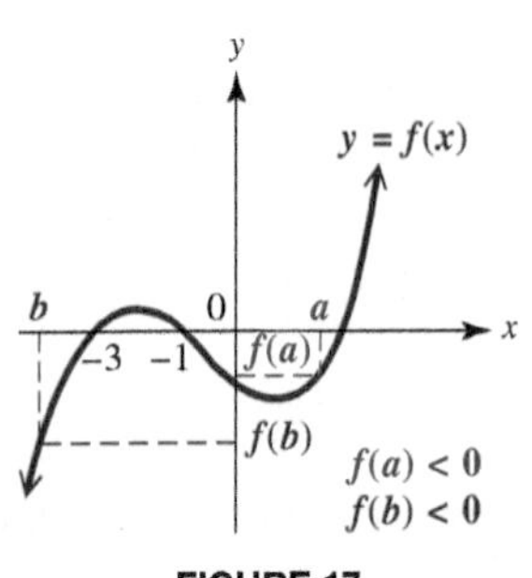

FIGURE 17

CAUTION ***Be careful when interpreting the intermediate value theorem.*** If $f(a)$ and $f(b)$ are *not* opposite in sign, it does not necessarily mean that there is no zero between a and b. For example, in **FIGURE 17,** $f(a)$ and $f(b)$ are both negative, but -3 and -1, which are between a and b, are zeros of $f(x)$.

NOW TRY EXERCISE 5
Show that

$$f(x) = x^3 + x^2 - 6x$$

has a real zero between -4 and -2.

EXAMPLE 5 Using the Intermediate Value Theorem to Locate a Zero

Show that $f(x) = x^3 - 2x^2 - x + 1$ has a real zero between 2 and 3.

Use synthetic division to find $f(2)$ and $f(3)$.

$$\begin{array}{r|rrrr} 2 & 1 & -2 & -1 & 1 \\ & & 2 & 0 & -2 \\ \hline & 1 & 0 & -1 & -1 = f(2) \end{array} \qquad \begin{array}{r|rrrr} 3 & 1 & -2 & -1 & 1 \\ & & 3 & 3 & 6 \\ \hline & 1 & 1 & 2 & 7 = f(3) \end{array}$$

The results show that $f(2) = -1$ and $f(3) = 7$. Because $f(2)$ is negative and $f(3)$ is positive, by the intermediate value theorem there must be a real zero between 2 and 3.

NOW TRY 5

The intermediate value theorem for polynomials helps limit the search for real zeros to smaller and smaller intervals. In **Example 5**, we used the theorem to verify that there is a real zero between 2 and 3. We could use the theorem repeatedly to locate the zero more accurately.

The next theorem, the **boundedness theorem,** shows how the bottom row of a synthetic division can be used to place upper and lower bounds on the possible real zeros of a polynomial function.

Boundedness Theorem

Let $f(x)$ be a polynomial function of degree $n \geq 1$ with *real coefficients* and with a *positive* leading coefficient. Suppose $f(x)$ is divided synthetically by $x - c$.

(a) If $c > 0$ and all numbers in the bottom row of the synthetic division are nonnegative, then $f(x)$ has no real zero greater than c.

(b) If $c < 0$ and the numbers in the bottom row of the synthetic division alternate in sign (with 0 considered positive or negative, as needed), then $f(x)$ has no real zero less than c.

PROOF We outline the proof of part (a). The proof for part (b) is similar.

By the division algorithm, if $f(x)$ is divided by $x - c$, then for some $q(x)$ and r,

$$f(x) = (x - c)q(x) + r,$$

where all coefficients of $q(x)$ are nonnegative, $r \geq 0$, and $c > 0$. If $x > c$, then $x - c > 0$. Because $q(x) > 0$ and $r \geq 0$, $f(x) = (x - c)q(x) + r > 0$. This means that $f(x)$ will never be 0 for $x > c$.

EXAMPLE 6 Using the Boundedness Theorem

Show that the real zeros of $f(x) = 2x^4 - 5x^3 + 3x + 1$ satisfy the following.

(a) No real zero is greater than 3.

Because $f(x)$ has real coefficients and the leading coefficient, 2, is positive, the boundedness theorem can be used. Divide $f(x)$ synthetically by $x - 3$.

$$\begin{array}{r|rrrrr} 3 & 2 & -5 & 0 & 3 & 1 \\ & & 6 & 3 & 9 & 36 \\ \hline & 2 & 1 & 3 & 12 & 37 \end{array}$$

Divide $f(x)$ by $x - 3$.

All are nonnegative.

Here $3 > 0$ and all numbers in the bottom row of the synthetic division are nonnegative, so $f(x)$ has no real zero greater than 3.

NOW TRY ANSWER

5. $f(-4) = -24 < 0$ and $f(-2) = 8 > 0$

NOW TRY EXERCISE 6

Show that

$f(x) = 2x^4 + 21x^3 - 36x^2 + x + 12$

has no real zero greater than 2.

(b) No real zero is less than -1.

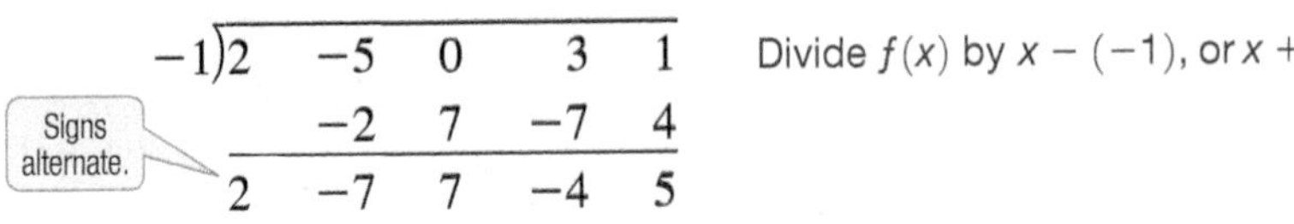

$$\begin{array}{r|rrrrr} -1 & 2 & -5 & 0 & 3 & 1 \\ & & -2 & 7 & -7 & 4 \\ \hline & 2 & -7 & 7 & -4 & 5 \end{array}$$

Divide $f(x)$ by $x - (-1)$, or $x + 1$.

Here $-1 < 0$ and the numbers in the bottom row alternate in sign, so $f(x)$ has no real zero less than -1. **NOW TRY**

OBJECTIVE 4 Approximate real zeros of polynomial functions.

NOW TRY EXERCISE 7

Use a graphing calculator to approximate the real zeros of

$f(x) = x^3 - 3x^2 - 3x + 2,$

to the nearest hundredth.

EXAMPLE 7 Approximating Real Zeros of a Polynomial Function

Use a graphing calculator to approximate the real zeros of the function

$$f(x) = x^4 - 6x^3 + 8x^2 + 2x - 1.$$

The dominating term is x^4, so the graph will have end behavior similar to the graph of $f(x) = x^4$, which is positive for all values of x with large absolute values. That is, the end behavior is up at the left and the right, ↖↗. There are at most four real zeros because the polynomial is fourth-degree.

Here $f(0) = -1$, so the y-intercept is $(0, -1)$. The end behavior is positive on the left and the right, and the y-value of the y-intercept is negative, so by the intermediate value theorem, f has at least one zero on either side of $x = 0$. The calculator graph in **FIGURE 18(a)** supports these facts. We can see that there are four real zeros, and the table in **FIGURE 18(b)** indicates that they are between

$$-1 \text{ and } 0, \quad 0 \text{ and } 1, \quad 2 \text{ and } 3, \quad \text{and} \quad 3 \text{ and } 4,$$

because there is a change of sign in $f(x) = Y_1$ in each case.

Using the capability of the calculator, we can find the zeros to a great degree of accuracy. **FIGURE 18(c)** shows that the negative zero is approximately -0.414214. To the nearest hundredth, the four real zeros are

$$-0.41, \quad 0.27, \quad 2.41, \quad \text{and} \quad 3.73.$$

NOW TRY

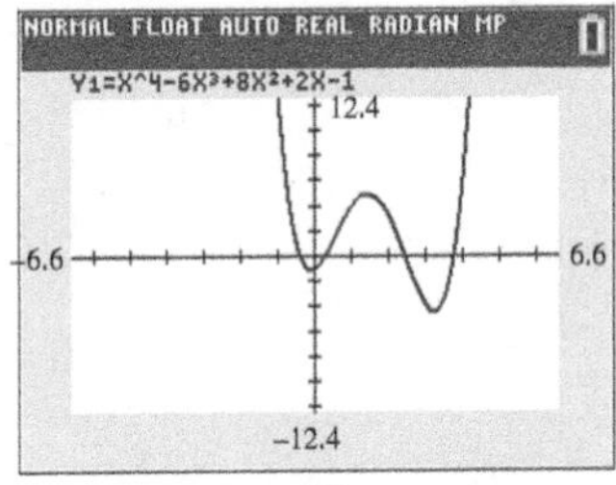

(a)

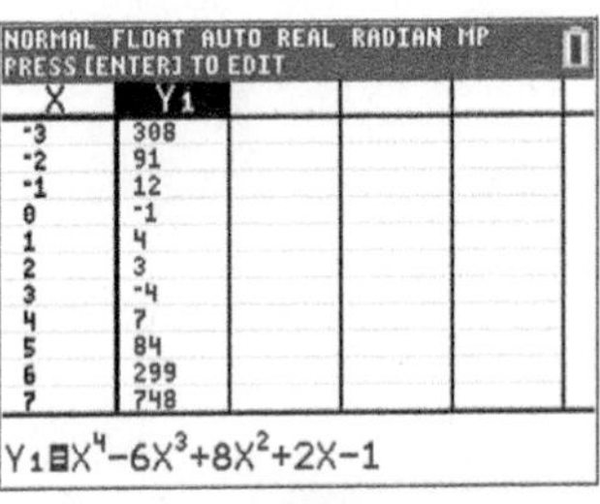
NORMAL FLOAT AUTO REAL RADIAN MP
PRESS [ENTER] TO EDIT

X	Y1
-3	308
-2	91
-1	12
0	-1
1	4
2	3
3	-4
4	7
5	84
6	299
7	748

$Y_1 = X^4 - 6X^3 + 8X^2 + 2X - 1$

(b)

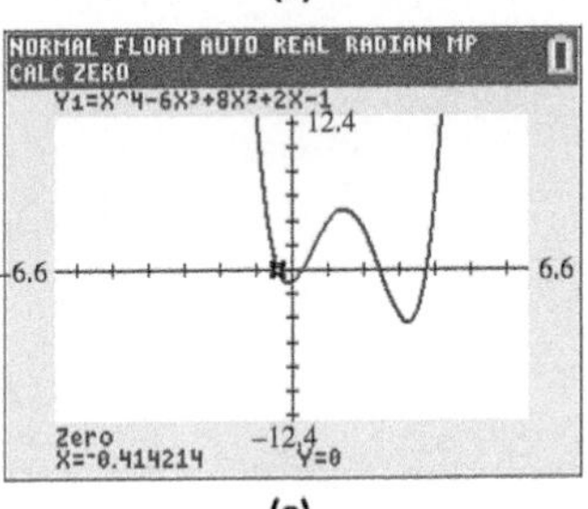

(c)

FIGURE 18

OBJECTIVE 5 Use a polynomial function to model data.

Year	Dollars (in billions)
2010	802.6
2011	865.4
2012	958.6
2013	1069.4
2014	1153.7
2015	1230.7

Data from 2016 *Student Loan Update.*

EXAMPLE 8 Examining a Polynomial Model

The table shows U.S. student loan balances, in billions of dollars, for selected years.

(a) Using $x = 0$ to represent 2010, $x = 1$ to represent 2011, and so on, use the regression feature of a calculator to determine the quadratic function that best fits the data. Give coefficients to four significant digits. Plot the data and the graph.

The best-fitting quadratic function for the data is

$$y = 0.6321x^2 + 85.87x + 792.9.$$

NOW TRY ANSWERS

6. Because $2 > 0$ and all numbers in the bottom row of the synthetic division are nonnegative, $f(x)$ has no real zero greater than 2.
7. $-1.15, 0.48, 3.67$

The student loan balance in 2009 was $713.7 billion. Repeat **Example 8** after adding these data to the table, using $x = 0$ to represent 2009 rather than 2010.

The regression coordinates screen and the graph are shown in **FIGURE 19.**

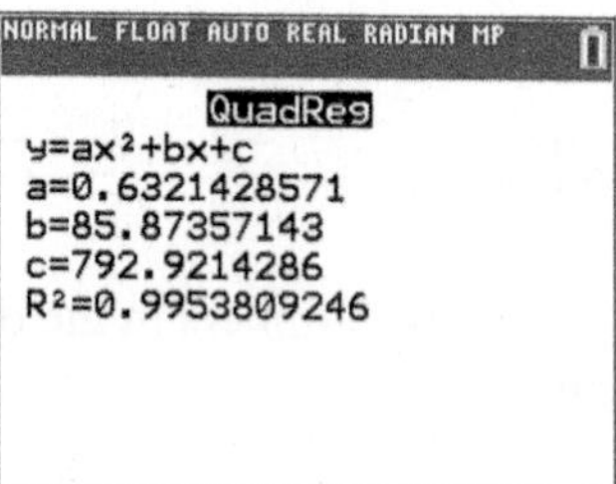

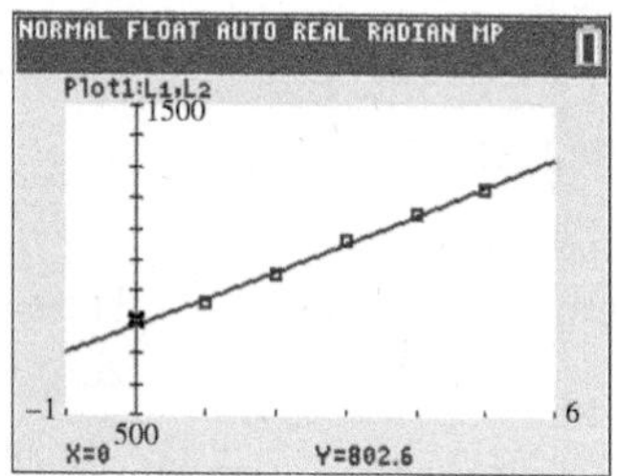

FIGURE 19

(b) Repeat part (a) for a cubic function (degree 3).

The best-fitting cubic function is shown in **FIGURE 20** and is

$$y = -2.970x^3 + 22.91x^2 + 45.18x + 801.8.$$

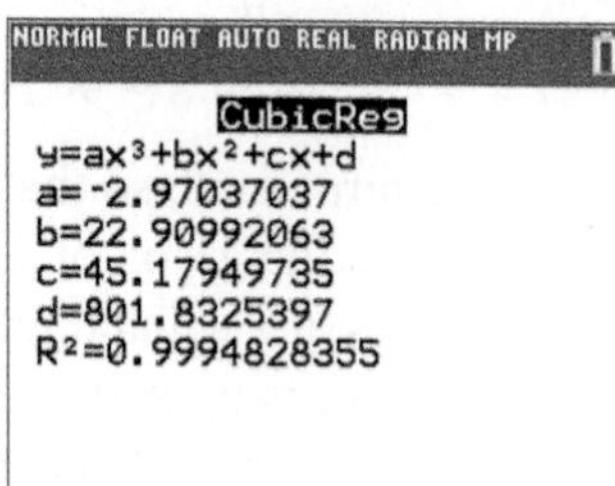

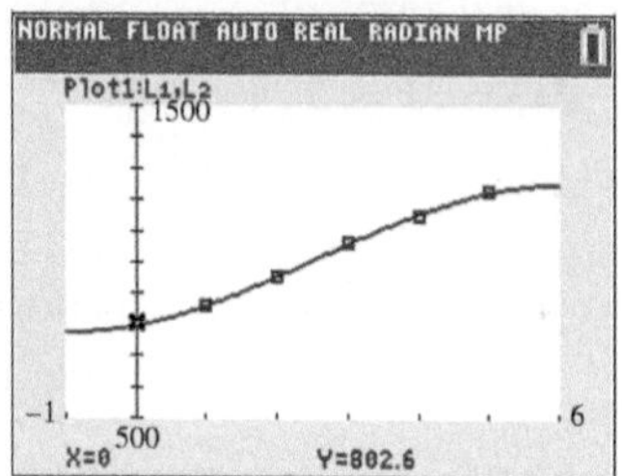

FIGURE 20

(c) Repeat part (a) for a quartic function (degree 4).

The best-fitting quartic function is shown in **FIGURE 21** and is

$$y = 0.6667x^4 - 9.637x^3 + 43.39x^2 + 26.13x + 803.0.$$

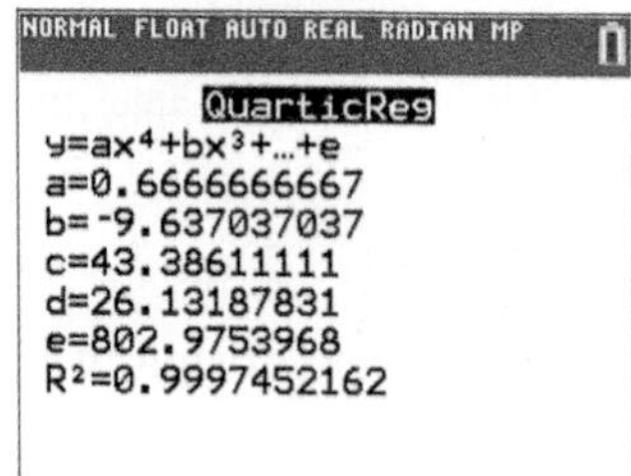

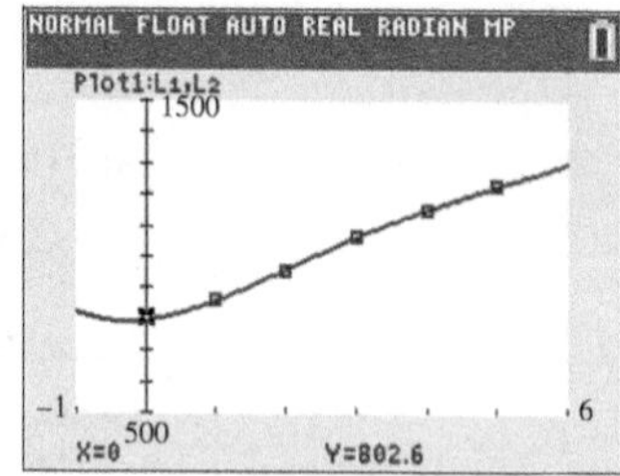

FIGURE 21

(d) The **correlation coefficient,** R, is a measure of the strength of the relationship between two variables. The values of R and R^2 are used to determine how well a regression model fits a set of data. The closer the value of R^2 is to 1, the better the fit. Compare R^2 for the three functions found in parts (a)–(c) to determine which function best fits the data.

With the statistical diagnostics turned on, the value of R^2 is displayed with the regression results on the TI 84 Plus C each time that a regression model is executed. By inspecting the R^2 value for each model, we see that the quartic function provides the best fit—it has the largest R^2 value of 0.9997452162.

NOW TRY 

NOW TRY ANSWERS

8. (a) $y = 0.9905x^2 + 81.81x + 712.3$

(b) $y = -1.058x^3 + 10.52x^2 + 60.65x + 718.6$

(c) $y = -0.6610x^4 + 6.874x^3 - 18.85x^2 + 94.07x + 715.2$

(d) The quartic function, with $R^2 = 0.9990165737$, provides the best fit.

NOTE In **Example 8(d),** we selected the quartic function as the best model based on a comparison of R^2 values of the models. In practice, however, the best choice of a model also depends on the data being analyzed, as well as on analysis of its trends and attributes.

11.3 Exercises

FOR EXTRA HELP  MyLab Math

Video solutions for select problems available in MyLab Math

Concept Check *The graphs of four polynomial functions are shown in A–D. They represent the graphs of functions defined by these four equations, but not necessarily in the order listed.*

$$y = x^3 - 3x^2 - 6x + 8 \qquad y = x^4 + 7x^3 - 5x^2 - 75x$$

$$y = -x^3 + 9x^2 - 27x + 17 \qquad y = -x^5 + 36x^3 - 22x^2 - 147x - 90$$

Apply the concepts of this section to answer each question.

A.

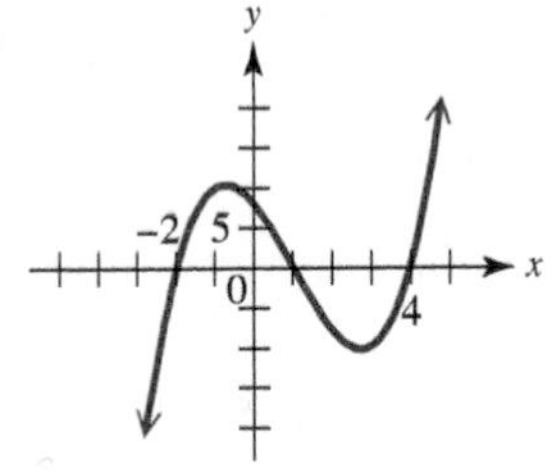

B.

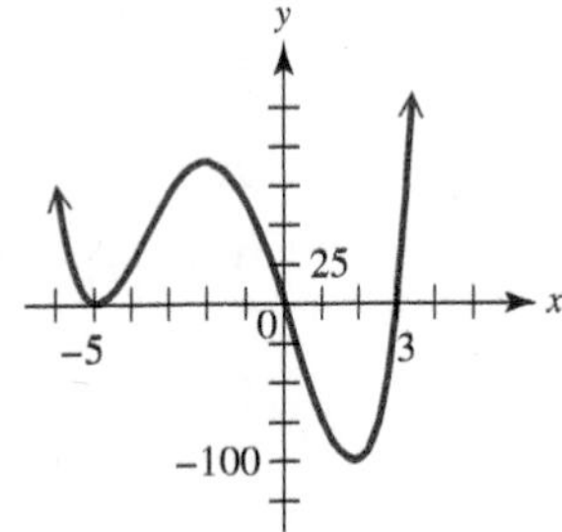

C.

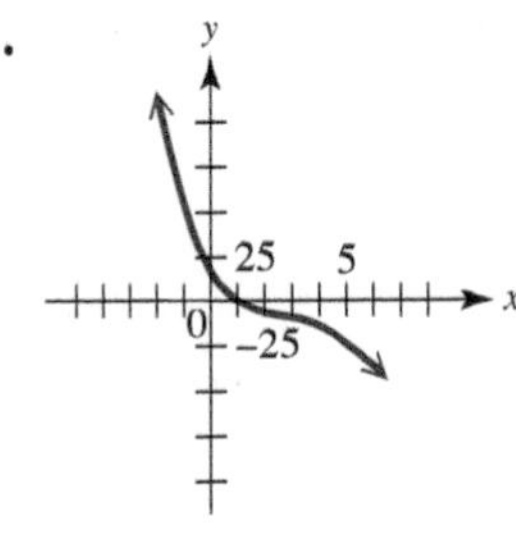

D.

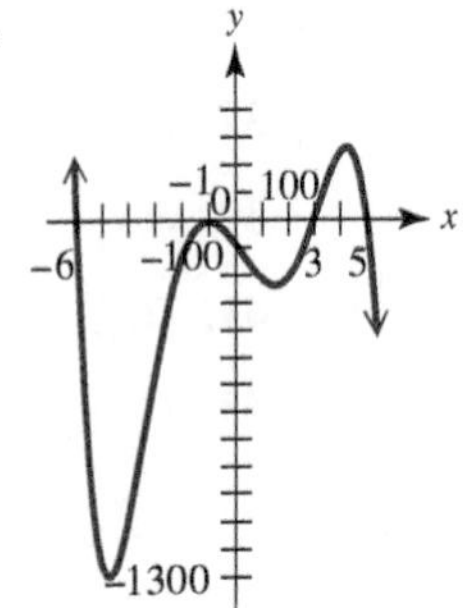

1. Which one of the graphs is that of $y = x^3 - 3x^2 - 6x + 8$?
2. Which one of the graphs is that of $y = x^4 + 7x^3 - 5x^2 - 75x$?
3. How many real zeros does the graph in C have?
4. Which one of C and D is the graph of $y = -x^3 + 9x^2 - 27x + 17$? (*Hint:* Look at the y-intercept.)
5. Which of the graphs cannot be that of a cubic polynomial function?
6. How many positive real zeros does the function graphed in D have?
7. How many negative real zeros does the function graphed in A have?
8. Which one of the graphs is that of a function whose range is *not* $(-\infty, \infty)$?

Graph each function. ***See Examples 1 and 2.***

9. $f(x) = \frac{1}{4}x^6$ **10.** $f(x) = 2x^4$ **11.** $f(x) = -\frac{5}{4}x^5$ **12.** $f(x) = -\frac{2}{3}x^5$

13. $f(x) = \frac{1}{2}x^3 + 1$ **14.** $f(x) = -x^4 + 2$ **15.** $f(x) = -(x + 1)^3$

16. $f(x) = \frac{1}{3}(x + 3)^4$ **17.** $f(x) = (x - 1)^4 + 2$ **18.** $f(x) = (x + 2)^3 - 1$

Determine the maximum possible number of turning points of the graph of each polynomial function. ***See the discussion following*** **FIGURE 10.**

19. $f(x) = x^3 - 3x^2 - 6x + 8$

20. $f(x) = x^3 + 4x^2 - 11x - 30$

21. $f(x) = -x^4 - 4x^3 + 3x^2 + 18x$

22. $f(x) = -x^4 + 2x^3 + 8x^2$

23. $f(x) = 2x^4 - 9x^3 + 5x^2 + 5x - 4$

24. $f(x) = 4x^4 + 2x^3 - 4x^2 - 4x - 3$

Use an end behavior diagram, ↖ ↗, ↙ ↘, ↖ ↘, *or* ↙ ↗, *to describe the end behavior of the graph of each polynomial function.* ***See*** **FIGURES 11 AND 12.**

25. $f(x) = 5x^5 + 2x^3 - 3x + 4$

26. $f(x) = -x^3 - 4x^2 + 2x - 1$

27. $f(x) = -4x^3 + 3x^2 - 1$

28. $f(x) = 4x^7 - x^5 + x^3 - 1$

29. $f(x) = 9x^6 - 3x^4 + x^2 - 2$

30. $f(x) = 10x^6 - x^5 + 2x - 2$

31. $f(x) = 3 + 2x - 4x^2 - 5x^{10}$

32. $f(x) = 7 + 2x - 5x^2 - 10x^4$

Graph each polynomial function. Factor first if the expression is not in factored form. Use the rational zeros theorem as necessary. ***See Examples 3 and 4.***

33. $f(x) = x^3 + 5x^2 + 2x - 8$

34. $f(x) = x^3 + 3x^2 - 13x - 15$

35. $f(x) = 2x(x - 3)(x + 2)$

36. $f(x) = x^2(x + 1)(x - 1)$

37. $f(x) = x^2(x - 2)(x + 3)^2$

38. $f(x) = x^2(x - 5)(x + 3)(x - 1)$

39. $f(x) = (3x - 1)(x + 2)^2$

40. $f(x) = (4x + 3)(x + 2)^2$

41. $f(x) = x^3 + 5x^2 - x - 5$

42. $f(x) = x^3 + x^2 - 36x - 36$

43. $f(x) = x^3 - x^2 - 2x$

44. $f(x) = 3x^4 + 5x^3 - 2x^2$

45. $f(x) = 2x^3(x^2 - 4)(x - 1)$

46. $f(x) = x^2(x - 3)^3(x + 1)$

47. $f(x) = 2x^3 - 5x^2 - x + 6$

48. $f(x) = 3x^3 + x^2 - 10x - 8$

49. $f(x) = x^3 + x^2 - 8x - 12$

50. $f(x) = x^3 + 6x^2 - 32$

51. $f(x) = -x^3 - x^2 + 8x + 12$

52. $f(x) = -x^3 + 10x^2 - 33x + 36$

53. $f(x) = x^4 - 18x^2 + 81$

54. $f(x) = x^4 - 8x^2 + 16$

For each of the following, ***(a)*** *show that the polynomial function has a zero between the two given integers and* ***(b)*** *approximate all real zeros to the nearest thousandth.* ***See Examples 5 and 7.***

55. $f(x) = x^4 + x^3 - 6x^2 - 20x - 16$; between -2 and -1

56. $f(x) = x^4 - 2x^3 - 2x^2 - 18x + 5$; between 0 and 1

57. $f(x) = x^4 - 4x^3 - 20x^2 + 32x + 12$; between -4 and -3

58. $f(x) = x^4 - 4x^3 - 44x^2 + 160x - 80$; between 2 and 3

Show that the real zeros of each polynomial function satisfy the given conditions. ***See Example 6.***

59. $f(x) = x^4 - x^3 + 3x^2 - 8x + 8$; no real zero greater than 2

60. $f(x) = 2x^5 - x^4 + 2x^3 - 2x^2 + 4x - 4$; no real zero greater than 1

61. $f(x) = x^4 + x^3 - x^2 + 3$; no real zero less than -2

62. $f(x) = x^5 + 2x^3 - 2x^2 + 5x + 5$; no real zero less than -1

63. $f(x) = 3x^4 + 2x^3 - 4x^2 + x - 1$; no real zero greater than 1

64. $f(x) = 3x^4 + 2x^3 - 4x^2 + x - 1$; no real zero less than -2

Extending Skills *Find a polynomial function $f(x)$ of least possible degree having the graph shown.*

65.

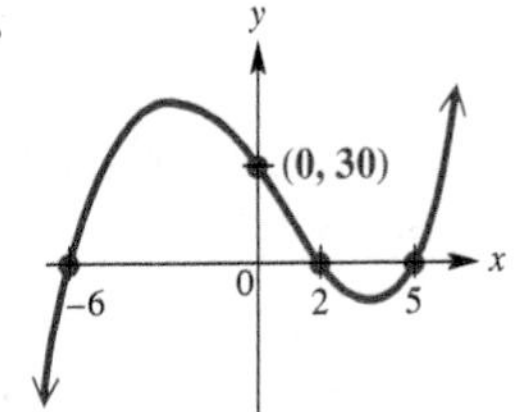

66.

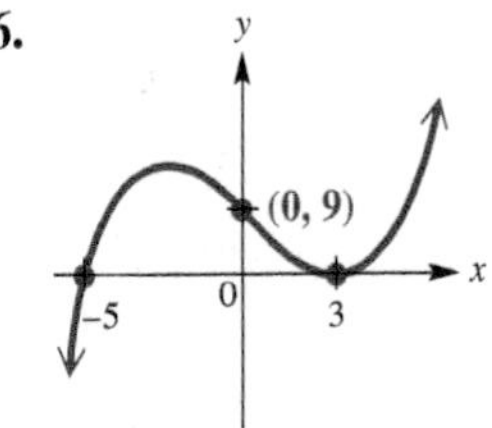

67.

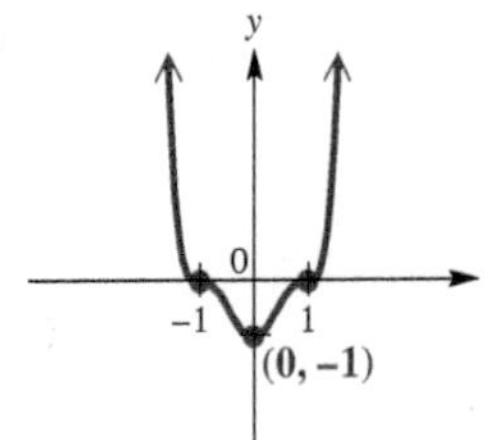

68.

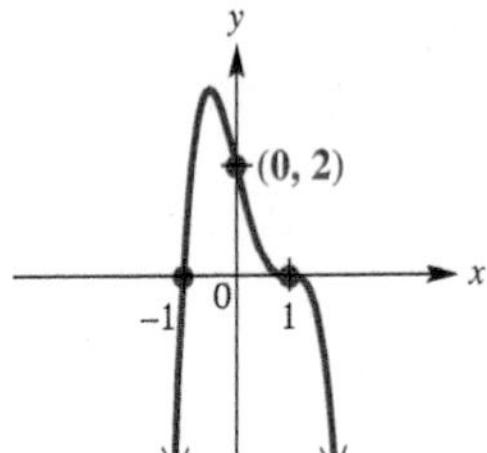

69.

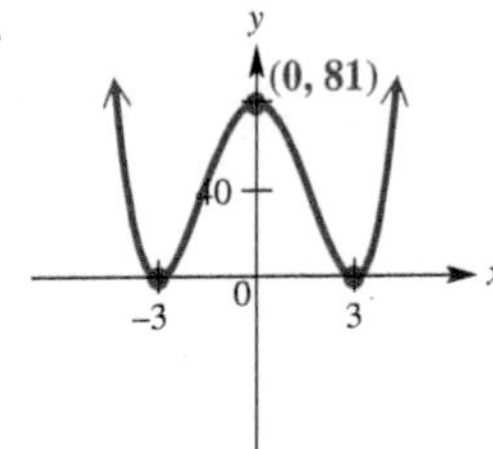

70.

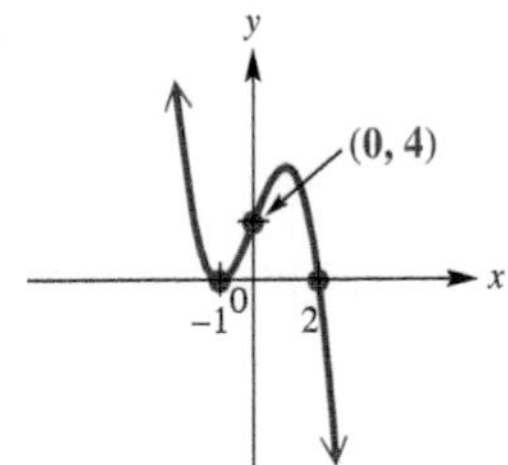

Approximate all real zeros of each function to the nearest hundredth. ***See Example 7.***

71. $f(x) = 0.86x^3 - 5.24x^2 + 3.5x + 7.8$

72. $f(x) = -2.47x^3 - 6.58x^2 - 3.3x + 0.1$

73. $f(x) = \sqrt{7}x^3 + \sqrt{5}x^2 + \sqrt{17}$

74. $f(x) = \sqrt{10}x^3 - \sqrt{11}x - \sqrt{8}$

75. $f(x) = -\sqrt{15}x^4 - \sqrt{3}x^2 + 7$

76. $f(x) = -\sqrt{17}x^4 + \sqrt{22}x^2 - 1$

Approximate to the nearest hundredth the coordinates of the turning point in the given interval of the graph of each polynomial function.

77. $f(x) = x^3 + 4x^2 - 8x - 8, \quad [-3.8, -3]$

78. $f(x) = x^3 + 4x^2 - 8x - 8, \quad [0.3, 1]$

79. $f(x) = 2x^3 - 5x^2 - x + 1, \quad [-1, 0]$

80. $f(x) = 2x^3 - 5x^2 - x + 1, \quad [1.4, 2]$

81. $f(x) = -x^4 + x + 3, \quad [0, 1]$

82. $f(x) = -2x^4 - 3x + 5, \quad [-1, 0]$

Solve each problem. ***See Example 8.***

83. The table shows the worldwide number of Amazon.com employees (in thousands) from 2009 to 2017.

Year	Amazon.com employees (in thousands)
2009	24.3
2010	33.7
2011	56.2
2012	88.4
2013	117.3
2014	154.1
2015	230.8
2016	341.4
2017	541.9

Data from www.amazon.com

(a) Using $x = 0$ to represent 2009, $x = 1$ to represent 2010, and so on, use the regression feature of a calculator to determine the quadratic function that best fits the data. Give coefficients to the nearest hundredth.

(b) Repeat part (a) for a cubic function (degree 3). Give coefficients to the nearest hundredth.

(c) Repeat part (a) for a quartic function (degree 4). Give coefficients to the nearest hundredth.

(d) Compare the correlation coefficient R^2 for the three functions in parts (a)–(c) to determine which function best fits the data. Give its value to the nearest ten-thousandth.

84. The table shows the total (cumulative) number of ebola cases reported in Sierra Leone during a serious West African ebola outbreak in 2014–2015. The total number of cases is reported x months after the start of the outbreak in May 2014.

x Months after May 2014	*Total Ebola Cases*
0	16
2	533
4	2021
6	7109
8	10,518
10	11,841
12	12,706
14	13,290
16	13,823
18	14,122

Data from World Health Organization.

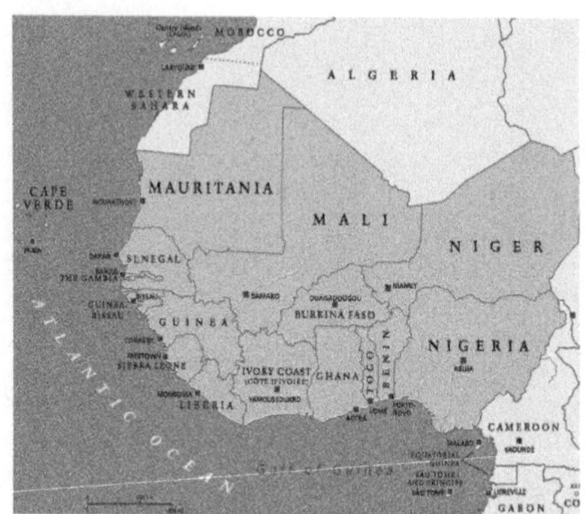

(a) Use the regression feature of a calculator to determine the quadratic function that best fits the data. Let x represent the number of months after May 2014, and let y represent the total number of ebola cases. Give coefficients to the nearest hundredth.

(b) Repeat part (a) for a cubic function (degree 3). Give coefficients to the nearest hundredth.

(c) Repeat part (a) for a quartic function (degree 4). Give coefficients to the nearest hundredth.

(d) Compare the correlation coefficient R^2 for the three functions in parts (a)–(c) to determine which function best fits the data. Give its value to the nearest ten-thousandth.

85. A rectangular piece of cardboard measuring 12 in. by 18 in. is to be made into a box with an open top by cutting equal-size squares from each corner and folding up the sides. Let x represent the length of a side of each such square in inches. Give approximations to the nearest hundredth.

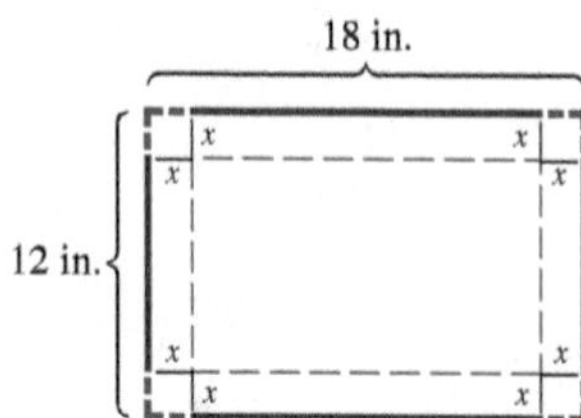

(a) Give the restrictions on x.

(b) Determine a function V that gives the volume of the box as a function of x.

(c) For what value of x will the volume be a maximum? What is this maximum volume? (*Hint:* Use the function of a graphing calculator that enables us to determine a maximum point within a given interval.)

(d) For what values of x will the volume be greater than 80 in.^{3}?

86. A certain right triangle has area 84 in.^{2} One leg of the triangle measures 1 in. less than the hypotenuse. Let x represent the length of the hypotenuse.

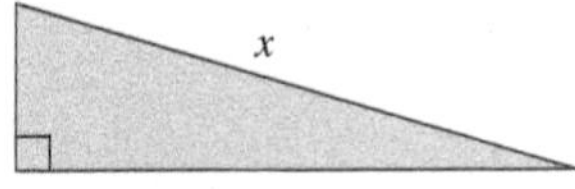

(a) Express the length of the leg mentioned above in terms of x.

(b) Express the length of the other leg in terms of x.

(c) Write an equation based on the information determined thus far. Square each side and then write the equation with one side as a polynomial with integer coefficients, in descending powers, and the other side equal to 0.

(d) Solve the equation in part (c) graphically. Find the lengths of the three sides of the triangle.

87. A piece of rectangular sheet metal is 20 in. wide. It is to be made into a rain gutter by turning up the edges to form parallel sides. Let x represent the length of each of the parallel sides.

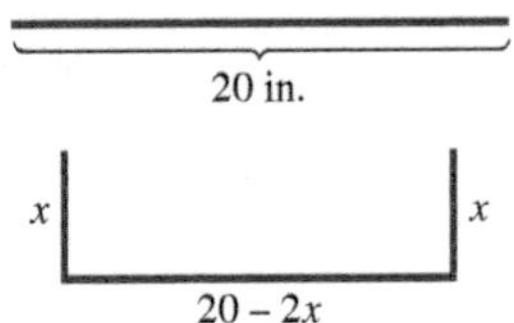

(a) Give the restrictions on x.

(b) Determine a function $\mathscr{A}$ that gives the area of a cross section of the gutter as a function of x.

(c) For what value of x will $\mathscr{A}$ be a maximum (and thus maximize the amount of water that the gutter will hold)? What is this maximum area?

(d) For what values of x will the area of a cross section be less than 40 in.^{2}?

88. A storage tank for butane gas is to be built in the shape of a right circular cylinder of altitude 12 ft, with a half sphere attached to each end. If x represents the radius of each half sphere, what radius should be used to cause the volume of the tank to be 144π ft^3?

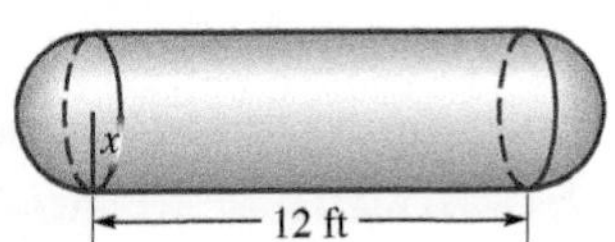

89. Find the value of x in the figure that will maximize the area of rectangle $ABCD$. Round to the nearest thousandth.

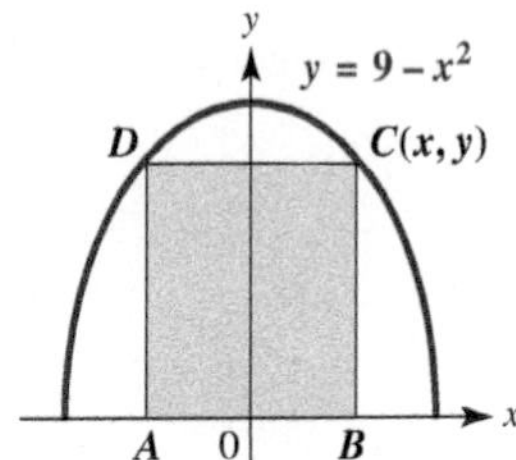

90. Grandfather clocks use pendulums to keep accurate time. The relationship between the length of a pendulum L and the time T for one complete oscillation can be expressed by the equation

$$L = kT^n,$$

where k is a constant and n is a positive integer to be determined. The data in the table were taken for different lengths of pendulums.

L (ft)	T (sec)
1.0	1.11
1.5	1.36
2.0	1.57
2.5	1.76
3.0	1.92
3.5	2.08
4.0	2.22

(a) As the length of the pendulum increases, what happens to T?

(b) Discuss how n and k can be found.

(c) Use the data to approximate k and determine the best value for n.

(d) Using the values of k and n from part (c), predict T for a pendulum having length 5 ft. Round to the nearest hundredth.

(e) If the length L of a pendulum doubles, what happens to the period T?

RELATING CONCEPTS For Individual or Group Work (Exercises 91–100)

Some polynomial functions are examples of even or odd functions.

A function $y = f(x)$ is an **even function** if $f(-x) = f(x)$ for all x in the domain of f.

A function $y = f(x)$ is an **odd function** if $f(-x) = -f(x)$ for all x in the domain of f.

For example, $f(x) = x^2$ *is an even function, because*

$$f(-x) = (-x)^2 = x^2 = f(x).$$

Also, $f(x) = x^3$ *is an odd function, because*

$$f(-x) = (-x)^3 = -x^3 = -f(x).$$

Determine whether each polynomial function is even, odd, *or* neither.

91. $f(x) = 2x^3$

92. $f(x) = -4x^5$

93. $f(x) = 0.2x^4$

94. $f(x) = -x^6$

95. $f(x) = -x^5$

96. $f(x) = -5x^7$

97. $f(x) = 2x^3 + 3x^2$

98. $f(x) = 4x^5 - x^4$

99. $f(x) = x^4 + 3x^2 + 5$

100. Fill in the blanks with the correct responses:

By the definition of an even function, if (a, b) lies on the graph of an even function, then so does $(-a, b)$. Therefore, the graph of an even function is symmetric with respect to the ______.

If (a, b) lies on the graph of an odd function, then by definition, so does $(-a, -b)$. Therefore, the graph of an odd function is symmetric with respect to the ______.

SUMMARY EXERCISES Examining Polynomial Functions and Graphs

For each polynomial function, do the following in order.

(a) *(Optional) Use Descartes' rule of signs to find the possible number of positive and negative real zeros.* ***(See the Exercises in the second section of this chapter.)***

(b) *Use the rational zeros theorem to determine the possible rational zeros of the function.*

(c) *Find the rational zeros, if any.*

(d) *Find all other real zeros, if any.*

(e) *Find any other complex zeros (that is, zeros that are not real), if any.*

(f) *Find the x-intercepts of the graph, if any.*

(g) *Find the y-intercept of the graph.*

(h) *Use synthetic division to find* $f(4)$, *and give the coordinates of the corresponding point on the graph.*

(i) *Determine the end behavior of the graph.*

(j) *Sketch the graph of the function.*

1. $f(x) = x^4 + 3x^3 - 3x^2 - 11x - 6$

2. $f(x) = -2x^5 + 5x^4 + 34x^3 - 30x^2 - 84x + 45$

3. $f(x) = 2x^5 - 10x^4 + x^3 - 5x^2 - x + 5$

4. $f(x) = 3x^4 - 4x^3 - 22x^2 + 15x + 18$

5. $f(x) = -2x^4 - x^3 + x + 2$

6. $f(x) = 4x^5 + 8x^4 + 9x^3 + 27x^2 + 27x$ (*Hint:* Factor out x first.)

7. $f(x) = 3x^4 - 14x^2 - 5$ (*Hint:* Factor the polynomial.)

8. $f(x) = -x^5 - x^4 + 10x^3 + 10x^2 - 9x - 9$

9. $f(x) = -3x^4 + 22x^3 - 55x^2 + 52x - 12$

10. For the polynomial functions in **Exercises 1–9** that have irrational zeros, find their approximations to the nearest thousandth.

11.4 Graphs and Applications of Rational Functions

OBJECTIVES

1. Graph rational functions using reflection and translation.
2. Find asymptotes of the graph of a rational function.
3. Graph rational functions.
4. Graph a rational function that is not in lowest terms.
5. Solve an application using a rational function.

VOCABULARY

- ☐ rational function
- ☐ discontinuous graph
- ☐ reciprocal function
- ☐ vertical asymptote
- ☐ horizontal asymptote
- ☐ oblique (slant) asymptote
- ☐ point of discontinuity (hole)

A rational expression is a fraction that is the quotient of two polynomials. A *rational function* is defined by a quotient of two polynomial functions.

Rational Function

A **rational function** is defined by a quotient of polynomials and has the form

$$f(x) = \frac{P(x)}{Q(x)}, \quad \textbf{where} \quad Q(x) \neq 0.$$

The domain of a rational function includes all real numbers except those that make $Q(x)$— that is, the denominator—equal to 0.

Examples: $f(x) = \frac{1}{x}$, $f(x) = \frac{x+1}{2x^2+5x-3}$, $f(x) = \frac{3(x+1)(x-2)}{(x+4)^2}$

Any values of x for which $Q(x) = 0$ are excluded from the domain of a rational function, so this type of function often has a **discontinuous graph**—that is, the graph has one or more breaks in it.

OBJECTIVE 1 Graph rational functions using reflection and translation.

The simplest rational function with a variable denominator is the **reciprocal function.**

$$f(x) = \frac{1}{x} \qquad \text{Reciprocal function}$$

The domain of this function is the set of all real numbers except 0. The number 0 cannot be used as a value of x, although for graphing it is helpful to find the values of $f(x)$ for some values of x close to 0. The table shows what happens to $f(x)$ as x gets closer and closer to 0 from either side.

x approaches 0.

x	−1	−0.1	−0.01	−0.001	0.001	0.01	0.1	1
$f(x)$	−1	−10	−100	−1000	1000	100	10	1

$|f(x)|$ increases without bound.

The table suggests that $|f(x)|$ gets larger and larger as x gets closer and closer to 0, which is written in symbols as

$$|f(x)| \to \infty \text{ as } x \to 0.$$

(The symbol $x \to 0$ means that x approaches as close as desired to 0, without necessarily ever being equal to 0.) Because x cannot equal 0, the graph of $f(x) = \frac{1}{x}$ will never intersect the vertical line $x = 0$. This line is a **vertical asymptote.**

As $|x|$ increases without bound, the values of $f(x) = \frac{1}{x}$ approach 0.

$x \to -\infty$ $\qquad$ $x \to \infty$

x	−10,000	−1000	−100	−10	10	100	1000	10,000
$f(x)$	−0.0001	−0.001	−0.01	−0.1	0.1	0.01	0.001	0.0001

$|f(x)| \to 0$

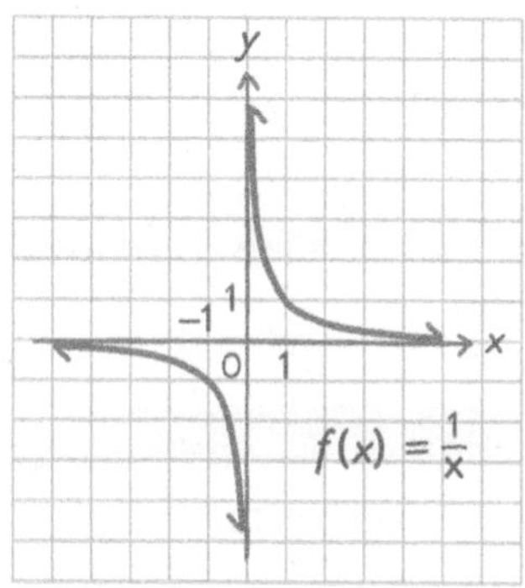

FIGURE 22

Letting $|x|$ increase without bound (which is written $|x| \to \infty$) causes the graph of $y = f(x) = \frac{1}{x}$ to approach the horizontal line $y = 0$. That is, $|f(x)| \to 0$. This line is a **horizontal asymptote.**

If the point (a, b) lies on the graph of $f(x) = \frac{1}{x}$, then so does the point $(-a, -b)$. Therefore, the graph of f is symmetric with respect to the origin. Choosing some positive values of x and finding the corresponding values of $f(x)$ gives the first-quadrant part of the graph shown in **FIGURE 22**. The other part of the graph (in the third quadrant) can be found by symmetry.

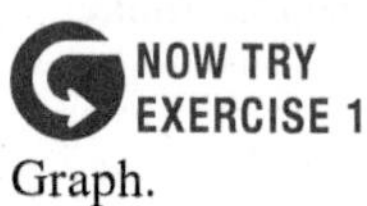
NOW TRY EXERCISE 1

Graph.

$$f(x) = -\frac{5}{x}$$

EXAMPLE 1 Graphing a Rational Function Using Reflection

Graph $f(x) = -\frac{2}{x}$.

The expression on the right side of the equation can be rewritten so that

$$f(x) = -2 \cdot \frac{1}{x}.$$

Compared to $f(x) = \frac{1}{x}$, the graph will be reflected about the x-axis (because of the negative sign), and the factor of 2 makes each point twice as far from the x-axis. The x-axis and y-axis remain the horizontal and vertical asymptotes. See **FIGURE 23**.

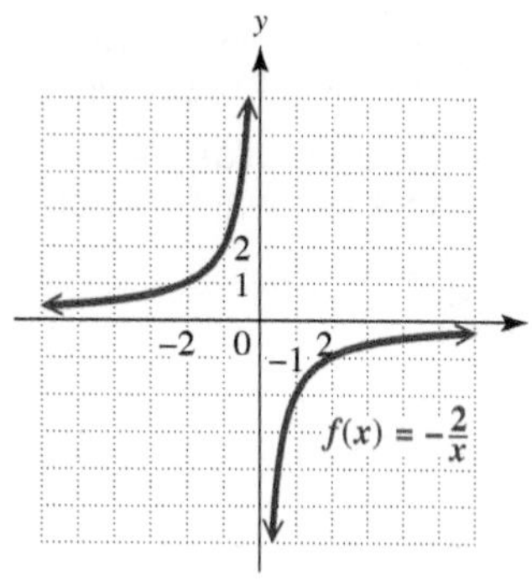

FIGURE 23

NOW TRY

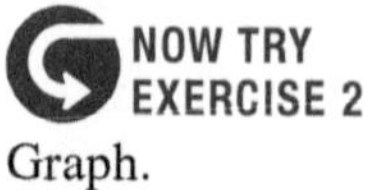
NOW TRY EXERCISE 2

Graph.

$$f(x) = \frac{1}{x+3}$$

EXAMPLE 2 Graphing a Rational Function Using Translation

Graph $f(x) = \frac{2}{x+1}$.

The domain of this function is the set of all real numbers except -1. As shown in **FIGURE 24(a)**, the graph is that of $f(x) = \frac{1}{x}$, translated to the left 1 unit, with each y-value doubled because of the numerator 2.

$$f(x) = 2 \cdot \frac{1}{x - (-1)}$$ Write in the form $x - k$.

The line $x = -1$ is the vertical asymptote, and the line $y = 0$ (the x-axis) remains the horizontal asymptote. **FIGURE 24(b)** shows a graphing calculator depiction.

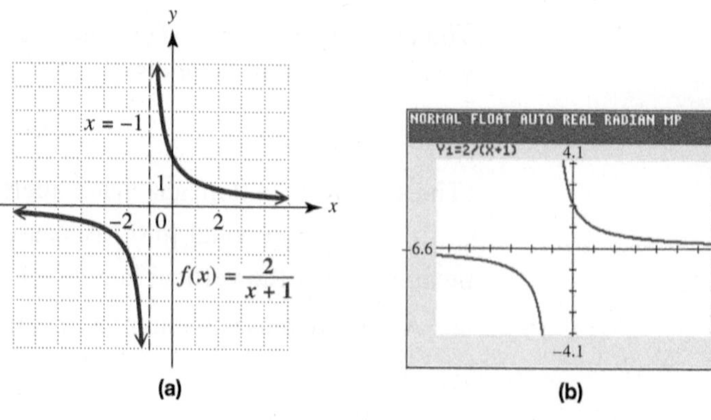

(a) (b)

FIGURE 24

NOW TRY

NOW TRY ANSWERS

1. 2.

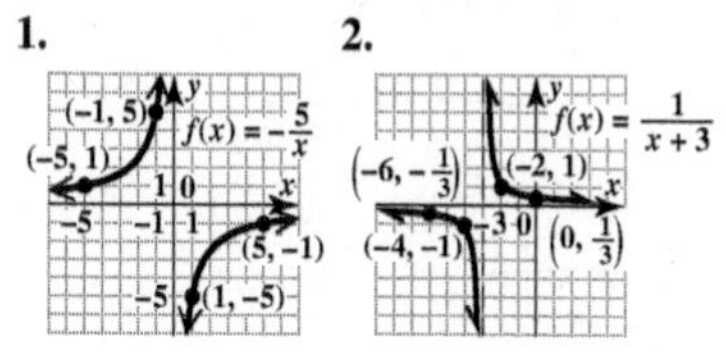

OBJECTIVE 2 Find asymptotes of the graph of a rational function.

The preceding examples suggest the definitions of vertical and horizontal asymptotes.

Vertical and Horizontal Asymptotes

Let $P(x)$ and $Q(x)$ define polynomials. For the rational function

$$f(x) = \frac{P(x)}{Q(x)},$$

written in lowest terms, and for real numbers a and b:

1. If $|f(x)| \to \infty$ as $x \to a$, then the line $x = a$ is a **vertical asymptote.**
2. If $f(x) \to b$ as $|x| \to \infty$, then the line $y = b$ is a **horizontal asymptote.**

Locating asymptotes is important when graphing rational functions.

- We find *vertical asymptotes* by determining the values of x that make the denominator equal TO 0.
- We find *horizontal asymptotes* (and, in some cases, *oblique asymptotes*) by considering what happens to $f(x)$ as $|x| \to \infty$. These asymptotes determine the end behavior of the graph.

Determining Asymptotes

To find asymptotes of a rational function defined by a rational expression *written in lowest terms,* use the following procedures.

1. **Vertical Asymptotes**

   Find any vertical asymptotes by setting the denominator equal to 0 and solving for x. If a is a zero of the denominator, then the line $x = a$ **is a vertical asymptote.**

2. **Other Asymptotes**

   Determine any other asymptotes by considering three possibilities:

   (a) If the numerator has lesser degree than the denominator, then there is a **horizontal asymptote, $y = 0$** (the x-axis).

   (b) If the numerator and denominator have the same degree, and the function is of the form

   $$f(x) = \frac{a_n x^n + \cdots + a_0}{b_n x^n + \cdots + b_0}, \quad \text{where } b_n \neq 0,$$

   then dividing by x^n in the numerator and denominator produces the **horizontal asymptote $y = \frac{a_n}{b_n}$.**

   (c) If the numerator is of degree exactly one more than the denominator, then there is **an oblique (slant) asymptote.** To find it, divide the numerator by the denominator and disregard any remainder. Set the rest of the quotient equal to y to obtain the equation of the asymptote.

EXAMPLE 3 Finding Equations of Asymptotes

For each rational function, find all asymptotes.

(a) $f(x) = \dfrac{x+1}{2x^2+5x-3}$

To find the vertical asymptotes, set the denominator equal to 0 and solve.

$$2x^2 + 5x - 3 = 0$$
$$(2x-1)(x+3) = 0 \quad \text{Factor.}$$
$$2x - 1 = 0 \quad \text{or} \quad x + 3 = 0 \quad \text{Zero-factor property}$$
$$x = \frac{1}{2} \quad \text{or} \quad x = -3 \quad \text{Solve each equation.}$$

The equations of the vertical asymptotes are $x = \frac{1}{2}$ and $x = -3$.

To find the horizontal asymptote, divide each term by the greatest power of x in the expression. Here, we divide each term by x^2 because 2 is the greatest exponent on x.

$$f(x) = \frac{\frac{x}{x^2} + \frac{1}{x^2}}{\frac{2x^2}{x^2} + \frac{5x}{x^2} - \frac{3}{x^2}} = \frac{\frac{1}{x} + \frac{1}{x^2}}{2 + \frac{5}{x} - \frac{3}{x^2}}$$

Simplify each term as needed.

As $|x| \to \infty$, the quotients $\frac{1}{x}$, $\frac{1}{x^2}$, $\frac{5}{x}$, and $\frac{3}{x^2}$ all approach 0, and

$$f(x) \to \frac{0+0}{2+0-0} = \frac{0}{2} = 0.$$

Therefore the line $y = 0$ (that is, the x-axis) is the horizontal asymptote.

(b) $f(x) = \dfrac{2x+1}{x-3}$

Set the denominator $x - 3$ equal to 0 to find that the vertical asymptote has the equation $x = 3$. To find the horizontal asymptote, divide each term in the rational expression by x because the greatest power of x in the expression is 1.

$$f(x) = \frac{2x+1}{x-3} = \frac{\frac{2x}{x} + \frac{1}{x}}{\frac{x}{x} - \frac{3}{x}} = \frac{2 + \frac{1}{x}}{1 - \frac{3}{x}}$$

As $|x| \to \infty$, both $\frac{1}{x}$ and $\frac{3}{x}$ approach 0, and

$$f(x) \to \frac{2+0}{1-0} = \frac{2}{1} = 2.$$

Therefore the line $y = 2$ is the horizontal asymptote.

(c) $f(x) = \dfrac{x^2+1}{x-2}$

Setting the denominator $x - 2$ equal to 0 shows that the vertical asymptote has the equation $x = 2$. If we divide by the greatest power of x as before (x^2 in this case), we see that there is no horizontal asymptote.

$$f(x) = \frac{\frac{x^2}{x^2} + \frac{1}{x^2}}{\frac{x}{x^2} - \frac{2}{x^2}} = \frac{1 + \frac{1}{x^2}}{\frac{1}{x} - \frac{2}{x^2}}$$

This approaches 0 as $|x| \to \infty$.

NOW TRY EXERCISE 3

For each rational function, find all asymptotes.

(a) $f(x) = \dfrac{x-2}{2x^2-3x-9}$

(b) $f(x) = \dfrac{x-2}{x+1}$

(c) $f(x) = \dfrac{x^2+3x+3}{x+2}$

Thus, $f(x)$ does not approach any real number as $|x| \to \infty$ because $\frac{1}{0}$ is undefined. This happens whenever the degree of the numerator is greater than the degree of the denominator. In such cases we divide the denominator into the numerator to write the expression in another form. Using synthetic division gives the following.

$$\begin{array}{r|rrr} 2 & 1 & 0 & 1 \\ & & 2 & 4 \\ \hline & 1 & 2 & 5 \end{array} \quad \leftarrow (x^2+1) \div (x-2)$$

The function can now be written as follows.

$$f(x) = \frac{x^2+1}{x-2} = x + 2 + \frac{5}{x-2}$$

As $|x| \to \infty$, $\frac{5}{x-2} \to 0$, and the graph approaches the line $y = x + 2$. This line is an **oblique asymptote** (neither vertical nor horizontal) for the function. NOW TRY

OBJECTIVE 3 Graph rational functions.

Graphing a Rational Function

Let $f(x) = \frac{P(x)}{Q(x)}$ define a rational function, where $P(x)$ and $Q(x)$ are polynomials and the rational expression is *written in lowest terms*. Sketch its graph as follows.

Step 1 **Find any vertical asymptotes.**

Step 2 **Find any horizontal or oblique asymptote.**

Step 3 **Find the *y*-intercept** by evaluating $f(0)$.

Step 4 **Find the *x*-intercepts,** if any, by solving $f(x) = 0$. (These will be the zeros of the numerator, $P(x)$.)

Step 5 **Determine whether the graph will intersect its nonvertical asymptote** $y = b$ or $y = mx + b$ by solving $f(x) = b$ or $f(x) = mx + b$.

Step 6 **Plot a few selected points, as necessary.** Choose an x-value in each interval of the domain as determined by the vertical asymptotes and x-intercepts.

Step 7 **Complete the sketch.**

NOTE The graph of a rational function may have more than one vertical asymptote, or it may have none at all. ***The graph cannot intersect any vertical asymptote. There can be only one other (nonvertical) asymptote, and the graph may intersect that asymptote.*** See Example 6.

EXAMPLE 4 Graphing a Rational Function (*x*-Axis as Horizontal Asymptote)

Graph $f(x) = \dfrac{x+1}{2x^2+5x-3}$.

Step 1 As shown in **Example 3(a),** $2x^2 + 5x - 3 = (2x-1)(x+3)$ and the vertical asymptotes have equations $x = \frac{1}{2}$ and $x = -3$.

Step 2 Again, as shown in **Example 3(a),** the horizontal asymptote is the x-axis.

NOW TRY ANSWERS

3. **(a)** V.A.: $x = -\frac{3}{2}$, $x = 3$
H.A.: $y = 0$
(b) V.A.: $x = -1$
H.A.: $y = 1$
(c) V.A.: $x = -2$
O.A.: $y = x + 1$

NOW TRY EXERCISE 4

Graph.

$$f(x) = \frac{x-2}{2x^2-3x-9}$$

Step 3 Evaluate $f(0)$ to find that the y-intercept is $\left(0, -\frac{1}{3}\right)$.

$$f(0) = \frac{0+1}{2(0)^2+5(0)-3} = -\frac{1}{3}$$ The y-intercept corresponds to the ratio of the constant terms.

Step 4 Find the x-intercept by solving $f(x) = 0$.

$$\frac{x+1}{2x^2+5x-3} = 0$$ Let $f(x) = 0$.

$$x + 1 = 0$$ If a rational expression equals 0, then its numerator must equal 0.

$$x = -1$$ The x-intercept is $(-1, 0)$.

Step 5 To determine whether the graph intersects its horizontal asymptote, solve

$$f(x) = 0.$$ y-value of horizontal asymptote

The horizontal asymptote is the x-axis, so the solution of this equation was found in Step 4. (This will not happen in all cases.) The graph intersects its horizontal asymptote at $(-1, 0)$.

Step 6 Plot a point in each of the intervals determined by the x-intercepts and the vertical asymptotes, $(-\infty, -3)$, $(-3, -1)$, $\left(-1, \frac{1}{2}\right)$, and $\left(\frac{1}{2}, \infty\right)$, to see how the graph behaves in each interval.

Interval	*Test Point x*	*Value of $f(x)$*	*Sign of $f(x)$*	*Graph Above or Below x-Axis*
$(-\infty, -3)$	-4	$-\frac{1}{3}$	Negative	Below
$(-3, -1)$	-2	$\frac{1}{5}$	Positive	Above
$\left(-1, \frac{1}{2}\right)$	0	$-\frac{1}{3}$	Negative	Below
$\left(\frac{1}{2}, \infty\right)$	2	$\frac{1}{5}$	Positive	Above

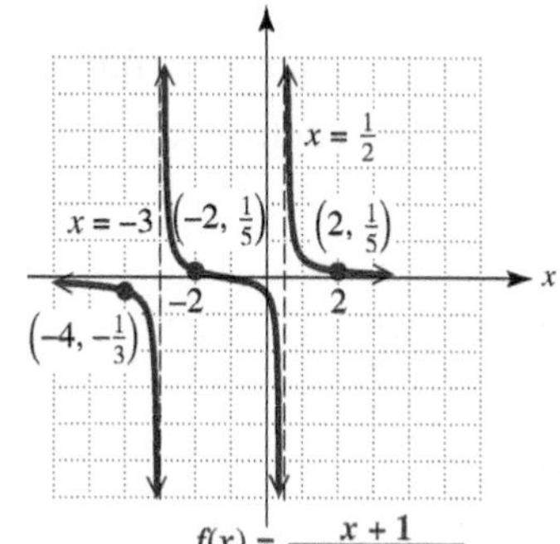

FIGURE 25

Step 7 Complete the sketch. The graph of this function, which decreases on every interval of its domain, is shown in **FIGURE 25**.

NOW TRY

EXAMPLE 5 Graphing a Rational Function (Does Not Intersect Its Horizontal Asymptote)

Graph $f(x) = \dfrac{2x+1}{x-3}$.

Steps 1 and 2 As determined in **Example 3(b),** the equation of the vertical asymptote is $x = 3$. The horizontal asymptote has equation $y = 2$.

Step 3 $f(0) = -\frac{1}{3}$, so the y-intercept is $\left(0, -\frac{1}{3}\right)$.

Step 4 Solve $f(x) = 0$ to find any x-intercepts.

$$\frac{2x+1}{x-3} = 0$$ Let $f(x) = 0$.

$$2x + 1 = 0$$ If a rational expression is equal to 0, then its numerator must equal 0.

$$x = -\frac{1}{2}$$ The x-intercept is $\left(-\frac{1}{2}, 0\right)$.

NOW TRY ANSWER

4. 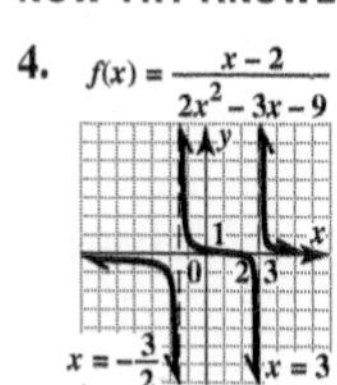

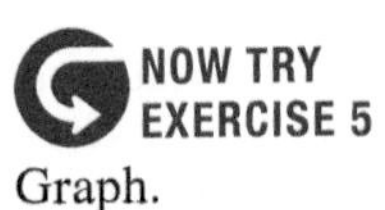

Graph.

$$f(x) = \frac{x-2}{x+1}$$

Step 5 The graph does not intersect its horizontal asymptote because $f(x) = 2$ has no solution.

$$\frac{2x+1}{x-3} = 2 \qquad \text{Let } f(x) = 2.$$

$$2x + 1 = 2x - 6 \qquad \text{Multiply each side by } x - 3.$$

$$1 = -6 \qquad \text{Subtract } 2x.$$

A false statement results.

Steps 6 and 7 The points $(-4, 1)$, $\left(1, -\frac{3}{2}\right)$, and $\left(6, \frac{13}{3}\right)$ are on the graph and can be used to complete the sketch of this function, which decreases on every interval of its domain. See **FIGURE 26.**

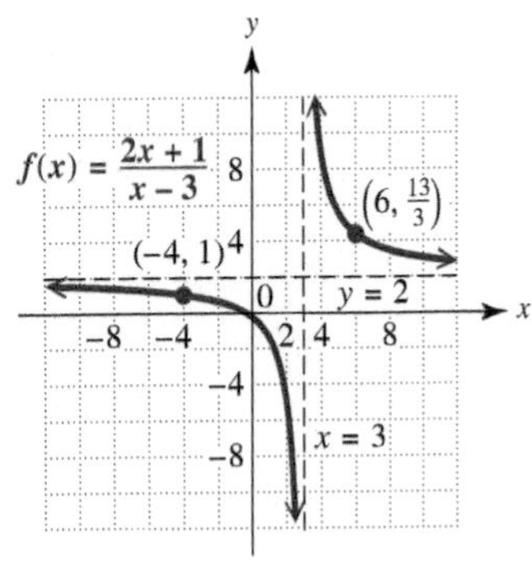

FIGURE 26

NOW TRY 

EXAMPLE 6 Graphing a Rational Function (Graph Intersects the Asymptote)

Graph $f(x) = \dfrac{3(x+1)(x-2)}{(x+4)^2}$.

Step 1 The only vertical asymptote is the line $x = -4$.

Step 2 To find any horizontal asymptotes, we multiply the factors in the numerator and denominator.

$$f(x) = \frac{3x^2 - 3x - 6}{x^2 + 8x + 16}$$

We divide all terms by x^2 and consider the behavior of each term as $|x|$ increases without bound to obtain the equation of the horizontal asymptote.

$$y = \frac{3}{1}, \quad \begin{matrix} \leftarrow \text{Leading coefficient of numerator} \\ \leftarrow \text{Leading coefficient of denominator} \end{matrix} \qquad \text{or} \qquad y = 3$$

Step 3 $f(0) = -\frac{3}{8}$, so the y-intercept is $\left(0, -\frac{3}{8}\right)$.

Step 4 Solve $f(x) = 0$ to find any x-intercepts.

$$\frac{3x^2 - 3x - 6}{x^2 + 8x + 16} = 0 \qquad \text{Let } f(x) = 0.$$

$$3x^2 - 3x - 6 = 0 \qquad \text{Set the numerator equal to 0.}$$

$$x^2 - x - 2 = 0 \qquad \text{Divide by 3.}$$

$$(x-2)(x+1) = 0 \qquad \text{Factor.}$$

$$x = 2 \quad \text{or} \quad x = -1 \qquad \text{Zero-factor property}$$

The x-intercepts are $(-1, 0)$ and $(2, 0)$.

NOW TRY ANSWER

5. 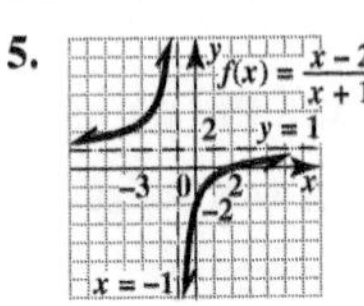

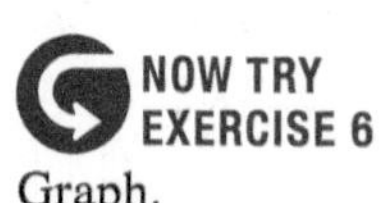
NOW TRY EXERCISE 6

Graph.

$$f(x) = \frac{2x(x+3)}{(x+2)^2}$$

Step 5 We let $f(x) = 3$ and solve to locate the point where the graph intersects the horizontal asymptote.

$$\frac{3x^2 - 3x - 6}{x^2 + 8x + 16} = 3 \quad \text{Let } f(x) = 3.$$

$$3x^2 - 3x - 6 = 3x^2 + 24x + 48 \quad \text{Multiply each side by } x^2 + 8x + 16.$$

$$-27x = 54 \quad \text{Subtract } 3x^2 \text{ and } 24x. \text{ Add } 6.$$

$$x = -2 \quad \text{Divide by } -27.$$

The graph intersects its horizontal asymptote at $(-2, 3)$.

Steps 6 and 7 Some other points that lie on the graph are $(-10, 9)$, $\left(-8, 13\frac{1}{8}\right)$, and $\left(5, \frac{2}{3}\right)$. These are used to complete the graph, as shown in **FIGURE 27**.

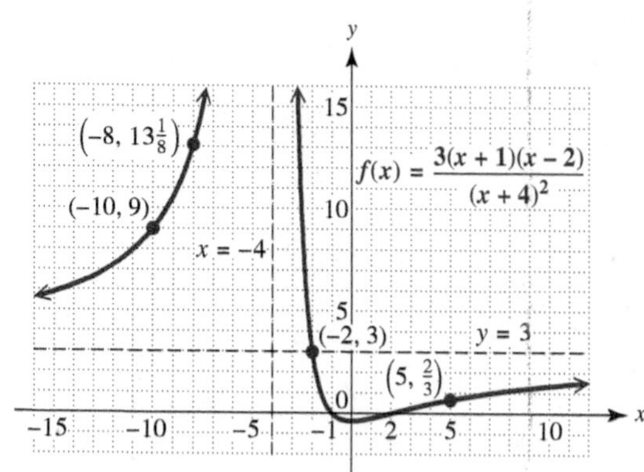

FIGURE 27

NOW TRY 

NOW TRY EXERCISE 7

Graph.

$$f(x) = \frac{x^2 + 3x + 3}{x + 2}$$

EXAMPLE 7 Graphing a Rational Function with an Oblique Asymptote

Graph $f(x) = \frac{x^2 + 1}{x - 2}$.

As shown in **Example 3(c),** the vertical asymptote has the equation $x = 2$, and the graph has an oblique asymptote with equation $y = x + 2$. The y-intercept is $\left(0, -\frac{1}{2}\right)$, and the graph has no x-intercepts because the numerator, $x^2 + 1$, has no real zeros. The graph does not intersect its oblique asymptote because the following has no solution.

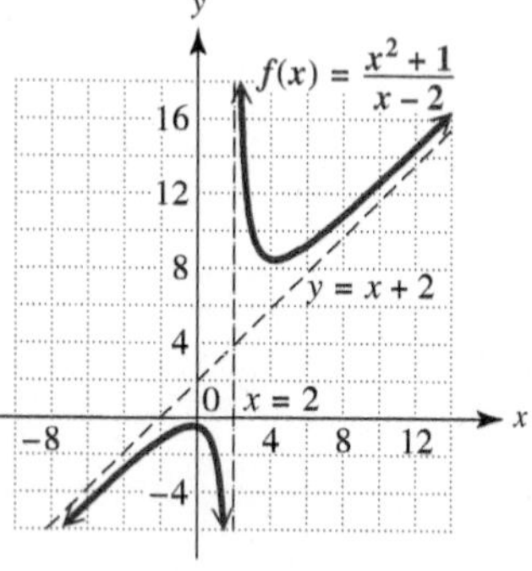

FIGURE 28

$$\frac{x^2 + 1}{x - 2} = x + 2 \quad \text{Set the expressions defining the function and the oblique asymptote equal.}$$

$$(x-2)\left(\frac{x^2+1}{x-2}\right) = (x-2)(x+2) \quad \text{Multiply each side by } x - 2.$$

$$x^2 + 1 = x^2 - 4 \quad (x-y)(x+y) = x^2 - y^2$$

$$1 = -4 \quad \text{False}$$

Using the intercepts, asymptotes, the points $\left(4, \frac{17}{2}\right)$ and $\left(-1, -\frac{2}{3}\right)$, and the general behavior of the graph near its asymptotes, we obtain the graph shown in **FIGURE 28**.

NOW TRY

NOW TRY ANSWERS

6.

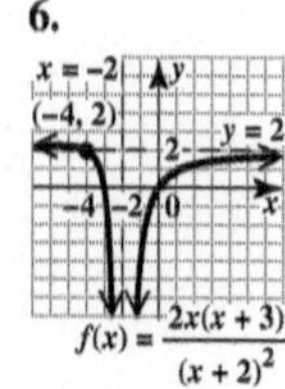

7.

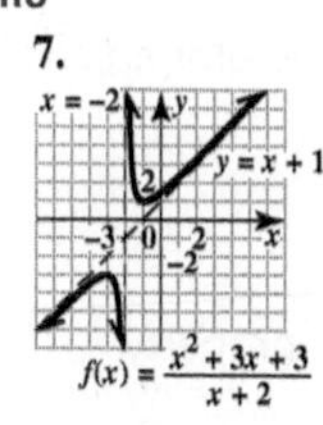

OBJECTIVE 4 Graph a rational function that is not in lowest terms.

A rational function must be in lowest terms before we can use the methods discussed thus far in this section to determine its graph. A rational function that has a common variable factor in the numerator and denominator will have a hole, or **point of discontinuity,** in its graph.

NOW TRY EXERCISE 8

Graph.

$$f(x) = \frac{2x^2 - 5x - 3}{x - 3}$$

EXAMPLE 8 Graphing a Rational Function (Not in Lowest Terms)

Graph $f(x) = \frac{x^2 - 4}{x - 2}$.

Start by noticing that the domain of this function does not include 2. The rational expression $\frac{x^2-4}{x-2}$ can be written in lowest terms by factoring the numerator and using the fundamental property.

$$f(x) = \frac{x^2 - 4}{x - 2} = \frac{(x + 2)(x - 2)}{x - 2} = x + 2 \qquad (x \neq 2)$$

Therefore, the graph of $f(x) = \frac{x^2-4}{x-2}$ will be the same as the graph of $y = x + 2$ (a straight line), with the exception of the point with x-value 2. Instead of an asymptote, there is a hole in the graph at $(2, 4)$. See **FIGURE 29(a)**. The calculator graph in **FIGURE 29(b)** confirms our work.

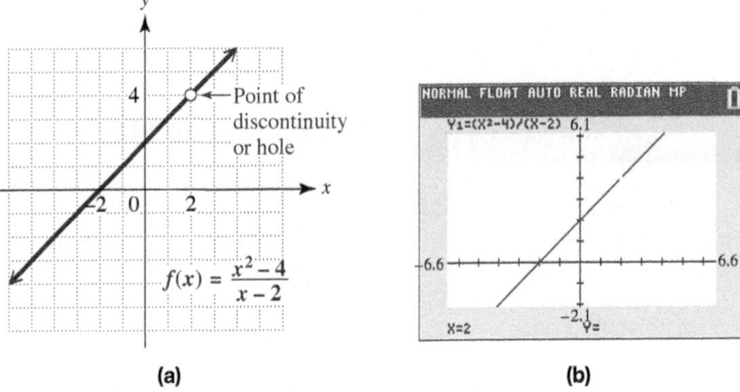

FIGURE 29

NOW TRY

OBJECTIVE 5 Solve an application using a rational function.

EXAMPLE 9 Applying a Rational Function to Traffic Intensity

Suppose that vehicles arrive randomly at a parking ramp with an average rate of 2.6 vehicles per minute. The parking attendant can admit 3.2 vehicles per minute. However, since arrivals are random, lines form at various times. (Data from Mannering, F., and W. Kilareski, *Principles of Highway Engineering and Traffic Control,* 2nd ed., John Wiley & Sons.)

(a) The **traffic intensity** x is defined as the ratio of the average arrival rate to the average admittance rate. Determine x for this parking ramp.

The average arrival rate is 2.6 vehicles, and the average admittance rate is 3.2 vehicles, so

$$x = \frac{2.6}{3.2} = 0.8125.$$

NOW TRY ANSWER

8.

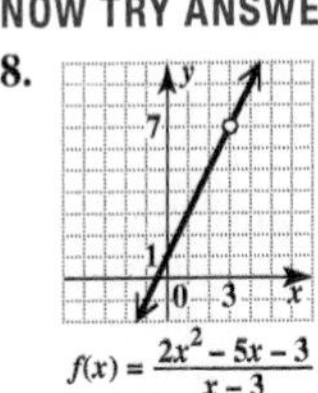

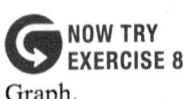

NOW TRY EXERCISE 9

In **Example 9,** suppose the vehicles arrive with an average rate of 3 vehicles per minute and the parking attendant can admit 3.2 vehicles per minute.

(a) Determine the traffic intensity x.

(b) Evaluate $f(x)$ to the nearest hundredth using the traffic intensity value from part (a).

(b) The average number of vehicles waiting in line to enter the ramp is given by

$$f(x) = \frac{x^2}{2(1 - x)},$$

where $0 \leq x < 1$ is the traffic intensity. Graph this function using a graphing calculator, and compute $f(0.8125)$ for this parking ramp.

FIGURE 30 shows the graph. Find $f(0.8125)$ by substitution.

$$f(0.8125) = \frac{0.8125^2}{2(1 - 0.8125)} \approx 1.76 \text{ vehicles}$$

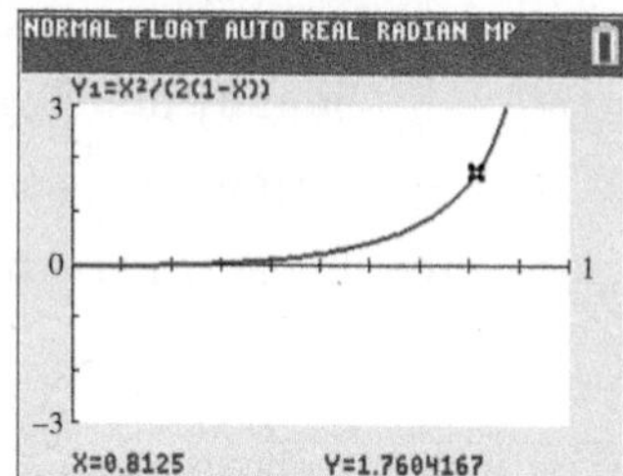

FIGURE 30

(c) What happens to the number of vehicles waiting in line as the traffic intensity approaches 1?

From the graph in **FIGURE 30**, we see that as x approaches 1, $y = f(x)$ gets very large. Thus, the average number of waiting vehicles gets very large. This is what we would expect.

NOW TRY 

NOW TRY ANSWERS

9. **(a)** 0.9375 **(b)** 7.03

11.4 Exercises

FOR EXTRA HELP MyLab Math

Video solutions for select problems available in MyLab Math

Concept Check *Use the graphs of the rational functions in A–D to answer each question. Give all possible answers. There may be more than one correct choice.*

A.

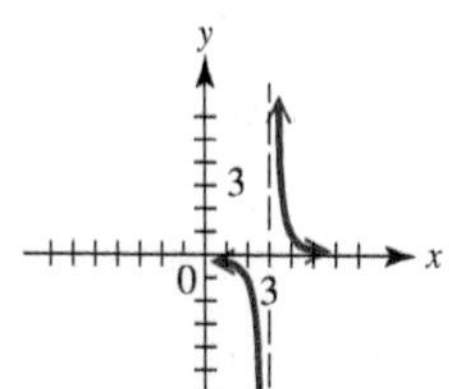

B.

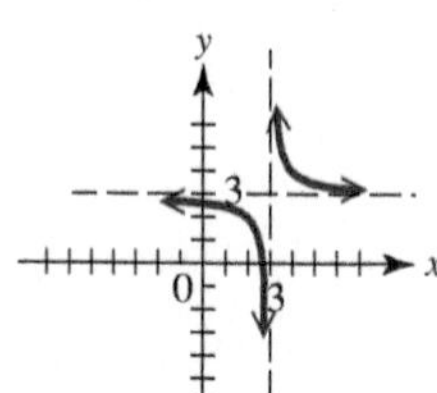

C.

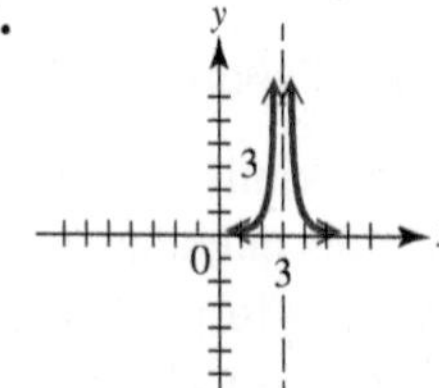

D.

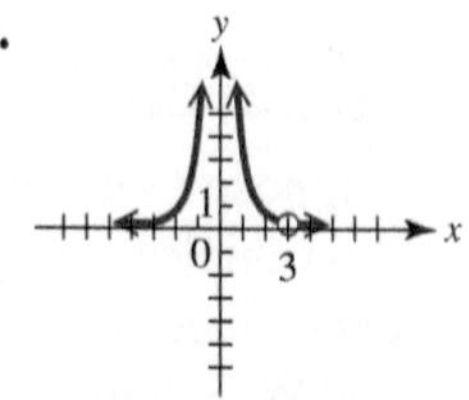

1. Which choices have domain $(-\infty, 3) \cup (3, \infty)$?

2. Which choices have range $(-\infty, 3) \cup (3, \infty)$?

3. Which choices have range $(-\infty, 0) \cup (0, \infty)$?

4. If f represents the function, only one choice has a single solution to the equation $f(x) = 3$. Which one is it?

5. Which choices have the x-axis as a horizontal asymptote?

6. Which choice is symmetric with respect to a vertical line?

Use reflections and/or translations to graph each rational function. ***See Examples 1 and 2.***

7. $f(x) = -\frac{3}{x}$ **8.** $f(x) = -\frac{2}{x}$ **9.** $f(x) = \frac{1}{x+2}$

10. $f(x) = \frac{1}{x-3}$ **11.** $f(x) = \frac{1}{x} + 1$ **12.** $f(x) = \frac{1}{x} - 2$

13. Concept Check A student claimed that the graphs of

$$f(x) = \frac{x^2 - 49}{x + 7} \quad \text{and} \quad g(x) = x - 7$$

are *exactly* the same. This is incorrect. **WHAT WENT WRONG?** Explain.

14. Concept Check A student claimed that the graph of a rational function cannot intersect an asymptote. This is incorrect. **WHAT WENT WRONG?** Explain.

Give the equations of any vertical, horizontal, or oblique asymptotes for the graph of each rational function. ***See Example 3.***

15. $f(x) = \frac{-8}{3x-7}$ **16.** $f(x) = \frac{-5}{4x-9}$ **17.** $f(x) = \frac{x+3}{3x^2+x-10}$

18. $f(x) = \frac{x-8}{2x^2-9x-18}$ **19.** $f(x) = \frac{2-x}{x+2}$ **20.** $f(x) = \frac{x-4}{5-x}$

21. $f(x) = \frac{3x-5}{2x+9}$ **22.** $f(x) = \frac{4x+3}{3x-7}$ **23.** $f(x) = \frac{2}{x^2-4x+3}$

24. $f(x) = \frac{-5}{x^2-3x-10}$ **25.** $f(x) = \frac{x^2-1}{x+3}$ **26.** $f(x) = \frac{x^2-4}{x-1}$

27. $f(x) = \frac{(x-3)(x+1)}{(x+2)(2x-5)}$ **28.** $f(x) = \frac{(x+2)(x-4)}{(6x-1)(x-5)}$

29. Concept Check Which choice has a graph that does not have a vertical asymptote?

A. $f(x) = \frac{1}{x^2+2}$ **B.** $f(x) = \frac{1}{x^2-2}$ **C.** $f(x) = \frac{3}{x^2}$ **D.** $f(x) = \frac{2x+1}{x-8}$

30. Concept Check Which choice has a graph that does not have a horizontal asymptote?

A. $f(x) = \frac{2x-7}{x+3}$ **B.** $f(x) = \frac{3x}{x^2-9}$

C. $f(x) = \frac{x^2-9}{x+3}$ **D.** $f(x) = \frac{x+5}{(x+2)(x-3)}$

Graph each rational function. ***See Examples 4–8.***

31. $f(x) = \frac{x+1}{x-4}$ **32.** $f(x) = \frac{x-5}{x+3}$

33. $f(x) = \frac{-x}{x^2-4}$ **34.** $f(x) = \frac{x}{x^2-9}$

35. $f(x) = \frac{3x}{(x+1)(x-2)}$ **36.** $f(x) = \frac{5x}{(x+1)(x-1)}$

37. $f(x) = \frac{2x+1}{(x+2)(x+4)}$ **38.** $f(x) = \frac{3x+2}{(x-2)(x-4)}$

39. $f(x) = \frac{3x}{x - 1}$

40. $f(x) = \frac{-4x}{3x - 1}$

41. $f(x) = \frac{1}{(x + 5)(x - 2)}$

42. $f(x) = \frac{1}{(x - 2)(x + 4)}$

43. $f(x) = \frac{3}{(x + 4)^2}$

44. $f(x) = \frac{-2}{(x + 3)^2}$

45. $f(x) = \frac{(x - 3)(x + 1)}{(x - 1)^2}$

46. $f(x) = \frac{x(x - 2)}{(x + 3)^2}$

47. $f(x) = \frac{x^2 + 1}{x + 3}$

48. $f(x) = \frac{x^2 + 2x}{2x - 1}$

49. $f(x) = \frac{2x^2 + 3}{x - 4}$

50. $f(x) = \frac{x(x - 1)}{x + 2}$

51. $f(x) = \frac{(x - 5)(x - 2)}{x^2 + 9}$

52. $f(x) = \frac{(x + 2)(x - 3)}{x^2 + 4}$

53. $f(x) = \frac{x^2 - 9}{x + 3}$

54. $f(x) = \frac{x^2 - 16}{x - 4}$

55. $f(x) = \frac{25 - x^2}{x - 5}$

56. $f(x) = \frac{36 - x^2}{x - 6}$

57. Concept Check The graphs in A–D show the four ways that a rational function can approach the vertical line $x = 2$ as an asymptote. Match each rational function in (a)–(d) with the appropriate graph in A–D.

(a) $f(x) = \frac{1}{(x - 2)^2}$ **(b)** $f(x) = \frac{1}{x - 2}$ **(c)** $f(x) = \frac{-1}{x - 2}$ **(d)** $f(x) = \frac{-1}{(x - 2)^2}$

A.

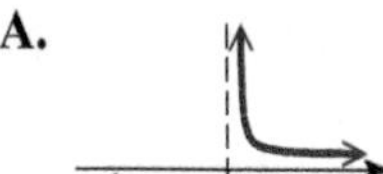

B.

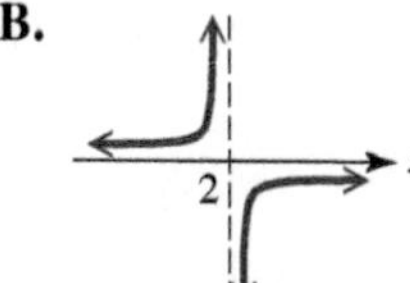

C.

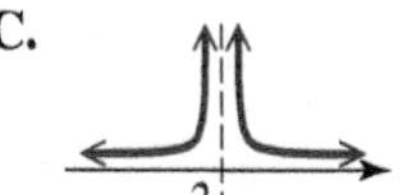

D.

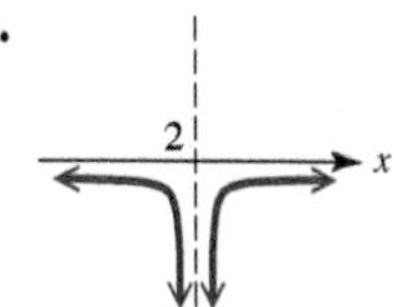

58. Concept Check Suppose that a friend explains that the graph of

$$f(x) = \frac{x^2 - 25}{x + 5}$$

has a vertical asymptote with equation $x = -5$. This is incorrect. **WHAT WENT WRONG?** Correctly describe the behavior of the graph at $x = -5$.

Solve each problem.

59. Refer to **Example 9.** Let the average number of vehicles arriving at the gate of an amusement park per minute equal k, and let the average number of vehicles admitted by the park attendants equal r. Then the average waiting time T (in minutes) for each vehicle arriving at the park is given by the rational function

$$T(r) = \frac{2r - k}{2r^2 - 2kr}, \quad \text{where } r > k.$$

(Data from Mannering, F., and W. Kilareski, *Principles of Highway Engineering and Traffic Control,* 2nd ed., John Wiley & Sons.)

(a) It is known from experience that on Saturday afternoon, $k = 25$. Use graphing to estimate the admittance rate r that is necessary to keep the average waiting time T for each vehicle to 30 sec.

(b) If one park attendant can serve 5.3 vehicles per minute, how many park attendants will be needed to keep the average wait to 30 sec?

60. The table contains incidence ratios by age for deaths due to coronary heart disease (CHD) and lung cancer (LC) when comparing smokers (21–39 cigarettes per day) to nonsmokers.

Age	*CHD*	*LC*
55–64	1.9	10
65–74	1.7	9

The incidence ratio of 10 means that smokers are 10 times more likely than nonsmokers to die of lung cancer between the ages of 55 and 64. If the incidence ratio is x, then the percent P (in decimal form) of deaths caused by smoking can be calculated using the rational function

$$P(x) = \frac{x - 1}{x}.$$

(Data from Walker, A., *Observation and Inference: An Introduction to the Methods of Epidemiology*, Epidemiology Resources Inc.)

(a) As x increases, what value does $P(x)$ approach?

(b) Why might the incidence ratios be slightly less for ages 65–74 than for ages 55–64?

61. Braking distance for automobiles traveling at x mph, where $20 \le x \le 70$, can be modeled by the rational function

$$d(x) = \frac{8710x^2 - 69{,}400x + 470{,}000}{1.08x^2 - 324x + 82{,}200}.$$

(Data from Mannering, F., and W. Kilareski, *Principles of Highway Engineering and Traffic Control*, 2nd ed., John Wiley & Sons.)

(a) Use graphing to estimate x to the nearest unit when $d(x) = 300$.

(b) Complete the table for each value of x, to the nearest unit.

x	20	25	30	35	40	45	50	55	60	65	70
d(x)											

(c) If a car doubles its speed, does the stopping distance double or more than double? Explain.

(d) **Concept Check** Suppose the stopping distance doubled whenever the speed doubled. What type of relationship would exist between the stopping distance and the speed?

62. The **grade** x of a hill is a measure of its steepness. For example, if a road rises 10 ft for every 100 ft of horizontal distance, then it has an uphill grade of

$$x = \frac{10}{100}, \quad \text{or} \quad 10\%.$$

Grades are typically kept quite small—usually less than 10%. The braking distance D for a car traveling at 50 mph on a wet, uphill grade is given by

$$D(x) = \frac{2500}{30(0.3 + x)}.$$

(Data from Haefner, L., *Introduction to Transportation Systems*, Holt, Rinehart and Winston.)

(a) Evaluate $D(0.05)$ and interpret the result.

(b) Describe what happens to braking distance as the hill becomes steeper. Does this agree with your driving experience?

(c) Estimate the grade associated with a braking distance of 220 ft.

RELATING CONCEPTS For Individual or Group Work **(Exercises 63–72)**

Consider the following "monster" rational function.

$$f(x) = \frac{x^4 - 3x^3 - 21x^2 + 43x + 60}{x^4 - 6x^3 + x^2 + 24x - 20}$$

Analyzing this function will synthesize many of the concepts of this and earlier sections. ***Work Exercises 63–72 in order.***

63. Find the equation of the horizontal asymptote.

64. Given that -4 and -1 are zeros of the numerator, factor the numerator completely.

65. Given that 1 and 2 are zeros of the denominator, factor the denominator completely.

66. Write the entire quotient for f so that the numerator and the denominator are in factored form.

Answer each question.

67. (a) What is the common factor in the numerator and the denominator?

(b) For what value of x will there be a point of discontinuity (a hole)?

68. What are the x-intercepts of the graph of f?

69. What is the y-intercept of the graph of f?

70. Find the equations of the vertical asymptotes.

71. Determine the point or points of intersection of the graph of f with its horizontal asymptote.

72. Sketch the graph of f.

Chapter 11 Summary

Key Terms

11.1
polynomial function of degree n
zero (of a function)
root (solution)

11.2
multiplicity of a zero
complex conjugate

11.3
turning points
end behavior
dominating term

11.4
rational function
discontinuous graph
reciprocal function
vertical asymptote
horizontal asymptote
oblique (slant) asymptote
point of discontinuity (hole)

New Symbols

$\bar{c}$ the complex conjugate of the complex number c

$x \to a$ x approaches a

Test Your Word Power

See how well you have learned the vocabulary in this chapter.

1. **Synthetic division** is
   - A. a method for dividing any two polynomials
   - B. a method for multiplying two binomials
   - C. a shortcut method for dividing a polynomial by a binomial of the form $x - k$
   - D. the process of adding to a binomial the number that makes it a perfect square trinomial.

2. A **polynomial function** is a function f of the form
   - A. $f(x) = \sqrt{x}$, for $x \geq 0$
   - B. $f(x) = |mx + b|$
   - C. $f(x) = a_n x^n + a_{n-1}x^{n-1} + \cdots + a_1 x + a_0$ for complex numbers $a_n, a_{n-1}, \ldots, a_1$, and a_0, where $a_n \neq 0$
   - D. $f(x) = \frac{P(x)}{Q(x)}$, where P and Q are polynomials, and $Q(x) \neq 0$.

3. A **zero of a function** f is
   - A. a point where the function changes from increasing to decreasing or from decreasing to increasing
   - B. a value of x that satisfies $f(x) = 0$
   - C. a value for which the function is undefined
   - D. a point where the graph of the function intersects the x-axis or the y-axis.

4. A **turning point** is
   - A. a point where the function changes from increasing to decreasing or from decreasing to increasing
   - B. a value of x that satisfies $f(x) = 0$
   - C. a value for which the function is undefined
   - D. a point where the graph of the function intersects the x-axis or the y-axis.

5. A **rational function** is a function f of the form
   - A. $f(x) = \sqrt{x}$, for $x \geq 0$
   - B. $f(x) = |mx + b|$
   - C. $f(x) = a_n x^n + a_{n-1}x^{n-1} + \cdots + a_1 x + a_0$ for complex numbers $a_n, a_{n-1}, \ldots, a_1$, and a_0, where $a_n \neq 0$
   - D. $f(x) = \frac{P(x)}{Q(x)}$, where P and Q are polynomials, and $Q(x) \neq 0$.

6. An **asymptote** is
   - A. a line that a graph intersects just once
   - B. the x-axis or y-axis
   - C. a line about which a graph is symmetric
   - D. a line that a graph approaches more closely as x approaches a certain value or as $|x| \to \infty$.

ANSWERS

1. C; *Example:* See the first section of the Quick Review. **2.** C; *Examples:* $f(x) = x^2 - 4x + 5$, $f(x) = 6x^3 - 31x^2 - 15x$, $f(x) = x^4 + 2$
3. B; *Example:* 3 is a zero of the polynomial function $f(x) = x^2 - 3x$ because $f(3) = 0$. **4.** A; *Example:* The graph of $f(x) = 2x^3 + 5x^2 - x - 6$ has two turning points. See **FIGURE 14** in this chapter. **5.** D; *Examples:* $f(x) = \frac{4}{x}$, $f(x) = \frac{1}{x-5}$, $f(x) = \frac{x^2 - 9}{x + 1}$
6. D; *Example:* The graph of $f(x) = \frac{2}{x+4}$ has $x = -4$ as a vertical asymptote.

Quick Review

CONCEPTS	EXAMPLES
11.1 Zeros of Polynomial Functions (I)	
Division Algorithm Let $f(x)$ and $g(x)$ be polynomials with $g(x)$ of lesser degree than $f(x)$, and $g(x)$ of degree 1 or more. There exist unique polynomials $q(x)$ and $r(x)$ such that $$f(x) = g(x) \cdot q(x) + r(x),$$ where either $r(x) = 0$ or the degree of $r(x)$ is less than the degree of $g(x)$.	Express the polynomial function $$f(x) = 2x^3 - 3x + 2$$ in the form $f(x) = (x - k)q(x) + r$ for $k = 1$. Use synthetic division to divide $f(x)$ by $x - 1$.
Synthetic Division Synthetic division is a shortcut procedure for polynomial division that eliminates writing the variable factors. It is used only when dividing a polynomial by a binomial of the form $x - k$.	$$\begin{array}{r\|rrrr} 1 & 2 & 0 & -3 & 2 \\ & & 2 & 2 & -1 \\ \hline & 2 & 2 & -1 & 1 \end{array} \leftarrow f(1) = 1$$ Coefficients of the quotient; Remainder
Remainder Theorem If the polynomial $f(x)$ is divided by $x - k$, then the remainder is equal to $f(k)$.	$$f(x) = (x - 1)(2x^2 + 2x - 1) + 1$$
11.2 Zeros of Polynomial Functions (II)	
Factor Theorem The polynomial $x - k$ is a factor of the polynomial $f(x)$ if and only if $f(k) = 0$.	For the polynomial function $$f(x) = x^3 + x + 2,$$ $f(-1) = 0$. Therefore, $x - (-1)$, or $x + 1$, is a factor of $f(x)$.
Rational Zeros Theorem For $a_n \neq 0$, let $$f(x) = a_nx^n + a_{n-1}x^{n-1} + \cdots + a_1x + a_0$$ define a polynomial function with integer coefficients. If $\frac{p}{q}$ is a rational number written in lowest terms, and if $\frac{p}{q}$ is a zero of f, then p is a factor of the constant term a_0, and q is a factor of the leading coefficient a_n.	The only rational numbers that can possibly be zeros of $$f(x) = 2x^3 - 9x^2 - 4x - 5$$ are ± 1, $\pm \frac{1}{2}$, ± 5, and $\pm \frac{5}{2}$. By synthetic division, it can be shown that the only rational zero of $f(x)$ is 5. $$\begin{array}{r\|rrrr} 5 & 2 & -9 & -4 & -5 \\ & & 10 & 5 & 5 \\ \hline & 2 & 1 & 1 & 0 \end{array} \leftarrow f(5)$$
Fundamental Theorem of Algebra Every polynomial function of degree 1 or more has at least one complex zero.	$f(x) = x^3 + x + 2$ has at least one and at most three distinct zeros.
Number of Zeros Theorem A polynomial function of degree n has at most n distinct zeros.	
Conjugate Zeros Theorem If $f(x)$ is a polynomial function *having only real coefficients* and if $a + bi$ is a zero of $f(x)$, where a and b are real numbers, then $a - bi$ is also a zero of $f(x)$.	Because $1 + 2i$ is a zero of $f(x) = x^3 - 5x^2 + 11x - 15$, its conjugate $1 - 2i$ is a zero as well.

CONCEPTS	EXAMPLES

11.3 Graphs and Applications of Polynomial Functions

Graphing Using Translations
The graph of a function of the form

$$f(x) = a(x - h)^n + k$$

can be found by considering the effects of the constants a, h, and k on the graph of $y = x^n$.

Graph $f(x) = -(x + 2)^4 + 1$.

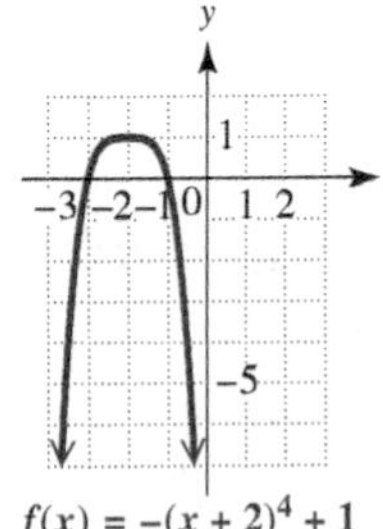

$f(x) = -(x + 2)^4 + 1$

The negative sign causes the graph to be reflected across the x-axis as compared with the graph of $y = x^4$.

The graph is also translated to the left 2 units and up 1 unit.

Turning Points
A polynomial function of degree n has at most $n - 1$ turning points, with at least one turning point between each pair of successive zeros.

The graph of

$$f(x) = 4x^5 - 2x^3 + 3x^2 + x - 10$$

has at most $5 - 1 = 4$ turning points.

End Behavior
The end behavior of the graph of a polynomial function $f(x)$ is determined by the dominating term, or term of greatest degree. If ax^n is the dominating term of $f(x)$, then the end behavior is as follows.

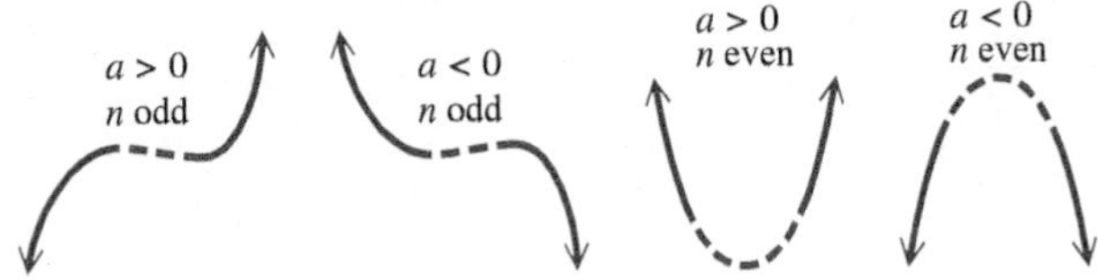

The end behavior of the graph of

$$f(x) = 3x^5 + 2x^2 + 7$$

is ↙↗.

The end behavior of the graph of

$$f(x) = -x^4 - 3x^3 + 2x - 9$$

is ↙↘.

Graphing a Polynomial Function

Step 1 Find the real zeros of f. Plot the corresponding x-intercepts.

Step 2 Find $f(0) = a_0$. Plot the corresponding y-intercept.

Step 3 Use end behavior, whether the graph crosses, bounces on, or wiggles through the x-axis at the x-intercepts, and selected points as necessary to complete the graph.

Graph $f(x) = (x + 2)(x - 1)(x + 3)$.

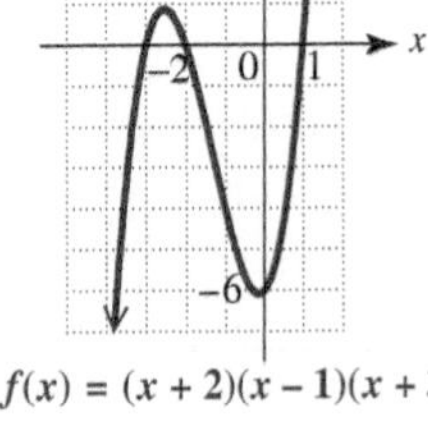

$f(x) = (x + 2)(x - 1)(x + 3)$

The x-intercepts correspond to the zeros of f, which are -2, 1, and -3.

Because $f(0) = 2(-1)(3) = -6$, the y-intercept is $(0, -6)$. The dominating term is $x(x)(x)$, or x^3,

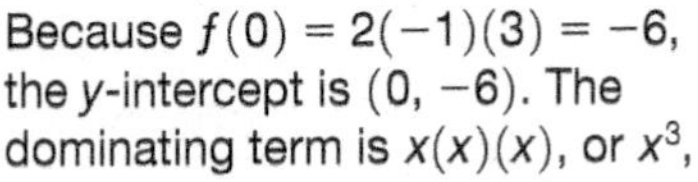

so the end behavior is ↙↗.

Begin at either end of the graph with the correct end behavior, and draw a smooth curve that crosses the x-axis at each zero, has a turning point between successive zeros, and passes through the y-intercept.

Intermediate Value Theorem
If $f(x)$ is a polynomial function *with only real coefficients,* and if for real numbers a and b the values of $f(a)$ and $f(b)$ are opposite in sign, then there exists at least one real zero between a and b.

For the polynomial function

$$f(x) = -x^4 + 2x^3 + 3x^2 + 6,$$

$$f(3.1) = 2.0599 \quad \text{and} \quad f(3.2) = -2.6016.$$

Because $f(3.1) > 0$ and $f(3.2) < 0$, there exists at least one real zero between 3.1 and 3.2.

CONCEPTS	EXAMPLES

Boundedness Theorem

Let $f(x)$ be a polynomial function of degree $n \geq 1$ *with real coefficients* and with a *positive* leading coefficient. Suppose $f(x)$ is divided synthetically by $x - c$.

(a) If $c > 0$ and all numbers in the bottom row of the synthetic division are nonnegative, then $f(x)$ has no real zero greater than c.

(b) If $c < 0$ and the numbers in the bottom row of the synthetic division alternate in sign (with 0 considered positive or negative, as needed), then $f(x)$ has no real zero less than c.

Show that $f(x) = x^3 - x^2 - 8x + 12$ has no real zero greater than 4 and no real zero less than -4.

$$\begin{array}{r|rrrr} 4 & 1 & -1 & -8 & 12 \\ & & 4 & 12 & 16 \\ \hline & 1 & 3 & 4 & 28 \end{array} \leftarrow \text{All signs positive}$$

$$\begin{array}{r|rrrr} -4 & 1 & -1 & -8 & 12 \\ & & -4 & 20 & -48 \\ \hline & 1 & -5 & 12 & -36 \end{array} \leftarrow \text{Alternating signs}$$

11.4 Graphs and Applications of Rational Functions

Graphing a Rational Function

To graph a rational function

$$f(x) = \frac{P(x)}{Q(x)},$$

where $P(x)$ and $Q(x)$ are polynomials and the rational expression is *written in lowest terms,* find asymptotes and intercepts. Determine whether the graph intersects the nonvertical asymptote. Plot a few points, as necessary, to complete the sketch.

Graph $f(x) = \dfrac{x^2 - 1}{(x + 3)(x - 2)}$.

$f(x) = \dfrac{x^2-1}{(x+3)(x-2)}$

The x-intercepts are $(-1, 0)$ and $(1, 0)$.

The y-intercept is $\left(0, \frac{1}{6}\right)$.

The vertical asymptotes are $x = -3$ and $x = 2$, and the horizontal asymptote is $y = 1$.

Point of Discontinuity

If a rational function is *not* written in lowest terms—that is, it has a common variable factor in the numerator and denominator—then there will be a hole in the graph instead of an asymptote.

Graph $f(x) = \dfrac{x^2 - 1}{x + 1}$.

$$f(x) = \frac{x^2 - 1}{x + 1} = \frac{(x - 1)(x + 1)}{x + 1} = x - 1 \qquad (x \neq -1)$$

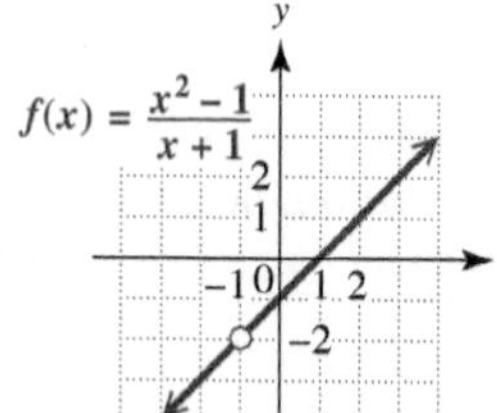

The graph is the same as that of $y = x - 1$, with a hole at $(-1, -2)$.

Chapter 11 Review Exercises

11.1 *Use synthetic division to divide.*

1. $\dfrac{3x^2 - x - 2}{x - 1}$

2. $\dfrac{10x^2 - 3x - 15}{x + 2}$

3. $\dfrac{2x^3 - 5x^2 + 12}{x - 3}$

4. $\dfrac{-x^4 + 19x^2 + 18x + 15}{x + 4}$

Use synthetic division to determine whether -5 is a zero of the polynomial function.

5. $f(x) = 2x^3 + 8x^2 - 14x - 20$

6. $f(x) = -3x^4 + 2x^3 + 5x^2 - 9x + 1$

For each polynomial function, use the remainder theorem and synthetic division to find $f(k)$.

7. $f(x) = 3x^3 - 5x^2 + 4x - 1;\quad k = -1$

8. $f(x) = x^4 - 2x^3 - 9x - 5;\quad k = 3$

11.2 *Find all rational zeros of each polynomial function.*

9. $f(x) = 2x^3 - 9x^2 - 6x + 5$

10. $f(x) = 3x^3 - 10x^2 - 27x + 10$

11. $f(x) = x^3 - \frac{17}{6}x^2 - \frac{13}{3}x - \frac{4}{3}$

12. $f(x) = 8x^4 - 14x^3 - 29x^2 - 4x + 3$

Determine whether $(x + 1)$ is a factor of $f(x)$. If it is, express it with linear factors.

13. $f(x) = x^3 + 2x^2 + 3x - 1$

14. $f(x) = 2x^3 + x^2 - 16x - 15$

Work each problem.

15. Find a polynomial function $f(x)$ of degree 3 with real coefficients, having -2, 1, and 4 as zeros, and $f(2) = 16$.

16. Find a polynomial function $f(x)$ of least possible degree with real coefficients, having 2, -2, and $-i$ as zeros, and $f(1) = -6$.

17. Find a polynomial function $f(x)$ of least possible degree with real coefficients, having 2, -3, and $5i$ as zeros, and $f(0) = -150$.

18. Find a polynomial function $f(x)$ of least possible degree with real coefficients, having -3 and $1 - i$ as zeros, and $f(1) = 4$.

19. Find all zeros of f, if $f(x) = x^4 - 3x^3 - 8x^2 + 22x - 24$, given that $1 - i$ is a zero, and factor $f(x)$.

20. Suppose that a polynomial function has six real zeros. Is it possible for the function to be of degree 5? Explain.

11.3 *Determine the maximum possible number of turning points of the graph of each polynomial function.*

21. $f(x) = x^3 - 9x$

22. $f(x) = 4x^4 - 6x^2 + 2$

Graph each polynomial function.

23. $f(x) = x^3 + 5$

24. $f(x) = 1 - x^4$

25. $f(x) = x^2(2x + 1)(x - 2)$

26. $f(x) = 2x^3 + 13x^2 + 15x$

27. $f(x) = 12x^3 - 13x^2 - 5x + 6$

28. $f(x) = x^4 - 2x^3 - 5x^2 + 6x$

Show that the real zeros of each polynomial function satisfy the given conditions.

29. $f(x) = 3x^3 - 8x^2 + x + 2;$ zeros in $[-1, 0]$ and $[2, 3]$

30. $f(x) = x^3 + 2x^2 - 22x - 8;$ zeros in $[-1, 0]$ and $[-6, -5]$

31. $f(x) = 6x^4 + 13x^3 - 11x^2 - 3x + 5;$ no zero greater than 1 or less than -3

32. Approximate the real zeros of each polynomial function. Round to the nearest thousandth for irrational zeros.

(a) $f(x) = 2x^3 - 11x^2 - 2x + 2$

(b) $f(x) = x^4 - 4x^3 - 5x^2 + 14x - 15$

(c) $f(x) = x^3 + 3x^2 - 4x - 2$

Graph each polynomial function.

33. $f(x) = 2x^3 - 11x^2 - 2x + 2$ (See **Exercise 32(a).**)

34. $f(x) = x^4 - 4x^3 - 5x^2 + 14x - 15$ (See **Exercise 32(b).**)

35. $f(x) = x^3 + 3x^2 - 4x - 2$ (See **Exercise 32(c).**)

36. $f(x) = 2x^4 - 3x^3 + 4x^2 + 5x - 1$

11.4 *Graph each rational function.*

37. $f(x) = \frac{8}{x}$

38. $f(x) = \frac{2}{3x - 1}$

39. $f(x) = \frac{4x - 2}{3x + 1}$

40. $f(x) = \frac{6x}{(x - 1)(x + 2)}$

41. $f(x) = \frac{2x}{x^2 - 1}$

42. $f(x) = \frac{x^2 + 4}{x + 2}$

43. $f(x) = \frac{x^2 - 1}{x}$

44. $f(x) = \frac{x^2 + 6x + 5}{x - 3}$

45. $f(x) = \frac{4x^2 - 9}{2x + 3}$

46. $f(x) = \frac{(x + 4)(2x + 5)}{x - 1}$

*The **Laffer curve,** an idealized version of which is shown here, suggests that increasing a tax rate (say from x_1 percent to x_2 percent on the graph) can actually lead to a decrease in government revenue. All economists agree on the endpoints, 0 revenue at tax rates of both 0% and 100%, but there is much disagreement on the location of the rate x_1 that produces maximum revenue.*

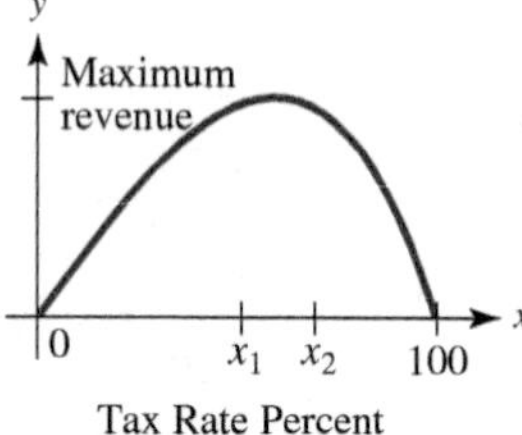

47. Suppose an economist studying the Laffer curve produces the rational function

$$R(x) = \frac{80x - 8000}{x - 110},$$

where $R(x)$ is government revenue, in tens of millions of dollars, from a tax rate of x percent, with $R(x)$ valid for $55 \leq x \leq 100$. Find the revenue for each tax rate.

(a) 55% **(b)** 60% **(c)** 70% **(d)** 90% **(e)** 100%

48. What are the equations of the vertical and horizontal asymptotes of the graph of $y = R(x)$ in **Exercise 47?**

Chapter 11 Mixed Review Exercises

1. Consider the polynomial function

$$f(x) = x^3 + 7x^2 + 7x - 15.$$

(a) Given that -3 is a zero of f, find all other zeros.

(b) Factor $f(x)$ into linear factors.

(c) What are the x-intercepts of the graph of f? What is the y-intercept?

(d) Sketch the graph.

2. Consider the rational function

$$f(x) = \frac{x^2 - 4}{x^2 - 9}.$$

(a) Give the equations of the vertical asymptotes of the graph of f.

(b) What are the x-intercepts of the graph of f?

(c) What is the equation of the horizontal asymptote of the graph of f?

(d) Sketch the graph.

Graph each function.

3. $f(x) = \dfrac{-4x + 3}{2x + 1}$

4. $f(x) = 3x^3 + 2x^2 - 27x - 18$

Solve each problem.

5. Express the polynomial function

$$f(x) = 2x^3 + x - 6$$

in the form $f(x) = (x - k)q(x) + r$ for $k = -2$.

6. Find a polynomial function $f(x)$ of degree 4 with real coefficients, having 1, -1, and $3i$ as zeros, and $f(2) = 39$.

7. For the polynomial function $f(x) = x^3 - 3x^2 - 7x + 12$, we have $f(4) = 0$. Therefore, we can say that 4 is a(n) ______ of the function, 4 is a(n) ______ of the equation $x^3 - 3x^2 - 7x + 12 = 0$, and $(4, 0)$ is a(n) ______ of the graph of the function.

8. Which one of the following is not a polynomial function?

A. $f(x) = x^2$ B. $f(x) = 2x + 5$ C. $f(x) = \dfrac{1}{x}$ D. $f(x) = x^{100}$

Chapter 11 Test

Step-by-step test solutions are found on the Chapter Test Prep Videos available in MyLab Math.

View the complete solutions to all Chapter Test exercises in MyLab Math.

Use synthetic division to work each problem.

1. Find the quotient when $2x^3 - 3x - 10$ is divided by $x - 2$.

2. Express the polynomial function

$$f(x) = x^4 + 2x^3 - x^2 + 3x - 5$$

in the form $f(x) = (x - k)q(x) + r$ for $k = -1$.

3. Determine whether $x + 3$ is a factor of $x^4 - 2x^3 - 15x^2 - 4x - 12$.

4. Evaluate $f(-4)$ if $f(x) = 3x^3 - 4x^2 - 5x + 9$.

5. Is 3 a zero of $f(x) = 6x^4 - 11x^3 - 35x^2 + 34x + 24$? Why or why not?

Work each problem.

6. Find a polynomial function $f(x)$ of degree 4 with real coefficients, having 2, -1, and i as zeros, and $f(3) = 80$.

7. Consider the polynomial function

$$f(x) = 6x^3 - 25x^2 + 12x + 7.$$

(a) List all possible rational zeros.

(b) Find all rational zeros.

8. Show that the polynomial function

$$f(x) = 2x^4 - 3x^3 + 4x^2 - 5x - 1$$

has no real zeros greater than 2 or less than -1.

9. Consider the polynomial function

$$f(x) = 2x^3 - x + 3.$$

(a) Use the intermediate value theorem to show that there is a zero between -2 and -1.

(b) Use a graphing calculator to approximate this zero to the nearest thousandth.

10. Consider the graph of the polynomial function

$$f(x) = 2x^3 - 9x^2 + 4x + 8.$$

Answer the following without actually graphing.

(a) Determine the maximum possible number of x-intercepts.

(b) Determine the maximum possible number of turning points.

Graph each polynomial function.

11. $f(x) = (x - 1)^4$

12. $f(x) = x(x + 1)(x - 2)$

13. $f(x) = 2x^3 - 7x^2 + 2x + 3$

14. $f(x) = x^4 - 5x^2 + 6$

15. Suppose that the polynomial function

$$f(x) = -0.184x^3 + 1.45x^2 + 10.7x - 27.9$$

models the average temperature in degrees Fahrenheit, where $x = 1$ corresponds to January and $x = 12$ to December. What is the average temperature in June?

Graph each rational function.

16. $f(x) = \dfrac{-2}{x + 3}$

17. $f(x) = \dfrac{3x - 1}{x - 2}$

18. $f(x) = \dfrac{x^2 - 1}{x^2 - 9}$

19. $f(x) = \dfrac{2x^2 + x - 6}{x - 1}$

20. Which one of the following functions has a graph with no x-intercepts?

A. $f(x) = (x - 2)(x + 3)^2$

B. $f(x) = \dfrac{x + 7}{x - 2}$

C. $f(x) = x^3 - x$

D. $f(x) = \dfrac{1}{x^2 + 4}$

Chapters R–11 Cumulative Review Exercises

Solve each problem.

1. Write $\frac{16}{64}$ in lowest terms.

2. Multiply $\frac{8}{17} \cdot \frac{51}{4}$. Write the answer in lowest terms.

3. Add $\frac{9}{4} + \frac{2}{3}$. Write the answer in lowest terms.

4. Write 0.473 as a percent.

5. Write 326% as a decimal.

6. Write $\frac{3}{4}$ as a percent.

Solve each equation or inequality.

7. $4x + 8x = 17x - 10$

8. $0.10(x - 6) + 0.05x = 0.06(50)$

9. $\frac{x+1}{3} + \frac{2x}{3} = x + \frac{1}{3}$

10. $4(x + 2) \geq 6x - 8$

11. $5 < 3x - 4 < 9$

12. $|x + 2| - 3 > 2$

Solve each problem.

13. A jar of 38 coins contains only nickels and dimes. The amount of money in the jar is \$2.50. How many nickels and how many dimes are in the jar?

14. Find an equation of the line passing through the point $(5, -3)$ and perpendicular to the graph of $y = \frac{1}{3}x - 6$. Give it in slope-intercept form.

Graph each equation or inequality.

15. $-3x + 5y = -15$

16. $y \leq -2x + 7$

17. $f(x) = x^2$

Solve each system.

18. $3x + 2y = -4$
$y = 2x + 5$

19. $-2x + y - 4z = 2$
$3x + 2y - z = -3$
$-x - 4y - 2z = -17$

Perform the indicated operations.

20. $(r^4 - 2r^3 + 6)(3r - 1)$

21. $[(k - 5h) + 2]^2$

22. Solve $x^3 + 3x^2 - x - 3 = 0$. (*Hint:* Factor by grouping.)

Factor.

23. $6x^2 - 15x - 9$

24. $729 + 8y^6$

25. Consider the rational expression $\frac{x^2 - 36}{4x^2 - 21x - 18}$.

(a) For what values of x is this expression undefined?

(b) Express it in lowest terms.

26. Simplify the complex fraction $\frac{\frac{1}{y} + \frac{1}{y-1}}{\frac{1}{y} - \frac{2}{y-1}}$.

Perform each operation and express the answer in lowest terms.

27. $\frac{2r+4}{5r} \cdot \frac{3r}{5r+10}$

28. $\frac{y^2 - 2y - 3}{y^2 + 4y + 4} \div \frac{y^2 - 1}{y^2 + y - 2}$

29. $\frac{3x+12}{2x+7} + \frac{-7x-26}{2x+7}$

30. $\frac{2}{r-2} - \frac{r+3}{r-1}$

Solve each equation.

31. $\frac{10}{x^2} - \frac{3}{x} = 1$

32. $\frac{3}{x+1} = \frac{1}{x-1} - \frac{2}{x^2-1}$

Simplify each radical expression.

33. $\sqrt{32} - \sqrt{128} + \sqrt{162}$

34. $\frac{3 - 4\sqrt{2}}{1 - \sqrt{2}}$

Consider the quadratic equation $x^2 + 4x + 2 = 0$.

35. What is the value of the discriminant?

36. Based on the answer to **Exercise 35,** are the solutions *rational, irrational,* or *not real*?

37. Solve the equation.

Decide whether the graph of the relation is symmetric with respect to the x-axis, the y-axis, the origin, or none of these.

38. $x = y^2 + 3$

39. $y = x^2 + 3$

40. $y = 5x^3$

Work each problem.

41. Find $f(3)$ if $f(x) = \begin{cases} x^2 - 6 & \text{if } x \leq 3 \\ 2x & \text{if } x > 3. \end{cases}$

42. Find $f^{-1}(x)$ if $f(x) = \sqrt[3]{3x + 5}$.

43. Solve $\log_2 x + \log_2 (x + 2) - 3 = 0$.

44. Use synthetic division to divide.

$$\frac{x^4 + 7x^3 - 5x^2 + 2x + 13}{x + 1}$$

Graph each function.

45. $f(x) = -\frac{1}{x^3}$

46. $f(x) = \frac{x^3 + 1}{x + 1}$

STUDY SKILLS **REMINDER**

Have you begun to prepare for your final exam? **Review Study Skill 10, *Preparing for Your Math Final Exam.***

12

CONIC SECTIONS AND NONLINEAR SYSTEMS

An *ellipse,* one of a group of curves known as *conic sections,* has a special reflecting property responsible for "whispering galleries" like that in the Old House Chamber of the U.S. Capitol. We investigate ellipses in this chapter.

12.1 Circles Revisited and Ellipses

OBJECTIVES

1. Graph circles.
2. Write an equation of a circle given its center and radius.
3. Determine the center and radius of a circle given its equation.
4. Recognize the equation of an ellipse.
5. Graph ellipses.

VOCABULARY

- ☐ conic section
- ☐ circle
- ☐ center (of a circle)
- ☐ radius
- ☐ ellipse
- ☐ foci (singular *focus*)
- ☐ center (of an ellipse)

When an infinite cone is intersected by a plane, the resulting figure is called a **conic section.** The parabola is one example of a conic section. Circles, ellipses, and hyperbolas may also result. See **FIGURE 1.**

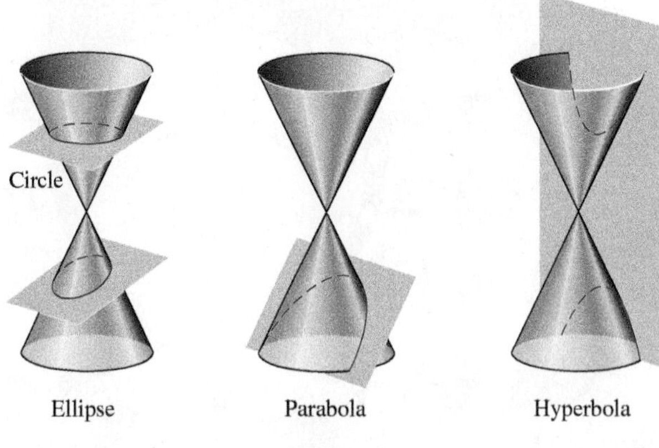

FIGURE 1

OBJECTIVE 1 Graph circles.

Recall from our earlier work that a **circle** is a set of all points in a plane that lie a fixed distance from a fixed point. This fixed point is the **center** of the circle. The fixed distance is its **radius.**

Suppose that the point (h, k) is the center of a circle with radius r. We let (x, y) represent any point on the circle. See **FIGURE 2.** By definition, the distance between (h, k) and (x, y) must be r.

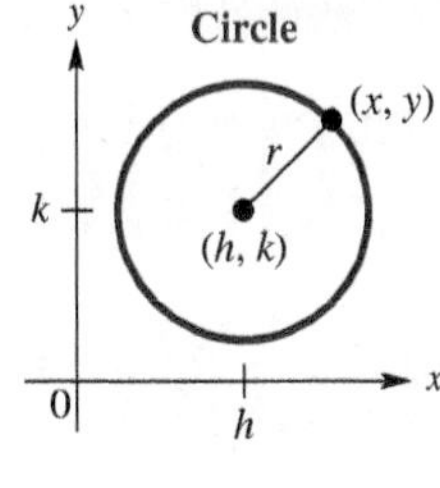

FIGURE 2

$$\sqrt{(x_2 - x_1)^2 + (y_2 - y_1)^2} = r \quad \text{Distance formula}$$

$$\sqrt{(x - h)^2 + (y - k)^2} = r \quad \text{Let } (h, k) = (x_1, y_1) \text{ and } (x, y) = (x_2, y_2).$$

$$(x - h)^2 + (y - k)^2 = r^2 \quad \text{Square each side.}$$

This result is the *center-radius form* of the equation of a circle.

Equation of a Circle (Center-Radius Form)

A circle with center (h, k) and radius r has an equation that can be written in the form

$$(x - h)^2 + (y - k)^2 = r^2 \quad \textbf{(where } r > 0\textbf{).}$$

If a circle has its center at the origin $(0, 0)$, then its equation becomes

$$x^2 + y^2 = r^2. \quad \text{Here, } h = 0 \text{ and } k = 0.$$

Find the center and radius of each circle. Then graph the circle.

(a) $(x-2)^2 + (y-3)^2 = 9$

(b) $x^2 + y^2 = 49$

EXAMPLE 1 Graphing Circles

Find the center and radius of each circle. Then graph the circle.

(a)
$$(x-3)^2 + (y+1)^2 = 16$$

The radius r is the positive square root of 16.

$$(x-3)^2 + [y-(-1)]^2 = 4^2 \qquad (x-h)^2 + (y-k)^2 = r^2$$

(h, k, r indicated by arrows under 3, −1, and 4.)

The center (h, k) is $(3, -1)$, and the radius r is 4.

To graph the circle, plot the center $(3, -1)$, and then move right, left, up, and down 4 units from the center to plot four points on the circle. Draw a smooth curve through the four points. See **FIGURE 3.** (The center is *not* part of the graph.)

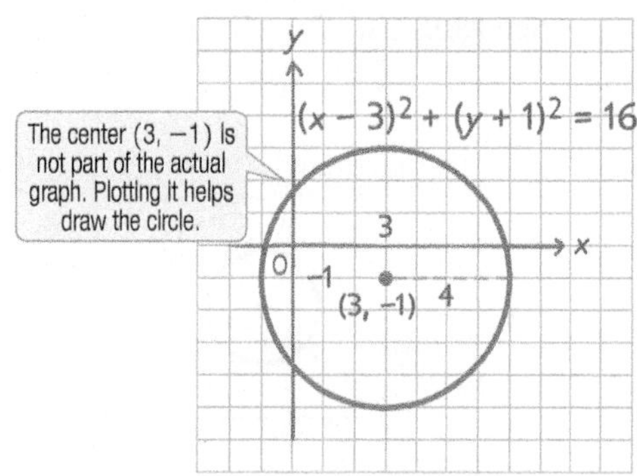

FIGURE 3

(b)
$$x^2 + y^2 = 64$$

$$(x-0)^2 + (y-0)^2 = 8^2 \qquad \text{Here, } h = 0 \text{ and } k = 0.$$

The center is $(0, 0)$ and the radius is 8. The graph, which uses a scale of 2 on the axes, is shown in **FIGURE 4.**

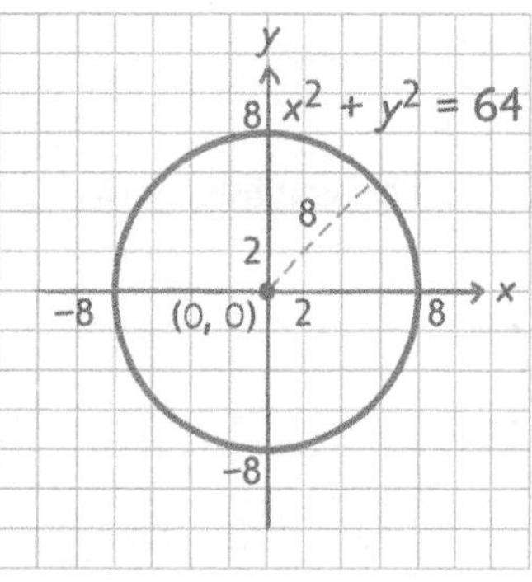

FIGURE 4

NOW TRY 

NOTE The x- and y-coordinates of the center of a circle are the values of x and y that make each squared binomial in its center-radius form equal 0. Consider the equation

$$(x-3)^2 + (y+1)^2 = 16. \qquad \textbf{See Example 1(a).}$$

If we let $x = 3$, then $(x-3)^2$ equals 0. If we let $y = -1$, then $(y+1)^2$ equals 0. Thus, the center of the circle is $(3, -1)$.

NOW TRY ANSWERS

1. (a) center: $(2, 3)$; radius: 3

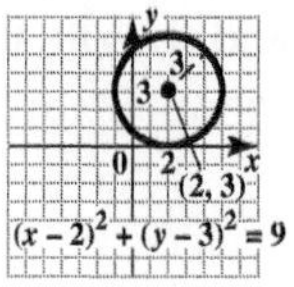

(graph: $(x-2)^2 + (y-3)^2 = 9$, center (2, 3))

(b) center: $(0, 0)$; radius: 7

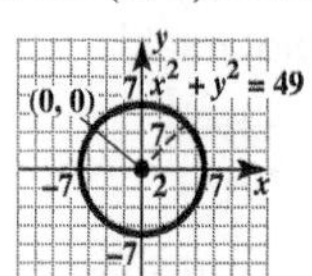

NOW TRY EXERCISE 2

Write the center-radius form of each circle described. Then graph the circle.

(a) Center: $(1, -2)$; radius: 3

(b) Center: $(0, 0)$; radius: $\sqrt{7}$

OBJECTIVE 2 Write an equation of a circle given its center and radius.

EXAMPLE 2 Writing Equations of Circles and Graphing Them

Write the center-radius form of each circle described. Then graph the circle.

(a) Center: $(-3, 4)$; radius: 2

Here, $(h, k) = (-3, 4)$ and $r = 2$.

$$(x - h)^2 + (y - k)^2 = r^2 \quad \text{Center-radius form}$$

$$[x - (-3)]^2 + (y - 4)^2 = 2^2 \quad \text{Substitute for } h, k, \text{ and } r.$$

Watch signs here.

$$(x + 3)^2 + (y - 4)^2 = 4 \quad \text{Simplify.}$$

See **FIGURE 5.**

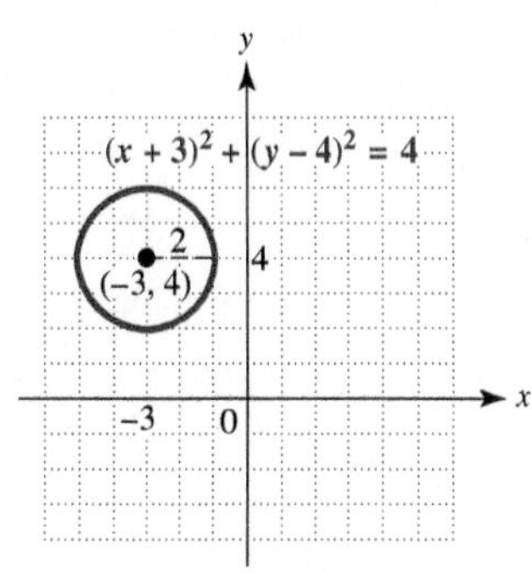

FIGURE 5

FIGURE 6

(b) Center: $(2, 0)$; radius: $\sqrt{10}$

$$(x - h)^2 + (y - k)^2 = r^2 \quad \text{Center-radius form}$$

$$(x - 2)^2 + (y - 0)^2 = (\sqrt{10})^2 \quad \text{Let } h = 2, k = 0, \text{ and } r = \sqrt{10}.$$

$$(x - 2)^2 + y^2 = 10 \quad \text{Simplify; } (\sqrt{a})^2 = a.$$

The graph is shown in **FIGURE 6.** For the radius,

$$r = \sqrt{10} \approx 3.16. \quad \text{Evaluate with a calculator.}$$

NOW TRY 

NOW TRY ANSWERS

2. (a) $(x - 1)^2 + (y + 2)^2 = 9$

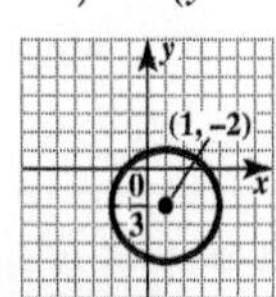

(b) $x^2 + y^2 = 7$

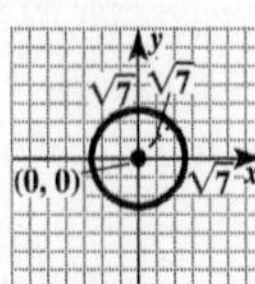

OBJECTIVE 3 Determine the center and radius of a circle given its equation.

If we begin with the center-radius form of the equation of a circle and rewrite it so that the binomials are expanded and the right side equals 0, we obtain another form of the equation. Consider the following.

$$(x + 3)^2 + (y - 4)^2 = 4 \quad \text{See Example 2(a).}$$

Remember the middle term when squaring each binomial.

$$x^2 + \mathbf{6x} + 9 + y^2 - \mathbf{8y} + 16 = 4 \quad \text{Square each binomial.}$$

$$x^2 + y^2 + 6x - 8y + 21 = 0 \quad \text{Subtract 4. Combine and rearrange terms.}$$

This result is a different, yet equivalent, form of the equation of the circle.

Equation of a Circle (General Form)

For some real numbers c, d, and e, an equation of the form

$$x^2 + y^2 + cx + dy + e = 0$$

may represent a circle.

In the above general form, both x^2- and y^2-terms have equal coefficients, here 1. If the coefficients are equal but *not* 1, the above form can be obtained by dividing through by that coefficient.

NOW TRY EXERCISE 3
Write the center-radius form of the circle with equation

$$x^2 + y^2 - 8x + 10y - 8 = 0.$$

Give the center and radius.

EXAMPLE 3 Completing the Square to Write the Center-Radius Form

Write the center-radius form of the circle with equation

$$x^2 + y^2 + 2x + 6y - 15 = 0.$$

Give the center and radius, and graph the circle.

To write the center-radius form, we complete the squares on x and y.

$$x^2 + y^2 + 2x + 6y - 15 = 0$$

$$x^2 + y^2 + 2x + 6y = 15 \quad \text{Transform so that the constant is on the right.}$$

$$(x^2 + 2x \qquad) + (y^2 + 6y \qquad) = 15 \quad \text{Write in anticipation of completing the square.}$$

$$\left[\tfrac{1}{2}(2)\right]^2 = 1 \qquad \left[\tfrac{1}{2}(6)\right]^2 = 9 \quad \text{Square half the coefficient of each middle term.}$$

$$(x^2 + 2x + 1) + (y^2 + 6y + 9) = 15 + 1 + 9 \quad \text{Complete the squares on both } x \text{ and } y.$$

Add 1 and 9 on *both* sides of the equation.

$$(x + 1)^2 + (y + 3)^2 = 25 \quad \text{Factor. Add. Center-radius form}$$

$$[x - (-1)]^2 + [y - (-3)]^2 = 5^2 \quad \text{Rewrite to identify the center and radius.}$$

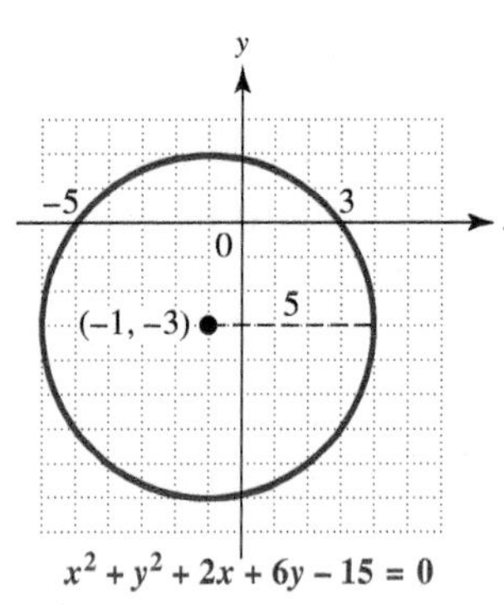

FIGURE 7

The circle has center $(-1, -3)$ and radius 5. See FIGURE 7. NOW TRY 

NOTE There are three possibilities for the graph of an equation of the form

$$(x - h)^2 + (y - k)^2 = m, \quad \text{where } m \text{ is a constant.}$$

1. If $m > 0$, then $r^2 = m$, and the graph of the equation is a circle with radius $\sqrt{m}$.
2. If $m = 0$, then the graph of the equation is the single point (h, k).
3. If $m < 0$, then no points satisfy the equation and the graph is nonexistent.

OBJECTIVE 4 Recognize the equation of an ellipse.

An **ellipse** is the set of all points in a plane the *sum* of whose distances from two fixed points is constant. These fixed points are the **foci** (singular: *focus*). The ellipse in FIGURE 8 has foci $(c, 0)$ and $(-c, 0)$, with x-intercepts $(a, 0)$ and $(-a, 0)$ and y-intercepts $(0, b)$ and $(0, -b)$. It is shown in more advanced courses that $c^2 = a^2 - b^2$ for an ellipse of this type. The origin is the **center** of the ellipse.

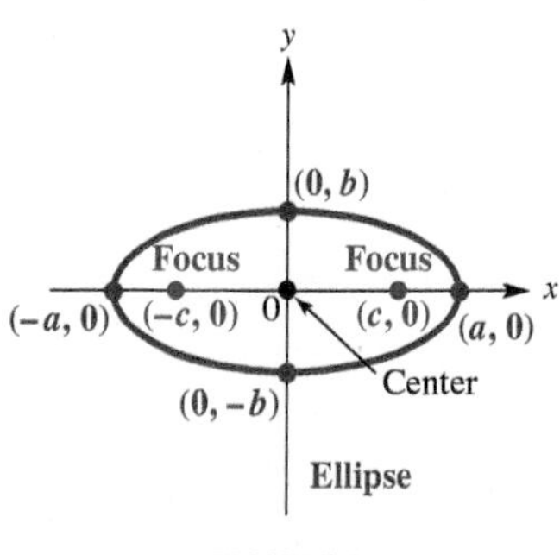

FIGURE 8

NOW TRY ANSWER
3. $(x - 4)^2 + (y + 5)^2 = 49$; center: $(4, -5)$; radius: 7

An ellipse has the following equation.

Equation of an Ellipse

An ellipse with x-intercepts $(a, 0)$ and $(-a, 0)$ and y-intercepts $(0, b)$ and $(0, -b)$ has an equation that can be written in the form

$$\frac{x^2}{a^2} + \frac{y^2}{b^2} = 1.$$

NOTE A circle is a special case of an ellipse, where $a^2 = b^2$.

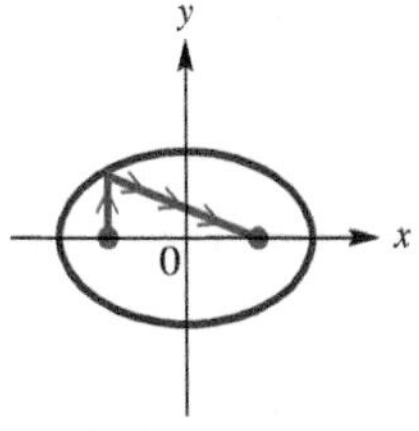

Reflecting property of an ellipse

FIGURE 9

When a ray of light or sound emanating from one focus of an ellipse bounces off the ellipse, it passes through the other focus. See **FIGURE 9**. This reflecting property is responsible for whispering galleries. In the Old House Chamber of the U.S. Capitol, John Quincy Adams was able to listen in on his opponents' conversations—his desk was positioned at one of the foci beneath the ellipsoidal ceiling, and his opponents were located across the room at the other focus. (See the chapter opener.)

Elliptical bicycle gears are designed to respond to the legs' natural strengths and weaknesses. At the top and bottom of the powerstroke, where the legs have the least leverage, the gear offers little resistance, but as the gear rotates, the resistance increases. This allows the legs to apply more power where it is most naturally available. See **FIGURE 10**.

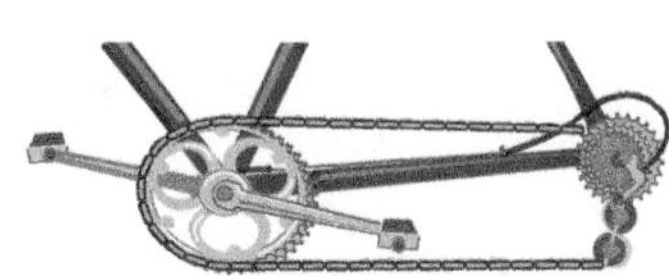

FIGURE 10

OBJECTIVE 5 Graph ellipses.

To graph an ellipse centered at the origin, we plot the four intercepts and sketch the ellipse through those points.

EXAMPLE 4 Graphing Ellipses

Graph each ellipse.

(a) $\frac{x^2}{49} + \frac{y^2}{36} = 1$

Here, $a^2 = 49$, so $a = 7$, and the x-intercepts are $(7, 0)$ and $(-7, 0)$. Similarly, $b^2 = 36$, so $b = 6$, and the y-intercepts are $(0, 6)$ and $(0, -6)$. Plotting the intercepts and sketching the ellipse through them gives the graph in **FIGURE 11**.

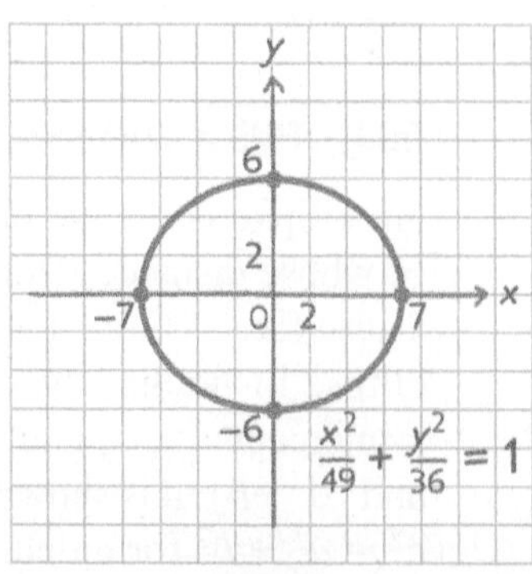

FIGURE 11

NOW TRY EXERCISE 4

Graph $\frac{x^2}{16} + \frac{y^2}{25} = 1$.

(b) $\frac{x^2}{36} + \frac{y^2}{121} = 1$

The x-intercepts are $(6, 0)$ and $(-6, 0)$, and the y-intercepts are $(0, 11)$ and $(0, -11)$. Join these intercepts with the smooth curve of an ellipse. See **FIGURE 12**. (Note the scale of 2 on the axes.)

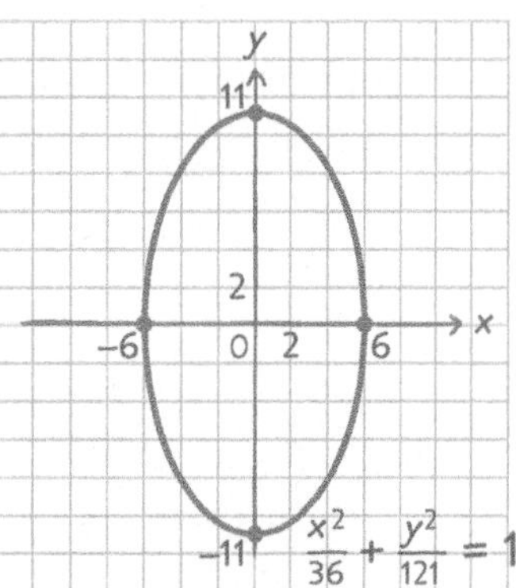

FIGURE 12

NOW TRY 

CAUTION Hand-drawn graphs of ellipses are smooth curves that show symmetry with respect to the center. Again, the center is not part of the actual graph. It just provides help in drawing a more accurate graph.

NOW TRY EXERCISE 5

Graph

$$\frac{(x-3)^2}{36} + \frac{(y-4)^2}{4} = 1.$$

EXAMPLE 5 Graphing an Ellipse Shifted Horizontally and Vertically

Graph $\frac{(x-2)^2}{25} + \frac{(y+3)^2}{49} = 1$.

Just as $(x-2)^2$ and $(y+3)^2$ would indicate that the center of a circle would be $(2, -3)$, so it is with this ellipse. **FIGURE 13** shows that the graph goes through the four points

$$(2, 4), \quad (7, -3), \quad (2, -10), \quad \text{and} \quad (-3, -3).$$

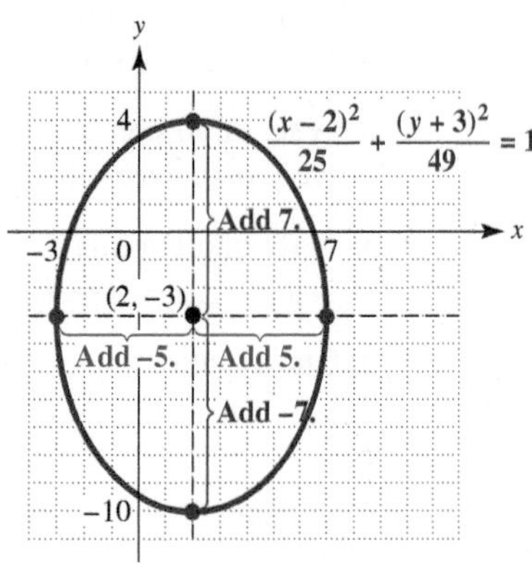

FIGURE 13

The x-values of these points are found by adding $\pm a = \pm 5$ to 2 (the x-value of the center). The y-values are found by adding $\pm b = \pm 7$ to -3 (the y-value of the center).

NOW TRY 

NOW TRY ANSWERS

4.

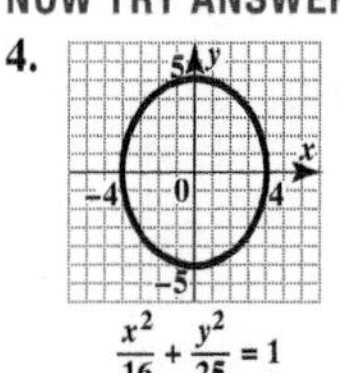

5.

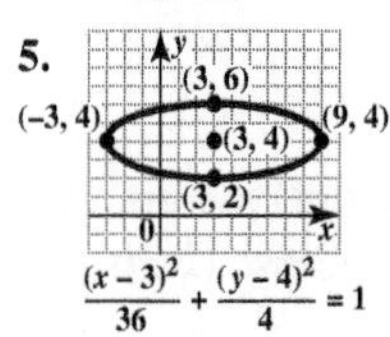

Graphs of circles and ellipses are not graphs of functions—they fail the conditions of the vertical line test. The only conic section whose graph represents a function is the vertical parabola with equation $f(x) = ax^2 + bx + c$.

NOTE A graphing calculator in function mode cannot directly graph a circle or an ellipse because they do not represent functions. We must first solve the equation for y to obtain two functions y_1 and y_2. The union of these two graphs is the graph of the entire figure. Consider the following equation of a circle.

$$(x+3)^2 + (y+2)^2 = 25$$ Solve the equation for y.

$$(y+2)^2 = 25 - (x+3)^2$$ Subtract $(x+3)^2$.

$$y + 2 = \pm\sqrt{25 - (x+3)^2}$$ Take square roots.

$$y = -2 \pm \sqrt{25 - (x+3)^2}$$ Add -2.

Remember both roots.

The two functions to be graphed are

$$y_1 = -2 + \sqrt{25 - (x+3)^2} \quad \text{and} \quad y_2 = -2 - \sqrt{25 - (x+3)^2}.$$

To get an undistorted screen, a **square viewing window** must be used. See **FIGURE 14**. The two semicircles seem to be disconnected because the calculator cannot show a true picture of the behavior at these points.

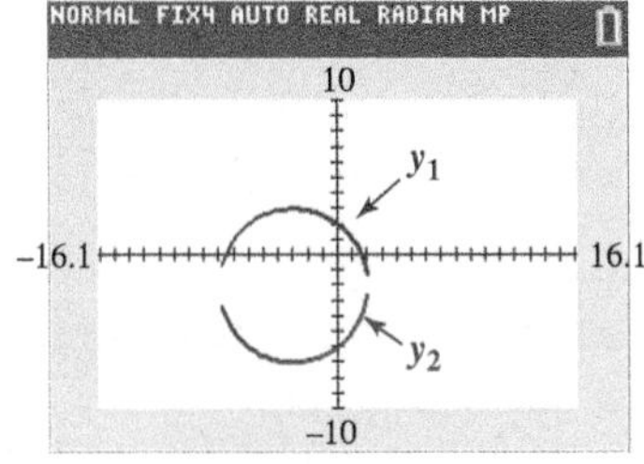

Square Viewing Window

FIGURE 14

12.1 Exercises

FOR EXTRA HELP MyLab Math

Video solutions for select problems available in MyLab Math

Concept Check *Complete each statement. Choices may be used once, more than once, or not at all.*

circle	parabolas	radius	intercepts
foci	center	ellipse	conic sections

1. When a plane intersects an infinite cone at different angles, ______ such as ______, circles, ellipses, and hyperbolas result.

2. A set of all points in a plane that lie a fixed distance from a fixed point is a(n) ______. The fixed distance is the ______ and the fixed point is the ______.

3. A set of all points in a plane the sum of whose distances from two fixed points is constant is a(n) ______. The fixed points are the ______.

4. The equation $(x+1)^2 + (y-4)^2 = 4$ represents a(n) ______ with center

   $[(1, -4)/(1, 4)/(-1, 4)]$ and ______ equal to 2.

5. **Concept Check** Consider the circle whose equation is $x^2 + y^2 = 25$.

   (a) What are the coordinates of its center? **(b)** What is its radius?

   (c) Sketch its graph.

6. **Concept Check** Complete the following: The graph of a circle (*is*/*is not*) the graph of a function because it fails the conditions of the ______. The graph of an ellipse (*is*/*is not*) the graph of a function.

Concept Check *Match each equation with the correct graph.*

7. $(x-3)^2+(y-1)^2=25$

8. $(x+3)^2+(y-1)^2=25$

9. $(x-3)^2+(y+1)^2=25$

10. $(x+3)^2+(y+1)^2=25$

A.

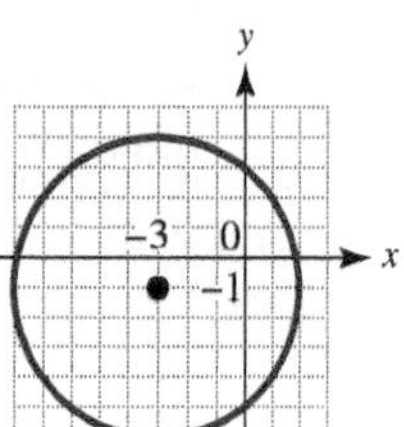

B.

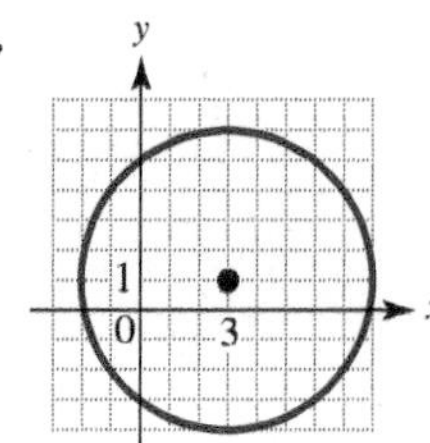

C.

D.

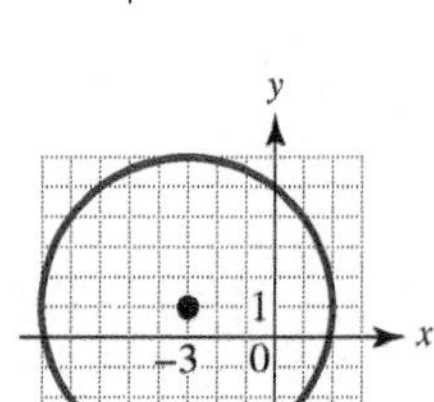

Find the center and radius of each circle. Then graph the circle. ***See Example 1.***

11. $x^2+y^2=4$

12. $x^2+y^2=1$

13. $x^2+y^2=81$

14. $x^2+y^2=36$

15. $(x-5)^2+(y+4)^2=49$

16. $(x+2)^2+(y-5)^2=16$

17. $(x+1)^2+(y+3)^2=25$

18. $(x-3)^2+(y-2)^2=4$

Write the center-radius form of each circle described. Then graph the circle. ***See Examples 1 and 2.***

19. Center: $(0, 0)$; radius: 4

20. Center: $(0, 0)$; radius: 3

21. Center: $(-3, 2)$; radius: 3

22. Center: $(1, -3)$; radius: 4

23. Center: $(4, 3)$; radius: 5

24. Center: $(-3, -2)$; radius: 6

25. Center: $(-2, 0)$; radius: $\sqrt{5}$

26. Center: $(3, 0)$; radius: $\sqrt{13}$

27. Center: $(0, -3)$; radius: 7

28. Center: $(0, 4)$; radius: 4

Write the center-radius form of the circle with the given equation. Give the center and radius. (Hint: Divide each side by a common factor as needed.) ***See Example 3.***

29. $x^2+y^2+4x+6y+9=0$

30. $x^2+y^2-8x-12y+3=0$

31. $x^2+y^2+10x-14y-7=0$

32. $x^2+y^2-2x+4y-4=0$

33. $3x^2+3y^2-12x-24y+12=0$

34. $2x^2+2y^2+20x+16y+10=0$

Write the center-radius form of the circle with the given equation. Give the center and radius, and graph the circle. ***See Example 3.***

35. $x^2+y^2-4x-6y+9=0$

36. $x^2+y^2+8x+2y-8=0$

37. $x^2+y^2+6x-6y+9=0$

38. $x^2+y^2-4x+10y+20=0$

Concept Check *Answer each question, and give a short explanation.* ***See the Note following Example 3.***

39. How many points are there on the graph of $(x-4)^2+(y-1)^2=0$?

40. How many points are there on the graph of $(x-4)^2+(y-1)^2=-1$?

41. **Concept Check** A circle can be drawn on a piece of posterboard by fastening one end of a string with a thumbtack, pulling the string taut with a pencil, and tracing a curve, as shown in the figure. Why does this method work?

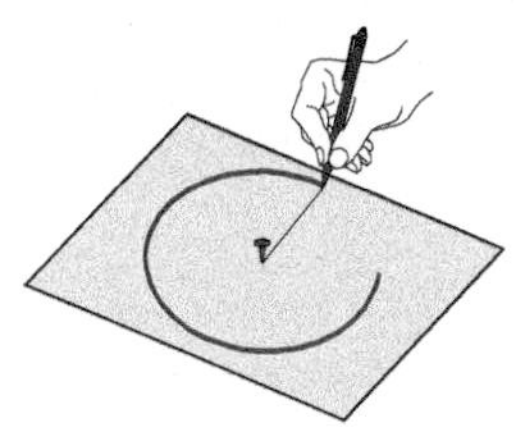

42. **Concept Check** An ellipse can be drawn on a piece of posterboard by fastening two ends of a length of string with thumbtacks, pulling the string taut with a pencil, and tracing a curve, as shown in the figure. Why does this method work?

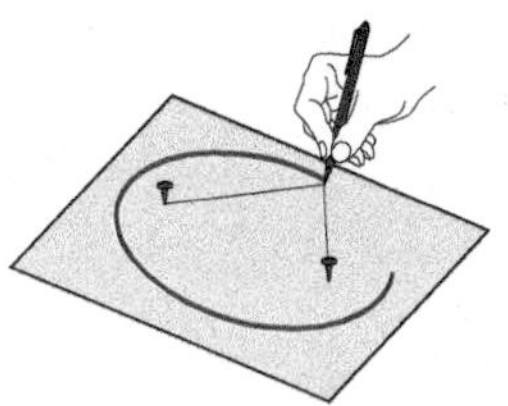

Graph each ellipse. ***See Examples 4 and 5.***

43. $\frac{x^2}{9} + \frac{y^2}{25} = 1$

44. $\frac{x^2}{9} + \frac{y^2}{16} = 1$

45. $\frac{x^2}{36} + \frac{y^2}{16} = 1$

46. $\frac{x^2}{9} + \frac{y^2}{4} = 1$

47. $\frac{x^2}{16} + \frac{y^2}{4} = 1$

48. $\frac{x^2}{49} + \frac{y^2}{81} = 1$

49. $\frac{y^2}{25} = 1 - \frac{x^2}{49}$

50. $\frac{y^2}{9} = 1 - \frac{x^2}{16}$

51. $\frac{(x+1)^2}{64} + \frac{(y-2)^2}{49} = 1$

52. $\frac{(x-4)^2}{9} + \frac{(y+2)^2}{4} = 1$

53. $\frac{(x-2)^2}{16} + \frac{(y-1)^2}{9} = 1$

54. $\frac{(x+3)^2}{25} + \frac{(y+2)^2}{36} = 1$

Extending Skills *A* ***lithotripter*** *is a machine used to crush kidney stones using shock waves. The patient is in an elliptical tub with the kidney stone at one focus of the ellipse. Each beam is projected from the other focus to the tub so that it reflects to hit the kidney stone. See the figure.*

55. Suppose a lithotripter is based on an ellipse with equation

$$\frac{x^2}{36} + \frac{y^2}{9} = 1.$$

How far from the center of the ellipse must the kidney stone and the source of the beam be placed? (*Hint:* Use the fact that $c^2 = a^2 - b^2$ because $a > b$ here.)

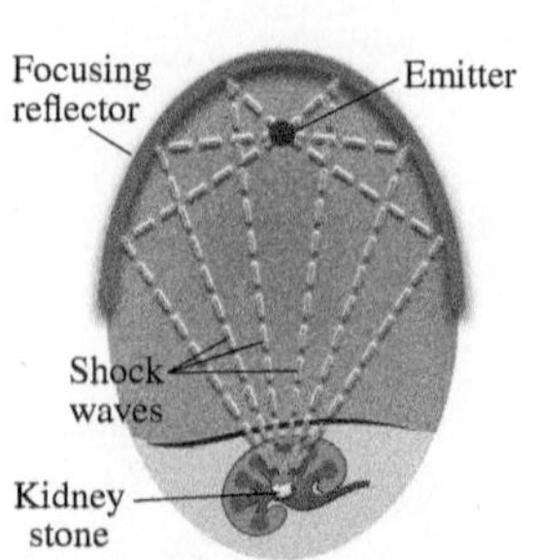

The top of an ellipse is illustrated in this depiction of how a lithotripter crushes a kidney stone.

56. Rework **Exercise 55** if the equation of the ellipse is

$$9x^2 + 4y^2 = 36.$$

(*Hint:* Write the equation in fractional form by dividing each term by 36, and use $c^2 = b^2 - a^2$ because $b > a$ here.)

Extending Skills *Solve each problem.*

57. An arch has the shape of half an ellipse. The equation of the ellipse, where x and y are in meters, is

$$100x^2 + 324y^2 = 32{,}400.$$

(a) How high is the center of the arch?

(b) How wide is the arch across the bottom?

Not to scale

58. A one-way street passes under an overpass, which is in the form of the top half of an ellipse, as shown in the figure. Suppose that a truck 12 ft wide passes directly under the overpass. What is the maximum possible height of this truck?

Not to scale

59. Work each of the following.

(a) The Roman Colosseum is an ellipse with $a = 310$ ft and $b = \frac{513}{2}$ ft. Find the distance, to the nearest tenth, between the foci of this ellipse.

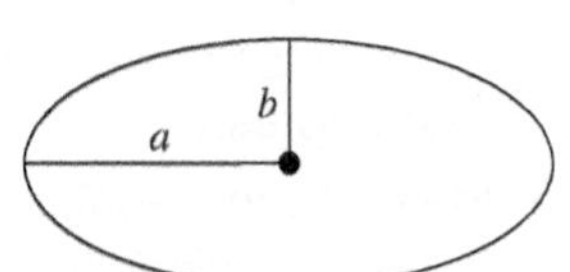

(b) The approximate perimeter of an ellipse is given by

$$P \approx 2\pi\sqrt{\frac{a^2 + b^2}{2}},$$

where a and b are the lengths given in part (a). Use this formula to find the approximate perimeter, to the nearest tenth, of the Roman Colosseum.

60. A satellite is in an elliptical orbit around Earth with least distance (**perigee**) of 160 km and greatest distance (**apogee**) of 16,000 km. Earth is located at a focus of the elliptical orbit. See the figure. (Data from *Space Mathematics,* Kastner, B., NASA.)

Find the equation of the ellipse. (*Hint:* Use the fact that $c^2 = a^2 - b^2$ here.)

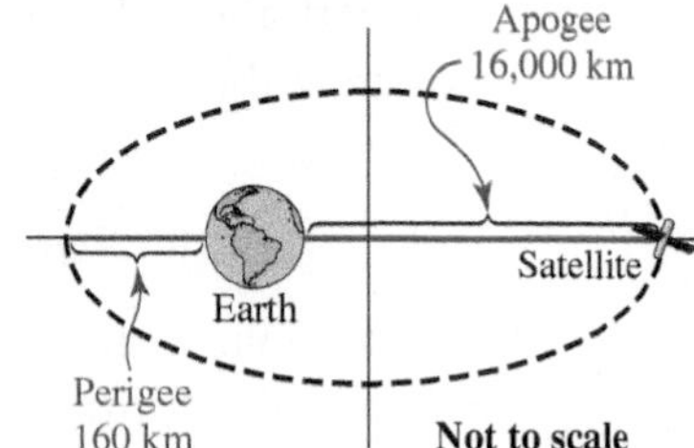

Not to scale

Extending Skills *Solve each problem. See* **FIGURE 8** *and use the fact that* $c^2 = a^2 - b^2$, *where* $a^2 > b^2$. *Round answers to the nearest tenth. (Data from* Astronomy!, *Kaler, J. B., Addison-Wesley.)*

61. The orbit of Mars is an ellipse with the sun at one focus. For x and y in millions of miles, the equation of the orbit is

$$\frac{x^2}{141.7^2} + \frac{y^2}{141.1^2} = 1.$$

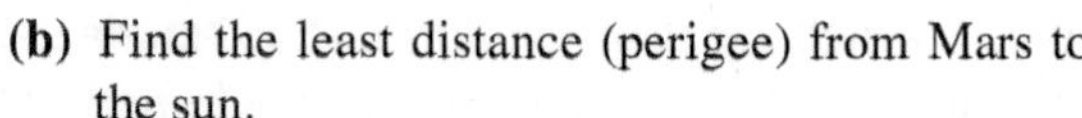

(a) Find the greatest distance (apogee) from Mars to the sun.

(b) Find the least distance (perigee) from Mars to the sun.

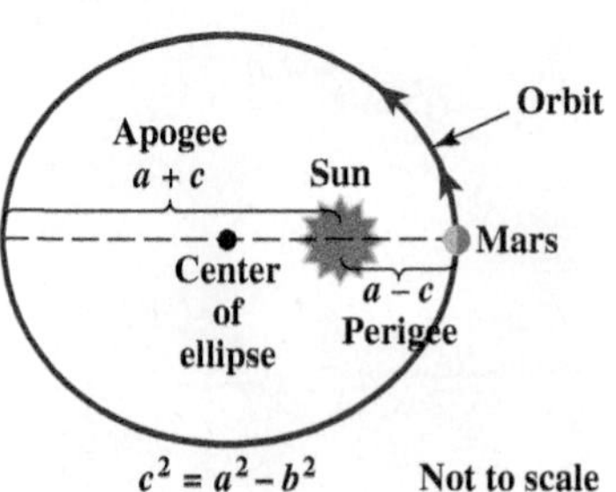

Not to scale

62. The orbit of Venus around the sun (one of the foci) is an ellipse with equation

$$\frac{x^2}{5013} + \frac{y^2}{4970} = 1,$$

where x and y are measured in millions of miles.

(a) Find the greatest distance (apogee) between Venus and the sun.

(b) Find the least distance (perigee) between Venus and the sun.

12.2 Hyperbolas and Functions Defined by Radicals

OBJECTIVES

1 Recognize the equation of a hyperbola.
2 Graph hyperbolas using asymptotes.
3 Identify conic sections using their equations.
4 Graph generalized square root functions.

VOCABULARY

☐ hyperbola
☐ transverse axis
☐ fundamental rectangle
☐ asymptotes (of a hyperbola)
☐ generalized square root function

OBJECTIVE 1 Recognize the equation of a hyperbola.

A **hyperbola** is the set of all points in a plane such that the absolute value of the *difference* of the distances from two fixed points (the *foci*) is constant. The graph of a hyperbola has two parts, or *branches,* and two intercepts (or *vertices*) that lie on its axis, called the **transverse axis.**

The hyperbola in FIGURE 15 has a horizontal transverse axis, with foci $(c, 0)$ and $(-c, 0)$ and x-intercepts $(a, 0)$ and $(-a, 0)$. (A hyperbola with vertical transverse axis would have its intercepts on the y-axis.)

A hyperbola centered at the origin has one of the following equations. (It is shown in more advanced courses that for a hyperbola, $c^2 = a^2 + b^2$.)

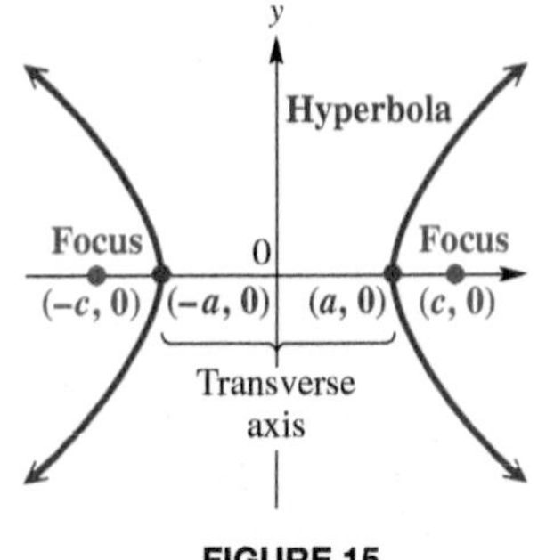

FIGURE 15

Equations of Hyperbolas

A hyperbola with x-intercepts $(a, 0)$ and $(-a, 0)$ has an equation that can be written in the form

$$\frac{x^2}{a^2} - \frac{y^2}{b^2} = 1. \quad \text{Transverse axis on } x\text{-axis}$$

A hyperbola with y-intercepts $(0, b)$ and $(0, -b)$ has an equation that can be written in the form

$$\frac{y^2}{b^2} - \frac{x^2}{a^2} = 1. \quad \text{Transverse axis on } y\text{-axis}$$

If we were to throw two stones into a pond, the ensuing concentric ripples would be shaped like a hyperbola. A cross-section of the cooling towers for a nuclear power plant is hyperbolic, as shown in the photo.

OBJECTIVE 2 Graph hyperbolas using asymptotes.

The two branches of the graph of a hyperbola approach a pair of intersecting straight lines, which are its *asymptotes.* (See FIGURE 16 on the next page.) The asymptotes are useful for sketching the graph of the hyperbola.

Asymptotes of Hyperbolas

The extended diagonals of a rectangle, called the **fundamental rectangle,** with vertices (corners) at the points (a, b), $(-a, b)$, $(-a, -b)$, and $(a, -b)$ are the **asymptotes** of the hyperbolas

$$\frac{x^2}{a^2} - \frac{y^2}{b^2} = 1 \quad \text{and} \quad \frac{y^2}{b^2} - \frac{x^2}{a^2} = 1.$$

Using previous methods, we could show that the equations of these asymptotes are

$$y = \frac{b}{a}x \quad \text{and} \quad y = -\frac{b}{a}x.$$ Equations of the asymptotes of a hyperbola

Graphing a Hyperbola

Step 1 **Find and locate the intercepts.**

- At $(a, 0)$ and $(-a, 0)$ if the x^2-term has a positive coefficient
- At $(0, b)$ and $(0, -b)$ if the y^2-term has a positive coefficient

Step 2 **Find the fundamental rectangle.** Its vertices will be located at the points (a, b), $(-a, b)$, $(-a, -b)$, and $(a, -b)$.

Step 3 **Sketch the asymptotes.** The extended diagonals of the fundamental rectangle are the asymptotes of the hyperbola. They have equations $y = \pm\frac{b}{a}x$.

Step 4 **Draw the graph.** Sketch each branch of the hyperbola through an intercept, approaching (but not touching) the asymptotes.

NOW TRY EXERCISE 1

Graph $\frac{x^2}{25} - \frac{y^2}{9} = 1$.

EXAMPLE 1 **Graphing a Horizontal Hyperbola**

Graph $\frac{x^2}{16} - \frac{y^2}{25} = 1$.

Step 1 Here $a = 4$ and $b = 5$. The x-intercepts are $(4, 0)$ and $(-4, 0)$.

Step 2 The vertices of the fundamental rectangle are the four points

$$\begin{array}{cccc} (a, b) & (-a, b) & (-a, -b) & (a, -b) \\ \downarrow\ \downarrow & \downarrow\ \downarrow & \downarrow\ \downarrow & \downarrow\ \downarrow \\ (4, 5), & (-4, 5), & (-4, -5), \text{ and} & (4, -5). \end{array}$$

Steps 3 and 4 The equations of the asymptotes are $y = \pm\frac{b}{a}x$, or $y = \pm\frac{5}{4}x$. The hyperbola approaches these lines as x and y get larger in absolute value. See **FIGURE 16.**

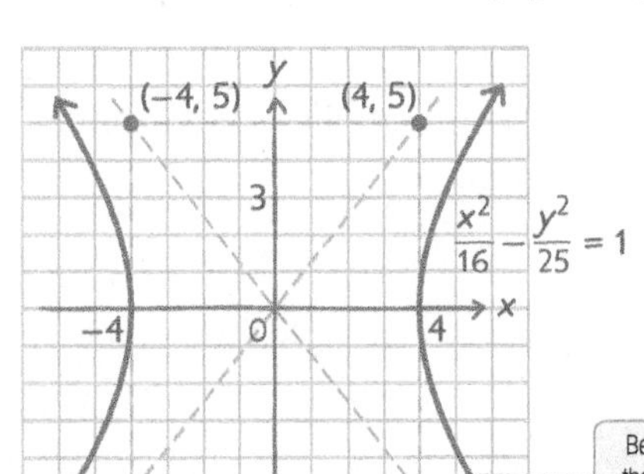

FIGURE 16

When graphing a hyperbola, the fundamental rectangle and the asymptotes are not part of the actual graph. They provide help in drawing a more accurate graph.

NOW TRY

NOW TRY ANSWER

1.

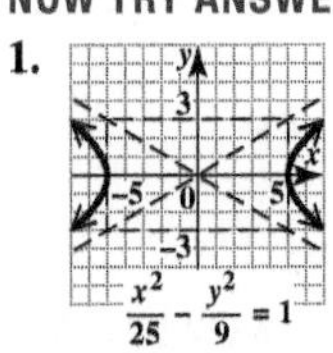

Graph $\frac{y^2}{9} - \frac{x^2}{16} = 1$.

EXAMPLE 2 Graphing a Vertical Hyperbola

Graph $\frac{y^2}{49} - \frac{x^2}{16} = 1$.

This hyperbola has y-intercepts $(0, 7)$ and $(0, -7)$. The asymptotes are the extended diagonals of the fundamental rectangle with vertices at

$$(4, 7), \quad (-4, 7), \quad (-4, -7), \quad \text{and} \quad (4, -7).$$

Their equations are $y = \pm\frac{7}{4}x$. See **FIGURE 17.**

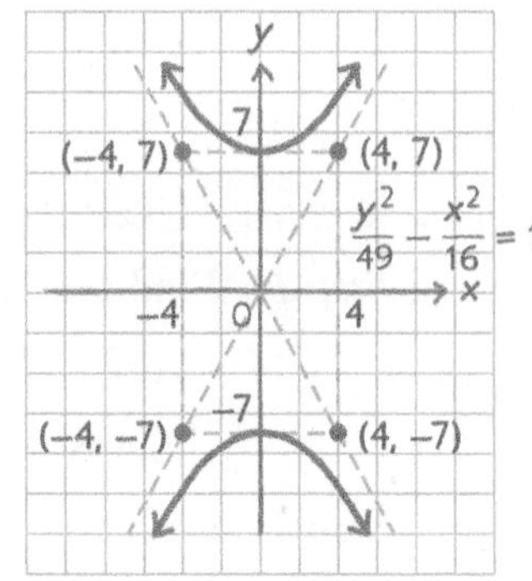

FIGURE 17

NOW TRY 

OBJECTIVE 3 Identify conic sections using their equations.

Rewriting a second-degree equation in one of the forms given for ellipses, hyperbolas, circles, or parabolas makes it possible to identify the graph of the equation.

NOW TRY EXERCISE 3

Identify the graph of each equation.

(a) $y^2 - 10 = -x^2$

(b) $y - 2x^2 = 8$

(c) $3x^2 + y^2 = 4$

EXAMPLE 3 Identifying the Graphs of Equations

Identify the graph of each equation.

(a) $9x^2 = 108 + 12y^2$

Both variables are squared, so the graph is either an ellipse or a hyperbola. (This situation also occurs for a circle, which is a special case of an ellipse.) To see which conic section it is, rewrite the equation so that the x^2- and y^2-terms are on one side of the equation and 1 is on the other side.

$$9x^2 - 12y^2 = 108 \quad \text{Subtract } 12y^2.$$

$$\frac{x^2}{12} - \frac{y^2}{9} = 1 \quad \text{Divide by 108.}$$

The subtraction symbol indicates that the graph of this equation is a hyperbola.

(b) $x^2 = y - 3$

Only one of the two variables, x, is squared, so this is the vertical parabola $y = x^2 + 3$.

(c) $x^2 = 9 - y^2$

Write the variable terms on the same side of the equation.

$$x^2 + y^2 = 9 \quad \text{Add } y^2.$$

The graph of this equation is a circle with center at the origin and radius 3.

NOW TRY

NOW TRY ANSWERS

2.

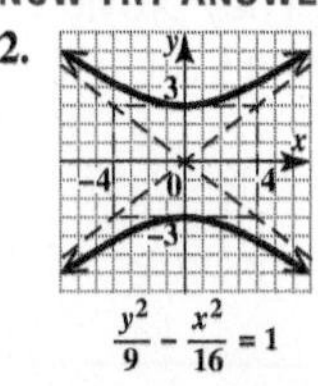

3. **(a)** circle **(b)** parabola **(c)** ellipse

▼ SUMMARY OF CONIC SECTIONS

Equation	Graph	Description	Identification
$y = ax^2 + bx + c$ or $y = a(x - h)^2 + k$	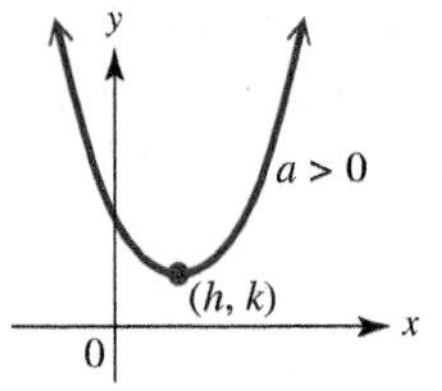 Parabola	It opens up if $a > 0$, down if $a < 0$. The vertex is (h, k).	It has an x^2-term. y is not squared.
$x = ay^2 + by + c$ or $x = a(y - k)^2 + h$	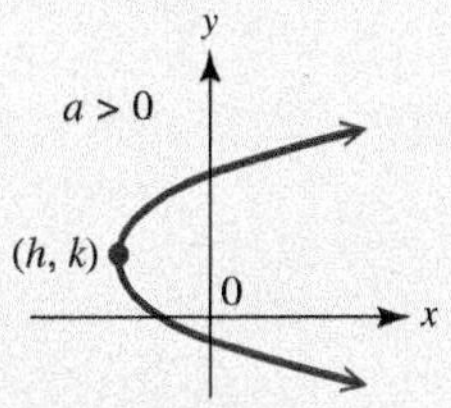 Parabola	It opens to the right if $a > 0$, to the left if $a < 0$. The vertex is (h, k).	It has a y^2-term. x is not squared.
$(x - h)^2 + (y - k)^2 = r^2$	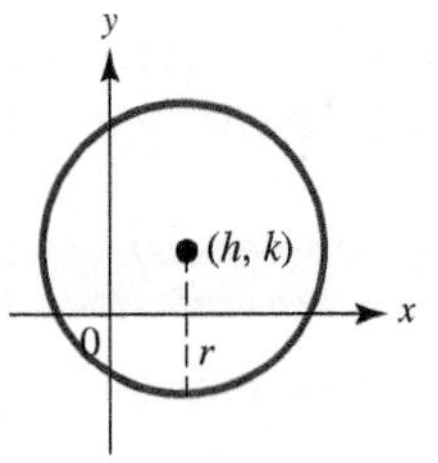 Circle	The center is (h, k), and the radius is r.	x^2- and y^2-terms have the same positive coefficient.
$\frac{x^2}{a^2} + \frac{y^2}{b^2} = 1$	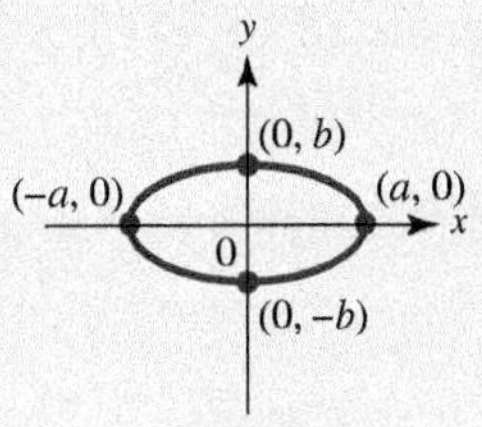 Ellipse	The x-intercepts are $(a, 0)$ and $(-a, 0)$. The y-intercepts are $(0, b)$ and $(0, -b)$.	x^2- and y^2-terms have different positive coefficients.
$\frac{x^2}{a^2} - \frac{y^2}{b^2} = 1$	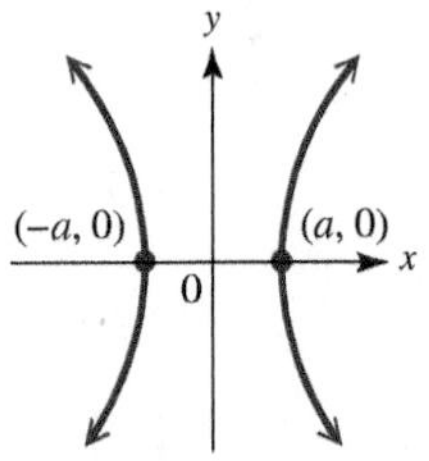 Hyperbola	The x-intercepts are $(a, 0)$ and $(-a, 0)$. The asymptotes are found from (a, b), $(a, -b)$, $(-a, -b)$, and $(-a, b)$.	x^2 has a positive coefficient. y^2 has a negative coefficient.
$\frac{y^2}{b^2} - \frac{x^2}{a^2} = 1$	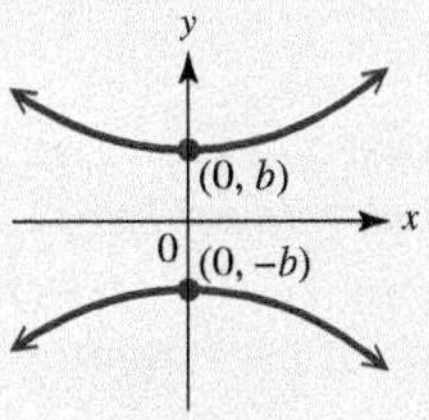 Hyperbola	The y-intercepts are $(0, b)$ and $(0, -b)$. The asymptotes are found from (a, b), $(a, -b)$, $(-a, -b)$, and $(-a, b)$.	y^2 has a positive coefficient. x^2 has a negative coefficient.

OBJECTIVE 4 Graph generalized square root functions.

Because they do not satisfy the conditions of the vertical line test, the graphs of horizontal parabolas and all circles, ellipses, and hyperbolas with horizontal or vertical axes do not satisfy the conditions of a function. However, by considering only a part of the graph of each of these, we have the graph of a function, as seen in **FIGURE 18.**

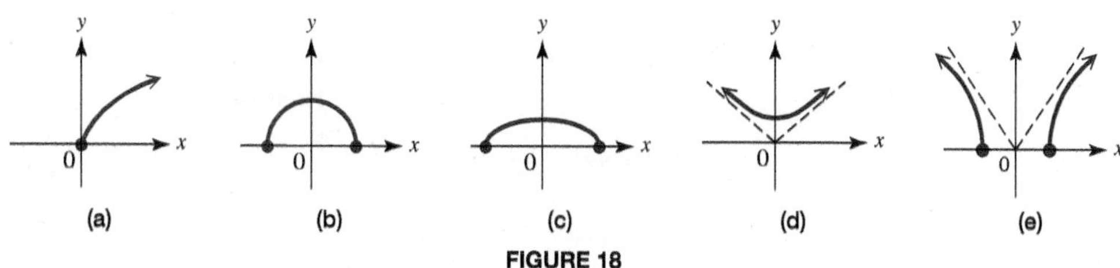

FIGURE 18

In parts (a)–(d) of **FIGURE 18**, the top portion of a conic section is shown (parabola, circle, ellipse, and hyperbola, respectively). In part (e), the top two portions of a hyperbola are shown. In each case, the graph is that of a function because the graph satisfies the conditions of the vertical line test.

Earlier in this chapter, we worked with the square root function $f(x) = \sqrt{x}$. To find equations for the types of graphs shown in **FIGURE 18**, we extend its definition.

Generalized Square Root Function

For an algebraic expression in x defined by u, where $u \geq 0$, a function of the form

$$f(x) = \sqrt{u}$$

is a **generalized square root function.**

NOW TRY EXERCISE 4

Graph $f(x) = \sqrt{64 - x^2}$. Give the domain and range.

EXAMPLE 4 Graphing a Semicircle

Graph $f(x) = \sqrt{25 - x^2}$. Give the domain and range.

$f(x) = \sqrt{25 - x^2}$	Given function
$y = \sqrt{25 - x^2}$	Replace $f(x)$ with y.
$y^2 = \left(\sqrt{25 - x^2}\right)^2$	Square each side.
$y^2 = 25 - x^2$	$(\sqrt{a})^2 = a$
$x^2 + y^2 = 25$	Add x^2.

This is the equation of a circle with center at $(0, 0)$ and radius 5.

Because $f(x)$, or y, represents a principal square root in the original equation, $f(x)$ must be nonnegative. This restricts the graph to the upper half of the circle.

See **FIGURE 19.** Use the graph and the vertical line test to verify that it is indeed a function. The domain is $[-5, 5]$, and the range is $[0, 5]$.

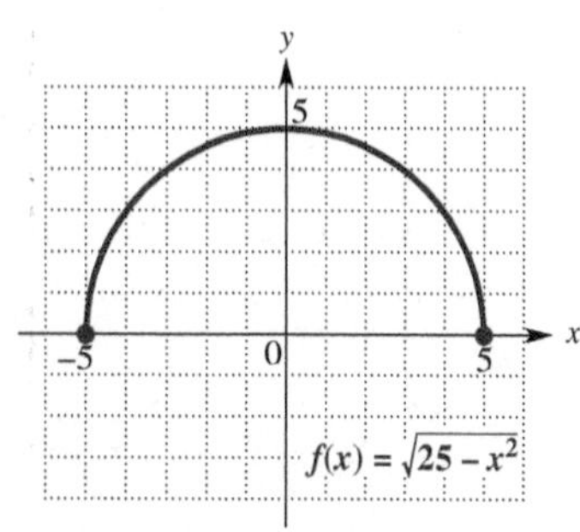

FIGURE 19

NOW TRY 

NOW TRY ANSWER

4. $f(x) = \sqrt{64 - x^2}$

domain: $[-8, 8]$; range: $[0, 8]$

NOW TRY EXERCISE 5

Graph $\frac{y}{4} = -\sqrt{1 - \frac{x^2}{9}}$.

Give the domain and range.

EXAMPLE 5 Graphing a Portion of an Ellipse

Graph $\frac{y}{6} = -\sqrt{1 - \frac{x^2}{16}}$. Give the domain and range.

$$\frac{y}{6} = -\sqrt{1 - \frac{x^2}{16}} \quad \text{Given equation}$$

$$\left(\frac{y}{6}\right)^2 = \left(-\sqrt{1 - \frac{x^2}{16}}\right)^2 \quad \text{Square each side.}$$

$$\frac{y^2}{36} = 1 - \frac{x^2}{16} \quad \text{Apply the exponents.}$$

$$\frac{x^2}{16} + \frac{y^2}{36} = 1 \quad \text{Add } \frac{x^2}{16}.$$

This is the equation of an ellipse with center at $(0, 0)$. The x-intercepts are $(4, 0)$ and $(-4, 0)$. The y-intercepts are $(0, 6)$ and $(0, -6)$.

Because $\frac{y}{6}$ equals a negative square root in the original equation, y must be nonpositive, restricting the graph to the lower half of the ellipse.

See **FIGURE 20**. The graph is that of a function. The domain is $[-4, 4]$, and the range is $[-6, 0]$.

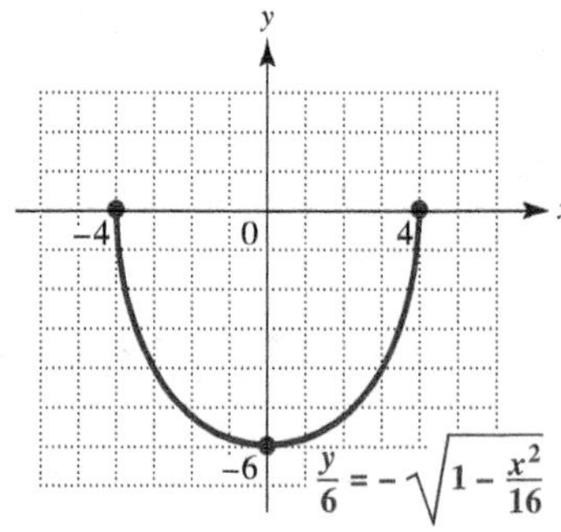

FIGURE 20

NOW TRY 5

NOW TRY ANSWER

5.

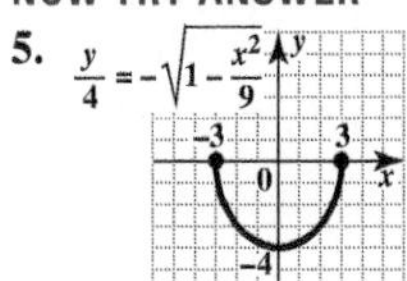

domain: $[-3, 3]$; range: $[-4, 0]$

12.2 Exercises

FOR EXTRA HELP

MyLab Math

Video solutions for select problems available in MyLab Math

Concept Check *Based on the discussions of ellipses in the previous section and of hyperbolas in this section, match each equation with its graph.*

1. $\frac{x^2}{25} + \frac{y^2}{9} = 1$

2. $\frac{x^2}{9} + \frac{y^2}{25} = 1$

3. $\frac{x^2}{9} - \frac{y^2}{25} = 1$

4. $\frac{x^2}{25} - \frac{y^2}{9} = 1$

A.

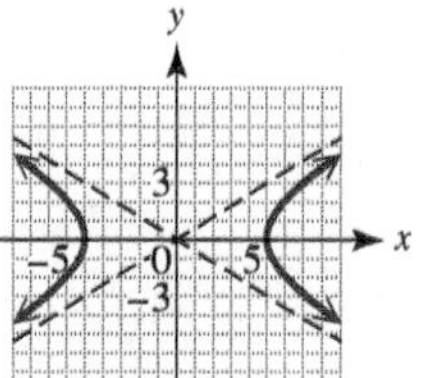

B.

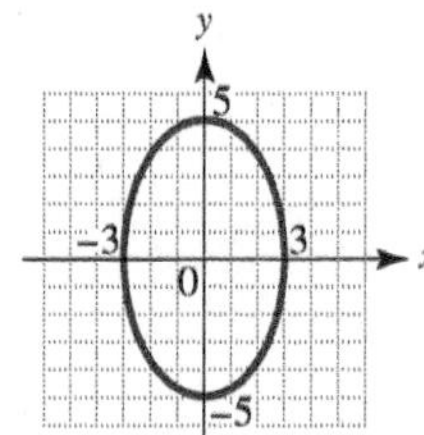

C.

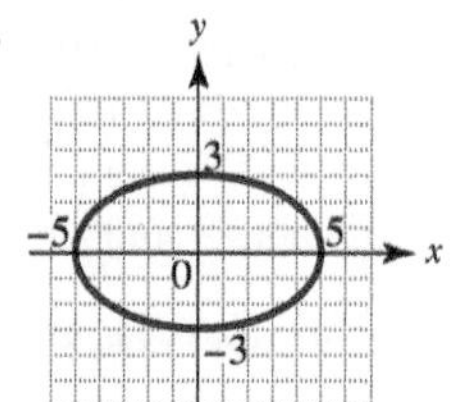

D.

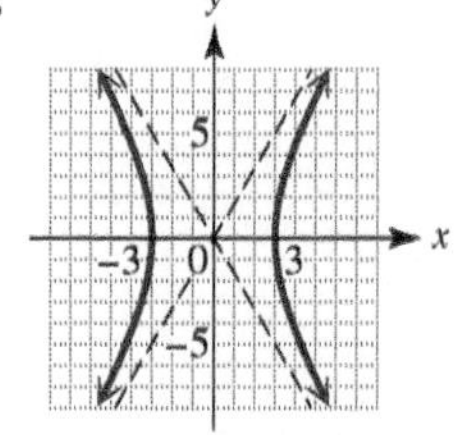

5. **Concept Check** A student incorrectly described the graph of the equation

$$\frac{x^2}{9} - \frac{y^2}{16} = 1$$

as a vertical hyperbola with y-intercepts $(0, -4)$ and $(0, 4)$. **WHAT WENT WRONG?** Give the correct description of the graph.

6. **Concept Check** A student incorrectly described the graph of the function

$$f(x) = \sqrt{49 - x^2}$$

as a circle centered at the origin with radius 7. **WHAT WENT WRONG?** Give the correct description of the graph.

Graph each hyperbola. ***See Examples 1 and 2.***

7. $\frac{x^2}{16} - \frac{y^2}{9} = 1$

8. $\frac{x^2}{25} - \frac{y^2}{9} = 1$

9. $\frac{y^2}{4} - \frac{x^2}{25} = 1$

10. $\frac{y^2}{9} - \frac{x^2}{4} = 1$

11. $\frac{x^2}{25} - \frac{y^2}{36} = 1$

12. $\frac{x^2}{49} - \frac{y^2}{16} = 1$

13. $\frac{y^2}{9} - \frac{x^2}{9} = 1$

14. $\frac{y^2}{16} - \frac{x^2}{16} = 1$

Identify the graph of each equation as a parabola, circle, ellipse, *or* hyperbola, *and then sketch the graph.* ***See Example 3.***

15. $x^2 - y^2 = 16$

16. $x^2 + y^2 = 16$

17. $4x^2 + y^2 = 16$

18. $9x^2 = 144 + 16y^2$

19. $y^2 = 36 - x^2$

20. $9x^2 + 25y^2 = 225$

21. $x^2 + 9y^2 = 9$

22. $x^2 - 2y = 0$

23. $x^2 = 4y - 8$

24. $y^2 = 4 + x^2$

Graph each generalized square root function. Give the domain and range. ***See Examples 4 and 5.***

25. $f(x) = \sqrt{16 - x^2}$

26. $f(x) = \sqrt{9 - x^2}$

27. $f(x) = -\sqrt{36 - x^2}$

28. $f(x) = -\sqrt{25 - x^2}$

29. $\frac{y}{3} = \sqrt{1 + \frac{x^2}{9}}$

30. $\frac{y}{2} = \sqrt{1 + \frac{x^2}{4}}$

31. $y = -2\sqrt{1 - \frac{x^2}{9}}$

32. $y = -3\sqrt{1 - \frac{x^2}{25}}$

Extending Skills *Solve each problem.*

33. Two buildings in a sports complex are shaped and positioned like a portion of the branches of the hyperbola with equation

$$400x^2 - 625y^2 = 250{,}000,$$

where x and y are in meters.

(a) How far apart are the buildings at their closest point?

(b) Find the distance d in the figure to the nearest tenth of a meter.

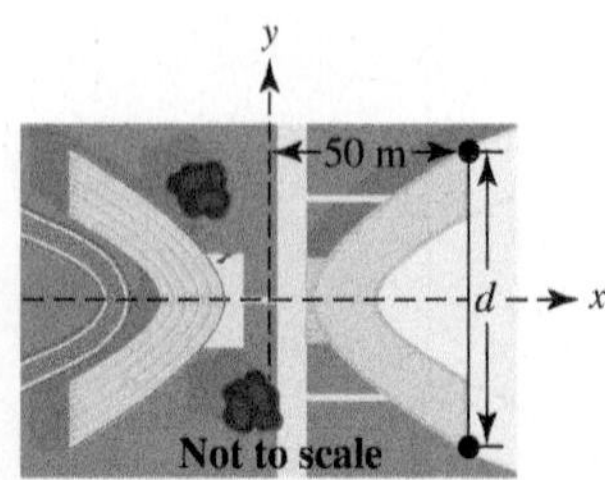

34. Using LORAN, a location-finding system, a radio transmitter at M sends out a series of pulses. When each pulse is received at transmitter S, it then sends out a pulse. A ship at P receives pulses from both M and S. A receiver on the ship measures the difference in the arrival times of the pulses. A special map gives hyperbolas that correspond to the differences in arrival times (which give the distances d_1 and d_2 in the figure). The ship can then be located as lying on a branch of a particular hyperbola.

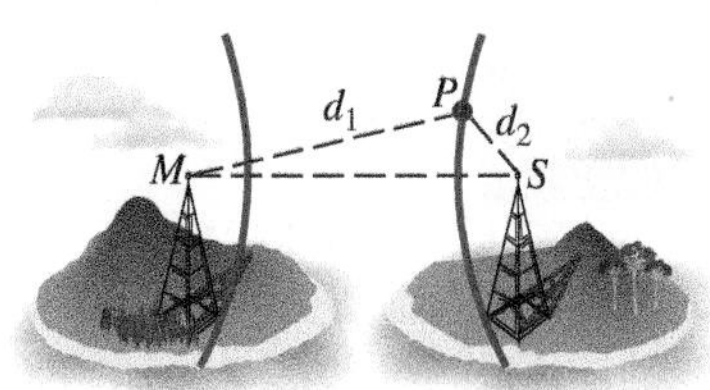

Suppose $d_1 = 80$ mi and $d_2 = 30$ mi, and the distance between transmitters M and S is 100 mi. Use the definition to find an equation of the hyperbola on which the ship is located.

Extending Skills *In rugby, after a* try *(similar to a touchdown in American football) the scoring team attempts a kick for extra points. The ball must be kicked from directly behind the point where the try was scored. The kicker can choose the distance but cannot move the ball sideways. It can be shown that the kicker's best choice is on the hyperbola with equation*

$$\frac{x^2}{g^2} - \frac{y^2}{g^2} = 1,$$

where 2g is the distance between the goal posts. Since the hyperbola approaches its asymptotes, it is easier for the kicker to estimate points on the asymptotes instead of on the hyperbola. (Data from Isaksen, Daniel C., "How to Kick a Field Goal." The College Mathematics Journal.)

35. What are the asymptotes of this hyperbola?

36. Why is it relatively easy to estimate the asymptotes?

RELATING CONCEPTS For Individual or Group Work (Exercises 37–40)

We have seen that the center of an ellipse may be shifted away from the origin. The same process applies to hyperbolas. For example, the hyperbola

$$\frac{(x+5)^2}{4} - \frac{(y-2)^2}{9} = 1,$$

has the same graph as

$$\frac{x^2}{4} - \frac{y^2}{9} = 1,$$

but it is centered at $(-5, 2)$, as shown at the right.

Graph each hyperbola with center shifted away from the origin.

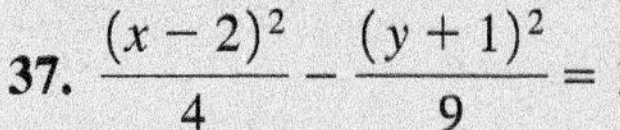

37. $\frac{(x-2)^2}{4} - \frac{(y+1)^2}{9} = 1$

38. $\frac{(x+3)^2}{16} - \frac{(y-2)^2}{25} = 1$

39. $\frac{y^2}{36} - \frac{(x-2)^2}{49} = 1$

40. $\frac{(y-5)^2}{9} - \frac{x^2}{25} = 1$

12.3 Nonlinear Systems of Equations

OBJECTIVES

1 Solve a nonlinear system using substitution.
2 Solve a nonlinear system with two second-degree equations using elimination.
3 Solve a nonlinear system that requires a combination of methods.

VOCABULARY

☐ nonlinear equation
☐ nonlinear system of equations

An equation in which some terms have more than one variable or a variable of degree 2 or greater is called a **nonlinear equation.** A **nonlinear system of equations** includes at least one nonlinear equation.

When solving a nonlinear system, it helps to visualize the types of graphs of the equations of the system to determine the possible number of points of intersection. For example, if a system includes two equations where the graph of one is a circle and the graph of the other is a line, then there may be zero, one, or two points of intersection, as illustrated in **FIGURE 21**.

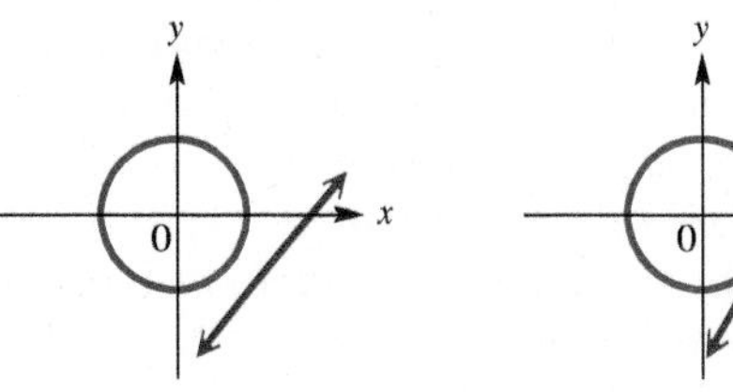

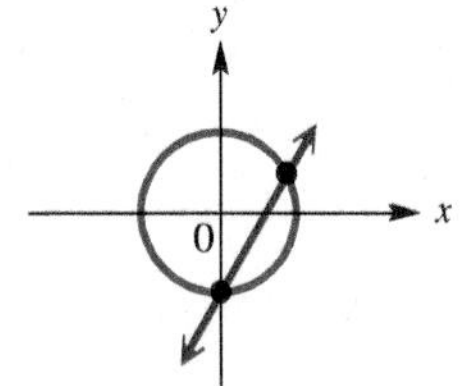

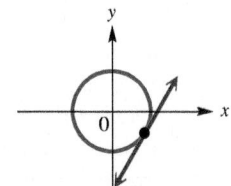

No points of intersection One point of intersection Two points of intersection

FIGURE 21

OBJECTIVE 1 Solve a nonlinear system using substitution.

We can usually solve a nonlinear system using the substitution method if one equation is linear.

EXAMPLE 1 Solving a Nonlinear System Using Substitution

Solve the system.

$$x^2 + y^2 = 9 \quad (1)$$
$$2x - y = 3 \quad (2)$$

The graph of equation (1) is a circle and the graph of equation (2) is a line, so the graphs could intersect in zero, one, or two points, as in **FIGURE 21**. We begin by solving the linear equation (2) for one of the two variables.

$$2x - y = 3 \quad (2)$$
$$y = 2x - 3 \quad \text{Solve for } y. \quad (3)$$

Then we substitute $2x - 3$ for y in the nonlinear equation (1).

$$x^2 + y^2 = 9 \quad (1)$$
$$x^2 + (2x - 3)^2 = 9 \quad \text{Let } y = 2x - 3.$$
$$x^2 + 4x^2 - 12x + 9 = 9 \quad \text{Square } 2x - 3.$$
$$5x^2 - 12x = 0 \quad \text{Combine like terms. Subtract 9.}$$
$$x(5x - 12) = 0 \quad \text{Factor. The GCF is } x.$$
$$x = 0 \quad \text{or} \quad 5x - 12 = 0 \quad \text{Zero-factor property}$$

Set *both* factors equal to 0.

$$x = \frac{12}{5} \quad \text{Solve for } x.$$

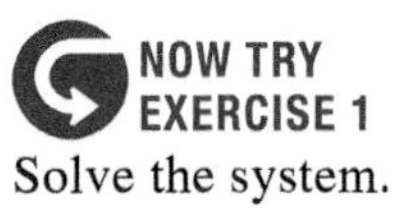

NOW TRY EXERCISE 1

Solve the system.

$$4x^2 + y^2 = 36$$
$$x - y = 3$$

Let $x = 0$ in equation (3), $y = 2x - 3$, to obtain $y = -3$. Then let $x = \frac{12}{5}$ in equation (3).

$$y = 2\left(\frac{12}{5}\right) - 3 \quad \text{Let } x = \frac{12}{5}.$$

$$y = \frac{24}{5} - \frac{15}{5} \quad \text{Multiply; } 3 = \frac{15}{5}$$

$$y = \frac{9}{5} \quad \frac{a}{c} - \frac{b}{c} = \frac{a-b}{c}$$

The solution set of the system is $\left\{(0, -3), \left(\frac{12}{5}, \frac{9}{5}\right)\right\}$. The graph in **FIGURE 22** confirms the two points of intersection.

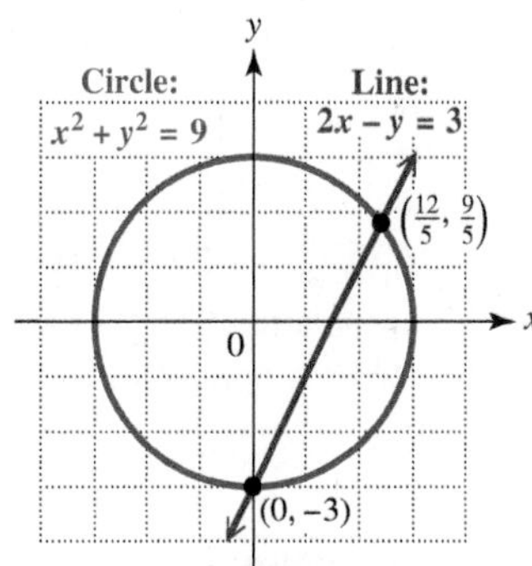

FIGURE 22

NOW TRY 

EXAMPLE 2 Solving a Nonlinear System Using Substitution

Solve the system.

$$6x - y = 5 \quad (1)$$
$$xy = 4 \quad (2)$$

The graph of (1) is a line. It can be shown by plotting points that the graph of (2) is a hyperbola. Visualizing a line and a hyperbola indicates that there may be zero, one, or two points of intersection.

Since neither equation has a squared term, we can solve either equation for one of the variables and then substitute the result into the other equation. Solving $xy = 4$ for x gives $x = \frac{4}{y}$. We substitute $\frac{4}{y}$ for x in equation (1).

$$6x - y = 5 \quad (1)$$

$$6\left(\frac{4}{y}\right) - y = 5 \quad \text{Let } x = \frac{4}{y}.$$

$$\frac{24}{y} - y = 5 \quad \text{Multiply.}$$

$$24 - y^2 = 5y \quad \text{Multiply by } y,\ y \neq 0.$$

$$y^2 + 5y - 24 = 0 \quad \text{Standard form}$$

$$(y - 3)(y + 8) = 0 \quad \text{Factor.}$$

$$y - 3 = 0 \quad \text{or} \quad y + 8 = 0 \quad \text{Zero-factor property}$$

$$y = 3 \quad \text{or} \quad y = -8 \quad \text{Solve each equation.}$$

NOW TRY ANSWER

1. $\left\{(3, 0), \left(-\frac{9}{5}, -\frac{24}{5}\right)\right\}$

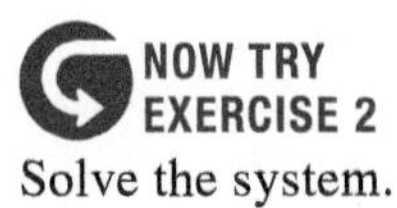

Solve the system.

$$xy = 2$$
$$x - 3y = 1$$

We substitute these values of y into $x = \frac{4}{y}$ to obtain the corresponding values of x.

If $y = 3$, then $x = \frac{4}{3}$.

If $y = -8$, then $x = -\frac{1}{2}$.

The solution set of the system is

$$\left\{\left(\frac{4}{3}, 3\right), \left(-\frac{1}{2}, -8\right)\right\}.$$

Write the x-coordinates first in the ordered-pair solutions.

See the graph in **FIGURE 23.**

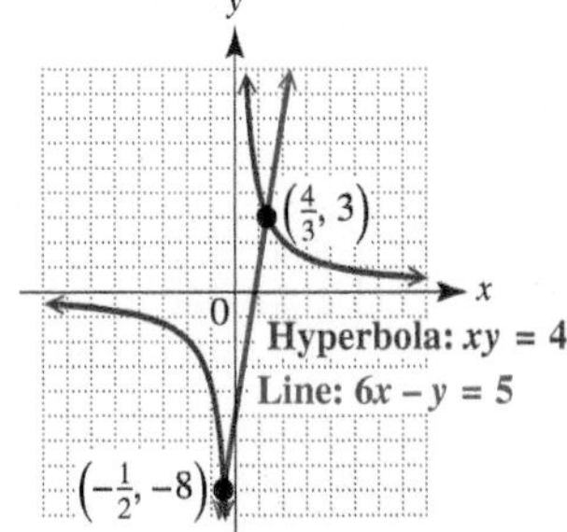

FIGURE 23

NOTE In **Example 2,** we could solve the *linear* equation for one of its variables and substitute for this variable in the *nonlinear* equation. There is often more than one way to solve a nonlinear system of equations.

OBJECTIVE 2 Solve a nonlinear system with two second-degree equations using elimination.

If a system consists of two second-degree equations, then there may be zero, one, two, three, or four solutions. **FIGURE 24** shows a case where a system consisting of a circle and a parabola has four solutions, all made up of ordered pairs of real numbers.

The elimination method is often used when both equations of a nonlinear system are second degree.

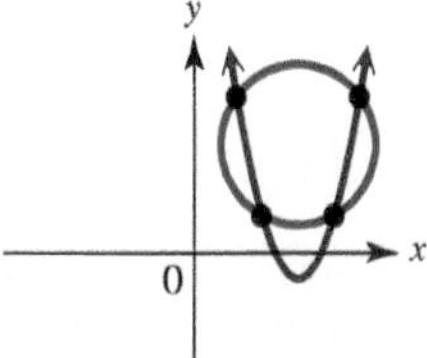

This system has four solutions because there are four points of intersection.

FIGURE 24

EXAMPLE 3 Solving a Nonlinear System Using Elimination

Solve the system.

$$x^2 + y^2 = 9 \quad (1)$$
$$2x^2 - y^2 = -6 \quad (2)$$

The graph of (1) is a circle, while the graph of (2) is a hyperbola. By analyzing the possibilities, we conclude that there may be zero, one, two, three, or four points of intersection. Adding the two equations will eliminate y.

$x^2 + y^2 = 9$	(1)
$2x^2 - y^2 = -6$	(2)
$3x^2 = 3$	Add.
$x^2 = 1$	Divide by 3.
$x = 1$ or $x = -1$	Square root property

Each value of x gives corresponding values for y when substituted into one of the original equations. Using equation (1) is easier because the coefficients of the x^2- and y^2-terms are 1.

NOW TRY ANSWER

2. $\left\{(-2, -1), \left(3, \frac{2}{3}\right)\right\}$

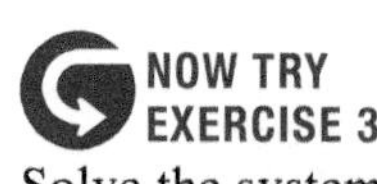

NOW TRY EXERCISE 3

Solve the system.

$$x^2 + y^2 = 16$$
$$4x^2 + 13y^2 = 100$$

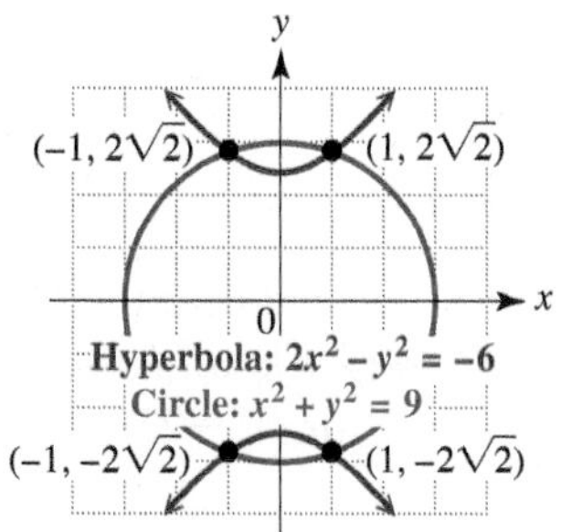

FIGURE 25

$$x^2 + y^2 = 9 \quad (1)$$
$$1^2 + y^2 = 9 \quad \text{Let } x = 1.$$
$$y^2 = 8$$
$$y = \sqrt{8} \quad \text{or} \quad y = -\sqrt{8}$$
$$y = 2\sqrt{2} \quad \text{or} \quad y = -2\sqrt{2}$$

$$x^2 + y^2 = 9 \quad (1)$$
$$(-1)^2 + y^2 = 9 \quad \text{Let } x = -1.$$
$$y^2 = 8$$
$$y = \sqrt{8} \quad \text{or} \quad y = -\sqrt{8}$$
$$y = 2\sqrt{2} \quad \text{or} \quad y = -2\sqrt{2}$$

The solution set is

$$\{(1, 2\sqrt{2}), (1, -2\sqrt{2}), (-1, 2\sqrt{2}), (-1, -2\sqrt{2})\}.$$

FIGURE 25 shows the four points of intersection. **NOW TRY** 

OBJECTIVE 3 Solve a nonlinear system that requires a combination of methods.

EXAMPLE 4 Solving a Nonlinear System Using a Combination of Methods

Solve the system.

$$x^2 + 2xy - y^2 = 7 \quad (1)$$
$$x^2 - y^2 = 3 \quad (2)$$

While we have not graphed equations like (1), its graph is a hyperbola. The graph of (2) is also a hyperbola. Two hyperbolas may have zero, one, two, three, or four points of intersection. We use the elimination method here in combination with the substitution method.

$$x^2 + 2xy - y^2 = 7 \quad (1)$$
$$-x^2 \qquad + y^2 = -3 \quad \text{Multiply (2) by } -1.$$
$$2xy = 4 \quad \text{Add.}$$

The x^2- and y^2-terms were eliminated.

Next, we solve $2xy = 4$ for one of the variables. We choose y.

$$2xy = 4$$
$$y = \frac{2}{x} \quad \text{Divide by } 2x. \quad (3)$$

Now, we substitute $y = \frac{2}{x}$ into one of the original equations.

$$x^2 - y^2 = 3 \quad \text{The substitution is easier in (2).}$$
$$x^2 - \left(\frac{2}{x}\right)^2 = 3 \quad \text{Let } y = \tfrac{2}{x}.$$
$$x^2 - \frac{4}{x^2} = 3 \quad \text{Square } \tfrac{2}{x}.$$
$$x^4 - 4 = 3x^2 \quad \text{Multiply by } x^2,\ x \neq 0.$$
$$x^4 - 3x^2 - 4 = 0 \quad \text{Subtract } 3x^2.$$
$$(x^2 - 4)(x^2 + 1) = 0 \quad \text{Factor.}$$
$$x^2 - 4 = 0 \quad \text{or} \quad x^2 + 1 = 0 \quad \text{Zero-factor property}$$
$$x^2 = 4 \quad \text{or} \quad x^2 = -1 \quad \text{Solve each equation.}$$
$$x = 2 \quad \text{or} \quad x = -2 \quad \text{or} \quad x = i \quad \text{or} \quad x = -i$$

NOW TRY ANSWER

3. $\{(2\sqrt{3}, 2), (2\sqrt{3}, -2), (-2\sqrt{3}, 2), (-2\sqrt{3}, -2)\}$

NOW TRY EXERCISE 4

Solve the system.

$$x^2 + 3xy - y^2 = 23$$
$$x^2 - y^2 = 5$$

Substituting these four values of x into $y = \frac{2}{x}$ (equation (3)) gives the corresponding values for y.

If $x = 2$, then $y = \frac{2}{2} = 1$.

If $x = -2$, then $y = \frac{2}{-2} = -1$.

If $x = i$, then $y = \frac{2}{i} = \frac{2}{i} \cdot \frac{-i}{-i} = -2i$.

Multiply by the complex conjugate of the denominator. $i(-i) = 1$

If $x = -i$, then $y = \frac{2}{-i} = \frac{2}{-i} \cdot \frac{i}{i} = 2i$.

If we substitute the x-values we found into equation (1) or (2) instead of into equation (3), we get extraneous solutions. ***It is always wise to check all solutions in both of the given equations.*** We show a check for the ordered pair $(i, -2i)$.

CHECK Let $x = i$ and $y = -2i$ in both equations (1) and (2).

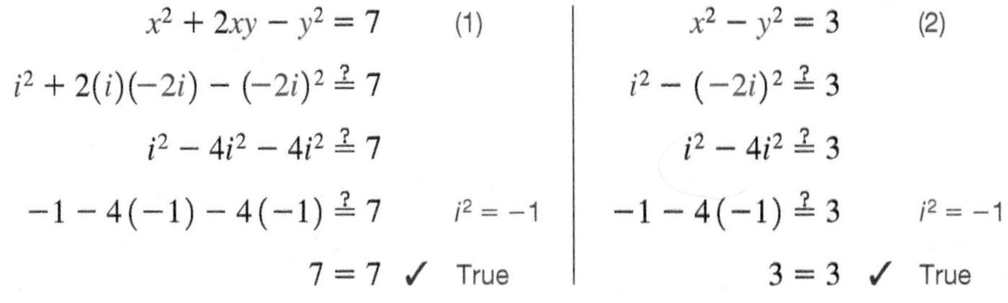

$x^2 + 2xy - y^2 = 7$	(1)	$x^2 - y^2 = 3$	(2)
$i^2 + 2(i)(-2i) - (-2i)^2 \stackrel{?}{=} 7$		$i^2 - (-2i)^2 \stackrel{?}{=} 3$	
$i^2 - 4i^2 - 4i^2 \stackrel{?}{=} 7$		$i^2 - 4i^2 \stackrel{?}{=} 3$	
$-1 - 4(-1) - 4(-1) \stackrel{?}{=} 7$	$i^2 = -1$	$-1 - 4(-1) \stackrel{?}{=} 3$	$i^2 = -1$
$7 = 7$ ✓	True	$3 = 3$ ✓	True

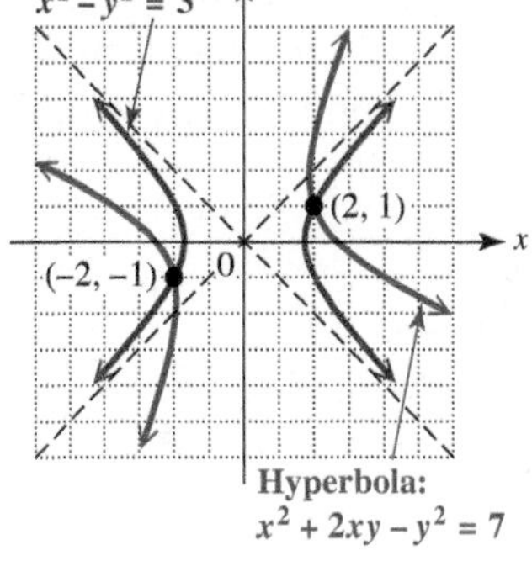

FIGURE 26

The other ordered pairs would be checked similarly. There are four ordered pairs in the solution set, two with real values and two with pure imaginary values. The solution set is

$$\{(2, 1), (-2, -1), (i, -2i), (-i, 2i)\}.$$

The graph of the system, shown in **FIGURE 26**, shows only the two real intersection points because the graph is in the real number plane. In general, if solutions contain nonreal complex numbers as components, they do not appear on the graph.

NOW TRY 

NOTE It is not essential to visualize the number of points of intersection of the graphs in order to solve a nonlinear system. Sometimes we are unfamiliar with the graphs or, as in **Example 4,** there are nonreal complex solutions that do not appear as points of intersection in the real plane. Visualizing the graphs is only an aid to solving these systems.

NOW TRY ANSWER

4. $\{(3, 2), (-3, -2), (2i, -3i), (-2i, 3i)\}$

12.3 Exercises

FOR EXTRA HELP  **MyLab Math**

Video solutions for select problems available in MyLab Math

Concept Check *Each sketch represents the graphs of a pair of equations in a system. How many ordered pairs of real numbers are in each solution set?*

1.

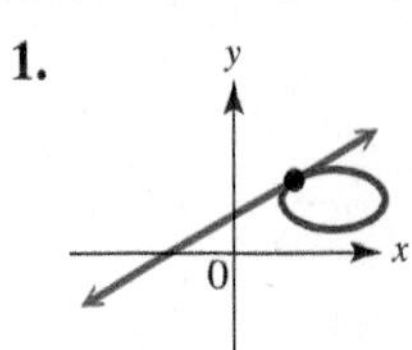

2.

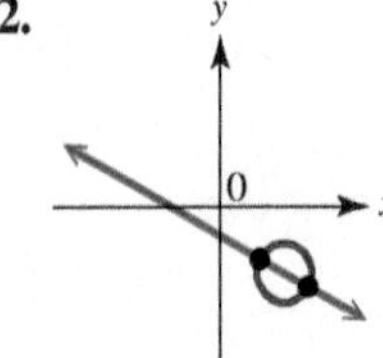

3.

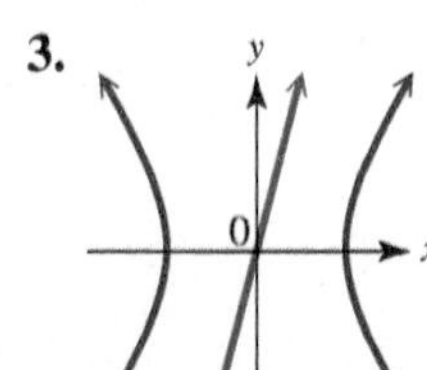

4.

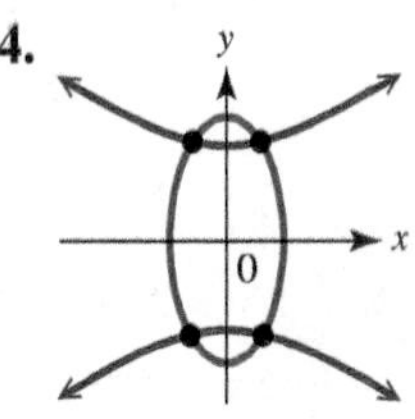

Concept Check *Suppose that a nonlinear system is composed of equations whose graphs are those described, and the number of points of intersection of the two graphs is as given. Make a sketch satisfying these conditions. (There may be more than one way to do this.)*

5. A line and a circle; no points

6. A line and a circle; one point

7. A line and a hyperbola; one point

8. A line and an ellipse; no points

9. A circle and an ellipse; four points

10. A parabola and an ellipse; one point

11. A parabola and an ellipse; four points

12. A parabola and a hyperbola; two points

Solve each system using the substitution method. **See Examples 1 and 2.**

13. $y = 4x^2 - x$
$y = x$

14. $y = x^2 + 6x$
$3y = 12x$

15. $y = x^2 + 6x + 9$
$x + y = 3$

16. $y = x^2 + 8x + 16$
$x - y = -4$

17. $x^2 + y^2 = 2$
$2x + y = 1$

18. $2x^2 + 4y^2 = 4$
$x = 4y$

19. $xy = 4$
$3x + 2y = -10$

20. $xy = -5$
$2x + y = 3$

21. $xy = -3$
$x + y = -2$

22. $xy = 12$
$x + y = 8$

23. $y = 3x^2 + 6x$
$y = x^2 - x - 6$

24. $y = 2x^2 + 1$
$y = 5x^2 + 2x - 7$

25. $2x^2 - y^2 = 6$
$y = x^2 - 3$

26. $x^2 + y^2 = 4$
$y = x^2 - 2$

27. $x^2 - xy + y^2 = 0$
$x - 2y = 1$

28. $x^2 - 3x + y^2 = 4$
$2x - y = 3$

Solve each system using the elimination method or a combination of the elimination and substitution methods. **See Examples 3 and 4.**

29. $3x^2 + 2y^2 = 12$
$x^2 + 2y^2 = 4$

30. $x^2 + 6y^2 = 9$
$4x^2 + 3y^2 = 36$

31. $5x^2 - 2y^2 = -13$
$3x^2 + 4y^2 = 39$

32. $x^2 + y^2 = 41$
$x^2 - y^2 = 9$

33. $2x^2 + 3y^2 = 6$
$x^2 + 3y^2 = 3$

34. $3x^2 + y^2 = 15$
$x^2 + y^2 = 5$

35. $2x^2 + y^2 = 28$
$4x^2 - 5y^2 = 28$

36. $x^2 + 3y^2 = 40$
$4x^2 - y^2 = 4$

37. $xy = 6$
$3x^2 - y^2 = 12$

38. $xy = 5$
$2y^2 - x^2 = 5$

39. $2x^2 = 8 - 2y^2$
$3x^2 = 24 - 4y^2$

40. $5x^2 = 20 - 5y^2$
$2y^2 = 2 - x^2$

41. $x^2 + xy + y^2 = 15$
$x^2 + y^2 = 10$

42. $2x^2 + 3xy + 2y^2 = 21$
$x^2 + y^2 = 6$

43. $x^2 + xy - y^2 = 11$
$x^2 - y^2 = 8$

44. $x^2 + xy - y^2 = 29$
$x^2 - y^2 = 24$

45. $3x^2 + 2xy - 3y^2 = 5$
$-x^2 - 3xy + y^2 = 3$

46. $-2x^2 + 7xy - 3y^2 = 4$
$2x^2 - 3xy + 3y^2 = 4$

Extending Skills *Solve each problem using a nonlinear system.*

47. The area of a rectangular rug is 84 ft^2 and its perimeter is 38 ft. Find the length and width of the rug.

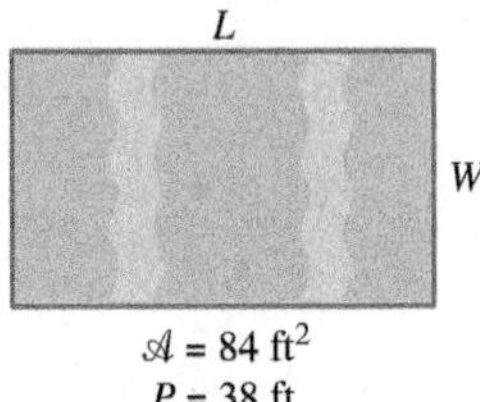

48. Find the length and width of a rectangular room whose perimeter is 50 m and whose area is 100 m^2.

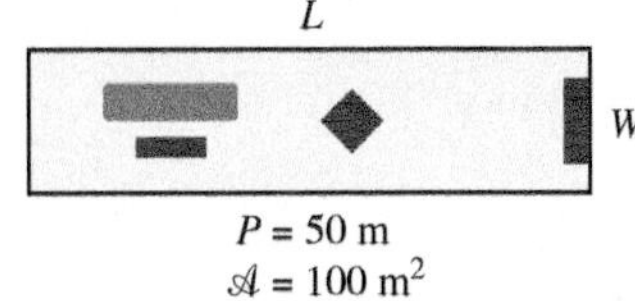

49. A company has found that the price p (in dollars) of its scientific calculator is related to the supply x (in thousands) by the equation

$$px = 16.$$

The price is related to the demand x (in thousands) for the calculator by the equation

$$p = 10x + 12.$$

The **equilibrium price** is the value of p where demand equals supply. Find the equilibrium price and the supply/demand at that price. (*Hint:* Demand, price, and supply must all be positive.)

50. A company has determined that the cost y to make x (thousand) computer tablets is

$$y = 4x^2 + 36x + 20,$$

and that the revenue y from the sale of x (thousand) tablets is

$$36x^2 - 3y = 0.$$

Find the **break-even point,** where cost equals revenue.

12.4 Second-Degree Inequalities, Systems of Inequalities, and Linear Programming

OBJECTIVES

1 Graph second-degree inequalities.
2 Graph the solution set of a system of inequalities.
3 Solve linear programming problems by graphing.

OBJECTIVE 1 Graph second-degree inequalities.

A **second-degree inequality** is an inequality with at least one variable of degree 2 and no variable of degree greater than 2.

Examples: $x^2 + y^2 > 9$, $y \leq 2x^2 - 4$, $x \geq y^2$ Second-degree inequalities

To graph a second-degree inequality, we graph the related second-degree equation and then determine the region to shade.

VOCABULARY

- ☐ second-degree inequality
- ☐ system of inequalities
- ☐ constraints
- ☐ objective function
- ☐ region of feasible solutions
- ☐ vertices (singular *vertex*) (corner points)

NOW TRY EXERCISE 1

Graph $x^2 + y^2 \geq 9$.

EXAMPLE 1 Graphing a Second-Degree Inequality

Graph $x^2 + y^2 \leq 36$.

The boundary of the inequality $x^2 + y^2 \leq 36$ is the graph of the equation $x^2 + y^2 = 36$, a circle with radius 6 and center at the origin, as shown in **FIGURE 27**.

The inequality $x^2 + y^2 \leq 36$ includes the points of the boundary (because the symbol $\leq$ includes equality) and either the points "outside" the circle or the points "inside" the circle. To decide which region to shade, we substitute any test point not on the circle.

$x^2 + y^2 < 36$ — We are testing the region.

$0^2 + 0^2 \overset{?}{<} 36$ — Use (0, 0) as a test point.

$0 < 36$ — True

Because a true statement results, the original inequality includes the points *inside* the circle, the shaded region in **FIGURE 27**, and the boundary.

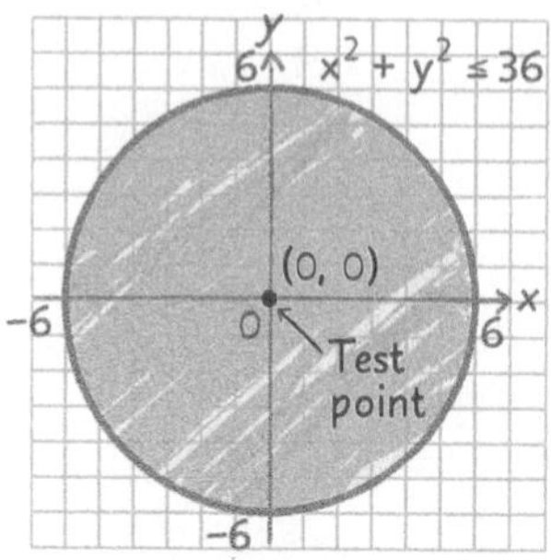

FIGURE 27

NOTE Because the substitution is easy, the origin is the test point of choice unless the graph actually passes through (0, 0).

NOW TRY EXERCISE 2

Graph $y \geq -(x + 2)^2 + 1$.

EXAMPLE 2 Graphing a Second-Degree Inequality

Graph $y < -2(x - 4)^2 - 3$.

The boundary, $y = -2(x - 4)^2 - 3$, is a parabola that opens down with vertex $(4, -3)$. We must decide whether to shade the region "inside" or "outside" the parabola.

$y < -2(x - 4)^2 - 3$ — Original inequality

$0 \overset{?}{<} -2(0 - 4)^2 - 3$ — Use (0, 0) as a test point.

$0 \overset{?}{<} -32 - 3$ — Simplify.

$0 < -35$ — False

Because the final inequality is a false statement, the points in the region containing $(0, 0)$ do not satisfy the inequality. As a result, the region inside (or below) the parabola is shaded in **FIGURE 28**. The parabola is drawn as a dashed curve since the points of the parabola itself do not satisfy the inequality.

CHECK As additional confirmation, select a test point in the shaded region, such as $(4, -7)$, and substitute it into the original inequality.

$y < -2(x - 4)^2 - 3$

$-7 \overset{?}{<} -2(4 - 4)^2 - 3$ — Test (4, −7).

$-7 < -3$ ✓ — True

A true statement results, so the correct region is shaded.

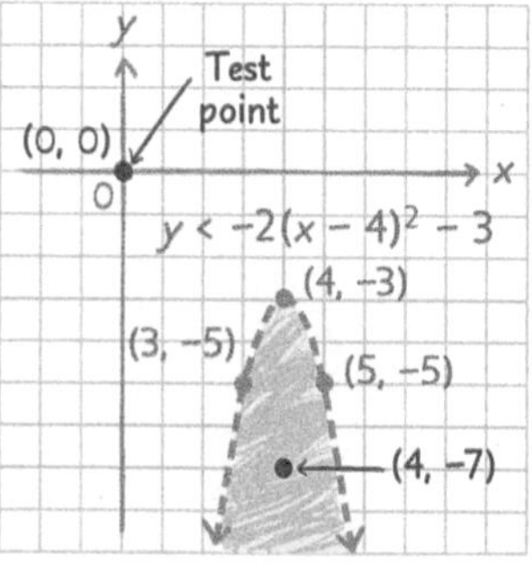

FIGURE 28

NOW TRY

NOW TRY ANSWERS

1.

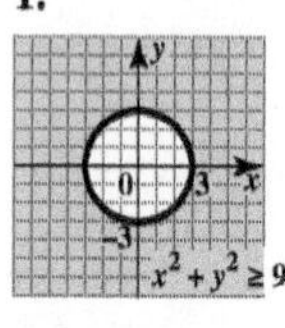

2.

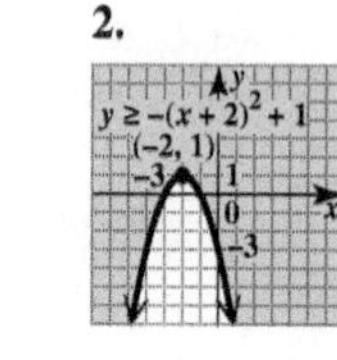

NOW TRY EXERCISE 3

Graph $25x^2 - 16y^2 > 400$.

EXAMPLE 3 Graphing a Second-Degree Inequality

Graph $16y^2 \le 144 + 9x^2$.

$$16y^2 \le 144 + 9x^2$$

$$16y^2 - 9x^2 \le 144 \quad \text{Subtract } 9x^2.$$

$$\frac{y^2}{9} - \frac{x^2}{16} \le 1 \quad \text{Divide by 144.}$$

This form shows that the boundary is the hyperbola given by

$$\frac{y^2}{9} - \frac{x^2}{16} = 1.$$

Because the graph is a vertical hyperbola, the desired region will be either the region "between" the branches or the regions "above" the top branch and "below" the bottom branch. We choose $(0, 0)$ as a test point.

$$16y^2 < 144 + 9x^2$$

$$16(0)^2 \stackrel{?}{<} 144 + 9(0)^2 \quad \text{Test } (0, 0).$$

$$0 < 144 \quad \text{True}$$

Because a true statements results, we shade the region between the branches containing $(0, 0)$. See **FIGURE 29**.

NOW TRY

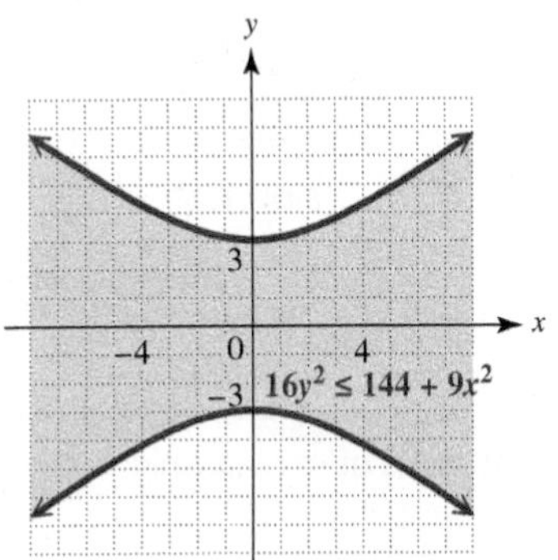

FIGURE 29

OBJECTIVE 2 Graph the solution set of a system of inequalities.

If two or more inequalities are considered at the same time, we have a **system of inequalities.** To find the solution set of the system, we find the intersection of the graphs (solution sets) of the inequalities in the system.

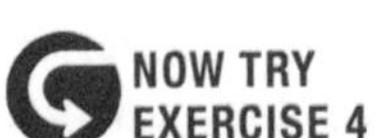

NOW TRY EXERCISE 4

Graph the solution set of the system.

$$2x + 3y \ge 6$$
$$x - 5y \ge 5$$

EXAMPLE 4 Graphing a System of Two Inequalities

Graph the solution set of the system.

$$x - y \le 4$$
$$x + 2y < 2$$

Both inequalities in the system are linear. We begin by graphing the solution set of $x - y \le 4$. The inequality includes the points of the boundary line $x - y = 4$ because the symbol $\le$ includes equality. The test point $(0, 0)$ leads to a true statement in $x - y < 4$, so we shade the region above the line. See **FIGURE 30**.

The graph of the solution set of $x + 2y < 2$ does not include the boundary line $x + 2y = 2$. Testing the point $(0, 0)$ leads to a true statement, indicating that we should shade the region below the dashed boundary line. See **FIGURE 31**.

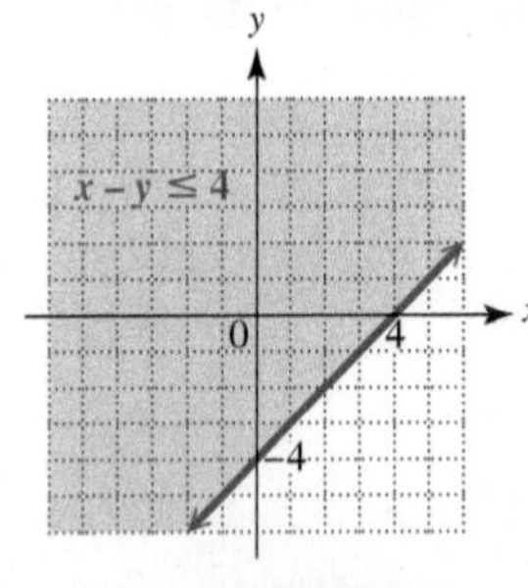

FIGURE 30

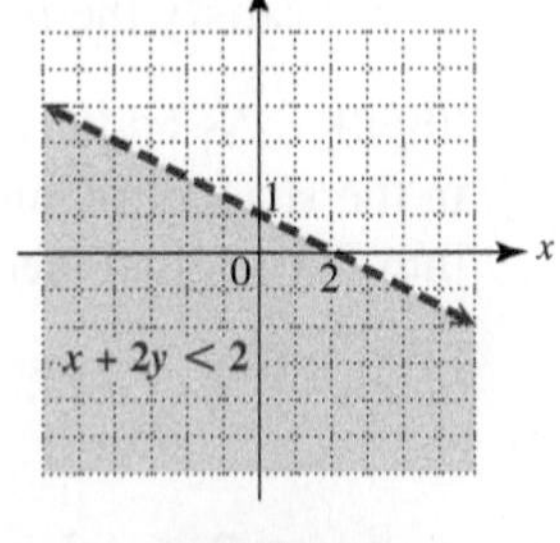

FIGURE 31

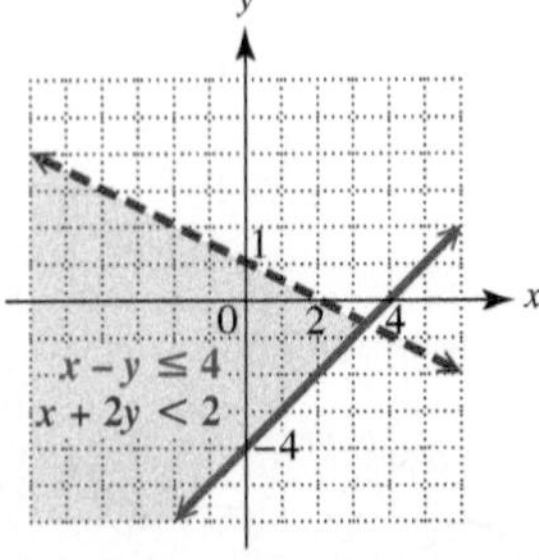

FIGURE 32

The solution set is the intersection of the graphs of the two inequalities. See the overlapping region in **FIGURE 32**, which includes one boundary line.

NOW TRY

NOW TRY ANSWERS

3. 4.

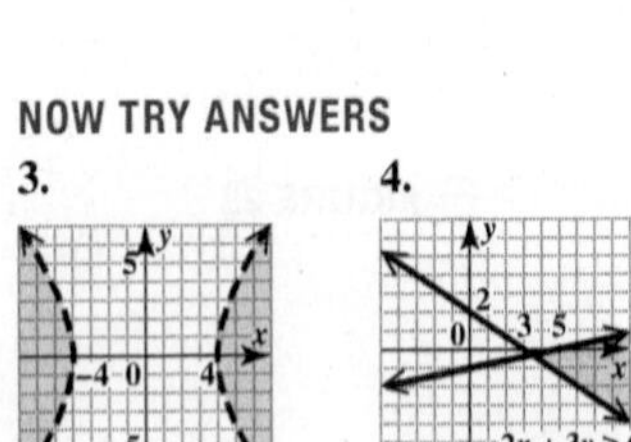

NOW TRY EXERCISE 5

Graph the solution set of the system.

$$x^2 + y^2 \le 25$$
$$x + y \le 3$$

EXAMPLE 5 Graphing a System of Two Inequalities

Graph the solution set of the system.

$$2x + 3y > 6$$
$$x^2 + y^2 < 16$$

We begin by graphing the solution set of the linear inequality $2x + 3y > 6$. The boundary line is the graph of $2x + 3y = 6$ and is a dashed line because the symbol $>$ does not include equality. The test point $(0, 0)$ leads to a false statement in $2x + 3y > 6$, so we shade the region above the line, as shown in **FIGURE 33.**

The graph of $x^2 + y^2 < 16$ is the region inside of a dashed circle centered at the origin with radius 4. This is shown in **FIGURE 34.**

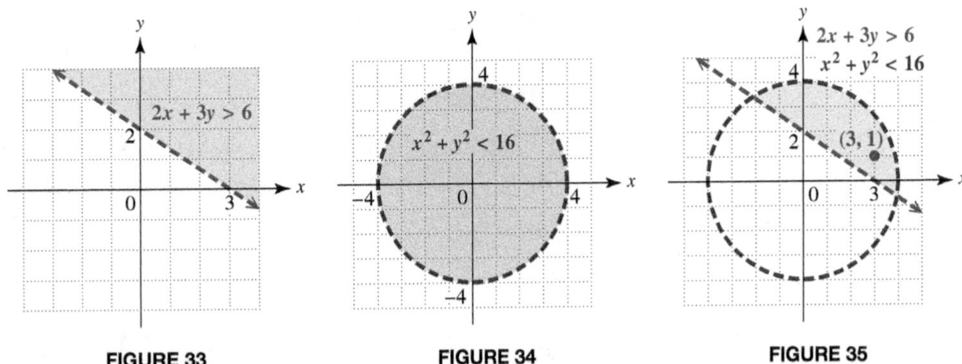

FIGURE 33 **FIGURE 34** **FIGURE 35**

The graph of the solution set of the system is the intersection of the graphs of the two inequalities. The overlapping region in **FIGURE 35** is the solution set.

CHECK As additional confirmation, select a test point in the shaded region, such as $(3, 1)$, and substitute it into *both* inequalities.

$2x + 3y > 6$		$x^2 + y^2 < 16$	
$2(3) + 3(1) \overset{?}{>} 6$	Test (3, 1).	$3^2 + 1^2 \overset{?}{<} 16$	Test (3, 1).
$9 > 6$ ✓	True	$10 < 16$ ✓	True

True statements result, so the correct region is shaded in **FIGURE 35.** NOW TRY 

NOW TRY EXERCISE 6

Graph the solution set of the system.

$$3x + 2y > 6$$
$$y \ge \frac{1}{2}x - 2$$
$$x \ge 0$$

EXAMPLE 6 Graphing a System of Three Inequalities

Graph the solution set of the system.

$$x + y < 1$$
$$y \le 2x + 3$$
$$y \ge -2$$

We graph each linear inequality on the same axes.

- The graph of $x + y < 1$ is the region that lies below the dashed line $x + y = 1$.
- The graph of $y \le 2x + 3$ is the region that lies below the solid line $y = 2x + 3$.
- The graph of $y \ge -2$ is the region above the solid horizontal line $y = -2$.

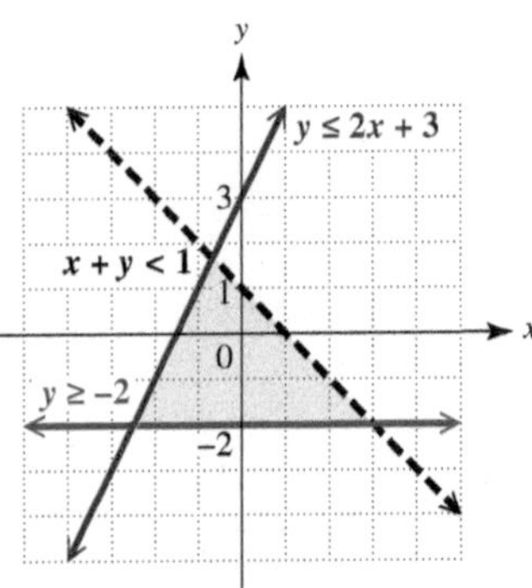

FIGURE 36

The test point $(0, 0)$ satisfies all three inequalities. The graph of the system, the intersection of these three graphs, is the triangular region enclosed by the three boundary lines in **FIGURE 36,** including two of its boundary lines. NOW TRY

NOW TRY ANSWERS

5. 6.

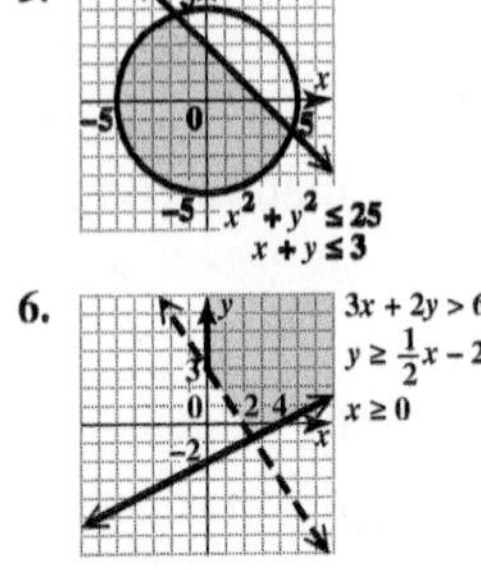

NOW TRY EXERCISE 7

Graph the solution set of the system.

$$\frac{x^2}{4} + \frac{y^2}{16} \le 1$$
$$y \le x^2 - 2$$
$$y + 3 > 0$$

EXAMPLE 7 Graphing a System of Three Inequalities

Graph the solution set of the system.

$$y \ge x^2 - 2x + 1$$
$$2x^2 + y^2 > 4$$
$$y < 4$$

We graph each inequality on the same axes.

- The graph of $y = x^2 - 2x + 1$ is a parabola with vertex at $(1, 0)$. Those points inside (or above) the parabola satisfy the condition $y > x^2 - 2x + 1$. Thus, the solution set of $y \ge x^2 - 2x + 1$ includes points on or inside the parabola.
- The graph of the equation $2x^2 + y^2 = 4$ is an ellipse. We draw it as a dashed curve. To satisfy the inequality $2x^2 + y^2 > 4$, a point must lie outside the ellipse.
- The graph of $y < 4$ includes all points below the dashed line $y = 4$.

The graph of the system is the shaded region in **FIGURE 37**, which lies outside the ellipse, inside or on the boundary of the parabola, and below the line $y = 4$.

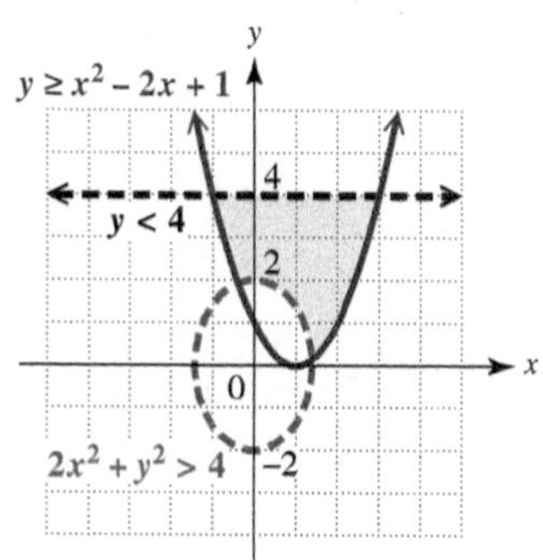

FIGURE 37

NOW TRY 

OBJECTIVE 3 Solve linear programming problems by graphing.

An important application of mathematics is *linear programming*. We use **linear programming** to find an optimum value such as minimum cost or maximum profit. It was first developed to solve problems in allocating supplies for the U.S. Air Force during World War II.

EXAMPLE 8 Finding a Maximum Profit Model

A company makes two models of cellular phones: model A and model B. Each model A phone produces a profit of \$30, while each model B phone produces a \$70 profit. The company must manufacture at least 10 model A phones per day, but no more than 50 because of production restrictions. The number of model B phones produced cannot exceed 60 per day, and the number of model A phones cannot exceed the number of model B phones.

How many of each model of cellular phone should the company manufacture to obtain maximum profit?

First, we translate the statement of the problem into symbols.

Let x = number of model A phones to be produced daily,

and y = number of model B phones to be produced daily.

NOW TRY ANSWER

7.
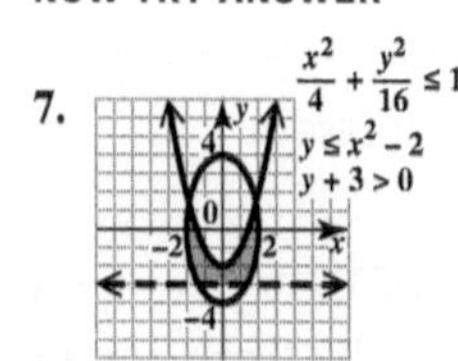

NOW TRY EXERCISE 8

Hazel has \$21,000 to invest. Her financial advisor recommends that she place her money in Treasury bills yielding 3% per year and corporate bonds yielding 4% per year. Both investments earn simple interest for one year. To balance these investments, she invests no more than twice as much money in bonds as she does in Treasury bills.

What is Hazel's maximum annual income from these two investments?

The company must produce at least 10 model A phones (10 or more), so

$$x \geq 10.$$

Because no more than 50 model A phones may be produced,

$$x \leq 50.$$

No more than 60 model B phones may be made in one day, so

$$y \leq 60.$$

The number of model A phones may not exceed the number of model B phones, which translates as

$$x \leq y.$$

The numbers of model A and model B phones cannot be negative, so

$$x \geq 0 \quad \text{and} \quad y \geq 0.$$

These restrictions, or **constraints,** form the following system of inequalities.

$$x \geq 10, \quad x \leq 50, \quad y \leq 60, \quad x \leq y, \quad x \geq 0, \quad y \geq 0$$

Each model A phone gives a profit of \$30, so the daily profit from production of x model A phones is $30x$ dollars. Also, the profit from production of y model B phones will be $70y$ dollars per day. The total daily profit is

$$\text{profit} = 30x + 70y.$$

This equation defines the **objective function** to be maximized.

To find the maximum possible profit, subject to these constraints, we sketch the graph of each constraint. The only feasible values of x and y are those which satisfy *all* constraints—that is, the values that lie in the intersection of the graphs.

The intersection is shown in **FIGURE 38**. Any point lying inside the shaded region or on the boundary satisfies the restrictions as to numbers of model A and model B phones that may be produced. (For practical purposes, however, only points with integer coefficients are useful.) This region is the **region of feasible solutions.**

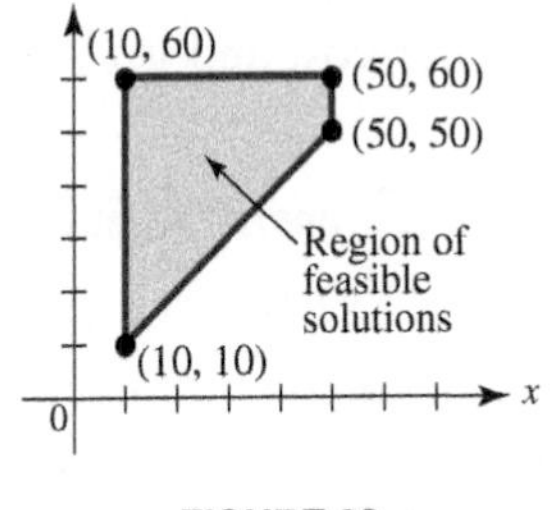

FIGURE 38

The **vertices** (singular *vertex*), or **corner points,** of the region of feasible solutions have coordinates

$$(10, 10), \quad (10, 60), \quad (50, 50), \quad \text{and} \quad (50, 60).$$

We must find the value of the objective function $30x + 70y$ for each vertex. We want the vertex that produces the maximum possible value of this function.

Vertex	$30x + 70y =$ *Profit*	
(10, 10)	$30(10) + 70(10) = 1000$	
(10, 60)	$30(10) + 70(60) = 4500$	
(50, 50)	$30(50) + 70(50) = 5000$	
(50, 60)	$30(50) + 70(60) = 5700$	← Maximum

The maximum profit, obtained when 50 model A phones and 60 model B phones are produced each day, will be

$$30(50) + 70(60) = \$5700 \text{ per day.}$$

NOW TRY 

NOW TRY ANSWER

8. \$770 (with \$7000 in Treasury bills and \$14,000 in bonds)

Example 8 illustrates an application of the method used to solve linear programming problems in general. It is based on the following theorem.

Fundamental Theorem of Linear Programming

If the optimal value for a linear programming problem exists, then it occurs at a vertex of the region of feasible solutions.

To solve such problems, follow these steps.

Solving a Linear Programming Problem

Step 1 Write the objective function and all necessary constraints.

Step 2 Graph the region of feasible solutions.

Step 3 Identify all vertices (corner points).

Step 4 Find the value of the objective function at each vertex.

Step 5 Choose the required maximum or minimum value, given by the vertex that produces the optimal value of the objective function.

EXAMPLE 9 Finding a Minimum Cost Model

Robin takes vitamin pills each day. She wants at least 16 units of vitamin A, at least 5 units of vitamin B_1, and at least 20 units of vitamin C. Capsules, costing \$0.10 each, contain 8 units of A, 1 of B_1, and 2 of C. Chewable tablets, costing \$0.20 each, contain 2 units of A, 1 of B_1, and 7 of C.

How many of each should she take each day to minimize her cost and yet fulfill her daily requirements?

Step 1 Let x represent the number of capsules to take, and let y represent the number of chewable tablets to take. Then the cost in *pennies* per day is

$$\text{cost} = 10x + 20y. \quad \text{Objective function}$$

Robin takes x of the \$0.10 capsules and y of the \$0.20 tablets, and she gets 8 units of vitamin A from each capsule and 2 units of vitamin A from each tablet. She takes $8x + 2y$ units of A per day. She wants at least 16 units, so

$$8x + 2y \geq 16.$$

Each capsule and each tablet supplies 1 unit of vitamin B_1. Robin wants at least 5 units per day, so

$$x + y \geq 5.$$

For vitamin C, the inequality is

$$2x + 7y \geq 20.$$

Robin cannot buy negative numbers of pills, so $x \geq 0$ and $y \geq 0$.

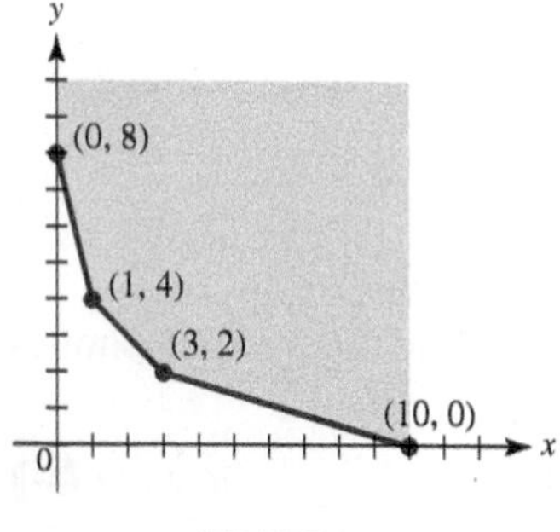

FIGURE 39

Step 2 FIGURE 39 shows the intersection of the graphs of the following system.

$$8x + 2y \geq 16, \quad x + y \geq 5, \quad 2x + 7y \geq 20, \quad x \geq 0, \quad \text{and} \quad y \geq 0$$

NOW TRY EXERCISE 9

A manufacturer must ship at least 100 refrigerators to two warehouses. Each warehouse holds a maximum of 100 refrigerators. Warehouse A holds 25 refrigerators already, while warehouse B has 20 on hand. It costs \$12 to ship a refrigerator to warehouse A and \$10 to ship one to warehouse B.

How many refrigerators should be shipped to each warehouse to minimize cost? What is the minimum cost?

NOW TRY ANSWER

9. 20 to A and 80 to B; \$1040

Step 3 The vertices are $(0, 8)$, $(1, 4)$, $(3, 2)$, and $(10, 0)$.

Steps 4 and 5 See the table. The minimum cost occurs at $(3, 2)$.

Vertex	$10x + 20y =$ *Cost*	
(0, 8)	10(0) + 20(8) = 160	
(1, 4)	10(1) + 20(4) = 90	
(3, 2)	10(3) + 20(2) = 70	⟵ Minimum
(10, 0)	10(10) + 20(0) = 100	

Robin's best choice is to take 3 capsules and 2 tablets each day, for a minimum cost of \$0.70 per day. She receives the minimum amounts of vitamins B_1 and C and an excess of vitamin A.

NOW TRY

12.4 Exercises

MyLab Math

Video solutions for select problems available in MyLab Math

1. Concept Check Match each inequality in parts (a)–(d) with its graph in choices A–D.

(a) $y \geq x^2 + 4$ **(b)** $y \leq x^2 + 4$ **(c)** $y < x^2 + 4$ **(d)** $y > x^2 + 4$

A.

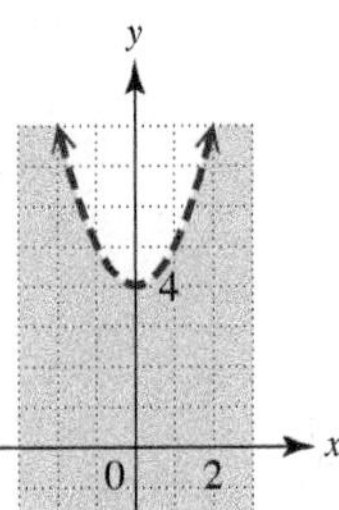

B.

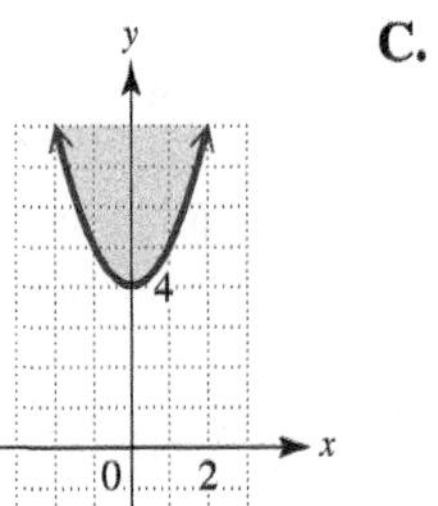

C.

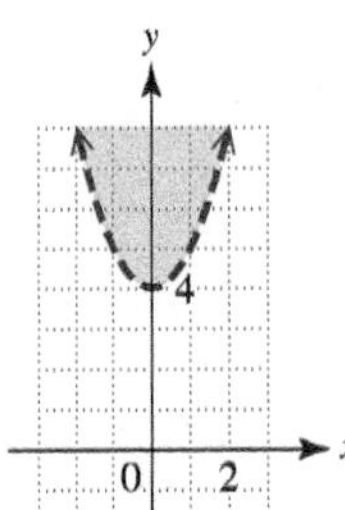

D.

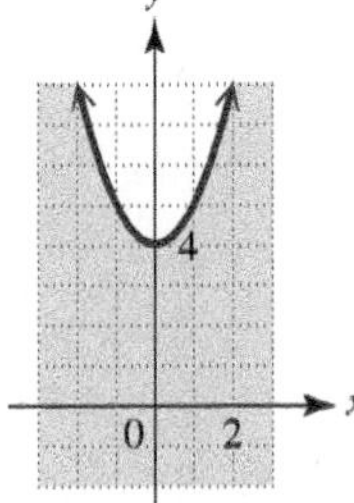

2. Concept Check Match each system of inequalities in parts (a)–(d) with its graph in choices A–D.

(a) $x > -2$, $y \leq 2$ **(b)** $x < -2$, $y \geq 2$ **(c)** $x \geq 2$, $y < -2$ **(d)** $x \leq 2$, $y > -2$

A.

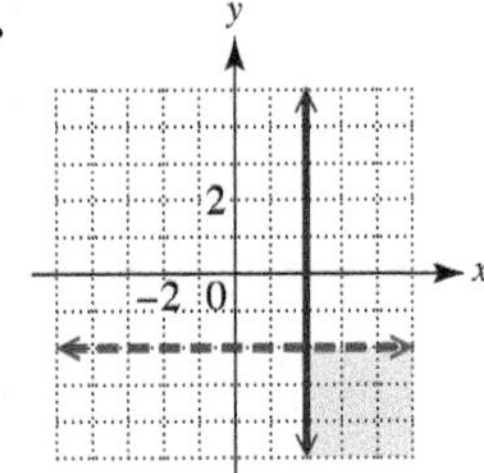

B.

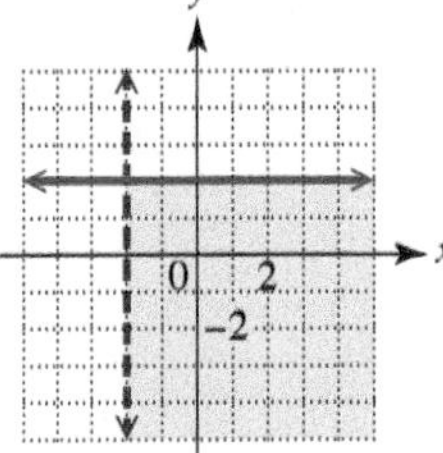

C.

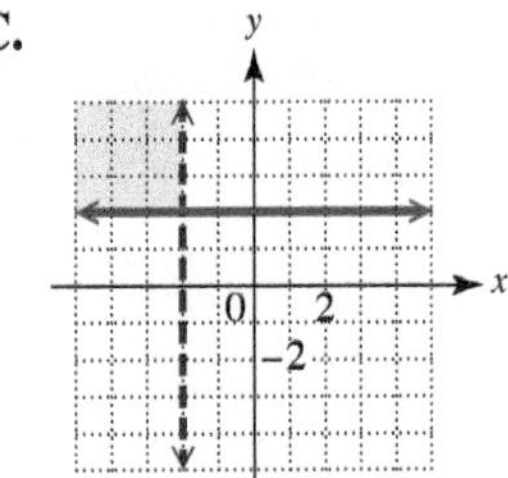

D.

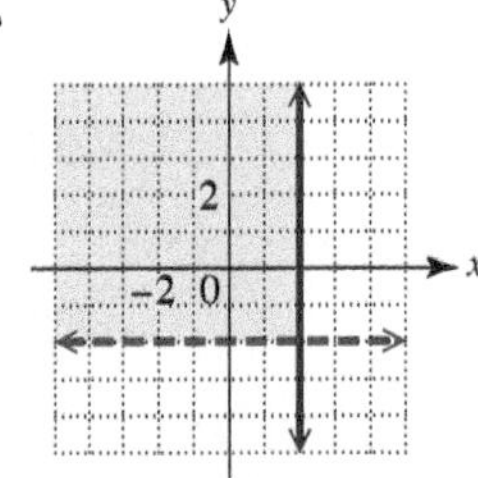

Concept Check *Write a system of inequalities for the indicated region.*

3. Quadrant I, including the x- and y-axes

4. Quadrant IV, not including the x- and y-axes

5. Quadrant III, including the x-axis but not including the y-axis

6. Quadrant II, not including the x-axis but including the y-axis

7. Concept Check Which one of the following is a description of the graph of the solution set of the following system?

$$x^2 + y^2 < 25$$
$$y > -2$$

A. All points outside the circle $x^2 + y^2 = 25$ and above the line $y = -2$

B. All points outside the circle $x^2 + y^2 = 25$ and below the line $y = -2$

C. All points inside the circle $x^2 + y^2 = 25$ and above the line $y = -2$

D. All points inside the circle $x^2 + y^2 = 25$ and below the line $y = -2$

8. Concept Check Fill in each blank with the appropriate response. The graph of the solution set of the system

$$y > x^2 + 1$$
$$\frac{x^2}{9} + \frac{y^2}{4} > 1$$
$$y < 5$$

consists of all points ________ (above / below) the parabola $y = x^2 + 1$, ________ (inside / outside) the ellipse $\frac{x^2}{9} + \frac{y^2}{4} = 1$, and ________ (above / below) the line $y = 5$.

Graph each inequality. **See Examples 1–3.**

9. $y \geq x^2 - 2$

10. $y > x^2 - 1$

11. $2y^2 \geq 8 - x^2$

12. $y^2 \leq 4 - 2x^2$

13. $x^2 > 4 - y^2$

14. $x^2 \leq 16 - y^2$

15. $9x^2 > 16y^2 + 144$

16. $x^2 \leq 16 + 4y^2$

17. $9x^2 < 16y^2 - 144$

18. $y^2 > 4 + x^2$

19. $x^2 - 4 \geq -4y^2$

20. $4y^2 \leq 36 - 9x^2$

21. $x \leq -y^2 + 6y - 7$

22. $x \geq y^2 - 8y + 14$

23. $25x^2 \leq 9y^2 + 225$

24. $y^2 - 16x^2 \leq 16$

Graph each system of inequalities. **See Examples 4–7.**

25. $2x + 5y < 10$
$x - 2y < 4$

26. $3x - y > -6$
$4x + 3y > 12$

27. $5x - 3y \leq 15$
$4x + y \geq 4$

28. $4x - 3y \leq 0$
$x + y \leq 5$

29. $x \leq 5$
$y \leq 4$

30. $x \geq -2$
$y \leq 4$

31. $x^2 + y^2 > 9$
$y > x^2 - 1$

32. $x^2 - y^2 \geq 9$
$\frac{x^2}{16} + \frac{y^2}{9} \leq 1$

33. $y > x^2 - 4$
$y < -x^2 + 3$

34. $y \leq -x^2 + 5$
$y \leq x^2 - 3$

35. $x + y > 1$
$y \geq 2x - 2$
$y \leq 4$

36. $3x - 4y \geq 12$
$x + 3y > 6$
$y \leq 2$

37. $x^2 + y^2 \geq 4$
$x + y \leq 5$
$x \geq 0$
$y \geq 0$

38. $y^2 - x^2 \geq 4$
$-5 \leq y \leq 5$

39. $y \leq -x^2$
$y \geq x - 3$
$y \leq -1$
$x < 1$

40. $y < x^2$
$y > -2$
$x + y < 3$
$3x - 2y > -6$

Extending Skills *The nonlinear inequality*

$$x^2 + y^2 \leq 4, \quad x \geq 0$$

is graphed in the figure. Only the right half of the interior of the circle and its boundary is shaded because of the restriction that x must be nonnegative. Graph each nonlinear inequality with the given restrictions.

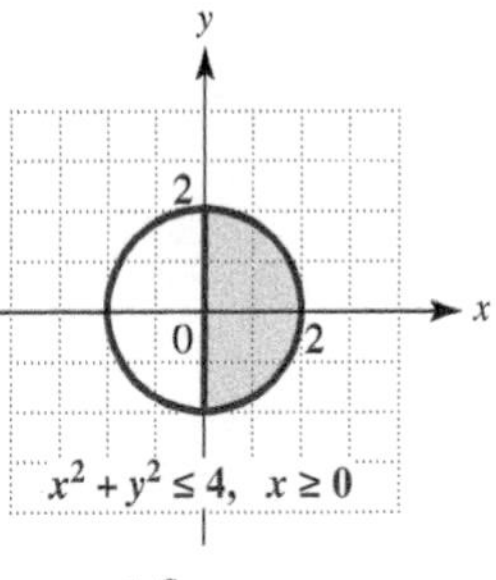

41. $x^2 + y^2 > 36, \quad x \geq 0$

42. $4x^2 + 25y^2 < 100, \quad y < 0$

43. $x < y^2 - 3, \quad x < 0$

44. $x^2 - y^2 < 4, \quad x < 0$

45. $4x^2 - y^2 > 16, \quad x < 0$

46. $x^2 + y^2 > 4, \quad y < 0$

47. $x^2 + 4y^2 \geq 1, \quad x \geq 0, y \geq 0$

48. $2x^2 - 32y^2 \leq 8, \quad x \leq 0, y \geq 0$

The graphs show regions of feasible solutions. Find the maximum and minimum values of each objective function. ***See Examples 8 and 9.***

49. $3x + 5y$

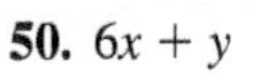
50. $6x + y$

51. $40x + 75y$

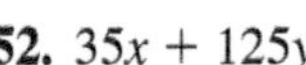
52. $35x + 125y$

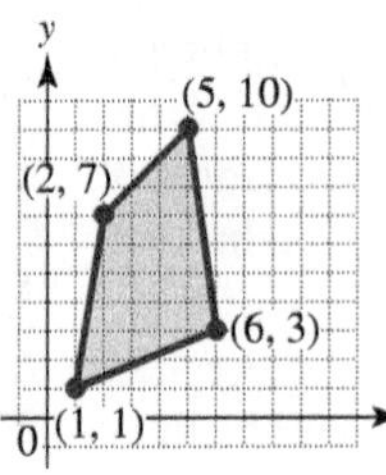

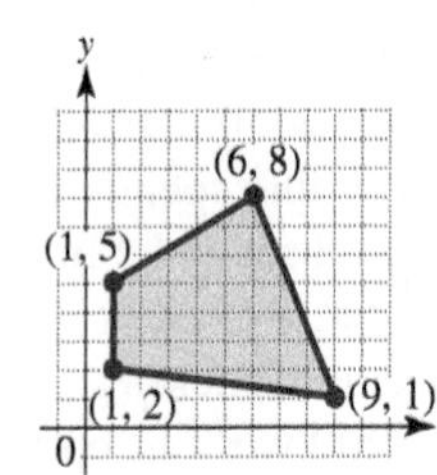

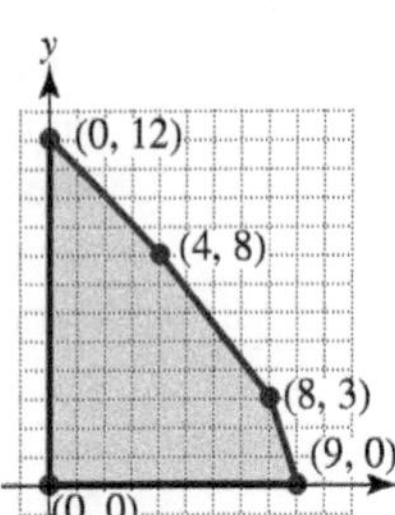

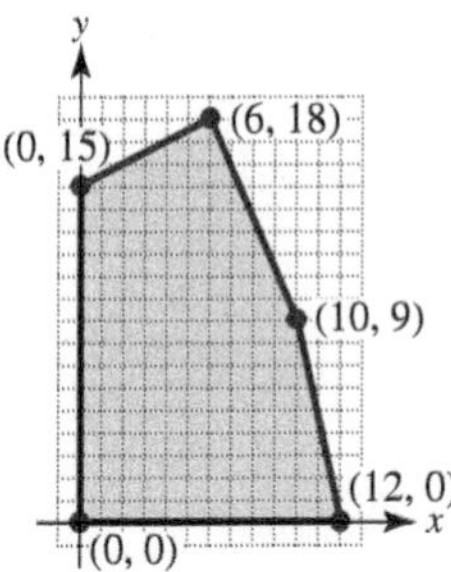

Use graphical methods on the given constraints to find the indicated optimal value of the given objective function. (Hint: It may be necessary to solve a system of equations in order to find vertices.) ***See Examples 8 and 9.***

53. $2x + 3y \leq 6$
$4x + y \leq 6$
$x \geq 0$
$y \geq 0$
Maximize $5x + 2y$.

54. $x + y \leq 10$
$5x + 2y \geq 20$
$x - 2y \leq 0$
$x \geq 0$
$y \geq 0$
Minimize $x + 3y$.

55. $3x - y \geq 12$
$x + y \leq 15$
$x \geq 2$
$y \geq 5$
Minimize $2x + y$.

56. $2x + 3y \leq 100$
$5x + 4y \leq 200$
$x \geq 10$
$y \geq 20$
Maximize $x + 3y$.

Solve each problem. ***See Examples 8 and 9.***

57. A 4-H member raises pigs and geese. He wants to raise no more than 16 animals with no more than 12 geese. He spends \$50 to raise a pig and \$20 to raise a goose. He has \$500 available for this purpose.

If the 4-H member makes a profit of \$80 per goose and \$40 per pig, how many of each animal should he raise in order to maximize profit? What is the maximum profit?

58. A wholesaler of sporting goods wishes to display hats and uniforms at a convention. Her booth has 12 m^2 of floor space to be used for display purposes. A display unit for hats requires 2 m^2, and a display unit for uniforms requires 4 m^2. She never wants to have more than a total of 5 units of uniforms and hats on display at one time.

If she receives three inquiries for each unit of hats and two inquiries for each unit of uniforms on display, how many of each should she display in order to receive the maximum number of inquiries? What is the maximum number of inquaries?

59. An office manager wants to buy some filing cabinets. She knows that cabinet A costs \$20 each, requires 6 ft^2 of floor space, and holds 8 ft^3 of files. Cabinet B costs \$40 each, requires 8 ft^2 of floor space, and holds 12 ft^3. She can spend no more than \$280 due to budget limitations, and there is room for no more than 72 ft^2 of cabinets.

To maximize storage capacity within the limits imposed by funds and space, how many of each type of cabinet should she buy? What is the maximum storage capacity?

60. A certain manufacturing process requires that oil refineries manufacture at least 2 gal of gasoline for each gallon of fuel oil. To meet the winter demand for fuel oil, at least 3 million gal per day must be produced. The demand for gasoline is no more than 6.4 million gal per day.

If the price of gasoline is \$2.90 per gal and the price of fuel oil is \$2.50 per gal, how much of each should be produced to maximize revenue? What is the maximum revenue?

61. The GL Company makes color television sets. It produces a bargain set that sells for \$100 profit and a deluxe set that sells for \$150 profit. On the assembly line, the bargain set requires 3 hr, while the deluxe set takes 5 hr. The finishing line spends 1 hr on the finishes for the bargain set and 3 hr on the finishes for the deluxe set. Both sets require 2 hr of time for testing and packing. The company has available 3900 work hr on the assembly line, 2100 work hr on the finishing line, and 2200 work hr for testing and packing.

How many sets of each type should the company produce to maximize profit? What is the maximum profit?

62. The Schwab Company designs and sells two types of rings: the VIP and the SST. The company can produce up to 24 rings each day using up to 60 total hours of labor. It takes 3 hr to make one VIP ring and 2 hr to make one SST ring. The profit on one VIP ring is \$30, and the profit on one SST ring is \$40.

How many of each type of ring should be made daily to maximize profit? What is the maximum profit?

63. Tsunami victims in Southeast Asia need medical supplies and bottled water. Each medical kit measures 1 cubic foot and weighs 10 lb. Each container of water is also 1 cubic foot, but weighs 20 lb. The plane can carry only 80,000 lb, with total volume of 6000 cubic feet. Each medical kit will aid 6 people, while each container of water will serve 10 people.

How many of each should be sent in order to maximize the number of people assisted? How many people will be assisted?

64. Refer to **Exercise 63.** If each medical kit could aid 4 people instead of 6, how many of each should be sent in order to maximize the number of people assisted? How many people would be assisted?

Chapter 12 Summary

Key Terms

12.1

conic section
circle
center (of a circle)
radius
ellipse
foci (singular *focus*)
center (of an ellipse)

12.2

hyperbola
transverse axis
fundamental rectangle
asymptotes (of a hyperbola)
generalized square root function

12.3

nonlinear equation
nonlinear system of equations

12.4

second-degree inequality
system of inequalities
constraints
objective function
region of feasible solutions
vertices (singular *vertex*) (corner points)

Test Your Word Power

See how well you have learned the vocabulary in this chapter.

1. Conic sections are
A. graphs of first-degree equations
B. the result of two or more intersecting planes
C. graphs of first-degree inequalities
D. figures that result from the intersection of an infinite cone with a plane.

2. A **circle** is the set of all points in a plane
A. such that the absolute value of the difference of the distances from two fixed points is constant
B. that lie a fixed distance from a fixed point
C. the sum of whose distances from two fixed points is constant
D. that make up the graph of any second-degree equation.

3. An **ellipse** is the set of all points in a plane
A. such that the absolute value of the difference of the distances from two fixed points is constant
B. that lie a fixed distance from a fixed point
C. the sum of whose distances from two fixed points is constant
D. that make up the graph of any second-degree equation.

4. A **hyperbola** is the set of all points in a plane
A. such that the absolute value of the difference of the distances from two fixed points is constant
B. that lie a fixed distance from a fixed point
C. the sum of whose distances from two fixed points is constant
D. that make up the graph of any second-degree equation.

ANSWERS

1. D; *Example:* Parabolas, circles, ellipses, and hyperbolas are conic sections. **2.** B; *Example:* The graph of $x^2 + y^2 = 9$ is a circle centered at the origin with radius 3. **3.** C; *Example:* The graph of $\frac{x^2}{49} + \frac{y^2}{36} = 1$ is an ellipse centered at the origin with x-intercepts $(7, 0)$ and $(-7, 0)$ and y-intercepts $(0, 6)$ and $(0, -6)$. **4.** A; *Example:* The graph of $\frac{x^2}{16} - \frac{y^2}{25} = 1$ is a horizontal parabola centered at the origin with vertices $(4, 0)$ and $(-4, 0)$.

Quick Review

CONCEPTS | **EXAMPLES**

12.1 Circles Revisited and Ellipses

Circle

A circle with center (h, k), and radius $r > 0$ has an equation that can be written in the form

$$(x - h)^2 + (y - k)^2 = r^2. \quad \text{Center-radius form}$$

If its center is $(0, 0)$, then this equation becomes

$$x^2 + y^2 = r^2.$$

For some real numbers c, d, and e, an equation of the form

$$x^2 + y^2 + cx + dy + e = 0 \quad \text{General form}$$

may represent a circle. The x^2- and y^2-terms have equal coefficients, here 1.

Graph $(x + 2)^2 + (y - 3)^2 = 25$.

This equation, which can be written

$$[x - (-2)]^2 + (y - 3)^2 = 5^2,$$

represents a circle with center $(-2, 3)$ and radius 5.

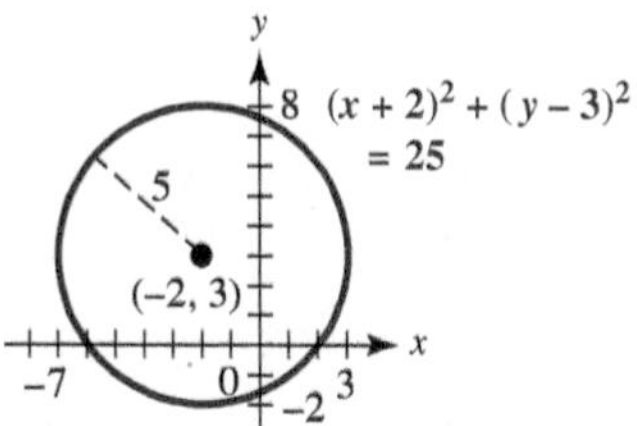

The general form of the above equation is found as follows.

$(x + 2)^2 + (y - 3)^2 = 25$	Center-radius form
$x^2 + 4x + 4 + y^2 - 6y + 9 = 25$	Square each binomial.
$x^2 + y^2 + 4x - 6y - 12 = 0$	General form

Ellipse

An ellipse with x-intercepts $(a, 0)$ and $(-a, 0)$ and y-intercepts $(0, b)$ and $(0, -b)$ has an equation that can be written in the form

$$\frac{x^2}{a^2} + \frac{y^2}{b^2} = 1.$$

Graph $\frac{x^2}{9} + \frac{y^2}{4} = 1$.

x-intercepts: $(3, 0)$ and $(-3, 0)$

y-intercepts: $(0, 2)$ and $(0, -2)$

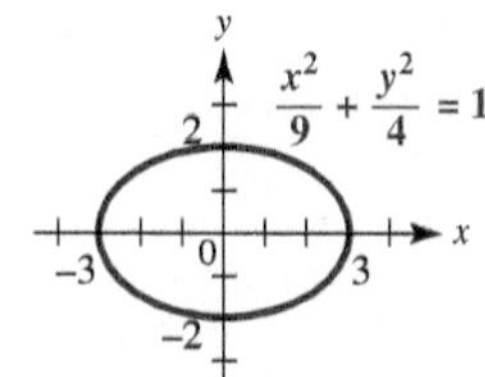

12.2 Hyperbolas and Functions Defined by Radicals

Hyperbola

A hyperbola centered at the origin has an equation that can be written in one of the following forms.

$$\frac{x^2}{a^2} - \frac{y^2}{b^2} = 1 \quad \text{or} \quad \frac{y^2}{b^2} - \frac{x^2}{a^2} = 1$$

x-intercepts $(a, 0)$ and $(-a, 0)$ | y-intercepts $(0, b)$ and $(0, -b)$

The extended diagonals of the fundamental rectangle with vertices at the points (a, b), $(-a, b)$, $(-a, -b)$, and $(a, -b)$ are the asymptotes of these hyperbolas.

Graph $\frac{x^2}{4} - \frac{y^2}{4} = 1$.

x-intercepts: $(2, 0)$ and $(-2, 0)$

Vertices of the fundamental rectangle:

$(2, 2)$, $(-2, 2)$, $(-2, -2)$, and $(2, -2)$

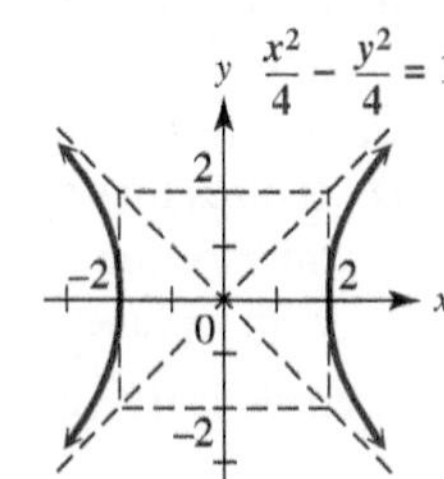

Generalized Square Root Function

For an algebraic expression in x defined by u, where $u \geq 0$, a function of the form

$$f(x) = \sqrt{u}$$

is a generalized square root function.

Graph $y = -\sqrt{4 - x^2}$.

Square each side and rearrange terms.

$$x^2 + y^2 = 4$$

This equation has a circle as its graph. However, graph only the lower half of the circle because the original equation indicates that y cannot be positive.

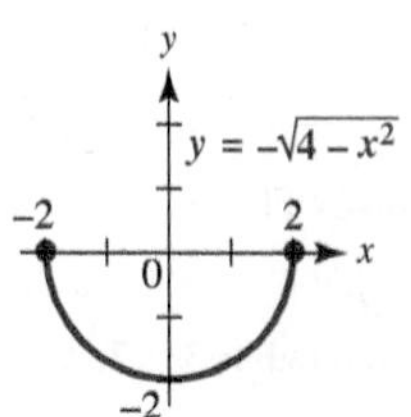

CONCEPTS | **EXAMPLES**

12.3 Nonlinear Systems of Equations

Solving a Nonlinear System

A nonlinear system can be solved by the substitution method, the elimination method, or a combination of the two.

Geometric Interpretation

If, for example, a nonlinear system includes two second-degree equations, then there may be zero, one, two, three, or four solutions of the system.

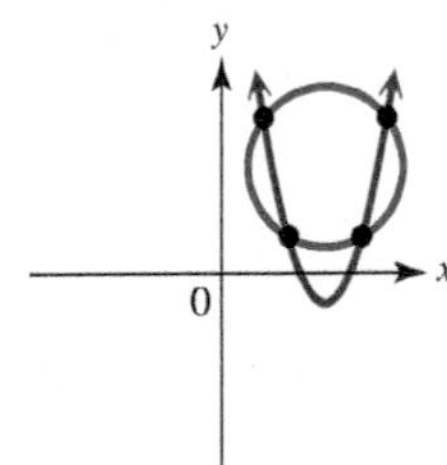

This system has four solutions because there are four points of intersection.

Solve the system.

$$x^2 + 2xy - y^2 = 14 \qquad (1)$$
$$x^2 - y^2 = -16 \qquad (2)$$

Multiply equation (2) by -1 and use the elimination method.

$$\begin{aligned} x^2 + 2xy - y^2 &= 14 \\ -x^2 \qquad + y^2 &= 16 \\ \hline 2xy \qquad &= 30 \\ xy &= 15 \end{aligned}$$

Solve $xy = 15$ for y to obtain $y = \frac{15}{x}$, and substitute into equation (2).

$$x^2 - y^2 = -16 \qquad (2)$$
$$x^2 - \left(\frac{15}{x}\right)^2 = -16 \qquad \text{Let } y = \tfrac{15}{x}.$$
$$x^2 - \frac{225}{x^2} = -16 \qquad \text{Apply the exponent.}$$
$$x^4 + 16x^2 - 225 = 0 \qquad \text{Multiply by } x^2. \text{ Add } 16x^2.$$
$$(x^2 - 9)(x^2 + 25) = 0 \qquad \text{Factor.}$$
$$x^2 - 9 = 0 \quad \text{or} \quad x^2 + 25 = 0 \qquad \text{Zero-factor property}$$
$$x = \pm 3 \quad \text{or} \quad x = \pm 5i \qquad \text{Solve each equation}$$

Substitute these values of x into $y = \frac{15}{x}$ to obtain the corresponding values of y.

If $x = 3$, then $y = \frac{15}{3} = 5$.

If $x = -3$, then $y = \frac{15}{-3} = -5$.

If $x = 5i$, then $y = \frac{15}{5i} = \frac{3}{i} \cdot \frac{-i}{-i} = -3i$.

If $x = -5i$, then $y = \frac{15}{-5i} = \frac{3}{-i} \cdot \frac{i}{i} = 3i$.

Solution set: $\{(3, 5), (-3, -5), (5i, -3i), (-5i, 3i)\}$

12.4 Second-Degree Inequalities, Systems of Inequalities, and Linear Programming

Graphing a Second-Degree Inequality

To graph a second-degree inequality, graph the corresponding equation as a boundary, and use test points to determine which region(s) form the solution set. Shade the appropriate region(s).

Graph $y \geq x^2 - 2x + 3$.

The boundary is a parabola that opens up with vertex $(1, 2)$. Use $(0, 0)$ as a test point. Substituting into the inequality $y > x^2 - 2x + 3$ gives a false statement,

$$0 > 3. \qquad \text{False}$$

Shade the region inside (or above) the parabola.

CONCEPTS	EXAMPLES

Graphing a System of Inequalities

The solution set of a system of inequalities is the intersection of the graphs (solution sets) of the inequalities in the system.

Graph the solution set of the system.

$$3x - 5y > -15$$
$$x^2 + y^2 \leq 25$$

Graph each inequality separately, on the same axes. The graph of the system is the region that lies below the dashed line *and* inside the solid circle.

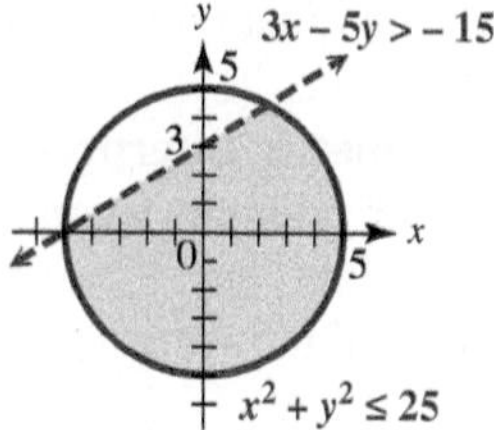

Solving a Linear Programming Problem

Step 1 Write the objective function and all constraints.

Step 2 Graph the region of feasible solutions.

Step 3 Identify all vertices (corner points).

Step 4 Find the value of the objective function at each vertex.

Step 5 Choose the required maximum or minimum value, given by the vertex that produces the optimal value of the objective function.

The feasible region for

$$x + 2y \leq 14$$
$$3x + 4y \leq 36$$
$$x \geq 0$$
$$y \geq 0$$

is given in the figure. Maximize the objective function $8x + 12y$.

Vertex	*Value of* $8x + 12y$	
(0, 0)	$8(0) + 12(0) = 0$	
(0, 7)	$8(0) + 12(7) = 84$	
(12, 0)	$8(12) + 12(0) = 96$	
(8, 3)	$8(8) + 12(3) = 100$	← Maximum

The objective function is maximized at 100 for $x = 8$ and $y = 3$.

Chapter 12 Review Exercises

12.1 *Write the center-radius form of each circle described.*

1. Center: $(0, 0)$; radius: 7

2. Center $(-2, 4)$; radius: 3

3. Center $(4, 2)$; radius: 6

4. Center: $(-5, 0)$; radius: $\sqrt{2}$

Write the center-radius form of the circle with the given equation. Give the center and radius.

5. $x^2 + y^2 + 6x - 4y - 3 = 0$

6. $x^2 + y^2 - 8x - 2y + 13 = 0$

7. $2x^2 + 2y^2 + 4x + 20y = -34$

8. $4x^2 + 4y^2 - 24x + 16y = 48$

9. Identify the graph of each equation as a *circle* or an *ellipse*.

(a) $x^2 + y^2 = 121$

(b) $x^2 + 4y^2 = 4$

(c) $\frac{x^2}{16} + \frac{y^2}{25} = 1$

(d) $3x^2 + 3y^2 = 300$

10. Match each equation of an ellipse in Column I with the appropriate intercepts in Column II.

I	II
(a) $36x^2 + 9y^2 = 324$	**A.** $(-3, 0), (3, 0), (0, -6), (0, 6)$
(b) $9x^2 + 36y^2 = 324$	**B.** $(-4, 0), (4, 0), (0, -5), (0, 5)$
(c) $\frac{x^2}{25} + \frac{y^2}{16} = 1$	**C.** $(-6, 0), (6, 0), (0, -3), (0, 3)$
(d) $\frac{x^2}{16} + \frac{y^2}{25} = 1$	**D.** $(-5, 0), (5, 0), (0, -4), (0, 4)$

Graph each equation.

11. $x^2 + y^2 = 16$

12. $(x + 3)^2 + (y - 2)^2 = 9$

13. $\frac{x^2}{16} + \frac{y^2}{9} = 1$

14. $\frac{x^2}{49} + \frac{y^2}{25} = 1$

12.2 *Graph each equation.*

15. $\frac{x^2}{16} - \frac{y^2}{25} = 1$

16. $\frac{y^2}{25} - \frac{x^2}{4} = 1$

17. $f(x) = -\sqrt{16 - x^2}$

18. Match each equation of a hyperbola in Column I with the appropriate intercepts in Column II.

I	II
(a) $\frac{x^2}{25} - \frac{y^2}{9} = 1$	**A.** $(-3, 0), (3, 0)$
(b) $\frac{x^2}{9} - \frac{y^2}{25} = 1$	**B.** $(0, -3), (0, 3)$
(c) $\frac{y^2}{25} - \frac{x^2}{9} = 1$	**C.** $(-5, 0), (5, 0)$
(d) $\frac{y^2}{9} - \frac{x^2}{25} = 1$	**D.** $(0, -5), (0, 5)$

Identify the graph of each equation as a parabola, circle, ellipse, *or* hyperbola.

19. $x^2 + y^2 = 64$

20. $y = 2x^2 - 3$

21. $y^2 = 2x^2 - 8$

22. $y^2 = 8 - 2x^2$

23. $x = y^2 + 4$

24. $x^2 - y^2 = 64$

12.3 *Answer each question.*

25. How many solutions are possible for a system of two equations whose graphs are a circle and a line?

26. How many solutions are possible for a system of two equations whose graphs are a parabola and a hyperbola?

Solve each system.

27. $2y = 3x - x^2$
$x + 2y = -12$

28. $y + 1 = x^2 + 2x$
$y + 2x = 4$

29. $x^2 + 3y^2 = 28$
$y - x = -2$

30. $xy = 8$
$x - 2y = 6$

31. $x^2 + y^2 = 6$
$x^2 - 2y^2 = -6$

32. $3x^2 - 2y^2 = 12$
$x^2 + 4y^2 = 18$

12.4 *Graph each inequality.*

33. $9x^2 \geq 16y^2 + 144$

34. $4x^2 + y^2 \geq 16$

35. $y < -(x + 2)^2 + 1$

36. Which one of the following is a description of the graph of $x^2 + y^2 > 4$?

A. The region inside a circle with radius 4

B. The region outside a circle with radius 4

C. The region inside a circle with radius 2

D. The region outside a circle with radius 2

Graph each system of inequalities.

37. $2x + 5y \leq 10$
$3x - y \leq 6$

38. $9x^2 \leq 4y^2 + 36$
$x^2 + y^2 \leq 16$

39. $x + y \geq -2$
$y < x + 4$
$x < 1$

40. Match each system of inequalities in Column I with the quadrant that represents its graph in Column II.

I		II
(a) $x < 0$, $y > 0$	**(b)** $x > 0$, $y < 0$	**A.** Quadrant I
		B. Quadrant II
(c) $x < 0$, $y < 0$	**(d)** $x > 0$, $y > 0$	**C.** Quadrant III
		D. Quadrant IV

Solve each problem.

41. A bakery makes both cakes and cookies. Each batch of cakes requires 2 hr in the oven and 3 hr in the decorating room. Each batch of cookies needs $1\frac{1}{2}$ hr in the oven and $\frac{2}{3}$ hr in the decorating room. The oven is available no more than 15 hr per day, while the decorating room can be used no more than 13 hr per day.

How many batches of cakes and cookies should the bakery make in order to maximize profits if cookies produce a profit of $20 per batch and cakes produce a profit of $30 per batch? What is the maximum profit?

42. A company makes two kinds of pizza: basic and plain. Basic contains cheese and beef, while plain contains onions and beef. The company sells at least 3 units per day of basic and at least 2 units of plain. The beef costs $5 per unit for basic and $4 per unit for plain. The company can spend no more than $50 per day on beef. Dough for basic is $2 per unit, while dough for plain is $1 per unit. The company can spend no more than $16 per day on dough.

How many units of each kind of pizza should the company make in order to maximize revenue if basic sells for $20 per unit and plain for $15 per unit? What is the maximum revenue?

Chapter 12 Mixed Review Exercises

1. Find the center and radius of the circle with equation

$$x^2 + y^2 + 4x - 10y - 7 = 0.$$

2. *True* or *false:* The graph of $y = -\sqrt{1 - x^2}$ is the graph of a function.

Graph.

3. $x^2 + 9y^2 = 9$

4. $x^2 - 9y^2 = 9$

5. $f(x) = \sqrt{4 - x}$

6. $x^2 + y^2 = 25$

7. $\frac{y^2}{4} - 1 \leq \frac{x^2}{9}$

8. $4y > 3x - 12$
   $x^2 < 16 - y^2$

Solve each system.

9. $y = x^2 - 84$
   $4y = 20x$

10. $x^2 - xy - y^2 = -5$
    $x^2 - y^2 = -3$

Consider the given constraints.

$$\begin{aligned} 3x + 2y &\leq 6 \\ -2x + 4y &\leq 8 \\ x &\geq 0 \\ y &\geq 0 \end{aligned}$$

11. Find the values of x and y that maximize the objective function $2x + 5y$.

12. What is the maximum value?

Chapter 12 Test

Step-by-step test solutions are found on the Chapter Test Prep Videos available in MyLab Math.

View the complete solutions to all Chapter Test exercises in MyLab Math.

1. Find the center and radius of the circle with equation $(x - 2)^2 + (y + 3)^2 = 16$. Sketch the graph.

2. Which one of the following equations is represented by the graph of a circle?

   A. $x^2 + y^2 = 0$ **B.** $x^2 + y^2 = -1$

   C. $x^2 + y^2 = x^2 - y^2$ **D.** $x^2 + y^2 = 1$

3. For the equation in **Exercise 2** that is represented by a circle, what are the coordinates of the center? What is the radius?

4. Find the center and radius of the circle with equation $x^2 + y^2 + 8x - 2y = 8$.

Graph.

5. $f(x) = \sqrt{9 - x^2}$

6. $4x^2 + 9y^2 = 36$

7. $16y^2 - 4x^2 = 64$

8. $\frac{y}{2} = -\sqrt{1 - \frac{x^2}{9}}$

Identify the graph of each equation as a parabola, hyperbola, ellipse, *or* circle.

9. $6x^2 + 4y^2 = 12$

10. $16x^2 = 144 + 9y^2$

11. $y^2 = 20 - x^2$

12. $4y^2 + 4x = 9$

Solve each system.

13. $2x - y = 9$
$xy = 5$

14. $x - 4 = 3y$
$x^2 + y^2 = 8$

15. $x^2 + y^2 = 25$
$x^2 - 2y^2 = 16$

Graph.

16. $y < x^2 - 2$

17. $x - y \leq 2$
$x + 3y > 6$

18. $x^2 + 25y^2 \leq 25$
$x^2 + y^2 \leq 9$

19. The graph shows a region of feasible solutions.

(a) Find the values of x and y that maximize the objective function $2x + 4y$.

(b) What is the maximum value of $2x + 4y$?

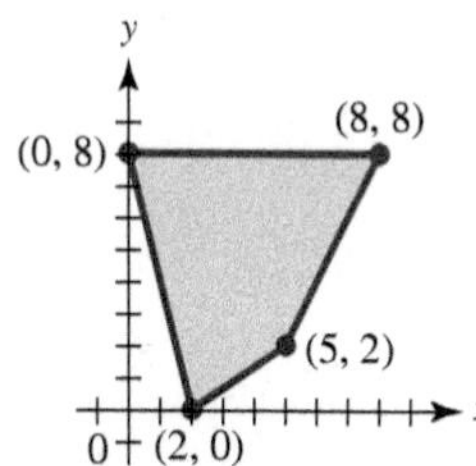

20. The Alessi company manufactures two products: radios and DVD players. Each radio results in a profit of \$15, and each DVD player gives a profit of \$35. Due to demand, the company must produce at least 5, and not more than 25, radios per day. The number of radios cannot exceed the number of DVD players, and the number of DVD players cannot exceed 30.

How many of each should the company manufacture to obtain maximum profit? What will that profit be?

Chapters R–12 Cumulative Review Exercises

1. Complete the table of fraction, decimal, and percent equivalents.

	Fraction in Lowest Terms (or Mixed Number)	*Decimal*	*Percent*
(a)	$\frac{1}{25}$		
(b)			15%
(c)		0.8	
(d)	$1\frac{1}{4}$		

2. Multiply or divide as indicated.

(a) 2.45×10 **(b)** 2.45×1000 **(c)** $2.45 \div 10$ **(d)** $2.45 \div 100$

3. Evaluate each expression in the complex number system.

(a) 4^2 **(b)** $(-4)^2$ **(c)** -4^2 **(d)** $-(-4)^2$ **(e)** $\sqrt{4}$ **(f)** $-\sqrt{4}$ **(g)** $\sqrt{-4}$

4. Simplify $3(6 + 7) + 6 \cdot 4 - 3^2$.

Solve.

5. $4 - (2x + 3) + x = 5x - 3$

6. $-4x + 7 \geq 6x + 1$

7. $|5x| - 6 = 14$

8. $|2p - 5| > 15$

9. Find the slope of the line passing through the points $(2, 5)$ and $(-4, 1)$.

10. Find the equation of the line passing through the point $(-3, -2)$ and perpendicular to the graph of $2x - 3y = 7$.

Solve each system.

11. $3x - y = 12$
$2x + 3y = -3$

12. $x + y - 2z = 9$
$2x + y + z = 7$
$3x - y - z = 13$

13. $xy = -5$
$2x + y = 3$

14. Al and Bev traveled from their apartment to a picnic 20 mi away. Al traveled on his bike while Bev, who left later, took her car. Al's average rate was half of Bev's average rate. The trip took Al $\frac{1}{2}$ hr longer than Bev. What was Bev's average rate?

	d	*r*	*t*
Al	20		
Bev			

Perform the indicated operations.

15. $(5y - 3)^2$

16. $\dfrac{8x^4 - 4x^3 + 2x^2 + 13x + 8}{2x + 1}$

17. $\dfrac{y^2 - 4}{y^2 - y - 6} \div \dfrac{y^2 - 2y}{y - 1}$

18. $\dfrac{5}{c + 5} - \dfrac{2}{c + 3}$

19. $\dfrac{p}{p^2 + p} + \dfrac{1}{p^2 + p}$

20. $\dfrac{1}{\frac{1}{x} - \frac{1}{y}}$

21. $4\sqrt[3]{16} - 2\sqrt[3]{54}$

22. $\dfrac{5 + 3i}{2 - i}$

Factor.

23. $12x^2 - 7x - 10$

24. $z^4 - 1$

25. $a^3 - 27b^3$

26. Henry and Lawrence want to clean their office. Henry can do the job alone in 3 hr, and Lawrence can do it alone in 2 hr. How long will it take them if they work together?

	Rate	Time Working Together	Fractional Part of the Job Done
Henry	$\frac{1}{3}$		
Lawrence			

Simplify. Assume all variables represent positive real numbers.

27. $\dfrac{(2a)^{-2}a^4}{a^{-3}}$

28. $\dfrac{3\sqrt{5x}}{\sqrt{2x}}$

Solve.

29. $2\sqrt{x} = \sqrt{5x + 3}$

30. $10q^2 + 13q = 3$

31. $3x^2 - 3x - 2 = 0$

32. $2(x^2 - 3)^2 - 5(x^2 - 3) = 12$

33. $F = \dfrac{kwv^2}{r}$ for v

34. $\log(x + 2) + \log(x - 1) = 1$

Work each problem.

35. If $f(x) = x^2 + 2x - 4$ and $g(x) = 3x + 2$, find the following.

(a) $(g \circ f)(1)$ **(b)** $(f \circ g)(x)$

36. Give the domain and range of the function defined by $f(x) = |x - 3|$.

37. If $f(x) = x^3 + 4$, find $f^{-1}(x)$.

38. Evaluate. **(a)** $3^{\log_3 4}$ **(b)** $e^{\ln 7}$

39. Use properties of logarithms to write $2 \log (3x + 7) - \log 4$ as a single logarithm.

40. Use synthetic division to determine $f(3)$, if $f(x) = 2x^3 - 4x^2 + 5x - 10$.

41. Use the factor theorem to determine whether $x + 2$ is a factor of

$$5x^4 + 10x^3 + 6x^2 + 8x - 8.$$

If it is, what is the other factor? If it is not, explain why.

42. Find all zeros of $f(x) = 3x^3 + x^2 - 22x - 24$, given that one zero is -2.

Graph.

43. $f(x) = -3x + 5$

44. $f(x) = \sqrt{x - 2}$

45. $f(x) = -2(x - 1)^2 + 3$

46. $f(x) = 3^x$

47. $\frac{x^2}{4} - \frac{y^2}{16} = 1$

48. $\frac{x^2}{25} + \frac{y^2}{16} \le 1$

STUDY SKILLS **REMINDER**

Have you begun to prepare for your final exam? **Review Study Skill 10, *Preparing for Your Math Final Exam.***

13

FURTHER TOPICS IN ALGEBRA

The number of ancestors in the family tree of a male honeybee forms the *Fibonacci sequence*

$$1, 1, 2, 3, 5, 8, \ldots,$$

where each term after the second is found by adding the two previous terms. This chapter covers *sequences, series,* and other topics in algebra.

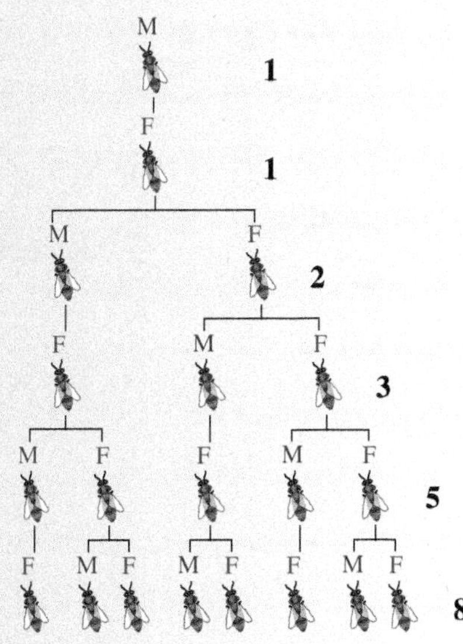

13.1 Sequences and Series

OBJECTIVES

1 Define infinite and finite sequences.
2 Find the terms of a sequence, given the general term.
3 Find the general term of a sequence.
4 Use sequences to solve applied problems.
5 Use summation notation to evaluate a series.
6 Write a series using summation notation.
7 Find the arithmetic mean (average) of a group of numbers.

VOCABULARY

- ☐ infinite sequence
- ☐ finite sequence
- ☐ terms of a sequence
- ☐ general term
- ☐ series
- ☐ summation notation
- ☐ index of summation
- ☐ arithmetic mean (average)

NOW TRY EXERCISE 1

Given an infinite sequence with $a_n = 5 - 3n$, find a_3.

NOW TRY ANSWER

1. $a_3 = -4$

OBJECTIVE 1 Define infinite and finite sequences.

In the Palace of the Alhambra, residence of the Moorish rulers of Granada, Spain, the Sultana's quarters feature an interesting architectural pattern.

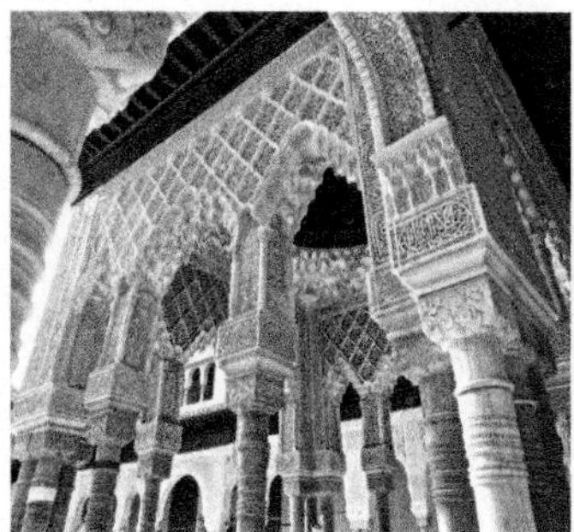

There are 2 matched marble slabs inlaid in the floor, 4 walls, an octagon (8-sided) ceiling, 16 windows, 32 arches, and so on.

If this pattern is continued indefinitely, the set of numbers forms an *infinite sequence* whose *terms* are powers of 2.

Sequence

An **infinite sequence** is a function with the set of positive integers (natural numbers)* as domain.

A **finite sequence** is a function with domain of the form $\{1, 2, 3, \ldots, n\}$, where n is a positive integer.

For any positive integer n, the function value of a sequence is written as a_n (read "***a* sub-*n***"). The function values

$$a_1, a_2, a_3, \ldots,$$

written in order, are the **terms** of the sequence, with a_1 the first term, a_2 the second term, and so on. The expression a_n, which defines the sequence, is its **general term.**

In the Palace of the Alhambra example, the first five terms of the sequence are

$$a_1 = 2, \quad a_2 = 4, \quad a_3 = 8, \quad a_4 = 16, \quad \text{and} \quad a_5 = 32.$$

The general term for this sequence is $a_n = 2^n$.

OBJECTIVE 2 Find the terms of a sequence, given the general term.

EXAMPLE 1 Writing the Terms of Sequences from the General Term

Given an infinite sequence with $a_n = n + \frac{1}{n}$, find the following.

(a) The second term of the sequence

$$a_2 = 2 + \frac{1}{2} = \frac{5}{2} \qquad \text{Replace } n \text{ in } a_n = n + \tfrac{1}{n} \text{ with 2.}$$

(b) $a_{10} = 10 + \frac{1}{10} = \frac{101}{10}$ $\qquad 10 + \frac{1}{10} = \frac{10}{1} \cdot \frac{10}{10} + \frac{1}{10} = \frac{100}{10} + \frac{1}{10} = \frac{101}{10}$

(c) $a_{12} = 12 + \frac{1}{12} = \frac{145}{12}$ $\qquad 12 + \frac{1}{12} = \frac{12}{1} \cdot \frac{12}{12} + \frac{1}{12} = \frac{144}{12} + \frac{1}{12} = \frac{145}{12}$

NOW TRY 

*In this chapter we will use the terms *positive integers* and *natural numbers* interchangeably.

Graphing calculators can generate and graph sequences. See **FIGURE 1**. The calculator must be in dot mode so that the discrete points on the graph are not connected. ***Remember that the domain of a sequence consists only of positive integers.***

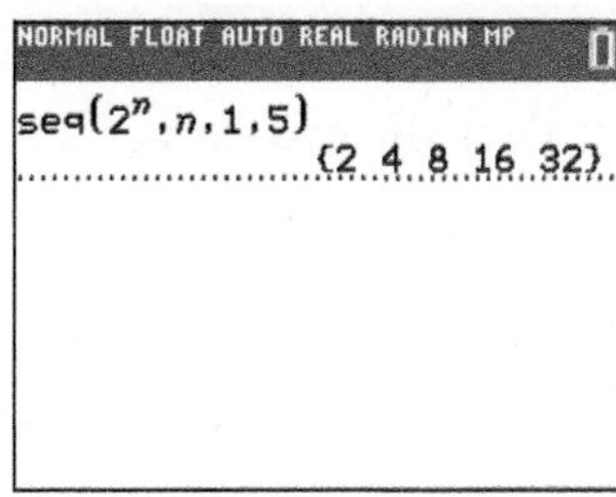

The first five terms of the sequence $a_n = 2^n$

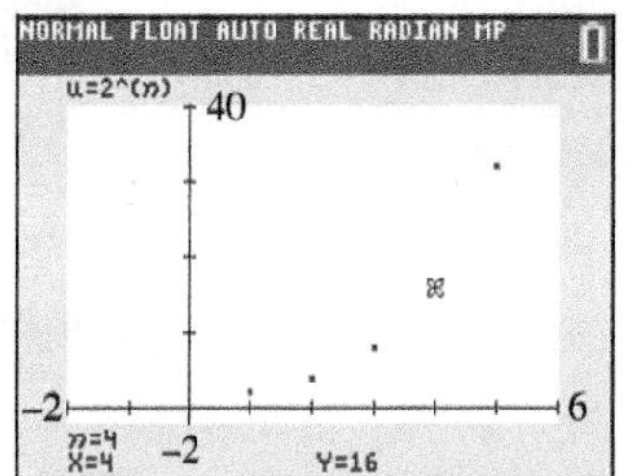

The first five terms of $a_n = 2^n$ are graphed here. The display indicates that the fourth term is 16; that is, $a_4 = 2^4 = 16$.

FIGURE 1

OBJECTIVE 3 Find the general term of a sequence.

NOW TRY EXERCISE 2

Determine an expression for the general term a_n of the sequence.

$$-3, 9, -27, 81, \ldots$$

EXAMPLE 2 Finding the General Term of a Sequence

Determine an expression for the general term a_n of the sequence.

$$5, 10, 15, 20, 25, \ldots$$

Notice that the terms are all multiples of 5. The first term is 5(1), the second is 5(2), and so on. The general term that will produce the given first five terms is

$$a_n = 5n.$$

NOW TRY 

OBJECTIVE 4 Use sequences to solve applied problems.

Applied problems may involve *finite sequences.*

NOW TRY EXERCISE 3

Chase borrows \$8000 and agrees to pay \$400 monthly, plus interest of 2% on the unpaid balance from the beginning of the first month. Find the payments for the first four months and the remaining debt at the end of that period.

EXAMPLE 3 Using a Sequence in an Application

Saad borrows \$5000 and agrees to pay \$500 monthly, plus interest of 1% on the unpaid balance from the beginning of the first month. Find the payments for the first four months and the remaining debt at the end of that period.

The payments and remaining balances are calculated as follows.

First month Payment: \$500 + 0.01(\$5000) = \$550
Balance: \$5000 − \$500 = \$4500

Second month Payment: \$500 + 0.01(\$4500) = \$545
Balance: \$5000 − 2 · \$500 = \$4000

Third month Payment: \$500 + 0.01(\$4000) = \$540
Balance: \$5000 − 3 · \$500 = \$3500

Fourth month Payment: \$500 + 0.01(\$3500) = \$535
Balance: \$5000 − 4 · \$500 = \$3000

The payments for the first four months are

\$550, \$545, \$540, \$535

and the remaining debt at the end of the period is \$3000.

NOW TRY 

NOW TRY ANSWERS

2. $a_n = (-3)^n$
3. payments: \$560, \$552, \$544, \$536; balance: \$6400

OBJECTIVE 5 Use summation notation to evaluate a series.

By adding the terms of a sequence, we obtain a *series.*

Series

A **series** is the indicated sum of the terms of a sequence.

For example, if we consider the sum of the payments listed in **Example 3,** namely,

$$550 + 545 + 540 + 535,$$

we have a series that represents the total payments for the first four months. Because a sequence can be finite or infinite, there are both finite and infinite series.

We use a compact notation, called **summation notation,** to write a series from the general term of the corresponding sequence. In mathematics, the Greek letter **Σ (sigma)** is used to denote summation. For example, the sum of the first six terms of the sequence with general term

$$a_n = 3n + 2$$

is written as

$$\sum_{i=1}^{6} (3i + 2).$$

The letter i is the **index of summation.** We read this as "*the sum from* $i = 1$ *to* 6 *of* $3i + 2$." To find this sum, we replace the letter i in $3i + 2$ with 1, 2, 3, 4, 5, and 6, and add the resulting terms.

CAUTION This use of i as the index of summation has no connection with the complex number i.

EXAMPLE 4 Evaluating Series Written Using Summation Notation

Write each series as a sum of terms and then find the sum.

(a) $\sum_{i=1}^{6} (3i + 2)$

Multiply and then add.

$= (3 \cdot 1 + 2) + (3 \cdot 2 + 2) + (3 \cdot 3 + 2)$
$+ (3 \cdot 4 + 2) + (3 \cdot 5 + 2) + (3 \cdot 6 + 2)$ Replace i with 1, 2, 3, 4, 5, 6.

$= 5 + 8 + 11 + 14 + 17 + 20$ Simplify inside the parentheses.

$= 75$ Add.

(b) $\sum_{i=1}^{5} (i - 4)$

$= (1 - 4) + (2 - 4) + (3 - 4) + (4 - 4) + (5 - 4)$ $i = 1, 2, 3, 4, 5$

$= -3 - 2 - 1 + 0 + 1$ Subtract.

$= -5$ Simplify.

NOW TRY EXERCISE 4

Write the series as a sum of terms and then find the sum.

$$\sum_{i=1}^{5}(i^2 - 4)$$

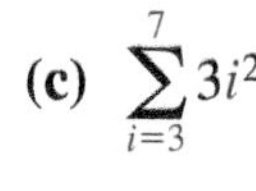

(c) $\sum_{i=3}^{7} 3i^2$

$= 3(3)^2 + 3(4)^2 + 3(5)^2 + 3(6)^2 + 3(7)^2$ $\quad i = 3, 4, 5, 6, 7$

$= 27 + 48 + 75 + 108 + 147$ $\quad$ Square, and then multiply.

$= 405$ $\quad$ Add. **NOW TRY**

OBJECTIVE 6 Write a series using summation notation.

In **Example 4,** we started with summation notation and wrote each series using + signs. Given a series, we can write it using summation notation by observing a pattern in the terms and writing the general term accordingly.

NOW TRY EXERCISE 5

Write each series using summation notation.

(a) $3 + 5 + 7 + 9 + 11$

(b) $-1 - 4 - 9 - 16 - 25$

EXAMPLE 5 Writing Series Using Summation Notation

Write each series using summation notation.

(a) $2 + 5 + 8 + 11$

First, find a general term a_n that will give these four terms for a_1, a_2, a_3, and a_4, respectively. Each term is one less than a multiple of 3, so try $3i - 1$.

$3(1) - 1 = 2 \quad i = 1$

$3(2) - 1 = 5 \quad i = 2$

$3(3) - 1 = 8 \quad i = 3$

$3(4) - 1 = 11 \quad i = 4$

(There may be other expressions that also work.) Because i ranges from 1 to 4,

$$2 + 5 + 8 + 11 = \sum_{i=1}^{4}(3i - 1).$$

(b) $8 + 27 + 64 + 125 + 216$

These numbers are the cubes of 2, 3, 4, 5, and 6, so the general term is i^3.

$$8 + 27 + 64 + 125 + 216 = \sum_{i=2}^{6} i^3$$

NOW TRY 

OBJECTIVE 7 Find the arithmetic mean (average) of a group of numbers.

Arithmetic Mean or Average

The **arithmetic mean,** or **average,** of a set of values is symbolized $\bar{x}$ and is found by dividing their sum by the number of values.

$$\bar{x} = \frac{\sum_{i=1}^{n} x_i}{n}$$

The values of x_i represent the individual numbers in the group, and n represents the number of values.

NOW TRY ANSWERS

4. $-3 + 0 + 5 + 12 + 21 = 35$

5. (a) $\sum_{i=1}^{5}(2i + 1)$ **(b)** $\sum_{i=1}^{5} -i^2$

NOW TRY EXERCISE 6

The table shows the top five American Quarter Horse States in 2016 based on the total number of registered Quarter Horses. To the nearest whole number, what is the average number of Quarter Horses registered per state in these top five states?

State	Number of Registered Quarter Horses
Texas	418,249
Oklahoma	164,265
California	114,623
Missouri	95,877
Montana	87,418

Data from American Quarter Horse Association.

NOW TRY ANSWER

6. 176,086 quarter horses

EXAMPLE 6 Finding the Arithmetic Mean (Average)

The table shows the number of FDIC-insured financial institutions for each year during the period from 2011 through 2017. What was the average number of institutions per year for this 7-yr period?

Year	Number of Institutions
2011	7523
2012	7255
2013	6950
2014	6669
2015	6358
2016	6068
2017	5797

Data from U.S. Federal Deposit Insurance Corporation.

$$\bar{x} = \frac{\sum_{i=1}^{7} x_i}{7}$$ Let $x_1 = 7523$, $x_2 = 7255$, and so on. There are 7 values in the group, so $n = 7$.

$$\bar{x} = \frac{7523 + 7255 + 6950 + 6669 + 6358 + 6068 + 5797}{7}$$

$$\bar{x} = 6660$$

The average number of institutions per year for this 7-yr period was 6660.

NOW TRY 

13.1 Exercises

  MyLab Math

Video solutions for select problems available in MyLab Math

STUDY SKILLS REMINDER

Make study cards to help you learn and remember the material in this chapter. **Review Study Skill 5, *Using Study Cards.***

Concept Check *Fill in each blank with the correct response.*

1. The domain of an infinite sequence is ______.
2. In the sequence 3, 6, 9, 12, the term $a_3 =$ ______.
3. If $a_n = 2n$, then $a_{40} =$ ______.
4. If $a_n = (-1)^n$, then $a_{115} =$ ______.
5. The value of the sum $\sum_{i=1}^{3}(i+2)$ is ______.
6. The arithmetic mean of -4, -2, 0, 2, and 4 is ______.

Write the first five terms of each sequence. ***See Example 1.***

7. $a_n = n + 1$ **8.** $a_n = n + 4$ **9.** $a_n = \frac{n+3}{n}$

10. $a_n = \frac{n+2}{n}$ **11.** $a_n = 3^n$ **12.** $a_n = 2^n$

13. $a_n = -\frac{1}{n^2}$ **14.** $a_n = -\frac{2}{n^2}$ **15.** $a_n = 5(-1)^{n-1}$

16. $a_n = 6(-1)^{n+1}$ **17.** $a_n = n - \frac{1}{n}$ **18.** $a_n = n + \frac{4}{n}$

Find the indicated term for each sequence. **See Example 1.**

19. $a_n = -9n + 2;\quad a_8$

20. $a_n = -3n + 7;\quad a_{12}$

21. $a_n = \frac{3n + 7}{2n - 5};\quad a_{14}$

22. $a_n = \frac{5n - 9}{3n + 8};\quad a_{16}$

23. $a_n = (n + 1)(2n + 3);\quad a_8$

24. $a_n = (5n - 2)(3n + 1);\quad a_{10}$

Determine an expression for the general term a_n of each sequence. **See Example 2.**

25. 4, 8, 12, 16, . . .

26. 7, 14, 21, 28, . . .

27. −8, −16, −24, −32, . . .

28. −10, −20, −30, −40, . . .

29. $\frac{1}{3}, \frac{1}{9}, \frac{1}{27}, \frac{1}{81}, \ldots$

30. $\frac{2}{5}, \frac{2}{25}, \frac{2}{125}, \frac{2}{625}, \ldots$

31. $\frac{2}{5}, \frac{3}{6}, \frac{4}{7}, \frac{5}{8}, \ldots$

32. $\frac{1}{2}, \frac{2}{3}, \frac{3}{4}, \frac{4}{5}, \ldots$

Solve each applied problem by writing the first few terms of a sequence. **See Example 3.**

33. Horacio borrows \$1000 and agrees to pay \$100 plus interest of 1% on the unpaid balance each month. Find the payments for the first six months and the remaining debt at the end of that period.

34. Leslie is offered a new job with a salary of 20,000 + 2500*n* dollars per year at the end of the *n*th year. Write a sequence showing her salary at the end of each of the first 5 yr. If she continues in this way, what will her salary be at the end of the tenth year?

35. Suppose that an automobile loses $\frac{1}{5}$ of its value each year; that is, at the end of any given year, the value is $\frac{4}{5}$ of the value at the beginning of that year. If a car costs \$20,000 new, what is its value at the end of 5 yr, to the nearest dollar?

36. A certain car loses $\frac{1}{2}$ of its value each year. If this car cost \$40,000 new, what is its value at the end of 6 yr?

Write each series as a sum of terms and then find the sum. **See Example 4.**

37. $\sum_{i=1}^{5} (i + 3)$

38. $\sum_{i=1}^{6} (i + 9)$

39. $\sum_{i=1}^{3} (i^2 + 2)$

40. $\sum_{i=1}^{4} (i^3 + 3)$

41. $\sum_{i=1}^{6} (-1)^i \cdot 2$

42. $\sum_{i=1}^{5} (-1)^i \cdot i$

43. $\sum_{i=3}^{7} (i - 3)(i + 2)$

44. $\sum_{i=2}^{6} (i + 3)(i - 4)$

Write each series using summation notation. **See Example 5.**

45. 3 + 4 + 5 + 6 + 7

46. 7 + 8 + 9 + 10 + 11

47. −2 + 4 − 8 + 16 − 32

48. −1 + 2 − 3 + 4 − 5 + 6

49. 1 + 4 + 9 + 16

50. 1 + 16 + 81 + 256

51. Concept Check When asked to write the series −1 − 4 − 9 − 16 − 25 using summation notation, a student incorrectly wrote the following.

$$\sum_{i=1}^{5} (-i)^2$$

WHAT WENT WRONG? Give the correct summation notation.

52. Concept Check When asked to write the first five terms of the sequence defined by $a_n = 5n - 1$, a student incorrectly wrote the following.

$$4 + 9 + 14 + 19 + 24$$

WHAT WENT WRONG? Give the correct answer.

Find the arithmetic mean for each set of values. ***See Example 6.***

53. 8, 11, 14, 9, 7, 6, 8

54. 10, 12, 8, 19, 23, 12

55. 5, 9, 8, 2, 4, 7, 3, 2, 0

56. 2, 1, 4, 8, 3, 7, 10, 8, 0

Solve each problem. ***See Example 6.***

57. The number of mutual funds operating in the United States each year during the period 2012 through 2016 is given in the table. To the nearest whole number, what was the average number of mutual funds operating per year during the given period?

Year	*Number of Mutual Funds*
2012	8744
2013	8972
2014	9258
2015	9517
2016	9511

Data from Investment Company Institute.

58. The total assets of mutual funds operating in the United States, in billions of dollars, for each year during the period 2012 through 2016 are shown in the table. What were the average assets per year during this period?

Year	*Assets (in billions of dollars)*
2012	13,054
2013	15,049
2014	15,873
2015	15,650
2016	16,344

Data from Investment Company Institute.

13.2 Arithmetic Sequences

OBJECTIVES

1. Find the common difference of an arithmetic sequence.
2. Find the general term of an arithmetic sequence.
3. Use an arithmetic sequence in an application.
4. Find any specified term or the number of terms of an arithmetic sequence.
5. Find the sum of a specified number of terms of an arithmetic sequence.

OBJECTIVE 1 Find the common difference of an arithmetic sequence.

In this section, we introduce *arithmetic sequences.*

Arithmetic Sequence

An **arithmetic sequence, or arithmetic progression,** is a sequence in which each term after the first is found by adding a constant number to the preceding term.

For example, $\quad 6, \overset{+5}{\frown} 11, \overset{+5}{\frown} 16, \overset{+5}{\frown} 21, \overset{+5}{\frown} 26, \ldots$ $\quad$ Arithmetic sequence

is an arithmetic sequence in which the difference between any two adjacent terms is always 5. The number 5 is the **common difference** of the arithmetic sequence. The common difference d is found by subtracting a_n from a_{n+1} in any such pair of terms.

$$\boldsymbol{d = a_{n+1} - a_n} \quad \text{Common difference}$$

VOCABULARY

□ arithmetic sequence (arithmetic progression)

□ common difference

NOW TRY EXERCISE 1

Determine the common difference d for the arithmetic sequence.

$-4, -13, -22, -31, -40, \ldots$

EXAMPLE 1 Finding the Common Difference

Determine the common difference d for the arithmetic sequence.

$$-11, \ -4, \ 3, \ 10, \ 17, \ 24, \ldots$$

The sequence is arithmetic, so d is the difference between any two adjacent terms. We arbitrarily choose the terms 10 and 17.

$$d = a_{n+1} - a_n$$

$$d = 17 - 10 \qquad a_5 - a_4$$

$$d = 7 \qquad \text{Common difference}$$

Verify that *any* two adjacent terms would give the same result.  

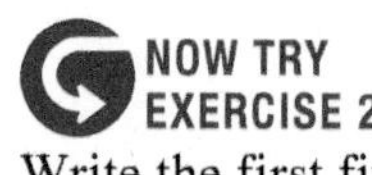

NOW TRY EXERCISE 2

Write the first five terms of the arithmetic sequence with first term 10 and common difference -8.

EXAMPLE 2 Writing the Terms of an Arithmetic Sequence

Write the first five terms of the arithmetic sequence with first term 3 and common difference -2.

$$a_1 = 3 \qquad \text{First term}$$

$$a_2 = 3 + (-2) = 1 \qquad \text{Add } d = -2.$$

$$a_3 = 1 + (-2) = -1 \qquad \text{Add } -2.$$

$$a_4 = -1 + (-2) = -3 \qquad \text{Add } -2.$$

$$a_5 = -3 + (-2) = -5 \qquad \text{Add } -2.$$

The first five terms are $3, \ 1, \ -1, \ -3, \ -5$. NOW TRY 

OBJECTIVE 2 Find the general term of an arithmetic sequence.

Generalizing from **Example 2,** if we know the first term a_1 and the common difference d of an arithmetic sequence, then the sequence is completely defined as

$$a_1, \quad a_2 = a_1 + d, \quad a_3 = a_1 + 2d, \quad a_4 = a_1 + 3d, \ldots.$$

Writing the terms of the sequence in this way suggests the following formula for a_n.

General Term of an Arithmetic Sequence

The general term a_n of an arithmetic sequence with first term a_1 and common difference d is given by the following.

$$\boldsymbol{a_n = a_1 + (n - 1)d}$$

Because $\quad a_n = a_1 + (n - 1)d \qquad$ Formula for a_n

$$a_n = dn + (a_1 - d) \qquad \text{Properties of real numbers}$$

is a linear function in n, any linear expression of the form $dn + c$, where d and c are real numbers, defines an arithmetic sequence.

NOW TRY ANSWERS

1. $d = -9$

2. $10, 2, -6, -14, -22$

NOW TRY EXERCISE 3

Determine an expression for the general term of the arithmetic sequence. Then find a_{20}.

$-5, 0, 5, 10, 15, \ldots$

EXAMPLE 3 Finding the General Term of an Arithmetic Sequence

Determine an expression for the general term of the arithmetic sequence. Then find a_{20}.

$$-9, -6, -3, 0, 3, 6, \ldots$$

The first term is $a_1 = -9$. Begin by finding d.

$$d = -6 - (-9) \quad \text{Let } d = a_2 - a_1.$$

$$d = 3 \quad \text{Subtract.}$$

Now find a_n.

$$a_n = a_1 + (n - 1)d \quad \text{Formula for } a_n$$

$$a_n = -9 + (n - 1)(3) \quad \text{Let } a_1 = -9, \text{ and } d = 3.$$

$$a_n = -9 + 3n - 3 \quad \text{Distributive property}$$

$$a_n = 3n - 12 \quad \text{Combine like terms.}$$

The general term is $a_n = 3n - 12$. Now find a_{20}.

$$a_{20} = 3(20) - 12 \quad \text{Let } n = 20.$$

$$a_{20} = 60 - 12 \quad \text{Multiply.}$$

$$a_{20} = 48 \quad \text{Subtract.}$$

NOW TRY 

OBJECTIVE 3 Use an arithmetic sequence in an application.

NOW TRY EXERCISE 4

Ginny is saving money for her son's college education. She makes an initial contribution of \$1000 and deposits an additional \$120 each month for the next 96 months. Disregarding interest, how much money will be in the account after 96 months?

EXAMPLE 4 Applying an Arithmetic Sequence

Leonid's uncle decides to start a fund for Leonid's education. He makes an initial contribution of \$3000 and deposits an additional \$500 each month. How much will be in the fund after 24 months? (Disregard any interest.)

After n months, the fund will contain

$$a_n = 3000 + 500n \text{ dollars.} \quad \text{Use an arithmetic sequence.}$$

To find the amount in the fund after 24 months, find a_{24}.

$$a_{24} = 3000 + 500(24) \quad \text{Let } n = 24.$$

$$a_{24} = 3000 + 12{,}000 \quad \text{Multiply.}$$

$$a_{24} = 15{,}000 \quad \text{Add.}$$

The fund will contain \$15,000 (disregarding interest) after 24 months. NOW TRY 

OBJECTIVE 4 Find any specified term or the number of terms of an arithmetic sequence.

The formula for the general term of an arithmetic sequence has four variables:

$$a_n, \quad a_1, \quad n, \quad \text{and} \quad d.$$

If we know the values of any three of these variables, we can use the formula to find the value of the fourth variable.

NOW TRY ANSWERS

3. $a_n = 5n - 10$; $a_{20} = 90$

4. \$12,520

Find the indicated term for each arithmetic sequence.

(a) $a_1 = 21, d = -3;\quad a_{22}$

(b) $a_7 = 25, a_{12} = 40;\quad a_{19}$

EXAMPLE 5 Finding Specified Terms in Arithmetic Sequences

Find the indicated term for each arithmetic sequence.

(a) $a_1 = -6, d = 12;\quad a_{15}$

$$a_n = a_1 + (n-1)d \qquad \text{Formula for } a_n$$
$$a_{15} = a_1 + (15-1)d \qquad \text{Let } n = 15.$$
$$a_{15} = -6 + 14(12) \qquad \text{Let } a_1 = -6,\ d = 12.$$
$$a_{15} = 162 \qquad \text{Multiply, and then add.}$$

(b) $a_5 = 2, a_{11} = -10;\quad a_{17}$

Any term can be found if a_1 and d are known. Use the formula for a_n.

$a_n = a_1 + (n-1)d$		$a_n = a_1 + (n-1)d$	
$a_5 = a_1 + (5-1)d$	Let $n = 5$.	$a_{11} = a_1 + (11-1)d$	Let $n = 11$.
$a_5 = a_1 + 4d$	Simplify.	$a_{11} = a_1 + 10d$	Simplify.
$2 = a_1 + 4d$	$a_5 = 2$	$-10 = a_1 + 10d$	$a_{11} = -10$

This gives a system of two equations in two variables, a_1 and d.

$$a_1 + 4d = 2 \qquad (1)$$
$$a_1 + 10d = -10 \qquad (2)$$

Multiply equation (2) by -1 and add to equation (1) to eliminate a_1.

$$a_1 + 4d = 2 \qquad (1)$$
$$-a_1 - 10d = 10 \qquad \text{Multiply (2) by } -1.$$
$$-6d = 12 \qquad \text{Add.}$$
$$d = -2 \qquad \text{Divide by } -6.$$

Now find a_1 by substituting -2 for d into either equation.

$$a_1 + 10d = -10 \qquad (2)$$
$$a_1 + 10(-2) = -10 \qquad \text{Let } d = -2 \text{ in (2).}$$
$$a_1 - 20 = -10 \qquad \text{Multiply.}$$
$$a_1 = 10 \qquad \text{Add 20.}$$

Use the formula for a_n to find a_{17}.

$$a_n = a_1 + (n-1)d \qquad \text{Formula for } a_n$$
$$a_{17} = a_1 + (17-1)d \qquad \text{Let } n = 17.$$
$$a_{17} = a_1 + 16d \qquad \text{Subtract.}$$
$$a_{17} = 10 + 16(-2) \qquad \text{Let } a_1 = 10,\ d = -2.$$
$$a_{17} = -22 \qquad \text{Multiply, and then add.}$$

NOW TRY

NOW TRY ANSWERS

5. **(a)** -42 **(b)** 61

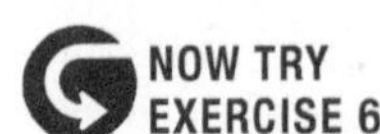
NOW TRY EXERCISE 6

Find the number of terms in the arithmetic sequence.

$$1, \frac{4}{3}, \frac{5}{3}, 2, \ldots, 11$$

EXAMPLE 6 Finding the Number of Terms in an Arithmetic Sequence

Find the number of terms in the arithmetic sequence.

$$-8, -2, 4, 10, \ldots, 52$$

Let n represent the number of terms in the sequence. Here $a_n = 52$, $a_1 = -8$, and $d = -2 - (-8) = 6$, so we use the formula for a_n to find n.

$$a_n = a_1 + (n-1)d \quad \text{Formula for } a_n$$

$$52 = -8 + (n-1)(6) \quad \text{Let } a_n = 52, a_1 = -8, d = 6.$$

$$52 = -8 + 6n - 6 \quad \text{Distributive property}$$

$$52 = 6n - 14 \quad \text{Simplify.}$$

$$66 = 6n \quad \text{Add 14.}$$

$$n = 11 \quad \text{Divide by 6.}$$

The sequence has 11 terms. NOW TRY 

OBJECTIVE 5 Find the sum of a specified number of terms of an arithmetic sequence.

To find a formula for the sum S_n of the first n terms of a given arithmetic sequence, we can write out the terms in two ways. We start with the first term, and then with the last term. Then we add the terms in columns.

$$S_n = a_1 + (a_1 + d) + (a_1 + 2d) + \cdots + [a_1 + (n-1)d]$$

$$S_n = a_n + (a_n - d) + (a_n - 2d) + \cdots + [a_n - (n-1)d]$$

$$S_n + S_n = (a_1 + a_n) + (a_1 + a_n) + (a_1 + a_n) + \cdots + (a_1 + a_n)$$

$$2S_n = n(a_1 + a_n) \quad \text{There are } n \text{ terms of } a_1 + a_n \text{ on the right.}$$

Formula for S_n

$$\mathbf{S_n = \frac{n}{2}(a_1 + a_n)} \quad \text{Divide by 2.}$$

NOW TRY EXERCISE 7

Evaluate the sum of the first seven terms of the arithmetic sequence in which $a_n = 5n - 7$.

EXAMPLE 7 Finding the Sum of the First n Terms of an Arithmetic Sequence

Evaluate the sum of the first five terms of the arithmetic sequence in which $a_n = 2n - 5$.

Begin by evaluating a_1 and a_5.

$a_1 = 2(1) - 5$	$a_5 = 2(5) - 5$
$a_1 = -3$	$a_5 = 5$

Now evaluate the sum using $a_1 = -3$, $a_5 = 5$, and $n = 5$.

$$S_n = \frac{n}{2}(a_1 + a_n) \quad \text{Formula for } S_n$$

$$S_5 = \frac{5}{2}(-3 + 5) \quad \text{Substitute.}$$

$$S_5 = 5 \quad \text{Add, and then multiply.}$$

NOW TRY

NOW TRY ANSWERS

6. 31
7. 91

It is possible to express the sum S_n of an arithmetic sequence in terms of a_1 and d, the quantities that define the sequence. We have established that

$$S_n = \frac{n}{2}(a_1 + a_n) \quad \text{and} \quad a_n = a_1 + (n-1)d.$$

We substitute the expression for a_n into the expression for S_n.

$$S_n = \frac{n}{2}(a_1 + [a_1 + (n-1)d]) \qquad \text{Substitute for } a_n.$$

$$\mathbf{S_n = \frac{n}{2}[2a_1 + (n-1)d]} \qquad \text{Combine like terms.}$$

Sum of the First n Terms of an Arithmetic Sequence

The sum S_n of the first n terms of the arithmetic sequence with first term a_1, nth term a_n, and common difference d is given by either of the following.

$$S_n = \frac{n}{2}(a_1 + a_n) \quad \text{or} \quad S_n = \frac{n}{2}[2a_1 + (n-1)d]$$

The first formula is used when the first and last terms are known; otherwise, the second formula is used.

NOW TRY EXERCISE 8
Evaluate the sum of the first nine terms of the arithmetic sequence having first term -8 and common difference -5.

EXAMPLE 8 Finding the Sum of the First n Terms of an Arithmetic Sequence

Evaluate the sum of the first eight terms of the arithmetic sequence having first term 3 and common difference -2.

The known values $a_1 = 3$, $d = -2$, and $n = 8$ appear in the second formula for S_n, so we use it.

$$S_n = \frac{n}{2}[2a_1 + (n-1)d] \qquad \text{Second formula for } S_n$$

$$S_8 = \frac{8}{2}[2(3) + (8-1)(-2)] \qquad \text{Let } a_1 = 3,\ d = -2,\ n = 8.$$

$$S_8 = 4[6 - 14] \qquad \text{Work inside the brackets.}$$

$$S_8 = -32 \qquad \text{Subtract, and then multiply.}$$

NOW TRY

As mentioned earlier, a linear expression of the form

$$dn + c, \quad \text{where } d \text{ and } c \text{ are real numbers,}$$

defines an arithmetic sequence. For example, the sequences defined by $a_n = 2n + 5$ and $a_n = n - 3$ are arithmetic sequences. For this reason,

$$\sum_{i=1}^{n} (di + c)$$

represents the sum of the first n terms of an arithmetic sequence having first term $a_1 = d(1) + c = d + c$ and general term $a_n = d(n) + c = dn + c$. We can find this sum with the first formula for S_n, as shown in the next example.

NOW TRY ANSWER
8. -252

NOW TRY EXERCISE 9

Evaluate $\sum_{i=1}^{11} (5i - 7)$.

EXAMPLE 9 Using S_n to Evaluate a Summation

Evaluate $\sum_{i=1}^{12} (2i - 1)$.

This is the sum of the first 12 terms of the arithmetic sequence having $a_n = 2n - 1$. We find this sum, S_{12}, using the first formula for S_n.

$$S_n = \frac{n}{2}(a_1 + a_n)$$ First formula for S_n

$$S_{12} = \frac{12}{2}[(2(1) - 1) + (2(12) - 1)]$$ Let $n = 12$, $a_1 = 2(1) - 1$, and $a_{12} = 2(12) - 1$.

$$S_{12} = 6(1 + 23)$$ Evaluate a_1 and a_{12}.

$$S_{12} = 6(24)$$ Add.

$$S_{12} = 144$$ Multiply. **NOW TRY**

NOW TRY ANSWER

9. 253

13.2 Exercises

MyLab Math

Video solutions for select problems available in MyLab Math

STUDY SKILLS REMINDER

Have you begun to prepare for your final exam? **Review Study Skill 10, *Preparing for Your Math Final Exam.***

Concept Check *Fill in each blank with the correct response.*

1. In an arithmetic sequence, if any term is subtracted from the term that follows it, the result is the common ______ of the sequence.

2. For the arithmetic sequence having $a_n = 2n + 4$, the term $a_3 =$ ______.

3. The sum of the first five terms of the arithmetic sequence 1, 6, 11, . . . is ______.

4. The number of terms in the arithmetic sequence 2, 4, 6, . . . , 100 is ______.

Concept Check *Fill in the blanks to complete the terms of each arithmetic sequence.*

5. $1, \frac{3}{2}, 2,$ ______, ______, ______

6. ______, ______, 1, 5, 9

7. 11, ______, 21, ______, 31

8. 7, ______, ______, −17, ______

If the given sequence is arithmetic, find the common difference d. If the sequence is not arithmetic, say so. **See Example 1.**

9. 1, 2, 3, 4, 5, . . .

10. 2, 5, 8, 11, . . .

11. 2, −4, 6, −8, 10, −12, . . .

12. 1, 2, 4, 7, 11, 16, . . .

13. 10, 5, 0, −5, −10, . . .

14. −6, −10, −14, −18, . . .

Write the first five terms of each arithmetic sequence. **See Example 2.**

15. $a_1 = 5, d = 4$

16. $a_1 = 6, d = 7$

17. $a_1 = -2, d = -4$

18. $a_1 = -3, d = -5$

19. $a_1 = \frac{1}{3}, d = \frac{2}{3}$

20. $a_1 = 0.25, d = 0.55$

Determine an expression for the general term of each arithmetic sequence. Then find a_{25}. **See Example 3.**

21. $a_1 = 2, d = 5$

22. $a_1 = 5, d = 3$

23. $3, \frac{15}{4}, \frac{9}{2}, \frac{21}{4}, \ldots$

24. $1, \frac{5}{3}, \frac{7}{3}, 3, \ldots$

25. $-3, 0, 3, \ldots$

26. $-10, -5, 0, \ldots$

Find the indicated term for each arithmetic sequence. **See Example 5.**

27. $a_1 = 4, d = 3;\quad a_{25}$

28. $a_1 = 1, d = -3;\quad a_{12}$

29. $2, 4, 6, \ldots;\quad a_{24}$

30. $1, 5, 9, \ldots;\quad a_{50}$

31. $a_{12} = -45, a_{10} = -37;\quad a_1$

32. $a_{10} = -2, a_{15} = 13;\quad a_3$

Find the number of terms in each arithmetic sequence. **See Example 6.**

33. $3, 5, 7, \ldots, 33$

34. $4, 8, 12, \ldots, 204$

35. $-7, -13, -19, \ldots, -157$

36. $4, 1, -2, \ldots, -32$

37. $\frac{3}{4}, 3, \frac{21}{4}, \ldots, 12$

38. $2, \frac{3}{2}, 1, \frac{1}{2}, \ldots, -5$

39. Concept Check A student incorrectly claimed that the common difference for the arithmetic sequence

$$-15, -10, -5, 0, 5, \ldots$$

is -5. **WHAT WENT WRONG?** Find the correct common difference.

40. Concept Check A student incorrectly claimed that there are 100 terms in the arithmetic sequence

$$2, 4, 6, 8, \ldots, 100.$$

WHAT WENT WRONG? How many terms are there?

Evaluate S_6 for each arithmetic sequence. **See Examples 7 and 8.**

41. $a_n = 3n - 8$

42. $a_n = 5n - 12$

43. $a_n = 4 + 3n$

44. $a_n = 9 + 5n$

45. $a_n = -\frac{1}{2}n + 7$

46. $a_n = -\frac{2}{3}n + 6$

47. $a_1 = 6, d = 3$

48. $a_1 = 5, d = 4$

49. $a_1 = 7, d = -3$

50. $a_1 = -5, d = -4$

51. $a = 0.9, d = -0.5$

52. $a_1 = 2.4, d = 0.7$

Use a formula for S_n to evaluate each series. **See Example 9.**

53. $\sum_{i=1}^{10} (8i - 5)$

54. $\sum_{i=1}^{17} (3i - 1)$

55. $\sum_{i=1}^{20} \left(\frac{3}{2}i + 4\right)$

56. $\sum_{i=1}^{11} \left(\frac{1}{2}i - 1\right)$

57. $\sum_{i=1}^{250} i$

58. $\sum_{i=1}^{2000} i$

Solve each problem. **See Examples 4, 7, 8, and 9.**

59. Nancy's aunt has promised to deposit $1 in her account on the first day of her birthday month, $2 on the second day, $3 on the third day, and so on for 30 days. How much will this amount to over the entire month?

60. Billy's aunt deposits \$2 in his savings account on the first day of February, \$4 on the second day, \$6 on the third day, and so on for the entire month. How much will this amount to? (Assume that it is a leap year.)

61. Suppose that Cherian is offered a job at \$1600 per month with a guaranteed increase of \$50 every six months for 5 yr. What will Cherian's salary be at the end of that time?

62. Malik is offered a job with a starting salary of \$2000 per month with a guaranteed increase of \$100 every four months for 3 yr. What will Malik's salary be at the end of that time?

63. A seating section in a theater-in-the-round has 20 seats in the first row, 22 in the second row, 24 in the third row, and so on for 25 rows. How many seats are there in the last row? How many seats are there in the section?

64. Constantin has started on a fitness program. He plans to jog 10 min per day for the first week and then to add 10 min per day each week until he is jogging an hour each day. In which week will this occur? What is the total number of minutes he will run during the first four weeks?

65. A child builds with blocks, placing 35 blocks in the first row, 31 in the second row, 27 in the third row, and so on. Continuing this pattern, can he end with a row containing exactly 1 block? If not, how many blocks will the last row contain? How many rows can he build this way?

66. A stack of firewood has 28 pieces on the bottom, 24 on top of those, then 20, and so on. If there are 108 pieces of wood, how many rows are there?

13.3 Geometric Sequences

OBJECTIVES

1. Find the common ratio of a geometric sequence.
2. Find the general term of a geometric sequence.
3. Find any specified term of a geometric sequence.
4. Find the sum of a specified number of terms of a geometric sequence.
5. Apply the formula for the future value of an ordinary annuity.
6. Find the sum of an infinite number of terms of a geometric sequence.

OBJECTIVE 1 **Find the common ratio of a geometric sequence.**

In an arithmetic sequence, each term after the first is found by *adding* a fixed number to the previous term. A *geometric sequence* is defined using multiplication.

Geometric Sequence

A **geometric sequence,** or **geometric progression,** is a sequence in which each term after the first is found by multiplying the preceding term by a nonzero constant.

We find the constant multiplier r, called the **common ratio,** by dividing any term a_{n+1} by the preceding term, a_n.

$$r = \frac{a_{n+1}}{a_n} \quad \text{Common ratio}$$

For example,

$$\overset{\times 3 \quad \times 3 \quad \times 3 \quad \times 3}{2, \; 6, \; 18, \; 54, \; 162, \ldots} \quad \text{Geometric sequence}$$

is a geometric sequence in which the first term is 2 and the common ratio is 3.

$$r = \frac{6}{2} = \frac{18}{6} = \frac{54}{18} = \frac{162}{54} = 3 \leftarrow \frac{a_{n+1}}{a_n} = 3 \text{ for all } n.$$

VOCABULARY

- ☐ geometric sequence (geometric progression)
- ☐ common ratio
- ☐ annuity
- ☐ ordinary annuity
- ☐ payment period
- ☐ term of an annuity
- ☐ future value of an annuity

EXAMPLE 1 Finding the Common Ratio

Determine the common ratio r for the geometric sequence.

$$15, \frac{15}{2}, \frac{15}{4}, \frac{15}{8}, \ldots$$

To find r, choose any two successive terms and divide the second one by the first. We choose the second and third terms of the sequence.

NOW TRY EXERCISE 1

Determine the common ratio r for the geometric sequence.

$$\frac{1}{4}, -1, 4, -16, 64, \ldots$$

$r = \frac{a_3}{a_2}$ Common ratio $r = \frac{a_{n+1}}{a_n}$

$r = \frac{\frac{15}{4}}{\frac{15}{2}}$ Substitute.

$r = \frac{15}{4} \div \frac{15}{2}$ Write as a division problem.

$r = \frac{15}{4} \cdot \frac{2}{15}$ Definition of division

$r = \frac{1}{2}$ Multiply. Write in lowest terms.

Any other two successive terms could have been used to find r. Additional terms of the sequence can be found by multiplying each successive term by $\frac{1}{2}$. NOW TRY

OBJECTIVE 2 Find the general term of a geometric sequence.

The general term a_n of a geometric sequence $a_1, a_2, a_3, \ldots$ is expressed in terms of a_1 and r by writing the first few terms.

$$a_1, \quad a_2 = a_1r, \quad a_3 = a_1r^2, \quad a_4 = a_1r^3, \ldots, \quad a_n = a_1r^{n-1}.$$

General Term of a Geometric Sequence

The general term a_n of the geometric sequence with first term a_1 and common ratio r is given by the following.

$$a_n = a_1r^{n-1}$$

NOW TRY EXERCISE 2

Determine an expression for the general term of the geometric sequence.

$$\frac{1}{4}, -1, 4, -16, 64, \ldots$$

EXAMPLE 2 Finding the General Term of a Geometric Sequence

Determine an expression for the general term of the geometric sequence.

$$15, \frac{15}{2}, \frac{15}{4}, \frac{15}{8}, \ldots$$ See Example 1.

The first term is $a_1 = 15$ and the common ratio is $r = \frac{1}{2}$.

$a_n = a_1r^{n-1}$ Formula for the general term a_n

$a_n = 15\left(\frac{1}{2}\right)^{n-1}$ Let $a_1 = 15$ and $r = \frac{1}{2}$.

It is not possible to simplify further because the exponent must be applied before the multiplication can be done. NOW TRY

NOW TRY ANSWERS

1. $r = -4$
2. $a_n = \frac{1}{4}(-4)^{n-1}$

OBJECTIVE 3 Find any specified term of a geometric sequence.

NOW TRY EXERCISE 3

Find the indicated term for each geometric sequence.

(a) $a_1 = 3, r = -2;\quad a_8$

(b) $10, 2, \frac{2}{5}, \frac{2}{25}, \ldots;\quad a_7$

EXAMPLE 3 Finding Specified Terms in Geometric Sequences

Find the indicated term for each geometric sequence.

(a) $a_1 = 4, r = -3;\quad a_6$

$$a_n = a_1 r^{n-1} \qquad \text{Formula for the general term } a_n$$

$$a_6 = a_1 \cdot r^{6-1} \qquad \text{Let } n = 6.$$

$$a_6 = 4(-3)^5 \qquad \text{Let } a_1 = 4, r = -3.$$

$$a_6 = -972$$

Evaluate $(-3)^5$ and then multiply.

(b) $\frac{3}{4}, \frac{3}{8}, \frac{3}{16}, \ldots;\quad a_7$

$$a_7 = \frac{3}{4}\left(\frac{1}{2}\right)^{7-1} \qquad \text{Let } a_1 = \tfrac{3}{4}, r = \tfrac{1}{2}, n = 7.$$

$$a_7 = \frac{3}{4}\left(\frac{1}{64}\right) \qquad \text{Apply the exponent.}$$

$$a_7 = \frac{3}{256} \qquad \text{Multiply.}$$

NOW TRY

NOW TRY EXERCISE 4

Write the first five terms of the geometric sequence having first term 25 and common ratio $-\frac{1}{5}$.

EXAMPLE 4 Writing the Terms of a Geometric Sequence

Write the first five terms of the geometric sequence having first term 5 and common ratio $\frac{1}{2}$.

$$a_1 = 5, \quad a_2 = 5\left(\frac{1}{2}\right) = \frac{5}{2}, \quad a_3 = 5\left(\frac{1}{2}\right)^2 = \frac{5}{4},$$

$$a_4 = 5\left(\frac{1}{2}\right)^3 = \frac{5}{8}, \quad a_5 = 5\left(\frac{1}{2}\right)^4 = \frac{5}{16}$$

Use $a_n = a_1 r^{n-1}$, with $a_1 = 5$, $r = \frac{1}{2}$, and $n = 1, 2, 3, 4, 5$.

NOW TRY

OBJECTIVE 4 Find the sum of a specified number of terms of a geometric sequence.

It is convenient to have a formula for the sum S_n of the first n terms of a geometric sequence. We can develop such a formula by first writing out S_n.

$$S_n = a_1 + a_1 r + a_1 r^2 + a_1 r^3 + \cdots + a_1 r^{n-1}$$

Next, we multiply both sides by $-r$.

$$-rS_n = -a_1 r - a_1 r^2 - a_1 r^3 - a_1 r^4 - \cdots - a_1 r^n$$

Now add.

$$\begin{aligned} S_n &= a_1 + a_1 r + a_1 r^2 + a_1 r^3 + \cdots + a_1 r^{n-1} \\ -rS_n &= - a_1 r - a_1 r^2 - a_1 r^3 - \cdots - a_1 r^{n-1} - a_1 r^n \\ \hline S_n - rS_n &= a_1 \phantom{+ a_1 r + a_1 r^2 + a_1 r^3 + \cdots + a_1 r^{n-1}} - a_1 r^n \end{aligned}$$

$$S_n(1 - r) = a_1 - a_1 r^n \qquad \text{Factor on the left.}$$

$$\mathbf{S_n = \frac{a_1(1 - r^n)}{1 - r}} \qquad \text{Factor on the right. Divide each side by } 1 - r.$$

NOW TRY ANSWERS

3. (a) -384 (b) $\frac{2}{3125}$

4. $25, -5, 1, -\frac{1}{5}, \frac{1}{25}$

Sum of the First n Terms of a Geometric Sequence

The sum S_n of the first n terms of the geometric sequence having first term a_1 and common ratio r is given by the following.

$$S_n = \frac{a_1(1 - r^n)}{1 - r} \quad \text{(where } r \neq 1\text{)}$$

If $r = 1$, then $S_n = a_1 + a_1 + a_1 + \cdots + a_1 = na_1$.

Multiplying the formula for S_n by $\frac{-1}{-1}$ gives an alternative form.

$$S_n = \frac{a_1(1 - r^n)}{1 - r} \cdot \frac{-1}{-1} = \frac{a_1(r^n - 1)}{r - 1} \qquad \text{Alternative form}$$

NOW TRY EXERCISE 5
Evaluate the sum of the first six terms of the geometric sequence with first term 4 and common ratio 2.

EXAMPLE 5 **Finding the Sum of the First n Terms of a Geometric Sequence**

Evaluate the sum of the first six terms of the geometric sequence having first term -2 and common ratio 3.

$$S_n = \frac{a_1(1 - r^n)}{1 - r} \qquad \text{Formula for } S_n$$

$$S_6 = \frac{-2(1 - 3^6)}{1 - 3} \qquad \text{Let } n = 6,\ a_1 = -2,\ r = 3.$$

$$S_6 = \frac{-2(1 - 729)}{-2} \qquad \text{Evaluate } 3^6\text{. Subtract in the denominator.}$$

$$S_6 = -728 \qquad \text{Simplify.}$$

NOW TRY

A series of the form

$$\sum_{i=1}^{n} a \cdot b^i$$

represents the sum of the first n terms of a geometric sequence having first term $a_1 = a \cdot b^1 = ab$ and common ratio b. The next example illustrates this form.

NOW TRY EXERCISE 6

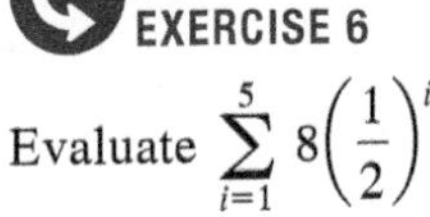

Evaluate $\sum_{i=1}^{5} 8\left(\frac{1}{2}\right)^i$.

EXAMPLE 6 **Using the Formula for S_n to Evaluate a Summation**

Evaluate $\sum_{i=1}^{4} 3 \cdot 2^i$.

Because the series is in the form $\sum_{i=1}^{n} a \cdot b^i$, it represents the sum of the first n terms of the geometric sequence with $a_1 = a \cdot b^1 = 3 \cdot 2^1 = 6$ and $r = b = 2$.

$$S_n = \frac{a_1(1 - r^n)}{1 - r} \qquad \text{Formula for } S_n$$

$$S_4 = \frac{6(1 - 2^4)}{1 - 2} \qquad \text{Let } n = 4,\ a_1 = 6,\ r = 2.$$

$$S_4 = \frac{6(1 - 16)}{-1} \qquad \text{Evaluate } 2^4\text{. Subtract in the denominator.}$$

$$S_4 = 90 \qquad \text{Simplify.}$$

NOW TRY

NOW TRY ANSWERS

5. 252

6. $\frac{31}{4}$, or 7.75

```
NORMAL FLOAT AUTO REAL RADIAN MP
seq(3*2^I,I,1,4)→L1
                    {6 12 24 48}
sum(L1)
                              90
```

FIGURE 2

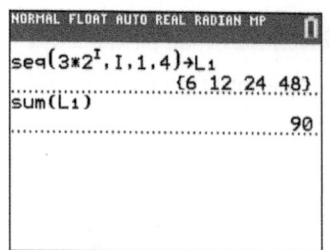

FIGURE 2 shows how a graphing calculator can store the terms in a list and then find the sum of these terms. The figure supports the result of **Example 6.**

OBJECTIVE 5 Apply the formula for the future value of an ordinary annuity.

A sequence of equal payments made over equal periods is an **annuity.** If the payments are made at the end of the period, and if the frequency of payments is the same as the frequency of compounding, the annuity is an **ordinary annuity.** The time between payments is the **payment period,** and the time from the beginning of the first payment period to the end of the last is the **term of the annuity.** The **future value of the annuity,** the final sum on deposit, is defined as the sum of the compound amounts of all the payments, compounded to the end of the term.

We state the following formula without proof.

Future Value of an Ordinary Annuity

The future value of an ordinary annuity is given by

$$S = R\left[\frac{(1+i)^n - 1}{i}\right],$$

where S is the future value,

R is the payment at the end of each period,

i is the interest rate per period,

and n is the number of periods.

NOW TRY EXERCISE 7

Work each problem. Give answers to the nearest cent.

(a) Billy deposits \$600 at the end of each year into an account paying 2% per yr, compounded annually. Find the total amount on deposit after 18 yr.

(b) How much will be in Billy's account after 18 yr if he deposits \$100 at the end of each month at 1.5% interest compounded monthly?

EXAMPLE 7 Applying the Formula for Future Value of an Annuity

Work each problem. Give answers to the nearest cent.

(a) Igor is an athlete who believes that his playing career will last 7 yr. He deposits \$22,000 at the end of each year for 7 yr in an account paying 2% compounded annually. How much will he have on deposit after 7 yr?

Igor's payments form an ordinary annuity with $R = 22{,}000$, $n = 7$, and $i = 0.02$.

$$S = 22{,}000\left[\frac{(1+0.02)^7 - 1}{0.02}\right] \quad \text{Substitute into the formula for } S.$$

$$S = 163{,}554.23 \quad \text{Evaluate with a calculator.}$$

The future value of this annuity is \$163,554.23.

(b) Amy has decided to deposit \$200 at the end of each month in an account that pays interest of 2.8% compounded monthly for retirement in 20 yr. How much will be in the account at that time?

Because the interest is compounded monthly, $i = \frac{0.028}{12}$. Also, $R = 200$ and $n = 12(20)$. Substitute into the formula for S to find the future value.

$$S = 200\left[\frac{\left(1+\frac{0.028}{12}\right)^{12(20)} - 1}{\frac{0.028}{12}}\right] = 64{,}245.50, \quad \text{or} \quad \$64{,}245.50$$

NOW TRY

NOW TRY ANSWERS

7. (a) \$12,847.39
(b) \$24,779.49

OBJECTIVE 6 Find the sum of an infinite number of terms of a geometric sequence.

Consider an infinite geometric sequence such as

$$\frac{1}{3}, \frac{1}{6}, \frac{1}{12}, \frac{1}{24}, \frac{1}{48}, \ldots, \quad \text{with first term } \frac{1}{3} \text{ and common ratio } \frac{1}{2}.$$

The sum of the first two terms is

$$S_2 = \frac{1}{3} + \frac{1}{6} = \frac{1}{2}.$$

In a similar manner, we can find additional "partial sums."

$$S_3 = S_2 + \frac{1}{12} = \frac{1}{2} + \frac{1}{12} = \frac{7}{12} \approx 0.583$$

$$S_4 = S_3 + \frac{1}{24} = \frac{7}{12} + \frac{1}{24} = \frac{15}{24} = 0.625$$

$$S_5 = \frac{31}{48} \approx 0.64583$$

$$S_6 = \frac{21}{32} = 0.65625$$

$$S_7 = \frac{127}{192} \approx 0.6614583, \quad \text{and so on.}$$

Each term of the geometric sequence is less than the preceding one, so each additional term is contributing less and less to the partial sum. In decimal form (to the nearest thousandth), the first 7 terms and the 10th term are given in the table.

Term	a_1	a_2	a_3	a_4	a_5	a_6	a_7	a_{10}
Value	0.333	0.167	0.083	0.042	0.021	0.010	0.005	0.001

As the table suggests, the value of a term gets closer and closer to 0 as the number of the term increases. To express this idea, we say that as n increases without bound (written $n \to \infty$), the limit of the term a_n is 0, which is written

$$\lim_{n \to \infty} a_n = 0.$$

A number that can be defined as the sum of an infinite number of terms of a geometric sequence is found by starting with the expression for the sum of a finite number of terms.

$$S_n = \frac{a_1(1 - r^n)}{1 - r}$$

If $|r| < 1$, then as n increases without bound, the value of r^n gets closer and closer to 0. As r^n approaches 0, $1 - r^n$ approaches $1 - 0 = 1$. S_n approaches the quotient $\frac{a_1}{1 - r}$.

$$\lim_{n \to \infty} S_n = \lim_{n \to \infty} \frac{a_1(1 - r^n)}{1 - r} = \frac{a_1(1 - 0)}{1 - r} = \frac{a_1}{1 - r}$$

This limit is defined to be the *sum* of the terms of the infinite geometric sequence.

$$a_1 + a_1r + a_1r^2 + a_1r^3 + \cdots = \frac{a_1}{1 - r}, \quad \text{if } |r| < 1$$

Sum of the Terms of an Infinite Geometric Sequence

The sum S of the terms of an infinite geometric sequence having first term a_1 and common ratio r, where $|r| < 1$, is given by the following.

$$S = \frac{a_1}{1 - r}$$

If $|r| \geq 1$, then the sum does not exist.

Now consider $|r| > 1$. For example, suppose the sequence is

$$6,\ 12,\ 24,\ \ldots,\ 3(2)^n,\ \ldots. \qquad \text{Here, } a_1 = 6 \text{ and } r = 2.$$

In this kind of sequence, as n increases, the value of r^n also increases and so does the sum S_n. Each new term adds a greater and greater amount to the sum, so there is no limit to the value of S_n. The sum S does not exist. Recall that if $r = 1$, then

$$S_n = a_1 + a_1 + a_1 + \cdots + a_1 = na_1.$$

NOW TRY EXERCISE 8

Evaluate the sum of the terms of the infinite geometric sequence having $a_1 = -4$ and $r = \frac{2}{3}$.

EXAMPLE 8 Finding the Sum of the Terms of an Infinite Geometric Sequence

Evaluate the sum of the terms of the infinite geometric sequence having $a_1 = 3$ and $r = -\frac{1}{3}$.

$$S = \frac{a_1}{1 - r} \qquad \text{Infinite sum formula}$$

$$S = \frac{3}{1 - \left(-\frac{1}{3}\right)} \qquad \text{Let } a_1 = 3,\ r = -\tfrac{1}{3}.$$

$$S = \frac{3}{\frac{4}{3}} \qquad \text{Subtract in the denominator; } 1 - \left(-\tfrac{1}{3}\right) = 1 + \tfrac{1}{3} = \tfrac{3}{3} + \tfrac{1}{3} = \tfrac{4}{3}.$$

$$S = 3 \div \frac{4}{3} \qquad \text{Write as a division problem.}$$

$$S = 3 \cdot \frac{3}{4} \qquad \text{Definition of division}$$

$$S = \frac{9}{4} \qquad \text{Multiply.}$$

NOW TRY 

In summation notation, the sum of an infinite geometric sequence is written as

$$\sum_{i=1}^{\infty} a_i.$$

For instance, the sum in **Example 8** would be written

$$\sum_{i=1}^{\infty} 3\left(-\frac{1}{3}\right)^{i-1}.$$

NOW TRY ANSWER

8. -12

NOW TRY EXERCISE 9

Evaluate $\sum_{i=1}^{\infty} \left(\frac{5}{8}\right)\left(\frac{3}{4}\right)^i$.

NOW TRY ANSWER

9. $\frac{15}{8}$

EXAMPLE 9 Evaluating an Infinite Geometric Series

Evaluate $\sum_{i=1}^{\infty} \left(\frac{2}{3}\right)\left(\frac{1}{2}\right)^i$.

This is the infinite geometric series

$$\frac{1}{3} + \frac{1}{6} + \frac{1}{12} + \cdots,$$

with $a_1 = \frac{1}{3}$ and $r = \frac{1}{2}$. Because $|r| < 1$, we find the sum as follows.

$$S = \frac{a_1}{1 - r}$$ Infinite sum formula

$$S = \frac{\frac{1}{3}}{1 - \frac{1}{2}}$$ Let $a_1 = \frac{1}{3}$, $r = \frac{1}{2}$.

$$S = \frac{\frac{1}{3}}{\frac{1}{2}}$$ Subtract in the denominator; $1 - \frac{1}{2} = \frac{2}{2} - \frac{1}{2} = \frac{1}{2}$.

$$S = \frac{2}{3}$$ Divide; $\frac{\frac{1}{3}}{\frac{1}{2}} = \frac{1}{3} \cdot \frac{2}{1} = \frac{2}{3}$.

NOW TRY 

13.3 Exercises

FOR EXTRA HELP

MyLab Math

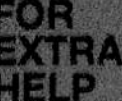 *Video solutions for select problems available in MyLab Math*

Concept Check *Fill in each blank with the correct response.*

1. In a geometric sequence, if any term after the first is divided by the term that precedes it, the result is the common _____ of the sequence.

2. For the geometric sequence having $a_n = (-2)^n$, the term $a_5 =$ _____.

3. The sum of the first five terms of the geometric sequence 1, 2, 4, . . . is _____.

4. The number of terms in the geometric sequence 1, 2, 4, . . . , 2048 is _____.

Concept Check *Fill in the blanks to complete the terms of each geometric sequence.*

5. $\frac{1}{3}, -\frac{1}{9}, \frac{1}{27}$, _____, _____, _____

6. _____, _____, 1, 5, 25

7. 7, _____, 28, _____, 112

8. −2, _____, _____, −54, _____

If the given sequence is geometric, find the common ratio r. If the sequence is not geometric, say so. ***See Example 1.***

9. 4, 8, 16, 32, . . .

10. 5, 15, 45, 135, . . .

11. $\frac{1}{3}, \frac{2}{3}, \frac{3}{3}, \frac{4}{3}, \ldots$

12. $\frac{5}{7}, \frac{8}{7}, \frac{11}{7}, 2, \ldots$

13. 1, −3, 9, −27, 81, . . .

14. 2, −8, 32, −128, . . .

15. $1, -\frac{1}{2}, \frac{1}{4}, -\frac{1}{8}, \ldots$

16. $\frac{2}{3}, -\frac{2}{15}, \frac{2}{75}, -\frac{2}{375}, \ldots$

Determine an expression for the general term of each geometric sequence. **See Example 2.**

17. $-5, -10, -20, \ldots$

18. $-2, -6, -18, \ldots$

19. $-2, \frac{2}{3}, -\frac{2}{9}, \ldots$

20. $-3, \frac{3}{2}, -\frac{3}{4}, \ldots$

21. $10, -2, \frac{2}{5}, \ldots$

22. $8, -2, \frac{1}{2}, \ldots$

Find the indicated term for each geometric sequence. **See Example 3.**

23. $a_1 = 2, r = 5; \quad a_{10}$

24. $a_1 = 1, r = 3; \quad a_{15}$

25. $\frac{1}{2}, \frac{1}{6}, \frac{1}{18}, \ldots; \quad a_{12}$

26. $\frac{2}{3}, \frac{1}{3}, \frac{1}{6}, \ldots; \quad a_{18}$

27. $a_2 = 18, a_5 = -486; \quad a_7$

28. $a_5 = 48, a_8 = -384; \quad a_{10}$

Write the first five terms of each geometric sequence. **See Example 4.**

29. $a_1 = 2, r = 3$

30. $a_1 = 4, r = 2$

31. $a_1 = 5, r = -\frac{1}{5}$

32. $a_1 = 6, r = -\frac{1}{3}$

33. $a_1 = -4, r = 0.5$

34. $a_1 = -40, r = 0.25$

Evaluate the sum of the terms of each geometric sequence. In Exercises 39–44, give answers to the nearest thousandth. **See Examples 5 and 6.**

35. $a_1 = -3, r = 4;$ Find S_{10}.

36. $a_1 = -5, r = 7;$ Find S_9.

37. $\frac{1}{3}, \frac{1}{9}, \frac{1}{27}, \frac{1}{81}, \frac{1}{243}$

38. $\frac{1}{4}, \frac{1}{16}, \frac{1}{64}, \frac{1}{256}, \frac{1}{1024}$

39. $-\frac{4}{3}, -\frac{4}{9}, -\frac{4}{27}, -\frac{4}{81}, -\frac{4}{243}, -\frac{4}{729}$

40. $\frac{5}{16}, -\frac{5}{32}, \frac{5}{64}, -\frac{5}{128}, \frac{5}{256}$

41. $\sum_{i=1}^{7} 4\left(\frac{2}{5}\right)^i$

42. $\sum_{i=1}^{8} 5\left(\frac{2}{3}\right)^i$

43. $\sum_{i=1}^{10} (-2)\left(\frac{3}{5}\right)^i$

44. $\sum_{i=1}^{6} (-2)\left(-\frac{1}{2}\right)^i$

Solve each problem. Round answers to the nearest cent. **See Example 7.**

45. A father opened a savings account for his daughter on her first birthday, depositing \$1000. Each year on her birthday he deposits another \$1000, making the last deposit on her 21st birthday. If the account pays 1.5% interest compounded annually, how much is in the account at the end of the day on the daughter's 21st birthday?

46. A teacher puts \$1000 in a retirement account at the end of each quarter $\left(\frac{1}{4}\text{ of a year}\right)$ for 15 yr. If the account pays 2.2% annual interest compounded quarterly, how much will be in the account at that time?

47. At the end of each quarter, a 50-year-old woman puts \$1200 in a retirement account that pays 2% interest compounded quarterly. When she reaches age 60, she withdraws the entire amount and places it in a fund that pays 1% annual interest compounded monthly. From then on, she deposits \$300 in the fund at the end of each month. How much is in the account when she reaches age 65?

48. At the end of each quarter, a 45-year-old man puts \$1500 in a retirement account that pays 1.5% interest compounded quarterly. When he reaches age 55, he withdraws the entire amount and places it in a fund that pays 0.75% annual interest compounded monthly. From then on, he deposits \$400 in the fund at the end of each month. How much is in the account when he reaches age 60?

Find the sum, if it exists, of the terms of each infinite geometric sequence. **See Examples 8 and 9.**

49. $a_1 = 6,\ r = \frac{1}{3}$

50. $a_1 = 10,\ r = \frac{1}{5}$

51. $a_1 = 1000,\ r = -\frac{1}{10}$

52. $a_1 = 8800,\ r = -\frac{3}{5}$

53. $\sum_{i=1}^{\infty} \frac{9}{8}\left(-\frac{2}{3}\right)^i$

54. $\sum_{i=1}^{\infty} \frac{3}{5}\left(\frac{5}{6}\right)^i$

55. $\sum_{i=1}^{\infty} \frac{12}{5}\left(\frac{5}{4}\right)^i$

56. $\sum_{i=1}^{\infty} \left(-\frac{16}{3}\right)\left(-\frac{9}{8}\right)^i$

Extending Skills *Solve each application.*

57. When dropped from a certain height, a ball rebounds to $\frac{3}{5}$ of the original height. How high will the ball rebound after the fourth bounce if it was dropped from a height of 10 ft? Round to the nearest tenth.

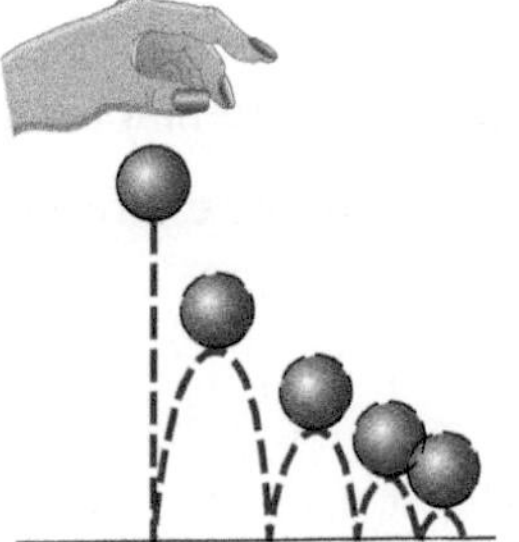

58. A fully wound yo-yo has a string 40 in. long. It is allowed to drop, and on its first rebound it returns to a height 15 in. lower than its original height. Assuming that this "rebound ratio" remains constant until the yo-yo comes to rest, how far does it travel on its third trip up the string? Round to the nearest tenth.

59. A ball is dropped from a height of 20 m, and on each bounce it returns to $\frac{3}{4}$ of its previous height. How far will the ball travel before it comes to rest? (*Hint:* Consider the sum of two sequences.)

60. A fully wound yo-yo is dropped the length of its 30-in. string. Each time it drops, it returns to $\frac{1}{2}$ of its original height. How far does it travel before it comes to rest? (*Hint:* Consider the sum of two sequences.)

61. A particular substance decays in such a way that it loses half its weight each day. In how many days will 256 g of the substance be reduced to 32 g? How much of the substance is left after 10 days?

62. A tracer dye is injected into a system with an ingestion and an excretion. After 1 hr, $\frac{2}{3}$ of the dye is left. At the end of the second hour, $\frac{2}{3}$ of the remaining dye is left, and so on. If one unit of the dye is injected, how much is left after 6 hr?

63. In a certain community, the consumption of electricity has increased about 6% per yr.

(a) If the community uses 1.1 billion units of electricity now, how much will it use 5 yr from now? Round to the nearest tenth.

(b) Find the number of years (to the nearest year) it will take for the consumption to double.

64. A growing community increases its consumption of electricity 2% per yr.

(a) If the community uses 1.1 billion units of electricity now, how much will it use 5 yr from now? Round to the nearest tenth.

(b) Find the number of years (to the nearest year) it will take for the consumption to double.

65. A machine depreciates by $\frac{1}{4}$ of its value each year. If it cost \$50,000 new, what is its value after 8 yr?

66. A vehicle depreciates by 20% of its value each year. If it cost \$35,000 new, what is its value after 6 yr?

67. The repeating decimal 0.99999 . . . can be written as the sum of the terms of a geometric sequence with $a_1 = 0.9$ and $r = 0.1$.

$$0.99999\ldots = 0.9 + 0.9(0.1) + 0.9(0.1)^2 + 0.9(0.1)^3 + 0.9(0.1)^4 + 0.9(0.1)^5 + \cdots$$

Because $|0.1| < 1$, this sum can be found from the formula $S = \frac{a_1}{1-r}$. Use this formula to find a more common way of writing the decimal 0.99999 . . . .

68. If the result of **Exercise 67** seems hard to believe, look at it this way: Use long division to find the repeating decimal representations of $\frac{1}{3}$ and $\frac{2}{3}$. Then line up the decimals vertically and add them. The result must equal

$$\frac{1}{3} + \frac{2}{3} = 1.$$

13.4 The Binomial Theorem

OBJECTIVES

1 Expand a binomial raised to a power.

2 Find any specified term of the expansion of a binomial.

VOCABULARY

☐ Pascal's triangle

OBJECTIVE 1 Expand a binomial raised to a power.

Observe the expansion of the expression $(x + y)^n$ for the first six nonnegative integer values of n.

Expansions of $(x + y)^n$

$$(x + y)^0 = 1$$
$$(x + y)^1 = x + y$$
$$(x + y)^2 = x^2 + 2xy + y^2$$
$$(x + y)^3 = x^3 + 3x^2y + 3xy^2 + y^3$$
$$(x + y)^4 = x^4 + 4x^3y + 6x^2y^2 + 4xy^3 + y^4$$
$$(x + y)^5 = x^5 + 5x^4y + 10x^3y^2 + 10x^2y^3 + 5xy^4 + y^5$$

By identifying patterns, we can write a general expansion for $(x + y)^n$.

First, if n is a positive integer, each expansion after $(x + y)^0$ begins with x raised to the same power to which the binomial is raised. That is, the expansion of $(x + y)^1$ has a first term of x^1, the expansion of $(x + y)^2$ has a first term of x^2, and so on. Also, the last term in each expansion is y to this same power, so the expansion of

$$(x + y)^n$$

should begin with the term x^n and end with the term y^n.

The exponents on x decrease by 1 in each term after the first, while the exponents on y, beginning with y in the second term, increase by 1 in each succeeding term. Thus, the *variables* in the expansion of $(x + y)^n$ have the following pattern.

$$x^n, \quad x^{n-1}y, \quad x^{n-2}y^2, \quad x^{n-3}y^3, \quad \ldots, \quad xy^{n-1}, \quad y^n$$

This pattern suggests that the sum of the exponents on x and y in each term is n. For example, in the third term shown, the variable part is $x^{n-2}y^2$ and the sum of the exponents is $n - 2 + 2 = n$.

Now examine the pattern for the *coefficients* of the terms of the preceding expansions. Writing the coefficients alone in a triangular pattern gives **Pascal's triangle,** named in honor of the 17th-century mathematician Blaise Pascal.

Blaise Pascal (1623–1662)

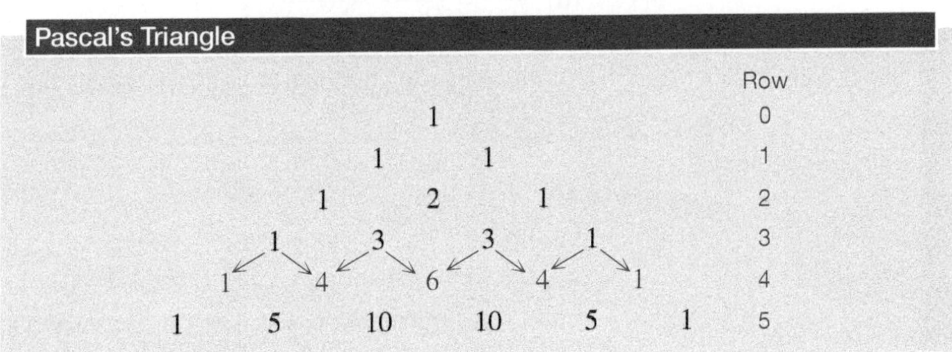

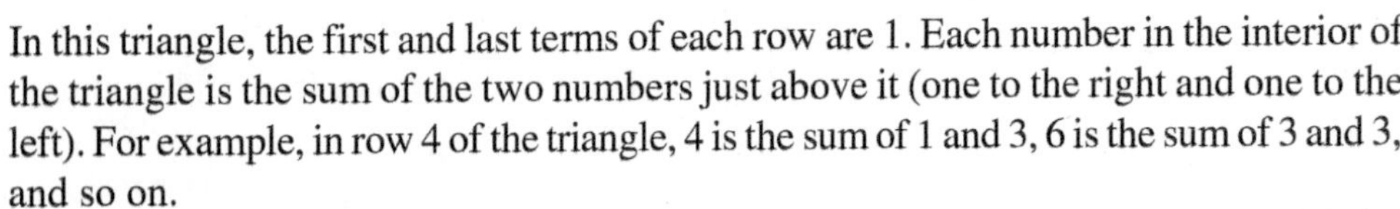

In this triangle, the first and last terms of each row are 1. Each number in the interior of the triangle is the sum of the two numbers just above it (one to the right and one to the left). For example, in row 4 of the triangle, 4 is the sum of 1 and 3, 6 is the sum of 3 and 3, and so on.

To obtain the coefficients for $(x + y)^6$, we need to attach row 6 to the table by starting and ending with 1, and adding pairs of numbers from row 5.

1 6 15 20 15 6 1 Row 6

We then use these coefficients to expand $(x + y)^6$.

$$(x + y)^6 = x^6 + 6x^5y + 15x^4y^2 + 20x^3y^3 + 15x^2y^4 + 6xy^5 + y^6$$

Although it is possible to use Pascal's triangle to find the coefficients in $(x + y)^n$ for any positive integer value of n, it is impractical for large values of n. A more efficient way to determine these coefficients uses the symbol $n!$ (read "***n* factorial**").

NORMAL FLOAT AUTO REAL RADIAN MP
3! 6
5! 120
0! 1

FIGURE 3

n Factorial ($n!$)

For any positive integer n,

$$n! = n(n-1)(n-2)(n-3)\cdots(3)(2)(1).$$

By definition, $\mathbf{0! = 1}$.

EXAMPLE 1 Evaluating Factorials

Evaluate each factorial.

(a) $3! = 3 \cdot 2 \cdot 1 = 6$ **(b)** $5! = 5 \cdot 4 \cdot 3 \cdot 2 \cdot 1 = 120$ **(c)** $0! = 1$

FIGURE 3 shows how a graphing calculator computes factorials. NOW TRY 

NOW TRY EXERCISE 1
Evaluate 7!.

NOW TRY ANSWER
1. 5040

NOW TRY EXERCISE 2

Find the value of each expression.

(a) $\frac{8!}{6!2!}$ (b) $\frac{8!}{5!3!}$

(c) $\frac{6!}{6!0!}$ (d) $\frac{6!}{5!1!}$

EXAMPLE 2 Evaluating Expressions Involving Factorials

Find the value of each expression.

(a) $\frac{5!}{4!1!} = \frac{5 \cdot 4 \cdot 3 \cdot 2 \cdot 1}{(4 \cdot 3 \cdot 2 \cdot 1)(1)} = 5$

(b) $\frac{5!}{3!2!} = \frac{5 \cdot 4 \cdot 3 \cdot 2 \cdot 1}{(3 \cdot 2 \cdot 1)(2 \cdot 1)} = \frac{5 \cdot 4}{2 \cdot 1} = 10$

(c) $\frac{6!}{3!3!} = \frac{6 \cdot 5 \cdot 4 \cdot 3 \cdot 2 \cdot 1}{(3 \cdot 2 \cdot 1)(3 \cdot 2 \cdot 1)} = \frac{6 \cdot 5 \cdot 4}{3 \cdot 2 \cdot 1} = 20$

(d) $\frac{4!}{4!0!} = \frac{4 \cdot 3 \cdot 2 \cdot 1}{(4 \cdot 3 \cdot 2 \cdot 1)(1)} = 1$

NOW TRY 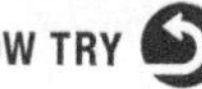

Now look again at the coefficients of the expansion

$$(x + y)^5 = x^5 + 5x^4y + 10x^3y^2 + 10x^2y^3 + 5xy^4 + y^5.$$

The coefficient of the second term is 5, and the exponents on the variables in that term are 4 and 1. From **Example 2(a),** $\frac{5!}{4!1!} = 5$. The coefficient of the third term is 10, and the exponents are 3 and 2. From **Example 2(b),** $\frac{5!}{3!2!} = 10$. Similar results are true for the remaining terms. The first term can be written as $1x^5y^0$, and the last term can be written as $1x^0y^5$. Then the coefficient of the first term should be $\frac{5!}{5!0!} = 1$, and the coefficient of the last term would be $\frac{5!}{0!5!} = 1$. (This is why 0! is *defined* to be 1.)

The coefficient of a term in $(x + y)^n$ in which the variable part is x^ry^{n-r} is

$$\frac{n!}{r!(n - r)!}.$$ This is the binomial coefficient.

The **binomial coefficient** $\frac{n!}{r!(n-r)!}$ is often represented by the symbol $_nC_r$. This notation comes from the fact that if we choose *combinations* of *n* things taken *r* at a time, the result is given by that expression. We read $_nC_r$ as ***"combinations of n things taken r at a time."*** Another common representation is $\binom{n}{r}$.

Formula for the Binomial Coefficient $_nC_r$

For nonnegative integers n and r, where $r \leq n$, $_nC_r$ is defined as follows.

$$_nC_r = \frac{n!}{r!(n - r)!}$$

EXAMPLE 3 Evaluating Binomial Coefficients

Evaluate each binomial coefficient.

(a) $_5C_4 = \frac{5!}{4!(5 - 4)!}$ Let $n = 5$, $r = 4$.

$= \frac{5!}{4!1!}$ Subtract.

$= \frac{5 \cdot 4 \cdot 3 \cdot 2 \cdot 1}{4 \cdot 3 \cdot 2 \cdot 1 \cdot 1}$ Definition of n factorial

$= 5$ Lowest terms

NOW TRY ANSWERS

2. (a) 28 (b) 56 (c) 1 (d) 6

NOW TRY EXERCISE 3
Evaluate ${}_7C_2$.

(b) $${}_5C_3 = \frac{5!}{3!(5-3)!} = \frac{5!}{3!2!} = \frac{5 \cdot 4 \cdot 3 \cdot 2 \cdot 1}{3 \cdot 2 \cdot 1 \cdot 2 \cdot 1} = 10$$

Binomial coefficients will always be whole numbers.

(c) $${}_6C_3 = \frac{6!}{3!(6-3)!} = \frac{6!}{3!3!} = \frac{6 \cdot 5 \cdot 4 \cdot 3 \cdot 2 \cdot 1}{3 \cdot 2 \cdot 1 \cdot 3 \cdot 2 \cdot 1} = 20$$

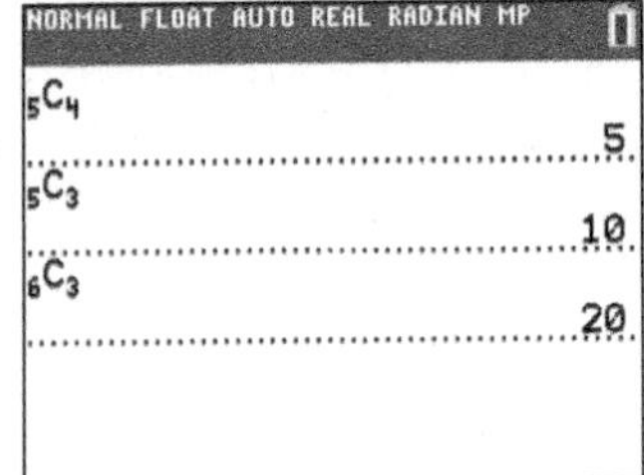

FIGURE 4

FIGURE 4 displays the binomial coefficients computed here.

NOW TRY 

Our observations about the expansion of $(x+y)^n$ are summarized as follows.

- There are $n+1$ terms in the expansion.
- The first term is x^n, and the last term is y^n.
- In each succeeding term, the exponent on x decreases by 1 and the exponent on y increases by 1.
- The sum of the exponents on x and y in any term is n.
- The coefficient of the term with $x^r y^{n-r}$ or $x^{n-r}y^r$ is ${}_nC_r$.

We now state the **binomial theorem,** or the **general binomial expansion.**

Binomial Theorem

For any positive integer n, $(x+y)^n$ is expanded as follows.

$$(x+y)^n = x^n + \frac{n!}{1!(n-1)!}x^{n-1}y + \frac{n!}{2!(n-2)!}x^{n-2}y^2 + \frac{n!}{3!(n-3)!}x^{n-3}y^3 + \cdots + \frac{n!}{(n-1)!1!}xy^{n-1} + y^n$$

The binomial theorem can be written using summation notation as follows.

$$(x+y)^n = \sum_{r=0}^{n} \frac{n!}{r!(n-r)!}x^{n-r}y^r$$

NOW TRY EXERCISE 4
Expand $(a+3b)^5$.

EXAMPLE 4 Using the Binomial Theorem

Expand $(2m+3)^4$.

$$(2m+3)^4$$
$$= (2m)^4 + \frac{4!}{1!3!}(2m)^3(3) + \frac{4!}{2!2!}(2m)^2(3)^2 + \frac{4!}{3!1!}(2m)(3)^3 + 3^4$$
$$= 16m^4 + 4(8m^3)(3) + 6(4m^2)(9) + 4(2m)(27) + 81$$
$$= 16m^4 + 96m^3 + 216m^2 + 216m + 81$$

Remember: $(ab)^m = a^m b^m$.

NOW TRY 

NOTE The binomial coefficients for the first and last terms in **Example 4** are both 1. These values can be taken from Pascal's triangle or computed as follows.

$\frac{4!}{0!4!} = 1$ Binomial coefficient of the first term

$\frac{4!}{4!0!} = 1$ Binomial coefficient of the last term

NOW TRY ANSWERS

3. 21
4. $a^5 + 15a^4b + 90a^3b^2 + 270a^2b^3 + 405ab^4 + 243b^5$

NOW TRY EXERCISE 5

Expand $\left(\frac{x}{3} - 2y\right)^4$.

EXAMPLE 5 Using the Binomial Theorem

Expand $\left(a - \frac{b}{2}\right)^5$.

$$\left(a - \frac{b}{2}\right)^5$$

$$= a^5 + \frac{5!}{1!4!}a^4\left(-\frac{b}{2}\right) + \frac{5!}{2!3!}a^3\left(-\frac{b}{2}\right)^2 + \frac{5!}{3!2!}a^2\left(-\frac{b}{2}\right)^3$$

$$+ \frac{5!}{4!1!}a\left(-\frac{b}{2}\right)^4 + \left(-\frac{b}{2}\right)^5$$

$$= a^5 + 5a^4\left(-\frac{b}{2}\right) + 10a^3\left(\frac{b^2}{4}\right) + 10a^2\left(-\frac{b^3}{8}\right)$$

$$+ 5a\left(\frac{b^4}{16}\right) + \left(-\frac{b^5}{32}\right)$$

Notice that signs alternate positive and negative.

$$= a^5 - \frac{5}{2}a^4b + \frac{5}{2}a^3b^2 - \frac{5}{4}a^2b^3 + \frac{5}{16}ab^4 - \frac{1}{32}b^5$$

NOW TRY 

CAUTION When the binomial is a *difference* of two terms, as in **Example 5,** the signs of the terms in the expansion will alternate.

- Those terms with odd exponents on the second variable expression $\left(-\frac{b}{2}\text{ in } \textbf{Example 5}\right)$ will be negative.
- Those terms with even exponents on the second variable expression will be positive.

OBJECTIVE 2 Find any specified term of the expansion of a binomial.

Any single term of a binomial expansion can be determined without writing out the whole expansion. For example, if $n \geq 10$, then the 10th term of $(x + y)^n$ has y raised to the ninth power (because y has the power of 1 in the second term, the power of 2 in the third term, and so on).

The exponents on x and y in any term must have a sum of n, so the exponent on x in the 10th term is $n - 9$. The quantities 9 and $n - 9$ determine the factorials in the denominator of the coefficient. Thus, the 10th term of $(x + y)^n$ is

$$\frac{n!}{9!(n-9)!}x^{n-9}y^9.$$

kth Term of the Binomial Expansion

If $n \geq k - 1$, then the kth term of the expansion of $(x + y)^n$ is given by the following.

$$\frac{n!}{(k-1)![n-(k-1)]!}x^{n-(k-1)}y^{k-1}$$

NOW TRY ANSWER

5. $\frac{x^4}{81} - \frac{8x^3y}{27} + \frac{8x^2y^2}{3} - \frac{32xy^3}{3} + 16y^4$

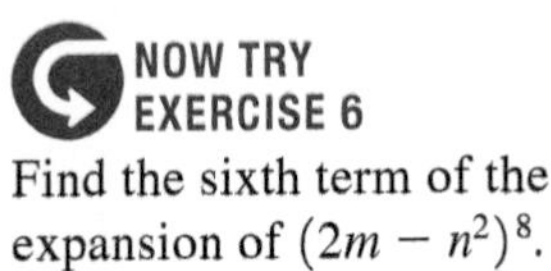

Find the sixth term of the expansion of $(2m - n^2)^8$.

NOW TRY ANSWER

6. $-448m^3n^{10}$

EXAMPLE 6 Finding a Single Term of a Binomial Expansion

Find the fourth term of the expansion of $(a + 2b)^{10}$.

In the fourth term, $2b$ has an exponent of $4 - 1 = 3$, and a has an exponent of $10 - 3 = 7$. The fourth term is determined as follows.

$$\frac{10!}{3!7!}(a^7)(2b)^3$$ Parentheses MUST be used with $2b$. Let $n = 10$, $x = a$, $y = 2b$, and $k = 4$.

$$= 120(a^7)(2^3b^3)$$ Simplify the factorials; $(ab)^m = a^mb^m$

$$= 120a^7(8b^3)$$ Simplify.

$$= 960a^7b^3$$ Multiply.

NOW TRY

13.4 Exercises

FOR EXTRA HELP MyLab Math

Video solutions for select problems available in MyLab Math

Concept Check *Fill in each blank with the correct response.*

1. In each row of Pascal's triangle, the first and last terms are ______, and each number in the interior of the triangle is the ______ of the two numbers just above it (one to the right and one to the left).
2. The complete row of Pascal's triangle that begins with the terms 1, 4, is ______.
3. The first term in the expansion of $(x + y)^3$ is ______.
4. The last term in the expansion of $(x - y)^4$ is ______.
5. The value of 3! is ______.
6. The value of 0! is ______.
7. For any nonnegative integer n, the binomial coefficient ${}_nC_0$ is equal to ______.
8. For any nonnegative integer n, the binomial coefficient ${}_nC_n$ is equal to ______.

Evaluate each expression. **See Examples 1–3.**

9. $6!$	**10.** $4!$	**11.** $8!$	**12.** $9!$
13. $\frac{6!}{4!2!}$	**14.** $\frac{7!}{3!4!}$	**15.** $\frac{4!}{0!4!}$	**16.** $\frac{5!}{5!0!}$
17. $4! \cdot 5$	**18.** $6! \cdot 7$	**19.** ${}_6C_2$	**20.** ${}_7C_4$
21. ${}_{13}C_{11}$	**22.** ${}_{13}C_2$	**23.** ${}_{10}C_7$	**24.** ${}_{10}C_3$

Use the binomial theorem to expand each binomial. **See Examples 4 and 5.**

25. $(m + n)^4$	**26.** $(x + r)^5$	**27.** $(a - b)^5$
28. $(p - q)^4$	**29.** $(2x + 3)^3$	**30.** $(4x + 2)^3$
31. $\left(\frac{x}{2} - y\right)^4$	**32.** $\left(\frac{x}{3} - 2y\right)^5$	**33.** $(x^2 + 1)^4$
34. $(y^3 + 2)^4$	**35.** $(3x^2 - y^2)^3$	**36.** $(2p^2 - q^2)^3$

Write the first four terms of each binomial expansion. **See Examples 4 and 5.**

37. $(r + 2s)^{12}$ **38.** $(m + 3n)^{20}$ **39.** $(3x - y)^{14}$

40. $(2p - 3q)^{11}$ **41.** $(t^2 + u^2)^{10}$ **42.** $(x^2 + y^2)^{15}$

Find the indicated term of each binomial expansion. **See Example 6.**

43. $(2m + n)^{10}$; fourth term

44. $(a + 3b)^{12}$; fifth term

45. $\left(x + \frac{y}{2}\right)^8$; seventh term

46. $\left(a + \frac{b}{3}\right)^{15}$; eighth term

47. $(k - 1)^9$; third term

48. $(r - 4)^{11}$; fourth term

49. The middle term of $(x^2 + 2y)^6$

50. The middle term of $(m^3 + 2y)^8$

51. The term with x^9y^4 in $(3x^3 - 4y^2)^5$

52. The term with x^8y^2 in $(2x^2 + 3y)^6$

13.5 Mathematical Induction

OBJECTIVES

1. Learn the principle of mathematical induction.
2. Use mathematical induction to prove a statement.

OBJECTIVE 1 Learn the principle of mathematical induction.

Many statements in mathematics are claimed true for any positive integer. Any of these statements can be checked for $n = 1, n = 2, n = 3$, and so on, but because the set of positive integers is infinite, it is impossible to check every possible case.

For example, let S_n represent the statement that the sum of the first n positive integers is $\frac{n(n+1)}{2}$.

$$S_n\text{: } 1 + 2 + 3 + \cdots + n = \frac{n(n+1)}{2}$$

The truth of this statement can be checked quickly for the first few values of n.

If $n = 1$, S_1 is $\quad 1 = \frac{1(1+1)}{2}$. This is true because $1 = 1$.

If $n = 2$, S_2 is $\quad 1 + 2 = \frac{2(2+1)}{2}$. This is true because $3 = 3$.

If $n = 3$, S_3 is $\quad 1 + 2 + 3 = \frac{3(3+1)}{2}$. This is true because $6 = 6$.

If $n = 4$, S_4 is $\quad 1 + 2 + 3 + 4 = \frac{4(4+1)}{2}$. This is true because $10 = 10$.

We cannot conclude that the statement is true for all positive integers simply by observing a finite number of examples. To prove that a statement is true for *every* positive integer value of n, we use the following principle.

Principle of Mathematical Induction

Let S_n be a statement concerning the positive integer n. Suppose that both of the following are satisfied.

1. S_1 is true.

2. For any positive integer k, $k \leq n$, if S_k is true, then S_{k+1} is also true.

Then S_n is true for every positive integer value of n.

A proof by mathematical induction can be explained as follows. By assumption (1) above, the statement is true when $n = 1$. If (2) has been proven, the fact that the statement is true for $n = 1$ implies that it is true for $n = 1 + 1 = 2$. Using (2) again reveals that the statement is thus true for

$$2 + 1 = 3, \quad \text{for } 3 + 1 = 4, \quad \text{for } 4 + 1 = 5, \quad \text{and so on.}$$

FIGURE 5

Continuing in this way shows that the statement must be true for *every* positive integer.

The situation is similar to that of an infinite number of dominoes lined up as suggested in **FIGURE 5**. If the first domino is pushed over, it pushes the next, which pushes the next, and so on, continuing indefinitely.

Another example of the principle of mathematical induction might be an infinite ladder. Suppose the rungs are spaced so that, whenever we are on a rung, we know we can move to the next rung. Thus *if* we can get to the first rung, then we can go as high up the ladder as we wish.

OBJECTIVE 2 Use mathematical induction to prove a statement.

Two separate steps are required for a proof by mathematical induction.

Procedure for Proof by Mathematical Induction

Step 1 Prove that the statement is true for $n = 1$.

Step 2 Show that for any positive integer k, $k \leq n$, if S_k is true, then S_{k+1} is also true.

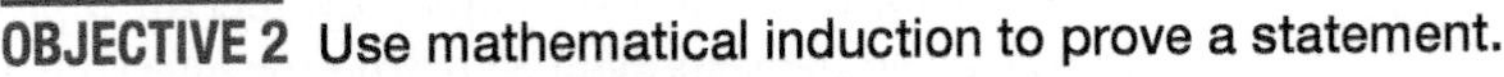

EXAMPLE 1 Proving an Equality Statement by Mathematical Induction

Let S_n represent the statement

$$1 + 2 + 3 + \cdots + n = \frac{n(n+1)}{2}.$$

Prove that S_n is true for every positive integer value of n.

PROOF The proof by mathematical induction is as follows.

Step 1 Show that the statement is true when $n = 1$. If $n = 1$, then S_1 becomes

$$1 = \frac{1(1+1)}{2}, \quad \text{which is true.}$$

Let S_n represent the statement

$$6 + 12 + 18 + \cdots + 6n = 3n(n + 1).$$

Prove that S_n is true for every positive integer value of n.

Step 2 Show that if S_k is true, then S_{k+1} is also true, where S_k is the statement

$$1 + 2 + 3 + \cdots + k = \frac{k(k + 1)}{2},$$

and S_{k+1} is the statement

$$1 + 2 + 3 + \cdots + k + (k + 1) = \frac{(k + 1)[(k + 1) + 1]}{2}.$$

Start with S_k and assume it is a true statement.

$$1 + 2 + 3 + \cdots + k = \frac{k(k + 1)}{2}$$

$$1 + 2 + 3 + \cdots + k + (k + 1) = \frac{k(k + 1)}{2} + (k + 1)$$ Add $k + 1$ to each side to obtain S_{k+1}.

$$= (k + 1)\left(\frac{k}{2} + 1\right)$$ Factor out the common factor $k + 1$ on the right.

$$= (k + 1)\left(\frac{k + 2}{2}\right)$$ $\frac{k}{2} + 1 = \frac{k}{2} + \frac{2}{2} = \frac{k + 2}{2}$

$$1 + 2 + 3 + \cdots + k + (k + 1) = \frac{(k + 1)[(k + 1) + 1]}{2}$$ $k + 2 = (k + 1) + 1$

This is the original statement with $k + 1$ substituted for n.

This final result is the statement for $n = k + 1$. We have shown that if S_k is true, then S_{k+1} is also true. The two steps required for a proof by mathematical induction are complete, so the statement S_n is true for every positive integer value of n.

NOW TRY 

CAUTION Notice that the left side of the statement always includes *all* the terms up to the nth term, as well as the nth term.

EXAMPLE 2 Proving an Equality Statement by Mathematical Induction

Prove that $4 + 7 + 10 + \cdots + (3n + 1) = \frac{n(3n + 5)}{2}$ for every positive integer value of n.

PROOF

Step 1 Show that the statement is true when $n = 1$. If $n = 1$, then S_1 becomes

$$4 = \frac{1(3 \cdot 1 + 5)}{2}.$$

The right side equals 4, so S_1 is a true statement.

Step 2 Show that if S_k is true, then S_{k+1} is true, where S_k is the statement

$$4 + 7 + 10 + \cdots + (3k + 1) = \frac{k(3k + 5)}{2},$$

NOW TRY EXERCISE 2

Prove that

$$8 + 11 + 14 + \cdots + (3n + 5) = \frac{n(3n + 13)}{2}$$

for every positive integer value of n.

and S_{k+1} is the statement

$$4 + 7 + 10 + \cdots + (3k + 1) + [3(k + 1) + 1] = \frac{(k + 1)[3(k + 1) + 5]}{2}.$$

Start with S_k and assume it is a true statement.

$$4 + 7 + 10 + \cdots + (3k + 1) = \frac{k(3k + 5)}{2}$$

To transform the left side of S_k to become the left side of S_{k+1}, we must add the $(k + 1)$st term. Add $[3(k + 1) + 1]$ to each side of S_k.

$$4 + 7 + 10 + \cdots + (3k + 1) + [3(k + 1) + 1]$$

$$= \frac{k(3k + 5)}{2} + [3(k + 1) + 1]$$

$$= \frac{k(3k + 5)}{2} + 3k + 3 + 1 \quad \text{Clear parentheses on the right side.}$$

$$= \frac{k(3k + 5)}{2} + 3k + 4$$

Use parentheses in the numerator.

$$= \frac{k(3k + 5)}{2} + \frac{2(3k + 4)}{2} \quad \text{The LCD is 2.}$$

$$= \frac{k(3k + 5) + 2(3k + 4)}{2} \quad \text{Add.}$$

$$= \frac{3k^2 + 5k + 6k + 8}{2} \quad \text{Distributive property}$$

$$= \frac{3k^2 + 11k + 8}{2} \quad \text{Combine like terms.}$$

$$= \frac{(k + 1)(3k + 8)}{2} \quad \text{Factor.}$$

Because $3k + 8$ can be written as $3(k + 1) + 5$, the following is true.

$$4 + 7 + 10 + \cdots + (3k + 1) + [3(k + 1) + 1] = \frac{(k + 1)[3(k + 1) + 5]}{2}$$

The final result is the statement for $n = k + 1$. Therefore, if S_k is true, then S_{k+1} is true. The two steps required for a proof by mathematical induction are complete, so the general statement S_n is true for every positive integer value of n.

NOW TRY EXERCISE 3
Prove that for every positive integer n,

$$2^n > n.$$

EXAMPLE 3 Proving an Inequality Statement by Mathematical Induction

Prove that if x is a real number between 0 and 1, then for every positive integer n,

$$0 < x^n < 1.$$

PROOF

Step 1 Let S_n represent the given statement. Here S_1 is the statement

if $0 < x < 1$, then $0 < x^1 < 1$, which is true.

Step 2 S_k is the statement

if $0 < x < 1$, then $0 < x^k < 1$.

To show that this implies that S_{k+1} is true, multiply the three parts of $0 < x^k < 1$ by x.

$$x \cdot 0 < x \cdot x^k < x \cdot 1 \quad \text{Use the fact that } 0 < x.$$

$$0 < x^{k+1} < x \quad \text{Simplify.}$$

Because $x < 1$,

$$x^{k+1} < x < 1 \quad \text{and} \quad 0 < x^{k+1} < 1.$$

By this work, if S_k is true, then S_{k+1} is true. The given statement S_n is true for every positive integer n.

NOW TRY

13.5 Exercises

FOR EXTRA HELP

MyLab Math

Video solutions for select problems available in MyLab Math

Concept Check *Write out S_4 for each of the following, and determine whether it is* true *or* false.

1. S_n: $3 + 6 + 9 + \cdots + 3n = \frac{3n(n+1)}{2}$

2. S_n: $1^2 + 2^2 + 3^2 + \cdots + n^2 = \frac{n(n+1)(2n+1)}{6}$

3. S_n: $\frac{1}{2} + \frac{1}{2^2} + \frac{1}{2^3} + \cdots + \frac{1}{2^n} = \frac{2^n - 1}{2^n}$

4. S_n: $6 + 12 + 18 + \cdots + 6n = 3n^2 + 3n$

5. S_n: $2^n < 2n$

6. S_n: $n! > 6n$

7. Concept Check A proof by mathematical induction allows us to prove that a statement is true for all (*positive/negative*) ______.

8. Write out in full and verify the statements S_1, S_2, S_3, S_4, and S_5 for the following statement.

$$2 + 4 + 6 + \cdots + 2n = n(n+1)$$

Then use mathematical induction to prove that the statement is true for every positive integer value of n.

Use mathematical induction to prove that each statement is true for every positive integer value of n. ***See Examples 1–3.***

9. $3 + 6 + 9 + \cdots + 3n = \frac{3n(n+1)}{2}$

10. $5 + 10 + 15 + \cdots + 5n = \frac{5n(n+1)}{2}$

11. $1 + 3 + 5 + \cdots + (2n - 1) = n^2$

12. $2 + 4 + 8 + \cdots + 2^n = 2^{n+1} - 2$

13. $1^3 + 2^3 + 3^3 + \cdots + n^3 = \dfrac{n^2(n+1)^2}{4}$

14. $1^2 + 2^2 + 3^2 + \cdots + n^2 = \dfrac{n(n+1)(2n+1)}{6}$

15. $7 \cdot 8 + 7 \cdot 8^2 + 7 \cdot 8^3 + \cdots + 7 \cdot 8^n = 8(8^n - 1)$

16. $5 \cdot 6 + 5 \cdot 6^2 + 5 \cdot 6^3 + \cdots + 5 \cdot 6^n = 6(6^n - 1)$

17. $\dfrac{1}{1 \cdot 2} + \dfrac{1}{2 \cdot 3} + \dfrac{1}{3 \cdot 4} + \cdots + \dfrac{1}{n(n+1)} = \dfrac{n}{n+1}$

18. $\dfrac{1}{1 \cdot 4} + \dfrac{1}{4 \cdot 7} + \dfrac{1}{7 \cdot 10} + \cdots + \dfrac{1}{(3n-2)(3n+1)} = \dfrac{n}{3n+1}$

19. $x^{2n} + x^{2n-1}y + \cdots + xy^{2n-1} + y^{2n} = \dfrac{x^{2n+1} - y^{2n+1}}{x - y}$

20. $x^{2n-1} + x^{2n-2}y + \cdots + xy^{2n-2} + y^{2n-1} = \dfrac{x^{2n} - y^{2n}}{x - y}$

21. $\dfrac{4}{5} + \dfrac{4}{5^2} + \dfrac{4}{5^3} + \cdots + \dfrac{4}{5^n} = 1 - \dfrac{1}{5^n}$

22. $\dfrac{1}{2} + \dfrac{1}{2^2} + \dfrac{1}{2^3} + \cdots + \dfrac{1}{2^n} = 1 - \dfrac{1}{2^n}$

23. $(a^m)^n = a^{mn}$
(Assume that a and m are constant.)

24. $(ab)^n = a^n b^n$
(Assume that a and b are constant.)

25. If $a > 1$, then $a^n > 1$.

26. If $a > 1$, then $a^n > a^{n-1}$.

27. If $0 < a < 1$, then $a^n < a^{n-1}$.

28. $3^n > n$

Extending Skills *Work each problem.*

29. Suppose that n straight lines (with $n \geq 2$) are drawn in a plane, where no two lines are parallel and no three lines pass through the same point. Show that the number of points of intersection of the lines is $\dfrac{n^2 - n}{2}$.

30. The series of sketches below starts with an equilateral triangle having sides of length 1. In the following steps, equilateral triangles are constructed on each side of the preceding figure. The lengths of the sides of these new triangles are $\frac{1}{3}$ the length of the sides of the preceding triangles. Develop a formula for the number of sides of the nth figure. Use mathematical induction to prove your answer.

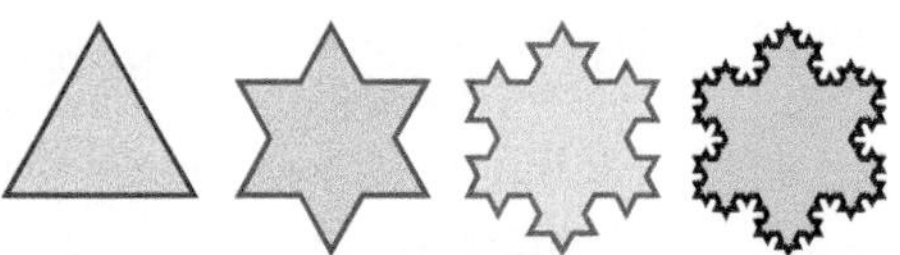

31. Find the perimeter of the nth figure in **Exercise 30.**

32. Show that the area of the nth figure in **Exercise 30** is

$$\sqrt{3}\left[\frac{2}{5} - \frac{3}{20}\left(\frac{4}{9}\right)^{n-1}\right].$$

13.6 Counting Theory

OBJECTIVES

1 Use the fundamental principle of counting.
2 Learn the formula ${}_nP_r$ for permutations.
3 Use the permutations formula to solve counting problems.
4 Review the formula ${}_nC_r$ for combinations.
5 Use the combinations formula to solve counting problems.
6 Distinguish between permutations and combinations.

OBJECTIVE 1 Use the fundamental principle of counting.

If there are 3 roads from Albany to Baker and 2 roads from Baker to Creswich, in how many ways can one travel from Albany to Creswich by way of Baker?

For each of the 3 roads from Albany to Baker, there are 2 different roads from Baker to Creswich. There are

$$3 \cdot 2 = 6 \text{ different ways}$$

to make the trip, as shown in the **tree diagram** in FIGURE 6 at the right.

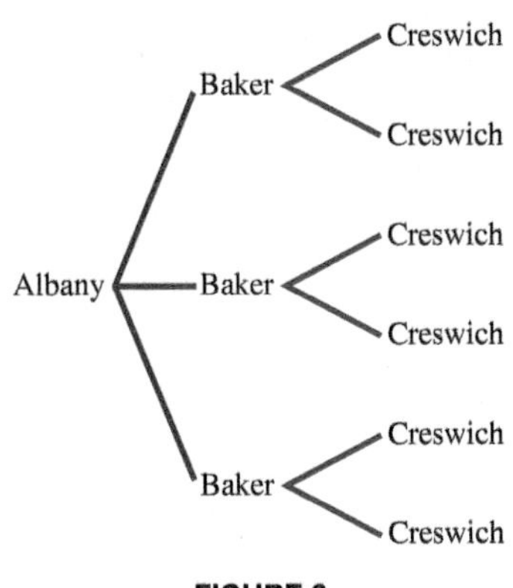

FIGURE 6

This example from *counting theory* illustrates the following property.

Fundamental Principle of Counting

If one event can occur in m ways and a second event can occur in n ways, then both events can occur in mn ways, provided that the outcome of the first event does not influence the outcome of the second.

The fundamental principle of counting can be extended to any number of events, provided that the outcome of no one event influences the outcome of another. Such events are **independent events.**

VOCABULARY

- ☐ tree diagram
- ☐ independent events
- ☐ permutation
- ☐ combination

EXAMPLE 1 Using the Fundamental Principle of Counting

A restaurant offers a choice of 3 salads, 5 main dishes, and 2 desserts. Use the fundamental principle of counting to find the number of different 3-course meals that can be selected.

Three independent events are involved: selecting a salad, selecting a main dish, and selecting a dessert. The first event can occur in 3 ways, the second event can occur in 5 ways, and the third event can occur in 2 ways.

$$3 \cdot 5 \cdot 2 = 30 \text{ possible meals}$$

NOW TRY 

NOW TRY EXERCISE 1

A wireless cell phone company offers 21 cell phone styles, 5 messaging packages, and 4 data services. How many different options are there for choosing a phone along with a messaging package and a data service?

EXAMPLE 2 Using the Fundamental Principle of Counting

Eli has 5 different books that he wishes to arrange on his desk. How many different arrangements are possible?

Five events are involved: selecting a book for the first spot, selecting a book for the second spot, and so on. Here the outcome of the first event *does* influence the outcome of the other events (because one book has already been chosen). For the first spot Eli has 5 choices, for the second spot 4 choices, for the third spot 3 choices, and so on. We use the fundamental principle of counting.

$$5 \cdot 4 \cdot 3 \cdot 2 \cdot 1 = 120 \text{ different arrangements}$$

NOW TRY

NOW TRY EXERCISE 2

A store manager wishes to display 7 brands of toothpaste in a row. In how many ways can this be done?

NOW TRY ANSWERS

1. 420
2. 5040

When using the fundamental principle of counting, we encounter products such as

$$5 \cdot 4 \cdot 3 \cdot 2 \cdot 1. \quad \text{See Example 2.}$$

To write these products, we use the symbol $n!$ (read "*n factorial*").

n Factorial (n!)

For any positive integer n,

$$n! = n(n-1)(n-2)(n-3)\cdots(3)(2)(1).$$

Example: $5 \cdot 4 \cdot 3 \cdot 2 \cdot 1$ is written as $5!$.

The definition of $n!$ means that $n[(n-1)!] = n!$ for all positive integers $n \geq 2$. This relationship also holds for $n = 1$, so, by definition, $\mathbf{0! = 1}$.

NOW TRY EXERCISE 3
An instructor wishes to select 4 of 7 exercises to present to students in a lecture. In how many ways can she arrange the 4 selected exercises?

EXAMPLE 3 Arranging r of n Items $(r < n)$

Suppose that Eli (from **Example 2**) wishes to place only 3 of the 5 books on his desk. How many arrangements of 3 books are possible?

He still has 5 ways to fill the first spot, 4 ways to fill the second spot, and 3 ways to fill the third. He wants to use 3 books, so there are only 3 spots to be filled (3 events) instead of 5. Again, we use the fundamental principle of counting.

$$5 \cdot 4 \cdot 3 = 60 \text{ arrangements}$$

NOW TRY 

OBJECTIVE 2 Learn the formula $_nP_r$ for permutations.

In **Example 3,** 60 is the number of *permutations* of 5 things taken 3 at a time, written

$$_5P_3 = 60.$$

A **permutation** of n elements taken r at a time is one of the *arrangements* of r elements taken from a set of n elements $(r \leq n)$. Generalizing, the number of permutations of n elements taken r at a time, denoted by $_nP_r$, is given as follows.

$$_nP_r = n(n-1)(n-2)\cdots(n-r+1)$$

$$_nP_r = \frac{n(n-1)(n-2)\cdots(n-r+1)(n-r)(n-r-1)\cdots(2)(1)}{(n-r)(n-r-1)\cdots(2)(1)}$$

$$_nP_r = \frac{n!}{(n-r)!}$$

This proves the following result.

Permutations of n Elements Taken r at a Time

If $_nP_r$ represents the number of **permutations** of n elements taken r at a time, for $r \leq n$, then the following holds true.

$$_nP_r = \frac{n!}{(n-r)!}$$

Alternative notations for $_nP_r$ are $P(n, r)$ and P_r^n.

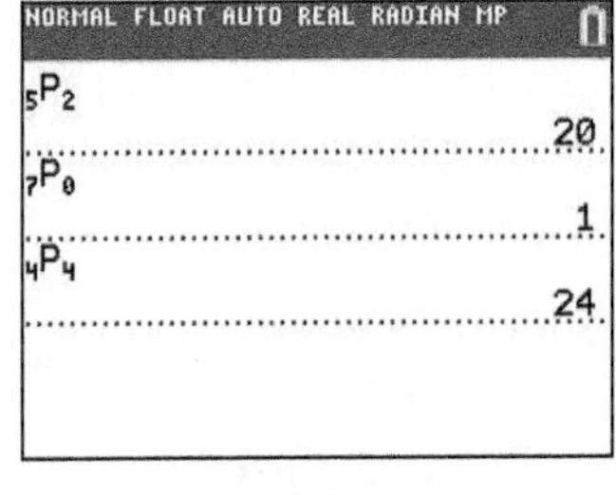

FIGURE 7

NOW TRY ANSWER
3. 840

FIGURE 7 shows how a calculator evaluates $_5P_2$, $_7P_0$, and $_4P_4$.

OBJECTIVE 3 Use the permutations formula to solve counting problems.

NOW TRY EXERCISE 4
There are 7 dance crews in a competition. In how many ways can the judges choose first, second, and third place winners?

EXAMPLE 4 Using the Permutations Formula

Suppose 8 people enter an event in a swim meet. Assuming that there are no ties, in how many ways could the gold, silver, and bronze medals be awarded?

Using the fundamental principle of counting, there are 3 choices to be made, giving

$$8 \cdot 7 \cdot 6 = 336.$$

However, we can also use the formula for ${}_nP_r$ and obtain the same result.

$${}_nP_r = \frac{n!}{(n-r)!} \qquad \text{Permutations formula}$$

$${}_8P_3 = \frac{8!}{(8-3)!} \qquad \text{Let } n = 8 \text{ and } r = 3.$$

$${}_8P_3 = \frac{8!}{5!} \qquad \text{Subtract in the denominator.}$$

$${}_8P_3 = \frac{8 \cdot 7 \cdot 6 \cdot 5 \cdot 4 \cdot 3 \cdot 2 \cdot 1}{5 \cdot 4 \cdot 3 \cdot 2 \cdot 1} \qquad \text{Definition of } n!$$

$${}_8P_3 = 8 \cdot 7 \cdot 6 \qquad \text{Divide out the common factors.}$$

$${}_8P_3 = 336 \text{ ways} \qquad \text{Multiply.}$$

NOW TRY

NOW TRY EXERCISE 5
There are 10 players on a women's slow-pitch softball team. How many possible batting lineups are there for these 10 players?

EXAMPLE 5 Using the Permutations Formula

In how many ways can 6 students be seated in a row of 6 desks?

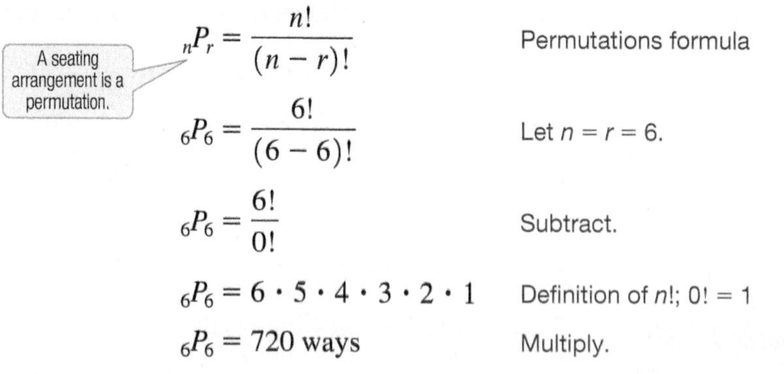
A seating arrangement is a permutation.

$${}_nP_r = \frac{n!}{(n-r)!} \qquad \text{Permutations formula}$$

$${}_6P_6 = \frac{6!}{(6-6)!} \qquad \text{Let } n = r = 6.$$

$${}_6P_6 = \frac{6!}{0!} \qquad \text{Subtract.}$$

$${}_6P_6 = 6 \cdot 5 \cdot 4 \cdot 3 \cdot 2 \cdot 1 \qquad \text{Definition of } n!;\ 0! = 1$$

$${}_6P_6 = 720 \text{ ways} \qquad \text{Multiply.}$$

NOW TRY

NOW TRY EXERCISE 6
A company provides employee identification numbers that consist of 3 letters followed by 2 digits. How many identification numbers are possible if the first letter must be an S or F, and repetitions are allowed?

EXAMPLE 6 Using the Fundamental Principle of Counting with Restrictions

In how many ways can three letters of the alphabet be arranged if a vowel cannot be used in the middle position, and repetitions of the letters are allowed?

We cannot use ${}_{26}P_3$ here, because of the restriction on the middle position, and because repetition is allowed. In the first and third positions, we can use any of the 26 letters of the alphabet. In the middle position, we can use only one of $26 - 5 = 21$ letters (because there are 5 vowels).

$$26 \cdot 21 \cdot 26 = 14{,}196 \text{ ways} \qquad \text{Use the fundamental principle of counting.}$$

NOW TRY

NOW TRY ANSWERS
4. 210
5. 3,628,800
6. 135,200

OBJECTIVE 4 Review the formula $_nC_r$ for combinations.

We have discussed a method for finding the number of ways to arrange r elements taken from a set of n elements. Sometimes, however, the arrangement (or order) of the elements is not important. The discussion that follows leads to the same results seen with binomial coefficients studied earlier in the chapter.

For example, suppose three people (Ms. Opelka, Mr. Adams, and Ms. Jacobs) apply for 2 identical jobs. *Ignoring all other factors, in how many ways can 2 people be selected from the 3 applicants?* Here the arrangement or order of the people is unimportant. Selecting Ms. Opelka and Mr. Adams is the same as selecting Mr. Adams and Ms. Opelka. Therefore, there are only 3 ways to select 2 of the 3 applicants.

Ms. Opelka and Mr. Adams
Ms. Opelka and Ms. Jacobs
Mr. Adams and Ms. Jacobs

Ways to select 2 applicants from a pool of 3

These three choices are *combinations* of 3 elements taken 2 at a time. A **combination** of n elements taken r at a time is one of the ways in which r elements can be chosen from n elements. Each combination of r elements forms $r!$ permutations. Therefore, the number of combinations of n elements taken r at a time is found by dividing the number of permutations $_nP_r$ by $r!$ to obtain

$$\frac{_nP_r}{r!} \text{ combinations.}$$

This expression can be rewritten as follows.

$$\frac{_nP_r}{r!} = \frac{\frac{n!}{(n-r)!}}{r!} = \frac{n!}{(n-r)!} \cdot \frac{1}{r!} = \frac{n!}{r!(n-r)!}$$

Combinations of n Elements Taken r at a Time

If $_nC_r$ represents the number of **combinations** of n elements taken r at a time, for $r \le n$, then the following holds true.

$$_nC_r = \frac{n!}{r!(n-r)!}$$

Alternative notations for $_nC_r$ are $C(n, r)$, C_r^n, and $\binom{n}{r}$.

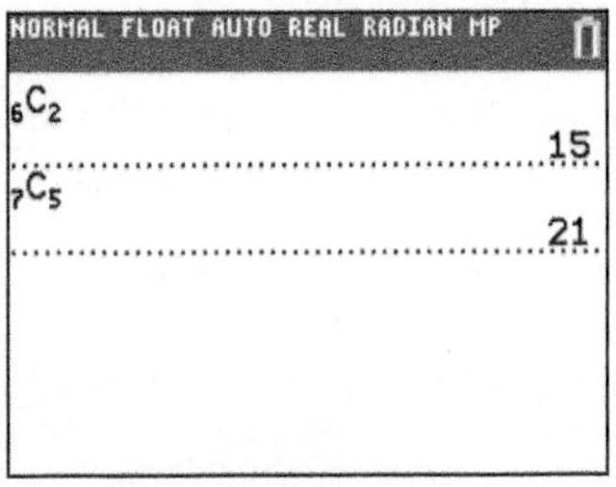

NORMAL FLOAT AUTO REAL RADIAN MP
$_6C_2$ 15
$_7C_5$ 21

FIGURE 8

FIGURE 8 shows how a calculator evaluates $_6C_2$ and $_7C_5$.

OBJECTIVE 5 Use the combinations formula to solve counting problems.

In the preceding discussion on job applications, it was shown that $_3C_2 = 3$. This can also be found using the combinations formula.

$$_nC_r = \frac{n!}{r!(n-r)!} \qquad \text{Combinations formula}$$

$$_3C_2 = \frac{3!}{2!(3-2)!} \qquad \text{Let } n = 3 \text{ and } r = 2.$$

$$_3C_2 = \frac{3 \cdot 2 \cdot 1}{2 \cdot 1 \cdot 1} \qquad \text{Definition of } n!$$

$$_3C_2 = 3 \text{ ways} \qquad \text{Divide out the common factors.}$$

NOW TRY EXERCISE 7

In how many ways can a committee of 4 be selected from a group of 10 people?

EXAMPLE 7 Using the Combinations Formula

How many different committees of 3 people can be chosen from a group of 8 people?

Because the order in which the members of the committee are chosen does not affect the result, use combinations.

$${}_nC_r = \frac{n!}{r!(n-r)!} \qquad \text{Combinations formula}$$

$${}_8C_3 = \frac{8!}{3!(8-3)!} \qquad \text{Let } n = 8 \text{ and } r = 3.$$

$${}_8C_3 = \frac{8!}{3!5!} \qquad \text{Subtract in the denominator.}$$

$${}_8C_3 = \frac{8 \cdot 7 \cdot 6 \cdot 5 \cdot 4 \cdot 3 \cdot 2 \cdot 1}{3 \cdot 2 \cdot 1 \cdot 5 \cdot 4 \cdot 3 \cdot 2 \cdot 1} \qquad \text{Definition of } n!$$

$${}_8C_3 = 56 \text{ committees} \qquad \text{Divide out the common factors. Multiply.}$$

NOW TRY EXERCISE 8

A committee of 4 faculty members must be selected from a department of 12 members.

(a) In how many ways can the committee be selected?

(b) In how many ways can the committee be selected if the department chair, who is one of the 12 faculty members, must serve on the committee?

EXAMPLE 8 Using the Combinations Formula

From a group of 30 bank employees, 3 are to be selected to work on a special project.

(a) In how many different ways can the employees be selected?

The number of 3-element combinations from a set of 30 elements must be found. (Use combinations because order within the group of 3 does not affect the result.)

$${}_{30}C_3 = \frac{30!}{3!(30-3)!} = \frac{30!}{3!27!} = 4060 \text{ ways} \qquad \text{Use a calculator.}$$

(b) In how many different ways can the group of 3 be selected if it has already been decided that a certain employee must work on the project?

One employee has already been selected to work on the project, so the problem is reduced to selecting 2 more employees from the 29 employees that are left.

$${}_{29}C_2 = \frac{29!}{2!(29-2)!} = \frac{29!}{2!27!} = 406 \text{ ways} \qquad \text{Use a calculator.}$$

NOW TRY EXERCISE 9

A student organization consists of 5 freshmen and 8 sophomores. A group of 4 students is selected to serve on an advisory panel. In how many ways can this advisory panel be selected if it must include exactly 3 sophomores?

EXAMPLE 9 Using Combinations

A congressional committee consists of 4 senators and 6 representatives. A delegation of 5 members is to be chosen. In how many ways could this delegation include exactly 3 senators?

"Exactly 3 senators" implies that there must be $5 - 3 = 2$ representatives as well. The 3 senators could be chosen in

$${}_4C_3 = 4 \text{ ways.}$$

The 2 representatives could be chosen in

$${}_6C_2 = 15 \text{ ways.}$$

We apply the fundamental principle of counting.

$$4 \cdot 15 = 60 \text{ ways}$$

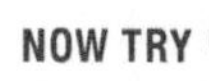

NOW TRY ANSWERS

7. 210
8. (a) 495 (b) 165
9. 280

OBJECTIVE 6 Distinguish between permutations and combinations.

Consider the following table, which can be used to determine whether permutations or combinations should be used in an application.

▼ **Similarities and Differences between Permutations and Combinations**

Permutations	*Combinations*
Number of ways of selecting r items out of n items Repetitions are not allowed.	
Order is important.	Order is not important.
They are *arrangements* of r items from a set of n items.	They are *subsets* of r items from a set of n items.
${}_nP_r = \frac{n!}{(n-r)!}$	${}_nC_r = \frac{n!}{r!(n-r)!}$
Clue words: *arrangement, schedule, order*	Clue words: *group, committee, sample, selection*

NOW TRY EXERCISE 10

Determine whether *permutations* or *combinations* should be used to solve each problem.

(a) In how many different ways can a security guard visit 5 buildings on a campus?

(b) In how many ways can a jury of 6 men and 6 women be selected from 8 men and 10 women?

(c) In a board of directors of 8 people, how many ways can a president, secretary, and treasurer be selected?

EXAMPLE 10 Distinguishing between Permutations and Combinations

Determine whether *permutations* or *combinations* should be used to solve each problem.

(a) How many 4-digit codes are possible if no digits are repeated?

Changing the order of the 4 digits results in a different code, so *permutations* should be used.

(b) A sample of 3 light bulbs is randomly selected from a batch of 15 bulbs. How many different samples are possible?

The order in which the 3 light bulbs are selected is not important. The sample is unchanged if the items are rearranged, so *combinations* should be used.

(c) In a basketball tournament with 8 teams, how many games must be played so that each team plays every other team exactly once?

Selection of 2 teams for a game is an *unordered* subset of 2 from the set of 8 teams. Use *combinations.*

(d) In how many ways can 4 stockbrokers be assigned to 6 offices so that each broker has a private office?

The office assignments are an *ordered* selection of 4 offices from the 6 offices. Exchanging the offices of any 2 stockbrokers within a selection of 4 offices gives a different assignment, so *permutations* should be used.

NOW TRY

NOW TRY ANSWERS

10. (a) permutations
(b) combinations
(c) permutations

13.6 Exercises

MyLab Math

Video solutions for select problems available in MyLab Math

Concept Check *Fill in each blank with the correct response.*

1. From the two choices *permutation* and *combination,* a computer password is an example of a ______, and a hand of cards is an example of a ______.

2. If there are 3 ways to choose a salad, 5 ways to choose an entrée, and 4 ways to choose a dessert, then there are ______ ways to form a meal consisting of these three choices.

3. There are ______ ways to form a three-digit number consisting of the digits 4, 5, and 9.

4. If there are 3 people to choose from, there are ______ ways to choose a pair of them.

5. **Concept Check** Padlocks with digit dials are often referred to as "combination locks." According to the mathematical definition of combination, is this an accurate description? Why or why not?

6. **Concept Check** What is the difference between a permutation and a combination? What should we look for in a problem to decide which of these is an appropriate solution?

Evaluate each expression.

7. $_6P_4$
8. $_7P_5$
9. $_9P_2$
10. $_6P_5$
11. $_5P_0$
12. $_6P_0$
13. $_4C_2$
14. $_9C_3$
15. $_6C_1$
16. $_8C_1$
17. $_3C_0$
18. $_9C_0$

Use the fundamental principle of counting or permutations to solve each problem. ***See Examples 1–6.***

19. How many different types of homes are available if a builder offers a choice of 6 basic plans, 4 roof styles, and 2 exterior finishes?

20. A menu offers a choice of 4 salads, 8 main dishes, and 5 desserts. How many different 3-course meals (salad, main dish, dessert) are possible?

21. In an experiment on social interaction, 8 people will sit in 8 seats in a row. In how many different ways can the 8 people be seated?

22. In how many ways can 7 of 10 mice be arranged in a row for a genetics experiment?

23. For many years, the state of California used 3 letters followed by 3 digits on its automobile license plates.

(a) How many different license plates are possible with this arrangement?

(b) When the state ran out of new plates, the order was reversed to 3 digits followed by 3 letters. How many additional plates were then possible?

(c) Eventually, the plates described in part (b) were also used up. The state then issued plates with 1 letter followed by 3 digits and then 3 letters. How many plates does this scheme provide?

24. How many 7-digit telephone numbers are possible if the first digit cannot be 0, and

(a) only odd digits may be used?

(b) the telephone number must be a multiple of 10 (that is, it must end in 0)?

(c) the first three digits must be 456?

25. If a college offers 400 courses, 20 of which are in mathematics, and a counselor arranges a schedule of 4 courses by random selection, how many schedules are possible that do not include a mathematics course?

26. In how many ways can 5 players be assigned to the 5 positions on a basketball team, assuming that any player can play any position? In how many ways can 10 players be assigned to the 5 positions?

27. In a sales force of 35 people, how many ways can 3 salespeople be selected for 3 different leadership jobs?

28. In how many ways can 6 bank tellers be assigned to 6 different windows? In how many ways can 10 tellers be assigned to the 6 windows?

Use combinations to solve each problem. ***See Examples 7–9.***

29. A professional stockbrokers' association has 50 members. If a committee of 6 is to be selected at random, how many different committees are possible?

30. A group of 5 financial planners is to be selected at random from a professional organization with 30 members to participate in a seminar. In how many ways can this be done? In how many ways can the group that will not participate be selected?

31. Harry's Hamburger Heaven sells hamburgers with cheese, relish, lettuce, tomato, onion, mustard, and/or ketchup. How many different hamburgers can be concocted using any 4 of the extras?

32. At a soda machine, there are 7 different soft drink choices, with additional flavor options of vanilla, cherry, raspberry, lemon, and lime. How many different drinks can be created using 1 soft drink choice and 2 additional flavor options?

33. How many different 5-card poker hands can be dealt from a deck of 52 playing cards?

34. Seven cards are marked with the numbers 1 through 7 and are shuffled, and then 3 cards are drawn. How many different 3-card combinations are possible?

35. A bag contains 18 marbles. How many samples of 3 marbles can be drawn from it? How many samples of 5 marbles?

36. If a bag of 18 marbles contains 5 purple, 4 green, and 9 black marbles, how many samples of 3 can be drawn in which all the marbles are black? How many samples of 3 can be drawn in which exactly 2 marbles are black?

37. How many different samples of 4 light bulbs can be selected from a carton of 2 dozen bulbs?

38. In a carton of 2 dozen light bulbs, 5 are defective. How many samples of 4 can be drawn in which all are defective? How many samples of 4 can be drawn in which there are 2 good bulbs and 2 defective bulbs?

Determine whether each of the following is a permutation *or a* combination. ***See Example 10.***

39. Your 5-digit postal zip code

40. Your 9-digit Social Security number

41. A particular 5-card hand in a poker game

42. A committee of school board members

Solve each problem using any method. ***See Examples 1–10.***

43. From a pool of 7 secretaries, 3 are selected to be assigned to 3 managers, with 1 secretary for each manager. In how many ways can this be done?

44. In a game of musical chairs, 12 children, staying in the same order, circle around 11 chairs. Each child who is next to a chair must sit down when the music stops. (One will be left out.) How many seatings are possible?

45. From 10 names on a ballot, 4 will be elected to a political party committee. How many different committees are possible? In how many ways can the committee of 4 be formed if each person will have a different responsibility?

46. In how many ways can 5 of 9 plants be arranged in a row on a windowsill?

47. Hazel specializes in making different vegetable soups with carrots, celery, onions, beans, peas, tomatoes, and potatoes. How many different soups can she make using any 4 ingredients?

48. How many 4-letter radio-station call letters can be made if the first letter must be K or W, and no letter may be repeated? How many if repeats are allowed? How many of the call letters with no repeats can end in K?

49. In an office with 8 men and 11 women, how many 5-member groups with the following compositions can be chosen for a training session?

(a) All men **(b)** All women

(c) 3 men and 2 women **(d)** No more than 3 women

50. In an experiment on plant hardiness, a researcher gathers 6 wheat plants, 3 barley plants, and 2 rye plants. Four plants are to be selected at random.

(a) In how many ways can this be done?

(b) In how many ways can this be done if exactly 2 wheat plants must be included?

51. A group of 12 workers decides to send a delegation of 3 to their supervisor to discuss their work assignments.

(a) How many delegations of 3 are possible?

(b) How many are possible if one of the 12, the foreman, must be in the delegation?

(c) If there are 5 women and 7 men in the group, how many possible delegations would include exactly 1 woman?

52. The Riverdale board of supervisors is composed of 2 liberals and 5 conservatives. Three members are to be selected randomly as delegates to a convention.

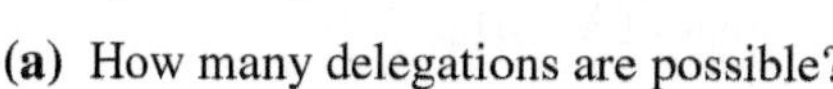

(a) How many delegations are possible?

(b) How many delegations could have all liberals?

(c) How many delegations could have 2 conservatives and 1 liberal?

(d) If the supervisor who serves as chair of the board must be included, how many delegations are possible?

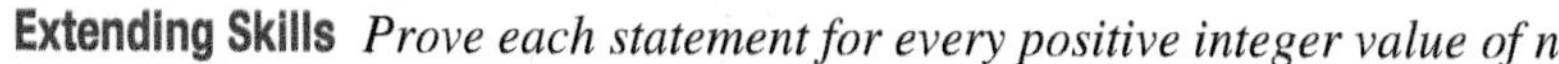

Extending Skills *Prove each statement for every positive integer value of n.*

53. ${}_nP_{n-1} = {}_nP_n$ **54.** ${}_nP_1 = n$ **55.** ${}_nP_0 = 1$

56. ${}_nC_n = 1$ **57.** ${}_nC_0 = 1$ **58.** ${}_nC_{n-1} = n$

13.7 Basics of Probability

OBJECTIVES

1 Learn the terminology of probability theory.
2 Find the probability of an event.
3 Find the probability of the complement of *E*, given the probability of *E*.
4 Find the odds in favor of an event.
5 Find the probability of a compound event.

OBJECTIVE 1 Learn the terminology of probability theory.

In probability, each repetition of an experiment is a **trial.** The possible results of each trial are **outcomes** of the experiment. In this section, we are concerned with outcomes that are equally likely to occur.

For example, the experiment of tossing a coin has two equally likely outcomes:

$$\text{landing heads up } (H) \quad \text{or} \quad \text{landing tails up } (T).$$

Also, the experiment of rolling a fair die has six equally likely outcomes:

landing so the face that is up shows 1, 2, 3, 4, 5, or 6 dots.

The set S of all possible outcomes of a given experiment is the **sample space** for the experiment. (In this section, all sample spaces are finite.) A sample space S can be written in set notation.

$S = \{H, T\}$ Sample space for the experiment of tossing a coin

$S = \{1, 2, 3, 4, 5, 6\}$ Sample space for the experiment of rolling a single die

VOCABULARY

- ☐ trial
- ☐ outcome
- ☐ sample space
- ☐ event
- ☐ probability
- ☐ certain event
- ☐ impossible event
- ☐ complement
- ☐ Venn diagram
- ☐ odds
- ☐ compound event
- ☐ mutually exclusive events

Any subset of the sample space is an **event.** In the experiment with the die, for example, "the number showing is a three" is an event, say E_1, such that $E_1 = \{3\}$. "The number showing is greater than three" is also an event, say E_2, such that $E_2 = \{4, 5, 6\}$. To represent the number of outcomes that belong to event E, the notation $\boldsymbol{n(E)}$ is used. Then

$$n(E_1) = 1 \quad \text{and} \quad n(E_2) = 3.$$

OBJECTIVE 2 Find the probability of an event.

The notation $\boldsymbol{P(E)}$ is used to designate the *probability* of event E.

Probability of Event E

In a sample space with equally likely outcomes, the **probability** of an event E, written $P(E)$, is the ratio of the number of outcomes in sample space S that belong to event E, $n(E)$, to the total number of outcomes in sample space S, $n(S)$.

$$P(E) = \frac{n(E)}{n(S)}$$

NOW TRY EXERCISE 1

A single die is rolled. Write the following events in set notation, and give the probability for each event.

(a) E_7: the number showing is less than 4

(b) E_8: the number showing is greater than 0

EXAMPLE 1 Finding Probabilities of Events

A single die is rolled. Write the following events in set notation, and give the probability for each event.

(a) E_3: the number showing is even

Because event $E_3 = \{2, 4, 6\}$, $n(E_3) = 3$. Also, because $S = \{1, 2, 3, 4, 5, 6\}$, $n(S) = 6$.

$$P(E_3) = \frac{n(E_3)}{n(S)} = \frac{3}{6} = \frac{1}{2}$$

(b) E_4: the number showing is greater than 4

Again, $n(S) = 6$. Event $E_4 = \{5, 6\}$, with $n(E_4) = 2$.

$$P(E_4) = \frac{n(E_4)}{n(S)} = \frac{2}{6} = \frac{1}{3}$$

(c) E_5: the number showing is less than 7

$$E_5 = \{1, 2, 3, 4, 5, 6\} \quad \text{and} \quad P(E_5) = \frac{6}{6} = 1$$

(d) E_6: the number showing is 7

$$E_6 = \varnothing \quad \text{and} \quad P(E_6) = \frac{0}{6} = 0$$

NOW TRY 

NOW TRY ANSWERS

1. (a) $E_7 = \{1, 2, 3\}$; $P(E_7) = \frac{1}{2}$
   (b) $E_8 = \{1, 2, 3, 4, 5, 6\}$; $P(E_8) = 1$

In **Example 1(c),** $E_5 = S$. Therefore, the event E_5 is *certain* to occur every time the experiment is performed. On the other hand, in **Example 1(d),** $E_6 = \varnothing$ and $P(E_6) = 0$, so E_6 is *impossible.*

Probability Values and Terminology

- A **certain event**—that is, an event that is certain to occur—always has probability 1.
- The probability of an **impossible event** is always 0, because none of the outcomes in the sample space satisfy the event.
- For any event E, **$P(E)$ is between 0 and 1 inclusive of both.**

OBJECTIVE 3 Find the probability of the complement of E, given the probability of E.

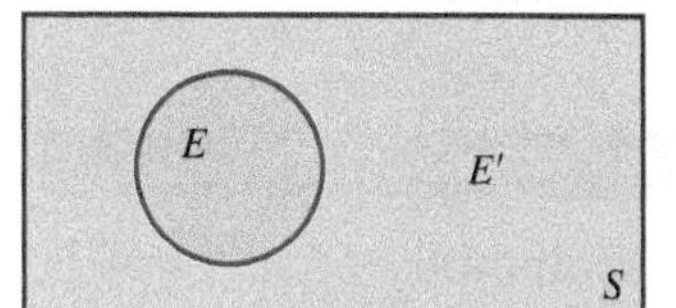

FIGURE 9

The set of all outcomes in the sample space S that do *not* belong to event E is the **complement** of E, written E'.

For example, in the experiment of drawing a single card from a standard deck of 52 cards, let E be the event "the card is an ace." Then E' is the event "the card is not an ace." From the definition of E', for any event E,

$$E \cup E' = S \quad \text{and} \quad E \cap E' = \varnothing.$$

Probability concepts can be illustrated using **Venn diagrams,** as shown in FIGURE 9. The rectangle in FIGURE 9 represents the sample space S in an experiment. The area inside the circle represents event E, and the area inside the rectangle, but outside the circle, represents event E'.

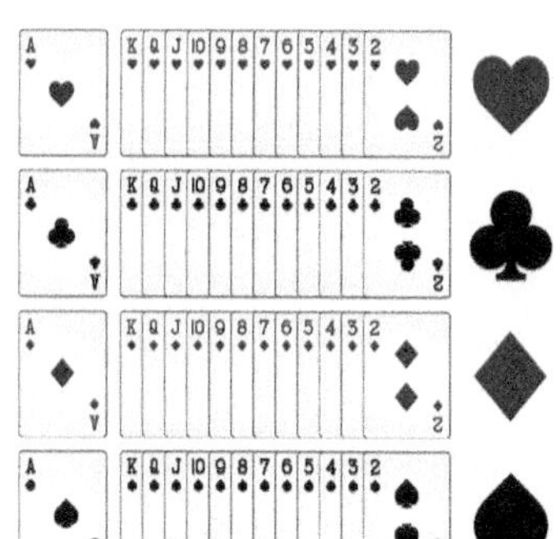
Standard deck of 52 cards

NOTE A standard deck of 52 cards, pictured in the margin, has four suits:

hearts ♥, clubs ♣, diamonds ♦, and spades ♠.

There are 13 cards in each suit, including a jack, a queen, and a king (sometimes called the "face cards"), an ace, and cards numbered from 2 to 10. The hearts and diamonds are red, and the spades and clubs are black. We refer to this standard deck of cards in this section.

NOW TRY EXERCISE 2

In the experiment of drawing a card from a well-shuffled standard deck, find the probability of event E, "the card is a red card," and of event E'.

EXAMPLE 2 Using the Complement in a Probability Problem

In the experiment of drawing a card from a well-shuffled standard deck, find the probability of event E, "the card is an ace," and of event E'.

There are four aces in a standard deck of 52 cards, so $n(E) = 4$ and $n(S) = 52$.

$$P(E) = \frac{n(E)}{n(S)} = \frac{4}{52} = \frac{1}{13}$$

Of the 52 cards, 48 are not aces, so $n(E') = 48$.

$$P(E') = \frac{n(E')}{n(S)} = \frac{48}{52} = \frac{12}{13} \quad \text{Lowest terms}$$

In **Example 2,** $P(E) + P(E') = \frac{1}{13} + \frac{12}{13} = 1$. This is always true for any event E and its complement E'.

NOW TRY ANSWER

2. $P(E) = \frac{1}{2}$; $P(E') = \frac{1}{2}$

Properties of $P(E)$ and $P(E')$

For any event E, the following holds true.

$$P(E) + P(E') = 1$$

This can be restated as

$$P(E) = 1 - P(E') \quad \text{or} \quad P(E') = 1 - P(E).$$

The last two equations suggest an alternative way to compute the probability of an event. For example, if it is known that $P(E) = \frac{1}{10}$, then $P(E') = 1 - \frac{1}{10} = \frac{9}{10}$.

OBJECTIVE 4 Find the odds in favor of an event.

Sometimes probability statements are expressed in terms of *odds,* a comparison of $P(E)$ with $P(E')$. The **odds** *in favor of* an event E are expressed as

$$\textbf{the ratio of } P(E) \textbf{ to } P(E'), \quad \text{or as the fraction} \quad \frac{P(E)}{P(E')}.$$

For example, if the probability of rain can be established as $\frac{1}{3}$, then the odds that it will rain are

$$P(\text{rain}) \text{ to } P(\text{no rain}) = \frac{1}{3} \text{ to } \frac{2}{3} = \frac{\frac{1}{3}}{\frac{2}{3}} = \frac{1}{3} \div \frac{2}{3} = \frac{1}{3} \cdot \frac{3}{2} = \frac{1}{2}, \quad \text{or} \quad 1 \text{ to } 2.$$

On the other hand, the odds *against* rain are $\frac{2}{3}$ to $\frac{1}{3}$ (or 2 to 1). If the odds in favor of an event are, say, 3 to 5, then the probability of the event is $\frac{3}{8}$, and the probability of the complement of the event is $\frac{5}{8}$.

Probability and Odds

If the odds favoring event E are m to n, then the following hold true.

$$P(E) = \frac{m}{m+n} \quad \text{and} \quad P(E') = \frac{n}{m+n}$$

EXAMPLE 3 Finding Odds in Favor of or against an Event

Solve each problem involving odds.

(a) A manager is to be selected at random from 6 sales managers and 4 office managers. Find the odds in favor of a sales manager being selected.

Let E represent the event "a sales manager is selected." Then

$$P(E) = \frac{6}{10} = \frac{3}{5} \quad \text{and} \quad P(E') = 1 - \frac{3}{5} = \frac{2}{5}.$$

Therefore, the odds in favor of a sales manager being selected are

$$P(E) \text{ to } P(E') = \frac{3}{5} \text{ to } \frac{2}{5} = \frac{\frac{3}{5}}{\frac{2}{5}} = \frac{3}{2}, \quad \text{or} \quad 3 \text{ to } 2.$$

Solve each problem involving odds.

(a) Find the odds in favor of rolling a die and getting a 4.

(b) If the probability that 3-D Design will win his next horserace is 0.2, find the odds against 3-D Design winning his next race.

(b) Suppose that in a recent year, the probability that corporate stock was owned by a pension fund was 0.227. Find the odds that year against a corporate stock being owned by a pension fund.

Let E represent the event "corporate stock is owned by a pension fund." Then

$$P(E) = 0.227 \quad \text{and} \quad P(E') = 1 - 0.227 = 0.773.$$

Because $\quad \dfrac{P(E')}{P(E)} = \dfrac{0.773}{0.227} \approx 3.4, \quad$ Evaluate with a calculator.

the odds against a corporate stock being owned by a pension fund were about

3.4 to 1, or 34 to 10, or 17 to 5. NOW TRY 

OBJECTIVE 5 Find the probability of a compound event.

A **compound event** involves an *alternative,* such as "E or F," where E and F are events.

For example, in the experiment of rolling a die, suppose H is the event "the result is a 3," and K is the event "the result is an even number." *What is the probability of the compound event "the result is a 3 or an even number"?* Here,

$$H = \{3\} \qquad K = \{2, 4, 6\} \qquad H \text{ or } K = \{2, 3, 4, 6\}$$

$$P(H) = \frac{1}{6} \qquad P(K) = \frac{3}{6} = \frac{1}{2} \qquad P(H \text{ or } K) = \frac{4}{6} = \frac{2}{3}.$$

Notice that $P(H) + P(K) = P(H \text{ or } K)$.

Before assuming that this relationship is true in general, consider another event G for this experiment, "the result is a 2."

$$G = \{2\} \qquad K = \{2, 4, 6\} \qquad K \text{ or } G = \{2, 4, 6\}$$

$$P(G) = \frac{1}{6} \qquad P(K) = \frac{3}{6} = \frac{1}{2} \qquad P(K \text{ or } G) = \frac{3}{6} = \frac{1}{2}$$

In this case $P(K) + P(G) \neq P(K \text{ or } G)$.

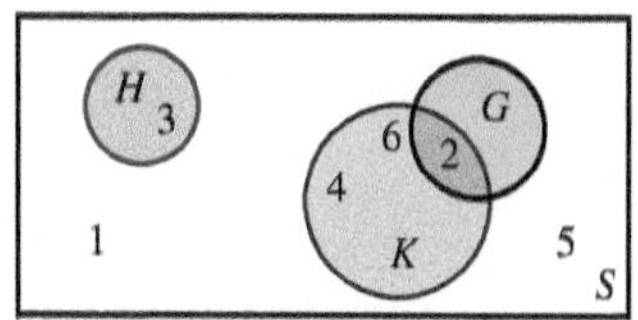

FIGURE 10

As FIGURE 10 shows, the difference in the two examples above comes from the fact that events H and K cannot occur simultaneously. Such events are **mutually exclusive events.** In fact,

$$H \cap K = \varnothing, \quad \text{which is true for any two mutually exclusive events.}$$

Events K and G, however, can occur simultaneously. Both are satisfied if the result of the roll is a 2, the element in their intersection $(K \cap G = \{2\})$.

Probability of a Compound Event

For any events E and F, the following holds true.

$$P(E \text{ or } F) = P(E) + P(F) - P(E \text{ and } F),$$

or

$$P(E \cup F) = P(E) + P(F) - P(E \cap F)$$

NOW TRY ANSWERS

3. (a) 1 to 5 **(b)** 4 to 1

NOW TRY EXERCISE 4

One card is drawn from a well-shuffled standard deck of 52 cards. What is the probability of each compound event?

(a) The card is a red card or a spade.

(b) The card is a red card or a king.

EXAMPLE 4 Finding the Probability of Compound Events

One card is drawn from a well-shuffled standard deck of 52 cards. What is the probability of each compound event?

(a) The card is an ace or a spade.

The events "drawing an ace" and "drawing a spade" are not mutually exclusive because it is possible to draw the ace of spades, an outcome satisfying both events.

$$P(\text{ace or spade}) = P(\text{ace}) + P(\text{spade}) - P(\text{ace and spade})$$ Rule for compound events

There are 4 aces, 13 spades, and 1 ace of spades.

$$= \frac{4}{52} + \frac{13}{52} - \frac{1}{52}$$ Find and substitute each probability.

$$= \frac{16}{52}$$ Add and subtract fractions.

$$= \frac{4}{13}$$ Write in lowest terms.

(b) The card is a 3 or a king.

"Drawing a 3" and "drawing a king" are mutually exclusive events because it is impossible to draw one card that is both a 3 and a king.

$$P(3 \text{ or } K) = P(3) + P(K) - P(3 \text{ and } K)$$ Rule for compound events

$$= \frac{4}{52} + \frac{4}{52} - 0$$ Find and substitute each probability.

$$= \frac{2}{13}$$ Add and subtract. Write in lowest terms: $\frac{8}{52} = \frac{2}{13}$.

NOW TRY

EXAMPLE 5 Finding the Probability of a Compound Event

For the experiment consisting of one roll of a pair of dice, find the probability that the sum of the dots showing is at most 4.

"At most 4" can be rewritten as "2 or 3 or 4." (A sum of 1 is meaningless.)

$$P(\text{at most } 4) = P(2 \text{ or } 3 \text{ or } 4)$$
$$= P(2) + P(3) + P(4) \quad (1)$$

The events represented by "2," "3," and "4" are mutually exclusive.

The sample space includes $6 \cdot 6 = 36$ possible pairs of numbers. See **FIGURE 11**.

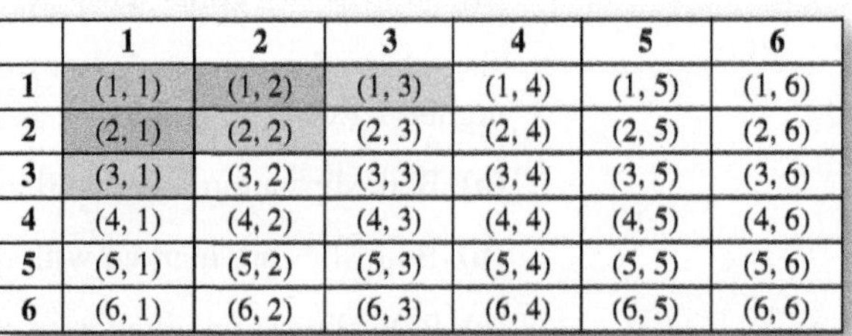

	1	2	3	4	5	6
1	(1, 1)	(1, 2)	(1, 3)	(1, 4)	(1, 5)	(1, 6)
2	(2, 1)	(2, 2)	(2, 3)	(2, 4)	(2, 5)	(2, 6)
3	(3, 1)	(3, 2)	(3, 3)	(3, 4)	(3, 5)	(3, 6)
4	(4, 1)	(4, 2)	(4, 3)	(4, 4)	(4, 5)	(4, 6)
5	(5, 1)	(5, 2)	(5, 3)	(5, 4)	(5, 5)	(5, 6)
6	(6, 1)	(6, 2)	(6, 3)	(6, 4)	(6, 5)	(6, 6)

FIGURE 11

NOW TRY ANSWERS

4. (a) $\frac{3}{4}$ **(b)** $\frac{7}{13}$

NOW TRY EXERCISE 5

For the experiment consisting of one roll of a pair of dice, find the probability that the sum of the dots showing is 7 or 11.

The pair $(1, 1)$ is the only one with a sum of 2, so $P(2) = \frac{1}{36}$. Also $P(3) = \frac{2}{36}$ because both $(1, 2)$ and $(2, 1)$ give a sum of 3. The pairs $(1, 3)$, $(2, 2)$, and $(3, 1)$ have a sum of 4, so $P(4) = \frac{3}{36}$.

$$P(\text{at most } 4) = \frac{1}{36} + \frac{2}{36} + \frac{3}{36} \quad \text{Substitute into equation (1).}$$

$$= \frac{1}{6} \quad \text{Add fractions. Write in lowest terms: } \frac{6}{36} = \frac{1}{6}.$$

NOW TRY 

Summary of Properties of Probability

For any events E and F, the following hold true.

1. $0 \le P(E) \le 1$
2. $P(\text{a certain event}) = 1$
3. $P(\text{an impossible event}) = 0$
4. $P(E') = 1 - P(E)$
5. $P(E \cup F) = P(E) + P(F) - P(E \cap F)$

NOW TRY ANSWER

5. $\frac{2}{9}$

13.7 Exercises

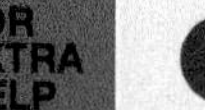
FOR EXTRA HELP

MyLab Math

Video solutions for select problems available in MyLab Math

Concept Check *Write a sample space with equally likely outcomes for each experiment.*

1. Two ordinary coins are tossed.
2. Three ordinary coins are tossed.
3. Five slips of paper marked with the numbers 1, 2, 3, 4, and 5 are placed in a box. After mixing well, two slips are drawn.
4. A die is rolled and then a coin is tossed.
5. **Concept Check** A student gives the answer to a problem requiring a probability as $\frac{6}{5}$. **WHAT WENT WRONG?** Why is this answer incorrect?
6. **Concept Check** If the probability of an event is 0.857, what is the probability that the event will not occur?

Write each event in set notation, and give its probability. ***See Examples 1–5.***

7. Two ordinary coins are tossed.
   (a) Both coins show the same face.
   (b) At least one coin turns up heads.
8. Three ordinary coins are tossed.
   (a) The result of the toss is exactly 2 heads and 1 tail.
   (b) At least two coins show tails.
9. Two slips of paper are drawn from a box containing five slips of paper marked with the numbers 1, 2, 3, 4, and 5.
   (a) Both slips are marked with even numbers.
   (b) Both slips are marked with odd numbers.
   (c) Both slips are marked with the same number.
   (d) One slip is marked with an odd number and the other with an even number.

10. A die is rolled and then a coin is tossed.

(a) The die shows an even number. **(b)** The coin shows heads.

(c) The die shows 6.

(d) The die shows 2 and the coin shows tails.

Solve each problem. ***See Examples 2–5.***

11. A marble is drawn at random from a box containing 3 yellow, 4 white, and 8 blue marbles. Find each probability.

(a) A yellow marble is drawn. **(b)** A blue marble is drawn.

(c) A black marble is drawn. **(d)** A white marble is drawn.

12. A marble is drawn from a bag containing 5 yellow, 6 white, and 10 blue marbles.

(a) What are the odds in favor of drawing a yellow marble?

(b) What are the odds against drawing a blue marble?

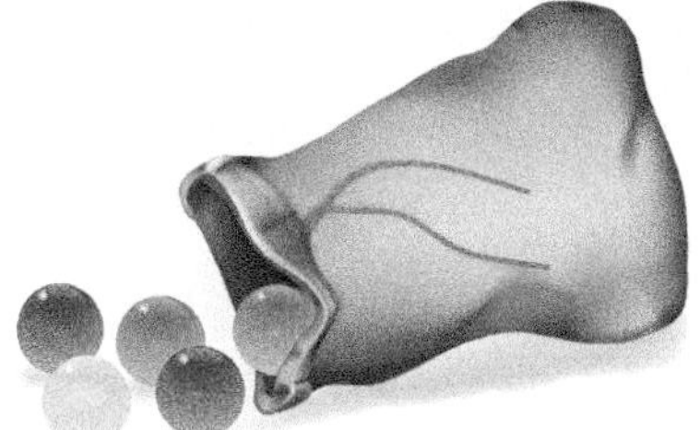

13. A baseball player with a batting average of .300 comes to bat. What are the odds in favor of his getting a hit?

14. Suppose that the probability that a bank with assets greater than or equal to \$30 billion will make a loan to a small business is 0.002. What are the odds against such a bank making a small business loan?

15. If the odds that it will rain are 4 to 5, what is the probability of rain? Against rain?

16. In the experiment of drawing a card from a well-shuffled standard deck, find the probability of the events E, "the card is a face card (K, Q, J of any suit)," and E'.

17. Ms. Bezzone invites 10 relatives to a party: her mother, 2 uncles, 3 brothers, and 4 cousins. If the chances of any one guest arriving first are equally likely, find each probability.

(a) The first guest is an uncle or a cousin.

(b) The first guest is a brother or a cousin.

(c) The first guest is an uncle or her mother.

(d) The first guest is an aunt.

18. A card is drawn from a well-shuffled standard deck of 52 cards. Find the probability that the card is each of the following.

(a) A queen **(b)** Red **(c)** A black 3 **(d)** A club or red

19. A card is drawn from a well-shuffled standard deck of 52 cards. Find the probability that the card is each of the following.

(a) A face card (K, Q, J of any suit) **(b)** Red or a 3

(c) Less than a 4 (Consider an ace as a 1.)

20. Two dice are rolled. Find the probability of each event.

(a) The sum of the dots is at least 10.

(b) The sum of the dots is either 7 or at least 10.

(c) The sum of the dots is 3 or the dice both show the same number.

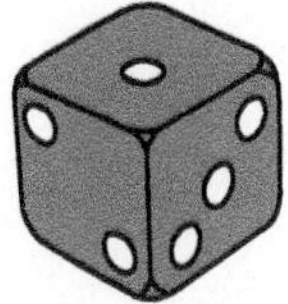

21. If a marble is drawn from a bag containing 2 yellow, 5 red, and 3 blue marbles, what is the probability of each event?

(a) The marble is yellow or blue.

(b) The marble is yellow or red.

(c) The marble is green.

22. The law firm of Alam, Bartolini, Chinn, Dickinson, and Ellsberg has two senior partners, Alam and Bartolini. Two of the attorneys are to be selected to attend a conference. Assuming that all are equally likely to be selected, find each probability.

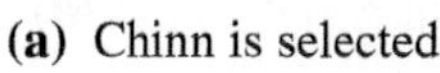

(a) Chinn is selected.

(b) Alam and Dickinson are selected.

(c) At least one senior partner is selected.

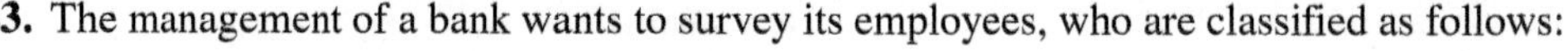

23. The management of a bank wants to survey its employees, who are classified as follows:

30% have worked for the bank more than 5 years;

28% are female;

65% contribute to a voluntary retirement plan;

Half of the female employees contribute to the retirement plan.

Find each probability.

(a) A male employee is selected.

(b) An employee is selected who has worked for the bank for 5 years or less.

(c) An employee is selected who contributes to the retirement plan or is female.

24. Suppose that a game requires a player to pick 6 different numbers from 1 to 99.

(a) How many ways are there to choose 6 numbers if order is not important?

(b) How many ways are there if order is important?

(c) Assume order is unimportant. What is the probability of picking all 6 numbers correctly?

25. The table shows the probabilities of a person accumulating specific amounts of credit card charges over a 12-month period. Find the probabilities that a person's total charges during the period are the following.

Charges	*Probability*
Under \$ 100	0.31
\$ 100–\$ 499	0.18
\$ 500–\$ 999	0.18
\$1000–\$1999	0.13
\$2000–\$2999	0.08
\$3000–\$4999	0.05
\$5000–\$9999	0.06
\$10,000 or more	0.01

(a) \$500–\$999 **(b)** \$500–\$2999

(c) \$5000–\$9999 **(d)** \$3000 or more

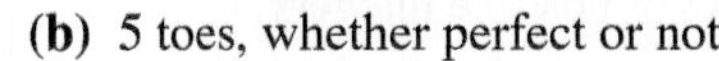

26. In most animals and plants, it is very unusual for the number of main parts of the organism (arms, legs, toes, flower petals, etc.) to vary from generation to generation. Some species, however, have **meristic variability,** in which the number of certain body parts varies from generation to generation. One researcher studied the front feet of certain guinea pigs and produced the following probabilities. (Data from "Analysis of Variability in Number of Digits in an Inbred Strain of Guinea Pigs," by S. Wright in *Genetics*, v. 19, 506–36.)

$$P(\text{only four toes, all perfect}) = 0.77$$

$$P(\text{one imperfect toe and four good ones}) = 0.13$$

$$P(\text{exactly five good toes}) = 0.10$$

Find the probability of each event.

(a) No more than 4 good toes

(b) 5 toes, whether perfect or not

Extending Skills *The table shows the probabilities for the outcomes of an experiment having the following sample space.*

$$S = \{s_1, s_2, s_3, s_4, s_5, s_6\}$$

Outcomes	s_1	s_2	s_3	s_4	s_5	s_6
Probability	0.17	0.03	0.09	0.46	0.21	0.04

Let $E = \{s_1, s_2, s_5\}$, *and let* $F = \{s_4, s_5\}$. *Find each probability.*

27. $P(E)$

28. $P(F)$

29. $P(E \cap F)$

30. $P(E \cup F)$

31. $P(E' \cup F')$

32. $P(E' \cap F)$

Chapter 13 Summary

Key Terms

13.1

infinite sequence
finite sequence
terms of a sequence
general term
series
summation notation
index of summation
arithmetic mean (average)

13.2

arithmetic sequence (arithmetic progression)
common difference

13.3

geometric sequence (geometric progression)
common ratio
annuity
ordinary annuity
payment period
term of an annuity
future value of an annuity

13.4

Pascal's triangle

13.6

tree diagram
independent events
permutation
combination

13.7

trial
outcome
sample space
event
probability
certain event
impossible event
complement
Venn diagram
odds
compound event
mutually exclusive events

New Symbols

a_n nth term of a sequence	$\lim_{n\to\infty} a_n$ limit of a_n as n increases without bound	${}_nP_r$ permutation of n elements taken r at a time	$n(E)$ number of outcomes that belong to event E
$\sum_{i=1}^{n} a_i$ summation notation with index i	$\sum_{i=1}^{\infty} a_i$ sum of an infinite number of terms	${}_nC_r$ binomial coefficient; combination of n elements taken r at a time	$P(E)$ probability of event E
S_n sum of first n terms of a sequence	$n!$ n factorial		E' complement of event E

Test Your Word Power

See how well you have learned the vocabulary in this chapter.

1. An **infinite sequence** is
   A. the values of a function
   B. a function whose domain is the set of positive integers
   C. the sum of the terms of a function
   D. the average of a group of numbers.

2. A **series** is
   A. the sum of the terms of a sequence
   B. the product of the terms of a sequence
   C. the average of the terms of a sequence
   D. the function values of a sequence.

3. An **arithmetic sequence** is a sequence in which
   A. each term after the first is a constant multiple of the preceding term
   B. the numbers are written in a triangular array
   C. the terms are added
   D. each term after the first differs from the preceding term by a common amount.

4. A **geometric sequence** is a sequence in which
   A. each term after the first is a constant multiple of the preceding term
   B. the numbers are written in a triangular array
   C. the terms are multiplied
   D. each term after the first differs from the preceding term by a common amount.

5. A **permutation** is
   A. the ratio of the number of outcomes in an equally likely sample space that satisfy an event to the total number of outcomes in the sample space
   B. one of the ways r elements taken from a set of n elements can be arranged
   C. one of the (unordered) subsets of r elements taken from a set of n elements
   D. the ratio of the probability that an event will occur to the probability that it will not occur.

6. A **combination** is
   A. the ratio of the number of outcomes in an equally likely sample space that satisfy an event to the total number of outcomes in the sample space
   B. one of the ways r elements taken from a set of n elements can be arranged
   C. one of the (unordered) subsets of r elements taken from a set of n elements
   D. the ratio of the probability that an event will occur to the probability that it will not occur.

ANSWERS

1. B; *Example:* The ordered list of numbers 3, 6, 9, 12, 15, . . . is an infinite sequence. **2.** A; *Example:* $3 + 6 + 9 + 12 + 15$, written in summation notation as $\sum_{i=1}^{5} 3i$, is a series. **3.** D; *Example:* The sequence $-3, 2, 7, 12, 17, \ldots$ is arithmetic. **4.** A; *Example:* The sequence 1, 4, 16, 64, 256 , . . . is geometric. **5.** B; *Example:* The permutations of the three letters *m*, *n*, and *t* taken two at a time are *mn*, *mt*, *nt*, *nm*, *tm*, and *tn*. **6.** C; *Example:* The combinations of the letters in Answer 5 are *mn*, *mt*, and *nt*. (Order does not matter here.)

Quick Review

CONCEPTS	EXAMPLES
13.1 Sequences and Series	
An infinite sequence is a function with the set of positive integers as domain.	The infinite sequence $1, \frac{1}{2}, \frac{1}{3}, \frac{1}{4}, \ldots, \frac{1}{n}$ has general term $a_n = \frac{1}{n}$.
A finite sequence is a function with domain of the form $\{1, 2, 3, \ldots, n\}$, where n is a positive integer.	The corresponding series is the *sum*
A series is the indicated sum of the terms of a sequence.	$1 + \frac{1}{2} + \frac{1}{3} + \frac{1}{4} + \cdots + \frac{1}{n}$.

CONCEPTS	EXAMPLES

13.2 Arithmetic Sequences

Assume that a_1 is the first term, a_n is the nth term, and d is the common difference in an arithmetic sequence.

Common Difference

$$d = a_{n+1} - a_n$$

***n*th Term**

$$a_n = a_1 + (n - 1)d$$

Sum of the First *n* Terms

$$S_n = \frac{n}{2}(a_1 + a_n)$$

or $$S_n = \frac{n}{2}[2a_1 + (n - 1)d]$$

Consider the following arithmetic sequence.

$$2, 5, 8, 11, \ldots$$

$a_1 = 2$ $\quad a_1$ is the first term.

$d = 3$ $\quad d = a_2 - a_1$

(Any two successive terms could have been used.)

The tenth term is found as follows.

$a_{10} = 2 + (10 - 1)(3)$ $\quad$ Let $a_1 = 2$, $d = 3$, and $n = 10$.

$a_{10} = 2 + 9 \cdot 3$ $\quad$ Subtract inside the parentheses.

$a_{10} = 29$ $\quad$ Multiply, and then add.

The sum of the first ten terms can be found in either of two ways.

$S_{10} = \frac{10}{2}(2 + a_{10})$	$S_{10} = \frac{10}{2}[2(2) + (10 - 1)(3)]$
$S_{10} = 5(2 + 29)$	$S_{10} = 5(4 + 9 \cdot 3)$
$S_{10} = 5(31)$	$S_{10} = 5(4 + 27)$
$S_{10} = 155$	$S_{10} = 5(31)$
	$S_{10} = 155$

13.3 Geometric Sequences

Assume that a_1 is the first term, a_n is the nth term, and r is the common ratio in a geometric sequence.

Common Ratio

$$r = \frac{a_{n+1}}{a_n}$$

***n*th Term**

$$a_n = a_1 r^{n-1}$$

Sum of the First *n* Terms

$$S_n = \frac{a_1(1 - r^n)}{1 - r} \quad \text{(where } r \neq 1\text{)}$$

Consider the following geometric sequence.

$$1, 2, 4, 8, \ldots$$

$a_1 = 1$ $\quad a_1$ is the first term.

$r = 2$ $\quad r = \frac{a_4}{a_3}$

(Any two successive terms could have been used.)

The sixth term is found as follows.

$a_6 = 1(2)^{6-1}$ $\quad$ Let $a_1 = 1$, $r = 2$, and $n = 6$.

$a_6 = 1(2)^5$ $\quad$ Subtract in the exponent.

$a_6 = 32$ $\quad$ Apply the exponent and multiply.

The sum of the first six terms is found as follows.

$S_6 = \frac{1(1 - 2^6)}{1 - 2}$ $\quad$ Substitute for a_1, r, and n.

$S_6 = \frac{1 - 64}{-1}$ $\quad$ Evaluate 2^6. Subtract in the denominator.

$S_6 = 63$ $\quad$ Simplify.

CONCEPTS	EXAMPLES
Future Value of an Ordinary Annuity $$S = R\left[\frac{(1+i)^n - 1}{i}\right],$$ where S is the future value, R is the payment at the end of each period, i is the interest rate per period, and n is the number of periods.	If \$5800 is deposited into an ordinary annuity at the end of each quarter for 4 yr and interest is earned at 2.4% compounded quarterly, how much will be in the account at that time? $$R = \$5800, \quad i = \frac{0.024}{4} = 0.006, \quad n = 4(4) = 16$$ $$S = 5800\left[\frac{(1+0.006)^{16} - 1}{0.006}\right] = \$97{,}095.24 \quad \text{Nearest cent}$$
Sum of the Terms of an Infinite Geometric Sequence $$S = \frac{a_1}{1-r} \quad (\text{where } \lvert r \rvert < 1)$$	Evaluate the sum S of the terms of an infinite geometric sequence with $a_1 = 1$ and $r = \frac{1}{2}$. $$S = \frac{1}{1-\frac{1}{2}} = \frac{1}{\frac{1}{2}} = 1 \div \frac{1}{2} = 1 \cdot \frac{2}{1} = 2$$

13.4 The Binomial Theorem

CONCEPTS	EXAMPLES
***n* Factorial ($n!$)** For any positive integer n, $$n! = n(n-1)(n-2)(n-3)\cdots(3)(2)(1).$$ By definition, $0! = 1$.	Evaluate 4!. $$4! = 4 \cdot 3 \cdot 2 \cdot 1 = 24$$
Binomial Coefficient $${}_nC_r = \frac{n!}{r!(n-r)!} \quad (\text{where } r \le n)$$	Evaluate ${}_5C_3$. $${}_5C_3 = \frac{5!}{3!(5-3)!} = \frac{5!}{3!2!} = \frac{5 \cdot 4 \cdot 3 \cdot 2 \cdot 1}{3 \cdot 2 \cdot 1 \cdot 2 \cdot 1} = 10$$
Binomial Theorem For any positive integer n, $(x+y)^n$ is expanded as follows. $$\begin{aligned}(x+y)^n &= x^n + \frac{n!}{1!(n-1)!}x^{n-1}y + \frac{n!}{2!(n-2)!}x^{n-2}y^2 \\ &\quad + \frac{n!}{3!(n-3)!}x^{n-3}y^3 + \cdots + \frac{n!}{(n-1)!1!}xy^{n-1} \\ &\quad + y^n\end{aligned}$$	Expand $(2m+3)^4$. $$\begin{aligned}(2m+3)^4 &= (2m)^4 + \frac{4!}{1!3!}(2m)^3(3) + \frac{4!}{2!2!}(2m)^2(3)^2 \\ &\quad + \frac{4!}{3!1!}(2m)(3)^3 + 3^4 \\ &= 2^4m^4 + 4(2)^3m^3(3) + 6(2)^2m^2(9) + 4(2m)(27) + 81 \\ &= 16m^4 + 12(8)m^3 + 54(4)m^2 + 216m + 81 \\ &= 16m^4 + 96m^3 + 216m^2 + 216m + 81\end{aligned}$$
***k*th Term of the Binomial Expansion of $(x+y)^n$** If $n \ge k-1$, then the kth term of the expansion of $(x+y)^n$ is given by the following. $$\frac{n!}{(k-1)![n-(k-1)]!}x^{n-(k-1)}y^{k-1}$$	Find the eighth term of the expansion of $(a-2b)^{10}$. $\frac{10!}{7!3!}(a^3)(-2b)^7$ Let $n = 10$, $x = a$, $y = -2b$, and $k = 8$. $= 120(a^3)(-2)^7b^7$ Simplify the factorials; $(ab)^m = a^mb^m$ $= 120a^3(-128b^7)$ Simplify. $= -15{,}360a^3b^7$ Multiply.

CONCEPTS	EXAMPLES

13.5 Mathematical Induction

Principle of Mathematical Induction

Let S_n be a statement concerning the positive integer n. Suppose that both of the following are satisfied.

1. S_1 is true.
2. For any positive integer k, $k \le n$, if S_k is true, then S_{k+1} is also true.

Then S_n is true for every positive integer value of n.

See **Examples 1–3** in **Section 13.5.**

13.6 Counting Theory

Fundamental Principle of Counting

If one event can occur in m ways and a second event can occur in n ways, then both events can occur in mn ways, provided that the outcome of the first event does not influence the outcome of the second.

If there are 2 ways to choose a pair of socks and 5 ways to choose a pair of shoes, then how many ways are there to choose socks and shoes?

$$2 \cdot 5 = 10 \text{ ways}$$

Permutations Formula

If ${}_nP_r$ represents the number of permutations of n elements taken r at a time, for $r \le n$, then the following holds true.

$$_nP_r = \frac{n!}{(n-r)!}$$

How many ways are there to arrange the letters of the word *triangle* using 5 letters at a time?

$$_8P_5 = \frac{8!}{(8-5)!} = \frac{8!}{3!} = 6720 \text{ ways} \qquad \text{Let } n = 8 \text{ and } r = 5.$$

Combinations Formula

If ${}_nC_r$ represents the number of combinations of n elements taken r at a time, for $r \le n$, then the following holds true.

$$_nC_r = \frac{n!}{r!(n-r)!}$$

How many committees of 4 senators can be formed from a group of 9 senators?

The arrangement of senators does not matter, so this is a combinations problem.

$$_9C_4 = \frac{9!}{4!(9-4)!} = \frac{9!}{4!5!} = 126 \text{ committees} \qquad \text{Let } n = 9 \text{ and } r = 4.$$

13.7 Basics of Probability

Probability of an Event E

In a sample space with equally likely outcomes, the probability of an event E, written $P(E)$, is the ratio of the number of outcomes in sample space S that belong to event E, $n(E)$, to the total number of outcomes in sample space S, $n(S)$.

$$P(E) = \frac{n(E)}{n(S)}$$

Properties of Probability

For any events E and F, the following hold true.

1. $0 \le P(E) \le 1$
2. $P(\text{a certain event}) = 1$
3. $P(\text{an impossible event}) = 0$
4. $P(E') = 1 - P(E)$
5. $P(E \cup F) = P(E) + P(F) - P(E \cap F)$

A number is chosen at random from the set

$$S = \{1, 2, 3, 4, 5, 6\}.$$

What is the probability that the number is less than 3?

The event is $E = \{1, 2\}$. We have $n(S) = 6$ and $n(E) = 2$.

$$P(E) = \frac{2}{6} = \frac{1}{3}$$

What is the probability that the number is 3 or more?

This event is E'.

$$P(E') = 1 - \frac{1}{3} = \frac{2}{3}$$

Chapter 13 Review Exercises

13.1 *Write the first four terms of each sequence.*

1. $a_n = 2n - 3$

2. $a_n = \frac{n-1}{n}$

3. $a_n = n^2$

4. $a_n = \left(\frac{1}{2}\right)^n$

5. $a_n = (n+1)(n-1)$

6. $a_n = n(-1)^{n-1}$

Write each series as a sum of terms and then find the sum.

7. $\sum_{i=1}^{5} i^2$

8. $\sum_{i=1}^{6} (i+1)$

9. $\sum_{i=3}^{6} (5i-4)$

10. $\sum_{i=1}^{4} (i+2)^2$

11. $\sum_{i=1}^{6} 2^i$

12. $\sum_{i=4}^{7} \frac{i}{i+1}$

13. The table shows the worldwide number of electric vehicles, in thousands, in use from 2012 to 2017. To the nearest tenth, what was the average number of electric vehicles for this period?

Year	*Electric Vehicles (in thousands)*
2012	110
2013	220
2014	409
2015	727
2016	1186
2017	1928

Data from www.statista.com

13.2, 13.3 *Determine whether each sequence is* arithmetic, geometric, *or* neither. *If the sequence is arithmetic, find the common difference d. If it is geometric, find the common ratio r.*

14. 2, 5, 8, 11, . . .

15. −6, −2, 2, 6, 10, . . .

16. $\frac{2}{3}, -\frac{1}{3}, \frac{1}{6}, -\frac{1}{12}, \ldots$

17. −1, 1, −1, 1, −1, . . .

18. $64, 32, 8, \frac{1}{2}, \ldots$

19. 64, 32, 8, 1, . . .

20. The *Fibonacci sequence* begins 1, 1, 2, 3, 5, 8, ... What is the eleventh term of this sequence?

13.2 *Find the indicated term for each arithmetic sequence.*

21. $a_1 = -2, d = 5;\quad a_{16}$

22. $a_6 = 12, a_8 = 18;\quad a_{25}$

Determine an expression for the general term of each arithmetic sequence.

23. $a_1 = -4, d = -5$

24. 6, 3, 0, −3, . . .

Find the number of terms in each arithmetic sequence.

25. 7, 10, 13, . . . , 49

26. 5, 1, −3, . . . , −79

Evaluate S_8 for each arithmetic sequence.

27. $a_1 = -2, d = 6$

28. $a_n = -2 + 5n$

13.3 *Determine an expression for the general term of each geometric sequence.*

29. $-1, -4, -16, \ldots$

30. $\frac{2}{3}, \frac{2}{15}, \frac{2}{75}, \ldots$

Find the indicated term for each geometric sequence.

31. $2, -6, 18, \ldots;\ a_{11}$

32. $a_3 = 20, a_5 = 80;\ a_{10}$

Evaluate each sum if it exists.

33. $\sum_{i=1}^{5} \left(\frac{1}{4}\right)^i$

34. $\sum_{i=1}^{8} \frac{3}{4}(-1)^i$

35. $\sum_{i=1}^{\infty} 4\left(\frac{1}{5}\right)^i$

36. $\sum_{i=1}^{\infty} 2(3)^i$

13.4 *Use the binomial theorem to expand each binomial.*

37. $(2p - q)^5$

38. $(x^2 + 3y)^4$

39. $(3t^3 - s^2)^4$

40. Write the fourth term of the expansion of $(3a + 2b)^{19}$.

13.5 *Use mathematical induction to prove that each statement is true for every positive integer value of n.*

41. $2 + 6 + 10 + 14 + \cdots + (4n - 2) = 2n^2$

42. $2^2 + 4^2 + 6^2 + \cdots + (2n)^2 = \frac{2n(n+1)(2n+1)}{3}$

43. $2 + 2^2 + 2^3 + \cdots + 2^n = 2(2^n - 1)$

44. $1^3 + 3^3 + 5^3 + \cdots + (2n - 1)^3 = n^2(2n^2 - 1)$

13.6 *Evaluate each expression.*

45. ${}_5P_5$

46. ${}_9P_2$

47. ${}_7C_3$

48. ${}_8C_5$

Solve each problem.

49. Two people are planning their wedding. They can select from 2 different chapels, 4 soloists, 3 organists, and 2 ministers. How many different wedding arrangements will be possible?

50. John is furnishing his apartment and wants to buy a new couch. He can select from 5 different styles, each available in 3 different fabrics, with 6 color choices. How many different couches are available?

51. Four students are to be assigned to 4 different summer jobs. Each student is qualified for all 4 jobs. In how many ways can the jobs be assigned?

52. How many different license plates can be formed with a letter followed by 3 digits and then 3 letters? How many such license plates have no repeats?

13.7 *Solve each problem.*

53. A marble is drawn at random from a box containing 4 green, 5 black, and 6 white marbles. Find the probability of each event.

(a) A green marble is drawn.

(b) A marble that is not black is drawn.

(c) A blue marble is drawn.

(d) A black or a white marble is drawn.

54. A marble is drawn from a bag containing 5 green, 4 black, and 10 white marbles.

(a) What are the odds in favor of drawing a green marble?

(b) What are the odds against drawing a white marble?

A card is drawn from a standard deck of 52 cards. Find the probability that each card described is drawn.

55. A black king

56. A face card or an ace

57. An ace or a diamond

58. A card that is not a diamond

59. A card that is not a diamond or not black

60. An 8, 9, or 10

Chapter 13 Mixed Review Exercises

Find the indicated term and evaluate S_{10} for each sequence.

1. a_{10}: geometric; $-3, 6, -12, \ldots$

2. a_{40}: arithmetic; $1, 7, 13, \ldots$

3. a_{15}: arithmetic; $a_1 = -4, \quad d = 3$

4. a_9: geometric; $a_1 = 1, \quad r = -3$

Determine an expression for the general term of each arithmetic or geometric sequence.

5. $2, 8, 32, \ldots$

6. $2, 7, 12, \ldots$

7. $12, 9, 6, \ldots$

8. $27, 9, 3, \ldots$

Solve each problem.

9. When Faith's sled goes down the hill near her home, she covers 3 ft in the first second. Then, for each second after that, she goes 4 ft more than in the preceding second. If the distance she covers going down is 210 ft, how long does it take her to reach the bottom?

10. An ordinary annuity is set up so that \$672 is deposited at the end of each quarter for 7 yr. The money earns 4.5% annual interest compounded quarterly. What is the future value of the annuity?

11. The school population in Middleton has been dropping 3% per yr. The current population is 50,000. If this trend continues, what will the population be in 6 yr?

12. A pump removes $\frac{1}{2}$ of the liquid in a container with each stroke. What fraction of the liquid is left in the container after seven strokes?

13. A student council consists of a president, vice-president, secretary/treasurer, and 3 representatives at large. Three members are to be selected to attend a conference.

(a) How many different such delegations are possible?

(b) How many are possible if the president must attend?

14. Nine football teams are competing for first-, second-, and third-place titles in a statewide tournament. In how many ways can the winners be determined?

15. A sample shipment of 5 swimming pool filters is chosen. The probability of exactly 0, 1, 2, 3, 4, or 5 filters being defective is given in the following table.

Number Defective	0	1	2	3	4	5
Probability	0.31	0.25	0.18	0.12	0.08	0.06

Find the probability that the given number of filters are defective.

(a) No more than 3

(b) At least 2

Chapter 13 Test

Step-by-step test solutions are found on the Chapter Test Prep Videos available in MyLab Math.

View the complete solutions to all Chapter Test exercises in MyLab Math.

Write the first five terms of each sequence described.

1. $a_n = (-1)^n + 1$

2. arithmetic, with $a_1 = 4$ and $d = 2$

3. geometric, with $a_4 = 6$ and $r = \frac{1}{2}$

Determine a_4 for each sequence described.

4. arithmetic, with $a_1 = 6$ and $d = -2$

5. geometric, with $a_5 = 16$ and $a_7 = 9$

Evaluate S_5 for each sequence described.

6. arithmetic, with $a_2 = 12$ and $a_3 = 15$

7. geometric, with $a_5 = 4$ and $a_7 = 1$

Solve each problem.

8. The first eight time intervals, in minutes, between eruptions of the Old Faithful geyser in Yellowstone National Park on July 19, 2018, are given here. Calculate the average number of minutes between eruptions to the nearest tenth. (Data from www.geysertimes.org)

89, 94, 99, 85, 101, 89, 97, 86

9. If \$4000 is deposited in an ordinary annuity at the end of each quarter for 7 yr and earns 6% interest compounded quarterly, how much will be in the account at the end of this term?

10. Under what conditions does an infinite geometric series have a sum?

Evaluate each sum if it exists.

11. $\sum_{i=1}^{5} (2i + 8)$

12. $\sum_{i=1}^{6} (3i - 5)$

13. $\sum_{i=1}^{500} i$

14. $\sum_{i=1}^{3} \frac{1}{2}(4^i)$

15. $\sum_{i=1}^{\infty} \left(\frac{1}{4}\right)^i$

16. $\sum_{i=1}^{\infty} 6\left(\frac{3}{2}\right)^i$

17. Expand $(3k - 5)^4$.

18. Write the fifth term of the expansion of $\left(2x - \frac{y}{3}\right)^{12}$.

Solve each problem.

19. Christian bought a new bicycle for \$300. He agreed to pay \$20 per month for 15 months, plus interest of 1% each month, on the unpaid balance. Find the total cost of the bicycle.

20. During the summer months, the population of a certain insect colony triples each week. If there are 20 insects in the colony at the end of the first week in July, how many are present by the end of September? (Assume exactly four weeks in a month.)

21. Use mathematical induction to prove that this statement is true for every positive integer value of n.

$$8 + 14 + 20 + 26 + \cdots + (6n + 2) = 3n^2 + 5n$$

Evaluate each expression.

22. ${}_{11}P_3$

23. ${}_{45}C_1$

Solve each problem.

24. A clothing manufacturer makes women's coats in 4 different styles. Each coat can be made from one of 3 fabrics. Each fabric comes in 5 different colors. How many different coats can be made?

25. A club with 30 members is to elect a president, secretary, and treasurer from its membership. If a member can hold at most one position, in how many ways can the offices be filled?

26. In how many ways can a committee of 3 representatives be chosen from a group of 9 representatives?

A card is drawn from a standard deck of 52 cards. Find the probability that each card described is drawn.

27. A red 3

28. A card that is not a face card

29. A king or a spade

30. If a card is drawn from a standard deck, what are the odds in favor of drawing a face card?

Chapters R–13 Cumulative Review Exercises

Simplify each expression.

1. $|-7| + 6 - |-10| - (-8 + 3)$

2. $-\frac{7}{30} + \frac{11}{45} - \frac{3}{10}$

Solve each equation or inequality.

3. $9 - (5 + 3x) + 5x = -4(x - 3) - 7$

4. $7x + 18 \le 9x - 2$

5. $2x > 8$ or $-3x > 9$

6. $|4x - 3| = 21$

7. $|2x - 5| \ge 11$

8. $2x^2 + x = 10$

9. $6x^2 + 5x = 8$

10. $\frac{4}{x - 3} - \frac{6}{x + 3} = \frac{24}{x^2 - 9}$

11. $3^{2x-1} = 81$

12. $\log_8 x + \log_8 (x + 2) = 1$

Solve each problem.

13. Find the slope of the line passing through $(4, -5)$ and $(-12, -17)$.

14. Find the standard form of the equation of the line passing through $(-2, 10)$ and parallel to the line with equation $3x + y = 7$.

15. Write the equation of a circle with center $(-5, 12)$ and radius 9.

Solve each system of equations.

16. $y = 5x + 3$
$2x + 3y = -8$

17. $x + 2y + z = 8$
$2x - y + 3z = 15$
$-x + 3y - 3z = -11$

18. $xy = -5$
$2x + y = 3$

19. Nuts worth \$3 per lb are to be mixed with 8 lb of nuts worth \$4.25 per lb to obtain a mixture that will be sold for \$4 per lb. How many pounds of the \$3 nuts should be used?

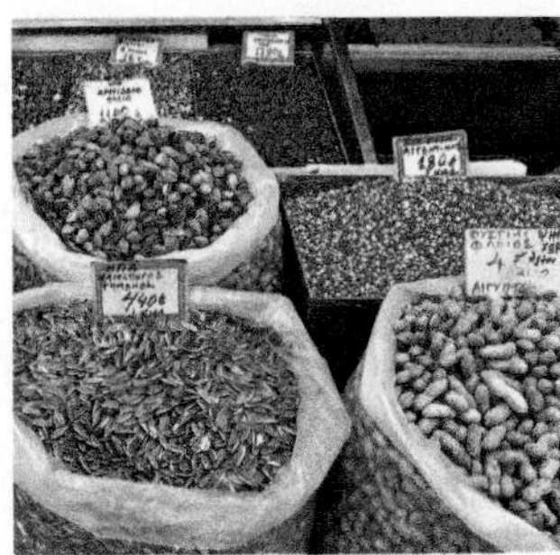

Simplify. Assume that all variables represent nonzero real numbers.

20. $\left(\frac{2}{3}\right)^{-2}$

21. $\frac{(3p^2)^3(-2p^6)}{4p^3(5p^7)}$

Perform the indicated operations.

22. $(4p + 2)(5p - 3)$

23. $(2m^3 - 3m^2 + 8m) - (7m^3 + 5m - 8)$

24. $(6t^4 + 5t^3 - 18t^2 + 14t - 1) \div (3t - 2)$

25. $(8 + 3i)(8 - 3i)$

Factor.

26. $6z^3 + 5z^2 - 4z$

27. $49a^4 - 9b^2$

28. $c^3 + 27d^3$

Simplify.

29. $\frac{x^2 - 16}{x^2 + 2x - 8} \div \frac{x - 4}{x + 7}$

30. $\frac{5}{p^2 + 3p} - \frac{2}{p^2 - 4p}$

31. $5\sqrt{72} - 4\sqrt{50}$

Graph.

32. $x - 3y = 6$

33. $4x - y < 4$

34. $f(x) = 2(x - 2)^2 - 3$

35. $g(x) = \left(\frac{1}{3}\right)^x$

36. $y = \log_{1/3} x$

37. $f(x) = \frac{1}{x - 3}$

38. $\frac{x^2}{9} + \frac{y^2}{25} = 1$

39. $x^2 - y^2 = 9$

Solve each problem.

40. Find $f^{-1}(x)$ if $f(x) = 9x + 5$.

41. Factor $f(x) = 2x^3 + 9x^2 + 3x - 4$ into linear factors given that $f(-4) = 0$.

42. Write the first five terms of the sequence with general term $a_n = 5n - 12$.

43. Find each of the following.

(a) The sum of the first six terms of the arithmetic sequence with $a_1 = 8$ and $d = 2$

(b) The sum of the geometric series $15 - 6 + \frac{12}{5} - \frac{24}{25} + \cdots$

44. Find the sum $\sum_{i=1}^{4} 3i$.

45. Use the binomial theorem to expand $(2a - 1)^5$.

46. Find the fourth term in the expansion of $\left(3x^4 - \frac{1}{2}y^2\right)^5$.

47. Use mathematical induction to prove that this statement is true for every positive integer value of n.

$$4 + 8 + 12 + 16 + \cdots + 4n = 2n(n + 1)$$

48. Evaluate. **(a)** $9!$ **(b)** ${}_7P_3$ **(c)** ${}_{10}C_4$

49. Find the probability of rolling a sum of 11 with two dice.

50. If the odds that it will rain are 3 to 7, what is the probability of rain?

Solving Systems of Linear Equations by Matrix Methods

APPENDIX A

OBJECTIVES

1 Define a matrix.
2 Write the augmented matrix of a system.
3 Use row operations to solve a system with two equations.
4 Use row operations to solve a system with three equations.
5 Use row operations to solve special systems.

VOCABULARY

☐ matrix (plural *matrices*)
☐ elements
☐ rows
☐ columns
☐ dimensions
☐ square martix

OBJECTIVE 1 Define a matrix.

An ordered array of numbers is a **matrix.**

$$\text{Rows}\rightarrow\begin{bmatrix} 2 & 3 & 5 \\ 7 & 1 & 2 \end{bmatrix}\ \text{Matrix}$$

(Columns indicated by arrows above each column.)

The numbers are **elements** of the matrix. *Matrices* (the plural of *matrix*) are named according to the number of **rows** and **columns** they contain. The rows are read horizontally, and the columns are read vertically. This matrix is a 2×3 (read "*two by three*") matrix, because it has 2 rows and 3 columns. The number of rows followed by the number of columns gives the **dimensions** of the matrix.

$$\begin{bmatrix} -1 & 0 \\ 1 & -2 \end{bmatrix}\ \begin{matrix}2 \times 2 \\ \text{matrix}\end{matrix} \qquad \begin{bmatrix} 8 & -1 & -3 \\ 2 & 1 & 6 \\ 0 & 5 & -3 \\ 5 & 9 & 7 \end{bmatrix}\ \begin{matrix}4 \times 3 \\ \text{matrix}\end{matrix}$$

A **square matrix** is a matrix that has the same number of rows as columns. The 2×2 matrix above is a square matrix. **FIGURE 1** shows how a graphing calculator displays the preceding two matrices. Consult your owner's manual for details for using matrices.

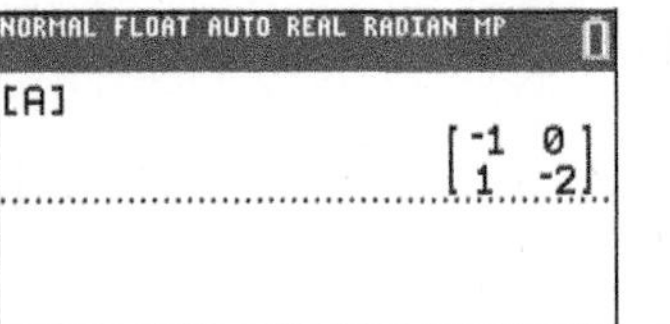

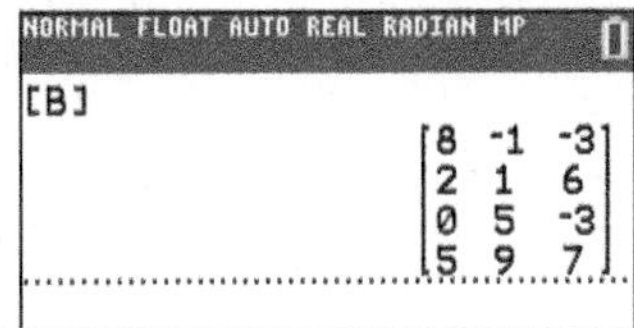

FIGURE 1

In this section, we discuss a matrix method of solving linear systems that is a structured way of using the elimination method. The advantage of this new method is that it can be done by a graphing calculator or a computer.

OBJECTIVE 2 Write the augmented matrix of a system.

To solve a linear system using matrices, we begin by writing an *augmented matrix* of the system. An **augmented matrix** has a vertical bar that separates the columns of the matrix into two groups.

$$\begin{aligned} x - 3y &= 1 \\ 2x + y &= -5 \end{aligned} \quad \text{System of equations} \qquad \left[\begin{array}{cc|c} 1 & -3 & 1 \\ 2 & 1 & -5 \end{array}\right] \quad \text{Corresponding augmented matrix}$$

Place the coefficients of the variables to the left of the bar, and the constants to the right. The bar separates the coefficients from the constants.

An augmented matrix is simply a shorthand way of writing a system of equations. The rows of the augmented matrix can be treated the same as the equations of the corresponding system.

Exchanging the positions of two equations in a system does not change the system. Also, multiplying any equation in a system by a nonzero number does not change the system. Comparable changes to the augmented matrix of a system of equations produce new matrices that correspond to systems with the same solutions as the original system.

The following **row operations** produce new matrices that lead to systems having the same solutions as the original system.

Matrix Row Operations

1. Any two rows of the matrix may be interchanged.
2. The elements of any row may be multiplied by any nonzero real number.
3. Any row may be changed by adding to the elements of the row the product of a real number and the corresponding elements of another row.

Example of row operation 1:

$$\begin{bmatrix} 2 & 3 & 9 \\ 4 & 8 & -3 \\ 1 & 0 & 7 \end{bmatrix} \text{ becomes } \begin{bmatrix} 1 & 0 & 7 \\ 4 & 8 & -3 \\ 2 & 3 & 9 \end{bmatrix}$$ Interchange row 1 and row 3.

Example of row operation 2:

$$\begin{bmatrix} 2 & 3 & 9 \\ 4 & 8 & -3 \\ 1 & 0 & 7 \end{bmatrix} \text{ becomes } \begin{bmatrix} 6 & 9 & 27 \\ 4 & 8 & -3 \\ 1 & 0 & 7 \end{bmatrix}$$ Multiply the numbers in row 1 by 3.

Example of row operation 3:

$$\begin{bmatrix} 2 & 3 & 9 \\ 4 & 8 & -3 \\ 1 & 0 & 7 \end{bmatrix} \text{ becomes } \begin{bmatrix} 0 & 3 & -5 \\ 4 & 8 & -3 \\ 1 & 0 & 7 \end{bmatrix}$$ Multiply the numbers in row 3 by −2. Add them to the corresponding numbers in row 1.

The third row operation corresponds to the way a variable can be eliminated from a pair of equations.

OBJECTIVE 3 Use row operations to solve a system with two equations.

Row operations can be used to rewrite a matrix until it is the matrix of a system whose solution is easy to find. The goal is a matrix in the form

$$\left[\begin{array}{cc|c} 1 & a & b \\ 0 & 1 & c \end{array}\right] \quad \text{or} \quad \left[\begin{array}{ccc|c} 1 & a & b & c \\ 0 & 1 & d & e \\ 0 & 0 & 1 & f \end{array}\right]$$

for systems with two and three equations, respectively. Notice that there are 1's down the diagonal from upper left to lower right and 0's below the 1's. A matrix written this way is in **row echelon form.**

Use row operations to solve the system.

$$x + 3y = 3$$
$$2x - 3y = -12$$

EXAMPLE 1 Using Row Operations to Solve a System with Two Variables

Use row operations to solve the system.

$$x - 3y = 1$$
$$2x + y = -5$$

We start by writing the augmented matrix of the system.

$$\left[\begin{array}{cc|c} 1 & -3 & 1 \\ 2 & 1 & -5 \end{array}\right] \quad \text{Write the augmented matrix.}$$

Our goal is to use the various row operations to change this matrix into one that leads to a system that is easier to solve. It is best to work by columns.

We start with the first column and make sure that there is a 1 in the "first row, first column" position. There already is a 1 in this position.

Next, we obtain 0 in every position below the first. To introduce 0 in row two, column one, we add to the numbers in row two the result of multiplying each number in row one by -2. (We abbreviate this as $-2R_1 + R_2$.) Row one remains unchanged.

$$\left[\begin{array}{cc|c} 1 & -3 & 1 \\ 2 + 1(-2) & 1 + (-3)(-2) & -5 + 1(-2) \end{array}\right]$$

Original number from row two ↑   ↑ −2 times number from row one

$$\left[\begin{array}{cc|c} 1 & -3 & 1 \\ 0 & 7 & -7 \end{array}\right] \quad -2R_1 + R_2$$

1 in the first position of column one → (row one); 0 in every position below the first → (row two)

Now we go to column two. The number 1 is needed in row two, column two. We use the second row operation, multiplying each number in row two by $\frac{1}{7}$.

$$\left[\begin{array}{cc|c} 1 & -3 & 1 \\ 0 & 1 & -1 \end{array}\right] \quad \tfrac{1}{7}R_2$$

Stop here—this matrix is in row echelon form.

This augmented matrix leads to the system of equations

$$\begin{aligned} 1x - 3y &= 1 \\ 0x + 1y &= -1, \end{aligned} \quad \text{or} \quad \begin{aligned} x - 3y &= 1 \\ y &= -1. \end{aligned}$$

From the second equation, $y = -1$, we substitute -1 for y in the first equation to find x.

$$x - 3y = 1$$
$$x - 3(-1) = 1 \quad \text{Let } y = -1.$$
$$x + 3 = 1 \quad \text{Multiply.}$$
$$x = -2 \quad \text{Subtract 3.}$$

Check these values by substitution in both equations of the system. The solution set of the system is $\{(-2, -1)\}$.

The calculator screen in **FIGURE 2** shows how a graphing calculator evaluates the augmented matrix.

NOW TRY

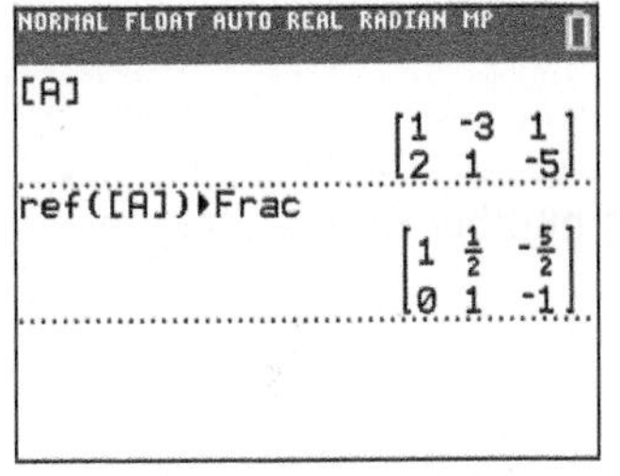

Even though this system looks different from the one we obtained in **Example 1,** it is equivalent, because its solution set is also $\{(-2, -1)\}$.

FIGURE 2

NOW TRY ANSWER

1. $\{(-3, 2)\}$

OBJECTIVE 4 Use row operations to solve a system with three equations.

EXAMPLE 2 Using Row Operations to Solve a System with Three Variables

Use row operations to solve the system.

$$\begin{aligned} x - y + 5z &= -6 \\ 3x + 3y - z &= 10 \\ x + 3y + 2z &= 5 \end{aligned}$$

Start by writing the augmented matrix of the system.

$$\left[\begin{array}{ccc|c} 1 & -1 & 5 & -6 \\ 3 & 3 & -1 & 10 \\ 1 & 3 & 2 & 5 \end{array}\right] \quad \text{Write the augmented matrix.}$$

This matrix already has 1 in row one, column one. Now obtain 0's in the rest of column one. First, add to row two the results of multiplying each number in row one by -3.

$$\left[\begin{array}{ccc|c} 1 & -1 & 5 & -6 \\ 0 & 6 & -16 & 28 \\ 1 & 3 & 2 & 5 \end{array}\right] \quad -3R_1 + R_2$$

Now add to the numbers in row three the results of multiplying each number in row one by -1.

$$\left[\begin{array}{ccc|c} 1 & -1 & 5 & -6 \\ 0 & 6 & -16 & 28 \\ 0 & 4 & -3 & 11 \end{array}\right] \quad -1R_1 + R_3$$

Introduce 1 in row two, column two, by multiplying each number in row two by $\frac{1}{6}$.

$$\left[\begin{array}{ccc|c} 1 & -1 & 5 & -6 \\ 0 & 1 & -\frac{8}{3} & \frac{14}{3} \\ 0 & 4 & -3 & 11 \end{array}\right] \quad \frac{1}{6}R_2$$

To obtain 0 in row three, column two, add to row three the results of multiplying each number in row two by -4.

$$\left[\begin{array}{ccc|c} 1 & -1 & 5 & -6 \\ 0 & 1 & -\frac{8}{3} & \frac{14}{3} \\ 0 & 0 & \frac{23}{3} & -\frac{23}{3} \end{array}\right] \quad -4R_2 + R_3$$

Obtain 1 in row three, column three, by multiplying each number in row three by $\frac{3}{23}$.

This matrix is in row echelon form.

$$\left[\begin{array}{ccc|c} 1 & -1 & 5 & -6 \\ 0 & 1 & -\frac{8}{3} & \frac{14}{3} \\ 0 & 0 & 1 & -1 \end{array}\right] \quad \frac{3}{23}R_3$$

The final matrix gives this system of equations.

$$\begin{aligned} x - y + 5z &= -6 \\ y - \frac{8}{3}z &= \frac{14}{3} \\ z &= -1 \end{aligned}$$

NOW TRY EXERCISE 2

Use row operations to solve the system.

$$\begin{aligned} x + y - 2z &= -5 \\ -x + 2y + z &= -1 \\ 2x - y + 3z &= 14 \end{aligned}$$

Substitute -1 for z in the second equation, $y - \frac{8}{3}z = \frac{14}{3}$, to find that $y = 2$. Finally, substitute 2 for y and -1 for z in the first equation,

$$x - y + 5z = -6,$$

to determine that $x = 1$. The solution set of the original system is $\{(1, 2, -1)\}$. Check these values by substitution. **NOW TRY**

OBJECTIVE 5 Use row operations to solve special systems.

NOW TRY EXERCISE 3

Use row operations to solve each system.

(a) $\begin{aligned} 3x - y &= 8 \\ -6x + 2y &= 4 \end{aligned}$

(b) $\begin{aligned} x + 2y &= 7 \\ -x - 2y &= -7 \end{aligned}$

EXAMPLE 3 Recognizing Inconsistent Systems or Dependent Equations

Use row operations to solve each system.

(a) $\begin{aligned} 2x - 3y &= 8 \\ -6x + 9y &= 4 \end{aligned}$

$$\left[\begin{array}{rr|r} 2 & -3 & 8 \\ -6 & 9 & 4 \end{array}\right] \quad \text{Write the augmented matrix.}$$

$$\left[\begin{array}{rr|r} 1 & -\frac{3}{2} & 4 \\ -6 & 9 & 4 \end{array}\right] \quad \tfrac{1}{2}R_1$$

$$\left[\begin{array}{rr|r} 1 & -\frac{3}{2} & 4 \\ 0 & 0 & 28 \end{array}\right] \quad 6R_1 + R_2$$

The corresponding system of equations can be written as follows.

$$x - \frac{3}{2}y = 4$$

$$0 = 28 \quad \text{False}$$

This system has no solution and is inconsistent. The solution set is $\varnothing$.

(b) $\begin{aligned} -10x + 12y &= 30 \\ 5x - 6y &= -15 \end{aligned}$

$$\left[\begin{array}{rr|r} -10 & 12 & 30 \\ 5 & -6 & -15 \end{array}\right] \quad \text{Write the augmented matrix.}$$

$$\left[\begin{array}{rr|r} 1 & -\frac{6}{5} & -3 \\ 5 & -6 & -15 \end{array}\right] \quad -\tfrac{1}{10}R_1$$

$$\left[\begin{array}{rr|r} 1 & -\frac{6}{5} & -3 \\ 0 & 0 & 0 \end{array}\right] \quad -5R_1 + R_2$$

The corresponding system of equations can be written as follows.

$$x - \frac{6}{5}y = -3$$

$$0 = 0 \quad \text{True}$$

This system has dependent equations. We use the second equation of the given system, which is in standard form, to express the solution set.

$$\{(x, y) \mid 5x - 6y = -15\} \quad \text{Set-builder notation}$$

NOW TRY

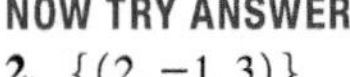

NOW TRY ANSWERS

2. $\{(2, -1, 3)\}$

3. (a) $\varnothing$

(b) $\{(x, y) \mid x + 2y = 7\}$

A Exercises

MyLab Math

Video solutions for select problems available in MyLab Math

Concept Check *Answer the following for each matrix.*

(a) What are the elements of the second row?

(b) What are the elements of the third column?

(c) Is this a square matrix? Explain why or why not.

(d) Give the matrix obtained by interchanging the first and third rows.

(e) Give the matrix obtained by multiplying the first row by $-\frac{1}{2}$.

(f) Give the matrix obtained by multiplying the third row by 3 and adding to the first row.

1. $\begin{bmatrix} -2 & 3 & 1 \\ 0 & 5 & -3 \\ 1 & 4 & 8 \end{bmatrix}$

2. $\begin{bmatrix} -7 & 0 & 1 \\ 3 & 2 & -2 \\ 0 & 1 & 6 \end{bmatrix}$

Concept Check *Give the dimensions of each matrix.*

3. $\begin{bmatrix} 3 & -7 \\ 4 & 5 \\ -1 & 0 \end{bmatrix}$

4. $\begin{bmatrix} 4 & 9 & 0 \\ -1 & 2 & -4 \end{bmatrix}$

5. $\begin{bmatrix} 6 & 3 \\ -2 & 5 \\ 4 & 10 \\ 1 & -1 \end{bmatrix}$

6. $\begin{bmatrix} 8 & 4 & 3 & 2 \end{bmatrix}$

Use row operations to solve each system. ***See Examples 1 and 3.***

7. $\begin{aligned} x + y &= 5 \\ x - y &= 3 \end{aligned}$

8. $\begin{aligned} x + 2y &= 7 \\ x - y &= -2 \end{aligned}$

9. $\begin{aligned} 2x + 4y &= 6 \\ 3x - y &= 2 \end{aligned}$

10. $\begin{aligned} 4x + 5y &= -7 \\ 3x - y &= 9 \end{aligned}$

11. $\begin{aligned} 3x + 4y &= 13 \\ 2x - 3y &= -14 \end{aligned}$

12. $\begin{aligned} 5x + 2y &= 8 \\ 3x - 5y &= 11 \end{aligned}$

13. $\begin{aligned} -4x + 12y &= 36 \\ x - 3y &= 9 \end{aligned}$

14. $\begin{aligned} 2x - 4y &= 8 \\ -x + 2y &= 1 \end{aligned}$

15. $\begin{aligned} 2x + y &= 4 \\ 4x + 2y &= 8 \end{aligned}$

16. $\begin{aligned} -3x - 4y &= 1 \\ 6x + 8y &= -2 \end{aligned}$

17. $\begin{aligned} -3x + 2y &= 0 \\ 2x - 2y &= 0 \end{aligned}$

18. $\begin{aligned} -5x + 3y &= 0 \\ 7x + 2y &= 0 \end{aligned}$

Use row operations to solve each system. ***See Examples 2 and 3.***

19. $\begin{aligned} x + y - 3z &= 1 \\ 2x - y + z &= 9 \\ 3x + y - 4z &= 8 \end{aligned}$

20. $\begin{aligned} 2x + 4y - 3z &= -18 \\ 3x + y - z &= -5 \\ x - 2y + 4z &= 14 \end{aligned}$

21. $\begin{aligned} x + y - z &= 6 \\ 2x - y + z &= -9 \\ x - 2y + 3z &= 1 \end{aligned}$

22. $\begin{aligned} x + 3y - 6z &= 7 \\ 2x - y + 2z &= 0 \\ x + y + 2z &= -1 \end{aligned}$

23. $\begin{aligned} x - y &= 1 \\ y - z &= 6 \\ x + z &= -1 \end{aligned}$

24. $\begin{aligned} x + y &= 1 \\ 2x - z &= 0 \\ y + 2z &= -2 \end{aligned}$

25. $\begin{aligned} x - 2y + z &= 4 \\ 3x - 6y + 3z &= 12 \\ -2x + 4y - 2z &= -8 \end{aligned}$

26. $\begin{aligned} x + 3y + z &= 1 \\ 2x + 6y + 2z &= 2 \\ 3x + 9y + 3z &= 3 \end{aligned}$

27. $\begin{aligned} x + 2y + 3z &= -2 \\ 2x + 4y + 6z &= -5 \\ x - y + 2z &= 6 \end{aligned}$

28. $\begin{aligned} 4x + 8y + 4z &= 9 \\ x + 3y + 4z &= 10 \\ 5x + 10y + 5z &= 12 \end{aligned}$

Determinants and Cramer's Rule

APPENDIX B

OBJECTIVES

1. Evaluate 2 × 2 determinants.
2. Use expansion by minors to evaluate 3 × 3 determinants.
3. Understand the derivation of Cramer's rule.
4. Apply Cramer's rule to solve linear systems.

Recall that an ordered array of numbers within square brackets is a **matrix** (plural *matrices*). Matrices are named according to the number of rows and columns they contain. A **square matrix** has the same number of rows and columns.

Columns

Rows $\begin{bmatrix} 2 & 3 & 5 \\ 7 & 1 & 2 \end{bmatrix}$ 2 × 3 matrix $\qquad \begin{bmatrix} -1 & 0 \\ 1 & -2 \end{bmatrix}$ 2 × 2 square matrix

Associated with every *square matrix* is a real number called the **determinant** of the matrix. A determinant is symbolized by the entries of the matrix placed between two vertical lines.

$\begin{vmatrix} 2 & 3 \\ 7 & 1 \end{vmatrix}$ 2 × 2 determinant $\qquad \begin{vmatrix} 7 & 4 & 3 \\ 0 & 1 & 5 \\ 6 & 0 & 1 \end{vmatrix}$ 3 × 3 determinant

Like matrices, determinants are named according to the number of rows and columns they contain.

VOCABULARY

- ☐ matrix (plural *matrices*)
- ☐ square matrix
- ☐ determinant
- ☐ minor
- ☐ array of signs

OBJECTIVE 1 Evaluate 2 × 2 determinants.

The value of a determinant is a *real number*. We use the following rule to evaluate a 2 × 2 determinant.

Value of a 2 × 2 Determinant

$$\begin{vmatrix} a & b \\ c & d \end{vmatrix} = ad - bc$$

NOW TRY EXERCISE 1

Evaluate the determinant.

$$\begin{vmatrix} 6 & 3 \\ 4 & -5 \end{vmatrix}$$

EXAMPLE 1 Evaluating a 2 × 2 Determinant

Evaluate the determinant.

$$\begin{vmatrix} -1 & -3 \\ 4 & -2 \end{vmatrix}$$

Use the rule stated above. Here $a = -1$, $b = -3$, $c = 4$, and $d = -2$.

$$\begin{vmatrix} -1 & -3 \\ 4 & -2 \end{vmatrix} = -1(-2) - (-3)4$$
$$= 2 + 12$$
$$= 14$$

NOW TRY 

NOW TRY ANSWER

1. −42

Value of a 3 × 3 Determinant

$$\begin{vmatrix} a_1 & b_1 & c_1 \\ a_2 & b_2 & c_2 \\ a_3 & b_3 & c_3 \end{vmatrix} = (a_1b_2c_3 + b_1c_2a_3 + c_1a_2b_3) - (a_3b_2c_1 + b_3c_2a_1 + c_3a_2b_1)$$

To calculate a 3 × 3 determinant, we rearrange terms using the distributive property.

$$\begin{vmatrix} a_1 & b_1 & c_1 \\ a_2 & b_2 & c_2 \\ a_3 & b_3 & c_3 \end{vmatrix} = a_1(b_2c_3 - b_3c_2) - a_2(b_1c_3 - b_3c_1) + a_3(b_1c_2 - b_2c_1) \qquad (1)$$

Each quantity in parentheses represents a 2 × 2 determinant that is the part of the 3 × 3 determinant remaining when the row and column of the multiplier are eliminated, as shown below.

$$a_1(b_2c_3 - b_3c_2) \begin{vmatrix} a_1 & b_1 & c_1 \\ a_2 & b_2 & c_2 \\ a_3 & b_3 & c_3 \end{vmatrix}$$ Eliminate the 1st row and 1st column.

$$a_2(b_1c_3 - b_3c_1) \begin{vmatrix} a_1 & b_1 & c_1 \\ a_2 & b_2 & c_2 \\ a_3 & b_3 & c_3 \end{vmatrix}$$ Eliminate the 2nd row and 1st column.

$$a_3(b_1c_2 - b_2c_1) \begin{vmatrix} a_1 & b_1 & c_1 \\ a_2 & b_2 & c_2 \\ a_3 & b_3 & c_3 \end{vmatrix}$$ Eliminate the 3rd row and 1st column.

These 2 × 2 determinants are **minors** of the elements in the 3 × 3 determinant. In the determinant above, the minors of a_1, a_2, and a_3 are, respectively,

$$\begin{vmatrix} b_2 & c_2 \\ b_3 & c_3 \end{vmatrix},$$

$$\begin{vmatrix} b_1 & c_1 \\ b_3 & c_3 \end{vmatrix}, \quad \text{Minors}$$

and

$$\begin{vmatrix} b_1 & c_1 \\ b_2 & c_2 \end{vmatrix}.$$

OBJECTIVE 2 **Use expansion by minors to evaluate 3 × 3 determinants.**

We evaluate a 3 × 3 determinant by multiplying each element in the first column by its minor and combining the products as indicated in equation (1) above. This procedure is called **expansion of the determinant by minors** about the first column.

NOW TRY EXERCISE 2

Evaluate the determinant using expansion by minors about the first column.

$$\begin{vmatrix} 0 & -2 & 3 \\ 4 & 1 & -5 \\ 6 & -1 & 5 \end{vmatrix}$$

EXAMPLE 2 Evaluating a 3 × 3 Determinant

Evaluate the determinant using expansion by minors about the first column.

$$\begin{vmatrix} 1 & 3 & -2 \\ -1 & -2 & -3 \\ 1 & 1 & 2 \end{vmatrix}$$

In this determinant, $a_1 = 1$, $a_2 = -1$, and $a_3 = 1$. Multiply each of these numbers by its minor, and combine the three terms using the definition.

Notice that the second term in the definition is *subtracted.*

$$\begin{vmatrix} 1 & 3 & -2 \\ -1 & -2 & -3 \\ 1 & 1 & 2 \end{vmatrix} = 1\begin{vmatrix} -2 & -3 \\ 1 & 2 \end{vmatrix} - (-1)\begin{vmatrix} 3 & -2 \\ 1 & 2 \end{vmatrix} + 1\begin{vmatrix} 3 & -2 \\ -2 & -3 \end{vmatrix}$$

$$= 1[-2(2) - (-3)1] + 1[3(2) - (-2)1]$$

Use parentheses and brackets to avoid errors.

$$+ 1[3(-3) - (-2)(-2)]$$

$$= 1(-1) + 1(8) + 1(-13)$$

$$= -1 + 8 - 13$$

$$= -6$$

NOW TRY 

To obtain equation (1) on the preceding page, we could have rearranged terms in the definition of the determinant and used the distributive property to factor out the three elements of the second or third column or of any of the three rows.

Expanding by minors about any row or any column results in the same value for a 3 × 3 determinant.

Array of Signs for a 3 × 3 Determinant

+	−	+
−	+	−
+	−	+

To determine the correct signs for the terms of other expansions, the **array of signs** in the margin is helpful. The signs alternate for each row and column beginning with a + in the first row, first column position. For example, if the expansion is to be about the second column, the first term would have a minus sign associated with it, the second term a plus sign, and the third term a minus sign.

NOW TRY EXERCISE 3

Evaluate the determinant using expansion by minors about the second column.

$$\begin{vmatrix} 0 & -2 & 3 \\ 4 & 1 & -5 \\ 6 & -1 & 5 \end{vmatrix}$$

EXAMPLE 3 Evaluating a 3 × 3 Determinant

Evaluate the determinant of **Example 2** using expansion by minors about the second column.

$$\begin{vmatrix} 1 & 3 & -2 \\ -1 & -2 & -3 \\ 1 & 1 & 2 \end{vmatrix} = -3\begin{vmatrix} -1 & -3 \\ 1 & 2 \end{vmatrix} + (-2)\begin{vmatrix} 1 & -2 \\ 1 & 2 \end{vmatrix} - 1\begin{vmatrix} 1 & -2 \\ -1 & -3 \end{vmatrix}$$

$$= -3(1) - 2(4) - 1(-5)$$

$$= -3 - 8 + 5$$

$$= -6$$ The result is the same as in **Example 2.**

NOW TRY 

NOW TRY ANSWERS

2. 70

3. 70

NORMAL FLOAT AUTO REAL RADIAN MP

[A]

$\begin{bmatrix} -1 & -3 \\ 4 & -2 \end{bmatrix}$

det([A])

14

NORMAL FLOAT AUTO REAL RADIAN MP

[B]

$\begin{bmatrix} 1 & 3 & -2 \\ -1 & -2 & -3 \\ 1 & 1 & 2 \end{bmatrix}$

det([B])

-6

The TI-84 Plus C gives the results found in **Examples 1–3.**

OBJECTIVE 3 Understand the derivation of Cramer's rule.

We can use determinants to solve a system of equations of the form

$$a_1x + b_1y = c_1 \quad (1)$$
$$a_2x + b_2y = c_2. \quad (2)$$

The result is a formula to solve any system of two equations with two variables.

$$a_1b_2x + b_1b_2y = c_1b_2 \quad \text{Multiply equation (1) by } b_2.$$
$$-a_2b_1x - b_1b_2y = -c_2b_1 \quad \text{Multiply equation (2) by } -b_1.$$
$$(a_1b_2 - a_2b_1)x = c_1b_2 - c_2b_1 \quad \text{Add.}$$

Solve for x.

$$x = \frac{c_1b_2 - c_2b_1}{a_1b_2 - a_2b_1} \quad (\text{if } a_1b_2 - a_2b_1 \neq 0)$$

To solve for y, we multiply each side of equation (1) by $-a_2$ and each side of equation (2) by a_1 and add.

$$-a_1a_2x - a_2b_1y = -a_2c_1 \quad \text{Multiply equation (1) by } -a_2.$$
$$a_1a_2x + a_1b_2y = a_1c_2 \quad \text{Multiply equation (2) by } a_1.$$
$$(a_1b_2 - a_2b_1)y = a_1c_2 - a_2c_1 \quad \text{Add.}$$

$$y = \frac{a_1c_2 - a_2c_1}{a_1b_2 - a_2b_1} \quad (\text{if } a_1b_2 - a_2b_1 \neq 0)$$

We can write both numerators and the common denominator of these values for x and y as determinants because

$$a_1c_2 - a_2c_1 = \begin{vmatrix} a_1 & c_1 \\ a_2 & c_2 \end{vmatrix}, \quad c_1b_2 - c_2b_1 = \begin{vmatrix} c_1 & b_1 \\ c_2 & b_2 \end{vmatrix}, \quad \text{and} \quad a_1b_2 - a_2b_1 = \begin{vmatrix} a_1 & b_1 \\ a_2 & b_2 \end{vmatrix}.$$

Using these results, the solutions for x and y become

$$x = \frac{\begin{vmatrix} c_1 & b_1 \\ c_2 & b_2 \end{vmatrix}}{\begin{vmatrix} a_1 & b_1 \\ a_2 & b_2 \end{vmatrix}} \quad \text{and} \quad y = \frac{\begin{vmatrix} a_1 & c_1 \\ a_2 & c_2 \end{vmatrix}}{\begin{vmatrix} a_1 & b_1 \\ a_2 & b_2 \end{vmatrix}}, \quad \text{where} \quad \begin{vmatrix} a_1 & b_1 \\ a_2 & b_2 \end{vmatrix} \neq 0.$$

For convenience, we denote the three determinants in the solution as

$$\begin{vmatrix} a_1 & b_1 \\ a_2 & b_2 \end{vmatrix} = D, \quad \begin{vmatrix} c_1 & b_1 \\ c_2 & b_2 \end{vmatrix} = D_x, \quad \text{and} \quad \begin{vmatrix} a_1 & c_1 \\ a_2 & c_2 \end{vmatrix} = D_y.$$

The elements of D are the four coefficients of the variables in the given system. The elements of D_x and D_y are obtained by replacing the coefficients of x and y by the respective constants. These results are summarized as **Cramer's rule.**

Cramer's Rule for 2 × 2 Systems

For the system $\begin{matrix} a_1x + b_1y = c_1 \\ a_2x + b_2y = c_2 \end{matrix}$, the values of x and y are given by

$$x = \frac{\begin{vmatrix} c_1 & b_1 \\ c_2 & b_2 \end{vmatrix}}{\begin{vmatrix} a_1 & b_1 \\ a_2 & b_2 \end{vmatrix}} = \frac{D_x}{D} \quad \text{and} \quad y = \frac{\begin{vmatrix} a_1 & c_1 \\ a_2 & c_2 \end{vmatrix}}{\begin{vmatrix} a_1 & b_1 \\ a_2 & b_2 \end{vmatrix}} = \frac{D_y}{D}.$$

$a_1b_2 - a_2b_1 = D$, and $D \neq 0$

OBJECTIVE 4 Apply Cramer's rule to solve linear systems.

To use Cramer's rule to solve a system of equations, we find the three determinants, D, D_x, and D_y, and then write the necessary quotients for x and y.

CAUTION As indicated in the box on the preceding page, ***Cramer's rule does not apply if***

$$D = a_1b_2 - a_2b_1 = 0.$$

When $D = 0$, the system is inconsistent or has dependent equations.

NOW TRY EXERCISE 4

Use Cramer's rule to solve the system.

$$3x - 2y = -33$$
$$2x + 3y = -9$$

EXAMPLE 4 Using Cramer's Rule to Solve a 2 × 2 System

Use Cramer's rule to solve the system.

$$5x + 7y = -1$$
$$6x + 8y = 1$$

It is a good idea to evaluate D first in case it equals 0.

$$D = \begin{vmatrix} 5 & 7 \\ 6 & 8 \end{vmatrix} = 5(8) - 7(6) = -2 \qquad \text{Find } D.$$

$$D_x = \begin{vmatrix} -1 & 7 \\ 1 & 8 \end{vmatrix} = -1(8) - 7(1) = -15$$

$D \neq 0$, so find D_x and D_y.

$$D_y = \begin{vmatrix} 5 & -1 \\ 6 & 1 \end{vmatrix} = 5(1) - (-1)6 = 11$$

From Cramer's rule, $x = \frac{D_x}{D} = \frac{-15}{-2} = \frac{15}{2}$ and $y = \frac{D_y}{D} = \frac{11}{-2} = -\frac{11}{2}$.

A check confirms that the solution set is $\left\{\left(\frac{15}{2}, -\frac{11}{2}\right)\right\}$. **NOW TRY**

Cramer's Rule for 3 × 3 Systems

For the system

$$a_1x + b_1y + c_1z = d_1$$
$$a_2x + b_2y + c_2z = d_2$$
$$a_3x + b_3y + c_3z = d_3,$$

with

$$D_x = \begin{vmatrix} d_1 & b_1 & c_1 \\ d_2 & b_2 & c_2 \\ d_3 & b_3 & c_3 \end{vmatrix}, \quad D_y = \begin{vmatrix} a_1 & d_1 & c_1 \\ a_2 & d_2 & c_2 \\ a_3 & d_3 & c_3 \end{vmatrix},$$

$$D_z = \begin{vmatrix} a_1 & b_1 & d_1 \\ a_2 & b_2 & d_2 \\ a_3 & b_3 & d_3 \end{vmatrix}, \quad D = \begin{vmatrix} a_1 & b_1 & c_1 \\ a_2 & b_2 & c_2 \\ a_3 & b_3 & c_3 \end{vmatrix} \neq 0,$$

the values of x, y, and z are given by

$$x = \frac{D_x}{D}, \quad y = \frac{D_y}{D}, \quad \text{and} \quad z = \frac{D_z}{D}.$$

NOW TRY ANSWER

4. $\{(-9, 3)\}$

NOW TRY EXERCISE 5

Use Cramer's rule to solve the system.

$$\begin{aligned} 4x + 2y + z &= 15 \\ -2x + 5y - 2z &= -6 \\ x - 3y + 4z &= 0 \end{aligned}$$

EXAMPLE 5 Using Cramer's Rule to Solve a 3 × 3 System

Use Cramer's rule to solve the system.

$$\begin{aligned} x + y - z &= -2 \\ 2x - y + z &= -5 \\ x - 2y + 3z &= 4 \end{aligned}$$

We expand by minors about row 1 to find D.

$$D = \begin{vmatrix} 1 & 1 & -1 \\ 2 & -1 & 1 \\ 1 & -2 & 3 \end{vmatrix}$$

$$= 1\begin{vmatrix} -1 & 1 \\ -2 & 3 \end{vmatrix} - 1\begin{vmatrix} 2 & 1 \\ 1 & 3 \end{vmatrix} + (-1)\begin{vmatrix} 2 & -1 \\ 1 & -2 \end{vmatrix}$$

$$= 1(-1) - 1(5) - 1(-3)$$

$$= -3$$

Verify that $D_x = 7$, $D_y = -22$, and $D_z = -21$. Thus,

$$x = \frac{D_x}{D} = \frac{7}{-3} = -\frac{7}{3}, \quad y = \frac{D_y}{D} = \frac{-22}{-3} = \frac{22}{3}, \quad z = \frac{D_z}{D} = \frac{-21}{-3} = 7.$$

Check that the solution set is $\left\{\left(-\frac{7}{3}, \frac{22}{3}, 7\right)\right\}$.

NOW TRY 

NOW TRY EXERCISE 6

Use Cramer's rule, if possible, to solve the system.

$$\begin{aligned} 5x + 3y + z &= 1 \\ x - 2y + 3z &= 6 \\ 10x + 6y + 2z &= 3 \end{aligned}$$

EXAMPLE 6 Determining When Cramer's Rule Does Not Apply

Use Cramer's rule, if possible, to solve the system.

$$\begin{aligned} 2x - 3y + 4z &= 8 \\ 6x - 9y + 12z &= 24 \\ x + 2y - 3z &= 5 \end{aligned}$$

First, find D.

$$D = \begin{vmatrix} 2 & -3 & 4 \\ 6 & -9 & 12 \\ 1 & 2 & -3 \end{vmatrix}$$

$$= 2\begin{vmatrix} -9 & 12 \\ 2 & -3 \end{vmatrix} - 6\begin{vmatrix} -3 & 4 \\ 2 & -3 \end{vmatrix} + 1\begin{vmatrix} -3 & 4 \\ -9 & 12 \end{vmatrix}$$

$$= 2(3) - 6(1) + 1(0)$$

$$= 0$$

Because $D = 0$ here, Cramer's rule does not apply and we must use another method to solve the system. Multiplying each side of the first equation by 3 shows that the first two equations have the same solution set, so this system has dependent equations and an infinite solution set.

NOW TRY

NOW TRY ANSWERS

5. $\{(4, 0, -1)\}$
6. Cramer's rule does not apply because $D = 0$.

B Exercises

FOR EXTRA HELP  MyLab Math

Video solutions for select problems available in MyLab Math

Concept Check *Determine whether each statement is* true *or* false. *If false, explain why.*

1. A matrix is an array of numbers, while a determinant is a single number.

2. A square matrix has the same number of rows as columns.

3. The determinant $\begin{vmatrix} a & b \\ c & d \end{vmatrix}$ is equal to $ad + bc$.

4. The value of $\begin{vmatrix} 0 & 0 \\ x & y \end{vmatrix}$ is 0 for any replacements for x and y.

Concept Check *Which choice is an expression for each determinant?*

A. $-2(-6) + (-3)4$ **B.** $-2(-6) - 3(4)$

C. $-3(4) - (-2)(-6)$ **D.** $-2(-6) - (-3)4$

5. $\begin{vmatrix} -2 & -3 \\ 4 & -6 \end{vmatrix}$ **6.** $\begin{vmatrix} -2 & 4 \\ 3 & -6 \end{vmatrix}$

Evaluate each determinant. ***See Example 1.***

7. $\begin{vmatrix} -2 & 5 \\ -1 & 4 \end{vmatrix}$ **8.** $\begin{vmatrix} 3 & -6 \\ 2 & -2 \end{vmatrix}$ **9.** $\begin{vmatrix} 1 & -2 \\ 7 & 0 \end{vmatrix}$

10. $\begin{vmatrix} -5 & -1 \\ 1 & 0 \end{vmatrix}$ **11.** $\begin{vmatrix} 0 & 4 \\ 0 & 4 \end{vmatrix}$ **12.** $\begin{vmatrix} 8 & -3 \\ 0 & 0 \end{vmatrix}$

Evaluate each determinant by expansion by minors about the first column. ***See Example 2.***

13. $\begin{vmatrix} -1 & 2 & 4 \\ -3 & -2 & -3 \\ 2 & -1 & 5 \end{vmatrix}$ **14.** $\begin{vmatrix} 2 & -3 & -5 \\ 1 & 2 & 2 \\ 5 & 3 & -1 \end{vmatrix}$

15. $\begin{vmatrix} 1 & 0 & -2 \\ 0 & 2 & 3 \\ 1 & 0 & 5 \end{vmatrix}$ **16.** $\begin{vmatrix} 2 & -1 & 0 \\ 0 & -1 & 1 \\ 1 & 2 & 0 \end{vmatrix}$

Evaluate each determinant using expansion by minors about any row or column. (Hint: The work is easier if a row or a column with 0s is used.) ***See Example 3.***

17. $\begin{vmatrix} 3 & -1 & 2 \\ 1 & 5 & -2 \\ 0 & 2 & 0 \end{vmatrix}$ **18.** $\begin{vmatrix} 4 & 4 & 2 \\ 1 & -1 & -2 \\ 1 & 0 & 2 \end{vmatrix}$ **19.** $\begin{vmatrix} 0 & 0 & 3 \\ 4 & 0 & -2 \\ 2 & -1 & 3 \end{vmatrix}$

20. $\begin{vmatrix} 3 & 5 & -2 \\ 1 & -4 & 1 \\ 3 & 1 & -2 \end{vmatrix}$ **21.** $\begin{vmatrix} 1 & 1 & 2 \\ 5 & 5 & 7 \\ 3 & 3 & 1 \end{vmatrix}$ **22.** $\begin{vmatrix} 3 & 3 & -2 \\ 1 & 1 & 1 \\ 3 & 3 & -2 \end{vmatrix}$

23. Concept Check For the following system, $D = -43$, $D_x = -43$, $D_y = 0$, and $D_z = 43$. What is the solution set of the system?

$$\begin{aligned} x + 3y - 6z &= 7 \\ 2x - y + z &= 1 \\ x + 2y + 2z &= -1 \end{aligned}$$

24. Concept Check Consider this system.

$$\begin{aligned} 4x + 3y - 2z &= 1 \\ 7x - 4y + 3z &= 2 \\ -2x + y - 8z &= 0 \end{aligned}$$

Match each determinant in parts (a)–(d) with its correct representation from choices A–D.

(a) D **(b)** D_x **(c)** D_y **(d)** D_z

A. $\begin{vmatrix} 1 & 3 & -2 \\ 2 & -4 & 3 \\ 0 & 1 & -8 \end{vmatrix}$ **B.** $\begin{vmatrix} 4 & 3 & 1 \\ 7 & -4 & 2 \\ -2 & 1 & 0 \end{vmatrix}$

C. $\begin{vmatrix} 4 & 1 & -2 \\ 7 & 2 & 3 \\ -2 & 0 & -8 \end{vmatrix}$ **D.** $\begin{vmatrix} 4 & 3 & -2 \\ 7 & -4 & 3 \\ -2 & 1 & -8 \end{vmatrix}$

Use Cramer's rule to solve each system. **See Example 4.**

25. $5x + 2y = -3$
$4x - 3y = -30$

26. $3x + 5y = -5$
$-2x + 3y = 16$

27. $3x - y = 9$
$2x + 5y = 8$

28. $8x + 3y = 1$
$6x - 5y = 2$

29. $4x + 5y = 6$
$7x + 8y = 9$

30. $2x + 3y = 4$
$5x + 6y = 7$

Use Cramer's rule, if possible, to solve each system. **See Examples 5 and 6.**

31. $x - y + 6z = 19$
$3x + 3y - z = 1$
$x + 9y + 2z = -19$

32. $2x + 3y + 2z = 15$
$x - y + 2z = 5$
$x + 2y - 6z = -26$

33. $7x + y - z = 4$
$2x - 3y + z = 2$
$-6x + 9y - 3z = -6$

34. $2x - 3y + 4z = 8$
$6x - 9y + 12z = 24$
$-4x + 6y - 8z = -16$

35. $-x + 2y = 4$
$3x + y = -5$
$2x + z = -1$

36. $3x + 5z = 0$
$2x + 3y = 1$
$-y + 2z = -11$

37. $-5x - y = -10$
$3x + 2y + z = -3$
$-y - 2z = -13$

38. $x - 3y = 13$
$2y + z = 5$
$-x + z = -7$

Extending Skills *Solve each equation by finding an expression for the determinant on the left, and then solving using earlier methods.*

39. $\begin{vmatrix} 4 & x \\ 2 & 3 \end{vmatrix} = 8$ **40.** $\begin{vmatrix} -2 & 10 \\ x & 6 \end{vmatrix} = 0$ **41.** $\begin{vmatrix} x & 4 \\ x & -3 \end{vmatrix} = 0$ **42.** $\begin{vmatrix} 5 & 3 \\ x & x \end{vmatrix} = 20$

Properties of Matrices

APPENDIX C

OBJECTIVES

1 Apply the basic definitions for matrices.
2 Add and subtract matrices.
3 Multiply a matrix by a scalar.
4 Multiply matrices.
5 Use matrices in applications.

OBJECTIVE 1 Apply the basic definitions for matrices.

Recall that we use capital letters to name matrices. Subscript notation is used to name the elements of a matrix, as in the following matrix A.

$$A = \begin{bmatrix} a_{11} & a_{12} & a_{13} & \cdots & a_{1n} \\ a_{21} & a_{22} & a_{23} & \cdots & a_{2n} \\ a_{31} & a_{32} & a_{33} & \cdots & a_{3n} \\ \vdots & \vdots & \vdots & & \vdots \\ a_{m1} & a_{m2} & a_{m3} & \cdots & a_{mn} \end{bmatrix}$$

The first-row, first-column element is a_{11} (read "a-sub-one-one"). The second-row, third-column element is a_{23}. In general, the ith-row, jth-column element is a_{ij}.

An $n \times n$ matrix is a **square matrix of order *n*.** Also, a matrix with just one row is a **row matrix,** and a matrix with just one column is a **column matrix.**

Two matrices are equal if they have the same dimensions and if corresponding elements, position by position, are equal. Using this definition, the following matrices are *not* equal (even though they contain the same elements and have the same dimensions) because the corresponding elements differ.

$$\begin{bmatrix} 2 & 1 \\ 3 & -5 \end{bmatrix} \quad \text{and} \quad \begin{bmatrix} 1 & 2 \\ -5 & 3 \end{bmatrix} \quad \text{Not equal}$$

NOW TRY EXERCISE 1

Find the values of the variables for which each statement is true.

(a) $\begin{bmatrix} 9 & a \\ b & 13 \end{bmatrix} = \begin{bmatrix} c & 4 \\ -2 & d \end{bmatrix}$

(b) $\begin{bmatrix} x & y & z \end{bmatrix} = \begin{bmatrix} 1 \\ 2 \\ 3 \end{bmatrix}$

EXAMPLE 1 Deciding Whether Two Matrices Are Equal

Find the values of the variables for which each statement is true.

(a) $\begin{bmatrix} 2 & 1 \\ p & q \end{bmatrix} = \begin{bmatrix} x & y \\ -1 & 0 \end{bmatrix}$ The only way this statement can be true is if $2 = x$, $1 = y$, $p = -1$, and $q = 0$.

(b) $\begin{bmatrix} x \\ y \end{bmatrix} = \begin{bmatrix} 1 \\ 4 \\ 0 \end{bmatrix}$ This statement can never be true. The two matrices have different dimensions. (One is 2×1 and the other is 3×1.)

NOW TRY

VOCABULARY

- ☐ square matrix of order n
- ☐ row matrix
- ☐ column matrix
- ☐ zero matrix
- ☐ additive inverse (negative)
- ☐ scalar

OBJECTIVE 2 Add and subtract matrices.

Addition of Matrices

To add two matrices of the same dimensions, add corresponding elements. Only matrices of the same dimensions can be added.

NOW TRY ANSWERS

1. **(a)** $a = 4, b = -2, c = 9, d = 13$
(b) This can never be true.

NOW TRY EXERCISE 2

Find each sum, if possible.

(a) $\begin{bmatrix} 6 & -3 & 5 \\ 8 & 8 & 10 \end{bmatrix} + \begin{bmatrix} -9 & 12 & -14 \\ 3 & 5 & 4 \end{bmatrix}$

(b) $\begin{bmatrix} 2 & 4 \end{bmatrix} + \begin{bmatrix} 9 \\ -6 \end{bmatrix}$

EXAMPLE 2 Adding Matrices

Find each sum, if possible.

(a) $\begin{bmatrix} 5 & -6 \\ 8 & 9 \end{bmatrix} + \begin{bmatrix} -4 & 6 \\ 8 & -3 \end{bmatrix} = \begin{bmatrix} 5+(-4) & -6+6 \\ 8+8 & 9+(-3) \end{bmatrix} = \begin{bmatrix} 1 & 0 \\ 16 & 6 \end{bmatrix}$

(b) $\begin{bmatrix} 2 \\ 5 \\ 8 \end{bmatrix} + \begin{bmatrix} -6 \\ 3 \\ 12 \end{bmatrix} = \begin{bmatrix} -4 \\ 8 \\ 20 \end{bmatrix}$ These matrices have the same dimension, 3×1. Add corresponding elements.

(c) $\begin{bmatrix} 5 & 8 \\ 6 & 2 \end{bmatrix} + \begin{bmatrix} 3 & 9 & 1 \\ 4 & 2 & 5 \end{bmatrix}$ Because these matrices have different dimensions, the sum cannot be found.

NOW TRY 

A matrix containing only zero elements is a **zero matrix.**

$$O = \begin{bmatrix} 0 & 0 & 0 \end{bmatrix} \quad 1 \times 3 \text{ zero matrix} \qquad O = \begin{bmatrix} 0 & 0 & 0 \\ 0 & 0 & 0 \end{bmatrix} \quad 2 \times 3 \text{ zero matrix}$$

By the additive inverse property, each real number has an additive inverse: If a is a real number, then there is a real number $-a$ such that

$$a + (-a) = 0 \quad \text{and} \quad -a + a = 0.$$

Similarly, given the matrix

$$A = \begin{bmatrix} -5 & 2 & -1 \\ 3 & 4 & -6 \end{bmatrix},$$

for example, there is a matrix $-A$ such that

$$A + (-A) = O, \quad \text{where } O \text{ is the } 2 \times 3 \text{ zero matrix.}$$

$-A$ has as elements the additive inverses of the elements of A. (Remember, each element of A is a real number and therefore has an additive inverse.)

$$-A = \begin{bmatrix} 5 & -2 & 1 \\ -3 & -4 & 6 \end{bmatrix}$$

To check, test that $A + (-A)$ equals the zero matrix, O.

$$A + (-A) = \begin{bmatrix} -5 & 2 & -1 \\ 3 & 4 & -6 \end{bmatrix} + \begin{bmatrix} 5 & -2 & 1 \\ -3 & -4 & 6 \end{bmatrix} = \begin{bmatrix} 0 & 0 & 0 \\ 0 & 0 & 0 \end{bmatrix} = O$$

Matrix $-A$ is the **additive inverse,** or **negative,** of matrix A. Every matrix has an additive inverse.

The real number b is subtracted from the real number a, written $a - b$, by adding a and the additive inverse of b. That is, $a - b = a + (-b)$.

The same definition applies to subtraction of matrices.

Subtraction of Matrices

To subtract two matrices of the same dimensions, use the usual definition of subtraction.

$$A - B = A + (-B)$$

NOW TRY ANSWERS

2. **(a)** $\begin{bmatrix} -3 & 9 & -9 \\ 11 & 13 & 14 \end{bmatrix}$

(b) The sum cannot be found.

NOW TRY EXERCISE 3

Find each difference, if possible.

(a) $\begin{bmatrix} -3 & 7 \\ -4 & 12 \end{bmatrix} - \begin{bmatrix} -5 & 8 \\ 12 & 0 \end{bmatrix}$

(b) $\begin{bmatrix} 1 \\ 3 \end{bmatrix} - \begin{bmatrix} 3 & 1 \end{bmatrix}$

EXAMPLE 3 Subtracting Matrices

Find each difference, if possible.

(a) $\begin{bmatrix} -5 & 6 \\ 2 & 4 \end{bmatrix} - \begin{bmatrix} -3 & 2 \\ 5 & -8 \end{bmatrix} = \begin{bmatrix} -5-(-3) & 6-2 \\ 2-5 & 4-(-8) \end{bmatrix} = \begin{bmatrix} -2 & 4 \\ -3 & 12 \end{bmatrix}$

(b) $\begin{bmatrix} 8 & 6 & -4 \end{bmatrix} - \begin{bmatrix} 3 & 5 & -8 \end{bmatrix} = \begin{bmatrix} 5 & 1 & 4 \end{bmatrix}$ Subtract corresponding elements.

(c) $\begin{bmatrix} -2 & 5 \\ 0 & 1 \end{bmatrix} - \begin{bmatrix} 3 \\ 5 \end{bmatrix}$ These matrices have different dimensions, so the difference cannot be found.

NOW TRY 

OBJECTIVE 3 Multiply a matrix by a scalar.

In work with matrices, a real number is called a **scalar** to distinguish it from a matrix.

The product of a scalar k and a matrix X is the matrix kX, each of whose elements is k times the corresponding element of X.

NOW TRY EXERCISE 4

Find the product.

$$-\frac{2}{3}\begin{bmatrix} 6 & -18 \\ 12 & 9 \end{bmatrix}$$

EXAMPLE 4 Multiplying a Matrix by a Scalar

Find each product.

(a) $5\begin{bmatrix} 2 & -3 \\ 0 & 4 \end{bmatrix} = \begin{bmatrix} 5(2) & 5(-3) \\ 5(0) & 5(4) \end{bmatrix} = \begin{bmatrix} 10 & -15 \\ 0 & 20 \end{bmatrix}$

(b) $\frac{3}{4}\begin{bmatrix} 20 & 36 \\ 12 & -16 \end{bmatrix} = \begin{bmatrix} \frac{3}{4}(20) & \frac{3}{4}(36) \\ \frac{3}{4}(12) & \frac{3}{4}(-16) \end{bmatrix} = \begin{bmatrix} 15 & 27 \\ 9 & -12 \end{bmatrix}$

NOW TRY 

Properties of Scalar Multiplication

If A and B are matrices that have the same dimensions, and c and d are real numbers, then the following hold true.

$$(c+d)A = cA + dA \qquad c(A)d = cd(A)$$
$$c(A+B) = cA + cB \qquad (cd)A = c(dA)$$

OBJECTIVE 4 Multiply matrices.

Multiplication of matrices is defined in such a way that it is appropriate in certain types of applications. To illustrate, we find the product of

$$A = \begin{bmatrix} -3 & 4 & 2 \\ 5 & 0 & 4 \end{bmatrix} \quad \text{and} \quad B = \begin{bmatrix} -6 & 4 \\ 2 & 3 \\ 3 & -2 \end{bmatrix}.$$

First locate *row* one of A and *column* one of B, shown shaded below. Multiply corresponding elements, and find the sum of the products.

$$\begin{bmatrix} -3 & 4 & 2 \\ 5 & 0 & 4 \end{bmatrix}\begin{bmatrix} -6 & 4 \\ 2 & 3 \\ 3 & -2 \end{bmatrix} \qquad -3(-6) + 4(2) + 2(3) = 32$$

The result 32 is the element for row one, column one of the product matrix.

NOW TRY ANSWERS

3. (a) $\begin{bmatrix} 2 & -1 \\ -16 & 12 \end{bmatrix}$

(b) The difference cannot be found.

4. $\begin{bmatrix} -4 & 12 \\ -8 & -6 \end{bmatrix}$

Now use *row* one of A and *column* two of B (shaded below) to determine the element in row one, column two of the product matrix.

$$\begin{bmatrix} -3 & 4 & 2 \\ 5 & 0 & 4 \end{bmatrix} \begin{bmatrix} -6 & 4 \\ 2 & 3 \\ 3 & -2 \end{bmatrix} \qquad -3(4) + 4(3) + 2(-2) = -4$$

Next, use *row* two of A and *column* one of B. This will give the row two, column one entry of the product matrix.

$$\begin{bmatrix} -3 & 4 & 2 \\ 5 & 0 & 4 \end{bmatrix} \begin{bmatrix} -6 & 4 \\ 2 & 3 \\ 3 & -2 \end{bmatrix} \qquad 5(-6) + 0(2) + 4(3) = -18$$

Finally, use *row* two of A and *column* two of B to find the entry for row two, column two of the product matrix.

$$\begin{bmatrix} -3 & 4 & 2 \\ 5 & 0 & 4 \end{bmatrix} \begin{bmatrix} -6 & 4 \\ 2 & 3 \\ 3 & -2 \end{bmatrix} \qquad 5(4) + 0(3) + 4(-2) = 12$$

The product matrix can now be written using the four entries just found.

$$\begin{bmatrix} -3 & 4 & 2 \\ 5 & 0 & 4 \end{bmatrix} \begin{bmatrix} -6 & 4 \\ 2 & 3 \\ 3 & -2 \end{bmatrix} = \begin{bmatrix} 32 & -4 \\ -18 & 12 \end{bmatrix}$$

This is the product AB.

We see that the product of a 2×3 matrix and a 3×2 matrix is a 2×2 matrix.

By definition, the product AB of an $m \times n$ matrix A and an $n \times p$ matrix B is found as follows:

> To find the ith-row, jth-column element of AB, multiply each element in the ith row of A by the corresponding element in the jth column of B. (Note the shaded areas in the matrices below.) The sum of these products will give the element of row i, column j of AB.

$$A = \begin{bmatrix} a_{11} & a_{12} & a_{13} & \cdots & a_{1n} \\ a_{21} & a_{22} & a_{23} & \cdots & a_{2n} \\ \vdots & & & & \\ a_{i1} & a_{i2} & a_{i3} & \cdots & a_{in} \\ \vdots & & & & \\ a_{m1} & a_{m2} & a_{m3} & \cdots & a_{mn} \end{bmatrix} \qquad B = \begin{bmatrix} b_{11} & b_{12} & \cdots & b_{1j} & \cdots & b_{1p} \\ b_{21} & b_{22} & \cdots & b_{2j} & \cdots & b_{2p} \\ \vdots & & & & & \\ b_{n1} & b_{n2} & \cdots & b_{nj} & \cdots & b_{np} \end{bmatrix}$$

Matrix Multiplication

If the number of columns of matrix A is the same as the number of rows of matrix B, then entry c_{ij} of the product matrix $C = AB$ is found as follows.

$$c_{ij} = a_{i1}b_{1j} + a_{i2}b_{2j} + \cdots + a_{in}b_{nj}$$

The final product C will have as many rows as A and as many columns as B.

NOW TRY EXERCISE 5

Determine whether the product can be found. If so, give its dimensions.

(a) $\begin{bmatrix} 0 & -1 \\ 5 & 12 \end{bmatrix} \begin{bmatrix} 6 \\ 8 \end{bmatrix}$

(b) $\begin{bmatrix} 6 \\ 8 \end{bmatrix} \begin{bmatrix} 0 & -1 \\ 5 & 12 \end{bmatrix}$

EXAMPLE 5 Deciding Whether Two Matrices Can Be Multiplied

Given $A = \begin{bmatrix} 3 & 1 \\ 4 & 1 \\ 5 & 9 \end{bmatrix}$ and $B = \begin{bmatrix} 2 & 7 & 1 & 8 \\ 2 & 8 & 1 & 8 \end{bmatrix}$, can the product AB be calculated? If so, what are the dimensions of AB? Can the product BA be calculated? If so, what are the dimensions of BA?

The following diagram helps answer the questions about the product AB.

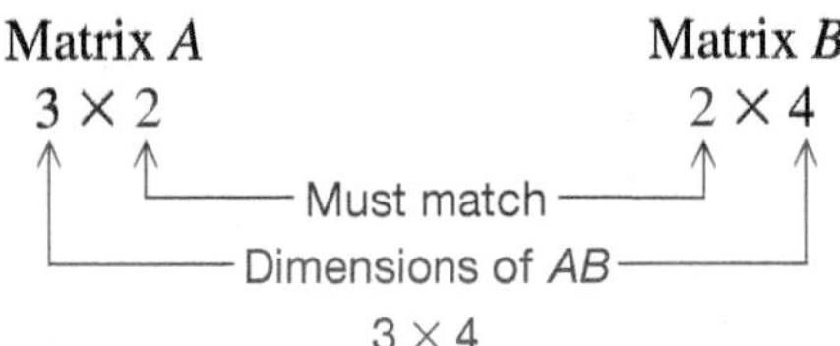

The product AB exists because the number of columns of A equals the number of rows of B. (Both are 2.) The product is a 3×4 matrix.

Matrix B 2×4 — Different — Matrix A 3×2

Make a similar diagram for BA.

The product BA is not defined because B has 4 columns and A has only 3 rows.

NOW TRY EXERCISE 6

Find AB and BA, if possible.

$A = \begin{bmatrix} 1 & -2 & 3 \\ 5 & 0 & 4 \\ -8 & 7 & -7 \end{bmatrix}$,

$B = \begin{bmatrix} 1 \\ -2 \\ 3 \end{bmatrix}$

EXAMPLE 6 Multiplying Two Matrices

Find AB and BA, if possible.

$$A = \begin{bmatrix} 1 & -3 \\ 7 & 2 \end{bmatrix}, \quad B = \begin{bmatrix} 1 & 0 & -1 & 2 \\ 3 & 1 & 4 & -1 \end{bmatrix}$$

$$AB = \begin{bmatrix} 1 & -3 \\ 7 & 2 \end{bmatrix} \begin{bmatrix} 1 & 0 & -1 & 2 \\ 3 & 1 & 4 & -1 \end{bmatrix}$$

A is 2×2 and *B* is 2×4, so the product will be a 2×4 matrix.

$$= \begin{bmatrix} 1(1) + (-3)3 & 1(0) + (-3)1 & 1(-1) + (-3)4 & 1(2) + (-3)(-1) \\ 7(1) + 2(3) & 7(0) + 2(1) & 7(-1) + 2(4) & 7(2) + 2(-1) \end{bmatrix}$$

$$= \begin{bmatrix} -8 & -3 & -13 & 5 \\ 13 & 2 & 1 & 12 \end{bmatrix}$$

B is a 2×4 matrix and A is a 2×2 matrix, so the number of columns of B (4) does not equal the number of rows of A (2). The product BA is not defined. NOW TRY

NOW TRY EXERCISE 7

Find AB and BA.

$A = \begin{bmatrix} -1 & 0 \\ 2 & 5 \end{bmatrix}, B = \begin{bmatrix} -3 & 5 \\ 2 & 1 \end{bmatrix}$

EXAMPLE 7 Multiplying Square Matrices in Different Orders

Find AB and BA.

$$A = \begin{bmatrix} 1 & 3 \\ -2 & 5 \end{bmatrix}, \quad B = \begin{bmatrix} -2 & 7 \\ 0 & 2 \end{bmatrix}$$

$$AB = \begin{bmatrix} 1 & 3 \\ -2 & 5 \end{bmatrix} \begin{bmatrix} -2 & 7 \\ 0 & 2 \end{bmatrix}$$

$$AB = \begin{bmatrix} -2 & 13 \\ 4 & -4 \end{bmatrix}$$

$$BA = \begin{bmatrix} -2 & 7 \\ 0 & 2 \end{bmatrix} \begin{bmatrix} 1 & 3 \\ -2 & 5 \end{bmatrix}$$

$$BA = \begin{bmatrix} -16 & 29 \\ -4 & 10 \end{bmatrix}$$

NOW TRY

NOW TRY ANSWERS

5. (a) yes; 2×1 (b) no

6. $AB = \begin{bmatrix} 14 \\ 17 \\ -43 \end{bmatrix}$; BA is not defined.

7. $AB = \begin{bmatrix} 3 & -5 \\ 4 & 15 \end{bmatrix}$; $BA = \begin{bmatrix} 13 & 25 \\ 0 & 5 \end{bmatrix}$

Examples 5 and 6 showed that the order in which two matrices are to be multiplied may determine whether their product can be found. **Example 7** showed that even when both products AB and BA can be found, they may not be equal. ***In general, for matrices A and B, $AB \neq BA$. Matrix multiplication is not commutative.***

Matrix multiplication does satisfy the associative and distributive properties.

Properties of Matrix Multiplication

If A, B, and C are matrices such that all the following products and sums exist, then the following hold true.

$$(AB)C = A(BC)$$
$$A(B + C) = AB + AC$$
$$(B + C)A = BA + CA$$

All of the operations on matrices illustrated in this section can be performed by graphing calculators. As an example, the screens in **FIGURE 1** support the results of matrix multiplication seen in **Example 7.**

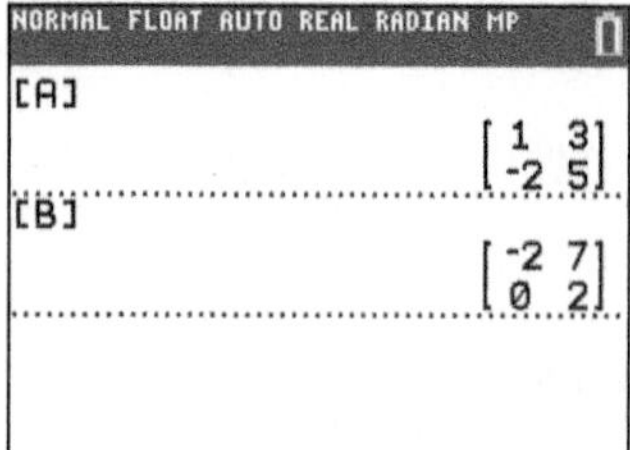

FIGURE 1

OBJECTIVE 5 Use matrices in applications.

EXAMPLE 8 Applying Matrix Multiplication

A contractor builds three kinds of houses, models A, B, and C, with a choice of two styles, colonial or ranch. Matrix P below shows the number of each kind of house the contractor is planning to build for a new 100-home subdivision. The amounts for each of the main materials used depend on the style of the house. These amounts are shown in matrix Q below, and matrix R gives the cost in dollars for each kind of material. Concrete is measured here in cubic yards, lumber in 1000 board feet, brick in 1000s, and shingles in 100 square feet.

$$\begin{matrix} \\ \text{Model A} \\ \text{Model B} \\ \text{Model C} \end{matrix} \begin{matrix} \text{Colonial} \quad \text{Ranch} \\ \begin{bmatrix} 0 & 30 \\ 10 & 20 \\ 20 & 20 \end{bmatrix} \end{matrix} = P$$

$$\begin{matrix} \\ \text{Colonial} \\ \text{Ranch} \end{matrix} \begin{matrix} \text{Concrete} \quad \text{Lumber} \quad \text{Brick} \quad \text{Shingles} \\ \begin{bmatrix} 10 & 2 & 0 & 2 \\ 50 & 1 & 20 & 2 \end{bmatrix} \end{matrix} = Q \qquad \begin{matrix} \text{Concrete} \\ \text{Lumber} \\ \text{Brick} \\ \text{Shingles} \end{matrix} \begin{matrix} \text{Cost per unit} \\ \begin{bmatrix} 20 \\ 180 \\ 60 \\ 25 \end{bmatrix} \end{matrix} = R$$

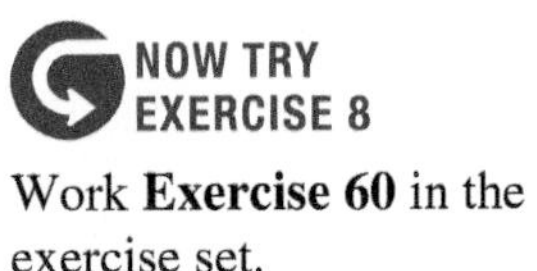

Work **Exercise 60** in the exercise set.

(a) What is the total cost of materials for all houses of each model?

To find the materials cost of each model, first find matrix PQ, which will show the total amount of each material needed for all houses of each model.

$$PQ = \begin{bmatrix} 0 & 30 \\ 10 & 20 \\ 20 & 20 \end{bmatrix} \begin{bmatrix} 10 & 2 & 0 & 2 \\ 50 & 1 & 20 & 2 \end{bmatrix} = \begin{matrix} \text{Concrete} & \text{Lumber} & \text{Brick} & \text{Shingles} & \\ \left[\begin{matrix} 1500 \\ 1100 \\ 1200 \end{matrix}\right. & \begin{matrix} 30 \\ 40 \\ 60 \end{matrix} & \begin{matrix} 600 \\ 400 \\ 400 \end{matrix} & \left.\begin{matrix} 60 \\ 60 \\ 80 \end{matrix}\right] & \begin{matrix} \text{Model A} \\ \text{Model B} \\ \text{Model C} \end{matrix} \end{matrix}$$

Multiplying PQ and the cost matrix R gives the total cost of materials for each model.

$$(PQ)R = \begin{bmatrix} 1500 & 30 & 600 & 60 \\ 1100 & 40 & 400 & 60 \\ 1200 & 60 & 400 & 80 \end{bmatrix} \begin{bmatrix} 20 \\ 180 \\ 60 \\ 25 \end{bmatrix} = \begin{matrix} \text{Cost} & \\ \begin{bmatrix} 72{,}900 \\ 54{,}700 \\ 60{,}800 \end{bmatrix} & \begin{matrix} \text{Model A} \\ \text{Model B} \\ \text{Model C} \end{matrix} \end{matrix}$$

(b) How much of each of the four kinds of material must be ordered?

The totals of the columns of matrix PQ will give a matrix whose elements represent the total amounts of each material needed for the subdivision. Call this matrix T and write it as a row matrix.

$$T = \begin{bmatrix} 3800 & 130 & 1400 & 200 \end{bmatrix}$$

(c) What is the total cost of the materials?

The total cost of all the materials is given by the product of matrix R, the cost matrix, and matrix T, the total amounts matrix. To multiply these and obtain as the product a 1×1 matrix, representing the total cost, requires multiplying a 1×4 matrix and a 4×1 matrix. This is why in part (b) a row matrix was written rather than a column matrix. The total materials cost is given by TR.

$$TR = \begin{bmatrix} 3800 & 130 & 1400 & 200 \end{bmatrix} \begin{bmatrix} 20 \\ 180 \\ 60 \\ 25 \end{bmatrix} = \begin{matrix} \text{Total} \\ \text{cost} \\ \begin{bmatrix} 188{,}400 \end{bmatrix} \end{matrix}$$

The total cost of the materials is \$188,400.

NOW TRY 

NOW TRY ANSWER

8. See the answer section at the back of the book.

C Exercises

  MyLab Math

Video solutions for select problems available in MyLab Math

Concept Check *Fill in each blank with the correct response.*

1. Only matrices of the same ______ can be added or subtracted.

2. To multiply a matrix by a scalar, multiply each ______ of the matrix by the scalar.

3. If matrix A has dimensions 3×2, in order to find the product AB, matrix B must have ______ rows.

4. If A and B are square matrices of order n, then AB and BA both have order ______ and are not necessarily ______.

Find the values of the variables for which each statement is true. ***See Example 1.***

5. $\begin{bmatrix} w & x \\ y & z \end{bmatrix} = \begin{bmatrix} 3 & 2 \\ -1 & 4 \end{bmatrix}$

6. $\begin{bmatrix} -3 & a \\ b & 5 \end{bmatrix} = \begin{bmatrix} c & 0 \\ 4 & d \end{bmatrix}$

7. $\begin{bmatrix} 0 & 5 & x \\ 1 & 3 & y+2 \\ 4 & 1 & z \end{bmatrix} = \begin{bmatrix} 0 & w+3 & 6 \\ 1 & 3 & 0 \\ 4 & 1 & 8 \end{bmatrix}$

8. $\begin{bmatrix} 5 & x-4 & 9 \\ 2 & 3 & 8 \\ 6 & 0 & 5 \end{bmatrix} = \begin{bmatrix} y+3 & 2 & 9 \\ z+4 & 3 & 8 \\ 6 & 0 & w \end{bmatrix}$

9. $\begin{bmatrix} -7+z & 4r & 8s \\ 6p & 2 & 5 \end{bmatrix} + \begin{bmatrix} -9 & 8r & 3 \\ 2 & 5 & 19 \end{bmatrix} = \begin{bmatrix} 2 & 36 & 27 \\ 20 & 7 & 12a \end{bmatrix}$

10. $\begin{bmatrix} a+2 & 3z+1 & 5m \\ 8k & 0 & 3 \end{bmatrix} + \begin{bmatrix} 3a & 2z & 5m \\ 2k & 5 & 6 \end{bmatrix} = \begin{bmatrix} 10 & -14 & 80 \\ 10 & 5 & 9 \end{bmatrix}$

Concept Check *Find the dimensions of each matrix. Identify any square, column, or row matrices.*

11. $\begin{bmatrix} -4 & 8 \\ 2 & 3 \end{bmatrix}$

12. $\begin{bmatrix} 7 & -4 \\ 4 & 2 \end{bmatrix}$

13. $\begin{bmatrix} 2 & 1 & 6 & 8 & 3 \end{bmatrix}$

14. $\begin{bmatrix} 8 & -2 & 4 & 6 & 3 \end{bmatrix}$

15. $\begin{bmatrix} 2 \\ 4 \end{bmatrix}$

16. $\begin{bmatrix} 9 \\ 13 \end{bmatrix}$

17. $\begin{bmatrix} 5 & 5 & -2 \\ -4 & -8 & 3 \\ 1 & 1 & 6 \end{bmatrix}$

18. $\begin{bmatrix} 6 & 1 & 5 \\ -9 & 6 & 2 \\ 4 & 1 & 8 \end{bmatrix}$

19. $\begin{bmatrix} -6 & 8 & 0 & 0 \\ 4 & 1 & 9 & 2 \\ 3 & -5 & 7 & 1 \end{bmatrix}$

20. $\begin{bmatrix} 9 & 8 & 0 & 4 \\ 2 & 1 & 9 & -1 \\ 5 & 3 & 3 & 2 \end{bmatrix}$

Find each sum or difference, if possible. ***See Examples 2 and 3.***

21. $\begin{bmatrix} 6 & -9 & 2 \\ 4 & 1 & 3 \end{bmatrix} + \begin{bmatrix} -8 & 2 & 5 \\ 6 & -3 & 4 \end{bmatrix}$

22. $\begin{bmatrix} 4 & -3 & 8 \\ 2 & 3 & 5 \end{bmatrix} + \begin{bmatrix} 0 & 6 & -5 \\ 2 & 1 & 8 \end{bmatrix}$

23. $\begin{bmatrix} 8 & -1 \\ 5 & 7 \end{bmatrix} - \begin{bmatrix} -4 & 2 \\ 2 & -7 \end{bmatrix}$

24. $\begin{bmatrix} 9 & 4 \\ -8 & 2 \end{bmatrix} - \begin{bmatrix} -3 & 2 \\ -4 & 7 \end{bmatrix}$

25. $\begin{bmatrix} 8 & -3 \\ 7 & 4 \\ 10 & 12 \end{bmatrix} - \begin{bmatrix} -1 & 6 \\ 4 & -1 \\ 5 & -8 \end{bmatrix}$

26. $\begin{bmatrix} 1 & -4 \\ 2 & -3 \\ -8 & 4 \end{bmatrix} - \begin{bmatrix} -6 & 9 \\ -2 & 5 \\ -7 & -12 \end{bmatrix}$

27. $\begin{bmatrix} 3 \\ 2 \end{bmatrix} + \begin{bmatrix} 2 & 3 \end{bmatrix}$

28. $\begin{bmatrix} 4 \\ 9 \end{bmatrix} + \begin{bmatrix} 0 & 8 \end{bmatrix}$

29. $\begin{bmatrix} 3x + y & x - 2y & 2x \\ 5x & 3y & x + y \end{bmatrix} + \begin{bmatrix} 2x & 3y & 5x + y \\ 3x + 2y & x & 2x \end{bmatrix}$

30. $\begin{bmatrix} 9x + 8y & 2x - 4y & 3x \\ 7x & 5y & x - 3y \end{bmatrix} + \begin{bmatrix} -2x + 7y & 8x - 8y & -3x \\ 5x & 2y & 3x - y \end{bmatrix}$

31. $\begin{bmatrix} 2r - 5s \\ 6a - 9b \\ -5m + 6n \\ -8x + 4y \end{bmatrix} - \begin{bmatrix} -8r + 2s \\ 7a + 7b \\ -2m - 3n \\ -6x - 4y \end{bmatrix}$

32. $\begin{bmatrix} 4k - 8y \\ 6z - 3x \\ 2k + 5a \\ -4m + 2n \end{bmatrix} - \begin{bmatrix} 5k + 6y \\ 2z + 5x \\ 4k + 6a \\ 4m - 2n \end{bmatrix}$

Let $A = \begin{bmatrix} -2 & 4 \\ 0 & 3 \end{bmatrix}$ *and* $B = \begin{bmatrix} -6 & 2 \\ 4 & 0 \end{bmatrix}$. *Find each of the following.* ***See Example 4.***

33. $2A$

34. $-3B$

35. $2A - B$

36. $-2A + 4B$

37. $-A + \frac{1}{2}B$

38. $\frac{3}{4}A - B$

Decide whether each product can be found. Give the dimensions of each product if it exists. ***See Example 5.***

$$A = \begin{bmatrix} 3 & 7 & 1 \\ -2 & 4 & 0 \end{bmatrix}, \quad B = \begin{bmatrix} 5 & 2 & 9 & 6 & 7 \\ 1 & 1 & 0 & 0 & 4 \\ 8 & 3 & 6 & 5 & 2 \end{bmatrix}, \quad C = \begin{bmatrix} -3 & 6 \\ 5 & 5 \\ 0 & 18 \\ -1 & 9 \\ 7 & -13 \end{bmatrix}$$

39. AB

40. CA

41. BA

42. AC

Find each product, if possible. ***See Examples 6 and 7.***

43. $\begin{bmatrix} 1 & 2 \\ 3 & 4 \end{bmatrix} \begin{bmatrix} -1 \\ 7 \end{bmatrix}$

44. $\begin{bmatrix} -1 & 5 \\ 7 & 0 \end{bmatrix} \begin{bmatrix} 6 \\ 2 \end{bmatrix}$

45. $\begin{bmatrix} 3 & -4 & 1 \\ 5 & 0 & 2 \end{bmatrix} \begin{bmatrix} -1 \\ 4 \\ 2 \end{bmatrix}$

46. $\begin{bmatrix} -6 & 3 & 5 \\ 2 & 9 & 1 \end{bmatrix} \begin{bmatrix} -2 \\ 0 \\ 3 \end{bmatrix}$

47. $\begin{bmatrix} 5 & 2 \\ -1 & 4 \end{bmatrix} \begin{bmatrix} 3 & -2 \\ 1 & 0 \end{bmatrix}$

48. $\begin{bmatrix} -4 & 0 \\ 1 & 3 \end{bmatrix} \begin{bmatrix} -2 & 4 \\ 0 & 1 \end{bmatrix}$

49. $\begin{bmatrix} 2 & 2 & -1 \\ 3 & 0 & 1 \end{bmatrix} \begin{bmatrix} 0 & 2 \\ -1 & 4 \\ 0 & 2 \end{bmatrix}$

50. $\begin{bmatrix} -9 & 2 & 1 \\ 3 & 0 & 0 \end{bmatrix} \begin{bmatrix} 2 \\ -1 \\ 4 \end{bmatrix}$

51. $\begin{bmatrix} -1 & 2 & 0 \\ 0 & 3 & 2 \\ 0 & 1 & 4 \end{bmatrix} \begin{bmatrix} 2 & -1 & 2 \\ 0 & 2 & 1 \\ 3 & 0 & -1 \end{bmatrix}$

52. $\begin{bmatrix} -2 & -3 & -4 \\ 2 & -1 & 0 \\ 4 & -2 & 3 \end{bmatrix} \begin{bmatrix} 0 & 1 & 4 \\ 1 & 2 & -1 \\ 3 & 2 & -2 \end{bmatrix}$

53. $\begin{bmatrix} -2 & 4 & 1 \end{bmatrix} \begin{bmatrix} 3 & -2 & 4 \\ 2 & 1 & 0 \\ 0 & -1 & 4 \end{bmatrix}$

54. $\begin{bmatrix} 0 & 3 & -4 \end{bmatrix} \begin{bmatrix} -2 & 6 & 3 \\ 0 & 4 & 2 \\ -1 & 1 & 4 \end{bmatrix}$

55. $\begin{bmatrix} -3 & 0 & 2 & 1 \\ 4 & 0 & 2 & 6 \end{bmatrix}\begin{bmatrix} -4 & 2 \\ 0 & 1 \end{bmatrix}$

56. $\begin{bmatrix} -1 & 2 & 4 & 1 \\ 0 & 2 & -3 & 5 \end{bmatrix}\begin{bmatrix} 1 & 2 & 4 \\ -2 & 5 & 1 \end{bmatrix}$

Solve each problem. ***See Example 8.***

57. A hardware chain does an inventory of a particular size of screw and finds that its Adelphi store has 100 flat-head and 150 round-head screws, its Beltsville store has 125 flat and 50 round, and its College Park store has 175 flat and 200 round. Write this information first as a 3×2 matrix and then as a 2×3 matrix.

58. At the grocery store, Liam bought 4 quarts of milk, 2 loaves of bread, 4 potatoes, and an apple. Mary bought 2 quarts of milk, a loaf of bread, 5 potatoes, and 4 apples. Write this information first as a 2×4 matrix and then as a 4×2 matrix.

59. Yummy Yogurt sells three types of yogurt, nonfat, regular, and super creamy, at three locations. Location I sells 50 gal of nonfat, 100 gal of regular, and 30 gal of super creamy each day. Location II sells 10 gal of nonfat and Location III sells 60 gal of nonfat each day. Daily sales of regular yogurt are 90 gal at Location II and 120 gal at Location III. At Location II, 50 gal of super creamy are sold each day, and 40 gal of super creamy are sold each day at Location III.

(a) Write a 3×3 matrix that shows the sales figures for the three locations.

(b) The income per gallon for nonfat, regular, and super creamy is \$12, \$10, and \$15, respectively. Write a 1×3 matrix or a 3×1 matrix displaying the income.

(c) Find a matrix product that gives the daily income at each of the three locations.

(d) What is Yummy Yogurt's total daily income from the three locations?

60. **(Now Try Exercise 8)** The Bread Box, a neighborhood bakery, sells four main items: sweet rolls, bread, cakes, and pies. The amount of each ingredient (in cups, except for eggs) required for these items is given by matrix A.

	Eggs	Flour	Sugar	Shortening	Milk
Rolls (dozen)	1	4	$\frac{1}{4}$	$\frac{1}{4}$	1
Bread (loaves)	0	3	0	$\frac{1}{4}$	0
Cakes	4	3	2	1	1
Pies (crust)	0	1	0	$\frac{1}{3}$	0

$= A$

The cost (in cents) for each ingredient when purchased in large lots or small lots is given in matrix B.

	Cost: Large lot	Small lot
Eggs	5	5
Flour	8	10
Sugar	10	12
Shortening	12	15
Milk	5	6

$= B$

(a) Use matrix multiplication to find a matrix giving the comparative cost per item for the two purchase options.

(b) A day's orders consist of 20 dozen sweet rolls, 200 loaves of bread, 50 cakes, and 60 pies. Write the orders as a 1×4 matrix. Using matrix multiplication, write as a matrix the amount of each ingredient needed to fill the day's orders.

(c) Use matrix multiplication to find a matrix giving the costs under the two purchase options to fill the day's orders.

Extending Skills *Matrices A, B, and C are defined as follows, where all the elements are real numbers.*

$$A = \begin{bmatrix} a_{11} & a_{12} \\ a_{21} & a_{22} \end{bmatrix}, \quad B = \begin{bmatrix} b_{11} & b_{12} \\ b_{21} & b_{22} \end{bmatrix}, \quad \textit{and} \quad C = \begin{bmatrix} c_{11} & c_{12} \\ c_{21} & c_{22} \end{bmatrix},$$

Use these matrices to show that each statement is true for 2×2 *matrices.*

61. $A + B = B + A$ (commutative property)

62. $A + (B + C) = (A + B) + C$ (associative property)

63. $(AB)C = A(BC)$ (associative property)

64. $A(B + C) = AB + AC$ (distributive property)

65. $c(A + B) = cA + cB$, for any real number c

66. $(c + d)A = cA + dA$, for any real numbers c and d

67. $c(A)d = cd(A)$

68. $(cd)A = c(dA)$

Matrix Inverses

APPENDIX D

OBJECTIVES

1 Define and apply identity matrices.
2 Find multiplicative inverse matrices.
3 Use inverse matrices to solve systems of linear equations.

VOCABULARY

☐ identity matrix
☐ multiplicative inverse
☐ square system

OBJECTIVE 1 Define and apply identity matrices.

By the identity property for real numbers, $\boldsymbol{a} \cdot \mathbf{1} = \boldsymbol{a}$ and $\mathbf{1} \cdot \boldsymbol{a} = \boldsymbol{a}$ for any real number a. If there is to be a multiplicative **identity matrix** $\boldsymbol{I}$, such that

$$AI = A \quad \text{and} \quad IA = A,$$

for any matrix A, then A and I must be square matrices of the same dimensions. Otherwise it would not be possible to find both products.

2 × 2 Identity Matrix

I_2 represents the 2×2 identity matrix.

$$I_2 = \begin{bmatrix} 1 & 0 \\ 0 & 1 \end{bmatrix}$$

To verify that I_2 is the 2×2 identity matrix, we must show that $AI_2 = A$ and $I_2A = A$ for any 2×2 matrix. Let

$$A = \begin{bmatrix} x & y \\ z & w \end{bmatrix}.$$

Then

$$AI_2 = \begin{bmatrix} x & y \\ z & w \end{bmatrix}\begin{bmatrix} 1 & 0 \\ 0 & 1 \end{bmatrix} = \begin{bmatrix} x \cdot 1 + y \cdot 0 & x \cdot 0 + y \cdot 1 \\ z \cdot 1 + w \cdot 0 & z \cdot 0 + w \cdot 1 \end{bmatrix} = \begin{bmatrix} x & y \\ z & w \end{bmatrix} = A,$$

and

$$I_2A = \begin{bmatrix} 1 & 0 \\ 0 & 1 \end{bmatrix}\begin{bmatrix} x & y \\ z & w \end{bmatrix} = \begin{bmatrix} 1 \cdot x + 0 \cdot z & 1 \cdot y + 0 \cdot w \\ 0 \cdot x + 1 \cdot z & 0 \cdot y + 1 \cdot w \end{bmatrix} = \begin{bmatrix} x & y \\ z & w \end{bmatrix} = A.$$

Generalizing from this example, an $n \times n$ identity matrix has 1's on the main diagonal and 0's elsewhere.

***n × n* Identity Matrix**

I_n represents the $n \times n$ identity matrix.

$$I_n = \begin{bmatrix} 1 & 0 & \cdots & 0 \\ 0 & 1 & \cdots & 0 \\ \vdots & \vdots & a_{ij} & \vdots \\ 0 & 0 & \cdots & 1 \end{bmatrix}$$

Element $a_{ij} = 1$ when $i = j$ (the diagonal elements), and $a_{ij} = 0$ otherwise.

NOW TRY EXERCISE 1

Let $A = \begin{bmatrix} 2 & 1 & 3 & 9 \\ 5 & -2 & 8 & 6 \\ 6 & 2 & 1 & 0 \\ 5 & 3 & 7 & 7 \end{bmatrix}$.

Give the 4×4 identity matrix and show that $AI_4 = A$.

EXAMPLE 1 Stating and Verifying the 3 × 3 Identity Matrix

Let $A = \begin{bmatrix} -2 & 4 & 0 \\ 3 & 5 & 9 \\ 0 & 8 & -6 \end{bmatrix}$. Give the 3×3 identity matrix I_3 and show that $AI_3 = A$.

The 3×3 identity matrix is

$$I_3 = \begin{bmatrix} 1 & 0 & 0 \\ 0 & 1 & 0 \\ 0 & 0 & 1 \end{bmatrix},$$

so $$AI_3 = \begin{bmatrix} -2 & 4 & 0 \\ 3 & 5 & 9 \\ 0 & 8 & -6 \end{bmatrix}\begin{bmatrix} 1 & 0 & 0 \\ 0 & 1 & 0 \\ 0 & 0 & 1 \end{bmatrix} = \begin{bmatrix} -2 & 4 & 0 \\ 3 & 5 & 9 \\ 0 & 8 & -6 \end{bmatrix} = A.$$ Definition of matrix multiplication

NOW TRY 

The graphing calculator screen in **FIGURE 1(a)** shows the identity matrices for $n = 2$ and $n = 3$. The screen in **FIGURE 1(b)** supports the result in **Example 1.**

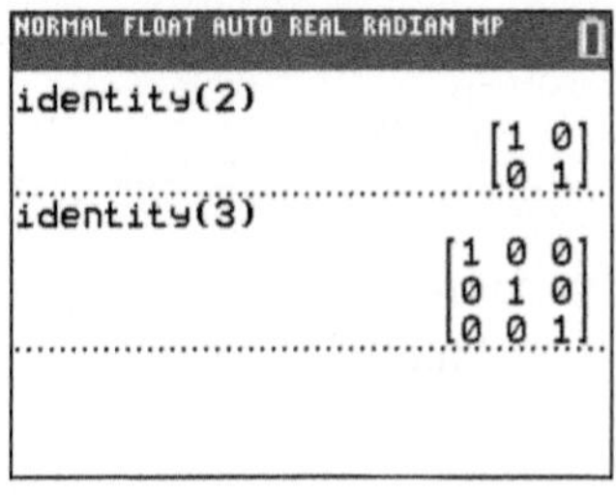

(a)

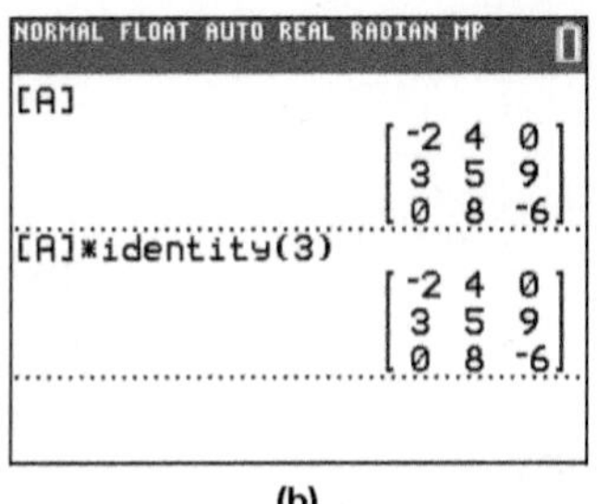

(b)

FIGURE 1

OBJECTIVE 2 Find multiplicative inverse matrices.

For every nonzero real number a, there is a multiplicative inverse $\frac{1}{a}$ such that

$$a \cdot \frac{1}{a} = 1 \quad \text{and} \quad \frac{1}{a} \cdot a = 1.$$

Recall that $\frac{1}{a}$ is also written a^{-1}. In a similar way, if A is an $n \times n$ matrix, then its **multiplicative inverse,** written A^{-1}, must satisfy both

$$AA^{-1} = I_n \quad \text{and} \quad A^{-1}A = I_n.$$

This means that only a square matrix can have a multiplicative inverse.

EXAMPLE 2 Finding the Inverse of a 2 × 2 Matrix

Find the inverse of

$$A = \begin{bmatrix} 2 & 4 \\ 1 & -1 \end{bmatrix}.$$

Let the unknown inverse matrix be

$$A^{-1} = \begin{bmatrix} x & y \\ z & w \end{bmatrix}.$$

By the definition of matrix inverse, $AA^{-1} = I_2$.

$$AA^{-1} = \begin{bmatrix} 2 & 4 \\ 1 & -1 \end{bmatrix}\begin{bmatrix} x & y \\ z & w \end{bmatrix} = \begin{bmatrix} 1 & 0 \\ 0 & 1 \end{bmatrix}$$

NOW TRY ANSWER

1. $I_4 = \begin{bmatrix} 1 & 0 & 0 & 0 \\ 0 & 1 & 0 & 0 \\ 0 & 0 & 1 & 0 \\ 0 & 0 & 0 & 1 \end{bmatrix}$;

$$\begin{bmatrix} 2 & 1 & 3 & 9 \\ 5 & -2 & 8 & 6 \\ 6 & 2 & 1 & 0 \\ 5 & 3 & 7 & 7 \end{bmatrix}\begin{bmatrix} 1 & 0 & 0 & 0 \\ 0 & 1 & 0 & 0 \\ 0 & 0 & 1 & 0 \\ 0 & 0 & 0 & 1 \end{bmatrix} = \begin{bmatrix} 2 & 1 & 3 & 9 \\ 5 & -2 & 8 & 6 \\ 6 & 2 & 1 & 0 \\ 5 & 3 & 7 & 7 \end{bmatrix}$$

By matrix multiplication, we have the following.

$$\begin{bmatrix} 2x + 4z & 2y + 4w \\ x - z & y - w \end{bmatrix} = \begin{bmatrix} 1 & 0 \\ 0 & 1 \end{bmatrix}$$

Set corresponding elements equal to give a system of equations.

$$\begin{aligned} 2x + 4z &= 1 \quad (1) \\ 2y + 4w &= 0 \quad (2) \\ x - z &= 0 \quad (3) \\ y - w &= 1 \quad (4) \end{aligned}$$

Because equations (1) and (3) involve only x and z, while equations (2) and (4) involve only y and w, these four equations lead to two systems of equations.

$$\begin{aligned} 2x + 4z &= 1 \\ x - z &= 0 \end{aligned} \quad \text{and} \quad \begin{aligned} 2y + 4w &= 0 \\ y - w &= 1 \end{aligned}$$

We use row operations to solve this system.

$$\left[\begin{array}{cc|c} 2 & 4 & 1 \\ 1 & -1 & 0 \end{array}\right] \quad \text{and} \quad \left[\begin{array}{cc|c} 2 & 4 & 0 \\ 1 & -1 & 1 \end{array}\right]$$

Write augmented matrices for the systems.

The elements to the left of the vertical bar are identical, so the two systems can be combined into one matrix.

$$\left[\begin{array}{cc|cc} 2 & 4 & 1 & 0 \\ 1 & -1 & 0 & 1 \end{array}\right]$$

We need to change the numbers on the left of the vertical bar to the 2×2 identity matrix.

$$\left[\begin{array}{cc|cc} 1 & -1 & 0 & 1 \\ 2 & 4 & 1 & 0 \end{array}\right] \quad \text{Interchange } R_1 \text{ and } R_2.$$

$$\left[\begin{array}{cc|cc} 1 & -1 & 0 & 1 \\ 0 & 6 & 1 & -2 \end{array}\right] \quad -2R_1 + R_2$$

$$\left[\begin{array}{cc|cc} 1 & -1 & 0 & 1 \\ 0 & 1 & \frac{1}{6} & -\frac{1}{3} \end{array}\right] \quad \tfrac{1}{6}R_2$$

$$\left[\begin{array}{cc|cc} 1 & 0 & \frac{1}{6} & \frac{2}{3} \\ 0 & 1 & \frac{1}{6} & -\frac{1}{3} \end{array}\right] \quad R_2 + R_1$$

The numbers in the first column to the right of the vertical bar give the values of x and z. The second column gives the values of y and w.

$$\left[\begin{array}{cc|cc} 1 & 0 & x & y \\ 0 & 1 & z & w \end{array}\right] = \left[\begin{array}{cc|cc} 1 & 0 & \frac{1}{6} & \frac{2}{3} \\ 0 & 1 & \frac{1}{6} & -\frac{1}{3} \end{array}\right],$$

so

$$A^{-1} = \begin{bmatrix} x & y \\ z & w \end{bmatrix} = \begin{bmatrix} \frac{1}{6} & \frac{2}{3} \\ \frac{1}{6} & -\frac{1}{3} \end{bmatrix}.$$

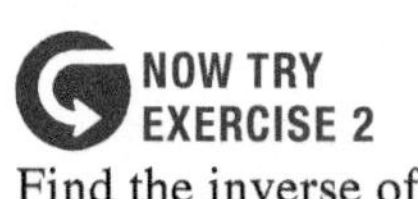

NOW TRY EXERCISE 2
Find the inverse of

$$A = \begin{bmatrix} -6 & 5 \\ -5 & 4 \end{bmatrix}.$$

CHECK Multiply A by A^{-1}. The result should be I_2.

$$AA^{-1} = \begin{bmatrix} 2 & 4 \\ 1 & -1 \end{bmatrix} \begin{bmatrix} \frac{1}{6} & \frac{2}{3} \\ \frac{1}{6} & -\frac{1}{3} \end{bmatrix} = \begin{bmatrix} \frac{1}{3}+\frac{2}{3} & \frac{4}{3}-\frac{4}{3} \\ \frac{1}{6}-\frac{1}{6} & \frac{2}{3}+\frac{1}{3} \end{bmatrix} = \begin{bmatrix} 1 & 0 \\ 0 & 1 \end{bmatrix} = I_2 \checkmark$$

Use the definition.

Therefore,
$$A^{-1} = \begin{bmatrix} \frac{1}{6} & \frac{2}{3} \\ \frac{1}{6} & -\frac{1}{3} \end{bmatrix}.$$

NOW TRY 

NOTE See **Example 2.** To confirm that two $n \times n$ matrices A and B are inverses of each other, it is sufficient to show that $AB = I_n$. It is not necessary to show also that $BA = I_n$.

Finding an Inverse Matrix

To obtain A^{-1} for any $n \times n$ matrix A for which A^{-1} exists, follow these steps.

Step 1 Form the augmented matrix $[A \mid I_n]$, where I_n is the $n \times n$ identity matrix.

Step 2 Perform row operations on $[A \mid I_n]$ to obtain a matrix of the form $[I_n \mid B]$.

Step 3 Matrix B is A^{-1}.

EXAMPLE 3 Finding the Inverse of a 3 × 3 Matrix

Find A^{-1} if $A = \begin{bmatrix} 1 & 0 & 1 \\ 2 & -2 & -1 \\ 3 & 0 & 0 \end{bmatrix}$.

Step 1 Write the augmented matrix $[A \mid I_3]$.

$$\left[\begin{array}{ccc|ccc} 1 & 0 & 1 & 1 & 0 & 0 \\ 2 & -2 & -1 & 0 & 1 & 0 \\ 3 & 0 & 0 & 0 & 0 & 1 \end{array}\right]$$

Step 2 Because 1 is already in the upper left-hand corner as desired, begin by using a row operation that will result in a 0 for the first element in the second row. Multiply the elements of the first row by -2, and add the results to the second row.

$$\left[\begin{array}{ccc|ccc} 1 & 0 & 1 & 1 & 0 & 0 \\ 0 & -2 & -3 & -2 & 1 & 0 \\ 3 & 0 & 0 & 0 & 0 & 1 \end{array}\right] \quad -2R_1 + R_2$$

To obtain 0 for the first element in the third row, multiply the elements of the first row by -3, and add the results to the third row.

$$\left[\begin{array}{ccc|ccc} 1 & 0 & 1 & 1 & 0 & 0 \\ 0 & -2 & -3 & -2 & 1 & 0 \\ 0 & 0 & -3 & -3 & 0 & 1 \end{array}\right] \quad -3R_1 + R_3$$

NOW TRY ANSWER

2. $A^{-1} = \begin{bmatrix} 4 & -5 \\ 5 & -6 \end{bmatrix}$

Find A^{-1} if

$$A = \begin{bmatrix} -4 & 2 & 0 \\ 1 & -1 & 2 \\ 0 & 1 & 4 \end{bmatrix}.$$

We want 1 for the second element in the second row, so multiply the elements of the second row by $-\frac{1}{2}$.

$$\left[\begin{array}{ccc|ccc} 1 & 0 & 1 & 1 & 0 & 0 \\ 0 & 1 & \frac{3}{2} & 1 & -\frac{1}{2} & 0 \\ 0 & 0 & -3 & -3 & 0 & 1 \end{array}\right] \quad -\tfrac{1}{2}R_2$$

To obtain 1 for the third element in the third row, multiply the elements of the third row by $-\frac{1}{3}$.

$$\left[\begin{array}{ccc|ccc} 1 & 0 & 1 & 1 & 0 & 0 \\ 0 & 1 & \frac{3}{2} & 1 & -\frac{1}{2} & 0 \\ 0 & 0 & 1 & 1 & 0 & -\frac{1}{3} \end{array}\right] \quad -\tfrac{1}{3}R_3$$

Now we need 0 as the third element in the first row. Multiply the elements of the third row by -1, and add to the first row.

$$\left[\begin{array}{ccc|ccc} 1 & 0 & 0 & 0 & 0 & \frac{1}{3} \\ 0 & 1 & \frac{3}{2} & 1 & -\frac{1}{2} & 0 \\ 0 & 0 & 1 & 1 & 0 & -\frac{1}{3} \end{array}\right] \quad -1R_3 + R_1$$

Finally, to introduce 0 as the third element in the second row, multiply the elements of the third row by $-\frac{3}{2}$, and add to the second row.

$$\left[\begin{array}{ccc|ccc} 1 & 0 & 0 & 0 & 0 & \frac{1}{3} \\ 0 & 1 & 0 & -\frac{1}{2} & -\frac{1}{2} & \frac{1}{2} \\ 0 & 0 & 1 & 1 & 0 & -\frac{1}{3} \end{array}\right] \quad -\tfrac{3}{2}R_3 + R_2$$

Step 3

$$A^{-1} = \begin{bmatrix} 0 & 0 & \frac{1}{3} \\ -\frac{1}{2} & -\frac{1}{2} & \frac{1}{2} \\ 1 & 0 & -\frac{1}{3} \end{bmatrix}$$

The last operation gives A^{-1}. Confirm this by forming the product $A^{-1}A$ or AA^{-1}, each of which should equal the matrix I_3.

NOW TRY

To perform the row operations, make changes column by column from left to right, so for each column the required 1 is the result of the first change. Next, perform the operations that obtain the 0's in that column. Then proceed to another column.

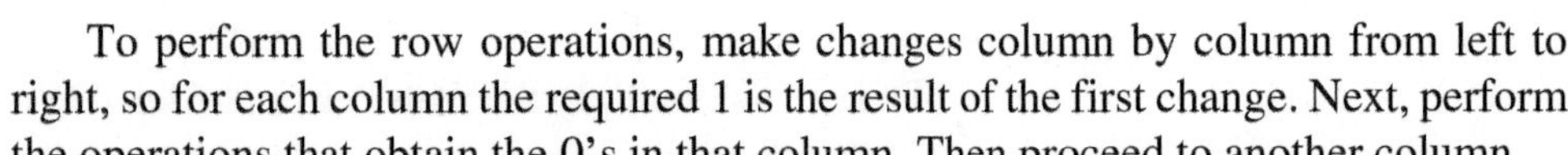

EXAMPLE 4 Identifying a Matrix with No Inverse

Find A^{-1}, if possible, given $A = \begin{bmatrix} 2 & -4 \\ 1 & -2 \end{bmatrix}$.

Using row operations to change the first column of the augmented matrix

$$\left[\begin{array}{cc|cc} 2 & -4 & 1 & 0 \\ 1 & -2 & 0 & 1 \end{array}\right]$$

results in the following matrices.

$$\left[\begin{array}{cc|cc} 1 & -2 & \frac{1}{2} & 0 \\ 1 & -2 & 0 & 1 \end{array}\right] \quad \tfrac{1}{2}R_1 \qquad \text{and} \qquad \left[\begin{array}{cc|cc} 1 & -2 & \frac{1}{2} & 0 \\ 0 & 0 & -\frac{1}{2} & 1 \end{array}\right] \quad R_2 - R_1$$

NOW TRY ANSWER

3. $A^{-1} = \begin{bmatrix} -\frac{3}{8} & -\frac{1}{2} & \frac{1}{4} \\ -\frac{1}{4} & -1 & \frac{1}{2} \\ \frac{1}{16} & \frac{1}{4} & \frac{1}{8} \end{bmatrix}$

NOW TRY EXERCISE 4

Find A^{-1} if possible, given

$$A = \begin{bmatrix} 4 & -2 & 5 \\ 0 & 1 & 0 \\ -8 & 4 & -10 \end{bmatrix}.$$

At this point, the matrix should be changed so that the second-row, second-column element will be 1. Because that element is now 0, there is no way to complete the desired operations, so A^{-1} does not exist for this matrix A.

Just as there is no multiplicative inverse for the real number 0, not every matrix has a multiplicative inverse. Matrix A is an example of such a matrix.

NOW TRY 

If the inverse of a matrix exists, it is unique. That is, any given square matrix has no more than one inverse.

A graphing calculator can find the inverse of a matrix. The screens in **FIGURE 2** support the result of **Example 3.** The elements of the inverse are expressed as fractions.

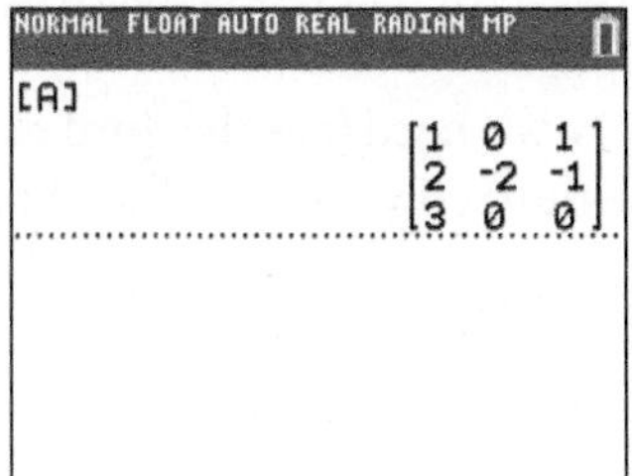

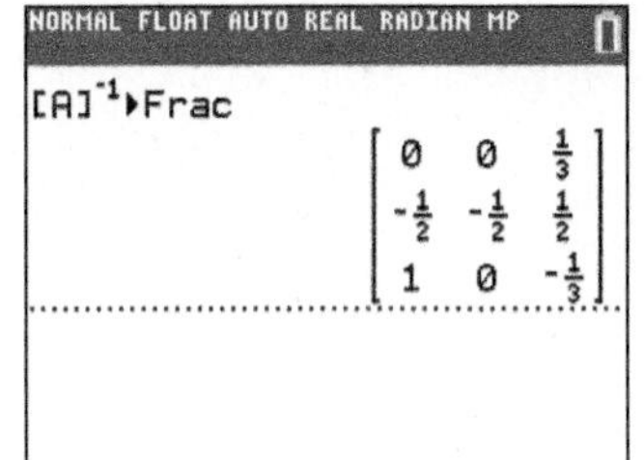

FIGURE 2

OBJECTIVE 3 Use inverse matrices to solve systems of linear equations.

Matrix inverses can be used to solve *square* linear systems of equations. A **square system** has the same number of equations as variables. Consider this linear system.

$$a_{11}x + a_{12}y + a_{13}z = b_1$$
$$a_{21}x + a_{22}y + a_{23}z = b_2$$
$$a_{31}x + a_{32}y + a_{33}z = b_3$$

The definition of matrix multiplication can be used to rewrite the system.

$$\begin{bmatrix} a_{11} & a_{12} & a_{13} \\ a_{21} & a_{22} & a_{23} \\ a_{31} & a_{32} & a_{33} \end{bmatrix} \cdot \begin{bmatrix} x \\ y \\ z \end{bmatrix} = \begin{bmatrix} b_1 \\ b_2 \\ b_3 \end{bmatrix} \qquad (1)$$

(To see this, multiply the matrices on the left.) We now define A, X, and B.

$$A = \begin{bmatrix} a_{11} & a_{12} & a_{13} \\ a_{21} & a_{22} & a_{23} \\ a_{31} & a_{32} & a_{33} \end{bmatrix}, \quad X = \begin{bmatrix} x \\ y \\ z \end{bmatrix}, \quad \text{and} \quad B = \begin{bmatrix} b_1 \\ b_2 \\ b_3 \end{bmatrix}$$

The system given in (1) becomes $AX = B$. *If* A^{-1} *exists*, then we can multiply both sides of $AX = B$ by A^{-1} on the left.

$$A^{-1}(AX) = A^{-1}B \quad \text{Assume } A^{-1} \text{ exists.}$$
$$(A^{-1}A)X = A^{-1}B \quad \text{Associative property}$$
$$I_3X = A^{-1}B \quad \text{Inverse property}$$
$$X = A^{-1}B \quad \text{Identity property}$$

$A^{-1}B$ gives the solution of the system.

NOW TRY ANSWER

4. A^{-1} does not exist.

Solution of a Linear System by the Inverse Matrix Method

If A is an $n \times n$ matrix with inverse A^{-1}, X is an $n \times 1$ matrix of variables of a system, and B is an $n \times 1$ matrix of constants of the system, then a matrix equation for the system can be written as follows.

$$AX = B$$

The solution is given by X.

$$X = A^{-1}B$$

NOW TRY EXERCISE 5

Use the inverse of the coefficient matrix to solve the system.

$$5x + 2y = -1$$
$$2x + 3y = 15$$

EXAMPLE 5 Solving Systems of Equations Using Matrix Inverses

Use the inverse of the coefficient matrix to solve each system.

(a) $2x - 3y = 4$
$x + 5y = 2$

To represent the system as a matrix equation, use one matrix for the coefficients, one for the variables, and one for the constants, as follows.

$$A = \begin{bmatrix} 2 & -3 \\ 1 & 5 \end{bmatrix}, \quad X = \begin{bmatrix} x \\ y \end{bmatrix}, \quad \text{and} \quad B = \begin{bmatrix} 4 \\ 2 \end{bmatrix}$$

Write the system in matrix form as the equation $AX = B$.

$$AX = \begin{bmatrix} 2 & -3 \\ 1 & 5 \end{bmatrix}\begin{bmatrix} x \\ y \end{bmatrix} = \begin{bmatrix} 2x - 3y \\ x + 5y \end{bmatrix} = \begin{bmatrix} 4 \\ 2 \end{bmatrix} = B$$

To solve the system, first find A^{-1}. Then find the product $A^{-1}B$.

$$A^{-1} = \begin{bmatrix} \frac{5}{13} & \frac{3}{13} \\ -\frac{1}{13} & \frac{2}{13} \end{bmatrix} \quad \text{Use the method of Example 2.} \quad A^{-1}B = \begin{bmatrix} \frac{5}{13} & \frac{3}{13} \\ -\frac{1}{13} & \frac{2}{13} \end{bmatrix}\begin{bmatrix} 4 \\ 2 \end{bmatrix} = \begin{bmatrix} 2 \\ 0 \end{bmatrix}$$

Because $X = A^{-1}B$, we can determine X.

$$X = \begin{bmatrix} x \\ y \end{bmatrix} = \begin{bmatrix} 2 \\ 0 \end{bmatrix}$$

The solution set of the system is $\{(2, 0)\}$.

(b) $2x - 3y = 1$
$x + 5y = 20$

This system has the same matrix of coefficients as the system in part (a). Only matrix B is different. Use A^{-1} from part (a) and multiply by B.

$$X = A^{-1}B = \begin{bmatrix} \frac{5}{13} & \frac{3}{13} \\ -\frac{1}{13} & \frac{2}{13} \end{bmatrix}\begin{bmatrix} 1 \\ 20 \end{bmatrix} = \begin{bmatrix} 5 \\ 3 \end{bmatrix}$$

The solution set is $\{(5, 3)\}$.

NOW TRY 

NOW TRY ANSWER
5. $\{(-3, 7)\}$

D Exercises

FOR EXTRA HELP  MyLab Math

Video solutions for select problems available in MyLab Math

Concept Check *Fill in each blank with the correct response.*

1. Any identity matrix is a square matrix with all entries along the diagonal from upper left to lower right equal to ________, and all other entries equal to ________.

2. If a square matrix A is multiplied by its inverse A^{-1} in either order, the product is the ________ matrix with the same dimensions.

3. The matrix $\begin{bmatrix} -9 & 8 & 0 \\ 1 & 2 & -4 \end{bmatrix}$ cannot have an inverse because the number of ________ does not equal the number of ________.

4. Any given square matrix has no more than one ________.

5. Let $A = \begin{bmatrix} 4 & -2 \\ 3 & 1 \end{bmatrix}$. Give the 2×2 identity matrix I_2 and show that $AI_2 = A$. **See Example 1.**

6. In **Exercise 5,** verify that $I_2A = A$.

Decide whether the given matrices are inverses of each other. See the check in **Example 2.**

7. $\begin{bmatrix} 5 & 7 \\ 2 & 3 \end{bmatrix}, \begin{bmatrix} 3 & -7 \\ -2 & 5 \end{bmatrix}$

8. $\begin{bmatrix} 2 & 3 \\ 1 & 1 \end{bmatrix}, \begin{bmatrix} -1 & 3 \\ 1 & -2 \end{bmatrix}$

9. $\begin{bmatrix} -1 & 2 \\ 3 & -5 \end{bmatrix}, \begin{bmatrix} -5 & -2 \\ -3 & -1 \end{bmatrix}$

10. $\begin{bmatrix} 2 & 1 \\ 3 & 2 \end{bmatrix}, \begin{bmatrix} 2 & 1 \\ -3 & 2 \end{bmatrix}$

11. $\begin{bmatrix} 0 & 1 & 0 \\ 0 & 0 & -2 \\ 1 & -1 & 0 \end{bmatrix}, \begin{bmatrix} 1 & 0 & 1 \\ 1 & 0 & 0 \\ 0 & -1 & 0 \end{bmatrix}$

12. $\begin{bmatrix} 1 & 2 & 0 \\ 0 & 1 & 0 \\ 0 & 1 & 0 \end{bmatrix}, \begin{bmatrix} 1 & -2 & 0 \\ 0 & 1 & 0 \\ 0 & -1 & 1 \end{bmatrix}$

13. $\begin{bmatrix} -1 & -1 & -1 \\ 4 & 5 & 0 \\ 0 & 1 & -3 \end{bmatrix}, \begin{bmatrix} 15 & 4 & -5 \\ -12 & -3 & 4 \\ -4 & -1 & 1 \end{bmatrix}$

14. $\begin{bmatrix} 1 & 3 & 3 \\ 1 & 4 & 3 \\ 1 & 3 & 4 \end{bmatrix}, \begin{bmatrix} 7 & -3 & -3 \\ -1 & 1 & 0 \\ -1 & 0 & 1 \end{bmatrix}$

Find the inverse, if it exists, for each matrix. **See Examples 2–4.**

15. $\begin{bmatrix} -1 & -2 \\ 3 & 5 \end{bmatrix}$

16. $\begin{bmatrix} 3 & -1 \\ -5 & 2 \end{bmatrix}$

17. $\begin{bmatrix} 5 & 10 \\ -3 & -6 \end{bmatrix}$

18. $\begin{bmatrix} -6 & 4 \\ -3 & 2 \end{bmatrix}$

19. $\begin{bmatrix} 1 & 0 & 1 \\ 0 & -1 & 0 \\ 2 & 1 & 1 \end{bmatrix}$

20. $\begin{bmatrix} 1 & 0 & 0 \\ 0 & -1 & 0 \\ 1 & 0 & 1 \end{bmatrix}$

21. $\begin{bmatrix} 3 & 3 & -1 \\ 2 & 6 & 0 \\ -6 & -6 & 2 \end{bmatrix}$

22. $\begin{bmatrix} 5 & -3 & 2 \\ -5 & 3 & -2 \\ 1 & 0 & 1 \end{bmatrix}$

23. $\begin{bmatrix} 2 & 2 & -4 \\ 2 & 6 & 0 \\ -3 & -3 & 5 \end{bmatrix}$

24. $\begin{bmatrix} 2 & 4 & 6 \\ -1 & -4 & -3 \\ 0 & 1 & -1 \end{bmatrix}$

25. $\begin{bmatrix} 1 & 1 & 0 & 2 \\ 2 & -1 & 1 & -1 \\ 3 & 3 & 2 & -2 \\ 1 & 2 & 1 & 0 \end{bmatrix}$

26. $\begin{bmatrix} 1 & -2 & 3 & 0 \\ 0 & 1 & -1 & 1 \\ -2 & 2 & -2 & 4 \\ 0 & 2 & -3 & 1 \end{bmatrix}$

Solve each system by using the inverse of the coefficient matrix. **See Example 5.**

27. $-x + y = 1$
$2x - y = 1$

28. $x + y = 5$
$x - y = -1$

29. $2x - y = -8$
$3x + y = -2$

30. $x + 3y = -12$
$2x - y = 11$

31. $2x + 3y = -10$
$3x + 4y = -12$

32. $2x - 3y = 10$
$2x + 2y = 5$

33. $x + 2y - z = 2$
$-3x + y + z = 3$
$-x + z = 1$

34. $-2x + 2y + 4z = 3$
$-3x + 4y + 5z = 1$
$x + 2z = 2$

35. $2x + 2y - 4z = 12$
$2x + 6y = 16$
$-3x - 3y + 5z = -20$

36. $2x + 4y + 6z = 4$
$-x - 4y - 3z = 8$
$y - z = -4$

37. $x + y + 2w = 3$
$2x - y + z - w = 3$
$3x + 3y + 2z - 2w = 5$
$x + 2y + z = 3$

38. $x - 2y + 3z = 1$
$y - z + w = -1$
$-2x + 2y - 2z + 4w = 2$
$2y - 3z + w = -3$

Extending Skills *Solve each problem.*

39. The amount of plate-glass sales S (in millions of dollars) can be affected by the number of new building contracts B issued (in millions) and automobiles A produced (in millions). A plate-glass company in California wants to forecast future sales by using the past three years of sales. The totals for three years are given in the table.

S	A	B
602.7	5.543	37.14
656.7	6.933	41.30
778.5	7.638	45.62

To describe the relationship among these variables, we use the equation

$$S = a + bA + cB,$$

where the coefficients a, b, and c are constants that must be determined. (Data from Makridakis, S., and S. Wheelwright, *Forecasting Methods for Management,* John Wiley & Sons.)

(a) Substitute the values for S, A, and B for each year from the table into the equation $S = a + bA + cB$, and obtain three linear equations involving a, b, and c.

(b) Use a graphing calculator to solve this linear system for a, b, and c. Use matrix inverse methods.

(c) Write the equation for S using these values for the coefficients.

(d) It is estimated that for the next year, $A = 7.752$ and $B = 47.38$. Predict S to the nearest tenth. (The actual value for S was 877.6.)

(e) It is predicted that in six years, $A = 8.9$ and $B = 66.25$. Find the value of S in this situation and discuss its validity.

40. The number of automobile tire sales is dependent on several variables. In one study the relationship among annual tire sales S (in thousands of dollars), automobile registrations R (in millions), and personal disposable income I (in millions of dollars) was investigated. The results for three years are given in the table.

S	R	I
10,170	112.9	307.5
15,305	132.9	621.63
21,289	155.2	1937.13

To describe the relationship between these variables, we use the equation

$$S = a + bR + cI,$$

where the coefficients a, b, and c are constants that must be determined. (Data from Jarrett, J., *Business Forecasting Methods,* Basil Blackwell.)

(a) Substitute the values for S, R, and I for each year from the table into the equation $S = a + bR + cI$, and obtain three linear equations involving a, b, and c.

(b) Use a graphing calculator to solve this linear system for a, b, and c. Use matrix inverse methods.

(c) Write the equation for S using these values for the coefficients.

(d) If $R = 117.6$ and $I = 310.73$, predict S to the nearest tenth. (The actual value for S was 11,314.)

(e) If $R = 143.8$ and $I = 829.06$, predict S to the nearest tenth. (The actual value for S was 18,481.)

ANSWERS TO SELECTED EXERCISES

In this section, we provide the answers that we think most students will obtain when they work the exercises using the methods explained in the text. If your answer does not look exactly like the one given here, it is not necessarily wrong. In many cases, there are equivalent forms of the answer that are correct. For example, if the answer section shows $\frac{3}{4}$ and your answer is 0.75, you have obtained the right answer, but written it in a different (yet equivalent) form. Unless the directions specify otherwise, 0.75 is just as valid an answer as $\frac{3}{4}$.

In general, if your answer does not agree with the one given in the text, see whether it can be transformed into the other form. If it can, then it is the correct answer. If you still have doubts, talk with your instructor.

R REVIEW OF THE REAL NUMBER SYSTEM

Section R.1

1. true **3.** false; This is an improper fraction. Its value is 1. **5.** C **7.** $\frac{1}{2}$ **9.** $\frac{5}{6}$ **11.** $\frac{3}{5}$ **13.** $\frac{1}{5}$ **15.** $\frac{6}{5}$ **17.** $1\frac{5}{7}$ **19.** $6\frac{5}{12}$ **21.** $\frac{13}{5}$ **23.** $\frac{38}{3}$ **25.** $\frac{24}{35}$ **27.** $\frac{1}{20}$ **29.** $\frac{6}{25}$ **31.** $\frac{6}{5}$, or $1\frac{1}{5}$ **33.** 9 **35.** $\frac{65}{12}$, or $5\frac{5}{12}$ **37.** $\frac{14}{27}$ **39.** $\frac{10}{3}$, or $3\frac{1}{3}$ **41.** 12 **43.** $\frac{1}{16}$ **45.** 10 **47.** 18 **49.** $\frac{35}{24}$, or $1\frac{11}{24}$ **51.** $\frac{84}{47}$, or $1\frac{37}{47}$ **53.** $\frac{11}{15}$ **55.** $\frac{2}{3}$ **57.** $\frac{8}{9}$ **59.** $\frac{29}{24}$, or $1\frac{5}{24}$ **61.** $\frac{107}{144}$ **63.** $\frac{43}{8}$, or $5\frac{3}{8}$ **65.** $\frac{5}{9}$ **67.** $\frac{2}{3}$ **69.** $\frac{1}{4}$ **71.** $\frac{17}{36}$ **73.** $\frac{67}{20}$, or $3\frac{7}{20}$ **75.** $\frac{32}{9}$, or $3\frac{5}{9}$ **77.** **(a)** $\frac{1}{2}$ **(b)** $\frac{1}{4}$ **(c)** $\frac{1}{3}$ **(d)** $\frac{1}{6}$ **79.** **(a)** 6 **(b)** 9 **(c)** 1 **(d)** 7 **(e)** 4 **81.** **(a)** 46.25 **(b)** 46.2 **(c)** 46 **(d)** 50 **83.** $\frac{4}{10}$ **85.** $\frac{64}{100}$ **87.** $\frac{138}{1000}$ **89.** $\frac{43}{1000}$ **91.** $\frac{3805}{1000}$ **93.** 143.094 **95.** 25.61 **97.** 15.33 **99.** 21.77 **101.** 81.716 **103.** 15.211 **105.** 116.48 **107.** 739.53 **109.** 0.006 **111.** 7.15 **113.** 2.8 **115.** 2.05 **117.** 1232.6 **119.** 5711.6 **121.** 94 **123.** 0.162 **125.** 1.2403 **127.** 0.02329 **129.** 1% **131.** $\frac{1}{20}$ **133.** $12\frac{1}{2}$%, or 12.5% **135.** 0.25; 25% **137.** $\frac{1}{2}$; 0.5 **139.** $\frac{3}{4}$; 75% **141.** 4.2 **143.** 2.25 **145.** 0.375 **147.** $0.\overline{5}$; 0.556 **149.** $0.1\overline{6}$; 0.167 **151.** 0.54 **153.** 0.07 **155.** 1.17 **157.** 0.024 **159.** 0.0625 **161.** 79% **163.** 2% **165.** 0.4% **167.** 128% **169.** 40% **171.** $\frac{51}{100}$ **173.** $\frac{3}{20}$ **175.** $\frac{1}{50}$ **177.** $\frac{7}{5}$, or $1\frac{2}{5}$ **179.** $\frac{3}{40}$ **181.** 80% **183.** 14% **185.** $18.\overline{18}$% **187.** 225% **189.** $216.\overline{6}$%

Section R.2

1. yes **3.** $\{1, 2, 3, 4, 5\}$ **5.** $\{5, 6, 7, 8, \ldots\}$ **7.** $\{\ldots, -1, 0, 1, 2, 3, 4\}$ **9.** $\{10, 12, 14, 16, \ldots\}$ **11.** $\{-4, 4\}$ **13.** $\varnothing$

In Exercises 15 and 17, we give one possible answer.

15. $\{x \mid x$ is an even natural number less than or equal to $8\}$
17. $\{x \mid x$ is a positive multiple of $4\}$

19. 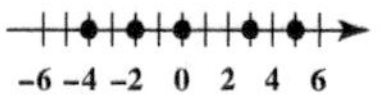

21. 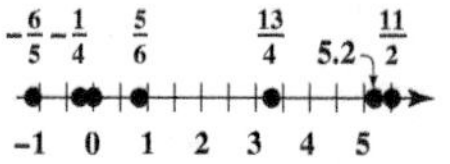

23. **(a)** 8, 13, $\frac{75}{5}$ (or 15) **(b)** 0, 8, 13, $\frac{75}{5}$ **(c)** $-9, 0, 8, 13, \frac{75}{5}$ **(d)** $-9, -0.7, 0, \frac{6}{7}, 4.\overline{6}, 8, \frac{21}{2}, 13, \frac{75}{5}$ **(e)** $-\sqrt{6}, \sqrt{7}$ **(f)** All are real numbers. **25.** false; Some are whole numbers, but negative integers are not. **27.** false; No irrational number is an integer. **29.** true **31.** true **33.** true **35.** **(a)** A **(b)** A **(c)** B **(d)** B **37.** **(a)** -6 **(b)** 6 **39.** **(a)** 12 **(b)** 12 **41.** **(a)** $-\frac{6}{5}$ **(b)** $\frac{6}{5}$ **43.** 8 **45.** $\frac{3}{2}$ **47.** -5 **49.** -2 **51.** -4.5 **53.** 5 **55.** -1 **57.** 6 **59.** 0 **61.** **(a)** New Orleans; The population increased by 13.9%. **(b)** Toledo; The population decreased by 3.0%. **63.** Pacific Ocean, Indian Ocean, Caribbean Sea, South China Sea, Gulf of California **65.** true **67.** true **69.** false **71.** true **73.** false **75.** $2 < 6$ **77.** $4 > -9$ **79.** $-10 < -5$ **81.** $7 > -1$ **83.** $5 \geq 5$ **85.** $13 - 3 \leq 10$ **87.** $5 + 0 \neq 0$ **89.** false **91.** true **93.** $-6 < 10$; true **95.** $10 \geq 10$; true **97.** $-3 \geq -3$; true **99.** $-8 > -6$; false **101.** Iowa (IA), Ohio (OH), Pennsylvania (PA) **103.** $x < y$

Section R.3

1. positive; $18 + 6 = 24$ **3.** greater; $-14 + 9 = -5$ **5.** the number with greater absolute value is subtracted from the one with lesser absolute value; $5 - 12 = -7$ **7.** positive; $-2(-8) = 16$ **9.** undefined; 0; $\frac{-17}{0}$ is undefined; $\frac{0}{42} = 0$ **11.** -19 **13.** -9 **15.** 9 **17.** 5 **19.** $-\frac{19}{12}$ **21.** 1.7 **23.** -5 **25.** -11 **27.** 21 **29.** -13 **31.** -10.18 **33.** $\frac{67}{30}$ **35.** 14 **37.** -5 **39.** -6 **41.** -11 **43.** 0 **45.** 12 **47.** -4 **49.** 3 **51.** $-\frac{7}{4}$ **53.** $-\frac{7}{8}$ **55.** 2 **57.** 1 **59.** 6 **61.** $\frac{13}{2}$, or $6\frac{1}{2}$ **63.** It is true for multiplication (and division). It is false for addition and false for subtraction when the number to be subtracted has the lesser absolute value. A more precise statement is "The product or quotient of two negative numbers is positive." **65.** 40 **67.** -35 **69.** 0 **71.** 0 **73.** -12 **75.** 2 **77.** $\frac{6}{5}$ **79.** 1 **81.** 0.4 **83.** -0.024 **85.** -7 **87.** 6 **89.** -4 **91.** 0 **93.** undefined **95.** $\frac{25}{102}$ **97.** $-\frac{9}{13}$ **99.** -9 **101.** 3.2 **103.** $\frac{17}{18}$ **105.** $\frac{17}{36}$ **107.** $-\frac{19}{24}$ **109.** $-\frac{22}{45}$ **111.** $-\frac{2}{15}$ **113.** $\frac{3}{5}$ **115.** $-\frac{20}{7}$ **117.** $-\frac{1}{3}$ **119.** $-\frac{4}{3}$ **121.** -32.351 **123.** -3 **125.** -4.14 **127.** 4800 **129.** 10,000 **131.** -4.39 **133.** 180°F **135.** \$30.13 **137.** **(a)** \$466.02 **(b)** \$190.68 **139.** **(a)** 31.33% **(b)** -36.87% **(c)** 58.64% **141.** 2001: \$128 billion; 2006: $-$\$248 billion; 2011: $-$\$1300 billion; 2016: $-$\$585 billion

Section R.4

1. false; $-7^6 = -(7^6)$ **3.** true **5.** true **7.** true **9.** false; The base is 2. **11.** 10^4 **13.** $\left(\frac{3}{4}\right)^5$ **15.** $(-9)^3$ **17.** $(0.8)^2$ **19.** z^7

21. (a) 64 (b) −64 (c) 64 (d) −64 **23.** 16 **25.** 0.027 **27.** $\frac{1}{125}$ **29.** $\frac{256}{625}$ **31.** −125 **33.** 256 **35.** −729 **37.** −4096 **39.** (a) negative (b) positive (c) negative (d) negative (e) negative (f) positive **41.** 9 **43.** 13 **45.** −20 **47.** $\frac{10}{11}$ **49.** −0.7 **51.** not a real number **53.** not a real number **55.** (a) The grandson's answer was correct. (b) The reasoning was *incorrect*. The operation of division must be done first, and then the addition follows. The grandson's "Order of Process rule" is **not correct.** It happened coincidentally, in this problem, that he obtained the correct answer the wrong way. **57.** 24 **59.** 15 **61.** 55 **63.** −91 **65.** −8 **67.** −48 **69.** −2 **71.** −97 **73.** 11 **75.** −10 **77.** 2 **79.** 1 **81.** undefined **83.** −1 **85.** 17 **87.** −96 **89.** 180 **91.** $\frac{5}{9}$ **93.** $-\frac{15}{238}$ **95.** 8 **97.** $-\frac{5}{16}$ **99.** $-\frac{11}{4}$, or −2.75 **101.** \$2434 **103.** (a) $48 \times 3.2 \times 0.075 \div 190 - 2 \times 0.015$ (b) 0.031 **105.** (a) 0.024; 0.023; Increased weight results in lower BACs. (b) Decreased weight will result in higher BACs; 0.040; 0.053 **107.** (a) \$31.5 billion (b) \$48.9 billion (c) \$63.9 billion (d) The amount spent on pets roughly doubled from 2003 to 2016.

Section R.5

1. B **3.** A **5.** $ab + ac$ **7.** grouping **9.** like **11.** $2m + 2p$ **13.** $-12x + 12y$ **15.** $8k$ **17.** $-2r$ **19.** cannot be simplified **21.** $8a$ **23.** $2x$ **25.** $-2d + f$ **27.** $x + y$ **29.** $2x - 6y + 4z$ **31.** $-3y$ **33.** $p + 11$ **35.** $-2k + 15$ **37.** $-4y + 2$ **39.** $30xy$ **41.** $-56wz$ **43.** $m - 14$ **45.** −1 **47.** $2p + 7$ **49.** $-6z - 39$ **51.** $(5 + 8)x = 13x$ **53.** $(5 \cdot 9)r = 45r$ **55.** $9y + 5x$ **57.** 7 **59.** 0 **61.** $8(-4) + 8x = -32 + 8x$ **63.** 0 **65.** Answers will vary. One example is washing your face and brushing your teeth. **67.** 1900 **69.** 75 **71.** 87.5 **73.** associative property **74.** associative property **75.** commutative property **76.** associative property **77.** distributive property **78.** arithmetic facts

Chapter R Test

[R.1] **1.** $\frac{11}{12}$ **2.** $\frac{2}{3}$ **3.** 6.833 **4.** 0.028 **5.** $\frac{1}{25}$; 0.04 **6.** $0.8\overline{3}$; $83\frac{1}{3}\%$, or $83.\overline{3}\%$ **7.** $\frac{3}{2}$, or $1\frac{1}{2}$; 150% [R.2] **8.**

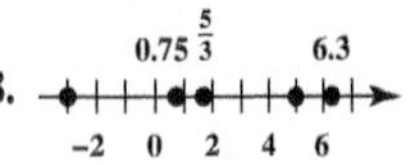

9. 0, 3, $\sqrt{25}$ (or 5), $\frac{24}{2}$ (or 12) **10.** −1, 0, 3, $\sqrt{25}$ (or 5), $\frac{24}{2}$ (or 12) **11.** −1, −0.5, 0, 3, $\sqrt{25}$ (or 5), 7.5, $\frac{24}{2}$ (or 12) **12.** All are real numbers except $\sqrt{-4}$. [R.3, R.4] **13.** 0 **14.** $-\frac{17}{63}$ **15.** −26 **16.** 19 **17.** $\frac{16}{7}$ **18.** undefined [R.4] **19.** 14 **20.** −15 **21.** not a real number **22.** $-\frac{6}{23}$ [R.5] **23.** $10k - 10$ **24.** $7r + 2$ [R.3, R.5] **25.** (a) B (b) D (c) A (d) F (e) C (f) C (g) E

1 LINEAR EQUATIONS, INEQUALITIES, AND APPLICATIONS

Section 1.1

1. algebraic expression; does; is not **3.** true; solution; solution set **5.** identity; all real numbers **7.** A, C **9.** equation **11.** expression **13.** equation **15.** A sign error was made when the distributive property was applied. The left side of the second line should be $8x - 4x + 6$. The correct solution is 1. **17.** (a) B (b) A (c) C **19.** $\{-1\}$ **21.** $\{-4\}$ **23.** $\{-7\}$ **25.** $\{0\}$ **27.** $\{4\}$ **29.** $\{-\frac{7}{8}\}$ **31.** $\emptyset$; contradiction **33.** $\{-9\}$ **35.** $\{-\frac{1}{2}\}$ **37.** $\{2\}$ **39.** $\{-2\}$ **41.** {all real numbers}; identity **43.** $\{-1\}$ **45.** $\{7\}$ **47.** $\{2\}$ **49.** {all real numbers}; identity **51.** 12 **53.** (a) 10^2, or 100 (b) 10^3, or 1000 **55.** $\{-\frac{18}{5}\}$ **57.** $\{-\frac{5}{6}\}$ **59.** $\{6\}$ **61.** $\{4\}$ **63.** $\{3\}$ **65.** $\{3\}$ **67.** $\{0\}$ **69.** $\{35\}$ **71.** $\{-10\}$ **73.** $\{2000\}$ **75.** $\{25\}$ **77.** $\{40\}$ **79.** $\{3\}$

Section 1.2

1. formula **3.** (a) 35% (b) 18% (c) 2% (d) 7.5% (e) 150%

There may be other acceptable forms of the answers in Exercises 5–35.

5. $r = \frac{I}{pt}$ **7.** (a) $W = \frac{\mathcal{A}}{L}$ (b) $L = \frac{\mathcal{A}}{W}$ **9.** $L = \frac{P - 2W}{2}$, or $L = \frac{P}{2} - W$ **11.** (a) $W = \frac{V}{LH}$ (b) $H = \frac{V}{LW}$ **13.** $r = \frac{C}{2\pi}$ **15.** (a) $b = \frac{2\mathcal{A} - hB}{h}$, or $b = \frac{2\mathcal{A}}{h} - B$ (b) $B = \frac{2\mathcal{A} - hb}{h}$, or $B = \frac{2\mathcal{A}}{h} - b$ **17.** $C = \frac{5}{9}(F - 32)$ **19.** (a) $x = -\frac{b}{a}$ (b) $a = -\frac{b}{x}$ **21.** $t = \frac{A - P}{Pr}$ **23.** D **25.** $y = -4x + 1$ **27.** $y = \frac{1}{2}x + 3$ **29.** $y = -\frac{4}{9}x + \frac{11}{9}$ **31.** $y = \frac{3}{2}x + \frac{5}{2}$ **33.** $y = \frac{6}{5}x - \frac{7}{5}$ **35.** $y = \frac{3}{2}x - 3$ **37.** 3.492 hr **39.** 52 mph **41.** 108°F **43.** 230 m **45.** (a) 240 in. (b) 480 in. **47.** 2 in. **49.** 25% **51.** 20% **53.** 75% **55.** 75% water; 25% alcohol **57.** 3% **59.** (a) .494 (b) .481 (c) .481 (d) .463 **61.** 19% **63.** 6.5 million **65.** \$72,324 **67.** \$14,465 **69.** 12% **71.** 5.1% **73.** 10.1% **75.** (a) 0.79 m^2 (b) 0.96 m^2 **77.** (a) 116 mg (b) 141 mg **79.** (a) $7x + 8 = 36$ (b) $ax + k = tc$ **80.** (a) $7x + 8 - 8 = 36 - 8$ (b) $ax + k - k = tc - k$ **81.** (a) $7x = 28$ (b) $ax = tc - k$ **82.** (a) $x = 4$; $\{4\}$ (b) $x = \frac{tc - k}{a}$ **83.** $a \neq 0$; If $a = 0$, the denominator is 0. **84.** $\{-5\}$ **85.** $a = \frac{2S - n\ell}{n}$, or $a = \frac{2S}{n} - \ell$; $n \neq 0$

Section 1.3

1. (a) $x + 15$ (b) $15 > x$ **3.** (a) $x - 8$ (b) $8 < x$ **5.** Because the unknown number comes first in the verbal phrase, it must come first in the mathematical expression. The correct expression is $x - 7$. **7.** $2x - 13$ **9.** $12 + 4x$ **11.** $8(x - 16)$ **13.** $\frac{3x}{10}$ **15.** $x + 6 = -31$; −37 **17.** $x - (-4x) = x + 9$; $\frac{9}{4}$ **19.** $12 - \frac{2}{3}x = 10$; 3 **21.** expression; $-11x + 63$ **23.** equation; $\{\frac{51}{11}\}$ **25.** expression; $\frac{1}{3}x - \frac{13}{2}$ **27.** *Step 1:* the number of patents each corporation secured; *Step 2:* patents that Samsung secured; *Step 3:* x; $x - 2250$; *Step 4:* 7309; *Step 5:* 7309; 5059; *Step 6:* 2250; IBM patents; 5059; 12,368

29. width: 165 ft; length: 265 ft **31.** 24.34 in. by 29.88 in. **33.** 850 mi; 925 mi; 1300 mi **35.** Walmart: \$486.9 billion; Berkshire Hathaway: \$223.6 billion **37.** Eiffel Tower: 984 ft; Leaning Tower: 180 ft **39.** National Geographic: 83.4 million followers; Nike: 75.4 million followers **41.** 19.0% **43.** 1300 million users **45.** \$49.86 **47.** \$225 **49.** We cannot expect the final mixture to be worth more than the more expensive of the two ingredients. **51.** \$4000 at 3%; \$8000 at 4% **53.** \$10,000 at 4.5%; \$19,000 at 3% **55.** \$24,000 **57.** 5 L **59.** 4 L **61.** 1 gal **63.** 150 lb **65.** **(a)** $800 - x$ **(b)** $800 - y$
66. **(a)** $0.03x$; $0.06(800 - x)$ **(b)** $0.03y$; $0.06(800 - y)$
67. **(a)** $0.03x + 0.06(800 - x) = 800(0.0525)$
(b) $0.03y + 0.06(800 - y) = 800(0.0525)$
68. **(a)** \$200 at 3%; \$600 at 6% **(b)** 200 L of 3% acid; 600 L of 6% acid **(c)** The processes are the same. The amounts of money in Problem A correspond to the amounts of solution in Problem B.

Section 1.4

1. \$5.70 **3.** 30 mph **5.** 81 in.^{2} **7.** The right side of the equation should be 180, not 90. The correct equation is $x + (x + 1) + (x + 2) = 180$. The angles measure 59°, 60°, and 61°. **9.** 17 pennies; 17 dimes; 10 quarters **11.** 23 loonies; 14 toonies **13.** 28 \$10 coins; 13 \$20 coins **15.** 872 adult tickets **17.** 8.01 m per sec **19.** 10.11 m per sec **21.** $2\frac{1}{2}$ hr **23.** 7:50 P.M. **25.** 45 mph **27.** $\frac{1}{2}$ hr **29.** 60°, 60°, 60° **31.** 40°, 45°, 95° **33.** 41°, 52°, 87° **35.** Both measure 122°. **37.** 64°, 26° **39.** 65°, 115° **41.** 19, 20, 21 **43.** 27, 28, 29, 30 **45.** 61 yr old **47.** 28, 30, 32 **49.** 21, 23, 25 **51.** 40°, 80° **52.** 120° **53.** The sum is equal to the measure of the angle found in **Exercise 52.** **54.** The sum of the measures of angles ① and ② is equal to the measure of angle ③.

SUMMARY EXERCISES Applying Problem-Solving Techniques

1. length: 8 in.; width: 5 in. **2.** length: 60 m; width: 30 m **3.** \$279.98 **4.** \$425 **5.** \$800 at 4%; \$1600 at 5% **6.** \$12,000 at 3%; \$14,000 at 4% **7.** Westbrook: 2558 points; Harden: 2356 points **8.** 2016: 718 films; 2009: 557 films **9.** 5 hr **10.** $1\frac{1}{2}$ cm **11.** $13\frac{1}{3}$ L **12.** $53\frac{1}{3}$ kg **13.** 84 fives; 42 tens **14.** 1650 tickets at \$9; 810 tickets at \$7 **15.** 20°, 30°, 130° **16.** 107°, 73° **17.** 31, 32, 33 **18.** 9, 11 **19.** 6 in., 12 in., 16 in. **20.** 23 in.

Section 1.5

1. D **3.** B **5.** F **7.** **(a)** $x < 100$ **(b)** $100 \le x \le 129$ **(c)** $130 \le x \le 159$ **(d)** $160 \le x \le 189$ **(e)** $x \ge 190$
9. The student divided by 4, a *positive* number. Reverse the symbol only when multiplying or dividing by a *negative* number. The solution set is $[-16, \infty)$.

11. $[16, \infty)$ **13.** $(7, \infty)$

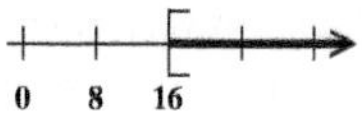

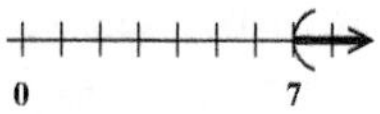

15. $(-\infty, -4)$ **17.** $(-4, \infty)$

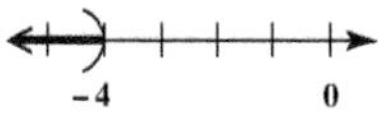

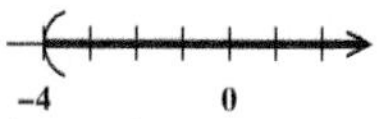

19. $(-\infty, -40]$ **21.** $(-\infty, 4]$

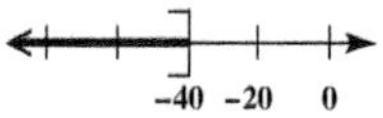

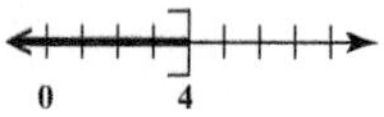

23. $(-\infty, -10]$ **25.** $(7, \infty)$

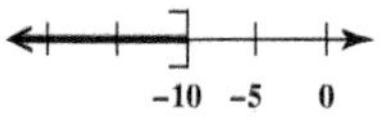

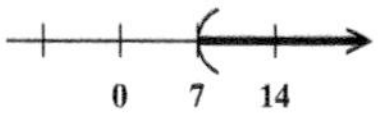

27. $\left(-\infty, -\frac{15}{2}\right)$ **29.** $(-\infty, -7)$

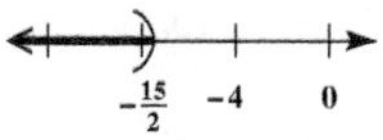

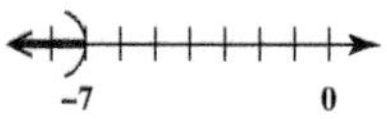

31. $\left[\frac{1}{2}, \infty\right)$ **33.** $[2, \infty)$

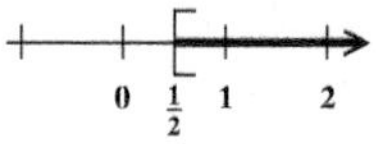

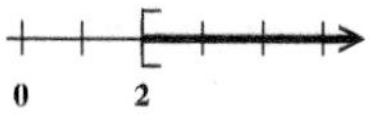

35. $(3, \infty)$ **37.** $(-\infty, 4)$

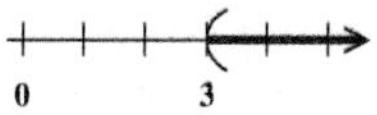

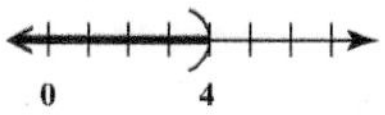

39. $\left(-\infty, \frac{23}{6}\right]$ **41.** $\left(-\infty, \frac{76}{11}\right)$

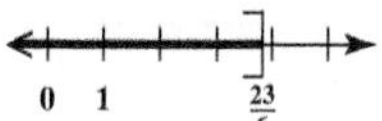

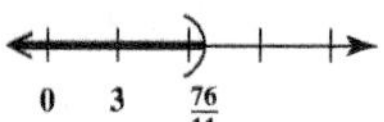

43. $(-\infty, \infty)$ **45.** $\emptyset$

47. $(1, 11)$ **49.** $[-14, 10]$

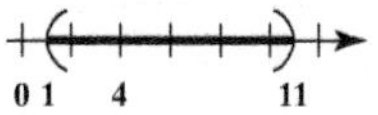

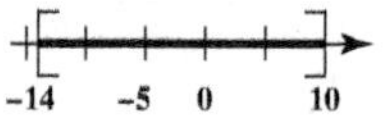

51. $[-5, 6]$ **53.** $\left[-\frac{14}{3}, 2\right]$

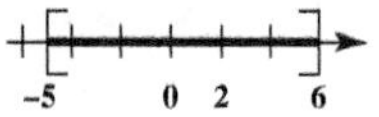

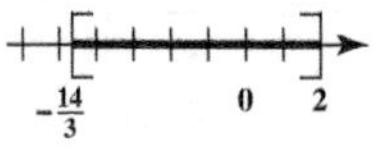

55. $\left(-\frac{1}{3}, \frac{1}{9}\right]$ **57.** $\left[-1, \frac{5}{2}\right]$

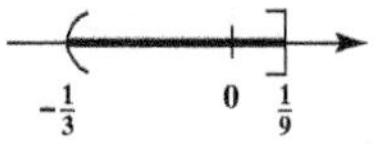

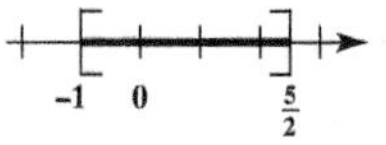

59. $(-4, 8)$ **61.** $\left[-\frac{1}{2}, \frac{35}{2}\right]$

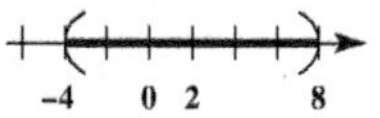

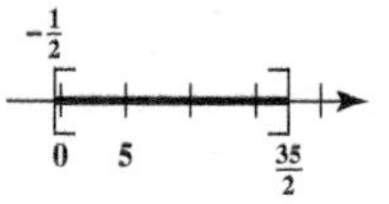

63. $\left[\frac{6}{5}, \frac{16}{5}\right]$ **65.** $(0, 1)$ **67.** $(-2, 2)$ **69.** $[3, \infty)$ **71.** $[-9, \infty)$ **73.** at least 80

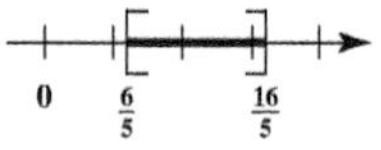

75. at least 78 **77.** 26 months **79.** more than 4.4 in. **81.** **(a)** 140 to 184 lb **(b)** 107 to 141 lb **(c)** Answers will vary. **83.** 26 DVDs

85. $\{-9\}$ **86.** $(-9, \infty)$

87. $(-\infty, -9)$

88. We obtain the set of all real numbers.

89. $\{-3\}$; $(-3, \infty)$; $(-\infty, -3)$ **90.** It is the set of all real numbers; $(-\infty, \infty)$

Section 1.6

1. true **3.** false; The union is $(-\infty, 7) \cup (7, \infty)$. **5.** $\{4\}$, or D **7.** $\{1, 3, 5\}$, or B **9.** $\varnothing$ **11.** $\{1, 2, 3, 4, 5, 6\}$, or A **13.** $\{1, 3, 5, 6\}$

15.

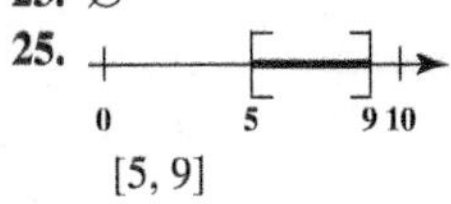

17.

19. $(-3, 2)$

21. $(-\infty, 2]$

23. $\varnothing$

25. $[5, 9]$

27. $(-3, -1)$

29. $(-\infty, 4]$

31.

33.

35. $(-\infty, 8]$

37. $[-2, \infty)$

39. $(-\infty, \infty)$

41. $(-\infty, -5) \cup (5, \infty)$

43. $(-\infty, -1) \cup (2, \infty)$

45. $(-\infty, \infty)$

47. $[-4, -1]$ **49.** $[-9, -6]$ **51.** $(-\infty, 3)$ **53.** $[3, 9)$

55. intersection; $(-5, -1)$

57. union; $(-\infty, 4)$

59. union; $(-\infty, 0] \cup [2, \infty)$

61. intersection; $[4, 12]$

63. {Tuition and fees} **65.** {Tuition and fees, Board rates, Dormitory charges} **67.** 160 ft; 150 ft; 220 ft; 120 ft **69.** Maria, Joe **71.** none of them **73.** Maria, Joe **75.** **(a)** A: $185 + x \geq 585$; B: $185 + x \geq 520$; C: $185 + x \geq 455$ **(b)** A: $x \geq 400$; 89%; B: $x \geq 335$; 75%; C: $x \geq 270$; 60% **76.** $520 \leq 185 + x \leq 584$; $335 \leq x \leq 399$; $75\% \leq \text{average} \leq 89\%$ **77.** A: $105 + x \geq 585$; $x \geq 480$; impossible; B: $105 + x \geq 520$; $x \geq 415$; 93%; C: $105 + x \geq 455$; $x \geq 350$; 78% **78.** $455 \leq 105 + x \leq 519$; $350 \leq x \leq 414$; $78\% \leq \text{average} \leq 92\%$

Section 1.7

1. E; C; D; B; A **3.** **(a)** one **(b)** two **(c)** none **5.** $\{-12, 12\}$ **7.** $\{-5, 5\}$ **9.** $\{-6, 12\}$ **11.** $\{-5, 6\}$ **13.** $\left\{-3, \frac{11}{2}\right\}$ **15.** $\left\{-\frac{19}{2}, \frac{9}{2}\right\}$ **17.** $\left\{\frac{7}{3}, 3\right\}$ **19.** $\{12, 36\}$ **21.** $\{-12, 12\}$ **23.** $\{-10, -2\}$ **25.** $\left\{-\frac{32}{3}, 8\right\}$ **27.** $\{-75, 175\}$

29.

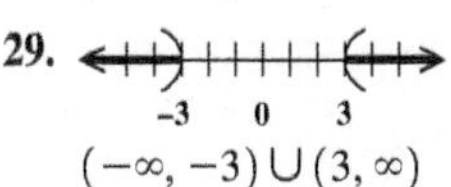

$(-\infty, -3) \cup (3, \infty)$

31. $(-\infty, -4] \cup [4, \infty)$

33. $(-\infty, -25] \cup [15, \infty)$

35. $\left(-\infty, -\frac{12}{5}\right) \cup \left(\frac{8}{5}, \infty\right)$

37. $(-\infty, -2) \cup (8, \infty)$

39. $\left(-\infty, -\frac{9}{5}\right] \cup [3, \infty)$

41. **(a)** **(b)**

43. $[-3, 3]$

45. $(-4, 4)$

47. $(-25, 15)$

49. $\left[-\frac{12}{5}, \frac{8}{5}\right]$

51. $[-2, 8]$

53.

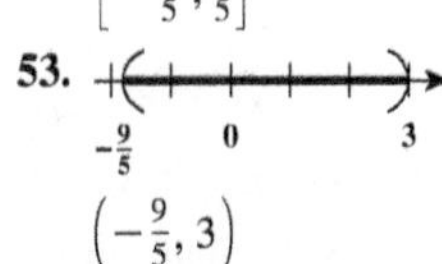

$\left(-\frac{9}{5}, 3\right)$

55. $(-\infty, -5) \cup (13, \infty)$

57. $\left(-\frac{13}{3}, 3\right)$

59. $\{-6, -1\}$

61.

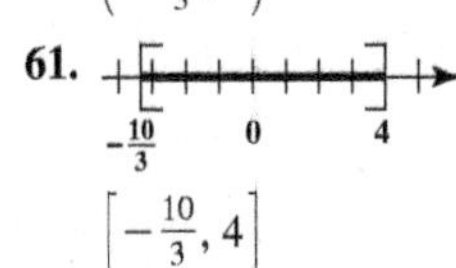

$\left[-\frac{10}{3}, 4\right]$

63.

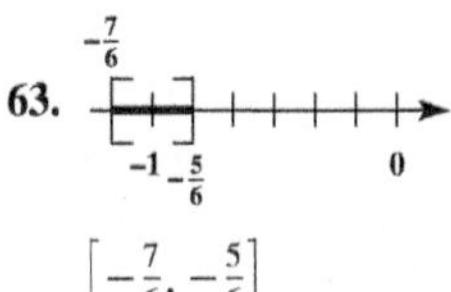

$\left[-\frac{7}{6}, -\frac{5}{6}\right]$

65.

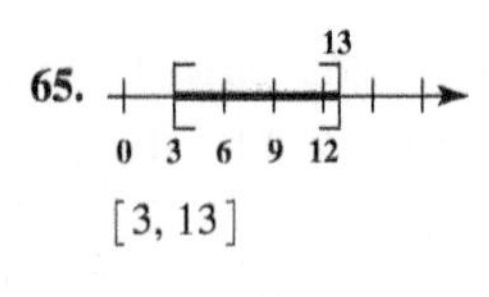

$[3, 13]$

67. $\left(-\infty, \frac{1}{2}\right] \cup \left[\frac{7}{6}, \infty\right)$ **69.** $\left\{-\frac{5}{3}, \frac{11}{3}\right\}$ **71.** $(-\infty, -20) \cup (40, \infty)$ **73.** $\{-5, 1\}$ **75.** $\{3, 9\}$ **77.** $\{0, 20\}$ **79.** $\{-5, 5\}$ **81.** $\{-5, -3\}$ **83.** $(-\infty, -3) \cup (2, \infty)$ **85.** $[-10, 0]$ **87.** $(-\infty, 20] \cup [30, \infty)$ **89.** $\left\{-\frac{5}{3}, \frac{1}{3}\right\}$ **91.** $\{-1, 3\}$ **93.** $\left\{-3, \frac{5}{3}\right\}$ **95.** $\left\{-\frac{1}{3}, -\frac{1}{15}\right\}$ **97.** $\left\{-\frac{5}{4}\right\}$ **99.** $(-\infty, \infty)$ **101.** $\varnothing$ **103.** $\left\{-\frac{1}{4}\right\}$ **105.** $\varnothing$ **107.** $(-\infty, \infty)$ **109.** $\left\{-\frac{3}{7}\right\}$ **111.** $\left\{\frac{2}{5}\right\}$ **113.** $(-\infty, \infty)$ **115.** $\varnothing$ **117.** between 60.8 and 67.2 oz, inclusive **119.** between 31.36 and 32.64 oz, inclusive **121.** $(-0.05, 0.05)$ **123.** $(2.74975, 2.75025)$ **125.** between 6.8 and 9.8 lb **127.** $|x - 1000| \leq 100$; $900 \leq x \leq 1100$ **129.** 814.8 ft **130.** Bank of America Center

131. Williams Tower, Bank of America Center, Texaco Heritage Plaza, 609 Main at Texas, Enterprise Plaza, Centerpoint Energy Plaza, 1600 Smith St., Fulbright Tower **132. (a)** $|x - 814.8| \geq 95$ **(b)** $x \geq 909.8$ or $x \leq 719.8$ **(c)** JPMorgan Chase Tower, Wells Fargo Plaza **(d)** It makes sense because it includes all buildings *not* listed in the answer to **Exercise 131.**

SUMMARY EXERCISES Solving Linear and Absolute Value Equations and Inequalities

1. $\{12\}$ **2.** $\{-5, 7\}$ **3.** $\{7\}$ **4.** $\{-\frac{2}{5}\}$ **5.** $\emptyset$ **6.** $(-\infty, -1]$ **7.** $[-\frac{2}{3}, \infty)$ **8.** $\{-1\}$ **9.** $\{-3\}$ **10.** $\{1, \frac{11}{3}\}$ **11.** $(-\infty, 5]$ **12.** $(-\infty, \infty)$ **13.** $\{2\}$ **14.** $(-\infty, -8] \cup [8, \infty)$ **15.** $\emptyset$ **16.** $(-\infty, \infty)$ **17.** $(-5.5, 5.5)$ **18.** $\{\frac{13}{3}\}$ **19.** $\{-\frac{96}{5}\}$ **20.** $(-\infty, 32]$ **21.** $(-\infty, -24)$ **22.** $[\frac{3}{4}, \frac{15}{8}]$ **23.** $\{\frac{7}{2}\}$ **24.** $\{60\}$ **25.** {all real numbers} **26.** $(-\infty, 5)$ **27.** $[-\frac{9}{2}, \frac{15}{2}]$ **28.** $\{24\}$ **29.** $\{-\frac{1}{5}\}$ **30.** $(-\infty, -\frac{5}{2}]$ **31.** $[-\frac{1}{3}, 3]$ **32.** $[1, 7]$ **33.** $\{-\frac{1}{6}, 2\}$ **34.** $\{-3\}$ **35.** $(-\infty, -1] \cup [\frac{5}{3}, \infty)$ **36.** $\{\frac{3}{8}\}$ **37.** $\{-\frac{5}{2}\}$ **38.** $(-6, 8)$ **39.** $(-\infty, -4) \cup (7, \infty)$ **40.** $(1, 9)$ **41.** $(-\infty, \infty)$ **42.** $\{\frac{1}{3}, 9\}$ **43.** {all real numbers} **44.** $\{-\frac{10}{9}\}$ **45.** $\{-2\}$ **46.** $\emptyset$ **47.** $(-\infty, -1) \cup (2, \infty)$ **48.** $[-3, -2]$

Chapter 1 Review Exercises

1. $\{-\frac{9}{5}\}$ **2.** $\{16\}$ **3.** $\{-\frac{7}{5}\}$ **4.** $\emptyset$ **5.** {all real numbers}; identity **6.** $\emptyset$; contradiction **7.** $\{0\}$; conditional equation **8.** {all real numbers}; identity **9.** $L = \frac{V}{HW}$ **10.** $b = \frac{2\mathcal{A} - Bh}{h}$, or $b = \frac{2\mathcal{A}}{h} - B$ **11.** $c = P - a - b - B$ **12.** $y = -\frac{4}{7}x + \frac{9}{7}$ **13.** 6 ft **14.** 14.4% **15.** 3.25% **16.** 25° **17.** 138.6 million subscriptions **18.** 67.3 million subscriptions **19.** $9 - \frac{1}{3}x$ **20.** $\frac{4x}{x+9}$ **21.** length: 13 m; width: 8 m **22.** 17 in., 17 in., 19 in. **23.** 12 kg **24.** 30 L **25.** 10 L **26.** \$10,000 at 6%; \$6000 at 4% **27.** 15 dimes; 8 quarters **28.** 7 nickels; 12 dimes **29.** C **30.** 530 mi **31.** 328 mi **32.** 2.2 hr **33.** 50 km per hr; 65 km per hr **34.** 46 mph **35.** 40°, 45°, 95° **36.** 150°, 30° **37.** 48, 49, 50 **38.** 38, 40, 42 **39.** $(-9, \infty)$ **40.** $(-\infty, -3]$ **41.** $(\frac{3}{2}, \infty)$ **42.** $[-3, \infty)$ **43.** $[3, 5)$ **44.** $(\frac{59}{31}, \infty)$ **45. (a)** $x < 6.75$ **(b)** $6.75 \leq x \leq 7.25$ **(c)** $x > 7.25$ **46.** 38 m or less **47.** 31 tickets or less (but at least 15) **48.** any score greater than or equal to 61 **49.** {a, c} **50.** {a} **51.** {a, c, e, f, g} **52.** {a, b, c, d, e, f, g}

53. 0 3 6 9
$(6, 9)$

54. 0 8 14 16
$(8, 14)$

55. −3 0 5
$(-\infty, -3] \cup (5, \infty)$

56. −2 0 2
$(-\infty, \infty)$

57. $\emptyset$

58. −2 0 7
$(-\infty, -2] \cup [7, \infty)$

59. $(-3, 4]$ **60.** $(-\infty, 2]$ **61.** $(4, \infty)$ **62.** $(1, \infty)$ **63.** $\{-7, 7\}$ **64.** $\{-11, 7\}$ **65.** $\{-\frac{1}{3}, 5\}$ **66.** $\emptyset$ **67.** $\{0, 7\}$ **68.** $\{-\frac{3}{2}, \frac{1}{2}\}$ **69.** $\{-\frac{3}{4}, \frac{1}{2}\}$ **70.** $\{-\frac{1}{2}\}$ **71.** $(-14, 14)$ **72.** $[-1, 13]$ **73.** $[-3, -2]$ **74.** $(-\infty, \infty)$ **75.** $\emptyset$ **76.** $(-\infty, \infty)$ **77.** between 46.56 and 49.44 oz, inclusive **78.** between 31.68 and 32.32 oz, inclusive

Chapter 1 Mixed Review Exercises

1. $(-2, \infty)$ **2.** $k = \frac{6r - bt}{a}$ **3.** $[-2, 3)$ **4.** $\{0\}$ **5.** $(-\infty, \infty)$ **6.** $(-\infty, 2]$ **7.** $\{-\frac{7}{3}, 1\}$ **8.** $\{300\}$ **9.** $[-16, 10]$ **10.** $(-\infty, \frac{14}{17})$ **11.** $(-3, \frac{7}{2})$ **12.** $(-\infty, 1]$ **13.** $(-\infty, -\frac{13}{5}) \cup (3, \infty)$ **14.** $(-\infty, \infty)$ **15.** $\{-4, -\frac{2}{3}\}$ **16.** $\{30\}$ **17.** $\{1, \frac{11}{3}\}$ **18.** $\emptyset$ **19.** $(6, 8)$ **20.** $(-\infty, -2] \cup [7, \infty)$ **21.** 10 ft **22.** 46, 47, 48 **23.** any amount greater than or equal to \$1100 **24.** 5 L

Chapter 1 Test

[1.1] **1.** $\{-19\}$ **2.** $\{5\}$ **3. (a)** $\emptyset$; contradiction **(b)** {all real numbers}; identity **(c)** $\{0\}$; conditional equation **(d)** {all real numbers}; identity [1.2] **4.** $h = \frac{3V}{b}$ **5.** $y = \frac{3}{2}x + 3$ [1.3, 1.4] **6.** 3.497 hr **7.** 1.25% **8. (a)** .562 **(b)** .494 **(c)** .469 **(d)** .463 **9.** \$8000 at 1.5%; \$20,000 at 2.5% **10.** 13.3% **11.** 80 L **12.** faster car: 60 mph; slower car: 45 mph **13.** 40°, 40°, 100°

[1.5]

14. 0 1
$[1, \infty)$

15. 0 10 28
$(-\infty, 28)$

16. 0 1 2
$(1, 2)$

17. −3 0 3
$[-3, 3]$

18. C **19.** 82 or more [1.6] **20. (a)** $\{1, 5\}$ **(b)** $\{1, 2, 5, 7, 9, 12\}$ **21.** $[2, 9)$ **22.** $(-\infty, 3) \cup [6, \infty)$ [1.7] **23.** $[-\frac{5}{2}, 1]$ **24.** $(-\infty, -\frac{7}{6}) \cup (\frac{17}{6}, \infty)$ **25.** $\{1, 5\}$ **26.** $(\frac{1}{3}, \frac{7}{3})$ **27.** $\emptyset$ **28.** $\{-\frac{5}{3}, 3\}$ **29.** $\{-\frac{5}{7}, \frac{11}{3}\}$ **30. (a)** $\emptyset$ **(b)** $(-\infty, \infty)$ **(c)** $\emptyset$

Chapters R and 1 Cumulative Review Exercises

[R.1] **1. (a)** 0.35 **(b)** 35% **2. (a)** 66% **(b)** $\frac{33}{50}$ **3.** $\frac{37}{60}$ **4.** −7.865 [R.3, R.4] **5.** −8 **6.** 10 [R.4] **7.** −125 **8.** $\frac{81}{16}$ **9.** −34 **10.** $\frac{3}{16}$ [R.5] **11.** $-20r + 17$ **12.** $13k + 42$ [1.1] **13.** $\{-12\}$ **14.** $\{26\}$ [1.7] **15.** $\{\frac{3}{4}, \frac{7}{2}\}$ [1.2] **16.** $n = \frac{A - P}{iP}$

[1.5–1.7]

17. −14 −2 0
$[-14, \infty)$

18.
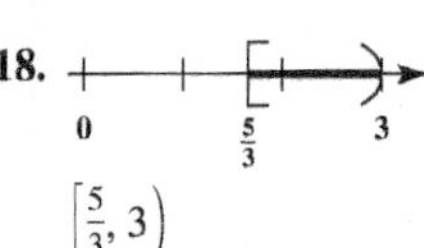

$[\frac{5}{3}, 3)$

19.
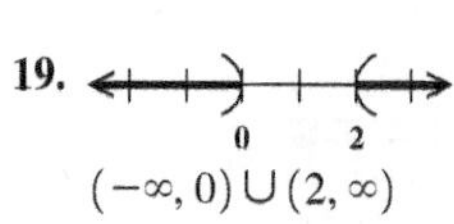

$(-\infty, 0) \cup (2, \infty)$

20.
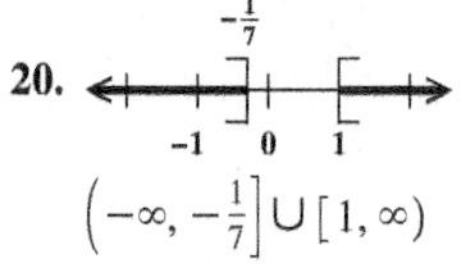

$(-\infty, -\frac{1}{7}] \cup [1, \infty)$

[1.3] **21.** \$5000 [1.5] **22.** $6\frac{1}{3}$ g **23.** 74 or greater [R.3] **24.** 11,331 ft [1.3] **25.** 4 cm; 9 cm; 27 cm

2 LINEAR EQUATIONS, GRAPHS, AND FUNCTIONS

Section 2.1

1. origin **3.** y; x; x; y **5.** two; 2; 6 **7.** **(a)** x represents the year; y represents personal spending on medical care in billions of dollars. **(b)** about \$2830 billion **(c)** (2016, 2830) **(d)** 2010

9. **(a)** I **(b)** III **(c)** II **(d)** IV **(e)** none **(f)** none

11. **(a)** I or III **(b)** II or IV **(c)** II or IV **(d)** I or III

13. The student interchanged the x- and y-coordinates. To plot this point correctly, move from 0 to the left 4 units on the x-axis and then up 2 units parallel to the y-axis.

15.–23.

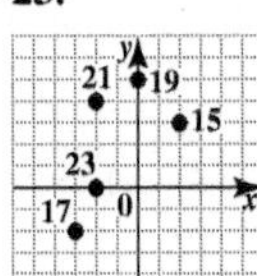

25. -4; -3; -2; -1; 0

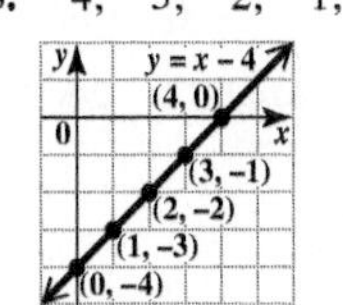

27. -3; 3; 2; -1 **29.** $\frac{5}{2}$; 5; $\frac{3}{2}$; 1 **31.** -4; 5; $-\frac{12}{5}$; $\frac{5}{4}$

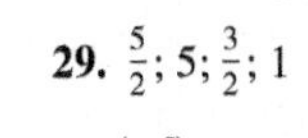

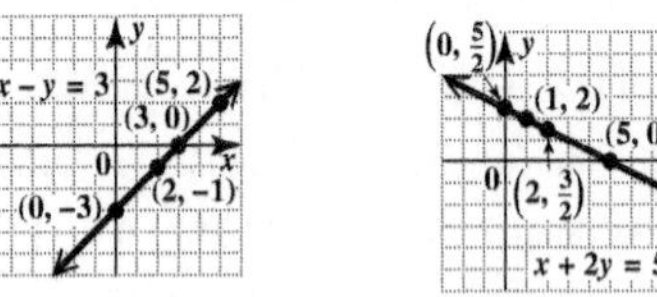

33. **(a)** C **(b)** D **(c)** B **(d)** A

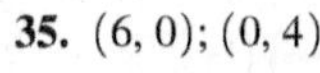

35. (6, 0); (0, 4) **37.** (6, 0); (0, -2) **39.** (-2, 0); $\left(0, -\frac{5}{3}\right)$

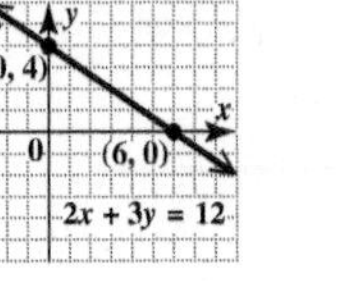

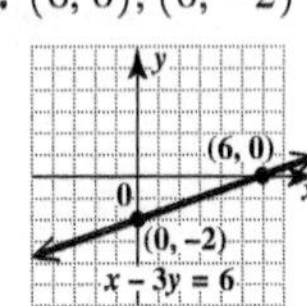

41. $\left(\frac{21}{2}, 0\right)$; $\left(0, -\frac{7}{3}\right)$ **43.** (0, 0); (0, 0) **45.** (0, 0); (0, 0)

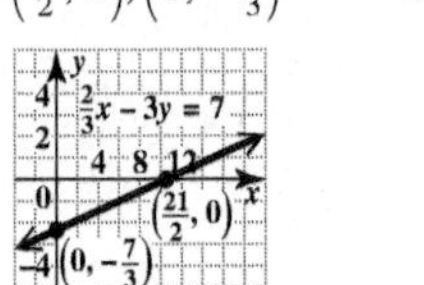

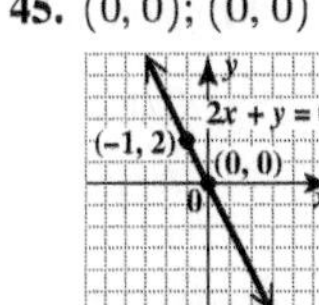

47. (0, 0); (0, 0) **49.** (0, 0); (0, 0) **51.** none; (0, 5)

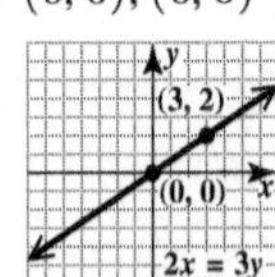

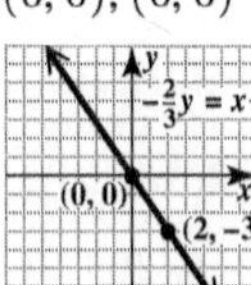

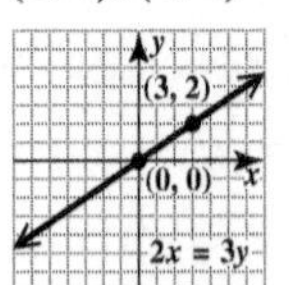

53. (2, 0); none **55.** (-4, 0); none **57.** none; (0, -2)

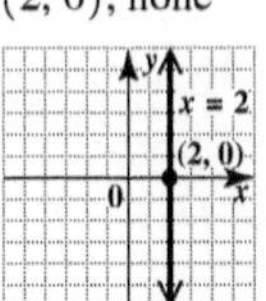

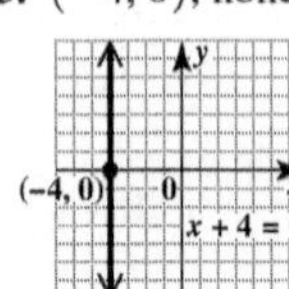

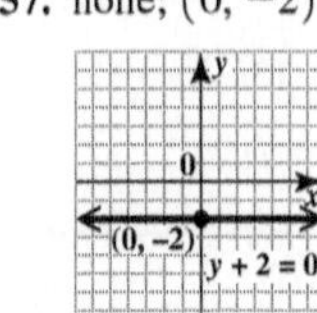

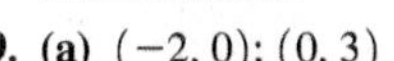

59. **(a)** (-2, 0); (0, 3) **(b)** B **61.** **(a)** none; (0, -1) **(b)** A

(c) **(c)**

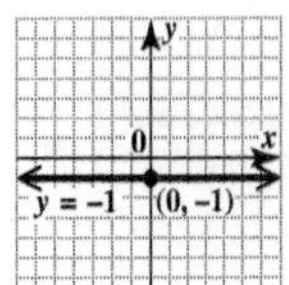

63. (-5, -1) **65.** $\left(\frac{9}{2}, -\frac{3}{2}\right)$ **67.** $\left(0, \frac{11}{2}\right)$ **69.** (2.1, 0.9) **71.** (1, 1)

73. $\left(-\frac{5}{12}, \frac{5}{28}\right)$ **75.** $Q(11, -4)$ **77.** $Q(4.5, 0.75)$

Section 2.2

1. A, B, D, F **3.** 5 ft **5.** **(a)** C **(b)** A **(c)** D **(d)** B **7.** 2

9. undefined **11.** 1 **13.** -1 **15.** **(a)** B **(b)** C **(c)** A **(d)** D **17.** The x-values in the denominator should be subtracted in the same order as the y-values in the numerator. The denominator should be $2 - (-4)$. The correct slope is $\frac{2}{6} = \frac{1}{3}$. **19.** 2 **21.** 0

23. undefined **25.** **(a)** 8 **(b)** rises **27.** **(a)** $\frac{5}{6}$ **(b)** rises

29. **(a)** 0 **(b)** horizontal **31.** **(a)** $-\frac{1}{2}$ **(b)** falls **33.** **(a)** undefined **(b)** vertical **35.** **(a)** -1 **(b)** falls **37.** 6 **39.** -3 **41.** -2

43. $\frac{4}{3}$ **45.** $-\frac{5}{2}$ **47.** undefined

In part (a) of Exercises 49–53, we used the intercepts. Other points can be used.

49. **(a)** (4, 0) and (0, -8); 2 **(b)** $y = 2x - 8$; 2 **(c)** $A = 2$, $B = -1$; 2

51. **(a)** (4, 0) and (0, 3); $-\frac{3}{4}$ **(b)** $y = -\frac{3}{4}x + 3$; $-\frac{3}{4}$ **(c)** $A = 3$, $B = 4$; $-\frac{3}{4}$ **53.** **(a)** (-3, 0) and (0, -3); -1 **(b)** $y = -x - 3$; -1 **(c)** $A = 1$, $B = 1$; -1

55. $-\frac{1}{2}$ **57.** $\frac{5}{2}$ **59.** 4

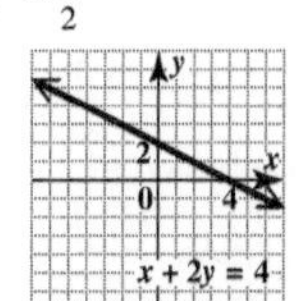

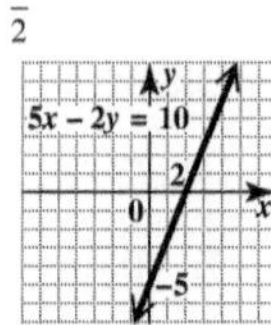

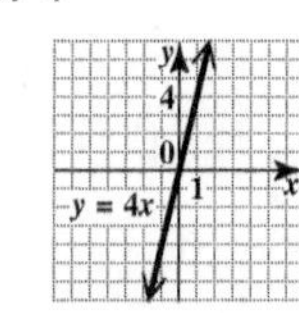

61. undefined **63.** 0 **65.** 0

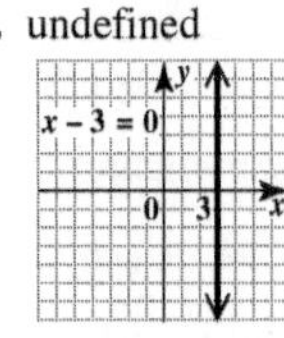

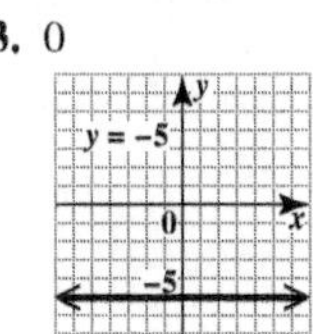

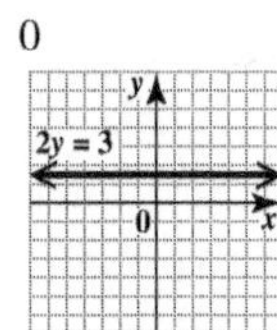

67. **69.** **71.**

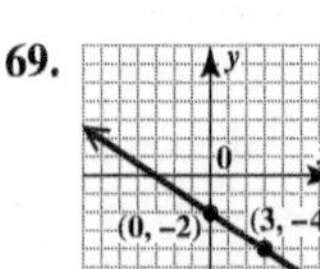

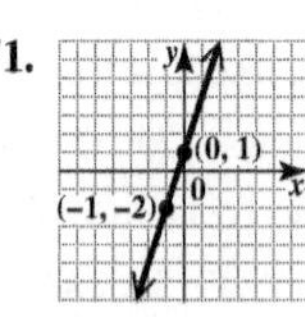

73. **75.** **77.**

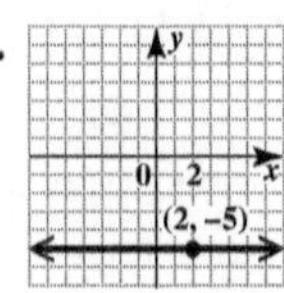

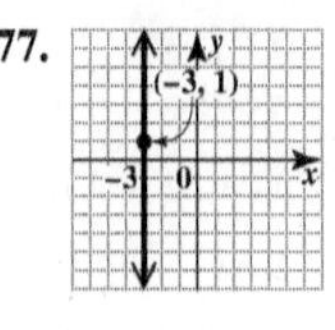

79. $-\frac{4}{9}$; $\frac{9}{4}$ **81.** parallel **83.** perpendicular **85.** neither

87. parallel **89.** neither **91.** perpendicular **93.** perpendicular

95. $\frac{7}{10}$ **97.** $-\$4000$ per year; The value of the machine is decreasing \$4000 per year during these years. **99.** 0% per year (or no change); The percent of pay raise is not changing—it is 3% per year during these years. **101.** **(a)** In 2016, there were 396 million wireless subscriber connections in the U.S.

101. **(b)** 16 **(c)** The number of subscribers increased by an average of 16 million per year from 2011 to 2016. **103.** **(a)** -9 theaters per yr **(b)** The negative slope means that the number of drive-in theaters decreased by an average of 9 per year from 2010 to 2017. **105.** \$0.04 per year; The price of a gallon of gasoline increased by an average of \$0.04 per year from 2000 to 2016. **107.** -16 million digital cameras per year; The number of digital cameras sold decreased by an average of 16 million per year from 2010 to 2016.
109. Because the slopes of both pairs of opposite sides are equal, the figure is a parallelogram. **111.** $\frac{1}{3}$ **112.** $\frac{1}{3}$ **113.** $\frac{1}{3}$
114. $\frac{1}{3} = \frac{1}{3} = \frac{1}{3}$ is true. **115.** collinear **116.** not collinear

Section 2.3

1. A **3.** A **5.** $3x + y = 10$ **7.** A **9.** C **11.** H **13.** B
15. $y = 5x + 15$ **17.** $y = -\frac{2}{3}x + \frac{4}{5}$ **19.** $y = x - 1$
21. $y = \frac{2}{5}x + 5$ **23.** $y = \frac{2}{3}x + 1$ **25.** $y = -x - 2$ **27.** $y = 2x - 4$
29. $y = -\frac{3}{5}x + 3$

31. **(a)** $y = x + 4$ **(b)** 1 **(c)** $(0, 4)$
(d)

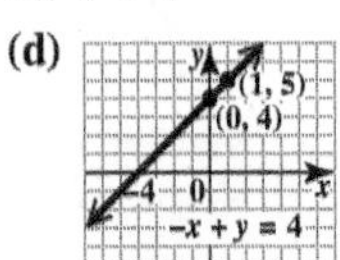

33. **(a)** $y = -\frac{6}{5}x + 6$ **(b)** $-\frac{6}{5}$ **(c)** $(0, 6)$
(d)

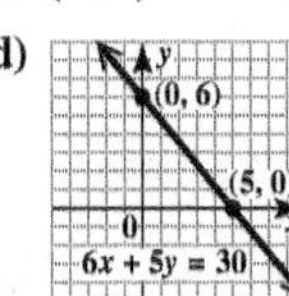

35. **(a)** $y = \frac{4}{5}x - 4$ **(b)** $\frac{4}{5}$ **(c)** $(0, -4)$
(d)

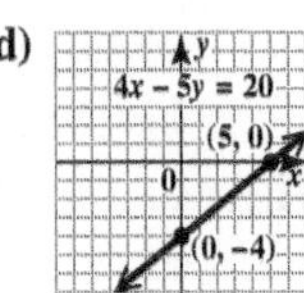

37. **(a)** $y = -\frac{1}{2}x - 2$ **(b)** $-\frac{1}{2}$ **(c)** $(0, -2)$
(d)

39. **(a)** $y = -2x + 18$ **(b)** $2x + y = 18$ **41.** **(a)** $y = -\frac{3}{4}x + \frac{5}{2}$
(b) $3x + 4y = 10$ **43.** **(a)** $y = \frac{1}{2}x + \frac{13}{2}$ **(b)** $x - 2y = -13$
45. **(a)** $y = 4x - 12$ **(b)** $4x - y = 12$ **47.** **(a)** $y = 1.4x + 4$
(b) $7x - 5y = -20$ **49.** $2x - y = 2$ **51.** $x + 2y = 8$ **53.** $y = 5$
55. $x = 7$ **57.** $y = -3$ **59.** $2x - 13y = -6$ **61.** $y = 5$
63. $x = 9$ **65.** $y = -\frac{3}{2}$ **67.** $y = 8$ **69.** $x = 0.5$ **71.** $y = 0$
73. **(a)** $y = 3x - 19$ **(b)** $3x - y = 19$ **75.** **(a)** $y = \frac{1}{2}x - 1$
(b) $x - 2y = 2$ **77.** **(a)** $y = -\frac{1}{2}x + 9$ **(b)** $x + 2y = 18$
79. **(a)** $y = 7$ **(b)** $y = 7$ **81.** $y = 45x$; $(0, 0)$, $(5, 225)$, $(10, 450)$
83. $y = 3.75x$; $(0, 0)$, $(5, 18.75)$, $(10, 37.50)$ **85.** $y = 150x$; $(0, 0)$, $(5, 750)$, $(10, 1500)$ **87.** **(a)** $y = 140x + 18.50$ **(b)** $(5, 718.50)$; The cost for 5 tickets and the delivery fee is \$718.50. **(c)** \$298.50
89. **(a)** $y = 41x + 99$ **(b)** $(5, 304)$; The cost for a 5-month membership is \$304. **(c)** \$591 **91.** **(a)** $y = 90x + 25$ **(b)** $(5, 475)$; The cost of the plan for 5 months is \$475. **(c)** \$2185
93. **(a)** $y = 6x + 30$ **(b)** $(5, 60)$; It costs \$60 to rent the saw for 5 days. **(c)** 18 days **95.** **(a)** $y = -79.7x + 716$; Sales of e-readers in the United States decreased by \$79.7 million per year from 2013 to 2016.
(b) \$636.3 million **97.** **(a)** The slope would be positive because spending on home health care is increasing over these years.
(b) $y = 3.5x + 36$ **(c)** \$88.5 billion; It is very close to the actual value. **99.** 32; 212 **100.** **(a)** $(0, 32)$ and $(100, 212)$
(b) $\frac{9}{5}$ **101.** $F = \frac{9}{5}C + 32$ **102.** $C = \frac{5}{9}(F - 32)$; $-40°$
103. 60° **104.** 59°; They differ by 1°. **105.** 90°; 86°; They differ by 4°. **106.** Because $\frac{9}{5}$ is a little less than 2, and 32 is a little more than 30, $\frac{9}{5}C + 32 \approx 2C + 30$.

SUMMARY EXERCISES Finding Slopes and Equations of Lines

1. $-\frac{3}{5}$ **2.** 0 **3.** 1 **4.** $\frac{3}{7}$ **5.** undefined **6.** $-\frac{4}{7}$ **7.** **(a)** B
(b) F **(c)** A **(d)** C **(e)** E **(f)** D **8.** C is in standard form;
A: $4x + y = -7$; B: $3x - 4y = -12$; D: $x + 2y = 0$;
E: $3x - y = 5$; F: $5x - 3y = 15$ **9.** **(a)** $y = -3x + 10$
(b) $3x + y = 10$ **10.** **(a)** $y = \frac{2}{3}x + 8$ **(b)** $2x - 3y = -24$
11. **(a)** $y = -\frac{5}{6}x + \frac{13}{3}$ **(b)** $5x + 6y = 26$ **12.** **(a)** $y = -\frac{5}{2}x + 2$
(b) $5x + 2y = 4$ **13.** **(a)** $y = 3x + 11$ **(b)** $3x - y = -11$
14. **(a)** $y = -\frac{5}{2}x$ **(b)** $5x + 2y = 0$ **15.** **(a)** $y = -8$
(b) $y = -8$ **16.** **(a)** $y = -\frac{7}{9}$ **(b)** $9y = -7$
17. **(a)** $y = 2x - 10$ **(b)** $2x - y = 10$ **18.** **(a)** $y = \frac{2}{3}x + \frac{14}{3}$
(b) $2x - 3y = -14$ **19.** **(a)** $y = \frac{1}{5}x - \frac{7}{5}$ **(b)** $x - 5y = 7$
20. **(a)** $y = -\frac{3}{4}x - 6$ **(b)** $3x + 4y = -24$

Section 2.4

1. **(a)** yes **(b)** yes **(c)** no **(d)** yes **3.** **(a)** no **(b)** no **(c)** no **(d)** yes **5.** solid; below **7.** dashed; above **9.** $\leq$ **11.** $>$

13.
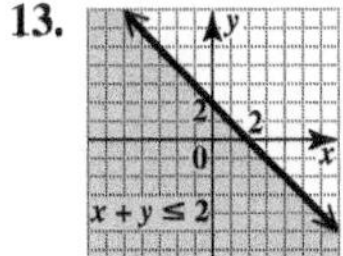

15.
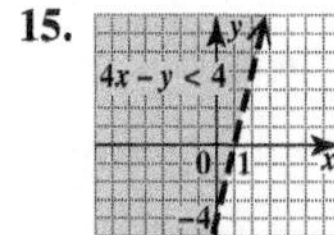

17.
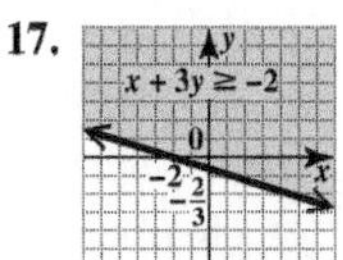

19.
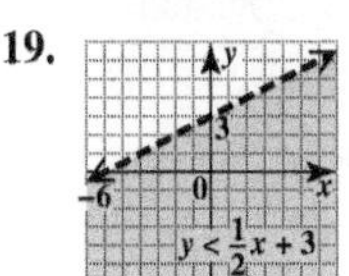

21.
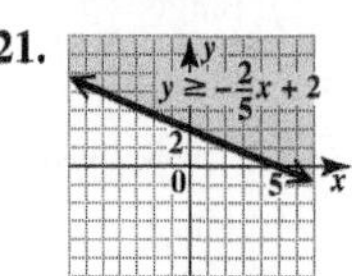

23.
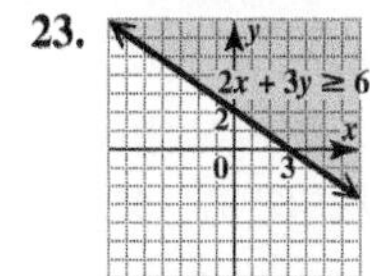

25.
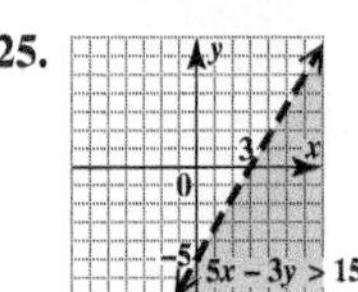

27.
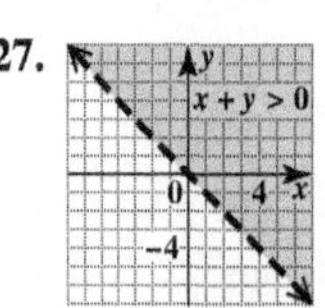

29.
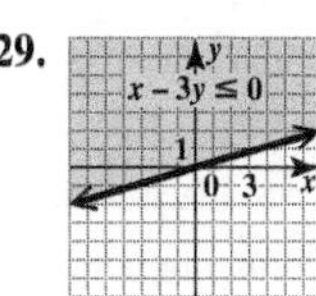

31.
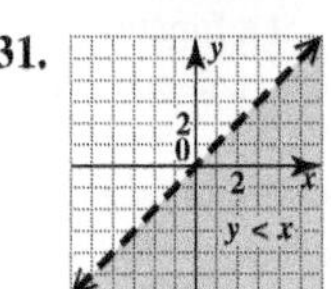

33.
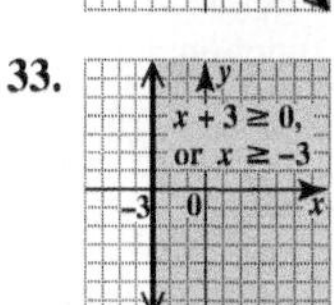

35.
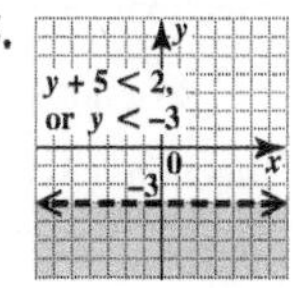

37. 2; $(0, -4)$; $2x - 4$; solid; above; $\geq$; $\geq 2x - 4$

39. 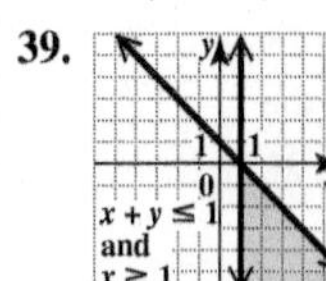

41. 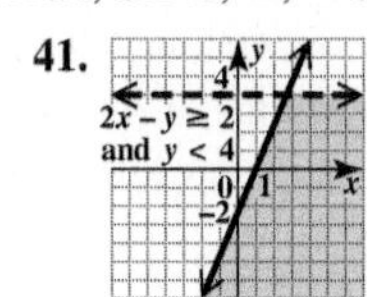

43. 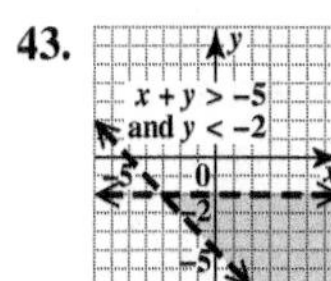

45. $-3 < x < 3$

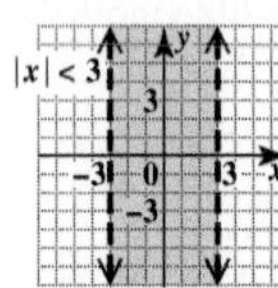

47. $-2 < x + 1 < 2$
$-3 < x < 1$

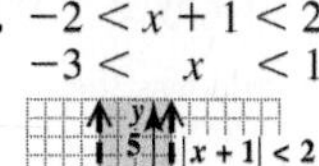
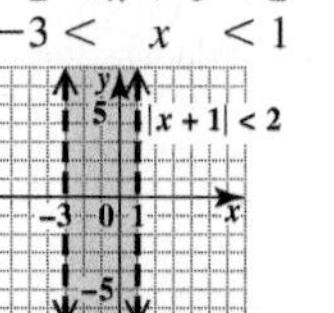

49. 
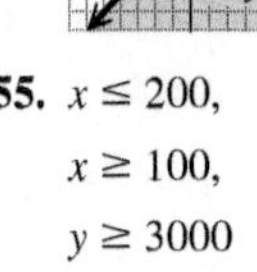

51. 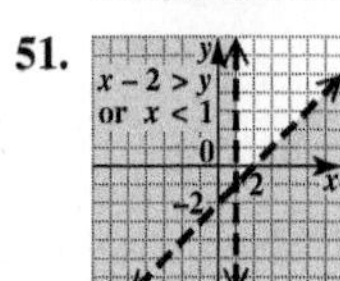

53. 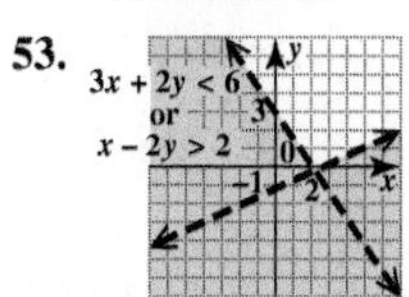

55. $x \leq 200$, $x \geq 100$, $y \geq 3000$

56. 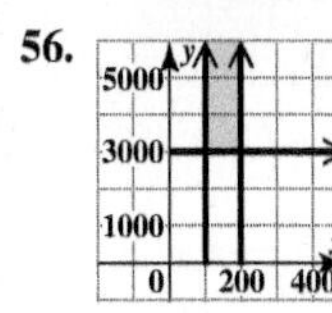

57. $C = 50x + 100y$ **58.** Some examples are $(100, 5000)$, $(150, 3000)$, and $(150, 5000)$. The corner points are $(100, 3000)$ and $(200, 3000)$. **59.** The least value occurs when $x = 100$ and $y = 3000$. **60.** The company should use 100 workers and manufacture 3000 units to achieve the least possible cost.

Section 2.5

1. relation; ordered pairs **3.** domain; range **5.** independent variable; dependent variable **7.** $\{(2, -2), (2, 0), (2, 1)\}$ **9.** $\{(1960, 0.76), (1980, 2.69), (2000, 5.39), (2016, 8.65)\}$ **11.** $\{(\text{A}, 4), (\text{B}, 3), (\text{C}, 2), (\text{D}, 1), (\text{F}, 0)\}$

In Exercises 13–17, answers will vary.

13. 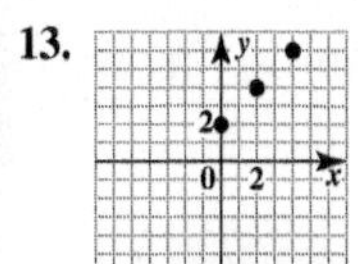

15. $\{(-1, -3), (0, -1), (1, 1), (3, 3)\}$ **17.** $y = 2$ **19.** function; domain: $\{5, 3, 4, 7\}$; range: $\{1, 2, 9, 6\}$ **21.** not a function; domain: $\{2, 0\}$; range: $\{4, 2, 5\}$

23. function; domain: $\{-3, 4, -2\}$; range: $\{1, 7\}$ **25.** not a function; domain: $\{1, 0, 2\}$; range: $\{1, -1, 0, 4, -4\}$ **27.** not a function; domain: $\{1\}$; range: $\{5, 2, -1, -4\}$ **29.** function; domain: $\{4, 2, 0, -2\}$; range: $\{-3\}$ **31.** function; domain: $\{2, 5, 11, 17, 3\}$; range: $\{1, 7, 20\}$ **33.** function; domain: $\{-2, 0, 3\}$; range: $\{2, 3\}$ **35.** function; domain: $(-\infty, \infty)$; range: $(-\infty, \infty)$ **37.** not a function; domain: $\{-2\}$; range: $(-\infty, \infty)$ **39.** function; domain: $(-\infty, \infty)$; range: $\{-2\}$ **41.** not a function; domain: $(-\infty, 0]$; range: $(-\infty, \infty)$ **43.** function; domain: $(-\infty, \infty)$; range: $(-\infty, 4]$ **45.** not a function; domain: $[-4, 4]$; range: $[-3, 3]$ **47.** not a function; domain: $(-\infty, \infty)$; range: $[2, \infty)$ **49.** function; $(-\infty, \infty)$ **51.** function; $(-\infty, \infty)$ **53.** function; $(-\infty, \infty)$ **55.** not a function; $[0, \infty)$ **57.** not a function; $(-\infty, \infty)$ **59.** function; $[0, \infty)$ **61.** function; $[3, \infty)$ **63.** function; $\left[-\frac{1}{2}, \infty\right)$ **65.** function; $(-\infty, \infty)$ **67.** function; $(-\infty, 0) \cup (0, \infty)$ **69.** function; $(-\infty, 4) \cup (4, \infty)$

71. function; $(-\infty, 0) \cup (0, \infty)$ **73.** **(a)** yes **(b)** domain: $\{2013, 2014, 2015, 2016, 2017\}$; range: $\{52.8, 52.6, 53.2, 53.7\}$ **(c)** 52.6; 2017 **(d)** Answers will vary. Two possible answers are $(2014, 52.6)$ and $(2017, 53.7)$.

Section 2.6

1. $f(x)$; function; domain; x; f of x (or "f at x") **3.** line; -2; linear; $-2x + 4$; -2; 3; -2 **5.** 4 **7.** 13 **9.** -11 **11.** 4 **13.** -296 **15.** 3 **17.** 2.75 **19.** $-3p + 4$ **21.** $3x + 4$ **23.** $-3x - 2$ **25.** $-6t + 1$ **27.** $-\pi^2 + 4\pi + 1$ **29.** $-3x - 3h + 4$ **31.** $-\frac{p^2}{9} + \frac{4p}{3} + 1$ **33.** **(a)** -1 **(b)** -1 **35.** **(a)** 2 **(b)** 3 **37.** **(a)** 15 **(b)** 10 **39.** **(a)** 4 **(b)** 1 **41.** **(a)** 3 **(b)** -3 **43.** **(a)** -3 **(b)** 2 **45.** **(a)** 2 **(b)** 0 **(c)** -1 **47.** **(a)** $f(x) = -\frac{1}{3}x + 4$ **(b)** 3 **49.** **(a)** $f(x) = 3 - 2x^2$ **(b)** -15 **51.** **(a)** $f(x) = \frac{4}{3}x - \frac{8}{3}$ **(b)** $\frac{4}{3}$

53. 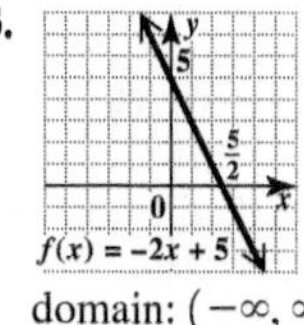

domain: $(-\infty, \infty)$; range: $(-\infty, \infty)$

55. 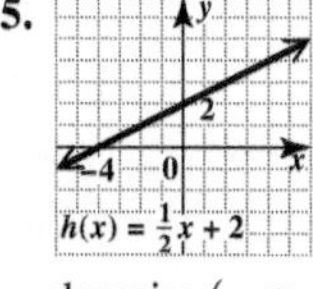

domain: $(-\infty, \infty)$; range: $(-\infty, \infty)$

57. 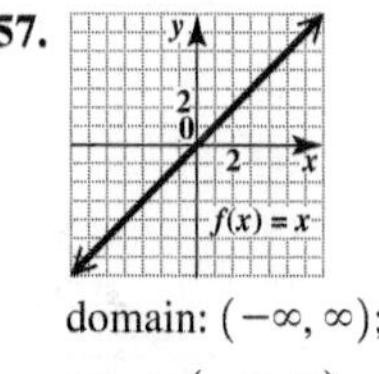

domain: $(-\infty, \infty)$; range: $(-\infty, \infty)$

59. 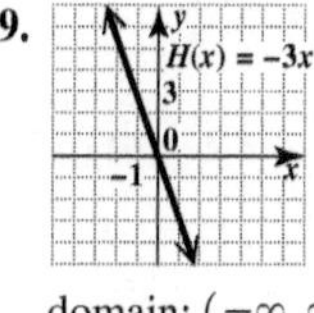

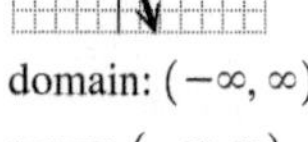
domain: $(-\infty, \infty)$; range: $(-\infty, \infty)$

61. 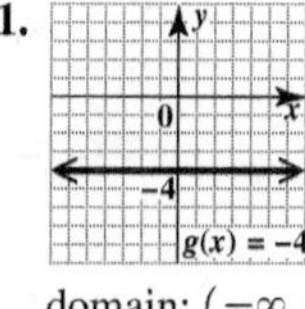

domain: $(-\infty, \infty)$; range: $\{-4\}$

63. 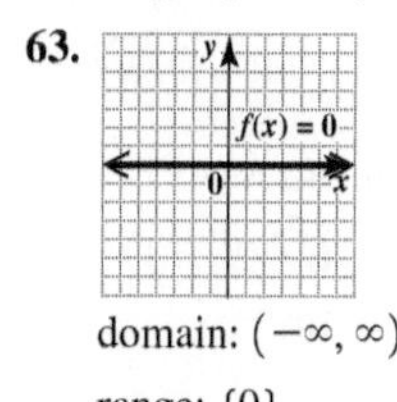

domain: $(-\infty, \infty)$; range: $\{0\}$

65. x-axis **67.** **(a)** 11.25 (dollars) **(b)** 3 is the value of the independent variable, which represents a package weight of 3 lb. $f(3)$ is the value of the dependent variable, which represents the cost to mail a 3-lb package. **(c)** \$18.75; $f(5) = 18.75$ **69.** **(a)** $f(x) = 12x + 100$ **(b)** 1600; The cost to print 125 t-shirts is \$1600. **(c)** 75; $f(75) = 1000$; The cost to print 75 t-shirts is \$1000. **71.** **(a)** 1.1 **(b)** 4 **(c)** -1.2 **(d)** $(0, 3.5)$ **(e)** $f(x) = -1.2x + 3.5$ **73.** **(a)** $[0, 100]$; $[0, 3000]$ **(b)** 25 hr; 25 hr **(c)** 2000 gal **(d)** $f(0) = 0$; The pool is empty at time 0. **(e)** $f(25) = 3000$; After 25 hr, there are 3000 gal of water in the pool. **75.** **(a)** 194.53 cm **(b)** 177.29 cm **(c)** 177.41 cm **(d)** 163.65 cm **77.** Because it falls from left to right, the slope is negative. **78.** $-\frac{3}{2}$ **79.** $-\frac{3}{2}; \frac{2}{3}$ **80.** $\left(\frac{7}{3}, 0\right)$ **81.** $\left(0, \frac{7}{2}\right)$ **82.** $f(x) = -\frac{3}{2}x + \frac{7}{2}$ **83.** $-\frac{17}{2}$ **84.** $\frac{23}{3}$ **85.** 2; $f(1) = -\frac{3}{2}(1) + \frac{7}{2}$, which simplifies on the right to $f(1) = -\frac{3}{2} + \frac{7}{2}$. This gives $f(1) = 2$.

Chapter 2 Review Exercises

1. 5; $\frac{10}{3}$; 2; $\frac{14}{3}$

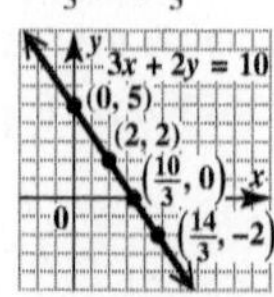

2. -6; 5; -5; 6

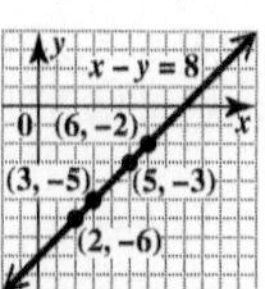

3. $(3, 0)$; $(0, -4)$

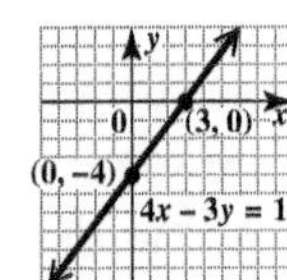

4. $\left(\frac{28}{5}, 0\right)$; $(0, 4)$

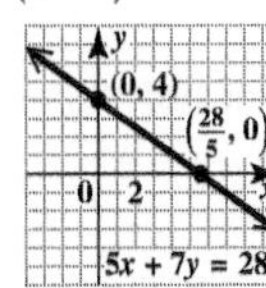

5. $(10, 0)$; $(0, 4)$

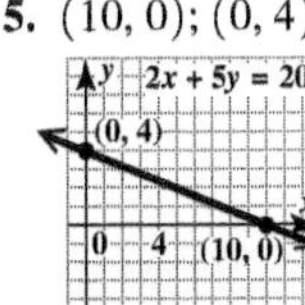

6. $(8, 0)$; $(0, -2)$

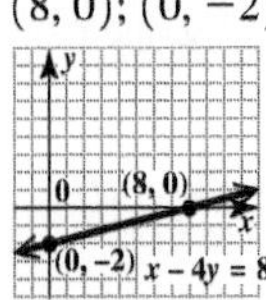

7. $(0, 2)$ **8.** $\left(-\frac{9}{2}, \frac{3}{2}\right)$ **9.** $-\frac{7}{5}$ **10.** $-\frac{1}{2}$ **11.** undefined **12.** 2 **13.** $\frac{3}{4}$ **14.** 0 **15.** $-\frac{1}{3}$ **16.** $\frac{2}{3}$ **17.** $-\frac{1}{3}$ **18.** -1 **19.** -3 **20.** **(a)** positive **(b)** negative **(c)** undefined **(d)** 0 **21.** perpendicular **22.** parallel **23.** 12 ft **24.** \$1057 per year **25.** **(a)** $y = \frac{3}{5}x - 8$ **(b)** $3x - 5y = 40$ **26.** **(a)** $y = -\frac{1}{3}x + 5$ **(b)** $x + 3y = 15$ **27.** **(a)** $y = -9x + 13$ **(b)** $9x + y = 13$ **28.** **(a)** $y = \frac{7}{5}x + \frac{16}{5}$ **(b)** $7x - 5y = -16$ **29.** **(a)** $y = 4x - 26$ **(b)** $4x - y = 26$ **30.** **(a)** $y = -\frac{5}{2}x + 1$ **(b)** $5x + 2y = 2$ **31.** $y = 12$ **32.** $x = 2$ **33.** $x = 0.3$ **34.** $y = 4$ **35.** **(a)** $y = 229x + 2829$ **(b)** \$11,760 **36.** **(a)** $y = 1007.5x + 29{,}822.5$; The cost at private 4-year universities increased by an average of \$1007.50 per year from 2013 to 2017. **(b)** \$44,935

37.

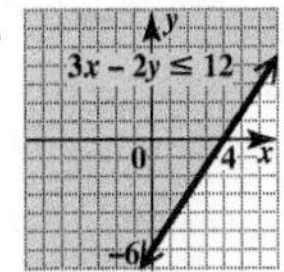

38.

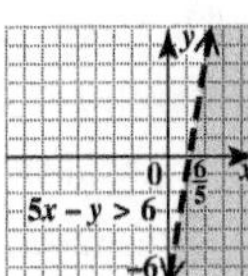

39.

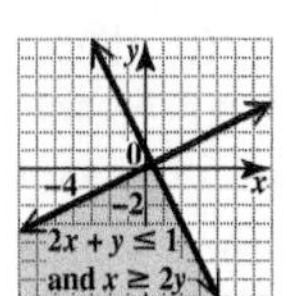

40.

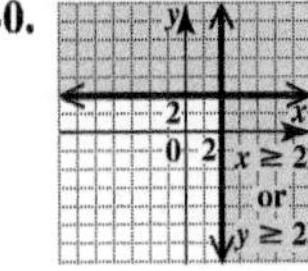

41. not a function; domain: $\{-4, 1\}$; range: $\{2, -2, 5, -5\}$ **42.** function; domain: $\{9, 11, 4, 17, 25\}$; range: $\{32, 47, 69, 14\}$ **43.** function; domain: $[-4, 4]$; range: $[0, 2]$ **44.** not a function; domain: $(-\infty, 0]$; range: $(-\infty, \infty)$ **45.** function; domain: $(-\infty, \infty)$; linear function **46.** not a function; domain: $(-\infty, \infty)$ **47.** function; domain: $(-\infty, \infty)$ **48.** function; domain: $\left[-\frac{7}{4}, \infty\right)$ **49.** not a function; domain: $[0, \infty)$ **50.** function; domain: $(-\infty, 6) \cup (6, \infty)$ **51.** -6 **52.** -15 **53.** -8 **54.** $-2k^2 + 3k - 6$ **55.** $f(x) = 2x^2$; 18 **56.** C **57.** It is a horizontal line. **58.** **(a)** yes **(b)** domain: $\{1960, 1970, 1980, 1990, 2000, 2010, 2015\}$; range: $\{69.7, 70.8, 73.7, 75.4, 76.8, 78.7, 78.8\}$ **(c)** Answers will vary. Two possible answers are $(1960, 69.7)$ and $(2010, 78.7)$. **(d)** 73.7; In 1980, life expectancy at birth was 73.7 yr. **(e)** 2000

Chapter 2 Mixed Review Exercises

1. parallel **2.** perpendicular **3.** -0.575 lb per year; Per capita consumption of potatoes decreased by an average of 0.575 lb per year from 2008 to 2016. **4.** $y = -0.575x + 37.8$ **5.** $y = 3x$ **6.** $x + 2y = 6$ **7.** $y = -3$ **8.** A, B, D **9.** D **10.** **(a)** -1 **(b)** -2 **(c)** 2 **(d)** $(-\infty, \infty)$; $(-\infty, \infty)$

Chapter 2 Test

[2.1] **1.** $-\frac{10}{3}$; -2; 0

2. $(-3, 0)$; $(0, 4)$

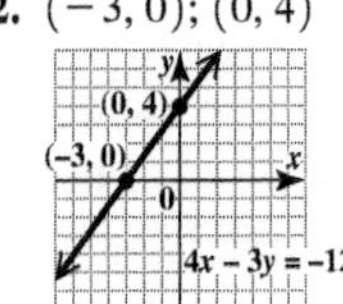

3. $(2, 0)$; none

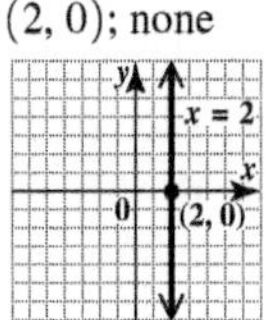

4. $(0, 0)$; $(0, 0)$

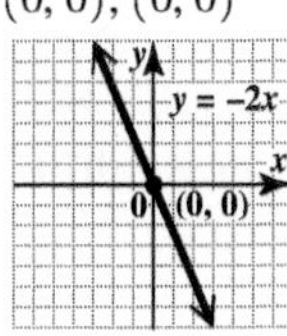

[2.2] **5.** $\frac{1}{2}$ **6.** It is a vertical line. [2.1, 2.2] **7.** $\frac{3}{2}$; $\left(\frac{13}{3}, 0\right)$; $\left(0, -\frac{13}{2}\right)$ **8.** 0; none; $(0, 5)$ [2.2] **9.** B **10.** perpendicular **11.** neither **12.** -438 farms per yr; The number of farms decreased an average of 438 farms per year from 2000 to 2016. [2.3] **13.** **(a)** $y = -\frac{2}{5}x + 3$ **(b)** $2x + 5y = 15$ **14.** **(a)** $y = -\frac{1}{2}x + 2$ **(b)** $x + 2y = 4$ **15.** **(a)** $y = -5x + 19$ **(b)** $5x + y = 19$ **16.** $y = 14$ **17.** $x = 5$ **18.** $y = 0$ **19.** **(a)** $y = -\frac{3}{5}x - \frac{11}{5}$ **(b)** $y = -\frac{1}{2}x - \frac{3}{2}$ **20.** **(a)** $y = 142.75x + 45$ **(b)** \$901.50

[2.4] **21.**

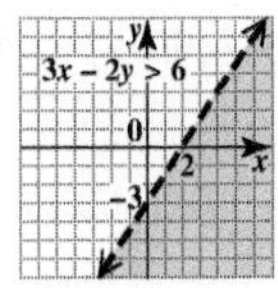

22.

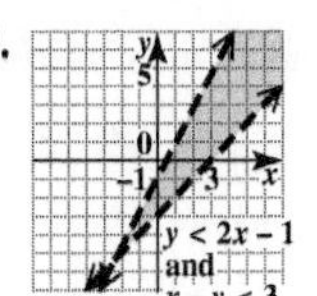

[2.5] **23.** D; domain: $(-\infty, \infty)$; range: $[0, \infty)$

24. D; domain: $\{0, 3, 6\}$; range: $\{1, 2, 3\}$ [2.6] **25.** 0; $-a^2 + 2a - 1$

26. domain: $(-\infty, \infty)$; range: $(-\infty, \infty)$

Chapters R–2 Cumulative Review Exercises

[R.1] **1.** **(a)** $\frac{11}{12}$ **(b)** $\frac{2}{3}$ **(c)** 4.75 **2.** **(a)** 8 **(b)** 150 **(c)** 0.125 **3.** **(a)** 0.01; 1% **(b)** $\frac{1}{2}$; 50% **(c)** $\frac{3}{4}$; 0.75 **(d)** 1; 100% [R.2] **4.** always true **5.** never true **6.** sometimes true; For example, $3 + (-3) = 0$, but $3 + (-1) = 2$ and $2 \neq 0$. [R.3] **7.** -3 [R.4] **8.** 0.64 **9.** -9 **10.** not a real number [R.4, R.5] **11.** -39 **12.** undefined **13.** $4x - 3$ **14.** 0 [1.1] **15.** $\left\{\frac{7}{6}\right\}$ **16.** $\{-1\}$ [1.4] **17.** 6 in. **18.** 2 hr [1.5] **19.** $\left(-3, \frac{7}{2}\right)$ **20.** $(-\infty, 1]$ [1.6] **21.** $(6, 8)$ **22.** $(-\infty, -2] \cup (7, \infty)$ [1.7] **23.** $\{0, 7\}$ **24.** $(-\infty, \infty)$

[2.1] **25.** $(4, 0)$; $\left(0, \frac{12}{5}\right)$

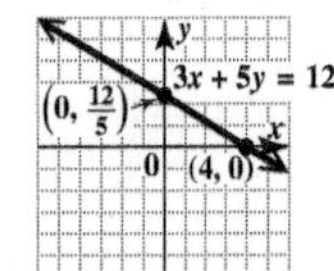

[2.2] **26.** **(a)** $-\frac{6}{5}$ **(b)** $\frac{5}{6}$

[2.4] **27.**

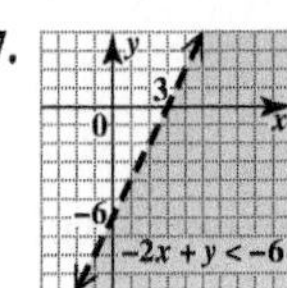

[2.3] **28.** **(a)** $y = -\frac{3}{4}x - 1$ **(b)** $3x + 4y = -4$ **29.** **(a)** $y = -\frac{4}{3}x + \frac{7}{3}$ **(b)** $4x + 3y = 7$ [2.5] **30.** domain: $\{14, 91, 75, 23\}$; range: $\{9, 70, 56, 5\}$; not a function

3 SYSTEMS OF LINEAR EQUATIONS

Section 3.1

1. 4; −3 **3.** ∅ **5.** no **7.** D; The ordered-pair solution must be in quadrant IV because that is where the graphs of the equations intersect. **9.** **(a)** B **(b)** C **(c)** A **(d)** D **11.** solution **13.** not a solution
15. $\{(-2, -3)\}$ **17.** $\{(0, 1)\}$ **19.** $\{(2, 0)\}$

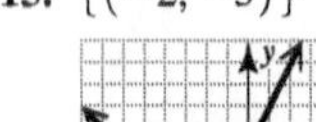
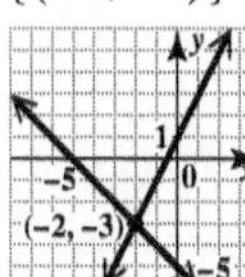

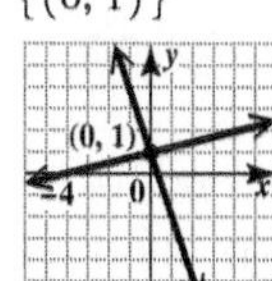

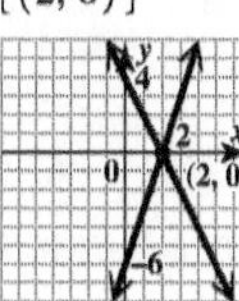

21. $\{(1, 2)\}$ **23.** $\{(2, 3)\}$ **25.** $\{(1, 3)\}$ **27.** $\{(0, -4)\}$ **29.** $\left\{\left(\frac{22}{9}, \frac{22}{3}\right)\right\}$ **31.** $\{(5, 4)\}$ **33.** $\left\{\left(-5, -\frac{10}{3}\right)\right\}$ **35.** $\{(2, 6)\}$ **37.** $\{(4, 2)\}$ **39.** $\{(-8, 4)\}$ **41.** $\{(x, y) \mid 2x - y = 0\}$; dependent equations **43.** ∅; inconsistent system **45.** $\{(2, -4)\}$ **47.** $\{(3, -1)\}$ **49.** $\{(2, -3)\}$ **51.** $\{(x, y) \mid 7x + 2y = 6\}$; dependent equations **53.** $\left\{\left(\frac{3}{2}, -\frac{3}{2}\right)\right\}$ **55.** ∅; inconsistent system **57.** $\{(0, 0)\}$ **59.** $\{(0, -4)\}$ **61.** $\left\{\left(6, -\frac{5}{6}\right)\right\}$ **63.** **(a)** Use elimination because the coefficients of the y-terms are opposites. **(b)** Use substitution because the second equation is solved for y. **(c)** Use elimination because the equations are in standard form with no coefficients of 1 or −1. Solving by substitution would involve fractions. **65.** $\{(-3, 2)\}$ **67.** $\left\{\left(\frac{1}{3}, \frac{1}{2}\right)\right\}$ **69.** $\{(-4, 6)\}$ **71.** $\{(5, 0)\}$ **73.** $\left\{\left(1, \frac{1}{2}\right)\right\}$ **75.** $\{(x, y) \mid 4x - y = -2\}$ **77.** $y = -\frac{3}{7}x + \frac{4}{7}$; $y = -\frac{3}{7}x + \frac{3}{14}$; no solution **79.** Both are $y = -\frac{2}{3}x + \frac{1}{3}$; infinitely many solutions **81.** **(a)** \$4 **(b)** 300 half-gallons **(c)** supply: 200 half-gallons; demand: 400 half-gallons **83.** **(a)** 2011–2015 **(b)** 2015 and 2016 **(c)** 2013; 260 thousand students **(d)** (2013, 260) **(e)** Enrollment in Kentucky was decreasing, and enrollment in Utah remained constant. **85.** Netflix **87.** (4.4, 48.7) **89.** $\{(2, 4)\}$ **91.** $\left\{\left(\frac{1}{2}, 2\right)\right\}$ **93.** $\left\{\left(\frac{c}{a}, 0\right)\right\}$ **95.** $\left\{\left(-\frac{1}{a}, -5\right)\right\}$

Section 3.2

1. Answers will vary. Some possible answers are **(a)** two perpendicular walls and the ceiling in a normal room, **(b)** the floors of three different levels of an office building, and **(c)** three pages of a book (because they intersect in the spine). **3.** The statement means that when −1 is substituted for x, 2 is substituted for y, and 3 is substituted for z in the three equations, the resulting three statements are true. **5.** 4 **7.** $\{(3, 2, 1)\}$ **9.** $\{(1, 4, -3)\}$ **11.** $\{(0, 2, -5)\}$ **13.** $\{(1, 0, 3)\}$ **15.** $\left\{\left(1, \frac{3}{10}, \frac{2}{5}\right)\right\}$ **17.** $\left\{\left(-\frac{7}{3}, \frac{22}{3}, 7\right)\right\}$ **19.** $\{(-12, 18, 0)\}$ **21.** $\{(0.8, -1.5, 2.3)\}$ **23.** $\{(4, 5, 3)\}$ **25.** $\{(2, 2, 2)\}$ **27.** $\left\{\left(\frac{8}{3}, \frac{2}{3}, 3\right)\right\}$ **29.** $\{(-1, 0, 0)\}$ **31.** $\{(-4, 6, 2)\}$ **33.** $\{(-3, 5, -6)\}$ **35.** ∅; inconsistent system **37.** $\{(x, y, z) \mid x - y + 4z = 8\}$; dependent equations **39.** $\{(3, 0, 2)\}$ **41.** $\{(x, y, z) \mid 2x + y - z = 6\}$; dependent equations **43.** $\{(0, 0, 0)\}$ **45.** ∅; inconsistent system **47.** $\{(2, 1, 5, 3)\}$ **49.** $\{(-2, 0, 1, 4)\}$ **51.** $2a + b + c = -5$ **52.** $a - c = 1$ **53.** $3a + 3b + c = -18$ **54.** $a = 1, b = -7, c = 0$ **55.** $x^2 + y^2 + x - 7y = 0$ **56.** The relation is not a function because it fails the vertical line test—that is, a vertical line can intersect its graph more than once.

Section 3.3

1. **(a)** 6 oz **(b)** 15 oz **(c)** 24 oz **(d)** 30 oz **3.** \$1.99x **5.** **(a)** $(10 - x)$ mph **(b)** $(10 + x)$ mph **7.** **(a)** 220 ft **(b)** $\frac{d}{44}$ sec **9.** length: 78 ft; width: 36 ft **11.** wins: 102; losses: 60 **13.** AT&T: \$163.8 billion; Verizon: \$125.1 billion **15.** $x = 40$ and $y = 50$, so the angles measure 40° and 50°. **17.** Red Sox: \$360.66; Indians: \$179.44 **19.** ribeye: \$30.30; salmon: \$21.60 **21.** Busch Gardens: \$90; Universal Studios: \$110 **23.** general admission: 76; with student ID: 108 **25.** 25% alcohol: 6 gal; 35% alcohol: 14 gal **27.** nuts: 14 kg; cereal: 16 kg **29.** 2%: \$1000; 4%: \$2000 **31.** pure acid: 6 L; 10% acid: 48 L **33.** \$1.75-per-lb candy: 7 lb; \$1.25-per-lb candy: 3 lb **35.** train: 60 km per hr; plane: 160 km per hr **37.** freight train: 50 km per hr; express train: 80 km per hr **39.** boat: 21 mph; current: 3 mph **41.** plane: 300 mph; wind: 20 mph **43.** $x + y + z = 180$; 70°, 30°, 80° **45.** 20°, 70°, 90° **47.** 12 cm, 25 cm, 33 cm **49.** gold: 7; silver: 6; bronze: 6 **51.** upper level: 1170; center court: 985; floor: 130 **53.** bookstore A: 140; bookstore B: 280; bookstore C: 380 **55.** first chemical: 50 kg; second chemical: 400 kg; third chemical: 300 kg **57.** wins: 46; losses: 23; overtime losses: 13 **59.** box of fish: 8 oz; box of bugs: 2 oz; box of worms: 5 oz

Chapter 3 Review Exercises

1. $\{(2, 2)\}$ **2.** $\{(-4, 0)\}$ **3.** $\{(1, -1)\}$

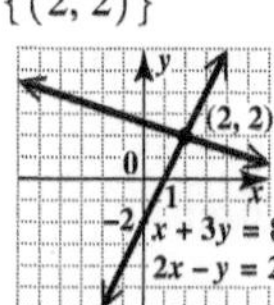

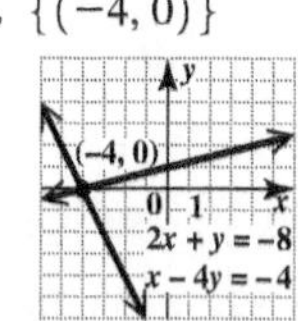

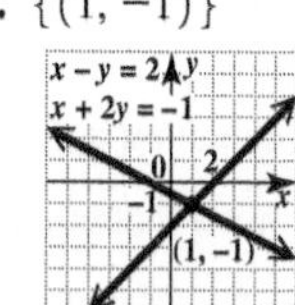

4. D **5.** $\left\{\left(-\frac{8}{9}, -\frac{4}{3}\right)\right\}$ **6.** $\{(0, 4)\}$ **7.** $\{(2, 4)\}$ **8.** $\{(2, 2)\}$ **9.** $\{(2, -4)\}$ **10.** $\{(0, 1)\}$ **11.** $\{(-1, 2)\}$ **12.** $\{(-6, 3)\}$ **13.** $\{(x, y) \mid 3x - y = -6\}$; dependent equations **14.** ∅; inconsistent system **15.** $\{(1, -5, 3)\}$ **16.** ∅; inconsistent system **17.** $\{(1, 2, 3)\}$ **18.** $\{(5, -1, 0)\}$ **19.** $\{(x, y, z) \mid 3x - 4y + z = 8\}$; dependent equations **20.** $\{(0, 0, 0)\}$ **21.** length: 200 ft; width: 85 ft **22.** plane: 300 mph; wind: 20 mph **23.** gold: 12; silver: 8; bronze: 21 **24.** \$6-per-lb nuts: 30 lb; \$3-per-lb candy: 70 lb **25.** 8% solution: 5 L; 20% solution: 3 L **26.** 85°, 35°, 60° **27.** 10%: \$40,000; 6%: \$100,000; 5%: \$140,000 **28.** Mantle: 54; Maris: 61; Berra: 22

Chapter 3 Mixed Review Exercises

1. Answers will vary.

(a) 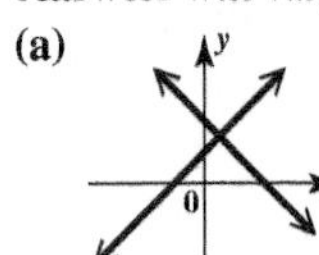 **(b)** 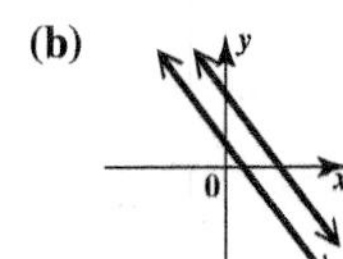 **(c)** 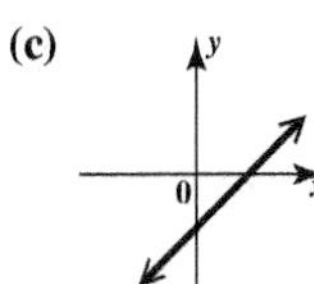

2. The two lines have the same slope, 3, but the y-intercepts, $(0, 2)$ and $(0, -4)$, are different. Therefore, the lines are parallel, do not intersect, and have no common solution. **3.** $\{(0, 4)\}$ **4.** $\left\{\left(\frac{82}{23}, -\frac{4}{23}\right)\right\}$ **5.** $\{(x, y) \mid x + 2y = 48\}$ **6.** $\{(5, 3)\}$ **7.** $\{(12, 9)\}$ **8.** $\{(3, -1)\}$ **9.** $\{(1, 2, 3)\}$ **10.** $\{(1, 0, -1)\}$ **11.** $\varnothing$ **12.** *B;* The second equation is already solved for y. **13.** 20 L **14.** United States: 121; China: 70; Great Britain: 67

Chapter 3 Test

[3.1] **1. (a)** Houston, Phoenix, Dallas **(b)** Philadelphia **(c)** Dallas, Phoenix, Philadelphia, Houston **2. (a)** 2010; 1.45 million **(b)** $(2025, 2.8)$

3. $\{(6, 1)\}$

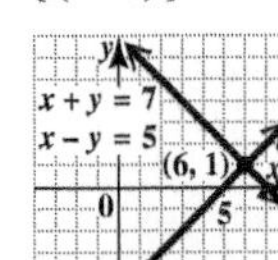

4. $\{(6, -4)\}$ **5.** $\left\{\left(-\frac{9}{4}, \frac{5}{4}\right)\right\}$ **6.** $\{(x, y) \mid 12x - 5y = 8\}$; dependent equations **7.** $\{(3, 3)\}$ **8.** $\{(0, -2)\}$ **9.** $\varnothing$; inconsistent system [3.2] **10.** $\left\{\left(-\frac{2}{3}, \frac{4}{5}, 0\right)\right\}$ **11.** $\{(3, -2, 1)\}$ **12.** $\varnothing$ [3.3] **13.** *Captain America:* \$408.1 million; *Deadpool:* \$363.1 million **14.** slower car: 45 mph; faster car: 75 mph **15.** 20% solution: 4 L; 50% solution: 8 L **16.** AC adaptor: \$8; rechargeable flashlight: \$15 **17.** 25°, 55°, 100° **18.** Orange Pekoe: 60 oz; Irish Breakfast: 30 oz; Earl Grey: 10 oz

Chapters R–3 Cumulative Review Exercises

[R.4] **1.** 81 **2.** −81 **3.** −81 **4.** 0.7 **5.** −0.7 **6.** It is not a real number. [R.3] **7.** −7.17 **8.** $-\frac{37}{24}$, or $-1\frac{13}{24}$ **9.** $-\frac{28}{33}$ [R.4] **10.** −30 **11.** −199 **12.** 455 [1.1] **13.** $\left\{-\frac{15}{4}\right\}$ [1.7] **14.** $\left\{\frac{2}{3}, 2\right\}$ [1.2] **15.** $x = \frac{d - by}{a}$ [1.1] **16.** $\{11\}$ [1.5] **17.** $\left(-\infty, \frac{240}{13}\right]$ [1.7] **18.** $\left[-2, \frac{2}{3}\right]$ **19.** $(-\infty, \infty)$ [1.6] **20.** $(-\infty, \infty)$ [1.3, 1.4] **21.** pennies: 35; nickels: 29; dimes: 30 **22.** 46°, 46°, 88° [R.1, 1.2] **23.** 2010; 1813; 62.8%; 57.2% [2.1] **24.** $y = 6$ **25.** $x = 4$ [2.2] **26.** $-\frac{4}{3}$ **27.** $\frac{3}{4}$ [2.3] **28.** $4x + 3y = 10$

[2.2] **29.**

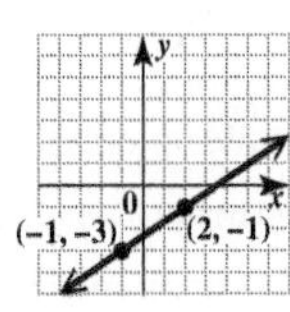

[2.4] **30.** $-3x - 2y \le 6$

[2.6] **31. (a)** −6 **(b)** $a^2 + 3a - 6$ [3.1] **32.** $\{(3, -3)\}$ **33.** $\{(x, y) \mid x - 3y = 7\}$ [3.2] **34.** $\{(5, 3, 2)\}$ [3.1] **35. (a)** $x = 8$, or 800 items; \$3000 **(b)** about \$400

4 EXPONENTS, POLYNOMIALS, AND POLYNOMIAL FUNCTIONS

Section 4.1

1. false; $(ab)^2 = a^2b^2$ **3.** false; $\left(\frac{4}{a}\right)^3 = \frac{4^3}{a^3}$ **5.** false; $xy^0 = x \cdot 1 = x$ **7.** false; $-(-10)^0 = -(1) = -1$ **9.** The bases should not be multiplied; $4^5 \cdot 4^2 = 4^7$ **11.** 13^{12} **13.** 8^{10} **15.** x^{17} **17.** The product rule does not apply. **19.** $-27w^8$ **21.** $18x^3y^8$ **23. (a)** B **(b)** C **(c)** B **(d)** C **25.** 1 **27.** −1 **29.** 1 **31.** −1 **33.** 1 **35.** 2 **37.** 0 **39.** −2 **41. (a)** B **(b)** D **(c)** B **(d)** D **43.** $\frac{1}{5^4}$ **45.** $\frac{1}{3^5}$ **47.** $\frac{1}{9}$ **49.** $\frac{1}{(4x)^2}$ **51.** $\frac{4}{x^2}$ **53.** $-\frac{1}{a^3}$ **55.** $\frac{1}{(-a)^4}$, or $\frac{1}{a^4}$ **57.** $\frac{1}{(-3x)^3}$ **59.** $\frac{11}{30}$ **61.** $\frac{9}{20}$ **63.** $-\frac{5}{24}$ **65.** 16 **67.** $\frac{27}{4}$ **69.** $\frac{16}{25}$ **71.** 4^2 **73.** x^4 **75.** $\frac{1}{r^3}$ **77.** 6^6 **79.** $\frac{1}{6^{10}}$ **81.** 7^2 **83.** r^3 **85.** The quotient rule does not apply. **87.** x^{18} **89.** $\frac{27}{125}$ **91.** $64t^3$ **93.** $216x^6$ **95.** $-\frac{64m^6}{t^3}$ **97.** $\frac{s^{12}}{t^{20}}$ **99. (a)** B **(b)** D **(c)** D **(d)** B **101.** 64 **103.** $\frac{27}{8}$ **105.** $\frac{25}{16}$ **107.** $\frac{81}{16t^4}$ **109.** $16x^2$ **111.** $\frac{32}{x^5}$ **113.** $\frac{1}{3}$ **115.** $\frac{1}{a^5}$ **117.** 5^6 **119.** $\frac{1}{x^{12}}$ **121.** $\frac{1}{k^2}$ **123.** $-4r^6$ **125.** $\frac{625}{a^{10}}$ **127.** $\frac{z^4}{x^3}$ **129.** $\frac{m^8}{n^{11}}$ **131.** $\frac{1}{5p^{10}}$ **133.** $\frac{1}{2pq}$ **135.** $\frac{4}{a^2}$ **137.** $\frac{1}{6y^{13}}$ **139.** $\frac{4k^5}{m^2}$ **141.** $\frac{8}{3pq^{10}}$ **143.** $\frac{y^9}{8}$ **145.** $\frac{n^{10}}{25m^{18}}$ **147.** $-\frac{125y^3}{x^{30}}$ **149.** $\frac{4k^{17}}{125}$ **151.** $-\frac{3}{32m^8p^4}$

Section 4.2

1. after; power; a; 10^n **3.** A number in scientific notation should have only one nonzero digit to the left of the decimal point. The correct answer is 9.275×10^6. **5.** 1×10^6 **7.** 1×10^{12} **9.** 5.3×10^2 **11.** 8.3×10^{-1} **13.** 6.92×10^{-6} **15.** -3.85×10^4 **17.** $\$1 \times 10^9$ (or $\$10^9$); $\$1 \times 10^{12}$ (or $\$10^{12}$); $\$4.094 \times 10^{12}$; 2.77843×10^5 **19.** 72,000 **21.** 0.00254 **23.** −60,000 **25.** 0.000012 **27.** −0.0289 **29.** 8.761 **31.** 0.06 **33.** 0.0000025 **35.** 2,700,000 **37.** 200,000 **39.** 10 **41.** 3000 **43. (a)** 3.231×10^8 **(b)** $\$1 \times 10^{12}$ (or $\$10^{12}$) **(c)** \$3095 **45.** \$42,500 **47. (a)** 5.95×10^2 **(b)** 998 mi² **49.** 300 sec **51.** approximately 5.87×10^{12} mi **53.** 7.5×10^9 **55.** 4×10^{17} **57.** The 2002 quake was 100 times as intense as the 2018 quake. **58.** The 1960 quake was 316 times as intense as the 2010 quake. **59.** The 2010 quake was 25 times as intense as the 2005 quake. **60.** $10^3 = 1000$ times as intense; $10^{-1} = \frac{1}{10}$ (one-tenth) as intense.

Section 4.3

1. **(a)** D **(b)** A **(c)** A **(d)** E **(e)** B **3.** Answers will vary. One example is $7x^5 + 2x^3 - 6x^2 + 9x$. **5.** To identify the leading coefficient, it is necessary to write the polynomial in descending powers of the variable as $-8x^4 - 5x^3 + 2x^2$. The leading term is $-8x^4$, so the leading coefficient is -8. **7.** 7; 1 **9.** -15; 2 **11.** 1; 4 **13.** $\frac{1}{6}$; 1 **15.** 8; 0 **17.** -1; 3 **19.** $2x^3 - 3x^2 + x + 4$; $2x^3$; 2 **21.** $p^7 - 8p^5 + 4p^3$; p^7; 1 **23.** $-3m^4 - m^3 + 10$; $-3m^4$; -3 **25.** monomial; 0 **27.** binomial; 1 **29.** binomial; 8 **31.** monomial; 6 **33.** trinomial; 3 **35.** none of these; 5 **37.** $8z^4$ **39.** $7m^3$ **41.** $5x$ **43.** already simplified **45.** $-t + 13s$ **47.** $8k^2 + 2k - 7$ **49.** $-2n^4 - n^3 + n^2$ **51.** $-2ab^2 + 20a^2b$ **53.** Only the coefficients of the like terms should be added, *not* the exponents. The correct sum is $6x^2 - 4x - 1$. **55.** $8x^2 + x - 2$ **57.** $-t^4 + 2t^2 - t + 5$ **59.** $5y^3 - 3y^2 + 5y + 1$ **61.** $r + 13$ **63.** $-2a^2 - 2a - 7$ **65.** $-3z^5 + z^2 + 7z$ **67.** $12p - 4$ **69.** $-9p^2 + 11p - 9$ **71.** $5a + 18$ **73.** $14m^2 - 13m + 6$ **75.** $13z^2 + 10z - 3$ **77.** $10y^3 - 7y^2 + 5y + 8$ **79.** $-5a^4 - 6a^3 + 9a^2 - 11$ **81.** $3y^2 - 4y + 2$ **83.** $-4m^2 + 4n^2 - 7n$ **85.** $y^4 - 4y^2 - 4$ **87.** $10z^2 - 16z$ **89.** $12x^2 + 8x + 5$

Section 4.4

1. polynomial; one; terms; powers **3.** C **5.** 0; 1; 2; (0, 0), (1, 1), (2, 2) **7.** The student either did not substitute correctly or did not apply the exponent correctly. Here $f(-2) = -(-2)^2 + 4 = -(4) + 4 = 0$, so $f(-2) = 0$. **9.** **(a)** -10 **(b)** 8 **(c)** -4 **11.** **(a)** 8 **(b)** -10 **(c)** 0 **13.** **(a)** 8 **(b)** 2 **(c)** 4 **15.** **(a)** 7 **(b)** 1 **(c)** 1 **17.** **(a)** 8 **(b)** 74 **(c)** 6 **19.** **(a)** -11 **(b)** 4 **(c)** -8 **21.** **(a)** 59.08 million lb **(b)** 162.20 million lb **(c)** 152.62 million lb **23.** **(a)** \$45.2 billion **(b)** \$84.3 billion **(c)** \$163.4 billion **25.** **(a)** $8x - 3$ **(b)** $2x - 17$ **27.** **(a)** $-x^2 + 12x - 12$ **(b)** $9x^2 + 4x + 6$ **29.** $f(x)$ and $g(x)$ can be any two polynomials that have a sum of $3x^3 - x + 3$, such as $f(x) = 3x^3 + 1$ and $g(x) = -x + 2$. **31.** $x^2 + 2x - 9$ **33.** 6 **35.** $x^2 - x - 6$ **37.** 6 **39.** -33 **41.** 0 **43.** $-\frac{9}{4}$ **45.** $-\frac{9}{2}$ **47.** **(a)** $P(x) = 8.49x - 50$ **(b)** \$799

49.

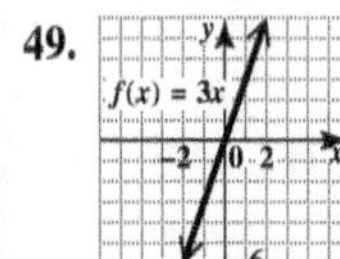

domain: $(-\infty, \infty)$; range: $(-\infty, \infty)$

51.

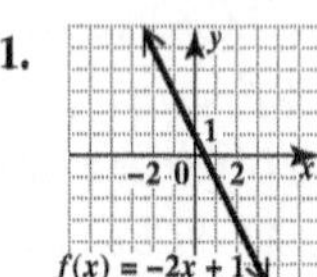

domain: $(-\infty, \infty)$; range: $(-\infty, \infty)$

53.

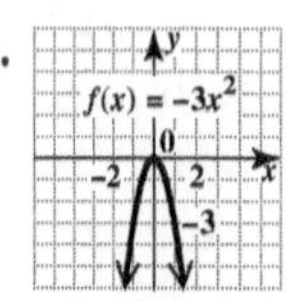

domain: $(-\infty, \infty)$; range: $(-\infty, 0]$

55.

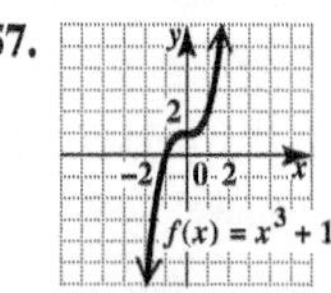

domain: $(-\infty, \infty)$; range: $[-2, \infty)$

57.

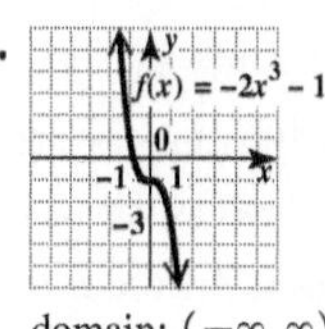

domain: $(-\infty, \infty)$; range: $(-\infty, \infty)$

59. $f(x) = -2x^3 - 1$

domain: $(-\infty, \infty)$; range: $(-\infty, \infty)$

61. B **63.** A **65.** The student multiplied the functions instead of composing them. The correct answer is $(f \circ g)(x) = -6$. **67.** 6 **69.** 83 **71.** 53 **73.** 13 **75.** $2x^2 + 11$ **77.** $2x - 2$ **79.** $\frac{97}{4}$ **81.** 8 **83.** 1 **85.** 9 **87.** 1 **89.** $(f \circ g)(x) = 63{,}360x$; It computes the number of inches in x miles. **91.** **(a)** $s = \frac{x}{4}$ **(b)** $y = \frac{x^2}{16}$ **(c)** 2.25 **93.** **(a)** $g(x) = \frac{1}{2}x$ **(b)** $f(x) = x + 1$ **(c)** $(f \circ g)(x) = \frac{1}{2}x + 1$ **(d)** $(f \circ g)(60) = 31$; The sale price is \$31. **95.** $(\mathscr{A} \circ r)(t) = 4\pi t^2$; This is the area of the circular layer as a function of time. **97.** $f(1) = 1$; $f(2) = 2$ **98.** $f(3) = 4$; $f(4) = 8$; $f(5) = 16$ **99.** $f(3) = 4$; $f(4) = 8$; $f(5) = 16$ **100.** The pattern 1, 2, 4, 8, 16 emerges, so most students predict 32 because the terms are doubling each time. However, $f(6) = 31$ (not 32). See the figure.

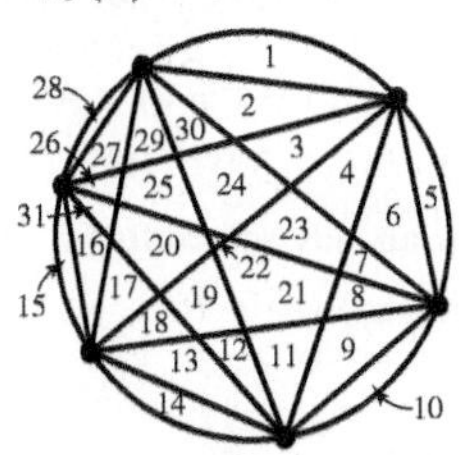

Section 4.5

1. C **3.** D **5.** $-24m^5$ **7.** $-28x^7y^4$ **9.** $-6x^2 + 15x$ **11.** $-2q^3 - 3q^4$ **13.** $18k^4 + 12k^3 + 6k^2$ **15.** $15x^3 + 8x^2 + x$ **17.** $6y^2 + y - 12$ **19.** $m^3 - 3m^2 - 40m$ **21.** $24z^3 - 20z^2 - 16z$ **23.** $4x^5 - 4x^4 - 24x^3$ **25.** $6t^3 + t^2 - 14t - 3$ **27.** $25m^2 - 9n^2$ **29.** $-2b^3 + 2b^2 + 18b + 12$ **31.** $6p^4 + p^3 + 4p^2 - 27p - 6$ **33.** $8z^4 - 14z^3 + 17z^2 + 20z - 3$ **35.** $m^2 - 3m - 40$ **37.** $12k^2 + k - 6$ **39.** $15x^2 - 13x + 2$ **41.** $20x^2 + 23xy + 6y^2$ **43.** $3z^2 + zw - 4w^2$ **45.** $12c^2 + 16cd - 3d^2$ **47.** The student squared each term instead of multiplying the binomial times itself. The correct product is $x^2 + 8x + 16$. **49.** $x^2 - 81$ **51.** $4p^2 - 9$ **53.** $25m^2 - 1$ **55.** $9a^2 - 4c^2$ **57.** $16m^2 - 49n^4$ **59.** $75y^7 - 12y$ **61.** $y^2 - 10y + 25$ **63.** $x^2 + 2x + 1$ **65.** $4p^2 + 28p + 49$ **67.** $16n^2 - 24nm + 9m^2$ **69.** $k^2 - \frac{10}{7}kp + \frac{25}{49}p^2$ **71.** $0.04x^2 - 0.56xy + 1.96y^2$ **73.** $16x^2 - \frac{4}{9}$ **75.** $0.1x^2 + 0.63x - 0.13$ **77.** $3w^2 - \frac{23}{4}wz - \frac{1}{2}z^2$ **79.** $25x^2 + 10x + 1 + 60xy + 12y + 36y^2$ **81.** $4a^2 + 4ab + b^2 - 12a - 6b + 9$ **83.** $4a^2 + 4ab + b^2 - 9$ **85.** $4h^2 - 4hk + k^2 - j^2$ **87.** $y^3 + 6y^2 + 12y + 8$ **89.** $125r^3 - 75r^2s + 15rs^2 - s^3$ **91.** $q^4 - 8q^3 + 24q^2 - 32q + 16$ **93.** $6a^3 + 7a^2b + 4ab^2 + b^3$ **95.** $4z^4 - 17z^3x + 12z^2x^2 - 6zx^3 + x^4$ **97.** $m^4 - 4m^2p^2 + 4mp^3 - p^4$ **99.** $a^4b - 7a^2b^3 - 6ab^4$ **101.** $\frac{9}{2}x^2 - 2y^2$ **103.** $15x^2 - 2x - 24$ **105.** $10x^2 - 2x$ **107.** $2x^2 - x - 3$ **109.** $8x^3 - 27$ **111.** $2x^3 - 18x$ **113.** -20 **115.** 32 **117.** 36 **119.** 0 **121.** $\frac{35}{4}$ **123.** $a - b$ **124.** $\mathscr{A} = s^2$; $(a - b)^2$ **125.** $(a - b)b$, or $ab - b^2$; $2ab - 2b^2$ **126.** b^2 **127.** a^2; a **128.** $a^2 - (2ab - 2b^2) - b^2 = a^2 - 2ab + b^2$ **129.** **(a)** They must be equal. **(b)** $(a - b)^2 = a^2 - 2ab + b^2$

130.

	a	b
a	Area: a^2	Area: ab
b	Area: ab	Area: b^2

The large square is made up of two smaller squares and two congruent rectangles. The sum of the areas is $a^2 + 2ab + b^2$. Because $(a + b)^2$ must represent the same quantity, they must be equal. Thus, $(a + b)^2 = a^2 + 2ab + b^2$.

Section 4.6

1. quotient; exponents **3.** 0 **5.** $3x^3 - 2x^2 + 1$ **7.** $3y + 4 - \frac{5}{y}$

9. $3 + \frac{5}{m} + \frac{6}{m^2}$ **11.** $\frac{2}{7n} - \frac{3}{2m} + \frac{9}{7mn}$ **13.** $\frac{2y}{x} + \frac{3}{4} + \frac{3w}{x}$

15. $r^2 - 7r + 6$ **17.** $y - 4$ **19.** $q + 8$ **21.** $t + 5$

23. $p - 4 + \frac{44}{p + 6}$ **25.** $m^2 + 2m - 1$ **27.** $x^2 + 2x - 3 + \frac{-3}{4x + 1}$

29. $2x - 5 + \frac{-4x + 5}{3x^2 - 2x + 4}$ **31.** $m^2 + m + 3$ **33.** $x^2 + x + 3$

35. $2x^2 - x - 5$ **37.** $3x^2 + 6x + 11 + \frac{26}{x - 2}$ **39.** $2k^2 + 3k - 1$

41. $z^2 + 3$ **43.** $2y^2 + 2$ **45.** $p^2 + p + 1$ **47.** $x^2 - 4x + 2 + \frac{9x - 4}{x^2 + 3}$

49. $p^2 + \frac{5}{2}p + 2 + \frac{-1}{2p + 2}$ **51.** $\frac{3}{2}a - 10 + \frac{77}{2a + 6}$ **53.** $\frac{2}{3}x - 1$

55. $\frac{3}{4}a - 2 + \frac{1}{4a + 3}$ **57.** $(2p + 7)$ feet **59.** $Q(x) = 2x^2 - 3x + 5$; $R(x) = 3$ **61.** $5x - 1; 0$ **63.** $2x - 3; -1$ **65.** $4x^2 + 6x + 9; \frac{3}{2}$

67. $\frac{x^2 - 9}{2x}$, $x \neq 0$ **69.** $-\frac{5}{4}$ **71.** $\frac{x - 3}{2x}$, $x \neq 0$ **73.** 0

75. $-\frac{35}{4}$ **77.** $\frac{7}{2}$

Chapter 4 Review Exercises

1. 64 **2.** $\frac{1}{81}$ **3.** -125 **4.** 18 **5.** $\frac{81}{16}$ **6.** $\frac{16}{25}$ **7.** $\frac{1}{30}$ **8.** $\frac{3}{4}$ **9.** 0

10. $\frac{1}{3^8}$ **11.** x^8 **12.** $\frac{y^6}{x^2}$ **13.** $\frac{1}{z^{15}}$ **14.** $\frac{25}{m^{18}}$ **15.** $4k^{11}$ **16.** $\frac{25}{z^4}$

17. $\frac{1}{96m^7}$ **18.** $\frac{2025}{8r^4}$ **19.** $-12x^2y^8$ **20.** $-\frac{2n}{m^5}$ **21.** $\frac{10p^8}{q^7}$

22. In $(-6)^0$, the base is -6 and the expression simplifies to 1. In -6^0, the base is 6 and the expression simplifies to -1.

23. 1.345×10^4 **24.** 7.65×10^{-8} **25.** -1.38×10^{-1} **26.** 3.214×10^8; 7.2×10^4; 1×10^2 (or 10^2) **27.** 1,210,000 **28.** $-267{,}000{,}000$

29. 0.0058 **30.** 2×10^{-4}; 0.0002 **31.** 1.5×10^3; 1500

32. 4.1×10^{-5}; 0.000041 **33.** 2.7×10^{-2}; 0.027

34. (a) 20,000 hr (b) 833 days **35.** 14; 5 **36.** -1; 2

37. $\frac{1}{10}$; 1 **38.** 504; 8 **39.** (a) $11k^3 - 3k^2 + 9k$ (b) trinomial (c) 3 **40.** (a) $9m^7 + 14m^6$ (b) binomial (c) 7

41. (a) $-5y^4 + 3y^3 + 7y^2 - 2y$ (b) none of these (c) 4

42. (a) $-7q^5r^3$ (b) monomial (c) 8 **43.** $-x^2 - 3x + 1$

44. $-5y^3 - 4y^2 + 6y - 12$ **45.** $6a^3 - 4a^2 - 16a + 15$

46. $8y^2 - 9y + 5$ **47.** $20x^2 - 6x + 6$ **48.** One example is $x^5 + 2x^4 - x^2 + x + 2$. **49.** (a) -11 (b) 4 (c) 7

50. (a) 132,976 twin births (b) 133,786 twin births (c) 135,689 twin births **51.** (a) $5x^2 - x + 5$ (b) $-5x^2 + 5x + 1$ (c) 11 (d) -9

52. $f(x) = -2x + 5$
domain: $(-\infty, \infty)$;
range: $(-\infty, \infty)$

53. $f(x) = x^2 - 6$
domain: $(-\infty, \infty)$;
range: $[-6, \infty)$

54. $f(x) = -x^3 + 1$
domain: $(-\infty, \infty)$;
range: $(-\infty, \infty)$

55. (a) 167 (b) 1495 (c) 20 (d) 42 (e) $75x^2 + 220x + 160$ (f) $15x^2 + 10x + 2$ **56.** $-12k^3 - 42k$ **57.** $2x^3 + 22x^2 + 56x$

58. $15m^2 - 7m - 2$ **59.** $6w^2 - 13wt + 6t^2$

60. $10p^4 + 30p^3 - 8p^2 - 24p$ **61.** $3q^3 - 13q^2 - 14q + 20$

62. $36r^4 - 1$ **63.** $16m^2 + 24m + 9$ **64.** $9t^2 - 12ts + 4s^2$

65. $y^2 - 3y + \frac{5}{4}$ **66.** $x^2 - 4x + 6$ **67.** $p^2 + 6p + 9 + \frac{54}{2p - 3}$

68. $p^2 + 3p - 6$ **69.** (a) $36x^3 - 9x^2$ (b) -45

70. (a) $4x - 1$, $x \neq 0$ (b) 7

Chapter 4 Mixed Review Exercises

1. (a) A (b) G (c) C (d) C (e) A (f) E (g) B (h) H (i) F (j) I **2.** 102,129 mi^2 **3.** $\frac{y^4}{36}$ **4.** $\frac{1}{125}$ **5.** -9

6. $-\frac{1}{5z^9}$ **7.** $21p^9 + 7p^8 + 14p^7$ **8.** $4x^2 - 36x + 81$ **9.** $\frac{1}{16y^{18}}$

10. $\frac{4w^6}{z^{18}}$ **11.** $8x^2 - 10x - 3$ **12.** $8x + 1 + \frac{5}{x - 3}$

13. $9m^2 - 30mn + 25n^2 - p^2$ **14.** $2y^2x + \frac{3y^3}{2x} + \frac{5x^2}{2}$

15. $-3k^2 + 4k - 7$

Chapter 4 Test

[4.1] **1.** (a) C (b) A (c) D (d) A (e) E (f) F (g) B (h) G (i) I (j) C **2.** $\frac{4x^7}{9y^{10}}$ **3.** $\frac{6}{r^{14}}$ **4.** $\frac{16}{9p^{10}q^{28}}$ **5.** $\frac{16}{x^6y^{16}}$

[4.2] **6.** 0.00000091 **7.** 3×10^{-4}; 0.0003 [4.4] **8.** (a) -18 (b) $-2x^2 + 12x - 9$ (c) $-2x^2 - 2x - 3$ (d) -7

9.

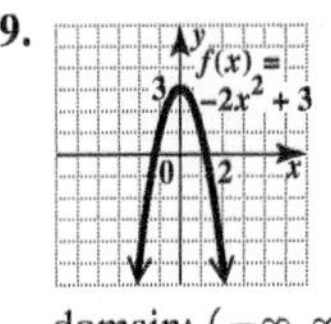

domain: $(-\infty, \infty)$;
range: $(-\infty, 3]$

10.

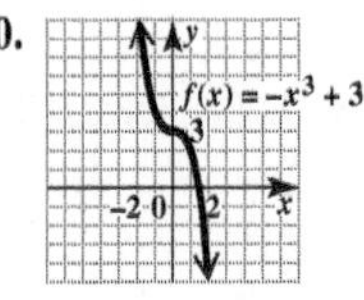

domain: $(-\infty, \infty)$;
range: $(-\infty, \infty)$

11. (a) 23 (b) $3x^2 + 11$ (c) $9x^2 + 30x + 27$

[4.5] **12.** $10x^2 - x - 3$ **13.** $6m^3 - 7m^2 - 30m + 25$ **14.** $36x^2 - y^2$

15. $9k^2 + 6kq + q^2$ **16.** $4y^2 - 9z^2 + 6zx - x^2$

[4.6] **17.** $4p - 8 + \frac{6}{p}$ **18.** $x^2 + 4x + 4$ [4.3] **19.** $x^3 - 2x^2 - 10x - 13$

[4.5, 4.6] **20.** (a) $x^3 + 4x^2 + 5x + 2$ (b) 0 (c) $x + 2$, $x \neq -1$ (d) 0

Chapters R–4 Cumulative Review Exercises

[R.1] **1.** (a) D, I (b) C, H (c) G (d) A, F (e) B (f) E

[R.2] **2.** (a) A, B, C, D, F (b) B, C, D, F (c) D, F (d) C, D, F (e) E, F (f) D, F [R.4] **3.** 32 **4.** $-\frac{1}{72}$ **5.** 0 [R.5] **6.** -7

[1.1] **7.** $\{-65\}$ **8.** {all real numbers} [1.2] **9.** $t = \frac{A - p}{pr}$

[1.5] **10.** $(-\infty, 6)$ [1.7] **11.** $\left\{-\frac{1}{3}, 1\right\}$ **12.** $\left(-\infty, -\frac{8}{3}\right] \cup [2, \infty)$

[R.1, 1.2] **13.** 3500; 1950; 36%; 31% [1.4] **14.** 15°, 35°, 130°

[2.2] **15.** $-\frac{4}{3}$ **16.** 0 [2.3] **17.** **(a)** $y = -4x + 15$

(b) $4x + y = 15$ **18.** **(a)** $y = 4x$ **(b)** $4x - y = 0$

[2.1] **19.**

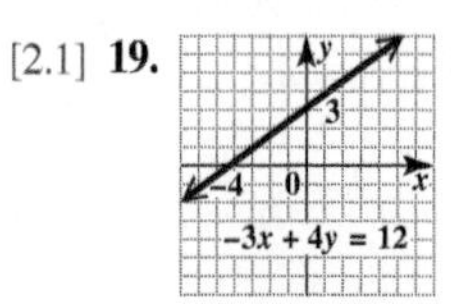

[2.4] **20.**

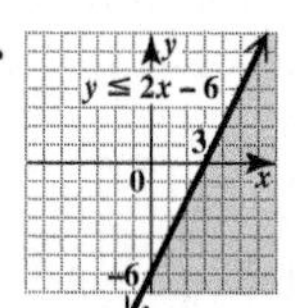

21.

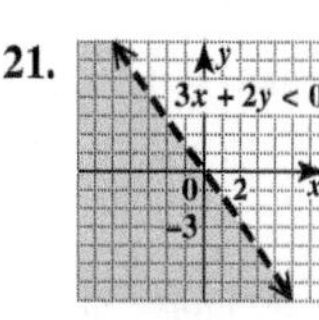

[2.2, 2.3] **22.** **(a)** -0.44 gal per yr; Per capita consumption of whole milk decreased by an average of 0.44 gal per year from 1970 to 2015. **(b)** $y = -0.44x + 25.3$ **(c)** 12.1 gal

[2.5] **23.** domain: $\{-4, -1, 2, 5\}$; range: $\{-2, 0, 2\}$; function

[2.6] **24.** -9 [3.1] **25.** $\{(3, 2)\}$ **26.** $\varnothing$ [3.2] **27.** $\{(1, 0, -1)\}$

[3.3] **28.** length: 42 ft; width: 30 ft **29.** 15% solution: 6 L; 30% solution: 3 L [4.1] **30.** $\frac{y^7}{x^{13}z^2}$ [4.3] **31.** $2x^2 - 5x + 10$

[4.5] **32.** $x^3 + 8y^3$ **33.** $15x^2 + 7xy - 2y^2$

[4.6] **34.** $4xy^4 - 2y + \frac{1}{x^2y}$ **35.** $m^2 - 2m + 3$

5 FACTORING

Section 5.1

1. C **3.** The student stopped too soon. This is a sum of two terms, not a product. $w^2 - 8$ must be factored out to obtain $(w^2 - 8)(2t + 5)$. **5.** $3m$ **7.** $8xy$ **9.** $3(r + t)^2$ **11.** $12(m - 5)$ **13.** $8(s + 2t)$ **15.** $4(1 + 5z)$ **17.** cannot be factored **19.** $8k(k^2 + 3)$ **21.** $-2p^2q^4(2p + q)$ **23.** $7x^3(1 + 5x - 2x^2)$ **25.** $2t^3(5t^2 - 1 - 2t)$ **27.** $5ac(3ac^2 - 5c + 1)$ **29.** $16zn^3(zn^3 + 4n^4 - 2z^2)$ **31.** $7ab(2a^2b + a - 3a^4b^2 + 6b^3)$ **33.** $(m - 4)(2m + 5)$ **35.** $11(2z - 1)$ **37.** $(2 - x)^2(1 + 2x)$ **39.** $(3 - x)(6 + 2x - x^2)$ **41.** $20z(2z + 1)(3z + 4)$ **43.** $5(m + p)^2(m + p - 2 - 3m^2 - 6mp - 3p^2)$ **45.** $r(-r^2 + 3r + 5)$; $-r(r^2 - 3r - 5)$ **47.** $12s^4(-s + 4)$; $-12s^4(s - 4)$ **49.** $2x^2(-x^3 + 3x + 2)$; $-2x^2(x^3 - 3x - 2)$ **51.** $(m + q)(x + y)$ **53.** $(5m + n)(2 + k)$ **55.** $(2 - q)(2 - 3p)$ **57.** $(p + q)(p - 4z)$ **59.** $(2x + 3)(y + 1)$ **61.** $(m + 4)(m^2 - 6)$ **63.** $(a^2 + b^2)(-3a + 2b)$ **65.** $(y - 2)(x - 2)$ **67.** $(3y - 2)(3y^3 - 4)$ **69.** $(1 - a)(1 - b)$ **71.** $2(m - 3q)(x + y)$ **73.** $4(a^2 + 2b)(a - b^2)$ **75.** $y^2(2x + 1)(x^2 - 7)$ **77.** $m^{-5}(3 + m^2)$ **79.** $p^{-3}(3 + 2p)$

Section 5.2

1. D **3.** C **5.** The factor $(4x + 10)$ can be factored further into $2(2x + 5)$, giving the *completely* factored form as $2(2x + 5)(x - 2)$. **7.** $x + 3$ **9.** $m - 7$ **11.** $r - 5$ **13.** $x - 2a$ **15.** $2x - 3$ **17.** $2u + v$ **19.** $(y - 3)(y + 10)$ **21.** $(p + 8)(p + 7)$ **23.** prime **25.** $(a + 5b)(a - 7b)$ **27.** $(a - 6b)(a - 3b)$ **29.** prime **31.** $-(6m - 5)(m + 3)$ **33.** $(5x - 6)(2x + 3)$ **35.** $(4k + 3)(5k + 8)$ **37.** $(3a - 2b)(5a - 4b)$ **39.** $(6m - 5)^2$ **41.** prime **43.** $(2xz - 1)(3xz + 4)$ **45.** $3(4x + 5)(2x + 1)$ **47.** $-5(a + 6)(3a - 4)$ **49.** $-11x(x - 6)(x - 4)$ **51.** $2xy^3(x - 12y)^2$ **53.** $6a(a - 3)(a + 5)$ **55.** $13y(y + 4)(y - 1)$ **57.** $3p(2p - 1)^2$ **59.** $(6p^3 - r)(2p^3 - 5r)$ **61.** $(5k + 4)(2k + 1)$ **63.** $(3m + 3p + 5)(m + p - 4)$ **65.** $(p^2 - 8)(p^2 - 2)$ **67.** $(2x^2 + 3)(x^2 - 6)$ **69.** $(4x^2 + 3)(4x^2 + 1)$ **71.** $(z - x)^2(z + 2x)$ **73.** $(a + b)^2(a - 3b)(a + 2b)$ **75.** $(p + q)^2(p + 3q)$ **77.** $\frac{36x^2 + 42x - 120}{6}$ **78.** $\frac{t^2 + 7t - 120}{6}$ **79.** $\frac{(t + 15)(t - 8)}{6}$ **80.** $\frac{(6x + 15)(6x - 8)}{6} = \frac{3(2x + 5) \cdot 2(3x - 4)}{6} = (2x + 5)(3x - 4)$ **81.** $(x - 4)(3x + 8)$ **82.** $(4x + 1)(3x + 8)$

Section 5.3

1. A, D **3.** B, C **5.** $x^2 + 4$ is a sum of squares with GCF = 1. It cannot be factored and is prime. Also, $(x + 2)^2 = x^2 + 4x + 4$, not $x^2 + 4$. In general, $x^2 + y^2 \neq (x + y)^2$. **7.** $(p + 4)(p - 4)$ **9.** $(5x + 2)(5x - 2)$ **11.** $2(3a + 7b)(3a - 7b)$ **13.** $4(4m^2 + y^2)(2m + y)(2m - y)$ **15.** $(y + z + 9)(y + z - 9)$ **17.** $(4 + x + 3y)(4 - x - 3y)$ **19.** $(p^2 + 16)(p + 4)(p - 4)$ **21.** $(k - 3)^2$ **23.** $(2z + w)^2$ **25.** $(4m - 1 + n)(4m - 1 - n)$ **27.** $(2r - 3 + s)(2r - 3 - s)$ **29.** $(x + y - 1)(x - y + 1)$ **31.** $2(7m + 3n)^2$ **33.** $(p + q + 1)^2$ **35.** $(a - b + 4)^2$ **37.** $(x - 3)(x^2 + 3x + 9)$ **39.** $(6 - t)(36 + 6t + t^2)$ **41.** $(x + 4)(x^2 - 4x + 16)$ **43.** $(10 + y)(100 - 10y + y^2)$ **45.** $(2x + 1)(4x^2 - 2x + 1)$ **47.** $(5x - 6)(25x^2 + 30x + 36)$ **49.** $(x - 2y)(x^2 + 2xy + 4y^2)$ **51.** $(4g - 3h)(16g^2 + 12gh + 9h^2)$ **53.** $(7p + 5q)(49p^2 - 35pq + 25q^2)$ **55.** $3(2n + 3p)(4n^2 - 6np + 9p^2)$ **57.** $(y + z + 4)(y^2 + 2yz + z^2 - 4y - 4z + 16)$ **59.** $(m^2 - 5)(m^4 + 5m^2 + 25)$ **61.** $(3 - 10x^3)(9 + 30x^3 + 100x^6)$ **63.** $(5y^2 + z)(25y^4 - 5y^2z + z^2)$ **65.** $k(5 - 4k)(25 + 20k + 16k^2)$ **67.** $(5p + 2q)(25p^2 - 10pq + 4q^2 + 5p - 2q)$ **69.** $(3a - 4b)(9a^2 + 12ab + 16b^2 + 5)$ **71.** $(t - 3)(2t + 1)(4t^2 - 2t + 1)$ **73.** $(8m - 9n)(8m + 9n - 64m^2 - 72mn - 81n^2)$ **75.** $(x^3 - y^3)(x^3 + y^3)$; $(x - y)(x^2 + xy + y^2)(x + y)(x^2 - xy + y^2)$ **76.** $(x^2 + xy + y^2)(x^2 - xy + y^2)$ **77.** $(x^2 - y^2)(x^4 + x^2y^2 + y^4)$; $(x - y)(x + y)(x^4 + x^2y^2 + y^4)$ **78.** $x^4 + x^2y^2 + y^4$ **79.** The product must equal $x^4 + x^2y^2 + y^4$. Multiply $(x^2 + xy + y^2)(x^2 - xy + y^2)$ to verify this. **80.** Start by factoring as a difference of squares.

Section 5.4

1. **(a)** B **(b)** D **(c)** A **(d)** A, C **(e)** A, B **3.** $3p^2(p - 6)(p + 5)$ **5.** $3pq(a + 6b)(a - 5b)$ **7.** prime **9.** $(6b + 1)(b - 3)$ **11.** $(x - 10)(x^2 + 10x + 100)$ **13.** $(p + 2)(4 + m)$ **15.** $9m(m - 5 + 2m^2)$ **17.** $2(3m - 10)(9m^2 + 30m + 100)$ **19.** $(3m - 5n)^2$ **21.** $(k - 9)(q + r)$ **23.** $16z^2x(zx - 2)$ **25.** $(r + 4)(2r - 3)$ **27.** $(25 + x^2)(5 - x)(5 + x)$ **29.** $(p + 1)(p^2 - p + 1)$ **31.** $(8m + 25)(8m - 25)$ **33.** $6z(2z^2 - z + 3)$ **35.** $16(4b + 5c)(4b - 5c)$ **37.** $8(4 + 5z)(16 - 20z + 25z^2)$

39. $(5r - s)(2r + 5s)$ **41.** $4pq(2p + q)(3p + 5q)$
43. $3(4k^2 + 9)(2k + 3)(2k - 3)$ **45.** $(m - n)(m^2 + mn + n^2 + m + n)$
47. $(x - 2m - n)(x + 2m + n)$ **49.** $6p^3(3p^2 - 4 + 2p^3)$
51. $2(x + 4)(x - 5)$ **53.** $8mn$ **55.** $2(5p + 9)(5p - 9)$
57. $4rx(3m^2 + mn + 10n^2)$ **59.** $(7a - 4b)(3a + b)$ **61.** prime
63. $(p + 8q - 5)^2$ **65.** $(7m^2 + 1)(3m^2 - 5)$
67. $(2r - t)(r^2 - rt + 19t^2)$ **69.** $(x + 3)(x^2 + 1)(x + 1)(x - 1)$
71. $(m + n - 5)(m - n + 1)$

Section 5.5

1. B, C, E, F **3.** The zero-factor property can be applied only with a product that equals 0. Multiply on the left, write the equation in standard form $x^2 - x - 12 = 0$, factor, and then apply the zero-factor property. The solution set is $\{-3, 4\}$. **5.** By dividing each side by a variable expression, here $3x$, the student "lost" the solution 0. On the left, set $3x = 0$ when the zero-factor property is applied. The solution set is $\{-4, 0\}$.
7. $\{-10, 5\}$ **9.** $\{-\frac{8}{3}, \frac{5}{2}\}$ **11.** $\{-2, 5\}$ **13.** $\{-6, -3\}$
15. $\{-\frac{1}{2}, 4\}$ **17.** $\{-\frac{1}{3}, \frac{4}{5}\}$ **19.** $\{-4, 0\}$ **21.** $\{0, 6\}$ **23.** $\{-2, 2\}$
25. $\{-3, 3\}$ **27.** $\{3\}$ **29.** $\{-\frac{4}{3}\}$ **31.** $\{-4, 2\}$ **33.** $\{-\frac{1}{2}, 6\}$
35. $\{-3, 4\}$ **37.** $\{-5, -\frac{1}{5}\}$ **39.** $\{1, 6\}$ **41.** $\{-\frac{1}{2}, 0, 5\}$
43. $\{-1, 0, 3\}$ **45.** $\{-\frac{4}{3}, 0, \frac{4}{3}\}$ **47.** $\{-\frac{5}{2}, -1, 1\}$ **49.** $\{-3, 3, 6\}$
51. $\{-\frac{1}{2}, 6\}$ **53.** $\{-\frac{2}{3}, \frac{4}{15}\}$ **55.** $\{-\frac{3}{2}, \frac{1}{2}\}$ **57.** width: 16 ft; length: 20 ft
59. base: 12 ft; height: 5 ft **61.** 50 ft by 100 ft **63.** −6 and −5 or 5 and 6
65. length: 15 in.; width: 9 in. **67.** 5 sec **69.** 6 sec **71.** $F = \frac{k}{d - D}$
73. $P = \frac{A}{1 + rt}$ **75.** $r = \frac{-2k - 3y}{a - 1}$, or $r = \frac{2k + 3y}{1 - a}$ **77.** $y = \frac{-x}{w - 3}$, or $y = \frac{x}{3 - w}$

Chapter 5 Review Exercises

1. $6p(2p - 1)$ **2.** $7x(3x + 5)$ **3.** $4qb(3q + 2b - 5q^2b)$
4. $6rt(r^2 - 5rt + 3t^2)$ **5.** $(x + 3)(x - 3)$ **6.** $(z + 1)(3z - 1)$
7. $(m + q)(4 + n)$ **8.** $(x + y)(x + 5)$ **9.** $(m + 3)(2 - a)$
10. $(x + 3)(x - y)$ **11.** $(3p - 4)(p + 1)$ **12.** $(3k - 2)(2k + 5)$
13. $(3r + 1)(4r - 3)$ **14.** $(2m + 5)(5m + 6)$ **15.** $(2k - h)(5k - 3h)$
16. prime **17.** $2x(4 + x)(3 - x)$ **18.** $3b(2b - 5)(b + 1)$
19. $(y^2 + 4)(y^2 - 2)$ **20.** $(2k^2 + 1)(k^2 - 3)$
21. $(p + 2)^2(p + 3)(p - 2)$ **22.** $(3r + 16)(r + 1)$
23. $3x^4(2x^2 - 9)(4x^2 - 3)$ **24.** $5y^3(3y^3 + 1)(2y^3 - 11)$
25. $(4x + 5)(4x - 5)$ **26.** $(3t + 7)(3t - 7)$
27. $(6m - 5n)(6m + 5n)$ **28.** $(x + 7)^2$ **29.** $(3k - 2)^2$
30. $(r + 3)(r^2 - 3r + 9)$ **31.** $(5x - 1)(25x^2 + 5x + 1)$
32. $(m + 1)(m^2 - m + 1)(m - 1)(m^2 + m + 1)$
33. $(x^4 + 1)(x^2 + 1)(x + 1)(x - 1)$ **34.** $(x + 3 + 5y)(x + 3 - 5y)$
35. $2b(3a^2 + b^2)$ **36.** $(x + 1)(x - 1)(x - 2)(x^2 + 2x + 4)$
37. $\{4\}$ **38.** $\{-1, -\frac{2}{5}\}$ **39.** $\{2, 3\}$ **40.** $\{-4, 2\}$ **41.** $\{-\frac{5}{2}, \frac{10}{3}\}$
42. $\{-\frac{3}{2}, \frac{1}{3}\}$ **43.** $\{-\frac{3}{2}, -\frac{1}{4}\}$ **44.** $\{-3, 3\}$ **45.** $\{-\frac{3}{2}, 0\}$
46. $\{\frac{1}{2}, 1\}$ **47.** $\{1, 4\}$ **48.** $\{-\frac{7}{2}, 0, 4\}$ **49.** $\{-3, -2, 2\}$
50. $\{-2, -\frac{6}{5}, 3\}$ **51.** 3 ft **52.** length: 60 ft; width: 40 ft **53.** 16 sec
54. 1 sec and 15 sec **55.** 8 sec **56.** The rock reaches a height of 240 ft once on its way up and once on its way down. **57.** $k = \frac{-3s - 2t}{b - 1}$, or $k = \frac{3s + 2t}{1 - b}$ **58.** $w = \frac{7}{z - 3}$, or $w = \frac{-7}{3 - z}$

Chapter 5 Mixed Review Exercises

1. $(x + y)(x + 5)$ **2.** $-(x + 2)(x - 5)$ **3.** $16(x^2 + 9)$
4. $a(6 - m)(5 + m)$ **5.** $(2 - a)(4 + 2a + a^2)$ **6.** $(9k + 4)(9k - 4)$
7. prime **8.** $5y^2(3y + 4)$ **9.** $(5z - 3m)^2$ **10.** D; The polynomial is not factored. **11.** $\{-\frac{3}{5}, 4\}$ **12.** $\{-1, 0, 1\}$ **13.** $\{0, 3\}$
14. $h = \frac{2\mathcal{A}}{b + B}$ **15.** $H = \frac{S - 2LW}{2W + 2L}$ **16.** width: 25 ft; length: 110 ft
17. 6 in. **18.** 5 sec and 9 sec

Chapter 5 Test

[5.1–5.4] **1.** $11z(z - 4)$ **2.** $5x^2y^3(2y^2 - 1 - 5x^3)$
3. $(x + y)(3 + b)$ **4.** $-(2x + 9)(x - 4)$ **5.** $(3x - 5)(2x + 7)$
6. $(4p - q)(p + q)$ **7.** $(4a + 5b)^2$ **8.** $(x + 1 + 2z)(x + 1 - 2z)$
9. $(a + b)(a - b)(a + 2)$ **10.** $(3k + 11j)(3k - 11j)$
11. $(y - 6)(y^2 + 6y + 36)$ **12.** $(2k^2 - 5)(3k^2 + 7)$
13. $(3x^2 + 1)(9x^4 - 3x^2 + 1)$ **14.** $(t^2 + 8)(t^2 + 2)$
15. prime **16.** $-(x + 5)(x - 6)$ **17.** $(5 - 6y^2)(25 + 30y^2 + 36y^4)$
18. D [5.5] **19.** $\{-2, -\frac{2}{3}\}$ **20.** $\{0, \frac{5}{3}\}$ **21.** $\{-\frac{2}{5}, 1\}$
22. $r = \frac{-2 - 6t}{a - 3}$, or $r = \frac{2 + 6t}{3 - a}$ **23.** length: 8 in.; width: 5 in.
24. 2 sec and 4 sec

Chapters R–5 Cumulative Review Exercises

[R.1] **1.** $\frac{13}{12}$ **2.** 16.1 [R.5] **3.** $-2m + 6$ **4.** $2x^2 + 5x + 4$
[R.4] **5.** 10 **6.** undefined [1.1] **7.** $\{\frac{7}{6}\}$ **8.** $\{-1\}$
[1.5] **9.** $(-\frac{1}{2}, \infty)$ [1.6] **10.** $(2, 3)$ **11.** $(-\infty, 2) \cup (3, \infty)$
[1.7] **12.** $\{-\frac{16}{5}, 2\}$ **13.** $(-11, 7)$ **14.** $(-\infty, -2] \cup [7, \infty)$
[1.4] **15.** 2 hr [2.2] **16.** 0 **17.** −1
[2.1] **18.**

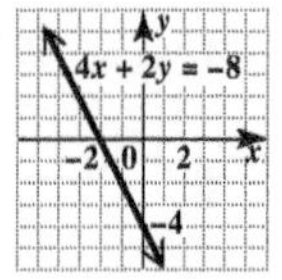

[2.6] **19.** −1 **20.** $(-\frac{7}{2}, 0)$ **21.** $(0, 7)$
[3.1] **22.** $\{(1, 5)\}$ **23.** $\varnothing$
[3.2] **24.** $\{(1, 1, 0)\}$
[4.5] **25.** $49x^2 + 42xy + 9y^2$ [4.6] **26.** $2x^2 + x + 3$
[4.3] **27.** $x^3 + 12x^2 - 3x - 7$ [4.2] **28.** **(a)** 4.638×10^{-4} **(b)** 566,000
[5.1–5.4] **29.** $(2w + 7z)(8w - 3z)$ **30.** $(2x - 1 + y)(2x - 1 - y)$
31. $(10x^2 + 9)(10x^2 - 9)$ **32.** $(2p + 3)(4p^2 - 6p + 9)$
[5.5] **33.** $\{-4, -\frac{3}{2}, 1\}$ **34.** $\{\frac{1}{3}\}$ **35.** 4 ft **36.** longer sides: 18 in.; distance between: 16 in.

6 RATIONAL EXPRESSIONS AND FUNCTIONS

Section 6.1

1. integers; 0; rational; polynomials; 0
3. (a) $\{x \mid x \text{ is a real number}, x \neq 7\}$ **(b)** $(-\infty, 7) \cup (7, \infty)$
5. (a) $\left\{x \mid x \text{ is a real number}, x \neq -\frac{1}{7}\right\}$ **(b)** $\left(-\infty, -\frac{1}{7}\right) \cup \left(-\frac{1}{7}, \infty\right)$
7. (a) $\{x \mid x \text{ is a real number}, x \neq 0\}$ **(b)** $(-\infty, 0) \cup (0, \infty)$
9. (a) $\left\{x \mid x \text{ is a real number}, x \neq -2, \frac{3}{2}\right\}$
(b) $(-\infty, -2) \cup \left(-2, \frac{3}{2}\right) \cup \left(\frac{3}{2}, \infty\right)$ **11. (a)** $\{x \mid x \text{ is a real number}\}$ **(b)** $(-\infty, \infty)$ **13. (a)** $\{x \mid x \text{ is a real number}\}$ **(b)** $(-\infty, \infty)$
15. $\frac{2}{15}$ **17.** $\frac{9}{10}$ **19.** $\frac{3}{4}$ **21.** B, E, F **23.** B **25.** The expression $\frac{2x}{4}$ has 2 as a factor of both the numerator and the denominator. It can be written in lowest terms as $\frac{x}{2}$. **27.** The student tried to simplify before factoring the numerator and denominator. The correct answer is $\frac{1}{2}$.
29. x **31.** $\frac{x-3}{x+5}$ **33.** $\frac{x+3}{2x(x-3)}$ **35.** It is already in lowest terms.
37. $\frac{6}{7}$ **39.** $\frac{z}{6}$ **41.** $\frac{t-3}{3}$ **43.** $\frac{2}{t-3}$ **45.** $\frac{x-3}{x+1}$ **47.** $\frac{4x+1}{4x+3}$
49. $a^2 - ab + b^2$ **51.** $\frac{c+6d}{c-d}$ **53.** $\frac{a+b}{a-b}$ **55.** -1
57. $-(x+y)$, or $-x-y$ **59.** $-(x+2)$, or $-x-2$ **61.** $-\frac{x+y}{x-y}$ (There are other answers.) **63.** $-\frac{1}{2}$ **65.** It is already in lowest terms. **67.** $\frac{x+4}{x-2}$ **69.** $\frac{2x+3}{x+2}$ **71.** $\frac{7x}{6}$ **73.** $-\frac{p+5}{2p}$ (There are other answers.) **75.** $\frac{35}{4}$ **77.** $-(z+1)$, or $-z-1$ **79.** $\frac{14x^2}{5}$
81. $\frac{-m(m+7)}{m+1}$ (There are other answers.) **83.** -2 **85.** $\frac{x+4}{x-4}$
87. $\frac{a^2+ab+b^2}{a-b}$ **89.** $\frac{2x-3}{2(x-3)}$ **91.** $\frac{a^2+2ab+4b^2}{a+2b}$ **93.** $\frac{2x+3}{2x-3}$
95. $\frac{27}{2mn^7}$ **97.** $\frac{k+5p}{2k+5p}$ **99.** $(k-1)(k-2)$

Section 6.2

1. product; factors; greatest; denominator **3.** C **5.** $\frac{4}{5}$ **7.** $-\frac{1}{18}$
9. $\frac{31}{36}$ **11.** $\frac{9}{t}$ **13.** $\frac{6x+y}{7}$ **15.** $\frac{2}{x}$ **17.** $-\frac{2}{x^3}$ **19.** 1 **21.** $x-5$
23. $\frac{1}{p+3}$ **25.** $\frac{5}{x-2}$ **27.** $a-b$ **29.** $72x^4y^5$ **31.** $z(z-2)$
33. $2(y+4)$ **35.** $(x+9)^2(x-9)$ **37.** $(m+n)(m-n)$
39. $x(x-4)(x+1)$ **41.** $(t+5)(t-2)(2t-3)$
43. $2y(y+3)(y-3)$ **45.** $2(x+2)^2(x-3)$ **47.** The expression $\frac{x-4x-1}{x+2}$ is incorrect. The third term in the numerator should be $+1$ because the $-$ sign should be distributed over both $4x$ and -1. The correct answer is $\frac{-3x+1}{x+2}$. **49.** $\frac{31}{3t}$ **51.** $\frac{5-22x}{12x^2y}$
53. $\frac{16b+9a^2}{60a^4b^6}$ **55.** $\frac{4pr+3sq^3}{14p^4q^4}$ **57.** $\frac{a^2b^5-2ab^6+3}{a^5b^7}$
59. $\frac{1}{x(x-1)}$ **61.** $\frac{5a^2-7a}{(a+1)(a-3)}$ **63.** 4 **65.** 3
67. $\frac{3}{x-4}$, or $\frac{-3}{4-x}$ **69.** $\frac{w+z}{w-z}$, or $\frac{-w-z}{z-w}$ **71.** $\frac{-2}{(x+1)(x-1)}$
73. $\frac{2(2x-1)}{x-1}$ **75.** $\frac{7}{y}$ **77.** $\frac{6}{x-2}$ **79.** $\frac{3x-2}{x-1}$
81. $\frac{4x-7}{x^2-x+1}$ **83.** $\frac{2x+1}{x}$ **85.** $\frac{4p^2-21p+29}{(p-2)^2}$
87. $\frac{x}{(x-2)^2(x-3)}$ **89.** $\frac{10x+23}{(x+2)^2(x+3)}$
91. $\frac{2x^2+24xy}{(x+2y)(x-y)(x+6y)}$ **93.** $\frac{2x^2+21xy-10y^2}{(x+2y)(x-y)(x+6y)}$
95. $\frac{3r-2s}{(2r-s)(3r-s)}$ **97. (a)** $C(x) = \frac{8000+10x}{49(101-x)}$
(b) 30.44 thousand dollars

Section 6.3

1. complex; numerator; both **3.** $\frac{1}{6}$ **5.** $\frac{9}{5}$ **7.** $\frac{4}{15}$ **9.** $\frac{7}{17}$
11. $\frac{2x}{x-1}$ **13.** $\frac{2(k+1)}{3k-1}$ **15.** $\frac{5x^2}{9z^3}$ **17.** $\frac{1+x}{-1+x}$
19. $\frac{6x+1}{7x-3}$ **21.** $\frac{y+x}{y-x}$ **23.** $4x$ **25.** $\frac{y+4}{2}$ **27.** $x+4y$
29. $\frac{a+b}{ab}$ **31.** xy **33.** $\frac{3y}{2}$ **35.** $\frac{y^2+yx+x^2}{xy(y+x)}$ **37.** $\frac{x^2+5x+4}{x^2+5x+10}$
39. The negative exponents are on terms, not factors. Terms with negative exponents cannot be simply moved across a fraction bar.
41. $\frac{x^2y^2}{y^2+x^2}$ **43.** $\frac{y^2+x^2}{xy^2+x^2y}$, or $\frac{y^2+x^2}{xy(y+x)}$ **45.** $\frac{p^2+k}{p^2-3k}$
47. $\frac{1}{2xy}$ **49.** $\frac{m^2+6m-4}{m(m-1)}$ **50.** $\frac{m^2-m-2}{m(m-1)}$ **51.** $\frac{m^2+6m-4}{m^2-m-2}$
52. $m(m-1)$ **53.** $\frac{m^2+6m-4}{m^2-m-2}$ **54.** Answers will vary.

Section 6.4

1. (a) equation **(b)** expression **(c)** expression **(d)** equation
3. The proposed solution 3 must be rejected because it causes a denominator to equal 0. The solution set is $\varnothing$.
In Exercises 5–15, we give the domains using set-builder notation.
5. $\{x \mid x \text{ is a real number}, x \neq 0\}$ **7.** $\{x \mid x \text{ is a real number}, x \neq -1, 2\}$
9. $\left\{x \mid x \text{ is a real number}, x \neq 4, \frac{7}{2}\right\}$ **11.** $\left\{x \mid x \text{ is a real number}, x \neq -\frac{7}{4}, 0, \frac{13}{6}\right\}$ **13.** $\{x \mid x \text{ is a real number}, x \neq \pm 4\}$
15. $\{x \mid x \text{ is a real number}, x \neq 0, 1, -3, 2\}$ **17.** $\{1\}$ **19.** $\{-6, 4\}$
21. $\{-7, 3\}$ **23.** $\left\{-\frac{7}{12}\right\}$ **25.** $\varnothing$ **27.** $\{-3\}$ **29.** $\{5\}$ **31.** $\{0\}$
33. $\left\{\frac{27}{56}\right\}$ **35.** $\varnothing$ **37.** $\{5\}$ **39.** $\varnothing$ **41.** $\varnothing$ **43.** $\{-6, 3\}$
45. $\{0\}$ **47.** $\{-3, -1\}$ **49.** $\{-10\}$ **51.** $\{-1\}$ **53.** $\{13\}$
55. $\{x \mid x \text{ is a real number}, x \neq \pm 3\}$ **57. (a)** D **(b)** C **(c)** A **(d)** B

59.

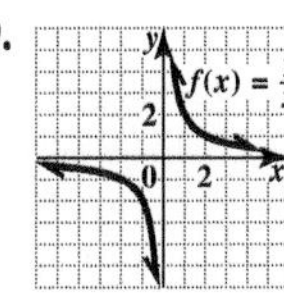

$x = 0; y = 0$

61.

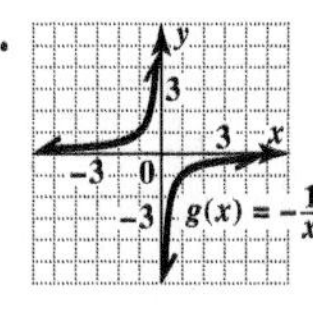

$x = 0; y = 0$

63.

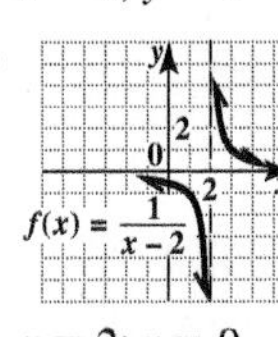

$x = 2; y = 0$

65.

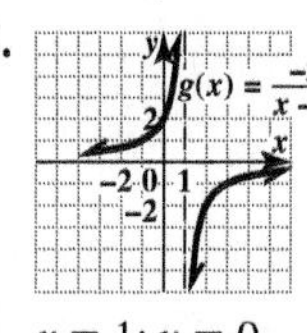

$x = 1; y = 0$

67. (a) 0 **(b)** 1.6 **(c)** 4.1 **(d)** The waiting time also increases. **69. (a)** 500 ft **(b)** It decreases.

SUMMARY EXERCISES Simplifying Rational Expressions vs. Solving Rational Equations

1. equation; $\{20\}$ **2.** expression; $\frac{2(x+5)}{5}$ **3.** expression; $-\frac{22}{7x}$ **4.** expression; $\frac{y+x}{y-x}$ **5.** equation; $\{\frac{1}{2}\}$ **6.** equation; $\{7\}$ **7.** expression; $\frac{43}{24x}$ **8.** equation; $\{1\}$ **9.** expression; $\frac{5x-1}{-2x+2}$, or $\frac{5x-1}{-2(x-1)}$ **10.** expression; $\frac{25}{4(r+2)}$ **11.** expression; $\frac{x^2+xy+2y^2}{(x+y)(x-y)}$ **12.** expression; $\frac{24p}{p+2}$ **13.** expression; $-\frac{5}{36}$ **14.** equation; $\{0\}$ **15.** expression; $\frac{2x+10}{x(x-2)(x+2)}$ **16.** equation; $\{\frac{1}{7}, 2\}$ **17.** expression; $\frac{-x}{3x+5y}$ **18.** expression; $\frac{t-2}{8}$ **19.** equation; $\emptyset$ **20.** expression; $\frac{13x+28}{2x(x+4)(x-4)}$ **21.** expression; $\frac{b+3}{3}$ **22.** expression; $\frac{5}{3z}$ **23.** expression; $\frac{3y+2}{y+3}$ **24.** equation; $\{-13\}$ **25.** expression; $\frac{-1}{x-3}$, or $\frac{1}{3-x}$ **26.** equation; $\{\frac{5}{4}\}$ **27.** equation; $\{-10\}$ **28.** expression; $\frac{2z-3}{2z+3}$ **29.** equation; $\emptyset$ **30.** expression; $\frac{k(2k^2-2k+5)}{(k-1)(3k^2-2)}$

Section 6.5

1. A **3.** D **5.** 24 **7.** $\frac{25}{4}$ **9.** The variable a cannot appear on both sides of the final answer. The correct answer is $a = \frac{mb}{m-b}$. **11.** $G = \frac{Fd^2}{Mm}$ **13.** $a = \frac{bc}{c+b}$ **15.** $v = \frac{PVt}{pT}$ **17.** $r = \frac{nE - IR}{In}$, or $r = \frac{IR - nE}{-In}$ **19.** $b = \frac{2\mathcal{A}}{h} - B$, or $b = \frac{2\mathcal{A} - hB}{h}$ **21.** $r = \frac{eR}{E-e}$ **23.** $R = \frac{D}{1-DT}$, or $R = \frac{-D}{DT-1}$ **25.** 21 girls, 7 boys **27.** 1.75 in. **29.** 5.4 in. **31.** 7.6 in. **33.** 56 teachers **35.** 210 deer **37.** 25,000 fish **39.** 6.6 more gallons **41.** $x = \frac{7}{2}$; $AC = 8$; $DF = 12$ **43.** 2.4 mL **45.** $\frac{1}{3}$ job per hr **47. (a)** $4 - x$ **(b)** $4 + x$ **49.** 3 mph **51.** 10 mph **53.** 1020 mi **55.** 1750 mi **57.** 190 mi **59.** 4 hr **61.** $6\frac{2}{3}$ min, or 6 min, 40 sec **63.** 30 hr **65.** $2\frac{1}{3}$ hr, or 2 hr, 20 min **67.** 20 hr **69.** $2\frac{4}{5}$ hr, or 2 hr, 48 min

Section 6.6

1. increases; decreases **3.** direct **5.** direct **7.** inverse **9.** inverse **11.** inverse **13.** direct **15.** joint **17.** combined **19.** The perimeter of a square varies directly as the length of its side. **21.** The surface area of a sphere varies directly as the square of its radius. **23.** The area of a triangle varies jointly as the lengths of its base and height. **25.** 4; 2; 4π; $\frac{4}{3}\pi$; $\frac{1}{2}$; $\frac{1}{3}\pi$ **27.** $A = kb$ **29.** $h = \frac{k}{t}$ **31.** $M = kd^2$ **33.** $I = kgh$ **35.** 36 **37.** $\frac{16}{9}$ **39.** 0.625 **41.** $\frac{16}{5}$ **43.** $222\frac{2}{9}$ **45.** \$3.92 **47.** 8 lb **49.** 450 cm^3 **51.** 256 ft **53.** $106\frac{2}{3}$ mph **55.** 100 cycles per sec **57.** $21\frac{1}{3}$ foot-candles **59.** \$420 **61.** 11.8 lb **63.** 448.1 lb **65.** 68,600 calls **67.** Answers will vary. **69.** $(0, 0)$, $(1, 3.75)$ **70.** 3.75 **71.** $y = 3.75x + 0$, or $y = 3.75x$ **72.** $a = 3.75$, $b = 0$ **73.** It is the price per gallon and the slope of the line. **74.** It can be written in the form $y = kx$ (where $k = a$). The value of a is the constant of variation.

Chapter 6 Review Exercises

1. (a) $\{x \mid x \text{ is a real number}, x \neq -6\}$ **(b)** $(-\infty, -6) \cup (-6, \infty)$ **2. (a)** $\{x \mid x \text{ is a real number}, x \neq 2, 5\}$ **(b)** $(-\infty, 2) \cup (2, 5) \cup (5, \infty)$ **3. (a)** $\{x \mid x \text{ is a real number}, x \neq 9\}$ **(b)** $(-\infty, 9) \cup (9, \infty)$ **4.** $\frac{x}{2}$ **5.** $\frac{5m+n}{5m-n}$ **6.** $\frac{-1}{2+r}$ **7.** $\frac{3y^2(2y+3)}{2y-3}$ **8.** $\frac{-3(w+4)}{w}$ **9.** $x - 4$ **10.** $\frac{z(z+2)}{z+5}$ **11.** $\frac{-2}{x+8}$ **12.** 1 **13.** $96b^5$ **14.** $9r^2(3r+1)$ **15.** $(3x-1)(2x+5)(3x+4)$ **16.** $3(x-4)^2(x+2)$ **17.** $\frac{16z-3}{2z^2}$ **18.** 12 **19.** $\frac{71}{30(a+2)}$ **20.** $\frac{13r^2+5rs}{(5r+s)(2r-s)(r+s)}$ **21.** $\frac{3+2t}{4-7t}$ **22.** -2 **23.** $\frac{1}{3q+2p}$ **24.** $\frac{y+x}{xy}$ **25.** $\{-3\}$ **26.** $\{-2\}$ **27.** $\emptyset$ **28.** $\{0\}$ **29.** $\{-5\}$ **30.** $\emptyset$ **31. (a)** equation; $\{-24\}$ **(b)** expression; $\frac{24+x}{6x}$

32.

$x = -1; y = 0$

33. $\frac{15}{2}$ **34.** 2 **35.** $m = \frac{Fd^2}{GM}$ **36.** $M = \frac{m\mu}{\nu - \mu}$ **37.** 6000 passenger-km per day **38.** 12.2 more gallons **39.** 16 km per hr **40.** $4\frac{4}{5}$ min, or 4 min, 48 sec **41.** $3\frac{3}{5}$ hr, or 3 hr, 36 min **42.** C **43.** 430 mm **44.** 5.59 vibrations per sec **45.** 22.5 ft^3

Chapter 6 Mixed Review Exercises

1. $\frac{1}{x-2y}$ **2.** $\frac{x+5}{x+2}$ **3.** $\frac{6m+5}{3m^2}$ **4.** $\frac{11}{3-x}$, or $\frac{-11}{x-3}$ **5.** $\frac{x^2-6}{2(2x+1)}$

6. $\frac{3-5x}{6x+1}$ **7.** $\frac{1}{3}$ **8.** $\frac{s^2+t^2}{st(s-t)}$ **9.** $\frac{k-3}{36k^2+6k+1}$ **10.** $\frac{x(9x+1)}{3x+1}$

11. $\frac{5a^2+4ab+12b^2}{(a+3b)(a-2b)(a+b)}$ **12.** $\frac{acd+b^2d+bc^2}{bcd}$ **13.** $\left\{\frac{1}{3}\right\}$

14. $\left\{-\frac{14}{3}\right\}$ **15.** $\{1, 4\}$ **16.** $r = \frac{AR}{R-A}$, or $r = \frac{-AR}{A-R}$

17. (a) 8.32 mm **(b)** 44.87 diopters **18.** \$53.17 **19.** $4\frac{1}{2}$ mi

20. 12 ft^2

Chapter 6 Test

[6.1] **1. (a)** $\left\{x \mid x \text{ is a real number}, x \neq -2, \frac{4}{3}\right\}$

(b) $(-\infty, -2) \cup \left(-2, \frac{4}{3}\right) \cup \left(\frac{4}{3}, \infty\right)$ **2.** $\frac{2x-5}{x(3x-1)}$ **3.** $\frac{3(x+3)}{4}$

4. $\frac{y+4}{y-5}$ **5.** -2 **6.** $\frac{x+5}{x}$ [6.2] **7.** $t^2(t+3)(t-2)$

8. $\frac{7-2t}{6t^2}$ **9.** $\frac{13x+35}{(x-7)(x+7)}$ **10.** $\frac{11x+21}{(x-3)^2(x+3)}$ **11.** $\frac{4}{x+2}$

[6.3] **12.** $\frac{72}{11}$ **13.** $-\frac{1}{a+b}$ **14.** $\frac{2y^2+x^2}{xy(y-x)}$ [6.4] **15. (a)** expression; $\frac{11(x-6)}{12}$ **(b)** equation; $\{6\}$ **16.** $\left\{\frac{1}{2}\right\}$ **17.** $\{5\}$

18. $\ell = \frac{2S}{n} - a$, or $\ell = \frac{2S - na}{n}$

19.

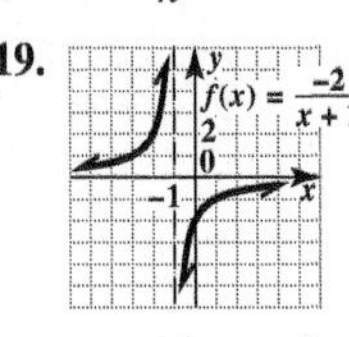

$x = -1$; $y = 0$

[6.5] **20.** $3\frac{3}{14}$ hr **21.** 15 mph **22.** 48,000 fish **23. (a)** 3 units **(b)** 0 [6.6] **24.** 200 amps **25.** 0.8 lb

Chapters R–6 Cumulative Review Exercises

[R.1, R.3] **1.** $-\frac{11}{12}$ **2.** $\frac{5}{8}$ **3.** $-\frac{6}{5}$ **4.** -5.67 **5.** 0.0525 **6.** 4360 [R.4] **7. (a)** 25 **(b)** -25 **(c)** 25 **(d)** 5 **(e)** -5 **(f)** not a real number **8.** -199 [1.1] **9.** $\left\{-\frac{15}{4}\right\}$ [1.7] **10.** $\left\{\frac{2}{3}, 2\right\}$

[1.5] **11.** $\left(-\infty, \frac{240}{13}\right]$ [1.7] **12.** $(-\infty, -2] \cup \left[\frac{2}{3}, \infty\right)$

[1.3] **13.** \$4000 at 4%; \$8000 at 3% **14.** 6 m

[2.1] **15.** $-4x + 2y = 8$

x-intercept: $(-2, 0)$;
y-intercept: $(0, 4)$

[2.2] **16.** $-\frac{3}{2}$ **17.** $-\frac{3}{4}$ [2.3] **18.** $y = -\frac{3}{2}x + \frac{1}{2}$

[2.4] **19.**

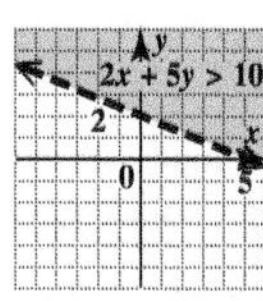

20. $x - y \ge 3$ and $3x + 4y \le 12$

[2.6] **21. (a)** $f(x) = \frac{5}{3}x - \frac{8}{3}$ **(b)** -1 [2.5] **22. (a)** yes **(b)** domain: $[-2, \infty)$; range: $(-\infty, 0]$ [3.1] **23.** $\{(-1, 3)\}$ [3.2] **24.** $\{(-2, 3, 1)\}$ **25.** $\varnothing$ [4.1] **26.** $\frac{m}{n}$ [4.3] **27.** $4y^2 - 7y - 6$ [4.5] **28.** $12f^2 + 5f - 3$ **29.** $\frac{1}{16}x^2 + \frac{5}{2}x + 25$ [4.6] **30.** $x^2 + 4x - 7$ [4.4] **31. (a)** $2x^3 - 2x^2 + 6x - 4$ **(b)** $2x^3 - 4x^2 + 2x + 2$ **(c)** -14 **(d)** $x^4 + 2x^2 - 3$ [5.2] **32.** $(2x+5)(x-9)$ [5.3] **33.** $25(2t^2+1)(2t^2-1)$ **34.** $(2p+5)(4p^2-10p+25)$ [6.1] **35.** $\frac{a(a-b)}{2(a+b)}$ **36.** $\frac{2(x+3)}{(x+2)(x^2+3x+9)}$ [6.2] **37.** 3 [5.5] **38.** $\left\{-\frac{7}{3}, 1\right\}$ [6.4] **39.** $\{-4\}$ [6.5] **40.** $1\frac{1}{5}$ hr, or 1 hr, 12 min

7 ROOTS, RADICALS, AND ROOT FUNCTIONS

Section 7.1

1. (a) 5 **(b)** 9 **(c)** 12 **3.** $\sqrt{16}$ is a number whose square equals 16, not a number that gives 16 when doubled. Thus, $\sqrt{16} = 4$. **5.** E **7.** D **9.** C **11.** C **13.** A **15. (a)** not a real number **(b)** negative **(c)** 0 **17.** -9 **19.** 6 **21.** -4 **23.** -8 **25.** 6 **27.** -2 **29.** It is not a real number. **31.** 2 **33.** It is not a real number. **35.** $\frac{8}{9}$ **37.** $\frac{4}{3}$ **39.** $-\frac{1}{2}$ **41.** 3 **43.** 0.5 **45.** -0.7 **47.** 0.1

In Exercises 49–55, we give the domain and then the range.

49.

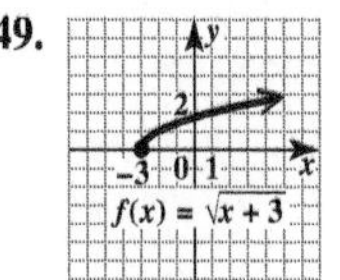

$[-3, \infty)$; $[0, \infty)$

51. $f(x) = \sqrt{x} - 2$

$[0, \infty)$; $[-2, \infty)$

53.

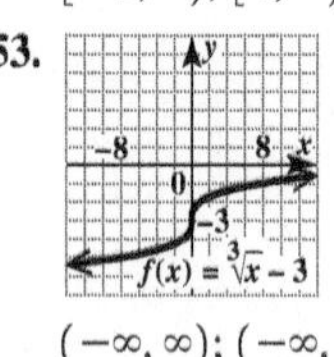

$(-\infty, \infty)$; $(-\infty, \infty)$

55. $f(x) = \sqrt[3]{x-3}$

$(-\infty, \infty)$; $(-\infty, \infty)$

57. 12 **59.** 10 **61.** 2 **63.** -9 **65.** -5 **67.** $|x|$ **69.** $|z|$ **71.** x **73.** x^5 **75.** $|x|^5$ (or $|x^5|$) **77.** 97.381 **79.** 16.863 **81.** -9.055 **83.** 7.507 **85.** 3.162 **87.** 1.885 **89. (a)** 1,183,000 cycles per sec **(b)** 118,000 cycles per sec **91.** 10 mi **93.** 1.732 amps **95.** 437,000 mi^2 **97.** 330 m^2

Section 7.2

1. C **3.** A **5.** H **7.** B **9.** D **11.** In $27^{1/3}$, the base and exponent should not be multiplied. The denominator of the rational exponent is the index of the radical, $\sqrt[3]{27}$. The correct answer is 3. **13.** 13 **15.** 9

17. 2 **19.** $\frac{8}{9}$ **21.** -3 **23.** It is not a real number. **25.** 1000

27. 27 **29.** -1024 **31.** 16 **33.** $\frac{1}{8}$ **35.** $\frac{1}{512}$ **37.** $\frac{9}{25}$ **39.** $\frac{27}{8}$

41. $\sqrt{10}$ **43.** $(\sqrt[4]{8})^3$ **45.** $5(\sqrt[3]{x})^2$ **47.** $9(\sqrt[8]{q})^5 - (\sqrt[3]{2x})^2$

49. $\frac{1}{(\sqrt[5]{x})^3}$ **51.** $(\sqrt[3]{2y + x})^2$ **53.** $15^{1/2}$ **55.** 64 **57.** 64

59. x **61.** x^{10} **63.** 9 **65.** 4 **67.** y **69.** $x^{5/12}$ **71.** $k^{2/3}$

73. x^3y^8 **75.** $\frac{1}{x^{10/3}}$ **77.** $\frac{1}{m^{1/4}n^{3/4}}$ **79.** $\frac{y^{17/3}}{x^8}$ **81.** $\frac{c^{11/3}}{b^{11/4}}$ **83.** $\frac{q^{5/3}}{9p^{7/2}}$

85. $p + 2p^2$ **87.** $k^{7/4} - k^{3/4}$ **89.** $6 + 18a$ **91.** $-5x^2 + 5x$

93. $x^{17/20}$ **95.** $t^{8/15}$ **97.** $\frac{1}{x^{3/2}}$ **99.** $y^{5/6}z^{1/3}$ **101.** $m^{1/12}$ **103.** $x^{1/24}$

105. $\sqrt{a^2 + b^2} = \sqrt{3^2 + 4^2} = 5$; $a + b = 3 + 4 = 7$; $5 \neq 7$

107. 4.5 hr **109.** 19.0°; The table gives 19°. **111.** 4.2°; The table gives 4°.

Section 7.3

1. D **3.** B **5.** D **7.** The student "dropped" the index, 3. The correct product is $\sqrt[3]{65}$. **9.** $\sqrt{9}$, or 3 **11.** $\sqrt{36}$, or 6 **13.** $\sqrt{30}$ **15.** $\sqrt{14x}$

17. $\sqrt{42pqr}$ **19.** $\sqrt[3]{10}$ **21.** $\sqrt[3]{14xy}$ **23.** $\sqrt[4]{33}$ **25.** $\sqrt[4]{6xy^2}$

27. This expression cannot be simplified by the product rule. **29.** $\frac{8}{11}$

31. $\frac{\sqrt{3}}{5}$ **33.** $\frac{\sqrt{x}}{5}$ **35.** $\frac{p^3}{9}$ **37.** $-\frac{3}{4}$ **39.** $\frac{\sqrt[3]{r^2}}{2}$ **41.** $-\frac{3}{x}$ **43.** $\frac{1}{x^3}$

45. $\sqrt{12}$ can be simplified further. The *greatest* perfect square factor that divides into 48 is 16, not 4. Thus, $\sqrt{48} = \sqrt{16 \cdot 3} = \sqrt{16} \cdot \sqrt{3} = 4\sqrt{3}$.

47. $2\sqrt{3}$ **49.** $12\sqrt{2}$ **51.** $-4\sqrt{2}$ **53.** $-2\sqrt{7}$ **55.** This radical cannot be simplified further. **57.** $4\sqrt[3]{2}$ **59.** $2\sqrt[3]{5}$ **61.** $-2\sqrt[3]{2}$

63. $-4\sqrt[4]{2}$ **65.** $2\sqrt[5]{2}$ **67.** $-3\sqrt[5]{2}$ **69.** $2\sqrt[6]{2}$ **71.** $6k\sqrt{2}$

73. $12xy^4\sqrt{xy}$ **75.** $11x^3$ **77.** $-3t^4$ **79.** $-10m^4z^2$ **81.** $5a^2b^3c^4$

83. $\frac{1}{2}r^2t^5$ **85.** $5x\sqrt{2x}$ **87.** $-10r^5\sqrt{5r}$ **89.** $x^3y^4\sqrt{13x}$ **91.** $2z^2w^3$

93. $-2zt^2\sqrt[3]{2z^2t}$ **95.** $3x^3y^4$ **97.** $-3r^3s^2\sqrt[4]{2r^3s}$ **99.** $\frac{y^5\sqrt{y}}{6}$

101. $\frac{x^5\sqrt[3]{x}}{3}$ **103.** $4\sqrt{3}$ **105.** $\sqrt{5}$ **107.** $x^2\sqrt{x}$ **109.** $x\sqrt[5]{x^3}$

111. $\sqrt[6]{432}$ **113.** $\sqrt[12]{6912}$ **115.** $\sqrt[6]{x^5}$ **117.** 5 **119.** $8\sqrt{2}$

121. $2\sqrt{14}$ **123.** 13 **125.** $9\sqrt{2}$ **127.** $\sqrt{17}$ **129.** 5

131. $6\sqrt{2}$ **133.** $\sqrt{5y^2 - 2xy + x^2}$ **135.** **(a)** B **(b)** C **(c)** D **(d)** A **137.** $x^2 + y^2 = 144$ **139.** $(x + 4)^2 + (y - 3)^2 = 4$

141. $(x + 8)^2 + (y + 5)^2 = 5$

143.

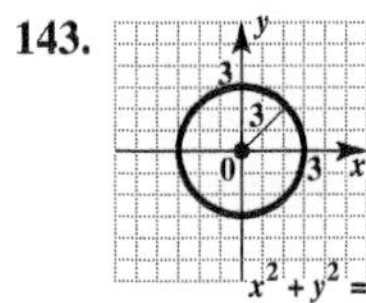

center: $(0, 0)$; radius: 3

145.

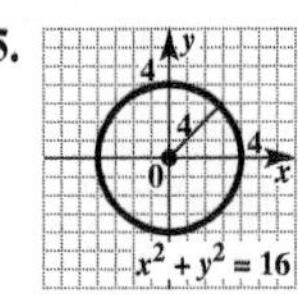

center: $(0, 0)$; radius: 4

147.

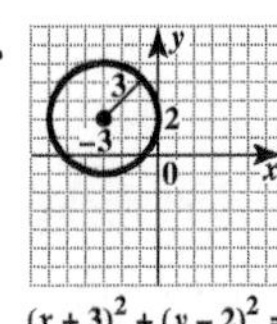

center: $(-3, 2)$; radius: 3

149. $(x - 2)^2 + (y - 3)^2 = 4$

center: $(2, 3)$; radius: 2

151. 42.0 in. **153.** 581

155. **(a)** $d = 1.224\sqrt{h}$ **(b)** 15.3 mi

157. $s = 13$ units **158.** $6\sqrt{13}$ sq. units

159. $h = \sqrt{13}$ units **160.** $3\sqrt{13}$ sq. units

161. $6\sqrt{13}$ sq. units **162.** The answers are equal, both $6\sqrt{13}$ sq. units, as expected.

Section 7.4

1. B **3.** The terms 3 and $3xy$ are not like terms and cannot be combined. The expression $(3 + 3xy)\sqrt[3]{xy^2}$ cannot be simplified further. **5.** -4 **7.** $8\sqrt{10}$ **9.** $-\sqrt{5}$ **11.** $7\sqrt{3}$

13. The expression cannot be simplified further. **15.** $14\sqrt[3]{2}$

17. $24\sqrt{2}$ **19.** $20\sqrt{5}$ **21.** $4\sqrt{2x}$ **23.** $-11m\sqrt{2}$ **25.** $5\sqrt[4]{2}$

27. $7\sqrt[3]{2}$ **29.** $2\sqrt[3]{x}$ **31.** $-7\sqrt[3]{x^2y}$ **33.** $-x\sqrt[3]{xy^2}$ **35.** $19\sqrt[4]{2}$

37. $x\sqrt[4]{xy}$ **39.** $9\sqrt[4]{2a^3}$ **41.** $(4 + 3xy)\sqrt[3]{xy^2}$ **43.** $4t\sqrt[3]{3st} - 3s\sqrt{3st}$

45. $4x\sqrt[3]{x} + 6x\sqrt[4]{x}$ **47.** $2\sqrt{2} - 2$ **49.** $\frac{5\sqrt{5}}{6}$ **51.** $\frac{7\sqrt{2}}{6}$

53. $\frac{5\sqrt{2}}{3}$ **55.** $5\sqrt{2} + 4$ **57.** $\frac{5 + 3x}{x^4}$ **59.** $\frac{30\sqrt{2} - 21}{14}$

61. $\frac{m\sqrt[3]{m^2}}{2}$ **63.** $\frac{3x\sqrt[3]{2} - 4\sqrt[3]{5}}{x^3}$ **65.** **(a)** $\sqrt{7}$ **(b)** 2.645751311 **(c)** 2.645751311 **(d)** equal **67.** A; 42 m **69.** $(12\sqrt{5} + 5\sqrt{3})$ in.

71. $(24\sqrt{2} + 12\sqrt{3})$ in. **73.** $2\sqrt{106} + 4\sqrt{2}$

Section 7.5

1. E **3.** A **5.** D **7.** $3\sqrt{6} + 2\sqrt{3}$ **9.** $6 - 4\sqrt{3}$

11. $20\sqrt{2}$ **13.** $6 - \sqrt{6}$ **15.** $\sqrt{6} + \sqrt{2} + \sqrt{3} + 1$

17. -1 **19.** 6 **21.** $\sqrt{22} + \sqrt{55} - \sqrt{14} - \sqrt{35}$ **23.** $8 - \sqrt{15}$

25. $9 + 4\sqrt{5}$ **27.** $26 - 2\sqrt{105}$ **29.** $4 - \sqrt[3]{36}$

31. 10 **33.** $6x + 3\sqrt{x} - 2\sqrt{5x} - \sqrt{5}$ **35.** $9r - s$

37. $4\sqrt[3]{4y^2} - 19\sqrt[3]{2y} - 5$ **39.** $3x - 4$ **41.** Because 6 and $4\sqrt{3}$ are not like terms, they cannot be combined. The expression $6 - 4\sqrt{3}$ cannot be simplified further. **43.** $\sqrt{7}$ **45.** $5\sqrt{3}$ **47.** $\frac{\sqrt{6}}{2}$

49. $\frac{9\sqrt{15}}{5}$ **51.** $-\frac{7\sqrt{3}}{12}$ **53.** $\frac{\sqrt{14}}{2}$ **55.** $-\frac{\sqrt{14}}{10}$ **57.** $\frac{2\sqrt{6x}}{x}$

59. $\frac{-8\sqrt{3k}}{k}$ **61.** $\frac{-5m^2\sqrt{6mn}}{n^2}$ **63.** $\frac{12x^3\sqrt{2xy}}{y^5}$ **65.** $\frac{5\sqrt{2my}}{y^2}$

67. $-\frac{4k\sqrt{3z}}{z}$ **69.** $\frac{\sqrt[3]{18}}{3}$ **71.** $\frac{\sqrt[3]{12}}{3}$ **73.** $\frac{\sqrt[3]{18}}{4}$ **75.** $-\frac{\sqrt[3]{2pr}}{r}$

77. $\frac{x^2\sqrt[3]{y^2}}{y}$ **79.** $\frac{2\sqrt[4]{x^3}}{x}$ **81.** $\frac{\sqrt[4]{2yz^3}}{z}$ **83.** $\frac{3(4 - \sqrt{5})}{11}$

85. $3(\sqrt{5} - \sqrt{3})$ **87.** $\frac{6\sqrt{2} + 4}{7}$ **89.** $\frac{2(3\sqrt{5} - 2\sqrt{3})}{33}$

91. $2\sqrt{3} + \sqrt{10} - 3\sqrt{2} - \sqrt{15}$ **93.** $\sqrt{m} - 2$

95. $\dfrac{4(\sqrt{x}+2\sqrt{y})}{x-4y}$ **97.** $\dfrac{x-2\sqrt{xy}+y}{x-y}$ **99.** $\dfrac{5\sqrt{k}(2\sqrt{k}-\sqrt{q})}{4k-q}$

101. $3+2\sqrt{6}$ **103.** $1-\sqrt{5}$ **105.** $\dfrac{4-2\sqrt{2}}{3}$ **107.** $\dfrac{6+2\sqrt{6p}}{3}$

109. $\dfrac{3\sqrt{x+y}}{x+y}$ **111.** $\dfrac{p\sqrt{p+2}}{p+2}$ **113.** Each expression is approximately equal to 0.2588190451. **115.** $\dfrac{33}{8(6+\sqrt{3})}$ **116.** $\dfrac{11}{2(2\sqrt{5}+3)}$

117. $\dfrac{4x-y}{3x(2\sqrt{x}+\sqrt{y})}$ **118.** $\dfrac{p-9q}{4q(\sqrt{p}+3\sqrt{q})}$

SUMMARY EXERCISES Performing Operations with Radicals and Rational Exponents

1. The radicand is a fraction, $\frac{2}{5}$. **2.** The exponent in the radicand and the index of the radical have greatest common factor 5. **3.** The denominator contains a radical, $\sqrt[3]{10}$. **4.** The radicand has two factors, x and y, that are raised to powers greater than the index, 3. **5.** $-6\sqrt{10}$ **6.** $7-\sqrt{14}$ **7.** $2+\sqrt{6}-2\sqrt{3}-3\sqrt{2}$ **8.** $4\sqrt{2}$ **9.** $73+12\sqrt{35}$ **10.** $\dfrac{-\sqrt{6}}{2}$ **11.** $4(\sqrt{7}-\sqrt{5})$ **12.** $-3+2\sqrt{2}$ **13.** -44 **14.** $\dfrac{\sqrt{x}+\sqrt{5}}{x-5}$ **15.** $2abc^3\sqrt[3]{b^2}$ **16.** $5\sqrt[3]{3}$ **17.** $3(\sqrt{5}-2)$ **18.** $\dfrac{\sqrt{15x}}{5x}$ **19.** $\frac{8}{5}$ **20.** $\dfrac{\sqrt{2}}{8}$ **21.** $-\sqrt[3]{100}$ **22.** $11+2\sqrt{30}$ **23.** $-3\sqrt{3x}$ **24.** $52-30\sqrt{3}$ **25.** $\dfrac{\sqrt[3]{117}}{9}$ **26.** $3\sqrt{2}+\sqrt{15}+\sqrt{42}+\sqrt{35}$ **27.** $2\sqrt[4]{27}$ **28.** $\dfrac{x\sqrt[3]{x^2}}{y}$ **29.** $-4\sqrt{3}-3$ **30.** $7+4\cdot 3^{1/2}$, or $7+4\sqrt{3}$ **31.** $3\sqrt[3]{2x^2}$ **32.** -2 **33.** $3^{5/6}$ **34.** $\dfrac{x^{5/3}}{y}$ **35.** $xy^{6/5}$ **36.** $x^{10}y$ **37.** $\dfrac{1}{25x^2}$ **38.** $\dfrac{-6y^{1/6}}{x^{1/24}}$

Section 7.6

1. (a) yes **(b)** no **3. (a)** yes **(b)** no **5.** There is no solution. The radical expression, which is nonnegative, cannot equal a negative number. The solution set is $\varnothing$. **7.** $\{11\}$ **9.** $\left\{\frac{1}{3}\right\}$ **11.** $\varnothing$ **13.** $\{5\}$ **15.** $\{18\}$ **17.** $\{5\}$ **19.** $\{4\}$ **21.** $\{17\}$ **23.** $\{5\}$ **25.** $\varnothing$ **27.** $\{0\}$ **29.** $\{1\}$ **31.** $\{-1, 3\}$ **33.** $\{0\}$ **35.** $\varnothing$ **37.** $\{1\}$ **39.** We cannot just square each term. The right side should be $(8-x)^2 = 64-16x+x^2$. The correct first step is $3x+4 = 64-16x+x^2$. The solution set is $\{4\}$. **41.** $\{7\}$ **43.** $\{7\}$ **45.** $\{4, 20\}$ **47.** $\varnothing$ **49.** $\left\{\frac{5}{4}\right\}$ **51.** $\{9\}$ **53.** $\{1\}$ **55.** $\{14\}$ **57.** $\{-1\}$ **59.** $\{8\}$ **61.** $\{0\}$ **63.** $\varnothing$ **65.** $\{-4\}$ **67.** $\{9, 17\}$ **69.** $\left\{\frac{1}{4}, 1\right\}$ **71.** $L = CZ^2$ **73.** $K = \dfrac{V^2m}{2}$ **75.** $M = \dfrac{r^2F}{m}$ **77. (a)** $r = \dfrac{a}{4\pi^2N^2}$ **(b)** $a = 4\pi^2N^2r$

Section 7.7

1. nonreal complex, complex **3.** real, complex **5.** pure imaginary, nonreal complex, complex **7.** i **9.** -1 **11.** $-i$ **13.** $13i$ **15.** $-12i$ **17.** $i\sqrt{5}$ **19.** $4i\sqrt{3}$ **21.** It is incorrect to use the product rule for radicals before using the definition of $\sqrt{-b}$. The correct product is -15. **23.** -15 **25.** $-\sqrt{105}$ **27.** -10 **29.** $i\sqrt{33}$ **31.** $5i\sqrt{6}$ **33.** $\sqrt{3}$ **35.** $5i$ **37.** -2 **39.** $-1+7i$ **41.** 0 **43.** $7+3i$ **45.** -2 **47.** $1+13i$ **49.** $6+6i$ **51.** $4+2i$ **53.** -81 **55.** -16 **57.** $-10-30i$ **59.** $10-5i$ **61.** $-9+40i$ **63.** $-16+30i$ **65.** 153 **67.** 97 **69.** 4 **71. (a)** $a-bi$ **(b)** a^2; b^2 **73.** $1+i$ **75.** $2+2i$ **77.** $-1+2i$ **79.** $-\frac{5}{13}-\frac{12}{13}i$ **81.** $1-3i$ **83.** $1+3i$ **85.** -1 **87.** i **89.** -1 **91.** $-i$ **93.** $-i$ **95.** 1 **97.** Because $i^{20} = (i^4)^5 = 1^5 = 1$, the student multiplied by 1, which is justified by the identity property for multiplication. **99.** $\frac{1}{2}+\frac{1}{2}i$ **101.** Substitute $1+5i$ for x in the equation. A true statement results—that is, $(1+5i)^2 - 2(1+5i) + 26$ will simplify to 0 when the operations are applied. Thus, $1+5i$ is a solution. **102.** Substituting $1-5i$ for x in the equation results in a true statement, indicating that $1-5i$ is a solution. **103.** They are complex conjugates. **104.** Substituting $3+2i$ for x in the equation results in a true statement, indicating that $3+2i$ is a solution. **105.** $3-2i$; Substituting $3-2i$ for x in the equation results in a true statement, indicating that $3-2i$ is a solution.

Chapter 7 Review Exercises

1. 42 **2.** -17 **3.** 6 **4.** -5 **5.** -3 **6.** -2 **7.** $|x|$ **8.** x **9.** $|x|^5$ (or $|x^5|$) **10.** $\sqrt[n]{a}$ is not a real number if n is even and a is negative.

11.

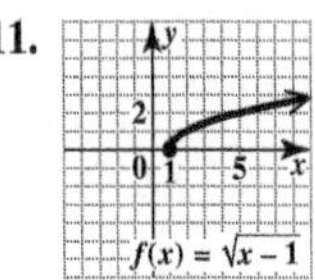

domain: $[1, \infty)$; range: $[0, \infty)$

12.

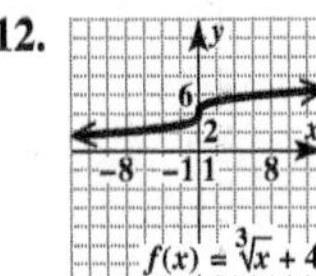

domain: $(-\infty, \infty)$; range: $(-\infty, \infty)$

13. -6.856 **14.** -5.053 **15.** 4.960 **16.** 4.729 **17.** -7.937 **18.** -5.292 **19.** 1.9 sec **20.** 66 in.^{2} **21.** 7 **22.** -11 **23.** 32 **24.** -4 **25.** $-\frac{216}{125}$ **26.** -32 **27.** $\frac{1000}{27}$ **28.** It is not a real number. **29.** $10\sqrt{x}$ **30.** $\dfrac{1}{(\sqrt[3]{3a+b})^5}$ **31.** $7^{9/2}$ **32.** 1331 **33.** r^4 **34.** z **35.** 25 **36.** 96 **37.** $a^{2/3}$ **38.** $\dfrac{1}{y^{1/2}}$ **39.** $\dfrac{z^{1/2}x^{8/5}}{4}$ **40.** $r^{1/2}+r$ **41.** $y^{8/15}$ **42.** $\dfrac{1}{x^{1/2}}$ **43.** $p^{1/2}$ **44.** $k^{9/4}$ **45.** $m^{13/3}$ **46.** $t^{3/2}$ **47.** $x^{1/8}$ **48.** $x^{1/15}$ **49.** $x^{1/36}$ **50.** The product rule for exponents applies only if the bases are the same. **51.** $\sqrt{66}$ **52.** $\sqrt{5r}$ **53.** $\sqrt[3]{30}$ **54.** $\sqrt[4]{21}$ **55.** $2\sqrt{5}$ **56.** $5\sqrt{3}$ **57.** $-5\sqrt{5}$ **58.** $-3\sqrt[3]{4}$ **59.** $10y^3\sqrt{y}$ **60.** $4pq^2\sqrt[3]{p}$ **61.** $3a^2b\sqrt[3]{4a^2b^2}$

62. $2r^2t\sqrt[3]{79r^2t}$ **63.** $\frac{y\sqrt{y}}{12}$ **64.** $\frac{m^5}{3}$ **65.** $\frac{\sqrt[3]{r^2}}{2}$ **66.** $\frac{a^2\sqrt[4]{a}}{3}$

67. $\sqrt{15}$ **68.** $p\sqrt{p}$ **69.** $\sqrt[12]{2000}$ **70.** $\sqrt[10]{x^7}$ **71.** 10

72. $\sqrt{197}$ **73.** $x^2 + y^2 = 121$ **74.** $(x+2)^2 + (y-4)^2 = 9$

75. $(x+1)^2 + (y+3)^2 = 25$ **76.** $(x-4)^2 + (y-2)^2 = 36$

77.

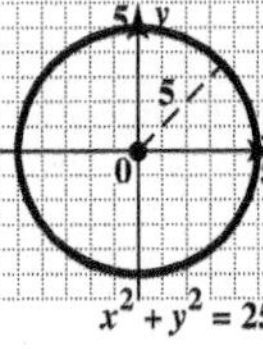

center: $(0, 0)$; radius: 5

78.

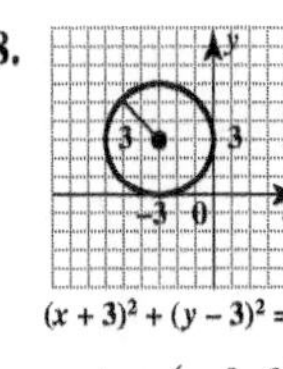

center: $(-3, 3)$; radius: 3

79.

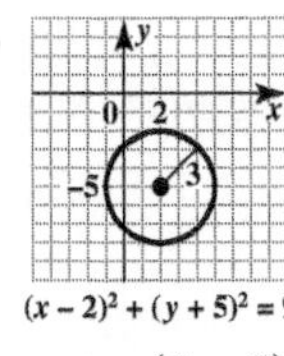

center: $(2, -5)$; radius: 3

80. It is impossible for the sum of the squares of two real numbers to be negative. **81.** $-11\sqrt{2}$ **82.** $23\sqrt{5}$ **83.** $7\sqrt{3y}$ **84.** $26m\sqrt{6m}$

85. $19\sqrt[3]{2}$ **86.** $-8\sqrt[4]{2}$ **87.** $1 - \sqrt{3}$ **88.** 2 **89.** $9 - 7\sqrt{2}$

90. $15 - 2\sqrt{26}$ **91.** 29 **92.** $2\sqrt[3]{2y^2} + 2\sqrt[3]{4y} - 3$ **93.** $\frac{\sqrt{30}}{5}$

94. $-3\sqrt{6}$ **95.** $\frac{3\sqrt{7py}}{y}$ **96.** $\frac{\sqrt{22}}{4}$ **97.** $-\frac{\sqrt[3]{45}}{5}$ **98.** $\frac{3m\sqrt[3]{4n}}{n^2}$

99. $\frac{\sqrt{2} - \sqrt{7}}{-5}$ **100.** $\frac{5(\sqrt{6} + 3)}{3}$ **101.** $\frac{1 - \sqrt{5}}{4}$ **102.** $\frac{-6 + \sqrt{3}}{2}$

103. $\{2\}$ **104.** $\{6\}$ **105.** $\{0, 5\}$ **106.** $\{9\}$ **107.** $\{3\}$ **108.** $\{7\}$

109. $\{-\frac{1}{2}\}$ **110.** $\{14\}$ **111.** $\varnothing$ **112.** $\{7\}$ **113.** $H = \sqrt{L^2 - W^2}$

114. 7.9 ft **115.** $4i$ **116.** $10i\sqrt{2}$ **117.** $-10 - 2i$ **118.** $14 + 7i$

119. $-\sqrt{35}$ **120.** -45 **121.** 3 **122.** $5 + i$ **123.** $32 - 24i$

124. $1 - i$ **125.** $-i$ **126.** 1 **127.** -1 **128.** 1

Chapter 7 Mixed Review Exercises

1. $-13ab^2$ **2.** $\frac{1}{100}$ **3.** $\frac{1}{z^{3/5}}$ **4.** $3z^3t^2\sqrt[3]{2t^2}$ **5.** $7i$ **6.** $-\frac{\sqrt{3}}{6}$

7. $\frac{\sqrt[3]{60}}{5}$ **8.** 1 **9.** $57\sqrt{2}$ **10.** $5 - 11i$ **11.** $-5i$ **12.** $\frac{1 + \sqrt{6}}{2}$

13. $5 + 12i$ **14.** $6x\sqrt[3]{y^2}$ **15.** $35 + 15i$ **16.** $\sqrt[12]{2000}$

17. $\frac{2\sqrt{z}(\sqrt{z} + 2)}{z - 4}$ **18.** $(12\sqrt{3} + 5\sqrt{2})$ ft **19.** $\{5\}$

20. $\{-4\}$ **21.** $\{\frac{3}{2}\}$ **22.** $\{7\}$

Chapter 7 Test

[7.1] **1.** -29 **2.** -8 [7.2] **3.** 5 [7.1] **4.** 21.863 **5.** -9.405

6.

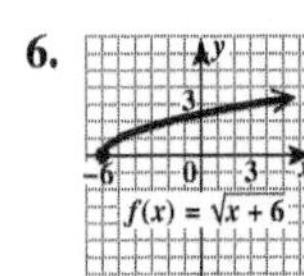

domain: $[-6, \infty)$; range: $[0, \infty)$

[7.2] **7.** $\frac{125}{64}$ **8.** $\frac{1}{256}$ **9.** $\frac{9y^{3/10}}{x^2}$ **10.** $x^{4/3}y^6$

11. $7^{1/2}$ **12.** $a^{11/3}$ [7.3] **13.** $\sqrt{145}$

14. 10 **15.** $(x+4)^2 + (y-6)^2 = 25$

16. $(x-2)^2 + (y+3)^2 = 16$; center: $(2, -3)$; radius: 4 **17.** $3x^2y^3\sqrt{6x}$ **18.** $2ab^3\sqrt[4]{2a^3b}$ **19.** $\sqrt[6]{200}$

[7.4] **20.** $26\sqrt{5}$ **21.** $(2ts - 3t^2)\sqrt[3]{2s^2}$

[7.5] **22.** $66 + \sqrt{5}$ **23.** $23 - 4\sqrt{15}$

24. $-\frac{\sqrt{10}}{4}$ **25.** $\frac{2\sqrt[3]{25}}{5}$

26. $-2(\sqrt{7} - \sqrt{5})$ **27.** $3 + \sqrt{6}$ [7.6] **28.** **(a)** 59.8

(b) $T = \frac{V_0^2 - V^2}{-V^2k}$, or $T = \frac{V^2 - V_0^2}{V^2k}$ **29.** $\{-1\}$ **30.** $\{3\}$

31. $\{-3\}$ [7.7] **32.** $-5 - 8i$ **33.** $-2 + 16i$ **34.** $3 + 4i$

35. i **36.** **(a)** true **(b)** true **(c)** false **(d)** true

Chapters R–7 Cumulative Review Exercises

[R.4] **1.** -47 **2.** $\frac{43}{10}$ [R.3] **3.** $\frac{5}{24}$ [1.1] **4.** $\{-4\}$ **5.** $\{-12\}$

6. $\{6\}$ [1.7] **7.** $\{-\frac{10}{3}, 1\}$ [1.5] **8.** $(-6, \infty)$ **9.** $(-2, 1)$

[1.3] **10.** 36 nickels; 64 quarters **11.** $2\frac{2}{39}$ L

[2.1] **12.**

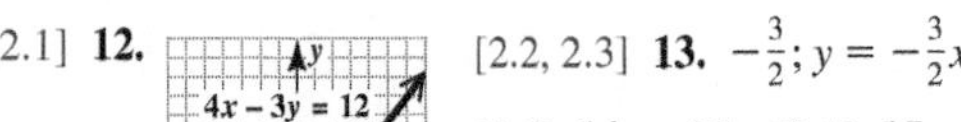

[2.2, 2.3] **13.** $-\frac{3}{2}$; $y = -\frac{3}{2}x$

[2.6] **14.** -37 [3.1] **15.** $\{(7, -2)\}$

[3.2] **16.** $\{(-1, 1, 1)\}$

[3.3] **17.** 2-oz letter: $0.70; 3-oz letter: $0.91

[4.3] **18.** $-k^3 - 3k^2 - 8k - 9$ [4.5] **19.** $8x^2 + 17x - 21$

[4.6] **20.** $3y^3 - 3y^2 + 4y + 1 + \frac{-10}{2y + 1}$

[5.2, 5.3] **21.** $(2p - 3q)(p - q)$ **22.** $(3k^2 + 4)(k - 1)(k + 1)$

23. $(x + 8)(x^2 - 8x + 64)$ [6.1] **24.** $\{x \mid x \text{ is a real number}, x \neq \pm 3\}$

25. $\frac{y}{y + 5}$ [6.2] **26.** $\frac{4x + 2y}{(x + y)(x - y)}$ [6.3] **27.** $-\frac{9}{4}$ **28.** $\frac{-1}{a + b}$

29. $\frac{1}{xy - 1}$ [5.5] **30.** $\{-3, -\frac{5}{2}\}$ **31.** $\{-\frac{2}{5}, 1\}$ [6.4] **32.** $\varnothing$

[7.6] **33.** $\{3, 4\}$ [6.5] **34.** Danielle: 8 mph; Richard: 4 mph

[7.2] **35.** $\frac{1}{9}$ [7.3] **36.** $2x\sqrt[3]{6x^2y^2}$ [7.4] **37.** $7\sqrt{2}$

[7.5] **38.** $\frac{\sqrt{10} + 2\sqrt{2}}{2}$ [7.3] **39.** $\sqrt{29}$ [7.7] **40.** $4 + 2i$

8 QUADRATIC EQUATIONS AND INEQUALITIES

Section 8.1

1. quadratic; second; two **3.** The zero-factor property requires a product equal to 0. The first step should have been to write the equation with 0 on one side. The solution set is $\{-2, 3\}$. **5.** $\{-2, -1\}$

7. $\{-3, \frac{1}{3}\}$ **9.** $\{-\frac{3}{5}, 2\}$ **11.** $\{2, 4\}$ **13.** $\{\frac{1}{2}, 4\}$ **15.** $\{-\frac{3}{2}, -\frac{1}{4}\}$

17. $\{\pm 9\}$ **19.** $\{\pm\sqrt{17}\}$ **21.** $\{\pm 4\sqrt{2}\}$ **23.** $\{\pm 2\sqrt{5}\}$

25. $\{\pm 2\sqrt{6}\}$ **27.** $\{\pm 3\sqrt{3}\}$ **29.** $\{\pm 4\sqrt{2}\}$ **31.** $\{-7, 3\}$

33. $\{-1, 13\}$ **35.** $\{4 \pm \sqrt{3}\}$ **37.** $\{-5 \pm 4\sqrt{3}\}$

39. $\left\{\frac{1 \pm \sqrt{7}}{3}\right\}$ **41.** $\left\{\frac{-1 \pm 2\sqrt{6}}{4}\right\}$ **43.** $\left\{\frac{2 \pm 2\sqrt{3}}{5}\right\}$

45. 5.6 sec **47.** Solve $(2x + 1)^2 = 5$ by the square root property. Solve $x^2 + 4x = 12$ by completing the square. **49.** 9; $(x + 3)^2$

51. 36; $(p - 6)^2$ **53.** $\frac{81}{4}$; $\left(q + \frac{9}{2}\right)^2$ **55.** $\{-4, 6\}$

57. $\{-2 \pm \sqrt{6}\}$ **59.** $\{-5 \pm \sqrt{7}\}$ **61.** $\left\{-\frac{8}{3}, 3\right\}$

63. $\left\{\frac{-7 \pm \sqrt{53}}{2}\right\}$ **65.** $\left\{\frac{-5 \pm \sqrt{41}}{4}\right\}$ **67.** $\left\{\frac{5 \pm \sqrt{15}}{5}\right\}$

69. $\left\{\frac{4 \pm \sqrt{3}}{3}\right\}$ **71.** $\left\{\frac{2 \pm \sqrt{3}}{3}\right\}$ **73.** $\{1 \pm \sqrt{2}\}$

75. $\{\pm 10i\}$ **77.** $\{\pm 2i\sqrt{3}\}$ **79.** $\{-3 \pm 2i\}$

81. $\{5 \pm i\sqrt{3}\}$ **83.** $\left\{\frac{1}{6} \pm \frac{\sqrt{2}}{3}i\right\}$ **85.** $\{-2 \pm 3i\}$

87. $\{-3 \pm i\sqrt{3}\}$ **89.** $\left\{-\frac{2}{3} \pm \frac{2\sqrt{2}}{3}i\right\}$ **91.** $\left\{-\frac{5}{2} \pm \frac{\sqrt{15}}{2}i\right\}$

93. $\{\pm\sqrt{b}\}$ **95.** $\left\{\pm\frac{\sqrt{b^2 + 16}}{2}\right\}$ **97.** $\left\{\frac{2b \pm \sqrt{3a}}{5}\right\}$

99. x^2 **100.** x **101.** $6x$ **102.** 1 **103.** 9 **104.** $(x + 3)^2$, or $x^2 + 6x + 9$

Section 8.2

1. No. The fraction bar should extend under the term $-b$. The correct formula is $x = \frac{-b \pm \sqrt{b^2 - 4ac}}{2a}$. **3.** The last step is wrong. Because 5 is not a common factor in the numerator, the fraction cannot be simplified. The solution set is $\left\{\frac{5 \pm \sqrt{5}}{10}\right\}$. **5.** $\{3, 5\}$ **7.** $\left\{-\frac{5}{2}, \frac{2}{3}\right\}$

9. $\left\{-\frac{3}{2}\right\}$ **11.** $\left\{\frac{1}{6}\right\}$ **13.** $\left\{\frac{-2 \pm \sqrt{2}}{2}\right\}$ **15.** $\left\{\frac{1 \pm \sqrt{3}}{2}\right\}$

17. $\{5 \pm \sqrt{7}\}$ **19.** $\left\{\frac{-1 \pm \sqrt{2}}{2}\right\}$ **21.** $\left\{\frac{-1 \pm \sqrt{7}}{3}\right\}$

23. $\{1 \pm \sqrt{5}\}$ **25.** $\left\{\frac{-2 \pm \sqrt{10}}{2}\right\}$ **27.** $\{-1 \pm 3\sqrt{2}\}$

29. $\left\{\frac{1 \pm \sqrt{29}}{2}\right\}$ **31.** $\left\{\frac{-4 \pm \sqrt{91}}{3}\right\}$ **33.** $\left\{\frac{-3 \pm \sqrt{57}}{8}\right\}$

35. $\left\{\frac{3}{2} \pm \frac{\sqrt{15}}{2}i\right\}$ **37.** $\{3 \pm i\sqrt{5}\}$ **39.** $\left\{\frac{1}{2} \pm \frac{\sqrt{6}}{2}i\right\}$

41. $\left\{-\frac{2}{3} \pm \frac{\sqrt{2}}{3}i\right\}$ **43.** $\left\{\frac{1}{2} \pm \frac{1}{4}i\right\}$ **45.** $\left\{\frac{4}{5} \pm \frac{2\sqrt{6}}{5}i\right\}$

47. 0; B; zero-factor property **49.** 8; C; quadratic formula **51.** 49; A; zero-factor property **53.** -80; D; quadratic formula

55. **(a)** 25; zero-factor property; $\left\{-3, -\frac{4}{3}\right\}$ **(b)** 44; quadratic formula; $\left\{\frac{7 \pm \sqrt{11}}{2}\right\}$ **57.** -10 or 10 **59.** 16 **61.** 25 **63.** $b = \frac{44}{5}; \frac{3}{10}$

Section 8.3

1. Multiply by the LCD, x. **3.** Substitute a variable for $x^2 + x$. **5.** The proposed solution -1 does not check. The solution set is $\{4\}$. **7.** $\{-2, 7\}$ **9.** $\{-4, 7\}$ **11.** $\left\{-\frac{2}{3}, 1\right\}$ **13.** $\left\{-\frac{14}{17}, 5\right\}$

15. $\left\{-\frac{11}{7}, 0\right\}$ **17.** $\left\{\frac{-1 \pm \sqrt{13}}{2}\right\}$ **19.** $\left\{-\frac{8}{3}, -1\right\}$

21. $\left\{\frac{2 \pm \sqrt{22}}{3}\right\}$ **23.** $\left\{\frac{-1 \pm \sqrt{5}}{4}\right\}$ **25.** **(a)** $(20 - t)$ mph **(b)** $(20 + t)$ mph **27.** the rate of her boat in still water; $x - 5$; $x + 5$; row 1 of table: 15, $x - 5$, $\frac{15}{x - 5}$; row 2 of table: 15, $x + 5$, $\frac{15}{x + 5}$; $\frac{15}{x - 5} + \frac{15}{x + 5} = 4$; 10 mph **29.** 25 mph **31.** 50 mph

33. 3.6 hr **35.** Rusty: 25.0 hr; Nancy: 23.0 hr **37.** 9 min

39. $\{2, 5\}$ **41.** $\{3\}$ **43.** $\left\{\frac{8}{9}\right\}$ **45.** $\{9\}$ **47.** $\left\{\frac{2}{5}\right\}$ **49.** $\{-2\}$

51. $\{\pm 2, \pm 5\}$ **53.** $\left\{\pm 1, \pm\frac{3}{2}\right\}$ **55.** $\{\pm 2, \pm 2\sqrt{3}\}$

57. $\{-6, -5\}$ **59.** $\left\{-\frac{16}{3}, -2\right\}$ **61.** $\{-8, 1\}$ **63.** $\{-64, 27\}$

65. $\left\{\pm 1, \pm\frac{27}{8}\right\}$ **67.** $\left\{-\frac{1}{3}, \frac{1}{6}\right\}$ **69.** $\left\{-\frac{1}{2}, 3\right\}$ **71.** $\left\{\pm\frac{\sqrt{6}}{3}, \pm\frac{1}{2}\right\}$

73. $\{3, 11\}$ **75.** $\{25\}$ **77.** $\left\{-\sqrt[3]{5}, -\frac{\sqrt[3]{4}}{2}\right\}$ **79.** $\left\{\frac{4}{3}, \frac{9}{4}\right\}$

81. $\left\{\pm\frac{\sqrt{9 + \sqrt{65}}}{2}, \pm\frac{\sqrt{9 - \sqrt{65}}}{2}\right\}$ **83.** $\left\{\pm 1, \pm\frac{\sqrt{6}}{2}i\right\}$

SUMMARY EXERCISES Applying Methods for Solving Quadratic Equations

1. square root property **2.** zero-factor property **3.** quadratic formula **4.** quadratic formula **5.** zero-factor property **6.** square root property **7.** $\{\pm\sqrt{7}\}$ **8.** $\left\{-\frac{3}{2}, \frac{5}{3}\right\}$ **9.** $\{-3 \pm \sqrt{5}\}$ **10.** $\{-2, 8\}$

11. $\left\{-\frac{3}{2}, 4\right\}$ **12.** $\left\{-3, \frac{1}{3}\right\}$ **13.** $\left\{\frac{2 \pm \sqrt{2}}{2}\right\}$ **14.** $\{\pm 2i\sqrt{3}\}$

15. $\left\{\frac{1}{2}, 2\right\}$ **16.** $\{\pm 1, \pm 3\}$ **17.** $\left\{\frac{-3 \pm 2\sqrt{2}}{2}\right\}$ **18.** $\left\{\frac{4}{5}, 3\right\}$

19. $\{\pm\sqrt{2}, \pm\sqrt{7}\}$ **20.** $\left\{\frac{1 \pm \sqrt{5}}{4}\right\}$ **21.** $\left\{-\frac{1}{2} \pm \frac{\sqrt{3}}{2}i\right\}$

22. $\left\{-\frac{\sqrt[3]{175}}{5}, 1\right\}$ **23.** $\left\{\frac{3}{2}\right\}$ **24.** $\left\{\frac{2}{3}\right\}$ **25.** $\{\pm 6\sqrt{2}\}$

26. $\left\{-\frac{2}{3}, 2\right\}$ **27.** $\{-4, 9\}$ **28.** $\{\pm 13\}$ **29.** $\left\{1 \pm \frac{\sqrt{3}}{3}i\right\}$

30. $\{3\}$ **31.** $\left\{\frac{1}{6} \pm \frac{\sqrt{47}}{6}i\right\}$ **32.** $\left\{-\frac{1}{3}, \frac{1}{6}\right\}$

Section 8.4

1. Find a common denominator, and then multiply both sides by the common denominator. **3.** Write it in standard form (with 0 on one side, in decreasing powers of w). **5.** $m = \sqrt{p^2 - n^2}$ **7.** $t = \frac{\pm\sqrt{dk}}{k}$

9. $r = \frac{\pm\sqrt{S\pi}}{2\pi}$ **11.** $d = \frac{\pm\sqrt{skI}}{I}$ **13.** $v = \frac{\pm\sqrt{kAF}}{F}$

15. $r = \frac{\pm\sqrt{3\pi Vh}}{\pi h}$ **17.** $t = \frac{-B \pm \sqrt{B^2 - 4AC}}{2A}$ **19.** $h = \frac{D^2}{k}$

21. $\ell = \frac{p^2 g}{k}$ **23.** $R = \frac{E^2 - 2pr \pm E\sqrt{E^2 - 4pr}}{2p}$

25. $r = \frac{5pc}{4}$ or $r = -\frac{2pc}{3}$ **27.** $I = \frac{-cR \pm \sqrt{c^2R^2 - 4cL}}{2cL}$

29. 7.9, 8.9, 11.9 **31.** eastbound ship: 80 mi; southbound ship: 150 mi **33.** 8 in., 15 in., 17 in. **35.** length: 24 ft; width: 10 ft **37.** 2 ft **39.** 7 m by 12 m **41.** 20 in. by 12 in. **43.** 1 sec and 8 sec **45.** 2.4 sec and 5.6 sec **47.** 9.2 sec **49.** It reaches its *maximum* height at 5 sec because this is the only time it reaches 400 ft. **51.** \$1.50 **53.** 0.035, or 3.5% **55.** 5.5 m per sec **57.** 5 or 14 **59. (a)** \$600 billion **(b)** \$597 billion; They are about the same. **61.** 2008 **63.** isosceles triangle; right triangle **64.** no; The formula for the Pythagorean theorem involves the *squares* of the sides, not the square roots.

65.

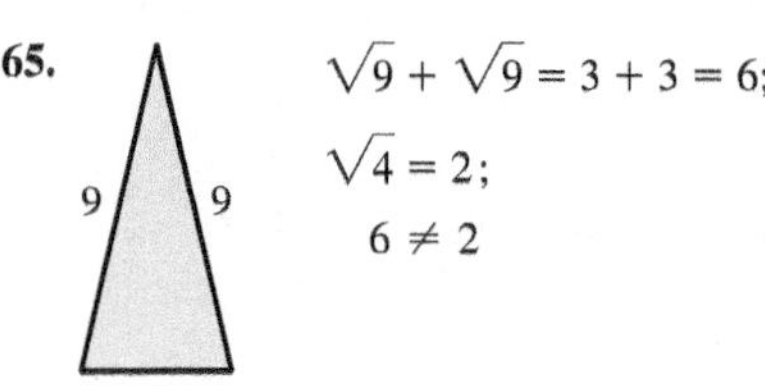

$\sqrt{9} + \sqrt{9} = 3 + 3 = 6;$
$\sqrt{4} = 2;$
$6 \neq 2$

66. The sum of the squares of the two shorter sides (legs) of a right triangle is equal to the square of the longest side (hypotenuse).

Section 8.5

1. B, C **3.** false **5.** true

7. $(-\infty, -1) \cup (5, \infty)$ **9.** $(-4, 6)$

11. $(-\infty, 1] \cup [3, \infty)$ **13.** $\left(-\infty, -\frac{3}{2}\right] \cup \left[\frac{3}{5}, \infty\right)$

15. $\left[-\frac{3}{2}, \frac{3}{2}\right]$ **17.** $\left(-\infty, -\frac{1}{2}\right] \cup \left[\frac{1}{3}, \infty\right)$

19. $(-\infty, 0] \cup [4, \infty)$ **21.** $\left[0, \frac{5}{3}\right]$

23. $(-\infty, 3 - \sqrt{3}] \cup [3 + \sqrt{3}, \infty)$ **25.** $(-\infty, \infty)$ **27.** $\varnothing$ **29.** $\varnothing$ **31.** $(-\infty, \infty)$

33. $(-\infty, 1) \cup (2, 4)$ **35.** $\left[-\frac{3}{2}, \frac{1}{3}\right] \cup [4, \infty)$

37. $(-\infty, 1) \cup (4, \infty)$ **39.** $\left[-\frac{3}{2}, 5\right)$

41. $(2, 6]$ **43.** $\left(-\infty, \frac{1}{2}\right) \cup \left(\frac{5}{4}, \infty\right)$

45. $[-7, -2)$ **47.** $(-\infty, 2) \cup (4, \infty)$

49. $\left(0, \frac{1}{2}\right) \cup \left(\frac{5}{2}, \infty\right)$ **51.** $\left[\frac{3}{2}, \infty\right)$

53. $\left(-2, \frac{5}{3}\right) \cup \left(\frac{5}{3}, \infty\right)$ **55.** 3 sec and 13 sec **56.** between 3 sec and 13 sec **57.** at 0 sec (the time when it is initially projected) and at 16 sec (the time when it hits the ground) **58.** between 0 and 3 sec and between 13 and 16 sec

Chapter 8 Review Exercises

1. $\{\pm 11\}$ **2.** $\{\pm\sqrt{3}\}$ **3.** $\left\{-\frac{15}{2}, \frac{5}{2}\right\}$ **4.** $\left\{\frac{2}{3} \pm \frac{5}{3}i\right\}$ **5.** $\{-2 \pm \sqrt{19}\}$ **6.** $\left\{\frac{1}{2}, 1\right\}$ **7.** 7.8 sec **8.** 5.9 sec **9.** $\left\{-\frac{7}{2}, 3\right\}$ **10.** $\left\{\frac{-5 \pm \sqrt{53}}{2}\right\}$ **11.** $\left\{\frac{1 \pm \sqrt{41}}{2}\right\}$ **12.** $\left\{-\frac{3}{4} \pm \frac{\sqrt{23}}{4}i\right\}$ **13.** $\left\{\frac{2}{3} \pm \frac{\sqrt{2}}{3}i\right\}$ **14.** $\left\{\frac{-7 \pm \sqrt{37}}{2}\right\}$ **15.** 17; C; quadratic formula **16.** 64; A; zero-factor property **17.** -92; D; quadratic formula **18.** 0; B; zero-factor property **19.** $\left\{-\frac{5}{2}, 3\right\}$ **20.** $\left\{-\frac{1}{2}, 1\right\}$ **21.** $\{-4\}$ **22.** $\left\{-\frac{11}{6}, -\frac{19}{12}\right\}$ **23.** $\left\{-\frac{343}{8}, 64\right\}$ **24.** $\{\pm 1, \pm 3\}$ **25.** 7 mph **26.** 40 mph **27.** 4.6 hr **28.** Zoran: 2.6 hr; Claude: 3.6 hr **29.** $v = \frac{\pm\sqrt{rFkw}}{kw}$ **30.** $y = \frac{6p^2}{z}$ **31.** $t = \frac{3m \pm \sqrt{9m^2 + 24m}}{2m}$

32. 9 ft, 12 ft, 15 ft **33.** 12 cm by 20 cm **34.** 1 in. **35.** 18 in. **36.** 3 min

37. 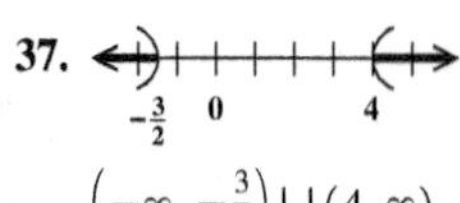

$\left(-\infty, -\frac{3}{2}\right) \cup (4, \infty)$

38. $[-4, 3]$

39. $(-\infty, -5] \cup [-2, 3]$

40. $\varnothing$

41. $\left(-\infty, \frac{1}{2}\right) \cup (2, \infty)$

42. $[-3, 2)$

Chapter 8 Mixed Review Exercises

1. $R = \dfrac{\pm\sqrt{Vh - r^2h}}{h}$ **2.** $\left\{1 \pm \dfrac{\sqrt{3}}{3}i\right\}$ **3.** $\left\{\dfrac{-11 \pm \sqrt{7}}{3}\right\}$ **4.** $d = \dfrac{\pm\sqrt{SkI}}{I}$ **5.** $(-\infty, \infty)$

6. $\{4\}$ **7.** $\{\pm\sqrt{4 + \sqrt{15}}, \pm\sqrt{4 - \sqrt{15}}\}$

8. $\left(-5, -\frac{23}{5}\right]$ **9.** $\left\{-\frac{5}{3}, -\frac{3}{2}\right\}$ **10.** $\{-2, -1, 3, 4\}$

11. $(-\infty, -6) \cup \left(-\frac{3}{2}, 1\right)$ **12.** 10 mph **13.** length: 2 cm; width: 1.5 cm **14.** 5.2 sec

Chapter 8 Test

[8.1] **1.** $\left\{-3, \frac{2}{5}\right\}$ **2.** $\{\pm 3\sqrt{6}\}$ **3.** $\left\{-\frac{8}{7}, \frac{2}{7}\right\}$ **4.** $\{-1 \pm \sqrt{5}\}$

[8.2] **5.** $\left\{\dfrac{3 \pm \sqrt{17}}{4}\right\}$ **6.** $\left\{\dfrac{2}{3} \pm \dfrac{\sqrt{11}}{3}i\right\}$ [8.3] **7.** $\left\{\frac{2}{3}\right\}$

[8.1] **8.** A [8.2] **9.** discriminant: 88; There are two irrational solutions.

[8.1–8.3] **10.** $\left\{-\frac{2}{3}, 6\right\}$ **11.** $\left\{\dfrac{-7 \pm \sqrt{97}}{8}\right\}$ **12.** $\left\{\pm\frac{1}{3}, \pm 2\right\}$

13. $\left\{-\frac{5}{2}, 1\right\}$ [8.4] **14.** $r = \dfrac{\pm\sqrt{\pi S}}{2\pi}$ [8.3] **15.** Terry: 11.1 hr; Callie: 9.1 hr **16.** 7 mph [8.4] **17.** 2 ft **18.** 16 m

[8.5] **19.** $(-\infty, -5) \cup \left(\frac{3}{2}, \infty\right)$ **20.** $(-\infty, 4) \cup [9, \infty)$

Chapters R–8 Cumulative Review Exercises

[R.1–R.3] **1.** −5.38 **2.** 2 **3.** 25 **4.** 0.72 [R.4, R.5] **5.** −45

[R.2, 7.7] **6.** **(a)** −2, 0, 7 **(b)** $-\frac{7}{3}, -2, 0, 0.7, 7, \frac{32}{3}$

(c) All are real except $\sqrt{-8}$. **(d)** All are complex numbers.

[1.1] **7.** $\left\{\frac{4}{5}\right\}$ [1.7] **8.** $\left\{\frac{11}{10}, \frac{7}{2}\right\}$ [7.6] **9.** $\left\{\frac{2}{3}\right\}$ [6.4] **10.** $\varnothing$

[8.1, 8.2] **11.** $\left\{\dfrac{7 \pm \sqrt{177}}{4}\right\}$ [8.3] **12.** $\{\pm 1, \pm 2\}$

[1.5] **13.** $[1, \infty)$ [1.7] **14.** $\left[2, \frac{8}{3}\right]$ [8.5] **15.** $(1, 3)$ **16.** $(-2, 1)$

[2.1, 2.5] **17.** function; domain: $(-\infty, \infty)$; range: $(-\infty, \infty)$; $f(x) = \frac{4}{5}x - 3$

[2.4, 2.5] **18.** not a function

[2.1, 2.2] **19.** $m = \frac{2}{7}$; x-intercept: $(-8, 0)$; y-intercept: $\left(0, \frac{16}{7}\right)$

[2.3] **20.** **(a)** $y = -\frac{5}{2}x + 2$ **(b)** $y = \frac{2}{5}x + \frac{13}{5}$ [3.1] **21.** $\{(1, -2)\}$

22. $\varnothing$ [3.2] **23.** $\{(3, -4, 2)\}$ [3.3] **24.** *Star Wars: The Last Jedi:* \$545 million; *Beauty and the Beast:* \$504 million [4.1] **25.** $\dfrac{x^8}{y^4}$

26. $\dfrac{4}{xy^2}$ [4.5] **27.** $\frac{4}{9}t^2 + 12t + 81$ [4.6] **28.** $4x^2 - 6x + 11 + \dfrac{4}{x + 2}$

[5.1–5.3] **29.** $(4m - 3)(6m + 5)$ **30.** $4(t + 5)(t - 5)$

31. $(2x + 3y)(4x^2 - 6xy + 9y^2)$ **32.** $(3x - 5y)^2$ [6.1] **33.** $-\frac{5}{18}$

[6.2] **34.** $-\dfrac{8}{x}$ [6.3] **35.** $\dfrac{r - s}{r}$ [8.4] **36.** southbound car: 57 mi; eastbound car: 76 mi [7.3] **37.** $\dfrac{3\sqrt[3]{4}}{4}$ [7.5] **38.** $\sqrt{7} + \sqrt{5}$

9 ADDITIONAL GRAPHS OF FUNCTIONS AND RELATIONS

Section 9.1

1. 8; 15; 23 (all in billions of dollars) **3.** 19; It represents the dollars in billions spent for space/other technologies. **5.** $P(x) = 50x - 800$ **7.** 24 **9.** −6 **11.** −1 **13.** 3 **15.** 364 **17.** −112 **19.** 1 **21.** $\frac{28}{13}$ **23.** 94 **25.** 13 **27.** 46 **29.** **(a)** $10x + 2$ **(b)** $-2x - 4$ **(c)** $24x^2 + 6x - 3$ **(d)** $\dfrac{4x - 1}{6x + 3}$; All domains are $(-\infty, \infty)$, except for $\frac{f}{g}$, which is $\left(-\infty, -\frac{1}{2}\right) \cup \left(-\frac{1}{2}, \infty\right)$. **31.** **(a)** $4x^2 - 4x + 1$ **(b)** $2x^2 - 1$ **(c)** $(3x^2 - 2x)(x^2 - 2x + 1)$ **(d)** $\dfrac{3x^2 - 2x}{x^2 - 2x + 1}$; All domains are $(-\infty, \infty)$, except for $\frac{f}{g}$, which is $(-\infty, 1) \cup (1, \infty)$.

33. **(a)** $2x + 5 + \sqrt{4x + 3}$ **(b)** $2x + 5 - \sqrt{4x + 3}$ **(c)** $(2x + 5)\sqrt{4x + 3}$ **(d)** $\dfrac{2x + 5}{\sqrt{4x + 3}}$; All domains are $\left[-\frac{3}{4}, \infty\right)$, except for $\frac{f}{g}$, which is $\left(-\frac{3}{4}, \infty\right)$. **35.** **(a)** 5 **(b)** 5 **(c)** 0 **(d)** undefined **37.** **(a)** $6h$ **(b)** 6 **39.** **(a)** $2xh + h^2$ **(b)** $2x + h$ **41.** **(a)** $4xh + 2h^2$ **(b)** $4x + 2h$ **43.** **(a)** $2xh + h^2 + 4h$ **(b)** $2x + h + 4$ **45.** −5 **47.** 7 **49.** 6 **51.** −1

In Exercises 53–63, we give $(f \circ g)(x)$, $(g \circ f)(x)$, and the domains.

53. $-30x - 33$; $-30x + 52$; Both domains are $(-\infty, \infty)$.

55. $-5x^2 + 20x + 18$; $-25x^2 - 10x + 6$; Both domains are $(-\infty, \infty)$.
57. $\sqrt{x+3}$; $\sqrt{x} + 3$; domain of $f \circ g$: $[-3, \infty)$; domain of $g \circ f$: $[0, \infty)$
59. $\frac{1}{x^2}$; $\frac{1}{x^2}$; Both domains are $(-\infty, 0) \cup (0, \infty)$. **61.** $2\sqrt{2x-1}$; $8\sqrt{x+2} - 6$; domain of $f \circ g$: $\left[\frac{1}{2}, \infty\right)$; domain of $g \circ f$: $[-2, \infty)$
63. $\frac{x}{2-5x}$; $2(x-5)$; domain of $f \circ g$: $(-\infty, 0) \cup \left(0, \frac{2}{5}\right) \cup \left(\frac{2}{5}, \infty\right)$; domain of $g \circ f$: $(-\infty, 5) \cup (5, \infty)$ **65.** 4 **67.** -3 **69.** 0 **71.** 1 **73.** 3 **75.** 1 **77.** 1 **79.** 9 **81.** 1 **83.** $g(1) = 9$, and $f(9)$ cannot be determined from the table.

Other correct answers are possible in Exercises 85–91.

85. $f(x) = x^2$; $g(x) = 6x - 2$ **87.** $f(x) = \sqrt{x}$; $g(x) = x^2 + 3$
89. $f(x) = \frac{1}{x+2}$; $g(x) = x^2$ **91.** $f(x) = 4x^3 + x + 5$; $g(x) = 2x - 3$
93. $(f \circ g)(x) = 5280x$; It computes the number of feet in x miles.
95. (a) $g(x) = 0.5x$ **(b)** $f(x) = x - 0.1x$ **(c)** $(f \circ g)(x) = 0.5x - 0.1(0.5x) = 0.45x$ **(d)** $(f \circ g)(80) = 36$; The sale price of the \$80 coat is \$36.
97. $D(c) = \frac{-c^2 + 10c - 25}{25} + 500$ **99.** 0; 0; 0 **100.** 1; 1; 1
101. $(f \circ g)(x) = g(x)$ and $(g \circ f)(x) = g(x)$. In each case, we obtain $g(x)$. **102.** $f(x) = x$ is called the identity function. **103.** $-a$; $-a$; $-a$
104. $\frac{1}{a}$; $\frac{1}{a}$; $\frac{1}{a}$ **105.** $(f \circ g)(x) = x$ and $(g \circ f)(x) = x$. In each case, we obtain x. **106.** f and g are inverses.

Section 9.2

1. (a) B **(b)** C **(c)** A **(d)** D **3. (a)** D **(b)** B **(c)** F **(d)** C **(e)** A **(f)** E **5.** $(0, 0)$ **7.** $(0, 0)$ **9.** $(0, 4)$ **11.** $(1, 0)$ **13.** $(-3, -4)$ **15.** $(5, 6)$ **17.** down; wider **19.** up; narrower **21.** down; narrower

23.

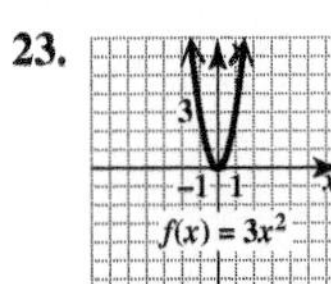

vertex: $(0, 0)$; axis: $x = 0$; domain: $(-\infty, \infty)$; range: $[0, \infty)$

25.

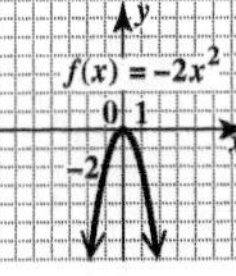

vertex: $(0, 0)$; axis: $x = 0$; domain: $(-\infty, \infty)$; range: $(-\infty, 0]$

27.

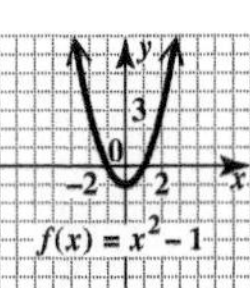

vertex: $(0, -1)$; axis: $x = 0$; domain: $(-\infty, \infty)$; range: $[-1, \infty)$

29.

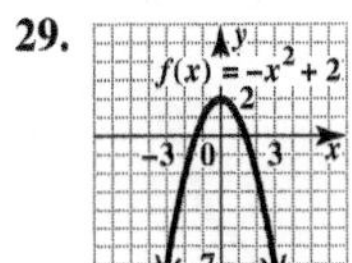

vertex: $(0, 2)$; axis: $x = 0$; domain: $(-\infty, \infty)$; range: $(-\infty, 2]$

31. $f(x) = (x - 4)^2$

vertex: $(4, 0)$; axis: $x = 4$; domain: $(-\infty, \infty)$; range: $[0, \infty)$

33.

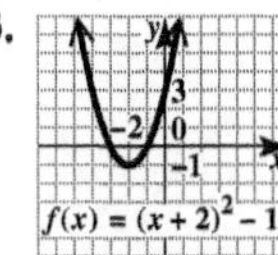

vertex: $(-2, -1)$; axis: $x = -2$; domain: $(-\infty, \infty)$; range: $[-1, \infty)$

35.

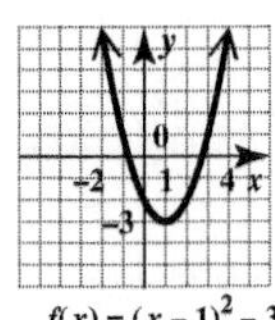

vertex: $(1, -3)$; axis: $x = 1$; domain: $(-\infty, \infty)$; range: $[-3, \infty)$

37. $f(x) = 2(x - 2)^2 - 4$

vertex: $(2, -4)$; axis: $x = 2$; domain: $(-\infty, \infty)$; range: $[-4, \infty)$

39. $f(x) = \frac{1}{2}(x - 2)^2 - 3$

vertex: $(2, -3)$; axis: $x = 2$; domain: $(-\infty, \infty)$; range: $[-3, \infty)$

41.

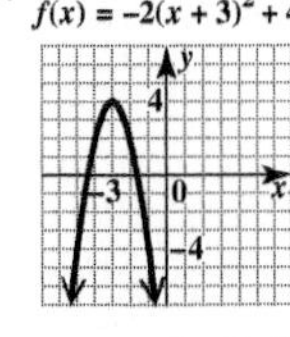

vertex: $(-3, 4)$; axis: $x = -3$; domain: $(-\infty, \infty)$; range: $(-\infty, 4]$

43.

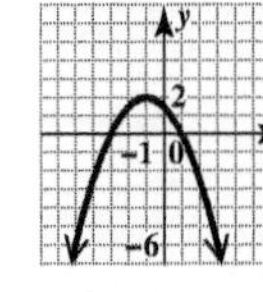

vertex: $(-1, 2)$; axis: $x = -1$; domain: $(-\infty, \infty)$; range: $(-\infty, 2]$

45. linear function; positive **47.** quadratic function; positive **49.** quadratic function; negative

51. (a)

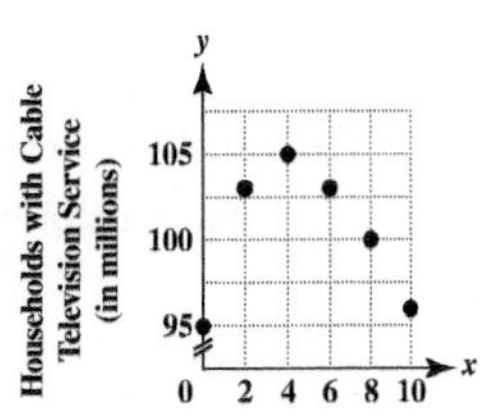

(b) quadratic function **(c)** negative **(d)** $y = -0.4x^2 + 4.1x + 95$ **(e)** 2009: 101.6 million households; 2015: 102.2 million households; The model approximates the data fairly well.

Section 9.3

1. If x is squared, it has a vertical axis. If y is squared, it has a horizontal axis. **3.** Find the discriminant of the function. If it is positive, there are two x-intercepts. If it is 0, there is one x-intercept (at the vertex), and if it is negative, there is no x-intercept. **5.** A, D are vertical parabolas. B, C are horizontal parabolas. **7.** $(-4, -6)$ **9.** $(1, -3)$ **11.** $\left(-\frac{1}{2}, -\frac{29}{4}\right)$ **13.** $(-1, 3)$; up; narrower; no x-intercepts **15.** $\left(\frac{5}{2}, \frac{37}{4}\right)$; down; same shape; two x-intercepts **17.** $(-3, -9)$; to the right; wider **19.** F **21.** C **23.** D

25.

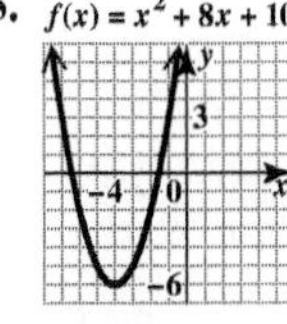

vertex: $(-4, -6)$; axis: $x = -4$; domain: $(-\infty, \infty)$; range: $[-6, \infty)$

27. $f(x) = x^2 + 4x + 3$

vertex: $(-2, -1)$; axis: $x = -2$; domain: $(-\infty, \infty)$; range: $[-1, \infty)$

29.

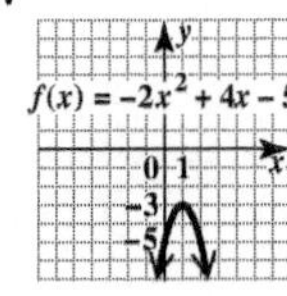

vertex: $(1, -3)$; axis: $x = 1$; domain: $(-\infty, \infty)$; range: $(-\infty, -3]$

31.

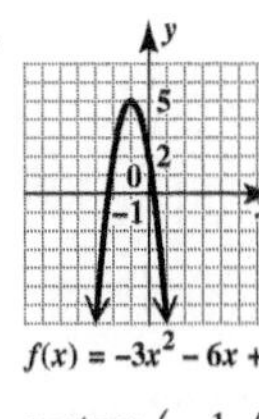

$f(x) = -3x^2 - 6x + 2$

vertex: $(-1, 5)$;
axis: $x = -1$;
domain: $(-\infty, \infty)$;
range: $(-\infty, 5]$

33.

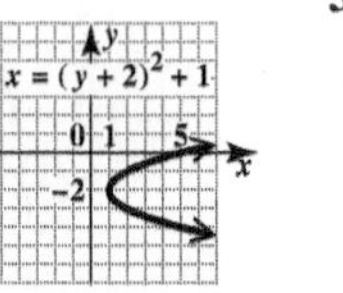

vertex: $(1, -2)$;
axis: $y = -2$;
domain: $[1, \infty)$;
range: $(-\infty, \infty)$

35. $x = -(y - 3)^2 - 1$

vertex: $(-1, 3)$;
axis: $y = 3$;
domain: $(-\infty, -1]$;
range: $(-\infty, \infty)$

37. $x = -\frac{1}{5}y^2 + 2y - 4$

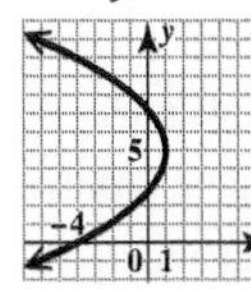

vertex: $(1, 5)$;
axis: $y = 5$;
domain: $(-\infty, 1]$;
range: $(-\infty, \infty)$

39.

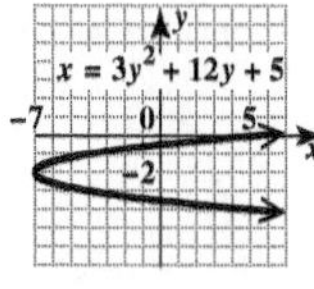

vertex: $(-7, -2)$;
axis: $y = -2$;
domain: $[-7, \infty)$;
range: $(-\infty, \infty)$

41. 20 and 20 **43.** 140 ft by 70 ft; 9800 ft^2 **45.** 2 sec; 65 ft **47.** 20 units; \$210 **49.** 16 ft; 2 sec **51.** The coefficient of x^2 is negative because a parabola that models the data must open down.

53. **(a)** $R(x) = (100 - x)(200 + 4x)$
$= 20{,}000 + 200x - 4x^2$

(b)

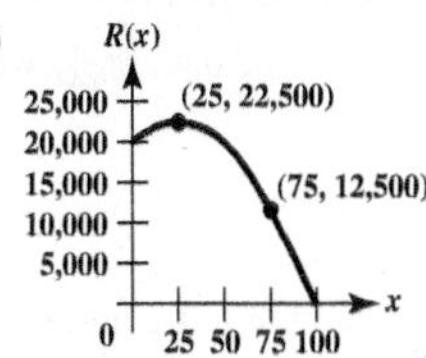

(c) 25 **(d)** \$22,500

Section 9.4

1. shrunken; $\frac{1}{2}$; 1; 2 **3.** reflected; x-axis; stretched; 2; −2; −4

5.

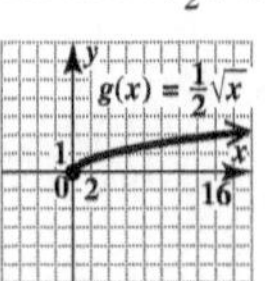

7.

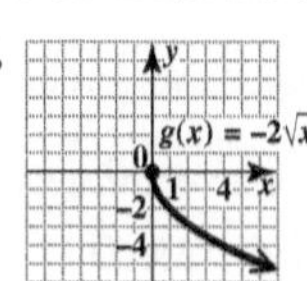

9. **(a)**

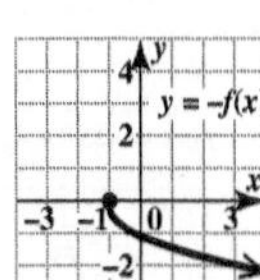

The graph of $f(x)$ is reflected across the x-axis.

(b)

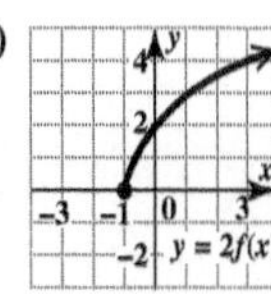

The graph is the same shape as that of $f(x)$, but stretched vertically by a factor of 2.

9. **(c)**

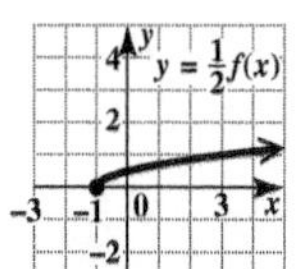

The graph is the same shape as that of $f(x)$, but shrunken vertically by a factor of $\frac{1}{2}$.

11. **(a)** symmetric **(b)** symmetric
13. **(a)** not symmetric **(b)** symmetric

15. (−4, 2) (a); (4, 2) (c); (−4, −2); (4, −2) (b)

17. (−8, 0) (a); (8, 0) (b),(c)

19. x-axis **21.** origin **23.** x-axis, y-axis, origin **25.** None of the symmetries apply. **27.** y-axis **29.** origin **31.** None of the symmetries apply. **33.** x-axis **35.** y-axis **37.** origin **39.** If the graph of a relation is symmetric with respect to the x-axis, it does not satisfy the conditions of the vertical line test for a function. **41.** $f(-2) = -3$ **43.** $f(4) = 3$ **45.** **(a)** $(-\infty, -3)$ **(b)** $(0, \infty)$ **(c)** $(-3, 0)$ **47.** **(a)** $(-\infty, -3)$; $(-1, \infty)$ **(b)** $(-3, -1)$ **(c)** none **49.** **(a)** $(-\infty, 2)$ **(b)** none **(c)** $(2, \infty)$ **51.** **(a)** $(-\infty, -2)$; $(1, \infty)$ **(b)** $(-2, 1)$ **(c)** none **53.** **(a)** none **(b)** $(-\infty, \infty)$ **(c)** none **55.** **(a)** none **(b)** $(-\infty, -2)$; $(3, \infty)$ **(c)** $(-2, 3)$ **57.** $(2009, 2012)$ **59.** In all cases, f is an even function. **60.** In all cases, f is an odd function. **61.** An even function has its graph symmetric with respect to the y-axis. **62.** An odd function has its graph symmetric with respect to the origin.

63. $y = f(x)$

64.

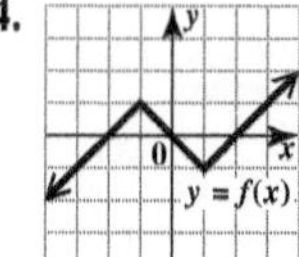

Section 9.5

1. E; $(0, \infty)$ **3.** A; 1 **5.** B; It does not satisfy the conditions of the vertical line test. **7.** B **9.** A

11. $f(x) = |x + 1|$

13.

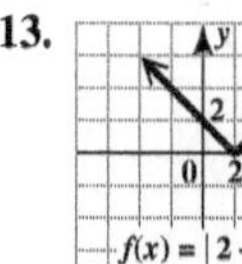

15.

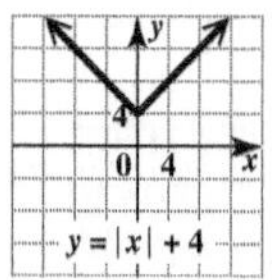

17. $y = 3|x - 2| - 1$

19. **(a)** −10 **(b)** −2 **(c)** −1 **(d)** 2 **(e)** 4
21. **(a)** 2 **(b)** 2 **(c)** 2 **(d)** −6 **(e)** −6

23. $f(x) = \begin{cases} 4 - x & \text{if } x < 2 \\ 1 + 2x & \text{if } x \geq 2 \end{cases}$

25.

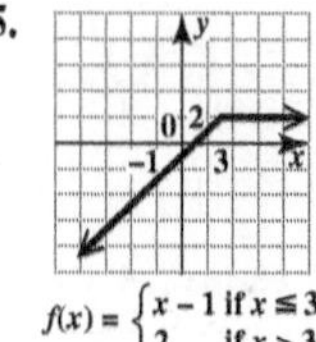

$f(x) = \begin{cases} x - 1 & \text{if } x \leq 3 \\ 2 & \text{if } x > 3 \end{cases}$

27.

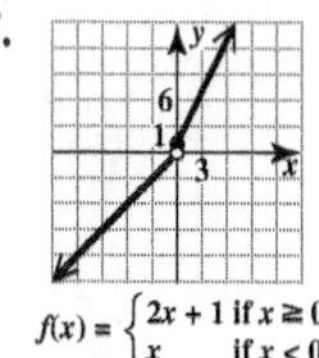

$f(x) = \begin{cases} 2x + 1 & \text{if } x \geq 0 \\ x & \text{if } x < 0 \end{cases}$

29.

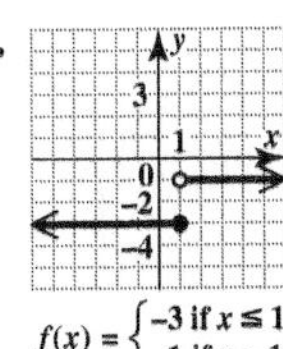

$f(x) = \begin{cases} -3 \text{ if } x \le 1 \\ -1 \text{ if } x > 1 \end{cases}$

31.

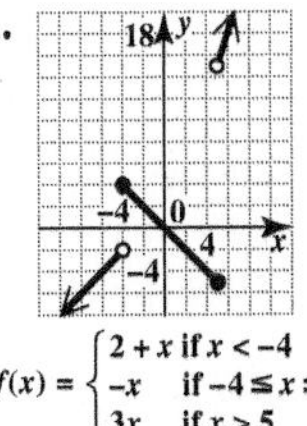

$f(x) = \begin{cases} 2 + x \text{ if } x < -4 \\ -x \quad \text{if } -4 \le x \le 5 \\ 3x \quad \text{if } x > 5 \end{cases}$

33.

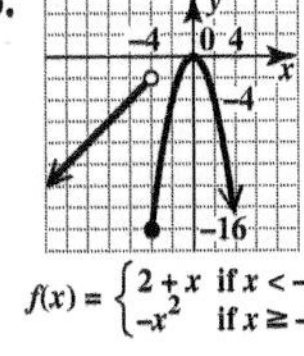

$f(x) = \begin{cases} 2 + x \text{ if } x < -4 \\ -x^2 \quad \text{if } x \ge -4 \end{cases}$

35.

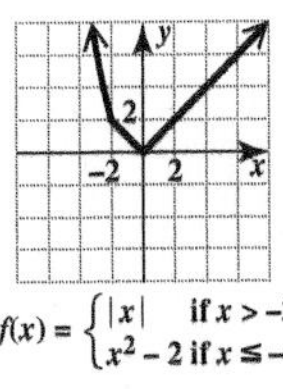

$f(x) = \begin{cases} |x| \quad \text{if } x > -2 \\ x^2 - 2 \text{ if } x \le -2 \end{cases}$

37.

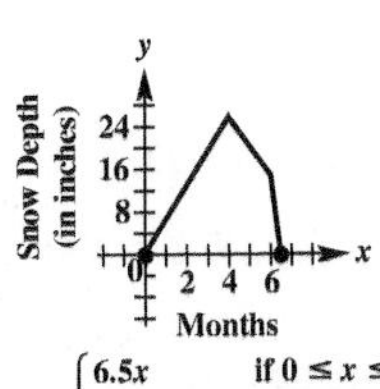

$f(x) = \begin{cases} 6.5x \quad \text{if } 0 \le x \le 4 \\ -5.5x + 48 \text{ if } 4 < x \le 6 \\ -30x + 195 \text{ if } 6 < x \le 6.5 \end{cases}$

39. 19.5 in.; 15 in. **41. (a)** $f(x) = \begin{cases} 27x + 1368 & \text{if } 0 \le x \le 5 \\ -34.6x + 1676 & \text{if } 5 < x \le 10 \end{cases}$

(b) 2010: 1449 news/talk stations; 2016: 1365 news/talk stations

43. $[\![x]\!]$ means the greatest integer *less than* or equal to x. Because -6 is less than -5.1, the greatest integer less than -5.1 is -6. **45.** 3

47. 4 **49.** 0 **51.** 2 **53.** -14 **55.** -11

We do not show ellipsis points for the step functions graphed over $(-\infty, \infty)$ in Exercises 57–65.

57.

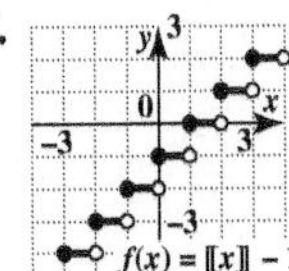

$f(x) = [\![x]\!] - 1$

59.

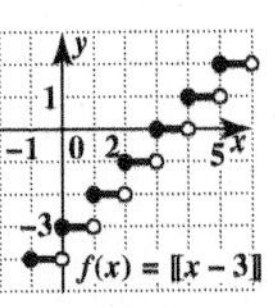

$f(x) = [\![x - 3]\!]$

61.

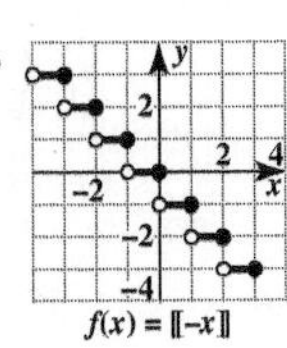

$f(x) = [\![-x]\!]$

63.

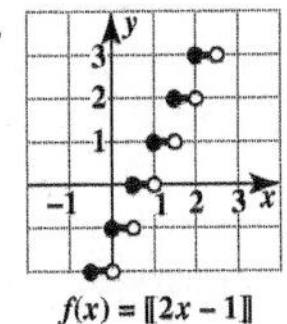

$f(x) = [\![2x - 1]\!]$

65.

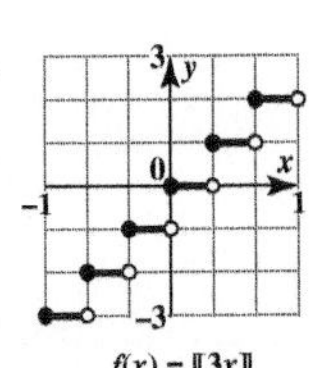

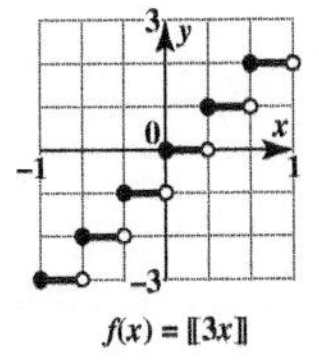

$f(x) = [\![3x]\!]$

67. (a) \$11 **(b)** \$18 **(c)** \$32 **(d)**

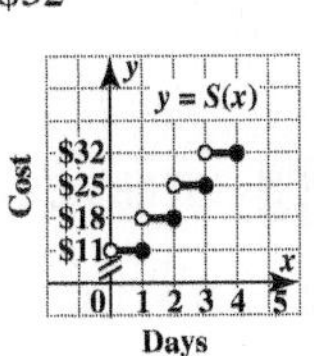

69.

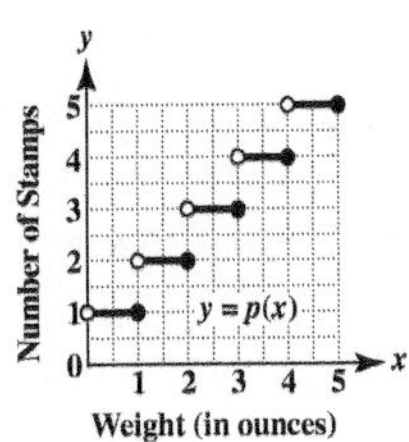

71. \$2.75

Chapter 9 Review Exercises

1. $x^2 + 3x + 3$; $(-\infty, \infty)$ **2.** $x^2 - 7x - 3$; $(-\infty, \infty)$

3. $(x^2 - 2x)(5x + 3)$; $(-\infty, \infty)$ **4.** $\frac{x^2 - 2x}{5x + 3}$; $\left(-\infty, -\frac{3}{5}\right) \cup \left(-\frac{3}{5}, \infty\right)$

5. $5x^2 - 10x + 3$; $(-\infty, \infty)$ **6.** $25x^2 + 20x + 3$; $(-\infty, \infty)$

7. 3 **8.** 5 **9.** $7\sqrt{5}$ **10.** $7 + \sqrt{5}$ **11.** -1 **12.** 1

13. $(f \circ g)(2) = 2\sqrt{2} - 3$; The answers are not equal, so composition of functions is not commutative. **14.** $f(5) = \sqrt{9 - 2(5)} = \sqrt{-1}$, which is not a real number. Thus, 5 is not in the domain of $f(x)$. **15.** $(1, 0)$

16. $(3, 7)$ **17.** $(-4, 3)$ **18.** $\left(\frac{2}{3}, -\frac{2}{3}\right)$

19.

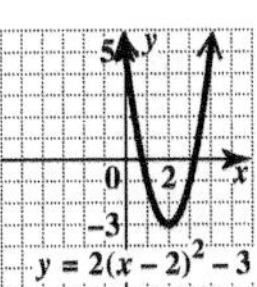

$y = 2(x - 2)^2 - 3$

vertex: $(2, -3)$;
axis: $x = 2$;
domain: $(-\infty, \infty)$;
range: $[-3, \infty)$

20. $f(x) = -2x^2 + 8x - 5$

vertex: $(2, 3)$;
axis: $x = 2$;
domain: $(-\infty, \infty)$;
range: $(-\infty, 3]$

21.

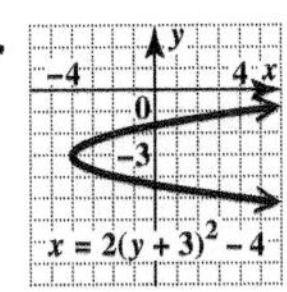

$x = 2(y + 3)^2 - 4$

vertex: $(-4, -3)$;
axis: $y = -3$;
domain: $[-4, \infty)$;
range: $(-\infty, \infty)$

22.

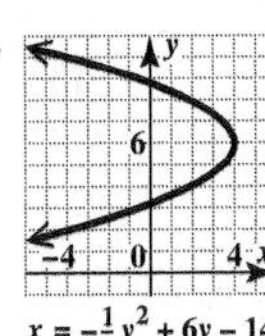

$x = -\frac{1}{2}y^2 + 6y - 14$

vertex: $(4, 6)$;
axis: $y = 6$;
domain: $(-\infty, 4]$;
range: $(-\infty, \infty)$

23. 5 sec; 400 ft **24.** length: 50 m; width: 50 m; maximum area: 2500 m^2

25. x-axis, y-axis, origin **26.** x-axis, y-axis, origin **27.** x-axis

28. y-axis **29.** None of the symmetries apply. **30.** y-axis

31. (a) yes **(b)** yes **(c)** yes **32.** decreasing

33. (a) $(-\infty, 2)$ **(b)** $(2, \infty)$ **(c)** none **34. (a)** $(4, \infty)$

(b) $(-\infty, -4)$ **(c)** $(-4, 4)$ **35. (a)** $(-\infty, -3)$; $(-1, \infty)$

(b) $(-3, -1)$ **(c)** none **36. (a)** $(1, \infty)$ **(b)** $(-\infty, 1)$

(c) none

37.

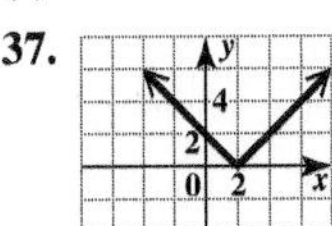

$f(x) = |x - 2|$

38.

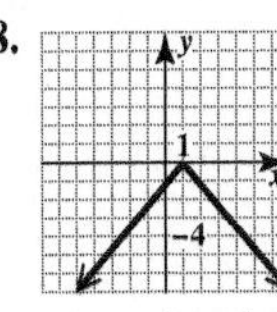

$f(x) = -|x - 1|$

39.

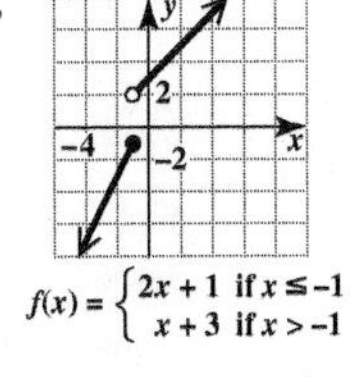

$f(x) = \begin{cases} 2x + 1 \text{ if } x \le -1 \\ x + 3 \text{ if } x > -1 \end{cases}$

40.

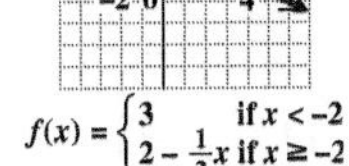

$f(x) = \begin{cases} 3 \quad \text{if } x < -2 \\ 2 - \frac{1}{2}x \text{ if } x \ge -2 \end{cases}$

41.

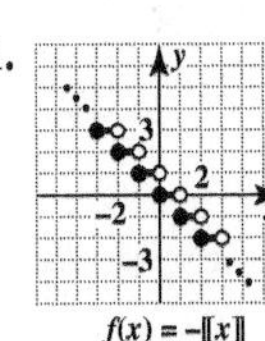

$f(x) = -[\![x]\!]$

42.

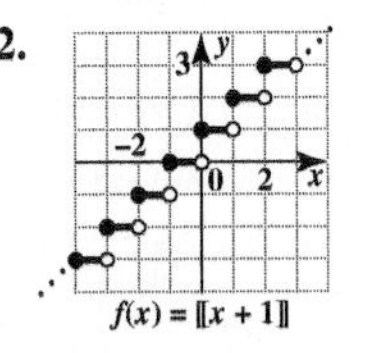

$f(x) = [\![x + 1]\!]$

43. The graph is narrower than the graph of $y = |x|$, and it is shifted (translated) to the left 4 units and down 3 units. **44. (a)** \$0.90

(b) \$1.10 **(c)** \$1.60 **(d)**

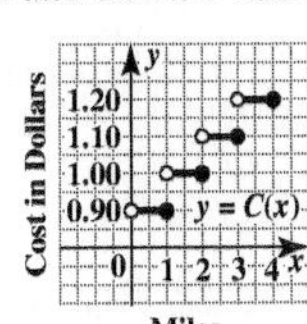

Chapter 9 Mixed Review Exercises

1. (a) $2x^2 - x + 1$ **(b)** $\frac{2x^2 - 3x + 2}{-2x + 1}$ **(c)** $\left(-\infty, \frac{1}{2}\right) \cup \left(\frac{1}{2}, \infty\right)$

(d) $4x + 2h - 3$ **2. (a)** 0 **(b)** -12 **(c)** 1 **3. (a)** F

(b) B **(c)** C **(d)** A **(e)** E **(f)** D

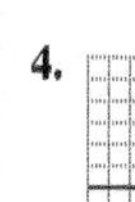

4. 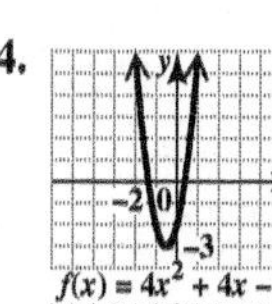

vertex: $\left(-\frac{1}{2}, -3\right)$; axis: $x = -\frac{1}{2}$; domain: $(-\infty, \infty)$; range: $[-3, \infty)$

5. (a) $(-\infty, -3)$ (b) $(4, \infty)$ (c) $(-3, 4)$

6. 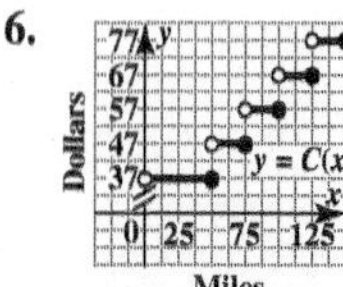

7. $f(x) = -3\sqrt{x}$

8. $f(x) = |2x + 1|$

9. 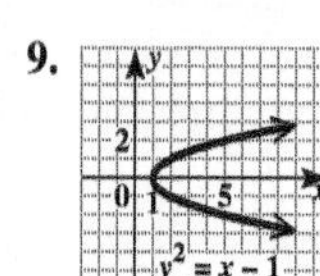

10. 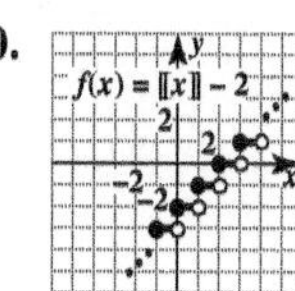

Chapter 9 Test

[9.1] **1.** 2 **2.** −7 **3.** $-\frac{7}{3}$ **4.** −2 **5.** $-16x^2 - 16x - 1$ **6.** $x^2 + 4x - 1$; $(-\infty, \infty)$

[9.2] **7.** A **8.** 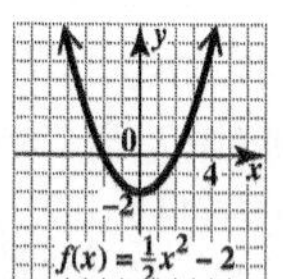

vertex: $(0, -2)$; axis: $x = 0$; domain: $(-\infty, \infty)$; range: $[-2, \infty)$

[9.3] **9.** 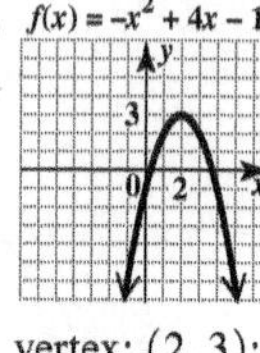

vertex: $(2, 3)$; axis: $x = 2$; domain: $(-\infty, \infty)$; range: $(-\infty, 3]$

10. 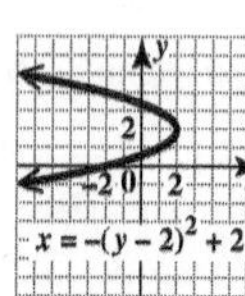

vertex: $(2, 2)$; axis: $y = 2$; domain: $(-\infty, 2]$; range: $(-\infty, \infty)$

11. 160 ft by 320 ft; 51,200 ft^2 [9.4] **12.** y-axis **13.** x-axis **14.** x-axis, y-axis, origin [9.5] **15.** (a) C (b) A (c) D (d) B [9.4] **16.** (a) $(-1, 1)$ (b) $(-\infty, -1)$ (c) $(1, \infty)$

[9.5] **17.** 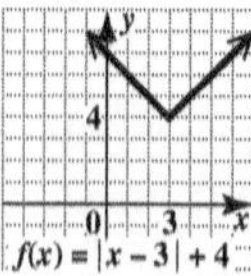

18. 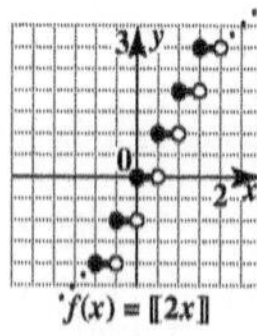

19. 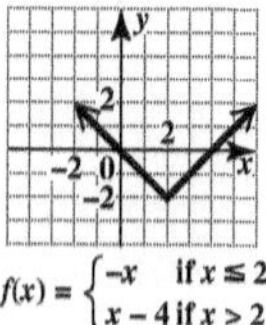

20. $\{\ldots, -2, -1, 0, 1, 2, \ldots\}$

Chapters R–9 Cumulative Review Exercises

[R.1] **1.** 0.01; 1% [R.3] **2.** 0 [R.2] **3.** (a) A, B, C, D, F (b) B, C, D, F (c) D, F (d) E, F (e) C, D, F (f) D, F [R.4] **4.** (a) 36 (b) −36 (c) 36 (d) 6 (e) −6 (f) not a real number [1.1] **5.** $\{-6\}$ [1.2] **6.** $p = \frac{q^2}{3 - q}$ [1.5] **7.** $(-\infty, 2]$ [1.6] **8.** $(-\infty, 0] \cup (2, \infty)$ [1.7] **9.** $\left\{-\frac{7}{3}, \frac{17}{3}\right\}$ **10.** $(-\infty, \infty)$ [2.2] **11.** $-\frac{5}{7}$ **12.** $\frac{2}{3}$ [2.3] **13.** (a) $y = 2x + 4$ (b) $2x - y = -4$ [2.5] **14.** $\left(-\infty, \frac{1}{4}\right]$ [2.6] **15.** 8 [3.1] **16.** $\{(4, 3)\}$ [3.2] **17.** $\{(1, 2, -1)\}$ [4.6] **18.** $4x^2 + x + 3$ [6.2] **19.** $\frac{3p - 2}{p - 1}$ [6.1] **20.** $\frac{2}{3x}$ [5.4] **21.** $(x - y)(x + y) \cdot (x^2 + xy + y^2)(x^2 - xy + y^2)$ **22.** $(2k^2 + 3)(k + 1)(k - 1)$ [4.1] **23.** $\frac{2}{9}m^{10}$ [7.2] **24.** $k\sqrt[6]{k}$ [5.5] **25.** $\left\{-\frac{7}{2}, 0, 4\right\}$ [6.4] **26.** $\left\{-\frac{1}{11}\right\}$ [7.6] **27.** $\{3\}$ [8.2] **28.** $\left\{\frac{-3 \pm \sqrt{69}}{10}\right\}$ [8.4] **29.** $r = \frac{-\pi h \pm \sqrt{\pi^2 h^2 + \pi S}}{\pi}$ [8.5] **30.** $\left(-\infty, -\frac{2}{3}\right) \cup (4, \infty)$ [5.5] **31.** (a) after 16 sec (b) 4 sec and 12 sec

[2.1] **32.** 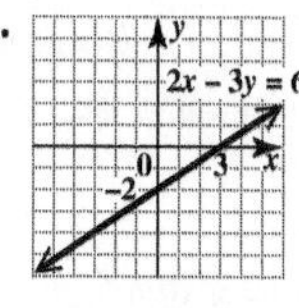

[2.4] **33.** 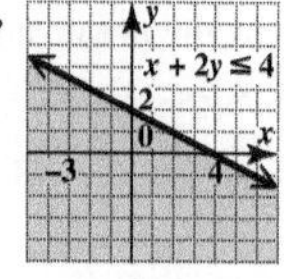

[9.3] **34.** $f(x) = -2x^2 + 5x + 3$

[9.5] **35.** 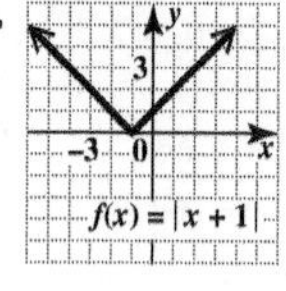

[9.4] **36.** x-axis **37.** x-axis, y-axis, origin [9.5] **38.** (a) 5 (b) −2

10 INVERSE, EXPONENTIAL, AND LOGARITHMIC FUNCTIONS

Section 10.1

1. B **3.** A **5.** It is not one-to-one. The Big Breakfast with Egg Whites and the Bacon Clubhouse Burger are paired with the same fat content, 41 g. **7.** Yes, it is one-to-one. Adding 1 to 1058 would make two distances be the same, so the function would not be one-to-one. **9.** $\{(6, 3), (10, 2), (12, 5)\}$ **11.** not one-to-one **13.** $\{(4.5, 0), (8.6, 2), (12.7, 4)\}$ **15.** $f^{-1}(x) = x - 3$ **17.** $f^{-1}(x) = -2x - 4$ **19.** $f^{-1}(x) = \frac{x - 4}{2}$, or $f^{-1}(x) = \frac{1}{2}x - 2$ **21.** $g^{-1}(x) = \frac{-x + 3}{4}$, or $g^{-1}(x) = -\frac{1}{4}x + \frac{3}{4}$ **23.** not one-to-one **25.** $f^{-1}(x) = x^2 + 3$, $x \ge 0$ **27.** $f^{-1}(x) = x^2 - 6$, $x \ge 0$ **29.** not one-to-one **31.** $g^{-1}(x) = \sqrt[3]{x} - 1$ **33.** $f^{-1}(x) = \sqrt[3]{x + 4}$ **35.** $f^{-1}(x) = \frac{-2x + 4}{x - 1}$, $x \ne 1$ **37.** $f^{-1}(x) = \frac{-5x - 2}{x - 4}$, $x \ne 4$

39. $f^{-1}(x) = \frac{5x+1}{2x+2}$, $x \neq -1$ **41.** (a) 8 (b) 3 **43.** (a) 1 (b) 0

45. (a) one-to-one **47.** (a) not one-to-one **49.** (a) one-to-one

(b) (b)

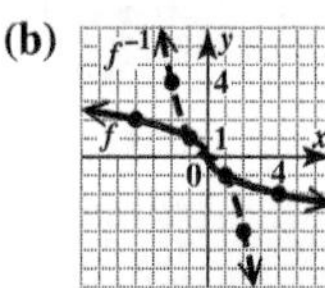

51. **53.**

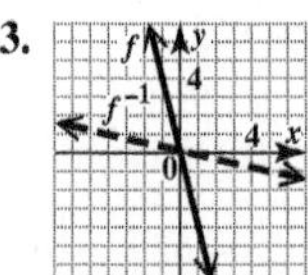

55.

x	f(x)
0	0
1	1
4	2

57.

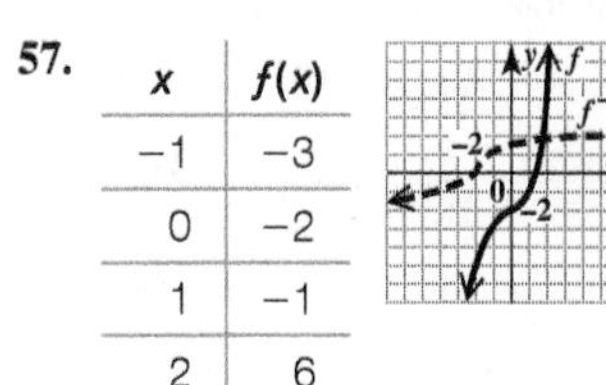

x	f(x)
−1	−3
0	−2
1	−1
2	6

59. $f^{-1}(x) = \frac{x+5}{4}$, or $f^{-1}(x) = \frac{1}{4}x + \frac{5}{4}$

60. MY CALCULATOR IS THE GREATEST THING SINCE SLICED BREAD. **61.** If the function were not one-to-one, there would be ambiguity in some of the characters, as they could represent more than one letter. **62.** Answers will vary. For example, Jane Doe is 1004 5 2748 129 68 3379 129.

Section 10.2

1. rises **3.** C **5.** A **7.** 3.732 **9.** 0.344 **11.** 1.995 **13.** 1.587 **15.** 0.192 **17.** 73.517

19.

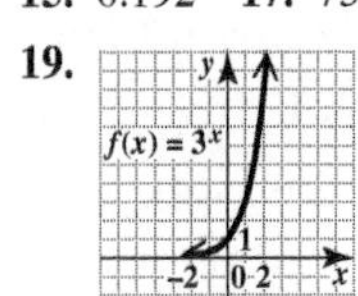

21.

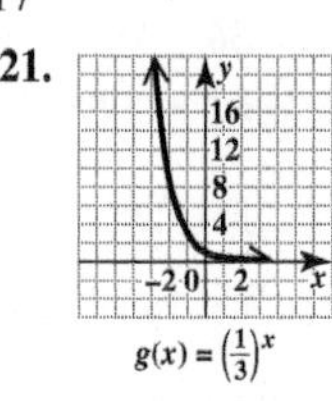

23.

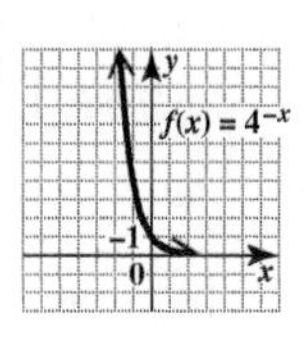

25.

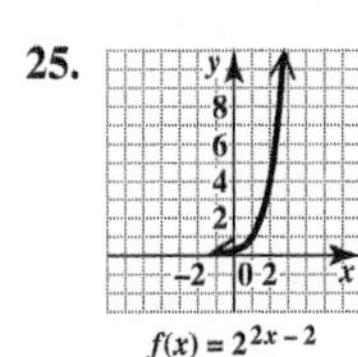

27. The division in the second step does not lead to x on the left side. By expressing each side using the same base, $2^x = 2^5$, we obtain the correct solution set, $\{5\}$. **29.** $\{2\}$ **31.** $\left\{\frac{3}{2}\right\}$ **33.** $\left\{\frac{3}{2}\right\}$

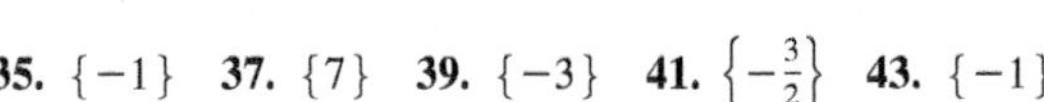

35. $\{-1\}$ **37.** $\{7\}$ **39.** $\{-3\}$ **41.** $\left\{-\frac{3}{2}\right\}$ **43.** $\{-1\}$

45. $\{-3\}$ **47.** $\{-2\}$ **49.** 100 g **51.** 0.30 g **53.** (a) 0.6°C (b) 0.3°C **55.** (a) 1.4°C (b) 0.5°C **57.** (a) 80.599 million users (b) 165.912 million users (c) 341.530 million users

59. (a) \$5000 (b) \$2973 (c) \$1768

(d)

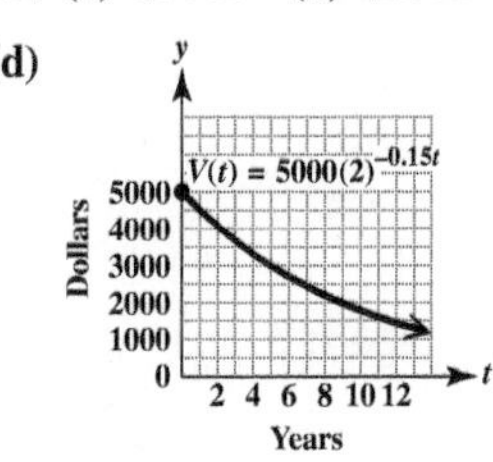

Section 10.3

1. (a) B (b) E (c) D (d) F (e) A (f) C **3.** $(0, \infty)$; $(-\infty, \infty)$

5. $\log_4 1024 = 5$ **7.** $\log_{1/2} 8 = -3$ **9.** $\log_{10} 0.001 = -3$

11. $\log_{625} 5 = \frac{1}{4}$ **13.** $\log_8 \frac{1}{4} = -\frac{2}{3}$ **15.** $\log_5 1 = 0$ **17.** $4^3 = 64$

19. $12^1 = 12$ **21.** $6^0 = 1$ **23.** $9^{1/2} = 3$ **25.** $\left(\frac{1}{4}\right)^{1/2} = \frac{1}{2}$

27. $5^{-1} = 5^{-1}$ **29.** (a) C (b) B (c) B (d) C **31.** 3.1699

33. 1.7959 **35.** −1.7925 **37.** −1.5850 **39.** 1.9243 **41.** 1.6990

43. $\left\{\frac{1}{3}\right\}$ **45.** $\left\{\frac{1}{125}\right\}$ **47.** $\{81\}$ **49.** $\left\{\frac{1}{5}\right\}$ **51.** $\{1\}$

53. $\{x \mid x > 0, x \neq 1\}$ **55.** $\{5\}$ **57.** $\left\{\frac{5}{3}\right\}$ **59.** $\{4\}$ **61.** $\left\{\frac{3}{2}\right\}$

63. $\{30\}$ **65.** $\left\{\frac{37}{9}\right\}$ **67.** 1 **69.** 0 **71.** 9 **73.** −1 **75.** 9

77. 5 **79.** 6 **81.** 4 **83.** −1 **85.** $\frac{1}{3}$

87.

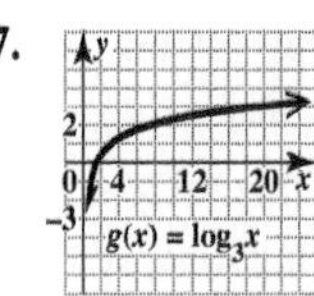

89.

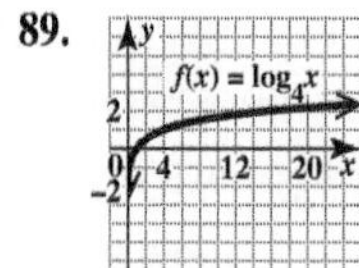

91.

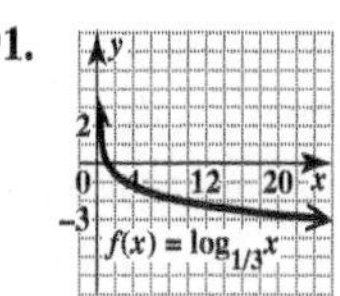

93.

95. 8 **97.** 24 **99.** Because every real number power of 1 equals 1, if $y = \log_1 x$, then $x = 1^y$ and so $x = 1$ for every y. This contradicts the definition of a function.

101. $x = \log_a 1$ is equivalent to $a^x = 1$. The only value of x that makes $a^x = 1$ is 0. (Recall that $a \neq 1$.)

103. (a) 130 thousand units (b) 190 thousand units

(c)

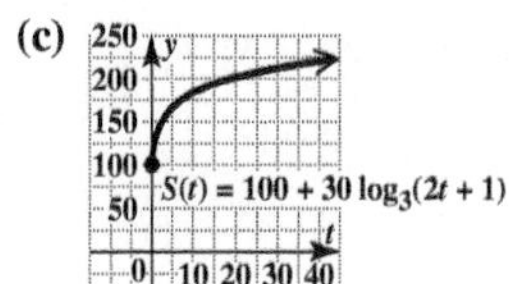

105. 5 advertisements

107. $\frac{1}{100}$; $\frac{1}{10}$; 1; 10; 100 **108.** −2; −1; 0; 1; 2

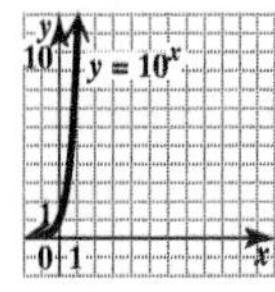

domain: $(-\infty, \infty)$; range: $(0, \infty)$

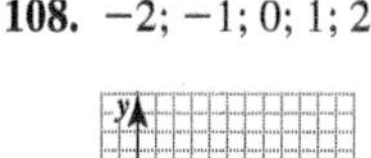

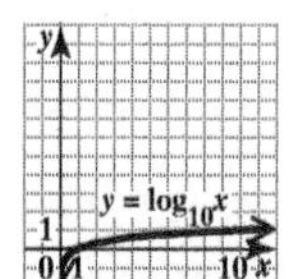

domain: $(0, \infty)$; range: $(-\infty, \infty)$

109. The graphs have symmetry across the line $y = x$.

110. They are inverses.

Section 10.4

1. false; $\log_b x + \log_b y = \log_b xy$ **3.** true **5.** true **7.** In the notation $\log_a (x + y)$, the parentheses do not indicate multiplication. They indicate that $x + y$ is the result of raising a to some power.

9. $\log_{10} 7 + \log_{10} 8$ **11.** 4 **13.** 9 **15.** $\log_7 4 + \log_7 5$

17. $\log_5 8 - \log_5 3$ **19.** $2 \log_4 6$ **21.** $\frac{1}{3}\log_3 4 - 2\log_3 x - \log_3 y$

23. $\frac{1}{2}\log_3 x + \frac{1}{2}\log_3 y - \frac{1}{2}\log_3 5$ **25.** $\frac{1}{3}\log_2 x + \frac{1}{5}\log_2 y - 2\log_2 r$

27. $\log_b xy$ **29.** $\log_a \frac{m}{n}$ **31.** $\log_a \frac{rt^3}{s}$ **33.** $\log_a \frac{125}{81}$

35. $\log_{10}(x^2 - 9)$ **37.** $\log_{10}(x^2 + 8x + 15)$ **39.** $\log_p \frac{x^3y^{1/2}}{z^{3/2}a^3}$

41. 1.2552 **43.** −0.6532 **45.** 1.5562 **47.** 0.2386 **49.** 0.4771

51. 4.7710 **53.** false **55.** true **57.** true **59.** false

Section 10.5

1. C **3.** B **5.** 9.6421 **7.** $\sqrt{3}$ **9.** −11.4007 **11.** 1.6335 **13.** 2.5164 **15.** −1.4868 **17.** 9.6776 **19.** 2.0592 **21.** −2.8896 **23.** 5.9613 **25.** 4.1506 **27.** 2.3026 **29.** **(a)** 2.552424846 **(b)** 1.552424846 **(c)** 0.552424846 **(d)** The whole number parts will vary, but the decimal parts are the same. **31.** poor fen **33.** bog **35.** rich fen **37.** 11.6 **39.** 8.4 **41.** 4.3 **43.** 4.0×10^{-8} **45.** 1.0×10^{-2} **47.** 2.5×10^{-5} **49.** **(a)** 142 dB **(b)** 126 dB **(c)** 120 dB **51.** **(a)** 800 yr **(b)** 5200 yr **(c)** 11,500 yr **53.** **(a)** 189.53 million monthly active Twitter users **(b)** 2015 **55.** **(a)** \$54 per ton **(b)** If $x = 0$, then $\ln(1 - x) = \ln 1 = 0$, so $T(x)$ would be negative. If $x = 1$, then $\ln(1 - x) = \ln 0$, but the domain of $\ln x$ is $(0, \infty)$. **57.** 2.2619 **59.** 0.6826 **61.** −0.0947 **63.** −2.3219 **65.** 0.3155 **67.** 0.8736

Section 10.6

1. common logarithms **3.** natural logarithms **5.** $\{0.827\}$ **7.** $\{0.833\}$ **9.** $\{1.201\}$ **11.** $\{2.269\}$ **13.** $\{15.967\}$ **15.** $\{-6.067\}$ **17.** $\{261.291\}$ **19.** $\{-10.718\}$ **21.** $\{3\}$ **23.** $\{5.879\}$ **25.** $\{-\pi\}$, or $\{-3.142\}$ **27.** $\{1\}$ **29.** $\{4\}$ **31.** $\{\frac{2}{3}\}$ **33.** $\{\frac{33}{2}\}$ **35.** $\{-1 + \sqrt[3]{49}\}$ **37.** $\{\pm 3\}$ **39.** 2 cannot be a solution because $\log(2 - 3) = \log(-1)$, and −1 is not in the domain of $\log x$. **41.** $\{\frac{1}{3}\}$ **43.** $\{2\}$ **45.** $\emptyset$ **47.** $\{8\}$ **49.** $\{\frac{4}{3}\}$ **51.** $\{8\}$ **53.** **(a)** \$2539.47 **(b)** 10.2 yr **55.** \$4934.71 **57.** 27.73 yr **59.** **(a)** \$11,260.96 **(b)** \$11,416.64 **(c)** \$11,497.99 **(d)** \$11,580.90 **(e)** \$11,581.83 **61.** \$137.41 **63.** \$210 billion **65.** **(a)** 1.62 g **(b)** 1.18 g **(c)** 0.69 g **(d)** 2.00 g **67.** **(a)** 179.73 g **(b)** 21.66 yr **69.** $\log 5^x = \log 125$ **70.** $x \log 5 = \log 125$ **71.** $x = \frac{\log 125}{\log 5}$ **72.** $\frac{\log 125}{\log 5} = 3$; $\{3\}$

Chapter 10 Review Exercises

1. not one-to-one **2.** one-to-one **3.** not one-to-one **4.** $\{(-8, -2), (-1, -1), (0, 0), (1, 1), (8, 2)\}$ **5.** $f^{-1}(x) = \frac{x - 7}{-3}$, or $f^{-1}(x) = -\frac{1}{3}x + \frac{7}{3}$ **6.** $f^{-1}(x) = \frac{x^3 + 4}{6}$ **7.** not one-to-one

8. This function is not one-to-one because two states in the list have minimum wage \$10.50.

9.

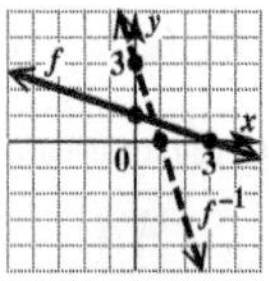

10.

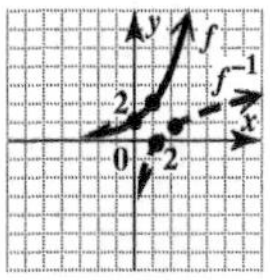

11. 172.466 **12.** 0.034 **13.** 0.079

14.

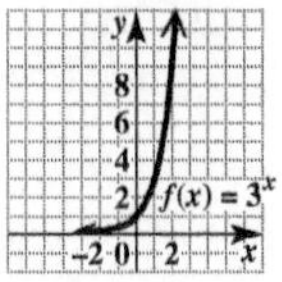

15.

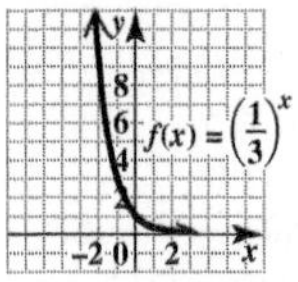

16.

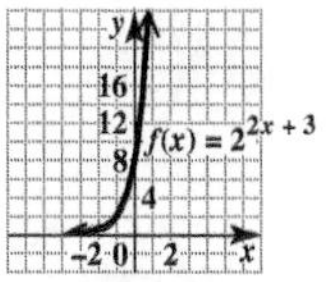

17. $\{\frac{1}{2}\}$ **18.** $\{4\}$ **19.** $\{\frac{3}{7}\}$ **20.** **(a)** 55.8 million people **(b)** 81.0 million people **21.** **(a)** $5^4 = 625$ **(b)** $\log_5 0.04 = -2$ **22.** 4; 3; fourth; 81 **23.** 2.5850 **24.** 0.5646 **25.** 1.7404

26.

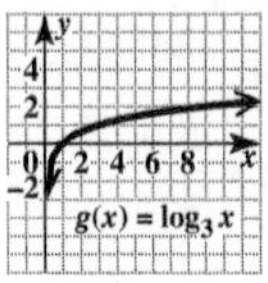

27.

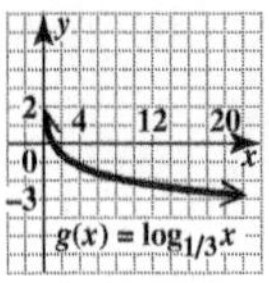

28. **(a)** 12 **(b)** 13 **(c)** 4 **29.** $\{2\}$ **30.** $\{\frac{3}{2}\}$ **31.** $\{7\}$ **32.** $\{8\}$ **33.** $\{4\}$ **34.** $\{\frac{1}{36}\}$

35. \$300,000

36.

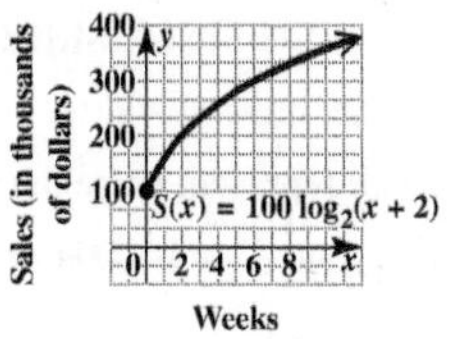

37. $\log_4 3 + 2 \log_4 x$ **38.** $3 \log_5 a + 2 \log_5 b - 4 \log_5 c$ **39.** $\frac{1}{2}\log_4 x + 2 \log_4 w - \log_4 z$

40. $2 \log_2 p + \log_2 r - \frac{1}{2}\log_2 z$ **41.** $\log_a \frac{49}{16}$ **42.** $\log_a 250$ **43.** $\log_b \frac{3x}{y^2}$ **44.** $\log_3 \frac{x + 7}{4x + 6}$ **45.** 1.4609 **46.** −0.5901 **47.** 3.3638 **48.** −1.3587 **49.** 0.9251 **50.** 1.7925 **51.** −2.0437 **52.** 0.3028 **53.** 6.4 **54.** 8.4 **55.** 2.5×10^{-5} **56.** Magnitude 1 is about 6.3 times as intense as magnitude 3. **57.** every 2 hr **58.** $\{2.042\}$ **59.** $\{4.907\}$ **60.** $\{18.310\}$ **61.** $\{\frac{1}{9}\}$ **62.** $\{-6 + \sqrt[3]{25}\}$ **63.** $\{2\}$ **64.** $\{\frac{3}{8}\}$ **65.** $\{4\}$ **66.** $\{1\}$ **67.** \$24,403.80 **68.** \$11,190.72 **69.** Plan A is better because it would pay \$2.92 more. **70.** 13.9 days

Chapter 10 Mixed Review Exercises

1. 7 **2.** 0 **3.** −3 **4.** 36 **5.** 4 **6.** e **7.** −5 **8.** 5.4 **9.** $\{72\}$ **10.** $\{5\}$ **11.** $\{\frac{1}{9}\}$ **12.** $\{\frac{4}{3}\}$ **13.** $\{3\}$ **14.** $\{0\}$ **15.** $\{\frac{1}{8}\}$ **16.** $\{\frac{11}{3}\}$ **17.** $\{\frac{2}{63}\}$ **18.** $\{-2, -1\}$ **19.** **(a)** \$4267 **(b)** 11% **20.** **(a)** 0.325 **(b)** 0.673 **21.** **(a)** 1.209061955 **(b)** 7 **(c)** $\{1.209061955\}$ **22.** Answers will vary. Suppose the name is Jeffery Cole, with $m = 7$ and $n = 4$. **(a)** $\log_7 4$ is the exponent to which 7 must be raised to obtain 4. **(b)** 0.7124143742 **(c)** 4

Chapter 10 Test

[10.1] **1.** **(a)** not one-to-one **(b)** one-to-one **2.** $f^{-1}(x) = x^3 - 7$

3.

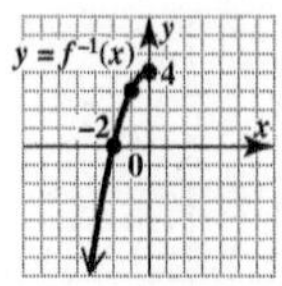

[10.2] **4.**

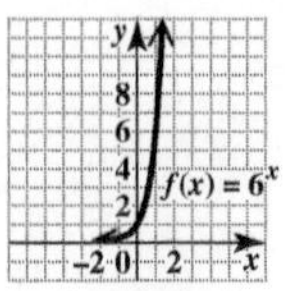

[10.3] **5.**

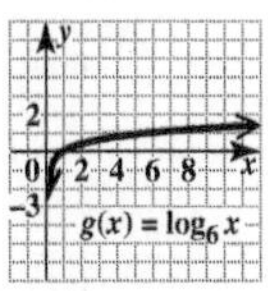

6. Interchange the x- and y-values of the ordered pairs, because the functions are inverses. **7.** 9 **8.** 6 **9.** 0 [10.2] **10.** $\{-4\}$ **11.** $\left\{-\frac{13}{3}\right\}$ **12.** **(a)** 775 millibars **(b)** 265 millibars [10.3] **13.** $\log_4 0.0625 = -2$ **14.** $7^2 = 49$ **15.** $\{32\}$ **16.** $\left\{\frac{1}{2}\right\}$ **17.** $\{2\}$ **18.** 5; 2; fifth; 32 [10.4] **19.** $2\log_3 x + \log_3 y$ **20.** $\frac{1}{2}\log_5 x - \log_5 y - \log_5 z$ **21.** $\log_b \frac{s^3}{t}$ **22.** $\log_b \frac{r^{1/4}s^2}{t^{2/3}}$ [10.5] **23.** 1.3636 **24.** −0.1985 **25.** 2.1245 **26.** **(a)** $\frac{\log 19}{\log 3}$ **(b)** $\frac{\ln 19}{\ln 3}$ **(c)** 2.6801 [10.6] **27.** $\{3.966\}$ **28.** $\{3\}$ **29.** **(a)** \$11,903.40 **(b)** 19.9 yr **30.** **(a)** \$17,427.51 **(b)** 23.1 yr

Chapters R–10 Cumulative Review Exercises

[R.1] **1.** **(a)** 0.05; 5% **(b)** 1.25; 125% [R.1, R.3] **2.** **(a)** $-\frac{6}{5}$ **(b)** $-\frac{4}{5}$ **3.** **(a)** 3750 **(b)** 0.0375 [R.2] **4.** **(a)** $-2, 0, 6, \frac{30}{3}$ (or 10) **(b)** $-\frac{9}{4}, -2, 0, 0.6, 6, \frac{30}{3}$ (or 10) **(c)** $-\sqrt{2}, \sqrt{11}$ [R.3, R.4] **5.** 16 **6.** −39 [1.1] **7.** $\left\{-\frac{2}{3}\right\}$ [1.5] **8.** $[1, \infty)$ [1.7] **9.** $\{-2, 7\}$ **10.** $(-\infty, -3) \cup (2, \infty)$ [2.1] **11.**

$5x + 2y = 10$

[2.4] **12.**

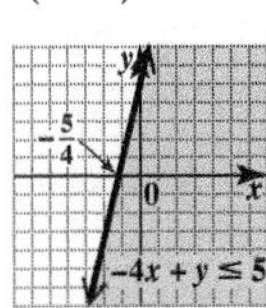

[2.2, 2.5] **13.** **(a)** yes **(b)** 1.955; The number of travelers shows an increase of an average of 1.955 million per year during these years. [2.3] **14.** $y = \frac{3}{4}x - \frac{19}{4}$ [3.1, 3.2] **15.** $\{(4, 2)\}$ **16.** $\{(x, y) \mid 3x - 2y = 3\}$ **17.** $\{(1, -1, 4)\}$ [3.3] **18.** 6 lb [4.5] **19.** $6p^2 + 7p - 3$ **20.** $16k^2 - 24k + 9$ [4.3] **21.** $-5m^3 + 2m^2 - 7m + 4$ [4.6] **22.** $5x^2 - 2x + 8$ [5.1] **23.** $x(8 + x^2)$ [5.2] **24.** $(3y - 2)(8y + 3)$ **25.** $z(5z + 1)(z - 4)$ [5.3] **26.** $(4a + 5b^2)(4a - 5b^2)$ **27.** $(2c + d)(4c^2 - 2cd + d^2)$ **28.** $(4r + 7q)^2$ [4.1] **29.** $-\frac{1875p^{13}}{8}$ [6.1] **30.** $\frac{x + 5}{x + 4}$ [6.2] **31.** $\frac{-3k - 19}{(k + 3)(k - 2)}$ [7.3] **32.** $12\sqrt{2}$ [7.4] **33.** $-27\sqrt{2}$ [7.7] **34.** 41 [7.6] **35.** $\{0, 4\}$ [8.1, 8.2] **36.** $\left\{\frac{1 \pm \sqrt{13}}{6}\right\}$ [8.5] **37.** $(-\infty, -4) \cup (2, \infty)$ [8.3] **38.** $\{\pm 1, \pm 2\}$ [10.2] **39.** $\{-1\}$ [10.6] **40.** $\{1\}$ [9.2] **41.** $f(x) = \frac{1}{3}(x - 1)^2 + 2$ [10.2] **42.**

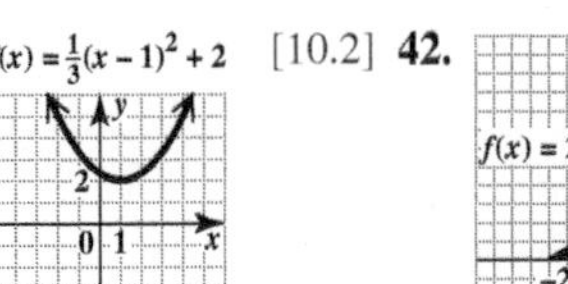

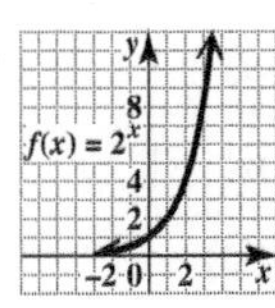

[10.3] **43.**

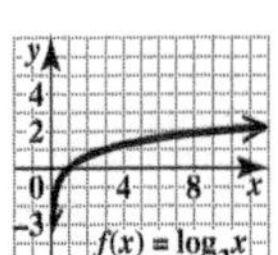

[9.4] **44.** x-axis, y-axis, origin **45.** y-axis [9.1] **46.** **(a)** $2x + 1$ **(b)** $3x^2 - 3x$ **(c)** $\frac{3x}{x - 1}$; The domains of $f - g$ and fg are $(-\infty, \infty)$. The domain of $\frac{f}{g}$ is $(-\infty, 1) \cup (1, \infty)$. [10.4] **47.** $3\log x + \frac{1}{2}\log y - \log z$ [10.6] **48.** **(a)** 25,000 bacteria **(b)** 30,500 bacteria **(c)** 37,300 bacteria **(d)** in 3.5 hr, or at 3:30 P.M.

11 POLYNOMIAL AND RATIONAL FUNCTIONS

Section 11.1

1. C **3.** 3; 4 **5.** 3; 6; 0 **7.** The student neglected to include 0 as the coefficient of the linear term. The correct setup is $-2\overline{)4 \quad 2 \quad 0 \quad 6}$, and the correct answer is $4x^2 - 6x + 12 + \frac{-18}{x + 2}$. **9.** $x - 5$ **11.** $4x - 1$ **13.** $2x + 4 + \frac{5}{x + 2}$ **15.** $4x^2 + x + 3$ **17.** $x^2 + 6x + 20 + \frac{83}{x - 4}$ **19.** $2x^4 + 4x^3 + 6x^2 + 15x + 6 + \frac{10}{x - 2}$ **21.** $-3x^4 + 5x^2 - 11x + 11 + \frac{-8}{x + 1}$ **23.** $x^4 + x^2 + 1 + \frac{2}{x + 1}$ **25.** $f(x) = (x + 1)(2x^2 - x + 2) + (-10)$ **27.** $f(x) = (x + 2)(-x^2 + 4x - 8) + 20$ **29.** $f(x) = (x - 3)(4x^3 + 9x^2 + 7x + 20) + 60$ **31.** 2 **33.** −1 **35.** −6 **37.** 45 **39.** $-6 - i$ **41.** 0 **43.** yes **45.** yes **47.** no **49.** no **51.** $(2x - 3)(x + 4)$ **52.** $\left\{-4, \frac{3}{2}\right\}$ **53.** 0 **54.** 0 **55.** a **56.** Yes, $x - 3$ is a factor; $g(x) = (x - 3)(3x - 1)(x + 2)$

Section 11.2

1. true **3.** false **5.** no **7.** yes **9.** yes **11.** $f(x) = (x - 2) \cdot (2x - 5)(x + 3)$ **13.** $f(x) = (x + 3)(3x - 1)(2x - 1)$ **15.** $-1 \pm i$ **17.** $3, 2 + i$ **19.** $-i, 7 \pm i$ **21.** **(a)** $\pm 1, \pm 2, \pm 5, \pm 10$ **(b)** $-1, -2, 5$ **(c)** $f(x) = (x + 1)(x + 2)(x - 5)$ **23.** **(a)** $\pm 1, \pm 2, \pm 3, \pm 5, \pm 6, \pm 10, \pm 15, \pm 30$ **(b)** $-5, -3, 2$ **(c)** $f(x) = (x + 5)(x + 3)(x - 2)$ **25.** **(a)** $\pm 1, \pm 2, \pm 3, \pm 4, \pm 6, \pm 12, \pm\frac{1}{2}, \pm\frac{3}{2}, \pm\frac{1}{3}, \pm\frac{2}{3}, \pm\frac{4}{3}, \pm\frac{1}{6}$ **(b)** $-4, -\frac{1}{3}, \frac{3}{2}$ **(c)** $f(x) = (x + 4)(3x + 1)(2x - 3)$ **27.** **(a)** $\pm 1, \pm 2, \pm 3, \pm 6, \pm\frac{1}{2}, \pm\frac{3}{2}, \pm\frac{1}{3}, \pm\frac{2}{3}, \pm\frac{1}{6}, \pm\frac{1}{12}, \pm\frac{1}{4}, \pm\frac{3}{4}$ **(b)** $-\frac{3}{2}, -\frac{2}{3}, \frac{1}{2}$ **(c)** $f(x) = (2x + 3)(3x + 2)(2x - 1)$ **29.** −4 (mult. 2), $\pm\sqrt{7}$, −1 (mult. 4) **31.** 0 (mult. 3), 2, −3, ±1 **33.** $-\frac{7}{9}$ (mult. 2), $4i$ (mult. 2), $-4i$ (mult. 2) **35.** $f(x) = x^2 - 6x + 10$ **37.** $f(x) = x^3 - 5x^2 + 5x + 3$ **39.** $f(x) = x^4 + 4x^3 - 4x^2 - 36x - 45$ **41.** $f(x) = x^3 - 2x^2 + 9x - 18$ **43.** $f(x) = x^4 - 6x^3 + 17x^2 - 28x + 20$ **45.** $f(x) = -3x^3 + 6x^2 + 33x - 36$ **47.** $f(x) = -\frac{1}{2}x^3 - \frac{1}{2}x^2 + x$ **49.** $f(x) = -4x^3 + 20x^2 - 4x + 20$ **51.** −1; 3; $f(x) = (x + 2)^2(x + 1)(x - 3)$ **53.** 2 or 0 positive; 1 negative **55.** 1 positive; 1 negative **57.** 2 or 0 positive; 3 or 1 negative **59.** −0.88, 2.12, 4.86 **61.** 0.44, 1.81 **63.** 1.40

Section 11.3

1. A **3.** one **5.** B and D **7.** one

9.
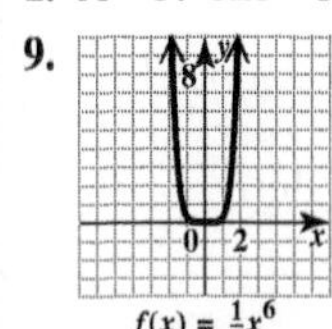

11.
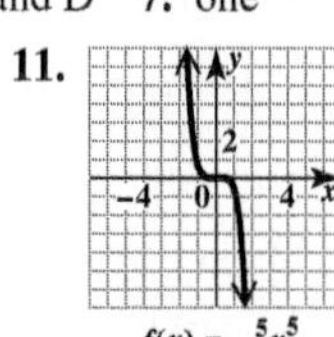

13.
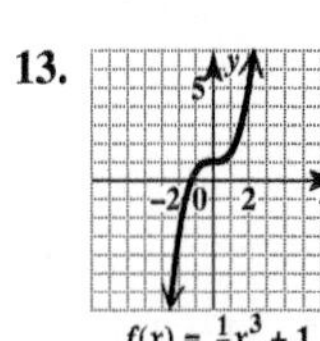

15.
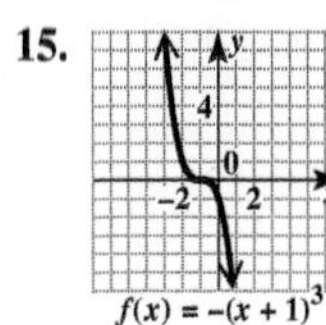

17.
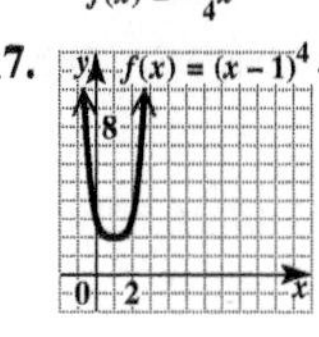

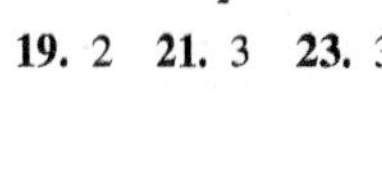

19. 2 **21.** 3 **23.** 3

25.
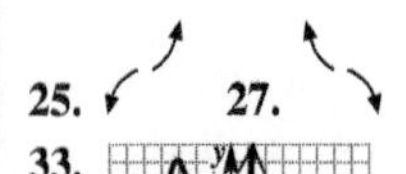
27.
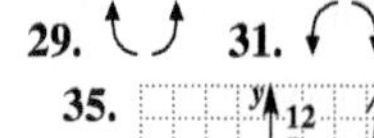
29. **31.**

33.
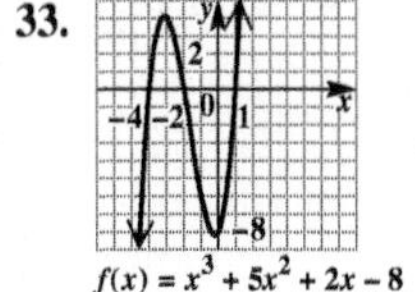

35.
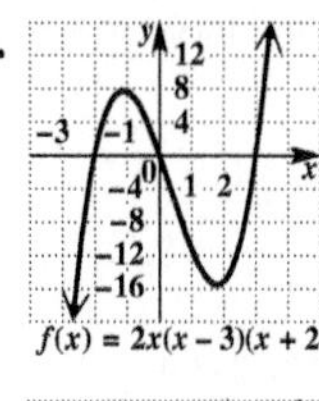

37.
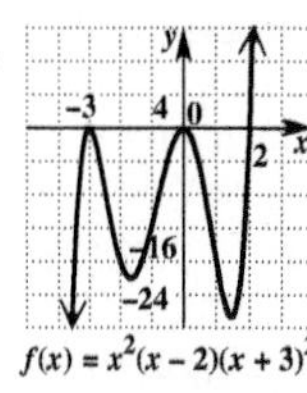

39.
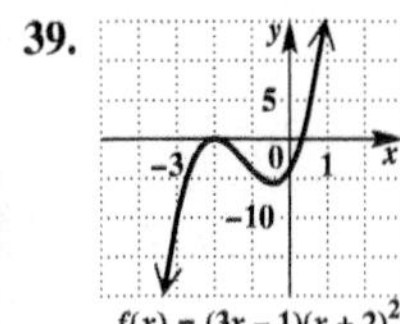

41.
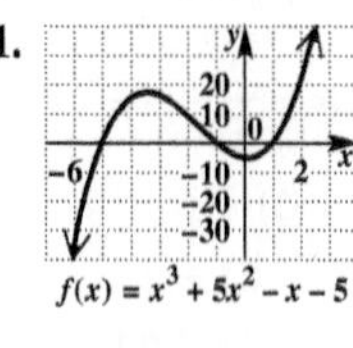

43.
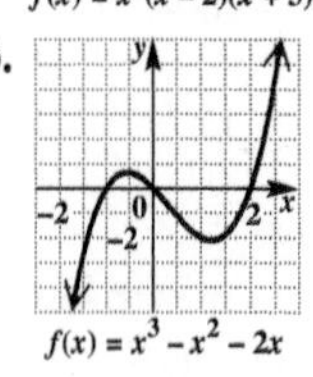

45.
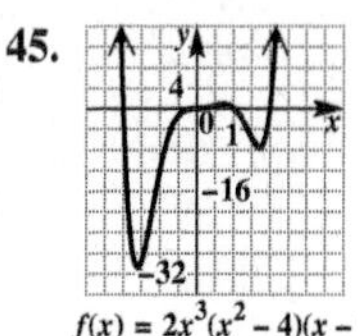

47.
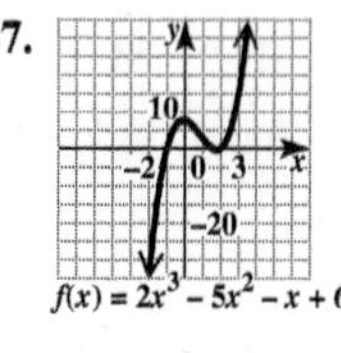

49.
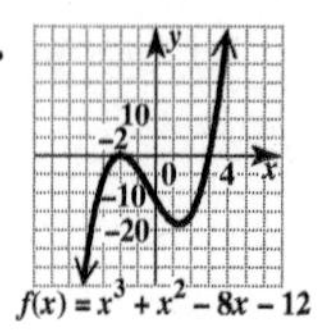

51.
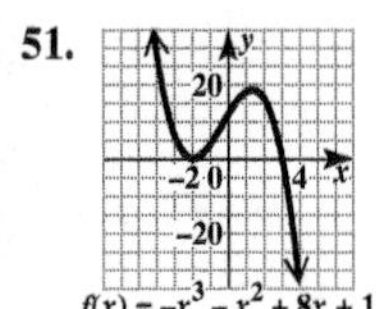

53.
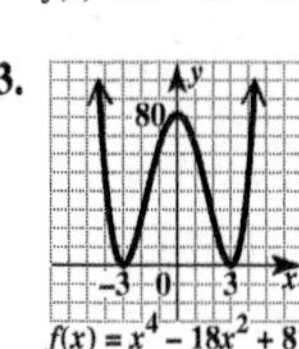

55. (a) $f(-2) = 8 > 0$ and $f(-1) = -2 < 0$ **(b)** $-1.236, 3.236$ **57. (a)** $f(-4) = 76 > 0$ and $f(-3) = -75 < 0$ **(b)** -3.646, $-0.317, 1.646, 6.317$ **65.** $f(x) = \frac{1}{2}(x + 6)(x - 2)(x - 5)$, or $f(x) = \frac{1}{2}x^3 - \frac{1}{2}x^2 - 16x + 30$ **67.** $f(x) = (x - 1)^3(x + 1)^3$, or $f(x) = x^6 - 3x^4 + 3x^2 - 1$ **69.** $f(x) = (x - 3)^2(x + 3)^2$, or $f(x) = x^4 - 18x^2 + 81$ **71.** $-0.89, 2.10, 4.88$ **73.** -1.52 **75.** $-1.07, 1.07$ **77.** $(-3.44, 26.15)$ **79.** $(-0.09, 1.05)$ **81.** $(0.63, 3.47)$ **83. (a)** $y = 10.51x^2 - 27.30x + 47.36$ **(b)** $y = 1.88x^3 - 12.03x^2 + 40.69x + 15.80$ **(c)** $y = 0.35x^4 - 3.65x^3 + 15.49x^2 - 2.38x + 24.10$ **(d)** The quartic function in part (c) best fits the data, with $R^2 = 0.9998$. **85. (a)** $0 < x < 6$ **(b)** $V(x) = x(18 - 2x)(12 - 2x)$, or $V(x) = 4x^3 - 60x^2 + 216x$ **(c)** $x \approx 2.35$; 228.16 in.^{3} **(d)** $0.42 < x < 5$ **87. (a)** $0 < x < 10$ **(b)** $\mathcal{A}(x) = x(20 - 2x)$, or $\mathcal{A}(x) = -2x^2 + 20x$ **(c)** $x = 5$; maximum cross section area: 50 in.^{2} **(d)** between 0 and 2.76 in. or between 7.24 and 10 in. **89.** 1.732 **91.** odd **92.** odd **93.** even **94.** even **95.** odd **96.** odd **97.** neither **98.** neither **99.** even **100.** y-axis; origin

SUMMARY EXERCISES Examining Polynomial Functions and Graphs

1. (a) positive zeros: 1; negative zeros: 3 or 1 **(b)** $\pm 1, \pm 2, \pm 3, \pm 6$ **(c)** $-3, -1$ (mult. 2), 2 **(d)** no other real zeros **(e)** no other complex zeros **(f)** $(-3, 0), (-1, 0), (2, 0)$ **(g)** $(0, -6)$ **(h)** $f(4) = 350$; $(4, 350)$ **(i)** **(j)**
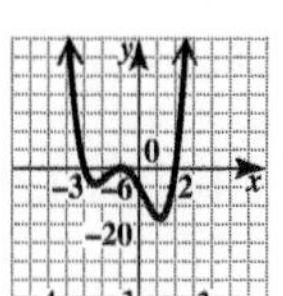

2. (a) positive zeros: 3 or 1; negative zeros: 2 or 0 **(b)** $\pm 1, \pm 3, \pm 5, \pm 9, \pm 15, \pm 45, \pm\frac{1}{2}, \pm\frac{3}{2}, \pm\frac{5}{2}, \pm\frac{9}{2}, \pm\frac{15}{2}, \pm\frac{45}{2}$ **(c)** $-3, \frac{1}{2}, 5$ **(d)** $\pm\sqrt{3}$ **(e)** no other complex zeros **(f)** $(-3, 0), \left(\frac{1}{2}, 0\right), (5, 0), (-\sqrt{3}, 0), (\sqrt{3}, 0)$ **(g)** $(0, 45)$ **(h)** $f(4) = 637$; $(4, 637)$ **(i)** **(j)**
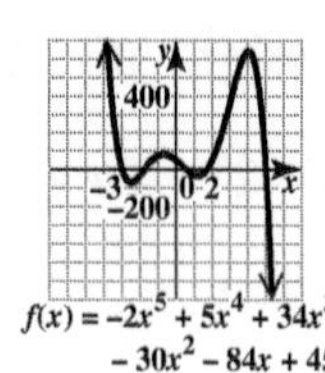

$f(x) = -2x^5 + 5x^4 + 34x^3 - 30x^2 - 84x + 45$

3. (a) positive zeros: 4, 2, or 0; negative zeros: 1 **(b)** $\pm 1, \pm 5, \pm\frac{1}{2}, \pm\frac{5}{2}$ **(c)** 5 **(d)** $\pm\frac{\sqrt{2}}{2}$ **(e)** $\pm i$ **(f)** $\left(-\frac{\sqrt{2}}{2}, 0\right), \left(\frac{\sqrt{2}}{2}, 0\right), (5, 0)$ **(g)** $(0, 5)$ **(h)** $f(4) = -527$; $(4, -527)$ **(i)** **(j)**
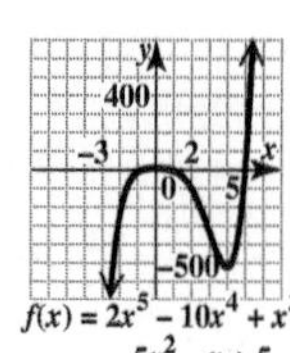

4. (a) positive zeros: 2 or 0; negative zeros: 2 or 0 **(b)** $\pm 1, \pm 2, \pm 3, \pm 6, \pm 9, \pm 18, \pm\frac{1}{3}, \pm\frac{2}{3}$ **(c)** $-\frac{2}{3}, 3$ **(d)** $\frac{-1 \pm \sqrt{13}}{2}$ **(e)** no other complex zeros **(f)** $\left(-\frac{2}{3}, 0\right), (3, 0), \left(\frac{-1 + \sqrt{13}}{2}, 0\right), \left(\frac{-1 - \sqrt{13}}{2}, 0\right)$ **(g)** $(0, 18)$ **(h)** $f(4) = 238$; $(4, 238)$ **(i)** **(j)**
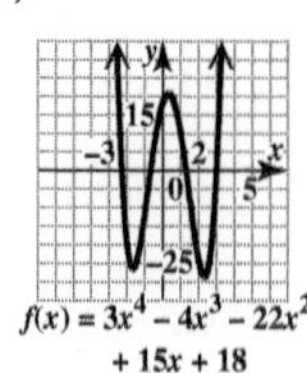

5. (a) positive zeros: 1; negative zeros: 3 or 1 **(b)** $\pm 1, \pm 2, \pm\frac{1}{2}$ **(c)** ± 1 **(d)** no other real zeros **(e)** $-\frac{1}{4} \pm \frac{\sqrt{15}}{4}i$ **(f)** $(-1, 0), (1, 0)$ **(g)** $(0, 2)$ **(h)** $f(4) = -570$; $(4, -570)$ **(i)** **(j)**
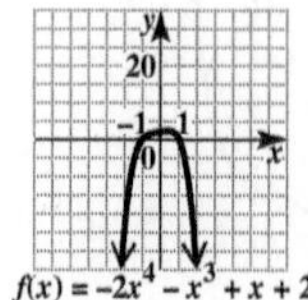

6. (a) positive zeros: 0; negative zeros: 4, 2, or 0

(b) $0, \pm 1, \pm 3, \pm 9, \pm 27, \pm\frac{1}{2}, \pm\frac{3}{2}, \pm\frac{9}{2}, \pm\frac{27}{2}, \pm\frac{1}{4}, \pm\frac{3}{4}, \pm\frac{9}{4}, \pm\frac{27}{4}$

(c) $0, -\frac{3}{2}$ (mult. 2) **(d)** no other real zeros **(e)** $\frac{1}{2} \pm \frac{\sqrt{11}}{2}i$

(f) $(0, 0), \left(-\frac{3}{2}, 0\right)$ **(g)** $(0, 0)$ **(h)** $f(4) = 7260$; $(4, 7260)$

(i) 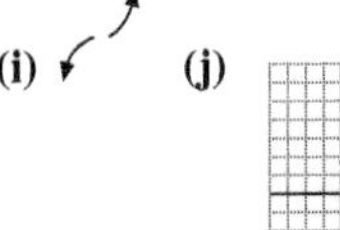 **(j)** 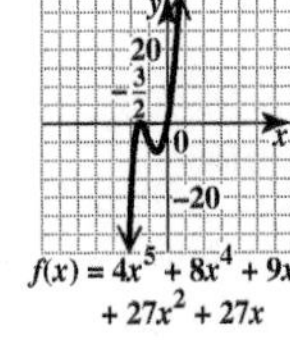

7. (a) positive zeros: 1; negative zeros: 1 **(b)** $\pm 1, \pm 5, \pm\frac{1}{3}, \pm\frac{5}{3}$

(c) no rational zeros **(d)** $\pm\sqrt{5}$ **(e)** $\pm\frac{\sqrt{3}}{3}i$

(f) $(-\sqrt{5}, 0), (\sqrt{5}, 0)$ **(g)** $(0, -5)$ **(h)** $f(4) = 539$; $(4, 539)$

(i) **(j)**

$f(x) = 3x^4 - 14x^2 - 5$

8. (a) positive zeros: 2 or 0; negative zeros: 3 or 1 **(b)** $\pm 1, \pm 3, \pm 9$

(c) $-3, -1$ (mult. 2), 1, 3 **(d)** no other real zeros

(e) no other complex zeros **(f)** $(-3, 0), (-1, 0), (1, 0), (3, 0)$

(g) $(0, -9)$ **(h)** $f(4) = -525$; $(4, -525)$ **(i)** **(j)** $f(x) = -x^5 - x^4 + 10x^3 + 10x^2 - 9x - 9$

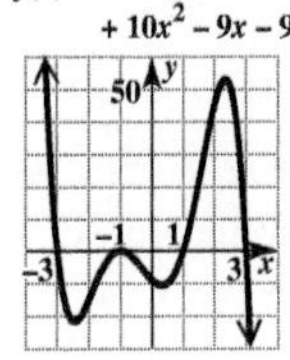

9. (a) positive zeros: 4, 2, or 0; negative zeros: 0 **(b)** $\pm 1, \pm 2, \pm 3, \pm 4, \pm 6, \pm 12, \pm\frac{1}{3}, \pm\frac{2}{3}, \pm\frac{4}{3}$ **(c)** $\frac{1}{3}$, 2 (mult. 2), 3 **(d)** no other real zeros **(e)** no other complex zeros **(f)** $\left(\frac{1}{3}, 0\right), (2, 0), (3, 0)$ **(g)** $(0, -12)$

(h) $f(4) = -44$; $(4, -44)$ **(i)** **(j)** $f(x) = -3x^4 + 22x^3 - 55x^2 + 52x - 12$

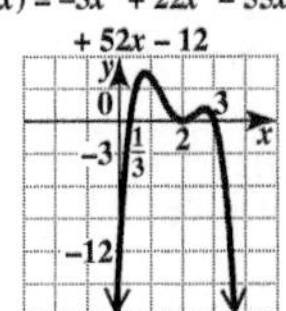

10. For the function in **Exercise 2:** ± 1.732; for the function in **Exercise 3:** ± 0.707; for the function in **Exercise 4:** $-2.303, 1.303$; for the function in **Exercise 7:** ± 2.236

Section 11.4

1. A, B, C **3.** A **5.** A, C, D

7. 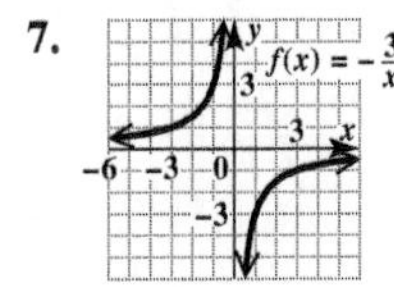

9. 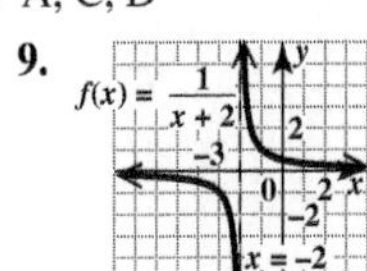

11. 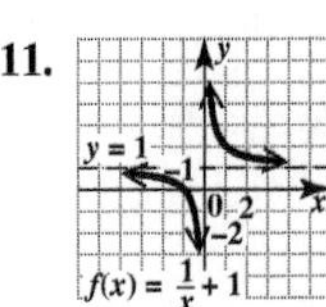

13. The graph of $f(x) = \frac{x^2 - 49}{x + 7}$ is a straight line with hole at $x = -7$. The graph of $g(x) = x - 7$ is the same line but contains the point $(-7, -14)$.

15. V.A.: $x = \frac{7}{3}$; H.A.: $y = 0$ **17.** V.A.: $x = \frac{5}{3}, x = -2$; H.A.: $y = 0$

19. V.A.: $x = -2$; H.A.: $y = -1$ **21.** V.A.: $x = -\frac{9}{2}$; H.A.: $y = \frac{3}{2}$

23. V.A.: $x = 3, x = 1$; H.A.: $y = 0$ **25.** V.A.: $x = -3$; O.A.: $y = x - 3$

27. V.A.: $x = -2, x = \frac{5}{2}$; H.A.: $y = \frac{1}{2}$ **29.** A

31. $f(x) = \frac{x+1}{x-4}$

33. 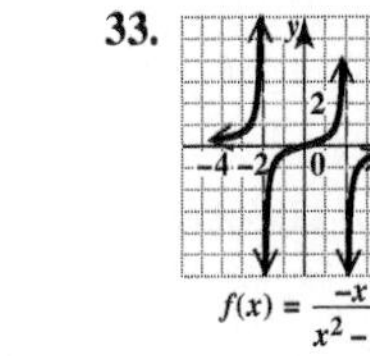

35. 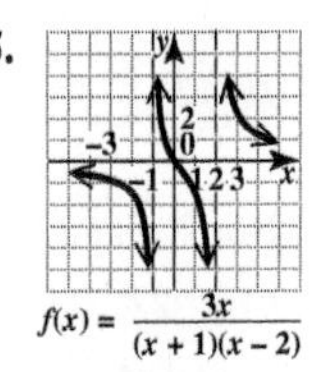

37. 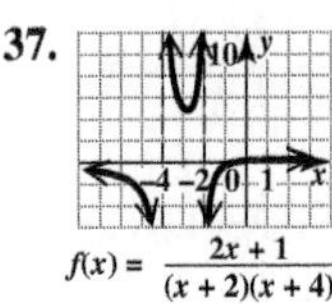

39. 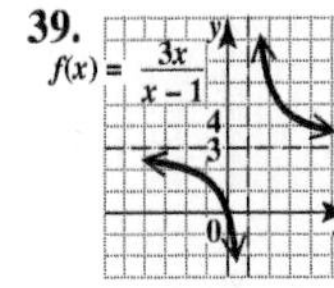

41. 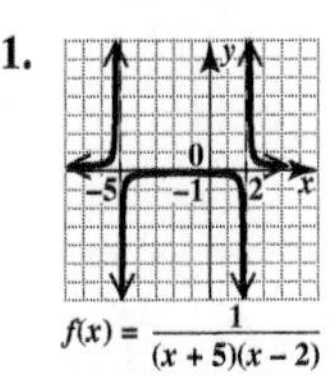

43. 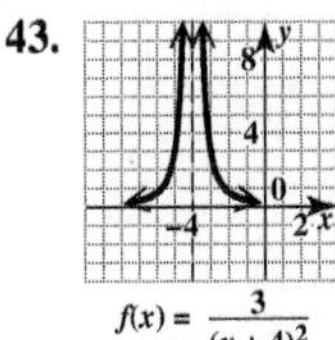

45. 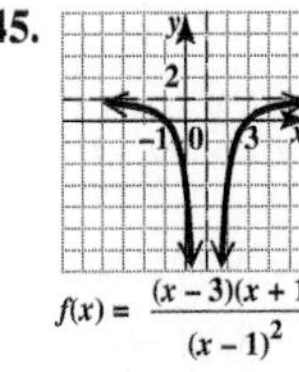

47. 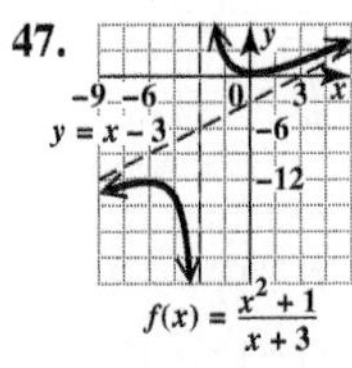

49. 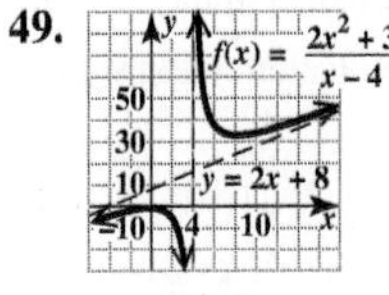

51. 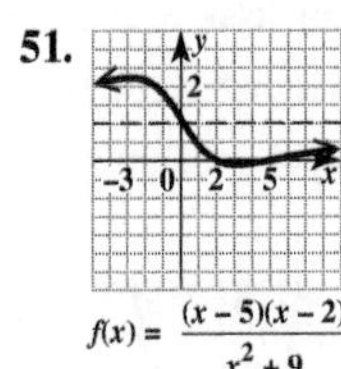

53. 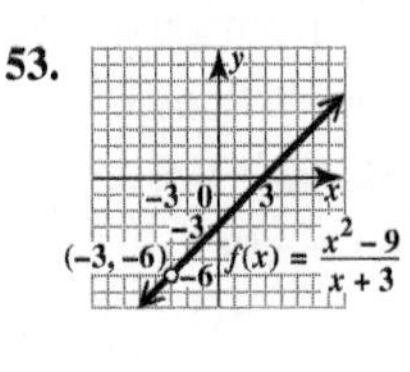

55. 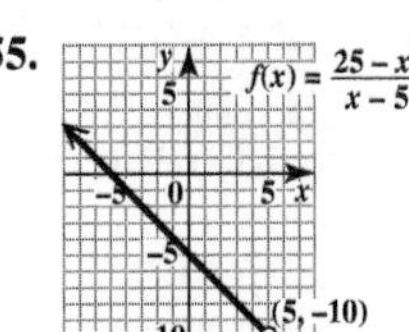

57. (a) C **(b)** A **(c)** B **(d)** D

59. (a) 26 per min **(b)** 5

61. (a) 52 mph **(b)** 34, 56, 85, 121, 164, 215, 273, 340, 415, 499, 591 **(c)** From the table we see that if the speed doubles, the stopping distance more than doubles. **(d)** linear

63. $y = 1$ **64.** $(x + 4)(x + 1)(x - 3)(x - 5)$

65. $(x - 1)(x - 2)(x + 2)(x - 5)$

66. $f(x) = \frac{(x + 4)(x + 1)(x - 3)(x - 5)}{(x - 1)(x - 2)(x + 2)(x - 5)}$ **67. (a)** $x - 5$ **(b)** 5

68. $(-4, 0), (-1, 0), (3, 0)$ **69.** $(0, -3)$ **70.** $x = 1, x = 2, x = -2$

71. $\left(\frac{7 \pm \sqrt{241}}{6}, 1\right)$ **72.**

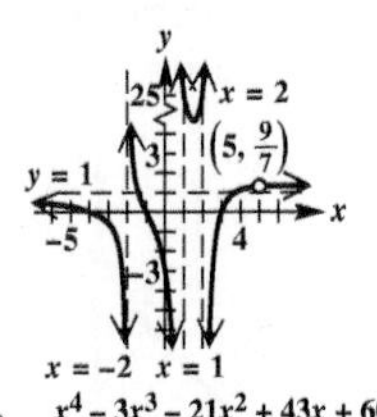

Chapter 11 Review Exercises

1. $3x+2$ **2.** $10x-23+\frac{31}{x+2}$ **3.** $2x^2+x+3+\frac{21}{x-3}$

4. $-x^3+4x^2+3x+6+\frac{-9}{x+4}$ **5.** yes **6.** no **7.** -13

8. -5 **9.** $\frac{1}{2}, -1, 5$ **10.** $\frac{1}{3}, -2, 5$ **11.** $4, -\frac{1}{2}, -\frac{2}{3}$

12. $3, -1, \frac{1}{4}, -\frac{1}{2}$ **13.** no **14.** yes; $f(x)=(x+1)(x-3)(2x+5)$

15. $f(x)=-2x^3+6x^2+12x-16$ **16.** $f(x)=x^4-3x^2-4$

17. $f(x)=x^4+x^3+19x^2+25x-150$ **18.** $f(x)=x^3+x^2-4x+6$

19. $1-i, 1+i, 4, -3$; $f(x)=(x-1+i)(x-1-i)(x-4)(x+3)$

20. No. The number of real zeros cannot exceed the degree.

21. two **22.** three

23.

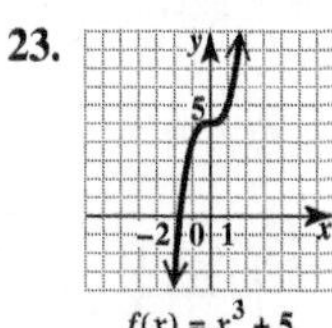

$f(x)=x^3+5$

24.

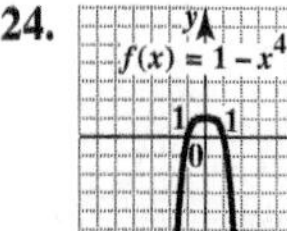

$f(x)=1-x^4$

25.

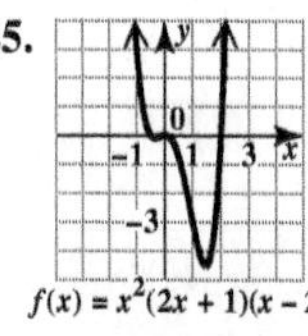

$f(x)=x^2(2x+1)(x-2)$

26.

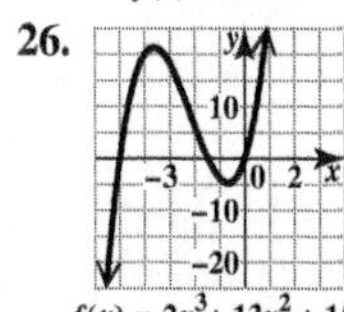

$f(x)=2x^3+13x^2+15x$

27.

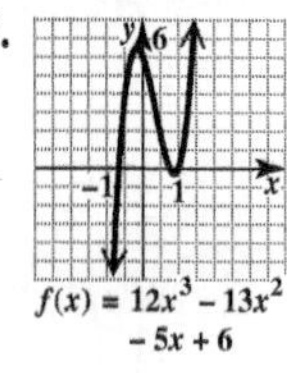

$f(x)=12x^3-13x^2-5x+6$

28.

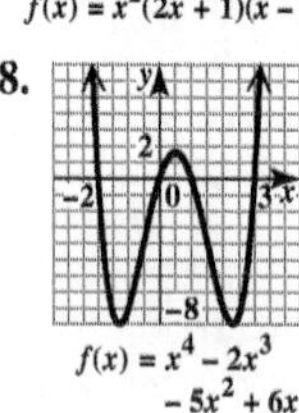

$f(x)=x^4-2x^3-5x^2+6x$

29. $f(-1)=-10$ and $f(0)=2$; $f(2)=-4$ and $f(3)=14$

30. $f(-1)=15$ and $f(0)=-8$; $f(-6)=-20$ and $f(-5)=27$

31. $1>0$ and synthetic division leads to all positive numbers in the bottom row; $-3<0$ and synthetic division leads to numbers with alternating signs.

32. **(a)** $-0.5, 0.354, 5.646$ **(b)** $-2.259, 4.580$ **(c)** $-3.895, -0.397, 1.292$

33.

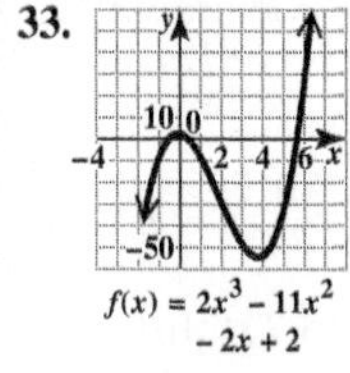

$f(x)=2x^3-11x^2-2x+2$

34.

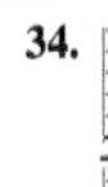

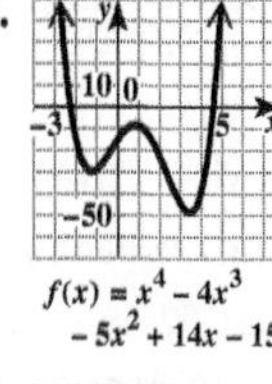

$f(x)=x^4-4x^3-5x^2+14x-15$

35.

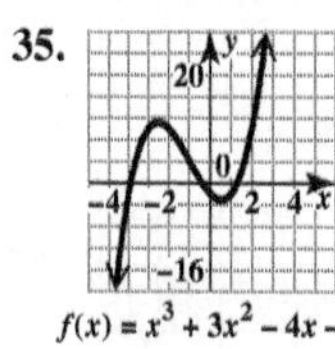

$f(x)=x^3+3x^2-4x-2$

36.

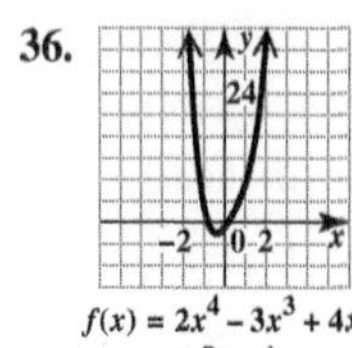

$f(x)=2x^4-3x^3+4x^2+5x-1$

37.

$f(x)=\frac{8}{x}$

38.

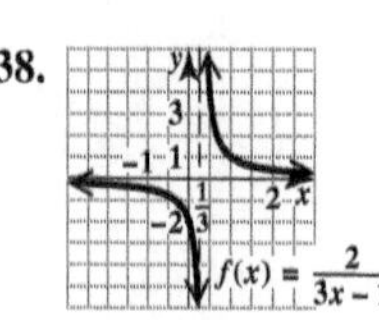

$f(x)=\frac{2}{3x-1}$

39.

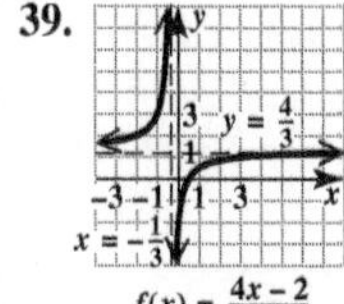

$f(x)=\frac{4x-2}{3x+1}$

40.

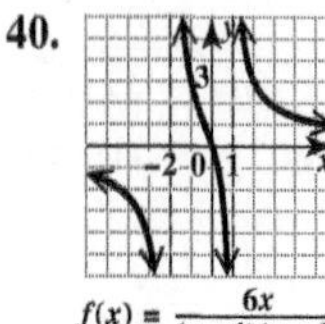

$f(x)=\frac{6x}{(x-1)(x+2)}$

41.

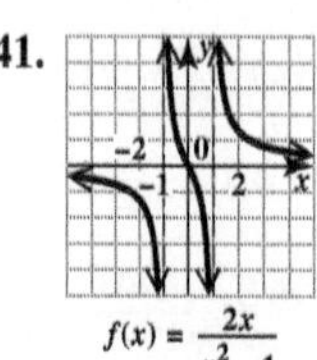

$f(x)=\frac{2x}{x^2-1}$

42.

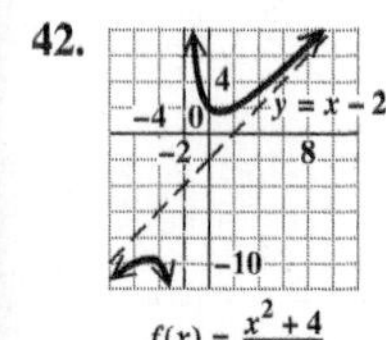

$f(x)=\frac{x^2+4}{x+2}$

43.

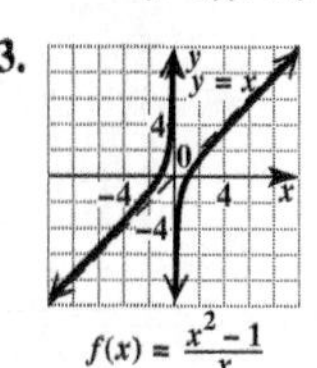

$f(x)=\frac{x^2-1}{x}$

44.

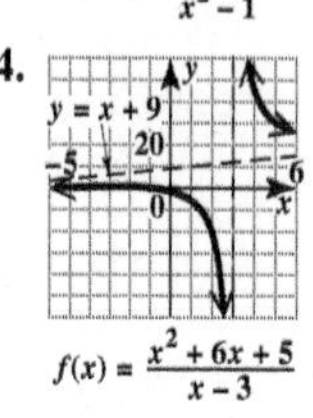

$f(x)=\frac{x^2+6x+5}{x-3}$

45. $f(x)=\frac{4x^2-9}{2x+3}$

46.

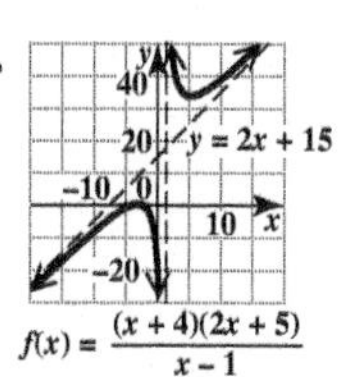

$f(x)=\frac{(x+4)(2x+5)}{x-1}$

47. All answers are given in tens of millions. **(a)** \$65.5 **(b)** \$64 **(c)** \$60 **(d)** \$40 **(e)** \$0 **48.** V.A.: $x=110$; H.A.: $y=80$

Chapter 11 Mixed Review Exercises

1. **(a)** $-5, 1$ **(b)** $f(x)=(x+3)(x+5)(x-1)$ **(c)** $(-3,0), (-5,0), (1,0)$; y-intercept: $(0,-15)$ **(d)**

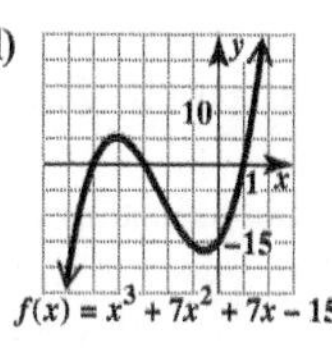

$f(x)=x^3+7x^2+7x-15$

2. **(a)** $x=-3, x=3$ **(b)** $(-2,0), (2,0)$ **(c)** $y=1$ **(d)**

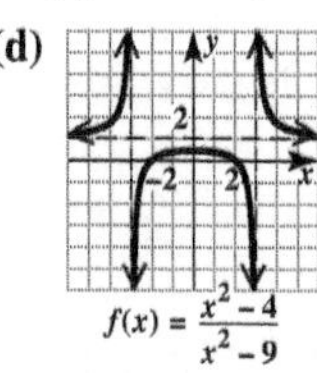

$f(x)=\frac{x^2-4}{x^2-9}$

3.

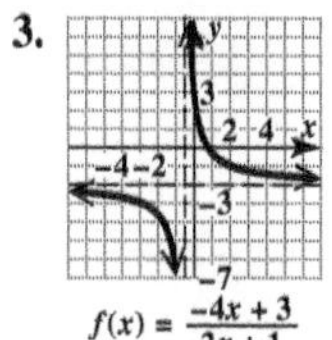

$f(x)=\frac{-4x+3}{2x+1}$

4.

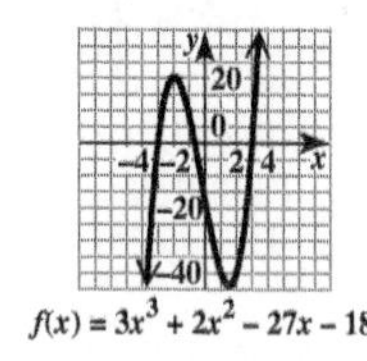

$f(x)=3x^3+2x^2-27x-18$

5. $f(x)=(x+2)(2x^2-4x+9)+(-24)$

6. $f(x)=x^4+8x^2-9$ **7.** zero; solution; x-intercept **8.** C

Chapter 11 Test

[11.1] **1.** $2x^2+4x+5$ **2.** $f(x)=(x+1)(x^3+x^2-2x+5)+(-10)$ **3.** yes **4.** -227 **5.** Yes, 3 is a zero because the last term in the bottom row of the synthetic division is 0.

[11.2] **6.** $f(x)=2x^4-2x^3-2x^2-2x-4$ **7.** **(a)** $\pm 1, \pm\frac{1}{2}, \pm\frac{1}{3}, \pm\frac{1}{6}, \pm 7, \pm\frac{7}{2}, \pm\frac{7}{3}, \pm\frac{7}{6}$ **(b)** $-\frac{1}{3}, 1, \frac{7}{2}$ [11.3] **8.** $2>0$ and synthetic division leads to all positive numbers in the bottom row; $-1<0$ and synthetic division leads to numbers with alternating signs. **9.** **(a)** $f(-2)=-11<0$ and $f(-1)=2>0$ **(b)** -1.290 **10.** **(a)** 3 **(b)** 2

11. $f(x)=(x-1)^4$

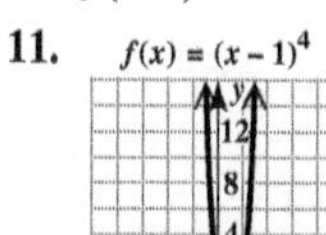

12. $f(x)=x(x+1)(x-2)$

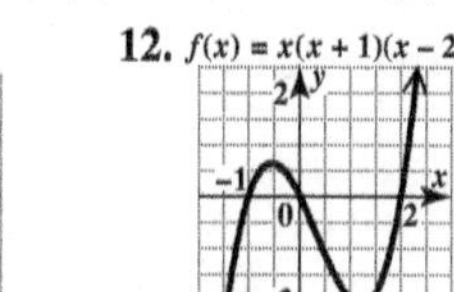

13.

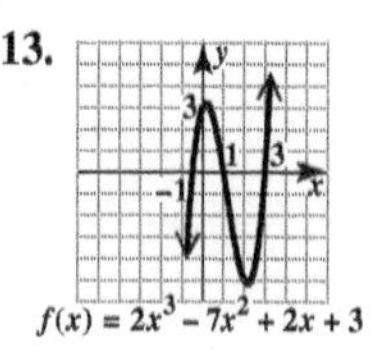

$f(x)=2x^3-7x^2+2x+3$

14.

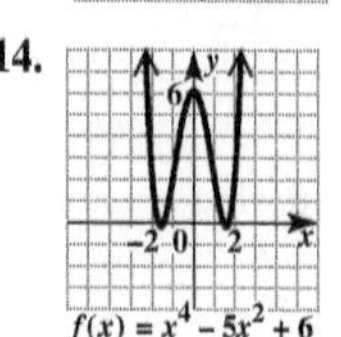

$f(x)=x^4-5x^2+6$

15. 49°F [11.4] **16.** $f(x)=\frac{-2}{x+3}$

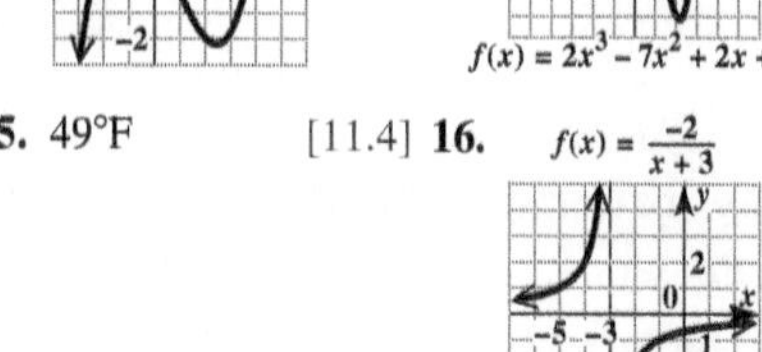

17. $f(x)=\frac{3x-1}{x-2}$ **18.** $f(x)=\frac{x^2-1}{x^2-9}$

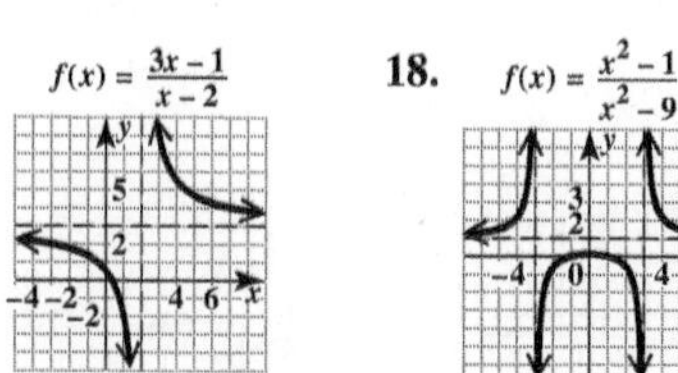

19. $f(x)=\frac{2x^2+x-6}{x-1}$

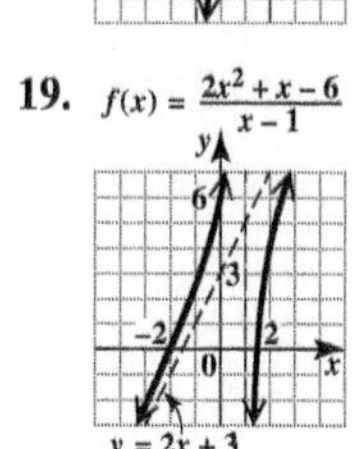

$y=2x+3$

20. D

Chapters R–11 Cumulative Review Exercises

[R.1] **1.** $\frac{1}{4}$ **2.** 6 **3.** $\frac{35}{12}$ **4.** 47.3% **5.** 3.26 **6.** 75% [1.1] **7.** $\{2\}$ **8.** $\{24\}$ **9.** {all real numbers} [1.5] **10.** $(-\infty, 8]$ **11.** $\left(3, \frac{13}{3}\right)$ [1.7] **12.** $(-\infty, -7) \cup (3, \infty)$ [1.3] **13.** 26 nickels and 12 dimes [2.3] **14.** $y = -3x + 12$ [2.1] **15.**

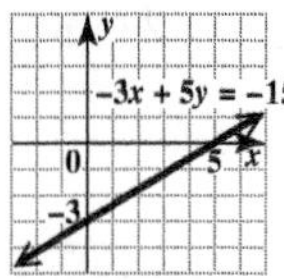

[2.4] **16.**

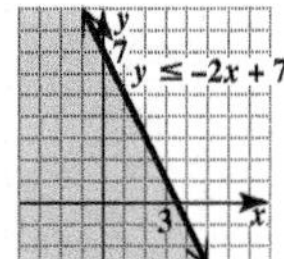

[4.4] **17.**

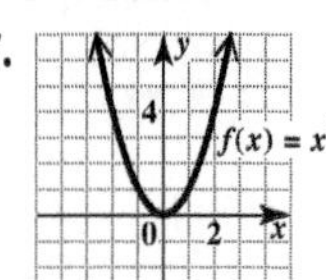

[3.1] **18.** $\{(-2, 1)\}$ [3.2] **19.** $\{(-3, 4, 2)\}$ [4.5] **20.** $3r^5 - 7r^4 + 2r^3 + 18r - 6$ **21.** $k^2 - 10kh + 25h^2 + 4k - 20h + 4$ [5.1, 5.5] **22.** $\{-3, -1, 1\}$ [5.2] **23.** $3(x - 3)(2x + 1)$ [5.3] **24.** $(9 + 2y^2)(81 - 18y^2 + 4y^4)$ [6.1] **25.** **(a)** $-\frac{3}{4}, 6$ **(b)** $\frac{x + 6}{4x + 3}$ [6.3] **26.** $\frac{2y - 1}{-y - 1}$, or $\frac{1 - 2y}{y + 1}$ [6.1] **27.** $\frac{6}{25}$ **28.** $\frac{y - 3}{y + 2}$ [6.2] **29.** -2 **30.** $\frac{-r^2 + r + 4}{(r - 2)(r - 1)}$ [6.4] **31.** $\{2, -5\}$ **32.** $\varnothing$ [7.3] **33.** $5\sqrt{2}$ [7.5] **34.** $5 + \sqrt{2}$ [8.2] **35.** 8 **36.** irrational **37.** $\{-2 \pm \sqrt{2}\}$ [9.4] **38.** x-axis **39.** y-axis **40.** origin [9.5] **41.** 3 [10.1] **42.** $f^{-1}(x) = \frac{x^3 - 5}{3}$ [10.6] **43.** $\{2\}$ [11.1] **44.** $x^3 + 6x^2 - 11x + 13$ [11.4] **45.**

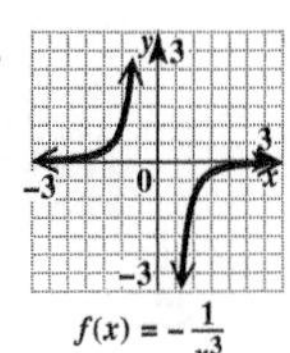

$f(x) = -\frac{1}{x^3}$

46.

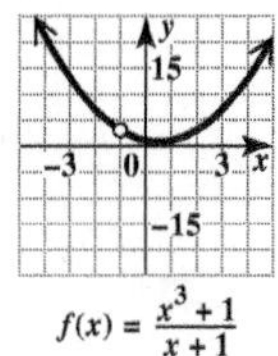

$f(x) = \frac{x^3 + 1}{x + 1}$

12 CONIC SECTIONS AND NONLINEAR SYSTEMS

Section 12.1

1. conic sections; parabolas **3.** ellipse; foci **5.** **(a)** $(0, 0)$ **(b)** 5 **(c)**

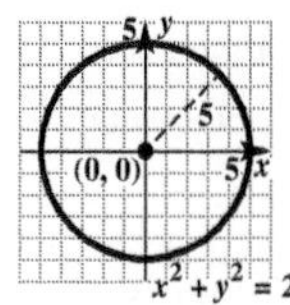

$x^2 + y^2 = 25$

7. B **9.** C

11. $(0, 0)$; $r = 2$

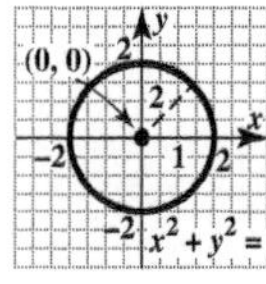

$x^2 + y^2 = 4$

13. $(0, 0)$; $r = 9$

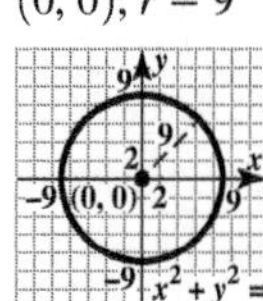

$x^2 + y^2 = 81$

15. $(5, -4)$; $r = 7$

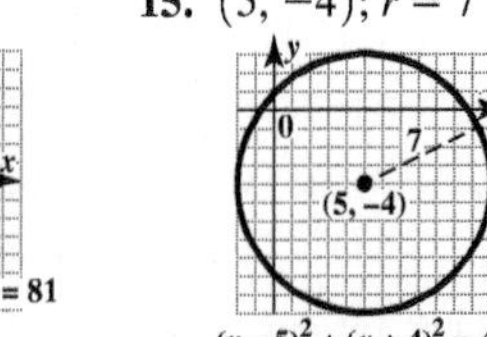

$(x - 5)^2 + (y + 4)^2 = 49$

17. $(-1, -3)$; $r = 5$

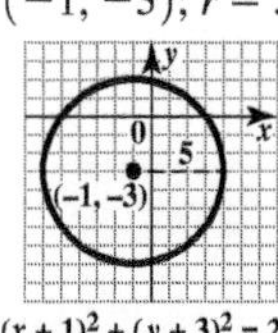

$(x + 1)^2 + (y + 3)^2 = 25$

19. $x^2 + y^2 = 16$

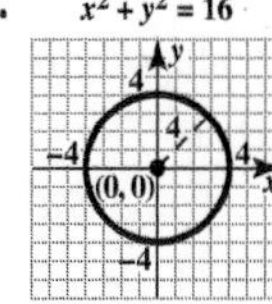

21. $(x + 3)^2 + (y - 2)^2 = 9$

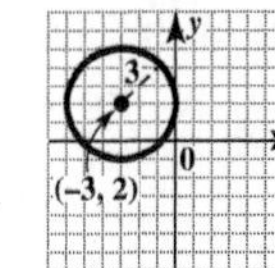

23. $(x - 4)^2 + (y - 3)^2 = 25$

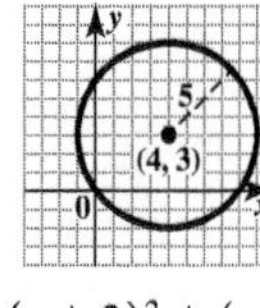

25. $(x + 2)^2 + y^2 = 5$

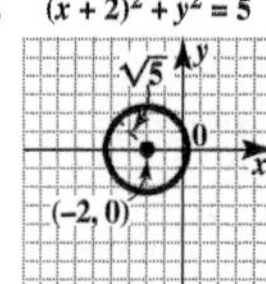

27. $x^2 + (y + 3)^2 = 49$

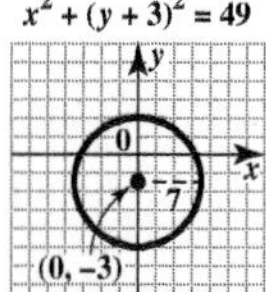

29. $(x + 2)^2 + (y + 3)^2 = 4$; $(-2, -3)$; $r = 2$ **31.** $(x + 5)^2 + (y - 7)^2 = 81$; $(-5, 7)$; $r = 9$ **33.** $(x - 2)^2 + (y - 4)^2 = 16$; $(2, 4)$; $r = 4$ **35.** $(x - 2)^2 + (y - 3)^2 = 4$; $(2, 3)$; $r = 2$

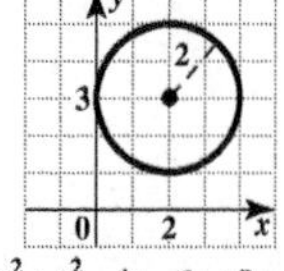

$x^2 + y^2 - 4x - 6y + 9 = 0$

37. $(x + 3)^2 + (y - 3)^2 = 9$; $(-3, 3)$; $r = 3$

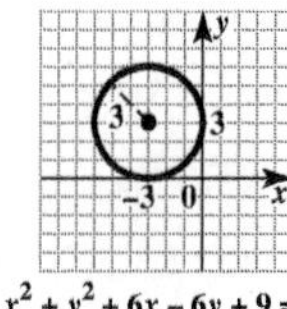

$x^2 + y^2 + 6x - 6y + 9 = 0$

39. one point; The only ordered pair that satisfies the equation is $(4, 1)$. **41.** The thumbtack acts as the center, and the length of the string acts as the radius.

43.

$\frac{x^2}{9} + \frac{y^2}{25} = 1$

45.

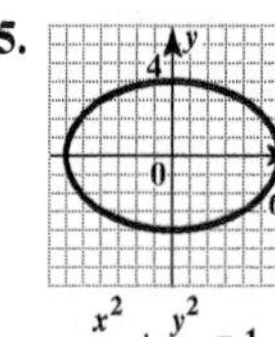

$\frac{x^2}{36} + \frac{y^2}{16} = 1$

47.

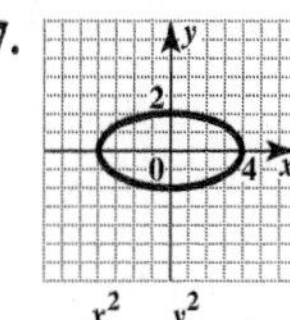

$\frac{x^2}{16} + \frac{y^2}{4} = 1$

49.

$\frac{y^2}{25} = 1 - \frac{x^2}{49}$

51.

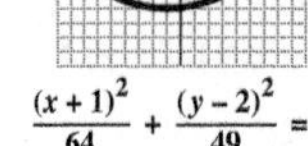

$\frac{(x + 1)^2}{64} + \frac{(y - 2)^2}{49} = 1$

53.

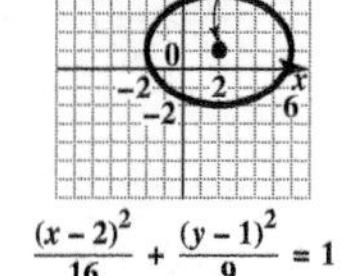

$\frac{(x - 2)^2}{16} + \frac{(y - 1)^2}{9} = 1$

55. $3\sqrt{3}$ units **57.** **(a)** 10 m **(b)** 36 m **59.** **(a)** 348.2 ft **(b)** 1787.6 ft **61.** **(a)** 154.7 million mi **(b)** 128.7 million mi

Section 12.2

1. C **3.** D **5.** Because the coefficient of the x^2-term is positive, this is a horizontal hyperbola with x-intercepts $(-3, 0)$ and $(3, 0)$.

7.

$\frac{x^2}{16} - \frac{y^2}{9} = 1$

9.

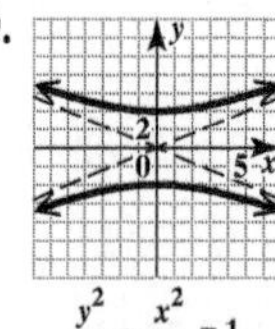

$\frac{y^2}{4} - \frac{x^2}{25} = 1$

11.

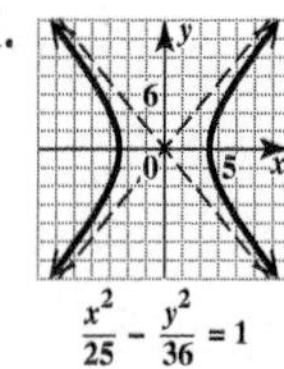

$\frac{x^2}{25} - \frac{y^2}{36} = 1$

13.

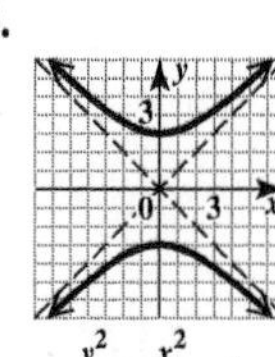

$\frac{y^2}{9} - \frac{x^2}{9} = 1$

15. hyperbola

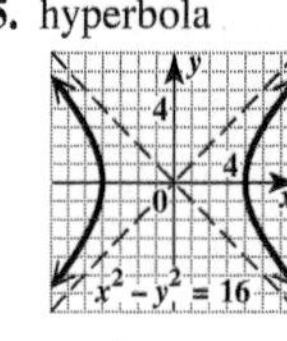

17. ellipse

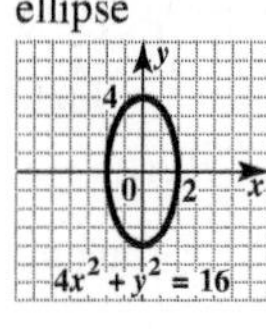

19. circle

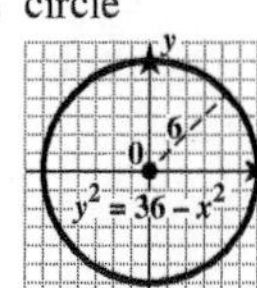

21. ellipse

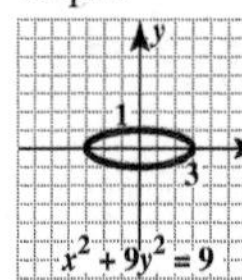

23. parabola

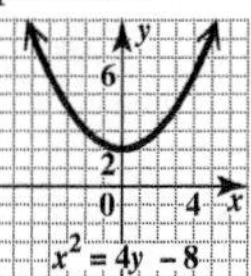

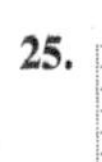

25. 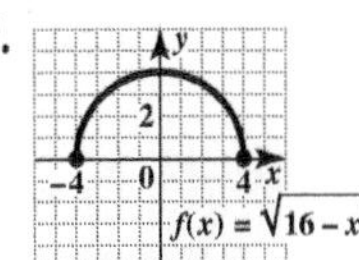

domain: $[-4, 4]$; range: $[0, 4]$

27. 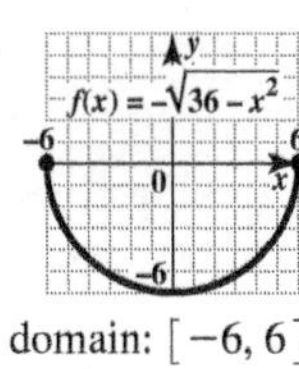

domain: $[-6, 6]$; range: $[-6, 0]$

29. 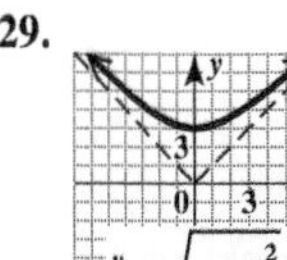

domain: $(-\infty, \infty)$; range: $[3, \infty)$

31. 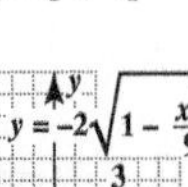 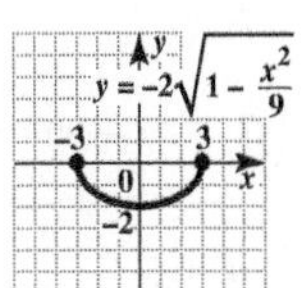 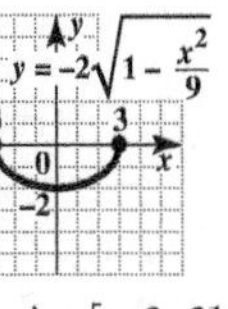

domain: $[-3, 3]$; range: $[-2, 0]$

33. **(a)** 50 m **(b)** 69.3 m

35. $y = \pm x$

37. $\frac{(x-2)^2}{4} - \frac{(y+1)^2}{9} = 1$

38. $\frac{(x+3)^2}{16} - \frac{(y-2)^2}{25} = 1$

 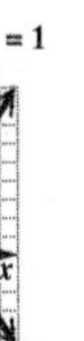

39. $\frac{y^2}{36} - \frac{(x-2)^2}{49} = 1$

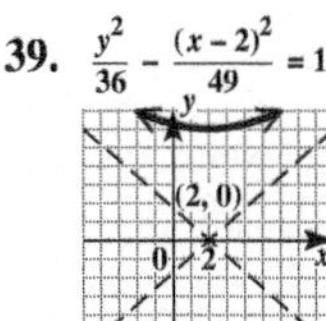 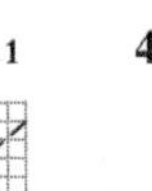

40. $\frac{(y-5)^2}{9} - \frac{x^2}{25} = 1$

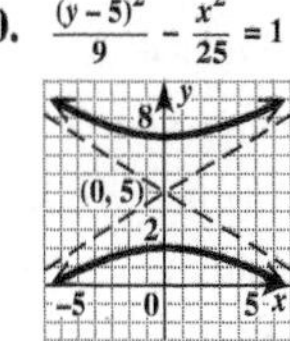

Section 12.3

1. one **3.** none

5. 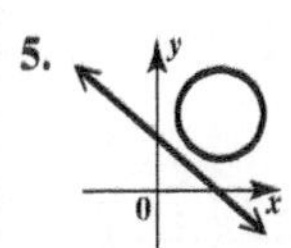 **7.** 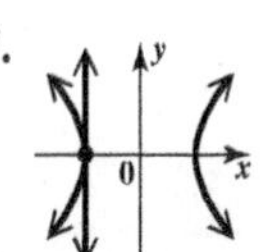 **9.** 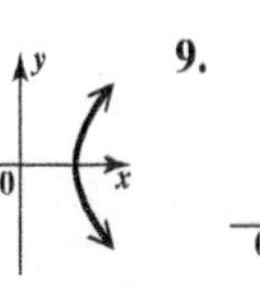 **11.** 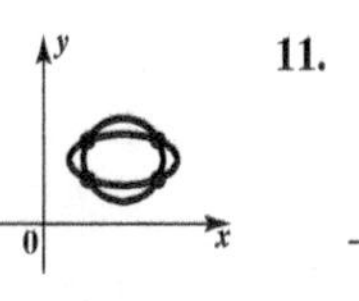

13. $\left\{(0, 0), \left(\frac{1}{2}, \frac{1}{2}\right)\right\}$ **15.** $\{(-6, 9), (-1, 4)\}$

17. $\left\{\left(-\frac{1}{5}, \frac{7}{5}\right), (1, -1)\right\}$ **19.** $\left\{(-2, -2), \left(-\frac{4}{3}, -3\right)\right\}$

21. $\{(-3, 1), (1, -3)\}$ **23.** $\left\{\left(-\frac{3}{2}, -\frac{9}{4}\right), (-2, 0)\right\}$

25. $\{(-\sqrt{3}, 0), (\sqrt{3}, 0), (-\sqrt{5}, 2), (\sqrt{5}, 2)\}$

27. $\left\{\left(\frac{\sqrt{3}}{3}i, -\frac{1}{2} + \frac{\sqrt{3}}{6}i\right), \left(-\frac{\sqrt{3}}{3}i, -\frac{1}{2} - \frac{\sqrt{3}}{6}i\right)\right\}$

29. $\{(-2, 0), (2, 0)\}$ **31.** $\{(1, 3), (1, -3), (-1, 3), (-1, -3)\}$

33. $\{(-\sqrt{3}, 0), (\sqrt{3}, 0)\}$

35. $\{(-2\sqrt{3}, -2), (-2\sqrt{3}, 2), (2\sqrt{3}, -2), (2\sqrt{3}, 2)\}$

37. $\{(i\sqrt{2}, -3i\sqrt{2}), (-i\sqrt{2}, 3i\sqrt{2}), (-\sqrt{6}, -\sqrt{6}), (\sqrt{6}, \sqrt{6})\}$

39. $\{(-2i\sqrt{2}, -2\sqrt{3}), (-2i\sqrt{2}, 2\sqrt{3}), (2i\sqrt{2}, -2\sqrt{3}), (2i\sqrt{2}, 2\sqrt{3})\}$

41. $\{(-\sqrt{5}, -\sqrt{5}), (\sqrt{5}, \sqrt{5})\}$

43. $\{(-3, -1), (3, 1), (-i, 3i), (i, -3i)\}$

45. $\{(i, 2i), (-i, -2i), (2, -1), (-2, 1)\}$ **47.** length: 12 ft; width: 7 ft

49. $20; $\frac{4}{5}$ thousand or 800 calculators

Section 12.4

1. **(a)** B **(b)** D **(c)** A **(d)** C **3.** $x \geq 0$, $y \geq 0$ **5.** $x < 0$, $y \leq 0$ **7.** C

9. 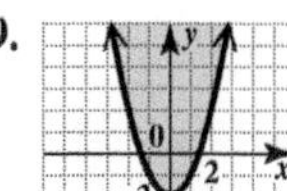 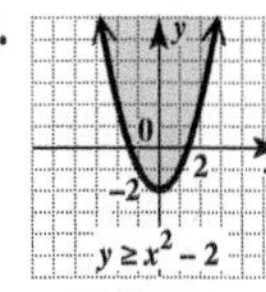 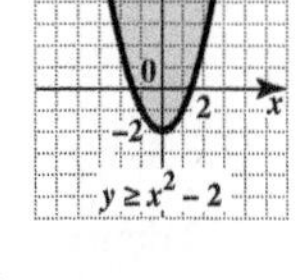

11. 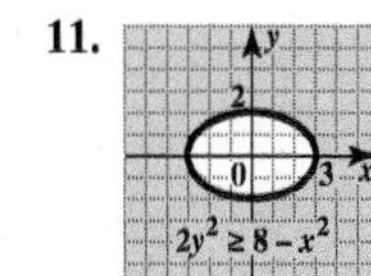

13. 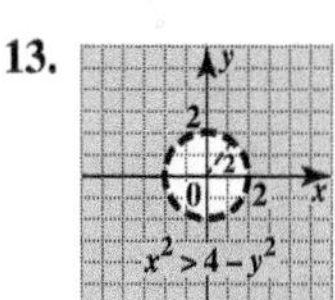

15. 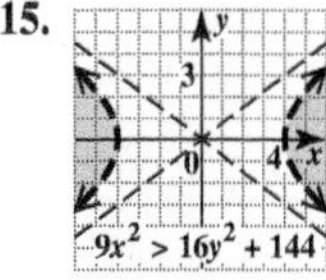

17. 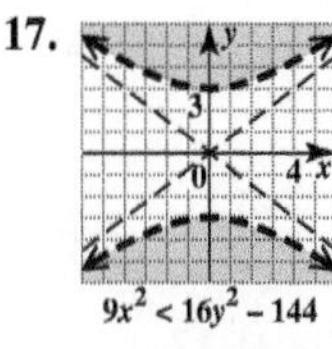

19. 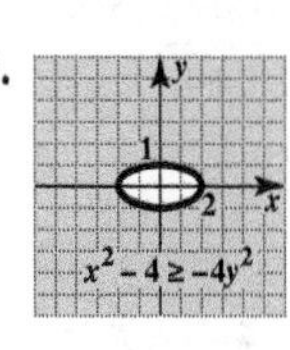

21. 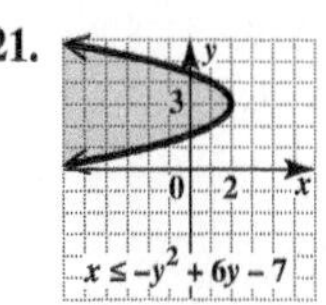

23. 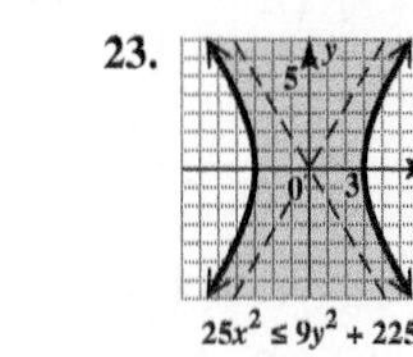

25. 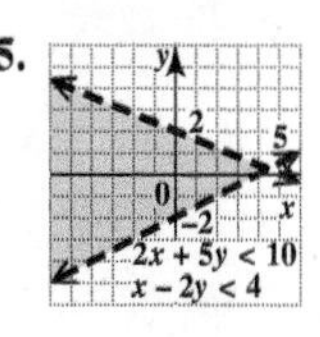

27. 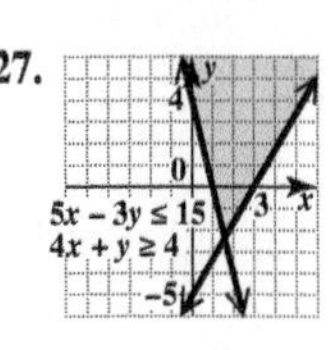

29. 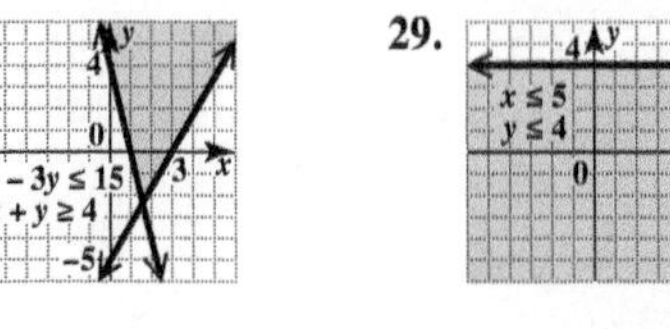

31. 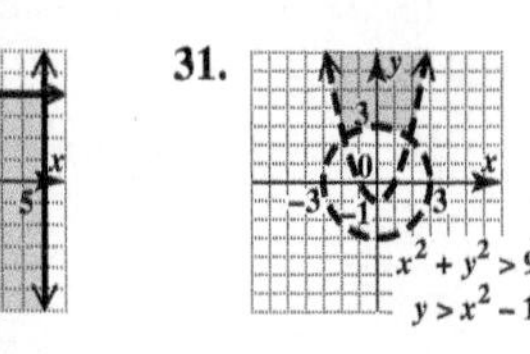

33. 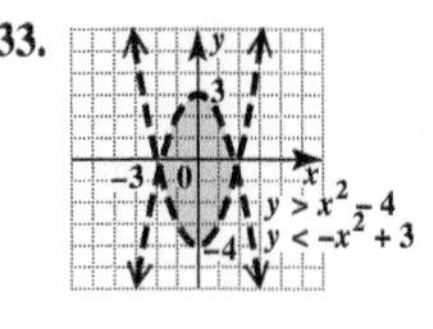

35. 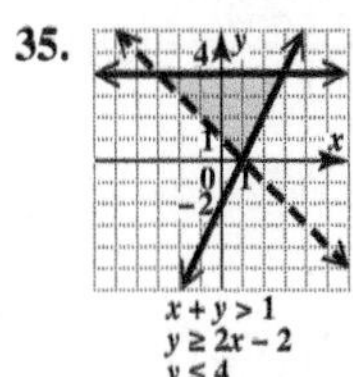

37. 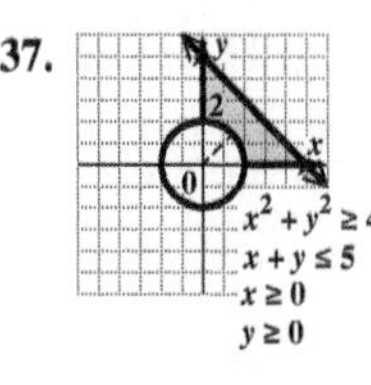

39. 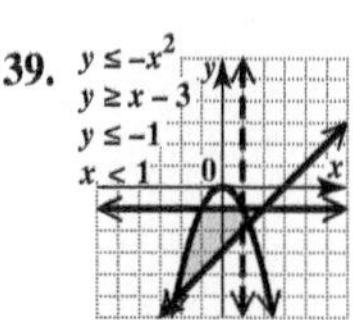

41. 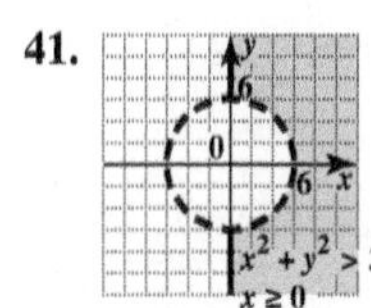

43. 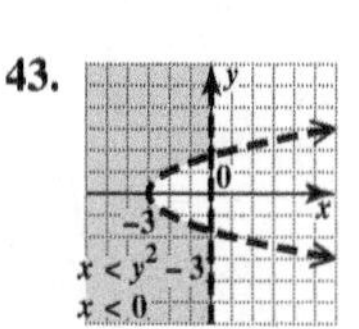

45. 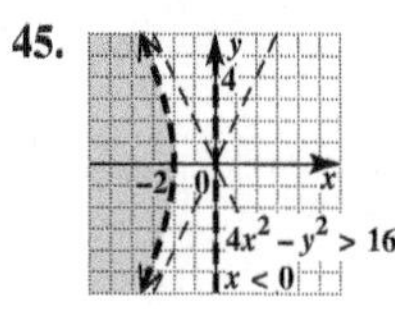

47. 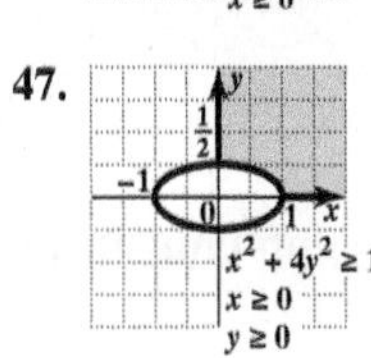

49. maximum: 65; minimum: 8

51. maximum: 900; minimum: 0

53. maximum of $\frac{42}{5}$ at $\left(\frac{6}{5}, \frac{6}{5}\right)$ **55.** minimum of $\frac{49}{3}$ at $\left(\frac{17}{3}, 5\right)$

57. 4 pigs, 12 geese; $1120 **59.** 8 of cabinet A, 3 of cabinet B; 100 ft^3 of storage **61.** 800 bargain sets, 300 deluxe sets; $125,000

63. 4000 medical kits, 2000 containers of water; 44,000 people

Chapter 12 Review Exercises

1. $x^2 + y^2 = 49$ **2.** $(x + 2)^2 + (y - 4)^2 = 9$

3. $(x - 4)^2 + (y - 2)^2 = 36$ **4.** $(x + 5)^2 + y^2 = 2$

5. $(x + 3)^2 + (y - 2)^2 = 16$; $(-3, 2)$; $r = 4$

6. $(x - 4)^2 + (y - 1)^2 = 4$; $(4, 1)$; $r = 2$

7. $(x + 1)^2 + (y + 5)^2 = 9$; $(-1, -5)$; $r = 3$

8. $(x - 3)^2 + (y + 2)^2 = 25$; $(3, -2)$; $r = 5$

9. **(a)** circle **(b)** ellipse **(c)** ellipse **(d)** circle

10. **(a)** A **(b)** C **(c)** D **(d)** B

11. 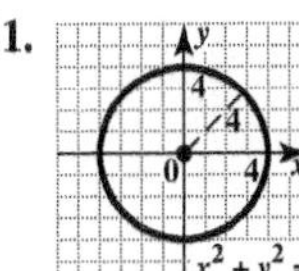

12. 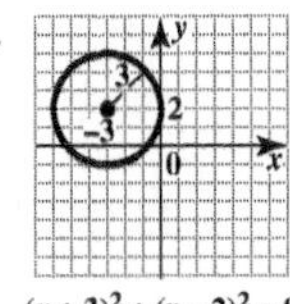

13. 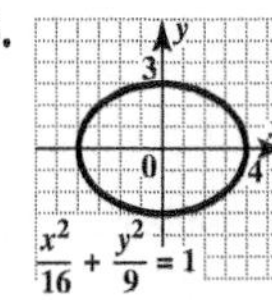

14. 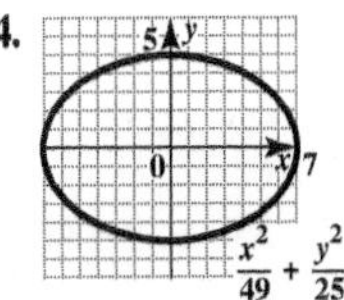

15. 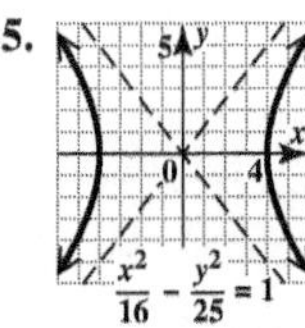

16. 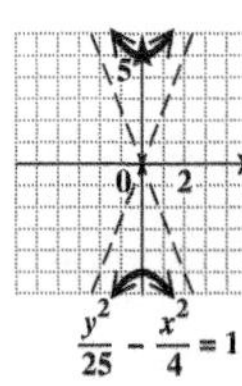

17. 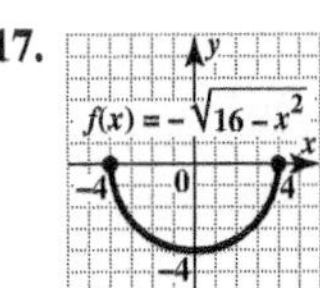

18. (a) C (b) A (c) D (d) B **19.** circle **20.** parabola **21.** hyperbola **22.** ellipse **23.** parabola **24.** hyperbola **25.** 0, 1, or 2 **26.** 0, 1, 2, 3, or 4

27. $\{(6, -9), (-2, -5)\}$ **28.** $\{(1, 2), (-5, 14)\}$ **29.** $\{(4, 2), (-1, -3)\}$ **30.** $\{(-2, -4), (8, 1)\}$ **31.** $\{(-\sqrt{2}, 2), (-\sqrt{2}, -2), (\sqrt{2}, -2), (\sqrt{2}, 2)\}$ **32.** $\{(-\sqrt{6}, -\sqrt{3}), (-\sqrt{6}, \sqrt{3}), (\sqrt{6}, -\sqrt{3}), (\sqrt{6}, \sqrt{3})\}$

33. 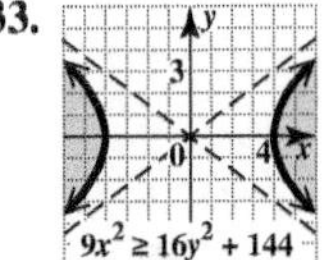

34. 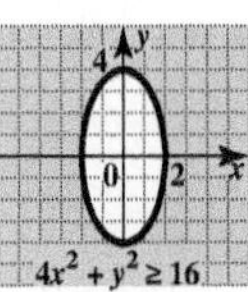

35. 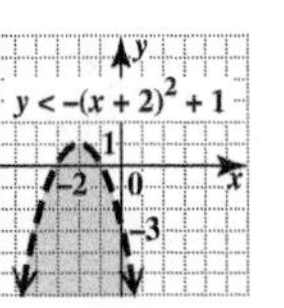

36. D

37. 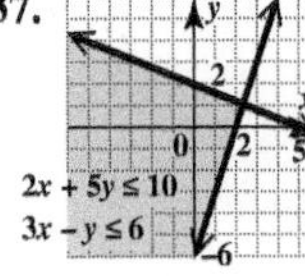

38. 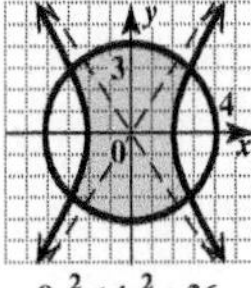

39. 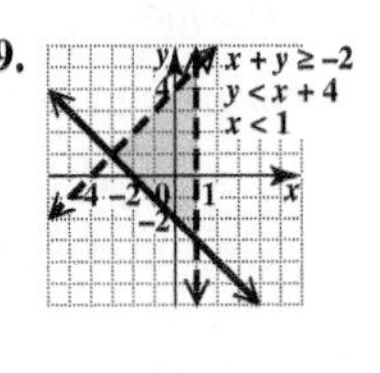

40. (a) B (b) D (c) C (d) A **41.** 3 batches of cakes, 6 batches of cookies; $210 **42.** $\frac{14}{3}$ units of basic, $\frac{20}{3}$ units of plain; $193.33

Chapter 12 Mixed Review Exercises

1. $(-2, 5)$; $r = 6$ **2.** true

3. 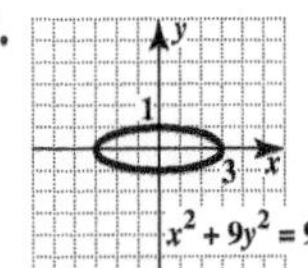

4. 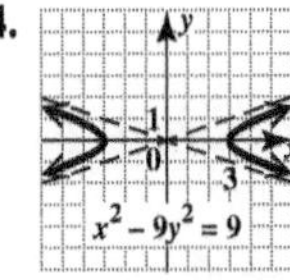

5. 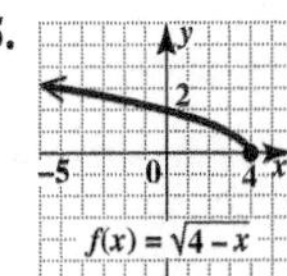

6. 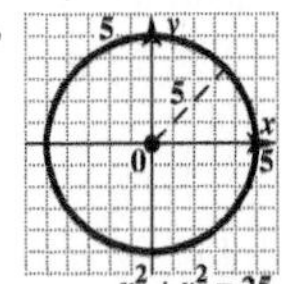

7. 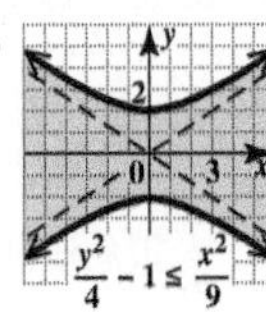

8. 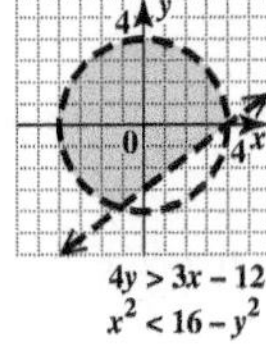

9. $\{(-7, -35), (12, 60)\}$ **10.** $\{(-1, -2), (1, 2), (-2i, i), (2i, -i)\}$ **11.** $\left(\frac{1}{2}, \frac{9}{4}\right)$ **12.** $12\frac{1}{4}$

Chapter 12 Test

[12.1] **1.** 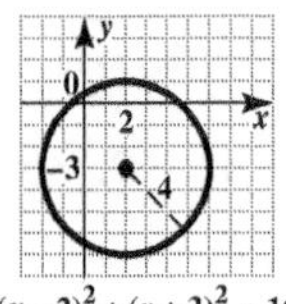

$(2, -3)$; $r = 4$

2. D **3.** $(0, 0)$; $r = 1$ **4.** $(-4, 1)$; $r = 5$

[12.2] **5.** 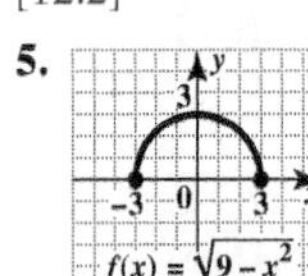

[12.1] **6.** 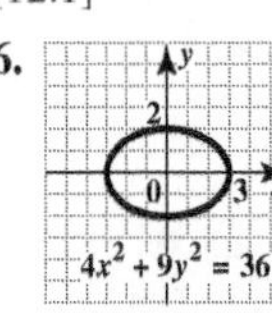

[12.2] **7.** 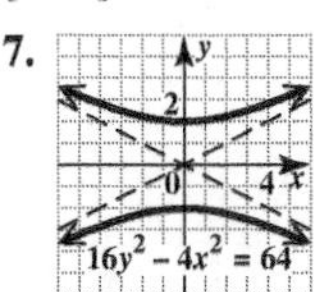

8. $\frac{y}{2} = -\sqrt{1 - \frac{x^2}{9}}$

9. ellipse **10.** hyperbola **11.** circle **12.** parabola [12.3] **13.** $\left\{\left(-\frac{1}{2}, -10\right), (5, 1)\right\}$ **14.** $\left\{(-2, -2), \left(\frac{14}{5}, -\frac{2}{5}\right)\right\}$

15. $\{(-\sqrt{22}, -\sqrt{3}), (-\sqrt{22}, \sqrt{3}), (\sqrt{22}, -\sqrt{3}), (\sqrt{22}, \sqrt{3})\}$

[12.4]

16. 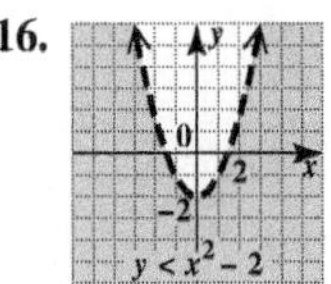

17. 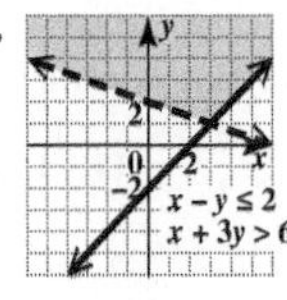

18. 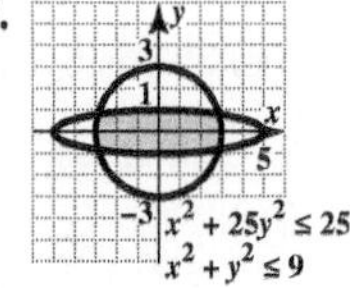

19. (a) $(8, 8)$ (b) 48 **20.** 25 radios, 30 DVD players; $1425

Chapters R–12 Cumulative Review Exercises

[R.1] **1.** (a) 0.04; 4% (b) $\frac{3}{20}$; 0.15 (c) $\frac{4}{5}$; 80% (d) 1.25; 125% **2.** (a) 24.5 (b) 2450 (c) 0.245 (d) 0.0245 [R.4, 4.1, 7.1, 7.7] **3.** (a) 16 (b) 16 (c) -16 (d) -16 (e) 2 (f) -2 (g) $2i$ [R.4] **4.** 54 [1.1] **5.** $\left\{\frac{2}{3}\right\}$ [1.5] **6.** $\left(-\infty, \frac{3}{5}\right]$ [1.7] **7.** $\{-4, 4\}$ **8.** $(-\infty, -5) \cup (10, \infty)$ [2.2] **9.** $\frac{2}{3}$ [2.3] **10.** $3x + 2y = -13$ [3.1] **11.** $\{(3, -3)\}$ [3.2] **12.** $\{(4, 1, -2)\}$ [12.3] **13.** $\left\{(-1, 5), \left(\frac{5}{2}, -2\right)\right\}$ [3.3] **14.** 40 mph [4.5] **15.** $25y^2 - 30y + 9$ [4.6] **16.** $4x^3 - 4x^2 + 3x + 5 + \frac{3}{2x + 1}$ [6.1] **17.** $\frac{y - 1}{y(y - 3)}$ [6.2] **18.** $\frac{3c + 5}{(c + 5)(c + 3)}$ **19.** $\frac{1}{p}$ [6.3] **20.** $\frac{xy}{y - x}$ [7.4] **21.** $2\sqrt[3]{2}$ [7.7] **22.** $\frac{7}{5} + \frac{11}{5}i$ [5.2] **23.** $(3x + 2)(4x - 5)$ [5.3] **24.** $(z^2 + 1)(z + 1)(z - 1)$ **25.** $(a - 3b)(a^2 + 3ab + 9b^2)$ [6.5] **26.** $1\frac{1}{5}$ hr [4.1] **27.** $\frac{a^5}{4}$ [7.5] **28.** $\frac{3\sqrt{10}}{2}$ [7.6] **29.** $\varnothing$ [5.5] **30.** $\left\{\frac{1}{5}, -\frac{3}{2}\right\}$ [8.1, 8.2] **31.** $\left\{\frac{3 \pm \sqrt{33}}{6}\right\}$ [8.3] **32.** $\left\{\pm\frac{\sqrt{6}}{2}, \pm\sqrt{7}\right\}$ [8.4] **33.** $v = \frac{\pm\sqrt{rFkw}}{kw}$ [10.6] **34.** $\{3\}$ [4.4, 9.1] **35.** (a) -1 (b) $9x^2 + 18x + 4$ [9.5] **36.** domain: $(-\infty, \infty)$; range: $[0, \infty)$

[10.1] **37.** $f^{-1}(x) = \sqrt[3]{x-4}$ [10.4, 10.5] **38. (a)** 4 **(b)** 7

[10.4] **39.** $\log \frac{(3x+7)^2}{4}$ [11.1] **40.** 23 [11.2] **41.** Yes, $x + 2$ is a factor of $f(x)$. The other factor is $5x^3 + 6x - 4$. **42.** $-2, -\frac{4}{3}, 3$

[2.6] **43.**

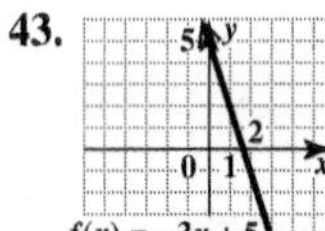

[7.1] **44.**

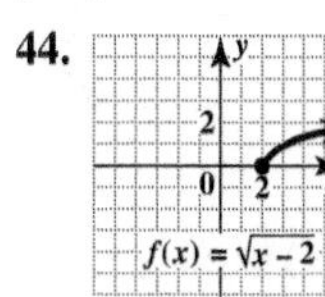

[9.2] **45.**

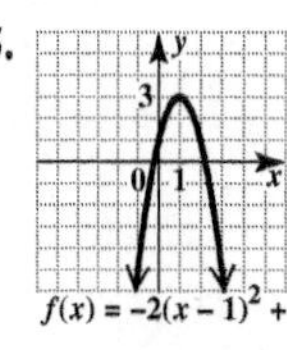

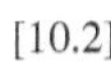

[10.2] **46.**

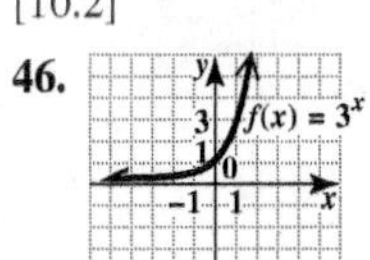

[12.2] **47.**

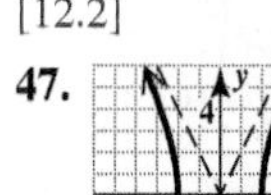

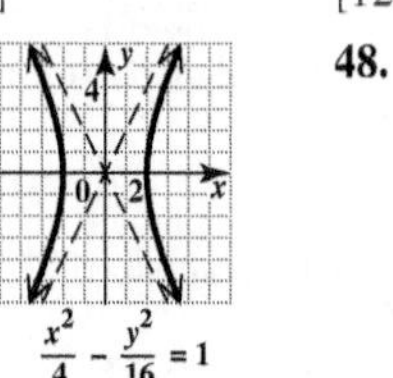

[12.4] **48.**

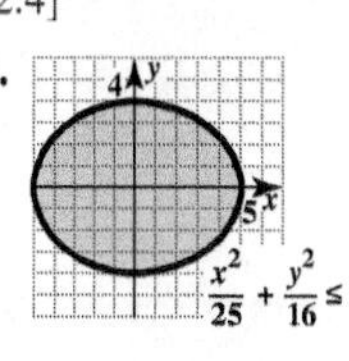

13 FURTHER TOPICS IN ALGEBRA

Section 13.1

1. the set of all positive integers (natural numbers) **3.** 80 **5.** 12 **7.** 2, 3, 4, 5, 6 **9.** $4, \frac{5}{2}, 2, \frac{7}{4}, \frac{8}{5}$ **11.** 3, 9, 27, 81, 243 **13.** $-1, -\frac{1}{4}, -\frac{1}{9}, -\frac{1}{16}, -\frac{1}{25}$ **15.** 5, −5, 5, −5, 5 **17.** $0, \frac{3}{2}, \frac{8}{3}, \frac{15}{4}, \frac{24}{5}$ **19.** −70 **21.** $\frac{49}{23}$ **23.** 171 **25.** $4n$ **27.** $-8n$ **29.** $\frac{1}{3^n}$ **31.** $\frac{n+1}{n+4}$ **33.** \$110, \$109, \$108, \$107, \$106, \$105; \$400 **35.** \$6554 **37.** $4 + 5 + 6 + 7 + 8 = 30$ **39.** $3 + 6 + 11 = 20$ **41.** $-2 + 2 - 2 + 2 - 2 + 2 = 0$ **43.** $0 + 6 + 14 + 24 + 36 = 80$

Answers may vary for Exercises 45–49.

45. $\sum_{i=1}^{5} (i+2)$ **47.** $\sum_{i=1}^{5} 2^i(-1)^i$ **49.** $\sum_{i=1}^{4} i^2$

51. In summation notation, each replacement for i should be squared first before the negative symbol is applied. The correct answer is $\sum_{i=1}^{5} -i^2$. **53.** 9 **55.** $\frac{40}{9}$ **57.** 9200 mutual funds

Section 13.2

1. difference **3.** 55 **5.** $\frac{5}{2}; 3; \frac{7}{2}$ **7.** 16; 26 **9.** $d = 1$ **11.** not arithmetic **13.** $d = -5$ **15.** 5, 9, 13, 17, 21 **17.** −2, −6, −10, −14, −18 **19.** $\frac{1}{3}, 1, \frac{5}{3}, \frac{7}{3}, 3$ **21.** $a_n = 5n - 3$; 122 **23.** $a_n = \frac{3}{4}n + \frac{9}{4}$; 21 **25.** $a_n = 3n - 6$; 69 **27.** 76 **29.** 48 **31.** −1 **33.** 16 **35.** 26 **37.** 6 **39.** The student subtracted terms in the wrong order. Subtracting the first term from the second term, we find that $d = -10 - (-15) = -10 + 15 = 5$. **41.** 15 **43.** 87 **45.** $\frac{63}{2}$ **47.** 81 **49.** −3 **51.** −2.1 **53.** 390 **55.** 395 **57.** 31,375 **59.** \$465 **61.** \$2100 per month **63.** 68 seats; 1100 seats **65.** no; 3 blocks; 9 rows

Section 13.3

1. ratio **3.** 31 **5.** $-\frac{1}{81}; \frac{1}{243}; -\frac{1}{729}$ **7.** 14; 56 **9.** $r = 2$ **11.** not geometric **13.** $r = -3$ **15.** $r = -\frac{1}{2}$

There are alternative forms of the answers in Exercises 17–21.

17. $a_n = -5(2)^{n-1}$ **19.** $a_n = -2\left(-\frac{1}{3}\right)^{n-1}$ **21.** $a_n = 10\left(-\frac{1}{5}\right)^{n-1}$ **23.** 3,906,250 **25.** $\frac{1}{354{,}294}$ **27.** −4374 **29.** 2, 6, 18, 54, 162 **31.** $5, -1, \frac{1}{5}, -\frac{1}{25}, \frac{1}{125}$ **33.** −4, −2, −1, −0.5, −0.25 **35.** −1,048,575 **37.** $\frac{121}{243}$ **39.** −1.997 **41.** 2.662 **43.** −2.982 **45.** \$24,470.52 **47.** \$74,156.06 **49.** 9 **51.** $\frac{10{,}000}{11}$ **53.** $-\frac{9}{20}$ **55.** The sum does not exist. **57.** 1.3 ft **59.** 140 m **61.** 3 days; $\frac{1}{4}$ g **63. (a)** 1.5 billion units **(b)** 12 yr **65.** \$5005.65 **67.** $\frac{a_1}{1-r} = \frac{0.9}{1-0.1} = \frac{0.9}{0.9} = 1$; Therefore, $0.99999\ldots = 1$.

Section 13.4

1. 1; sum **3.** x^3 **5.** 6 **7.** 1 **9.** 720 **11.** 40,320 **13.** 15 **15.** 1 **17.** 120 **19.** 15 **21.** 78 **23.** 120 **25.** $m^4 + 4m^3n + 6m^2n^2 + 4mn^3 + n^4$ **27.** $a^5 - 5a^4b + 10a^3b^2 - 10a^2b^3 + 5ab^4 - b^5$ **29.** $8x^3 + 36x^2 + 54x + 27$ **31.** $\frac{x^4}{16} - \frac{x^3y}{2} + \frac{3x^2y^2}{2} - 2xy^3 + y^4$ **33.** $x^8 + 4x^6 + 6x^4 + 4x^2 + 1$ **35.** $27x^6 - 27x^4y^2 + 9x^2y^4 - y^6$ **37.** $r^{12} + 24r^{11}s + 264r^{10}s^2 + 1760r^9s^3$ **39.** $3^{14}x^{14} - 14(3^{13})x^{13}y + 91(3^{12})x^{12}y^2 - 364(3^{11})x^{11}y^3$ **41.** $t^{20} + 10t^{18}u^2 + 45t^{16}u^4 + 120t^{14}u^6$ **43.** $120(2^7)m^7n^3$ **45.** $\frac{7x^2y^6}{16}$ **47.** $36k^7$ **49.** $160x^6y^3$ **51.** $4320x^9y^4$

Section 13.5

1. $3 + 6 + 9 + 12 = \frac{3(4)(4+1)}{2}$; true **3.** $\frac{1}{2} + \frac{1}{2^2} + \frac{1}{2^3} + \frac{1}{2^4} = \frac{2^4 - 1}{2^4}$; true **5.** $2^4 < 2(4)$; false **7.** positive; integers

Although we do not usually give proofs, the answers to Exercises 9 and 17 are shown here.

9. ***Step 1:*** $3(1) = 3$ and $\frac{3(1)(1+1)}{2} = \frac{6}{2} = 3$, so S_n is true for $n = 1$.

Step 2: S_k: $3 + 6 + 9 + \cdots + 3k = \frac{3k(k+1)}{2}$;

S_{k+1}: $3 + 6 + 9 + \cdots + 3(k+1) = \frac{3(k+1)[(k+1)+1]}{2}$;

Add $3(k+1)$ to each side of S_k and simplify until S_{k+1} is obtained. Because S_n is true for $n = 1$, and S_n is true for $n = k + 1$ when it is true for $n = k$, S_n is true for every positive integer value of n.

17. ***Step 1:*** $\frac{1}{1 \cdot (1+1)} = \frac{1}{2}$ and $\frac{1}{1+1} = \frac{1}{2}$, so S_n is true for $n = 1$.

Step 2: S_k: $\frac{1}{1 \cdot 2} + \frac{1}{2 \cdot 3} + \frac{1}{3 \cdot 4} + \cdots + \frac{1}{k(k+1)} = \frac{k}{k+1}$;

S_{k+1}: $\frac{1}{1 \cdot 2} + \frac{1}{2 \cdot 3} + \cdots + \frac{1}{(k+1)[(k+1)+1]} = \frac{k+1}{(k+1)+1}$;

Add $\frac{1}{(k+1)[(k+1)+1]}$ to each side of S_k and simplify until S_{k+1} is obtained. Because S_n is true for $n = 1$, and S_n is true for $n = k + 1$ when it is true for $n = k$, S_n is true for every positive integer value of n.

31. $\frac{4^{n-1}}{3^{n-2}}$, or $3\left(\frac{4}{3}\right)^{n-1}$

Section 13.6

1. permutation; combination **3.** $3 \cdot 2 \cdot 1 = 6$
5. No, this is a misnomer. Rearranging the numbers or letters in this interpretation of "combination" produces a different "combination." For example, 1, 4, 3 is a different "combination" than 4, 1, 3 or 3, 4, 1.
7. 360 **9.** 72 **11.** 1 **13.** 6 **15.** 6 **17.** 1 **19.** 48 **21.** 40,320 **23.** **(a)** 17,576,000 **(b)** 17,576,000 **(c)** 456,976,000 **25.** $2.052371412 \times 10^{10}$ **27.** 39,270 **29.** 15,890,700 **31.** 35 **33.** 2,598,960 **35.** 816; 8568 **37.** 10,626 **39.** permutation **41.** combination **43.** 210 **45.** 210; 5040 **47.** 35 **49.** **(a)** 56 **(b)** 462 **(c)** 3080 **(d)** 8526 **51.** **(a)** 220 **(b)** 55 **(c)** 105

Section 13.7

1. $S = \{HH, HT, TH, TT\}$ **3.** $S = \{(1, 2), (1, 3), (1, 4), (1, 5), (2, 3), (2, 4), (2, 5), (3, 4), (3, 5), (4, 5)\}$ **5.** A probability cannot be greater than 1. **7.** **(a)** $\{HH, TT\}$; $\frac{1}{2}$ **(b)** $\{HH, HT, TH\}$; $\frac{3}{4}$
9. **(a)** $\{(2, 4)\}$; $\frac{1}{10}$ **(b)** $\{(1, 3), (1, 5), (3, 5)\}$; $\frac{3}{10}$ **(c)** $\emptyset$; 0 **(d)** $\{(1, 2), (1, 4), (2, 3), (2, 5), (3, 4), (4, 5)\}$; $\frac{3}{5}$ **11.** **(a)** $\frac{1}{5}$ **(b)** $\frac{8}{15}$ **(c)** 0 **(d)** $\frac{4}{15}$ **13.** 3 to 7 **15.** $\frac{4}{9}$; $\frac{5}{9}$ **17.** **(a)** $\frac{3}{5}$ **(b)** $\frac{7}{10}$ **(c)** $\frac{3}{10}$ **(d)** 0 **19.** **(a)** $\frac{3}{13}$ **(b)** $\frac{7}{13}$ **(c)** $\frac{3}{13}$ **21.** **(a)** $\frac{1}{2}$ **(b)** $\frac{7}{10}$ **(c)** 0 **23.** **(a)** 0.72 **(b)** 0.70 **(c)** 0.79 **25.** **(a)** 0.18 **(b)** 0.39 **(c)** 0.06 **(d)** 0.12 **27.** 0.41 **29.** 0.21 **31.** 0.79

Chapter 13 Review Exercises

1. $-1, 1, 3, 5$ **2.** $0, \frac{1}{2}, \frac{2}{3}, \frac{3}{4}$ **3.** 1, 4, 9, 16 **4.** $\frac{1}{2}, \frac{1}{4}, \frac{1}{8}, \frac{1}{16}$
5. 0, 3, 8, 15 **6.** $1, -2, 3, -4$ **7.** $1 + 4 + 9 + 16 + 25 = 55$
8. $2 + 3 + 4 + 5 + 6 + 7 = 27$ **9.** $11 + 16 + 21 + 26 = 74$
10. $9 + 16 + 25 + 36 = 86$ **11.** $2 + 4 + 8 + 16 + 32 + 64 = 126$
12. $\frac{4}{5} + \frac{5}{6} + \frac{6}{7} + \frac{7}{8} = \frac{2827}{840}$ **13.** 763.3 thousand electric vehicles
14. arithmetic; $d = 3$ **15.** arithmetic; $d = 4$ **16.** geometric; $r = -\frac{1}{2}$
17. geometric; $r = -1$ **18.** neither **19.** neither **20.** 89 **21.** 73
22. 69 **23.** $a_n = -5n + 1$ **24.** $a_n = -3n + 9$ **25.** 15 **26.** 22
27. 152 **28.** 164 **29.** $a_n = -1(4)^{n-1}$ **30.** $a_n = \frac{2}{3}\left(\frac{1}{5}\right)^{n-1}$
31. 118,098 **32.** 2560 or -2560 **33.** $\frac{341}{1024}$ **34.** 0 **35.** 1
36. The sum does not exist. **37.** $32p^5 - 80p^4q + 80p^3q^2 - 40p^2q^3 + 10pq^4 - q^5$ **38.** $x^8 + 12x^6y + 54x^4y^2 + 108x^2y^3 + 81y^4$
39. $81t^{12} - 108t^9s^2 + 54t^6s^4 - 12t^3s^6 + s^8$ **40.** $7752(3)^{16}a^{16}b^3$
45. 120 **46.** 72 **47.** 35 **48.** 56 **49.** 48 **50.** 90 **51.** 24
52. 456,976,000; 258,336,000 **53.** **(a)** $\frac{4}{15}$ **(b)** $\frac{2}{3}$ **(c)** 0 **(d)** $\frac{11}{15}$
54. **(a)** 5 to 14 **(b)** 9 to 10 **55.** $\frac{1}{26}$ **56.** $\frac{4}{13}$ **57.** $\frac{4}{13}$ **58.** $\frac{3}{4}$
59. 1 **60.** $\frac{3}{13}$

Chapter 13 Mixed Review Exercises

1. $a_{10} = 1536$; $S_{10} = 1023$ **2.** $a_{40} = 235$; $S_{10} = 280$
3. $a_{15} = 38$; $S_{10} = 95$ **4.** $a_9 = 6561$; $S_{10} = -14{,}762$ **5.** $a_n = 2(4)^{n-1}$
6. $a_n = 5n - 3$ **7.** $a_n = -3n + 15$ **8.** $a_n = 27\left(\frac{1}{3}\right)^{n-1}$
9. 10 sec **10.** \$21,973.00 **11.** approximately 42,000 people
12. $\frac{1}{128}$ **13.** **(a)** 20 **(b)** 10 **14.** 504 **15.** **(a)** 0.86 **(b)** 0.44

Chapter 13 Test

[13.1] **1.** 0, 2, 0, 2, 0 [13.2] **2.** 4, 6, 8, 10, 12 [13.3] **3.** 48, 24, 12, 6, 3
[13.2] **4.** 0 [13.3] **5.** $\frac{64}{3}$ or $-\frac{64}{3}$ [13.2] **6.** 75 [13.3] **7.** 124 or 44
[13.1] **8.** 92.5 min [13.3] **9.** \$137,925.91 **10.** It has a sum if $|r| < 1$.
[13.2] **11.** 70 **12.** 33 **13.** 125,250 [13.3] **14.** 42
15. $\frac{1}{3}$ **16.** The sum does not exist.
[13.4] **17.** $81k^4 - 540k^3 + 1350k^2 - 1500k + 625$ **18.** $\frac{14{,}080x^8y^4}{9}$
[13.1] **19.** \$324 [13.3] **20.** 3,542,940 insects
[13.6] **22.** 990 **23.** 45 **24.** 60 **25.** 24,360 **26.** 84
[13.7] **27.** $\frac{1}{26}$ **28.** $\frac{10}{13}$ **29.** $\frac{4}{13}$ **30.** 3 to 10

Chapters R–13 Cumulative Review Exercises

[R.1–R.3] **1.** 8 **2.** $-\frac{13}{45}$ [1.1] **3.** $\left\{\frac{1}{6}\right\}$ [1.5] **4.** $[10, \infty)$
[1.6] **5.** $(-\infty, -3) \cup (4, \infty)$ [1.7] **6.** $\left\{-\frac{9}{2}, 6\right\}$
7. $(-\infty, -3] \cup [8, \infty)$ [5.5] **8.** $\left\{-\frac{5}{2}, 2\right\}$
[8.2] **9.** $\left\{\frac{-5 \pm \sqrt{217}}{12}\right\}$ [6.4] **10.** $\emptyset$ [10.2] **11.** $\left\{\frac{5}{2}\right\}$
[10.6] **12.** $\{2\}$ [2.2] **13.** $\frac{3}{4}$ [2.3] **14.** $3x + y = 4$
[7.3, 12.1] **15.** $(x + 5)^2 + (y - 12)^2 = 81$
[3.1] **16.** $\{(-1, -2)\}$ [3.2] **17.** $\{(2, 1, 4)\}$
[12.3] **18.** $\left\{(-1, 5), \left(\frac{5}{2}, -2\right)\right\}$ [3.3] **19.** 2 lb [4.1] **20.** $\frac{9}{4}$
21. $-\frac{27p^2}{10}$ [4.5] **22.** $20p^2 - 2p - 6$ [4.3] **23.** $-5m^3 - 3m^2 + 3m + 8$
[4.6] **24.** $2t^3 + 3t^2 - 4t + 2 + \frac{3}{3t - 2}$ [7.7] **25.** 73
[5.2] **26.** $z(3z + 4)(2z - 1)$ [5.3] **27.** $(7a^2 + 3b)(7a^2 - 3b)$

28. $(c + 3d)(c^2 - 3cd + 9d^2)$ [6.1] **29.** $\frac{x+7}{x-2}$

[6.2] **30.** $\frac{3p - 26}{p(p+3)(p-4)}$ [7.4] **31.** $10\sqrt{2}$

[2.1] **32.** [2.4] **33.**

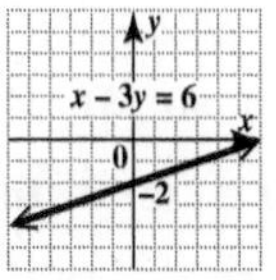

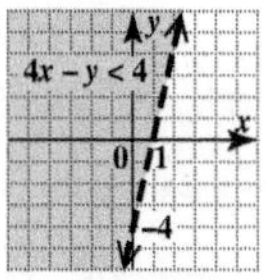

[9.2] **34.**

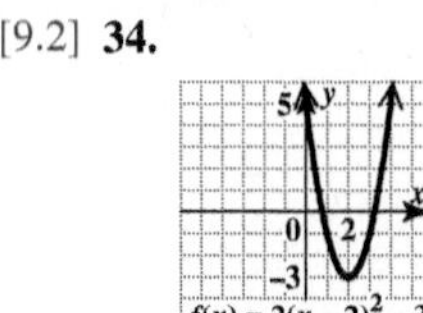

[10.2] **35.** $g(x) = \left(\frac{1}{3}\right)^x$

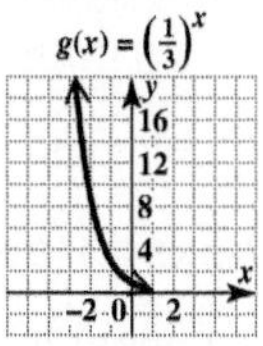

[10.3] **36.**

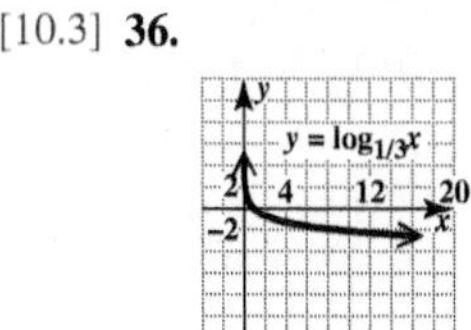

[11.4] **37.** $f(x) = \frac{1}{x-3}$

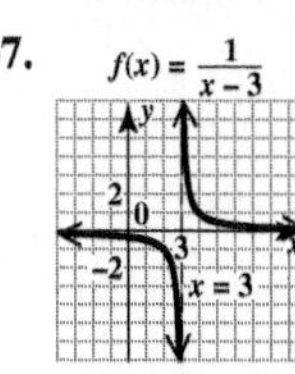

[12.1] **38.**

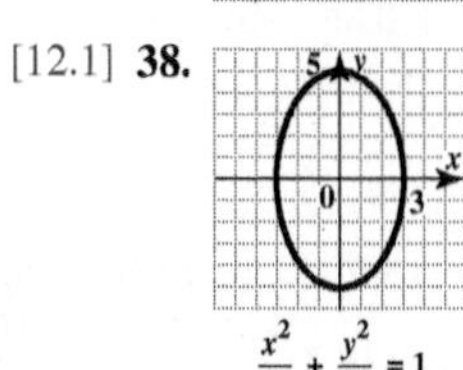

[12.2] **39.**

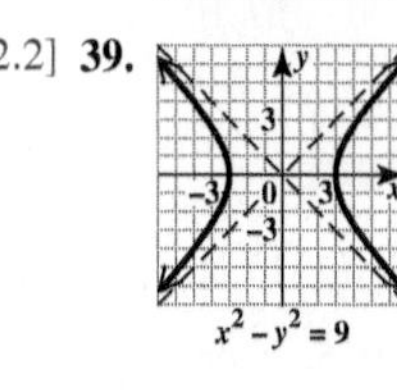

[10.1] **40.** $f^{-1}(x) = \frac{x-5}{9}$, or $f^{-1}(x) = \frac{1}{9}x - \frac{5}{9}$

[11.2] **41.** $f(x) = (2x - 1)(x + 4)(x + 1)$

[13.1] **42.** $-7, -2, 3, 8, 13$ [13.2, 13.3] **43. (a)** 78 **(b)** $\frac{75}{7}$

[13.1] **44.** 30 [13.4] **45.** $32a^5 - 80a^4 + 80a^3 - 40a^2 + 10a - 1$

46. $-\frac{45x^8y^6}{4}$ [13.4, 13.6] **48. (a)** 362,880 **(b)** 210 **(c)** 210

[13.7] **49.** $\frac{1}{18}$ **50.** $\frac{3}{10}$

APPENDICES

Appendix A

1. (a) $0, 5, -3$ **(b)** $1, -3, 8$ **(c)** yes; The number of rows is the same as the number of columns (three). **(d)** $\begin{bmatrix} 1 & 4 & 8 \\ 0 & 5 & -3 \\ -2 & 3 & 1 \end{bmatrix}$

(e) $\begin{bmatrix} 1 & -\frac{3}{2} & -\frac{1}{2} \\ 0 & 5 & -3 \\ 1 & 4 & 8 \end{bmatrix}$ **(f)** $\begin{bmatrix} 1 & 15 & 25 \\ 0 & 5 & -3 \\ 1 & 4 & 8 \end{bmatrix}$ **3.** 3×2 **5.** 4×2

7. $\{(4, 1)\}$ **9.** $\{(1, 1)\}$ **11.** $\{(-1, 4)\}$ **13.** $\varnothing$

15. $\{(x, y) \mid 2x + y = 4\}$ **17.** $\{(0, 0)\}$ **19.** $\{(4, 0, 1)\}$

21. $\{(-1, 23, 16)\}$ **23.** $\{(3, 2, -4)\}$

25. $\{(x, y, z) \mid x - 2y + z = 4\}$ **27.** $\varnothing$

Appendix B

1. true **3.** false; The determinant equals $ad - bc$. **5.** D **7.** -3

9. 14 **11.** 0 **13.** 59 **15.** 14 **17.** 16 **19.** -12 **21.** 0

23. $\{(1, 0, -1)\}$ **25.** $\{(-3, 6)\}$ **27.** $\left\{\left(\frac{53}{17}, \frac{6}{17}\right)\right\}$ **29.** $\{(-1, 2)\}$

31. $\{(4, -3, 2)\}$ **33.** Cramer's rule does not apply.

35. $\{(-2, 1, 3)\}$ **37.** $\left\{\left(\frac{49}{9}, -\frac{155}{9}, \frac{136}{9}\right)\right\}$ **39.** $\{2\}$ **41.** $\{0\}$

Appendix C

1. dimension **3.** 2 **5.** $w = 3, x = 2, y = -1, z = 4$

7. $w = 2, x = 6, y = -2, z = 8$ **9.** $z = 18, r = 3, s = 3, p = 3, a = 2$

11. 2×2; square **13.** 1×5; row **15.** 2×1; column

17. 3×3; square **19.** 3×4 **21.** $\begin{bmatrix} -2 & -7 & 7 \\ 10 & -2 & 7 \end{bmatrix}$

23. $\begin{bmatrix} 12 & -3 \\ 3 & 14 \end{bmatrix}$ **25.** $\begin{bmatrix} 9 & -9 \\ 3 & 5 \\ 5 & 20 \end{bmatrix}$ **27.** The matrices cannot be added.

29. $\begin{bmatrix} 5x + y & x + y & 7x + y \\ 8x + 2y & x + 3y & 3x + y \end{bmatrix}$ **31.** $\begin{bmatrix} 10r - 7s \\ -a - 16b \\ -3m + 9n \\ -2x + 8y \end{bmatrix}$

33. $\begin{bmatrix} -4 & 8 \\ 0 & 6 \end{bmatrix}$ **35.** $\begin{bmatrix} 2 & 6 \\ -4 & 6 \end{bmatrix}$ **37.** $\begin{bmatrix} -1 & -3 \\ 2 & -3 \end{bmatrix}$ **39.** yes; 2×5

41. no **43.** $\begin{bmatrix} 13 \\ 25 \end{bmatrix}$ **45.** $\begin{bmatrix} -17 \\ -1 \end{bmatrix}$ **47.** $\begin{bmatrix} 17 & -10 \\ 1 & 2 \end{bmatrix}$ **49.** $\begin{bmatrix} -2 & 10 \\ 0 & 8 \end{bmatrix}$

51. $\begin{bmatrix} -2 & 5 & 0 \\ 6 & 6 & 1 \\ 12 & 2 & -3 \end{bmatrix}$ **53.** $\begin{bmatrix} 2 & 7 & -4 \end{bmatrix}$ **55.** The matrices cannot be multiplied. **57.** $\begin{bmatrix} 100 & 150 \\ 125 & 50 \\ 175 & 200 \end{bmatrix}$; $\begin{bmatrix} 100 & 125 & 175 \\ 150 & 50 & 200 \end{bmatrix}$

59. (a) $\begin{bmatrix} 50 & 100 & 30 \\ 10 & 90 & 50 \\ 60 & 120 & 40 \end{bmatrix}$ **(b)** $\begin{bmatrix} 12 \\ 10 \\ 15 \end{bmatrix}$ (If the rows and columns are interchanged in part (a), this should be a 1×3 matrix.)

(c) $\begin{bmatrix} 2050 \\ 1770 \\ 2520 \end{bmatrix}$ (This may instead be a 1×3 matrix.) **(d)** \$6340

60. (Now Try Exercise 8) (a) $\begin{bmatrix} 47.5 & 57.75 \\ 27 & 33.75 \\ 81 & 95 \\ 12 & 15 \end{bmatrix}$

(b) $\begin{bmatrix} 20 & 200 & 50 & 60 \end{bmatrix}$; $\begin{bmatrix} 220 & 890 & 105 & 125 & 70 \end{bmatrix}$

(c) $\begin{bmatrix} 11{,}120 & 13{,}555 \end{bmatrix}$

Appendix D

1. 1; 0 **3.** rows; columns (These may be interchanged.)

5. $I_2 = \begin{bmatrix} 1 & 0 \\ 0 & 1 \end{bmatrix}$; $AI_2 = \begin{bmatrix} 4 & -2 \\ 3 & 1 \end{bmatrix}\begin{bmatrix} 1 & 0 \\ 0 & 1 \end{bmatrix} = \begin{bmatrix} 4 & -2 \\ 3 & 1 \end{bmatrix} = A$

7. yes **9.** no **11.** no **13.** yes **15.** $\begin{bmatrix} 5 & 2 \\ -3 & -1 \end{bmatrix}$

17. The inverse does not exist. **19.** $\begin{bmatrix} -1 & 1 & 1 \\ 0 & -1 & 0 \\ 2 & -1 & -1 \end{bmatrix}$

21. The inverse does not exist. **23.** $\begin{bmatrix} -\frac{15}{4} & -\frac{1}{4} & -3 \\ \frac{5}{4} & \frac{1}{4} & 1 \\ -\frac{3}{2} & 0 & -1 \end{bmatrix}$

25. $\begin{bmatrix} \frac{1}{2} & 0 & \frac{1}{2} & -1 \\ \frac{1}{10} & -\frac{2}{5} & \frac{3}{10} & -\frac{1}{5} \\ -\frac{7}{10} & \frac{4}{5} & -\frac{11}{10} & \frac{12}{5} \\ \frac{1}{5} & \frac{1}{5} & -\frac{2}{5} & \frac{3}{5} \end{bmatrix}$ **27.** $\{(2, 3)\}$ **29.** $\{(-2, 4)\}$

31. $\{(4, -6)\}$ **33.** $\left\{\left(-\frac{1}{4}, \frac{3}{2}, \frac{3}{4}\right)\right\}$ **35.** $\{(11, -1, 2)\}$

37. $\{(1, 0, 2, 1)\}$

39. **(a)** $602.7 = a + 5.543b + 37.14c$

$656.7 = a + 6.933b + 41.30c$

$778.5 = a + 7.638b + 45.62c$

(b) $a \approx -490.547$, $b = -89$, $c = 42.71875$

(c) $S = -490.547 - 89A + 42.71875B$ **(d)** 843.5

(e) $S \approx 1547.5$; Using only three consecutive years to forecast six years into the future is probably not wise.

PHOTO CREDITS

STUDY SKILLS
p. S-1 Borchee/E+/Getty Images; **p. S-2** Originoo stock/123RF; **p. S-4** G-stockstudio/Shutterstock; **p. S-6** Rawpixel/123RF; **p. S-8** Studio DMM Photography/Designs & Art/Shutterstock; **p. S-10** Doyeol (David) Ahn/Alamy Stock Photo

CHAPTER R
p. 17 Terry McGinnis; **p. 25** Callie Daniels; **p. 35** Proxima Studio/Shutterstock; **p. 44** (both) Terry McGinnis; **p. 45** Maximilian Laschon/Shutterstock; **p. 47** Tatyana Tomsickova/123RF

CHAPTER 1
p. 55 Callie Daniels; **p. 70** John Hornsby; **p. 75** Anne Richard/Shutterstock; **p. 76** Dinodia/123RF; **p. 82** Pedro Monteiro/Shutterstock; **p. 89** (top) Michele Perbellini/Shutterstock, (bottom) Hornsby stamp; **p. 90** Rob/Fotolia; **p. 91** Spotmatik Ltd/Shutterstock, **p. 97** Pearson Education, Inc.; **p. 98** (top) Josefauer/Shutterstock, (bottom) Pete Niesen/Shutterstock; **p. 102** Oles Ishchuk/123RF; **p. 114** (top) Fredrick Kippe/Alamy Stock Photo, (bottom left) Duplass/Shutterstock, (bottom right) John Hornsby; **p. 121** Jeayesy/123RF; **p. 123** Michaeljung/Shutterstock; **p. 124** Chad McDermott/ Shutterstock; **p. 131** Leonid Shcheglov/Shutterstock; **p. 135** Katherine Martin/123RF; **p. 144** John Hornsby

CHAPTER 2
p. 149 Michelangeloop/Shutterstock; **p. 151** Library of Congress; **p. 182** Kazzland Inc./Shutterstock; **p. 184** Podfoto/Shutterstock; **p. 187** (top) Siraphol/123RF, (bottom) Karen roach/123RF; **p. 188** (top) Wavebreakmedia/Shutterstock, (bottom) Goran Bogicevic/123RF; **p. 190** Barry Blackburn/Shutterstock; **p. 198** Derek Meijer/Alamy Stock Photo; **p. 216** Rido/Shutterstock; **p. 218** Neelsky/Shutterstock

CHAPTER 3
p. 229 Graham Moore/123RF; **p. 254** Pavol Kmeto/Shutterstock; **p. 255** Blickwinkel/Alamy Stock Photo; **p. 262** Marie C Fields/Shutterstock; **p. 264** AndreAnita/Shutterstock; **p 265** (top) Chris Whitehead/Cultura Creative (RF)/Alamy Stock Photo, (bottom) Thomas M Perkins/Shutterstock; **p. 267** Callie Daniels; **p. 268** Dpa picture alliance/Alamy Stock Photo; **p. 269** Callie Daniels

CHAPTER 4
p. 279 Rawpixel/123RF; **p. 294** (top) Vovan/Shutterstock, (bottom) Terry McGinnis; **p. 295** Doug McLean/Shutterstock; **p. 303** Gareth Boden/Pearson Education Ltd; **p. 308** Gorbelabda/Shutterstock; **p. 310** Fair Trade USA; **p. 311** Topseller/Shutterstock; **p. 312** Coprid/Shutterstock; **p. 313** Terry MGinnis; **p. 335** Lucky Dragon/Fotolia; **p. 336** Lopolo/Shutterstock; **p. 340** National Atomic Museum Foundation

CHAPTER 5
p. 341 Don Fink/Shutterstock; **p. 372** Konstantin Shaklein/123RF

CHAPTER 6
p. 385 Ballemans/Shutterstock; **p. 404** Terry McGinnis; **p. 418** Lars Christensen/Shutterstock; **p. 422** Disability Images/Alamy Stock Photo; **p. 424** Lucio pepi/Fotolia; **p. 426** Dmitry Kalinovsky/ Shutterstock; **p. 429** (top) John Hornsby, (center) Hurst Photo/Shutterstock, (bottom) CCimage/ Shutterstock; **p. 430** (top) Terry McGinnis, (bottom) Scott Rothstein/Shutterstock; **p. 438** Dana Bartekoske/123RF; **p. 441** Dmitry Deshevykh/Alamy Stock Photo; **p. 442** Cathy Yeulet/123RF

CHAPTER 7
p. 455 Giulio Meinardi/Fotolia; **p. 463** Stoupa/Shutterstock; **p. 472** Maria Moroz/Fotolia; **p. 485** (dog) Anneka/Shutterstock, (screen) Cobalt/Fotolia; **p. 486** Kevin Kipper/Alamy Stock Photo; **p. 509** Johnson Space Center/NASA; **p. 528** Scol22/Fotolia

CHAPTER 8

p. 531 Callie Daniels; **p. 534** Portrait of Galileo (ca. 1853), Charles Knight. Steel engraving. Library of Congress Prints and Photographs Division [LC-DIG-pga-38085]; **p. 540** (left) Piotrek Jastrzebski/Shutterstock, (right) Rudi1976/Alamy Stock Photo; **p. 551** Phovoir/Shutterstock; **p. 564** Micro10x/Shutterstock; **p. 569** H. & D. Zielske/LOOK Die Bildagentur der Fotografen GmbH/Alamy Stock Photo; **p. 570** Steven J. Everts/123RF; **p. 577** Hartrockets/iStock/Getty Images

CHAPTER 9

p. 587 Sarot Chamnankit/123RF; **p. 591** Terry McGinnis; **p. 598** Hxdbzxy/Shutterstock; **p. 603** Zurijeta/Shutterstock; **p. 607** Andriy Popov/123RF; **p. 608** Alexwhite/Shutterstock; **p. 617** Kaband/Shutterstock; **p. 618** Bonniej/E+/Getty Images; **p. 631** Africa Studio/Shutterstock; **p. 634** Samuel Acosta/Shutterstock; **p. 636** Terry McGinnis

CHAPTER 10

p. 649 Chien321/Shutterstock; **p. 651** Frontpage/Shutterstock; **p. 658** Mark Higgins/Shutterstock; **p. 667** Bernard Staehli/Shutterstock; **p. 668** Maksym Bondarchuk/123RF; **p. 673** Per Tillmann/Fotolia; **p. 676** Newman Mark/Prisma by Dukas Presseagentur GmbH/Alamy Stock Photo; **p. 685** Romrodphoto/Shutterstock; **p. 687** Georgios Kollidas/Fotolia; **p. 690** (top) Jim West/Alamy Stock Photo, (bottom) Silvano Rebai/Fotolia; **p. 691** WaterFrame_fba/Alamy Stock Photo; **p. 699** Sergey Nivens/Shutterstock; **p. 700** Armadillo Stock/Shutterstock; **p. 706** Shutterstock; **p. 708** Syda Productions/Shutterstock; **p. 709** IndustryAndTravel/Alamy Stock Photo

CHAPTER 11

p. 715 James Morgan/Shutterstock; **p. 726** Science History Images/Alamy Stock Photo; **p. 745** Matthew Horwood/Alamy Stock Photo; **p. 746** Peter Hermes Furian/123RF; **p. 761** Alice-photo/Shutterstock

CHAPTER 12

p. 773 Terry McGinnis; **p. 784** Petr Student/Shutterstock; **p. 791** Faiz Azizan/Shutterstock; **p. 808** (top) Thanest Sonmueng/123RF, (bottom) Rawpixel/123RF; **p. 809** A.S. Zain/Shutterstock; **p. 814** Ruth Black/123RF

CHAPTER 13

p. 819 Callie Daniels; **p. 820** Madeinitaly4k/Shutterstock; **p. 824** Frank11/Shutterstock; **p. 833** Blend Images/Shutterstock; **p. 834** Creativa Images/Shutterstock; **p. 842** Paul Hakimata Photography/Shutterstock; **p. 845** Atlaspix/Alamy Stock Photo; **p. 856** HDesert/Shutterstock; **p. 858** Mr.markin/Fotolia; **p. 860** Orhan Cam/Shutterstock; **p. 862** Swisshippo/Fotolia; **p. 863** Dolgachov/123RF; **p. 864** Shutterstock; **p. 872** Kurhan/123RF; **p. 873** Peter Kirillov/123RF; **p. 881** Jtbaskinphoto/Shutterstock; **p. 883** Baloncici/123RF

APPENDIX C

p. 904 Alfred Emmerichs/123RF; **p. 905** Jonathan Weiss/Shutterstock

APPENDIX D

p. 920 Belchonock/123RF

INDEX

D

H

I

J

L

T

Triangles and Angles

Right Triangle

Triangle has one 90° (right) angle.

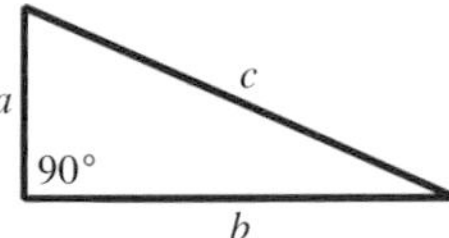

Pythagorean Theorem (for right triangles)

$a^2 + b^2 = c^2$

Isosceles Triangle

Two sides are equal.

$AB = BC$

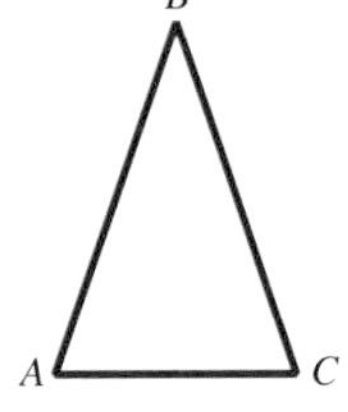

Equilateral Triangle

All sides are equal.

$AB = BC = CA$

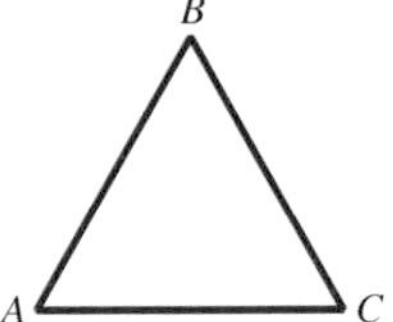

Sum of the Angles of Any Triangle

$A + B + C = 180°$

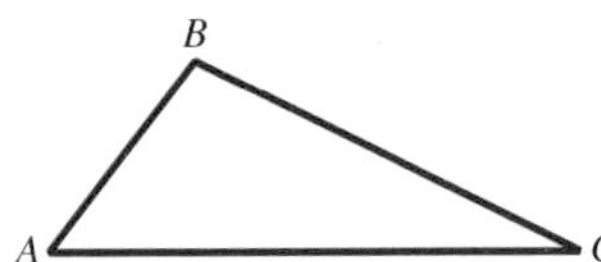

Similar Triangles

Corresponding angles are equal. Corresponding sides are proportional.

$A = D, B = E, C = F$

$$\frac{AB}{DE} = \frac{AC}{DF} = \frac{BC}{EF}$$

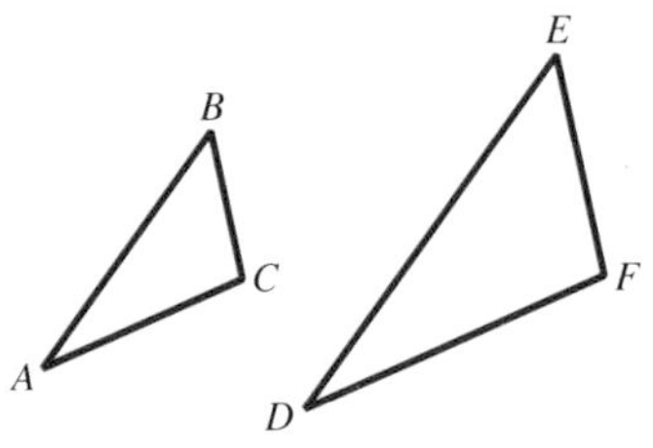

Right Angle

Measure is 90°.

Straight Angle

Measure is 180°.

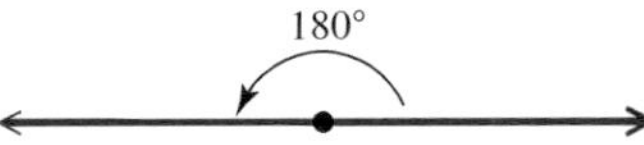

Complementary Angles

The sum of the measures of two complementary angles is 90°.

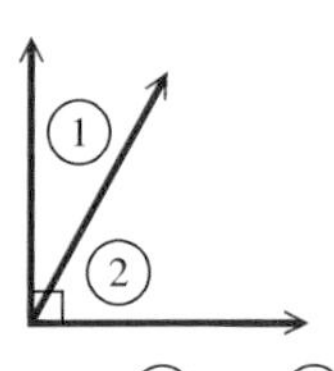

Angles (1) and (2) are complementary.

Supplementary Angles

The sum of the measures of two supplementary angles is 180°.

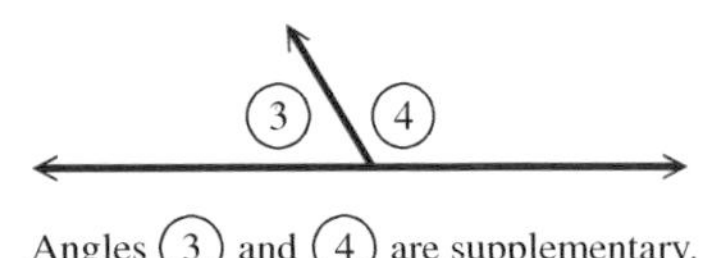

Angles (3) and (4) are supplementary.

Vertical Angles

Vertical angles have equal measures.

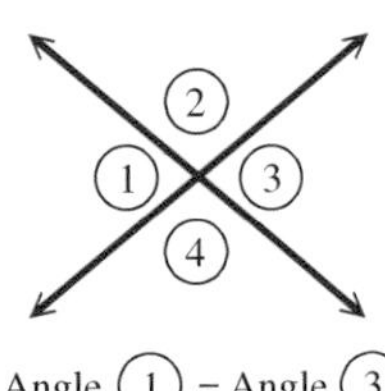

Angle (1) = Angle (3)

Angle (2) = Angle (4)

Geometry Formulas

Figure	*Formulas*	*Illustration*
Square	Perimeter: $P = 4s$ Area: $\mathscr{A} = s^2$	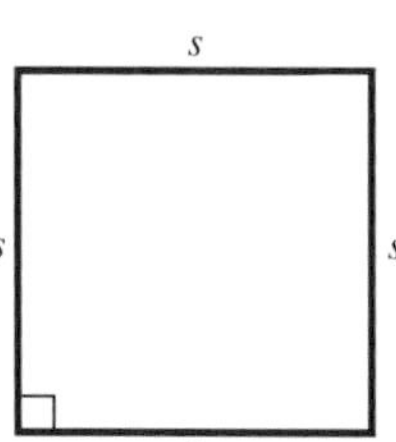
Rectangle	Perimeter: $P = 2L + 2W$ Area: $\mathscr{A} = LW$	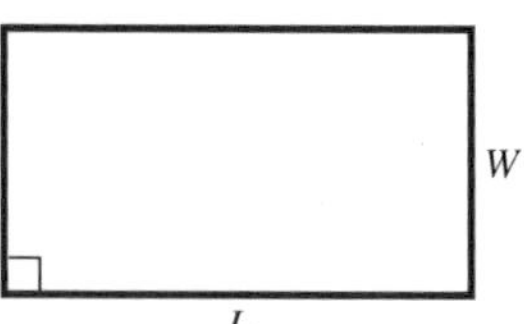
Triangle	Perimeter: $P = a + b + c$ Area: $\mathscr{A} = \frac{1}{2}bh$	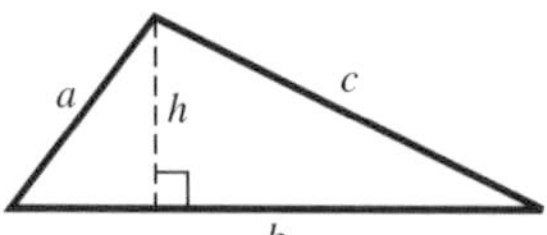
Parallelogram	Perimeter: $P = 2a + 2b$ Area: $\mathscr{A} = bh$	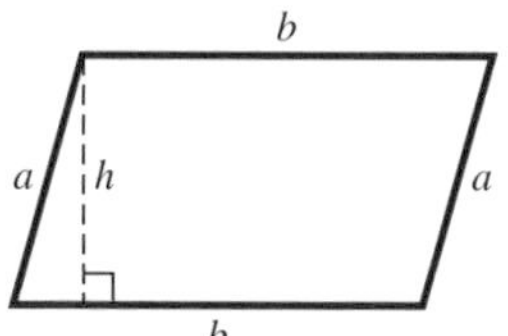
Trapezoid	Perimeter: $P = a + b + c + B$ Area: $\mathscr{A} = \frac{1}{2}h(b + B)$	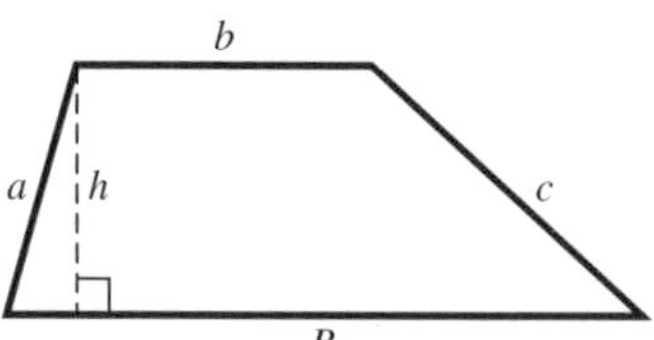
Circle	Diameter: $d = 2r$ Circumference: $C = 2\pi r$ $C = \pi d$ Area: $\mathscr{A} = \pi r^2$	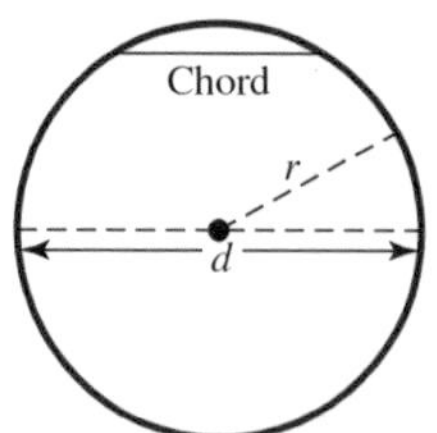